Tailoring Surfaces

Modifying Surface Composition and Structure for Applications in Tribology, Biology and Catalysis

IISc CENTENARY LECTURE SERIES

Series Editor: Anurag Kumar *(Indian Institute of Science, India)*

Description:

World Scientific Publishing Company Singapore and Indian Institute of Science (IISc), Bangalore, will co-publish a series of prestigious lectures delivered during IISc's centenary year (2008-09), and a series of textbooks and monographs, by prominent scientists and engineers from IISc and other institutions.

This pioneering collaboration will contribute significantly in disseminating current Indian scientific advancement worldwide. In addition, the collaboration also proposes to bring the best scientific ideas and thoughts across the world in areas of priority to India through specially designed India editions.

The "IISc Centenary Lecture Series" will comprise lectures by designated Centenary Professors – eminent teachers and researchers from all over the world.

Published

Vol. 1 Trends in Chemistry of Materials
Selected Research Papers of C N R Rao
by C N R Rao

Vol. 2 The Foundations of the Digital Wireless World
Selected Works of A J Viterbi
by A J Viterbi

Vol. 3 Science and Sustainable Food Security
Selected Papers of M S Swaminathan
by M S Swaminathan

Vol. 5 Tailoring Surfaces
Modifying Surface Composition and Structure for Applications in Tribology, Biology and Catalysis
by N D Spencer

Forthcoming

Vol. 4 Ultrafast 2D IR Vibrational Echo Spectroscopy
by M Fayer

IISc Centenary Lecture Series

Tailoring Surfaces

Modifying Surface Composition and Structure for Applications in Tribology, Biology and Catalysis

Nicholas D Spencer

ETH Zurich, Switzerland

IISc Press

World Scientific

NEW JERSEY • LONDON • SINGAPORE • BEIJING • SHANGHAI • HONG KONG • TAIPEI • CHENNAI

Published by

World Scientific Publishing Co. Pte. Ltd.
5 Toh Tuck Link, Singapore 596224
USA office: 27 Warren Street, Suite 401-402, Hackensack, NJ 07601
UK office: 57 Shelton Street, Covent Garden, London WC2H 9HE

British Library Cataloguing-in-Publication Data
A catalogue record for this book is available from the British Library.

IISc Centenary Lecture Series — Vol. 5
TAILORING SURFACES
Modifying Surface Composition and Structure for Applications in Tribology, Biology and Catalysis

ISBN-13 978-981-4289-42-9
ISBN-10 981-4289-42-6

Printed in Singapore by World Scientific Printers.

Dylan Spencer-Davidson

Professor Nicholas D. Spencer

Preface

At the time I began my PhD research, more than thirty years ago, surface chemistry involved "putting things on top of other things" in ultrahigh vacuum. Most papers in the field involved the adsorption of simple gases, favorites being CO, NO, or O_2, onto well-defined crystal faces of metals, such as Pt(111), Ni(100), or W(110). Analytical techniques were those that allowed structural and chemical analysis of such systems (e.g. low-energy electron diffraction, Auger spectroscopy, and X-ray photoelectron spectroscopy), and some means to measure bond energies to the surface (thermal desorption, for instance). The liquid-solid interface was rather neglected at that time, and the solid-solid interface hardly mentioned in polite company.

In the subsequent decades, the field has been transformed by new developments, both in instrumentation and in chemistry. Scanning-probe methods have not only made access to the nanometer scale far easier (and in real space) but, together with new spectroscopic methods, they have also brought surface science out of the vacuum. With improved ways of looking at liquid-solid interfaces on small scales, a bridge has been built to areas of technical importance such as corrosion and tribology, as well as to biology and hence to medicine. At the same time, developments in surface-chemical functionalization have meant that it has become increasingly feasible to modify, or tailor, the surfaces of materials with molecular precision outside of the vacuum in myriad different ways and for a vast array of applications.

I have had the pleasure of participating in this vibrant environment, as surface chemistry has broadened towards many other disciplines, and become central to a host of developments in medicine, semiconductors, chemical manufacturing, and many other technologies. In this book I have tried to convey the flavor of that development, through examples from my own work in surface functionalization, catalysis, tribology, and biointerfaces.

Nicholas D. Spencer
Zurich, Switzerland
2010

Biography

Nicholas Spencer was born in the suburbs of London on April 15th, 1955, attending local schools and then Dulwich College, where his primary interests were in music, science, and computers. After a brief stay in West Berlin in 1974, working as a systems analyst in the nascent computer industry, Spencer began his studies at King's College, Cambridge, first obtaining his BA in Natural Sciences in 1977 and then going on to do a PhD in the Department of Physical Chemistry, under the guidance of Richard Lambert, on the surface chemistry of silver and gold.

Excited by the possibilities of the United States, and tired of the biting east wind in Cambridge, Spencer moved to the University of California, Berkeley in 1980 for a two-year post-doctoral stint in the laboratory of Gabor Somorjai. Here he focused on catalysis, notably on ammonia synthesis over single crystals of iron, using the newly developed high-pressure-low-pressure apparatus. The structural sensitivity for this reaction was the highest that had been observed up to that time. There then followed an 11-year stay in industry, in the Research Division of W.R. Grace & Co, in Columbia, Maryland, where Spencer worked initially on methane partial oxidation catalysts and high-temperature superconductors, for which he developed new synthetic routes and founded an internal startup company to manufacture and market the materials. Subsequently Spencer ran the surface analysis, microscopy and spectroscopy sections of the Division.

The return to academia took place in 1993, with a call to the Chair of Surface Science and Technology ("Oberflächentechnik") at the Swiss Federal Institute of Technology (ETH) in Zurich, Switzerland. Spencer built up a sizeable research group at the ETH in the areas of surface modification, tribology, and biointerfaces. In addition, he co-founded, with Eddy Tysoe, University of Wisconsin, Milwaukee, the journal *Tribology Letters*, which they continue to edit, and he sits on the editorial boards of several other tribology journals. Spencer also co-edited the *Encyclopedia of Chemical Physics and Physical Chemistry* with the late Jack Moore, University of Maryland. He is the author of over 250 research articles and a dozen patents.

Nicholas Spencer has been Chairman of the ETH Department of Materials for a total of six years, founding Director of the ETH Materials Research Center, and he is currently President of the ETH Research Commission. He was cofounder of the Swiss Tribology Society and the International Nanotribology Forum (INF) (which he currently chairs). The INF has been active in organizing many conferences and workshops throughout Asia, with the goal of bringing students, especially those from developing countries, in contact with top scientists from all over the world in the general areas of nanotribology, surface mechanical properties, and cell-surface interactions.

Spencer has been awarded a number of distinctions, including the Lectureship Award of the Chemical Society of Japan, Surface and Colloid Division (1999), the G.S. Tendolkhar Memorial Lecture (Indian Institute of Technology, Bombay, 2002), the Hascoe Distinguished Lectureship (University of Connecticut, 2004), and the Royal Society of Chemistry Tribochemistry Lectureship, 2007. Spencer was also made a Fellow of the Royal Society of Chemistry in 2007, and a Centenary Visiting Professor of the Indian Institute of Science, Bangalore, in 2008, where he continues to be Visiting Professor at Large.

Spencer's current research interests continue to focus on surface modification and analysis, with a particular emphasis on gradients in surface chemistry and morphology, the fabrication and properties of surface-grafted polymers, and applications in lubrication and cell-surface interactions.

Acknowledgments

This book came into existence as a result of my year as Centenary Visiting Professor at the Indian Institute of Science, Bangalore, during which time I received extraordinary hospitality and was exposed both to exciting research activities and to innumerable cultural stimuli. The initial chapters were written while in Bangalore, and many of their themes were touched upon during my seminars given in various IISc departments throughout the year. I am deeply indebted to Professor Sanjay K. Biswas and the Director of the Institute, Professor P. Balaram, for making my stay possible, and helping to forge a connection to India that will remain very strong.

I am most grateful to the publishers who have granted copyright permission to reproduce the following articles in this book:

American Chemical Society (Articles 2.1–2.6, 2.8, 2.11, 2.16, 2.19, 2.21, 2.22, 2.24–2.27, 2.29, 2.30, 2.32, 2.33, 3.3, 3.4, 4.1–4.3, 4.6, 4.7, 4.9–4.13, 5.2, 5.4, 5.5)
Springer (Articles 2.7, 2.9, 2.10, 2.13, 2.17, 2.18, 2.20, 2.23, 2.34, 2.35, 2.36, 2.38, 2.39, 5.3)
Elsevier (Articles 2.12, 2.37, 2.40, 2.41, 3.2, 3.5, 3.6, 4.4, 5.1, 5.8)
Nature Publishing Group (Articles 3.1, 4.15)
Royal Society of Chemistry (Article 4.8)
American Academy for the Advancement of Science (Article 2.15)
American Institute of Physics (Articles 4.5, 5.6, 5.7)
Wiley (Articles 2.14, 2.31, 2.42, 2.43, 4.14, 4.16)
The Biophysical Society (Article 2.28)

The work presented in this book has all been the result of innumerable collaborations, with my students, postdocs, technical staff, colleagues in my own institution and around the world, and my mentors, Richard Lambert and Gabor Somorjai. I am most grateful to them all for helping make the pursuit of science so stimulating and enjoyable. I am also thankful to those who have provided me with the necessary means to carry out research, especially my current employer, the ETH Zurich, and the Swiss National Science Foundation. The ETH Zurich has also provided an excellent working environment for the last 17 years, and I am grateful to my colleagues, especially in the Department of Materials, for making it such a special place.

A number of people have made lasting and important contributions to our research group. I would like to make a special mention of Marcus Textor, who has been a constant supportive, stimulating and generous presence for my entire time in Zurich. Many of the papers reproduced in this volume are the result of collaborations with him and his students. Antonella Rossi has also been with me at the ETH for many years, and has made sure that our XPS measurements have been carried out in a most rigorous manner, as well as leading our efforts in antiwear additives. Georg Hähner was my first postdoc and contributed enormously to our research in chemical force microscopy and our early work in nanotribology. Manfred Heuberger brought us exceptional and lasting instrumental innovations in the surface forces area. Seunghwan Lee, whom I originally engaged to work on chocolate tribology, turned out to be the driving force behind much of our work on aqueous tribology. Venkataraman began as our infrared specialist, but went on to do much elegant work on gradients. Janos Vörös brought us up to speed in optical methods for measuring adsorption, as well as contributing greatly to our polymer brush and biosensor projects, before becoming an ETH professor himself. Samuele Tosatti first impressed us as a well-organized undergraduate with a skill for management, moved our alkanephosphate efforts a long way forward as PhD and postdoc, and went on to become CEO of SuSoS, a spin-off from our group. Stefan Zürcher is his business partner at SuSoS, its CTO, and the man who, as a postdoc, guided us (gently) into organic synthesis. Last but not least, Tanja Drobek gave us insights into many different areas, ranging from surface forces to wetting behavior and tribology.

Collaborations have been very important to our group, and I would particularly like to acknowledge the following colleagues with whom we have taken great pleasure in carrying out research and publishing a number of papers over the last few years: Prof. Don Brunette (University of British Columbia, Canada), Drs Rowena Crockett, Siegfried Derler, and Beat Keller (Empa, Switzerland), Dr Gabor Csucs (ETH, Switzerland), Prof. Andrew Gellman (Carnegie Mellon University, USA), Dr Hans Griesser (Ian Wark Research Institute, Australia), Prof. Jeff Hubbell (EPFL, Switzerland), Prof. Bengt Kasemo (Chalmers University, Sweden), Prof. Scott Perry (University of Florida, USA), Prof. Jeremy Ramsden (Cranfield University, UK), Prof. Hugh Spikes (Imperial College, UK), and Dr Heiko Wolf (IBM Zurich, Switzerland).

I would also like to express my thanks to my assistant, Ms Josephine Baer, who has provided invaluable help during the preparation of this book, and to Jennifer Davidson, Doris Spori, Rowena Crockett, Sanjay Biswas, and Scott Perry for their critical reading of the manuscript.

Finally, I would like to thank my wife Jennifer, and my children, Dylan, Xanthe, and Lucy for their constant love and support.

Contents

CHAPTER 1
Introduction

My aim in this introduction is to describe a number of fundamental aspects and techniques of surface modification that are built upon in subsequent chapters. Techniques range from the adsorption of small or large molecules onto surfaces, through the incorporation of layer-forming additives into a fluid layer above the surface, to the deliberate roughening or morphological modification of the surface. Gradual or abrupt spatial control of surface modification is also addressed and the impact of surface modification in application areas such as tribology, medicine, and catalysis is emphasized.

Materials properties can be broadly classified into those resulting from the nature of the bulk and those resulting from the characteristics of the surface. Examples of bulk properties include tensile strength, magnetic susceptibility, density, heat capacity, and even price (since materials are generally sold by the kg rather than the m^2). Examples of properties where surface effects dominate include wear-resistance, frictional behavior, wettability, paintability or printability, biocompatibility, corrosion resistance, and, to some extent, aesthetic appearance.

It often happens that the ideal properties for a particular application cannot be found in a single material, but that the best solution is to coat a material possessing ideal bulk properties with a substance that imparts the desirable surface performance. The application of coatings represents an important industrial activity that navigates the often-tortuous path between the optimization of bulk and surface properties and the realities of bulk-material cost and coating-process economics. Further constraints that add to the challenges facing the industrial coater include adhesion between coating and bulk, the speed of the process and evenness of coating, the temperature required during coating, and the toxicity of the materials used in the process.

Many industrial coatings can be micrometers thick, this book focuses on surface modification on the molecular scale. This is where the most drastic changes occur: Many of the surface properties described above are fully established after only a single molecular layer of the coating has been applied. Molecular-scale coating studies require a "surface-science approach" to the subject, since well-defined, clean surfaces of the bulk material, or "substrate" are essential, if the properties of the coated object are to be predictable and the coating is to be readily characterized.

Many coating processes, such as painting or galvanizing, have histories going back several centuries. However, the surface-science approach to surface modification really started to develop in the second half of the twentieth century, as ultrahigh vacuum (UHV) analytical techniques — such as Auger electron spectroscopy, X-ray photoelectron spectroscopy and low-energy electron diffraction — started to become available[1,2], enabling the quantitative characterization of monolayer coatings for the first time. In those early days of surface science, monomolecular layers of simple substances, such as oxygen, carbon monoxide, or halogens, were typically applied to well-characterized surfaces in the analytical UHV chamber[2]. The past half-century has witnessed a steady increase in the complexity and applicability of systems that can be studied by a rigorous, surface-science approach, which has grown from a field dealing with solid surfaces in vacuum, to encompass interfaces between solids and other condensed phases (see Chapter 5).

1a. Self-assembled monolayers

The invention of self-assembled monolayers (SAMs) (see Chapter 2a) represented a major step forward in the fabrication of monomolecular layers, since they allow a surface to be readily and reproducibly functionalized with a monolayer, without the use of a UHV chamber[3]. The essential components of SAMs (Figure 1) are the anchoring group, which attaches the molecules to the surface, the head group, which defines the state of functionalization of the new outer surface following modification, and the linking group, which, via van der Waals' interactions, provides an additional driving force for the adsorption reaction, and can create a certain degree of order in the system. Many head groups have been reported[4] including $-CH_3$, $-CF_3$, $-OH$, $-COOH$, $-NH_2$, and biotin. The linking groups generally consist of hydrocarbon (or fluorocarbon) chains with > 8 carbon atoms, at which point the total van der Waals interaction between the chains becomes large enough to lead to ordering phenomena. Anchoring groups depend on the substrate, since it is generally a covalent or coordination bond that is being formed, and include thiols[5] (on gold and other noble metals), silanes[6] (on hydroxyl-terminated oxides),

---

[1] *Surface Analysis: The Principal Techniques*, John C. Vickerman, Ian Gilmore (Editors), 2nd Ed., John Wiley & Sons, 2009.

[2] *Introduction to Surface Chemistry and Catalysis*, Gabor A. Somorjai, John Wiley & Sons, 1994.

[3] W.C. Bigelow, D.L. Pickett, W.A. Zisman, Oleophobic Monolayers 1. Films Adsorbed From Solution In Non-Polar Liquids. *J Coll Sci Imp U Tok* (1946) Vol. 1(6) pp. 513–538; R.G. Nuzzo, D.L. Allara, Adsorption of Bifunctional Organic Disulfides on Gold Surfaces. *J Am Chem Soc* (1983) Vol. 105(13) pp. 4481–4483; R. Maoz, J. Sagiv. On The Formation and Structure of Self-Assembling Monolayers A Comparative ATR-Wettability Study of Langmuir-Blodgett and Adsorbed Films on Flat Substrates and Glass Microbeads. *J Colloid Interf Sci* (1984) Vol. 100(2) pp. 465–496

[4] D. Witt, R. Klajn, P. Barski, B.A. Grzybowski, Applications, Properties, and Synthesis of Omega-Functionalized n-Alkanethiols and Disulfides – the Building Blocks of Self-Assembled Monolayers. *Curr Org Chem* (2004) Vol. 8(18) pp. 1763–1797.

[5] J.C. Love, L.A. Estroff, J.K. Kriebel, R.G. Nuzzo, G.M. Whitesides, Self-Assembled Monolayers of Thiolates on Metals as a Form of Nanotechnology. *Chem Rev* (2005) Vol. 105(4) pp. 1103–1169.

[6] A. Ulman, Formation and Structure of Self-Assembled Monolayers. *Chem Rev* (1996) Vol. 96(4) pp. 1533–1554.

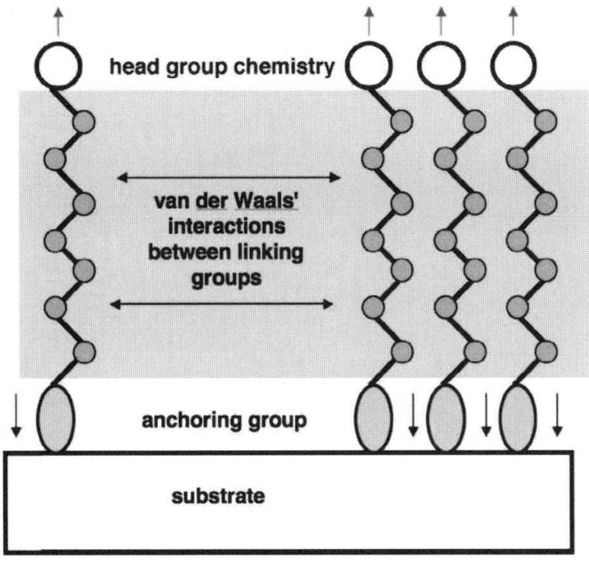

Figure 1: Diagram of the interactions involved in a self-assembled monolayer (courtesy of Dr. S. Tosatti, SuSoS AG).

phosphates[7] or phosphonates[8] (on many oxides) and catechols[9] (on transition metal oxides) (Figure 2). Self assembly of these systems occurs rapidly (typically minutes to hours), although the molecular order in the monolayer often develops with slower kinetics[10].

SAMs represent a powerful approach to covering a surface in a layer of a given functionality. In many cases moderate surface contamination is displaced by the SAM-forming reaction[5], meaning that surface functionalization with a high degree of perfection can be achieved under ambient conditions.

It is, of course possible to adsorb more than one single SAM-forming molecule (with different head groups) on a surface, in order to have more control over the precise functionality that is exposed. The *density* of the functionality can also be tuned by mixing the adsorbate containing the desired head group with an adsorbate terminated in an "inert" head group that has no function for the particular applica-

[7]D. Brovelli, G. Hähner, L. Ruiz, R. Hofer, G. Kraus, A. Waldner, J. Schlosser, P. Oroszlan, M. Ehrat, N.D. Spencer, Highly Oriented, Self-Assembled Alkanephosphate Monolayers on Tantalum(V) Oxide Surfaces. *Langmuir* (1999) Vol. 15 pp. 4324–4327.

[8]J.T. Woodward, A. Ulman, D.K. Schwartz, Self-Assembled Monolayer Growth of Octadecylphosphonic Acid on Mica. *Langmuir* (1996) Vol. 12(15) pp. 3626–3629; W. Gao, L. Dickinson, C. Grozinger, F.G. Morin, L. Reven, Self-Assembled Monolayers of Alkylphosphonic Acids on Metal Oxides. *Langmuir* (1996) Vol. 12(26) pp. 6429–6435.

[9]J.L. Dalsin, B.H. Hu, B.P. Lee, P.B. Messersmith, Mussel Adhesive Protein Mimetic Polymers for the Preparation of Nonfouling Surfaces. *J Am Chem Soc* (2003) Vol. 125(14) pp. 4253–4258; J.-Y. Wach, B. Malisova, S. Bonazzi, S. Tosatti, M. Textor, S. Zuercher, K. Gademann, Protein-Resistant Surfaces through Mild Dopamine Surface Functionalization. *Chem-Eur J* (2008) Vol. 14(34) pp. 10579–10584.

[10]D.K. Schwartz, Mechanisms and Kinetics of Self-Assembled Monolayer Formation. *Annu Rev Phys Chem* (2001) Vol. 52 pp. 107–137.

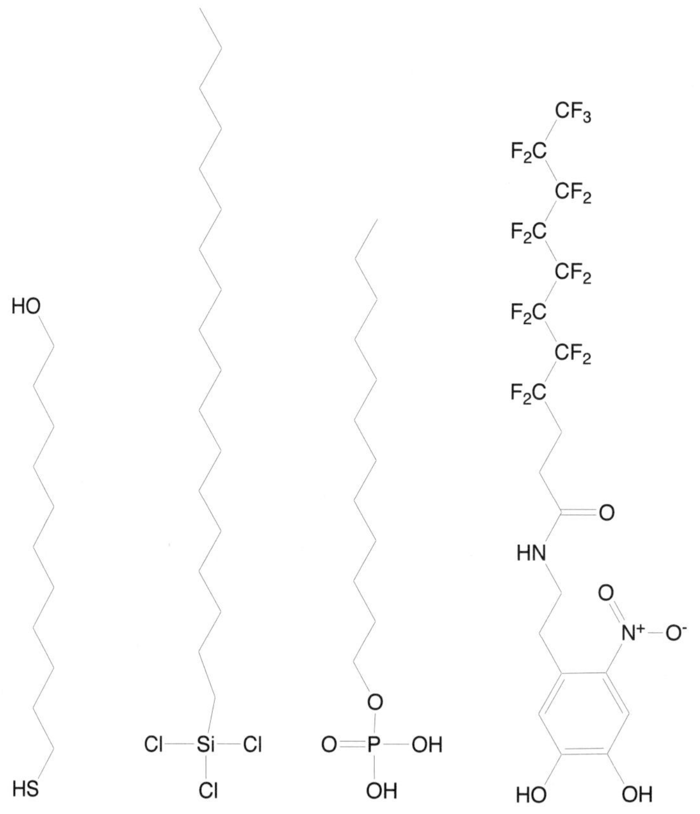

mercaptoundecanol | octadecyl trichlorosilane | dodecyl phosphoric acid | N-(4,5-dihydroxy-2-nitro-phenethyl)-2H,2H,3H,3H-perfluoro-undecanamide

a) thiols **b) silanes** **c) phosphates** **d) catechols**

Figure 2: Examples of molecules that have been used to produce self-assembled monolayers on various surfaces. a) Thiols have been used extensively to functionalize gold, silver, and other metals[5]. b) Trichloro- and trialkoxy silanes are used to functionalize many oxide surfaces[6] and are used industrially as adhesion promoters. c) Phosphates and phosphonates[7,8] and d) catechols[9] have been used to functionalize a number of transition metal oxides.

tion.[11] This dilution can also be spatially varied, leading to a concentration gradient in the functionality[12]. SAMs can also be printed in patterns on surfaces, for example by microcontact printing, which involves using an elastomeric stamp to print SAM-forming molecules onto substrates[13], or ink-jet printing, where the adsorbing

[11]C.D. Bain, G.M. Whitesides, Formation of 2-Component Surfaces by the Spontaneous Assembly of Monolayers on Gold from Solutions Containing Mixtures of Organic Thiols. *J Am Chem Soc* (1988) Vol. 110(19) pp. 6560–6561.

[12]S. Morgenthaler, C. Zink, N.D. Spencer, Surface-Chemical and -Morphological Gradients. *Soft Matter* (2008) Vol. 4 pp. 419–434.

[13]Y.N. Xia, G.M. Whitesides, Soft Lithography. *Angew Chem Int Edit* (1998) Vol. 37(5) pp. 551–575.

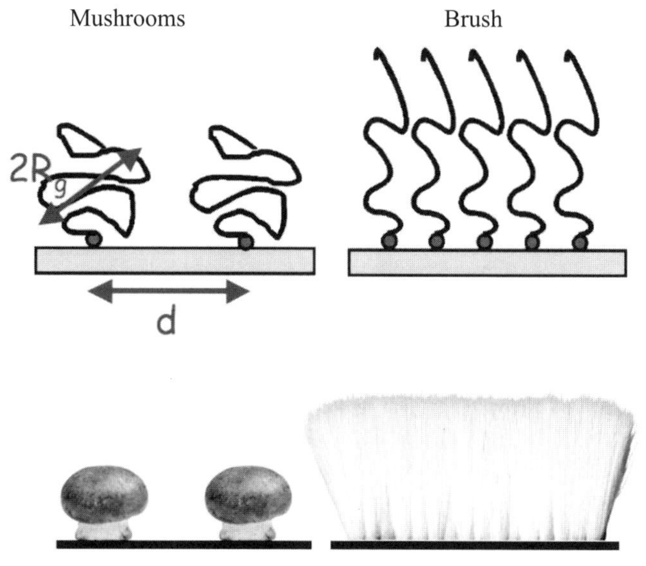

Figure 3: End-grafted polymer chains in a good solvent. If d > $2R_g$, the chains assume a conformation similar to that of free chains in solution, for which R_g is the radius of gyration. This is known as the "mushroom" conformation. If d < $2R_g$, the chains stretch out into the solvent to form a polymer "brush".

molecule is used as ink in a conventional ink-jet printer[14]. A very broad range of applications for SAMs has been reported, ranging from biosensors[15] to lubricants for microelectromechanical devices[16] and they have been crucial in the development of a number of new fields, from nanoparticles[17] to nanowires[18].

1b. Functionalizing surfaces with polymer brushes

When polymer chains are tethered to a surface in the presence of a good solvent, and are spaced closely together (separated by a distance, d, that is less than twice their radii of gyration[19], R_g, measured as the free molecule in a good solvent), they have a tendency to stretch out into the solvent in a brush-like configuration, rather than interacting closely with each other.

[14]A. Bietsch, J. Zhang, M. Hegner, H.P. Lang, C. Gerber, Rapid Functionalization of Cantilever Array Sensors by Inkjet Printing. *Nanotechnology* (2004) vol 15 pp. 873–880.
[15]N.K. Chaki, K. Vijayamohanan, Self-Assembled Monolayers as a Tunable Platform for Biosensor Applications. *Biosens Bioelectron* (2002) Vol. 17(1-2) pp. 1-12.
[16]W.R. Ashurst, C. Yau, C. Carraro, R. Maboudian, M.T. Dugger, Dichlorodimethylsilane as an Anti-Stiction Monolayer for MEMS: A Comparison to the Octadecyltrichlosilane Self-Assembled Monolayer. *J Microelectromech S* (2001) Vol. 10(1) pp. 41–49.
[17]G.H. Woehrle, L.O. Brown, J.E. Hutchison, Thiol-Functionalized, 1.5-nm Gold Nanoparticles Through Ligand Exchange Reactions: Scope and Mechanism of Ligand Exchange. *J Am Chem Soc* (2005) Vol. 127(7) pp. 2172–2183.
[18]C.N.R. Rao, G.U. Kulkarni, A. Govindaraj, B.C. Satishkumar, P.J. Thomas, Metal Nanoparticles, Nanowires, and Carbon Nanotubes. *Pure Appl Chem* (2000) Vol. 72(1-2) pp. 21–33.
[19]*Principles of Polymer Chemistry*, P.J. Flory, Cornell University Press (1953).

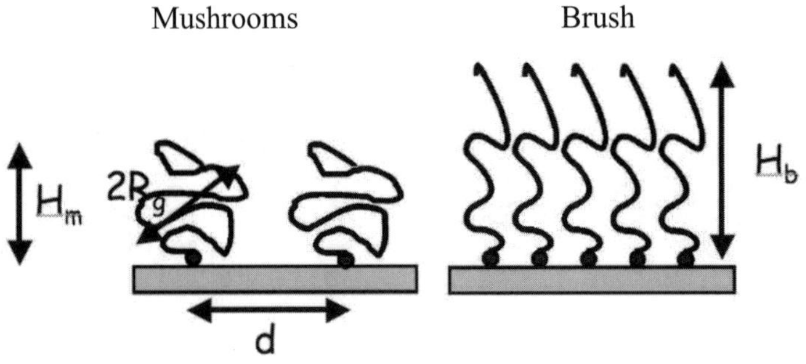

Figure 4: Scaling of polymer height from the substrate surface in both mushroom and brush conformations. The quantity $d/2R_g$ gives an indication of the degree of brush formation. In the mushroom conformation, the height, H_m is independent of the grafting density, while in a brush, the height, H_b scales as the cube root of the grafting density, σ, which is $1/d^2$.

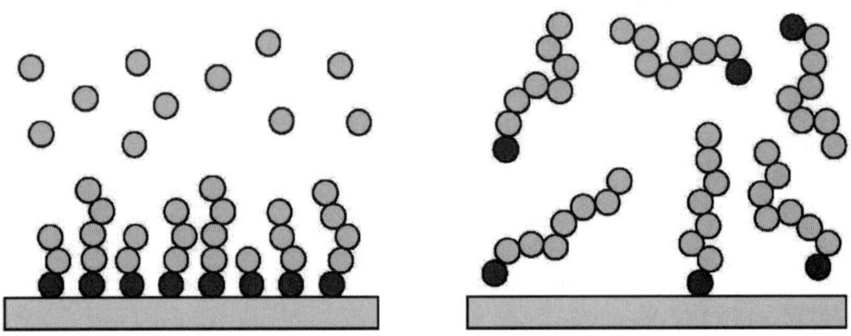

Figure 5: "Grafting-from" (left) and "grafting-to" (right) approaches for the synthesis of polymer brushes.

In other words, the free-energy cost incurred due to interaction between the chains exceeds the contribution due to the entropy elasticity of the chain (Figure 3). In the case of the widely spaced, (or "mushroom") conformation, the height of the polymer layer above the substrate is independent of the grafting density of the chains (number of attached chains per unit area). In the brush conformation, however, the height scales as the cube root of the grafting density (Figure 4)[20], σ, which itself is equivalent to $1/d^2$.[21]

A number of approaches have been described for the preparation of polymer brushes, and these can be broadly described as "grafting from" and "grafting to"[22] (Figure 5). In grafting from, an initiator is immobilized on a surface, and

[20]T. Wu, K. Efimenko, P. Vlcek, V. Subr, J. Genzer, Formation and Properties of Anchored Polymers with a Gradual Variation of Grafting Densities on Flat Substrates. *Macromolecules* (2003) Vol. 36 pp. 2448–2453.

[21]A. Halperin, M. Tirrell, T.P. Lodge, Tethered Chains in Polymer Microstructures. *Adv Polym Sci* (1992) Vol. 100 pp. 31–71; *Polymers at Interfaces*, G.J. Fleer, M.A. Cohen Stuart, J.M.H.M. Scheutjens, T. Cosgrove, B. Vincent, Chapman & Hall, London, 1993.

[22]*Polymer Brushes: Synthesis, Characterization, Applications*, R.C. Advincula, W.J. Brittain, K.C. Caster, J. Rühe, Wiley-VCH, 2004.

a polymerization reaction then takes place, with the polymer chain growing out from the surface. A number of different synthetic approaches have been taken, including atom-transfer radical polymerization (ATRP)[23] and reversible addition-fragmentation chain transfer (RAFT)[24]. The grafting-from approach has the advantage of producing a high density of polymer chains on the surface (high σ), but the disadvantage for applications that synthetic chemistry needs to be performed on the surface, whenever a brush is needed.

The grafting-to approach involves the adsorption of ready-synthesized, end-functionalized polymers onto a surface, for example using thiol-modified PEG on a gold surface[25]. While this approach is much more straightforward, once the end-functionalized polymer has been synthesized, it generally does not yield the high values of σ that are seen with grafting-from approaches, due to the steadily increasing steric inhibition of adsorption — as the coverage increases — by already-adsorbed polymer chains. An improvement in the σ obtainable by grafting-to methods can be achieved via the use of graft copolymers, where multiple brush-forming chains are grafted onto a backbone, to form a bottle-brush-like molecule. The backbone can then provide the adhesion to the surface[26] via coulombic, hydrophobic, or covalent interactions, for example. Many examples of this approach will be found throughout this book.

Polymer brushes have a number of fascinating properties that lead to many important applications including biocompatibility, colloidal stabilization, chromatography and control of wetting phenomena (see Chapter 2b). As with self-assembled monolayers, they represent a convenient way of covering a surface with a particular functionality, be it the polymer chains themselves, which may have useful properties related to adhesion or friction, or an end-functionalization on the brush that may be used for subsequent reactions. An example of the former is the widespread use of poly(ethylene glycol) (PEG) brushes to inhibit protein adsorption[27]; an example of the latter is the use of biotin-end-functionalized PEG brushes to serve as a platform for the immobilization of biomolecules for cells via integrin receptors[28], paving the

[23]M. Ejaz, S. Yamamoto, K. Ohno, Y. Tsujii, T. Fukuda, Controlled Graft Polymerization of Methyl Methacrylate on Silicon Substrate by the Combined Use of the Langmuir-Blodgett and Atom Transfer Radical Polymerization Techniques. *Macromolecules* (1998) Vol. 31(17) pp. 5934–5936.

[24]J. Chiefari, Y.K. Chong, F. Ercole, J. Krstina, J. Jeffery, T.P. Le, R.T.A Mayadunne, G.F. Meijs, C.L. Moad, G. Moad, E. Rizzardo, S.H. Thang. Living Free-Radical Polymerization by Reversible Addition-Fragmentation Chain Transfer: The RAFT Process. *Macromolecules* 31(16) 5559–5562, 1998.

[25]M. Himmelhaus, T. Bastuck, S. Tokumitsu, M. Grunze, L. Livadaru, H.J. Kreuzer, Growth of a Dense Polymer Brush Layer from Solution. *Europhys Lett* (2003) Vol. 64(3) pp. 378–384.

[26]G.L. Kenausis, J. Vörös, D.L. Elbert, N.P. Huang, R. Hofer, L. Ruiz, M. Textor, J.A. Hubbell, N.D. Spencer, Poly(L-Lysine)-g-Poly(Ethylene Glycol) Layers on Metal Oxide Surfaces: Attachment Mechanism and Effects of Polymer Architecture on Resistance to Protein Adsorption. *J. Phys. Chem. B* (2000) Vol. 104(14) pp. 3298–3309.

[27]S.I. Jeon, J.H. Lee, J.D. Andrade, P.G. De Gennes, Protein Surface Interactions in the Presence of Polyethylene Oxide. 1. Simplified Theory. *J Colloid Interf Sci* (1991) Vol. 142(1) pp. 149–158

[28]N.P. Huang, J. Vörös, S.M. De Paul, M. Textor, N.D. Spencer, Biotin-Derivatized Poly(L-Lysine)-g-Poly(Ethylene Glycol): A Novel Polymeric Interface for Bioaffinity Sensing. *Langmuir* (2002) Vol. 18(1) pp. 220–230.

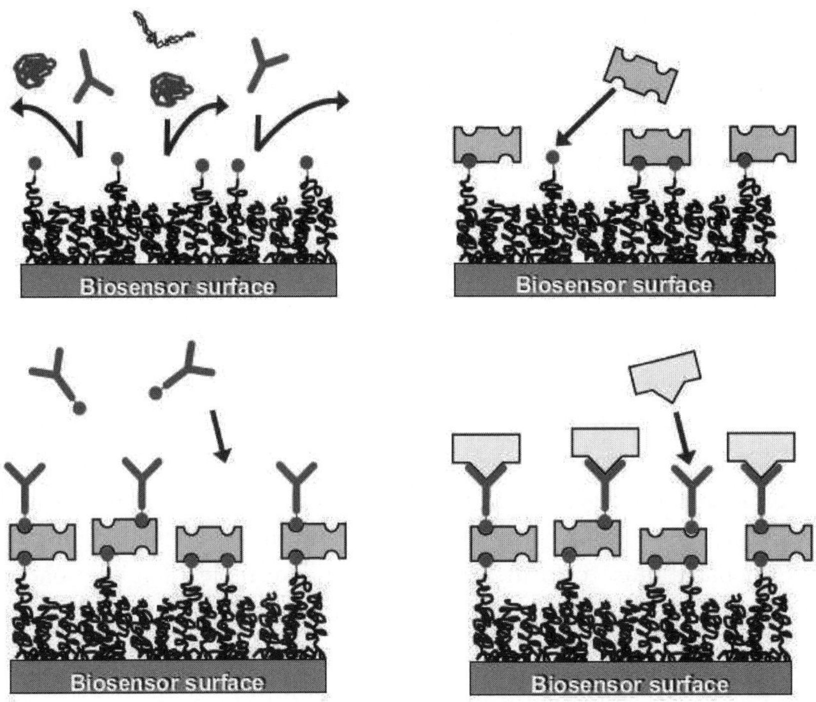

Figure 6: Schematic of the different steps in a model immunoassay. Upper left: a surface coated with a PEG brush, partially terminated with biotin, resists non-specific protein adsorption; upper right: the biotinylated PEG chains specifically adsorb streptavidin; lower left: an antibody attaches to the streptavidin and serves as a capture molecule; lower right: the target molecule binds to the biotinylated antibody through antibody-antigen interactions. Reproduced from Reference 28 with kind permission.

way for proteomics and immunoassay applications (Figure 6). A significant advantage of this approach is that the immobilized biomolecules, which can be used for specific protein or cell interactions, for example, are incorporated in a background of PEG, which suppresses non-specific adsorption to the surface. It is thus a very convenient way to steer the surface towards highly selective interactions with proteins and cells (see Chapter 2c).

The mechanical properties of polymer brushes are also of interest. Klein was among the first to recognize that polymer-brush-coated surfaces displayed exceedingly low friction coefficients under the appropriate good solvent[29], and explained this in terms of entropic effects that inhibit both compression and interdigitation of the brush-coated surfaces. Klein began his work with polystyrene brushes, toluene serving as a solvent, but later moved on to PEG/water systems[30]. The use of water-based brushes was significant, since it mirrors the way in which nature lubricates (see Chapter 2e). This involves the adsorption of glycoproteins to yield oligosaccharide-

[29] J. Klein, E. Kumacheva, D. Mahalu, D. Perahia, L.J. Fetters, Reduction of Frictional Forces Between Solid-Surfaces Bearing Polymer Brushes. *Nature* (1994) Vol. 370 pp. 634–636.

[30] U. Raviv, R. Tadmor, J. Klein, Shear and Frictional Interactions Between Adsorbed Polymer Layers in a Good Solvent. *J Phys Chem B* (2001) Vol. 105 pp. 8125–8134.

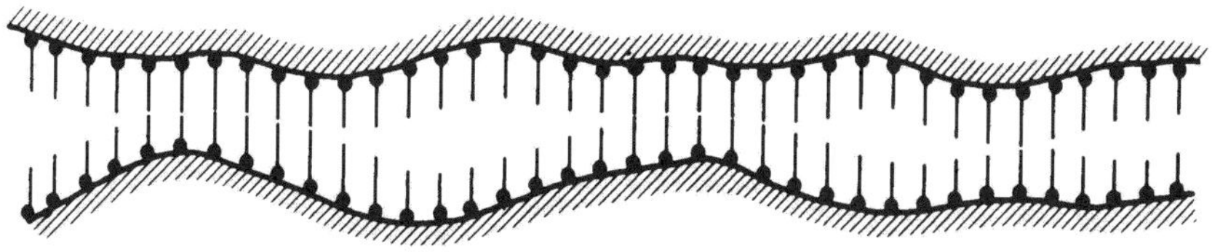

Figure 7: The adsorption of carboxylic acids onto sliding surfaces in a lubricated contact under oil, to produce a low-shear-strength, self-repairing interface. Reproduced from Reference 34 with kind permission.

covered surfaces, which bear a certain resemblance to the man-made, brush-covered systems described above[31].

1c. Using additives to modify surfaces in a self-repairing way

The traditional life cycle for coatings in most applications begins with the surface-modification process in a manufacturing environment. During use in the field, the coating suffers degradation due to processes such as wear and corrosion, until it reaches the end of its service life.

In contrast, in the case of lubricants, surface modification is frequently a continuous process that takes place as and when it is needed. A simple example is that of friction modifiers such as long-chain carboxylic acids, which are often used in motor oils, in order to reduce friction under conditions that are incapable of producing a hydrodynamic lubricating film. According to the well-established model that originates with Hardy[32], the additive molecules adsorb onto the opposed metallic or oxidic surfaces in monolayers (Figure 7), providing a low-shear-strength interface between the hydrocarbon chains, at which sliding can readily occur. This leads to a reduction in friction. Wear of the adsorbed layer does occur to a small extent, but this is rapidly compensated by re-adsorption from the additive-containing oil. Thus the protective system is self repairing. An alternative mechanism involving metal-soap formation appears to occur under certain circumstances, depending on the local water concentration[33].

In the case of antiwear additives and the so-called "extreme-pressure" additives, the situation is more complicated; the crucial reaction that leads to the production of a wear-protective layer may occur only in the case of high pressure and/or locally high temperatures. The precise modes of action of such additives, which are present in virtually every lubricating oil, have been the topic of intensive research for several decades (see Chapter 2d). A significant incentive for such research has been the search for alternative additives that are less damaging to the pollution-control devices on modern automobiles. The commonly used zinc dialkyldithiophosphates

[31] S. Lee, N.D. Spencer, Sweet, Hairy, Soft, and Slippery. *Science* (2008) Vol. 319 pp. 575–576.

[32] W.B. Hardy, I. Bircumshaw, Boundary Lubrication, Plane Surfaces and the Limitations of Amontons' Law. *Proc. R. Soc. Lond.* (1925) Vol. A108 pp. 1–27.

[33] M. Ratoi, V. Anghel, C.H. Bovington, H.A. Spikes, Mechanisms of Oiliness Additives. *Tribology Int.* (2000) Vol. 33 pp. 241–247.

(ZnDTPs) have been shown to react, under high tribological stress, to produce a glassy layer on the sliding surfaces, effectively protecting the underlying steel[34,35]. Worn areas of this layer are replaced by further surface-reactions of the additive, which is always present in the oil solution. Not only are the layers formed in this way self-repairing, but they also seem to display some behavior that can be described as "responsive": Nanoindentation measurements suggest that the higher the load to which the layers are subjected in tribological tests, the harder — and thus the more protective — the layer becomes[36].

An alternative method for lowering friction is the surface attachment of brush-forming polymer chains, as described in Chapter 1b, and a number of different approaches have been taken to realizing this goal. One method is to synthesize brush-forming molecules that spontaneously and reversibly adsorb to the surface from solution, i.e. when dissolved in the lubricant, they can function as friction-reducing additives in a similar way to that described above for the carboxylic acid additives[37]. These surface-grafted polymers are removed from the surface under extreme mechanical stress, but are then replaced by identical molecules that are present in solution. Thus, such brush systems can also be self-healing, provided that the time constants for removal are slower than those for diffusion to the surface and readsorption. A relatively weak interaction with the surface is actually an advantage in this case, since it ensures that the polymers are removed intact from the surface, leaving behind empty sites for replacement[38].

1d. Structure: a new dimension to surface tailoring

Chemical effects of structure

While the chemical composition of a surface is clearly important in determining its properties, the topographical structure of surfaces, from the atomic up to the micrometer scale, can have an additional, but in many cases very significant, effect on surface behavior.

On the atomic level, this is clearly observed in many catalytic reactions on metal surfaces, where the crystal face of the metal concerned can greatly influence the catalyzed rate of reaction (see Chapter 3a). One of the most significant effects of

[34] A.J. Gellman, N.D. Spencer, Surface Chemistry in Tribology. *P I Mech Eng J-J Eng* (2002) Vol. 216 pp. 443–461.
[35] H. Spikes. The History and Mechanisms of ZDDP. *Tribology Letters* (2004) Vol. 17(3) pp. 469–489.
[36] S. Bec, A. Tonck, J.M. Georges, R.C. Coy, J.C. Bell, G.W. Roper. Relationship Between Mechanical Properties and Structures of Zinc Dithiophosphate Anti-Wear Films. *P Roy Soc Lond A Mat* (1999) Vol. 455 (1992) pp. 4181–4203.
[37] S. Lee, M. Müller, M. Ratoi-Salagean, J. Vörös, S. Pasche, S.M. De Paul, H.A. Spikes, M. Textor, N.D. Spencer. Boundary Lubrication of Oxide Surfaces by Poly(L-Lysine)-g-Poly(Ethylene Glycol) (PLL-g-PEG) in Aqueous Media. *Tribology Letters* (2003) Vol. 15(3), pp. 231–239.
[38] S. Lee, M. Müller, R. Heeb, S. Zürcher, S. Tosatti, M. Heinrich, F. Amstad, S. Pechmann, N.D. Spencer. Self-Healing Behavior of a Polyelectrolyte-Based Lubricant Additive for Aqueous Lubrication of Oxide Materials. *Tribology Letters* (2006) Vol. 24(3), pp. 217–223.

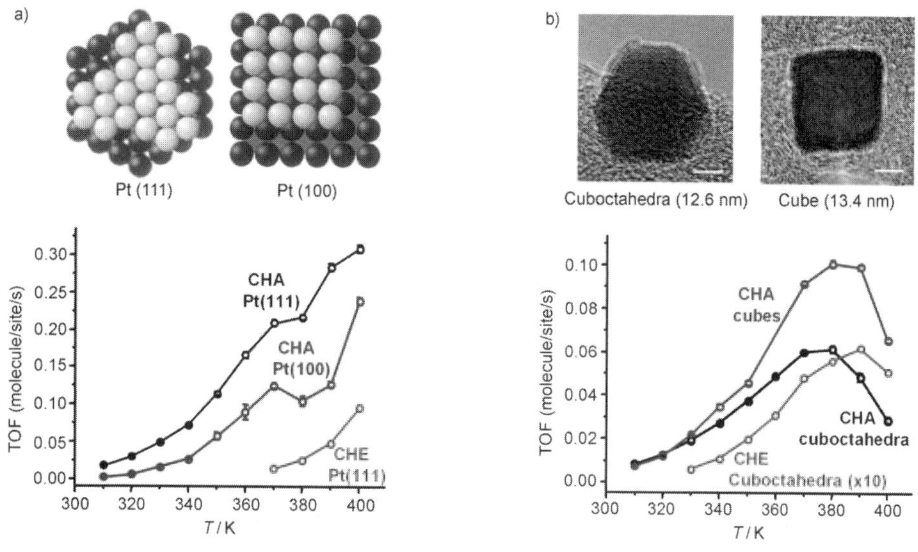

Figure 8: Turnover rates of cyclohexene and cyclohexane formation during the hydrogenation of benzene on Pt(111) and Pt(100) surfaces a), as well as the cubic and cuboctahedra platinum nanoparticles b), which demonstrate the similarities between the single-crystal and nanoparticle surfaces. CHA=cyclohexane, CHE=cyclohexene. Reproduced from Reference 42 with kind permission.

this kind is seen in the iron-catalyzed synthesis of ammonia, where more than a 400-fold difference in catalytic activity can be seen between the different crystal faces of iron[39]. This behavior can be ascribed to the presence of atoms with different coordination numbers in different crystal faces, those with seven coordinated atoms being only revealed in particularly open crystal faces, such as Fe(111) and constituting the most active sites for both nitrogen adsorption and ammonia synthesis[40].

In the case of benzene hydrogenation to cyclohexane or cyclohexene on platinum single crystals, a change in selectivity is observed from Pt(111), which catalyzes the formation of both products, to Pt(100), where only cyclohexane is produced. This effect is due to the structure sensitivity of the cyclohexene-forming reaction, while the cyclohexane-forming reaction is structure insensitive. Interestingly, the same effect can be observed on nanoparticles of different shapes; cuboctahedral particles, which contain (111) faces, catalyze the production of both products, whereas cubic nanoparticles (with (100) faces) produce only cyclohexane (Figure 8)[41,42].

[39]N.D. Spencer, R.C. Schoonmaker, G.A. Somorjai. Structure Sensitivity in the Iron Single-Crystal Catalyzed Synthesis of Ammonia. *Nature* (1981) Vol. 294 (5842) pp. 643–644.

[40]N.D. Spencer, R.C. Schoonmaker, G.A. Somorjai. Iron Single-Crystals as Ammonia-Synthesis Catalysts — Effect of Surface-Structure on Catalyst Activity. *J Catal* (1982) Vol. 74(1) pp. 129–135.

[41]K.M. Bratlie, H. Lee, K. Komvopoulos, P. Yang, G.A. Somorjai. Platinum Nanoparticle Shape Effects on Benzene Hydrogenation Selectivity, *Nano Lett.* (2007) Vol. 7 pp. 3097–3101.

[42]G.A. Somorjai, J.Y. Park. Molecular Factors of Catalytic Selectivity. *Angew. Chem. Int. Ed.* (2008) Vol. 47 pp. 9212–9228.

a) b)

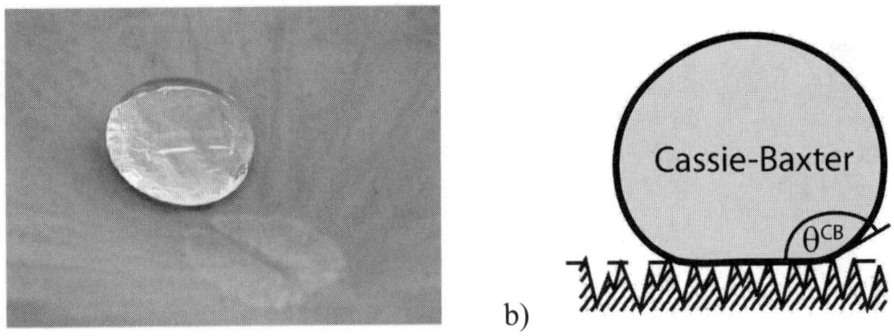

Figure 9: a) A drop of water on a lotus-leaf surface, b) "Cassie & Baxter" state of a droplet on a hydrophobic, rough surface. Note the air enclosed under the drop. Pictures courtesy of Doris Spori, ETH Zurich.

Physical effects of structure

In addition to its influence on surface reactivity, surface structure is also seen to affect wettability on the micrometer scale, as is best illustrated by the "lotus effect"[43] (see Chapter 3b). The lotus leaf is superhydrophobic, i.e. has a water contact angle of about 160°, thanks to the combination of the waxes on the surface with a characteristic dual micrometer- and nanometer-scale surface topography. Without the structure, the wax chemistry would only impart mild hydrophobicity to the surface. Superhydrophobicity comes about only when a water droplet is in contact with a rough surface with a substantial enclosure of air beneath the drop (Figure 9). This is the so-called "Cassie-Baxter" state, named after the authors of the work[44] that described the contact angle of water droplets in this state by means of the equation:

$$\cos(\theta^{CB}) = f_1^* \cos(\theta^Y) - f_2$$

where f_1 and f_2 are fractions of the drop area in contact with the surface and with air, respectively, and θ^Y is the contact angle on a flat surface of the same chemistry. Figure 10 illustrates the effect on the contact angle, of varying the f_1 parameter while maintaining constant surface chemistry[45].

Superhydrophobic surfaces such as the lotus leaf have a particularly interesting property of being "self-cleaning", since water rolling over the surface tends to remove traces of dirt. This phenomenon lends itself to a number of applications, ranging from self-cleaning textiles to self-cleaning buildings.

Adhesion and friction are also dependent on surface structure in the sense of roughness, although the effect can be difficult to predict. Frictional forces often contain an adhesive component, which tends to decrease as roughness increases, but

[43]W. Barthlott, C. Neinhuis. The Purity of Sacred Lotus or Escape From Contamination in Biological Surfaces, *Planta* (1997) Vol. 202 pp. 1–8.

[44]A.B.D. Cassie, S. Baxter. Wettability of Porous Surfaces, *Trans. Faraday Soc.* (1944) Vol. 40 pp. 546–550.

[45]D.M. Spori, T. Drobek, S. Zürcher, N.D. Spencer. Cassie-State Wetting Investigated by Means of a Hole-to-Pillar-Density Gradient. *Langmuir* (2010) Vol. 26(12) pp. 9465-9473.

Figure 10: Drops of water in contact with poly(dimethyl siloxane) surfaces of varying hole/pillar densities (shown beneath the drops), corresponding to different fractions of air enclosed beneath the drops (f_2), from Reference 45 with kind permission.

friction itself is also somewhat roughness dependent, especially when interlocking effects[46] and conformal contacts of elastomers[47] are involved. The effects are familiar to us in everyday life, but the science remains poorly understood.

In biological adhesion, roughness also plays an important role. For example, roughness is routinely used to enhance cell adhesion to titanium implants that are designed to integrate with bone, such as those in hip-joint or tooth replacements. However, it is also clear that roughness does not affect the adhesion of all cells in a similar manner[48], and the biochemical aspects of cell responses to roughness remain a much-explored research topic (see Chapter 3c).

Thus, surface structure is an important factor in determining surface properties, and represents an additional parameter that can be varied in order to tailor material surfaces for a particular application. Much research remains to be done in this area.

1e. Spatial distributions on surfaces: from patterns to gradients

There are many reasons why one might want to pattern a surface, chemically or morphologically. Patterns can be used to highlight the contrasting behavior of different surface-chemical modifications, in terms of their chemical reactions or interaction with living organisms, for example. This approach forms the basis of many diagnostic methods, such as gene or protein chips, where different biomolecules are distributed across a surface and their interactions with analytes are followed by fluorescence, for example. Patterned surfaces can also be used to explore biological

[46]P.V.K. Porgess, H. Wilman. The Dependence of Friction on Surface Roughness. *Proc. Roy. Soc. A* Vol. 252 (1959), pp. 35–44.

[47]G.A.D. Briggs, B.J. Briscoe. Surface Roughness and the Friction and Adhesion of Elastomers. *Wear* Vol. 57(2) pp. 269–280.

[48]T.P. Kunzler, T. Drobek, M. Schuler, N.D. Spencer. Systematic Study of Osteoblast and Fibroblast Response to Roughness by Means of Surface Morphology Gradients. *Biomaterials* (2007) Vol. 28 pp. 2175–2182.

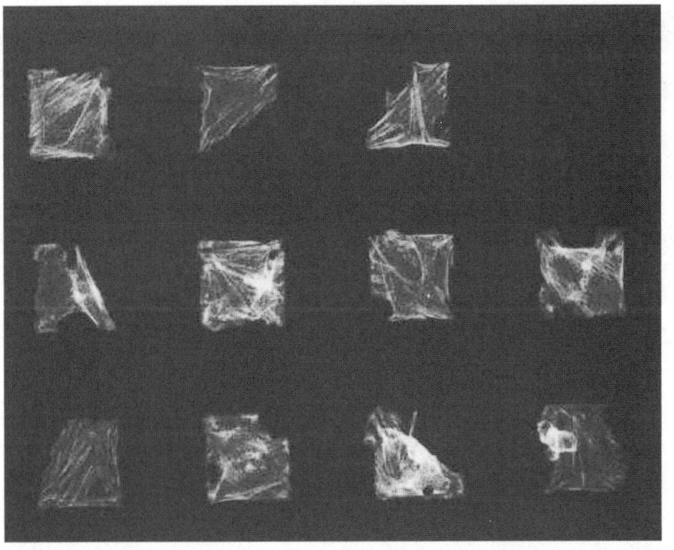

Figure 11: 60×60 μm^2 features of TiO_2 in a SiO_2 substrate, subsequently functionalized with alkane chains and poly(ethylene glycol) brushes, respectively, and then exposed to human foreskin fibroblasts. Fibroblasts are shown to spread out on the alkane chains to the border of the PEG brushes, and are visualized by immunostaining for f-actin. From Reference 50 with kind permission.

phenomena, such as neurite outgrowth on surfaces[49], and how adhesion molecules influence such processes. Morphological and chemical patterns have also been explored in terms of their influence on cell adhesion (Figure 11)[50] and alignment[51], drop dynamics[52] and wetting phenomena[53] (see Chapter 4b).

Approaches to patterning can be roughly divided into two categories: parallel and serial. Approaches such as dip-pen nanolithography (Figure 12)[54], which uses a modified atomic force microscope tip to write chemical patterns directly onto a surface, can produce features on the nanometer scale, but are inherently serial, one feature being written after another. Fabrication of a macro- or even microscale array is therefore an extremely slow process, unless multiple-tip systems are employed. Parallel approaches, such as photolithography and related techniques, or printing

[49]J.A. Hammarback, S.L. Palm, L.T. Furcht, P.C. Letourneau. Guidance of Neurite Outgrowth by Pathways of Substratum-Adsorbed Laminin. *J. Neurosci. Res.* (1985) Vol. 13(1-2) pp. 213–220.

[50]R. Michel, J.W. Lussi, G. Csúcs, I. Reviakine, G. Danuser, B. Ketterer, J.A. Hubbell, M. Textor, N.D. Spencer. Selective Molecular Assembly Patterning: A New Approach to Micro- and Nanochemical Patterning of Surfaces for Biological Applications. *Langmuir* (2002) Vol. 18(8) pp. 3281–3287.

[51]J.L. Charest, M.T. Eliason, A.J. García, W.P. King. Combined Microscale Mechanical Topography and Chemical Patterns on Polymer Cell Culture Substrates. *Biomaterials* (2006) Vol. 27(11) pp. 2487–2494.

[52]H. Kusumaatmaja, J. Léopoldès, A. Dupuis, J.M. Yeomans. Drop Dynamics on Chemically Patterned Surfaces. *Europhys. Lett.* (2006) Vol. 73(5) pp. 740–746.

[53]M. Morita, T. Koga, H. Otsuka, A. Takahara. Macroscopic-Wetting Anisotropy on the Line-Patterned Surface of Fluoroalkylsilane Monolayers. *Langmuir* (2005) Vol. 21 pp. 911–918.

[54]R.D. Piner, J. Zhu, F. Xu, S. Hong, C.A. Mirkin. "Dip-Pen" Nanolithography. *Science* (1999) vol 283 pp. 661–663.

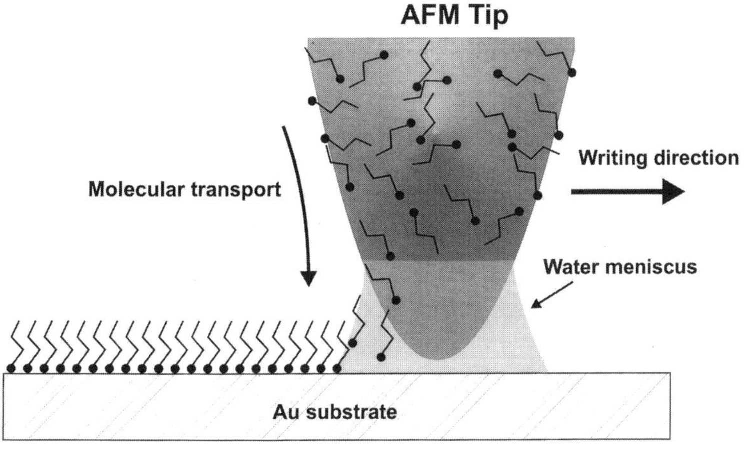

Figure 12: Principle of "dip-pen nanolithography", whereby SAM-forming molecules are transported to a specific position with nanometer resolution, by means of an AFM tip. The molecules are transported within the water meniscus at the tip-substrate interface. From Reference 54 with kind permission.

Figure 13: A gradient in hydrophobicity, formed by slow dipping of a gold-covered substrate into a dilute alkanethiol solution, and backfilling with a complementary, OH-terminated thiol. From Reference 60 with kind permission.

processes, such as microcontact printing (Figure 13)[55,56] or inkjet printing[57], have much greater potential for patterning large areas, but generally are not suited to nanometer-scale resolution.

Recently, not only molecules, but also particles, have been patterned onto surfaces, by means of a silicone-rubber-based printing step, opening the way to a patterning approach for the fabrication of submicron-scale devices[58].

Patterning leads to a set of discrete, chemically or morphologically distinct regions on a surface. Sometimes it is more useful to fabricate surfaces where a particular property changes gradually as a function of its spatial location. Such systems

[55] G.M. Whitesides, E. Ostuni, S. Takayama, X. Jiang, D.E. Ingber. Soft Lithography in Biology and Biochemistry. *Ann. Rev. Biomed. Eng.* (2001) Vol. 3 pp. 335–373.

[56] A.P. Quist, E. Pavlovic, S. Oscarsson. Recent Advances in Microcontact Printing. *Anal. Bioanal. Chem.* (2005) Vol. 381 pp. 591–600.

[57] N.E. Sanjana, S.B. Fuller. A Fast Flexible Ink-Jet Printing Method for Patterning Dissociated Neurons in Culture. *J. Neurosci. Meth.* (2004) Vol. 136(2) pp. 151–163.

[58] T. Kraus, L. Malaquin, H. Schmid, W. Riess, N.D. Spencer, H. Wolf. Nanoparticle Printing with Single-Particle Resolution. *Nature Nanotechnology* (2007) Vol. 2 pp. 570–576.

are known as surface gradients, and are finding increasing application in a number of different areas[59] (see Chapter 4a). On the one hand, gradients allow experiments to be carried out very rapidly as a function of a particular surface property, which is changed along a spatial dimension of the sample. This has the distinct advantage that all other conditions remain the same while this property is being explored. On the other hand, certain dynamic phenomena, such as cell mobility, can be explored as a function of gradient slope. Not only has chemical functionality (leading e.g. to hydrophobicity[60] (Figure 13) or charge gradients) been explored in this context, but so also have polymer properties (block structure, molecular weight)[61] and surface roughness have also been incorporated into gradients. A large number of different methods have been developed in recent years for the fabrication of chemical and morphological gradients, and their application as combinatorial research tools is now widespread.

[59]S. Morgenthaler, C. Zink, N.D. Spencer. Surface-Chemical and -Morphological Gradients. *Soft Matter* (2008) Vol. 4 pp. 419–434.

[60]S. Morgenthaler, S. Lee, S. Zürcher, N.D. Spencer. A Simple, Reproducible Approach to the Preparation of Surface-Chemical Gradients. *Langmuir* (2003) Vol. 19(25) pp. 10459–10462.

[61]J. Genzer, R.R. Bhat. Surface-Bound Soft Matter Gradients. *Langmuir* (2008) Vol. 24(6) pp. 2294–2317.

CHAPTER 2
Chemical Modification of Surfaces

2a. Self-assembled monolayers: new approaches

Commentary

As described in the introduction, self-assembled monolayers represented a breakthrough in surface functionalization, in that they permitted surfaces to be modified under ambient conditions in a reproducible manner. Efforts in our group have focused on extending the utility of this approach, both to new functionalities and to new functionalization chemistries that attach to a wide variety of substrates.

Our first foray in this direction, in 1996, was to collaborate with the renowned ETH sugar chemist, Andrea Vasella, who was synthesizing short oligosaccharides terminating in sulfur species. By using a thiol group attached to a hexasaccharide as an anchor onto gold, we were able to create a high surface concentration of sugars, creating a potential platform for sugar-based experiments in biorecognition, tribology, and other areas. The sugars were adsorbed in both free and protected form, leading to different, but controllable surface coverages. The adsorbed protected hexasaccharides could be deprotected on the surface — an early example of carrying out organic reactions on surfaces (2.1).

Alkanephosphates were of great interest to us, since they seemed to provide a means to functionalize oxide surfaces with monolayers analogously to the way in which thiols functionalize gold. Unlike other oxide-functionalizing methods such as silanes, however, they reliably produced monolayers. We performed a number of studies in this area, including detailed surface characterization of the monolayers adsorbed from organic solvents (2.2, 2.3), the use of ammonium salts of the phosphates to adsorb the layers from aqueous solutions (2.4), the use of the phosphates for the functionalization of titanium, and the influence of the roughness on that system (2.5), as well as the effect of chain-length on the ordering behavior of the alkanephosphates (2.6).

While alkanephosphates were found to be useful in a number of applications, the search continued for monolayer-forming systems that were effective on a variety of substrates and stable for extended periods in biological milieux. Contacts with Phil Messersmith's group (Northwestern University) led us to begin exploring catechol-based systems, and we found we were able to use nitrodopamine-based anchoring groups to produce robust monolayers and surface-chemical gradients with a number of useful functionalities (2.8).

In the area of tribology, we had unsuccessfully tried, in the mid 1990s, to explore the relationship between nanoscale (AFM) and macroscale (pin-on-disk) friction measurements, by the use of thiol-based self-assembled monolayers on gold. While being useful models in nanotribological experiments, the monolayers were found, despite our best efforts, to wear away very rapidly against a silicon counterface, under loads of the order of a Newton. The solution to this problem came to us a decade later. If a soft material, such as a silicone rubber, is used as a counterface, the local pressure and shear stress becomes much lower, thereby preserving the integrity of the monolayer during sliding. The larger contact area with the elastomer also has the advantage that measurements with moderate spatial resolution, such as infrared reflection-absorption spectroscopy, can be made in the wear track, allowing observations of molecular orientation and order to be made in the monolayer, both before and after tribological stress (2.7).

6074 *Langmuir* **1996,** *12,* 6074–6082

Self-Assembled Hexasaccharides: Surface Characterization of Thiol-Terminated Sugars Adsorbed on a Gold Surface

Michaela C. Fritz, Georg Hähner,* and Nicholas D. Spencer*

Laboratory for Surface Science and Technology, Department of Materials, ETH-Zürich, Sonneggstrasse 5, CH-8092 Zürich, Switzerland

Roland Bürli and Andrea Vasella

Laboratory of Organic Chemistry, Department of Chemistry, ETH-Zürich, Universitätsstrasse 16, CH-8092 Zürich, Switzerland

Received June 24, 1996. In Final Form: August 28, 1996®

A thiol-terminated hexasaccharide, protected with acetyl groups (AHS) was synthesized for the purpose of depositing self-assembled monolayers (SAMs) from solution onto gold surfaces. X-ray photoelectron spectroscopy (XPS), ellipsometry, contact angle measurements, and imaging time-of-flight secondary ion mass spectroscopy (iToF-SIMS) were used to determine coverage, homogeneity, chemical composition, film thicknesses, and kinetics of film growth. Deprotection of the molecules, i.e. replacing acetoxy groups by hydroxyl groups, was performed following adsorption of AHS onto the surface, as well as prior to adsorption from solution. The chemical composition of the resulting films, the film thickness, the density of molecules, and the nature of the surface functional groups were determined. Adsorption of the deprotected molecules (DHS) from solution was found to lead to a higher density of adsorbed species.

Introduction

Self-assembled monolayers (SAMs) have been an active research area for nearly two decades. A large number of studies has concerned thiol-terminated molecules that adsorb spontaneously from solution onto gold, silver, and copper surfaces, establishing self-assembled monolayer films.[1] Much of this research has been directed at the preparation of tailored organic surfaces; their importance has been steadily increasing in various applications, ranging from the construction of biosensors to the development of special electronic devices.[1] Films of ω-functionalized alkanethiols have facilitated fundamental studies of interfacial phenomena, such as adhesion,[2,3] corrosion protection,[4] electrochemistry,[5] wetting,[6] protein adsorption,[7,8] and molecular recognition.[9–13] The last two areas are of fundamental interest for biological applications. An understanding of the mechanism of protein adsorption, the interaction of proteins with "artificial" substrates, and the way in which these interactions determine the biological activity of these substrates is of immense biomedical significance.[14–18]

Self-assembled monolayers play a particularly important role here, since they can serve as models of polymer surfaces, allowing surface chemical properties to be investigated independent from the effects of surface morphology. In this way, macroscopic concepts such as hydrophobicity, hydrophilicity, wettability, and water content, which are crucial to understanding cell adhesion and anchorage-dependent cell behavior,[19–23] can be substituted by more fundamental, molecular-level concepts of surface organization, reactivity, and structure. Efforts have been undertaken to engineer gradients of surface hydrophobicity/hydrophilicity on polymeric surfaces[24–27] and, very recently, on SAMs prepared from thiols.[28]

Hydroxylated surfaces are of particular interest, due to the possibility of derivatizing the OH groups with biologically active moieties. The spatial arrangement and density of OH groups within a monolayer matrix are relevant, since they may regulate the accessibility of a specific functional group to biomolecules. One approach

* To whom correspondence should be addressed.

® Abstract published in *Advance ACS Abstracts,* November 15, 1996.

(1) Ulman, A. *Ultrathin Organic Films*; Academic Press: San Diego, CA, 1991.
(2) Stewart, K. R.; Whitesides, G. M.; Godfried, H. P.; Silvera, I. F. *Rev. Sci. Instrum.* **1986**, *57*, 1381.
(3) Young, J. T.; Boerio, F. J.; Zhang, Z.; Beck, T. L. *Langmuir* **1996**, *12*, 1219.
(4) Volmer, M.; Stratmann, M.; Viefhaus, H. *Surf. Interfacial Anal.* **1990**, *16*, 278.
(5) Stern, D. A.; Wellner, E.; Salaita, G. N.; Laguren-Davidson, L.; Lu, F.; Batina, N.; Frank, D. G.; Zapien, D. C.; Walton, N.; Hubbard, A. T. *J. Am. Chem. Soc.* **1988**, *110*, 4885.
(6) Bain, C. D.; Whitesides, G. M. *Langmuir* **1989**, *5*, 1370.
(7) Prime, K. L.; Whitesides, G. M. *Science* **1991**, *252*, 1164.
(8) Pale-Grosdemange, C.; Simon, E. S.; Prime, K. L.; Whitesides, G. M. *J. Am. Chem. Soc.* **1991**, *113*, 12.
(9) Häussling, L.; Michel, B.; Ringsdorf, H.; Rohrer, H. *Angew. Chem., Int. Ed. Engl.* **1991**, *30*, 569.
(10) Häussling, L.; Ringsdorf, H.; Schmitt, F.-J.; Knoll, W. *Langmuir* **1991**, *7*, 1837.
(11) Spinke, J.; Liley, M.; Schmitt, F.-J.; Guder, H.-J.; Angermaier, L.; Knoll, W. *J. Chem. Phys.* **1993**, *99*, 7012.
(12) Spinke, J.; Liley, M.; Guder, H.-J.; Angermaier, L.; Knoll, W. *Langmuir* **1993**, *9*, 1821.
(13) Schierbaum, K. D.; Weiss, T.; Thoden van Velzen, E. U.; Engbersen, J. F. J.; Reinhoudt, D. N.; Göpel, W. *Science* **1994**, *265*, 1413.
(14) Norde, W.; Lyklema, J. *J. Colloid Interface Sci.* **1979**, *71*, 350.
(15) Norde, W. *Adv. Colloid Interface Sci.* **1986**, *25*, 267.
(16) Andrade, J. D.; Hlady, V. *Ann. N. Y. Acad. Sci.* **1988**, 160.
(17) Brynda, E.; Hlady, V.; Andrade, J. D. *J. Colloid Interface Sci.* **1990**, *139*, 374.
(18) Golander, C.-G.; Lin, Y.-S.; Hlady, V.; Andrade, J. D. *Colloids Surf.* **1990**, *49*, 280.
(19) Yamada, K. M.; Kennedy, D. W. *J. Cell Biol.* **1978**, *99*, 29.
(20) Lewandowska, K.; Balachander, N.; Sukenik, C. N.; Culp, L. A. *J. Cell Physiol.* **1989**, *141*, 334.
(21) Grinnell, F.; Phan, T. V. *Thromb. Res.* **1985**, *39*, 65.
(22) Chinn, J. A.; Horbett, T. A.; Ratner, B. D.; Schway, M. B.; Haque, Y.; Hauschka, S. D. *J. Colloid Interface Sci.* **1989**, *127*, 67.
(23) Dekker, A.; Reitsma, K.; Beugling, T.; Bantjes, A.; Feijen, J.; van Aken, W. G. *Biomaterials* **1991**, *12*, 130.
(24) Maroudas, N. G. *J. Cell. Physiol.* **1977**, *90*, 511.
(25) van Wachem, P. B.; Beugeling, T.; Feijen, J.; Bantjes, A.; Detmers, J. P.; van Aken, W. G. *Biomaterials* **1985**, *6*, 403.
(26) Pratt, K. J.; Williams, S. K.; Jarrel, B. E. *J. Biomed. Mater. Res.* **1989**, *23*, 1131.
(27) Dekker, A.; Beugeling, T.; Wind, H.; Poot, A.; Bantjes, A.; Feijen, J.; van Aken, W. G. *J. Mater. Sci.* **1991**, *2*, 227.
(28) Liedberg, B.; Tengvall, P. *Langmuir* **1995**, *11*, 3821.

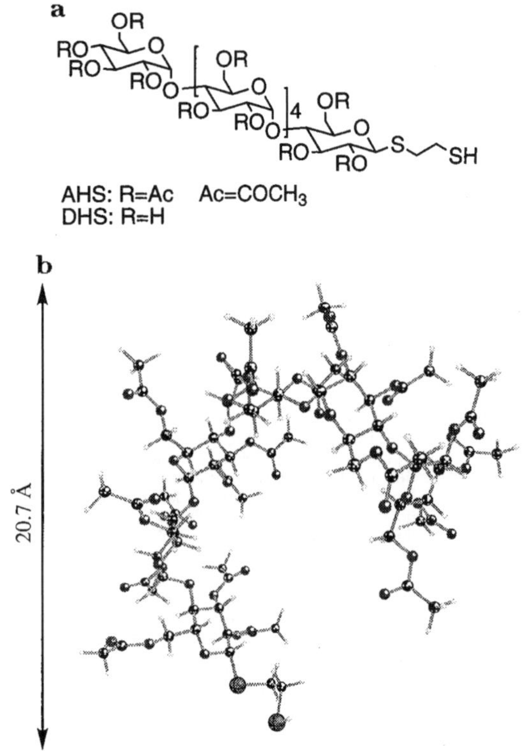

Figure 1. Schematic diagram of the hexasaccharide: (a) Chemical composition; (b) three-dimensional contorted structure of the protected molecule AHS. The conformation is likely to be close to equilibrium.

to controlling these properties uses mixed chain length CH_3- and OH-terminated alkanethiolate SAMs.[28–33]

We report on studies of a thiol-terminated hexasaccharide (AHS), corresponding to the first six units of amylose, that was synthesized in order to establish self-assembling monolayer films on gold surfaces. The acetoxy groups of the hexasaccharide, which is displayed in Figure 1, can be replaced by hydroxy groups, serving as a starting point for further transformations. We have studied the deprotection of adsorbed AHS on the surface as well as the adsorption of deprotected molecules (DHS) from solution. These species combine both hydrophobic and hydrophilic structural elements, obviating the need for adsorption from multicomponent solutions. We have used X-ray photoelectron spectroscopy (XPS), ellipsometry, contact angle measurements, and imaging time-of-flight secondary ion mass spectroscopy (iToF-SIMS) to characterize the adsorbed layers. This is essential before further optimization of these monolayers for biological applications can be undertaken.

Experimental Section

Materials. Solvents and chemicals were purchased from Fluka and Aldrich Chemical Co. and used without further purification.

Synthesis of Saccharides. *2-Mercaptoethyl O-(2,3,4,6-Tetra-O-acetyl-α-D-glucopyranosyl)-(1→4)-tetrakis[O-(2,3,6-tri-O-acetyl-α-D-glucopyranosyl)-(1→4)]-2,3,6-tri-O-acetyl-1-thio-β-D-glucopyranoside.* A suspension of (2,3,4,6-tetra-*O*-acetyl-α-D-glucopyranosyl)-(1→4)-tetrakis[*O*-(2,3,6-tri-*O*-acetyl-α-D-glucopyranosyl)-(1→4)]-1,2,3,6-tetra-*O*-acetyl-D-glucopyranoside[34] (0.400 g, 0.22 mmol), ZnI_2 (0.700 g, 2.2 mmol), and 3 Å molecular sieves (1 g) in $Cl(CH_2)_2Cl$ (14 mL, dest. CaH_2) was treated at room temperature under N_2 with bis(trimethylsilyl)ethanedithiol (842 μL, 3.3 mmol), stirred for 5 h, diluted with CH_2Cl_2, and filtered through Celite. The filtrate was washed with a 1 M HCl solution, a saturated $NaHCO_3$ solution, and H_2O, dried over $MgSO_4$, and evaporated. Flash chromatography (toluene/AcOEt, 4:1 → 1:1) gave AHS (274 mg, 67%) as a white solid. AHS was purified by HPLC (Nucleosil 5 CN 250 × 21 mm^2; hexane/AcOEt, 3:2) prior to adsorption onto gold. R_f (toluene/AcOEt, 2:3): 0.25. $[\alpha]^{25}_D$ = 115.6° (c = 0.52, $CHCl_3$). Mp 118–120°. IR: 2962w, 1753s (br), 1430w, 1370m, 1032s (br). ^{1}H-NMR (500 MHz, $CDCl_3$): 5.33–5.41 (m, 6 H); 5.27–5.30 (m, 5 H); 5.07 (t, J = 10.0, 1 H); 4.83–4.87 (m, 2 H); 4.71–4.78 (m, 4 H); 4.57 (d, J = 10.0, H–C(1)); 4.48–4.53 (m, 5 H); 4.15–4.36 (m, 6 H); 3.89–4.06 (m, 11 H); 3.70–3.75 (m, 1 H); 2.94 (td, J = 6.6, 13.5, C*H*CH_2SH); 2.73–2.83 (m, CHC*H*$_2$SH); 1.98–2.20 (several s, 19 Ac); 1.69 (t, exchange D_2O, J = 8.2 SH). ^{13}C-NMR (125 MHz, $CDCl_3$): 170.72, 170.70, 170.67, 170.65 (2×), 170.53, 170.44, 170.41, 170.35 (2×), 170.32, 169.99, 169.74, 169.65, 169.60, 169.54, 169.53, 169.47, 169.46 (17 s, 19 C=O); 95.76, 95.75 (2×), 95.74, 95.65, 83.62 (5 d, 6 C(1)); 76.27, 76.17, 73.70, 73.54, 73.37, 73.29, 72.39, 71.74, 71.70, 71.63, 71.61, 70.81, 70.51, 70.48, 70.45, 70.44, 70.06, 69.38, 69.07, 68.98 (2×), 68.97, 68.47, 67.97 (23 d, 6 C(2), 6 C(3), 6 C(4), 6 C(5)); 62.96, 62.52, 62.50, 62.38, 62.20, 61.39 (6 t, 6 C(6)); 34.44, 25.39 (2 t, CH_2-CH_2SH); 20.55–20.90 (several q, Me). FAB-MS: 1853 (<1, [M + 23]$^+$), 1771 (8, [M − 59]$^+$), 1483 (<1, [M − 347]$^+$), 1423 (<1, [M − 347 − 60]$^+$), 1195 (<1, [M − 347 − 288]$^+$), 1135 (<1, [M − 347 − 288 − 60]$^+$), 907 (<1, [M − 347 − 288 − 288]$^+$), 847 (2, [M − 347 − 288 − 288 − 60]$^+$), 619 (6, [M − 347 − 288 − 288 − 288]$^+$), 559 (7, [M − 347 − 288 − 288 − 288 − 60]$^+$), 331 (26, [M − 347 − 288 − 288 − 288 − 288]$^+$), 169 (100), 108 (70). Anal. Calc for $C_{76}H_{104}O_{49}Si_2$ (1865.76): C, 48.93; H, 5.62. Found: C, 48.79; H, 5.71.

2-Mercaptoethyl Pentakis[O-α-D-glucopyranosyl-(1→4)]-1-thio-β-D-glucopyranoside. A suspension of AHS (222 mg, 0.12 mmol) in MeOH (4 mL) was treated at room temperature under N_2 with NaOMe in MeOH (200 μL, 0.34 M), stirred for 15 h (clear solution after 30 min → suspension), diluted with H_2O (4 mL), neutralized with ion exchange (H$^+$), filtered, and evaporated to give DHS (120 mg, 94%) as a white solid. DHS was purified by HPLC (Nucleosil 5 CN 250 × 21 mm^2; H_2O) prior to adsorption onto gold. IR (KBr): 3405s (br), 2927w, 1654w (br), 1155m, 1023s (br). ^{1}H-NMR (300 MHz, D_2O): 5.31–5.36 (m, 5 H); 4.42 (d, J = 9.9, H–C(1)); 3.30–3.98 (m, 36 H); 2.59–2.86 (m, SCH_2CH_2S). ^{13}C-NMR (125 MHz, D_2O): 99.79, 99.65 (3×), 99.51, 85.33 (4 d, 6 C(1)); 76.77–78.4 (several d); 69.30–73.29 (several d); 60.79, 60.45 (2×), 60.39 (3×) (3 d, 6 C(6)); 38.29, 29.42 (2 t, CH_2CH_2SH). FAB-MS: 1067 (68, [M + H]$^+$), 297 (100).

Sample Preparation. Gold slides were prepared by evaporation of ca. 2000 Å of gold (99.99%, Balzers, Liechtenstein) with a deposition rate of ca. 15 Å/s onto silicon wafers, which had previously been coated to a depth of ca. 60 Å with chromium (99.99%, Balzers, Liechtenstein) as an adhesion promoter. This procedure was carried out in a Balzers MED 010 coater, operating at a pressure of ca. 10^{-5} mbar during evaporation. After the evaporation process was finished, the chamber was flushed with argon. Glassware was cleaned with methylene chloride or methanol. The slides were kept under argon.

(29) Bain, C. D.; Whitesides, G. M. *J. Am. Chem. Soc.* **1989**, *111*, 7164.
(30) Bain, C. D.; Evall, J.; Whitesides, G. M. *J. Am. Chem. Soc.* **1989**, *111*, 7155.
(31) Dubois, L. H.; Nuzzo, R. G. *Annu. Rev. Phys. Chem.* **1992**, *43*, 437.
(32) Bertilsson, L.; Liedberg, B. *Langmuir* **1993**, *9*, 141.
(33) Atre, S. V.; Liedberg, B.; Allara, D. L. *Langmuir* **1995**, *11*, 3882.

(34) (2,3,4,6-Tetra-*O*-acetyl-α-D-glucopyranosyl)-(1→4)-tetrakis[*O*-(2,3,6-tri-*O*-acetyl-α-D-glucopyranosyl)-(1→4)]-1,2,3,6-tetra-*O*-acetyl-D-glucopyranoside was prepared in a manner similar to a procedure reported by Sakairi et al.[35] The acetolytic cleavage of per-*O*-acetyloxy-α-cyclodextrin was performed in Ac_2O and $HClO_4$ 70% (0.06 M). The yield of the linear hexasaccharide was improved from 47% to 65%.[36] However, the separation (MPLC) of the hexasaccharide from the smaller homologues (pentasaccharide to monosaccharide) was tedious.
(35) Sakairi, N.; Wang, L.-X.; Kuzuhara, H. *J. Chem. Soc., Chem. Commun.* **1991**, 289.
(36) Rädler, U. Diploma Thesis, ETH-Zürich, 1996.

The adsorption experiments were performed by immersing the slides into the peracetylated or deacetylated saccharide solutions (10 mL). Following the immersion times indicated in the text, the slides were removed from the solution and rinsed with several milliliters of methylene chloride or in the case of the deprotected layers in a bath of methanol and methylene chloride. A third set of samples was made by adsorbing peracetylated saccharide from solution, followed by a deprotection of the adsorbed monolayers in place, on the substrates. The adsorbed monolayers were deprotected by immersing the sample in a 0.02% NaOMe/MeOH solution (10 mL), shaking for 24 h, and rinsing in a bath of methanol and methylene chloride.

The solubility of AHS in pure methanol was low, probably due to the relatively nonpolar acetoxy groups. For this reason a 1:1 mixture of methylene chloride and ethanol was used as solvent. In the deprotected form of the hexasaccharide, however, the acetoxy groups are substituted by hydroxyl groups, changing the solubility of the molecules drastically. Methanol was used as a solvent in this latter case, since the molecules were no longer soluble in the ethanol/ methylene chloride mixture.

Ellipsometry. Ellipsometric measurements were performed with a PLASMOS SD 2300 ellipsometer, equipped with a He–Ne laser (λ = 632.8 nm), at an angle of incidence of 70°. The beam diameter was ca. 1 mm. For each set of samples a bare gold sample was measured as a reference. Following determination of the optical constants of the substrate, the layer thickness was measured at 10 different points. The method for calculation of film thickness was based on a three-phase ambient/film/gold model, in which the film was assumed to be homogeneous and isotropic and was assigned a scalar refractive index of 1.45. The observed scatter in the data was typically ±3 Å, arising largely, we believe, from differences in the amount of adventitious material adsorbed on the bare gold substrates.

Contact Angle. Contact angle measurements were made using a Ramé-Hart 100–00 goniometer at room temperature and ambient humidity. Ultrapure water was used as a wetting liquid, and advancing angles, θ_a, are reported. For the measurements, 3 μL of water were put on the surface with a syringe, followed by adding another 3 μL to the first drop. Under these conditions, the contact angles were stable for several minutes.

XPS. X-ray photoelectron spectroscopy experiments were carried out using ESCA 5400 and ESCA 5600 instruments (Physical Electronics, Eden Prairie, MN), both of which were equipped with a Mg-Kα source, operating at 200 W. Survey spectra of a saccharide-covered sample were recorded both prior to and following exposure to x-ray radiation for 45 min, in order to check for any possible beam damage. No difference was found between spectra of exposed and nonexposed samples, indicating that X-ray beam damage is negligible under the conditions employed.

The energy resolution during acquisition of single-region scans was approximately 1 and 0.8 eV, on the 5400 and 5600 instruments, respectively. Electron binding energies were calibrated using the Au 4f (84.0 eV) and C 1s (284.6 eV) lines. The base pressure in the chamber was better than 10^{-8} mbar. Collected data were analyzed with a least-squares fit routine. Quantitative analysis was performed using a Shirley background subtraction method.

The ratio of the Au 4f signal from the substrate and the carbon signal from the SAMs was used to determine the thickness of the adsorbed layers. A dodecanethiol film was used as reference. Angle-resolved (AR) XPS measurements were performed on the 5600 system, while varying the angle between surface normal and analyzer from 15° to 75°.

Imaging ToF-SIMS. Imaging time-of-flight secondary ion mass spectroscopy measurements were carried out on a PHI 7200 ToF-SIMS system (Physical Electronics, Eden Prairie, MN). For performing spatially resolved experiments, this instrument is equipped with a Ga liquid metal ion (LMI) gun, which was operated at 25 keV during this study. For a probe beam size in the micrometer range it is necessary to use pulses of approximately 50 ns duration and therefore only moderate mass resolution may be achieved in this mode. The ion dose during measurement was kept in the static regime, well below the dynamic limit.

Results

General. The structure of the protected hexasaccharide AHS, displayed in Figure 1, is composed of six glucose units and a surface-active tail. Each glucose unit is protected by three acetyl groups, with the tail unit containing an additional (final) protective group. The head unit is connected via a sulfur atom to the surface-active entity, a short alkanethiol ($SH-(CH_2)_2$). In its deprotected form, all acetoxy groups are substituted by hydroxyl groups.

The molecule bears two chemically different sulfur atoms: one in the SH bond, which establishes the bonding to the gold surface, and a second one, which substitutes for an ether oxygen, connecting the hexasaccharide to the surface-active tail. The number of chemically different carbon atoms in the molecule is quite large, since within each glucose unit, six chemically distinct carbon species are to be found. The acetoxy groups contribute another two, in the case of AHS. Four different oxygen species are present in AHS and three in DHS.

Signals from C, O, and S can be detected in X-ray photoelectron spectroscopy (XPS) measurements. The binding energy difference of S in thiolate and sulfide forms can be up to approximately 2 eV,[37] and their ratio within the stoichiometric molecule is 1:1. Three carbon species are expected and can be classified as aliphatic, single-bonded C—O, and carbonyl (C=O). Their ratio is approximately 1:2:1 for AHS, while mainly C—O and only very small aliphatic amounts (C—C and C—S) should be present for DHS. In the case of oxygen, two contributions should be distinguishable, based on ether and carbonyl oxygen with a ratio of 3:2 for AHS and only a single peak (C—O) evident for DHS.

Table 1 summarizes the binding energies of the different chemical atomic species expected from the molecules and their stoichiometric ratios.

Protected Molecules. Gold substrates were immersed in a 5 μM 1:1 mixture of ethanol and methylene chloride for various periods between 2 min and 6 days. Survey and single-region spectra were recorded for these films. The carbon and oxygen peaks in the survey spectra indicate the presence of AHS on the surface. The strong Au 4f peaks are indicative of a 'thin' overlayer, i.e. the thickness is well below the mean free path of the Au 4f electrons (~35 Å).[38]

Figure 2 shows single-region scans of C 1s, O 1s, and S 2p for a sample that has been immersed for 5 min. In the C 1s spectrum (Figure 2a), three distinct species are clearly visible at binding energies of 284.9, 286.8, and 289.3 eV. The fwhm values for these peaks are 1.8 eV for the aliphatic, 1.9 eV for the single-bonded C—O, and 1.6 eV for the carbonyl carbon. The small expected chemical shift between the two single-bonded C—O species results in a single peak with a correspondingly broader fwhm. The stoichiometric ratio of the C=O and C—O species is approximately 1:2, as expected for the AHS molecule, although the aliphatic region shows a higher amount due to some residual contamination on the surface. The excess within the aliphatic carbon peak amounts to approximately 30%; i.e., the remaining contamination on a sample covered with saccharide can be estimated to be on the order of 10% of the overall carbon intensity, on the basis of the aliphatic peak in the carbon region. All samples immersed for periods of more than 2 min show similar spectra, indicating that the adsorption process is already

(37) *Handbook of X-Ray Photoelectron Spectroscopy*; Chastain, J., Ed.; Perkin Elmer Corp., Physical Electronics: Eden Praire, MN, 1992.
(38) Hansen, H. S.; Tougaard, S.; Biebuyck, H. *J. Electron Spectrosc.* **1992**, *58*, 141.

Table 1. XPS Binding Energies of the Different Elements Expected from the Molecule in its Protected Form (AHS) and Their Stoichiometric Ratios[a]

		expected distribution	observed distribution	expected position	observed position
C 1s	**CH_3**	25%		284.6	284.9
		}1.1	1.6		
	C–S	3%			
	C–O–**C**	22%		286.5	
		}1.9	2		286.8
	C–O–C=O	25%		286.9	
	C–O–**C**=O	25% }1	1	289.0	289.3
O 1s	C–**O**	61% }3	3	533.0	533.6
	C=**O**	39% }2	2	532.0	532.6
S 2p	C–**S**$^-$	50% }1	~2	162.0	162.1
	C–**S**–C	50% }1	1	~164	164.1
Au 4f				84.0	84.0

[a] For DHS only some of the indicated species are expected.

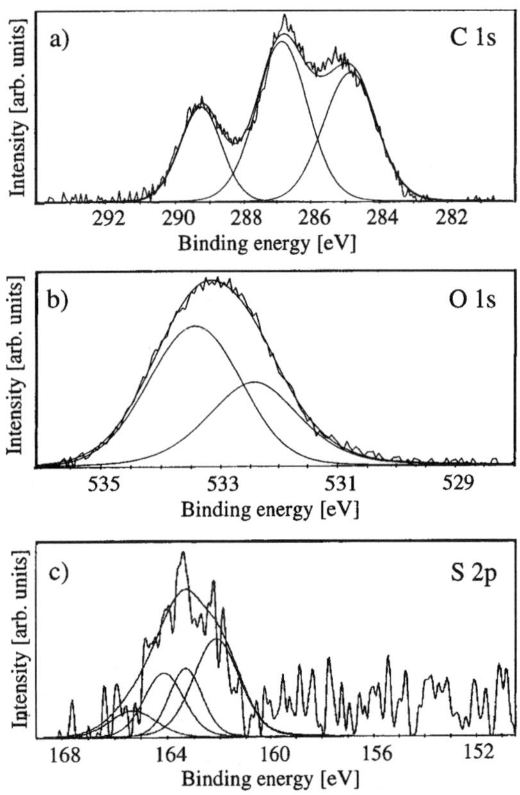

Figure 2. Single-region C 1s, O 1s, and S 2p XPS spectra of a sample covered with a 'monolayer' of AHS (immersion time, 5 min). The spectrum of the S 2p region has been slightly smoothed.

completed by this time, i.e. that saturation coverage has been reached.

The oxygen region shown in Figure 2b displays one broad and rather symmetric peak, which can be well modeled by two components, separated by approximately 1 eV, as expected for the two distinguishable oxygen species contained in the molecule, i.e. single- and double-bonded O. The binding energies are 533.6 and 532.6 eV with fwhm of 1.7 and 1.8 eV and a stoichiometric ratio of 3:2.

The S region consists of a rather broad and slightly asymmetric peak centered around 164 eV. It can be modeled by two components (each a doublet with an intensity ratio of 2:1). The S $2p_{3/2}$ and S $2p_{1/2}$ peaks at 162.1 and 163.3 eV correspond to a thiolate[39,40] while those at 164.1 and 165.3 eV are indicative of a sulfide. However, due to the low signal and large noise connected with the S region, it is hard to determine the intensity ratio of the thiolate and the sulfide unambiguously. In general, the deconvolution fits were better when higher amounts of thiolate were assumed. The absolute amount of sulfur found on the surface was found to be higher than expected in most cases.

We also measured bare gold substrates, in order to obtain information about the contamination present on the surface. Typically, aliphatic carbons at energies around 284.6 eV were found with peaks showing a shoulder at higher binding energy. As has been observed with other thiols, these species were partially removed by the adsorbing AHS.[41] Additionally, traces of oxygen were sometimes found at binding energies of 532.8 eV. However, as in the case of the C signal, these can be distinguished from the oxygen that is due to the saccharide or are negligible due their very small surface concentration.

For samples covered with a full monolayer, i.e. with immersion times above 2 min, the ratio found for C/O was typically 2.2, compared to an expected ratio of 1.6, that for C/S was 30 versus 38, and that for O/S was 13 in comparison with the theoretical value of 24.5. The largest scatter in the ratios is found for those where S is involved, i.e. C/S and O/S, due to the small signal itself and the low signal-to-noise ratio recorded for S.

Contact angle measurements performed directly after preparation of the films of the protected molecules yielded values around 55°. This value is significantly lower than that found for bare gold surfaces covered with some contamination prior to the immersion and is consistent with the displacement of the aliphatic contamination from the gold substrate by the saccharide. The contaminated surfaces typically display contact angle values around 75–80° and clean gold around 30°. The polar character of the surface increases upon the adsorption of the saccharide, as expected from the structure of the molecule.

Ellipsometry and XPS measurements were used to generate information about the film thickness of the prepared layers. The ratio of the C and Au signals from a dodecanethiol-covered Au substrate with a thickness of 16 Å was used as a reference system.[41] These films can be prepared in a highly reproducible manner. The thickness of one monolayer of AHS then corresponds to

(39) Nuzzo, R. G.; Zegarski, B. R.; Dubois, L. H. *J. Am. Chem. Soc.* **1987**, *109*, 733.

(40) Bain, C. D.; Biebuyck, H. A.; Whitesides, G. M. *Langmuir* **1989**, *5*, 723.

(41) Bain, C. D.; Troughton, E. B.; Tao, Y. T.; Evall, J.; Whitesides, G. M.; Nuzzo, R. G. *J. Am. Chem. Soc.* **1989**, *111*, 321.

20 Å, as determined by XPS.[42] With ellipsometry the same film thickness was obtained.

The amount of contamination present on the bare gold substrates before the adsorption of the saccharide, determined in a similar way, varied between 3 and 10 Å in height, assuming that the films were homogeneous. We determined this value separately for each set of samples prior to the adsorption of the saccharides.

The molecular concentration of hexasaccharide adsorbed on the surface at maximum coverage is roughly one order of magnitude lower than the corresponding amount of dodecanethiol (i.e. $\sim 5 \times 10^{13}$ cm^{-2}), as estimated from the respective C/Au ratios of samples covered with the alkanethiol and the hexasaccharide.

Angular-dependent XPS (take-off angle between 15° and 75°) measurements show that the sulfur signal increases with increasing angle, while the signal from carbon shows the opposite behavior. The variation of the oxygen signal with angle is similar to that of carbon, as expected from the structure of the molecules. This indicates that the sulfur atoms are in close proximity to the gold surface and that the glucose units of the saccharide are likely to be mainly pointing away from it, i.e. that they are "covering" the sulfur.

Kinetic Studies. To gain information about film growth and intermolecular interactions, the kinetics of film establishment was investigated with XPS and contact angle measurements. A series of samples was immersed in a 5 μM solution for various periods between 1 s and 5 min. Single-region spectra for different immersion times, recorded for the carbon region, are shown in Figure 3. Up to an immersion time of approximately 1 min the aliphatic peak shows the highest intensity. All other carbon contributions increase with increasing periods of immersion, as expected from the adsorption of the molecule. The intensities in the oxygen and sulfur regions also increase with increasing immersion time.

From the ratio of the C/Au and the O/Au signals, the coverage can be determined assuming a final film thickness of 20 Å.[42] Intermediate states therefore consist of areas that are covered with the saccharide (20 Å in height) and areas corresponding to the "bare gold" surface, i.e. where the surface is covered with contamination to an approximate depth of 7 Å. The immersion-time-dependence of the coverage, determined from the carbon region, is displayed in Figure 4, together with a fit based on Langmuir-type adsorption kinetics of the form

$$\theta(t) \propto 1 - \exp(-\lambda c t) \quad (1)$$

where λ is the rate of adsorption, c the concentration of the solution, and t the immersion time. For this type of adsorption, $d\theta/dt \propto -(1-\theta)$. The rate of change in coverage is thus simply proportional to the number of nonoccupied sites; i.e., there is no interaction between the molecules. Only contributions due to the saccharide, i.e. C=O and C—O, have been taken into consideration, and the signal from gold has been corrected in order to take into account the contamination present in the intermediate stages. The error bars shown in Figure 4 have been calculated by taking into consideration all errors from the XPS signals, the reference film thickness, and the mean free path of the electrons.

Other models have also been tested to fit the immersion-time-dependence of the coverage, on the basis of the intensities that had been measured for the C 1s region, taking into account various molecule–molecule interactions. We tried $d\theta/dt \propto -(1-\theta^2)$ to model attractive interactions and $d\theta/dt \propto -(1-\theta)^2$ to model repulsive ones, resulting in an immersion-time-dependence of the coverage of $\theta(t) \propto \tanh(\lambda c t)$ and $\theta(t) \propto 1 - 1/(\lambda c t + 1)$, respectively. However, deviation of our data from those expected from a simple Langmuir-type adsorption model is not significant in the crucial regime of $\theta \sim 0.7$, suggesting that intermolecular interactions do not play a significant role. A similar treatment of the oxygen signal showed the same time-dependence effects.

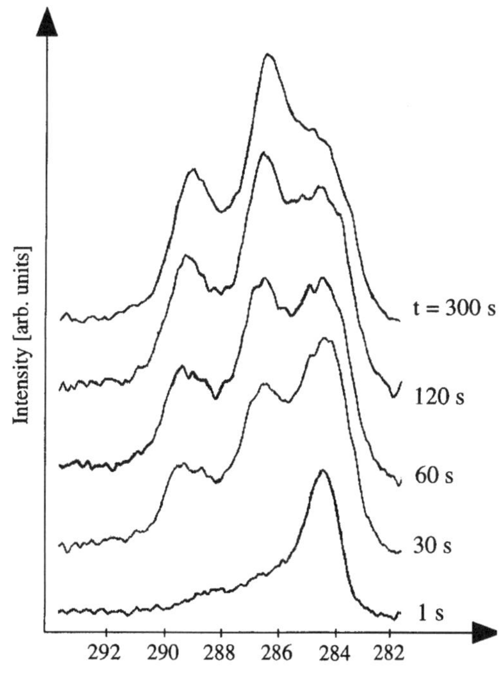

Figure 3. Single-region C 1s XPS spectra of samples that have been immersed for various periods ranging from 1 s to 5 min in a 5 μM solution of AHS. Contaminants are partially replaced by the adsorbing saccharide. The displayed spectra have been slightly smoothed.

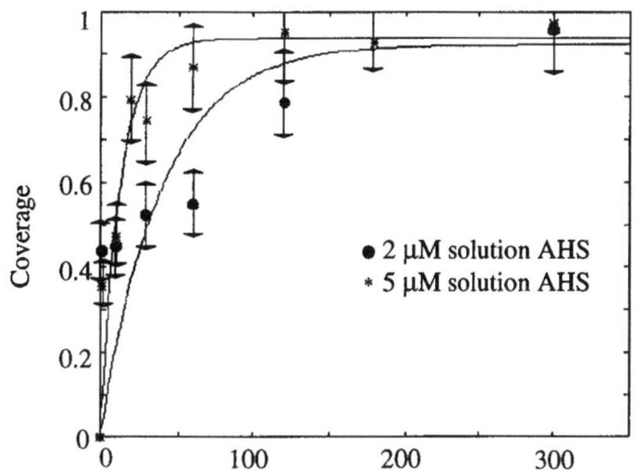

Figure 4. Experimentally determined coverages versus immersion time together with a fit according to a Langmuir-type adsorption for the adsorption of AHS. A rate constant for adsorption, λ, that is on the order of $\sim 10^7$ cm^3 mol^{-1} s^{-1} can be derived (see text for details).

(42) Ertl, G.; Küppers, J. *Low Energy Electrons and Surface Chemistry*, 2nd ed.; VCH: Weinheim, 1985.

By measuring the intensity of the XPS carbon signal at different concentrations (2 and 5 μM) as a function of immersion time, a rate constant for adsorption, λ, can be deduced from the parameters of the fit shown in Figure 4. This value corresponds to $17(\pm 5.5) \times 10^6$ cm^3 mol^{-1} s^{-1} if the Langmuir-type adsorption model (eq 1) is used. We have estimated the error of the rate of adsorption using the errors determined for the time-dependent coverages. Measurement of the oxygen region yielded a value of the same order of magnitude.

Imaging ToF-SIMS measurements were performed on samples that were covered with intermediate layers ($\theta \sim 0.7$). Mass peaks indicative of the saccharide (331 and 169 amu; see Experimental Section, Synthesis of Saccharides) and the bare gold ion at 197 amu were recorded within a square of 10×10 μm^2 and at micrometer resolution. The data revealed a uniform film growth on a micrometer scale. The uniform distribution of both saccharide and gold within this area supports our assertion that possible intermolecular interactions at the surface are weak.

Contact angle measurements on these samples as immersion time is increased show a shift from higher values (77°) to lower ones (53°), indicating the adsorption of the saccharide and partial removal of contamination with time. However, the data are not sufficiently noise-free to allow us to extract an adsorption rate constant within reasonable error bars.

Deprotected Molecules. *Deprotection at the Surface.* Following adsorption of a monolayer of AHS, the molecules at the surface were deprotected by immersing the sample in a 0.02% NaOMe/MeOH solution for 15 h, followed by extensive rinsing with MeOH. The resulting XPS single-region spectra of C, O, and S are displayed in Figure 5. The carbon region still shows a small amount of carbonyl and a larger peak due to aliphatic species. Their combined fraction of the total carbon signal amounts to ~23%. The highest contribution stems, however, from C—O, as expected. The oxygen signal also implies the presence of a small amount of residual carbonyl oxygen. The ratio of 'ether' and carbonyl oxygen is approximately 6:1. The sulfur peak is similar to that measured for AHS. The overall intensity in the S region is again higher than expected from the intact molecule. Once more, better fits in modeling the sulfur region were obtained by assigning a significantly higher signal to the thiolate species than to the sulfide.

The ratios found for C/O , O/S, and C/S are 2.4, 8.1, and 13.2 compared to the expected ratios of 1.3, 14.8, and 18.7, respectively. After correction for the contribution from the aliphatic carbon, the C/O and C/S ratios amount to 1.8 and 14.8, respectively. The signals from C and S are thus higher than those expected from pure DHS.

The film thickness, as determined with XPS and using a dodecanethiol monolayer as reference, corresponds to 16 Å and is thus slightly smaller than that of the AHS monolayer. The film thickness determined with ellipsometry was 19 Å. The contact angle after deprotection of the layer is 35°, i.e. significantly lower than the value observed for AHS.

Deprotection in Solution before Adsorption. Adsorption of the deacetylated saccharide (DHS) from solution (methanol) was performed by immersing a gold substrate for 3 h in a 40 μM solution. This higher concentration was necessary for adsorption to occur within a reasonable time. For a 5 μM solution we found that immersion times in excess of 22 h are necessary to reach saturation coverage; i.e., the rate of adsorption is dramatically lower than that of AHS. The molecules were no longer soluble in ethanol and methylene chloride.

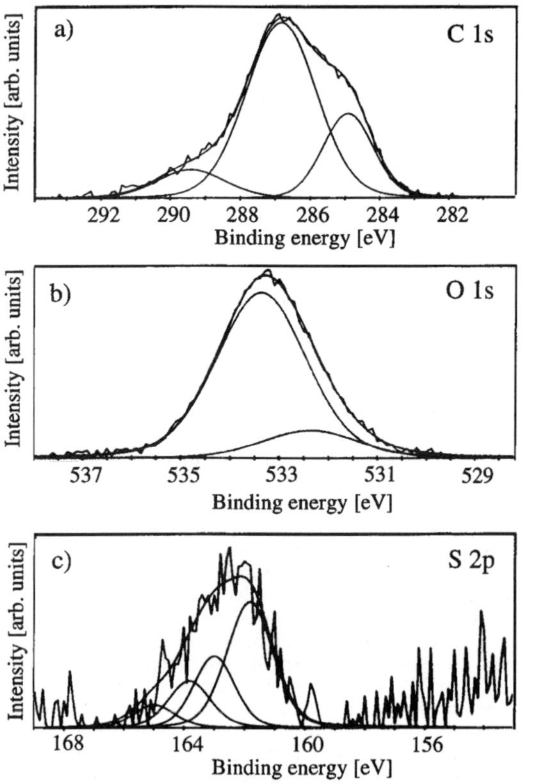

Figure 5. Single-region XPS spectra of a monolayer of AHS that has been deacetylated at the surface. Small amounts of contamination are still present, as can be seen by the aliphatic carbon peak. The spectrum of the S 2p region has been slightly smoothed.

The XPS spectra are very similar to those found for a layer that has been deprotected at the surface (see Figure 6). Single-region spectra recorded for C also indicate that carbonyl and aliphatic species are still present, with these species amounting to 19% of total C. The ratio of ether and carbonyl oxygen is approximately 9:1.

The experimentally determined ratios of the various chemical species are C/O = 2.6, O/S = 15.5, and C/S = 41.2. After correcting for the contamination, the C/O ratio is 2.1 and the C/S ratio is 33.6. These ratios are indicative of excess surface carbon. The O/S ratio is close to the expected value, although the sulfur peak itself shows a higher amount of thiolate than expected.

The film thickness determined with XPS indicates 24 Å, which is slightly higher than the value derived for a monolayer of AHS. Ellipsometry yields values around 25 Å, in good agreement with the value found with XPS. The measured contact angle was less than 20°. This value is also lower than that of a film that has been deacetylated at the surface (35°) and suggests that the density of OH groups present in the film is higher than that of a deacetylated AHS layer.

Table 2 summarizes the chemical composition, contact angle, and film thickness data determined for the monolayers of AHS, AHS deprotected at the surface, and adsorbed DHS.

Discussion

Protected Molecules. AHS adsorbs spontaneously on a gold surface from solution, establishing self-assembled

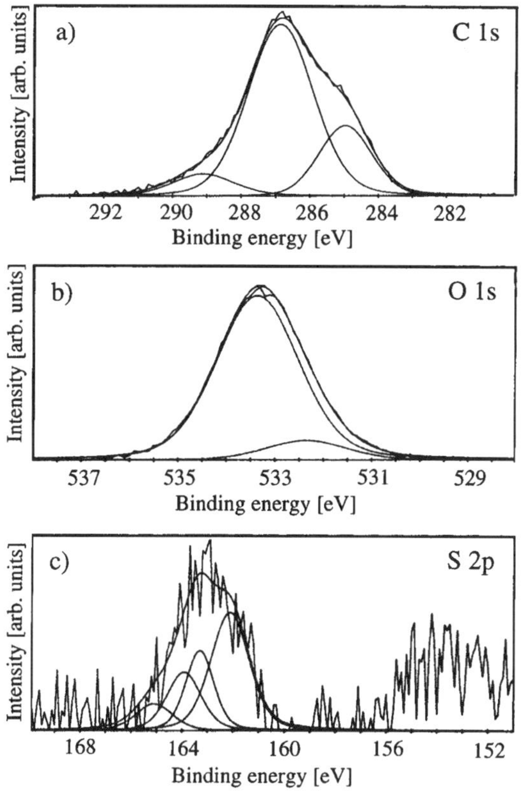

Figure 6. Single-region XPS spectra of a monolayer of DHS that has been adsorbed directly from solution. Small amounts of contamination are still present, as can be seen by the aliphatic carbon peak. The spectrum of the S 2p region has been slightly smoothed.

monolayers akin to those formed from long-chain alkanethiols. The film thicknesses of ~20 Å, determined with XPS for saturation coverage, are in good agreement with those expected for a monolayer, given the dimensions of the molecule (see Figure 1). However, the density of carbon atoms in the saccharide layers is likely to be smaller than that of a densely packed alkane chain layer such as dodecanethiol, considering the expected contorted configuration, depicted in Figure 1, which shows that the equilibrium structure may well correspond to the well-known helical arrangement seen for the corresponding polysaccharide, amylose.

The peak shapes measured for C, O, and S with XPS, together with the fitted peaks, are in good agreement with those expected for AHS. However, the ratios of C/O of 2.2 compared to an expected value of 1.6, C/S of typically 30 versus 38, and O/S of 13 in comparison to a value of 24.5 indicate deviations from the theoretical values. The sulfur signal leads to some uncertainty, due to its low intensity and the significant noise connected with it. This might explain the large range of ratios and the deviation from the theoretical values found with S. The C/O ratio, which is slightly too high, even when taking contamination into account, might partly be due to the different mean free paths of the detected electrons of the two species. The mean free path of the O 2s electrons is shorter than that of C 1s, since the kinetic energy is smaller. Using an energy dependence of $E^{0.7}$,[43] it can be estimated that an apparent lack of oxygen on the order of 7% compared to the signal from C 1s is expected, while S 2p (with higher kinetic energy) should show 3% higher intensity when compared to C due to its larger mean free path and 11% when compared to O. However, this does not fully account for the observed discrepancy. Hence, the observed ratios with S and C must be interpreted in terms of excess surface carbon and sulfur. This phenomenon has been observed throughout nearly all experiments performed with AHS.

The excess surface sulfur might be due to bond cleavage at the ether sulfur. This is supported by the XPS spectra of the S 2p region, which show a higher amount of thiolate species present (up to a factor of 2). After cleavage of the ether bond, both sulfur atoms may form a thiolate on the gold substrate. However, the mechanism responsible for this is not clear, and we have no final explanation for this phenomenon. One possibility might be an interaction between the Au atoms and the C−S bond, where electron density is transferred from the metal into the bond, leading to subsequent cleavage. The former sulfide may then also establish a bond with the gold substrate. This would explain the large thiolate signal. Such a mechanism may also be responsible for parts of the signal detected for the aliphatic carbon. The remaining excess aliphatic carbon signal has, nevertheless, to be assigned to contamination.

The density of molecules determined on the surface in a saturation monolayer of AHS was 5×10^{13} cm^{-2}, corresponding to an area of 200 Å^2/molecule. In view of the molecular size, this can be classified as a densely packed layer. It is not surprising that some contamination is still present, since the conformation of the molecules in such a layer leaves some vacant sites where 'small' molecules may readily penetrate or remain adsorbed. This is consistent with the signal from the aliphatic carbon being slightly higher than that expected for a pure saccharide layer.

The angle-dependent XPS measurements, together with the measured contact angles, strongly suggest a contorted conformation of the molecules in the adsorbed state with the glucose units covering the sulfur and stretching away from the surface. The replacement of a contaminant layer by the saccharide can be directly monitored by following the aliphatic carbon signal with immersion time. The coverage reached when the peak from the aliphatic carbon and the signal from C−O are comparable in intensity is ~0.7. This observation allows the amount of contamination present at the surface to be determined. For a coverage of 0.7, i.e., ~3.5×10^{13} saccharides/cm^2, both aliphatic contributions are comparable. This observation is consistent with a contamination layer some 3−10 Å in height.

The adsorption kinetics experiments, as well as ToF-SIMS studies, imply that no significant intermolecular interactions are present, i.e., that the resulting films are homogeneous on a micrometer scale.

The contact angle determined for a monolayer of AHS of 55° is significantly lower than the value observed for bare (contaminated) gold substrates (~75°). It is likely that the glucose units, in connection with the protecting acetyl groups, determine the wetting properties of the prepared layer. Since our ToF-SIMS experiments indicate that no macroscopic islands are formed (>1 μm), the observed wetting angle is likely to represent an average value of a film that consists of homogeneously distributed glucose units over the surface.

(43) Penn, L. R. *J. Electron Spectrosc.* **1976**, *9*, 40.

Table 2. Chemical Composition, Thickness and Contact Angles of the Layers Formed from AHS, after Deprotection of AHS at the Surface and from Direct Adsorption of DHS from Solution

	AHS	deac. AHS	DHS
contact angle	55°	35°	<20°
film thickness			
XPS	20 Å	16 Å	24 Å
ellipsometry	20 Å	19 Å	25 Å
chemical composition			
C/O expected/observed/corrected[a]	1.6/2.2	1.3/2.4/1.8	1.3/2.6/2.1
C/S expected/observed/corrected[a]	38/30	18.7/13.2/14.8	18.7/41.2/33.6
O/S expected/observed	24.5/13	14.8/8.1	14.8/15.5

[a] Corrected means after substraction of the contributions from contamination.

The experimentally determined rate of adsorption of $17(\pm 5.5) \times 10^6$ cm^3 mol^{-1} s^{-1} is of the same order of magnitude as that of alkanethiols adsorbed from ethanol (3.3×10^6 cm^3 mol^{-1} s^{-1}).[41,44] However, the solvent plays a major role in the adsorption kinetics, as is evident when comparing the results obtained at different immersion times and concentrations with the pre-deacetylated saccharide (DHS).

Deprotected Molecules. For future applications, especially further bonding of proteins to the prepared layer or a derivatization of the hydroxyl groups in the film, the deprotection of the molecules is of principal interest. Deprotection on the substrate and in solution prior to the adsorption led to comparable chemical compositions, as is evident from the XPS spectra. The slight difference found in the ratio of ether/carbonyl signal might indicate that a smaller percentage of OH groups can be converted under in-situ conditions at the surface. However, the carbon concentrations found in both layers are too high. The sulfur content within the layer prepared from DHS is close to the theoretical one, although too much thiolate is present. Possibly both sulfur atoms in the molecule establish bonds with the gold substrate, without cleavage of the C–S–C bond. However, this issue is not resolved.

The carbonyl peaks, which contain approximately 10% of the whole carbon signal for both layers, may partly stem from residual protective acetyloxy groups. From the NMR spectra of the solution, the amount can be estimated as less than 5% in the provided DHS. The remaining 5% then has to be assigned to contamination. By far the majority of the contamination shows aliphatic character.

The loss of carbon atoms, which is correlated with the removal of acetyl groups, explains the film thickness of 17 Å, as determined with XPS after deprotection of the molecules at the surface. The higher value found for the film that was deprotected prior to adsorption can be understood in terms of a higher density of saccharide molecules on the surface after adsorption from solution. Steric reasons are probably responsible for the higher space requirement per molecule in its protected form (AHS) compared to a deacetylated molecule (DHS) due to the bulky protective acetoxy groups.

The difference in absolute coverage between the surface-deprotected AHS and the adsorbed DHS leads, of course, to a difference in surface OH concentration and thus to a change in hydrophilicity, as seen by the different measured contact angles. The rather long reaction time (15 h) and the high concentration of NaOMe necessary to deacetylate the preadsorbed films is consistent with steric hindrance in the contorted molecules but could also be due to a slow migration of AcO groups along the molecule. In other words, the slow reaction rate could be caused by poor accessibility to the adsorbed species by the reagent or by the hindered diffusion of the MeOAc reaction product away from the saccharide, possibly resulting in intramolecular AcO group transfer. The difference observed between the two layers in film thickness and contact angles can be explained by the individual space requirements necessary for the adsorption of an AHS and a DHS molecule.

On the basis of the film thickness of 24 Å compared to 20 Å for AHS, an increase in the density of molecules of ~20% can be estimated. Hence the density of DHS after adsorption from the solution is approximately 6×10^{13}/cm^2, with a corresponding space requirement of 167 Å^2/molecule. The surface-projected space requirement per OH group then amounts to an area of approximately ~11 Å^2/OH for a layer that has been fully deprotected at the surface and of ~9 Å^2/OH for a DHS layer that has been adsorbed from solution. However, it has to be borne in mind that the functional groups are homogeneously distributed through the adsorbed layer. For comparison, a densely packed alkane chain layer corresponds to approximately ~22 Å^2/molecule.[45]

As mentioned above, in all cases the XPS carbon signals show the presence of small amounts of carbonyl and larger amounts of aliphatic species. This underlines the fact that even for saturation coverage a significant amount of contamination is present that cannot be displaced by the saccharide, probably due to its large size, compared to that of a surface adsorption site. Moreover we found that the rate of aliphatic carbon contamination following film formation was larger for layers of deacetylated saccharides than for layers of AHS. This might be due to the polar hydroxyl groups and their strong interaction with contaminated water from the atmosphere.

The adsorption kinetics are found to be completely different for the two molecules, AHS and DHS. The rate of dissolution of molecules into solution from fully formed monolayers at room temperature is negligible, so equilibrium cannot be established by desorption and readsorption of monolayer components in the complete monolayer. Equilibration could proceed through the physisorbed thiol. Rapid equilibration between the physisorbed molecule and the molecules in solution would be followed by relatively slow conversion of the physisorbed thiols to chemisorbed thiolates. If the rate constant for conversion of thiol to surface thiolate is independent of the structure of the thiol, which is likely, a chemisorbed layer would be kinetically trapped. The adsorption rate constant would then be determined by the equilibration between the physisorbed thiol and the thiols in solution, which might explain the observed large differences between AHS and DHS. A similar argument has been used by Bain et al. to explain the observed composition of mixed monolayers.[30] This would explain the major role of the solvent in the adsorption kinetics and possibly also in the resulting film structure. A further consideration is the expected hy-

(44) Buck, M.; Eisert, F.; Fischer, J.; Grunze, M.; Träger, F. *Appl. Phys. A* **1991**, *53*, 552.

(45) Strong, L.; Whitesides, G. M. *Langmuir* **1988**, *4*, 546.

drogen bonding from the SH moiety to OH groups on the unprotected DHS. This would most likely affect the concentration of the intermediate physisorbed thiol species and therefore also alter the observed overall adsorption kinetics.

Although the hydroxyl groups in the films we have described are not all located in the outermost layer, their homogeneous distribution makes the molecules promising for further studies in biological applications. The hydrophobic/hydrophilic character of the prepared surface may be varied by varying the density of the adsorbed molecule. This, in turn, can be tailored by varying the immersion time or the number of glucose units in the adsorbate. Moreover, the number of hydroxyl groups present in the adsorbed layer could also be tailored by varying the time allowed for deprotection. It has previously been shown that, within a set of SAMs derived from similar components, greater quantities of proteins are adsorbed on the more hydrophobic surfaces.[7]

The final (deprotected) monolayers consist of a still undefined fraction of water-accessible hydrophobic spots, consisting of contaminated gold, which are separated by hydrophilic areas corresponding to the saccharide-covered substrate. When the monolayers are employed in complex biofluids, the hydrophobic spots might react as adsorption (aggregation) sites for proteins and other macromolecules. This type of aggregation phenomena is undesirable in many biological applications. However, the removal of residual contamination is very challenging: A possible solution might involve subsequent adsorption of short-chain thiols for example, although this might also remove a fraction of the adsorbed saccharide molecules. It should be borne in mind that, although the hydrophobic spots are accessible by water, as is evident from the contact angle measurements, they might well be inaccessible to larger molecules. Further investigations concerning the homogeneity of the resulting films below the micrometer scale are necessary to tackle this problem.

Conclusion

A hexasaccharide synthesized with a surface-active tail was used to establish self-assembling monolayers on gold surfaces. These were characterized with XPS, contact angle measurements, ellipsometry, and iToF-SIMS.

It was found that the protected hexasaccharide forms homogeneous monolayer films with a density of molecules at the surface that is roughly one order of magnitude lower than that of alkanethiols. The concentration and time necessary to deprotect the layer at the surface, as well as the observed contact angle and angle-dependent XPS measurements, indicate contorted molecules at the surface (possibly the onset of the helical amylose structure) with consequent limited accessibility to the protective groups.

The preparation of a deprotected layer is possible by performing the deprotection on the surface as well as by adsorption of deprotected molecules from solution. The degree of deprotection was found to be comparable in both cases; however, the degree observed for deprotection on the surface was slightly lower. The density of molecules and hence of functionalized groups was found to be highest for molecules that had been deprotected before adsorption. It is expected that the apparent ease with which the hydrophobicity/hydrophilicity of these (outer) surfaces can be controlled will be of significance in future applications of these molecules.

Acknowledgment. A.V. and R.B. thank F. Hoffmann La Roche, Basel, for generous support.

LA960623J

4324 *Langmuir* **1999,** *15,* 4324–4327

Highly Oriented, Self-Assembled Alkanephosphate Monolayers on Tantalum(V) Oxide Surfaces

Dorothee Brovelli,[†] Georg Hähner,[†] Laurence Ruiz,[†] Rolf Hofer,[†,‡] Gerolf Kraus,[‡] Adrian Waldner,[‡,§] Johanna Schlösser,[‡] Peter Oroszlan,[‡] Markus Ehrat,[‡] and Nicholas D. Spencer*,[†]

Laboratory for Surface Science and Technology, Department of Materials, ETH-Zürich, CH-8092 Zürich, Switzerland, and Novartis Pharma AG, CH-4002 Basel, Switzerland

Received December 30, 1998. In Final Form: April 7, 1999

Octadecyl phosphoric acid ester has been found to produce oriented, well-ordered monolayers on a flat tantalum(V) oxide surface, via self-assembly from a heptane/propan-2-ol solution. By means of contact angle, optical waveguide lightmode spectroscopy (OWLS), near-edge X-ray absorption fine structure spectroscopy (NEXAFS), and X-ray photoelectron spectroscopy (XPS) measurements, it has been shown that these layers closely resemble those formed by the corresponding thiol–gold system, with respect to packing density, inclination, and order. The system shows promise as an approach to functionalizing oxide surfaces with well-ordered organic monolayers, with potential applications in the fields of biochemical analysis and sensors.

Introduction

The development of methods for the generation of self-assembled monolayers (SAMs)[1,2] has presented the surface scientist and engineer with a highly flexible approach for the creation of concentrated planes of functionality. While this methodology has tremendous possibilities for applications in such varied areas as biosensors,[3] corrosion-resistant systems,[4] adhesion promotion,[5] and so forth, the specific chemistries generally employed have led to certain limitations. The largest classes of SAMs investigated until now have been based either on the interaction of chlorosilanes[6] with OH-terminated oxide surfaces or on the adsorption of thiols on gold.[1] While the former approach offers great flexibility, it has the disadvantage of frequently producing ill-defined surfaces due to the onset of uncontrolled polymerization reactions. The latter (thiol–gold) approach can produce monolayer films with a high degree of perfection, but the necessity for a gold (or in certain circumstances silver[7]) surface all but rules it out in many applications, especially those in which optical transmission is a requirement for the system.

A smaller number of publications has appeared where alternative chemistries have been employed to coat oxide surfaces with SAMs. These have included hydroxamic,[8] carboxylic,[9] and phosphonic acids[10,11] and, to a very limited extent, phosphoric acids.[11] In addition, several papers have appeared using alkyl phosphates and phosphonates in a tails-down configuration, for the purpose of building zirconium phosphate layer structures.[12]

In this paper we describe, for the first time, a self-assembly technique that employs alkyl phosphoric acid esters to produce dense, highly ordered monolayers in a "tails-up" configuration, on a Ta_2O_5 surface. Tantalum oxide was chosen because of its high refractive index, which renders it ideal for application in a planar-waveguide-based bioaffinity sensor.[13] Upon appropriate ω-functionalization, alkanephosphate-based SAMs have the potential to be used as the interface that anchors active sensing elements or as the basis of passive, biomolecule-resistant regions on the sensor surface.

Experimental Section

Synthesis of *n*-Octadecyl Phosphoric Acid Esters. *n*-Octadecyl phosphoric acid ester (ODP) was prepared according to the protocol reported by Okamoto[14] (Scheme 1). ODP is a stable, waxy solid and was recrystallized from hot *n*-hexane. ^1H NMR spectra ($CDCl_3$) δ: 0.83–0.93 (CH_3, 3H, t), 1.18–1.45 ($(CH_2)_{15}$, 30H, m), 1.62–1.76 (P(O) CH_2CH_2, 2H, quintet), 3.97–4.13 (P(O) CH_2, 2H, quintet). Elemental anal (C, H, N: Leco CHN-900. P via photometry). Calc for $C_{18}H_{39}O_4P$: C, 61.69; H, 11.22; P, 8.84. Found: C, 61.72; H, 11.02; P, 8.82. The C/P ratio estimated from the elemental analysis results is 18.05, in agreement with expectations.

Cleaning of Materials and Containers for Substrate Handling. Glass bottles were used for storing the amphiphile

* To whom correspondence should be addressed.

† ETH-Zürich.

‡ Novartis Pharma AG.

§ Current Address: Novartis Animal Health, Ltd., CH-4002 Basel, Switzerland.

(1) Bain, C. D.; Troughton, E. B.; Tao, Y.-T.; Evall, J.; Whitesides, G. M.; Nuzzo, R. G. *J. Am. Chem. Soc.* **1989**, *111*, 321.

(2) Nuzzo, R. G.; Allara, D. L. *J. Am. Chem. Soc.* **1983**, *105*, 4481. Sellers, H.; Ulman, A.; Schnidman, Y.; Eilers, J. E. *J. Am. Chem. Soc.* **1993**, *115*, 9389. Bishop, A. R.; Nuzzo, R. G. *Curr. Opin. Colloid Interface Sci.* **1996**, *1*, 127. Ulman, A. *Chem. Rev.* **1996**, *96*, 1533.

(3) Hickman, J. J.; Ofer, D.; Laibinis, P. E.; Whitesides, G. M.; Wrighton, M. S. *Science* **1991**, *252*, 688. Swalen, J. D.; Allara, D. L.; Andrade, J. D.; Chandross, E. A.; Garoff, S.; Israelachvili, J.; McCarthy, T. J.; Murray, R.; Pease, R. F.; Rabolt, J. F.; Wanne, K. J.; Yu, H. *Langmuir* **1987**, *3*, 932.

(4) Bram, Ch.; Jung, Ch.; Stratmann, M. *Fres. J. Anal. Chem.* **1997**, *358*, 108.

(5) Maoz, R.; Netzer, L.; Gun, J.; Sasgiv, J. *J. Chim. Phys.* **1988**, *85*, 1059.

(6) Ulman, A. *Adv. Mater.* **1990**, *2*, 573.

(7) Porter, M. D.; Bright, T. B.; Allara, D. L.; Chidsey, C. E. D. *J. Am. Chem. Soc.* **1987**, *109*, 3559.

(8) Folkers, J. P.; Gorman, C. B.; Laibinis, P. E.; Buchholz, S.; Whitesides, G. M. *Langmuir* **1995**, *11*, 813–824.

(9) Aronoff, Y. G.; Chen, B.; Lu, G.; Seto, C.; Schwartz, J.; Bernasek, S. L. *J. Am. Chem. Soc.* **1997**, *119*, 259. Laibinis, P. E.; Hickman, J. J.; Wrighton, M. S.; Whitesides, G. M. *Science* **1989**, *245*, 845.

(10) Woodward, J. T.; Ulman, A.; Schwartz, D. K. *Langmuir* **1996**, *12*, 3626. Gao, W.; Dickinson, L.; Grozinger, C.; Morin, F. G.; Reven, L. *Langmuir* **1996**, *12*, 6429.

(11) Maege, I.; Jaehne, E.; Henke, A.; Adler, H.-J. P.; Bram, C.; Jung, C.; Stratmann, M. *Macromol. Symp.* **1997**, *126*, 7–24.

(12) Lee, H.; Kepley, L. J.; Hong, H. G.; Akhter, S.; Mallouk, T. E. *J. Phys. Chem.* **1988**, *92*, 2597–2601. Lee, H.; Hong, H. G.; Mallouk, T. E.; Kepley, L. J. *J. Am. Chem. Soc.* **1988**, *110*, 618–620.

(13) Duveneck, G. L.; Pawlak, M.; Neuschäfer, D.; Bär, E.; Budak, W.; Pieles, U.; Ehrat, M. *Sens. Actuators B* **1997**, *38–39*, 88.

(14) Okamoto, Y. *Bull. Chem. Soc. Jpn.* **1985**, *58*, 3393.

Published on Web 05/22/1999

Scheme 1

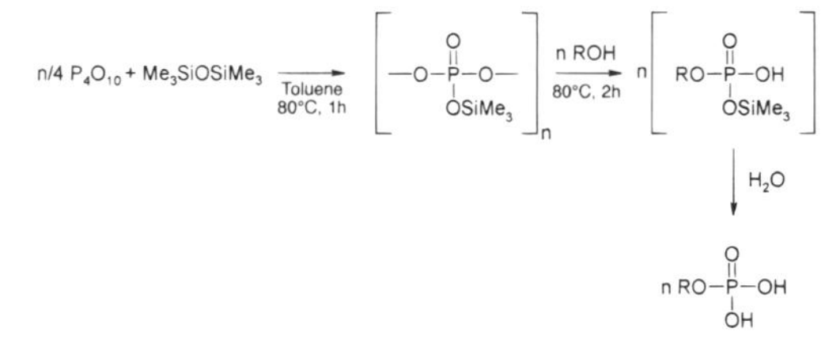

stock solutions. SAM formation was performed in PTFE containers. Glass bottles and PTFE containers were immersed for 2 h in piranha solution (a 1:2 mixture of 30% H_2O_2 and 98% H_2SO_4—*Caution, this solution can explode in contact with organic matter!*), followed by extensive rinsing in ultrapure water prior to first use. During the investigation, all materials were cleaned with propan-2-ol after use and stored in a class-1000 laminar flow hood.

Preparation of Stock Solutions of Amphiphiles. Stock solutions for SAM formation were prepared by dissolving the ODP in a 100:0.4 (v/v) mixture of *n*-heptane and propan-2-ol at concentrations of 5, 50, and 500 μM, followed by sonication for 10 min. For comparison purposes, a 500 μM solution was also prepared in pure propan-2-ol (ODP was found to be insoluble in pure *n*-heptane). The resulting solutions were filtered through a 200 nm cellulose nitrate membrane and stored at ambient temperature in glass bottles until use. No deleterious effects were observed upon storage of these solutions for up to several weeks.

Substrate Cleaning and Formation of Self-Assembled Monolayers. SAM formation was studied on tantalum(V) oxide films deposited by physical vapor deposition (ion plating) on Corning glass 7059 (16 mm × 16 mm, 150 nm thickness, <1 nm roughness (AFM), Balzers AG, Liechtenstein). A grating of 320 nm period was etched into those substrates destined for OWLS measurements (see below). All substrates were cleaned in an ultrasonic bath of propan-2-ol for 15 min and dried in helium (5.0 purity). Surfaces were then cleaned using a UV cleaner (model 135500, Boekel Ind. Inc., PA) for 30 min. Representative substrates were checked for cleanliness by XPS and advancing water contact angle measurements (<10°). The substrates were transferred immediately after cleaning into a PTFE container and immersed in a stock solution of the ODP (1 mL of stock solution per 250 mm^2 of metal oxide surface area). SAM formation was conducted in the closed container at ambient temperature. The duration of SAM formation was varied systematically during this study. Substrates were kept in the SAM solution up to 48 h. After SAM formation, the substrates were removed from the solution and immediately rinsed with 1.5 mL of propan-2-ol followed by blow drying in helium. Contact angle measurements were carried out immediately afterward. The functionalized substrates were stored in a closed PTFE container prior to characterization and were found to be stable over many weeks.

Characterization Methods. Measurements of advancing contact angles were performed in ambient air using automatic drop shape analysis (contact angle measuring system G2/G40, Krüss GmbH, Hamburg, Germany). The values reported here are mean values derived from three identically prepared samples and at least four positions on each sample.

SAM adsorption was also followed in situ by optical waveguide lightmode spectroscopy (OWLS)[15] by means of a grating coupler system (GKR 401, Fraunhofer-Institut für Physikalische Messtechnik, Freiburg, Germany). Prior to starting adsorption experiments, the baseline was stabilized with the pure solvent. For kinetic measurements, a solution of ODP was pumped at 1 mL/s through the fluid cell at room temperature and the adsorption of ODP was followed in real time.

NEXAFS experiments were carried out at the National Synchrotron Light Source (NSLS), Brookhaven National Lab (New York), on Exxon Beamline U1A in the partial electron yield mode for various irradiation angles between normal and grazing incidence. To compensate for the severe charging of the samples,

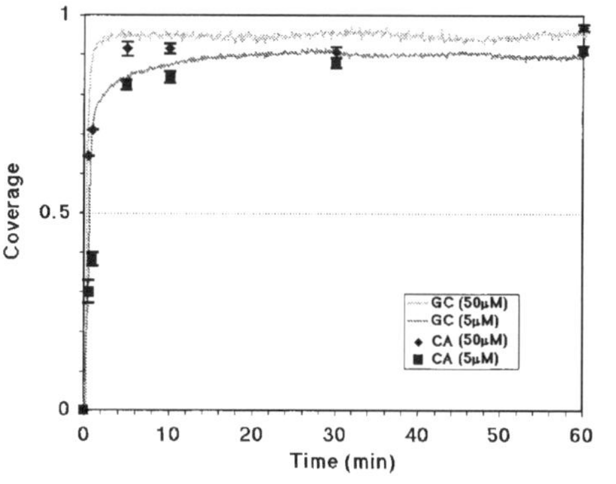

Figure 1. Coverage variation with assembly time derived from contact angle (CA) and grating coupler (GC) results, for two different concentrations of ODP in *n*-heptane/propan-2-ol (100/0.4 v/v). Error bars show standard deviations for repeated measurements.

a low-electron-energy flood gun was used.[16] Data analysis was carried out by means of difference spectra according to the analysis procedure described previously.[17–19] NEXAFS spectroscopy probes transitions of core electrons into unoccupied molecular orbitals. Since these excitations are governed by dipole selection rules, the polarization dependence of their signal intensities can be used to extract information on the orientation of molecular adsorbates.[17]

X-ray photoelectron spectroscopy (XPS) analyses were performed using a PHI 5700 spectrometer equipped with a concentric hemispherical analyzer (CHA) and a non-monochromatized Al Kα source (200 W) (Physical Electronics, Eden Prairie, MN). Pass energies used for survey and detailed scans (for Ta_{4f}, C_{1s}, O_{1s}, and P_{2p}) were 187.85 and 23.5 eV, respectively. For survey and individual detailed scans, the acquisition times were 4.17 and 2.5 min, and the measured resolutions (fwhm Ag $3d_{5/2}$) were 2.7 eV and 1.1 eV, respectively. Sample charging was minimized by means of a low-energy electron flood gun (<15 eV). Data were fitted following a Shirley iterative background subtraction. Spectra were referenced to the aliphatic hydrocarbon C_{1s} signal at 285.0 eV. No evidence of sample degradation was observed during the course of XPS measurements.

Results and Discussion

Film Growth and Wettability. The formation of adsorbed overlayers from a heptane/propan-2-ol solution of ODP on tantalum(V) oxide films was studied as a function of assembly time by measuring both the advancing water contact angle and the OWLS incoupling angle for different concentrations of the amphiphile solution. The data are compiled in Figure 1. The maximum contact angle reached, after 48 h of adsorption from a 50 μM (or higher) amphiphile solution onto tantalum (V) oxide, was 113 ± 2°. This value was taken to be equivalent to a coverage of unity, and contact angle data were converted into coverages accordingly, using the Cassie equation[20] to account for the nonuniformity of the partially covered surfaces. OWLS results were normalized to the contact angle data. The high-contact-angle value strongly suggests a "tails-up" orientation of the adsorbed molecules and a high degree of order. The self-assembly process appears to occur rapidly, with near-monolayer coverages and high contact angles being obtained within several minutes of immersion in the 50 μM amphiphile solution.

(15) Ramsden, J. J. *Phys. Rev. Lett.* **1993**, *71*, 295.
(16) Hähner, G.; Marti, A.; Caseri, W. R.; Spencer, N. D. *J. Chem. Phys.* **1996**, *104*, 7749.
(17) Stöhr, J. *NEXAFS Spectroscopy*; Springer: Heidelberg, 1992; and references therein.

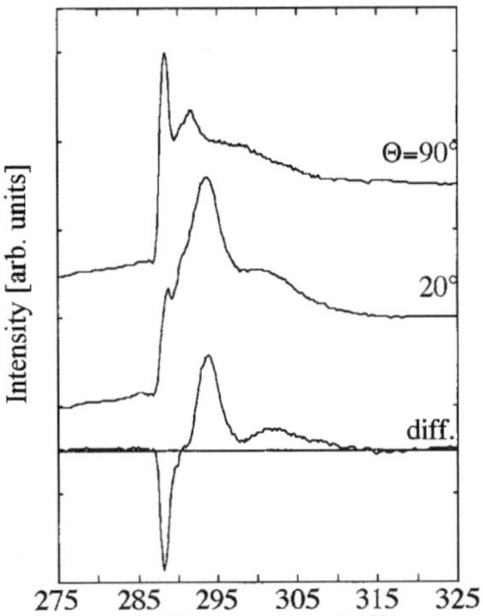

Figure 2. NEXAFS spectra of ODP adsorbed on tantalum(V) oxide from a 500 μM solution in n-heptane/propan-2-ol (100/0.4 v/v) for 48 h at normal and grazing incidence.

An adsorption experiment was also carried out using pure propan-2-ol as an amphiphile solvent. After 48 h of adsorption, the maximum contact angle reached was 95° ± 2°, suggesting a much lower coverage and/or degree of order than those with the mixed n-heptane/propan-2-ol solvent.

Characterization of the Adsorbed Film. *NEXAFS.* Figure 2 displays NEXAFS spectra of ODP adsorbed on Ta_2O_5 from a 0.5 mM solution in n-heptane/propan-2-ol for 48 h at normal and grazing incidence of the incoming X-rays. The sharp transition situated at 287.6 eV close to the absorption step can be assigned to transitions into C−H* valence orbitals.[21] Its intensity is a maximum at normal X-ray incidence (θ = 90°). The two broader resonances above the step at around 293.1 and 301 eV stem from transitions into C−C σ^* orbitals.[21] They are strongest for grazing incidence (θ = 20°). To emphasize this angular variation, the difference spectrum between grazing and normal incidence has been included in the figure. The transitions into C−C σ^* and C−H* orbitals exhibit an opposite polarization dependence, since the corresponding molecular orbitals are perpendicular for alkyl chains in an all-trans conformation.[21,22] A comparison with the spectra of a well-ordered SAM of hexadecanethiol on gold (reference sample) reveals virtually no difference, indicating a very similar degree of order in the two samples.

A more quantitative analysis[17−19] of the order in the films yields a similar average tilt angle to that obtained for hexadecanethiol monolayers on gold (i.e. 33° between the alkyl chain axis and the surface normal). An error of less than 3° was calculated (for details see ref 19).

An additional adsorption experiment was carried out using pure propan-2-ol as a solvent. In this case, analysis of the angular-dependent NEXAFS data revealed a larger average tilt angle (48°) or lower degree of order. (It should be borne in mind that the signal of randomly oriented transition dipole moments cannot be distinguished from the signal of dipole moments with a specific tilt angle of 54.7° ("magic angle"), both being invariant with the angle of incidence.) For alkanethiols on gold it has been previously reported that increasing the surface coverage of the thiols leads to an increase in the degree of order in these SAMs.[23]

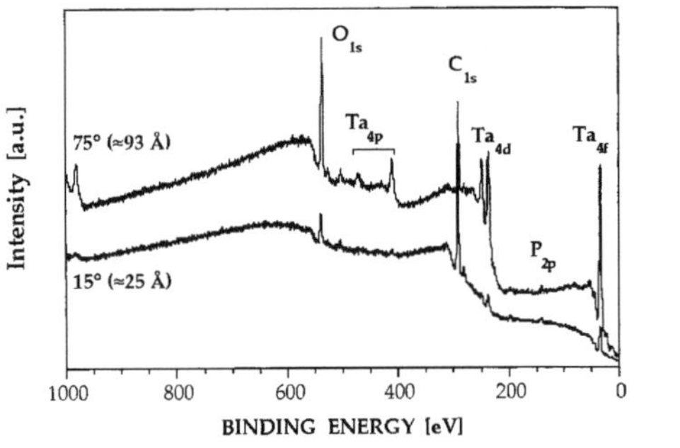

Figure 3. Survey spectra of ODP overlayer on tantalum(V) oxide obtained at different emission angles (15° and 75°).

The NEXAFS spectra of SAMs of ODP on Ta_2O_5 therefore indicate a strong preferential orientation of the alkyl chains on the surface. Although the degree of order in the films is strongly dependent on the solvent used in the adsorption solution, NEXAFS spectra of films produced from the n-heptane/propan-2-ol mixture are virtually indistinguishable from those of octadecanethiol SAMs on gold, indicating a similar electronic structure, an equivalent degree of order, and a similar average orientation. Since the length of the ODP molecule can be estimated from known bond lengths and angles to be ~25 Å, the 33° tilt angle yields a film thickness of ~21 Å (assuming monolayer coverage).

Since the tilt angle arises from the system's attempt to maximize interchain van der Waals interaction for a given anchor-group spacing, the similar tilt angles of ODP on Ta_2O_5 and hexadecanethiol on gold also indicate that the density of ODP anchoring groups on the surface is comparable to the density of sulfur atoms in the ordered $(\sqrt{3}\times\sqrt{3})$R30° structure of alkanethiols on gold(111), that is, 1 per 21.4 Å^2.[24]

XPS. Overlayers deposited from the ODP/(n-heptane/propan-2-ol) solution were also analyzed by XPS at two emission angles, 15° and 75°, corresponding to an information depth ($Z = 3\lambda \sin\theta$) of ≈25 Å (approximately one monolayer thickness) and ≈93 Å, respectively, for phosphorus and tantalum. λ-values were derived from the Seah and Dench relation,[25] assuming an organic matrix. Figure 3 displays the survey spectra obtained for both takeoff angles.

Atomic concentrations and elemental ratios, derived from detailed elemental scans, are reported in Table 1. Such estimations are, however, only strictly valid when considering a homogeneous slab—not the case for the SAM layer. Nevertheless, it is clear that, in the case of grazing incidence detection (information depth ≈25 Å), the calculated elemental C/P ratio is much higher than that

(18) Kinzler, M.; Schertel, A.; Hähner, G.; Wöll, Ch.; Grunze, M.; Albrecht, H.; Holzhüter, G.; Gerber, Th. *J. Chem. Phys.* **1994**, *100*, 7722.
(19) Fischer, D.; Marti, A.; Hähner, G. *J. Vac. Sci. Technol., A* **1997**, *15*, 2173.
(20) Cassie, A. B. D. *Discuss. Faraday Soc.* **1948**, *3*, 11.
(21) Outka, D. A.; Stöhr, J.; Rabe, J. P.; Swalen, J. D. *J. Chem. Phys.* **1988**, *88*, 4076.
(22) Hähner, G.; Kinzler, M.; Wöll, Ch.; Grunze, M.; Scheller, M. K.; Cederbaum, L. S. *Phys. Rev. Lett.* **1991**, *67*, 851; see also **1991**, *69*, 694 (Erratum).
(23) Hähner, G.; Wöll, Ch.; Buck, M.; Grunze, M. *Langmuir* **1993**, *9*, 1955.
(24) Ulman, A. *Ultrathin Organic Films*; Academic Press: San Diego, CA, 1991.
(25) Seah, M. P.; Dench, W. A. *Surf. Interface Anal.* **1979**, *1*, 2.

Table 1. Atomic Concentration (%) and Elemental Ratio of a Self-Assembled ODP Overlayer on Ta_2O_5, Based on a Simple, Homogeneous Slab Model

	atomic conc and C/P ratio at emission angle (info depth) =	
	15° (≈25 Å)	75° (≈93 Å)
C	88	49
P	1.6	2.2
O	8.8	37
Ta	2.1	12
C/P	55	22

for the 75° case, for which a ratio close to the stoichiometric one is measured. This is presumably due to attenuation of the phosphorus signal intensity by inelastic scattering of the P_{2p} electrons in the hydrocarbon tail of the adsorbed ODP molecule. Similar observations have been reported for alkanethiols on gold.[1] This is completely consistent with the ordered, tails-up monolayer picture that emerges from both contact angle and NEXAFS results described above. A more rigorous three-layer XPS model is described elsewhere[26] and independently yields a layer thickness of 21 Å, in agreement with the NEXAFS results described above.

(26) Textor, M.; Ruiz, L.; Hofer, R.; Rossi, A.; Feldman, K.; Hähner, G.; Spencer, N. D. In preparation.

Conclusions

Octadecyl phosphoric acid ester appears to form an ordered, oriented, self-assembled monolayer on a flat tantalum(V) oxide surface, when deposited from an *n*-heptane/propan-2-ol solution. The adsorption rate is rapid, with a complete "tails-up" layer being formed within minutes, according to contact angle and grating-coupler results. An optimal combination of assembly rate and film quality appears to be achieved with 50 or 500 μM solutions of ODP in *n*-heptane/propan-2-ol (100/0.4 v/v).

NEXAFS indicates that the degree of order and the average orientation (approximately 33° inclination to the normal) of the molecular layer resemble those of hexadecanethiol on gold(111). XPS measurements are consistent with this picture and monolayer film thickness. The solvent appears to play a critical role in the formation of the highly ordered and oriented layer.

Acknowledgment. NEXAFS measurements were performed at the NSLS, Brookhaven, NY (USA), which is supported by the U.S. Department of Energy. We thank Dr. J. G. Chen and B. D. DeVries (Exxon) for their support during our beam time at beamline U1A at the NSLS. We acknowledge the Swiss National Science Foundation for financial support.

LA981758N

Langmuir **2000,** *16,* 3257–3271 3257

Structural Chemistry of Self-Assembled Monolayers of Octadecylphosphoric Acid on Tantalum Oxide Surfaces

Marcus Textor, Laurence Ruiz,† Rolf Hofer, Antonella Rossi,‡ Kirill Feldman, Georg Hähner, and Nicholas D. Spencer*

Laboratory for Surface Science and Technology, Department of Materials, ETH-Zürich, CH-8092 Zürich, Switzerland

Received July 16, 1999. In Final Form: December 2, 1999

Octadecylphosphoric acid ester is shown to self-assemble on amorphous/nanocrystalline tantalum oxide (Ta_2O_5) layers deposited by physical vapor deposition onto glass substrates. Three complementary surface-analytical techniques (angle-dependent X-ray photoelectron spectroscopy, time-of-flight secondary ion mass spectrometry, and atomic force microscopy in lateral force mode), showed that a 2.2 nm thick, "tails-up"-oriented adlayer is formed, which displays local near-hexagonal order, strong P–O–Ta bonding, and the presence of $(-P-O-)_2Ta$ species. A model for the binding and the structural organization of the octadecyl phosphate molecules on the tantalum oxide surface is proposed involving direct coordination of the terminal phosphate headgroup to Ta(V) cations forming a strong complexation bond, two types of bonding of the octadecyl phosphate with both monodentate and bidentate phosphate-Ta(V) coordinative interactions, and, locally, the formation of a coincidence lattice of approximately hexagonal structure defined by both the location of Ta(V) cation sites and an intermolecular spacing between the octadecyl phosphate ligands of approximately 0.5 nm. This is very similar to the self-assembled monolayer structure of long-chain alkanethiols on gold. The use of phosphoric acid ester derivatives is believed to have potential for designing specific interface architectures in sensor technology, in surface modification of oxide-passivated metallic biomaterials, and in composite metal (oxide)–polymer interfaces.

1. Introduction

Self-assembled monolayers (SAMs)[1,2] represent a powerful and highly flexible approach for the creation of concentrated planes of surface functionality. Although this methodology has the potential for applications in many areas, such as biosensors,[3] corrosion-resistant systems,[4] adhesion promotion,[5] etc., the specific chemistries that have been widely used to date have suffered from certain inherent restrictions. The principal classes of SAMs investigated and applied until now have been based either on the interaction of chlorosilanes[6] with OH-terminated oxide surfaces or on the adsorption of thiols on gold.[1] Silanes have a tendency to form films that are thicker than one monolayer on oxides, due to the onset of uncontrolled polymerization reactions. On the other hand, thiols can produce monolayer films with a high degree of perfection, but necessitate the use of a gold (or in certain circumstances silver)[7] surface. This can be a problem, particularly in cases where optical transmission is required. A smaller number of publications has appeared where alternative chemistries have been employed to coat *oxide* surfaces with SAMs. These have included hydroxamic,[8] carboxylic,[9] and phosphonic acid,[10,11] as well as, to a limited extent, phosphoric acids.[11]

In a previous paper[12] we described a self-assembly technique that employs octadecyl phosphoric acid ester to produce dense, highly ordered monolayers in a "tails-up" configuration on a Ta_2O_5 surface. Adsorbate orientation and thickness were determined by a combination of near-edge X-ray absorption fine structure (NEXAFS) data, contact angle measurements, grating coupler results, and simple, angular-dependent X-ray photoelectron spectroscopy (XPS) studies. In the present paper we detail further investigations of the same phosphate ester–tantalum oxide system and focus on the structural chemistry at the phosphate–oxide interface, as revealed by a much more detailed XPS study, together with time-of-flight secondary ion mass spectrometry (ToF-SIMS), and atomic force microscopy (AFM).

2. Material and Methods

2.1. Substrate. Surface modifications were studied on tantalum pentoxide films deposited via physical vapor (ion plating)

* To whom correspondence should be addressed.

† Present address: Zyomyx, Inc., 3911 Trust Way, Hayward, CA 94545.

‡ Department of Inorganic and Analytical Chemistry, University of Cagliari, S.S. 554 bivio per Sestu, I-09042 Monserrato (Cagliari), Italy.

(1) Bain, C. D.; Troughton, E. B.; Tao, Y.-T.; Evall, J.; Whitesides, G. M.; Nuzzo, R. G. *J. Am. Chem. Soc.* **1989**, *111*, 321.

(2) Nuzzo, R. G.; Allara, D. L. *J. Am. Chem. Soc.* **1983**, *105*, 4481. Sellers, H.; Ulman, A.; Schnidman, Y.; Eilers, J. E. *J. Am. Chem. Soc.* **1993**, *115*, 9389. Bishop, A. R.; Nuzzo, R. G. *Curr. Opin. Colloid Interface Sci.* **1996**, *1*, 127. Ulman, A. *Chem. Rev.* **1996**, *96*, 1533.

(3) Hickman, J. J.; Ofer, D.; Laibinis, P. E.; Whitesides, G. M.; Wrighton, M. S. *Science* **1991**, *252*, 688. Swalen, J. D.; Allara, D. L.; Andrade, J. D.; Chandross, E. A.; Garoff, S.; Israelachvili, J.; McCarthy, T. J.; Murray, R.; Pease, R. F.; Rabolt, J. F.; Wanne, K. J.; Yu, H. *Langmuir* **1987**, *3*, 932.

(4) Bram, Ch.; Jung, Ch.; Stratmann, M. *Fresenius J. Anal. Chem.* **1997**, *358*, 108.

(5) Maoz, R.; Netzer, L.; Gun, J.; Sasgiv, J. *J. Chim. Phys.* (*Paris*) **1988**, *85*, 1059.

(6) Ulman, A. *Adv. Mater.* **1990**, *2*, 573.

(7) Porter, M. D.; Bright, T. B.; Allara, D. L.; Chidsey, C. E. D. *J. Am. Chem. Soc.* **1987**, *109*, 3559.

(8) Folkers, J. P.; Gorman, C. B.; Laibinis, P. E.; Buchholz, S.; Whitesides, G. M. *Langmuir* **1995**, *11*, 813–824.

(9) Aronoff, Y. G.; Chen, B.; Lu, G.; Seto, C.; Schwartz, J.; Bernasek, S. L. *J. Am. Chem. Soc.* **1997**, *119*, 259. Laibinis, P. E.; Hickman, J. J.; Wrighton, M. S.; Whitesides, G. M. *Science* **1989**, *245*, 845.

(10) Woodward, J. T.; Ulman, A.; Schwartz, D. K. *Langmuir* **1996,** *12,* 3626. Gao, W.; Dickinson, L.; Grozinger, C.; Morin, F. G.; Reven, L. *Langmuir* **1996**, *12*, 6429.

(11) Maege, I.; Jaehne, E.; Henke, A.; Adler, H.-J. P.; Bram, C.; Jung, C.; Stratmann, M. *Macromol. Symp.* **1997**, *126*, 7–24.

(12) Brovelli, D.; Hähner, G.; Ruiz, L.; Hofer, R.; Kraus, G.; Waldner, A.; Schlösser, J.; Oroszlan, P.; Ehrat, M.; Spencer, N. D. *Langmuir* **1999**, *15*, 4324.

Published on Web 02/17/2000

deposition onto Corning glass substrates (150 nm oxide layer thickness, sub-nanometer average roughness, from Balzers AG, Liechtenstein). The cleanliness of the surface of the tantalum pentoxide layer was tested using XPS and ToF-SIMS.

2.2. Synthesis of Octadecylphosphoric Acid Ester. Octadecylphosphoric acid ester ($C_{18}H_{37}OPO(OH)_2$) was prepared according to the protocol reported by Okamoto.[13] It is a stable, waxy solid and was recrystallized from hot *n*-hexane. Details of the synthesis protocol and of the chemical analysis have been published separately.[12] Elemental analysis (weight %), data are given again as they are needed for the discussion of the XPS quantification: C, 61.72; H, 11.02; P, 8.82; O, 18.44 (remaining amount added up to 100%). The atomic ratios of H/C, C/P, and O/P of 2.13, 18.04, and 4.05, respectively, calculated from these elemental analysis data, are in good agreement with the values expected for the formal stoichiometry of the compound (2.17, 18.00, and 4.00).

For the sake of simplicity, the immobilized molecule will simply be referred to as octadecyl phosphate or ODP in the following text.

2.3. Self-Assembly Protocols. Octadecylphosphoric acid ester ($C_{18}H_{37}OPO(OH)_2$) was dissolved in *n*-heptane (Uvasol)/2-propanol (Uvasol) from MERCK in a 100/0.4 (v/v) solvent mixture at a 500 μM concentration. Solutions were filtered using a 0.2 μm cellulose nitrate filter and stored until use.

Ta_2O_5-coated glass substrates were cleaned in an ultrasonic bath (BRANSON 3200) in 2-propanol for 15 min followed by UV/ozone cleaning (BOEKEL model 135500, Boekel Ind. Inc., PA) for 30 min. A SAM was formed by a subsequent immersion in the octadecylphosphoric acid ester solution for up to 48 h. Following immersion, the substrates were removed from the solution and rinsed with 2-propanol, blow-dried with He, and stored in air until analysis. The ODP SAM is stable for several hours in the *n*-heptane/2-propanol mixed solution, as well as in pure 2-propanol, and for weeks if stored in air.

The quality and uniformity of the self-assembled monolayers of ODP were checked by water contact-angle measurements and microdroplet imaging techniques,[14] as contact angle and hydrophobicity are known to be extremely sensitive to the degree of coverage and order in SAMs with nonpolar, hydrophobic tails.

2.4. Surface Investigation Techniques. 2.4.1. Atomic Force Microscopy. AFM measurements were performed with a commercial scanning probe microscope (Nanoscope E, Digital Instruments, Santa Barbara, CA). Measurements of surface topography and lateral force were made simultaneously by operating the instrument in the contact mode while scanning the cantilever laterally. For AFM probes, we used sharpened Si_3N_4 microlevers (Park Scientific, Sunnyvale, CA) with a nominal probe tip radius of 20 nm and a force constant of 0.03 N/m. Only those probes that provided good quality in the high-resolution imaging of freshly cleaved mica were employed for the imaging of the ODP samples. The applied load during scanning was in all cases below 0.5 nN and generally a slight "pulling" of the tip was necessary, i.e., applying negative loads, to get best resolution. All measurements were performed in ambient air.

2.4.2. X-ray Photoelectron Spectroscopy. XPS analyses were performed using a PHI 5700 spectrophotometer equipped with a concentric hemispherical analyzer in the standard configuration (Physical Electronics, Eden Prairie, MN). Spectra were acquired at a base pressure of 10^{-9} mbar using a nonmonochromatic Al Kα source operating at 200 W and positioned ~13 mm away from the sample. The instrument was run in the minimum-area mode using an aperture of 0.8 mm diameter. The CHA was used in the fixed analyzer transmission mode. Pass energies used for survey scans and detailed scans for tantalum Ta4f, carbon C1s, oxygen O1s, and phosphorus P2p were 187.85 and 23.5 eV, respectively. Under these conditions, the energy resolution (full width at half maximum height, fwhm) measured on silver $Ag3d_{5/2}$ is 2.7 and 1.1 eV, respectively. Acquisition times were approximately 5 min for survey scans and 9 min (total) for high-energy resolution elemental scans. These experimental conditions were chosen in order to have an adequate signal-to-noise ratio in a minimum time and to limit beam-induced damage. Under these conditions, sample damage was negligible, even after 90 min of X-ray exposure, and reproducible analyzing conditions were obtained on all samples. In addition, only one sample was introduced into the analyzing chamber at a time.

Angle-resolved XPS (AR-XPS) measurements were conducted at different takeoff angles (detection angle), namely 15, 45, 75, and 90° with respect to the surface plane, to obtain depth-dependent information and to determine the octadecylphosphoric acid ester monolayer thickness deposited on the tantalum oxide substrate.

Spectra were referenced to the aliphatic hydrocarbon C1s signal at 285.0 eV.

Data were analyzed using a least-squares fit routine following Shirley iterative background subtraction. Atomic concentrations were calculated using published ionization cross sections[15] and calculated attenuation length values.[16] Intensities were also corrected for the energy dependence of the transmission function. Spectra were fitted using Gaussian–Lorentzian functions.

As a reference, octadecylphosphoric acid bulk powder pressed onto an indium foil was analyzed with a takeoff angle of 45° with respect to the surface. Both as-received and sputter-cleaned bare tantalum pentoxide substrates were analyzed as reference substrates.

2.4.3. Time-of-Flight Secondary Ion Mass Spectrometry (ToF-SIMS). Secondary ion mass spectra of the bare Ta_2O_5 and the ODP-treated Ta_2O_5 surface were recorded on a PHI 7200 time-of-flight secondary ion mass spectrometer in the mass range 0–1000 *m/e*. The total ion dose of the 8 kV Cs^+ primary ion beam (200 μm diameter) was typically 9×10^{11} ions·cm^{-2}, corresponding to a value below the static limit. Time per data point was 1.25 ns. Due to the low conductivity of the Ta_2O_5/glass substrate, intermittent, pulsed electron-beam neutralization had to be used during measurement of both positive and negative secondary ion mass spectrometry (SIMS) spectra. Mass resolution $M/\Delta M$ was typically 5500 in the positive and 2000 in the negative mode (mass 43 and 17, respectively). To calibrate the mass scale, the whole mass range was first calibrated using a single standard set of low ion masses followed by the assignment of species in the whole mass range using the PHI software TOFPAK. To improve the quality of the mass calibration at higher masses, different sets of ion species were used due to the large mass range analyzed and the very different nature of the observed secondary ion species (likely to leave the surface with varying velocities): (a) low mass range (< ca. 200), C_nH_m species (positive ions) and OH^-, PO_2^-, PO_3^- (negative ions); (b) medium mass range, $Ta_aO_b^{\pm}$ and $TaP_aO_bH_c^{\pm}$ species (both types of ions); (c) high mass range (> ca. 400), $TaP_aO_bC_cH_d^+$ species (positive ions) and $Ta_2O_aH_b^-$ species (negative ions).

The mean calibration deviations from the exact mass of the assigned species were always below 50 ppm, in most cases below 20 ppm.

2.4.4. Contact Angle Measurements. Surface wettability was investigated by measuring the *advancing water contact angle* (contact angle measuring system, G2/G40 2.05-D, Krüss GmbH, Hamburg, Germany). The contact angle measurement (20 volume pulses of 0.22 μL each with a pulse frequency of 1 s) was performed at five different places on each chip, and the average contact angle value was determined.

3. Results

3.1. Atomic Force Microscopy. Figure 1 shows a lateral force image of an ODP layer on Ta_2O_5. Clearly visible is a certain degree of ordering in the layer in the form of small patterned areas. However, apart from these "ordered" regions, there are regions where no structure is clearly observable and the images are "blurred". Both structured and nonstructured regions are on the order of a few nanometers in diameter. The small "structured" regions display a roughly hexagonal pattern with an average nearest-neighbor distance of 0.49 ± 0.01 nm (see

(13) Okamoto, Y. *Bull. Chem. Soc. Jpn.* **1985**, *58*, 3393.
(14) Hofer, R.; Textor, M.; Spencer, N. D. In preparation.
(15) Scofield, J. H. *J. Electron Spectrosc. Relat. Phenom.* **1976**, *8*, 129.
(16) Seah, M. P.; Dench, W. A. *Surf. Interface Anal.* **1979**, *1*, 1.

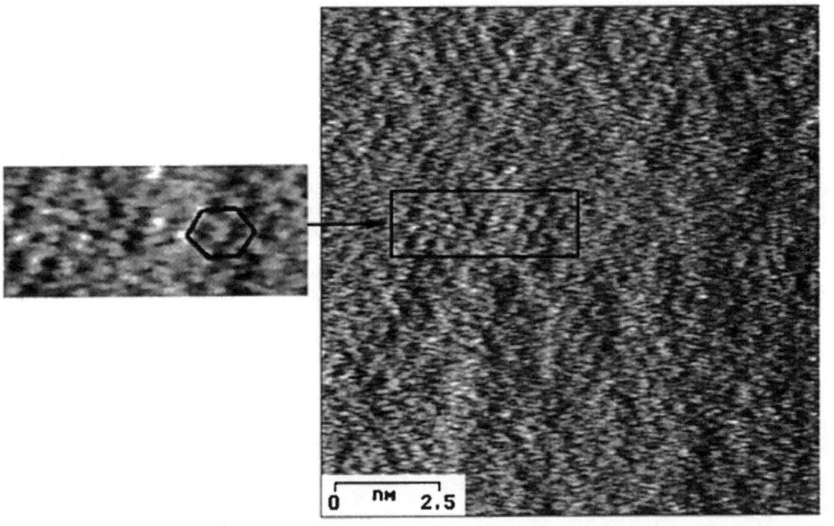

Figure 1. Lateral force image of an ODP layer on a Ta_2O_5 surface. The left image shows a slightly enlarged and smoothed version of the small inset shown in the image on the right-hand side. Also indicated is the local hexagonal structure.

Figure 1). Assuming this arrangement and published[17] bond lengths (0.16 nm for the O–P, 0.14 nm for O–C, and 0.17 nm for O–Ta), densities of the phosphate and hydrocarbon regions of 2.0 and 1.1 g/cm^3, respectively were calculated (this information is used later in the calculations of layer thicknesses and composition from XPS data). The pattern is similar to that found for alkanethiols on gold,[2] and we interpret the resolved "bumps" as being the terminal methyl groups of the alkane chains. AFM images over 1 μm^2 regions of the untreated Ta_2O_5 surface revealed a very flat, featureless topography with a roughness (R_a) of 0.2 nm.

3.2. X-ray Photoelectron Spectroscopy. 3.2.1. Reference Materials. Bulk Tantalum Oxide. Tantalum pentoxide was analyzed as a reference material to obtain the curve-fitting parameters for both O1s and Ta4f signals. The spectra were collected both on "as received" samples and following argon ion etching (sputtering) to remove the contamination layer. The curve-fitted spectra of Ta4f and O1s before (a) and after ion etching (b) are shown in Figure 2.

Curve fitting of the tantalum signal, Ta4f7/2 and Ta4f5/2, was performed using Gaussian–Lorentzian curves with $\Delta BE = 1.9$ eV and a branching ratio of 0.75. These two values were kept constant during the curve-fitting procedure of the SAM on tantalum pentoxide. The binding energy of the Ta4f was found at 26.4 ± 0.1 eV and the full width at half-maximum (fwhm) was 1.6 ± 0.1 eV. The oxygen signal, O1s, of the as-received sample (Figure 2a) showed contributions at 530.5 ± 0.1 eV, assigned to oxygen bonded to the tantalum ion, and at 531.9 ± 0.1 eV and 533.3 ± 0.1 attributable to hydroxides and water adsorbed on the contamination layer. The fwhm values were 1.7, 1.8, and 1.8 eV, respectively. After ion etching the component at 533.3 eV disappeared and the other minor component was greatly reduced (Figure 2b).

From the integrated intensities of Ta4f and O1s (530.5 eV) the composition of the tantalum pentoxide has been calculated with a three-layer model (for reasons of consistency with the calculations for the adsorbate system: see below) to be 81.4 ± 0.3 wt % Ta and 18.6 ± 0.3 wt % O, in very good agreement with the expected values of 81.9 wt % Ta and 18.1 wt % O.

Octadecylphosphoric Acid (Powder). The detailed spectra of C1s, O1s, and P2p obtained on bulk ODP powder (free acid) are shown in Figure 3. The C1s signal of the ODP powder is asymmetric, containing a contribution at 285.0 eV and one at 286.8 ± 0.2 eV. The first component is assigned to the carbon of the aliphatic chain and the second to the carbon covalently bond to one oxygen of the phosphate group (**C**-O–P). Two Gaussian–Lorentzian curves have been used in the curve-fitting routine applied to the O1s signal, which itself has been resolved into two signals: one at 532.1 ± 0.1 eV and the other at 533.6 ± 0.1 eV. The assignments have been carried out taking only initial chemical state effects into account and on the basis of literature data on sodium phosphate glasses.[18]

The P2p signal is a doublet with $2p_{1/2}$ and $2p_{3/2}$ components: their theoretical energy separation (0.9 eV) and the intensity ratio of 0.5 have been fixed for the curve-fitting analyses of all P2p spectra. The binding energy of the $P2p_{3/2}$ component was at 134.7 ± 0.1 eV. The results of qualitative and quantitative analysis are summarized in Table 1.

The carbon content (81.6 atom %) is higher compared to the expected values from elemental analysis (C = 78.3 atom %). This is probably due to a slight hydrocarbon surface contamination.

3.2.2. Qualitative Analysis of Self-Assembled ODP Monolayers on Tantalum Oxide. The self-assembled monolayers deposited from the *n*-heptane + 0.4 vol % 2-propanol solution were analyzed at four different takeoff angles: 15°, 45°, 75°, and 90°. Comparison of the survey spectra taken at various takeoff angles (Figure 4) shows a strong attenuation of the signals due to tantalum, oxygen, and phosphorus at the most grazing takeoff angle (15°).

The detailed spectra of C1s, O1s, P2p, and Ta4f taken at several different takeoff angles were resolved into their

(17) *Handbook of Chemistry and Physics*; CRC Press: Boca Raton, FL, 1997.

(18) Gresch, R.; Müller-Warmuth, W.; Dutz, H. *J. Non-Cryst. Solids* **1979**, *34*, 127.

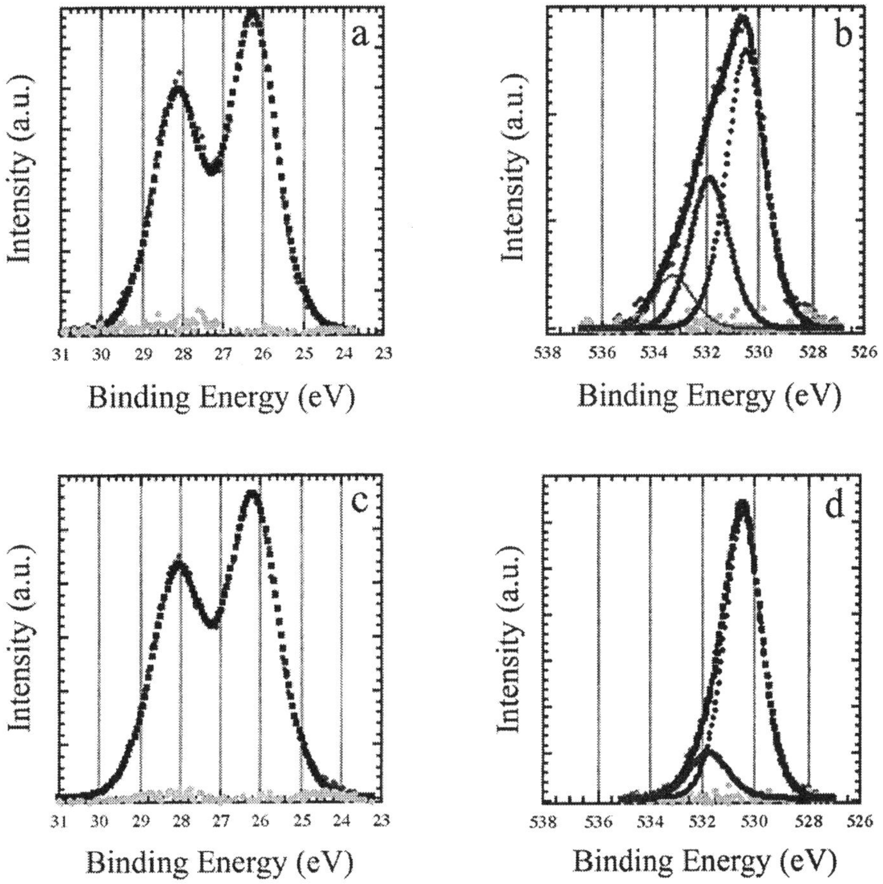

Figure 2. Curve-fitted Ta4f and O1s XP spectra collected on Ta_2O_5 before (a and b) and after (c and d) removing the contamination layer by ion etching.

Table 1. XPS Binding Energies (E_B ±0.1 eV), Full Width at Half-Maximum Height (fwhm), and Experimental Results of Quantitative Analysis of Bulk ODP (Free Acid)[a]

	C1s(1)	C1s(2)	O1s(1)	O1s(2)	P2p
assignments	**CH_2, CH_3**	**C**−O−P	P=**O**	P−**O**H, P−**O**C	C−O−**P**(O)(OH)$_2$
BE (eV)	285.0	286.8	532.1	533.6	134.7
fwhm	1.4	1.4	1.8	1.9	1.7
atomic ratio	17:1		1:3		
atom % (calcd)[b]	78.3		17.4		4.3
atom % (exptl)[c]	81.6 ± 1		14.5 ± 1		3.9 ± 0.5
atom ratio O/P (exptl)					3.7 ± 0.6
atom ratio O/P (ODP)[d]					4.05

[a] Comparison to atomic concentrations ("calcd"), calculated from the formal stoichiometry of $C_{18}H_{37}PO_4$. Photoelectron takeoff angle $\theta = 45°$. [b] Expected concentration based on formal stoichiometry. [c] Based on the use of photoelectron ionization cross sections[15] and attenuation length[16] values from the literature. [d] Based on elemental analysis of ODP powder (see section 2.2).

Table 2. XPS Binding Energy (BE) Values (±0.1 eV) of a Self-Assembled ODP Monolayer on Ta_2O_5 and Chemical Shifts (ΔE) Referred to the ODP Powder and Tantalum Oxide Substrate Energy Levels, Respectively[a]

	photoelectron emission peaks						
	C1s(1)	C1s(2)	O1s(1)	O1s(2)	P2p	O1s(3)	Ta4f$_{7/2}$
assignment	aliphatic	**C**−O−P				Ta_2O_5	
E_B (eV)	285.0	286.2	531.8	533.1	134.2	530.7	26.8
ΔE vs ref (eV)		−0.6	−0.3	−0.5	−0.5	+0.1	+0.4
intensity ratio O1s(1)/O1s(2)			1.6 ± 0.2				

[a] Average of data taken at different takeoff angles.

components by the curve-fitting procedure as described in section 2.4.1. The data are presented in Table 2.

The C1s signal was fitted with contributions from the aliphatic chain (285.0 eV) and the carbon bonded to the

SAMs on Tantalum Oxide *Langmuir, Vol. 16, No. 7, 2000* 3261

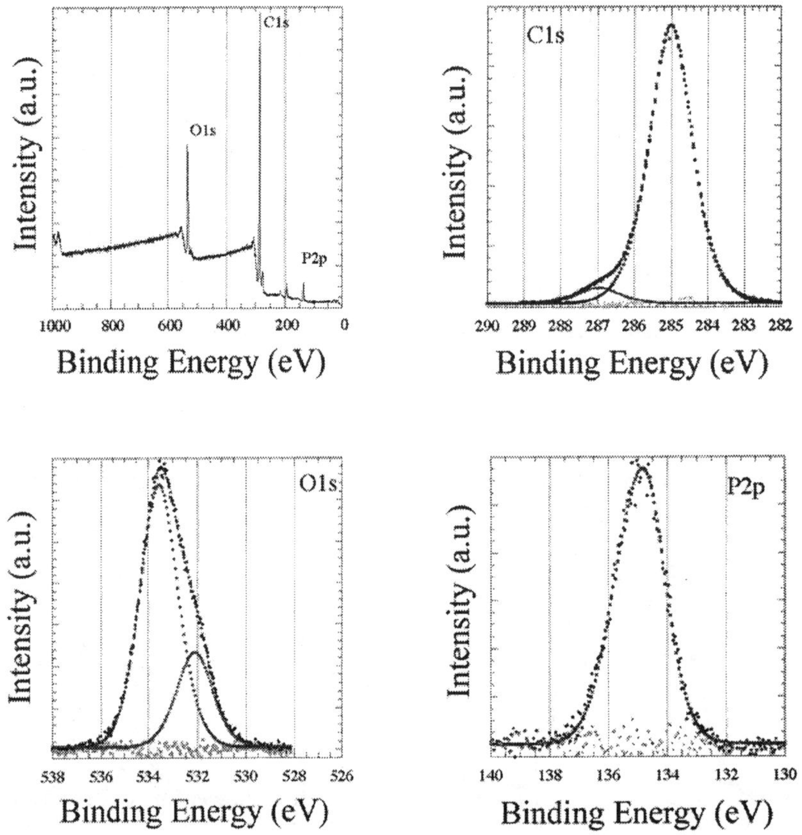

Figure 3. XPS survey spectrum and C1s, O1s, and P2p high-resolution spectra after background subtraction and curve fitting of ODP powder pressed onto indium foil. The difference between the original and the curve fitted spectrum is shown.

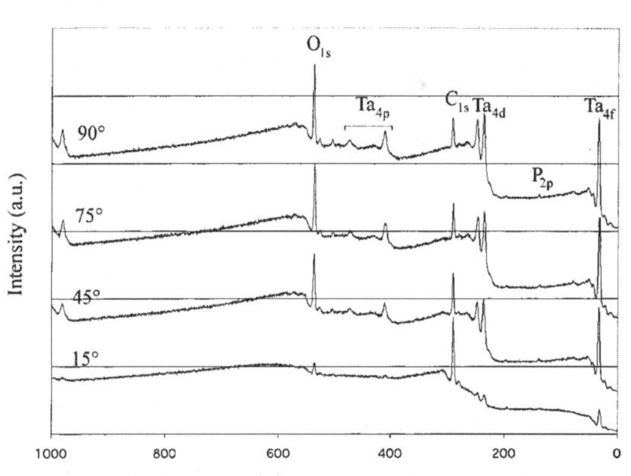

Figure 4. XPS survey spectra of self-assembled ODP monolayer on tantalum pentoxide obtained at different takeoff angles (15 and 75°, corresponding to approximate sampling depths of 2.5 and 9 nm).

phosphate group (286.2 ± 0.2 eV). The O1s signal is of particular interest regarding the assessment of the type of bond between ODP and the substrate. This signal was fitted with three contributions (Figure 5): the oxygen from the tantalum oxide was found at 530.7 ± 0.1 eV (O1s (3)) and the two oxygen components from the ODP molecule were found at 531.8 ± 0.1 eV (O1s (1)) and at 533.1 ± 0.1 eV (O1s (2)). The energy separation between O1s(2) and O1s(3) was 2.4 ± 0.1 eV for all data, independent of the emission angle. The energy separation O1s(2) − O1s(1) is found to be 1.35 ± 0.1 eV at the 15° takeoff angle. The intensity of the O1s(3) from the Ta_2O_5 substrate is a strong function of the chosen detection angle. The intensity decreases as the takeoff angle is reduced from 90° to 15° (grazing exit). The intensity ratio O1s(1):O1s(2) was determined from independent measurements taken at both takeoff angles; at 15°, the mean value was found to be 1.2 ± 0.2. At emission angles of 75° the intensity ratio was 1.9 ± 0.2. This difference may be attributed either to the influence in the curve fitting of the strong oxygen signal at 530.7 eV due to the substrate or to the fact that at this angle the contribution of the oxygen which faces the tantalum pentoxide is higher. Therefore, it has been decided to take the average over all the 22 measurements, equal to 1.6 ± 0.2.

The P2p signal showed a comparatively poor signal-to-noise ratio and was fitted with a single 2p doublet. The binding energy of the $P2p_{3/2}$ component was 134.2 ± 0.1 eV, independent of takeoff angle. The binding energy of the Ta4f was found to be 26.8 ± 0.1 eV.

The binding energy data in Table 2 show a shift of +0.4 eV for Ta4f and ca. −0.5 eV in C1s(2), O1s(1), O1s(2), and P2p, relative to the free acid (ODP powder) and the Ta_2O_5 substrate, respectively. The shift to lower binding energies of the ODP-related signals is likely to be due to a (full or partial) deprotonation of the phosphate acid head upon

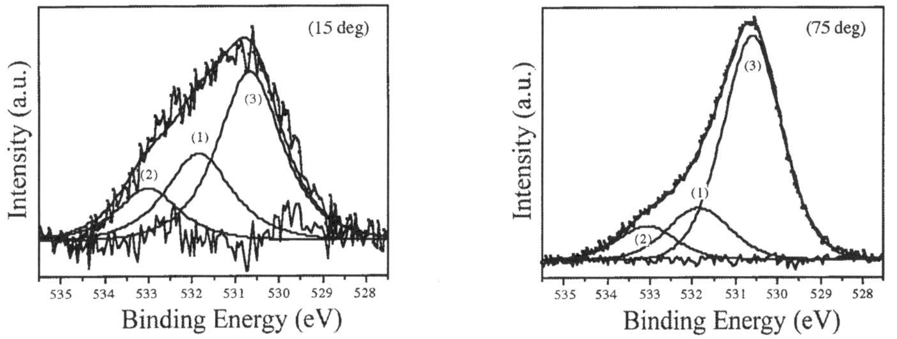

Figure 5. Curve fitting of the XPS O1s signal of an ODP SAM on Ta_2O_5 at 15 and 75° detection (takeoff) angle.

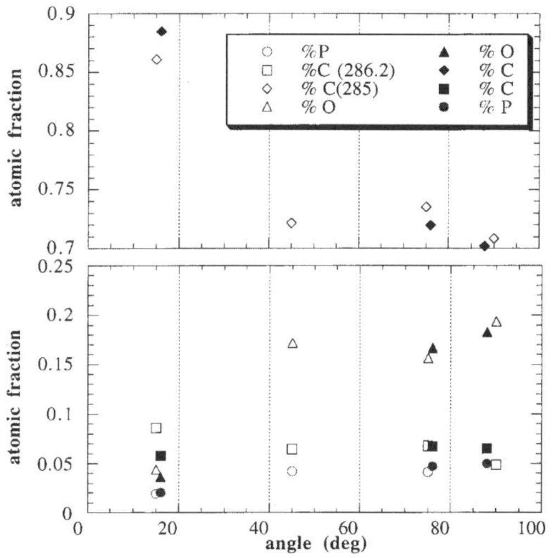

Figure 6. Angle-resolved XPS measurements of the elements C, O, and P in the ODP–SAM on Ta_2O_5. The atomic fractions were calculated from the intensities of the curve fitting procedure corrected for the sensitivity factors given by Scofield.[15] Results of two independent measurements (full and open symbols).

coordination to the surface and the corresponding formation of a negative charge on the phosphate headgroup. The positive shift of the Ta_2O_5 substrate suggests a charge transfer from the substrate to the ODP.

3.2.3. Angle-Resolved XPS: Quantitative Analysis of the ODP SAM on Tantalum Pentoxide. The results of the quantitative analysis of four independent angle-resolved XPS (AR-XPS) measurements are summarized in Figure 6. As the takeoff angle is reduced to 15°, the amount of the C1s assigned to the aliphatic chain (285.0 eV) increases and the contributions of the phosphorus, oxygen, and carbon components of the C–O–P head of the molecule decrease. This indicates that the polar head of the ODP molecule is located in the inner part of the ODP SAM.

In the evaluation of the thickness and composition of the ODP monolayer on the Ta_2O_5, the layered structure of the system under investigation has to be considered; it is therefore not particularly useful to calculate an overall percentage composition that does not take the different depth location of the various elements and groups into account. Instead, the depth origin of the different signals has to be borne in mind and the composition and thickness of each layer calculated within a multiple layer model as follows:

(a) First, all signals from Ta4f and the O1s(3) oxygen at 530.7 eV originate exclusively from the Ta_2O_5 substrate and are thus attenuated by the ODP–SAM.

(b) Second, the results of the angle-resolved XPS measurements (see Figure 6) indicate that the P2p signals originate from the inner part of the ODP layer. In other words, the P2p and O1s at 531.9 and at 533.2 eV are located on top of the Ta_2O_5. The phosphorus and the oxygen components of the polar head of the molecule can be considered as a very thin film containing only P and O atoms. Their XPS signal intensities are attenuated by the carbon chain of the ODP located at the top.

(c) Finally, the C1s contribution of the $C_{18}H_{37}$ hydrocarbon chain (and possibly of contamination) originates from the outermost part of the layer system.

On the basis of these observations, a "three-layer model" known from XPS analysis of thin oxide films on metallic substrates[19–21] has been applied to the ODP–SAM on tantalum oxide. This model allows the composition of (a) the substrate, (b) the P–O interfacial layer, and (c) the hydrocarbon top layer to be calculated, as well as the thicknesses of layers b and c, based on the intensities and the origin of the individual components of the elements as defined above. The density of the Ta_2O_5 substrate was taken as 8.6 g/cm^3, that of the hydrocarbon chains in the film as 1.1 g/cm^3, and that of the thin P–O polar head layer as 2.0 g/cm^3. The latter two values were estimated from AFM results, as outlined in section 3.1. Cross sections used for the calculations were taken from ref 15. The attenuation length values $\lambda(E_{kin})$ of the photoelectrons were calculated as $\lambda_I = B\sqrt{E_{kin}}$, with $B = 0.096$ for the inorganic compounds and $B = 0.087$ for the organic layer.[16] The parameters utilized in the three-layer model are listed in Table 3.

The composition and thickness of the Ta_2O_5 substrate and the ODP self-assembled monolayer, calculated with the three-layer model, are summarized in Table 4. The results are averaged over all experimental measurements, and no significant differences were found when analyzing data taken at different emission angles. For both the ODP layer and the substrate, good agreement is found between the expected composition and that calculated within the three-layer model. The thickness of the ODP layer (including both the polar head and the hydrocarbon chain)

(19) Asami, K., Hashimoto K. *Corros. Sci.* **1984**, *24*, 83 and references quoted therein.
(20) Rossi, A., Elsener, B. *Surf. Interface Anal.* **1992**, *18*, 499.
(21) Elsener, B., Rossi, A. *Electrochim. Acta* **1992**, *37*, 2269.

Table 3. Photoelectron Ionization Cross Sections σ and Electron Attenuation Length λ of Self-Assembled ODP Monolayer on Ta_2O_5, from Data Given in Refs 15 and 16

	photoelectron emission peaks						
	C1s(1)	C1s(2)	O1s(1)	O1s(2)	P2p	O1s(3)	$Ta4f_{7/2}$
assignment	aliphatic	C–O–P	see text	see text		Ta_2O_5	Ta_2O_5
σ (Scofield)	1	1	2.93	2.93	1.192	2.93	8.62
λ (nm) inorg			2.97	2.97	3.53	2.97	3.67
λ (nm) org	3.32	3.32	2.69	2.69	3.20	2.69	3.32

Table 4. Thickness and Composition of Self-Assembled ODP Monolayer on Ta_2O_5, Based on Evaluation of the XPS Data within a "Three-Layer Model"[19–21]

Thickness					
physical parameters	elements				
	carbon	oxygen	phosphorus	oxygen	tantalum
layers 1–3	ODP chain	polar head	polar head	substrate Ta_2O_5	substrate Ta_2O_5
thickness (nm)	1.4 ± 0.1	0.8 ± 0.1	0.8 ± 0.1	semi-infinite	semi-infinite

Composition				
	composition, atom %			
	oxygen	phosphorus	oxygen	tantalum
theoretical	80	20	71.4	28.6
experimental	77 ± 4	23 ± 4	70 ± 2	30 ± 2

was found to be 2.2 ± 0.2 nm at all emission angles investigated.

3.3. Time-of-Flight Mass Spectrometry. The most prominent secondary ion masses observed in the static SIMS spectra are listed in Table 5 (negative ions) and Table 6 (positive ions), together with their intensities relative to the most intense peak within each class of fragment species. SIMS spectra of the Ta_2O_5–ODP surface are only selectively shown in Figure 7 for the most informative mass range $m/z = 200–400$ (negative secondary ions), since all data are presented in Tables 5 and 6. However, spectra across the full mass range are available as Supporting Information.

The assignment of the masses to molecular ion species is discussed on the basis of the classes of fragments:

Class 0: C_mH_n Ions. These fragments occur on both the untreated and treated Ta_2O_5 surface and do not carry specific information about the nature and structure of the adlayer. They originate both from fragmentation of the ODP and from the naturally adsorbed (contaminant) hydrocarbons on the treated and the bare surface, respectively. They are not included in Tables 5 and 6.

Class I: $C_aH_bPO_c$ Ions. Ions of this type are only observed on the Ta_2O_5–ODP surface. Several species are observed in both modes, starting from the pure phosphate fragments (e.g., PO_2^-, PO_3^-, HPO_4^-, $H_2PO_4^-$, $H_4PO_4^+$) through various partly fragmented phosphoric ester species up to the molecular masses ($[M - H]^-$ and $[M \pm H]^+$). As expected, a certain amount of reduction of the phosphate (oxidation state +V) to the phosphonate (oxidation state +III) is taking place.

Class II: $Ta_aO_bH_c$. Fragments of this type occur in both the negative and positive modes and on both the treated and the bare Ta_2O_5 surfaces. Species corresponding to a formal stoichiometry of Ta(+V) and Ta (+III)—the most stable oxidation states in inorganic tantalum compounds—tend to have higher intensities compared to the others. This is a frequent observation in SIMS spectra of metal oxides and hydroxides. The general pattern of the $Ta_aO_bH_c^{\pm}$ ion fragments is similar on both the bare and the ODP-modified surface, although the distribution of intensities is somewhat different.

Class III: $Ta_aP_bO_cH_d$. The presence of strong peaks characteristic of tantalum phosphate and phosphonate species in both the positive and negative spectra of the

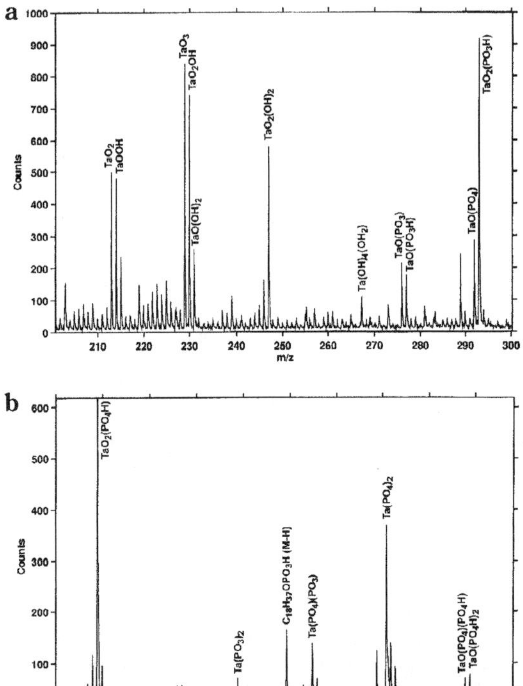

Figure 7. Selected ToF-SIMS spectra (negative secondary ions) of the ODP/Ta_2O_5 surface in the mass range $m/z = 200–300$ (a) and 300–400 (b). Spectra across the whole mass range investigated (0–600) are available as Supporting Information.

ODP-modified surface strongly suggests that the tantalum ions of the Ta_2O_5 oxide layer are actually directly bound to the chelating phosphoric acid group. It is unlikely that complex tantalum oxide phosphate species would be detected with high intensity if there were no direct complexation between the phosphate group and the tantalum(V) cation. In particular, assuming that the phosphoric acid binds to the oxide surface via hydrogen

Table 5. ToF-SIMS: List of the Most Prominent *Negative* Secondary Ion Masses (*m/e*) and Assignment to Molecular Fragment Species (M = Molecular Mass $C_{18}H_{37}OPO_3H_2$)

secondary ion mass [*m/e*] (obsd)	species charge: −1	formal oxidation state of Ta	formal oxidation state of P	dev of obsd mass[a] (ppm)	rel intens[b] (%)
	Class I: $C_aH_bPO_c$				
62.963	PO_2		+III	+4	23
78.958	PO_3		+V	+4	100
95.957	PO_4H		+V	+43	4
96.968	PO_4H_2		+V	+13	7
109.97	CH_3OPO_3		+V	+47	3
122.99	$C_2H_3OPO_3H$		+V	−19	2
165.07	$C_6H_{13}OPO_2H$		+III	−13	0.7
181.06	$C_6H_{13}OPO_3H$		+V	+14	3
349.25	$C_{18}H_{37}OPO_3H$ (M − H)		+V	−1	0.6
	Class II: $Ta_aO_bH_c$				
197.95	TaOH	0		+9	0.5
198.98	$TaOH_2$	-I		+95	0.8
212.94	TaO_2	+III		+11	1.3
213.95	TaOOH	+II		+31	1.3
228.94	TaO_3	+V		+13	2
229.95	TaO_2OH	+IV		+33	2
230.95	$TaO(OH)_2$	+III		0	0.8
246.96	$TaO_2(OH)_2$	+V		+51	1.6
266.97	$Ta(OH)_4(OH_2)$	+III		0	0.4
410.91	Ta_2O_2OH	+II		−54	0.5
411.87	$Ta_2O(OH)_2$	+I/II		+3	0.1
450.94	$Ta_2OH(OH_2)_4$	0		+3	0.04
458.87	Ta_2O_5OH	+V		+3	0.04
	Class III: $Ta_aP_bO_cH_d$				
275.90	$TaO(PO_3)$	+IV	+III	+9	0.5
276.91	$TaO(PO_3H)$	+III	+III	−11	0.4
291.90	$TaO(PO_4)$	+IV	+V	−16	1.0
292.91	$TaO(PO_4H)$ or $TaO_2(PO_3H)$	+III or +V	+V +III	+8	3
308.90	$TaO_2(PO_4H)$	+V	+V	+2	3
338.86	$Ta(PO_3)_2$	+V	+III	+27	0.2
354.85	$Ta(PO_4)(PO_3)$	+V	+V/+III	+28	0.4
370.86	$Ta(PO_4)_2$	+V	+V	+19	1.5
387.86	$TaO(PO_4)(PO_4H)$	+VI	+V	+4	0.2
388.88	$TaO(PO_4H)_2$	+V	+V	−48	0.2
504.84	$Ta_2(PO_2)(PO_3H)$	+II	+III	+25	0.2
520.83	$Ta_2(PO_3)(PO_3H)$	+II	+III	+13	0.05
538.83	$Ta_2O_5(PO_4H_2)$	+V	+V	+10	0.1
584.91	$Ta_2O_2(PO_4)(PO_4H)$	+III/+V	+V	−85	0.2
600.79	$Ta_2(PO_3)(PO_3H)_2$	+III	+V	+1	0.3
662.75	$Ta_2O(PO_4)_3$	+V	+V	+1	0.3
681.01	$Ta_2O(OH_2)(PO_4)_3$	+V	+V	−360	0.2
892.31	$Ta_3O_4(PO_4)_2(PO_4H)$	+V	+V	+419	0.2
	Class IV: $Ta_aP_bO_cC_dH_e$				
545.16	$TaO(O_3POC_{18}H_{37})$ = TaO·(M − 2H)	+III	+V	+53	0.06
561.18	$TaO{\cdot}(O_3POC_{18}H_{37})$ = TaO_2·(M − 2H)	+V	+V	+2	0.2

[a] Deviation of experimentally observed mass from exact mass of assigned species in ppm. [b] Intensity of the secondary ion peaks relative to the most intense peak (PO_3^- = 100%) of the whole spectrum.

bonding, species consisting of both tantalum oxides and the phosphoric acid group are unlikely to survive the emission process without further fragmentation into pure tantalum oxide and phosphate species, respectively. The fact that complex fragments such as $TaO(PO_4)(PO_4H)^-$, $Ta_2O_5(PO_4H_2)^-$, $Ta_2O_2(PO_4)(PO_4H)^-$, $Ta_2(PO_3)(PO_3H)_2^-$, $Ta_2O(OH_2)(PO_4)_3^-$, $Ta_3O_4(PO_4)_2(PO_4H)^-$, $Ta(PO_4H)_2^+$, or $Ta(OH)_3(PO_4H_2)^+$ in the negative spectra and $Ta(PO_4H)_2^+$ or $TaOH(PO_4H)(PO_4H_2)^+$ in the positive spectra are observed provides evidence for the close molecular packing of the self-assembled ODP molecules on the Ta_2O_5 surface and a binding scheme that involves, at least to some extent, more than one phosphate headgroup coordinated to one Ta ion.

Class IV: $Ta_aP_bO_cC_dH_e$. Prominent peaks corresponding to $TaO(O_3POC_{18}H_{37})^-$ = $TaO{\cdot}(M - 2H)^-$, $TaO{\cdot}(O_3POC_{18}H_{37})^-$ = $TaO_2{\cdot}(M - 2H)$ and fragmented species, such as $Ta(OH)_3(HO_3POC_2H_5)^+$, $Ta(OH)_2(O_3POC_6H_{13})^+$, or $Ta(OH)_3(HO_3POC_6H_{13})^+$, again support the presence of a strong bond between the tantalum (oxide) and the ODP molecules. The majority of these class IV species correspond to the formal, most stable oxidation of tantalum, i.e., Ta(+V). The observation of $Ta(PO_4H)(O_3POC_7H_{15})^+$ positive ions, albeit of weak intensity, suggests a close spacing of the phosphoric acid headgroups on the surface and, again, the coordination of two SAM molecules to the same Ta ion.

4. Discussion

4.1. Ta_2O_5 Substrate and Bulk ODP. XPS spectra of the bare Ta_2O_5 substrate showed, as expected, a 2:5 stoichiometry of Ta/O. In addition, the surface shows small peaks that can be attributed to −OH (hydroxide) and H_2O, as well as adventitious carbon contamination.

The interpretation and quantitative results of the XPS spectra obtained from bulk ODP are straightforward. The assignment of the curve-fitted C1s at binding energies of

Table 6. ToF-SIMS: List of the Most Prominent *Positive* Secondary Ion Masses (*m/e*) and Assignment to Molecular Fragment Species (M = Molecular Mass $C_{18}H_{37}OPO_3H_2$)

secondary ion mass [*m/e*] (obsd)	species charge: +1	formal oxidation state of Ta	formal oxidation state of P	dev of obsd mass[a] (ppm)	rel intens[b] (%)
	Class I: $C_aH_bPO_c$				
98.984	H_4PO_4		+V	+6	100
120.97	$C_2H_2OPO_3$		+V	+6	8
125.00	$C_2H_2OPO_3H_3$		+V	0	13
207.08	$C_8H_{13}OPO_3H_3$		+V	+20	10
214.12	$C_{11}H_{19}OPO$		+I	+30	10
219.08	$C_9H_{13}OPO_3H_3$		+V	+15	5
220.11	$C_{10}H_{18}OPO_2H_3$		+III	+53	9
223.12	$C_9H_{17}OPO_3H_3$		+V	+37	5
237.14	$C_{10}H_{19}OPO_3H_3$		+V	−76	5
249.16	$C_{12}H_{23}OPO_2H_3$		+III	+2	7
263.18	$C_{13}H_{25}OPO_2H_3$		+III	−5	5
275.18	$C_{14}H_{25}OPO_2H_3$		+III	−3	3
277.20	$C_{14}H_{27}OPO_2H_3$		+III	−35	4
349.25	$C_{18}H_{35}OPO_3H_3$ (M−H)		+V	−3	0.3
351.27	$C_{18}H_{37}OPO_3H_3$ (M+H)		+V	−2	0.2
	Class II: $Ta_aO_bH_c$				
180.92	Ta	+I		+182	26
181.93	TaH	0		+162	30
196.94	TaO	+III		0	30
197.96	TaOH	+II		+25	24
212.95	TaO_2	+V		+43	5
230.94	$TaO(OH)_2$	+V		−20	6
231.95	$Ta(OH)_3$	+IV		−17	0.4
232.95	$Ta(OH)_2(OH_2)$	+III		−44	7
	Class III: $Ta_aP_bO_cH_d$				
310.93	$Ta(OH)_2(PO_4H)$	+V	+V	−51	0.8
328.94	$Ta(OH)_3(PO_4H_2)$	+V	+V	−31	1.3
372.90	$Ta(PO_4H)_2$	+V	+V	−10	0.8
390.86	$TaOH(PO_4H)(PO_4H_2)$	+V	+V	+59	2
	Class IV: $Ta_aP_bO_cC_dH_e$				
356.96	$Ta(OH)_3(HO_3POC_2H_5)$	+V	+V	+2	1.0
395.01	$Ta(OH)_2(O_3POC_6H_{13})$	+V	+V	0	1.0
412.98	$Ta(OH)_3(HO_3POC_6H_{13})$	+V	+V	+85	0.6
470.98	$Ta(PO_4H)(O_3POC_7H_{15})$	+V	+V	−1	0.4

[a] Deviation of experimentally observed mass from exact mass of assigned species in ppm. [b] Relative intensity of the secondary ion peak relative to the most intense peak ($H_4PO_4^+$ = 100%).

285.0 and 286.8 eV to hydrocarbon and **C**−O−P, respectively, and of O1s at 532.1 and 533.6 eV to P=**O** (O type 1) and P−**O**−R (O type 2), respectively, are in agreement with expectations based on published reference data.[18,22] The experimentally determined O/P atomic ratio of 3.7 is consistent with the stoichiometry of the phosphate functional group (Table 1).

4.2. ODP on Ta_2O_5. 4.2.1. Investigation of Order by Atomic Force Microscopy. We found it challenging to obtain high-resolution AFM images of the ODP monolayer compared to that of dodecanethiols on gold, for example. Small regions showing periodic structures were observed (Figure 1), predominantly in friction mode and with low-pass filtering, and only under very moderate or negative applied loads. In contrast, high-resolution imaging of dodecanethiol on gold is straightforward: imaging is insensitive to the applied load and does not require filtering, and good quality images are easily obtained in lateral as well as in the height mode. Difficulty in imaging the periodic structure of the ODP monolayer might be due to partially weak bonds of the alkanephosphates with the substrate or the absence of strong cohesive forces in the film, due to partially aperiodic adsorption sites for the phosphates on the amorphous substrate. Slightly differing distances between neighboring molecules on the amorphous substrate might cause the disruption of the molecular order in the layer when the latter is subjected to the external pressure and torsional forces exerted by the AFM probe This could be the reason applied loads above 1 nN did not yield good resolution, although the *average* nearest neighbor distance and the chain length are similar to those of a thiol film on gold.

4.2.2. Orientation, Stoichiometry, and Thickness of the ODP Adlayer. The results of the angle-dependent XPS investigation on the ODP/Ta_2O_5 surface demonstrate that the terminal phosphate groups are oriented toward the Ta_2O_5 substrate surface. This is strongly suggested by the evolution of the carbon, oxygen, phosphorus, and tantalum XPS intensities as a function of electron emission angle (Figure 6). A three-layer model has been applied to calculate the thickness and stoichiometry within each layer (hydrocarbon, phosphate headgroup, and tantalum substrate) (Table 4). Assuming densities of each individual layer that were calculated from AFM data, and the molecular orientation discussed above, thickness and stoichiometry values for the individual layers could be calculated:

(a) The adlayer thickness 2.2 ± 0.2 nm calculated using the three-layer model is in excellent agreement with the value of 2.1 ± 0.05 nm calculated from the total length of the ODP molecule (2.5 nm) and from the (average) tilt angle of 30−35° determined experimentally from NEXAFS measurements.[12]

(b) The O/P atomic ratio of 3.4 ± 0.8, calculated from the curve-fitted O1s and from the P2p XPS peak areas,

(22) Moulder, J. F.; Stickle, W. F.; Sobol, P. E.; Bomben, K. D. *Handbook of X-ray photoelectron spectroscopy;* Perkin-Elmer Corporation/PHI Division: Eden Prairie, MN, 1992.

Table 7. XPS Binding Energies in Reference Compounds Depending on the Ratio of "Free" O Ligands (n) to Covalently Bound OR Ligands (m, R = H and P)[a]

$[PO_4]^{3-}$ $n = 4, m = 0$ $-y/4 = -0.75$	$[PO_3(OR)]^{2-}$ $n = 3, m = 1$ $-y/4 = -0.5$	$[PO_2(OR)_2]^-$ $n = 2, m = 2$ $-y/4 = -0.25$	$[PO(OR)_3]^0$ $n = 1, m = 3$ $-y/4 = 0$
PO_4^{3-} in Na_3PO_4: 132.4 eV PO_4^{3-} in Na_3PO_4: 132.4−132.5 eV	HPO_4^{2-} in Na_2HPO_4: 133.1 eV $P_2O_7^{4-}$ ($PO_{3.5}^{2-}$) in $Na_4P_2O_7$: 133.2−133.4 eV	$H_2PO_4^-$ in NaH_2PO_4: 134.2 eV PO_3^- in $NaPO_3$: 134.2−134.4 eV	H_3PO_4: ≈135 eV $PO_{2.5}$ in P_4O_{10}: 135.3−135.8 eV

[a] XPS binding energies (eV) of P2p in structures of the type $[PO_n(OR)_m]^{y-}$ with mean formal charge per O atom ($-y/4$). Data are from the literature[18,22] and referenced to the C(1s) of aliphatic hydrocarbons at 285.0 eV.

Table 8. XPS Binding Energies E_B for ODP in Bulk Form (Free Acid) and as a SAM on Ta_2O_5 Substrate[a]

experimental data	O1s E_B for O of type 1 (eV)	O1s E_B for O of type 2 (eV)	O1s E_B for O of type 3 (eV)	atomic ratio of O(1)/O(2)
ODP bulk powder ($\theta = 45°$)	532.1	533.6		0.33 ± 0.03 (theor. 0.333)
ODP SAM on Ta_2O_5 (θ = 15 and 75°)	531.8	533.1	530.7	1.6 ± 0.2
reference data[b]	531.7−532.1	533.1−534.3	530.6−530.8	
Model Calculation for Different Coordination Regimes				
ODP type A coordination (bidentate)[c]				3:1 = 3.0
ODP type B coordination (monodentate)[c]				2:2 = 1.0
ODP type C coordination (tridentate)[c]				3:1 = 3.0
ODP type A (1 mol) *plus* type B (2 mol) coordination[c]				7:5 = 1.40

[a] Comparison with different theoretical models for the coordination of the phosphate head groups to tantalum cations at the oxide surface. [b] From literature data[18,22] and references therein. [c] See Figure 8.

is consistent with the expected ratio of 4:1 for the phosphate group. The relatively high standard deviation is due to the low signal-to-noise ratio for the P2p emission.

(c) The O/Ta atomic ratio of 2.3 ± 0.3 is close to the expected value for the Ta_2O_5 stoichiometry.

4.2.3. ODP Substrate Bonding. In this section we discuss the observations from ToF-SIMS and XPS related to the question of the bonding mechanism of the phosphate headgroup of the ODP molecule to the tantalum oxide substrate.

Time-of-Flight Secondary Ion Mass Spectrometry (ToF-SIMS). The positive and negative ToF-SIMS spectra showing a variety of fragments corresponding to the classes I to IV (see section 3.3) provide conclusive information as regards the different potential models for the coordination of the phosphate headgroup to the Ta cations:

(a) The fact that prominent peaks of class III ($Ta_aP_bO_cH_d$) and IV ($Ta_aP_bO_cC_dH_e$) are observed gives strong evidence for a direct coordination of the phosphate group to Ta cations. If the binding of the phosphate group were weak, such as hydrogen bonding to tantalum oxides and hydroxides at the surface (e.g., $ROPO(OH)_2\cdots O{-}Ta$), it would be unlikely that complex fragments of types III and IV would survive during the fragmentation and detection process and lead to mass peaks of relatively high intensity. Moreover, many of the observed fragments such as m/e = 371 (negative), 391 (positive), 561 (negative), etc. cannot readily be explained assuming indirect coordination via P−O· · ·H−O−Ta or P−O−H· · ·O−Ta bonding.

(b) The fact that a very particular pattern of fragment stoichiometries is observed supports the findings of a preferred model of coordination, discussed in detail in section 4.3. The fragmentation types displayed in Table 9 are experimentally observed in the positive and/or negative SIMS spectra.

The findings are exactly what one would expect for a model of coordination of ODP on Ta_2O_5 involving the presence of both bidentate ODP bound to one Ta ion and of monodentate coordination of two ODP molecules to one Ta ion. The assumption is that the secondary fragments in (static) ToF-SIMS do reflect the original molecular structure of the surface and that recombination of fragments that are originally apart from each other is not a likely process during secondary ion formation.

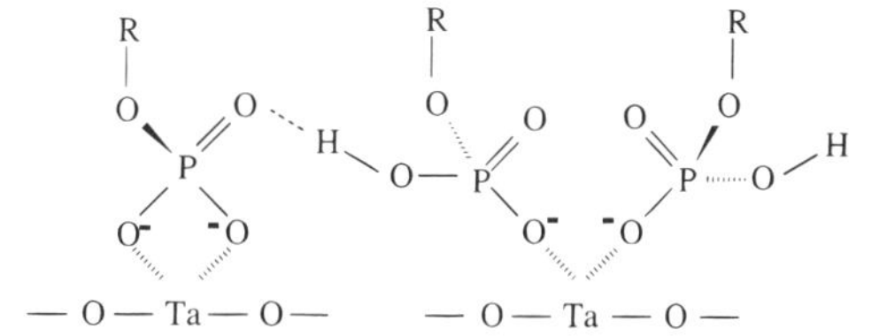

Figure 8. Bidentate (type A, left) and monodentate (type B, right) phosphate coordination to tantalum ions, with the possibility for the formation of intermolecular hydrogen bonding.

Table 9. Observed Combinations m/n for Secondary Fragments of the Type $Ta_n(PO_x)_m$ in the Positive and Negative ToF-SIMS Spectra of ODP SAM on Ta_2O_5

no. (n) of Ta atoms in secondary ion fragment	no. (m) of PO_x groups in secondary ion fragment
1	1
1	2
2	1
2	2
2	3
3	3

On the basis of the structure and preferred coordination of tantalum(V) in oxides, structures of the molecular ion species corresponding to some of the more interesting fragments of Table 5 and Table 6 are proposed in Figure 9.

The predominant structural pattern in the crystalline state of Ta_2O_5 is characterized by 6-fold coordination of Ta and by edge-sharing octahedra, leading to the 2:5 stoichiometry.[23] Although the sputtered Ta_2O_5 used as substrate in this study is amorphous to nanocrystalline, it is still highly likely that the local, short-range order environment is similar to that of the crystalline state. The species proposed in Figure 9 can be seen as fragments of the original Ta_2O_5 polymeric structure with coordinatively bound phosphate moieties. The fact that the coordination number of the Ta atoms in these fragments is generally lower than the preferred value of six is likely

(23) Wells, A. F. *Structural Inorganic Chemistry*; Clarendon Press: Oxford, 1991.

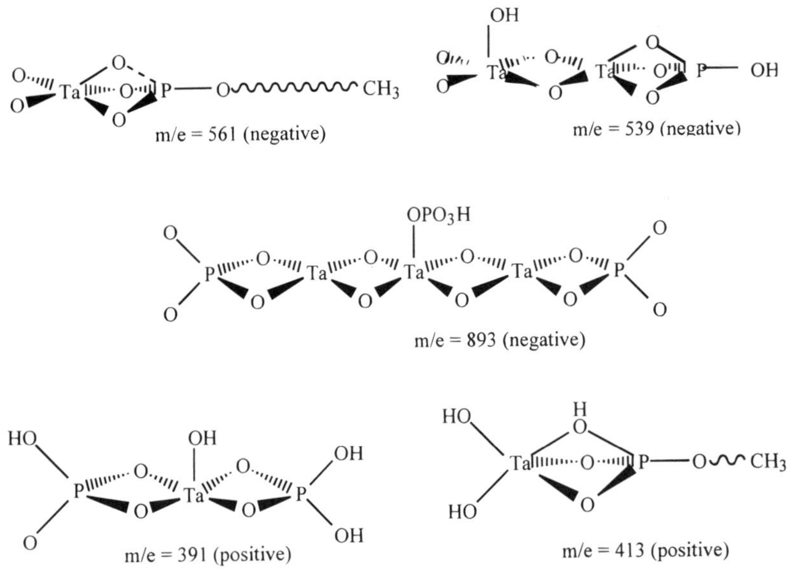

Figure 9. Proposed structures for some of the observed fragments in the negative and positive ToF-SIMS spectra of the ODP SAM monolayer on tantalum oxide (Ta_2O_5).

to be a consequence of the preferred oxidation state of Ta being +V, +IV, or +III. In the proposals of Figure 9, the coordination number of tantalum in the fragments has been maximized due to physicochemical considerations, although it is proposed that in the original SAM layer only mono- and bidentate phosphate coordination occurs (as opposed to 3-fold coordination).

XPS Binding Energies and Chemical Shifts. The interpretation of the binding energies of the P2p and of the curve-fitted O1s spectra provides an insight into the binding of the phosphate headgroup of the ODP molecule at the surface.

P2p Binding Energy. The experimental binding energies E_B of chemical moieties of type $[MO_n(OR)_m]^{y-}$ generally follow a systematic rule: an increase of E_B as the ratio $n{:}m$ of "free" O ligands (n) to covalently bound OR ligands (m) is stepwise increased from 4:0 to 3:1, to 2:2, etc. This is a consequence of a systematic dependence of the partial charge of the M atom on the chemical environment (nearest and next-nearest neighbor atoms). In our case, M corresponds to P and R is either H or C. This incremental increase for metal phosphate[18] and PO_x structures[23] is typically about 1 eV (Table 7).

The experimental value of $P2p_{3/2}$ at 134.7 eV (Table 1) for the *free acid (bulk ODP powder)* agrees with the general trend in Table 7, being closest to the $[PO(OR)_3]^0$ ("free acid") case. The corresponding value of 134.2 eV for the *ODP SAM*, however, is definitely lower, suggesting a change of the chemical structure of the phosphate headgroup upon coordination to the surface. The value of 134.2 eV lies close to the reference values for $[PO_2(OR)_2]^-$. However, the P2p binding energy is further affected by the observed charge transfer of approximately 0.5 eV between substrate and adlayer (see Table 2). Assuming no charge transfer (i.e., $Ta4f_{7/2}$ at 26.4 eV as in the case of the bare Ta_2O_5), the position of the P2p binding energy would be shifted in the direction of the reference value for $[PO_3(OR)]^{2-}$. It cannot be excluded therefore that both $[PO_2(OR)_2]^-$ and $[PO_3(OR)]^{2-}$ may coexist at the surface and in that respect this is not in contradiction to our preferred model of both monodentate and bidentate coordination of the alkane phosphoric acid headgroup to tantalum cations (Figure 8) as presented and discussed in detail in the following subsections. However our view is that it is not possible to draw a final conclusion just based on the XPS P2p signal and further experimental evidence for one or the other binding model is needed (XPS O1s binding energies and ToF-SIMS data, see below). What can be definitely excluded from the observed P2p binding energy of the ODP adlayer is the presence of substantial concentrations of either the free acid or of "free" phosphate $[PO_4]^{3-}$. In such a case the P2p signals would have to be clearly different from the experimentally observed value.

The P2p signal does not, in fact, show any evidence of asymmetry due to different chemical states, although the P2p binding energies in the type A and type B environments would be expected to be different by 1 eV or so. We believe that the reason is intermolecular hydrogen bonding within the SAM layer, leading to partial charge transfer between adjacent phosphate groups and a leveling of the differences in partial charge on the P atom in the type A and type B coordination situation.

O1s Binding Energies. While the O1s signal of the ODP bulk powder shows two different chemical states at E_B = 532.1 (O type 1) and 533.6 eV (O type 2), respectively (Table 1), the O1s spectrum of the ODP SAM (Figure 4, Table 2) shows a third component, due to O from the Ta_2O_5 substrate at $E_B = 530.7 \pm 0.1$ eV (O type 3). Table 8 summarizes the quantitative XPS O1s data obtained. Different models of the ODP complexation to the surface have been tested with respect to how closely they reflect the observed experimental data. These include monodentate coordination of two ODP units (two $C_{18}H_{37}OP(OH)O_2^-$) to one Ta cation (type B bond in Figure 8), for which the expected atomic ratio O(1)/O(2) is 2:2; bidentate coordination of one ODP ($C_{18}H_{37}OPO_3^{2-}$) to one Ta cation (type A bond in Figure 8), for which the expected atomic ratio O(1)/O(2) is 3:1; and finally, tridentate coordination, previously proposed for the coordination of phosphoric acid carriers to titania by Busca et al.[24] and for which the expected atomic ratio O(1)/O(2) is again 3:1. None of the simple models assuming a *single* type of coordination of the phosphate headgroup to Ta(V) cations turns out to be in close agreement with the experimental findings (Table 8).

The data listed in Table 8 show the best agreement with a coordination regime based on a mixed monodentate and bidentate binding of the headgroup to the Ta(V) cations. The experimental atomic ratio of the two different O atoms O(1)/O(2) is 1.60 ± 0.20 (average of 11 individual measurements at emission angle of 15 and 75°) and thus is consistent with the mixed model, for which the theoretical value is 7:5 or 1.40. Another possibility that would formally be consistent with the observed O(1)/O(2) ratio is a mixture of tridentate (type C) and monodentate (type B) coordination. This, however, is less likely from the point of view of the preferred coordination number of Ta(V) being 6 or 7, rather than 8.

The difference in E_B between O(1) and O(2) is somewhat smaller than expected from the reference data[18] (1.7–1.9 eV); again we believe this to be due to intermolecular hydrogen bonding as already discussed above in the context of the P2p signals.

4.3. Structure and Formation of the ODP/Ta_2O_5 Surface. To construct a reasonable model of the ODP–Ta_2O_5 system, the following observations need to be accounted for: (a) the NEXAFS[12] evidence for chain order and an average tilt angle of 30–35°; (b) the AFM evidence of local nearly hexagonal order; (c) the ToF-SIMS evidence for P–O–Ta bonding; (d) the ToF-SIMS evidence for coordination of more than one phosphate to a single tantalum; (e) the XPS evidence for tails-up orientation, possible charge transfer from substrate to ODP, an adsorbed layer thickness of 2.2 nm, the possible presence of both $[PO_3(OR)]^{2-}$ and $[PO_2(OR)_2]^{-}$ species, and the inability of a single type of coordination to account for the observed ratio between different oxygen environments.

4.3.1. Molecular Model. The Ta_2O_5 coating was deposited by physical vapor deposition and is—according to the manufacturer—nanocrystalline to amorphous. However, even for an XRD amorphous structure, short-range order is likely to be present, with the Ta cations in preferred oxide coordination symmetries. The further discussion is based on the assumption of preferred coordination of Ta cations and a short-range order deduced from considerations of the structure in crystalline Ta_2O_5.

The low-temperature form of Ta_2O_5 (L-Ta_2O_5, below 1360 °C) is characterized by chains built from octahedral (and partly bipyramidal) coordination groups.[23] These chains are linked to each other by edge and vertex sharing to form the 3D network and satisfy the overall stoichiometry $TaO_{2.5}$. The coordination number in the bulk structure of Ta_2O_5 is 6 and 7. The relative proportion of the different structural elements, however, varies depending on the synthesis conditions; Ta_2O_5 (as well as other M_2O_5 oxides such as Nb_2O_5) shows a pronounced tendency to polymorphism or polytypism.[25]

The combination of XPS, ToF-SIMS, AFM, and NEXAFS[12] results is believed to constitute conclusive evidence for the presence of ordered monolayers of ODP tilted by an angle of 30–35° relative to the surface normal and for a coordination regime that involves both unidentate and bidentate direct binding of the phosphate headgroup to the tantalum cations at the surface of the Ta_2O_5 substrate.

(24) Busca, G.; Ramis, G.; Lorenzelli, V.; Rossi, P. F.; Ginestra, A. L.; Patrono, P. *Langmuir* **1989**, *5*, 911.

(25) Cotton, F. A.; Wilkinson, G. *Advanced Inorganic Chemistry*, 5th ed.; Wiley: New York, 1988.

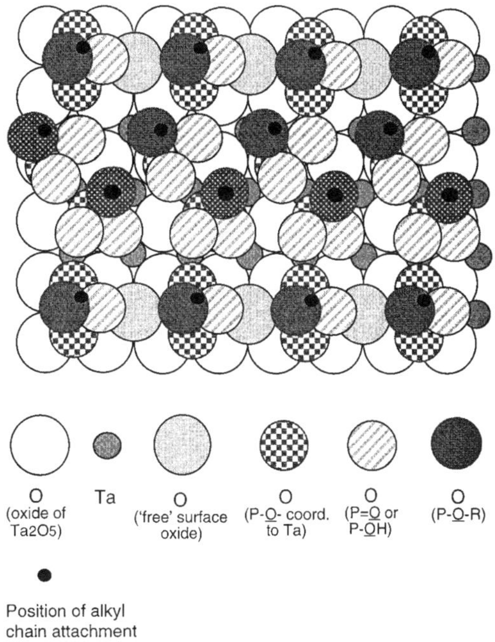

Figure 10. Schematic, idealized view of the arrangement and orientation of phosphate groups of ODP at a Ta_2O_5 surface (with square substrate lattice). The phosphates are bound to the Ta ions through either unidentate or bidentate coordination. The P–O–R groups form a nearly perfect hexagonal lattice with a mean distance between the hydrocarbon (R) chains of approximately 0.49 nm (corresponding to a specific area of 0.21 nm^2 per molecule).

Assuming a certain degree of (short-range) order in the oxide substrate, a model of packing of the phosphate groups on top of the octahedral Ta ion sites with hexagonal structure is proposed. Ionic radii of 0.14 nm for O(–II) and 0.064 nm for Ta(+V), an O covalent radius of 0.125 nm, and bond lengths of 0.15–0.16 nm for P=O and *P–O–*R(H) have been assumed. There is a particular arrangement of the phosphate headgroups on the square tantalum oxide lattice that satisfies the assumption of monodentate and bidentate ODP coordination (in the molecular ratio of 2:1) and at the same time leads to an approximately close-packed phosphate ligand ordering at the surface as shown in Figure 10. The AFM study provides direct evidence for such a nearly hexagonal, nearly close-packed adlayer, although the order is localized to rather small regions, possibly due to the generally noncrystalline nature of the surface, where only local order can be expected. Tantalum cation sites not coordinated to phosphate are likely to be linked to oxygen, hydroxide, or water molecules to complete the coordination sphere. For the sake of clarity this is only partly shown in Figure 10 (as "free" surface oxide).

A possible arrangement of a row of ODP molecules is shown in Figure 11, based on a ball-and-stick model. There is a particular arrangement that allows hydrogen bonding between two adjacent ODP molecules (types A and B respectively, see Figure 8) while keeping the hydrocarbon chains at the distance of approximately 0.5 nm, known to be a favorable distance for strong intermolecular van der Waals interactions in long-chain, alkane-based SAMs. The angle of the hydrocarbon chain in the molecular model arrangement of Figure 11 is approximately 30°, as calculated from experimental NEXAFS measurements.[12]

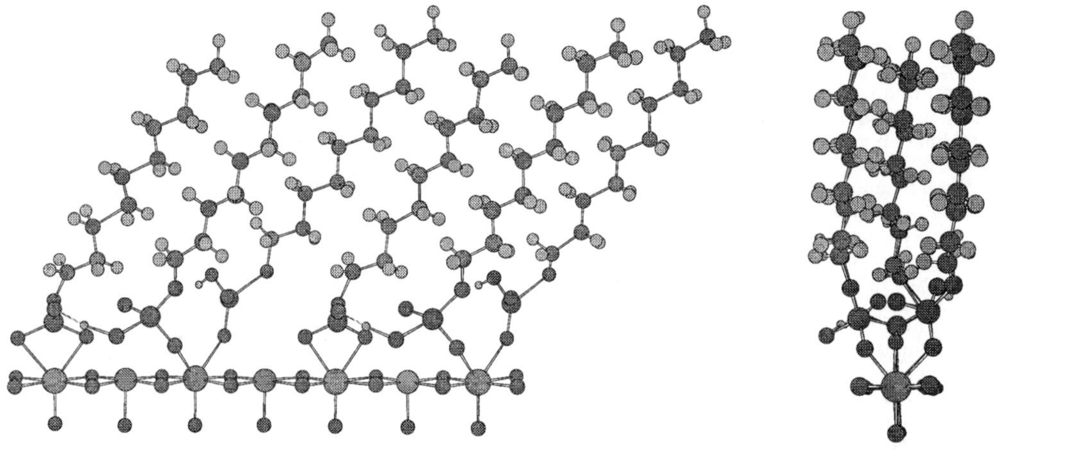

Figure 11. Ball-and-stick model of six adjacent ODP molecules with monodentate and bidentate coordination (in the ratio of 2:1) to the Ta_2O_5 substrate surface (see Figure 8 for binding of the phosphate group to the substrate). Only 10 carbon atoms of the alkane chain are shown for the sake of simplicity. The tilt angle of the hydrocarbon chain is approximately 30° relative to the surface normal.

Table 10. Comparison of Structural Parameters of Octadecylphosphoric Acid on Tantalum Oxide, Octadecylphosphonic Acid on Mica and Alkanethiols on Gold

SAM on substrate	length, l (nm)[a]	thickness, d (nm)[b]	angle, v (deg)[c]	area, A (nm^2)[d]	distance, d' (nm)[e]	technique (ref)
octadecylphosphoric acid on Ta_2O_5	2.5	2.2 ± 0.3	30–35	0.209 ± 0.04	0.49 ± 0.01	NEXAFS (ref 12); AFM, molecular model, XPS, SIMS (this work)
octadecylphosphonic acid on mica	2.5	1.8 ± 0.2	(~40)	0.188–0.220	0.47–0.5[f]	isotherm, AFM (ref 10)
alkanethiols on gold (C6–C22)	1.0 (C6) 1.5 (C10) 1.7 (C12) 2.2 (C16) 2.5 (C18) 3.0 (C22)	~0.9 ~1.3 ~1.5 ~1.9 ~2.2 ~2.6	~30 (for all alkanethiols)	0.214 (for all alkanethiols)	0.50[f]	contact angle, FTIRAS, XPS (ref 1)

[a] Total length l of extended molecule. [b] Thickness d of SAM layer d (measured perpendicular to surface). [c] Average angle v between axis of molecule and surface normal. [d] Average area A occupied per molecule. [e] Average intermolecular spacing d' between adjacent molecules, measured parallel to the surface. [f] Assuming a hexagonal arrangement of assembled molecules.

The parallel ordering of the hydrocarbon chain can be achieved by a single gauche $O-CH_2-CH_2$ conformation of the bidentate ODP molecule, while the adjacent monodentate ODP molecule (type B) has exclusively trans conformations. A consequence of such a conformational arrangement would be a slight difference in the height level of the terminal methyl group of ODP molecules A and B, respectively. This effect is likely to be too small to be detected by AFM, however.

The hydrocarbon chains—attached to the phosphate group at the dark-colored spots in Figure 10—form an approximately hexagonal pattern. On the basis of an assumed, idealized square substrate (Ta_2O_5) lattice of dimensions 0.28×0.28 nm^2, a $[4 \times 2]$-overlayer coincidence lattice can be formed. Within this overlayer lattice of 0.63 nm^2 dimension, 3 ODP molecules can be accommodated with each ODP molecule formally occupying an average area of 0.209 nm^2 at an average intermolecular distance (parallel to the surface) of 0.49 nm. The intermolecular distances and area of occupancy per SAM molecule according to AFM and model calculations are listed in Table 10, together with corresponding literature values for alkanethiols on gold and octadecylphosphonic acid on mica.

The structural geometric parameters found for the ODP/Ta_2O_5 system are indeed very close to those reported for alkanethiols on gold[26] as well as for octadecylphosphoric acid on mica[10] (the latter adlayer is, however, chemically not stable). For the alkanethiol/gold system there is general agreement that the lateral periodicity of the SAM is directly linked to the periodicity of the gold surface structure with the terminal sulfur occupying hollow sites in the gold substrate lattice.[26,27] Since the size of the alkanethiol molecule is too large to occupy every hollow site, an overlayer is formed that conforms with the steric requirements of the molecule. On Au(111), the most extensively studied single crystal plane, a $(\sqrt{3} \times \sqrt{3})R30°$ overlayer structure is formed. The observed specific tilt angle of the alkanethiol molecule is believed to result from a maximization of the van der Waals attraction between adjacent alkane chains within the SAM, leading to the energetically most favorable conformation.

It is tempting to use an argumentation based on the same or similar principle as in the gold/thiol system to explain the (local) order observed for the ODP–Ta_2O_5 system. In our proposed model, the periodicity of the Ta cations is again believed to be the prime factor determining

(26) Ulman, A. In *An introduction to organic thin films; from Langmuir–Blodgett to Self-Assembly*; Academic Press: San Diego, CA, 1991; p 279.
(27) Feuter, P.; Eisenberger, P.; Liang, K. S. *Phys. Rev. Lett.* **1993**, *70*, 2447.

order in the ODP adlayer. In view of the extremely high strength of the Ta−O bond (approximately 800 kJ/mol) and the generally high strength of transition metal−phosphate bonds, one expects deep potential wells for the phosphate at the Ta cation coordination sites. There are, however, two main differences between the two SAM systems: (a) there is no *simple* coincidence lattice for phosphate groups positioned right on top of the cation sites that results in a close-packed adlayer and at the same time maintains a realistic intermolecular phosphate−phosphate distance; (b) for the phosphate group, on the other hand, there are several possible geometries to coordinate to cation,: one-, two-, and 3-fold, which offers more flexibility as regards geometric orientation of the phosphate group relative to the cation lattice. Our preferred model of an adlayer binding based on both monodentate and bidentate phosphate coordination results in a hexagonal overlayer lattice geometry with an intermolecular spacing that is almost exactly the same as the one in the thiol/gold system. It is then straightforward to assume that the experimentally observed mean tilt angle of the ODP molecules of 30−35° (relative to the surface normal) again, as in the gold/thiol case, results from a maximization of the van der Waals attraction of the hydrocarbon chains.

Order in the ODP−Ta_2O_5 system, as seen in our AFM results, is restricted to areas of a few nm^2—much smaller than is observed in the case of alkanethiols on gold. This may be due to the fact that the Ta_2O_5 film is of nanocrystalline to amorphous nature. If the assumption that the Ta cation lattice geometry is responsible for adlayer order is correct, then it is straightforward to assume that order in the tantala−ODP system can only extend over areas comparable to the size of the nanocrystals. Areas that show no order in the AFM investigation would then correspond to entirely amorphous Ta_2O_5 regions, completely lacking periodicity in the cation lattice, or to nanocrystalline patches that are too small to induce a measurable order in the adlayer. Nevertheless, NEXAFS data[12] suggest that the overlayer maintains its order over larger distances, essentially bridging the gap between the ordered tantala patches. A second possibility could be linked to the different crystallographic planes of the nanocrystals exposed at the surface, not all of which would be expected to have the appropriate symmetry to induce order in the ODP SAM. Still another explanation could be the local presence of strongly coordinated ODP molecules with a direct phosphate−Ta(V) bond in the ordered areas surrounded by areas with more weakly bound ODP molecules, e.g., weakly bound to the oxide surface through hydrogen bonding rather than chemisorbed by direct coordination. AFM and NEXAFS studies on different single crystal planes of transition metal oxides would be needed to determine which of these factors is chiefly responsible for limiting the AFM-observed order to small areas.

Regarding our proposal for the structure in the ordered ODP areas, other models, which may be in accord with part of the experimental findings, are, of course, feasible. Our belief in the proposed model is not based on single observations but on the sum of the information from the various techniques applied to characterize the surface and interface composition and structure. Also from purely chemical considerations, the two proposed coordination structures of the phosphate/Ta_2O_5 interface may actually make more sense than alternative models, since:

(a) Both types of coordination proposed (A and B, see Figure 8) satisfy formal charge considerations, in the sense that formally an oxide O^{2-} is replaced by either two single-coordinated anions of charge −1 or by one bidentate anion of charge −2. In view of the high O−Ta bond strength (mostly ionic character) of ca. 800 kJ/mol, it seems very important that a similar gain in enthalpy is achieved through replacement of the oxide and coordination of the phosphate in order to get a thermodynamically stable adlayer. Therefore coordination involving anions of the same total negative charge should be preferred over, e.g., single coordination to an anion with charge −1.

(b) Taking the simplified model of an octahedral structure of Ta_2O_5 (edge- and vertex-shared octahedra), the coordination number of the coordinated surface Ta ion turns out to be 7 for both proposed types A and B of surface bonding. Seven is—in addition to six—a preferred coordination number for Ta(V). The often proposed tridentate coordination of the phosphate group would, on the other hand, lead to a coordination number of 8, which we believe is less likely, in particular for steric reasons.

(c) The combination of types A and B coordination bonding at adjacent sites allows for the formation of hydrogen bonding, which may be important for further stabilization of the monolayer. Most other models, such as pure tridentate phosphate coordination or pure type A (bidentate) bonding, do not provide the possibility for stabilization through hydrogen bonding.

(d) Polymerization across the phosphate intermediate layer through condensation of phosphate groups to form structures such as $(R-OPO_2)_\infty$ (similar to $NaPO_3$) is unlikely for both thermodynamic and kinetic reasons, since (a) there would be no strong bond of sufficient ionic character (only P=O· · ·Ta type of coordination) and (b) in contrast to silanes, which very easily form polymeric siloxane structures through condensation, phosphates are kinetically much more inert. Furthermore, no fragments indicative of polyphosphates were observed in the ToF−SIMS data.

(e) Indirect (weak) coordination of the phosphate to the tantalum oxide or hydroxide surface via hydrogen bonding is believed to be less likely than the presence of strong phosphate−Ta(V) complexes. In the former case the surface bonds are likely to be too weak to survive as complex secondary ion fragments (see section 4.2.2). Moreover, the SIMS fragmentation pattern can only be fully understood when assuming direct complex coordination. However it cannot be excluded that—*in localized regions*—less strongly bound states are also present, e.g., hydrogen-bonded phosphoric acid molecules. In fact, one possible explanation for the local lability of the SAM order as observed by AFM may be the presence of such weaker surface interactions in localized areas (see section 4.2.3).

4.3.2. Surface Reaction Mechanism. In terms of reaction mechanism, we have no direct evidence for a particular mechanism of surface complex formation. However, in view of the very strong O−Ta bond, it is not likely that "free" Ta ions are available at the surface under ambient conditions. Rather, we have to assume that an oxide ion has to be replaced by phosphate(s). Again this is not likely to be directly possible, and we assume that a low activation energy path is only feasible through intermediate structures such as those proposed in Figure 12.

In fact, hydroxylation (or protonation of oxide) has been shown to be an important initial reaction step prior to the SAM formation of phosphonic acid esters on aluminum oxide. It has been demonstrated that fast adsorption in this system only takes place if a minimum fraction of hydroxide is present on the aluminum oxide surface.[11,28,29] The fact that solvent and pH are important factors for the kinetics of the SAM formation reaction may be related to

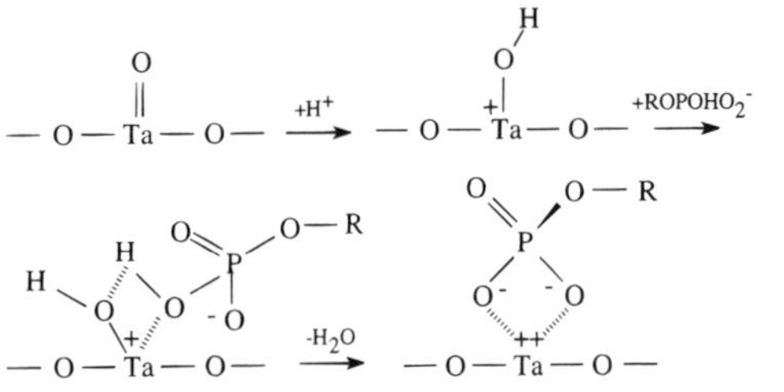

Figure 12. Proposed reaction sequence in the displacement of oxide ligands at the Ta_2O_5 surface by alkyl phosphates through intermediate hydroxide formation.

the question of protonation or hydroxylation and the existence of intermediate surface-coordinated species as discussed above.

5. Conclusions and Outlook

Octadecyl phosphate (ODP) has been shown to adsorb onto tantalum oxide (Ta_2O_5) surfaces forming a monolayer with direct, coordinative complexation of the phosphate headgroup to Ta(V) cations. This direct coordination is believed to be one of the reasons for the stability of the ODP adlayer on Ta_2O_5 surfaces. The molecular order within the overlayer—proven by NEXAFS and AFM studies—is proposed to be the consequence of a nearly hexagonal coincidence overlayer lattice of the phosphate groups on the Ta_2O_5 substrate with a packing density that turns out to be very close to that observed in ordered thiol SAMs on gold surfaces. Stability and order can therefore be explained by a combination of complex coordination $PO_4\cdots Ta$ and van der Waals interactions between the hydrocarbon chains at a separation of about 0.5 nm. The proposed model assumes that—on a very local scale—the nanocrystalline Ta_2O_5 substrate is ordered with a structure similar to bulk, crystalline tantalum oxide, i.e., with TaO_6 octahedra forming a network by sharing edges and vertices. The fact that nearly hexagonal molecular order in the adlayer is seen with AFM only over distances of a few nanometers is believed to be related to the lack of long-range order in the tantalum oxide substrate and/or the local presence of weakly adsorbed ODP molecules.

On the basis of the combined results of X-ray photoelectron spectroscopy, time-of-flight secondary ion mass spectrometry, and atomic force microscopy, the ODP molecules are proposed to preferentially coordinate to tantalum cations via both monodentate and bidentate complexation, leading to 7-fold site coordination, a preferred coordination number for Ta(V). A further stabilization of the ODP layer by intermolecular hydrogen bonding between monodentate and bidentate molecules is also likely to occur in such a model.

The proposed model implies that the crystallographic structure (symmetry and cation–cation distances) of the oxide substrate determines whether order can be achieved in the case of a particular phosphoric acid ester SAM. Studying single-crystal metal oxide surfaces with different Miller indices will be used in the future to test this assumption.

Regarding applications, long-chain phosphoric (and also phosphonic) acid esters are expected to have potential for applications in areas where the specific surface functionalization of oxides by extremely thin films is essential to the quality of a product. One area where application of octadecyl phosphate on tantalum oxide has already proven to be successful is optical biosensor technology, where the use of well-controlled ODP self-assembled monolayers on optical waveguide chips has been demonstrated to increase both the specificity and selectivity when sensing extremely low quantities of biomolecules using evanescent field and fluorescence techniques. The application potential will be further increased if a second functionality can be introduced in the terminal, ω-position of phosphoric acid esters.

Supporting Information Available: ToF-SIMS spectra (negative secondary ions) of the ODP/Ta_2O_5 surface in the mass range $m/z = 0-600$. This material is available free of charge via the Internet at http://pubs.acs.org.

LA990941T

(28) Bram, C. *Oberflächenanalytische Untersuchungen zur Selbstorganisation von aliphatischen Phosphonsäuren auf Aluminium.* PhD Dissertation, Erlangen, 1998.

(29) Jung, C. *Alkylphosphonsäuren als molekulare Haftvermittler für Aluminium und Zinkwerkstoff.* PhD Dissertation, Erlangen, 1998.

4014 *Langmuir* **2001,** *17,* 4014–4020

Alkyl Phosphate Monolayers, Self-Assembled from Aqueous Solution onto Metal Oxide Surfaces

R. Hofer, M. Textor, and N. D. Spencer*

Laboratory for Surface Science and Technology, Department of Materials, ETH-Zürich, CH-8092 Zürich, Switzerland

Received December 15, 2000. In Final Form: March 16, 2001

Metal oxide surfaces have been coated by self-assembled monolayers (SAMs) of dodecyl phosphate ($DDPO_4$) and 12-hydroxy dodecyl phosphate (OH-$DDPO_4$), by means of a novel surface-modification protocol based on the adsorption of the alkyl phosphate ammonium salts from aqueous solution. Formation of $DDPO_4$ SAMs has been successfully demonstrated on anodized aluminum (Al_2O_3) as well as on smooth, thin films of tantalum oxide (Ta_2O_5), niobium oxide (Nb_2O_5), zirconium oxide (ZrO_2), and titanium oxide (TiO_2) deposited on glass substrates, resulting in highly hydrophobic surfaces with advancing water contact angles $\geq 110°$. SAM formation does not occur on silica surfaces under the same conditions. The formation of SAMs based on hydroxy-terminated alkyl phosphates (OH-$DDPO_4$) onto Ta_2O_5 and Nb_2O_5 has also been observed. $DDPO_4$ and OH-$DDPO_4$ were codeposited onto Ta_2O_5 from aqueous solutions at different concentration ratios. It is shown that the water contact angle can be precisely controlled within the range of 110–50° by adjusting the molar ratio of the two different molecules in the aqueous SAM forming solution. The SAMs were characterized by advancing water contact angle (wettability), microdroplet density measurements (condensation figures to judge homogeneity), and X-ray photoelectron spectroscopy (coverage and orientation of molecules).

1. Introduction

The formation of self-assembled monolayers (SAMs) on metal[1] or metal oxide surfaces[2] is widely employed for the fabrication of model surfaces with highly controlled chemical properties. SAMs can be used in the modification of metal oxide surfaces for the investigation of protein adsorption,[3,4] for the study of cell behavior,[5] or for the fabrication of tailored sensor surfaces.[6] Passivation of metal surfaces, adhesion promotion, and interface corrosion protection in metal/lacquer systems are other examples of industrial applications of SAMs.[7]

To date, the best-studied SAM system has been that of self-assembled alkanethiolates on gold, where thiols adsorb onto the gold surface in an initial "lying-down" phase followed by rearrangement into a "standing-up" phase, which completes the monolayer and results in highly ordered, two-dimensional structures.[1,8–12] Alkyl phosphates and phosphonates constitute two further systems that have been shown to form ordered SAMs on metal oxide surfaces.[1,13–16] Transition metal oxides such as tantalum oxide (Ta_2O_5), niobium oxide (Nb_2O_5), and titanium oxide (TiO_2), in particular, are known to interact strongly with phosph(on)ates and to form relatively stable interfacial bonds. Both systems, thiol–gold and phosph(on)ate–metal oxide, form monolayers with a "tail-up" orientation and a tilt angle of the hydrocarbon chains of about 30° with respect to the surface normal.[17,18] Such amphiphile adlayers are generally produced from solutions in organic solvents.

The use of organic solvents in the deposition process has three main disadvantages: (1) In biomaterial applications, organic solvent molecules may be trapped within the adlayer and reduce the biocompatibility of the surface. (2) For the formation of mixed adlayer systems, it can be difficult to find an organic solvent that is suitable for both adlayer components. (3) On an industrial scale, organic solvents are increasingly falling into disfavor because of both air-pollution and water-pollution issues. We demonstrate in this paper that both pure and mixed alkyl phosphates with different terminal functionalities can be deposited from aqueous solution by converting the free (water-insoluble) alkyl phosphoric acids into the corresponding water-soluble salts. Mixed SAMs on metal oxide surfaces are of particular interest to the biosensor and biomaterial field, because they allow surface properties such as wettability, polarity, surface charge, and so forth to be tailored in a precise manner. Such surface tailoring may prove to be highly relevant for controlling the interaction between the SAM-modified surface and biological systems, such as proteins, antibodies, and cells.

(1) Folkers, J. P.; Gorman, Ch. B.; Laibinis, P. E.; Buchholz, S.; Whitesides, G. M.; Nuzzo, R. G. *Langmuir* **1995**, *11*, 813.
(2) Gao, W.; Dickinson, L.; Grozinger, Ch.; Morin, F. G.; Reven, L. *Langmuir* **1997**, *13*, 115.
(3) Shumaker-Parry, J. S.; Campbell, Ch. T.; Stormo, G. D.; Silbaq, F. S.; Aebersold, R. H. *Proceedings of Medical and Fiber Optic Sensors and Delivery Systems*; SPIE: San Jose, CA, 2000; Vol. 3922, pp 158–166.
(4) Harder, P.; Grunze, M.; Dahint, R.; Whitesides, G. M.; Laibinis, P. E. *J. Phys. Chem. B* **1998**, *102*, 426.
(5) Kingshott, P.; Griesser, H. J. *Curr. Opin. Solid State Mater. Sci.* **1999**, *4*, 403.
(6) Nyquist, R. M.; Eberhardt, A. S.; Silks, L. A., III; Li, Z.; Yang, X.; Swanson, B. I. *Langmuir* **2000**, *16*, 1793.
(7) Van Alsten, J. G. *Langmuir* **1999**, *15*, 7605.
(8) Chidsey, C. E. D. *Science* **1991**, *251*, 919.
(9) Chechik, V.; Schonherr, H. *Langmuir* **1998**, *14* (11), 3003.
(10) Cotton, C.; Glidle, A. *Langmuir* **1998**, *14* (18), 5139.
(11) Ishida, T.; Yamamoto, S. *Langmuir* **1997**, *13* (13), 3261.
(12) Tamada, K.; Hara, M. *Langmuir* **1997**, *13* (6), 1558.
(13) Gao, W.; Dickinson, L.; Grozinger, Ch.; Morin, F. G.; Reven, L. *Langmuir* **1996**, *12*, 6429.
(14) Brovelli, D.; Hähner, G.; Ruiz, L.; Hofer, R.; Kraus, G.; Waldner, A.; Schlösser, J.; Oroszlan, P.; Ehrat, M.; Spencer, N. D. *Langmuir* **1999**, *15*, 4324.
(15) Bram, Ch.; Jung, Ch.; Stratmann, M. *Fresenius' J. Anal. Chem.* **1997**, *358*, 108.
(16) Maege, I.; Jaehne, E.; Henke, A.; Adler, H.-J. P.; Bram, C.; Jung, C.; Stratmann, M. *Macromol. Symp.* **1997**, *126*, 7.
(17) Textor, M.; Ruiz, L.; Hofer, R.; Rossi, A.; Feldman, K.; Hähner, G.; Spencer, N. D. *Langmuir* **2000**, *16*, 3257.
(18) Hofer, R. Surface Modification for Optical Biosensor Applications. Ph.D. Thesis 13873, ETH Zurich, Zurich, Switzerland, 2000.

Published on Web 05/23/2001

The present paper describes two novel approaches to modifying metal oxide surfaces: the use of aqueous solutions of alkane phosphates in the self-assembly process and the tailoring of surface wettability based on mixed alkane phosphate/hydroxy-terminated alkane phosphate SAMs.

2. Materials and Methods

2.1. Substrates. *Samples with a Ta_2O_5 or Nb_2O_5 Surface.* Glass slides (15 × 15 × 1 mm) were coated with 150 nm thick layers of Ta_2O_5 or Nb_2O_5 by physical vapor deposition (sputter coating) (Balzers AG, Balzers, Liechtenstein).

Samples with a ZrO_2 or SiO_2 Surface. Commercial optical waveguide substrates were used, consisting of AF45 glass slides (8 × 12 × 1 mm), which were coated by a 200 nm thick $Ti_{0.4}Si_{0.6}O_2$ sol–gel waveguiding layer, followed by a 14 nm thick top layer of ZrO_2 or SiO_2, using a magnetron sputtering unit (Microvacuum, Ltd., Budapest, Hungary).

TiO_2. Glass slides were sputter coated with a 100 nm thick layer of TiO_2 (Paul Scherrer Institut, Villigen, Switzerland).

Al_2O_3. Aluminum (Al 99.99) thin strip samples (1 mm in thickness) were electrolytically polished in phosphoric acid/sulfuric acid electrolyte and anodized in sulfuric acid electrolyte (200 g/L H_2SO_4, 20 °C, 4 min), producing an anodic oxide layer of about 2 μm thickness (Alusuisse, Neuhausen am Rheinfall, Switzerland).

The Nb_2O_5 and Ta_2O_5 surfaces are reported by the manufacturer to be nanocrystalline (crystallite size in low-nanometer range). The TiO_2 and Al_2O_3 are largely amorphous, as judged by X-ray diffraction. The cleanliness of the samples prior to functionalization was reported in a previous paper.[17]

2.2. Alkyl Phosphates. *Precipitation of Ammonium Salt $DDPO_4(NH_4)_2$.* A quantity (2.00 g) of dodecyl phosphate ($DDPO_4$) (technical grade, Aldrich) was dissolved in 200 mL of 2-propanol (UVASOL, Merck), heated to 82 °C, and refluxed. Subsequently, 6 mL of ammonia (25% aq, p.a., Merck) was added. After the reaction mixture was cooled with ice water, the precipitated ammonium salt of $DDPO_4$ was filtered. The product was washed with ice-cold 2-propanol and dried at 60 °C and 10 mbar in a vacuum oven for 20 h. A white powdery product (1.61 g; mp, 225°) could be isolated, corresponding to a yield of 71%. ^{1}H NMR (DMSO or CD_3OD): 0.88 ppm (t, 3H, $-(CH_2)_nCH_3$), 1.28 ppm (m, 18H, $-CH_2CH_2(CH_2)_9CH_3$), 1.54 ppm (m, 2H, $-CH_2CH_2(CH_2)_9CH_3$), 3.72 ppm (q, 2H, $-OCH_2CH_2(CH_2)_9CH_3$), 4.9 ppm (s, 8H, NH_4). The ^{31}P NMR showing one single peak suggests that the product consists of a uniform alkyl phosphate species. Elemental analysis calculated: [C] 50.87%, [H] 10.67%, [N] 4.94%, [O] 22.59%, [P] 10.93%. Analysis: [C] 50.61%, [H] 10.94%, [N] 4.95%, [O] 22.75%, [P] 10.69%.

Precipitation of Ammonium Salt OH-$DDPO_4(NH_4)_2$. A quantity (500 mg) of hydroxy dodecyl phosphate (OH-$DDPO_4$) was dissolved in 20 mL of 2-propanol (UVASOL, Merck), and NH_3 was bubbled through the solution for 5 min. NH_3 was obtained from a hot 25% aqueous ammonia solution and transferred into the amphiphile solution via a glass pipet fixed onto a plastic tube. The precipitated ammonium salt of OH-$DDPO_4$ was separated from the solvent by centrifugation and removal of the solvent. The product was dried in a flow of dry nitrogen at room temperature using an evaporation system (EVAPOR) followed by drying at 10 mbar for 2 days. A white powdery product (515 mg; mp, 165°) could be isolated, corresponding to a yield of 91%. ^{1}H NMR (DMSO or CD_3OD): 1.24 ppm (m, 16H, $-CH_2CH_2(CH_2)_8CH_2CH_2OH$), 1.38 ppm (m, 2H, $-CH_2CH_2(CH_2)_8CH_2CH_2OH$), 1.44 ppm (m, 2H, $-CH_2CH_2(CH_2)_{10}OH$), 3.35 ppm (t, 2H, $-(CH_2)_nCH_2OH$), 3.57 ppm (q, 2H, $-OCH_2CH_2(CH_2)_{10}OH$), 5.2 ppm (s, 8H, NH_4). Elemental analysis calculated as diammonium salt: [C] 45.6%, [H] 10.5%, [N] 8.9%. Analysis: [C] 44.1%, [H] 9.6%, [N] 5.7%.

The NMR results of the obtained product suggest that the substance is pure. The somewhat low nitrogen content found experimentally is likely to reflect a partial loss of ammonia and partial formation of the monoammonium salt. To be consistent with $DDPO_4(NH_4)_2$, we abbreviate the ammonium salt of hydroxy dodecyl phosphate with OH-$DDPO_4(NH_4)_2$.

Preparation of Amphiphile Solutions. (a) A quantity (15.0 mg) of $DDPO_4(NH_4)_2$ was dissolved in 5 mL of high-purity water by heating to ca. 50 °C, and the volume was subsequently adjusted to 100 mL with water. (b) A quantity (15.8 mg) of OH-$DDPO_4(NH_4)_2$ was dissolved in 5 mL of high-purity water by heating to ca. 80 °C. The solution was cooled to room temperature, filtered through a 0.22 μm filter (MILLEX-GV, MILLIPORE, Bedford, MA), and adjusted to a volume of 100 mL.

The alkyl phosphate and hydroxyalkyl phosphate solutions, (a) and (b), were mixed in different ratios from 0 to 100 vol % with respect to the amount of OH-$DDPO_4(NH_4)_2$, in steps of 10% resulting in 11 different solutions at constant total phosphate concentration of 0.5 mM.

2.3. Sample Preparation. The substrates were sonicated in high-purity water for 15 min, followed by a second sonication in 2-propanol (UVASOL, Fluka) for another 15 min. The slides were removed from the cleaning solvent, blow-dried with nitrogen, and transferred to an oxygen-plasma cleaner (Harrick Plasma Cleaner/Sterilizer PDC-32G, Ossining, NY). After 3 min of oxygen-plasma cleaning, the chips were transferred to 5 ml glass vials and amphiphile solution was subsequently added. The chips were immersed for 48 h in the solution for self-assembly and removed; each was rinsed with 10 mL of high-purity water and blow-dried with nitrogen.

2.4. Investigation of Surface Properties. The wettability of surfaces was investigated by measuring the advancing water contact angle (Contact-Angle Measuring System, with Kontaktwinkel Messsystem G2/G40 2.05-D software, Krüss GmbH, Hamburg, Germany).

Surface homogeneity was determined by means of microdroplet density, μdd, measurements.[19] Microdroplet density data from condensation figures were obtained by means of an apparatus consisting of a metal table placed in a transparent humidity chamber and a CCD camera (Panasonic, model WV-BP 310/6, Matsushita Communication Deutschland GmbH, Germany) fixed on a microscope stage (Zeiss, Carl Zeiss AG, Switzerland). Images of the pattern of growing droplets on the surfaces were made after cooling the samples, which were placed on the metal table in the humidity chamber.[18,19]

X-ray photoelectron spectroscopy (XPS) analyses were performed using a PHI 5700 spectrometer equipped with a standard Al Kα source (200 W) (Physical Electronics, Eden Prairie, MN). Photoelectron intensities were normalized using the standard sensitivity factors provided in the PHI quantification software. The details of the curve-fitting procedures and of the quantification of peak intensities have been published elsewhere.[18]

3. Results and Discussion

3.1. $DDPO_4$ on Different Metal Oxides. Self-assembled monolayers of dodecyl phosphate were produced on Al_2O_3, Ta_2O_5, Nb_2O_5, ZrO_2, TiO_2, and SiO_2 by immersion in 0.5 mM $DDPO_4(NH_4)_2$, as described in section 3.3.

Contact Angle. Advancing water contact angles (CA) were measured immediately following sample preparation (Table 1). The results show that highly hydrophobic SAMs form on all metal oxide surfaces investigated in this work, with the exception of silica. The contact angle values of $\geq 110°$ are typical for well-defined alkyl phosphate SAMs.[14] The isoelectric points (IEP) of the different bare metal oxides, which vary from 2.7–3.0 for Ta_2O_5 to 7.5–8.0 for Al_2O_3, do not seem to be an important parameter for SAM formation with this system (Table 1). On SiO_2, the contact angle remains in the same range as it was prior to immersion in the amphiphile solution. This suggests that $DDPO_4$ does not adsorb on SiO_2 with this approach.

Microdroplet Density (Condensation Figures). To remove any particles that had adsorbed to the surfaces, the samples were rinsed with high-purity water just before measuring the microdroplet density (μdd). $DDPO_4$ SAMs on pure metal oxide surfaces show a low μdd value of 150–260 droplets/mm^2 (Table 1) and a homogeneous

(19) Hofer, R.; Textor, M.; Spencer, N. D. *Langmuir*, **2001**, *17*, 4123.

Table 1. Literature Values of IEP[a] of the Investigated Bare Oxides and Experimental Advancing Water Contact Angles, θ_a, (± Standard Deviation for 3 Probing Drops in Different Areas of the Sample, 10 Measurements per Drop) and Microdroplet Density, μdd, (± Standard Deviation for 4 Independent Measurements, Each on an Area of 0.25 mm²) Following Self-Assembly of $DDPO_4$ on the Oxide Surfaces[b]

substrate	IEP	θ_a (± st dev)/deg	μdd (± st dev)/ droplets mm^{-2}
Ta_2O_5	2.7−3.0	114.6 ± 0.5	129 ± 19
Al_2O_3	7.5−8.0	111.4 ± 1.0	152 ± 22
Nb_2O_5	3.4−3.6	109.7 ± 0.6	120 ± 18
ZrO_2	4.0	110.1 ± 0.6	258 ± 39
TiO_2	4.7−6.2	111.4 ± 1.2	221 ± 33
SiO_2	1.8−2.2	10.0 ± 3.4	>3000

[a] References 15−22. [b] References 20−27.

distribution of the growing droplets of condensed water (Figure 1). SiO_2 remains hydrophilic (as mentioned above), and microdroplets coalesce to a complete layer of condensed water before they can be detected as single droplets. The maximum value that can be detected with the used setup corresponds to about 3000 droplets/mm². The μdd value of >3000 droplets/mm² is consistent with the suggestion of the contact angle data, that $DDPO_4$ does not adsorb onto SiO_2 from aqueous amphiphile solution.

X-ray Photoelectron Spectroscopy. After immersion of the metal oxide and silica samples in a 0.5 mM $DDPO_4$-$(NH_4)_2$ solution for 48 h, the surfaces were investigated by XPS at two different takeoff angles Θ: at a takeoff angle of 15° (with respect to the surface plane), the analysis is highly surface sensitive, whereas a takeoff angle of 75° yields more information on the substrate.

Photoelectron intensities (peak areas) were normalized by dividing them by the standard sensitivity factors and converting them to a percent of the sum of all observed normalized intensity values (Table 2). We deliberately do not call these values "atomic concentrations" to take into account the fact that SAM surfaces are typically not homogeneous in the *z* direction (i.e., perpendicular to the surface). These normalized intensities are, however, useful for the comparison of coverages between different surfaces. The relative differences in the normalized intensities between the 15° (grazing angle) and the 75° electron takeoff angle provide qualitative information about the location of the hydrocarbon chain relative to the phosphate headgroup and the oxide substrate, respectively.

The comparison of the data at 15° and 75° electron takeoff angles supports our model of a surface architecture where the alkyl phosphates adsorb onto metal oxide substrate surfaces with the phosphate attached to the oxide surface and a hydrocarbon tail-up orientation. This can be deduced qualitatively from the fact that the carbon intensities are much higher and the metal intensities are much lower at 15° (i.e., in the more surface-sensitive mode) compared to 75° (i.e., in the higher sampling depth mode). This observation is true for all transition metal surfaces investigated as well as for aluminum oxide. Extensive quantitative evaluation of the XPS data using a multilayer model has been discussed in an earlier paper describing octadecyl phosphate self-assembled onto tantalum oxide surfaces from organic solvents. The results of these calculations were in excellent agreement with the assumption of a monolayer of ordered, oriented, and close-packed alkyl phosphate molecules.[17]

The amount of adsorbed amphiphile at the surfaces is in the same range for all investigated metal oxide substrates and points to a coverage close to one monolayer as discussed in an earlier publication.[17] Consistent with the CA and μdd results, no adsorbed alkyl phosphate could be detected on the SiO_2 substrate (no phosphorus and very low carbon concentration, typical of the clean substrate, Table 2). The absence of $DDPO_4$ SAM formation from aqueous solutions on silicon oxide surfaces is likely to be due to a much lower affinity of phosphate for Si(IV) in comparison to Al(III), Ti(IV), Nb(V), and Ta(V), which all form highly stable phosphate−metal cation coordination bonds, that is, a direct bond of the phosphate headgroup to the metal cation through either one oxygen atom (monodentate) or two oxygen atoms (bidentate).[13]

3.2. OH-$DDPO_4$ on Ta_2O_5 and Nb_2O_5. 12-Hydroxy dodecyl phosphate (OH-$DDPO_4$) was self-assembled on Ta_2O_5 and Nb_2O_5 slides by immersion in a 0.5 mM OH-$DDPO_4(NH_4)_2$ aqueous solution for 48 h, as described in section 3.3.

Contact Angle and XPS. The advancing water contact angle was measured immediately after the self-assembly process. The contact angle was approximately 50°; that is, the surface is much more hydrophilic than in the case of the methyl-terminated $DDPO_4$ SAMs. This is evidence that the terminal hydroxy groups are indeed exposed at the SAM surface. However, in comparison to hydroxy-terminated alkane thiol SAMs (advancing water contact angle of typically 20°[28]), the value of 50° of the OH-$DDPO_4$ SAMs seems high. This may indicate a comparatively higher degree of heterogeneity, a lower packing density, and/or a lesser degree of order in case of the alkane phosphates. A near-edge X-ray absorption fine structure (NEXAFS) study to determine orientation and order in a number of alkane phosphate SAMs is planned.

To prove this hypothesis, XPS spectra were recorded at different takeoff angles. The variation of the signal intensity as a function of the takeoff angle may give information about the vertical position of the corresponding element within the depth sampled by XPS. The O(1s) signals of the spectra at the two grazing angles 11.5° and 20.5° are shown in Figure 2. Compared to the nonfunctionalized, methyl-terminated SAM, the hydroxy-terminated SAM shows an additional shoulder at approximately 534 eV in the O(1s) XPS spectrum (Figure 2) with a relative intensity that is significantly higher at the electron takeoff angle of 11.5° compared to 20.5°. This is good evidence that the terminal hydroxy groups in the OH-$DDPO_4$ SAM are indeed exposed at the surface, because the O(1s) signal from such hydroxy groups is not affected by inelastic scattering within the alkyl monolayer, whereas the O(1s) signal intensities from the interfacial phosphate groups and from the metal oxide substrate are strongly reduced at the grazing takeoff angles because of inelastic scattering within the organic overlayer.

No nitrogen could be detected by XPS, demonstrating that the ammonium cations are not incorporated in the self-assembled adlayer.

It has been shown that water contact angle data of alkyl phosphate SAMs reflect the degree of coverage of SAMs of long-chain alkyl phosphates.[14] Contact angle values of complete monolayers from various types of hydroxy-

(20) Bousse, L. J. *J. Colloid Interface Sci.* **1991**, *147*, 22.
(21) Mattson, S. *Soil Sci.* **1930**, *30*, 459.
(22) Verwey, E. J. W. *Recl. Trav. Chim. Pays-Bas* **1941**, *60*, 625.
(23) Hazel, F.; Ayres, G. H. *J. Phys. Chem.* **1931**, *35*, 2930.
(24) Mattson, S.; Pugh, A. J. *Soil Sci.* **1934**, *38*, 229.
(25) Kenausis, G.; Vörös, J.; Elbert, D. L.; Huang, N.; Hofer, R.; Ruiz, L.; Textor, M.; Hubbell, J. A.; Spencer, N. D. *J. Phys. Chem. B* **2000**, *104*, 3298.
(26) Johannsen, P. G.; Buchanan, A. S. *Aust. J. Chem.* **1957**, *10*, 398.
(27) Parks, G. A. *Chem. Rev.* **1965**, *65*, 177.
(28) Tam-Chang, S.-W.; Biebuyck, H. A.; Whitesides, G. M.; Jeon, N.; Nuzzo, R. G. *Langmuir* **1995**, *11*, 4371.

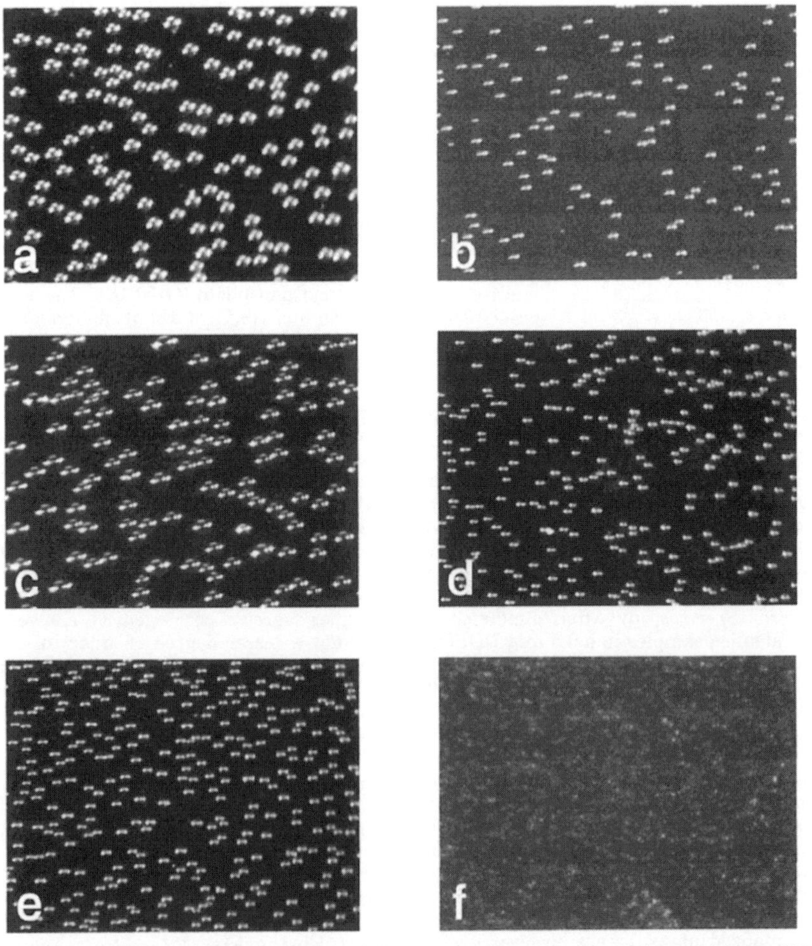

Figure 1. Microdroplet density images of $DDPO_4$ SAMs on different smooth metal oxide surfaces: (a) Al_2O_3, (b) Ta_2O_5, (c) Nb_2O_5, (d) ZrO_2, (e) TiO_2, and (f) SiO_2. The homogeneous distribution of growing water droplets reflects the homogeneity of the hydrophobic SAM surfaces on a microscopic scale. SiO_2 (f) does not adsorb DDP under the same self-assembly conditions; it remains hydrophilic and is completely covered by a layer of condensed water during the water condensation test. Image size: 1 mm^2.

Table 2. Normalized XPS Intensities (Intensities of C(1s), O(1s), Ta(4f), Al(2p), Nb(3d), Zr(4f), Ti(2p), (Si2p), and P(2p) Divided by the Corresponding Elemental Sensitivity Factors and Normalized to 100%) of $DDPO_4$, Self-Assembled onto Different Metal Oxide Substrates[a]

substrate	normalized intensity				substrate	normalized intensity			
MO_x	% C	% O	% M	% P	MO_x	% C	% O	% M	% P
15° Electron Takeoff Angle					75° Electron Takeoff Angle				
Ta_2O_5	67.8	23.2	6.71	2.36	Ta_2O_5	27.3	53.7	17.4	1.61
Al_2O_3	67.6	21.1	8.85	2.53	Al_2O_3	25.3	47.1	25.5	2.10
Nb_2O_5	59.7	29.6	8.5	2.26	Nb_2O_5	31.9	49.4	17.3	1.30
ZrO_2	72.2	22.0	3.06	2.77	ZrO_2	35.9	49.0	12.9	2.28
TiO_2	68.6	23.0	6.13	2.23	TiO_2	31.2	49.4	17.6	1.83
SiO_2	6.2	70.8	22.9	0	SiO_2	nd[b]	nd[b]	nd[b]	nd[b]

[a] The measurements were performed at takeoff angles of 15° and 75°, with respect to the surface plane. [b] nd = not determined.

terminated amphiphiles have been reported to be 50–80°.[1] Although contact angles on hydrophobic, methyl-terminated SAMs provide a very sensitive indicator of complete coverage and order, no straightforward prediction can be made from contact angles of hydroxy (or other polar group) ω-functionalized alkyl phosphates. Therefore, to test whether OH-$DDPO_4$ forms densely packed SAMs, we first compared the coverage-dependent XPS C(1s) signals from $DDPO_4$ SAMs with those of OH-$DDPO_4$ measured by XPS at different takeoff angles (Figure 3). The carbon concentration is plotted in Figure 3 as a function of the sine of the takeoff angle Θ, because the sampling depth scales linearly with the sine of the takeoff angle. The carbon signal intensities for both amphiphiles on Ta_2O_5 as well as on Nb_2O_5 are very similar, suggesting that hydroxy-terminated dodecyl phosphate and dodecyl phosphate indeed form SAMs with comparable molecular surface densities.

Normalized XPS intensities of OH-$DDPO_4$ self-assembled onto Ta_2O_5 and Nb_2O_5 are summarized in Table 3 as a function of the electron takeoff angle. The data are again qualitatively consistent with our model of a surface architecture where the alkyl phosphates adsorb onto metal oxide substrate surfaces with a tail-up orientation of the alkane chain.

The O(1s) signal was curve-fitted into three different O(1s) subpeaks representing five different oxygen species: (a) the metal oxide (530.2 eV), (b) the phosphate oxygen (531.4 eV for P–O–Metal and P=O and 532.6 eV for R–O–P and P–OH), and (c) the hydroxy group

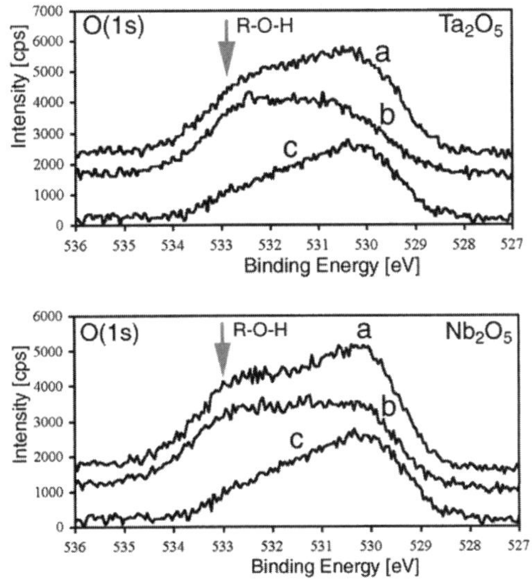

Figure 2. XPS O(1s) spectra of *hydroxy-terminated* dodecyl phosphate (OH-$DDPO_4$), self-assembled onto Ta_2O_5 and Nb_2O_5 (curves a and b). The spectra were recorded at two different electron takeoff angles Θ (curve a, 20.5°, and curve b, 11.5°, relative to the surface plane) and are compared to the O(1s) spectrum of a *methyl-terminated* dodecyl phosphate ($DDPO_4$) SAM on the corresponding substrate, measured at a takeoff angle Θ of 15° (curve c). The electron emission at a binding energy of approximately 534 eV is assigned to the terminal R−OH group in the OH-$DDPO_4$ SAM. The observed increase of the intensity with decreasing electron takeoff angle provides evidence for the location of the hydroxy group at the outermost surface.

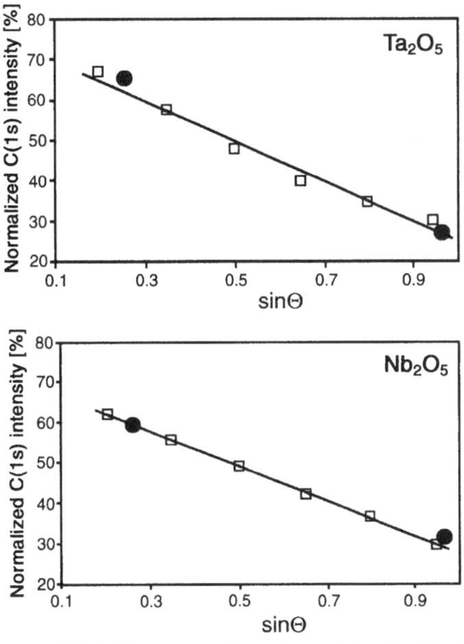

Figure 3. XPS C(1s) normalized intensities (in %) of OH-$DDPO_4$ and $DDPO_4$ SAMs on Ta_2O_5 (top) and Nb_2O_5 (bottom), measured at different electron takeoff angles Θ and plotted as a function of sin Θ. C(1s) intensity values of the OH-$DDPO_4$ SAM are represented by squares; those of the $DDPO_4$ SAM (only two measurements) are represented by circles. A linear trend line is added.

(approximately 534 eV, position can only be approximately determined because of the comparatively weak signal). The details of the curve-fitting procedures have been published elsewhere.[18] The assignment of different binding energies to the different species in the substrate and coordinated phosphate groups has been described in a previous paper.[17] The angle-dependent intensity data (peak areas) for the Nb(3d), C(1s), and P(2p) photoelectrons and for the three deconvoluted contributions to the O(1s) signal are plotted in Figure 4 for the case of the OH-$DDPO_4$-coated niobium oxide surface.

The increase of the substrate *oxygen* signal as the takeoff angle approaches the surface normal is due to the increase of the sampling depth and the consequent increased detection of the metal oxide layer, which can be considered as semi-infinite for XPS measurements. The phosphate oxygen O(1s) signal decreases slightly upon increasing the takeoff angle. This can be explained by our model where the phosphate groups form an interfacial monolayer between the substrate and the alkane chain layer. The signal from the ω-terminal hydroxy group (ROH) decreases only slightly with increasing takeoff angle Θ.

The strong decrease of the *metal* (Ta, Nb) signal originating from the cations in the semi-infinite oxide substrate with decreasing takeoff angle Θ is again, as in the case of the oxygen substrate signal, clearly due to the corresponding decrease of the sampling depth, depending on the sine of the takeoff angle Θ. This is paralleled by a corresponding increase of the *carbon* signal of the organic adlayer.

All signals, on the other hand, that originate from a purely two-dimensional, roughly monatomic surface or interface layer (O(1s) and P(2p) of the PO_4 interface and O(1s) of the terminal ROH group) give rise to intensities that show comparatively less dependence on the takeoff angle. The same observation has been reported for the interfacial phosphate in the closely related system of octadecyl phosphate, self-assembled on tantalum oxide and discussed in terms of a quantitative evaluation of the angle-dependent XPS intensities assuming a three-layer model.[17]

Table 3. Normalized XPS Intensities (Intensities of Ta(4f), Nb(3d), O(1s), P(2p), and C(1s) Divided by the Corresponding Elemental Sensitivity Factors and Normalized to 100%) of OH-$DDPO_4$ Self-Assembled onto Ta_2O_5 and Nb_2O_5[a]

takeoff angle Θ	sin Θ	normalized XPS intensities OH-$DDPO_4$ on Ta_2O_5			
		% Ta	% O	% P	% C
11.5	0.20	3.05	26.6	3.69	66.72
20.5	0.35	6.27	33.0	3.91	56.79
30.0	0.50	9.21	39.8	3.12	47.78
40.5	0.65	11.9	45.8	2.83	39.46
53.1	0.80	13.42	50.3	2.00	34.29
71.8	0.95	14.88	53.3	2.12	29.65

takeoff angle Θ	sin Θ	normalized XPS intensities OH-$DDPO_4$ on Nb_2O_5			
		% Nb	% O	% P	% C
11.5	0.20	4.64	29.3	3.86	62.21
20.5	0.35	6.74	33.9	3.32	55.97
30.0	0.50	9.53	38.5	3.07	48.62
40.5	0.65	11.97	43.6	2.78	41.97
53.1	0.80	14.05	47.7	2.01	36.24
71.8	0.95	16.38	51.6	2.07	29.98

[a] The measurements were performed at different electron takeoff angles Θ. The sine of the takeoff angle (Θ) was increased linearly corresponding to a linear increase in the effective sampling depth.

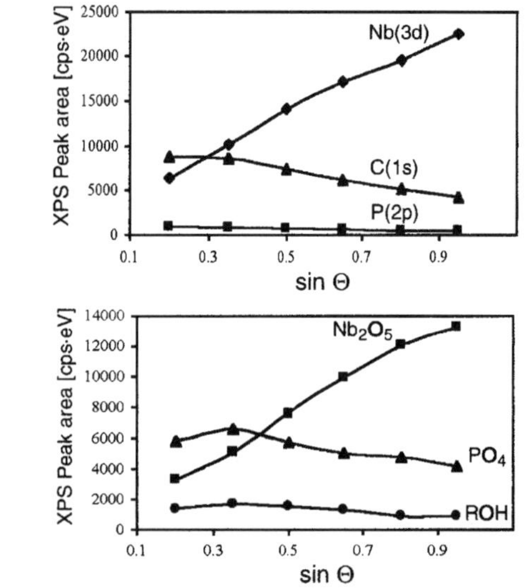

Figure 4. Angle-dependent XPS peak areas of Nb(3d), C(1s), and P(2p) (top) and of O(1s) of different oxygen types (bottom) for the OH-DDPO$_4$ SAM on Nb$_2$O$_5$. The O(1s) detail spectrum was deconvoluted in three different subpeaks representing three types of oxygen [O(Nb$_2$O$_5$), O(PO$_4$), and O(ROH)]. The peak areas are shown as a function of the sine of the electron takeoff angle Θ, basically reflecting the sin Θ dependence of the effective sampling depth. The results of OH-DDPO$_4$ SAMs on Ta$_2$O$_5$ are very similar and not shown.

3.3. Mixed SAMs of OH-DDPO$_4$/DDPO$_4$ on Ta$_2$O$_5$ and TiO$_2$. Mixed solutions of 0.5 mM OH-DDPO$_4$(NH$_4$)$_2$ and 0.5 mM DDPO$_4$(NH$_4$)$_2$ were prepared according to section 2.2. Ta$_2$O$_5$ samples were cleaned and immersed in the mixed amphiphile solutions, as described in section 2.3.

Contact Angle and Microdroplet Density. The advancing water contact angle was measured immediately following sample preparation. The results show that the wettability of the resulting SAM correlates with the molar ratio OH-DDPO$_4$(NH$_4$)$_2$/DDPO$_4$(NH$_4$)$_2$ of the amphiphile solution and spans a range of approximately 110° (pure DDPO$_4$) to 50° (pure OH-DDPO$_4$) (Figure 5).

The relationship between amphiphile solution composition and molecular ratio of the two alkane phosphate molecules in the SAM based on XPS and time-of-flight secondary ion mass spectrometry (ToF-SIMS) measurements will be published in a separate paper.

The microdroplet density images appear homogeneous on the micrometer scale, suggesting that hydroxy-functionalized and nonfunctionalized alkyl phosphates in the monolayer are both equally homogeneously distributed at this scale. The large difference in wettability across the range of SAM composition ($\Delta = 50°$) does not significantly affect the μdd (Figure 5). In separate publications,[18,19] we have shown that the microdroplet density value does not primarily depend on the wettability or water contact angle of a surface but on the concentration of condensation nucleation sites, believed to be induced by features such as surface roughness, molecular defects, and/or disorder in surfaces and adlayers. The experimental μdd value of 90−230 droplets/mm^2 is low when compared to more disordered SAMs, for which this value is typically in the range of 1000−3000 droplets/mm^2.[18,19] From the condensation figure evaluation, we deduce the presence of rather

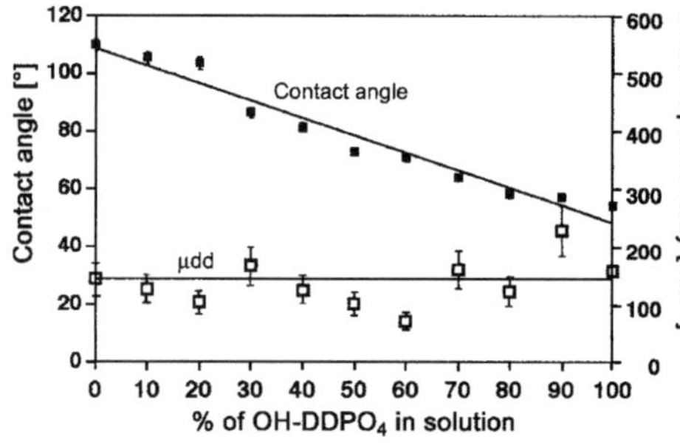

Figure 5. Advancing water contact angle θ_a (sessile drop) and microdroplet density (μdd) data of mixed OH-DDPO$_4$/DDPO$_4$ SAMs on Ta$_2$O$_5$. The results are presented as a function of the mole percentage of OH-DDPO$_4$ in the amphiphile solutions, defined as 100% × [OH-DDPO$_4$]/([OH-DDPO$_4$] + [DDPO$_4$]). The total alkyl phosphate concentration [OH-DDPO$_4$] + [DDPO$_4$] is kept constant at 0.5 mM. 0% corresponds to the pure DDPO$_4$ SAM, and 100% to the pure OH-DDPO$_4$ SAM. Error bars correspond to ± standard deviation (contact angle, θ_a: 3 probing drops in different areas of the sample, 10 measurements per drop; microdroplet density, μdd: 4 independent measurements, each on an area of 0.25 mm^2).

well-defined, smooth adlayers, independent of the composition of the SAM within the investigated OH-DDPO$_4$/DDPO$_4$ system.[19] To what extent these SAM layers are ordered on a molecular level, however, cannot be concluded from the results of the experimental techniques used in this work. NEXAFS and atomic force microscopy (AFM) would be the methods of choice to unambiguously comment on structural order and molecular orientation, and such a study is currently underway.

4. Conclusion and Outlook

It has been shown that it is possible to self-assemble alkyl phosphates from aqueous solutions of their ammonium salts, on a wide range of different metal oxides. XPS analysis did not reveal the presence of nitrogen, suggesting that ammonium cations do not adsorb on the substrate surface but serve only to increase the solubility of the alkyl phosphates in water. Highly hydrophobic surfaces are readily formed by self-assembly from aqueous solutions of the ammonium salt of dodecyl phosphate. On all investigated transition metal oxides as well as on aluminum oxide, defined SAMs are formed with approximately the same molecular coverage and correspondingly similar properties (e.g., advancing contact angles in the range of 110−115°). Silicon dioxide, on the other hand, does not form adlayers from aqueous alkyl phosphate solutions, probably because of the far lower affinity of phosphate for Si(IV) compared to Al(III), Ti(IV), Nb(V), and Ta(V), which are known to form strong phosphate coordination complexes.

It is also possible to produce more hydrophilic SAMs using aqueous solutions of the ammonium salt of hydroxy-terminated dodecyl phosphate. Furthermore, the wettability (water contact angle) imposed onto a metal oxide surface can be varied in a controlled, predictable way by the formation of mixed methyl-/hydroxy-terminated dodecyl phosphate SAMs, self-assembled from aqueous solutions containing both molecules in an appropriate ratio. Investigations of the relationship between adsorbate composition and solution amphiphile concentration ratio for mixed SAM systems (by XPS and ToF-SIMS) as well as amphiphile orientation and order in the different SAMs (by NEXAFS and AFM) are currently under way.

Acknowledgment. We acknowledge financial support provided by the International Team of Oral Implantology, ITI Foundation Research Committee, Waldenburg, Switzerland, and the Commission for Technology and Innovation (CTI), Berne, Switzerland. We thank Professor H.-J. Adler and Dr. E. Jähne, Institut für Makromolekulare Chemie und Textilchemie, University of Dresden, Germany, for the synthesis of the 12-hydroxy dodecyl phosphate and the Paul Scherrer Institut (PSI), Villigen, Switzerland, for the preparation of the metal oxide coated surfaces.

LA001756E

Langmuir **2002,** *18,* 3537−3548 3537

Self-Assembled Monolayers of Dodecyl and Hydroxy-dodecyl Phosphates on Both Smooth and Rough Titanium and Titanium Oxide Surfaces

S. Tosatti, R. Michel, M. Textor,* and N. D. Spencer

Laboratory for Surface Science and Technology, Department of Materials, Swiss Federal Institute of Technology (ETHZ), CH-8092 Zürich, Switzerland

Received September 21, 2001. In Final Form: February 1, 2002

Dodecyl phosphate and hydroxy-terminated dodecyl phosphate are shown to spontaneously assemble on smooth titanium oxide and titanium metal coated glass and silicon substrates, as well as on rough titanium metal implant surfaces. The surfaces were dipped in aqueous solutions of the corresponding ammonium salts for 48 h. The molecules are shown by X-ray photoelectron spectroscopy (XPS) to form densely packed, self-assembled monolayers (SAMs) on all surfaces investigated. The phosphate headgroups are believed to attach to the titanium (oxide) surface with the terminal end group (either methyl or hydroxy) pointing toward the ambient environment (air, vacuum, or water). Mixed SAMs are shown to be formed from mixed aqueous solutions of the two amphiphiles, with the hydroxy-terminated dodecyl phosphate adsorbing more favorably than the methyl-terminated molecule. The advancing water contact angles can be easily tailored via the composition of the self-assembly solution in the range of 110° (pure methyl) to 55° (pure hydroxy) on flat, smooth titanium surfaces. Surface roughness strongly modifies the wetting properties, with advancing contact angles in the range of 150−100° being observed, as well as the degree of hysteresis (difference between advancing and receding angles). Model calculations based on XPS intensities have been successfully used to quantify the adlayer composition and molecular surface densities across the whole range of mixed adlayer chemistry. The organophosphate monolayers on titanium are believed to have a significant potential for precise control of the surface chemistry and interfacial tension on both smooth and rough titanium surfaces in application areas such as medical implants and other devices where independent control of surface chemistry and topography is essential to performance.

1. Introduction

Self-assembled monolayers (SAMs) have attracted a lot of attention, thanks to their ability to form chemically well-controlled, structurally ordered surfaces. Thiols on gold have been particularly well studied and used as model systems for a variety of applications including biomaterial and biosensor surfaces.[1,2] Recently, octadecyl phosphate has been shown to form adlayers spontaneously from heptane/2-propanol solutions on tantalum oxide (Ta_2O_5) surfaces, resulting in highly hydrophobic surfaces with structural and chemical properties resembling those of long-chain alkanethiols on gold.[3,4] In particular, the average tilt angle of 30° (relative to the surface normal) determined by near-edge X-ray absorption fine structure (NEXAFS) spectroscopy and the detection by atomic force microscopy (AFM) of (local) two-dimensional, hexagonal patterns of the terminal methyl group with a characteristic intermolecular spacing of 0.49 nm (measured parallel to the surface) are similar to the corresponding characteristic values reported for gold (111)/alkanethiol SAMs.[1,5] Tantalum oxide was chosen as a substrate in view of its use in transparent, high-refractive-index waveguiding layers in optical sensor applications.[6,7] The production of bioaffinity sensor surfaces with controlled chemical properties is of great relevance in the areas of DNA/RNA and protein sensing, genomics, and proteomics.[8]

The use of organic solvents to produce surface layers has clear disadvantages if the aim is to use such techniques on an industrial scale (due to both environmental emission and disposal issues). Moreover, for applications in areas such as medical devices and implants, the presence of even minor organic solvent residues in the adlayers cannot be tolerated, in view of potential cell-toxicity effects and other biological risks. On the other hand, self-assembled monolayers, in particular of molecules with functional terminal groups, are of great interest for the modification and functionalization of biomaterials and medical devices. Therefore, a technique based on the deposition of SAMs from aqueous alkyl phosphate solutions has been developed and successfully applied to a variety of metal oxide substrates.[9]

This paper covers the deposition and characterization of monolayers of dodecyl phosphate, 12-hydroxy-dodecyl phosphate, and a mixture of the two from aqueous solutions of their ammonium salts onto three different substrates: titanium oxide and titanium metal films, deposited by physical vapor deposition onto glass and silicon substrates (serving as smooth, flat model surfaces), and a special titanium dental implant surface with a rough, highly corrugated surface. The aim of the present investigation is twofold: first, to describe the physicochemical and structural properties of the single-component and mixed SAMs on titanium (oxide), and second, to test the

* To whom correspondence should be addressed.

(1) Bain, C. D.; Troughton, E. B.; Tao, Y. T.; Evall, J.; Whitesides, G. M.; Nuzzo, R. G. *J. Am. Chem. Soc.* **1989**, *111*, 321.
(2) Nuzzo, R.; Allara, D. L. *J. Am. Chem. Soc.* **1983**, *105*, 4481.
(3) Brovelli, D.; Hähner, D.; Ruiz, L.; Hofer, R.; Kraus, G.; Waldner, A.; Schlösser, J.; Oroszlan, P.; Ehrat, M.; Spencer, N. D. *Langmuir* **1999**, *15*, 4324.
(4) Textor, M.; Ruiz, L.; Hofer, R.; Rossi, A.; Feldman, K.; Hähner, G.; Spencer, N. D. *Langmuir* **2000**, *16*, 3257.
(5) Folkers, J. P.; Gorman, C. B.; Laibinis, P. E.; Buchholz, S.; Whitesides, G. M.; Nuzzo, R. G. *Langmuir* **1995**, *11*, 813.
(6) Duveneck, G. L.; Pawlak, M.; Neuschäfer, D.; Bär, E.; Budak, W.; Pieles, U.; Ehrat, M. *Sens. Actuators, B* **1997**, *38−39*, 88.
(7) Budach, W.; Abel, A. P.; Bruno, A. E.; Neuschafer, D. *Anal. Chem.* **1999**, *71*, 3347.
(8) Duveneck, G. L.; Abel, A. P. *Proc. SPIE (Prepr.)* **1999**, Vol. 3858.
(9) Hofer, R.; Textor, M.; Spencer, N. D. *Langmuir* **2001**, *17*, 4014.

Published on Web 04/05/2002

feasibility of using alkyl phosphate based SAMs to tailor the surface properties of titanium medical devices.

2. Materials and Methods

2.1. Substrates. Three different substrates were used. Thin films of titanium dioxide (TiO_2) (20 nm thick) and of titanium (Ti(metal)) (100 nm thick) were deposited by physical vapor deposition (PVD, reactive magnetron sputtering, Paul Scherrer Institut, Villigen, Switzerland) onto silicon wafers and flat glass slides, respectively. Before the coating, the uncoated substrates were ultrasonically cleaned in toluene (Uvasol, Merck, Dietikon, Switzerland) for 15 min and dried under flowing nitrogen gas (5.0). Commercially pure titanium (CP Ti) disks, 15 mm in diameter and 1 mm in thickness, were grit-blasted with alumina beads (average particle size, 250 μm) under low-impact-energy, industrial particle-blasting conditions and subsequently etched in a hot solution of HCl/H_2SO_4 (Institut Straumann AG, Waldenburg, Switzerland). This surface, known commercially as "SLA", is characterized by a duplex surface topography with roughness contributions in the range of typically 20–50 μm (originating from the blasting process) and 0.5–2 μm (from the chemical etching process).

2.2. Self-Assembled Monolayers. Ammonium salts of dodecyl ($DDPO_4$) and 12-hydroxy dodecyl phosphate ($OH-DDPO_4$) were precipitated from a 2-propanol (UVASOL, Merck) solution of the free acid molecules using ammonia (25% aq, p.a., Merck). Details are given in refs 9 and 10. Equimolar solutions (0.5 mM) of both amphiphiles were prepared by dissolving their corresponding ammonium salts in high-purity water (HPLC, Fluka, Buchs, Switzerland) at 50 °C and subsequently cooling to room temperature. The alkyl phosphate and hydroxyalkyl phosphate aqueous solutions were mixed in different ratios from 0:100 to 100:0 vol % in steps of either 10% or 25% with respect to the amount of $OH-DDPO_4$. Cleaning and SAM formation for TiO_2 and Ti(metal) samples were carried out according to details given in previous papers.[9,10] The samples were immersed for 48 h in the amphiphile solution and then removed, each being rinsed with 10 mL of high-purity water (HPLC, Fluka) and finally blown dry with nitrogen (5.0 purity).

2.3. X-ray Photoelectron Spectroscopy (XPS). XPS spectra were recorded with a SAGE 100 system (Specs, Berlin, Germany) using nonmonochromatized Al Kα radiation at 320 W (13 kV) and an electron-energy analyzer pass energy of 50 eV for low-resolution surveys and of 14 eV for high-resolution detailed scans. The analyzed area was 6 mm^2, and the results therefore represent a laterally averaged chemical composition. To determine the quantitative surface composition from XPS data, the sensitivity factors of Scofield[11] are used. All binding energies are referenced relative to the main hydrocarbon peak (from residual contamination in the case of the clean surfaces and the SAM hydrocarbon chain in the case of the SAM-coated surfaces), set at a binding energy of 285.0 eV.

2.4. Atomic Force Microscopy (AFM). AFM measurements were performed with a commercial scanning probe microscope (Nanoscope III, Digital Instruments, Santa Barbara, CA). Measurements of surface topography and lateral force were made simultaneously by operating the instrument in contact mode while scanning the cantilever laterally. All measurements were performed in ambient air on freshly plasma cleaned samples.

2.5. Contact-Angle Measurements. Surface wettability was investigated by measuring advancing and receding contact angles in a sessile water drop experiment (Contact Angle Measuring System, G2/G40 2.05-D, Krüss GmbH, Hamburg, Germany). The measurements were performed in an automated way by increasing and decreasing the water drop size stepwise. Averaged data and error bars refer to six advancing-angle and three receding-angle measurements at different locations on each sample.

2.6. Time-of-Flight Secondary Ion Mass Spectrometry (ToF-SIMS) Measurements. Secondary ion mass spectra were recorded on a PHI 7200 time-of-flight secondary ion mass spectrometer (Physical Electronics, Eden Prairie, MN) in the mass range 0–1000 *m/e*. The total ion dose of the 8 kV Cs^+ primary ion beam (100 μm diameter) was below the static limit (<1.0 × 10^{13} ions/cm^{-2}). Time per data point was 1.25 ns. Three positive and three negative spectra were taken across the sample to ensure reproducibility and determine homogeneity of the monolayers. Mass resolution $M/\Delta M$ was typically around 5000 in the positive and 3000 in the negative mode. The PHI Tofpack software was used to calibrate the entire mass range from a single standard set of low ion masses. To improve the quality of mass calibration in the higher mass range, a higher-mass metal ion peak was used in addition.

Table 1. Roughness Values R_a and R_q for the Different Substrates

substrate	PVD coating		roughness values
glass		R_q	5.65 ± 0.28 nm[a]
		R_a	3.78 ± 0.19 nm[a]
	TiO_2	R_q	3.63 ± 0.18 nm[a]
		R_a	2.05 ± 0.10 nm[a]
	Ti(metal)	R_q	3.85 ± 0.19 nm[a]
		R_a	2.48 ± 0.12 nm[a]
Si wafer		R_q	0.11 ± 0.01 nm[a]
		R_a	0.07 ± 0.01 nm[a]
	TiO_2	R_q	0.60 ± 0.03 nm[a]
		R_a	0.44 ± 0.02 nm[a]
	Ti(metal)	R_q	1.67 ± 0.08 nm[a]
		R_a	1.32 ± 0.07 nm[a]
Ti(metal)	blasted, acid-etched (SLA)	R_q	2.01 ± 0.22 μm[b]
		R_a	2.56 ± 0.28 μm[b]

[a] Measured with AFM over a 10 × 10 μm area and calculated from lines of different scan lengths. [b] Measured with the noncontact laser profilometer technique (ref 12) in a scan range of 10–50 μm.

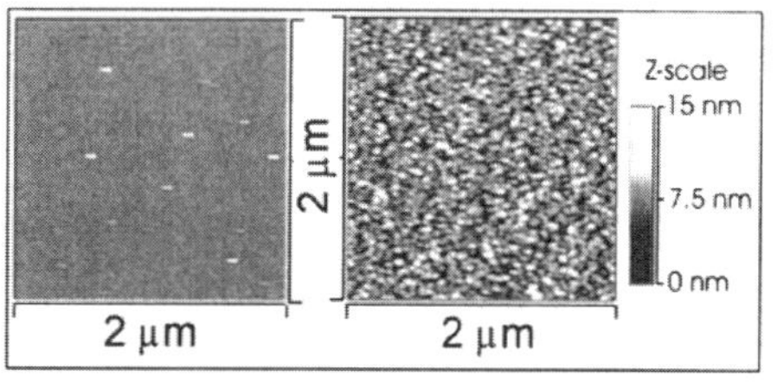

Figure 1. AFM images of TiO_2 (left) and Ti(metal) (right) on a silicon wafer. Scan area: 2 × 2 μm.

3. Results

3.1. Properties of the Clean Titanium (Oxide) Surfaces. The surfaces of all three types of substrates were cleaned according to the protocol given in refs 9 and 10 and characterized by XPS, AFM, and scanning electron microscopy (SEM). Roughness parameters of the different substrates were determined using AFM line scans in characteristic areas along distances of 10, 2, and 0.5 μm. They are reported as R_a and R_q values in Table 1. In the case of the Si wafer, the titanium oxide surface is very smooth with a roughness that is only slightly higher than that of the wafer substrate surface, while the titanium metal surface is significantly rougher. These findings can be explained by the fact that the PVD-deposited oxide film is amorphous, while the titanium metal film is expected, at least partially, to recrystallize following deposition, inducing grain-size- and grain-orientation-related topographic features at the surface (Figure 1). Roughness measurements on the blasted, acid-etched surface (SLA) were performed using the noncontact laser profilometry technique.[12,13] The roughness parameters are summarized in Table 1. The experimental values are characteristic of the complex topography of alumina-blasted, acid-etched

(10) Hofer, R. Surface Modification for Optical Biosensor Application. Ph.D. Thesis No. 13873, ETH Zürich, Zürich, Switzerland, 2000.

(11) Scofield, J. H. *J. Electron Spectrosc. Relat. Phenom.* **1976**, *8*, 129.

(12) Wieland, M. Ph.D. Thesis No. 13247, ETH Zürich, Zürich, Switzerland, 1999.

Table 2. XPS Binding Energies ($E_B \pm 0.1$ eV) of the Different Peak Components and Normalized Intensities[a] of Cleaned Titanium Oxide and Titanium Metal Films on Si Wafer Substrates

element	assignment	energy [eV]	normalized intensities[a] TiO$_2$ sample	normalized intensities[a] Ti(metal) sample	normalized intensities[a] Ti(SLA) sample
C(1s)(1)	**C**H$_2$, **C**H$_3$ contamination	285.0	6.9	6.4	8.8
C(1s)(2)	−**C**−O−H, −**C**−OOH contamination	286.5, 289.0			
O(1s)(1)	Ti**O**$_2$	530.1	64.3	62.0	61.9
Ti(2p1/2)	**Ti(metal)**	454.1		8.3	4.6
Ti(2p3/2)	**Ti(metal)**	459.7			
Ti(2p1/2)	**Ti**O$_2$	458.7	28.8	23.3	24.7
Ti(2p3/2)	**Ti**O$_2$	464.4			

[a] Experimental (overall) peak areas divided by the corresponding elemental sensitivity factors and expressed as a percentage of the summed normalized peak areas.

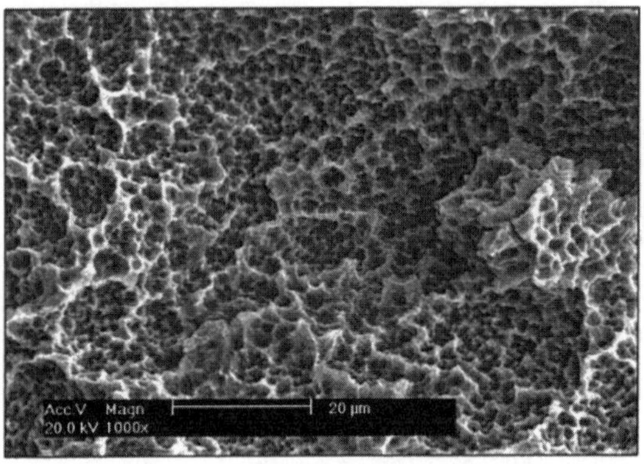

Figure 2. SEM image of the topography of the particle-blasted and acid-etched CP titanium (SLA) surface.

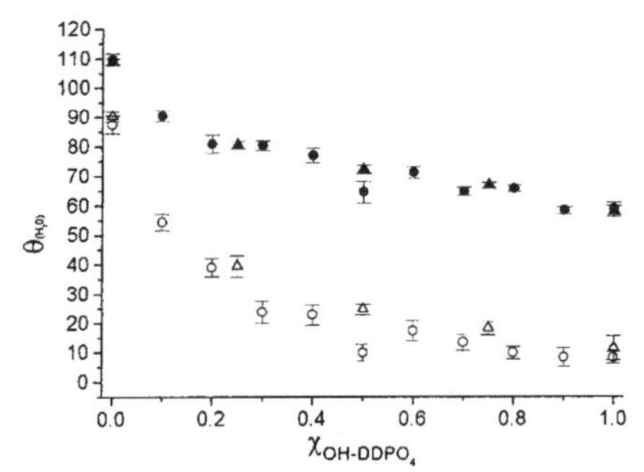

Figure 3. Advancing and receding contact angles of water on smooth titanium dioxide coated glass (adv = ●; rec = ○) and silicon wafer (adv = ▲; rec = △) surfaces as a function of the OH−DDPO$_4$(NH$_4$)$_2$ concentration in the SAM-forming solution, expressed as a mole fraction, $\chi_{OH-DDPO_4}$ = [OH−DDPO$_4$(NH$_4$)$_2$]/{[OH−DDPO$_4$(NH$_4$)$_2$] + [DDPO$_4$(NH$_4$)$_2$]}.

titanium surfaces.[12−14] A SEM image of the SLA surface taken at a magnification of 1000 is shown in Figure 2.

The XPS results (Table 2) show the presence of titanium, oxygen, and carbon, as expected for a well-cleaned titanium (oxide) surface.[15] The carbon is due to adventitious contamination that cannot be prevented in practice for samples exposed even briefly to ambient air. However, in our experience, this degree of contamination is low enough not to interfere adversely with the subsequent self-assembly process.

The water contact angles are typically below 10° for the flat titanium oxide and titanium metal surfaces when measured immediately (<5 min) after UV/ozone cleaning.

Positive ion ToF-SIMS spectra of the bare, smooth TiO$_2$ surfaces exhibit peaks resulting from hydrocarbon contamination, including the peak series $C_nH_{2n+1}^+$ and $C_nH_{2n-1}^+$, and from a few further contaminants, such as NH_4^+ and Na^+, present at very low concentrations (not detectable by XPS). The main peaks, however, are due to mono- or multinuclear oxide species of the general type $Ti_aO_bH_c$, with Ti^+ and TiO^+ displaying the highest intensities.[4] The negative ion spectra exhibit O^- and OH^- as the most prominent peaks observed from the oxide surface. Smaller peaks due to hydrocarbon contamination and some recombinant $Ti_aO_bC_cH_d$ peaks at higher mass ranges are also detected.[16,17] These data are consistent with results published previously on clean titanium oxide surfaces.[4,15]

3.2. Formation of Self-Assembled Monolayers and Contact-Angle Measurement. The alkyl phosphate and hydroxyalkyl phosphate solutions were mixed in different ratios from 0 to 100 vol %, leading to solutions with different mole fractions $\chi_{OH-DDPO_4}$ = [OH−DDPO$_4$(NH$_4$)$_2$]/{[OH−DDPO$_4$(NH$_4$)$_2$] + [DDPO$_4$(NH$_4$)$_2$]} at a constant total phosphate concentration of 0.5 mM. Following the protocols given in section 2.2, self-assembled monolayers were prepared on the smooth and structured samples. The water sessile drop contact-angle data are plotted in Figures 3−5 for the three different types of surfaces exposed to the amphiphile solution at different values of $\chi_{OH-DDPO_4}$ for 48 h. The advancing/receding contact angles on the SAM-coated, flat titanium metal and titanium oxide substrates show a relatively smooth, continuous change from around 110°/85° for the pure methyl-terminated SAM to around 55°/10° for the pure hydroxy-terminated SAM layers, with a clear trend toward increasing hysteresis (difference between advancing and receding angles) as the OH−DDPO$_4$ content ($\chi_{OH-DDPO_4}$) of the SAM solution is increased. The corresponding experimental data for the rough SLA surface, on the other hand, are quite different, with comparatively much higher advancing contact angles of between ca. 150° (DDPO$_4$) and 100° (OH−DDPO$_4$) being

(13) Wieland, M.; Textor, M. Measurement and Evaluation of the Chemical Composition and Topography of Titanium Implant Surfaces. In *Bone Engineering*; Davies, J. E., Ed.; em squared: Toronto, 2000; pp 163−182.
(14) Vörös, J.; Wieland, M.; Ruiz-Taylor, L.; Textor, M.; Brunette, D. M. Characterization of Titanium Surfaces. In *Titanium in Medicine: Material Science, Surface Science, Engineering, Biological Responses and Medical Applications*; Springer-Verlag: Berlin, 2001; pp 87−144.
(15) Textor, M.; Sittig, C.; Frauchiger, V.; Tosatti, S.; Brunette, D. M. Properties and Biological Significance of Natural Oxide Films on Titanium and Its Alloys. In *Titanium in Medicine: Material Science, Surface Science, Engineering, Biological Responses and Medical Applications*; Springer-Verlag: Berlin, 2001; pp 171−230.
(16) Lazzeri, P.; Lui, A.; Moro, L.; Vanzetti, L. *Surf. Interface Anal.* **2000**, *29* (11), 798.
(17) Michiels, F.; Adams, J. *Anal. At. Spectrosc.* **1987**, *2* (8), 773.

Table 3. XPS Binding Energies ($E_B \pm 0.1$ eV) of the Different Peak Components and Normalized Intensities of the Smooth Titanium Oxide (TiO_2 on Glass) Surface Covered with Self-Assembled Monolayers of Methyl- or Hydroxy-Terminated Dodecyl Phosphate

element	assignment	energy [eV]	normalized intensities I' [a] DDPO$_4$ SAM	normalized intensities I' [a] OH−DDPO$_4$ SAM
C(1s)(1)	**CH_2, CH_3**	285.0	27.0	29.3
C(1s)(2)	**C**−O−P, **C**−O−H (only OH−DDPO$_4$)	286.5		
O(1s)(1)	**TiO$_2$**	530.1	50.2	49.4
O(1s)(2)	Ti−**O**−P, P=**O**	531.3		
O(1s)(3)	P−**O**−H, P−**O**−R	532.6		
O(1s)(4)	C−**O**−H (only OH−DDPO$_4$)	533.4		
P(2p)	C−O−**P**(=O)(O$^-$)$_2$	134.0	2.5	2.5
Ti(2p1/2)	**TiO$_2$**	458.7	20.3	18.8
Ti(2p3/2)	**TiO$_2$**	464.4		

[a] Experimental (overall) peak areas divided by the corresponding elemental Scofield sensitivity factors.

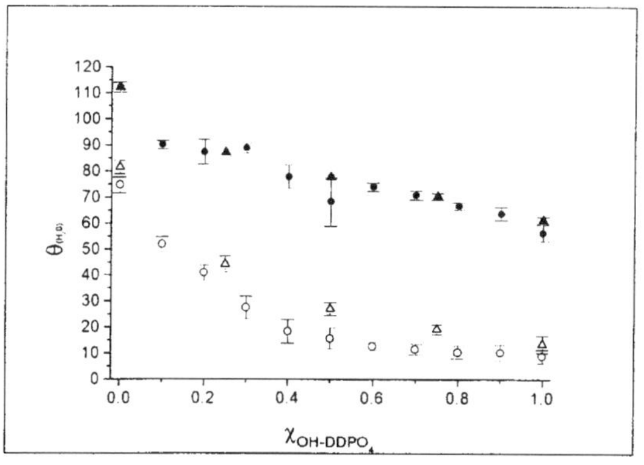

Figure 4. Advancing and receding contact angles of water on smooth titanium metal coated glass (adv = ●; rec = ○) and silicon wafer (adv = ▲; rec = △) surfaces as a function of the OH−DDPO$_4$(NH$_4$)$_2$ concentration in the SAM-forming solution, expressed as a mole fraction, $\chi_{OH-DDPO_4}$ = [OH−DDPO$_4$(NH$_4$)$_2$]/{[OH−DDPO$_4$(NH$_4$)$_2$] + [DDPO$_4$(NH$_4$)$_2$]}.

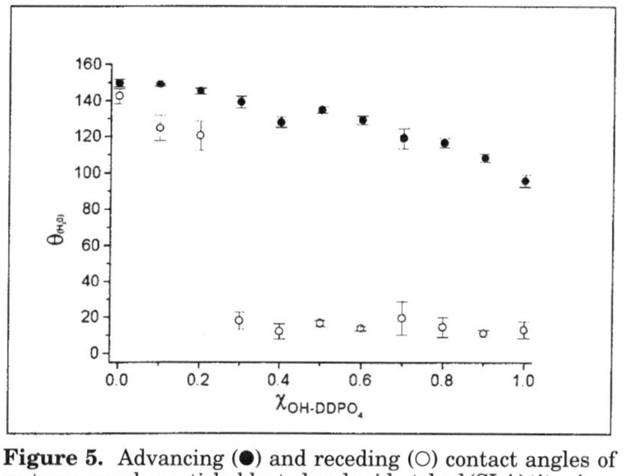

Figure 5. Advancing (●) and receding (○) contact angles of water on rough, particle-blasted and acid-etched (SLA) titanium metal surfaces as a function of the OH−DDPO$_4$(NH$_4$)$_2$ concentration in the SAM-forming solution, expressed as a mole fraction, $\chi_{OH-DDPO_4}$ = [OH−DDPO$_4$(NH$_4$)$_2$]/{[OH−DDPO$_4$(NH$_4$)$_2$] + [DDPO$_4$(NH$_4$)$_2$]}.

observed, as well as a characteristic sharp drop in the receding angle at $\chi_{OH-DDPO_4}$ values of 0.2−0.3 and a strong increase in hysteresis for $\chi_{OH-DDPO_4}$ = 0.3.

3.3. XPS Analysis of the SAM-Treated Surfaces. Table 3 summarizes the experimental binding energies of the C(1s), O(1s), P(2p), and Ti(2p) XPS peaks for a smooth (TiO_2 on glass) surface. The peak fitting (deconvolution) of the C(1s) and O(1s) peaks and assignment to particular species were performed according to protocols and argumentation discussed in detail in reference 4. Table 3 also contains the normalized intensities, that is, the experimental intensities (peak areas) divided by the corresponding sensitivity factors and expressed as percentages of the summed normalized peak areas. We deliberately refer to these values as normalized intensities rather than atomic concentrations to make clear that the type of surfaces under investigation should not be treated quantitatively using a homogeneous surface model, that is, assuming that the elemental concentrations are uniformly distributed across the sampling depth of the XPS method. Nevertheless, the data are useful for drawing some general conclusions and for semiquantitative comparisons of the different clean and modified surfaces.

Table 4 lists in detail the position, full width at half-maximum intensity (fwhm), and relative peak areas of the deconvolved C(1s) and O(1s) signals for both the smooth titanium oxide and smooth titanium metal surfaces treated in solutions with $\chi_{OH-DDPO_4}$ = 0, 0.5, and 1.0. In Figures 6 and 7, the intensity ratios of C(1s)/Ti(2p), O(1s)/Ti(2p), C(1s)/O(1s), and P(2p)/C(1s) are plotted as a function of $\chi_{OH-DDPO_4}$.

There are relatively minor changes in the intensity ratios (Figures 6 and 7) across the range of $\chi_{OH-DDPO_4}$ values, pointing to relatively constant densities of the assembled molecules independent of the $\chi_{OH-DDPO_4}$ ratio. There seems to be a slight tendency for the C/Ti ratio to increase with increasing $\chi_{OH-DDPO_4}$. Considering the error bars, however, this is not statistically relevant. Two observations are clearly significant: first, the proportion of C−O−R (R = H or C) increases with increasing $\chi_{OH-DDPO_4}$ (Table 4), pointing to an increase of the fraction of OH−DDPO$_4$ in the adlayer, as expected from the contact-angle measurements; second, the C/Ti intensity ratio is much higher in the case of the rough SLA titanium surface in comparison to the flat case (titanium metal or titanium oxide). This is consistent with expectations, since on average the electrons have to travel a longer distance through the SAM layer for a rough surface with a mean angle of inclination, α (relative to the surface plane), that is different from zero (see discussion section). The C(1s) intensity is expected to increase, while the Ti(2p) and O(1s) (oxide) intensities decrease with increasing α, making the C(1s)/Ti(2p) ratio a sensitive measure of the average tilt angle, α, of the surface. Corresponding quantitative analysis will be presented in the discussion section.

3.4. ToF-SIMS of SAM Surfaces. For the series of mixed self-assembled monolayer surfaces ($\chi_{OH-DDPO_4}$ = 0, 0.25, 0.5, 0.75, 1), two negative ion spectra per sample were recorded for three independent series. The dodecyl phosphate molecular ion peak, with the loss of one hydrogen $(M-H)^-$ in the negative secondary ion spectrum,

SAMs of Dodecyl Phosphates on Titanium *Langmuir, Vol. 18, No. 9, 2002* 3541

Table 4. Differences in the XPS Binding Energies ($E_B \pm 0.1$ eV), Full Widths at Half-Maximum, and Relative Peak Areas of Oxygen and Carbon Peaks after Curve Fitting for SAMs Assembled from Solutions with Different Molar Fractions, $\chi_{OH-DDPO_4}$[a]

				relative element areas for $\chi_{OH-DDPO_4} = 0$			relative element areas for $\chi_{OH-DDPO_4} = 0.5$			relative element areas for $\chi_{OH-DDPO_4} = 1$		
peak	species	ΔE^b [eV]	fwhm	calc	expt TiO_2	expt Ti(metal)	calc	expt TiO_2	expt Ti(metal)	calc	expt TiO_2	expt Ti(metal)
C(1s)(1)	CH_2, CH_3		1.55	90.9	91.4	89.0	86.4	85.0	85.9	81.8	82.2	81.5
C(1s)(2)	C–O–P, C–O–H	1.60	1.75	9.1	8.6	10.9	13.6	15.0	14.1	18.2	17.8	18.5
O(1s)(1)	TiO_2		1.55		79.5	70.1		76.1	69.4		73.9	67.7
O(1s)(2)	Ti–O–P, P=O	1.25	1.70		15.6	22.4		14.9	20.2		15.8	19.3
O(1s)(3)	P–O–H, P–O–R	2.55	1.70		4.9	7.5		6.3	8.4		5.0	8.9
O(1s)(4)	C–O–H	3.35	1.70					2.7	2.0		5.3	4.1

[a] Substrates: titanium oxide and titanium metal coatings on silicon substrates. [b] Binding energy shift relative to main peak component.

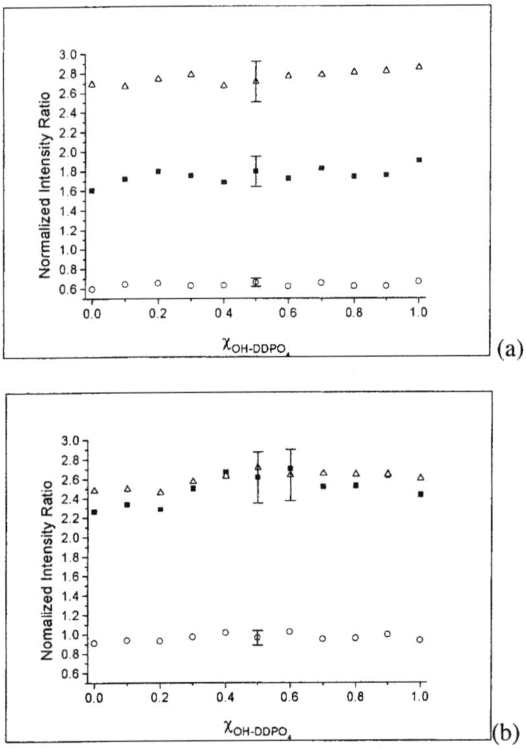

Figure 6. XPS normalized intensity ratios C/Ti (■), C/O (○), and O/Ti (△) for the smooth titanium oxide on glass (a) and rough (SLA) titanium (b), after SAM formation in solutions of different molar fractions, $\chi_{OH-DDPO_4}$ = [OH–DDPO$_4$(NH$_4$)$_2$]/{[OH–DDPO$_4$(NH$_4$)$_2$] + [DDPO$_4$(NH$_4$)$_2$]}.

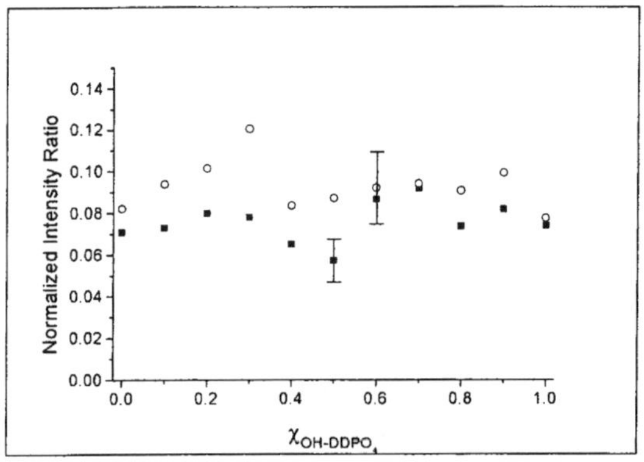

Figure 7. XPS normalized intensity ratios P/C on the smooth titanium oxide surface (TiO_2 on glass) (■) and the rough (SLA) titanium metal (○) surface, after SAM formation in solutions of different molar fractions, $\chi_{OH-DDPO_4}$ = [OH–DDPO$_4$(NH$_4$)$_2$]/{[OH–DDPO$_4$(NH$_4$)$_2$] + [DDPO$_4$(NH$_4$)$_2$]}.

was used to semiquantitatively evaluate the composition of the pure and mixed alkane phosphate adlayers. The comparison between the spectra of the clean TiO_2 surface and the $DDPO_4$ monolayer is in good agreement with the results of a previous study,[4] which described the adsorption of octadecyl phosphate onto Ta_2O_5 surfaces.

In general, SIMS cannot be considered a quantitative analytical technique due to strong matrix effects on secondary ion intensities.[16–18] However, under favorable conditions, for example, when a series of closely chemically related samples are measured under the same experimental conditions, *semi*quantitative results can be achieved.[19,20] We used sets of SAM samples that were prepared according to the same protocol and measured using the same experimental parameters. In the *positive* ion mode, the $DDPO_4$ self-assembled surface displays similar hydrocarbon ion peaks to that of the bare oxide (in this case due to adventitious hydrocarbon contamination), but with strongly decreased contaminant peak intensity (Na^+, NH_4^+, etc.), as well as decreased intensities of the Ti^+ and TiO^+ peaks. Fragments of the type $C_aH_bP_cO_d$, $Ti_aP_bO_cH_d$, and $Ti_aP_bO_cC_dH_e$ are indicative of the formation of a SAM and of a direct bond between the phosphate headgroup and the metal cations at the oxide surface. The *negative* ion spectra displayed strong PO_2^- and PO_3^- fragments that were not observed on the clean surfaces, as well as the molecular mass peaks $(M - H)^-$ characteristic of the two types of alkane phosphates ($DDPO_4$ at 265.16, OH–$DDPO_4$ at 281.15), providing good evidence that alkane phosphate molecules chemisorb in molecular form (i.e., without decomposition) onto the TiO_2 surface.

The molecular ion $(M - H)^-$ intensity of the methyl-terminated phosphate is 1 order of magnitude greater than that of the hydroxy-terminated phosphate. It has been reported that the detection of molecular mass ions M – H is correlated with the degree of order in self-assembled monolayers, resulting from energy transfer and stability effects.[21] We assume that the higher count rate

(18) Vickerman, J. C.; Brown, A.; Reed, N. *Secondary ion mass spectrometry: principles and applications*; Clarendon Press: Oxford, 1989.

(19) Chakraborty, P. *Pramana* **1998**, *50* (6), 617.

(20) Thompson, P. M. *Anal. Chem.* **1991**, *63* (21), 2447.

(21) Graham, D. J.; Ratner, B. D. Multivariate Analysis of TOF-SIMS Spectra from Dodecanethiol SAM Assembly on Gold: Spectral interpretation and TOF-SIMS fragmentation processes. *Langmuir*, submitted.

Table 5. Thicknesses of the Different Layers Used for Quantitative XPS-Based Calculations Assuming the Model of Figure 9 with an Ordered SAM and Molecular Tilt Angle of 30° Relative to the Surface Normal

layer no.	chemistry	atom code[a]	parameter code	thickness[b] [nm] for SAM on TiO_2 surface	thickness[b] [nm] for SAM on Ti(metal) surface
1	Ti(metal)	Ti(metal)	$d_{Ti(metal)}$		semi-infinite
2	TiO_2	TiO_2 or O1	d_{TiO_2}	semi-infinite	5.0[c]
3	phosphate group	P, O2, O3	d_{PO_4}	0.37	0.37
4	carbon chain for $DDPO_4$ and OH-$DDPO_4$	C1, C2	$d_{(CH_2)12}$	1.3	1.3
5	hydroxy group (OH, only OH-$DDPO_4$)	O4	d_{OH}	0.11	0.11

[a] See Table 4. [b] Measured perpendicular to the surface plane. [c] Natural oxide film thickness calculated from XPS intensity ratio I(Ti(2p), oxide)/I(Ti(2p), metal) of a clean surface (see text).

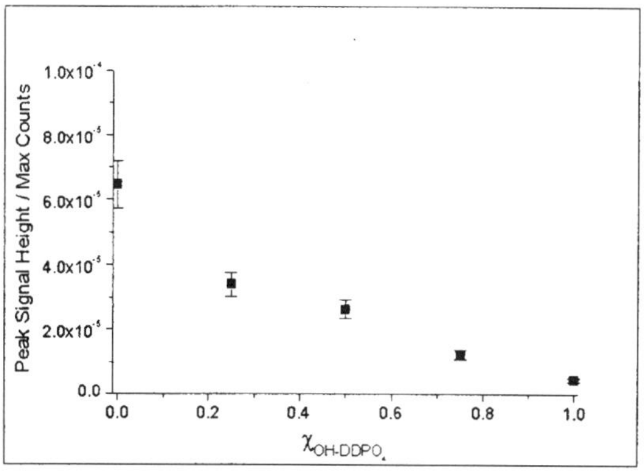

Figure 8. ToF-SIMS normalized peak signal heights of the $DDPO_4$ negative ion $(M - H)^-$ on smooth titanium oxide surfaces (TiO_2 on Si wafer) (■) after SAM formation in solutions of different molar fractions, $\chi_{OH-DDPO_4} = [OH-DDPO_4(NH_4)_2]/\{[OH-DDPO_4(NH_4)_2] + [DDPO_4(NH_4)_2]\}$.

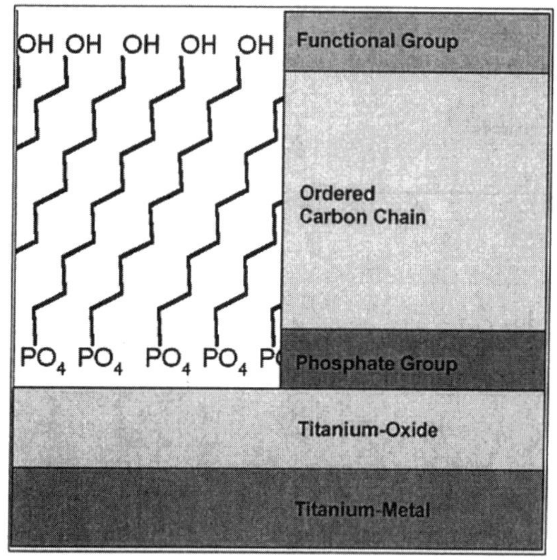

Figure 9. Idealized view and structural model of the alkane phosphate SAM on the titanium/titanium oxide substrate.

of the methyl-terminated phosphate is due to higher order in the self-assembled monolayer as well as to lower cohesion effects that may hinder the desorption of a hydrogen-bonded network of hydroxyl-terminated phosphates.

Therefore, the methyl-terminated phosphate signal can be used to show the dependency of the surface composition of the mixed self-assembled monolayers on the molar concentrations of the amphiphile in solution. Figure 8 shows the relative counts for the $DDPO_4$ ion expressed as a function of the solution molar fraction of the OH-terminated molecule ($\chi_{OH-DDPO_4}$). If taken quantitatively, this would represent a nonlinear dependence similar to the observations based on our other measurement techniques and in that way supports our view of the dependence of adlayer composition on assembly solution composition. However, it has to be admitted that the situation is rather complex in the sense that the adlayer composition and order are changing at the same time. More information on the order of these mixed adlayers is presented elsewhere.[22]

3.5. Model Calculations Based on XPS Results. This section concerns a multilayer model for the SAM-covered surface and corresponding calculations of XPS intensities based on standard equations and depending on the molecular orientation and packing density within the adlayer. The calculations are first applied to the less complex, smooth surfaces and then to the rough (SLA) case.

3.5.1. Smooth Case. *Model.* Based on earlier studies of molecular order and orientation in alkane phosphate SAMs on metal oxide surfaces,[4,9,10] a model assuming a uniform, dense packing of molecules tilted by an angle of 30° to the surface normal was used (Figure 9).

Based on this idealized model and using standard bond lengths,[23] the thicknesses of the different layers were estimated (Table 5). These layer thicknesses are used as input parameters for the calculations discussed below.

The following standard formulas for the intensity of the various photoelectron peaks were used:

Layer 1:

$$I^{Ti}_{Ti(metal)} \propto n^{Ti}_{Ti(metal)}\, \sigma_{Ti}\, T_{Ti}\, \lambda^{Ti}_{Ti(metal)} \exp\left(\frac{-d_{SAM+Oxide}}{\lambda^{Ti}_{SAM+Oxide}\sin(\Theta)}\right) \quad (1)$$

Layer 2:

Natural TiO_2 layer for Ti(metal) surfaces:

$$I^{Ti}_{TiO_2;Natural-Oxide} \propto n^{Ti}_{TiO_2}\, \sigma_{Ti}\, T_{Ti}\, \lambda^{Ti}_{TiO_2} \exp\left(\frac{-d_{SAM}}{\lambda^{Ti}_{SAM}\sin(\Theta)}\right) \times \left(1 - \exp\left(\frac{-d_{TiO_2}}{\lambda^{Ti}_{TiO_2}\sin(\Theta)}\right)\right) \quad (2)$$

(22) Zwahlen, M.; Tosatti, S.; Textor, M.; Hähner, G. *Langmuir*, in press.

(23) *CRC Handbook of Chemistry and Physics (1913–1995)*, 75th ed.; Lide, D. R., Ed.; CRC Press: Boca Raton, FL, 1994.

Table 6. Equations Used for the Calculation of the C(1s)/Ti(2p) Ratio for $DDPO_4$ and OH−$DDPO_4$ SAMs on Smooth TiO_2 and on Ti(metal) Surfaces Covered by a Natural Oxide Film

substrate	$DDPO_4$ SAM	OH−$DDPO_4$ SAM
TiO_2	$\frac{I^{C}_{DDPO_4}}{I^{Ti}_{TiO_2}}$ (6)	$\frac{I^{C}_{OH-DDPO_4}}{I^{Ti}_{TiO_2}}$ (7)
Ti(metal)	$\frac{I^{C}_{DDPO_4}}{I^{Ti}_{TiO_2;Natural-Oxide} + I^{Ti}_{Ti(metal)}}$ (8)	$\frac{I^{C}_{OH-DDPO_4}}{I^{Ti}_{TiO_2;Natural-Oxide} + I^{Ti}_{Ti(metal)}}$ (9)

PVD TiO_2 layer for TiO_2 surfaces:

$$I^{Ti}_{TiO_2} \propto n^{Ti}_{TiO_2}\, \sigma_{Ti}\, T_{Ti}\, \lambda^{Ti}_{TiO_2} \exp\left(\frac{-d_{SAM}}{\lambda^{Ti}_{SAM} \sin(\Theta)}\right) \quad (3)$$

Layer 4:

$$I^{C}_{DDPO_4} \propto n^{C}_{Layer4}\, \sigma_{C}\, T_{C}\, \lambda^{C}_{SAM}\left(1 - \exp\left(\frac{-d_{(CH_2)12}}{\lambda^{C}_{SAM} \sin(\Theta)}\right)\right) \quad (4)$$

Layer 5:

$$I^{C}_{OH-DDPO_4} \propto n^{C}_{Layer4}\, \sigma_{C}\, T_{C}\, \lambda^{C}_{SAM} \exp\left(\frac{-d_{OH}}{\lambda^{C}_{SAM} \sin(\Theta)}\right) \times \left(1 - \exp\left(\frac{-d_{(CH_2)12}}{\lambda^{C}_{SAM} \sin(\Theta)}\right)\right) \quad (5)$$

Definitions of the variables are presented in Table 7. The ratios were calculated by combining eqs 1−5 in an appropriate manner for each case (Table 6).

Several parameters are needed to estimate the corresponding XPS intensity ratios of eqs 1−9 within our model and to compare them to the experimental observations:

(1) The electron attenuation length (mean free path length λ) in the various layers was calculated using the general expressions given by Seah:[24]

Organic layer: $$\lambda = B\sqrt{E_{kin}} \quad (10a)$$

TiO_2 and Ti(metal) layer:

$$\lambda = \frac{E}{E_p\left[\beta \ln(\gamma E) - \frac{C}{E} + \frac{D}{E}\right]} \quad (10b)$$

with $B = 0.087$ and E = electron energy; for E_p, γ, C, and D, see eqs 5.23 and 5.24 of ref 24.

(2) The thickness d_{TiO_2} of the oxide film was calculated using eq 11:[25]

$$d_{TiO_2} = \lambda^{Ti}_{TiO_2} \cos(\theta) \ln\left(\frac{I^{Ti}_{TiO_2}}{I^{Ti}_{Ti(metal)}} \frac{n^{Ti}_{Ti(metal)}\, \lambda^{Ti}_{(metal)t}}{n^{Ti}_{TiO_2}\, \lambda^{Ti}_{TiO_2}} + 1\right) \quad (11)$$

(3) Atomic densities, n, for the titanium oxide and titanium metals were taken from ref 23.

(4) Photoelectron cross sections, σ, were taken from tabulations published by Scofield.[11]

(5) The density of alkane phosphate molecules at the surface and carbon atoms in layer 4 was calculated assuming a hexagonal arrangement of the molecules in the SAM layer with an intermolecular distance (parallel to the surface) of 0.49 nm as has been reported for octadecyl phosphate SAMs on tantalum oxide.[4] Such a layer has a molecular density of 4.8×10^{14} molecules/cm^2 and a carbon atomic density of $n^{C}_{Layer4} = 4.5 \times 10^{22}$ atoms/cm^3, calculated according to eq 12:

$$n^{C}_{Layer4} = \frac{n^{SAM-molecules}_{Surface}\, N^{C}_{SAM-molecule}}{d_{Layer4}} \quad (12)$$

The parameters and values used as input data for the calculations are summarized in Table 7.

The result of the calculation is the ratio between the C(1s) peak intensity of the alkane chains in the SAM and the Ti(2p) peak intensity of the TiO_2 or Ti(metal) substrate. To correct for a small contribution from hydrocarbon contamination, the C(1s) signal intensities (peak areas) of the SAM samples have been reduced by 7.5%, a value deduced from the XPS spectra of the corresponding bare (cleaned) titanium (oxide) surfaces.

The results of the experimental and calculated C(1s)/Ti(2p) intensity ratios are plotted in Figure 10 for the mixed SAMs (a) on the smooth TiO_2 and (b) on the smooth Ti(metal) substrate as a function of the solution molar fraction, $\chi_{OH-DDPO_4}$.

Figure 10 demonstrates that the agreement between the experimental ratios and the corresponding calculated values is quite satisfactory, in view of the experimental uncertainty (the C(1s)/Ti(2p) intensity ratio is a parameter that is highly sensitive to variations in SAM film thickness or density) and in view of the crude approximations of parameters such as λ. Two conclusions can be drawn. First, the density of the self-assembled molecules at the two titanium (oxide) surfaces appears to be comparable to that found experimentally for octadecyl phosphate on tantalum oxide as evidenced by the fairly good agreement between experimental and theoretical data based on the assumption of the same molecular density in both cases. Second, there seems to be no statistically significant trend for changes in the density of the organophosphate molecules (or of the effective adlayer thickness) within the SAM as one moves from the pure $DDPO_4$ SAM through the mixed adlayer case to the pure OH−$DDPO_4$ SAM.

The generally observed higher C/Ti ratios for the OH−$DDPO_4$ in comparison to the $DDPO_4$ SAM is likely to be related to one or both of two factors: First, the increase in surface energy that accompanies the increase in OH surface density is likely to increase the susceptibility to hydrocarbon surface contamination.[26] Second, differences in the structural order of the adlayer can also decrease the resistance of the surface to recontamination.[21] Such small differences in molecular order probably remain undetected by XPS but seem to affect the results of other techniques, such as ToF-SIMS.

(24) *Practical Surface Analysis*, 2nd ed.; Briggs, D., Seah, M. P., Eds.; John Wiley & Sons: New York, 1990; Vol. 1, Chapter 5.

(25) Sittig, C. E. Ph.D. Thesis No. 12657, ETH Zürich, Zürich, Switzerland, 1998.

(26) Barthlott, W.; Neinhuis, C. *Planta* **1997**, *202*, 1.

Table 7. List of Parameters, Definitions, and Values Used for the Calculations of XPS Intensity Ratios (Equations 1–9) within the Model

parameter	description	value
$n^{Ti}_{TiO_2}$	titanium atom density in TiO_2	2.9×10^{22} atoms/cm^3
$n^{Ti}_{Ti(metal)}$	titanium atom density in Ti(metal)	5.6×10^{22} atoms/cm^3
$\lambda^{Ti}_{TiO_2}$	electron attenuation length of Ti(2p) in TiO_2	1.86 nm
$\lambda^{Ti}_{Ti(metal)}$	electron attenuation length of Ti(2p) in Ti(metal)	2.11 nm
λ^{Ti}_{SAM}	electron attenuation length of Ti(2p) in the organic adlayer	2.75 nm
λ^{C}_{SAM}	electron attenuation length of C(1s) in the organic adlayer	3.32 nm
$\lambda^{Ti}_{SAM+Oxide} = X_{SAM}\,\lambda^{Ti}_{SAM} + X_{TiO_2}\,\lambda^{Ti}_{TiO_2}$ where X_i = fractional layer i with $X_{SAM} + X_{TiO_2} = 1$	electron attenuation length of Ti(2p) through the oxide film and the organic adlayer	2.08 nm
σ_C	C(1s) photoelectron cross section	1
σ_{Ti}	Ti(2p) photoelectron cross section	7.81
Θ	takeoff angle relative to the surface plane	90°

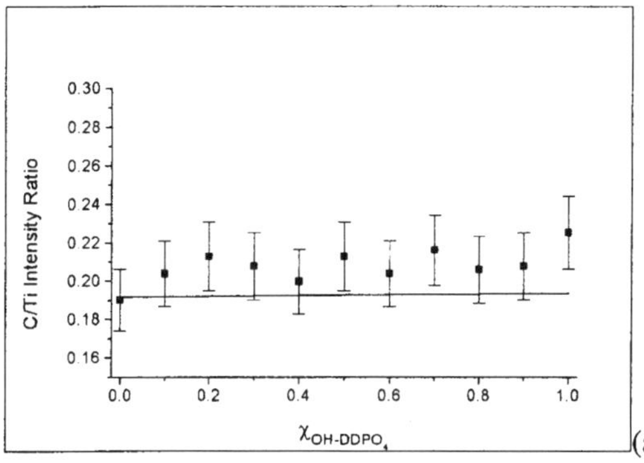

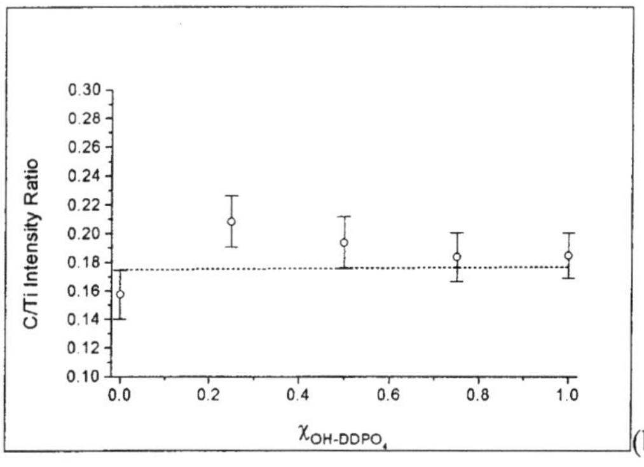

Figure 10. C(1s)/Ti(2p) XPS intensity ratios as a function of the composition of the self-assembly solution, $\chi_{OH-DDPO_4} = [OH-DDPO_4(NH_4)_2]/\{[OH-DDPO_4(NH_4)_2] + [DDPO_4(NH_4)_2]\}$, on smooth TiO_2 (a) and Ti(metal) (b) on Si wafers. Experimental values are given as symbols together with the estimated error range. The line refers to the calculated ratios assuming a density of 4.8×10^{14} molecules/cm^2 in the SAM layer, corresponding to a hexagonal close-packed SAM at an intermolecular distance of 0.49 nm (see text).

One question remains unanswered at this point: how does the molecular composition of the SAM layer compare with that of the amphiphile solution from which the SAMs have been produced? XPS offers the possibility to deconvolute signals that are characteristic for the $OH-DDPO_4$ molecule and which may be used to quantitatively determine the ratio of $OH-DDPO_4/DDPO_4$ within the organophosphate film. Two different peak components were employed for this purpose: the O(1s) signal of the –C–O–H terminal group (O4) with a binding energy of 533.4 ± 0.1 eV and the C(1s) signal of the same group with a binding energy of 286.5 ± 0.1 eV. In the first case, the hydroxy (O4) component was directly used to calculate the surface concentration of $OH-DDPO_4$. In the second case, the C(1s) binding energy (C2) of –C–O–H is very close (or identical) to that of –C–O–P–. Taking the difference of the C2 peak area between the pure $DDPO_4$ SAM and the mixed SAM therefore eliminated the latter contribution, which was present in the spectra of both $DDPO_4$ and $OH-DDPO_4$. Within a series of samples, the normalized intensity of COH (I'_{COH}) was normalized to the total concentration of C atoms (C1 + C2) in order to account for potential instrumental variations (such as X-ray flux) by using eq 13:

$$I'_{COH} = I'_{C2} - \left(\frac{I^{DDPO_4}_{COH}}{I^{DDPO_4}_{C-total}}\right) I'_{C-total} \qquad (13)$$

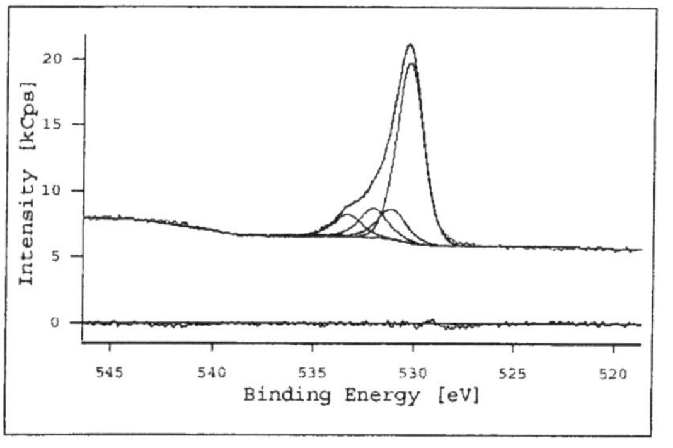

Figure 11. XPS oxygen peak (O(1s)) and resolved contributions of O1–O4 (see text), determined for a $OH-DDPO_4$ adlayer on TiO_2.

Figure 11 shows the O(1s) spectrum of the pure $OH-DDPO_4$ SAM on the TiO_2 surface together with the resolved contributions from the different oxygen (O1–O4) species present (see also Table 4).

In Figure 12, the normalized concentration of the COH terminal groups relative to the sum of all other C atoms is plotted as a function of the mole fraction, $\chi_{OH-DDPO_4}$.

We conclude from Figure 12 that there are statistical uncertainties when estimating the surface concentration of the terminal C–O–H group as a function of $\chi_{OH-DDPO_4}$. However, it is also obvious that the relationship is not a simple linear one, and a trend toward an increased proportion of $HO-DDPO_4$ in the surface layer in comparison to the solution concentration can be observed,

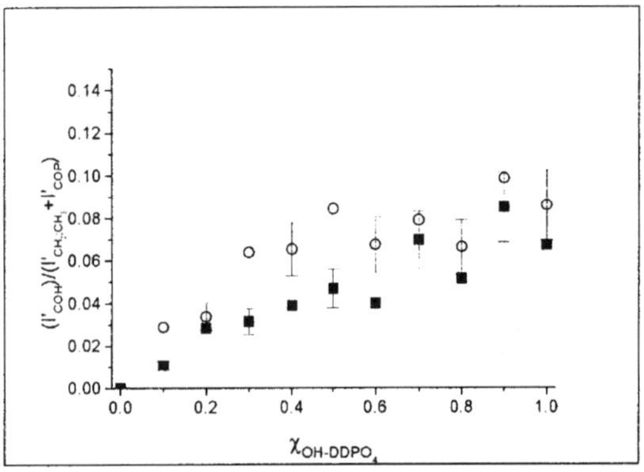

Figure 12. Ratio of the normalized intensity COH/(COH + CH_2 + CH_3 + COP) as a function of $\chi_{OH-DDPO_4}$ calculated from the resolved O(1s) peak contribution C−*O*−H (O4) (■) and the resolved C(1s) peak contribution *C*−O−H (○).

pointing to a preferential adsorption of HO−$DDPO_4$. This will be further discussed in section 4 together with information from contact-angle measurements.

3.5.2. Rough Surface. The XPS intensity ratios C(1s)/Ti(2p) for the SAM-coated, rough titanium metal surfaces (SLA) are significantly higher than on the flat titanium (oxide) surfaces. This is to be expected, since the effective local photoelectron takeoff angle is quite different from 90° on a highly rough, high-aspect-ratio surface, such as SLA. Correspondingly, the calculation procedure has to be adapted to account for the topographical effects. The most simple correction procedure was chosen by applying the same formulas as discussed for the flat case but introducing an effective takeoff angle corresponding to an idealized representation of the SLA surface with a single value for the local slope of the surface (Figure 13).

Using eq 11, the effective path length, $d_{TiO_2(SLA)}$, through the oxide layer at an angle of 90° relative to the macroscopic surface plane was calculated to be 6.35 nm. Assuming a 5.0 nm natural oxide layer for CP Ti samples,[25] the effective takeoff angle, α, calculated with the help of eq 14 corresponds to 51.9°.

$$\alpha = \frac{\pi}{2} - \arccos\left(\frac{d_{TiO_2(\mathrm{flat})}}{d_{TiO_2(\mathrm{SLA})}}\right) \tag{14}$$

The intensity ratios, C(1s)/Ti(2p), were now calculated again using eqs 1, 2−12 but replacing the thickness, *d*, of each surface layer by the effective path length $d' = d/\cos[(\pi/2) - \alpha]$. Within the experimental uncertainty, the experimental C(1s)/Ti(2p) intensity ratios agree with the calculated curve (Figure 14).

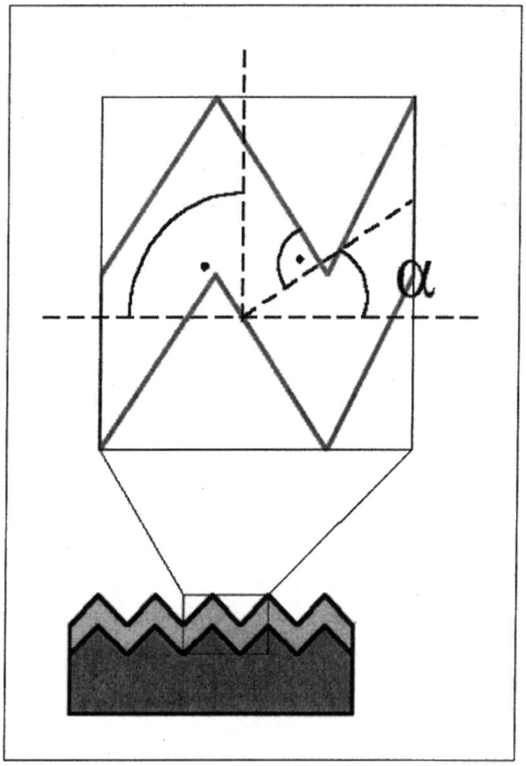

Figure 13. Idealized representation of the rough titanium (SLA) surface with a uniform local slope (angle α) of the surface relative to the average surface plane.

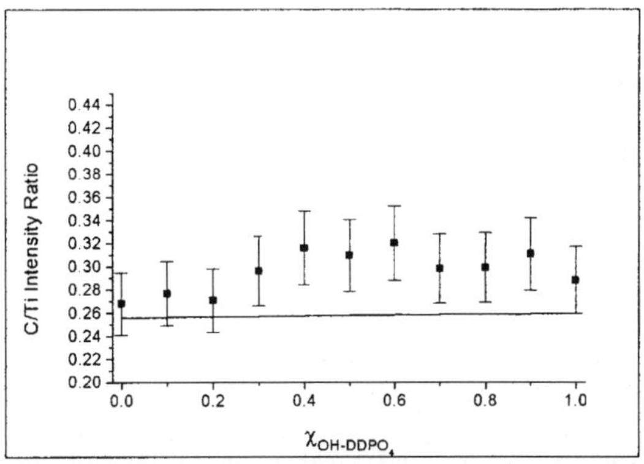

Figure 14. C(1s)/Ti(2p) XPS intensity ratios as function of the composition of the self-assembly solution, $\chi_{OH-DDPO_4}$ = [OH−$DDPO_4(NH_4)_2$]/{[OH−$DDPO_4(NH_4)_2$] + [$DDPO_4(NH_4)_2$]}, for the rough titanium metal (SLA) substrate. Experimental values are given as symbols together with the estimated error range. The line refers to calculated ratios assuming a density of 4.8 × 10^{14} molecules/cm^2 (effective surface) in the SAM layer (see text).

4. Discussion

4.1. Formation and Properties of Organophosphate SAMs on Smooth Surfaces. The combined results of XPS, ToF-SIMS, and contact-angle investigations provide good evidence for the formation of spontaneously formed monomolecular adlayers of dodecyl phosphate ($DDPO_4$) and hydroxy-terminated dodecyl phosphate (OH−$DDPO_4$) on titanium oxide, as well as on titanium metal coated glass and silicon substrates. The oxide and metal substrate surfaces behave similarly in terms of the properties of the SAMs, which is to be expected, since on the metallized surface the outermost layer of ca. 4−5 nm is composed of an amorphous, naturally formed oxide. The advancing contact angles of 110−112° for the methyl-terminated self-assembled monolayer and of 55° for the hydroxy-terminated surfaces support the model of an organized layer with the $-CH_3$ and −OH end groups of the amphiphilic molecules being exposed to the ambient environment (vacuum, air, or water) and with the phosphate groups being oriented toward the substrate.

However, the experimentally observed advancing (but not the receding) contact angle of 50−55° for the hydroxy-terminated alkane phosphate SAM is significantly higher compared to that of a well-ordered hydroxyl-terminated alkanethiol film on gold or other hydrophilic surfaces, for which water contact angles below 20° have been reported.[5] Our hypothesis was that this observation is indicative of a comparably lower order for the hydroxy-terminated alkane phosphate SAM, thus partly exposing more hydrophobic $-CH_2-$ groups at the surface. In a companion paper,[22] the authors have determined order in the same type of SAMs using NEXAFS. Indeed, there is a clear trend to a continuously lower degree of molecular order in these SAMs (increased angle of orientation of the alkane chains relative to the surface normal) as the proportion of the hydroxy-terminated alkane phosphate in the mixed adlayer is increased.

The ToF-SIMS spectra show a clear difference between bare oxide and self-assembled substrates, and the observed secondary ion fragments of the SAM surfaces provide further evidence for this type of molecular organization at the titanium oxide surface. The XPS intensity-ratio calculations are in reasonably good agreement with the assumed model of the organophosphate molecules assembled in the form of a densely packed monolayer with an intermolecular distance of 0.49 nm and an average density of 4.8×10^{14} molecules/cm^2 or 21 Å^2 per molecule, similar values having been reported for octadecyl phosphate monolayers on (amorphous) tantalum oxide surfaces[3,4] as well as for alkanethiols on gold (111).[1,5]

The water contact angles of the SAMs formed in the mixed OH−DDPO$_4$/DDPO$_4$ solutions provide evidence that there is a gradual change in the composition of the adlayer as the ratio of concentration in solution is changed. The quantification of the terminal −C−O−H functional group (XPS) and of the (M − H) (M = DDPO$_4$) mass fragment at the surface (ToF-SIMS), however, points to an adlayer composition that is richer in OH−DDPO$_4$ than would be expected from the solution composition (Figures 8 and 12).

The evolution of the advancing contact angles of SAMs deposited from mixed solutions may be compared with predictions from model equations describing micro- to macroscopically heterogeneous, two-component surface systems. Cassie has proposed the following equation for such systems:[27]

$$\cos \theta_{\text{Mixed-SAM}} = f_{\text{DDPO}_4} \cos \theta_{\text{DDPO}_4} + f_{\text{OH-DDPO}_4} \cos \theta_{\text{OH-DDPO}_4} \quad (15)$$

where f_{DDPO_4} and $f_{\text{OH-DDPO}_4}$ denote the relative surface coverage due to DDPO$_4$ and OH−DDPO$_4$, respectively ($f_{\text{DDPO}_4} + f_{\text{OH-DDPO}_4} = 1$).

Israelachvili and Gee[28] have proposed a modified equation that accounts for surfaces that exhibit heterogeneities on a scale close to molecular dimensions. This is likely the case for the SAMs under investigation:

$$(1 + \cos \theta_{\text{Mixed-SAM}})^2 = f_{\text{DDPO}_4}(1 + \cos \theta_{\text{DDPO}_4})^2 + f_{\text{OH-DDPO}_4}(1 + \cos \theta_{\text{OH-DDPO}_4})^2 \quad (16)$$

Both theoretical equations have been used to estimate the deviations of the surface composition from that of the solution, by plotting the mole fraction $f_{\text{OH-DDPO}_4}$ at the surface, as calculated by the Cassie and Israelachvili & Gee equations, against the solution mole fraction $\chi_{\text{OH-DDPO}_4}$. This is shown in Figure 15, together with the corresponding values based on the experimentally measured surface concentration of the −C−O−H terminal groups (Figure 12) and of the normalized peak signal of (M − H) (Figure 8). The different approaches all point to a preferential adsorption and assembly of the OH−DDPO$_4$ molecules.

(27) Cassie, A. B. D. *Discuss. Faraday Soc.* **1952**, *4*, 5041.
(28) Israelachvili, J. N.; Gee, M. L. *Langmuir* **1989**, *5*, 288.

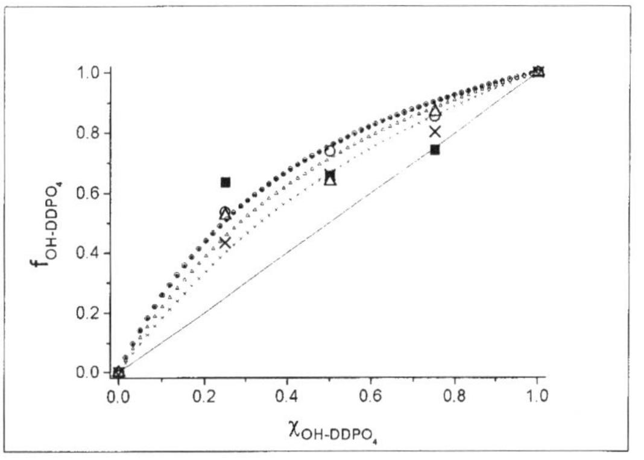

Figure 15. Plot of the molar fraction of OH−DDPO$_4$ in the SAM, $f_{\text{OH-DDPO}_4}$, against that of the self-assembly solution, $\chi_{\text{OH-DDPO}_4}$, using four different approaches: (a) Cassie equation for the contact angles of heterogeneous, patchy surfaces (○), (b) Israelachvili & Gee equation for the contact angle of heterogeneous systems at the molecular scale (×), (c) surface composition estimated via the concentration of the −C−O−H terminal groups at the surface (XPS, oxygen type 4) (■), and (d) surface composition estimated via the number of DDPO$_4$ molecules (ToF-SIMS, counts of (M − H) (M = DDPO$_4$)) (△). The straight line assumes a linear dependency between surface, $f_{\text{OH-DDPO4}}$, and solution, $\chi_{\text{OH-DDPO}_4}$, concentrations. Large markers are used for experimental data; small markers are calculated using eq 19 as a basis for the fitting function (see text). Surface: smooth titanium oxide coated Si wafer.

The data can be related to a SAM-formation model using the simplest approach, that is, a "Langmuir type" model where there is no interaction between the molecules, independent adsorption, and first-order kinetics. The kinetics of adsorption for DDPO$_4$ and OH−DDPO$_4$ can then be expressed as in eqs 17 and 18:

$$\frac{df_{\text{DDPO}_4}(t)}{dt} = k_{\text{DDPO}_4} C^{\text{Solution}}_{\text{DDPO}_4}(1 - f_{\text{DDPO}_4}(t) - f_{\text{OH-DDPO}_4}(t)) \quad (17)$$

$$\frac{df_{\text{OH-DDPO}_4}(t)}{dt} = k_{\text{OH-DDPO}_4} C^{\text{Solution}}_{\text{OHDDPO}_4}(1 - f_{\text{DDPO}_4}(t) - f_{\text{OH-DDPO}_4}(t)) \quad (18)$$

where $C^{\text{Solution}}_{\text{Molecule}}$ is the concentration of the molecule in the solution and f_{molecule} is the partial coverage of the surface by the corresponding amphiphile. Since all of the solutions have a phosphate concentration of 0.5 mM, it is possible to substitute the concentration $C^{\text{Solution}}_{\text{Molecule}}$ by the corresponding mole fraction $\chi_{\text{OH-DDPO}_4}$. Solving the set of two differential equations and calculating the ratio $f_{\text{DDPO}_4}/f_{\text{OH-DDPO}_4}$, we obtain eq 19 for $t \to \infty$:

$$\frac{1 - f_{\text{OH-DDPO}_4}}{f_{\text{OH-DDPO}_4}} = k\frac{1 - \chi_{\text{OH-DDPO}_4}}{\chi_{\text{OH-DDPO}_4}} \quad (19)$$

Table 8. Ratio of the First-Order Adsorption Rate Constants $k = k_{DDPO_4}/k_{OH-DDPO_4}$ Estimated by Fitting the Kinetic Model Equation to XPS Measurements, Contact-Angle Data, and ToF-SIMS Data

experimental method	evaluation technique	k
ToF-SIMS	signal peak of (M − H) (M = $DDPO_4$)	0.40
XPS	quantification of COH using O(1s) peak component	0.33
contact-angle	Israelachvili equation	0.33
contact-angle	Cassie equation	0.50

With $k = k_{DDPO_4}/k_{OH-DDPO_4}$, eq 19 can be solved for $f_{OH-DDPO_4}$ and used as a physical model based relation to fit the $f_{surface} = f_{surface}(\chi_{solution})$ curves presented in Figure 15, thus providing values for the ratio k of the first-order adsorption rate constants. The k values for the three different procedures to estimate $f_{surface}(\chi_{solution})$ are presented in Table 8. The fact that the values differ is not surprising, in view of the quite different experimental and theoretical approaches taken. The k value based on the Cassie equation may be expected to be less appropriate than that of Israelachvili and Gee since the mixed SAMs are likely not to show any substantial segregation and formation of extended patches with different surface chemistries. Preliminary AFM studies have indeed not yielded any evidence of segregation. We may therefore conclude that the formation of mixed SAMs on flat titanium oxide and titanium metal surfaces from aqueous solutions of the two amphiphiles can be described by a simple first-order adsorption model with the adsorption rate for OH−$DDPO_4$ being 2−3 times higher than that of $DDPO_4$.

4.2. Formation and Properties of Organophosphate SAMs on Rough (SLA) Surfaces. The estimation of surface coverage and properties on rough surfaces, in comparison to the ideal smooth-surface case, is complicated by the influence of surface roughness on both the quantitative XPS intensities and the contact angles.

In terms of the XPS investigation, the intensity ratio Ti(2p)(oxide)/Ti(2p)(metal) (a sensitive measure of oxide film thickness) and the ratio C(1s)/Ti(2p) (a sensitive function of the self-assembly coverage) depend on the distribution of local slopes of the surface relative to the mean (macroscopic) surface plane (Figure 13). The SLA surface topography profiles are too complex to be used directly within the XPS model. Therefore, the assumption was made that a mean slope may be a reasonable first approximation to be introduced as a parameter in both intensity ratios mentioned above.

Assuming that the natural oxide film thickness of a flat CP titanium metal surface and of the rough SLA surface is the same, an approximate (mean) slope of 51.9° was calculated using eq 14. This value was then applied in the calculation of the C(1s)/Ti(2p) intensity ratio assuming the same specific SAM coverage as for the flat case, that is, a close-packed arrangement of the organophosphate molecules along the corrugated surface (corresponding to a coverage per geometric (flat) unit area that is higher than that of the smooth surface by the effective surface area factor). The calculated intensity ratios are plotted in Figure 14. Despite the crudeness of the assumption, the consistency of the results is promising, providing good evidence that the assumed model of a close-packed assembly of alkane phosphate molecules is indeed reasonable for the corrugated SLA surface.

The water contact angle is shown in Figure 5 as a function of the molar ratio $\chi_{OH-DDPO_4}$ of the solution. In comparison to the smooth surfaces, the advancing contact angles are significantly higher, by roughly 40°, on the SLA surfaces. Roughness is well-known to influence static contact angles. The vectors describing the interfacial tension are locally changed in direction, and the magnitudes of the vectors composing the well-known Young equation are correspondingly modified. The Young equation, which strictly holds only for ideally homogeneous and smooth surfaces, is therefore no longer applicable. Wenzel has proposed the use of a modified equation that considers the effective surface area,[29] also called roughness factor, r, as an explicit parameter describing the real, rough surface:

$$r(\gamma_S - \gamma_{SL}) = \gamma_{LV} \cos \theta_W \tag{19}$$

where θ_W is now called the Wenzel angle and r = specific surface/geometric surface ≥ 1. γ_S is defined as the surface energy of the solid, γ_{SL} is defined as the surface energy at the interface between solid and liquid, and γ_{LV} is defined as the surface energy at the interface between liquid and vapor. The relation to the Young angle θ_Y is

$$\cos \theta_W = r \cos \theta_Y \tag{20}$$

This approach predicts that the Wenzel contact angle should increase with roughness for Young angles greater than 90° and decrease for angles lower than 90°. Since in our experiments all advancing angles are higher for the rough case, even when the Young angle (smooth surface value) is below 90°, this model does not explain the evolution of the contact angles with $\chi_{solution}$ in our case.

A very simple, purely geometric argument is that for a surface with a pronounced local inclination of the surface relative to the average surface plane, the advancing water drop would preferentially wet the local surface regions pointing away from the droplet center, microscopically jumping from slope to adjacent slope as a consequence. This would mean that the advancing contact angle would be larger by approximately the mean angle of inclination of the surface. The 40° difference observed in advancing contact angle between SLA and the smooth surface is actually close to the value determined by XPS intensity-ratio considerations (52°, see above). However, the difference between advancing and receding contact angle (hysteresis) cannot be explained by this oversimplified model. Hysteresis may provide very useful information about the surface chemistry and structure if it can be treated with a suitable model. It has been discussed in terms of surface heterogeneities (chemical and morphological) and the existence of many metastable energy minima in the free energy versus contact angle dependence.[30] In addition, hysteresis information may contain kinetic contributions if surface chemical changes occur during the course of the wettability studies.

The behavior of very rough, high-aspect-ratio surfaces (such as the SLA) is most likely governed by pronounced heterogeneities. In the $DDPO_4$-rich case in particular, this likely means that the fine structure of the surface is not completely wetted, water being prevented from penetrating into the topographical features of the surface. The Cassie and Baxter equation[31,32] may describe such surfaces:

(29) Wenzel, R. N. *Ind. Eng. Chem.* **1936**, *28*, 988.
(30) Garbassi, F.; Morra, M.; Occhiello, E. *Polymer Surfaces*; John Wiley & Sons: New York, 1994; Chapter 4.
(31) Cassie, A. B. D.; Baxter, S. *Trans. Faraday Soc.* **1944**, *40*, 549.
(32) Cassie, A. B. D.; Baxter, S. *J. Text. Inst.* **1945**, *36*, T67.

$$\cos \theta_{CB} = f_1 \cos \theta_1 - f_2 \quad (21)$$

where the heterogeneity is due to the local water–surface contact area f_1 (with contact angle θ_1) and local water/air contact areas f_2 (with θ_2 assumed to be 180°, $\cos \theta_2 = -1$). For a pure $DDPO_4$ SAM on Ti(SLA), the effective surface contact area may thus be only 20% of the total area. In the hydrophobic case, the general behavior of such hydrophobic and rough surfaces is characterized by extremely high (ca. 140–170°) contact angles and, if studied as a function of surface roughness, a sudden decrease of hysteresis above a critical roughness, due to a sudden decrease of the energy barrier between metastable states.[31,32,33] This is the situation for highly nonwettable surfaces, where the water droplet can actually move almost unhindered along the surface, a concept used by nature in leaves ("Lotus effect"), feathers, and so forth, to prevent wetting and to reduce the risk of adhesion of particles, bacteria, and fungi ("self-cleaning" properties).[26] The left-hand side of Figure 5 certainly corresponds to this case with a high advancing contact angle of ca. 140–150° and a relatively small hysteresis, similar to that observed in etched, microscopically rough poly(tetrafluoroethylene) surfaces.[33] On surfaces with Young contact angles below 90°, there is a rapid increase of hysteresis as roughness is increased up to the point where penetration of water into the microscopic surface texture ("wicking") occurs, at which point the advancing angle decreases again and the receding angle essentially goes to zero. The mixed SAMs on SLA do indeed show an extreme jump from low (about 5–10°) to very pronounced hysteresis (about 110°) as the molar fraction $\chi_{OH-DDPO_4}$ reaches 0.2–0.25. In that sense, the observation of much higher hysteresis as the adlayer becomes more hydrophilic is in agreement with the general observations reported in the literature.[29] However, the reason for the sharp discontinuity is not known to us. It is rather unlikely that this discontinuity reflects the onset of wetting of the surface structures (at least not of the fine etch structure of pits with diameters in the range 0.5–2 μm). First, this would also have to result in abrupt changes of the advancing contact angle, and second, it is known that even quite low Young contact angles (i.e., moderately hydrophilic surfaces) are still effective in preventing wicking.

More work, including surface-tension measurements (e.g., via Wilhelmy balance investigations), will be needed to better explain the details of the behavior of mixed organophosphate SAMs on rough (e.g., SLA type) titanium (oxide) surfaces. The motivation for further investigation is twofold: we believe that the interfacial energetics is important in the context of protein adsorption and conformational effects and therefore highly relevant to the biomaterials area; second, the mixed $DDPO_4$/OH–$DDPO_4$ system, together with well-defined surface structures, is expected to be a useful model system for investigating surface/interface energetics with truly independent control of surface chemistry and topography.

5. Conclusion

The technique of spontaneous organization of organophosphate molecules on titanium (oxide) surfaces is believed to have potential for the modification of titanium-based medical implants and devices with the aim of tailoring the surface chemistry (chemical or biological functionalities) to the particular needs of the application. It is a cost-effective technique (simple dipping process) but needs careful control of all relevant parameters, in particular in terms of cleanliness or purity of surfaces and materials used. Methyl- and hydroxy-terminated dodecyl phosphate were used as model systems to investigate self-assembly processes from aqueous solution onto titanium (oxide) surfaces. By using the technique of coadsorbed self-assembled monolayers, it was demonstrated that the technique allows the wettability of the surface to be tailored, not only in the case of smooth surfaces but also for high-surface-area materials, such as the SLA dental implant surface.

The technique also has applicability for basic studies. Wettability phenomena on surfaces of defined topography or roughness (including the observation of contamination-resistant, self-cleaning natural surfaces (Lotus effect)) could be studied much more systematically by having truly independent control over surface chemistry and topography.

In addition to tailoring wettability with alkyl phosphates, one could also impart further properties of interest in the fields of biomaterials and cell–surface interactions, including electric charges (e.g., by means of terminal amine, carboxylate, or phosphate functionalities) and biologically active groups such as poly(ethylene glycol), cell-adhesive peptides, or growth factors.

Further investigation should lead to a better understanding of the influence that minor changes in surface order have on adlayer properties such as resistance to recontamination or stability on contact with solutions of different compositions (water, buffer, etc.).

Acknowledgment. We acknowledge financial support by the International Team of Oral Implantology, ITI Foundation Research Committee, Waldenburg, Switzerland, N.P. Huang for her careful AFM measurements on the cleaned samples, and R. Hofer and S. DePaul for the scientific support.

LA011459P

(33) Lupis, H. C. O. *Chemical Thermodynamics of Materials*; North-Holland: New York, 1983; Chapter 23.

Langmuir **2007,** *23,* 8053–8060 8053

Influence of Alkyl Chain Length on Phosphate Self-Assembled Monolayers

Doris M. Spori, Nagaiyanallur V. Venkataraman, Samuele G. P. Tosatti, Firat Durmaz, Nicholas D. Spencer,* and Stefan Zürcher

Laboratory for Surface Science and Technology, Department of Materials, ETH Zurich, Wolfgang-Pauli-Strasse 10, CH-8093 Zurich, Switzerland

Received February 17, 2007. In Final Form: May 8, 2007

A series of alkyl phosphates with alkyl chain lengths ranging from C_{10} to C_{18} have been synthesized. Self-assembled monolayers (SAMs) of these molecules were prepared on titanium oxide surfaces by immersion of the substrates in alkyl phosphate solutions of 0.5 mM concentration in *n*-heptane/isopropanol. The SAMs were characterized by means of dynamic water contact angle (dCA) measurements, variable-angle spectroscopic ellipsometry (VASE), X-ray photoelectron spectroscopy (XPS), and polarization-modulated infrared reflection–absorption spectroscopy (PM-IRRAS). A higher degree of order and packing density within the monolayers was found for alkyl phosphates with alkyl chain lengths exceeding 15 carbon atoms. This is reflected in a lower dCA hysteresis, as well as a film thickness measured by VASE and XPS close to the expected values for SAMs with an average alkyl chain tilt angle of 30° to the surface normal. Additionally a shift of the symmetric and antisymmetric C–H stretching modes in the PM-IRRAS spectra to lower wave numbers was observed. These findings imply a higher two-dimensional crystallinity of the films derived from alkyl phosphates with a longer alkyl chain length.

1. Introduction

Self-assembled monolayers (SAMs) represent an easy, accurate and precise approach to the modification of surface properties.[1–3] Consequently, a significant amount of research has been dedicated to the investigation of the fundamentals associated with the spontaneous adsorption and assembly of monomolecular layers.[1–6] Most work has been carried out with alkyl silanes adsorbing on silicon oxide[7] or thiols or disulfides adsorbing on gold.[5,8,9] A further class of self-assembling materials—phosphonic and phosphoric acids—has gained interest due to their ability to bind to a wide range of metal oxide surfaces and form robust SAMs of a similar quality to that of thiol SAMs on gold. This has been demonstrated for the system of octadecyl phosphate deposited from a heptane/2-propanol solution on tantalum oxide,[10–13] for example. It is known that long-chain phosphonic and phosphoric acids lead to dense, well-ordered SAMs.[10,14–17] Some comparison studies between different lengths of alkyl phosph(on)ates have been carried out,[18–21] but there are still many open questions with respect to order, packing density, simplicity, and reproducibility of adsorption and stability, as well as binding mechanism and resulting bond-architecture to the corresponding metal oxide surfaces.

The aim of this paper was to investigate the SAM quality, in terms of binding mechanisms, density, and order, of alkyl phosphoric acids with alkyl chains ranging from C_{10} to C_{18} deposited from heptane/2-propanol solution on titanium oxide, in order to determine if there is a transition from liquid-like SAMs to more ordered and crystalline structures, as has been found for silanes and thiols on silicon or gold, respectively.[8,22,23] Research on aspects such as film thickness, tilt angle, and water contact angle as a function of the chain length has been carried out on various systems,[19,24–28] but, so far, no research has been carried out systematically for alkyl phosphates on titanium oxide. Changing from the commonly used octadecyl phosphate to shorter chain length improves the solubility of the molecule,[19] but, at the same time, the van der Waals interactions between the chains—a major stabilizing factor within the monolayer—are reduced. Titanium oxide was chosen as the substrate due to its biocompatibility and its potential for use in biosensor applications. For

* To whom correspondence should be addressed. E-mail: spencer@mat.ethz.ch, fax: +41 44 633 10 27.

(1) Schreiber, F. *Prog. Surf. Sci.* **2000**, *65*, 151–256.
(2) Smith, R. K.; Lewis, P. A.; Weiss, P. S. *Prog. Surf. Sci.* **2004**, *75*, 1–68.
(3) Ulman, A. *Chem. Rev.* **1996**, *96*, 1533–1554.
(4) Aswal, D. K.; Lenfant, S.; Guerin, D.; Yakhmi, J. V.; Vuillaume, D. *Anal. Chim. Acta* **2006**, *568*, 84–108.
(5) Love, J. C.; Estroff, L. A.; Kriebel, J. K.; Nuzzo, R. G.; Whitesides, G. M. *Chem. Rev.* **2005**, *105*, 1103–1169.
(6) Schwartz, D. K. *Annu. Rev. Phys. Chem.* **2001**, *52*, 107–137.
(7) Sagiv, J. *J. Am. Chem. Soc.* **1980**, *102*, 92–98.
(8) Bain, C. D.; Troughton, E. B.; Tao, Y. T.; Evall, J.; Whitesides, G. M.; Nuzzo, R. G. *J. Am. Chem. Soc.* **1989**, *111*, 321–335.
(9) Nuzzo, R. G.; Allara, D. L. *J. Am. Chem. Soc.* **1983**, *105*, 4481–4483.
(10) Brovelli, D.; Hähner, G.; Ruiz, L.; Hofer, R.; Kraus, G.; Waldner, A.; Schlosser, J.; Oroszlan, P.; Ehrat, M.; Spencer, N. D. *Langmuir* **1999**, *15*, 4324–4327.
(11) Hähner, G.; Hofer, R.; Klingenfuss, I. *Langmuir* **2001**, *17*, 7047–7052.
(12) Hofer, R.; Textor, M.; Spencer, N. D. *Langmuir* **2001**, *17*, 4014–4020.
(13) Textor, M.; Ruiz, L.; Hofer, R.; Rossi, A.; Feldman, K.; Hähner, G.; Spencer, N. D. *Langmuir* **2000**, *16*, 3257–3271.
(14) Foster, T. T.; Alexander, M. R.; Leggett, G. J.; McAlpnie, E. *Langmuir* **2006**, *22*, 9254–9259.
(15) Gao, W.; Dickinson, L.; Grozinger, C.; Morin, F. G.; Reven, L. *Langmuir* **1996**, *12*, 6429–6435.
(16) Pawsey, S.; Yach, K.; Reven, L. *Langmuir* **2002**, *18*, 5205–5212.
(17) Helmy, R.; Fadeev, A. Y. *Langmuir* **2002**, *18*, 8924–8928.
(18) Chen, Y. X.; Liu, W. M.; Ye, C. F.; Yu, L. G.; Qi, S. K. *Mater. Res. Bull.* **2001**, *36*, 2605–2612.
(19) Pellerite, M. J.; Dunbar, T. D.; Boardman, L. D.; Wood, E. J. *J. Phys. Chem. B* **2003**, *107*, 11726–11736.
(20) Tosatti, S.; Michel, R.; Textor, M.; Spencer, N. D. *Langmuir* **2002**, *18*, 3537–3548.
(21) Zwahlen, M.; Tosatti, S.; Textor, M.; Hähner, G. *Langmuir* **2002**, *18*, 3957–3962.
(22) Troughton, E. B.; Bain, C. D.; Whitesides, G. M.; Nuzzo, R. G.; Allara, D. L.; Porter, M. D. *Langmuir* **1988**, *4*, 365–385.
(23) Wasserman, S. R.; Tao, Y. T.; Whitesides, G. M. *Langmuir* **1989**, *5*, 1074–1087.
(24) Barrena, E.; Palacios-Lidon, E.; Munuera, C.; Torrelles, X.; Ferrer, S.; Jonas, U.; Salmeron, M.; Ocal, C. *J. Am. Chem. Soc.* **2004**, *126*, 385–395.
(25) Fenter, P.; Eberhardt, A.; Liang, K. S.; Eisenberger, P. *J. Chem. Phys.* **1997**, *106*, 1600–1608.
(26) Frey, S.; Shaporenko, A.; Zharnikov, M.; Harder, P.; Allara, D. L. *J. Phys. Chem. B* **2003**, *107*, 7716–7725.
(27) Tao, Y. T. *J. Am. Chem. Soc.* **1993**, *115*, 4350–4358.
(28) Zharnikov, M.; Kuller, A.; Shaporenko, A.; Schmidt, E.; Eck, W. *Langmuir* **2003**, *19*, 4682–4687.

Published on Web 06/15/2007

a complete understanding of the structure of the deposited films, a combination of surface-characterization techniques was applied: dynamic water contact angle (dCA), variable-angle spectroscopic ellipsometry (VASE), X-ray photoelectron spectroscopy (XPS), and polarization-modulated infrared reflection−absorption spectroscopy (PM-IRRAS). It was found that, for chain lengths above 15 carbon atoms, alkyl phosphates form crystalline structures.

2. Materials and Methods

General Considerations. All reactions were carried out under a N_2 atmosphere using standard Schlenk techniques. The used alcohols ($H(CH2)_nOH$, $n = 10-18$) and $POCl_3$ were obtained from Fluka (Switzerland) in a purity equal to or higher than 95%. Petrolether for recrystallizations (bp fraction 80−110 °C) was freshly distilled before use. Routine ^1H, ^{31}P, and ^{13}C NMR spectra were recorded with a Bruker 300 MHz spectrometer. Elemental analyses were performed with a Leco CHN-900 at the Laboratory for Organic Chemistry, ETH Zurich.

2.1. Phosphate Synthesis. The alkyl phosphates were all synthesized according to a slightly adjusted protocol developed in the late 1970s by Imokawa and Tsutsumi (Scheme 1).[29] Briefly, an

Scheme 1. Synthesis of Alkyl Phosphates from Alcohols and Phosphorus Oxychloride.

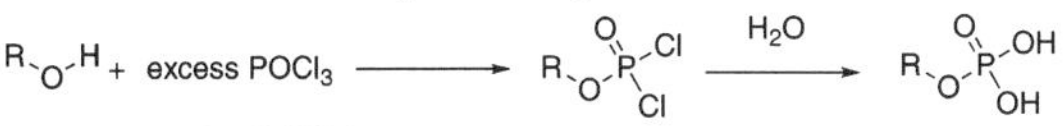

R = H-(CH$_2$)$_n$- n = 10, 11, 12, 13, 14, 15, 16, 17, 18

alcohol with the desired chain length was added directly to a 1.5-fold excess of neat phosphorus oxychloride under an atmosphere of nitrogen. The resulting mixture was warmed to 50 °C for 5 h, and the evolving hydrogen chloride was removed from the flask by a stream of nitrogen. After cooling to room temperature, the final mixture was poured into a water/ice mixture and stirred for several hours. The products were isolated by extraction into diethyl ether and purified by recrystallization from freshly distilled petroleum ether (bp 80 °C). This procedure readily allows several grams of product to be synthesized in good quality in a single step. Using an excess of phosphorus oxychloride reduces the formation of dialkyl phosphates, which, together with the phosphoric acid anhydride, constitute the two main side products and can be easily detected by ^{31}P NMR. Several recrystallizations from petroleum ether remove the side-products to a level below the NMR detection limit. The purity of the final compounds was determined by ^1H, ^{13}C, and ^{31}P NMR and elemental analysis.

2.2. Substrates. We used 8 × 6 mm^2 single-side-polished silicon wafers coated by physical vapor deposition (PVD) sputtering of 20 nm TiO_2 for XPS, VASE, and CA studies. PM-IRRAS substrates were obtained by PVD sputtering of 8 nm TiO_2 onto flat gold samples prepared as reported previously.[30]

2.3. SAMs. SAMs were prepared by immersion of the wafers into 0.5 mM solutions of the corresponding phosphate dissolved in a heptane/isopropanol mixture (99.2/0.8 v/v for C_{10} to C_{17} and 99.4/0.6 for C_{18}). After 48 h of immersion, they were removed from the solution, rinsed thoroughly with isopropanol, and blown dry with a stream of nitrogen.

2.4. XPS. XPS analyses were performed using either a PHI 5700 spectrophotometer equipped with a concentric hemispherical analyzer (CHA) (Physical Electronics, Eden Prairie, MN) or a SAGE 100 system (Specs, Berlin, Germany) in the standard configurations. On the PHI 5700, spectra were acquired at a base pressure of 10^{-9} mbar or below, using a non-monochromatic Al−Kα source operating at 350 W and positioned 10 mm away from the sample. The instrument was run in the minimum-area mode using an aperture of 0.4 mm diameter. The CHA was used in the fixed-analyzer-transmission mode. The pass energies used for survey scans and detailed scans were 187.85 and 46.95 eV, respectively, for titanium Ti2p, carbon C1s, oxygen O1s, and phosphorus P2p. Under these conditions, the energy resolution (full width at half-maximum height, fwhm) measured on silver Ag3d5/2 is 2.7 and 1.1 eV, respectively. Acquisition times were approximately 5 min for survey scans and 15 min (total) for high-energy-resolution elemental scans. These experimental conditions were chosen in order to have an adequate signal-to-noise ratio in a minimum time and to limit beam-induced damage. Under these conditions, sample damage was negligible, and reproducible analyzing conditions were obtained on all samples. In addition, only one sample was introduced into the analyzing chamber at a time. The measurements were performed with a takeoff angle (detection angle to the surface) of 45° with respect to the surface plane. In addition and for qualitative comparison, a series of samples were also measured on a SAGE 100 system (Specs, Berlin, Germany) using recently published conditions.[20] The main difference between these measurements and those recorded on the PHI 5700 is the 90° takeoff angle (detection angle to the surface) with a concomitantly lower surface sensitivity and lower signal-to-noise ratio. Data acquired with this instrument are included in the Supporting Information. All recorded spectra were referenced to the aliphatic hydrocarbon C1s signal at 285.0 eV. Data were analyzed using the program CasaXPS [version 2.3.5, www.casaxps.com]. The signals were fitted using Gaussian−Lorentzian functions and least-squares-fit routines following Shirley iterative background subtraction. Sensitivity factors were calculated using published ionization cross sections[31] corrected for the energy dependence of the transmission function of the instrument and the attenuation-length dependence on kinetic energy.

2.5. VASE. The monolayer thickness was measured, using a VAS ellipsometer (M-2000FTM, J.A. Wollam, Inc., Lincoln, NE), and the data were evaluated using the software WVASE32 (WexTech Systems, Inc., New York). The measurement was conducted in the spectral range of 370−1000 nm at three angles of incidence (65, 70, and 75°) under ambient conditions immediately before and after monolayer formation. The parameters for data evaluation are summarized in the supporting materials.

2.6. dCA Measurements. Surface wettability was investigated by measuring advancing and receding contact angles in a sessile drop (water) experiment (contact angle measuring system, G2/G40 2.05-D, Krüss GmbH, Hamburg, Germany). The measurements were performed in an automated way by increasing and decreasing the drop size at a speed of 15 μL per minute. Averaged data and error bars refer to a total of 480 points for the advancing angle and 240 points for the receding angle, resulting from measurements taken at three different locations on each sample. Data were analyzed using the *tangent method 2* fit routine of the Drop-Shape Analysis program (DSA version 1.80.0.2 for Windows 9x/NT4/2000, (c) 1997−2002 KRUESS)

2.7. PM−IRRAS. PM-IRRAS measurements were carried out on a Bruker IFS66v spectrometer equipped with a PMA50 photoelastic modulator accessory. The interferogram from the spectrometer was modulated with a ZnSe photoelastic modulator (Hinds Co.) at a frequency of 50 kHz and analyzed with a lock-in amplifier (SR380, Stanford Research, USA). Typically 1024 scans were acquired at a resolution of 8 cm^{-1}. The sample chamber was purged continuously with dry air during the measurements. The resulting interferograms were processed using OPUS (Bruker Optics) software, baseline corrected with a polynomial background, and normalized to the highest intensity band. The spectra showed strong bands in the frequency region below 1000 cm^{-1} arising from the TiO_2 layer. However, the C−H stretching region, which is of interest in this study, was free of any interfering features from the TiO_2 layer underneath.

3. Results

3.1. dCAs. One of the most sensitive techniques to probe the outermost surface of a sample is the measurement of dynamic

(29) Imokawa, G.; Tsutsumi, H. *J. Am. Oil Chem. Soc.* **1978**, *55*, 839−843.
(30) Venkataraman, N. V.; Zürcher, S.; Spencer, N. D. *Langmuir* **2006**, *22*, 4184−4189.
(31) Scofield, J. H. *J. Electron Spectrosc. Relat. Phenom.* **1976**, *8*.

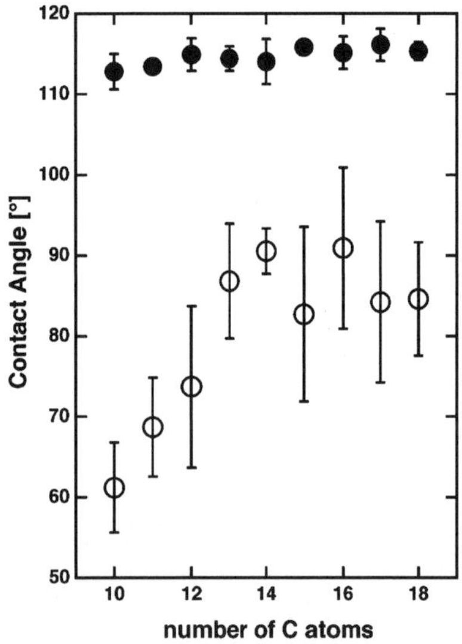

Figure 1. dCA measurements on the C_{10} to C_{18} series of phosphate SAMs. Only a small increase in the advancing CA (closed symbols) can be observed by increasing the chain length. In comparison, the receding CA (open symbols) is very low for C_{10} and increases rapidly with chain length up to C_{14}. For the longer-chain phosphates, the receding CA remains more or less constant at around 90°.

contact angles. This technique gives information not only on surface energy but also on the degree of order of the molecules directly located at the solid−air interface.[20,32] Advancing water contact angles for all measured SAMs were found to be above 110°—a clear sign that the hydrophobic alkyl chains are exposed at the SAM−air interface (Figure 1). A small but significant increase was observed in going from C_{10} to C_{18}. The receding contact angles, in contrast, show a large increase from approximately 60° for C_{10} to above 85° for SAMs having alkyl chains longer than C_{13}. The decreasing hysteresis between advancing and receding contact angles as the carbon chain length is increased is an indication of the increasing order of the SAMs.[20,21,32] An influence of substrate roughness on hysteresis can be excluded, since all samples were manufactured under the same conditions and possessed a highly reproducible subnanometer morphology, similar to that of ultraflat gold substrates.[20,21]

3.2 XPS Analysis. *Qualitative XPS Analysis of Self−Assembled Alkyl Phosphates on Titanium Oxide.* The SAMs for all nine investigated phosphates C_{10} to C_{18} were analyzed at a 45° takeoff angle. In the survey spectra, the only elements that could be detected were titanium, oxygen, carbon, and phosphorus, as expected for these SAMs on TiO_2 (Figure 2a). The high-resolution detail spectra for O 1s, Ti 2p, C 1s, and P 2p were resolved into their components using a fitting procedure as described below. The spectra are displayed in Figure 2, and the parameters used are presented in Table 1. In respect to each element, it was possible to observe that

P 2p: The phosphorus signal (Figure 2b) was fitted as a single doublet with P $2p_{3/2}$ at 134.5 ± 0.1 eV and P $2p_{1/2}$ with a binding-energy difference of 0.81 eV and a fixed-area ratio of 2:1. On a Ta_2O_5 substrate, the binding energy of a octadecylphosphoric acid SAM was found to be 134.2 ± 0.1 eV.[13] The P 2p binding energy for the free, nonbonded acid (octadecylphosphoric acid powder) was detected at a slightly higher binding energy of 134.7 ± 0.1 eV.[13]

Ti 2p: Only one chemical species could be detected for the Ti 2p signal (Figure 2c) with a binding energy corresponding to that of TiO_2 at 458.6 eV for Ti $2p_{3/2}$.[33] With increasing length of the phosphates, the intensity of Ti 2p decreases due to greater attenuation by the thicker organic overlayer.

C 1s: The main contribution of the C 1s signal originates from aliphatic carbon with a binding energy of 285 eV. A shoulder at a higher binding energy of 286.6 eV can be attributed to the CH_2 group connected to the phosphate. Whereas this higher binding energy contribution remains more or less constant in intensity (the signal being too small to observe attenuation effects), the aliphatic carbon contribution increases with increasing chain length, as expected (Figure 2d).

O 1s: At a takeoff angle of 45°, the main contribution to the O 1s signal originates from the oxygen in the TiO_2 substrate (binding energy 530.1 eV), (Figure 2e). The remaining oxygen contributions should all therefore originate at the substrate-SAM interface and can be assigned to alkyl−O−P, P═O, P−OH, P−O−Ti, and Ti−O−H. Since the amount of Ti can be measured from the Ti 2p peak, the amount of oxygen from the substrate can be calculated and subtracted from the O 1s signal. The remaining peak clearly shows two shoulders on each side (Figure 2f). It was therefore modeled using three contributions assigned to R−O−P at 532.8 eV—P−O−H, P−O−Ti, and Ti−OH—which are assumed to have very similar binding energies and are fitted with a peak centered at 531.4 eV and an fwhm of 2.0 eV, and P═O at 530.8 eV. Since the resolution of the O 1s peak does not allow us to distinguish between all these contributions, some additional constraints had to be implemented in the fitting routine to obtain consistent results. Knowing that R−O−P and P═O are located at a depth in the structure similar to that of phosphorus, and since the amount of phosphorus can be measured independently, the areas of these two peaks were also calculated and constrained. This then only allows the area of the component assigned to P−O−H, P−O−Ti, and Ti−OH to vary freely. For this contribution, a clear trend can be observed: It decreases more with increasing chain length than would be expected from attenuation effects alone. This observation can be explained with a decreasing amount of free Ti−OH and an increasing phosphate density. As expected, the relative amount of the substrate oxygen contribution decreases with increasing carbon chain length, which is a clear sign that the overlayer thickness is increasing.

The first indications of a structural change with increasing chain length can be clearly observed in Table 2. Indeed, although the C/Ti ratio increases with increasing chain length, a change in the (interface O)/P ratio, as well as a change in the P/Ti ratio with increasing alkyl chain length, suggests a change in the surface coverage and packing density of the alkyl chains in addition to a simple increase in the adlayer thickness. Finally, the oxygen found at the SAM substrate interface also decreases with increasing carbon chain length. If one calculates the excess oxygen at the interface, it can be attributed to the number of Ti−OH species per phosphate group. There is clearly more Ti−OH present in the case of the shorter molecules than for the longer ones. This suggests that, for longer chain lengths, the phosphate density of the monolayers is increasing at the cost of either the surface-bound water or hydroxyl groups.

Quantitative XPS Analysis of Self−Assembled Alkyl Phosphates on Titanium Dioxide. Since the studied SAMs on titanium dioxide

(32) Garbassi, F.; Morra, M.; Occhiello, E. *Polymer Surfaces*; John Wiley & Sons: New York, 1994.

(33) NIST Data Gateway. http://srdata.nist.gov.

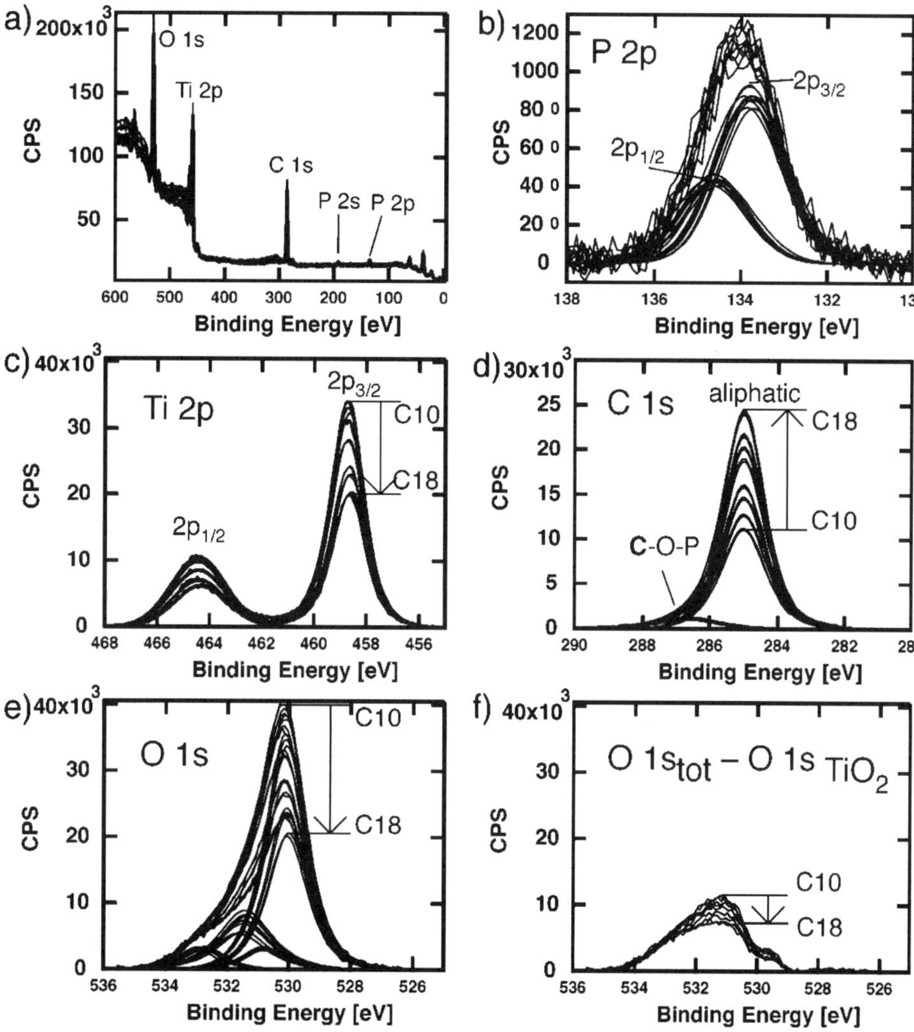

Figure 2. (a) XPS survey spectra for the series of C_{10} to C_{18} phosphate SAMs on TiO_2. (b) P 2p detail spectra with their deconvolution into P $2p_{3/2}$ and P $2p_{1/2}$. (c) Ti 2p detail spectra with deconvolution into Ti $2p_{3/2}$ and Ti $2p_{1/2}$. With increasing chain length of the phosphate SAMs, the Ti 2p intensity decreases due to higher attenuation. (d) C 1s detail spectra. An opposite trend compared to that of Ti 2p with increasing chain length is observed. (e) O 1s detail spectra. For an explanation of the different components, see text. (f) O 1s spectra after subtraction of the oxygen contribution originating from the TiO_2 substrate. All detail spectra are displayed after Shirley background subtraction.

are not homogeneous in the z-direction, the layered structure of the system has to be taken into account, in order to carry out quantitative XPS analysis. The expected layered structure involves the polar phosphate groups located at the TiO_2−SAM interface on top of the bulk TiO_2, and the nonpolar alkyl chains constituting the top surface. The contact-angle data are the most direct indication of this structure (see above). The attenuation of electrons emitted from buried layers by inelastic scattering in layers above leads to a lower detected intensity and accounts for (a) the lower calculated atomic percentage for phosphorus and oxygen and (b) the higher atomic percentage for carbon, in comparison to the values calculated according to the stoichiometry of the adsorbates (Table 3).

Three-Layer Model. Data from an XPS experiment are usually specified as atomic concentrations, calculated using an equation of the form

$$X_A = \frac{I_A/R_A}{\sum_j I_j/R_j} \tag{1}$$

where X_A is the atomic composition of element A in a sample containing j components, I_A is the measured spectral intensity for element A, and R_A is the relative sensitivity factor for element A.[34] However, these coefficients are only strictly correct for homogeneous samples. For nonhomogeneous samples in the z-direction, one needs a method to calculate the effective intensity of the emitted electrons after being attenuated by all the overlayers. This depends on the structure of the sample as well as the

(34) Smith, G. C.; Livesey, A. K. *Surf. Interface Anal.* **1992**, *19*, 175−180.

Table 1. XPS Binding Energies and Fitting Parameters[a]

	binding energy (BE) [eV]	fwhm [eV]	constraints[b]	line shape %Gauss−Lorentz and asymmetry [lit. Casa]
O 1s R−**O**−P	532.8 ± 0.1	1.53 ± 0.02	fwhm = fwhm(O 1s Ti**O**$_2$); area = area(O 1s P=**O**)	GL(78)
O 1s P−**O**−Ti P−**O**−H Ti−**O**−H	531.4 ± 0.1	1.98 ± 0.06	none	GL(78)
O 1s P=**O**	530.8 ± 0.1	1.53 ± 0.02	fwhm = fwhm(O 1s TiO2);area = area(P 2p) × RSF	GL(78)
O 1s Ti**O**$_2$	530.1 ± 0.1	1.53 ± 0.02	area = area(Ti 2p) × RSF × 2	GL(78)
Ti 2p$_{1/2}$	464.4 ± 0.1	2.45 ± 0.04	none	GL(1)
Ti 2p$_{3/2}$	458.6 ± 0.1	1.33 ± 0.02	none	GL(82)T(2.1)
C 1s **C**−O−P	286.65	1.51 ± 0.05	fwhm = fwhm(C 1s aliphatic); BE = BE(C 1s aliphatic) +1.65	GL(75)
C 1s aliphatic	285.0	1.51 ± 0.05	none	GL(75)
P 2p$_{1/2}$	134.6 ± 0.1	1.90 ± 0.08	fwhm = fwhm(P 2p$_{3/2}$); area = 0.5 × area(P 2p$_{3/2}$);BE = BE(P 2p$_{3/2}$) + 0.81	GL(30)
P 2p$_{3/2}$	133.8 ± 0.1	1.90 ± 0.08	none	GL(30)

[a] Data measured on PHI 5700. [b] RSF = relative sensitivity factor.

Table 2. XPS Intensity Ratios Measured at a 45° Takeoff Angle on a PHI 5700 XPS Machine[a]

	C/Ti	P/Ti	interface O/P	Ti−OH/P
$C_{10}H_{21}PO_4$	1.8	0.16	6.0	2.0
$C_{11}H_{23}PO_4$	2.1	0.17	6.0	2.0
$C_{12}H_{25}PO_4$	2.4	0.19	5.3	1.3
$C_{13}H_{27}PO_4$	2.6	0.19	5.1	1.1
$C_{14}H_{29}PO_4$	3.6	0.22	5.0	1.0
$C_{15}H_{31}PO_4$	4.0	0.24	4.8	0.8
$C_{16}H_{33}PO_4$	4.9	0.26	4.5	0.5
$C_{17}H_{35}PO_4$	5.5	0.28	4.4	0.4
$C_{18}H_{37}PO_4$	6.4	0.30	4.5	0.5

[a] All numbers are averaged over at least 3 samples with the same chain length and have a relative accuracy of ± 10%.

Table 3. Atomic Concentrations of Elements in the Nine SAMs Compared to Calculated Values for the Stoichiometry of the Molecules[a]

	At% C	At% P	At% O (PO$_4$)
$C_{10}H_{21}PO_4$	69.5 ± 1.4 (66.7)	6.1 ± 0.3 (6.7)	24.4 ± 1.1 (26.7)
$C_{11}H_{23}PO_4$	68.1 ± 4.5 (68.8)	6.4 ± 0.9 (6.3)	25.5 ± 3.6 (25.0)
$C_{12}H_{25}PO_4$	72.3 ± 2.1 (70.6)	5.5 ± 0.4 (5.9)	22.1 ± 1.7 (23.5)
$C_{13}H_{27}PO_4$	70.2 ± 2.3 (72.2)	6.0 ± 0.5 (5.6)	23.9 ± 1.8 (22.2)
$C_{14}H_{29}PO_4$	75.7 ± 2.3 (73.7)	4.9 ± 0.5 (5.3)	19.4 ± 1.9 (21.1)
$C_{15}H_{31}PO_4$	76.5 ± 1.6 (75.0)	4.7 ± 0.3 (5.0)	18.8 ± 1.3 (20.0)
$C_{16}H_{33}PO_4$	78.8 ± 0.5 (76.2)	4.2 ± 0.1 (4.8)	16.9 ± 0.4 (19.1)
$C_{17}H_{35}PO_4$	80.0 ± 0.4 (77.3)	4.0 ± 0.1 (4.6)	16.0 ± 0.3 (18.2)
$C_{18}H_{37}PO_4$	80.8 ± 1.7 (78.3)	3.8 ± 0.3 (4.4)	15.4 ± 1.4 (17.4)

[a] The numbers are average values of at least three samples. Calculated values are written in brackets.

attenuation length of the electrons in each layer. To do so, we use the method described by Smith and Livesey.[34] A detailed description of our calculations with all the parameters listed can be found in the Supporting Information for this article. In brief, we used a three-layer model similar to that used by Textor et al. on tantalum oxide,[13,35,36] with a layer of titanium oxide, a layer containing all the phosphate and surface oxygen, and a final layer of the aliphatic carbon (Figure 3). With this model, where only the thickness of the aliphatic layer was allowed to vary, we calculated calibration curves for the takeoff angles of the electrons (45° and 90°) in the two spectrometers used (for data using the 90° angle, see Supporting Information). The thickness d of the alkyl part of the SAM was obtained by finding the best fit of the measured average atomic composition for each chain length (Figure 3). As one can see, the layer thickness observed for the series of alkyl phosphate SAMs is not simply proportional to their chain length.

The biggest deviations from the model occur for the short alkyl phosphates and show up mainly as an increased value for the interface oxygen, suggesting again a higher amount of free Ti−OH per phosphate molecule at the TiO_2−SAM interface for shorter alkyl chains. Plotting the measured thickness versus the number of carbon atoms, an S-shaped trend is obtained with the measured thicknesses (XPS) of the alkyl phosphates C_{10} to C_{13} following a calculated thickness for a mean tilt angle of 45°. This value has, of course, no physical meaning for poorly ordered monolayers but is similar to that previously published for poorly ordered SAMs,[21] where the chains are in a liquid-like conformation. For the longer molecules, the measured thickness more closely follows the calculated value with a tilt angle of 30° or below, which is again an indication for more densely packed and well-ordered layers (Figure 4). A similar trend can be found for the measured thickness using VASE.

3.3. VASE. As in XPS, the data obtained from VASE depend on the model parameters used to fit the measured data. Since VASE is very sensitive to small deviations in the layer thickness and optical parameters of the substrates used, each sample was measured individually before and after adsorption of the SAM. The difference between these two measurements was fitted as a Cauchy layer (see Supporting Information) with the only parameter to be fitted being d, the layer thickness on top of a layered structure of $TiO_2/SiO_2/Si$. For the underlying substrate the parameters have been fitted individually for each sample measured before adsorption and then fixed. Since the refractive index for the phosphate SAM is not known, the obtained layer thicknesses are not absolute. Nevertheless, the value used is unlikely to deviate from the true refractive index by more than

(35) Elsener, B.; Rossi, A. *Electrochim. Acta* **1992**, *37*, 2269−2276.
(36) Rossi, A.; Elsener, B. *Surf. Interface Anal.* **1992**, *18*, 499−504.

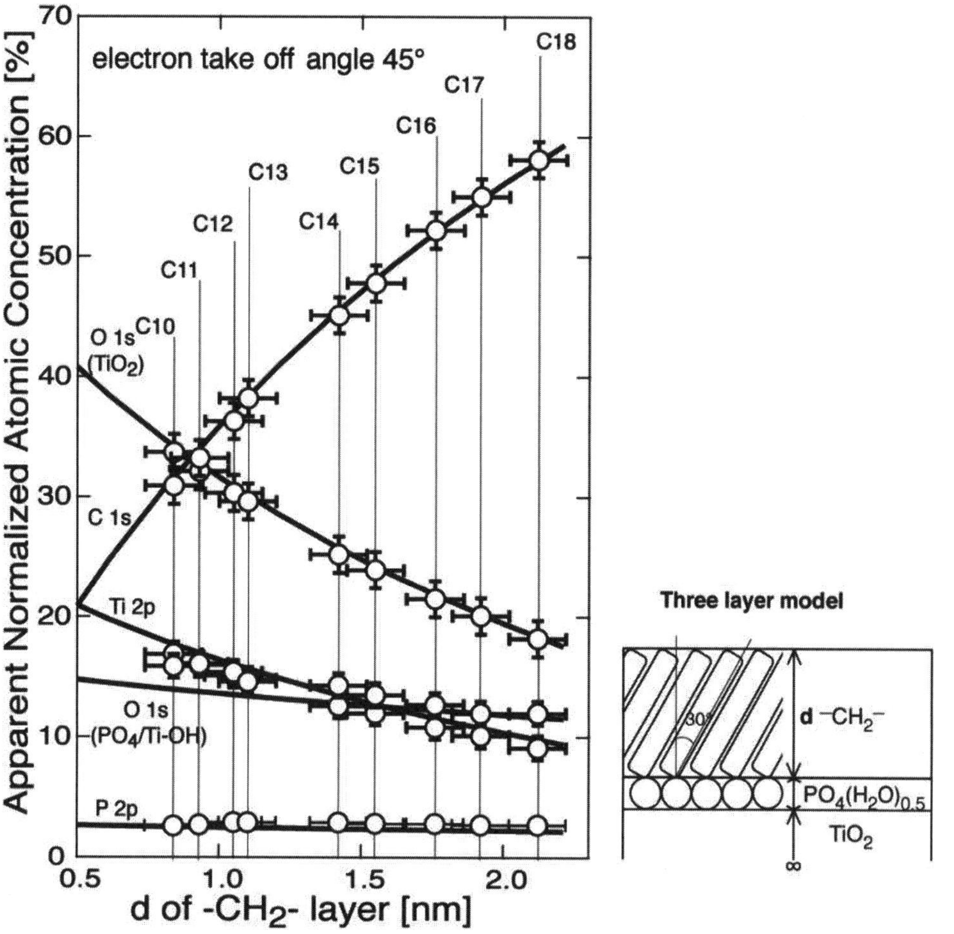

Figure 3. Calculated calibration curves of apparent normalized atomic concentrations as a function of alkyl chain overlayer thickness using a three-layer model and a 45° electron takeoff angle. Measured apparent atomic concentrations for each chain length were then fitted to these curves in order to measure the alkyl layer thickness *d*.

10%, and therefore the measured thicknesses should be close to true values. This assumption has been confirmed by the close match to the XPS data (Figure 4). For SAMs obtained with C_{18}, C_{17}, and somewhat less for C_{15}, quite large deviations from sample to sample were observed. Especially in the case of C18, it turned out to be quite difficult to obtain a reproducible thickness. Small deviations in solvent composition seemed to have an especially large influence on measured layer thickness. A possible explanation for this behavior could be coadsorption of solvent, leading to a higher apparent thickness. This is not found in XPS since the unbound solvent will evaporate during pumping down to ultrahigh vacuum.

3.4. PM-IRRAS. Infrared spectroscopic measurements on phosphate SAMs were carried out to gain further insight into the structure of these SAMs, especially the chain-length dependence of alkyl chain conformational order within the monolayers. PM-IRRA spectra were carried out on phosphate monolayers deposited onto TiO_2-covered gold substrates, owing to the poor quality of the infrared spectra obtainable on bare TiO_2 substrates. Such an optically thin layer of TiO_2 is essentially transparent in the wavelength region of interest, and thereby allows for the measurement of the spectrum of the adsorbed phosphate

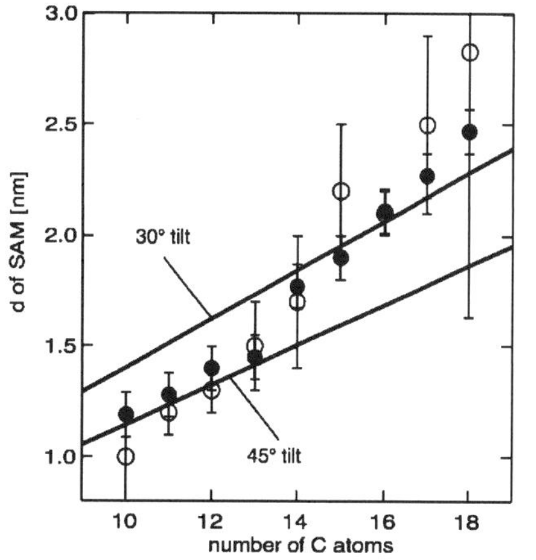

Figure 4. SAM thickness versus number of carbon atoms measured for the alkyl phosphate monolayers. The data are compared with a theoretical thickness calculated for 30° and 45° tilt angles of the alkyl chains. The data are obtained by XPS (filled symbols) (using a three-layer model) and VASE (open symbols) measurements.

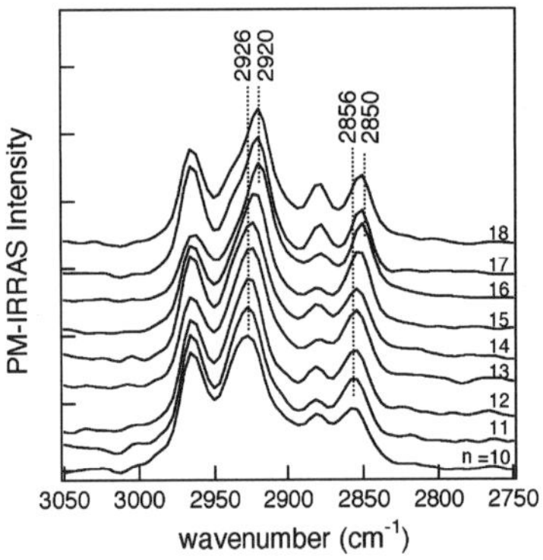

Figure 5. PM-IRRAS spectra in the C−H stretching region of the series of phosphate SAMs. The number of carbon atoms (n) in the alkyl chain of the phosphates is indicated alongside the spectra.

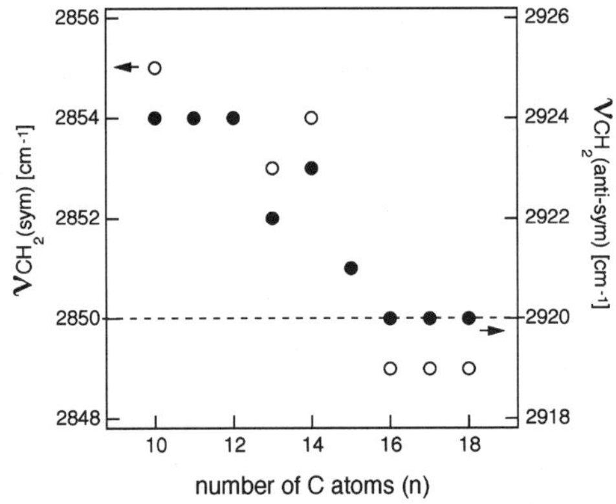

Figure 6. Positions of the methylene symmetric (open symbols) and antisymmetric (filled symbols) stretching modes in the PM-IRRA spectrum of the series of phosphates. The horizontal dotted line corresponds to the typical values of these modes observed in crystalline ordered alkyl chain assemblies.

monolayers with practically the same sensitivity as for monolayers adsorbed onto a metallic substrate. Such methods using a thick metallic layer underneath thin layers of a dielectric, often referred to as buried metal layer[37] substrates, have been used in the studies of adsorbates on silicon oxide surfaces with a metallic Ni or Cu layer below.[38,39] The SAMs on TiO_2/Au substrates used in IR measurements were thoroughly characterized by means of XPS, VASE, and dCA measurements as described above, and were found to be identical in film composition, thickness, and order to those formed on TiO_2/SiO_2/Si substrates.

The PM-IRRAS spectra, in the wavelength region of 2800−3000 cm^{-1}, of the series of phosphates are shown in Figure 5. The spectra show four prominent bands assignable to the C−H stretching modes of the methylene (CH_2) chain and that of the terminal methyl (CH_3) group. The asymmetric and symmetric stretching modes of the methyl groups appear at 2960 and 2870 cm^{-1}, respectively. The symmetric and antisymmetric stretching modes of the methylene chains appear in the regions of 2850 and 2920 cm^{-1}, respectively. The most interesting feature of the series of spectra shown in Figure 5 is the shift in the positions of these bands. These bands show a clear shift to lower frequencies with increasing chain length, whereas the positions of the methyl stretching bands remain almost invariant. The methylene stretching modes are known to be sensitive to the conformational order of alkyl chains shifting to higher frequencies with increasing conformational disorder.[40−42] For the series of phosphates studied here, the positions of these two bands are plotted as a function of alkyl chain length in Figure 6. The positions of the methylene stretching modes show a sharp transition at chain lengths above 15. For the shorter chains ($n < 14$) these bands appear at 2927 and 2856 cm^{-1}, indicating that the alkyl chains adopt a disordered, liquid-like structure, whereas, for higher chain lengths ($n > 15$) they appear at 2920 and 2950 cm^{-1}, indicating that they adopt a more ordered crystalline structure. For intermediate chain lengths, these bands appear at a value between these two extremes ($n = 14,15$). These results are in agreement with the XPS and ellipsometric measurements described above. As mentioned in the discussions above, the SAMs formed from $n = 18$ alkyl phosphates showed some inconsistencies in the position and the intensity of the stretching bands, presumably due to solvent incorporation. However, the spectra of alkyl phosphates of other chain lengths remained consistent in their peak positions.

4. Discussion and Conclusions

All measurement methods show a clear difference between short (C_{10} to C_{13}) and long (C_{16} to C_{18}) alkyl phosphate SAMs. We believe that the shorter molecules assemble into a less dense, liquid-like structure. This is supported by the fact that there is an excess of oxygen found for those alkyl phosphates with XPS, which we interpreted as free surface Ti−OH groups. The phosphate heads therefore must be assumed to be more widely spaced, leading to a model of the surface where the average tilt angle is higher than in a densely packed layer, as measured by XPS and VASE. PM-IRRAS also supports this widely spaced structure, in that more "liquid-like" C−H stretching vibrations are observed for the alkyl phosphates with shorter chain lengths. Increasing the chain length increases the van der Waals interaction between the alkyl chains, which leads to a more crystalline, all-trans conformation of the chains. At the same time, the packing density increases and therefore more surface OH groups are replaced with phosphate groups. The higher order is also evident in the smaller hysteresis of the dCAs and the shift of the C−H stretching frequency to values that are similar to those of crystalline aliphatic chains. A sketch of the proposed structures for short versus long alkyl phosphates on TiO_2 is depicted in Figure 7. Since we do not exactly know the atomic surface structure of our amorphous or nano crystalline TiO_2 films, the exact binding mode of the phosphate groups to the TiO_2 surface remains speculative. Nevertheless, we can propose a surface structure, which is in accordance to our findings. In the case of thiol SAMs on gold, even short decanethiols form densely packed ordered monolayers.[5,8,30] Whereas thiols have only one possible binding interaction with gold surfaces, phosphate groups can, in principle, bind in a mono-, bi- or even tridentate manner to the TiO_2 surface. In single-crystal structures of layered alkyl phosphates, the phosphate groups bind almost exclusively in a bridged binding mode with each oxygen atom binding to a different metal ion. Two- and 3-fold coordination can be found, while a chelating binding mode with two oxygen atoms binding to the same metal is very uncommon. Therefore, we can most probably exclude such a chelating binding mode, since this would

(37) Gardner, P.; LeVent, S.; Pilling, M. *Surf. Sci.* **2004**, *559*, 186−200.
(38) Finke, S. J. S., G. L. *Spectrochem. Acta* **1990**, *46A*, 91−96.
(39) McGonigal, M. B. V. M.; Butler, J. E. *J. Electron Spectrosc. Relat. Phenom.* **1990**, *54/55*, 1033−1044.
(40) Snyder, R. G.; S., H. L.; Elliger, C. A. *J. Phys. Chem.* **1982**, *86*, 5145−5150.
(41) Macphail, R. A.; Strauss, H. L.; Snyder, R. G.; Elliger, C. A. *J. Phys. Chem.* **1984**, *88*, 334−341.
(42) Nuzzo, R. G.; Dubois, L. H.; Allara, D. L. *J. Am. Chem. Soc.* **1990**, *112*, 558−569.

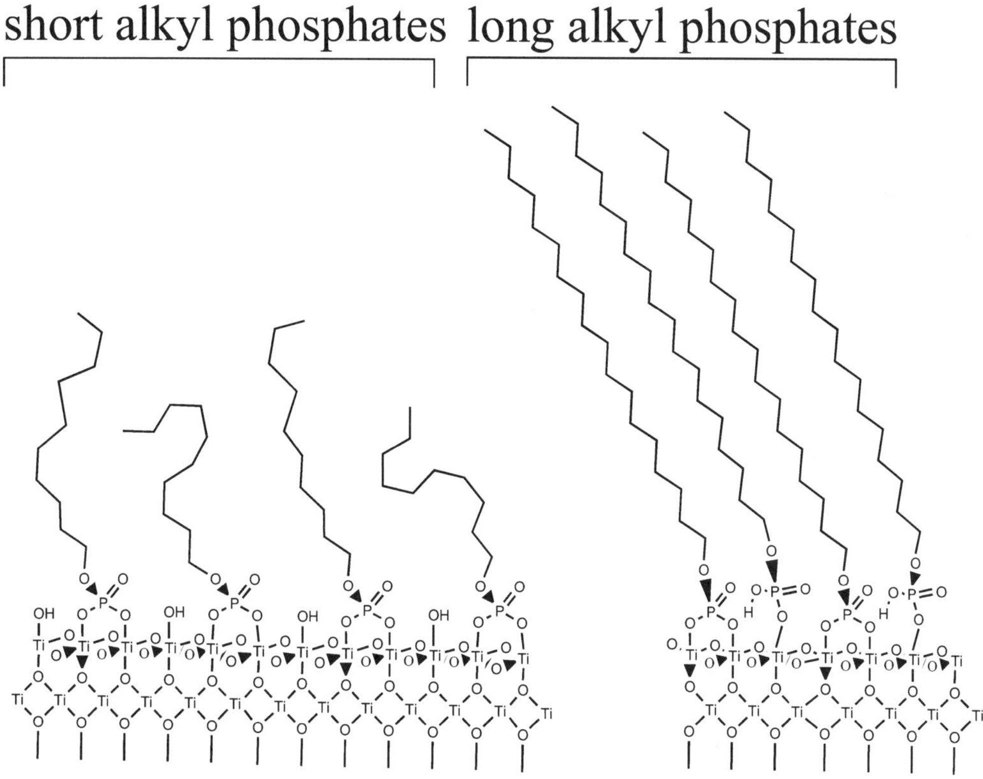

Figure 7. Sketch of the proposed binding structure for short (C_{10}) and long (C_{18}) alkyl phosphates self-assembled on TiO_2.

lead to coordination numbers for the titanium of larger than six, and a highly strained four-numbered Ti−O−P−O cycle, which should not be energetically favorable. We believe, in view of the coverages observed, that a realistic picture involves a mixture of single and double, but not triple coordination, especially for the long alkyl phosphates. This is in accordance with our XPS data evaluation, which implies the presence of a free P=O unit. The short alkyl phosphates (C_{10}−C_{13}) are mainly stabilized due to their interaction with the surface. This might be the reason that they do not pack as densely in order to bind at an optimal place in a strong bridging bidentate manner, which can be further stabilized by hydrogen bonds to free surface Ti−OH groups. For the longer alkyl chains, the enthalpy gain due to van der Waals interactions between the alkyl chains becomes more important and also makes monodentate binding energetically favorable. Therefore additional molecules can squeeze between the bidentate bonded molecules and almost double the final surface density (Figure 7). This study has shown for the first time that there is a strong influence of the chain length on the final structure in alkyl phosphate SAMs. While we have chosen identical adsorption protocols for all chain lengths in order to be able to compare the spectroscopic data, an optimization of the adsorption parameters for each molecule could potentially further improve the layer quality, order, and reproducibility of SAM formation.

Acknowledgment. We thank Michael Horisberger of the PSI (Villigen, Switzerland) for support of substrate coating.

Supporting Information Available: A table with the parameters used for the fitting of the VASE data as well as a detailed description of the calculation of the calibration curves for the XPS evaluation. This information is available free of charge via the Internet at http://pubs.acs.org.

LA700474V

Tribol Lett (2007) 28:229–239
DOI 10.1007/s11249-007-9266-1

ORIGINAL PAPER

Macroscopic Tribological Testing of Alkanethiol Self-assembled Monolayers (SAMs): Pin-on-disk Tribometry with Elastomeric Sliding Contacts

Seunghwan Lee · Raphael Heeb · Nagaiyanallur V. Venkataraman · Nicholas D. Spencer

Received: 12 July 2007 / Accepted: 24 August 2007 / Published online: 26 September 2007

Abstract We demonstrate that the frictional properties of alkanethiol self-assembled monolayers (SAMs) with various surface-chemical and structural features can be investigated on a macroscopic scale by employing an elastomer as the sliding partner in pin-on-disk tribometry. The mild contact conditions at the elastomeric tribological interface allow the SAM films to remain virtually intact despite the tribological stress. Sliding contact between SAMs and elastomers over the speed range available from an ordinary tribometer in a liquid environment induced a broad range of lubrication mechanisms, ranging from boundary to fluid-film lubrication regimes. Thus, the impact of both the chemical and structural characteristics of SAMs on the formation of fluid films and interfacial friction forces could be probed in the absence of wear processes. Given the large SAM "toolbox" that is readily available for the modification of surface-chemical characteristics, this approach provides an opportunity to investigate the influence of surface chemistry on the frictional properties of elastomeric tribological contacts.

Keywords Self-assembled monolayers (SAMs) · Elastomer · Pin-on-disk tribometry · Boundary lubrication · Soft elastohydrodynamic lubrication (soft EHL)

S. Lee · R. Heeb · N. V. Venkataraman · N. D. Spencer (✉)
Laboratory for Surface Science and Technology, Department of Materials, ETH Zurich, Wolfgang-Pauli-Strasse 10, CH-8093 Zurich, Switzerland
e-mail: spencer@mat.ethz.ch

1 Introduction

In the past two decades, self-assembled monolayers (SAMs) generated from alkanethiols and alkylsilanes have been extensively studied as model systems for boundary lubricants [1–24]. Since the structure and chemistry of SAM films can be systematically tailored at the molecular level, the majority of reports have focused on the study of friction and lubricating properties of these systems under nanoscale contact conditions [1–18]. Technically, this has become possible for two reasons; firstly, SAMs can be readily generated on substrates displaying extremely smooth surfaces, such as monocrystalline gold, mica, and highly polished silicon, and thus boundary lubricant films with ideally smooth morphology have become available. As a comparison, it is worth noting that more classical boundary lubricant films, such as fatty acids and alcohols [25–28], have generally been formed on polished metal/metal oxide surfaces, which display much higher surface roughness. Secondly, as the counterface to these boundary lubricant films, a slider consisting of a single asperity, such as the atomic force microscope (AFM) tip [29, 30], has become available. Thus, the control of contact pressure within the contact area became feasible by simply controlling the external load. This is important in that tribological stress could be applied exclusively onto the SAM films, i.e., without penetrating to the underlying substrate, and thus their tribological properties as boundary-lubricant films can be probed. In contrast, tribological contacts involving SAM films probed by more conventional, macroscale approaches, such as pin-on-disk tribometry [19–24], have generally resulted in irreversible damage to the samples, both the SAM films and the supporting substrates, which renders the interpretation of the observed tribological properties far more complicated.

Clearly, this is due to the multi-asperity contacts arising from the counterpart roughness, e.g., the macroscale slider (pin), and the resulting uncontrolled, extremely high local contact pressures.

The poor mechanical durability of SAM films, and often of the supporting substrates themselves in addition, is a major barrier to fundamental tribological studies of SAM films on a macroscopic scale. This problem, however, can be readily overcome if a highly compliant material, such as a rubber, is employed as the sliding counterpart to the SAM films. Under ordinary tribometer experiment conditions, the apparent contact pressure applied by an elastomer can be easily maintained below the MPa range, and due to its high compliance, the surface asperities of the elastomer can readily flatten out under pressure. Thus, the wear problems arising from uncontrolled asperity contact can be avoided. In fact, this configuration, either 'soft' slider on 'rigid' track or 'rigid' slider on 'soft' track, has routinely been employed in tribological studies of elastomers [31–33], especially in soft elastohydrodynamic lubrication (soft EHL) studies [34–43]. As schematically depicted in Fig. 1, the effect of using an elastomeric slider for pin-on-disk tribometric studies of SAM films is to reduce the contact pressure to a level typically available from nanotribological contacts or lower, while the contact area and sliding speed are maintained at levels typical for macrotribological studies. Thus, the tribological properties of SAM films can be investigated in a wear-less regime, as is the case with nanotribological approaches, yet with the possibility of both *macroscopic contact area* and *high sliding speeds*. Furthermore, the contact area is large enough that standard surface-analytical approaches can readily access it to characterize the structural and chemical changes present after tribological stress; currently, no generally available surface-analytical approach can access the contact area generated by an AFM probe other than AFM itself.

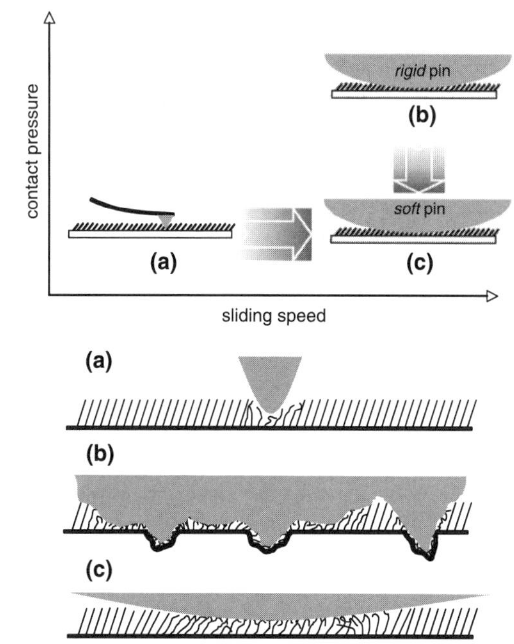

Fig. 1 A pressure-versus-sliding-speed diagram showing the contact configuration of an elastomeric slider on self-assembled monolayer (SAM) films: schematic illustrations for (**a**) single-asperity contacts on a nanoscopic scale as in AFM (low contact pressure, low speed), (**b**) multi-asperity contacts on a macroscopic scale as in conventional pin-on-disk tribometry employing rigid sliders (high contact pressure, high speed) and (**c**) soft contacts on a macroscopic scale by employing an elastomeric slider in pin-on-disk tribometry (low contact pressure, high speed)

We consider that this approach—soft slider on SAM films/rigid substrate—is particularly useful to investigate the role of surface chemistry on the efficacy of lubricant-film formation, which is crucial in the soft elastohydrodynamic lubrication (soft EHL) regime. Our recent studies, [34, 35] involving water and poly(dimethylsiloxane) (PDMS) as lubricant and elastomeric tribopair, respectively, have shown that the formation of a soft EHL film is determined not merely by the bulk mechanical and rheological properties of the tribosystem alone as predicted by classical theory [44, 45], but that the surface-chemical properties of tribopairs also play a significant role; when the surface of the tribopair is insufficiently wetted by the lubricant, the predicted formation of an EHL film is significantly retarded, even if the bulk mechanical properties meet the conditions for the formation of a soft-EHL film. Given the wealth of useful approaches that have been established in self-assembly techniques to modify surface-chemical characteristics, the effects of a broad range of surface-chemical modifications on soft EHL can be conveniently explored. In the present article, we aim to establish this methodology by employing SAM systems whose tribological properties have been well established, allowing us to focus on testing the proposed methodology.

2 Materials and Methods

2.1 Self-assembled Monolayers (SAMs)

Three types of SAM films have been prepared by spontaneous adsorption from ethanolic solutions of 11-mercaptoundecanol ($HS(CH_2)_{11}OH$), 1-dodecanethiol ($HS(CH_2)_{11}CH_3$), and 1-hexanethiol ($HS(CH_2)_5CH_3$) onto polycrystalline gold surfaces deposited on glass. For simplicity, these SAMs are abbreviated as $\mathbf{C_{11}OH}$, $\mathbf{C_{11}CH_3}$, and $\mathbf{C_5CH_3}$, respectively, throughout this article. A schematic illustration of the SAM films on gold substrates generated from these molecules is

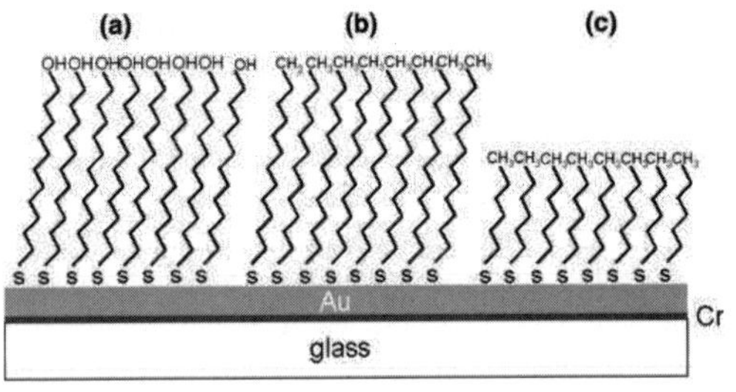

Fig. 2 Schematic representation of the alkanethiol SAM films on gold substrate used in this study: (**a**) 11-mercaptoundecanol ($HS(CH_2)_{11}OH$), (**b**) 1-dodecanethiol ($HS(CH_2)_{11}CH_3$) and (**c**) 1-hexanethiol ($HS(CH_2)_5CH_3$)

presented in Fig. 2. The alkanethiols (Sigma-Aldrich, Switzerland) were used as received. Ethanol (Fluka, Switzerland) was used as solvent for alkanethiol solutions (1 mM). Soda-lime glass microscope slides (SuperFrost, Menzel-Gläser, Braunschweig, Germany) were cut into $\sim 2.5 \times 2.5$ cm^2 pieces and used as substrates for the gold films. The glass substrates were cleaned according to the following procedure; (a) ultrasonication in ethanol four times for 15 min—the solvent was exchanged each time (b) O_2 plasma for 2 min (high power, Harrick Plasma Cleaner/Sterilizer, Ossining, NY, USA). After cleaning, the substrates were coated with a 10-nm adhesive layer of chromium, followed by 100-nm layer of gold by thermal evaporation (MED020 coating system, BALTEC, Balzers, Lichtenstein). The SAM films were generated by immersion of the substrates in the relevant 1 mM ethanolic alkanethiol solution overnight. Prior to the immersion step, the gold substrates were rinsed with pure ethanol and subsequently dried with nitrogen.

2.2 Poly(dimethylsiloxane)(PDMS)

Poly(dimethylsiloxane) (PDMS) was employed as an elastomeric counterpart (pin) against the SAM films/gold/glass in pin-on-disk tribometry experiments. The PDMS pins were prepared from a commercial silicone elastomer kit (SYLGARD® 184 silicone elastomer, base and curing agent, Dow Corning, Midland, MI, USA). In order to prepare PDMS pins with hemispherical ends, a commercial polystyrene cell-culture plate with round-shaped wells (radius 3 mm, 96 MicroWell Plates™, NUNCLON™ Delta Surface, Denmark) was employed as master. The PDMS pins were prepared according to a conventional recipe [34, 35]. Briefly, the base and crosslinker of the SYLGARD® 184 elastomer kit were mixed at 10:1 ratio (by weight). The foams generated during mixing were removed by gentle vacuum. The mixture was transferred into the master and incubated in an oven ($\sim$70 °C) overnight. Planar PDMS sheets were also prepared for the characterization of surface hydrophilicity according to the same procedure.

The surface of the PDMS pins was hydrophilized by means of air plasma treatment for 1 min, and thus they are denoted as '*ox*-PDMS' throughout this article. For a control experiment, the PDMS pins were immersed in *n*-hexane, to extract uncrosslinked monomer species prior to air-plasma treatment; the solvent, *n*-hexane, was exchanged twice during 24 h.

2.3 Water Contact Angle Measurements

The surface hydrophilicity of the tribopairs, including the SAM films and *ox*-PDMS, was characterized by measuring static water contact angles, θ_w, by employing a contact-angle goniometer (Ramé-Hart model 100). All contact-angle measurements were averaged over five runs.

2.4 Ellipsometry

The dry thicknesses of the SAM films were determined by variable angle spectroscopic ellipsometry (VASE, M-2000F, L.O.T. Oriel GmbH, Germany). Measurements were conducted under ambient conditions at three angles of incidence (65°, 70°, and 75°) in the spectral range of 370–995 nm. Measurements were fitted with the WVASE32 analysis software using a three-layer model for an organic layer on a gold/glass substrate.

2.5 Polarization–Modulation Infrared Reflection–Absorption Spectroscopy (PM-IRRAS)

Polarization–modulation infrared reflection–absorption spectra (PM-IRRAS) were recorded on a Bruker IFS 66v IR spectrometer, equipped with a PMA37 polarization-modulation accessory (Bruker Optics, Germany). The interferogram from the spectrometer's external beam port was passed through a KRS-5 wire-grid polarizer and a ZnSe photoelastic modulator before reflecting off the sample surface at an angle of 80° and being detected with a liquid-nitrogen-cooled MCT detector. The sample compartment was continuously purged with dry air. The sample holder was suitably modified to be able to record spectra of different regions of the sample with a 2-mm aperture. Typically, 1,024 scans of multiplexed interferograms were collected with 4-cm^{-1} resolution and processed with the OPUS software (Bruker Optics, Germany). The spectra were background corrected with a polynomial.

2.6 Atomic Force Microscopy (AFM)

Atomic force microscopy (AFM, Dimension 3000, Veeco Metrology Group, Santa Barbara, CA, USA) was employed to characterize the surface morphological and nanotribological properties of the samples. The surface morphology was characterized by TappingMode™ AFM; the R_a values of the SAM films were obtained by directly imaging the samples, whereas the R_a values of the PDMS pin were obtained by imaging the polystyrene master. A tapping-mode silicon cantilever and tip (PointProbe® Plus, k = 92 N/m, f_0 = 330 kHz, Germany) was used as a probe. Nanotribological properties of the SAM films were characterized by contact-mode AFM under distilled water (>18 MΩ). Based on the conventional beam-deflection and four-quadrant photodetector method [1, 2, 29], the interfacial friction force between a sample and a tip was measured as a function of applied load. Under a fixed load, the friction force was obtained by subtracting the lateral force obtained during retrace from that obtained during trace in a 'friction loop' [29, 44]. The friction force under a given load was obtained at more than 10 different positions on the samples for statistical evaluation. The scan length was 1 μm and the sliding speed 2 μm/s. This procedure was repeated by varying the applied load. Since the purpose of the AFM nanotribological studies in this work is to provide a reference for the relative order of frictional forces of the SAM films on a nanoscopic contact scale, the normal spring constant value of the commercial silicon nitride cantilevers indicated by the manufacturer (Nanoprobes, Veeco Metrology Group, Santa Barbara, CA, USA), k_N (0.12 N/m), was employed without further calibration. In order to ensure valid comparison of the frictional data, however, all measurements were performed with an identical tip/cantilever assembly. The applied load varied between 10 nN and 80 nN. Friction forces were expressed in arbitrary units, as received from the photodiode detector without further conversion. Along with friction measurements, the adhesive properties of the SAM films were also characterized by acquiring force-versus-distance curves using the identical tip/cantilever assembly. Prior to the measurements, the tip/cantilever assembly was immersed in a commercial acidic cleaning solution (Cleaner, COBAS INTEGRA (HCl 300 mM, detergent 1%), Roche, Germany) for 5 min, followed by immersion in distilled water for 10 min, and finally oxygen-plasma cleaning for 30 s.

2.7 Pin-on-disk Tribometry

The macroscopic-scale tribological properties of the SAM films were characterized by means of pin-on-disk tribometry (CSM Instruments SA, Peseux, Switzerland). In this approach, the load was controlled by dead weights and the friction forces were measured by a strain gauge. After forming a contact between the loaded pin and the disk, the latter was rotated at a controlled speed by a motor, thus generating sliding friction forces. PDMS with an end-radius of 3 mm was employed as a standard pin material, but for comparison, a stainless steel ball (DIN 5401-20 G20, Hydrel AG, Romanshorn, Switzerland) with a radius of 3 mm was also employed as pin. The stainless steel ball was rinsed with ethanol, blown dry with nitrogen, and oxygen-plasma cleaned for 30 s. The raw data for the friction forces were recorded as a function of time (or the number of rotations) over a fixed track, using the software (InstrumX version 2.5A) provided by the manufacturer. All measurements were carried out in distilled water (>18 MΩ).

The standard protocol for the friction measurements involves the acquisition of μ-versus-speed plots (μ = friction force/load) and friction-versus-load plots. For μ-versus-speed plots, the speed was varied from 0.25 mm/s to 100 mm/s under a fixed load (1 N), and for friction-versus-load plots, the load was varied from 0.5 N to 5 N under two fixed speeds (1 mm/s and 50 mm/s). The number of rotations was 20 for each measurement. The average friction force from the latter half of these (11–20th) was obtained to avoid "running-in" effects. For control experiments involving the examination of the sliding track by PM-IRRAS, the number of rotations was extended to 500 under 1 N.

3 Results and Discussion

3.1 Initial Characterization of the SAM Films

Before the start of the tribological experiments, the SAM films formed on thermally evaporated gold substrates were characterized by ellipsometry, water-contact-angle measurements, AFM, and PM-IRRAS. Ellipsometry revealed monolayer thicknesses of 1.57 ± 0.11 nm for the $\mathbf{C_{11}OH}$, 1.43 ± 0.13 nm for the $\mathbf{C_{11}CH_3}$, and 0.47 ± 0.52 nm for the $\mathbf{C_5CH_3}$ SAM films. The refractive index for the organic layers was assumed to be 1.45. The water contact angle, θ_w, for the $\mathbf{C_{11}OH}$ SAM film, 20 ± 5°, was substantially lower than those for the $\mathbf{C_{11}CH_3}$, 109 ± 2°, and the $\mathbf{C_5CH_3}$, 106 ± 2°, films as well as that of bare gold substrate, 70 ± 5°, which indicates the exposure of –OH ($\mathbf{C_{11}OH}$) and –CH_3 ($\mathbf{C_{11}CH_3}$ and $\mathbf{C_5CH_3}$) groups on the surfaces of the corresponding SAM films. The average values and the standard deviation of water contact angles were obtained from more than 10 different measurements on three different samples for each SAM film. The slightly lower water contact angle of $\mathbf{C_5CH_3}$, 106°, than that of $\mathbf{C_{11}CH_3}$, 109°, on average, is generally ascribed to more

Tribol Lett (2007) 28:229–239 233

probable exposure of methylene groups (–CH_2–) arising from the less well-packed backbone structure of $\mathbf{C_5CH_3}$ than $\mathbf{C_{11}CH_3}$ SAM film [45, 46]. The θ_w values of the *ox*-PDMS surfaces were <3°.

AFM revealed homogenously distributed grainy features of the gold substrates (diameter: a few tens of nm scale), which are typical morphological characteristics of thermally evaporated polycrystalline gold surfaces on glass or silicon oxide. The morphologies of the SAM films on the gold substrates were indistinguishable from those of the substrates, which supports the formation of homogenous, monolayer films. The R_a values obtained from the gold substrates over 1 μm × 1 μm, 10 μm × 10 μm, and 50 μm × 50 μm, were 1.3, 1.4, and 1.6 nm, respectively (the error bars are ±0.60 nm), and those for the three SAM films were virtually identical. In parallel, the topographic images of the master for the PDMS pins were also obtained. Compared with the SAM films on gold substrates, slightly rougher surfaces, including some random features a few tens of nm in height, were observed. The R_a values obtained over 1 μm × 1 μm, 10 μm × 10 μm, and 50 μm × 50 μm, were 1.2, 2.4, and 3.4 nm, respectively (the error bars are ±1.10 nm).

Analysis by PM-IRRAS revealed more detailed information on the chemical and structural features of the SAM films. The asymmetric and symmetric methyl C–H stretches, $\nu_a(CH_3)$ at 2,962 cm^{-1} and $\nu_s(CH_3)$ at 2,877 cm^{-1}, respectively, and the band at 2,936 cm^{-1} assigned to a Fermi resonance (FR) between the CH_3 symmetric stretching mode and overtones of bending modes, appear only for $\mathbf{C_{11}CH_3}$ and $\mathbf{C_5CH_3}$ SAM films, but are absent for $\mathbf{C_{11}OH}$ SAM films, as expected. The position and peak width of the methylene asymmetric C–H stretching modes are particularly sensitive to the conformational order of the alkyl chains [15, 16, 45–48]. As seen in Fig. 3, the $\nu_a(CH_2)$ peak of the $\mathbf{C_{11}CH_3}$ film appears at 2,919 cm^{-1}, indicating a highly ordered hydrocarbon backbone structure. The $\nu_a(CH_2)$ peak of the $\mathbf{C_{11}OH}$ film was observed at a slightly higher value of 2,921 cm^{-1}. In contrast, the $\nu_a(CH_2)$ peak of the $\mathbf{C_5CH_3}$ film is poorly defined and appears much broader than those of the $\mathbf{C_{11}CH_3}$ and $\mathbf{C_{11}OH}$ SAM films, which indicates that the conformation of the $\mathbf{C_5CH_3}$ film is very disordered and liquid-like.

As a whole, these data verified that the three SAM films displaying (a) varying surface hydrophilicity ($\mathbf{C_{11}OH}$ versus $\mathbf{C_{11}CH_3/C_5CH_3}$ SAM films) and (b) varying conformational order of the alkyl chains ($\mathbf{C_{11}CH_3}$ versus $\mathbf{C_5CH_3}$ SAM films) have been successfully generated on polycrystalline gold substrates.

3.2 Nanotribological Properties of the SAM Films by Means of AFM

The nanotribological properties of the SAM films generated from $\mathbf{C_{11}OH}$, $\mathbf{C_{11}CH_3}$, and $\mathbf{C_5CH_3}$ SAM films were characterized through the measurement of frictional forces with a plasma-cleaned silicon nitride tip as a function of load under distilled water. Representative friction-versus-load plots obtained by AFM are presented in Fig. 4a. The force-versus-distance curves obtained from the same three SAM films are shown in Fig. 4b. In order to ensure a statistically valid comparison, the measurements were repeated several times in varying order using the same tip/

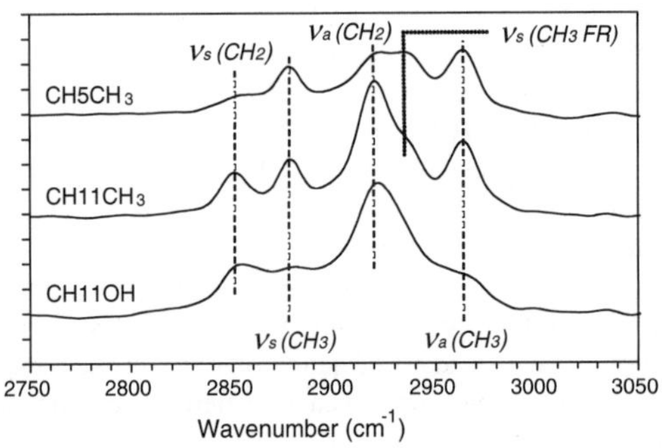

Fig. 3 PM-IRRA spectra obtained from the three alkanethiol SAM films

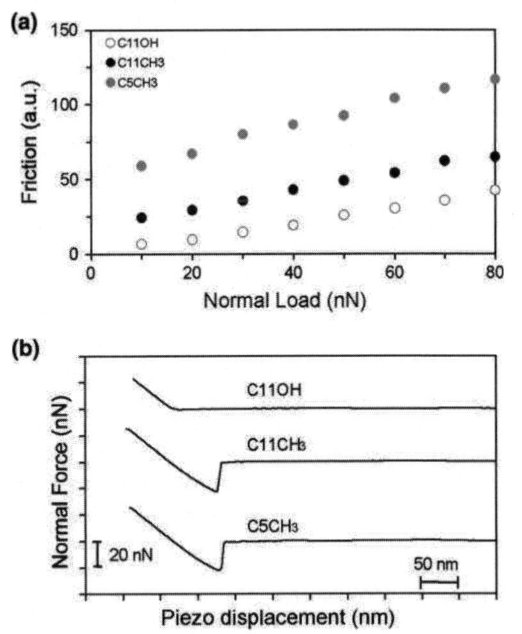

Fig. 4 (**a**) Friction force versus load plots and (**b**) force-versus-distance plots obtained from the interaction between an AFM tip and the three SAM films in distilled water

cantilever assembly. Although minor differences were observed in the frictional and adhesive properties of the samples, the relative order of friction forces of the SAM films were consistently observed as in Fig. 4a: **C_5CH_3 > $C_{11}CH_3$ > $C_{11}OH$**. This order is highly correlated with the order of 'pull-off forces' obtained from force-versus-distance curves as shown in Fig. 4b: **$C_5CH_3 \approx C_{11}CH_3 > C_{11}OH$**.

As mentioned in Sect. 1, choice of the three SAMs in this work was motivated by an attempt to employ standard SAM systems whose tribological properties have been well established and, therefore, we can focus on the proposed test methodology. For the CH_3-terminated normal alkanethiol or alkylsilane SAM films, it has been well known that longer hydrocarbon chains usually contribute to enhancing both the packing density and order of the backbone and, therefore, tend to lower the interfacial frictional properties [3–9]. While most previous nanotribological studies of these systems have been carried out under ambient conditions, we have now shown the same trend in distilled water: Alkyl-chain deformation still appears to be the dominant energy-dissipation mechanism for the two CH_3-terminated SAM films.

$C_{11}OH$ and **$C_{11}CH_3$** SAM films display similarly well-ordered backbones, are constructed from identical hydrocarbon chains, and yet present different terminal groups (OH– versus CH_3–). The tribological properties of SAM films displaying chemically different terminal groups are known to be closely associated with their adhesive properties, which are, in turn, dependent on both the chemical identity of the counterface and the medium in which the tribological interaction takes place [9–18]. Since the oxygen-plasma-treated silicon nitride AFM probe in this work displays hydrophilic surface characteristics (OH-groups), the work of adhesion between two hydrophilic surfaces (the AFM probe versus **$C_{11}OH$**) is apparently lower than that involving a hydrophobic surface (the AFM probe versus **$C_{11}CH_3$**) in aqueous media [11], as shown in Fig. 4b. The difference in the adhesive forces for these two SAM films with the AFM probe is thus mainly responsible for the difference in the nanotribological properties.

Overall, the nanotribological properties of the SAM films revealed by AFM in this work serve as a reference for the intrinsic tribological properties of the films imparted by their structural and chemical features in a wear-less contact regime.

3.3 Elastomeric Sliding Contact on the SAM Films on a Macroscopic Scale

Our primary interest in employing an elastomeric (PDMS) pin as the sliding partner for these three SAM films is to test if the alkyl-chain-dependent (**C_5CH_3** versus **$C_{11}CH_3$** SAMs) and terminal-functional-group-dependent (**$C_{11}CH_3$** versus **$C_{11}OH$** SAMs) frictional properties, as observed in AFM experiment (Fig. 4), can be reproduced on a macroscopic scale. Figure 5a shows that this is indeed the case; When the friction-versus-load plots were obtained from the sliding contacts between *ox*-PDMS pins, which are chemically similar to the oxidized silicon nitride AFM tip, and the three SAM films in distilled water; the same order of the interfacial frictional properties, **C_5CH_3 > $C_{11}CH_3$ > $C_{11}OH$**, was measured at 1 mm/s. This trend was consistently observed from the friction-versus-load plot obtained at much higher sliding speed, 50 mm/s, as shown in Fig. 5b. However, the μ values at 50 mm/s are significantly lower than those at 1 mm/s for all three SAMs: for

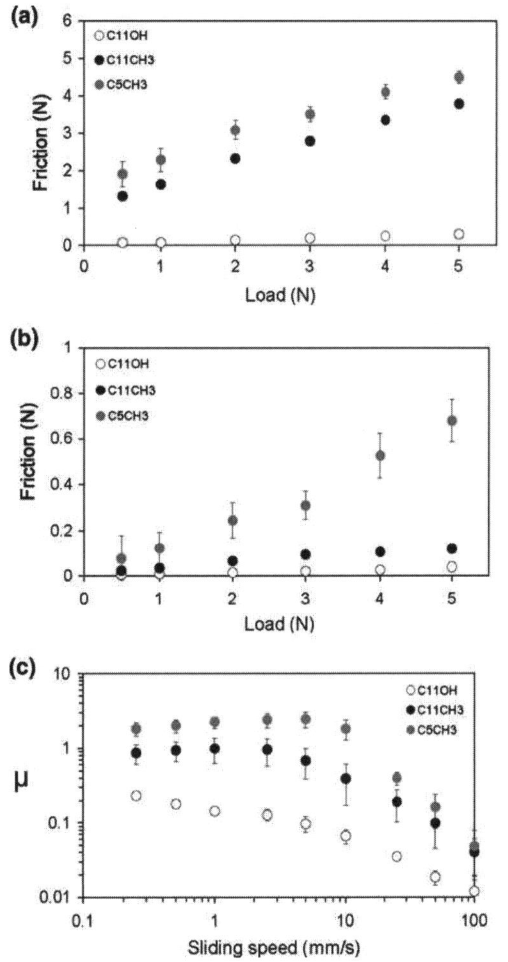

Fig. 5 (**a**) Friction-force-versus-load plots (sliding speed = 1 mm/s), (**b**) friction-force-versus-load plots (sliding speed = 50 mm/s) and (**c**) μ-versus-speed plots (load = 1 N) obtained from the tribopair of *ox*-PDMS/SAM films in distilled water

instance, 0.008 (50 mm/s) vs. 0.061 (1 mm/s) for the **C11OH** film, 0.024 vs. 0.76 for the **$C_{11}CH_3$** film, and 0.14 vs. 0.90 for the **C_5CH_3** film, under the load of 5 N. The speed-dependent frictional behavior of the sliding contacts of the *ox*-PDMS/SAM films is more clearly seen in the μ-versus-speed plots (load = 1 N), as shown in Fig. 5c. While the μ values were generally decreasing with increasing speed for all cases, the onset of the decrease in μ appears to be dependent on the hydrophilicity of the SAM films. For the two CH_3-terminated SAM films, for instance, the μ values started to decrease from ca. 5 mm/s, reaching 0.04–0.05 at the highest speed, whereas the decrease of μ for the **$C_{11}OH$** SAM started from the lowest speed, 0.25 mm/s, gradually to the highest speed, 100 mm/s, reaching a lowest value of $\mu \sim 0.01$. These observations suggest that the lubrication mechanisms in the low- and high-speed regimes might be different, despite the same relative order of the frictional properties of the three SAM films. This issue will be discussed in detail in Sect. 3.5.

When the SAM films were slid against a hydrophilic (oxygen-plasma treated), yet rigid slider, a stainless steel pin, wear of the film and the substrate became so significant that the structural and chemical characteristics of SAM films were not discernable any more, as shown in Fig. 6. Due to the apparent damage on the sliding track for each measurement, the sliding track was changed for each speed condition, and the μ values remained constant at ca. 0.15 over the entire range of speed for all three samples (two SAM films and bare gold substrate). Although some previous macrotribological studies using a rigid slider have revealed alkyl-chain-length-dependent frictional properties of SAMs, this distinction was possible only under a load of a few mN [20].

The pin-on-disk tribometer experiments in this work have thus revealed that the frictional properties of the SAM films observed by AFM could be reproduced on a macroscopic contact scale by employing a hydrophilic elastomeric slider.

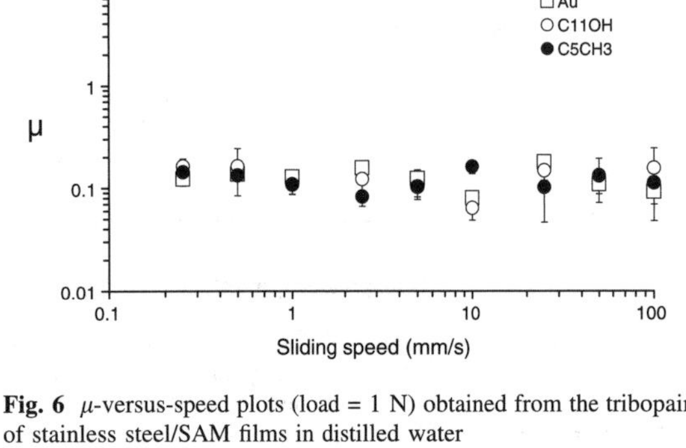

Fig. 6 μ-versus-speed plots (load = 1 N) obtained from the tribopair of stainless steel/SAM films in distilled water

3.4 PM-IRRAS Studies on the Sliding Track

One distinct advantage of the macrotribological versus nanotribological approach is that the sliding tracks are generally wide enough to be accessed by standard surface-spectroscopic tools. Given that the tribological properties of the three SAM films characterized by pin-on-disk tribometer using an elastomer slider, *ox*-PDMS, revealed the same order of frictional properties as were observed by AFM, it is of interest to examine if the SAM films retain their structural integrity following tribological stress. It is noted that most tribological experiments on SAM films by AFM are believed to take place in a wear-less regime. However, no direct spectroscopic evidence is currently available due to the very small contact area involved. In this work, we have employed PM-IRRAS to examine the molecular structure of the SAM films following tribological stress applied by the elastomeric slider. For these experiments, the sliding of *ox*-PDMS pin against the **$C_{11}OH$** and **$C_{11}CH_3$** SAM films was extended to 500 rotations at 1 mm/s and 1 N. The PM-IRRA spectra obtained from inside and outside the sliding tracks are compared in Fig. 7. In Fig. 7a, the spectral region sensitive to the characteristic structural features of the SAM films is shown. Firstly, it is notable that the peak positions for the asymmetric methylene C–H stretch, $\nu_a(CH_2$, 2,919–2,921 cm^{-1}), are virtually identical for inside (*solid line*) and outside (*dotted line*) the sliding track, which indicates that the SAM films remain intact and retain a nearly well-ordered backbone structure despite the tribological stress. Secondly, the intensity of the asymmetric methyl C–H stretch, $\nu_s(CH_3$, ca. 2,962 cm^{-1}) has somewhat increased inside the sliding track, and this change is more significant for the cases of **$C_{11}CH_3$** SAM than of **$C_{11}OH$** SAM. The increase in the intensity of the asymmetric methyl C–H stretch peak coincides with the occurrence of the peaks at 1,110 and 1,265 cm^{-1} (arising from Si–O asymmetric stretching and Si–CH_3 asymmetric bending modes, respectively [48]) inside the sliding track, as shown in Fig. 7b, which suggests that the transfer of uncrosslinked monomer species from the pin to the SAM films might have occurred. In order to test this hypothesis, we have repeated the same experiments after extracting uncrosslinked monomer species from the PDMS pin network (see the Sect. 2.2 for extraction process). As shown in Fig. 7a, the PM-IRRA spectra corresponding to the C–H stretching modes, from 2,750 cm^{-1} to 3,050 cm^{-1}, obtained from inside and outside the sliding track are almost identical. A slight difference in peak intensity for some cases shown in Fig. 7a was statistically not significant. The occurrence of Si–O–Si peaks in the low wavenumber region, Figure 7b, has also been substantially suppressed for all cases. However, the transfer of a trace amount of the monomer species

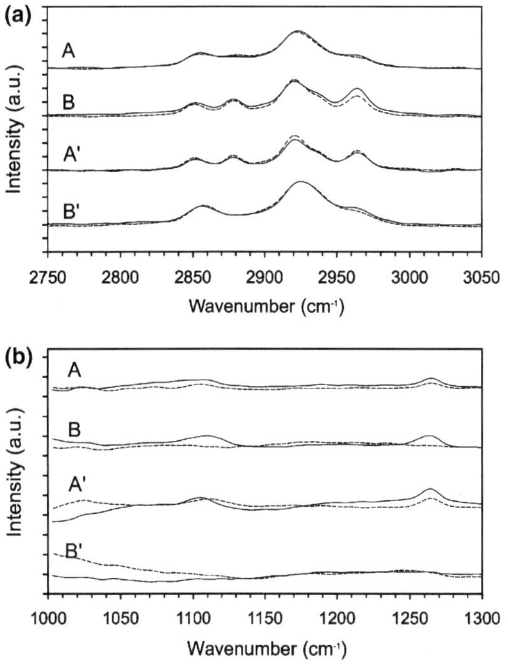

Fig. 7 PM-IRRA spectra obtained from inside and outside the sliding track of the SAM films. (**a**) the spectral region sensitive to the characteristic structural features of the SAM films, (**b**) the spectral region sensitive to the Si–O–Si stretching region (A: *ox*-PDMS/$\mathbf{C_{11}OH}$, B: *ox*-PDMS/$\mathbf{C_{11}CH_3}$, C: *ox*-PDMS (extracted)/$\mathbf{C_{11}OH}$, D: *ox*-PDMS (extracted)/$\mathbf{C_{11}CH_3}$). The tribostress was applied by generating the sliding contacts between *ox*-PDMS/SAM films for 500 rotations (load = 1 N and sliding speed = 1 mm/s)

to $\mathbf{C_{11}OH}$ SAM film by *ox*-PDMS pins was observed to be not completely avoidable. The extraction of the monomer species from the PDMS pins, however, did not influence on the relative frictional properties of the SAM films shown in Fig. 5.

Spectroscopic examination of the sliding track by PM-IRRAS clearly verified that the SAM films retain their ordered structure following the tribostress provided by the elastomeric slider. Analysis of the sliding track that has received the tribological stress is common practice in macrotribological studies of SAM films [20, 21, 24]; however, it was even easier when an elastomer was employed as the sliding partner because of the larger contact area achieved at the elastomeric interface. It should be emphasized that the spatially resolved PM-IRRAS measurements presented here were simply measured by positioning different regions of the sample in the beam path under normal sampling conditions and, therefore, the spectra presented above as "inside" the wear track do not completely exclude the beam sampling some of the area outside the track, as well. As a rough estimate, the fraction of the wear track probed by the beam in our measurement, assuming an ideally aligned beam path and sample positioning, is about 80%. However, the reasonable resolution between inside and outside the sliding track by PM-IRRAS obtained even with such a simple set-up suggests that this approach holds potential as powerful research methodology. Although PM-IRRAS was employed mainly to verify the mechanical durability of the SAM films in this work, more diverse and systematic spectroscopic studies associated with tribological contacts are expected; the detection of the monomer transfer following the sliding contact between *ox*-PDMS/SAM films can be one example.

3.5 Lubrication Mechanisms: The Transition from Boundary Lubrication to Soft EHL

Another important aspect of macrotribometric approaches is that they can easily explore the high-speed regime. This feature is particularly advantageous when the tribological testing is carried out in a liquid medium as in this work; with increasing sliding speed, the entrainment of the liquid into the contact zone and the formation of the fluid film starts to occur, and thus the transition from boundary lubrication to fluid-film lubrication is also observed with increasing speed. Despite its extremely low pressure-coefficient of viscosity, water can also form a fluid-film lubricant layer when one or both sides of the tribopair is comprised of an elastomer, i.e., in the soft-EHL regime (also known as isoviscous-elastic lubrication regime) [34–43]. The likelihood of the soft-EHL mechanism occurring and the corresponding fluid-film thickness for a given elastomeric tribosystem can be readily predicted according to the theoretical model described by Hamrock and Dowson [42] (later revised by Esfahanian and Hamrock [43]) based upon the bulk mechanical properties of the tribopair and the rheological properties of the lubricant. When expressed in terms of the material and measurement parameters, the minimum film thickness is

$$h_{\min} \approx 2.1 \cdot R^{0.77} \cdot (\eta \cdot u_s)^{0.66} \cdot E'^{-0.45} \cdot w^{-0.21} \quad (1)$$

where R is the radius of the slider for a sphere-on-plane contact (3 mm), η is the viscosity of the lubricant (9×10^{-4} Pa s), u_s is the sliding speed (variable), E' is the reduced contact modulus, $E' = 2((1-\nu_1^2)/E_1) + ((1-\nu_2^2)/E_2)$ where ν is Poisson ratio (E_{PDMS} = 2 MPa, E_{glass} = 270 GPa, ν_{PDMS} = 0.5, ν_{glass} = 0.2) and w is the applied load (variable). We have carried out the calculation of the film thickness by employing the parameters employed for the tribometer experiments (Figs. 4, 5). The results are shown in Fig. 8a and b. Since the film thickness predicted from Eq. 1 is that for ideally smooth surfaces (R_a = 0), the

Tribol Lett (2007) 28:229–239 237

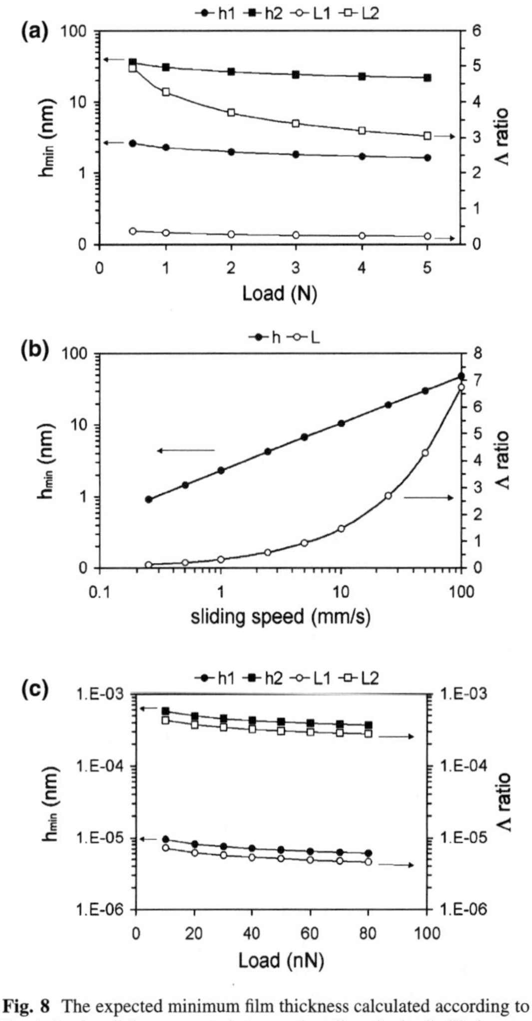

Fig. 8 The expected minimum film thickness calculated according to Eq. 1 for the sliding contacts between the tribopairs in distilled water and the Λ ratio between the expected film thickness and the surface roughness ($\Lambda = h_{min}/\sigma$, where h_{min} is the minimum film thickness and σ is the combined surface roughness of the tribopair, $\sigma = \sqrt{R_a^{pin} + R_a^{disk}}$) (**a**) for the tribopair of *ox*-PDMS/glass pair as a function of load (h_1 and L_1: sliding speed, 1 mm/s and h_2 and L_2: sliding speed, 50 mm/s) in water, (**b**) for the tribopair of *ox*-PDMS and glass as a function of speed in water (h: minimum film thickness, L: Λ ratio) and (c) for the tribopair of silicon nitride AFM probe and SAM film as a function of load in water (h_1 and L_1: sliding speed, 2 μm/s and h_2 and L_2: sliding speed, 1,000 μm/s)

relative magnitude of the film thickness compared to the surface roughness can be taken into account by estimating the ratio between them, i.e., Λ ratio ($\Lambda = h_{min}/\sigma$, where σ is the combined surface roughness of the tribopair, $\sigma = \sqrt{R_a^{pin} + R_a^{disk}}$, and h_{min} is the minimum film thickness). This Λ ratio is often employed to estimate the lubrication mechanism for rough surfaces; generally, fluid-film lubrication is expected when $\Lambda \geq 3$, and boundary lubrication is expected when $\Lambda \leq 1$, and mixed lubrication is expected when $1 \leq \Lambda \leq 3$ [49]. It is noted that due to the larger area of contact for the tribometer experiments in this work (mm-range in contact radius), the R_a values of the sample surfaces characterized by AFM (Sect. 3.1) are extrapolated to the corresponding scale.

Figure 8a shows that the Λ ratio for the friction-versus-load plots at 1 mm/s ranged from 0.37 (0.5 N) to 0.23 (5 N), whereas those at 50 mm/s ranged from 4.95 (0.5 N) to 3.05 (5 N), which suggests that the sliding contacts of *ox*-PDMS/SAM films might have occurred in different lubrication regimes in distilled water, i.e., boundary lubrication for 1 mm/s and fluid-film lubrication for 50 mm/s. Although the experimental data for the fluid-film thickness are currently not available, significantly lower μ values observed from the friction-versus-load plots at 50 mm/s (Fig. 5b) compared with those at 1 mm/s (Fig. 5a) are consistent with this argument. The Λ-ratio-versus-speed plots in Fig. 8b shows that the transition to fluid-film lubrication, i.e., $\Lambda \geq 3$, is expected to occur from ca. 30 mm/s in the case of an applied load of 1 N.

It is emphasized that for the experiments carried out in distilled water, the occurrence of different lubrication mechanisms at different speeds was only to be expected for elastomeric contact on the macroscopic scale, such as the *ox*-PDMS/SAM film tribopairs tested by means of pin-on-disk tribometry. When the same calculation was carried out under the same experimental conditions for the AFM probe/SAM films tribopair ($E_{AFM\ tip}$ = 140 GPa, E_{SAM} = 9.3 GPa, $v_{AFM\ tip}$ = 0.25, v_{SAM} = 0.35, R = 50 nm, all the parameters quoted from the references [50] and [51]), the expected film thickness is in the range of ca. 10^{-5} nm, and the Λ ratio is smaller than 7.14×10^{-6} as shown in Fig. 8c. Even if 1,000 μm/s, upper speed limit of an ordinary AFM, is employed for the calculation of the film thickness, the expected film thickness is not higher than 5.71×10^{-4} nm and the Λ ratio is also not higher than 4.23×10^{-4} as also shown in Figure 8(c). In other words, the transition from boundary lubrication to fluid-film lubrication is unlikely to occur for typical tribological contacts on SAM films by AFM probes in distilled water.

3.6 Lubrication Mechanisms: the Role of Surface Chemistry

The three SAM films employed in this work contributed to the modification of the interfacial frictional properties of the sliding contacts of elastomer/rigid substrate in two ways. Firstly, in the context of activating the fluid-film lubrication mechanism, the SAM films provided different surface hydrophilicities of the sliding track to vary the interfacial hydrophilicity of the tribopair; as addressed in a previous study involving self-mated PDMS sliding contacts in water [34], the bulk mechanical properties of the tribopairs and the surface roughness, discussed in the previous section, are, in practice, not the sole parameters that determine the activation of the soft-EHL mechanism: The surface chemistry of the tribopairs also plays a very significant role. Although the **$C_{11}OH$** SAM is the most hydrophilic of the three SAM films used in this work with $\theta_w \approx 25°$, it is less hydrophilic than *ox*-PDMS, $\theta_w < 3°$, and thus the μ values for *ox*-PDMS/**$C_{11}OH$** were observed to be generally higher than those of *ox*-PDMS/*ox*-PDMS [34]. Secondly, in the boundary-lubrication regime, where the formation of the soft-EHL film is not expected (i.e., $\Lambda < 1$), both surface chemistry and the structural integrity of the SAM films influenced the interfacial frictional properties in a similar fashion to that observed in AFM experiments. In terms of surface chemistry, the hydrophilic **$C_{11}OH$** SAM showed lower frictional forces than those observed for the two CH_3-terminated SAMs when sliding against an *ox*-PDMS pin in the low-speed and/or high-load regime; under such conditions, however, this trend was not due to the higher feasibility of forming aqueous fluid-film, but rather due to the lower work of adhesion for the hydrophilic interface in an aqueous environment, similarly to the AFM experiments shown in the Sect. 3.2. In terms of the structural integrity, the **$C_{11}CH_3$** SAM revealed lower frictional forces than the **C_5CH_3** SAM for all cases, again mirroring the AFM experiments.

4 Conclusions

In this work, we have demonstrated that the intrinsic frictional properties of three alkanethiol SAM films, previously assessed by nanotribological approaches only, can be manifested on a macroscopic scale by employing an elastomer, PDMS, as the sliding partner in pin-on-disk tribometery. This approach provides unique opportunities to investigate the tribological properties of SAM films that have hitherto not been accessible by either conventional macrotribological or nanotribological approaches. Firstly, while the contact pressure can be maintained low enough to retain the integrity of SAM films, as with nanotribological approaches, high-speed, macroscale tribological contacts can be achieved by employing a pin-on-disk tribometer with an elastomeric counterface. Secondly, the wide sliding tracks generated in this approach—even wider than those obtained in conventional tribometry using rigid sliders—allow for standard spectroscopic approaches to access the contact area and characterize the influence of the tribological contacts on the SAM films. Thirdly, by running the measurements in a liquid medium, it is possible to induce a range of lubrication regimes, from boundary lubrication to fluid-film lubrication, over the speed range available from an ordinary pin-on-disk tribometer. Given that a large toolbox of self-assembly approaches to controlling surface chemistry has been established, this approach possesses the potential for systematic studies to investigate the influence of surface chemistry on elastomer-based tribological systems.

References

1. Carpick, R.W., Salmeron, M.: Scratching the surface: fundamental investigations of tribology with atomic force microscopy. Chem. Rev. **97**(4), 1163–1194 (1997)
2. Leggett, G.J., Brewerb, N.J., Chonga, K.S.L.: Friction force microscopy: towards quantitative analysis of molecular organisation with nanometre spatial resolution. Phys. Chem. Chem. Phys. **7**, 1107–1120 (2007)
3. Xiao, X., Hu, J., Charych, D.H., Salmeron, M.: Chain length dependence of the frictional properties of alkylsilane molecules self-assembled on mica studied by atomic force microscopy. Langmuir **12**, 235–237 (1996)
4. Lio, A., Charych, D.H., Salmeron, M.: Comparative atomic force microscopy study of the chain length dependence of frictional properties of alkanethiols on gold and alkylsilanes on mica. J. Phys. Chem. B **101**, 3800–3805 (1997)
5. Wong, S.-S., Takano, H., Porter, M.D.: Mapping orientation differences of terminal functional groups by friction force microscopy. Anal. Chem. **70**, 5209–5212 (1998)
6. Lee, S., Shon, Y.-S., Colorado, R. Jr., Guenard, R.L., Lee, T.R., Perry, S.S.: The influence of packing densities and surface order on the frictional properties of alkanethiol self-assembled monolayers (SAMs) on gold: a comparison of SAMs derived from normal and spiroalkanedithiols. Langmuir **16**, 2220–2224 (2000)
7. van der Vegte, E.W., Subbotin, A., Hadziioannou, G.: Nanotribological properties of unsymmetrical *n*-dialkyl sulfide monolayers on gold: effect of chain length on adhesion, friction, and imaging. Langmuir **16**, 3249–3256 (2000)
8. Beake, B.D., Leggett, G.J.: Variation of frictional forces in air with the composition of heterogeneous organic surfaces. Langmuir **16**, 735–739 (2000)
9. Brewer, N.J., Beake, B.D., Leggett, G.J.: Friction force microscopy of self-assembled monolayers: influence of adsorbate alkyl chain length, terminal group chemistry, and scan velocity. Langmuir **17**, 1970–1974 (2001)
10. Frisbie, C.D., Rozsnyai, L.F., Noy, A., Wrighton, M.S., Lieber, C.M.: Functional group imaging by chemical force microscopy. Science **265**, 2071–2074 (1994)
11. Sinniah, S.K., Steel, A.B., Miller, C.J., Reutt-Robey, J.E.: Solvent exclusion and chemical contrasts in scanning force microscopy. J. Am. Chem. Soc. **118**, 8925–8931 (1996)
12. Vezenov, D.V., Noy, A., Rozsnyai, L.F., Lieber, C.M.: Force titration and ionization state sensitive imaging of functional groups in aqueous solution by chemical force microscopy. J. Am. Chem. Soc. **119**, 2006–2015 (1997)
13. Noy, A., Vezenov, D.V., Lieber, C.M.: Chemical force microscopy. Annu. Rev. Mater. Sci. **27**, 381–421 (1997)

14. Kim, H.I., Koini, T., Lee, T.R., Perry, S.S.: Systematic studies of the frictional properties of fluorinated monolayers with atomic force microscopy: comparison of CF_3- and CH_3-terminated films. Langmuir **13**, 7192–7196 (1997)
15. Kim H.I., Graupe, M., Oloba, O., Koini, T., Imaduddin, S., Lee, T.R., Perry, S.S.: Molecularly specific studies of the frictional properties of monolayer films: a systematic comparison of CF_3-, $(CH_3)_2CH$-, and CH_3-terminated films. Langmuir **15**, 3179–3185 (1999)
16. Lee, S., Puck, A., Graupe, M., Colorado, R. Jr., Shon, Y.-S., Lee, T.R., Perry, S.S.: Structure, wettability, and frictional properties of phenyl-terminated self-assembled monolayers on gold. Langmuir **17**, 7364–7370 (2001)
17. Clear, S.C., Nealey, P.F.: Chemical force microscopy study of adhesion and friction between surfaces functionalized with self-assembled monolayers and immersed in solvents. J. Colloid Interface Sci. **213**, 238–250 (1999)
18. Clear, S.C., Nealey, P.F.: Lateral force microscopy study of the frictional behavior of self-assembled monolayers of octadecyltrichlorosilane on silicon/silicon dioxide immersed in *n*-alcohols. Langmuir **17**, 720–732 (2001)
19. Ren, S., Yang, S., Zhao, Y., Zhou, J., Xu, T., Liu, W.: Friction and wear studies of octadecyltrichlorosilane SAM on silicon. Tribol. Lett. **13**, 233–239 (2002)
20. Nakano, M., Ishida, T., Numata, T., Ando, Y., Sasaki, S.: Alkyl chain length effect on tribological behavior of alkanethiol self-assembled monolayers on Au. Jpn. J. Appl. Phys. **42**, 4734–4738 (2003)
21. Nakano, M., Ishida, T., Numata, T., Ando, Y., Sasaki, S.: Tribological behavior of terphenyl self-assembled monolayer studied by a pin-on-plate method and friction force microscopy. Jpn. J. Appl. Phys. **43**, 4619–4623 (2004)
22. Sung, I.H., Kim, D.-E.: Surface damage characteristics of self-assembled monolayers of alkanethiols on metal surfaces. Tribol. Lett. **17**, 835–844 (2004)
23. Ishida, H., Koga, T., Morita, M., Otsuka, H., Takahara, A.: Macro- and nanotribological properties of organosilane monolayers prepared by a chemical vapor adsorption method on silicon substrates. Tribol. Lett. **19**, 3–8 (2005)
24. Nicholas, A. Jr., Street, S.C.: Spectroscopic analysis of the tribological behavior of a model boundary layer lubricant. Analyst **126**, 1269–1273 (2001)
25. Hardy, W.B., Bircumshaw, I.: Boundary lubrication-plane surfaces and the limitations of Amonton's law. Proc. R. Soc. (Lond.) A **108**, 1–27 (1925)
26. Tabor, D.: Mechanism of boundary lubrication. Proc. R. Soc. Ser. A **212**, 498–505 (1952)
27. Levine, O., Zisman, W.A.: Physical properties of monolayers adsorbed at the solid-air interface. I. Friction and wettability of aliphatic polar compounds and effect of halogenation. J. Phys. Chem. **61**, 1068–1077 (1957)
28. Levine, O., Zisman, W.A.: Physical properties of monolayers adsorbed at the solid-air interface. II. Mechanical durability of aliphatic polar compounds and effect of halogenation. J. Phys. Chem. **61**, 1188–1196 (1957)
29. Mate, C.M., McClelland, G.M., Erlandsson, R., Chiang, S.: Atomic-scale friction of a tungsten tip on a graphite surface. Phys. Rev. Lett. **59**, 1942–1945 (1987)
30. Carpick, R.W., Agrait, N., Ogletree, D.F., Salmeron, M.: Measurement of interfacial shear (friction) with an ultrahigh vacuum atomic force microscope. J. Vac. Sci. Technol. B **14**, 1289-1295 (1996)
31. Brown, H.R.: Chain pullout and mobility effects in friction and lubrication. Science **263**, 1411–1413 (1994)
32. Greenwood, J.A., Minshall, H., Tabor, D.: Hysterersis losses in rolling and sliding friction. Proc. R. Soc. Ser. A **259**, 480–507 (1961)
33. Schallamach, A.: How does rubber slide? Wear **17**, 301–312 (1971)
34. Lee, S., Spencer, N.D.: Aqueous lubrication of polymers: influence of surface chemical modification. Tribol. Int. **38**, 922–930 (2005)
35. Lee, S., Iten, R., Müller, M., Spencer, N.D.: Influence of molecular architecture on the adsorption of poly(ethylene oxide)-poly(propylene oxide)-poly(ethylene oxide) (PEO-PPO-PEO) on PDMS surfaces and implications for aqueous lubrication. Macromolecules **37**, 8349–8356 (2004)
36. de Vincente, J., Strokes, J.R., Spikes, H.A.: Lubrication properties of non-absorbing polymer solutions in soft-elastohydrodynamic (EHD) contacts. Tribol. Int. **38**, 515–526 (2005)
37. Roberts, A.D., Tabor, D.: The extrusion of liquids between highly elastic solids. Proc. R. Soc. Lond. A **325**, 323–345 (1971)
38. Moore, D.F.: The elastohydrodynamic transition speed for spheres sliding on lubricated rubber. Wear **35**, 159–170 (1975)
39. Medley, J.B., Strong A.B., Pilliar, R.M., Wong, E.W.: The breakdown of fluid film lubrication in elastic-isoviscous point contacts. Wear **63**, 25–40 (1980)
40. Richards, S.C., Roberts, A.D.: Boundary lubrication of rubber by aqueous surfactant. J. Phys. D **25**, A76–A80 (1992)
41. Dowson, D., Jin, Z.M.: Microelastohydrodynamic lubrication of low-elastic-modulus solids on rigid substrates. J. Phys. D 25, A116–A123 (1992)
42. Hamrock, B.J., Dowson, D.: Minimum film thickness in elliptical contacts for different regimes of fluid-film lubrication. Proceedings of the 5th Leeds-Lyon Symposium on Tribology, pp. 22–27. Mech. Eng. Publ. Bury St. Edmunds, Suffolk (1979)
43. Esfahanian, M., Hamrock, B.J.: Fluid-film lubrication regimes revisited. Tribol. Trans. **34**, 628–632 (1991)
44. Marti, A., Hähner, G., Spencer, N.D.: Sensitivity of frictional forces to pH on a nanometer scale: a lateral force microscopy study. Langmuir **11**, 4632–4635 (1995)
45. Nuzzo, R.G., Dubois, L.H., Allara, D.L.: Fundamental studies of microscopic wetting on organic surfaces. 1. Formation and structural characterization of a self-consistent series of polyfunctional organic monolayers. J. Am. Chem. Soc. **112**, 558–569 (1990)
46. Nuzzo, R.G., Fusco, F.A., Allara, D.L.: Spontaneously organized molecular assemblies. 3. Preparation and properties of solution adsorbed monolayers of organic disulfides on gold surfaces. J. Am. Chem. Soc. **109**, 2358–2368 (1987)
47. Bensebaa, F., Voicu, R., Huron, L., Ellis, T.H.: Kinetics of formation of long-chain *n*-alkanethiolate monolayers on polycrystalline gold. Langmuir **13**, 5335–5340 (1997)
48. Truong, K.D., Rowntree, P.A.: Formation of self-assembled butanethiol monolayers of Au substrate: spectroscopic evidence for highly ordered island formation in sub-monolayer films. J. Phys. Chem. **100**, 19917–19926 (1996)
49. Hutchings, I.M.: Ch.4 lubricants and lubrication In: Tribology; Friction and Wear of Engineering Materials. Edward Arnold (1992)
50. Barrena, E., Kopta, S., Ogletree, D.F., Charych, D.H., Salmeron, M.: Relationship between friction and molecular structure: alkylsilane lubricant films under pressure. Phys. Rev. Lett. **82**, 2880–2883 (1999)
51. Weihs, T.P., Nawaz, Z., Jarvis, S.P., Pethica, J.B.: Limits of imaging resolution for atomic force microscopy of molecules. Appl. Phys. Lett. **59**, 3536–3538 (1991)

Langmuir
Article
pubs.acs.org/Langmuir

Fabricating Chemical Gradients on Oxide Surfaces by Means of Fluorinated, Catechol-Based, Self-Assembled Monolayers[†]

Mathias Rodenstein,[‡] Stefan Zürcher,[‡,§] Samuele G. P. Tosatti,[‡,§] and Nicholas D. Spencer*[,‡]

[‡]*Laboratory for Surface Science and Technology, Department of Materials, ETH Zurich, Wolfgang-Pauli-Strasse 10, CH-8093 Zurich, Switzerland, and* [§]*SuSoS AG, Lagerstrasse 14, CH-8600 Dübendorf, Switzerland*

Received February 24, 2010. Revised Manuscript Received April 21, 2010

Catechols bind strongly to several metal oxides and can thus be used as a binding group for generating self-assembled monolayers. Furthermore, their derivatives can be used to produce well-defined, centimeter-scale surface-chemical gradients on technologically relevant surfaces, such as titanium dioxide (TiO_2). A simple dip-and-rinse gradient-preparation technique was utilized to produce surface-hydrophobicity gradients from perfluoro-alkyl catechols and nitrodopamine (ND). Chemical composition, quality, and properties of the functionalized surfaces were determined by means of X-ray photoelectron spectroscopy (XPS), variable-angle spectroscopic ellipsometry (VASE), and static water contact angle (sCA) measurements. Contact angles were found to be in the range of 30°−95°, correlating well with the determined surface chemical composition and adlayer thickness.

Introduction

The use of monolayer coatings that are spontaneously generated by immersion into solutions of adsorbing molecules, so-called self-assembled monolayers (SAMs), has gained a lot of popularity over the past 20 years.[1] SAMs allow facile and economical tailoring of surface properties, such as hydrophobicity,[2−4] protein resistance,[5,6] or specific biological activity.[7] Depending on the substrate to be modified, different anchoring chemistries (e.g., thiols for noble metals,[2−4] silanes for silica,[8] or phosph(on)ates for metal oxides[9,10]) have been established.[11−15] In general, a limitation has been that the binding chemistry to be applied is determined by the surface chemistry of the substrate material. Hence, the synthesis and combination of different anchor and functional groups have been necessary. Recently, it was reported that mussels can stick to virtually any surface under the harshest of conditions, by means of catechol-based adhesive compounds.[16−18] Catechols represent a very promising, biomimetic approach to extend the field of surface functionalization. Their adhesion properties on a wide variety of substrates, including metals (Ag, Au),[19,20] native oxide surfaces (on Ti, Cu, Fe),[16,21,22] semiconductors (Si),[23] and polymers such as polystyrene,[18] have been reported.[18] The study of their adsorption mechanisms, theoretically[24,25] and experimentally,[26] the influence of parameters such as pH,[21] as well as their stability under a variety of conditions (solvents, temperature, and time) is thus of fundamental interest.[27]

The fact that, in addition to mussels, cyanobacteria also employ a catechol-based molecule, the siderophore anachelin, to covalently bind to metal ions, such as Fe^{3+},[28−30] further suggests the flexibility of this chemistry as a potential coating platform. Wach et al.,[30] for example, have used a hybrid catechol-based molecule to generate a functional coating with active antibacterial properties.

In the case of anachelin, it is possible to observe how one of the general limitations of catechol chemistry (its low stability toward oxidation, especially at elevated pH) is circumvented by the introduction of the electron-withdrawing dimethyl-ammonium group, bound to the aromatic ring of the catechol.[28] To the same end, we have introduced the nitro ($-NO_2$) group in the para-position to the hydroxide, due to synthetic ease and relatively small additional steric bulk.[27]

In the present work, we further functionalize the nitro-containing species, nitrodopamine (ND), with a perfluorinated alkyl

[†] Part of the Molecular Surface Chemistry and Its Applications special issue.

*To whom correspondence should be addressed. E-mail: nspencer@ethz.ch.

(1) Ulman, A. *Chem. Rev.* **1996**, *96*, 1533.
(2) Nuzzo, R. G.; Fusco, F. A.; Allara, D. L. *J. Am. Chem. Soc.* **1987**, *109*, 2358.
(3) Nuzzo, R. G.; Zegarski, B. R.; Dubois, L. H. *J. Am. Chem. Soc.* **1987**, *109*, 733.
(4) Troughton, E. B.; Bain, C. D.; Whitesides, G. M.; Nuzzo, R. G.; Allara, D. L.; Porter, M. D. *Langmuir* **1988**, *4*, 365.
(5) Lin, Y. S.; Hlady, V.; Gölander, C.-G. *Colloids Surf., B* **1994**, *3*, 49.
(6) Tosatti, S. G. P. Ph.D. Dissertation, ETH Zurich, Switzerland, 2003.
(7) Michel, R.; Lussi, J. W.; Csucs, G.; Reviakine, I.; Danuser, G.; Ketterer, B.; Hubbell, J. A.; Textor, M.; Spencer, N. D. *Langmuir* **2002**, *18*, 3281.
(8) Chaudhury, M. K.; Whitesides, G. M. *Science* **1992**, *255*, 1230.
(9) Hofer, R.; Textor, M.; Spencer, N. D. *Langmuir* **2001**, *17*, 4014.
(10) Tosatti, S. G. P.; Michel, R.; Textor, M.; Spencer, N. D. *Langmuir* **2002**, *18*, 3537.
(11) Liedberg, B.; Tengvall, P. *Langmuir* **1995**, *11*, 3821.
(12) Guerrero, G.; Mutin, P.; Vioux, A. *Chem. Mater.* **2001**, *13*, 4367.
(13) Bhat, R.; Fischer, D.; Genzer, J. *Langmuir* **2002**, *18*, 5640.
(14) Carbonell, L.; Whelan, C.; Kinsella, M.; Maex, K. *Superlattices Microstruct.* **2004**, *36*, 149.
(15) Love, J. C.; Estroff, L. A.; Kriebel, J. K.; Nuzzo, R. G.; Whitesides, G. M. *Chem. Rev.* **2005**, *105*, 1103.
(16) Dalsin, J. L.; Lin, L.; Tosatti, S. G. P.; Voros, J.; Textor, M.; Messersmith, P. B. *Langmuir* **2005**, *21*, 640.
(17) Dalsin, J. L.; Messersmith, P. B. *Mater. Today* **2005**, *8*, 38.
(18) Lee, H.; Dellatore, S. M.; Miller, W. M.; Messersmith, P. B. *Science* **2007**, *318*, 426.
(19) Kawabata, T. *Biochem. Pharmacol.* **1996**, *51*, 1569.
(20) Brooksby, P. A.; Schiel, D. R.; Abell, A. D. *Langmuir* **2008**, *24*, 9074.
(21) Araujo, P. Z.; Morando, P. J.; Blesa, M. A. *Langmuir* **2005**, *21*, 3470.
(22) Creutz, C.; Chou, M. H. *Inorg. Chem.* **2008**, *47*, 3509.
(23) Lambert, J.; Singer, S. *J. Organomet. Chem.* **2004**, *689*, 2293.
(24) Terranova, U.; Bowler, D. R. *J. Phys. Chem. C.* **2010**, *114*, 6491.
(25) Li, S.-C.; Wang, J.-G.; Jacobson, P.; Gong, X.-Q.; Selloni, A.; Diebold, U. *J. Am. Chem. Soc.* **2009**, *131*, 980.
(26) Lee, H.; Lee, K. D.; Pyo, K. B.; Park, S. Y.; Lee, H. *Langmuir* **2010**, *26*, 3790.
(27) Malisova, B.; Tosatti, S.; Textor, M.; Gademan, K.; Zürcher, S. *Langmuir* **2010**, *26*, 4018.
(28) Zürcher, S.; Wäckerlin, D.; Bethuel, Y.; Malisova, B.; Textor, M.; Tosatti, S. G. P.; Gademann, K. *J. Am. Chem. Soc.* **2006**, *128*, 1064.
(29) Wach, J.-Y.; Bonazzi, S.; Gademann, K. *Angew. Chem., Int. Ed.* **2008**, *47*, 7123.
(30) Wach, J.-Y.; Malisova, B.; Bonazzi, S.; Tosatti, S. G. P.; Textor, M.; Zürcher, S.; Gademann, K. *Chem.—Eur. J.* **2008**, *14*, 10579.

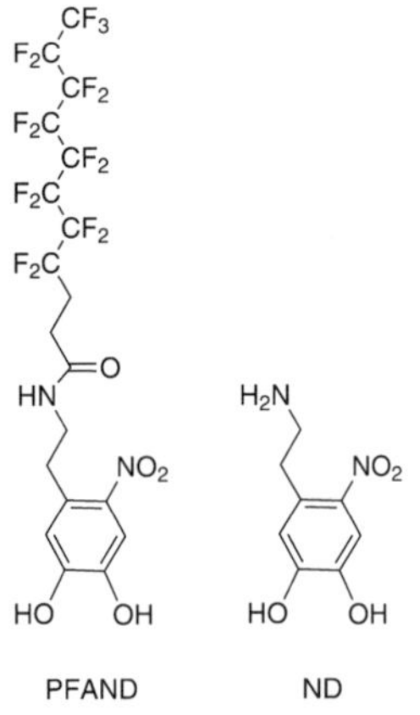

Figure 1. Structures of the two molecules used for self-assembly: perfluoro-alkyl-nitrodopamine (PFAND) and nitrodopamine (ND).

chain (yielding perfluoro-alkyl-nitrodopamine (PFAND)), shown in Figure 1, in order to obtain a molecule that combines the functional properties of perfluoro-alkyls with the anchoring mechanisms of catechols in an architecture that can generate self-assembled monolayers.

Fluoropolymers, such as poly(tetrafluoroethylene) (PTFE), and SAM-forming molecules, such as perfluorinated thiols, silanes, and *n*-alkanoic acids, are widely used for surface fluorination in academia and industry (e.g., as easy-to-clean surfaces).[31−34] The modification of surfaces with PTFE is chemically challenging because of its low reactivity and solubility. Thiols, on the other hand, generally form stable layers on (noble) metals only and are thus not applicable to the functionalization of most materials. Utilizing catechol derivatives to fluorinate a wide variety of surfaces in order to render them hydrophobic or nonadhesive is therefore a promising alternative.

A recent development has been the replacement of sets of discrete sample arrays of varying surface parameters, such as substrate roughness, hydrophobicity, protein resistance, or (bio)ligand densities by a continuum of these properties.[35,36] Preparation and comparison of discrete arrays can be very time-consuming and yield only fragmentary results. Samples that exhibit a continuous spectrum of a surface-chemical functionality along one defined spatial dimension on a single substrate are known as surface-chemical gradients. Using surfaces that incorporate such a continuous change of a defined property can result in a multiplication and parallelization of experiments and thus a radical increase in research speed. The approach can also yield insights into phenomena that are not otherwise accessible, for example, taxis of cells on various scales.[35−44]

Numerous kinds of surface-chemical gradients have been reported, including one-, two-, or three-dimensional gradients and gradients on the micrometer or centimeter scales. Fabrication approaches have involved polymer brushes, alkanethiols on Au, micropillars, and poly(dimethylsiloxane) (PDMS) microfluidic channel systems among others.[11,35,41,45,46]

In the present work, in order to demonstrate the controllable adsorption and application of PFAND and ND, we have adapted the linear motion drive (LMD) gradient-fabrication protocol developed by Morgenthaler et al.[42] to the technologically relevant TiO_2 substrate. Changes in the adlayer thickness, its chemical composition, and surface energy have been measured for the gradient and compared to corresponding values for homogeneous adlayers.

Materials and Methods

Substrates and Chemicals. Solvents and chemicals were obtained from Sigma-Aldrich, Buchs, Switzerland unless otherwise stated. Silicon wafers with a natural SiO_2 layer were obtained from Si-Mat Silicon Materials, Landsberg/Lech, Germany. Entire wafers were coated with TiO_2 by physical vapor deposition (PVD, reactive magnetron sputtering) at the Paul Scherer Institute, Villigen, Switzerland, to obtain TiO_2 films of about 20 nm in thickness (verified by ellipsometry). The wafers were diced into pieces of 1×1 cm^2 (homogeneous samples) and 1×4 cm^2 (gradients). Prior to functionalization, the substrates were sonicated twice for 7 min in toluene and then twice for 7 min in 2-propanol, and dried under a stream of nitrogen (5.0). The substrates were further cleaned for 30 min in a UV/O_3 cleaner (Boekel Scientific, Feasterville, PA, model 135500), followed by overnight immersion in ultrapure water (TOC < 5 ppb, R = 18.2 MΩcm).

Nitrodopamine (ND) and perfluoro-alkyl-nitrodopamine (PFAND) used for self-assembly in this study were synthesized according to the method of Tosatti and Zürcher.[47]

Synthesis of 2H,2H,3H,3H-Perfluoro-undecanoic-acid-*N*-succinimidyl-ester. 2H,2H,3H,3H-Perfluoroundecanoic acid (1.354 g, 2.75 mmol, Fluorous Technologies Inc., Pittsburgh, PA), *N*-hydroxysuccinimide (348 mg, 3.02 mmol), and dicyclohexylcarbodiimide (622 mg, 3.02 mmol) were dissolved in ethyl acetate (120 mL) and stirred for 18 h at room temperature. The white precipitate formed (dicyclohexyl urea, DCU) was filtered off, and the remaining solution evaporated to dryness. The residue was recrystallized twice from ethyl acetate. Yield: 1.00 g (62%), containing some traces of DCU. ^{1}H NMR ($CDCl_3$, 300 MHz, ppm): 3.0 (m, 2H CH_2), 2.88 (s, 4H CH_2 NHS), 2.6 (m, 2H CH_2).

Synthesis of 6-Nitro-3-hydroxytyramine Hemisulfate Salt (Nitrodopamine, ND). 3-Hydroxytyramine hydrochloride (1.90 g, 10 mmol) and sodium nitrite (1.52 g, 22 mmol) were dissolved in water (25 mL) and cooled to 0 °C. Sulfuric acid (17.4 mmol in 10 mL of water) was added slowly to the mixture, and a yellow precipitate was formed. After stirring at room temperature overnight, the precipitate was filtered and recrystallized from water. The product was dried under high vacuum to yield ND as a hemisulfate salt.[48] Yield: 1.389 g (58%). ^{1}H NMR (D_2O, 300 MHz, ppm): 7.62 (s, 1H ND), 6.83 (s, 1H ND), 3.24 (t, 2H CH_2), 3.12 (t, 2H CH_2). Elemental Analysis for $C_{16}H_{22}N_4O_{12}S$ in %

(31) Wallace, R.; Chen, P.; Henck, S.; Webb, D. *J. Vac. Sci. Technol., A* **1995**, *13*, 1345.
(32) Evans, S. D.; Flynn, T.; Ulman, A.; Beamson, G. *Surf. Interface Anal.* **1996**, *24*, 187.
(33) Geer, R.; Stenger, D.; Chen, M.; Calvert, J. M.; Shashidhar, R.; Jeong, Y.; Pershan, P. *Langmuir* **1994**, *10*, 1171.
(34) Barriet, D.; Lee, T. R. *Curr. Opin. Colloid Interface Sci.* **2003**, *8*, 236.
(35) Genzer, J.; Bhat, R. *Langmuir* **2008**, *24*, 2294.
(36) Morgenthaler, S.; Zink, C.; Spencer, N. D. *Soft Matter* **2008**, *4*, 419.
(37) Lee, S.-W.; Laibinis, P. E. *J. Am. Chem. Soc.* **2000**, *122*, 5395.
(38) Lahann, J.; Mitragotri, S.; Tran, T.-N.; Kaido, H.; Sundaram, J.; Choi, I. S.; Hoffer, S.; Somorjai, G.; Langer, R. *Science* **2003**, *299*, 371.
(39) Daniel, S.; Chaudhury, M. K. *Langmuir* **2002**, *18*, 3404.
(40) Jeon, N. L.; Baskaran, H.; Dertinger, S. K. W.; Whitesides, G. M.; Water, L. V. D.; Toner, M. *Nat. Biotechnol.* **2002**, *20*, 826.
(41) Wu, T.; Efimenko, K.; Vlcek, P.; Subr, V.; Genzer, J. *Macromolecules* **2003**, *36*, 2448.
(42) Morgenthaler, S.; Lee, S.; Zürcher, S.; Spencer, N. D. *Langmuir* **2003**, *19*, 10459.
(43) Yu, X.; Wang, Z.; Jiang, Y.; Zhang, X. *Langmuir* **2006**, *22*, 4483.
(44) Morgenthaler, S.; Zink, C.; Städler, B.; Vörös, J.; Lee, S.; Spencer, N. D.; Tosatti, S. G. P. *Biointerphases* **2007**, *1*, 156.
(45) Yoshimitsu, Z.; Nakajima, A.; Watanabe, T.; Hashimoto, K. *Langmuir* **2002**, *18*, 5818.
(46) Dertinger, S. K. W.; Chiu, D.; Jeon, N.; Whitesides, G. M. *Anal. Chem.* **2001**, *73*, 1240.
(47) Tosatti, S. G. P., Zürcher, S., Patent Appl. PCT/CH2007/000603 **2007**.
(48) Ranganathan, S.; Tamilarasu, N. *J. Indian Chem. Soc.* **1999**, *76*, 727.

found (calculated): C, 38.01 (38.87); H, 4.48 (4.48); N, 11.10 (11.33); S, 6.89 (6.49).

Synthesis of *N*-(4,5-Dihydroxy-2-nitro-phenethyl)-2H,2H,3H,3H-perfluoro-undecanamide (Perfluoro-alkyl-nitrodopamine, PFAND). 6-Nitro-3-hydroxytyramine hemisulfate salt (170 mg, 0.84 mmol) and *N*-methylmorpholine (150 μL) were dissolved in dimethylformamide (DMF, 5 mL). 2H,2H,3H,3H-Perfluoro-undecanoic-acid-*N*-succinimidyl-ester (500 mg, 0.85 mmol) was added, and the mixture was stirred under an atmosphere of nitrogen at room temperature overnight. Hydrochloric acid (1 N, 25 mL) was added, and the yellow precipitate formed was filtered off and washed with water. The solid was dissolved in ethyl acetate, and the organic phase dried over magnesium sulfate. The solvent was evaporated, and the residue recrystallized from a 1:1 chloroform/ethyl acetate mixture (20 mL, 4 °C). Yield: 279 mg (48%). ^{1}H NMR (DMSO-d_6, 300 MHz, ppm): 11−9 (broad, 2H OH), 8.2 (t, 1H NH), 7.48 (s, 1H nitro-dopamine), 6.7 (s, 1H nitro-dopamine), 2.9 (2H CH_2), 2.7−2.2 (m, 6H CH_2). Elemental Analysis for $C_{19}H_{13}N_2F_{17}$ in % found (calculated): C, 33.79 (33.94); H, 2.08 (1.95); N, 4.07 (4.17).

Solutions for Self-Assembly. ND is very hydrophilic due to its hydroxyl and amine groups and therefore soluble in pure H_2O (TKA-GenPure UV-TOC/UF, Huber & Co AG, Reinach BL, Switzerland). PFAND is strongly hydrophobic because of its perfluorinated alkyl tail. It was dissolved in 2-propanol (Sigma-Aldrich, Switzerland) first and then diluted with H_2O in a 2:1 ratio of H_2O/2-propanol.

Sample Preparation. *Homogeneous Samples.* Substrates were immersed in 1.5 mL of solution of different concentrations (between 0.4 and 440 μM) for different durations (2 min, 4 min, 8 min, 16 min, 32 min, 64 min, and 48 h). ND samples were subsequently immersed in pure H_2O and sonicated for 1 min, while PFAND samples were immersed and sonicated in 2-propanol for 3 min after adsorption, to remove loosely bound molecules. Subsequently, samples were blown dry under a stream of nitrogen.

Gradient Samples. The gradient-preparation method was adapted from that of Morgenthaler et al.[42] The immersion protocol was a logarithmic *z*-position-versus-time program (see the Supporting Information, part 1). Bare substrates were thus immersed in PFAND solution in a controlled manner. The following parameters were chosen for the preparation of gradient samples: (i) concentration of PFAND, 220 μM; (ii) immersion time at high-coverage end, 1 h; low coverage end, 4 s; (iii) logarithmic immersion depth versus time dependence, immediate sonication of samples for 3 min in 2-propanol after gradual immersion, followed by blow-drying with N_2 gas.

Backfilling was carried out by immersing one-component PFAND gradients in ND (500 μM) solution for 1 h and then rinsing them thoroughly with ultrapure water, followed by blow-drying with N_2.

Variable-Angle Spectroscopic Ellipsometry (VASE). VASE was used to measure surface coverage/layer thickness. We utilized a Woollam MF-2000F (WOOLLAM Co., Inc., Lincoln, NE) ellipsometer at three incident angles of 65°, 70°, and 75° with respect to the surface normal, averaging 50 measurements at each point. The spectral range was λ = 350−750 nm. The calculation of the thickness of the surface layer was performed using a four-layer model (Si/SiO_2/TiO_2/Cauchy), where Si and SiO_2 were assumed to be constant for all wafers. The TiO_2 layer was fitted before adsorption, and the Cauchy layer (organic layer) after adsorption with $n_{org} = n + ik = 1.45 + 0.01i$, where n is the refractive index and k is the extinction coefficient. Refractive index variations depending on the order of the SAM (estimated to be approximately 10%) were not taken into account. The positioning of the gradient samples was adjusted manually with an estimated error of ±0.5 mm. Due to the elliptical beam size, the measured thickness corresponds to an average over 2 mm × 5 mm (65°) to 2 mm × 8 mm (75°). If not stated otherwise, the error bars in VASE graphs are standard deviations from repeated experiments (three or more).

X-ray Photoelectron Spectroscopy (XPS). A Sigma2 instrument (Thermo Fisher Scientific, Loughborough, Great Britain) was utilized for routine XPS experiments. The Sigma2 is equipped with a UHV chamber (pressure < 10^{-6} Pa during measurements). The X-ray source is a non-monochromated 300 W Al Kα source ($h\nu$ = 1486.6 eV) that illuminates the sample at an angle of 54° to the surface normal. The hemispherical analyzer is mounted at 0° with respect to the surface normal, thus operating at the magic source-analyzer angle, which eliminates the need for angular-distribution correction.[49] A detector consisting of seven channeltrons is used. The spot size of the analyzed area (large-area mode) was 400 μm, and the results therefore represent a laterally averaged chemical composition. For this setup, the full width at half-maximum (FWHM) of Ag 3d$_{5/2}$ is 1.4 eV using a pass energy of 25 eV. Standard measurements comprised averages over nine (for C, F, and N) or three (for Ti and O) scans for each element plus survey scans with pass energies of 25 and 50 eV, respectively. The dwell time was left at 100 ms at all times, resulting in 3−5 min measurement time per spot for each element, accumulating to about 30 min for a complete elemental scan on each measuring position. In order to assess whether it was necessary to account for the degradation of the coatings during XPS measurements, the degradation kinetics on homogeneous samples were iteratively measured (C, F, and N spectra only (see Results section)).

For peak modeling, 30 (PFAND) or 10 (ND) high-resolution XPS spectra were used, recorded using a PHI5000 Versa probe (ULVAC-PHI, INC., Chigasaki, Japan) on homogeneous samples. The spectrometer is equipped with a 180° spherical capacitor energy analyzer and a multichannel detection system with 16 channels. Spectra were acquired at a base pressure of 5 × 10^{-8} Pa using a focused scanning monochromatic Al Kα source (1486.6 eV) with a spot size of 200 μm and 47.6 W power. The instrument was run in the FAT analyzer mode with electrons emitted at 45° to the surface normal. Pass energies used for survey scans was 187.85 and 46.95 eV for detail spectra. The FWHM of this setup is < 0.8 eV for Ag 3d$_{5/2}$.

All XPS spectra were evaluated using CasaXPS (version 2.3.12 and later). All binding energies are referenced relative to the main hydrocarbon peak (from residual contamination in the case of the clean surfaces and the SAM hydrocarbon chain in case of the SAM-coated surfaces), set at a binding energy of 285.0 eV.

Atomic percent (atom %) concentrations were calculated from detail spectra of each element present on the surface (region over all species), corrected by the appropriate relative sensitivity factors (RSFs), transmission function of the spectrometer, and inelastic mean free paths (IMFPs). The photoionization cross sections are normalized to C 1s according to Scofield.[50] The instrumental transmission function was calculated according to the formula:

$$T(E_k) = E_p(a^2/(a^2+R^2))^b \tag{1}$$

where E_k is the kinetic energy of the electrons, R is the retard ratio (E_p/E_k), E_p is the pass energy, and a and b are parameters. For the calculation, all available peaks of Au, Ag, and Cu reference samples were used. For the Sigma2, a best data fit yielded a = 21.75, b = 0.44. The IMFPs were calculated according to Cumpson and are comparable to those in PTFE.[51] In order to account for the adlayer structure, we assumed a stacked model (see Supporting Information, part 3) and adjusted the general attenuation formula:

$$I_s = I_0 \exp(-d/\lambda) \tag{2}$$

(49) Reilman, R.; Msezane, A.; Manson, S. *J. Electron Spectrosc. Relat. Phenom.* **1976**, *8*, 389.
(50) Scofield, J. H. *J. Electron Spectrosc. Relat. Phenom.* **1976**, *8*, 129.
(51) Cumpson, P. J. *Surf. Interface Anal.* **2001**, *31*, 23.

Table 1. Apparent Normalized Atomic Concentration (Determined with XPS, Analysis Depth ≈ 10 nm), Adlayer Thickness (VASE), and Static Contact Angle for Homogenous Monolayers Assembled from PFAND (440 μM in 1:2 Propanol/H_2O) and ND (500 μM in Pure H_2O) Solutions after 48 h of Immersion[a]

sample		XPS apparent atom % for layers on TiO_2 carbon	nitrogen	oxygen	fluorine	titanium	VASE *d* [nm]	CA [deg]
blank reference		5.1	0.2	66.0	0.0	28.7	0.0	< 10
PFAND	incl. substrate	18.5	1.9	44.8	16.0	18.9	1.3 ± 0.1	105 ± 2
	overlayer only (without TiO_2)	44.5	4.6	12.4	38.5	–		
	calculated	45.7	4.7	11.6	39.5		2.1 ± 0.1[b]	
ND	incl. substrate	17.6	3.0	56.2	0.0	23.2	0.5 ± 0.1	< 10
	overlayer only (without TiO_2)	61.8	10.5	27.7	0.0			
	calculated	57.1	14.3	28.6	0.0		0.6 ± 0.1[b]	

[a] As a comparison, the calculated values for pure PFAND and pure ND are given. [b] Calculated longest intramolecular distance (O–C of catechol versus F in CF_3 (PFAND) and H in NH_2 (ND)).

to

$$I_c = I_0(\lambda/d)(1 - \exp(-d/\lambda)) \tag{3}$$

where I is the attenuated intensity I_0 of an element with IMFP λ, after passing through a layer of thickness d. I_s is the intensity measured if the electrons originate below the layer, I_c if they are emitted within the layer itself. In the intensity corrections, we assumed that all electrons emitted from fluorine originate from the top $(CF_2)_8$ layer (eq 2), those from carbon originate from the entire PFAND/ND layer (eq 2), and those from nitrogen and oxygen (from molecules) originate from just below the $(CF_2)_8$ layer (eq 1). Electrons coming from TiO_2 were assumed to pass through the whole PFAND/ND layer (eq 1).

Ratios of elements are useful for the comparison of measurements on different samples. Using ratios instead of raw-intensity data eliminates errors arising from fluctuations in the photon flux.

Contributions to signals in the C 1s region arise from both the adlayer and potential surface contamination, the O 1s signal originates from the adlayer, contamination, surface hydroxyls and TiO_2, the Ti 2p from TiO_2 only, N 1s from adventitious contamination and the adlayer, and F 1s from the adlayer only (see Figure 1). The ratios F/Ti and N/Ti, assuming a homogeneous distribution of potential N-contaminants along the gradients, are thus directly related to the surface coverages of the adlayers.

Contact Angle Measurements. Static contact angles (sCA) were measured (with a NRL C.A. goniometer, model 100-00-230, Ramé-Hart, Netcong, NJ) for one-component as well as two-component homogeneous samples. The measurements for the two-component gradients were repeated after 1 and 14 h of immersion in H_2O to test their stability in aqueous environments; water droplets of 2 μL were placed on the sample, and the contact angles on both sides of the drops recorded. In order to achieve good correlation between XPS, VASE, and sCA measurements on gradient samples, the first drop was always placed 2 mm from the edge of the hydrophobic end. Subsequent drops were placed at a distance of 4 mm from the center of every preceding drop.

Statistics and Errors. The main positional error in the gradients (and thus the main error in correlations between XPS, VASE, and CA measurements) originates from the sample-preparation process, because of the difficulty in precise placing of the substrates above the solution before immersion and removing them quickly afterward.

Due to the VASE spot size, ellipsometric measurements are averages over 2 × 0.5 mm^2 with the longer axis of the beam spot perpendicular to the gradient. Only a negligible error is induced, provided the coverage gradient is linear. The main error (0.5 mm) for gradients is intrinsic to the measurement and results from the beam spot size.

For static contact angle measurements, errors occur because of size, evaporation, and creeping of the 2 μL water drop between deposition on the sample and measurement. The manual positioning along the gradient samples can be performed with an accuracy of 0.1 mm. Based on these estimates, a total positional error of ±1 mm is assumed.

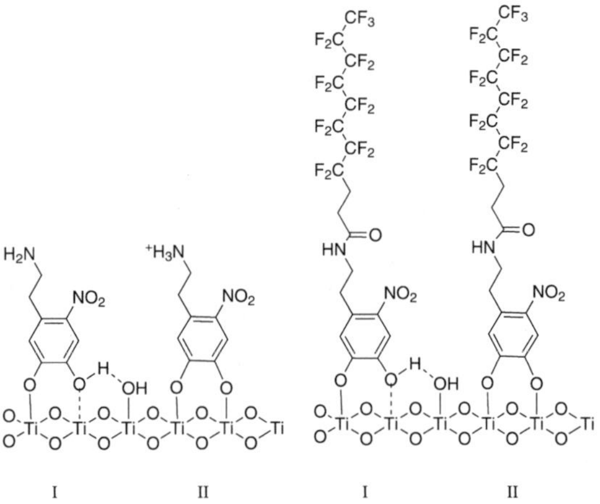

Figure 2. Two possible configurations of ND and PFAND when adsorbed on TiO_2. (I) Monodentate binding with a hydrogen bridge to a neighboring surface hydroxide. (I) Bidentate binding, in which both catechol oxygens are deprotonated.

Results

1. Homogenous Coatings and Investigation of Adlayer Degradation under X-ray Beam. Hydrophobic coatings with maximal surface coverage corresponding to one monomolecular layer were obtained by immersing the cleaned substrates into PFAND solutions (440 μM) for 48 h. The layer thickness of such samples measured by VASE was 1.3 ± 0.1 nm (the length of a fully stretched PFAND molecule is ~2.1 nm) and sCA 105 ± 2°. Such coatings served as references when determining X-ray degradation of the coating as a function of time (see the Supporting Information, part 2).[52–54] Homogeneous layers of ND are completely hydrophilic (sCA < 10°), and their thickness was determined to be 0.5 ± 0.1 nm (see Table 1).

Table 1 shows the surface-chemical composition (atom %), adlayer thickness, and static contact angle for different layers

(52) Beard, B. *Propellants* **2004**, *16*, 81.
(53) Mendes, P.; Belloni, M.; Ashworth, M.; Hardy, C.; Nikitin, K.; Fitzmaurice, D.; Critchley, K.; Evans, S. D.; Preece, J. *ChemPhysChem* **2003**, *4*, 884.
(54) Graham, R. L.; Bain, C. D.; Biebuyck, H. A.; Laibinis, P. E. *J. Phys. Chem.* **1993**, *97*, 9456.

Table 2. Binding Energies, FWHM, Assignment, and Measured Areas for a Homogeneous PFAND Film on TiO_2 Relative to the Carbon Area of the CF_3 Group[a]

Element	Peak	binding energy [eV] (PHI5000)	FWHM [eV] (PHI5000)	assignment (atoms/molecule)	RSF corrected measured area relative to CF_3 (PHI5000)	RSF corrected measured area relative to CF_3 (Sigma2[b])
carbon (1s)	C1	285	1.63 ± 0.08	C-**C**-C (4)	3.3 ± 0.2	2.6 ± 0.1
	C2	286.18 ± 0.02	= FWHM(C1)	**C**−O; **C**−N; **C**−CON; **C**−CF_2−(5)	5.5 ± 0.2	4.9 ± 0.3
	C3	288.22 ± 0.03	1.35 ± 0.04	−**C**(=O)−N (1)	1.1 ± 0.1	0.9 ± 0.1
	C4	289.22 ± 0.04	= FWHM(C3)	−**C**−NO_2 (1)	1.1 ± 0.1	0.9 ± 0.1
	C5	290.93 ± 0.08	= FWHM(C3)	C−**C**F_2−CF_2 (1)	1.0 ± 0.1	0.9 ± 0.1
	C6	292.00 ± 0.09	= FWHM(C3)	CF_2−**C**F_2−CF_x (6)	5.5 ± 0.4	5.1 ± 0.2
	C7	294.24 ± 0.09	1.17 ± 0.07	CF_2−**C**F_3 (1)	1.0 ± 0.1	1.0 ± 0.0
nitrogen (1s)	N1	400.42 ± 0.05	1.78 ± 0.12	−C(=O)−**N** (1)	1.06 ± 0.08	1.05 ± 0.07
	N2	405.09 ± 0.12	= FHWM(N1)	−C−**N**O_2 (II) (*x*)	0.20 ± 0.04	0.26 ± 0.04
	N3	406.56 ± 0.12	= FHWM(N1)	−C−**N**O_2 (I) (1 − *x*)	0.31 ± 0.03	0.23 ± 0.06
fluorine (1s)		689.27 ± 0.1	1.88 ± 0.03	−C**F**$_x$ (17)	18.8 ± 0.9	16.7 ± 0.8
oxygen (1s)		530.29 ± 0.05	1.27 ± 0.03	Ti**O**$_2$ (2·Ti)	19.18 ± 0.82	26.01 ± 1.63
		531.59 ± 0.19	2.00 ± 0.23	C−**O**−H; C−**O**−Ti; −C(=**O**)−N; −C−N**O**$_2$; Ti−**O**H (total 5 from PFAND)	4.67 ± 0.36	3.82 ± 0.21

[a] Comparison between experimental and stoichiometrically calculated intramolecular ratios, measured using monochromatic (high resolution, PHI5000) or non-monochromatic X-rays (Sigma2 Probe). [b] In the case of the Sigma2 Probe, in addition to the FWHM constraints, peak positions were also constrained relative to the C−**C**−C component as found in the high-resolution spectra (PHI5000).

assembled from a 440 μM and 500 μM solution of PFAND and ND, respectively. For the XPS data, the calculations are based on a stack model for the adlayer and include corrections for transmission function and IMFPs associated with the XPS measurements. To discriminate substrate from overlayer oxygen, we assigned all oxygen species except that of TiO_2 to the adlayer, assuming that all surface (hydroxyl) groups participate in the binding of catechol. This assumption might lead to an overestimation of surface coverage because not all surface hydroxyl groups are occupied. PFAND and ND are assumed to bind via the catechol group to the TiO_2 surface in two possible configurations. Figure 2 shows monodentate (I) and bidentate (II) configurations of ND molecules adsorbed on TiO_2, as described for catechol on rutile by Li et al. The different binding states of the catechol's hydroxyl groups lead to changes in the electron distribution within the benzene ring and thus also in the attached nitro group. These differences become detectable in the nitrogen XP spectra for the nitro group as a binding-energy shift. The presence of both binding modes causes the appearance of two nitrogen peaks for the NO_2 group. The results are comparable in terms of layer thickness/surface coverage, composition, and contact angles to other surfaces that have been hydrophobically functionalized by SAMs.[55−58]

The theoretical ratios for atomic species present in ND and PFAND (C/N/F) are as follows: 8:2:0 (= 0.250 N/C) and 19:2:17 (= 0.105 N/C, 0.895 F/C), respectively. Calculating the ratios from the atom % shown in Table 1 (C/N/F) yields 0.06 ± 0.01 N/C for the clean TiO_2, 0.179 ± 0.011 N/C for ND, 0.105 ± 0.007 N/C, and 0.874 ± 0.021 F/C for PFAND after 48 h of immersion. After 1 h of immersion, TiO_2 in ND yields 0.17 ± 0.01 N/C, and 1 h in PFAND, 0.089 ± 0.005 N/C and 0.855 ± 0.022 F/C (Table 4). This suggests a displacement mechanism of the contaminants, as previously reported in the literature.[57,59,60] Nevertheless, the measured ratios are higher than the theoretical values. This also suggests the unavoidable presence of carbon-containing adventitious contamination when working under ambient conditions.

Tables 2 and 3 show the analysis of PFAND and ND layers on TiO_2 as determined from high-resolution XPS (PHI5000) spectra. Figure 3 shows the corresponding normalized high-resolution C 1s and N 1s spectra of PFAND and ND on TiO_2, respectively. All identified components are marked C1−C7 and N1−N4.

For C 1s, positions of the fitted components are as expected from the intramolecular bonds of the molecules. Experimental area ratios are higher than the theoretical values, suggesting the presence of contamination. Differences between the FWHM of identified C 1s and O 1s components are due to the presence of contaminants as well as the slightly different electronic configurations or hybridization of bonds associated with a single binding energy (Tables 2 and 3). For N 1s, PFAND shows 3 peaks and ND 4 shows peaks, although only two kinds of nitrogen (NO_2, NH_2) were expected from the stoichiometry. The amine-to-nitro peak area ratio is not 1, in contrast to the molecular stoichiometry. Possible explanations for this observation are presented in the Discussion.

X-ray Induced Degradation of the SAM Layers. The effects of X-ray exposure on the adlayers have been investigated in detail (see the Supporting Information, part 2). Repeated measurement of the same spot results in (i) <10% of the carbon present on the surface being lost due to degradation; (ii) continuous decrease of fluorine over the whole 20 h of X-ray exposure; and (iii) decrease and eventually disappearance of the nitro peak, while the amine/amide signal increases. Figure 4 compares the N 1s signals for PFAND and ND after short and long X-ray exposure. For PFAND, the nitro peak is significantly reduced below the noise level after about 4.5 h, while for ND only a small change, within the error margin, is noticeable.

2. ND and PFAND Self-Assembly Kinetics. In order to be able to apply our gradient-deposition protocol, it is necessary to determine and control the kinetics of molecular adsorption.

(55) Venkataraman, N. V.; Zürcher, S.; Rossi, A.; Lee, S. *J. Phys. Chem. C* **2009**, *113*, 5620.
(56) Spori, D. M.; Venkataraman, N. V.; Tosatti, S. G. P.; Durmaz, F.; Spencer, N. D.; Zürcher, S. *Langmuir* **2007**, *23*, 8053.
(57) Hoque, E.; DeRose, J. A.; Hoffmann, P.; Mathieu, H. J.; Bhushan, B.; Cichomski, M. *J. Chem. Phys.* **2006**, *124*, 174710.
(58) Suzuki, S.; Whittaker, M. R.; Wentrup-Byrne, E.; Monteiro, M. J.; Grondahl, L. *Langmuir* **2008**, *24*, 13075.
(59) Dannenberger, O.; Buck, M.; Grunze, M. *J. Phys. Chem. B* **1999**, *103*, 2202.
(60) Himmelhaus, M.; Eisert, F.; Buck, M.; Grunze, M. *J. Phys. Chem. B* **2000**, *104*, 576.

Table 3. Binding Energies, FWHM, Assignment, and Measured Areas for a Homogeneous ND Film on TiO_2 Relative to the Nitrogen Area[a]

element	peak	binding energy [eV] (PHI5000)	FWHM [eV] (PHI5000)	assignment (atoms/molecule)	RSF corrected measured area relative to N_{tot} = 2 (PHI5000)	RSF corrected measured area relative to N_{tot} = 2 (Sigma2[b])
carbon (1s)	C1	285	1.58 ± 0.02	C−**C**−C (4)	5.80 ± 0.24	3.6 ± 0.3
	C2	286.19 ± 0.03	= FWHM(C1)	**C**−O; **C**−N (3)	3.56 ± 0.12	3.6 ± 0.3
	C3	287.47 ± 0.11	= FWHM(C1)	−**C**−NO_2 (1 − x)	1.07 ± 0.05	0.7 ± 0.1
	C4	288.90 ± 0.06	= FWHM(C1)	−**C**−NO_2 (x)	1.19 ± 0.05	0.5 ± 0.1
nitrogen (1s)	N1	400.14 ± 0.03	1.96 ± 0.03	−C−NH_2 (1 − y)	0.73 ± 0.02	0.5 ± 0.0
	N2	401.88 ± 0.03	= FHWM(N1)	−C−NH_3^+ (y)	0.62 ± 0.01	0.5 ± 0.0
	N3	405.19 ± 0.06	= FHWM(N1)	−C−NO_2 (II) (x)	0.22 ± 0.01	0.3 ± 0.0
	N4	406.54 ± 0.04	= FHWM(N1)	−C−NO_2 (I) (1 − x)	0.43 ± 0.01	0.2 ± 0.0
oxygen (1s)		530.14 ± 0.02	1.28 ± 0.02	Ti**O**$_2$ (2·Ti)	16.08 ± 0.49	15.7 ± 0.7
		531.47 ± 0.19	2.23 ± 0.00	C−**O**−H; C−**O**−Ti;	6.63 ± 0.31	7.5 ± 0.5
		532.95 ± 0.34	1.95 ± 0.39	−C−N**O**$_2$; Ti−**O**H (total 4 from ND)	0.90 ± 0.81	0.0 ± 0.0

[a] Comparison between experimental and stoichiometrically calculated intramolecular ratios, measured using monochromatic (high resolution, PHI5000) or non-monochromatic X-rays (Sigma2 Probe). [b] In the case of the Sigma2 Probe, in addition to the FWHM constraints, peak positions were also constrained relative to the C−C−C component, as found in the high-resolution spectra (PHI5000).

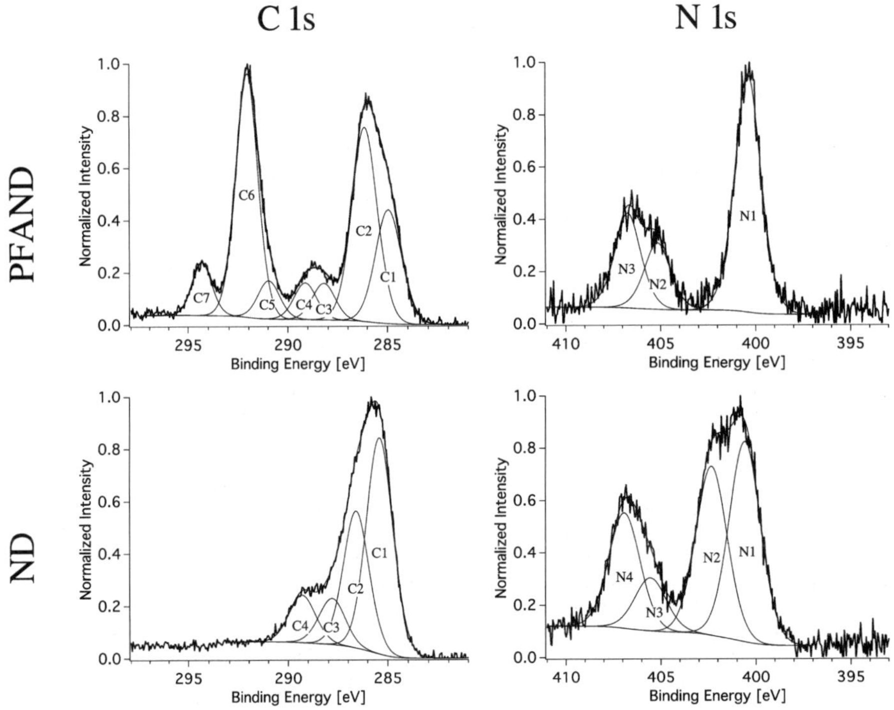

Figure 3. XPS C 1s and N 1s spectra of TiO_2 substrates homogenously coated with PFAND and ND, respectively (obtained with PHI5000). C1−C7 and N1−N4 are curve-fitted carbon and nitrogen components (details in Table 2).

Therefore, the concentration has to be chosen such that a full monolayer is assembled within several minutes, in order to practicably achieve gradients of a few centimeters in length.

In the case of PFAND (Figure 5), the dependence of surface coverage on assembly time as well as on solution concentration, with standard deviations below 10%, is clearly visible. While for a concentration of 7 μM the layer thickness shows no time dependence, values of concentration in the range of 44−440 μM were found to be suitable for the fabrication of gradients by the method used in this study. A value of 220 μM was chosen as the standard concentration. XPS data confirm the (logarithmic) trend observed by ellipsometry for the N/Ti and F/Ti ratios measured for PFAND layers after different immersion times (Figure 5). The data indicate the presence of about 100% more adsorbed molecules after 1 h than after 1 min.

In the case of ND, solutions with concentrations below 10 μM yield coverages comparable to contamination levels. Higher solution concentrations result in higher coverage. A simple time dependence for different concentrations could not be deduced. The highest thickness measured was about 0.5 nm (48 h in 500 μM ND), but substantial standard deviations of up to 20% have been recorded (see the Supporting Information, part 4), likely due to the hydrophilic nature of ND (static water contact angle < 10°; see Table 1) resulting in varying degrees of wetting and conta-

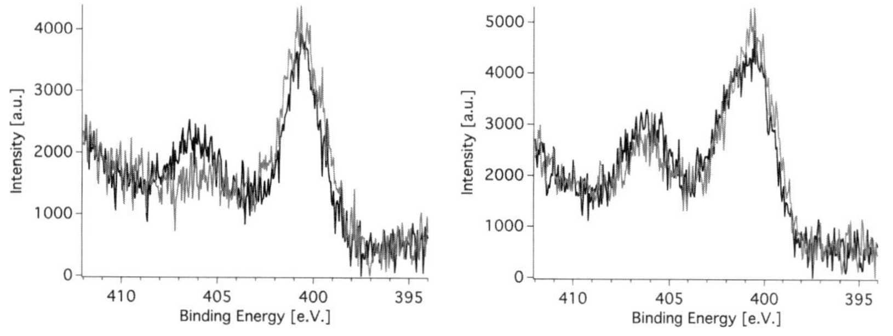

Figure 4. Nitrogen 1s spectra at beginning (black line) and after 4.5 h (gray line) of XPS measurement (on Sigma2) of PFAND (left) and ND (right).

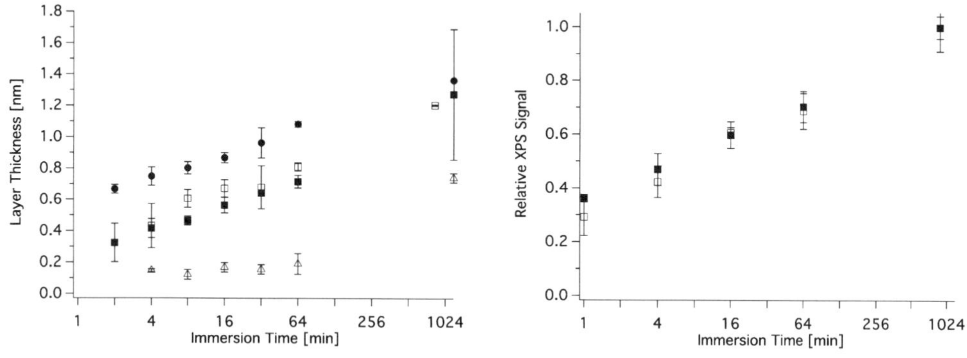

Figure 5. (a) Logarithmic plot of adsorption kinetics of PFAND from 1:2 2-propanol/H_2O with concentrations of (●) 220 μM, (□) 110 μM, (■) 44 μM, and (△) 7 μM. The higher the concentration, the higher the adsorption rate and the higher the "initial coverage" at the short immersion times. A value of 220 μM was determined as the ideal concentration for making gradients. (b) Logarithmic plot of normalized F/Ti (■) and N/Ti (□) XPS ratios for PFAND (sub)monolayers adsorbed from 220 μM 1:2 2-propanol/H_2O. Time dependence is in agreement with that of the measured surface coverage.

Table 4. Apparent Atomic % (atom %) of Mixed PFAND (220 μM) versus ND (500 μM) Layers on TiO_2 as Determined by XPS, Film Thickness as Determined Using VASE, and Static Water Contact Angles

	XPS apparent atom % for layers on TiO_2						
sample	carbon	nitrogen	oxygen	fluorine	titanium	VASE *d* [nm]	CA [deg]
clean TiO_2	6.9	0.4	65.5	0.0	27.2	0.0	< 10
PFAND 1 h	15.0	1.2	45.8	18.6	19.4	1.1 ± 0.1	90 ± 2
ND 1 h	12.2	1.9	61.4	0.0	24.4	0.4 ± 0.1	< 10
PFAND 1 h + ND 1 h	17.7	1.9	49.9	10.7	19.8	1.0 ± 0.1	82 ± 2

mination under ambient laboratory conditions but also due to the resolution of the VASE instrument (±0.1 nm). High fluctuations in detected layer thickness suggest that ND would not be a good choice as a first component for LMD-fabricated gradients.

3. Gradients. The gradients in surface coverage can be detected by measuring film thickness (VASE) as well as F/Ti and N/Ti XPS signal ratios. While PFAND forms hydrophobic layers, ND results in an amine-terminated hydrophilic surface. Thus, water contact angles should also show gradients in hydrophobicity. The extreme ends of the gradients were tested in the form of homogeneous samples (Table 4) with different coating steps: different timings and combinations of PFAND adsorption followed by ND backfilling.

It must be noted that ND adsorbs onto and possibly even partially replaces PFAND adlayers, especially for incomplete PFAND monolayers, resulting in higher nitrogen (+60%) and lower fluorine (−40%) apparent atom % for mixed layers, as outlined in Table 4. Such replacement is also observed for gradients, but can readily be compensated for during the fabrication process.

Figure 6 presents the film thickness (VASE), relative F/Ti ratio and N/Ti ratio (XPS), and static water contact angles of the gradients immersed in 220 μM PFAND solution, before and after backfilling, as determined by VASE, CA, and XPS. A surface-coverage gradient going from one end to the other end of the sample is observable in all cases. Before backfilling, the measured layer thickness varies in a linear manner by a factor of 7 from 0.15 to 1.05 nm (Figure 6a). Similar changes and position dependencies are observed for the F/Ti and N/Ti ratios measured for the same gradients while contact angles range from 35° to 95° (Figure 6b–d).

After backfilling the gradients with ND for 1 h, the surface coverage levels out to be almost homogeneous, with the lower-coverage end of PFAND being about 15% of the high-coverage end. Standard deviation becomes noticeably higher along the whole sample (up to 20%). The same observations hold true for

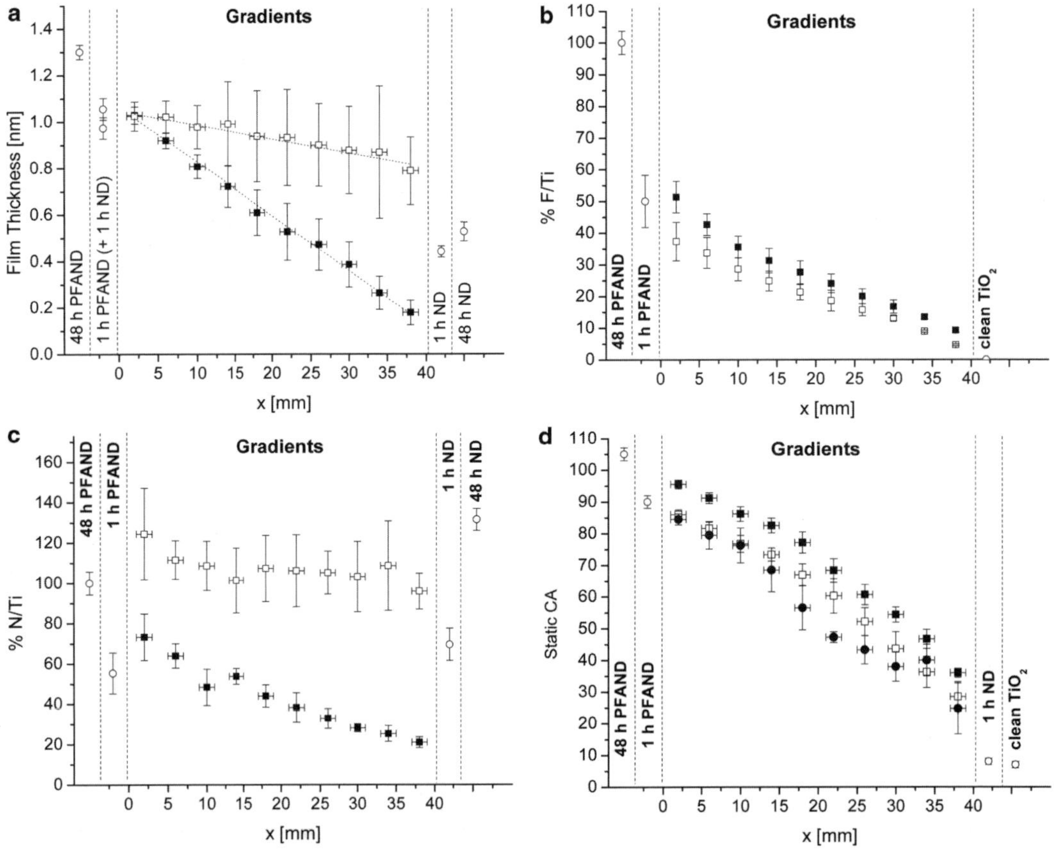

Figure 6. (a) Film thickness (ellipsometry) depends linearly (dotted lines are linear fits to the data sets) on the position along the PFAND one-component gradient (■). After ND-backfilling (□), the film thickness becomes more homogeneous. ND coverage in the two-component (difference to one-component) gradient is also linear. Gradients are in range of homogeneous layers (○). (b) F/Ti XPS ratios for homogeneous PFAND layers (○) and gradients (■) and PFAND-ND (□) gradients normalized to 48 h homogeneous layer. Dependence is linear, as for VASE, along immersion direction. High-PFAND-coverage end (left) and low-coverage end (right) differ by a factor of 5. (c) Relative N/Ti XPS ratios for homogeneous PFAND and ND layers (○), PFAND (■) and PFAND-ND (□) gradients. Dependence is linear, as for VASE, along immersion direction. High-PFAND-coverage end (left) and low-coverage end (right) differ by a factor of about 4 for one-component. Two-component gradients show more homogeneous N/Ti ratio. (d) Static water contact angles for PFAND one-component (■), ND backfilled gradients (□) after their additional immersion in H_2O for 14 h (●). Homogeneous samples (○).

the N/Ti ratio (Figure 6c). Additionally, the data show that the N/Ti signal of backfilled gradients is at least 50% higher than the one-component ones, meaning that there is additional ND adsorbed along the whole gradient, also on the higher PFAND coverage side. This also explains that backfilling PFAND gradients with ND causes a reduction of contact angles for the whole gradient by approximately 10° (Figure 6d). On the other hand, H_2O conditioning for 14 h does not change the sCA position dependence significantly, suggesting that the backfilled gradients remain stable in water for 14 h.

Discussion

We utilized the time-dependency of the Langmuir-type adsorption kinetics of perfluoro-alkyl-nitrodopamine (PFAND) with the linear-motion-drive technique (LMD) to make one-component surface-coverage gradients, which, once backfilled with ND, yielded catechol-based hydrophobicity gradients on TiO_2. In order to better understand the backfilling process, we considered the surface coverage of the two adsorbates.

VASE measurements reveal a thickness of 1.3−1.4 nm for densely packed monolayers of PFAND. The difference between molecular length (≈ 2.1 nm) and measured adlayer thickness is attributed to three considerations: (i) relaxed PFAND molecules are most likely not straight but helical and bent as in the case of fluorinated alkyl thiols,[34] (ii) adsorbed PFAND molecules, as fluorinated alkyls, might be tilted with respect to the surface normal by 30°−35° to optimize packing and intermolecular interactions,[34,61] and (iii) VASE-determined thickness is systematically underestimated by the layer model that was employed, because of adventitious carbon/contamination, which is constantly replaced by PFAND during adsorption, a process referred to as "self-cleaning" by Buck et al. for thiols on gold.[62] Comparison of C 1s XP spectra allows an estimation of the replacement of surface contamination: the amount of contamination on clean TiO_2 surfaces is equivalent to a fraction (20−25%) of what is

(61) Schonherr, H.; Ringsdorf, H. *Langmuir* **1996**, *12*, 3891.
(62) Buck, M.; Eisert, F.; Fischer, J.; Grunze, M.; Träger, F. *Appl. Phys. A: Mater. Sci. Process.* **1991**, *53*, 552.

measured on adlayers. XPS C 1s spectra indicate (complete) replacement of this contamination during (PFAND) adlayer formation. Hence, VASE-measured thickness of a PFAND monolayer might be compensated by 25% yielding 1.7 nm. We assume that the packing of PFAND and ND is mainly governed by the adsorption kinetics of catechols on TiO_2 surfaces, which have been addressed extensively in the literature, as well as the interaction between the perfluoro-alkyl chains.[24,25,63,64] While it is not clear how much the different properties of perfluoro-alkyl and hydrocarbon alkyl chains (the radius of $-CF_2-$ links is about 0.04 nm larger than that of CH_2)[33] influence ordering, an average tilt angle of ~32° is assumed for a densely packed PFAND monolayer in analogy to thiol−gold[65] and phosphate−metal-oxide[56,66] systems. Such a tilt angle and comparison with the literature for perfluorinated thiol SAMs on gold and density of binding sites for catechols on TiO_2 yield a molecular density in the range of 2−4 molecules/nm^2.[64,67] An assessment for ND proved difficult and likely erroneous because of the higher degree of hydrophilicity and thus affinity for contamination. Comparing N spectra of PFAND and ND monolayers indicate a molecular surface density only 10% higher for ND. Both results are consistent with values reported before for other systems.[21,64] Araujo et al.[21] and Rodriguez et al.[64] investigated the adsorption kinetics of catechol and gallic acid on TiO_2. They reported areas of 0.3−0.8 nm^2 (depending on adsorbate concentration) per molecule for monolayers (of catechol and gallic acid) on TiO_2.[21]

The higher surface density of ND versus PFAND adlayers suggests the presence of unoccupied binding sites within PFAND monolayers. This general abundance of adsorption sites and the sub-monolayer state of PFAND one-component gradients can explain the additional adsorption of ND everywhere within the PFAND gradients, as can be observed for VASE (Figure 6a) and N/Ti XPS (Figure 6c) data.

XPS data (Table 2 and 3) show good agreement between molecular stoichiometry and elemental distribution in surface adlayers. Deviations between the experimental and theoretical data are attributed to contamination.

The different mechanisms by which catechols can bind to TiO_2 have been described previously.[64] Monodentate and bidentate configurations are possible (see Figure 2), leading to different bond strengths and electron distributions in the adsorbate (particularly the benzene). Such different electron distributions lead to the appearance of neighboring peaks for the $-NO_2$ groups in the N 1s XP spectra of PFAND and ND as shown in Figure 3. PFAND and ND show two peaks at 405.1 and 406.6 eV associated to the nitro group. The peak splitting arises from the nitro group, which is conjugated to the benzene ring and therefore sensitive to the binding state of the catechol (see Figure 2). For ND, there are two peaks at 400.1 and 401.9 eV, which we attribute to partial protonation of the amine group. This splitting is not observed in the case of the amide nitrogen of PFAND at 400.4 eV, since protonation is not possible. The N 1s XPS amine/amide-to-nitro area ratio is greater than 1 for both molecules, which implies a possible reduction of NO_2 to NO_x or loss of NO_2 induced by the X-ray radiation or emitted photoelectrons. Such an effect could also be the reason for the disappearance of the PFAND nitro peak during X-ray exposure (Figure 4).

(63) Vasudevan, D.; Stone, A. *Environ. Sci. Technol.* **1996**, *30*, 1604.
(64) Rodriguez, R.; Blesa, M.; Regazzoni, A. *J. Colloid Interface Sci.* **1996**, *177*, 122.
(65) Chidsey, C. E. D.; Loiacono, D. *Langmuir* **1990**, *6*, 682.
(66) Pellerite, M.; Dunbar, T.; Boardman, L.; Wood, E. *J. Phys. Chem. B* **2003**, *107*, 11726.
(67) Wasserman, S.; Whitesides, G. M.; Tidswell, I.; Ocko, B.; Pershan, P.; Axe, J. D. *J. Am. Chem. Soc.* **1989**, *111*, 5852.

From VASE (Figure 6a) and XPS (Figure 6b, c) results, one can see that the two-component gradients apparently consist of complete mixed monolayers. In Figure 6a, the difference between the one- and two-component gradient is a linear curve ranging from 0.0 nm to about 0.6 nm (not shown), which essentially represents the adsorbed ND (concentration) in the two-component gradient but also includes contamination and potential replacement of PFAND by ND during backfilling. Yet, one has to be aware that up to 10% of PFAND remains (at the lower end of the gradient), as can be seen in Figure 6b. Accordingly, the N/Ti ratio, being equally representative for PFAND and ND, in Figure 6c almost reaches the ND monolayer level along the gradient (only slightly attenuated by the lower coverage end PFAND density), indicating complete surface coverage by catechols. The high N/Ti ratios versus the small ND coverage as measured by VASE can be explained by molecular displacement. The sub-monolayers of PFAND might not be stable with respect to substitution. Yet, most likely the abundance in Ti−OH binding sites allows plentiful ND coadsorption with airborne contamination, impeding an unambiguous interpretation of the obtained data. Steric hindrance of PFAND molecules might impede formation of strong bonds between molecules and substrate. Subsequent backfilling with ND can lead to replacement of PFAND because ND favors stronger bonds due to its size and possible electrostatic interaction of protonated amines with the negatively charged surface.

Both XPS and VASE show that PFAND coverage at the opposing low and high PFAND ends of one- and two-component gradients differs by a factor of 5−7. The high-end coverage, as deduced from XPS (F 1s spectra) data, corresponds to only about 50% of maximum coverage (48 h in 440 μM), while VASE yields about 75%. This difference can be explained by high levels of airborne contamination, as is expected for the long time of sample exposure to air (more than 1.5 h during adsorption and VASE measurements). The process of immersing the clean TiO_2 substrate in a normal lab environment over 1 h and the time for performing VASE measurements allows for contamination of one-component gradients. Backfilling one-component PFAND gradients with ND replaces this contamination and creates a well-defined surface-chemical composition. The gradient in hydrophobicity is still intact (Figure 6d) after backfilling, since PFAND coverage is only slightly influenced (Figure 6b) while the whole sample surface is covered by catechols (Figure 6c).

Conclusion and Perspectives

The perfluoro-alkyl-nitrodopamine (PFAND) and nitrodopamine (ND) catechol derivatives can be adsorbed onto TiO_2 from aqueous solution to produce either hydrophobically (sCA = 105°) or hydrophilically (sCA = 10°) functionalized surfaces. The adlayers formed in this way are stable for at least 12 h of water exposure, as confirmed by XPS and VASE. Static water contact angle measurements show that the adlayer functionality is maintained after the exposure.

PFAND adsorption onto TiO_2 follows clearly reproducible Langmuir-type adsorption kinetics, such that PFAND-based one- and (ND-backfilled) two-component hydrophobicity gradients on TiO_2 can be produced by utilizing a simple dip-and-rinse technique.[42] This is the first example of catechol-based surface-chemical gradients. XPS and VASE data show that surface coverage on the opposing ends of such gradients differs by a factor of 5−7. Static water contact angles are linearly dependent on the position along the gradients over a range of 35°−95° when a logarithmic time-dependence immersion protocol is employed.

Nitrogen 1s XP spectra indicate that PFAND as well as ND bind to the TiO_2 in more than a single configuration. From the literature, it is known that the different binding configurations of catechol derivatives to (amorphous) TiO_2[64,68] vary in bond strength and can thus lead to different stabilities of the adlayers.

The adsorption of catechol-bound self-assembled monolayers appears to be a powerful approach to the fabrication of water-stable surface-chemical gradients. Of particular technological significance is the compatibility of these systems with oxide substrates, such as titanium dioxide.

Acknowledgment. We thank Michael Horisberger, PSI Villigen, for providing substrate coatings. The work was funded by the Swiss National Science Foundation (Project 200020-116150).

Supporting Information Available: Additional experimental results and figures. This material is available free of charge via the Internet at http://pubs.acs.org.

(68) Sever, M.; Wilker, J. *Dalton Trans.* **2004**, 1061.

2b. Surfaces functionalized with polymer brushes for lubrication

Commentary

We spent a number of years investigating the tribological consequences of using poly(L-lysine)-*g*-polyethylene glycol (PLL-*g*-PEG) to attach chains of polyethylene glycol to a surface in an aqueous environment, forming a polymer brush. The motivation was to imitate the kind of lubricating structures that are often encountered in nature (2.15). The molecules attach to negatively charged surfaces (such as oxide layers on metals or ceramics (2.18)) via the positive charges on the protonated amine groups on the polylysine backbone. The adsorption of PLL-*g*-PEG turned out to be a very effective way of lubricating in aqueous environments, despite water's intrinsically poor lubricating properties, especially at low speeds. We used macroscopic methods to examine the lubricating effects of the PLL-*g*-PEG, collaborating with Hugh Spikes and his colleagues at Imperial College to make measurements under rolling conditions (2.9). The system worked particularly well under rolling, presumably due to the lower shear stresses leading to less removal of the polymer from the surface than under sliding conditions. While this removal happens under most conditions, it appears to be immediately followed by a rehealing process (2.16), as the molecules in solution readsorb on the vacated surface adsorption sites. This could be readily followed by fluorescence microscopy (2.13). In collaboration with Scott Perry, at the University of Florida, we also found that by changing the architectural parameters of the polymer — such as backbone length, PEG-chain length, and chain spacing — changes in the lubricating properties could be observed. It was found that longer PEG chains appeared to lubricate better, but that changing the spacing between the chains on the backbone had the effect of both increasing chain density and reducing the efficacy of chain attachment (since anchoring groups were replaced by PEG chains), and thus had to be undertaken with caution (2.10, 2.21).

Measurements of the PEG brush thickness in an aqueous environment by optical and gravimetric methods enabled us to determine that they typically contain around 80% water, with this proportion decreasing as the quality of the solvent is decreased (replacing water with methanol or higher alcohols, for example). A good correlation between the solvent quality and the resulting friction (as measured by AFM) was found (2.11).

We also found that the multiple lysine side chains in the PLL-*g*-PEG molecule contained sufficient methylene groups in total, to allow a substantial hydrophobic interaction with many polymer surfaces, allowing PEG brushes to be formed and effective lubrication to take place (2.12, 2.14). Furthermore, we found that by adding glycerol to the PLL-*g*-PEG solution, we obtained a lubricant with useful properties over a much broader range of speed and load conditions than that obtainable with the polymer alone (2.20).

While PLL-*g*-PEG proved to be an interesting and potentially useful lubricant additive, we began to examine some alternatives that overcame some of the drawbacks of this molecule. One option was to replace the PEG with dextran, which appeared to work well and had a significant economic advantage (2.17). Another was to move away from grafting molecules to the surface in the direction of surface-induced polymerization (SIP) (2.19), which has opened up a rich new area of investigation for us.

Tribology Letters, Vol. 15, No. 3, October 2003 (© 2003) 231

Boundary lubrication of oxide surfaces by Poly(L-lysine)-*g*-poly(ethylene glycol) (PLL-*g*-PEG) in aqueous media

Seunghwan Lee[a], Markus Müller[a], Monica Ratoi-Salagean[b], Janos Vörös[a], Stéphanie Pasche[a], Susan M. De Paul[a], Hugh A. Spikes[b], Marcus Textor[a], and Nicholas D. Spencer[a,*]

[a]*Laboratory for Surface Science and Technology, Department of Materials, ETH-Zürich, Sonneggstrasse 5, Zürich, CH-8092, Switzerland*
[b]*Tribology Section, Department of Mechanical Engineering, Imperial College, London, SW7 2BX U.K.*

Received 15 December 2002; accepted 10 March 2003

In this work, we have explored the application of poly(L-lysine)-*g*-poly(ethylene glycol) (PLL-*g*-PEG) as an additive to improve the lubricating properties of water for metal-oxide-based tribo-systems. The adsorption behavior of the polymer onto both silicon oxide and iron oxide has been characterized by optical waveguide lightmode spectroscopy (OWLS). Several tribological approaches, including ultra-thin-film interferometry, the mini traction machine (MTM), and pin-on-disk tribometry, have been employed to characterize the frictional properties of the oxide tribo-systems in various contact regimes. The polymer appears to form a protective layer on the tribological interface in aqueous buffer solution and improves both the load-carrying and boundary-layer-lubrication properties of water.

KEY WORDS: poly(L-lysine)-*g*-poly(ethylene glycol) (PLL-*g*-PEG), aqueous lubrication, boundary lubrication

1. Introduction

Water possesses several characteristics that are desirable for its use as a lubricant, such as being environmentally friendly, advantageous for heat management and economical. Synovial joints, such as hip, knee, shoulder, ankle and finger joints, can display friction coefficients, μ, that are less than 0.003 [1–4]. However, extremely low friction in natural joint systems cannot be achieved by water alone, due to its inability to form useful boundary films and its extremely low pressure-coefficient of viscosity [1–4]. The latter has serious consequences for elastohydrodynamic lubrication with water and thus limits the load-carrying capacity of water-lubricated tribo-systems. Nature deals with these issues by using "smart", pressure-responsive cartilage surfaces as sliding partners [4]. In efforts to use water as a lubricant in engineering systems, several approaches are employed to modify the lubricating properties of water, including oil-in-water (O/W) emulsions [5–7] or aqueous surfactants [8–11]. In both approaches, the formation of a protective film, either instantaneously or irreversibly, is known to be responsible for a lubrication effect.

In recent years, some attention has been paid to end-grafted polymers that exhibit unique tribological properties in organic media [12–18]. In "good" solvents, end-grafted polymers are strongly stretched and keep the polymer-bearing interfaces apart during sliding, while maintaining a relatively fluid layer at the interface. This mechanism strongly contrasts with the sliding of a rigid interface where surfaces are brought into close contact and the solvent molecules squeezed out during sliding. Coefficients of friction as low as 0.001 have been observed under light compression of polymer-grafted interfaces, although higher friction forces were observed in a higher-pressure regime [13,14]. These observations hint that brush-like polymers may provide an alternative approach for improving the lubrication properties in an aqueous environment, providing that their chemical and structural properties are properly designed.

We employed poly(L-lysine)-*g*-poly(ethylene glycol) (PLL-*g*-PEG) as an additive to improve the lubricating properties of water for metal-oxide-based tribo-systems. PLL-*g*-PEG has been extensively investigated in previous studies for use in a wide array of biomedical applications [19–23]. As shown in figure 1, this co-polymer is composed of a polycationic PLL backbone and a non-reactive PEG side chain. The PLL backbone is positively charged at pH $\leq$ 10 due to protonation of the primary amine groups, and readily adsorbs onto a negatively charged surface, mainly through electrostatic interactions. Many metal oxide surfaces display iso-electric points (IEP) that lie below pH 10 [24], and thus can be effectively coated by this polyelectrolyte in aqueous conditions. Some of the NH_2 functional groups of the PLL are used for coupling PEG side-chains, thus resulting in hydrophilic, flexible, brush-like side chains.

In this work, we explore the potential use of PLL-*g*-PEG as an additive to improve the lubrication properties of water for tribo-systems composed of metal

* To whom correspondence should be addressed. E-mail: nspencer @surface.mat.ethz.ch

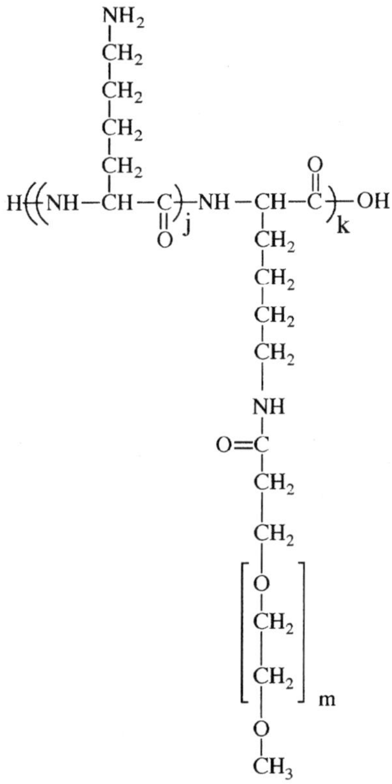

Figure 1. A schematic illustration of PLL-*g*-PEG.

oxides. For this purpose, several macroscopic tribological approaches, including ultra-thin film interferometry, the mini traction machine (MTM), and pin-on-disk tribometry, representing a pure rolling, mixed sliding/rolling, and pure sliding respectively, all in a non-conformal sphere-on-plane geometry, have been employed.

2. Materials and methods

2.1. PLL(10)-g[2.9]-PEG(2)

PLL(10)-*g*[2.9]-PEG(2) denotes a graft co-polymer with a PLL backbone of molecular weight 10 kDa, a grafting ratio of (lysine-mer)/(PEG side chains) of 2.9, and a PEG molecular weight 2 kDa.

The PLL-*g*-PEG co-polymer has been synthesized according to a method described in previous publications [22,23]. Briefly, poly(L-lysine) hydrobromide (16 kDa including HBr, Sigma, St. Louis, MO, USA) was dissolved at a concentration of 100 mM in sodium borate buffer solution (50 mM) adjusted to pH 8.5. The solution was filter sterilized (0.22 μm pore-size filter). For the grafting of PEG onto PLL, the *N*-hydroxysuccinimidyl ester of methoxypoly(ethylene glycol) propionic acid (mol. wt. 2 kDa, mPEG-SPA, Shearwater Corporation, Huntsville, AL, USA) was added to PLL solution. The reaction was allowed to proceed for 6 h at room temperature, after which the reaction mixture was dialyzed (Spectra-Por, mol. wt. cutoff size 6–8 kDa, Spectrum, Houston, TX, USA) for 48 h against deionized water. The product was freeze-dried and stored at −20 °C. Detailed analytical information of the product produced using this method is available in previous publications [22,23].

The PLL(10)-*g*[2.9]-PEG(2) was dissolved in 10 mM HEPES (4-(2-hydroxyethyl)piperazine-1-ethanesulfonic acid (Sigma, St. Louis, MO, USA), adjusted to pH 7.4 with 1.0 M NaOH solution) with a concentration of 0.25 mg/ml and this solution was used for all subsequent experiments. For cases where the comparison of the PLL(10)-*g*[2.9]-PEG(2) with PLL(10) alone is needed, PLL(10) solution was prepared by dissolving the poly(L-lysine) hydrobromide (mol. wt. 16 kDa including HBr, Sigma, St. Louis, MO, USA) in 10 mM HEPES with a concentration of 0.045 mg/ml to maintain the concentration of PLL constant in both PLL and PLL-*g*-PEG experiments.

2.2. Optical waveguides lightmode spectroscopy (OWLS)

Optical Waveguide Lightmode Spectroscopy (OWLS) was employed to investigate the adsorption behavior of the polymers on the tribo-pair surfaces. OWLS is based on grating-assisted in-coupling of a He-Ne laser into a planar waveguide, and allows the direct online monitoring of macromolecule adsorption. This method is highly sensitive up to a distance of 100 nm above the surface of the waveguide (sensitivity limit, ~1 ng/cm^2). Furthermore, a measurement-time resolution of 3 s allows for the *in situ*, real-time study of adsorption kinetics. The waveguide chips used for OWLS measurements were purchased from MicroVacuum Ltd. (Budapest, Hungary) and consisted of a 1-mm-thick AF 45 glass substrate and a 200-nm-thick $Si_{0.75}Ti_{0.25}O_2$ waveguiding layer at the surface. A silicon oxide (ca. 12 nm) or iron oxide layer (ca. 1 nm) was sputter coated on top of the waveguiding layer in a Leybold dc-magnetron Z600 sputtering unit. The coating conditions and the principles of OWLS investigations have been described in detail elsewhere [25–27].

All OWLS experiments were carried out in a BIOS-I instrument (ASI AG, Zürich, Switzerland) using a Kalrez (Dupont, Wilmington, DE, USA) flow-through cell (8 × 2 × 1 mm) [25]. The solution exchange was carried out by syringe injection (1 ml within 10 s). Adsorbed mass density data were calculated according to de Feijter's formula from the adsorbed layer thickness and refractive index values from the mode equations [28]. A refractive index increment (dn/dc) value of 0.169 cm^3/g, as determined in a refractometer (Carl

Zeiss, Jena, Germany), was used for the calculation of PLL(10)-*g*[2.9]-PEG(2) adsorption.

2.3. Ultra-thin film interferometry

Ultra-thin film interferometry is a unique approach to measuring the film thickness under a lubricated contact [29]. The set-up is schematically illustrated in figure 2. A lubricated contact is formed between a loaded steel ball and the flat surface of a glass disk. The disk is driven with respect to the nominal axis in contact with the ball and thus can drive the ball in pure rolling contact. The surface of the glass disk is coated, first with a very thin, semi-reflecting layer of chromium, on top of which is a 400-nm layer of transparent silica, as a spacer layer. Thus, the ultra-thin film interferometry tribo-pair in this work represents an FeO_x/SiO_x interface. White light is shone through the glass into the contact, where a proportion is reflected back from the chromium layer while the remainder passes through the silica layer and the polymer-containing water before being reflected from the steel ball surface. Since the two beams have traveled different distances, they interfere at wavelengths that depend on the path difference, i.e. on the sum of silica and lubricant film thickness. The interference pattern from the contact is passed to a spectrometer and the resultant dispersed light analyzed to determine the precise wavelength of maximum constructive interference from the center of the contact. This yields an accurate measure of the composite film thickness, from which the spacer layer can be subtracted.

The refractive index of the lubricant solution is a necessary input for the calculation of the film thickness. The refractive index of the polymer-containing buffer solution is expected to be different from that of pure water (1.33) for two reasons. Firstly, addition of polymers into water results in an increase of the refractive index of the lubricant solution, which was already described in section 2.2. The same incremental value of $0.169\,cm^3/g$ (dn/dc) was considered here. Secondly, the increased pressure at the contact point also results in an increase of the refractive index of the lubricant solution. However, in previous studies, measurement by two-angle interferometry showed that an increase in pressure from atmospheric to 1 GPa alters the refractive index of oil by less than 10% [29–31]. We assume that this is applicable to polymer solutions as well. The refractive index of the polymer solutions is thus calculated to lie between 1.34 (negligible pressure effect) and 1.47 (10% increase by pressure effect). As a compromise, a refractive index of 1.4 has been used for the calculation. As the film thickness is inversely proportional to the refractive index [29], the uncertainty of the film thickness in this work corresponds to $\pm 5\%$.

All the measurements were carried out under a fixed load (20.0 N) and temperature ($\sim 25\,°C$) and along the same track of the disk. The radii of the ball and disk tracks are 9.5 and 17.75 mm, respectively. The roughness values (R_a) of the ball and disk tracks are 11 and 5 nm respectively. For smooth surface contact, the maximum contact pressure in this configuration is, according to Hertz equations, 0.51 GPa (see table 1). The separation of the surfaces by a thin lubricating film will not significantly change this pressure. In practice, however, because of the finite roughness of the surfaces, there will be local variations from the Hertz pressure distribution at an asperity scale. A thin boundary or EHL film may significantly reduce these variations.

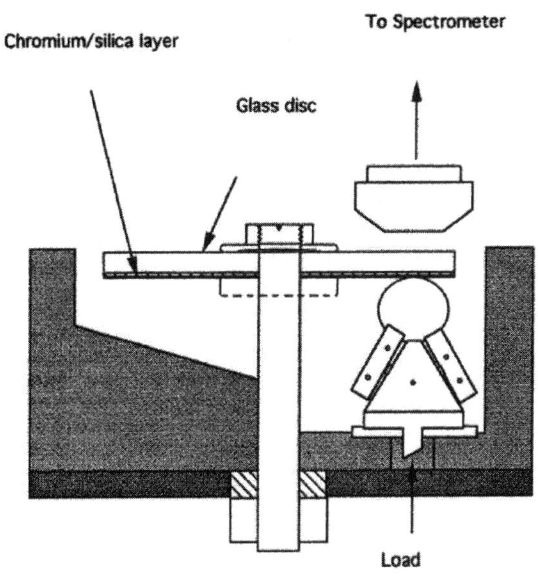

Figure 2. A schematic set-up of ultra-thin film interferometry.

2.4. Mini traction machine (MTM)

A mini traction machine (MTM, PCS Instruments, London, U.K.) was employed to characterize the lubrication properties of the polymers in a mixed sliding/rolling contact regime. The set-up of the instrument is schematically illustrated in figure 3. A lubricated contact is formed between a polished steel ball and a flat steel surface, thus representing an FeO_x/FeO_x interface. In contrast to the other instruments employed in this work, which provide either a pure rolling contact (ultra-thin film interferometry, section 2.3) or a pure sliding contact (pin-on-disk tribometer, section 2.5), MTM provides a mixed sliding/rolling contact through the independent control of the ball and disk velocities. The slide/roll ratio, SRR, is defined as the percentage ratio of the *difference* and the *mean* of the ball velocity (u_{ball}) and disk velocity (u_{disk}), i.e.

$$SRR = \frac{|u_{\text{ball}} - u_{\text{disk}}|}{\left(\frac{u_{\text{ball}} + u_{\text{disk}}}{2}\right)} \times 100\%.$$

Table 1
The calculated maximum contact pressure (Hertzian) for the configuration (sphere on plane) used in this work (maximum contact pressure, $P_0 = 1/\pi \cdot \sqrt[3]{6WE^{*2}/R^2}$ where W = load, R = radius of the balls and $E^* = \left((1-\nu_1^2)/E_1 + (1-\nu_2^2)/E_2\right)^{-1}$ where E_1 and ν_1 are Young's modulus and Poisson ratio of the ball and E_2 and ν_2 are Young's modulus and the Poisson ratio of the disk respectively. E_{steel} = 203 GPa, E_{silica} = 72 GPa, ν_{steel} = 0.3 and ν_{silica} = 0.2 [32]).

Instrument	Tribo-pair (sphere/plane)	R	W	P_0
Ultra-thin film interferometry	silica/steel	9.5 mm	20.0 N	0.51 GPa
MTM	steel/steel	9.5 mm	3.0 N	0.04 GPa
	steel/steel	9.5 mm	20.0 N	0.81 GPa
Pin-on-disk	silica glass/steel	3 mm	0.5 N	0.32 GPa
	silica glass/steel	3 mm	1.0 N	0.41 GPa
	silica glass/steel	3 mm	2.0 N	0.51 GPa
	silica glass/steel	3 mm	5.0 N	0.70 GPa
	steel/steel	3 mm	2.0 N	0.81 GPa

Thus, SRR = 0% (i.e. $u_{ball} = u_{disk}$) represents a pure rolling contact and SRR = 200% (i.e. either u_{ball} or $u_{disk} = 0$) represents a pure sliding contact, while the values between 0 and 200% represent a mixed sliding/rolling contact. With the given software (PCS Instruments, MTM version 1.0), values of 1 to 200% of SRR were accessible. Under a given SRR, the coefficient of friction (dF/dN) is measured as a function of *mean* velocity. To investigate the lubrication behaviors of the polymers at different slide/roll ratios, the coefficient of friction versus mean velocity plots were obtained at different SRRs.

All the friction measurements were performed under a fixed load (3.0 or 20.0 N) and temperature (25 °C) and along the same ball and disk tracks; the radii of the ball and disk tracks are 9.5 and 20.7 mm, respectively. The roughness values (R_a) of the ball and disk are 11 and 10 nm respectively. However, different pairs of the ball/disk tracks were used for the comparison of the friction measurement in the absence versus presence of the polymers in buffer solution, due to significant wear of the tribo-pair in the former case. The calculated maximum contact pressures are 0.86 and 0.46 GPa for 20.0 and 3.0 N, respectively (see table 1).

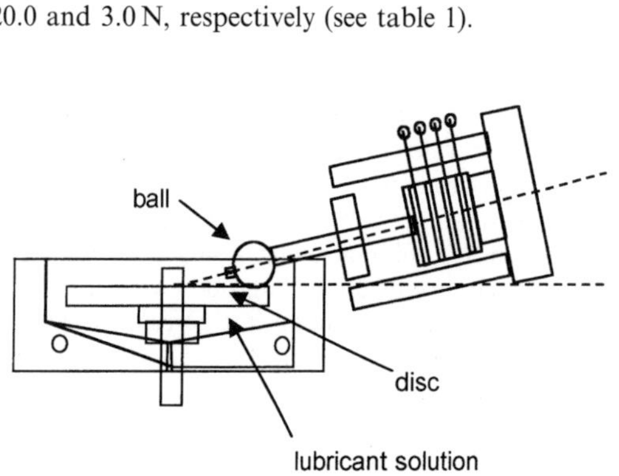

Figure 3. A schematic set-up of the mini traction machine (MTM).

2.5. Pin-on-disk tribometry

A pin-on-disk tribometer (CSM, Neuchâtel, Switzerland) was employed to characterize the friction and lubrication properties of aqueous PLL(10)-*g*[2.9]-PEG(2) solutions in the pure sliding contact regime. In this set-up, a fixed, spherical steel ball (6 mm in diameter, DIN 17 230, Hydrel, Romanshorn, Switzerland) is in contact with a flat disk. Both glass (SuperFrost, soda-glass, Menzel-Gläser, Braunschweig, Germany) and steel (DIN 17 230, XXX, Switzerland) were selected for the disk, thus representing the FeO_x/SiO_x and FeO_x/FeO_x pairings. The roughness values (R_a) for steel ball, steel disk, and glass disk are 32, 5, and 5 nm respectively. The load was controlled by placing dead weights on top of the ball holder (0.5 to 5.0 N) and the friction forces were measured by a strain gauge. The friction signals were recorded with a Macintosh Power PC using Labview and an ADC card of the MIO family (both from National Instruments, Austin, TX, USA). To investigate the lubricating properties of the polymer, both load and velocity dependence of the frictional properties were explored. For each friction measurement, different fresh tracks were selected. For this reason, the comparison of the frictional properties with or without polymers could be performed with exactly the same set-up of the tribo-pair. All the measurements were performed at room temperature (~25 °C). The calculated maximum contact pressures are presented in table 1.

2.6. Cleaning of the tribo-pairs

All the instrumental parts that are expected to be coated with the polymer, including the waveguides for the OWLS and the balls/disks of all tribological instruments, were oxygen-plasma cleaned for ca. 2 min in a Harrick Plasma Cleaner/Sterilizer PDC-32G instrument (Ossining, NY, USA) prior to the measurements. Before the plasma cleaning, the instrumental parts were rinsed with copious amounts of water or organic solvents as follows: The waveguides of the OWLS were sonicated in 0.1 M HCl for 10 min, extensively rinsed with ultra-high-purity water, and dried under nitrogen; the ball/disk and the assembly parts of ultra-thin film interferometry and MTM were degreased by sonication in toluene for 10 min, followed by sonication in iso-propanol for 10 min, and subsequently dried under nitrogen; the balls and disks of pin-on-disk and assembly parts were sonicated in ethanol for 10 min and dried under nitrogen.

3. Results and discussion

3.1. *Adsorption behavior of PLL(10)-g[2.9]-PEG(2) on silicon oxide and iron oxide substrates*

In figure 4, representative OWLS adsorption measurements for the PLL(10)-*g*[2.9]-PEG(2) on (a) silicon oxide and (b) iron oxide are presented. All measurements were carried out *in situ* in a flow-through cell without an intermittent drying stage.

As expected, the results of OWLS experiments indicate that the PLL(10)-*g*[2.9]-PEG(2) spontaneously adsorbed from a pH 7.4 HEPES (10 mM) buffered aqueous solution (0.25 mg/ml) onto both oxide surfaces. A PLL-*g*-PEG layer of areal density of approximately 120 ng/cm^2 and 60 ng/cm^2 was formed on silicon oxide and iron oxide, respectively. The difference in areal density for the two substrates is understood by considering the difference in IEPs of the two oxide substrates and thus the corresponding differences in charge density at a given pH. In previous studies involving several other metal oxides, a lower IEP for a given substrate has been correlated with a higher areal density of PLL-*g*-PEG [22,23]. The lower IEP of silicon oxide (~ 2) when compared with iron oxide (6–9) is consistent with the higher areal density of PLL-*g*-PEG adsorbed on the silicon oxide. In both cases, 95% of the final mass was reached within the first 5 min. There was no apparent desorption following rinsing with buffer solution on the timescale of the experiment. All tribological experiments involving the polymers were carried out 30 min after the tribo-pair was immersed in polymer-containing solution, to ensure adequate time for adsorption.

As described in the experimental section, the tribo-pairs used in this work were prepared in different ways from the substrates (waveguides) used in the OWLS experiment, although all are either silicon oxide- or iron oxide-based materials. It is thus noted that the actual amount of the adsorbed polymers on each tribo-pair may be slightly different from those shown in figure 4.

3.2. *Lubrication properties in rolling contact*

The lubrication properties of the PLL(10)-*g*[2.9]-PEG(2) were characterized under pure rolling contact conditions by employing ultra-thin-film interferometry. In contrast to the other tribological approaches in this work, which determine the interfacial friction forces, ultra-thin film interferometry uniquely measures the film thickness of a lubricated contact. In this work, the film-thickness measurements were first performed under polymer-free buffer solution. No meaningful data were obtained, due to significant wear of the contact in this condition. Only in a high-velocity regime (> 1000 mm/s) has an interfacial film involving pure water as a lubricant been reported [8]. However, following addition of the polymers to the buffer solution, the measurement of the lubricant-film thickness was reproducibly achieved over a wide range of

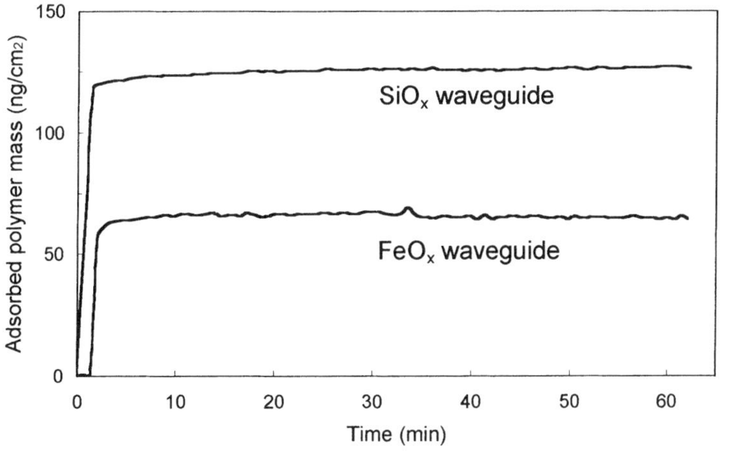

Figure 4. Representative adsorption profiles of PLL(10)-*g*[2.9]-PEG(2) onto (a) silicon oxide and (b) iron oxide substrates as measured by OWLS (buffer solution = 10 mM HEPES (pH 7.4), concentration of the polymer = 0.25 mg/ml; T = 25 °C).

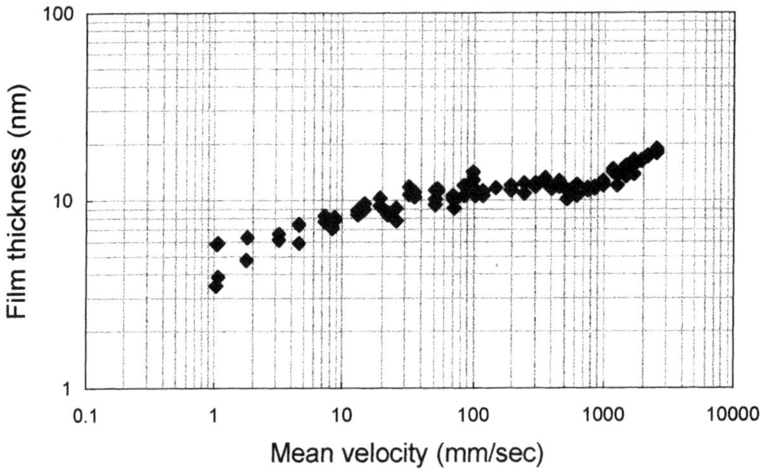

Figure 5. Log (film thickness) versus log (velocity) plot by ultra-thin-film interferometery (ball = AISI 52100 steel (19 mm in diameter), disk surface = sputtered silica, buffer solution = 10 mM HEPES (pH 7.4), concentration of the polymers = 0.25 mg/ml, load = 20.0 N, T = 25 °C).

velocities. The results are shown as a log (film thickness) versus log (velocity) plot in figure 5.

The lubricant-film formation in the presence of the polymer in figure 5 exhibits three velocity regimes: (i) velocity range lower than ~30 mm/s, where the film thickness gradually decreases with decreasing velocity, (ii) velocity range from ~30 to ~1000 mm/s, where the measured film thickness is nearly constant (thickness 11.4 ± 1.0 nm in average), and (iii) velocity range higher than ~1000 mm/s, where the film thickness increases with increasing velocity (slope ~0.4 in log–log plot). The nearly constant film thickness obtained over a wide range of velocities (~30 to 1000 mm/s) suggests that a boundary-layer-lubrication mechanism is active in this regime. The high-velocity region where film thickness increases with increasing velocity is similar to that previously observed for pure water and probably represents the hydrodynamic entrainment of water into the contact. The velocity-dependent behavior of the lubricant-film thickness for the PLL(10)-*g*[2.9]-PEG(2)-containing buffer solution in the low-velocity regime is qualitatively similar to that of surfactant (sodium olefin (C16) sulfonate)-containing water, reported in a previous study [8]. However, a relatively thicker layer is observed in this work (~ 11 nm versus ~3 nm), attributed to the brush-like structure of this polymer.

3.3. *Lubrication properties in mixed sliding/rolling contact*

The lubrication properties of PLL(10)-*g*[2.9]-PEG(2) solutions have been characterized in a mixed sliding/rolling contact regime by means of MTM. As described in the experimental section, a mixed sliding/rolling contact was achieved by independent driving of the ball and disk, to form a lubricated contact. In this work, the measurement was initially performed under 20.0 N load. Regardless of the slide/rolling ratio, the friction forces were very high and irreproducible. Furthermore, no noticeable reduction of friction forces was observed by addition of the polymers into buffer solution, which is in contrast to the results shown in section 3.2. This may be due partly to the higher contact pressure for the FeO_x/FeO_x pair (max. 0.86 GPa) compared to the FeO_x/SiO_x pair (max. 0.51 GPa), and partly to the lower areal density of PLL-*g*-PEG on the iron oxide surface, as measured by OWLS in section 3.1. Thus, the measurements were carried out under a lower-pressure regime by applying a load of 3.0 N (max. 0.46 GPa). The coefficient of friction was measured as a function of velocity at three different slide/roll ratios, SRR = 50 and 100% for comparison. Due to an instrumental resolution issue, measurements were only able to be made in the velocity range shown in figure 6.

For SRR = 50%, apparently higher values of friction were observed in pure buffer solution than the polymer-containing solution. Furthermore, monotonically decreasing frictional properties (from $\mu = 0.43$ to $\mu = 0.22$) as a function of increasing velocity were observed. The polymer-containing solution, on the other hand, showed similar velocity-dependent frictional behavior, but with clearly lower coefficients of friction ($\mu = 0.25$ to $\mu = 0.12$) within the same velocity range. The frictional properties at SRR = 100% were observed to be similar to those at SRR = 50%. The coefficient of friction reduced from 0.42 to 0.26 in buffer solution to 0.26 to 0.09 in the presence of the polymers. It is noted that all measurements were performed on the same disk track, although different balls and disks were used for measurements with or without polymers.

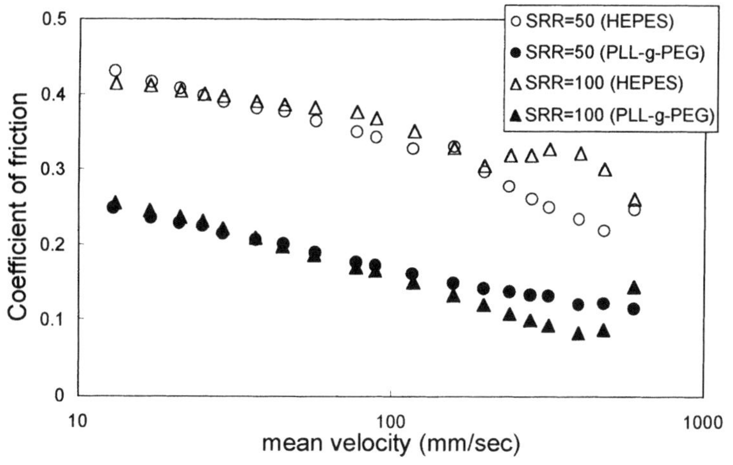

Figure 6. Log (coefficient of friction) versus log (velocity) plot by MTM slide/roll ratio (SRR) = 50% and 100%. Empty symbols are for buffer solution only and filled symbols are for the polymer in buffer solution (ball = AISI 52100 steel (17.4 mm in diameter), substrate = AISI 52100 steel, buffer solution = 10 mM HEPES (pH 7.4), concentration of the polymer = 0.25 mg/ml, load = 3.0 N, T = 25 °C).

Due to the relatively narrow range of velocities available, it is not trivial to determine what lubrication regime is activated for the mixed sliding/rolling contact in this work. However, the results above show that the PLL(10)-*g*[2.9]-PEG(2) can effectively reduce the frictional properties at mixed sliding/rolling contact of FeO_x/FeO_x in buffer solution.

3.4. Lubrication properties in sliding contact

Finally, the lubrication properties of the PLL(10)-*g*[2.9]-PEG(2) were characterized in a pure sliding-contact regime by employing pin-on-disk tribometry. Stainless steel was selected as ball and both glass and steel were selected as disks. The lubrication properties of the PLL(10)-*g*[2.9]-PEG(2) were investigated by comparing the frictional properties of the tribo-pair in the presence and absence of the polymers in buffer solution. In this experiment, the lubrication properties of PLL(10) were also included for comparison.

Firstly, the load dependence of the frictional properties for the steel/glass tribo-pair was investigated. In this experiment, the friction forces were measured on fresh tracks for each load (0.5 to 5.0 N), while keeping the rpm constant and the total number of revolutions below 5. The resultant velocity was within the range of ~ 0.2 to ~ 0.8 mm/s, which is close to the lowest value achieved with the tribometer. As will be shown below, the frictional properties were virtually independent of the slight variation of the velocity in that range. The results are shown in figure 7. In all cases, the frictional forces are fairly linear with increasing applied load. It is also noticeable that the friction forces of the steel/glass pair are significantly reduced by addition of the polymers into the buffer solution. The coefficient of friction of the steel/glass tribo-pair, obtained by taking the slope of friction versus load plots ($\Delta F/\Delta L$ ($L = 0.5$ to 5.0 N)) decreased from 0.34 to 0.15 by the addition of PLL(10), and further decreased to 0.09 by the addition of PLL(10)-*g*-[2.9]-PEG(2).

Secondly, the velocity dependence of the frictional properties for the steel/glass tribo-pair was investigated. For this experiment, the friction forces were also measured on fresh tracks for each velocity in the range of ~ 0.1 to ~ 400 mm/s. All the measurements were performed under a fixed load (2.0 N) and the total number of revolutions was extended to 100. The results are shown in figure 8. The friction versus velocity plots in figure 8 reveal that there are two distinct velocity regimes in which the frictional properties of the tribo-pair in the presence versus absence of the polymers in buffer solution can be compared. In the velocity regime lower than ~40 mm/s, the frictional properties of the tribo-pair can clearly be distinguished in magnitude in decreasing order, *buffer solution only* ≥ *PLL(10)* > *PLL(10)*-g*[2.9]-PEG(2)*, which is consistent with the load-dependent frictional properties shown in figure 7. In the velocity regime higher than ~40 mm/s, however, the difference between the lubricants diminishes and the frictional properties of the tribo-pair in buffer solution is not significantly influenced by the presence of the polymers. Throughout the entire velocity range, the sliding of the tribo-pair in the presence of buffer solution only or with PLL(10) appears to decrease with an increase of the velocity. Meanwhile, the corresponding sliding in the presence of PLL(10)-*g*[2.9]-PEG(2) appears to exhibit a virtually constant coefficient of friction (0.062 ± 0.019 on average). This is highly indicative of boundary-layer formation by the PLL(10)-*g*[2.9]-PEG(2) on these oxide surfaces in aqueous buffer solution. For both load- and velocity-

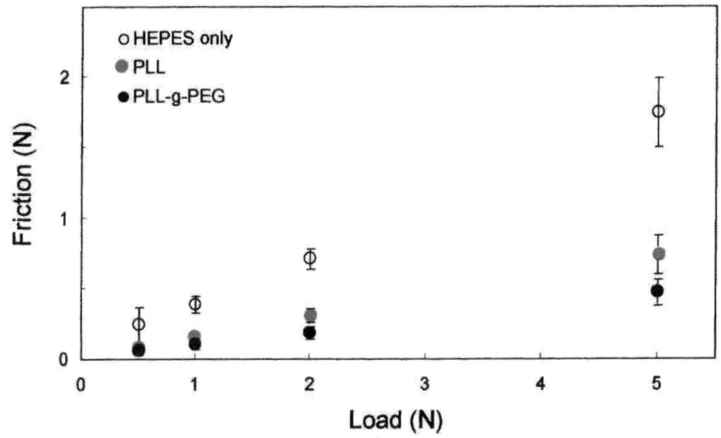

Figure 7. Friction force versus load plot (pin-on-disk tribometer) of buffer solution only (○), PLL(10) in buffer solution (◉), and PLL(10)-*g*[2.9]-PEG(2) in buffer solution (●) (ball = steel (6 mm in diameter), disk = glass, buffer solution = 10 mM HEPES (pH 7.4), concentration of the polymer = 0.25 mg/ml, load = 0.5 – 5.0 N, T = 25 °C).

dependence measurements, it appears that the lower frictional properties of the PLL(10)-*g*[2.9]-PEG(2) compared with PLL(10) indicate a significant role of the brush-like PEG side chains in lubrication of the tribo-interfaces, as suggested in studies on end-grafted polymers in organic solvents [12–18].

Finally, the load and velocity dependence for pure sliding of the FeO_x/FeO_x tribo-pair has been investigated. The coefficient of friction obtained from the slope of the friction versus load plots ($\Delta F/\Delta L$ (L = 0.5 to 5.0 N)) after 5 rotations decreased from 0.30 to 0.14 upon addition of polymer. However, the coefficient of friction, dF/dN (N = 2.0 N) that was obtained after 100 rotations for each velocity, which remained virtually constant over the range of ~1 to ~100 mm/s, was not significantly reduced upon addition of polymer ($\mu = 0.31 \pm 0.05$ without polymer to $\mu = 0.27 \pm 0.03$ with polymer on average). The reduced lubrication effect of PLL(10)-*g*[2.9]-PEG(2) for the sliding of a FeO_x/FeO_x tribo-pair (with respect to FeO_x/SiO_x tribo-pair) is thought to be mainly correlated with the relatively lower amount of polymer adsorption (see figure 4).

4. Summary

The lubrication properties of PLL(10)-*g*[2.9]-PEG(2) (PLL-*g*-PEG) as an additive for tribo-systems composed of silicon oxide and iron oxide in an aqueous environment have been characterized employing several macroscopic tribological approaches. The tribological properties of tribo-pairs lubricated by a PLL-*g*-PEG-containing aqueous buffer solution have been investigated in various dynamic contact regimes encountered

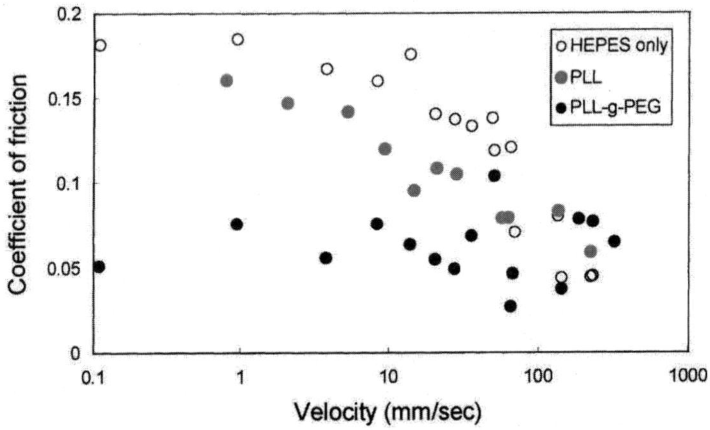

Figure 8. Log (coefficient of friction) versus log (velocity) plot by pin-on-disk tribometer of buffer solution only (○), PLL(10) in buffer solution (◉), and PLL(10)-*g*[2.9]-PEG(2) in buffer solution (●) (ball = steel (6 mm in diameter), disk = glass, buffer solution = 10 mM HEPES (pH 7.4), concentration of the polymer = 0.25 mg/ml, load = 2.0 N, T = 25 °C). In this plot, the error bars, typically 0.05 for buffer solution only and 0.02 for the polymers, are omitted for clarity.

in engineering systems: (1) pure rolling contact of a FeO_x/SiO_x tribo-pair, as measured by ultra-thin film interferometry, showed that the polymer-containing buffer solution forms a stable lubricant film (11.4±1.0 nm on average) over a wide velocity range (~ 30 to 1000 mm/s); (2) the mixed sliding/rolling contact of an FeO_x/FeO_x tribo-pair measured by MTM showed that the friction forces were reduced by approximately one-half upon addition of PLL-*g*-PEG when the slide/roll ratio is 50 or 100% in a low-pressure regime; (3) pure sliding contact of an FeO_x/SiO_x tribo-pair, measured by a pin-on-disk tribometer, showed a significant reduction of friction (remains constant at 0.06±0.019 on average for ~ 0.1 to 400 mm/s), while less effective lubrication was observed upon the sliding contact of an FeO_x/FeO_x tribo-pair.

The effectiveness of boundary lubrication by PLL(10)-*g*[2.9]-PEG(2) in aqueous buffer solution is very apparent in relatively low velocity regimes, where lubrication by water alone is practically impossible due to its extremely low pressure-coefficient of viscosity and poor film-forming properties. The relative adsorption behavior of the polymer onto SiO_x and FeO_x surfaces, as investigated by OWLS (~ 120 ng/cm^2 for SiO_x and ~ 60 ng/cm^2 for FeO_x surfaces), seems to explain the relatively less effective lubrication for FeO_x/FeO_x compared with the FeO_x/SiO_x tribo-pair. In summary, the PLL(10)-*g*[2.9]-PEG(2) appears to form a protective layer both on silicon oxide and iron oxide surfaces, thus effectively improving load-carrying and boundary lubrication properties of water for a variety of dynamic contact regimes.

Acknowledgments

This work was financially supported by the Council of the Swiss Federal Institutes of Technology (ETH-Rat TopNano 21) and National Research Program 47 of the Swiss National Science Foundation. We are also grateful to Dr. Philippa Cann and Ms. Ksenia Toplovec Miklozic of Imperial College for their valuable advice and assistance, as well as Michael Horisberger (Paul Scherrer Institute, Villigen, Switzerland) for the sputter coating.

References

[1] G.D. Jay, K. Haberstroh and C.-J. Cha, J. Biomed. Mat. Res. 40 (1998) 414.

[2] I.M. Schwarz and B.A. Hills, British J. Rheum. 37 (1998) 21.

[3] B.A. Hills, Proc. Instrn. Mech. Engrs. 214 (2000) 83.

[4] M. Scherge and S.N. Gorb, *Biological Micro- and Nano-tribology* (Springer, Berlin, 2001).

[5] M. Ratoi-Salagean, H.A. Spikes and R. Higendoorn, Proc. Instrn. Mech. Engrs. 211 (1997) 195.

[6] M. Ratoi-Salagean, H.A. Spikes and H.L. Rieffe, Tribol. Trans. 40 (1997) 569.

[7] S.R. Schmid and W.R.D. Wilson, Lub. Eng. 52 (1995) 168.

[8] M. Ratoi-Salagean and H.A. Spikes, Tribol. Trans. 42 (1999) 479.

[9] S. Plaza, L. Margielewski, G. Celichowski, R.W. Wesolowski and R. Stanecka, Wear 249 (2001) 1077.

[10] S. Hollinger, J.-M. Georges, D. Mazuyer, G. Lorentz, O. Aguerre and N. Du, Tribol. Lett. 9 (2000) 143.

[11] B. Duan and H. Lei, Wear 249 (2001) 528.

[12] J. Klein, Annu. Rev. Mater. Sci. 26 (1996) 581.

[13] J. Klein, E. Kumacheva, D. Perahia and L.J. Fetters, Acta Polym. 49 (1998) 617.

[14] U. Raviv, R. Tadmor and J. Klein J. Phys. Chem. B. 105 (2001) 8125.

[15] S. Granick, A.L. Demirel, L.L. Cai and J. Peanasky, Isr. J. Chem. 35 (1995) 75.

[16] J. Klein, D. Peraha and S. Warburg, Nature 352 (1991) 143.

[17] G.S. Grest, Phys. Rev. Lett. 76 (1996) 4979.

[18] T. Kreer, M.H. Müser, K. Binder and J. Klein, Langmuir 17 (2001) 7804.

[19] A.S. Sawhney and J.A. Hubbell, Biomaterials 13 (1992) 863.

[20] D.L. Elbert and J.A. Hubbell, Chem. Biol. 5 (1998) 177.

[21] D.L. Elbert and J.A. Hubbell, J. Biomed. Mater. Res. 42 (1998) 55.

[22] G.L. Kenausis, J. Vörös, D.L. Elbert, N.-P. Huang, R. Hofer, L. Ruiz-Taylor, M. Textor, J.A. Hubbell and N.D. Spencer, J. Phys. Chem. B 104 (2000) 3298.

[23] N.-P. Huang, R. Michel, J. Vörös, M. Textor, R. Hofer, A. Rossi, D.L. Elbert, J.A. Hubbell and N.D. Spencer, Langmuir 17 (2001) 489.

[24] G.A. Parks, Chem. Rev. 65 (1965) 177.

[25] R. Kurrat, M. Textor, J. J. Ramsden, P. Böni and N.D. Spencer, Rev. Sci. Instrum. 68 (1997) 2172.

[26] R. Kurrat, B. Walivaara, A. Marti, M. Textor, P. Tengvall, J. J. Ramsden and N.D. Spencer, Colloids Surf. B 11 (1998) 187.

[27] J. Vörös, J.J. Ramsden, G. Csucs, I. Szendrö, S.M. De Paul, M. Textor and N.D. Spencer, Biomaterials 23 (2002) 3699.

[28] J.J. Ramsden, J. Stat. Phys. 73 (1993) 853.

[29] G.J. Johnston, R. Wayte and H.A. Spikes, Tribol. Trans. 34 (1991) 187.

[30] G.R. Paul and A. Cameron, Proc. Roy. Soc. Lond. A331 (1972) 171.

[31] M. Smeeth and H.A. Spikes, J. Tribol. 119 (1997) 291.

[32] T.H. Courtney, *Mechanical Behavior of Materials*, 2nd ed. (McGraw-Hill/Higher Education, International Edition, 2000).

Tribology Letters, Vol. 15, No. 4, November 2003 (© 2003) 395

The influence of molecular architecture on the macroscopic lubrication properties of the brush-like co-polyelectrolyte poly(L-lysine)-*g*-poly(ethylene glycol) (PLL-*g*-PEG) adsorbed on oxide surfaces

Markus Müller[a], Seunghwan Lee[a], Hugh A. Spikes[b] and Nicholas D. Spencer[a,*]

[a]*Laboratory for Surface Science and Technology, Department of Materials, ETH-Zürich, Sonneggstrasse 5, Zürich, CH-8092, Switzerland*
[b]*Tribology Section, Department of Mechanical Engineering, Imperial College, London, SW7 2BX UK*

Received 21 February 2003; accepted 29 May 2003

The co-polymer poly(L-lysine)-*g*-poly(ethylene glycol) (PLL-*g*-PEG) has been investigated as a potential biomimetic boundary-lubrication additive for aqueous lubrication systems. In this work, the influence of the co-polymer's architecture on its tribological performance has been investigated. The architectural parameters investigated comprise side-chain (PEG) length, Lys/PEG grafting ratio and backbone chain (PLL) length. The tribological approaches applied in this work include ultra-thin-film interferometry, the mini-traction machine (MTM), and pin-on-disk tribometry. Both an increase in the molecular weight of the PEG side chains and a reduction in the grafting ratio result in an improvement in the lubricating properties of aqueous PLL-*g*-PEG solution at low speeds. MTM measurements show that an increase in the molecular weight of the PLL backbone results in an increase of the coefficient of friction.

KEY WORDS: poly(L-lysine)-*g*-poly(ethylene glycol) (PLL-*g*-PEG), aqueous lubrication, polymer architecture, boundary lubrication

1. Introduction

In a previous paper [1] we explored the application of the poly(L-lysine)-*g*-poly(ethylene glycol) (PLL-*g*-PEG) as a potential biomimetic boundary-lubrication additive for aqueous lubrication systems. We showed that, for oxide tribo-systems, the boundary-lubrication properties of aqueous buffer solution are significantly improved upon addition of PLL-*g*-PEG. On the molecular scale, Yan *et al.* [2] recently investigated the influence of the architectural parameters of PLL-*g*-PEG on its lubrication properties in aqueous solution by means of atomic force microscopy (AFM). The results revealed that the adsorption of PLL-*g*-PEG onto silicon oxide surfaces significantly reduces friction, with the lowest friction being observed when polymer adsorption occurs on both surfaces. As shown in figure 1, PLL-*g*-PEG is composed of a polyionic PLL backbone with covalently bound PEG side chains. The enhanced lubricating effect of PLL-*g*-PEG in aqueous solution was ascribed to its unique adsorption behavior: the positively charged amino groups of the PLL backbone are electrostatically attracted to negatively charged oxide surfaces, forcing the PEG side chains into a dense, brush-like structure. Polymer "brushes", i.e., systems of polymer chains end-grafted onto surfaces, are often used to modify surface forces and have important technological applications in many areas, including colloidal stabilization and adherence [3,4]. Over the past ten years, attention has been paid to end-grafted polymers that exhibit unique tribological properties in "good" solvents, and it is suggested that brush-like polymer layers may be useful as lubricants. Models for the shear interactions between polymer-bearing surfaces have been described by both experimental [5–9] and computer-simulation studies [10,11]. The origin of the low frictional forces between brush-bearing surfaces has been attributed both to steric repulsion between the polymers supporting high normal loads and to intermolecular interactions between the polymer brushes and the solvent molecules which maintain a lubricating fluid layer at the sheared interfacial region [11]. Extremely low coefficients of friction (COF) (as low as the detection limit <0.001) have been reported in nanotribological studies performed by means of the surface forces apparatus (SFA) [7].

Most tribological studies concerned with investigating the lubrication properties of end-grafted polymers have been performed at the nanometer scale in a very-low-load regime. However, little is known about the lubrication properties of end-grafted polymers on a macroscopic scale, which is more relevant when considering technical and engineering applications such as polymer-bearing interfaces. In our previous work [1], a significant reduction of the macroscopic

*To whom correspondence should be addressed. E-mail: nicholas.spencer@mat.ethz.ch

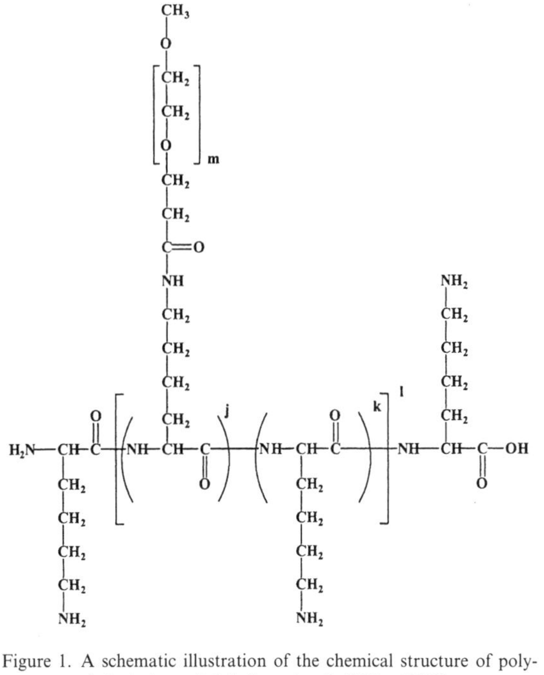

Figure 1. A schematic illustration of the chemical structure of poly-(L-lysine)-g-poly(ethylene glycol) (PLL-g-PEG).

COF for oxide-based tribo-systems was observed when adding PLL-g-PEG to aqueous buffer solution. The present study is a continuation of this work, focusing on the architectural parameters of the PLL-g-PEG co-polymer and its influence on the macroscopic lubrication properties. The architectural parameters investigated in this work include side-chain (PEG) length, Lys/PEG grafting ratio and backbone (PLL) length. The experimental techniques employed are ultra-thin-film interferometry, the mini-traction machine (MTM), and pin-on-disk tribometry.

2. Materials and methods

2.1. Synthesis of poly(L-lysine) grafted poly(ethylene glycol)

The following abbreviations are used for the various polymers discussed in this paper, to indicate the average molecular weight and the grafting ratio: PLL(x)-g[y]-PEG(z) signifies that the graft co-polymer has a PLL backbone of mol. wt x kDa, a graft ratio, y, of lysine unit/PEG side chain, and PEG side chains of mol. wt z kDa.

Table 1 shows the details of the reagent masses that were used to synthesize the different polymers according to a method described in previous publications [12]. Briefly, poly(L-lysine) hydrobromide (Fluka, Switzerland) was dissolved at a concentration of 100 mM in SBB (sodium borate buffer solution, 50 mM) adjusted to pH 8.5. The solution was filter sterilized (0.22 μm pore-size filter). For the grafting of PEG onto PLL, the *N*-hydroxysuccinimidyl ester of methoxypoly (ethylene glycol) propionic acid (mPEG-SPA, Nektar, Huntsville, AL, USA) was added to PLL-HBr solution. The reaction was allowed to proceed for 6 h at room temperature, after which the reaction mixture was dialyzed (Spectra-Por, mol. wt cutoff size 6–8 kDa, Spectrum, Houston, TX, USA) for 48 h against deionized water. The product was freeze-dried and stored in powder form at −20 °C. Detailed analytical information concerning the product produced by this method is available in previous publications [12,13].

Table 1.
Details of the masses that were used to synthesize the polymers possessing different architecture. PLL(x)-g[y]-PEG(z) signifies that the graft copolymer has a PLL backbone of mol. wt x kDa; a graft ratio, y, of lysine unit/PEG side chain, and PEG side chains of mol. wt z kDa.

Polymer PLL(x)-g[y]-PEG(z)	PLL	PEG
PLL(20)-g[3.4]-PEG(2)	245.0 mg	689.6 mg
PLL(20)-g[3.4]-PEG(5)	106.9 mg	751.9 mg
PLL(20)-g[2.2]-PEG(2)	107.5 mg	543.9 mg
PLL(20)-g[5.7]-PEG(2)	278.2 mg	433.2 mg
PLL(350)-g[3.4]-PEG(2)	106.9 mg	751.9 mg

For all subsequent experiments, PLL(x)-g[y]-PEG(z) was dissolved at a concentration of 0.25 mg/ml in 10 mM HEPES (4-(2-hydroxyethyl)piperazine-1-ethanesulfonic acid (Fluka, Switzerland), adjusted to pH 7.4 with 1.0 M NaOH solution).

2.2 Ultra-thin-film interferometry

The lubricant film-forming ability of the test solutions in a high-pressure rolling contact was measured by means of ultra-thin-film interferometry. The detailed method has been described elsewhere [1,14]. A lubricated contact is formed between a loaded stainless steel ball (radius $R = 9.5$ mm, AISI 440, RMS surface roughness 11 nm, PCS Instruments, London, UK) and the flat surface of a silicate glass disk (RMS surface roughness 2 nm). The disk drives the ball in nominally pure rolling and although the precise motion of the ball is not externally controlled, it is believed to roll against the disk with less than 2% sliding at speeds up to 1 m/s. The surface of the glass disk in contact with the ball is coated with a very thin, semi-reflecting layer of chromium, on top of which is a 400-nm layer of transparent silica which acts as a spacer layer. Thus the tribological pair consists of FeO_x/SiO_x. White light is shone through the glass into the contact, where a proportion is reflected back from the chromium layer while the remainder passes through the silica layer and the polymer-containing water before being reflected from the steel ball surface. Since the two reflected beams have traveled different distances, they interfere with a relative phase shift that depends on the path difference, i.e., on the sum of silica and lubricant film thickness. The interference pattern from the contact is passed to a spectrometer and the resultant dispersed light is analyzed to determine the precise wavelength of maximum constructive interference from the centre of the contact. This yields an accurate measure of the composite film thickness, from which the thickness of the spacer layer can be subtracted. For the calculation of the film thickness, the refractive index of the lubricant has to be known. Based on our previous work [1], a uniform refractive index of 1.4 has been used for the calculations of the film thickness for all PLL-*g*-PEG architectures. All measurements were carried out at room temperature ~25 °C and a fixed load of 10 N, which corresponds to a maximum Hertzian contact pressure of 0.41 GPa. The maximum rotation speed of the disk track was limited to 1 m/s. Each measurement was performed on a fresh disk track by varying the radii of measurement within 35 to 42 mm and a fresh stainless steel ball was used for each test.

2.3. Mini-traction machine (MTM)

A mini-traction machine (MTM, PCS Instruments, London, UK) was employed to characterize the lubrication properties of the polymer solutions in a mixed sliding/rolling contact regime. The set-up of the instrument is described in ref. [1]. A lubricated contact is formed between a polished stainless steel ball (radius $R = 9.5$ mm, AISI 440, RMS surface roughness 11 nm, PCS Instruments, London, UK) and a flat silicate glass disk (RMS surface roughness 2 nm, PCS Instruments, London, UK), thus representing an FeO_x/SiO_x interface. The MTM provides a mixed sliding/rolling contact through the independent control of the ball and disk velocities. The slide/roll ratio, SRR, is defined as the percentage ratio of the *difference* and the *mean* of the ball speed (u_{ball}) and disk speed (u_{disk}), i.e., $SRR = [|u_{ball} - u_{disk}|/(u_{ball} + u_{disk})/2] \times 100\%$. Thus, $SRR = 0\%$ (i.e. $u_{ball} = u_{disk}$) represents a pure rolling contact and higher SRR values represent a higher portion of the sliding character. With the software provided (PCS Instruments, MTM version 1.0, London, UK), values in the range SRR = 1 to 200% were accessible. At a given SRR, the COF ($\mu = F/N$) is measured as a function of mean speed. For the current work, MTM tests were initially conducted at several different SRR values between 5 to 25% but no dependence of friction on SRR in this range was detected. Therefore a standard SRR value of 10% was chosen, which was large enough to provide reliable friction measurement but sufficiently small so as to be comparable with the nominally pure rolling conditions of the ultra-thin-film interferometry work. It has been shown that fluid entrainment and thus film thickness is independent of slide/roll ratio within the range SRR = 0 to 50% [15]. MTM tests on polymer solutions were performed at a fixed load of 10 N and a controlled temperature of 25 °C, the same conditions as in the ultra-thin-film interferometry. The disk track radius was fixed at 20.7 mm and for each measurement, a new stainless steel ball and a new glass disk were used.

2.4. Pin-on-disk tribometry

A pin-on-disk tribometer (CSEM, Neuchâtel, Switzerland) was employed in this work to characterize the influence of the architectural parameters of PLL-*g*-PEG on lubrication properties in pure sliding conditions as a function of load. There were two reasons for using pin-on-disk tribometry for pure sliding tests, rather than the MTM. One is the availability of a wide range of substrate materials (e.g., ceramics) for pin-on-disk, which was important in determining the ideal settings and material pairings at the beginning of the experimental work. More importantly, the pin-on-disk was able to operate at slower sliding speeds than the MTM and thus ensures full boundary-lubrication conditions, as discussed later in this paper. In the pin-on-disk setup, a fixed, spherical steel pin (6 mm in diameter, DIN 5401-20 G20, Hydrel AG, Romanshorn, Switzerland) was loaded against a flat glass disk (RMS surface roughness 2 nm). The load was controlled by placing dead weights on top of the pin holder (1 to 5 N) and the friction forces were measured by a strain gauge at a constant speed of 0.005 m/s at room temperature ~25 °C. The maximum Hertzian contact pressure in this configuration is calculated to be 0.41, 0.51, and 0.70 GPa at applied loads of 1, 2, and 5 N, respectively. The number of rotations of the disk was limited to 100. For each friction measurement, a new steel ball and a fresh track on the glass disk was used. The friction signals were recorded with a Macintosh Power PC using Labview and an ADC card of the MIO family (both from National Instruments, Austin, TX, USA).

2.5. Cleaning of the tribo-pairs

All the instrumental parts that are expected to be in contact with the polymer solution, including the balls and disks of all tribological instruments, were oxygen-plasma cleaned for ca. 2 min in a Harrick Plasma Cleaner/Sterilizer PDC-32G instrument (Ossining, NY, USA) prior to the measurements. Before the plasma cleaning, the instrumental parts and the tribo-pair of ultra-thin-film interferometry and MTM were degreased by sonication in toluene for 10 min, followed by sonication in isopropanol for 10 min, and dried in nitrogen atmosphere. The pins, disks and assembly parts

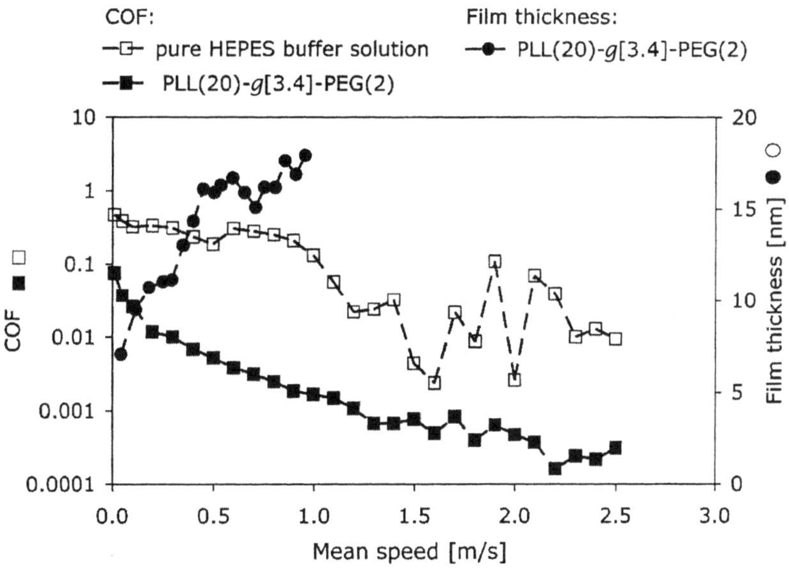

Figure 2. A direct comparison of the coefficient of friction (y-axis on the left-hand-side) and the lubricant film thickness (y-axis on the right-hand-side) for both polymer-free and PLL(20)-*g*[3.4]-PEG(2)-containing HEPES buffer solution, measured by MTM and ultra-thin-film interferometry. Squares are for the coefficient of friction and circles for the lubricant film thickness. The lines between data points serve as a guide for the eye. Ball = stainless steel (19 mm in diameter), substrate = silica, buffer solution = 10 mM HEPES (pH 7.4), polymer concentration = 0.25 mg/ml, load = 10 N, T = 25 °C.

of pin-on-disk were sonicated in ethanol for 10 min and dried in nitrogen atmosphere.

3. Results

*3.1. Lubrication properties of PLL-*g*-PEG in aqueous buffer solution compared with pure buffer solution*

Figure 2 compares the coefficient of friction (COF) of PLL(20)-*g*[3.4]-PEG(2)-containing HEPES buffer solution and pure (polymer-free) HEPES buffer solution as a function of mean rolling speed, in a glass/steel tribopair measured using the MTM. Also shown in this figure is the film thickness of the polymer solution measured as a function of rolling speed using ultra-thin-film interferometry. As shown in our previous work [1], the film thickness of pure HEPES buffer solution could not be measured reliably below 1 m/s because of significant wear and resulting damage to the coatings on the glass disk. However, upon addition of PLL(20)-*g*[3.4]-PEG(2) to the pure HEPES buffer solution, no such damage occurred and film thickness was found to increase from 7 to 18 nm as the speed was increased from 0.01 to 1 m/s. At speeds above 1 m/s an erratic and irreproducible decrease of the film thickness was seen, possibly due to damage to the disk coating. Hence, the speed range investigated by ultra-thin-film interferometry was limited to the range of 0.01 to 1 m/s.

The friction-speed curve obtained for pure HEPES buffer solution shows a constant COF of around 0.5 for a speed below 1 m/s. However, at higher speed, the COF became unstable and oscillated with a high irregular amplitude between 0.1 and 0.01. On the other hand, PLL(20)-*g*[3.4]-PEG(2) solution displayed a quite uniform decrease of the COF over the whole speed range, to reach values at least an order of magnitude lower than those obtained with pure HEPES buffer solution.

Figure 3 shows the results of pin-on-disk tribometry, also performed on a steel/glass tribo-pair, at a constant, pure sliding speed of 0.005 m/s at three applied loads. For the PLL(20)-*g*[3.4]-PEG(2) solution, the COF (~0.15) was about half that of pure HEPES buffer solution (~0.3) at all loads investigated. No reduction of the COF was observed upon addition of free PEG of similar molecular weight to the pure HEPES solution (data not shown). The COF for all test fluids decreased with increasing the load from 1 to 2 N, but remained broadly constant as the load was further increased to 5 N.

*3.2. The influence of the PLL-*g*-PEG architecture on the tribological properties*

The influence of the architectural parameters of the PLL-*g*-PEG polymers on film formation and friction was investigated by varying the side chain (PEG) length, the Lys/PEG grafting ratio, and the backbone (PLL) length.

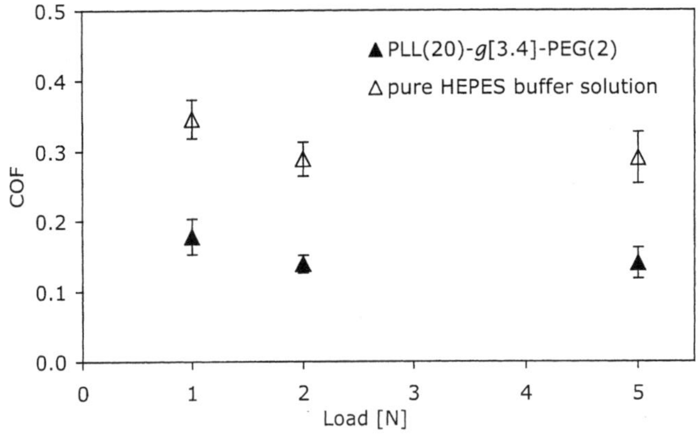

Figure 3. Coefficient of friction versus load determined by pin-on-disk tribometry for both polymer-free (white triangles) and PLL(20)-*g*[3.4]-PEG(2)-containing (black triangles) HEPES buffer solution. Mean values of three experiments ± standard deviation. Ball = steel (6 mm in diameter), disk = glass, buffer solution 10 mM HEPES (pH 7.4), polymer concentration = 0.25 mg/ml, load = 10 N, T = 25 °C.

3.2.1 Influence of the molecular weight of the PEG chain

Two PLL-*g*-PEG polymers with different PEG side-chain molecular weights (M_w = 2 kDa and 5 kDa) but the same molecular weight of the PLL backbone (M_w = 20 kDa) and the same Lys/PEG grafting ratio (3.4), were synthesized. Figure 4 shows the results of the MTM and ultra-thin-film interferometry experiments on solutions of these polymers. Over the entire range of speed (0.01 to 1 m/s), the polymers with 5 kDa PEG side chains form thicker lubricant films than those with 2 kDa chains. Under the same conditions, the COF is consistently lower for the polymer with 5 kDa PEG side chains than for the 2 kDa side-chain one. Figure 5 shows the COF measured as a function of load by means of pin-on-disk tribometry. Again, the polymer with the 5 kDa side chain reveals a significantly reduced COF compared to the polymer with 2 kDa chains.

3.2.2 Influence of the Lys/PEG grafting ratio

The influence of the Lys/PEG grafting ratio on frictional properties and interfacial film formation of PLL-*g*-PEG was investigated by varying grafting ratio while keeping molecular weight of the PLL backbone and PEG chain constant at 20 kDa and 2 kDa, respectively. Grafting ratios of $g = 2.2$, 3.4, and 5.7 were investigated. Figure 6 shows the results of both

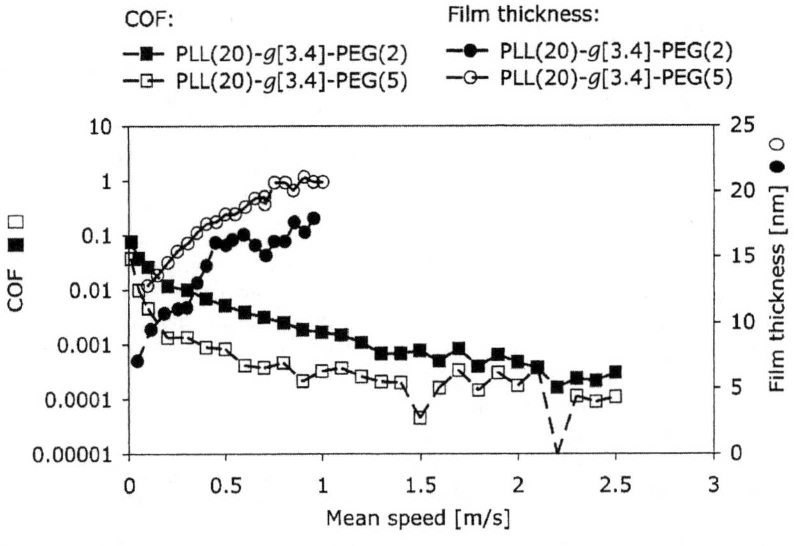

Figure 4. The influence of the molecular weight of the PEG side chains on the coefficient of friction (squares; y-axis on the left-hand-side) and lubricant film thickness (circles; y-axis on the right-hand-side) was measured as a function of speed by means of MTM and ultra-thin-film interferometry. The test lubricants were aqueous buffer solutions containing either PLL(20)-*g*[3.4]-PEG(2) (black symbols) or PLL(20)-*g*[3.4]-PEG(5) (white symbols). The lines between data points serve as a guide for the eye. Ball = stainless steel (19 mm in diameter), substrate = silica, buffer solution = 10 mM HEPES (pH 7.4), polymer concentration = 0.25 mg/ml, load = 10 N, T = 25 °C.

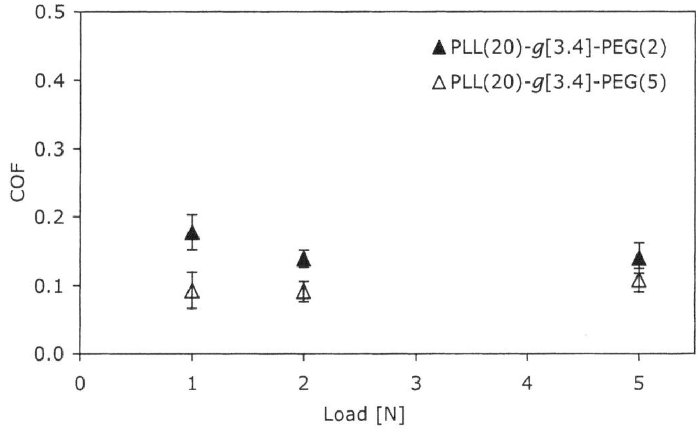

Figure 5. The influence of the molecular weight of the PEG side chains on the coefficient of friction was measured as a function of load by means of pin-on-disk tribometry. The test lubricants were aqueous buffer solutions containing either PLL(20)-g[3.4]-PEG(2) (black symbols) or PLL(20)-g[3.4]-PEG(5) (white symbols). Mean values of three experiments ± standard deviation. Ball = steel (6 mm in diameter), disk = glass, buffer solution 10 mM HEPES (pH 7.4), polymer concentration = 0.25 mg/ml, load = 10 N, T = 25 °C.

ultra-thin-film interferometry and MTM. Both film thickness and COF values for the two lower grafting ratios are very similar, although $g = 2.2$ gives slightly lower friction, especially at low speed. However film thickness is significantly lower and COF higher for the highest grafting ratio, $g = 5.7$. Figure 7 shows pin-on-disk measurements which reveal a similar COF dependence on grafting ratio in the decreasing order of $g = 5.7 > g = 3.4 > g = 2.2$. In other words, the more PEG chains grafted on the PLL backbone, the lower the resulting COF.

3.2.3. Influence of the PLL backbone length

The third structural parameter investigated was PLL backbone molecular weight. Two polymers having 20 kDa and 350 kDa PLL backbone molecular weight were compared: PLL(20)-g[3.4]-PEG(5) and PLL(350)-g[3.4]-PEG(5). The molecular weight of the PEG-side-chain (5 kDa) and the PEG grafting ratio (3.4), were kept constant. Figure 8 shows the results of ultra-thin-film interferometry and MTM. The higher molecular weight polymer shows both poorer film-forming properties and greater COF than the lower molecular weight

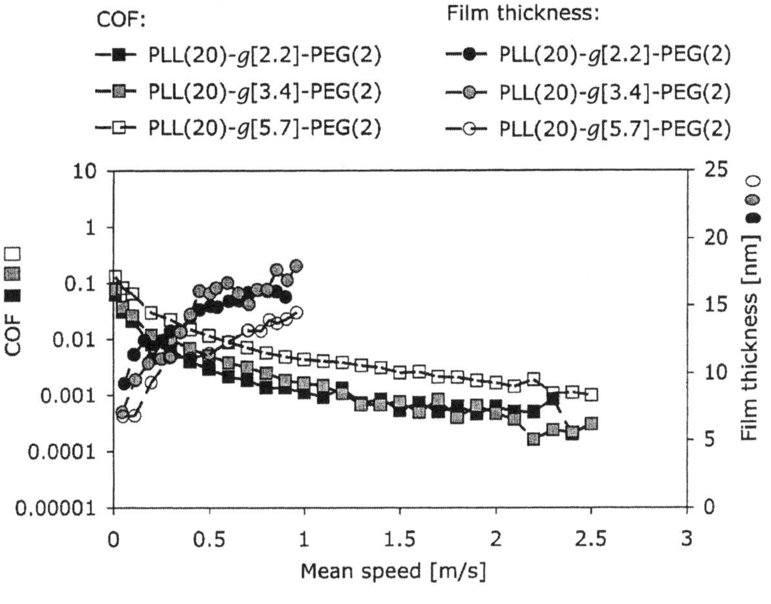

Figure 6. The influence of the Lys/PEG grafting ratio on the coefficient of friction (squares; y-axis on the left-hand-side) and lubricant film thickness (circles; y-axis on the right-hand-side) was measured as a function of speed by means of MTM and ultra-thin-film interferometry. The test lubricants were aqueous buffer solutions containing PLL(20)-g[2.2]-PEG(2) (black symbols), PLL(20)-g[3.4]-PEG(2) (grey symbols) or PLL(20)-g[5.7]-PEG(2) (white symbols). The lines between data points serve as a guide for the eye. Ball = stainless steel (19 mm in diameter), substrate = silica, buffer solution = 10 mM HEPES (pH 7.4), polymer concentration = 0.25 mg/ml, load = 10 N, T = 25 °C.

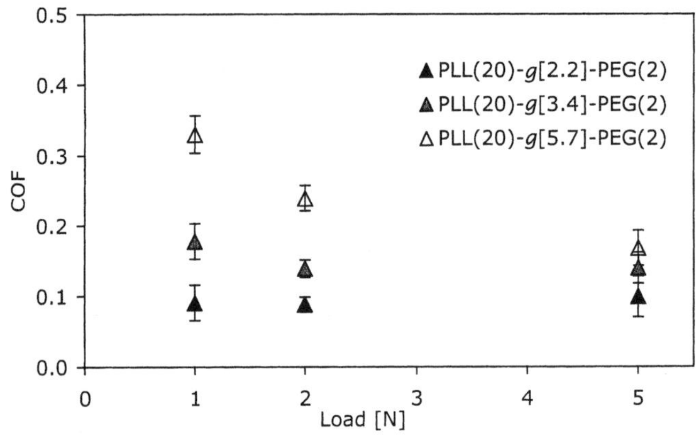

Figure 7. The influence of the Lys/PEG grafting ratio on the coefficient of friction was measured as a function of load by means of pin-on-disk tribometry. The test lubricants were aqueous buffer solutions containing PLL(20)-*g*[2.2]-PEG(2) (black triangles), PLL(20)-*g*[3.4]-PEG(2) (grey triangles) or PLL(20)-*g*[5.7]-PEG(2) (white triangles). Mean values of three experiments ± standard deviation. Ball = steel (6 mm in diameter), disc = glass, buffer solution 10 mM HEPES (pH 7.4), polymer concentration = 0.25 mg/ml, load = 10 N, T = 25 °C.

one. The difference in film thickness is most marked at low speed and tends to be lost at high speed. However the difference in COF is very striking, with the lower molecular weight polymer giving friction at least an order of magnitude lower than the higher molecular weight one over almost the whole speed range. Interestingly, as shown in figure 9, this difference between the two polymers was not seen in pin-on-disk tests.

4. Discussion

As shown above, the polymer PLL-*g*-PEG functions in aqueous solution as an effective boundary-lubricating additive and this effectiveness depend upon the polymer architecture. The focus of this discussion is to consider the mechanism of this boundary-lubricating performance and thus to determine how the polymer

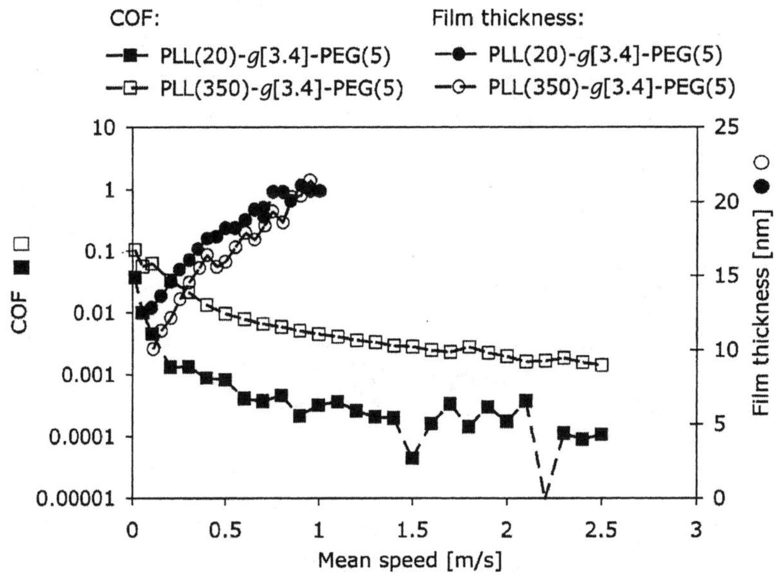

Figure 8. The influence of the molecular weight of the PLL backbone on the coefficient of friction (squares; y-axis on the left-hand-side) and lubricant film thickness (circles; y-axis on the right-hand-side) was measured as a function of speed by means of MTM and ultra-thin-film interferometry. The test lubricants were aqueous buffer solutions containing either PLL(350)-*g*[3.4]-PEG(5) (white symbols) or PLL(20)-*g*[3.4]-PEG(5) (black symbols). The lines between data points serve as a guide for the eye. Ball = stainless steel (19 mm in diameter), substrate = silica, buffer solution = 10 mM HEPES (pH 7.4), polymer concentration = 0.25 mg/ml, load = 10 N, T = 25 °C.

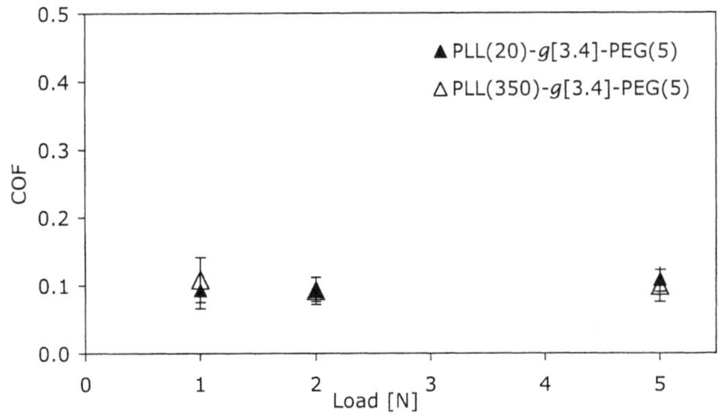

Figure 9. The influence of the molecular weight of the PLL backbone on the coefficient of friction was measured as a function of load by means of pin-on-disk tribometry. The test lubricants were aqueous buffer solutions containing PLL(350)-*g*[3.4]-PEG(5) (white triangles) and PLL(20)-*g*[3.4]-PEG(5) (black triangles). Mean values of three experiments ± standard deviation. Ball = steel (6 mm in diameter), disc = glass, buffer solution 10 mM HEPES (pH 7.4), polymer concentration = 0.25 mg/ml, load = 10 N, T = 25 °C.

molecules behave within a high-pressure contact to form a film and reduce friction.

4.1. *Lubrication regimes and lambda ratio*

According to elastohydrodynamic theory, in high-pressure lubricated contacts, the motion of the surfaces entrains lubricant into the contact to form a film whose thickness is dependent on the mean speed, U, the dynamic viscosity, η, of the fluid in the contact inlet and the pressure viscosity coefficient, α, of this fluid according to:

$$h \propto (U\eta)^{a}(\alpha)^{b} \qquad (1)$$

where a and b are generally 0.6–0.7 and 0.5 respectively [16].

This means that, at very low speeds, negligible fluid film is formed and the applied load is supported wholly by asperity–asperity contact between the solid surfaces. The friction is then determined by the shear strength properties of any thin adsorbed or reacted "boundary film" on these asperities. As the speed is raised, however, a fluid film gradually develops until, at high speeds, the surfaces are fully separated by a low-shear-strength, and thus low-friction, liquid lubricating film and the contact operates in full elastohydrodynamic lubrication.

One way of quantifying the extent of fluid film lubrication is in terms of the "lambda ratio", Λ, which is defined as the ratio between the film thickness h and the out-of-contact, composite surface roughness of the sliding bodies [16]. In the current study, it was possible to calculate the lambda ratio directly, since the film thickness was measured. Therefore, since the contact conditions and tribo-pair were the same in both film thickness and MTM friction measurements, it was possible to determine the dependence of COF on lambda ratio for the fluids tested. This is shown in figure 10, where the COF (from MTM) is plotted against Λ (calculated from h measured by ultra-thin-film interferometry). Figure 10 illustrates two important features. Firstly, there is a rapid decrease in friction coefficient with increasing lambda ratio, which indicates that in all of the MTM measurements the contact is operating either in boundary or in mixed-film lubrication, depending on the experimental settings. In this regime, some of the load is supported by asperity contact and some by fluid film pressure, so that the friction arises from a combination of asperity and fluid-film friction. As Λ increases, the fluid-film component of the load support increases and thus the friction correspondingly decreases. A second feature of figure 10 is that the COF versus lambda ratio plots collapse onto one single line for all different PLL-*g*-PEG architectures, except for the single set synthesized from the high-molecular-weight PLL. This would be expected if the polymers were to have the same boundary friction at very low speed and the same fluid-film friction at higher speed. At intermediate speeds, friction would then depend solely on the polymer's contribution to fluid-film formation and thus on the Λ ratio. The anomalous behavior of the high-molecular-weight PLL-containing polymer will be discussed later.

4.2. *Film formation by HEPES and polymer solutions*

The tests on polymer-free HEPES showed negligible film formation below 1 m/s. This is primarily because HEPES, like pure water, has an extremely low pressure viscosity coefficient ($\alpha \sim 0.36\,\text{GPa}^{-1}$ at 55 °C) compared to that of oil-based lubricants ($\alpha = 10\text{–}20\,\text{GPa}^{-1}$ at 55 °C) [17]. This means that, in accord with equation (1), it forms negligible fluid-film lubrication up to a mean speed of 1 m/s [18]. The very high friction coefficient of

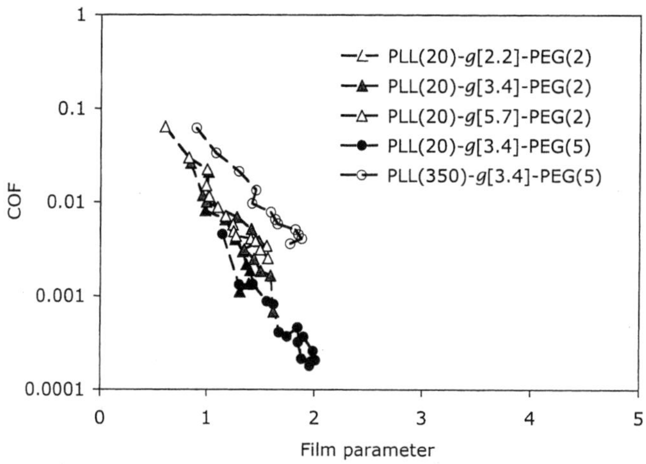

Figure 10. Friction coefficient as a function of the ratio of film thickness to surface roughness, Λ, showing regimes of lubrication [16]. The dimensionless film parameter Λ is defined as the ratio between the film thickness h_c and the composite surface roughness R_c of the sliding bodies, $\Lambda = h_c/R_c$, where $R_c = \sqrt{R_{q,\text{ball}}^2 + R_{q,\text{disk}}^2}$ and $R_{q,\text{ball}}$ and $R_{q,\text{disk}}$ are the RMS roughness of the ball (11 nm) and the disk (2 nm) respectively. Thus $R_c = 11.2$ nm.

HEPES at low speeds also indicates that the fluid possesses very poor boundary-film-forming properties.

In contrast to HEPES, the polymer solutions clearly show both a low-speed boundary-film formation (as evidenced by the pin-on-disk tests) and some degree of fluid-film entrainment. In conventional elastohydrodynamic (EHD) lubrication, as shown in equation (1), the film thickness should depend upon the mean speed raised to a constant exponent of between $a = 0.6$ and 0.7. Figure 11 shows a plot of log(film thickness) versus log(mean speed) for the polymer solution of PLL(20)-*g*[3.4]-PEG(2). The build-up of film thickness with speed is quite irregular but has gradient of $a \approx 0.3$. A similar gradient was found for all the polymers tested. This indicates that, although a film is clearly being entrained by the rolling/sliding motion of the surfaces, this entrainment does not conform with EHD theory. In figure 11 it can be seen that at a mean speed of above 0.8 m/s, the film thickness-speed curve appears to have become more regular, with a gradient approaching 0.6. This was also seen more clearly in previous work, which studied film thickness up to a speed of 3 m/s [1] and is believed to represent the entrainment of bulk-water phase into the contact in a conventional EHD fashion. Previous work has shown that a significant EHD film starts to be formed by water at speeds above about 0.8 m/s [18].

*4.3. Mechanism of film formation by PLL-*g*-PEG*

From a combination of the film thickness and friction results it is possible to arrive at a coherent model for the molecular behavior of the PLL-*g*-PEG polymer in thin-film, high-pressure contact. In this model we consider that outside the contact region, the PLL chains are partially or wholly adsorbed on the surfaces with the hydrophilic PEG chains oriented outward into the buffer solution, building a brush-like structure that incorporates a large amount of coordinated water molecules [12,19]. Due to the strong hydrogen bonding interactions, the effective viscosity of the confined water will be increased several times. Within the contact zone, at zero or very slow speeds, the polymer brushes are effectively squashed down under load to a thickness of a few nm. The friction coefficient under these conditions thus represents the shear strength of the squashed polymer and has a value of around 0.1, typical of most adsorbed organic boundary-lubricating films. As the rolling/sliding speed is raised, however, the viscous film in the inlet, consisting of water confined in polymer, will begin to be entrained into the contact against the contact pressure to generate a quasi-elastohydrodynamic film. This will be promoted by an osmotic force that will tend to induce coordinated water molecules to move into the squashed polymer film [3,20,21]. As the speed is increased, the squashed PEG chains will respond to the entrained water by stretching and swelling, resulting in an increasing height of the brush layer with an associated increase of the normal force acting between the surfaces of the tribo-pair [22,23]. The thickness of this water/polymer film will continue to increase with entrainment speed until it reaches the uncompressed thickness of the brush layer upstream of the inlet and the polymer film can contribute no further to water entrainment. Thereafter, any further entrainment will be governed by the viscosity of bulk water and no further film growth occurs until the speed is sufficient

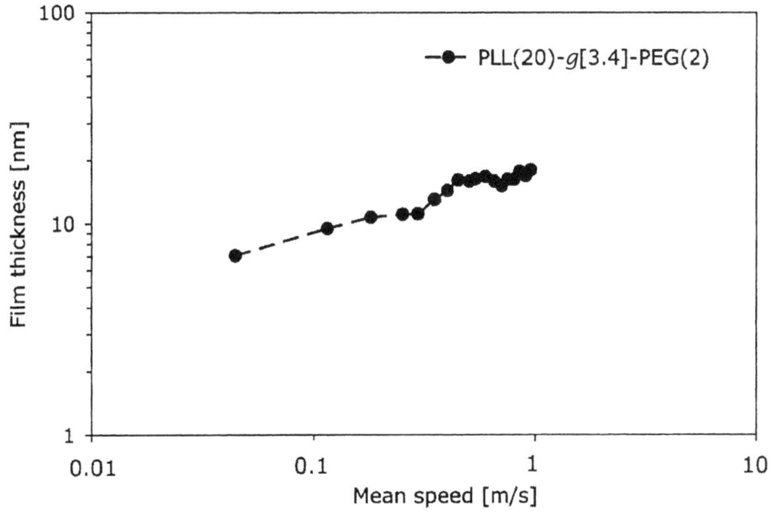

Figure 11. Log(film thickness) versus log(mean speed) for PLL(20)-g[3.4]-PEG(2)-containing HEPES buffer solution, measured by ultra-thin-film interferometry. The lines between data points serve as a guide for the eye. Ball = stainless steel (19 mm in diameter), substrate = silica, buffer solution = 10 mM HEPES (pH 7.4), polymer concentration = 0.25 mg/ml, load = 10 N, T = 25 °C.

for bulk water to lubricate via a conventional EHD mechanism. The leveling out of film thickness at around 20 nm noted in figure 11 (and figure 5 in [1]) thus represents the essentially full extension of the polymer PEG brushes.

4.4. *Influence of PLL-g-PEG architecture*

The architectural parameters investigated in this study were PEG side-chain length, Lys/PEG grafting ratio and backbone (PLL) length. First, the influence of the PEG chain length ($M_w = 2$ and 5 kDa) is discussed.

The radii of gyration (R_g) of the unperturbed PEG chains in aqueous solution have been estimated in other work by static light-scattering measurements to be about 1.65 and 2.82 nm for PEG side chains of 2 and 5 kDa, respectively [11]. However, if the average spacing between two grafted PEG side chains is smaller than $2R_g$, the coiled PEG side chains overlap, resulting in a stretching of the chains, resulting in a brush structure [19]. According to Alexander [24] and De Gennes [25], the brush height grows linearly with N under stretching conditions, while the nonstretched chain dimension R_g only grows with $N^{1/2}$. A stretching of the PEG side chains is thus expected to lead to a brush thickness that is much larger than R_g. As can be seen in figures 4 and 5, an increase in the molecular weight of the PEG chain from 2 to 5 kDa results in an increase in the interfacial film thickness by a factor of ~ 1.2 and in a significant reduction of the COF by a factor of ~ 3.2. It is suggested that the higher brush thickness associated with a higher water uptake, which results in an increase in Λ ratio, is responsible for the superior lubrication properties of the polymer having higher-molecular-weight PEG side chains.

The influence of the Lys/PEG grafting ratio on the lubrication properties was investigated by varying the grafting ratios (2.2, 3.4, and 5.7). The results are shown in figure 6 and 7. By changing the Lys/PEG grafting ratio, both the number of PEG chains and the segmental charge of the polyelectrolyte molecule are changed, since the PEG chains are covalently bound via the NH_2-terminated side chains of the poly(L-lysine) backbone, eliminating the charge contribution of the PEG-ylated amine groups. In other words, an increase in the Lys/PEG grafting ratio leads to both a decreased number of PEG side chains per PLL-g-PEG molecule and an increase in the segmental charge on the PLL backbone. The latter leads to an additional decrease of the adsorbed PEG side-chain density on the negatively charged native oxide surface, since fewer polymer molecules are required to absorb in order to compensate the surface charge [26]. Hence, a reduction in the Lys/PEG grafting ratio should lead to an increasing PEG density on a negatively charged oxide surface. This has been confirmed by optical waveguide lightmode spectroscopy (OWLS) and time-of-flight secondary ion mass spectrometry (ToF-SIMS) in a separate study [19]. As shown in figures 6 and 7, the COF for different Lys/PEG grafting ratios, measured by MTM and pin-on-disk tribometry, decreased in the order of $g = 5.7 > g = 3.4 > g = 2.2$, which is consistent with the very reasonable conclusion that an increase in the PEG density on the interface of the tribo-pair results in an improvement of the lubrication properties of PLL-g-PEG.

The influence of the PLL backbone length was investigated by examining two polymers with very different PLL molecular weights (20 and 350 kDa). The influence of the PLL backbone length on the film thickness was found to be quite small, but its effect on MTM COF was considerable, with a significantly higher COF being found for the higher-molecular-weight PLL backbone. This difference is seen very clearly in figure 10, where the high-molecular-weight PLL polymer lies on a markedly different COF versus Λ line than the low molecular weight PLL polymers. This difference in COF must arise from the higher molecular weight PLL polymer having greater friction either in the very low speed, "squashed polymer boundary lubrication" condition or in the high speed, "extended brush fluid film" condition. However, the very low speed pin-on-disk test showed no influence of the molecular weight of the PLL backbone on the COF. Thus it seems likely that the larger COF for the high-molecular-weight PLL backbone polymer results from this polymer forming a more viscous water/polymer film in the fluid-film regime. The reason for this is not clear, but possibly the fluid film contains a significant proportion of unbound polylysine chain, which has a high resistance to shear within the contact.

5. Conclusions

The co-polymer poly (L-lysine)-*g*-poly(ethylene glycol) (PLL-*g*-PEG) has been investigated as a potential biomimetic boundary-lubrication additive for aqueous lubrication systems. In this work, the PLL-*g*-PEG polymers were found to provide both a boundary lubricating film at very slow sliding speeds and also to promote the entrainment of a fluid, separating film of up to 20 nm thickness at intermediate-to-high speeds. The development of a fluid film resulted in a very low friction coefficient. A model for the development of this fluid film has been proposed. Further, the influence of the co-polymer architecture on the macroscopic tribological performance has been investigated. Both increasing the molecular weight of the PEG side chains and reducing the grafting ratio were found to result in an improvement in the lubricating properties of aqueous PLL-*g*-PEG solutions, in terms of increasing film thickness and friction reduction. Increasing the molecular weight of the PLL backbone resulted in an increase of the coefficient of friction at intermediate-to-high speeds but had little effect on film thickness.

The results of this study reveal the importance of the microscopic design of brush-like co-polyelectrolytes to control and optimize the lubrication capabilities of such polymer systems.

Acknowledgments

This work was financially supported by the Council of the Swiss Federal Institutes of Technology (ETH-Rat TopNano 21), the Swiss National Science Foundation NRP 47 and the US Air Force Office of Scientific Research (Contract No. F49620-02-1-0346). We are also grateful to Dr. Rowena Crockett of EMPA, Dübendorf, Switzerland, Monica Ratoi-Salagean of Imperial College, London, UK, and Stéphanie Pasche, BioInterface Group, LSST, Department of Materials, ETH-Zürich, for their valuable advice and assistance.

References

[1] S. Lee, M. Müller, J. Vörös, M. Ratoi-Salagean, H.A. Spikes and N.D. Spencer, Tribol. Lett., in press (2003).
[2] X. Yan, M.S. Lim, S. Pasche, S.M. De Paul, M. Textor, N.D. Spencer and S.S. Perry, submitted to Langmuir (2003).
[3] S.T. Milner, Science 251 (1991) 905.
[4] K. Kato, E. Uchida, E.-T. Kang, Y. Uyama and Y. Ikada, Progress in Polymer Science 28 (2003) 209.
[5] J.F. Joanny, Langmuir 8 (1992) 989.
[6] T.A. Witten, L. Leibler and P.A. Pincus, Macromolecules 23 (1990) 824.
[7] J. Klein, E. Kumacheva, D. Mahalu, D. Perahia and L.J. Fetters, Nature 370 (1994) 634.
[8] U. Raviv, R. Tadmor and J. Klein, J. Phys. Chem. B 105 (2001) 8125.
[9] T. Bouhacina, J.P. Aime, S. Gauthier, D. Michel and V. Heroguez, Phys. Rev. B 56 (1997) 7694.
[10] G.S. Grest, Current Opinion in Colloid & Interface Science 2 (1997) 271.
[11] G.S. Grest, Adv. Polymer Sci. 138 (1999) 149.
[12] G.L. Kenausis, J. Vörös, D.L. Elbert, N.P. Huang, R. Hofer, L. Ruiz-Taylor, M. Textor, J.A. Hubbell and N.D. Spencer, J. Phys. Chem. B 104 (2000) 3298.
[13] N.P. Huang, R. Michel, J. Vörös, M. Textor, R. Hofer, A. Rossi, D.L. Elbert, J.A. Hubbell and N.D. Spencer, Langmuir 17 (2001) 489.
[14] G.J. Johnston, R. Wayte and H.A. Spikes, Tribol. Trans. 34 (1991) 187.
[15] M. Smeeth and H. Spikes, *22nd Leeds/Lyon Symposium on Tribology, The Third Body Concept Proc.* (1996) p. 695.
[16] B.J. Hamrock, *Fundamentals of Fluid Film Lubrication* (McGraw-Hill, 1994).
[17] P. Vergne, M. Kamel and M. Querry, ASME J. Tribol. 119 (1997) 250.
[18] M. Ratoi and H.A. Spikes, Tribol. Trans. 42 (1999) 479.
[19] S. Pasche, S.M. De Paul, J. Vörös, N.D. Spencer and M. Textor, submitted to Langmuir (2003).
[20] J.L. Barrat, Macromolecules 25 (1992) 832.
[21] Y. Rabin and S. Alexander, Europhysics Lett. 13 (1990) 49.
[22] J. Klein, E. Kumacheva, D. Perahia, D. Mahalu and S. Warburg, Faraday Disc. (1994) 173.
[23] B. Bhushan, J.N. Israelachvili and U. Landman, Nature 374 (1995) 607.
[24] S. Alexander, Le Journal de Physique 38 (1977) 977.
[25] P.G. De Gennes, Macromolecules 13 (1980) 1069.
[26] N.G. Hoogeveen, M.A.C. Stuart and G.J. Fleer, J. Coll. Inter. Sci. 182 (1996) 133.

5706 *Macromolecules* **2005,** *38,* 5706–5713

Lubrication Properties of a Brushlike Copolymer as a Function of the Amount of Solvent Absorbed within the Brush

Markus T. Müller,[†] Xiaoping Yan,[‡] Seunghwan Lee,[†] Scott S. Perry,[‡] and Nicholas D. Spencer*[,†]

Laboratory for Surface Science and Technology, Department of Materials, Swiss Federal Institute of Technology, ETH-Hönggerberg, Wolfgang-Pauli-Strasse 10, CH-8093 Zürich, Switzerland, and Department of Chemistry, University of Houston, Texas 77204-5003

Received January 25, 2005; Revised Manuscript Received May 2, 2005

ABSTRACT: The shear forces between poly(L-lysine)-*graft*-poly(ethylene glycol) (PLL-*g*-PEG)-modified SiO_2 tribopairs have been measured with colloidal-probe, lateral force microscopy (LFM) and related to the mass of solvent absorbed within the brushlike structure of immobilized PEG chains. The amount of solvent (per unit substrate area) absorbed within the tethered, brushlike polymer, referred to as areal solvation, Ψ, appears to be of importance in determining the lubrication properties of the tethered polymers. In this study, the degree of solvation was varied by choosing different solvents (aqueous buffer solution, methanol, ethanol, and 2-propanol) and was determined by a technique that combines the results of quartz crystal microbalance (QCM-D) experiments and optical waveguide lightmode spectroscopy (OWLS). The highest degree of solvation was measured for aqueous buffer solutions, and a progressive decrease in solvation of PLL-*g*-PEG was observed in moving from methanol to ethanol to 2-propanol. A concomitant increase in the measured shear force was observed with this decrease in solvation. The lubrication mechanism of the PLL-*g*-PEG-coated SiO_2 tribopair is discussed in terms of solvation and solvent quality and compared with the lubrication mechanism of the corresponding tribopair where only one surface is coated with the polymer brush.

Introduction

The lubrication properties of polymer "brushes", i.e., systems of polymers end-grafted to a flat surface, have been investigated in numerous experimental and theoretical studies.[1–14] Vanishingly low friction forces have been reported for solid surfaces bearing polymer brushes under good solvent conditions and in the low-contact-pressure regime of the surface forces apparatus (SFA).[1] When brushlike polymers are brought into contact under good solvent conditions, long-ranged repulsive forces of osmotic origin act to keep the surfaces apart and entropic effects restrict mutual interpenetration of opposing polymer chains to a narrow interfacial zone, thus maintaining a highly fluid layer at the interface between them.[4,7,8,11] The effective viscosity in that zone differs only slightly from the solvent viscosity prevailing outside of the contact area. Hence, in the low-pressure regime of the SFA, the force required to maintain sliding is extraordinarily low and corresponds solely to the viscous drag of the polymer chain ends in the narrow interpenetration zone.[4] However, under the high-pressure regime of some 100 MPa, exerted by a colloidal-probe lateral force microscope (LFM), compression of the brush becomes more significant and the resistance to shear increases somewhat; nevertheless, friction still remains low.[15]

In this work, poly(ethylene glycol) brushes were formed from the comblike graft copolymer poly(L-lysine)-*graft*-poly(ethylene glycol) (PLL-*g*-PEG). This was synthesized by covalently linking PEG to a cationic poly(L-lysine) backbone. The structure of PLL-*g*-PEG is depicted in Figure 1. It consists of PEG chains that are attached to a PLL backbone, which is positively charged at pH

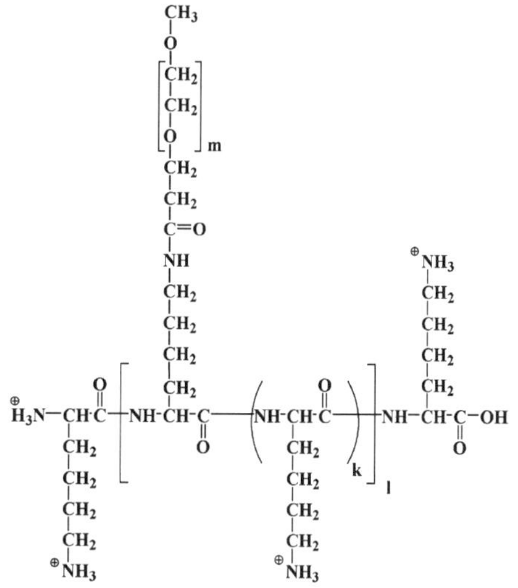

Figure 1. Structure of PLL-*g*-PEG.

≤ 10 due to protonation of the primary amine groups. In aqueous buffer solution, the PLL backbone readily adsorbs, for electrostatic reasons, onto a negatively charged surface, such as an oxide, forcing the PEG side chains into a dense, brushlike structure as depicted schematically in Figure 2A. Replacement of the good solvent (e.g., aqueous buffer solution) for a solvent of inferior quality (e.g., organic solvents) causes a contraction of the PEG brush to a more random, coillike conformation (Figure 2B).[16,17]

PEG has been extensively investigated for application in a wide array of biomedical devices, since immobilization of PEG onto surfaces has long been known to

[†] Swiss Federal Institute of Technology.
[‡] University of Houston.

Published on Web 06/04/2005

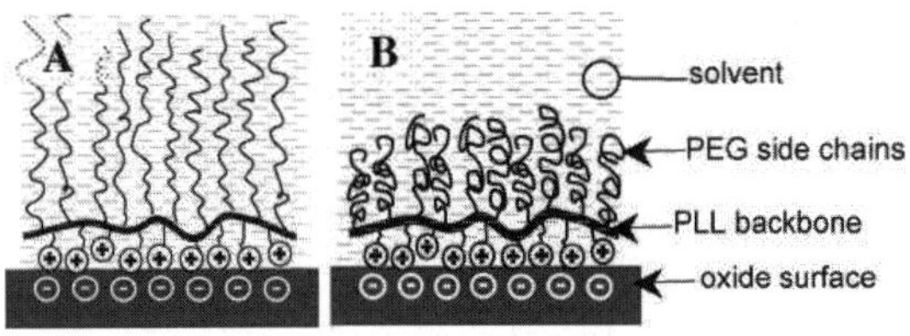

Figure 2. Schematic diagram of PLL-*g*-PEG adsorbed onto a negatively charged surface in (A) good solvent and (B) bad solvent.

decrease protein adsorption.[18–23] Very recently, PLL-*g*-PEG has also attracted interest in the field of tribology. Both macro- and nanoscale tribological studies have shown a significant reduction of interfacial friction between two PLL-*g*-PEG-coated oxide surfaces in an aqueous environment.[15,24,25]

The present study aims to understand the influence of solvent quality on the molecular-level friction mechanism of tethered, brushlike polymers. It involves complementary adsorption studies of PLL-*g*-PEG by means of optical waveguide lightmode spectroscopy (OWLS) and quartz crystal microbalance with dissipation (QCM-D) as well as friction studies performed on the nanoscale using colloidal-probe lateral force microscopy (LFM). The adsorbed mass measured by QCM-D includes a contribution from solvent molecules absorbed within the surface-bound polymer film.[23,26,27] This is in contrast to optical techniques, such as OWLS, which are sensitive only to the "dry mass" of a polymer adsorbed onto the surface of the waveguide.[28,29] By subtracting the "dry mass", derived from OWLS measurements, from the "wet mass", derived from QCM-D measurements, it is therefore possible to determine the mass of the solvent per unit substrate area absorbed in the brushlike structure of PLL-*g*-PEG, expressed as areal solvation, Ψ. Areal solvation was varied by choosing solvents (aqueous buffer solution, methanol, ethanol, and 2-propanol) of different quality with respect to the PEG brush. The solvents were characterized in terms of the three-component Hansen solubility parameters,[30] and these values were compared with measured areal solvation of the PEG brush.

The amount of solvent absorbed within a polymer brush is of great significance because it is regarded as a key parameter in determining the lubrication properties of brushlike polymers.[11,12] In this study we examined the relation between areal solvation of an unperturbed PLL-*g*-PEG-bearing SiO_2 surface and the force needed to shear two PLL-*g*-PEG-bearing SiO_2 surfaces past each other by the use of colloidal-probe LFM in liquid. The molecular friction mechanism is discussed in the context of solvation and on the basis of experiments performed on PLL-*g*-PEG-bearing tribopairs and on bare silica sliding against a flat PLL-*g*-PEG-bearing silica counter surface.

Materials and Methods

Synthesis of Poly(L-lysine)-*graft*-poly(ethylene glycol) (PLL-*g*-PEG). The synthesis of poly(L-lysine)-*graft*-poly(ethylene glycol) (PLL-*g*-PEG) molecules has been previously described.[20,31] PLL–HBr (Fluka, Switzerland) was dissolved at a concentration of 100 mM in SBB (sodium borate buffer solution, 50 mM) adjusted to pH 8.5. The solution was filter-sterilized (0.22 μm pore size filter). For the grafting of PEG onto PLL, the *N*-hydroxysuccinimidyl ester of methoxypoly(ethylene glycol) propionic acid (mPEG-SPA, Nektar, Huntsville, AL) was added to PLL–HBr solution. The reaction was allowed to proceed for 6 h at room temperature, after which the reaction mixture was dialyzed (Spectra-Por, mol wt cutoff size 6–8 kDa, Spectrum, Houston, TX) for 48 h against deionized water. The PEG graft ratio y, the number of lysine monomers per PEG chain, was determined using ^{1}H NMR. The product was freeze-dried and stored in powder form at −20 °C. The nomenclature of these polymers takes the form of PLL-(x)-g[y]-PEG(z) and signifies that the graft copolymer has a PLL backbone derived from PLL–HBr of molecular weight x kDa, a graft ratio y = number of lysine units/PEG side chain, and PEG side chains of molecular weight z kDa. In the present study, we used PLL(20)-*g*[3.5]-PEG(5): a PLL backbone of molecular weight 12 kDa, corresponding to 20 kDa PLL-HBr, was used for synthesis, and the polydispersity was M_w/M_n = 1.2. The grafting ratio was g = 3.5, and the PEG side chains were of molecular weight 5 kDa with a polydispersity of M_w/M_n = 1.1. The structure of the PLL-*g*-PEG polymer is shown in Figure 1, and the adsorption of PLL-*g*-PEG onto a negatively charged oxide surface is depicted in Figure 2.

Surface Preparation. PLL(20)-*g*[3.5]-PEG(5) adlayers were prepared by immersion of SiO_2 films into an aqueous buffer solution of the polymer. The films were sputter-coated onto quartz crystals for QCM-D or onto optical waveguides for OWLS. For LFM measurements, thermally oxidized Si (100) wafers and colloidal SiO_2 probes (diameter = 5.1 μm) were employed. Prior to immobilization of the polymer onto the surface, the oxidized Si (100) wafers and SiO_2 colloidal probe were treated by the following procedure: sonication in toluene for 2 min and then in 2-propanol for 10 min followed by extensive rinsing with ultrapure water (18.3 MΩ cm) (EM Science, Gibbstown, NJ), drying in a nitrogen flow, and exposure to an oxygen plasma PDC-32G (Harrick Scientific Corp., Ossining, NY) for 2 min (10 s in the case of the colloidal SiO_2 probes). The power applied to the rf coil of the plasma cleaner was set to 18 W. The oxidized substrate and the colloidal probe were immediately transferred to a 0.25 mg/mL solution of PLL(20)-*g*[3.5]-PEG(5) in 10 mM HEPES (4-[2-hydroxyethyl]piperazine-1-[2-ethanesulfonic acid], pH 7.4) buffer solution for 30 min and subsequently rinsed with polymer-free aqueous HEPES buffer solution to remove unbound PLL-*g*-PEG from the surface. The SiO_2 sputter-coated quartz crystal sensors and the optical waveguides were treated in a manner similar to the Si (100) wafers, but the plasma cleaning was reduced to 10 s in order to prevent damage to the thin SiO_2 coating. In the case of OWLS and QCM-D, the polymer was allowed to adsorb onto the sample directly in the liquid cell of the analytical devices.

Optical Waveguide Lightmode Spectroscopy (OWLS). Optical waveguide lightmode spectroscopy (OWLS) was carried out on a BIOS-I instrument (ASI AG, Zürich, Switzerland) using a Kalrez (Dupont, Wilmington, DE) flow-through cell with a volume of 16 μL. The waveguide chips used for OWLS measurements (MicroVacuum Ltd. Budapest, Hungary) consisted of a 1 mm thick glass substrate and a 200 nm thick $Si_{0.75}Ti_{0.25}O_2$ waveguiding layer at the surface. A silica layer (ca. 12 nm) was sputter-coated on top of the waveguiding layer in a Leybold dc-magnetron Z600 sputtering unit. The coating conditions and the principles of OWLS investigations have been described in detail elsewhere.[28,29] Briefly, in optical waveguide lightmode spectroscopy, the adsorbed mass is calculated from the change of the refractive index in the vicinity of the surface upon adsorption of molecules from solution. The refractive index is a linear function of the concentration over a wide range, and thus the absolute amount of the adsorbed molecules can be determined using de Feijter's formula.[32] A refractive index increment (dn/dc) value of 0.134 cm^3/g for PLL(20)-*g*[3.5]-PEG(5) was derived from a linear interpolation between 0.13 cm^3/g (pure PEG) and 0.18 cm^3/g (pure PLL). It is important to note that the surface-adsorbed areal mass density determined by OWLS is regarded as a "dry areal mass density" due to the fact that solvent molecules coupled to the adsorbate will not contribute to a change in the refractive index and thus do not contribute to the detected adsorbate mass. The reported dry areal mass density, m_{dry}, represents the average of three individual experiments.

Quartz Crystal Microbalance with Dissipation (QCM-D). All QCM-D measurements were performed with a commercial quartz crystal microbalance with dissipation monitoring from Q-Sense, Gothenburg, Sweden.[33] The instrument was equipped with a home-built laminar flow cell with a glass window that allows visual monitoring of the injection and exchange of the liquids. The sensor crystals used in the measurements were 5 MHz AT-cut quartz, sputter-coated with SiO_2 (Q-Sense). All measurements were recorded at four different frequencies (5, 15, 25, 35 MHz). The QCM-D liquid chamber was temperature-stabilized to 25 ± 0.02 °C. The QCM-D response, i.e. changes in both resonant frequency, Δf_0, and the dissipation factor, ΔD, at different overtones, is, in contrast to OWLS, sensitive to viscoelastic properties and density of any mass coupled to the mechanical oscillation of the quartz crystal. In our case, the adsorbed mass consists of the copolymer, PLL-*g*-PEG, along with solvent molecules coupled to the polymer. The positively charged PLL backbone is strongly attached to the negatively charged SiO_2 sensor surface whereas, depending on the solvent quality, the PEG side chains form a solvated brushlike structure, extending into the liquid phase, thereby incorporating solvent molecules within the brush (see schematic representation in Figure 2). These absorbed solvent molecules may be loosely associated (i.e., hydrodynamically) or strongly attracted (e.g., via hydrogen bonding) to the brush but do not behave like bulk liquid above the film when probed by the crystal's oscillation. Hence, it is important to note that the mass sensed by QCM-D is the mass of polymer plus the mass of absorbed solvent molecules and is referred to as wet mass, m_{wet}, in contrast to the dry mass, m_{dry}, obtained by OWLS.[34]

In QCM-D, the change in resonance frequency is often directly related to the mass of the adsorbed layer according to the Sauerbrey equation (eq 1):[35]

$$m_{wet(Sauerbrey)} = -C\frac{\Delta f}{n} \tag{1}$$

where $m_{wet(Sauerbrey)}$ is the adsorbed mass, C is a constant characteristic of the crystal, Δf is the change in frequency, and n is the shear wavenumber. However, the Sauerbrey equation holds strictly only in a vacuum and gaseous environments.[36] Thus, in this study, we used a Voigt-based model (software: Q-tools, version 2.0.1) where the adsorbed layer is represented by a homogeneous viscoelastic film characterized by a shear viscosity, η_{shear}, a shear modulus, E_{shear}, and a film thickness, h_{film}.[36–38] The input parameters include the viscosity and the density of the solvent and the density of the polymer film, ρ_{film}. While the viscosity and density of the solvents used in the present study are well-known, ρ_{film} remains unknown. However, ρ_{film} can be approximated fairly well by combining the mass fraction of its contributory terms, $\rho_{solvent}$ and $\rho_{polymer}$ (eq 2). The relative weighting of the two terms, $\rho_{solvent}$ and $\rho_{polymer}$, can be calculated using both m_{dry} and m_{wet}. While m_{dry} is known from OWLS experiments, m_{wet} must be estimated by applying eq 1 using the Δf obtained by QCM-D experiments.

$$\rho_{film} = \frac{m_{wet(Sauerbrey)} - m_{dry}}{m_{wet(Sauerbrey)}}\rho_{solvent} + \frac{m_{dry}}{m_{wet(Sauerbrey)}}\rho_{polymer} \tag{2}$$

where $m_{wet(Sauerbrey)}$ is the mass derived from the Sauerbrey equation. $\rho_{polymer}$ stands for the density of PLL-*g*-PEG (≈1 g/cm^3) and $\rho_{solvent}$ for the densities of the liquids, being $\rho_{HEPES} = 1.0$, $\rho_{methanol} = 0.791$, $\rho_{ethanol} = 0.787$, and $\rho_{2\text{-propanol}} = 0.786$ g/cm^3 for aqueous HEPES buffer solution, methanol, ethanol, and 2-propanol, respectively.[39] By multiplying the layer thickness h_{film}, obtained by the Voigt-based model, by ρ_{film}, we obtain the wet areal mass density, m_{wet}, which in our case is some 5–10% higher than the wet mass density, $m_{wet(Sauerbrey)}$, derived from the Sauerbrey equation. In agreement with Larson et al.,[38] we observed that, unlike h_{film}, m_{wet} does not depend on the choice of the input parameter ρ_{film}. The value of m_{wet} is preserved as it results from the multiplication of h_{film} and the input parameter ρ_{film} which is obtained from eq 2. The reported wet mass density, m_{wet}, represents the average of three individual experiments performed on the same quartz crystal.

Atomic Force Microscopy. A home-built atomic force microscope was used to probe frictional forces of the polymer-modified SiO_2 substrates in liquid environments. The microscope was equipped with a liquid cell/tip holder (Digital Instruments, Santa Barbara, CA) and controlled by SPM 1000 electronics and SPM 32 software (RHK Technology, Inc., Troy, MI). A silica microsphere (diameter = 5.1 μm), attached to an AFM cantilever (Novascan Technologies, Ames, IA), was used as the counterface to the polymer-modified SiO_2 surface. Kinetic friction data were acquired by monitoring the lateral deflection of the cantilever as a function of position across the sample surface while sliding. While the sample was rastered in a line-scan mode, the load was first increased and then decreased. During this procedure, frictional forces and normal forces were measured simultaneously at a scan speed of ≈1.4 μm/s over a distance of 0.1 μm. The reported frictional data represent the average of at least six results obtained at different locations as a function of decreasing load across the surface. Normal loads were limited to ≈35 nN in order to avoid wear of the tip and surface, which would invalidate the comparison of frictional data. AFM measurements were carried out in the sequence of aqueous HEPES buffer solution, methanol, ethanol, and 2-propanol. The composition of the liquid environment encompassing the tip–sample interface was controlled by transferring aliquots of solvents in and out of the liquid cell through the use of two 5 mL syringes.

In the reported AFM measurements, normal loads have been calibrated directly from the reported spring constant of AFM cantilever (0.58 N/m), while the friction forces have been calibrated through an improved wedge calibration method. Briefly, the force calibration was experimentally carried out by sliding the tip across a silicon grating surface containing two known slopes (TGG01, MikroMasch, Narva mnt 13, 10151 Tallinn, Estonia) as a function of applied load. More details of this calibration method have been reported elsewhere.[40,41]

Results

Comparative Adsorption Measurements Using QCM-D and OWLS. In a QCM-D experiment the resonance frequency, f_0, and the dissipation factor, D, of a single quartz crystal were measured first in one of the organic solvents (methanol, ethanol, or 2-propanol) and subsequently in aqueous HEPES buffer solution to set baselines. After recording the baselines, the polymer-free aqueous HEPES buffer solution was replaced by PLL(20)-*g*[3.5]-PEG(5)-containing aqueous HEPES buffer solution (0.25 mg/mL). After adsorption for 30 min, the liquid cell was rinsed with polymer-free aqueous HEPES buffer solution, and f_0 and D were recorded. Subsequently, the aqueous HEPES buffer solution was exchanged by one of the organic solvents, and both f_0 and D were recorded again. Note that PLL-*g*-PEG was allowed to adsorb only from aqueous HEPES buffer solution. To ascertain that no surface-bound polymer desorbed due to the solvent exchange, the organic solvent was once again replaced by aqueous HEPES buffer solution. The shift in both f_0 and D upon this control measurement was highly reproducible with Δf_0 and ΔD measured for the solvent exchange directly after the polymer adsorption showing that no surface-bound polymer was lost during solvent exchange. The raw data from a typical QCM-D measurement for the HEPES–methanol system are displayed in Figure 3. The difference in both f_0 and D before and after polymer adsorption is a measure of the mass of the adsorbate in the corresponding solvent and is converted to a mass per

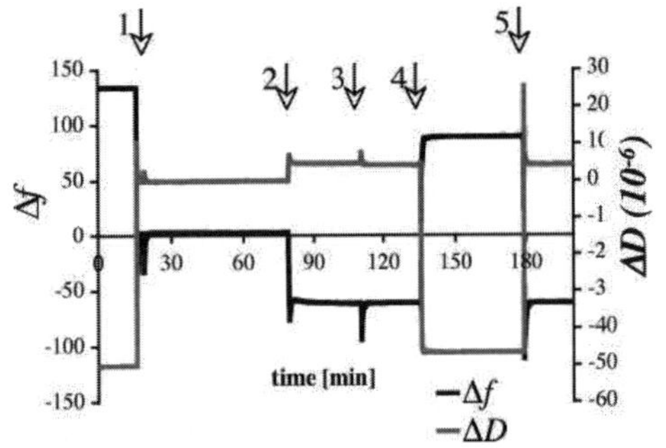

Figure 3. Changes in the normalized third overtone resonance frequency, Δf (black line), and dissipation, ΔD (gray line), during adsorption of PLL(20)-*g*[3.5]-PEG(5) onto a SiO_2 sputter-coated surface for the HEPES−methanol system (polymer injection at arrow number 2). Before injection of the polymer the baseline of the SiO_2-coated quartz crystal was measured in methanol and subsequently in HEPES buffer solution. The exchange of methanol for HEPES buffer solution is indicated by the arrow number 1. The measurement chamber was rinsed with polymer-free aqueous HEPES buffer solution 30 min after polymer injection (arrow number 3). Subsequently, the aqueous HEPES buffer solution was replaced by methanol, and the resonance frequency f_0 and the dissipation factor D were measured again (arrow number 4). The reproducibility of the Δf_0 and ΔD shifts upon solvent changes was tested by replacing methanol by aqueous HEPES buffer solution (arrow number 5). This measurement protocol for the methanol solvent system was repeated and applied to the other two solvent systems, ethanol and 2-propanol.

unit area using a Voigt-based model described in the Experimental Section. The obtained masses, referred to as wet mass, m_{wet}, are $m_{wet(HEPES)} = 1240.6$ ng/cm^2, $m_{wet(methanol)} = 876$ ng/cm^2, $m_{wet(ethanol)} = 637$ ng/cm^2, and $m_{wet(2-propanol)} = 402$ ng/cm^2 for aqueous HEPES buffer solution, methanol, ethanol, and 2-propanol, respectively. To obtain the dry mass of the surface adsorbed copolymer, we utilized *in situ* OWLS. The advantage of using OWLS instead of QCM-D for determining the adsorbed mass is that OWLS is an optical technique, which allows the quantification of the "dry" polymer mass in a liquid environment. The average dry mass value of PLL(20)-*g*[3.4]-PEG(5) adsorbed on a SiO_2 coated waveguide is $m_{dry} = 198$ ng/cm^2. By subtracting the dry mass from the wet mass, the solvent mass per unit substrate area, or areal solvation, Ψ, is obtained. It was found that the solvation is highest for aqueous HEPES buffer solution ($\Psi = 1043$ ng/cm^2) and decreases in the order of methanol ($\Psi = 678$ ng/cm^2), ethanol ($\Psi = 439$ ng/cm^2), and 2-propanol ($\Psi = 204$ ng/cm^2). From the experimental results for m_{dry} and m_{wet} it is feasible to calculate the number of solvent molecules per EG monomer averaged over the cross section of the PEG brush, $\Lambda_{sol/EG}$, by applying the following equations:

$$\Sigma_{sol} = \frac{N_A(m_{wet} - m_{dry})}{Mw_{sol}} \tag{2}$$

$$\Sigma_{EG} = \frac{Mw_{PLL}Mw_{PEG}m_{dry}}{N_A Mw_{EG} Mw_{LL} g\left(Mw_{PLL} + \frac{Mw_{PLL}Mw_{PEG}}{Mw_{LL}g}\right)} \tag{3}$$

$$\Lambda_{sol/EG} = \frac{\Sigma_{sol}}{\Sigma_{EG}} \tag{4}$$

where Σ_{sol} and Σ_{EG} are the areal density of the solvent molecules and ethylene glycol monomer units, EG, respectively, N_a is the Avogadro constant, Mw_{sol} is the molecular weight of the solvent, Mw_{PLL} is the molecular weight of the PLL backbone, Mw_{LL} is the molecular weight of a L-lysine monomer unit, Mw_{PEG} is the molecular weight of a PEG side chain, Mw_{EG} is the molecular weight of an ethylene glycol monomer unit, and g is the grafting ratio of the copolymer PLL-*g*-PEG. The calculated average ratio of EG monomers to solvent molecules of PLL-*g*-PEG immersed in HEPES aqueous buffer solutions is $\Lambda_{water/EG} = 14$, in methanol $\Lambda_{methanol/EG} = 5.1$, in ethanol $\Lambda_{ethanol/EG} = 2.3$, and in 2-propanol $\Lambda_{2-propanol/EG} = 0.8$.

Another technique that was used to estimate the solvent content and the number of solvent molecules per EG monomers of PLL-*g*-PEG coatings was recently developed by Pasche et al. and involves colloidal-probe AFM surface force measurements.[42] The main assumption made in this technique is that the unperturbed PEG layer is compressed by the colloidal probe from a fully solvated state to a solvent-free, dry state. Thus, the decrease in the layer thickness upon compression is likely to reflect the amount of solvent absorbed within the polymer brush. The results of that study are in reasonable agreement with the findings of the present work.

Table 1 provides an overview of the results of the comparative adsorption measurements using QCM-D and OWLS together with the coefficients of friction, which will be discussed below.

Solvent Characteristics. In an attempt to rationalize the effect of solvent characteristics on solvation, we make use of the three-component Hansen solubility parameter model.[30,43] These solubility parameters are commonly used in polymer chemistry to predict the solubility of a polymer in a solvent.[44] The Hansen solubility parameters are defined in terms of the cohesive energy density that relates to the amount of energy required to vaporize 1 mol of the solvent.

$$\delta = \left(\frac{-E}{V}\right)^{0.5} \tag{5}$$

where $-E$ is the internal energy change of vaporization and V is the molar volume of the solvent at the temperature of vaporization. Each solvent is characterized by a set of three parameters δ (Hansen solubility parameters, HSP) designated by subscripts d (dispersion), p (polar), and h (hydrogen bonding), according to the nature of the intermolecular cohesion energy they reflect. Total cohesion energy, $-E$, is then the sum of the individual energies that comprise it.

$$-E = -E_d - E_p - E_h \tag{6}$$

Dividing this by the molar volume, V, gives the total Hansen solubility parameter δ_0 (eq 7).

$$\delta_0^{\,2} = -\frac{E}{V} = -\frac{E_d}{V} - \frac{E_p}{V} - \frac{E_h}{V} \tag{7}$$

This parameter can also be stated in a form in which the individual terms refer to the dispersion, polar, and hydrogen-bonding parameters δ_d, δ_p, and δ_h, respectively (eq 8).

$$\delta_0^{\,2} = \delta_d^{\,2} + \delta_p^{\,2} + \delta_h^{\,2} \tag{8}$$

Table 1. Average Values (± Standard Deviation) for PLL(20)-*g*[3.5]-PEG(5), Adsorbed on a Silica Substrate Surface, of the Wet Mass Density, m_{wet}, the Dry Mass Density, m_{dry}, the Solvation, Ψ, the Number of Solvent Molecules per EG Monomer Averaged over the Cross Section of the PEG Brush, $\Lambda_{sol/EG}$, and the Coefficient of Friction for Symmetric (Silicon Wafer and Colloidal Probe Coated with PLL-*g*-PEG), and Asymmetric (Silicon Wafer Coated with PLL-*g*-PEG, Colloidal Probe Uncoated) Tribopairs

polymer–solvent system	h_{film} [nm]	m_{wet} [ng/cm²]	m_{dry} [ng/cm²]	ψ [ng/cm²]	$\Lambda_{sol/EG}$	μ (symmetric)	μ (asymmetric)
HEPES	12.4	1240.6 ± 62.9	197.8 ± 12.2	1042.7 ± 61.7	14	0.035 ± 0.001	0.220 ± 0.008
methanol	10.5	875.8 ± 52.5		678.0 ± 51.0	5.1	0.081 ± 0.001	0.306 ± 0.007
ethanol	7.5	636.7 ± 43.3		438.8 ± 41.6	2.3	0.159 ± 0.013	0.450 ± 0.01
2-propanol	4.6	401.5 ± 21.9		203.7 ± 18.2	0.8	0.321 ± 0.002	0.572 ± 0.007

Table 2. Hansen Solubility Parameters[30] (δ_0 = Total Hansen Solubility Parameter, δ_d = Dispersion Parameter, δ_p = Polar Parameter, δ_h = Hydrogen-Bonding Parameter)

Hansen no.	solvent	ψ [ng/cm²]	δ_0 [MPa$^{1/2}$]	δ_d [MPa$^{1/2}$]	δ_p [MPa$^{1/2}$]	δ_h [MPa$^{1/2}$]
696	water	1042.7 ± 61.7	47.9	15.5	16	42.4
570	methanol	678.0 ± 51.0	29.7	15.1	12.3	22.3
325	ethanol	438.8 ± 41.6	26.6	15.8	8.8	19.4
456	2-propanol	203.7 ± 18.2	23.5	15.8	6.1	16.9

Table 2 summarizes the values of solvation of the different solvent/PLL-*g*-PEG systems as a function of the total Hansen solubility parameters as well as a function of the single Hansen solubility parameters of the solvents, while Figure 4 represents these data in a graphical format. The total solubility parameter of the solvents decreases in the order water (δ_0 = 47.9 MPa$^{1/2}$), methanol (δ_0 = 29.7 MPa$^{1/2}$), ethanol (δ_0 = 26.6 MPa$^{1/2}$), and 2-propanol (δ_0 = 23.5 MPa$^{1/2}$). Solvation decreases in the same order; thus, a strong correlation between the Hansen solubility parameters and the measured solvation of the PEG brush is observed in Figure 4A. The values of the dispersion parameter of the solvents all fall within the very narrow range of 15.1 MPa$^{1/2}$ < δ_d < 15.8 MPa$^{1/2}$, and no systematic dependency between the dispersion parameter and solvation is observed in Figure 4B. However, when the solvation is plotted as a function of either the polar solubility parameter (Figure 4C) or the hydrogen-bonding solubility parameter (Figure 4D), a strong positive correlation is observed.

Frictional Properties of PLL-*g*-PEG in Different Solvents Using LFM. Figure 5 shows friction vs load measurements for "symmetric" (both the silicon wafer and the colloidal probe are coated with PLL-*g*-PEG) and "asymmetric" (only the silicon wafer is coated with PLL-*g*-PEG and the colloidal probe remained uncoated) interfaces under the different solvents (aqueous HEPES buffer solution, methanol, ethanol, 2-propanol). In both cases all friction curves show a linear dependence on load and go through the origin, indicating that the interfacial contact between colloidal probe and substrate is of a nonadhesive nature. Further, a clear dependence of the friction force on the type of solvent is observed over the entire range of loads investigated in this study. The slopes of the friction–load plots for both tribosystems decrease steeply in the order 2-propanol, ethanol, methanol, and aqueous HEPES buffer solution. Sliding contact of the asymmetric tribosystem showed higher friction forces than the symmetric case throughout the study. Table 1 lists the coefficients of friction, μ, obtained from the linear portion of the friction vs load plots, whereas in Figure 6, the coefficients of friction are plotted against areal solvation, Ψ. Both tribosystems, asymmetric and symmetric, experience a significant increase in the coefficient of friction with decreasing solvation. Interestingly, the shapes of the two plots are very similar. The absence of a qualitative difference for

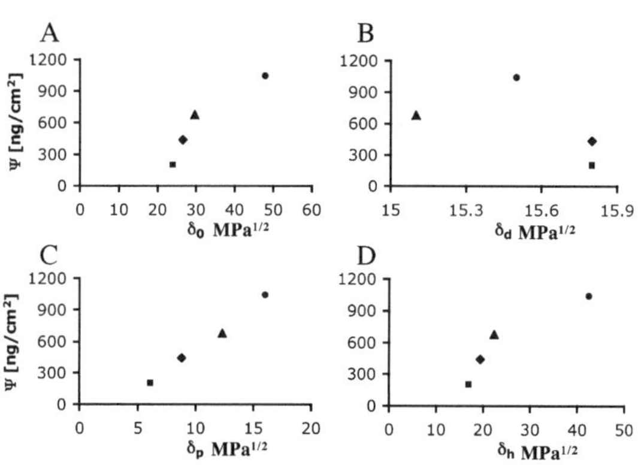

Figure 4. Effect of the total, δ_0, and the individual Hansen solubility parameters, δ_d, δ_p, δ_h, on the solvation, Ψ, of the PEG side chains. The study investigated the following solvents: ●, water; ▲, methanol; ◆, ethanol; ■, 2-propanol.

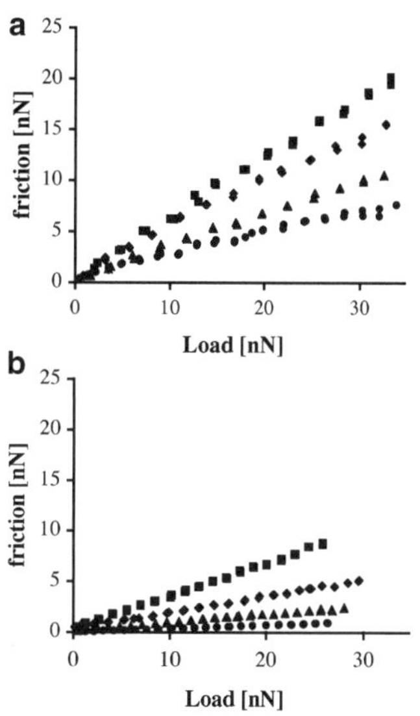

Figure 5. (●) Water, (▲) methanol, (◆) ethanol, and (■) 2-propanol. Interfacial friction measured as a function of decreasing load for the contact of (a) a bare and (b) a PLL-(20)-*g*[3.5]-PEG(5)-modified SiO_2 colloidal LFM-probe (diameter = 5.1 μm) and SiO_2 substrates coated with PLL(20)-*g*[3.5]-PEG(5). A single colloidal LFM probe was employed throughout the series of measurements.

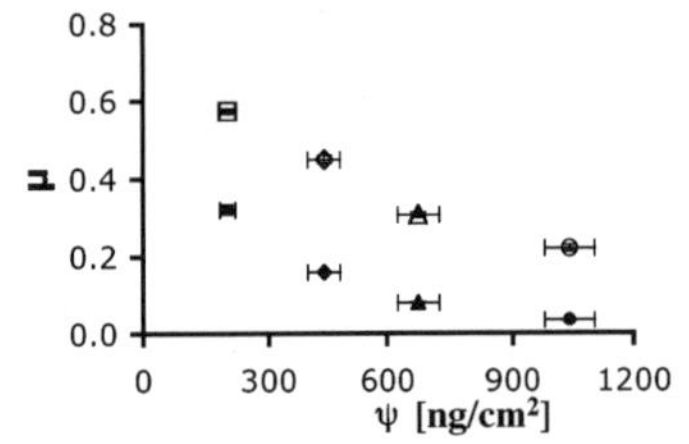

Figure 6. Coefficient of friction, μ, vs solvation, Ψ, for both asymmetric (open symbols) and symmetric (filled symbols) PLL(20)-*g*[3.5]-PEG(5) coated tribointerfaces (●, water; ▲, methanol; ◆, ethanol; ■, 2-propanol). Coefficients of friction were derived from a linear regression of the friction–load plots shown in Figure 5 and represent the mean values of three experiments (± standard deviation).

the two cases (symmetric and asymmetric) suggests that the molecular friction mechanism is not substantially different for these tribopairs.

Discussion

Previous tribological studies on PLL-*g*-PEG bearing tribopairs revealed a significant reduction of interfacial friction in an aqueous environment on both the macro- and the nanoscale.[15,24,25] The excellent lubrication properties of PLL-*g*-PEG have been ascribed to its distinctive adsorption behavior: the positively charged amino groups of the PLL backbone are electrostatically attracted to negatively charged oxide surfaces, forcing the PEG chains into a dense, brushlike structure that is highly solvated, storing a significant amount of solvent, when under good solvent conditions (Figure 2). To elucidate the relationship between solvent quality, conformation of the PEG brush, and the lubrication properties, we have quantified the amount of solvent absorbed in the PEG brush of surface-bound PLL-*g*-PEG in different solvents that vary in solvent quality with respect to PEG.

The wet mass obtained by QCM-D and the dry mass obtained by OWLS for PLL-*g*-PEG adsorbed from aqueous HEPES buffer solution on a flat SiO_2 substrate surface are $m_{wet} = 1241$ ng/cm^2 and $m_{dry} = 198$ ng/cm^2, respectively, resulting in a solvation of $\Psi = 1043$ ng/cm^2. This means that in the case where the PLL-*g*-PEG-bearing interface is immersed in aqueous HEPES buffer solutions, the mass of the hydrated PLL-*g*-PEG adlayer is more than 80% aqueous buffer solution and less than 20% polymer. However, in the case where the aqueous HEPES buffer solution is exchanged for an organic solvent, the mass of solvent molecules absorbed in the PEG brush reduces in the order $\Psi = 678$ ng/cm^2 for methanol, $\Psi = 438$ ng/cm^2 for ethanol, and $\Psi = 203$ ng/cm^2 for 2-propanol. It is important to note that the PLL-*g*-PEG layers were only adsorbed from pure aqueous HEPES buffer solution, and in the case of the organic solvents, a repeated exchange of the aqueous HEPES buffer solution for one of the organic solvents was carried out. To ensure a complete exchange of the solvent, the liquid cell was purged with an excess amount of liquid. We conclude that the quality of solvent, with respect to the PEG brush, decreases in the order aqueous HEPES buffer solution, methanol, ethanol, and 2-propanol.

The friction force measurements were performed in two different configurations: asymmetric, where only the flat substrate surface was coated by PLL-*g*-PEG and the colloidal probe of the LFM remains bare, and symmetric, where both surfaces were coated by PLL-*g*-PEG. As shown in Figure 5, all the friction–load plots obtained in this study show an Amontons-type behavior, and thus the coefficient of friction can be expressed through linear regressions of the friction–load plots. The coefficients of friction for all the measurements performed in this study are listed in Table 1. Raviv et al.[45] have reported on the properties and interactions of end-functionalized PEG (M_w = 3400 Da) layers physically grafted on mica and measured in aqueous solution by means of surface force apparatus (SFA). For the case of a symmetrically coated tribopair in the present study, the measured coefficient of friction, $\mu = 0.035 \pm 0.001$, is comparable to the value reported by Raviv et al.,[45] $\mu = 0.03$.

As expected, the coefficient of friction for both asymmetric and symmetric interfaces showed a clear dependence on the type of solvent. For both types of interface, the coefficient of friction increases in the same order as the solvent quality decreases: aqueous HEPES buffer solution, methanol, ethanol, and 2-propanol.

A comparison of the coefficient of friction with the amount of solvent molecules absorbed in the brushlike structure of PLL-*g*-PEG is shown in Figure 6. A strong correlation between coefficient of friction and solvation is observed for both symmetric and asymmetric tribointerfaces; i.e., the higher the solvation, the lower the coefficient of friction.

It is generally accepted that PEG has a strong tendency to hydrogen bond with surrounding water molecules, thus forming PEG–water complexes with a composition of approximately three water molecules directly associated with each EG monomer unit.[46] However, a closer examination of the experimental results of this study reveals that in aqueous HEPES buffer solution every EG monomer of the surface-bound PLL-*g*-PEG brush is, on the average over the whole interface of the polymer brush layer, surrounded by 14 water molecules, $\Lambda_{water/EG} = 14$. Thus, it can be concluded that, in addition to the three hydrogen-bonded water molecules, on average there exist another 11 water molecules per EG monomer within the brush structure immersed in aqueous HEPES solution. In comparison, the ratio of EG monomers to solvent molecules in methanol, ethanol, and 2-propanol is approximately $\Lambda_{methanol/EG} = 5.1$, $\Lambda_{ethanol/EG} = 2.3$, and $\Lambda_{2-propanol/EG} = 0.8$, respectively. However, it is not likely that hydrogen-bonded water molecules will easily be removed by a solvent exchange. Thus, in the case of the organic solvents, the ratio of EG monomers to solvent molecules, stated above, represents an overestimation by the fact that a small portion of water molecules will possibly remain in the brush structure following the exchange of the solvent. From the analysis of the solvents in terms of the Hansen solubility parameters, a fair correlation of the brush solvation with the total solubility parameter is observed in Figure 4. A detailed analysis of the different solubility parameters as a function of solvation reveals that there is no correlation between solvation and the dispersion parameter (Figure 4B). However, when solvation is plotted as a function of either the polar or the hydrogen-bonding interaction parameters, a strong correlation is observed (Figure 4C,D). The ability of the PEG brush to absorb solvent molecules from the bulk solution increases significantly with the increase of both the solvent's polar and hydrogen-bonding interaction parameters. This is not unexpected; the formation of structures between the solvent molecules and the monomer units of PEG is progressively less pronounced in the case of the organic solvents because the organic solvents show a significantly weaker ability to form either dipole–dipole interactions or hydrogen bonds, when compared to water. Thus, the interactions of the EG/EG monomer units become relatively more attractive with the deterioration of the solvent quality, and the remaining organic solvent molecules absorbed in the structure of the PEG chains experience a progressively smaller retaining force compared to the water molecules, as hydrogen-bonding and polar interactions decrease. As a result, solvent molecules are squeezed out more easily when the PLL-*g*-PEG interface comes under compression under poor solvent conditions. The case of a PLL-*g*-PEG-bearing tribopair sliding in a solvent of inferior quality, therefore, differs significantly from that of a PLL-*g*-PEG bearing tribopair sliding in good solvents as both long-ranged repulsive forces of osmotic origin and the entropic effects restricting mutual interpenetration are reduced under conditions of poor solvent quality. This leads to a less fluid and less mobile interface where more energy is dissipated during sliding motion, resulting in a higher coefficient of friction for brush-bearing tribopairs sliding in solvents of inferior quality.

Comparing the results of the friction force measurements for the two different tribogeometries, asymmetric and symmetric, it is important to note that the sliding contact of the asymmetric tribosystem shows significantly higher friction forces compared to that of the symmetric tribosystem throughout this work. However, the dependence of the coefficient of friction on areal solvation for the asymmetric case mirrors that of the symmetric interface (Figure 6). This behavior strongly suggests that the molecular friction mechanism is not intrinsically different for the two different tribointerfaces although the contact zone of the asymmetric interface differs from that of the symmetric interface in a number of important aspects.[10] In the former case, the polymer brush encounters the rigid surface of the SiO_2 colloidal probe of the LFM, whereas in the latter case, surface-bound PLL-*g*-PEG contacts an opposing PLL-*g*-PEG brush layer. In both cases, attractive forces must also be considered within the contact zone of the two opposing tribopairs due to potential bridging forces in the asymmetric case and due to both entanglement and bridging forces of opposing PEG chains in the symmetric case.[4,7,10] Bridging forces in the asymmetric case would arise if the PEG chains had a preference to adsorb onto the opposing silicon substrate surface. However, in the present case, force–distance curves (data not shown) have shown that no adhesion is detected for either asymmetric or symmetric interfaces. This observation indicates that the bridging between PEG and bare SiO_2 colloidal probe and the entanglement of opposing polymer brushes can either be ruled out or are of minor importance. In line with the findings for the effect of the solvent quality, discussed previously in this paper, the differences in the friction forces for the asymmetric and symmetric tribointerfaces may rather be explained by differences in both the fluidity and the shear mobility of the brush interfaces. Notably, and as shown by Brown,[47] the shear stress associated with the brushes depends crucially on the segment mobility of the materials on both sides, and when the materials on both sides of the interface are highly mobile, as is the case in the brush–brush interaction (symmetric tribopair), lower stress is produced.

However, considering that the silicon wafer and the SiO_2 colloidal probe possess a finite surface roughness in the nanometer range, the interfacial film thickness is a potential parameter contributing to the differences in the coefficients of friction of the two tribosystems. Asperity contacts are less effectively shielded in the case of lower film thickness, and thus a higher additional shear force is likely to be required in the case of the asymmetrically coated tribointerface to overcome surface irregularities than for the symmetric tribopair.

In summary, we suggest that both the degree of interfacial fluidity and the polymer layer thickness play key roles in determining the lubrication performance of surface-bound brushlike polymers in general and of PLL-*g*-PEG in particular. Both solvent quality and the nature of the counterface can influence these parameters.

Conclusions

The friction forces between "symmetric" (both sides coated with PLL(20)-*g*[3.5]-PEG(5)) and "asymmetric" (Si wafer coated with PLL(20)-*g*[3.5]-PEG(5), silica probe unmodified) interfaces have been measured by means of colloidal probe LFM in relation to the mass of solvent absorbed in the unperturbed brushlike structure of the PEG chains. The mass of solvent absorbed in the PEG brush per unit area, referred to as solvation, was varied by choosing different solvents. The solvent environment plays a key role in determining both the conformational and the frictional properties of surface-bound brushlike polymers. Notably, both solvation and the coefficient of friction are strongly dependent on the nature of the solvent, characterized in terms of the three-component Hansen solubility parameters. It was found that solvation of the PEG brush decreases with both decreasing polar and decreasing hydrogen-bonding solubility parameters causing an increase of the coefficient of friction. No dependency of friction on the dispersion parameter was observed. The energy-dissipating components of the molecular interaction mechanism of surface-bound PLL-*g*-PEG have been discussed on the basis of the comparison of the frictional behavior of symmetric vs asymmetric coated tribointerfaces. In line with previous theoretical and experimental work, we conclude that both the interfacial fluidity and the segment mobility of the brush chains play a prominent role in determining the lubrication properties of brush-like polymers. Additionally, in cases where the substrate surfaces are not atomically smooth, a finite surface roughness may contribute to the observed tribological properties.

Acknowledgment. This work was financially supported by the Council of the Swiss Federal Institutes of Technology (ETH-Rat TopNano 21) and the US Air Force Office of Scientific Research (Contract F49620-02-1-0346). We are grateful to Dr. Janos Vörös of the Laboratory of Surface Science and Technology, ETH Zürich, Switzerland, for his valuable advice and assistance in the QCM-D and OWLS experiments.

References and Notes

(1) Klein, J.; Kumacheva, E.; Mahalu, D.; Perahia, D.; Fetters, L. J. *Nature (London)* **1994**, *370*, 634–636.

(2) Tadmor, R.; Janik, J.; Klein, J.; Fetters, L. J. *Phys. Rev. Lett.* **2003**, *91*, art. no.-115503.
(3) Raviv, U.; Tadmor, R.; Klein, J. *J. Phys. Chem. B* **2001**, *105*, 8125–8134.
(4) Klein, J.; Kumacheva, E.; Perahia, D.; Mahalu, D.; Warburg, S. *Faraday Discuss.* **1994**, *98*, 173–188.
(5) Fredrickson, G. H.; Pincus, P. *Langmuir* **1991**, *7*, 786–795.
(6) Kreer, T.; Müser, M. H. *Wear* **2003**, *254*, 827–831.
(7) Grest, G. S. *Adv. Polym. Sci.* **1999**, *138*, 149–183.
(8) Grest, G. S. *Phys. Rev. Lett.* **1996**, *76*, 4979–4982.
(9) Kreer, T.; Binder, K.; Muser, M. H. *Langmuir* **2003**, *19*, 7551–7559.
(10) Kampf, N.; Gohy, J.-F.; Jerome, R.; Klein, J. *J. Polym. Sci., Part B: Polym. Phys.* **2005**, *43*, 193–204.
(11) Klein, J. *J. Annu. Rev. Mater. Sci.* **1996**, *26*, 581–612.
(12) Schorr, P. A.; Kwan, T. C. B.; Kilbey, S. M.; Shaqfeh, E. S. G.; Tirrell, M. *Macromolecules* **2003**, *36*, 389–398.
(13) Granick, S.; Kumar, S. K.; Amis, E. J.; Antonietti, M.; Balazs, A. C.; Chakraborty, A. K.; Grest, G. S.; Hawker, C.; Janmey, P.; Kramer, E. J.; Nuzzo, R.; Russell, T. P.; Safinya, C. R. *J. Polym. Sci., Part B: Polym. Phys.* **2003**, *41*, 2755–2793.
(14) Raviv, U.; Giasson, S.; Kampf, N.; Gohy, J.-F.; Jérôme, R.; Klein, J. *Nature (London)* **2003**, *425*, 163–165.
(15) Yan, X. P.; Perry, S. S.; Spencer, N. D.; Pasche, S.; De Paul, S. M.; Textor, M.; Lim, M. S. *Langmuir* **2004**, *20*, 423–428.
(16) Lai, P. Y.; Binder, K. *J. Chem. Phys.* **1992**, *97*, 586–595.
(17) Roters, A.; Schimmel, M.; Ruhe, J.; Johannsmann, D. *Langmuir* **1998**, *14*, 3999–4004.
(18) Harris, M. J. E. *Poly(ethylene glycol) Chemistry: Biomedical and Biomedical Applications;* Plenum Press: New York: 1992.
(19) Elbert, D. L.; Hubbell, J. A. *Chem. Biol.* **1998**, *5*, 177–183.
(20) Kenausis, G. L.; Vörös, J.; Elbert, D. L.; Huang, N. P.; Hofer, R.; Ruiz-Taylor, L.; Textor, M.; Hubbell, J. A.; Spencer, N. D. *J. Phys. Chem. B* **2000**, *104*, 3298–3309.
(21) Pasche, S.; De Paul, S. M.; Vörös, J.; Spencer, N. D.; Textor, M. *Langmuir* **2003**, *19*, 9216–9225.
(22) Wagner, M. S.; Pasche, S.; Castner, D. G.; Textor, M. *Anal. Chem.* **2004**, *76*, 1483–1492.
(23) Heuberger, M.; Drobek, T.; Vörös, J. *Langmuir* **2004**, *20*, 9445–9448.
(24) Lee, S.; Müller, M.; Ratoi-Salagean, M.; Vörös, J.; Pasche, S.; De Paul, S. M.; Spikes, H. A.; Textor, M.; Spencer, N. D. *Tribol. Lett.* **2003**, *15*, 231–239.
(25) Müller, M.; Lee, S.; Spikes, H. A.; Spencer, N. D. *Tribol. Lett.* **2003**, *15*, 395–405.
(26) Stalgren, J. J. R.; Eriksson, J.; Boschkova, K. *J. Colloid Interface Sci.* **2002**, *253*, 190–195.
(27) Craig, V. S. J.; Plunkett, M. *J. Colloid Interface Sci.* **2003**, *262*, 126–129.
(28) Kurrat, R.; Textor, M.; Ramsden, J. J.; Böni, P.; Spencer, N. D. *Rev. Sci. Instrum.* **1997**, *68*, 2172–2176.
(29) Vörös, J.; Ramsden, J. J.; Csúcs, G.; Szendrö, I.; De Paul, S. M.; Textor, M.; Spencer, N. D. *Biomaterials* **2002**, *23*, 3699–3710.
(30) Hansen, C. M. *Hansen Solubility Parameters: A User's Handbook*, CRC Press: Boca Raton, FL, 2000; pp 3–208.
(31) Huang, N. P.; Michel, R.; Vörös, J.; Textor, M.; Hofer, R.; Rossi, A.; Elbert, D. L.; Hubbell, J. A.; Spencer, N. D. *Langmuir* **2001**, *17*, 489–498.
(32) Defeijter, J. A.; Benjamins, J.; Veer, F. A. *Biopolymers* **1978**, *17*, 1759–1772.
(33) Rodahl, M.; Höök, F.; Krozer, A.; Brzezinski, P.; Kasemo, B. *Rev. Sci. Instrum.* **1995**, *66*, 3924–3927.
(34) Vörös, J. *Biophys. J.* **2004**, *87*, 553–561.
(35) Rodahl, M.; Kasemo, B. *Sens. Actuators, A* **1996**, *54*, 448–456.
(36) Voinova, M. V.; Rodahl, M.; Jonson, M.; Kasemo, B. *Phys. Scr.* **1999**, *59*, 391–396.
(37) Bandey, H. L.; Hillman, A. R.; Brown, M. J.; Martin, S. J. *Faraday Discuss.* **1997**, *107*, 105–121.
(38) Larsson, C.; Rodahl, M.; Höök, F. *Anal. Chem.* **2003**, *75*, 5080–5087.
(39) Lide, D. R. *Handbook of Chemistry and Physics*, 72nd ed.; CRC Press: London, 1992; pp 15.43–15.50.
(40) Ogletree, D. F.; Carpick, R. W.; Salmeron, M. *Rev. Sci. Instrum.* **1996**, *67*, 3298–3306.
(41) Varenberg, M.; Etsion, I.; Halperin, G. *Rev. Sci. Instrum.* **2003**, *74*, 3362–3367.
(42) Pasche, S.; Textor, M.; Meagher, L.; Spencer, N. D.; Griesser, H. J., submitted to *Langmuir*.
(43) Barton, A. F. M. *Handbook of Polymer-Liquid Interaction Parameters and Solubility Parameters*, 2nd ed.; CRC Press: Boca Raton, FL, 1991; pp 11–15.
(44) Zhao, L. Y.; Choi, P. *Polymer* **2004**, *45*, 1349–1356.
(45) Raviv, U.; Frey, J.; Sak, R.; Laurat, P.; Tadmor, R.; Klein, J. *Langmuir* **2002**, *18*, 7482–7495.
(46) Tasaki, K. *J. Am. Chem. Soc.* **1996**, *118*, 8459–8469.
(47) Brown, H. R. *Science* **1994**, *263*, 1411–1413.

MA0501545

ELSEVIER

Tribology International 38 (2005) 922–930

TRIBOLOGY INTERNATIONAL

www.elsevier.com/locate/triboint

Aqueous lubrication of polymers: Influence of surface modification

Seunghwan Lee, Nicholas D. Spencer*

Laboratory for Surface Science and Technology, Department of Materials, Swiss Federal Institute of Technology, ETH-Hönggerberg, Wolfgang-Pauli-Strasse 10, CH-8093 Zürich, Switzerland

Available online 10 August 2005

Abstract

We have investigated the influence of surface modification of an elastomer, poly(dimethylsiloxane) (PDMS), on its aqueous lubrication properties. A dramatic reduction in frictional forces has been observed upon hydrophilization by oxygen-plasma treatment or by surface coating with amphiphilic co-polymers, when PDMS was slid against PDMS in an aqueous environment. This effect is attributed to the removal of the strong hydrophobic interaction between PDMS surfaces in water, thereby enabling the isoviscous-elastic lubrication (or soft EHL) mechanism to predominate. This study demonstrates the significance of surface modification in allowing effective soft EHL of an elastomer.

Keywords: Aqueous lubrication; Poly(dimethylsiloxane) (PDMS); Oxygen-plasma-treatment; Hydrophilization; Soft EHL; PLL-*g*-PEG; PEO-*b*-PPO-*b*-PEO

1. Introduction

Many theoretical models have been developed to predict the film-thickness for fluid-film lubrication of non-conformal contacts [1–4]. In these models, the elasticity of the tribopairs and the viscosity of the lubricant are typically considered as the two main parameters that determine the fluid-film lubrication regime, such as isoviscous-rigid, piezoviscous-rigid, isoviscous-elastic, and piezoviscous-elastic lubrication. According to a model proposed by Hamrock and Dowson [1] (later revised by Esfahanian and Hamrock [2]), for instance, a specific fluid-film lubrication regime can be easily identified based upon the elasticity and viscosity parameters of a given tribosystem, and the fluid-film-thickness for a specific condition (such as load, mean speed) can be predicted. In reality, however, the theoretically predicted lubricating films are frequently not observed, and consequently the desired reduction of friction is not achieved. This is mainly because many other parameters than those used to model the fluid-film-thickness, such as the surface roughness of contacting bodies, also influence the interfacial friction forces. It is noted that most film-thickness models assume non-interacting and ideally smooth tribopair surfaces.

Among the aforementioned fluid-film lubrication regimes, isoviscous-elastic lubrication is often known as soft elastohydrodynamic lubrication (soft EHL) and typically occurs for lubricated contact involving an elastomer. In this lubrication regime, deformation of the tribopair plays a significant role in determining film-thickness due to the low elasticity moduli involved, whereas the increase of the lubricant viscosity with increasing pressure is negligible. Since an increase of lubricant viscosity is not a prerequisite for this lubrication mechanism, even liquids with extremely low pressure-coefficients of viscosity, such as water, can be used as lubricants when the tribopair is constructed from elastomeric materials. Some examples for this type of lubrication include human and animal synovial joints, tyres on wet roads, and various machines that are constructed from elastomeric materials [5–11]. While for rigid materials surface roughness is a common cause of premature breakdown of lubricating films, for elastomeric materials, surface interactions (e.g. hydrophobic) between the tribopairs often dominate, hindering the formation of stable lubricating films. For soft EHL of elastomers by water, for instance, the failure of the fluid film is often related to an increase in the adhesive component of friction, since most rubber-like materials tend to display surface hydrophobicity [11,12]. Thus, the control of the surface properties appears to be of prime importance to

* Corresponding author. Fax: +41 44 633 10 27.
E-mail address: nicholas.spencer@mat.ethz.ch (N.D. Spencer).

0301-679X/$ - see front matter © 2005 Elsevier Ltd. All rights reserved.
doi:10.1016/j.triboint.2005.07.017

ensure effective aqueous lubrication of rubber-like materials.

In this work, we have employed a silicone elastomer, poly(dimethylsiloxane) (PDMS), to investigate the influence of surface modification on its aqueous lubrication properties. PDMS was chosen for these studies, since it not only possesses a low elasticity modulus, but also can be readily fabricated into (hemi)spherical and plane shapes with a smooth surface finish. Thus, the influence of surface roughness can be ruled out in achieving a lubricating fluid-film, leaving the surface interaction as the dominant factor. We thus demonstrate that surface modification of PDMS, either by oxygen-plasma-treatment or by surface coating with amphiphilic co-polymers, can significantly modify its frictional properties in an aqueous environment, by facilitating the formation of a lubricating film.

2. Materials and methods

2.1. Tribopairs

Poly(dimethylsiloxane) (PDMS) has been employed as an elastomer and was used for both slider (pin) and track (disk). A commercial silicone elastomer kit (SYLGARD® 184 silicone elastomer, base and curing agent, Dow Corning, Midland, MI, USA) was purchased to prepare both PDMS pin and disk. For the preparation of hemispherical pin and flat disk, corresponding molds were fabricated; a commercial polystyrene cell-culture plate with round-shaped wells (96 MicroWell Plate™, NUCLON™ Delta Surface, Roskilde, Denmark) was used as a pin master (radius 3 mm) and a home-machined aluminum plate with flat wells (diameter 30 mm and depth 5 mm) was used as a disk master. The PDMS pin and disk were prepared according to a conventional recipe [12]. Briefly, the base and the curing agent of SYLGARD® 184 silicone elastomer kit were mixed at 10:1 ratio by weight. After removing the foams generated during mixing by a gentle vacuum, the mixture was transferred into the masters and incubated in an oven (~70 °C) overnight. The water-contact angle for the PDMS prepared according to this method was 110° (±2°). The elasticity modulus and Poisson ratio of PDMS were ca. 2 and 0.5 MPa, respectively. The surface roughness (Ra) of the PDMS disk surface was 0.5 nm over $10\times10\ \mu m^2$ (for the side exposed to ambient during the curing process) as measured by atomic force microscopy (Dimension 3000, Digital Instruments, Santa Barbara, CA, USA). The surface roughness (Ra) of the PDMS pin was estimated as ca. 2 nm over $10\times10\ \mu m^2$ by measuring the morphology of the polystyrene master by AFM. Polypropylene (PP) and polyamide 6,6 (PA-6,6) have also been employed for tribopairs. Both pin and disk materials were purchased from Maagtechnic (Dübendorf, Switzerland). The PP and PA-6,6 sphere pins (3 mm in radius) were polished by the manufacturer and were used as received. Five millimetre thick disks of both polymers were machined from 30 mm-diameter rods. The surfaces of the disks were also polished with silica paper on a rotating polishing wheel (grade P600 followed by grade P1200). The surface roughness (Ra) of the polymer pins characterized by AFM was 80 ± 10 and 250 ± 20 nm over $10\times10\ \mu m^2$ for PP and PA-6,6, respectively. The surface roughness (Ra) of the polymer disks was characterized as 30 ± 10 nm over $10\times10\ \mu m^2$ by AFM. The surface and mechanical properties of the tribopairs are summarized in Table 1.

2.2. Pin-on-disk tribometry

The frictional properties of the tribopairs during aqueous-lubricated sliding have been characterized by means of conventional pin-on-disk tribometry (CSM, Neuchâtel, Switzerland). In this approach, the load was determined by dead weights (0.5–5 N) and the frictional forces measured by a strain gauge. After forming a contact between loaded pin and disk, the disk was rotated at a controlled speed by a motor, thus generating sliding friction forces. The raw data for the friction forces were recorded as a function of time (or the number of rotations) over a fixed track, using a Macintosh Power PC with a Labview program and an ADC card of the MIO family (both from National Instruments, Austin, TX, USA).

To characterize the lubricating properties of water for a variety of tribopairs, load- and speed-dependent frictional data have been measured. Frictional forces were measured as a function of load at a fixed sliding speed (at 0.005 m/s unless otherwise mentioned), and μ ($\mu=F/w$, where F is friction and w is load) was measured as a function of speed (from 0.00025 to 0.1 m/s) at a fixed load (1 N unless otherwise mentioned). For both types of measurement,

Table 1
Some mechanical and surface properties of the tribopairs employed in this work (viscosity of water, $\eta_0=9\times10^{-4}$ Pa s and pressure-coefficient of viscosity of water, $\xi=3.6\times10^{-10}\ Pa^{-1}$)

	Elasticity modulus (MPa)[a]	Poisson ratio	Surface roughness (nm/$10\times10\ \mu m^2$) pin	disk	Static water contact angle (°) as is	O_2 plasma
PDMS	2	0.5	2 ± 0.5	0.5 ± 0.2	110 ± 2	<3
PP	1100-1300	0.3	80 ± 10	30 ± 10	107 ± 2	48 ± 2
PA-6,6	2000	0.3	250 ± 20	30 ± 10	82 ± 2	20 ± 2

[a] The elasticity moduli of PP and PA-6,6 are from the manufacturer, and the mid-values were used for the calculation shown in this table.

the average frictional forces were obtained during the 11–20th rotations out of a total of 20 rotations, in order to rule out running-in effects.

2.3. Hydrophilization of surfaces

In most cases, the hydrophilization of the tribopair was achieved by means of oxygen-plasma-treatment for 1 min in a Harrick Plasma Cleaner/Sterilizer PDC-32G instrument (Ossining, NY, USA). Following oxygen-plasma-treatment, the water-contact angle of the PDMS surface (ox-PDMS) was measured to be lower than 3°. The static water contact angles (±2°) for PP and PA-6,6 disk surfaces before (and after) oxygen-plasma-treatment were 107 (48°) and 82 (20°), respectively.

Amphiphilic copolymers have also been employed for the hydrophilization of PDMS surfaces. Two types of copolymers have been employed; poly(ethylene oxide)-*block*-poly(propylene oxide)-*block*-poly(ethylene oxide) (PEO-*b*-PPO-*b*-PEO) and poly(L-lysine)-graft-poly(ethylene glycol) (PLL-*g*-PEG). PEO-*b*-PPO-*b*-PEO (Pluronic®) was a commercially available, standard material that was kindly supplied by BASF (Mt Olive, NJ, USA). The molecular weight of the PEO and PPO blocks of the PEO-*b*-PPO-*b*-PEO used in this work were ca. 3250 (1625 for each PEO side) and 3250 g/mol, respectively (denoted as 'P105' according to the manufacturer [13,14]). The PEO-*b*-PPO-*b*-PEO was dissolved in distilled water at 2 mg/ml concentration and used as an aqueous lubricant. Since PEO-*b*-PPO-*b*-PEO is well known to adsorb onto various hydrophobic surfaces through hydrophobic interactions between the PPO block and the surface, no treatment was carried out other than immersing the tribopair into the polymer solution (15 min prior to the friction measurements).

The PLL-*g*-PEG was synthesized by the reaction of *N*-hydroxysucinimidyl-ester-functionalized PEG (*N*-hydroxysuccinimidyl ester of methoxy(ethylene glycol) propionic acid, mPEG-SPA, Nektar AL, Huntsville, AL, USA) with the amino groups of PLL (poly(L-lysine) hydrobromide, Sigma, St Louis, MO, USA) at a controlled stoichiometric ratio. More details on the synthesis and analytical information are available in previous publications [15–17]. The PLL-*g*-PEG employed in this work, denoted as PLL(20)-*g*[3.4]-PEG(5), possesses the molecular weight of 20,000 g/mol for the PLL backbone (including HBr as a precursor), and 5000 g/mol for PEG side chains, and 3.4 for the graft ratio (lysine-mer/PEG side chains). Since PLL-*g*-PEG has been characterized to form a stable layer through the electrostatic attraction between polycationic PLL and negatively charged oxide surfaces in aqueous environment [15–17], the PDMS surface was oxygen-plasma-treated to generate a negatively charged surface [12,18]. The copolymer was dissolved in 10 mM HEPES (4-(2-hydroxyethyl)-1piperaine-ethanesulfonic acid (Sigma, St Louis, MO, USA), adjusted to pH 7.4 with 1.0 M NaOH solution) at 0.25 mg/ml concentration and used as an aqueous lubricant.

3. Results and discussion

3.1. Theoretically predicted film thickness

As mentioned in Section 1, the two main parameters determining the full-fluid-film lubrication regime for non-conformal contact are the combined elasticity of the tribopair and the viscosity of the lubricant. These parameters are often made dimensionless for simplification and generalization. For instance, according to Hamrock and Dowson [1], the dimensionless viscosity and elasticity parameters are defined as follows

$$\text{Dimensionless viscosity parameter,} \quad g_V = \frac{GW^3}{U^2} \tag{1}$$

$$\text{Dimensionless elasticity parameter,} \quad g_E = \frac{W^{8/3}}{U^2} \tag{2}$$

where G, W, and U represent dimensionless materials parameter, dimensionless load parameter, and dimensionless speed parameter, respectively, and are defined as

$$\text{Dimensionless materials parameter,} \quad G = \xi E' \tag{3}$$

$$\text{Dimensionless load parameter,} \quad W = \frac{w}{E'R_x^2} \tag{4}$$

$$\text{Dimensionless speed parameter,} \quad U = \frac{\eta_0 u}{E'R_x} \tag{5}$$

where ξ is the pressure-coefficient of viscosity, E' is the effective elastic modulus, w is the applied load, η_0 is the viscosity at atmospheric pressure, u is the mean speed, and R_x is the effective radius in the x-direction (which is the same as R_y in sphere/plane contact).

Finally, the dimensionless film-thickness parameter is defined as

Dimensionless film thickness parameter,

$$\hat{H} = H\left(\frac{W}{U}\right)^2 \tag{6}$$

where H is the dimensionless film-thickness and is, in turn, defined as

$$\text{Dimensionless film thickness,} \quad H = \frac{h}{R_x} \tag{7}$$

where h is film thickness. Given the ellipticity parameter $k=1$ for sphere/plane contact, as in the present study, the lubrication regime can be identified from the map with the film-thickness contours on a log–log grid of viscosity and elasticity parameters [1,2].

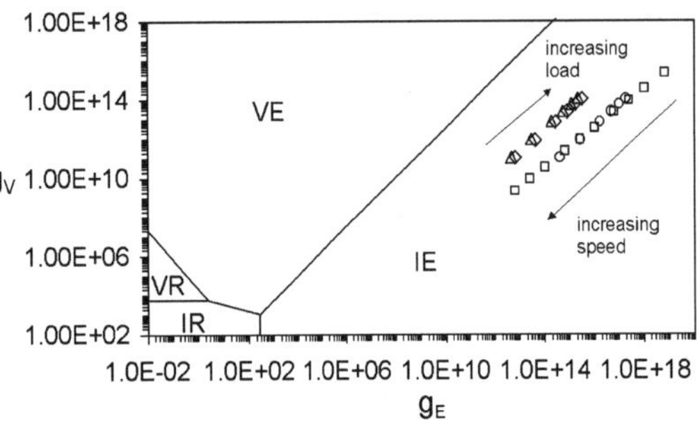

Fig. 1. The map of lubrication regimes for the non-conformal fluid-film contact (ellipticity parameter $k=1$) and the location of the tribo-contacts employed in this work (○ for PDMS in load-dependence and □ for PDMS in speed-dependence measurements, and ◇ for PP and △ for PA-6,6 in load-dependence measurements). IR, VR, IE, and VE stand for isoviscous-rigid, piezoviscous-rigid, isoviscous-elastic, piezoviscous-elastic, respectively.

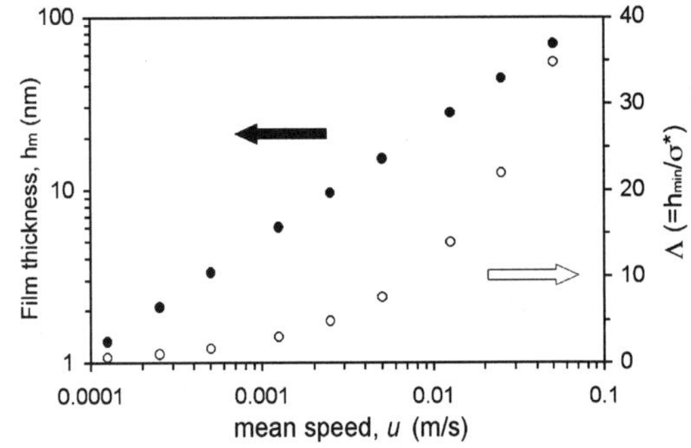

Fig. 3. The theoretically predicted minimum film-thickness (●) (Hamrock and Dowson model [1,2]) and corresponding Λ ratio (○) ($\Lambda = h_{min}/\sigma^*$, where σ^* is mean surface roughness, $\sigma^* = \sqrt{\sigma_{pin}^2 + \sigma_{disk}^2} \cong 2$ nm for PDMS surface (see Table 1)) for the water-lubricated sliding of PDMS vs. PDMS as a function of speed ($w=1$ N).

The tribopairs in this work include self-mated PDMS, PP, and PA-6,6 and the lubricant is pure water. The bulk mechanical properties of the tribopairs and the viscosity properties of water are summarized in Table 1. Three dimensionless parameters, i.e. those for material (G), load (W), and speed (U), for each tribopair were calculated from Eqs. (3)–(5), respectively, employing loads (w) and speeds (u) that were employed for the experimental measurements (see Fig. 4–9). It is noted that the mean speed u is defined as the half of the sliding speed for pure sliding contact. The dimensionless parameters for viscosity, g_V, and elasticity, g_E, for each tribopair and the experimental conditions were then obtained using Eqs. (1) and (2) and are plotted on the map of $\hat{H}$ contours on a g_V vs. g_E grid. The results are shown in Fig. 1. It is noted that we have employed the revised model by Esfahanian and Hamrock [2] to reproduce the $\hat{H}$ contour map in Fig. 1. For all cases, the lubrication regime for these g_V and g_E values was identified as belonging to the isoviscous-elastic lubrication regime (or soft EHL). In this lubrication regime, the minimum film-thickness is expressed as [1,2]

$$\hat{H}_{min} = 8.70 g_E^{0.67}(1 - 0.85\, e^{-0.31k}) \quad (8)$$

By equating Eqs. (6) and (8), and then using Eq. (7), the minimum film-thickness, h_{min}, can be obtained as a function of load or speed for each tribopair. The calculated results for the load-dependent and speed-dependent aqueous sliding of the tribopairs are presented in Figs. 2 and 3, respectively. In Fig. 3, the Λ ratio, defined as the ratio between the minimum film-thickness and the mean surface roughness, $\Lambda = h_{min}/\sigma^*$ (σ^* is mean surface roughness, $\sigma^* = \sqrt{\sigma_{pin}^2 + \sigma_{disk}^2}$) for the PDMS case is also presented as a function of mean speed.

3.2. Influence of oxygen-plasma treatment for PDMS sliding

Fig. 4 shows the influence of oxygen-plasma treatment of PDMS surface on the measured friction vs. load plots ($u=0.0025$ m/s). The friction forces and error bars represent the average kinetic friction forces and standard deviations from the friction vs. time (or number of rotation) plots for each measurement. A few characteristics are observed. Firstly, the sliding contact of untreated PDMS

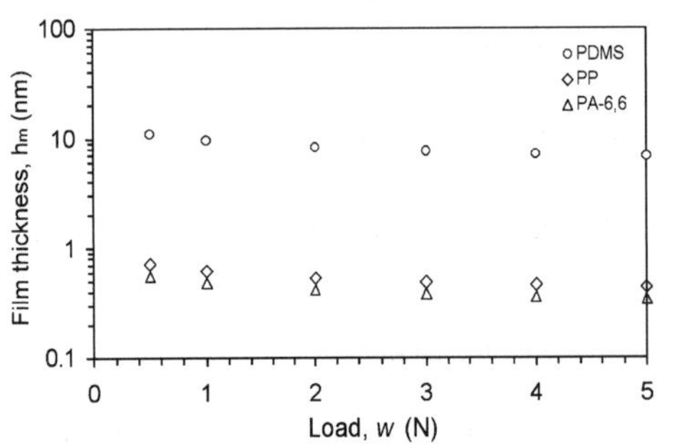

Fig. 2. The theoretically predicted minimum film-thickness (Hamrock and Dowson model [1,2]) for the water-lubricated sliding of PDMS (○), PP (◇), and PA-6,6 (△) as a function of load (from 0.5 to 5.0 N) ($u=0.0025$ m/s)

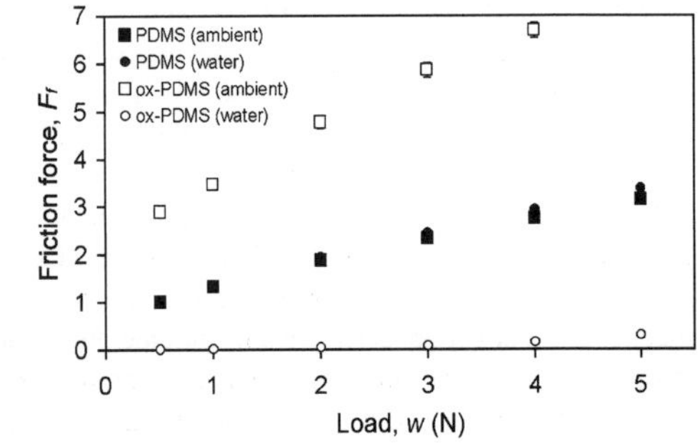

Fig. 4. The influence of oxygen-plasma treatment on the friction vs. load plots of PDMS sliding in ambient and water. 'PDMS' represents the untreated PDMS surface and 'ox-PDMS' represents oxygen-plasma-treated PDMS surface ($u=0.0025$ m/s).

exhibited very high μ values in ambient (for instance, μ(PDMS)ambient = 2.04 at 0.5 N and 0.63 at 5.0 N) and is indistinguishable from that in water (for instance, μ(PDMS)water = 2.09 at 0.5 N and 0.66 at 5.0 N). Secondly, oxygen-plasma treatment of both surfaces resulted in a significant reduction in μ when the tribopair is slid in water (for instance, μ(ox-PDMS)water = 0.03 at 0.5 N and 0.06 at 5.0 N). On the other hand, the corresponding sliding contact in ambient resulted in a significant increase in μ (for instance, μ(ox-PDMS)ambient = 5.80 at 0.5 N and 1.59 at 5.0 N). Thirdly, since the friction forces showed an excellent linear dependence as a function of load ($R^2 = 0.99$ for all cases, except for ox-PDMS in water, where $R^2 = 0.86$), the friction vs. load plots can be extrapolated to zero load, and thus the friction force at zero load, F_0, can be estimated. While the sliding contact of ox-PDMS in water showed a value close to zero ($F_0 = -0.05$ N), all the other cases showed finite F_0 values, namely, 0.85 N (PDMSambient), 0.86 N (PDMSwater), and 0.99 N (ox-PDMSambient), respectively.

The high μ values observed from the untreated PDMS sliding in water implies that the lubricant film, which is theoretically predicted to range from 11.14 nm (0.5 N) to 6.81 nm (5.0 N) (Fig. 2), is not formed in practice. As mentioned in Section 1, surface roughness is the most common cause for the breakdown of the lubricating film. However, since the mean surface roughness, $\sigma^* = \sqrt{\sigma_{\text{pin}}^2 + \sigma_{\text{disk}}^2} \cong 2$ nm, the Λ ratios for this case (data not shown) ranged from 5.57 (0.5 N) to 3.41 (5.0 N), which certainly implies a separation of the two surfaces by a full, fluid film [3,4]. The origin of the high μ values or the prevention of aqueous film formation for untreated PDMS surface thus needs to be explained by factors other than surface roughness. The presence of finite and high friction force at zero load, $F_0 = 0.85$ N, suggests that the adhesion between the two untreated PDMS surfaces contributes to the friction forces to a significant degree. A simplistic analytical approach to a tribological contact involving an adhesion component but displaying linear dependence of friction forces upon load change is to express the observed friction forces by the combination of load-dependent and adhesion-dependent factors; $F_f = sw + F_0$ [19] (F_f = friction force, s = slope of friction vs. load plot, w = load, and F_0 = y-intercept of friction vs. load plot). The four tribological systems explored in Fig. 4 revealed that the surface treatment leads to significant changes in F_0 in different environments, although it causes changes in s to a certain degree as well. Thus, the change in the frictional properties of PDMS sliding as a result of oxygen-plasma treatment can be partially attributed to the change in adhesion properties in each environment, even though the exact relationship needs to be further investigated. The hydrophobic interaction between PDMS surfaces has been extensively studied [12,20–22], and the work of adhesion has been determined to lie between 42.2 and 50.7 mJ m^{-2} in ambient [12,20–22], and 75 mJ m^{-2} in water [12]. More importantly, oxygen-plasma treatment of the PDMS surface did not induce any significant changes in the bulk properties of the PDMS network, e.g. elasticity modulus [12], yet it induces a hydrophilization by the generation of surface hydroxyl groups [12,18]. The hydrophilization of PDMS surface via oxygen-plasma-treatment increases the work of adhesion in ambient (by nearly three times [12]) due to strong electrostatic and H-bonding interactions in the hydrophobic air environment, and consequently increases the frictional forces, as observed in this work. In contrast, the hydrophilization of the PDMS surface reduces the friction forces observed in water, due to the elimination of unfavorable hydrophobic interactions.

The influence of oxygen-plasma treatment on the aqueous lubrication of PDMS surfaces has been further investigated by varying the mean speed from 0.000125 to 0.05 m/s. Within this speed range, the fluid-film-thickness is expected to increase exponentially from 0.73 nm (lowest mean speed $u = 0.000125$ m/s) to 69.44 nm (highest mean speed $u = 0.05$ m/s) according to the Hamrock and Dowson model [1,2]. As shown in Fig. 3, the variation of the film-thickness within this speed range corresponds to $\Lambda \leq 1$ for $u \leq 0.000231$ m/s, $1 \leq \Lambda \leq 3$ for 0.000231 m/s $\leq u \leq$ 0.00123 m/s, and $\Lambda \geqq 3$ for $u \geqq 0.00123$ m/s. In other words, a transition from boundary lubrication ($u \leq 0.000231$ m/s) through mixed lubrication (0.000231 m/s $\leq u \leq$ 0.00123 m/s), and finally to fluid-film lubrication ($u \geqq 0.00123$ m/s) is expected if the expected lubricating films were formed in practice. As shown in Fig. 5 ($w = 1$ N), however, the experimental measurements on untreated PDMS sliding partners showed only a gradual increase of μ with increasing speed (from 0.89 at the lowest speed to 2.10 at the highest speed). This observation also implies that the predicted lubricating films (Fig. 3) may not be readily formed for the speed range selected in this work, when the PDMS surface remained hydrophobic. Upon oxygen-plasma treatment of PDMS, however, μ showed a systematic decrease with increasing speed and this variation is fairly well correlated with that of Λ ratio; $\mu > 0.1$ for $u \leq 0.000125$ m/s ($\Lambda < 1$), $0.071 \leq \mu \leq 0.092$ for 0.00025 m/s $\leq u \leq$ 0.0005 m/s ($1 < \Lambda < 3$), and finally

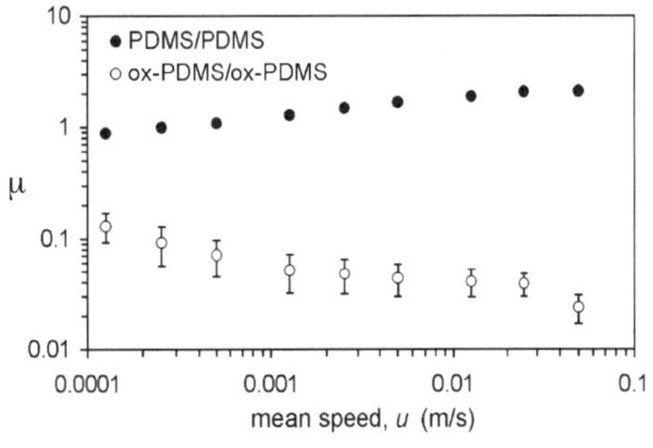

Fig. 5. The influence of oxygen-plasma-treatment on the μ vs. speed plots for the aqueous-lubricated sliding of PDMS contact ($w = 1$ N).

$\mu < 0.05$ for $u \geqq 0.00125$ m/s ($\Lambda > 3$). In other words, the influence of oxygen-plasma treatment is, in essence, to enable the formation of aqueous lubricating films that are theoretically expected in isoviscous-elastic lubrication regime.

This load- and speed-dependent variation of the frictional properties of self-mated sliding PDMS surfaces shows that oxygen-plasma treatment is an effective way of removing hydrophobic interactions and consequently allows effective aqueous lubrication. Given that it is only the hydrophilicity of the PDMS surface that is modified by oxygen-plasma treatment, this study clearly demonstrates that the control of surface properties is a crucial factor that determines the modality of lubricating-film formation for this tribosystem.

3.3. Influence of oxygen-plasma-treatment on sliding of PP/PP and PA-6,6/PA-6,6

Given the effect of oxygen-plasma-treatment on the aqueous lubrication of PDMS sliding as shown above, it is of interest to test whether a similar effect can be observed for other commonly studied hydrophobic materials of engineering importance. To this end, two other hydrophobic polymers PP and PA-6,6, have been studied. For these, much harder tribopairs, considerably smaller film-thicknesses (smaller than 1 nm) are expected than with PDMS, as shown in Fig. 2. However, the lubrication regime is still identified as belonging to the isoviscous-elastic regime, according to the Hamrock and Dowson model (see Fig. 1)

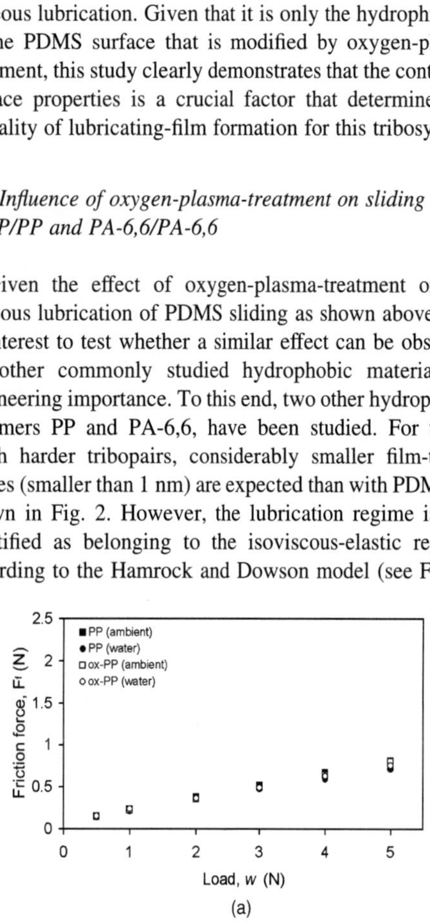

Fig. 6. The influence of oxygen-plasma-treatment on the friction vs. load plots of (a) PP, and (b) PA-6,6 in ambient and water ($u = 0.0025$ m/s).

[1,2]. The influence of oxygen-plasma treatment on the friction vs. load plots obtained for the self-mated sliding of these tribopairs is presented in Fig. 6. The methods of data acquisition and assessment of the friction forces were identical with those in Fig. 3. The mean speed was also 0.0025 m/s. While the μ values for the untreated polymers in ambient are lower than those of PDMS (for instance, μ(PP)ambient = 0.30 and 0.16, and μ(PA-6,6)ambient = 0.18 and 0.08 at 0.5 and 5.0 N, respectively), the oxygen-plasma treatment resulted in either no noticeable decrease or rather an increase of μ in water (for instance, μ(ox-PP)water = 0.31 and 0.15, and μ(ox-PA-6,6)water = 0.64 and 0.24 at 0.5 and 5.0 N, respectively).

The failure of effective lubrication for these polymers could have been anticipated, since the predicted lubricating film-thickness is lower than 1 nm, as shown in Fig. 2. Furthermore, oxygen-plasma treatment of these polymers did not lead to as hydrophilic surfaces as observed with PDMS, as confirmed by water contact angle measurements (see Table 1). To expect any lubricating effect at all from such small film-thicknesses, the tribopair surface would have to be extremely smooth to obtain as high Λ ratio as possible. For instance, to obtain a Λ ratio higher than three under the conditions given above, the surface roughness of both pin and disk should be lower than 3×10^{-3} nm, which is clearly impossible to achieve in practice. Consequently, asperity contacts appear to prevail for these polymers, even in the presence of water as lubricant.

A clear contrast between the two types of polymers, PDMS vs. PP and PA-6,6, in terms of aqueous lubricating effect is primarily due to the difference in their elasticity modulus; 2 MPa for PDMS and in the GPa range for the other two (see Table 1). However, the difference in the surface roughness of the two types of polymers needs to be considered as an additional factor in determining aqueous lubricating properties.

3.4. Oxygen-plasma-treatment vs. amphiphilic copolymers

An alternative to oxygen-plasma-treatment for hydrophilizing the PDMS surface is to coat the surface with amphiphilic copolymers such as PEO-*b*-PPO-*b*-PEO or PLL-*g*-PEG. In an aqueous environment, PEO-*b*-PPO-*b*-PEO copolymers are known to adsorb onto PDMS surfaces through hydrophobic interactions between the PPO block and the surface [14,23,24], and thus no pretreatment of the PDMS surface is needed. PLL-*g*-PEG copolymers are known to adsorb onto negatively charged surfaces most effectively [15–17], and thus the PDMS tribopair was oxygen-plasma treated to generate a negatively charged surface in water [12,18]. The optimum solution concentrations of 2.0 mg/ml for PEO-*b*-PPO-*b*-PEO and 0.25 mg/ml for PLL-*g*-PEG were selected based upon results from previous studies [14,16,17]. Optical waveguide lightmode spectroscopy (OWLS) studies, which detect the adsorbate mass based upon refractive-index change [25,26], have shown that the adsorbed masses of the PEO-*b*-PPO-*b*-PEO and the PLL-*g*-PEG employed in this work (see Section 2.3 for the details of molecular structure) are ca. 164 ng/cm^2 [14] and ca. 146 ng/cm^2, respectively, which implies a high density of EO or EG moieties on the surface. Although the driving force of anchoring the PPO block or PLL backbone onto the PDMS surface is different, i.e. hydrophobic vs. electrostatic interactions, both copolymers exhibit an extended conformation of PEO or PEG brushes into the bulk water due to strong brush–water interactions [14–17]. A schematic illustration of this conformation is shown in Fig. 7.

The surface modification of PDMS by adsorption of PEO-*b*-PPO-*b*-PEO or PLL-*g*-PEG showed an indistinguishably effective lubricating effect compared with oxygen-plasma-treatment in friction vs. load plots (data not shown); for instance, the μ for the PEO-*b*-PPO-*b*-PEO-coated PDMS was 0.032 and 0.025, and the μ for the PLL-*g*-PEG-coated PDMS was 0.028 and 0.027 at 0.5 and 5.0 N, respectively. It is clear that the adsorption of the amphiphilic copolymers does not induce any significant change in the bulk mechanical properties of PDMS either, while significantly modifying the surface properties. As mentioned above, the densely packed PEO or PEG brushes have been known to exhibit a fully extended conformation due to favorable interaction with the solvent, and thus facilitate the entrainment of lubricant, i.e. water, during sliding contact [14,16,17].

An interesting difference was observed when comparing the three PDMS hydrophilization approaches by acquiring μ vs. speed plots, as shown in Fig. 8 ($w=1$ N). While the μ

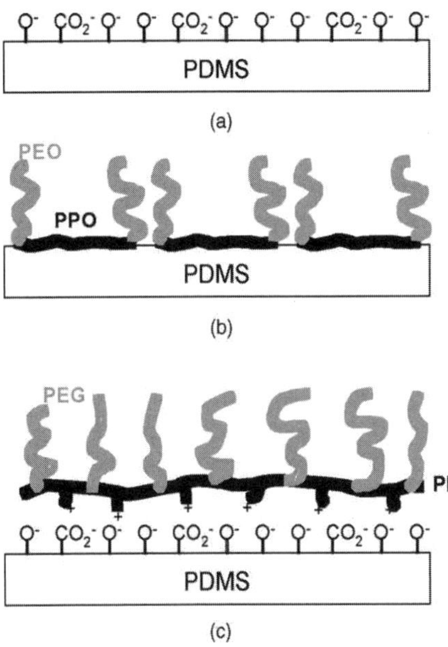

Fig. 7. A schematic illustration of (a) bare ox-PDMS, (b) PDMS coated with PEO-*b*-PPO-*b*-PEO, (c) PDMS coated with PLL-*g*-PEG in water.

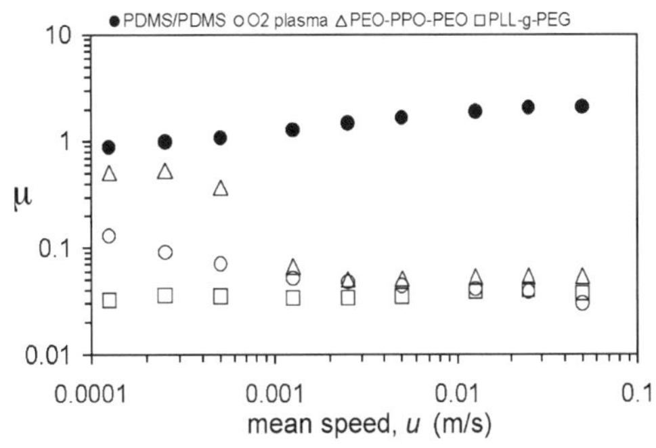

Fig. 8. μ vs. speed plots for the water-lubricated sliding of PDMS/PDMS contact in the absence of treatment, and hydrophilized by means of oxygen-plasma-treatment, adsorption of PEO-*b*-PPO-*b*-PEO co-polymer, and adsorption of PLL-*g*-PEG co-polymer ($w=1$ N).

again showed indistinguishably low values ($<\sim 0.05$) at speeds higher than ~ 0.025 m/s for all treated cases, μ started to increase with decreasing speed at speeds lower than ~ 0.025 m/s to distinctly different degrees. The order of μ in the low-speed regime is μ(PEO-*b*-PPO-*b*-PEO-coated PDMS) > μ(oxygen-plasma treated PDMS) > μ(PLL-*g*-PEG-coated PDMS, following oxygen-plasma treatment). The variation of μ as a function of speed was observed to be independent of whether the speed was increased or decreased, and thus the increasing μ with decreasing speed cannot be ascribed simply to the removal of polymers. Furthermore, it is important to note that the μ values for all cases are still lower than that of untreated PDMS sliding, i.e. some lubrication mechanism is still active in the low-speed regime even though it is not as effective as in high-speed regime. Considering that the transition speed, ~ 0.00125 m/s, coincides with the transition of Λ ratio from higher than to lower than three as discussed in Section 3.1, the increasing μ with decreasing speed in the low-speed regime is associated with the initiation of asperity contact or boundary lubrication. Thus, the order of the μ values for the three different approaches of hydrophilizing the PDMS surface at low speed regime may an indication of their effectiveness for boundary lubrication rather than being influenced by full-fluid lubrication behavior. A few points deserve further discussion. Firstly, the highest μ, i.e. the poorest boundary lubricating effect observed with PEO-*b*-PPO-*b*-PEO-coated PDMS may be related to the fact that the substrate for this coating is untreated, hydrophobic PDMS, while the other cases consist of oxygen-plasma-treated PDMS, with and without a PLL-*g*-PEG coating. When the asperity contact starts to be relevant in the low-speed regime, hydrophobic–hydrophobic interactions (PDMS vs. PDMS), which have been shown in the previous sections to be much stronger than interactions of hydrophilic surfaces (ox-PDMS vs. ox-PDMS) in an aqueous environment, start to become important, in the PEO-*b*-PPO-*b*-PEO-coated case. The superior boundary lubricating properties of PLL-*g*-PEG-coating compared with those of PEO-*b*-PPO-*b*-PEO-coating can also be associated with the higher molecular weight of PEG compared with PEO brushes (ca. three times higher), which has been found to be crucial for boundary lubrication in previous studies [15,16, 27]. However, the more effective boundary lubricating properties of bare ox-PDMS compared with PEO-*b*-PPO-*b*-PEO-coated PDMS can be explained only by the difference in the hydrophilicity of the substrates. Secondly, for the two tribosystems employing ox-PDMS surfaces, the presence of polymer brushes, PLL-*g*-PEG, resulted in an enhanced boundary lubricating properties over the bare substrate. In fact, the increase of μ in the low-speed regime even appeared to be absent for the PLL-*g*-PEG-coated ox-PDMS case (see Fig. 8). While the increase of lubricant viscosity due to the increase of pressure is not expected for soft EHL, the increase of polymer concentration at the confined interface in the low-speed regime has often been shown to increase the effective viscosity for various tribosystems involving viscosity-index improvers [5,28,29]. This effect typically leads to a higher film-thickness than is expected according to elastohydrodynamic theory in the low-speed regime [28,29].

As shown above, while the three approaches to hydrophilize PDMS surfaces exhibited equally effective fluid-film lubricating properties in the high-speed regime, PLL-*g*-PEG-coating of an oxygen-plasma treated PDMS surface also provided distinctly enhanced boundary-lubrication properties in low-speed regime as well. Another benefit of PLL-*g*-PEG-coating an oxygen-plasma treated PDMS surface has been observed when the modified PDMS surfaces were aged in ambient. As shown in Fig. 9, high-speed friction vs. load plots obtained immediately after surface modification showed a similar lubricating effect, whether ox-PDMS surface was further modified with PLL-*g*-PEG or not. This lubricating effect was observed to remain virtually unchanged when both modified surfaces were aged in an aqueous environment for a few days (data not shown). However, PLL-*g*-PEG-coated ox-PDMS also

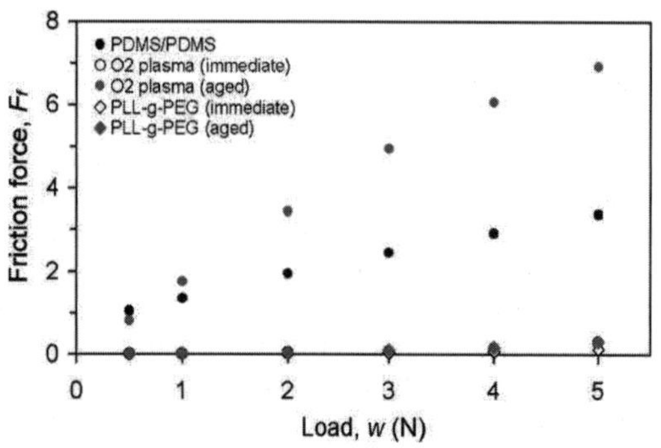

Fig. 9. Friction vs. load plots showing the effect of oxygen-plasma treatment and the adsorption of PLL-*g*-PEG upon aqueous lubrication of PDMS/PDMS sliding (u=0.0025 m/s). 'immediate' represents the measurements immediately after the surface modification and 'aged' represents the measurements after aging of the surface-modified PDMS pins and disks in ambient for 1 week.

maintained its aqueous lubricating properties after aging in ambient for 1 week, whereas bare ox-PDMS showed a clear deterioration in aqueous lubricating properties. This is attributed to the well-known hydrophobic recovery of ox-PDMS (and many other plasma-treated polymers) in ambient. It is well known that the hydrophobic recovery of the ox-PDMS surface involves a restructuring of hydrophilic (OH–) and hydrophobic (CH_3– or CH_2–) moieties at the PDMS surface, and is often accompanied by fracturing of the surface [18]. In other words, the coating of the ox-PDMS surface with PLL-*g*-PEG appears to prevent or delay this process for the duration of 1 week in ambient. Thus, further modification of ox-PDMS with PLL-*g*-PEG led to a preservation of surface hydrophilicity (water contact angle ~30°), which has been found to be essential to ensure effective aqueous lubrication.

4. Summary

In this work, we have investigated the influence of hydrophilization on the aqueous lubrication of a silicone elastomer, PDMS. According to the theoretical model of Hamrock and Dowson [1,2], film-thickness on the tens-of-nanometers scale was expected for the aqueous sliding of the tribopair within the load and speed range employed in this work. In contrast to the theoretical predictions, extremely high μ values were observed when the tribopair was slid in water without any surface treatment. Upon oxygen-plasma treatment, however, a very significant reduction in μ was observed. This is attributed to the removal of a strong hydrophobic interaction by hydrophilization of the PDMS surfaces, which thus facilitates the formation of aqueous lubricating films. Meanwhile, the influence of oxygen-plasma treatment on the aqueous lubrication properties of other hydrophobic polymers, including PP and PA-66, was observed to be negligible. This is mainly due to the much higher elasticity moduli of these polymers, which leads to the prediction of much thinner lubricating films (less than 1 nm). The surface modification of PDMS with amphiphilic copolymers, PEO-*b*-PPP-*b*-PEO and PLL-*g*-PEG, also exhibited lubricating properties as effective aqueous as those of oxygen-plasma-treated PDMS in the high-speed regime. This observation confirms that the hydrophilization of PDMS surface is an effective, or even a necessary, procedure to ensure a full-fluid lubrication of this elastomer by water. In addition, the surface coating of PDMS with PLL-*g*-PEG provided an additional benefit of distinctively superior boundary-lubricating properties in the low-speed regime, which is attributed to the role of the PLL-*g*-PEG copolymer as a viscosity-index improver at the confined interface in the low-speed regime. Furthermore, the surface coating of PDMS with PLL-*g*-PEG showed significantly enhanced maintenance of hydrophilicity, even in ambient for 1 week, over that obtained from oxygen-plasma-treatment only.

This study demonstrates that isoviscous-elastic lubrication by water-based lubricants may not be readily achieved simply by employing a smooth elastomer as a tribopair. The control of surface properties can be of great significance in ensuring effective aqueous lubrication of elastomeric materials.

Acknowledgements

This work was financially supported by the US Air Force Office of Scientific Research under Contract No. F49620-02-0346. The authors are grateful to Dr Markus Müller for the synthesis of PLL-*g*-PEG copolymers and Dr Kirill Feldman for the elasticity measurements of PDMS and to both for helpful discussions.

References

[1] Hamrock BJ, Dowson D. Minimum film thickness in elliptical contacts for different regimes of fluid-film lubrication Proceedings of the 5th leeds-lyon symposium on tribology. 1979. Bury St Edmunds, Suffolk: Mechanical Engineering Publication; 1979 p. 22–27.

[2] Esfahanian M, Hamrock BJ. Fluid-film lubrication regimes revisited. Tribol Trans 1991;34(4):628–32.

[3] Stachowiak GW, Batchelor AW. Ch.7 Elastohydrodynamic lubrication In: Engineering tribology, Elsevier, 1993.

[4] Hutchings IM, Ch.4 Lubricants and lubrication In: Tribology; Friction and wear of engineering materials. Edward Arnold, 1992.

[5] Roberts AD, Tabor D. The extrusion of liquids between highly elastic solids. Proc Roy Soc Lond A 1971;325:323–45.

[6] Moore DF. The elastohydrodynamic transition speed for spheres sliding on lubricated rubber. Wear 1975;35:159–70.

[7] Medley JB, Strong AB, Pilliar RM, Wong EW. The breakdown of fluid film lubrication in elastic-isoviscous point contacts. Wear 1980; 63:25–40.

[8] Richards SC, Roberts AD. Boundary lubrication of rubber by aqueous surfactant. J Phys D 1992;25:A76–A80.

[9] Dowson D, Jin ZM. Microelastohydrodynamic lubrication of low-elastic-modulus solids on rigid substrates. J Phys D 1992;25: A116–A23.

[10] Drews MJ, LaBerge M. An investigation of the fatigue induced failure modes of filber/elastomer composites as bearing surfaces in total hip joint prosthesis In: National Textile Center Annual report: November 1996. p. 71–9.

[11] Mansour, JM. Ch.5 Biomechanics of cartilage In: Applied kinesiology, Lippincott Williams and Wilkins, 2003.

[12] Chaudhury MK, Whitesides GM. Direct measurement of interfacial interaction between semispherical lenses and flat sheets of poly(-dimethylsiloxane) and their chemical derivatives. Langmuir 1991;7: 1013–25.

[13] BASF Technical Brochure. Parsippany, NJ: BASF Co.; 1989.

[14] Lee S, Iten R, Müller M, Spencer ND. Influence of molecular architecture on the adsorption of poly(ethylene oxide)-poly(propylene oxide)-poly(ethylene oxide) (PEO–PPO–PEO) on PDMS surfaces and implications for aqueous lubrication. Macromolecules 2004;(37): 8349–56.

[15] Kenausis GL, Vörös J, Elbert DL, Huang N-P, Hofer R, Ruiz-Taylor L, et al. Poly(L-lysine)-*g*-poly(ethylene glycol) layers on metal oxide surfaces: attachment mechanism and effects of polymer architecture on resistance to protein adsorption. J Phys Chem B 2000;104:3298–309.

[16] Lee S, Müller M, Ratoi-Salagean M, Vörös J, Pasche S, De Paul SM, et al. Boundary lubrication of oxide surfaces by poly(L-lysine)-*g*-poly(ethylene glycol) (PLL-*g*-PEG) in aqueous media. Tribol Lett 2003;15:231–9.

[17] Müller M, Lee S, Spikes HA, Spencer ND. The influence of molecular architecture on the macroscopic lubrication properties of the brush-like co-polyelectrolyte poly(L-lysine)-*g*-poly(ethylene glycol) (PLL-*g*-PEG) adsorbed on oxide surfaces. Tribol Lett 2003;15:395–405.

[18] Makamba H, Kim JH, Lim K, Park N, Hahn J. Surface modification of poly(dimethylsiloxane) microchannels. Electrophoresis 2003;24: 3607–19.

[19] Derjaguin BV, Karassev VV, Zakhavaeva NN, Lazarev VP. The mechanism of boundary lubrication and the properties of the lubricating film short- and long-range action in the theory of boundary lubrication. Wear 1957;1:277–90.

[20] Tirrell M. Measurement of interfacial energy at solid polymer surfaces. Langmuir 1996;12:4548–51.

[21] Choi GY, Kim S, Ulman A. Adhesion hysteresis studies of extracted poly(dimethylsiloxane) using contact mechanics. Langmuir 1997;13: 6333–8.

[22] Rundlöf M, Karlsson M, Wågberg L, Poptoshev E, Rutland M, Claesson PJ. Application of the JKR method to the measurement of adhesion to Langmuir-Blodgett cellulose surfaces. J Colloid Interf Sci 2000;230:441–7.

[23] Alexandridis P, Hatton TA. Poly(ethylene oxide)-poly(propylene oxide)-poly(ethylene oxide) block copolymer surfactants in aqueous solutions and at interfaces: thermodynamics, structure, dynamics, and modeling. Colloids Surf A 1995;96:1–46.

[24] Tiberg F, Malmsten M, Linse P, Lindman B. Kinetic and equilibrium aspects of block copolymer adsorption. Langmuir 1991;7:2723–30.

[25] Kurrat R, Textor M, Ramsden JJ, Böni P, Spencer ND. Instrumental improvements in optical waveguide lightmode spectroscopy for the study of biomolecule adsorption. Rev Sci Instrum 1997;68:2172–6.

[26] Vörös J, Ramsden JJ, Csúcs G, Szendrő I, De Paul SM, Textor M, et al. Optical grating coupler biosensors. Biomaterials 2002;23: 3699–710.

[27] Yan X, Perry SS, Spencer ND, Pasche S, De Paul SM, Textor M, et al. Reduction of friction at oxide interfaces upon polymer adsorption from aqueous solutions. Langmuir 2004;20:423–8.

[28] Smeeth M, Spikes HA, Gunsel S. The formation of viscous surface films by polymer solutions: boundary or elastohydrodynamic lubrication? Tribol Trans 1996;39:720–5.

[29] Smeeth M, Spikes HA, Gunsel S. Boundary film formation by viscosity index improvers. Tribol Trans 1996;39:726–34.

Tribology Letters, Vol. 24, No. 3, December 2006 (© 2006) 217
DOI: 10.1007/s11249-006-9121-9

Self-healing behavior of a polyelectrolyte-based lubricant additive for aqueous lubrication of oxide materials

S. Lee, M. Müller, R. Heeb, S. Zürcher, S. Tosatti, M. Heinrich, F. Amstad, S. Pechmann and N.D. Spencer*

Laboratory for Surface Science and Technology, Department of Materials, ETH Zurich, Wolfgang-Pauli-Strasse 10, CH-8093 Zurich, Switzerland

Received 17 July 2005; accepted 6 November 2005; published online 14 November 2006

We report on the self-healing behavior of a polyelectrolyte-based aqueous lubricant additive, poly(L-lysine)-*graft*-poly(ethylene glycol) (PLL-*g*-PEG), during aqueous lubrication of an oxide-based tribosystem. Combined pin-on-disk tribometry and fluorescence microscopy experiments have shown that stable lubricating performance was enabled by means of rapid healing of the worn tribopair surface by polymers dissolved in the adjoining bulk lubricant. This rapid 'self-healing' of PLL-*g*-PEG is attributed to electrostatic interactions between the polycationic poly(L-lysine) (PLL) backbone of the polymer and negatively charged oxide surface. In contrast, a similar healing effect was not readily achievable in the case of methoxy-poly(ethylene glycol)-trimethylsilylether (Sil-PEG), a lubricant additive that is covalently bonded to the surface prior to tribological stress.

KEY WORDS: self-healing, poly(L-lysine)-*graft*-poly(ethylene glycol), methoxy-poly(ethylene glycol)-trimethylsilylether, aqueous lubrication, fluorescence microscopy

1. Introduction

Boundary lubricants (or lubricant additives) are designed and employed to lubricate tribological contacts where asperity–asperity contact is dominant, such as under high-load and/or low-speed conditions. The mechanism of boundary lubrication is thus based upon the modification of shear strength of the tribological interface, rather than a formation of a hydrodynamic film through the entrainment of lubricants. For this reason, the continuous presence of boundary lubricants on the tribopair surface is critical in determining lubrication performance. In general, boundary-lubricating performance has been observed to improve with increasing adsorption energy [1–5]. On the other hand, if the tribostressed boundary lubricant film can heal itself following the tribostress-induced damage, it can offer an alternative efficient solution to the issue of durability under boundary lubrication.

While boundary lubricant additives are present in virtually every oil-based lubricant, the concept of adding boundary lubricants to water, in order to improve its lubrication properties, has been relatively little explored. Recently, it has been shown that a polycationic-based copolymer, poly(L-lysine)-*graft*-poly(ethylene glycol) (PLL-*g*-PEG) can be used as an effective boundary lubricant additive for the aqueous lubrication of oxide-based tribosystems [6–8]. As schematically depicted in figure 1, this copolymer is composed of a polycationic poly(L-lysine) (PLL) backbone and non-reactive poly(ethylene glycol) (PEG) side chains. The adsorption of PLL-*g*-PEG onto oxide substrates is mainly driven by electrostatic interactions between the positively charged PLL backbone and a negatively charged oxide surface in an aqueous environment at neutral pH [6–11]. As will be detailed below in this paper, we have observed that the lubricating performance of PLL-*g*-PEG as an aqueous lubricant additive is significantly enhanced by the presence of excess copolymers in the bulk lubricant; while excess copolymers appear to cure or prevent the degradation of the lubricating performance, a monolayer of PLL-*g*-PEG on the tribopair surface in a polymer-free aqueous environment results in rapid loss of its lubricating capabilities. In this study, in parallel with several control pin-on-disk tribometry experiments, we have employed a novel fluorescence microscopy approach to determine if exchange occurs, during the tribological contact, between the PLL-*g*-PEG copolymers initially adsorbed on the tribopair surface and those dissolved in the adjoining bulk lubricant. In addition, we have also compared PLL-*g*-PEG with its analog possessing covalent-bond-based anchoring groups, methoxy-poly(ethylene glycol)-trimethylsilylether (Sil-PEG), with respect to its 'self-healing' capacity under tribological stress.

*To whom correspondence should be addressed.
E-mail: spencer@mat.ethz.ch

2. Materials and methods

2.1. Polymers

The molecular structure of PLL-*g*-PEG and its conformation at the water/oxide interface are depicted in figure 1. The PLL-*g*-PEG was synthesized by the reaction of *N*-hydroxysuccinimidyl-ester-functionalized PEG (N-hydroxysuccinimidyl ester of methoxy (ethylene glycol) propionic acid, mPEG-SPA, Nektar AL, Huntsville, AL, USA) with the amino groups of PLL hydrobromide (Sigma, St. Louis, MO, USA) at a controlled stoichiometric ratio. Details on the synthesis and analytical information are available in previous publications [10,11]. The PLL-*g*-PEG employed in this work, denoted as "PLL(20)-*g*[3.4]-PEG(5)" [10,11], possesses the molecular weight of 20,000 g/mol for the PLL backbone (including HBr as a precursor), 5,000 g/mol for PEG side chains, and 3.4 for the graft ratio (lysine-mer/PEG side chains). This means that, on average, every 3.4th lysine unit carries a PEG side chain with 113 EG units. The non-functionalized lysine chains will be protonated at pH 7 and therefore responsible for binding to the surface. The polydispersities of the reagents, M_w/M_n, are 1.2 and 1.1 for the PLL backbone and the PEG side chains, respectively.

In order to investigate the tribostress-induced detachment and re-adsorption, i.e. "self-healing", of PLL-*g*-PEG during the sliding contact in pin-on-disk tribometry, PLL-*g*-PEG was labeled by either of the two fluorescence markers: fluorescein 5-isothyocyanate (5-FITC) and rhodamine B isothiocyanate (RBITC) (Fluka, Deisenhofen, Switzerland). Both fluorescence markers are isothiocyanate-activated chromophores and react covalently with the PLL backbone through a thiourea linkage. The fluorescence markers were added in a ratio of roughly one chromophore per 25 lysine monomers, i.e. 3.7 per PLL-*g*-PEG on average.

For comparison, Sil-PEG possessing the identical PEG molecular weight, 5,000 g/mol, as that of PLL-*g*-PEG was purchased (Nektar, Huntsville, AL, USA) and used as received.

2.2. Preparation of substrates and polymer coating

Polymer adlayers were prepared on top of silicon oxide (SiO_x) substrates; thermally oxidized Si(100) wafers were used for ellipsometry (ELM) and soda-glass slides (SuperFrost, Menzel-Gläser, Braunschweig, Germany) were used for pin-on-disk tribometry. Prior to polymer coating, the substrates were first ultrasonicated in ethanol for 5 min. and blown dry with nitrogen. Then, they were cleaned in piranha solution (H_2SO_4:$H_2O_2 \approx 3$:1) for 10 min., rinsed with a copious amount of distilled water, and dried with nitrogen again. The substrates were then oxygen-plasma treated for 1 min. in a plasma cleaner (Harrick, Ossining, NY, USA, PDC-32) to remove contaminants and to obtain a well-defined oxide surface.

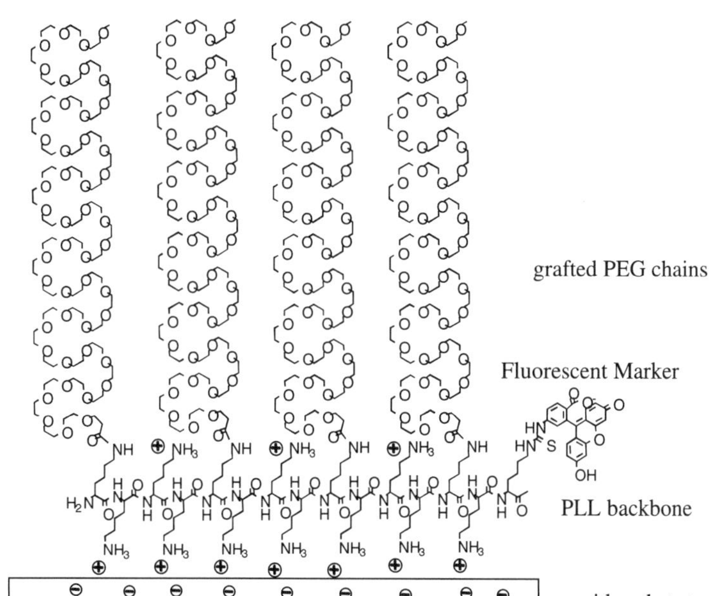

Figure 1. A schematic illustration of PLL-*g*-PEG/FITC adsorbed onto an oxide substrate surface in an aqueous environment (pH 7). Note that the ratio between the lysine monomers grafted to PEG side chains and those available for interaction with the oxide surface in this figure is 1:3 (lysine-mer/PEG side chain = 4), which is slightly different from that of the PLL-*g*-PEG employed throughout the experiments, 1:2.4 (g = 3.4). The FITC label is also not necessarily at the end of the PLL chain in reality.

Surface modification of the cleaned substrates with polymers was carried out either *ex situ* (pre-coating) or *in situ*. The pre-coating was achieved by incubating the substrates in the polymer solutions; in PLL-*g*-PEG solution (0.25 mg/ml in HEPES aqueous buffer (10 mM, 4-(2-hydroxyethyl)piperazine-1-ethanesulfonic acid (Sigma) for 30 min. or in Sil-PEG solution (1 mg/ml in toluene or 1 mg/ml in distilled water with 0.05% (v/v) of hydrochloric acid (36–38%) [12]) overnight, respectively. The modified substrates were then rinsed with the respective solvent to remove the unbound or loosely bound polymers. This approach was employed for all of the ELM measurements and some of the pin-on-disk tribometry measurements. For some control pin-on-disk tribometry measurements, *in situ* polymer coating was carried out either by employing polymer solutions as lubricants or injecting the polymer solution into the pin-on-disk tribometer cup during the measurements (see figure 3 for the schematic of the pin-on-disk tribometer).

2.3. Ellipsometry

A Variable Angle Spectroscopic Ellipsometer of the type M200-F (J.A. Woollam Co. Inc., Lincoln, USA) with a spectral range from 245 to 995 nm was used to determine the thickness of the adsorbed polymer layers. Measurements were performed in ambient at three different angles (65, 70, and 75° with respect to the surface normal). For each polymer adlayer, i.e. Sil-PEG (from toluene), Sil-PEG (from acidic aqueous solution), and PLL-*g*-PEG (from aqueous HEPES buffer), five samples were prepared to obtain statistical data. The measurements were fitted with multilayer models using WVASE32 analysis software. The analysis of optical constants was based on a bulk silicon/SiO_x layer, fitted in accordance with the Jellison model. After adsorption of the molecules, the adlayer thickness was determined using a Cauchy model ($A = 1.45$, $B = 0.01$, $C = 0$).

2.4. Pin-on-disk tribometry

Tribological measurements were performed by means of pin-on-disk tribometry (CSEM, Neuchâtel, Switzerland) [6, 7]. In this setup, a fixed, spherical steel pin (6 mm in diameter, DIN 5401-20, G100, Hydrel AG, Romanshorn, Switzerland) was loaded against a flat glass disk by placing dead weights on top of the pin holder. The sliding speed was controlled with a motor underneath the disk. Friction signals were recorded as a function of the number of revolutions over a fixed sliding track with a Macintosh Power PC using Labview 5.0 (National instruments, Austin, TX, USA). All the measurements were performed in aqueous solution, employing a load and sliding speed of 2 N and 5 mm/s, respectively. The temperature was ca. 20°C. The mean Hertzian contact pressure for this pair is estimated to be 0.51 GPa under the experimental conditions used in this work. The surface roughness of the steel pin and disk were 10 and 2 nm, respectively.

Generally, the revolution of a pin against a disk was allowed to proceed for up to 50 cycles. For measurements involving fluorescence-labeled PLL-*g*-PEG and subsequent fluorescence microscopy, however, the number of revolutions was carefully controlled over the four adjacent tracks on a single disk.

The steel pins were cleaned by consecutive sonication in toluene for 5 min., in ethanol for 10 min., and distilled water for 5 min. Then, they were oxygen-plasma cleaned for 2 min.

2.5. Fluorescence microscopy

Fluorescence microscopy (Zeiss Axiovert 135 TV, Carl Zeiss, Jena, Germany, equipped with a CCD camera (ORCA-ER, Hamamatsu, Japan)) was employed to visualize the exchange between the pre-adsorbed PLL-*g*-PEG polymers and those dissolved in the adjoining solution during pin-on-disk tribometry measurements. As described above, the PLL-*g*-PEG copolymers from the same batch were marked either by fluorescein isothyocyanate (FITC) or by RBITC. The FITC absorbs light at 490 and 494 nm, and re-emits light at 520 and 525 nm (green). These wavelengths were filtered out with the Zeiss Filter-set No. 10. RBITC absorbs light at 555 nm and emits at 576 nm (red). These wavelengths were filtered out with the Zeiss Filter-set No. 15. In pin-on-disk tribometry, the tribopair was pre-coated with PLL-*g*-PEG labeled with a different chromophore FITC than the excess PLL-*g*-PEG dissolved in the solution RBITC. By changing the wavelength of excitation in the fluorescence microscopy, it was possible to distinguish the two differently labeled polymers and, therefore, verify the exchange between the two in the course of the sliding contacts. The intensity of each fluorescence signal was quantified by means of a brightness profile measured with ImageJ 1.33U (National Institute of Health, Washington, DC, USA) from bitmap files exported from AxioVision 4.2 (Carl Zeiss Vision GmbH).

3. Results and discussion

3.1. Characterization of polymer film thickness

Table 1 shows the results of thickness measurements of the three polymer adlayers by ELM: Sil-PEG from toluene, Sil-PEG from acidic aqueous solution, and PLL-*g*-PEG from HEPES buffer. The highest adlayer thickness was obtained from Sil-PEG adsorbed from toluene (2.21 ± 0.25 nm), followed by PLL-*g*-PEG (1.41 ± 0.05 nm) adsorbed from aqueous HEPES buffer

Table 1.
The film thickness of Sil-PEG (adsorbed from toluene), Sil-PEG (adsorbed from acidic aqueous solution), and PLL-*g*-PEG (adsorbed from HEPES buffer solution) determined by means of ELM.

Sil-PEG (adsorbed from toluene)	Sil-PEG (adsorbed from acidic aqueous solution)	PLL-g-PEG (adsorbed from HEPES buffer solution)
2.21 ± 0.25 nm	0.49 ± 0.06 nm	1.41 ± 0.05 nm

solution, and finally Sil-PEG adsorbed from aqueous buffer solution (0.49 ± 0.06 nm).

Previous studies have shown that the film thickness of PLL-*g*-PEG adlayer is strongly influenced by its architectural features [10, 11], including grafting ratio, (number of lysine units per PEG chain), which simultaneously determines the spacing of PEG chains along the PLL backbone and the number of free protonated amine groups that can serve as surface-anchoring units. Generally, a higher film thickness is expected for smaller grafting ratios, unless the number of free amino groups is too small (e.g. *grafting ratio* ≤ 2.0) [11]. The higher film thickness obtained from Sil-PEG (from toluene) than PLL-*g*-PEG suggests that the lateral spacing between PEG chains of Sil-PEG on the substrate surface is probably closer than that of the PLL-*g*-PEG employed in the present work. Considering that most organic solvents, including toluene, are not 'good' solvents for PEG [9], the collapsed PEG structure of Sil-PEG in toluene can be advantageous for compact adsorption onto substrate surfaces. Although densely packed PEG films can be readily achieved on SiO_x surfaces by adsorption of Sil-PEG from organic solvents, as shown in this work, its usage as an aqueous lubricant additive can be rather inconvenient since the film preparation must always be carried out outside the tribosystem. For this reason, the adsorption of Sil-PEG from aqueous solution, following the method suggested by Papra *et al.* [12], has been employed as an alternative in this work. However, the film thickness of Sil-PEG generated from this approach is clearly smaller than for the other two protocols and molecules considered, while still being readily detectable. This is attributed to uncontrolled polymerization of siloxanes in the presence of water [13–15].

3.2. Pin-on-disk tribometry: poly(L-lysine)-graft-poly(ethylene glycol)

In figure 2(a), the pin-on-disk tribometry measurements for steel/glass tribopairs employing PLL-*g*-PEG as the aqueous lubricant additive are presented. Three sets of experiments with varying surface modification of the tribopair and composition of lubricant are included: (i) an unmodified steel/glass pair in HEPES (○); (ii) a steel/glass pair in PLL-*g*-PEG solution (0.25 mg/ml in HEPES); (□) (iii) a steel/glass pair pre-coated with PLL-*g*-PEG, then transferred into HEPES (●). For the case of the steel/glass pair in PLL-*g*-PEG solution, i.e. case (ii), the measurement was started 30 min. after immersion of the tribopair into the lubricant (PLL-*g*-PEG solution), to provide sufficient time for the adsorption of polymers onto the tribopair surface. Thus, the difference between the cases (ii) and (iii) is only the presence of the solution-phase PLL-*g*-PEG in the case of (ii).

For the bare steel/glass pair in HEPES, (i), the friction coefficient μ (= friction/load) increased erratically during the initial few revolutions ("running-in" process) and then settled down to a constant value of $\mu \approx 0.5$. On the other hand, the steel/glass pair in PLL-*g*-PEG solution, (ii), showed $\mu \approx 0.1$ from the initial contact, and this value persisted to the end of the measurement without significant variation. Finally, the tribopair pre-coated with PLL-*g*-PEG but tested in polymer-free HEPES aqueous buffer, (iii), exhibited a significant change of μ over the initial ca. five revolutions, then

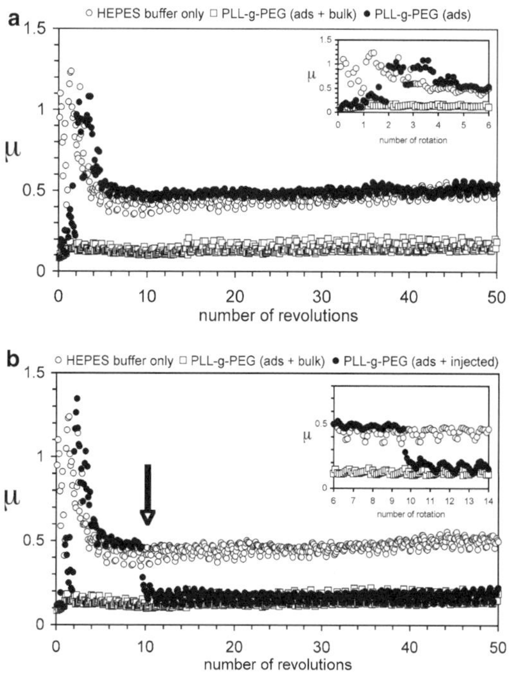

Figure 2. μ versus number of revolutions for sliding contact of a steel/glass (pin/disk) tribopair, by means of pin-on-disk tribometry: (a) in HEPES buffer solution (○); in PLL-*g*-PEG solution (0.25 mg/ml in HEPES buffer solution) (□); and in HEPES buffer solution with PLL-*g*-PEG coating at surfaces (●); (load: 2 N, sliding speed: 5 mm/s, radius of pin: 3 mm). Inset: expanded display of the data for the initial six revolutions, (b) in HEPES buffer solution (○); in PLL-*g*-PEG solution (0.25 mg/ml in HEPES buffer solution) (□); and in HEPES buffer solution with PLL-*g*-PEG coating at surfaces, followed by the injection of concentrated PLL-*g*-PEG solution (●); (final concentration: 0.25 mg/ml) at ~10th rotation, as indicated by the arrow (load: 2 N, sliding speed: 5 mm/s, radius of pin: 3 mm). Inset: expanded display of the data for the sixth–14th revolutions.

displaying a constant value, similar to that observed for the bare tribopair in HEPES, (i), until the end of the measurement. The detailed variation of μ in the initial testing period is displayed in the inset of figure 2(a); while the μ values for the first one or two revolutions are as low as with those for the steel/glass in PLL-*g*-PEG solution, (ii), those during the third to the fifth revolutions are much higher, and are similar to those of the bare steel/glass pair in HEPES, (i). This observation suggests that some apparently irreversible change in the pre-coated PLL-*g*-PEG film, such as removal or permanent disruption of the film structure, appears to have occurred during the initial sliding period for case (ii). Meanwhile, the consistently low μ values observed in case (iii) imply that such effects can be either prevented or instantaneously repaired by the presence of PLL-*g*-PEG polymers in the bulk lubricant.

To investigate this behavior in more detail, another pin-on-disk tribometry measurement involving a steel/glass tribopair pre-coated with PLL-*g*-PEG, i.e. case (iii), was carried out; for this run, however, an aliquot (5 ml) of concentrated PLL-*g*-PEG solution (1 mg/ml) was injected into the tribometer cup in which a pin-on-disk test was in progress under HEPES (15 ml). Thus, the final concentration was 0.25 mg/ml. The results are shown in figure 2(b). In this plot, the data obtained from two other tribopairs in figure 2(a), (i) and (ii), are also presented for comparison. The initial 10 revolutions are basically a reproduction of the data shown in figure 2(a). However, upon injection of the PLL-*g*-PEG solution at ~10th revolution (as indicated by the arrow), the μ values immediately decreased to the level of the steel/glass pair in PLL-*g*-PEG solution, the case (ii). As shown in the inset of figure 2(b), where the data for the sixth–14th revolutions are magnified, the reduction of μ upon injection of the PLL-*g*-PEG solution is completed even before a single revolution has elapsed. Another injection measurement was carried out, starting with a bare steel/glass tribopair; as in figure 2(b), upon injection of the polymer solution, a rapid reduction of μ was again observed to the level of the tribopair in PLL-*g*-PEG solution (data not shown).

The instantaneous reduction of μ upon injection of PLL-*g*-PEG solution into the tribometer is ascribed to the fast adsorption kinetics of PLL-*g*-PEG onto oxide surfaces at neutral pH. Previous optical waveguide lightmode spectroscopy (OWLS) studies have already revealed that the adsorption of PLL-*g*-PEG onto oxide surfaces, including SiO_x [6, 10] and iron oxide [6], which comprise the tribopair of this work, starts to occur within seconds, upon exposure of an oxide substrate to the polymer solution, and is virtually completed within a few minutes. However, fast adsorption of PLL-*g*-PEG is feasible only when the concentration of the polymer solution is sufficiently high, e.g. > 0.1 mg/ml. In fact, in the low-concentration regime, the lubricating capabilities of PLL-*g*-PEG solution are clearly inferior, even if excess polymers are present in the bulk solution. Thus, the 'irreversible' change of μ in the initial few revolutions for the tribopair with the pre-coated PLL-*g*-PEG alone (the case (iii) in figure 2(a)) can be understood in terms of the concentration of the PLL-*g*-PEG in bulk solution, which may come from the tribostressed detachment from the tribopair surface, being too low. In other words, at such low concentrations, the tribostress-induced desorption of PLL-*g*-PEG polymers is always faster than their re-adsorption onto the substrate under the experimental condition employed. In the same context, the stable lubricating performance of the PLL-*g*-PEG solution from the initial tribological contact (case (ii) in figure 2(a)) can be attributed to the solution concentration of PLL-*g*-PEG being sufficiently high to replenish, or repair, the worn area instantaneously. Based upon the results in figure 2(a) and (b), it can be inferred that the 'irreversible' change observed for the tribopair only possessing the surface PLL-*g*-PEG film (case (iii) in figure 2(a)) in the initial contact stage is, in fact, simple removal as entire molecules, rather than irreversible disruption of the film structure. Since PLL-*g*-PEG is mainly adsorbed onto SiO_x surfaces through electrostatic attraction in a neutral aqueous environment, adsorption is rapid when bare oxide surfaces are exposed.

3.3. *Fluorescence microscopy: poly(L-lysine)-graft-poly(ethylene glycol)*

The tribostress-induced desorption of PLL-*g*-PEG and exchange with excess polymers dissolved in the bulk solution has been manifested in a more striking manner by means of fluorescence microscopy. As schematically shown in figure 3, a steel/glass tribopair was pre-coated with PLL-*g*-PEG labeled with FITC (green), and was then transferred into the tribometer cup containing a solution of PLL-*g*-PEG labeled with RBITC (red). On the same disk, four adjacent tribostressed tracks with

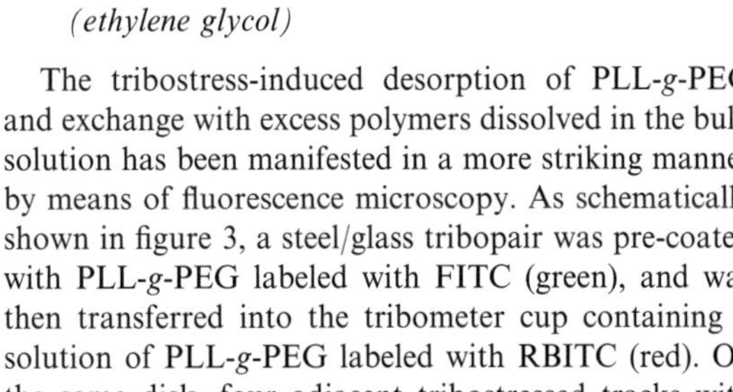

Figure 3. A schematic illustration of the pin-on-disk tribometer and the location of the two different fluorescently labeled PLL-*g*-PEG copolymers (fluorescein isothiocyanate (FITC-PLL-*g*-PEG), pre-coated onto both pin and disk, RBITC-PLL-*g*-PEG, in the lubricant solution, dissolved in HEPES buffer).

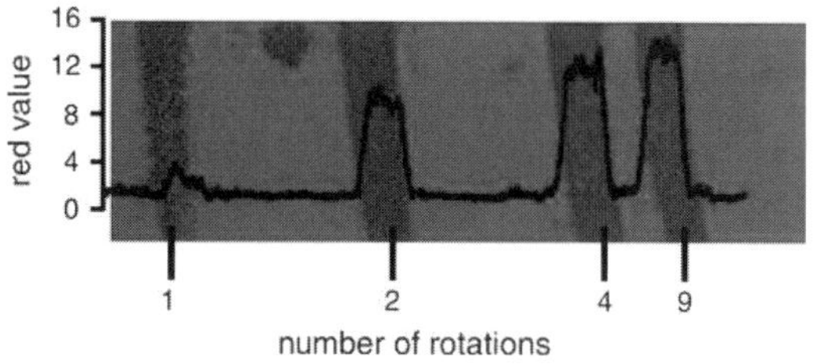

Figure 4. Fluorescence microscopy image of the disk following pin-on-disk tribometry experiment. The number of revolutions was 9, 4, 2, and 1 (in the order of the experiment) in the tracks of radius 1.6, 2.0, 2.4, and 2.8 mm. The load and sliding speed were fixed at 2 N and 5 mm/s, respectively.

varying numbers of revolutions, 1, 2, 4, 9, were generated. Because of their low concentration and small size, the chromophores were assumed to have negligible influence on the tribological properties. Fluorescence microscopy was performed *ex situ* after a drying stage in ambient following pin-on-disk tribometry. The results are shown in figure 4.

It can be clearly seen that, as a function of the number of rotations, part of the FITC-labeled (green) PLL-*g*-PEG, initially adsorbed on the glass disk, has been exchanged by the RBITC-labeled (red) PLL-*g*-PEG during the course of the pin-on-disk tribometry experiment. Since the choromophores are located exclusively on the PLL backbone, this exchange supports the idea that the tribostress applied to the sliding track mainly induces simple, molecular desorption of the entire PLL-*g*-PEG rather than disruption of the film structure or partial removal of the polymer involving the breaking of covalent bonds. It is also noted that the intensity of the red color in the sliding tracks increases upon increasing the number of revolutions. The fluorescence image shown in figure 4 confirms that the exchange of the pre-adsorbed polymer with polymer molecules dissolved in the bulk volume does occur, and that it accounts for the stable lubricating performance of aqueous PLL-*g*-PEG solution, as shown in figure 2(a) and (b). From the absence of color change in the regions between the test tracks, it can readily be observed that exchange of PLL-*g*-PEG between surface and solution does not occur to a detectable extent in the absence of tribological stress.

3.4. *Pin-on-disk tribometry: methoxy-poly(ethylene glycol)-trimethylsilylether*

In figure 5(a), the pin-on-disk tribometry measurements employing a steel pin and a glass disk pre-coated with Sil-PEG are presented. The data for the unmodified steel/disk in HEPES are also plotted as a reference in this figure. In a similar way to the steel/glass tribopair pre-coated with PLL-*g*-PEG ((iii) in figure 2(a)), the steel/glass tribopair involving the Sil-PEG adlayer on the disk

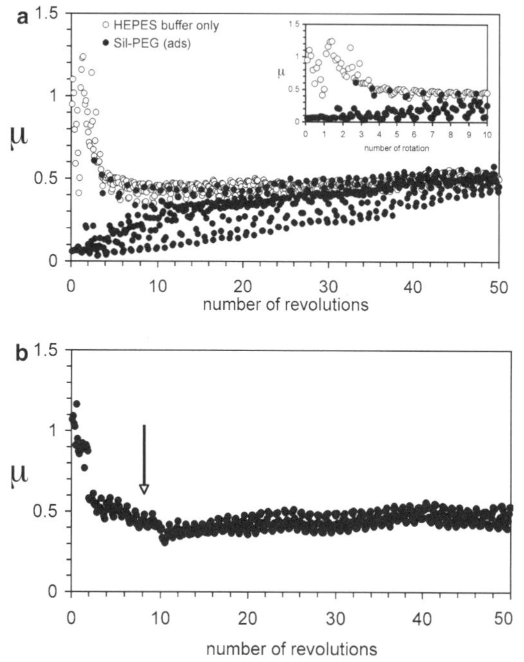

Figure 5. μ versus number of revolutions plots for sliding contact of a steel/glass (pin/disk) tribopair by means of pin-on-disk tribometry: (a) in HEPES buffer solution (○); and in HEPES buffer solution with Sil-PEG coating at surfaces (●); (load: 2 N, sliding speed: 5 mm/s, radius of pin: 3 mm). Inset: enlarged display of the data for the initial 10 revolutions, (b) an aliquot (5 ml) of concentrated PLL-*g*-PEG solution was injected (final concentration, 1.0 mg/ml) at the ~10th rotation, as indicated by the arrow (load: 2N, sliding speed: 5 mm/s, radius of pin: 3 mm).

also exhibited much lower μ values (≈ 0.08) than the *bare* steel/glass tribopair in HEPES, yet only in the initial revolutions. With increasing number of revolutions, the μ values gradually increased, reaching the level of the bare steel/glass tribopair by the end of the measurement. In this plot, 10 data points correspond to one revolution of the sliding contact. Thus, the scatter in friction signals within a single cycle in the majority of the data, ca. fifth–45th revolutions, suggests that the damage to the Sil-PEG coating may be initially occurring due to local asperities, eventually spreading over the entire track.

Although both of the pre-coated polymer films had completely lost their lubricating properties by the 50th rotation, the rate of increase in μ is much slower for the case of Sil-PEG. This difference is clearly demonstrated in the enlarged display of the initial 10 revolutions for the Sil-PEG coating, as shown in the inset of figure 5(a). While the removal of the PLL-*g*-PEG film appears to be virtually completed within the first five revolutions (see the inset of figure 2(a)), it took almost the entire 50 revolutions for the μ of Sil-PEG to reach the same level; the average μ values for the five revolutions are 0.58 ± 0.31 and 0.13 ± 0.13 for the PLL-*g*-PEG and Sil-PEG coatings, respectively. This difference implies that

Sil-PEG adlayer may be more durable than the PLL-*g*-PEG adlayer upon encountering the tribostress applied in this study, at least for the initial revolutions. This is presumably due to the stronger, covalent bonds that join the Sil-PEG coating to the surface as well as the higher initial thickness, leading to a better shielding of small asperities [16].

Previous studies on alkylsilane or alkanethiol self-assembled monolayers (SAMs) have shown that the chemical and structural properties of covalently bound films that have undergone tribostress in macroscopic contacts are diverse, and depend on the detailed experimental conditions, such as load, sliding speed, accumulated number of tribological interactions, and the details of film preparation, as well as bonding characteristics, etc. [17–20]. In the present work, the nature of the damage that occurred to the Sil-PEG films has not been investigated in detail. We were particularly interested in whether the 'self-healing' process, which was observed for PLL-*g*-PEG, could also occur for a covalently bound lubricant additive, such as Sil-PEG. In this sense, even though the film thickness of Sil-PEG adsorbed from acidic aqueous solution was smaller than that adsorbed from toluene (see table 1), an aqueous Sil-PEG solution has the potential to 'heal' the worn surface *in situ* since the liquid phases for both the film preparation and operation are aqueous. However, as shown in figure 5(b), when an aliquot (5 ml) of concentrated acidic aqueous Sil-PEG solution (4 mg/ml) was added into the tribometer cup during a steel/glass tribological test (at ~10th revolution, the final concentration being 1 mg/ml), no noticeable reduction of μ occurred, except for a very transient and minor change at the moment of polymer solution injection. This can be attributed to slower adsorption kinetics of Sil-PEG, as well as a lower adsorption amount from aqueous solution, or a combination of both effects.

4. Summary

We have shown that a polyelectrolyte-based copolymer, PLL-*g*-PEG, displays the unique feature of forming a self-healing lubricating layer for oxide tribosystems in an aqueous environment. When a steel/pin tribopair, pre-coated with a PLL-*g*-PEG adlayer, was subjected to tribological contact by means of pin-on-disk tribometry in polymer-free aqueous buffer solution, the μ values quickly increased to the level of the unmodified tribopair ($\approx$0.5) within only a few revolutions. Meanwhile, the same measurement repeated in PLL-*g*-PEG solution revealed consistently low μ values ($\approx$0.1) from the very initial contact to the end of measurement (50 revolutions). This was attributed to the tribostress-desorbed PLL-*g*-PEG polymers being quickly replaced by the polymers from the bulk lubricant. The exchange between the initially adsorbed PLL-*g*-PEG and that dissolved in the bulk lubricant was most directly manifested by functionalizing the adsorbed and solution-phase polymers with different fluorescing moeities. This behavior is chiefly ascribed to the unique adsorption kinetics of the PLL-*g*-PEG; the polycationic PLL backbone is rapidly attracted to negatively charged SiO_x surface in aqueous solution at neutral pH, due to electrostatic interactions. Although a covalently bonded analogue of PLL-*g*-PEG, Sil-PEG, showed higher durability in the initial stages of sliding, if the film was prepared from toluene, a complete loss of lubricating ability was observed by the end of the measurement. Moreover, *in situ* healing of the worn tribopair did not appear to be feasible when Sil-PEG was used as an aqueous lubricant.

Acknowledgments

This research was financially supported by the US Air Force Office of Scientific Research under Contract No. F49620-02-0346. The authors are grateful to Dr. Sonika Sharma for her help in the synthesis of fluorescently labeled PLL-*g*-PEG.

References

[1] E.P. Kingsbury, J. Appl. Phys. 29 (1958) 888.
[2] E.P. Kingsbury, ASLE Trans. 3 (1960) 30.
[3] H. Okabe, M. Masuko and K. Sakurai, ASLE Trans. 24 (1981) 467.
[4] S. Jahanmir and M. Beltzer, ASLE Trans. 29 (1985) 423.
[5] S. Jahanmir and M. Beltzer, J. Tribol. 108 (1986) 109.
[6] S. Lee, M. Müller, M. Ratoi-Salagean, J. Vörös, S. Pasche, S.M. De Paul, H.A. Spikes, M. Textor and N.D. Spencer, Tribol. Lett. 15 (2003) 231.
[7] M. Müller, S. Lee, H.A. Spikes and N.D. Spencer, Tribol. Lett. 15 (2003) 395.
[8] .D. Spencer, S.S. Perry, S. Lee, M. Müller, S. Pasche, S. de Paul, M. Textor, X. Yan and M. S. Lim, *Proc 29th Leeds-Lyon Symposium on Tribology* (*Tribological Research and Design for Engineering Systems*), University of Leeds, Leeds, United Kingdom, 2003, p. 411.
[9] M. Müller, X. Yang, S. Lee, S.S. Perry and N.D. Spencer, Macromolecules 38 (2005) 3861.
[10] G.L. Kenausis, J. Vörös, D.L. Elbert, N.P. Huang, R. Hofer, L. Ruiz, M. Textor, J.A. Hubbell and N.D. Spencer, J. Phys. Chem. B 104 (2000) 3298.
[11] S. Pasche, S.M. De Paul, J. Vörös, N.D. Spencer and M. Textor, Langmuir 19 (2003) 9216.
[12] A. Papra, A. Bernard, D. Juncker, N.B. Larsen, B. Michel and E. Delamarche, Langmuir 17 (2001) 4090.
[13] K.C. Popat and T.A. Desai, Biosens Bioelectron 19 (2004) 1037.
[14] H.-G. Hong, M. Jiang, S.G. Sligar and P.W. Bohn, Langmuir 10 (1994) 153.
[15] I. Haller, J. Am. Chem. Soc. 100 (1978) 8050.
[16] R.F. Boyer, *Concepts in Biochemistry* 2nd edn (California Brooks/ Cole Thomson Learning, Pacific Grove, CA, 2002).
[17] A. Nichols Jr and S.C. Street, Analyst 126 (2001) 1269.
[18] S. Ren, S. Yang, Y. Zhao, J. Zhou, T. Xu and W. Liu, Tribol. Lett. 13 (2002) 233.
[19] M. Nakano, T. Ishida, T. Numata, Y. Ando and S. Sasaki, Jpn. J. Appl. Phys. Part I 42 (2003) 4734.
[20] H. Ishida, T. Koga, M. Morita, H. Otsuka and A. Takahara, Tribol. Lett. 19 (2005) 3.

LUBRICATION SCIENCE
Lubrication Science 2008; **20**: 21–34
Published online 21 September 2007 in Wiley InterScience
(www.interscience.wiley.com) DOI: 10.1002/ls.50

Poly(L-lysine)-*graft*-poly(ethylene glycol): a versatile aqueous lubricant additive for tribosystems involving thermoplastics

Seunghwan Lee and Nicholas D. Spencer*,†

Laboratory for Surface Science and Technology, Department of Materials, ETH Zurich, Wolfgang-Pauli-Strasse 10, CH-8093 Zurich, Switzerland

ABSTRACT

The adsorption and aqueous lubricating behaviour of poly(L-lysine)-*graft*-poly(ethylene glycol) (PLL-*g*-PEG) have been investigated for tribopairs involving thermoplastic materials, including polypropylene, polyamide-6,6 and polyethylene. A major finding is that PLL-*g*-PEG adsorbs onto both hydrophobic, non-polar surfaces and hydrophilic, polar (negatively charged) surfaces from aqueous solution, and thus plays as a very unique and effective aqueous boundary lubricant additive for the sliding contact of thermoplastics against themselves as well as against many hydrophilic, polar materials, including metals (e.g. stainless steel) or ceramics (e.g. zirconia, ZrO_2). Copyright © 2007 John Wiley & Sons, Ltd.

KEY WORDS: aqueous lubrication; thermoplastics; poly(L-lysine)-*graft*-poly(ethylene glycol)

INTRODUCTION

Thermoplastics have long attracted attention from the tribology community as alternative bearing materials to metals due to a few distinctive advantages, such as low cost, low density and high resistance to corrosion and oxidation.[1–6] In particular, their resistance to corrosion in aqueous environments allows for the use of water as a lubricant for thermoplatics.[1–6] Some early studies have shown that thermoplastics possessing hygroscopic properties, such as polyamides, exhibit dramatic lubrication effects in the presence of water, especially when they are slid against polar materials, such as glass.[3] Absorption of water and consequent swelling of these materials, however, can also cause unfavourable tribological effects, such as an increased wear rate compared to sliding under dry conditions.[1,2,5] On the other hand, most thermoplastics are not wetted by water due to their extreme surface hydrophobicity,[2] and thus are not effectively lubricated by water alone. Previous research on these issues has tended to focus on the improvement of the material properties themselves, e.g. synthesis of composite

*Correspondence to: Nicholas D. Spencer, Laboratory for Surface Science and Technology, Department of Materials, ETH Zurich, Wolfgang-Pauli-Strasse 10, CH-8093 Zurich, Switzerland.
†E-mail: nicholas.spencer@mat.ethz.ch

thermoplastics,[1,2,5] and yet, tribological approaches to improve the lubricating performance, e.g. development of suitable aqueous lubricant additives, have been relatively rare, with the exception of the use of aqueous solutions of some simple surfactants as lubricants for thermoplastics.[3]

Recently, the authors have reported the tribological behaviour of various poly(ethylene glycol) (PEG)-based copolymers as aqueous lubricant additives.[7–14] Due to the unique affinity of ethylene glycol (EG) for water, PEG polymer brushes have been observed to greatly enhance the formation of aqueous lubricant films when grafted onto the surfaces of tribopairs.[7–14] Among the many different approaches to grafting PEG onto surfaces, initial attachment of PEG chains onto a polycationic backbone, namely poly(L-lysine), thus generating a poly(L-lysine)-*graft*-PEG (PLL-*g*-PEG) copolymer (see Figure 1 for the molecular structure of PLL-*g*-PEG), followed by a simple immersion of the tribopairs in an aqueous solution of the copolymer, has proven to be an exceedingly efficient adsorption method, leading to good lubricating properties.[7–12,14] This is due to stable attachment onto the surface through multiple interactions via anchoring groups. The choice of a polycation, poly(L-lysine), as a backbone onto which PEG chains are to be grafted, was originally made with the aim of lubricating negatively charged surfaces in aqueous environments, such as metal oxides; the adsorption of PLL-*g*-PEG onto metal oxide surfaces is thus driven by the electrostatic attraction between positively charged protonated primary amine groups, —NH_3^+, of PLL-*g*-PEG and the negatively charged surfaces.

The grafting of PEG polymers onto hydrophobic, non-polar surfaces can be readily achieved by means of hydrophobic anchoring groups in an aqueous environment; an outstanding example is the aqueous lubrication of the self-mated sliding of poly(dimethylsiloxane) (PDMS) by means of an amphiphilic tri-block copolymer, poly(ethylene oxide)-*block*-poly(propylene oxide)-*block*-poly(ethylene oxide) (PEO-*b*-PPO-*b*-PEO, also known as 'Pluronic'), as recently reported by the authors.[13,14]

In this work, we report that the aforementioned graft copolymer, PLL-*g*-PEG, readily adsorbs onto hydrophobic surfaces as well, and thus also behaves as an effective aqueous lubricant additive for hydrophobic surfaces. As will be elaborated later in this work, the C_4 hydrocarbon chain, —$CH_2CH_2CH_2CH_2$—, of the L-lysine monomer (dark grey area (in Figure 1(a))), is believed to play a role as an anchoring unit onto hydrophobic, non-polar surfaces. The versatile adsorption behaviour of PLL-*g*-PEG onto both hydrophilic, polar (negatively charged) surfaces as well as onto hydrophobic surfaces makes it particularly suitable for the aqueous lubrication of tribosystems involving thermoplastics; while thermoplastic-based bearing systems can employ self-mated sliding contacts, they often require polar surfaces, such as metals and ceramics, as sliding partners.[1,2] In the following, we demonstrate the unique efficacy of PLL-*g*-PEG as an aqueous lubricant additive for tribosystems involving thermoplastics by comparison with other aqueous lubricant additives.

MATERIALS AND METHODS

Tribopairs

The materials employed for tribopairs can be grouped into: (i) hydrophobic: thermoplastics, including polypropylene (PP) (Maagtechnic, Dübendorf, Switzerland), polyamide-6,6 (PA-6,6) (Maagtechnic) and polyethylene (PE) (for disk, Maagtechnic, and for pin, DSM Sittard, The Netherlands Stamylan, UH210, Switzerland), and (ii) hydrophilic: metal or ceramics, including stainless steel (Hydrel AG, Romanshorn, Switzerland), soda glass (SuperFrost, Menzel-Gläser, Braunschweig, Germany;

DOI: 10.1002/ls

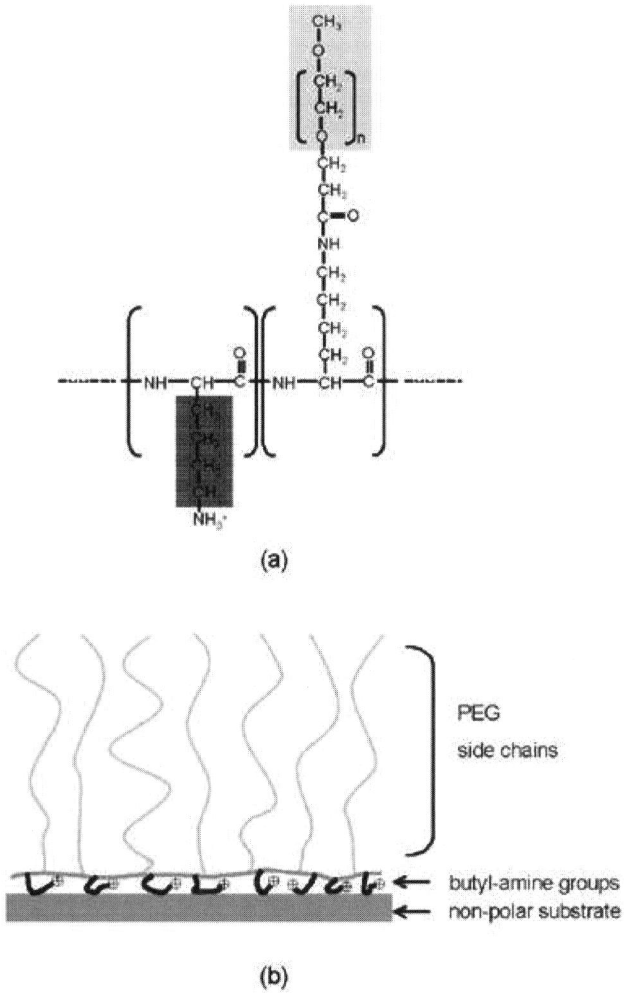

Figure 1. (a) A schematic of poly(L-lysine)-*graft*-poly(ethylene glycol) (PLL-g-PEG) and (b) its conformation at the interface of water/non-polar surface in an aqueous environment. The molecular structure of PLL-g-PEG displaying two L-lysine monomer units with (right bracket) and without (left bracket) a grafted poly(ethylene glycol) (PEG) side chain (light grey area). The C4 hydrocarbon chain, —$CH_2CH_2CH_2CH_2$—, of the ungrafted L-lysine monomer (dark grey area), is believed to play a role as an anchoring unit onto hydrophobic, non-polar surfaces.

approximate composition: SiO_2 72.2%, Na_2O 14.3%, CaO 6.4%, MgO 4.3%, Al_2O_3 1.2%, K_2O 1.2%, other impurities, Fe_2O_3 and SO_3, below 1%) and zirconia (ZrO_2, Saphirwerk, Brügg Switzerland). These materials have been employed for both pin and disk, except for soda glass (used as disk only). Two different pin geometries have been employed: flat-ended puck with a diameter of 3 mm for PE pin, and sphere with a diameter of 6 mm for the rest. All pin materials and the ZrO_2 disk (5 mm in thickness and 30 mm in diameter) had been supplied by the manufacturers with submicron roughness (see

DOI: 10.1002/ls

Table I. Basic information on the mechanical and surface properties of the materials employed for this study.

	Young's modulus (GPa)*	Poisson's ratio	Surface roughness/R_a (nm over 100 μm^2)		Static water contact angle (°)**
			Pin	Disk	
PP	1.1–1.3	0.3	77	32	106.8 ± 2.59
PA-6,6	2	0.3	239	30	81.2 ± 0.84
PE	0.7–1.4	0.3	125	42	90.8 ± 3.96
PDMS	—	—	—	—	109.4 ± 1.34
Glass	72	0.2	—	2.2	<5
ZrO_2	200	0.27	9.6	7.4	<5
Stainless steel	203	0.3	32	5	<5

*The elasticity moduli of PP and PA-6,6 are from the manufacturer, and the mid-values were used for the calculation shown in Table II.
**Error bars come from the standard deviation over five measurements.
PP, polypropylene; PA-6,6, polyamide-6,6; PE, polyethylene; PDMS, poly(dimethylsiloxane).

Table I) and were used without further polishing. Disks (5 mm thick) of the thermoplastics and stainless steel were machined from 30-mm-diameter rods, and their surfaces were polished with silica paper on a rotating polishing wheel (grade P600 followed by grade P1200) in our laboratory. The surface roughness (R_a) of the pins and disks was characterised by Atomic Force Microscopy (AFM), and the results are shown in Table 1. Prior to tribological measurements, all materials had been ultrasonicated in ethanol for 5 min and blown dry with nitrogen. Hydrophilic materials, including glass, ZrO_2 and stainless steel, were further cleaned by air plasma treatment for 1 min in a plasma cleaner (Harrick PDC-32, Ossining, NY, USA) to remove contaminants and to obtain well-defined oxide surfaces.

Lubricant and Lubricant Additives

Neutral aqueous buffer solution, prepared by dissolving KH_2PO_4 in water (1 mM) and adjusting pH to 7.0 by 6 M KOH, was used as the lubricant for all measurements in this work.

As aqueous lubricant additives, sodium dodecylsulphate (SDS) ($CH_3(CH_2)_{11}SO_4Na$), PEO-*b*-PPO-*b*-PEO, also known as 'Pluronic' and PLL-*g*-PEG have been employed. SDS (Sigma Aldrich, Buchs Switzerland) and PEO-*b*-PPO-*b*-PEO (BASF, Mt. Olive, NJ, USA) were purchased, and PLL-*g*-PEG was synthesised in our laboratory according to a method described in previous publications.[7,8,15] The molecular weight of the PEO and PPO blocks of the PEO-*b*-PPO-*b*-PEO used in this work is ca. 3250 (1625 g mol^{-1} for each PEO block) and 3250 g mol^{-1}, respectively (denoted as 'P105' according to the manufacturer[16] or can be denoted as EO_{37}-PO_{56}-EO_{37} in which 37 and 56 represent the numbers of EO and PO monomer units, respectively). The PLL-*g*-PEG copolymer employed in this work, denoted as PLL(375)-*g*[3.4]-PEG(5), possesses the molecular weights of 375 000 g mol^{-1} for the PLL backbone (including HBr in the precursor) and 5000 g mol^{-1} for PEG side chains, and a graft ratio of 3.4 (lysine-mer/PEG side chains). Thus, PLL(375)-*g*[3.4]-PEG(5), on average, carries 1793 lysine monomer units, of which 527 units are used to graft PEG side chains (each with 113 EG monomer

DOI: 10.1002/ls

Table II. Mechanical contact conditions for the tribopairs employed in this study (the elasticity modulus and Poisson's ratio of each material comes from Table I, and load, W, is 10 N, and the radius of pin, R, is 3 mm).

Tribopairs (pin/disk)	Reduced elasticity modulus, E' (GPa) $E' = 2\left(\frac{1-\nu_1^2}{E_1} + \frac{1-\nu_2^2}{E_2}\right)^{-1}$	Contact radius, a (mm) $a = \sqrt[3]{\frac{3WR}{4E'}}$	Maximum contact pressure, P_0 (MPa) $P_0 = \frac{1}{\pi}\sqrt[3]{\frac{6WE'^2}{R^2}}$	Mean contact pressure, P_m (MPa) $P_m \approx \frac{2}{3}P_0$
PP/PP	1.32	2.6×10^{-4}	45.4	30.3
PA-6,6/PA-6,6	2.20	2.2×10^{-4}	63.8	42.5
PE/PE*	—	1.5*	1.4*	1.4*
PP/Glass	2.59	2.1×10^{-4}	71.2	47.5
PA-6,6/Glass	4.27	1.7×10^{-4}	99.3	66.2
PE/Glass*	—	1.5*	1.4*	1.4*
PP/ZrO_2	2.75	2.0×10^{-4}	74.1	49.4
PA-6,6/ZrO_2	4.72	1.7×10^{-4}	106.2	70.8
PE/ZrO_2*	—	1.5*	1.4*	1.4*
ZrO_2/PP	2.75	2.0×10^{-4}	74.1	49.4
ZrO_2/PA-6,6	4.72	1.7×10^{-4}	106.2	70.8
ZrO_2/PE	2.39	2.1×10^{-4}	67.5	45.0
PP/Steel	2.62	2.0×10^{-4}	71.8	47.9
PA-6,6/Steel	4.35	1.7×10^{-4}	106.5	71.0
PE/Steel*	—	3*	1.4*	1.4*
Steel/PP	2.62	2.0×10^{-4}	71.8	47.8
Steel/PA-6,6	4.35	1.7×10^{-4}	106.5	71.0
Steel/PE	2.30	2.1×10^{-4}	65.7	43.8

*The contact radius for the tribopairs involving PE pin, 1.5 mm, was estimated from the radius of the PE pin, and the contact pressure is simply $W/\pi a^2$, due to flat-on-flat contact geometry.
PP, polypropylene; PA-6,6, polyamide-6,6; PE, polyethylene.

units on average), and the other 1266 units are left un-reacted. The PEO-*b*-PPO-*b*-PEO and PLL-*g*-PEG were dissolved in the aqueous buffer solution at 2 and 0.25 mg mL^{-1} concentration, respectively.

Optical Waveguide Lightmode Spectroscopy (OWLS)

OWLS has been employed to characterise the adsorption behaviour of lubricant additives onto the tribopair surfaces. OWLS is based upon grating-assisted in-coupling of a He–Ne laser into a planar waveguide coating (200-nm-thick $Si_{0.75}Ti_{0.25}O_2$ waveguiding layer on a 1-mm-thick AF 45 glass, MicroVacuum Ltd., Budapest, Hungary), and allows the direct online monitoring of the 'dry' mass of adsorption, in that water that is hydrodynamically coupled into adsorbates is not taken into account in mass detection. More detailed information on the operational principles of OWLS is available in previous publications.[17–20] All OWLS experiments were carried out in a BIOS-I instrument (ASI AG, Zurich, Switzerland) using a Kalrez (Dupont, Wilmington, DE, USA) flow-through cell ($8 \times 2 \times 1$ mm). The OWLS waveguides were first exposed to buffer solution until a stable baseline was obtained.

DOI: 10.1002/ls

Then, the solution containing lubricant additives was injected into the flow cell. It is emphasised that the change in refractive index at this stage reflects the adsorption of additives, yet includes the contribution from the refractive index change of bulk solution itself as well. The adsorbed mass was thus measured only after the flow cell was rinsed with buffer solution. The refractive index increment (dn/dc) values of 0.151 and 0.169 $cm^3 g^{-1}$ were used for PEO-*b*-PPO-*b*-PEO and PLL-*g*-PEG, respectively.

To apply the OWLS approach to surfaces other than the waveguide coating itself ($Si_{0.75}Ti_{0.25}O_2$), a thin layer of the materials of interest can be deposited on top of the waveguide coating. The ceramic materials — glass (SiO_2) and ZrO_2 — and stainless steel in this work were sputtered (ca. 12 nm) onto the waveguides in a Leybold dc-magnetron Z600 sputtering unit (Paul Scherrer Institute, Villigen, Switzerland). However, it is technically much more challenging to deposit thin layers of the relevant thermoplastics onto the waveguides. Therefore, we have employed PDMS as a *model* hydrophobic, non-polar thin polymer layer for OWLS studies. This has been successfully applied in previous studies, the film being ca. 30 nm in thickness and displaying a static water contact angle of $100 \pm 2°$.[13,21,22] Briefly, the base and curing agent of a commercial silicon elastomer, Sylgard 184 (Dow Corning, Midland, MI, USA), were dissolved in hexane at a ratio of 10 : 3, and the total concentration was 0.5% (w/w). A mixture of base and curing agent in hexane were spin coated onto waveguides at 2000 rpm for 40 s, followed by curing in an oven at 70°C for 24 h.

While the two coating materials, SiO_2 and PDMS, have different thicknesses and refractive indexes, the *change* of the refractive index, which is the basis for the determination of adsorbed mass, was observed to be nearly identical by testing with polymer-free aqueous solutions with different bulk refractive indexes, which validates the comparison of the adsorbed mass of the polymers onto both surfaces.

Pin-on-disk Tribometry

The frictional properties of the tribopairs during aqueous-lubricated sliding have been characterised by means of conventional pin-on-disk tribometry (CSM, Neuchâtel, Switzerland).[7–9,13,14] In this approach, the load was determined by dead weights, and the frictional forces were measured by a strain gauge. After forming a contact between the loaded pin and the disk, the disk was rotated at a controlled speed by a motor, thus generating sliding friction forces. The raw data for the friction forces were recorded as a function of time (or the number of rotations) over a fixed track, using a Macintosh Power PC with a Labview program and an ADC card of the MIO family (both from National Instruments, Austin, TX, USA). Throughout this work, the load, speed, sliding track radius and number of rotations were fixed at 10 N, 5.1 $mm\,s^{-1}$, 8 mm and 1000, respectively. The mechanical contact conditions for the tribopairs employed in this study are summarised in Table 2. The coefficients of friction, μ, were obtained from the average value from 501st to 1000th rotations only, to exclude any contribution from running-in effects.

RESULTS AND DISCUSSION

Adsorption of Lubricant Additives onto Surfaces

The adsorption behaviour of the lubricant additives onto a hydrophilic, polar surface (SiO_2) and a hydrophobic, non-polar surface (PDMS) has been characterised by means of OWLS, and the results

DOI: 10.1002/ls

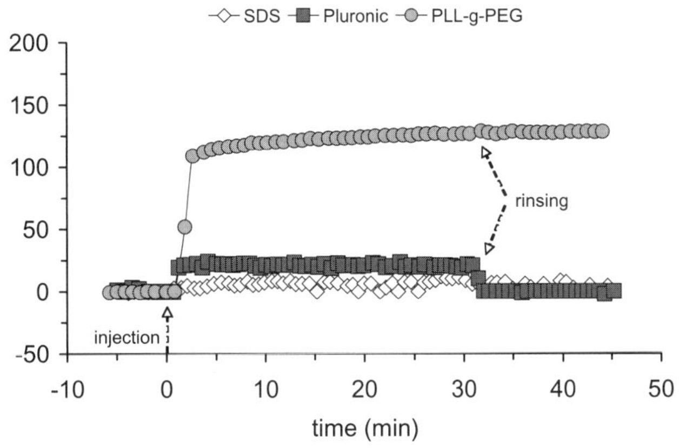

Figure 2. Adsorption behaviour of the three aqueous lubricant additives, sodium dodecylsulphate (SDS), poly(ethylene oxide)-*block*-poly(propylene oxide)-*block*-poly(ethylene oxide) and poly(L-lysine)-*graft*-poly(ethylene glycol) (PLL-*g*-PEG), onto a SiO_2 surface from aqueous solution (pH 7.0), as determined by means of optical waveguide lightmode spectroscopy.

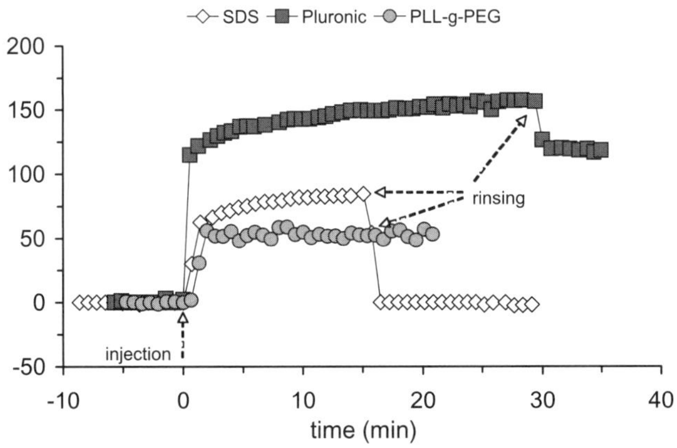

Figure 3. Adsorption behaviour of the three aqueous lubricant additives, sodium dodecylsulphate (SDS), poly(ethylene oxide)-*block*-poly(propylene oxide)-*block*-poly(ethylene oxide) and poly(L-lysine)-*graft*-poly(ethylene glycol) (PLL-*g*-PEG), onto poly(dimethylsiloxane) surface from aqueous solution (pH 7.0), as determined by means of optical waveguide lightmode spectroscopy.

are presented in Figures 2 and 3, respectively. These plots show the adsorbed mass, determined from the change in refractive index in the vicinity of the surface, as a function of time. The time $t = 0$ corresponds to the injection of the lubricant additive-containing solution into the flow cell, and thus the data prior to injection ($t < 0$) serve as a reference line. For all the lubricant additives tested, the adsorbed mass appears to increase upon injection of the solution and quickly reaches a pseudo-equilibrium value (within ca. 10 min). The data at this stage reflect the change in refractive index due to both surface-adsorbed species and bulk solution. Thus, an assessment of adsorbed mass is only possible after the flow cell has been rinsed with buffer solution.

DOI: 10.1002/ls

Among the three lubricant additives, PLL-*g*-PEG was observed to be the only one that adsorbed onto a SiO_2 surface (ca. $128\,ng\,cm^{-2}$). As was shown in previous studies,[11,12,15] the adsorption of PLL-*g*-PEG onto SiO_2 is driven by the electrostatic interaction between positively charged primary amine groups ($—NH_3^+$) of ungrafted lysine units (see Figure 1) and the negatively charged substrate (IEP of $SiO_2 \sim 2$)[23] in a neutral aqueous environment. In this context, SDS, carrying negative charges ($—SO_4^-$), and PEO-*b*-PPO-*b*-PEO, carrying no net charge, lack the corresponding driving force to adsorb onto a SiO_2 substrate, accounting for their negligible adsorption.

On the PDMS surface, PEO-*b*-PPO-*b*-PEO showed the highest amount of adsorption, ca. $118\,ng\,cm^{-2}$, followed by PLL-*g*-PEG, ca. $58\,ng\,cm^{-2}$, and SDS again showed virtually no adsorption. The adsorption behaviour of PEO-*b*-PPO-*b*-PEO onto hydrophobic surfaces, including PDMS,[13] from aqueous solution has been extensively investigated in previous studies,[24,25] and its driving force is attributed to a hydrophobic interaction between the PPO block and the surface.

It is noteworthy that PLL-*g*-PEG, carrying positively charged species in its anchoring units, displays distinct adsorption behaviour onto a hydrophobic, non-polar PDMS surface. Similarly to PEO-*b*-PPO-*b*-PEO, the adsorption of PLL-*g*-PEG onto the PDMS surface is thought to occur through hydrophobic interactions between a hydrophobic component of the lysine units — the C_4H_8 hydrocarbon chains, $\mathbf{CH_2CH_2CH_2CH_2}—NH_3^+$ — and the surface. While the interaction of an individual C_4H_8 hydrocarbon chain with the surface is weak, there are 1266 such groups per single copolymer molecule (see Lubricant and Lubricant Additives) that are available to collectively participate in the interaction with the surface. It should also be noted that the only structural difference between a PPO block, $—[CH(CH_3)CH_2O]_m—$, and a PEO block, $—[CH_2CH_2O]_n—$, in PEO-*b*-PPO-*b*-PEO is the replacement of a hydrogen atom, —H, with one methyl unit, CH_3, and yet, a distinctive hydrophobicity is imparted to PPO block because of the collective interaction of methyl groups with hydrophobic surfaces ($m = 56$ for the PEO-*b*-PPO-*b*-PEO employed in this work, see Lubricant and Lubricant Additives). The presence of positively charged primary amine groups next to the C_4 hydrocarbon chain, $—CH_2CH_2CH_2CH_2—\mathbf{NH_3^+}$, for PLL-*g*-PEG, however, might hinder anchoring onto non-polar surfaces due to charge accumulation at the surface. This accounts for the observation that the adsorption of the PEO-*b*-PPO-*b*-PEO onto the PDMS surface is more effective than that of PLL-*g*-PEG, when judged by the amount of adsorption (Figure 3), despite the significantly smaller number of anchoring units of the former, i.e. the number of PO monomer units, 56, compared to the corresponding anchoring units of the latter, i.e. the number of ungrafted lysine units, 1266. In the same context, SDS, which possesses much longer C_{12} hydrocarbon chains, ($\mathbf{CH_3(CH_2)_{11}—}$), displays a much weaker interaction with hydrophobic surfaces because it occurs only through individual molecules and therefore lacks the collective effects seen with the polymeric adsorbates.

The net effect of adsorbing either PEO-*b*-PPO-*b*-PEO or PLL-*g*-PEG onto hydrophobic surfaces is to cover the surface with hydrophilic PEO (or PEG) chains. With respect to aqueous lubrication of the thermoplastics in this work, this PEO (or PEG) layer is expected to make it feasible to form aqueous boundary lubricant films due to improved affinity with water, as will be shown in the following section.

It is important to clarify that the PDMS surface employed for adsorption studies in this work serves only as a model hydrophobic, non-polar surface. While all of the thermoplastics investigated in this study also display hydrophobic surface characteristics, the magnitude is notably different for each material, as demonstrated by water contact angle measurements (see Table I). For instance, PA-6,6 is well known for its net hydrophobic, yet slightly polar surface characteristics due to amide bonds.[1,2] A detailed characterisation of the adsorption behaviour of the lubricant additives onto each

DOI: 10.1002/ls

thermoplastic material will be required in the future for the following two reasons. Firstly, because PLL-*g*-PEG has revealed very versatile adsorption properties onto both hydrophilic, polar surfaces (e.g. SiO_2) and hydrophobic, non-polar surfaces (e.g. PDMS), the adsorption of this copolymer onto an inhomogeneous surface displaying both non-polar and polar (to be more precise, negatively charged) patches, for instance, is expected to be particularly effective. Secondly, previous tribological studies involving rubber materials by Roberts *et al.*[26–28] have shown that the addition of anionic surfactants, such as SDS, into water, facilitates the formation of a thicker aqueous lubricating film than water alone. Because this effect was more pronounced when rubber was slid against glass than the self-mated rubber–rubber pair, its origin was attributed to electrostatic double-layer repulsion; if SDS adsorbs onto rubber surfaces, which are usually hydrophobic, through its hydrocarbon chains, the negative charges carried by sulphate groups, —SO_4^-, can repel opposing negatively charged surfaces, either another rubber surface, or more effectively, a glass surface, which displays more distinctive negative charges in aqueous environment. In other words, it cannot be excluded that even the most weakly interacting SDS may adsorb onto hydrophobic surfaces other than PDMS, such as the thermoplastic materials employed in this work, to an appreciable magnitude. Nevertheless, no direct evidence for the adsorption of SDS onto rubber materials has been provided to date.

Influence of the Lubricant Additives on the Aqueous Lubrication Properties of Tribopairs Involving Thermoplastics

The influence of the lubricant additives on the aqueous lubricating properties of tribocontacts involving thermoplastics has been characterised firstly by employing the self-mated pairs of PP, PA-6,6 and PE. In these measurements, the tribological contact condition was selected such that the lubrication by water lies within the boundary lubrication regime (applied load, 10 N; sliding speed, 5.1 mm s^{-1}). It is noted that the contact pressure for PE/PE pair is much lower than the others, by more than an order of magnitude, due to the selected flat-on-flat contact geometry (see Table II for details). An example of raw friction data as a function of rotation is presented in Figure 4 (the case for self-mated PP sliding lubricated by aqueous solution with and without PLL-*g*-PEG). With the exception of a slight

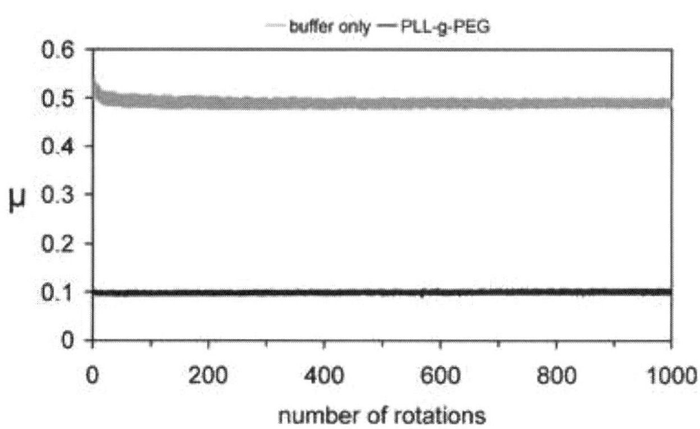

Figure 4. Raw friction data as a function of rotation for the self-mated polypropylene sliding contact lubricated by aqueous solution with and without poly(L-lysine)-*graft*-poly(ethylene glycol) (PLL-*g*-PEG) (load, 10 N; diameter of the pin, 6 mm; sliding speed, 5.1 mm s^{-1}; number of rotations, 1000).

DOI: 10.1002/ls

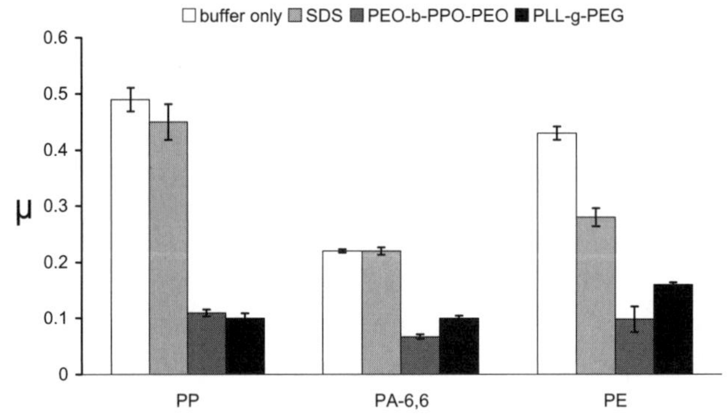

Figure 5. Comparison of the coefficients of friction (μ) values for the self-mated sliding of the thermoplastics, polypropylene (PP), polyamide-6,6 (PA-6,6) and polyethylene (PE), lubricated by aqueous lubricants containing sodium dodecylsulphate (SDS), poly(ethylene oxide)-*block*-poly(propylene oxide)-*block*-poly(ethylene oxide) (PEO-*b*-PPO-*b*-PEO) and poly(L-lysine)-*graft*-poly(ethylene glycol) (PLL-*g*-PEG) (load, 10 N; diameter of the pin, 6 mm; sliding speed, 5.1 mm s^{-1}; number of rotations, 1000).

variation in μ in the initial running-in stage, the lubrication of the tribopair with water, with or without lubricant additives, revealed very stable μ values over 1000 rotations, for all cases.

For the self-mated thermoplastic tribopairs, attention was paid to the relative effectiveness of the lubricant additives, and the results are presented in Figure 5. While all three lubricant additives have been observed to generally improve the lubricating properties of aqueous buffer solution, the magnitude is significantly higher for PEO-*b*-PPO-*b*-PEO and PLL-*g*-PEG than for SDS. This is thought to be closely associated with the distinctly higher amount of adsorption onto a model hydrophobic PDMS surface for the two polymers than for SDS, as described in the previous section. PEO-*b*-PPO-*b*-PEO and PLL-*g*-PEG were comparable in their lubricating properties for PP, but the former was slightly better than the latter in its lubrication capabilities for PA-6,6 and PE. The contribution of SDS in improving the lubricating capabilities of PP and PA-6,6, was negligible, whereas an appreciable reduction in μ was observed for PE case. As mentioned in the previous section, this difference might arise from different adsorption behaviour on different thermoplastics, but it should also be noted that the contact pressure for PE/PE sliding was much lower than in the other cases. If the double-layer repulsion mechanism is relevant at the contacts involving the thermoplastics, as it is for the tribological contact of rubbers, the degree of electrostatic repulsion would be inversely proportional to contact pressure.[26–28]

Secondly, the same measurements shown in Figure 5 were repeated, replacing the thermoplastic disks with glass. The tribopairs thus consisted of hydrophobic surfaces (sliders) and a hydrophilic, polar surface (track). The comparative lubricating properties of the three aqueous lubricant additives for these tribopairs are presented in Figure 6. First of all, it is notable that the μ values in the absence of the lubricant additives are generally smaller than those observed for the self-mated sliding contacts of the thermoplastics. This is attributed primarily to the hydrophilic characteristics of the glass surface, which enable the maintenance of a water boundary lubricant film at the tribological interface. The lubricating capabilities of additives have been observed to be in the order of PLL-*g*-PEG < SDS < PEO-*b*-PPO-*b*-PEO. The superior lubricating properties of PLL-*g*-PEG are unsurprising, considering its unique adsorption properties onto both hydrophobic and hydrophilic, polar surfaces (Figures 2 and

DOI: 10.1002/ls

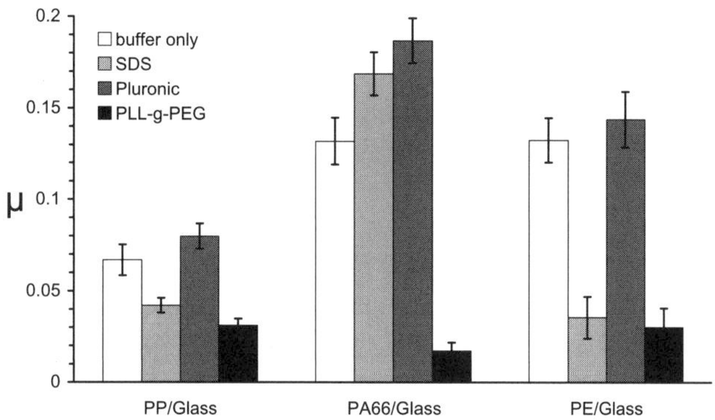

Figure 6. Comparison of the coefficients of friction (μ) values for the sliding of the thermoplastics, polypropylene (PP), polyamide-6,6 (PA-6,6) and polyethylene (PE), against a glass substrate lubricated by aqueous lubricants containing sodium dodecylsulphate (SDS), poly(ethylene oxide)-*block*-poly(propylene oxide)-*block*-poly(ethylene oxide) and poly(L-lysine)-*graft*-poly(ethylene glycol) (PLL-*g*-PEG) (load, 10 N; diameter of the pin, 6 mm; sliding speed, 5.1 mm s^{-1}; number of rotations, 1000).

3). SDS showed most effective lubrication again for the tribopair involving PE. Compared with the PE/PE pair (Figure 5), the degree of reduction in μ is enhanced when the sliding partner is replaced with glass, similarly to the aqueous lubrication of rubber.[26–28] Because SDS showed virtually no adsorption onto a SiO_2 surface (Figure 2), this effect should be ascribed to genuine negative charges provided by the glass surface as a sliding partner, rather than any active role of SDS at the glass surface, which supports the arguments proposed by Roberts *et al.*[26–28] It is notable that PEO-*b*-PPO-*b*-PEO, the best-performing lubricant additive at the self-mated sliding contacts (Figure 5), does not contribute to improving the lubricating properties at all. This is thought to be associated with the fact that PEO-*b*-PPO-*b*-PEO is only adsorbing onto hydrophobic surfaces. Exacerbating this effect, the thermoplastics have been employed as the pins (sliders) in the pin-on-disk experiments; in contrast to the disk (track), which experiences tribological contacts over a contact area that is periodically changing with time, the pin (slider) experiences continuous tribological contacts within a confined area, and thus, any adsorbates may be more easily removed, and less easily readsorbed.

Encouraged by the unique adsorption properties of PLL-*g*-PEG onto both hydrophobic and hydrophilic surfaces, we have tested its efficacy as an aqueous lubricant additive for tribopairs of relevance to engineering applications, such as thermoplastics/ceramic (ZrO_2) and thermoplastics/metal (stainless steel). The adsorption properties of PLL-*g*-PEG onto an FeO_x surface have already been reported in a previous study,[7] and very efficient adsorption behaviour of PLL-*g*-PEG onto ZrO_2 surfaces has also been confirmed in this study (ca. 190 ng cm^{-2}; raw data not shown). The tribological test results for the thermoplastics/ZrO_2 and the thermoplastics/stainless steel are presented in Figures 7 and 8, respectively. In these measurements, both thermoplastics and ZrO_2 or stainless steel have been employed as both pin (slider) and disk (track), in order to cover different geometries that may be relevant for bearing design. In the absence of the lubricant additive, soft slider (thermoplastic pins)/rigid track (ZrO_2 or stainless steel disks) pairs generally revealed lower friction forces than their counterparts — rigid slider (ZrO_2 or stainless steel pins)/soft track (thermoplastic disks) pairs. While an elastic deformation of the side with lower elasticity modulus is common for both contact configurations, rigid pin/soft disk pair experiences a continuous deformation–restoration cycle along the sliding track of

DOI: 10.1002/ls

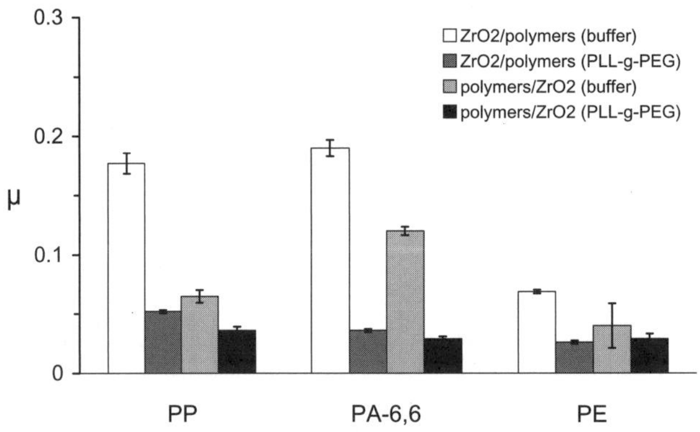

Figure 7. Comparison of the coefficients of friction (μ) values for the sliding of the thermoplastics, polypropylene (PP), polyamide-6,6 (PA-6,6) and polyethylene (PE), against ZrO_2 substrate lubricated by aqueous lubricants containing sodium dodecylsulphate, poly(ethylene oxide)-*block*-poly(propylene oxide)-*block*-poly(ethylene oxide) and poly(L-lysine)-*graft*-poly(ethylene glycol) (PLL-*g*-PEG). The results for both thermoplastic/ZrO_2 and ZrO_2/thermoplastic (pin/disk) pairs are shown (load, 10 N; diameter of the pin, 6 mm; sliding speed, 5.1 mm s^{-1}; number of rotations, 1000).

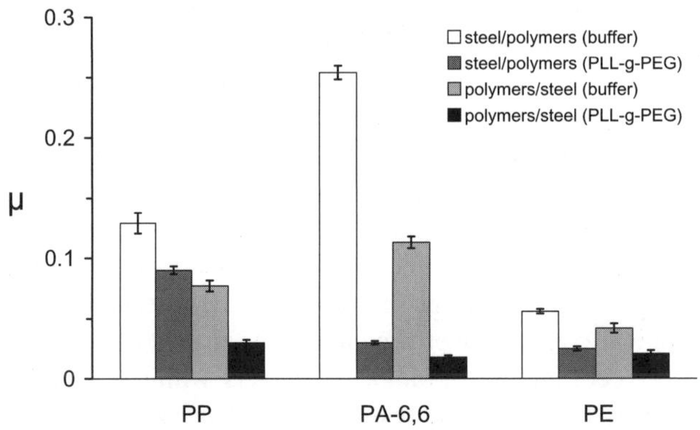

Figure 8. Comparison of the coefficients of friction (μ) values for the sliding of the thermoplastics, polypropylene (PP), polyamide-6,6 (PA-6,6) and polyethylene (PE), against a stainless steel substrate lubricated by the aqueous lubricants containing sodium dodecylsulphate, poly(ethylene oxide)-*block*-poly(propylene oxide)-*block*-poly(ethylene oxide) and poly(L-lysine)-*graft*-poly(ethylene glycol) (PLL-*g*-PEG). The results for both thermoplastic/stainless steel and stainless steel/thermoplastic (pin/disk) pairs are shown (load, 10 N; diameter of the pin, 6 mm; sliding speed, 5.1 mm s^{-1}; number of rotations, 1000).

the disk, for which the resulting energy dissipation is greater than its counterpart, for which elastic deformation is limited to a confined area of the pin and does not change during the rotation cycle. Nevertheless, as with the thermoplastics/glass pair case shown in Figure 6, the μ values of both configurations are generally lower than those of the self-mated thermoplastic tribopairs (Figure 5), due to the hydrophilic surface characteristics of the rigid sliding partners employed in this work (see Table 1 for water contact angles). As expected, the addition of PLL-*g*-PEG into the aqueous buffer

DOI: 10.1002/ls

induced distinctive improvement in lubricating properties. Overall, the lowest μ values were observed from the soft slider/rigid track tribopairs lubricated by PLL-*g*-PEG-containing aqueous buffer solutions.

CONCLUSIONS

In this work, we have investigated the adsorption and lubrication behaviour of a surfactant, SDS, and two PEG (or PEO)-based copolymers, PEO-*b*-PPO-*b*-PEO and PLL-*g*-PEG, for tribopairs comprised of hydrophobic thermoplastics, such as PP, PA-6,6 and PE, sliding against themselves or against hydrophilic, polar surfaces, such as glass, stainless steel or ZrO_2. While SDS and PEO-*b*-PPO-*b*-PEO led to an improvement in aqueous lubrication properties only for thermoplastics/hydrophilic tribopairs or self-mated thermoplastic tribopairs, respectively, PLL-*g*-PEG caused an improvement in the aqueous lubrication properties for both cases. This is thought to originate from PLL-*g*-PEG's unique, versatile adsorption behaviour from aqueous solution onto both hydrophilic, polar (negatively charged) surfaces as well as hydrophobic, non-polar surfaces. Because the engineering tribological applications of thermoplastic materials require not only self-mated sliding contacts, but often contacts of thermoplastics with polar surfaces, such as metals and ceramics, this study shows that PLL-*g*-PEG holds great potential to be used as a versatile aqueous lubricant additive for thermoplastic materials.

ACKNOWLEDGEMENT

This research was financially supported by the US Air Force Office of Scientific Research under contract no. F49620-02-0346.

REFERENCES

1. Yamaguchi Y. Application to sliding machine parts, in *Tribology of Plastic Materials* (ed.), Elsevier, Amsterdam 1990: 203–362.
2. Stachowiak GW, Batchelor AW. Wear of non-metallic materials, in *Engineering Tribology* (ed.), Elsevier, Amsterdam, The Netherlands 1993: 715–771.
3. Cohen SC, Tabor D. The friction and lubrication of polymers. *Proc. R. Soc. London, Ser. A* 1966; **291**(1425):186–207.
4. Mens JWM, de Gee AWJ. Friction and wear behavior of 18 polymers in contact with steel in environments of air and water. *Wear* 1991; **149**:255–268.
5. Lutton MD, Stolarski TA. The effect of water lubrication on polymer wear under rolling contact conditions. *J. Appl. Polym. Sci.* 1994; **54**:771–782.
6. Unal H, Mimaroglu A. Friction and wear characteristics of PEEK and its composite under water lubrication. *J. Reinf. Plast. Compos.* 2006; **25**(16):1659–1667.
7. Lee S, Müller M, Ratoi-Salagean M, Vörös J, Pasche S, De Paul SM, Spikes HA, Textor M, Spencer ND. Boundary lubrication of oxide surfaces by poly(L-lysine)-*g*-poly(ethylene glycol) (PLL-*g*-PEG) in aqueous media. *Tribol. Lett.* 2003; **15**:231–239.
8. Müller M, Lee S, Spikes HA, Spencer ND. The influence of molecular architecture on the macroscopic lubrication properties of the brush-like co-polyelectrolyte poly(L-lysine)-*g*-poly(ethylene glycol) (PLL-*g*-PEG) adsorbed on oxide surfaces. *Tribol. Lett.* 2003; **15**:395–405.
9. Lee S, Müller M, Heeb R, Zürcher S, Tosatti S, Heinrich M, Amstad F, Pechmann S, Spencer ND. Self-healing behavior of a polyelectrolyte-based lubricant additive for aqueous lubrication of oxide materials. *Tribol. Lett.* 2006; **24**(3):217–223.
10. Yan X, Perry SS, Spencer ND, Pasche S, De Paul SM, Textor M, Lim MS. Reduction of friction at oxide interfaces upon polymer adsorption from aqueous solutions. *Langmuir* 2004; **20**:423–428.

11. Müller M, Yan X, Lee S, Perry SS, Spencer ND. Lubrication properties of a brushlike copolymer as a function of the amount of solvent absorbed within the brush. *Macromolecules* 2005; **38**:5706–5713.
12. Müller M, Yan X, Lee S, Perry SS, Spencer ND. Preferential solvation and its effect on the lubrication properties of a surface-bound, brushlike copolymer. *Macromolecules* 2005; **38**:3861–3866.
13. Lee S, Iten R, Müller M, Spencer ND. Influence of molecular architecture on the adsorption of poly(ethylene oxide)-poly(propylene oxide)-poly(ethylene oxide) (PEO-PPO-PEO) on PDMS surfaces and implications for aqueous lubrication. *Macromolecules* 2004; **37**:8349–8356.
14. Lee S, Spencer ND. Aqueous lubrication of polymers: influence of surface modification. *Tribol. Int.* 2005; **38**:922–930.
15. Kenausis GL, Vörös J, Elbert DL, Huang NP, Hofer R, Ruiz-Taylor L, Textor M, Hubbell JA, Spencer ND. Poly(L-lysine)-*g*-poly(ethylene glycol) layers on metal oxide surfaces: attachment mechanism and effects of polymer architecture on resistance to protein adsorption. *J. Phys. Chem. B* 2000; **104**:3298–3309.
16. BASF. *BASF Technical Brochure*, BASF Co., Parsippany, NJ 1989.
17. Kurrat R, Textor M, Ramsden JJ, Böni P, Spencer ND. Instrumental improvements in optical waveguide lightmode spectroscopy for the study of biomolecule adsorption. *Rev. Sci. Instrum.* 1997; **68**:2172–2176.
18. Vörös J, Graf R, Kenausis GL, Bruinink A, Mayer J, Textor M, Wintermantel E, Spencer ND. Feasibility study of an online toxicological sensor based on the optical waveguide technique. *Biosens. Bioelectron.* 2000; **15**:423–429.
19. Vörös J, Ramsden JJ, Csúcs G, Szendro I, De Paul SM, Textor M, Spencer ND. Optical grating coupler biosensors. *Biomaterials* 2002; **23**:3699–3710.
20. Höök F, Vörös J, Rodahl M, Kurrat R, Böni P, Ramsden JJ, Textor M, Spencer ND, Tengvall P, Gold J, Kasemo. A comparative study of protein adsorption on titanium oxide surfaces using in situ ellipsometry, optical waveguide lightmode spectroscopy and quartz microbalance/dissipation. *Colloids Surf., B* 2002; **24**:155–170.
21. Lee S, Vörös J. An aqueous-based surface modification of poly(dimethylsiloxane) with poly(ethylene glycol) to prevent biofouling. *Langmuir* 2005; **21**:11957–11962.
22. Lee S, Müller M, Rezwan K, Spencer ND. Porcine gastric mucin (PGM) at the water/poly(dimethylsiloxane) (PDMS) interface: influence of pH and ionic strength on its conformation, adsorption, and aqueous lubrication properties. *Langmuir* 2005; **21**:8344–8353.
23. Parks GA. The isoelectric points of solid oxides, solid hydroxides, and aqueous hydroxo complex systems. *Chem. Rev.* 1965; **65**:177–198.
24. Alexandridis P, Hatton TA. Poly(ethylene oxide)-poly(propylene oxide)-poly(ethylene oxide) block copolymer surfactants in aqueous solutions and at interfaces: thermodynamics, structure, dynamics, and modeling. *Colloids Surf., A* 1995; **96**:1–46.
25. Tiberg F, Malmsten M, Linse P, Lindman B. Kinetic and equilibrium aspects of block copolymer adsorption. *Langmuir* 1991; **7**:2723–2730.
26. Roberts AD, Tabor D. The extrusion of liquids between highly elastic solids. *Proc. R. Soc. London, Ser. A* 1971; **325**:323–345.
27. Roberts AD. The shear of thin liquid films. *J. Phys. D* 1971; **4**:433–441.
28. Richards SC, Roberts, AD. Boundary lubrication of rubber by aqueous surfactant. *J. Phys. D* 1992; **25**:A76–A80.

DOI: 10.1002/ls

MATERIALS SCIENCE

Sweet, Hairy, Soft, and Slippery

Synthetic polymer lubricants inspired by biological systems may be the key to water-based lubrication.

Seunghwan Lee and Nicholas D. Spencer

Water, together with surfaces containing sugar chains, forms the basis of all biological lubrication systems, from the slithering of a snail to the passage of food along the digestive tract. Yet humans have typically lubricated their machines with oils and fats. Understanding of biological lubrication has now advanced to the point where these principles can be applied to systems of technological importance using synthetic polymers.

Water lubrication is used in some niche applications such as reservoir pumps, and oil-water emulsions are often used for metal cutting, taking advantage of the effectiveness of water as a coolant. In the mining industry, hydraulic fluids are frequently based on water so as to exclude flammable materials from underground working areas. Interest in water lubrication is also high in the food, textile, and pharmaceutical industries, where product contamination by oil is a concern (*1*).

Water on its own is, however, generally a poor lubricant, and unlike oil, its viscosity does not rise substantially with pressure. This property is essential to the mechanism by which oils can form a lubricating film in high-pressure, nonconformal contacts of hard materials such as gears or ball bearings (*1*).

The low viscosity of water at high pressures can be overcome by biological lubricant additives, usually glycoproteins, in which large numbers of sugar chains are bound along a protein backbone. For example, mucins are found in most parts of the human body that need lubricating, such as eyes and knees (*2*). These molecules probably aid lubrication both via their intrinsic viscoelastic properties in solution (*3*) and via their behavior when adsorbed on the sliding surfaces. The characteristic bottlebrush structure of the molecules is crucial to this mechanism: The hydrophilic sugars immobilize large amounts of water within the contact region, while the backbone interconnects to other bottlebrushes or to a surface. Hierarchically structured, sugar-based bottlebrushes also play a key role in the mechanical properties of cartilage (see the figure) (*4*).

One biomimetic approach is to decorate the sliding surfaces with a high density of brush-forming polymer chains. Klein *et al.* have shown that when two mica surfaces bearing polymer brushes are rubbed past each other under compression in "good solvents," the interfacial friction forces lie below the detection limit (*5*, *6*). The remarkable lubricating effect of such hairy polymer layers is ascribed to interchain repulsion, which leads to the incorporation of large quantities of solvent. The resulting fluid-like cushioning layer on the surface can sustain the externally applied pressure, thereby lowering the friction forces (*7*). This behavior, first observed for polystyrene chains in toluene, is a result of the interplay between the polymer and solvent rather than an intrinsic property of either component. Thus, it can also be observed for hydrophilic polymers such as polyethylene glycol (PEG) in water (*8*–*10*).

How high a pressure can brush-like polymer layers withstand? With increasing pressure, the compression and interpenetration of opposing polymer brushes also increase, even in good solvents; the disruption of the brushes, and thus the onset of substantial friction forces, occurs at pressures between 0.01 and 1 MPa, corresponding to pressures in contacts between internal organs or non–load-bearing mammalian joints (*11*). Nonetheless, Müller *et al.* attached dense PEG brushes to steel and glass surfaces sliding in water to investigate their lubricating properties under conditions approaching those of industrial bearings (*12*). They found that the formation of a lubricating film was substantially enhanced by the brushes at contact pressures as high as 0.5 GPa (conditions where water alone cannot form a lubricating film), but the brush layer became detached during sliding contact, and direct contacts between peaks in the surface roughness could not be suppressed completely.

Another important characteristic of natural tribological systems is that they usually involve soft surfaces, as exemplified by slugs, eyes, tongues, and cartilage-coated articular joints. In response to external loads, such soft surfaces deform elastically and increase the contact area, resulting in a relatively low contact pressure. This is why liquids whose viscosity increases only slightly with pressure, such as water, can form lubricating films in soft contacts. This property sparked an extensive study of the aqueous lubrication of elastic polymers (rubbers or elastomers) (*13*) and has led to applications, for example, in tires, seals, windshield wipers, and biomedical implants.

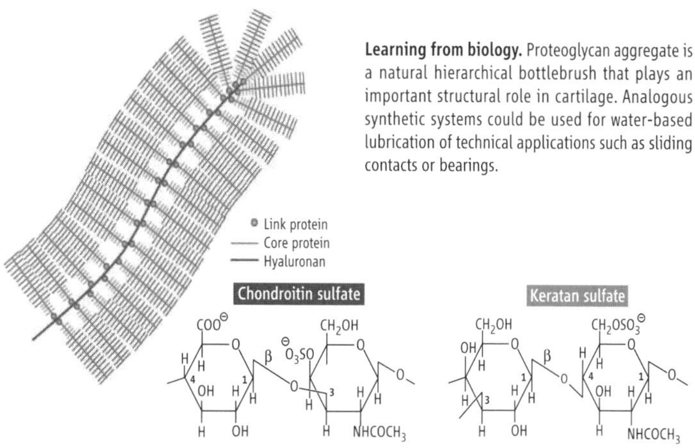

Learning from biology. Proteoglycan aggregate is a natural hierarchical bottlebrush that plays an important structural role in cartilage. Analogous synthetic systems could be used for water-based lubrication of technical applications such as sliding contacts or bearings.

The synergistic combination of hairy polymers and soft surfaces for water lubrication has, however, been investigated only recently. For example, end-grafted PEG chains can greatly enhance the water lubrication of silicone-rubber surfaces, especially at low speeds (*14*). In contrast, short-chain surfactants only led to minor reductions in friction in comparison to brush-like systems (*15*). Gong *et al.* have gone a step further in biomimicry by using hydrogels for aqueous lubrication. In addition to being soft, hydrogels allow water

The authors are in the Laboratory for Surface Science and Technology, Department of Materials, ETH Zürich, 8093 Zürich, Switzerland. E-mail: spencer@mat.ethz.ch

Published by AAAS

to permeate into the network, as does cartilage. Among the many hydrogels that have been investigated, those with brush-like polymer chains at their surfaces are the most effective for aqueous lubrication (*16*).

The combination of brushes with soft surfaces is clearly a key aspect of biological lubrication. However, it is less clear whether carbohydrates possess any specific or unique properties that are absent in other, synthetic brush-forming hydrophilic polymer chains. Moreover, natural lubricant additives appear to form hierarchical bottlebrush structures, such as that shown in the figure, more readily than the synthetic water-soluble brushes that have been investigated to date (*9*, *12*, *14*). The role of both the composition and structure of sugar-based, bottlebrush-structured molecules in natural lubrication thus needs to be clarified.

Although the feasibility of aqueous lubrication with elastomers and hydrogels has been established, the poor mechanical properties and wear resistance of soft materials have been limiting factors for applications. However, it has been shown that the mechanical properties of hydrogels can be improved to a similar level as those of elastomers (*17*). Thus, there is hope that hydrogels can be used in applications where mechanical strength is required. The practical implementation of biomimetic, aqueous lubrication approaches may become a reality in the not-too-distant future.

References

1. D. Dowson, *History of Tribology* (Professional Engineering Publishing, London, 1998).
2. R. Bansil *et al.*, *Annu. Rev. Physol.* **57**, 635 (1995).
3. J. Celli *et al.*, *Biomacromolecules* **6**, 1329 (2005).
4. L. Han *et al.*, *Biophys. J.* **92**, 1384 (2007).
5. J. Klein *et al.*, *Nature* **370**, 634 (1994).
6. J. Klein, *Proc. Inst. Mech. Eng. Part J* **220**, 691 (2006).
7. P. G. De Gennes, *Macromolecules* **13**, 1069 (1980).
8. U. Raviv *et al.*, *Langmuir* **18**, 7482 (2002).
9. T. Drobek, N. D. Spencer, *Langmuir* 10.1021/la702289n (2007).
10. M. Kobayashi *et al.*, *Soft Matter* **3**, 740 (2007).
11. J. Klein *et al.*, *Acta Polym.* **49**, 617 (1998).
12. M. Müller *et al.*, *Tribol. Lett.* **15**, 395 (2003).
13. D. F. Moore, *The Friction and Lubrication of Elastomers* (Pergamon, Oxford, 1972).
14. S. Lee, N. D. Spencer, *Tribol. Int.* **38**, 922 (2005).
15. S. Lee, N. D. Spencer, *Lubric. Sci.* **20**, 21 (2008).
16. J. P. Gong, *Soft Matter* **2**, 544 (2006).
17. Y. Tanaka *et al.*, *Prog. Polym. Sci.* **30**, 1 (2005).

10.1126/science.1153273

Published by AAAS

1484 *Langmuir* **2008,** *24,* 1484–1488

Nanotribology of Surface-Grafted PEG Layers in an Aqueous Environment†

Tanja Drobek and Nicholas D. Spencer*

Laboratory for Surface Science and Technology, Department of Materials, ETH Zurich, Wolfgang-Pauli-Strasse 10, CH-8093 Zurich, Switzerland

Received July 28, 2007. In Final Form: September 5, 2007

The lubrication properties of adsorbed poly(L-lysine)-*graft*-poly(ethylene glycol) in aqueous buffer solution were studied with the surface forces apparatus. In general, the polymer brushes revealed extremely low friction forces. Two distinct regimes could be identified. In response to lateral shear, the friction forces of intact polymer films at moderate loads were below the detection limit. At high loads, when the films were compressed to about 10% of the original equilibrium film thickness, the friction showed a reversible increase with load. Under certain conditions, film destruction was observed, immediately followed by a dramatic increase in the frictional force and an expansion of the adsorbed brush layer. By the addition of free polymer to the buffer solution, the resistance of the polymer brushes to abrasion was dramatically increased by readsorption of the polymer following friction-induced desorption. This self-healing capacity and the extremely low friction of the adsorbed copolymer films contribute to their excellent properties as lubricant additives for water-based lubrication under boundary conditions.

Introduction

Surface-grafted, brushlike polymers can dramatically modify the lubricious properties of surfaces.[1−3] The ability to bind a significant amount of solvent in a surface layer is thought to be one of the key mechanisms for low-friction, polymer-brush films. A brush composed of water-soluble, biocompatible polymers, such as poly(ethylene glycol), in an aqueous environment can provide an oil-free, environmentally friendly, food-compatible lubricious surface.

Poly(L-lysine)-*graft*-poly(ethylene glycol) (PLL-*g*-PEG) is a water-soluble copolymer consisting of a poly(L-lysine) backbone and poly(ethylene glycol) side chains.[4] The PLL chain carries multiple positive charges and spontaneously adsorbs onto negatively charged surfaces, such as many metal oxides or mica. This leads to a straightforward, reliable method for coating surfaces with a dense PEG brush simply by immersing them in an aqueous PLL-*g*-PEG solution.[5,6] The adsorption of PLL-*g*-PEG onto a substrate is primarily controlled by the balance between the attractive electrostatic interaction of the backbone and the steric repulsion of the PEG side chains.[5] In macroscopic tribological contacts, layers of PLL-*g*-PEG can reduce the coefficient of friction under sliding conditions and even more effectively under rolling conditions.[7,8] As long as the polymer is present in the solution, damage to the film is easily repaired by readsorption.[9,10] This self-healing capacity makes the polyelectrolyte-anchoring approach attractive for tribological applications in which the contacting surfaces are immersed in a reservoir of lubricant solution.

In previous studies, we have investigated the properties of PLL-*g*-PEG films with the surface forces apparatus (SFA) under compression.[11−13] The molecules form brushlike homogeneous films on the surface, exhibiting predominantly repulsive, nearly elastic interaction forces upon compression. The equilibrium film thickness is dependent on the polymer architecture, adsorption conditions, and temperature. A comparison of brush−brush and brush−hard wall experiments revealed a significant overlap of the two opposing films.

In this article, we report an investigation of the tribological properties of PLL-*g*-PEG films carried out with a surface forces apparatus. Although the conditions applied in the experiments (moderate surface pressures, low velocities) are far removed from those of many practical macrotribological applications, they help to elucidate the underlying mechanisms of film structure, lubrication, and repair.

Experimental Section

Mica Samples. Thin mica sheets (1.5−5 μm) were prepared by the manual cleavage of optical-quality ruby mica blocks (Spruce Pine Mica Company, Spruce Pine, NC). To avoid the presence of nanoparticles,[14−16] the individual pieces were cut with surgical scissors. The mica samples were covered on one side with a 30−40 nm silver film by thermal evaporation. To glue the mica to cylindrical

† Part of the Molecular and Surface Forces special issue.

* Corresponding author. Tel: +41 44 63 25850. Fax: +41 44 63 31027. E-mail: spencer@mat.ethz.ch.

(1) Klein, J. *Ann. Rev. Mater. Sci.* **1996**, *26*, 581−612.
(2) Raviv, U.; Frey, J.; Sak, R.; Laurat, P.; Tadmor, R.; Klein, J. *Langmuir* **2002**, *18*, 7482−7495.
(3) Lee, S.; Spencer, N. D. Achieving Ultralow Friction by Aqueous, Brush-Assisted Lubrication. In *Superlubricity*; Erdemir A., Jean-Michel Martin, J.-M., Eds.; Elsevier: Amsterdam, 2007; Chapter 21.
(4) Sawhney, A. S.; Hubbell, J. A. *Biomaterials* **1992**, *13*, 863−870.
(5) Kenausis, G. L.; Vörös, J.; Elbert, D. L.; Huang, N. P.; Hofer, R.; Ruiz-Taylor, L.; Textor, M.; Hubbell, J. A.; Spencer, N. D. *J. Phys. Chem B* **2000**, *104*, 3298−3309.
(6) Huang, N. P.; Michel, R.; Voros, J.; Textor, M.; Hofer, R.; Rossi, A.; Elbert, D. L.; Hubbell, J.; Spencer, N. D. *Langmuir* **2001**, *17*, 489−498.
(7) Lee, S.; Müller, M.; Ratoi-Salagean, M.; Voros, J.; Pasche, S.; De Paul, S. M.; Spikes, H. A.; Textor, M.; Spencer, N. D. *Tribol. Lett.* **2003**, *15*, 231−239.
(8) Müller, M.; Lee, S.; Spikes, H. A.; Spencer, N. D. *Tribol. Lett.* **2003**, *15*, 395−405.
(9) Müller, M. Aqueous Lubrication by Means of Surface-Bound Brush-Like Copolymers. Doctoral Thesis No. 16030, Swiss Federal Institute of Technology (ETH) Zürich, Zürich, Switzerland, 2005.
(10) Lee, S.; Müller, M.; Heeb, R.; Zürcher, S.; Tosatti, S.; Heinrich, M.; Amstad, F.; Pechmann, S.; Spencer, N. D. *Tribol. Lett.* **2006**, *24*, 217−223.
(11) Heuberger, M.; Drobek, T.; Spencer, N. D. *Biophys. J.* **2005**, *88*, 495−504.
(12) Heuberger, M.; Drobek, T.; Vörös, J. *Langmuir* **2004**, *20*, 9445−9448.
(13) Drobek, T.; Spencer, N. D.; Heuberger, M. *Macromolecules* **2005**, *38*, 5254−5259.
(14) Ohnishi, S.; Hato, M.; Tamada, K.; Christenson, H. K. *Langmuir* **1999**, *15*, 3312−3316.
(15) Kohonen, M. M.; Meldrum, F. C.; Christenson, H. K. *Langmuir* **2003**, *19*, 975−976.
(16) Heuberger, M.; Zäch, M. *Langmuir* **2003**, *19*, 1943−1947.

Published on Web 10/16/2007

glass lenses of 20 mm radius, the glass was spin coated with a film of pure epoxy resin (EPON 1004, Shell Chemicals) and heated to 140 °C. The mica sheets were glued with the silver-coated side on the glass lenses. After mounting into the SFA and purging for at least 1 h with dry nitrogen, the thickness of the mica samples was determined in a dry nitrogen atmosphere in unloaded adhesive contact prior to the adsorption of the polymer.

Polymer Film. The molecules used in this study were synthesized and characterized according to standardized procedures, which have been previously described in great detail.[6,17,18] In the experiments described here, PLL(20)-*g*[3.5]-PEG(2) and PLL(20)-*g*[2.9]-PEG(5) were used. The molecular weight of the backbone as a hydrobromide salt and the molecular weight of the side chains are in parentheses, and the grafting (Lys to PEG) ratio is in square brackets. A chain of 96 lysine units was used as a backbone. Yielding grafting ratios of 2.9 and 3.5, 32 grafted PEG 44-mers (PEG(2)) or 27 grafted PEG 114-mers (PEG(5)) were used as side chains, respectively. All PEGs were terminated with methoxy groups.

The polymer was dissolved in 10 mM 4-[2-hydroxyethyl]-piperazine-1-[2-ethanesulfonic acid] (HEPES) buffer (Fluka, Switzerland), adjusted to pH 7.4 with NaOH. After determining the thickness of the mica sheets in the SFA, the samples were demounted and completely immersed in solutions containing 0.5 mg/mL PLL-*g*-PEG at room temperature for more than 30 min. After removal from the polymer solution, the samples were thoroughly rinsed with a jet of ultrapure water (Fluka, Switzerland) and mounted in the SFA with a drop of either HEPES buffer or polymer solution placed between the mica surfaces. In one experiment, the surfaces were rinsed with 150 mM KCl, followed by rinsing with 10 mM HEPES buffer. The surfaces were mounted with a drop of polymer solution between them prior to the friction experiments.

Surface Forces Apparatus. The experiments were carried out with a modified version of the Mk 3 SFA (Surforce, Santa Barbara, CA).[19−21] The instrument is equipped with a fully automated data acquisition and evaluation program based on the fast spectral correlation algorithm and is mounted in a box with precise temperature control.[22] All experiments were carried out at 25 °C.

For the friction experiments, the lower surface was moved by means of a piezo bimorph with an amplitude of 50 μm and velocities in the range of 0.5 to 5 μm/s. The bimorph was mounted on a spring with a spring constant of 551 N/m. The shear forces acting on the upper surface were measured with a strain gauge mounted on a measuring spring (for technical details see Luengo et al.[23]) with a measuring rate of 300 Hz. The approach control mechanism of the SFA contains a stepper motor under computer control and a manual micrometer screw. A schematic view of the experimental setup is shown in Figure 1. An experimental setup with only a drop of liquid was necessary to avoid immersing the piezo and the strain gauge in the liquid.

Low-load approach and retraction measurements (up to 4 mN) were carried out by continuous motor travel, in part simultaneously with the friction measurements. For the high-load experiments described below, a manual approach with the micrometer screw was employed. The average contact pressure was calculated by dividing the load by the contact area. This latter parameter changed considerably during loading because of surface deformation, which varied from experiment to experiment as a result of differences in the thickness of mica and glue. Because of this deformation, the pressure does not increase linearly with load but shows a substantial lower-order dependency.

The drop configuration used in the experiments led to additional mechanical drifts of the surfaces as a result of the evaporation of the solvent. For this reason, approach and retract measurements had to be carried out at higher velocities (typically 10 nm/s) than in previously reported experiments, in which the surfaces were immersed in a bath of conductivity water.[11−13]

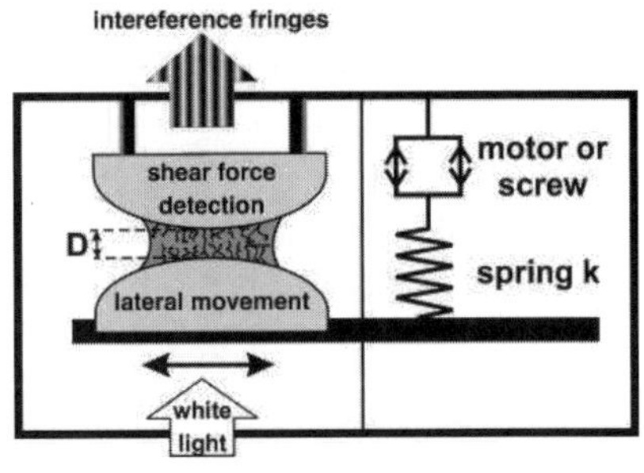

Figure 1. Schematic drawing of the SFA setup for friction measurements. The two mica surfaces with the grafted polymer films are separated by a drop of aqueous polymer solution. The distance between the two surfaces is measured by means of thin-film interferometry.

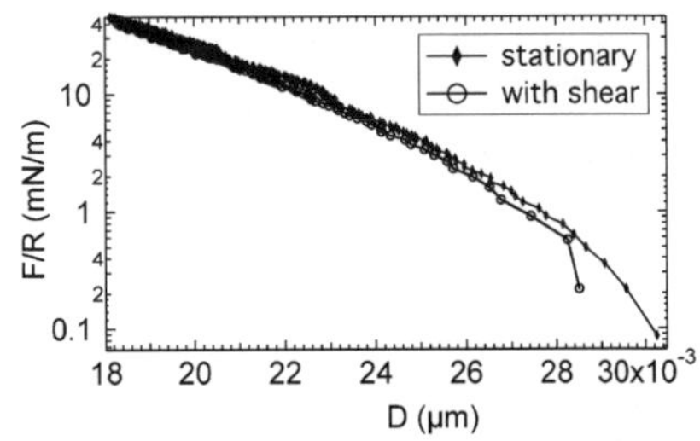

Figure 2. Compression isotherms measured upon approach of two opposing layers of PLL(20)-*g*[2.9]-PEG(5) in HEPES buffer (10 mM, pH 7.4). One data set (filled symbols) was taken without lateral movement of the surfaces, whereas a shear motion of 5 μm/s and an amplitude of 50 μm were applied during the measurement of the other data set (open symbols).

Results

In Figure 2, compression isotherms measured on two surfaces with adsorbed PLL(20)-*g*[2.9]-PEG(5) are shown. Data were taken upon approaching the surfaces with 10 nm/s, once without shearing and once with a shear rate of 5 μm/s. The difference between the two data sets lies within the scatter of this kind of experiment, which is caused by changes in the mechanical drift due to the evaporation of the solvent drop. In this set of experiments as well as in the others reported in this article, no change in the film thickness caused by the shear motion was observed. The equilibrium film thickness is slightly higher than previously reported in a publication for a comparable copolymer.[13] This difference is caused by a small difference in the grafting ratio and the use of 10 mM HEPES buffer instead of conductivity water. Both of these parameters affect the amount of polymer adsorbing onto the surface. The amount is determined by the balance between the electrostatic interaction of the polyelectrolyte backbone with the oppositely charged surface of the mica and the steric repulsion of the PEG side chains, which are forced into a brushlike configuration.

During the compression isotherms, refractive index data were collected. The refractive index, $n(D)$, is modeled by[2]

(17) Pasche, S.; De Paul, S. M.; Vörös, J.; Spencer, N. D.; Textor, M. *Langmuir* **2003**, *19*, 9216−9225.
(18) Pasche, S. Mechanisms of Protein Resistance of Adsorbed PEG-Graft Copolymers. Doctoral Thesis No. 15712, Swiss Federal Institute of Technology (ETH) Zürich, Zürich, Switzerland, 2004.
(19) Israelachvili, J. N.; McGuiggan, P. M. *J. Mater. Res.* **1990**, *5*, 2223−2231.
(20) Heuberger, M. *Rev. Sci. Instrum.* **2001**, *72*, 1700−1707.
(21) Zäch, M.; Heuberger, M. *Rev. Sci. Instrum.* **2003**, *74*, 260−266.
(22) Heuberger, M.; Vanicek, J.; Zäch, M. *Rev. Sci. Instrum.* **2001**, *72*, 3556−3560.
(23) Luengo, G.; Schmitt, F. J.; Hill, R.; Israelachvili, J. *Macromolecules* **1997**, *30*, 2482−2494.

$$n(D) = n_{H_2O} + \frac{m\zeta(n_{PEG} + n_{H_2O})}{D} \text{ for } D > \zeta \qquad (1)$$

Here, $n_{H_2O} = 1.337$ is the refractive index of the aqueous solution, and $n_{PEG} = 1.51$ is the refractive index of the dry PEG polymer. The prefactor $m = 2$ accounts for the brush–brush configuration. The equivalent dry thickness $\zeta = \Gamma/\rho$ of a dried polymer layer is the ratio of the adsorbed polymer mass, Γ, and the dry polymer density, $\rho \approx 1.12$ g/cm^3.

Fitting the data shown in Figure 3 reveals a dry polymer thickness of 2.6 nm, which corresponds to an adsorbed mass of 290 ± 100 ng/cm^2. This implies an average distance of grafting points of about $d_g = 3.5$ nm and an average PEG monomer volume fraction of $\phi = 0.17$ of the uncompressed, solvated polymer layer.

In the friction experiments, two distinct regimes could be identified. An intact polymer film at moderate loads (several tens of micronewtons corresponding to average pressures on the order of 10–20 MPa) always showed extremely low frictional forces, below the detection limit of 20 μN, in response to lateral shear. At high loads of about 200 mN, when the film was compressed to about 10% of the original equilibrium film thickness, the friction showed a reversible increase with load (i.e., upon reducing the load, the friction decreased again). Under some experimental conditions, film destruction was observed, immediately followed by a dramatic increase in the frictional force.

Intact Polymer Films. Under a variety of experimental conditions, the friction force between the two polymer-coated surfaces remained below the detection and cross-talk limit of 20 μN. An example of the data measured between brushes of PLL-(20)-*g*[2.9]-PEG(5) in HEPES buffer with a sliding velocity of 0.5 μm/s is shown in Figure 4. In this experiment, the load was gradually increased up to 30 mN without a significant increase in the friction signal being detected. This corresponds to a friction coefficient of <0.001.

Similar observations were also made on PLL(20)-*g*[3.5]-PEG(2) brushes mounted with a drop of polymer solution. Here, no friction could be detected up to loads of about 60 mN. At this load (corresponding to an average contact pressure of about 10 MPa), the film was compressed to 5.4 nm, which is one-third of its equilibrium film thickness. Film destruction, as described below, was never observed in the presence of excess polymer solution in the load range covered in this experiment.

A different result was obtained on PLL(20)-*g*[3.5]-PEG(2) brushes after a rinse with 150 mM KCl solution (Figure 5). The ionic strength of the rinsing solution was chosen to match biological conditions (e.g., the salt concentration in human body fluids). (This information is important if the polymer is to be considered as a lubricant for biomedical devices such as endoscopes.) The stability of the polymer layer on the surface is reduced by the ionic strength of the rinsing solution, and therefore a fraction of the polymer chains (presumably those attached by only a fraction of the backbone) is removed from the surface and the remaining molecules rearrange. For a detailed discussion of the effect of ionic strength on the stability of the PLL-g-PEG layers, see Pasche.[18]

Surfaces treated in this way showed a small but significant friction signal that increased with load. A linear fit of the data reveals a friction coefficient of 0.003. This result can be attributed to the reduced layer density and thickness.

Film Destruction. A different type of behavior was observed when the PLL-*g*-PEG films became damaged during the friction experiment. This was always accompanied by a dramatic increase

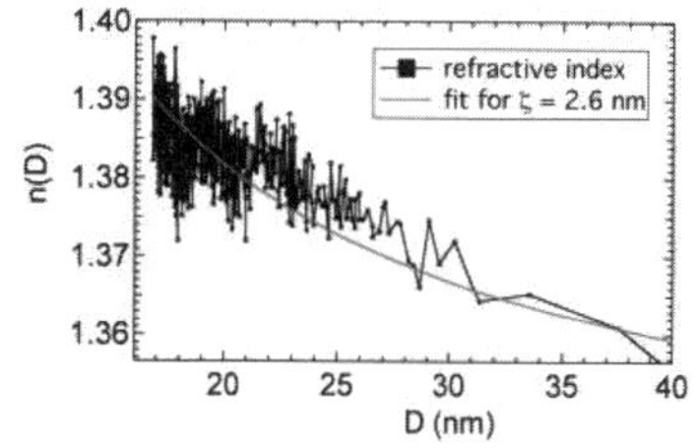

Figure 3. Refractive index measured upon compression of two opposing PLL(20)-*g*[2.9]-PEG(5) layers in HEPES buffer. The line shows a fit according to eq 1, revealing an adsorbed mass of 290 ± 100 ng/cm^2.

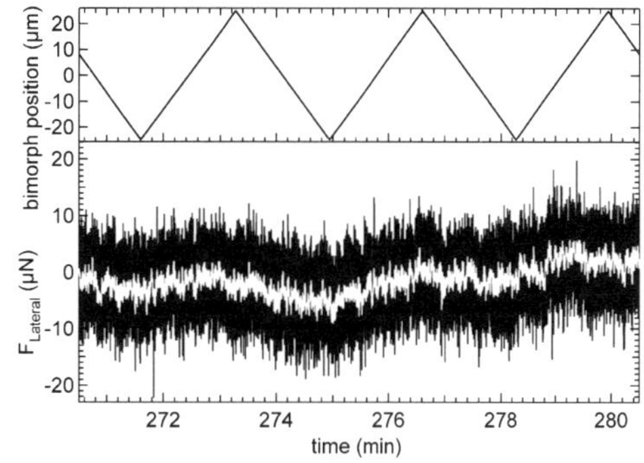

Figure 4. Friction signal measured with both surfaces coated with PLL(20)-*g*[2.9]-PEG(5) in HEPES buffer at a load of 15 mN and a sliding velocity of 0.5 μm/s. The upper part of the graph shows the position of the piezo bimorph. The lower part shows the corresponding strain-gauge (friction) signal. The white line in the middle shows the data after considerable smoothing. The small correlation between the bimorph movement and the friction signal is caused by cross talk and is also visible when the two surfaces are separated. It therefore corresponds to the minimum detectable friction signal.

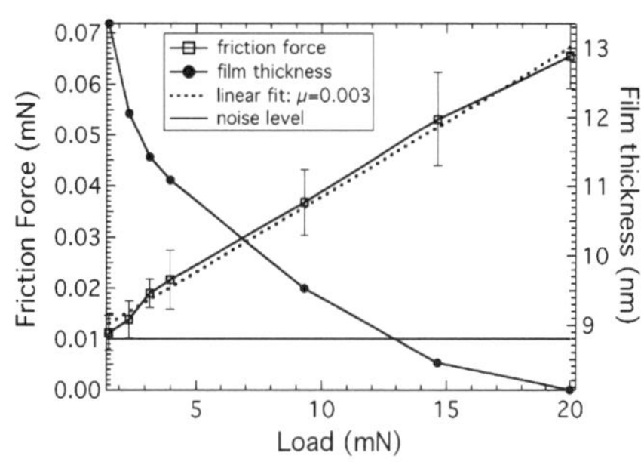

Figure 5. Friction force measured between two brushes of PLL-(20)-*g*[3.5]-PEG(2) after rinsing with 150 mM KCl, mounted with a drop of polymer solution. The sample showed a moderate increase in friction with load, leading to a friction coefficient of $\mu = 0.003$.

in friction. An example is shown in Figure 6. The data were collected between two brushes of PLL(20)-*g*[3.5]-PEG(2) in HEPES buffer solution in the absence of unbound polymer molecules. The load was increased by a continuous motor movement of 1 nm/s, and the two surfaces were sheared with $v_s = 0.5$ μm/s with an amplitude of 50 μm. Above a load of about 0.5 mN, the friction force increased linearly with load. A linear fit of the load dependence leads to a coefficient of friction of 0.35. The load dependence appeared to be reversible during subsequent unloading, but with small hysteresis. However, the extremely low friction measured at the beginning of the loading

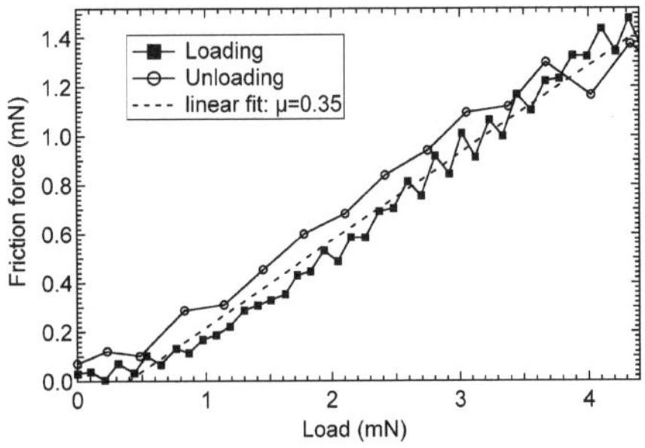

Figure 6. Friction vs load data for contacting PLL(20)-*g*[3.5]-PEG(2) brushes in pure HEPES buffer measured upon a loading–unloading cycle. At a load of about 0.5 mN, the friction started to increase.

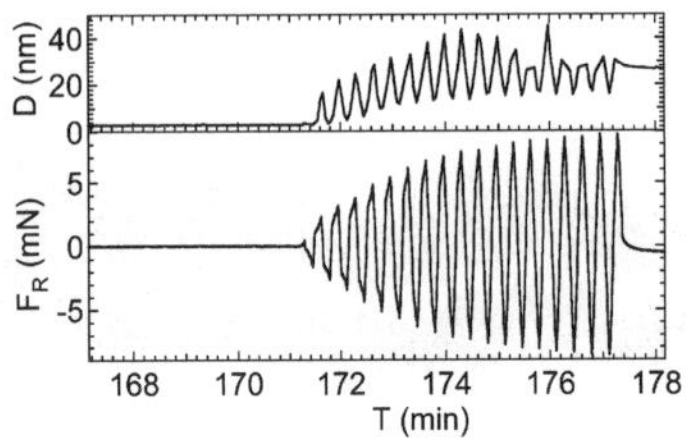

Figure 7. Onset of high friction on a single PLL(20)-*g*[2.9]-PEG-(5) brush sliding against a bare mica surface in pure HEPES buffer at a load of 23 mN. The surfaces were moved at 5 μm/s with an amplitude of 50 μm. Simultaneous to the buildup of friction, surface separation *D* increased. After the lateral motion was stopped, there was an exponential decrease in the lateral force with a decay rate of 480 s^{-1}.

was not reestablished upon unloading. The load and shear at the onset of the friction increase correspond to an average contact pressure of roughly 3 MPa and a shear rate $g = v_s/D$ of about 50 s^{-1}.

Typically, the increase in the friction coefficient was accompanied by an increase in the surface separation. This is illustrated in Figure 7. Here, a friction experiment was carried out on a single PLL(20)-*g*[2.9]-PEG(5) film opposing a bare mica surface in HEPES buffer without excess polymer in solution. In contrast to the situation described above for PLL(20)-*g*[3.5]-PEG(2) in pure HEPES buffer, a load of 23 mN (roughly a 10 MPa average contact pressure) had to be applied before the breakdown of the film occurred. Again, a corresponding film thickness increase was observed. The higher-frequency structure in the film thickness is caused by the relative lateral movement of the two surfaces. The data indicate that the well-ordered, homogeneous layer became seriously disrupted by tribologically induced desorption of the polymer. In vertical-approach experiments following such a film breakdown, repulsive forces were even detectable at separation distances of about 100–200 nm, providing further evidence of the destruction of the ordered surface layer and possible damage of the mica substrates.

High-Load Regime. When high loads are applied to the polymer brush, the film becomes increasingly compressed, and the water is gradually squeezed out of the contact region. In Figure 8, friction data obtained in a series of high-loading steps on two contacting PLL(20)-*g*[2.9]-PEG(5) brushes in pure HEPES buffer are shown. At a load of 200–250 mN (corresponding to an average contact pressure on the order of 25 MPa), the film was compressed to about 10% of its original equilibrium film thickness. At this load, the friction force increased. An unloading–loading cycle revealed this friction increase still to be reversible.

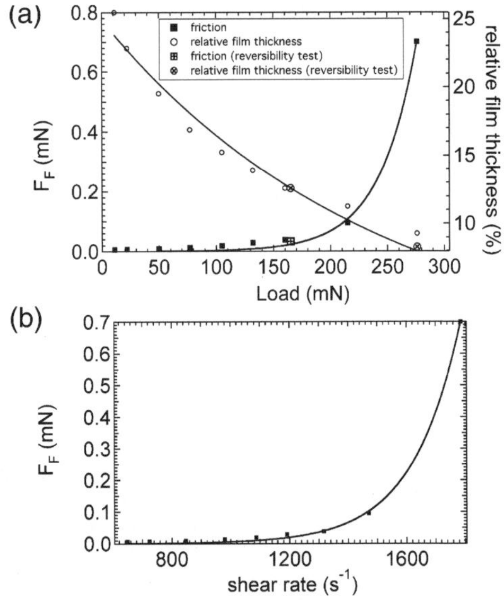

Figure 8. (a) Friction and surface separation between two PLL-(20)-*g*[2.9]-PEG(5) brushes in pure HEPES buffer. The surfaces were moved at 5 μm/s with an amplitude of 50 μm. When the film was compressed to about 10% of its equilibrium film thickness, the friction force increased. Unloading and loading showed this increase to be reversible. The two lines represent exponential fits of the data sets ($f(x) = Ae^{bx}$) with parameters $A = 1.5 \times 10^{-4}$ mN and $b = 3.0 \times 10^{-2}$ mN^{-1} for the friction and $A = 25$ mN and $b = -4.1 \times 10^{-3}$ mN^{-1} for the relative film thickness. (b) Friction force vs shear rate $g = v_s/D$ (data from part a). Squares show the experimental data, and the line is an exponential fit $f(x) = Ae^{bg}$ with $A = 1.3 \times 10^{-5}$ mN and $b = 0.006$ s.

Also, from the optical data there was no indication of film disruption, as had been observed in the experiments described above on PLL(20)-*g*[3.5]-PEG(2) in pure HEPES buffer.

Discussion

During adsorption from aqueous solution, the SFA results indicate that the copolymer forms moderately dense, brushlike layers. The adsorption itself is a random process, and the polymer chains have a finite length. Results confirm previous reports that the layer thickness depends on the length of the PEG side chains, adsorption conditions, and surface concentration of adsorbed molecules.[18]

The extremely low friction achieved on PLL(20)-*g*[2.9]-PEG(5) films in SFA experiments is not a unique property of this specific copolymer but has also been observed for other grafted poly(ethylene glycol) films[2] as well as for other polymer-brush systems in a good solvent, such as grafted polystyrene in toluene.[1,24] The osmotic pressure, which leads to strong repulsive forces as the polymer brushes are compressed, effectively prevents the direct contact of the solid surfaces. At comparable solid surface separations, thin films of water or aqueous salt solutions have been shown to retain a shear fluidity characteristic of the bulk liquid.[25] In a control experiment, this was also observed in 10 mM HEPES buffer solution (data not shown).

(24) Klein, J.; Kumacheva, E.; Perahia, D.; Mahalu, D.; Warburg, S. *Faraday Discuss.* **1994**, *98*, 173–188.
(25) Klein, J.; Raviv, U.; Perkin, S.; Kampf, N.; Chai, L.; Giasson, S. *J Phys.: Condens. Matter* **2004**, *16*, S5437–S5448.
(26) Neelov, I. M.; Borisov, O. V.; Binder, K. *J. Chem. Phys.* **1998**, *108*, 6973–6988.
(27) Grest, G. S. *Adv. Polym. Sci.* **1999**, *138*, 149–183.

The stationary brushes are likely to interpenetrate each other. On similar PLL-*g*-PEG layers, a comparison of brush−brush and brush−hard-wall experiments revealed a significant overlap of the opposing brushes at the onset of measurable repulsion.[13] On the basis of stochastic dynamics simulations, Neelov et al.[28] predict that at low compressions the polymer brushes interpenetrate only in the outermost region next to the interface. At high compressions and zero shear rates, the polymer chains completely interpenetrate each other. However, a shear motion of the two brushes will result in a shear alignment of the chains and a reduced interpenetration of the brushes.[26−28]

When a lateral motion is applied, two effects contribute to the friction force. First, there is the friction of the polymer chain and the solvent. This mainly affects the end segments of the chains because the flow field of the solvent does not range far into the brush. Second, there is the friction between chains of opposing brushes in the overlap region.

At moderate loads, the interfacial layer of the polymer brushes, where the sliding occurs, is still fluid. At shear rates of 0.5 to 5 μm/s, as reported in this article, the time scale of one polymer chain traveling a grafting distance—one bristle crossing another in a macroscopic view—is milliseconds, much longer than the relaxation time of the polymer chain, which is on the picosecond time scale, although the mobility of the water in the hydration layer is slower than in bulk water.[29] Therefore, the contribution of polymer−polymer friction should be small at moderate compression. Also, the viscosity of the solvent in the layer is still relatively small. The situation is different in the high-load regime. At compressions to about 1/10 of the equilibrium film thickness, where we observed an increase in friction, most of the water is squeezed out of the contact region. When the amount of water is reduced to two to three water molecules per EG unit, it is not sufficient for complete hydration shells of the polymer, so each water molecule participates in the shells of adjacent chains.[30] Therefore, with decreasing water content, the viscosity of the layer increases,[31] resulting in higher friction forces between the brushes.

A comparison of friction data measured with different PLL-g-PEG architectures and experimental setups (brush−wall, brush−brush, and rinsing) shows a general trend: thicker layers of PLL-*g*-PEG show a higher resistance to tribo-stress-induced abrasion of the polymer film. Films of PLL(20)-*g*[3.5]-PEG(2) have been shown to be destroyed by shear, even at moderate loads, when no self-healing was possible. The failure of a single PLL(20)-*g*[2.9]-PEG(5) brush sheared against a bare mica surface fits into the same framework because the total amount of PEG and water is comparable to a brush−brush configuration of PLL(20)-*g*[3.5]-PEG(2). The small but significant friction measured on PLL(20)-*g*[3.5]-PEG(2) after rinsing with 150 mM KCl solution is due to a less dense film, caused by partial removal of the polymer. These results are in good agreement with colloidal-probe AFM experiments on different PLL-*g*-PEG architectures, which revealed a systematically greater reduction of friction with increasing PEG length.[31] Similar experiments with different solvents as well as mixtures of HEPES buffer and 2-propanol[32,33] have shown that there is a strong correlation between the friction properties of the polymer film and the amount of solvent absorbed within the polymer brush.

Pin-on-disk experiments on fluorescently labeled PLL-*g*-PEG films suggest that the tribo-stress-induced desorption of the PLL-*g*-PEG films is mainly due to the removal of complete molecules rather than their partial removal.[10] In the presence of sufficient concentrations of free polymer in the solution, a replacement of the desorbed polymer was observed. This self-healing capacity is assumed to be the reason for the enhanced resistance to abrasion that we observed with PLL(20)-*g*[3.5]-PEG(2) in the presence of excess polymer in the solution.

An important requirement for a lubrication system based on self-healing is the kinetics of the adsorption process being faster than that of the shear-induced desorption. For PLL-*g*-PEG, this can be achieved by the presence of sufficiently high polymer concentrations. Additionally, the region of desorption has to be exposed to the free polymer. The high amplitude of 50 μm for the shear motion applied in the SFA experiments reported here is on the order of the contact region between the two mica samples, which is typically about 20 to 30 μm in diameter for a moderately loaded contact and increases up to about 100 μm for high loads as a result of elastic and plastic deformations of mica and glue. When the sheared region of the polymer film is exposed to the polymer solution prior to the next shearing cycle, defects eventually caused by the shear can be patched by readsorption of the polymer. Such a self-healing mechanism cannot take place if the sliding amplitude is smaller than the contact diameter because practically no free polymer is present in the shearing region as a result of the solution being squeezed out of the contact region during compression.

Conclusions

In the surface-forces experiments reported here, PLL-*g*-PEG brushes showed a remarkably low frictional response to shear. Two opposing brushes of PLL(20)-*g*[2.9]-PEG(5) were able to glide against each other without measurable friction forces, even at considerable compressions. The increase in friction observed at compressions to less than 10% of the equilibrium film thickness can be attributed to an increase of viscosity when the solvent is squeezed out of the contact.

The resistance of the brushes to tribologically induced desorption depends on the architecture of the polymer and therefore on the amount of PEG in the contact region. In a brush-against-wall experiment on PLL(20)-*g*[2.9]-PEG(5), this film was able to resist shear at intermediate compression but lost its integrity upon higher loading. Even very low loads were sufficient to destroy films of PLL(20)-*g*[3.5]-PEG(2). With the addition of free polymer to the solution, the resistance of the PLL(20)-*g*[3.5]-PEG(2) film to abrasion was dramatically increased. This enhancement of the loading capacity appears to be due to readsorption of the polymer following friction-induced desorption. This self-healing mechanism is a major factor in the efficacy of PLL-*g*-PEG and related water-soluble-brush-bearing polyelectrolytes as lubricant additives for water-based boundary lubrication.

Acknowledgment. We thank Dr. F. Durmaz (ETH Zurich) and Dr. S. Zürcher (Susos AG, Zurich) for their polymer synthesis efforts and Dr. M. Heuberger (Empa, St. Gallen) for useful discussions.

LA702289N

(28) Kreer, T.; Müser, M. H. *Wear* **2003**, *254*, 827−831.
(29) Tasaki, K. *J. Am. Chem. Soc.* **1996**, *118*, 8459−8469.
(30) Kingman, N. G.; Rosenberg, A.; Bastos, M.; Wadso, I. *Thermochim. Acta* **1990**, *169*, 339−346.
(31) Yan, X. P.; Perry, S. S.; Spencer, N. D.; Pasche, S.; De Paul, S. M.; Textor, M.; Lim, M. S. *Langmuir* **2004**, *20*, 423−428.
(32) Müller, M. T.; Yan, X. P.; Lee, S. W.; Perry, S. S.; Spencer, N. D. *Macromolecules* **2005**, *38*, 3861−3866.
(33) Müller, M. T.; Yan, X. P.; Lee, S. W.; Perry, S. S.; Spencer, N. D. *Macromolecules* **2005**, *38*, 5706−5713.

Tribol Lett (2009) 33:83–96
DOI 10.1007/s11249-008-9402-6

ORIGINAL PAPER

End-grafted Sugar Chains as Aqueous Lubricant Additives: Synthesis and Macrotribological Tests of Poly(L-lysine)-*graft*-Dextran (PLL-*g*-dex) Copolymers

Chiara Perrino · Seunghwan Lee · Nicholas D. Spencer

Received: 19 September 2008 / Accepted: 8 December 2008 / Published online: 30 December 2008

Abstract Comb-like graft copolymers with carbohydrate side chains have been developed as aqueous lubricant additives for oxide-based tribosystems, in an attempt to mimic biological lubrication systems, whose surfaces are known to be covered with sugar-rich layers. As adopted in the previous studies of the graft copolymer poly(L-lysine)-*graft*-poly(ethylene glycol) (PLL-*g*-PEG), which showed both excellent lubricating and antifouling properties, a similar approach was chosen to graft dextran chains onto the same backbone, thus generating PLL-*g*-dex. PLL-*g*-dex copolymers readily adsorb from aqueous solution onto negatively charged oxide surfaces. Tribological characterization at the macroscopic scale, either under pure sliding conditions or a mixed sliding/rolling contact regime, shows that PLL-*g*-dex is very effective for the lubrication of oxide-based tribosystems. The relative lubricating capabilities of PLL-*g*-dex copolymers compared with PLL-*g*-PEG copolymers were observed to be highly dependent on the molecular structure of the copolymers (in particular, side-chain density along the backbone) and the measurement conditions (in particular, time between tribocontacts); the PLL-*g*-dex copolymers with a low degree of grafted side chains ($\leq$20% grafting of available protonated primary amine groups along the backbone) showed better lubricating performance than their PLL-*g*-PEG counterparts at high tribocontact frequency ($\geq$ca. 0.32 Hz).

Keywords Aqueous lubrication · Boundary lubrication · Biomimetic lubricant additives · Poly(L-lysine)-*g*-dextran

C. Perrino · S. Lee · N. D. Spencer (✉)
ETH Zurich, Wolfgang-Pauli-Strasse 10,
CH 8093 Zurich, Switzerland
e-mail: nicholas.spencer@mat.ethz.ch

1 Introduction

Water-based lubricants possess distinctive advantages over oil-based lubricants (such as environmental compatibility, biocompatibility, availability, cost effectiveness), and could represent a viable alternative in certain applications. Their environmental compatibility makes them suitable for a number of industrial applications, such as food processing or textile and pharmaceutical manufacturing, where the use of oil-based lubricants can be problematic due to contamination issues [1, 2]. In the biomedical field, aqueous lubrication is particularly important, since water is virtually the only acceptable base lubricant. Any attempts to improve lubrication properties in biomedical applications (such as lubricious coatings for catheters) should therefore take both water compatibility and biocompatibility into account [3].

Poly(L-lysine)-*graft*-poly(ethylene glycol) (PLL-*g*-PEG), a graft copolymer consisting of a polycationic PLL backbone and PEG side chains, has proven to be highly effective on oxide surfaces with regard to both preventing non-specific adsorption of proteins and lubricating in an aqueous environment [4–11]. PLL-*g*-PEG has been shown to spontaneously adsorb from aqueous solutions onto negatively charged surfaces via electrostatic interactions; the positive charges present on the protonated primary amine groups of the PLL backbone in a neutral aqueous environment lead to its rapid immobilization onto negatively charged surfaces. PEG side chains, which are radially distributed along the backbone in bulk solution in order to minimize steric repulsion, stretch out into the solution just after the copolymers have adsorbed on the surface, forming a polymer brush, providing the inter-PEG spacing is sufficiently close. PEG, however, has some disadvantages, such as lower solubility with increasing temperature and oxidation to peroxides in air [12, 13], which could limit the usage of PLL-*g*-PEG in certain applications. An alternative approach is to mimic biological membrane surfaces, which are typically covered with carbohydrate-rich layers. In vivo cellular interactions can be strongly influenced by pendant oligosaccharides chains, the principal components of the highly hydrated *glycocalyx*, which surrounds certain kinds of cell, and is known to prevent undesirable, non-specific adsorption of proteins [14–18]. Carbohydrates are also known to play an important role in natural lubrication mechanisms, which often involve a brush-like structure of sugar chains, anchored to a protein backbone. Proteoglycans, for instance, have been suggested to play a role in natural joint lubrication [19, 20], and mucins, large glycoproteins that make up the mucosal hydrogel coating covering epithelial cell surfaces, act as both protective and lubricious layers [21–24].

Based on this natural model, sugar-based copolymers have been developed for use as aqueous lubricant additives. In this study, encouraged by the successful lubricating performance of PLL-*g*-PEG in previous studies [6–8], we have chosen to maintain the comb-like graft polymer structure, selecting the same PLL backbone, and replacing the PEG side chains with dextran chains (Fig. 1). Dextran is a natural and neutral polysaccharide consisting of an α (1$\rightarrow$6)-linked glucan with side chains attached to the 3-positions of the backbone glucose units. Our choice of dextran for the carbohydrate chains was motivated by several factors. First, there are already reports on the synthesis and applications for drug/gene delivery of poly (L-lysine)-*g*-dextran (PLL-*g*-dex) in the literature [25–31]. Second, since the dextran is neutral, it does not present problems due to electrostatic interactions when grafted onto a polycationic backbone (PLL). Finally, previous studies have shown the effectiveness of dextran coatings at preventing non-specific protein adsorption [14–18, 32–40], this being a property that often seems to be closely associated with good aqueous lubricating properties [5, 6, 41, 42].

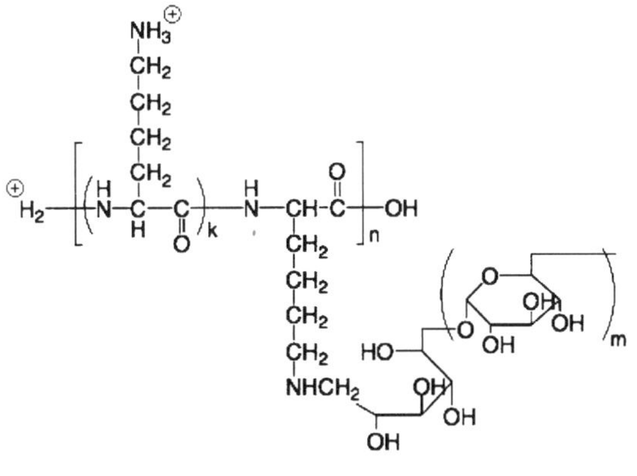

Fig. 1 A schematic illustration of the chemical structure of PLL-*g*-dextran. k should be taken as an average value. k + 1 represents the grafting ratio of the polymer

We have focussed on the characterization of adsorption and aqueous lubricating properties of PLL-*g*-dex copolymers on oxide surfaces under standard, macroscopic tribological conditions at ambient temperatures, including sliding and mixed sliding/rolling contacts. In particular, we have synthesized and employed various PLL-*g*-dex copolymers with varying architectural features, including the molecular weight and the grafting ratio of the dextran side chains along the backbone, to characterize the relative lubricating capabilities compared with counterpart PLL-*g*-PEG copolymers.

2 Experimental

2.1 Poly(L-lysine)-*graft*-dextran (PLL-*g*-dex)

Poly(L-lysine)-*graft*-dextran (PLL-*g*-dex) copolymers were synthesized by a reductive amination reaction of PLL HBr (20 kDa, polydispersity 1.1, Sigma-Aldrich, Switzerland) with dextran (dextran T5, 5.2 and T10, 10 kDa, polydispersity 1.8, Pharmacosmos A/S, Denmark). Borate buffer (50 mM, pH 8.5) was used as solvent for the reaction. The reaction proceeded in two steps: first, the formation of a Schiff base between the terminal dextran aldehyde group and primary amine groups of PLL, and second, the reduction of the unstable Schiff base to secondary amines. PLL was dissolved in borate buffer solution at 40 °C, and dextran was subsequently added to the solution, and the mixture was incubated with stirring overnight for the formation of the Schiff base. Sodium cyanoborohydride ($NaBH_3CN$, Fluka Chemika, Switzerland) was then added to the mixture (10 × molar excess to dextran was used) and it was kept under stirring for two more days. The resulting copolymers were isolated by ultracentrifugation (Vivaspin 15R centrifugation tubes, 30000 MWCO, Sartorius AG, Switzerland) to remove the unreacted starting materials. Varying the ratio of Lys/dex allowed us to control the degree of grafting of dextran chains onto the PLL backbone (see below for the definition of the grafting ratio, *g*). A 2–3 × molar excess of dextran was required for the synthesis to obtain a given grafting degree. PLL-*g*-dex molecules were characterized by ^{1}H-NMR spectroscopy and elemental analysis (EA). ^{1}H-NMR spectra of the copolymers in D_2O were recorded on a Bruker spectrometer (300 MHz) and both NMR spectra and elemental analysis data were used to determine the grafting ratio.

The notation PLL(*x*)-*g*[*y*]-dex(*z*) for the copolymers was used to represent the molar mass of PLL in kDa (*x*) (including the counterions, Br-, as precursor), the molar mass of dextran in kDa (z), and the grafting ratio $g[y]$ (defined as the number of lysine monomers/dextran side chain). Although it is possible, in principle, to control all three structural parameters, we have varied only the grafting ratio, y, and the molar mass of dextran, z, whereas the molecular weight of PLL was kept constant at 20 kDa, in this study. For comparison purposes, PLL(20)-*g*-PEG(5) copolymers provided by SuSoS AG (Dübendorf, Switzerland) have also been investigated, with PEG(5) side chains (molecular weight, 5 kDa) on a PLL(20) backbone and covering a similar range of grafting ratios to that of the PLL-*g*-dex copolymers employed [5].

2.2 Optical Waveguide Lightmode Spectroscopy (OWLS)

Optical waveguide lightmode spectroscopy (OWLS) was used to characterize the adsorption properties of the polymers. Experiments were performed using an OWLS 110 instrument (Microvacuum, Budapest, Hungary).

OWLS is an optical biosensing technique for the in situ label-free analysis of adsorption processes [43]. In brief, linearly polarized light (He–Ne laser) is coupled with a diffraction grating into a planar waveguide, provided that the in-coupling condition is fulfilled. The light is guided by total internal reflection to the ends of the waveguide layer, where it is detected by a photodiode detector. The adsorbed mass is calculated from the change in the refractive index in the vicinity of the waveguide surface upon adsorption of molecules [43–45]. The refractive index increment (dn/dc) of dextran was measured by means of a refractometer and a value of 0.131 was used for all measurements to calculate the mass of polymer adsorbed. For the measurement of the mass of adsorbed PLL-*g*-PEG copolymer, a value of 0.139 was always used.

Prior to the experiments, optical waveguides chips (standard: $Si_{0.75}Ti_{0.25}O_2$ on glass, 1.2×0.8 cm^2, Microvacuum, Budapest, Hungary) were ultrasonicated in 0.1 M HCl for 10 min, rinsed with Millipore water, ultrasonicated in 2-propanol for 10 min, rinsed again with Millipore water, and then dried under a dry nitrogen stream. The substrates were subsequently cleaned by UV/ozone cleaner (Boeckel industries Inc., model 135500) for 30 min.

The cleaned waveguides were assembled into the OWLS flow cell and equilibrated by exposing to HEPES buffer solution (10 mM 4-(2-hydroxyethyl)piperazine-1-ethanesulfonic acid (Sigma, St. Louis, MO, USA), adjusted to pH 7.4 with 6.0 M NaOH solution) overnight in order to obtain a stable baseline. The waveguides were then exposed to a polymer solution (0.25 mg mL^{-1} in HEPES buffer) for at least 30 min, resulting in the formation of a polymer adlayer, and rinsed three times by soaking them in buffer solution for another 30 min.

2.3 Pin-on-disk Tribometry

A pin-on-disk tribometer (CSM Instruments, Switzerland) was employed to investigate the lubricating properties of the polymer solutions under pure sliding conditions.

As a tribopair, we chose steel balls (6 mm in diameter, DIN 5401-20 G20, Hydrel AG, Romanshorn, Switzerland) against flat glass squares (2.5×2.5 cm^2, 1-mm thick), cut from microscope slides (Medite Medizintechnik AG, Switzerland, approximate chemical composition according to the manufacturer: 72.2% SiO_2, 14.3% Na_2O, traces of K_2O, CaO, MgO, Al_2O_3, Fe_2O_3, SO_3). The lubricating properties of the polymers were evaluated by acquiring the coefficient of friction, μ (= friction force/load), as a function of speed under a fixed load (2 N, dead weights) at room temperature. The maximum Hertzian contact pressure in this configuration is estimated to be 0.34 GPa. The sliding speed was controlled to vary from 1 to 19 mm s^{-1} using the instrument's software (InstrumX version 2.5A, CSM Instruments, Switzerland). The weight concentration of the polymer solutions was chosen in order to attain similar molar concentrations for all copolymers (0.25 mg mL^{-1} for PLL-*g*-dex(5.2) and PLL-*g*-peg(5) copolymers and 0.5 mg mL^{-1} for PLL-*g*-dex(10)) with a similar range of grafting ratios as used in the OWLS experiments. The number of rotations was fixed at 200. For each measurement with a given copolymer, the same pair of pin and disk was used. However, the position of the ball and the radius of the track on the disk were changed for each speed, to provide fresh tribocontact points. Balls and disks were cleaned immediately before each measurement as follows: they were ultrasonicated in ethanol for 10 min, dried under a nitrogen stream and then oxygen-plasma cleaned for 2 min. All instrumental parts expected to be in contact with the polymer solution were cleaned by rinsing with ethanol and water.

2.4 Mini-traction Machine

A mini-traction machine (MTM, PCS Instruments, London, UK) was employed to characterize the lubricating behavior of the copolymer solutions in a mixed sliding/rolling contact regime over a speed range of 0–2,500 mm s^{-1}. AISI 52100 steel balls (9.5 mm in radius, PCS instruments, London, UK) and flat silicate glass disks (46 mm in diameter, PCS Instruments, London, UK) were used as tribopairs. The ball is loaded against the disk and ball and disk driven independently to create a mixed sliding/rolling contact. The slide/roll ratio (SRR) is defined as the percentage ratio of the difference to the mean of the ball speed (u_{ball}) and disk speed (u_{disk}); i.e., SRR = $[|u_{ball} - u_{disc}|/(u_{ball} + u_{disc})/2] \times 100\%$. Thus, SRR = 0% (i.e., $u_{ball} = u_{disk}$) represents a pure rolling contact and higher SRR values represent a higher portion of the sliding character. With the software provided by the manufacturer (PCS Instruments, MTM version 1.0, London, UK), values in the range of SRR 1–200% are accessible. For this study, an SRR value of 10% was chosen, which enables friction measurements in conditions close to pure rolling. Speed-dependence tests on polymer solutions (0.25 or 0.5 mg mL^{-1} depending on molecular weight) were carried out at a constant load of 10 N (Hertzian contact pressure, 0.42 GPa) and at a controlled temperature (25 °C). The disk track radius was fixed at 21 mm, and for each measurement a new ball and a new disk were used. Balls and disks were cleaned according to the same procedure used for pin-on-disk tribometry experiments, except that air-plasma was used instead of oxygen-plasma. The instrumental parts expected to be in contact with the polymer solution were cleaned with a commercial cleaner (Hydrochloric acid 300 mmol/L and detergent 1%, Roche Diagnostics GmbH, Mannheim, Germany).

3 Results and Discussion

3.1 Synthesis and Structural Features of PLL-*g*-dex Copolymers

PLL-*g*-dextran copolymers with different molecular weights and grafting ratios of dextran side chains were successfully synthesized and characterized by ^{1}H-NMR. The molecular weights of the selected dextrans were 5.2 and 10 kDa (denoted as dex(5.2) and dex(10)). Grafting ratio was varied between roughly 3 and 9 and evaluated by means of NMR and EA. As a reference, a series of PLL (20)-*g*-PEG(5) (20 kDa for the molecular weight of PLL backbone and 5 kDa for the molecular weight of PEG side chains) with varying grafting ratios, roughly *g* [3] to *g* [11], were purchased from SuSoS AG (Dübendorf, Switzerland). In Table 1, the structural features of the synthesized copolymers and the yield of the synthesis are presented in detail.

It should be noted that the two dextran polymers with different molecular weights in this study, dex(5.2) and dex(10), were selected to maintain the structural features that may critically influence adsorption and lubricating properties comparable with those of PLL(20)-*g*-PEG(5). Firstly, dex(5.2) is nearly identical with PEG(5) in molecular weight, and thus can be employed to generate PLL-*g*-dex copolymers with similar molecular weights to those of PLL-*g*-PEG at the same grafting ratio. As will be addressed below, the molecular weights of the copolymers need to be comparable when the diffusion properties of copolymers through the liquid medium play an important role in the lubricating behavior.

PEG(5) is composed of 113.6 monomer units (ethylene glycol (EG)), whose fully extended chain length is estimated to be 40.5 nm (based on the molecular length of EG, 0.358 nm [46]), whereas dex(5.2) is composed of 31.8 monomer units (sugar rings), whose fully extended chain length is estimated to be only 22.3 nm (based on the molecular length of dextran: 0.7 nm [47]). While the lubricating film thickness generated by the adsorption of comb-like copolymer will be determined by many

Table 1 Synthesized PLL-*g*-dex copolymers and PLL-g-PEG copolymers used for comparison purposes

Polymer	Synthesis yield [%]	No. of grafted side chains per PLL	No. of free lysines per PLL	Percentage of side-chain grafting (%)	M.W. of copolymer (kDa)
PLL(20)-g[3.4]-dex(5.2)	30	28.15	67.5	29.4	157.4
PLL(20)-g[5.3]-dex(5.2)	60	18.06	77.6	18.9	105.4
PLL(20)-g[6.6]-dex(5.2)	56	14.50	81.2	15.2	87.0
PLL(20)-g[7.3]-dex(5.2)	27	13.11	82.6	13.7	80.0
PLL(20)-g[8.7]-dex(5.2)	60	11.00	84.7	11.5	69.0
PLL(20)-g[3.7]-dex(10)	46	25.86	69.8	27.0	270.1
PLL(20)-g[4.8]-dex(10)	40	19.94	75.8	20.8	211.6
PLL(20)-g[6.5]-dex(10)	59	14.72	81.0	15.4	159.5
PLL(20)-g[8.6]-dex(10)	58	11.13	84.6	11.6	123.5
PLL(20)-g[3]-PEG(5)	–	31.90	63.8	33.3	171.8
PLL(20)-g[4.4]-PEG(5)	–	21.75	73.9	22.7	121.0
PLL(20)-g[6.6]-PEG(5)	–	14.50	81.2	15.2	84.8
PLL(20)-g[11.2]-PEG(5)	–	8.54	87.1	8.9	55.0

The numbers of lysine units for the PLL-HBr (20 kDa), before grafting with dextran or PEG, is 95.7. The numbers of monomer units for the PEG (5 kDa) and dex (5.2 kDa) are 113.6 and 31.8, respectively parameters, the fully extended length of side chains will set the upper bound of the film thickness. For this reason, dex(10) (fully extended length, 44.6 nm) was employed to generate PLL-*g*-dex copolymers that can potentially generate comparable film thicknesses with those of PLL-*g*-PEG(5) copolymers. The molecular weights of PLL-*g*-dex(10) are, of course, significantly higher than those of PLL-*g*-dex(5.2) or PLL-*g*-PEG(5) for the same grafting ratio (see Table 1).

3.2 Adsorption Behavior of PLL-*g*-dex: OWLS

A representative adsorption profile for PLL-*g*-dex copolymer onto the oxide surface as characterized by OWLS is presented in Fig. 2 (for the case of PLL(20)-*g*[3.4]-dex(5.2)). Upon exposure of a waveguide surface to the polymer solution (after ca. 60 min exposure to HEPES buffer to achieve the baseline), the adsorption of PLL-*g*-dex proceeded very rapidly such that more than the 90% of the final mass of adsorbed polymer was realized within the first 5 min. Because the raw data signals (change in the refractive index) at this stage reflect the contributions from both the surface-bound polymers and the bulk polymer solution, the final adsorbed mass was determined after the flow cell was rinsed with buffer solution (after ca. 50 min exposure to the polymer solution) to exclude the contribution from the bulk solution and loosely bound polymers. The negligible reduction in the adsorbed mass after the rinse indicates that no noticeable polymer desorption occurred during rinsing and assures the formation of a stable polymer adlayer. The same procedure was repeated for all the PLL-*g*-dex and PLL-*g*-PEG copolymers, as well as dex(5.2). The resulting adsorbed masses per unit area are presented as a function of grafting ratio for the series of PLL-*g*-dex(5.2), PLL-*g*-dex(10), and PLL-*g*-PEG(5) copolymers, as shown in Fig. 3.

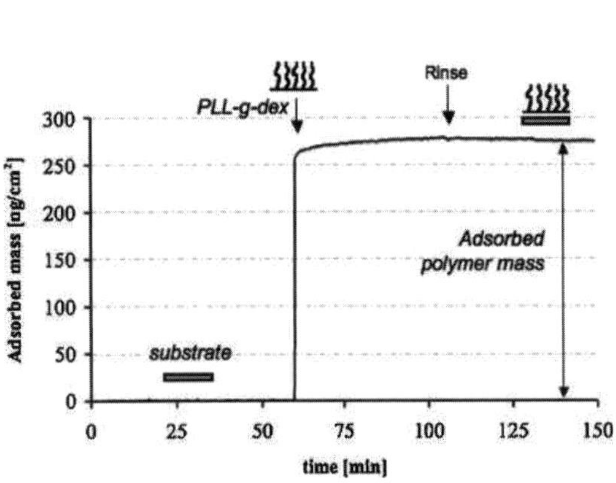

Fig. 2 In situ OWLS measurement of the polymer adsorption from solution onto uncoated waveguide chips

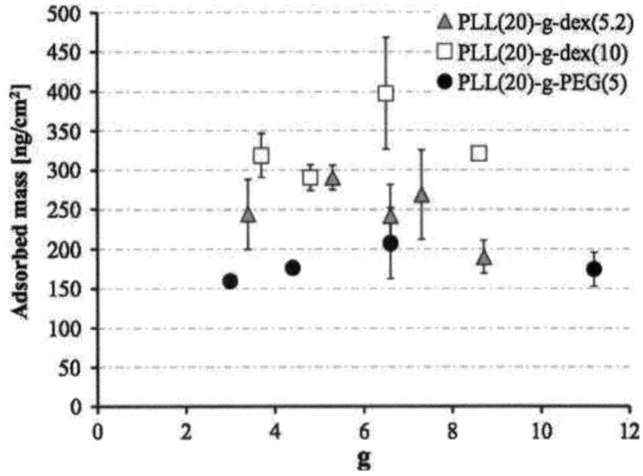

Fig. 3 OWLS results for the adsorption of PLL-*g*-dex copolymers and the counterpart PLL-*g*-PEG copolymers employed in this work. For some data points, error bars are smaller than the symbols

All the copolymers employed in this study showed a significant amount of adsorption onto the OWLS waveguide surfaces, ranging from ca. 200 to 290 ng cm^{-2} for PLL-*g*-dex(5.2), ca. 290 to 390 ng cm^{-2} for PLL-*g*-dex(10), and ca. 160 to 200 ng cm^{-2} for PLL-*g*-PEG(5) on average (see Table 2 and Fig. 3). Meanwhile, dextran (dex(5.2)) alone revealed negligible adsorption onto the surfaces (5.6 ± 3.7 ng cm^{-2}). This observation supports the hypothesis that both the adsorption mechanism and the conformation of the PLL-*g*-dex copolymer on negatively charged surfaces are very similar to those of PLL-*g*-PEG copolymer [5, 6, 9]; the adsorption of the copolymer proceeds through an electrostatic interaction between the PLL backbone and the surface in an aqueous environment, and thus the PLL backbone generally lies flat on the substrate surface, whereas the dextran side chains stretch into bulk aqueous solution because of their hydrophilic characteristics, forming brush layers. As shown in Fig. 3, the PLL-*g*-dex(10) copolymers employed in this study lead to either comparable or slightly higher adsorbed masses compared with PLL(20)-*g*-dex(5.2) copolymers, whereas both of them show somewhat higher adsorbed masses compared to PLL(20)-*g*-PEG(5) copolymers. In addition, for all three groups of copolymers reveal that the adsorbed mass is not significantly influenced by the variation in grafting ratio within the range selected in this study. The adsorbed mass is the product of the molecular weight of each copolymer molecule and the number of moles adsorbed on the surface. Because the (average) molecular weights of the copolymers are precisely known from their structures, more insights into the adsorption behavior of the copolymers can be obtained by decomposing their adsorbed mass into various coverage parameters, as shown in Fig. 4.

First, the molecular weights of the copolymers are shown as a function of the inverse of grafting ratio, $1/g$, in

Table 2 Summary of the adsorption data determined by OWLS for the synthesized PLL-*g*-dex copolymers

Surface	m_{pol} [ng/cm^2]	n_{Lys} [1/nm^2]	$n_{dex\ or\ PEG}$ [1/nm^2]	$n_{monomer\ units\ dex\ or\ PEG}$ [1/nm^2]	L [nm]	L/2Rg
PLL(20)-g[3.4]-dex(5.2)	244 ± 44	0.89 ± 0.20	0.26 ± 0.05	8.36 ± 1.51	2.12 ± 0.21	0.45 ± 0.04
PLL(20)-g[5.3]-dex(5.2)	290 ± 16	1.59 ± 0.09	0.30 ± 0.02	9.53 ± 0.52	1.96 ± 0.05	0.42 ± 0.01
PLL(20)-g[6.6]-dex(5.2)	241 ± 40	1.60 ± 0.26	0.24 ± 0.04	7.71 ± 1.27	2.19 ± 0.18	0.47 ± 0.04
PLL(20)-g[7.3]-dex(5.2)	269 ± 57	1.94 ± 0.41	0.27 ± 0.06	8.45 ± 1.80	2.10 ± 0.22	0.45 ± 0.05
PLL(20)-g[8.7]-dex(5.2)	190 ± 21	1.59 ± 0.17	0.18 ± 0.02	5.80 ± 0.62	2.52 ± 0.1	0.54 ± 0.03
PLL(20)-g[3.7]-dex(10)	319 ± 28	0.68 ± 0.06	0.18 ± 0.02	10.16 ± 0.58	2.51 ± 0.11	0.36 ± 0.02
PLL(20)-g[4.8]-dex(10)	290 ± 17	0.79 ± 0.05	0.16 ± 0.01	10.16 ± 0.49	2.65 ± 0.08	0.38 ± 0.01
PLL(20)-g[6.5]-dex(10)	394 ± 100	1.43 ± 0.36	0.22 ± 0.06	13.53 ± 3.43	2.32 ± 0.30	0.33 ± 0.04
PLL(20)-g[8.6]-dex(10)	321 ± 2	1.50 ± 0.01	0.17 ± 0.00	10.75 ± 0.07	2.57 ± 0.01	0.37 ± 0.00
PLL(20)-g(3)-PEG(5)	160 ± 3	0.54 ± 0.00	0.18 ± 0.00	20.49 ± 0.07	2.53 ± 0.00	0.45 ± 0.00
PLL(20)-g[4.4]-PEG(5)	180 ± 8	0.86 ± 0.04	0.19 ± 0.01	22.14 ± 1.04	2.44 ± 0.06	0.43 ± 0.01
PLL(20)-g[6.6]-PEG(5)	218 ± 58	1.48 ± 0.39	0.22 ± 0.06	25.47 ± 6.71	2.30 ± 0.31	0.41 ± 0.05
PLL(20)-g[11.2]-PEG(5)	184 ± 20	1.92 ± 0.20	0.17 ± 0.02	19.52 ± 2.03	2.60 ± 0.14	0.46 ± 0.02

m_{pol} adsorbed polymer mass, m_{serum} mass of serum adsorbed, n_{lys} surface density of lysine monomers, $n_{dex\ or\ PEG}$ surface density of dextran or PEG, $n_{monomer\ units\ dex\ or\ PEG}$ surface density of the monomer units of dextran or PEG, *L* spacing between grafted dextran/PEG chains, $L/2R_g$ degree of overlap of dextran/PEG chains

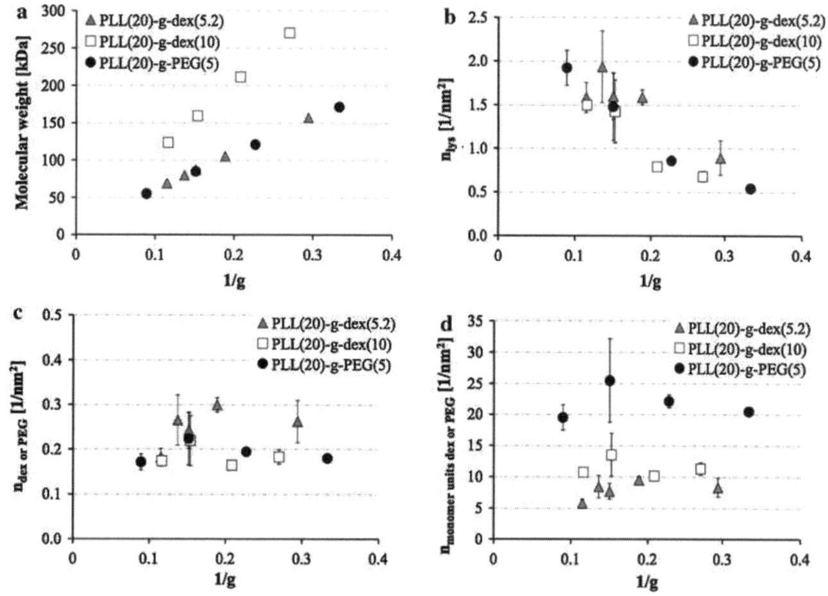

Fig. 4 Molecular weight (**a**), lysine monomer density (n_{lys}) (**b**), dextran (n_{dex}) or PEG (n_{PEG}) chain density (**c**), and density of the monomer units of dextran ($n_{monomer\ units\ dex}$) or PEG ($n_{monomer\ units\ PEG}$) (**d**) of the synthesized PLL-*g*-dex copolymers and the counterpart PLL-*g*-PEG copolymers on the silica-titania surface. For some data points, error bars are smaller than the symbols

Fig. 4(a); in this plot, increasing the value of $1/g$ represents the increase in the number of grafted side chains along the PLL backbone. In general, the molecular weight of the copolymer increases linearly with increasing $1/g$ values, due to increasing number of grafted side chains. The variation of molecular weight as a function of grafting ratio is nearly identical for PLL-*g*-PEG(5) and PLL-*g*-dex(5.2), while the molecular weights of PLL-*g*-dex(10) copolymers are much higher. In turn, from the adsorbed mass and the compositional features of the copolymers, it is possible to calculate the surface concentration of lysine monomers, n_{lys}, and dextran or PEG chains, $n_{dex\ or\ PEG}$. The surface concentration of lysine monomers, n_{lys}, reflects the number of copolymer molecules on the surface, whereas the surface density of dextran or PEG chains reflects the efficacy of the copolymer in grafting the hydrophilic polymer chains (either dextran or PEG) onto the surface. For the case of dextrans, because two different molecular weights (dextran chain lengths) were employed to generate PLL-*g*-dex copolymers in this study, the efficacy of grafting hydrophilic moieties can be expressed by the surface concentration of dextran monomer units (sugar rings) or

EG units, $n_{\text{monomer units dex or PEG}}$ [9]. The surface concentration of lysine monomers, n_{Lys}, dextran or PEG chains, n_{dex} or n_{PEG}, and monomer units of them, $n_{\text{monomer units dex}}$ or $n_{\text{monomer units PEG}}$, are plotted as a function of the inverse of grafting ratio, $1/g$, in Fig. 4b, c, and d, respectively. Finally, the mean spacing between the dextran chains, L, as well as the ratio between the spacing and the radius of gyration of dextran chains, $L/2R_g$, can be obtained to estimate the conformation of the surface-grafted dextran or PEG chains [9]. The results of these calculations are summarized in Table 2.

As shown in Fig. 4b, the surface concentrations of lysine monomers, n_{Lys}, for both PLL(20)-*g*-dex(5.2) copolymers show roughly constant values until $1/g$ reaches ca. 0.2, and yet rapidly decrease at higher $1/g$ values. Degraded adsorption with increasing number of side chains (increasing $1/g$) may be due either to the decreasing number of anchoring units (free NH_3^+) and/or increasing steric hindrance between grafted side chains. A previous study [9] involving PLL(20)-*g*-PEG and varying the molecular weight of PEG side chains, including PEG(1), PEG(2), and PEG(5), has, however, confirmed that this trend is mainly driven by the steric hindrance between side chains; a decrease in n_{Lys} with increasing $1/g$ was observed in the cases of PLL(20)-*g*-PEG(2) and PLL(20)-*g*-PEG(5), but not in the case of PLL-*g*-PEG(1), which has smaller steric hindrance between side chains as a consequence of the PEG(1)'s smaller radius of gyration. For PLL(20)-*g*-dex(10) and PLL(20)-*g*-PEG(5), a linear decrease of n_{Lys} is observed from $1/g \approx 0.1$ to the highest value of ≈ 0.28, which implies that the steric hindrance between dex(10) is more serious than in the other two cases. The radii of gyration for the three hydrophilic polymers are in the order dex(5.2) (2.35 nm [48]) <PEG(5) (2.82 nm [49]) <dex(10) (3.49 nm [48]).

The surface density of dextran or PEG chains, Fig. 4c, is thus proportional to the product of the n_{Lys} and the number of side chains per molecule, which also linearly increases with increasing $1/g$ values (see the Table 1). For PLL(20)-*g*-dex(5.2) copolymers, the increase in number of grafted side chains on the backbone leads to a proportional increase in the grafted side chains on the surface in the initial $1/g$ range of 0.1 $\sim$ 0.2, and yet, n_{dex} or n_{PEG} reaches a plateau at higher $1/g$ values. Apparently, this plateau is a result of opposing effects with increasing $1/g$—the decreasing number of copolymers on the surface, as shown by the decrease of n_{Lys}, and increasing number of grafted side chain on the backbone. The radii of gyration for dex(10) and PEG(5) are larger than that of dex(5), and the plateau for the corresponding copolymers is already reached at $1/g = 0.1$. Finally, the monomer densities, as shown in Fig. 4d, reveal nearly the same trends with those of n_{dex} or n_{PEG}, the relative magnitudes between dex(5.2) and dex(10) being reversed.

3.3 Lubrication Properties of the Copolymers: General Considerations

To evaluate the lubricating properties of PLL-*g*-dex copolymer solutions and compare them with other standard aqueous solutions, including HEPES buffer solution, dextran solutions (dex(5.2)), as well as PLL(20)-*g*-PEG(5) solutions, coefficient of friction (μ) versus speed plots were acquired under both sliding and mixed sliding/rolling conditions using a pin-on-disk tribometer and MTM, respectively.

The lubricating capabilities of the copolymers in this study are considered to be directly associated with their adsorption properties onto the tribopair surfaces. Although the adsorption properties of the copolymers onto the pins (stainless steel) remain uncharacterized, the OWLS measurements shown in the previous section can represent the adsorption properties of the copolymers on the disk (glass) because silicon oxides prevail on both the waveguide and the glass disk surfaces. Given that not only adsorbed masses but also lysine monomer densities, n_{Lys}, hydrophilic polymer chain densities, n_{dex} or n_{PEG}, as well as the hydrophilic monomer unit densities, $n_{\text{monomer units dex}}$ or $n_{\text{monomer units PEG}}$, can be deduced from OWLS measurements (Table 2 and Fig. 4), it is of interest to see with which structural parameter the lubricating properties are best correlated.

In macroscopic sliding contacts between two rigid surfaces, as in this study, the lubricating capabilities of the copolymers are expected to be further influenced by their adsorption kinetics. As was previously addressed in a study employing fluorescently labeled PLL-*g*-PEG copolymers, [50] the electrostatically adsorbed polymer layers are easily scraped away from the surfaces during sliding contacts in pin-on-disk tribometry, particularly due to asperity contacts arising from surface roughness and consequently high local contact pressures. Nevertheless, the excess polymers in bulk solution, if present, re-adsorb onto the area where the initially adsorbed copolymers have been desorbed, and the tribostressed area 'self-heals'. In other words, effective lubrication of electrostatically adsorbed copolymers over a long period in the presence of excess polymer in bulk solution is rather due to their fast adsorption kinetics than their adsorption strength. An important prerequisite to test the efficacy of the copolymers as lubricant additives and to take into account their surface adsorption kinetics behavior is to keep the number of excess polymer molecules constant, i.e., their molar concentration available to 'heal' the tribostressed area. For this reason, we have doubled the mass concentration of PLL(20)-*g*-dex(10) copolymers, 0.5 mg/mL, with respect to those of PLL(20)-*g*-dex(5.2) and PLL(20)-*g*-PEG(5) copolymers, 0.25 mg/mL, for the tribological tests. If the mass concentrations were kept constant for all the copolymers, the molar concentrations of

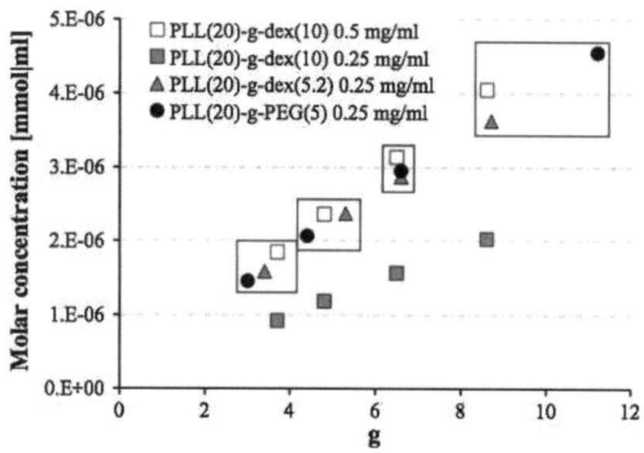

Fig. 5 Molar concentration [mmol/ml] of the copolymer solutions employed in this study (highlighted in the squares the concentrations chosen for the characterization of the lubricating properties of PLL-*g*-dex and the counterpart PLL-*g*-PEG)

PLL-*g*-dex(10) copolymers would be significantly lower than those of the other two types of copolymers, as shown in Fig. 5.

3.3.1 *Lubrication at Sliding Contacts: Pin-on-disk Tribometry*

Sliding-contact experiments, carried out by pin-on-disk tribometry, were intended to assess the lubricating capabilities of the polymer solutions in the boundary-lubrication regime. To compare the relative lubricating properties of three kinds of copolymers, PLL(20)-*g*-dex(5.2), PLL(20)-*g*-PEG(5), and PLL(20)-*g*-dex(10), they were grouped mainly according to grafting ratio (*x*-axis) as shown in Fig. 5. As mentioned above, this was to keep the molar concentration (*y*-axis) nearly constant for the copolymers employed for the comparison. As shown in Fig. 6, all the copolymer solutions employed in this study led to apparent reduction in μ values compared to those of HEPES buffer solution (as much as 50% for most copolymers at speeds above 10 mm s^{-1}, generally ca. 30% reduction at lower speeds). In contrast, the dextran (5.2) solution revealed negligible improvement in the lubricating properties. This behavior is consistent with the very low adsorption of dex(5.2) onto the oxide surfaces, as determined by OWLS measurements (Table 2). To highlight the comparison of the lubricating properties of PLL-*g*-PEG and PLL-*g*-dex, however, the data for the HEPES buffer solutions and dex(5.2) are omitted in subsequent figures.

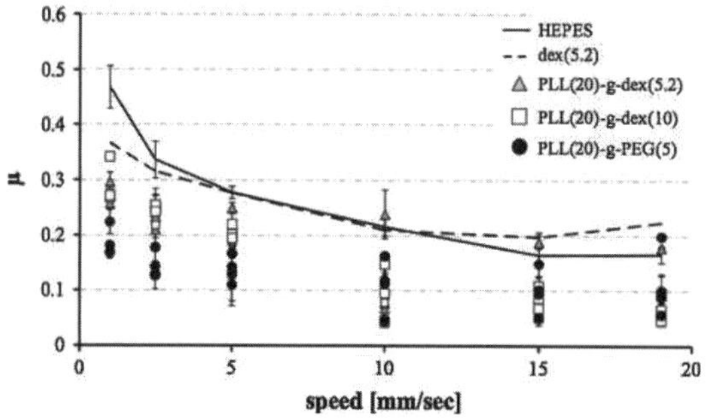

Fig. 6 Sliding pin-on-disk results of all PLL-*g*-dex and PLL-*g*-PEG copolymers (with grafting ratio varying from 3 to 11) investigated in this study (tribopair: steel balls versus glass disks, load: 2 N, room temperature): coefficients of friction (μ) as a function of the sliding speed

The μ vs. speed plots for the copolymers with *g*[3.5], *g*[4.8], *g*[6.5], and high *g*[y], between ca. *g* [8] and *g* [11] (average values) are plotted in Figs. 7, 8, 9, and 10, respectively. Generally, μ values (load = 2 N) of the

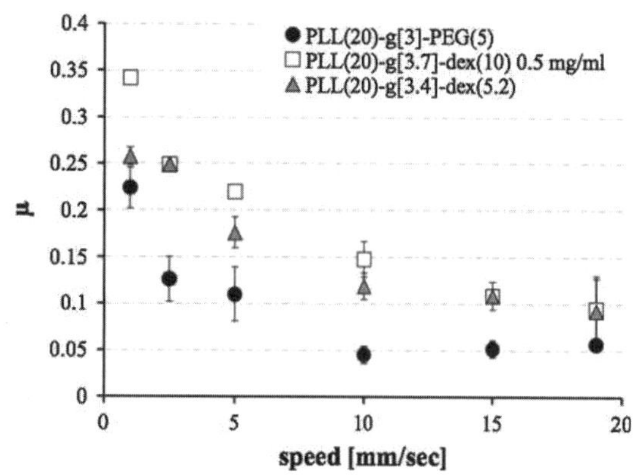

Fig. 7 Sliding pin-on-disk results of the PLL-*g*-dex copolymers and PLL-*g*-PEG counterpart with ca. *g*[3.5]: coefficients of friction as a function of the sliding speed. For some data points, error bars are smaller than the symbols

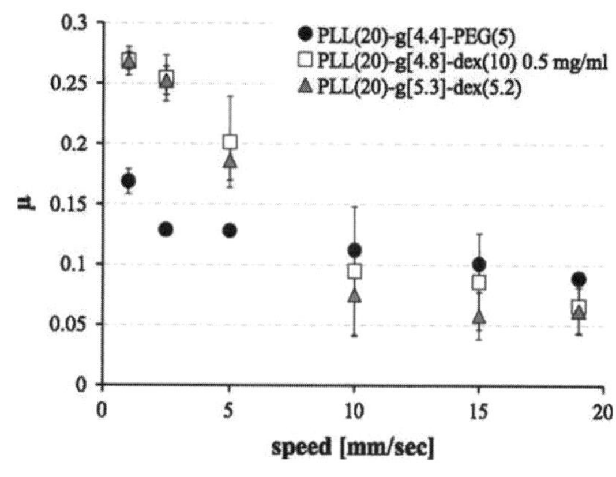

Fig. 8 Sliding pin-on-disk results of the PLL-*g*-dex copolymers and PLL-*g*-PEG counterpart with ca. *g*[4.8]: coefficients of friction as a function of the sliding speed. For some data points, error bars are smaller than the symbols

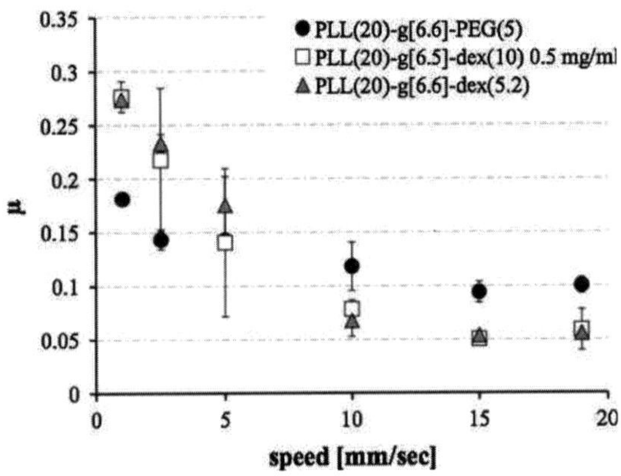

Fig. 9 Sliding pin-on-disk results of the PLL-*g*-dex copolymers and PLL-*g*-PEG counterpart with ca. *g*[6.5]: coefficients of friction as a function of the sliding speed. For some data points, error bars are smaller than the symbols

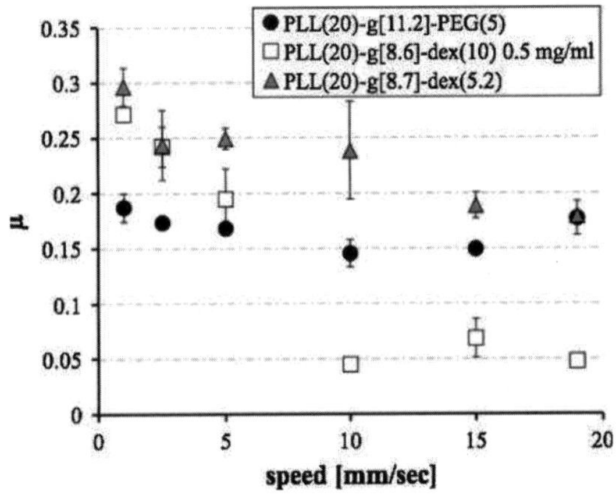

Fig. 10 Sliding pin-on-disk results of the PLL-*g*-dex copolymers and PLL-*g*-PEG counterpart with high grafting ratio, between ca. *g*[8] and *g*[11]: coefficients of friction as a function of the sliding speed. For some data points, error bars are smaller than the symbols

copolymer solutions revealed an initial reduction with increasing sliding speed (corresponding to increasing frequency of tribocontact), but they started to level off from approximately 10 mm s^{-1} to the highest speed investigated, reaching approximately $\mu = 0.05$ for the best-behaving copolymer in each comparison set. As shown in Fig. 7, for the copolymers possessing the grafting ratio of $\sim g$ [3.5], the μ values of the two PLL-*g*-dex solutions were observed to be generally very similar, whereas those of PLL(20)-*g*-PEG(5) are slightly, yet noticeably, lower than the two PLL-*g*-dex solutions over the entire speed regime. For all other cases, as shown in Figs. 8, 9, and 10, the relative lubricating efficacy of PLL-*g*-dex copolymers compared with PLL(20)-*g*-PEG(5) was highly dependent upon the sliding speed; in the low-speed regime ($\leq$10 mm s^{-1}), the lubricating performance of PLL(20)-*g*-PEG(5) was consistently superior to those of PLL-*g*-dex copolymers. In the high-speed regime ($\geq$10 mm s^{-1}), however, PLL-*g*-dex copolymers reveal lower friction forces than PLL(20)-*g*-PEG(5) in most cases. Exceptions are the cases of $\sim g$[3.5] (Fig. 7) and PLL(20)-*g*[8.6]-dex(5.2) (Fig. 10), which have too high a grafting ratio to be in a brush conformation.

To more clearly visualize the relative magnitude of lubricating efficacy of the copolymers employed, and also to visualize the variation of the lubricating properties as a function of architectural features of the copolymers, the μ values representing the characteristics in the low-speed regime (1 mm s^{-1}) and the high-speed regime (19 mm s^{-1}) are plotted against the inverse of grafting ratio, 1/*g*, in Figs. 11 and 12, respectively. It should be borne in mind that all speeds investigated in the pin-on-disk studies correspond to the boundary lubrication regime, and that the frequency of tribocontact is the parameter that is being investigated via speed changes.

In the low-speed regime, (Fig. 11), the μ values of PLL(20)-*g*-PEG(5) compared with PLL-*g*-dex copolymers are, again, consistently lower over the entire range of grafting ratio. Because the magnitudes of extension of the PEG and dextran chains, as quantified by $L/2R_g$, are comparable for the PLL(20)-*g*-PEG(5) and PLL(20)-*g*-dex(5.2) copolymers (see the Table 2), and in fact, those of PLL(20)-*g*-dex(10) are even greater (lower values of $L/2R_g$) than PLL(20)-*g*-PEG(5), it is unlikely that the difference in the conformation of the two different hydrophilic polymer chains on the surface is the principal cause of the

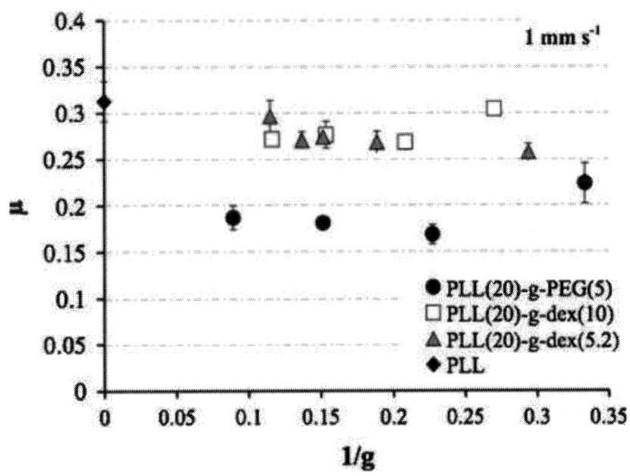

Fig. 11 Sliding pin-on-disk results of all PLL-*g*-dex and PLL-*g*-PEG copolymers investigated in this study at the lowest speed investigated (1 mm s^{-1}): coefficients of friction as a function of the degree of grafting of the side chains, 1/*g*. For some data points, error bars are smaller than the symbols

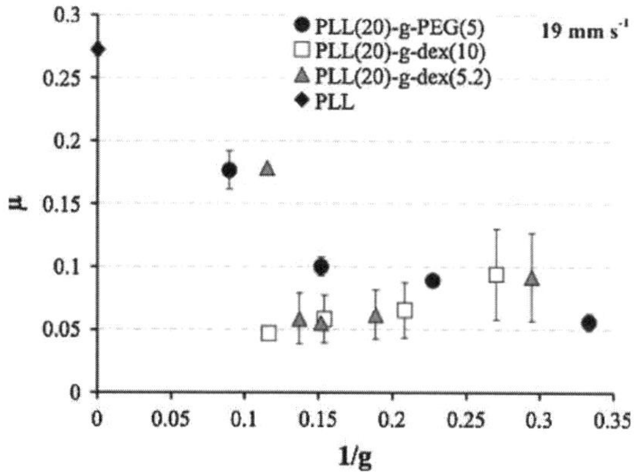

Fig. 12 Sliding pin-on-disk results of all PLL-*g*-dex and PLL-*g*-PEG copolymers investigated in this study at the highest speed investigated (19 mm s^{-1}): coefficients of friction as a function of the degree of grafting of the side chains, $1/g$. For some data points, error bars are smaller than the symbols

superior lubricating properties of PLL(20)-*g*-PEG(5) under these conditions. Furthermore, based upon $L/2R_g$ values and the fully extended lengths of PEG(5) and dex(10), the film thicknesses are also estimated to be comparable for the two hydrophilic polymer chains. We thus speculate that dextran and PEG chains may display different innate water-based lubricating capabilities, mainly arising from different degrees and mechanisms of hydration and/or differences in the stiffness and/or dynamical behavior of the chains [51]. Quantitative measurements of the hydration for PLL-*g*-PEG and PLL-*g*-dex copolymers are currently in progress.

From Fig. 11, it is also noticeable that in the low-speed regime, the μ values of all the copolymers remain nearly constant over the entire range of grafting ratio, except for the onset of an increase in μ for PLL-*g*-PEG(5) and PLL-*g*-dex(10) at the highest $1/g$ value, i.e., the highest number of grafted side chains along the PLL backbone. It is important to note that although the molar concentrations of the copolymers were controlled to be constant for each set of three copolymers with similar grafting ratio, the molar concentrations of the copolymers do, of course, increase with increasing grafting ratio, (i.e., decrease with increasing $1/g$) as shown in Fig. 4a. Thus, the molar concentrations of the copolymers shown in Fig. 11 are different as much as a factor of three. In turn, generally constant μ values observed over the entire $1/g$ range in Fig. 11 suggest that the lubricating properties in the low-speed regime are not sensitively influenced by the variation of the molar concentration. As will be discussed in more detail below, this feature is in stark contrast to the lubricating behavior in the high-speed regime, and indicates that the lubricating properties of the copolymers in the low-speed regime are mainly governed by the adsorption properties in equilibrium, since the solution concentration is considerably in excess of that necessary to achieve monolayer coverage on the sliding surfaces.

Comparing the various parameters representing adsorption properties of the copolymers, including adsorbed mass, the surface densities of lysine units, n_{Lys}, surface densities of dextran or PEG chains, n_{dex} or n_{PEG}, surface densities of monomer units, $n_{\mathrm{monomer\ units\ dex}}$ or $n_{\mathrm{monomer\ units\ PEG}}$, and the magnitude of brush extension, $L/2R_g$ (Table 2 and Fig. 4), the lubricating properties appear to be most highly correlated with the surface densities of dextran or PEG chains, albeit not in a simple, linear fashion; both lubricating properties and surface densities were usually insensitive to the variation of grafting ratio within the selected range. The onset of the increase in μ values observed from PLL-*g*-PEG(5) and PLL-*g*-dex(10), but not from PLL-*g*-dex(5.2), is associated with the radius of gyration for the side chains changing in the order of dex(5.2) < PEG(5) < dex(10). As was previously stated, too high a number of the grafted side chains along the PLL backbone tends to degrade the adsorption properties of comb-like graft copolymers, and consequently their lubricating properties. Because the steric hindrance between neighboring side chains plays a major role in this trend, larger radii of gyration of the side chains are expected to exacerbate such behavior.

In the high-speed regime, (Fig. 12), μ values of all copolymers employed in this work fall in the same range, between 0.05 and 0.1, with the exception of the two values for PLL(20)-*g*[11.2]-PEG(5) and PLL(20)-*g*[8.7]-dex(5.2), which show significantly higher μ values (close to that of the PLL backbone alone ($1/g = 0$)) and are presumably not in a brush conformation.

PLL(20)-*g*-dex(5.2) and PLL(20)-*g*-dex(10) copolymers behave similarly in the regime where $1/g \geq$ ca. 0.15, with their coefficients of friction slightly increasing in a fairly linear fashion. In the same regime ($1/g \geq$ ca. 0.15), PLL(20)-*g*-PEG(5) copolymers, instead, show a small decrease in the μ values. A general increase in μ values for PLL-*g*-dex copolymers with high $1/g$ values (high number of grafted side chains) can be readily understood in terms of smaller molar concentrations. The fact that the variation of molar concentration (from ca. 1.5 to 4×10^{-6} mmol/mL) influences the lubricating properties of the PLL-*g*-dex copolymers exclusively in high-speed regime is closely related to the time intervals between the cycles of tribostress. Under the experimental conditions (pin-on-disk tribometry), the tribostress is applied in a cyclic fashion, the sliding speed (or rpm, to be more precise) setting the time scale during which excess polymers can replenish the area where initially adsorbed polymers may have been detached by tribostress.

While with increasing $1/g$, PLL-*g*-dex copolymers become heavier (higher molecular weight), and therefore, slower in returning to the surface after being scraped away from it during sliding contact, the molar concentration also decreases as the degree of grafted side chains increases, meaning that the excess polymer in bulk solution available to re-adsorb, i.e., "self-heal" the tribostressed area, decreases at the same time.

The slight decrease of μ values for PLL(20)-*g*-PEG(5) with increasing $1/g$ suggests that the adsorption kinetics of PLL(20)-*g*-PEG(5) copolymers, in contrast to that of PLL-*g*-dex copolymers, are sufficiently fast such that the variation of the molar concentration within the range of $1.5–4 \times 10^{-6}$ mmol/mL does not lead to any substantial disadvantages in the lubricating performance. Furthermore, the higher flexibility of PEG chains compared to that of dextran's stiffer sugar units might facilitate the rearrangement of the polymer molecules on the surface after desorption.

3.3.2 *Lubrication at Mixed Sliding/Rolling Contacts: MTM*

The lubricating properties of PLL-*g*-dex copolymer solutions under mixed sliding/rolling contact conditions have been characterized by means of MTM. The MTM experiments serve primarily to assess the lubricating properties of the copolymers under milder contact conditions than the pure sliding contacts characteristic of pin-on-disk tribometry. In addition, owing to the much higher speed range available from MTM (up to 2,500 mm s^{-1}), the formation of a full fluid-film lubricant layer by the polymer solutions can be investigated. Based on the pin-on-disk tribometry results above, two PLL-*g*-dex(5) copolymers, PLL(20)-*g*[3.4]-dex(5.2) and PLL(20)-*g*[5.3]-dex(5.2), which showed slightly different lubricating behavior in sliding contact, have been employed, together with other lubricant solutions. The results are shown in Fig. 13.

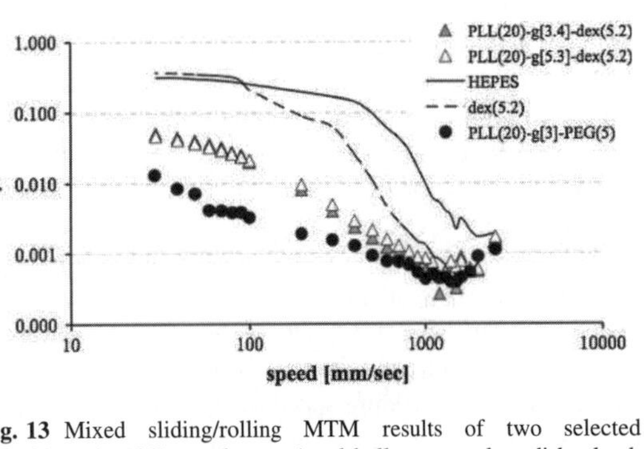

Fig. 13 Mixed sliding/rolling MTM results of two selected PLL(20)-*g*-dex(5.2) copolymers (steel balls versus glass disks, load: 10 N, room temperature): coefficients of friction as a function of the sliding speed

The μ vs. speed plots shown in Fig. 13 revealed significant reduction of μ values in the high-speed regime (starting from ca. 200 mm s^{-1}) for all aqueous lubricants employed, including polymer solutions and polymer-free HEPES buffer solution. The lowest μ value for HEPES buffer solution, 1.9×10^{-3}, which was achieved at the highest speed (2,500 mm s^{-1}), for instance, represents approximately two orders of magnitude reduction compared with the highest μ value, 0.3, observed at the lowest speed (10 mm s^{-1}). This observation suggests that the entrainment of the base lubricant, water, is starting to contribute to the lubrication of this mixed sliding/rolling contacts (SRR = 10%) in the high-speed regime. Nevertheless, all the polymer solutions investigated have been observed to further improve the lubricating properties, either in the high-speed regime alone or in both low- and high-speed regimes, depending on the type of polymers. For instance, the dextran solution revealed virtually identical μ vs. speed behavior to that of HEPES buffer in the low-speed regime ($\leq$100 mm s^{-1}), whereas a significant reduction in μ was observed in the high-speed regime, as mentioned above. It should be noted that at the dex(5.2) concentrations employed, no detectable increase in viscosity over that of pure HEPES solution could be detected in shear rheometry measurements (data not shown). Although dex(5.2) chains do not display noticeable surface-adsorption properties (Fig. 3), and thus are poor boundary-lubricating additives, they appear to facilitate the entrainment of aqueous lubricant due to their hydrophilic characteristics [52].

The two PLL-*g*-dex copolymer solutions showed a distinct reduction in μ values, both in low- and high-speed regimes, reflecting their adsorption onto the tribopair surfaces; compared with the dextran solution, the μ values are observed to be consistently lower over nearly the entire speed range investigated. This observation implies that the reduction in μ values in the high-speed regime is due not entirely to the formation of fluid-film, but also due partially to the improvement in the boundary lubrication properties. In fact, previous combined studies of the film thickness and frictional properties by means of ultra-thin film interferometry and MTM, respectively, have shown that the mixed lubrication mechanism is dominant for tribological contacts lubricated by PLL-*g*-PEG aqueous solutions [8]. In contrast to the results for sliding contacts, however, the two PLL-*g*-dex copolymers showed no noticeable difference in their lubricating properties, presumably because of the characteristics of the tribological contacts provided by MTM; first, the contact is much milder due to the dominance of rolling contact (SRR = 10%) under these conditions, and thus the readsorption properties of the molecules are less critical. Second, the entrainment of the base lubricant, water, becomes more likely with increasing speed, even in the absence of additives.

Finally, the lubricating performance of the PLL-*g*-dex copolymers under mixed sliding/rolling contact conditions was observed to be inferior to that of PLL(20)-*g*[3.0]-PEG(5). This difference is more noticeable under low-speed conditions, and starts to diminish with increasing speed. PLL(20)-*g*[6.5]-dex(10) and its PLL-*g*-PEG counterpart were also investigated by MTM, but they showed no difference in terms of the μ-value trend, when compared to the PLL-*g*-dex(5) copolymers employed. These results have therefore been omitted in Fig. 13 for clarity purposes, to highlight the results discussed above.

4 Conclusions

Comb-like graft copolymers with dextran side chains (PLL-*g*-dex) have been developed and investigated as biomimetic lubricant additives in an aqueous environment. The synthesized copolymers have been shown to adsorb spontaneously onto negatively charged surfaces, similarly to PLL-*g*-PEG, and to behave as boundary lubricants, both in the pure sliding and in mixed sliding/rolling contact regimes.

Two series of PLL-*g*-dex copolymers, with varying grafting ratios and molecular weights of the dextran side chains (dex(5.2) and dex(10)), were employed in this study and their lubricating capabilities compared to the PLL-*g*-PEG counterpart copolymers. The selection of the two dextrans was motivated by the desire to perform a fair comparison with the PLL-*g*-PEG copolymers: dex(5.2) is nearly identical with PEG(5) in terms of molecular weight (diffusion properties) and dex(10) and PEG(5) are comparable for the fully extended chain length (generable film thickness). The lubricating properties of the involved copolymers in the low-speed regime ($\leq$ca. 10 mm s^{-1}) appeared to be related to the adsorption properties in equilibrium and not to be significantly influenced by the variation of molar concentration. Lubricating performances of PLL(20)-*g*-PEG(5) compared with PLL-*g*-dex copolymers are, in this regime, better over the entire range of grafting ratio: this might be attributed to the higher degree of hydration and flexibility of PEG chains compared to that of the sugar units in dextran.

In the high-speed regime ($\geq$ca. 10 mm s^{-1}), all the PLL-*g*-dex copolymers, except for the case of *g*[3.5], showed better lubricating performance than the PLL-*g*-PEG counterpart copolymers, but with a slightly different trend in the μ values with increasing $1/g$; the self-healing behavior of PLL-*g*-dex copolymers seems to be diminished, in fact, by decreasing molar concentration and increasing molecular weight. Nevertheless, such changes do not influence the performance of the PLL-*g*-PEG copolymers, presumably due to the greater flexibility of the PEG chains.

Acknowledgments We would like to thank Dr. Rowena Crockett of EMPA, Dübendorf, who made the Mini Traction Machine available to us. Effort sponsored by the Air Force Office of Scientific Research, Air Force Material Command, USAF, under grant number FA8655-05-1-3042. The U.S. Government is authorized to reproduce and distribute reprints for Government purpose notwithstanding any copyright notation thereon. The views and conclusions contained herein are those of the authors and should not be interpreted as necessarily representing the official policies or endorsements, either expressed or implied, of the Air Force Office of Scientific Research or the U.S. Government.

References

1. Havet, L., Blouet, J., Robbe Valloire, F., Brasseur, E., Slomka, D.: Tribological characteristics of some environmentally friendly lubricants. Wear **248**, 140–146 (2001). doi:10.1016/S0043-1648(00)00550-0
2. Willing, A.: Lubricants based on renewable resources—an environmentally compatible alternative to mineral oil products. Chemosphere **43**, 89–98 (2001). doi:10.1016/S0045-6535(00)00328-3
3. Uyama, Y., Tadokoro, H., Ikada, Y.: Low-frictional catheter materials by photo-induced graft polymerization. Biomaterials **12**, 71–75 (1991). doi:10.1016/0142-9612(91)90135-W
4. Heuberger, M., Drobek, T., Spencer, N.D.: Interaction forces and morphology of a protein-resistant Poly(ethylene glycol) layer. Biophys. J. **88**, 495–504 (2005). doi:10.1529/biophysj.104.045443
5. Kenausis, G.L., Vörös, J., Elbert, D.L., Huang, N., Hofer, R., Ruiz-Taylor, L., Textor, M., Hubbell, J., Spencer, N.D.: Poly(L-lysine)-g-Poly(ethylene glycol) layers on metal oxide surfaces: attachment mechanism and effects of polymer architecture on resistance to protein adsorption. J. Phys. Chem. B **104**, 3298–3309 (2000). doi:10.1021/jp993359m
6. Lee, S., Müller, M., Ratoi-Salagean, M., Vörös, J., Pasche, S., De Paul, S.M., Spikes, H.A., Textor, M., Spencer, N.: Boundary lubrication of oxide surfaces by Poly(L-lysine)-g-poly(ethylene glycol) (PLL-g-PEG) in aqueous media. Tribol. Lett. **15**, 231–239 (2003). doi:10.1023/A:1024861119372
7. Müller, M.: Aqueous Lubrication by Means of Surface-Bound Brush-Like Copolymers. ETH Zurich, Zurich (2005)
8. Müller, M., Lee, S., Spikes, H.A., Spencer, N.D.: The influence of molecular architecture on the macroscopic lubrication properties of the brush-like co-polyelectrolyte poly: (L-lysine)-g-poly(ethylene glycol) (PLL-g-PEG) adsorbed on oxide surfaces. Tribol. Lett. **15**, 395–405 (2003). doi:10.1023/B:TRIL.0000003063.98583.bb
9. Pasche, S., Paul, S.M., Vörös, J., Spencer, N.D., Textor, M.: Poly(L-lysine)-graft-poly(ethylene glycol) assembled monolayers on niobium oxide surfaces: a quantitative study of the influence of polymer interfacial architecture on resistance to protein adsorption by ToF-SIMS and in situ OWLS. Langmuir **19**, 9216–9225 (2003). doi:10.1021/la034111y
10. Pasche, S., Textor, M., Meagher, L., Spencer, N.D., Griesser, H.J.: Relationship between interfacial forces measured by colloid-probe atomic force microscopy and protein resistance of poly/ethylen glycol)-grafted poly(L-lysine) adlayers on niobia surfaces. Langmuir **21**, 6508–6520 (2005). doi:10.1021/la050386x
11. Blaettler, T.M., Pasche, S., Textor, M., Griesser, H.J.: High salt stability and protein resistance of poly(L-lysine)-g-poly(ethylene glycol) copolymers covalently immobilized via aldehyde plasma polymer interlayers on inorganic and polymeric substrates. Langmuir **22**, 5760–5769 (2006). doi:10.1021/la034111y

12. Han, S., Kim, C., Kwon, D.: Thermal degradation of poly(ethylenelycol). Polym. Degrad. Stab. **47**, 203–208 (1995). doi:10.1016/0141-3910(94)00109-L
13. Han, S., Kim, C., Kwon, D.: Thermal/oxidative degradation and stabilization of polyethylene glycol. Polymer (Guildf.) **38**, 317–323 (1997). doi:10.1016/S0032-3861(97)88175-X
14. Holland, N.B., Qiu, Y., Ruegsegger, M., Marchant, R.E.: Biomimetic engineering of non-adhesive glycocalyx-like surfaces using oligosaccharide surfactant polymers. Nature **392**, 799–801 (1998). doi:10.1038/33894
15. McArthur, S.L., McLean, K.M., Kingshott, P., John, H.A.W., Chatelier, R.C., Griesser, H.J.: Effect of polysaccharide structure on protein adsorption. Colloids Surf. B Biointerfaces **17**, 37–48 (2000). doi:10.1016/S0927-7765(99)00086-7
16. Piehler, J., Brecht, A., Hehl, K., Gauglitz, G.: Protein interactions in covalently attached dextran layers. Colloids Surf. B Biointerfaces **13**, 325–336 (1999). doi:10.1016/S0927-7765(99)00046-6
17. Qiu, Y., Zhang, T., Ruegsegger, M., Marchant, R.E.: Novel nonionic oligosaccharide surfactant polymers derived from poly(vinylamine) with pendant dextran and hexanoyl groups. Macromolecules **31**, 165–171 (1998). doi:10.1021/ma9707401
18. Sen Gupta, A., Wang, S., Link, E., Anderson, E.H., Hofmann, C., Lewandowski, J., Kottke-Marchant, K., Marchant, R.E.: Glycocalyx-mimetic dextran-modified poly(vinyl amine) surfactant coating reduces platelet adhesion on medical-grade polycarbonate surface. Biomaterials **27**, 3084–3095 (2006). doi:10.1016/j.biomaterials.2006.01.002
19. Naka, M., Morita, Y., Ikeuchi, K.: Influence of proteoglycan contents and of tissue hydration on the frictional characteristics of articular cartilage. Proc. Inst. Mech. Eng. [H] **219**, 175–182 (2005). doi:10.1243/095441105X34220
20. Pickard, J., Ingham, E., Egan, J., Fisher, J.: Investigation into the effect of proteoglycan molecules on the tribological properties of cartilage joint tissues. Proc. Inst. Mech. Eng. **212**, 177–182 (1998)
21. Bansil, R., Turner, B.S.: Mucin structure, aggregation, physiological functions and biomedical applications. Curr. Opin. Colloid Interface Sci. **11**, 164–170 (2006). doi:10.1016/j.cocis.2005.11.001
22. Dalsin, J.L.: Mussel adhesive-inspired surface modification for the preparation of nonfouling biomaterials. Ph.D. Dissertation, Northwestern University (2004)
23. Hahn Berg, C.I., Lindh, L., Arnebrant, T.: Intraoral lubrication of PRP-1, Statherin and Mucin as Studied by AFM. Biofouling **20**, 65–70 (2004). doi:10.1080/08927010310001639082
24. Lee, S., Müller, M., Rezwan, K., Spencer, N.D.: Porcine gastric mucin (PGM) at the water/poly(dimethylsiloxane) (PDMS) interface: influence of pH and ionic strength on its conformation, adsorption, and aqueous lubrication properties. Langmuir **21**, 8344–8353 (2005). doi:10.1021/la050779w
25. Choi, S.W., Sato, Y., Akaike, T., Maruyama, A.: Preparation of cationic comb-type copolymer having guanidino moieties and its interaction with DNAs. J. Biomater. Sci. Polym. Ed. **15**, 1099–1110 (2004). doi:10.1163/1568562041753089
26. Ferdous, A., Akaike, T., Maruyama, A.: Inhibition of sequence-specific protein-DNA interaction and restriction endonuclease cleavage via triplex stabilization by poly(L-lysine)-graft-dextran copolymer. Biomacromolecules **1**, 186–193 (2000). doi:10.1021/bm9900141
27. Ferdous, A., Watanabe, H., Akaike, T., Maruyama, A.: Comb-type copolymer: stabilization of triplex DNA and possible application in antigene strategy. J. Pharm. Sci. **87**, 1400–1405 (1998). doi:10.1021/js980066g
28. Ferdous, A., Watanabe, H., Akaike, T., Maruyama, A.: Poly(L-lysine)–graft–dextran copolymer: amazing effects on triplex stabilization under physiological pH and ionic conditions (in vitro). Nucleic Acids Res. **26**, 3949–3954 (1998). doi:10.1093/nar/26.17.3949
29. Maruyama, A., Watanabe, H., Ferdous, A., Katoh, M., Ishihara, T., Akaike, T.: Characterization of interpolyelectrolyte complexes between double-stranded DNA and polylysine comb-ype copolymers having hydrophilic side chains. Bioconjugate Chem. **9**, 292–299 (1998)
30. Sato, Y., Kobayashi, Y., Kamiya, T., Watanabe, H., Akaike, T., Yoshikawa, K., Maruyama, A.: The effect of backbone structure on polycation comb-type copolymer/DNA interactions and the molecular assembly of DNA. Biomaterials **26**, 703–711 (2005). doi:10.1016/j.biomaterials.2004.03.018
31. Torigoe, H., Ferdous, A., Watanabe, H., Akaike, T., Maruyama, A.: Poly(L-lysine)-graft-dextran copolymer promotes pyrimidine motif triplex DNA formation at physiological pH. J. Biol. Chem. **274**, 6161–6167 (1999). doi:10.1074/jbc.274.10.6161
32. Frazier, R.A., Matthijs, G., Davies, M.C., Roberts, C.J., Schacht, E., Tendler, S.J.B.: Characterization of protein-resistant dextran monolayers. Biomaterials **21**, 957–966 (2000). doi:10.1016/S0142-9612(99)00270-7
33. Griesser, H.J., Hartley, P.G., McHarthur, S.L., McLean, K.M., Meagher, L., Thissen, H.: Interfacial properties and protein resistance of nano-scale polysaccharide coatings. Smart Mater. Struct. **11**, 652–661 (2002). doi:10.1088/0964-1726/11/5/305
34. Martwiset, S.: Nonfouling characteristics of dextran-containing surfaces. Langmuir **22**, 8192–8196 (2006). doi:10.1021/la061064b
35. Mc Lean, K.M., Johnson, G., Chatelier, R.C., Beumer, G.J., Steele, G., Griesser, H.J.: Method of immobilization of carboxymethyl-dextran affects resistance to tissue and cell colonization. Colloids Surf. B Biointerfaces **18**, 221–234 (2000). doi:10.1016/S0927-7765(99)00149-6
36. Österberg, E., Bergström, K., Holmberg, K., Schuman, T.P., Riggs, J.A., Burns, N.L., Van Alstine, J.M., Harris, J.M.: Protein-rejecting ability of surface-bound dextran in end-on and side-on configurations: comparison to PEG. J. Biomed. Mater. Res. **29**, 741–747 (1995). doi:10.1002/jbm.820290610
37. Snabre, P., Grossmann, G.H., Mills, P.: Effects of dextran polydispersity on red blood cell aggregation. Colloid Polym. Sci. **263**, 478–483 (1985). doi:10.1007/BF01458338
38. Wang, M.S., Palmer, L.B., Schwartz, J.D., Razatos, A.: Evaluating protein attraction and adhesion to biomaterials with the atomic force microscope. Langmuir **20**, 7753–7759 (2004). doi:10.1021/la049849+
39. Liu, W., Yu, D., Yang, M.: Blood compatibility of thermoplastic polyurethane membrane immobilized with water-soluble chitosan/dextran sulfate. Colloids Surf. B Biointerfaces **44**, 82–92 (2005). doi:10.1016/j.colsurfb.2005.05.015
40. Perrino, C., Lee, S., Spencer, N.D.: A biomimetic alternative to PEG as an antifouling coating: resistance to non-specific protein adsorption of poly(L-lysine)-graft-dextran. Langmuir **24**, 8850–8856 (2008). doi:10.1021/la800947z
41. Green, R.J., Davies, M.C., Roberts, C.J., Roberts, S.J.B.J.: J. Biomed. Mater. Res. **42**, 165–171 (1998). doi :10.1002/(SICI)1097-4636(199811)42:2<165::AID-JBM1>3.0.CO;2-N
42. Lee, S., Iten, R., Müller, M., Spencer, N.D.: Influence of molecular architecture on the adsorption of poly(ethylene oxide)-poly(propylene oxide)-poly(ethylene oxide) (PEO-PPO-PEO) on PDMS surfaces and implications for aqueous lubrication. Macromolecules **37**, 8349 (2004). doi:10.1021/ma049076w
43. Vörös, J., Ramsden, J.J., Csucs, G., Szendro, I., De Paul, S.M., Textor, M., Spencer, N.D.: Optical grating coupler biosensors. Biomaterials **23**, 3699–3710 (2002). doi:10.1016/S0142-9612(02)00103-5
44. Hook, F., Vörös, J., Rodahl, M., Kurrat, R., Böni, P., Ramsden, J.J., Textor, M., Spencer, N.D., Tengvall, P., Gold, J.,

Kasemo, B.: A comparative study of protein adsorption on titanium oxide surfaces using in situ ellipsometry, optical waveguide lightmode spectroscopy, and quartz crystal microbalance/dissipation. Colloids Surf. B Biointerfaces **24**, 155–170 (2002). doi:10.1016/S0927-7765(01)00236-3

45. Kurrat, R., Textor, M., Ramsden, J.J., Böni, P., Spencer, N.D.: Instrumental improvements in optical waveguide light mode spectroscopy for the study of biomolecule adsorption. Rev. Sci. Instrum. **68**, 2172–2176 (1997). doi:10.1063/1.1148069
46. Rixman, M.A., Dean, D., Ortiz, C.: Nanoscale intermolecular interactions between human serum albumin and low grafting density surfaces of poly(ethylene oxide). Langmuir **19**, 9357–9372 (2003). doi:10.1021/la0340571
47. Kawaguchi, T., Hasegawa, M.: Structure of dextran-magnetite complex: relation between conformation of dextran chains covering core and its molecular weight. J. Mater. Sci. Mater. Med. **11**, 31–35 (2000). doi:10.1023/A:1008933601813
48. Görisch, S.M., Wachsmuth, M., Fejes Toth, K., Lichter, P., Rippe, K.: Histone acetylation increases chromatin accessibility. J. Cell Sci. **118**, 5825–5834 (2005). doi:10.1242/jcs.02689
49. Kawaguchi, S., Imai, G., Suzuki, J., Miyahara, A., Kitano, T., Ito, K.: Aqueous solution properties of oligo- and poly(ethylene oxide) by static light scattering and intrinsic viscosity. Polymer (Guildf.) **38**, 2885–2891 (1997). doi:10.1016/S0032-3861(96)00859-2
50. Lee, S., Müller, M., Heeb, R., Zürcher, S., Tosatti, S., Heinrich, M., Amstad, F., Pechmann, S., Spencer, N.D.: Self-healing behavior of a polyelectrolyte-based lubricant additive for aqueous lubrication of oxide materials. Tribol. Lett. **24**, 217–223 (2006). doi:10.1007/s11249-006-9121-9
51. Tyrode, E., Johnson, C., Kumpulainen, A., Rutland, M.W., Claesson, P.M.: Hydration state of nonionic surfactant monolayers at the liquid/vapor interface: structure determination by vibrational sum frequency spectroscopy. J. Am. Chem. Soc. **127**, 16848–16859 (2005). doi:10.1021/ja053289z
52. Smeeth, M., Spikes, H.A., Gunsel, S.: The formation of viscous surface films by polymer solutions: boundary or elastohydrodynamic lubrication? Tribol. Trans. **39**, 720–725 (1996). doi:10.1080/10402009608983589

Tribol Lett (2009) 34:201–210
DOI 10.1007/s11249-009-9424-8

ORIGINAL PAPER

Aqueous Lubrication of SiC and Si_3N_4 Ceramics Aided by a Brush-like Copolymer Additive, Poly(L-lysine)-*graft*-poly(ethylene glycol)

Whitney Hartung · Antonella Rossi · Seunghwan Lee · Nicholas D. Spencer

Received: 8 December 2008 / Accepted: 19 February 2009 / Published online: 10 March 2009

Abstract We have examined the adsorption properties of poly(L-lysine)-*graft*-poly(ethylene glycol) (PLL-*g*-PEG)—a brush-like polymer—on Si_3N_4 and SiC surfaces and determined its impact on the aqueous lubrication of Si_3N_4 and SiC at various speeds and applied loads. The addition of PLL-*g*-PEG in aqueous solution reduces the interfacial friction forces significantly for self-mated sliding contacts of these two ceramics, as compared to lubrication with water or buffer solution alone. For SiC, the improved lubricating performance by addition of PLL-*g*-PEG was apparent for all tested speeds (from 1.4 to 185 mm/s under 2 N load). For Si_3N_4, the effect was more apparent in the slow-speed regime ($\leq$20 mm/s under 2 N load) than in the high-speed regime (>100 mm/s), where extremely low coefficients of friction ($\mu \leq 0.006$) are readily achieved by aqueous buffer solution alone. It was further observed that the optimal lubricating effect with Si_3N_4 is achieved when the tribopairs are first run-in in polymer-free aqueous buffer to render the sliding surfaces smooth, after which the PLL-*g*-PEG copolymer is added to the buffer solution.

Keywords Lubrication · Ceramics · Silicon nitride · Silicon carbide · Aqueous lubrication · Polymer brushes · PLL-*g*-PEG

W. Hartung · A. Rossi · S. Lee · N. D. Spencer (✉)
Laboratory for Surface Science and Technology, Department of Materials, ETH Zurich, Wolfgang-Pauli-Strasse 10, 8093 Zurich, Switzerland
e-mail: spencer@mat.ethz.ch

A. Rossi
Dipartimento di Chimica Inorganica ed Analitica, Università degli Studi di Cagliari, Cittadella Universitaria di Monserrato, 09100 Cagliari, Italy

1 Introduction

Si_3N_4 and SiC ceramics display excellent tribological properties, such as high abrasive wear resistance and high critical loads to seizure [1], and are used in applications such as abrasives, mechanical seals, and various types of bearings. In particular, their resistance to corrosion [2] and the extremely low friction forces observed at high speeds in an aqueous environment [1–13], even lower than some metal–oil systems [2], have led to an interest in tribological applications of Si_3N_4 and SiC in water, such as in water pumps without separate oil lubrication.

The aqueous lubrication properties of Si_3N_4 are particularly interesting because very low coefficients of friction ($\mu < 0.001$ [2, 3, 11, 12]) can be achieved in water alone. These low μ values can only be achieved at high speeds (>40 mm/s) [3, 12] with an estimated water film thickness of 5–70 nm [3, 10]. However, if these conditions are not met, water does not act as a good base lubricant and μ can become as high as 0.7 [3, 14]. The unique self-lubricating properties of Si_3N_4 in water have been attributed to the tribochemically activated hydrolysis of SiO_2 present on the Si_3N_4 surface [13], but the precise cause of the low friction has been the subject of debate. One argument is that the dissolution of asperities, via SiO_2 hydrolysis, leads to extremely smooth surfaces that are conducive to hydrodynamic lubrication [3, 5–9, 12, 13]. Alternatively, it has also been argued that the readily sheared products of the tribochemical reaction remain on the surface, acting as a boundary lubricant [1, 4, 10].

SiC, which is generally harder, more brittle, more wear resistant, and chemically more inert than Si_3N_4 [9], has shown more consistent μ values (around 0.01) in water over a broad range of speeds and loads [2, 3, 11]. This higher degree of chemical inertness of SiC can be attributed to the difference in electronic properties of Si_3N_4 and SiC. Water has a band gap of 8.0 eV, and Si_3N_4 and SiC have band gaps of 5.0 and 2.8 eV, leading to differences compared to the band gap of water of 3.0 and 5.2 eV, respectively [11]. Consistent with the theory of Klopman, who proposed that a smaller band gap difference among reagents leads to a higher reaction rate [15], SiC requires a much longer running-in time and much higher sliding speeds (500 mm/s) [12] to yield μ values that are as low as those of Si_3N_4.

Despite the extremely favorable tribological properties of SiC and Si_3N_4 under high-speed conditions, a reduction of the coefficients of friction under low-speed conditions would be highly advantageous for further applications of the materials in water-lubricated bearings. Past approaches have included modifying the surface charge by adjusting the pH of the lubricant [16] or using additives such as ionic liquids [17]. An alternative, promising approach for aqueous lubrication of ceramics is to use water-soluble brush-like copolymers, which have recently been applied to several oxide-based tribosystems [18–30]. Poly(L-lysine)-*graft*-poly(ethylene glycol) (PLL-*g*-PEG) has a polycationic backbone that adsorbs onto negatively charged surfaces, such as SiO_2 under neutral pH conditions. Upon adsorption, the hydrated PEG side chains stretch out into the water, due to steric effects at high grafting densities, forming a brush-like layer over the entire surface. Under certain types of abrasive tribocontact, such as sliding contact in the presence of macroscopic asperities, PLL-*g*-PEG performs best when an excess amount of copolymer is present in the base aqueous lubricant. Although the copolymer layer is easily rubbed away during tribocontact, due to its relatively weak but reversible electrostatic attachment, the excess copolymer in the vicinity of the surface can rapidly readsorb onto the surface due to fast surface adsorption kinetics, reforming, or "self-healing" the lubricating brush film [18–20, 22–29].

The surfaces of SiC and Si_3N_4 tend to spontaneously form an oxide layer in air [8, 31]. Past work on the adsorption [18–20, 24, 32–35] and tribological properties [18–20, 22–29] of PLL-*g*-PEG on oxide surfaces, especially SiO_2, suggest that PLL-*g*-PEG is a promising candidate as an aqueous lubricant additive. To investigate the possible benefits of PLL-*g*-PEG as an aqueous lubricant additive, adsorption tests using ellipsometry (ELM) and X-ray photoelectron spectroscopy, as well as tribological tests at a variety of speeds, using a pin-on-disc tribometer, have been carried out.

2 Experimental

2.1 Materials

2.1.1 Polymer Synthesis

For the experiments in this study, PLL(20)-*g*[2.9]-PEG(5) and PLL(20)-*g*[3.6]-PEG(5), indicating a poly(L-lysine) backbone polymer of 20 kDa (including Br^- counterions) and poly(ethylene glycol) side chains of 5 kDa grafted onto every 2.9th or 3.6th lysine unit, were employed. Little difference in the adsorption and tribological performance is expected in these two polymers [34]. Complete details of the synthesis of PLL-*g*-PEG can be found in previous publications [32, 36, 37]. Briefly, 20 kDa poly(L-lysine) hydrobromide (Fluka, Switzerland) was dissolved at a concentration of 100 mM in a 50 mM sodium borate buffer solution adjusted to pH 8.5. The solution was filtered through a 0.22 μm filter. To graft PEG onto PLL, the *N*-hydroxysuccinimidyl ester of methoxypoly(ethylene glycol) propionic acid (mPEGSPA, Nektar, Huntsville, AL) was added to the PLL–HBr solution. The reaction was allowed to proceed for 6 h at room temperature, after which the reaction mixture was dialyzed (Spectra-Por, molecular weight cutoff size 6–8 kDa, Spectrum, Houston, TX) for 48 h against deionized water. The product was freeze dried and stored in powder form at −20 °C.

For all tribometry and adsorption experiments described in this article, PLL-*g*-PEG was dissolved at a concentration of 0.25 mg/mL in an aqueous buffer solution consisting of 1 mM 4-(2-hydroxyethyl)piperazine-1-ethanesulfonic acid (abbreviated HEPES 0, Fluka, Switzerland), adjusted to pH 7. For the adsorption measurements described in Sects. 2.2.1 and 2.2.2, samples were allowed to incubate for 30 min in the PLL-*g*-PEG solution to allow for polymer adsorption.

2.1.2 Preparation of Tribopairs

Discs of SiC (Hexaloy SA, Saint-Gobain, Northboro, MA) and Si_3N_4 (approximately 25 mm in diameter) were polished to a 1 μm finish with diamond paste. Along with 6-mm ball bearings of the same materials (polished by the manufacturer Saphirwerk AG, Brügg, Switzerland, and used as received), they were ultrasonicated in a series of solvents (heptane or pentane, acetone, ethanol, water, and ethanol again) for 10 min each to remove all possible contaminants from the polishing process. They were subsequently plasma cleaned in air plasma (Harrick Scientific, Ithaca, NY) for 2 min before use in the tribometer or the adsorption experiments with ELM.

2.2 Experimental Procedures

2.2.1 Polymer Adsorption Measurements with Ellipsometry (ELM)

ELM was performed with an M200-F Variable Angle Spectroscopic Ellipsometer (J. A. Woollam Co. Inc., Lincoln, NE) with a spectral range of 245–995 nm. Measurements were performed in air at angles of 65°,

70°, and 75° from the surface normal. Optical constants of clean, bare surfaces of Si_3N_4, SiC, and SiO_2 (silicon wafers with thermal oxide) were measured first, after which samples were incubated in a PLL-*g*-PEG solution (described in Sect. 2.1.1) for 30 min and rinsed with HEPES 0 and water, and dried in a stream of N_2. WVASE 32 analysis software was used to fit the data to multilayer models, using a Cauchy model ($A = 1.45$, $B = 0.01$, $C = 0$) [38].

2.2.2 Adsorption Measurements with XPS

The surface chemistry of the samples before and after polymer adsorption was investigated with X-ray photoelectron spectroscopy (XPS). The X-ray source of the PHI Quantera SXM (ULVAC-PHI, Chanhassen, MN) is a focused and scanned monochromatic AlKα beam with a diameter that can be chosen between 9 and 200 μm. The emitted electrons are collected and retarded with a lens system similar to the Omega lens available from Physical Electronics Inc. After passing a spherical capacitor energy analyzer, the electrons are detected by a 32-channel detector. The system is also equipped with a high-performance, floating-column ion gun and an electron neutralizer for charge compensation.

Small-area XPS spectra were collected with a beam size of 100 μm diameter with a power of 24.5 W in the constant analyzer energy mode using a pass energy of 26 eV and a step size of 0.05 eV. Under these conditions, the full-width-at-half-maximum (fwhm) for $Ag3d_{5/2}$ is 0.55 eV. Survey spectra were acquired with 280 eV pass energy and a step size of 1 eV. The whole set of spectra (detailed and survey spectra) was acquired within 30 min/area.

The residual pressure was always below 5×10^{-7} Pa. The system was calibrated according to ISO 15472:2001 with an accuracy of ±0.05 eV.

Measurements were taken on one clean, polished (not tribostressed) sample each of Si_3N_4 and SiC and one of each incubated for 30 min in a PLL-*g*-PEG solution (described in Sect. 2.1.1) for 30 min and rinsed with HEPES 0 and water, and dried in a stream of N_2. Three measurements were taken for the SiC coated with PLL-*g*-PEG.

Detailed spectral analyses were processed using Casa-XPS software (V2.3.12, Casa Software Ltd., UK). An iterated Shirley–Sherwood background subtraction was applied before peak fitting using a linear-least-squares algorithm. Minor charging was observed and corrected by referencing, for SiC, to the carbide component of the Si 2p 3/2 peak 100.8, as per Contarini et al. [39] and for Si_3N_4, to the hydrocarbon component of the C1s peak at 284.6, as per Bertoti et al. [40].

2.2.3 Pin-on-Disc Tribometry

All pin-on-disc experiments were performed on a CSM tribometer (Peseux, Switzerland) with self-mated tribopairs of either SiC or Si_3N_4.

2.2.3.1 Change of Lubricant A series of experiments was performed on Si_3N_4 by changing the lubricant during the test to examine the effects of tribological contact history on the value of μ. To change the lubricant from PLL(20)-*g*[2.9]-PEG(5) solution (see Sect. 2.1.1) to HEPES 0, the PLL-*g*-PEG solution was removed from the tribological testing cup, the sample and cup were dried with hot air, and HEPES 0 was added to the cup. As indicated in a previous study [24], an excess of PLL-*g*-PEG should be present in the lubricant to insure the self-healing of the PLL-*g*-PEG coating and thus effective lubrication. Therefore, a superficial layer remaining on the surface, if present, should be removed within a few rotations and have no persistent lubricating effect. To change from HEPES 0 to PLL-*g*-PEG solution, the HEPES 0 was removed from the cup and the PLL-*g*-PEG solution was added. These experiments were performed at 120 mm/s under 5 N applied load with no running-in procedure.

2.2.3.2 Speed Dependence A series of measurements was performed in which the speed dependence of the friction of self-mated pairs of Si_3N_4 and SiC was tested in solutions of HEPES 0 and PLL(20)-*g*[2.9]-PEG(5), as further described in Sect. 3.1. For Si_3N_4, experiments were performed under an applied load 2 N and a speed of 10 mm/s on different tracks with no running-in procedure. Another series of experiments was performed on samples of both Si_3N_4 and SiC. First, a running-in procedure of 100 rotations at a speed of 2 mm/s was performed, and the same ball position was used for all experiments on the same sample to insure a consistent pin geometry. Experiments on each sample were first performed in HEPES 0, then in PLL-*g*-PEG solution, at speeds of 185, 17, and 1.4 mm/s, with a different track for each speed and lubricant. Each pair of HEPES 0 and PLL-*g*-PEG experiments was performed on tracks of similar radii (±0.5 mm) to maintain similar rotational speeds and linear speeds among the different lubricants. These experiments were performed under an applied load of 2 N, and each group of experiments was repeated three to four times.

2.2.3.3 Wear Track Morphology The final series of experiments was performed to compare the surface topographies of SiC and Si_3N_4 after tribological contact in HEPES 0 buffer solution and in PLL-*g*-PEG solution. With a fresh area of the pin and fresh track for each test, 30,000 rotations were performed at 185 mm/s under 2 N applied load in either HEPES 0 or PLL(20)-*g*[3.6]-PEG(5) solution. The resulting surface topography was examined by AFM (Dimension II, Digital Instruments, Santa Barbara, CA) using a sharp Si_3N_4 tip with a triangular cantilever of spring constant 0.12 N/m. Wear tracks on the pins and discs were measured with an Axioskop 2 plus optical microscope and AxioVision software (Carl Zeiss Inc., Jena, Germany).

3 Results

3.1 Adsorption Measurements

3.1.1 XPS

The data collected for Si_3N_4 can be seen in Fig. 1 and Table 1, with the C–C component of the C1s peak at 284.6 eV used as an internal reference. The Si 2p3/2 peaks at 101.6 and 103.5 eV closely match the analysis of Bertoti, who measured a Si–N component at 101.8 eV and a Si–O component at 103.5 eV [40]. This differs from the analysis of Hah, who reports Si–O around 101 eV, a SiO_xN_y at 102 eV, and a SiO_aN_b/SiO_2 species at 103 eV [40]. The O1s signal consisted of a major contribution at 532.1 eV, corresponding to an O–Si bond, measured at 532.3 eV by Bertoti. An additional minor contribution at 530.9 eV, too low to be an SiO_xN_y species [41], which can be attributed to hydroxide or carbonate bonded to a NH_3 species, corresponds to the value of 530.8 eV measured by Bertoti. A final contribution to the O1s spectrum at 533.1 eV can be ascribed to adsorbed water or ethanol remaining from the cleaning process or possibly the Na Auger contribution, which was present in the full spectra as a minor contaminant. The N1s spectrum contained a NSi_3 contribution at 397.5 eV, but a shoulder corresponding to the NH_2 contribution at 399.3 eV, as reported by Bertoti, was not found.

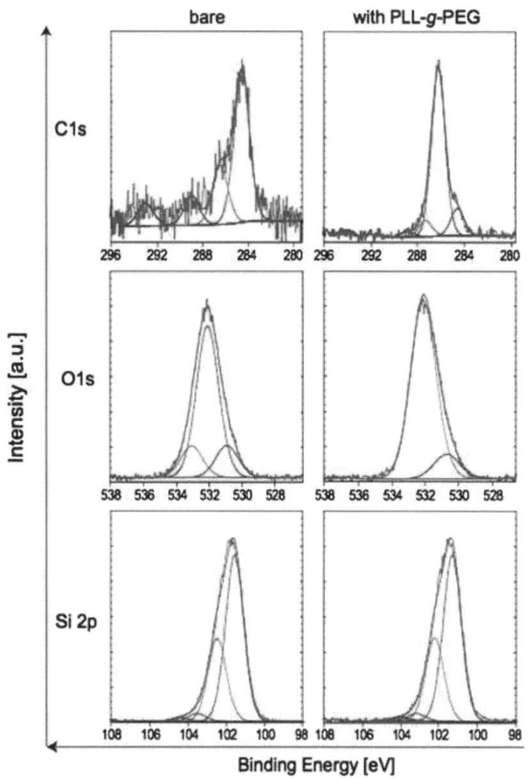

Fig. 1 XPS measurements of Si_3N_4 before and after adsorption of PLL-*g*-PEG

After PLL-*g*-PEG adsorption there is good agreement with the values of Huang et al. [33] (referred to C1s at 284.6 eV). The O1s-Si peak overlaps with the NHC=O,

Table 1 XPS analysis of PLL-*g*-PEG adsorbed onto Si_3N_4 with 0.4 eV subtracted from the values of Huang et al. [33] to be consistent with a binding energy of 284.6 eV for hydrocarbon C1s

Element	Bond	Bare Si_3N_4, measured	Si_3N_4 with PLL-*g*-PEG, measured	Si_3N_4 by Bertoti et al. [40]	PLL-*g*-PEG referred to Cls at 284.6 eV [33]
C1s	C–C/C–H	284.6	284.6	284.6	284.6
	C–O/C–N		286.2		286.1/285.8
	NHC=O		287.3		287.7
O1s	O–Si and NHC=O	532.1	532.2	532.3	531
	OH/CO_3 to NH_4 and C–O	530.9	530.8	530.8	532.4
	Contamination (H_2O, ethanol, or Na Auger)	533.1			
Si 2p 3/2	Si–N	101.6	101.4	101.8	
	Si–O	103.5	103.2	103.5	

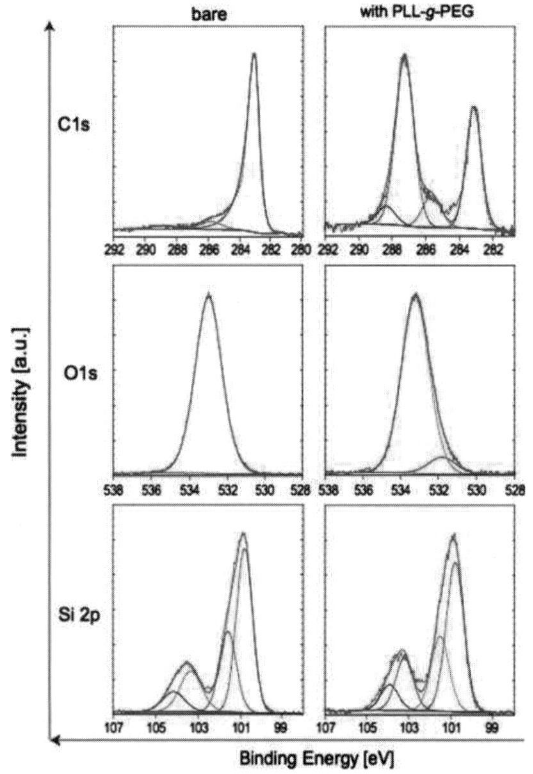

Fig. 2 XPS measurements of SiC before and after adsorption of PLL-*g*-PEG

which, at 532.2 eV, still matches well with that of the bare Si_3N_4 (532.1 eV) and the measurements of Bertoti (532.3 eV), but is higher than that of the theoretical value of 531.0 eV of Huang. In Fig. 1, a clear C–O peak appears upon adsorption of PLL-*g*-PEG. Notable shifts include a shift to lower binding energies in the C1s contribution from the NHC=O, from 287.3 to 287.7 eV, and small chemical shifts in the Si 2p 3/2 of 0.4 eV for the SiN_4 component and 0.3 eV for the SiO_4 component.

The results of the SiC analysis are provided in Fig. 2 and Table 2. In the case of SiC, the Si–C component of the Si 2p 3/2 at 100.8 eV was used for internal calibration, yielding the Si–O component of the Si 2p 3/2 at 103.3 eV, correlating well with the results of Contarini et al. [39]. However, slight discrepancies exist between the present analysis with those of Contarini, including a slightly higher value (285.6 eV) for the C–C/C–H components of the C1s signal than Contarini (285 eV), as well as a slightly higher O1s signal corresponding to the O–Si bond (533 eV compared to 532.3 eV). Both of these discrepancies can probably be attributed to the different forms of SiC used in the analysis—sintered discs in this study and powders by Contarini.

Following adsorption of PLL-*g*-PEG, the value of the Si–O component of the Si 3p 2/3 peak remains within 0.1 eV of the previously measured value of the bare SiC as well as the value of Contarini. The C–Si component of the C1s peak shows a slight shift to 283.2 ± 0.03 eV. The C–C/C–H component is found at 285.6 ± 0.05 eV, which is 0.6 eV higher than the bare SiC value reported by Contarini et al. [39] or the theoretical PLL-*g*-PEG value reported by Huang et al. [33], and is shifted by about 1.6 eV from the value measured on bare SiC. Additionally, a C–O/C–N component at 287.2 ± 0.05 eV is shifted 1.4 eV from the measurement on bare SiC, and is around 0.7 eV higher than the theoretical value of the C–O component for PLL-*g*-PEG. Small N1s peaks are measured at 400.7 and 402.4 eV.

3.1.2 ELM

ELM results, in general, show that dry film thicknesses of adsorbed PLL-*g*-PEG on Si_3N_4 and SiC are similar to the amounts seen on silicon wafers. Si_3N_4 and SiC have adsorption amounts of 15.7 ± 1.6 and 16.8 ± 0.7 nm,

Table 2 XPS analysis of PLL-*g*-PEG adsorbed onto SiC (using an internal calibration of Si 2p 3/2 to 100.8 eV binding energy)

Element	Bond	Bare SiC, measured	SiC with PLL-*g*-PEG, measured (average of three measurements)	SiC, bare by Contarini et al. [39]	Theoretical PLL-*g*-PEG by Huang et al. [33]
C1s	C–Si	283	283.2 ± 0.03	282.8	
	C–C/C–H	284	285.6 ± 0.05	285	285
	C–O/C–N	285.8	287.2 ± 0.05		286.5(/286.2)
	NHC=O		288.3 ± 0.06		288.1
O1s	O–Si	533	533.2 ± 0.02	532.3	
	C/NHC=O		531.8 ± 0.02		531.4/532.8
Si 3p 3/2	Si–C	100.8	100.8	100.8	
	Si–O	103.3	103.2 ± 0.02	103.3	

respectively, which is within the same range of adsorption as for PLL-*g*-PEG on SiO_2 (17.5 ± 0.7 nm).

3.2 Pin-on-Disc

3.2.1 Change of Lubricant

Figure 3 details an example of the evolution of the coefficient of friction (μ) on a self-mated sliding tribopair of Si_3N_4 at 120 mm/s under an applied load of 5 N. After an initial running-in period in HEPES 0 (<10,000 rotations), μ stabilized at around 0.02. After the addition of PLL-*g*-PEG solution, μ immediately decreased to about 0.003. Upon subsequent removal of the PLL-*g*-PEG, μ increased to around 0.009. In short, low μ obtained in buffer solution can be further lowered with the addition of PLL-*g*-PEG. However, when the experiments were performed with the changing of the lubricants in reverse order, the results are somewhat different. Figure 4 shows an initial μ value of a self-mated sliding contact of Si_3N_4 tribopair in PLL-*g*-PEG solution at 5 N applied load and 120 mm/s of around 0.03 (from ca. 5,000 to 20,000 rotation). Upon removal of the PLL-*g*-PEG solution and addition of HEPES 0 (at ca. 21,000 rotations), the μ is reduced to 0.01. After changing back to PLL-*g*-PEG solution (ca. at 40,000 rotations), μ is slightly reduced to 0.008.

3.2.2 Speed Dependence

Figure 5 details μ of Si_3N_4 at a slower speed of 10 mm/s with and without PLL-*g*-PEG in the aqueous lubricant at 2 N, with no running-in procedure performed. At this slow speed, μ was reduced tenfold from 0.4 with buffer alone to 0.04 with PLL-*g*-PEG solution. As shown in Fig. 6, the series of Si_3N_4 measurements conducted after running in reveals a clear reduction in μ in PLL-*g*-PEG solution under loads of 2 N at lower speeds. The dramatic reduction of μ is observed at speeds of 1.4 mm/s (0.91–0.20) and 17 mm/s (0.87–0.17). However, Si_3N_4 lubricated with buffer solution at high speed yielded a slightly lower μ value (0.02) than that of Si_3N_4 in PLL-*g*-PEG (0.04); this observation is consistent with Fig. 4, in which the μ values are slightly higher in PLL-*g*-PEG solution if the tribological contact occurs first in PLL-*g*-PEG solution, followed by contact in HEPES solution. PLL-*g*-PEG is thus most effective at low-to-medium speeds, which occur during start-up and in reciprocal motion.

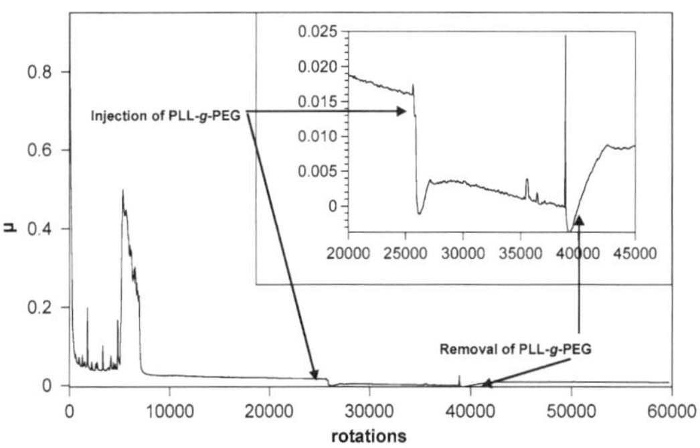

Fig. 3 Pin-on-disc experiment of Si_3N_4 pin against a Si_3N_4 disc at 185 mm/s under 5 N applied load. The experiment began with HEPES 0 buffer as the lubricant. The buffer was then replaced with PLL-*g*-PEG solution, after which an immediate reduction in friction can be seen. After the PLL-*g*-PEG is replaced with HEPES 0, the friction increases slightly

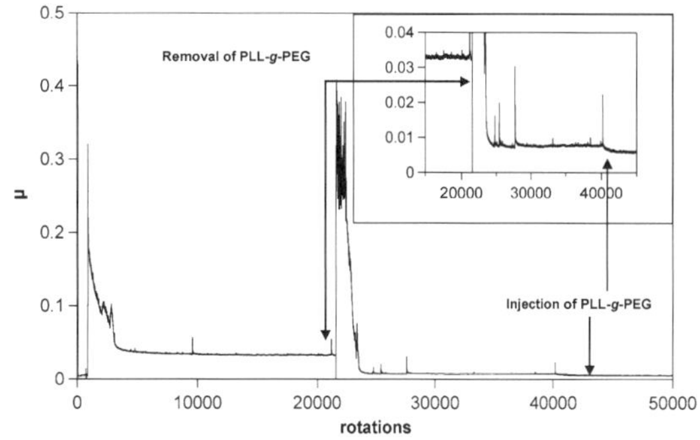

Fig. 4 Pin-on-disc experiment of a Si_3N_4 pin against a Si_3N_4 disc at 185 mm/s under 5 N applied load. The experiment began with PLL-*g*-PEG as the lubricant. When the PLL-*g*-PEG was replaced with HEPES 0, after an initial period of running-in, μ decreased. After the HEPES 0 was then replaced with PLL-*g*-PEG solution, μ decreased further

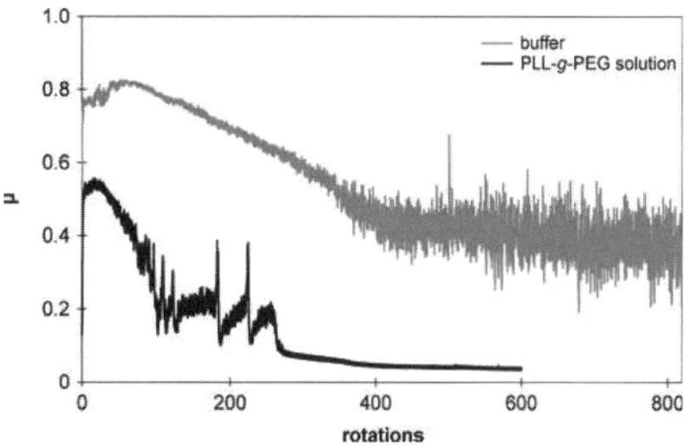

Fig. 5 Coefficient of friction (μ) of self-mated Si_3N_4 at a speed of 10 mm/s and applied force of 2 N lubricated with buffer and with PLL-*g*-PEG solution

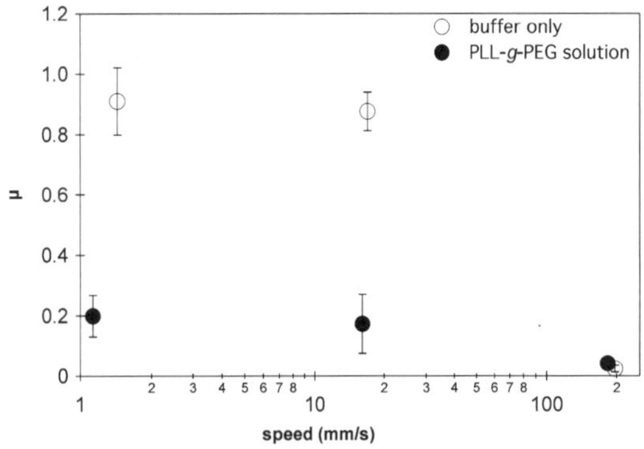

Fig. 6 Average coefficient of friction (μ) of self-mated Si_3N_4 under 2 N applied load at various speeds in buffer and in PLL-*g*-PEG solution

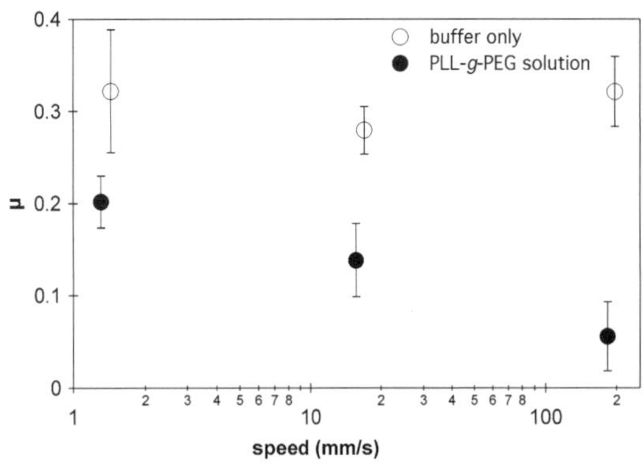

Fig. 7 Average coefficient of friction (μ) of SiC against SiC under 2 N applied load at various speeds in buffer and in PLL-*g*-PEG solution

PLL-*g*-PEG reduces μ in each of the studied speed regimes on SiC, as can be observed in Fig. 7. Although the μ values for SiC in buffer show no apparent speed dependence, a clearly decreasing trend of μ (from 0.20 to 0.06) with increasing speed (from 1.4 to 185 mm/s) is observed for the samples lubricated with PLL-*g*-PEG solution.

3.2.3 Wear-Track Morphology

A dramatic difference in the appearance of the wear track can be seen in the AFM images in Fig. 8. In comparison to the as-polished surface in Fig. 8 (left), the surface of the sliding track in HEPES 0 buffer is noticeably smoother; the microcracks, presumably created during the polishing process prior to testing, appear to have been significantly polished away, tribochemically. The topography of the Si_3N_4 surface lubricated with PLL-*g*-PEG shows a number of relatively large pits, larger than those of the as-polished surface, with smooth areas in between, suggesting both tribochemical polishing of the surface as well as microfracture induced by the tribocontact. Figure 9 (left) details the values for average roughness (R_a) and root-mean-square roughness (RMS) of the wear scars of the Si_3N_4 discs. In both cases, the roughness values for Si_3N_4 lubricated with PLL-*g*-PEG are approximately twice as high as the values for Si_3N_4 lubricated in HEPES 0. Finally, in Fig. 9 (middle), Si_3N_4 lubricated with HEPES 0 has wear scars approximately twice the width or diameter of Si_3N_4 lubricated with PLL-*g*-PEG. For SiC, little difference can be seen in the morphology of the wear tracks lubricated with HEPES 0 solution and with PLL-*g*-PEG solution, as seen in Fig. 10. No difference was seen in the roughness for Fig. 10 (middle) and (right), in which R_a was approximately 52 nm and RMS was around 65 nm.

4 Discussion

According to the ELM (Sect. 3.1.2) and XPS measurements (Figs. 1, 2; Tables 1, 2), PLL-*g*-PEG adsorbs onto both Si_3N_4 and SiC surfaces. Film thicknesses on both ceramics are similar to that measured on a Si wafer, and the typical components of PLL-*g*-PEG are visible within the XPS spectra. Possible chemical shifts of 0.2–0.3 eV in the SiN_4 and SiO_4 components of the Si 2p 3/2 peak may indicate that the adsorption of PLL-*g*-PEG occurs not only through the interaction with oxygen but with the nitrogen as well, suggesting that PLL-*g*-PEG could be used on other nonoxide surfaces, not only for lubrication but also for

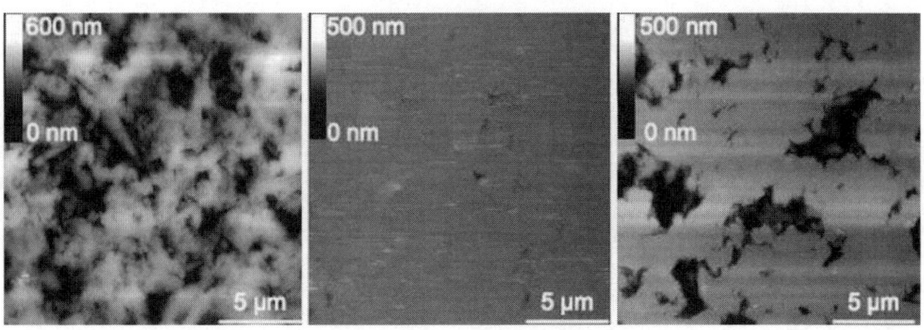

Fig. 8 Topography of the as-polished Si_3N_4 disc (*left*) as well as the wear track surface of Si_3N_4 after 30,000 rotations with 5 N applied force and 185 mm/s linear speed in HEPES 0 (*middle*) and PLL-*g*-PEG solution (*right*); larger fracture pits can be seen in the wear track of the sample lubricated with PLL-*g*-PEG solution

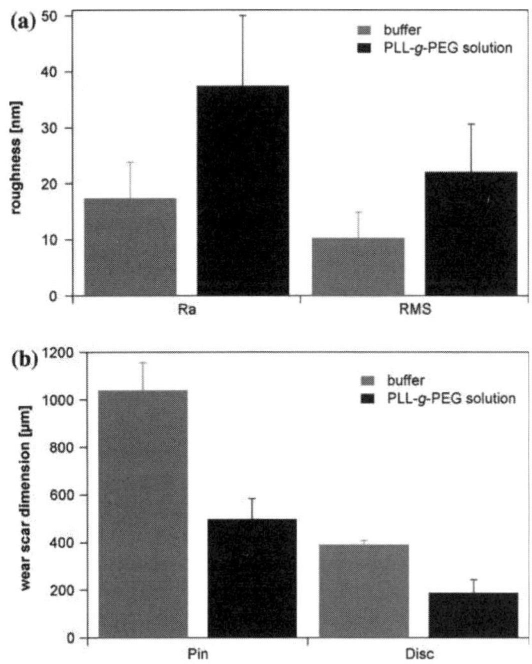

Fig. 9 Comparison of (**a**) R_a and RMS of the wear tracks and (**b**) dimensions of the wear scars (pin diameter and disc scar width) of Si_3N_4 pin-on-disc experiments of 30,000 rotations at speeds of 185 mm/s and 5 N applied load, lubricated with buffer and with PLL-*g*-PEG solution

prevention of nonspecific protein adsorption. However, it was not possible to detect any chemical shifts for SiC with the present analysis.

The morphological features of wear scars examined with AFM (Figs. 8, 9) suggest that tribochemical polishing occurs to a much lesser degree when the ceramic surfaces are lubricated with PLL-*g*-PEG than with HEPES 0; since former studies have observed extremely smooth surfaces when tribochemical polishing occurs [3, 5, 6, 10, 12, 13], it can be assumed that the smoother, lower-friction surfaces must be the result of the tribochemical polishing reactions. Below a certain threshold speed, which is around 40 mm/s for Si_3N_4 [3, 12], wear occurs predominately via microfracture, whether it is in HEPES or in a PLL-*g*-PEG solution. At high speeds, the essential tribochemical reaction is believed to be partially inhibited by the PLL-*g*-PEG layer, causing wear to occur via both microfracture and tribochemical reaction. This is consistent with a previous study [42], in which the tribochemical reactions of Si_3N_4 with water to produce a smooth morphology are inhibited by the addition of NaCl, due to the strong interaction of Na^+ ions with the surface. The narrower wear scars observed for PLL-*g*-PEG-lubricated Si_3N_4 may further indicate an inhibition of tribochemical wear. Tribochemical wear occurs mostly at asperities; wear scar edges are also more chemically active than other surfaces, allowing the wear scars to propagate laterally. In the case of wear via microfracture, wear would only take place under the contact area vertically, and propagation at the edges would be slower.

In contrast to the distinct topographies of the differently lubricated Si_3N_4 surfaces, little difference in the wear tracks of SiC lubricated with HEPES 0 and PLL-*g*-PEG solution can be seen in the AFM images in Fig. 10. Similarly, both R_a and RMS roughness are around 90 nm for both HEPES 0-lubricated and PLL-*g*-PEG-lubricated samples. Since tribochemical wear is not expected at speeds lower than 500 mm/s [12], no difference is expected in the topography or the wear mechanisms of SiC lubricated with water or with PLL-*g*-PEG solution. Since tribochemical smoothening of SiC is only observed at speeds higher than 500 mm/s, and wear occurs predominantly via microfracture, it is not expected that the behavior of SiC observed is similar to that of Si_3N_4 at low speeds, where PLL-*g*-PEG is most effective as an aqueous lubricant additive.

The adsorption of PLL-*g*-PEG as characterized by XPS (Tables 1, 2), the dry film thickness (Sect. 3.1.2), and the reduction of friction at lower speeds for Si_3N_4 (Fig. 6) and at all tested speeds for SiC (Fig. 7) are all consistent with the view that the lubrication mechanism of PLL-*g*-PEG on these ceramic surfaces is similar to those discussed in previous investigations on oxide surfaces [18, 19, 22–25,

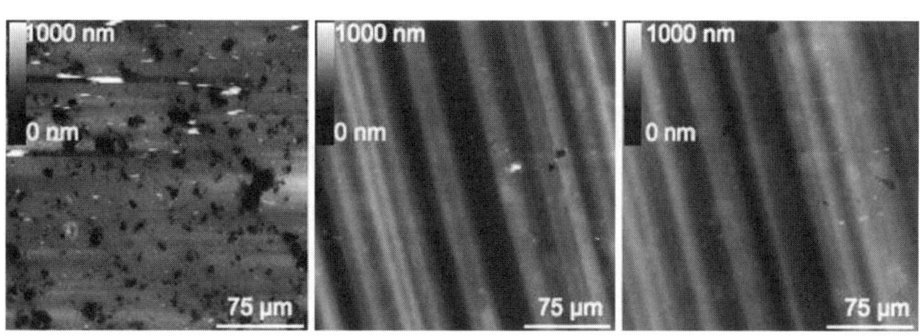

Fig. 10 Topography of the as-polished SiC disc (*left*) as well as the wear track surface of SiC after 30,000 rotations with 5 N applied force and 185 mm/s linear speed in HEPES 0 (*middle*) and PLL-g-PEG solution (*right*)

27–29, 43, 44]; PLL-*g*-PEG acts as a boundary lubricant additive via adsorption onto the ceramic surfaces and maintains the lubricating effect through a "self-healing" mechanism [24]. The only exception to this apparent boundary lubrication is in the case of sliding contact at high speed for Si_3N_4, as shown in Fig. 2. The magnitude of reduction in μ values by PLL-*g*-PEG under these conditions is less pronounced, mainly because the extremely low friction forces are already obtained upon lubrication in aqueous buffer solution alone. On the other hand, the fact that PLL-*g*-PEG further lowers the already-low friction achieved with buffer may suggest that a purely hydrodynamic lubrication regime is not reached with water alone under the conditions employed in this study.

5 Conclusions

PLL-*g*-PEG is an effective aqueous lubricant additive that can adsorb onto Si_3N_4 and SiC and lower the coefficient of friction at all speeds with SiC and at moderate-to-low speeds on Si_3N_4. At higher speeds on Si_3N_4, PLL-*g*-PEG partially inhibits the tribochemical reactions that lead to smooth surfaces, resulting in a combination of wear via microfracture and tribochemical polishing, and leading to rougher surfaces. However, coefficients of friction measured on Si_3N_4 in PLL-*g*-PEG solution are nearly as low as those in water alone, and can be further lowered at high speeds with PLL-*g*-PEG after performing a running-in procedure in water.

Acknowledgment The authors would like to thank the Research Commission of the ETH Zurich for funding.

References

1. Wong, H.C., Umehara, N., Kato, K.: The effect of surface roughness on friction of ceramics sliding in water. Wear **218**, 237–243 (1998). doi:10.1016/S0043-1648(98)00208-7
2. Wang, X., Kato, K., Adachi, K.: The critical condition for the transition from HL to ML in water-lubricated SiC. Tribol. Lett. **16**, 253–258 (2004). doi:10.1023/B:TRIL.0000015200.37141.f5
3. Tomizawa, H., Fischer, T.E.: Friction and wear of silicon nitride and silicon carbide in water: hydrodynamic lubrication at low sliding speed obtained by tribochemical wear. Tribol. Trans. **30**, 41–46 (1987). doi:10.1080/05698198708981728
4. Sasaki, S.: The effects of the surrounding atmosphere on the friction and wear of alumina, zirconia, silicon-carbide and silicon-nitride. Wear **134**, 185–200 (1989). doi:10.1016/0043-1648(89)90068-9
5. Hah, S.R., Fischer, T.E.: Tribochemical polishing of silicon nitride. J. Electrochem. Soc. **145**, 1708–1714 (1998). doi:10.1149/1.1838544
6. Muratov, V.A., Luangvaranunt, T., Fischer, T.E.: The tribochemistry of silicon nitride: effects of friction, temperature and sliding velocity. Tribol. Int. **31**, 601–611 (1998). doi:10.1016/S0301-679X(98)00081-4
7. Muratov, V.A., Olsen, J.E., Gallois, B.M., Fischer, T.E., Bean, J.C.: Tribochemical reactions of silicon: an in situ infrared spectroscopy characterization. J. Electrochem. Soc. **145**, 2465–2470 (1998). doi:10.1149/1.1838661
8. Hah, S.R., Burk, C.B., Fischer, T.E.: Surface quality of tribochemically polished silicon nitride. J. Electrochem. Soc. **146**, 1505–1509 (1999). doi:10.1149/1.1391795
9. Zhu, Z.Z., Muratov, V., Fischer, T.E.: Tribochemical polishing of silicon carbide in oxidant solution. Wear **229**, 848–856 (1999). doi:10.1016/S0043-1648(98)00392-5
10. Xu, J.G., Kato, K.: Formation of tribochemical layer of ceramics sliding in water and its role for low friction. Wear **245**, 61–75 (2000). doi:10.1016/S0043-1648(00)00466-X
11. Chen, M., Kato, K., Adachi, K.: The difference in running-in period and friction coefficient between self-mated Si_3N_4 and SiC under water lubrication. Tribol. Lett. **11**, 23–28 (2001). doi:10.1023/A:1016621929078
12. Jordi, L., Iliev, C., Fischer, T.E.: Lubrication of silicon nitride and silicon carbide by water: running in, wear and operation of sliding bearings. Tribol. Lett. **17**, 367–376 (2004). doi:10.1023/B:TRIL.0000044485.77019.fb
13. Muratov, V.A., Fischer, T.E.: Tribochemical polishing. Annu. Rev. Mater. Sci. **30**, 27–51 (2000). doi:10.1146/annurev.matsci.30.1.27
14. Jahanmir, S., Fischer, T.E.: Friction and wear of silicon nitride lubricated by humid air, water, hexadecane and hexadecane + 0.5 percent stearic acid. Tribol. Trans. **31**, 32–43 (1988). doi:10.1080/10402008808981795
15. Klopman, G.: Chemical reactivity and concept of charge- and frontier-controlled reactions. J. Am. Chem. Soc. **90**, 223–234 (1968). doi:10.1021/ja01004a002
16. Kalin, M., Novak, S., Vizintin, J.: Surface charge as a new concept for boundary lubrication of ceramics with water. J. Phys. D Appl. Phys. **39**, 3138–3149 (2006). doi:10.1088/0022-3727/39/15/S03
17. Phillips, B.S., Zabinski, J.S.: Ionic liquid lubrication effects on ceramics in a water environment. Tribol. Lett. **17**, 533–541 (2004). doi:10.1023/B:TRIL.0000044501.64351.68
18. Lee, S., Muller, M., Ratoi-Salagean, M., Voros, J., Pasche, S., De Paul, S.M., Spikes, H.A., Textor, M., Spencer, N.D.: Boundary lubrication of oxide surfaces by poly(L-lysine)-*g*-poly(ethylene glycol) (PLL-*g*-PEG) in aqueous media. Tribol. Lett. **15**, 231–239 (2003). doi:10.1023/A:1024861119372
19. Muller, M., Lee, S., Spikes, H.A., Spencer, N.D.: The influence of molecular architecture on the macroscopic lubrication properties of the brush-like co-polyelectrolyte poly(L-lysine)-*g*-poly(ethylene glycol) (PLL-*g*-PEG) adsorbed on oxide surfaces. Tribol. Lett. **15**, 395–405 (2003). doi:10.1023/B:TRIL.0000003063.98583.bb
20. Yan, X., Perry, S.S., Spencer, N.D., Pasche, S., De Paul, S.M., Textor, M., Lim, M.S.: Reduction of friction at oxide interfaces upon polymer adsorption from aqueous solutions. Langmuir **20**, 423–428 (2004). doi:10.1021/la035785b
21. Lee, S., Voros, J.: An aqueous-based surface modification of poly(dimethylsiloxane) with poly(ethylene glycol) to prevent biofouling. Langmuir **21**, 11957–11962 (2005). doi:10.1021/la051932p
22. Muller, M.T., Yan, X., Lee, S., Perry, S.S., Spencer, N.D.: Preferential solvation and its effect on the lubrication properties of a surface-bound, brushlike copolymer. Macromolecules **38**, 3861–3866 (2005). doi:10.1021/ma047468x
23. Muller, M.T., Yan, X., Lee, S., Perry, S.S., Spencer, N.D.: Lubrication properties of a brushlike copolymer as a function of the amount of solvent absorbed within the brush. Macromolecules **38**, 5706–5713 (2005). doi:10.1021/ma0501545
24. Lee, S., Muller, M., Heeb, R., Zurcher, S., Tosatti, S., Heinrich, M., Amstad, F., Pechmann, S., Spencer, N.D.: Self-healing

behavior of a polyelectrolyte-based lubricant additive for aqueous lubrication of oxide materials. Tribol. Lett. **24**, 217–223 (2006). doi:10.1007/s11249-006-9121-9

25. Lee, S., Spencer, N.D.: Aqueous lubrication of polymers: influence of surface modification. Tribol. Int. **38**, 922–930 (2006). doi:10.1016/j.triboint.2005.07.017
26. Drobek, T., Spencer, N.D.: Nanotribology of surface-grafted PEG layers in an aqueous environment. Langmuir **24**, 1484–1488 (2007). doi:10.1021/la702289n
27. Lee, S., Spencer, N.D.: Poly(L-lysine)-graft-poly(ethylene glycol): a versatile aqueous lubricant additive for tribosystems involving thermoplastics. Lubr. Sci. **20**, 21–34 (2007). doi:10.1002/ls.50
28. Lee, S., Spencer, N.D.: Achieving ultralow friction by aqueous, brush-assisted lubrication. In: Erdemir, A., Martin, J.M. (eds.) Superlubricity, pp. 365–396. Elsevier, Amsterdam (2007)
29. Lee, S., Spencer, N.D.: Materials science. Sweet, hairy, soft, and slippery. Science **319**, 575–576 (2008). doi:10.1126/science.1153273
30. Hartung, W., Drobek, T., Lee, S., Zürcher, S., Spencer, N.: The influence of anchoring-group structure on the lubricating properties of brush-forming graft copolymers in an aqueous medium. Tribol. Lett. **31**, 119–128 (2008). doi:10.1007/s11249-008-9344-z
31. Du, H., Tressler, R.E., Spear, K.E.: Thermodynamics of the Si-N-O system and kinetic modeling of oxidation of Si_3N_4. J. Electrochem. Soc. **136**, 3210–3215 (1989). doi:10.1149/1.2096427
32. Kenausis, G.L., Voros, J., Elbert, D.L., Huang, N., Hofer, R., Ruiz-Taylor, L., Textor, M., Hubbell, J.A., Spencer, N.D.: Poly(L-lysine)-*g*-poly(ethylene glycol) layers on metal oxide surfaces: attachment mechanism and effects of polymer architecture on resistance to protein adsorption. J. Phys. Chem. B **104**, 3298–3309 (2000). doi:10.1021/jp993359m
33. Huang, N.P., Michel, R., Voros, J., Textor, M., Hofer, R., Rossi, A., Elbert, D.L., Hubbell, J.A., Spencer, N.D.: Poly(L-lysine)-g-poly(ethylene glycol) layers on metal oxide surfaces: surface-analytical characterization and resistance to serum and fibrinogen adsorption. Langmuir **17**, 489–498 (2001). doi:10.1021/la000736+
34. Pasche, S., De Paul, S.M., Voros, J., Spencer, N.D., Textor, M.: Poly(L-lysine)-graft-poly(ethylene glycol) assembled monolayers on niobium oxide surfaces: a quantitative study of the influence of polymer interfacial architecture on resistance to protein adsorption by ToF-SIMS and in situ OWLS. Langmuir **19**, 9216–9225 (2003). doi:10.1021/la034111y
35. Wagner, M.S., Pasche, S., Castner, D.G., Textor, M.: Characterization of poly(L-lysine)-*graft*-poly(ethylene glycol) assembled monolayers on niobium pentoxide substrates using time-of-flight secondary ion mass spectrometry and multivariate analysis. Anal. Chem. **76**, 1483–1492 (2004). doi:10.1021/ac034873y
36. Elbert, D.L., Hubbell, J.A.: Self-assembly and steric stabilization at heterogeneous, biological surfaces using adsorbing block copolymers. Chem. Biol. **5**, 177–183 (1998). doi:10.1016/S1074-5521(98)90062-X
37. Sawhney, A.S., Hubbell, J.A.: Poly(ethylene oxide)-graft-poly(L-lysine) copolymers to enhance the biocompatibility of poly(L-lysine)-alginate microcapsule membranes. Biomaterials **13**, 863–870 (1992). doi:10.1016/0142-9612(92)90180-V
38. Azzam, R.M.A., Bashara, N.M.: Ellipsometry and Polarized Light. North-Holland, Amsterdam (1987)
39. Contarini, S., Howlett, S.P., Rizzo, C., De Angelis, B.A.: XPS study on the dispersion of carbon additives in silicon carbide powders. Appl. Surf. Sci. **51**, 177–183 (1991). doi:10.1016/0169-4332(91)90400-E
40. Bertoti, I., Varsanyi, G., Mink, G., Szekely, T., Vaivads, J., Millers, T., Grabis, J.: XPS characterization of ultrafine Si_3N_4 powders. Surf. Interface Anal. **12**, 527–530 (1988). doi:10.1002/sia.740121004
41. Raider, S.I., Flitsch, R., Aboaf, J.A., Pliskin, W.A.: Surface oxidation of silicon nitride films. J. Electrochem. Soc. **123**, 560–565 (1976). doi:10.1149/1.2132877
42. Saito, T., Imada, Y., Honda, F.: An analytical observation of the tribochemical reaction of silicon nitride sliding with low friction in aqueous solutions. Wear **205**, 153–159 (1997). doi:10.1016/S0043-1648(96)07287-0
43. Lee, S., Iten, R., Muller, M., Spencer, N.D.: Influence of molecular architecture on the adsorption of poly(ethylene oxide)-poly(propylene oxide)-poly(ethylene oxide) on PDMS surfaces and implications for aqueous lubrication. Macromolecules **37**, 8349–8356 (2004). doi:10.1021/ma049076w
44. Drobek, T., Spencer, N.D., Heuberger, M.: Compressing PEG brushes. Macromolecules **38**, 5254–5259 (2005). doi:10.1021/ma0504217

9124 *Macromolecules* **2009**, *42*, 9124–9132
DOI: 10.1021/ma901607w

Macromolecules
ARTICLE

Room-Temperature, Aqueous-Phase Fabrication of Poly(methacrylic acid) Brushes by UV-LED-Induced, Controlled Radical Polymerization with High Selectivity for Surface-Bound Species

Raphael Heeb, Robert M. Bielecki, Seunghwan Lee, and Nicholas D. Spencer*

Laboratory for Surface Science and Technology, Department of Materials, ETH Zurich, Wolfgang-Pauli-Strasse 10, CH-8093 Zurich, Switzerland

Received July 22, 2009; Revised Manuscript Received September 4, 2009

ABSTRACT: Poly(methacrylic acid) (PMAA) brushes were grafted from Si/SiO_2 substrates by means of immobilized-photoiniferter-mediated controlled radical polymerization. The employed UV setup was based on ultraviolet light-emitting diodes (UV-LEDs), which allowed for a precise control of the brush height with irradiation time, as observed by *in situ* quartz crystal microbalance experiments with dissipation monitoring (QCM-D). In contrast to many alternative approaches, it was shown that the novel UV source in combination with a photoiniferter renders lengthy postcleaning steps of the synthesized brushes unnecessary. Following characterization of the polymer layers by means of variable-angle spectroscopic ellipsometry (VASE) and static contact angle measurements, the lubrication properties of the PMAA brushes were investigated in macroscopic tribological experiments under low-contact-pressure, aqueous conditions. Results indicated that PMAA brushes have the potential to dramatically reduce sliding friction in an aqueous environment.

Introduction

Surface modifications by means of polymer brushes represent a very attractive tool for the tailoring and control of interfacial properties such as adhesion, friction, wettability, or biocompatibility.[1−3] Two principal experimental approaches are available to generate polymer brushes on solid substrates, namely "grafting to" and "grafting from".[4,5] While both approaches have their own unique characteristics, "grafting from" methods are known to be more suited to producing a high grafting density of chains because the polymers are generated *in situ* from a surface with a high density of initiating species. While "grafting to" approaches are more experimentally straightforward, they are restricted in terms of grafting densities, mainly due to the diffusion-limited adsorption of preformed polymers.[6] In recent years, numerous experimental strategies have been developed for the preparation of polymer brushes by means of "grafting from" methods.[4,7,8] Among them, controlled radical polymerization (CRP) strategies such as atom-transfer radical polymerization (ATRP),[9] nitroxide-mediated polymerization (NMP),[10−12] reversible addition−fragmentation transfer polymerization (RAFT),[13,14] and photoiniferter-mediated photopolymerization (PMP)[15,16] have gained considerable attention. The common feature of these methods is that the propagating chain continuously experiences activation−deactivation cycles to maintain a low radical concentration, which in turn minimizes irreversible chain termination. Therefore, CRP methods allow for the precise control of the polymer molecular weight and usually yield polymer brushes with low polydispersity. In addition, the preparation of block copolymer brushes becomes feasible as the active chain ends are typically preserved when a polymerization step is interrupted.

In this work, we have adapted the PMP method originally developed by Otsu et al.[15,16] to prepare poly(methacrylic acid) (PMAA) brushes on silicon surfaces, which are covered with native SiO_2. The PMP method is based on dithiocarbamate derivatives, which act as *ini*tiator, trans*fer* agent, and *ter*minating species (*iniferter*). Upon ultraviolet (UV) irradiation, the iniferter dissociates into a highly reactive radical as well as a noninitiating counter radical that reversibly terminates the propagation reaction. Advantages of the iniferter concept include the facile control of the polymerization reaction by means of irradiation time and UV intensity, the comparatively fast reaction kinetics compared to other CRP methods, and the fact that the polymerization can be easily performed at room temperature or below to avoid thermal polymerization of heat-sensitive monomers. Furthermore, the photoiniferter technique is compatible with aqueous media, it is suitable for micropatterning,[17,18] and it does not require any sacrificial initiator in the monomer solution. The latter is significant since it limits the formation of free polymers in the bulk solution and eliminates the need for extensive cleaning steps after brush formation.

The UV source employed in this work consisted of a recently developed, commercial high-power ultraviolet light-emitting diode,[19−22] and it was chosen for its distinct advantages compared to conventional UV sources. The sharp spectrum of these novel UV-LEDs (which are available in several different wavelengths) allows for a specific selection of the wavelength region according to the employed photoinitiating system and the reactive monomer. In comparison to conventional mercury lamps, the necessity for optical filters to block irradiation of undesired regions in the lamp spectrum is greatly reduced. This is advantageous since such filters frequently reduce the intensity in the desired wavelength region, leading to very poor energy efficiency and long irradiation times. The narrow spectrum of UV-LEDs is highly beneficial for controlled, surface-initiated polymerization reactions since the low-wavelength region of unfiltered mercury lamps generally induces polymerization of the monomer in the bulk solution. As a consequence, free polymer chains can become entangled within the growing polymer brush, which makes their controlled growth difficult and lengthy postcleaning processes unavoidable.

*Corresponding author. E-mail: nspencer@ethz.ch.

 Published on Web 09/23/2009

The polymer brushes prepared by means of UV-LED-induced photopolymerization were intended for the reduction of interfacial friction in an aqueous environment. While covalently attached polymer chains are readily removed during macroscopic sliding friction under contact pressures above 300–400 MPa,[23] the tribological experiments in this work aimed at aqueous lubrication under low contact pressures, for medical applications, for example. We have recently shown that strongly attached self-assembled monolayers (SAMs) and polymer brushes prepared by a "grafting to" approach represent promising aqueous lubrication additives under mild contact-pressure conditions.[23] While the previously employed surface modifications consisted of neutral, hydrophilic molecules with limited molecular weights and/or grafting densities, the introduction of charge, by means of weak polyelectrolyte brushes prepared with the PMP method, was expected to further enhance the lubricating ability of polymer brushes in an aqueous environment. In comparison to neutral hydrophilic brushes, polyelectrolyte brushes exhibit a very high osmotic pressure in aqueous environments of low ionic strengths, which renders them highly suitable for water-based lubrication purposes. For this reason, methacrylic acid (MAA) was chosen as a monomer, since it shows a high affinity toward water and because direct photopolymerization to PMAA brushes is feasible under carefully controlled reaction conditions. The preparation of PMAA brushes on Si/SiO_2 surfaces has been performed with a silane-derivatized dithiocarbamate iniferter developed by de Boer et al.[24] A number of polyelectrolyte brushes have been synthesized by "grafting from" approaches.[2,5,25–30] Since the carboxyl groups of acrylic acid-based monomers are prone to interact with metal catalysts, the synthesis of polyelectrolyte brushes by means of ATRP usually involves a hydrolysis step after the preparation of a neutral brush.[2,27,29,30] Therefore, the direct polymerization of polyelectrolyte brushes has typically been limited to either thermally induced approaches that require thorough cleaning steps after the brush synthesis, especially if a sacrificial initiator or free radicals are present in the monomer solution,[25,26] or the UV-induced photoiniferter approach.

In this work, a silane-derivatized dithiocarbamate iniferter was utilized to prepare PMAA brushes on Si/SiO_2 surfaces under UV irradiation. The combination of the photoiniferter-mediated photopolymerization with a UV-LED source appears to be ideally suited to the direct preparation of polyelectrolyte brushes with minimal free polymer formation during brush synthesis. Following characterization of the PMAA brushes by means of surface-analytical techniques, such as quartz crystal microbalance with dissipation monitoring (QCM-D), spectroscopic ellipsometry, and static contact-angle measurements, the PMAA brushes were demonstrated to enhance aqueous lubrication of Si/SiO_2 under low-contact-pressure conditions.

Experimental Section

Materials. *p*-(Chloromethyl)phenyltrimethoxysilane (ABCR, Germany), tetrahydrofuran (THF, 99.5% extra dry, Acros, Germany), methanol (Fluka, Switzerland), 2-propanol (Fluka, Switzerland), sulfuric acid (95–97%, Sigma-Aldrich, Germany), and hydrogen peroxide (30 wt % in water, VWR, Germany) were used as received. Sodium *N,N*-diethyldithiocarbamate (97%, Fluka, Switzerland) was recrystallized from methanol. Water was deionized with a GenPure filtration system (18.2 MΩ cm, TKA, Switzerland), and methacrylic acid (98%, Fluka, Switzerland) was first distilled under vacuum and subsequently passed through an alumina column (Sigma-Aldrich, Germany). The aqueous buffer solution employed for tribological experiments was prepared by adding 1 mM 4-(2-hydroxyethyl)-1-piperazineethanesulfonic acid (HEPES, BioChemika Ultra, Fluka, Switzerland) in pure water, and the solution pH was adjusted to a value of 7.4 by the addition of sodium hydroxide

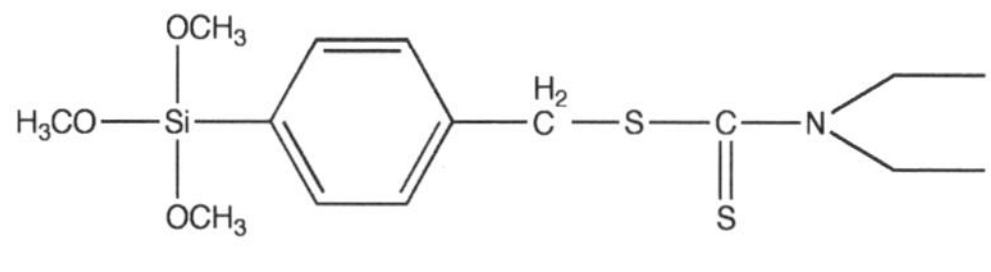

Figure 1. Chemical structure of the synthesized *N,N*-(diethylamino)dithiocarbamoylbenzyl(trimethoxy)silane (SBDC) photoiniferter.

(NaOH, Fluka, Switzerland). The buffer is abbreviated as HEPES 0 throughout this article.

Synthesis of the Photoiniferter, SBDC. The photoiniferter employed in this work (Figure 1) was synthesized according to a previously described protocol.[24] Briefly, *p*-(chloromethyl)-phenyltrimethoxsilane (CMPTMS) (1.48 g, 6 mmol) and sodium *N,N*-diethyldithiocarbamate (STC) (1.02 g, 6 mmol) were dissolved separately in 10 mL of dry THF before the STC solution was added slowly to the dissolved CMPTMS solution. After 3 h of stirring at room temperature, the solution was passed through a glass filter to remove the white NaCl precipitate, and THF was evaporated under reduced pressure. Prior to use, the photoiniferter (SBDC) was connected to a vacuum line for further drying ($p = 10^{-2}$ mbar, 4 h) and stored at −20 °C. The purified photoiniferter was obtained as a yellow viscous liquid and was characterized by ^{1}H NMR, and the data were in accordance with the values obtained by deBoer et al.[24]

Ultraviolet–Visible (UV–vis) Spectroscopy. The UV absorption spectra of both the SBDC photoiniferter and methacrylic acid were measured with a Cary 1 UV–vis spectrophotometer (Varian, Germany). The obtained spectra were taken as a reference for the selection of the spectral UV range in order to ensure effective initiation while avoiding polymerization of the monomer in solution.

Vapor Deposition of the Photoiniferter onto Si/SiO_2. Silicon wafers (P/B ⟨100⟩, Si-Mat Silicon Wafers, Germany) were cut into 20 mm × 20 mm pieces and ultrasonicated in 2-propanol twice for 15 min each time. Surface hydroxyl groups were generated on the silicon substrates by immersing the samples in a solution of concentrated sulfuric acid and 30 wt % hydrogen peroxide (H_2SO_4:H_2O_2 = 7:3), also known as piranha solution, for 60 min. (WARNING: piranha solution is very corrosive and must be handled with extreme caution; it reacts violently with organic materials and may not be stored in tightly closed vessels.) After rinsing the substrates with copious amounts of deionized water, they were dried with N_2 gas and immediately employed for surface modification.

In order to deposit the SBDC photoiniferter on the hydroxylated Si/SiO_2 substrates from the vapor phase, a 10 μL drop of the photoiniferter was placed in a desiccator, which was evacuated for 60 s with a rotary vane pump to evaporate residual solvent. Thereafter, the freshly cleaned Si/SiO_2 substrates were placed around the SBDC drop, and the desiccator was evacuated again, this time for 60 min ($p \approx 10^{-2}$ mbar). After closing the valve to the vacuum pump, the photoiniferter was allowed to adsorb onto the silicon oxide substrates for > 48 h until atmospheric pressure was reached. Prior to UV-induced polymerization reactions, the photoiniferter-modified substrates were ultrasonicated in toluene for 2 min to remove physisorbed initiator before the initiator layer was characterized by variable angle spectroscopic ellipsometry and static contact-angle measurements.

This deposition method was chosen because the formation of a homogeneous SBDC layer, which is essential for the controlled and uniform growth of polymer brushes over large areas, is more readily achievable by the vapor-deposition approach than by adsorption from solution.[31]

Controlled Radical Photopolymerization by Means of a UV-LED. The grafting solution consisted of 10 vol % MAA in distilled water. Prior to polymerization, the solution was degassed in a glass flask by three alternating ultrasonication

(5 min) and vacuum (5 min) cycles. The photoiniferter-modified substrates were placed in a round-bottom Schlenk flask and continuously purged with N_2 gas for 5 min. Then, the previously degassed monomer solution was transferred to the sample-containing glass flask via a syringe under nitrogen atmosphere.

The high-power UV-LED (NCSU033A, NICHIA Corp., Japan) with a narrow emission spectrum at 365 ± 5 nm was mounted onto a printed circuit board (PCB) in series with a 5 Ω resistor and operated with a laboratory power supply. During operation, two axial fans were placed in front of the UV-LED to avoid a thermally related lifetime degradation of the diode. The distance from the LED surface to the sample was typically 25 mm, and the intensity at 365 nm was measured with a radiometer (UVX radiometer with UVX-36 sensor, UVP, Upland, CA). The LED-to-sample distance of 25 mm was determined by the size and the angle-dependent light intensity of the UV-LED to ensure a uniform exposure over the whole sample area (20 mm × 20 mm). As previously mentioned, the optical power output of the LED is conveniently related to the forward current. We applied a voltage of 8.5 V, resulting in a fixed forward current of 500 mA, which is the maximum for long-term operation recommended by the manufacturer. The photopolymerization reaction was initiated by turning the power supply up to this maximum value.

After the polymerization reaction, the samples were taken out of the monomer solution, rinsed with water for 60 s, and blown dry with N_2 gas before they were characterized by means of surface-analytical techniques.

Characterization of PMAA Brushes. The brush thickness in a dry state was determined with a variable-angle spectroscopic ellipsometer (VASE) (M-2000F, LOT Oriel GmbH, Darmstadt, Germany) at three different angles of incidence (65°, 70°, 75°). In order to ensure the formation of homogeneous polymer brushes over the whole sample area of 400 mm^2, five different spots were measured on each sample, from which the thickness values were determined via the analysis of a three-layer model (software WVASE32, LOT Oriel GmbH, Darmstadt, Germany). The spectral range considered was from 370 to 995 nm, and the dry film thickness of the PMAA brush layer was assumed to have a refractive index of 1.45.

Static water contact angles were determined by the sessile-drop method employing a Ramé-Hart goniometer (Ramé-Hart Instrument Co., Model-100, Netcong, NJ) at all stages of the surface-modification process, i.e., after cleaning of the SiO_2 substrates, after vapor deposition of the photoiniferter, and after photopolymerization of the PMAA brushes.

The photopolymerization kinetics of PMAA brushes on SiO_2 surfaces were monitored *in situ* by means of quartz-crystal microbalance experiments with dissipation monitoring (QCM-D) using a Q-Sense E4 instrument (Q-Sense, Västra Frölunda, Sweden). As substrates, AT-cut quartz crystals (fundamental resonance frequency = 5 MHz) with a sensor area of 1.54 cm^2 and a 50 nm silicon dioxide top layer (QSX 303, Q-Sense, Västra Frölunda, Sweden) were employed. The flexibility of the UV-LED setup allowed for the positioning of the LED at a 20 mm distance from the QCM cell with a transparent quartz glass window. The photoiniferter-modified quartz crystal resonator was immobilized inside the cell, and the sample-to-LED distance was maintained at 25 mm. The degassed monomer solution was injected into the QCM cell with a syringe.

Macroscopic tribological experiments were performed by recording the coefficient of friction (μ) at different sliding speeds with a conventional pin-on-disk tribometer (CSM Instruments, Peseux, Switzerland). The sliding partner of the PMAA brush-bearing substrates was a spherical-ended elastomeric poly-(dimethylsiloxane) (PDMS) pin (Young's modulus = 1.4 MPa) with a diameter of 6 mm. Its high elasticity, its physiological inertness, and the straightforward fabrication make cross-linked PDMS ideally suitable as a model elastomer. In addition, it represents a technologically important material which displays a hydrophilic SiO_x surface layer after a 60 s air-plasma treatment (Harrick plasma cleaner/sterilizer, Ossining, NY).[32,33] The plasma-treated and thus oxidized PDMS pins are denoted as "ox-PDMS" pins throughout the article. The basic frictional properties of the PMAA brushes were tested at sliding speeds ranging from 0.25 to 10 mm/s. For each tribopair, 20 rotations were carried out at six different sliding speeds on a fixed sliding track (radius = 6 mm). The normal load was kept constant at 1 N. In order to analyze the stability of PMAA brush-bearing substrates under tribological stress, long-term pin-on-disk experiments (1000 rotations) were performed under 1 N normal load and at a sliding speed of 1 mm/s (Hertzian contact pressure ≈ 0.28 MPa).

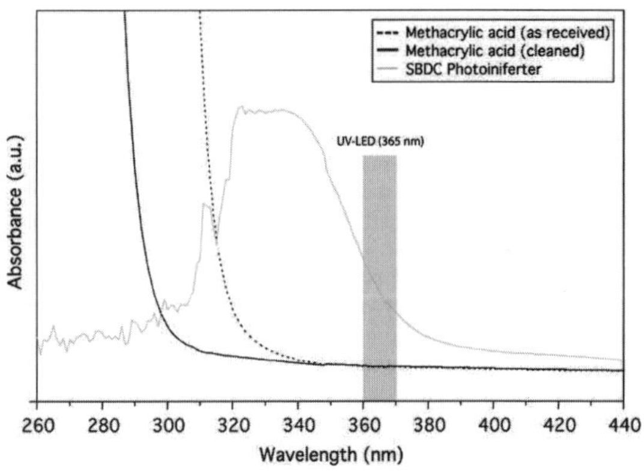

Figure 2. UV−vis spectra recorded from the SBDC photoiniferter and from the as-received as well as from the cleaned MAA monomer. The emission spectrum of the 365 nm UV-LED is indicated.

Results and Discussion

Ultraviolet−Visible (UV−vis) Absorption of Photoiniferter and Monomer. Figure 2 shows the UV−vis absorption spectra recorded from the SBDC photoiniferter as well as from as-received and cleaned methacrylic acid. While 10 μL of SBDC was dissolved in 1 mL of acetone and measured against an acetone reference, MAA was measured against air. As can be seen from Figure 2, significant UV absorption of the inhibitor-free, clean MAA starts around 300 nm, which is significantly lower than that of the as-received, stabilized MAA at around 320 nm. The SBDC photoiniferter shows an absorption maximum at ~340 nm. Importantly, the monomer and the SBDC photoiniferter have distinct regions of UV absorption that do not overlap significantly, which is considered essential for a controlled surface-initiated polymerization process. In order to avoid polymerization of the monomer in solution, the spectral region of the UV source has to be matched with the absorption band of both the photoiniferter and the monomer; i.e., the UV source should have minimal emission in the region where the monomer shows significant UV absorption. Hence, careful selection of photoiniferter, monomer, and UV source are believed to significantly enhance the degree of control over the polymerization reaction. In this respect, the narrow emission spectrum of the 365 nm high-power UV-LED employed in this work is highly advantageous. The indicated emission spectrum of the UV-LED in Figure 2 shows that UV-induced polymerization of MAA in solution is unlikely to occur since the narrow emission spectrum of the UV-LED does not overlap with the UV absorption region of the monomer.

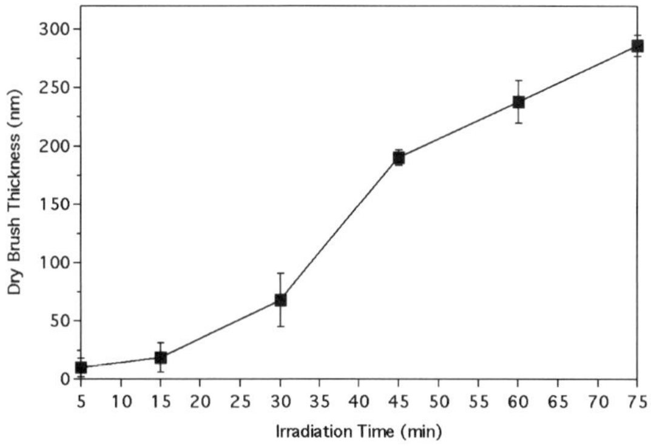

Figure 3. Characterization of the dry PMAA brush thickness as a function of irradiation time. The monomer solution consisted of 10 vol % MAA in water, and the 365 nm UV-LED intensity at the sample surface was 12 mW/cm^2. The lines serve as a guide to the eye.

Characterization of the Photoiniferter-Modified Substrates. After vapor deposition of the SBDC photoiniferter onto the Si/SiO_2 substrates and removal of any physisorbed material, the average static water contact angle was measured to be 68 ± 3°. This value is significantly higher than the contact angle of the cleaned SiO_2 surfaces (5 ± 3°) and is consistent with the successful formation of a monomolecular photoiniferter layer on the silicon substrate. The obtained water contact angles are also in good agreement with previously reported values that were obtained following SBDC adsorption from solution.[34] In order to verify the homogeneity of the vapor-deposited photoiniferter layers, the ellipsometric thickness was determined at three distinct spots on each sample by assuming a refractive index of 1.45.[34] The average thickness obtained from at least 10 samples was determined to be 0.75 ± 0.06 nm. This value is significantly lower than the theoretical maximum thickness (1.3 nm) calculated by Rahane et al.[34] and suggests that the surface coverage is below that of a full monolayer. However, it has been previously shown that too high a concentration of surface radicals can lead to extensive termination reactions in surface-initiated polymerization approaches, thus favoring initiator densities below full-monolayer coverage.[35−37] The employed vapor deposition method can be applied to materials that are not compatible with organic solvents, such as polymeric substrates, and the probability of multilayer formation is significantly reduced when adsorbing trifunctional silanes from the vapor phase.[31,38]

Photopolymerization of Methacrylic Acid to Form PMAA Brushes. The photoinitiated grafting of PMAA brushes from SBDC-modified Si/SiO_2 substrates was performed with a 365 nm UV intensity of 12 mW/cm^2, measured at the sample surface. Figure 3 shows the dry ellipsometric thickness of the PMAA brushes obtained from 10 vol % monomer solutions as a function of irradiation time. As is visible from the dry thickness of the PMAA brushes, the UV-LED-initiated polymerization method allows for an effective control of the brush thickness with irradiation time. The PMAA brushes reach high dry thickness values after short irradiation times and with low monomer concentrations. Immersion of the samples in water for 24 h and subsequent drying did not noticeably change the brush thickness. The initial slow growth of the PMAA brushes is followed by a regime, in which the layer thickness increases very rapidly with irradiation time. For exposure times beyond 45 min, however, the brush growth was found to slow down. Similar growth characteristics have been found in earlier investigations using the same[34] or different[28,35−37] SIP methods. It is generally agreed that, in comparison to controlled/"living" polymerization reactions in solution, controlled SIP methods do not exhibit a "living" character.[34,36,37,39] The relatively low concentration of deactivating species, i.e., dithiocarbamyl radicals in this work, facilitates irreversible termination reactions, leading to a loss of reactive, surface-bound radicals. Hence, a saturation of the brush thickness at longer irradiation times is expected eventually. Since previous studies with the identical SBDC photoiniferter showed that the saturation of poly(methyl methacrylate) (PMMA) brush growth occurs significantly faster for higher monomer concentrations,[34] aqueous solutions with a fixed monomer concentration (10 vol % MAA) were employed in this work in order to ensure effective control over the PMAA brush thickness with irradiation time.

Many of the attractive properties of polymer brushes are closely related to the molecular weight of the individual polymer chains as well as to their proximity to each other on the surface, i.e., the surface grafting density. According to a simple model developed by Sofia et al.,[40] the surface grafting density can be estimated from ellipsometry data.

As illustrated in Figure 4, this model is based on close-packed unit cells with side lengths L and height d, each bearing one polymer chain. The volume V occupied by a single chain can be calculated from

$$V = L^2 d = \frac{M_w}{\rho_{dry} N_A} \tag{1}$$

where L^2 is the surface area occupied by a single chain, d is the dry thickness measured by ellipsometry, M_w is the molecular weight of a single chain, ρ_{dry} is the density of the dry monolayer, and N_A is Avogadro's number. In order to obtain the surface grafting density of the SBDC photoiniferter layer, for instance, eq 1 can be transformed to

$$\frac{1}{L_{PI}^2} = \left(\frac{d_{PI}\rho_{dry,PI}N_A}{M_{w,PI}}\right) \tag{2}$$

In addition to the ellipsometric thickness of the SBDC monolayer ($d_{PI} = 0.75 \pm 0.06$ nm) as well as the molecular weight ($M_{w,PI} = 317.5$ g/mol) of the individual chains, a constant value has to be assumed for the dry density ($\rho_{dry,PI}$) of the photoiniferter layer. Hence, the validity of this model is restricted to polymer monolayers in the brush regime that do not possess a significant density gradient along the chain. Assuming a value of 1 g/cm^3 for the density of the SBDC monolayer, its surface grafting density was estimated to be 1.42 chains/nm^2. On the basis of this data, it is possible to calculate the lower molecular weight limit of the PMAA chains after photopolymerization. Provided that all photoiniferter chains induce polymerization and grow with an identical rate, i.e., $L_{PI} = L_{PMAA}$, and by further assuming a dry density ($\rho_{dry,PMAA} = 1.12$ g/cm^3)[26] for the PMAA chains, eq 1 can be transformed to

$$M_{w,PMAA} = L_{PMAA}{}^2 d_{PMAA}\rho_{dry,PMAA}N_A \tag{3}$$

Table 1 shows the calculated minimum molecular weights of the PMAA brushes that were obtained from average dry ellipsometry thickness data of the photoiniferter monolayer and the PMAA brushes. Since the fraction of photoiniferter molecules that induces simultaneous polymer growth is

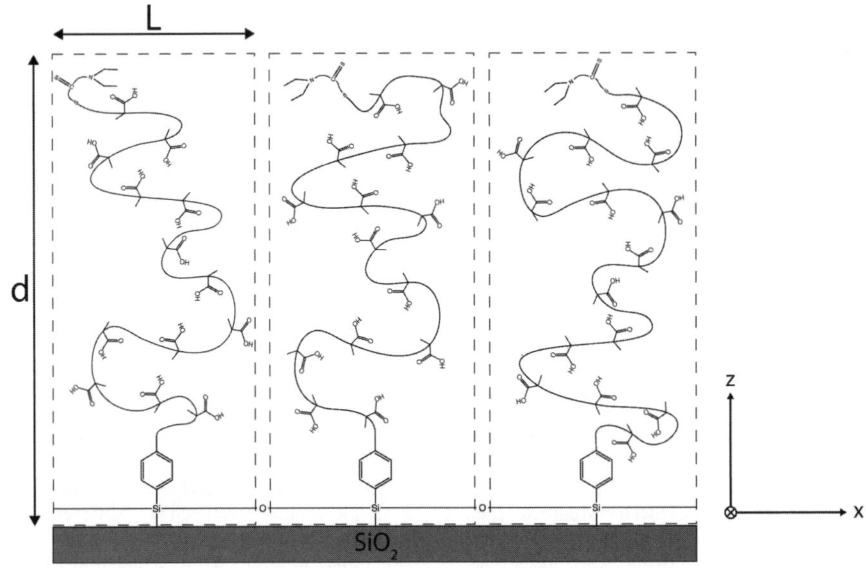

Figure 4. Two-dimensional model consisting of close-packed unit cells for the calculation of the surface grafting density and/or the molecular weight of surface-tethered polymer chains.

Table 1. Minimum Molecular Weights of the PMAA Brushes Prepared in This Work Based on Calculations from Eqs 1−3

irradiation time (min)	dry PMAA thickness (nm)	molecular weight (g/mol)
5	10	4 700
15	20	9 500
30	70	33 200
45	190	90 100
60	240	113 800
75	290	137 500

undoubtedly less than unity, it is clear that the molecular weight values of the PMAA brushes in Table 1 represent a lower limit.

Investigation of the *in Situ* PMAA Brush Growth by Means of QCM-D. The *in situ* observation of surface-initiated polymer brush growth by means of QCM has previously been reported in the literature.[35,41−44] It is a method particularly suited to UV-initiated polymerization. In order to ensure that the monomer solution does not polymerize in the absence of initiator, a blank SiO_2 crystal was first exposed to 10 vol % MAA in water and UV-irradiated for 30 min in a QCM-D setup. Figure 5 shows an increase of 6−7 Hz in the resonance frequency of the third overtone of the SiO_2 crystal upon UV exposure. This behavior has been observed previously[41] and is probably attributed to photoinduced noise. After the UV-LED was switched off after 30 min, the original resonance frequency values were retrieved, confirming that the 365 nm UV irradiation did not induce polymerization of the MAA monomer.

The *in situ* growth characteristics of PMAA brushes, determined by means of QCM-D experiments, are shown in Figure 6. After the UV-LED was switched on, the resonance frequency (black circles) increased slightly before a continuous decrease was observed. The onset of the negative frequency shift is believed to mark the start of PMAA brush polymerization since the decrease in resonance frequency can be attributed to an increase in the mass that is coupled to the QCM crystal. Up to ∼25 min of brush growth, i.e., until ca. 40 min in Figure 6, the frequency of the third

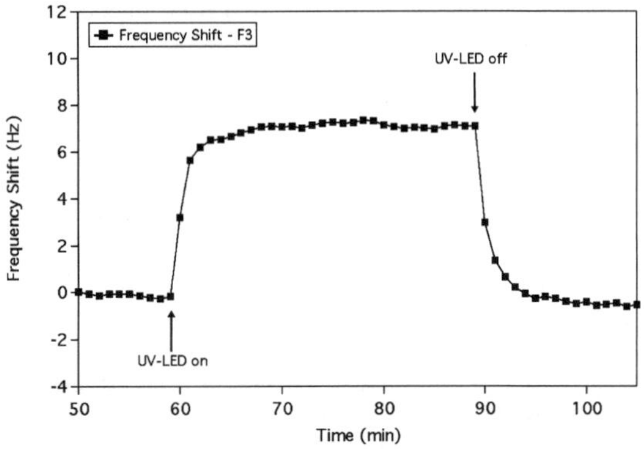

Figure 5. Shift in the resonance frequency of the third overtone of a SiO_2 QCM crystal in 10 vol % MAA solution upon exposure to 365 nm UV-LED irradiation.

overtone of the QCM crystal decreases almost linearly, suggesting a steady, continuous polymer growth. Prolonged irradiation was shown to lead to a higher polymerization rate. In contrast to the growth characteristics derived from ellipsometry data, the QCM-D experiment did not show a slower growth beyond 45 min of UV irradiation. This difference could be arising from that the difference in the physical quantity being probed by the two approaches, i.e., the total mass of the solvated polymer for QCM vs the thickness of the dry polymer for ellipsometry. QCM also detects incorporated solvent molecules and is therefore sensitive to configurational effects, i.e., the incorporation of more solvent in the brush than in mushroom configuration. Ellipsometry, on the other hand, simply provides the mass of polymer produced. The precise molecular nature of the transition in the QCM graph is, however, currently under investigation.

Our results suggest that a very slight degree of chain transfer, and consequent loss from the surface, occurs after longer irradiation times, since rinsing with fresh monomer

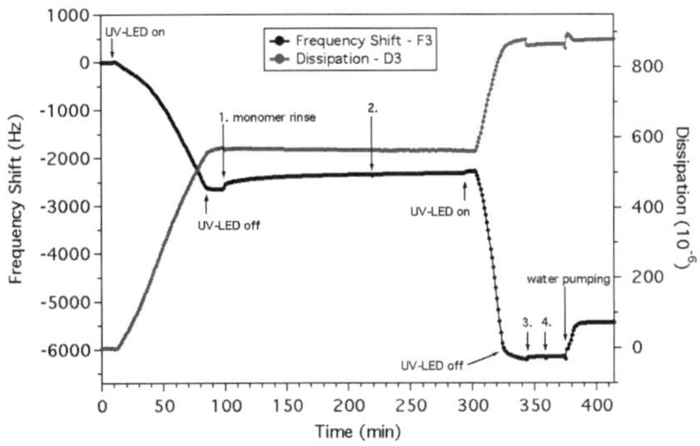

Figure 6. Frequency shift (black circles) and dissipation changes (gray circles) observed in QCM-D experiments during surface-initiated, *in situ* polymerization of PMAA brushes from 10 vol % monomer solutions in water. After a first polymerization step, the QCM cell was rinsed with fresh monomer solution (after 100 min) before the UV-LED was switched on again (after 290 min) to induce a second polymerization step. After the UV source was switched off (after 325 min), the cell was rinsed with fresh monomer twice before pure water was continuously pumped through the system in order to remove unbound polymer.

solution after the UV-LED was switched off (after 85 min in Figure 6) led to a minor decrease in the mass coupled to the QCM crystal from −2700 Hz after 100 min to −2380 Hz after 290 min in Figure 6. In contrast, the dissipation, which represents the sum of all energy losses per oscillation cycle, was almost constant during that period. It is believed that unbound polymer chains, which evolved from chain transfer reactions, can become entangled inside the brush and consequently reduce the mobility of individual brush chains. From the point of view of tribology, the viscoelastic properties of polymer brushes are extremely important since the solvation of the brush leads to a fluidlike cushion layer, which promotes facile sliding.

With the SI-PMP technique employed in this work, the active chain ends should be preserved after the polymerization step. In order to verify that, the UV-LED was switched on again (after 290 min) after stable resonance frequencies were obtained. As visible from the frequency and dissipation shifts after 300 min in Figure 6, the photopolymerization of PMAA brushes continues ~10 min after the UV-LED was switched on for the second time. Compared to the initial step, the second photopolymerization reaction occurs at a significantly faster rate, indicated by the steep decrease of the resonance frequency from −2300 to −6000 Hz within 25 min. After the UV-LED was switched off, the frequency continued to show a slight shift toward lower values, suggesting that the polymerization did not stop completely after the UV source was turned off. This is in contrast to the observations from the first polymerization step and indicates that the concentration of chain-terminating species is somewhat lower. It is thus very likely that fewer polymer chains grow much faster compared to the initial polymerization step. This is also reflected in the dissipation curve of the third overtone frequency (gray circles), which only increases by $\approx 290 \times 10^{-6}$ whereas the first polymerization led to a 560×10^{-6} increase in dissipation. The relatively low increase in dissipation compared to the high frequency shift during the second polymerization step also suggests that the growing chains are entangled inside the existing brush.

The generation of polymer in solution is very limited during the second polymerization step, as indicated by the small increase in the resonance frequency and decrease in the dissipation, respectively, after rinsing with fresh monomer (rinsing steps 3 and 4 in Figure 6). After ~375 min, deionized water was continuously pumped through the QCM cell in order to remove potential unanchored PMAA chains as well as to see differences in the hydration properties of the brush. The fast saturation of both the resonance frequency and the dissipation values suggest that the presence of unbound polymer chains is very limited during the UV-LED-induced photopolymerization method employed in this work. Hence, simple rinsing steps after the polymerization reaction appear to be sufficient to obtain well-defined PMAA brushes. While the frequency shifts toward higher values under a continuous flow of water, the dissipation was found to increase slightly. The higher frequency, indicative of a lower mass coupled to the QCM crystal, is associated with the exchange of the heavier methacrylic acid (M_w = 86.10 g/mol) with water (M_w = 18.02 g/mol), rather than with the removal of polymer chains.

Thermally induced polymerization of the monomer could be excluded during photopolymerization, since the UV-LED did not induce any temperature increase in the QCM-D setup. The temperature which was set to 25 °C varied by less than 0.01 °C in the course of the experiment. Furthermore, from the fact that UV exposure of the monomer itself did not induce polymerization (Figure 5), we suspect that chain transfer to monomer is occurring at later stages of the polymerization. Free polymer formed in the course of polymerization cannot be distinguished in a QCM-D experiment if the chains are entangled inside the brush or if the polymer significantly increases the viscosity of the monomer solution. The rinsing step after switching off the UV-LED, however, showed that some free polymer chains are present in solution. Unfortunately, most other studies presenting QCM data do not show such a rinsing step, although this would help to effectively distinguish between surface-tethered and unbound polymer.

Hydrophilicity of PMAA Brushes. Since the PMAA brushes prepared in this work were intended for aqueous tribology, the water compatibility of the surfaces is of particular interest. Therefore, the hydrophilicity of the PMAA brushes was measured by means of static contact-angle experiments. In comparison to the photoiniferter-modified substrates with a static water contact angle of 68 ± 3°, the PMAA brush-bearing samples exhibited average contact-angle values of 49 ± 7°, irrespective of the brush lengths. The water contact angles of hydrophilic polymer brushes have previously been found to be finite and nearly independent of the molecular weight, which has been attributed to the fact that the polymer chains can bridge the solvent−vapor interface.[45] Since the active chain ends were preserved during the PMAA polymerization reaction (Figure 6), the diethyldithiocarbamate-terminating groups are likely to contribute to the hydrophobic character of the PMAA-modified substrates in air. Nevertheless, it is expected that PMAA brushes can provide a lubricious interface in an aqueous environment because they exhibit a large number of ionizable carboxyls along their backbone.

Macroscopic Aqueous Lubrication Properties of PMAA Brushes. If the relative sliding speed between two contacting bodies is low or if the applied normal load is high, the lubricant is squeezed out of the contact area. This behavior is even more pronounced for low-viscosity fluids such as water and normally results in high interfacial friction and wear. The presence of a protecting polymer brush has been shown to greatly reduce the friction in this so-called "boundary-lubrication regime".[1,46,47] While direct contact and adhesion between asperities can be avoided for polymer brush-bearing surfaces, the incorporation of lubricant molecules

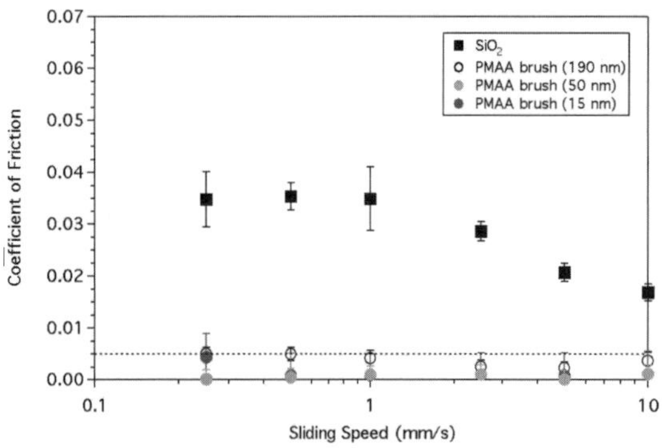

Figure 7. Pin-on-disk speed-dependence measurements obtained from PMAA brushes and from bare Si/SiO_2 substrates against an oxidized PDMS pin in HEPES 0 (normal load: 1 N). Dotted line indicates the sensitivity limit of the tribometer.

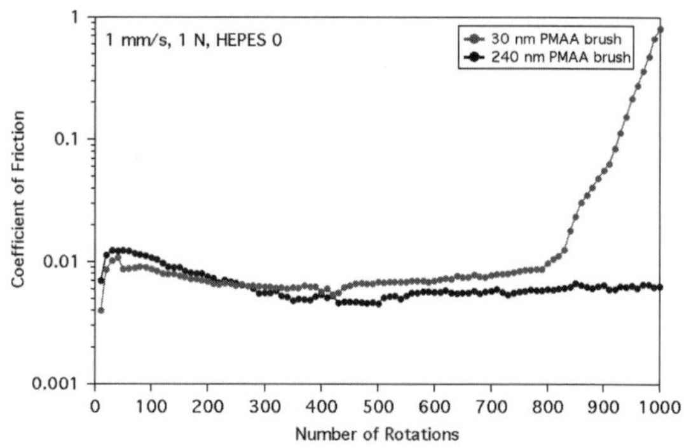

Figure 8. Long-term pin-on-disk experiments involving PMAA brushes with two distinct brush heights (30 and 240 nm). The shorter brush (gray circles) did not sustain the applied tribological stress (1 N normal load, 1 mm/s sliding speed) for 1000 rotations, while the 240 nm PMAA samples displayed very low friction coefficients throughout the entire experiment.

inside the brush additionally creates a fluidlike layer of low shear strength and thus facilitates sliding.[46]

We have previously demonstrated that hydrophilic polymer brushes can effectively reduce the interfacial friction in an aqueous environment under low-sliding-speed conditions.[48,49] In those studies, we employed "grafting to" approaches to generate polymer brushes on a variety of substrates. By applying the "grafting from" method described in this work, the formation of high-surface-density polyelectrolyte brushes became feasible. To date, only little experimental work has been dedicated to the macroscopic lubrication properties of polymer brushes prepared with a "grafting from" method.[50,51]

In many practical lubrication applications, high contact pressures on the order of 1 GPa are routinely encountered. Under such conditions, even strongly attached polymer brushes are easily removed during macroscopic sliding experiments. The reduction of the contact pressure by means of a soft rather than a rigid slider is usually sufficient for enabling the polymer brushes to sustain the tribological stress. The macroscopic pin-on-disk data obtained from different PMAA brushes against oxidized PDMS pins in HEPES 0 are presented in Figure 7. For comparison purposes, results from a bare Si/SiO_2 sample are included. From the coefficient of friction (μ) vs sliding speed plots, it is clear that clean, hydrophilic silica surfaces exhibit moderately good aqueous lubrication properties under low contact pressure conditions, especially at relatively high sliding speeds ($\mu \approx 0.017$ at 10 mm/s). Toward lower sliding speeds, i.e., in the boundary regime, these surfaces are not capable of maintaining a sufficiently thick lubricant film, which is manifested in the increase of the coefficient of friction by a factor of 2 ($\mu \approx 0.035$ at 0.25 mm/s).

In this respect, the PMAA brush-bearing surfaces were found to serve as very effective surfaces under boundary-lubrication conditions in water. All tested brushes with a dry ellipsometric thickness ranging from 15 to 190 nm showed friction coefficients below the sensitivity limit ($\mu = 0.005$) of the employed macrotribometer at the load employed. This resolution limit is indicated with a dotted line in Figure 7, and the plot shows that these low friction values were maintained over the entire sliding-speed range. Therefore, it was not possible to effectively distinguish the different PMAA brushes by their lubricating properties in HEPES 0. The results suggest that all PMAA brushes were able to create a highly hydrated, fluidlike interface, rendering these surface modifications highly compatible for aqueous lubrication purposes at neutral pH.

In order to test the stability of the PMAA brushes in longer term pin-on-disk experiments, two samples with different brush heights (30 and 240 nm dry thickness) were exposed to tribological stress for 1000 rotations under 1 N normal load and a sliding speed of 1 mm/s. The brush-bearing samples were immersed in HEPES 0 for 10 min prior to the pin-on-disk experiment. Figure 8 shows that the μ values from the shorter brush reach a maximum of 0.01 during initial rotations before the friction decreases to $\mu \approx 0.006$ after 200 rotations. The frictional response during that period is explained by the relatively slow complete hydration of the brush. Since the static water contact angles of the dry brushes are rather high, it seems likely that the complete hydration of PMAA chains takes some time. Apparently, the immersion of the brushes into the aqueous lubricant prior to pin-on-disk experiments is not sufficient for a complete hydration, but the samples become more hydrated under the influence of tribological stress. Between 200 and 450 rotations, the coefficient of friction from the 30 nm PMAA brush is nearly constant, before a slight but continuous increase in friction could be observed. After 790 rotations, the μ values increase drastically up to $\mu \approx 1.0$. Hence, the short PMAA brushes did not sustain the tribological stress for 1000 rotations.

When the identical long-term pin-on-disk experiment was performed with the longer PMAA brush (240 nm dry thickness), the coefficient of friction decreases from initially $\mu = 0.01$ to values below the detection limit of the pin-on-disk tribometer ($\mu = 0.005$) within the first 400 rotations. This behavior was again attributed to a continuous hydration process of the thick PMAA brush, the highest lubricity being expected after maximum swelling. In comparison to the short brush, the friction coefficients from the 240 nm PMAA sample were found to decrease over a longer period of time, which is believed to be due to a prolonged hydration process of brushes with a higher thickness. The next 200 rotations are characterized by undetectably low friction coefficients, after which the μ values seem to increase slightly and stabilize at around $\mu = 0.006$ for the remaining 400 rotations. In comparison to the speed-dependence experiments in Figure 7, where the undetectably low friction was observed at the initial sliding speed of 10 mm/s, the swelling of both short and long brushes, i.e., the fluid uptake, appears to be significantly slower at a sliding speed of 1 mm/s. Nonetheless, the

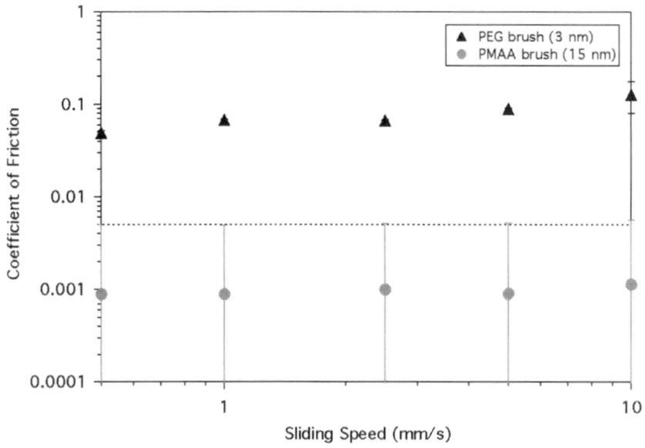

Figure 9. Comparison of μ vs sliding speed plots between 15 nm PMAA brushes and 3 nm PEG (5000) monolayers, which have been prepared via "grafting from" and "grafting to" methods, respectively. The molecular weight of individual polymer chains was expected to be of the same order of magnitude for both brushes.

obtained friction coefficients from both brushes were well below $\mu = 0.01$ over several hundred rotations, with the difference that the 30 nm PMAA brush did not show the superior long-term stability of the 240 nm PMAA samples.

In Figure 9, a comparison is made between the frictional properties of a 15 nm PMAA brush and a 3 nm PEG (5000) monolayer. On the basis of the ellipsometric dry thickness values from the 15 nm PMAA brush, its molecular weight was determined to be 6350 g/mol, which is in the range of the employed PEG chains (5000 g/mol). The fact that the dry thickness of the PMAA brush is 5 times higher than that of the PEG monolayer can be explained by the differences in surface grafting density due to the different preparation methods; i.e., the PEG monolayer was obtained by the adsorption of PEG-silane (5000) molecules on Si/SiO_2 substrates from toluene solution—a "grafting to" method. Since PEG-based polymer brushes have previously proven to be very effective aqueous boundary lubricant additives,[48,49,52] it was interesting to compare them directly with PMAA brushes. Even though the PEG (5000) monolayer showed very low and extremely stable friction coefficients of around 0.05 over the tested speed range (10−0.5 mm/s), the values obtained from the 15 nm PMAA brush were again found to be below the detection limit ($\mu = 0.005$) of the tribometer, indicated by the dotted line in Figure 9. This also explains the apparently high standard deviations in the μ values obtained from PMAA brushes compared to those from PEG monolayers, for which friction was 1 order of magnitude *above* the sensitivity limit. The superior lubrication properties of PMAA brushes compared to PEG monolayers at neutral solution pH are attributed to a number of factors. First, the PMAA brushes prepared by the "grafting from" method show a significantly higher grafting density than that of the PEG monolayers. Second, the density of hydrophilic moieties (−COOH) inside the brush is higher for PMAA brushes, and their deprotonation at neutral solution pH is expected to enhance the swelling of polyelectrolyte brushes. The hydrated thickness of the boundary lubricant is also important with regard to the surface roughness of the tribopair. If the hydrated brush length is significantly larger than the roughness of the tribopair, direct asperity contact between the sliding surfaces can be avoided, even if only one surface is bearing a polymer brush. A further effect that may reduce the friction between PMAA brushes against oxidized PDMS pins results from electrostatic repulsion between the two sliding surfaces at neutral pH. The schematic in Figure 10 illustrates the presumed tribological interface formed by a PMAA brush against an ox-PDMS in an aqueous environment at neutral pH.

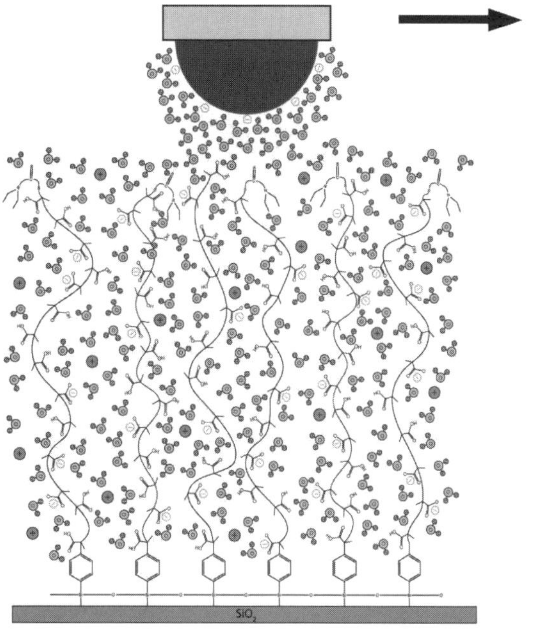

Figure 10. Presumed tribological interface between a highly hydrated PMAA brush and an ox-PDMS slider in an aqueous environment of neutral pH.

Conclusions

We reported the controlled growth of poly(methacrylic acid) brushes on Si/SiO_2 surfaces via a photoinduced "grafting from" approach. The employment of a relatively novel UV-LED setup allowed for the preparation of polymer brushes with a high dry thickness within comparatively short reaction times and low monomer concentrations. By careful purification of monomer and choice of photoinitiating system and UV source, the polymerization of monomer in solution could be suppressed, which rendered lengthy cleaning steps of the formed polymer brushes unnecessary. In this context, we consider the utilization of UV-LEDs to be very advantageous, since the narrow emission spectrum of the LED could be selected in a region where the monomer does not absorb UV irradiation. In conventional mercury arc UV lamps, for instance, the lower regions of the emission spectrum have to be filtered out in order to avoid polymerization of the monomer in solution. Besides the fact that optical filters often drastically reduce the intensity in the desired wavelength region, such lamps have to be effectively cooled to avoid heating of the monomer solution and associated thermally induced polymerization.

After the preparation of the PMAA brushes, their lubrication ability under low contact pressures was tested in a neutral aqueous solution. It was shown that the macroscopic friction between polyelectrolyte brushes of different molecular weights and soft, hydrophilic ox-PDMS pins was below the detection limit of the employed pin-on-disk tribometer over the entire speed range tested. While the PMAA brushes could not be distinguished with μ vs sliding speed plots, the long-term stability of short 15 nm PMAA brushes was shown to be inferior to long brushes (240 nm dry thickness). A further comparison between

PMAA brushes and PEG monolayers, of which the latter represent well-known aqueous boundary lubricants that are generally adsorbed on surfaces via "grafting to" methods, showed that PMAA brushes display significantly lower friction under aqueous lubrication conditions. Besides the higher grafting densities of the PMAA compared to PEG layers, enhanced swelling of the polyelectrolyte brushes in neutral aqueous media and additional electrostatic repulsion against the oxidized PDMS slider are presumably responsible for the significantly decreased frictional response.

In summary, we expect that strongly attached polyelectrolyte brushes hold great potential as effective surfaces for aqueous boundary lubrication under low contact pressures. To this end, the controlled polymerization of dense polymer brushes via the employed photopolymerization method is believed to serve as a versatile tool for the specific fabrication of ultralow-friction surfaces.

Acknowledgment. The authors are grateful to Drs. Andreas Mühlebach (Ciba Specialty Chemicals), Rupert Konradi (BASF), and Erik Reimhult (ETH Zurich) for useful discussions. The financial assistance of the ETH Research Commission is greatly appreciated.

References and Notes

(1) Raviv, U.; Giasson, S.; Kampf, N.; Gohy, J. F.; Jerome, R.; Klein, J. *Nature* **2003**, *425* (6954), 163–165.
(2) Dong, R.; Krishnan, S.; Baird, B. A.; Lindau, M.; Ober, C. K. *Biomacromolecules* **2007**, *8* (10), 3082–3092.
(3) Limpoco, F. T.; Advincula, R. C.; Perry, S. S. *Langmuir* **2007**, *23* (24), 12196–12201.
(4) Zhao, B.; Brittain, W. J. *Prog. Polym. Sci.* **2000**, *25* (5), 677–710.
(5) Ruhe, J.; Ballauff, M.; Biesalski, M.; Dziezok, P.; Grohn, F.; Johannsmann, D.; Houbenov, N.; Hugenberg, N.; Konradi, R.; Minko, S.; Motornov, M.; Netz, R. R.; Schmidt, M.; Seidel, C.; Stamm, M.; Stephan, T.; Usov, D.; Zhang, H. N. *Polyelectrolytes Defined Mol. Archit. I* **2004**, *165*, 79−150.
(6) Ruhe, J. *Polymer Brushes: On the Way to Tailor-Made Surfaces*; Wiley-VCH: New York, 2004; pp 1−31.
(7) Edmondson, S.; Osborne, V. L.; Huck, W. T. S. *Chem. Soc. Rev.* **2004**, *33* (1), 14–22.
(8) Tsujii, Y.; Ohno, K.; Yamamoto, S.; Goto, A.; Fukuda, T. *Surf.-Initiated Polym. I* **2006**, *197*, 1−45.
(9) Wang, J. S.; Matyjaszewski, K. *Macromolecules* **1995**, *28* (23), 7901–7910.
(10) Hawker, C. J.; Bosman, A. W.; Harth, E. *Chem. Rev.* **2001**, *101* (12), 3661–3688.
(11) Benoit, D.; Chaplinski, V.; Braslau, R.; Hawker, C. J. *J. Am. Chem. Soc.* **1999**, *121* (16), 3904–3920.
(12) Hawker, C. J.; Barclay, G. G.; Orellana, A.; Dao, J.; Devonport, W. *Macromolecules* **1996**, *29* (16), 5245–5254.
(13) Chiefari, J.; Chong, Y. K.; Ercole, F.; Krstina, J.; Jeffery, J.; Le, T. P. T.; Mayadunne, R. T. A.; Meijs, G. F.; Moad, C. L.; Moad, G.; Rizzardo, E.; Thang, S. H. *Macromolecules* **1998**, *31* (16), 5559–5562.
(14) Moad, G.; Rizzardo, E.; Thang, S. H. *Aust. J. Chem.* **2005**, *58* (6), 379–410.
(15) Takayuki Otsu, M. Y. *Makromol. Chem., Rapid Commun.* **1982**, *3* (2), 127–132.
(16) Takayuki Otsu, M. Y. T. T. *Makromol. Chem., Rapid Commun.* **1982**, *3* (2), 133–140.
(17) Nakayama, Y.; Matsuda, T. *Macromolecules* **1996**, *29* (27), 8622–8630.
(18) Luo, N.; Metters, A. T.; Hutchison, J. B.; Bowman, C. N.; Anseth, K. S. *Macromolecules* **2003**, *36* (18), 6739–6745.
(19) Morita, D.; Sano, M.; Yamamoto, M.; Matoba, K.; Yasutomo, K.; Akaishi, K.; Kasai, Y.; Nagahama, S.; Mukai, T. In 365-mn ultraviolet light emitting diodes with an output power of over 400 mW, Conference on Quantum Sensing and Nanophotonic Devices, San Jose, CA, Jan 25−29, 2004; Razeghi, M., Brown, G. J., Eds.; **2004**; pp 415−421.
(20) Morita, D.; Yamamoto, M.; Akaishi, K.; Matoba, K.; Yasutomo, K.; Kasai, Y.; Sano, M.; Nagahama, S.; Mukai, T. *Jpn. J. Appl. Phys., Part 1* **2004**, *43* (9A), 5945–5950.
(21) Morita, D.; Sano, M.; Yamamoto, M.; Nonaka, M.; Yasutomo, K.; Akaishi, K.; Nagahama, S.; Mukai, T. In Over 200 mW on 365 nm ultraviolet light emitting diode of GaN-free structure, 5th International Conference on Nitride Semiconductors (ICNS-5), Nara, Japan, May 25−30, 2003; **2003**, pp 114−117.
(22) Morita, D.; Sano, M.; Yamamoto, M.; Murayama, T.; Nagahama, S.; Mukai, T. *Jpn. J. Appl. Phys., Part 2* **2002**, *41* (12B), L1434–L1436.
(23) Lee, S.; Heeb, R.; Venkataraman, N. V.; Spencer, N. D. *Tribol. Lett.* **2007**, *28* (3), 229–239.
(24) de Boer, B.; Simon, H. K.; Werts, M. P. L.; van der Vegte, E. W.; Hadziioannou, G. *Macromolecules* **2000**, *33* (2), 349–356.
(25) Biesalski, M.; Johannsmann, D.; Ruhe, J. *J. Chem. Phys.* **2002**, *117* (10), 4988–4994.
(26) Konradi, R.; Ruhe, J. *Macromolecules* **2004**, *37* (18), 6954–6961.
(27) Parnell, A. J.; Martin, S. J.; Dang, C. C.; Geoghegan, M.; Jones, R. A. L.; Crook, C. J.; Howse, J. R.; Ryan, A. J. *Polymer* **2009**, *50* (4), 1005–1014.
(28) Jain, P.; Dai, J. H.; Baker, G. L.; Bruening, M. L. *Macromolecules* **2008**, *41* (22), 8413–8417.
(29) Matyjaszewski, K.; Miller, P. J.; Shukla, N.; Immaraporn, B.; Gelman, A.; Luokala, B. B.; Siclovan, T. M.; Kickelbick, G.; Vallant, T.; Hoffmann, H.; Pakula, T. *Macromolecules* **1999**, *32* (26), 8716–8724.
(30) Wu, T.; Gong, P.; Szleifer, I.; Vlcek, P.; Subr, V.; Genzer, J. *Macromolecules* **2007**, *40* (24), 8756–8764.
(31) Ashurst, W.R.; Carraro, C.; Maboudian, R. *IEEE Trans. Device Mater. Reliab.* **2003**, *3* (4), 173–178.
(32) Fritz, J. L.; Owen, M. J. *J. Adhes.* **1995**, *54* (1−2), 33–45.
(33) Efimenko, K.; Wallace, W. E.; Genzer, J. *J. Colloid Interface Sci.* **2002**, *254* (2), 306–315.
(34) Rahane, S. B.; Kilbey, S. M.; Metters, A. T. *Macromolecules* **2005**, *38* (20), 8202–8210.
(35) Niwa, M.; Date, M.; Higashi, N. *Macromolecules* **1996**, *29* (11), 3681–3685.
(36) Xiao, D. Q.; Wirth, M. J. *Macromolecules* **2002**, *35* (8), 2919–2925.
(37) Kim, J. B.; Huang, W. X.; Miller, M. D.; Baker, G. L.; Bruening, M. L. *J. Polym. Sci., Part A: Polym. Chem.* **2003**, *41* (3), 386–394.
(38) Jung, G. Y.; Li, Z. Y.; Wu, W.; Chen, Y.; Olynick, D. L.; Wang, S. Y.; Tong, W. M.; Williams, R. S. *Langmuir* **2005**, *21* (4), 1158–1161.
(39) Rahane, S. B.; Kilbey, S. M.; Metters, A. T. *Macromolecules* **2008**, *41* (24), 9612–9618.
(40) Sofia, S. J.; Premnath, V.; Merrill, E. W. *Macromolecules* **1998**, *31* (15), 5059–5070.
(41) Nakayama, Y.; Matsuda, T. *Macromolecules* **1999**, *32* (16), 5405–5410.
(42) Benetti, E. M.; Reimhult, E.; de Bruin, J.; Zapotoczny, S.; Textor, M.; Vancso, G. J. *Macromolecules* **2009**, *42* (5), 1640–1647.
(43) Moya, S. E.; Brown, A. A.; Azzaroni, O.; Huck, W. T. S. *Macromol. Rapid Commun.* **2005**, *26* (14), 1117–1121.
(44) Kurosawa, S.; Aizawa, H.; Talib, Z. A.; Atthoff, B.; Hilborn, J. *Biosens. Bioelectron.* **2004**, *20* (6), 1165–1176.
(45) Stuart, M. A. C.; de Vos, W. M.; Leermakers, F. A. M. *Langmuir* **2006**, *22* (4), 1722–1728.
(46) Klein, J. *Annu. Rev. Mater. Sci.* **1996**, *26*, 581–612.
(47) Lee, S.; Spencer, N. D. *Science* **2008**, *319* (5863), 575–576.
(48) Lee, S.; Muller, M.; Heeb, R.; Zurcher, S.; Tosatti, S.; Heinrich, M.; Amstad, F.; Pechmann, S.; Spencer, N. D. *Tribol. Lett.* **2006**, *24* (3), 217–223.
(49) Lee, S.; Iten, R.; Muller, M.; Spencer, N. D. *Macromolecules* **2004**, *37* (22), 8349–8356.
(50) Kobayashi, M.; Terayama, Y.; Hosaka, N.; Kaido, M.; Suzuki, A.; Yamada, N.; Torikai, N.; Ishihara, K.; Takahara, A. *Soft Matter* **2007**, *3* (6), 740–746.
(51) Sakata, H.; Kobayashi, M.; Otsuka, H.; Takahara, A. *Polym. J.* **2005**, *37* (10), 767–775.
(52) Lee, S.; Spencer, N. D. Achieving Ultralow Friction by Aqueous, Brush-Assisted Lubrication. In *Superlubricity*; Erdemir, A., Martin, J.-M., Eds.; Elsevier: Amsterdam, 2007; pp 365−396.

Tribol Lett (2010) 37:541–552
DOI 10.1007/s11249-009-9549-9

ORIGINAL PAPER

Macrotribological Studies of Poly(*L*-lysine)-*graft*-Poly(ethylene glycol) in Aqueous Glycerol Mixtures

Prathima C. Nalam · Jarred N. Clasohm · Alireza Mashaghi · Nicholas D. Spencer

Received: 25 September 2009 / Accepted: 19 November 2009 / Published online: 11 December 2009

Abstract We have investigated the tribological properties of surfaces with adsorbed poly(*L*-lysine)-*graft*-poly(ethylene glycol) (PLL-*g*-PEG) sliding in aqueous glycerol solutions under different lubrication regimes. Glycerol is a polar, biocompatible liquid with a significantly higher viscosity than that of water. Macrotribological performance was investigated by means of pin-on-disk and mini-traction-machine measurements in glycerol-PLL-*g*-PEG-aqueous buffer mixtures of varying compositions. Adsorption studies of PLL-*g*-PEG from these mixtures were conducted with the quartz-crystal-microbalance technique. The enhanced viscosity of the glycerol-containing lubricant reduces the coefficient of friction due to increased hydrodynamic forces, leading to a more effective separation of the sliding partners, while the presence of hydrated polymer brushes at the interface leads to an entropically driven repulsion, which also helps mitigate direct asperity–asperity contact between the solid surfaces under boundary-lubrication conditions. The combination of polymer layers on surfaces with aqueous phases of enhanced viscosity thus enables the friction to be reduced by several orders of magnitude, compared to the behavior of pure water, over a large range of sliding speeds. The individual contributions of the polymer and the aqueous glycerol solutions in reducing the friction have been studied across different lubrication regimes.

Keywords Boundary lubrication · Aqueous lubrication · Glycerol · Polymer brushes · Viscosity

P. C. Nalam · J. N. Clasohm · A. Mashaghi ·
N. D. Spencer (✉)
Laboratory for Surface Science and Technology, Department of Materials, ETH Zurich, HCI H523, Wolfgang-Pauli-Strasse 10, 8093 Zurich, Switzerland
e-mail: spencer@mat.ethz.ch

1 Introduction

The study of macromolecules at the solid–liquid interface has led to improved understanding and new technologies in many fields including colloid science, biomedicine, and tribology [1]. Klein et al. have studied the shear forces between polymer-bearing surfaces with the surface-forces apparatus, to understand the frictional forces at the interface. These studies show that when two surfaces covered with a high density of terminally attached polymers are immersed in a good solvent and brought into contact, the swollen polymer brushes reduce interfacial frictional forces [2]. As they approach each other, opposing polymer brushes exhibit repulsive forces due to osmotic effects on the one hand and the free-energy penalty (due to reduced configurational entropy) resulting from the overlap of the brush layers on the other. There have been several studies, both theoretical [3] and experimental [4, 5] that have investigated the origin of frictional forces between contacting brushes at different shear rates. For water-soluble polymer brushes in aqueous environments, the presence of bound (or 'hydration') water surrounding the polymer chains can result in structural forces between the hydrated brushes [6]. Strongly hydrated polymers, together with a continuous rapid exchange of bound water with other free water molecules, keep the surfaces separated while maintaining a high fluidity at the brush–brush interface at high compressions, thus leading to a very low coefficient of friction [7, 8].

Aqueous lubrication is of interest in a number of technological applications where lubrication with oil presents contamination problems. This is the case, for example, in the food, textile, and pharmaceutical industries. The adsorption of synthetic, hydrated polymer brushes at the interface overcomes the drawback of the low viscosity of water and, to some extent, mimics the situation encountered in nature [9]. Adsorption of the poly(*L*-lysine)-*graft*-poly(ethylene glycol) (PLL-*g*-PEG) copolymer has been extensively used [10, 11] as a facile approach for the attachment of water-compatible polymer brushes to surfaces. Studies have been conducted to understand the lubrication properties of these polymer brushes both at macroscopic [11, 12] and at nanoscopic scales [13, 14] in aqueous environments. PLL-*g*-PEG contains a positively charged polypeptide backbone that adsorbs spontaneously via electrostatic interactions onto several metal oxide surfaces, such as TiO_2, Nb_2O_5, and SiO_2 at neutral pH. In aqueous media, PEG chains become hydrated to form "brush-like" structures at the interface, which reduce the frictional forces when surfaces are rubbed against each other. If sheared off under tribological stress, the electrostatically attached polymers can immediately be replaced by molecules readsorbing from solution and thus can act as better lubricants when compared to covalently attached polymers [15], since the latter generally require specific, non-aqueous reaction conditions for reattachment.

Along with polymer architecture [11], the quality of the solvent surrounding the polymer brush is an important parameter for determining both adsorption kinetics and lubrication properties [16–18]. For end-grafted polymers in poor solvents, the cohesive forces between polymer molecules (both inter- and intrachain) or polymers and surface dominate, resulting in a dense collapsed structure of the polymer when adsorbed on the surface (pancake structure). In contrast, good solvents can induce, at low surface coverages, a structure resembling that of the free polymer chains in solution (mushroom structure), or, at high coverages, a significant stretching of the polymers leading to a polymer brush. Several studies have been conducted to understand the effect of solvent quality on the structure and stability of brushes [18, 19]. The structural changes and preferential solvation of polymer brushes have been studied in detail [20–22] by varying the solvent quality using binary solutions, which contain varying volume fractions of good and bad solvents in the solution. Müller et al. [22] studied the frictional properties of adsorbed PLL-*g*-PEG polymers on silica surfaces using colloidal-probe AFM for binary mixtures of water and 2-propanol. They observed little or no variation in the frictional properties of the brushes until the critical volume fraction of $\phi = 0.85$ (2-propanol) is reached, beyond which the friction increases remarkably with even a slight increase in the volume fraction of the solvent.

In this study, we have investigated the tribological properties of PLL-*g*-PEG copolymer brushes in binary mixtures of buffer solution and glycerol. Studies of the fluidity of water, when it is confined as a molecularly thin film between two solid surfaces, show that there is only a nominal increase in the viscosity of the confined water at the interface [23]. In contrast to the behavior of oils, the low pressure-viscosity coefficient of water can impose a major constraint for aqueous tribology at high loads, since the boundary regime is extended to higher speeds. Increasing viscosity by the addition of water-compatible viscous fluids is an alternative approach to rectifying this situation. Glycerol is a polar, biocompatible, and highly viscous liquid, which readily dissolves in water. As PEG does not dissolve in glycerol, glycerol behaves as a poor solvent in the buffer-solution–glycerol binary mixture. We have conducted tribological tests at various speeds and loads with buffer-solution–glycerol solutions of different compositions and viscosities and explored the effect of polymer brushes across different lubrication regimes. It was found that a combination of polymer brushes and the enhanced viscosity obtained by glycerol addition provided effective lubrication over a wide range of speeds, and therefore lubrication regimes. While the enhanced viscosity fluids were highly effective in extending the hydrodynamic regime to lower speeds, it was clear that the polymer brushes enhanced lubrication within the boundary and mixed regimes. The adsorption kinetics of polymers from viscous binary solutions has also been investigated. Lastly, a calculation of the lubricating film thickness at the interface determines the importance of polymer brushes at the interface under different lubrication regimes.

2 Materials and Methods

All tribological experiments were conducted with a steel ball loaded against a glass disk. HEPES [10 mM of 4-(2-hydroxyethyl)-1-piperazine-1-ethanesulfonic acid (Sigma, St. Louis, MO, USA), with 6.0 M NaOH solution] was used as the aqueous buffer to maintain the pH at 7.4. Due to the low isoelectric point of silicon dioxide ($\sim$2), negative charges reside on the surface at neutral pH. These negatively charged surfaces adsorb the positively charged backbones of PLL-*g*-PEG copolymers to form brush-like structures spontaneously upon immersion in the aqueous-polymer-containing solution.

2.1 Materials Used

PLL-*g*-PEG copolymer was purchased from SuSoS AG (Dübendorf, Switzerland). The specific copolymer used, PLL(20)-*g*(3.6)-PEG(5), with a PLL molecular weight of 20 kDa, PEG side chains of molecular weight 5 kDa, and a grafting ratio (number of lysine units/number of PEG chains) of 3.6 shows maximum adsorption on the surface and optimum brush density to maintain a hydrated and stretched brush structure [11]. For all tribological experiments, a polymer concentration of 0.25 mg/ml in HEPES was used, which is a sufficient concentration for the rapid re-healing of the polymers following tribocontact [15]. Different volume percentages of either glycerol or ethylene glycol in HEPES (0, 25, 50, and 75% v/v) were used as binary mixtures to vary the viscosity of the solution at the tribological contact. Glycerol (ABCR GmbH, Karlsruhe, Germany) and ethylene glycol (Aldrich-Sigma, Steinheim, Germany) were used without further processing. PLL-*g*-PEG does not dissolve either in pure glycerol or pure ethylene glycol, and thus experiments with the undiluted liquids were not conducted. Viscosities of the mixtures according to their volume fractions are given in Table 1. All polymer solutions were freshly prepared just before the experiment and were homogenized in a sonicator for 15–20 min before use.

2.2 Tribological Experiments

Disks and balls used for the tribological tests were sonicated in ethanol absolute (Scharlau, Analytical grade, ACS, Sentmenat, Spain) in Teflon boxes for 30 min. N_2-dried samples were then plasma-treated in an oxygen environment (for pin-on-disk) and in air (for MTM) for 90 s to remove adventitious organic matter. Treated disks and balls were transferred to the polymer solution and the experiments conducted after soaking for a minimum of 30 min.

2.3 Pin-on-disk Measurements

Pin-on-disk tribometers (CSEM, Neuchâtel, Switzerland) were used to measure macroscopic frictional forces under pure sliding conditions. Two tribometers operating in different speed ranges were employed to enable the sliding speed to be varied over a wide range. The slower tribometer measures frictional forces in the speed range of 0.1–20 mm/s and the faster tribometer from 25 to 400 mm/s. A fixed pin that holds the steel ball (diameter = 6 mm, DIN 5401-20 G20, Hydrel AG, Romanshorn, Switzerland) was brought into contact with the flat, rotating glass slide (2.5 × 2.5 cm^2, 1-mm thick; Super Frost microscope slides, Menzel Gläser, Braunschweig, Germany with a composition as specified by the manufacturer: 72.2% SiO_2, 14.3% Na_2O, 1.2% K_2O, 6.4% CaO, 4.3% MgO, 1.2% Al_2O_3, 0.03% Fe_2O_3, and 0.3% SO_3). RMS roughness values of the steel ball and the glass disk were measured by AFM as 32 and 5 nm, respectively. A stainless steel cup held the polymer solution (capacity ~20 ml) such that the pin and disk were completely immersed in the solution. The coefficient of friction (μ) is plotted as a function of number of laps under a normal load of 2 N for all experiments (Hertzian contact pressure = 0.34 GPa). Experiments were conducted at ambient temperature and a fresh track and new pin were used for every measurement. Data acquisition and operating speeds were controlled by means of *Tribo X* software (InstrumX version 2.5A, CSM Instruments, Switzerland). Friction coefficients were averaged over 200 laps for speeds above 20 mm/s and over 50 laps for speeds below 20 mm/s.

Table 1 Dynamic viscosities (mPa s) for different percentages of glycerol and ethylene glycol in water at 25 °C [33, 34]

Concentration of glycerol/ethylene glycol in water (vol.%)	Dynamic viscosity of glycerol–water mixture at 25 °C (mPa s)	Dynamic viscosity of ethylene-glycol–water mixture at 25°C (mPa s)
0	0.893	0.893
25	1.81	1.5
50	5.04	2.8
75	27.7	7.75
100	945	14

2.4 Mini-Traction-Machine Measurements

The mini-traction-machine (MTM, PCS instruments, London, UK) was used to measure frictional forces in rolling contact between PEG-coated surfaces immersed in the copolymer solutions. In the experimental setup, a 9.5-mm radius steel ball (AISI 52100, RMS roughness = 11 nm, PCS Instruments, London, UK) was brought into contact with a 46-mm diameter glass disk (RMS roughness = 2 nm, PCS Instruments, London, UK). Only one track with a radius of 20.7 mm per disk was used. The rotation of the ball and the disk can be independently controlled and thus a mixture of sliding and rolling can be achieved. The slide/roll ratio (SRR) is defined as the percentage ratio of the difference between the ball and the disk speed to the mean of ball (u_{ball}) and disk speed (u_{disk}); SRR = ($u_{ball} - u_{disk}$)/[($u_{ball} + u_{disk}$)/2]. The SRR varies from 0 to 200% with SRR = 0% ($u_{ball} = u_{disk}$) representing pure rolling and SRR = 200% for complete sliding conditions. A SRR of 10% was used for all experiments to maintain the conditions of near-pure rolling. Using the manufacturer's software (PCS Instruments, MTM version 1.0, London, UK) the speed can be varied from 0 to 2500 mm/s. A load of 10 N was applied (Hertzian contact pressure = 0.42 GPa) and the coefficient of friction measured as a function of the mean speed of the disk and the ball. A temperature of 25 °C was maintained by means of a water bath. New disks and balls were used for every measurement.

2.5 Quartz Crystal Microbalance

The quartz crystal microbalance (QCM) is a mass-sensing device. In contrast to many other mass-sensing techniques that work under liquids, QCM is sensitive to the mass of both the adsorbed polymer layer and the mass of the solvent associated with it. The measurements were performed with a commercially available QCM with dissipation monitoring (Q-sense, Gothenburg, Sweden).

AT-cut quartz crystals (diameter = 25 mm) coated with SiO_2 (QSX 303, LOT Oriel Group, Germany) with a fundamental frequency of 5 MHz were used to study the mass of the polymer adsorbed from different HEPES–glycerol mixtures. QCM oscillators can be used to monitor the change in mass (Δm) by measuring the change in resonance frequency (Δf) of the crystal resonator during polymer adsorption. Frequencies can be measured over different overtones ($n = 1$ to $n = 13$), which have different surface-sensitivities associated with them. According to the Sauerbrey equation, the mass of the adsorbed polymer along with its associated water molecules can be calculated by:

$$m_{\text{Sauerbrey}} = -C\frac{\Delta f}{n}$$

where $m_{\text{Sauerbrey}}$ is the wet mass of the polymer adsorbed, Δf is the change in frequency of the quartz crystal upon adsorption, C is the characteristic constant of the instrument and n is the shear wave number.

Quartz crystals were cleaned in ethanol for 30 min and then ozone-treated for 30 min before placing them in the QCM chamber. Inlet and outlet tubing and the QCM chambers were rinsed with ultra pure water (GenPure UV, TKA GmbH, Niederelbert, Germany) before use. The fundamental frequencies were characterized in pure water. The chamber is designed to provide a non-perturbing exchange of liquids over the quartz crystal by means of a pump. A flow rate of 20 μl/min was used and the chamber temperature was maintained at 25 °C during all of the measurements.

3 Results and Discussion

3.1 Tribological Studies of PLL-*g*-PEG in Aqueous Glycerol Solutions

The coefficient of friction (μ) under sliding, measured with the pin-on-disk tribometer, has been plotted as a function of speed (0.1–400 mm/s) for different HEPES–glycerol mixtures in Fig. 1. μ was observed to decrease with increasing sliding velocity, implying the onset of the mixed lubrication regime. Since hydrodynamic forces are viscosity-dependent, increasing the volume fraction of glycerol in the mixture leads to an additional decrease in μ (within the mixed lubrication regime), for a given speed range. A polymer-free HEPES–glycerol solution with a glycerol 75% v/v showed a 55% (from 0.45 to 0.2)

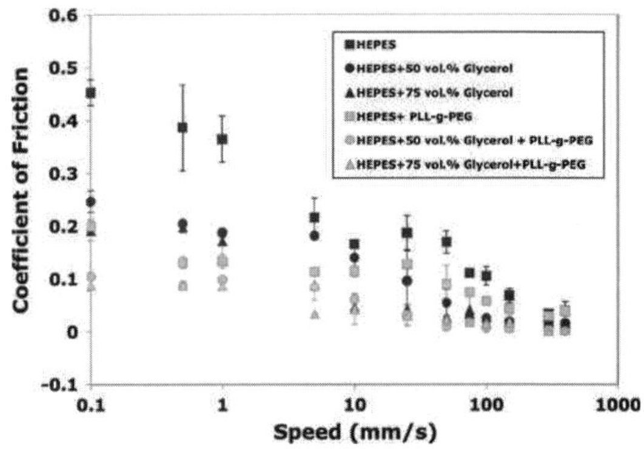

Fig. 1 Speed dependence of coefficient of friction (μ) from pin-on-disk measurements for different compositions of HEPES–glycerol mixtures with (■) and without (■) polymer

reduction in friction at a sliding velocity of 0.1 mm/s in comparison to pure HEPES solution at the same speed. The coefficient of friction values in the absence of polymer appear to converge at 150 mm/s irrespective of the viscosity of the operating fluid, suggesting that asperity contacts between surfaces are no longer occurring; the frictional forces originate only from viscous dissipation within the fluid film, and thus the coefficient of friction at the interface diminishes to a very low value.

PLL-*g*-PEG, when adsorbed from HEPES onto glass and steel surfaces, has been observed to reduce the coefficient of friction at the sliding interface [12]. The adsorption of PLL-*g*-PEG polymers from viscous HEPES–glycerol solutions has been observed by means of the QCM technique (Fig. 2). The change in the fundamental frequency of the crystal measured in real time provides an indication of the adsorption kinetics of the polymer along with the total adsorbed polymer mass (including that of the associated solvent). Since the adsorbed polymer will undergo a mushroom-brush transition during the adsorption process, the mass fraction of solvent associated with the polymer changes. Figure 2 shows the changes in frequency of the crystals in the 7th overtone while adsorbing PLL-*g*-PEG from different binary mixtures of HEPES and glycerol (0, 25, and 50% v/v of glycerol in HEPES). The percentage of glycerol in HEPES never exceeded 50%, as the free flow of higher concentration solutions through the thin inlet tubing was restricted due to the high viscous drag present.

In the polymer-adsorption studies carried out with different HEPES–glycerol solutions, the baseline was first obtained in HEPES buffer and the solution subsequently exchanged with the HEPES–glycerol mixture (arrow marked 'A' in Fig. 2). A decrease in the resonant frequency of the crystal was observed due to the drag forces

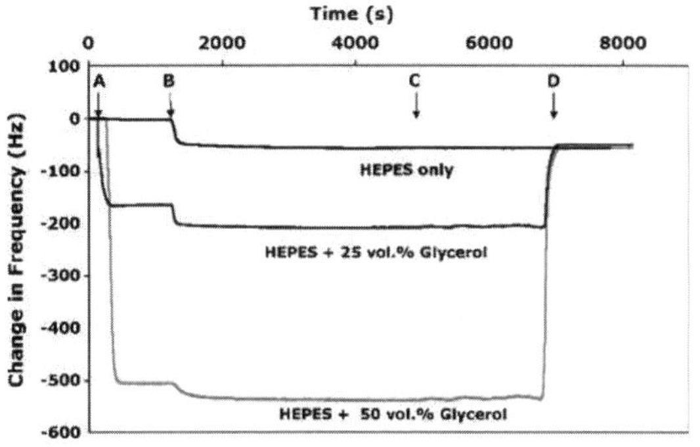

Fig. 2 Change in oscillating frequency of QCM crystal coated with SiO_2 with time during the adsorption of PLL-*g*-PEG from different compositions of HEPES–glycerol mixtures. The baseline was obtained in HEPES for all experiments. *A* Injection of HEPES–glycerol mixtures without polymer, *B* injection of PLL-*g*-PEG dissolved in corresponding HEPES–glycerol mixture, *C* rinsing of the physisorbed polymer with HEPES–glycerol solutions, *D* exchanging the solution back to HEPES

on the crystal originating from the viscosity of the solution. PLL-*g*-PEG dissolved in a HEPES–glycerol solution of the same mixing ratio, at a polymer concentration of 0.25 mg/ml, was injected into the cell (B). The kinetics of adsorption of the copolymers onto the bare SiO_2 surface are observed. The adsorption was carried out in flow mode until no further decrease in the frequency was seen. The crystal was then rinsed with the HEPES–glycerol mixture to remove any physisorbed or loosely bound copolymers from the crystal (C). The solution was finally exchanged back to HEPES solution (D) and the change in the frequency with respect to the baseline noted.

The total mass of copolymer adsorbed on the SiO_2 surfaces from HEPES–glycerol mixtures is comparable to that adsorbed from pure HEPES solution. The changes in frequency and the corresponding mass of the polymer for different HEPES–glycerol ratios have been tabulated (Table 2). In calculating the adsorbed polymer mass on the crystal, the effect of dissipative forces arising from the viscosity of the HEPES–glycerol solutions is eliminated by taking the difference in frequencies measured in pure HEPES solution before and after adsorption for all HEPES–glycerol mixtures. We also note that the viscoelastic nature of the polymer is not considered in the mass calculations from the Sauerbrey equation. The polymer on the surface is considered to be stiff and rigid to calculate the wet mass of the polymer. This is a gross approximation, and thus the calculated mass can only be considered for comparing the amount of mass adsorbed from different solutions. The calculated masses show a comparable adsorption of the polymer onto the SiO_2-coated quartz surface from different HEPES–glycerol solution compositions.

Table 2 The change in frequency and the corresponding mass of PLL-*g*-PEG adsorbed on the SiO_2-coated QCM crystal from HEPES–glycerol mixtures

Concentration of glycerol in water (vol.%)	Change in frequency upon polymer adsorption relative to baseline (Hz)	Mass of the adsorbed polymer, including solvent (from Sauerbrey equation) (ng/cm^2)
0	−55	135
25	−54	133
50	−55	135

The wet mass of the polymer is calculated from the Sauerbrey equation. The visco-elastic properties of the polymer in the solution are neglected ($C = 17.7$ ng cm^{-2} Hz^{-1} for a 5 MHz crystal and $n = 7$th overtone)

The polymer, when adsorbed from viscous solutions assists in the reduction in the coefficient of friction (Fig. 1) under the boundary- and mixed-lubrication conditions of pin-on-disk tribometry. The reduction in the friction was observed to be about 60% (from 0.25 to 0.1) at 0.1 mm/s when the polymer was adsorbed at the interface from 50% v/v HEPES–glycerol solution in comparison to a similar system with no polymer at the interface. The coefficient of friction also appears to converge at high speed (150 mm/s) in the presence of polymer for all HEPES–glycerol mixtures, indicating no effect of the polymer on friction as the surfaces become completely separated by a fluid film.

A MTM, in nearly pure rolling contact, can be operated at higher speeds in comparison to pin-on-disk experiments. Thus, MTM can be used to characterize the lubrication behavior of PLL-*g*-PEG copolymer in HEPES–glycerol mixtures in lubrication regimes beyond boundary lubrication. Milder shear stresses are applied to the adsorbed copolymer in rolling contact as compared to the sliding pin-on-disk experiments. Thus, the values of coefficient of friction for MTM measurements are observed to be much lower than for pin-on-disk experiments for the same speeds and solvent conditions, since the removal of polymer from the surface by shear (itself a dissipative process) is a less frequent occurrence. Coefficients of friction for different percentages of glycerol in HEPES (0, 50, and 75% v/v) are plotted against mean speed, the average speed of the ball and the disk at a constant SRR of 10%, in Fig. 3. In MTM measurements, the coefficient of friction was obtained for contact speeds ranging from 10 to 2500 mm/s. At lower operating speeds, MTM results show behavior similar to those observed under sliding conditions with pin-on-disk (Fig. 1). Again there is an initial decrease in the friction coefficient with increasing speed in the presence of viscous lubricant. In pure HEPES solution, the coefficient of friction between bare surfaces decreases by three orders of magnitude as the contact speed is increased from the lowest

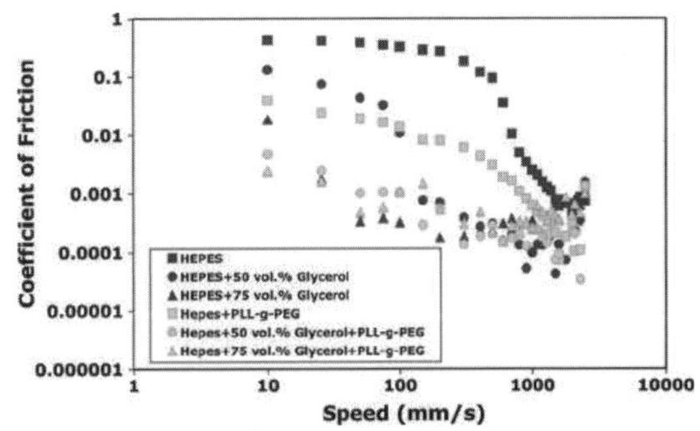

Fig. 3 Speed dependence of coefficient of friction (μ) from mini-traction-machine measurements for different compositions of HEPES–glycerol mixtures with (■) and without (■) polymer. A rotating steel ball is brought into contact with a rotating glass disk under an applied load of 10 N and with a track radius of 20.7 mm. The mean speed of the ball and disk varied from 0 to 2500 mm/s with a SRR of 10%

(10 mm/s) to the highest value (2500 mm/s). At 10 mm/s with MTM, the addition of 50% v/v glycerol to the pure HEPES solution reduces the coefficient of friction from 0.5 to 0.2 and addition of polymer to the 50% HEPES–glycerol solution further reduces the friction by more than an order of magnitude (from 0.2 to 0.006). As the viscosity of the lubricants is increased, the onset of hydrodynamic lubrication is shifted to lower speeds (Fig. 3). In addition to viscosity-related effects, the adsorption of polymer at the interface assists in the further reduction of the friction in the boundary and the mixed regime, by reducing asperity contact, as also observed in sliding contact (pin-on-disk).

At higher operating speeds, the viscous solutions form a complete lubricating film. For 50% v/v glycerol in HEPES solution, the coefficient of friction increases with increasing speed above 1000 mm/s, indicating the onset of full-fluid-film lubrication. Upon complete film formation between the contacts, the presence of polymer at the interface no longer has any effect on the frictional properties and thus the frictional curves (Fig. 3) show similar behavior with and without polymer. At high lubricant viscosity (i.e., 75% v/v glycerol in HEPES) the increase in friction forces occurs at speeds as low as 100 mm/s due to the onset of the hydrodynamic regime at much lower contact speeds.

The effect of polymer brushes in different lubrication regimes is seen in Fig. 4a and b, in which the Stribeck curves obtained from pin-on-disk and MTM measurements are plotted, respectively. The coefficient of friction is plotted against speed multiplied by viscosity for all HEPES–glycerol mixtures both in the presence and the absence of the polymer. As expected, the effect of polymer in reducing the friction is predominantly seen in the boundary-lubrication regime. The effect of polymer on the friction is also extended to the mixed-lubrication regime

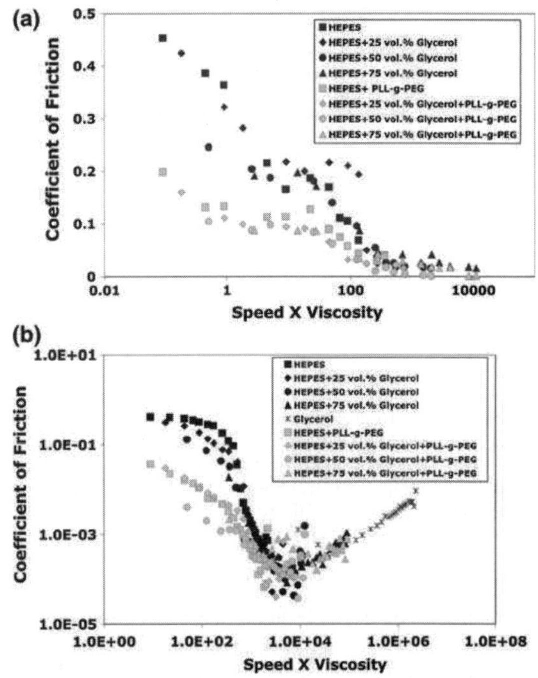

Fig. 4 **a** (Speed × viscosity) dependence of coefficient of friction (μ) from pin-on-disk measurements for different HEPES–glycerol mixtures with (■) and without (■) polymer, **b** (Speed × viscosity) dependence of coefficient of friction (μ) from mini-traction-machine measurements for different HEPES–glycerol mixtures with (■) and without (■) polymer

where there still exists a partial contact between the surfaces. In the hydrodynamic regime, the presence of polymer did not affect the frictional properties.

Similar studies have been conducted with different ratios of ethylene glycol in HEPES solution. Figure 5

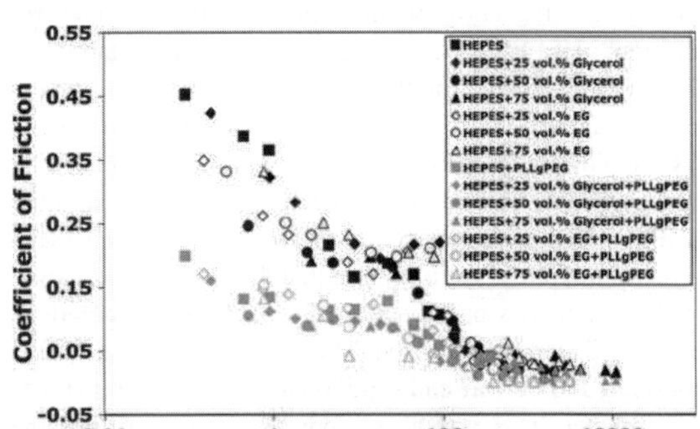

Fig. 5 (Speed × viscosity) dependence of coefficient of friction (μ) from pin-on-disk measurements for different compositions of HEPES–glycerol and HEPES–ethylene glycol (EG) mixtures with (■, □) and without (■, □) polymer

shows the results of pin-on-disk studies for both HEPES–glycerol and HEPES–EG solutions and the similarity of the curves indicates that it is purely the effect of viscosity of the lubricant that is being observed.

3.2 "Rehealing" Studies of PLL-*g*-PEG in Aqueous Glycerol Mixtures

There is a continuous shear of the polymer at the interface when the two contacting surfaces slide against each other. Under the conditions used in the pin-on-disk tribological experiments, the shear forces exerted on the polymer chains in the vicinity of underlying asperities were comparable to the binding strength of the polymer to the substrate and thus the polymer was partially removed from the interface after each cycle. In order to maintain a low coefficient of friction over a large number of rotations there is a need for either a strong polymer attachment with the surface or replacement of the sheared polymer by fresh polymer from the bulk solution. PLL-*g*-PEG interacts with the surface via a weak electrostatic attraction and thus the polymer is partially sheared from the interface under tribological stress. Lee et al. [15] have studied the rehealing of the tribo-stressed contact by diffusion of the polymer to the surface by means of tribometry and fluorescence microscopy. It was shown that a high concentration of polymer in the bulk lubricant (0.25 mg/ml) will provide sufficient polymer in the vicinity of the stressed area and thus rehealing of the sheared area can occur before the onset of the next rotation. The diffusivity of PLL-*g*-PEG and the concentration of the polymer in the vicinity of the contact are important parameters that influence the rehealing process. By maintaining the same concentration as used by Lee et al., we have explored the effect of solution viscosity (and thus the diffusion rate of the PLL-*g*-PEG) on the rehealing mechanism. Viscosity was varied by adding different volume fractions of either glycerol or ethylene glycol to HEPES solution.

Figure 6 shows the rehealing properties of adsorbed PLL-*g*-PEG in the presence of PLL-*g*-PEG-containing viscous solutions, as measured with the pin-on-disk tribometer. Figure 6a (0% v/v glycerol) and b (50% v/v glycerol) are representative graphs for the different viscosities tested. Four sets of experiments were conducted for each mixture of HEPES with glycerol or ethylene glycol. All experiments were conducted at a sliding velocity of 0.5 mm/s on a track of radius 3 mm under a 2-N load in the presence of 20 ml of the lubricant solution. The number of rotations used for the experiment was increased with increasing lubricant viscosity, i.e., 50 revolutions for 0% glycerol and 200 for 50% v/v glycerol in HEPES. For *Case I*, the bare steel/glass pair was brought into contact in the presence of solutions of different viscosity (data not

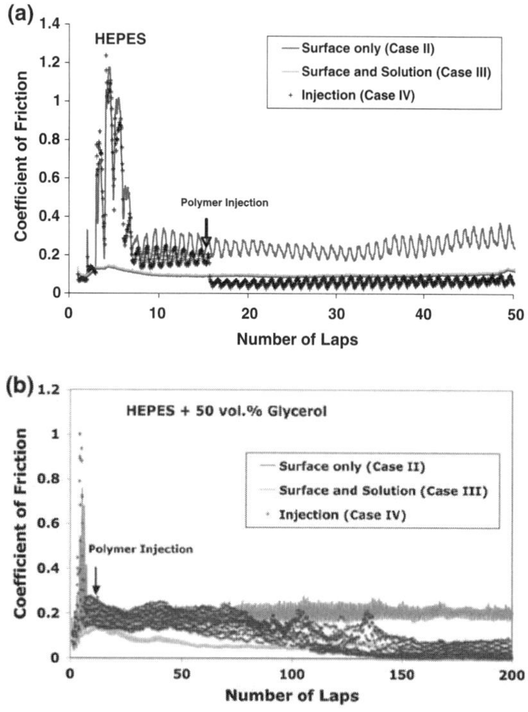

Fig. 6 Coefficient of friction (μ) versus number of laps for sliding contact between a steel pin and a glass disk in a pin-on-disk tribometer. **a** For HEPES solution as lubricant fluid **b** For 50 vol.% glycerol in HEPES as lubricant fluid. Case II concerns the tribopair pre-incubated in an aqueous-polymer-containing glycerol solution, while the solution in the cup does not contain any dissolved polymer. In Case III, the surface was also pre-coated with polymer in a similar way, and also the solution in the cup contains the polymer at a concentration of 0.25 mg/ml. Case IV represents surfaces that are similarly pre-coated with polymer but the polymer concentration of the solution in the cup is changed from 0 to 0.25 mg/ml following the injection of the polymer solution into the cup at the 13th lap (load = 2 N, sliding speed = 5 mm/s and track radius = 3 mm)

shown). A running-in effect was observed within the first few laps and the coefficient of friction achieved a steady value. The friction at a bare steel/glass contact decreased with increasing viscosity of the solution, as previously described (Fig. 1). For *Case II*, both the pin and the disk were pre-incubated in an aqueous glycerol or ethylene glycol solution containing polymer at a concentration of 0.25 mg/ml for 30 min. The solution in the cup during tribological measurements had the same percentage of glycerol or ethylene glycol in HEPES as used for the adsorption, but with no dissolved polymer. Due to the prior adsorption of the polymer at the interface, the first 3–5 laps show a low coefficient of friction. However, due to high tribological stresses the polymer was sheared away from the contact—with no polymer in the solution to heal the track, the coefficient of friction reached the friction coefficient of bare contact as observed for Case I. In *Case III*, the pin and the disk were pre-incubated in the polymer solution for 30 min and the cup contained a HEPES–glycerol solution with PLL-*g*-PEG polymer at a concentration of 0.25 mg/ml. The graph shows a low value of μ for all laps due to continuous replacement of polymer, implying rapid rehealing of the sheared contact area during the tribological test. Rapid rehealing of the contact by the polymer from the bulk solution to the surface was observed for all cases, irrespective of the viscosity of the solution. Finally, for *Case IV*, samples incubated in HEPES–glycerol–PLL-*g*-PEG for 30 min were first run in polymer-free HEPES–glycerol or HEPES–ethylene glycol mixtures (15 ml). The system displayed a low coefficient of friction in the initial laps but soon the value of μ reached that of the bare contact, as seen for Case II. The contact was run for several laps under these conditions to ensure complete removal of polymer from the track. After lap 13, a 5-ml mixture of aqueous glycerol or ethylene glycol with a polymer concentration of 1 mg/ml was injected into the cup, so that the overall concentration of the solution in the cup was brought to 0.25 mg/ml. The μ value decreased from that seen in the later stages of Case II to that obtained for the polymer-lubricated contact (Case III). At low viscosities, the decrease in friction was observed immediately—within one rotation of the tribo pair. With higher viscosities, however, the number of rotations required to observe the onset of rapid rehealing was increased.

Figure 7 shows the time required to establish rapid rehealing of the contact for different percentages of glycerol or ethylene glycol in HEPES buffer. The rate of diffusion of PLL-*g*-PEG to the surface to establish the rapid rehealing process is seen to be highly viscosity-dependent, taking less than one lap for pure HEPES solution but more than 125 laps (500 s) for 50% glycerol in HEPES. The effect of viscosity of the solution on the diffusion of the polymer to the interface is shown in Fig. 7 (inset) which plots the time required to establish the rapid rehealing process against the dynamic viscosity of the glycerol or ethylene glycol solution. The rate of diffusion of the polymer through the solution is a function of the size of the polymer and the viscosity of the solution. With the help of the Einstein–Stokes equation the diffusion rate of the polymer through a solution can be expressed as [24]:

$$D = \frac{kT}{6\pi\eta R}$$

where D is the Fickian diffusion constant of the polymer, η is the dynamic viscosity, and R is the effective radius of the molecule. Making the simplifying assumption that the extended volume of a PLL-*g*-PEG copolymer molecule is similar in all the HEPES–glycerol mixtures, the rate of diffusion should be inversely proportional to the viscosity of the solution. A straight line through all the data points (Fig. 7 inset) shows the linear dependence of the time to onset of rapid rehealing on the viscosity of the solution and hence its dependence on the (rate of diffusion)$^{-1}$ of PLL-*g*-PEG in the viscous solution.

According to Lee et al. [15], the rapid rehealing properties observed for PLL-*g*-PEG can be attributed to the fast adsorption kinetics of the polymer through a low-viscosity bulk solution to the surface. The present rehealing studies of PLL-*g*-PEG in viscous solutions also show a rapid replacement of the polymer when the surfaces are significantly covered with polymer, i.e., after a fraction has been removed by shear (Case III). However, similar kinetics were not observed for surfaces from which polymer had been completely tribologically removed (Case IV). This suggests that the time required to reestablish a monolayer of polymer on these essentially bare contact regions is noticeably increased with increasing viscosity. In the case of the largely covered surface (Case III), the need for polymer adsorption to reestablish the monolayer is much less pronounced, and therefore the effects of viscosity less noticeable on the timescales probed in these experiments. The experiments of Case IV show that adsorption is slowed down by diffusion of molecules from the bulk solution, which, in turn, is slowed down at higher viscosities.

The adsorption kinetics of the PLL-*g*-PEG from HEPES–glycerol mixtures, as monitored by QCM measurements, also show that the rate of adsorption of the polymer onto the surface has a clear dependence on the concentration of the glycerol in the solution (Fig. 2). The time required to form a fully covered polymer film on the surface increases with increasing glycerol content in the HEPES–glycerol solution.

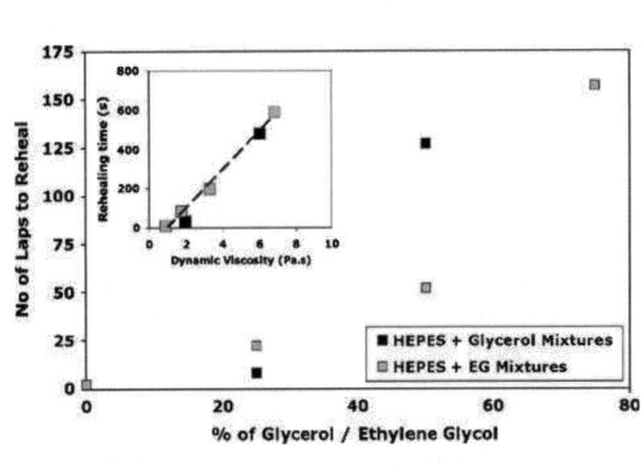

Fig. 7 Number of laps or time required for the onset of rapid rehealing of the PLL-*g*-PEG layer in the contact for different compositions of HEPES–glycerol (■) and HEPES–ethylene glycol (EG) (■) solutions are shown. *Inset* rehealing time has a linear relationship with the bulk dynamic viscosity of aqueous glycerol and EG solutions, irrespective of the volume fractions

Tribol Lett (2010) 37:541–552 549

3.3 Film-thickness Calculations for Lubricants Consisting of PLL-*g*-PEG in Aqueous Glycerol Mixtures

While solutions of PLL-*g*-PEG in water or buffer solution have previously been shown to lubricate a variety of contacting materials in the boundary-lubrication regime [11, 25–27], hydrodynamic lubrication was not observed with these solutions, except at very high speeds. The use of aqueous glycerol solutions of PLL-*g*-PEG means that while PLL-*g*-PEG can still function as a boundary lubrication additive, the intrinsically higher viscosity of the base fluid leads to hydrodynamic lubrication at lower speeds than observed with water or HEPES. Calculations of lubrication regimes and film-thickness values for hard contacts in aqueous glycerol mixtures are presented in this section. In addition to their intrinsically higher viscosities at ambient pressures, the pressure-viscosity coefficients, α, of aqueous glycerol mixtures, which are higher than that of water, can further improve the load-bearing capacity of hard contacts and can thus modify the operative lubrication regime. Theoretical models suggest that the elasticity of the contact material and the viscosity of the lubricant are the two main parameters determining the nature of the fluid film formed between the contacting surfaces [28]. Depending on the relative magnitudes of these quantities, fluid-film lubrication can be divided into four regimes, namely iso-viscous rigid (IR), piezo-viscous rigid (VR), iso-viscous elastic (IE), and piezo-viscous elastic (VE). The equations for calculating the dimensionless viscosity and elasticity parameters for the contact are given by Hamrock and Dowson [28, 29].

Dimensionless viscosity parameter, $g_v = \dfrac{GW^3}{U^2}$

Dimensionless elasticity parameter, $g_E = \dfrac{W^{8/3}}{U^2}$

where G, W, and U represent the dimensionless material parameter, dimensionless load parameter, and dimensionless speed parameter, respectively. These parameters are generalized for different contact geometries (for example, elliptical or line contact) and contacting materials (such as elastomers to steel) and are defined as given below:

Table 3 Pressure-viscosity coefficient values (α) for different concentrations of glycerol in water (vol.%)

Concentration of glycerol in water (vol.%)	0	25	50	75	100
Pressure-viscosity coefficient (α) ($\times 10^{-9}$ m^2 N^{-1})	0.75	1.5	2.25	3	3.74

The intermediate pressure viscosity values are approximated by linear interpolation of the pure water and glycerol α values [30, 31]

Dimensionless material parameter, $G = \alpha E'$

Dimensionless load parameter, $W = \dfrac{w}{E'R_x^2}$

Dimensionless speed parameter, $U = \dfrac{\eta_0 u}{E'R_x}$

where α is the pressure-viscosity coefficient of the lubricant, E' the effective elastic modulus, w the applied load, η_0 the viscosity of the lubricant, u the mean speed of the contact, and R_x the effective radius of the contact in the sliding direction.

The calculated values are plotted on the lubrication maps in Fig. 8. The pressure-viscosity coefficients for pure water and pure glycerol have been taken from the literature [30, 31] and linear interpolation of these values approximates the values for intermediate aqueous glycerol mixtures (Table 3). Operating conditions of POD and MTM experiments are used as input parameters to calculate the non-dimensional elastic (g_E) and viscosity (g_V) parameters. The contact area is considered to be circular and thus an ellipticity parameter (k) equal to 1 is used for the steel ball in contact with the glass (silica) disk ($E' = 112$ GPa, $R_x = 9.53 \times 10^{-3}$ m). Values of g_V plotted against g_E lie in the isoviscous-elastic regime (Fig. 8). The thickness of the lubricating film is calculated from the dimensionless film thickness parameter derived for the isoviscous-elastic regime [29]:

Dimensionless film thickness parameter,

$$(\hat{H}_{\min}) = 8.70 g_E^{0.67}\left(1 - 0.85e^{-0.31k}\right)$$

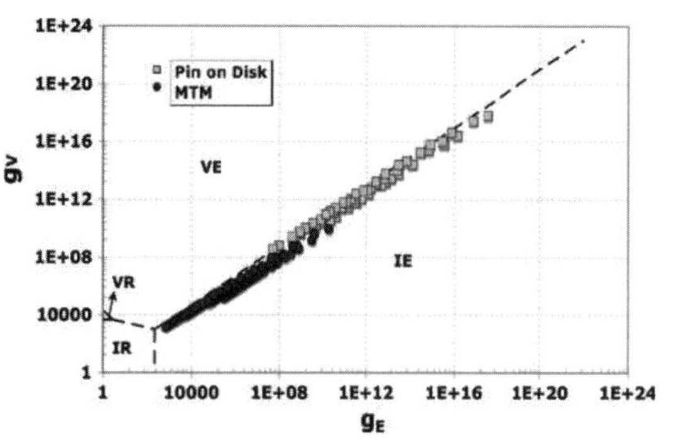

Fig. 8 All MTM (●) and pin-on-disk (■) data at different concentrations of aqueous glycerol mixtures, plotted on a lubrication-regime map, obtained from the Esfahanian–Hamrock–Dowson equations [28] for a circular contact (ellipticity parameter $k = 1$). The four different regimes in the dimensionless viscosity (g_V) versus elastic (g_E) parameter plot are iso-viscous rigid (IR), iso-viscous elastic (IE), piezo-viscous rigid (VR), and piezo-viscous elastic (VE). All the values reported in this study lie in the iso-viscous elastic regime. It should be noted that while the equations in [28] apply to rolling contact, the model can also be used for sliding geometries at the low speeds used in our pin-on-disk experiments [35]

also, $(\hat{H}_{\min}) = H\left(\frac{W}{U}\right)^2$ where H is the dimensionless film thickness, $H = \frac{h}{R_x}$ and h is the film thickness of the lubricating film between the contacts. The values of h are plotted against speed for different HEPES–glycerol mixtures (Fig. 9a).

In actual tribological contacts, the surfaces coming into contact will have a non-negligible roughness in comparison to the lubricating film thickness formed at the interface. The relative magnitude of the film thickness in comparison to the surface roughness is given by the λ ratio [32].

$$\lambda = \frac{h_{\min}}{\sqrt{\left(\sigma_{\text{pin}}^2 + \sigma_{\text{disk}}^2\right)}}$$

where σ_{pin} is the RMS surface roughness of the pin, σ_{disk} is the RMS surface roughness of the disk, $h_{\min}$ is the minimum film thickness, and the λ ratio estimates the lubrication regime for rough surfaces.

The RMS roughness values for steel ball and silica disk used for MTM are 11 and 2 nm, respectively (obtained from PCS instruments) and those for the steel ball and glass wafer used for POD experiments are 32 and 5 nm, respectively (measured by AFM). Figure 9b plots the λ values against speed for all of the POD and MTM experiments. The curves show that all the contacts during POD measurements were in the boundary-lubrication regime. Though there was an increase in the λ value with addition of glycerol, the values still lie below 1, indicating a high probability for asperity contact. For MTM measurements, on the other hand, although contacts tested at high speeds and in viscous lubricant (0.5 or 0.75 volume fraction of glycerol in HEPES) showed λ values above 3, for other operating conditions the λ values indicated either the boundary- or mixed-lubrication regime. Thus, there is a need for adsorbed copolymers of PLL-*g*-PEG on the sliding surfaces to reduce the interfacial friction generated at the asperity contacts, even in the presence of aqueous viscous lubricants.

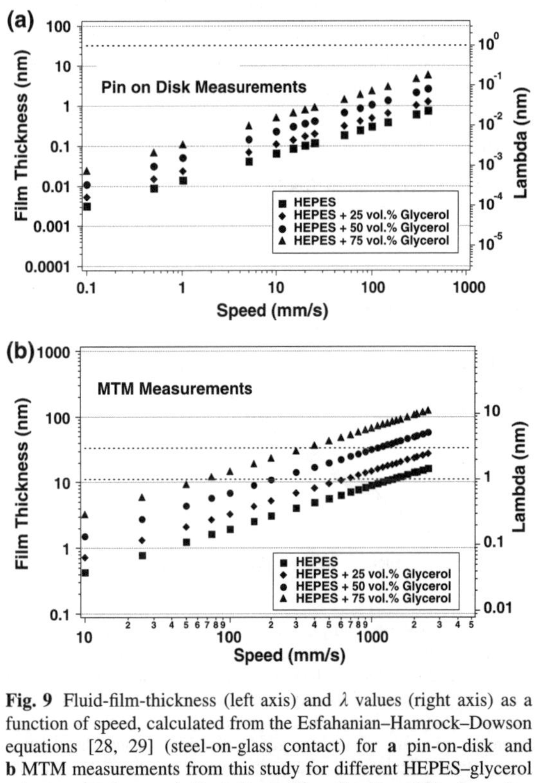

Fig. 9 Fluid-film-thickness (left axis) and λ values (right axis) as a function of speed, calculated from the Esfahanian–Hamrock–Dowson equations [28, 29] (steel-on-glass contact) for **a** pin-on-disk and **b** MTM measurements from this study for different HEPES–glycerol mixtures. λ values are indicated to enable estimation of film formation on the rough surfaces. $\lambda < 1$ represents boundary lubrication, $\lambda > 3$ represents full-fluid-film lubrication and $1 \leq \lambda \leq 3$ represents the mixed regime

4 Conclusions

It is has been shown that when poly(*L*-lysine)-*g*-poly(ethylene glycol) is dissolved in aqueous glycerol solutions, the tribological properties can be improved both in the boundary- and the mixed-lubrication regimes. Different percentages of glycerol in HEPES have been used to vary the viscosity of the solvent, which enables hydrodynamic lubrication to take place over a wider speed range than for the pure HEPES case. The effect of the polymer (in glycerol-containing solution) in different lubrication regimes was demonstrated by means of a Stribeck plot, which shows that the presence of polymer at the interface can reduce the friction as long as there exists asperity–asperity contact between the tribo pair; the viscous solvents separate the contacting surfaces due to hydrodynamic forces and the presence of hydrated polymer brushes reduces the interfacial friction between contacting surface asperities. Also, Esfahanian–Hamrock–Dowson film-thickness calculations show that the lubricating HEPES–glycerol films formed in our pin-on-disk experiments are very thin, so that numerous asperity–asperity contacts are expected, and thus the presence of copolymer at the surface is necessary to further reduce the friction. In mini-traction-machine measurements in rolling contact, on the other hand, the full-fluid-film-lubricated region could be examined, in which the presence of the polymer was found to have negligible effect.

Quartz-crystal microbalance measurements showed that the total amount of adsorbed polymer appeared unaffected by the presence of glycerol. The kinetics of adsorption of PLL-*g*-PEG from the HEPES–glycerol solution to the interface was investigated to help understand the effect of the increased viscosity on the rehealing of the tribo-stressed contact. Although there is an inverse relation between the viscosity, the solvent, and the rate of diffusion of the polymer to the interface, the time required for the polymer to reheal wear-damaged polymer layers appears to be less than that between contacts in successive laps, and thus re-healing of the contact remains independent of the viscosity of the solvent under the conditions investigated. These results show that the use of HEPES–glycerol solutions as a viscosity-enhanced solvent for PLL-*g*-PEG can expand the applicability of aqueous lubrication to a significantly larger range of operating conditions.

Acknowledgments The financial assistance of the European Science Foundation, through their Eurocores (FANAS) program is gratefully acknowledged. We would also like to thank Dr. Rowena Crockett of EMPA, Dübendorf, Switzerland for allowing us to use her mini-traction-machine and Prof. Hugh Spikes of Imperial College, London for his valuable suggestions.

References

1. Granick, S., Kumar, S.K., Amis, E.J., Antonietti, M., Balazs, A.C., Chakraborty, A.K., Grest, G.S., Hawker, C., Janmey, P., Kramer, E.J., Nuzzo, R., Russell, T.P., Safinya, C.R.: Macromolecules at surfaces: research challenges and opportunities from tribology to biology. J. Polym. Sci. B **41**, 2755–2793 (2003). doi: 10.1002/Polb.10669
2. Klein, J., Kumacheva, E., Mahalu, D., Perahia, D., Fetters, L.J.: Reduction of frictional forces between solid-surfaces bearing polymer brushes. Nature **370**, 634–636 (1994). doi: 10.1038/370634a0
3. Grest, G.S.: Interfacial sliding of polymer brushes a molecular dynamics simulation. Phys. Rev. Lett. **76**, 4979–4982 (1996). doi: 10.1103/PhysRevLett.76.4979
4. Klein, J., Perahia, D., Warburg, S.: Forces between polymer-bearing surfaces undergoing shear. Nature **352**, 143–145 (1991). doi:10.1038/352143a0
5. Klein, J., Kamiyama, Y., Yoshizawa, H., Israelachvili, J.N., Fredrickson, G.H., Pincus, P., Fetters, L.J.: Lubrication forces between surfaces bearing polymer brushes. Macromolecules **26**, 5552–5560 (1993). doi:10.1021/ma00073a004
6. Heuberger, M., Drobek, T., Spencer, N.D.: Interaction forces and morphology of a protein-resistant poly(ethylene glycol) layer. Biophys. J. **88**, 495–504 (2005). doi:10.1529/Biophysj.104.045443
7. Irfachsyad, D., Tildesley, D., Malfreyt, P.: Dissipative particle dynamics simulation of grafted polymer brushes under shear. Phys. Chem. Chem. Phys. **4**, 3008–3015 (2002). doi:10.1039/B110738k
8. Chen, M., Briscoe, W.H., Armes, S.P., Klein, J.: Lubrication at physiological pressures by polyzwitterionic brushes. Science **323**, 1698–1701 (2009). doi:10.1126/Science.1169399
9. Lee, S., Spencer, N.D.: Materials science—sweet, hairy, soft, and slippery. Science **319**, 575–576 (2008). doi:10.1126/Science.1153273
10. Pasche, S., Textor, M., Meagher, L., Spencer, N.D., Griesser, H.J.: Relationship between interfacial forces measured by colloid-probe atomic force microscopy and protein resistance of poly(ethylene glycol)-grafted poly(L-lysine) adlayers on niobia surfaces. Langmuir **21**, 6508–6520 (2005). doi:10.1021/La050386x
11. Müller, M., Lee, S., Spikes, H.A., Spencer, N.D.: The influence of molecular architecture on the macroscopic lubrication properties of the brush-like co-polyelectrolyte poly(L-lysine)-g-poly(ethylene glycol) (PLL-*g*-PEG) adsorbed on oxide surfaces. Tribol. Lett. **15**, 395–405 (2003). doi:10.1023/B:TRIL.0000003063.98583.bb
12. Lee, S., Müller, M., Ratoi-Salagean, M., Voros, J., Pasche, S., De Paul, S.M., Spikes, H.A., Textor, M., Spencer, N.D.: Boundary lubrication of oxide surfaces by poly(L-lysine)-*g*-poly(ethylene glycol) (PLL-*g*-PEG) in aqueous media. Tribol. Lett. **15**, 231–239 (2003). doi:10.1023/A:1024861119372
13. Yan, X.P., Perry, S.S., Spencer, N.D., Pasche, S., De Paul, S.M., Textor, M., Lim, M.S.: Reduction of friction at oxide interfaces upon polymer adsorption from aqueous solutions. Langmuir **20**, 423–428 (2004). doi:10.1021/La035785b
14. Drobek, T., Spencer, N.D.: Nanotribology of surface-grafted PEG layers in an aqueous environment. Langmuir **24**, 1484–1488 (2008). doi:10.1021/La702289n
15. Lee, S., Muller, M., Heeb, R., Zurcher, S., Tosatti, S., Heinrich, M., Amstad, F., Pechmann, S., Spencer, N.D.: Self-healing behavior of a polyelectrolyte-based lubricant additive for aqueous lubrication of oxide materials. Tribol. Lett. **24**, 217–223 (2006). doi:10.1007/S11249-006-9121-9
16. Raviv, U., Tadmor, R., Klein, J.: Shear and frictional interactions between adsorbed polymer layers in a good solvent. J. Phys. Chem. B **105**, 8125–8134 (2001). doi:10.1021/Jp0041860
17. Müller, M.T., Yan, X.P., Lee, S.W., Perry, S.S., Spencer, N.D.: Lubrication properties of a brushlike copolymer as a function of the amount of solvent absorbed within the brush. Macromolecules **38**, 5706–5713 (2005). doi:10.1021/Ma0501545
18. Ross, R.S., Pincus, P.: The polyelectrolyte brush—poor solvent. Macromolecules **25**, 2177–2183 (1992). doi:10.1021/ma00034a018
19. Soga, K.G., Guo, H., Zuckermann, M.J.: Polymer brushes in a poor solvent. Europhys. Lett. **29**, 531–536 (1995)
20. Auroy, P., Auvray, L.: Collapse-stretching transition for polymer brushes—preferential solvation. Macromolecules **25**, 4134–4141 (1992). doi:10.1021/ma00042a014
21. Roters, A., Schimmel, M., Ruhe, J., Johannsmann, D.: Collapse of a polymer brush in a poor solvent probed by noise analysis of a scanning force microscope cantilever. Langmuir **14**, 3999–4004 (1998). doi:10.1021/la971409d
22. Müller, M.T., Yan, X.P., Lee, S.W., Perry, S.S., Spencer, N.D.: Preferential solvation and its effect on the lubrication properties of a surface-bound, brushlike copolymer. Macromolecules **38**, 3861–3866 (2005). doi:10.1021/Ma047468x
23. Raviv, U., Klein, J.: Fluidity of bound hydration layers. Science **297**, 1540–1543 (2002). doi:10.1126/science.1074481
24. Carslaw, H.S., Jaeger, J.C.: Conduction of heat in solids, 2nd edn. Oxford University Press, London (1959)
25. Lee, S., Spencer, N.D.: Aqueous lubrication of polymers: Influence of surface modification. Tribol. Int. **38**, 922–930 (2005). doi: 10.1016/J.Triboint.2005.07.017
26. Hartung, W., Rossi, A., Lee, S.W., Spencer, N.D.: Aqueous lubrication of SiC and Si_3N_4 ceramics aided by a brush-like copolymer additive, poly(L-lysine)-graft-poly(ethylene glycol). Tribol. Lett. **34**, 201–210 (2009). doi:10.1007/S11249-009-9424-8
27. Erdemir, A., Martin, J.-M.: Superlubricity. Elsevier, Amsterdam (2007)
28. Hamrock, B.J, Dowson, D.: Minimum film thickness in elliptical contacts for different regimes of fluid-film lubrication. In: Proceedings of the 5th Leeds-Lyon Symposium on Tribology, pp 22–27, 1979
29. Esfahanian, M., Hamrock, B.J.: Fluid-film lubrication regimes revisited. Tribol. Trans. **34**, 628–632 (1991). doi:10.1080/10402009108982081

30. Totten, G.E.: Handbook of Hydraulic Fluid Technology. Marcel Dekker, Inc, New York (2000)
31. Ohno, N., Ziaur Rahman, M.D., Tsutsumi, H.: High-pressure short time behavior of traction fluids. Lubr. Sci. **18**, 25–36 (2006). doi:10.1002/1s.3
32. Stachowiak, G.W., Batchelor, A.W.: Engineering Tribology. Elsevier, Amsterdam (1993)
33. Lide, D.R.: Handbook of Chemistry and Physics, 20th edn. Chemical Rubber Publishing Co, Cleveland (1948)
34. Corradini, F., Marchetti, A., Tagliazucchi, M., Tassi, L., Tosi, G.: Thermodynamics of viscous-flow in ethane-1, 2-diol+water binary-mixtures. Aust. J. Chem. **48**, 103–113 (1995)
35. Smeeth, M., Spikes, H.A.: The influence of slide/roll ratio on the film thickness of an EHD contact operating within the mixed lubrication regime. In: Proceedings of the 22nd Leeds-Lyon Symposium on Tribology, pp 695–703 (1996)

ARTICLE

Tribological Properties of Poly(L-lysine)-*graft*-poly(ethylene glycol) Films: Influence of Polymer Architecture and Adsorbed Conformation

Scott S. Perry,*,† Xiaoping Yan,‡ F. T. Limpoco,§ Seunghwan Lee,‖ Markus Müller,‖ and Nicholas D. Spencer‖

Departments of Materials Science & Engineering and of Chemistry, University of Florida, Gainesville, Florida 32611, Seagate Technology, 1251 Waterfront Place, Pittsburgh, Pennsylvania 15222, and Laboratory for Surface Science and Technology, Department of Materials, Swiss Federal Institute of Technology, ETH-Hönggerberg, Wolfgang-Pauli-Strasse 10, CH-8093 Zürich, Switzerland

ABSTRACT The tribological properties of poly(L-lysine)-*graft*-poly(ethylene glycol) (PLL-*g*-PEG)-coated oxide interfaces have been investigated with atomic force microscopy (AFM) as a function of the molecular structure. Polymer-bearing surfaces were obtained via spontaneous adsorption of the polymer onto the oxide substrate from a buffered solution of physiological pH. Interfacial friction of these PLL-*g*-PEG-coated surfaces was found to be highly dependent on the duration of deposition and the architecture of PLL-*g*-PEG. In terms of the architecture, the PEG *chain length* and the *grafting ratio* (i.e., the molar ratio of L-lysine monomer to PEG side chain) of adsorbed PLL-*g*-PEG significantly influence the interfacial friction; specifically, friction is reduced as the PEG chain length increases and as the molar ratio of L-lysine monomer to PEG side chain decreases. The characteristics of the polymer deposition time and the influence of the lysine/PEG grafting ratio are rationalized in terms of spatial packing density considerations.

KEYWORDS: atomic force microscopy • friction • lubrication • thin films • polymer coatings • poly(L-lysine)-*graft*-poly(ethylene glycol) • polymer architecture • grafting ratio • silicon oxide

1. INTRODUCTION

Poly(L-lysine)-*graft*-poly(ethylene glycol) (PLL-*g*-PEG) is a member of a family of polycationic PEG-grafted copolymers that have been shown to adsorb on negatively charged surfaces, including various metal oxides, providing a high degree of resistance to protein adsorption (1−3). As a result, PLL-*g*-PEG-modified surfaces have received strong interest in a variety of applications including sensor chips for bioaffinity assays and blood-contacting biomedical devices (4, 5). Similarly, making use of its spontaneous adsorption onto metal oxide surfaces, PLL-*g*-PEG is expected to significantly lubricate these interfaces in physiological aqueous environments by forming a boundary polymer brush layer. As with protein resistance, the lubrication of metal oxide surfaces is of great significance in medical applications.

The structure of PLL-*g*-PEG, depicted in Figure 1, consists of a PLL backbone with multiple PEG side chains that have been grafted onto the backbone via amino groups on a fraction of the lysine units. The remaining amino groups provide the pathway for electrostatic interaction of the PLL backbone with appropriately charged surfaces (1).

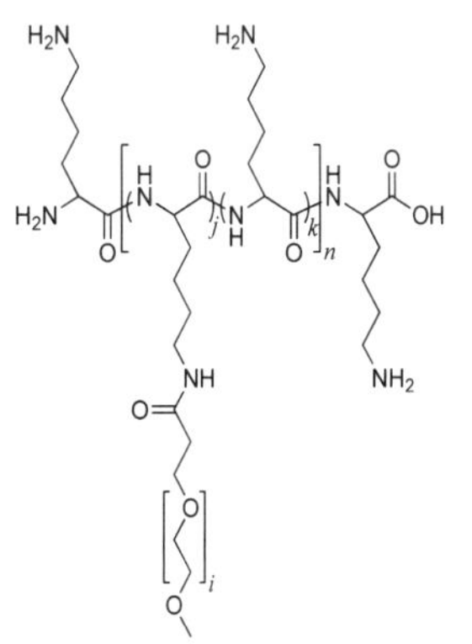

FIGURE 1. Schematic structure of a PLL-*g*-PEG copolymer consisting of a PLL backbone and randomly grafted PEG side chains. In this scheme, $(k+j)/j$ represents the *grafting ratio*, the fraction of lysine units to PEG side chains, for very large n.

In earlier nanotribological investigations using atomic force microscopy (AFM) (6, 7), a significant reduction in the interfacial friction measured between silicon oxide substrates and a sodium borosilicate microsphere was observed upon adsorption of PLL-*g*-PEG on either one or both sides of the interface. In addition, an investigation of PLL-*g*-PEG polymers, differing only in the PEG side-chain length, revealed that interfacial forces measured under aqueous con-

* Author to whom correspondence should be addressed.

Received for review February 16, 2009 and accepted May 5, 2009

† Department of Materials Science & Engineering, University of Florida.

‡ Seagate Technology.

§ Department of Chemistry, University of Florida.

‖ Swiss Federal Institute of Technology.

DOI: 10.1021/am900101m

Published on Web 05/22/2009

ditions were reduced with increasing PEG chain length, indicating that interfacial adhesion and friction can be modified through control of the polymer molecular structure (7). Macroscopic investigations showed that the boundary-lubrication properties of aqueous buffer solutions were significantly improved upon addition of PLL-*g*-PEG, with, again, a strong polymer architecture dependence (8, 9).

In this report, we present a systematic investigation of the dependence of interfacial friction on the molecular architecture of PLL-*g*-PEG brush polymers adsorbed on oxide surfaces. In brush systems, the grafting ratio of backbone to side-chain entities represents one of the most significant synthetic design opportunities. While the influence of the grafting ratio on the solution properties is well understood, less is known regarding how this parameter influences interfacial properties such as adsorption and interfacial friction. To this end, we have employed AFM, a quartz crystal microbalance (QCM), and optical waveguide lightmode spectroscopy (OWLS) to methodically explore a matrix of polymer brushes in which the grafting ratio is systematically varied for a series of side-chain molecular weights. In these studies, we have used a sodium borosilicate microsphere, attached to the end of an AFM cantilever, as the probe to avoid substantial deformation of the polymer layer that would occur under the high contact pressures common in the use of conventional (<100 nm radius) Si_3N_4 tips. Interfacial friction measurements were carried out in a *N*-(2-hydroxyethyl)piperazine-*N*'-2-ethanesulfonic acid (HEPES) buffer solution on polymer-coated silicon oxide substrates with bare microsphere probes. Results for three series of polymer samples, distinguished by the PEG molecular weight, are reported; within each series, the polymers differed only in the lysine/PEG ratio (i.e., the *grafting ratio* or the molar ratio of L-lysine-mers to PEG side chains). On the basis of friction measurements on a series of silicon oxide substrates coated with PLL-*g*-PEG that varied only in their grafting ratios, it was observed that the lysine/PEG fraction substantially influences the interfacial friction of these polymer-associated interfaces in a manner intimately tied to the conformation of PEG side chains extending into the solution. These results have been rationalized in terms of the PEG packing density on the surface using scaling arguments.

2. EXPERIMENTAL SECTION

2.1. Synthesis of PLL-*g*-PEG. In this paper, polymer substrates are designated as PLL(*x*)-*g*[*y*]-PEG(*z*), where the copolymer consists of a PLL backbone of molecular weight *x* kDa and a grafted PEG side chain of molecular weight *z* kDa with a grafting ratio [(lysine-mers)/(PEG side chains)] of *y*.

The PLL-*g*-PEG copolymers were synthesized according to a previously described method (1, 2). Briefly, poly(L-lysine) hydrobromide (PLL-HBr, MW 20 kDa; Sigma, St. Louis, MO) was dissolved in a 50 mM sodium borate buffer solution, followed by filter sterilization of the solution (0.22 μm pore-size filter). The *N*-hydroxysuccinimidyl ester of methoxypoly(ethylene glycol)propionic acid (SPA-PEG; Shearwater Polymers, Inc., Huntsville, AL) was then added to the dissolved PLL-HBr. The reaction was allowed to proceed for 6 h at room temperature, after which the reaction mixture was dialyzed (SpectraPor, MW cutoff size 6–8 kDa, Spectrum, Houston, TX) against deionized water for 48 h. The product was freeze-dried and stored at −20 °C.

Through variation of the molecular weight and the amount of starting material (PLL-HBr and SPA-PEG) as well as effective control of the reaction progress, a series of PLL-*g*-PEG graft copolymers of varying PEG side-chain length and grafting ratios were prepared. Detailed preparation procedures and analytical information of the product obtained via this method have been reported elsewhere (1, 2). In this study, three series of polymers synthesized using different PEG molecular weights were employed. The synthesis approach produced a series of polymers of the general composition PLL(20)-*g*-PEG(2), PLL(20)-*g*-PEG(5), and PLL(20)-*g*-PEG(10), with each series consisting of three or four polymer samples differing only in their grafting ratio.

2.2. Preparation of PLL-*g*-PEG-Coated Metal Oxide Substrates. For polymer deposition on oxide substrates, a given PLL(*x*)-*g*[*y*]-PEG(*z*) was dissolved in 10 mM HEPES [Sigma-Aldrich Inc., St. Louis, MO] at a concentration of 1.0 mg/mL. Unless otherwise noted, all HEPES solutions in this study were adjusted to pH 7.4 with 1.0 M NaOH.

Silicon (100) wafers, passivated with silica, were employed as substrates. Prior to immobilization of PLL-*g*-PEG onto the oxide surface, the wafers (0.5 cm × 0.5 cm) were prepared by sonication in toluene (2 min) and in 2-propanol (10 min), extensively rinsed with ultrapure water (EM SCIENCE, Gibbstown, NJ), dried under a gaseous nitrogen flow, and exposed for 2 min to an oxygen plasma (PDC-32G, Harrick Scientific Corp., Ossining, NY). The oxidized substrates were immediately transferred to a 1.0 mg/mL solution of PLL-*g*-PEG in a 10 mM HEPES buffer and incubated there for 40 min. The polymer-coated substrates were then stored in a HEPES solution (in the absence of PLL-*g*-PEG) until use in AFM experiments. Prior to AFM measurements, the polymer-coated substrates were withdrawn from solution, rinsed with a HEPES buffer and ultrapure water to remove free PLL-*g*-PEG, and then dried under a nitrogen flow.

2.3. Friction Measurements with AFM. AFM was used to probe friction forces at the interfaces of polymer-modified substrates under physiological pH solutions. The microscope was equipped with a liquid cell/tip holder (Digital Instruments, Santa Barbara, CA) and controlled by AFM100/SPM 1000 electronics and software (RHK Technology, Inc., Troy, MI). The microscope makes use of a single-tube scanner, on which substrates are rastered with respect to a fixed tip position, and a beam deflection technique, in which light from a laser diode is reflected from the back of a microfabricated cantilever onto a four-quadrant photodetector. Greater details of this instrumental design have been reported previously (10, 11).

Deflection of the cantilever normal to the surface served to monitor surface topograghy and interfacial adhesion. Torsion or twisting of the cantilever was indicative of frictional forces at the tip–sample interface. Kinetic friction data were acquired by monitoring the lateral deflection of the cantilever as a function of the position across the sample surface and applied normal load. This was accomplished by rastering the sample in a line-scan mode while first increasing and then decreasing the applied load. During this procedure, friction and normal forces were measured simultaneously with a scan speed of 1400 nm/s over a distance of 100 nm. Normal loads were determined from the cantilever's nominal spring constant (k = 0.58 N/m, manufacturer's reported value) and direct measurements of sample displacement. Friction forces were calibrated through an improved wedge calibration method (12, 13).

AFM measurements were carried out in aqueous HEPES solutions; the composition of the liquid environment encompassing the tip–sample interface was controlled by the transfer of aliquots of solution in and out of the liquid cell through the use of two 5 mL syringes. Sodium borosilicate microspheres (Novascan Technologies, Inc., Ames, IA) with 5.1 μm diameter

ARTICLE

ARTICLE

affixed to the end of cantilevers were employed as bare sliding counterfaces against the polymer-coated silicon substrates. These colloidal probes were rinsed with dilute HCl (pH 1) and exposed to oxygen plasma for 15 s between measurements to remove any adhering polymer. Normal loads were limited in order to avoid wear of both the probe and polymer-coated surface. A valid comparison of the friction data was enabled by employing the same tip–cantilever assembly throughout a series of friction measurements, while systematically varying other parameters such as the PEG chain length, grafting ratio, or deposition time. Reported friction data represent the average of at least six results obtained at different locations across the surface. Generally, the measurement of at least one tip–sample condition was repeated at the end of a series to ensure that significant wear had not occurred during the course of measurements.

2.4. OWLS. OWLS was carried out on a BIOS-I instrument (ASI AG, Zürich, Switzerland) using a Kalrez (Dupont, Wilmington, DE) flow-through cell with a volume of 16 μL. The waveguide chips (MicroVacuum Ltd., Budapest, Hungary) consisted of a 1-mm-thick glass substrate and a 200-nm-thick $Si_{0.75}Ti_{0.25}O_2$ waveguiding layer at the surface. A silica layer (ca. 12 nm) was sputter-coated on top of the waveguiding layer in a Leybold direct-current magnetron Z600 sputtering unit. Coating conditions and the principles of OWLS investigations have been described in detail elsewhere (14–16). It is important to note that the surface-adsorbed areal mass density determined by OWLS is regarded as a "dry" areal mass density because of the fact that solvent molecules coupled to the adsorbate will not contribute to a change in the refractive index and, thus, do not contribute to the detected adsorbate mass. The reported "dry" areal mass density (m_{dry}) represents the average of three individual experiments. Because this technique is highly sensitive (detection limit $\approx$ 1 ng/cm^2) and allows for the direct online monitoring of macromolecular adsorption (16), a measurement error of less than 1 % is expected.

2.5. QCM with Dissipation (QCM-D). All QCM-D measurements were performed with a commercial QCM with dissipation monitoring (Q-Sense, Gothenburg, Sweden) equipped with a home-built laminar flow cell, with a glass window allowing visual monitoring of injection and the exchange of liquids (17). All experiments employed 5 MHz AT-cut quartz sensor crystals, sputter-coated with SiO_2 (also Q-Sense). Details of this setup and measurements have been reported elsewhere (18, 19).

The QCM-D response to mass uptake on the crystal oscillator is reflected in the changes in both the resonant frequency (Δf) and dissipation factor (ΔD) at different overtones. In contrast to OWLS, the QCM-D approach is sensitive to viscoelastic properties and the density of *any* mass coupled to the mechanical oscillation of the quartz crystal. In this case, the adsorbed mass consists of the PLL-*g*-PEG copolymer along with solvent molecules associated with it. A Voigt-based model was therefore used in the analysis (software: *Q-tools*, version 2.0.1), where the adsorbed layer was represented as a homogeneous, viscoelastic film characterized by shear viscosity (η_{shear}), shear modulus (E_{shear}), and film thickness (h_{film}) (20–22).

3. RESULTS

3.1. Influence of the Duration of Polymer Deposition. Using the same microsphere–cantilever assembly throughout, the frictional properties of several PLL(20)-*g*[3.5]-PEG(2)-coated SiO_2 substrates were evaluated as a function of the polymer deposition times: 0.5, 1, 5, 10, 30, and 60 min. The data of Figure 2 portray the distinct influence of the duration of the polymer deposition on interfacial friction, indicating that longer deposition times result in lower friction. This is reflected in the rapid reduction in the coefficient of friction (Figure 2b), which is defined as the slope of the friction–load plot (Figure 2a). Within the first 5 min of deposition, the value of the coefficient of friction is reduced to 90 % of that measured for the interface of a bare microsphere against a bare oxidized silicon substrate. This reduction in the coefficient of friction was observed to reach a steady-state value after 1 h. The reduction in friction occurring more slowly over the latter portion of this period likely entails the reorganization of the polymer brush at the substrate–solution interface. This behavior was generally observed for all polymer architectures explored in this study.

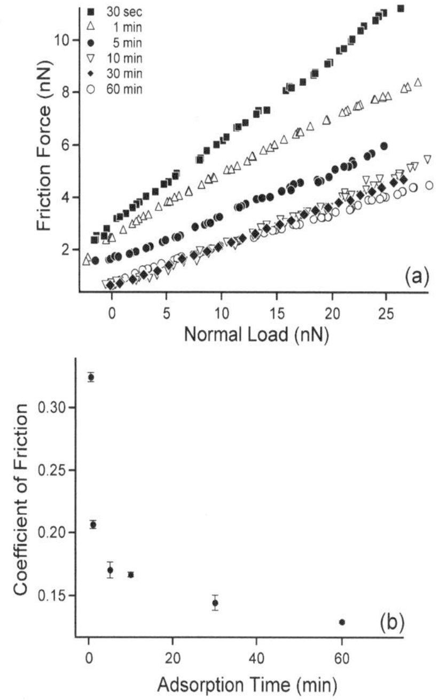

FIGURE 2. (a) Friction versus decreasing load plots of a bare sodium borosilicate microsphere against PLL(20)-*g*[3.5]-PEG(2)-coated SiO_2 substrates prepared at different polymer deposition times. (b) Plot of the coefficient of friction (i.e., slope of the friction–load plots) as a function of the duration of polymer deposition.

3.2. Influence of the Polymer Architecture. Interfacial friction was measured for the contact of a 5.1 μm bare borosilicate probe sliding against silicon oxide substrates coated with a series of PLL-*g*-PEG polymer brushes of different PEG molecular weights and lysine/PEG grafting ratios (Figure 3). Three general effects of the polymer architecture on interfacial friction were apparent. First, lower interfacial friction was observed for tribosystems with increased PEG molecular weight (side-chain length), as is exhibited through the maximum friction forces measured at a given normal load. For example, at 30 nN, friction forces decrease with respect to the PEG molecular weight (for a given approximate grafting ratio) in the order of ~10 nN for PLL(20)-*g*(5.7)-PEG(2), ~8 nN PLL(20)-*g*(5.2)-PEG(5), and ~6 nN PLL(20)-*g*(5.8)-PEG(10), in good agreement with prior reports for this polymer brush system (7).

Second, for each PLL-*g*-PEG series, the interfacial friction decreases with decreasing lysine/PEG grafting ratio, as

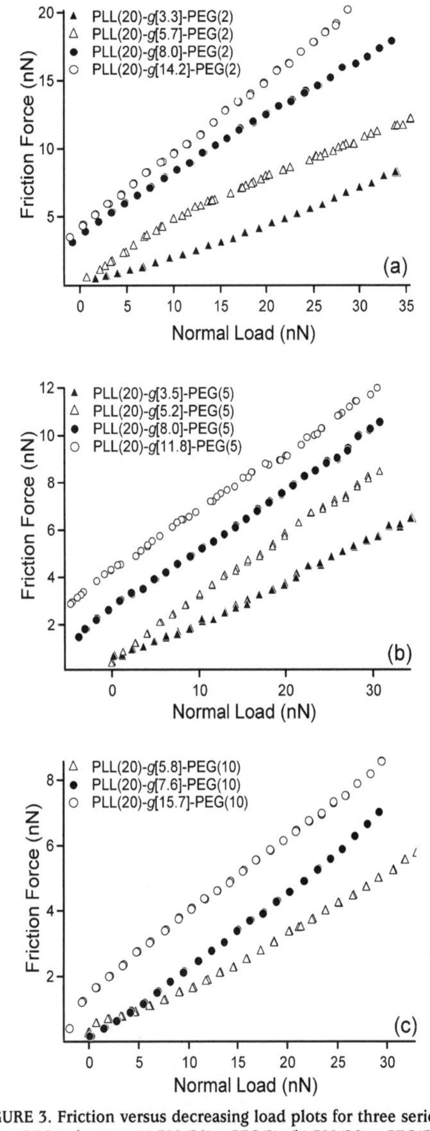

FIGURE 3. Friction versus decreasing load plots for three series of PLL-*g*-PEG polymers: (a) PLL(20)-*g*-PEG(2); (b) PLL(20)-*g*-PEG(5); (c) PLL(20)-*g*-PEG(10). In each series, the polymers vary only in their lysine/PEG grafting ratios. All measurements have been performed using the same AFM microsphere–cantilever assembly for the asymmetrically coated (i.e., bare microsphere/coated substrate) tribointerface. Negative normal loads correspond to adhesive forces existing between the polymer brush and the sodium borosilicate microsphere.

evidenced through both a reduction in the *magnitude* of the friction forces at a specific load and in the coefficient of friction (*slope* of the friction–load plot). For each series, the lowest coefficients of friction are observed for the lowest grafting ratio, corresponding to the greatest density of PEG chains grafted to the PLL backbone.

Third, interfacial adhesion between the bare sodium borosilicate microsphere and the adsorbed polymer film is reduced/eliminated at respectively lower grafting ratios for the systems composed of higher molecular weight PEG. For nonadhesive contacts, the friction force is expected to be zero at zero normal load; adhesion increases the effective contact area between the probe and substrate as a function of the normal load and results in finite contact areas, and corresponding finite friction forces, at zero and negative normal loads. In the PLL-*g*-PEG system, the friction–load plots approach the nonadhesive limit as the number and size of PEG chains increases. Where adhesive forces are detected in the case of high grafting ratios, the interaction likely entails partially exposed charges on the PLL backbone and surface charges on the bare oxide probe tip.

ARTICLE

Table 1 summarizes the data for the coefficient of friction (μ) versus the PEG molecular weight (z) and lysine/PEG grafting ratio (y). Additionally, data characterizing the adsorption of PLL-*g*-PEG from an aqueous solution composed of the physiological buffer, HEPES (a good solvent), is included in terms of the "dry" polymer mass (m_{dry}) and "wet" polymer mass measured with OWLS and QCM-D, respectively. From m_{dry} and the parameters of the polymer architecture (x, y, and z), the PEG surface packing density (σ), or the number of PEG chains per unit area of substrate, may be calculated. An effective solvated film thickness (h_{film}) can also be estimated from this areal PEG density and the "wet" mass of the solvated polymer. In each case, the reported values of the effective film thickness systematically scale with the polymer architecture, increasing for greater PEG chain lengths (z) and decreasing for higher grafting ratios (y).

4. DISCUSSION

4.1. Influence of the Duration of Polymer Deposition. The kinetics and thermodynamics of polyelectrolyte adsorption on oxide surfaces have been extensively studied (23, 24). In general, polyelectrolytes will adsorb spontaneously in order to neutralize charges on the surface. The initial rate of adsorption is diffusion-limited and, therefore, a function of the degree of swelling (size) of the polyelectrolyte in the bulk solution (23). Low ionic strengths preclude the screening of electrostatic interactions, thus causing the polyelectrolytes to adsorb irreversibly. Moreover, once on the surface, spreading and reconformation is slow, resulting in dangling loops and tails as well as conformational heterogeneity and overcompensation of surface charge (24). The adsorption of comb copolymers, in particular, has been studied using self-consistent-field methods (25). For comb copolymers with an "adsorbing backbone" (e.g., a polyelectrolyte) and "nonadsorbing teeth", the teeth will tend to protrude into the solution to compensate for the decrease in entropy, resulting in confinement and thus decreasing the critical adsorption energy. Moreover, the volume fraction profiles of adsorbed comb copolymers with narrow spacing between teeth are expected to exhibit brushlike behavior.

The adsorption performance of PLL-based polymers has also been studied, with PLL-*g*-PEG found to adsorb spontaneously from aqueous solution onto many oxide surfaces (1, 2). Electrostatic interaction between cations on the PLL backbone and the negative charge on the oxide surface lead to strong attraction under appropriate solution conditions.

Table 1. Summary of Data for the Three PEG Chain Length Series of PLL(x)-g[y]-PEG(z), Varying Only in Lysine/PEG Grafting Ratios (y)[a]

polymer architecture PLL(x)-g[y]-PEG(z)	y (lysine/PEG)	m_{dry}, ng/cm^2	σ, nm^{-2}	h_{film}, nm	μ
PLL(20)-g-PEG(2)	3.3	75.183	0.18	5.91	0.20 ± 0.04
	5.7	55.45	0.12	5.01	0.431 ± 0.005
	8.0	45.06	0.09	4.19	0.75 ± 0.03
	14.2	36.76	0.05	3.09	0.88 ± 0.04
PLL(20)-g-PEG(5)	3.5	147.57	0.16	11.18	0.199 ± 0.006
	5.2	111.95	0.12	9.75	0.308 ± 0.006
	8.0	88.27	0.09	9.35	0.46 ± 0.02
	11.8	59.81	0.05	6.98	0.58 ± 0.03
PLL(20)-g-PEG(10)	5.8	133.76	0.07	15.6	0.162 ± 0.003
	7.6	109.14	0.06	12.9	0.226 ± 0.001
	15.7	55.76	0.03	10.01	0.35 ± 0.01

[a] x and z are PLL and PEG molecular weights in kDa, respectively. Note: m_{dry}, "dry" mass measured by OWLS; h_{film}, thickness of the "wet" brush determined from QCM-D measurements; σ, calculated PEG surface packing density; μ, coefficient of friction.

In general, the pH of the medium must be above the isoelectric point (IEP) of the oxide, where it will be negatively charged, and below the pK_a of the primary amines on PLL, where it will be positively charged (1). In the present study, the adsorption of PLL(20)-g[3.5]-PEG(2) onto SiO_2-passivated surfaces has been performed from 1.0 mg/mL polymer solutions in 10 mM HEPES, adjusted to the physiological pH (7.4). Because the IEP of SiO_2 is ~2.0 (26, 27), its surfaces are negatively charged under these conditions, while the amino groups ($pK_a \approx 10$) would be positively charged.

The kinetics of adsorption of PLL-g-PEG has been previously reported from OWLS measurements for various oxide surfaces, including Nb_2O_5, $Si_{0.4}Ti_{0.6}O_2$, and TiO_2 (1, 2). It has been observed that adsorption takes place rapidly and irreversibly, with 95% of the final adsorbed mass reached within the first 5 min, followed by a stable plateau after 20 min. The plot of the coefficient of friction versus deposition time in Figure 2 essentially tracks the OWLS kinetic plot, with 90% of the reduction in friction occurring within the first 5 min, followed by a slow leveling off after 30 min. These friction measurements were performed using the same tip−cantilever assembly on several samples prepared at different deposition times, in contrast to the OWLS measurements, which were performed continuously over time.

The presence of a solvated polymer layer, especially the outer layer composed of water-soluble, flexible PEG side chains, proves to be favorable to the reduction in friction. The observed lubricity with the duration of polymer deposition is, therefore, a function of the development of this solvated polymer layer on the surface. Not only is the coverage increased over time, but as discussed below, the increase in the packing density drives the PEG chains to form more extended conformations as well.

4.2. Influence of the Polymer Architecture. The time dependence of the friction reduction with PLL-g-PEG adsorption suggests that the frictional properties of these interfaces are closely related to the areal density of the PEG chains immobilized near the surface. The full effect of this areal density is revealed through an analysis of the coupled contribution of the PEG chain length and grafting ratio to the interfacial friction. To determine the number density of PEG chains on the surface, it is assumed that the PLL backbone lies nearly flat on the surface and that the PEG chains are protruding into the solution in a brushlike fashion, as predicted for comb copolymers with "adsorbing backbones" and "nonadsorbing teeth". Table 2 presents the data for the "wet" thickness of the polymer film (from Table 1), comparing it with the theoretical thickness (h_0) predicted from brush scaling laws: $h_0 = aN\sigma^{1/3}$, where a relates to the monomer length, N the number of monomer units, and σ the PEG packing density (28, 29).

Table 2. "Wet" Brush Thickness Measured by QCM-D (h_{film}) Compared with the Brush Thickness Calculated from Scaling Laws ($h_0 = aN\sigma^{1/3}$)

polymer architecture PLL(x)-g[y]-PEG(z)	y (lysine/PEG)	h_{film}, nm	h_0, nm
PLL(20)-g-PEG(2)	3.3	5.91	4.66
	5.7	5.01	4.03
	8.0	4.19	3.63
	14.2	3.09	3.12
PLL(20)-g-PEG(5)	3.5	11.18	11.18
	5.2	9.75	10.05
	8.0	9.35	9.08
	11.8	6.98	7.75
PLL(20)-g-PEG(10)	5.8	15.6	17.27
	7.6	12.9	16.01
	15.7	10.01	12.38

In bulk solution, the radius of gyration (R_g) reflects the conformation of a linear polymer that follows the statistics of a self-avoiding random walk; this is considered to be its unperturbed dimension. Upon confinement to a surface, adsorbed polymers can be described in the form of a two-dimensional lattice with an average distance between graft points (L), which scales with the packing density, as shown below. If L is larger than R_g, the polymer molecules maintain their unperturbed dimension on the surface. However, when L becomes smaller than R_g, adjacent polymer chains begin to overlap and assume more stretched conformations due to repulsive (excluded volume) interactions (28−30). The ratio $L/2R_g$ can be used as a gauge for the extent of stretching of a polymer grafted onto a surface (1, 31, 32); it includes information about both the polymer's *dimensions* and *packing density*. For the series of PLL-*g*-PEG polymers, the polymer dimension would be that of the PEG side chains ($R_g \propto z$) that are expected to protrude perpendicular to the surface, while the polymer packing density will be closely related to the grafting ratio ($\sigma \propto y^{-1}$).

Estimates of the radius of gyration can be made using an empirical equation derived from static light scattering experiments (31, 33),

$$R_g = 0.181N^{0.58}\ \text{(nm)} \tag{1}$$

where N is the degree of polymerization. From this equation, the unperturbed dimensions of the PEG side chains of different molecular weights are estimated to be 1.65 nm for 2 kDa PEG, 2.82 nm for 5 kDa PEG, and 4.21 nm for 10 kDa PEG. The prefactor in the above equation relates to the PEG's monomer unit length. In turn, the average grafting distance can be estimated from the surface packing density, assuming a hexagonal close-packed lattice arrangement (31, 32).

$$L = \left(\frac{4}{3}\right)^{1/4}\sigma^{-1/2} \approx \sigma^{-1/2} \tag{2}$$

Table 1 presents values for σ that are equivalent to the number of PEG chains per unit area (nm^{-2}), as calculated from the "dry" mass (from OWLS) of PLL-*g*-PEG adsorbed on the surface and the parameters (x, y, and z) of the polymer architecture. As such, L relates to both the lysine/PEG grafting ratio and how the PLL-*g*-PEG copolymer stacks side-by-side on the surface during adsorption.

On the basis of these values, the coefficient of friction, determined from data in Figure 3 and presented in Table 1, can be plotted as a function of $L/2R_g$, a parameter that expresses the relative degree of extension of the PEG brush on the surface in terms of its chain length and packing density.

$L/2R_g$ values below 1 represent the regime where lateral interaction between PEG chains begins to occur and the polymer begins to stretch relative to its unperturbed dimensions. As architectures are modified such that $L/2R_g$ values approach 0.5, scaling theory predicts that strong stretching occurs and the adsorbed polymer switches from the mushroom to the brush regime (28−30). The data of Figure 4 portray a reduction in friction with decreasing values of $L/2R_g$, with the lowest friction observed at $L/2R_g \approx 0.5$, supporting the idea that the conformational state of the PEG chains contributes to the lubricity of the PLL-*g*-PEG copolymer system. At higher grafting ratios, changes in the adhesive component of friction also contribute to the observed trend; however, the conformational state represents the primary influence because the trend is clearly propagated to lower grafting ratios.

This observation is underscored by the extent of the data derived from a matrix of architectures varying in both the PEG molecular weight and grafting density. The coefficient of friction values of PLL-*g*-PEG polymers with 2 kDa PEG side

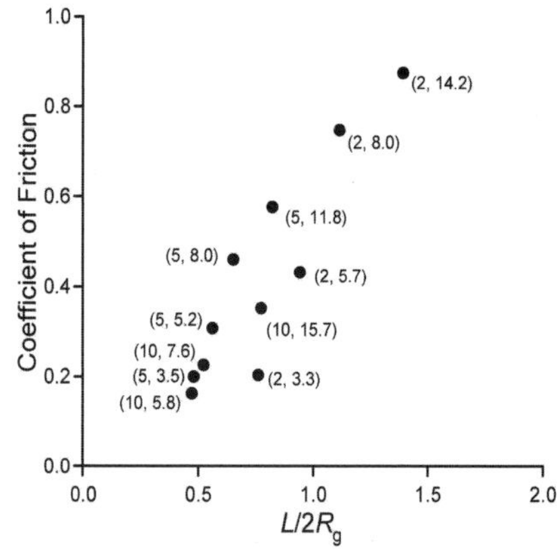

FIGURE 4. Plot of the coefficient of friction versus $L/2R_g$, estimated from eqs 1 and 2. The data labels (z, y) indicate the PEG molecular weight in kDa (z) and the lysine/PEG grafting ratio (y).

chains are observed to lay slightly outside the predicted relationship between friction and strong stretching. This result can be rationalized through the potential for a closer proximity of the PLL backbone with shorter side chains to the surface, thus producing a greater effective surface packing density and leading to an overestimation of L in these cases. In general, it is seen that brush architectures possessing longer PEG side chains and lower grafting ratios (relatively more PEG chains attached to the PLL backbone) exhibit the lowest frictional forces.

It is also noteworthy to observe that the tribological behavior of the PLL-*g*-PEG system, namely, the lowest friction corresponding to $L/2R_g \approx 0.5$, exhibits a trend remarkably similar to that observed for the resistance to protein adsorption for PLL-*g*-PEG-coated surfaces (1, 31). In general, reductions in the surface energy and increases in steric repulsion are known to enhance protein resistance; in addition, the amount of bound water at the interacting interface has also been considered in the context of PLL-*g*-PEG-coated surfaces. Similarly here, the reductions in friction observed for the series of thin polymer films, essentially equivalent in chemical composition at the sliding interface, suggest an important role of the solvent molecules. It is surmised that the conformational changes of the polymer associated with an extended or brushlike state in which water is effectively complexed also produce low shear strength at the sliding interface.

Finally, it is important to note that the results and discussion presented here correspond to tribological properties of adsorbed bottle brush systems, measured on a microscopic scale under boundary conditions in the absence of interfacial wear. In macroscopic tribological settings in which the polymer is removed at the sliding interface through shearing action, a more complex relationship between friction and molecular architecture is anticipated because of the contribution of interfacial binding and adsorption kinetics. Additional differences in the relationship between friction and structure are also likely for surfaces at which PEG moieties have been directly attached.

ARTICLE

5. CONCLUSION

Frictional properties of PLL-*g*-PEG-coated silicon oxide surfaces have been systematically studied by AFM, as a function of the deposition time and polymer architecture, employing a 5.1-μm-diameter sodium borosilicate probe under aqueous media at physiological pH. The most significant reduction in friction occurred within the first 5 min of substrate exposure to the polymer solution, after which a more moderate reduction was observed owing to the conformational reorganization of the polymer film. Friction between the polymer-coated substrate and colloidal probe was observed to systematically vary with the polymer architecture, specifically, the PEG chain length and the lysine/PEG grafting ratio. The coefficient of friction decreased with respect to increased PEG molecular weight and decreased lysine/PEG ratio. The general friction response of PLL-*g*-PEG as a function of the polymer architecture has been rationalized in terms of the spatial density of the PEG side chains. This areal density has been characterized in terms of the distance between PEG chains on the surface (L), related to the grafting ratio and coverage, and the radius of gyration (R_g) of the side chains, related to the PEG molecular weight. It was observed that the lowest friction was encountered at $L/2R_g \approx 0.5$, the point where brush scaling theory predicts the onset of strong stretching. This trend is analogous to the increase in protein resistance previously observed (1, 31) for PLL-*g*-PEG-coated surfaces within the same strong stretching regime. The observed trends in frictional properties for the matrix of polymer brush architectures are highly relevant to the design of future applications entailing brush structures as boundary layer lubricants or as biomimetic lubricants (34). Specifically the results highlight the need for tailoring the local conformation of brush side chains while simultaneously maintaining the backbone functionality required for adsorption at interfaces.

Acknowledgment. This work has been supported by the U.S. Air Force Office of Scientific Research under Contract F49620-02-1-0346. The authors are also grateful for financial support received from the TopNano21 Program of the Council of the Swiss Federal Institutes of Technology (ETH-Rat).

REFERENCES AND NOTES

(1) Kenausis, G. L.; Vörös, J.; Elbert, D. L.; Huang, N.; Hofer, R.; Ruiz-Taylor, L.; Textor, M.; Hubbell, J. A.; Spencer, N. D. *J. Phys. Chem. B* **2000**, *104*, 3298.

(2) Huang, N.; Michel, R.; Voros, J.; Textor, M.; Hofer, R.; Rossi, A.; Elbert, D. L.; Hubbell, J. A.; Spencer, N. D. *Langmuir* **2001**, *17*, 489.

(3) Elbert, D. L.; Hubell, J. A. *Chem. Biol.* **1998**, *5*, 177.

(4) Ruiz-Taylor, L. A.; Martin, T. L.; Wagner, P. *Langmuir* **2001**, *17*, 7313.

(5) Huang, N.; Vörös, J.; De Paul, S. M.; Textor, M.; Spencer, N. D. *Langmuir* **2002**, *18*, 220.

(6) Spencer, N. D.; Perry, S. S.; Lee, S.; Müller, M.; Pasche, S.; De Paul, S. M.; Textor, M.; Yan, X.; Lim, M. S. In *Tribological Research and Design for Engineering Systems: Proceedings of the 29th Leeds-Lyon Symposium (Tribology and Interface Engineering)*; Dowson, D., Lubrecht, A. A., Dalmaz, D., Priest, M., Eds.; Elsevier Science: New York, 2003; pp 411−416.

(7) Yan, X.; Perry, S. S.; Spencer, N. D.; Pasche, S.; De Paul, S. M.; Textor, M.; Lim, M. S. *Langmuir* **2004**, *20*, 423.

(8) Lee, S.; Müller, M.; Ratoi-Salagean, M.; Vörös, J.; Pasche, S.; De Paul, S. D.; Spikes, H. A.; Textor, M.; Spencer, N. D. *Tribol. Lett.* **2003**, *15*, 231.

(9) Müller, M.; Lee, S.; Spikes, H. A.; Spencer, N. D. *Tribol. Lett.* **2003**, *15*, 395.

(10) Perry, S. S. *MRS Bull.* **2004**, *29*, 478.

(11) Perry, S. S.; Somorjai, G. A.; Mate, C. M.; White, R. L. *Tribol. Lett.* **1995**, *1*, 233.

(12) Ogletree, D. F.; Carpick, R. W.; Salmeron, M. *Rev. Sci. Instrum.* **1996**, *67*, 3298.

(13) Varenberg, M.; Etsion, I.; Halperin, G. *Rev. Sci. Instrum.* **2003**, *74*, 3362.

(14) Kurrat, R.; Textor, M.; Ramsden, J. J.; Böni, P.; Spencer, N. D. *Rev. Sci. Instrum.* **1997**, *68*, 2172.

(15) Vörös, J.; Ramsden, J. J.; Csúcs, G.; Szendrö, I.; De Paul, S. M.; Textor, M.; Spencer, N. D. *Biomaterials* **2002**, *23*, 3699.

(16) Vörös, J. *Biophys. J.* **2004**, *87*, 553.

(17) Rodahl, M.; Höök, F.; Krozer, A.; Brzezinski, P.; Kasemo, B. *Rev. Sci. Instrum.* **1995**, *66*, 3924.

(18) Müller, M. T.; Yan, X.; Lee, S.; Perry, S. S.; Spencer, N. D. *Macromolecules* **2005**, *38*, 3861.

(19) Müller, M. T.; Yan, X.; Lee, S.; Perry, S. S.; Spencer, N. D. *Macromolecules* **2005**, *38*, 5706.

(20) Voinova, M. V.; Rodahl, M.; Jonson, M.; Kasemo, B. *Phys. Scr.* **1999**, *59*, 391.

(21) Bandey, H. L.; Hillman, A. R.; Brown, M. J.; Martin, S. J. *Faraday Discuss.* **1997**, *107*, 105.

(22) Larsson, C.; Rodahl, M.; Höök, F. *Anal. Chem.* **2003**, *75*, 5080.

(23) Hoogeveen, N. G.; Stuart, M. A. C.; Fleer, G. J. *J. Colloid Interface Sci.* **1996**, *182*, 133.

(24) Hoogeveen, N. G.; Stuart, M. A. C.; Fleer, G. J. *J. Colloid Interface Sci.* **1996**, *182*, 146.

(25) van der Linden, C. C.; Leermakers, F. A. M.; Fleer, G. J. *Macromolecules* **1996**, *29*, 1000.

(26) Lin, X.; Creuzet, F.; Arribart, H. *J. Phys. Chem.* **1993**, *97*, 7272.

(27) Parks, G. A. *Chem. Rev.* **1965**, *65*, 177.

(28) de Gennes, P. G. *Macromolecules* **1980**, *13*, 1069.

(29) de Gennes, P. G. *Adv. Colloid Interface Sci.* **1987**, *27*, 189.

(30) Hansen, P. L.; Cohen, J. A.; Podgornik, R.; Parsegian, V. A. *Biophys. J.* **2003**, *84*, 350.

(31) Pache, S.; De Paul, S. M.; Vörös, J.; Spencer, H. D.; Textor, M. *Langmuir* **2003**, *19*, 9216.

(32) Henn, G.; Bucknall, D. G.; Stamm, M.; Vanhoorne, P.; Jérôme, R. *Macromolecules* **1996**, *29*, 4305.

(33) Kawaguchi, S.; Imai, G.; Suzuki, J.; Miyahara, A.; Kitano, T.; Ito, K. *Polymer* **1997**, *38*, 2885.

(34) Lee, S.; Spencer, N. D. *Science* **2008**, *319*, 575.

AM900101M

2c. Surface modification with biomolecules and its control

Commentary

Our introduction to PLL-*g*-PEG came from one of its inventors, Jeff Hubbell, with whom we collaborated and who was a professor at the ETH at that time. As well as its applications as a lubricant additive, as described above, PLL-*g*-PEG changed our views of surface functionalization completely. Up to that time, we had carried out complex step-wise organic reactions, in order to modify surfaces with biomolecules, for example (2.22, 2.23). This was effective, but very time consuming. Given that many applications of surface-bound biomolecules were in our minds (implants, biosensors, diagnostics), we wanted the process of surface functionalization to be as simple as possible. PLL-*g*-PEG is attractive as a surface modifier in that the molecule contains PEG-brush-forming functionality, with all its useful properties such as resistance to protein adsorption (2.24, 2.25), as well as charged anchoring groups. PLL-*g*-PEG also self assembles in the right configuration on a wide variety of surfaces (2.27). The level of functionality can also be increased, by adding functional groups, such as biotin (2.26) or nitriloacetic acid (2.31), to the ends of the PEG chains. These can be used to link to a variety of biomolecules via streptavidin or his-tags, respectively. When a mixture of PLL-*g*-PEG and PLL-*g*-PEG-biotin is adsorbed on a surface, for example, the possibility exists of immobilizing biomolecules in a non-adsorptive sea of PEG chains — a very useful platform for sensors and diagnostics, where non-specific adsorption is generally to be avoided at all costs. Thus a simple dipping process can impart complex functionality to a surface, because the pre-synthesized multifunctional molecules assemble, right-side up, onto the surface, and no further surface organic chemistry is necessary.

The protein-non-adhesive properties of PEG brushes have been examined by many authors, but to a certain extent remain something of a mystery. We have tried to shed some light on the issue by the application of both AFM (in collaboration with Hans Griesser at the Ian Wark Research Institute in Australia), and surface-force apparatus approaches (2.28, 2.29, 2.30). One issue that needed tackling was the surface charge, which was necessary for the adhesion of the charged PLL-*g*-PEG molecules. Charge was also found to influence protein resistance to an extent determined by the effects of ionic strength and PEG molecular weight.

While PEG brushes have been useful for many applications, we also applied dextran brushes as a protein-non-adhesive surface, with some success (2.32), although it does not appear to be quite as effective as PEG, for reasons that we are still trying to understand.

Langmuir **1998,** *14,* 5507–5516 5507

Covalent Attachment of Cell-Adhesive, (Arg-Gly-Asp)-Containing Peptides to Titanium Surfaces

Shou-Jun Xiao, Marcus Textor, and Nicholas D. Spencer*

Laboratory for Surface Science and Technology, Department of Materials, Swiss Federal Institute of Technology, ETH-Zürich, CH-8092 Zürich, Switzerland

Hans Sigrist

CSEM (Centre Suisse d'Electronique et de Microtechnique SA), Jaquet-Droz 1, CH-2007 Neuchâtel, Switzerland

Received March 3, 1998. In Final Form: June 18, 1998

A three-step reaction procedure was applied to introduce RGD-containing peptides on the titanium surface. Water–vapor–plasma-pretreated titanium surfaces were first silanized with (3-aminopropyl)-triethoxysilane, resulting in a multilayer film of poly(3-aminopropyl)siloxane. In a second reaction step, the free primary amino groups were linked to one of the three hetero-cross-linkers: *N*-succinimidyl-6-maleimidylhexanoate, *N*-succinimidyl-3-maleimidylpropionate, and *N*-succinimidyl *trans*-4-(maleimidyl-methyl)cyclohexane-1-carboxylate. Onto the resulting terminal-maleimide surface, two model, cell-adhesive peptides, H-Gly-Arg-Gly-Asp-Ser-Pro-Cys-OH and H-Arg-Gly-Asp-Cys-OH were immobilized through covalent addition of the cysteine thiol (–SH) group. X-ray photoelectron spectroscopy, infrared reflection absorption spectroscopy, and radiolabeling techniques were applied to characterize the surfaces. From independent quantitative analysis, an approximate coverage of 0.2~0.4 peptides/nm^2 was calculated.

1. Introduction

Much attention has recently been directed toward the development of bioactive and biocompatible material surfaces for a variety of technological applications, among them biosensors,[1] bioreactors,[2] chromatographic supports,[3] and functionalized building blocks for biomaterials.[4] A major challenge in the development of functional surfaces is to devise strategies, employing either existing synthetic technologies or novel fabrication methods, to assemble various complex molecular species on material surfaces, such as organized thin organic films functionalized with peptides, proteins, or DNA/RNA strands. To select appropriate synthetic routes, the following restrictions have to be considered: (1) The attachment site and chemistry must not interfere with the functional structure or the active site of the biomolecule, (2) the attached biomolecule must not be denatured or inactivated at the surface during or following attachment, and (3) the attached biomolecule should be stably bound at the surface through linkages that are not susceptible to disruption by hydrolysis or other interactions with species in the environment.

Titanium is a successful biocompatible material that is extensively used today for manufacturing bone-anchoring systems, such as dental implants or hip-joint fixation and replacement, as well as for pacemakers, heart valves, and ear-drum drainage tubes. It has advantageous bulk and surface properties: in particular, a low modulus of elasticity, a high strength-to-weight ratio, excellent resistance to corrosion, and an inert, biocompatible surface oxide film.[5] The surface chemistry and structure are prime factors governing bone integration[6] and there is—from the standpoint of both surgeon and patient—considerable interest in increasing both speed of formation (healing time) and degree (long-term success) of close bone aposition for cement-free implantation. To further improve the biocompatibility of titanium, many physical and chemical surface modification methods such as electrochemical oxidation, plasma coating with titanium or hydroxyapatite, and ion implantation are in use.[7] Reports concerning biochemical modification to change a bioinert surface to a bioactive surface, however, are rare. One example is a paper by Sukenik et al.[8] describing the modification of the titanium surface with terminal groups such as CH_3, OH, and Br through silanization and their biological effect in neural cell culture tests.

The amino acid sequence RGD (Arg-Gly-Asp) is present in many extracellular matrix (ECM) proteins and has been found to play an important role in cellular growth, differentiation, proliferation, and regulation of overall cell function.[9] Immobilized, synthetic, RGD-containing peptides on solid supports such as polymers[10] and silicon oxide surfaces[11] have been reported to mediate specific surface–cell interactions and to promote cell organization. Studies of cell migration suggest the importance of the surface density of RGD sequences for efficient cell–matrix interaction.[12]

* To whom correspondence should be addressed.

(1) Schultz, J. S. *Sci. Am.* **1991**, *265*, 64.
(2) Fodor, S. P. A.; Read, L.; Pirrung, M. C.; Stryer, L.; Lu, A. T.; Solas, D. *Science* **1991**, *251*, 767.
(3) Nawrocki, J.; Dunlap, J.; Carr, P. W.; Blackwell, J. A. *Biotechnol. Prog.* **1994**, *10*, 561.
(4) Ratner, B. D. *J. Mol. Rec.* **1996**, *9*, 617.
(5) Brown, S. A.; Lemons, J. E. *Medical Application of Titanium and Its Alloys: The Material and Biological Issues;* ASTM: Philadelphia, PA, 1996.
(6) Doherty, P. J.; Williams, D. F. *Biomaterial-Tissue Interfaces: Advances in Biomaterials;* Elsevier: New York, 1992; Vol. 10.
(7) Geckeler, K. E.; Rupp, F.; Geis-Gerstorfer, J. *J. Adv. Mater.* **1997**, *9*, 513 and references therein.
(8) Sukenik, C. N.; Balachander, N.; Culp, L. A.; Lewandowska, K.; Merritt, K. *J. Biomed. Mater. Res.* **1990**, *24*, 1307.
(9) Ruoslahti, E.; Pierschbacher, M. D. *Science* **1987**, *238*, 491.
(10) Glass, J. R.; Dickerson, K. T.; Stecker, K.; Polarek, J. W. *Biomaterials* **1996**, *17*, 1101.
(11) Massia, S. P.; Hubbell, J. A. *Anal. Biochem.* **1990**, *187*, 292.
(12) Brandly, B. K.; Schaar, R. L. *Anal. Biochem.* **1988**, *172*, 270.

Published on Web 08/21/1998

The most commonly used immobilization methods on inorganic oxide surfaces involve reactively deposited silane films with terminal functional groups that can be further modified with different linking moieties. For aminosilanes, a variety of routes for further modification have been applied, for example, (1) reaction with glutaraldehyde,[13] yielding an aldehyde that can form an imine linkage with primary amines on the peptides, or (2) reaction with a mixture of peptides and carbodiimides, yielding an amide linkage with carboxyl groups on the peptides.[14] The two methods utilize the primary amino groups and carboxyl groups that occur with high frequency in peptides and proteins, but specific attachment at a defined site is very difficult with these functionalities. It is apparent that these approaches must produce highly heterogeneous surfaces.

The most recently examined method employs heterobifunctional cross-linkers with both thiol- and amino-reactive moieties,[15,17–21] which have been widely used in protein conjugation and cross-linking.[16] Bathia et al.[17] linked the maleimidyl group onto thiol-terminal silanized silica surfaces, followed by immobilization of *anti*-IgG antibody through the reaction of the primary amine moieties with the succinimidyl ester. Hong et al.[18] achieved the attachment of cytochrome by reaction of a succinimidyl ester with amine-terminated, silanized glass surfaces, followed by covalent binding of a single unique cysteine thiol on the cytochrome through maleimide. Using similar chemistry, Heyse et al.[19] attached thiol-bearing phospholipids onto optical waveguide surfaces (SiO_2-TiO_2 waveguide layer); Chrisey et al.[20] immobilized DNA on silica surfaces; Matsuzawa et al.[21] attached a synthetic peptide on glass for neuron culture tests. This functionalization technique satisfies to some extent the aforementioned requirements: production of a single functional site and mild reaction conditions. A particular advantage is the much higher surface coverage of biomolecules with this method compared to other coupling approaches. Although the surfaces were characterized by radiolabeling and UV−vis spectroscopy and subjected to biological activity assays, our knowledge regarding the physicochemical properties of these surfaces is still rudimentary.

In a recent publication,[22] we described the binding of H-Arg-Gly-Asp-Cys-OH (RGDC) onto a maleimide-functionalized surface. The present paper extends the studies to the binding of H-Gly-Arg-Gly-Asp-Ser-Pro-Cys-OH (GRGDSPC) as well as the use of a variety of surface functionalization and characterization techniques. The aim of the work is to covalently bind cell-adhesive RGD-containing peptides onto Ti surfaces and to evaluate the biocompatibility of these novel surfaces. Both surface-specific techniques (X-ray photoelectron spectroscopy (XPS), infrared reflection absorption spectroscopy (IRAS), ellipsometry, and contact angle goniometer) and radiolabeling methods were used to qualitatively and quantitatively characterize the surfaces after each reaction step. Good agreement is achieved between XPS and radiolabeling techniques in regard to the quantitative estimation of surface coverages. The results of cell culture tests and the biocompatibility performance of these novel surfaces will be published separately.

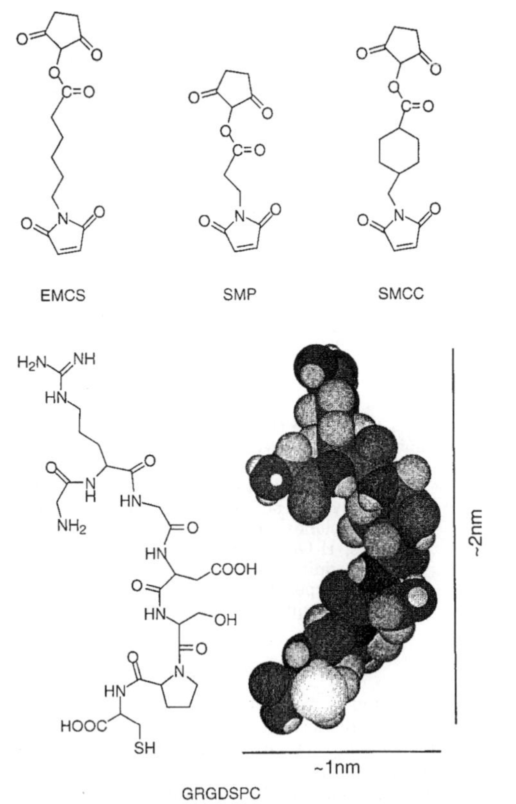

Figure 1. Molecular structures of the three cross-linkers: EMCS, SMP, SMCC, and the peptide GRGDSPC. The space-filling model of GRGDSPC with the minimum energy conformation calculated from MM2 is also shown.

2. Materials and Methods

2.1. Materials. The 100-nm-thick Ti coatings were produced on both sides of round glass cover slips (diameter = 15 mm, thickness = 0.16 mm) (Huber & Co. AG, Reinach/Switzerland) in a Leybold Z600 DC-magnetron sputtering facility.[23] Under the same conditions, larger Ti substrates (2×5 cm^2) were also prepared for IRAS measurements. For the plasma treatment, a plasma cleaner/sterilizer PDC-32G (Harrick, New York) was used. Pure water was obtained from an EASYpure device, Barnstead, USA. (3-Aminopropyl)triethoxysilane (APTES) was bought from Fluka, Buchs, Switzerland, distilled, stored, and used under N_2. The hetero-cross-linkers, *N*-succinimidyl-6-maleimidylhexanoate (EMCS), *N*-succinimidyl-3-maleimidylpropionate (SMP) (Fluka), *N*-succinimidyl *trans*-4-(maleimidylmethyl)cyclohexane-1-carboxylate (SMCC) (Molecular Probe,

(13) Puleo, D. A. *Biomaterials* **1996**, *17*, 217.
(14) (a) Dee, K. C.; Andersen, T. T.; Bizios, R. *Tissue Eng.* **1995**, *1*, 135. (b) Dee, K. C.; Rueger, D. C.; Andersen, T. T.; Bizios, R. *Biomaterials* **1996**, *17*, 209.
(15) Jonsson, U.; Malmqvist, M.; Olofsson, G.; Ronnberg, I. In *Methods in Enzymology*; Mosbach, K., Ed.; Academic Press: San Diego, 1988; Vol. 137, p 381.
(16) Wong, S. S. *Chemistry of Protein Conjugation and Cross-linking*; CRC Press: Boca Raton, FL, 1991.
(17) Bhatia, S. K.; Shriver-Lake, L. C.; Prior, K. J.; Georger, J. H.; Calvert, J. M.; Bredehorst, R.; Ligler, F. S. *Anal. Biochem.* **1989**, *178*, 408.
(18) (a) Hong, H.-G.; Bohn, P. W.; Sligar, S. G. *Anal. Chem.* **1993**, *65*, 1635. (b) Hong, H.-G.; Jiang, M.; Sliger, S. G.; Bohn, P. W. *Langmuir* **1994**, *10*, 153.
(19) Heyse, S.; Vogel, H.; Sanger, M.; Sigrist, H. *Protein Chem.* **1995**, *4*, 2532.
(20) (a) Chrisey, L. A.; Lee, G. U.; O'Ferrall, C. E. *Nucleic Acids Res.* **1996**, *24*, 3031. (b) Lee, G. U.; Chrisey, L. A.; O'Ferrall, C. E.; Pilloff, D. E.; Turner, N. H.; Colton, R. J. *Israel J. Chem.* **1996**, *36*, 81.
(21) Matsuzawa, M.; Umemura, K.; Beyer, D.; Sugioka, K.; Knoll, W. *Thin Solid Films* **1997**, *305*, 74.
(22) Xiao, S. J.; Textor, M.; Spencer, N. D.; Wieland, M.; Keller, B.; Sigrist, H. *J. Mater. Sci. Mater. Med.* **1997**, *8*, 867.

(23) Kurrat, R.; Textor, M.; Ramsden, J. J.; Böni, P.; Spencer, N. D. *Rev. Sci. Instrumen.* **1997**, *68*, 2172.

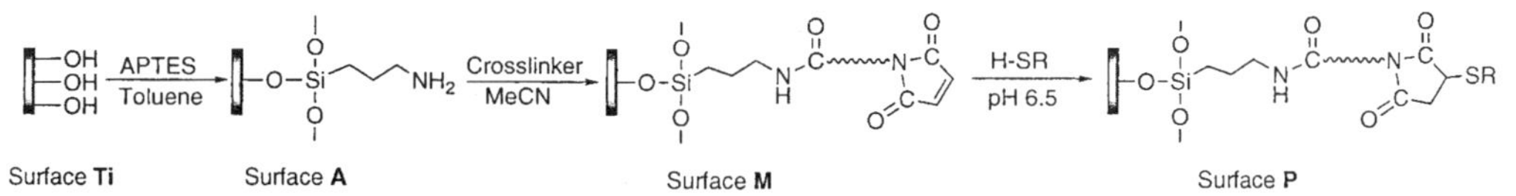

Figure 2. Schematic representation of the modification route. Surface **Ti**: water–vapor–plasma-pretreated titanium. Surface **A**: poly(3-aminopropyl)siloxane pendant surface. Surface **M**: maleimide-modified surfaces with different alkyl chains. Surface **P**: peptide- or L-cysteine-modified surfaces. H–SR: L-cysteine, RGDC, and GRGDSPC.

Netherlands), were stored as recommended by the supplier. The radiolabeling reagents, [^{14}C]-formaldehyde ([^{14}C]-FA) with specific radioactivity 54.0 mCi/mmol, [^{35}S]-L-cysteine ([^{35}S]-Cys) with 20–150 mCi/mmol, and [^{14}C]-phenylglyoxal ([^{14}C]-PG) with 27.0 mCi/mmol, were purchased from Amersham, Buckinghamshire, U.K., stored, and used as indicated. RGDC (65.2%) and GRGDSPC (65%) were purchased from Bachem AG, Bubendorf, Switzerland, and all other chemicals were obtained from Fluka.

The molecular structures of cross-linkers and GRGDSPC are illustrated in Figure 1.

2.2. Surface Modification Route. The surface modification route is shown in Figure 2. Water–vapor–plasma-pretreated titanium surfaces were first activated by APTES, followed by reaction of terminal amines with succinimidyl esters of the cross-linkers, and finally by the covalent binding of the thiol-bearing RGD-containing peptides through maleimidyl groups.

The different surfaces are defined as follows: **Ti**, water–vapor–plasma-pretreated titanium surface; **A**, poly(3-aminopropyl)siloxane-modified surface; **^{14}CA**, [^{14}C]-formaldehyde + **A**; **M**, maleimide-modified surface; **MH**, *N*-maleimidyl-6-hexanoyl (MH) pendant surface; **^{35}SMH**, [^{35}S]-L-cysteine + **MH**; **MP**, *N*-maleimidyl-3-propanoyl (MP) pendant surface; **MC**, *trans*-4-(maleimidylmethyl)cyclohexane-1-carbonyl (MC) pendant surface; **P**, peptide- or l-cysteine-modified surface; **CMH**, L-cysteine + **MH**; **CMP**, L-cysteine + **MP**; **CMC**, L-cysteine + **MC**; **GMH**, GRGDSPC + **MH**; **RMH**, RGDC + **MH**; **GMP**, GRGDSPC + **MP**; **RMP**, RGDC + **MP**; **GMC**, GRGRSPC + **MC**; **RMC**, RGDC + **MC**; **^{14}CGMH**, [^{14}C]-phenylglyoxal + **GMH**.

2.3. Chemical Functionalization. Preparation of Silanized Samples. Prior to silanization, the titanium-coated glass cover slips were first pretreated by water–vapor–plasma cleaning (0.42 mbar, 2 min) and then dried in a vacuum for 1 h. Three hundred samples of **Ti** were incubated using a self-fabricated glass holder in 200 mL of dry toluene containing 2 mL of APTES (43 mM) at 80 °C for 48 h. After reaction, the substrates were ultrasonically washed with chloroform five times, acetone twice, and methanol five times, and were extensively rinsed with water, then dried in a vacuum, and cured at 100 °C under N_2 for 1 h. The treated samples (**A**) were stored under argon for further reactions as detailed below.

Five samples of type **A** were acetylated by overnight treatment at room temperature with a mixture of toluene (1 mL), acetic acid anhydride (100 μL), and dry pyridine (100 μL). After 12 h, the samples were rinsed with toluene, CH_2Cl_2, and ethanol.

Preparation of Maleimide- (M) and Peptide- (or L-Cysteine-) Modified (P) Samples. Twenty silanized titanium samples (**A**) were placed in a self-fabricated polypropylene vessel with 2 mL of acetonitrile (CH_3CN) containing 5.0 mM cross-linkers. After incubation at 20 °C for 30 min, with 30 s of sonication every 10 min, the excess cross-linker solution was removed and the samples were washed extensively with acetonitrile, acetone, and hexane. After they were dried with N_2, the samples were subject to the next chemical step. Twenty maleimide-grafted substrates were incubated at 20 °C for 1 h in 2 mL of water containing 2.0 mM peptide or L-cysteine, with 30 s of sonication every 10 min. The pH of the peptide or l-cysteine solution was adjusted to 6.5 with 0.1 M NaOH before injection. The peptide-grafted substrates were washed thoroughly with water, dried with N_2, and stored in argon.

2.4. Surface Analysis Methods. X-ray Photoelectron Spectroscopy (XPS) Measurements. XPS spectra were recorded using a Specs SAGE 100 system with unmonochromatized Mg Kα radiation at 300 W (12 kV). Measurements were carried out using a takeoff angle of 90° with respect to the sample surface. The analyzed area was typically 9 × 9 mm^2. Survey scans over a binding energy range of 0–1150 eV were taken for each sample with a constant detector pass energy range of 50 eV, followed by a high-resolution XPS measurement (pass energy 14 eV) for quantitative determination of binding energy and atomic concentration. Background subtraction, peak integration, and fitting were carried out using SpecsLab software. Electron binding energies were calibrated to the hydrocarbon C 1s at 284.6 eV on pure titanium surfaces. To convert peak areas to surface concentration, sensitivity factors published by Evans et al. were used.[24]

Infrared Reflection Absorption Spectroscopy (IRAS) Measurements. The IRAS measurements were performed on a Bruker IFS 66V spectrometer operating at approximately 100 Pa. A mercury–cadmium–telluride (MCT) detector was used to collect spectra with a resolution of 2 cm^{-1}. The angle of incidence was 80° from the surface normal. A water–vapor–plasma-pretreated Ti mirror was used as the reference. For both sample and reference, 500 scans were collected.

Ellipsometry Measurements. Film thickness was measured by means of a Gaertner L-116C ellipsometer and software. The angle of incidence was set at 70°. The optical constants of surface **Ti** were determined in at least four different areas on each individual substrate. The thickness of the modified organic films was calculated using the software for a single organic thin film and a refractive index of 1.40. They are about 2-, 3-, and 4-nm thick for surface **A**, **MH**, and **GMH**, respectively.

Contact Angle Measurements. Advancing contact angles were measured on a Ramé-Hart NRL model goniometer at room temperature and ambient humidity. For the measurements, 6 μL of water was put on the surface, followed by adding another 6 μL to the first drop. The contact angle was measured immediately within 1 min. They are 30°, 44°, 60°, and 45° for surfaces **Ti**, **A**, **MH**, and **GMH**, respectively. The experimental error of this method is estimated to be ±3°.

Radiolabeling Procedure. [^{14}C]-Formaldehyde labeling of surface **A** was performed in a self-fabricated metal sample holder with four wells (diameter = 7 mm). Acetonitrile (50 μL) containing 10.0 mM $NaBH_3CN$ and 2.0 mM [^{14}C]-formaldehyde was injected into each well. After incubation for 4 h at 20 °C, the excess radioactive solution was removed and the exposed surface washed with CH_3CN 10 times and water 10 times. The demounted samples were extensively washed again with water and then dried with N_2. The same procedure was also applied to the control surfaces of **Ti** and acetylated **A**. The radioactivity of each sample was measured by scintillation counting in 5 mL of scintillation fluid (1080 mL of toluene, 920 mL of Triton X-100, 5.4 g of 2,5-diphenyloxazole, 0.2 g of 1,4-bis-2-(5-phenyloxazolyl) benzene, and 40 mL of acetic acid) on a Tri-Carb 2300TR liquid scintillation counter (Packard Instrument Co., USA).

Maleimide-functionalized substrates (**M**), kept between two metal plates, were covered with 50 μL of 0.11 μCi/μL [^{35}S]-cysteine aqueous solution in each well. The pH of the solution was adjusted to 6.5 by 0.1 mM NaOH before use. After incubation for 1 h at 20 °C, the excess radioactive solution was rinsed off 10 times with water. The substrates were then removed from the metal plates, washed again with excess water, and dried with N_2, and the radioactivity was determined. Control experiments for the covalent binding of cysteine to maleimide-functionalized substrates were carried out as above by applying the same [^{35}S]-cysteine solution to surface **A** and unlabeled-L-cysteine-reacted surface **M**.

Radiolabeling of RGD-grafted surfaces was carried out by injecting 50 μL of sodium phosphate (25 mM) buffer (pH 7.4) containing 5 mM [^{14}C]-PG to each well as described above,

(24) Evans, S.; Pritchard, R. G.; Thomas, J. M. *J. Electron Spectrosc. Relat. Phenom.* **1978**, *14*, 341.

Table 1. XPS Binding Energies C 1s, N 1s, Ti 2p, and O 1s for the Surfaces Ti, A, MH, and GMH and the Proposed Assignments to Surface Functionalities

surface	C 1s region BE (eV), %, assignments	N 1s region BE (eV), %, assignments	Ti 2p region[b] BE (eV), %, assignments	O 1s region BE (eV), %, assignments
Ti	284.6 (100) hydrocarbon[a]		453.3 (5) $2p_{3/2}$ Ti metal 458.5 (63) $2p_{3/2}$ TiO_2 464.3 (32) $2p_{1/2}$ TiO_2	530.0 (84) TiO_2 531.7 (16) OH
A	285.0 (72) C−C 286.4 (22) C−N 288.3 (6) C=O[a]	399.6 (75) NH_2 401.7 (25) NH_3^+	453.3 (1) $2p_{3/2}$ Ti metal 458.5 (66) $2p_{3/2}$ TiO_2 464.3 (33) $2p_{1/2}$ TiO_2	530.0 (58) TiO_2 532.3 (42) SiO
MH	285.1 (56) C−C 286.3 (26) C−N 288.2 (6) amide-*C* 289 (12) imide-*C*	399.6 (21) NH_2 400.1 (25) amide-*N* 400.7 (30) imide-*N* 401.7 (24) NH_3^+	458.5 (67) $2p_{3/2}$ TiO_2 464.3 (33) $2p_{1/2}$ TiO_2	530.0 (48) TiO_2 532 3 (52) SiO, C=O
GMH	285.1 (51) C−C 286.3 (29) C−N,C−O 288.2 (9) amide-*C* 289.0 (11) imide-, carboxyl-, guanidinyl-*C*	399.6 (14) NH_2 400.1 (39) amide-*N* 400.7 (26) imide-*N* 401.7 (21) NH_3^+, guanidinyl-*N*	458.5 (67) $2p_{3/2}$ TiO_2 464.3 (33) $2p_{1/2}$ TiO_2	530.0 (45) TiO_2 532.3 (55) SiO, C=O, COOH, C−OH

[a] Trace contaminations. [b] The detailed deconvolution of Ti 2p to different titanium oxidation states; see ref 29.

incubating for 24 h at 20 °C, then washing thoroughly with water, drying with N_2, and counting as detailed above.

The degradation experiments were carried out with the radiolabeled surfaces, **^{14}CA**, **^{35}SMH**, and **^{14}CGMH** by measuring their specific radioactivity loss as a function of storage time in 5 mL of water at room temperature.

3. Results and Discussion

3.1. Silanization Procedure. Silanization is the crucial step in regard to subsequent reproducibility of the chemical functionalization. Although silanization has been extensively studied, the resulting structure, coverage, orientation, and organization of the layers have not yet been satisfactorily determined and are still the subject of controversy.[25] It is generally accepted that silanization on inorganic surfaces occurs by reaction of silanol groups with hydroxyl groups present on oxide surfaces. Due to the ability of the silane to polymerize at the surface, the siloxane film can be produced as a monolayer or multilayer, depending on substrates, silane properties, and reaction conditions. In general, long-alkyl-chain silanes tend to form monolayers, while short bifunctional silanes, such as aminosilanes, tend to form multilayers.

Pretreatment of the titanium surface is an essential prerequisite for reproducible silanization. Untreated titanium films (as sputtered) gave unsatisfactory results. Optimum surface cleanliness and reactivity were achieved through water−vapor−plasma treatment, although other pretreatments such as H_2O_2 or HCl also gave satisfactory results.

APTES was chosen to silanize the titanium surface. From our experience, the reaction media (aqueous or organic), temperature, concentration, ratio of reactants, incubation time, washing steps, and soforth influence the amount of APS on titanium surfaces, to different degrees. Silanization in aqueous media resulted in a low surface concentration of amines (atomic ratio N/Si < 0.5 from XPS), which was not beneficial for further modification. However, silanization in toluene produced an APS film with a consistently high concentration of amines (N/Si = 0.7, see section 3.2). Thin and smooth APS films can be produced by extended silanization (toluene, 80 °C, and 48 h), followed by a three-step washing procedure: apolar organic solvent, polar organic solvent, and water. Immersion in water is a very efficient way to obtain thin, smooth films.[26] To exclude batch-to-batch variations and get reproducible data for further modification, a large number of samples (300 pieces) were produced in one batch with the same reaction conditions, washing steps, and curing. Curing has been widely used to stabilize the siloxane films because it drives the surface derivatization reaction further to completion. As a result, the cured films are more resistant to hydrolysis. However, a long time (12 h) curing at 100 °C in air resulted in a pronounced reduction of the surface amine concentration, as a result of oxidation of primary amines to imines and nitriles.[27] This significantly reduced the reaction yield of the subsequent reaction step with succinimidyl ester. Curing for 1 h at 100 °C under N_2 gas was employed in the present study.

The surface coverage of amino groups on samples of this batch was determined as 6.0 NH_2 groups/nm^2 via reaction of the amine with [^{14}C]-formaldehyde. The surface coverage of hydroxyl groups on the TiO_2 surface was estimated to be 6 OH groups/nm^2.[28] One aminosilane molecule reacts with 2 or 3 OH groups on the TiO_2 surface. Considering the N/Si ratio of 0.7 (Table 2), 9 silicon atoms/nm^2 would be present for every 4−6 APS layers on surface **A**. However, this does not mean that there are less than 25% primary amines exposed to the surface, since the surface roughness enhances the total proportion of exposed amines.

3.2. Monitoring Reaction Steps by XPS. XPS was used to monitor each reaction step as it can provide information on chemical structure, atomic concentration, and surface contamination. Figure 3 shows the evolution of the XPS signals of C 1s, N 1s, Ti 2p, and O 1s from **Ti** to **A**, **MH**, and **GMH**. Table 1 lists the experimental XPS binding energies of the deconvoluted detailed spectra and the proposed assignments to chemical bonds/oxidation states based on chemical shifts. Figure 4 illustrates the changes of the atomic concentrations on the above surfaces. Surfaces modified with other cross-linkers and peptides gave results consistent with the proposed interpretation and are not presented here.

Water−Vapor−Plasma-Pretreated Titanium Surface (Ti). Spectra of surface **Ti** show three elements: Ti, O, and C. Ti $2p_{3/2}$ at 458.5 eV and Ti $2p_{1/2}$ at 464.3 eV are assigned to TiO_2; Ti $2p_{3/2}$ at 453.3 eV is due to Ti metal because the XPS information depth is greater than the

(25) Ulman, A. *Chem. Rev.* **1996**, *96*, 1533.
(26) Haller, I. *J. Am. Chem. Soc.* **1978**, *100*, 8050.
(27) Ondrus, D. J.; Boerio, F. J. *J. Colloid Interface Sci.* **1988**, *124*, 349.
(28) Yates, D. E.; James, R. O.; Healy, T. W. *J. Chem. Soc., Faraday Trans.* **1980**, *76*, 1.

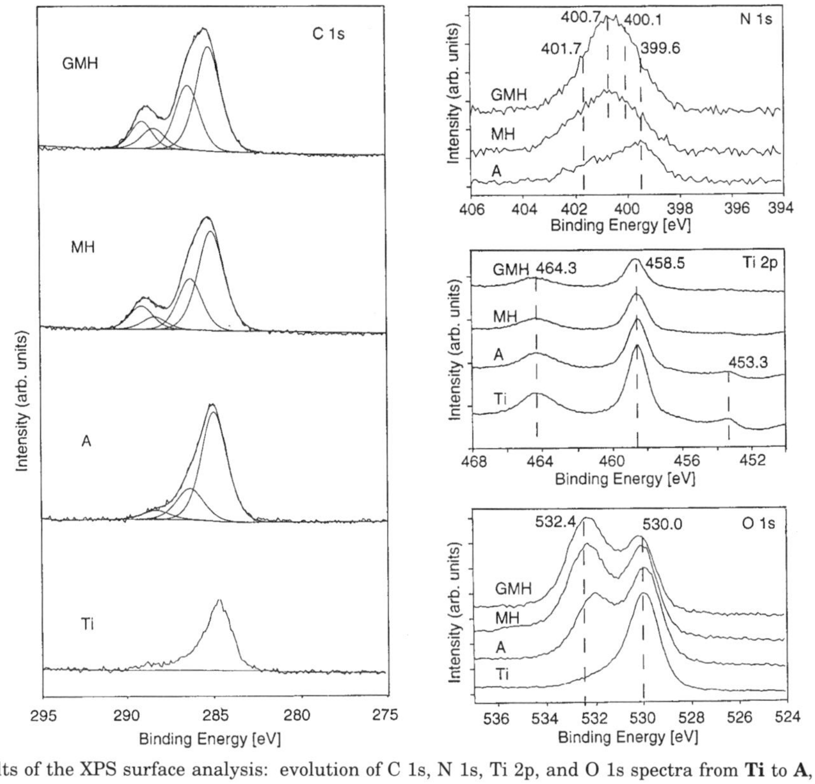

Figure 3. Results of the XPS surface analysis: evolution of C 1s, N 1s, Ti 2p, and O 1s spectra from **Ti** to **A**, **MH**, and **GMH**.

Table 2. XPS Atomic Ratios N/Si for APS-, Maleimide-, and Peptide- (or L-Cysteine-) Modified Surfaces and the Reaction Yields ($R_{X\to Y}$) Estimated Using Formulas 1 and 2 (See Text, Section 3.3)

surface	N/Si (±s.d.)	reaction ($X\to Y$)	$R_{X\to Y}$ % (± calculated error)
MH	0.95 ± 0.02	A → MH	36 ± 5
MP	0.90 ± 0.02	A → MP	29 ± 5
MC	0.87 ± 0.02	A → MC	24 ± 5
CMH	1.15 ± 0.04	MH → CMH	80 ± 20
CMP	1.05 ± 0.04	MP → CMP	75 ± 25
CMC	1.00 ± 0.04	MC → CMC	76 ± 32
RMH	1.26 ± 0.03	MH → RMH	15 ± 3
RMP	1.15 ± 0.04	MP → RMP	18 ± 4
RMC	1.05 ± 0.03	MC → RMC	15 ± 4
GMH	1.40 ± 0.04	MH → GMH	18 ± 3
GMP	1.26 ± 0.03	MP → GMP	18 ± 3
GMC	1.20 ± 0.03	MC → GMC	19 ± 4

thickness of the native TiO_2 layer (~5 nm).[29] O 1s at 530.0 eV is typical for TiO_2, while C 1s at 284.6 eV is due to the environmental hydrocarbon contamination.

Silanized Surface (A). XPS has shown[22] to be a powerful tool for following the silanization reaction from surface **Ti** to **A** via the appearance of N (N 1s at around 400 eV) and Si (Si 2p at 102.0 and 153.0 eV), as well as the new high binding energy peak of O 1s at 532.3 eV due to Si–O. The broad N 1s peak can be deconvoluted into two peaks 399.6 eV (75%) due to free amines and 401.7 eV (25%) assigned to protonated amines.[30]

(29) Sittig, C.; Wieland, M.; Vallotton, P.-H.; Textor, M.; Spencer, N. D. *J. Mater. Sci. Mater. Med.*, in press.

(30) Kallury, K. M. R.; Macdonald, P. M.; Thompson, M. *Langmuir* **1994**, *10*, 492.

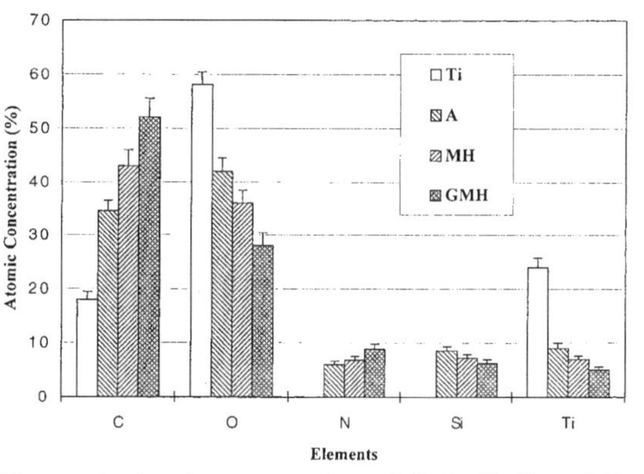

Figure 4. Atomic concentrations of C, O, N, Si, and Ti calculated from XPS intensities on surfaces **Ti**, **A**, **MH**, and **GMH**.

Maleimide-Modified Surface. The C 1s spectra of surface **MH** can be deconvoluted into four peaks with those at binding energies 289.0 and 288.2 eV being assigned to the newly introduced imide and amide functionalities, respectively. The N 1s emission has a new predominant contribution between 400 and 401 eV, again indicative of amide and imide functional groups reported to have binding energies of 400.1 and 400.7 eV, respectively.[31] The disappearance of the Ti $2p_{3/2}$ (metal) peak at 453.3 eV demonstrates that the total thickness of the oxide plus organic surface layer now exceeds the information depth of XPS (ca. 8 nm).

(31) Beamson, G.; Briggs, D. *High-Resolution XPS of Organic Polymers*; John Wiley & Sons Ltd.: New York, 1992.

5512 *Langmuir, Vol. 14, No. 19, 1998* *Xiao et al.*

Table 3. Surface Concentrations and Surface Coverages of Amino Group on Surface A, of Maleimidyl Group on Surfaces MH, MP, and MC, of RGD Group on Surfaces RMH, RMP, RMC, GMH, GMP, and GMC

surface	radiolabeling reagents	radioactivity (nCi/cm^2 ± s.d.)	surface concentration[a] (nmol/cm^2 ± s.d.)	surface coverage[a] (group/nm^2 ± s.d.)	calculated surface coverage[b] (group/nm^2)
A	[^{14}C]-FA	53.3 ± 5.4	1.0 ± 0.2	6.0 ± 1.2	
MH	[^{35}S]-Cys	8.2 ± 1.0	0.05–0.77	0.4–4.6	2.2 ± 0.5
MP	[^{35}S]-Cys	6.3 ± 0.8	0.04–0.71	0.2–4.3	1.7 ± 0.5
MC	[^{35}S]-Cys	5.6 ± 0.7	0.03–0.72	0.2–4.3	1.4 ± 0.4
RMH	[^{14}C]-PG	2.81 ± 0.43	0.052 ± 0.008	0.31 ± 0.05	0.3 ± 0.1
RMP	[^{14}C]-PG	2.17 ± 0.37	0.040 ± 0.007	0.24 ± 0.04	0.3 ± 0.1
RMC	[^{14}C]-PG	1.84 ± 0.27	0.034 ± 0.005	0.20 ± 0.03	0.2 ± 0.1
GMH	[^{14}C]-PG	3.27 ± 0.48	0.061 ± 0.009	0.37 ± 0.05	0.4 ± 0.1
GMP	[^{14}C]-PG	2.45 ± 0.40	0.045 ± 0.007	0.27 ± 0.04	0.3 ± 0.1
GMC	[^{14}C]-PG	2.33 ± 0.32	0.043 ± 0.006	0.26 ± 0.04	0.3 ± 0.1

[a] To calculate the surface concentration and surface coverage from the radioactivity, we assume the following molar ratios: [FA]/[NH_2] = 1, [Cys]/[Maleimidyl group] = $R_{X\to Y}$ in Table 2, respectively, [PG]/[RGD] = 2 (see text in section 3.4). The specific radioactivity of [^{14}C]-FA is 54.0 nCi/nmol, of [^{35}S]-Cys is between 20 and 150 *n*Ci/nmol, and of [^{14}C]-PG is 27.0 nCi/nmol. [b] The values are calculated by a combination of the NH_2 surface coverage on **A** and the reaction yields from Table 2. The errors correspond to calculated values.

MP and **MC** show similar XPS chemical shifts and the spectra are not shown here. The C 1s intensities above 288.0 eV are somewhat lower compared to those of **MH** and exhibit shoulders rather than obvious peaks (not shown here). This implies that the largest surface coverage occurs in the case of **MH**, which is consistent with results discussed below.

Peptide-Modified Surface. On the peptide-modified surface, **GMH**, no significant chemical shifts from **MH** were observed. The concentration of the newly introduced thiolether group is too low to be detected with confidence. The guanidinyl C 1s (289.0 eV) and amide C 1s (288.2 eV) overlap the imide and amide C 1s. Carboxyl, hydroxyl, and amide O 1s peaks at around 532 eV overlap Si–O and carbonyl O 1s. The guanidinyl (C=NH) N 1s at around 402 eV overlaps the protonated amines. However, the significant increase of the relative peak areas of N 1s, of O 1s above 532.0 eV, and of C 1s above 288.0 eV, as well as the decrease of the Ti 2p and Si 2p, are consistent with the presence of peptides.

3.3. Estimation of Reaction Yields. Atomic concentrations calculated from XPS intensities depend on the measured volume and the chemical components within this volume. Quantitative statements in regard to reaction yields for the subsequent modification reactions are difficult because of the inevitable carbon contamination and oxygen content on the starting surface **Ti**, and the changes of the chemical components and the film thickness on different surfaces. Although the average atomic concentrations reflect the sequential reactions well, the above-mentioned factors, as well as normal quantitative error, can result in a considerable deviation in the case of the low-concentration elements such as N and Si. Fortunately, silicon, which is not present on the starting surface, **Ti**, can be used as an internal reference. The absolute silicon surface content from **A** through **P** remains constant. Since the whole organic surface layer thickness is below the information depth of XPS, the atomic ratio of N to Si is a more representative parameter than the atomic concentration to follow quantitative surface changes. The reaction yields can be deduced from the following simple formula:

Reaction yield (**A** → **M**) = (imide-*N* content on **M**)/(*N* content on **A**)

$$R_{A\to M} = \frac{n_{N(M)} - n_{N(A)}}{n_{N(A)}} = \frac{(N/Si)_M - (N/Si)_A}{(N/Si)_A} \quad (1)$$

Reaction yield (**M** → **P**) = [(peptide-*N* content on **P**)/*m*]/(imide-*N* content on **M**)

$$R_{M\to P} = \frac{(n_{N(P)} - n_{N(M)})/m}{n_{N(M)} - n_{N(A)}} = \frac{[(N/Si)_P - (N/Si)_M]/m}{(N/Si)_M - (N/Si)_A} \quad (2)$$

where $R_{X\to Y}$ is the reaction yield from *X* to *Y*, $n_{N(i)}$ is the absolute nitrogen content of surface *i*, and $(N/Si)_i$ is the atomic ratio of N to Si on surface *i*. The divisor *m* takes account of the *m* nitrogen atom(s) in one peptide molecule (1 for cysteine, 7 for RGDC, and 10 for GRGDSPC).

The calculated reaction yields are summarized in Table 2. EMCS in the reaction **A** → **MH** shows the highest reaction yield among the three cross-linkers, suggesting the importance of steric hindrance. SMP has a short chain of three methylene groups between succinimidyl ester and maleimide, SMCC a bulky cyclohexane, while EMCS has a longer chain of six methylene groups. The former two have more rigid structures than the latter. One possible explanation is that EMCS, due to its molecular flexibility, has a higher probability of reaching partially occupied, reactive sites on the surface. Another factor could be the molecular organization on the surface: Because of the shorter spacer and molecular configuration, MP and MC groups are bound close to the binding sites, leading to increased disorder on a macroscale, and occupying more space. Particularly in the valley areas, the steric hindrance from the occupied maleimidyl group could largely inhibit the accessibility of the surface. However, for the MH group, the relatively longer spacer with its potential for greater interchain van der Waals interactions may result in some long-range ordering.

For thioether formation, the molecular size seems to be the key factor affecting the reaction yields. This might be the reason for the high yield (up to 80%) for cysteine and the generally low yields (about 18%) for RGDC and GRGDSPC.

3.4. Surface Coverage. Quantitative surface coverages were determined using radiolabeling techniques. [^{14}C]-Formaldehyde, [^{35}S]-cysteine, and [^{14}C]-phenylglyoxal were employed to measure the surface coverage of primary amine in **A**, maleimide in **M**, and peptide in **P**, respectively.

The surface coverages of maleimide and peptide were additionally estimated on the basis of the radiolabeling-derived amine surface concentration on **A** and the XPS-derived reaction yields shown in Table 2. These data are in quite good agreement with the directly measured values (see Table 3).

[^{14}C]-Formaldehyde Radiolabeling on Silanized Surfaces. [^{14}C]-Formaldehyde with $NaBH_3CN$ has been used as a standard procedure to label the free amino groups of proteins.[32] The primary amine reacts first with formaldehyde to form a Schiff base, which is then reduced by $NaBH_3CN$ to the secondary amine. Although the secondary amine can again react and form the tertiary amine, the stoichiometry of $[NH_2]/[H_2CO]$ can be fairly well-controlled to 1 under appropriate reaction conditions. We carried out the reaction in CH_3CN solution at 20 °C for 4 h. It is assumed that the small formaldehyde molecule can diffuse into the open-structure APS film and react with all the primary amine groups. With this assumption, the surface concentration (1.0 nmol/cm^2) and coverage (6.0 group/nm^2) of primary amines can be simply calculated from the specific radioactivity and the radiolabeled area. To exclude contributions from physically adsorbed ^{14}C-labeled formaldehyde, the surfaces of **Ti** and acetylated **A** were used as controls. Only negligible amounts of about 1 nCi/cm^2 (0.02 nmol/cm^2) were detected in these cases.

[^{35}S]-Cysteine Radiolabeling on Maleimide-Modified Surfaces. [^{35}S]-Cysteine was chosen to label the maleimidyl groups because of two reasons: (1) To mimic the reaction of terminal cysteine on peptides with maleimidyl groups; (2) to evaluate the surface coverage and reactivity of maleimide since maleimidyl groups may hydrolyze to maleamic acid. A still open question is whether cysteine reacted completely with maleimide or not. The estimated reaction yields from XPS were used to calculate the maleimide coverage. The estimated surface coverages of maleimide are only semiquantitative, however, since the specific radioactivity is specified by the supplier as 20−150 mCi/mmol. Surface **A** and the unlabeled-cysteine-reacted surface **M** were used as the control surfaces. Only negligible amounts of around 1 nCi/cm^2 on control surfaces were detected and thus supported the covalent nature of the bond.

[^{14}C]-Phenylglyoxal Radiolabeling on Peptide-Modified Surfaces. A standard target reaction of the guanidinyl moiety of arginine residues is with 1,2-dicarbonyl reagents.[33] Under mild alkaline conditions, these compounds condense with the guanidinyl group in an initial reaction very similar to the Schiff base formation, which is followed by further rearrangement to form different products. Arginine-containing peptides or proteins can form adducts with phenylglyoxal in a stoichiometric ratio (phenyglyoxal/arginine) of 1 or 2 depending on the peptides or proteins involved, and the reaction conditions. Following a published procedure,[34] we carried out the reaction in 25 mM sodium phosphate buffer (pH 7.4) at 20 °C for 24 h. A stoichiometry of 2 phenylglyoxal to 1 arginine is assumed here to calculate the peptide surface coverages. Both the direct radiolabeling measurements and the indirect calculation from the estimated reaction yields indicate surface coverages in the same range of 0.2−0.4 molecules/nm^2. According to a molecular modeling study, the minimum energy conformation of GRGDSPC (Figure 1) has dimensions of about 2×1 nm^2. Assuming equidistant attachment sites, this would correspond to an average separation of 2.2 nm of GRGDSPC on surface **GMH**.

3.5. Optimization of Conjugation Reactions. Surface A to Surface M. The linkers serve two purposes: (1) To covalently bind two distinct chemical entities that otherwise would remain unreactive toward each other; (2) to provide a physical spacer allowing for greater accessibility and/or orientational freedom of the attached biomolecules.

(32) Jentoft, N.; Dearborn, D. G. *J. Biol. Chem.* **1979**, *254*, 4359.

(33) Yankeelov, J. A., Jr. *Methods Enzymol.* **1972**, *25*, 566.

(34) Liao, T.-H.; Ho, H.-C.; Abe, A. *Biochim. Biophys. Acta* **1991**, *1079*, 335.

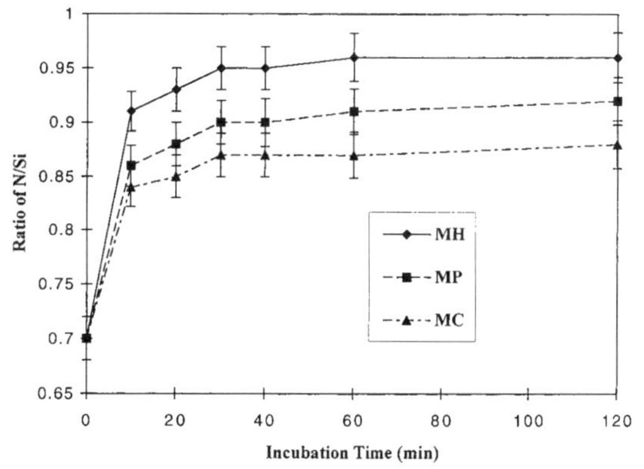

Figure 5. The relationship between the atomic ratios N/Si (XPS) on the maleimide-modified surfaces and the incubation time.

The succinimidyl ester reacts preferentially with amino groups, eliminating *N*-hydrosuccinimide as the leaving group. The reaction in solution is complete within 10−20 min at pH 6−9.[35] The competing reaction is the hydrolysis of the succinimidyl ester. The reaction was carried out in different solvents (e.g., aqueous buffer, *N,N*-dimethyl formamide (DMF), dimethyl sulfoxide (DMSO), and acetonitrile). The highest yield was achieved in CH_3CN, while an aqueous buffer is least suitable due to the low solubility of the cross-linkers and the hydrolysis of both the succinimidyl ester and the maleimide.

The effect of cross-linker concentration in the range of 1.0−10.0 mM was checked by XPS; no obvious differences were observed with a molar ratio of 2 cross-linkers to 1 primary amine. Thus, a 5.0 mM solution was used as the standard. All the experiments were carried out at 20 °C under an ambient atmosphere. The main variable is the reaction time, which is an important parameter, controlling both the extent of reaction and the surface uniformity. Figure 5 shows the relationship of the atomic ratio N/Si (XPS) with the incubation time. The reaction is very fast within the first 10 min, reaching its maximum after approximately 30 min.

Surface M to Surface P. Maleimides are quite specific to the thiol group, especially at pH < 7 where other nucleophiles are protonated. In acidic and near-neutral solutions, the reaction rate with simple thiols is about 1000-fold faster than with the corresponding simple amines.[36] The other major competing reaction is the hydrolysis of maleimide to maleamic acid. However, this reaction is much slower than the thiolether formation in near-neutral solutions. At pH 7, it is estimated that the half-life reaction time between millimolar concentrations of mercaptan and maleimide is of the order of 1 s.[36] The resulting thioether bond is very stable and cannot be cleaved under physiological conditions. The peptide modification was carried out in an aqueous solution. Samples treated in different concentrations and incubation times were evaluated by XPS. The effect of varying the peptide concentration in the range of 0.1−5 mM was not significant. Figure 6 shows the effect of incubation time on the atomic ratio N/Si (XPS). A sharp increase in the peptide surface concentration takes place in the first 20

(35) Lindsay, D. G. *FEBS Lett.* **1972**, *21*, 105.

(36) Glaser, A. N. In *The Proteins*, 3rd ed.; Neurath, H., Hill, R. L., Eds.; Academic Press: New York, 1976; Chapter 2.

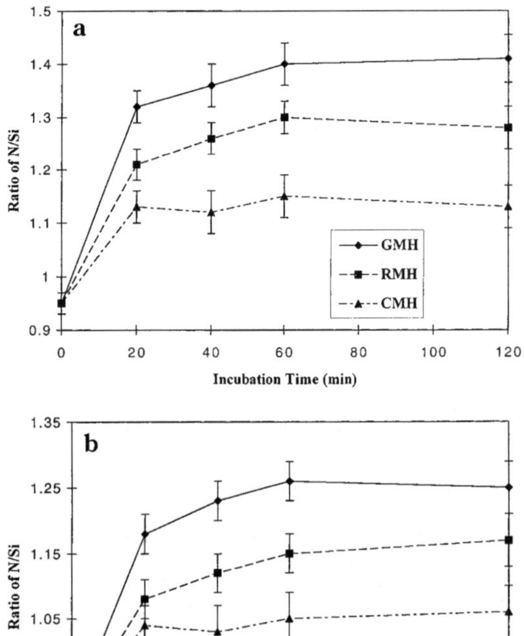

Figure 6. The dependence of the atomic ratios N/Si (XPS) of the peptide- or L-cysteine-modified surfaces on the incubation time.

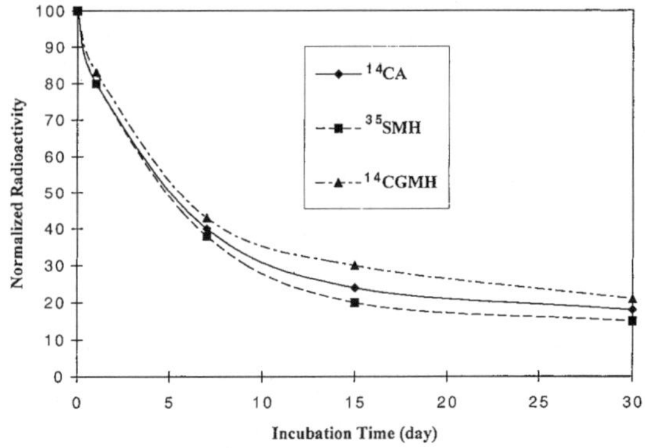

Figure 7. The degradation of **^{14}CA**, **^{35}SMH**, and **^{14}CGMH** in water measured through the loss of radioactivity. The data are normalized to the initial value.

min, followed by a slow increase. A reaction time of 1 h was chosen for the standard protocol. Although N/Si slowly increases with longer incubation times, this could be due to hydrolysis of siloxane films.

Degradation in Water. A problem related to the application of immobilized biomolecules via silanization techniques is the bioactivity loss due to hydrolysis of the siloxane films. Degradation was followed by measuring the radioactivity loss of the radiolabeled samples immersed in water (Figure 7). Comparing degradation kinetics of the different surfaces, **^{14}CA**, **^{35}SMH**, and **^{14}CGMH**, the main origin for the loss of peptide functionality appears to be the hydrolysis of siloxane films. A loss of about 50% occurs in the first week, followed by a gradual loss over several months.

3.6. Infrared Reflection Absorption Spectroscopy Measurements. Infrared reflection absorption spectroscopy (IRAS) has been thoroughly described.[37] It relies on reflecting an infrared beam at near-grazing incidence from the mirrorlike metallic surface on which the thin film of interest has been deposited. Since the surface roughness of the modified samples is much smaller ($R_a \approx$ 1.7 nm from AFM data) than the wavelength of the IR radiation, the electric field vector of the incident and reflected beams in our IRAS studies is expected to be along the normal of the surface plane. Hence, only the surface normal component of the dipole moment change can interact with the IR standing wave electric field at the surface, since the field vector and dipole moment derivative vector must be parallel. Thus, information on the orientation of the IRAS-active functional groups can be obtained.

Different infrared techniques have been applied to study the APS structure on silica[38] and metal surfaces.[27] Generally, the broad strong bands occur around 1140 cm^{-1} and are due to the Si−O−Si antisymmetric stretching mode. Absorption bands around 1570 cm^{-1} are caused by amine groups, but these were not detected in our case because of the low thickness of APS films (Figure 8).

The cyclic imide carbonyl groups in the crystalline state or in liquid solution exhibit at least two bands in the 1700−1800-cm^{-1} region.[39] One band is located between 1800 and 1740 cm^{-1} (symmetric stretch) and a more intense band between 1740 and 1700 cm^{-1} (antisymmetric stretch). The separation of the two bands is due to vibrational coupling of the carbonyls, leading to symmetric and antisymmetric stretching modes (Figure 9). Various explanations have been offered[39] for the symmetric and antisymmetric carbonyl stretch being assigned at the higher and lower frequency, respectively. These explanations include mechanical coupling, hydrogen bonding, and electronic effects. The antisymmetric vibration of the O=C−N−C=O group alternatively stabilizes the two resonance structures with a positive coupling constant lowering the frequency of the antisymmetric vibration.

(37) Allara, D. L.; Nuzzo, R. G. *Langmuir* **1985**, *1*, 52.
(38) Vandenberg, E. T.; Bertilsson, L.; Liedberg, B.; Uvdal, K.; Erlandsson, R.; Elwing, H.; Lundström, I. *J. Colloid Interface Sci.* **1991**, *147*, 103.
(39) (a) Mckittrick, P. T.; Katon, J. E. *Appl. Spectrosc.* **1990**, *44*, 812. (b) Parker, S. F. *Spectrochim. Acta, Part A* **1995**, *51*, 2067. (c) Matsuo, T. *Bull. Chem. Soc. Jpn.* **1964**, *37*, 1844.

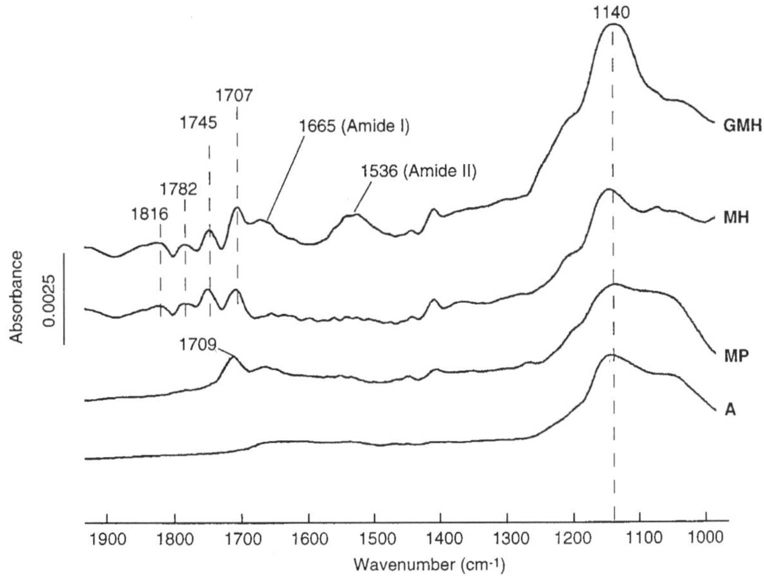

Figure 8. Infrared reflection spectra (IRAS) of surfaces **A**, **MP**, **MH**, and **GMH**.

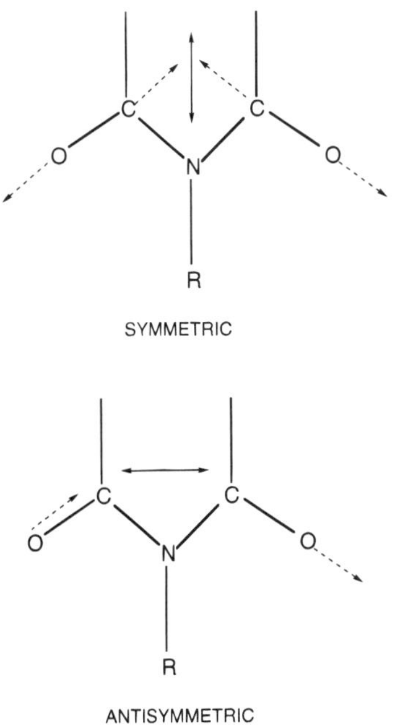

Figure 9. Symmetric and antisymmetric stretching vibrations of the imide moiety. Both-end-arrow lines represent the directions of the transition dipole moments.

On the **MP** and **MC** surfaces, only the antisymmetric vibration at 1709 cm^{-1} is clearly observed, while the symmetric one above 1740 cm^{-1} is barely detected, similar to the intensity distribution of polycrystalline maleimide derivatives with random orientation. However, on the **MH** surface, in addition to the antisymmetric vibration of the maleimidyl group at 1707 cm^{-1}, there are three other peaks at 1745 (strong), 1782 (weak), and 1816 (weak) cm^{-1}. The reason for these differences is not yet clearly understood and is under further investigation. A possible explanation is the presence of byproducts of an unknown side reaction. Another, more likely, explanation is a statistically preferential orientation of the molecules on surface **MH**. In this case, we tentatively assign 1745 cm^{-1} to the symmetric stretching mode of maleimide. The stronger intensity of this peak compared to those of the polycrystalline maleimide derivatives with random orientation may be attributed to a preferred orientation or aggregation of the maleimidyl groups, since the surface selection rule requires that the transition dipole moment of the symmetric stretching mode (Figure 9) has a substantial component along the substrate surface normal.

After the peptide attachment step, part of the maleimide is converted to succinimide. The bands of cyclic imide carbonyl groups appear virtually unchanged (Figure 8, **GMH**). The main difference between maleimide and succinimide is the fact that succinimide has a much stronger band around 1200 cm^{-1} (C−N−C).[39] The increased absorption intensity around 1200 cm^{-1} is attributed to the stretching vibration of C−N−C of succinimide and/or amide III (ca. 1200 cm^{-1}). Another obvious character on **GMH** is the enhanced amide groups. Additional information can be gained in the amide vibration regions: amide I at around 1660 cm^{-1} and amide II at around 1540 cm^{-1} are the characteristic absorption bands for peptides and proteins. Although an amide group is already present on **MH** before coupling the peptides, the strong increase in intensity at 1665 cm^{-1} (amide I) and at 1536 cm^{-1} (amide II) on surface **GMH** reflects the increased number of amide functionalities.

Conclusions

In summary, a three-step reaction procedure was employed to attach RGD-containing peptides onto a titanium surface. First, water−vapor−plasma-pretreated titanium surfaces were silanized with (3-aminopropyl)-triethoxysilane in dry toluene, resulting in a multilayer film of poly(3-aminopropyl)siloxane. Second, the free primary amino groups were linked to one of the three hetero-cross-linkers: *N*-succinimidyl-6-maleimidylhexanoate, *N*-succinimidyl-3-maleimidylpropionate, and *N*-

succinimidyl *trans*-4-(maleimidylmethyl)cyclohexane-1-carboxylate. Finally, onto the resulting terminal-maleimide surface, two model, cell-adhesive peptides, H-Gly-Arg-Gly-Asp-Ser-Pro-Cys-OH (GRGDSPC) and H-Arg-Gly-Asp-Cys-OH (RGDC), were immobilized through covalent addition of the cysteine thiol (−SH) group. X-ray photoelectron spectroscopy (XPS), infrared reflection absorption spectroscopy (IRAS), and radiolabeling techniques were applied to characterize the surfaces. The main results are as follows: (1) Silanization on Ti surfaces is shown to be the key step in terms of reproducibility in the subsequent modification steps. Samples produced in one batch with a surface coverage of 6 amino groups/nm^2 were used for further surface reactions. (2) The maleimidyl group introduced in the second-step reaction has been characterized using XPS (binding energy of C 1s at 289.0 eV), IRAS (band at around 1707 cm^{-1}), and radiolabeling techniques with [^{35}S]-cysteine. (3) The atomic ratio N/Si determined by XPS is used to estimate the reaction yields and to follow the reaction kinetics. The reaction yields are estimated to be about 30% for the reaction step of the aminosiloxane to the maleimide surface and about 18% for the conversion of the maleimide to the peptide surface. Optimal incubation times are 30 min for the former reaction and 1 h for the latter. (4) The grafted peptides, RGDC and GRGDSPC, have been qualitatively and quantitatively characterized with XPS, IRAS, and [^{14}C]-phenylglyoxal radiolabeling techniques. The surface coverage is estimated to be 0.2−0.4 peptide molecules/nm^2.

Acknowledgment. The authors would like to thank Dr. P. Böni and Mr. M. Horisberger of the Paul Scherrer Institute (PSI), CH-5232 Villigen PSI, for Ti coating, Mr. S. Brunner of the Department of Materials at ETH Zurich for technical help in IRAS measurements, Dr. M. Morstein of the Laboratory for Surface Science and Technology for generating the space-filling model of GRGDSPC, Dr. H. Chai-Gao and Ms. O. Bucher of CSEM for technical help at the beginning of this work, Dr P.-H. Vallotton of Institut Straumann, CH-4437 Waldenburg, and Mr. M. Windler of Sulzer Orthopedics, CH-8404 Winterthur, for their support. We are grateful to the Swiss Priority Program on Materials (PPM) (Council of the Swiss Federal Institutes of Technology) for their generous financial assistance.

LA980257Z

JOURNAL OF MATERIALS SCIENCE: MATERIALS IN MEDICINE **10** (1999) 255–263

Microstructured bioreactive surfaces: covalent immobilization of proteins on Au(1 1 1)/silicon via aminoreactive alkanethiolate self-assembled monolayers

F. G. ZAUGG, N. D. SPENCER*
Laboratory for Surface Science and Technology, Department of Materials, Swiss Federal Institute of Technology (ETHZ), CH-8092 Zurich, Switzerland

P. WAGNER
Department of Biochemistry, Beckman Center, Stanford University, Stanford CA 94305-5307, USA

P. KERNEN
EMPA, Lerchenfeldstrasse 5, CH-9014 St. Gallen, Switzerland

A. VINCKIER
Department of Biochemistry II, Universitätsstrasse 16, Swiss Federal Institute of Technology (ETHZ), CH-8092 Zurich, Switzerland

P. GROSCURTH
Institute of Anatomy, University of Zurich-Irchel, CH-8057, Switzerland

G. SEMENZA
Department of Biochemistry II, Swiss Federal Institute of Technology (ETHZ), and Dipartimento di Chimica e Biochimica Medica, Università di Milano, Via Saldini 50, 1-20133 Milano, Italy

Micrometer-scale patterns of a defined surface chemistry and structure were produced on both ultraflat Au(1 1 1) and on gold-coated monocrystalline silicon surfaces by a method combining microcontact printing, wet chemical etching and the replacement of etch-resist self-assembled monolayers (SAMs) by functionalized or reactive SAMs. Key steps in this methodology were characterized by X-ray photoelectron spectroscopy (XPS), ellipsometry and contact angle measurements. The covalent immobilization of (functional) biological systems on these surfaces was tested using an *N*-hydroxysuccinimide ester ω-functionalized disulphide (DSU), which covalently binds primary amines without the need for further activation steps. Atomic force microscope images of native collagen V single molecules immobilized on these patterned surfaces revealed both high spatial resolution and strong attachment to the monolayer/gold surface. Microcontact printing of DSU is shown to be feasible on specially prepared, ultraflat Au(1 1 1) surfaces providing a valuable tool for scanning probe experiments with biomolecules. The retention of enzymatic activity upon immobilization of protein was demonstrated for the case of horseradish peroxidase. The described approach can thus be used to confine biological activity to predetermined sites on microstructured gold/silicon devices – an important capability in biomedical and biomolecular research.

1. Introduction

The covalent immobilization of functional, biological molecules onto a defined and conductive surface provides the basis for sophisticated biomolecular architectures with numerous applications for *in vitro* studies on the behavior of biological structures such as proteins and cells, for implant "surface tailoring" or for biosensor devices. In recent years, much work has been devoted to the development of new techniques for protein immobilization on different substrates [1]. These

*Author to whom correspondence should be addressed. E-mail: nspencer@surface.mat.ethz.ch

developments have taken place in parallel with rapid progress in protein engineering technology. The immobilization of small, protein-containing structures such as cells or bilayer membranes can be facilitated by patterned surface chemistries or even topographically structured surfaces. Our ultimate goal is to build such structures mimicking parts of biological organelles, which are amenable to scanning probe microscopy. The goal of the present paper is to describe the foundations of such structures, based on functionalized alkanethiols or disulphides chemisorbed on gold surfaces.

The structure and formation of self-assembled monolayers (SAMs) of alkanethiols and dialkyldisulphides on gold surfaces has been intensively studied since 1983 [2–5] (for recent reviews see: [6, 7]). A broad range of applications of such systems has arisen from the synthesis of alkanethiols having functional groups in the ω-position. These make it possible to impart virtually any chemical property to a gold surface, and thus, parameters such as wettability [8, 9], and reactivity [10–14] can be controlled on such a surface. Monolayers on gold have been used to selectively and/or covalently bind cells [15–18], proteins [11, 13, 19–21], membranes [22], and other subcellular structures [23] (for a review see: [24]).

Micro- and nano-meter patterning of such alkanethiolate SAMs on gold can be achieved by the recently developed microcontact printing (μCP) method [25–27], providing monolayer surfaces with spatially resolved head-group functionalities. This method was favored in the present study over photolithographic techniques [28], because of its simplicity and wide applicability. In addition, this technique can be directly combined with wet-etching procedures to achieve topographical patterns [29].

We previously developed a procedure for producing Au(111) substrates for these monolayers, having extreme flatness over very large areas [30, 31]. More recently, we have shown that an amino-reactive, ω-substituted alkanethiolate self-assembled monolayer could be used to immobilize and study proteins in their native state [13, 23].

In this paper we report the application of the μCP technique to a functionalized alkanedisulphide – 11,11′-dithio-bis(succinimidyl-undecanoate) (DSU). We also describe a strategy for rendering etched silicon/gold structures accessible to this reactive crosslinker SAM. We subsequently present the site-specific immobilization of proteins onto such surfaces (Fig. 1) demonstrating the retained functionality of the immobilized protein. The atomic force microscope (AFM) was used to image the immobilized proteins and to determine their amount and spatial distribution on the patterned, template-stripped gold (TSG) surface.

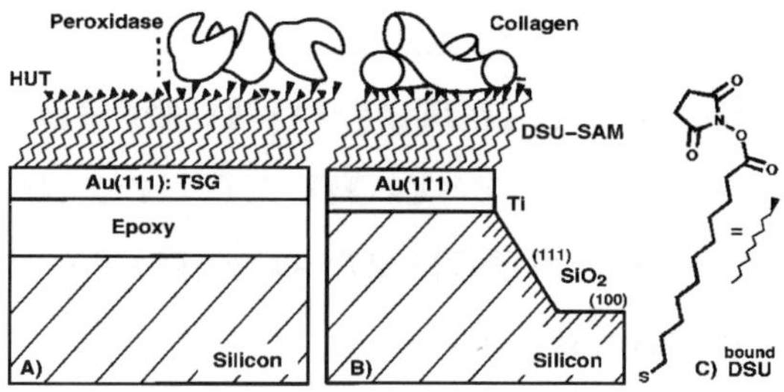

Figure 1 (A) Schematic section through a chemically microstructured TSG sample after μCP with reactive DSU, formation of a HUT SAM and immobilization of proteins. Primary amines of proteins react with the end groups of the DSU crosslinker (11, 11′-dithio-bis(succinimidylundecanoate), Fig. 1C: bound DSU (thiolate form) with the formation of an amide bond. (B) Schematic section through a microstructured silicon/gold structure after μCP, etching the gold, titanium and silicon, removal of the etch-resistant monolayer, formation of a reactive monolayer and finally reaction with protein.

2. Materials and methods

All chemicals were of the highest available purity. Ultrapure water was prepared by passage through a Barnstead purification system. 11-Hexadecanethiol (95%) and *N*-succinimidyl palmitate (X-ray photo electron spectroscopy; XPS reference) were purchased from Fluka (Buchs, Switzerland), silicon wafers (Si(100), *n*-type, 2–50 Ω cm) from Faselec AG (Zürich, Switzerland) and gold (99.99%) from Cendres et Metaux SA (Biel, Switzerland). Lyophilized bovine uterus collagen V (800 $\mu g\, ml^{-1}$ in 0.1 M acetic acid, purified according to a slightly modified protocol from Miller and Rhodes [32] was a gift from Dr Beat Trüeb, Maurice E. Mueller Institute for Biomechanics, Berne. Horseradish peroxidase (POD, type II, EC 1.11.1.7) was purchased from Sigma (Buchs, Switzerland). The synthesis of DSU was carried out as previously described [13]. 11-Hydroxyundecanethiol (HUT) was synthesized according to Bain *et al.* [2].

2.1. Preparation of gold surfaces

Two different gold surfaces have been used in this study:

(a) A 200-nm Au layer was deposited by thermal evaporation using a BAE 370 vacuum coating system from Bal-Tec (Liechtenstein) onto silicon wafers (Faselec AG, Zürich, Switzerland) coated with an adhesion layer consisting of 8 nm of Ti (e-beam). These substrates were placed under an argon atmosphere immediately after the preparation in order to avoid contamination of the gold from the ambient. In the present work we always refer to this type of gold, if not otherwise stated.

(b) Ultraflat, template-stripped gold (TSG) with a mean roughness of 0.2–0.5 nm over 25 μm^2 was prepared as described previously [14, 31]. These surfaces are protected from contamination by the mica template itself, which can easily be lifted off mechanically with the help of tweezers immediately before use. One can reproducibly obtain bare TSG samples without remaining mica pieces by cutting the edges of the glued sample with a sharp knife prior to the mica template removal.

2.2. Chemically patterned surfaces

Monolayer patterns of different alkanethiolates (dodecanethiol, hexadecanethiol (HDT), HUT, DSU) were prepared according the ''microcontact printing'' method [25, 33]. Briefly, an elastomeric stamp with the desired pattern was made by pouring poly(dimethylsiloxane) (PDMS) onto a silicon chip master, with subsequent curing at 80 °C for 5 h. For route A, the stamp was exposed to a solution of DSU (1 mM in dioxane). After drying with nitrogen and one minute in air, the stamp was lightly pressed onto the gold sample: alkanethiol was thereby transferred to the gold, forming an adsorbed layer. The DSU-pattern was then rinsed with HUT (5 mM in ethanol) for 30 s in order to cover the residual bare gold surface with a defined chemical functionality. For route B, the gold was patterned with a monolayer of hexadecanethiolate (1 mM in ethanol) by μCP and then subjected to chemical etching to create topographically patterned surfaces.

2.3. Physically patterned surfaces

Gold areas not covered by monolayers were selectively etched by immersion in an aqueous solution of 2 mM $K_3Fe(CN)_6$, 20 mM KCN and 1 M KOH at 3 °C for 10 min under stirring. The titanium layer and the underlying thin silicon oxide layer were then removed by etching with 1% HF. Subsequent boiling in 4 M KOH in 15% v/v isopropanol for 4 min at 60 °C resulted in V-shaped, anisotropic etching of the silicon. These substrates were then washed with 1 M KOH and water, dried under nitrogen and exposed to "piranha" solution consisting of 30% H_2O_2/conc. H_2SO_4 (3 : 7), at 60 °C for 2 min, to remove the organic monolayer etch-resist from the gold surface. Finally, after washing with water and drying, these substrates were incubated in a solution of 1 mM DSU in dioxane for 2 h, washed with dioxane and were then ready for immobilization of protein.

2.4. Immobilization of collagen

The acidic collagen V solution (800 μg/ml) was diluted 1 : 4000 with high-salt buffer (phosphate-buffered saline; PBS, pH 7.9, supplemented with 500 mM NaCl) and then applied to a 1 cm^2 HUT/DSU patterned surface for 2 h at room temperature. After washing with 15 ml of the buffer, the surface was again incubated in high-salt buffer for 2 h under gentle shaking. In some cases, non-specifically adsorbed protein had to be removed by brief sonication in NaCl-supplemented buffer with 0.05% Tween-20, followed by rinsing with phosphate buffered saline (PBS). The collagen pattern was examined by atomic force microscopy AFM either in PBS or after air-drying. This strategy was also used for the immobilization of collagen V on the 3D-patterned surfaces.

2.5. Peroxidase activity test

One milliliter of a solution of horseradish peroxidase (POD type II, 200 U mg^{-1}) in PBS (pH 7.5, activity 30 mU ml^{-1}) was applied to different DSU- and/or HUT-monolayer substrates and kept for 4 h at room temperature under gentle shaking. These surfaces were washed three times with a stream of PBS containing 0.1% (v/v) Tween-20 and further soaked for 20 min in the same buffer. The enzymic activity of the chemisorbed peroxidase was quantified as follows: after washing (as described) the samples were incubated individually at 24 °C in 1 ml of a 0.42 mM solution of 3,3′-5,5′-tetramethylbenzidine (TMB) in PBS (pH 6, containing 1% DMSO and 0.004% H_2O_2). After 45 min the blue supernatants were transferred into disposable cuvettes, the reaction stopped by adding 100 μl of 2 M H_2SO_4 and the absorption measured at 450 nm (Jasco 7800 UV/Vis Spectrophotometer). For the high-resolution AFM samples, POD (12.5 U ml^{-1}, 10 h at 4 °C) was used and the sample dried under nitrogen after rinsing with water.

2.6. Surface analysis

The AFM images were obtained either on a Bioscope or a NanoScope III equipped with a fluid cell and a "J-type" piezoscanner with a maximum scan range of 140 μm (Digital Instruments, Santa Barbara, CA). Monocrystalline silicon cantilevers (Lot, Darmstadt, Germany) were used with force constants ranging from 0.06 to 0.17 N m^{-1} (13–100 N m^{-1} for tapping-mode). In contact mode, the z-feedback loop was frequently adjusted to guarantee minimal forces (the force, just before the tip jumps out of contact if further retracted) between tip and substrate.

Prior to scanning electron microscopy (Philips SEM 505, 30 kV), an 8-nm thick gold layer was deposited on the etched sample in a sputtering device (Balzers Union, Balzers, Liechtenstein) to avoid electrostatic charging of the surface.

The XPS spectra were recorded using an ESCA 5400 instrument (Physical Electronics, Eden Prairie, MN) with MgK_α radiation at 300 W (15 kV). Measurements were taken at a take-off angle of 45° with respect to the sample surface. The analyzed area was 3.5 mm^2. Survey scans were taken for each sample with a constant detector pass energy range of 50 eV, followed by high-resolution XPS measurements (pass energy 17.9 eV) between 528 and 738 eV. Electron binding energies were calibrated to the Au 4f (84.0 eV) line.

The thickness of the stamped DSU layers was mesured using a Plasmos SD 2300 ellipsometer at 623.8 nm. The angle of incidence was set to 70° and the layer thickness measured relative to non-modified areas of bare gold on the same sample, assuming a refractive index of 1.44 for organic thin films. DSU (20 mM) in dioxane was used for stamping.

Advancing contact angle measurements were carried out on a G-I contact angle meter (Krüss, Hamburg, Germany) by applying a 3-μl drop of ultrapure water to a freshly prepared surface. A second drop was centered on the first, and the advancing contact angle measured within 30 s. A set of six locations was averaged per sample.

3. Results and discussion

Many aspects have to be taken into account when developing interfacial devices involving native biological molecules and inorganic, microstructured surfaces: biological objects, because of their sensitivity and complexity, need a well-defined, mild "physiological" environment for their structural integrity and activity. On the other hand, these systems have to be rendered convenient, making them accessible to experimental research or applications in artificial devices. In this regard, an important tool is the spatial direction of the site of immobilization of biomolecules on microstructured surfaces. However, care has to be taken not to denature the biological systems upon immobilization and to make sure that the attachment is strong enough to resist changes in the chemical and physical composition of the liquid environment, which can substantially impair the strength of immobilization of, for example, physisorbed proteins. Structured and chemically defined surfaces based on the alkanethiolate-SAM chemistry offer several advantages for the purposes mentioned above: alkanethiolate monolayers on gold are stable under aqueous conditions and they are ideal crosslinkers between biomolecules and the inorganic substrate. Moreover, the densely packed arrangement of the hydrocarbon chains with a defined spacer length provides a mechanical as well as a chemical barrier protecting the underlying substrate from direct contact with the environment. There are also no chemical limitations with respect to the synthesis of alkanethiols having complex end-group functionalities.

We have produced (by μCP) and then tested two different sets of structured surfaces for the covalent immobilization of biomolecules via SAMs: on ultraflat gold (i.e. on TSG) we prepared patterned DSU/HUT monolayers and tested them using proteins such as collagen V and POD (Fig. 1a). In a second strategy, we prepared structured silicon/gold surfaces with immobilized collagen (Fig. 1b).

3.1. Immobilization of biomolecules on chemically patterned TSG surfaces

While hexadecanethiol has been used in innumerable μCP studies and has been shown to be readily printable as a complete monolayer, it is not suited to the applications described in this paper, as its hydrophobic character is likely to induce protein adsorption with subsequent denaturation. Stamping the reactive SAM first, appropriate monolayer end-group chemistries can be chosen at will for the remaining, free gold areas, and the reactivity of the surface can either be tuned towards the immobilization of a specific ligand or to minimize non-specific adsorption of contaminants. We have concentrated on functionalization by stamping the reactive DSU disulphide. The higher polarity and increased size of this molecule compared to the *n*-alkanethiols used in standard μCP techniques also changes the stamping performance; ellipsometric measurements show that after stamping, a mean layer thickness of 0.48 ± 0.19 nm is obtained, compared to 1.7 nm for a fully formed DSU SAM after incubation [13]. Thus, only 1/4 of a monolayer is transferred to the substrate, on average. Subsequent rinsing with HUT, which adsorbs between the DSU molecules, gives rise to a SAM composed of mixed DSU and HUT molecules. The smaller amount of reactive sites available in the resulting mixed SAM may be counterbalanced by an increased reactivity of the DSU molecules – because of a reduced steric hindrance by neighboring DSU molecules towards nucleophilic attack [13, 34]. Additionally, it is beneficial to embed the reactive species into a matrix with non-adsorbing surface chemistry, leading to reduced non-specific adsorption on the reactive areas.

3.1.1. Collagen V immobilization

Collagen V is a typical extracellular matrix protein with a triple helical, 300-nm long and 1.4-nm wide structure consisting of 3042 amino acids. One hundred and fifty-two lysine residues are fairly homogeneously spread over its surface, enabling the protein to develop multiple bonds with the NHS-ester groups of a DSU monolayer.

Individual (i.e. three sub-units) molecules of collagen V were applied to a pattern of DSU and HUT monolayers. Fig. 2 shows an AFM image at high magnification of collagen V molecules immobilized only on those areas covered by DSU – resulting in a network of single collagen monomers – while no protein is observed on ultraflat-gold areas covered by HUT molecules. The amount of covalently immobilized collagen monomers on this system can be controlled by using different pH values and collagen concentrations in the buffer solution [13]. Densities ranging from a few molecules per square micron to dense networks of collagen can be obtained. The very characteristic filamentous shape of these proteins makes them clearly distinguishable from the template-stripped gold morphology in the background of the AFM images. They are visibly flexible and often follow the boundaries of HUT/DSU patterns. Interestingly, despite its size and shape, collagen V forms very sharp boundaries between patterned DSU and the protein-resistant, OH-terminated

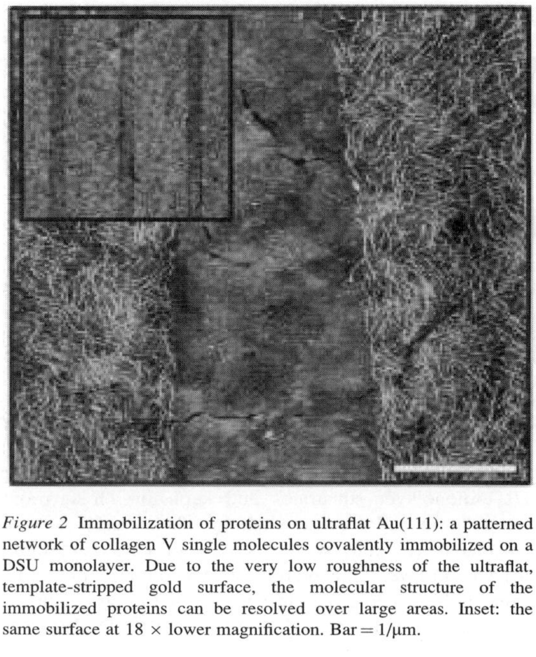

Figure 2 Immobilization of proteins on ultraflat Au(111): a patterned network of collagen V single molecules covalently immobilized on a DSU monolayer. Due to the very low roughness of the ultraflat, template-stripped gold surface, the molecular structure of the immobilized proteins can be resolved over large areas. Inset: the same surface at 18 × lower magnification. Bar = 1/μm.

monolayer. Conceivably, once a collagen molecule has bound with one lysine to a monolayer-site close to the DSU/HUT-SAM border, the non-bound triple-helical strand may be induced to zip-like binding also. In other words, being unable to create covalent bonds with HUT-SAM regions, the protein is thermodynamically forced to find its final location upon reaction with the NHS headgroups of the DSU monolayer. Contact-mode AFM revealed the very strong binding of the covalent (amide) bond between the proteins and the monolayer: an increased loading force on the scanning tip did not sweep aside the collagen until the protein chain itself was ruptured by the strong lateral forces applied. Physisorption of collagen on a HUT monolayer is also fairly stable, but can be overcome by special washing procedures, as described in Materials and Methods. Other ω-terminated alkanethiols (i.e. amino-, carboxyl-, or methyl-terminated) were also tried as fill-in monolayers for DSU patterns, but none proved superior to HUT. We are currently exploring the use of a polyethylene-glycol-terminated SAM, which was shown to resist physical adsorption of many proteins and even of cells [19, 35, 36], as an alternative protein-resistant monolayer coating.

3.1.2. Immobilization of peroxidase

In order to investigate the influence of the monolayer immobilization method on the activity of a model-protein, we adsorbed horseradish peroxidase on a similar pattern as described above and detected its enzymatic activity. Survival of the enzymatic activity at the solid–liquid interface cannot be taken for granted, since proteins often suffer large conformational changes – with partial or complete loss of their activity – when adsorbed at surfaces [37]. Non-covalently adsorbed peroxidase could be very efficiently removed by washing with a buffer solution containing 0.1% (v/v) Tween-20. This mild, non-polar detergent is known not to interfere with the activity of most proteins. The AFM image in Fig. 3a demonstrates that peroxidase molecules were exclusively adsorbed on those areas containing DSU in the monolayer. The height of these regions determined by contact mode AFM was 3.3 ± 0.6 nm, which is a realistic size if compared with X-ray structural data of the enzyme [38], taking into account some degree of compression and convolution caused by the interaction of the tip with the sample. A quantification of bound peroxidase can be attempted on the basis of tapping mode AFM images, on which individual immobilized peroxidase molecules can be distinguished as globular structures which can be counted and thence an average surface-coverage determined. Fig. 3b shows a typical TSG surface with individual POD molecules. A peroxidase surface density of 0.13 pmol cm^{-2} peroxidase could be calculated for an immobilization time of 10 h in PBS pH 7.5 at 4 °C, on ultraflat gold. X-ray data show [38] that the size of the enzyme is approximately $4 \times 4 \times 6$ nm. A maximal density of 3 pmol cm^{-2} can thus be expected for a tightly packed protein monolayer on ultraflat gold, assuming an exclusion radius of half the diameter around each protein.

The enzymatic activity of the covalently bound

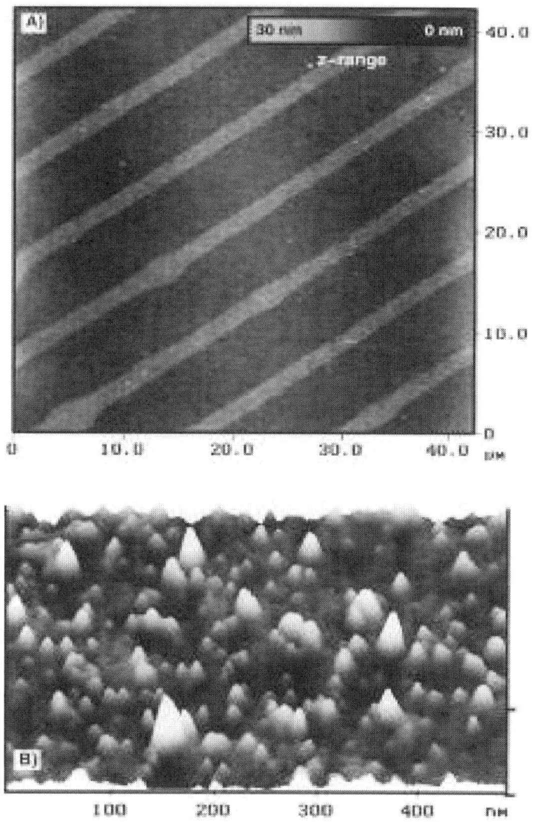

Figure 3 AFM image of bands of covalently immobilized horseradish peroxidase (A, light, higher areas) on ultraflat Au(111). Prior to the attachment of peroxidase, the gold was microcontact printed with a solution containing DSU and then rinsed with a solution of HUT, which covers the areas that appear dark in the image. (B) Individual POD molecules can be distinguished -and counted- at higher magnification on the tapping-mode AFM image. z-Bar = 5 nm.

peroxidase was estimated using 3,3′-5,5′-tetramethylbenzidine (TMB) (see Materials and Methods and Fig. 4). Three different surfaces have been examined for activity of the attached peroxidase after removing physisorbed enzyme by washing with PBS containing detergent (Tween): (i) surfaces totally covered by DSU binding the enzyme covalently; (ii) surfaces covered by a HUT monolayer; (iii) surfaces with DSU/HUT monolayer patterns. In three control experiments, the peroxidase activity was always the highest in the first case (i) 6.2 mU cm^{-2} activity on TSG and the lowest in the second case (ii) 1.4 mU cm^{-2} activity on TSG, while the patterned samples (iii) 3.3 mU cm^{-2} activity on TSG displayed intermediate activities. As expected, the enzyme activity correlates at least roughly with the amount of DSU present on the respective gold surface and, thus, with the number of bound enzyme molecules.

The influence of roughness of the gold surface on the amount of bound enzyme was investigated by comparing two types of surfaces: (i) a substrate prepared by thermal

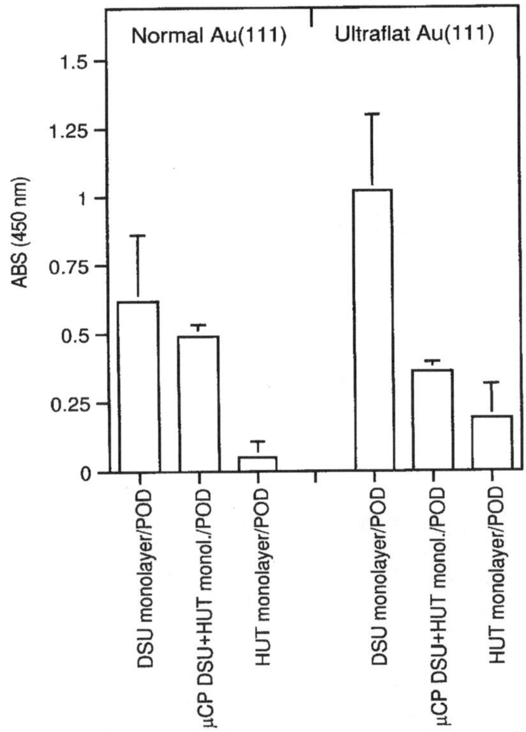

Figure 4 The relative activity of covalently immobilized peroxidase on two different Au(111) substrates precoated with three different SAM compositions: Au(111) normal = gold thermally evaporated on a Ti-primed Si(1 0 0) wafer; ultraflat Au(111) = ultraflat, TSG gold (see Materials and Methods) with a mean roughness of 0.2–0.5 nm/25 μm^2; DSU monolayer = the gold was incubated with the aminoreactive DSU-SAM prior to the immobilization of peroxidase (POD); HUT monolayer = the gold was incubated with 11-hydroxyundecanethiol prior to the immobilization of POD; μCP DSU + HUT monol. = the gold was first microcontact printed with DSU followed by rinsing with a solution of HUT, prior to POD incubation.

evaporation of gold on silicon wafers having a roughness of approx. 2–6 nm and (ii) ultraflat gold surfaces having mean roughness values one order of magnitude lower. Bound POD proved to retain enzymatic activity in both cases (Fig. 4), but only half as much activity was detected on the rather rough surface of the first substrate compared to the second type consisting only of monoatomic steps over cm^2 areas. Increased peroxidase activity on smooth surfaces was observed for the DSU-, HUT- and DSU/HUT-monolayer samples mentioned above. This may be explained by preferential adsorption of the protein in the (more reactive) grooves of the rougher substrate, which might lead to a different protein conformation caused by multiple binding with concomitant loss of activity and/or to unstirred layers during testing, which are known to result in an apparent increase of K_m and decrease in V_{max}. An equivalent activity of 0.08 pmol dissolved POD was found immobilized on 1 cm^2 of a TSG surface covered with DSU.

3.2. Generation of etch-patterned surfaces with immobilized biomolecules

The creation of etch-structured gold/silicon surfaces was achieved by a multistep procedure. First, a hexadecanethiolate monolayer pattern resistant to a cyanide-based gold etch solution was prepared on a 200-nm thick film of gold (deposited on a silicon substrate with a 8-nm titanium layer as an adhesion mediator) by microcontact printing. Non-protected, i.e. monolayer-free, gold areas were etched by subsequent treatment with a cyanide solution. A second etching step with 1% hydrofluoric acid was carried out to remove the titanium and the underlying thin native silicon oxide layer, followed by a third bath to create several-micron-deep, V-shaped grooves (cavities) in the monocrystalline Si(100) substrate by anisotropic etching. This resulted in flat, elevated micrometer-sized gold islands on a silicon support. No changes in the Au(111) morphology could be observed as a result of the etching. The hexadecanethiolate monolayer still left on these gold areas was finally removed by exposing the surface to hot "piranha" solution (30% H_2O_2/conc. $H_2SO_4 = 3:7$; 60 °C) for 5 min (*caution: this solution can explode in contact with organic matter*). Contact angle measurements on non-etched test Au samples (Fig. 5) confirm the removal of the hydrophobic etch-resist SAM upon 2 min exposure to piranha. This treatment leads to a bare gold surface which is now again accessible to the adsorption

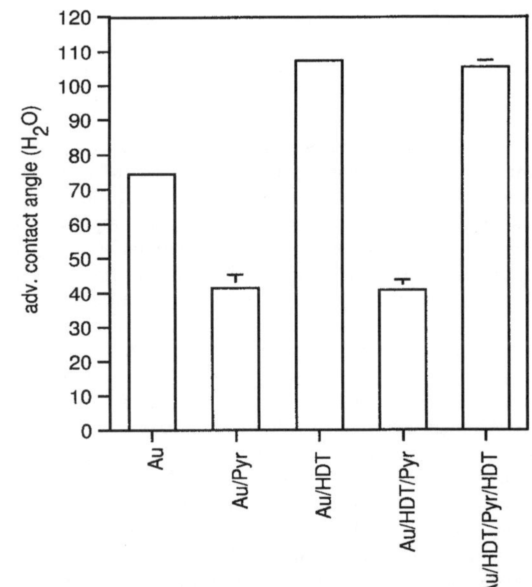

Figure 5 Changes in contact angle related to the process of oxidative SAM removal with piranha solution. Legend: Au: pure Au(111) with natural contamination layer; Au/Pyr: Au treated with hot Piranha solution for 2 min. Au/HDT: Au(111) after 2 h exposure to hexadecanethiol (5 mM in ethanol); Au/HDT/Pyr: Au after the removal of the hexadecanethiolate-SAM (hot piranha for 2 min); Au/HDT/Pyr/HDT: new formation of a hexadecanethiolate-SAM on Au(111) after piranha removal of a previous SAM (5 mM in ethanol).

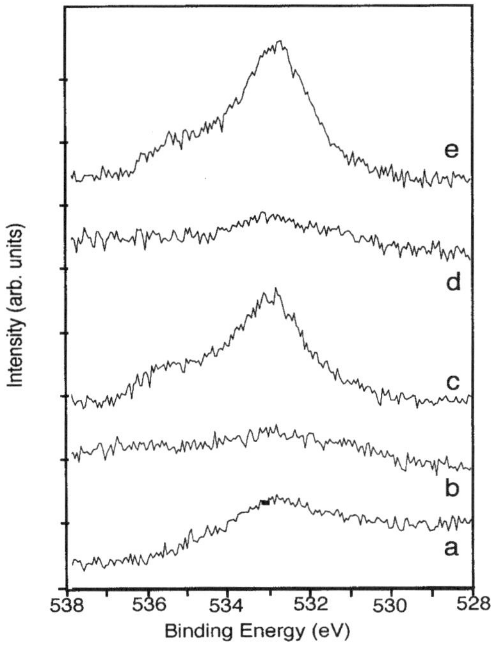

Figure 6 O(1s) X-ray photoelectron spectrum progression of a gold sample at different stages of functionalized-SAM replacement. (a) Fresh, naturally contaminated gold; (b) gold after exposure to 2 min piranha (60 °C); (c) DSU-SAM on gold (0.5 mM in dioxane for 18 h); (d) sample c exposed to 2 min piranha (60 °C), (e) sample d upon repeated exposure to DSU.

of a new SAM, as can be seen from the recovery of hydrophobicity upon exposure to a 5 mM hexadecanethiol solution. The same behavior can also be demonstrated for functionalized SAMs, such as DSU: Fig. 6 shows the progression of the O1s XPS spectra after the different reaction steps: only a very weak oxygen contamination peak can be detected on freshly deposited gold, which then disappears after inmersion in hot piranha solution. After formation of a DSU SAM, the O1s envelope of the succinimidyl-ester functionality appears as a strong band showing two shoulders at 532.2 ± 0.5 eV $(C = O)$ and 534.3 ± 0.13 eV $(C - O - C)$. The peak shape is similar for DSU chemisorbed on Au(111) as for a powder-spectrum of *N*-succinimidyl palmitate, a commercial compound containing the same succinimidyl-ester group (data not shown). Nevertheless, in no case could a good 1/1/2 curve-fitting for the oxygen peak be obtained, probably due to partial hydrolysis of the succinimidylester, which leads to the appearance of carboxylate and hydroxy peaks within the same binding-energy region. Exposure of the DSU-covered substrate to piranha solution leads to a complete disappearance of the oxygen peaks, whereas upon repeated formation of a DSU SAM, these peaks reappear in the original shape, demonstrating the reversability of the SAM formation. After removal of the hexadecanethiolate etch-resist and formation of a DSU SAM, the quality and specificity of this amino-reactive DSU pattern was again examined by immobilization of collagen V molecules. Fig. 7 shows a SEM image of such a topographically and chemically patterned sample. AFM imaging of the monolayer-free silicon oxide areas reveals no physisorbed collagen, while a network of single, covalently bound collagen monomers can be distinguished on the Au(111) islets. Collagen immobilized on the gold tracks of etch-patterned surfaces shows the same characteristics as that on chemically patterned surfaces. However, the apparent edge definition of the collagen on this kind of surface (Fig. 7) is much lower than on TSG (Fig. 2), because AFM imaging is negatively affected by the rough substrate morphology of Au(111) surfaces prepared by standard thermal evaporation of gold on titanium-primed silicon.

Many materials are processable by the sequence described above, since gold layers of well-defined thickness and topography can easily be deposited on a large number of substrates by vacuum-evaporation techniques. Subsequent structuring steps in the micrometer range can be performed by applying traditional wet-etching procedures. The possibility of removing a SAM from gold either by means of an electrical field, or heating above 80 °C in a suitable solvent [2], or by oxidation [39] allows the specific replacement of one monolayer by another. Interesting applications of this include the described possibility of using inert alkanethiols to pattern and etch a surface by μCP and then – through SAM exchange – to impart a desired chemical reactivity to it. Additionally, the process allows monolayer-immobilized proteins that have become non-functional to be replaced by fresh ones: silicon devices containing enzymes on small gold areas could be regenerated through a simple three-step procedure: removing the damaged monolayer/protein layer by photo-oxidation; formation of a new monolayer by rinsing with reactive compounds, and finally binding native enzyme. This kind of surface recycling can be of use in biophysical research applications where complex devices can only be produced in small numbers.

4. Conclusions and outlook

In this study, we have presented a mild procedure for covalent immobilization of proteins on microstructured gold surfaces. The use of ultraflat, TSG as a substrate opens the possibility of making patterned biomolecule structures accessible to scanning probe methods on a background with defined surface chemistry. On microstructured $Si/SiO_2/Ti/Au$ surfaces produced by μCP and etching, the immobilization site of proteins could also be spatially controlled using an amino-reactive SAM. The production of these micropatterned substrates including SAM exchange and protein immobilization, can all be performed in one day, thus providing a high degree of flexibility in experimental work. Although most biological samples have amino groups on their surfaces and can thus be immobilized via DSU, the need may arise for other ω-substituted alkanethiols with chemical reactivity directed towards other functional groups. For example, aryldiazonium-terminated alkanethiols [14] selectively

Figure 7 Covalently immobilized proteins on microstructured Au/Si surfaces: (right), SEM image of a structured Au/Si surface produced by μCP, etching, SAM replacement and covalent collagen immobilization (top, left; bar = 200 nm; z-range = 30 nm): An AFM (tapping mode in air) zoom of the Au/DSU-SAM areas reveals the presence of a two-dimensional network of covalently bound collagen V molecules. The resolution is lower than in Fig. 2 because the round-shaped (rough) morphology of this Au(111) impairs AFM measurements. (bottom, left): no collagen is visible on the silicon areas on the surface (same scales as top, left).

bind activated aromatic hydrogens of, e.g. histidine and tyrosine residues; Ni^{2+}-chelating monolayer-headgroups bind histidine-tagged fusion proteins [12]; (Wagner *et al.*, manuscript in preparation). We are currently exploring the use of this technology for the immobilization of cells and biological bilayer membranes.

Acknowledgment

We are grateful to Dr Beat Trueb for donating the collagen V. Many thanks also to Dr Urs Ziegler for his help on many occasions and Dr Antonella Rossi for helpful discussions. This research was supported in parts by the Swiss National Science Foundation Projects NFP-36 and Nr. 31-43459.95 and the O. Mayenfisch Stiftung Zürich. P. Wagner was supported by a Humboldt fellowship.

References

1. R. F. TAYLOR, in "Protein immobilization: fundamentals and applications," (Marcel Dekker, Inc., New York, 1991).
2. C. D. BAIN, E. B. TROUGHTON, Y. -T. TAO, J. EVALL, G. M. WHITESIDES and G. NUZZO, *J. Amer. Chem. Soc.* **111** (1989) 321.
3. R. G. NUZZO and D. L. ALLARA, *ibid.* **105** (1983) 4481.
4. M. D. PORTER, T. B. BRIGHT, D. L. ALLARA and C. E. D. CHIDSEY, *ibid.* **109** (1987) 3559.
5. H. SELLERS, A. ULMAN, Y. SHNIDMAN and J. E. EILERS, *ibid.* **115** (1993) 9389.
6. A. R. BISHOP and R. G. NUZZO, *Curr. Opin. Coll. Interf. Sci.* **1** (1996) 127.
7. A. ULMAN, *Chem. Rev.* **96** (1996) 1533.
8. C. D. BAIN and G. M. WHITESIDES, *Angew. Chem. Int. Ed. Engl.* **28** (1989) 506.
9. J. DRELICH, J. D. MILLER, A. KUMAR and G. M. WHITESIDES, *Coll. Surf. A* **93** (1994) 1.
10. L. F. ROZSNYAI and M. S. WRIGHTON, *Langmuir* **11** (1995) 3913.
11. E. DELAMARCHE, G. SUNDARABABU, H. BIEBUYCK, B. MICHEL, C. GERBER, H. SIGRIST, H. RINGSDORF, N. XANTHOPOULOS and H.J. MATHIEU, *ibid.* **12** (1996) 1997.
12. G. B. SIGAL, C. BAMDAD, A. BARBERIS, J. STROMINGER and G. M. WHITESIDES, *Anal. Chem.* **68** (1996) 490.
13. P. WAGNER, M. HEGNER, P. KERNEN, F. G. ZAUGG and G. SEMENZA, *Biophys. J.* **70** (1996) 2052.
14. P. WAGNER, F. Z. ZAUGG, P. KERNEN, M. HEGNER and G. SEMENZA, *J. Vac. Sci. Technol. B* **14** (1996) 1466.
15. G. P. LOPEZ, M. W. ALBERS, S. L. SCHREIBER, R. CARROLL, E. PERALTA and G. M. WHITESIDES, *J. Am. Chem. Soc.* **115** (1993) 5877.
16. R. SINGHVI, A. KUMAR, G. P. LOPEZ, G. N. STEPHANOPOULOS, D. I. C. WANG, G. .M WHITESIDES and D.E. INGBER *Science* **264** (1994) 696.
17. C. B. HERBERT, T. L. MCLERNON, C. L. HYPOLITE, D. N. ADAMS, L. PIKUS, C. C. HUANG, G. B. FIELDS, P. C. LETOURNEAU, M. D. DISTEFANO and W. S. HU *Chem. Biol.* **4** (1997) 731.
18. M. MRKSICH, L. E. DIKE, J. TIEN, D. E. INGBER and G. M. WHITESIDES, *Exp. Cell Res.* **235** (1997) 305.
19. K. L. PRIME and G. M. WHITESIDES, *Science* **252** (1991) 1164.
20. G. P. LOPEZ, H. A. BIEBUYCK, R. HAERTER, A. KUMAR and G. M. WHITESIDES, *J. Amer. Chem. Soc.* **115** (1993) 10774.
21. G. B. SIGAL, M. MRKSICH and G. M. WHITESIDES, *ibid.* **120** (1998) 3464.
22. H. LANG, C. DUSCHL and H. VOGEL, *Langmuir* **10** (1994) 197.
23. P. WAGNER, P. KERNEN, M. HEGNER, E. UNGEWICKELL and G. SEMENZA, *FEBS Lett.* **356** (1994) 267.
24. M. MRKSICH and G. M. WHITESIDES, *Annu. Rev. Biophys. Biomol. Struct.* **25** (1996) 55.

25. J. L. WILBUR, E. KUMAR and G. M. WHITESIDES, *Adv. Mater.* **6** (1994) 600.
26. M. MRKSICH and G. M. WHITESIDES, *Trends Biotechnol.* **13** (1995) 228.
27. Y. XIA and G. M. WHITESIDES, *Angew. Chem. Int. Ed.* **37** (1998) 550.
28. V. W. JONES, J. R. KENSETH, M. D. PORTER, C. L. MOSHER and E. HENDERSON, *Anal. Chem.* **70** (1998) 1233.
29. Y. XIA, X. -M. ZHAO and G. M. WHITESIDES, *Microelectr. Enging* **32** (1996) 255.
30. P. WAGNER, M. HEGNER, H.-J. GÜNTHERODT and G. SEMENZA, *Langmuir* **11** (1995) 3867.
31. M. HEGNER, P. WAGNER and G. SEMENZA, *Surf. Sci.* **291** (1993) 39.
32. E. J. MILLER and R. K. RHODES, *Methods Enzymol.* **82** (1982) 33.
33. A. KUMAR, H. A. BIEBUYCK and G. M. WHITESIDES, *Langmuir* **10** (1994) 1498.
34. J. H. WANG, J. R. KENSETH, V. W. JONES, J. B. D. GREEN, M. T. MCDERMOTT, and M. D. PORTER, *J. Amer. Chem. Soc.* **119** (1997) 12796.
35. J. M. HARRIS and S. ZALIPSKY, in ''Poly(ethylene glycol),'' (ACS Symposium Series, Vol. 680, edited by the American Chemical Society, Washington DC, 1997).
36. H. DU, P. CHANDAROY and S. W. HUI, *Biochim. Biophys. Acta Biomembranes* **1326** (1997) 236.
37. G. RIALDI and E. BATTISTEL, *J. Thermal Anal.* **47** (1996) 17.
38. L. BANCI, P. CARLONI and G. G. SAVELLINI, *Biochemistry* **33** (1994) 12356.
39. J. HUANG, D. A. DAHLGREN and J. C. HEMMINGER, *Langmuir* **10** (1994) 626.

Received 1 June
and accepted 20 July 1998

3298 *J. Phys. Chem. B* **2000,** *104,* 3298–3309

Poly(L-lysine)-*g*-Poly(ethylene glycol) Layers on Metal Oxide Surfaces: Attachment Mechanism and Effects of Polymer Architecture on Resistance to Protein Adsorption[†]

Gregory L. Kenausis,[‡] Janos Vörös,[‡] Donald L. Elbert,[§] Ningping Huang,[‡] Rolf Hofer,[‡] Laurence Ruiz-Taylor,[‡,||] Marcus Textor,[‡] Jeffrey A. Hubbell,[§] and Nicholas D. Spencer*,[‡]

Laboratory for Surface Science and Technology and Institute for Biomedical Engineering, Department of Materials, ETH Zurich, CH-8092 Zurich, Switzerland

Received: September 20, 1999; In Final Form: January 21, 2000

The generation of surfaces and interfaces that are able to withstand protein adsorption is a major challenge in the design of blood-contacting materials for both medical implants and bioaffinity sensors. Poly(ethylene glycol)-derived materials are generally considered to be particularly effective candidates for the fabrication of protein-resistant materials. Most metallic biomaterials are covered by a protective, stable oxide film; converting such oxide surfaces, which are known to strongly interact with proteins, into noninteractive surfaces requires a specific design of the surface/interface architecture. A class of copolymers based on poly(L-lysine)-*g*-poly(ethylene glycol) (PLL-*g*-PEG) was found to spontaneously adsorb from aqueous solutions onto several metal oxide surfaces, such as TiO_2, $Si_{0.4}Ti_{0.6}O_2$, and Nb_2O_5, as measured by the in situ optical waveguide lightmode spectroscopy technique and by ex situ X-ray photoelectron spectroscopy. The resulting adsorbed layers are highly effective in reducing the adsorption both of blood serum and of individual proteins such as fibrinogen, which is known to play a major role in the cascade of events that lead to biomaterial-surface-induced blood coagulation and thrombosis. Adsorbed protein levels as low as <5 ng/cm^2 could be achieved for an optimized polymer architecture. The modified surfaces are stable to desorption under flow conditions at 37 °C and pH 7.4 in HEPES [4-(2-hydroxyethyl)piperazine-1-ethanesulfonic acid] and PBS (phosphate-buffered saline) buffers. The adsorbed layer of copolymer is thought to form a comblike structure at the surface, with positively charged primary amine groups of the PLL bound to the negatively charged metal oxide surface, while the hydrophilic and uncharged PEG side chains are exposed to the solution phase. Copolymer architecture is an important factor in the resulting protein resistance; it is discussed on the basis of packing-density considerations and the corresponding radii of gyration of the different PEG chain lengths studied. This surface functionalization technology is believed to be of value for use in both the biomaterial and biosensor areas, as the chosen macromolecules are biocompatible and the application is straightforward and cost-effective.

1. Introduction

1.1. General Background. Nonspecific protein adsorption plagues a wide array of biomedical devices, such as catheters, heart valves, and stents, as well as blood-, serum-, or plasma-contacting sensors.[1] Protein adsorption onto the implant surface is the first stage in the series of events that, depending on the type and nature of the adsorbed proteins, may lead to a deleterious response. In the case of blood-contacting devices such as stents and catheters, in particular, protein adsorption is the first step in a cascade of surface processes that induce platelet deposition and thrombus formation. Because metals such as titanium or steel, covered by their related surface oxides, are often applied in blood-contacting devices, a method for reducing protein adsorption on oxides would be an important development in biomaterials technology. Moreover, oxides of transition metals, such as titanium, tantalum, or niobium, are preferred surface coatings for optical waveguides because of their high optical transparency and high index of refraction. Optical waveguide lightmode spectroscopy (OWLS) is a sensitive detection approach for biomolecule sensing because of its high interfacial sensitivity. In general, a technique for reducing nonspecific adsorption in a consistent manner is required to improve the performance of analyte-specific sensors in terms of both selectivity and sensitivity.

Poly(ethylene glycol) (PEG) has been extensively investigated for use in a wide array of biomedical applications, and immobilization of PEG on surfaces has long been known to decrease protein adsorption. Of the many models proposed to explain this effect, steric stabilization and excluded-volume effects are the most commonly cited.[2–4] A general approach to the immobilization of PEG on surfaces involves the coupling of PEG to a functional group that has an affinity for the target surface. Processes ranging from chemical surface derivatization with reactively functionalized PEG[5–7] to adsorption of PEG-containing surfactants such as Pluronics[8] have been studied extensively. In the latter case, treatment of hydrophobic surfaces with Pluronics resulted in surfaces that were resistant to protein adsorption and platelet adhesion.[9–11]

Another approach to PEG immobilization involves the grafting of PEG side chains onto a polymer backbone, resulting

[†] Part of the special issue "Gabor Somorjai Festschrift".

* Corresponding author. E-mail: nspencer@surface.mat.ethz.ch. Tel.: +41 1 632 5850. Fax: +41 1 633 10 27.

[‡] Laboratory for Surface Science and Technology.

[§] Institute for Biomedical Engineering.

[||] Present address: Zyomyx, Inc., 3911 Trust Way, Hayward, CA 94545.

Published on Web 03/15/2000

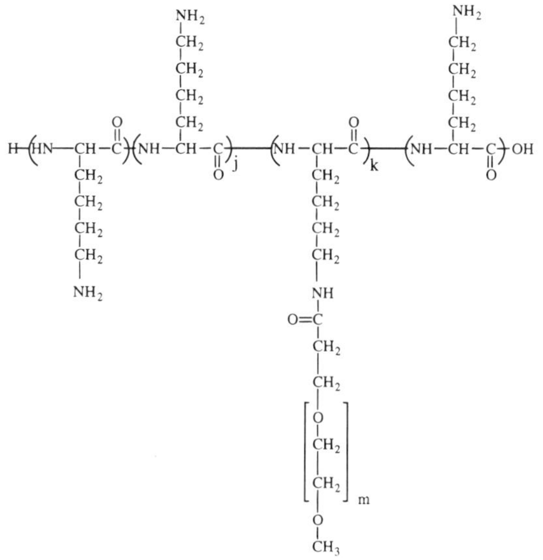

Figure 1. Chemical structure of the poly(L-lysine)-*g*-poly(ethylene glycol) (PLL-*g*-PEG) polymer. (See the first paragraph of section 2 for a comprehensive description of the specific structures of the various PLL-*g*-PEG polymers discussed in this paper.)

in the formation of a comblike structure.[2,12,13] The adsorption behavior of graft copolymers, in general, has been modeled by several authors[14−20] and clearly depends on the interaction between the polymer backbone and the target surface.

The aim of the present study was to immobilize PEG on metal oxide surfaces such as silicon dioxide, titanium dioxide, and niobium pentoxide. The adsorption behavior of polyelectrolytes on such metal oxide surfaces has been characterized,[21−27] and polycations, in particular, were found to form stable adsorbed layers on negatively charged oxides such as silicon dioxide and titanium dioxide.

The polymer system investigated in the present study, poly(L-lysine)-*g*-poly(ethylene glycol) (PLL-*g*-PEG), consists of a poly(L-lysine) (PLL) backbone that has been grafted with PEG side chains (see Figure 1).[51] PLL was chosen as the polymer backbone because it is highly cationic at physiological pH and because PLL is a polypeptide with well-understood toxicology.[28] Furthermore, graft copolymers of PLL and PEG and similar, related polymers have been evaluated for toxicity, immunogenicity, immunomodulatory potential, pyrogenicity, and biodegradation, largely in the context of drug delivery[29−33] and tissue adhesion suppression.[24−27,34]

In the present study, the assembly of an adsorbed layer of the PLL-*g*-PEG copolymer onto several metal oxide surfaces is shown to be caused by the electrostatic interaction of the positively charged PLL backbone with the negatively charged metal oxide surfaces. This, then, presumably creates a surface-bound comb of PEG. The resultant surfaces have been found to exhibit drastically reduced protein adsorption relative to the untreated metal oxide control surfaces. Because this immobilization relies on electrostatic interactions in an electrolyte, pH and ionic strength are clearly important parameters in determining the extent and stability of such adsorption. Because of its versatility, high sensitivity, and suitability for in situ measurements,[35−38] the optical waveguide lightmode spectroscopy (OWLS) technique was employed, as it is well-suited for the quantitative determination of copolymer adsorption and stability, as well as inhibition of protein adsorption. Additionally, X-ray photoelectron spectroscopy was used to assess the properties of the adsorbed layer of PLL-*g*-PEG and its interaction with the underlying metal oxide surface.

1.2. Review of the Surface Adsorption of Graft Copolymers and Polyelectrolytes. An understanding of the ability of adsorbed copolymers to passivate a surface against the adsorption of proteins will rely heavily on previous work concerning the steric stabilization of colloids because of the colloidal nature of individual protein molecules. From studies of polymeric steric stabilization, it is clear that the ability to stabilize a suspension of colloids will depend on the ability to immobilize an otherwise soluble polymer on the colloid surface. This can be achieved through chemical grafting of the polymer, through initiation of polymerization at the colloid surface, or through adsorption. Of these techniques, adsorption is perhaps the simplest means for immobilizing polymer on the colloid surface. Compared with homopolymers and random copolymers, block copolymers will typically be more useful for achieving adsorption and steric stabilization on a surface. Block copolymers of different architectures have been utilized in this regard, such as AB, ABA, BAB, $(AB)_n$, and $A(B)_n$ (that is, comb graft copolymer), as well as dendrimeric-type structures.[39]

In aqueous solution, the great majority of surfactants for the dispersion of colloids are not polymers, but rather charged long-chain hydrocarbons, with stability imparted by electrostatic stabilization. However, electrostatic stabilization is not generally feasible for the prevention of protein adsorption onto surfaces, because proteins will generally present a balance of positive and negative residues. Polymeric steric stabilization is much more appropriate, and in aqueous media, this is most often accomplished with nonionic surfactants. These are most often poly(ethylene glycol)s, attached to various hydrophobic components, such as alkyl phenols, alkyl alcohols, fatty acids, and water-insoluble polyalkyl glycols. Because the adsorption of the polymer will generally be irreversible, the thermodynamics and kinetics of adsorption will be important, as these will determine the surface density of poly(ethylene glycol) segments. To achieve efficient steric stabilization, the surface spacing of bound poly(ethylene glycol) units should be less than the radius of gyration of poly(ethylene glycol) in solution.[20]

An important theoretical result is that, as the polymer chains adsorb onto the surface, the amount of space left for other chains decreases, and thus, an adsorption isotherm will be observed. This is particularly important with block copolymers, because of the large volume of the adsorbing segment and because of the steric effects of the poly(ethylene glycol) segments. A fast adsorption step is observed, followed by slow surface rearrangements that allow the adsorption of more polymer.[40−42] To achieve efficient protein repulsion, a balance must be found so as to maximize the ratio of the radius of gyration of poly(ethylene glycol) to the distance between attachment sites of the poly(ethylene glycol) segments, while still adsorbing the copolymer to the surface essentially irreversibly.

Numerous examples of these systems have been studied experimentally. Tirrell et al.[43] measured the adsorption of AB copolymers of polystyrene and polyvinylpyridine on mica. A wide range of molecular weights was tested, and it was found that, for most samples, about 200 ng/cm^2 of polymer adsorbed to the surface, and that the buoy segments easily formed "polymer combs", unless the size of the anchoring segment was too large. The maximum surface density was found with small

TABLE 1: Details of the Synthesis of Different Types of PLL-*g*-PEG

polymer PLL(*x*)-*g*[*y*]-PEG(*z*)[a]	PLL	PEG
PLL(375)-*g*[5.6]-PEG(5)	500 mg in 10 mL of SBB[b]	2.0 g in 2.5 mL of SBB[b]
PLL(20)-*g*[6.0]-PEG(5)	500 mg in 10 mL of SBB[b]	2.0 g in 2.5 mL of SBB[b]
PLL(20)-*g*[3.5]-PEG(2)	83.6 mg in 1.05 mL of STBB[c]	215.7 mg of solid
PLL(20)-*g*[5.5]-PEG(2)	106.8 mg in 1.34 mL of STBB[c]	193.2 mg of solid

[a] PLL(*x*)-*g*[*y*]-PEG(*z*) signifies that the graft copolymer has a PLL backbone of mol wt *x* kDa; a graft ratio, *y*, of lysine-mer/PEG side chain; and PEG side chains of mol wt *z* kDa. [b] SBB is sodium borate buffer (pH 8.5). [c] STBB is sodium tetraborate buffer (pH 8.5).

anchor segments and intermediate-sized buoy segments, as would be expected for a copolymer containing anchor segments with strong binding capabilities.[43]

The adsorption of polyelectrolytes, in particular, onto surfaces has been reviewed,[20,44] and it has been found that, in general, the polyelectrolyte will attempt to neutralize any opposite charge present on the surface. Furthermore, the presence of polyelectrolyte loops and tails leads to an overcompensation and a reversal of surface charge. A polyelectrolyte will generally remain very close to the surface, orienting parallel to the surface and producing very thin layers, often less than the radius of gyration of the polymer, with the layer thickness increasing as the concentration of salt is decreased.[20] The concentration of salt is important and may aid or hinder adsorption of the polyelectrolyte. An increase in salt concentration decreases interpolyelectrolyte repulsion, leading to enhanced adsorption of the polymer, but also compacts the electric double layer, leading to repulsion and competition for surface ions, which, in turn, lead to desorption of the polymer. The adsorption of polylysine on silica was found to be maximal at 10−100 mM NaCl (physiological NaCl concentration is 150 mM), and the adsorbed amount of the polymer increased almost linearly with an increase in surface charge.[44] The kinetics of adsorption and desorption of polycations on titanium dioxide and silicon dioxide have also been studied.[22,23] The initial adsorption was found to be transport-limited, and a maximum in the amount adsorbed was found at a salt concentration of about 100 mM. Desorption of polycations after the polycation solution was replaced with 100 mM NaCl was slow even under flow. However, changing the pH so as to neutralize the charge of the polymer or the surface led to rapid desorption of the polymer.

2. Materials and Methods

For clarity, the following system of abbreviations will be used when referring to the various polymers discussed in this paper: PLL(*x*)-*g*[*y*]-PEG(*z*) signifies that the graft copolymer has a PLL backbone of molecular weight *x* kDa; a graft ratio, *y*, of lysine-mer/PEG side chain; and PEG side chains of mol wt *z* kDa.

2.1. Synthesis of PLL-*g*-PEG. Table 1 shows the details of the masses and solvents that were used to synthesize the different polymers according to the following procedure. Poly-L-lysine hydrobromide of mol wt 20 000 or 375 000 (Sigma, St. Louis, MO) was dissolved in 50 mM sodium borate buffer (SBB), pH 8.5. The solution was filter sterilized (0.2-μm pore-size filter). Monomethoxy PEG-nitrophenyl carbonate, mol wt 5000 (Shearwater Polymers, Huntsville, AL), or *N*-hydroxysuccinimidyl ester of methoxypoly(ethylene glycol) propionic acid, mol wt 2000 (SPA-PEG, Shearwater Polymers Europe, Inc.), was either quickly dissolved with stirring in 2.5 mL of 50 mM SBB, pH 8.5, or taken as a solid and added to the dissolved PLL. The reaction was allowed to proceed for 6 h at room temperature, after which the reaction mixture was dialyzed (Spectra-Por, mol wt cutoff 12 000−14 000; Spectrum, Houston, TX) for 24 h, first against phosphate-buffered saline (PBS; 0.2 g/L KCl, 0.2 g/L KH_2PO_4, 8 g/L NaCl, 1.15 g/L anhydrous Na_2HPO_4, pH 7.4 ± 0.1, 285 ± 5% mOsm/kg H_2O) and subsequently against deionized water. The product mixture was freeze-dried and stored at −20 °C under Ar. PLL(375)-*g*(5.6)-PEG(5): ^{1}H NMR (D_2O, ppm) 1.35, 1.60, 1.68 (−CH_2−), 2.88 (−CH_2−N−), 3.55 (PEG), 4.20 (−N−CHR−COO−). By NMR, the areas of the lysine side-chain peaks were compared with the area of the PEG peak to determine the graft ratio of the comb copolymer. Graft ratios of the PLL-*g*-PEG copolymers were also analyzed by aqueous size-exclusion chromatography using refractive index detection (Shodex OHpak column, SB-804HQ, Alltech, Deerfield, IL; Delory & King's carbonate−bicarbonate buffer eluent, 0.2 M anhydrous sodium carbonate and 0.2 M sodium bicarbonate mixed to make pH 10; sample was 1% in eluent buffer). Graft ratios were estimated by comparing the area of the free PEG peak with the area of the PLL-*g*-PEG peak, assuming that PEG and PLL-*g*-PEG have identical, linear refractive index increments in this concentration range. Because free mPEG is not removed during dialysis, the amount of PEG coupled to poly(L-lysine) can be calculated. The reported graft ratios were determined by the chromatographic method.

A dendron copolymer (Dendron-5) was prepared according to the following procedure. Monomethoxy PEG of mol wt 20 000 (PEG 20K; Shearwater Polymers, Huntsville, AL) was dried by azeotropic distillation from benzene. The hydroxyl terminus of the PEG was first esterified with Fmoc-Gly. The anhydride of Fmoc-Gly was produced by the reaction of Fmoc-Gly (1.78 g, 8 equiv; Novabiochem, San Diego, CA) with diisopropylcarbodiimide (DIPCDI; 0.469 mL, 4 equiv; Aldrich, Milwaukee, WI) in dimethylformamide (DMF; 10 mL, anhydrous; Aldrich) and dichloromethane (DCM; 4 mL, anhydrous; Aldrich) for 30 min at room temperature with stirring under argon. The PEG (15 g, 1 equiv) was dissolved in DCM (30 mL). (Dimethylamino)pyridine (91.6 mg, 1 equiv; Novabiochem) in DCM (5 mL) was added to the dissolved PEG, and the resulting solution was added to the Fmoc-Gly anhydride. The vessel that contained the PEG was washed with 5 mL of DCM, which was also added to the reaction mixture. The reaction proceeded for 6 h at room temperature with stirring under argon. The reaction mixture was filtered through paper under vacuum and then precipitated in cold, rapidly stirred ether and collected by vacuum filtration. This resulting product constituted the zeroth-generation dendron copolymer. The reaction sequence detailed below was repeated five times to yield Dendron-5, a fifth-generation dendron copolymer.

Fmoc-Lys(Fmoc)-OH was added to the PEG as follows. Fmoc groups were removed from the PEG by dissolving the Fmoc-protected, lysine-grafted PEG in 20% piperidine in DMF (4 mL/g PEG 20K; Perseptive Biosystems, Framingham, MA), heating with swirling at 45 °C until it dissolved, and then allowing the mixture to stand at room temperature for 30 min. These steps were followed by precipitation in cold, stirred ether, vacuum filtration, and drying under vacuum, with 500 mg of each PEG-amine product retained. The Fmoc-amino acid (Fmoc-L-Lys(Fmoc)-OH; 3 equiv; Bachem, King of Prussia, PA) and HOBT (3 equiv; Novabiochem) were dissolved in DMF (3 mL/g of amino acid) and DCM (2 mL/g of amino acid). DIPCDI

(3 equiv) was added, and after 15 min, the PEG (2 mL DCM/g of PEG) was added (the vessel that contained the PEG solution was washed two times with 5 mL of DCM). The reaction proceeded with stirring at room temperature for at least 6 h under argon. The reaction mixture was then precipitated in stirred, cold ether, vacuum filtered, and dried under vacuum.

The dendron product was analyzed by gel permeation chromatography (GPC) at 1% in DMF (Polymer Laboratories PL-EMD 950 evaporative mass detector; Polymer Laboratories 5 μm Mixed D 300 × 7.5 and 5 μm 500 Å 300 × 7.5 in series columns; Polymer Laboratories, Amherst, MA). The chromatograms from the mass detector, which allowed a measurement of the mass of polymer, were compared with the UV adsorption at 300 nm (Fmoc absorption $\epsilon_{300} = 6558$ L cm^{-1} mol^{-1}) in order to calculate the percentage of coupling.

Pluronics F-108 NF and F-68 NF were obtained from BASF (Mount Olive, NJ).

2.2. Substrates. All of the waveguides used in this study were purchased from Microvacuum, Ltd. (Budapest, Hungary) and consisted of a 1-mm-thick AF45 glass substrate and a 200-nm-thick $Si_{0.4}Ti_{0.6}O_2$ waveguiding layer at the surface. For experiments involving titanium oxide and niobium oxide surfaces, an additional 14-nm-thick oxide layer was sputter coated in a Leybold dc-magnetron Z600 sputtering unit onto the waveguiding layer. All of these surfaces were characterized by XPS, AFM, and ToF-SIMS as previously reported.[36,37,45]

Before each experiment, the waveguides were cleaned according to the following procedure: sonication in 0.1 M HCl for 10 min, extensive rinsing with ultra-high-purity water, and drying under nitrogen, followed by 2 min of oxygen-plasma cleaning in a Harrick Plasma Cleaner/Sterilizer PDC-32G instrument (Ossining, NY).

2.3. Experimental Techniques. *2.3.1. Optical Waveguide Lightmode Spectroscopy (OWLS).* The OWLS technique involves the incoupling of the evanescent field of a He−Ne laser into a planar waveguide that allows for the direct online monitoring of macromolecule adsorption. The method is highly sensitive (specifically, to ~1 ng/cm^2) up to a distance of 100 nm above the surface of the waveguide. Furthermore, a measurement time resolution of 3 s allows for the in situ, real-time study of adsorption kinetics.

Areal adsorbed mass density data were calculated from the thickness and refractive index values derived from the mode equations[35] according to Feijter's formula. A refractive index increment (dn/dc) value of 0.182 cm^3/g was used for the protein-adsorption calculations,[35,46] and a value of 0.202 cm^3/g, as determined in a Raleigh interferometer, was used for the PLL-*g*-PEG adsorption calculations. All OWLS experiments were conducted in a BIOS-I OWLS instrument (ASI AG, Zurich, Switzerland) using a Kalrez (Dupont, Wilmington, DE) flow-through cell as described previously.[36,45] The flow-though cell was used for studying both PLL-*g*-PEG adsorption and protein adsorption. The flow rate and wall shear rate were 1 mL/h and 0.83 s^{-1}, respectively.

2.3.2. X-ray Photoelectron Spectroscopy. X-ray photoelectron spectroscopy (XPS) analyses were performed using a PHI 5700 photoelectron spectrometer equipped with a concentric hemispherical analyzer in the standard configuration (Physical Electronics, Eden Prairie, MN). Spectra were acquired at a chamber pressure of 10^{-9} mbar using a non-monochromatized Al Kα source operating at 200 W and positioned ~13 mm away from the sample. The analyzer was used in the fixed-analyzer transmission mode. Pass energies used for survey scans and detailed scans were 187.85 and 23.5 eV, respectively. Under these conditions, the energy resolution [fwhm measured on silver Ag(3d5/2)] is 2.7 and 1.1 eV, respectively. Acquisition times were approximately 5 min for survey scans and 9 min for high-energy resolution elemental scans.

Angle-resolved XPS (AR-XPS) measurements were conducted at two different takeoff angles, namely 15° and 75° with respect to the surface plane, to obtain depth-dependent information on the molecular layers adsorbed onto the oxide substrate. (The typical sampling depths for 15° and 75° are 2−3 and 5−8 nm, respectively.) Spectra were referenced to the Ti(2p3/2) signal at 458.5 eV. Data were analyzed using a least-squares fitting routine following Shirley iterative background subtraction. Measured intensities (peak areas) were transformed into atomic concentrations by taking into account the respective atomic photoionization cross sections corresponding to PHI sensitivity factors[47] and by correcting for transmission functions and asymmetry factors. Spectra were fitted with the PC-Access V6.0F PHI software using the sum of a 90% Gaussian and 10% Lorentzian function.

2.4. Protocol for the Adsorption Experiments. *2.4.1. Protocol for XPS Sample Preparation.* Prior to XPS analysis, the samples were ultrasonically cleaned in 0.1 M HCl for 10 min, extensively rinsed with ultra-high-purity water, and dried in a nitrogen stream, followed by 2 min of oxygen-plasma cleaning, as described earlier. PLL-*g*-PEG-modified surfaces were prepared by dip coating for 10 min in a 1 mg/mL solution of PLL-*g*-PEG in 10-mM HEPES [4-(2-hydroxyethyl)piperazine-1-ethanesulfonic acid, adjusted to pH 7.4 with 1 M NaOH solution]. This buffer solution will be referred to as HEPES Z1 hereafter. Subsequently, the modified waveguides were rinsed immediately with ultrapure water and dried under nitrogen. Some samples were analyzed and used without the proper cleaning step described in the sample preparation procedure to test the effect of surface contamination on the adsorption and subsequent performance of PLL-*g*-PEG.

2.4.2. Protocol for OWLS Experiment and Sample Preparation. Samples with adsorbed PLL-*g*-PEG layers were prepared as described in section 2.4.1 and dried in flowing nitrogen. Their protein- and serum-adsorption performance was subsequently measured by OWLS as described later in section 2.4.3.

Samples modified with PLL-*g*-PEG in situ were initially placed in HEPES Z1 immediately following the cleaning procedure and allowed to soak overnight. Prior to assembly of the flow-through cuvette in the OWLS instrument, the samples were rinsed with ultrapure water and dried under nitrogen. These presoaked samples equilibrated and reached a flat baseline in HEPES Z1 in less than 1 h. Then, the samples were exposed, in situ, to the PLL-*g*-PEG solution (1 mg/mL in HEPES Z1). The adsorption was subsequently monitored for 30 min. The polymer solution was then replaced with HEPES Z1, and the protein- and serum-adsorption performance was measured as described later in section 2.4.3.

$Si_{0.4}Ti_{0.6}O_2$ waveguides were used for the pH-dependence measurements and were prepared as described in the immediately preceding procedure. However, in this case, the solution consisted of 10 mM HEPES, titrated to the predetermined pH by the addition of either 1 M NaOH or 1 M HCl. The polymer concentration was 0.1 mg/mL in the same pH-adjusted HEPES solution. After 1 h of polymer adsorption and 30 min of washing with the pH-adjusted solution, the solution was changed to the pH 7.4 buffer (HEPES Z1), and the protein-adsorption performance was measured, as described in section

3302 *J. Phys. Chem. B, Vol. 104, No. 14, 2000* Kenausis et al.

TABLE 2: XPS Binding Energies/Peak Areas and Surface Concentrations for Cleaned, Contaminated, and PLL(375)-*g*[5.6]-PEG(5)-Modified TiO_2 Surfaces

		atomic concentration (atom %)					binding energy (eV) (relative peak area (%))					
surface	angle of detection [°](deg)	C	O	Ti	N	Si	$C1s_A$ C–C, C–H	$C1s_B$ C–O	$C1s_C$ NHC(=O) O–C=O	$O1s_A$ TiO_2	$O1s_B$ COO*, OH	$O1s_C$ C–O, H_2O
clean TiO_2	15	34.5	49.8	15.7	–	–	284.9 (76.5)	286.6 (13.9)	288.7 (9.6)	530.0 (67.1)	531.5 (23.9)	532.7 (9.0)
	75	12.7	63.7	23.6	–	–	284.9 (70.2)	286.6 (16.7)	288.7 (13.1)	530.0 (77.3)	531.5 (17.1)	532.7 (5.6)
PLL(375)-*g*(5.6)-PEG(5)-covered TiO_2	15	52.5	41.8	5.0	0.7	–	285.1 (14.6)	286.7 (75.6)	288.5 (9.8)	530.0 (24.6)	531.4 (16.7)	532.9 (58.7)
	75	26.7	55.7	16.6	1.0	–	285.1 (16.1)	286.7 (72.5)	288.5 (11.4)	530.0 (62.0)	531.4 (16.2)	532.9 (21.8)
contaminated TiO_2	15	47.0	38.5	10.1	0.5	3.9	284.9 (89.5)	286.6 (7.4)	289.1 (3.1)	530.3 (66.7)	531.9 (26.0)	533.0 (7.3)
	75	15.6	59.0	21.5	0.4	3.5	284.9 (86.8)	286.6 (9.7)	289.1 (3.5)	530.3 (85.9)	531.9 (14.1)	–
PLL(375)-*g*(5.6)-PEG(5)-covered contaminated TiO_2	15	69.8	25.1	2.7	1.6	0.8	285.4 (59.9)	286.9 (33.9)	288.6 (6.2)	530.4 (26.7)	531.6 (5.5)	533.2 (67.8)
	75	35.7	46.0	15.1	1.7	1.5	285.4 (58.0)	286.9 (34.4)	288.6 (7.6)	530.4 (70.6)	531.6 (10.1)	533.2 (19.3)
theoretical PLL(375)-*g*(5.6)-PEG(5)		66.7	30.0	–	3.3	–	285.0 (13.3)	286.5 (83.8)	288.1, 289.3 (2.9)	–	531.6 (6.5)	532.8 (93.5)

2.4.3. The same procedure was used for the ionic-strength-dependence experiments, except that NaCl was used to generate the solution of predetermined ionic strength.

2.4.3. Protocol for Protein-Adsorption Experiments. The waveguides were exposed to a solution of human serum (Control Serum N, Art.# 07 3711 9, US# 42384, Roche, Basel, Switzerland) for 1 h at a temperature of 25 °C and subsequently washed for 30 min in HEPES Z1.

Human serum albumin (HSA) and human fibrinogen were obtained from Sigma Chemical Co., and their related antibodies were obtained from Dako A/S (Glostrup, Denmark). In the case of the single protein-adsorption experiments, the waveguide was exposed to a 1 mg/mL solution of the appropriate protein in HEPES Z1 for 1 h at a temperature of 25 °C and subsequently washed for 30 min in HEPES Z1.

In some cases, after the serum exposure, the waveguide was tested against solutions of 0.1 mg/mL rabbit anti-HSA and 0.28 mg/mL rabbit anti-human fibronectin for 30 min at a temperature of 25 °C and subsequently washed for 30 min in HEPES Z1. Adsorption of human fibrinogen was tested separately by exposure to a 1 mg/mL solution of fibrinogen for 1 h at 25 °C followed by exposure to a 0.1 mg/mL solution of rabbit anti-human fibrinogen for 30 min.

3. Results

3.1. Surface Characterization of Unmodified and Modified Waveguides. *3.1.1. X-ray Photoelectron Spectroscopy.* X-ray photoelectron spectroscopy was performed at two different detection angles (75° and 15° relative to the surface plane) on a variety of untreated and treated titanium oxide (TiO_2) surfaces, specifically, contaminated (not plasma-cleaned), plasma-cleaned, contaminated and treated with PLL(375)-*g*[5.6]-PEG(5), and plasma-cleaned and treated with PLL(375)-*g*[5.6]-PEG(5). The quantitative results of the XPS study are summarized in Tables 2 and 3. Figure 2 shows the most relevant detail spectra, that is, the C(1s) and O(1s) spectra of the clean TiO_2 surface and of the clean TiO_2 surface with adsorbed PLL-*g*-PEG at the two different detection angles.

TABLE 3: Atomic Ratios Calculated from XPS Data for PLL(375)-*g*[5.6]-PEG(5)-Modified TiO_2 Surfaces[a]

surface	angle of detection [°](deg)	atomic ratio				
		C_B/C_A[c]	O_C/O_A[c]	O_A/Ti[c]	C_B/O_C[c]	O_C/N[c]
PLL(375)-*g*(5.6)-PEG(5)-covered TiO_2	15	5.2	2.4	2.12	1.98	34.4
	75	4.5	0.35	2.06	1.95	10.5
PLL(375)-*g*(5.6)-PEG(5) covered contaminated TiO_2	15	0.57	2.5	2.4	1.53	14.0
	75	0.59	0.27	2.2	2.09	4.5
theoretical[b] PLL(375)-*g*(5.6)-PEG(5) on TiO_2		6.3	–	2.0	2.0	8.3

[a] A homogeneous elemental distribution is assumed. [b] Theoretical values expected for the stoichiometry of the polymer. [c] Contributions A, B, and C are defined in Table 2 above.

A rough quantification based on the overall detected intensities of the elements (when not distinguishing between different chemical states) shows the effect of the cleaning procedure on the remaining surface contamination (see Tables 2 and 3). At a 15° detection angle, the carbon concentration at the oxide surface (mainly adventitious hydrocarbon) clearly decreased from 47.0 to 34.5 atom % after plasma cleaning. The latter value is considered to be typical for well-cleaned surfaces.[37] After adsorption of PLL(375)-*g*[5.6]-PEG(5) on both types of surfaces, the carbon concentration significantly increased, as expected. The overall oxygen concentration showed less variation between uncoated and coated surfaces, as both the organic PLL(375)-*g*[5.6]-PEG(5) overlayer and the titanium oxide substrate contain high levels of oxygen. In comparing the data from spectra taken at 15° and 75° detection angles, one finds appreciably higher carbon concentrations in the case of the grazing exit angle (15°). This is a clear confirmation that both the adventitious contamination and the PLL(375)-*g*[5.6]-PEG(5) adlayer were located on top of the titanium oxide surface, because the information depth of the XPS technique is much lower for the 15° arrangement than for the 75° arrangement.

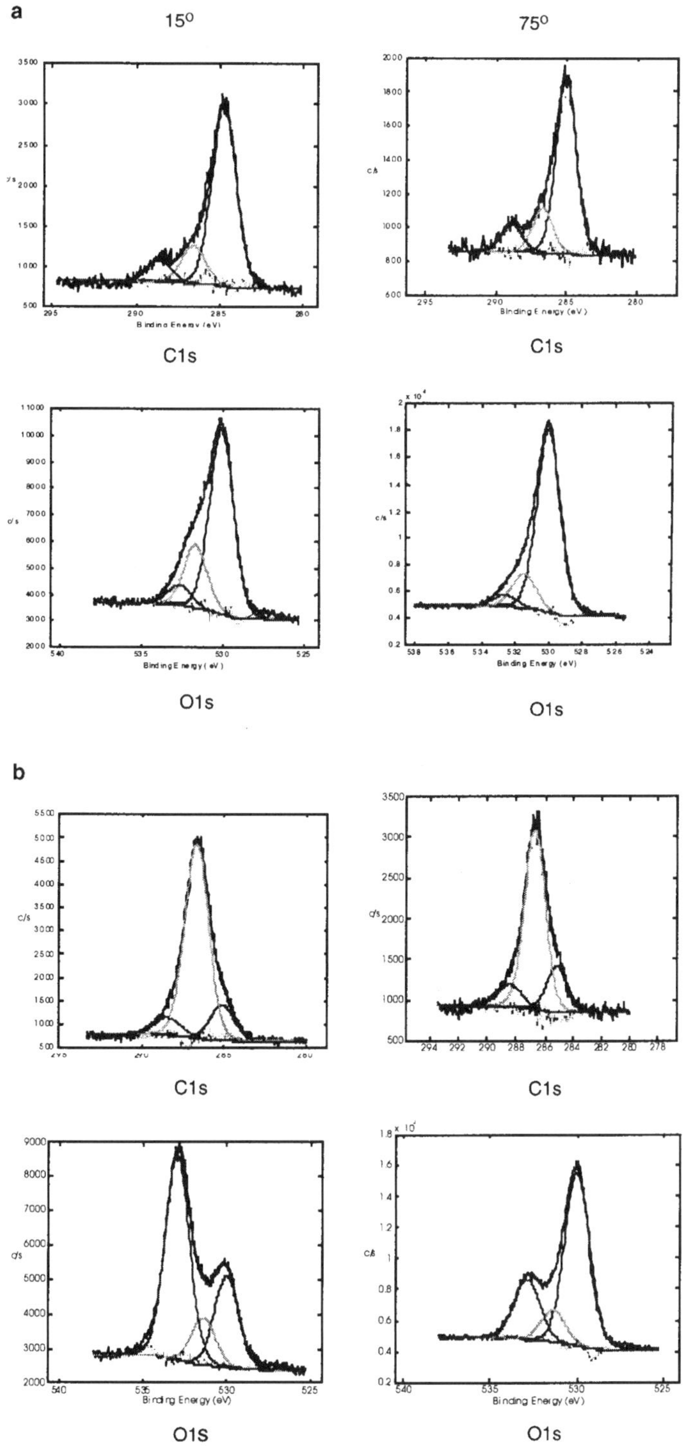

Figure 2. Detail XPS spectra in the C(1s) and O(1s) region at the two different detection angles (relative to the surface plane) of 15° (more surface-sensitive) and 75° (less surface-sensitive): (a) plasma-cleaned TiO_2 surface and (b) plasma-cleaned TiO_2 surface coated with PLL(375)-*g*[5.6]-PEG(5).

The data based on the high-resolution spectra, presented in Tables 2 and 3 and Figure 2, provide more detailed insight into the surface architecture of the PLL(375)-*g*[5.6]-PEG(5)-modified titanium oxide surfaces. Both carbon and oxygen were detected

3304 *J. Phys. Chem. B, Vol. 104, No. 14, 2000* Kenausis et al.

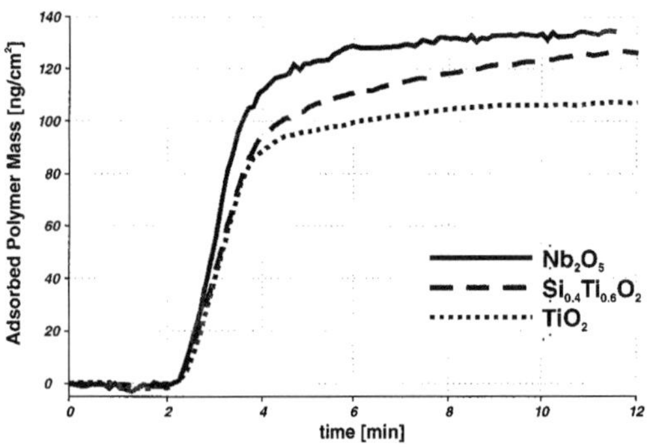

Figure 3. Examples of PLL(375)-*g*[5.6]-PEG(5) adsorption curves on three metal oxide surfaces, as measured by the OWLS technique [10 mM HEPES Z1 (pH 7.4), 1 mg/mL polymer, 1 mL/h, $T = 26$ °C].

in different chemical states, and the deconvoluted peaks were assigned to particular molecular functionalities on the basis of existing experience and published reference data.[47] Further discussion of the XPS results is presented in section 4.

3.2. PLL(375)-*g*[5.6]-PEG(5) Adsorption and Protein-Resistance Study. *3.2.1. PLL(375)-g[5.6]-PEG(5) Adsorption Characteristics.* All of the following measurements were carried out in a flow-through cell, and both the PLL-*g*-PEG pretreatment and the protein-adsorption tests were carried out in situ and consecutively without an intermittent drying stage unless otherwise noted.

The results of the OWLS experiments (see Figure 3) indicate that the PLL-*g*-PEG polymer spontaneously adsorbed from a pH 7.4 buffered aqueous solution onto metal oxide surfaces. The example shown in Figure 3 involved the adsorption of PLL-*g*-PEG onto three different metal oxide surfaces, specifically, titanium, niobium, and silicon/titanium. This adsorption process occurred rapidly and resulted in the formation of a layer of adsorbed polymer on the surface. Typically, for $Si_{0.4}Ti_{0.6}O_2$ surfaces, a layer with an adsorbed areal density of approximately 125 ng/cm^2 formed, and 95% of the final observed mass was reached within the first 5 min. Similar behavior was observed for the two other metal oxide surfaces investigated (that is, niobium pentoxide and titanium dioxide). Although the adsorption kinetics were quite similar, the resulting amount of PLL-*g*-PEG adsorbed to the surface was different and depended on the characteristic isoelectric point of the metal oxide, as shown in Figure 4 and Table 4. [37,38,46,48,49]

3.2.2. Protein-Adsorption Characteristics of PLL(375)-g[5.6]-PEG(5)-Modified Surfaces. Subsequent protein-adsorption experiments revealed that PLL-*g*-PEG modification of the metal oxide surfaces resulted in sharply reduced protein adsorption. Typically, the exposure of a metal oxide surface to serum

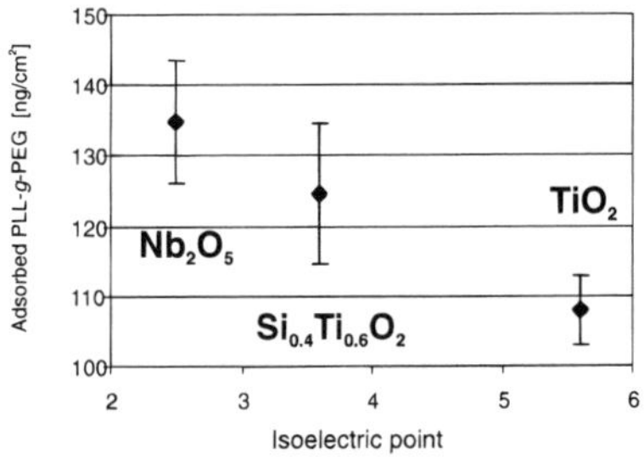

Figure 4. Dependence of the adsorbed amount of PLL(375)-*g*[5.6]-PEG(5) on the isoelectric point of the metal oxide surfaces, as measured by the OWLS technique[10 mM HEPES Z1 (pH 7.4), 1 mg/mL polymer, 1 mL/h, $T = 26$ °C].

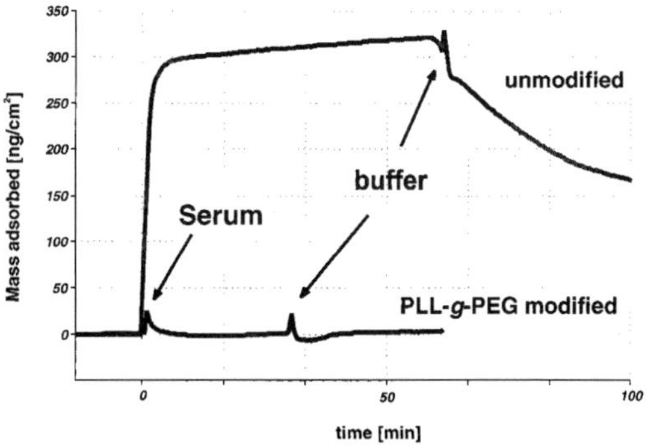

Figure 5. Serum-adsorption spectrum of an unmodified and a PLL-(375)-*g*[5.6]-PEG(5)-modified $Si_{0.4}Ti_{0.6}O_2$ surface, as measured by the OWLS technique [PBS (pH 7.4), 1 mg/mL polymer, $T = 26$ °C]. The baseline was achieved under PBS, and the spikes are due to temporary flow rate-related pressure changes.

produces a layer of adsorbed protein with an areal density of between 150 and 250 ng/cm^2 after the washing step.[50] However, waveguides that are precoated with a layer of PLL-*g*-PEG show a drastic reduction in subsequent serum protein adsorption. An example of this is shown in Figure 5. This experiment involved modified and unmodified silicon/titanium dioxide waveguides that were prepared according to the procedure described in section 1.4.1. Generally, PLL(375)-*g*[5.6]-PEG(5) pretreatment, regardless of the preparation procedure, caused an order-of-magnitude decrease in the areal density of adsorbed protein following serum exposure on titanium dioxide, silicon/titanium dioxide, and niobium oxide surfaces (see Figure 6). Moreover, the adsorption of human serum albumin (1 mg/mL, 10 mM HEPES-buffered solution, pH 7.4) was decreased by two orders of magnitude following the PLL-*g*-PEG pretreatment (see Figure

TABLE 4: Summary of the Isoelectric Points and the Charge Densities[46,49] of Investigated Surfaces and the Observed Adsorbed Areal Density of PLL(375)-*g*[5.6]-PEG(5), Serum, and Human Serum Albumin (HSA) on Unmodified Surfaces and of Serum and HSA on PLL(375)-*g*[5.6]-PEG(5)-Modified Surfaces[a]

substrate	isoelectric point	surface charge at pH 7.4 (uC/cm^2)	adsorbed mass of PLL(375)-*g*(5.6)-PLL(5) (ng/cm^2)	adsorbed mass of serum on an untreated surface (ng/cm^2)	adsorbed mass of serum on a PLL(375)-*g*(5.6)-PLL(5)-treated surface (ng/cm^2)	adsorbed mass of HSA on an untreated surface (ng/cm^2)	adsorbed mass of HSA on a PLL(375)-*g*(5.6)-PLL(5)-treated surface (ng/cm^2)
TiO_2	5.6	5	108 ± 9	320 ± 31	20 ± 40	200 ± 20	<1.0
$Si_{0.4}Ti_{0.6}O_2$	3.6	25	125 ± 13	596 ± 89	25 ± 5	176 ± 10	<1.0
Nb_2O_5	2.5	50	135 ± 13	445 ± 8	19 ± 22	96 ± 70	24 ± 16

[a] Polymer adsorption carried out under 10 mM HEPES Z1, 1 mg/mL polymer, 1 mL/h, $T = 26$ °C; serum adsorption, 1 mL/h, $T = 26$ °C; HSA adsorption, 10 mM HEPES Z1, 1 mg/mL HSA, 1 mL/h, $T = 26$ °C.

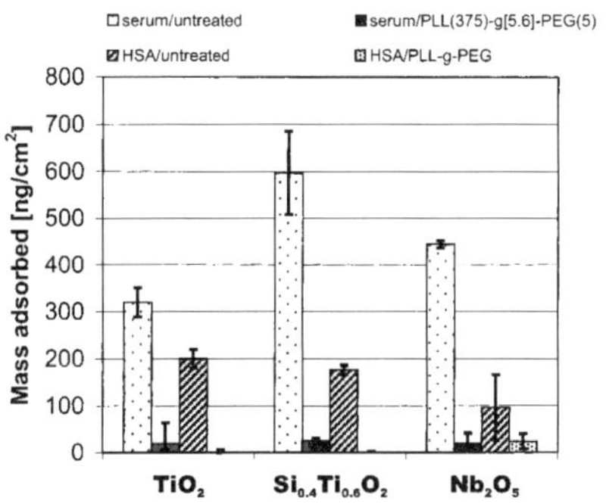

Figure 6. Adsorbed areal mass of serum and HSA on PLL(375)-*g*[5.6]-PEG(5)-modified and unmodified $Si_{0.4}Ti_{0.6}O_2$, TiO_2, and Nb_2O_5 surfaces, as measured by the OWLS technique [10 mM HEPES Z1 (pH 7.4), 1 mg/mL polymer, 1 mL/h, $T = 26$ °C].

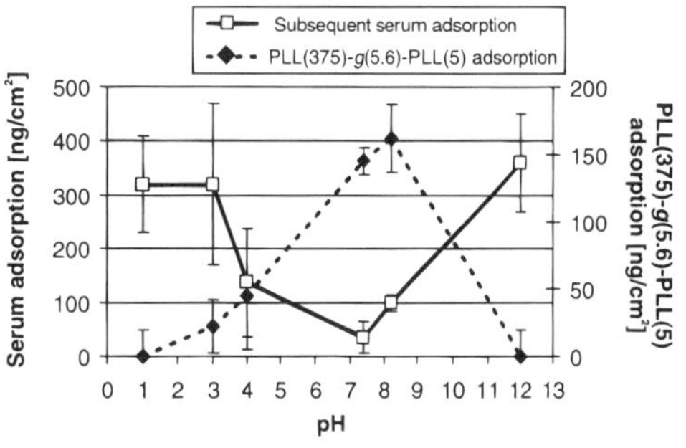

Figure 7. pH dependence of PLL(375)-*g*[5.6]-PEG(5) and subsequent serum adsorption on $Si_{0.4}Ti_{0.6}O_2$ surfaces, as measured by the OWLS technique [polymer adsorption carried out under titrated 10 mM HEPES Z1, 1 mg/mL polymer, 1 mL/h, $T = 26$ °C; serum adsorption, 1 mL/h, $T = 26$ °C].

6). The quantitative protein-adsorption data for all surfaces studied are listed in Table 4.

The residual material remaining on the surface after serum exposure was tested against antibodies of several common serum proteins, such as anti-albumin, anti-fibronectin, and anti-γ-globulin. Each of these antibodies exhibited an adsorbed areal density lower than the detection limit of the OWLS technique (that is, <2 ng/cm^2), which suggests that none of the proteins are present in their active conformations on the modified surfaces. It is likely that serum components other than proteins are picked up by the modified surfaces. Because serum is fibrinogen-depleted, fibrinogen was also tested separately from a 5 mg/mL solution in HEPES Z1, followed by an anti-fibrinogen assay. For a PLL(375)-*g*[5.6]-PEG(5)-modified silicon/titanium dioxide surface, 85 and 150 ng/cm^2 of adsorbed areal density was observed for fibrinogen and anti-fibrinogen, respectively. The antibody results are not shown.

3.2.3. Model of the PLL(375)-g[5.6]-PEG(5)/Surface Oxide Interaction. The adsorption of PLL(375)-*g*[5.6]-PEG(5) was found to be dependent on electrostatic interactions. The observation that the limiting mass of PLL-*g*-PEG adsorption increases for surfaces of decreasing isoelectric points provides evidence that the mechanism of surface adsorption is of an electrostatic nature (see Figures 3 and 4). By varying the pH of the solution in contact with the surface of the waveguide, one can control the sign and density of the surface charge. For example, at pH values lower than the isoelectric point of the metal oxide, the surface bears a positive charge, and at pH values higher than the isoelectric point, the surface bears a negative charge. Similarly, because the PLL backbone of this graft copolymer contains primary amines with a pK_a of about 10, pH values below 10 lead to a net positive charge on the polymer backbone, whereas pH values well above 10 render the polymer backbone essentially uncharged.

Maximal polymer adsorption occurs within a pH range between the isoelectric point of the surface and the pK_a of the polymer. Experiments performed at different pH values, as described in section 2.4.2, showed that negligible PLL-*g*-PEG adsorption is observed at pH values higher than the pK_a of the polymer (pH ~10) or lower than the isoelectric point of the metal oxide surface. The experimental data shown in Figure 7 demonstrate that, for a silicon/titanium dioxide surface, PLL-*g*-PEG adsorption takes place only in the pH range between 3 and 11. Outside of this range, negligible adsorption is observed.

The surfaces were found to exhibit suppression of protein adsorption only when an appreciable layer of PLL(375)-*g*[5.6]-PEG(5) was present at the surface. This behavior was observed for experiments involving the adsorption of human serum onto a modified silicon/titanium dioxide surface, as shown in Figure 7. Clearly, suppression of protein adsorption depended on the presence of a preadsorbed layer of PLL(375)-*g*[5.6]-PEG(5). The same general behavior was observed for human serum albumin adsorption onto modified titanium dioxide surfaces.

Experiments involving the adsorption of PLL(375)-*g*[5.6]-PEG(5) onto the surface of titanium oxide waveguides from solutions of varying ionic strength showed the adsorption to be ionic-strength dependent. An increase in the ionic strength of the PLL(375)-*g*[5.6]-PEG(5) solution brought about a decrease in the surface areal density of resulting adsorbed polymer. At ionic strengths above 2 M, negligible adsorption was observed. Importantly, an adsorbed areal density of 130 ng/cm^2 was observed at an ionic strength of 150 mM, the physiological ionic strength. Furthermore, this adsorbed layer of PLL(375)-*g*[5.6]-PEG(5) demonstrated an order-of-magnitude decrease in the subsequent serum protein adsorption (specifically, ~25 ng/cm^2), a performance equivalent to that of the polymer layer adsorbed from 10-mM ionic-strength solution.

Exposure of a preadsorbed layer of PLL(375)-*g*[5.6]-PEG-(5) to a solution with a pH value outside of the pH range where adsorption is observed or to a solution with a high ionic strength causes a gradual desorption of the polymer. Therefore, the adsorption of PLL(375)-*g*[5.6]-PEG(5) and the stability of this adsorbed polymer layer require that the contacting solution be limited to the pH range between the isoelectric point of the surface and the pK_a of the polymer, and to low ionic strengths.

3.2.4. Performance and Long-Term Stability of the Adsorbed PLL(375)-g[5.6]-PEG(5) Layer. Once established on the surface, the adsorbed layer of PLL(375)-*g*[5.6]-PEG(5) was found to be stable (that is, <5% loss in mass of the adsorbed layer after 1 week) and resistant to protein adsorption over 24 h at 37 °C under a flowing HEPES Z1 solution, as shown in Figure 8. This experiment involved the in situ deposition, within the first hour, of a PLL(375)-*g*[5.6]-PEG(5) layer on the surface of a silicon/titanium dioxide waveguide. Two subsequent exposures to serum produced less than 20 ng/cm^2 of surface-adsorbed protein. Eighteen hours later, two additional serum exposures similarly produced less than 20 ng/cm^2. Similar performance was observed when PBS was used as the buffer instead of HEPES Z1.

PLL(375)-*g*[5.6]-PEG(5)-modified waveguides that were stored dry were found to retain their protein-resistant properties after more than 3 months, and those stored in HEPES-buffered solution were found to retain their protein-resistant properties after more than 1 month.

3306 *J. Phys. Chem. B, Vol. 104, No. 14, 2000* Kenausis et al.

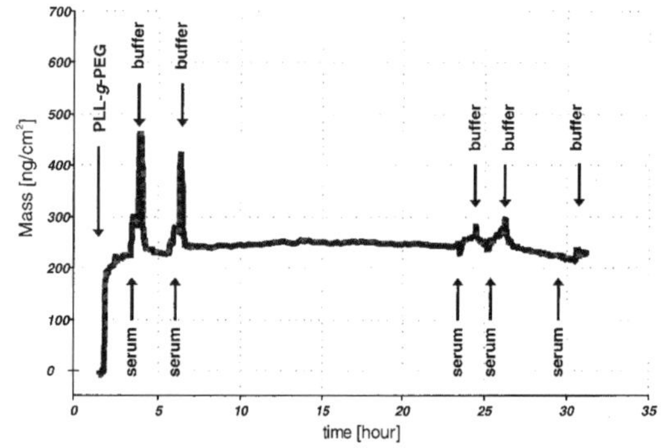

Figure 8. Long-term stability of the protein-adsorption suppression of a PLL(375)-*g*[5.6]-PEG(5)-modified $Si_{0.4}Ti_{0.6}O_2$ surface, as measured by the OWLS technique [polymer adsorption carried out under 10 mM HEPES Z1, 1 mg/mL polymer, 1 mL/h, $T = 26$ °C; serum adsorption, 1 mL/h, $T = 26$ °C].

Pluronics, diblock copolymers consisting of poly(propylene oxide) flanked by two poly(ethylene oxide) chains, are commonly used to immobilize PEG on hydrophobic surfaces. However, Pluronics F-108 and F-68 were found not to adsorb onto any of the metal oxide surfaces investigated here and, as a result, did not show any protein-adsorption-suppressing properties.

PLL(375)-*g*[5.6]-PEG(5) was also found to adsorb onto a preadsorbed layer of serum proteins. After a typical adsorption of serum protein (that is, about 260 ng/cm²), subsequent exposure to a 1 mg/mL solution of PLL(375)-*g*[5.6]-PEG(5) effected an overlayer of polymer with a surface areal density of approximately 90 ng/cm².

3.2.5. Effect of Precontamination on PLL(375)-g[5.6]-PEG-(5) Adsorption. Metal oxide surfaces that exhibited large amounts of hydrocarbon surface contamination nevertheless adsorbed a layer of PLL(375)-*g*[5.6]-PEG(5) that suppressed subsequent serum adsorption. Titanium dioxide waveguides that were not cleaned according to the procedure described in section 1.4.1 exhibited substantial hydrocarbon surface contamination (see Tables 2 and 3 and Figure 2). However, these XPS data also indicate that an additional layer of PLL(375)-*g*[5.6]-PEG-(5) does, indeed, adsorb onto this contaminated surface. Furthermore, OWLS experiments showed that the typical adsorbed areal density of 120 ng/cm² forms on contaminated titanium dioxide waveguides and that this adsorbed layer of polymer suppresses subsequent serum protein adsorption by about 95%. That is, the adsorption and performance characteristics of PLL(375)-*g*[5.6]-PEG(5) are identical in the case of both contaminated and cleaned titanium dioxide surfaces.

3.3. Effect of Polymer Architecture. Alternative architectures of the polymer were explored for the suppression of protein-adsorption performance. These architectures included comblike graft copolymers with differing PEG side-chain length, PLL backbone length, and PEG grafting ratio (see Table 1). The inverted tree-like dendrimeric PLL having a single PEG side chain attached at the base was also investigated.

3.3.1. Effect of the Backbone (PLL) Structure on Subsequent Protein Adsorption. The graft copolymer architecture was found to influence the subsequent serum-adsorption suppression. All of the polymers investigated demonstrate adsorption on silicon/titanium dioxide surfaces in the areal density range of about 150 ng/cm², as shown in Figure 9. Of the two comb copolymers with PEG side chains of mol wt 5000, PLL(375)-*g*[5.6]-PEG-(5) demonstrates only a slightly more pronounced suppression

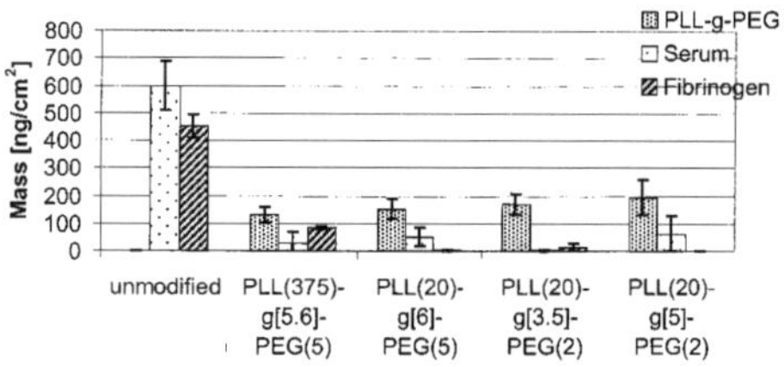

Figure 9. Effect of polymer architecture on the adsorbed areal density of the copolymer and subsequent serum and fibrinogen on $Si_{0.4}Ti_{0.6}O_2$ surfaces as measured by the OWLS technique [polymer adsorption carried out under 10 mM HEPES Z1, 1 mg/mL polymer, 1 mL/h, $T =$ 26 °C; serum adsorption, 1 mL/h, $T = 26$ °C].

TABLE 5: Observed Adsorbed Areal Density of Polymer, Serum, and Fibrinogen on $Si_{0.4}Ti_{0.6}O_2$ Surfaces Modified with Various PLL-*g*-PEG Copolymers[a]

	adsorbed mass of PLL-*g*-PEG polymer (ng/cm²)	adsorbed mass of serum (ng/cm²)	adsorbed mass of fibrinogen (ng/cm²)
unmodified $Si_{0.4}Ti_{0.6}O_2$	–	596 ± 89	451 ± 42
PLL(375)-*g*[5.6]-PEG(5)	133 ± 28	30 ± 42	85 ± 5
PLL(20)-*g*[6]-PEG(5)	152 ± 35	50 ± 33	3 ± 2.5
PLL(20)-*g*[3.5]-PEG(2)	169 ± 36	3 ± 4	15 ± 14
PLL(20)-*g*[5]-PEG(2)	194 ± 63	63 ± 63	1 ± 1

[a] Polymer adsorption carried out under 10 mM HEPES Z1, 1 mg/mL polymer, 1 mL/h, $T = 26$ °C; serum adsorption, 1 mL/h, $T = 26$ °C; fibrinogen adsorption, 10 mM HEPES Z1, 1 mg/mL fibrinogen, 1 mL/h, $T = 26$ °C.

of serum adsorption, decreasing the observed adsorption by about 95% to 30 ng/cm². The graft copolymer with the lower molecular weight, PLL(20)-*g*[6.0]-PEG(5), decreases the observed protein adsorption from serum exposure by about 90% to 50 ng/cm². Dendron-5, consisting of a 20 000 mol wt PEG with a terminus consisting of a fifth-generation lysine dendrimer, was also found to adsorb onto silicon/titanium dioxide surfaces in the areal density range of 120–150 ng/cm². Furthermore, this adsorbed polymer layer was found to decrease the subsequent protein adsorption due to serum exposure by about 85% to 80 ng/cm².

3.3.2. Effect of the Copolymer Grafting Ratio on Subsequent Protein Adsorption. The grafting ratios that were investigated were found to have only a slight influence on the subsequent serum-adsorption performance. To explore the effect of the grafting ratio on the polymer adsorption and subsequent protein-resistance behavior, PLL(20)-*g*[3.5]-PEG(2) and PLL(20)-*g*[5]-PEG(2) were synthesized. These two polymers differ only in their respective grafting ratios of 3.5/1 and 5.0/1 (where the ratio is for lysine monomer to PEG side chain). As shown in Figure 9, PLL(20)-*g*[3.5]-PEG(2) and PLL(20)-*g*[5]-PEG(2) exhibited adsorbed mass densities of 170 and 190 ng/cm², respectively. PLL(20)-*g*[3.5]-PEG(2) appears to have a slightly greater suppression of protein adsorption [specifically, a 99% decrease to 3 ng/cm² vs a 90% decrease to 63 ng/cm² for PLL-(20)-*g*[5]-PEG(2)], but the differences actually are not statistically significant. In many of the experiments involving both of these polymers, this amount of adsorbed mass was close to the detection limit of the OWLS technique (~2 ng/cm²) (see Table 5).

3.3.3. Coadsorption Effects on Overall Graft Copolymer Adsorption. Coadsorption effects were explored by varying the sequence of polymer types exposed to the surface, as well as by mixing the polymer types, and then observing the dependence of the behavior of the adsorbed polymer layer. No significant coadsorption effects were observed. For example, exposing a layer of preadsorbed high-molecular-weight graft copolymer to a solution of low-molecular-weight graft copolymer showed no increase in adsorbed polymer mass and no change in subsequent protein-adsorption behavior from that of the preadsorbed layer alone. Furthermore, exposing a layer of preadsorbed low-molecular-weight graft copolymer to a solution of high-molecular-weight graft copolymer showed no increase in adsorbed polymer mass and no change in subsequent protein-adsorption behavior from that of the preadsorbed layer alone. Exposing the surface to a solution containing a mixture of both polymers showed the same adsorbed areal density and protein-adsorption resistance as that of a layer of high-molecular-weight graft copolymer alone.

4. Discussion

In this study, the adsorption of polyelectrolytes, such as the class of PLL-*g*-PEG graft copolymers on metal oxide surfaces, was found to be consistent with the related observations from previous studies. In previous characterizations of the adsorption behavior of polyelectrolytes on such metal oxide surfaces, polycations in particular, such as PLL, were found to form stable adsorbed layers on negatively charged oxides such as silicon dioxide and titanium dioxide.[21−23] However, the class of PLL-*g*-PEG-based copolymers thus far had been evaluated only in the contexts of drug delivery[32,33] and the reduction of cell adhesion[24−27,34] without a thorough characterization of the related modified surfaces and their reduction of protein adsorption.

4.1. Adsorption and Structure of the PLL-*g*-PEG Adlayer. The adsorption behavior of the class of PLL-*g*-PEG graft copolymers investigated here was evaluated by means of XPS. In the case of the uncoated TiO_2 surfaces, the C(1s) spectrum is dominated by an adventitious hydrocarbon emission (binding energy of ~285.0 eV), whereas the O(1s) spectrum shows a contribution that is characteristic of the titanium oxide substrate (binding energy of ~530 eV). However, there are additional contributions at higher binding energies in both the C(1s) and the O(1s) spectra, which are believed to be derived from oxygen-containing organic contaminants, hydroxide groups, and water molecules that are present at the titanium surface. As expected, these additional surface species are more pronounced in the more surface-sensitive 15° spectra (see Figure 2), as is the case for all species that are presumably located at the outermost layer. The quantitative ratio of O(TiO_2)/Ti is slightly higher than expected but close to the stoichiometric value of 2.0 (Table 3).

In the case of PLL(375)-*g*[5.6]-PEG(5) coated on the clean TiO_2 surfaces, both the carbon and the oxygen signals appear quite different from the corresponding signals in the bare TiO_2 spectra. The C(1s) region is now dominated by a peak at ~286.7 eV, which is typical for the C−O−C entity that is present in PEG (see Figure 1). Similarly, the O(1s) spectra show substantial intensities at the binding energy of ~533 eV, which is assigned to the O atoms of the PEG chains. Both the O(TiO_2)/Ti and the C(PEG)/O(PEG) ratios are close to the expected ratio of 2.0, providing further evidence that the curve fitting and the assignment of spectral components to functionalities at the surface is, indeed, reasonable (Tables 2 and 3). The species assigned to −NHC(=O) in the C(1s) and O(1s) spectra have higher intensities than would be expected from the stoichiometry of the PLL-*g*-PEG (Tables 2 and 3) polymer. This discrepancy is primarily due to the fact that the bare TiO_2 surfaces also show emission peaks in the same specific energy ranges (288−289 and 531−532 eV) from O-containing natural organic contaminants and OH species at the surface of the titanium oxide. The ratio of O(C−O−C)/N determined at a detection angle of 75° (10.5) is only slightly higher than the theoretical value of 8.3 for the stoichiometric ratio of the PLL(375)-*g*[5.6]-PEG(5) molecule. A higher value would be expected if the nitrogen atoms were preferentially located at the surface of the TiO_2. Indeed, at the detection angle of 15°, this value is much higher (34), which would be expected for a polymer configuration with the amine groups of the PLL backbone located at or close to the interface and the PEG chains located preferentially above the PLL polymer.

The sample with PLL(375)-*g*[5.6]-PEG(5) adsorbed onto the contaminated TiO_2 surface shows a much higher proportion of hydrocarbon contamination than the sample with PLL-*g*-PEG adsorbed onto a clean oxide surface (see Tables 2 and 3, C(1s) contribution at a binding energy of 285 eV). The ratio of C(PEG)/C(C−C) in Table 3 can be taken as a rough measure of the ratio of the amount of PEG to the amount of adventitious hydrocarbon contamination; this ratio is approximately a factor 10 higher for the "clean case" than for the "contaminated case". This finding implies that these contaminants are not displaced by PLL-*g*-PEG during the adsorption step. Nevertheless, the amount of PLL-*g*-PEG that is adsorbed onto the contaminated oxide surface is comparable to the amount adsorbed in the clean case. This can be roughly deduced in Table 3 from the ratio of O(PEG)/O(TiO_2), which is a sensitive function of PEG coverage. This ratio shows a similar value for the two cases (2.4−2.5 at the 15° detection angle and 0.3−0.4 at the 75° detection angle). The conclusion is that the surface contaminants are not displaced by the PLL-*g*-PEG molecules, but that the polymer molecules bridge across the contaminants at the surface and adsorb at a density comparable to that obtained in the clean case.

Complementary to the XPS studies, the OWLS technique was used in the present study to quantify the polymer and biomolecule adsorption. This technique has a detection limit of about 2 ng/cm^2. Furthermore, the OWLS technique is direct in that it does not require labeling or any other potentially property-altering modifications to the adsorbing species. The areal density of PLL(375)-*g*[5.6]-PEG(5) adsorbed onto the oxide surfaces was found to be only weakly dependent on the oxide surface charge density. As shown in Table 4, the surface charge density of titanium at a pH of 7.4 and an ionic strength of 10 mM is an order of magnitude lower than that of niobium, although the adsorbed areal density of the polymer on titanium was only 20% lower than that on niobium. As expected, spatial packing considerations are likely to be a more important factor than surface charge density in determining the ultimate amount of polymer adsorbed.

The pH-range requirement for maximal polymer coverage confirms that the mechanism of polymer adsorption depends on an electrostatic interaction between the surface and the polymer. At pH values lower than the isoelectric point of the surface, both the polymer and the surface are positively charged, and no electrostatically driven adsorption is expected to take place. At pH values higher than the pK_a of the polymer, the surface is negatively charged and the polymer is uncharged, and likewise, no electrostatically driven adsorption is expected to take place. However, at intermediate pH values, the polymer is positively charged and the surface is negatively charged, allowing for the electrostatically driven adsorption of the polymer onto the surface.

Experiments exploring the dependence of the PLL(375)-*g*[5.6]-PEG(5) adsorption behavior on the ionic strength of the solution offer additional support to this adsorption model. Electrostatic shielding effects increase with increasing ionic strength, and therefore, electrostatically driven adsorption of polyelectrolytes, for example, is generally attenuated at high ionic strengths.

PLL(375)-*g*[5.6]-PEG(5) was also found to adsorb onto a preadsorbed layer of serum proteins. After a typical adsorption of serum protein (about 260 ng/cm^2), subsequent exposure to a 1 mg/mL solution of PLL(375)-*g*[5.6]-PEG(5) led to an overlayer of polymer with a surface areal density of approximately 90 ng/cm^2. Earlier studies showed that the treatment of living red blood cells with PLL(375)-*g*[5.6]-PEG(5) prevented their subsequent agglutination.[51] The observation of PLL(375)-*g*[5.6]-PEG(5) adsorption onto an adsorbed protein layer confirms the model proposed to explain this effect.

4.2. Reduction of Protein Adsorption. The protein-suppression performance of the adsorbed layer of copolymer was found to depend on the areal density of PEG that was immobilized on the surface, and this amount was found to be limited only by the spatial packing of the PEG side chains, regardless of the backbone length and grafting ratio. Approximate values for both the spacing and the radius of gyration of the PEG side chains are required in order to estimate the extent of packing at the surface. If one assumes that the side chains were arranged in a two-dimensional lattice at the surface, then the distance between each PEG side chain (L) (that is, the distance between each neighboring point within the lattice) can be calculated from the adsorbed copolymer areal density data. The radius of gyration (R_g) of the PEG side chains was estimated using an empirical equation based on static light-scattering measurements.[52]

$$R_g = 0.181N^{0.58}\ \text{(nm)} \qquad (1)$$

where N is the number of repeat units. Such an estimate provides radii of gyration of about 1.65 and 2.82 nm for PEG side chains of mol wt 2000 and 5000, respectively. Calculated from the spacing (L) and the radius of gyration (R_g), the quantity $L/2R_g$ provides a value that represents the extent of packing on the surface. That is, values of $L/2R_g$ that are less than unity imply that the radii of gyration of neighboring PEG side chains overlap and that the surface is densely packed. Conversely, values of $L/2R_g$ that are greater than unity imply that there is space between neighboring PEG side chains, as their radii of gyration do not overlap.[53]

The data presented in Figure 10 include protein-adsorption data from several different surface-immobilized PEG systems, including graft copolymers on metal oxide surfaces as described herein, self-assembled-monolayer-forming oligo(ethylene glycol) alkanethiolates[54] on gold surfaces, and poly(ethylene glycol) covalently bound to silanized silica.[53] Values from this work are compared to published data, for which both the areal mass density of proteins and the average PEG spacing (L) are taken from refs 53 and 54. The radii of gyration were calculated according to eq 1 for PEG, whereas the original value of 0.42 nm was used for the oligo(ethylene glycol). The evaluated data for the PLL-*g*-PEG graft copolymer system are in agreement with the evaluated data for the previously reported systems. The trend shown in Figure 10 suggests that the best protein-adsorption-suppression performance is observed for systems with the lowest $L/2R_g$ values and that, as $L/2R_g$ increases, the average spacing between the immobilized PEG chains increases and the suppression of protein adsorption decreases. Furthermore, the $L/2R_g$ value of 0.47 appears to be the limit of packing density, as $L/2R_g$ values lower than 0.47 were never observed

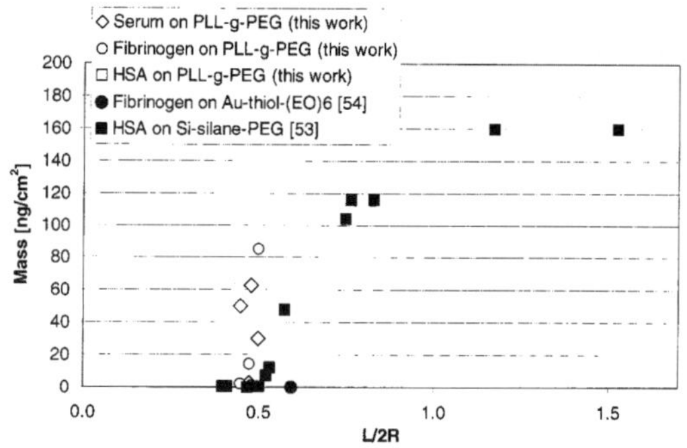

Figure 10. Dependence of serum adsorption on the extent of surface PEG side chain packing density. $L/2R_g$ is calculated as described in section 4.2. Values from this work are compared to published data, for which both the areal mass density of proteins and the average PEG spacing (L) are taken from refs 53 and 54. The radii of gyration were calculated according to eq 1 for PEG, whereas the original value of 0.42 nm was used for the oligo(ethylene glycol).

within our study and such values would imply an entropically unfavorable interaction between the PEG side chains.

The highest protein resistance among the PLL-*g*-PEG molecules investigated in this work was exhibited by PLL(20)-*g*[3.5]-PEG(2) and PLL(20)-*g*[5]-PEG(2), which often showed residual adsorbed masses of serum and fibrinogen, respectively, on the order of the detection limit of the OWLS technique (~2 ng/cm^2). This is an important finding because it has been shown that there is a lower limit of fibrinogen adsorbed mass (~5 ng/cm^2) below which the pathways leading to blood coagulation and thrombosis are not activated.[55]

5. Conclusions

All of the copolymers discussed here, all of which were based on poly(L-lysine)-*g*-poly(ethylene glycol) (PLL-*g*-PEG), were found to adsorb spontaneously in an electrostatically driven process onto several different negatively charged metal oxide surfaces. The adsorption process was rapid and was ultimately limited by the packing of the PEG side chains at the surface. The resulting adsorbed layer was found to drastically reduce the subsequent interfacial protein adsorption. Reduction of protein adsorption was observed for all of the different PLL-*g*-PEG copolymer architectures investigated. However, only in the case of optimized PLL-*g*-PEG architectures could extremely low levels of serum and fibrinogen (on the order of <5 ng/cm^2) be achieved. In addition, antibody experiments gave evidence that the adsorbed serum mass that was observed was due not to proteins but rather to other components of the serum. The combination of data obtained from X-ray photoelectron spectroscopy and optical waveguide measurements suggests that the backbone of the copolymer acts as an anchor to the solid metal oxide surface, thereby exposing the PEG side chains to the solution phase and yielding a "comblike" interface configuration. Furthermore, the anchoring backbone of the copolymer appears to be capable of bridging surface contamination, still producing a protein-resistant surface. This observation and the fact that devices with complex shapes can be easily treated by PLL-*g*-PEG make the PLL-*g*-PEG surface-treatment technology a candidate for future applications in both the biomaterial/implant and biosensor areas. As a next step, proof-of-concept studies are envisioned to investigate the long-term performance of the polymer adlayers under more application-relevant conditions, such as contact with blood flow.

References and Notes

(1) Leonard, E. F.; Turitto, V. T.; Vroman, L. *Blood in Contact with Natural and Artificial Surfaces*; New York Academy of Sciences: New York, 1987; Vol. 516.

(2) Jeon, S. I.; Lee, J. H.; Andrade, J. D.; de Gennes, P. G. *J. Colloid Interface Sci.* **1991**, *142*, 149−166.

(3) Isreals, R.; Leermakers, F. A. M.; Fleer, G. J. *Macromolecules* **1995**, *28*, 1626−1634.

(4) Bjorling, M. *Macromolecules* **1992**, *25*, 3956−3970.

(5) Claesson, P. *Colloids Surf., A* **1993**, *77*, 109−118.

(6) Gombotz, W. R. *Surfaces: Synthesis, Characterization and Biological Interaction Studies*; University of Washington: Seattle, WA, 1988.

(7) Gombotz, W. R.; Guanghui, W.; Hoffman, A. S. *J. Appl. Polym. Sci.* **1989**, *37*, 91−107.

(8) Schroen, C. G. P. H.; Stuart, M. A. C.; van der Voort Maarschaalk, K.; van der Padt, A.; van't Riet, K. *Langmuir* **1995**, *11*, 3068−3074.

(9) Freij-Larsson, C.; Nylander, T.; Jannasch, P.; Wesslen, B. *Biomaterials* **1996**, *17*, 2199−2207.

(10) Amiji, M. M.; Park, K. *J. Appl. Polym. Sci.* **1994**, *52*, 539−544.

(11) Amiji, M.; Park, K. *Biomaterials* **1992**, *13*, 682−692.

(12) Lee, J. H.; Kopeckova, P.; Kopecek, J.; Andrade, J. D. *Biomaterials* **1990**, *11*, 455−464.

(13) Lee, J. H.; Kopecek, J.; Andrade, J. D. *J. Biomed. Mater. Res.* **1989**, *23*, 351−368.

(14) Balazs, A. C.; Siemasko, C. P. *J. Chem. Phys.* **1991**, *95*, 3798−3803.

(15) van der Linden, C. C.; Leermakers, F. A. M.; Fleer, G. J. *Macromolecules* **1996**, *29*, 1000−1005.

(16) Kosmas, M. K. *Macromolecules* **1990**, *23*, 2061−2065.

(17) Marzio, E. A. D.; Gutmann, C. M.; Mah, A. *Macromolecules* **1995**, *28*, 2930−2937.

(18) Marques, C. M.; Joanny, J. F. *Macromolecules* **1990**, *23*, 268−276.

(19) Halperin, A.; Tirrell, M.; Lodge, T. P. *Adv. Polym. Sci.* **1992**, *100*, 31−71.

(20) Kawaguchi, M.; Takahashi, A. *J. Colloid Interface Sci.* **1992**, *37*, 219−317.

(21) Hoogeveen, N. G.; Stuart, M. A. C.; Fleer, G. J. *Faraday Discuss.* **1994**, *98*, 161−172.

(22) Hoogeveen, N. G.; Stuart, M. A. C.; Fleer, G. J. *J. Colloid Interface Sci.* **1996**, *182*, 146−157.

(23) Hoogeveen, N. G.; Stuart, M. A. C.; Fleer, G. J. *J. Colloid Interface Sci.* **1996**, *182*, 133−145.

(24) Hubbell, J. A.; Sawhney, A. S. Biocompatible Microcapsules. U.S. Patent 5,232,984, 1993.

(25) Hubbell, J. A.; Elbert, D. L.; Hill-West, J. L.; Drumheller, P. D.; Chowdhury, S.; Sawhney, A. S. Multifunctional Organic Polymers. U.S. Patent 5,462,990, 1995.

(26) Hubbell, J. A.; Sawhney, A. S. Biocompatible Microcapsules. U.S. Patent 5,380,536, 1995.

(27) Hubbell, J. A.; Donald, D. L. E.; Hill-West, J. L.; Drumheller, P. D.; Chowdhury, S.; Sawhney, A. S. Methods for Modifying Cell Contact with a Surface. U.S. Patent 5,567,440, 1996.

(28) Choksakulnimitr, S.; Musada, S.; Tokuda, H.; Takakura, Y.; Hashida, M. *J. Controlled Release* **1995**, *34*, 233−241.

(29) Clegg, J. A.; Hudecz, F.; Mezo, G.; Pimm, M. V.; Szekerke, M.; Baldwin, R. W. *Bioconjugate Chem.* **1990**, *1*, 425−430.

(30) Ryser, H. J. P.; Mandel, R.; Hacobian, A.; Chen, W. *J. Cell. Physiol.* **1988**, *135*, 277−284.

(31) Koyo, H.; Tsuruta, T.; Kataoka, K. *Polym. J.* **1993**, *25*, 141−152.

(32) Bogdanov, A. A.; Martin, C.; Bogdanova, A. V.; Brady, T. J.; Weissleder, R. *Bioconjugate Chem.* **1996**, *7*, 144−149.

(33) Bogdanov, A. A.; Weissleder, R.; Frank, H. W.; Bogdanova, A. V.; Nossif, N.; Schaffer, B.; Tsai, E.; Papisov, M. I.; Brady, T. J. *Radiology* **1993**, *187*, 701−706.

(34) Elbert, D. L.; Hubbell, J. A. *J. Biomed. Mater. Res.* **1998**, *42*, 55−65.

(35) Ramsden, J. J. *J. Stat. Phys.* **1993**, *73*, 853−877.

(36) Kurrat, R.; Textor, M.; Ramsden, J. J.; Boni, P.; Spencer, N. D. *Rev. Sci. Instrum.* **1997**, *68*, 2172−2176.

(37) Kurrat, R.; Walivaara, B.; Marti, A.; Textor, M.; Tengvall, P.; Ramsden, J. J.; Spencer, N. D. *Colloids Surf.* **1998**, *11*, 187−201.

(38) Ramsden, J. J.; Mate, M. *J. Chem. Soc., Faraday Trans.* **1998**, *94*, 783−788.

(39) Jukubauskas, H. L. *J. Coat. Technol.* **1986**, *58*, 71−82.

(40) Xu, R.; D'Unger, G.; Winnik, M. A.; Marinho, J. M. G.; d'Oliviera, J. M. R. *Langmuir* **1994**, *10*, 2977−2984.

(41) Tripp, C. P.; Hair, M. L. *Langmuir* **1996**, *12*, 2.

(42) Pefferkorn, E.; Elaissari, A.; Huguenard, C. *Macromol. Rep.* **1992**, *A29*, 147−153.

(43) Tirrell, M.; Parsonage, E.; Watanabe, H.; Dhoot, S. *Polym. J.* **1991**, *23*, 641−664.

(44) Stuart, M. A. C.; Fleer, G. J.; Lyklema, J.; Norde, W. *Adv. Colloid Interface Sci.* **1991**, *34*, 477−535.

(45) Kurrat, R. Adsorption of Biomolecules on Titanium Oxide Layers in Biological Model Solutions. Ph.D. Dissertation, ETH Zurich, Zurich, Switzerland, 1998.

(46) Ramsden, J. J.; Roush, D. J.; Gill, D. S.; Kurrat, R.; Willson, R. C. *J. Am. Chem. Soc.* **1995**, *117*, 8511−8516.

(47) Moulder, J. F.; Stickle, W. F.; Sobol, P. E.; Bomben, K. D. *Handbook of X-ray Photoelectron Spectroscopy*; Perkin-Elmer Corp.: Eden Prairie, MN, 1992.

(48) Marti, A. Untersuchung Lokaler Chemischer und Tribologischer Eigenschaften von Oxidischen und Organischen Oberflaechen in Elektrolytloesung mit dem Raterkraftmikroskop. Ph.D. Dissertation, ETH Zurich, Zurich, Switzerland, 1997.

(49) Healy, T. W.; White, L. R. *Adv. Colloid Interface Sci.* **1978**, *9*, 304−345.

(50) Vörös, J.; Höök, F.; Askendal, A.; Wälivaara, B.; Tengvall, P.; Kasemo, B.; Ramsden, J. J.; Böni, P.; Kurrat, R.; Textor, M.; Spencer, N. D. *Colloids Surf.* **1999**, submitted for publication.

(51) Elbert, D. L.; Hubbell, J. A. *Chem. Biol.* **1998**, *5*, 177−183.

(52) Kawaguchi, S.; Imai, G.; Suzuki, J.; Miyahara, A.; Kitano, T. *Polymer* **1997**, *38*, 2885−2891.

(53) Sofia, S. J.; Premnath, V.; Merrill, E. W. *Macromolecules* **1998**, *31*, 5059−5070.

(54) Harder, P.; Grunze, M.; Dahint, R.; Whitesides, G. M.; Laibinis, P. E. *J. Phys. Chem. B* **1998**, *102*, 426−436.

(55) Tsai, W. B.; Grunkemeier, J. M.; McFarland, C. D.; Horbett, T. A. *Selectively Depleted Plasmas To Study the Role of Four Plasma Proteins in Platelet Adhesion to Biomaterials*; LeBerge, M., Agrawal, C. M., Eds.; Society for Biomaterials: Providence, RI, 1999; Vol. 22, p 266.

Langmuir **2001,** *17,* 489–498 489

Poly(L-lysine)-*g*-poly(ethylene glycol) Layers on Metal Oxide Surfaces: Surface-Analytical Characterization and Resistance to Serum and Fibrinogen Adsorption

Ning-Ping Huang,[†] Roger Michel,[†] Janos Voros,[†] Marcus Textor,[†] Rolf Hofer,[†] Antonella Rossi,[†,§] Donald L. Elbert,[‡] Jeffrey A. Hubbell,[‡] and Nicholas D. Spencer*[,†]

Laboratory for Surface Science and Technology and Institute for Biomedical Engineering, Department of Materials, ETH Zurich, CH-8092 Zurich, Switzerland, and Dipartimento di Chimica Inorganica ed Analitica, University of Cagliari, Italy

Received May 30, 2000. In Final Form: October 5, 2000

Poly(L-lysine)-*g*-poly(ethylene glycol) (PLL-*g*-PEG) is a member of a family of polycationic PEG-grafted copolymers that have been shown to chemisorb on anionic surfaces, including various metal oxide surfaces, providing a high degree of resistance to protein adsorption. PLL-*g*-PEG-modified surfaces are attractive for a variety of applications including sensor chips for bioaffinity assays and blood-contacting biomedical devices. The analytical and structural properties of PLL-*g*-PEG adlayers on niobium oxide (Nb_2O_5), tantalum oxide (Ta_2O_5), and titanium oxide (TiO_2) surfaces were investigated using reflection–absorption infrared spectroscopy (RAIRS), angle-dependent X-ray photoelectron spectroscopy (XPS), and time-of-flight secondary ion mass spectrometry (ToF-SIMS). The combined analytical information provides clear evidence for an architecture with the cationic poly(L-lysine) attached electrostatically to the oxide surfaces (charged negatively at physiological pH) and the poly(ethylene oxide) side chains extending out from the surface. The relative intensities of the vibrational modes in the RAIRS spectra and the angle-dependent XPS data point to the PLL backbone being located directly at and parallel to the oxide/polymer interface, whereas the PEG chains are preferentially oriented in the direction perpendicular to the surface. Both positive and negative ToF-SIMS spectra are dominated by PEG-related secondary ion fragments with strongly reduced metal (oxide) intensities pointing to an (almost) complete coverage by the densely packed PEG comblike grafts. The three different transition metal oxide surfaces with isoelectric points well below 7 were found to behave very similarly, both in respect to the kinetics of the polymer adlayer adsorption and properties as well as in terms of protein resistance of the PLL-*g*-PEG-modified surface. Adsorption of serum and fibrinogen was evaluated using the OWLS optical planar waveguide technique. The amount of human serum adsorbed on the modified surfaces was consistently below the detection limit of the optical sensor technique used (<1–2 ng cm^{-2}), and fibrinogen adsorption was reduced by 96–98% in comparison to the nonmodified (bare) oxide surfaces.

1. Introduction

The generation of protein-resistant surfaces is a key element in the design of medical devices and implants in contact with blood, such as catheters, vascular stents, and in vivo sensors for long-term, real-time monitoring of biochemical variables (pO_2, pCO_2, pH, or glucose). Protein-resistant surfaces are needed to avoid or reduce nonspecific protein adsorption, platelet adhesion, and thrombus formation, to prevent undesirable responses of the living system to the device or implant. Because metals such as steel or titanium and its alloys, covered by their respective surface oxides (passive film), are often applied in blood-contacting devices, a reliable method for reducing protein adsorption on oxide surfaces in vivo would be highly beneficial to the biomaterials field. Moreover, protein-resistant surfaces are a necessity in bioaffinity sensing, where the selectivity and sensitivity of the antigen/antibody assay directly depends on how well nonspecific adsorption can be suppressed. Oxides of transition metals, such as niobium, tantalum, or titanium oxide, are preferred waveguiding layers on optical (grating coupler) waveguides because of their high optical transparency and high index of refraction.[1] Oxide-coated chips are used as sensing devices in optical sensor technology, based on the interaction of the evanescent field with analyte constituents adsorbed on or immobilized at the waveguide/fluid interface.[2]

A variety of methods and techniques have been investigated to design protein-resistant surfaces. They include anticoagulant coatings (such as heparin and nondenatured serum albumin),[3] phosphorylcholine-coated polymer surfaces,[4,5] and surfaces modified by grafting hydrogel or hydrophilic chains.[6] Surface modification by poly(ethylene glycol) (PEG) is a well-known strategy for rendering surfaces protein resistant, and the nonfouling properties

* To whom correspondence should be addressed. E-mail: nspencer@surface.mat.ethz.ch. Tel: +41 1 632 5850. Fax: +41 1 633 10 27.

[†] Laboratory for Surface Science and Technology, Department of Materials, ETH Zurich.

[‡] Institute for Biomedical Engineering, Department of Materials, ETH Zurich.

[§] Dipartimento di Chimica Inorganica ed Analitica, University of Cagliari.

(1) Broer, M. M.; Sigel, G. H.; Kersten, R. Th.; Kawazoe, H. *Optical Waveguide Materials*; Materials Research Society: Pittsburgh, PA, 1992.
(2) Duveneck, G. L.; Pawlak, M.; Neuschaefer, D.; Budach, W.; Ehrat, M. *Proc. Biomed. Syst. Technol.* **1996**, *2928*, 98–109.
(3) Rosenbery, R. *Semin. Hematol.* **1977**, *14*, 429.
(4) Ishihara, K.; Aragaki, R.; Ueda, T.; Watenabe, A.; Nakabayashi, N. *J. Biomed. Mater. Res.* **1990**, *24*, 1069–1077.
(5) Ruiz, L.; Fine, E.; Voros, J.; Makohliso, S. A.; Leonard, D.; Johnston, D. S.; Textor, M.; Mathieu, H. J. *J. Biomater. Sci., Polym. Ed.* **1999**, *10*, 931–955.
(6) Mori, Y.; Nagaoka, S.; Takinchi, H.; Kikuchi, T.; Noguchi, N.; Tanzawa, H.; Noishiki, Y. *Trans. Am. Soc. Artif. Intern. Organs* **1982**, *28*, 459.

Published on Web 12/21/2000

of such surfaces have been attributed to steric repulsion and excluded-volume effects between proteins in solution and the PEG-modified surface.[7,8] Despite many efforts, however, we still lack reproducible and cost-effective surface modifications to reduce protein adsorption to the very low levels needed for consistent blood compatibility (e.g., less than 5 ng/cm^2 fibrinogen adsorption[9]).

A general approach to the immobilization of PEG onto surfaces involves the coupling of PEG to functional groups that have an affinity for the target surface.[10,11] Another approach is grafting of PEG side chains to a polymer backbone, resulting in the formation of a comblike structure. In the latter case, the immobilization relies on a sufficiently strong interaction between the polymer backbone and the target surface.[7,12] Our objective is to immobilize PEG onto metal oxide surfaces that are negatively charged at physiological pH (TiO_2, Nb_2O_5, and Ta_2O_5 with isoelectric points (IEP) of 4.7−6.2, 3.4−3.8, and 2.7−3.0, respectively[13−15]). Poly(L-lysine) (PLL) is an obvious candidate for the backbone, in view of its biodegradability[16] and the positive charge on the amino groups of the pendant chain at pH values below 10 (pK_a of amino groups in the pendant chain is 10.5). PEG of different molecular weights can be grafted onto the poly-(L-lysine) at different PEG/PLL ratios, to optimize the polymer architecture for protein resistance.[17] A typical example for the molecular structure of a PLL-*g*-PEG copolymer is shown in Figure 1. PLL-*g*-PEG is believed to attach to negatively charged surfaces through electrostatic interaction of the multiple positive charges of the PLL backbone ($-NH_3^+$) with the negative charges on the metal oxide surfaces. The resulting PLL-*g*-PEG-modified surfaces have been found to display drastically reduced protein adsorption relative to unmodified metal oxide surfaces, both toward solutions of single proteins, such as albumin and fibrinogen, as well as toward serum.[18] The protein-suppression performance was found to depend on the molecular weight of PEG, the grafting ratio in the PLL-*g*-PEG copolymer, and the surface coverage of the polymer. The PLL-*g*-PEG-modified surfaces have, however, not yet been fully characterized. This is a prerequisite for an improved understanding of the relationship between polymer adlayer architecture and performance of these surfaces.

In this paper, we focus on the structure and properties of PLL-*g*-PEG-modified surfaces as well as on the adsorption kinetics of PLL-*g*-PEG onto metal oxide surfaces. We present the results of surface investigations using reflection−absorption infrared spectroscopy (RAIRS), X-ray photoelectron spectroscopy (XPS), and time-of-flight secondary ion mass spectrometry (ToF-SIMS) for surface-analytical characterization as well as optical waveguide lightmode spectroscopy (OWLS)[19] for the investigation of both the polymer assembly kinetics and protein adsorption onto the nonmodified and PLL-*g*-PEG-modified surfaces. The copolymer with a 20 000 Da PLL backbone and 2 000 Da PEG grafted at a ratio of lysine/PEG of ~3.5 is chosen as a model polymer on TiO_2, Nb_2O_5, and Ta_2O_5 surfaces in view of its optimum protein resistance.[18]

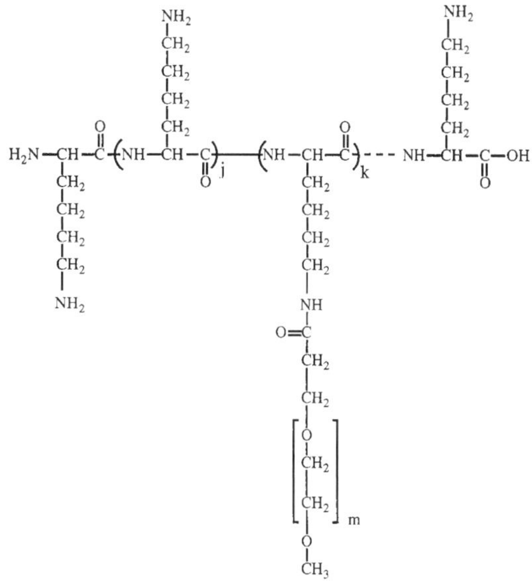

Figure 1. Molecular structure of the PLL-*g*-PEG copolymer. For the specific polymer investigated in this work, $j = 2.5$, $k = 1$, and $m = 41$.

2. Materials and Methods

2.1. Synthesis of PLL(20)-*g*[3.5]-PEG(2). PLL(20)-*g*[3.5]-PEG(2) denotes the graft copolymer with a PLL backbone of molecular weight 20 kDa, a grafting ratio of lysine-mer/PEG side chain of 3.5 and PEG side chains of molecular weight 2 kDa.

Poly-L-lysine hydrobromide (84 mg, mol wt 20 kDa, Sigma, USA) was dissolved in 1.05 mL of 50 mM sodium borate buffer solution (pH 8.5). The solution was filter sterilized (0.22 μm pore-size filter). The *N*-hydroxysuccinimidyl ester of methoxypoly-(ethylene glycol) propionic acid (216 mg, mol wt 2 kDa, SPA-PEG, Shearwater Polymers, Inc., USA) was added to the dissolved PLL. The reaction was allowed to proceed for 6 h at room temperature, after which the reaction mixture was dialyzed (Spectra-Por, mol wt cutoff size 6−8 kDa, Spectrum, USA) for 24 h, first against phosphate-buffered saline (PBS, pH 7.4) and subsequently against deionized water. The product was freeze-dried and stored at −20 °C. The product of PLL(20)-*g*[3.5]-PEG-(2) was characterized by a ^{1}H NMR (D_2O, ppm) spectrum: 1.35, 1.60, 1.68 ($-CH_2-$); 2.88 ($-CH_2-N-$); 3.55 (PEG); 4.20 (−N−CHR−COO−). The areas of the lysine side-chain peaks were compared with the area of the PEG peak to estimate the grafting ratio of the comb copolymer. The grafting ratio of the PLL-*g*-PEG copolymer was analyzed and determined by aqueous size exclusion chromatography using refractive index detection (Shodex OHpak column, SB-804HQ, Alltech, Deerfield, IL; Delory & King's carbonate−bicarbonate buffer eluent, 0.2 M anhydrous sodium carbonate and 0.2 M sodium bicarbonate mixed to get pH 10; sample was 1% in eluent buffer).

2.2. Substrates. Surface modification by PLL-*g*-PEG was studied on metal oxide films. Nb_2O_5 and Ta_2O_5 were sputter coated

(7) Jeon, S. I.; Lee, J. H.; Andrade, J. D.; de Gennes, P. G. *J. Colloid Interface Sci.* **1991**, *142*, 149−166.
(8) Bjorling, M. *Macromolecules* **1992**, *25*, 3956−3970.
(9) Tsai, W. B.; Grunkemeier, J. M.; McFarland, C. D.; Horbett, T. A. In *Selectively depleted plasmas to study the role of four plasma proteins in platelet adhesion to biomaterials*; LeBerge, M., Agrawal, C. M., Eds.; Society for Biomaterials: Providence, RI, 1999; Vol. 22, p 266.
(10) Claesson, P. *Colloids Surf., A* **1993**, *77*, 109−118.
(11) Gombotz, W. R.; Guanghui, W.; Hoffman, A. S. *J. Appl. Polym. Sci.* **1989**, *37*, 91−107.
(12) Lee, J. H.; Kopeckova, P.; Kopecek, J.; Andrade, J. D. *Biomaterials* **1990**, *11*, 455−464.
(13) Parks, G. A. *Chem. Rev.* **1965**, *65*, 177−198.
(14) Gonzalez, G.; Saraiva, S. M.; Aliaga, W. *J. Dispersion Sci. Technol.* **1994**, *15*, 123−132.
(15) Bousse, L.; Mostarshed, S.; Vandershoot, B.; Derooij, N. F.; Gimmel, P.; Gopel, W. *J. Colloid Interface Sci.* **1991**, *147*, 22−32.
(16) Kamath, K. R.; Park, K. *Adv. Drug Delivery Rev.* **1993**, *11*, 59−84.
(17) Elbert, D. L.; Hubbell, J. A. *Chem. Biol.* **1998**, *5*, 177−183.
(18) Kenausis, G. L.; Voros, J.; Elbert, D. L.; Huang, N. P.; Hofer, R.; Ruiz, L.; Textor, M.; Hubbell, J. A.; Spencer, N. D. *J. Phys. Chem. B* **2000** *104*, 3298.

(19) Kurrat, R.; Textor, M.; Ramsden, J. J.; Boni, P.; Spencer, N. D. *Rev. Sci. Instrum.* **1997**, *68*, 2172−2176.

onto Corning glass (150 nm oxide layer thickness, <1 nm average roughness (AFM), Balzers AG, Switzerland). TiO_2 (50 nm) was sputter coated onto commercial float glass using reactive magnetron sputtering (PSI, Villigen, Switzerland). Coatings of 100-nm-thick Ti metal were deposited onto silicon wafers using reactive magnetron sputtering, to be used as substrates for RAIRS measurements (PSI, Villigen, Switzerland). The waveguide chips used for OWLS measurements were purchased from Microvacuum Ltd. (Budapest, Hungary) and consisted of a 1-mm-thick AF45 glass substrate and a 200-nm-thick $Si_{0.75}Ti_{0.25}O_2$ waveguiding layer at the surface. For experiments involving titanium and niobium oxide surfaces, an additional 12-nm-thick oxide layer was sputter coated on top of the waveguiding layer in a Leybold dc-magnetron Z600 sputtering unit. The coating conditions and the principles of OWLS investigations have been given in detail elsewhere.[19,20]

2.3. Substrate Cleaning and Formation of PLL-*g*-PEG Monolayer. All substrates were sonicated in 2-propanol for 10 min, extensively rinsed with ultrapure water, and dried in a nitrogen stream, followed by 2 min of oxygen-plasma cleaning in a Plasma Cleaner/Sterilizer PDC-32G instrument (Harrick, Ossining, USA). Water used in the experiments was purified with an EASYpure device (Barnstead, USA). Representative substrates were checked for cleanliness by XPS and ToF-SIMS. The clean substrates were transferred immediately after oxygen-plasma treatment into a 1 mg/mL solution of PLL-*g*-PEG in 10 mM HEPES buffer solution (4-(2-hydroxyethyl)piperazine-1-ethanesulfonic acid, adjusted to pH 7.4 with 1 M NaOH solution). Following 30 min of immersion, the modified samples were withdrawn, rinsed with water to remove any excessively adsorbed PLL-*g*-PEG, and then dried under nitrogen. The samples were stored in an argon-filled chamber until analysis by XPS, ToF-SIMS, and RAIRS. For the purpose of standard experiments on adsorbed PLL-*g*-PEG by RAIRS, surfaces with physisorbed PLL-*g*-PEG were also prepared. This was achieved by dissolving the PLL-*g*-PEG powder in water. The clean titanium-covered wafer was soaked in PLL-*g*-PEG aqueous solution for 30 min and then removed without washing and dried in air.

2.4. Surface Analysis Methods. 2.4.1. Optical Waveguide Lightmode Spectroscopy (OWLS). OWLS is based on grating-assisted incoupling of a He−Ne laser into a planar waveguide that allows for the direct online monitoring of macromolecule adsorption. This method is highly sensitive (sensitivity limit ~1 ng/cm^2) up to a distance of 100 nm above the surface of the waveguide. Furthermore, a measurement-time resolution of 3 s allows for the in situ, real-time study of adsorption kinetics. All OWLS experiments were conducted in a BIOS-I instrument (ASI AG, Switzerland) using a Kalrez (Dupont, USA) flow-through cell (8 × 2 × 1 mm).[19] For all OWLS experiments (both PLL-*g*-PEG and protein adsorption), the solution change was carried out by syringe injection (1 mL within 10 s). Areal adsorbed mass density data are calculated from the adlayer thickness and refractive index values derived from the mode equations according to Feijter's formula.[21] A refractive index increment (dn/dc) value of 0.202 cm^3/g, as determined in a Raleigh interferometer, is used for the PLL-*g*-PEG-adsorption calculations, and a value of 0.182 cm^3/g is used for the protein-adsorption calculations.[22]

2.4.2. Reflection−Absorption Infrared Spectroscopy (RAIRS) Measurements. The RAIRS measurements were performed on a Bruker IFS 66V spectrometer operating at a pressure of approximately 100 Pa. A liquid nitrogen cooled mercury cadmium telluride (MCT) detector was used to collect spectra with a resolution of 2 cm^{-1}. The angle of incidence was 80° relative to the surface normal. An oxygen-plasma-pretreated titanium-coated Si wafer was used as the reference. For both samples and reference, 500 scans were collected. Pure PLL-*g*-PEG samples, ground with powdered KBr and then pressed into a disk, were measured with 32 scans without liquid nitrogen cooling.

2.4.3. X-ray Photoelectron Spectroscopy (XPS) Measurements. XPS analyses were performed using a PHI 5700 photoelectron spectrometer equipped with a concentric hemispherical analyzer in the standard configuration (Physical Electronics, Eden Prairie, USA). Spectra were acquired at a chamber pressure of 10^{-9} mbar using a nonmonochromatic Al Kα source operating at 200 W. The instrument was run in the minimum-area mode using an aperture of 0.8 mm diameter. The analyzer was used in the fixed analyzer transmission mode. Pass energies used for survey scans and detailed scans were 187.85 and 23.5 eV, respectively, the latter giving an experimental resolution of 1.0 eV for the Ag($3d_{5/2}$) reference peak. Acquisition times were approximately 5 min for survey scans and 9 min for high-energy-resolution scans.

Angle-resolved XPS measurements were conducted at two different takeoff angles, namely 15° and 75° relative to the surface plane, to obtain depth-dependent information on the molecular layers adsorbed onto the oxide substrate.

Spectra were referenced to the aliphatic hydrocarbon C1s signal at 285.0 eV. Data were analyzed using a least-squares fit routine following Shirley background subtraction. Measured intensities (peak areas) were transformed into normalized intensities using PHI sensitivity factors.[23] A three-layer model was also applied, using the parameters listed in ref 24. Spectra were fitted with the Multipak 6.0 software using the sum of an 80% Gaussian and 20% Lorentzian function.

2.4.4. Time-of-Flight Secondary Ion Mass Spectrometry (ToF-SIMS) Measurements. Secondary ion mass spectra were recorded on a PHI 7200 time-of-flight secondary ion mass spectrometer in the mass range 0−1000 *m/e*. The total ion dose of the 8 kV Cs^+ primary ion beam (200 μm diameter) was below the static limit (<1.0 × 10^{12} ions/cm^2). The time per data point was 1.25 ns. Because of the low conductivity of the metal oxide/glass substrate, intermittent, pulsed electron-beam neutralization had to be used during measurement of both positive and negative SIMS spectra. Mass resolution $M/\Delta M$ was typically 5500 in the positive mode and 2000 in the negative mode. To calibrate the mass scale, the whole mass range was first calibrated using a single standard set of low ion masses using the PHI software Tofpak. To improve the quality of the mass calibration in the higher mass range, different sets of ion species were used because of the large mass range analyzed and the very different nature of the observed secondary ion species.

2.5. Adsorption Kinetics of the PLL-*g*-PEG Monolayer by Optical Waveguide Lightmode Spectroscopy (OWLS). The protocol for in situ study of PLL-*g*-PEG adsorption kinetics by OWLS was as follows: Clean waveguides were initially placed in 10 mM HEPES buffer solution for 5 h. Prior to assembly of the flow-through cuvette in the OWLS instrument, the samples were rinsed with water and dried under nitrogen. These presoaked samples equilibrated and reached a flat baseline in HEPES solution in less than 1 h. The samples were then exposed in situ to the PLL-*g*-PEG solution (1 mg/mL in HEPES buffer solution). The adsorption was subsequently monitored for 1 h. Then the PLL-*g*-PEG solution was replaced with HEPES solution, and surface coverage was monitored for another 30 min.

2.6. Protein Adsorption on Clean Substrates and PLL-*g*-PEG-Modified Surfaces by OWLS Measurements. Protein-adsorption measurements on clean substrates (waveguides) and PLL-*g*-PEG-modified surfaces (waveguides) were carried out by following the same procedure as mentioned in section 2.5, except that after reaching a flat baseline the samples were separately exposed to a solution of human serum (Control Serum N, Roche, Switzerland) and a solution of 2.5 mg/mL human fibrinogen (Sigma Chemical Co., USA) in 10 mM HEPES buffer solution for 1 h at a temperature of 25 °C and subsequently rinsed for 30 min in HEPES solution.

3. Results and Discussion

3.1. Adsorption Kinetics of PLL-*g*-PEG on Metal Oxide Surfaces. In the present study, the molecular assembly of the PLL-*g*-PEG copolymer onto several metal

(20) Kurrat, R.; Walivaara, B.; Marti, A.; Textor, M.; Tengvall, P.; Ramsden, J. J.; Spencer, N. D. *Colloids Surf., B* **1998**, *11*, 187−201.
(21) Ramsden, J. J. *J. Stat. Phys.* **1993**, *73*, 853−877.
(22) Ramsden, J. J.; Roush, D. J.; Gill, D. S.; Kurrat, R.; Willson, R. C. *J. Am. Chem. Soc.* **1995**, *117*, 8511−8516.
(23) Moulder, J. F.; Stickle, W. F.; Sobol, P. E.; Bomben, K. D. *Handbook of X-Ray Photoelectron Spectroscopy*; Perkin-Elmer Corp.: Eden Prairie, MN, 1992.
(24) Textor, M.; Ruiz, L.; Hofer, R.; Rossi, A.; Feldman, K.; Hähner, G.; Spencer, N. D. *Langmuir* **2000** *16*, 3257.

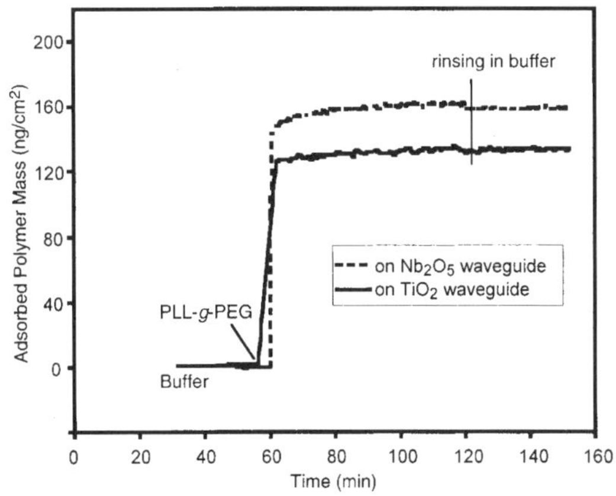

Figure 2. PLL-*g*-PEG adsorption versus time measured by the OWLS technique on metal oxide surfaces followed by rinsing in buffer (10 mM HEPES buffer solution (pH 7.4), 1 mg/mL polymer, $T = 25$ °C).

oxide surfaces is performed at pH 7.4 in HEPES buffer solution, followed by a washing step to remove excessively physisorbed polymer molecules. At pH 7.4, all the metal oxide surfaces investigated (TiO_2, Nb_2O_5, and Ta_2O_5) are negatively charged, whereas the PLL backbone of PLL-*g*-PEG is highly cationic. The electrostatic interaction between the multicharged adsorbate and the substrate is expected to lead to strong chemisorption of PLL-*g*-PEG on the three types of metal oxide surfaces. Figure 2 shows the adsorption kinetics measured in situ by the OWLS technique for a TiO_2- and a Nb_2O_5-coated waveguide chip. The adsorption is fast, as measured in situ by OWLS, with 95% of the total PLL-*g*-PEG coverage achieved within the first 5 min. After 20 min, the adsorbed mass of PLL-*g*-PEG reached a plateau that was stable even after the final washing step with HEPES solution, suggesting that the PLL-*g*-PEG adsorption process is essentially irreversible (providing the pH is held constant). The amount of PLL-*g*-PEG adsorbed onto the surface is found to be different, 123 ± 4 ng/cm^2 for TiO_2 and 148 ± 10 ng/cm^2 for Nb_2O_5, which most likely reflects differences in surface charge density of the two oxides at pH 7.4 (IEP of approximately 5.5 and 3.6 for TiO_2 and Nb_2O_5, respectively.) The adsorbed amounts of PLL-*g*-PEG correspond to a single adsorbed layer with different packing densities. The adsorbed layers of PLL-*g*-PEG were found to be quite stable from our previous study.[18] Less than 5% loss in mass of the adsorbed copolymer occurred after the adsorbed layer was exposed to flowing HEPES buffer solution for one week.

3.2. Characterization of the Unmodified and PLL-*g*-PEG-Modified Surfaces. 3.2.1. RAIRS Measurements of the PLL(20)-*g*[3.5]-PEG(2) Monolayer. Reflection–absorption infrared spectroscopy (RAIRS) is well suited to study adsorbates on metallic surfaces, which are highly reflective. It relies on reflecting an infrared beam at near-grazing incidence from the metallic surface on which the thin film of interest has been deposited. Only the component of the vibrational transition dipole moments perpendicular to the surface plane contributes to the absorption spectra. The intensity of an absorption band is proportional to the squared cosine of the angle between the transition dipole moment and the surface normal. Therefore, RAIRS provides information not only on functional groups but also on orientation and conformation of adsorbed molecules or molecular entities. Metal oxides

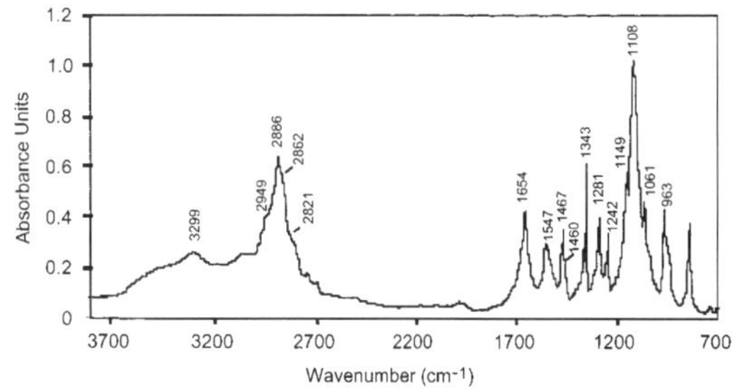

Figure 3. FTIR spectrum of PLL-*g*-PEG powder (pressed KBr pellet).

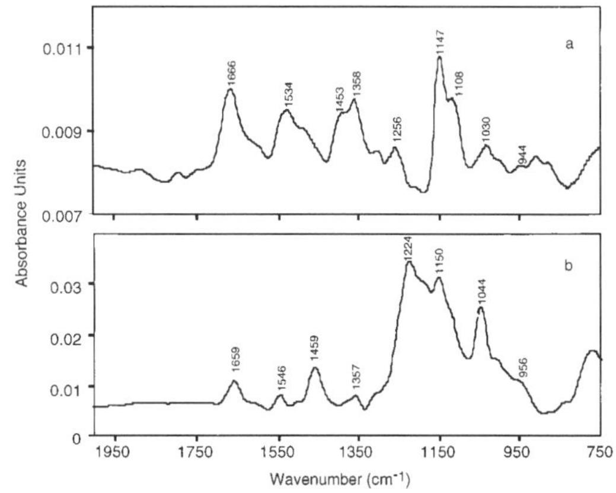

Figure 4. (a) FT-RAIRS spectrum of PLL-*g*-PEG chemisorbed as a monolayer on a natural-oxide-covered titanium surface. (b) FT-RAIRS spectrum of PLL-*g*-PEG physisorbed as multilayers on a natural-oxide-covered titanium surface.

(TiO_2, Nb_2O_5, and Ta_2O_5) are not suitable for RAIRS measurements because of their low reflectivity. Therefore, Si wafers coated with 100 nm of Ti metal instead of TiO_2 were used as substrates for RAIRS measurements. Ti metal surfaces always carry a native (passive) oxide film, whose IEP has been shown to be similar to that of "bulk" TiO_2 surfaces.[20] XPS analysis of the Ti-coated Si wafer indicates that there is an outermost ~5 nm thick native TiO_2 layer.[25] Therefore, the mechanism of PLL-*g*-PEG adsorption onto Ti metal surfaces is found to be very similar to that occurring on TiO_2 substrates.

As a basis to discuss the vibrational spectra of the PLL-*g*-PEG monolayer, IR spectra of pure PLL-*g*-PEG powder (KBr pellets) were measured first (Figure 3). The vibrational band frequencies and assignments referred to pure poly(ethylene glycol) and poly-L-lysine[26,27] are listed in Table 1. The RAIRS spectrum of the PLL-*g*-PEG monolayer on Ti is shown in Figure 4a and band assignments are summarized in Table 1. The bands at 2951 and 2887 cm^{-1} are ascribed to the asymmetric and symmetric CH_2 stretching band of the PEG chains, and the band at 2864 cm^{-1} is due to the symmetric CH_2 stretching band of the PLL units. The C–O–C stretching vibration gives very strong absorption bands at 1108 and 1147 cm^{-1}. The band at 1453 cm^{-1} is dominated by the CH_2 scissoring mode of the ether methylene units in PEG chains. Bands at 1358,

(25) Sittig, C.; Textor, M.; Spencer, N. D.; Wieland, M.; Vallotton, P. H. *J. Mater. Sci.: Mater. Med.* **1999**, *10*, 35–46.
(26) Smith, E. L.; Alves, C. A.; Anderegg, J. W.; Porter, M. D.; Siperko, L. M. *Langmuir* **1992**, *8*, 2707–2714.
(27) Duevel, R. V.; Corn, R. M. *Anal. Chem.* **1992**, *64*, 337–342.

Table 1. RAIRS Spectral Mode Assignments for PLL-*g*-PEG Prepared as Powder/KBr Pellet, Chemisorbed, and Physisorbed on Natural-Oxide-Covered Titanium Surface[a]

mode assignment	PLL-*g*-PEG (KBr)	chemisorbed PLL-*g*-PEG/Ti	physisorbed PLL-*g*-PEG/Ti
PLL amide NH stretching	3299		
PEG CH_3 asym stretching	2980[b]		
PEG CH_2 asym stretching	2949	2951	2941
PLL alkyl CH_2 asym stretching	2930[b]		
PEG CH_2 sym stretching	2886	2887	2887
PLL alkyl CH_2 sym stretching	2862	2864	2865
PEG CH_3 sym stretching	2821	2826	2823
PLL amide I (C=O stretching)	1654	1666	1659
PLL amide II (CNH stretch−bend)	1547	1534	1546
PEG CH_2 scissoring	1467, 1460	1453	1459
PEG CH_2 wagging	1343	1358	1357
PEG CH_2 twisting	1281, 1242	1256	1224
PEG C−O, C−C stretching	1149, 1108, 1061	1147, 1108, 1030	1150, 1102, 1044
PEG CH_2 rocking	963	944	956

[a] Values of the vibrational band frequency are in cm^{-1}. [b] Not clearly visible in PLL-*g*-PEG, but reported in PEG or PLL standards (refs 25 and 26).

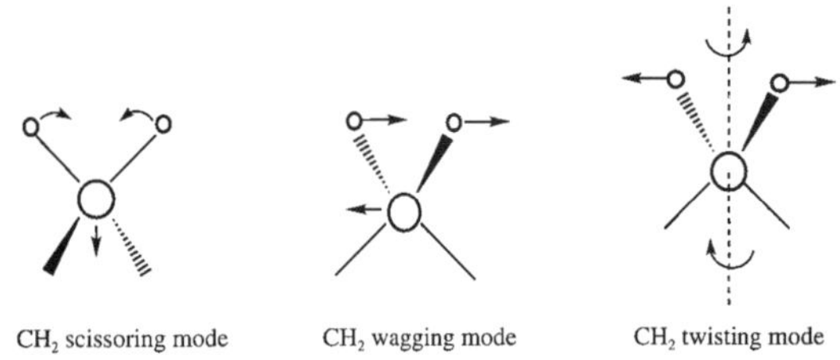

Figure 5. CH_2 vibrational modes. Small circles represent hydrogen, and large circles represent carbon. Wedge-shaped heavy lines represent the bond in front of the plane of the page, and wedge-shaped dashed lines represent the bond behind the plane of the page.

1256, and 944 cm^{-1} are associated with the ether CH_2 wagging, twisting, and rocking modes, respectively. In comparison to the IR spectra of PLL-*g*-PEG powder, these three band positions are somewhat shifted from the band positions at 1343, 1242, and 963 cm^{-1} typical for crystalline material, toward the band positions at 1352, 1249, and 945 cm^{-1} reported to be typical for PEG in amorphous environments.[28] The bands at 1666 and 1534 cm^{-1} are assigned to the amide I (C=O stretching) and amide II (CNH bend−stretch mode) of the PLL unit.

To investigate potential orientational effects of the chemisorbed PLL-*g*-PEG monolayer, a physisorbed (multilayer) PLL-*g*-PEG film on Ti was also measured by RAIRS (Figure 4b). These two spectra (4a,b) are qualitatively similar but show some specific relative intensity changes. In particular, both the CH_2 scissoring mode at 1453 cm^{-1} and the CH_2 twisting mode at 1256 cm^{-1} in spectrum 4a are weaker than the CH_2 wagging mode at 1358 cm^{-1}, whereas in spectrum 4b both the CH_2 scissoring mode (1459 cm^{-1}) and the CH_2 twisting mode (1224 cm^{-1}) are stronger than the CH_2 wagging mode (1357 cm^{-1}). This is likely to be due to different orientations of the PEG chains in the PLL-*g*-PEG layer obtained by chemisorption (monolayer) and physisorption (multilayer, likely to be randomly oriented), respectively. The scissoring, wagging, and twisting modes of the CH_2 group are shown in Figure 5.[29] Because only the component of the vibrational transition dipole moments perpendicular to the surface plane contributes to the absorption spectra, CH_2 chains parallel to the substrate surface will show strong scissoring and twisting bands, whereas CH_2 chains vertical to the surface will show a strong CH_2 wagging band. Therefore, we can deduce that the PEG chains in the chemisorbed PLL-*g*-PEG monolayer are on average preferentially oriented perpendicular to the surface in comparison with those in the physisorbed layer, which are probably randomly distributed. These observations are in agreement with the model of a chemisorbed PLL-*g*-PEG where the PLL chains are parallel to the surface and the PEG chains form a comblike structure extended perpendicular to the substrate surface. The majority of the PEG chains in a close-packed layer are assumed to be stretched out perpendicularly to the PLL backbone and located at the outermost surface.

3.2.2. XPS Analysis. *Model.* XPS measurements were performed at two different takeoff angles (75° and 15°) on TiO_2, Nb_2O_5, and Ta_2O_5 with both unmodified and PLL-*g*-PEG-modified surfaces. The quantitative results of the XPS study are summarized in Tables 2 and 3. Table 2a lists the normalized intensities (i.e., the measured peak areas divided by the corresponding sensitivity factors and normalized to 100% total intensity) of the elements on the different clean and modified surfaces, making the oversimplified assumption of a uniform distribution of the elements across the sampling depth of the XPS signals. More precise information on the PLL-*g*-PEG-modified oxide surfaces can be obtained, however, if one takes into account the layered structure found by RAIRS by using a three-layer model for quantitative analysis, which allows the simultaneous calculation of thickness and composition. This model was originally developed for thin oxide films[30] and has been successfully applied to self-assembled octadecyl phosphate monolayers on Ta_2O_5.[24]

The three layers are (1) the outermost surface layer, containing the PEG chains ($C1s_B$ and $O1s_B$ in Table 3) in addition to some hydrocarbon contamination, (2) an intermediate layer composed of the polylysine backbone ($C1s_A$, $C1s_C$, N1s, and $O1s_B$), and (3) the bulk oxide (metal Ti2p, Nb3d, Ta4f, and oxygen $O1s_A$). Integrated intensities of the individual peaks determined from curve fitting, corrected as reported in ref 30, were used in the calculations; the density of the intermediate layer was assumed to be 2 g cm^{-3}, the density of the outer layer was 1 g cm^{-3}, and the densities of the oxides were 4.26, 4.47, and 8.2 g

(28) Harder, P.; Grunze, M.; Dahint, R.; Whitesides, G. M.; Laibinis, P. E. *J. Phys. Chem. B* **1998**, *102*, 426−436.
(29) Colthup, N. B.; Daly, L. H.; Wiberley, S. E. *Introduction to Infrared and Raman Spectroscopy*, 3rd ed.; Academic Press: San Diego, CA, 1990.
(30) Rossi, A.; Elsener, B. *Surf. Interface Anal.* **1992**, *18*, 495. Elsener, B.; Rossi, A. *Electrochim. Acta* **1992**, *37*, 2269.

Table 2

(a) Normalized Intensities of Unmodified and PLL(20)-*g*[3.5]-PEG(2)-Modified Metal Oxide Surfaces Determined by XPS Analysis Using PHI Sensitivity Factors[a]

surface	takeoff angle [deg]	normalized intensities C	O	metal (Ti, Nb, Ta)	N
clean TiO_2	15	32.2	54.3	13.5	
	75	10.3	65.4	24.3	
clean Nb_2O_5	15	29.5	51.6	18.9	
	75	12.2	64.2	23.6	
clean Ta_2O_5	15	39.7	47.2	13.1	
	75	14.0	65.3	20.7	
PLL-*g*-PEG-modified TiO_2	15	57.3	37.2	3.3	2.2
	75	27.2	54.1	16.8	1.9
PLL-*g*-PEG-modified Nb_2O_5	15	56.6	36.7	4.2	2.5
	75	35.2	49.3	13.3	2.2
PLL-*g*-PEG-modified Ta_2O_5	15	57.5	36.8	3.7	2.0
	75	28.5	55.5	14.8	1.2

(b) Quantitative Analysis of the PLL-*g*-PEG-Modified Nb_2O_5 Surface According to the Three-Layer Model [24,30]

angle	thickness (nm) PEG	PLL	interface composition (PLL) (wt %) C1s	N1s	O1s	bulk composition (wt %) metal	O1s
15	0.9 ± 0.3	0.5 ± 0.2	63	20	17	73	27
75	1.4 ± 0.3	0.8 ± 0.2	59	21	20	72	28
theory			59	23	18	70	30

[a] Metal oxides: TiO_2, Nb_2O_5, and Ta_2O_5.

Table 3. Angle-Resolved XPS Quantitative Analysis of PLL(20)-*g*[3.5]-PEG(2) Monolayer on TiO_2, Nb_2O_5, and Ta_2O_5

surface	takeoff angle [deg]	binding energies [eV][a] (relative peak area [%]) $C1s_A$ C–C, C–H	$C1s_B$ C–O	$C1s_C$ OC(=O)	$O1s_A$ MO_x [b]	$O1s_B$ OH O–C=O*	$O1s_C$ C–O	atomic ratio C(PEG)[c] /C(C–C)	O(PEG) /N(PLL)	C(PEG)[c] /O(PEG)	$O(MO_x)/M$[b] (Ti, Nb, Ta)
clean TiO_2	75	285.0 (65.5)	286.6 (23.1)	288.9 (11.4)	530.2 (80.2)	531.7 (15.3)	532.8 (4.5)				2.1
clean Nb_2O_5	75	285.0 (70.3)	286.6 (22.3)	289.0 (7.4)	530.6 (88.3)	531.7 (6.6)	532.9 (5.1)				2.4
clean Ta_2O_5	75	285.0 (74.2)	286.7 (17.3)	289.0 (8.5)	530.6 (81.8)	531.7 (13.7)	532.8 (4.5)				2.6

surface	takeoff angle [deg]	$C1s_A$ C–C, C–H	$C1s_B$ C–O C–N	$C1s_C$ NHC(=O)	$O1s_A$ MO_x [b]	$O1s_B$ NHC(=O)	$O1s_C$ C–O	C(PEG)[c] /C(C–C)	O(PEG) /N(PLL)	C(PEG)[c] /O(PEG)	$O(MO_x)/M$[b] (Ti, Nb, Ta)
PLL-*g*-PEG-mod TiO_2	15	285.0 (12.1)	286.7 (81.4)	287.9 (6.5)	529.9 (17.9)	531.3 (13.0)	532.9 (69.1)	6.2	11.1	1.9	2.1
	75	285.0 (15.0)	286.7 (75.3)	287.9 (9.7)	529.9 (63.8)	531.3 (13.9)	532.9 (22.3)	4.6	5.7	1.9	2.1
PLL-*g*-PEG-mod Nb_2O_5	15	285.0 (15.3)	286.6 (79.5)	288.0 (5.2)	530.0 (24.3)	531.5 (7.0)	532.9 (68.7)	4.8	9.7	1.9	2.5
	75	285.0 (17.8)	286.6 (75.4)	288.0 (6.8)	530.0 (64.1)	531.5 (5.3)	532.9 (30.6)	3.9	6.2	2.0	2.4
PLL-*g*-PEG-mod Ta_2O_5	15	285.0 (12.8)	286.5 (82.6)	288.0 (4.6)	530.0 (18.0)	531.4 (8.9)	532.8 (73.1)	6.0	12.9	1.8	2.5
	75	285.0 (16.5)	286.5 (74.0)	288.0 (9.5)	530.0 (64.8)	531.4 (13.1)	532.8 (22.1)	4.2	9.2	1.9	2.5
theoretical PLL-*g*-PEG		285.0 (10.8)	286.5 (78.3) 286.2 (6.6)	288.1 (4.3)		531.4 (9.9)	532.8 (90.1)	7.3	5.9	2.0	(2.0, 2.5, 2.5)

[a] using C(1s) of hydrocarbon peak at 285.0 eV as calibration. [b] MO_x represents the metal oxides: TiO_2, Nb_2O_5, Ta_2O_5. [c] C(PEG) corresponds to the fraction of the $C1s_B$ peak due to PEG (C–O bands), i.e., 92.2% of the total $C1s_B$ peak as calculated from the stoichiometry of the polymer.

cm^{-3} for TiO_2, Nb_2O_5, and Ta_2O_5, respectively. The influence of the contamination layer was taken into account by assuming that it comprises 25% of the $C1s_A$ (C–C bonds) and 25% of the $O1s_B$ (O–H bonds) signals.

The results obtained for the Nb_2O_5-modified surfaces (Table 2b) show that for both takeoff angles the thickness of the outermost PEG layer is 1.1 ± 0.3 nm and that the interface layer (PLL) has a thickness of ca. 0.6 ± 0.2 nm, independent of the modified bulk oxide surface. This is fully consistent with monolayer coverage of PEG-*g*-PLL and with a surface coverage of 148 ng cm^{-2} measured by OWLS, (assuming the density of PEG-*g*-PLL to be intermediate between those of PEG and PLL). The composition of the bulk oxide beneath the modified surface layer of Nb_2O_5 corresponds to that expected from the stoichiometry. The composition of the interface layer corresponds, for all systems investigated, to that expected from the stoichiometry of the PLL part of the adsorbed molecules (59% C, 23% N, 18% O). In addition, the ratio of carbon to oxygen (corrected for the cross sections) for

PLL-g-PEG Layers on Metal Oxide Surfaces *Langmuir, Vol. 17, No. 2, 2001* 495

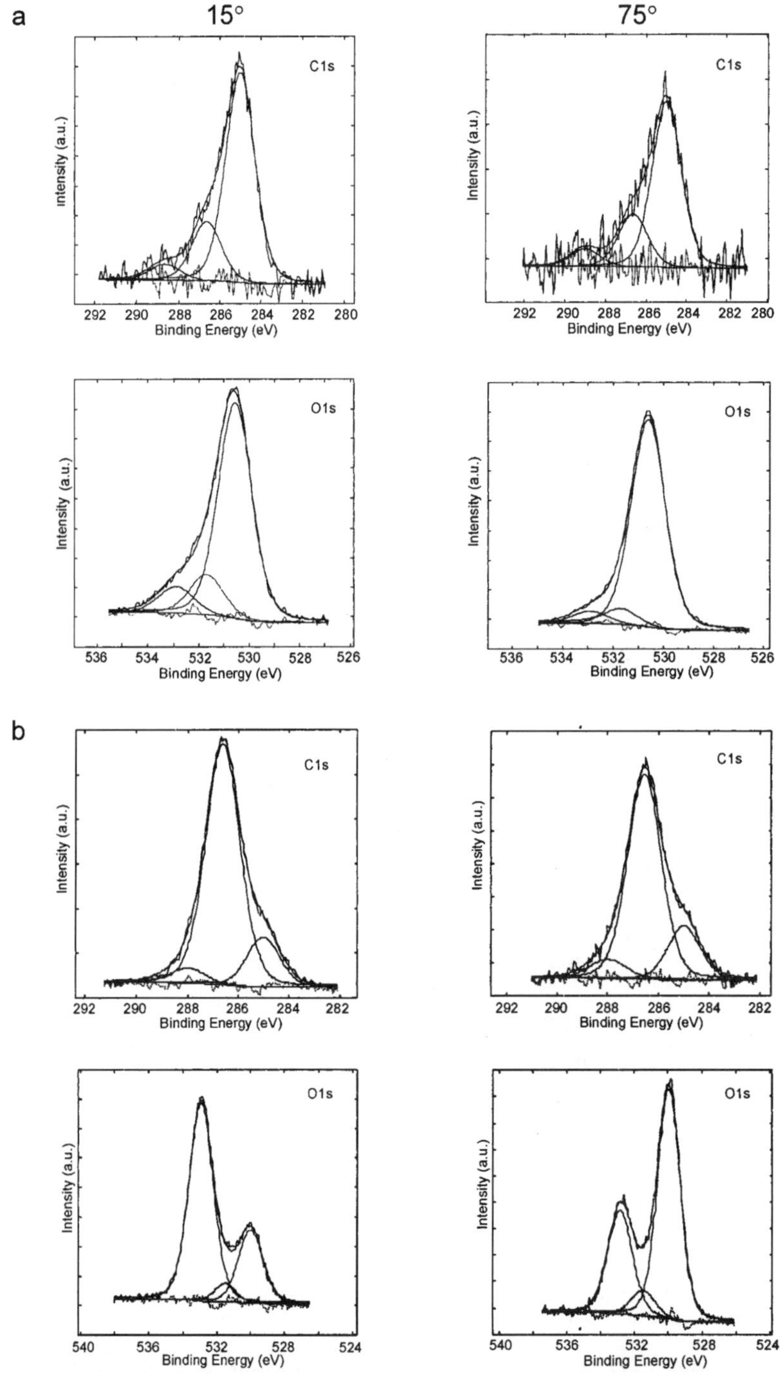

Figure 6. Detailed XPS spectra in the C(1s) and O(1s) region at the two different takeoff angles (relative to the surface plane) of 15° and 75°: (a) clean Nb_2O_5 surface and (b) clean Nb_2O_5 surface modified with a monolayer of chemisorbed PLL(20)-*g*[3.5]-PEG(2).

the PEG chains was found to be 1.95 ± 0.1, in very good agreement with the theoretical value of 2.00.

Chemical Inferences. Table 3 summarizes the experimental XPS binding energies of the deconvoluted detailed spectra from the different surfaces, together with the proposed assignments to chemical bonds/oxidation states based on observed chemical shifts. The most relevant detailed spectra, that is, the C1s and O(1s) spectra of the clean and PLL-*g*-PEG-modified surfaces, are shown in Figure 6 for the case of a Nb_2O_5 substrate. The detailed

C1s and O(1s) spectra for the corresponding TiO_2 and Ta_2O_5 surfaces are found to be very similar to those of Nb_2O_5 and are therefore not shown here.

Spectra of the clean surfaces of TiO_2, Nb_2O_5, and Ta_2O_5 show three elements: C, O, and the corresponding metal. The metal and most of the O emission intensity can be attributed to the metal oxide surfaces, whereas the C signal is due to unavoidable adventitious hydrocarbon contamination. After adsorption of PLL-*g*-PEG on all three types of clean oxide surfaces, the normalized carbon intensity significantly increases, as expected (Table 2a). The normalized overall oxygen intensity shows less variation between clean and modified surfaces, because both the organic PLL-*g*-PEG overlayer and the metal oxide substrate contain high concentrations of oxygen. In a comparison of the spectra taken at 15° and 75° takeoff angles, higher carbon intensities are always found in the case of the grazing exit angle (15°, analysis is more surface sensitive). This confirms that both the hydrocarbon contamination and the PLL-*g*-PEG adlayer are located on top of the metal oxide surface, because the sampling depth at the 15° takeoff angle is much lower than that at 75°.

In the case of the clean metal oxide surfaces (taking Nb_2O_5 as an example, Figure 6a), the C(1s) spectrum is dominated by a hydrocarbon emission peak (binding energy of 285.0 eV), whereas the O(1s) spectrum shows a contribution that is characteristic of the niobium pentoxide substrate (binding energy of 530.6 eV). There are two smaller, additional contributions at higher binding energies in both the C(1s) and O(1s) spectra, which are derived from oxygen-containing organic contaminants (C−O and OC=O) and hydroxide groups at the metal oxide surface. As expected, the additional O(1s) signal at higher binding energies is stronger in the 15° spectra compared to the 75° spectra, a typical observation for organic contaminants and hydroxides, which are expected to be located at the outermost surface. The quantitative ratio of O(Nb_2O_5)/Nb is 2.4, close to the stoichiometric value of 2.5 (Table 3). The same observation holds correspondingly for the other two metal oxides, Ta_2O_5 and TiO_2.

In the case of PLL-*g*-PEG-modified metal oxide surfaces (taking PLL-*g*-PEG-modified Nb_2O_5 as an example, Figure 6b), both carbon and oxygen signals are entirely different from those of the "clean" Nb_2O_5 spectra. The C(1s) region is now dominated by a peak at around 286.6 eV, typical for the C−O entity present in PEG, and a minor contribution to the same peak area (~7.8%, according to the stoichiometry of the PLL-*g*-PEG) originates from the carbon adjacent to the amino group in PLL (C−N bond, at E_B = 286.0 eV [$-C^*H_2-NH_2$], 286.1 eV [$-C^*H_2-NH_3^+$], and 286.3 eV [−NH−C*H−C=O]).[31] Two additional peaks are assigned to the methylene (285.0 eV) and amide (288.0 eV) groups in the PLL units. The O(1s) spectra are deconvoluted into three peaks: the peak with substantial intensity at the binding energy of 532.9 eV is attributed to the oxygen of the PEG chains, the peak at 531.5 eV is assigned to the oxygen of the amide group in PLL, and the oxygen from Nb_2O_5 is observed at the binding energy of 530.0 eV. The latter shows a shift of −0.6 eV relative to the unmodified Nb_2O_5 surface, and the same phenomenon is also observed in the case of TiO_2 and Ta_2O_5. This shift to lower binding energies of the O(1s) signal in metal oxide substrates is believed to reflect the buildup of additional negative charges at the interface upon adsorption of the positively charged PLL-*g*-PEG, for example, caused by deprotonation of surface hydroxides or partial hydrogen transfer from the hydroxide to the amino groups of the PLL.

Stoichiometry. The experimental intensity ratios of O(MO_x)/M (M = Ti, Nb, Ta) for all surfaces examined (Table 3), taking into account the corresponding sensitivity factors and the attenuation of the photoelectron signals due to the organic overlayer, are very close to the stoichiometric ratios (2.0, 2.5, 2.5). The C(PEG)/O(PEG) ratio is close to the expected value of 2.0 in all cases (Table 3). In the case of nitrogen, the simple use of sensitivity factors and the inherent assumption of a homogeneous overlayer breaks down, because of the nature of the nitrogen distribution within the organic layer. Here, the O(PEG)/N(PLL) ratio (Table 3) differs from the stoichiometric value. On the Nb_2O_5 surface, for example, the O(PEG)/N(PLL) ratio determined at the takeoff angle of 75° (6.2) is close to the theoretical value of 5.9 for the stoichiometric ratio of the PLL-*g*-PEG molecule. At the takeoff angle of 15°, this value increases (9.7), which is expected for a polymer configuration with the majority of the amino groups of the PLL backbone located close to the interface and the PEG chains extending further out of the surface.

The ratio of C(PEG)/C(C−C) in Table 3 represents the ratio of the amount of PEG to the amount of PLL and hydrocarbon contamination. At a 75° takeoff angle, the experimental ratio (3.9) is lower than the theoretical value for PLL-*g*-PEG (7.3), indicating that a small amount of hydrocarbon contamination probably still exists at the interface between Nb_2O_5 and the PLL-*g*-PEG monolayer. With the takeoff angle changed to 15°, this ratio (4.8) increases.

The XPS results support our view that the PLL backbone acts as an anchor to the metal oxide surface and that the PEG side chains are located further away from the interface.

3.2.3. ToF-SIMS Analysis. ToF-SIMS analysis was carried out on two samples of each of the three substrates (TiO_2, Nb_2O_5, and Ta_2O_5). One sample was cleaned followed by PLL-*g*-PEG adsorption, whereas the other sample was simply cleaned. Negative and positive ion ToF-SIMS was carried out on these 6 samples, totaling 12 spectra. All three substrates were equally representative in terms of observed species. Therefore, we concentrate upon the bare and PLL-*g*-PEG-modified Nb_2O_5 substrates and show both negative and positive secondary ion data. The most prominent peaks in the mass range of m/z = 0−1000 are summarized in Table 4 (positive and negative ion peaks) for the bare Nb_2O_5 and in Table 5 for the Nb_2O_5 modified by the PLL-*g*-PEG. Mass spectra at m/z = 43 for the clean and covered surfaces are shown in Figure 7.

Positive Ion ToF-SIMS: Clean Nb_2O_5. The positive ion spectrum of Nb_2O_5 exhibits peaks such as $(C_nH_{2n+1})^+$ and $(C_nH_{2n-1})^+$, typical of hydrocarbon contamination. Additional minor contaminants are observed, such as NH_4^+ and Na^+. It has been shown in our previous paper that the adsorption of PLL-*g*-PEG is not hindered by small amounts of surface contamination.[18] Nb^+ as well as $Nb_aO_bH_c^+$ are observed in the higher mass range (Table 4). Their stoichiometries often reflect, particularly in the high mass range, the preferred oxidation state +V for Nb, as has been found for other oxide systems.[24]

Positive Ion ToF-SIMS: PLL-g-PEG-Modified Nb_2O_5. In the positive ion spectrum, the PLL-*g*-PEG-covered Nb_2O_5 is dominated by peaks originating from the adsorbed polymeric molecule. Hydrocarbon fragment species are still detected and are likely to occur from PLL

(31) Beamson, G.; Briggs, D. *High-resolution XPS of organic polymers: the Scienta ESCA300 database*; John Wiley & Sons Ltd.: Chichester, U.K., 1992.

Table 4. List of the Most Prominent Positive and Negative Secondary Ions in the ToF-SIMS Spectra of Nb_2O_5, Cleaned Ultrasonically with 2-Propanol, Rinsed, and Dried Followed by Oxygen-Plasma Treatment

positive secondary ion mass	molecular species (charge 1+)	rel intens[a]	negative secondary ion mass	molecular species (charge 1−)	rel intens[a]
15.024	CH_3	0.17	13.008	CH	0.11
18.034	NH_4	0.34	15.995	O	1.00
27.023	C_2H_3	0.69	17.002	OH	0.60
29.024	C_2H_5	0.76	25.007	C_2H	0.12
31.019	CH_3O	0.08	26.003	CN	0.05
39.022	C_3H_3	0.47	31.988	O_2	0.07
41.039	C_3H_5	0.97	79.953	HPO_3	0.15
43.019	C_2H_3O	0.42	95.946	HPO_4	0.11
43.056	C_3H_7	0.66	96.954	H_2PO_4	0.10
55.056	C_4H_7	0.45	124.880	NbO_2	0.02
57.072	C_4H_9	0.34	125.891	NbO_2H	0.01
67.055	C_5H_7	0.11	140.875	NbO_3	0.20
69.072	C_5H_9	0.14	141.886	NbO_3H	0.04
77.039	C_6H_5	0.10	156.871	NbO_4	0.02
92.790	Nb	0.76	157.880	NbO_4H	0.04
108.862	NbO	1.00	158.890	NbO_4H_2	0.03
124.087	NbO_2	0.25	200.837	Nb_2CH_3	0.02
132.885	Cs	0.48	220.828	$Nb_2O_2H_3$	0.03
142.896	NbO_3H_2	0.06	265.756	Nb_2O_5	0.02
179.052	NbO_5H_6	0.04	282.763	Nb_2O_6H	0.05
233.937	Nb_2O_3	0.002	325.721	$Nb_3O_2CH_3$	0.002
249.951	Nb_2O_4	0.003	340.761	$Nb_3O_2C_2H_6$	0.002
266.763	Nb_2O_5H	0.019	406.646	Nb_3O_8	0.03
284.788	$Nb_2O_6H_3$	0.005	531.553	Nb_4O_{10}	0.003
374.903	Nb_3O_6	0.004	548.548	$Nb_4O_{11}H$	0.004
390.635	Nb_3O_7	0.007	672.438	Nb_5O_{13}	0.007
408.653	$Nb_3O_8H_2$	0.002			

[a] Fraction of peak with respect to the most intense peak set at 1.00.

fragmentation and some hydrocarbon contamination. The most intense peaks, however, are due to PEG-fragment species such as $C_2H_3O^+$ and $C_2H_5O^+$. A whole series of $C_aH_bO_c^+$ secondary ions is detected up to the C_6 species. The fragmentation series $C_aH_bO_c$ observed is typical for a PEG overlayer in SIMS.[32] Figure 7 shows the double peak at $m/z = 43$, which is indicative of two molecular ions: $C_2H_3O^+$, predominantly originating from PEG at $m/z = 43.018$, and $C_3H_7^+$, a typical hydrocarbon peak at $m/z = 43.055$. The $C_3H_7^+$ peak is still present with similar intensity after PLL-*g*-PEG adsorption, in contrast to the $C_2H_3O^+$, which shows a strong increase upon adsorption of PLL-*g*-PEG. Another indication for the presence of PLL-*g*-PEG is a dominant $C_5H_{10}N^+$ peak, identified in previous papers as a major lysine fragment.[33] The Nb^+ intensity drastically decreased compared to the bare sample. This is most likely due to two events. First of all, the niobium secondary ion ejection is presumably hindered by the PLL-*g*-PEG overlayer. Second, there is a high probability for the ejected niobium to recombine upon emission with fragments from the organic overlayer, in particular with PEG fragments. Some of these recombinant species are predominant in the high-mass range (m/z = 155−213, Table 5). The intensities of the $Nb_aO_bH_c^+$ ion species are notably smaller compared to the corresponding intensities found on the bare substrates, supporting the model of a "close-packed" PLL-*g*-PEG overlayer.

Negative Ion ToF-SIMS: Clean Nb_2O_5. The low-mass negative ion ToF-SIMS region is dominated by O^- and OH^-, typical for hydroxylated metal oxide surfaces. Some hydrocarbon peaks are observed, as well as a few contaminants such as HPO_3^- and $H_2PO_4^-$. In the higher mass range, $Nb_aO_bH_c^-$-type peaks are detected up to $m/z \approx 700$, again following the selection rule based on a preferred oxidation state +V of Nb. Furthermore, some

(32) Michel, R.; Luginbühl, R.; Graham, D. J.; Ratner, B. D. *Langmuir* **2000**, *16*, 6503.
(33) Mantus, D. S.; Ratner, B. D.; Carlson, B. A.; Moulder, J. F. *Anal. Chem.* **1993**, *65*, 1431−1438.

Table 5. List of the Most Prominent Positive and Negative Secondary Ions of Nb_2O_5, Surface Covered with Chemisorbed PLL-*g*-PEG

positive secondary ion mass	molecular species (charge 1+)	rel intens[a]	negative secondary ion mass	molecular species (charge 1−)	rel intens[a]
15.027	CH_3	0.08	13.008	CH	0.42
27.025	C_2H_3	0.12	15.995	O	1.00
29.041	C_2H_5	0.15	17.003	OH	0.79
31.021	CH_3O	0.21	25.0098	C_2H	0.31
41.041	C_3H_5	0.14	26.004	CN	0.25
43.019	C_2H_3O	0.36	41.999	CNO	0.26
45.034	C_2H_5O	1.00	43.019	C_2H_3O	0.23
59.051	C_3H_7O	0.34	44.997	CHO_2	0.25
73.028	$C_3H_5O_2$	0.39	58.005	$C_2H_2O_2$	0.11
84.083	$C_5H_{10}N$	0.19	61.031	$C_2H_5O_2$	0.06
87.044	$C_4H_7O_2$	0.10	79.955	HPO_3	0.05
91.094	C_7H_7	0.08	96.955	H_2PO_4	0.04
92.852	Nb	0.05	101.021	$C_4H_5O_3$	0.01
99.043	$C_5H_7O_2$	0.02	103.035	$C_4H_7O_3$	0.006
101.057	$C_5H_9O_2$	0.02	105.055	$C_4H_9O_3$	0.01
103.078	$C_5H_{11}O_2$	0.03	115.030	$C_5H_7O_3$	0.007
108.858	NbO	0.04	117.016	$C_5H_9O_3$	0.007
115.159	$C_6H_{11}O_2$	0.002	124.883	NbO_2	0.009
117.091	$C_6H_{13}O_2$	0.007	125.892	NbO_2H	0.01
124.896	NbO_2	0.01	126.900	NbO_2H_2	0.007
132.873	Cs	0.15	140.878	NbO_3	0.03
142.883	NbO_3H_2	0.008	142.895	NbO_3H_2	0.01
154.875	NbO_3CH_3	0.007	150.889	$NbOC_2H_2$	0.009
168.899	NbO_3CH_2	0.007	158.886	NbO_4H_2	0.03
182.908	$NbO_3C_2H_4$	0.003	167.887	$Nb_2O_2H_3$	0.01
212.924	$NbO_3C_3H_6$	0.004	184.917	$NbO_3C_2H_4$	0.01
			220.830	$Nb_2O_2H_3$	0.007
			265.762	Nb_2O_5	0.008
			282.763	Nb_2O_6H	0.03
			342.736	Nb_3O_4	0.004
			406.655	Nb_3O_8	0.02
			548.567	$Nb_4O_{11}H$	0.005
			672.481	Nb_5O_{13}	0.005

[a] Fraction of peak with respect to the most intense peak set at 1.00.

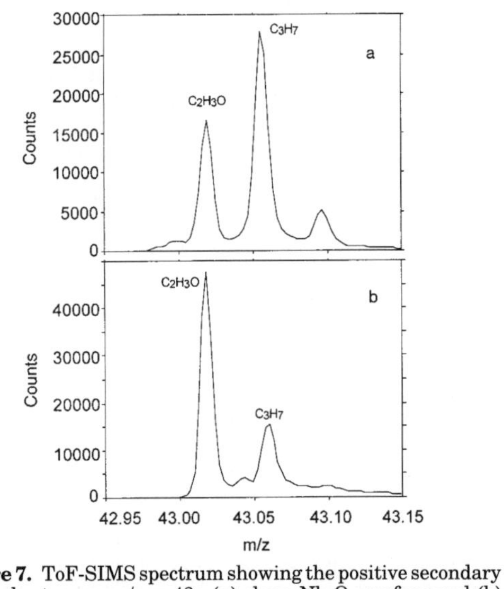

Figure 7. ToF-SIMS spectrum showing the positive secondary ion peak at mass $m/z = 43$: (a) clean Nb_2O_5 surface and (b) clean Nb_2O_5 surface modified with a monolayer of chemisorbed PLL-*g*-PEG.

Table 6. Adsorbed Mass of Human Serum and Human Fibrinogen on PLL(20)-*g*[3.5]-PEG(2)-Modified and Unmodified TiO_2 and Nb_2O_5 Surfaces, As Measured by the OWLS Technique[a]

	unmodified TiO_2	PLL(20)-*g*[3.5]-PEG(2)-modified TiO_2	unmodified Nb_2O_5	PLL(20)-*g*[3.5]-PEG(2)-modified Nb_2O_5
adsorbed mass of serum (ng/cm^2)[b]	378 ± 31	<2	471 ± 8	<2
adsorbed mass of fibrinogen (ng/cm^2)[b]	559 ± 15	22 ± 12	640 ± 13	13 ± 4

[a] Adsorption time 1 h, followed by 30 min rinsing with 10 mM HEPES solution (pH 7.4), T = 25 °C; human fibrinogen, 2.5 mg/mL in HEPES solution. [b] Average of three measurements and standard deviation.

recombinant peaks such as $Nb_aO_bC_cH_d^-$ are observed as well (Table 4).

Negative Ion ToF-SIMS: *PLL-g-PEG-modified Nb_2O_5*. The low-mass range of the negative ion spectrum reveals a large variety of peaks. O^- and OH^- are the dominating peaks again but are more likely to originate from the PEG chains rather than from the oxide substrate. CN^- and CNO^-, indicative of the PLL backbone, were not observed in the case of the bare substrate (Table 5). HPO_3^- and $H_2PO_4^-$ are still detected, suggesting that this type of contamination is not removed upon adsorption of PLL-*g*-PEG but rather covered and bridged by the assembled macromolecule. The $Nb_aO_bH_c^-$ peaks in the higher mass range have intensities reduced by nearly an order of magnitude, in comparison to the bare substrate. A large amount of recombinant species of the type $Nb_aO_bC_cH_d^-$ is observed as well.

In an attempt to obtain a semiquantitative estimate of relative surface coverage (RSC), we propose to use the ratio of overlayer/substrate related intensities as defined in eq 1. RSC is expected to be a steep function of surface coverage and may be of value for (future) correlations between PLL-PEG surface coverage and protein resistance.

$$\mathrm{RSC} = \frac{\sum \mathrm{C}_a\mathrm{H}_b\mathrm{O}_c}{\sum \mathrm{M}_m\mathrm{O}_n\mathrm{H}_o} \tag{1}$$

The corresponding RSC values for the three clean surfaces of Nb_2O_5 (0.225), Ta_2O_5 (0.81), and TiO_2 (0.77) compared to those of the PLL-*g*-PEG-modified substrates Nb_2O_5 (21.3), Ta_2O_5 (109.6), and TiO_2 (59.3) show an increase of 2 orders of magnitude for all three substrates. The change in the RSC ratio of 2 orders of magnitude for all three substrates is believed to provide good evidence for an overlayer that essentially covers the whole surface area.

3.3. Protein Adsorption Characteristics of Unmodified and PLL(20)-*g*[3.5]-PEG(2)-Modified Surfaces. Serum (containing albumin and IgG) and fibrinogen were used to measure the protein resistance of the PLL-*g*-PEG monolayer. They are the main components of blood plasma.

The experimental results of the in situ protein adsorption experiments using the OWLS technique are summarized in Table 6. For unmodified TiO_2 and Nb_2O_5 waveguides, exposure to the serum or fibrinogen solution produces a layer of adsorbed protein. After the subsequent buffer washing step, 378 ± 31 ng/cm^2 of serum and 559 ± 15 ng/cm^2 of fibrinogen irreversibly adsorbed on the TiO_2 surface, and 471 ± 8 and 640 ± 13 ng/cm^2 of serum and fibrinogen, respectively, adsorbed on the Nb_2O_5 surface. These values are in a range that is typical for a monolayer of adsorbed proteins.[20] However, waveguide surfaces modified with a layer of PLL-*g*-PEG show a drastic reduction in plasma-protein adsorption. For both the modified TiO_2 and Nb_2O_5 surfaces, the amount of serum that remains adsorbed to the surface after serum adsorption and subsequent buffer rinsing is lower than the detection limit of the OWLS technique, that is, <2 ng/cm^2. Under the same conditions, the adsorption of human fibrinogen (2.5 mg/mL, pH 7.4) is decreased by about 96% to 22 ng/cm^2 and by about 98% to 13 ng/cm^2 for the modified TiO_2 and Nb_2O_5 surfaces, respectively. The upper limits for the effective amount of adsorbed serum proteins and fibrinogen are lower than most of the values reported in the literature for other protein-resistant, nonfouling surfaces.[18]

4. Conclusions

The generation of surfaces and interfaces that are able to withstand protein adsorption is a major challenge in the use of blood-contacting materials both for medical implants (reduction of blood contact activation, platelet adhesion, and thrombus formation) and in the design of bioaffinity sensor surfaces with a reduced degree of nonspecific adsorption. Poly(ethylene glycol)-derived materials are generally considered to be particularly effective candidates for the fabrication of protein-resistant materials. Metallic biomaterials are typically covered by a protective, stable oxide film, and optical waveguide sensors often use highly transparent, high-refractive-index waveguiding oxide films. Because most of these oxide films (with the prominent exception of aluminum oxide) are negatively charged at or close to pH 7, they provide ideal substrates for the attachment of PEG-grafted poly(L-lysine) macromolecules through electrostatic interactions. The particular type of poly(L-lysine)-*g*-poly(ethylene glycol) investigated in this work {PLL(20 kDa)-*g*[3.5]-PEG(2 kDa)} was selected in view of its particularly effective protein resistance when adsorbed onto surfaces. The combined use of complementary surface-analytical and sensor techniques has provided an improved insight into the PLL-*g*-PEG polymeric adlayer: (a) the fast (minutes) formation of a monolayer with the cationic poly-L-lysine backbone in intimate contact with the negatively charged oxide surface; (b) the poly(ethylene glycol)-grafted side chains stretched out in a direction perpendicular to the surface; (c) an essentially close-packed, compressed PEG comblike structure with a very small proportion of "free" uncovered metal oxide surface. The amount of serum that remains adsorbed on the surface after buffer washing was found to be consistently below the detection method of the optical sensor system used, that is, below 1−2 ng cm^{-2} for all three modified oxide surfaces investigated (Nb_2O_5, Ta_2O_5, and TiO_2). Fibrinogen adsorption was reduced by almost 2 orders of magnitude at the PLL-*g*-PEG-treated surfaces in comparison to the bare oxide surfaces. This surface functionalization technology is believed to be of value for use in both the biomaterial and biosensor areas, because the chosen macromolecules are biocompatible and the application is fast and straightforward, even in the case of complex device shapes.

Acknowledgment. This work has been financially supported by the Swiss Federal Commission for Technology and Innovation (CTI). We also acknowledge support by the company Zeptosens of Witterswil, Switzerland.

LA000736+

220 *Langmuir* **2002,** *18,* 220–230

Biotin-Derivatized Poly(L-lysine)-*g*-poly(ethylene glycol): A Novel Polymeric Interface for Bioaffinity Sensing

Ning-Ping Huang, Janos Vörös, Susan M. De Paul, Marcus Textor,* and Nicholas D. Spencer

Laboratory for Surface Science and Technology, Department of Materials, Swiss Federal Institute of Technology, ETH-Zürich, CH-8952 Schlieren, Switzerland

Received June 18, 2001. In Final Form: October 4, 2001

A novel biosensor interface exploiting the spontaneous surface assembly of a polycationic, PEG-grafted, biotinylated copolymer was developed and tested on optical waveguide chips in a model immunoassay based on sequential immobilization of (strept)avidin and biotinylated goat antirabbit immunoglobulin (αRIgG–biotin) as a capture molecule to sense the rabbit immunoglobulin (RIgG) target molecule. Optical waveguide lightmode spectroscopy with niobium oxide waveguiding layers was used to monitor quantitatively and in situ the spontaneous adsorption of the (biotinylated) copolymer onto the waveguide surface, the resistance of the resulting adlayer to nonspecific protein adsorption, and the mass uptakes in each step of the model immunoassay. Poly(L-lysine)-*g*-poly(ethylene oxide) (PLL-*g*-PEG) is a polycationic copolymer that adsorbs spontaneously from aqueous solutions onto negatively charged surfaces via electrostatic interactions. It forms monolayers with densely packed PEG chains. PLL-*g*-PEG graft copolymers carrying terminal biotin groups on 0, 20, 30, or 50% of the PEG chains were synthesized and assembled onto the surface of niobium oxide (negatively charged at neutral pH). The surface concentration of biotin was tailored by adjusting the biotin grafting ratio in the polymeric molecule or by assembling mixed [PLL-*g*-PEG/PEGbiotin + PLL-*g*-PEG] adlayers from the corresponding mixed solutions. These biotinylated surfaces are shown to be highly resistant to nonspecific adsorption from serum while still allowing for the specific surface binding of the linkage proteins: streptavidin, avidin, or neutral avidin. The amount of immobilized linkage protein is shown to be closely related to the biotin surface concentration. The subsequent adsorption behavior of αRIgG–biotin and RIgG, however, depends in a more complex manner on each individual surface modification step and is discussed in the light of specific and nonspecific interactions, as well as of orientational and steric repulsion effects within the adlayers. In terms of the sensing signal-to-background ratio, the [PLL-*g*-PEG/PEGbiotin//NeutrAvidin//αRIgG–biotin] architecture demonstrated particularly promising performance as an interface architecture for bioaffinity sensing of proteins.

1. Introduction

Bioaffinity sensors are important tools in a variety of fields including immunoassays, toxicology analysis, forensics, drug screening, gene expression analysis, gene identification, agrodiagnostics, and pharmacogenetics.[1–3] To be quantitative and efficient, such sensors need to be rapid, specific, reproducible, and highly sensitive. This, in turn, necessitates controlled and optimized surface and interfacial chemistry.

Oxides of transition metals, such as niobium, tantalum, or titanium, are suitable as waveguiding layers on optical (grating coupler) waveguides due to their high refractive index and transparency.[4] Therefore, oxide-coated chips are used as sensing devices in technology based on the interaction of a laser-induced evanescent field with analyte constituents adsorbed on or immobilized at the waveguide/fluid interface.[5]

It has previously been shown that coating metal oxide surfaces with PEG-containing graft copolymers,[6–10] such as poly(L-lysine)-*g*-poly(ethylene glycol) (PLL-*g*-PEG), provides an attractive option for producing stable surfaces that are protein-resistant. At neutral pH, PLL-*g*-PEG (pK_a ~ 10) is positively charged and thus adsorbs readily (primarily via the electrostatic interaction) onto negatively charged metal oxide surfaces, such as Ta_2O_5, Nb_2O_5, TiO_2, and SiO_2.[9,10] By a proper choice of the grafting ratio of lysine units to poly(ethylene glycol) chains, excellent resistance to protein adsorption from human blood serum can be obtained.

Clearly, if some of the PEG chains in PLL-*g*-PEG could be end-functionalized such that they reacted selectively with a target molecule, a PLL-*g*-PEG-coated surface could be straightforwardly used as the basis for a specific biosensor. An obvious first choice for such a functionalization is the biotin group since biotinylated PEG is commercially available and one can immediately make use of the well-developed biotin–(strept)avidin technology to create model immunoassays.[3]

The interaction between the vitamin biotin and the protein avidin or streptavidin is one of the strongest

* To whom correspondence should be addressed. E-mail: textor@surface.mat.ethz.ch. Tel: +41 1 632 64 51. Fax: +41 1 633 10 48.

(1) Singh, P.; Sharma, B. P.; Tyle, P. *Diagnostics in the year 2000: antibody, biosensor, and nucleic acid technologies*; Van Nostrand Reinhold: New York, 1993.
(2) Blum, L. J.; Coulet, P. R. *Biosensor principles and applications*; Marcel Dekker: New York, 1991.
(3) Hermanson, G. T. *Bioconjugate techniques*; Academic Press: San Diego, 1996.
(4) Broer, M. M.; Sigel, G. H.; Kersten, R. T.; Kawazoe, H. *Optical Waveguide Materials*; Materials Research Society: Pittsburgh, PA, 1992.
(5) Duveneck, G. L.; Pawlak, M.; Neuschaefer, D.; Budach, W.; Ehrat, M. *Proc. Biomed. Syst. Technol.* **1996**, *2928*, 98–109.
(6) Claesson, P. M.; Blomberg, E.; Paulson, O.; Malmsten, M. *Colloids Surf., A* **1996**, *112*, 131–139.
(7) Malmsten, M.; Emoto, K.; Van Alstine, J. M. *J. Colloid Interface Sci.* **1998**, *202*, 507–517.
(8) Brink, C.; Osterberg, E.; Holmberg, K.; Tiberg, F. *Colloids Surf.* **1992**, *66*, 149–156.
(9) Kenausis, G. L.; Vörös, J.; Elbert, D. L.; Huang, N. P.; Hofer, R.; Ruiz, L.; Textor, M.; Hubbell, J. A.; Spencer, N. D. *J. Phys. Chem. B* **2000**, *104*, 3298–3309.
(10) Huang, N. P.; Michel, R.; Vörös, J.; Textor, M.; Hofer, R.; Rossi, A.; Elbert, D. L.; Hubbell, J. A.; Spencer, N. D. *Langmuir* **2001**, *17*, 489–498.

Published on Web 12/05/2001

noncovalent affinity interactions in nature ($K_{aff} = 10^{13-15}$ M^{-1} in solution), and as such it has been used for a wide range of applications including immunoassays, cytochemistry, protein purification, and diagnostics.[11] The strong interaction is due to the shape-specificity of the biotin-binding pocket, which allows the formation of multiple hydrogen bonds and van der Waals interactions.[12,13] Such binding is achieved without significant perturbation of the tertiary or quaternary structures of the protein and is stable over a wide range of pH values and temperatures.[12,14] Both avidin and streptavidin are homo-tetramers with four biotin-binding sites. The proteins show 222 symmetry, with pairs of biotin-binding sites on opposite sides of the protein, making them favorable for use as intermediate building blocks in immunoassays since two biotinylated ligands can be attached on each side.[12] One major difference between the proteins is that avidin carries a positive charge at neutral pH[15,16] whereas streptavidin is nearly neutral.[16,17] However, a neutral form of avidin can be produced by removing the carbohydrate groups.[18]

In this paper, we demonstrate that biotinylated PLL-g-PEG is a promising platform for immunoassays. By means of optical waveguide lightmode spectroscopy (OWLS),[19] streptavidin and avidin are shown to bind specifically to the biotin-functionalized PEG, while the resistance of the remaining PEG chains to protein adsorption yields a high specific binding to nonspecific binding ratio. Subsequent binding of biotinylated goat antirabbit immunoglobulin (αRIgG−biotin) to (strept)-avidin as a capture molecule allows the system to be used as an immunoassay for the target molecule, rabbit immunoglobulin (RIgG). The various components of this model immunoassay are shown to retain their biological activity, and the effects of protein charge and the ionic strength of the buffer are explored.

2. Materials and Methods

2.1. Materials. Poly(L-lysine) hydrobromide (mol wt ~ 20 kDa), streptavidin from *Streptomyces avidinii* (mol wt ~ 60 kDa), avidin from egg white (mol wt ~ 66 kDa), biotinylated goat antirabbit immunoglobulin (αRIgG−biotin, mol wt ~ 150 kDa), and biotinylated bovine serum albumin (BSA−biotin, mol wt ~ 66 kDa) were purchased from Sigma-Aldrich, Buchs, Switzerland. The *N*-hydroxysuccinimidyl ester of methoxy-poly(ethylene glycol) propionic acid (MeO−PEG−SPA, mol wt ~ 2 kDa) and the α-biotin-ω-*N*-hydroxysuccinimidyl ester of poly(ethylene glycol)-carbonate (biotin−PEG−CO_2−NHS, mol wt ~ 3.4 kDa) were obtained from Shearwater Polymers, Inc., Huntsville, AL. NeutrAvidin (mol wt ~ 60 kDa) was purchased from Molecular Probes, Leiden, The Netherlands. Rabbit immunoglobulin (antihuman albumin) (RIgG, mol wt ~ 150 kDa) and rabbit antibovine serum albumin (αBSA, mol wt ~ 150 kDa) were obtained from DAKO, Glostrup, Denmark. All antibody reagents used were polyclonal products. Control serum N (human) came from F. Hoffmann-La Roche AG, Basel, Switzerland. It consists of human serum components along with enzymes and other additives. The concentrations of the proteins are approximately equal to those found in clinically normal human serum. 4-(2-Hydroxyethyl)piperazine-1-ethane-sulfonic acid (HEPES) and other chemicals used for making buffers were purchased from Fluka, Buchs, Switzerland.

All aqueous solutions were formed using ultrapure water (18 MΩ cm) obtained from an EasyPure reverse osmosis system (Barnstead Thermolyne, Dubuque, IA).

2.2. Substrates. OWLS[19] experiments were performed using ASI2400μV sensor chips (Microvacuum Ltd., Budapest, Hungary), on which an additional 12-nm-thick niobium oxide (Nb_2O_5) layer was sputter-coated in a Leybold dc-magnetron Z600 sputtering unit at the Paul Scherrer Institute, Villigen, Switzerland. Before their initial use, the Nb_2O_5 waveguide chips were sonicated in 0.1 M HCl for 10 min, extensively rinsed with ultrapure water, dried in a nitrogen stream, and treated for 2 min in oxygen plasma in a Plasma Cleaner/Sterilizer PDC-32G (Harrick, Ossining, NY). After use in an experiment, sensor chips were regenerated by washing with "Cleaner" (a detergent from F. Hoffmann-La Roche, Basel, Switzerland) for 10 min and repeating the cleaning protocol described above.

2.3. Synthesis of PLL-*g*-PEG and Derivatives. The synthesis of PLL-g-PEG was first described by Sawhney and Hubbell.[20] The procedure used here is based on the protocol developed by Elbert and Hubbell[21] and described recently by our group.[9,10] All mixing and dialysis steps were carried out in a laminar flow box, and the lyophilized polymers were stored in a freezer at −25 °C until just before their use in an experiment.

On the basis of previous work, PLL-g-PEG coatings were found to reduce nonspecific protein adsorption from serum once the interchain spacing between PEG groups became smaller than the radius of gyration.[9] Thus, for 2 kDa PEG side chains, grafting ratios of lysine units/PEG side chains of 2.0 and 3.5 were found to resist adsorption of proteins more effectively than a grafting ratio of 5.0.[22] However, less of the polymer with the 2.0 grafting ratio adsorbed to the surface, most likely due to one of two reasons: (a) the strength of the electrostatic interaction between the PLL backbone and the negatively charged surface might have decreased due to the smaller number of underivatized, positively charged NH_3^+ groups available for anchoring; (b) steric repulsion between the PEG combs within the tightly packed adlayer could have resulted, for the same surface density of PEG chains, in a lower number of molecules adsorbed per unit area.

We therefore attempted to synthesize all biotinylated derivatives with a compromise value of the grafting ratio equal to 3.5, and the percentage of PEG chains that were biotinylated was then varied.

Figure 1 shows a schematic drawing of the synthesis of a biotinylated PLL-g-PEG derivative. *N*-Hydroxysuccinimidyl esters of both biotinylated and non-biotinylated poly(ethylene glycol) were reacted with poly(L-lysine) in stoichiometric ratios to produce the desired product. Details of the synthesis are provided in subsections 2.3.1 and 2.3.2. The longer chain length of the biotinylated PEG as compared to the methoxyPEG was hypothesized to have the advantage of improving the accessibility of the biotin to the (strept)avidin.

A useful nomenclature for describing the various PLL-g-PEG derivatives includes the molecular weights of the reagents, the grafting ratio, and the percentage of PEG chains that are biotinylated.[9,10] Thus, PLL(20)-g[3.5]-PEG(2)/PEGbiotin(3.4)30% describes a polymer formed from a 20 kDa poly(L-lysine) backbone with side chains of poly(ethylene glycol) that have a molecular weight of 2 kDa as well as side chains of biotinylated poly(ethylene glycol) that have a molecular weight of 3.4 kDa. The grafting ratio of 3.5 indicates that 2 out of every 7 lysine units are attached to PEG chains, and the designation 30% indicates that 3 out of 10 of those PEG chains are biotinylated. Since all of the polymers used in this paper were synthesized from similar precursors, simplified abbreviations such as PLL-g-PEG/PEGbiotin30% or PPB30 will also be used throughout the text.

(11) Wilchek, M.; Bayer, E. A. *Biomol. Eng.* **1999**, *16*, 1−4.
(12) Rosano, C.; Arioso, P.; Bolognesi, M. *Biomol. Eng.* **1999**, *16*, 5−12.
(13) Freitag, S.; Trong, I. L.; Klumb, L. A.; Chu, V.; Chilkoti, A.; Stayton, P. S.; Stenkamp, R. E. *Biomol. Eng.* **1999**, *16*, 13−19.
(14) González, M.; Argaraña, C.; Fidelio, G. D. *Biomol. Eng.* **1999**, *16*, 67−72.
(15) Woolley, D. W.; Longsworth, L. G. *J. Biol. Chem.* **1942**, *142*, 285−290.
(16) Green, N. M. *Adv. Protein Chem.* **1975**, *29*, 85−133.
(17) Diamandis, E. P.; Christopoulos, T. K. *Clin. Chem.* **1991**, *37*, 625−636.
(18) Hiller, Y.; Gershoni, J. M.; Bayer, E. A.; Wilchek, M. *Biochem. J.* **1987**, *248*, 167−171.
(19) Vörös, J.; Ramsden, J. J.; Csucs, G.; Szendrö, I.; De Paul, S. M.; Textor, M.; Spencer, N. D. *Biomaterials*, submitted.

(20) Sawhney, A. S.; Hubbell, J. A. *Biomaterials* **1992**, *13*, 863−870.
(21) Elbert, D. L.; Hubbell, J. A. *J. Biomed. Mater. Res.* **1998**, *42*, 55−65.
(22) Hofer, R. Surface Modification for Optical Biosensor Applications. Ph.D. Dissertation, ETH Zürich, 2000.

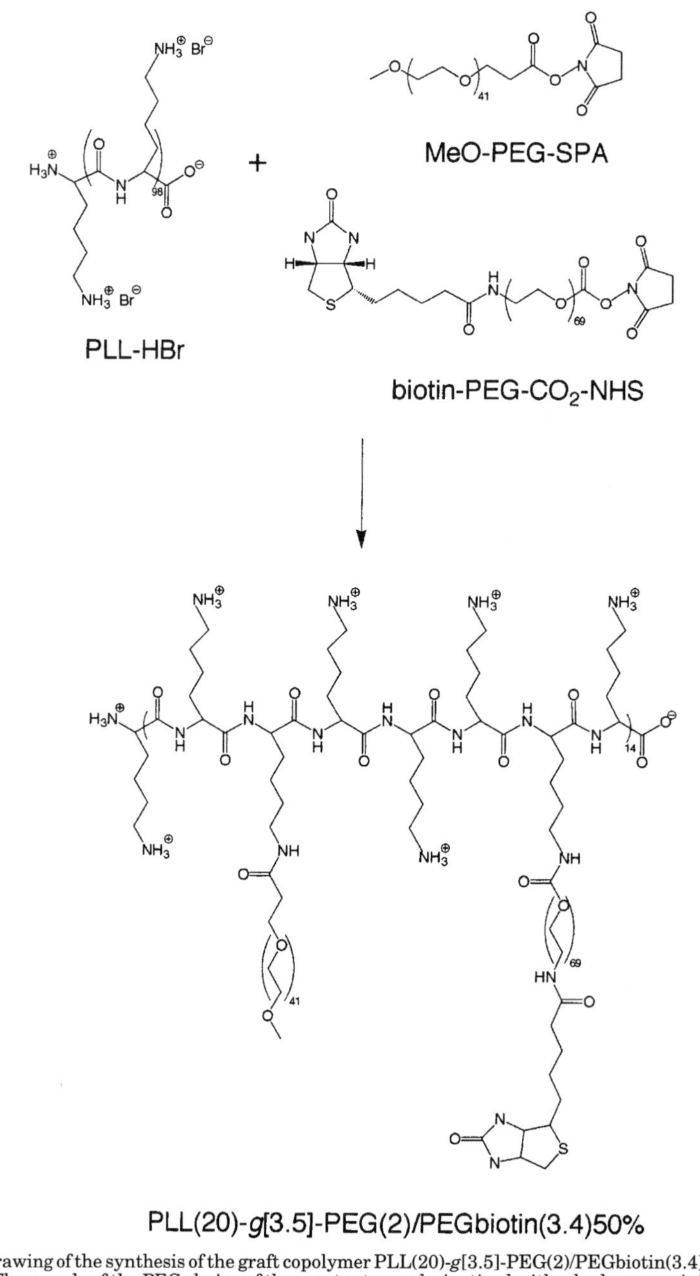

Figure 1. Schematic drawing of the synthesis of the graft copolymer PLL(20)-*g*[3.5]-PEG(2)/PEGbiotin(3.4)50% from commercially available components. The ω-ends of the PEG chains of the reactants are derivatized with a hydroxysuccinimidyl ester for reaction with the primary amine side chains of the lysine units. Details of the synthetic procedure are described in the text.

2.3.1. Synthesis of PLL(20)-*g*[3.5]-PEG(2). Poly-L-lysine hydrobromide (PLL–HBr) was dissolved in 25 mL of 50 mM sodium tetraborate buffer (STBB, pH 8.5) per gram of PLL–HBr. The resulting solution was stirred vigorously for approximately 30 min and subsequently filtered through a 0.22 μm Durapore membrane (sterile Millex GV, Sigma-Aldrich, Buchs, Switzerland) into a sterile culture tube. The appropriate stoichiometric amount of MeO–PEG–SPA powder was then slowly added to the solution while it was continuously stirred. After 6 more hours of vigorous stirring at room temperature, the solution was transferred to a dialysis tube (Spectr/Por dialysis tubing, molecular weight cutoff of 6–8 kDa, Spectrum Laboratories, Inc., Rancho Dominguez, CA). Dialysis was carried out for 24 h in 1 L of 10 mM phosphate-buffered saline (PBS, pH 7.0), followed by 24 additional hours of dialysis in 1 L of deionized water. The product was then freeze-dried for 48 h at a temperature of −50 °C and a pressure of 0.2 mbar.

2.3.2. Synthesis of Biotin-Derivatized PLL-*g*-PEG. The biotin-derivatized PLL-*g*-PEG was synthesized in a manner similar to that described in the previous section. A stoichiometric amount of biotin–PEG–CO_2–NHS powder was slowly added to the filtered PLL–HBr solution and stirred vigorously for 1 h. Addition of the corresponding stoichiometric amount of MeO–PEG–SPA powder then followed, and the resulting solution was

Table 1. Grafting Ratios and Percentages of Biotinylation for the Polymers Used in This Paper As Determined by Analysis of ^{1}H NMR Spectra[a]

polymer abbreviation	grafting ratio	% of PEG chains biotinylated
PLL-*g*-PEG	4.4 ± 1.5	0
PPB20	3.2 ± 1.0	23 ± 7
PPB30	3.9 ± 0.5	31 ± 5
PPB50	2.8 ± 0.5	51 ± 5

[a] The size of the error bars is significant due to partial peak overlap and imperfect baselines. In cases where several batches of polymer were synthesized, slightly larger error bars reflect the variability from batch to batch.

stirred vigorously for 5 additional hours at room temperature. Dialysis and recovery were the same as described in section 2.3.1.

2.3.3. Determination of Grafting Ratio and Biotin Percentage. Proton NMR was used to estimate the grafting ratio and percentage of biotinylated PEG in the derivatives. The lyophilized polymers were dissolved in D_2O, and the spectra were recorded on a 300 MHz spectrometer. The spectra were assigned based on comparison with previously published spectra of PLL-*g*-PEG[20] and of biotin[23] as well as with spectra of the individual reagents. Note that the PEG side chains described in an earlier publication[20] contained a carbamate ester linkage, which was not present in the methoxyPEG side chains in our polymer but was present in the biotinylated PEG chains (see Figure 1). This is due to differences in the structure of the reagents used and leads to differences in assignments of the peaks in the NMR spectra.

The ^{1}H NMR chemical shifts (in D_2O) were assigned as follows: 1.2–1.5 and 1.5–1.8 ppm (m, $-\mathbf{CH_2}-$, β–δ carbons of the lysine side chains and the three methylene groups in biotin that are nearest to the thiophene ring), 2.22 ppm (t, $-CH_2CH_2$-$\mathbf{CH_2}C(O)NH-$, biotin), 2.47 ppm (t, $-OCH_2\mathbf{CH_2}C(O)NH-$, methoxyPEG linked to lysine), 2.73 ppm (incompletely resolved q, $-CH_a\mathbf{H_b}S-$, biotin), 2.95 ppm (m, $-\mathbf{CH_a}H_bS-$ in biotin and $-\mathbf{CH_2}NH_3^+$ in ungrafted lysine chains), 3.05 and 3.12 ppm (incompletely resolved multiplets, $-\mathbf{CH_2}NHC(O)OCH_2-$ from biotinylated PEG linked to lysine and $-\mathbf{CH_2}NHC(O)CH_2-$ from methoxyPEG linked to lysine and from biotinylated PEG itself), 3.28 ppm (m, $-\mathbf{CH}S-$, biotin), 3.33 ppm (s, $\mathbf{CH_3}O-$, methoxy-PEG), 3.65 ppm (m, $-\mathbf{CH_2CH_2}O-$, ethylene glycol), 4.15 ppm (m, $-\mathbf{CH_2}OC(O)NH-$, biotinylated PEG linked to lysine), 4.25 ppm (m, $-NHC(O)\mathbf{CH}-$, lysine backbone), 4.37 and 4.56 ppm (m, $-\mathbf{CH}NHC(O)-$, biotin).

To estimate the grafting ratio and percent of biotinylation, it is necessary to use the integrated intensities of peaks which can be unambiguously assigned and have little or no overlap with neighboring peaks. The best peaks for these purposes are the peak at 2.47 ppm (methoxyPEG linked to lysine), the peak at 4.15 ppm (biotinylated PEG linked to lysine), and the peak at 4.25 ppm (lysine backbone). The grafting ratio is then given by the intensity of the peak at 4.25 ppm divided by half the sum of the intensities of the peaks at 2.47 and 4.15 ppm, whereas the percentage of biotinylation is given by the intensity of the peak at 4.15 ppm divided by the sum of the intensities of the peaks at 2.47 and 4.15 ppm. Calculated grafting ratios and percentages of biotinylation for the polymers used in this paper are summarized in Table 1. Evidence for small amounts of unreacted methoxyPEG can be seen in the spectra of PLL-*g*-PEG and in one batch of PPB20 by the presence of a small triplet at 2.41 ppm, but the remaining polymers did not show such peaks.

Due to inherent experimental error, we do not consider the values in Table 1 to be accurate to more than one significant figure. Although the observed variation of the grafting ratio is non-negligible, previous work has shown that differences in protein resistance between PLL(20)-*g*[3.5]-PEG(2) and PLL(20)-*g*[2.0]-PEG(2) are not statistically significant.[22] The labels in the leftmost column of Table 1 will, therefore, be used to refer to the polymers throughout the remainder of this paper.

2.4. Optical Waveguide Lightmode Spectroscopy. OWLS uses the evanescent field generated by the incoupling of a He–Ne laser via a grating into a planar waveguide to measure changes in the refractive index at the solid–liquid interface.[19] This technique enables the direct, on-line monitoring of the adsorption of macromolecules such as proteins, lipids, or polymers.[10,24–26] It is highly sensitive to adsorbed masses (i.e., as small as 1 ng/cm^2)[27] up to a distance of 200 nm above the surface of the waveguide. Furthermore, a measurement time resolution of 3 s permits real-time studies of the adsorption kinetics of macromolecules. OWLS experiments were conducted in a commercial IOS-1 instrument (Artificial Sensing Instruments ASI, Zurich, Switzerland) using a Kalrez (a perfluoronated elastomer, Dupont, Wilmington, DE) flow-through cell (8 × 2 × 1 mm).[28] Mass data are calculated from the thickness and refractive index values derived from the guided-mode equations according to de Feijter's formula.[24] Using a Raleigh interferometer, we measured a refractive index increment (dn/dc) of 0.158 cm^3/g for PLL-*g*-PEG-based polymers and used this value in subsequent calculations of adsorbed polymer mass. The literature value of 0.182 cm^3/g was used for the protein adsorption calculations.[19,29]

2.4.1. Formation of Polymeric Interfaces. A clean waveguide chip was soaked in HEPES-1 buffer (10 mM HEPES, pH 7.4) for a minimum of 5 h prior to each experiment.[30] After the presoaked sample was inserted in the OWLS instrument, the sample equilibrated (still under HEPES-1) and reached a flat baseline in approximately 1 h.

Solutions of 15 μM PLL-*g*-PEG and 15 μM PLL-*g*-PEG/PEGbiotin in HEPES-1 were filtered through 0.22 μm Durapore membranes and mixed directly before use. The polymeric interface was formed in situ by exposing the chip to such a mixed solution for 30 min at a flow rate of 1 mL/h. The sample was subsequently rinsed with HEPES-1 buffer for 30 min. If a higher ionic strength was desired for the protein-adsorption part of the experiment (see section 3.4 below), the HEPES-2 buffer (10 mM HEPES, 150 mM NaCl, pH 7.4) was introduced at this point and a new baseline was established (in approximately 30 min). All subsequent adsorption and rinsing steps were carried out in the same buffer.

2.4.2. Standard Bioaffinity Assay Protocol. For most biosensing experiments, the polymer-coated sensor chip was sequentially incubated with a continuous flow of 100 μg/mL streptavidin (or avidin or NeutrAvidin), 100 μg/mL biotinylated goat antirabbit immunoglobulin (αRIgG–biotin), and 200 μg/mL rabbit immunoglobulin (antihuman albumin) (RIgG). Each incubation period lasted 15 min (enough time to reach saturation in all cases) and was followed by a 30 min rinse with the chosen buffer to remove all molecules that were not adsorbed on the surface. The flow rate was 1 mL/h for all analytes.

2.4.3. Other Bioaffinity Assays. For experiments investigating the effect of streptavidin concentration on the above biosensing system, the protocol of section 2.4.2 was followed except that a streptavidin concentration of 2.5 μg/mL was used and an incubation time of 45 min was necessary to reach saturation.

For biotinylated albumin adsorption experiments, 100 μg/mL of biotinylated bovine serum albumin (BSA–biotin) and 200 μg/mL of rabbit antibovine serum albumin (αBSA) were used in place of αRIgG–biotin and RIgG, respectively.

2.4.4. Serum Adsorption Experiments. Nonspecific protein adsorption from human serum was tested in a series of experiments on Nb_2O_5-coated waveguides. Bioaffinity assays were carried out in HEPES-2 buffer, as described in sections 2.4.1 and 2.4.2, except that after a given step had been completed, the experiment was interrupted (and the flow stopped) and control human serum N was injected into the OWLS cuvette. After 30 min of equilibration, the cuvette was rinsed with 2 mL of HEPES-2 and allowed to equilibrate for 30 additional minutes.

For comparison, serum adsorption was also tested on a clean Nb_2O_5-coated waveguide and on a Nb_2O_5-coated waveguide that

(23) Ikura, M.; Hikichi, K. *Org. Magn. Reson.* **1982**, *20*, 266–273.
(24) Ramsden, J. J. *J. Stat. Phys.* **1993**, *73*, 853–877.
(25) Csucs, G.; Ramsden, J. J. *Biophys. J.* **1998**, *74*, A336.
(26) Kurrat, R.; Walivaara, B.; Marti, A.; Textor, M.; Tengvall, P.; Ramsden, J. J.; Spencer, N. D. *Colloids Surf., B* **1998**, *11*, 187–201.
(27) Lukosz, W. *Biosens. Bioelectron.* **1991**, *6*, 215–225.
(28) Kurrat, R.; Textor, M.; Ramsden, J. J.; Boni, P.; Spencer, N. D. *Rev. Sci. Instrum.* **1997**, *68*, 2172–2176.
(29) Ramsden, J. J.; Roush, D. J.; Gill, D. S.; Kurrat, R.; Willson, R. C. *J. Am. Chem. Soc.* **1995**, *117*, 8511–8516.
(30) Ramsden, J. J. *J. Mater. Chem.* **1994**, *4*, 1263–1265.

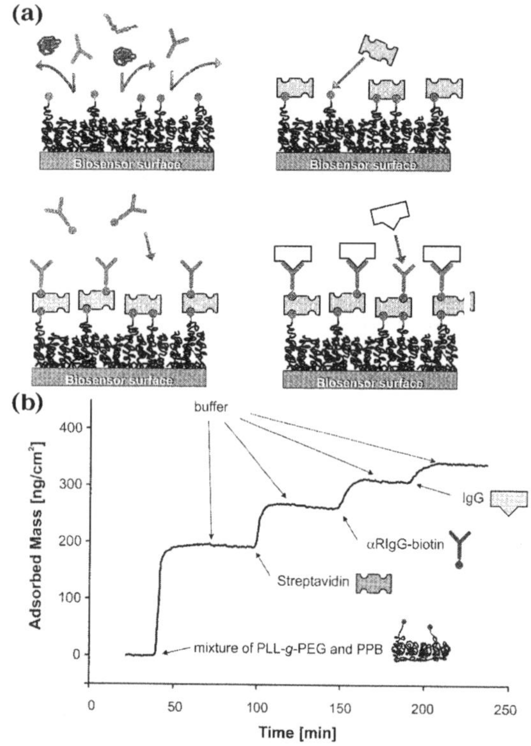

Figure 2. (a) Schematic of the different steps in the model immunoassay. Upper left: the PLL-*g*-PEG/PEGbiotin surface resists nonspecific protein adsorption; upper right: the biotinylated PEG chains specifically adsorb (strept)avidin; lower left: αRIgG–biotin attaches to the (strept)avidin and serves as a capture molecule; lower right: the target molecule RIgG selectively binds to αRIgG–biotin through antibody–antigen interactions. (b) A sample OWLS measurement of sequential adsorption of a mixed polymeric interface of PLL-*g*-PEG and PLL-*g*-PEG/PEGbiotin, streptavidin, biotinylated goat antirabbit IgG, and rabbit IgG. Each adsorption step was followed by rinsing with buffer.

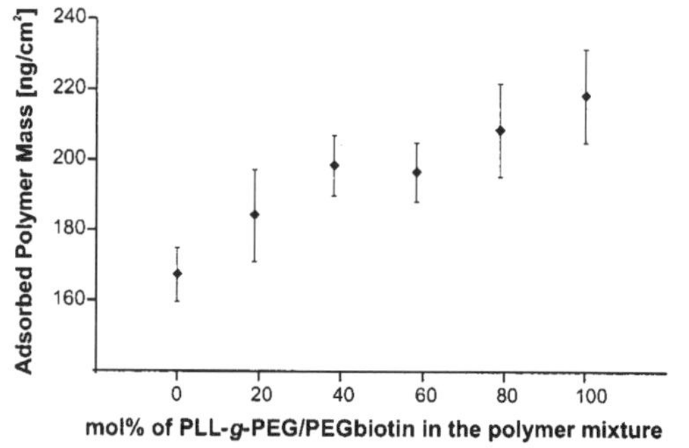

Figure 3. Polymer adsorption on Nb_2O_5 chips from mixed solutions of PLL-*g*-PEG and PPB20. The adsorbed masses were determined from OWLS experiments as described in the text.

was incubated for 60 min with 100 μg/mL of streptavidin to form a physisorbed protein layer.

3. Results and Discussion

3.1. Formation of Polymeric Interfaces on Nb_2O_5 Waveguide Chips. Figure 2a depicts a schematic of our model immunoassay, and Figure 2b shows a typical mass adsorption curve derived from an in situ OWLS measurement.

At pH 7.4, the Nb_2O_5 surface is negatively charged (isoelectric point, IEP ~ 3.6) whereas the PLL backbones of PLL-*g*-PEG and PLL-*g*-PEG/PEGbiotin are highly cationic, due to the protonated amine groups. The electrostatic interaction between the multicharged adsorbate and the substrate is likely the driving force for the strong chemisorption of PLL-based polymers onto the Nb_2O_5-coated waveguides.

Our motivation for mixing PLL-*g*-PEG and PLL-*g*-PEG/PEGbiotin was to tailor the chip surface in terms of specific analyte–surface interactions and to reduce or eliminate nonspecific adsorption. On the mixed polymeric interfaces, biotin sites from PLL-*g*-PEG/PEGbiotin were designed for the specific adhesion of proteins, such as streptavidin or avidin, which can specifically bind further biotinylated moieties (e.g., αRIgG–biotin), which in turn can capture target molecules (e.g., RIgG) as part of a bioaffinity sensor. The surface biotin sites, which can be considered as "docking sites" for (strept)avidin, are surrounded by nonadhesive PEG chains to prevent denaturing of the (strept)avidin and to reduce nonspecific adsorption of other proteins. Changing the ratio of PLL-*g*-PEG to PLL-*g*-PEG/PEGbiotin in the mixture will change the distribution of docking sites on the interface, allowing optimization of the sensor response (see section 3.2).

Figure 3 shows the mass of adsorbed polymer on Nb_2O_5 chips for different mixtures of PLL-*g*-PEG and PPB20. The amount of adsorbed polymer on Nb_2O_5 is 167 ± 8 ng/cm^2 for pure PLL-*g*-PEG and 218 ± 13 ng/cm^2 for pure PPB20. Most of this difference is due to the differences in molecular weight between PLL-*g*-PEG and PPB20. By using the molecular weights of the reagents and the grafting ratios determined by NMR, we estimated the average molecular weight for each polymer (~68 kDa for PLL-*g*-PEG and ~87 kDa for PPB20) and calculated the surface molecular concentration for each of the mixtures depicted in Figure 3. Within experimental error, all of the concentrations were the same, namely, 2.5 ± 0.1 pmol/cm^2 of polymer, which suggests that the ratio of the two polymers in the adsorbed polymer layer is the same as that in solution. Further evidence supporting this claim will be presented in section 3.6.

3.2. Bioaffinity Sensing Based on the Streptavidin–Biotin System. Figure 4 shows the results from the sequential adsorption of streptavidin, biotinylated goat antirabbit IgG, and rabbit IgG on mixed polymeric interfaces of PLL-*g*-PEG and PPB20. The amount of biotin at the interface (plotted on the abscissa) was calculated from the NMR-determined biotinylation percentages (see Table 1) and the ratios of the two polymers in the solution. The absolute error in these surface biotin concentrations could be quite large (up to ~30%) due to the inherent polydispersity of the polymers as well as to errors in the NMR-determined grafting ratio and biotinylation percentage. However, much of this error will be systematic since the same two batches of polymer are used for an entire concentration series. This allows us to be confident about the trends we observe, even if not about the absolute numbers.

When a 100 μg/mL solution of streptavidin was flowed through the cell chamber, the binding of streptavidin molecules to the biotinylated polymeric interface reached saturation in less than 15 min. Following the removal of the nonspecifically bound (loosely physisorbed) streptavidin by washing with HEPES-1 buffer, the amount of streptavidin that was strongly bound to a given interface

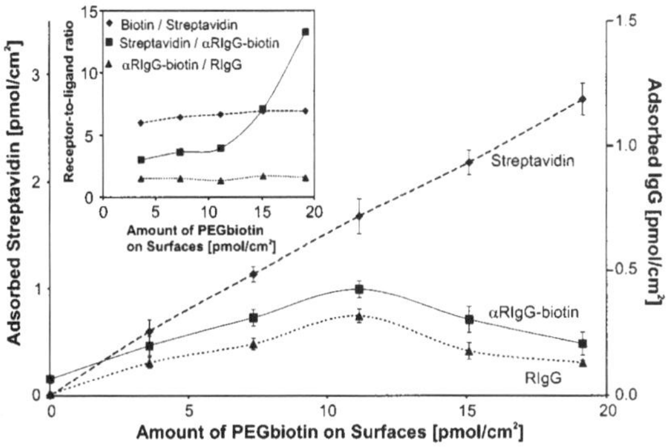

Figure 4. Results from OWLS measurements on the sequential adsorption of streptavidin, αRIgG−biotin, and RIgG on mixed polymeric interfaces. The surface concentration of adsorbed streptavidin is plotted on the left axis, whereas the concentration of adsorbed IgG is indicated on the right. The inset shows the receptor-to-ligand ratio for each step of the bioaffinity assay.

could be determined. This amount of bound streptavidin increased monotonically with the concentration of biotin on the surface. For example, no streptavidin was adsorbed on the pure PLL-*g*-PEG-modified surface (corresponding to 0 on the x-axis), which confirmed its protein resistance, but 2.77 pmol/cm^2 of streptavidin was bound to the pure PPB20 surface (corresponding to 19.15 pmol/cm^2 of biotin on the interface). From the two-dimensional crystal structure of streptavidin, it can be estimated that the individual molecular size of streptavidin is about (5.5 × 4.5) nm^2.[31] This means that a closely packed streptavidin monolayer would have a surface density of ~6.71 pmol/cm^2. Therefore, the range of surface densities of streptavidin obtained in the present system corresponds to approximately 0−41% of a closely packed arrangement. For a pure PPB50-modified surface (corresponding to 41.97 pmol/cm^2 of biotin at the interface), there is approximately 6.65 pmol/cm^2 of streptavidin bound to the interface (data not shown), which corresponds to nearly full coverage.

The ratio of interfacial biotin to streptavidin (see Figure 4, inset) is approximately equal to 6.5 for all concentrations studied. One possible explanation for this excess of PEGbiotin moieties relative to the number of biotin-binding sites on one side of streptavidin is that some of the biotin molecules are not accessible at the surface but rather are hidden within the PEG adlayer. Another possible contribution to the excess is that the flexibility of the biotinylated chains could allow the streptavidin to be tilted at some angle with respect to the surface, permitting more than two PEGbiotin chains to bind to a single streptavidin molecule (see Figure 5a).

Following the injection of biotinylated antirabbit immunoglobulin (αRIgG−biotin) to the streptavidin-modified sensor chip, the αRIgG−biotin molecules were rapidly bound to the surface through the remaining biotin-binding sites of streptavidin. Any loosely bound αRIgG−biotin was removed by rinsing with HEPES-1 buffer. Figure 4 indicates that the amount of adsorbed αRIgG−biotin increases as a function of surface biotin concentration (and, thus, surface streptavidin concentration) until it reaches a maximum of approximately 0.43 pmol/cm^2 of adsorbed αRIgG−biotin at a PEGbiotin surface concentration of

(31) Darst, S. A.; Ahlers, M.; Meller, P. H.; Kubalek, E. W.; Blankenburg, R.; Ribi, H. O.; Ringsdorf, H.; Kornberg, R. D. *Biophys. J.* **1991**, *59*, 387−396.

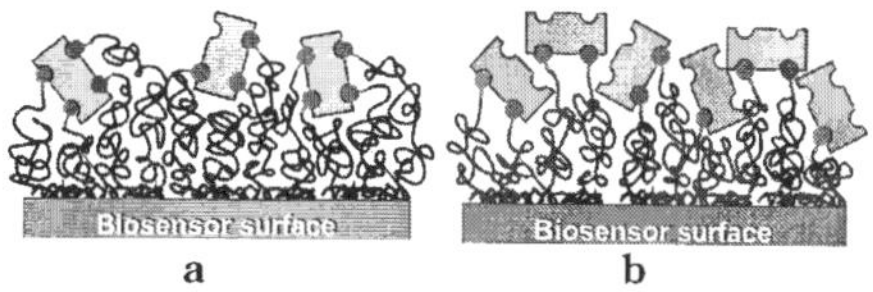

a b

Figure 5. Schematic depictions of possible streptavidin configurations on the polymer-coated surfaces: (a) binding of more than two PEG−biotin groups to a single (tilted) streptavidin molecule and (b) blocking of biotin-binding sites on streptavidin by a neighboring streptavidin molecule.

approximately 11.15 pmol/cm^2. Under these conditions, the amount of adsorbed streptavidin is 1.68 pmol/cm^2. If the surface PEGbiotin concentration, and thus the amount of adsorbed streptavidin, is further increased, the amount of adsorbed αRIgG−biotin begins to decrease. This is also reflected in the steep increase in the ratio of streptavidin to αRIgG−biotin as the polymer concentration exceeds 11.15 pmol/cm^2 (Figure 4, inset).

Two possible hypotheses could explain the decrease in the adsorbed mass of αRIgG−biotin at streptavidin concentrations higher than a certain value (e.g., 1.68 pmol/cm^2 for the present system):

First, an increase of the surface PEGbiotin concentration results not only in an increase in the number of bound streptavidin molecules (as expected) but also in an increase in the number of surface biotin molecules that could potentially block some of the remaining active sites on the adsorbed streptavidin (Figure 5a). Such additional binding of PEGbiotin to streptavidin would reduce the concentration of free sites available for subsequent biotin binding and lead to a reduction in the amount of immobilized αRIgG−biotin at high PEGbiotin surface coverages.

Second, the flexibility of the PEGbiotin chains suggests that it is realistic to assume that the streptavidin molecules adopt a variety of orientations relative to the surface plane rather than that they lie flat as in idealized models (see Figure 2a). When the coverage of streptavidin is increased to a certain level, it then becomes likely that some of these tilted streptavidin molecules will sterically block free biotin-binding sites on adjacent streptavidin molecules (Figure 5b). This effect would again lead to a decrease in the number of available binding sites for αRIgG−biotin at higher streptavidin concentrations.

To test these hypotheses, experiments were performed with a streptavidin solution concentration of 2.5 μg/mL (in place of the standard 100 μg/mL) on a surface with a PEGbiotin concentration of 11.15 pmol/cm^2. The experimental results indicated that after 45 min of incubation followed by 30 min of rinsing with buffer, the amount of adsorbed streptavidin reached, within experimental uncertainty, the same saturation value (1.60 pmol/cm^2) as in the experiments with the higher streptavidin solute concentration. However, the amount of subsequently captured αRIgG−biotin was only 35% (0.15 pmol/cm^2) of that obtained for the case of the experiment with the higher streptavidin concentration (0.43 pmol/cm^2). This is indicative of a kinetically controlled effect and suggests that there are two competing interfacial processes: (a) a fast adsorption of streptavidin from solution to one or two biotin moieties of the PEGbiotin surface and (b) a slower reconstruction process during which additional PEGbiotin groups may bind to streptavidin, thus reducing the average number of free sites available for subsequent αRIgG−biotin binding. We conclude from these observations that the PEGbiotin−streptavidin−biotinylated antibody architecture depends in a complex manner on the kinetics

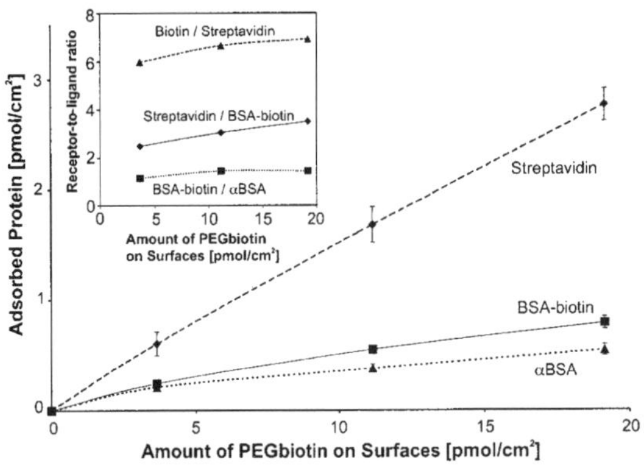

Figure 6. Results from OWLS measurements on the sequential adsorption of streptavidin, BSA–biotin, and αBSA on mixed polymeric interfaces. The inset shows the receptor-to-ligand ratio for each step of the bioaffinity assay.

of each individual surface modification step. Important parameters in this respect are believed to include the initial PEGbiotin surface concentration, the streptavidin solute concentration, and the streptavidin adsorption time.

An additional effect which could limit the maximum amount of adsorbed αRIgG–biotin is related to the molecular dimensions of αRIgG–biotin, which should be close to those of αRIgG, that is, $(14.3 \times 5.9 \times 13.1)$ nm^3.[32] The area that a single αRIgG–biotin molecule occupies on the surface is thus approximately 2.5 times larger than that occupied by streptavidin if we take the orientation of αRIgG–biotin to be that shown in Figure 2a, that is, assuming a footprint area of (14.3×5.9) nm^2. Therefore, when the coverage of streptavidin has reached a certain level, αRIgG–biotin moieties bound to streptavidin are likely to sterically block the access of other αRIgG–biotin molecules to nearby streptavidin sites. Note that this argument does not explain the observed decrease (as opposed to a leveling off) of the mass of adsorbed αRIgG–biotin. However, if this effect plays an important role, one would predict that the maximum coverage of biotinylated molecules, as well as the corresponding optimal surface streptavidin concentration, should depend on the size of the biotinylated capture molecule.

To test this hypothesis, the protein BSA–biotin (mol wt ~ 66 kDa) was used in place of the larger αRIgG–biotin (mol wt ~ 150 kDa), and the surface was subsequently exposed to the antibody αBSA (see section 2.4.3). Figure 6 shows OWLS results for the sequential adsorption of streptavidin, BSA–biotin, and αBSA on mixed polymeric interfaces. The amount of adsorbed BSA–biotin kept increasing beyond the point where the maximum was reached in the streptavidin/αRIgG–biotin system, and no maximum was seen up to surface streptavidin concentrations as high as 2.77 pmol/cm^2. In an additional experiment (data not shown), pure PPB50 was adsorbed onto the waveguide surface. This surface subsequently bound 6.65 pmol/cm^2 of streptavidin but only 0.12 pmol/cm^2 of BSA–biotin, thus suggesting that the BSA–biotin surface concentration also goes through a maximum but that this maximum is shifted to higher streptavidin surface concentrations (between 2.77 and 6.65 pmol/cm^2) in comparison with the αRIgG–biotin case. Correspondingly, the receptor-to-ligand ratios for streptavidin to BSA–biotin are smaller than those for streptavidin to αRIgG–biotin (Figure 6, inset), indicating that more BSA–biotin molecules can be accommodated at the streptavidin surface. The ratio of BSA–biotin to αBSA is approximately 1.5, suggesting that either one or two BSA–biotin molecules capture one αBSA molecule. Binding of αBSA also indicates that the BSA–biotin molecules remain biologically active after being captured by the streptavidin layer.

(32) Kratzin, H. D.; Palm, W.; Stangel, M.; Schmidt, W. E.; Friedrich, J.; Hilschmann, N. *Biol. Chem. Hoppe-Seyler* **1989**, *370,* 263–272.

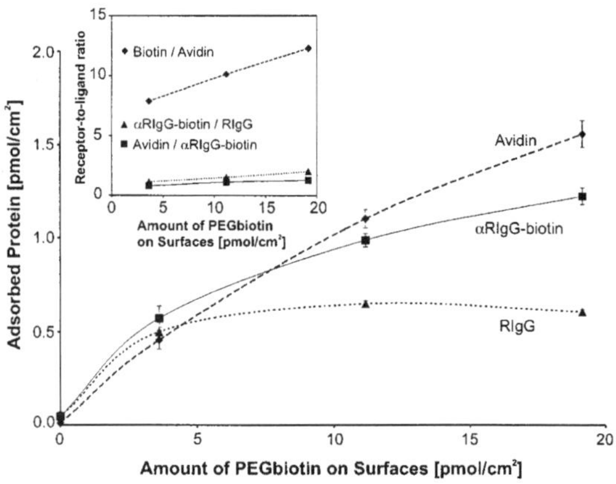

Figure 7. Results from OWLS measurements on the sequential adsorption of avidin, αRIgG–biotin, and RIgG on mixed polymeric interfaces based on mixtures of PLL-*g*-PEG and PPB20 with various ratios. The inset shows the receptor-to-ligand ratio for each step of the bioaffinity assay.

The last step in the model immunoassay depicted in Figure 2a was the exposure of the αRIgG–biotin-modified sensor chips to the target RIgG solution, followed by a rinsing with buffer. Figure 4 indicates that the amount of captured RIgG follows the same trend as the αRIgG–biotin. The ratio of bound αRIgG–biotin to bound RIgG was approximately 1.5 for all surfaces studied, suggesting that the αRIgG–biotin molecules captured by streptavidin on our polymeric interface remain immunologically active over the whole surface concentration range examined.

3.3. Bioaffinity Sensing Based on the Avidin–Biotin System. Avidin is another biotin-binding protein that has been widely used in assays. Its affinity for biotin ($K_{aff} = 1.7 \times 10^{15}$ M^{-1}) is even higher than that of streptavidin ($K_{aff} = 2.5 \times 10^{13}$ M^{-1}).[33] To determine whether streptavidin or avidin is the better protein to use in PLL-*g*-PEG/PEGbiotin-based immunoassays, the experiments described in section 3.2 were repeated with avidin taking the place of streptavidin.

Figure 7 shows the results from OWLS experiments that monitored the sequential adsorption of avidin, αRIgG–biotin, and RIgG on mixed polymeric interfaces. A comparison of Figure 7 with Figure 4 shows that the avidin–biotin bioaffinity sensor behaves quite differently from the streptavidin–biotin case. Although the PLL-*g*-PEG/PEGbiotin surfaces bind less avidin than streptavidin, the avidin-modified surfaces bind significantly more αRIgG–biotin, which in turn leads to more binding of RIgG. This behavior is probably dominated by electrostatic effects. At pH 7.4, the cationic polymer surfaces (pK_a ~ 10)[9] would be likely to interact repulsively with the positively charged avidin (IEP ~ 10.5).[15,16] In addition, any avidin molecules that did manage to bind to the polymeric surface could inhibit the binding of

(33) Bayer, E. A.; Wilchek, M. *Methods Biochem. Anal.* **1980**, *26,* 1–45.

additional avidin molecules to nearby biotin sites because of lateral repulsive effects. A comparison of the avidin adsorption curve (Figure 7) with the streptavidin adsorption curve (Figure 4) indicates similar initial slopes of 0.15 for avidin and 0.16 for streptavidin. However, the slope of the avidin adsorption curve decreases with increasing biotin surface concentration, showing a tendency to level off at the higher PEGbiotin surface concentrations, whereas the corresponding streptavidin curve (Figure 4) remains linear over the whole range investigated. The protein surface concentration at the highest biotin surface concentrations (see Figures 4 and 7) is therefore much higher for streptavidin (2.77 pmol/cm^2) than for avidin (1.56 pmol/cm^2). Our conclusion is that at low surface biotin concentrations (biotin surface sites well separated), repulsive effects between the charged avidin molecules are not so relevant but become increasingly important at the higher PEGbiotin surface concentrations, limiting the number of avidin molecules that can be accommodated at the surface.

The next assay step, the immobilization of αRIgG−biotin on the avidin-covered surface, again shows systematic differences in comparison to the streptavidin case. The adsorbed mass curve for αRIgG−biotin (Figure 7) has a general shape that is rather similar to that of the avidin curve. Correspondingly, the ratio of avidin to bound αRIgG−biotin is fairly constant and approximately equal to 1 across the whole range of PEGbiotin surface concentrations investigated (Figure 7, inset). Again, electrostatic effects are believed to play a major role. αRIgG−biotin is negatively charged at pH 7.4, and therefore, in addition to specific biotin−avidin binding events, nonspecific interactions of electrostatic origin between the oppositely charged avidin and αRIgG−biotin can be expected. Control experiments on a pure PPB20-coated surface showed that 0.61 ± 0.03 pmol/cm^2 of αRIgG (non-biotin-labeled) and 1.23 ± 0.05 pmol/cm^2 of αRIgG−biotin are adsorbed onto the same avidin-modified polymeric surface. We can therefore estimate that roughly 50% of the experimentally observed αRIgG−biotin adsorption on the avidin layer is due to nonspecific adsorption, mostly via the electrostatic interaction and about 50% due to specific binding. (Note that this holds for the buffer system HEPES-1 only; for further discussions on the aspects of nonspecific versus specific adsorption and on the influence of ionic strength, see sections 3.4 and 3.6.)

Two possible explanations for the constant, but lower, receptor-to-ligand ratio of avidin to bound αRIgG−biotin as compared to that of streptavidin to bound αRIgG−biotin can be given:

First, nonspecific interactions, which are clearly important for the avidin system, would be expected to depend to a lesser extent on orientational, packing, and steric exclusion considerations of the binding partners, in comparison to the specific biotin-based recognition and immobilization. There are likely many different orientations and conformations of both the surface avidin and the approaching antibody molecules that allow for nonspecific, electrostatic adsorption, in contrast to a situation (such as the streptavidin case) where the majority of the binding events require the formation of specific lock-and-key-type bonds.

Second, in view of the observed long-range repulsive interactions within the adlayer of charged avidin molecules, the average distance between the surface-immobilized avidin molecules is expected to be larger than for the uncharged streptavidin case, particularly since the rather flexible polymeric PEGbiotin chains in the adlayer are likely to allow for easy rearrangements of the surface avidin molecules and thus for minimization of the free energy. Independent of whether the subsequent adsorption of αRIgG−biotin is due to specific or nonspecific events, one would therefore expect a more efficient organization of the αRIgG−biotin adlayer and a lower avidin/αRIgG−biotin ratio when compared to the corresponding streptavidin/αRIgG−biotin ratio.

The subsequent binding of the target RIgG to the αRIgG−biotin capture molecule shows two different regimes. At low αRIgG−biotin concentrations (up to approximately 0.6 pmol/cm^2 αRIgG−biotin corresponding to 3.6 pmol/cm^2 PEGbiotin, Figure 7), the ratio of adsorbed molar concentration of RIgG to αRIgG−biotin is close to 1, implying an advantageous spatial organization and surface exposure of the αRIgG−biotin antibody and an efficient binding of RIgG to this immobilized antibody. At higher coverages of αRIgG−biotin, however, the amount of RIgG that can bind to the surface levels off at a value of about 0.65 pmol/cm^2 (Figure 7), although the amount of antibody still continues to increase. The amount of αRIgG−biotin immobilized at the surface at this point (0.99 pmol/cm^2) corresponds to roughly 50% of a close-packed monolayer (assumed to be 1.97 pmol/cm^2). It is therefore likely that steric repulsion among RIgG molecules begins to occur at this point and prevents further RIgG molecules from being accommodated at the sensing surface. Another contributing factor could be related to the observation that an important fraction of the αRIgG−biotin is immobilized nonspecifically at the avidin surface (see above). If nonspecifically adsorbed antibodies are less efficient in recognition events, this could explain the observed saturation in the RIgG adsorbed mass curve and the concomitant increase of the αRIgG−biotin/RIgG ratio in Figure 7 (inset).

3.4. Effects of Ionic Strength on Specific and Nonspecific Adsorption. From the experiments described in section 3.3, we observed that molecular charges can deleteriously affect the quality of bioaffinity sensing by introducing nonspecific interactions that can lead to elevated background signals in certain assay steps. To reduce unwanted electrostatic interactions, one can either use neutral molecules and/or increase the ionic strength of the medium.

As shown above (see section 3.3), the positively charged residues of avidin can interact nonspecifically with negatively charged molecules. However, by removing the carbohydrates and thereby lowering the isoelectric point of avidin, a modified avidin, NeutrAvidin, can be produced. As its name implies, this variant of avidin is nearly neutral at pH 7.4. NeutrAvidin retains the specific biotin-binding sites and the complement of amine-conjugation sites originally found in avidin.[18]

To test the effects of ionic strength, multistep immunoassay experiments were carried out based on all three proteins (streptavidin, avidin, and NeutrAvidin) for a single polymer surface composition in two buffer solutions: HEPES-1 (10 mM HEPES, pH 7.4) and HEPES-2 (10 mM HEPES, 150 mM NaCl, pH 7.4). The ionic strength of the HEPES-2 buffer is close to physiological.

Table 2 lists the masses of sequentially adsorbed streptavidin (or avidin or NeutrAvidin), αRIgG−biotin, and RIgG on a given polymer surface (pure PPB20), which corresponds to 19.15 pmol/cm^2 of biotin on the surface, for each of the two buffers. The nonspecific adsorption of αRIgG−biotin on the polymeric (PLL-*g*-PEG/PEGbiotin) interface and the nonspecific adsorption of RIgG on streptavidin (or avidin or NeutrAvidin) are also included, as well as the corresponding mass adsorption ratios, $R_{spec/nonspec}$, of specific-to-nonspecific adsorption, defined

228 *Langmuir, Vol. 18, No. 1, 2002* *Huang et al.*

Table 2. Specific and Nonspecific Adsorption in HEPES-1 and HEPES-2 Buffers for the Streptavidin, Avidin, and NeutrAvidin Systems[a]

	sequential adsorption ($pmol/cm^2$)			nonspecific adsorption ($pmol/cm^2$)		
buffer	SA on PPB20	αRIgG−biotin on SA	RIgG on αRIgG−biotin	αRIgG−biotin on PPB20	RIgG on SA	ratio specific/nonspecific SA−biotin system
HEPES-1	2.77 ± 0.15	0.21 ± 0.05	0.13 ± 0.01	0.09 ± 0.01	0.06 ± 0.01	2.5 ± 1.0
HEPES-2	2.20 ± 0.14	0.20 ± 0.01	0.10 ± 0.01	0.01 ± 0.003	0.03 ± 0.001	5.7 ± 0.4

	sequential adsorption ($pmol/cm^2$)			nonspecific adsorption ($pmol/cm^2$)		
buffer	A on PPB20	αRIgG−biotin on A	RIgG on αRIgG−biotin	αRIgG−biotin on PPB20	RIgG on A	ratio specific/nonspecific A−biotin system
HEPES-1	1.56 ± 0.07	1.23 ± 0.05	0.61 ± 0.01	0.09 ± 0.01	0.61 ± 0.03	1.0 ± 0.1
HEPES-2	2.04 ± 0.16	0.42 ± 0.02	0.18 ± 0.01	0.01 ± 0.003	0.17 ± 0.01	1.5 ± 0.2

	sequential adsorption ($pmol/cm^2$)			nonspecific adsorption ($pmol/cm^2$)		
buffer	NA on PPB20	αRIgG−biotin on NA	RIgG on αRIgG−biotin	αRIgG−biotin on PPB20	RIgG on NA	ratio specific/nonspecific NA−biotin system
HEPES-1	3.89 ± 0.25	0.53 ± 0.04	0.23 ± 0.02	0.09 ± 0.01	0.03 ± 0.002	16.7 ± 1.4
HEPES-2	3.29 ± 0.15	0.48 ± 0.03	0.23 ± 0.02	0.01 ± 0.003	<0.007	>68

[a] The nonspecific binding values refer to the adsorption of each analyte on the previous layer in the immunoassay. The labels of PPB20, SA, A, and NA listed in the table represent PLL-*g*-PEG/PEGbiotin20%, streptavidin, avidin, and NeutrAvidin, respectively.

as follows for streptavidin (and analogously for avidin and NeutrAvidin):

$$R_{\text{spec/nonspec}}\,(\text{SA}) = \frac{(\alpha\text{RIgG}-\text{biotin on SA}) - (\text{RIgG on SA})}{(\text{RIgG on SA})}$$

In calculating such ratios, we make the assumption that RIgG behaves in the same way as an unbiotinylated αRIgG, thus allowing the adsorption of RIgG on SA to serve as an estimate of nonspecific αRIgG adsorption. This assumption is reasonable because of the similar size, shape, and charge of these two molecules. The $R_{\text{spec/nonspec}}$ ratio provides a quantitative way of comparing the suitability of streptavidin, avidin, and NeutrAvidin for biosensor applications. Although this ratio does not directly include the binding of the target molecule, RIgG, to the capture molecule, we have already shown that the adsorption of RIgG follows the trends for αRIgG−biotin (see, for example, Figure 4). Thus, if we can optimize αRIgG−biotin binding, we will optimize the efficiency of our sensor.

For the streptavidin- and NeutrAvidin-based systems, the effect of the ionic strength of the buffer on the adsorbed masses is relatively minor, with a slight (0−20%) reduction of the adsorbed mass at the higher ionic strength. This is not surprising in view of the near neutrality of these proteins. However, the reduction in nonspecific adsorption is more significant (Table 2) and has a profound effect on the $R_{\text{spec/nonspec}}$ ratios. Particularly for the NeutrAvidin system, a significant improvement in the specific-to-nonspecific adsorption ratio is achieved by switching to the higher ionic strength buffer.

The absolute masses adsorbed are strongly buffer-dependent for the avidin system because the electrostatic effects caused by positively charged avidin are significantly reduced in HEPES-2. Reductions in both the total adsorbed mass (i.e., specific plus nonspecific events) and the nonspecific adsorbed mass of each immunoglobin are observed (Table 2). A slightly higher ratio of specific-to-nonspecific adsorption is obtained in the high ionic strength HEPES-2 buffer, but this improvement is not significant. We conclude that the PPB/avidin/αRIgG−biotin interface is clearly not a preferred architecture for a bioaffinity sensor chip surface.

Control experiments (not all data shown) were also carried out in HEPES-2 for the streptavidin and NeutrAvidin systems. They demonstrate that in a buffer with physiological ionic strength (1) streptavidin, NeutrAvidin, αRIgG−biotin, and RIgG do not adsorb (within the limits of detection of the OWLS technique, that is, 1 ng/cm^2) onto a layer of pure PLL-*g*-PEG; (2) αRIgG−biotin adsorbs only slightly onto a PLL-*g*-PEG/PEGbiotin layer; and (3) RIgG adsorbs only slightly onto a layer of streptavidin or NeutrAvidin. Such results indicate that both streptavidin and NeutrAvidin systems are promising for bioaffinity sensing applications.

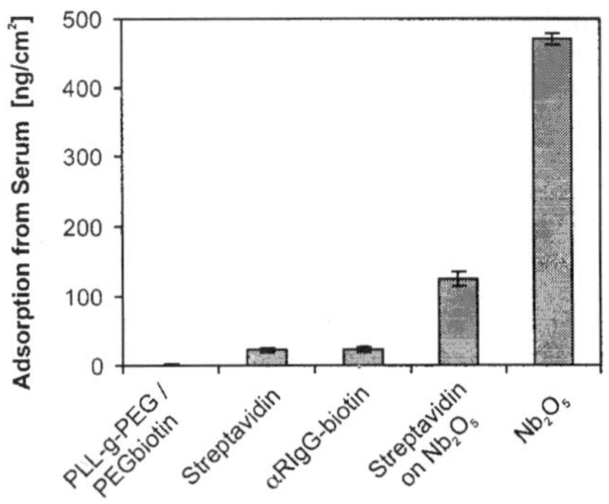

Figure 8. Nonspecific adsorption from human serum on different steps in the streptavidin−biotin immunoassay, on streptavidin physisorbed onto Nb_2O_5, and on bare Nb_2O_5.

3.5. Nonspecific Adsorption from Human Serum. Since human serum is commonly examined in clinical diagnosis, it is useful to test how resistant our proposed biosensor systems are to nonspecific adsorption from human serum. Figure 8 shows the results of adsorption experiments (see section 2.4.4) from human serum. Within the limits of detection of the OWLS technique, no human serum is detected on the PPB20-modified surface. A small amount of nonspecific adsorption was observed after each of the subsequent steps of the model bioaffinity experiment. However, the amounts of human serum adsorbed were considerably less than those observed on the physisorbed streptavidin layer and were only 5% of the amount adsorbed on a clean Nb_2O_5 surface, suggesting that the present biosensing system is highly selective and promising for biomedical applications.

3.6. Interface Architectures for Bioaffinity Sensor Surfaces Based on the NeutrAvidin−Biotin System.

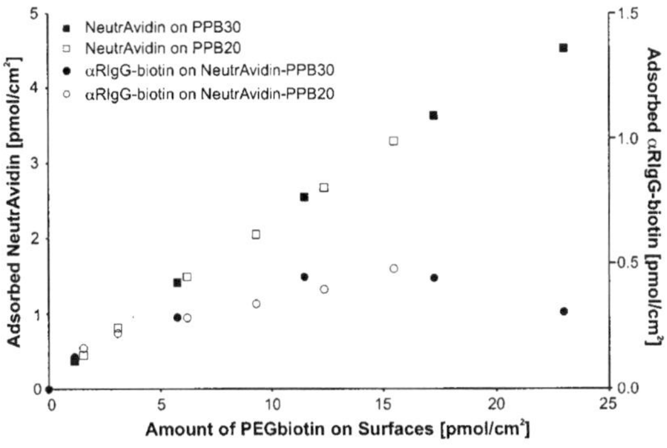

Figure 9. Results from OWLS measurements on the sequential adsorption of NeutrAvidin and αRIgG–biotin on mixed polymeric interfaces formed from mixtures of PLL-*g*-PEG and PPB20 and from mixtures of PLL-*g*-PEG and PPB30. The surface concentration of adsorbed NeutrAvidin is plotted on the left axis, whereas the concentration of adsorbed αRIgG–biotin is indicated on the right.

Since the experiments using NeutrAvidin in HEPES-2 (see Table 2) on PPB20 showed favorable specific binding to nonspecific binding ratios, we chose to further explore this system to find the optimal polymeric interface.

Two series of polymeric interfaces were prepared by exposing the Nb_2O_5 chips to different mixed polymer solutions. The first series was made by mixing PLL-*g*-PEG with 0, 10, 20, 40, 60, 80, and 100 mol % of PPB20. To simplify the discussion which follows, such mixtures shall be referred to using the nomenclature PPB20-mol %. The second series was made by mixing PLL-*g*-PEG with 0, 5, 25, 50, 75, and 100 mol % of PPB30; this series shall be designated using the nomenclature PPB30-mol %. Thus, the PPB20-75% solution, for example, should have the same amount of biotin groups as the PPB30-50% solution, simply distributed over a larger number of polymer molecules. The sequential adsorption of NeutrAvidin, αRIgG–biotin, and RIgG on both series of polymeric interfaces was carried out according to the protocol of section 2.4.2 using the HEPES-2 buffer, and the adsorption data are shown in Figure 9. Since the amount of adsorbed RIgG was proportional to the amount of adsorbed αRIgG–biotin in the present system (as was the case for the streptavidin-based system, see Figure 4), the RIgG data are not shown on the graph.

For the case of NeutrAvidin adsorption, it can clearly be seen that the data from both types of polymer mixtures fall onto a single curve. This agrees well with our previous assumption that the composition of the adsorbed polymer layer is the same as that in solution. The data for the binding of αRIgG–biotin to NeutrAvidin are also reasonably consistent between the two series. For instance, both the PPB30-50% and the PPB20-75% surface layers have comparable amounts of biotin (approximately 11.5 pmol/cm^2), and both have nearly the same amount of bound αRIgG–biotin (within experimental error). However, the OWLS results cannot tell us how the biotin sites are distributed on the surface or to what extent rearrangement of the polymers occurs after the (rapid) adsorption step. AFM studies that could potentially address such issues are currently under way in our group. Although the results of these studies were independent of the degree of biotinylation (20% vs 30%) of the polymers, some dependence on the degree of biotinylation is to be expected when higher biotin percentages are used.

Compared with the streptavidin–biotin system, NeutrAvidin behaves qualitatively like streptavidin because of its similar molecular charge and affinity to biotin; it is a particularly attractive immunoassay platform in view of its high ratio of specific detection to nonspecific adsorption events (Table 2). Since NeutrAvidin and streptavidin molecules have different primary structures, it is not surprising that they provide slightly different optimal layers for maximum binding of RIgG (via αRIgG–biotin). The percentage of PEGbiotin in the initial polymeric interface has been shown to be a crucial parameter. The optimal biotinylated polymeric surface varies with the size of biotinylated molecules introduced to NeutrAvidin or streptavidin. That is, the optimal surface concentration of PEGbiotin is not the same for all immunoassays.

4. Summary and Conclusions

PLL-*g*-PEG is shown to be a promising copolymer for the design of optical waveguide bioaffinity sensor surfaces. It spontaneously adsorbs from aqueous solution in the form of a molecular monolayer onto negatively charged oxide surfaces such as Nb_2O_5 (used as the waveguiding layer in this study) and provides, through formation of a dense layer of surface-exposed PEG chains, a highly protein-resistant surface. The nonspecific adsorption from serum was found to be typically below 1 ng/cm^2, the detection limit of the label-free OWLS instrument used in this study. PLL-*g*-PEG derivatives with 0, 20, 30, and 50% of the PEG chains carrying terminal biotin groups were synthesized, assembled onto the waveguide oxide surface, and used as a basis for testing different interface architectures in IgG-related immunoassays. Such surfaces with controlled surface densities of biotin were consecutively exposed to streptavidin (or avidin or NeutrAvidin), αRIgG–biotin, and RIgG. Streptavidin and NeutrAvidin, both nearly neutral at pH 7.4, adsorbed through specific binding to the PEGbiotin surface with adsorbed masses that depended linearly on the surface biotin concentration. Avidin, however, showed a nonlinear relationship with reduced amounts adsorbing at the higher biotin concentrations, which was likely due to repulsive interactions within the charged avidin adlayer. The subsequent immobilization of the αRIgG–biotin on all three types of protein adlayers followed a more complex adsorption pattern, reaching a maximum in adsorbed mass at a particular biotin surface concentration. The reduced availability of streptavidin (avidin, NeutrAvidin) to bind αRIgG–biotin at high surface biotin concentrations is believed to provide evidence for an increased blocking of the remaining sites of streptavidin by biotinylated PEG as a consequence of reorganization of the interface as well as evidence for hindered access to free sites in adjacent molecules due to steric effects in the more closely packed adlayers. The exact architecture and performance of the sensor interface depends in a complex manner on the kinetics of each individual surface modification step. The maximum achievable density of immobilized antibodies depends furthermore on the size of the biotinylated proteins, with a shift to higher coverages for the smaller albumin relative to the bulkier IgG. Finally, detection of the target RIgG molecule closely reflects the surface concentration of the corresponding immobilized antibody.

In terms of the ratio of specific-to-nonspecific detection, the positively charged avidin shows the lowest, least favorable values due to strong nonspecific adsorption of the (negatively charged) RIgG on the avidin interface, particularly when working with low ionic strength buffer solutions. For streptavidin and especially for NeutrAvidin, promising sensor performance in terms of this ratio could be observed due to the very low nonspecific background of the PLL-*g*-PEG/PEGbiotin adlayers and to the highly preserved immunological activity of proteins immobilized on a background of waterlike, nonadhesive PEG chains.

From an applied perspective, the novel bioaffinity sensor interface presented in this work displays a number of advantages:

Surface functionalization using PLL-*g*-PEG/PEGbiotin is a simple, reproducible, and cost-effective spontaneous adsorption process, not requiring complex, multistage surface chemistry.

The thickness of the interfacial layers remains small enough to guarantee high detection sensitivity of the target molecules well within the penetration depth of the evanescent field.

The biotin concentration at the interface can be quantitatively controlled in an elegant way by forming mixed adlayers of PLL-*g*-PEG/PEGbiotin and PLL-*g*-PEG at the oxide surface, using aqueous solutions of the two polymers in the appropriate ratio. This is an important advantage of this molecular assembly technology because optimal detection in terms of sensitivity and specificity requires biotin (and NeutrAvidin or streptavidin) surface concentrations that are different for different target molecules.

Acknowledgment. We gratefully acknowledge D. L. Elbert, J. A. Hubbell (Institute for Biomedical Engineering, ETH Zürich), R. Hofer (Laboratory for Surface Science and Technology, ETH Zürich), and M. Ehrat, A. Abel, M. Pawlak, and E. Schürmann (Zeptosens, Switzerland) for helpful discussions and suggestions; F. Bangerter and D. Suter (Department of Chemistry, ETH Zürich, Switzerland) for the NMR measurements; and M. Horisberger (Paul Scherrer Institute, Villingen, Switzerland) for the sputter coating. Financial contributions from the Swiss National Science Foundation (SNF, National Research Program NFP 47 "Supramolecular Functional Materials") and from the Swiss Federal Commission for Technology and Innovation (CTI, Project No. 4620.1 MTS) are gratefully acknowledged.

LA010913M

9216 *Langmuir* **2003,** *19,* 9216–9225

Poly(L-lysine)-*graft*-poly(ethylene glycol) Assembled Monolayers on Niobium Oxide Surfaces: A Quantitative Study of the Influence of Polymer Interfacial Architecture on Resistance to Protein Adsorption by ToF-SIMS and in Situ OWLS

Stéphanie Pasche, Susan M. De Paul,[†] Janos Vörös, Nicholas D. Spencer, and Marcus Textor*

BioInterfaceGroup, Laboratory for Surface Science and Technology, Swiss Federal Institute of Technology (ETH) Zurich, CH-8952 Schlieren, Switzerland

Received January 22, 2003. In Final Form: July 16, 2003

Poly(L-lysine) grafted with poly(ethylene glycol) (PLL-*g*-PEG), a polycationic copolymer that is positively charged at neutral pH, spontaneously adsorbs from aqueous solution onto negatively charged surfaces, resulting in the formation of stable polymeric monolayers and rendering the surfaces protein-resistant to a degree related to the PEG surface density. A set of PLL-*g*-PEG polymers with different architectures was synthesized. The grafting ratio, *g*, of the polymer, defined as the ratio of the number of lysine monomers to the number of PEG side chains, was systematically varied between 2 and 23, and PEG molecular weights of 1, 2, and 5 kDa were used. The polymers were adsorbed onto niobium oxide-coated substrates, leading to highly different but well-controlled PEG surface densities with maximal values of 0.9, 0.5, and 0.3 chains/nm^2 for the three PEG molecular weights, respectively. Time-of-flight secondary-ion mass spectrometry (ToF-SIMS) was used in conjunction with the in situ optical waveguide lightmode spectroscopy (OWLS) technique to investigate the interface architecture. While ToF-SIMS provided surface-analytical data on the polymeric adlayer, OWLS allowed the quantitative determination of the adsorbed polymer mass. Extremely good correlations were established between the ToF-SIMS data (obtained in UHV) and the in situ OWLS results. The amount of serum adsorbed, determined quantitatively by OWLS, was found to depend systematically on the surface coverage in terms of the ethylene glycol (EG) density, controlled by both PEG molecular weight and grafting ratio, *g*. Serum adsorption dropped gradually from 590 ng/cm^2 on bare Nb_2O_5 to <2 ng/cm^2 (=detection limit of the OWLS technique) for EG densities ≥ 20 nm^{-2}. The PLL-*g*-PEG technology shows itself to be an efficient, cost-effective, and robust tool for the immobilization of PEG chains onto metal oxide surfaces. The precise control over PEG surface density across a wide range allows for the production of tailored surfaces with controlled degrees of bio-interactiveness. Such surfaces are expected to have a substantial potential for applications in biomedical and bioanalytical devices.

1. Introduction

Adsorption of proteins from solution onto synthetic materials is a key factor in the response of a living body to artificial implanted materials and devices. Adsorbed proteins mediate cell attachment and spreading through specific peptide sequence–integrin receptor interactions and may therefore favorably influence the mechanical stability of the subsequently developed tissue–implant interface. However, the uncontrolled nonspecific adsorption of proteins from the extracellular matrix results in interfaces with many types of proteins in different conformations—a situation that is believed to cause deleterious reactions of the body, such as foreign-body response and fibrous encapsulation.[1]

Elimination of the nonspecific adsorption of biomolecules to artificial material surfaces is therefore an important issue in the design of biomedical and bioanalytical devices.[2–5] Applications of surfaces that are resistant to nonspecific protein adsorption ("protein-resistant" surfaces) include, among others, blood-contacting devices, such as stents and catheters, contact and intraocular lenses, and bacterial-resistant implants.[6] Moreover, bioligand (e.g., protein or oligonucleotide)-functionalized surfaces of biomaterials and implants,[7–9] tissue engineering scaffolds,[10,11] and bioaffinity sensors[12,13] require an inert, non-interactive background in order to elicit the desired biospecific responses.

* To whom correspondence should be addressed. E-mail: marcus.textor@mat.ethz.ch. Telephone: +41-1-6326451. Fax: +41-1-6331048.

[†] Current address: Solvias AG, Klybeckstrasse 191, Postfach, CH-4002 Basel, Switzerland. E-mail: susan.depaul@solvias.com. Telephone: +41-61-6866049. Fax: +41-61-6866565.

(1) Ratner, B. D. In *Titanium in Medicine: Material Science, Surface Science, Engineering, Biological Responses and Medical Applications*; Brunette, D. M., Tengvall, P., Textor, M., Thomsen, P., Eds.; Springer-Verlag: Heidelberg and Berlin, 2000.

(2) Kingshott, P.; Griesser, H. J. *Curr. Opin. Solid State Mater. Sci.* **1999**, *4*, 403–412.
(3) Ratner, B. D. *J. Biomed. Mater. Res.* **1993**, *27*, 837–850.
(4) Tirell, M.; Kokkoli, E.; Biesalski, M. *Surf. Sci.* **2002**, *500*, 61–83.
(5) Kasemo, B. *Surf. Sci.* **2002**, *500*, 656–677.
(6) Ostuni, E.; Chapman, R. G.; Liang, M. N.; Meluleni, G.; Pier, G.; Ingber, D. E.; Whitesides, G. M. *Langmuir* **2001**, *17*, 6336–6343.
(7) Rezania, A.; Johnson, R.; Lefkow, A. R.; Healy, K. E. *Langmuir* **1999**, *15*, 6931–6939.
(8) Drumheller, P. D.; Herbert, C. B.; Hubbell, J. A. In *Interfacial Phenomena and Bioproducts*; Brash, J. L., Wojciechowski, P. W., Eds.; Marcel Dekker Inc.: New York, 1996; pp 273–310.
(9) Textor, M.; Tosatti, S.; Wieland, M.; Brunette, D. M. In *Bio-Implant Interface: Improving Biomaterials and Tissue Reaction*; Lyngstadaas, E., Ed.; CRC Press: Boca Raton, FL (in press).
(10) Hubbell, J. A. *Curr. Opin. Solid State Mater. Sci.* **1998**, *3*, 246–251.
(11) Langer, R. *Acc. Chem. Res.* **2000**, *33*, 94–101.
(12) Duveneck, G. L.; Abel, A. P.; Bopp, M. A.; Kresbach, G. M.; Ehrat, M. *Anal. Chim. Acta* **2002**, *21771*, 1–13.
(13) Griesser, H. J.; Hartley, P. G.; McArthur, S. L.; McLean, K. M.; Meagher, L.; Thissen, H. *Smart Mater. Struct.* **2002**, *11*, 652–661.

Published on Web 09/13/2003

Besides the many applications of biomolecule-resistant devices, surfaces that show a tailored intermediate interactiveness have also attracted attention in, for example, the context of gentle surface immobilization of oligonucleotides[14] and proteins,[15] thus better preserving the natural conformation and functionality of the adsorbed biomolecular moiety.

A preferred strategy for blocking protein adsorption relies on the use of poly(ethylene glycol) (PEG) immobilized at surfaces. PEG's low toxicity and low immunogenicity makes it a suitable material for applications in the field of biomedical devices.[16] PEG has a number of outstanding properties and is a preferred material for imparting protein resistance to surfaces.[16,17] PEG is uncharged, hydrophilic, and soluble in water as well as in many organic solvents. Its hydrogen-bond acceptor character and the size compatibility of an EG monomer unit in a network of hydrogen-bonded water account for a higher-order intrachain structure of PEG in water.[18,19] Extensive hydration, good conformational flexibility, and high chain mobility contribute to a steric exclusion effect between PEG chains in water. Although not yet fully understood, the mechanism underlying the protein resistance of PEGylated, brush-type surfaces is frequently attributed to this "exclusion effect" or "steric stabilization effect" and to osmotic repulsion.[16,17]

Theoretical and experimental approaches[2,18] have demonstrated the relevance of chain length and surface coverage of PEG for imparting protein resistance to surfaces, with protein resistance improving as the length of the PEG chains increases from 20 to 120 EG units, resulting in larger excluded volumes, higher conformational entropy, and more pronounced steric repulsion.[20–26] However, short, oligo$(EG)_{2-7}$-terminated alkane thiol self-assembled monolayers on gold or silver can also be highly protein resistant, provided that the EG_n chains adopt certain particular conformations.[25,27,28] The protein resistance of these short chains has been attributed to the presence of stable water interfacial layers and long-range electrostatic repulsion, rather than conformational entropy or excluded volume effects, as is the case for longer PEG chains.[28–30]

Theoretical studies have aimed at elucidating quantitatively the thermodynamic and kinetic factors that govern protein interaction on surface-grafted polymer chains and brushes.[22,23] Two different types of protein adsorption states have been proposed:[23] (i) Primary adsorption at the substrate surface due to short-range attractive interaction, relevant for small (protein) molecules that can penetrate through the grafted chain interface to the substrate surface; in this case, increased polymer grafting density is reported to be the relevant factor to repress protein adsorption. (ii) Secondary adsorption at the outer edge of the brush, due to van der Waals and/or electrostatic attraction; this type of adsorption is particularly relevant for larger proteins and decreases as the brush thickness is increased. In practice, however, chain graft density and brush thickness are often not independent design parameters.[23]

Several experimental studies related to PEG surface coverage[31–34] indicate that both chain length and density affect the protein-resistant property of PEG-based coatings.[16,20,21,35,36] Long PEG chains (e.g., molecular weight 5000) as well as shorter chains (e.g., molecular weight 2000) rejected proteins, provided the chain density was sufficiently high, implying that protein resistance of PEGylated surfaces requires a sufficiently high surface density of ethylene glycol monomer units.[25,33] However, to be able to test some of the proposed theoretical models in a more quantitative manner, we believe there is still a need for an extended experimental, quantitative database, covering wide variations in PEG chain length and surface density, and corresponding protein adsorption data.

A range of techniques has been employed for the immobilization of PEG onto surfaces.[2,37] A frequently used approach is the covalent coupling of end-functionalized PEG to a surface ("grafting to") or the synthesis of PEG chains by polymerization of EG monomers at the surface ("grafting from"), providing stable coatings of medium grafting densities.[2] Higher PEG surface densities are difficult to produce by "grafting to" and "grafting from" techniques, unless special experimental conditions are used, such as carrying out the grafting reaction at or above the cloud point of the PEG solution.[36] Other approaches include the grafting of PEG using plasma polymerization of PEG precursors.[38,39] A frequent problem when using chemoreactive coupling schemes is irreproducibility due to the critical dependence of the reaction rate on the density of the reactive sites at the surface. An alternative simple and cost-effective approach for producing surfaces with dense arrays of PEG brushes relies on the spontaneous assembly of PEG-grafted copolymers. For example, amphiphilic PEG–PPO–PEG triblock copolymers (known as poloxamers or Pluronics)—amphiphilic copolymers with a hydrophobic poly(propylene oxide) (PPO) domain—have

(14) De Paul, S. M.; Falconnet, D.; Pasche, S.; Textor, M.; Abel, A. P.; Kauffmann, E.; Liedtke, R.; Ehrat, M. (to be submitted).
(15) Brenner, M. G. E. G. Immunoglobulin adsorption on modified surfaces. Ph.D. Thesis, University of Wageningen, The Netherlands, 2001.
(16) Harris, J. M. *Poly(ethylene glycol) Chemistry: Biotechnical and Biomedical Applications*; Plenum Press: New York, 1992.
(17) Harris, J. M. *Poly(ethylene glycol) Chemistry and Biological Applications*; American Chemical Society: Washington, DC, 1997; Vol. 680.
(18) Morra, M. *Water in Biomaterial Surface Science*; John Wiley & Sons Ltd.: Chichester, U.K., 2001.
(19) Israelachvili, J. *Proc. Natl. Acad. Sci. U.S.A.* **1997**, *94*, 8378–8379.
(20) Jeon, S. I.; Lee, J. H.; Andrade, J. D.; de Gennes, P. G. *J. Colloid Interface Sci.* **1991**, *142*, 149–158.
(21) Jeon, S. I.; Andrade, J. D. *J. Colloid Interface Sci.* **1991**, *142*, 159–166.
(22) Szleifer, I. *Biophys. J.* **1997**, *72*, 595–612.
(23) Halperin, A. *Langmuir* **1999**, *15*, 2525–2533.
(24) Zhu, B.; Eurell, T.; Gunawan, R.; Leckband, D. *J. Biomed. Mater. Res.* **2001**, *56*, 406–416.
(25) Prime, K. L.; Whitesides, G. M. *J. Am. Chem. Soc.* **1993**, *115*, 10714–10721.
(26) Nakao, M. D.; Nagaoka, S.; Mori, Y. *J. Biomater. Appl.* **1987**, *2*, 219–234.
(27) Harder, P.; Grunze, M.; Dahint, R.; Whitesides, G. M.; Laibinis, P. E. *J. Phys. Chem. B* **1998**, *102*, 426–436.
(28) Feldman, K.; Hahner, G.; Spencer, N. D.; Harder, P.; Grunze, M. *J. Am. Chem. Soc.* **1999**, *121*, 10134–10141.
(29) Dicke, C.; Hahner, G. *J. Am. Chem. Soc.* **2002**, *124*, 12619–12625.
(30) Wang, R. L. C.; Kreuzer, H. J.; Grunze, M. *J. Phys. Chem. B* **1997**, *101*, 9767–9773.
(31) McPherson, T.; Kidane, A.; Szleifer, I.; Park, K. *Langmuir* **1998**, *14*, 176–186.
(32) Malmsten, M.; Emoto, K.; Van Alstine, J. M. *J. Colloid Interface Sci.* **1998**, *202*, 507–517.
(33) Lin, Y. S.; Hlady, V.; Gölander, C. G. *Colloids Surf., B: Biointerfaces* **1994**, *3*, 49–62.
(34) Efremova, N. V.; Sheth, S. R.; Leckband, D. E. *Langmuir* **2001**, *17*, 7628–7636.
(35) Leckband, D.; Sheth, S.; Halperin, A. *J. Biomater. Sci., Polym. Ed.* **1999**, *10*, 1125–1147.
(36) Kingshott, P.; Thissen, H.; Griesser, H. J. *Biomaterials* **2002**, *23*, 2043–2056.
(37) Holmberg, K.; Tiberg, F.; Malmsten, M.; Brink, C. *Colloids Surf., A: Physicochem. Eng. Aspects* **1997**, *123*, 297–306.
(38) Kingshott, P.; McArthur, S.; Thissen, H.; Castner, D. G.; Griesser, H. J. *Biomaterials* **2002**, *23*, 4775–4785.
(39) Shen, M.; Martinson, L.; Wagner, M. S.; Castner, D. G.; Ratner, B. D.; Horbett, T. A. *J. Biomater. Sci., Polym. Ed.* **2002**, *13*, 367–390.

been used in conjunction with hydrophobic surfaces, exploiting strong hydrophobic–hydrophobic interactions between the PPO block and the surface.[31] Similarly, cationic polyelectrolytes, such as poly(ethylene imine) or poly(L-lysine) grafted with PEG side chains, have been shown to spontaneously adsorb from aqueous solutions onto negatively charged surfaces and to be stably immobilized, provided the electrostatic interaction between surface and polymer backbone is strong enough.[32,40–42]

Graft copolymers of poly(L-lysine) and poly(ethylene glycol) (PLL-*g*-PEG) have proven to be a particularly attractive system to render negatively charged surfaces highly resistant to nonspecific adsorption.[42,43] The interfacial architecture of the PLL-*g*-PEG adlayers has been studied by angle-dependent X-ray photoelectron spectroscopy (XPS), suggesting the existence of a monolayer with PLL bound to the metal oxide surface and densely packed PEG chains extending out of the surface.[43] A PLL-*g*-PEG polymer with a 20 kDa PLL backbone, 2 kDa PEG side chains, and a grafting ratio, *g* (ratio of the number of lysine monomers to the number of PEG side chains), of 3.5 was shown to render TiO_2 and Nb_2O_5 surfaces highly protein resistant in contact with human blood serum.[42,43] Other polymer architectures were less effective.[42] However, the influence of PLL-*g*-PEG structure on the physicochemical and biological properties of modified interfaces was not systematically evaluated.

This work comprises the systematic syntheses of PLL-*g*-PEG polymers with varying PEG molecular weights and grafting ratios, the quantitative evaluation of polymer surface coverage on Nb_2O_5-coated substrates, and the determination of the amount of serum that adsorbs onto the modified surfaces. The main aim is to develop a quantitative relationship between PEG surface coverage and the mass of subsequently adsorbed proteins, thus establishing criteria for the design and fabrication of PEGylated metal oxide surfaces with tailored interactiveness. The optical waveguide lightmode spectroscopy (OWLS) technique[44,45] was used to determine quantitatively and in situ the adsorbed masses of both polymer and serum, while time-of-flight mass spectrometry[46] provided surface-analytical data on the polymer adlayer and permitted the optical waveguide results to be correlated with the interfacial architecture of the various polymeric adlayers. Nb_2O_5 was used as the substrate surface in view of its excellent chemical stability, optical transparency, high refractive index, and low isoelectric point (IEP $\cong$ 4.3).[47]

2. Materials and Methods

2.1. Synthesis of Poly(L-lysine)-*graft*-poly(ethylene glycol). A systematic notation is used for the stoichiometry of PLL-*g*-PEG polymers, indicating the average molecular weights of both the PLL (molecular weight of PLL-HBr) and PEG and the grafting ratio. The grafting ratio is expressed as the number of lysine monomers divided by the number of PEG side chains (Lys/PEG ratio). For example, a polymer labeled PLL(20)-*g*[3.5]-PEG(2) has a PLL backbone corresponding to PLL-HBr of molecular weight ~ 20 kDa (96 Lys-mer), PEG side chains of molecular weight 2 kDa (45 EG-mer), and a grafting ratio of 3.5 (3.5 lysine units per PEG chain). The value of 20 kDa specified here for the average molecular weight of PLL-HBr is only an approximate figure in view of the substantial polydispersity and variations in PLL molecular weight from batch to batch (see below).

PLL-*g*-PEG polymers were synthesized from a stoichiometric mixture of poly(L-lysine) hydrobromide (PLL) (Sigma, USA, molecular weight 15–30 kDa, polydispersity 1.1–1.3) and an *N*-hydroxysuccinimidyl ester of methoxy-terminated poly(ethylene glycol) (mPEG-NHS) (Shearwater, USA, molecular weights 1, 2, and 5 kDa, polydispersity < 1.05). Poly(L-lysine) was dissolved in a 50 mM sodium tetraborate buffer solution (pH 8.5) in a concentration corresponding to 100 mM monomeric lysine. The solution was filter sterilized (0.22 μm pore size filter, Millex-GV, Sigma-Aldrich, Switzerland). The mPEG-NHS was then added to the solution, and the mixture was stirred for 6 h at room temperature. Subsequently, the reaction mixture was dialyzed (Spectra-Por, molecular weight cutoff size 6–8 kDa, Spectrum Laboratories, Inc., USA) for 48 h against deionized water, changing the water after 24 h. The product was freeze-dried and stored at −20 °C before use. Further details regarding the synthesis protocol have been published previously.[41,48,49]

The PLL-*g*-PEG product was characterized by ^1H NMR spectroscopy, using D_2O as a solvent on a 500-MHz Bruker instrument. A bulk value of the grafting ratio was determined from the NMR spectra.[49]

Choosing PEGs with molecular weights of 1, 2, and 5 kDa and different relative amounts of PLL and PEG allowed us to synthesize a matrix of different polymers with defined molecular architectures.

2.2. Substrates. PLL-*g*-PEG was used to modify the surface of metal oxide films. Nb_2O_5 (12 nm) was sputter-coated onto silicon wafers (WaferNet GmbH, Germany) using reactive magnetron sputtering (PSI, Villigen, Switzerland) to produce substrates for ToF-SIMS measurements. Optical waveguide chips for OWLS measurements were purchased from Microvacuum Ltd. (Budapest, Hungary) and consisted of a 1-mm-thick AF45 glass substrate and a 200-nm-thick $Si_{0.75}Ti_{0.25}O_2$ waveguiding surface layer. A 12-nm-thick Nb_2O_5 layer was sputter-coated on top of the waveguiding layer under the same deposition conditions as described above for the silicon wafers.

2.3. Surface Modification. Substrates used for ToF-SIMS experiments were sonicated in 2-propanol for 10 min, rinsed with ultrapure water, and dried under a nitrogen stream, followed by 2 min of oxygen-plasma cleaning in a plasma cleaner/sterilizer PDC-32G instrument (Harrick, Ossining, USA). The optical waveguide chips used for OWLS were cleaned and regenerated by washing with a "cleaner" solution (detergent from Roche Diagnostics, Switzerland), followed by 10 min of sonication in 0.1 M HCl, rinsing with ultrapure water, drying under a nitrogen stream, and exposure to 2 min of oxygen plasma. Water used in the experiments was purified with a Millipore water treatment apparatus (organic content less than 5 ppb).

For ToF-SIMS investigation of the modified oxide surfaces, the clean substrates were immediately transferred to a filtered 1 mg/mL solution of PLL-*g*-PEG in 10 mM HEPES buffer solution (4-(2-hydroxyethyl)piperazine-1-ethanesulfonic acid, adjusted to pH 7.4 with a 6 M NaOH solution). After 30 min of immersion, the modified samples were withdrawn, rinsed extensively with ultrapure water, and dried under nitrogen. Samples were analyzed immediately after adsorption, to avoid any uncertainty due to storage.

A similar surface modification protocol was adapted for the optical waveguide chips. For further details, the reader is referred to section 2.5.

2.4. Time-of-Flight Secondary-Ion Mass Spectrometry (ToF-SIMS). Secondary-ion mass spectra were recorded on a

(40) Claesson, P. M.; Blomberg, E.; Paulson, O.; Malmsten, M. *Colloids Surf., A: Physicochem. Eng. Aspects* **1996**, *112*, 131–139.
(41) Elbert, D. L.; Hubbell, J. A. *Chem. Biol.* **1998**, *5*, 177–183.
(42) Kenausis, G. L.; Voros, J.; Elbert, D. L.; Huang, N. P.; Hofer, R.; Ruiz-Taylor, L.; Textor, M.; Hubbell, J. A.; Spencer, N. D. *J. Phys. Chem. B* **2000**, *104*, 3298–3309.
(43) Huang, N. P.; Michel, R.; Voros, J.; Textor, M.; Hofer, R.; Rossi, A.; Elbert, D. L.; Hubbell, J. A.; Spencer, N. D. *Langmuir* **2001**, *17*, 489–498.
(44) Voros, J.; Ramsden, J. J.; Csucs, G.; Szendro, I.; De Paul, S. M.; Textor, M.; Spencer, N. D. *Biomaterials* **2002**, *23*, 3699–3710.
(45) Höök, F.; Vörös, J.; Rodahl, M.; Kurrat, R.; Böni, P.; Ramsden, J. J.; Textor, M.; Spencer, N. D.; Tengvall, P.; Gold, J.; Kasemo, B. *Colloids Surf., B: Biointerfaces* **2002**, *24*, 155–170.
(46) Castner, D. G.; Ratner, B. D. In *Surface Characterization of Biomaterials*; Ratner, B. D., Ed.; Elsevier Press: Amsterdam, 1988; pp 65–81.
(47) Kosmulki, M. *J. Colloid Interface Sci.* **2002**, *253*, 77–87.
(48) Sawhney, A. S.; Hubbell, J. A. *Biomaterials* **1992**, *13*, 863–870.
(49) Huang, N. P.; Vörös, J.; De Paul, S. M.; Textor, M.; Spencer, N. D. *Langmuir* **2002**, *18*, 220–230.

PHI 7200 time-of-flight secondary-ion mass spectrometer in the mass range 0−1000 *m/e*. The total ion dose of the 8 kV Cs^+ primary ion beam (200 μm beam diameter) was below the static limit (<1.0 × 10^{12} ions/cm^2). The time per data point was 1.25 ns. Using conductive silicon wafers as underlying substrates, measurements could be performed in the conductive mode. Mass resolution, $M/\Delta M$, was typically 6000 in the positive-ion mode and 2000 in the negative-ion mode. The positive spectra were calibrated using the secondary-ion peaks CH_3^+, $C_2H_3^+$, and $C_3H_5^+$, and the negative spectra using CH^-, OH^-, and C_2H^-.

2.5. Optical Waveguide Lightmode Spectroscopy (OWLS). In optical waveguide lightmode spectroscopy, the adsorbed mass is calculated from the change of the refractive index in the vicinity of the surface upon adsorption of molecules from solution. This change is monitored by the grating-induced incoupling of laser light into the waveguiding substrate, generating an evanescent wave. For each polarization mode of the light (transverse electric, T_E, and transverse magnetic, T_M) there is a discrete incoupling angle corresponding to maximum constructive interference. From the changes in incoupling angles for T_E and T_M upon molecular adsorption, the thickness and effective refractive index of the adlayer were calculated. Values of dn/dc were linearly interpolated between 0.13 cm^3/g (pure PEG) and 0.18 cm^3/g (pure PLL) for the different PLL-*g*-PEG polymers and used to calculate the adsorbed mass of polymer according to de Feijter's formula.[50] A published value of 0.182 cm^3/g was used for the proteins in serum.[50] The sensitivity of the OWLS technique is typically 1−2 ng/cm^2. Further details have been described elsewhere.[44]

In situ polymer adsorption was studied in stop-flow mode using a flow-through cell with a volume of 16 μL, typically replacing the solution every 15 min.[44] Nb_2O_5-coated waveguides were cleaned according to the protocol in section 2.3, then inserted into the OWLS flow-through cell, and equilibrated by immersing in HEPES buffer (pH 7.4) for at least 10 h in order to obtain a stable baseline. Subsequently, the buffer was exchanged in situ against a solution of 1 mg/mL of PLL-*g*-PEG in HEPES buffer. After 30 min of exposure to the polymer solution, resulting in the formation of a complete monolayer, the sample was rinsed with the HEPES buffer solution for another 30 min. To condition the polymer adlayer to the higher ionic strength used in the subsequent serum adsorption step and to remove any loosely attached PLL-*g*-PEG, the surface was first exposed to a higher ionic strength HEPES solution (10 mM HEPES with 150 mM NaCl) for 15 min before exchanging the buffer back to the original low ionic strength (10 mM) HEPES buffer. Protein adsorption onto bare and PLL-*g*-PEG-coated OWLS chips was studied in situ by exposure to a solution of human serum (Control Serum N, Roche Diagnostics, Switzerland) for 15 min and subsequent rinse with a 10 mM HEPES buffer solution (pH 7.4), typically every 15 min for about 1 h.

3. Results

3.1. Polymer Synthesis and Characterization. Poly(L-lysine)-*graft*-poly(ethylene glycol) copolymers (Figure 1a) with different copolymer architectures in terms of PEG molecular weight and grafting ratio were successfully synthesized and quantitatively characterized by ^{1}H NMR. The grafting ratio, *g* (lysine to PEG ratio), was systematically varied between 2 and 23 for each type of PLL-*g*-PEG copolymer with PEG molecular weights of 1, 2, and 5 kDa. The notation used for each polymer indicates the molecular weights of PLL and PEG, as well as the grafting ratio, *g* (see section 2.1). The average grafting ratio of the copolymer was calculated from ^{1}H NMR spectra, taking ratios of the areas of the lysine backbone peak (4.30 ppm; −NH−C**H**R−CO−), the lysine side chain peak (3.00 ppm; −C**H**$_2$−NH$_2$), and the lysine side chain peak grafted to PEG (3.18 ppm; −C**H**$_2$−NH−CO−).

Table 1 summarizes the polymers synthesized; their grafting ratios as well as their surface characteristics are discussed in the following subsections.

(50) de Feijter, J. A.; Benjamins, J.; Veer, F. A. *Biopolymers* **1978**, *17*, 1759−1772.

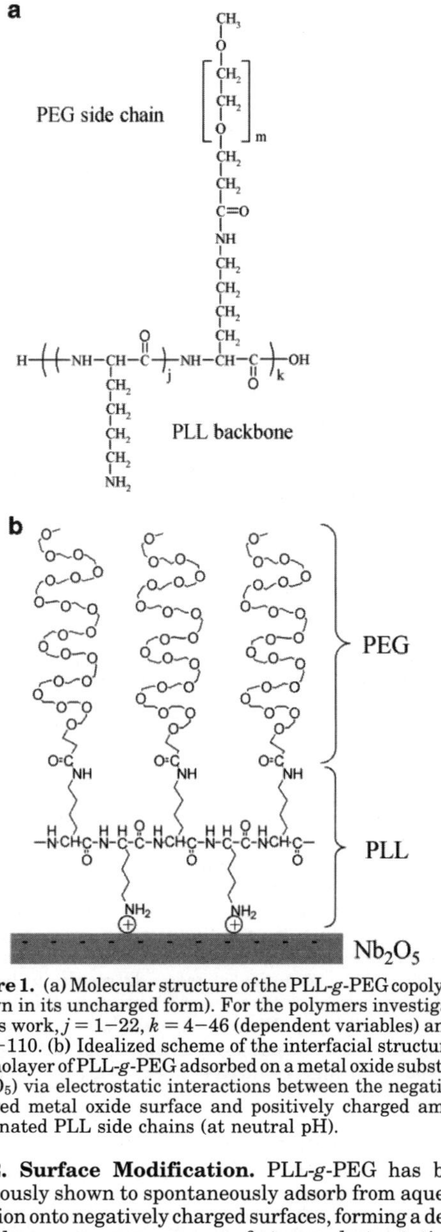

Figure 1. (a) Molecular structure of the PLL-*g*-PEG copolymer (shown in its uncharged form). For the polymers investigated in this work, j = 1−22, k = 4−46 (dependent variables) and m = 18−110. (b) Idealized scheme of the interfacial structure of a monolayer of PLL-*g*-PEG adsorbed on a metal oxide substrate (Nb_2O_5) via electrostatic interactions between the negatively charged metal oxide surface and positively charged amino-terminated PLL side chains (at neutral pH).

3.2. Surface Modification. PLL-*g*-PEG has been previously shown to spontaneously adsorb from aqueous solution onto negatively charged surfaces, forming a dense monolayer as a consequence of strong electrostatic interactions with the positively charged PLL backbone.[42,43] A schematic view of an adsorbed molecule is shown in Figure 1b. Each of the polymers was adsorbed onto the Nb_2O_5-coated substrates by immersing the samples in a solution of 1 mg/mL of PLL-*g*-PEG dissolved in 10 mM HEPES buffer at pH 7.4. An adsorption time of 30 min was sufficient to adsorb a monolayer of the polymer.[42,43]

3.3. ToF-SIMS Analysis. Positive and negative ToF-SIMS spectra were measured for the different polymers adsorbed onto the Nb_2O_5-coated wafer samples. Figure 2 shows a typical positive secondary-ion spectrum for a PLL-(20)-*g*[3.5]-PEG(2) monolayer on Nb_2O_5. All prominent peaks in the mass range m/z = 1−150 have been unambiguously assigned on the basis of the comparison of experimental and theoretical masses of the secondary

Table 1. Summary of the Adsorption Data Determined by OWLS for PLL-*g*-PEG Polymers with Different Architectures (see Figure 1) and Nb_2O_5 as the Substrate Surface[a]

polymer	*g*	m_{pol} (ng/cm²)	n_{Lys} (1/nm²)	n_{PEG} (1/nm²)	n_{EG} (1/nm²)	m_{serum} (ng/cm²)
uncoated Nb_2O_5						590 ± 57
PLL(20)		50 ± 5	2.11			280 ± 40
PLL(20)-*g*[2.7]-PEG(1)	2.7 ± 0.1	199 ± 20	2.39	0.89	20.12	1 ± 2
PLL(20)-*g*[4.1]-PEG(1)	4.1 ± 0.2	154 ± 14	2.48	0.60	13.74	10 ± 3
PLL(20)-*g*[6.5]-PEG(1)	6.5 ± 0.3	117 ± 12	2.49	0.38	8.69	65 ± 2
PLL(20)-*g*[11.7]-PEG(1)	11.7 ± 0.6	88 ± 1	2.48	0.21	4.81	210 ± 15
PLL(20)-*g*[14.2]-PEG(1)	14.2 ± 0.7	80 ± 2	2.43	0.17	3.88	293 ± 38
PLL(20)-*g*[2.2]-PEG(2)	2.2 ± 0.1	204 ± 10	1.18	0.54	24.36	6 ± 8
PLL(20)-*g*[3.5]-PEG(2)	3.5 ± 0.2	217 ± 14	1.86	0.53	24.14	2 ± 3
PLL(20)-*g*[5.7]-PEG(2)	5.7 ± 0.3	152 ± 13	1.91	0.33	15.22	10 ± 9
PLL(20)-*g*[10.1]-PEG(2)	10.1 ± 0.5	114 ± 11	2.10	0.21	9.44	86 ± 11
PLL(20)-*g*[22.6]-PEG(2)	22.6 ± 1.1	87 ± 3	2.42	0.11	4.86	195 ± 35
PLL(20)-*g*[2.1]-PEG(5)	2.1 ± 0.1	194 ± 30	0.46	0.22	25.05	85 ± 58
PLL(20)-*g*[3.5]-PEG(5)	3.5 ± 0.2	247 ± 77	0.95	0.27	30.85	0 ± 1
PLL(20)-*g*[5.3]-PEG(5)	5.3 ± 0.3	224 ± 24	1.25	0.24	26.88	0 ± 0
PLL(20)-*g*[7.9]-PEG(5)	7.9 ± 0.4	155 ± 6	1.22	0.15	17.60	23 ± 4
PLL(20)-*g*[18.7]-PEG(5)	18.7 ± 0.9	91 ± 17	1.38	0.07	8.36	116 ± 42

[a] *g* (±SE), grafting ratio of the polymer determined by ¹H NMR ± standard error; m_{pol} (±SD), adsorbed polymer mass from OWLS measurements ± standard deviation; n_{Lys}, number of lysine monomers per square nanometer; n_{PEG}, number of PEG chains per square nanometer; n_{EG}, number of ethylene glycol monomers per square nanometer; m_{serum} (±SD), adsorbed serum mass (from OWLS) upon exposure to human serum for 15 min and subsequent rinsing in 10 mM HEPES buffer (pH 7.4).

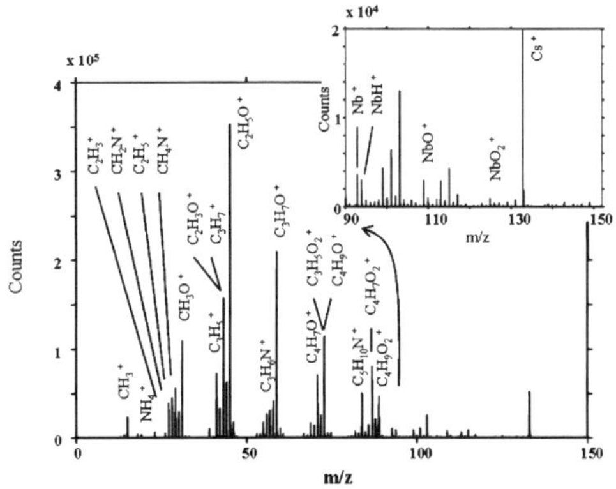

Figure 2. Positive ToF-SIMS spectrum of a PLL(20)-*g*[3.5]-PEG(2)-coated Nb_2O_5 substrate in the mass range $m/z = 0-150$. The more intense secondary-ion peaks were assigned to fragment species, based on the comparison of the theoretical to the experimental mass measured at high mass resolution.

species. Table 2 reports masses and relative intensities for the main peaks of an uncoated Nb_2O_5 surface, as well as a poly(L-lysine)- and a PLL(20)-*g*[3.5]-PEG(2)-coated Nb_2O_5 surface. The characteristic peaks that have been taken into account for quantitative data evaluation (discussed below) are shown in bold in Table 2 and represent fragments that originate exclusively from a specific component of the interfacial structure, that is, PLL or PEG. The peaks in the negative spectra appeared to be generally less specific; therefore, only the positive secondary-ion intensities were used for further quantitative evaluation of the relationship between SIMS intensities and polymer architecture.

3.4. Quantitative Analysis of Polymer Adsorption. The formation of the polymeric adlayer on the Nb_2O_5-coated optical waveguide chips and the subsequent adsorption of proteins onto the polymer-modified surfaces were monitored by the in situ, real-time OWLS technique. Nb_2O_5-coated optical chips were used to allow for unambiguous correlations with the ToF-SIMS data. A typical example of an adsorbed mass versus time curve is shown in Figure 3, demonstrating the changes of interfacial mass upon consecutive exposure of the chip to the polymer solution, buffer (rinse), serum, and buffer again. Since

Table 2. Positive Secondary-Ion Masses and Intensities from ToF-SIMS Measurements, Normalized to 1.0 for the Most Intense Peak in the Corresponding Spectrum[a]

positive secondary-ion mass	molecular species (charge +1)	Nb_2O_5	PLL(20) on Nb_2O_5	PLL(20)-*g*[3.5]-PEG(2) on Nb_2O_5
15.024	CH_3	0.12	0.08	0.06
18.034	NH_4	0.08	0.28	0.01
27.023	C_2H_3	0.65	0.37	0.10
28.019	CH_2N	-	0.32	-
29.039	C_2H_5	0.71	0.33	0.15
30.034	CH_4N	0.12	1.00	0.06
31.018	CH_3O	0.08	0.04	0.28
41.039	C_3H_5	1.00	0.67	0.24
43.018	C_2H_3O	0.24	0.10	0.41
43.055	C_3H_7	0.94	0.39	0.27
45.034	C_2H_5O	0.06	0.03	1.00
56.050	**C_3H_6N**	-	**0.72**	**0.08**
59.050	C_3H_7O	0.02	0.02	0.62
71.050	**C_4H_7O**	-	-	**0.21**
73.029	**$C_3H_5O_2$**	-	-	**0.30**
73.065	**C_4H_9O**	-	-	**0.10**
84.081	**$C_5H_{10}N$**	-	**0.99**	**0.17**
87.047	$C_4H_7O_2$	-	0.01	0.24
89.060	**$C_4H_9O_2$**	-	-	**0.15**
92.790	Nb	0.42	0.35	0.03
93.800	NbH	0.20	0.19	0.03
108.862	NbO	0.68	0.44	0.02
124.087	NbO_2	0.16	0.10	0.01
132.885	Cs	0.52	0.51	0.15

[a] Surfaces: Nb_2O_5 uncoated, Nb_2O_5 coated with PLL (20 kDa), and Nb_2O_5 coated with PLL(20)-*g*[3.5]-PEG(2). - refers to intensity values below the detection limit (normalized intensity < 0.001). Data in bold were chosen for quantitative correlations (see sections 3.3 and 4.1).

only a small change in mass is observed after the rinse after polymer adsorption, it can be assumed that no desorption occurs at this point and that the adlayer remains stable under these experimental conditions. Polymer polydispersity may contribute to scatter in the adsorbed mass measurements. Therefore, all measurements were repeated at least three times and the errors were calculated on the basis of statistical standard deviations over the different measurements.

Once the adsorbed polymer mass is known, it is straightforward to deduce the lysine (monomer) and the PEG chain surface densities expressed as [molecules per

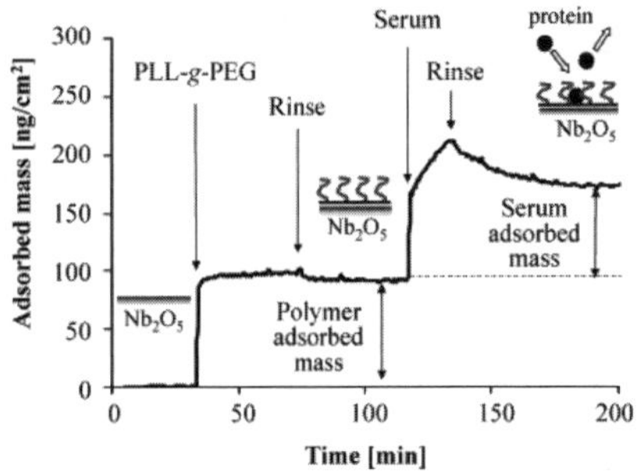

Figure 3. In situ measurement of the adsorption of PLL(20)-g[4.1]-PEG(1) from solution on a Nb_2O_5-coated waveguide chip and subsequent exposure to human blood serum, monitored by OWLS. A 10 mM HEPES buffer solution (pH 7.4) was used for both polymer adsorption and for the rinse between the different adsorption steps. (For simplification, the adsorption curve shown here does not include the additional rinse with a higher ionic strength solution, following polymer adsorption, as discussed in section 2.)

nm²] for the different polymers, since the corresponding grafting ratios, g, are known from NMR (eqs 1 and 2). Furthermore, to facilitate direct comparisons between the three polymer classes with PEG molecular weights of 1, 2, and 5 kDa, the PEG surface densities are converted to ethylene glycol (EG) monomer surface densities (eq 3).

$$n_{\mathrm{Lys}} = \frac{m_{\mathrm{pol}}}{M_{\mathrm{Lys}} + \dfrac{M_{\mathrm{PEG}}}{g}} \tag{1}$$

$$n_{\mathrm{PEG}} = \frac{m_{\mathrm{pol}}}{M_{\mathrm{Lys}}g + M_{\mathrm{PEG}}} \tag{2}$$

$$n_{\mathrm{EG}} = n_{\mathrm{PEG}}\frac{M_{\mathrm{PEG}}}{M_{\mathrm{EG}}} \tag{3}$$

In eqs 1–3, n_{Lys}, n_{PEG}, and n_{EG} denote the surface densities (expressed as number of molecules per unit area) of lysine-mer, PEG, and EG-mer, respectively, m_{pol} represents the mass of polymer adsorbed per unit area, g represents the grafting ratio, and M_{Lys}, M_{PEG}, and M_{EG} are the molecular weights of the lysine monomer, the PEG chain, and the EG monomer, respectively. The surface density data—n_{Lys}, n_{PEG}, and n_{EG}—are included in Table 1 and form the basis for the quantitative correlations between polymer architecture at the interface and protein adsorption data, as well as ToF-SIMS intensities (see Discussion section). Polymer adsorption data, expressed in terms of (total) polymer mass, m_{pol}, lysine surface density, n_{Lys}, and PEG chain surface density, n_{PEG}, are shown in Figure 4 as a function of the inverse of the grafting ratio g, that is, as a function of the PEG/Lys ratio. The curves demonstrate that the polymer surface coverage data depend in a systematic manner both on the polymer grafting ratio and on the molecular weight of PEG; this is discussed in detail in section 4.

3.5. Quantitative Analysis of Protein Adsorption. The adsorption of proteins from serum onto Nb_2O_5-coated optical chips without and with adsorbed PLL-g-PEG was studied in situ in the same way as described above and in section 2. Changes in the adsorbed mass were recorded for up to 15 min of exposure of the surfaces to human serum. This relatively short adsorption time has been proven to be sufficient to reach a kind of steady-state adsorbed serum mass when exposed to undiluted serum solutions; experiments with 2 h exposure resulted in the same values for the adsorbed serum mass (data not shown), implying that a type of "short-term" quasi-equilibrium state was reached after 15 min. Longer exposure times were not considered in this work, to avoid complications due to protein exchange phenomena ("Vroman effect"). On the other hand, exposure of PLL-g-PEG-covered, protein-resistant metal oxide and TCPS surfaces to serum for up to 2–6 days has been previously published, demonstrating long-term preservation of resistance to protein adsorption and cell adhesion.[42,51]

The rinse after serum adsorption results in a slow desorption process, as seen in Figure 3. After a rinse of approximately 1 h, the amount of adsorbed proteins, however, remained fairly stable; therefore, all adsorbed protein mass data refer to a standard 1 h rinse in 10 mM HEPES buffer. The masses of serum adsorbed onto the bare and polymer-modified surfaces (reported in Table 1) are presented in the bar diagram of Figure 5 and demonstrate a strong dependence on polymer architecture. The amount of serum adsorbed can be greatly reduced by the presence of a monolayer of PLL-g-PEG with optimum composition. While 590 ± 57 ng/cm² serum was found to adsorb onto bare Nb_2O_5 surfaces, no significant adsorption (<2 ng/cm²) could be detected on surfaces coated with, for example, PLL(20)-g[2.7]-PEG(1), PLL(20)-g[3.5]-PEG(2), and PLL(20)-g[5.3]-PEG(5). For each of the three polymer types with PEG molecular weights of 1, 2, and 5 kDa, a significant decrease in protein adsorption was observed with decreasing grafting ratio, g. As the molecular weight of PEG increases, the maximum g value required to achieve protein resistance shifts to higher values, demonstrating the importance of PEG surface coverage for achieving protein resistance, as reported in the literature.[24,32,33,52] However, there are exceptions to this generally observed pattern of reduced protein adsorption with increased PEG coverage. For very low grafting ratios (g = 2), proteins are found to adsorb again, especially for the polymer with 5 kDa PEG chains. This is further discussed in section 4.

4. Discussion

4.1. Interfacial Architecture. A matrix of PLL-g-PEG polymers with systematically varied molecular structure (grafting ratio, g, and PEG molecular weight) was synthesized and thoroughly characterized. The in situ OWLS technique was used to provide quantitative data on adsorbed mass for each of the different polymers. In view of the substantial polydispersity, one may argue that the polymer architecture at the interface does not necessarily reflect the average bulk polymer structure as determined by NMR. Therefore, ToF-SIMS was chosen as a technique that can provide information on the molecular structure of surface-adsorbed thin films and monolayers. Although generally considered to be a qualitative rather than quantitative technique, in view of the strong matrix effect on secondary-ion yield, secondary-ion intensities may provide semiquantitative compositional data when analyzing variations within a given, well-defined matrix system.

In Figure 6a the sum of the characteristic secondary-ion intensities of PLL (as discussed in section 3 and listed in Table 2), normalized to the sum of the characteristic

(51) VandeVondele, S.; Vörös, J.; Hubbell, J. A. *Biotechnol. Bioeng.* **2003**, *82*, 784–790.
(52) Sofia, S. J.; Premnath, V.; Merrill, E. W. *Macromolecules* **1998**, *31*, 5059–5070.

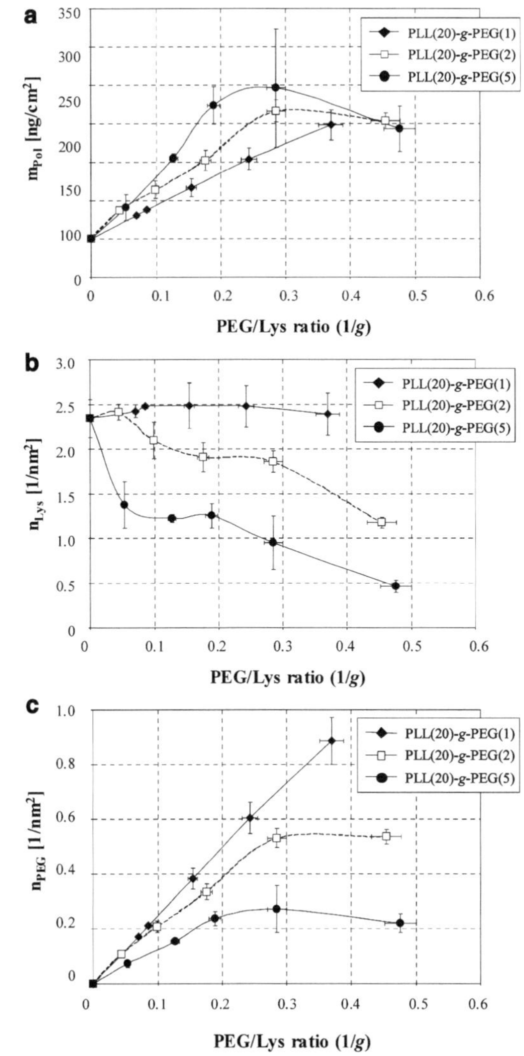

Figure 4. Adsorption data for PLL-*g*-PEG polymers with a 20 kDa PLL backbone, 1, 2, or 5 kDa PEG side chains, and NMR-derived grafting ratios *g* (=Lys-mers per PEG side chain) ranging from 2 to 23, plotted as a function of the inverse of the grating ratio *g*, that is, PEG/Lys, in the bulk polymer: (a) total polymer mass, m_{pol}, monitored by OWLS; (b) polymer surface density expressed as the number of lysine monomers per square nanometer, n_{Lys}; (c) PEG chain surface density expressed as the number of PEG chains per square nanometer, n_{PEG}.

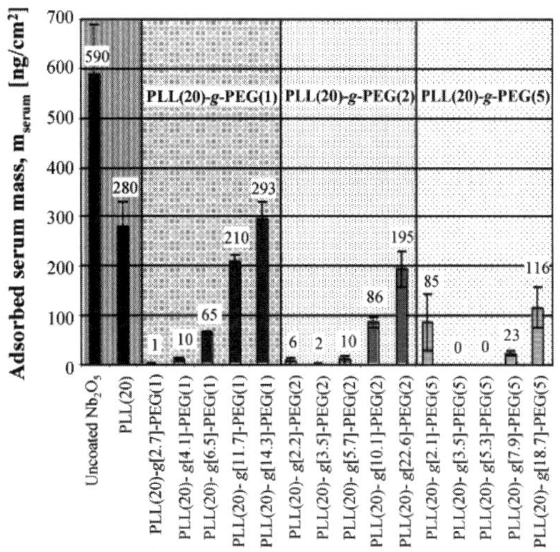

Figure 5. Mass of serum in nanograms per square centimeter, m_{serum}, that adsorbed onto uncoated and PLL-*g*-PEG-coated Nb_2O_5 surfaces upon exposure to human serum for 15 min and subsequent rinsing in 10 mM HEPES buffer (pH 7.4). The adsorbed serum mass was determined quantitatively by the OWLS technique. The polymers are based on 20 kDa PLL and PEG with molecular weights of 1, 2, or 5 kDa and have grafting ratios *g* (=Lys/PEG) varying from 2 to 23.

PLL and PEG intensities, is plotted for all polymers as a function of the ethylene glycol monomer density derived from OWLS (adsorbed mass) and NMR (grafting ratio) data, as described in section 3 (eqs 1−3). Figure 6b is the corresponding plot for the PEG-derived secondary-ion intensities. The correlation between the two data sets is not only surprisingly good but also linear with a R^2 value of 0.97. Parts a and b of Figure 6 provide very good evidence that the polymeric adlayers indeed reflect the composition of the corresponding bulk polymers in a systematic way. Moreover, the excellent quality of the fits demonstrates that ToF-SIMS intensities can be used for establishing semiquantitative correlations with surface-chemical composition within a given matrix. This is further demonstrated in Figure 6c, where the ratio of PLL to PEG ToF-SIMS intensities is plotted as a function of the grafting ratio, *g*, determined for the bulk polymers by NMR. Although the correlation is reasonable, it is not entirely linear; the Lys/PEG SIMS intensity ratios deviate increasingly with decreasing grafting ratio, *g*. This deviation from linearity is a clear consequence of the extreme surface sensitivity (ca. 1 nm) of the static SIMS technique: the high PEG densities achieved with polymers of low *g* value (high PEG/Lys ratio) "shield" the underlying PLL interface, resulting in Lys/PEG SIMS intensity ratios that are smaller than those expected from the bulk (average) composition. However, the linearity for $g > 5$ still suggests that ToF-SIMS is a suitable technique for determining grafting ratios of PLL-*g*-PEG polymers in monomolecular adlayers, both quantitatively and with excellent reproducibility once a corresponding calibration curve has been established.

The quality of the correlations demonstrated in Figure 6 provides strong evidence that the PLL-*g*-PEG technology is indeed suitable for producing PEGylated surfaces with a high degree of control over surface composition and with excellent reproducibility.

4.2. Polymer Adsorption. The systematic changes of adsorbed polymer mass with variations in the polymer architecture provide some useful insights into the factors that govern PLL-*g*-PEG adsorption onto negatively charged surfaces. The adsorbed polymer mass varies in a systematic manner with $1/g$ (=PEG/Lys ratio), as shown in Figure 4a, in the sense that it increases approximately linearly in the range between $1/g = 0$ (pure PLL) and an upper limit that depends on the PEG molecular weight. For PLL-*g*-PEG(1) this linear relationship extends across the entire range of polymer compositions investigated in this study, reflecting the continuously increasing average molecular weight of the polymer. However, for the polymers with 2 and 5 kDa PEG molecular weights, the adsorbed mass levels off (or even decreases) at higher values of $1/g$, implying a decrease in the polymer molar

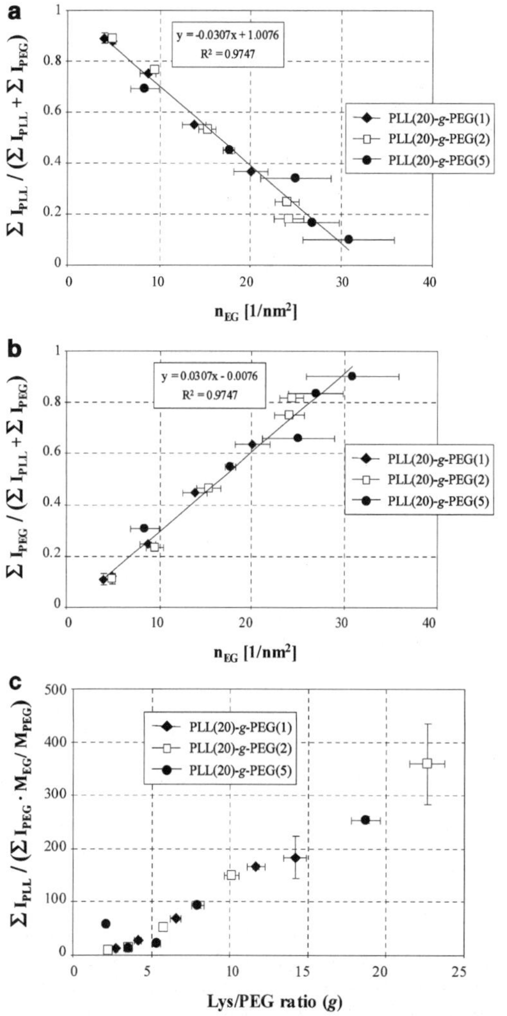

Figure 6. Positive ToF-SIMS intensity data of PLL-*g*-PEG-coated Nb_2O_5 surfaces (for a selection of secondary-ion fragments, see section 3.3). (a) Sum of lysine-specific secondary-ion intensities (ΣI_{PLL}) normalized to the sum of lysine- and PEG-specific intensities $\{\Sigma I_{PLL}/(\Sigma I_{PLL} + \Sigma I_{PEG})\}$, plotted as a function of the EG-monomer surface density, n_{EG}, calculated from OWLS polymer adsorption data and NMR grafting ratios of the bulk polymers. (b) Sum of PEG-related secondary-ion intensities (ΣI_{PEG}) normalized to the sum of lysine- and PEG-specific intensities $\{\Sigma I_{PEG}/(\Sigma I_{PLL} + \Sigma I_{PEG})\}$, plotted as a function of the EG-monomer surface density, n_{EG}. (c) Sum of lysine-specific positive-secondary-ion intensities divided by the sum of PEG-specific intensities (corrected for PEG molecular weight) $\Sigma I_{PLL}/(\Sigma I_{PEG} M_{EG}/M_{PEG})$, as a function of the grafting ratio *g* (Lys/PEG) of the bulk polymer, as determined by NMR.

density at the higher 1/*g* values. This is more directly obvious in Figure 4b: while the lysine surface density (and therefore the molar polymer surface density) is essentially constant for the 1 kDa PEG polymer, the corresponding curves for the 2 and 5 kDa PEG polymers have more complex shapes. For the latter two polymers the molar surface density shows a plateau at the lower 1/*g* values; above 1/*g* = 0.3 (2 kDa PEG) and 0.2 (5 kDa), however, the molar polymer surface density drops rapidly. The lysine surface density curves are paralleled by the corresponding PEG surface densities (Figure 4c), which increase linearly with the PEG/Lys ratio for the 1 kDa PEG polymer but level off at the higher 1/*g* values in the case of polymers with longer PEG chains.

The observation of at least two distinctly different adsorption regimes probably originates from an electrostatic and a steric effect, since the PLL-*g*-PEG adsorption is primarily controlled by the balance between the attractive electrostatic backbone–surface interaction and a steric repulsion between the PEG side chains.[42] Increasing the PEG/Lys ratio of the polymer is expected to result in both a weakening of the electrostatic bond to the substrate (due to the reduced number of "free", NH_3^+-terminated side chains of the PLL) and an increasing steric hindrance due to repulsive interactions between the densely packed PEG chains, both effects being unfavorable for the adsorption process. The fact that the transition from the adsorption regime of essentially constant lysine (or polymer) molar surface density to the one characterized by reduced lysine and constant PEG surface density shifts to lower PEG/Lys values with increasing PEG molecular weight points to the dominance of the steric effect. For the polymers with the higher PEG molecular weights (2 and 5 kDa) and the higher PEG/Lys ratios, PEG chain packing density considerations become the decisive factor, limiting the amount of polymer that can spontaneously adsorb onto the metal oxide surface.

It should, however, be pointed out that the level of self-limited PEG surface densities achieved with the PLL-*g*-PEG system is high: 0.9, 0.5, and 0.3 chains/nm² for PEGs with molecular weights of 1, 2, and 5 kDa, respectively—values that compare favorably with the maximal values of 0.3–0.4 and 0.1–0.2 for PEG molecular weights of 2 and 5 kDa, respectively, reported for end-grafted PEG.[24,32,33,52] According to Malmsten et al., the adsorption onto silica of a similar polycationic copolymer, poly(ethylene imine)-*graft*-poly(ethylene glycol) (PEI-*g*-PEG), with a PEG molecular weight of 6 kDa, resulted in a coverage at saturation of 0.15 PEG chains/nm².[32] The use of PEG-grafted polyelectrolytes therefore proves to be an elegant and efficient technique to reproducibly achieve very dense PEG brushes.

4.3. Protein Resistance. OWLS measurements provide good evidence that the mass of proteins adsorbed from serum strongly depends on the architecture of the preadsorbed polymer, varying systematically with the copolymer composition (Figure 5). To quantify this dependence for all polymer types in one graph, the adsorbed serum mass is expressed in Figure 7 as a function of the density of ethylene glycol monomeric units (EG) at the interface (eq 3). The amount of serum that adsorbs to the polymeric interface decreases systematically as the ethylene glycol monomer density at the surface, n_{EG}, increases, reaching values that typically lie below the detection limit of the OWLS technique (1–2 ng/cm²) for $n_{EG} \geq 20$ nm^{-2}. Since on the graph of Figure 7 all data points fall on the same curve (with one exception, see below), neither the grafting ratio nor the PEG chain length alone, but rather the product of the two, that is, the EG monomer surface density, determines the degree of protein resistance. Excellent protein resistance is observed for all three PEG chain lengths, as long as the PEG/Lys ratio is sufficiently large and results in a sufficiently high value for n_{EG} ($\geq$20 nm^{-2}).

The latter finding implies that the dominant contribution to the free energy of a protein molecule in contact with the PEG layer is a simple EG–protein interaction, resulting in a term proportional to the EG surface concentration. Thus, the product of PEG chain length and

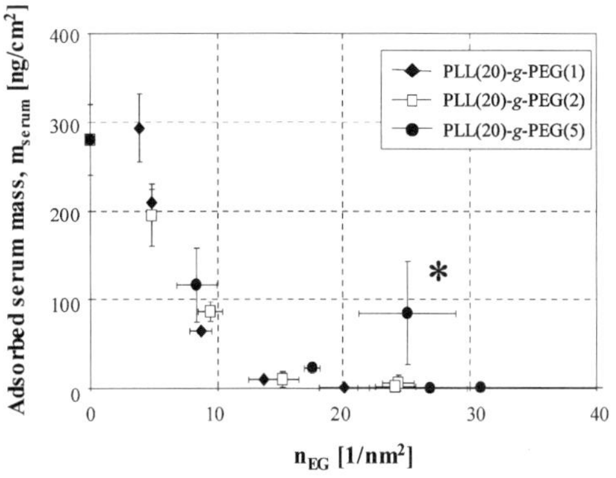

Figure 7. Mass of serum in nanograms per square centimeter, m_{serum}, that adsorbs onto PLL-*g*-PEG-coated Nb_2O_5 surfaces upon exposure to human serum for 15 min and subsequent rinsing in 10 mM HEPES buffer (pH 7.4) as a function of the EG monomer surface density, n_{EG}. The corresponding values of m_{serum} for the uncoated and PLL-coated Nb_2O_5 surfaces are 280 ± 40 and 590 ± 57 ng/cm², respectively. The adsorbed serum mass was determined by OWLS; the EG monomer surface density was determined from the adsorbed polymer mass (OWLS) and the NMR-determined grafting ratio, *g*, of the bulk polymer. (The outlier data point refers to polymer PLL(20)-*g*[2]-PEG(5), further discussed in section 4.3.)

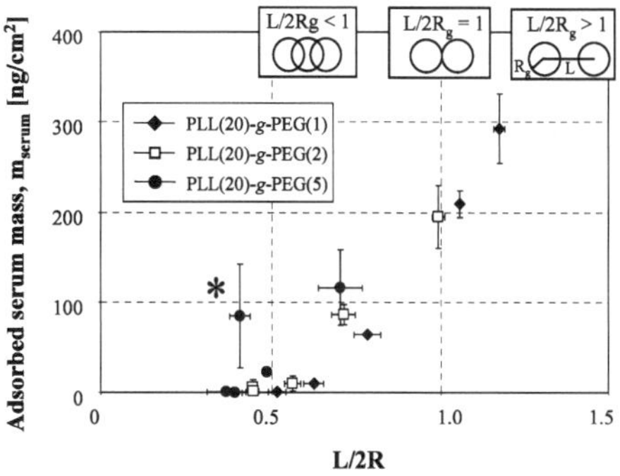

Figure 8. Adsorbed mass of serum in nanograms per square centimeter, m_{serum}, as a function of the ratio $L/2R_g$. L denotes the average mean distance between PEG chains of the PLL-*g*-PEG-coated Nb_2O_5 surface, calculated from the adsorbed polymer mass and the NMR grafting ratio of the bulk polymer. R_g is the radius of gyration of the "free" PEG as determined in aqueous solution[53] (see text). The PEG chains are assumed to be arranged in a close-packed (hexagonal) layer. Insets: $L/2R_g > 1$, PEG chains well separated; $L/2R_g = 1$, PEG spheres touch; $L/2R_g < 1$, PEG spheres overlap, that is, PEG chains become extended, forming a brush conformation. Protein resistance is only observed in the latter case ($L/2R_g < \sim 0.5$). (The outlier data point refers to polymer PLL(20)-*g*[2]-PEG(5), further discussed in section 4.3.)

density, a measure of the degree of overlap of the chains, seems to be the factor that largely determines the degree of protein (serum) adsorption or resistance for the PEG–copolymer system investigated in this study. The contributions of the thickness of the layer and the chain density are generally interrelated, however, and therefore cannot be independently investigated, as long as the polymer layer is thick enough to shield the surface charge.

It has been predicted by Szleifer on the basis of the single-chain mean-field (SCMF) approximation that protein resistance is expected to occur at PEG surface densities of 0.5 chains/nm² for PEG chains with 25 EG units and 0.4 chains/nm² for >50 EG units, corresponding to n_{EG} values of 13, 20, and 45 nm^{-2} for PEG molecular weights of 1, 2, and 5 kDa, respectively.[22] These theoretical values are somewhat different from but within the same order of magnitude as our experimental results. More closely related experimental studies reported by Malmsten et al.[32] for covalently linked PEG and physically adsorbed PEI-*g*-PEG copolymers on silica and by Sofia et al.[52] for covalently grafted PEG chains on silanized silica surfaces are in excellent and quantitative agreement with our findings: the data they report for PEGylated, protein-resistant surfaces correspond to EG monomer surface densities in the range 15–20 nm^{-2}.

There is basically only one outlier in Figure 7, a data point related to the polymer PLL(20)-*g*[2.1]-PEG(5), for which the amount of serum adsorbed is much higher than that expected from its n_{EG} value. The high molecular weight of PEG as well as the high PEG/Lys ratio must make this polymer rather bulky and stiff. One can imagine that thermodynamic and/or kinetic barriers prevent the molecule from achieving a conformation where the backbone lies flat on the surface. As a result, an inhomogeneous rather than a homogeneous high-density layer is formed. One possible conformation for the adlayer is that a fraction of the adsorbed molecules might have a tail-on conformation with part of the surface being uncovered. Although such a surface may have a sufficiently high EG surface density, it would not be expected to be protein-resistant. Furthermore, for this particular polymer, the standard deviations in the masses of the adsorbed polymer and the proteins are much higher in comparison to those of the other polymers, again pointing to pronounced heterogeneity of the adlayer structure. This observation suggests a limitation in the ability to repel proteins for high packing density of the PEG in the PLL-*g*-PEG molecule.

To visualize the steric constraints of the high-density PEG surface, a simple geometric packing model was applied on the basis of the experimentally determined quantitative PEG surface density data, following an approach published in a previous study (Figure 8).[42] We assume as a first approximation that the backbone lies flat on the surface and that the PEG chains retain a spherical shape (as free PEG has in solution) with a size given by the radius of gyration.[16] The radius of gyration (R_g) of the PEG side chains was calculated with an empirical formula based on static light-scattering measurements:[53]

$$R_g = 0.181N^{0.58}\ \text{(nm)} \tag{4}$$

where N is the number of ethylene glycol repeat units. Equation 4 results in radii of gyration of 1.1, 1.7, and 2.8 nm for PEGs of molecular weights 1, 2, and 5 kDa, respectively. The average spacing between the PEG side chains (L) was then calculated from the copolymer area density data, assuming a hexagonal (close-packed) pattern:

$$L = \left(\frac{2}{\sqrt{3}n_{PEG}}\right)^{0.5} \tag{5}$$

where the PEG chain surface density, n_{PEG}, is determined from the adsorbed mass of polymer and its grafting ratio (eq 2). Mean distances between PEG side chains were found to vary between 1.1 and 2.6 nm for 1 kDa PEG, 1.5 and 3.3 nm for 2 kDa PEG, and 2.3 and 4.0 nm for 5 kDa PEG for the polymer architectures investigated in this

(53) Kawaguchi, S.; Imai, G.; Suzuki, J.; Miyahara, A.; Kitano, T. *Polymer* **1997**, *38*, 2885–2891.

work. The radius of gyration of the PEG molecule can then be compared to the average spacing between two chains by using a characteristic parameter, $L/2R_g$, to describe the degree of overlap between the idealized spherical chains (Figure 8). If $L/2R_g$ is smaller than one, the PEG spheres overlap, meaning that the chains must get stretched out, toward a brush structure. A graphical representation of protein adsorption as a function of this characteristic parameter (Figure 8) shows that the PEG chain spheres need to overlap considerably in order to render the surface protein-resistant. Protein adsorption only drops to very small values for surfaces with $L/2R_g \leq$ 0.5−0.7, which is in good agreement with findings published formerly.[32,42,52] This confirms that PEG chains in the adsorbed layer have to be highly deformed from their original random coil conformation, forming a brush-type PEG structure, provided we assume PLL lies flat.[32] Furthermore, this deformation is likely to result in a significant rearrangement of those water molecules associated with the PEG film.

The use of serum is certainly relevant for biological (e.g., cell culture) studies and diagnostic and medical applications, but it is also complex in the sense of the presence of many different proteins with different molecular weights and properties. Therefore, we are not able to relate our protein adsorption data to models that cover, for example, the aspect of size-dependence of protein adsorption in PEG chain structures.[23] However, we are currently investigating single protein adsorption and AFM-based surface force measurements on the same type of PLL-*g*-PEG surfaces, expecting that these data will provide further insight into the role of steric and electrostatic effects for the adsorption of proteins to well-defined PEG chain interfaces.

5. Conclusions

In this study we synthesized a number of defined architectures of the copolymer poly(L-lysine)-*graft*-poly(ethylene glycol) (PLL-*g*-PEG) and functionalized niobium oxide surfaces in a highly controlled manner through spontaneous self-assembly from aqueous solution. Combining the ex situ ToF-SIMS and the in situ OWLS techniques permits quantitative information to be gained regarding polymer surface coverage, the surface density of the PEG chains, and the degree of protein resistance upon contact with serum. The formation of the polymeric monomolecular adlayer is shown to be governed by the balance between an attractive electrostatic force anchoring the copolymer backbone, poly(L-lysine), to the surface and a repulsive steric term originating from PEG−PEG side chain interactions. At low PEG/lysine ratios of the polymer, the adsorbed mass is essentially governed by the electrostatic interaction and space requirements of the PLL backbone, while, at high PEG/lysine ratios, PEG chain packing density limitations become the dominant factor that controls adsorbed protein mass.

Serum-exposure data indicated a close relationship between protein adsorption and the architecture of the polymeric interface. The ability of PLL-*g*-PEG coatings to resist protein adsorption could be attributed in a quantitative manner to the presence of a minimum ethylene glycol monomer surface density and the formation of a dense PEG-brush. The spontaneous assembly of PLL-*g*-PEG with different grafting ratios proved to be a very promising and efficient way to precisely control the interactiveness of the surface toward serum proteins. At ethylene glycol surface densities $\geq 20\ nm^{-2}$, highly protein resistant surfaces could be produced with adsorbed serum masses below the detection limit of the OWLS technique ($<2\ ng/cm^2$). Such stable, protein-resistant, high-PEG-density brush surfaces have numerous applications in bioaffinity sensors and blood-contacting devices, while surfaces with intermediate PEG densities may have potential for the controlled, physical adsorption of biomolecular entities, such as proteins, enzymes, and growth factors, in a biocompatible, PEG-chain-supported environment.

Acknowledgment. The authors gratefully acknowledge Drs. Paul Hug and Beat Keller at EMPA, Switzerland, for their help with the ToF-SIMS measurements, Dr. Michael Horrisberger at PSI for performing the niobium oxide coating, and Drs. Ning-Ping Huang (now at the Biozentrum, Basel) and Stephanie VandeVondele at ETH Zürich for their support in the synthesis of the polymers. Dr. Ilya Reviakine, now at the University of Houston, Prof. Jeffrey A. Hubbell, ETH Zürich, Prof. Hans J. Griesser, University of South Australia, Prof. David G. Castner, University of Washington (Seattle), Dr. Matthew S. Wagner, NIST (Gaithersburg), and Dr. Oleg Borisov, Université de Pau (France), are thanked for valuable discussions. This work was financially supported by EPF Lausanne and ETH Zürich (Project TH-33./01-3), the Swiss Commission for Technology and Innovation (CTI, Project 5439.1), and the International Team for Oral Implantology (ITI), Basel, Switzerland (Project 192).

Note Added after ASAP Posting. This article was posted ASAP on the Web on 9/13/2003. Changes were made to the ranges of the mean distances between PEG side chains. The correct version was posted on 9/24/2003.

LA034111Y

Biophysical Journal Volume 88 January 2005 495–504 495

Interaction Forces and Morphology of a Protein-Resistant Poly(ethylene glycol) Layer

M. Heuberger, T. Drobek, and N. D. Spencer
Laboratory for Surface Science and Technology, Department of Materials, ETH Zürich, Zürich, Switzerland

ABSTRACT The molecular interactions on a protein-resistant surface coated with low-molecular-weight poly(ethylene glycol) (PEG) copolymer brushes are investigated using the extended surface forces apparatus. The observed interaction force is predominantly repulsive and nearly elastic. The chains are extended with respect to the Flory radius, which is in agreement with qualitative predictions of scaling theory. Comparison with theory allows the determination of relevant quantities such as brush length and adsorbed mass. Based on these results, we propose a molecular model for the adsorbed copolymer morphology. Surface-force isotherms measured at high resolution allow distinctive structural forces to be detected, suggesting the existence of a weak equilibrium network between poly(ethylene glycol) and water—a finding in accordance with the remarkable solution properties of PEG. The occurrence of a fine structure is interpreted as a water-induced restriction of the polymer's conformational space. This restriction is highly relevant for the phenomenon of PEG protein resistance. Protein adsorption requires conformational transitions, both in the protein as well as in the PEG layer, which are energetically and kinetically unfavorable.

INTRODUCTION

Poly(ethylene glycol) (PEG) surface grafts have gained considerable attention as being stable films that provide resistance to nonspecific protein adsorption (Elbert and Hubbell, 1996; Harris and Zalipsky, 1997). This protein resistance is of practical importance for a number of surface-engineering applications including biomaterials, drug delivery, or biosensors.

The molecular mechanisms underlying protein resistance of grafted PEG have not yet been fully identified. Two different aspects of the problem are usually considered: first, the brush-induced "steric" repulsion that is thought to prevent direct contact between proteins and the underlying surface (Szleifer, 1997); and second, hydration shells, which energetically suppress adsorption of proteins onto the PEG layer. Recently, it has been argued that the protein resistance also observed on well-ordered model systems of self-assembled monolayers of oligo(ethylene glycol) may alternatively be due to charging effects (Chan et al., 2003; Herrwerth et al., 2003; Kreuzer et al., 2003).

The PEG-water interaction is truly remarkable, as one can see from its unusual and well-documented solution properties. It is commonly understood that PEG is an amphiphilic polymer, which is only structurally soluble in water, i.e., hydrated PEG preferentially resides in polar *gauche*-conformations (Björling et al., 1991) that closely match the structure of water. It is widely recognized that there must be a particularly enhanced water structure around PEG chains (Müller and Rasmussen, 1991) that gives rise to a solubility gap (phase separation) at higher solution temperatures (Karlström, 1985; Saeki et al., 1976). Based on thermodynamic considerations, a widely cited hydration model has been proposed (Kjellander and Florin, 1981), which portrays a helical PEG conformation encaged into a surrounding water structure. Indeed, the existence of extensive equilibrium hydration structures was postulated based on heat-capacity measurements (Kingman et al., 1990). A minimum of ~2–3 water molecules per PEG monomer seems to be required to complete basic hydration (Maisano et al., 1993).

Low-molecular-weight PEG does not readily adsorb onto surfaces. It must be grafted to the surface to form a brush layer of sufficiently high surface density to achieve protein resistance. Common grafting strategies include the use of block copolymers (Costello et al., 1993), electrostatic adsorption (Claesson and Gölander, 1987), chemical functionalization (Himmelhaus et al., 2003; Prime and Whitesides, 1993; Raviv et al., 2002), surface-induced polymerization (Ma et al., 2004), cross-linking of star-shaped PEG (Groll et al., 2004), or covalent grafting of PEG chains to phospholipids (Kuhl et al., 1994; Tirosh et al., 1998).

In this work, poly(ethylene glycol) was grafted onto a cationic poly-L-lysine backbone, forming the comb-like polymer, poly(L-lysine)-graft-poly(ethylene glycol) (PLL-*g*-PEG). The PLL backbone acts as an anchor onto anionic surfaces such as mica (Fig. 1 *a*). This type of copolymer exhibits an extraordinarily strong adsorption onto many different surfaces. Different PLL-*g*-PEG architectures are known to exhibit different degrees of protein resistance (Kenausis et al., 2000; Pasche et al., 2003). In this article, we have focused on the most efficient architecture for protein resistance.

Our motivation in studying these PEG films in the extended surface-forces apparatus (eSFA) is to obtain additional

Submitted May 4, 2004, and accepted for publication October 15, 2004.

Address reprint requests to M. Heuberger, E-mail: manfred.heuberger@mat.ethz.ch.

0006-3495/05/01/495/10
doi: 10.1529/biophysj.104.045443

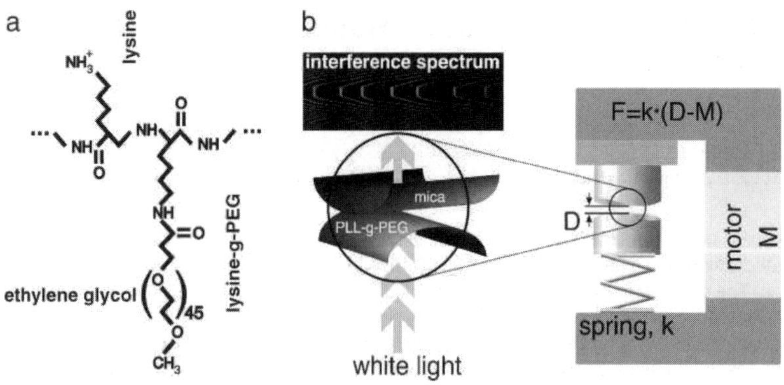

FIGURE 1 (*a*) Chemical structure of PLL-*g*-PEG used in this study. Before the grafting reaction in solution, the molecular weight of PLL was 12 kDa (polymerization: $\xi_{PLL} = 96$, $r_F \approx 5.4$ nm) and for PEG is was 2 kDa (polymerization: $\xi_{PEG} = 45$ and $r_F \approx 3.4$ nm). The PEG was then grafted to one lysine out of $g = 3.5$. (*b*) Schematic setup of the extended surface forces apparatus used to measure the surface force isotherms. The surface separation, *D*, is measured using thin-film interferometry in transmission (see interference pattern on *left*) while a calibrated motor, *M*, approaches and compresses the surfaces via a compliant spring, *k*. The mica surfaces are in crossed-cylinder geometry and optionally coated with the PLL-*g*-PEG copolymer.

insights into the molecular morphology and interactions of the adsorbed copolymer film and to identify the molecular mechanisms responsible for its excellent protein resistance.

MATERIALS AND METHODS

We have adsorbed PLL-*g*-PEG onto clean mica surfaces and measured the surface-force isotherms between them in aqueous solutions using the eSFA. Additionally we have measured the surface forces in the case where only one mica surface was covered with the adsorbed polymer.

Copolymer synthesis and chemical structure

The PLL-*g*-PEG used for this study was synthesized and characterized in our group according to procedures previously described in great detail (Huang et al., 2000; Pasche et al., 2003). The molecular weight of the poly(L-lysine) was 12 kDa, corresponding to 20 kDa PLL-HBr used for synthesis, and the polydispersity was $M_W/M_n = 1.2$. The molecular weight of the poly(ethylene glycol) was 2 kDa and the polydispersity $M_w/M_n = 1.1$. The structure of this copolymer is illustrated schematically in Fig. 1 *a*. The lysine side chains are positively charged in aqueous solutions, with a $pK = 10.5$. The grafting ratio, *g*, is defined as the total number of lysine units divided by the number of PEG-modified units. The grafting ratio of the PLL-*g*-PEG copolymer used in this study was $g = 3.5$. The degree of polymerization of PLL is $\xi_{PLL} = 96$ and the polymerization of PEG $\xi_{PEG} = 45$. Therefore, each PLL-*g*-PEG copolymer carries an average number of 27 grafted PEG side chains and 69 positively charged lysine side chains. The average distance between PEG grafting points along the PLL backbone is $d_0 \approx 1.2$ nm. The total molar mass of one PLL-*g*-PEG molecule is 67 kDa. This particular PLL-*g*-PEG architecture (i.e., grafting ratio and molecular weight) was chosen because it results in a highly protein-resistant adsorbed layer (Kenausis et al., 2000; Pasche et al., 2003).

Copolymer adsorption

We used the cleavage plane of mica crystals as a model substrate in this work. Ruby mica is known to expose negative surface sites on the basal cleavage plane that are neutralized by K^+ ions in the bulk crystal. The remaining K^+ ions can dissociate in an aqueous environment and a double layer is formed in the vicinity of this negatively charged surface. The surface density of such ion sites $(0.48\ nm^2)^{-1}$ is determined by the crystal structure.

The adsorption process of PLL-*g*-PEG on mica is controlled by electrostatic forces, Van der Waals attraction, and steric contributions that originate from the grafted PEG side chains. PEG_{2000} itself is of too low a molecular weight to adsorb efficiently on mica. The adsorption kinetics of the PLL-*g*-PEG copolymer have been investigated on metal oxide substrates in great detail and were found to be very fast and highly irreversible (Kenausis et al., 2000).

In our study, PLL(12 k)-*g*(3.5)-PEG(2 k) was adsorbed onto the mica substrates from a 1 mg/ml aqueous PLL-*g*-PEG solution ex situ with an incubation time of 30 min at room temperature. The substrates were thoroughly rinsed with ultrapure water to remove any possible excess copolymer or ions. After PLL-*g*-PEG adsorption, the substrates were kept wet at all times during handling.

Aqueous solutions used

All experiments were performed using ultrapure water (puriss. p.a., 0.2 μm membrane filtered; Fluka, Milwaukee, WI) as received. The fluid cell of the eSFA has a volume of ~80 ml. For the addition of salt, 10 ml of the water was drained from the SFA and replaced by 10 ml of a 0.2-μm-filtered solution of KCl (ultra, >99.0%; Sigma, St. Louis, MO) at the appropriate concentration.

Mica surface preparation

Thin mica sheets were prepared by manual cleavage of optical-quality ruby mica blocks in a class-100 laminar-flow cabinet. Sets of thin mica pieces of uniform thickness (in the range of 1.5–4.5 μm) were obtained at an approximate size of 20 × 20 mm. A number of smaller (≈8 × 8 mm) rectangular pieces were cut from the larger sheet.

To avoid potential complications due to the recently discovered nanoparticles on conventionally cut mica (Heuberger and Zäch, 2003; Kohonen et al., 2003; Ohnishi et al., 1999), we used surgical scissors for cutting. The cut mica sheets were deposited on a thicker, freshly cleaved mica block for intermediate storage and further handling. One edge of each mica piece was overlapped with a clean Teflon tape to facilitate subsequent liftoff.

The mica sheets were then coated with a silver film of 40 nm thickness by means of thermal evaporation in vacuum ($P_{base} < 5 \times 10^{-6}$ mbar). The silvered mica sheets were finally lifted off and glued onto cylindrical lenses using epoxy resin glue. After gluing, the samples were immediately (≈5 min) inserted into the sealed measurement apparatus. The fluid cell was then purged with dry nitrogen and the mica thickness determined using thin-film interferometry.

The extended surface-forces apparatus

The experimental data were obtained using the eSFA (Fig. 1 *b*), which is an enhanced and automated version of the SFA 3 (Surforce, Santa Barbara, CA)

(Israelachvili and McGuiggan, 1990). The eSFA is located inside a thermally insulated enclosure at a controlled temperature at 25.00(2)°C (Heuberger et al., 2001). The mica surfaces are arranged in crossed-cylinder geometry inside a fluid cell. The optical measurement of surface separation is based on thin-film interferometry (Israelachvili, 1973). The eSFA uses fully automated spectrum acquisition and evaluation, which permits measurements to be carried out rapidly and with great precision. Interference spectra are evaluated by means of the numerical fast spectral correlation algorithm (Heuberger, 2001), allowing the determination of both film thickness, D, and refractive index, n, at high acquisition rates (1–10 Hz). The instrumental precision of film-thickness measurement is typically $\sigma_D = \pm 25$ pm, and for the refractive index it is $\sigma_n < \pm 0.05$ (for $D > 1$ nm). A detailed error calculation is given elsewhere (Heuberger, 2001). In a typical experiment, the optical thickness of the mica substrates is first determined in a dry N_2 atmosphere while the mica surfaces are in unloaded adhesive contact.

All surface-force isotherms shown here were measured by approaching the surfaces toward each other with an actuator at a constant velocity of $v = 1.00(5)$ nm/s. Surface forces, $F(D)$, are calculated based on the difference between the fine-calibrated actuator movement, $M(t)$, and the optically measured film thickness, $D(t)$, measured in the point of closest approach. To obtain the normalized surface force, one can use the simple relationship

$$\frac{F(D)}{R} = \frac{k}{R}(D - M) = 2\pi E(D), \qquad (1)$$

where $R = (R_x \times R_y)^{0.5}$ is the effective radius representing the local surface curvature and $k = 1002(23)$ N/m is the compliance of the force-measuring spring (Fig. 1 *b*).

The calculated force, $F(D)$, is normalized by the effective local radius, $R \approx 18(2)$ mm of curvature. The mica surfaces showed only minor elastic flattening during compression of the polymer film, as seen from an analysis of the interference fringes. Using the Derjaguin approximation, one can thus estimate the free energy per unit area, $F/R \approx 2\pi E(D)$, which allows easy comparison between data and theory (Derjaguin, 1934; Israelachvili, 1991).

The calculation of this free energy (i.e., F/R) is subject to instrumental errors that are discussed in detail elsewhere (Zäch and Heuberger, 2000). Typical drift rates of the eSFA are 20 pm/min (Heuberger et al., 2001) and the cylinder radii, R_i, can be determined with <5% relative error using lateral scanning (Heuberger, 2001). In this work, the accuracy of $F(D)/R$ was typically ± 0.1 mN/m. The error bars shown in Figs. 2, 3, and 5 correspond to this instrumental error. Errors due to variations in sample preparation are discussed in the text.

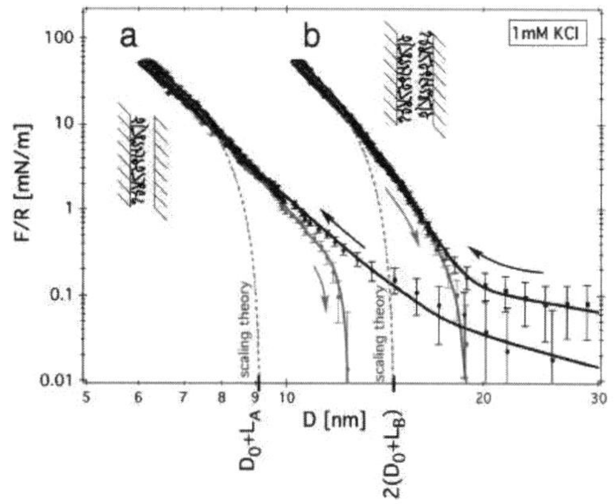

FIGURE 2 Surface-force isotherms (*black* is loading; *gray* is unloading) measured in two situations: (*a*) where only one, or (*b*) where both mica surfaces were covered with the PLL-*g*-PEG copolymer. The data points shown here were obtained in 1 mM KCl solution at 25°C. The data points were continuously acquired while the motor (c.f. Fig. 1 *b*) was operated at constant velocity, $dM/dt = -1$ nm/s; this corresponds to a compression rate of $d(F/R)/dt \approx 55\ \mu\text{Nm}^{-1}$/s. The marks, L_A and L_B, designate the PEG brush length obtained from a fit of the data (*dashed gray lines*) to the de Gennes scaling theory. Error bars represent instrumental errors, as discussed in the Methods section.

RESULTS

Surface-force isotherms

Typical surface-force isotherms, $F(D)/R$, for adsorbed PLL-*g*-PEG copolymer films measured in 1 mM KCl aqueous solution are shown in Fig. 2. The value $D = 0$ nm indicates the absolute location of the mica surfaces (± 0.2 nm, absolute). Under compression, D represents the polymer film thickness. Two curves are shown: in curve *a* the adsorbed copolymer covers only one mica surface; and in curve *b* both surfaces are covered by the adsorbed copolymer film. The measured force is chiefly repulsive, i.e., $F/R > 0$. The high-load regime can be well described by an exponential function. The film thickness of curve *b* is slightly less than twice that of curve *a* for a comparable compression. The exponential decay length is larger for the symmetric case. We note that the highly compressed copolymer film ($F/R > 1$ mN/m) deforms virtually free of hysteresis, which suggests an equilibrium measurement. The absence of hysteresis under compression is in agreement with previous force measurements of surface-grafted PEG chains (Efremova et al., 2001; Sheth and Leckband, 1997). In the highly compressed region, we found that the repulsive force was independent of the salt concentrations used here. This behavior is expected because of the nonionic nature of PEG and is in agreement with previous studies on grafted PEG brushes (Raviv et al., 2002). Only in the low-force regime, where the polymer is barely compressed, did we observe weak attractive forces (see below).

Refractive index and adsorbed mass

Simultaneously with the interferometric measurement of the surface separation, D, we have determined the refractive index, $n(D)$ (Fig. 3). This is a measure of the average density in a volume of thickness D and diameter $\varnothing \approx 1\ \mu$m. Because the refractive index of the adsorbed copolymer is higher than that of water, we expect to observe an increase of n as the surface separation is decreased. To estimate the total amount of PEG between the surfaces from the refractive index we can adopt a simple model (Raviv et al., 2002)

$$n(D) = n_{H_2O} + m\zeta(n_{PEG} - n_{H_2O})/D \quad \text{for} \quad D > \zeta. \qquad (2)$$

$\zeta = \Gamma/\rho$ is the equivalent thickness of the dried PLL-*g*-PEG layer; Γ is the adsorbed PLL-*g*-PEG mass per surface area, and $\rho \approx 1.12$ g/cm^3 is the dry polymer density. If we further assume two identical copolymer layers on both surfaces, we

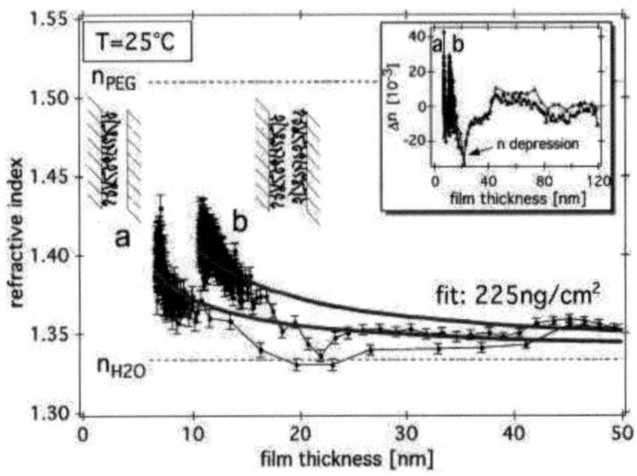

FIGURE 3 Refractive index measured simultaneously with the surface-force isotherms for the two separate situations: (*a*) where only one, or (*b*) where both mica surfaces were covered with the PLL-*g*-PEG copolymer. The medium was ultrapure water in equilibrium with atmospheric CO_2 at a temperature of 25°C. The solid gray lines correspond to the refractive index envelope predicted by the simple model described in Eq. 2. The error bars shown along the solid lines represent the estimated statistical errors based on the standard deviation of $\pm\ \sigma_\lambda = 10$ pm for interference peak detection, which is typical for this type of fast spectral correlation setup (Heuberger, 2001). The inset shows the difference between the simple model and the measurement; a significant depression of refractive index is visible at a surface separation of 20 nm. This *n*-depression is reproducible between different samples and in quantitative agreement with a recently reported ≈10% density decrease found in water adjacent to a PEG-ylated surface (Schwendel et al., 2002). Error bars represent instrumental errors, as discussed in the Methods section.

can set $m = 1$ for the asymmetric case and $m = 2$ for the symmetric case.

Fitting Eq. 2 to our data yields an adsorbed polymer mass $\Gamma \approx 200 \pm 100$ ng/cm^2, which is in good agreement with the typical adsorbed mass of this PLL-*g*-PEG architecture on various metal oxide substrates, as determined from an optical waveguide technique (180 ± 20 ng/cm^2) (Kenausis et al., 2000; Pasche et al., 2003).

The inset of Fig. 3 shows the difference between the model described by Eq. 2 and the measurement. We observe a significant depression of the refractive index at a separation of $D_{min} = 21 \pm 1$ nm. If interpreted in terms of water density, this is equivalent to a zone of 5–10% reduced density 5–10 nm above the PEG layers. We note that a quantitatively similarly density depression was recently suggested based on neutron reflectivity measurements (Schwendel et al., 2002).

Comparison with theory

To compare the surface-force isotherms with theory, we have to consider several different surface-force contributions. Van der Waals forces between mica surfaces can be neglected here because the closest mica separations are beyond the range of detectable force. Due to the charged nature of both mica and poly-lysine, we need to consider the presence of double-layer forces. We invoke the well-known double-layer model to describe the exponential force profiles at large separations (i.e., outside the polymer layer) with the characteristic Debye length, κ^{-1}, which depends on the ionic composition of the solution:

$$\text{Debye length in } [m]\kappa^{-1} = \sqrt{\frac{\varepsilon_0 \varepsilon_r k_B T}{10^3 N_A e^2 \sum_i C_i z_i^2}}, \tag{3}$$

where ε_0 is the dielectric field constant, ε_r the relative dielectric constant of the aqueous solution, k_B the Boltzmann constant, $T = 298.15$ K the absolute temperature, N_A the Avogadro number, e the electronic charge, C_i the molar concentration, and z_i the valency, of each ionic species, i, in the solution.

Assuming a constant dielectric permittivity for the aqueous solutions, $\varepsilon_r \approx 80$, the Debye lengths for ultrapure water (pH = 5.6, dissolved CO_2) and 1 mM KCl solution are $\kappa^{-1}(H_2O) \approx 190$ nm and κ^{-1}(1 mM KCl) ≈ 9.7 nm, respectively. The measured surface-force isotherms were fitted in the regime of sufficient surface separation, i.e., 50 nm $< D <$ 2.5 μm, using the well-known expression of double-layer forces between curved surfaces (Verwey and Overbeek, 1945):

$$\frac{F(D)}{R} = \frac{128\pi \times k_B T \times 10^3 N_A \times C}{\kappa} \times \tanh^2\left(\frac{e\Psi}{4kT}\right) \times e^{-\kappa D}$$
$$\text{for } D > \kappa^{-1}, \tag{4}$$

where C is the molar concentration of the salt solution and Ψ the surface potential in Volts.

There is no significant long-range force in 1 mM KCl ($\Psi = 20 \pm 20$ mV). In pure water the measurement was of similar magnitude, but less reliable due to the accumulation of small nonlinearities in the actuator over the longer distance range.

The surface-force isotherms presented in Fig. 2 reveal a strong exponential repulsion at small surface separations. This force is commonly assigned to the steric repulsion known for polymer-bearing surfaces. If the number of available conformers is restricted by an opposing surface, the associated entropy reduction gives rise to a repulsive force. Depending on the average distance, d, between PEG grafting sites, one can invoke different theoretical models to describe this steric force. For low-molecular-weight PEG, in particular, there is some controversy in the literature about the correct assignment of these regimes (Hansen et al., 2002). The so-called "brush" regime is often invoked for $d < 2r_F$, and the so-called "mushroom" regime for $d > 2r_F$. The quantity r_F is the Flory radius of the polymer random coil in a good solvent,

$$r_F = a \times \xi_{PEG}^{0.6} \approx 3.4\,\text{nm}, \tag{5}$$

where $a = 0.35$ nm is the PEG monomer size (Hansen et al., 2002), and $\xi_{PEG} = 45$ the degree of polymerization of PEG_{2000}.

The average distance, d, between PEG surface grafting sites determined from the optically measured adsorbed mass (200 ± 100 ng/cm^2) is $d_n = 1.4 \pm 0.4$ nm. Because this optically determined d_n is clearly smaller than the random coil size, $2r_F = 6.8$ nm, we conclude that the chains are extended with respect to their Gaussian dimension. The length, L, of such a brush can be predicted qualitatively from this grafting distance and the known number of monomer units. Using the scaling relation proposed by Alexander (1977) we have:

$$L = \xi_{PEG} \times \frac{a^{5/3}}{d^{2/3}} \approx 6.3\,\text{nm}, \quad \text{for} \quad PEG_{2000}. \tag{6}$$

Scaling theory strictly applies to highly extended (i.e., high lateral compression), end-grafted polymers of high molecular weight, so it can only be considered as an approximate guide here. We can also attempt to fit the parameters of the de Gennes scaling theory for brushes (deGennes, 1987) to reproduce the measured surface-force isotherm. After integration of the brush-induced pressure (Eq. 17 in deGennes, 1987), scaling theory predicts a surface force of the form:

$$\frac{F(D)}{R} = C_1\left[7\left(\frac{D - mD_0}{mL}\right)^{-5/4} + 5\left(\frac{D - mD_0}{mL}\right)^{7/4} - 12\right] \quad \text{for} \quad D < L, \tag{7a}$$

where

$$C_1 = \frac{8\pi}{35} m\,k_B T \frac{L}{d^3}\varepsilon \tag{7b}$$

with $\varepsilon \approx 1$ a dimensionless prefactor and $m = 1$ for the asymmetric experiment (Eq. 7a) and $m = 2$ for the symmetric experiment (Eq. 7b). The parameter D_0 is an offset that accounts for the finite thickness of the underlying PLL anchor.

Fitting of the surface-force isotherm (Fig. 2) gives a reasonable quantitative agreement in the regime of strong compression. Fixing the offset to $D_0 = 1$ nm gives the following brush lengths: $L_A = 8.2$ nm and $L_B = 6.3$ nm. The result, $L_b < L_a$, indicates the presence of interdigitation effects in case b as opposed to case a. Interdigitation is not included in this simple model and will be discussed in more detail in a later publication.

Weak attractive forces

A magnification of the weak-force regime reveals rather small attractive forces. In Fig. 4 *a*, two different isotherms are displayed for the case of polymer adsorbed onto one surface only. One curve is measured in pure water (*triangles*) and one is measured in 1 mM KCl solution (*squares*). The weak double-layer force contribution has been mathematically subtracted for better illustration of the hysteresis. The isotherms were fitted to Eq. 4 outside the range of the steric

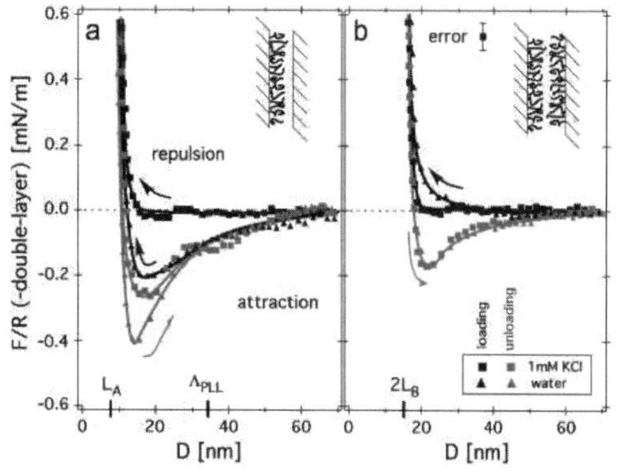

FIGURE 4 Magnification of surface-force isotherms in the low-force regime. The weak double-layer repulsion forces have been subtracted to illustrate the small hysteresis. Black symbols are loading cycles and gray symbols are unloading cycles; squares stand for 1 mM KCl solution and triangles for water; panel *a* illustrates the attractive forces in the asymmetric case, and panel *b* shows those observed in the symmetric case. The loading/unloading speed was 1 nm/s. The absolute errors in this graph are dominated by uncertainties of fitting (and subtracting) the weak double-layer force. For the sake of simplicity the estimated error bar is shown exemplarily for one point only.

repulsion (i.e., >30 nm) and then extrapolated into the range shown here. In the pure water, the Debye length is sufficiently large to justify a linear approximation of Eq. 4. The loading branch in pure water (*black triangles*) revealed a small attractive force that is strongest at $F/R(D = 16 \pm 2$ nm$) = -0.18 \pm 0.05$ mN/m. The maximum range of these small attractive forces is $\sim 48 \pm 5$ nm. These small attractive forces are not observed during approach in salt solution (*black squares*). The gray curves show the corresponding unloading branches. The adhesive minima are located in the same range of surface separation, i.e., $D = 16 \pm 2$ nm. The difference of the force between loading and unloading branches was comparable in both solutions. For comparison, the length of a fully stretched PLL backbone is $\Lambda_{PLL} \approx 34$ nm and a fully stretched PEG chain is $\Lambda_{PEG} \approx 17$ nm.

The curves shown in Fig. 4 *b* illustrate the weak-force regime for the case of polymer adsorbed onto both surfaces. As described above (Fig. 4 *a*), the weak double-layer force contribution has been subtracted. In both solutions we observed no attractive forces during loading and a weak attractive force during unloading at a surface separation of $D = 21 \pm 1$ nm. The small repulsion observed in pure water is at the significance limit of this measurement.

Embedded fine structure

When we magnify the D axis of the surface-force isotherms in the repulsive regime, as exemplified in Fig. 5, we find a ubiquitous fine structure in the form of distinctive film-thickness transitions. These transitions are embedded into the

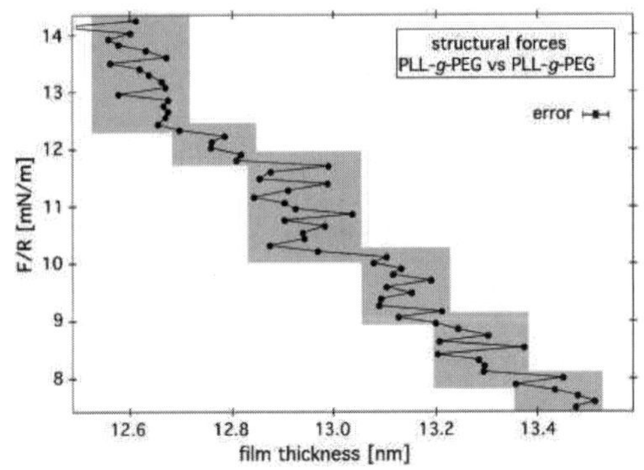

FIGURE 5 Magnification of the surface-force isotherm between two PLL-*g*-PEG covered surfaces in pure water. Discrete film-thickness transitions with a step size 125 ± 25 pm (statistics based on 40 isotherms) reveal the presence of long-lived, load-bearing equilibrium structures in the PLL-*g*-PEG film. Similar film-thickness transitions are also observed in the case where only one surface is decorated with PEG. The transitions are observed in sequences of 6–10 individual transitions, separated by regions of similar extension (≈1–1.5 nm) without clear transitions. The loading speed was held constant at 1 nm/s using a closed feedback loop. For the sake of simplicity, only one error bar is shown, which represents the instrumental precision of this measurement.

known steric repulsion curve described above (Fig. 1). The characteristic step size is 125 ± 25 pm. The data points were acquired at ≈1.9-s intervals, which means that some of the observed features effectively have a lifetime of at least 30 s. A similar fine structure could be observed for all salt concentrations used here, both for the symmetric and the asymmetric case, and, in a temperature range between 1 and 40°C.

Manual analysis of ~40 surface-force isotherms taken at 25°C reveals that these steps tend to occur in groups of 6–10 at diverse film compressions, starting around the nominal brush length. Similar groups of steps are also observed during unloading—often at comparable film compressions to those observed during loading. The surface-force isotherms are free of hysteresis in this load regime. This suggests that these fine transitions are caused by equilibrium structures that exist within this hydrated PLL-*g*-PEG layer.

The fine structure is not an instrumental artifact. The precision of the interferometric distance measurements used here is ±25 pm, as previously shown in detail (Heuberger, 2001). The actuator did not induce the steps because it moved at constant velocity (relative error <5%), as opposed to the stepwise measurement of force isotherms often used in other studies. More generally, we can exclude any instrumental instabilities (e.g., drift, vibrations, actuator irregularities) as the origin of the fine structure by the following argument. The sample surfaces are mechanically coupled to the eSFA body via the force-measuring spring (Fig. 1 *b*). The contact stiffness of the polymer film can be seen as a nonlinear spring in series with this force-measuring spring and the other much stiffer springs in the apparatus. At the compression where steps were observed, the contact stiffness varies from ~100 kN/m to 1000 kN/m (c.f. slope of isotherm in Fig. 2). This is 1000 times higher than the spring constant, $k = 1$ kN/m, of the force-measuring spring. To induce film transitions of the observed size in the contact zone, the eSFA frame would thus have to perform quasiperiodic jumps over distances of 125 nm. Careful analysis of high-resolution data both with and without actuator action at various surface separations and actuator speeds reveals that the largest instrumental instabilities are three orders of magnitude too small to produce such an artifact (Heuberger et al., 2000).

It is also important to note that the observed step size does not vary systematically with the amount of overall film compression. Because the effective stiffness of the polymer film varies over a wide range in this experiment, we can conclude that the transitions are indeed an intrinsic property of the polymer layer and reflect a characteristic structural length scale.

DISCUSSION

Based on the surface force and refractive index measurements, we propose the model of a well-organized film (Fig. 6). The PLL backbone is adsorbed to the mica surface. The in-plane order of the adsorbed PLL backbones cannot be determined here, but the high effective grafting density ($d_n \approx 1.4$ nm) compared to the PEG grafting distance along the PLL backbone, $d_0 \approx 1.2$ nm, indicates that the PEG side chains are indeed standing up and that the PLL backbones cover most of the available surface area. The PEG chains are in a brush-like configuration. For an adsorbed mass of 200 ng/cm^2 and a film thickness of 8.2 nm, the volume fraction of water in the film is 82%, which corresponds to a number of 10 water molecules per ethylene-oxide (ethylene oxide)-mer present in the film.

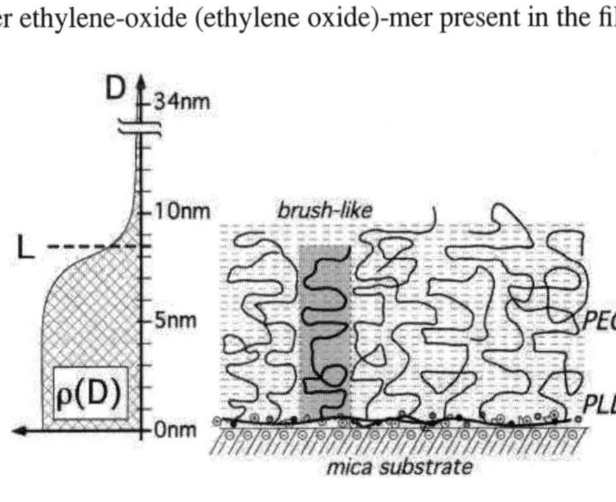

FIGURE 6 Schematic model exemplifying the molecular morphology of one PLL-*g*-PEG film adsorbed to mica based on the compression isotherms discussed in the text. The copolymer film has a high degree of molecular organization. The PLL backbone is strongly adsorbed to the mica surface and the PEG side chains extend away from the surface in a brush-like configuration. The proposed polymer density profile, $\rho(D)$, is shown on the left. The gray dashed lines in the background denote the underlying hydration structure, which gives rise to a quantized fine structure in the compression isotherm (c.f. Fig. 5).

A significant amount of PLL on the outside of the polymer film can be excluded due to the absence of strong electrostatic and adhesive surface forces against clean mica. However, the weak attractive forces shown in Fig. 4 *a* indicate that a very small number of PLL chain segments can still directly interact with the opposing surface. The differences of the weak attraction seen with different ionic concentrations can be explained as follows. The Debye length in 1 mM KCl aqueous solution is $\kappa^{-1} \approx 9.6$ nm, as opposed to $\kappa^{-1} \approx 190$ nm in pure water (pH 5.6 due to dissolved CO_2). As a consequence, extending poly-cationic PLL segments are expected to contract in salt solution. Additionally, electrostatic anchoring to the opposing negatively charged mica surface is less likely due to the enhanced double layer.

We have shown that the short-range repulsion can be reasonably well described by theories of grafted flexible polymer chains in a brush-like configuration (Alexander, 1977; deGennes, 1987). Analysis of the surface-force isotherms in terms of scaling theory at low forces indicate that the density profile is not rectangular, but must exhibit a finite polymer density beyond the length of the de Gennes brush (Fig. 2) (Hansen et al., 2002; Milner and Witten, 1988; Zhulina et al., 1989). Comparison of de Gennes brush lengths between the asymmetric (*a*) and symmetric (*b*) experiments yields $L_b < L_a$. This finding indicates that there is excess entropy of interdigitation in case *b* in contrast to case *a*. Interdigitation reduces the apparent brush length. This effect is not accounted for in the theoretical model, c.f. Eq. 7 and is sometimes neglected in comparable experimental studies.

The steric repulsion exhibited by the PLL-g-PEG film is in good qualitative agreement with previous direct force measurements reported for comparable grafted PEG systems (Claesson and Gölander, 1987; Costello et al., 1993; Efremova et al., 2001; Kenworthy et al., 1995; Kuhl et al., 1994; Needham et al., 1992; Raviv et al., 2002). However, the observation of fine film-thickness transitions is unprecedented. The question is, thus, how can the PEG layer consist of flexible chains and still exhibit quantized structural forces?

At a sufficiently high density, a surface-adsorbed polymer might be in a glass-like state (Kremer, 1986). It has been shown that many water-soluble polymers can exhibit a small but finite yield strength when spread and compressed at the water/air interface. However, in these studies, PEG was found to exhibit only a vanishingly small such effect, if any (Cohen-Stuart et al., 1986). Furthermore, the glass-transition temperature of PEG_{2000} is rather low, $T_g \approx -40°C$, and the PLL-g-PEG layer investigated in this study consists of 83% water. A glass-like polymer film is expected to exhibit a finite yield stress and plasticity over a wide temperature range, which is in contrast with the hysteresis-free (i.e., elastic) compression isotherm observed here. Some kind of glassy state as the source of the observed transitions can thus be excluded.

The characteristic step size of 125 ± 25 pm reported here is smaller than any molecular unit present in this system and thus cannot be attributed to a confinement-induced dynamic "layering" effect (Israelachvili and Pashley, 1983).

The change of free energy between two consecutive transitions can be estimated according to Eq. 1, $\Delta E \approx \Delta F / 2\pi R \approx 0.4 \pm 0.2$ mJ/m^2. Assuming that all PEG chains contribute to the transitions and taking the area per PEG chain determined above ($a_{PEG} \approx 2.0 \times 10^{-18}$ m^2), we obtain a change of free energy per PEG chain $\Delta G/PEG = 0.2 \pm 0.1$ kT. The change of free energy is small compared to kT and thus compatible with the chain flexibility paradigm. It is interesting to note that the quantum mechanically calculated change of free energy for conformational transitions in hydrated PEG is of comparable magnitude (Wang et al., 2000). The step size observed here (Fig. 5) is also comparable to the change of the oxygen-oxygen distance during a *trans-gauche* transition around the C-C bond in PEG.

The force isotherms measure the change of free energy upon compression. The observation of the fine structure in the absence of hysteresis thus implies two things: first, that the derivative of the entropy versus surface separation, dS/dD, is no longer a monotonous function; and secondly, a considerable fraction of the molecules in the contact region undergo the transition simultaneously, hence, there must be a molecular mechanism synchronizing these events.

To illustrate the first point, we note that a PEG chain can be in $\sim 10^{62}$ different conformations (random walk model). In the contact zone there are $\sim 10^8$ PEG chains, which results in a total of 10^{70} molecular configurations contributing to the eSFA measurement. Under confinement, the number of possible conformations is reduced continuously, and, due to the high number of possible states, no noticeable transitions are expected. The presence of fine structures must therefore be understood as a significant restriction or degeneration of the polymer conformational space.

Regarding the second point we note that the high degree of molecular synchronization is linked to the nonmonotonous change of free energy. It can be understood in terms of a critical stress that is locally needed to induce the transition. As soon as one PEG chain undergoes the transition, the load on the neighboring chains will increase above the critical stress, thus, triggering a cascade of similar transitions in its neighborhood. Considering the curved contact geometry in this experiment, one can readily show that a region of a few micrometers diameter can undergo the same transition simultaneously.

It is important to note that the transition discussed here is not an individual change of conformation, but rather a confinement-induced critical transition of the conformational space in a thermodynamic sense (i.e., a phase transition). Therefore, the measured barriers in the system free energy are not expected to produce appreciable loading speed dependence in this quasiequilibrium experiment. Indeed, an experimental variation of actuator speed in the range from 0.1 to 5 nm/s (data not shown) revealed no dependence on this parameter.

To this end, a comparison with the literature is made to identify possible molecular mechanisms that give rise to the observed fine structure. The existence of an extensive equilibrium hydration structure was postulated in numerous studies, namely in aqueous PEG solution experiments (Bieze et al., 1994; Janelli et al., 1994; Karlström, 1985; Kjellander and Florin, 1981), single-molecule force measurements (Oesterhelt et al., 1999), and recent surface-monolayer studies (Grunze and Pertsin, 2000) as well as theory (Wang et al., 1997, 2000). The extraordinary solution properties of PEG are reviewed in the introduction above. It is known that *gauche* conformers are more polar and offer two hydrogen bond acceptor sites with an oxygen-oxygen distance (2.88 Å) that is very close to that of bulk water (2.85 Å). For example, H-bonding with two successive oxygen atoms in PEG by one single water molecule forms a pentagonal bridge structure (Bandyopadhyay et al., 2000) reminiscent of the pentagonal equilibrium structure predominant in bulk water (Dougherty and Howard, 1998). Namely, it was shown that the *gauche* conformers are overpopulated in PEG in contact with water (Björling et al., 1991). Similarly, the existence of conformational substates and bound water beyond the first hydration shell was detected as an anomalous excess heat capacity (Kingman et al., 1990).

Although the influence of water on the PEG conformational structure is widely accepted, it is less clear to what extent the water structure is modified by the presence of PEG. The density depression illustrated in Fig. 3 would suggest that a modified water structure potentially exists several nanometers beyond the nominal brush length. This is comparable to recent neutron-scattering observations (Schwendel et al., 2002) on ethylene-oxide self-assembled monolayers. The general idea is that this effect be linked to the hydrophobic effect in the vicinity of the amphiphilic PEG layer.

The PLL(12 k)-*g*(3.5)-PEG(2 k) copolymer studied here is used as a very effective protein-resistant coating. It is thus interesting to discuss the role of the embedded fine structure in terms of protein adsorption. It is generally accepted that in a wide range of grafting densities, direct adsorption to the underlying substrate is prevented by the osmotic pressure (i.e., steric repulsion of the polymer film). Another important ingredient for protein resistance is that the proteins do not stick to the PEG. Recent SFA measurements suggest that there may be a rather high activation barrier of up to 44 kT, which kinetically hinders the proteins (e.g., streptdavidin) from adsorbing (Efremova et al., 2001). If correct, then protein resistance is ultimately a nonequilibrium effect. Indeed, adhesion energies on the order of ≈ -2 kT per protein were reported between strepdavidin and PEG brushes after external compression (Efremova et al., 2001; Sheth and Leckband, 1997).

Considering the size of a protein (footprint ≈ 20 nm^2) and the fact that proteins adsorb via multiple sites, we estimate that at least 10 PEG chains directly contribute to the adsorption barrier of a single protein. Therefore, it takes ~ 2 kT per protein to induce one transition of the kind observed here. It is important to note that to adsorb to PEG, a protein will restrict the conformational space of the PEG chains, even in the absence of a net compression. A simple thermodynamical analysis of this quasiequilibrium situation reveals that $\sim 99.8\%$ of the proteins are incapable of inducing more than three conformational transitions in PEG and thus will be repelled within the first 0.4 nm (i.e., three or less transitions).

With the possibility at hand to directly detect transitions in the conformational space of the hydrated PEG, it is now planned to apply this new tool to various other polymer architectures, particularly those that are known to be less protein resistant.

CONCLUSION

In quantitative agreement with previous studies, steric forces are found during compression of surfaces exposing grafted PEG_{2000} brushes on mica substrates. The compression isotherms can be fitted using a combination of double-layer repulsion and polymer-brush compression. The effective surface potential is small, 20 ± 20 mV, and the obtained polymer brush lengths agree with theoretical predictions based on chemical structure. The grafting density and film thickness were determined from refractive index and force measurements, respectively. A morphological model for the structure of the adsorbed copolymer film is presented. A significant density depression above the PEG layer is found from an analysis of the refractive index envelope. Multiple, partially coherent conformational transitions are resolved that are embedded into the well-known steric repulsion. The conformational space of PEG is significantly frustrated and quantized as a function of film thickness. This is in agreement with the presence of an equilibrium hydration structure, also postulated in the literature. The nonmonotonous change of free energy during each structural transition is ~ 0.2 kT per PEG chain. The resulting energy barrier is capable of inhibiting the adsorption of proteins within the first 0.4 nm of the polymer film.

We thank N. P. Huang for the synthesis of the PLL-*g*-PEG and S. Tosatti for providing unpublished data on refractive index for PEG_{400} solutions. Furthermore, we acknowledge invaluable discussions and advice from O. Borisov, J. Vörös, and M. A. Cohen-Stuart. This work profited from technical assistance provided by M. Elsener and J. Vanicek.

Financial support was provided by the TopNano21 Program of the Council of the Swiss Federal Institutes of Technology (ETH-Rat) as well as the US Air Force Office of Scientific Research under contract No. F49620-02-1-0346.

REFERENCES

Alexander, S. 1977. Adsorption of chain molecules with a polar head a scaling description. *Le Journal de Physique*. 38:983–987.

Bandyopadhyay, S., M. Tarek, M. L. Lynch, and M. L. Klein. 2000. Molecular dynamics study of the poly(oxyethylene) surfactant $C_{12}E_2$ and water. *Langmuir*. 16:2000.

Bieze, T. W. N., A. C. Barnes, C. J. M. Huige, J. E. Enderby, and J. C. Leyte. 1994. Distribution of water around poly(ethylene oxide): a neutron diffraction study. *J. Phys. Chem.* 98:6568–6576.

Björling, M., G. Karlström, and P. Linse. 1991. Conformational adaptation of poly(ethylene oxide). *J. Phys. Chem.* 95:6706–6709.

Chan, Y.-H. M., R. Schweiss, C. Werner, and M. Grunze. 2003. Electrokinetic characterization of oligo- and poly(ethylene glycol)-terminated self-assembled monolayers on gold and glass surfaces. *Langmuir*. 19: 7380–7385.

Claesson, P. M., and C.-G. Gölander. 1987. Direct measurements of steric interactions between mica surfaces covered with electrostatically bound low-molecular-weight polyethylene oxide. *J. Colloid Interface Sci.* 117: 366–374.

Cohen-Stuart, M. A., J. T. F. Keurentjes, B. C. Bonekamp, and J. G. E. M. Fraaye. 1986. Gelation of polymers adsorbed at a water-air interface. *Colloids and Surfaces*. 17:91–102.

Costello, B. A. De L., P. F. Luckham, and T. F. Tadros. 1993. Forces between adsorbed low molecular weight graft copolymers. *J. Colloid Interface Sci.* 156:72–77.

deGennes, P. G. 1987. Polymers at an interface: a simplified view. *Adv. Colloid Interface Sci.* 27:189–209.

Derjaguin, B. V. 1934. Untersuchungen über die Reibung und Adhäsion. *IV. Kolloid Zeitschrift*. 69:155–164. [in German].

Dougherty, R. C., and L. N. Howard. 1998. Equilibrium structural model of liquid water: evidence from heat capacity, spectra, density, and other properties. *J. Chem. Phys.* 109:7379–7393.

Efremova, N. V., S. R. Sheth, and D. E. Leckband. 2001. Protein-induced changes in poly(ethylene glycol) brushes: molecular weight and temperature dependence. *Langmuir*. 17:7628–7636.

Elbert, D. L., and J. Hubbell. 1996. Surface treatments of polymers for biocompatibility. *Annual Review of Materials Science*. 26:365–370.

Groll, J., E. V. Amirgoulova, T. Ameringer, C. D. Heyes, C. Röcker, G. U. Nienhaus, and M. Möller. 2004. Biofunctionalized, ultrathin coatings of cross-linked star-shaped poly(ethylene oxide) allow reversible folding of immobilized proteins. *J. Am. Chem. Soc.* 126:4234–4239.

Grunze, M., and A. Pertsin. 2000. Molecular Conformations in Organic Monolayers Affect their Ability to Resist Protein Adsorption. L. Fabbrizzi and A. Poggi, editors. Springer-Verlag, Berlin/Heidelberg, Germany.

Hansen, P. L., J. A. Cohen, R. Podgomik, and V. A. Parsegian. 2002. Osmotic properties of poly(ethylene glycols): quantitative features of brush and bulk scaling laws. *Biophys. J.* 84:350–355.

Harris, J. M., and S. Zalipsky. 1997. Poly(ethylene glycol). J. M. Harris and S. Zalipsky, editors. American Chemical Society, Washington, DC.

Herrwerth, S., W. Eck, S. Reinhardt, and M. Grunze. 2003. Factors that determine the protein resistance of oligoether self-assembled monolayers: internal hydrophilicity, terminal hydrophilicity, and lateral packing density. *J. Am. Chem. Soc.* 125:9359–9366.

Heuberger, M. 2001. The extended surface forces apparatus. Part I. Fast spectral correlation spectroscopy. *Rev. Sci. Instrum.* 72:1700–1707.

Heuberger, M., J. Vanicek, and M. Zäch. 2001. The extended surface forces apparatus. Part II. Precision temperature control. *Rev. Sci. Instrum.* 72: 3556–3560.

Heuberger, M., and M. Zäch. 2003. Nano-fluidics: structural forces, density anomalies and the pivotal role of nano-particles. *Langmuir*. 19:1943–1947.

Heuberger, M., M. Zäch, and N. D. Spencer. 2000. Sources and control of instrumental drift in the surface forces apparatus. *Rev. Sci. Instrum.* 71: 4502–4508.

Himmelhaus, M., T. Bastuck, S. Tokumitsu, M. Grunze, L. Livadaru, and H. J. Kreuzer. 2003. Growth of a dense polymer brush layer from solution. *Europhys. Lett.* 64:378–384.

Huang, N.-P., R. Michel, J. Voros, M. Textor, R. Hofer, A. Rossi, L. Elbert, J. Hubbell, and N. D. Spencer. 2000. Poly(L-lysine)-g-poly(ethylene glycol) layers on metal oxide surfaces: surface-analytical characterization and resistance to serum and fibrinogen adsorption. *Langmuir*. 17:489–498.

Israelachvili, J. N. 1973. Thin film studies using multiple beam interferometry. *J. Colloid Interface Sci.* 44:259–271.

Israelachvili, J. N. 1991. Intermolecular & Surface Forces. Academic Press, London, UK.

Israelachvili, J. N., and P. M. McGuiggan. 1990. Adhesion and short-range forces between surfaces. I. New apparatus for surface force measurements. *J. Mater. Res.* 5:2223–2231.

Israelachvili, J. N., and R. M. Pashley. 1983. Molecular layering of water at surfaces and origin of repulsive hydration forces. *Nature*. 306:249–250.

Janelli, M. P., S. Magazu, G. Maisano, D. Majolino, and P. Migliardo. 1994. Non-ideal compressibility in poly(ethylene oxide)-water solutions induced by H-bond interactions. *J. Mol. Struct.* 322:337–343.

Karlström, G. 1985. A new model for upper and lower critical solution temperatures in poly(ethylene oxide) solutions. *J. Phys. Chem.* 89:4962–4964.

Kenausis, G., J. Voros, D. L. Elbert, N.-P. Huang, R. Hofer, L. Ruiz-Taylor, M. Textor, J. A. Hubbell, and N. D. Spencer. 2000. Poly(L-lysine)-g-poly(ethylene glycol) layers on metal oxide surfaces: attachment mechanism and effects of polymer architecture on resistance to protein adsorption. *J. Phys. Chem. B*. 104:3298–3309.

Kenworthy, A. K., K. Hristova, D. Needham, and T. J. McIntosh. 1995. Range and magnitude of the steric pressure between bilayers containing phospholipids with covalently attached poly(ethylene glycol). *Biophys. J.* 68:1921–1936.

Kingman, N. G., A. Rosenberg, M. Bastos, and I. Wadsö. 1990. Heat capacity of poly(ethylene glycol)-water mixtures: poly(ethylene glycol)-water interactions. *Thermochimica Acta*. 169:339–346.

Kjellander, R., and E. Florin. 1981. Water structure and changes in thermal stability of the system poly(ethylene oxide)-water. *J. Chem. Soc. Faraday Trans. I*. 77:2053–2077.

Kohonen, M. M., F. C. Meldrum, and H. K. Christenson. 2003. Particles on melt-cut mica sheets are platinum. *Langmuir*. 19:975–976.

Kremer, K. 1986. Glassy states of adsorbed flexible polymers and spread polymer ''monolayers''. *Le Journal de Physique*. 47:1269–1271.

Kreuzer, H. J., R. L. C. Wang, and M. Grunze. 2003. Hydroxide ion adsorption on self-assembled monolayers. *J. Am. Chem. Soc.* 125:8384–8389.

Kuhl, T. L., D. E. Leckband, D. D. Lasic, and J. N. Israelachvili. 1994. Modulation of interaction forces between bilayers exposing short-chained ethylene oxide headgroups. *Biophys. J.* 66:1479–1488.

Ma, H., J. Hyun, P. Stiller, and A. Chilkoti. 2004. ''Non-fouling'' Oligo(ethylene glycol)- functionalized polymer brushes synthesized by surface-initiated atom transfer radical polymerization. *Adv. Mater.* 16: 338–341.

Maisano, G., D. Majolino, P. Migliardo, S. Venuto, F. Aliotta, and S. Magazu. 1993. Sound velocity and hydration phenomena in aqueous polymeric solutions. *Mol. Phys.* 78:421–435.

Milner, S. T., and T. A. Witten. 1988. Theory of the grafted polymer brush. *Macromolecules*. 21:2610–2619.

Müller, E. A., and P. Rasmussen. 1991. Densities and excess volumes in aqueous poly(ethylene glycol) solutions. *Journal of Chemical Engineering Data*. 36:214–217.

Needham, D., K. Hristova, T. J. McIntosh, M. Dewhirst, N. Wu, and D. D. Lasic. 1992. Polymer-grafted liposomes: physical basis for the ''stealth'' property. *J. Liposome Res.* 2:411–430.

Oesterhelt, F., M. Rief, H. E. Gaub. 1999. Single molecule force spectroscopy by AFM indicates helical structure of poly(ethylene-glycol) in water. *New J. Phys.* 1:6.1–6.11.

Ohnishi, S., M. Hato, T. Tamada, and H. K. Christenson. 1999. Presence of particles on melt-cut mica sheets. *Langmuir*. 15:3312–3316.

Pasche, S., S. M. De Paul, J. Voros, N. D. Spencer, and M. Textor. 2003. Poly(L-lysine-graft-poly(ethylene glycol) assembled monolayers on niobium oxide surfaces: a quantitative study of the influence of polymer interfacial architecture on resistance to protein adsorption by ToF-SIMS and in situ OWLS. *Langmuir*. 19:9216–9225.

Prime, K. L., and G. M. Whitesides. 1993. Adsorption of proteins onto surfaces containing end-attached oligo(ethylene oxide): a model system using self-assembled monolayers. *J. Am. Chem. Soc.* 115:10714–10721.

Raviv, U., J. Frey, R. Sak, P. Laurat, R. Tadmore, and J. Klein. 2002. Properties and interactions of physigrafted end-functionalized poly-(ethylene glycol) layers. *Langmuir*. 18:7482–7495.

Saeki, S., K. Nobuhiro, M. Nakata, and M. Kaneko. 1976. Upper and lower critical solution temperatures in poly (ethylene glycol) solutions. *Polymer*. 17:685–689.

Schwendel, D., T. Hayashi, R. Dahint, A. Pertsin, M. Grunze, R. Steitz, and F. Schreiber. 2002. Interaction of water with self-assembled monolayers: neutron reflectivity measurements of the water density in the interface region. *Langmuir*. 19:2284–2293.

Sheth, S. R., and D. Leckband. 1997. Measurements of attractive forces between proteins and end-grafted poly(ethylene glycol) chains. *Proc. Natl. Acad. Sci. USA*. 94:8399–8404.

Szleifer, I. 1997. Polymers and proteins: interactions at interfaces. *Curr. Opin. Colloid Interface Sci.* 2:337–344.

Tirosh, O., Y. Barenholz, J. Katzhendler, and A. Priev. 1998. Hydration of polyethylene glycol-grafted liposomes. *Biophys. J.* 74:1371–1379.

Verwey, E. J. W., and J. T. G. Overbeek. 1945. Theory of Stability of Lyophobic Colloids. Elsevier, Amsterdam, The Netherlands.

Wang, R. L. C., H. J. Kreuzer, and M. Grunze. 1997. Molecular conformation and solvation of oligo(ethylene glycol)-terminated self-assembled monolayers and their resistance to protein adsorption. *J. Phys. Chem. B*. 101:9767–9773.

Wang, R. L. C., H. J. Kreuzer, and M. Grunze. 2000. The interaction of oligo(ethylene oxide) with water: a quantum mechanical study. *Phys. Chem. Chem. Phys.* 2:3613–3622.

Zäch, M., and M. Heuberger. 2000. Statistical and systematic errors of the surface forces apparatus. *Langmuir*. 16:7309–7314.

Zhulina, E. B., O. V. Borisov, and V. A. Priamitsyn. 1989. Theory of steric stabilization of colloid dispersions by grafted polymers. *J. Colloid Interface Sci.* 137:495–511.

6508 *Langmuir* **2005,** *21,* 6508–6520

Relationship between Interfacial Forces Measured by Colloid-Probe Atomic Force Microscopy and Protein Resistance of Poly(ethylene glycol)-Grafted Poly(L-lysine) Adlayers on Niobia Surfaces

Stéphanie Pasche,[†,‡] Marcus Textor,[†] Laurence Meagher,[§] Nicholas D. Spencer,[†] and Hans J. Griesser[*,‡]

Laboratory for Surface Science and Technology, Department of Materials, Swiss Federal Institute of Technology, ETH Hönggerberg, Wolfgang-Pauli-Strasse 10, CH-8093 Zürich, Switzerland, Ian Wark Research Institute, University of South Australia, Mawson Lakes, SA 5095, Australia, and CSIRO Molecular Science, Bag 10, Clayton South, Vic 3169, Australia

Received February 11, 2005. In Final Form: May 9, 2005

Adsorbed layers of "comb-type" copolymers consisting of PEG chains grafted onto a poly(L-lysine) (PLL) backbone on niobium oxide substrates were studied by colloid-probe AFM in order to characterize the interfacial forces associated with coatings of varying architectures (PEG/PLL ratios and PEG chain lengths) and their relevance to protein resistance. The steric and electrostatic forces measured varied substantially with the architecture of the PLL-*g*-PEG copolymers. Varying the ionic strength of the buffer solutions enabled discrimination between electrostatic and steric–entropic contributions to the net interfacial force. For high PEG grafting densities the steric component was most prominent, but at low ionic strengths and high grafting densities, a repulsive electrostatic surface force was also observed; its origin was assigned to the niobia charges beneath the copolymer, as insufficient protonated amine groups in the PLL backbone were available for compensation of the oxide surface charges. For lower grafting densities and lower ionic strengths there was a substantial attractive electrostatic contribution arising from interaction of the electrical double layer arising from the protonated amine groups, with that of the silica probe surface (as under low ionic strength conditions, the electrical double layer was thicker than the PEG layer). For these PLL-*g*-PEG coatings the net interfacial force can thus be a markedly varying superposition of electrostatic and steric–entropic contributions, depending on various factors. The force curves correlate with protein adsorption data, demonstrating the utility of AFM colloid-probe force measurements for quantitative analysis of surface forces and how they determine interfacial interactions with proteins. Such characterization of the net interfacial forces is essential to elucidate the multiple types of interfacial forces relevant to the interactions between PLL-*g*-PEG coatings and proteins and to advance interpretation of protein adsorption or repellence beyond the oversimplified steric barrier model; in particular, our data demonstrate the importance of an ionic-strength-dependent minimum PEG layer thickness to screen the electrostatic interactions of charged interfaces.

1. Introduction

Control of the interfacial interactions between a synthetic material/device and the biological medium is a key aspect in the design of biomaterials and biosensor surfaces. Protein-resistant ("nonfouling") surfaces are particularly important in the context of blood-contacting devices and contact lenses and as a noninteractive background for biodiagnostic surfaces designed to elicit specific binding of target molecules to surface-grafted probes. Poly(ethylene glycol) (PEG) immobilized onto surfaces has been shown to confer high resistance to protein adsorption.[1,2] However, the extent of protein repellence varies considerably between studies. The reasons for variable performance and optimal protein repellence of PEG layers have been the subject of much discussion;[3–7] factors such as the PEG length and graft density are important,[3] and minimal attractive interactions with proteins, extensive hydration of PEG chains in water, good conformational flexibility, and high chain mobility are thought to be key properties of a "steric barrier" PEG layer capable of preventing adsorption of proteins.

The density of grafted chains must be sufficient to prevent proteins from diffusing through the grafted PEG layer and reaching the underlying substrate, onto which they can adsorb irreversibly.[3] In addition, proteins must not adsorb on top of the PEG layer; this would reduce the mobility of the hydrated grafted PEG chains. Hence the PEG layer provides a steric exclusion zone and also defines

* Author for correspondence. E-mail: hans.griesser@unisa.edu.au. Telephone: +61-8-83023703. Fax: +61-8-83023683.
† Swiss Federal Institute of Technology, ETH Hönggerberg.
‡ University of South Australia.
§ CSIRO Molecular Science.

(1) Harris, J. M. *Poly(Ethylene Glycol) Chemistry: Biotechnical and Biomedical Applications*; Plenum Press: New York, 1992.
(2) Harris, J. M. *Poly(ethylene glycol) Chemistry and Biological Applications*; ACS Symposium Series 680; American Chemical Society: Washington, DC, 1997.
(3) Kingshott, P.; Thissen, H.; Griesser, H. J. *Biomaterials* **2002**, *23*, 2043–2056.
(4) Kingshott, P.; Griesser, H. J. *Curr. Opin. Solid State Mater. Sci.* **1999**, *4*, 403–412.
(5) Blomberg, E.; Claesson, P. M. In *Proteins at Interfaces II: Fundamentals and Applications*; Horbett, T. A., Brash, J. L., Eds.; American Chemical Society: Washington, DC, 1995; ACS Symposium Series 602; pp 296–310.
(6) Morra, M. *Water in Biomaterial Surface Science*; John Wiley & Sons Ltd.: Chichester, England, 2001.
(7) Norde, W. In *Biopolymers at Interfaces*; Malmsten, M., Ed.; Surfactant Science Series 75; Marcel Dekker: New York, 1998; Vol. 75.

Published on Web 06/09/2005

an entropic barrier to protein adsorption.[8] Theoretical and experimental approaches have demonstrated that both PEG chain length and surface density affect protein adsorption. Protein resistance has specifically been shown to depend on the EG−monomer surface density defined as the product of both parameters.[4,6] However, there is no universal agreement yet on the mechanisms governing protein−PEG brush interactions.

Such "steric stabilization" has long been used in various industrial processes, for example to prevent the aggregation of colloidal particles, and theories of interfacial forces have proved useful for interpreting particle−particle interactions. In the biomaterials field, however, protein adsorption is not often discussed in terms of interfacial forces between synthetic surfaces and approaching proteins. Proteins can interact in various ways with surfaces; electrostatic and hydrophobic forces often dominate, and they usually combine with an entropy gain caused by conformational changes in the protein, as it loses its ordered structure upon adsorption.[7,9] Due to the heterogeneity and asymmetry of the protein surface, the types and magnitudes of interactions will also depend on the orientation of the molecule relative to the surface. Attractive and repulsive forces between PEG and proteins have been reviewed.[10]

A brush structure is considered essential for effective steric stabilization by a grafted layer. The ability of polymer brushes to prevent the adsorption of proteins has been studied theoretically.[10−13] Halperin expressed the interaction potential between proteins and brush-coated surfaces as the sum of the individual interactions of the protein with the bare substrate and the brush, predicting adsorbed protein mass as a function of grafting density. Three modes of adsorption have been considered: (a) primary adsorption due to proteins diffusing through the grafted layer to the underlying substrate and adsorbing to its surface (invasive mechanism); (b) secondary adsorption at the outer surface of the brush layer due to protein−brush interaction; (c) adsorption of proteins with compression of the brush layer (compressive mechanism).[13] Protein resistance requires exclusion of all three processes. The thickness of the grafted PEG layer must be sufficient to screen protein−substrate interactions, and the grafting density must block diffusion through the steric layer.[14] If primary adsorption is excluded, adsorption onto the brush layer or repellence is defined by the balance between attractive and repulsive interfacial forces experienced by an approaching protein molecule.

Measurement of interfacial forces thus offers the potential to study the factors involved in protein repellence or adsorption. Force measurements on adsorbed and grafted PEG layers have been reported, using both the surface forces apparatus (SFA)[15] and the colloid-probe atomic force microscopy (AFM) technique.[16,17] Adsorbed PEG layers show responses due to relaxation processes, hysteresis and bridging of loosely bound chains on the surface.[18−22] End-grafted PEG chains as well as PEG-copolymers have been studied with the SFA as well as with conventional and colloid-probe AFM to probe effects of solvent, ionic strength, pH and temperature, revealing the importance of different conformational regimes.[23−30] AFM has been used to probe interactions between various tips and oligo(ethylene glycol)-terminated alkanethiol monolayers on gold surfaces to study the effect of pH and ionic strength,[30−32] and interactions between PEG and proteins[32,33] and PEG and bacteria.[34]

AFM force measurements can be done with a range of surfaces. Conventional AFM tips, however, should not be used for quantitative interfacial force measurements because quantitative comparison with theoretical predictions requires that the radius of the approaching probe be much greater than the separation distance. The latter point has often been overlooked in surface force measurements with standard AFM tips. The colloid-probe AFM method[16,17,35] is appropriate for quantitative surface force measurements. A colloidal (usually silica) microsphere attached to the end of the AFM cantilever provides a well-defined, mathematically tractable sphere-vs-flat geometry for the scaling of forces, and allows the use of different colloid materials, or the surface modification of colloid spheres, for investigating interactions between surfaces with various physicochemical properties.

Among the various techniques for the immobilization of PEG on surfaces,[4,36,37] graft copolymers of poly(L-lysine) and poly(ethylene glycol) (PLL-*g*-PEG) are particularly attractive. PLL-*g*-PEG copolymers spontaneously adsorb from aqueous solutions as dense, monomolecular layers onto a range of negatively charged surfaces such as various metal oxides and tissue-culture polystyrene, protecting them against nonspecific protein adsorption; moreover,

(8) Heuberger, M.; Drobek, T.; Spencer, N. D. *Biophys. J.* **2004**, submitted.
(9) Rixman, M. A.; Dean, D.; Ortiz, C. *Langmuir* **2003**, *19*, 9357−9372.
(10) Jeon, S. I.; Andrade, J. D. *J. Colloid Interface Sci.* **1991**, *142*, 159−166.
(11) Jeon, S. I.; Lee, J. H.; Andrade, J. D.; de Gennes, P. G. *J. Colloid Interface Sci.* **1991**, *142*, 149−158.
(12) Szleifer, I. *Physica A* **1997**, *244*, 370−388.
(13) Halperin, A. *Langmuir* **1999**, *15*, 2525−2533.
(14) Israelachvili, J. N.; Adams, G. E. *J. Chem. Soc., Faraday Trans. 1* **1978**, *74*, 975.
(15) Israelachvili, J. *Acc. Chem. Res.* **1987**, *20*, 415−421.
(16) Ducker, W. A.; Senden, T. J.; Pashley, R. M. *Langmuir* **1992**, *8*, 1831−1836.
(17) Senden, T. J.; Ducker, W. A. *Langmuir* **1992**; *8*, 733−735.
(18) Luckham, P. F.; Klein, J. *J. Chem. Soc., Faraday Trans.* **1990**, *86*, 1363−1368.
(19) Braithwaite, G. J. C.; Howe, A.; Luckham, P. F. *Langmuir* **1996**, *12*, 4224−4237.
(20) Braithwaite, G. J. C.; Luckham, P. F. *J. Chem. Soc., Faraday Trans.* **1997**, *93*, 1409−1415.
(21) Giesbers, M.; Kleijn, J. M.; Fleer, G. J.; Stuart, M. A. C. *Colloids Surf. A−Physicochem. Eng. Asp.* **1998**, *142*, 343−353.
(22) Kuhl, T. L.; Leckband, D. E.; Lasic, D. D.; Israelachvili, J. N. *Biophys. J.* **1994**, *66*, 1479−1488.
(23) Lea, A. S.; Andrade, J. D.; Hlady, V. *Colloids Surf. A−Physicochem. Eng. Asp.* **1994**, *93*, 349−357.
(24) Sheth, S. R.; Efremova, N.; Leckband, D. E. *J. Phys. Chem. B* **2000**, *104*, 7652−7662.
(25) Efremova, N. V.; Bondurant, B.; O'Brien, D. F.; Leckband, D. E. *Biochemistry* **2000**, *39*, 3441−3451.
(26) Claesson, P. M.; Golander, C. G. *J. Colloid Interface Sci.* **1987**, *117*, 366−374.
(27) Claesson, P. *Colloids Surf. A−Physicochem. Eng. Asp.* **1993**, *77*, 109−118.
(28) Claesson, P. M.; Blomberg, E.; Paulson, O.; Malmsten, M. *Colloids Surf. A−Physicochem. Eng. Asp.* **1996**, *112*, 131−139.
(29) Butt, H. J.; Kappl, M.; Mueller, H.; Raiteri, R.; Meyer, W.; Ruhe, J. *Langmuir* **1999**, *15*, 2559−2565.
(30) Feldman, K.; Hahner, G.; Spencer, N. D.; Harder, P.; Grunze, M. *J. Am. Chem. Soc.* **1999**, *121*, 10134−10141.
(31) Hahner, G.; Dicke, C.; Feldman, K.; Eck, W.; Herrwerth, S. *Abstr. Pap. Am. Chem. Soc.* **2000**, *220*, 211-POLY.
(32) Sheth, S. R.; Leckband, D. *Proc. Natl. Acad. Sci. U.S.A.* **1997**, *94*, 8399−8404.
(33) Razatos, A.; Ong, Y. L.; Boulay, F.; Elbert, D. L.; Hubbell, J. A.; Sharma, M. M.; Georgiou, G. *Langmuir* **2000**, *16*, 9155−9158.
(34) Hartley, P. G. In *Colloid-Polymer Interactions: From Fundamentals to Practice*; Dubin, P. L., Ed.; John Wiley & Sons: New York, 1999; pp 253−286.
(35) Ducker, W. A.; Senden, T. J.; Pashley, R. M. *Nature (London)* **1991**, *353*, 239−241.
(36) Holmberg, K.; Tiberg, F.; Malmsten, M.; Brink, C. *Colloids Surf. A−Physicochem. Eng. Asp.* **1997**, *123*, 297−306.
(37) Vermette, P.; Meagher, L. *Colloids Surf. B−Biointerfaces* **2003**, *28*, 153−198.

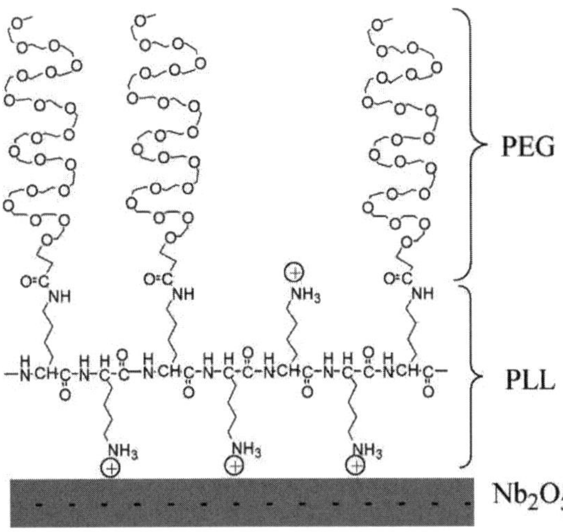

Figure 1. Idealized diagram of the interfacial structure of a monolayer of PLL-*g*-PEG adsorbed on a metal oxide substrate (Nb_2O_5) via electrostatic interactions between the negatively charged metal oxide surface and positively charged amino-terminated PLL side chains (at neutral pH).

they enable control of PEG length and surface density.[38-41] Protonated amine groups of the PLL backbone bind electrostatically to negatively charged surfaces. PEG chains are grafted, with controllable ratios, to PLL prior to adsorption, allowing tailoring of the PEG content in the copolymer and hence PEG surface coverage on adsorption. A schematic structure of the adlayer is shown in Figure 1. Protein adsorption studies on PLL-*g*-PEG-coated surfaces have shown the amounts of adsorbed protein to correlate with the EG−monomer surface density.[38]

In this study we investigated the interfacial forces emanating from adsorbed PLL-*g*-PEG layers on niobia substrates by the colloid-probe AFM technique with the aim of using the resultant understanding of the interfacial forces and characterization of the structure of the polymer coating (overlapping mushrooms vs brush) for interpretation of protein adsorption data previously obtained on the same layers.[38-40] Although the approach to, and compression of, a PEG layer by a colloid microsphere is physically somewhat different to the approach and adsorption of a protein, particularly by not being able to mimic the heterogeneity and structural flexibility of proteins, we undertook this work to assess whether such a simplified model of biointerfacial interactions could be useful toward improved understanding of the interfacial forces and PEG coating structures involved in protein adsorption or repellence.

2. Materials and Methods

2.1. Synthesis of Poly(L-lysine)-*graft*-poly(ethylene glycol). The notation used for PLL(x)-*g*[y]-PEG(z) copolymers indicates the average molecular weights (MWs) of PLL-hydrobromide (x) and PEG (z), and the grafting ratio, g (y). The grafting ratio is expressed as the number of lysine monomers divided by the number of PEG side chains (Lys/PEG ratio). For example, a copolymer labeled PLL(20)-*g*[3.5]-PEG(2) has a backbone of PLL-HBr of MW ~20 kDa (96 Lys-mers), PEG side chains of MW 2 kDa (45 EG-mers), and a grafting ratio of 3.5 (3.5 lysine units per PEG chain). The value of 20 kDa specified for the average MW of PLL-HBr is only approximate in view of the substantial polydispersity and variations in PLL molecular weight from batch to batch (see below).

PLL-*g*-PEG polymers were synthesized from a stoichiometric mixture of PLL-HBr (Sigma, MW 15−30 kDa, polydispersity 1.1−1.3) and a *N*-hydroxysuccinimidyl ester of methoxy-terminated poly(ethylene glycol) (mPEG-SPA) (Nektar, MWs 1, 2, and 5 kDa, polydispersity <1.05). PLL-HBr was dissolved in a 50 mM sodium borate buffer solution (pH 8.5) in a concentration corresponding to 100 mM monomeric lysine. The solution was filter sterilized (0.22 μm pore-size filter, Millex-GV, Sigma-Aldrich, Switzerland). The mPEG-SPA was then added to the solution, and the mixture was stirred for 6 h at room temperature. Subsequently, the reaction mixture was dialyzed (Spectra-Por, mol wt cutoff size 6−8 kDa, Spectrum Laboratories, Inc.) for 48 h against deionized water, changing the water after 24 h. The product was freeze-dried and stored at −20 °C until use. Further details regarding the synthesis protocol have been published previously.[42-44] Use of PEGs with molecular weights of 1, 2, and 5 kDa and different relative amounts of PLL and PEG allowed the synthesis of a matrix of different polymers with defined molecular architectures. The PLL-*g*-PEG product was characterized by ^{1}H NMR spectroscopy, using D_2O as the solvent, on a 500 MHz Bruker instrument. The grafting ratio was determined from the NMR spectra.[45]

2.2. Surface Preparation and Characterization. PLL-*g*-PEG was adsorbed onto niobium oxide films. 12 nm thick films of Nb_2O_5 were sputter coated onto silicon wafers ⟨110⟩ (WaferNet GmbH, Germany) using reactive magnetron sputtering (PSI, Villigen, Switzerland) to produce substrates for the AFM measurements. Substrates were sonicated in 2-propanol for 10 min, rinsed with ultrapure water, and dried under a nitrogen stream, followed by 2 min of oxygen plasma cleaning in a Plasma Cleaner/Sterilizer PDC-32G instrument (Harrick, Ossining, NY). Water used in the experiments was purified with a Millipore water treatment apparatus (organic content less than 5 ppb). The cleaned substrates were immediately transferred to a filtered 1 mg/mL solution of PLL-*g*-PEG in 10 mM HEPES buffer solution (4-(2-hydroxyethyl) piperazine-1-ethanesulfonic acid, adjusted to pH 7.4 with a 6 M NaOH solution). After 30 min immersion, the samples were withdrawn, rinsed extensively with water, and dried under nitrogen. Samples were analyzed immediately after PLL-*g*-PEG adsorption, to avoid uncertainties due to storage.

Polymer and protein adsorption was studied quantitatively with an optical waveguide sensing technique, optical waveguide lightmode spectroscopy (OWLS).[46,47]

2.3. Atomic Force Microscopy (AFM). A Nanoscope IIIA Multimode (Digital Instruments, Santa Barbara, CA) was used for atomic force microscopy measurements. The AFM was operated in force mode, with a scan rate of 1 Hz, and a z-piezo total displacement of 500 nm, without calibrating the scanner before each measurement (routine calibration only). Both approach/extension and retraction force curves of the cantilever were recorded.

Experiments were performed both with silica sphere tips and with PLL-*g*-PEG coated silica sphere tips, against uncoated and polymer-coated Nb_2O_5 surfaces. Silica microspheres (4−5 μm

(38) Pasche, S.; De Paul, S. M.; Voros, J.; Spencer, N. D.; Textor, M. *Langmuir* **2003**, *19*, 9216−9225.
(39) Huang, N. P.; Michel, R.; Voros, J.; Textor, M.; Hofer, R.; Rossi, A.; Elbert, D. L.; Hubbell, J. A.; Spencer, N. D. *Langmuir* **2001**, *17*, 489−498.
(40) Kenausis, G. L.; Voros, J.; Elbert, D. L.; Huang, N. P.; Hofer, R.; Ruiz-Taylor, L.; Textor, M.; Hubbell, J. A.; Spencer, N. D. *J. Phys. Chem. B* **2000**, *104*, 3298−3309.
(41) VandeVondele, S.; Voros, J.; Hubbell, J. A. *Biotechnol. Bioeng.* **2003**, *82*, 784−790.
(42) Elbert, D. L.; Hubbell, J. A. *Chem. Biol.* **1998**, *5*, 177−183.
(43) Sawhney, A. S.; Hubbell, J. A. *Biomaterials* **1992**, *13*, 863−870.
(44) Huang, N. P.; Vörös, J.; De Paul, S. M.; Textor, M.; Spencer, N. D. *Langmuir* **2002**, *18*, 220−230.
(45) Gibson, C. T.; Watson, G. S.; Myhra, S. *Nanotechnology* **1996**, *7*, 259−262.
(46) Voros, J.; Ramsden, J. J.; Csucs, G.; Szendro, I.; De Paul, S. M.; Textor, M.; Spencer, N. D. *Biomaterials* **2002**, *23*, 3699−3710.
(47) Hook, F.; Voros, J.; Rodahl, M.; Kurrat, R.; Boni, P.; Ramsden, J. J.; Textor, M.; Spencer, N. D.; Tengvall, P.; Gold, J.; Kasemo, B. *Colloids Surf. B−Biointerfaces* **2002**, *24*, 155−170.

Figure 2. Scanning electron micrograph of the Novascan AFM cantilever bearing a SiO_2 colloid microsphere as the tip, after use for force measurements.

diameter) (Bangs Laboratories) were attached to V-shaped Si_3N_4 AFM cantilevers with a nominal spring constant of 0.12 N/m (Novascan Technologies, Ames, IA) (Figure 2). The radius of the sphere was measured using an optical microscope. The bending stiffness of a number of colloid-probe-modified cantilevers was determined experimentally by deflecting the cantilever against a calibrated rectangular cantilever of known stiffness45 (0.14 N/m, Mikromash, Tallinn, Estonia); there was less than 10% variation between cantilevers (data not shown). Consequently an average spring constant of 0.11 N/m was used for all measurements; but to reduce uncertainty, related measurements were performed using the same cantilever. Prior to the force measurements the cantilevers were cleaned for 30 min in a UV/Ozone cleaner (UV/ Clean, model 135500, Boekel Industries, Inc., Feasterville, PA). Coating of the microsphere with PLL-*g*-PEG was done by immersing the cleaned cantilever in a 1 mg/mL filtered solution of PLL-*g*-PEG in 10 mM HEPES buffer (pH 7.4) for 30 min, followed by rinsing with ultrapure water, as described in section 2.2 for the flat substrates.

AFM experiments were performed in liquid using an open fluid cell, after equilibrating the system for 30−60 min. All buffers were filtered with a 0.22 μm pore-size filter. To avoid changes in the adsorbed polymer mass at the interface due to salt effects, all samples were immersed in the highest ionic strength buffer (10 mM HEPES + 150 mM NaCl) prior to each experiment.

Raw data were obtained using the Nanoscope V3 software (Digital Instruments, Santa Barbara, CA) and converted from cantilever deflection and z-piezo position into force vs distance curves.[16] The force was calculated from the cantilever deflection by Hooke's law, $F = k\delta$, where k is the spring constant and δ the deflection of the cantilever. The total force was held between 15 and 25 nN (no trigger), corresponding to a pressure of approximately 0.15−0.25 GPa. In this range the forces were observed not to be dependent on the load (data not shown). For comparison of the experimentally obtained forces to theory, the Derjaguin approximation provides a relationship between the force, $F(D)$, the radius of the sphere, R, and the calculated interaction energy per unit area between two flat plates, $W(D)_{\text{planes}}$.[49] Therefore, the data were represented as the force divided by the radius of the sphere, $F(D)/R$, as a function of the relative separation between the sphere and the surface, D. $F(D)/R > 0$ describes a repulsive interaction, while $F(D)/R < 0$ characterizes an attractive interaction. The zero separation ($D = 0$) between an approaching sphere and a soft coating such as our PLL-*g*-PEG layers is not accessible experimentally because the coating becomes compressed; accordingly, we define $D = 0$ experimentally as the "distance" at which linearity in the constant compliance region was observed.[37] At this point, the PEG coating has become maximally compressed but its residual thickness is not known. However, we believe the coatings are compressed to a small percentage of their equilibrium thickness.

Force curves recorded at different times and locations on a sample were reproducible, indicating homogeneous surface coverage. Considering the sphere diameter of 4−5 μm used in this study, more than 1000 PEG chains are expected to be compressed in the interaction area. The reproducibility of the force curves with time proved that the relaxation of the PEG chains after release of compression was a rapid process in comparison with the AFM cycle time of loading and unloading.[26]

Table 1. Polymer and Protein Adsorption Data Measured with OWLS, Expressed as the Surface Density of PEG Chains and EG Monomers, n_{PEG} [nm^{-2}], n_{EG} [nm^{-2}], and the Mass of Adsorbed Serum, m_{serum} [Ng/cm^2], ± Standard Deviation[38]

polymer	n_{PEG} [nm^{-2}]	n_{EG} [nm^{-2}]	m_{serum} [ng/cm^2]
uncoated Nb_2O_5			590 ± 57
PLL(20)			280 ± 40
PLL(20)-*g*[6.5]-PEG(1)	0.38 ± 0.04	8.7 ± 0.9	65 ± 2
PLL(20)-*g*[2.2]-PEG(2)	0.54 ± 0.03	24.4 ± 1.2	6 ± 8
PLL(20)-*g*[3.5]-PEG(2)	0.53 ± 0.04	24.1 ± 1.6	2 ± 3
PLL(20)-*g*[5.7]-PEG(2)	0.33 ± 0.03	15.2 ± 1.3	10 ± 9
PLL(20)-*g*[10.1]-PEG(2)	0.21 ± 0.02	9.4 ± 0.9	86 ± 11
PLL(20)-*g*[22.6]-PEG(2)	0.11 ± 0.01	4.9 ± 1.2	195 ± 35
PLL(20)-*g*[5.3]-PEG(5)	0.24 ± 0.03	26.9 ± 2.9	0 ± 0

3. Results and Analysis

3.1. Characterization of Surfaces and Protein Adsorption. PLL-*g*-PEG copolymers were synthesized, leading to a range of copolymers with PEG molecular weights of 1, 2, and 5 kDa and grafting ratios *g* varying from 2.1 to 22.6. Polymer and protein adsorption was studied quantitatively with OWLS.[46,47] Combined OWLS[38] and ToF−SIMS[48] results provided clear evidence for an interfacial polymer structure with the PLL backbone bound to the substrate surface (mostly by electrostatic interaction) and the PEG chains exposed at the surface with a surface density depending on the polymer architecture, in particular the grafting ratio *g* (Figure 1). Table 1 lists polymer and protein adsorption data, taken from ref 38, for the polymers used in this study. It was found in that study that the amount of adsorbed serum proteins depended only on the area density of ethylene glycol on the surface. Serum adsorption was reduced from 590 ng/cm^2 for uncoated Nb_2O_5 to <2 ng/cm^2 (detection limit of the OWLS) for EG surface densities >20 nm^{-2}. The protein resistance of such high-density PEG coatings has been attributed to the formation of PEG brushes at the interface, while lower PEG densities were believed to form a mushroom-type conformation that exhibited only partial protein resistance.[38]

While the molecular weight and PEG grafting density contribute mostly to the thickness of the layer, the grafting ratio not only influences the PEG surface density but also influences the amount of free positive charges in the PLL backbone, as shown in Figure 1. An increase in the grafting ratio (decreased PEG loading of the copolymer) both reduces the PEG surface density and increases the number of positive charges at the interface. Steric repulsion and electrostatic interaction effects are therefore intimately related for this copolymer system, and the interfacial force measurements reported below were designed to elucidate their contributions, which we expected to vary for the various PLL-*g*-PEG architectures.

3.2. Force Curve Analysis. Typical raw data are shown in Figure 3a,b, where the force is plotted as a function of the *z*-piezo position. Figure 3a displays the interaction between a bare SiO_2 microsphere and a PLL-

(48) Wagner, M.; Pasche, S.; Castner, D. G.; Textor, M. *Anal. Chem. A* **2004**, *76*, 1483−1492.

(49) Israelachvili, J. *Intermolecular & Surface Forces*, 2nd ed.; Academic Press: New York, 1992.

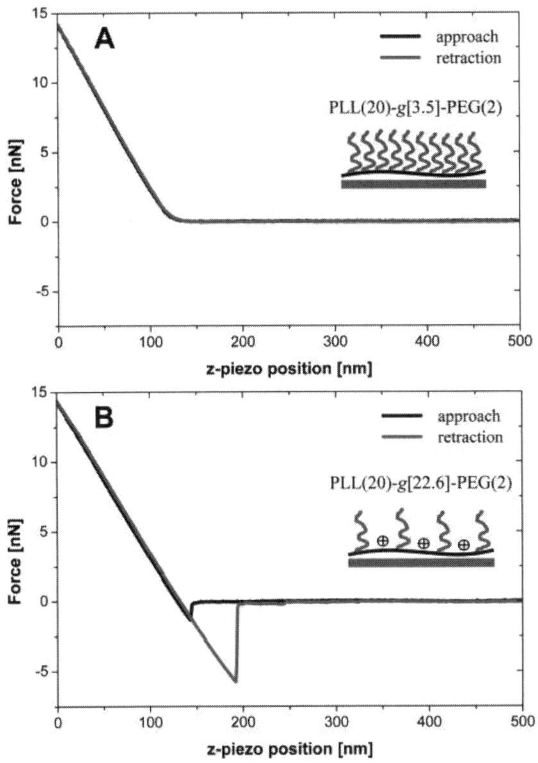

Figure 3. Raw AFM force data plots exhibiting the force experienced by the cantilever in nN as a function of the *z*-piezo position for both the approach and the retraction parts, for two polymers. PLL(20)-*g*[3.5]-PEG(2) (a) shows purely repulsive interaction, without hysteresis, while (b) PLL(20)-*g*[22.6]-PEG-(2) shows attraction on the approach curve and adhesion on the retraction.

(20)-*g*[3.5]-PEG(2)-coated Nb_2O_5 surface in 10 mM HEPES (pH 7.4); this sample has a high PEG density of 0.53 nm^{-2}. The interaction is purely repulsive, without hysteresis between the extension and retraction curves, indicating that the system is in equilibrium.[25] Such absence of hysteresis has been reported with other systems[18,26–28] and analogously suggests that the PLL-*g*-PEG chains are firmly anchored. The interaction observed in Figure 3b, on the other hand, is attractive, and shows hysteresis between extension and retraction curves, due to adhesion of the SiO_2 sphere on the sample surface. In this case, the PLL(20)-*g*[22.6]-PEG(2) coating has a low PEG density of 0.11 nm^{-2}, allowing the negatively charged SiO_2 microsphere to interact electrostatically with the positively charged PLL backbone.

3.3. Effects of the Ionic Strength. Parts a–d of Figure 4 display the measured interaction forces between a SiO_2 microsphere and a Nb_2O_5 flat surface that is uncoated, coated with PLL, and with PLL-*g*-PEG, respectively, at various ionic strengths. Figure 4a shows the effect of the ionic strength on the interaction between the negatively charged SiO_2 microsphere and the uncoated Nb_2O_5 surface, both on a linear and on a semilogarithmic scale, as the EDL interaction can be described by an exponential function.[49] We have also included theoretical predictions based on the DLVO theory[50,51] where the EDL contribution was calculated using the algorithm of McCormack et al.[52] In this case the fitting parameters were the Debye length, κ^{-1}, (described by eq 1)[49] and the diffuse double layer potential on the Nb_2O_5 surface (Table 2).

$$\kappa = \left(\sum_i \rho_\infty i e^2 z_i^2/\epsilon\epsilon_0 kT\right)^{1/2} [\mathrm{m}^{-1}] \qquad (1)$$

Literature values were used for the potential of the silica surface at the pH and ionic strength conditions of the measurements.[53] The Debye length was a fitting parameter here because the concentration of counterions for the negatively charged oxide surfaces was determined by the amount of NaOH added to the buffer to adjust the pH. HEPES is net negatively charged at the pH value used in these measurements and so acts as the co-ion. A nonretarded Hamaker constant of 2×10^{-20} J was used to calculate the van der Waals interaction. This value was chosen because the refractive index of Nb_2O_5 is similar to that of alumina and detailed Hamaker constant calculations for the interaction of silica and alumina surfaces across a water interlayer have been carried out.[54] Small variations in the Hamaker constant used generally do not significantly influence the fitted potential obtained at low ionic strengths, although for higher ionic strengths, retardation and salt screening effects can be apparent.

The force vs relative separation distance curves obtained between SiO_2 and Nb_2O_5 and presented in Figure 4a were purely repulsive and followed an exponentially decaying profile, with decay lengths corresponding to ~ 10 nm, 3.1 nm and finally to <1 nm as the ionic strength was increased from 1 mM HEPES buffer to a 10 mM HEPES buffer to a 10 mM HEPES + 150 mM NaCl buffer. This result, coupled with the reduction in the magnitude of the fitted potential for the Nb_2O_5 surfaces as the ionic strength of the solution was increased (i.e., from −46 to −19 mV), is as one would expect for the electrostatic interaction of two negatively charged surfaces in buffers of the ionic strength used in this study. Good agreement between the force data and theoretical predictions was obtained, except at small separation distances, where theory predicts the presence of an attractive force due to van der Waals interactions. The fact that the interactions were purely repulsive at all separation distances suggests the presence of an additional, short ranged repulsive interaction, presumably due to the presence of hydration/steric interactions, as are often obtained between silica surfaces.[16,55]

Figure 4b analogously shows the interaction between a SiO_2 microsphere and a PLL-coated Nb_2O_5 surface, exhibiting strong attractive forces in the low ionic strength 1 mM HEPES buffer and with the magnitude of the attractive force decreasing as the ionic strength of the buffer was increased. Also presented in Figure 4b are DLVO theoretical curves. As observed previously, the agreement between measured forces and theory was good, with the Debye length in the 1 and 10 mM HEPES buffers now being determined, at least in part, by the HEPES concentration. Adsorption of PLL onto the Nb_2O_5 surface reverses the sign of the surface charge from negative to positive, resulting in attractive electrostatic

(50) Derjaguin, B.; Landau, L. *Acta Physicochim. U.R.S.S.* **1941**, *14*, 633–662.
(51) Verwey, E. G. W.; Overbeek, J. T. G. *Theory of the Stability of Lyophobic Colloids*; Elsevier: Amsterdam, 1948.
(52) McCormack, D.; Carnie, S. L.; Chan, D. Y. C. *J. Colloid Interface Sci.* **1995**, *169*, 177–196.
(53) Meagher, L. Unpublished data 2004.
(54) Larson, I.; Drummond, C. J.; Chan, D. Y. C.; Grieser, F. *Langmuir* **1997**, *13*, 2109–2112.
(55) Grabbe, A.; Horn, R. G. *J. Colloid Interface Sci.* **1993**, *157*, 375–383.

Interfacial forces of PLL-g-PEG *Langmuir, Vol. 21, No. 14, 2005* 6513

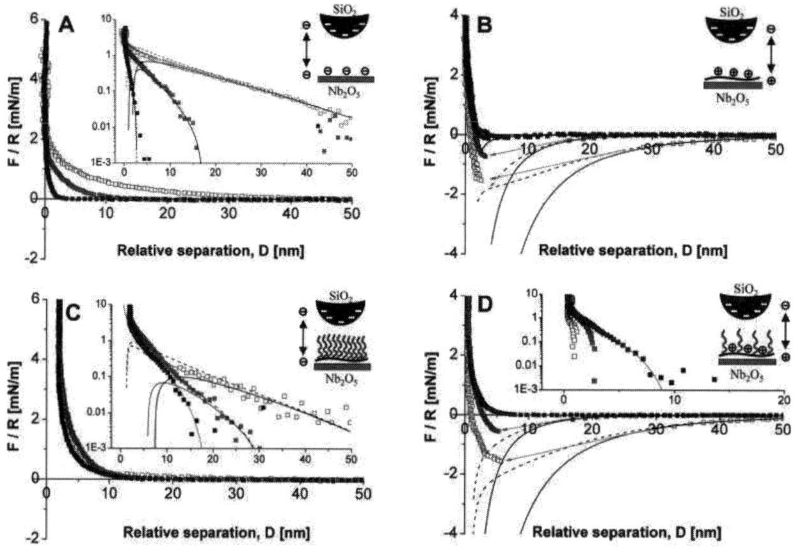

Figure 4. Force vs distance curves recorded upon the approach of a SiO_2 microsphere to (a) Nb_2O_5 bare substrate, (b) PLL(20)-coated Nb_2O_5, (c) PLL(20)-*g*[3.5]-PEG(2)-coated Nb_2O_5 (2 nm offset), and (d) PLL(20)-*g*[10.1]-PEG(2)-coated Nb_2O_5 (0.5 nm offset), measured at different ionic strengths: 1 mM (white squares), 10 mM (gray squares), and 160 mM (black squares). DLVO fits to the data are presented, calculated using the constant potential (solid black line) and constant charge (dashed line) boundary conditions, as well as steric MWC fits (solid gray line) for the curves in c) and (d) measured in 160 mM (see Table 2 for a summary of the fitting parameters used). In the DLVO fits, a Hamaker constant value, $A_H = 2 \times 10^{-20}$ J was used in all cases.

Table 2. Fitting Parameters for DLVO and MWC Theoretical Fits to Data in Figure 4a–d[a]

					MWC fit		
sample	C_0(buffer) [mM]	ψ_{d,SiO_2} [mV]	$\psi_{d,surface}$ [mV]	κ^{-1} [nm]	*s* [nm]	L_0 [nm]	*p*
Nb_2O_5	1	−60	−46	11.1			
	10	−35	−29	3.1			
	160	−19	−19	0.8			
PLL(20)-coated Nb_2O_5	1	−60	+100	13.6			
	10	−50	+50	5.6			
	160						
PLL(20)-*g*[3.5]-PEG(2)- coated Nb_2O_5	1	−60	−16	9.6			
	10	−50	−12	5.6			
	160				1.6	19.5	0.0055
PLL(20)-*g*[10.1]-PEG(2)- coated Nb_2O_5	1	−80	+70	15.2			
	10	−50	+35	5.6			
	160				2.5	10.0	0.0017

[a] For the DLVO fits, ψ_d is the diffuse double layer potential where $\psi_{d,\,SiO2}$ refers to the silica sphere, $\psi_{d,\,surface}$ refers to the opposing surface as defined in column 1 and κ^{-1} is the fitted Debye length. For the MWC fits, *s* is the spacing between PEG chains (obtained using $s = 2(\sigma/\pi)^{0.5}$, where σ is the surface area per PEG chain taken from Table 2), L_0 is the fitted thickness of the polymer layers and *p* is a prefactor.

interactions. Since the HEPES molecule is net negatively charged at the pH at which the measurements were carried out (~0.3 negative charges/molecule), it acts as the counterion for the PLL-Nb_2O_5 surface, while Na^+ ions act as the counterion for the negatively charged silica surface. Thus, the decay lengths obtained were longer than one would expect for a 1:1 electrolyte solution of the same concentration (i.e., in 1 mM HEPES the decay length obtained was 13.6 nm compared to 9.6 nm for a 1 mM solution of a 1:1 electrolyte). While the numerical solutions to the nonlinear Poisson–Boltzmann equation used to generate the theoretical curves presented here are valid strictly only for 1:1 electrolytes, the comparison between the force data and theoretical curves nevertheless strongly indicates that the attractive interactions obtained have an electrostatic origin. Currently, it is not possible to predict accurately the electrostatic forces resulting from the interaction of oppositely charged surfaces in an asymmetric electrolyte system such as in this study. At high ionic strengths, i.e., 10 mM HEPES + 150 mM NaCl, the interactions between the surfaces could be adequately described by a van der Waals interaction alone. There are also indications that the interactions at short range were due to compression of the adsorbed PLL layer, as the magnitude of the short-range repulsive force increased as the ionic strength of the buffer was increased, in line with expectations of the behavior of adsorbed polyelectrolyte layers.[56] Parts a and b of Figure 4 serve as reference curves for the experiments with PLL-*g*-PEG copolymer-coated surfaces.

The data presented in Figure 4c,d demonstrate the effect of the ionic strength on the interactions between silica and two PLL-*g*-PEG-coated Nb_2O_5 surfaces with different PEG surface densities. For the lower PEG density, PLL(20)-*g*[10.1]-PEG(2) (0.21 PEG nm^{-2}), attractive forces were observed at low ionic strength, revealing the

(56) Fleer, G. J.; Cohen Stuart, M. A.; Scheutjens, J. M. H. M.; Cosgrove, T.; Vincent, B. *Polymers at Interfaces*; Chapman & Hall: London, 1993.

presence of free positive charges in the PLL backbone (Figure 4d). The electrostatic origin of the longer ranged attractive force was verified by comparison with DLVO theoretical curves (the EDL fits were carried out at moderate to large separation distances, to avoid any contribution from steric forces, due to polymer layer compression, in the fits). As the polymer coated Nb_2O_5 surfaces were positively charged in this case, the arguments presented above regarding the fitted Debye length are relevant here also. Small offsets were applied to the scaled force data to try to account for the thickness of polymer present between the surfaces at high compressive loads, as this allowed a better fit to steric forces (high ionic strength) and did not change the potential fitted to the lower ionic strength data significantly. Although the offset did not change the fitted potential significantly, it allowed a better fit between the experimentally obtained force data and steric compressive force theoretical curves fitted to the data at high ionic strength (see below). The fitted potential for the polymer-coated Nb_2O_5 surface was lower than for PLL coated Nb_2O_5 (+70 vs +100 mV). Assuming that the PLL adsorbed amount was similar in both cases, the reduced positive diffuse double layer potential obtained by fitting the force data is consistent with some of the PLL amines being consumed in the grafting of PEG side chains, thus reducing the overall linear charge density of the adsorbing molecules. At smaller relative separation distances, a minimum in the magnitude of the force was present (i.e., the slope of the measured force changed from negative to positive) and at very small separation distances, the force was repulsive. This transition indicates a change in the mechanism of force generation between the surfaces.

As we increased the ionic strength to 10 mM HEPES, the magnitude and decay length of the longer ranged attractive forces was reduced. Again the correlation between the force data and the DLVO theoretical fits confirms the origin of the longer ranged forces as electrostatic in nature. The fitted potential in the 10 mM HEPES solution for the polymer coated Nb_2O_5 surface was +35 mV (compared to a value of +70 mV in 1 mM HEPES buffer). Finally in a 10 mM + 150 mM NaCl buffer, the forces were purely repulsive and could no longer be described by DLVO theoretical curves. Clearly here the onset of the repulsive force corresponds to the initial compression of the PEG layer (i.e., the approximate thickness of the adsorbed PLL-*g*-PEG layer), and under the ionic strength of the solution used, the predicted thickness of the EDL will be smaller than the thickness of the polymer layer. Hence for most of the range of the force, the PEG steric repulsive interactions prevail over any EDL attractive forces. We would, however, expect a small contribution from EDL interactions at relative separations < 3 nm.

The origin of the repulsive force in high ionic strength solution was verified by comparison with theoretical curves calculated using equations derived by Kenworthy et al.,[57] based on the mean field theory of Milner et al.[58,59] This theory (MWC), and the equations developed to describe the interaction energy between two opposing polymer brushes as they are compressed,[25,59] make use of a parabolic segment density profile and allows a more realistic representation of the structure of tethered polymer brushes than the uniform segment density layer originally proposed by de Gennes and Alexander.[60,61] The parameters used in the fitting routine were the distance between PEG molecules (s), the brush thickness (L_0), the number of EG monomers (45 for a 2 kDa chain), the monomer size (0.35 nm), and a prefactor (p) (Table 2). In addition, the force/radius as a function of separation distance predicted was divided by a factor of 2 to account for the fact that only one polymer brush was being compressed in the experiments reported here. The s value was calculated using the measured amount of adsorbed polymer and the average grafting (g) value for PEG chains along the PLL backbone, and the thickness of the brush was taken as the range of the repulsive forces due to compression of the adsorbed layer. In addition, a small offset was used to account for the thickness of compressed polymer between the surfaces at high loads. Inclusion of this offset also allowed a better fit over a larger range of relative separation distances. The value of the prefactor was allowed to vary as a dependent variable, and the L_0 value was adjusted within a limited range to obtain better fits. It is apparent that the agreement between the force data and theoretical curves was quite good.

At higher PEG density, for PLL(20)-*g*[3.5]-PEG(2) (0.53 PEG nm^{-2}), steric repulsive forces dominate the interaction even in the 10 mM HEPES buffer; the measured interaction forces are very similar for both 10 and 150 mM ionic strength buffers (Figure 4c), especially at high compressive loads, with only a small EDL contribution at larger separation distances in the 10 mM HEPES case. This is presumably due to the greater thickness of the PLL(20)-*g*[3.5]-PEG(2) layer compared to the PLL(20)-*g*[10.1]-PEG(2) layer. Here the EDL was only slightly thicker than the polymer layer. At the lowest ionic strength (1 mM HEPES), however, EDL *repulsive* forces were observed on approach at longer distances, until the sphere sensed a stronger steric repulsion from the PEG chains.[22,26] Therefore, with a high grafted loading of PEG side chains on the PLL backbone, which has consumed many of the amine groups, the remaining positive charges in the PLL backbone (<1.3 nm^{-2}) were not sufficient to compensate fully the negative charges of the underlying Nb_2O_5 substrate, resulting in an overall negative charge of the copolymer-coated surface. Also presented in the inset to Figure 4c is a steric force theoretical curve calculated using the same protocols as outlined above. Here the s spacing used was much smaller, due to the higher PEG grafting density, and the thickness of the brush layer was larger than that obtained for the PLL(20)-*g*[10.1]-PEG(2) case presented in Figure 4d. There is good agreement between experimental data and the theoretical curves, supporting the assumption that this force was steric in nature and due to compression of the PEG brush. In addition, the high load compression data from the three different ionic strength solutions used was very similar and could be described using the same polymer brush fitting parameters. This suggests that the structure of the polymer layer was independent of the solution ionic strength, as expected for neutral brushes and that the adsorbed amount did not change as the ionic strength of the solution was altered.

Parts a–d of Figure 4 demonstrate the effect of polymer architecture on the net charge at the interface, where an increase in the PEG surface density, determined by the grafting ratio, both enhances the steric repulsion and weakens the EDL attraction. These data also show that the net electrostatic charge, resulting from incomplete

(57) Kenworthy, A. K.; Hristova, K.; Needham, D.; McIntosh, T. J. *Biophys. J.* **1995**, *68*, 1921–1936.

(58) Milner, S.; Witten, T.; Cates, M. *Macromolecules* **1988**, *21*, 2610–2619.

(59) Milner, S. T.; Witten, T. A.; Cates, M. E. *Europhys. Lett.* **1988**, *5*, 413–418.

(60) de Gennes, P. G. *Adv. Colloid Interface Sci.* **1987**, *27*, 189–209.

(61) Alexander, S. *J. Phys. (Paris)* **1977**, *38*, 983–987.

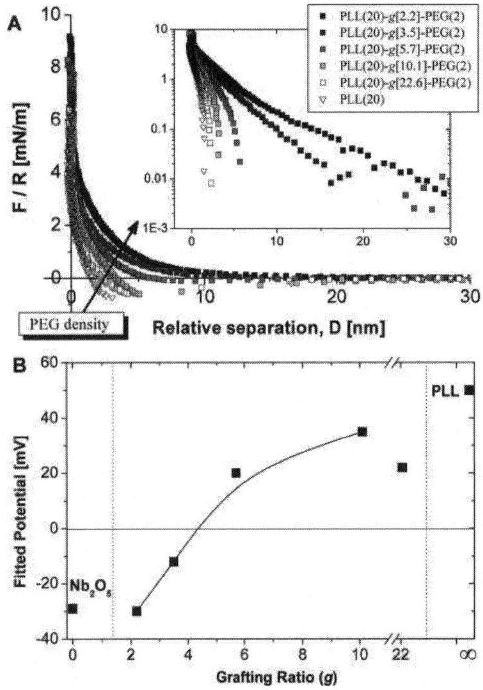

Figure 5. (a) Force–distance curves recorded upon compression of PLL(20)-*g*-PEG(2) polymer adlayers on niobia by a SiO_2 microsphere in 10 mM HEPES buffer solution (pH 7.4), with PEG chain densities from 0 to 0.54 PEG nm^{-2}. (b) Surface potential from DLVO fits (not shown) as a function of the grafting ratio, *g*, of the PLL-*g*-PEG polymer, highlighting charge reversal at $g = 4$.

association/compensation between negative metal oxide surface charges and protonated amine groups on the PLL, can have either sign, depending on the grafting ratio, and that the importance of the EDL contribution to the total interfacial force varies with the ionic strength. These effects are not readily predictable, and thus such measurements are invaluable and essential for taking into account possible electrostatic effects on the interactions between PLL-*g*-PEG coatings and proteins.

3.4. Effect of the PEG Surface Density. PLL and PLL-*g*-PEG copolymers, synthesized from 20 kDa PLL-HBr and 2 kDa PEG at grafting ratios between 2.1 and 22.6, were adsorbed onto Nb_2O_5, leading to PEG densities varying from 0 to 0.54 nm^{-2}. The PEG surface density was calculated from in situ quantitative measurements of adsorbed mass as described in section 3.1 and in.[38] Figure 5a presents the influence of the PEG chain surface density on the interaction forces with the SiO_2 microsphere, measured in 10 mM HEPES buffer at pH 7.4, where the EDL contribution to the force was still perceptible at longer relative separations for all of the polymers investigated. A higher PEG density has been shown to be related to an increase in both the compressed layer thickness and the magnitude of the steric force in a different PEG system.[28] The conformation is thought to be more brushlike for high PEG surface densities, leading to stronger and more extended steric repulsive forces.[22] In Figure 5a, as the PEG surface density was increased (from 0 to 0.33 nm^{-2}), the resulting force indeed became less attractive, as both the magnitude of the steric repulsion increased and the EDL attractive forces were weakened. At the highest PEG densities (>0.33 nm^{-2}), the observed interaction forces were due to a superposition of a repulsive EDL and steric interactions, the magnitudes of both the EDL and steric contributions becoming more repulsive as the PEG grafting density was increased.

Fitting of the EDL contribution was carried out (data not included for brevity) and a summary of the resulting fitted potential vs the grafting ratio (*g*) of the polymer is presented in Figure 5b. It shows that the sign and magnitude of the fitted potential varied as a function of the *g* value of the polymer. For *g* values between 22.6 and 5.7 (0.11–0.33 nm^{-2}), the longer ranged interactions measured were attractive and a positive fitted potential was obtained, due to overcompensation of the negative surface charge of the Nb_2O_5 surface by the positive charges along the PLL backbone. However, for the two higher grafting densities (*g* values of 3.5 and 2.2) the longer ranged EDL interactions were repulsive, resulting in a negative fitted potential, the magnitude of which increased as the grafting density of PEG chains along the PLL backbone increased. In these two cases, there were insufficient positive charges remaining along the PLL backbone to neutralize the negative charges on the Nb_2O_5 surface. Also included for comparison are the fitted diffuse double layer potentials obtained for bare Nb_2O_5 and Nb_2O_5 with an adsorbed PLL (i.e., with no grafted PEG). The *g* value where the net charge of the surface was zero is predicted by this relationship to be approximately 4.

3.5. Effect of the PEG Chain Length. To assess how the steric force and its range would vary with PEG chain length, the EDL contribution had to be minimized. Thus, the measurements reported in this section were performed in higher ionic strength solutions in order to reduce the range of the EDL force and thereby enhance the relative contribution of steric repulsion to the net interfacial force. This situation also corresponds to many potential applications, such as biomedical device coatings where the ionic strength is ~0.15 M, determined by physiological conditions.

Figure 6a reproduces force–distance curves recorded with a SiO_2 microsphere and Nb_2O_5 surfaces coated with PLL-*g*-PEGs of 1, 2, and 5 kDa PEG molecular weights and grafting ratios that resulted in similar EG surface densities. To extract the influence of the PEG chain length, PLL-*g*-PEGs with PEG MWs of 2 and 5 kDa and grafting ratios leading to a constant PEG surface density of 0.2 nm^{-2} were used for sample set 1. In addition, in a second set the influence of PEG MW for constant EG densities of 9 and 25 nm^{-2} was explored; these densities correspond to protein-interactive and protein-resistant situations, respectively. The measurements were performed in 10 mM HEPES + 150 mM NaCl buffer. This enabled us to (a) assess the contribution of the steric force, which dominates the observed forces at this high ionic strength, and (b) relate the measured forces to protein adsorption data from parallel experiments performed in the same buffer solution. The forces are plotted on a semilogarithmic scale for better visualization of the steric decay lengths obtained for compression of the various adsorbed polymer layers.

Increasing the PEG MW from 2 to 5 kDa while keeping the PEG surface density constant (sample set 1) resulted in an increase of the decay length, demonstrating the presence of a thicker but slightly softer PEG interfacial layer. At the same PEG surface density, longer PEG chains are likely to be more flexible than shorter ones unless the brushes are densely packed.

The experiments with sample set 2, comparing different PEG molecular weights and similar EG surface density values, reflecting a combined effect of PEG molecular

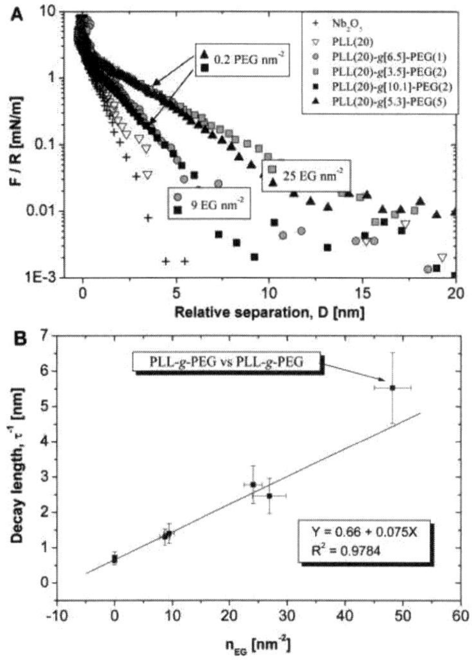

Figure 6. (a) Force–distance curves recorded upon compression of PLL(20)-*g*-PEG polymer adlayers on niobia by a SiO_2 microsphere in 10 mM HEPES + 150 mM NaCl buffer solution (pH 7.4), for different PEG MWs but constant PEG densities of 0.2 nm^{-2} or EG surface densities of 9 and 25 nm^{-2}. The curves are shown on a semilogarithmic plot in order to compare their decay lengths. (b) Decay length of the steric repulsive interaction as a function of the EG surface density.

weight and density, gave unexpected results. Independent of whether they resist protein adsorption or not, surfaces with the same EG density showed a similar response upon compression: PEGs of MWs 1 and 2 kDa with 9 EG nm^{-2}, and PEGs of MWs 2 and 5 kDa with 25 EG nm^{-2}, exhibited similar decay lengths independent of the PEG MW. Furthermore, as shown in Figure 6b, the decay length depends linearly on the EG surface density. This observation provides evidence that the EG–monomer density determines the stiffness of the layer as well as its thickness. Since protein adsorbed mass has been shown to depend primarily on EG density (and not on PEG MW or density per se), the observation of similar responses to compression at constant EG density validate AFM force measurements as a useful technique to provide fundamental understanding into the forces that govern protein adsorption and resistance.

3.6. PEG—PEG Interactions. To gain insight into the interactions between two PEG-copolymer coated surfaces, both the SiO_2 sphere and the flat niobia substrate were coated with a layer of protein-resistant PLL(20)-*g*[3.5]-PEG(2). As the spacing between the PEG chains (1–2 nm) is negligible compared to the radius of the sphere (4–5 μm), the PLL-*g*-PEG molecules should not sense any curvature. The measurements were performed in 10 mM HEPES buffer, pH 7.4. We will treat this as a symmetric system although the substrates for PLL-*g*-PEG adsorption are different, because under these solution conditions the interactions between the *uncoated* microsphere and the polymer-coated substrate were independent of whether SiO_2 or Nb_2O_5 was used as the substrate (data not shown),

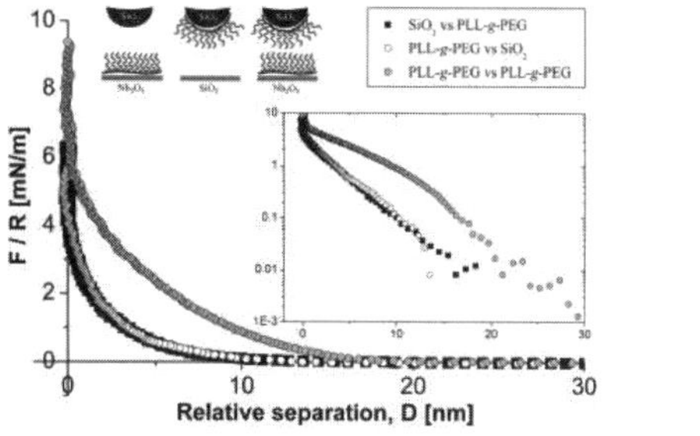

Figure 7. Force curves recorded for PEG-PEG interactions, upon coating both the SiO_2 microsphere and the Nb_2O_5 substrate with PLL(20)-*g*[3.5]-PEG(2), measured in 10 mM HEPES buffer (pH 7.4), both linear and semilogarithmic plots.

as both oxides are negatively charged at pH 7.4, with IEPs of ~2 for SiO_2 and 4.4 for Nb_2O_5.[62]

In Figure 7, force curves recorded with an uncoated SiO_2 colloid approaching PLL(20)-*g*[3.5]-PEG(2)-coated Nb_2O_5 are compared with those for a polymer-coated sphere against an uncoated SiO_2 surface. The two force profiles are very similar, indicating that the SiO_2 microsphere indeed carried a monolayer of PLL-*g*-PEG. [After many cycles, some desorption of the PLL-*g*-PEG layer adsorbed on the SiO_2 microsphere was observed. PLL-*g*-PEG is known to form a less stable layer on SiO_2 compared to Nb_2O_5, probably due to a lower charge density in the former case (data not shown). Thus, the polymer was adsorbed onto the silica sphere prior to each experiment, and the measurement time was kept short. A force curve was performed at the end against an uncoated Nb_2O_5 substrate, to verify the presence of the polymer layer.] The interaction between the two PLL-*g*-PEG-coated surfaces appears to be qualitatively additive (Figure 7). The film thickness and the magnitude of the force are, however, smaller than twice those of the asymmetric experiments in which only one interacting surface was coated with the polymer. This may be due to compression of two soft layers. If we assume that the density of chains at the extremity of the brush layer is low, as is the case with a parabolic brush profile, then to avoid interpenetration of the brushes, the thickness of the brush might reduce slightly, and this would be balanced by the concentration increase of polymer within the brush. Partial interpenetration of the PEG chains is entropically unfavorable. Quantitative interpretation is, however, hampered by the fact that the "zero" separations in the asymmetric and symmetric experiments correspond to one or both layers, respectively, being fully compressed, and as discussed above, there is some uncertainty as to the residual thickness of the compressed layers. The force curves show weak repulsion at longer range, and increasingly strong repulsion as the sphere approaches to within the PEG layer thickness and starts to compress the hydrated PEG layer. This apparent overlap might be related to the polydispersity of the PEG chains.[8]

The decay length of the steric force at separation distances <~12 nm for the symmetric case ($\tau^{-1} \approx 5.5 \pm 1.0$ nm) is approximately twice that when only one surface is coated ($\tau^{-1} \approx 2.8 \pm 0.5$ nm). This is consistent with the observation that the decay length depends on the EG density (section 3.5), as shown in Figure 6b. As no adhesion could be detected even when increasing the loads up to twice the original load (data not shown), there is no evidence in our experiments of bridging between PEG

(62) Kosmulski, M. *J. Colloid Interface Sci.* **2002**, *253*, 77–87.

brushes, in contrast to reported data from systems with loosely bound polymeric adlayers with much longer PEG chains of MW 56000 to 1.2×10^6 Da.[18-21] These findings provide further evidence that the PLL-*g*-PEG molecule is firmly attached and exhibits a stable brush conformation under the conditions (pH, ionic strength) of our experiments.

4. Discussion

The motivation behind this colloid-probe AFM study was to (a) investigate and distinguish between two types of interfacial forces that may be important in (bio)molecule-surface interactions, viz., electrical double layer interactions and steric–entropic repulsion, and to (b) correlate the experimental observations with previously reported quantitative data on protein (blood serum) adsorption to the same bare and polymer-functionalized surfaces.[38] In this way we aimed to understand the interfacial interaction forces between proteins and PLL-*g*-PEG coatings that lead to different behaviors of these coatings as the grafting ratio and the PEG chain length was varied. Thus, ultimately one may be able to rationalize protein adsorption in terms of interfacial forces and how these forces are determined by the composition and architecture of grafted layers and the solution conditions. Of course, simulating the approach of a protein to PLL-*g*-PEG coatings by a SiO_2 microsphere attached to an AFM cantilever is a rather simplified model experiment, as the sphere is considerably larger and lacks the heterogeneity and conformational mobility of proteins. Nevertheless we believe it is reasonable to assume that an approaching protein molecule is subject to the same types of surface forces emanating from the grafted PLL-*g*-PEG layer as are sensed by the colloid sphere probes, and hence the forces measured by colloid-probe AFM will assist in the interpretation of protein adsorption, though one will need to account for the effects of protein properties on the interaction forces. Clearly, interpretation of protein adsorption in terms of contact angles or the presence of specific chemical species is inadequate; to interpret the differences in protein adsorption shown by the various PLL-*g*-PEG copolymers, an approach based on EDL interactions and steric–entropic repulsion appears to be necessary.

One issue we must consider is that of surface homogeneity or heterogeneity: the colloid probe cannot sense nanoscopic heterogeneities of surfaces that might affect protein/surface interactions. Thus, colloid-probe force measurements are valid only if the surfaces are likely to be homogeneous across dimensions comparable to the sphere diameter. ToF–SIMS imaging and multiple force measurements (not shown) showed no detectable variations across coated samples; hence, we believe the coatings are sufficiently uniform and the measurements presented here are reasonable indicators of surface forces acting on approaching proteins. As shown by our results, the AFM colloid-probe method can provide, at the least, semiquantitative information on two inherently important types of surface forces and thus aid in establishing improved design criteria for PEGylated surfaces with controlled forces of interaction. The polycationic PLL-*g*-PEG copolymeric system is well suited to such studies as it forms monomolecular adlayers with sufficient stability for investigation at neutral pH and in the range of ionic strength values used in this study. In addition, systematic variations of the molecular structure (PEG chain length and grafting density) provide coatings with controlled PEG chain surface densities from isolated PEG chains to mushroom to brush conformations. Finally, changes in the PLL-*g*-PEG grafting ratio affect not only the PEG surface density but also the density of remaining charges associated with protonated free (non-PEGylated) amine groups in the PLL backbone, enabling us to systematically vary the resulting interfacial charge. The latter is determined by the sum of the negative charge of the Nb_2O_5 substrate surface and the positive charge of the polycationic copolymer backbone, and its surface coverage. Niobia was selected as the substrate surface in view of its well-defined isoelectric point (IEP ~ 4.4)[62] and its high surface charge density at physiological pH, which results in reproducible, stable binding of the polyelectrolyte adlayers through electrostatic attraction.

Among the interaction forces discussed in recent reviews,[9,37] electrical double layer forces and steric–entropic–osmotic interactions are expected to be main contributors to the total force between the silica microsphere and PLL-*g*-PEG-coated Nb_2O_5 surfaces, whereas van der Waals interactions are of short range and likely to be a minor contribution only, due to the low dielectric constant of PEG. As both the underlying substrate (Nb_2O_5) and the PLL backbone are charged, the charge distribution within the layer, and thus electrical double layer (EDL) interactions, are expected to be rather complex. The steric repulsive forces originating from the PEG side chains are most relevant, in view of their known role in imparting protein resistance. The two types of forces will, however, vary in different ways with the ionic strength of the solution, and hence force measurements at different ionic strengths should enable us to discriminate between electrostatic contributions and steric interactions. As the ionic strength is increased, the thickness of the EDL decreases to the point where it becomes significantly less than the PEG layer thickness; thus, we expect that above a certain concentration, dissolved ions will shield the surface charge and thus the steric force from the PEG brushes will dominate the net interfacial force.

The interaction of the SiO_2 microsphere with both the uncoated and PLL-coated Nb_2O_5 substrates (Figure 4, parts a and b) can be explained by the presence of a purely electrostatic double layer force over almost all separation distances, with some evidence of a small contribution from short-range hydration/steric repulsive forces for the Nb_2O_5 surface and a short-range repulsive force from compression of the PLL layer adsorbed onto the Nb_2O_5 surface, in line with expectations for these types of surfaces. Thus, as expected, the PLL is adsorbed essentially in a flat conformation. The EDL force depends on the ionic strength of the solution and the sign and magnitude of the diffuse double layer potentials associated with the surfaces, in agreement with Poisson–Boltzmann theory, and the experimental decay lengths match the Debye lengths expected under the solution conditions employed. These reference measurements authenticate the reliability of the experimental setup for measurements with more complex surfaces such as the PLL-*g*-PEG coatings.

The surface forces sensed by the approaching sphere can be discussed in terms of two regimes: (a) before the approaching sphere contacts the PEG chains, the forces are predominantly electrostatic in nature with a range, magnitude, and decay length depending on the sign and magnitude of diffuse double layer potentials associated with the surfaces, the ionic strength of the solution, and the thickness of the adsorbed polymer layer; (b) as the colloid sphere starts to contact and compress the PEG layer, steric repulsion forces arise which depend in a complex manner on chain density, conformation, and water content of the PLL-*g*-PEG layer. The commercial PEGs used for the synthesis of these copolymers are, however, somewhat polydisperse. For this reason, and because of the distribution of PEG grafting densities on the PLL backbone (the *g* value is an average determined by NMR) and conformational flexibility of the chains, the PEG surface is not to be viewed as a well-defined surface but as a rather gradual interface with increasing segment density. The sphere will at first sense only a sparse array of more protruding chains and gradually compress the layer. Thus, the distance at which the sphere "contacts" the PEG brushes is not distinct. The thickness of the adsorbed layer will be modified somewhat from the MWC picture of a polymer brush (the model assumes a single *s* value) because of the distribution of PEG interchain spacings and small polydispersity in PEG molecular weight. We expect, however, that the basic physics should be essentially the same. Both of these factors will increase the effective thickness of the PEG layer. For example, the effect of a polydispersity of 1.02 increased the layer thickness of grafted polystyrene layers by approximately 13%.[59]

4.1. Electrical Double Layer Interactions. The surface charge of PLL-*g*-PEG-coated surfaces varies in a systematic manner with the PEG surface density: the positive charge shown by the PLL-coated surface decreases continuously as the PEG surface density in the copolymer adlayers increases, ultimately leading to a surface charge reversal (at a *g* value of approximately 4) for the highest PEG densities as there are insufficient protonated amine groups to compensate the negatively charged oxide surface. The variations of the forces with PEG surface density and the associated fitted surface potentials (Figure 5) attest to the intimate relationship between PEG surface density, surface charge, and surface forces. High PEG surface densities, which are associated with low protein adsorption, generate repulsive interactions with the SiO_2 sphere, while lower PEG densities result in attractive forces between the flat surface and the negatively charged sphere, which can be assigned to the range of the EDL arising from remaining positive charges of the PLL backbone exceeding the thickness of the PEG layer. These observations reveal the existence of an electrical double layer force contribution between the SiO_2 microsphere and the polymer-coated surfaces. Yet, the interfacial forces recorded for PLL-*g*-PEG coated surfaces do not accord completely with DLVO theory, indicating that the EDL interaction is superimposed upon a steric repulsion force generated by the PEG chains as the sphere compresses the layer. While the compression of the PEG brushes by the sphere does not mimic events in protein adsorption, the sensing of the EDL force by the colloid probe is, we postulate, pertinent to interpretation of protein adsorption, since EDL forces, of varying magnitude and sign, are also experienced by approaching proteins. Clearly, EDL forces penetrating the PEG brush layer can markedly affect interfacial interactions with proteins, and their detection and quantification by colloid-probe AFM aids in the interpretation of protein adsorption patterns. In addition, our measurements demonstrate that the PLL-*g*-PEG coatings cannot be viewed simply as steric barrier coatings; the surface forces involved are more complex and depend both on the molecular architecture and the ionic strength of the solution.

The colloid-probe measurements may be most pertinent to the interpretation of interfacial interactions between these coatings and large, usually overall negatively charged proteins that would not be able to penetrate, by diffusion, into the steric barrier layer but are likely to be able to sense the EDL force. If such proteins sense an attractive EDL force originating from beneath the PEG layer, as is the case for lower PEG graft densities, they may adsorb on top of the PLL-*g*-PEG layer. Accordingly, the screening of residual "underlying" surface charges by sufficiently long and densely packed PEG chains and the charge neutrality of the "outer" surface are key parameters for preventing the adsorption of proteins. If the PEG chains are too short, or the ionic strength is decreased, proteins may be able to adsorb onto the PEG layer if the attractive EDL force outweighs the steric–entropic repulsion. This scenario of protein attraction by substrate charges applies not only to the PLL-*g*-PEG systems but also to other grafted PEG layers, such as covalently immobilized brushes.[3]

The relevance of the PEG layer thickness for the resulting surface forces and protein adsorption is demonstrated experimentally by the effect of the ionic strength on the forces (Figure 4a–d). In Figure 4c, no EDL force is observed at 160 mM salt concentration ($\kappa^{-1} < 1$ nm), while at 10 and 1 mM ($\kappa^{-1} \approx 3$ and 10 nm respectively) the bare SiO_2 probe senses either repulsive or attractive force upon approach before it reaches the onset of the steric repulsion. Therefore, a minimal layer thickness that depends on the ionic strength is required in order to shield the surface charge. As long as the PEG layer thickness exceeds the range of the EDL force, no electrostatic force is sensed by the probe (and, presumably, an approaching protein) before it reaches the outer reaches of the brush polymer layer and experiences steric–entropic repulsion. Indeed, the relevance of these considerations has been borne out experimentally by protein adsorption experiments performed at ionic strengths of 1, 10, and 160 mM, which demonstrated adsorption caused by ionic strength dependent attractive EDL forces; these results will be presented elsewhere.[63]

4.2. PEG Chains Conformation. Grafted PEG chains at an interface have been shown to adopt various conformations, depending on PEG surface density and chain length, which are referred to as pancakes, noninteracting or interacting mushrooms, and brushes.[22,37] Protein-resistant PEG coatings are thought to result from the formation of PEG brushes at the interface.[13] Both the response to compression and the observed layer thickness can provide information on conformation.[64] On the basis of the radius of gyration of free PEG in solution and the PEG surface coverage, one can estimate the conformation of the polymer adlayer, relating the surface area occupied by a single chain, σ, to the projected area of the unperturbed chain in solution, based on the radius of gyration, πR_g^2. To a first approximation $\sigma < \pi R_g^2$ corresponds to the regime of overlapping chains, in which the chains are extended into a brushlike structure.[25] On the basis of this criterion, the data in Table 3 suggest that most of the PLL-*g*-PEG copolymers used in this study form, upon adsorption onto niobia surfaces, PEG chains in the brush regime. Therefore, the layer thickness is expected to exceed the radius of gyration of "free" PEG in solution. This hypothesis was supported by comparison of the forces obtained at high ionic strength to theoretical curves for polymer brush compression (Figures 4c and d).

One challenge in AFM colloid-probe surface force measurements is the definition of the zero separation distance with compressible layers. The experimental zero distance is obtained by constant compliance; at this point no further compression of the PLL-*g*-PEG layer occurs. We do not, however, know the thickness of the compressed layer. Thus, approach force curves describe the transition

(63) Pasche, S.; Vörös, J.; Griesser, H. J.; Spencer, N. D.; Textor, M. *J. Phys. Chem. B* **2005**, submitted.

(64) Kidoaki, S.; Nakayama, Y.; Matsuda, T. *Langmuir* **2001**, *17*, 1080–1087.

Table 3. Estimation of the Conformation of the Polymers, Comparing R_g [nm], the Radius of Gyration of the Polymer in Solution, Estimated Empirically with $R_g = 0.181N^{0.58}$ (where N Is the Number of EG per PEG Chain),[70] σ [nm²], the Surface Area of One PEG Chain, Evaluated from NMR and OWLS Measurements,[38] and L [nm], a Qualitative Estimation for the Compressible Thickness of the Polymer Layer Determined from AFM Data

polymer	R_g [nm]	s [nm²]	$\sigma/\pi R_g^2$ [-]	L (±0.5) [nm]	conformation
PLL(20)-*g*[6.5]-PEG(1)	1.1	2.6 ± 0.3	0.7	5	brush
PLL(20)-*g*[2.2]-PEG(2)	1.7	1.9 ± 0.1	0.2	7	brush
PLL(20)-*g*[3.5]-PEG(2)	1.7	1.9 ± 0.1	0.2	7	brush
PLL(20)-*g*[5.7]-PEG(2)	1.7	3.0 ± 0.3	0.3	6	brush
PLL(20)-*g*[10.1]-PEG(2)	1.7	4.8 ± 0.5	0.6	5	brush
PLL(20)-*g*[22.6]-PEG(2)	1.7	9.4 ± 0.9	1.1	4	weak overlap
PLL(20)-*g*[5.3]-PEG(5)	2.8	4.2 ± 0.5	0.2	7	brush

from a hydrated to a fully compressed layer, and the distances over which such compression is observed in the figures do not correspond to the PEG layer thickness. The thickness of the compressed layers has not yet been experimentally determined. Assuming a mass density of dry PEG of 1.094 g/cm³ yields a "dry" thickness of 1–3 nm depending on molecular architecture. While the assumptions are untested, it appears that the sum of the dry thickness plus the compression distance significantly exceeds the radius of gyration except perhaps for the sparser PEG copolymers. Thus, PLL-*g*-PEGs with high PEG loading can reasonably be assumed to adopt a brush structure on adsorption.

Even though AFM measurements only display relative separation, the thickness of the *hydrated* layer can nevertheless be estimated semiquantitatively for some of the polymers investigated in this study, if full compression of the polymer is associated with expulsion of water molecules trapped between the PEG chains. Figure 4c shows that for a PLL(20)-*g*[3.5]-PEG(2)-coated surface the EDL contribution is overcompensated by the steric repulsion at a relative separation of about 7 nm, at which point the probe starts to compress the polymer layer. The PEG layer is then compressed over a z range of ~7 (±0.5) nm. In Figure 4d, assuming that the jump-to-contact occurs close to the edge of the polymer layer, the thickness of the adsorbed low-PEG-density copolymer PLL(20)-*g*[10.1]-PEG(2) can be estimated to be ~5 (±0.5) nm. Values for the compressible thickness of a selected number of different polymer layers are summarized in Table 3. Although these values are considered to be semiquantitative rather than quantitative (due to the superposition of EDL and steric forces and uncertainty around the compressed layer thickness), they are in surprisingly good agreement with the results of two other studies based on measurements with a surface force apparatus (SFA), a technique that allows absolute distances to be measured. The PEG layer thickness for the PLL(20)-*g*[3.5]-PEG(2) adlayer on mica was found to be 8.2 nm by comparison to theory,[8] which agrees well with the 7 (±0.5) nm over which colloid AFM compression is observed plus ~2 nm "dry" thickness estimated above. Corresponding thickness data for MW 2000 Da PEG surfaces of similar surface densities were in the range of 6.5–7 nm.[22,25]

Using higher ionic strength buffer enables study of the response of the chains toward compression with little interference from EDL contributions. In the case of PEG and other grafted chains, the repulsive force upon compression is often regarded as a steric force, and in our case was roughly exponentially decaying at larger separation distances, the decay length varying with the PEG chain length and density (Figure 6a). As shown in Figure 6b, the decay length of the steric repulsion increases linearly with the EG surface density. Since the EG monomer surface density has been shown to correlate quantitatively with the degree of protein resistance for PLL-*g*-PEG with PEG molecular weights of 1,2 and 5 kDa,[38] it is reasonable to assume that steric repulsion by densely packed PEG brush chains is essential for achieving nonfouling surface properties. In addition, the degree of stretching of the PEG molecules in the brushes (defined by $2R_{(F\ or\ g)}/s$, where R_F and R_g are the Flory radius and radius of gyration respectively and s is the distance between PEG grafting sites) also correlated with the degree of protein adsorption, indicating that the more densely packed brushes provide the best protection against protein adsorption.[38] It should also be noted that the interactions measured in high ionic strength buffer were slightly longer than that predicted by MWC theory, presumably due to the distribution of PEG spacings expected in the adsorbed layers and some polydispersity in the PEG chains themselves. The theory and equations used to fit the data do not take account of either of these factors. The presence of some highly stretched chains in the layers would increase the total range of the steric force, as would PEG polydispersity.

4.3. Hydration of PEG Layers. There is evidence, however, that steric repulsion by an array of surface-immobilized, flexible chains may not be sufficient to explain the observed protein resistance of polymer brush surfaces.[65,66] Theoretical models have identified both an osmotic repulsive term and an attractive elastic force, as shown in the scaling theory of de Gennes for polymer brushes.[60] Upon compression, the water trapped within the PEG chains is squeezed out of the layer. As the polymer concentration in the brush rises, the osmotic pressure increases, while the elastic restoring force decreases. The osmotic repulsive force on compression of the layer therefore depends on the water content within the layer. The amount of water (and probably its molecular structure) trapped in the PEG brush layer determines the osmotic repulsive effects. The water content of the PLL-*g*-PEG coatings can be estimated from our AFM force distance curves based on the following consideration: As the PEG layer is compressed from a fully hydrated state to a "dry" state, the decrease in thickness is likely to reflect the water of hydration. As an example, the PEG adsorbed mass in PLL(20)-*g*[3.5]-PEG(2) is 177 ± 11 ng/cm² as measured by OWLS, which corresponds to a PEG surface density of 0.53 nm^{-2} and a dry thickness of 1.6 nm assuming a specific density of dry PEG of 1.094 g/cm³. From the AFM compression distance of 7 ± 0.5 nm observed for the same polymer adlayer, the content of water in the polymer layer can be estimated to be 80 ± 6 wt %. Investigations with a quartz crystal microbalance, which is sensitive to the sum of adsorbed (dry) polymer mass and mechanically coupled surface water, and OWLS, which senses the dry polymer mass only, indicates a content of 78 wt % of trapped water,[67,68] which is in reasonable agreement with the colloid-probe AFM estimate.

(65) Prime, K. L.; Whitesides, G. M. *Science* **1991**, *252*, 1164–1167.
(66) Harder, P.; Grunze, M.; Dahint, R.; Whitesides, G. M.; Laibinis, P. E. *J. Phys. Chem. B* **1998**, *102*, 426–436.

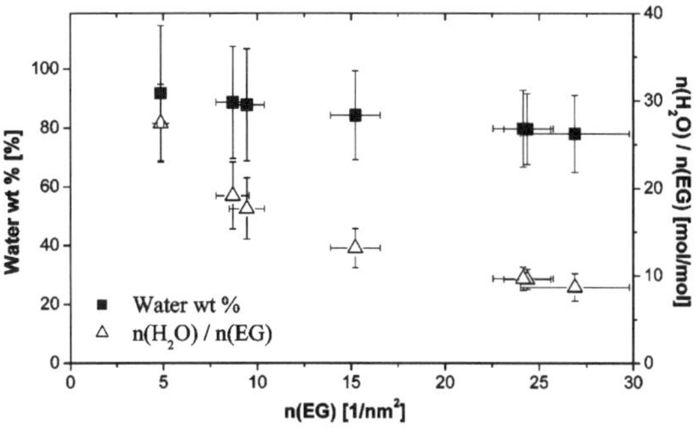

Figure 8. Semiquantitative plot of the water content in PLL-*g*-PEG layers in wt %, and the water to ethylene glycol unit molar ratios, as a function of the EG surface density.

In Figure 8, the water content (in wt %) of the various PLL-*g*-PEG layers (Table 2) and the corresponding molecular ratio H_2O/EG are plotted as a function of the EG monomer surface density. While within the error bars the water content in wt % does not show any clear dependence on the EG monomer surface density, the second representation showing the molecular ratio H_2O/EG shows an interesting trend. The water content decreases linearly with the EG monomer surface density, revealing the lowest hydration of the layer when a close-packed brush is formed. For comparison, the water content of "free" PEG in solution can be calculated from the difference between the dry PEG density and the hydrated PEG sphere. According to the radius of gyration (Table 2), PEGs with molecular weights 1, 2, and 5 kDa contain 81, 89, and 95 wt % water, corresponding to 6, 12, and 25 H_2O/EG, respectively. In the PLL-*g*-PEG adlayers, the water to EG ratio decreases from 27 to 9 water molecule per EG monomer as the EG surface density increases. These values appear to tend to a minimal value, associated with structural hydration water, reported as 3.5, 5, and 10 water molecules per EG for PEG 1, 2, and 5 kDa, respectively.[69] Thus, it appears that for PLL-*g*-PEG coatings with high PEG densities, most of the water in the layer is structural hydration water. Perhaps effective protein resistance of PEG coatings is associated with minimal "free" water between PEG chains through which proteins could diffuse to reach the substrate surface, but sufficient bound structural water of hydration to maintain a sufficient steric–entropic–osmotic character of the coating. This supposition then raises interesting questions of comparison with the effective protein repellence of oligo-EO–SAMs.[30–32]

5. Conclusions

Colloid-probe AFM surface force measurements on adsorbed PLL-*g*-PEG layers on niobia, with various PEG MWs and grafting ratios, have provided information on both the total surface forces and the components contributing to the net force. We found that the measured interfacial forces could be interpreted in terms of contributions that depended on the architecture of the PLL-*g*-PEG copolymer, in particular on the PEG chain length and density, as well as on the ionic strength of the solutions that measurements were taken in. Superimposed contributions, of varying magnitudes and range, arose from electrostatic double layer forces and steric–entropic repulsion upon compression of the hydrated PEG chains. van der Waals forces played a minor role only and did not significantly contribute to variations in force curves between different coatings. At low ionic strength, electrical double layer forces contribute significantly to the total force, whereas at high ionic strength steric repulsion dominated, provided the PEG grafting density was sufficiently high.

Our results show that the traditional interpretation of PEG coatings as steric barriers to protein adsorption is valid only if the PEG layer thickness is sufficient to shield EDL forces from the underlying substrate, and that reducing either the graft density or the ionic strength leads to the appearance of a significant EDL contribution to the total surface force. Such EDL forces will also be experienced by approaching proteins, and hence our surface force measurements assist in rationalizing observed differences in protein resistance of different PLL-*g*-PEG coatings as well as ionic strength effects. Clearly, EDL ranges exceeding the thickness of ("shining through") the PEG brush layer can markedly affect interfacial interactions with proteins, and their detection and quantification by colloid-probe AFM aids in the interpretation of protein adsorption patterns. Surface force measurements can also be used to assess whether steric barrier coatings are sufficiently thick to screen substrate charges for a given ionic strength of the solution.

Acknowledgment. The authors thank Dr. Graeme Gillies, Dr. Kristen Bremmell, and Dr. Fabiano Assi for assistance with AFM experiments, Dr Michael Horrisberger for performing the niobium oxide coating, and Christoph Huwiler for SEM imaging. Dr. Janos Vörös, Dr. Tanja Drobek, Dr. Manfred Heuberger, Dr. Georg Papastavrou, Dr. Priska Zammaretti, and Dr. Oleg Borisov are acknowledged for valuable discussions. This work was financially supported by EPF Lausanne and ETH Zürich (Project TH-33/01-3), the University of South Australia under the ARC Grant Special Research Centre for Particle and Material Interfaces, and the US Air Force Office of Scientific Research (Contract No. F49620-02-1-0346).

LA050386X

(67) Müller, M.; Yan, X.; Lee, S.; Perry, S. S.; Spencer, N. D. *Macromolecules* **2005**, to be submitted.

(68) Heuberger, M.; Drobek, T.; Vörös, J. *Langmuir* **2004**, *20*, 9445–9448.

(69) Branca, C.; Magazu, S.; Migliardo, F.; Romeo, G. *J. Mol. Liq.* **2003**, *103–104*, 181–185.

(70) Kawaguchi, S.; Imai, G.; Suzuki, J.; Miyahara, A.; Kitano, T. *Polymer* **1997**, *38*, 2885–2891.

J. Phys. Chem. B **2005,** *109,* 17545–17552 **17545**

Effects of Ionic Strength and Surface Charge on Protein Adsorption at PEGylated Surfaces

Stéphanie Pasche,*[†,‡] Janos Vörös,[†] Hans J. Griesser,[‡] Nicholas D. Spencer,[†] and Marcus Textor[†]

BioInterfaceGroup, Laboratory for Surface Science and Technology, Department of Materials, Swiss Federal Institute of Technology (ETH) Zurich, CH-8093 Zurich, Switzerland, and Ian Wark Research Institute, University of South Australia, Mawson Lakes, SA 5095, Australia

Received: January 25, 2005; In Final Form: May 17, 2005

PEGylated Nb_2O_5 surfaces were obtained by the adsorption of poly(L-lysine)-*g*-poly(ethylene glycol) (PLL-*g*-PEG) copolymers, allowing control of the PEG surface density, as well as the surface charge. PEG (MW 2 kDa) surface densities between 0 and 0.5 nm^{-2} were obtained by changing the PEG to lysine-mer ratio in the PLL-*g*-PEG polymer, resulting in net positive, negative and neutral surfaces. Colloid probe atomic force microscopy (AFM) was used to characterize the interfacial forces associated with the different surfaces. The AFM force analysis revealed interplay between electrical double layer and steric interactions, thus providing information on the surface charge and on the PEG layer thickness as a function of copolymer architecture. Adsorption of the model proteins lysozyme, α-lactalbumin, and myoglobin onto the various PEGylated surfaces was performed to investigate the effect of protein charge. In addition, adsorption experiments were performed over a range of ionic strengths, to study the role of electrostatic forces between surface charges and proteins acting through the PEG layer. The adsorbed mass of protein, measured by optical waveguide lightmode spectroscopy (OWLS), was shown to depend on a combination of surface charge, protein charge, PEG thickness, and grafting density. At high grafting density and high ionic strength, the steric barrier properties of PEG determine the net interfacial force. At low ionic strength, however, the electrical double layer thickness exceeds the thickness of the PEG layer, and surface charges "shining through" the PEG layer contribute to protein interactions with PLL-*g*-PEG coated surfaces. The combination of AFM surface force measurements and protein adsorption experiments provides insights into the interfacial forces associated with various PEGylated surfaces and the mechanisms of protein resistance.

1. Introduction

Spontaneous adsorption of proteins from biological fluids onto synthetic materials such as biomaterials and biomedical devices may induce undesirable reactions of the body to the foreign materials, such as immune responses, blood coagulation, or bacterial adhesion.[1,2] In the bioaffinity sensor field, suppression of nonspecific protein adsorption is crucial for achieving sufficient bioassay selectivity and sensitivity.[3] Elimination of protein adsorption requires the suppression of all attractive forces between proteins and the surface. Studies investigating the driving forces for protein adsorption have demonstrated the importance of enthalpic contributions, such as van der Waals, electrical double layer, and hydrophobic interactions. At the same time, entropically based mechanisms entail the release of counterions and/or solvation water, and the reduction of ordered structure due to adsorption-induced conformational changes.[4–6] Due to the diversity of the interactions between proteins and surfaces, a preferred strategy for blocking the adsorption of proteins is to immobilize polymers in the form of well-solvated brushes (e.g., poly(ethylene glycol), PEG) on the surface. The polymer layer shields the surface, introducing a high activation barrier for the proteins to adsorb.[7]

* To whom correspondence should be addressed. Present address: Centre Suisse d'Electronique et de Microtechnique SA (CSEM), Jaquet-Droz 1, CH-2007 Neuchâtal. E-mail: stephanie.pasche@csem.ch. Telephone: +41-32-7205540. Fax: +41-32-7205740.

[†] Swiss Federal Institute of Technology (ETH) Zurich.

[‡] University of South Australia.

The ability of polymer brushes to prevent the adsorption of proteins has been addressed theoretically, showing the importance of chain length and density.[8–12] Three modes of adsorption have been proposed: (i) primary adsorption due to proteins diffusing through the interfacial region to the underlying substrate and adsorbing onto the substrate (invasive mechanism); (ii) secondary adsorption at the outer surface of the brush due to protein-brush interactions; (iii) adsorption of proteins upon compression of the polymer film (compressive mechanism).[12] Protein resistance requires the exclusion of all three processes. The thickness of the grafted PEG layer must be sufficient to screen protein–substrate interactions, and the brush chain density must be high enough to block diffusion through this steric layer.

Poly(ethylene glycol) (PEG) has been shown to successfully confer protein resistance to a variety of surfaces.[13,14] Both PEG chain length and surface coverage have been demonstrated to play a key role in imparting protein resistance, with PEG chains of typically 1–10 kDa molecular weight providing protein resistance, provided that the chain density is sufficiently high, hence indicating the need for a high surface density of ethylene glycol monomer units.[7,15–20] Short, densely packed oligo-$(EG)_{2-7}$-terminated alkanethiol self-assembled monolayers on gold have also been shown to confer protein resistance to surfaces.[21–23] While there remains no general agreement, the protein resistance by PEG chains has been associated with two main mechanisms, steric repulsion and a hydration or water-structuring layer.[13,14]

Published on Web 08/24/2005

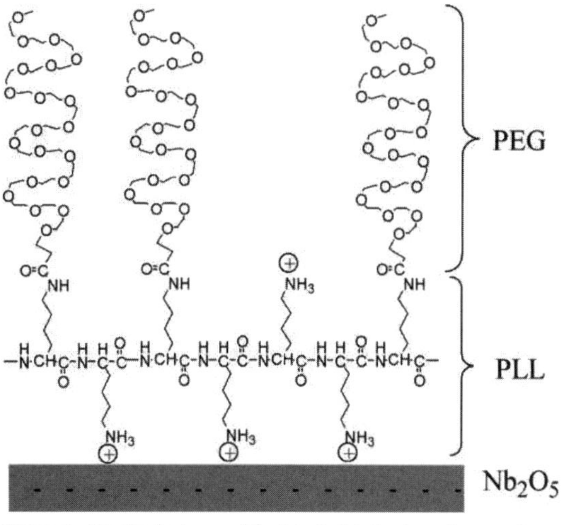

Figure 1. Idealized scheme of the interfacial structure of a monolayer of PLL-*g*-PEG adsorbed on a metal oxide substrate (Nb_2O_5) via electrostatic interactions between the negatively charged metal oxide substrate and positively charged amino-terminated PLL side chains (at neutral pH).

Self-organized monolayers of PEG-grafted polyelectrolytes such as poly(L-lysine)-*g*-poly(ethylene glycol) (PLL-*g*-PEG) have been shown to provide a means of quantitative control of PEG surface coverage, with the adsorbed protein (serum) mass correlating directly with the surface density of ethylene glycol monomer units.[20,24,25] PLL-*g*-PEG is a polycationic copolymer with PEG chains covalently grafted onto a positively charged (at neutral pH) PLL backbone. The copolymer spontaneously adsorbs from aqueous solution via electrostatic interactions onto negatively charged surfaces such as SiO_2, TiO_2, TCPS, and Nb_2O_5 (Figure 1).[26]

This study aims at an improved understanding of the mechanisms of protein resistance of PEG layers, using the comb-graft-copolymer poly(L-lysine)-*g*-poly(ethylene glycol) (PLL-*g*-PEG), and concerns the effects of systematic variations in the surface properties, in particular PEG chain length and density, and surface charge. PLL-*g*-PEG is used as a tool to decouple the different contributions to protein-surface interactions, by allowing surface charge and PEG surface density to be independently varied. Five surfaces with different surface charges and PEG surface densities have been analyzed by recording force−distance curves with the colloid probe atomic force microscope (AFM) at different ionic strengths, thus enabling identification and detailed analysis of the electrical double layer and steric repulsion contributions to the total interfacial force on each of the various PLL-*g*-PEG coated surfaces.[27] To elucidate how these interfacial forces and their relative magnitudes affect the adsorption of proteins, the results of the colloid probe AFM study are correlated with the adsorbed mass of three model proteins, lysozyme, myoglobin, and α-lactalbumin in buffers of different ionic strength, as measured by optical waveguide lightmode spectroscopy (OWLS). The proteins are selected on the basis of their similarity in size and structure, while showing variations in their net charge (Table 1).[28]

2. Materials and Methods

2.1. Materials. PLL-*g*-PEG polymers were synthesized from a stoichiometric mixture of poly(L-lysine) hydrobromide (PLL) (Sigma, USA, MW 15−30 kDa, polydispersity 1.3) and a *N*-hydroxysuccinimidyl ester of methoxy-terminated poly-(ethylene glycol) (mPEG-SPA) (Nektar, USA, MW 2 kDa, polydispersity <1.05). The notation used for PLL(*x*)-*g*[*y*]-PEG-(*z*) copolymers indicates the average molecular weights (MWs) of PLL-HBr (*x*) and PEG (*z*), and the grafting ratio, *g* (*y*). The grafting ratio, expressed as the number of lysine monomers divided by the number of PEG side chains (Lys/PEG ratio), was determined by ^{1}H NMR. The synthesis and characterization have been described elsewhere.[20]

Lysozyme from chicken egg white, myoglobin from horse heart, and α-lactalbumin from bovine milk (Sigma, Buchs, Switzerland) were used as model proteins for the protein adsorption assays. Lysozyme (LSZ), an enzyme (hydrolase) that attacks the protective cell walls of bacteria, myoglobin (MGB), an oxygen-binding haem protein active in the storage and transport of oxygen, and α-lactalbumin (α-LA), a calcium-binding protein involved in lactogenesis, are globular proteins with similar sizes and molecular weights, but different net charge (see Table 1).[29] At pH 7.4, lysozyme is positively charged, myoglobin neutral, and α-lactalbumin negatively charged. Protein solutions were obtained by dissolving the protein in buffer at a concentration of 0.5 mg/mL.

Adsorption of PLL-*g*-PEG as well as all in situ experiments and protein studies was conducted in a 4-(2-hydroxyethyl)-piperazine-1-ethane-sulfonic acid- (HEPES-) based buffer (Fluka, Buchs, Switzerland) adjusted to pH 7.4 with NaOH 6 M (Fluka, Buchs, Switzerland). The choice of the buffer system is important for the type of studies performed in this work: While phosphate-containing buffers result in strong binding of phosphate to transition metal cations and therefore change the surface properties in a time-dependent manner, HEPES is a buffer system with only weak interaction with metal oxide surfaces. The ionic strength of the buffer varied between 1 and 160 mM, either by dilution or by addition of NaCl (Fluka, Buchs, Switzerland). The following buffers were prepared: 1

TABLE 1: Selected Physical Properties of the Proteins Lysozyme (LSZ), Myoglobin (MGB), and a-Lactalbumin (α-LA) Investigated in This Study[28]

property	LSZ	MGB	α-LA
molar mass [g·mol^{-1}]	14 600	17 800	14 200
partial specific vol [cm^3·g^{-1}]	0.688	0.742	0.735
dimensions [nm^3]	4.5 × 3.0 × 3.0	4.5 × 3.5 × 2.5	3.7 × 3.2 × 2.5
diffusion coeff [m^2·s^{-1}]	$1.04 \cdot 10^{-10}$	$1.10 \cdot 10^{-10}$	$1.06 \cdot 10^{-10}$
isoelectric point [pH units]	11.1	7.0	4.3
no. of net charges at pH 7.4	+ 6.8	+ 0.6	- 11.2
overall hydrophobicity [J·g^{-1}]	−7.6	−4.1	−5.8
Gibbs energy of denaturation [J·g^{-1}][a]	−4.1	−2.8	−1.5
secondary structure			
percentage α-helix	42	75	26
percentage β-sheet			14

[a] Heat denaturation.

mM HEPES (H0), 10 mM HEPES (H1), and 150 mM NaCl in 10 mM HEPES (H2). Ultrapure Millipore water (organic content less than 5 ppb) was used for the preparation of all buffers and experiments.

2.2. Substrates. PLL-*g*-PEG was used to modify the surface of niobium oxide films. 12 nm thick films of Nb_2O_5 were sputter coated onto silicon wafers ⟨110⟩ (WaferNet GmbH, Eching, Germany) using reactive magnetron sputtering (PSI, Villigen, Switzerland) to produce substrates for the AFM measurements. Optical waveguide chips for OWLS measurements (Micro-vacuum Ltd., Budapest, Hungary) consisted of a 1 mm-thick AF45 glass substrate and a 200 nm-thick $Si_{0.75}Ti_{0.25}O_2$ waveguiding surface layer, with a total size of 1.2 × 0.8 cm. A 12 nm-thick Nb_2O_5 layer was sputter-coated on top of the waveguiding layer under the same deposition conditions as described above for the silicon wafers. Niobia was used as the substrate surface in view of its high negative charge density at neutral pH; in comparison to SiO_2, niobia surfaces proved to result in improved adlayer stability and consistency of the results.

The substrates used for AFM measurements were sonicated in 2-propanol for 10 min, rinsed with ultrapure water, dried under a nitrogen stream, followed by 2 min of oxygen-plasma cleaning in a plasma cleaner/sterilizer PDC-32G instrument (Harrick, Ossining, NY). The optical waveguide chips used for OWLS were cleaned and regenerated by washing with "Cleaner" solution (Roche, Basel, Switzerland), followed by 10 min sonication in 0.1 M HCl, rinsed with ultrapure water, dried under a nitrogen stream and exposed to 2 min of oxygen plasma.

2.3. Surface Modification. For AFM measurements, the clean substrates were immediately transferred to a filtered 1 mg/mL solution of PLL-*g*-PEG in 10 mM HEPES buffer solution (pH 7.4). After 30 min of immersion, resulting in the formation of a monolayer of PLL-*g*-PEG, the modified samples were withdrawn, rinsed extensively with ultrapure water and dried under nitrogen.[20,30] Deliberate desorption of part of the polymer monolayer was obtained through 10 min immersion of the samples in 10 mM HEPES with 2.8 M NaCl (pH 7.4), followed by rinsing with ultrapure water and drying under nitrogen. This resulted in the formation of a fractional monolayer of the polymer film with approximately 57% of the surface covered by the copolymer. Samples were analyzed with AFM immediately after adsorption, to avoid any uncertainty due to storage.

The surface modification process was monitored in situ with OWLS (see section 2.5).

2.4. Colloid Probe Atomic Force Microscopy. Colloid probe atomic force microscopy (AFM) uses a microsphere as a probe for the quantitative measurement of surface forces.[31−35] The main advantage of using a microsphere instead of a sharp tip is the improved definition of the contact geometry and thus the ability to perform quantitative comparisons with theoretical models of interfacial forces.

A Nanoscope IIIA Multimode (Digital Instruments, Santa Barbara, CA) was used for the atomic force microscopy measurements. The AFM was operated in the force mode, with a scan rate of 1 Hz and a *z*-piezo total displacement of 500 nm. Both approach/extension and retraction force curves of the cantilever were recorded. The experiments were performed in liquid, using an open fluid cell and letting the system equilibrate for 30−60 min. All buffers were filtered through a 0.22 μm pore size filter.

Experiments were performed with silica sphere tips and both uncoated and polymer-coated Nb_2O_5 surfaces. Silica microspheres (4−5 μm diameter) (Bangs Laboratories, USA) were attached to V-shaped Si_3N_4 AFM cantilevers with a nominal spring constant of 0.12 N/m (Novascan Technologies, Ames, IA). Prior to the force measurements the cantilevers were cleaned for 30 min in a UV/ozone cleaner (UV/ clean, model 135500, Boekel Industries, Inc., Feasterville, PA). Further details are provided elsewhere.[27]

Raw data were converted from cantilever deflection and z-piezo position into force-vs-distance curves. The force was calculated with Hooke's law ($F = k\delta$), where k is the spring constant and δ the deflection of the cantilever, and normalized to the radius of the SiO_2 sphere, R, relating the force to the interaction energy (Derjaguin approximation).[36] A positive force was associated with repulsion, while a negative force indicated attraction between the sphere and the surface. The point of zero separation, experimentally not accessible for a soft surface, was defined as the onset of the linear-constant-compliance region, where the polymer layer becomes maximally compressed for a given cantilever.[37] Data analysis was conducted for the approach/extension part of the curve.[27]

2.5. Optical Waveguide Lightmode Spectroscopy. In optical waveguide lightmode spectroscopy, the adsorbed mass is calculated from the change of the refractive index in the vicinity of the surface upon adsorption of molecules from solution. This change is monitored by the grating-induced incoupling of laser light into the waveguiding substrate, generating an evanescent wave.[38−40] The adsorbed mass was calculated from the thickness and refractive index of the adlayer according to de Feijter's formula, with dn/dc values of 0.18, 0.139, and 0.135 cm^3/g for pure PLL, PLL(20)-*g*[3.5]-PEG(2) and PLL(20)-*g*[10.1]-PEG(2) polymers, respectively, and 0.182 cm^3/g for the proteins.[20,41] The sensitivity limit of the OWLS technique is typically 1−2 ng/cm^2.

In situ polymer adsorption was studied using a flow-through cell with a volume of 16 μL. The solution was typically replaced every 10 min.[38] The cleaned Nb_2O_5-coated waveguides were inserted into the OWLS flow-through cell, and equilibrated by immersing in HEPES buffer (pH 7.4) for at least 10 h in order to obtain a stable baseline. Subsequently, the buffer was exchanged in situ against a filtered solution of PLL-*g*-PEG (1 mg/mL) in HEPES buffer. After 30 min exposure to the polymer solution, resulting in the formation of a complete monolayer, the sample was rinsed with the HEPES buffer solution for another 30 min. Polymer desorption was achieved after 10 min exposure of the polymer layer to 10 mM HEPES with 2.8 M NaCl (pH 7.4), and a subsequent rinse with buffer. Protein adsorption onto bare and PLL-*g*-PEG-coated OWLS chips was studied in situ after 1 h exposure to 0.5 mg/mL protein solution and a subsequent rinse with the same buffer for about 1 h.

3. Results and Discussion

3.1. Surfaces with Different Polymer Architectures. PLL and PLL-*g*-PEG copolymers with PLL MW of ~20 kDa (MW of PLL-HBr), PEG MW of 2 kDa and grafting ratios g of 3.5 and 10.1 were used to modify the Nb_2O_5-coated substrates, allowing the surface charge to be varied from negative (bare Nb_2O_5) to positive (PLL-coated Nb_2O_5), and the PEG surface density from 0 to 0.5 chains/nm^2, as measured by OWLS (eq 1).[20] While the PEG surface density is controlled through the PEG grafting ratio in the PLL-*g*-PEG polymer and the adsorbed polymer mass, the surface charge is determined by both the charge of the Nb_2O_5 substrate surface and the number of free amine groups in the lysine present in the polymeric monolayer. Only the lysine side chains that are not bound to a PEG chain carry a positive charge, as illustrated in the scheme of PLL-*g*-PEG in Figure 1. The number of free amines, n_{amine}(free) can

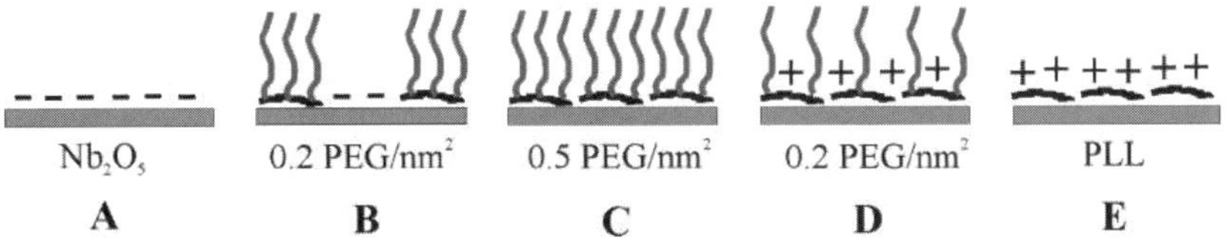

Figure 2. Surfaces used in the study that exhibit different surface charges and PEG surface densities: (A) bare niobium oxide substrate; (B) PLL(20)-*g*[3.5]-PEG(2)-coated Nb_2O_5 after partial desorption of the polymer layer with NaCl 2.8 M in 10 mM HEPES (half monolayer); (C) PLL(20)-*g*[3.5]-PEG(2)-coated Nb_2O_5 (full monolayer); (D) PLL(20)-*g*[10.1]-PEG(2)-coated Nb_2O_5 (monolayer); (E) PLL(20)-coated Nb_2O_5. The surface charge gradually changes from strongly negative (A) to strongly positive (E); the PEG surface density varies from 0 to a maximum of 0.5 chains per nm^2.

TABLE 2: Characteristics of the Surfaces Used for the Protein Adsorption Study: Surface Density of PEG Chains, n_{PEG}, and of Free Amine Groups in PLL, n_{amine}(free)[b], and "Surface Charge"

surface	n_{PEG} [nm^{-2}]	n_{amine}(free)[b] [nm^{-2}]	surface charge[c] [nm^{-2}]
Nb_2O_5	0		−1.30
PLL(20)	0	3.19 ± 0.52	+1.89
PLL(20)-*g*[10.1]-PEG(2)	0.27 ± 0.04	1.84 ± 0.28	+0.54
PLL(20)-*g*[3.5]-PEG(2)	0.51 ± 0.07	1.28 ± 0.17	−0.02
PLL(20)-*g*[3.5]-PEG(2) (1/2)[a]	0.22 ± 0.03	0.55 ± 0.08	−0.75

[a] (1/2) means that approximately half of the polymer monolayer has been desorbed by exposing the surface for 10 min to a 10 mM HEPES buffer solution with 2.8 M NaCl, leading to 57% desorption of the polymer (half monolayer coverage). [b] Amine groups of the lysine not bound to PEG chains and protonated at neutral pH. [c] Net "surface charge" calculated as the sum of the bare Nb_2O_5 surface charge (−1.30 nm^{-2}; see sections 3.1 and 3.2) and n_{amine}(free).

be calculated with eqs 1 and 2, where m_{pol} is the adsorbed polymer mass measured with OWLS [ng/cm^2], n_{Lys} and n_{PEG} are the numbers of adsorbed Lys monomers and PEG chains per nm^2, respectively, (eq 2), and g is the grafting ratio of the polymer, determined for the bulk polymer with^{1}H NMR.[20]

$$n_{PEG} = \frac{m_{pol}}{M_{Lys}g + M_{PEG}} \tag{1}$$

$$n_{amine}(\text{free}) = n_{Lys}\left(1 - \frac{1}{g}\right) = n_{PEG}(g - 1) \tag{2}$$

Assuming that every free lysine side chain is protonated at pH 7.4 (pK_a = 10.8)[42] we can estimate the remaining surface charge, provided that the surface charge density of the Nb_2O_5 substrate is known. As this is not the case, the total surface charge of the bare Nb_2O_5 surface was estimated based on the AFM force curves (section 3.2). The AFM data for the PLL-(20)-*g*[3.5]-PEG(2)-coated Nb_2O_5 substrate indicated that the net surface charge sensed by the colloidal AFM tip was very close to zero (in fact, very slightly negative), implying that the negative surface charge of the bare niobia surface and the positive charges of the protonated amine groups in the adsorbed polymeric monolayer almost canceled each other out. Since the latter is known from NMR of the polymer and the OWLS adsorbed mass data to be 1.28 ± 0.17 nm^{-2} (eq 2), the surface charge of bare niobia was set to ∼−1.30 nm^{-2} (or 0.21 C·m^{-2}), resulting in a very small negative surface charge for the PLL-(20)-*g*[3.5]-PEG(2)-coated Nb_2O_5 substrate, in agreement with the AFM information. Table 2 provides information on the surfaces used in this study, listing the PEG surface density, the number of ungrafted lysine-mers (i.e., non-PEGylated, protonated amine groups) per unit surface area, and the net surface charge estimated as discussed above.

As shown in Table 2, PLL(20)-*g*[3.5]-PEG(2) adsorption results in a nearly neutral surface with ∼0.5 PEG chains/nm^2, while PLL(20)-*g*[10.1]-PEG(2) shows a positive charge with ∼0.2 PEG chains/nm^2. To investigate the influence of the surface charge for a similar average PEG surface density, a negatively charged surface with ∼0.2 PEG/nm^2 was used. For this purpose, PLL(20)-*g*[3.5]-PEG(2) was adsorbed onto Nb_2O_5, and exposed to a 2.8 M NaCl in 10 mM HEPES buffer (pH 7.4) for 10 min, leading to 57% polymer desorption. As the resulting surface carries approximately half of a monolayer of PLL(20)-*g*-[3.5]-PEG(2), the notation "PLL(20)-*g*-[3.5]-PEG-(2) (1/2)" is used. Figure 2 illustrates the differences between the five surfaces used for the subsequent investigations, highlighting variations both in the PEG surface density and in the surface charge. The PLL-*g*-PEG technology therefore allows independent investigation of the influence of surface charge and PEG density on protein adsorption.

3.2. Surface Charge Measurement Using Colloid Probe AFM. Colloid probe AFM measurements were performed on all surfaces at different ionic strengths. The interaction between a 5 μm silica microsphere and the various surfaces contains both an electrostatic and a steric contribution to the total interfacial force; detailed analyses of the force curves including fits of the electrostatic forces to DLVO models are presented elsewhere.[27] While the electrostatic contribution provides information on the surface charge, the interplay with the steric force as a function of ionic strength demonstrated to what extent the PEG layer thickness is successful in screening the surface charge and eliminating the effects of the electrical double layer contribution beyond the uncompressed PEG layer, and it also provides an estimate for the hydrated thickness of the PEG layer.

Parts A−E of Figure 3 show the measured interfacial forces, normalized to the radius of the sphere, upon compression of the surface by the SiO_2 microsphere, plotted as a function of the apparent relative separation. The forces were measured at three ionic strengths, in 1 mM HEPES, 10 mM HEPES, and 10 mM HEPES with 150 mM NaCl buffer solutions (pH 7.4) (H0, H1, and H2), corresponding to Debye lengths of ∼10, ∼3, and <1 nm, respectively.[36]

The force between negatively charged silica and negatively charged niobia (Figure 3A) typically shows an electrostatic repulsion with a decay length increasing as the ionic strength decreases. Plotted on semilogarithmic axes (inset graph) the interactions at the three ionic strengths reflect the Debye lengths in the different solutions. Moving from A to E, the interaction gradually changes from repulsive to attractive, ending with a purely attractive interaction between the PLL-coated Nb_2O_5 and the silica microsphere. The data reflect the influence of the ionic strength and are characteristic of an electrical double layer force. As an attraction is associated with a negative force, the plots of Figure 3D,E are not shown on a semilogarithmic scale.

While the influence of the ionic strength probes the electrostatic nature of the interaction for the uncoated and PLL-coated Nb_2O_5 samples, it also provides relevant information for the

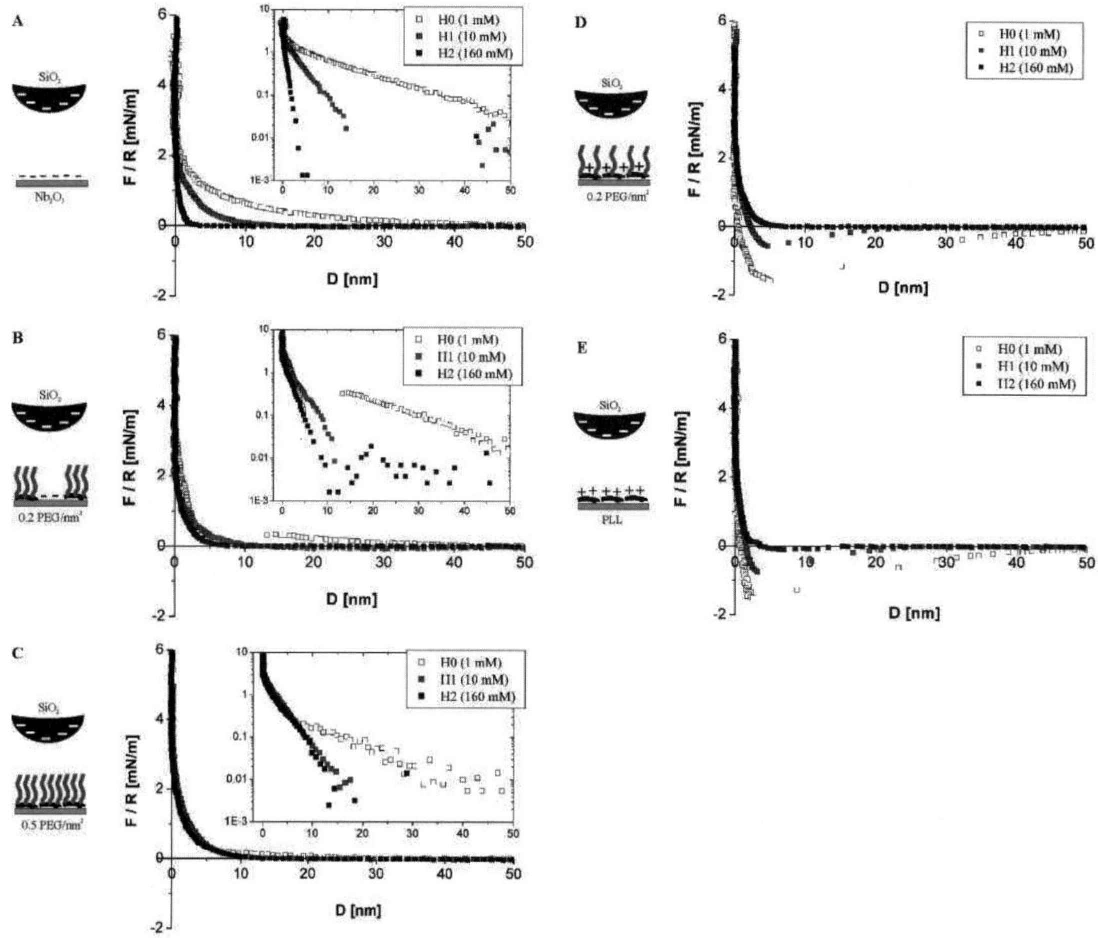

Figure 3. AFM force−distance curves between a negatively charged 4−5 μm SiO_2 microsphere probe and the following surfaces: (A) uncoated Nb_2O_5 substrate, (B) PLL(20)-*g*[3.5]-PEG(2) (1/2), (C) PLL(20)-*g*[3.5]-PEG(2), (D) PLL(20)-*g*[10.1]-PEG(2), and (E) PLL(20) adlayers on Nb_2O_5. From part A to part E, the surface charge gradually changes from negative to positive; the PEG surface density varies from 0 (A and E) to a maximum of 0.5 nm^{-2} (C). The force curves shown were determined in buffer solutions of three ionic strengths, 1, 10, and 160 mM (H0, H1, and H2). All measurements were performed at pH 7.4.

PLL-*g*-PEG-coated surfaces, as shown in Figure 3B−D. In view of the inherent interdependence between the PEG surface density/thickness and the surface charge for the chosen polymeric system, we explore the influence of the PEG layer thickness by varying the thickness of the diffuse ion layer (ionic strength). For the PLL-*g*-PEG surfaces, the measured interaction reflects both an electrostatic contribution and a steric repulsion. The steric repulsion is associated with the PEG side chains, while the electrostatic term arises from the remaining surface charge, where the initial negative charge of the Nb_2O_5 substrate is under- or overcompensated by the adsorbed polycationic polymer. As the ionic strength is associated with the thickness of the diffuse ion layer, the charge of the surface will only be perceptible from a certain distance to the surface. Therefore, the PEG layer shields the surface charge as long as the thickness of the PEG layer exceeds the effective distance of the electrical double layer force. The Debye lengths are ~10, ~3 and <1 nm for the 1, 10, and 160 mM buffer solutions, respectively. As there exists a diffuse ion layer on both the surface and the sphere, the minimum thickness of a PEG layer required to sufficiently screen the surface charges can be 2−3 times the Debye length. The graph of Figure 3C shows the force between the SiO_2 microsphere and the PLL(20)-*g*[3.5]-PEG(2)-coated Nb_2O_5, clearly illustrating the influence of the ionic strength. While similar forces in 10 mM and in 160 mM solutions are indicative of a purely steric repulsive interaction, the force measured in 1 mM solution exhibits two regimes. As the sphere approaches the surface, the net force at first is dominated by a distance dependence (slope) characteristic of the decay length of a repulsive electrostatic force in this medium, then shows a sharp transition to a stronger repulsion at the first contact with the PEG layer, i.e., at the onset of steric repulsion, thus providing information on the thickness of the layer. The sample with a half-monolayer coverage of PLL(20)-*g*[3.5]-PEG(2) shows an even sharper transition in 1 mM HEPES. The transition is also observed in 10 mM HEPES, however this is less obvious in this case as the decay length of the steric interaction is close to the Debye length. Finally, the PLL(20)-*g*[10.1]-PEG(2) polymer shows electrostatic attraction both in 1 and 10 mM HEPES, indicating that the PEG chains do not sufficiently screen the (positive) surface charge of these surfaces under these conditions.

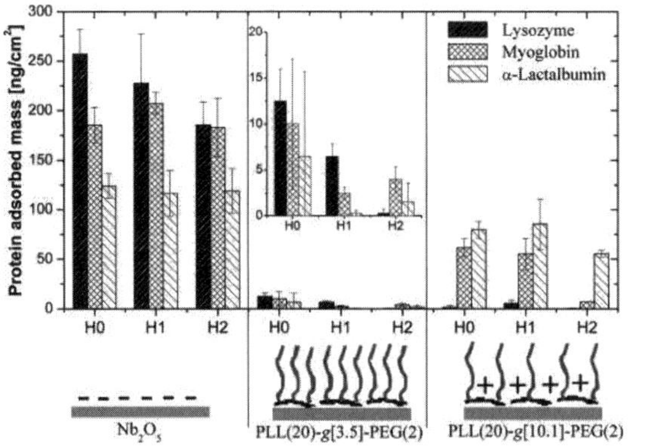

Figure 4. Adsorbed mass of the proteins lysozyme, myoglobin, α-lactalbumin on the five different surfaces, shown schematically in Figure 2, at increasing ionic strengths, measured by OWLS. The ionic strengths were 1 mM HEPES (H0), 10 mM HEPES (H1), and 10 mM HEPES + 150 mM NaCl (H2); all adsorption experiments were performed at pH 7.4.

AFM force measurements thus provide qualitative information on the charge of the different surfaces, as well as on the role of the PEG layer in screening the surface charge to an extent that depends on the ionic strength of the solution. Such measurements are useful as they can provide insight into the potential forces the proteins will sense upon approaching the surface (sections 3.3 and 3.4). For the present purpose, which is to correlate protein adsorption with the nature of the main interfacial forces, quantitative characterization of the forces is not necessary, but detailed analyses of force curves recorded with PLL-*g*-PEG coatings is presented elsewhere.[27]

3.3. Effect of the Ionic Strength on Protein Adsorption. The surfaces were exposed to three different proteins of similar size and shape, but carrying a different net charge at pH 7.4. Lysozyme (IEP 11.1), myoglobin (IEP 7.0), and α-lactalbumin (IEP 4.3) carry a positive, neutral, and negative overall charge, respectively (see Table 1). The adsorption of these proteins onto three surfaces was studied as a function of the ionic strength of the solution, to extract the electrostatic contribution to the interfacial interaction force between proteins and these surfaces. Protein adsorption at a charged surface involves overlap of the electrical double layers at the solvated substrate surface and the solvated protein surface. Figure 4 shows the protein adsorbed mass of the three proteins on uncoated, PLL(20)-*g*[3.5]-PEG(2)- and PLL(20)-*g*[10.1]-PEG(2)-coated Nb_2O_5, measured with OWLS at three ionic strengths, from 1 mM HEPES (H0) to 10 mM HEPES with 150 mM NaCl (H2), lowering the Debye length from ~10 nm down to <1 nm.

The adsorption of positively charged lysozyme onto Nb_2O_5 showed a dependence on the ionic strength, typical for an electrostatically dominated interaction, with protein adsorbed mass decreasing as the ionic strength increases.[43,44] Myoglobin and α-lactalbumin adsorbed in lesser amounts onto Nb_2O_5, and the adsorbed mass was independent of the ionic strength of the solution. This indicates that other mechanisms were responsible for the observed adsorption of these proteins, possibly accompanied by partial unfolding of the proteins at the interface.[45] The lower adsorbed mass of α-lactalbumin may be attributed in part to the effect of the net negative charge present on both protein and substrate. On the other hand, strong deformation of this soft protein upon adsorption would be expected to increase the footprint of the adsorbed protein, thus decreasing the number of protein molecules per unit area that can be accommodated on the surface.

The adsorption of all three proteins onto PLL(20)-*g*[3.5]-PEG(2)-coated Nb_2O_5 was significantly reduced in comparison to the corresponding behavior on the bare Nb_2O_5 surface (95–100% reduction in protein adsorption). This surface exhibits a dense PEG coating and only a very small negative charge is sensed at the lowest ionic strength (Table 2 and section 3.2).[20] However, a closer look (zoomed inset in Figure 3) reveals an influence of the ionic strength on the adsorption of the "hard" and positively charged lysozyme:[28] protein adsorption decreases with increasing ionic strength. The latter is associated with a diffuse ion layer of defined thickness, described by the Debye length. The influence of the ionic strength on protein adsorption demonstrates the role of PEG layer thickness in shielding the surface charge and thereby preventing strong electrostatic interaction between the surface and the protein. The minimal PEG thickness needed to hide the diffuse ion layer depends on the ionic strength of the solution.

While the positively charged PLL(20)-*g*[10.1]-PEG(2)-coated Nb_2O_5 surface did not adsorb a significant amount of positively charged lysozyme, due to a strong electrostatic repulsive force, the adsorption of myoglobin and α-lactalbumin decreased with increasing ionic strength, illustrating, as for the PLL(20)-*g*[3.5]-PEG(2) polymer, the interplay between the thickness of the polymer layer and that of the diffuse ion layer.

While for small proteins that can penetrate into the PEG layer (primary adsorption) the ionic strength is likely to have only a minor effect on the adsorbed mass, secondary adsorption on top of the polymer layer is expected to be highest at the lowest ionic strengths.[43] As a consequence, general resistance to protein adsorption requires the elimination of both primary and secondary adsorption. Reduction of primary adsorption is achieved primarily through the presence of a dense PEG brush, while secondary adsorption onto PEG graft layers on charged solid substrates due to electrostatic attraction of oppositely charged proteins can be reduced by charge neutralization at the interface and/or sufficiently thick PEG layers that screen the substrate interfacial charge.

3.4. Effect of the Surface Chemistry. To investigate the effects of the surface charge and chemistry on protein adsorption we chose a medium ionic strength with a 10 mM HEPES buffer solution ($\kappa^{-1} \sim 3$ nm), allowing better visualization of the electrostatic interactions compared to physiological conditions (~150 mM). At 10 mM ionic strength, the PLL(20)-*g*-[3.5]-PEG(2) surface was the only one *not* to display any charge in the colloid probe AFM measurements. The adsorbed mass values for the three proteins are presented in Figure 5 for five surfaces carrying different charges and PEG surface densities (Figure 2).

The surface properties significantly influenced the adsorption of the different proteins. To better discriminate between the effect of surface charge and PEG surface density, surfaces with both negative and positive charge at PEG densities of both 0 and 0.2 PEG/nm^2 were prepared as described in section 2.3 and 2.5 and exposed to the individual protein solutions. Again, as on the other surfaces, lysozyme adsorption showed a pronounced and systematic dependence on the surface charge, only adsorbing onto the negatively charged surfaces. This indicates a dominant electrostatic character of the interaction, as also demonstrated by the influence of the ionic strength (Figure 4). Lysozyme behaves in this respect similarly to a hard colloid particle. Indeed, lysozyme has been reported to be a stable protein that shows comparatively little tendency to lose its native conforma-

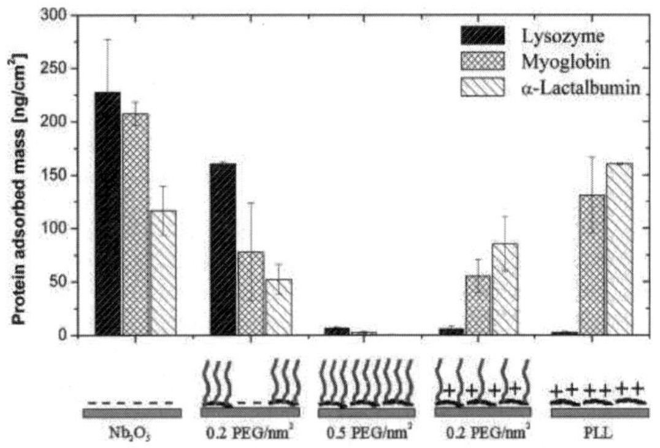

Figure 5. Protein adsorbed mass measured in 10 mM HEPES buffer (pH 7.4) by OWLS, on the five surfaces with various surface charges and PEG surface densities as shown schematically in Figure 2. From left to right the surface charge gradually changes from negative to positive; the PEG surface density passes through a maximum of 0.5 PEG/nm^2 for the PLL(20)-*g*[3.5]-PEG(2) polymer adlayer shown in the middle. Surfaces with a medium value for the average PEG surface density of 0.2 PEG/nm^2 were obtained with PLL(20)-*g*[3.5]-PEG(2) (1/2, i.e., half monolayer coverage), charged negatively, and a monolayer of PLL(20)-*g*[10.1]-PEG(2), charged positively.

tion.[28] In conclusion, controlling the adsorption of hard, charged proteins such as lysozyme therefore simply requires tuning of the surface charge.

The softer protein α-lactalbumin, on the other hand, showed a more complex response. While the adsorbed protein mass slightly increased as the surface became more positive, this overall negatively charged protein adsorbed in significant amounts also on the negatively charged bare (A in Figure 2) and PEG-coated (B) Nb_2O_5, independent of the ionic strength, indicating either the presence of other interaction mechanisms, deformation/unfolding of the protein upon adsorption, or a combination of both. While the overall negatively charged α-lactalbumin could bind with its local positive patches to the surface through electrostatic interaction, other driving forces for protein adsorption are likely be operative in this case including hydrophobic and entropic factors.[28] Indeed, α-lactalbumin has been shown to undergo conformational changes, going through a transition state characterized as "molten globule".[45] Therefore, the adsorption of soft, conformationally delicate proteins such as α-lactalbumin is much more complex and cannot be predicted based on a simple electrostatic model. The adsorbed mass data for the different surfaces suggest that elimination of the adsorption of a soft and "unpredictable" protein such as α-lactalbumin requires complete screening of electrostatic interactions with the underlying substrate by a sufficiently thick PEG layer as well as a repulsive regime exerted by densely packed PEG chains. Both criteria are met for the dense-brushed PLL(20)-*g*[3.5]-PEG(2) layer, the only surface among those tested in this work that resists adsorption of both lysozyme and the other proteins.

The adsorption of neutral myoglobin is mainly controlled by the PEG surface density. Independent of the surface charge, myoglobin adsorbs on surfaces to an extent that is inversely correlated with the PEG surface density (in the regime 0−0.2 nm^{-2}), the surface with the highest PEG surface density (0.5 nm^{-2}) preventing its adsorption almost completely.

In summary, the findings of this study are in general agreement with the conclusions from many other published studies regarding the mechanisms behind the protein resistance of PEGylated surfaces. Electrostatic interactions are clearly found to be important, and dominant in the case of lysozyme. As most proteins do not behave like the "hard" lysozyme, neutralization of the surface charge alone is not sufficient, however, to eliminate protein adsorption in general. Protein resistance is achieved by a combination of several effects: surface charge, surface energy, interfacial water structure, and, in the case of PEGylated surfaces, PEG layer thickness and PEG surface density. Charge neutralization is always effective in eliminating electrostatic interactions between a protein and the surface, but this is not always feasible; most interfaces in contact with aqueous media carry some surface charge. Sufficiently thick PEG layers are therefore needed in order to facilitate the screening of the interfacial charge, thus preventing (bio)-molecules from adsorbing on top of the polymer layer via electrical double layer forces transmitted through the PEG layer. Furthermore, dense PEG layers are essential as a steric barrier to diffusion of the protein between the PEG chains and subsequent adsorption onto the underlying substrate. In addition, the predominantly hydrophilic character of the PEG chains renders the formation of strong hydrophobic forces unlikely, and dense PEG brushes have the particular ability to act as a scaffold for structural water that prevents molecules from permanently interacting with the PEG layer.[46] Finally, the highly mobile and dynamic PEG chains are likely to resist the compression by large protein molecules as a consequence of steric-osmotic effects. Thus, all three modes of adsorption proposed in theoretical work (primary adsorption, secondary adsorption, and the compressive mechanism)[12] can be eliminated with a sufficiently thick, dense PEG layer.

An interesting outcome of this study is, however, that even for the best of the PLL-*g*-PEG coating, which resists protein adsorption under the physiological conditions of 150 mM ionic strength, protein adsorption can be observed as the ionic strength is lowered. Thus, a "nonfouling" coating can become fouling under some solution conditions, due to secondary, electrostatically induced adsorption onto the PEG layer. While an ionic strength of 1 mM is not relevant to biomedical implants, it may be important in some diagnostic applications to consider the possibility of protein/surface interactions arising from electrical double layer forces whose range exceeds the thickness of the repulsive PEG layer under the solution conditions applicable. This insight suggests that the length/molecular weight of the grafted PEG side chains should be tailored such as to match the dimensional requirements imposed by the electrical double layer force, defined by the ionic strength of the intended solution environment.

4. Conclusions

Five model surfaces with different surface charges and quantitatively controlled PEG surface densities were produced using PLL-*g*-PEG polymer adlayers on Nb_2O_5 substrates. Characterization of the surfaces with colloid probe AFM provided information on the surface charge and demonstrated the important role of PEG-layer thickness in shielding the electrical double layer forces. Although it is clear that a hard SiO_2 microsphere does not exactly mimic a soft, nanometer-sized protein, the colloid-probe AFM technique is nevertheless considered to be an excellent tool for independently sensing electrostatic and steric-repulsive forces, by means of ionic-strength variation. Both forces are important in biomolecule-surface interactions. While this approach clearly provides information that is laterally averaged over an area much larger than the size of a protein, and is therefore not sensitive to very local inhomogeneities such as nanodefects in coatings, the technique provides useful information on the interplay between averaged surface forces and overall levels of protein adsorption. Its usefulness is demonstrated in this study by the close correlation of observed surface forces with measured protein adsorption.

Adsorption of three small model proteins at different ionic strengths on the five types of surfaces revealed the importance of electrostatic forces in protein adsorption. Lysozyme behaved similarly to a hard, positively charged particle, being attracted and repelled by negatively and positively charged surfaces, respectively, but this "simple" behavior is probably rather the exception than the rule.

Myoglobin and α-lactalbumin showed a more complex interaction pattern with the same type of surfaces, likely a consequence of being rather soft proteins, exhibiting changes in conformation upon adsorption. Independent information on the influence of surface charge and PEG surface density highlighted major requirements that PEG graft layers have to fulfill in order to resist adsorption of proteins. While control of the surface charge is sufficient to prevent the adsorption of lysozyme, charge screening with a sufficiently thick and dense PEG layer (min 0.5 PEG (2 kDa) or 25 EG monomers per nm^2) is further needed for most other proteins, including those in serum.[20] Such a PEG layer is able to reduce or eliminate secondary adsorption on the top of the PEG layer, while a high PEG density (brush) is required to prevent primary adsorption by proteins diffusing through the layer and adsorbing onto the underlying surface. Finally, dense PEG chains in a brush conformation provide an entropic insurance preventing proteins from compressing the layer.

Acknowledgment. The authors thank Dr. Graeme Gillies and Dr. Fabiano Assi for assistance with AFM measurements, Michael Horrisberger for performing the niobium oxide coating, and Dr. Oleg Borisov for valuable discussions. This work was financially supported by EPF Lausanne and ETH Zürich (Project TH-33/01-3), the Swiss National Science Foundation, National Research Program NRP 47 (Project No 4047-057548), and by the University of South Australia under the ARC Grant Special Research Centre for Particle and Material Interfaces.

References and Notes

(1) Ratner, B. D. *J. Biomed. Mater. Res.* **1993**, *27*, 837−850.

(2) Ostuni, E.; Chapman, R. G.; Holmlin, R. E.; Takayama, S.; Whitesides, G. M. *Langmuir* **2001**, *17*, 5605−5620.

(3) Schneider, B. H.; Dickinson, E. L.; Vach, M. D.; Hoijer, J. V.; Howard, L. V. *Biosensors Bioelectron.* **2000**, *15*, 13−22.

(4) Norde, W. In *Biopolymers at Interfaces*; Malmsten, M., Ed.; Surfactant Science Series: Marcel Dekker: New York, 1998; Vol. 75.

(5) Norde, W. *Colloids and Interfaces in Life Sciences*; Marcel Dekker: New York, 2003.

(6) Malmsten, M. *J. Colloid Interface Sci.* **1998**, *207*, 186−199.

(7) Leckband, D.; Sheth, S.; Halperin, A. *J. Biomater. Sci.—Polym. Ed.* **1999**, *10*, 1125−1147.

(8) Jeon, S. I.; Andrade, J. D. *J. Colloid Interface Sci.* **1991**, *142*, 159−166.

(9) Jeon, S. I.; Lee, J. H.; Andrade, J. D.; de Gennes, P. G. *J. Colloid Interface Sci.* **1991**, *142*, 149−158.

(10) Szleifer, I. *Biophys. J.* **1997**, *72*, 595−612.

(11) Szleifer, I. *Physica A* **1997**, *244*, 370−388.

(12) Halperin, A. *Langmuir* **1999**, *15*, 2525−2533.

(13) Vermette, P.; Meagher, L. *Colloids Surf. B—Biointerfaces* **2003**, *28*, 153−198.

(14) Morra, M. *J. Biomater. Sci.—Polym. Ed.* **2000**, *11*, 547−569.

(15) Kingshott, P.; Griesser, H. J. *Curr. Opin. Solid State Mater. Sci.* **1999**, *4*, 403−412.

(16) McPherson, T.; Kidane, A.; Szleifer, I.; Park, K. *Langmuir* **1998**, *14*, 176−186.

(17) Efremova, N. V.; Sheth, S. R.; Leckband, D. E. *Langmuir* **2001**, *17*, 7628−7636.

(18) Malmsten, M.; Emoto, K.; Van Alstine, J. M. *J. Colloid Interface Sci.* **1998**, *202*, 507−517.

(19) Lin, Y. S.; Hlady, V.; Gölander, C. G. *Colloids Surf. B: Biointerfaces* **1994**, *3*, 49−62.

(20) Pasche, S.; De Paul, S. M.; Vörös, J.; Spencer, N. D.; Textor, M. *Langmuir* **2003**, *19*, 9216−9225.

(21) Feldman, K.; Hähner, G.; Spencer, N. D.; Harder, P.; Grunze, M. *J. Am. Chem. Soc.* **1999**, *121*, 10134−10141.

(22) Prime, K. L.; Whitesides, G. M. *J. Am. Chem. Soc.* **1993**, *115*, 10714−10721.

(23) Harder, P.; Grunze, M.; Dahint, R.; Whitesides, G. M.; Laibinis, P. E. *J. Phys. Chem. B* **1998**, *102*, 426−436.

(24) Kenausis, G. L.; Vörös, J.; Elbert, D. L.; Huang, N. P.; Hofer, R.; Ruiz-Taylor, L.; Textor, M.; Hubbell, J. A.; Spencer, N. D. *J. Phys. Chem. B* **2000**, *104*, 3298−3309.

(25) Huang, N. P.; Vörös, J.; De Paul, S. M.; Textor, M.; Spencer, N. D. *Langmuir* **2002**, *18*, 220−230.

(26) Elbert, D. L.; Hubbell, J. A. *Chem. Biol.* **1998**, *5*, 177−183.

(27) Pasche, S.; Textor, M.; Meagher, L.; Spencer, N. D.; Griesser, H. J. *Langmuir* **2005**, *21*. 6508−6520.

(28) Arai, T.; Norde, W. *Colloids Surf.* **1990**, *51*, 1−15.

(29) Berman, H. M.; Westbrook, J.; Zeng, Z.; Gilliland, G.; Bhat, T. N.; Weissig, H.; Shindyalov, I. N.; Bourne, P. E. *Nucleic Acids Res.* **2000**, *28*, 235−242.

(30) Huang, N. P.; Michel, R.; Vörös, J.; Textor, M.; Hofer, R.; Rossi, A.; Elbert, D. L.; Hubbell, J. A.; Spencer, N. D. *Langmuir* **2001**, *17*, 489−498.

(31) Ducker, W. A.; Senden, T. J.; Pashley, R. M. *Nature (London)* **1991**, *353*, 239−241.

(32) Ducker, W. A.; Senden, T. J.; Pashley, R. M. *Langmuir* **1992**, *8*, 1831−1836.

(33) Luckham, P. F.; Costello, B. A. D. *Adv. Colloid Interface Sci.* **1993**, *44*, 183−240.

(34) Meagher, L. *J. Colloid Interface Sci.* **1992**, *152*, 293.

(35) Hartley, P. G. In *Colloid-Polymer Interactions: From Fundamentals to Practice*; Farinato, R. S., Dubin, P. L., Eds.; John Wiley & Sons: New York, 1999; pp 253−286.

(36) Israelachvili, J. *Intermolecular & Surface Forces*, 2nd ed.; Academic Press: San Diego, CA, 1992.

(37) Seog, J.; Dean, D.; Plaas, A. H. K.; Wong-Palms, S.; Grodzinsky, A. J.; Ortiz, C. *Macromolecules* **2002**, *35*, 5601−5615.

(38) Vörös, J.; Ramsden, J. J.; Csucs, G.; Szendrő, I.; De Paul, S. M.; Textor, M.; Spencer, N. D. *Biomaterials* **2002**, *23*, 3699−3710.

(39) Kurrat, R.; Textor, M.; Ramsden, J. J.; Böni, P.; Spencer, N. D. *Rev. Sci. Instrum.* **1997**, *68*, 2172−2176.

(40) Höök, F.; Vörös, J.; Rodahl, M.; Kurrat, R.; Böni, P.; Ramsden, J. J.; Textor, M.; Spencer, N. D.; Tengvall, P.; Gold, J.; Kasemo, B. *Colloids Surf. B—Biointerfaces* **2002**, *24*, 155−170.

(41) de Feijer, J. A.; Benjamins, J.; Veer, F. A. *Biopolymers* **1978**, *17*, 1759−1772.

(42) Horbett, T. A. In *Biomaterials Science: An Introduction to Materials in Medicine*; Ratner, B. D., Hoffman, A. S., Schoen, F. J., Lemons, J. E., Eds.; Academic Press: New York, 1996; p 133.

(43) Carignano, M. A.; Szleifer, I. *Mol. Phys.* **2002**, *100*, 2993−3003.

(44) Su, T. J.; Lu, J. R.; Thomas, R. K.; Cui, Z. F.; Penfold, J. *J. Colloid Interface Sci.* **1998**, *203*, 419−429.

(45) Fink, A. L. In *Encyclopedia of Life Sciences*; Nature Publishing Group: London, 2001; http://www.els.net/.

(46) Heuberger, M.; Drobek, T.; Vörös, J. *Langmuir* **2004**, *20*, 9445−9448.

ADVANCED FUNCTIONAL MATERIALS

FULL PAPER

DOI: 10.1002/adfm.200500232

Nitrilotriacetic Acid Functionalized Graft Copolymers: A Polymeric Interface for Selective and Reversible Binding of Histidine-Tagged Proteins**

By *Guoliang Zhen*, *Didier Falconnet*, *Eva Kuennemann*, *Janos Vörös*, *Nicholas D. Spencer*, *Marcus Textor*, and *Stefan Zürcher**

A series of novel graft copolymers, poly(L-lysine)-*graft*-poly(ethylene glycol) (PLL-*g*-PEG) with 28, 46, and 96 % of the PEG chains carrying a terminal nitrilotriacetic acid (NTA) group, have been synthesized. Through electrostatic interactions, these polycationic graft copolymers assemble spontaneously from aqueous solution onto negatively charged surfaces, forming polymeric monolayers that present NTA groups at controlled surface densities. Such NTA-presenting surfaces on a highly PEGylated background are attractive sensor platforms for the reversible binding, through Ni^{2+} ions, of histidine (His)-tagged biomolecules such as proteins and antibodies. With these three polymers, NTA-ligand surface densities of 16, 23, and 37 pmol cm^{-2}, respectively, have been obtained. The surface assembly of the polymers, as well as their sensing performance, has been monitored quantitatively using in-situ optical-waveguide light-mode spectroscopy. The NTA-functionalized PLL-*g*-PEG surfaces prove to be highly resistant to non-specific adsorption in contact with human serum, while allowing the specific and reversible surface binding of the 6 × His-tagged green fluorescent protein (GFP)uv-6His in its native conformation. The amount of GFPuv-6His immobilized on the polymeric surface increases with increasing NTA surface density. Furthermore, micropatterns consisting of NTA-functionalized PLL-*g*-PEG in a background of PLL-*g*-PEG are produced using the "molecular assembly patterning by lift-off" technique. Exposure to Ni^{2+} and GFPuv-6His results in a protein pattern of excellent contrast, as judged by fluorescence microscopy. The NTA-functionalized PLL-*g*-PEG surface is considered to be a promising sensor platform for binding 6 × His-tagged proteins, thanks to the simplicity and cost-effectiveness of the surface modification protocol, the high specificity and nearly quantitative reversibility of the protein binding, and the potential for fabricating microarrays of multiple capture molecules.

1. Introduction

As a consequence of recent, large-scale genomic sequencing efforts, a strong interest has emerged in analyzing the function of DNA-encoded information on a similarly global scale. Since DNA/RNA microarray technology has been successfully used in functional genomic research, much attention has been devoted to the development of a similar technology to analyze proteins. Direct analysis of proteins is essential, because many aspects of modulation and regulation of cellular activity cannot be investigated on the level of nucleic acids. This is because the messenger RNA (mRNA) expression level is often poorly correlated with actual protein concentrations. A protein microarray technology would offer several distinct advantages, such as low sample consumption and compatibility with high-throughput analysis, thus greatly accelerating the understanding of complex biological systems and, hence, facilitating the diagnosis and/or prognosis of diseases.[1] Proteins, unlike DNA, are known to be susceptible to loss of activity upon immobilization on surfaces, owing to unfolding processes[2] and multiple-site chemical immobilization. Compared with DNA/RNA microarrays, which have been successfully applied in biomedical research, protein microarrays require much more sophisticated surface chemistry, both to inhibit non-specific interactions and to maintain the optimal conformation of immobilized protein molecules. Ideally, one would like the surface chemistry for protein immobilization in an array format to meet the following criteria: i) the surface should be inert, non-fouling, and resistant to non-specific adsorption; ii) the linking chemistry should allow control of the protein orientation and surface density, and provide a local chemical environment that favors the retention of the native conformation of the immobilized protein molecules; and iii) the tethering of the proteins should

[*] Dr. S. Zürcher, G. Zhen, Dr. D. Falconnet, Dr. J. Vörös, Prof. N. D. Spencer, Prof. M. Textor
Laboratory for Surface Science and Technology
Wolfgang-Pauli-Strasse 10, ETH-Hönggerberg
HCI, CH-8093 Zurich (Switzerland)
E-mail: stefan.zuercher@mat.ethz.ch

Dr. E. Kuennemann
Institute for Molecular Biology and Biophysics
Schafmattstrasse 20, ETH-Hönggerberg
HPK, CH-8093 Zurich (Switzerland)

[**] We thank Verena Eggli for her help in preparing the proteins and Doris Suter (Department of Chemistry and Applied Biosciences, ETH Zürich, Switzerland) for the NMR measurements. Financial support from the Swiss National Science Foundation (Swiss National Science Foundation, National Research Program NRP-47 "Supramolecular Functional Materials", Project No. 4047-057548/1) is gratefully acknowledged.

be highly specific and stable, yet reversible. For each of these individual criteria, solutions have already been described in the literature,[3–5] but their combination into a single device remains a major challenge in this field.

Coating of metal-oxide surfaces with poly(ethylene glycol) (PEG)-grafted polyelectrolytes, such as the brush copolymer poly(L-lysine)-*graft*-PEG (PLL-*g*-PEG), provides an attractive option for producing surfaces with a very low tendency towards unspecific protein adsorption. At neutral pH, PLL-*g*-PEG ($pK_a \approx 10$) has a positively charged backbone and, thus, adsorbs spontaneously (primarily via electrostatic interaction) onto negatively charged metal-oxide surfaces, such as Ta_2O_5, Nb_2O_5, TiO_2, and SiO_2,[6,7] and some other materials such as tissue culture polystyrene (TCPS).[8] The stability of such adlayers has been shown to be good for weeks to months, as long there is no large change in pH or ionic concentration.[6] By a proper choice of the grafting ratio of lysine units to PEG side chains, excellent resistance against non-specific adsorption of proteins can be achieved in contact with human blood serum at physiological conditions (less than 2 $ng\,cm^{-2}$).[6,9,10] Molecular adlayers of biotin-derivatized PLL-*g*-PEG (PLL-*g*-PEG/PEG-biotin) have been demonstrated to be a useful platform for the specific binding of avidin, neutral avidin, or streptavidin, and subsequent immobilization of biotinylated biomolecules, such as antibodies.[4] In a previous study, we also successfully used neutral avidin attached to biotinylated PLL-*g*-PEG polymers to immobilize biotinylated β-lactamase in a fully active form onto Nb_2O_5 surfaces with a defined, controlled enzyme density.[11] However, the application of biotinylated PLL-*g*-PEG has several limitations, such as non-specific adsorption due to avidin or streptavidin, non-reversibility of the immobilization procedure, and the necessity for biotinylation of proteins, a non-site-specific chemical modification that may affect the activity and native physicochemical properties.

A chelator-based immobilization strategy, very often used for chromatographic protein purification, exploits the use of oligohistidine tags ("His-tags") in combination with metal ions immobilized via chelators such as iminodiacetic acid[12] or nitrilotriacetic acid (NTA).[13] The chelating chemistry offers several distinct advantages over biochemical recognition elements such as antibodies or biotin/streptavidin (neutral avidin) systems: i) His-tags bound to either the C- or N-terminus are commercially available for a large number of proteins; ii) the formation of a histidine–Ni^{II} complex is a fast reaction; iii) the reaction is reversible when the surface is exposed to a stronger chelator, such as ethylenediaminetetraacetic acid (EDTA), or to an excess of a competitor ligand, such as imidazole; and iv) the spatially defined attachment of His-tags on proteins or antibodies offers the opportunity for their immobilization in controlled orientations. Large libraries of His-tagged receptor molecules, such as oligohistidine-tagged single-chain antibody variable-region fragments, are already available, and could be immobilized onto a microarray from expression supernatants.[14] Ni^{II}–NTA is the most popular system for chelation-based immobilization. NTA is a tetradentate chelating ligand occupying four of the six binding sites in the coordination sphere of the nickel ion, leaving two sites free to interact with the 6 × His-tag. NTA-containing resin binds metal ions far more strongly than other available chelating resins,[13] leading to a smaller loss of metal ions due to leakage. Many further applications using NTA–Ni^{II} to immobilize proteins on planar surfaces have been reported,[15] including NTA-functionalized dextran,[16,17] NTA-terminated (alkyl) thiol self-assembled monolayers (SAMs),[3,18] and solid-support lipid bilayers combined with NTA–lipids.[19–21] A critical, largely unsolved issue in the context of an NTA-based biosensor platform remains the elimination of non-specific protein adsorption.

Recently, the first approaches combining the advantage of using PEG-modified non-fouling surface chemistry with chelation-based immobilizations have been reported.[22,23] Chelation-based immobilization schemes were demonstrated to be compatible with high-density PEG-coated surfaces, resulting in surfaces that resist non-specific protein adsorption while preserving the highly specific interaction to His-tagged proteins. The major limitation of these approaches is the relatively complex, multistage surface chemistry preparation that makes it difficult to achieve defined surface properties and reproducibility from batch to batch.

Herein, we demonstrate that it is possible to combine the advantages of the PLL-*g*-PEG based polymeric assembly system with the well-established Ni^{II}–His-tag chemistry, thus extending the available toolbox of immobilization methods to cover different needs in protein sensing. The use of spontaneous assembly of multifunctional polymeric molecules has proven to be a highly promising approach, both in terms of basic research and in application-oriented developments. In basic research, it makes producing surfaces with very well-defined physicochemical properties by simple aqueous dip-and-rinse processes feasible. Concerning applications, spontaneous assembly of multifunctional molecules is a cost-effective, single-step process for functionalizing surfaces. It can be easily scaled-up to production scale, and is characterized by excellent reproducibility.

We describe here the synthesis of NTA-functionalized PLL-*g*-PEG polymers that spontaneously assemble on metal-oxide surfaces as a monolayer, which allows for the immobilization of proteins through NTA–Ni^{II}–histidine docking-site chemistry. The NTA–ligand surface density is tailored by adjusting the ratio of NTA-terminated PEG chains to inert methoxy (MeO)-terminated chains during the synthesis step. By means of optical-waveguide light-mode spectroscopy (OWLS)[24] with niobium-oxide-coated waveguide chips, we studied, in situ, both the absorption of the polymers and the ability of the modified chip to specifically and reversibly bind and sense 6 × His-tagged green fluorescent protein (GFPuv-6His). The surfaces were highly resistant to non-specific protein adsorption from human serum and GFPuv (without His tag). The effects of increased NTA density on protein resistance of different polymers were also investigated. Furthermore, we demonstrate that this new approach to functionalizing surfaces with NTA can be combined with the "molecular assembly patterning by lift-off" (MAPL) patterning technique,[25–27] providing a promising platform for the fabrication of protein microarrays to bind and study His-tagged recombinant proteins.

2. Synthesis of the Nitrilotriacetic-Acid-Functionalized Copolymer, PLL-*g*-PEG

For the synthesis of end-functionalized PEG with NTA and subsequent coupling of this polymer to PLL, one depends on α,ω-functionalized PEG. The main problem is that most commercially available compounds have functional groups that do not react quantitatively and selectively with reactive groups such as sulfhydryls or amines. This means that coupling of two different fragments on each end of the PEG chain requires optimum adjustment and exact control of the reaction conditions such as pH, concentrations of the reagents, and reaction times. Failure to do this leads to product mixtures, in which some PEG chains have the same group at both ends. One commercially available bifunctional PEG is the α-vinylsulfone-ω-*N*-hydroxysuccinimidyl ester of PEG-propionic acid, VS-PEG-NHS, which contains both an ω-vinylsulfone and an α-*N*-hydroxysuccinimidyl ester group. The vinylsulfone group is selective for reactions with sulfhydryl groups around pH 7, while reaction with amino groups and vinylsulfone proceed at higher pH.[28] However, even at pH 7, the reaction with amino groups is not completely suppressed. In a recently published work from Sagara and Kim,[29] it was shown that the ^{1}H NMR determination of the conjugated products of branched polyethyleneimine (PEI) with VS-PEG-NHS indicated that the VS groups were completely bound to the amine groups of PEI, as well as the NHS groups, at pH 6.0–8.0. At pH 7.0, all the VS groups reacted with the primary, secondary, or tertiary amine groups of PEI within 2 h. We have also found that the VS signals in the proton NMR disappeared completely after treating with PLL (molecular weight $MW = 20$ kDa; 1 Da = 1 g mol^{-1}) for 1 h at pH 8.5. Such different reaction behavior can be explained by the higher concentration of amine groups of PEI or PLL, compared to those of proteins. In the reaction with monoamines, the NHS groups can be selectively coupled with the amine groups, while the VS groups remain completely intact[28,29] if one uses the appropriate reaction conditions and stoichiometry. Since both of our targets (lysine-NTA and PLL) to be coupled to PEG contain amino groups (a monoamine and a polyamine, respectively), we decided to make use of this behavior and couple first the NTA–lysine to the NHS-side at low pH in methanol, i.e., under conditions that leave the VS group intact and, in a second step at increased pH, couple the VS to PLL in the desired ratio, followed by backfilling with MeO-terminated PEG-succinimidyl propionic acid to achieve the desired grafting ratio. This strategy proved successful. An outline of the synthesis is depicted in Scheme 1. The ^{1}H NMR spectrum after conjugating NTA–lysine with VS-PEG-NHS showed new peaks in the range 1.29 to 1.96 ppm (–$CHCH_2CH_2CH_2CH_2NH$–) that can be assigned to the coupled NTA–lysine. From the integration ratios between NTA–lysine and the PEG peak (–CH_2CH_2O–), the NTA–ly-

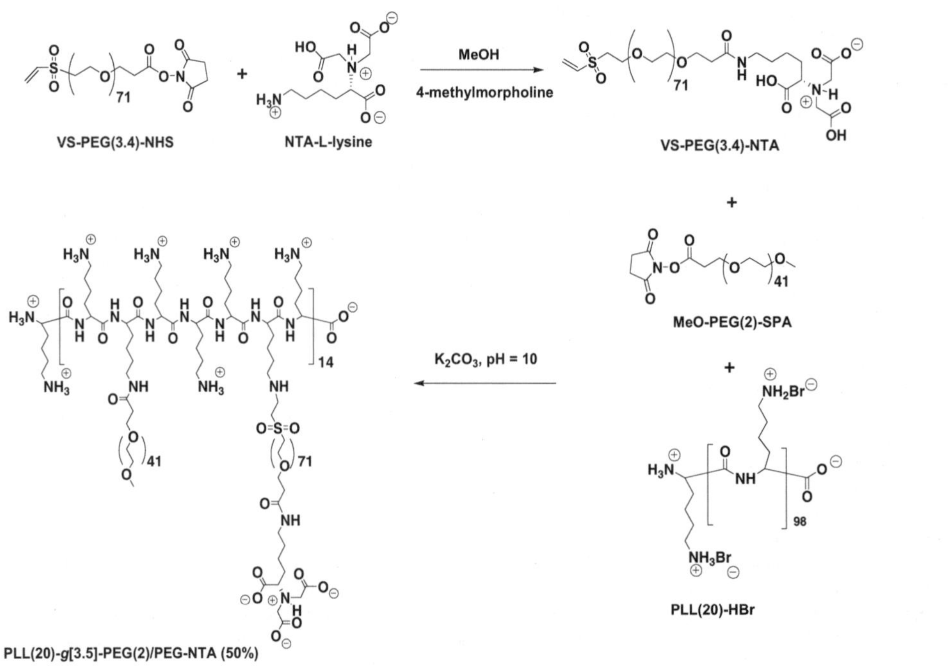

Scheme 1. Two-step synthesis of the sensing graft copolymer PLL(20)-*g*[3.5]-PEG(2)/PEG(3.4)-NTA50 %. The specific structure shown corresponds to a grafting ratio, *g* (lysine/PEG), of 3.5. Half of all PEG chains were functionalized with NTA.

ADVANCED FUNCTIONAL MATERIALS

sine conjugate percentage was calculated to be 96 %, assuming a PEG chain with 71 monomeric units. Comparison with the NMR spectrum of VS-PEG-NHS proved that the VS groups remained completely intact. This VS-PEG-NTA served as the starting material, together with PLL and the *N*-hydroxysuccinimidyl ester of MeO-poly(ethylene glycol) propionic acid (mPEG-SPA), to synthesize three graft copolymers with different amounts of NTA-functionalized end groups. The targeted grafting ratio was 3.5, known from previous studies[4,6] to be optimal for polymeric adlayers that impart resistance against non-specific protein adsorption. The final polymer architecture was quantitatively determined by proton NMR in terms of both the grafting ratio and the NTA-PEG/mPEG ratio. The results of these measurements are summarized in Table 1, and the procedure to calculate the numbers is described in detail in the Experimental section.

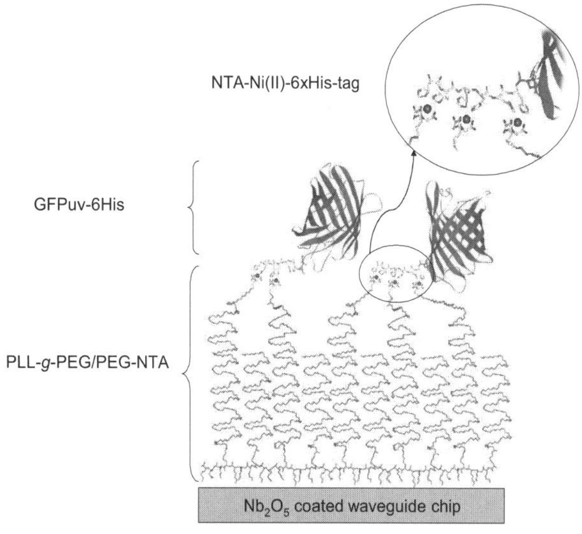

Figure 1. Schematic view of the molecular assembly monolayer of the polycationic polymer PLL-*g*-PEG/PEG-NTA on a negatively charged Nb_2O_5-coated waveguide chip. The GFPuv-6His is shown to be attached via the six adjacent histidine residues to three and two Ni-NTA complexes, with the NTA covalently bound to the end of the long PEG chains (molecular weight 3.4 kDa). The non-functionalized PEG chains have a molecular weight of 2 kDa. The magnified inset shows details of the NTA–Ni^{II}/6 × His-tag interaction.

3. Surface Preparation and Sensing by OWLS

The sensing surfaces were prepared by immersing niobium oxide coated waveguide chips in a buffer solution (4-(2-hydroxyethyl)piperazine-1-ethanesulfonic acid (HEPES-2); see Experimental) at physiological pH and salt concentration. After injection of a solution of the polymer in the same buffer, adsorption was monitored in situ, using OWLS in combination with a flow-through cell. The resulting surface was then activated by injection of a Ni^{2+} solution. Nickel ions were coordinated by the NTA ligands, resulting in complexes that bound 6 × His-tagged proteins. A sketch of such an interface is depicted in Figure 1. Nb_2O_5-coated chips were used in view of the high negative surface charge density (isoelectric point, IEP: ≈ 3.6) at neutral pH, and the resulting strong electrostatic binding of polycationic polymers. The high-density PEG chain brush renders the surface highly resistant to non-specific adsorption of single proteins and full human serum, whereas the PEG chains carrying NTA ligands can specifically bind to 6 × His-tagged recombinant proteins via the well-established NTA–Ni^{II}/6 × His-tag interaction. These long and flexible PEG chains (molecular weight 3.4 kDa) are presumed to allow more than two NTA–Ni^{II} complexes to bind to one 6 × His-tagged protein via multipoint attachment. Stable adsorption of 6 × His-tagged proteins has been reported to require two to three NTA–Ni^{II} complexes per protein entity. Local proximity of NTA functionalities is therefore essential for sufficiently strong (multivalent) binding.[20,23] We used GFPuv-6His as a model protein, since it is fluorescent and can also be used for the microarray study. Owing to its relatively high stability, results obtained in this study might not be translated directly to less-stable and more-sensitive proteins, but the PEGylated surface helps to prevent surface-induced denaturing.

Table 1. Polymer composition data determined from proton NMR and synthetic yield.

Polymer architecture [a]	Polymer code	Grafting ratio [b]	Percentage of NTA-functionalized PEG chains [c] [%]	Yield [%]
PLL(20)-*g*[3.5]-PEG(2)	PLL-*g*-PEG	3.5	0	75
PLL(20)-*g*[3.8]-PEG(2)/PEG(3.4)-NTA28 %	PLL-PEG-NTA28	3.8	28	78
PLL(20)-*g*[4.1]-PEG(2)/PEG(3.4)-NTA46 %	PLL-PEG-NTA46	4.1	46	72
PLL(20)-*g*[4.3]-PEG(3.4)-NTA 96%	PLL-PEG-NTA96	4.3	96	58

[a] The polymers are based on a PLL average molecular weight of 20 kDa (as the hydrobromide salt) and PEG molecular weights of 2 and 3.4 kDa for the unmodified and NTA-derivatized PEG chains, respectively. [b] The grafting ratio, g, denotes the number of lysine monomers divided by the number of PEG side chains, and was determined by ^{1}H NMR. [c] The fraction of PEG chains that carry a terminal NTA group as a percentage of the total number of PEG chains.

A typical curve obtained from OWLS of GFPuv-6His binding on the NTA polymeric surface is shown in Figure 2A. The proteins used (GFPuv and GFPuv-6His) were over-expressed in *E. coli* cytoplasm, purified by chromatographic methods, and showed at least 95 % purity. After exposure for 30 min to a HEPES-2 buffer solution containing 1 mg mL^{-1} PLL-PEG-NTA96, a stable layer formed with an

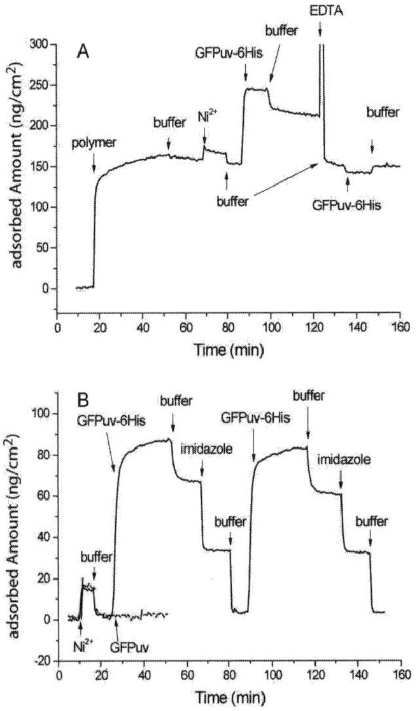

Figure 2. A) A typical OWLS adsorption curve of the polymer PLL-PEG-NTA96 on a Nb_2O_5-coated waveguide chip: activation of the surface with Ni^{2+}, adsorption of GFPuv-6His on the Ni^{2+} activated polymeric surface, and desorption of the GFPuv-6His from the surface by rinsing with EDTA solution. The non-specific adsorption of the polymeric surface was tested by adding GFPuv-6His after removing Ni^{2+} from the surface with EDTA. The observed, sharp "mass changes" (e.g., after EDTA injection) were due to refractive-index changes of the solutions. B) OWLS measurement of GFPuv-6His (line) and wild-type GFPuv (dotted line) on a Ni^{2+} charged PLL-PEG-NTA46 coated Nb_2O_5 surface. The binding of GFPuv-6His on the NTA-functionalized polymeric surface was repeated twice upon removal with 200 mM imidazole in HEPES-2 buffer.

adsorbed polymer mass of 154 ng cm^{-2} after washing with HEPES-2 buffer. After loading the surface with Ni^{2+}, it was possible to adsorb about 100 ng cm^{-2} GFPuv-6His on the surface. After rinsing of this initially formed surface with HEPES-2 buffer, approximately 35 % of the GFPuv-6His dissociated in a fast dissociation phase, which then slowed down and stabilized within 30 min. The OWLS measurement indicates that more than 65 % of the initially adsorbed protein mass remained on the surface. To test the reversibility of this system, the surface was rinsed with either an EDTA or imidazole solution. Because EDTA forms a more-stable complex with Ni^{2+} compared to NTA, the original PLL-PEG-NTA96-coated surface was regenerated. This was demonstrated by injecting GFPuv-6His a second time, which did not increase the total adsorbed mass (Fig. 2A). Non-specific adsorption of GFPuv-6His on the polymeric surface in the absence of Ni^{2+} was below 2 ng cm^{-2}. Imidazole, on the other hand, is only able to remove the GFPuv-6His (not the Ni^{2+}) owing to a competitive reaction. Since it is a much weaker ligand than the 6 × His-tag, a relatively high concentration (200 mM) had to be used. However, once a GFPuv-6 × His molecule had desorbed from the surface, the probability of its rebinding was close to zero in the presence of a high-concentration imidazole solution. To test the reusability of the surface after imidazole exposure, the reactivated surface was rinsed with buffer followed by injection of GFPuv-6His. The sensing capacity was reduced to 85 % compared with the first injection (Fig. 2B). The slightly reduced adsorption capability was probably due to leakage of some nickel ions during the imidazole rinse. Non-specific adsorption was also tested by injection of GPFuv (lacking the 6 × His-tag ligand) after activating the surface with nickel ions. The increase in adsorbed mass was below the detection limit of OWLS (2 ng cm^{-2}), which confirms the excellent specificity of this polymeric sensing surface (Fig. 2B, dotted line).

Knowing both the ratio between NTA-functionalized and non-functionalized PEG chains and the adsorbed mass of the polymers on the surface, the NTA ligand surface density can be calculated (Table 2). This method has been successfully used to estimate the density of biotin groups on biotinylated PLL-*g*-PEG-modified surfaces for bioaffinity assays[4] and enzyme immobilization.[11] Binding curves of GFPuv-6His on PLL-PEG-NTA28, PLL-PEG-NTA46, and PLL-PEG-NTA96 surfaces after being activated with Ni^{2+} measured by OWLS are plotted in Figure 3. During the washing with buffer step, a fast dissociation phase was observed during the first few minutes, which slowly stabilized at a level of approximately 65 % of the initially bound GFPuv-6His. This behavior was likely due to the presence of different surface-binding states of the GFPuv-6His

Table 2. Surface properties of the non-functionalized and the three NTA-functionalized copolymers.

Polymer code	Adsorbed polymer mass [a] [ng cm^{-2}]	Calculated NTA surface density [b] [pmol cm^{-2}]	Amount of stable, immobilized GFPuv-6His [c] [pmol cm^{-2}]	NTA per GFPuv-6His [d]	Protein adsorption from serum [e] [ng cm^{-2}]
PLL-g-PEG	164 ± 5	0	0	0	0
PLL-PEG-NTA28	169 ± 4	16 ± 1.6	1.1 ± 0.1	14	2 ± 1
PLL-PEG-NTA46	158 ± 4	23 ± 2.3	1.5 ± 0.1	15	5 ± 1
PLL-PEG-NTA96	149 ± 5	37 ± 3.7	2.1 ± 0.2	18	8 ± 2

[a] Measured by OWLS. [b] Calculated from the adsorbed mass and the average molecular weight of the polymer repeating unit per NTA-functionalized PEG chain. [c] After rinsing with HEPES-2. [d] Number of NTA ligands per bound GFPuv-6His protein. [e] Measured by OWLS on non-Ni^{2+}-activated surfaces.

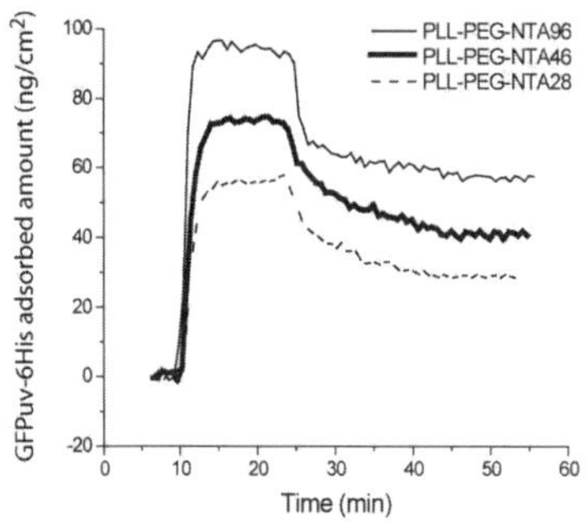

Figure 3. Binding curves of GFPuv-6His (1 μM) on the three polymeric surfaces with different NTA concentrations (see Table 1) after loading with Ni^{2+}.

molecules with one, two, or three NTA ligands involved, having highly different desorption kinetics. The single-NTA-bound proteins were expected to desorb quickly during the washing step, while immobilization stability increased progressively for proteins interacting with two or three NTA–Ni^{II} complexes. It can be seen from Table 2 that the amount of stable immobilized GFPuv-6His increased with increasing NTA-ligand density. If one assumes that three NTA ligands are necessary to interact with one GFPuv-6His for stable immobilization, the maximum protein-binding capability would be 5.2, 7.6, and 12.3 pmol cm^{-2} for PLL-PEG-NTA28, PLL-PEG-NTA46, and PLL-PEG-NTA96, respectively (found: 1.1, 1.5, and 2.1 pmol cm^{-2}, respectively, see Table 2). These numbers indicate that not all of the NTA ligands are accessible for binding GFPuv-6His. Assuming that the GFP "barrels" (≈4.2 nm height, ≈2.4 nm diameter) form a close-packed monolayer on the surface, one should arrive at a surface coverage of approximately 16 pmol cm^{-2}, corresponding to 440 ng cm^{-2}. We clearly are far below such a high-density layer. Therefore, we can rule out a steric reason for the adsorbing proteins being the limiting factor. Calculating the number of NTA ligands per stably bonded protein, one gets 14, 15, and 18 for the three polymers PLL-PEG-NTA28, PLL-PEG-NTA46, and PLL-PEG-NTA96, respectively (Table 2). It seems that the higher the NTA ligand density, the fewer ligands contribute to binding of GFPuv-6His. A possible explanation is that some ligands are hidden in the dense PEG brush, or even attracted by the positively charged PLL backbone, reducing the number of accessible ligands on the surface. This effect is more pronounced for PLL-PEG-NTA96, since in this case all PEG chains have the same length. In the other two cases, long, NTA-functionalized PEG chains are interspersed with shorter, MeO-terminated chains, therefore presenting relatively more NTA groups on the surface. We compared our results to a recently published similar system,[23] and found that approximately the same amount of stably bonded proteins per NTA ligand could be immobilized in both cases.

All four polymers (Table 2) were screened for non-specific adsorption of serum before Ni^{2+} activation. The non-functionalized PLL-*g*-PEG showed no serum adsorption, while, for the NTA-functionalized polymer layers, a small amount of adsorbed serum mass was observed that increased with the percentage of NTA functionalization. All four surfaces were calculated to have similar ethylene glycol densities of 17±1 ethylene glycol monomers per square nanometer, which is believed to be sufficient for protecting the surface against non-specific adsorption.[9] Therefore, the observed serum adsorption is likely to be a consequence of the negative charge on the NTA ligands, which is expected to interact with (positively charged) proteins, or due to metal ions present in human serum. Nevertheless, the measured non-specific serum adsorption was below 10 ng cm^{-2}, even for the highest NTA density surface (Table 2), compared to 450 ng cm^{-2} for a bare niobium-oxide surface.[4] If serum was adsorbed on Ni^{2+}-charged surfaces, a higher protein adsorption compared to non-activated surfaces was observed. This was probably due to the interactions between NTA–Ni^{II} and metal-coordinating amino acids (cysteine, glutamic acid, aspartic acid, arginine, lysine, histidine) in serum proteins. EDTA rinsing of such surfaces completely removes these adsorbed serum proteins, demonstrating that they were bound to the NTA–Ni^{II} ligands.

4. Microarray Using the MAPL Technique

We chose GFPuv-6His to test the specificity and reversibility of surface binding because of its robustness and the presence of an intrinsic chromophore, which allows us to easily verify that the immobilized protein has retained its native conformation.[30,31] We produced a MAPL array with PLL-PEG-NTA28 patterns (60 μm × 60 μm), surrounded by non-functionalized PLL-*g*-PEG. The polymer-coated patterned samples were loaded with Ni^{2+} ions, followed by incubation with GFPuv-6His. Figure 4A shows the specific coupling of GFPuv-6His to the NTA-functionalized patches, while the non-functional, inert PLL-*g*-PEG-coated background showed very low fluo-

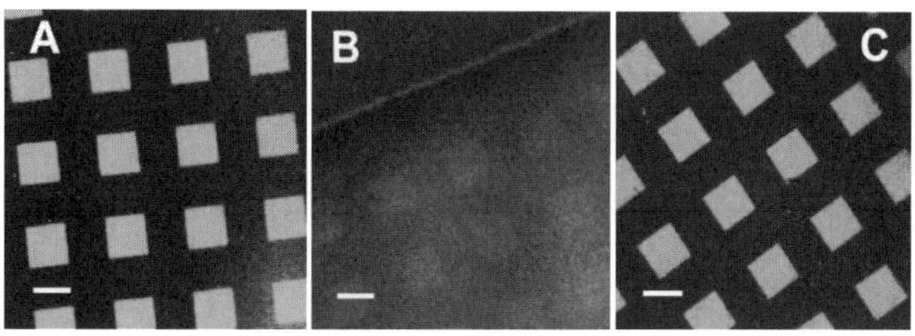

Figure 4. Patterns of GFPuv-6His on an Nb_2O_5 surface analyzed by confocal laser scanning microscopy (CLSM). The scale bars represent 60 μm. A) GFPuv-6His immobilized on a MAPL patterned substrate that has PLL-PEG-NTA28 micropatches surrounded by PLL-*g*-PEG after activation with Ni^{2+}. B) As for (A), fluorescence image after rinsing the surface with 200 mM imidazole for 2 min. C) As for (B), fluorescence image after reloading GFPuv-6His on the patterned substrate.

rescent intensity. Upon addition of imidazole, the immobilized protein was removed from the surface (Fig. 4B) because of competition between the His-tag and imidazole. Since this imidazole treatment preserves the majority of the nickel ions in complexed form, GFPuv-6His can simply be readsorbed on the surface and the pattern reappears, thus demonstrating the reversibility of the binding (Fig. 4C). The almost complete removal of the proteins with imidazole and the high protein resistance of the background illustrate the excellent specificity of the NTA–Ni^{II}/6 × His-tag interaction and the compatibility of this surface-immobilization scheme with the MAPL patterning technique. Because the 6 × His-tag in GFPuv-6His is selectively attached to the C-terminus of the GFP protein molecule, the surface-immobilized GFP should possess controlled orientation. The combination of this controlled orientation, together with the non-adhesive inert surface, can significantly prevent the denaturing of GFP on the surface. The intense fluorescence signal indicates that the surface-immobilized GFPuv-6His was in its active conformation.

5. Conclusions

We successfully synthesized and characterized a series of NTA-functionalized PLL-*g*-PEG polymers. Compared with other NTA–Ni^{II}/6 × His-tagged protein-binding systems, such as carboxy-methylated dextran matrix,[16,17] NTA lipid bilayers[19–21] and NTA SAMs,[3,18] the PLL-*g*-PEG/PEG-NTA polymers used in this study contain long PEG chains with about 71 EG units per chain that serve as spacers to connect the NTA–Ni^{II} binding site. This structure results in a flexible and highly hydrated polymer platform, which is believed to be able to facilitate more than one NTA–Ni^{II} binding site interacting with the 6 × His-tag, leading to increased stability of the interaction. Although not all of the NTA ligands seemed to be accessible for binding, the bound proteins were observed to be stable and did not detach within a reasonable time (up to 40 min). The surface NTA density was lower by a factor of about five compared to a recent report,[23] but our surface-modification protocol is far easier and, therefore, believed to be more reproducible. Approximately the same amount of protein, calculated per NTA ligand, could be immobilized in both cases. It is possible to fine-tune the surface NTA ligand density by simply changing the polymer architecture, or by coadsorption with non-functionalized PLL-*g*-PEG. The major advantage is that we can keep the excellent, very low non-specific adsorption properties in full serum, characteristic of non-functionalized PLL-*g*-PEG, at least for lower degrees of functionalization; this is a key parameter for potential applications. This and the excellent reversibility were proven using OWLS as well as the recently developed MAPL patterning technique. The combination of NTA-functionalized PLL-*g*-PEG polymers with MAPL is a promising technique for the development of new protein biosensors. To improve the stability of the protein binding further, a modified NTA ligand recently developed by Lata and Piehler[23] that has three NTAs in close proximity instead of lysine NTA could be coupled to the PLL-*g*-PEG polymer. With such a system, an additional reduction of the total NTA ligand density without loss of sensing capability and stability could probably be achieved.

6. Experimental

Materials: PLL hydrobromide (PLL HBr, $MW = 20$ kDa) was purchased from Sigma-Aldrich (Buchs, Switzerland). *N*-hydroxysuccinimidyl ester of methoxy-poly(ethylene glycol) propionic acid (MeO-PEG-SPA, molecular weight = 2 kDa), and α-vinylsulfone-ω-*N*-hydroxysuccinimidyl ester of PEG-propionic acid (VS-PEG-NHS; molecular weight: 3.4 kDa) were purchased from Nektar Therapeutics (Huntsville, AL). *N*,*N*-bis(carboxymethyl)-L-lysine hydrate, 4-methylmorpholine, 4-(2-hydroxyethyl)piperazine-1-ethanesulfonic acid (HEPES), EDTA, and imidazole were purchased from Fluka (Buchs, Switzerland). Unless stated otherwise, buffer components, organic solvents, and all other chemicals were from Fluka. HEPES-2 buffer (150 mM $NaCl_2$, 10 mM HEPES adjusted to pH 7.4) was prepared in our laboratory. Control human serum (Precinorm U, Roche Diagnostics GmbH, Mannheim, Germany) contained human serum components along with enzymes and other additives (≈ 70 mg mL^{-1}). The concentrations of the proteins were approximately equal to those found in clinically normal human serum.

Synthesis of VS-PEG-NTA (3.4 kDa): A slight excess of NTA-L-lysine (0.070 mmol, 19.3 mg) was dissolved in methanol (2 mL) containing 4-methylmorpholine (60 μL). Then the solution was added to VS-PEG-NHS (200 mg, 0.059 mmol), dissolved in methanol (2 mL), and stirred for 6 h at room temperature. The product (VS-PEG-NTA) was precipitated by adding diethyl ether (40 mL) and cooling to –20 °C, filtered, and vacuum dried to yield 191 mg (91 %) of a white powder. The molar ratios of NTA to PEG (91 %) and the vinylsulfone group to PEG (89 %) in the product were calculated from ^1H NMR measured in $CDCl_3$.

Synthesis of PLL-PEG-NTA28, PLL-PEG-NTA46, and PLL-PEG-NTA96: PLL-HBr (20 kDa) was dissolved in K_2CO_3 buffer (0.1 M; 47 mL g^{-1} of PLL-HBr, pH 10). The resulting solution was stirred vigorously for approximately 30 min and subsequently filtered through a Durapore membrane (0.22 μm; sterile Millex GV, Sigma-Aldrich, Buchs, Switzerland) into a sterile culture tube. Then a stoichiometric amount of VS-PEG-NTA powder was slowly added to the filtered PLL-HBr solution while it was vigorously stirred. After reacting for 12 h, the appropriate stoichiometric amount of MeO-PEG-SPA powder was slowly added to the solution while it was continuously stirred. After six additional hours of vigorous stirring at room temperature, the solution was transferred to a dialysis tube (Spectr/Por dialysis tubing, molecular weight cutoff of 6–8 kDa, Spectrum Laboratories, Inc., Rancho Dominguez, CA). Dialysis was carried out for 48 h in deionized water (1.5 L; the water was changed every 24 h). The product was then lyophilized at a pressure of 0.2 mbar (1 bar = 10^5 Pa). The products were recovered with yields of 78, 72, and 58 % for PLL-PEG-NTA28, PLL-PEG-NTA46, and PLL-PEG-NTA96, respectively.

Characterization by ^1H NMR: The dialyzed and lyophilized polymers were dissolved in D_2O and the spectra recorded on a 500 MHz spectrometer (Bruker). Spectral assignments were based on comparison with previously published spectra of PLL-*g*-PEG [6] and PLL-*g*-PEG/PEG-biotin [4], as well as with spectra of the individual reagents. ^1H NMR (D_2O) δ [ppm]: 2.47 (t, $-OCH_2CH_2C(O)NH-$, MeO-PEG linked to lysine), 2.95 (m, $-CH_2NH_3^+$ in free lysine chains), 3.05–3.10 (incompletely resolved multiplets, CH_2CH_2NH from VS-PEG-NTA-grafted lysine chains and $-CH_2NHC(O)OCH_2-$ from NTA–lysine linked to VS-PEG and $-CH_2NHC(O)CH_2-$ from MeO PEG linked to lysine), 3.33 (s, CH_3O-, methoxy PEG), 3.65 (broad m, $-CH_2CH_2O-$, ethylene glycol), 4.10 (t, $-S(O2)CH_2CH_2NH-$, vinylsulfone bonded to amine), 4.25 (m, $-NHC(O)CH-$, lysine backbone), 6.18 and 6.41 (d, $=CH_2$, vinylsulfone), 6.80 (dd, $-SO_2CH=$, vinylsulfone).

In order to calculate the grafting ratios of lysine units/PEG side chains and the percentage of NTA-derivatized PEG/PEG side chains, we used the integrated intensities of peaks that could be assigned and that have little or no overlap with neighboring peaks. We chose the peak at 4.25 ppm (m, –NHC(O)*CH*–, lysine backbone), the peak at 4.10 ppm (m, –$S(O_2)CH_2CH_2NH$–, vinylsulfone bonded to amine), the peak at 3.33 ppm (s, CH_3O–, methoxy PEG) and the peak at 2.95 ppm (m, –$CH_2NH_3^+$ in ungrafted lysine chains). The grafting ratio of PLL-*g*-PEG/PEG-NTA polymer was thus given by the intensity of the peak at 4.25 ppm divided by the difference in the intensity of the peak at 4.25 ppm and half of the intensity at 2.95 ppm. The percentage of NTA functionality was given by half of the intensity of the peak at 4.10 ppm divided by the sum of the half intensities of the peaks at 4.10 ppm and one third of the intensity of the peak at 3.33 ppm. The grafting ratios of the polymers and the percentage of NTA functionality are summarized in Table 1. Owing to inherent experimental and NMR spectrum integration errors, we do not consider these values to be accurate to more than one significant figure. However, these values confirm the synthesis strategy, target grafting ratio, and NTA percentage, which we could choose to modify the structure of the polymer. A spectrum of unreacted NHS-PEG-VS was recorded and showed two doublets at 6.18 and 6.41 ppm, and a doublet of doublets at 6.80 ppm. Therefore, the absence of these vinyl peaks in the final product also demonstrated that no detectable amount of free vinylsulfone remained.

Expression and Purification of GFPuv: The coding sequence of GFPuv [32] was cloned into a T7 expression plasmid [33] and GFPuv was highly over-expressed in the cytoplasm of *E. coli* BL21(DE3). Cells transformed with the corresponding plasmid were grown at 25 °C to an optical density at 600 nm of 0.75 in DYT (double yeast tryptone: 1 % Bacto yeast extract, 1.6 % Bacto tryptone, 0.5 % NaCl) medium containing ampicillin (100 µg mL^{-1}), followed by induction with IPTG (isopropyl-β-D-thiogalactopyranoside) (1 mM final concentration). The cells were harvested at 16 h after induction by centrifugation. The purification of GFPuv was performed according to Topell et al. [32]. Cytoplasmic extract (40 mL) was used in the purification step and, finally, around 20 mg of purified GFPuv (corresponding to 0.5 mg mL^{-1} extract) was obtained. The correct mass of GFPuv (26 649 Da) was verified by matrix-assisted laser desorption/ionization (MALDI) mass spectrometry.

Expression and Purification of GFPuv-6His: The expression plasmid for GFPuv-6His was constructed by Kunkel mutagenesis [33] of pGFPuv3 [32] with a primer introducing the DNA sequence for six histidines at the C-terminus of the coding sequence, resulting in the plasmid pGFP-6His. The plasmid was transformed into *E. coli* JM83 cells by electroporation, and cells were grown in DYT medium containing 100 µg mL^{-1} ampicillin to an optical density at 600 nm of 0.9 at 37 °C. After induction with IPTG (final concentration 1 mM) the temperature was reduced to 25 °C and growth was continued for a further 18 h. The cells were harvested by centrifugation (10 min, 2700 *g*, 4 °C), the pellet was resuspended in tris(hydroxymethyl)aminomethane (20 mM; Tris-HCl, pH 8.0) containing lysozyme (1 mg mL^{-1}) and DNaseI (10 µg mL^{-1}) at 4 °C. Cells were disrupted and the soluble extract cleared by ultracentrifugation (30 min, 180 000 *g*, 4 °C). After dialysis against Na phosphate (50 mM)/NaOH, pH 8.0, NaCl (1 M), the supernatant was applied on a Ni-NTA column (Qiagen) equilibrated with the same buffer. The column was washed with imidazole (20 and 40 mM). For elution, a linear gradient from 40 to 300 mM imidazole in Na phosphate (50 mM)/NaOH, pH 8.0, NaCl (1 M) was applied. The GFPuv-6His-containing fractions (between 120 and 200 mM imidazole) were combined and incubated at 60 °C for 30 min. Precipitated proteins were removed by ultracentrifugation (180 000 *g*, 1 h, 4 °C). The supernatant was adjusted with Tris-HCl (20 mM), pH 8.0, and ammonium sulfate (0.8 M) and loaded on a phenyl-Sepharose HP column (Pharmacia) equilibrated with Tris-HCl (20 mM), pH 8.0, ammonium sulfate (0.8 M) at room temperature. A gradient from 0.8 to 0 M ammonium sulfate led to the elution of GFPuv-6His. The protein was dialyzed against Tris-HCl (20 mM), pH 8.0, and applied to a MonoQ 10/10 anion-exchange column (Pharmacia) in the same buffer. GFPuv-6His was eluted by a gradient from 0 to 300 mM NaCl. Pure fractions were combined and dialyzed against HEPES (10 mM)/NaOH, pH 7.4. Cytoplasmic extract (40 mL) was used in the purification step and, finally, around 30 mg purified GFPuv-6His (corresponding to 0.75 mg mL^{-1} extract) was obtained. The correct mass of GFPuv-6His (27 472 Da) was verified by MALDI mass spectrometry.

OWLS Measurements; Substrate Preparation: OWLS measurements were carried out using an instrument described by Vörös et al. [24] and used in situ to monitor quantitatively the spontaneous adsorption of the NTA-functionalized copolymer on a niobium oxide coated waveguide chip. Optical waveguide chips for OWLS measurements were purchased from Microvacuum Ltd. (Budapest, Hungary) and consisted of a 1 mm thick AF45 glass substrate and a 200 nm thick $Si_{0.75}Ti_{0.25}O_2$ waveguiding surface layer. A 12 nm thick Nb_2O_5 layer was sputter-coated on top of the waveguiding layer in a Leybold direct-current magnetron Z600 sputtering unit at the Paul Scherrer Institute, Villigen, Switzerland. Nb_2O_5-coated waveguide chips were sonicated in HCl (0.1 M) for 10 min, extensively rinsed with ultrapure water, dried in a nitrogen stream, and treated for 1 min with oxygen plasma in a Plasma Cleaner/Sterilizer PDC-32G (Harrick, Ossining, NY). Used sensor chips were regenerated by washing with "Cleaner" (Cobas Integra, Roche Diagnostics GmbH, Mannheim, Germany) for 10 min and repeating the cleaning protocol described above. Nb_2O_5-coated waveguides were then inserted into the OWLS flow cell and equilibrated in HEPES-2 buffer for at least 6 h in order to establish a stable baseline.

Standard Assays: To prepare the polymeric adlayers by molecular assembly, solutions of PLL-*g*-PEG/PEG-NTA (1 mg mL^{-1}) in HEPES-2 buffer were injected into the flow cell for 30 min, followed by washing in HEPES-2 for another 15 min. Subsequently, this sensing surface was activated by coordination of $NiCl_2$ (5 mM) in HEPES-2 buffer to the surface-exposed NTA ligand. Protein attachment on the NTA polymeric surface was carried out by sequential injections of the 6 × His-tagged GFP solutions (1 µM) in HEPES-2. The surface was washed with HEPES-2 after each injection. In order to elute the protein, the surface was exposed to imidazole (200 mM) in HEPES-2 buffer or EDTA (200 mM) in HEPES-2 buffer. To determine binding of proteins in the absence of Ni^{II}, the $NiCl_2$ injection was omitted, or the surface was fully regenerated by injection of EDTA (200 mM) in HEPES-2 buffer. The adsorbed masses of polymers and proteins were calculated from the measured effective refractive indices and thickness values according to the formula of Defeijter et al. [34], using a refractive-index increment (dn/dc; n: refractive index; c: concentration) of 0.182 cm^3 g^{-1} for proteins [24] and 0.165 cm^3 g^{-1} for polymers [4].

Serum-Adsorption Experiments: Protein-resistance measurements were carried out by following the same procedure as above to form a PLL-*g*-PEG/PEG-NTA monolayer on the waveguide (or after the standard assay experiments of binding 6 × His-tagged proteins), and then the surface was fully regenerated by exposing it to EDTA (200 mM). After a flat baseline was reached, the control human serum was injected into the OWLS flow cell. After 15 min of equilibration, the flow cell was rinsed with HEPES-2 buffer and allowed to equilibrate for 30 additional minutes. The difference to the first baseline corresponds to the amount of non-specific adsorption of proteins.

Microarray of GFPuv-6His on a Nb_2O_5 Substrate Analyzed by Confocal Laser Scanning Microscopy (CLSM): The NTA-functionalized polymer was used to produce microarrays with interactive square micropatches (60 µm × 60 µm) in a non-interactive PLL-*g*-PEG background. The microarrays were prepared according to the recently published MAPL technique on Nb_2O_5-coated Pyrex surfaces [25]. The patterned surface was first activated with Ni^{2+} by dipping the sample into a $NiCl_2$ (10 mM) in HEPES-2 buffer for 10 min and washing with HEPES-2. Then the sample was dipped in GFPuv-6His (1 µM) in HEPES-2 buffer for 15 min, followed by a washing step with HEPES-2 buffer. To elute the GFPuv-6His from the surface, the sample was exposed to imidazole (200 mM) in HEPES-2 for 2 min. Then the sample was again dipped in GFPuv-6His solution to reload GFPuv-6His on the surface. The experimental steps were followed by CLSM (Zeiss LSM 510). CLSM was used to evaluate the quality of the GFPuv-6His patterns and the activity of immobilized GFP, as well as the specificity and reversibility of the PLL-*g*-PEG/PEG-NTA coated surface.

Received: April 18, 2005
Final version: July 3, 2005
Published online: September 29, 2005

[1] H. Zhu, M. Snyder, *Curr. Opin. Chem. Biol.* **2003**, *7*, 55.
[2] W. Norde, *Adv. Colloid Interface Sci.* **1986**, *25*, 267.
[3] G. B. Sigal, C. Bamdad, A. Barberis, J. Strominger, G. M. Whitesides, *Anal. Chem.* **1996**, *68*, 490.
[4] N. P. Huang, J. Vörös, S. M. De Paul, M. Textor, N. D. Spencer, *Langmuir* **2002**, *18*, 220.
[5] J. K. Lee, Y. G. Kim, Y. S. Chi, W. S. Yun, I. S. Choi, *J. Phys. Chem. B* **2004**, *108*, 7665.
[6] G. L. Kenausis, J. Vörös, D. L. Elbert, N. P. Huang, R. Hofer, L. Ruiz-Taylor, M. Textor, J. A. Hubbell, N. D. Spencer, *J. Phys. Chem. B* **2000**, *104*, 3298.
[7] N. P. Huang, R. Michel, J. Vörös, M. Textor, R. Hofer, A. Rossi, D. L. Elbert, J. A. Hubbell, N. D. Spencer, *Langmuir* **2001**, *17*, 489.
[8] S. VandeVondele, J. Vörös, J. A. Hubbell, *Biotechnol. Bioeng.* **2003**, *82*, 784.
[9] S. Pasche, S. M. De Paul, J. Vörös, N. D. Spencer, M. Textor, *Langmuir* **2003**, *19*, 9216.
[10] S. Tosatti, S. M. De Paul, A. Askendal, S. VandeVondele, J. A. Hubbell, P. Tengvall, M. Textor, *Biomaterials* **2003**, *24*, 4949.
[11] G. L. Zhen, V. Eggli, J. Vörös, P. Zammaretti, M. Textor, R. Glockshuber, E. Kuennemann, *Langmuir* **2004**, *20*, 10 464.
[12] J. Porath, J. Carlsson, I. Olsson, G. Belfrage, *Nature* **1975**, *258*, 598.
[13] E. Hochuli, H. Dobeli, A. Schacher, *J. Chromatogr.* **1987**, *411*, 177.
[14] C. Wingren, C. A. Borrebaeck, *Expert Rev. Proteomics* **2004**, *1*, 355.
[15] E. K. M. Ueda, P. W. Gout, L. Morganti, *J. Chromatogr. A* **2003**, *988*, 1.
[16] L. Nieba, S. E. NiebaAxmann, A. Persson, M. Hamalainen, F. Edebratt, A. Hansson, J. Lidholm, K. Magnusson, A. F. Karlsson, A. Pluckthun, *Anal. Biochem.* **1997**, *252*, 217.
[17] P. D. Gershon, S. Khilko, *J. Immunol. Methods* **1995**, *183*, 65.
[18] D. Kroger, M. Liley, W. Schiweck, A. Skerra, H. Vogel, *Biosens. Bioelectron.* **1999**, *14*, 155.
[19] L. Schmitt, C. Dietrich, R. Tampe, *J. Am. Chem. Soc.* **1994**, *116*, 8485.
[20] J. G. Altin, F. A. J. White, C. J. Easton, *Biochim. Biophys. Acta* **2001**, *1513*, 131.
[21] E. Gizeli, J. Glad, *Anal. Chem.* **2004**, *76*, 3995.
[22] T. Cha, A. Guo, Y. Jun, D. Q. Pei, X. Y. Zhu, *Proteomics* **2004**, *4*, 1965.
[23] S. Lata, J. Piehler, *Anal. Chem.* **2005**, *77*, 1096.
[24] J. Vörös, J. J. Ramsden, G. Csucs, I. Szendro, S. M. De Paul, M. Textor, N. D. Spencer, *Biomaterials* **2002**, *23*, 3699.
[25] D. Falconnet, A. Koenig, T. Assi, M. Textor, *Adv. Funct. Mater.* **2004**, *14*, 749.
[26] D. Falconnet, D. Pasqui, S. Park, R. Eckert, H. Schift, J. Gobrecht, R. Barbucci, M. Textor, *Nano Lett.* **2004**, *4*, 1909.
[27] R. N. Orth, M. Wu, D. A. Holowka, H. G. Craighead, B. A. Baird, *Langmuir* **2003**, *19*, 1599.
[28] M. Morpurgo, F. M. Veronese, D. Kachensky, J. M. Harris, *Bioconjugate Chem.* **1996**, *7*, 363.
[29] K. Sagara, S. W. Kim, *J. Controlled Release* **2002**, *79*, 271.
[30] M. Zimmer, *Chem. Rev.* **2002**, *102*, 759.
[31] M. Ormo, A. B. Cubitt, K. Kallio, L. A. Gross, R. Y. Tsien, S. J. Remington, *Science* **1996**, *273*, 1392.
[32] S. Topell, J. Hennecke, R. Glockshuber, *FEBS Lett.* **1999**, *457*, 283.
[33] S. Strobl, P. Muhlhahn, R. Bernstein, R. Wiltscheck, K. Maskos, M. Wunderlich, R. Huber, R. Glockshuber, T. A. Holak, *Biochemistry* **1995**, *34*, 8281.
[34] J. A. Defeijter, J. Benjamins, F. A. Veer, *Biopolymers* **1978**, *17*, 1759.

8850 *Langmuir* **2008,** *24,* 8850–8856

A Biomimetic Alternative to Poly(ethylene glycol) as an Antifouling Coating: Resistance to Nonspecific Protein Adsorption of Poly(L-lysine)-*graft*-dextran

Chiara Perrino,[†] Seunghwan Lee,[†] Sung Won Choi,[‡] Atsushi Maruyama,[‡] and Nicholas D. Spencer*[,†]

Laboratory for Surface Science and Technology, Department of Materials, ETH Zurich, Wolfgang-Pauli-Strasse 10, CH-8093 Zurich, Switzerland, and Institute for Materials Chemistry and Engineering, Kyushu University, 744-CE11 Motooka, Nishi, Fukuoka 819-0395, Japan

Received March 26, 2008. Revised Manuscript Received May 16, 2008

Poly(L-lysine)-*graft*-dextran (PLL-*g*-dex), graft copolymers with dextran side chains grafted onto a poly(L-lysine) backbone, previously shown to be effective as stabilizers of DNA triple helices and as carriers of functional genes to target cells or tissues, were employed in this work to prevent nonspecific adsorption of proteins, as determined by means of optical waveguide lightmode spectroscopy. PLL-*g*-dex copolymers readily adsorb from aqueous solution onto negatively charged oxide surfaces and significantly reduce nonspecific protein adsorption onto bare silica–titania surfaces. While effective and equivalent surface adsorption and antifouling properties were observed for PLL-*g*-dex copolymers in a variety of architectures, nanotribological analysis by atomic force microscopy was able to distinguish between the different brush densities produced.

Introduction

Control of biofouling is the motivation for a large body of ongoing research in many disparate fields. The design of antifouling surfaces for biomedical device applications would allow adverse biological reactions to be avoided, since these generally arise from the conformational transformations undergone by proteins adsorbed on biomaterials. Such reactions can be life threatening and can certainly limit the performance of the devices.[1–5] The undesirable adsorption of microorganisms and the formation of biofilms are also detrimental in several non-biomedical areas: the consequences of biofouling for water purification, transport or storage systems, heat-transfer components, ships' hulls, and static marine structures are considerable, both economically and ecologically.[6–11]

Poly(ethylene glycol) (PEG) has been of great interest in the preparation of protein-resistant surfaces, and PEG coatings have proven to successfully reduce or prevent nonspecific protein adsorption, especially when they are in a brush conformation.[12–23] PEG—a simple linear polyether—does, however, have some limitations: for instance it is susceptible to thermal and oxidative degradation.[24–28] The low long-term stability of PEG is one of its major limits and can be responsible for decreased performance in preventing nonspecific protein adsorption.

Mimicking biological membrane surfaces provides an alternative approach for conferring antifouling properties to surfaces: the highly hydrated glycocalyx, which surrounds certain kinds of cells, is known, for instance, to possess antiadhesive properties. Carbohydrates—the principal component of the glycocalyx—are thought to be mainly responsible for its ability to prevent undesirable nonspecific adsorption of proteins.[29–33] Carbohydrates may therefore constitute a

* Corresponding author. E-mail: spencer@mat.ethz.ch.
† ETH Zurich.
‡ Kyushu University.

(1) Anderson, J. M.; Cima, M. J.; Langer, R.; Shawgo, R. S.; Shive, M. S.; von Recum, H.; Voskerician, G. *Biomaterials* **2003**, *24*(11), 1959–1967.
(2) Dai, L.; St John, H. A. W.; Bi, J.; Zientek, P.; Chatelier, R. C.; Griesser, H. J. *Surf. Interface Anal.* **2000**, *29*, 46–55.
(3) Hook, F.; Vörös, J.; Rodahl, M.; Kurrat, R.; Böni, P.; Ramsden, J. J.; Textor, M.; Spencer, N. D.; Tengvall, P.; Gold, J.; Kasemo, B. *Colloids Surf., B* **2002**, (24), 155–170.
(4) Morra, M. *J. Biomater. Sci., Polym. Ed.* **2000**, *11*(6), 547–569.
(5) Wang, M. S.; Palmer, L. B.; Schwartz, J. D.; Razatos, A. *Langmuir* **2004**, *20*, 7753–7759.
(6) Griesser, H. J.; Hartley, P. G.; McHarthur, S. L.; McLean, K. M.; Meagher, L.; Thissen, H. *Smart Mater. Struct.* **2002**, *11*, 652–661.
(7) Brady, R. F. *J. Protective Coat. Linings* **2000**, *17*(6), 42–46.
(8) Rajagopal, S.; Sasikumar, N.; Jakapaul, A.; Nair, K. V. K. *Biofouling* **1991**, *3*(4), 311–324.
(9) Charmain, J. K.; Osborn, K. S.; Rickard, A. H.; Robson, G. D.; Handley, P. S. *Handb. Water Wastewater Microbiol.* **2003**, 757–775.
(10) Flemming, H.-C.; Griebe, T.; Schaule, G. *Water Sci. Technol.* **1996**, *34*(5), 517–524.
(11) Melo, L. F.; Bott, T. R. *Exp. Therm. Fluid Sci.* **1997**, *14*(4), 375–381.
(12) Yu, W. H.; Kang, E. T.; Neoh, K. G. *Langmuir* **2005**, *21*(1), 450–456.
(13) McGurk, S. L.; Green, R. J.; Sanders, G. H. W.; Davies, M. C.; Roberts, C. J.; Tendler, S. J. B.; Williams, P. M. *Langmuir* **1999**, *15*(15), 5136–5140.
(14) Lee, J. H.; Kopeckova, P.; Kopecek, J.; Andrade, J. D. *Biomaterials* **1990**, *11*, 455–464.
(15) Fan, X.; Lin, L.; Messersmith, P. B. *Biomacromolecules* **2006**, *7*(8), 2443–8.
(16) Otsuka, H.; Satomi, T.; Itadani, J. H.; Nagasaki, Y.; Okano, T.; Horiike, Y.; Kataoka, K. *Eur. Cells Mater.* **2003**, *6*(1), 102.
(17) Ma, H.; Hyun, J.; Stiller, P.; Chilkoti, A. *Adv. Mater.* **2004**, *16*(4), 338–341.
(18) Heuberger, M.; Drobek, T.; Spencer, N. D. *Biophys. J.* **2005**, *88*, 495–504.
(19) Kenausis, G. L.; Vörös, J.; Elbert, D. L.; Huang, N.; Hofer, R.; Ruiz-Taylor, L.; Textor, M.; Hubbell, J.; Spencer, N. D. *J. Phys. Chem. B* **2000**, *104*, 3298–3309.
(20) Lee, S.; Iten, R.; Müller, M.; Spencer, N. D. *Macromolecules* **2004**, *37*(22), 8349.
(21) Pasche, S.; De Paul, S. M.; Vörös, J.; Spencer, N. D.; Textor, M. *Langmuir* **2003**, *19*, 9216–9225.
(22) Pasche, S.; Vörös, J.; Griesser, H. J.; Spencer, N. D.; Textor, M. *J. Phys. Chem. B* **2005**, *109*, 17545–17552.
(23) Pasche, S.; Vörös, J.; Spencer, N. D.; Textor, M. *Langmuir* **2003**, *19*, 9216.
(24) Han, S.; Kim, C.; Kwon, D. *Polym. Degrad. Stab.* **1995**, *47*(2), 203–208.
(25) Han, S.; Kim, C.; Kwon, D. *Polymer* **1997**, *38*(2), 317–323.
(26) Madorsky, S. L.; Straus, S. *J. Polym. Sci.* **1959**, *26*, 183–194.
(27) Reich, L. *J. Appl. Polym. Sci.* **1969**, *13*, 977–988.
(28) Yang, L.; F., H.; Blease, T. G.; Thompson, R. I. G. *Eur. Polym. J.* **1996**, *32*(5), 535–547.
(29) Holland, N. B.; Qiu, Y.; Ruegsegger, M.; Marchant, R. E. *Nature* **1998**, *392*, 799–801.
(30) McArthur, S. L.; McLean, K. M.; Kingshott, P.; St John, H. A. W.; Chatelier, R. C.; Griesser, H. J. *Colloids Surf., B* **2000**, *17*(1), 37–48.
(31) Piehler, J.; Brecht, A.; Hehl, K.; Gauglitz, G. *Colloids Surf., B* **1999**, *13*, 325–336.

Published on Web 07/11/2008

biomimetic alternative to PEG as components of antifouling coatings.

Dextran is a natural polysaccharide consisting of an α(1→6)-linked glucan with side chains attached to the 3-positions of the backbone glucose units. It is nontoxic, water-soluble, and neutral—favorable characteristics for application in the biomedical field. Dextran has already been employed for the preparation of protein-resistant surfaces and revealed good antifouling properties.[29–31,33–39] Martwiset et al. compared the antifouling properties of dextrans of different molecular weights and different extents of oxidation grafted onto silicon surfaces and reported that the chemical structure of the grafted polymers (relative amount of oxidized groups, molecular weight of dextran) strongly affects their conformation and ability to prevent protein adsorption. They suggested that −OH groups of dextran chains, being both hydrogen bond acceptors and donors, can interact with each other strongly, leading to the collapse of pristine dextran chains, whereas −CHO groups of oxidized dextran only form hydrogen bonds with the surrounding water molecules, thus maintaining a relatively extended conformation.[36] Österberg et al. bound lightly oxidized dextran in a loops-and-trains configuration on polystyrene and compared its performance in preventing nonspecific protein adsorption to that of dextran/PEG bound in an end-on configuration. They reported that dextran chains grafted with the side-on configuration are much more effective than those with the end-on configuration in reducing protein adsorption.[39] However, for the end-on configuration, they did not study the influence of the grafting density on the protein-rejecting ability of the coatings and only a dilute density regime was investigated. In fact it is difficult to achieve high surface brush densities by covalent coupling of polymers onto surfaces ("grafting to") in an end-on configuration: polymer chains that form well-hydrated random coils in solution will not, in fact, spontaneously adsorb forming dense elongated brushes, due to their hydration shells, which cause mutual steric repulsion. Bosker et al. were the first to consider dextran brushes with high grafting densities: they demonstrated the almost complete protein resistance of densely packed polystyrene−dextran block copolymers (PS−dextran) adsorbed onto hydrophobic surfaces.[34]

An alternative approach to produce a dense polymer brush involves the grafting of polymer side chains onto a backbone, resulting in the formation of a comblike structure. When adsorbed onto a surface via the backbone, the side chains are forced away from the surface, spontaneously forming a brush, providing the surrounding medium constitutes a good solvent for the side chains. An example of such behavior is seen with poly(L-lysine)-*graft*-PEG (PLL-*g*-PEG),[19,40] a graft copolymer consisting of PEG chains grafted onto a polycationic PLL backbone, which has proven to be highly effective at preventing nonspecific adsorption of proteins onto oxide surfaces in an aqueous environment.[18,19,21–23,41] PLL-*g*-PEG copolymers have been shown to readily adsorb from aqueous solutions onto negatively charged surfaces via electrostatic interactions; the positive charges present on the protonated primary amine groups of the PLL backbone in neutral aqueous environment lead to its rapid immobilization onto negatively charged surfaces. The side chains, radially distributed along the backbone in bulk solution in order to minimize steric repulsion, stretch out into the solution once the copolymers have adsorbed on the surface, providing sufficiently close inter-PEG spacing.

The copolymers employed in the present study, poly(L-lysine)-*graft*-dextran (PLL-*g*-dex), have a very similar structure to PLL-*g*-PEG, with the same PLL backbone and dextran substituting for PEG as side chains (Figure 1). Dextran, like PEG, is neutral, and it therefore has no electrostatic interactions with the polycationic backbone. Furthermore it has multiple reactive sites,[42] facilitating subsequent functionalization, and is much cheaper than end-functionalized PEG. The undesirable oxidation to peroxides in air, a weakness of PEG-based systems, is not expected to be a problem with sugar chains (the relative long-term stability of dextran and PEG chains as antifouling coatings is currently under investigation). While the PS−dextran brushes studied by Bosker et al. are adsorbed onto hydrophobic substrates via hydrophobic interactions, PLL-*g*-dex copolymers, similarly to PLL-*g*-PEG, are expected to spontaneously and electrostatically adsorb onto metal oxide surfaces, thus broadening the range of potential applications of dextran brushes. PLL-*g*-dex copolymers have previously proven to be effective as stabilizers for DNA triple helices and as potential carriers of functional genes to target cells or tissues.[43–47] On the basis of the well-established excellent performance of PLL-*g*-PEG in preventing nonspecific protein adsorption and the above-mentioned studies of the protein-repelling capabilities of dextran coatings, the focus of this work is the characterization of both the adsorption and antifouling properties of PLL-*g*-dex graft-copolymers.

Previous nanotribological analysis of PEG brushes have allowed subtle differences in brush structure to be distinguished.[48] In this study, this approach has been employed to show differences in dextran brush structure as a consequence of modifications in the PLL-*g*-dex architecture.

Materials and Methods

Poly(L-lysine)-*graft*-Dextran (PLL-*g*-dex). Poly(L-lysine)-*graft*-dextran (PLL-*g*-dex) copolymers were synthesized by a reductive amination reaction of poly(L-lysine)-HBr (13 or 6 kDa, polydispersity 1.3 (for both), Sigma-Aldrich, St. Louis, MO) with dextran (5.9 kDa, polydispersity 1.6, Amersham Bioscience, Uppsala, Sweden). A sodium borate buffer (0.1 M, pH 8.5, 0.4 M NaCl) was used as solvent for the reaction. An approximately 10× molar excess of sodium cyanoborohydride ($NaBH_3CN$) to dextran was used to reduce the unstable Schiff base resulting

(32) Qiu, Y.; Zhang, T.; Ruegsegger, M.; Marchant, R. E. *Macromolecules* **1998**, *31*(1), 165–171.

(33) Sen Gupta, A.; Wang, S.; Link, E.; Anderson, E. H.; Hofmann, C.; Lewandowski, J.; Kottke-Marchant, K.; Marchant, R. E. *Biomaterials* **2006**, *27*(16), 3084–3095.

(34) Bosker, W. T. E.; Patzsch, K.; Cohen Stuart, M. A.; Norde, W. *Soft Matter* **2007**, *3*, 754–762.

(35) Frazier, R. A.; Matthijs, G.; Davies, M. C.; Roberts, C. J.; Schacht, E.; Tendler, S. J. B. *Biomaterials* **2000**, *21*, 957–966.

(36) Martwiset, S. *Langmuir* **2006**, *22*, 8192–8196.

(37) Mc Lean, K. M.; Johnson, G.; Chatelier, R. C.; Beumer, G. J.; Steele, G.; Griesser, H. J. *Colloids Surf., B* **2000**, *18*, 221–234.

(38) Österberg, E.; Bergström, K.; Holmberg, K.; Riggs, J. A.; Van Alstine, J. M.; Schuman, T. P.; Burns, N. L.; Harris, J. M. *Colloids Surf., A* **1993**, *77*, 159–169.

(39) Österberg, E.; Bergström, K.; Holmberg, K.; Schuman, T. P.; Riggs, J. A.; Burns, N. L.; Van Alstine, J. M.; Harris, J. M. *J. Biomed. Mater. Res.* **1995**, *29*, 741–747.

(40) Huang, N.; Michel, R.; Voros, J.; Textor, M.; Hofer, R.; Rossi, A.; Elbert, D.; Hubbell, J.; Spencer, N. D. *Langmuir* **2001**, *17*, 489–498.

(41) Pasche, S.; Textor, M.; Meagher, L.; Spencer, N. D.; Griesser, H. J. *Langmuir* **2005**, *21*, 6508–6520.

(42) Massia, S. P.; Stark, J. *J. Biomed. Mater. Res.* **2001**, *56*(3), 390–399.

(43) Choi, S. W.; Sato, Y.; Akaike, T.; Maruyama, A. *J. Biomater. Sci., Polym. Edn.* **2004**, *15*(9), 1099–1110.

(44) Ferdous, A.; Akaike, T.; Maruyama, A. *Biomacromolecules* **2000**, *1*, 186–193.

(45) Ferdous, A.; Watanabe, H.; Akaike, T.; Maruyama, A. *J. Pharm. Sci.* **1998**, *87*(11), 1400–1405.

(46) Ferdous, A.; Watanabe, H.; Akaike, T.; Maruyama, A. *Nucleic Acid Res.* **1998**, *26*(17), 3949–3954.

(47) Maruyama, A.; Watanabe, H.; Ferdous, A.; Katoh, M.; Ishihara, T.; Akaike, T. *Bioconjugate Chem.* **1998**, 9.

(48) Yan, X.; Perry, S. S.; Spencer, N. D.; Pasche, S.; De Paul, S. M.; Textor, M.; Lim, M. S. *Langmuir* **2004**, *20*, 423–428.

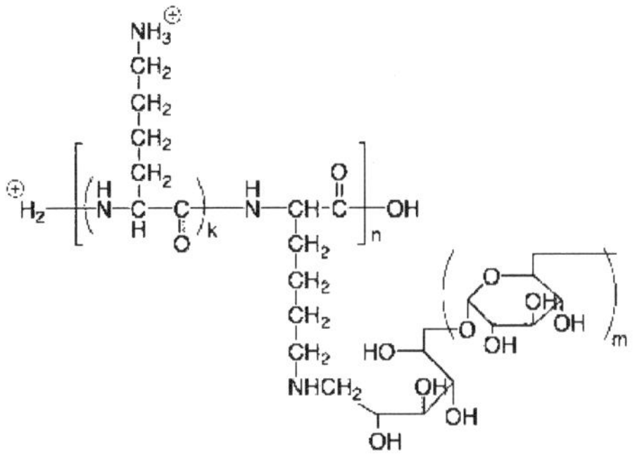

Figure 1. Structural formula of the PLL-*g*-dex copolymer. *k* should be taken as an average value. $k + 1$ represents the grafting ratio of the polymer.

from the reaction between the terminal dextran aldehyde group and primary amine groups of PLL.

The resulting graft copolymers (Figure 1) were isolated by dialysis, in order to free them from unreacted starting materials. The molecular weights of the starting materials, as well as of the graft copolymers, were determined by gel permeation chromatography (JASCO 880 model, Tokyo, Japan) with a multiangle light (GPC-MALS, Dawn-EOS, Wyatt Technology, Santa Barbara, CA). ^{1}H-NMR spectra were recorded in D_2O on a JEOL JNM-EX270 spectrometer (Tokyo, Japan) to determine the composition of the resulting copolymers and the grafting ratio.

The notation PLL(*x*)-*g*[*y*]-dex(*z*) for the copolymers was used to represent the molar mass of PLL in kilodaltons (*x*) (including the counterions, Br−, as precursor), the molar mass of dextran in kilodaltons (*z*), and the grafting ratio *g*[*y*] (defined as the number of lysine monomers/dextran side chain).

Four PLL-*g*-dex copolymers were employed in this work: while the molecular weight of dextran was kept constant at 5.9 kDa, the grafting ratio (from ca. *g*[3] to ca. *g*[10]) and the molecular weight of the PLL backbones (13 and 6 kDa) were varied, yielding four different types of PLL-*g*-dex (see Table 1).

For comparison purposes, a PLL(20)-*g*[3.0]-PEG(5) copolymer with PEG(5) side chains (molecular weight, 5 kDa) grafted onto a PLL(20) backbone at the grafting ratio of *g*[3.0], provided by SuSoS AG (Dübendorf, Switzerland), has also been employed.[19]

Optical Waveguide Lightmode Spectroscopy (OWLS). Optical waveguide lightmode spectroscopy (OWLS) was employed to characterize the adsorption properties of the polymers and to evaluate their ability to prevent nonspecific adsorption of proteins. Experiments were performed using an OWLS 110 instrument (Microvacuum, Budapest, Hungary).

OWLS is an optical biosensing technique for the in situ label-free analysis of adsorption processes.[49] The grating-assisted in-coupling of a He−Ne laser into a planar waveguide allows for a direct online monitoring of macromolecule adsorption. When the in-coupling condition is fulfilled, the light is guided by total internal reflection to the ends of the waveguiding layer, where it is detected by a photodiode detector. The adsorbed mass is calculated from the change in the refractive index in the vicinity of the waveguide surface upon adsorption of molecules.[3,49,50]

The refractive index increment (d*n*/d*c*) of dextran was measured by means of a refractometer and a value of 0.131 was used for all measurements. Since the d*n*/d*c* values of dextran and polylysine are very similar, no d*n*/d*c* correction was made for the different structures investigated. Prior to the experiments, optical waveguide chips (standard: $Si_{0.75}Ti_{0.25}O_2$ on glass, 1.2 × 0.8 cm^2, Microvacuum, Budapest, Hungary) were ultrasonicated in 0.1 M HCl for 10 min, rinsed with ultrapure water (Milli-Q Gradient A10, Millipore SA, Molsheim, France), ultrasonicated in 2-propanol for 10 min, rinsed again with Millipore water, and finally dried under a dry nitrogen stream. The substrates were subsequently cleaned in a UV/ozone cleaner (UV/Clean, model 135500, Boeckel industries Inc., Feasterville, PA) for 30 min.

A representative adsorption profile for PLL-*g*-dex (the case of PLL(6)-*g*[5]-dex(5.9)) is shown in Figure 2. The cleaned waveguides were placed into the OWLS flow cell and equilibrated by exposing to HEPES buffer solution (10 mM 4-(2-hydroxyethyl)piperazine-1-ethanesulfonic acid (Sigma), adjusted to pH 7.4 with 6.0 M NaOH solution) overnight, in order to obtain a stable baseline. The waveguides were then exposed to a polymer solution (0.25 mg mL^{-1} in a similar HEPES buffer) for at least 30 min, resulting in the formation of a polymer adlayer, and rinsed three times with buffer solution for another 30 min, to exclude the influence of the bulk solution and weakly bound polymer that might contribute to changes in the refractive index at the vicinity of the surface.

Low-ionic-strength HEPES was used during the polymer-adsorption step, in order to maximize the adsorbed mass. To condition the polymer layer to the ionic strength used in the subsequent serum-adsorption step for the evaluation of the protein resistance properties, the optical chips were exposed to a higher-ionic-strength HEPES solution (10 mM HEPES with 150 mM NaCl) before exchanging the buffer back to lower-ionic-strength HEPES. The waveguides were finally exposed to a solution of human serum (Control serum, Precinorm U, Roche, Basel, Switzerland) for 30 min and then rinsed with 10 mM HEPES solution every 15 min for 2 h to check the protein-desorption process.

Atomic Force Microscopy (AFM). A conventional beam-deflection-based atomic force microscope was employed to characterize the frictional properties of the PLL-*g*-dex copolymers adsorbed on SiO_2 surfaces on a nanometer scale in an aqueous environment. A commercial SPM scan head (Nanoscope IIIa, MultiMode, Veeco Instruments Inc., Santa Barbara, CA) equipped with a liquid cell/tip holder (Veeco Instruments Inc.) was used, and the fine movement of the sample, placed on top of the piezo scan tube, was controlled by SPM 1000 electronics and SPM 32 software (RHK Technology, Inc., Troy, MI), using an SPM Interface Module (RHK Technology) to interface the scan head and the controller. A commercial silicon nitride AFM tip−cantilever assembly (Veeco) was used as the counterface to the polymer-modified SiO_2 surfaces. The AFM tip−cantilever assembly was air-plasma cleaned (Plasma Cleaner/Sterilizer, PDC-32G instrument, Harrick, Ossining, NY) for 10 s immediately before the measurements and kept in distilled water prior to use. Silicon wafers (1.2 cm × 1.2 cm) were ultrasonicated in ethanol for 10 min, rinsed with Millipore water, dried under a nitrogen stream, and then oxygen-plasma cleaned for 2 min (Plasma Cleaner/Sterilizer, PDC-32G instrument, Harrick). After the cleaning procedure, the substrates were incubated in the polymer solution (0.1 mg/mL) for 30 min, rinsed with 10 mM HEPES buffer solution, and then dried under a nitrogen stream. All measurements were performed in 10 mM HEPES buffer solution.

The nanotribological properties of PLL-*g*-dex copolymers were characterized by the acquisition of "friction-vs-load" plots in a number of areas on each sample.[51] Briefly, the sample was laterally scanned relative to a fixed tip−cantilever assembly position in a line-scan mode while the sample was simultaneously ramped up and then ramped down in vertical direction to provide a variation in normal load. Both normal and lateral deflection of the tip−cantilever assembly generated from the interaction between the probe tip and the sample surface were detected by a four-quadrant photodiode and interpreted as the normal load (converted based on the manufacturers' normal spring constant value, $k_N = 0.58$ N m^{-1}) and the frictional forces (raw photodiode signals), respectively. With the friction forces plotted as a function of normal load, "friction-vs-load" plots were obtained. To ensure a valid comparison of the frictional properties

(49) Vörös, J.; Ramsden, J. J.; Csucs, G.; Szendro, I.; De Paul, S. M.; Textor, M.; Spencer, N. D. *Biomaterials* **2002**, *23*, 3699–3710.

(50) Kurrat, R.; Textor, M.; Ramsden, J. J.; Böni, P.; Spencer, N. D. *Rev. Sci. Instrum.* **1997**, *68*(5), 2172–2176.

(51) Müller, M. T.; Yan, X.; Lee, S.; Perry, S. S.; Spencer, N. D. *Macromolecules* **2005**, *38*(13), 5706–5713.

Table 1. Synthesized PLL-*g*-dex Copolymers in This Work (Data Relative to PLL(20)-*g*[3]-PEG(5) Shown for Comparison Purposes)[a]

polymer	no. of grafted side chains per PLL	no. of free lysines per PLL	percentage of side-chain grafting (%)	mol wt of copolymer (kDa)
PLL(13)-*g*[3.9]-dex(5.9)	16	46.3	25.6	88
PLL(6)-*g*[5]-dex(5.9)	5.7	23	20	28
PLL(13)-*g*[7.1]-dex(5.9)	8.8	53.4	14.1	78
PLL(6)-*g*[10.2]-dex(5.9)	2.8	25.9	9.8	17
PLL(20)-*g*[3]-PEG(5)	37.9	75.7	33.3	203.9

[a] The numbers of lysine units for the PLL−HBr (13 kDa), PLL−HBr (6 kDa), and PLL−HBr (20 kDa), before grafting with dextran or PEG, are 62.2, 28.7, and 95.7, respectively. The numbers of monomer units for the PEG (5 kDa) and dex (5.9 kDa) are 113.6 and 36.4, respectively.

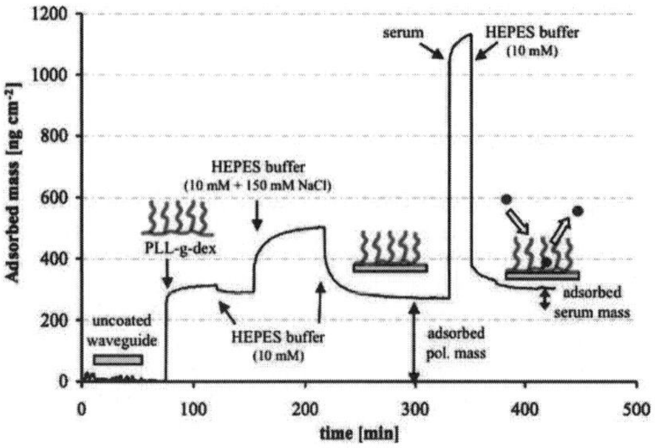

Figure 2. Representative adsorption profile for PLL-*g*-dex copolymer onto a silica−titania waveguide: adsorption of the copolymer (PLL(6)-*g*[5]-dex(5.9)) and subsequent exposure to serum.

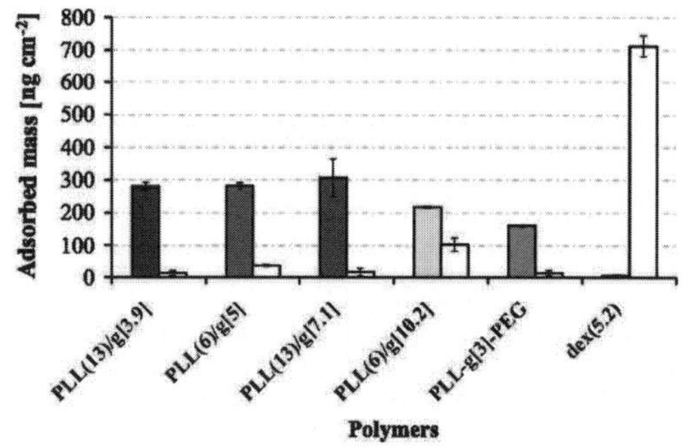

Figure 3. OWLS results for the polymer (filled histograms) and serum adsorption (empty histograms) onto bare silica−titania waveguides, including all the investigated PLL-*g*-dex copolymers (indicated with the length of the backbone and their grafting ratio), PLL(20)-*g*[3]-PEG(5) and dex.

of various samples, the same tip−cantilever assembly was used for all the measurements.

Results and Discussion

Synthesis and Structural Features. PLL-*g*-dextran copolymers with a variety of different architectures (different grafting ratios and lengths of the backbone) were successfully synthesized. Characterization by ^{1}H-NMR and GPC-MALS was performed to evaluate the grafting ratios and molecular weights, respectively. The detailed structural features of the synthesized copolymers are shown in Table 1.

The molecular weight of dextran chains (5.9 kDa) employed in this work was chosen to be comparable to that of PEG (5 kDa) in the PLL-*g*-PEG copolymers selected for comparison. The number of monomer units of dex(5.9), 36.4 sugar rings, is, however, significantly lower than that of corresponding EG units of PEG(5), 113.6, resulting in different chain lengths for the different side chains: the fully extended chain length of dextran(5.9), 25.5 nm, is significantly shorter than that of PEG(5), 40.5 nm (based on the molecular length of monomers: 0.7 nm for dextran[52] and 0.358 nm for ethylene glycol (EG)[53]).

Adsorption Properties: OWLS. *Quantitative Analysis of Polymer Adsorption.* Polymer adsorption properties were characterized by means of OWLS. The results of the OWLS experiments (the adsorption profile for PLL(6)-*g*[5]-dex(5.9) is shown in Figure 2 as example) indicate that the PLL-*g*-dex copolymers spontaneously adsorb from aqueous solution (10 mM HEPES buffer, pH 7.4) onto metal oxide surfaces. Upon exposure of a waveguide surface to the polymer solution (after 70 min of exposure to HEPES buffer to achieve the baseline), the adsorption process occurred rapidly, such that more than 90% of the final mass of adsorbed polymer was reached within the first 5 min, and resulted in the formation of a polymer adlayer on the waveguide surface, without significant polymer desorption upon rinsing with buffer solution.

The adsorbed mass of PLL-*g*-dex copolymers reported in Table 1 was measured, as well as that of dextran itself and PLL(20)-*g*[3]-PEG(5), for comparison purposes: the results are presented in Figure 3. From the adsorbed polymer mass and the compositional features of the copolymers, it is possible to calculate the surface density of dextran chains, n_{dex}, and lysine monomers, n_{lys}, expressed as molecules nm^{-2}, and the spacing between side chains on the surface, L, and finally to estimate the conformation of the surface-grafted dextran chains by comparing the spacing and the radius of gyration of dextran chains, $L/2R_g$. The results of these calculations are summarized in Table 2.

All the PLL-*g*-dex copolymers investigated showed significant adsorption on the OWLS waveguides, ranging from ca. 200 to ca. 300 ng cm^{-2} on average, whereas dextran alone revealed negligible adsorption onto the surfaces (5.6 ± 3.7 ng cm^{-2}). As with PLL-*g*-PEG copolymers, the adsorption of PLL-*g*-dex in an aqueous environment is thought to proceed through the electrostatic interactions between the polycationic PLL backbone and the surface: the dextran side chains stretch out toward the solution and the backbone lies flat on the substrate. In the case of dextran alone, no charges are available for electrostatic interactions, resulting in almost zero adsorption.

The molecular weight (and therefore the length) of the PLL backbone and the density of the grafted dextran side chains along the backbone were the two architectural parameters changed for the synthesis of the PLL-*g*-dex copolymers employed in this work: 6 kDa and 13 kDa PLL backbones were used and the grafting ratio was systematically varied from *g*[3.9] to *g*[10.2]. Despite the variation in these two architectural parameters, the adsorbed masses obtained from the PLL-*g*-dex copolymers employed are nearly constant within the error bars, except for PLL(6)-*g*[10.2]-dex(5.9), which presumably possesses too few

(52) Kawaguchi, T.; Hasegawa, M. *J. Mater. Sci.: Mater. Med.* **2000**, *11*, 31–35.

(53) Rixman, M. A.; Dean, D.; Ortiz, C. *Langmuir* **2003**, *19*, 9357–9372.

Table 2. Summary of the Adsorption Data Determined by OWLS for the PLL-*g*-dex Polymers. [a]

surface	m_{pol} (ng/cm^2)	m_{serum}	n_{Lys} (1/nm^2)	$n_{dex/PEG}$ (1/nm^2)	L (nm)	$L/2R_g$
PLL(13)-*g*[3.9]-dex(5.9)	281 ± 12	13 ± 8	1.03 ± 0.05	0.26 ± 0.01	2.09 ± 0.05	0.39 ± 0.01
PLL(6)-*g*[5]-dex(5.9)	281 ± 11	37 ± 4	1.29 ± 0.05	0.26 ± 0.01	2.11 ± 0.04	0.39 ± 0.01
PLL(13)-*g*[7.1]-dex(5.9)	307 ± 58	17 ± 12	1.90 ± 0.39	0.27 ± 0.06	2.10 ± 0.22	0.39 ± 0.04
PLL(6)-*g*[10.2]-dex(5.9)	217 ± 2	102 ± 20	1.85 ± 0.02	0.18 ± 0.00	2.52 ± 0.01	0.47 ± 0.00
PLL(20)-*g*(3)-PEG(5)	160 ± 3	13 ± 8	0.54 ± 0.00	0.18 ± 0.00	2.53 ± 0.00	0.45 ± 0.00

[a] m_{pol} = adsorbed polymer mass, m_{serum} = mass of serum adsorbed, n_{lys} = surface concentration of lysine monomers, $n_{dex/PEG}$ = surface concentration of dextran or PEG, L = spacing between grafted dextran or PEG chains, $L/2R_g$ = degree of overlap of dextran or PEG chains.

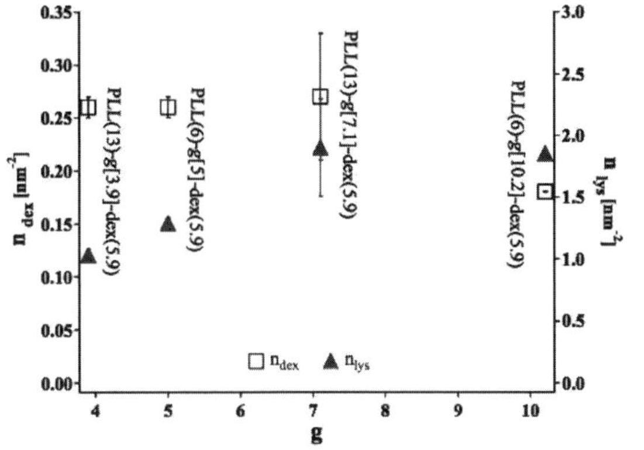

Figure 4. Lysine monomer density (n_{lys}) and dextran chain density (n_{dex}) as a function of grafting ratio (g) for the PLL(13)-*g*-dex(5.9) and PLL(6)-*g*-dex(5.9) copolymers.

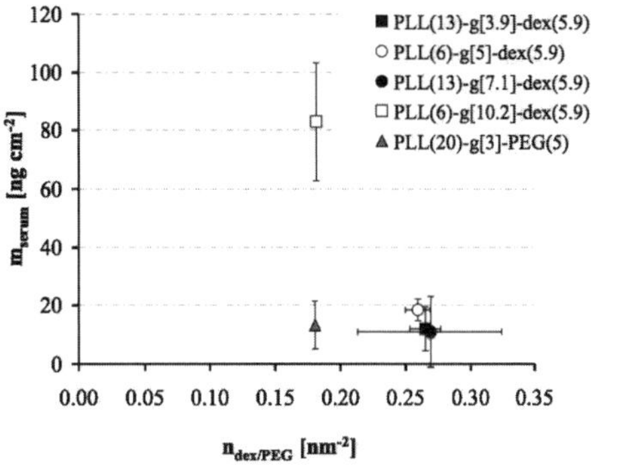

Figure 5. Adsorbed mass of serum, m_{serum}, as a function of the surface concentration of grafted dextran or PEG chains, $n_{dex/PEG}$.

dextran side chains to establish a brush, and they showed no clear trend as the molecular weight of the PLL backbone or grafting ratio was changed. The negligible influence of the PLL molecular weight on the adsorption properties is thought to be due to the fact that the length of PLL backbones selected in this work, 6 or 13 kDa, is sufficiently short that a "flat-lying" conformation of PLL is readily achieved in both cases. Second, the negligible influence of the grafting ratio can be attributed to the opposing effects of the molecular weight of a single PLL-*g*-dex copolymer and the probability of adsorption onto surfaces as a function of the grafting ratio. For instance, the grafting ratios *g*[3.9] and *g*[5] for the PLL(13)-*g*[3.9]-dex(5.9) and PLL(6)-*g*[5]-dex(5.9) represent very different total numbers of grafted dextran chains on a single copolymer molecule, 16 and 5.7, respectively, and different total molecular weights, 88 and 28 kDa, respectively (structural details are summarized in Table 1). In terms of molecular weight, the copolymer with the lowest grafting ratio, *g*[3.9], might show higher mass of surface adsorption due to the higher number of dextran chains per single molecule. The number of anchoring groups (free lysine groups) is, however, higher for higher grafting ratios, while at the same time the steric hindrance between neighboring dextran side chains is smaller, both of which are advantageous for polymer adsorption. The control of the amount of polymer adsorption by the balance between the attractive electrostatic backbone−surface interactions and a steric repulsion between dextran side chains is more directly manifested in the plot of the lysine monomer density (n_{lys}) and the dextran chain density (n_{dex}) of the copolymers as a function of the grafting ratio (Figure 4). If we still consider the PLL(6)-*g*[10.2]-dex(5.9) as an outlying exception, the lysine (or the polymer) molar surface density increases almost linearly with increasing grafting ratio, whereas n_{dex}, which represents the efficiency of the copolymers in grafting dextran chains onto the surface, is nearly constant. Among the four copolymers investigated, PLL(6)-*g*[10.2]-dex(5.9) is the one with the highest grafting ratio, i.e., the lowest molecular weight; in this case the effect of the molecular weight overwhelms that of the available anchoring points and/or steric hindrance in determining the amount of polymer adsorbed at the equilibrium that a smaller adsorbed mass was observed. Compared to PLL(20)-*g*[3]-PEG(5), all PLL-*g*-dex copolymers examined showed higher adsorbed masses (200−300 ng cm^{-2} vs ca. 160 ng cm^{-2}), although the molecular weight is definitely higher for PLL(20)-*g*[3]-PEG(5) than for all the PLL-*g*-dex copolymers. The difference in the radius of gyration (R_g) of the side chains, 2.68 nm for dextran(5.9)[54] vs 2.82 nm for PEG(5),[55] might explain the different amounts of polymer adsorbed: a smaller R_g of the side chains of the copolymer is, in fact, expected to enhance the surface adsorption due to the weaker shielding of the free lysine monomers (anchoring groups) by the side chains. In addition, the smaller the R_g, the weaker the steric interactions between neighboring side chains.

Quantitative Analysis of Protein Adsorption. OWLS measurements revealed that the amount of protein adsorbed can be greatly reduced by the presence of a PLL-*g*-dex adlayer on the surface. As summarized in Table 2 and shown in Figure 3, silica−titania surfaces exposed to a dextran solution, which showed negligible adsorption of dextran (5.6 ± 3.7 ng cm^{-2}) and can be therefore considered as bare substrates, adsorbed a significant amount of serum (756 ± 34 ng cm^{-2}), whereas all PLL-*g*-dex coated waveguides revealed a significant reduction in protein adsorption. Similarly to PLL-*g*-PEG, as shown in a previous study employing a broad range of polymer architectures (from ca. *g*[2] to *g*[20], PEG molecular weight = 1, 2, and 5 kDa),[21] the ability to resist nonspecific protein adsorption of the PLL-*g*-dex copolymers examined in this work was observed to be dependent on the surface density of the dextran chains (Figure 5): the amount of serum adsorbed decreases as the dextran density at the surface, n_{dex}, increases. As long as the grafting density is sufficiently low

(54) Görisch, S. M.; Wachsmuth, M.; Fejes Toth, K.; Lichter, P.; Rippe, K. *J. Cell Sci.* **2005**, *118*, 5825–5834.
(55) Kawaguchi, S.; Imai, G.; Suzuki, J.; Miyahara, A.; Kitano, T.; Ito, K. *Polymer* **1997**, *38*(12), 2885–2891.

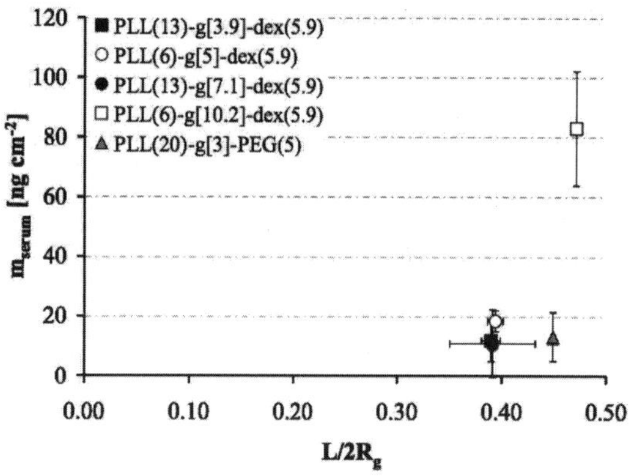

Figure 6. Adsorbed mass of serum, m_{serum}, as a function of the degree of overlap of dextran or PEG chains, $L/2R_g$. L denotes the average mean distance between dextran or PEG chains of the PLL-*g*-dex/PLL-*g*-PEG coated surfaces, calculated from the mass of polymer adsorbed and from the grafting ratios.

and thus a sufficiently high value for n_{dex} is obtained, PLL-*g*-dex copolymers greatly reduce the nonspecific adsorption of proteins, reaching a level comparable to that observed for PLL-*g*-PEG coatings.

The protein-resistance capabilities of PLL-*g*-PEG copolymers have previously proven to be dependent also on another factor: the degree of overlap between the PEG chains. This can be viewed as the product of the PEG chain length and density, and it is still related to the surface density of the surface-grafted chains.[21] It can be quantitatively estimated by using a characteristic parameter, $L/2R_g$, which compares the average spacing between neighboring polymer chains, L, to their radius of gyration, R_g. In the case of $L/2R_g$ being smaller than 1, the grafted polymer chains would overlap, were they to be in a mushroom configuration, and therefore they stretch out to form a brush structure. As shown in Figure 6, the dependence of the antifouling properties on $L/2R_g$ was also observed for the PLL-*g*-dex copolymers investigated. According to what was observed in this work, however, dextran chains need to overlap more ($L/2R_g \leq 0.4$) than PEG chains ($L/2R_g = 0.4-0.7$,[21]) to render the surface protein resistant. Compared to the PLL-*g*-PEG copolymer involved in this work, much higher values of surface density in the grafted chains and lower values of $L/2R_g$ are needed for PLL-*g*-dex copolymers to achieve comparable antifouling capabilities, as judged from the adsorbed amount of proteins from serum (Figure 3 and Table 2). For example, PLL(6)-*g*[10.2]-dex(5.9), which shows almost the same surface chain density and a similar value of $L/2R_g$ to those of PLL(20)-*g*[3]-PEG(5), shows markedly inferior protein-repelling properties. This might be attributed to the higher flexibility and degree of hydration of PEG chains compared to dextran's bulkier sugar units and, consequently, to the larger excluded volume of PEG in water, which allows PEG chains to form a more significant barrier against protein adsorption, even at smaller degree of side-chain overlap.

Nanotribological Analysis. Nanotribological measurements of the PLL-*g*-dex copolymers by means of AFM served as a further probe of the difference in the architectural features of the copolymers. Previous studies have shown that the tribological

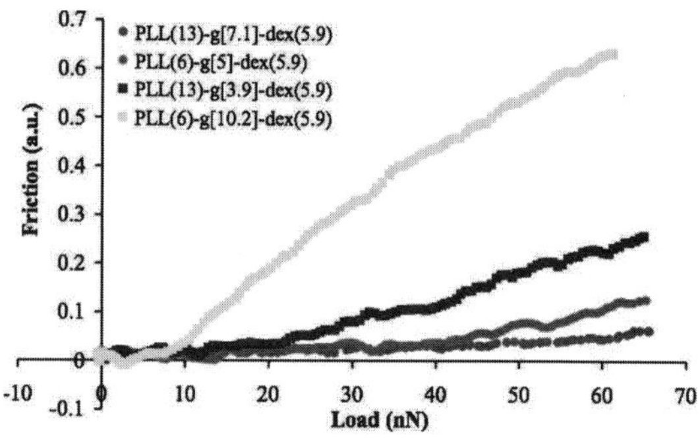

Figure 7. AFM results: friction force measured as a function of increasing load for the contact of a bare silicon nitride tip and silicon wafers coated with PLL-*g*-dex copolymers.

properties of graft copolymers, such as PLL-*g*-PEG, are significantly influenced by the architectural features, including side chain length,[48,56] grafting ratio,[48] and the backbone length.[56]

The nanotribological properties characterized by AFM (Figure 7) are indeed able to clearly distinguish between copolymers with different architectural features (grafting ratio and length of backbone), most of which were indistinguishable in their adsorption behavior and the protein-resistance properties. For instance, the two copolymers with 13 kDa PLL backbone and PLL(6)-*g*[5]-dex(5.9), which were only marginally different in their surface adsorption and antifouling properties, were observed to reveal distinctively different frictional behavior; the frictional forces being in the order of PLL(6)-*g*[10.2]-dex(5.9) ≫ PLL(13)-*g*[3.9]-dex(5.9) > PLL(6)-*g*[5]-dex(5.9) > PLL(13)-*g*[7.1]-dex(5.9).

Significantly higher frictional forces observed from the PLL(6)-*g*[10.2]-dex(5.9) can be attributed to the distinctly smaller surface-grafted dextran chain density, n_{dex} (Table 2), which, in turn, leads to its inferior ability to generate an aqueous lubricating film at the interface. On the other hand, since the n_{dex} and $L/2R_g$ values for the other copolymers are fairly similar, the improved lubricating behavior with increasing grafting ratio, from 3.9 to 7.1, might be associated with the stability of the polymer adlayer. For instance, the copolymer with higher grafting ratio, such as PLL(13)-*g*[7.1]-dex(5.9), may have stronger binding to the surface due to the higher number of available anchoring points, as well as lower steric hindrance between side chains, allowing easier access to the surface by the positively charged amine groups on the backbone. Finally, as with the surface adsorption and antifouling properties, the effect of the molecular weight of the PLL backbone between 6 and 13 kDa was observed to be inconsequential for the nanotribological properties.

Conclusions

The protein-repelling capabilities of a series of dextran-based graft copolymers (PLL-*g*-dex) with different grafting ratios and lengths of the PLL backbone, were investigated by means of OWLS and compared to PLL-*g*-PEG, a graft copolymer with the same PLL backbone but with PEG as side chains, which has been shown to be highly effective at preventing nonspecific protein adsorption.

Similarly to PLL-*g*-PEG, the amount of proteins adsorbed onto PLL-*g*-dex-coated silica−titania waveguides was observed to be dependent on the surface concentration (n_{dex}) and degree

(56) Müller, M.; Lee, S.; Spikes, H. A.; Spencer, N. D. *Tribol. Lett.* **2003**, *15*(4), 395–405.

of overlap ($L/2R_g$) of the surface-grafted polymer chains: PLL-*g*-dex copolymers with high values of n_{dex} (≥ 0.26 nm^{-2}) greatly reduce the nonspecific adsorption of proteins to a level comparable to those observed from PLL-*g*-PEG coatings. Nanotribological measurements by AFM were performed in order to further distinguish between the investigated copolymers: Those PLL-*g*-dex copolymers that showed the optimum, and yet indistinguishable, surface adsorption and antifouling properties resulted in notably different lubricating properties. For the PLL-*g*-dex copolymers employed in this work, the grafting ratio was observed to be the most important architectural parameter in determining the nanotribological properties. While the tribological studies in this work were motivated primarily by an attempt to better probe the architectural features of the PLL-*g*-dex copolymers, the combination of good lubricating and antifouling properties is highly desirable for some biomedical applications involving moving parts, e.g., coatings for stents, contact lenses, catheters, and endoscopes. The comparison of OWLS and AFM data can in this case be very helpful for the selection of the "ideal" polymer architecture for such applications.

LA800947Z

2d. Lubricant additives as surface modifiers

Commentary

Zinc dialkyldithiophosphates (ZnDTPs) are present in virtually every motor oil formulation, and function as both anti-oxidant and antiwear additives. Unfortunately their adverse effect on catalytic converters in automobile exhaust systems has made their (at least partial) replacement necessary. In order to find alternative compounds that protect engines as effectively against wear, the mechanism by which ZnDTP functions needed to be established. Our approach to this problem has been twofold. We wanted to adopt an *in situ* approach, such that the chemistry of forming tribofilms could be monitored while sliding was actually occurring at elevated temperatures. This was realized by means of in situ attenuated total reflectance infrared spectroscopy, coating the ATR element with iron, and sliding a cylinder on it while heating and taking spectra (2.33, 2.36). Also, we felt that one of the problems in the ZnDTP literature was that studies had come to different conclusions about the mechanism, based on data obtained under completely different conditions. In order to cover as large a tribological parameter space as possible, as efficiently as possible, we developed a combinatorial approach to tribological measurements. This allowed us to scan a region of load space and velocity space during a pin-on-disk experiment, separating the measurements spatially on the disk, such that the formed tribofilms could be individually examined by x-ray photoelectron spectroscopy and correlated with the input parameters and the tribological measurements made during the test (2.34, 2.35, 2.37).

One of the issues we wanted to explore was the relative influence of tribological conditions and temperature on the reactions occurring between ZnDTP and the steel surface. It appears that while polyphosphates can be formed by purely thermal treatment at elevated temperatures, tribological conditions lead to the formation of simple iron phosphates and tough, short-chain polyphosphates (2.38).

No effective replacement molecule for ZnDTP has been reported, to date. In an effort to understand why current replacement candidates are not sufficiently effective, we scrutinized the thermal chemistry of triphenyl phosphorothionate (TPPT) in oil by means of infrared and nuclear magnetic resonance spectroscopies. TPPT appeared to be both more thermally stable than ZnDTP (and therefore not activated under most conditions), as well as being a much less effective oxidation inhibitor — the original application of ZnDTP (2.39).

6606 *Langmuir* **2002,** *18,* 6606–6613

Growth of Tribological Films: In Situ Characterization Based on Attenuated Total Reflection Infrared Spectroscopy

Federica M. Piras,† Antonella Rossi,†,‡ and Nicholas D. Spencer*,†

Laboratory for Surface Science and Technology, Department of Materials, Swiss Federal Institute of Technology, ETH Zürich, CH-8092 Zürich, Switzerland, and Department of Inorganic and Analytical Chemistry, University of Cagliari, Cagliari, Italy

Received March 20, 2002. In Final Form: May 21, 2002

In this paper we describe the development of a new technique based on attenuated total reflection (ATR) Fourier transform infrared (FT-IR) spectroscopy and its application to the in situ study of lubricant system behavior under tribological conditions. The lubricated tribological system consisted of a fixed steel cylinder, sliding across the surface of a germanium ATR crystal, coated with a thin iron film in the presence of the well-studied secondary zinc dialkyl dithiophosphate (ZnDTP) lubricant additive. Using this approach, changes in the additive chemistry due to adsorption, as well as reaction film growth, can be studied as a function of sliding time and temperature. The ATR FT-IR spectra reported in this work are fully consistent with the existing ZnDTP literature and show the decomposition of ZnDTP with the formation of P–O–P species following thermal testing at 150 °C, while a simple phosphate film has been detected on the iron surface following tribological testing at the same temperature.

Introduction

In most mechanical systems (transport, energy production, manufacturing), lubricants are used to reduce friction and wear between moving parts. Lubricants generally consist of mineral or synthetic oils and contain low concentrations of different additives, including chemical compounds that adsorb on or react with the metallic surface to produce an organic and/or inorganic thin layer, which reduces wear and friction under conditions of boundary lubrication, where sliding speeds are too low or loads too high for a full fluid lubricating layer to be mantained.[1]

The analysis of these reaction films is of crucial importance for a better understanding of the mechanism of action of the additives. Most of the analytical techniques currently used for investigating the tribochemistry of additive-derived films that are formed in tribostressed systems are performed ex situ, i.e., *outside* the tribometer and *after* the friction test. Tribofilms analyzed in this way are not necessarily representative of the films in their active state, and monitoring changes with time, temperature, or other variables becomes more difficult. Although challenging, the in situ and in vivo[2] (also called in lubro[3]) surface chemical analysis of lubricated tribological systems during a friction test is extremely useful, since it provides information on the chemistry taking place under steady-state conditions.[4] The few reported studies involving in situ and in vivo analysis have involved optical methods and vibrational spectroscopy in particular.[3,5,6] These studies have several limitations: infrared microreflection absorption spectroscopy[3] is not a true surface-sensitive technique and is not effective for studies in the boundary lubrication regime, while the use of a diamond anvil cell[6] only allows static measurements to be made.

Attenuated total reflection (ATR) spectroscopy is one of the most widely used techniques for surface infrared analysis. Although the phenomenon of total internal reflection of light was described by Newton in the early 17th century, it was not until much later that Harrick[7] and, independently, Fahrenfort[8] were to exploit this phenomenon to obtain absorption spectra and develop the ATR technique. When applied to the study of in situ kinetics of adsorption and reaction of species at liquid/solid interfaces, ATR spectroscopy can yield valuable surface-chemical data. Such studies have been carried out in a variety of research and technological areas, including biomembranes,[9] biofilms,[10] thin film structure and reactivity,[11,12] and electrochemistry.[13]

The IR radiation propagating in the ATR element (the optically denser medium) undergoes total internal reflection at the interface with the sample (the optically rarer medium) because of the different refractive indices of the two media (Figure 1). There is, however, an exponentially decaying electromagnetic wave that penetrates into the rarer medium—the "evanescent wave"—interacting with

* To whom correspondence should be addressed. E-mail: nspencer@surface.mat.ethz.ch. Phone: +41-1-6325850. Fax: +41-1-6331027.

† ETH Zürich.

‡ University of Cagliari.

(1) Hähner, G.; Spencer, N. D. *Phys. Today* **1998**, *51*, *9*, 22.

(2) Donnet, C. *Handbook of Surface and Interface Analysis—Methods for Problem-Solving*; Rivière, J. C., Myhra, S., Eds.; Marcel Dekker Inc.: New York, 1998; Chapter 2.

(3) Cann, P. M.; Spikes, H. A. *Tribol. Trans.* **1991**, *34*, *2*, 248.

(4) Martin, J. M.; Le Mogne, Th.; Grossiord, C.; Palermo, Th. *Tribol. Lett.* **1996**, *2*, 313.

(5) Cann, P. M.; Spikes, H. A. *Lubr. Eng.* **1991**, *48*, 335.

(6) Westerfield, C.; Agnew, S. *Wear* **1995**, *181–183*, 805.

(7) Harrick, N. J. *Internal Reflection Spectroscopy*; Interscience Publishers, John Wiley & Sons Inc.: New York, 1967.

(8) Fahrenfort, J. *Spectrochim. Acta* **1961**, *17*, 698–709.

(9) Fringeli, U. P.; Goette, J.; Reiter, G.; Siam, M.; Baurecht, D. In *Fourier Transform Spectroscopy: 11th International Conference, AIP Conference Proceeding*; de Haseth, J. A., Ed.; American Institute of Physics: Woodbury, NY, 1998; Vol. 430, p 729. Wenzel, P.; Fringeli, M.; Goette, J.; Fringeli, U. P. *Langmuir* **1994**, *10*, 4253.

(10) Ishida, K. P.; Griffiths, P. R. *J. Colloid Interface Sci.* **1999**, *213*, 513.

(11) Chovelon, J. M.; Gaillard, F.; Wan, K.; Jaffrezic-Renault, N. *Langmuir* **2000**, *16*, 6228.

(12) Fanucci, G. E.; Talham, D. R. *Langmuir* **1999**, *15*, 3289.

(13) Zhu, Y.; Uchida, H.; Yajima, T.; Watanabe, M. *Langmuir* **2001**, *17*, 146.

Published on Web 07/24/2002

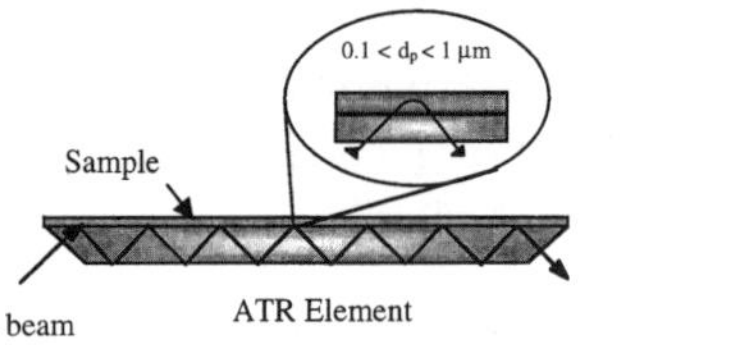

Figure 1. Scheme of the principle of ATR spectroscopy.

molecules in the vicinity of the surface, and thereby attenuating the reflected intensity and yielding an infrared spectrum similar to that obtained from a transmission experiment.[7,14] For radiation in the mid-IR range (4000–400 cm^{-1}), the penetration depth (d_p), which is a function of the wavelength, typically ranges from 0.1 to 1 μm, and thus standard ATR FT-IR spectroscopy can be regarded as a moderately surface-sensitive analytical technique. As Jakobsen has demonstrated,[15] ATR spectra can be obtained from metal-coated ATR crystals to study reactions at the liquid/metal interface. One of the applications reported in his work concerns the investigation of the adsorption of stearic acid onto a thin film of iron, sputter deposited onto a Ge ATR crystal. Applications of "metal-coated" ATR spectroscopy to study the adsorption process and structural changes at the iron/inhibitor interface[16] or to perform studies of thin copper films exposed to aqueous solutions[17] have also been reported.

In the present study, we used this ATR FT-IR approach with tribological experiments to probe the lubricant layer between sliding surfaces by measuring ATR spectra from the underside of a thin metallic film. In this way, the metal surface/lubricant interface can be continuously monitored, in situ, and in a nondestructive manner during a complete tribological sliding experiment.

Zinc dialkyl dithiophosphates (ZnDTPs) are widely used as extreme pressure and antiwear additives in many different kinds of engine and industrial lubricants. It is known that ZnDTP forms tribological films on rubbing metal surfaces; it has been proposed that these films consist of amorphous polyphosphates, but the exact chemical composition of the different polyphosphates in the ZnDTP tribofilm is not known, and a generally accepted reaction mechanism has not emerged to date. Most authors believe that thermal decomposition is the major mechanism of ZnDTP tribofilm formation; as a result only tribological experiments conducted at elevated temperatures (60–200 °C) are, typically, reported in the literature.[18–22] All the evidence obtained so far for substantiating the amorphous polyphosphate model has been based on ex situ experiments.

Our aim in this work was to develop in situ ATR spectroscopy as an analytical method for investigating tribological systems, thus using ATR FT-IR spectroscopy

(14) Mirabella, Jr., F. M. *Internal Reflection Spectroscopy—Theory and Applications*; Marcel Dekker: New York, 1993.

(15) Jakobsen, R. J. In *Fourier Transform Infrared Spectroscopy, Applications to Chemical Systems*; Ferraro, J. R., Basile, L. J., Eds.; Academic Press: New York, 1978; Vol. 2, p 165.

(16) Incorvia, M. J.; Haltmar, W. C. *J. Electrochem. Soc.* **1986**, *133*, 8, 41.

(17) Ishida, K. P.; Griffiths, P. R. *Anal. Chem.* **1994**, *66*, 522.

(18) Suominen Fuller, M. L.; Kasrai, M.; Bancroft, G. M.; Fyfe, K.; Tan, K. H. *Tribol. Int.* **1998**, *31, 10*, 627.

(19) Martin, J. M. *Tribol. Lett.* **1999**, *6*, 1.

(20) Bec, S.; Tonck, A.; Georges, J. M.; Coy, R. C.; Bell, J. C.; Roper, G. W. *Proc. R. Soc. London, A* **1999**, *455*, 4181.

(21) Bell, J. C.; Delargy, K. M.; Seeney, A. M. In *Proceedings of the 18th Leeds/Lyon Symposium, Wear Particles*; Dowson, D., et al., Eds.; Elsevier Science Publishers B.V.: New York, 1992; p 387.

(22) Choa, S. H.; Ludema, K. C.; Potter, G. E.; DeKoven, B. M.; Morgan, T. A.; Kar, K. K. *Wear* **1994**, *177*, 33.

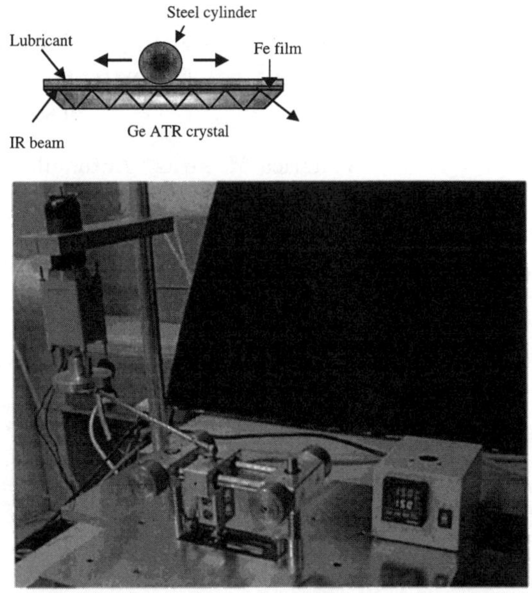

Figure 2. Schematic **(a, top)** and photograph **(b, bottom)** of the ATR tribometer for the in situ chemical analysis of tribological films.

to effectively be able to "see through" one of the rubbing surfaces (present as a thin iron film) and probe the lubricant layer. By means of this approach, changes in lubricant chemistry and the growth of reaction films have been studied in situ and as a function of sliding time and temperature.

To determine the usefulness of the in situ ATR tribometry, the experiments presented in this work have been performed in the presence of a well-studied system: ZnDTP on iron/steel.[18–22] Of particular interest in the ZnDTP lubricant–additive system studied is a greater understanding of the kinetics of formation of tribofilms and the relative roles of tribochemical and thermochemical processes.

Experimental Section

ATR Tribometer. An ATR FT-IR spectrometer has been equipped with a special tribometer (Figure 2), designed and constructed in our laboratory in Zürich. Trapezoid ATR elements of monocrystalline germanium with an angle of incidence of 45°, dimensions 72 × 10 × 6 mm (seven reflections), were used in this work. Germanium was chosen as the ATR element because of its favorable mechanical properties, high refractive index, and chemical resistance to many solvents. The crystal can be coated with different metals by magnetron sputtering, constituting one (fixed) sliding partner of the tribological system with a flat surface of 72 × 10 mm. The moving part of the tribological system is a fixed cylinder (diameter 10 mm, width 7 mm) sliding in a reciprocating motion across the metal-coated germanium crystal surface. The line contact between the cylinder and flat surface is thus swept back and forth along the entire length of the crystal. To move the cylinder along the crystal, an electric motor with an extremely high gear ratio (8640:1) is used. The resulting slow circular motion is transformed into sliding by an aluminum bar connected at one end to the plate of the motor and at the other end to the cylinder holder. The average sliding velocity of the cylinder is typically between 20 and 200 mm/min. The normal load on the contact region is applied by applying different weights on the aluminum rig. In this way, variations in the ATR FT-IR spectrum due to tribochemical reactions occurring between the lubricant and the contacting surfaces can be investigated as a function of tribological conditions.

Table 1. Transmission and ATR FT-IR Experimental Conditions

mode	detector	resolution (cm^{-1})	no. of scans	gain control	acquisition time	scan velocity
transmission	DTGS/KBr	2	32	1	1 min, 9 s	0.6329
ATR (Ge)	MCT/A	2	32	1	45 s	2.5317
ATR (Ge/Fe)	MCT/A	2	1024	8 (auto)	15 min	2.5317

Heating of the crystal surface of the ATR tribometer was performed by means of a heatable top-plate for the Benchmark-ATR (Portmann Instruments AG, Biel-Benken CH, Switzerland), heatable to 200 °C, with power provided by a temperature controller (FCS-23A, Shinko Technos Co., Ltd., Japan). The temperature was measured with a PT-100 thermocouple. In this way, both purely thermochemical reactions and tribochemical reactions can be studied at elevated temperatures.

The experiments in this work were performed with iron, sputter-deposited onto a Ge crystal, a steel cylinder as the sliding part, and commercial ZnDTP as a lubricant. The normal load applied onto the cylinder was 7 N, the Hertzian contact area 0.1 mm^2, the average pressure 34 MPa, and the mean sliding velocity 24 mm/min.

The tribological tests performed in standard tribometers and reported in the literature have been carried out under a range of tribological conditions, between 26 and 100 MPa of mean pressure, and between 100 and 750 mm/s sliding velocity.[18-21]

Thermal Tests. Thermal tests were performed in the ATR tribometer, without contact with the slider, under purely thermal conditions at room temperature, 80 °C, and 150 °C for up to 38 h. ATR FT-IR spectra were acquired periodically during the experiment, high-temperature thermal tests requiring cooling of the tribometer to room temperature prior to the spectra being acquired.

Tribological Tests. Tribological tests were carried out at room temperature, 80 °C, and 150 °C, while the steel cylinder was slid across the iron-coated germanium ATR crystal surface. Tribological tests at room temperature and 80 °C were performed for up to 90 h, whereas tribotests at 150 °C were performed for up to 20 h.

During the tribological tests the ATR spectra were collected periodically, after the tribometer was cooled to room temperature for measurements when the tribotest was carried out at high temperature.

Materials. The lubricant additive used was a commercial secondary ZnDTP (Hitec 7169, Ethyl Petroleum Additives International, England), purified by liquid chromatography. The composition of the purified secondary ZnDTP has been checked by both elemental analysis[23] and XPS quantitative analysis,[24] and found to correspond to the molecular formula $C_{18}H_{40}O_4P_2S_4$-Zn, suggesting a mixture of diisopropyl ZnDTP and hexyl ZnDTP. The thermal and tribological experiments reported in this work were carried out in the presence of the pure ZnDTP.

Transmission spectra of diisopropyl ZnDTP (pure, synthesized at the Institute Français du Petrole), $FePO_4 \cdot xH_2O$ (Alfa, Karlsruhe, Germany), and $Zn_3(PO_4)_2 \cdot xH_2O$ (Strem Chemicals, Newburyport, MA) were collected as references.

Cleaning of the ATR FT-IR tribometer was performed with cyclohexane p.a. (≥99.5%, Fluka, Buchs, Switzerland). The germanium crystals and the top-plates were cleaned with petroleum ether (technical grade) and ethanol p.a. (>99.8%) (Merck, Dietikon, Switzerland). To remove the Fe coating after each test, the germanium crystals were cleaned in a solution of 6 M HCl (HCl fuming 37%, puriss. p.a., Fluka), first with a soaked tissue and afterward by immersion in the HCl solution for at least 15 min.

Iron Coating. The iron was coated onto the germanium crystals, by means of magnetron sputtering, at the Paul Scherrer Institut (Villigen, Switzerland). A planar iron magnetron sputtering target (ISO 9001 Certified, target type PK 75) with a metallic purity >99.9% was used. The argon pressure during sputtering was 2.4×10^{-3} mbar, the sputtering rate being about 10 Å/s. The thickness and homogeneity of the iron coating were tested by ellipsometry prior to each run and found to be 12.0 ± 0.3 nm. The purity of the iron surface was checked by XPS, and the presence of a thin iron oxide film detected.

(23) Piras, F. M.; Rossi, A.; Spencer, N. D. In *Proceedings of the 28th Leeds/Lyon Symposium, Boundary and Mixed Lubrication: Science and Applications*; Dowson, D., et al., Eds.; 2002, p 72.

(24) Piras F. M.; Rossi, A.; Spencer, N. D., submitted.

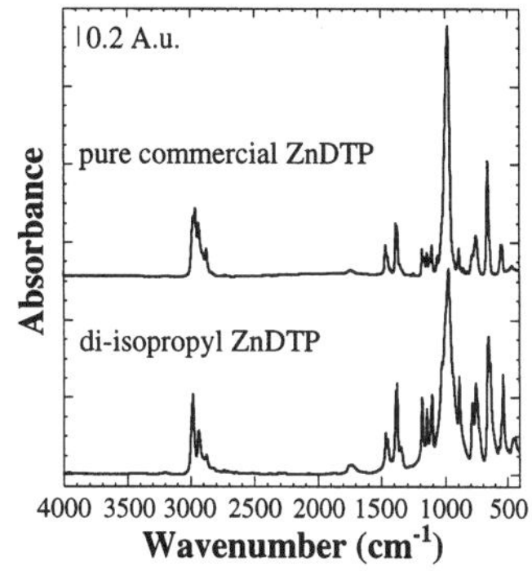

Figure 3. Transmission FT-IR spectra of diisopropyl ZnDTP and commercial ZnDTP.

XPS analysis was carried out on the iron-coated germanium crystals after each experiment. After sliding at 150 °C on the iron surface, three areas were visible (contact area/wear scar/noncontact area), differing both in morphology and in chemical composition, as indicated by imaging XPS.[24]

FT-IR Spectrometer. Transmission FT-IR and ATR spectra were obtained with a single-beam Nicolet Magna-IR System 550 Fourier transform spectrometer, equipped with a Greasby-Specac advanced overhead (specaflow) 1401 Series ATR system. The experimental conditions are listed in Table 1. Sampling in transmission was performed by placing one drop of the pure samples between KBr windows.

Data Processing of ATR FT-IR Spectra. A background correction always had to be applied to the experimental spectra, due to the single-beam acquisition mode. The background was acquired at the beginning of each experiment.

In the case of the ATR spectra that were to be compared with transmission spectra (only Figure 4), an ATR correction routine was used to allow for the variation in penetration depth by multiplying the sample spectrum by a wavelength-dependent factor to correct the relative peak intensities.[25] The standard ZnDTP ATR spectra collected on an iron-coated germanium crystal were corrected for the baseline after the ATR correction (only Figure 4). All the other ATR spectra presented in this work (Figures 5–7) are reported without ATR correction, as acquired.

Results

Transmission and ATR FT-IR Spectra. The transmission FT-IR spectra of the diisopropyl ZnDTP reference compound and the commercial ZnDTP (Figure 3) clearly reveal the characteristic peaks of the ZnDTP molecule. The IR peak assignments for the transmission spectra of ZnDTP are reported in Table 2. Four regions can be distinguished:

(1) the region at higher wavenumbers (around 2900 cm^{-1})—symmetric and asymmetric stretching vibrations of CH_3, CH_2, and CH;

(2) the region at around 1400 cm^{-1}—bending modes of CH_3 groups;

(3) the region at around 1000 cm^{-1} with the most intense peak at 978 cm^{-1} (commercial ZnDTP) and 971 cm^{-1} (diisopropyl ZnDTP) of the P–O–(C) bond;

(25) Nicolet's OMNIC ESP spectroscopy software, version 4.1a.

Table 2. IR Frequencies (cm^{-1}) and Functional Groups for the Transmission FT-IR Spectrum of Diisopropyl ZnDTP and for the Transmission and ATR FT-IR Spectra of Commercial ZnDTP

diisopropyl ZnDTP		commercial ZnDTP			
transmission	functional group	transmission	ATR (Ge)	ATR (Ge/Fe, 10 nm)	functional group
2978, 2931, 2872 s/m	$\nu_{as}(CH_3)$, $\nu(CH)$, $\nu_s(CH_3)$[26,31]	2976, 2958, 2934, 2871 m/s	2976, 2958, 2934, 2871 m/s	2974, 2955, 2930, 2870 w/m	$\nu_{as}(CH_3)$, $\nu_{as}(CH_2)$, $\nu(CH)$, $\nu_s(CH_3)$[26,31]
1466, 1452 m	$\delta_{as}(CH_3)$[29]	1468 m,1454 (sh)	1467 m, 1454 (sh)	1463, 1451 w	$\delta_{as}(CH_3)$[29]
1386,1375 s1347 m	$\delta_s(CH_3)$[26,29,31,a]	1386, 1373 m	1385, 1373 m	1383, 1371 w	$\delta_s(CH_3)$[26,29,31,a]
1180, 1142,1102 s	$\nu(CO(P))$[27,28,a]	1178, 1160[b], 1141, 1120, 1103[b] m	1178,1160,[b] 1141, 1120, 1103[b] m	1176, 1158,[b] 1140, 1120, 1099[b] w	$\nu(CO(P))$[27,28,a]
1024 (sh), 997(sh), 971 vs	$\nu(PO(C))$[26−28]	1022 (sh),978 vs	1022 (sh),969 vs	1022 vw,961 s	$\nu(PO(C))$[26−28]
887 s	$\nu(CC)$[28,33]	889 m	889 m	886w	$\nu(CC)$[28,33]
651, 636 s	$\nu as(PS)$[26,27,30]	658 s			$\nu as(PS)$[26,27,30]
546 (sh),532 s	$\nu s(PS)$[26,27,30]	539 m br			$\nu s(PS)$[26,27,30]

[a]Characteristics of the isopropyl group.[26,28] [b] Peaks not yet assigned.

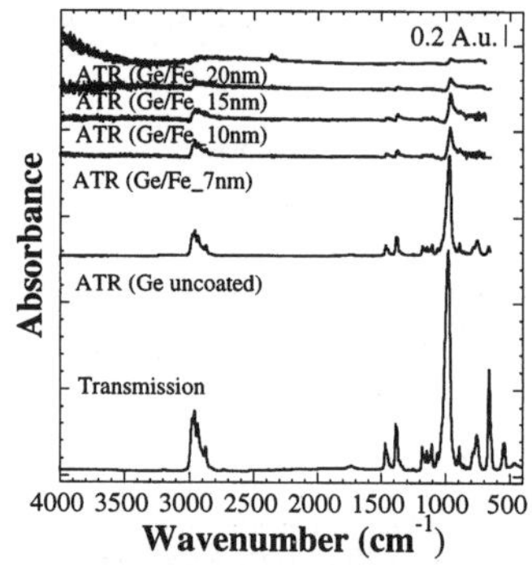

Figure 4. Transmission FT-IR and ATR spectra of commercial ZnDTP. The ATR spectra have been collected on an uncoated germanium crystal and on iron-coated germanium crystals; the thicknesses of the iron films were 7, 10, 15, and 20 nm.

(4) the region below 700 cm^{-1}—stretching P−S bands.

All vibrational frequencies assigned in Table 2 are in agreement with the literature.[26−31] Some minor peaks have not yet been assigned. By comparing the transmission spectra of the model compound diisopropyl ZnDTP with the commercial ZnDTP, it can be observed that all the characteristic peaks of the diisopropyl ZnDTP compound are also present in the spectrum of the commercial product (Table 2). However, differences are observed in the positions and intensities of the peaks, and the commercial ZnDTP spectrum shows more peaks in the regions around 2900 and 1200 cm^{-1}.

A comparison of the transmission spectra of commercial ZnDTP with the ATR FT-IR spectra on uncoated and on iron-coated (7−20 nm) germanium ATR crystals are shown in Figure 4. The IR peak assignment for commercial ZnDTP ATR FT-IR spectra are reported in Table 2 (columns 4 and 5). As can be seen in Figure 4, all the characteristic peaks of the ZnDTP transmission spectrum are also present in the ATR mode, with the exception of the peaks in the region below 650 cm^{-1}, where the strong absorbance of germanium obscures all other signals. Thus, the P−S region (Table 2) cannot be analyzed by ATR FT-IR if a germanium ATR element is involved. In addition, in the ATR spectrum collected with an uncoated germanium crystal, the peak assigned to the P−O−(C) stretching vibration is shifted to lower wavenumbers (969 cm^{-1}) compared to that in the transmission spectrum (978 cm^{-1}). This shift to lower wavenumbers is even more pronounced in the ATR spectrum collected with the (7 nm) iron-coated germanium crystal (963 cm^{-1}). Furthermore, this peak is skewed toward *low* wavenumbers in the ATR (Ge/Fe) spectrum, while it is skewed toward *high* wavenumbers in the other two spectra. All signals in the ATR spectrum collected on the iron-coated germanium crystal show a shoulder at lower wavenumbers, and their relative intensities are different with respect to the ATR spectrum collected on an uncoated germanium crystal. This effect

(26) Gallopoulos, N. E. *Prepr.—Am. Chem. Soc., Div. Pet. Chem.* **1966**, *11*, 21.
(27) Harrison, P. G.; Begley, M. J.; Kikabhai, T. *J. Chem. Soc., Dalton Trans.* **1986**, *5*, 929.
(28) Thomas, L. C.; Chittenden, R. A. *Spectrochim. Acta* **1964**, *20*, 489.
(29) Nadler, M. P.; Nissan, R. A.; Hollins, R. A. *Appl. Spectrosc.* **1988**, *42*, 634.
(30) Jiang, S.; Dasgupta, S.; Blanco, M.; Frazier, R.; Yamaguchi, E. S.; Tang, Y.; Goddar, W. A., III. *J. Phys. Chem.* **1996**, *100*, 15760.
(31) Lin-Vien, D.; Colthup, N. B.; Fateley, W. G.; Grasselli, J. G. *The Handbook of Infrared and Raman Characteristic Frequencies of Organic Molecules*; Academic Press Inc.: Boston, 1991; Chapter 16.

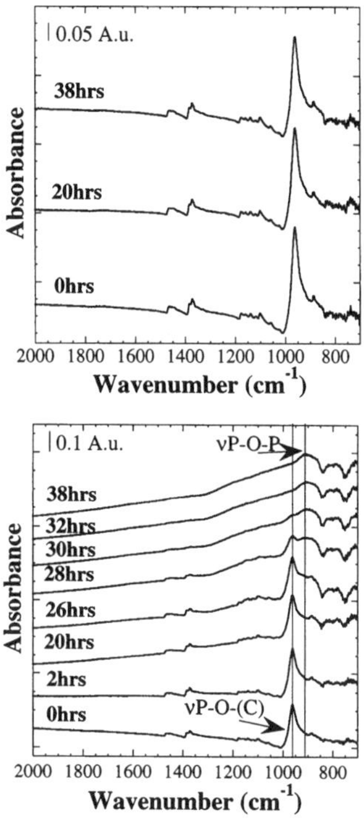

Figure 5. Commercial ZnDTP ATR spectra collected during thermal tests at room temperature (a, top) and at 150 °C (b, bottom) on germanium ATR crystals coated with iron (10 nm). ATR and baseline corrections have not been applied.

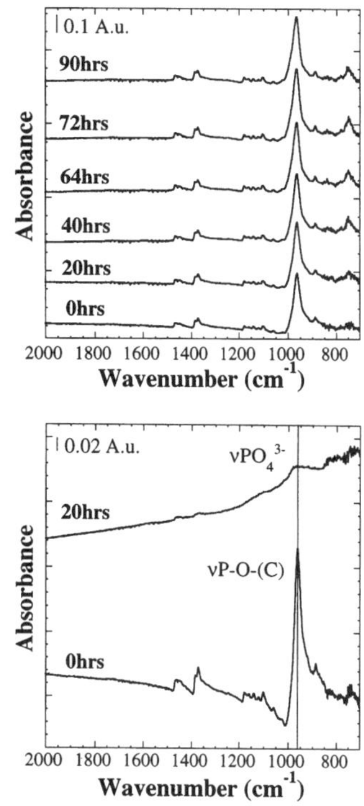

Figure 6. ATR spectra of commercial ZnDTP ATR acquired during tribological tests at room temperature (a, top) and at 150 °C (b, bottom) on iron-coated (10 nm) germanium ATR crystals. ATR and baseline corrections have not been applied.

has been confirmed by studying the influence of the thickness of the iron coating on the ATR FT-IR spectra of commercial ZnDTP (Figure 4). The intensities of the characteristic peaks of ZnDTP decrease with increasing thickness of the iron coating, and the peak assigned to the P–O–(C) stretching vibration (978 cm^{-1}) is shifted to progressively lower wavenumbers (960 cm^{-1} for a 20 nm iron coating). Germanium crystals coated with a 10 nm thick iron film have been used to perform the adsorption and tribotests reported in this work. It has been found that the thickness of 10 nm for the iron film is ideal, since it is thick enough to prevent damage to the germanium crystal surface during sliding, but thin enough to allow ATR measurements of the samples deposited on the iron surface.

Thermal Tests. ATR FT-IR spectra were collected during thermal tests of commercial ZnDTP on a Ge crystal, coated with 10 nm of iron. At room temperature (Figure 5a), no differences in the spectra in either intensity or peak position were found up to 38 h. The same behavior was found for a thermal test at 80 °C (not shown).

Working at higher temperatures becomes increasingly difficult due to the strong reduction in the transmittance of the Ge crystal, since the useful transmission range becomes narrower with increasing temperature. At 150 °C, germanium becomes opaque at wavenumbers below 1600 cm^{-1}. As is shown in Figure 7, the temperature effect is more pronounced when the Ge crystal is coated with an iron film; at 150 °C the transmittance of germanium is almost zero over the whole spectral range. Thus, when adsorption or tribological reactions are studied at temperatures above ca. 150 °C, the whole system has to be cooled to near room temperature to perform ATR FT-IR measurements.

In Figure 5b, the ZnDTP ATR FT-IR spectra collected at different times during a thermal test performed at a temperature of 150 °C on a Ge crystal coated with Fe (10 nm) are shown. The spectrum of commercial ZnDTP at the beginning of the experiment (0 h) shows the characteristic peaks of the ZnDTP molecule (Figure 4). The P–O–(C) stretching vibration peak (961 cm^{-1}) can be clearly seen. The spectra recorded after 26–38 h at 150 °C show clear differences: the strong ν(P–O–(C)) peak disappears, while new bands appear in the same wavenumber region. The spectrum after heating at $T = 150$ °C for 38 h exhibits a broad band in the region around 1100 cm^{-1}, indicating a modification of the molecule. In the functional group region (around 1400 cm^{-1}) the alkane group bending peaks are no longer detected. The peak at ~916 cm^{-1} is assigned to ν(P–O–P), and the band around 1100 cm^{-1} probably results from the overlap of the peaks assigned to ν(P=O) and ν(P–O), which, typically, fall in the regions 1320–1140 and 950–1060 cm^{-1}, respectively.[31–33] At this stage, a complete explanation for the change of the baseline slope over time has not been found. It is probably due to the formation of a further layer (thermal or tribological film) at the iron film/ZnDTP interface, since the same background spectrum was used for all spectra.

Tribological Tests. The ATR FT-IR spectra of ZnDTP on an iron-coated germanium crystal, recorded during tribotesting at room temperature, did not reveal any changes up to 90 h of sliding time (Figure 6a), the same

(32) Harrison, P. G.; Brown, P. *Wear* **1991**, *148*, 123.
(33) Thomas, L. C. *The Identification of functional Groups in Organophosphorus Compounds*; Academic Press: New York, 1974.

Table 3. IR Frequencies (cm^{-1}) and Functional Groups for the Transmission FT-IR Spectra of Iron Phosphate and Zinc Phosphate and for the ATR FT-IR Spectrum of Commercial ZnDTP after Tribochemical Reaction at 150 °C

$Zn_3(PO_4)_2{\cdot}xH_2O$	$FePO_4{\cdot}xH_2O$	ZnDTP after tribotest at 150 °C	functional group
		2955, 2925, 2868 m	ν(alkane groups)
		1466, 1452 w	$\delta_{as}(CH_3)$
		1375 vw	$\delta_s(CH_3)$
br band:1488−772,1105, 1067 (sh) vs	br band:1416−820,1117 (sh) s	br band:1241−860,1101 w	$\nu(PO_4^{3-})$
1015 vs	1019 vs		
938 vs		972 m	

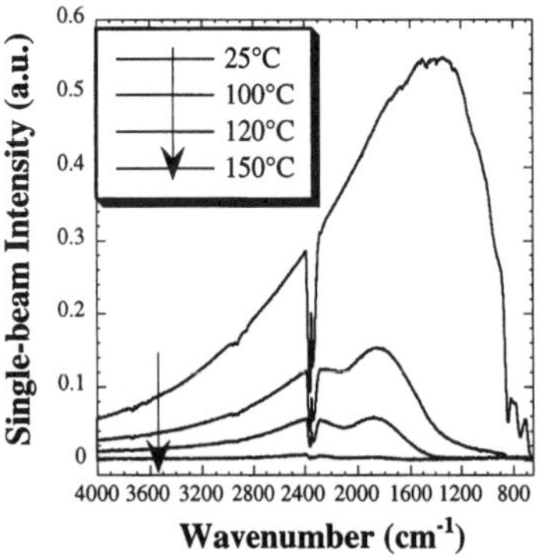

Figure 7. Temperature effect on the iron-coated (10 nm) germanium ATR spectrum.

result being found for tribological tests performed at 80 °C (spectra not shown). Only in the tribological test performed at 150 °C were clear differences observed (Figure 6b). The spectrum collected at the beginning of the test (0 h) shows the expected characteristic peaks of the ZnDTP molecule on iron (Table 2). After 20 h of sliding in the ATR tribometer at 150 °C the spectrum shows two new bands with maximum intensities at 1102 and 972 cm^{-1}, assigned to PO_4^{3-} stretching.[31−34] Transmission spectra collected on $FePO_4{\cdot}xH_2O$ and $Zn_3(PO_4)_2{\cdot}xH_2O$ confirmed this assignment (Table 3).

XPS analysis collected on the same sample supports the formation of phosphate in the contact area. Ge signals were neither detected in the contact area nor inside the wear scar.

Discussion

Spectra obtained with the in situ ATR FT-IR tribometer allow changes in the lubricant chemistry to be monitored upon adsorption of additives on iron, as well as the growth of reaction films to be followed as a function of sliding time and temperature in tribological experiments. To allow a correct interpretation, the spectra obtained in the in situ ATR FT-IR tribological apparatus were compared to traditional transmission spectra.

Comparison between Transmission and ATR FT-IR Spectra. By comparing the transmission and ATR FT-IR spectra of commercial ZnDTP reported in this work (Figure 4), it appears that the ATR spectra collected on the uncoated germanium ATR crystal closely resemble those obtained via transmission. Differences were noticed in the P−O−(C) peak position, however, this peak being shifted to lower wavenumbers in the ATR (Ge/Fe, 10 nm) spectrum (Table 2). A reason for this shift may be the adsorption of the ZnDTP onto the iron surface. Unfortunately the low-wavenumber region, where the Fe−S interaction may be detected to confirm this result, is not accessible. A shift of the peak positions in ATR spectra is expected to be caused by the wavelength dependence of the penetration depth (d_p) and the effective thickness (d_e) that is analyzed. The concept of d_e as the interaction of the evanescent field with the sample was introduced by Harrick.[7] The effective thickness represents the actual thickness of a film that would be required to obtain the same absorption in a transmission measurement as that obtained by means of an evanescent wave. For bulk materials in which the thickness of the rarer medium is much greater than the penetration depth of the evanescent field, d_e is proportional to d_p. Since d_p is in turn proportional to the wavelength, d_e also increases with wavelength. As a result of the consequent distorting effect on peak shapes, broad absorption bands in ATR spectra of bulk materials show a shift to longer wavelengths (lower wavenumbers).[7] The bands at longer wavelengths are also relatively more intense.

There is another effect that should be taken into account when ATR spectra are compared to transmission spectra: the variation of the refractive index across an absorption band. This phenomenon is known as dispersion, which affects the spectra for internal reflection, contributing to an increase in band intensity, a slight shift in the band position, and a distortion of the band shape.[35] The ATR spectra reported in this work should not be distorted by dispersion, however, since they were obtained at an incidence angle well above the critical angle.

Influence of the Thin Iron Coating. ZnDTP ATR spectra collected on a germanium crystal coated with Fe show a further shift of the P−O−(C) absorption band to lower wavenumbers, and the distortion of this band on the lower-wavenumber side. Furthermore, all the bands in these spectra show a distortion at lower wavenumbers and differences in relative peak intensities (Figure 4).

From data reported in the literature,[36] the optical properties of iron change over the wavelength (wavenumber) region investigated in this work. In particular, at low wavelengths (high wavenumbers), the refractive index of iron is close to that of germanium, while the refractive index increases as the wavelength increases. For wavelengths at which the refractive index of Fe and Ge are the same, these layers are optically indistinguishable. When the refractive indexes of Fe and Ge are different, a three-layered system has to be considered. More theoretical studies (e.g., electric-field analysis) are necessary to understand and determine how optical constants influence ATR spectra in the case of a stratified multilayered system where an iron film is the intermediate layer. In a very recent paper[37] on the influence of the optical properties of metals on ATR band shape at the metal−liquid interface, it has been reported that the

(34) Shagidullin, R. R.; Chernova, A. V.; Vinogradova, V. S.; Mukhametov, F. S. *Atlas of IR Spectra of Organophosphorus Compounds*; Kluwer Academic Publishers: Moscow, 1990.

(35) Belali, R.; Vigoreux, J. M.; Morvan, J. *J. Opt. Soc. Am. B* **1995**, *12*, 2377.

(36) Ordal, M. A.; Bell, R. J.; Alexander, R. W., Jr.; Nequist, L. A.; Querry, M. R. *Appl. Opt.* **1988**, *27*, *6*, 1203.

(37) Bürgi, T. *Phys. Chem. Chem. Phys.* **2001**, *3*, 2124.

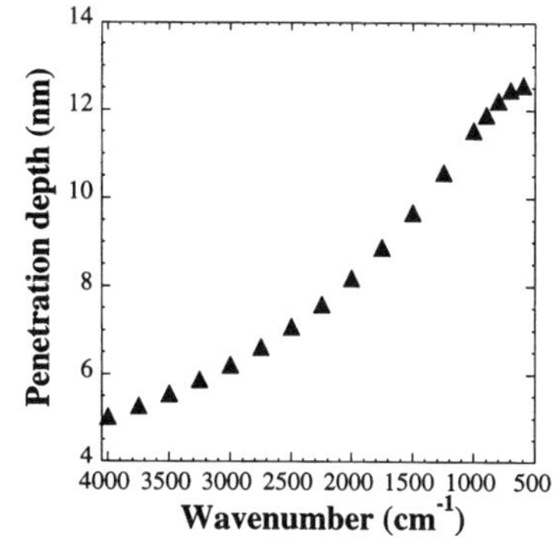

Figure 8. Penetration depth (nm) for the IR beam into the ZnDTP for the three-layered system Ge crystal/ Fe film/ZnDTP as a function of the wavenumber (cm^{-1}).

Chart 1

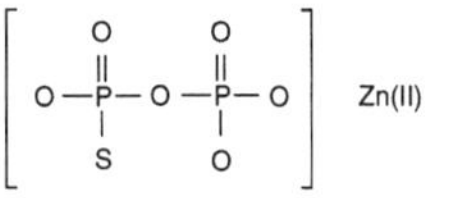

distortion of ATR increases with increasing thickness, increasing the refractive index and decreasing the absorption coefficient of the metal film. To have an initial estimate of the thickness of the film probed when a germanium crystal is coated with an iron film, the penetration depth (d_p) of the infrared beam into the ZnDTP for the three-layered system germanium crystal/iron coating/ZnDTP has been calculated according to the approach of Müller and Abraham-Fuchs.[37] The calculated penetration depth as a function of the wavenumber in the mid-IR range (4000–500 cm^{-1}) ranges from 5 to 13 nm (Figure 8).

Thermal vs Tribochemical Effect. Of particular interest in the interaction of ZnDTP with iron and steel are the relative roles of tribochemical and thermochemical processes. Some authors have reported that tribochemical reactions are simply due to the high temperature induced during metal–metal contact (by means of plastic deformation, etc.),[39] while other authors believe that there is something unique about the high-pressure–high-shear situation existing in tribological systems.[19] In fact, the ATR experiments performed in this work did not reveal chemical reactions of commercial ZnDTP with iron during simple adsorption at room temperature (Figure 5a) or at 80 °C. Reactions were neither detected during tribological tests (sliding) at room temperature (Figure 6a) nor at 80 °C despite test durations of up to 90 h. While in the case of the simple thermal experiment, this can be explained by a lack of the necessary activation energy; for the tribological sliding experiments this might be due to insufficient local pressure or an insufficient number of cycles.

During a *thermal test* at 150 °C, the ATR analysis of the ZnDTP surface reaction on an iron-coated germanium crystal indicates a rearrangement of the molecule with the formation of P–O–P bonds. The proposed structure for the thermochemical reaction product, supported by the XPS results collected on the same samples,[24] is given in Chart 1.

The molecular weight of the reaction product has not been determined at this stage, and thus a real chemical structure is not proposed, but the schematic of the reaction product contains all the functional groups detected by ATR and XPS analysis.[24] The decrease in intensity of the alkane C–H bend peaks, until their disappearance after 38 h of heating (Figure 5b), is presumably due to the elimination of alkenes during the thermal decomposition of ZnDTP. The results of the present study are supported by the mechanism proposed in the literature,[40,41] and the ATR results are in agreement with the infrared studies reported by Harrison and Brown.[32]

After the *tribological test* performed at 150 °C, bands with maximum intensities at 1102 and 972 cm^{-1} were present. They are assigned to PO_4^{3-} stretching, indicating the formation of phosphates (Table 3) as a product of the tribochemical reaction at 150 °C.

The tribological reaction seems to be faster than the thermal reaction, since after 20 h of heating only, no significant changes in the ZnDTP ATR spectrum were detected (Figure 5b). In contrast, the spectrum collected after the same time but during sliding at the same temperature shows the formation of an inorganic phosphate film (Figure 6b). These results, in agreement with the X-ray absorption and photoelectron spectroscopic studies of ZnDTP tribofilms formed at 150 °C,[42] indicate that the thermochemical and the tribochemical reactions follow two different mechanisms with different kinetics.

Some models for ZnDTP antiwear films reported in the literature[19–21] propose a multilayered structure where a polyphosphate film is responsible for the wear reduction. The results reported in this work do not indicate formation of P–O–P species, characteristic of polyphosphates, during up to 20 h of sliding at 150 °C. It has to be remembered, however, that the tribological conditions used in this work (normal load, mean pressure, and sliding velocity) are milder than those reported in the literature,[19–21] and that pure ZnDTP has been used instead of a solution in mineral oil.

It may be suggested that, at the extent of reaction probed in this study, the tribochemical reaction is not complete, and the formation of phosphate is at an intermediate stage. Further experiments at longer sliding times and/or higher normal loads are required to understand more fully the mechanisms of thermochemical and tribochemical processes and their relative roles.

Conclusions

ATR FT-IR spectroscopy appears to be a powerful and useful technique for the in situ monitoring of tribochemical film formation. By measuring ATR FT-IR spectra from the underside of a thin iron film, the chemistry of the iron/ZnDTP interface has been analyzed as a function of time under both purely thermal and tribological conditions.

In situ ATR thermal and tribological experiments have been performed in the presence of a pure secondary ZnDTP at room temperature, 80 °C, and 150 °C. No thermal or tribochemical reactions of ZnDTP with iron were detected during simple thermal tests at room temperature and at 80 °C.

The ATR results on the surface reaction of ZnDTP on iron under purely thermal conditions at 150 °C indicate a rearrangement of the molecule, with the formation of P–O–P species. The formation of a phosphate film from

(38) Müller, G. J.; Abraham-Fuchs, K. *Optik* **1991**, *88*, *3*, 83.
(39) Tysoe, W. T.; Surerus, K.; Lara, J.; Blunt, T. J.; Kotvis, P. V. *Tribol. Lett.* **1995**, *1*, 39.
(40) Dickert, J. J.; Rowe, C. N. *J. Org. Chem.* **1967**, *32*, 647.
(41) Coy, R. C.; Jones, R. B. *ASLE Trans.* **1980**, *24*, *1*, 77. Jones, B.; Coy, R. C. *ASLE Trans.* **1981**, *24*, *1*, 91.
(42) Kasrai, M.; Fuller, M.; Scaini, M.; Yin, Z.; Brunner, R. W.; Bancroft, G. M.; Fleet, M. E.; Fyfe, K.; Tan, K. H. In *Proceedings of the 23rd Leeds / Lyon Symposium, Lubricants and Lubrication*; Dowson, D., et al., Eds.; Elsevier Science Publishers B.V.: Amsterdam, 1995; p 659.

ZnDTP after friction at 150 °C (tribotest) is indicated by the ATR results.

In the present study we show the utility of the new approach for examining the interface in tribological systems in situ and for understanding the kinetics of formation of tribofilms, as well as the relative roles of tribochemical and thermochemical processes. Current work extends the study to diluted solutions of ZnDTP in mineral oil to simulate real working conditions.

Acknowledgment. Financial support of the ETH and Italian MURST (ex 40% grant to A.R.) and Regione Autonoma della Sardegna is gratefully acknowledged. Prof. J. M. Martin (Ecole Centrale de Lyon, France) and Dr. H. Camenzind (Ciba Specialty Chemicals, Switzerland) are thanked for supplying the pure additives. Mr. M. Horisberger kindly prepared the iron coatings by magnetron sputtering at PSI (Villigen, Switzerland). Prof. G. W. Stachowiak and Dr. C. Soto are thanked for their help in the mechanical design of the ATR tribometer. Finally, we thank Prof. J. R. Ferraro for his valuable advice and helpful discussions.

LA0202733

Tribology Letters, Vol. 15, No. 3, October 2003 (© 2003) 193

A combinatorial approach to elucidating tribochemical mechanisms

Michael Eglin[a,*], Antonella Rossi[a,b] and Nicholas D. Spencer[a]

[a]Laboratory for Surface Science and Technology, Department of Materials, Swiss Federal Institute of Technology, ETH Zürich, CH-8092 Zürich, Switzerland

[b]Department of Inorganic and Analytical Chemistry, University of Cagliari, INSTM, Cagliari, Italy

Received 15 December 2002; accepted 23 February 2003

A new type of combinatorial tribological experiment is presented, which explores a series of tribological conditions, such as load and relative velocity, spatially separated as a "library" on one single sample. As an example, a library displaying the results of tribological testing of an additive under a series of different loads has been prepared and analyzed. The tribological information acquired during the testing has been correlated with spectroscopic information from the tribologically stressed surface. The use of imaging and small-area X-ray photoelectron spectroscopy has allowed the identification of the different tribologically stressed areas and the acquisition of detailed spectroscopic information. The composition and the thickness of the tribofilm were found to be dependent on the applied load. The use of the combinatorial approach shows the potential to greatly facilitate rapid characterization of new lubricant additives.

KEY WORDS: tribochemistry, boundary lubrication, combinatorial tribochemistry

1. Introduction

The combinatorial synthetic approach, together with high-throughput screening of compounds, has been applied in pharmaceutical chemistry since the early eighties. A large number of molecules are synthesized in parallel and subsequently probed for chemical, physical and medicinal properties. This approach was first brought into materials science in the field of high-temperature superconductors [1], where a spatially addressable library of potential candidates for high-T_c was fabricated and tested and it has been adapted to different fields in materials science, such as semiconductors [2] or metallurgy [3]. The advantage of the combinatorial synthetic approach in both chemistry and materials science is the rapid production of many substances of varying compositions which are subsequently analyzed in a massively parallel way. This methodology speeds up the search for new substances with the desired properties.

In tribology, and especially in tribochemistry, one has to deal with a large parameter space, since the friction, wear and tribochemical reactions of a given tribological system have been shown to depend strongly on the applied conditions (e.g. relative velocity, contact pressure, sliding time, temperature) [4]. To understand the reactions that can occur in this system it is necessary to explore a significant portion of the parameter space. To date, this has been a very time-consuming process, involving an enormous number of experiments, suggesting that a variant of the combinatorial analytical approach could also be useful in a tribological context.

A wide variety of surface-analytical methods has been used to investigate tribochemical reaction products. X-ray photoelectron spectroscopy (XPS), scanning Auger microscopy (SAM), time-of-flight secondary ion mass spectroscopy (ToF-SIMS) and X-ray absorption near-edge structure (XANES) are frequently used methods for the chemical characterization of tribochemical reaction films. Most of these methods can combine imaging and spectroscopy. While scanning Auger electron microscopy (SAM) has been used to map elemental distribution of a contact area [5], the use of imaging XPS (i-XPS) has been shown recently to be of value in investigating the distribution of chemical species according to their chemical state [6,7]. So far, the imaging possibilities of these methods were used on relatively simple systems, allowing the contact and non-contact regions to be distinguished. By carrying out a combinatorial experiment, one can take further advantage of the imaging capability. In this type of experiment a parameter library is built, applying various tribological conditions (as a function of the lateral position on the disk) on a single sample, which is subsequently analyzed by an imaging surface analytical technique. The tribological information (coefficient of friction, wear) and the spectroscopic results can afterwards be mapped onto the parameter library (see figure 1). The correlation of this data should provide insight into the application range of a lubricant additive system and help uncover mechanistic reaction pathways.

Only a very little work has been done applying a combinatorial approach to tribological problems. Green

*To whom correspondence should be addressed. E-mail: michael.eglin@mat.ethz.ch

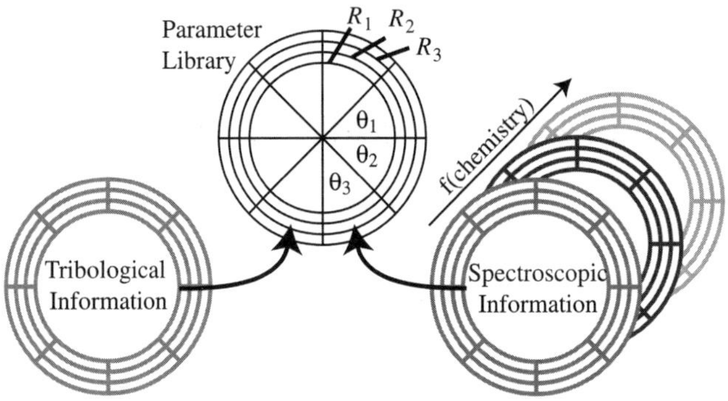

Figure 1. General principle of the combinatorial approach: a parameter library is produced by varying the tribological stress on a single sample depending on the lateral e.g. radial (R) and angular (θ) position on the disk. The tribological information acquired during or after the test (coefficient of friction, wear) and the spectroscopic information (chemistry of surface film) from surface-analytical examination of the sample can then be mapped onto this parameter library and correlated.

and Lee used an AFM with chemically patterned cantilevers and tip arrays to probe adhesive forces between carboxylic acid, alcohol and methyl groups [8]. This approach reveals tribological information on the nanometer scale, whereas the tribological load scanner introduced by Hogmark uses a macroscopic, crossed-cylinder configuration [9]. In the latter setup, two elongated cylinders repeatedly slide across each other with varying load in such a manner that each point along the sliding track of both cylinders experiences a unique load. This test was used in the evaluation of hard coatings.

A model system (di-isopropyl zinc dithiophosphate (i-ZnDTP) in decane) has been chosen as a lubricant-additive system in the present study for two reasons: firstly, the tribological system was to be kept as well-defined as possible for the first experiments with this new approach, and secondly because ZnDTP is a well-studied system, which readily allows comparison of our own results with the literature. The formation of tribofilms from ZnDTP has been studied extensively and has been described by several authors [4,10–12].

2. Experimental

Steel disks (AISI 52100) were polished to a final surface roughness (R_a) below 10 nm using silicon carbide paper and diamond paste. The samples were ultrasonically cleaned in ethanol and analyzed for surface contamination by XPS immediately prior to tribotesting. Commercial 4 mm ball-bearings (AISI 52100) were used as a counter-face.

A 1 wt.% solution of di-isopropyl zinc dithiophosphate (i-ZnDTP) in decane was used as a lubricant. To dissolve the additive in the solvent, the solution was stirred at 60 °C for 30 min. The tribotests were carried out at room temperature (24 ± 0.5 °C) and the relative humidity of the air was recorded during each test and found to be always between 22 and 38%. The sample was fully immersed in the lubricant during testing. A CETR 2 tribometer (Center for Tribology, Inc., Campbell, CA, USA) was used to run the tests using a ball-on-disk geometry. This tribometer allows the independent programming of normal load, velocity, radius and duration. Prior to the actual tests, running-in of the ball was performed at 5 N load with a speed of 31.4 mm/min for 2 h, outside the region to be used for XPS analysis and in the presence of the test lubricant. The running-in was performed in order to create a flat spot on the bottom of the ball. This flat spot was then placed in contact with the surface for the subsequent runs and defined the apparent contact area.

In figure 2, a schematic of the combinatorial tribotest is presented. The experiment consisted of producing five concentric tribologically stressed annuli on a single sample; in all cases a speed of 31.4 mm/s was used. In each annulus a different load (ranging from 0.05 to 5 N) was applied. Each annulus consisted of 11 overlapping wear tracks, with radii differing from each other by 25 μm (see insert of figure 2). The width of each wear track was defined by the flat spot produced during the running-in. It was found to have a diameter of approximately 150 μm. The 11 wear tracks created together an annular test region spanning more than 250 μm. Before the XPS measurements the sample was ultrasonically cleaned in cyclohexane. On the tribologically stressed sample, an O(1s) map was collected, which evidences the different wear regions. Within each tested area a small-area XPS analysis was carried out. The analyzed area consisted of a spot of 120 μm diameter, and thus completely inside the tribologically stressed area.

The XPS analyses were performed on a PHI 5700 system with an Omni Focus IV lens system. Spectroscopic maps were acquired using the imaging capabilities of the lens system. The analyzed spot (diameter 120 μm) was electrostatically rastered over the sample (typically 64×64 pixels, 2×2 mm). For each pixel a full spectrum of the selected energy region was acquired. The typical size of an XPS map with respect to the tested

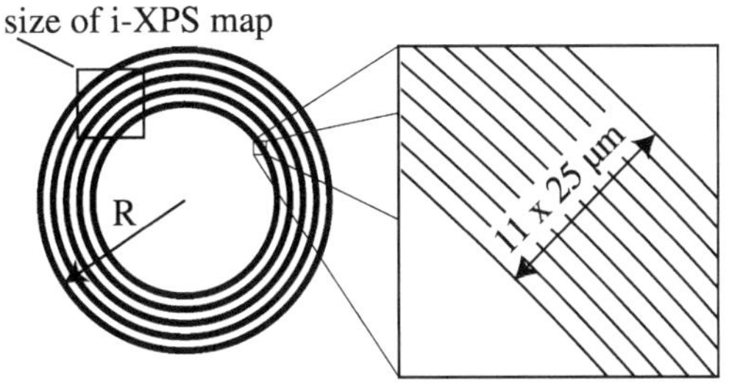

Figure 2. Schematic of the tribotest: the tests were performed in 5 concentric annuli consisting of 11 single tracks, each with a radius differing by 25 μm. The wear tracks are partially overlapping due to the width of the apparent contact area, which is defined by the flat spot (⌀150 μm) produced on the ball during the running-in of the tribotest. For each annulus, a different load (L) was applied ($R = 3.5$ mm, $L = 0.05$ N; $R = 4$ mm, $L = 0.1$ N; $R = 4.5$ mm, $L = 0.5$ N; $R = 5$ mm, $L = 1$ N; $R = 5.5$ mm, $L = 5$ N). The square in the top left corner represents the area analyzed by i-XPS.

areas is displayed in figure 2. The acquired spectroscopic maps were processed with PHI Multipak (V6.0) software.

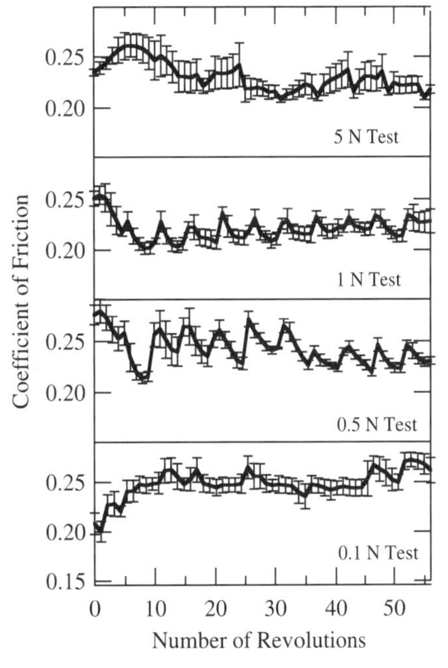

Figure 3. Coefficient of friction (COF) of a tribotest for the loads from 5 to 0.1 N. The COF is averaged over 1 full turn of the disk; the error bars represent the standard deviation during one turn. The increase of the COF after the 25 μm steps after each fifth turn can be seen (see text).

3. Results

In figure 3, the coefficient of friction (COF) acquired during a test is displayed versus the number of turns for loads from 5 to 0.1 N. The COF is averaged over one full turn of the disk and displayed versus the turn number. The error bars represent the standard deviation during one full turn. It can be noticed that at 0.5 and 1 N the COF decreases after the beginning of the test and shows a sudden increase after each fifth turn. This increase coincides with the 25 μm steps that occur after every five turns. After the increase, the COF decreases again and at the end of a five-turn cycle the COF has reached a steady state. This steady state is assumed to indicate that a tribofilm has been formed that is representative of the applied conditions. Thus the average of the COF during the last full turn of the test is given in table 1 as a representation of the tribological properties of the tribofilm. The COF shows a small decrease with increasing load.

The morphology of the wear tracks and the ball used for the tribotest were examined with an optical microscope. The ball shows a circular flat spot with a diameter of 150 μm, which was worn off mainly during the running-in period of the tribotest. It defines the apparent contact area during subsequent tests. On the tribologically stressed disk, the areas tested with 5 and 1 N are clearly visible by optical microscopy, but for lower loads no differences can be recognized between the contact and the non-contact area. Without additional information from imaging XPS it would be difficult to locate the areas tested at lower loads.

The total O(1s) intensity map of the tribologically stressed sample is presented in figure 4(a). Arcs of higher intensity can be seen running from the bottom left to top right, indicating the location of the tribologically stressed annuli. Each pixel of the O(1s) map contains the entire spectroscopic information for the O(1s) region. The spectra of the most intense and the least intense areas are extracted and displayed in figure 4(b). The dark areas (which correspond to a lower oxygen signal intensity) reveal a spectrum that is characteristic for the surface of an oxidized steel, showing a peak at 530 eV and a shoulder at a binding energy that is approximately 1.5 eV higher than the main peak. The bright areas show a different peak shape in the O(1s) region. The main peak is found at 531.7 eV plus a shoulder at 530 eV. The peak at 531.7 eV is characteristic for oxygen bound in a phosphate group, while the shoulder is due to contributing oxide. In the following, this shape of the O(1s) spectrum is referred to as being of the phosphate type.

The O(1s) map was further processed with the linear-least-squares (LLS) algorithm of the PHI Multipak (V6.0) software. This algorithm fits the spectra of the selected areas (shown in figure 4(b)) with the spectra in each pixel of the map. The correlation is shown in the chemical-state maps (figure 4(c) and (d)). Bright areas in figure 4(c) show areas with high correlation with the oxide-type spectra, while bright areas in figure 4(d) show high correlation with the phosphate-type spectra. The phosphate-type spectrum is most prominently repre-

Table 1
COF and elemental ratio for the various loads of two independent samples. The detailed analysis of the XP spectra is presented elsewhere [13].

Load	COF		P/S		$O_{phosphate}/P$	
0.05 N	0.25	0.31	0.66	0.68	5.1	7.0
0.1 N	0.26	0.31	0.71	1.13	5.0	4.7
0.5 N	0.23	0.21	0.94	1.30	4.1	4.5
1 N	0.22	0.20	0.95	1.42	4.0	4.5
5 N	0.20	0.19	0.91	1.45	3.6	4.2

sented in the outermost test area (5 N load) and decreases with decreasing load. The O(1s) peak shape characteristic of oxide ("oxide type") is almost absent in the area tested with 5 N but seems to be increasingly present at lower load and it is predominant in the non-contact area.

From the chemical-state map it is possible to select "areas of interest" for more detailed (small-area) XPS analysis. The results are summarized in table 1 and a more detailed spectral analysis can be found elsewhere [13]. No differences in the binding energies of the spectra taken in the various tested areas could be found, although changes in intensity ratios were observed. With increasing applied load, the phosphorus-to-sulfur ratio in the tribofilm was observed to increase. The ratio of the oxygen bound in a phosphate group (component at 531.7 eV, see figure 4(c)) to phosphorus was found be close to 4 : 1 for high loads, indicative of PO_4^{3-}. The binding-energy value of the P(2p) peak (133.6 eV) is also in agreement with phosphorus being bound within a phosphate group [14].

4. Discussion

The combinatorial approach to tribological testing presented here involves the creation of a set of spatially separated areas on a single sample that have undergone tribological tests under a variety of loads. The frictional information gathered during tribological tests as a function of test conditions can then be mapped onto subsequent, spatially resolved surface-analytical data. This methodology can be considered as the generation of a parameter library on the sample, which then serves as a means to relate tribological conditions with both friction and subsequent tribochemical reactions.

As described above, the COF increases after every fifth turn of the disk due to the 25 μm step that is performed to ensure a contact area wide enough for XPS analysis (see figure 2). At the end of a five-turn cycle, the COF reaches a steady-state value, indicating that a tribofilm characteristic of the applied load is formed and thus a representative surface analytical analysis can be carried out within these contact areas. This behavior of the COF is most evident in the 1 N and 0.5 N tests. It can be assumed that due to the high load, a surface film is formed more rapidly than at lower load.

The resolution of the load cell measuring both frictional force and normal load is ±5 mN. The friction force for the low-load experiments is on the order of 20

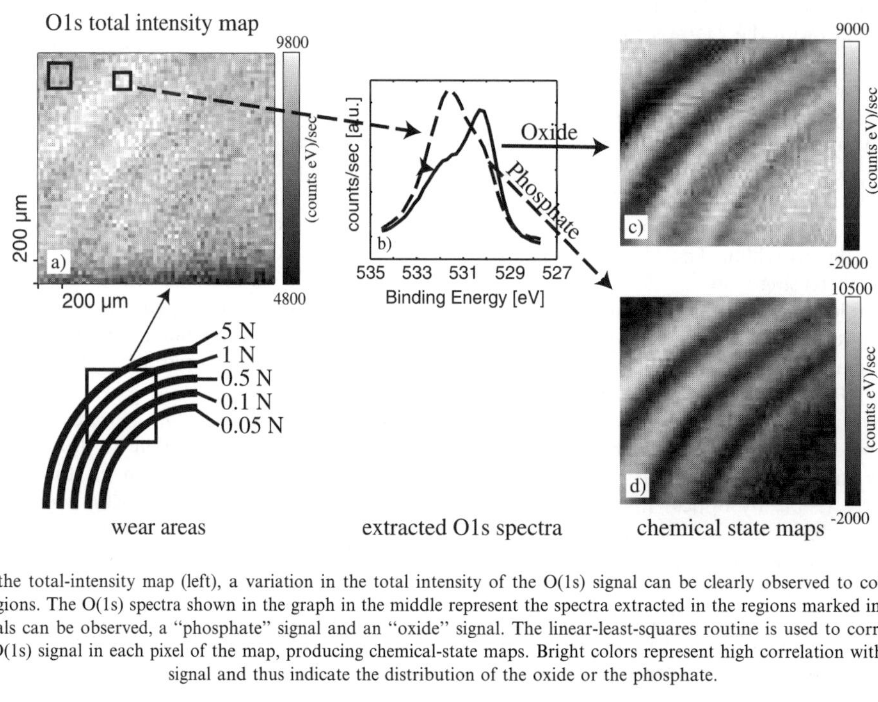

Figure 4. In the total-intensity map (left), a variation in the total intensity of the O(1s) signal can be clearly observed to coincide with the tribotested regions. The O(1s) spectra shown in the graph in the middle represent the spectra extracted in the regions marked in the map. Two different signals can be observed, a "phosphate" signal and an "oxide" signal. The linear-least-squares routine is used to correlate these two signals with O(1s) signal in each pixel of the map, producing chemical-state maps. Bright colors represent high correlation with the respective signal and thus indicate the distribution of the oxide or the phosphate.

and 10 mN for the 100 and 50 mN tests, respectively, and thus the experimental error for the friction force is on the order of 25–50% for these loads. Despite the high experimental error at low loads, a slight reduction of the COF at higher loads compared to the lower loads can be seen for both experiments reported in table 1.

A possible shortcoming of the described combinatorial test method, which involves using the same pin and disk for both running-in and load tests, can arise from the fact that the tribological conditions at the contact between the pin and disk result from the nature of the surface films present on both the pin and the disk and not just from the disk. Indeed, any change in the surface film composition of the pin might also affect the subsequent results. This needs to be determined in a controlled experiment for each system investigated. To ascertain whether any such interference was significant in the present study, oscillating load tests have been performed, during which the load is cycled from the minimum value to the maximum value, in synchrony with the angular position on the disk. These tests have shown that the coefficient of friction changes as a function of the applied load in a similar way during each cycle, as well as showing symmetric behavior during increasing- and decreasing-load phases [15]. In the presence of ZnDTP, the COF showed higher values at the smaller loads and decreased with increasing load. Results obtained performing traditional single-wear-track tests were in good agreement with those obtained during combinatorial tests for the same applied loads, confirming that the history of the pin did not influence subsequent tribological measurements in the system under investigation [15].

The decreasing oxygen-to-phosphorus ratio and the increasing phosphorus-to-sulfur ratio (see table 1) indicate that with increasing load an increasing amount of phosphate is formed in the contact area. Detailed analysis of the S(2p) spectra indicate that some of the sulfur is present in the sulfide state and some is present as organosulfur species [13]. Spectroscopic analysis of the Fe(2p) signals also indicates the presence of iron phosphate. Tribological films from ZnDTP have been reported to be polyphosphate films [4,10]. They are formed under pure thermal or combined thermal and tribological stress [11]. At temperatures above 100 °C, tribofilms with a thickness of a few tens of nanometers are found, while at lower temperatures thinner films are reported [16]. The absence of thermal activation in our experiments explains the lack of a thick polyphosphate film. Only the presence of a thin orthophosphate film is indicated.

5. Conclusion

A combinatorial experiment has been successfully applied to the investigation of a tribological system. The advantage of the combinatorial approach is that multiple experiments can be efficiently combined on a single sample and readily compared. Tribological and spectroscopic results could be acquired in one experiment for a number of conditions and mapped onto the parameter library.

Useful information concerning the reactions occurring in the tribological contact could be derived from the experiment: the i-ZnDTP molecule reacts under tribological stress with the steel surface and forms an iron phosphate-containing film. The amount of phosphate film formed and the composition is dependent on the applied load. Sulfides and organosulfur species are formed, but the amount of these species present is not as strongly load dependent as the amount of phosphate on the surface.

In future, the information density obtained in such experiments will need to be increased in order to realize the full potential of this type of experiment. The range of experimental parameters should also be enlarged in order to cover as many tribologically relevant regions of the parameter space as possible. Ideally, with one experiment the tribological conditions under which a given lubricant additive formulation shows desirable properties could be determined. This approach could therefore speed up the search for new lubricant-additive systems. The approach can also be extended to other (surface) analytical methods. Auger electron spectroscopy or time-of-flight secondary ion mass spectroscopy would have the advantage that the analysis could be performed with a higher lateral resolution and therefore allow a higher information density.

Acknowledgments

Financial support of the ETH Zurich and Italian MURST (ex 60% grant to A.R.) is gratefully acknowledged. Prof. J.M. Martin (Ecole Centrale de Lyon, France) is thanked for supplying the i-ZnDTP.

References

[1] X.D. Xiang, X.D. Sun, G. Briceno, Y.L. Lou, K.A. Wang, H.Y. Chang, W.G. Wallace-Freedman, S.W. Chen and P.G. Schultz, Science 268 (1995) 1738.

[2] R.B. van Dover, L.D. Schneemeyer and R.M. Fleming, Nature 392 (1998) 162.

[3] J.C. Zhao, J. Mater. Res. 16 (2001) 1565.

[4] A.J. Gellman and N.D. Spencer, J. Eng. Tribol. 216 (2002) 443.

[5] J.M. Martin, C. Grossiord, T. Le Mogne and J. Igarashi, Wear 245 (2000) 107.

[6] K. Matsumoto, A. Rossi and N.D. Spencer, *Proceedings of the International Tribology Conference Nagasaki, 2000 II* (2000) 1287.

[7] M. Eglin, A. Rossi and N.D. Spencer, in: *Proceedings of the 28th Leeds–Lyon Symposium on Tribology*, ed. D. Dowson *et al.* (Vienna, 2002) p. 49.

[8] J.-B.D. Green and G.U. Lee, Langmuir 16 (2000) 4009.

[9] S. Hogmark, S. Jacobson and O. Wänstrand, in: *Proceedings of the 21st IRG-OECD Meeting*, ed. D.J. Schipper (Amsterdam, 1999).
[10] J.M. Martin, Tribol. Lett. 6 (1999) 1.
[11] Z. Yin, M. Kasrai, M. Fuller, G.M. Bancroft, K. Fyfe and K.H. Tan, Wear 202 (1997) 172.
[12] M.L.S. Fuller, M. Kasrai, G.M. Bancroft, K. Fyfe and K.H. Tan, Tribol. Int. 31 (1998) 627.
[13] M. Eglin, A. Rossi and N.D. Spencer, Tribol. Lett. this issue.
[14] D. Schuetzle, R.O. Carter, J. Shyu, R.A. Dickie, J. Holubka and N.S. McIntyre, Appl. Spect. 40 (1986) 641.
[15] M. Eglin, PhD Thesis. Presented at the Department of Materials, ETH Zürich, 2003
[16] S.H. Choa, K.C. Ludema, G.E. Potter, B.M. Dekoven, T.A. Morgan and K.K. Kar, Wear 177 (1994) 33.

Tribology Letters, Vol. 15, No. 3, October 2003 (© 2003) 199

X-ray photoelectron spectroscopy analysis of tribostressed samples in the presence of ZnDTP: a combinatorial approach

Michael Eglin[a,*], Antonella Rossi[a,b] and Nicholas D. Spencer[a]

[a]*Laboratory for Surface Science and Technology, Department of Materials, Swiss Federal Institute of Technology, ETH Zürich, CH-8092 Zürich, Switzerland*

[b]*Department of Inorganic and Analytical Chemistry, University of Cagliari, INSTM, Cagliari, Italy*

Received 15 December 2002; accepted 12 April 2003

The influence of load on the chemistry of tribofilms formed on a steel surface in solution of pure di-isopropyl zinc dithiophosphate (i-ZnDTP) in *n*-decane has been investigated by means of a combinatorial tribological experiment involving X-ray photoelectron spectroscopy. The experiment consisted of the preparation of a set of spatially separated areas, produced under various tribological test conditions, and the subsequent spectroscopic probing of the chemical composition of the tribofilm. The experiment was carried out at room temperature under boundary-lubrication conditions and revealed a physically adsorbed layer of the additive in the non-contact area and a thin (ca. 5 nm), inhomogeneous phosphate film covering the tribostressed areas. The total amount of phosphate present in the tribostressed area was found to increase with increasing load. In the contact areas, iron oxides and metal sulfides have also been detected.

KEY WORDS: boundary lubrication, isopropyl ZnDTP, X-ray photoelectron spectroscopy, combinatorial tribochemistry

1. Introduction

In a previous paper [1] we presented tribological results obtained by applying a new combinatorial approach to the study of the interaction of a lubricant additive with tribostressed surfaces. In this paper, the combinatorial approach has been combined with imaging XPS (i-XPS) and small-area XPS (SAXPS) in order to characterize the tribochemistry of an additive interacting with steel surfaces. A model compound, di-isopropyl zinc dithiophosphate (i-ZnDTP), was chosen for investigation, since zinc dialkyl dithiophosphates (ZnDTPs) have been the most prominent antiwear additives in lubricating oils for over forty years [2]. ZnDTP was originally added to lubricating oil as an antioxidant, but it was rapidly discovered that it also functioned as a highly effective antiwear and extreme-pressure (EP) additive, and it is an essential ingredient in the vast majority of current commercial lubricant formulations [3]. Although it is one of the most studied additives, its outstanding properties in antiwear film formation are still not completely understood. An important role of ZnDTP in such lubricants is to act as an antiwear agent by forming a protective film on the rubbing surfaces [4]. Much information has been published on the thickness, structure and chemistry of the films as well as the mechanism of film formation under tribological stress. The ability of ZnDTP to function in different ways depending on the nature and severity of the tribological conditions may be one of the reasons for the often divergent results reported in the literature. Several authors have proposed a layered structure of a few 100 nm thickness, consisting of iron oxides, sulfides and poly (thio) phosphates or alkylphosphates [3,5–7]. It has also been reported that in the mild-wear regime (0.26 GPa contact pressure, 50 Hz, 15 mm stroke length, 60 °C), the wear-reducing action of ZnDTP is mainly due to the rheological properties of the thermally formed polyphosphate glassy films, whereas under severe conditions (1 GPa contact pressure, 0.5 m/s, 100 °C), short-chain mixed iron/zinc phosphate glasses are formed [5].

As far as the film-formation mechanism is concerned, it is generally agreed that the molecule interacts only weakly with the steel surface at room temperature, starting to decompose around 50 °C on iron substrate [8]. It has been suggested that upon reaching 100 °C ZnDTP undergoes a thermally activated rearrangement in solution to form a linkage isomer (LI-ZnDTP) [3,7,9] as a precursor to the formation of a long-chain glassy polyphosphate and poly-thiophosphate film on the surface. Due to the thermally activated formation of the LI-ZnDTP, as well as the thermo-oxidative processes and wear involved in the film formation, the composition of the generated films changes with temperature [3,6]. Without thermal activation, only very thin films are formed in the tribological contact, whereas at higher temperatures the film thickness increases [10]. Recent work from our laboratory using *in situ* attenuated total reflection infrared spectroscopy (ATR FT-IR) has allowed changes in the additive

*To whom correspondence should be addressed. E-mail: michael.eglin@mat.ethz.ch

1023-8883/03/1000–0199/0 © 2003 Plenum Publishing Corporation chemistry due to adsorption, as well as reaction film growth, to be followed as a function of temperature and time [11,12]. The reported ATR FT-IR spectra confirm thermal decomposition of ZnDTP with the formation of P–O–P species at 150 °C, while a short-chain phosphate film has been detected on the iron surface following tribotests at the same temperature [13].

The detailed chemistry of these films is still controversial: older works based on elemental analysis with SEM-EDX showed that the film contained Zn, S, P, C and O [14]. XPS [15] and AES [16,17] have also been used to provide chemical differentiation for the different elements; P was claimed to be present as a phosphate and S as a sulfide, the corresponding cation being zinc or iron. XANES spectroscopy revealed that the chemical nature of P and S in the tribofilms was different from that of the pure ZnDTP substances and depends on the alkyl group used in ZnDTP [18]. With isopropyl ZnDTP, phosphorus is found to correspond closely to metaphosphate (polyphosphate with a cyclic structure) whereas with *n*-butyl ZnDTP pyrophosphates are found. This is important when discussing the results obtained from commercial ZnDTP, which often consist of a blend of ZnDTP molecules containing different alkyl groups [3].

A detailed analysis of the high-resolution XP spectra allows further information to be obtained. From the oxygen O(1s) signal, not only oxide and phosphate can be distinguished but, due to the substantial chemical shifts in the signals of the phosphate group, non-bridging oxygen (NBO: PO^-, POH, P=O) at 532.2± 0.2 eV and bridging oxygen (BO: P–O–P) at 534.4± 0.2 eV can be differentiated [19–22], allowing the chain length of polyphosphates to be estimated [21,23]. The phosphorus P(2p) signal has been used to distinguish between orthophosphates, metaphosphates and polyphosphates [21,22]. After thermal decomposition of ZnDTP on an iron surface, zinc polyphosphates could be detected, whereas only simple phosphates were detectable after tribological testing at the same temperature [22].

The difficulty in assignment of chemical state based on binding energies alone should not be underestimated. While the sulfur S(2p) signal provides information about the oxidation state of sulfur, and thus sulfates and sulfide can be readily distinguished, it is much more difficult to identify the cation from the binding energy of the S(2p) peak alone. The chemical state of zinc cannot be identified from the Zn(2p) signal due to the very small chemical shifts between zinc signals in ZnO, ZnS or Zn-polyphosphates. However, the Zn(LMM) X-ray excited Auger signal shows a stronger dependency on the chemical environment [15,24]. It has been shown, for example, that a two-dimensional chemical-state plot of zinc can effectively distinguish between zinc sulfide and oxide [25]. This method can also be extended to differentiate between zinc phosphate or polyphosphate in the contact and non-contact areas of tribostressed surfaces [26].

Most of the published surface-analytical studies of ZnDTP have been based on the analysis of several mm^2, and thus an average chemical composition has been obtained, despite the well-known fact that the films at the micron level are inhomogeneous and "patchy" as revealed by SEM [27] and AFM [28]. Therefore, imaging spectroscopic methods were used to investigate films formed during tribological contact in the presence of lubricant additives. Imaging XPS provides quantitative elemental and chemical state information with a lateral resolution in the submillimeter range [29–31]. Synchrotron-based photoemission electron microscopy (PEEM) offers higher lateral resolution (micrometer range) with elemental and chemical state information [32], but it is less quantitative. Scanning Auger Microscopy allows elemental mapping with a lateral resolution in the sub-micrometer range [16,33], but chemical-state-induced Auger shifts are not easy to interpret, even though recent results suggest that it is possible to distinguish between oxygen in oxides and in phosphates [34].

So far, the imaging possibilities of these surface-analytical methods were applied to relatively simple systems, e.g. traditional pin-on-disk tests [32]. To gain a better insight into the possible tribochemical reactions of a given lubricant-additive system depending on the applied tribological conditions, imaging methods can be used in combination with a combinatorial experiment [1]. A combinatorial sample is produced by systematically varying the parameters defining the tribological stress, spatially separated on a single sample. Imaging spectroscopic methods are then used to analyze the combinatorial sample and extract spectroscopic information for the various conditions applied. One of the parameters studied is the load dependence of the film structure [6,10,35]. It has been reported that higher loads increase the ZnDTP decomposition rate and lead to long-chain polyphosphates [6]. The film thickness of the tribofilm increases to a maximum value with increasing load and may break down upon further increase [10]. The following describes an investigation into the load dependence of the chemical composition of the tribofilms formed in a 1 wt.% solution of i-ZnDTP in decane. These have been analyzed by means of i-XPS and SAXPS.

2. Experimental

2.1. Sample preparation

Steel disks (AISI 52100) were ground using P120, P600 (in tap water) and P1200, P2500 (in ethanol (p.a.)) silicon carbide paper, and polished in ethanol using 3, 1, and 1/4 μm diamond paste. The surface roughness (Ra) was measured by interferometric methods and always found to be less than 10 nm. After polishing, the samples were kept for three days in a desiccator to allow an oxide film to grow and thus to provide a reproducible surface composition. The samples were ultrasonically cleaned in ethanol and analyzed for surface contamination by XPS immediately prior to tribotesting. Commercial 4 mm ball-bearing balls (AISI 52100) were used as a counterface. They were ultrasonically cleaned in ethanol before the tribotest.

2.2. Tribotest

The steel disk was tribostressed in a ball-on-disk setup with a CETR UMT-2 tribometer (Center for Tribology, Inc., Campbell, CA, USA). This tribometer allows load, rotational velocity and duration of the tribotest to be programmed. The tribometer is equipped with a load cell with a maximum capacity of 5 N and a resolution of ±5 mN in two axes (normal load and friction force). Normal load is applied from the carriage via a spring. The normal load is constantly monitored and adjusted via a feedback loop moving the carriage up or down. The spring constant was determined to be 2.7 N/mm in the z-direction and 5.7 N/mm in the y-direction (friction force). The instrument is capable of recording normal load, friction force, rotational position of the disk and the y-position of the carriage versus time.

As a lubricant, a 1 wt% solution of di-isopropyl zinc dithiophosphate (i-ZnDTP) in decane was used. To dissolve the additive in the lubricant, the solution was stirred at 60 °C for 30 min. The tests were carried out at room temperature (24 ± 0.5 °C). The relative humidity of the air was recorded during each test and was determined to be between 22 and 38%. Decane has a boiling point of 174 °C. This fact hindered investigations at higher temperatures due to problems with evaporation.

Prior to the actual tests, the ball was run in at a load of 5 N with a speed of 31.4 mm/min for 2 h outside the region to be analyzed, in order to create a flat spot on the ball, defining the apparent contact area in the tribotest that followed.

A full tribotest consisted of the formation of five, individual, concentric test areas, each corresponding to a different load. The relative velocity between the ball and the disk surface was kept constant at 31.4 mm/min for all the tribostressed areas, while the load was changed from 0.05 to 5 N between the tested areas. The apparent contact pressure was calculated to be 2.8–280 MPa using the applied load and the apparent contact area defined by the flat spot produced during the running in. The number of turns of the ball on the disk was kept constant for each load, in order to impart comparable tribostress. The ball was moved radially by 25 μm after each fifth turn. Thus, for each load, 11 overlapping wear tracks formed a tribostressed annulus with a width of >250 μm. This allows XPS analysis to be performed completely within the tribostressed area. After the tribotest, the samples were cleaned in cyclohexane in an ultrasonic bath for 30 s and dried under an argon stream before being introduced into the analysis chamber.

2.3. XPS

The XPS analyses were performed on a PHI 5700 system with an Omni Focus IV lens system. The residual pressure in the spectrometer during the data acquisition was always below 5×10^{-7} Pa. The X-ray source was Al Kα (1486.6 eV) run at 350 W. The diameter of the analyzed area was 120 μm. The spectrometer was operated in the fixed analyzer transmission (FAT) mode with a pass energy of 46.95 eV (full width of half-maximum (FWHM) for Ag($3d_{5/2}$) = 1.1 eV). The instrument was calibrated using the spectral lines of Au($4f_{7/2}$) and Cu($2p_{3/2}$) at 83.95 eV and 932.63 eV, respectively [36]. Ag($3d_{5/2}$) and Cu(LMM) peak energies at 368.22 eV and 567.96 eV respectively, were used to verify the linearity of the binding-energy scale. The accuracy was ±0.05 eV.

Spectroscopic maps were acquired using the imaging capabilities of the Omni Focus IV lens system. The analyzed spot (diameter 120 μm) was electrostatically rastered over the sample (typically 64 × 64 pixels, 2 × 2 mm). For each pixel a full spectrum of the selected energy region was acquired. The acquired images were processed with the PHI Multipak (V6.0) software. Spectra can be extracted from the map by selecting a region of interest and can be used to reconstruct "chemical-state maps" with the linear-least-squares (LLS) routine. The correlation between the extracted spectra and the spectrum at each pixel generates a new map, which shows regions having similar chemical states of a given element.

From these maps, areas of interest were determined and analyzed by SAXPS. For curve fitting, CASA XPS software (V2.0, CASA Software Ltd, UK) was used. The spectra were resolved into their components after an integrated Shirley background subtraction. Curve-fitting parameters (FWHM and Gaussian/Lorenzian ratio) were determined from reference spectra. The spectra were fitted with Gaussian/Lorenzian peaks, keeping the FWHM and the Gaussian/Lorenzian ratio constant and fitting the energy of the peaks and their heights using a LLS algorithm. Quantitative analysis was performed using the photoionization cross-section according to Scofield [37] and correcting the intensities for the inelastic mean free path, according to Seah and Dench [38], as well as the analyzer transmission function [39]. The accuracy of the measurements was estimated to be ±0.1 for all measurements, unless otherwise stated. The results presented in this paper refer to two independent experiments.

2.4. Auger electron spectroscopy

The Auger mapping was performed on a VG ESCALAB 200 using an electron gun, which allows 2000 Å lateral resolution. The electron beam, of energy of 10 keV (inducing a specimen current of 20 nA), was rastered to produce 128 × 128 pixel maps, which were recorded on an IBM 486 computer.

The Auger maps were corrected for topographical effects using the (peak-background)/background algorithm.

2.5. Reference compounds

Reference compounds, $FePO_4.nH_2O$, ZnS and ZnO (Aldrich) were pressed into pellets and mounted in a standard PHI sample holder. i-ZnDTP was supplied by Prof. J.M. Martin. The samples were cooled with liquid nitrogen during measurements to reduce outgassing of the compounds. Reference compounds were analyzed by XPS under the same experimental conditions as reported in section 2.3. The results were confirmed by experiments performed with aperture #3 (0.4 mm diameter) and pass energy 23.5 eV with higher energy resolution.

3. Results

3.1. Tribological results

The coefficient of friction (COF) measured during the tribological test is presented in figure 1 as a function of the applied load. The two bars represent two independent samples. The reported COFs are the values averaged over the last full turn of the disk at the particular load. It is assumed that, at this time, a

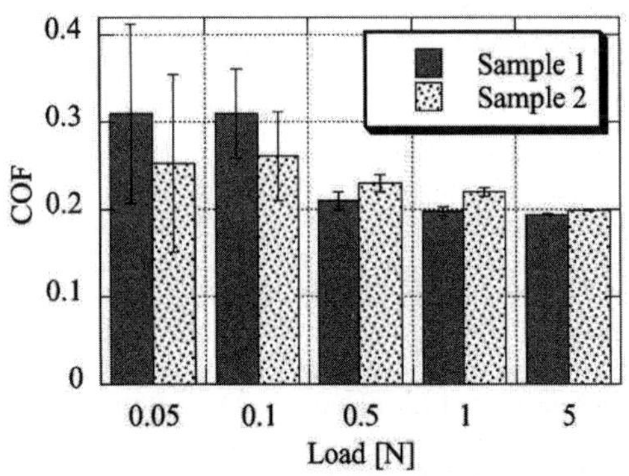

Figure 1. Coefficient of friction (COF) for various loads. The value presented is the average of the last full turn during the test at a particular load. It is assumed that at the end of the test the COF is representative for the protective film formed under the applied conditions. The error bars are calculated using error propagation from the load-cell resolution. The two samples represent two independent experiments.

tribological film that is representative for the particular load is formed and thus the COF is a representative value for the applied conditions. The error bars represent the experimental error, calculated from the error propagation of the load-cell resolution. Despite the large experimental error at low loads, a slight decrease in COF can be seen for larger loads.

The morphology of the wear tracks and the ball used for the tribotest were examined with an optical microscope. The ball shows a circular flat spot with a diameter of 150 μm. This spot was produced mainly during the running-in of the ball and defines the apparent contact area during the tests with different loads. In the 5 and 1 N regions the wear tracks are clearly visible on the disk, but for smaller loads no differences can be recognized between the contact and non-contact areas.

3.2. XPS imaging

An O(1s) chemical-state map is shown in figure 2. The chemical-state map is calculated using a linear-least-squares routine, fitting the peak shape of an O(1s) spectrum extracted from a contact region to the spectrum in each pixel of the O(1s) map. The spectrum used to calculate the chemical-state map is similar to that of the 5 N load spectrum shown on the right-hand side of figure 2, and shows a high correlation in the tribostressed areas, forming arcs of a width of 250 μm running across the chemical-state map. The tribostressed areas are labeled according to their applied loads.

On the right-hand side of figure 2, spectra extracted from different areas of the O(1s) map are shown. An increase in the peak intensity can be seen at 531.7 eV as the applied load increases. In the spectra extracted from the non-contact area, a shoulder can be noted at this binding-energy value, which is also present in spectra taken on the steel surface prior to tribotesting.

In order to provide more detail on the chemical species present on the surface, SAXPS analysis was performed in the areas marked with a circle in the O(1s) map.

3.3. Chemical states of the studied elements

3.3.1. Reference compounds

Reference materials, which include the steel surface after mechanical polishing, steel after argon-ion etching, the compounds iron phosphate, iron sulfide, zinc sulfide and zinc oxide, as well as the i-ZnDTP, were analyzed under the same analysis conditions used for obtaining the X-ray photoelectron spectra on the tribostressed samples, in order to derive the curve-fitting parameters to be used for monitoring the changes in the surface chemistry of the tribostressed samples. The peak

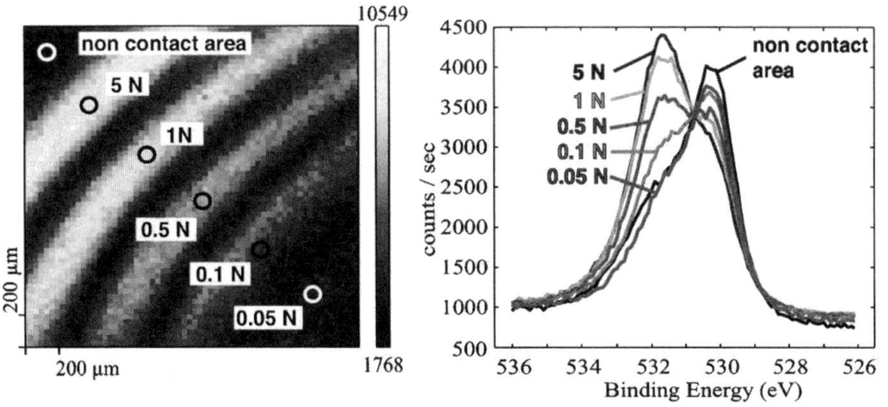

Figure 2. Left-hand side: O(1s) chemical-state map of the five contact areas with the different loads. The map shows the correlation with phosphate-type spectra (extracted from a contact area), and thus the distribution of phosphate-type species. Right-hand side: O(1s) spectra extracted at the points shown in the chemical-state map (circles). The spectra show an increasing peak at 531.7 eV with increasing load. This peak can be assigned to oxygen bound in a phosphate structure.

energies of the main components of the reference compounds are summarized in table 1.

Steel surface after mechanical polishing. The iron Fe($2p_{3/2}$) signal was fitted with Gaussian/Lorenzian peaks and its sum contains the contributions of four signals: The elemental iron Fe(0) signal at 706.8 ± 0.1 eV, Fe(II) at 708.9 ± 0.1 eV (including the Fe(II) shake-up calculated to be 5.5 eV higher in binding energy and 8% of the main peak intensity [40]), Fe(III) at 710.3 ± 0.1 eV and Fe(III) in the oxy-hydroxide layer at 711.9 ± 0.2 eV.

The oxygen signal (figure 3) is composed of three peaks: the first at 530.2 ± 0.1 eV can be assigned to the iron oxide; the second peak at 531.8 ± 0.1 eV is assigned to the iron oxy-hydroxides and the third, at 533.1 ± 0.1 eV, having the lower intensity in comparison with the others, is attributable to adsorbed water.

Steel surface after argon-ion etching. After ion etching, the oxy-hydroxide film, which is always present after mechanical polishing and air exposure, is removed and the spectrum of the iron signal exhibits a peak maximum at 706.8 ± 0.1 eV with an asymmetric tail on the higher-binding-energy side due to multielectron processes.

Iron phosphate. In this case, the P($2p_{3/2}$) signal is found at 133.7 ± 0.1 eV with a FWHM equal to 1.65. The oxygen O(1s) spectrum has been fitted with two Gaussian/Lorenzian curves. The resulting binding energy values are 531.7 ± 0.1 eV for the oxygen in the phosphate group with a FWHM of 1.8 eV and 533.2 ± 0.1 eV due to the water of crystallization. The iron peak has a maximum at 712.2 ± 0.2 eV and has been fitted using a Gaussian–Lorenzian peak with a FWHM of 3.0 ± 0.2 eV.

Zinc sulfide. The most intense zinc signal, Zn($2p_{3/2}$), is found at 1022.1 ± 0.1 eV. The sulfur S2p peak was fitted with the two contributions due to spin-orbit coupling: the binding energy value of the S($2p_{3/2}$) component was equal to 162.0 ± 0.1 eV. Quantitative surface analysis yielded 45 at.% Zn, 55 at.% S.

Table 1

Binding energies of the photoelectron peaks measured in the contact areas (mean value for all loads), the non-contact areas and for the reference compounds. The standard deviation is 0.1 eV for all values listed, except for S $2p_{3/2}$ (see text).

	O 1s (I)	O 1s (II)	O 1s (III)	P $2p_{3/2}$	S $2p_{3/2}$ (I)	S $2p_{3/2}$ (II)	Zn $2p_{3/2}$
			Tribostressed samples				
Contact area	530.2	531.6	532.9	133.6	161.9	163.2	1022.3
Non-contact area	530.2	531.7	532.9	133.4	162.0	163.4	1022.3
			Reference compounds				
Steel surface	530.2	531.8	533.1				
i-ZnDTP [41]			533.1	133.8	162.8		1022.7
$FePO_4$		531.7	533.2	133.7			
ZnS					162.0		1022.1
ZnO	530.2						1021.3

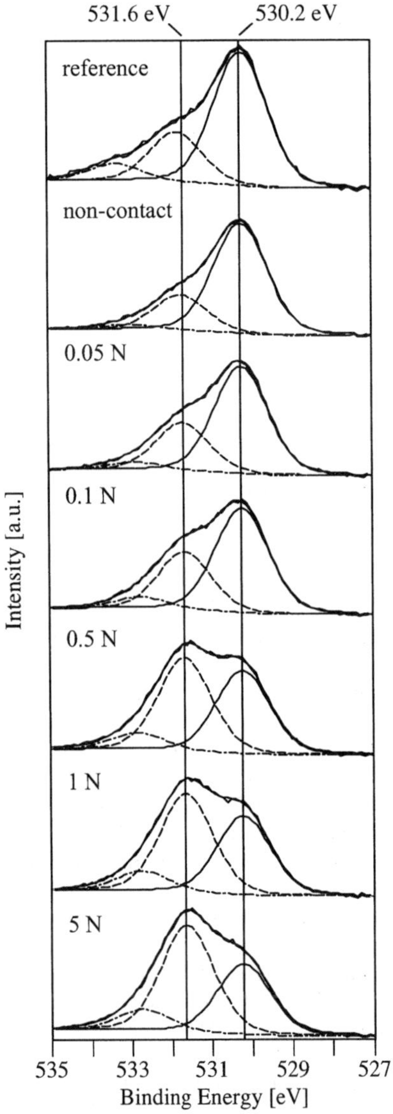

Figure 3. O(1s) peaks of the five different loads together with the non-contact area and the reference spectra of a nascent steel surface. The signals are fitted with three different peaks, a first at 530.2, a second at 531.6 and a third at 532.9 eV.

Zinc oxide. The most intense zinc signal is found at 1021.3 ± 0.1 eV and it exhibits a FWHM of 1.8 ± 0.1 eV; the oxygen O(1s) signal was measured at 530.2 ± 0.1 eV and its FWHM is 1.6 ± 0.1 eV. The quantitative analysis was checked and resulted in excellent agreement with the expected ratio of 1 : 1.

i-ZnDTP [41]. The zinc signal, $Zn(2p_{3/2})$, is found at 1022.7 ± 0.1 eV with a FWHM of 1.9 ± 0.1 eV. The oxygen signal comprises only one signal, found at 533.1 ± 0.1 eV. Phosphorus and sulfur were also fitted, taking into account the splitting and the branching ratio of the two components. $P(2p_{3/2})$ is found at 133.8 eV (FWHM = 1.55 eV) and $S(2p_{3/2})$ at 162.8 eV (FWHM = 1.7 eV). The atomic concentration was found to be 51% C, 15% O, 9% P, 19% S, 5% Zn being in good agreement with the calculated values (52.1% C, 17.4% O, 8.7% P, 17.4% S, 4.4% Zn).

3.3.2. Surface chemistry after tribological tests

High-resolution spectra of the O(1s), S(2p), P(2p), Zn(2p), Zn(3s) and Fe(2p) regions were acquired in the tribostressed areas (see circles in figure 2). For comparison, the non-contact area was analyzed as well. The size of the analyzed area is 120 μm, and thus smaller than the width of the contact region. This ensures that only the tribostressed area is analyzed and thus the spectroscopic information is not averaged with spectral contributions from the non-contact area. In figure 3, the O(1s) peaks for the various loads are displayed. They can be curve-fitted using three peaks. For the curve fitting, a Gaussian/Lorenzian function is used, the G/L ratio was kept fixed at 20% Lorenzian and the FWHM at 1.55 eV while the peak position (binding energy) and peak height were left free to converge. The peak positions do not shift within the experimental error (±0.1 eV) upon changing the applied load, although they were left free to vary during the curve fitting. The comparison of the spectra collected in the various contact areas corresponding to different applied loads shows that the peak at 531.6 eV increases with increasing load. In the same figure, the O1s spectra of the steel surface immediately prior to tribotesting, together with the oxygen spectra collected in the non-contact area, are also displayed. The binding-energy values of the three components are listed in table 1 with the binding-energy values of the $P(2p_{3/2})$, $S(2p_{3/2})$ and $Zn(2p_{3/2})$ peaks.

The P(2p) and S(2p) peaks were fitted using a doublet, taking into account the theoretical ratio of the spin-orbit coupling (1 : 2). The separations of the two peaks were determined by reference measurements and kept constant for data processing (1.2 eV for sulfur and 0.84 eV for phosphorus). The $P(2p_{3/2})$ is found at 133.6 ± 0.1 eV in the contact regions (independently of the applied load) and 133.4 ± 0.1 eV in the non-contact region.

The FWHM of the S(2p) peak in the contact areas were found to be broader (2.5 eV) than those of the reference spectra (ZnS: 2.2 eV). This suggested that more than one chemical state is present, and thus two doublets were used for the fit. The signal at lower binding energies was found at 161.9 ± 0.1 eV and the second has a peak maximum at 163.2 eV. Due to the low sulfur content in the non-contact area, the S(2p) spectra has a low S/N ratio, which makes it difficult to produce a proper peak fit. These values show a higher standard deviation (±0.2 eV).

In figure 4, an example of an iron $Fe(2p_{3/2})$ signal collected in an area tribostressed with 1 N is shown, following a Shirley–Sherwood background subtraction.

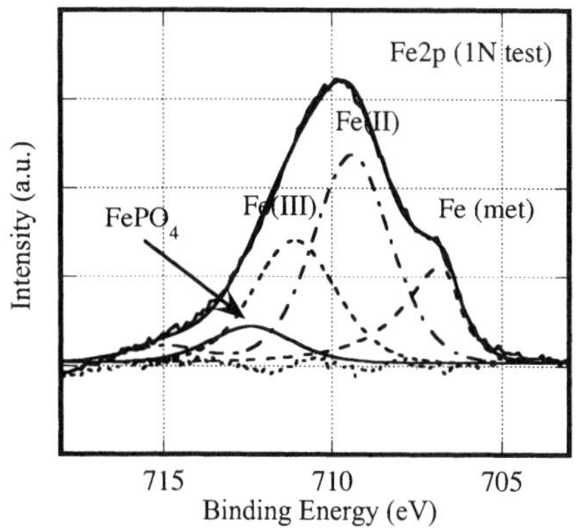

Figure 4. $Fe(2p_{3/2})$ spectra collected in an area tribostressed under 1 N load. It is possible to observe a contribution at 712.4 eV, which can be attributed to iron bound to phosphate.

The spectra is quite complex because of the superimposition of the contributions due to the different iron species present in the surface film and the substrate. To extract as much information as possible, a curve-fitting procedure based on the parameters obtained from the reference spectra and leaving only the heights of the curves free to vary was adopted. The contribution to the metallic iron (706.8 eV) is fitted with a tail function in order to avoid an overestimation of the intensity of the other signals. In this case, as for the reference compounds, the iron in the oxidation state (II), found at 709.4 eV, has been fitted together with its satellite (at a binding energy 5.5 eV higher than the main peak), which might interfere with the signals assigned to iron in a phosphate and/or in a hydroxide. The Fe (III) is present at 711.1 eV, whereas the signal at 712.4 eV might be assigned to the iron phosphate signal, according to the results obtained on the reference iron phosphate.

The Zn $2p_{3/2}$ signal was always a Gaussian/Lorenzian symmetric curve: the peak position was found at 1022.3 eV with a FWHM of 1.9 eV, independent of the applied load.

3.4. *Composition of the tribofilm*

In the following, quantitative results obtained from the measured XPS intensities (areas of the peaks) are presented for two independent samples.

In figure 5, the atomic ratio between phosphorus and sulfur is given for the different analyzed regions. The sum of the spin-orbit contributions (i.e. the total area under each signal) of the sulfur and the phosphorus peaks were used. At higher load the phosphorus-to-sulfur ratio increases. The theoretical ratio between phosphorus and sulfur in i-ZnDTP (0.5) is also given in figure 5. This ratio is close to that found in the non-contact area.

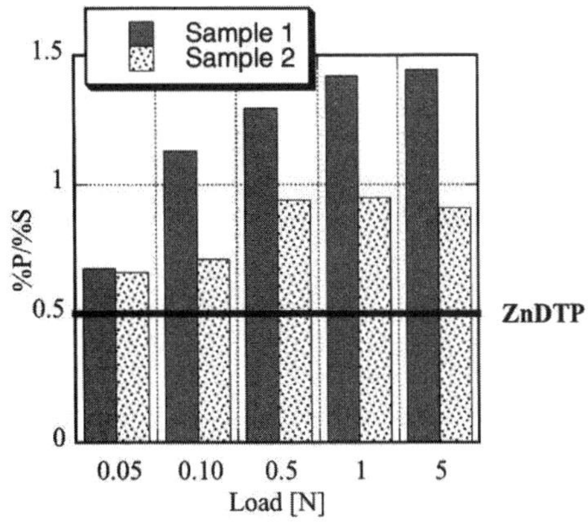

Figure 5. Atomic ratio of P and S for the various loads, calculated using the total area under the P and the S(2p) peaks, corrected with the appropriate sensitivity factors. A substantial increase in P can be seen with increasing load. The two samples represent two independent experiments.

As previously seen in figure 3, the O(1s) peak can be resolved into three different peaks. Figure 6 shows the ratio between the second component of the O(1s) spectra (at 531.6 eV) and the total amount of phosphorus. At high loads the ratio is close to 1 : 4, while at low loads it is higher. The difference between the two samples increases at low loads.

4. Discussion

Most of the work published so far concerning tribofilms formed from ZnDTPs has focused on experiments carried out above 100 °C. It is assumed that these films are a representation for the films produced under real conditions, since ZnDTPs are usually used in applications where elevated temperatures are present. However, although even at room temperature a positive effect of ZnDTP can be observed, very little has been published on the film structure/formation mechanism of

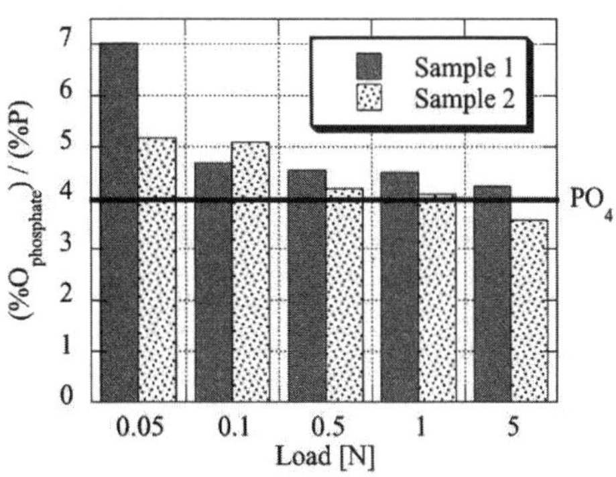

Figure 6. Atomic ratio of O to P for the various loads, calculated using the second component in the O(1s) spectra (at 531.6 eV, see figure 3), attributed to oxygen bound within phosphate and the total P(2p) peak. The two samples represent two independent experiments.

ZnDTP under ambient conditions [10]. Room-temperature measurements are of interest, however, since all machines in which lubricant additives are used start up from room temperature before reaching their operating conditions. It is also frequently the case that the most severe tribological conditions occur during the start-up phase of a machine.

4.1. Tribology and coefficient of friction

Coefficient of friction versus load. The COF values measured in this work are high at all applied loads. This fact substantiates the assumption that the experiments have been carried out in the boundary lubrication regime [42]. Despite the high experimental error in the COF at low loads, a slight decrease in COF can be seen.

Coefficient of friction versus time. A decrease in the COF is detected with time for any given load. This decrease is related to a change in the mechanical properties of the surface film and to a change in the surface composition of the two counterparts. However, with optical microscopy it is difficult to detect wear on the steel surface for applied loads lower than 1 N, although chemical-state maps clearly show the tribostressed areas. This indicates that when an effective additive is employed, it is difficult to observe surface changes with the traditional microscopic techniques but surface analytical methods can contribute to the characterization of the tribofilms. Many such studies have been carried out, but one of the limiting factors is the difficulty of collecting information over a wide-enough range of conditions in order to clarify the mechanism of the film formation. Here, the results of the combinatorial approach provide an insight into the influence of the applied load and other parameters on the chemical composition of the surface film. Differences revealed between the contact and the non-contact regions will also be discussed in the following.

4.2. The surface chemistry of the tribofilm

A chemical-state map (figure 2), reconstructed using the oxygen signal extracted from the contact area, allows the contact area to be distinguished from the non-contact area. Moreover, chemical differences can be detected by looking at the O(1s) signal extracted from areas within the contact regions that were tribostressed with different loads. This allows areas of interest to be selected and analyzed further by SAXPS. The results obtained reveal that in the regions with different loads, no differences in binding energies can be detected. However, differences in the *composition* of the surface film can be readily detected.

Non-contact area: Chemical state and composition. As was pointed out in a recent paper, i-ZnDTP is weakly adsorbed on the non-contact area of the sample surface [31]. The detailed O1s spectra show the main component to be at 530.2 eV with a second, less-intense peak at 531.7 eV, also in the presence of i-ZnDTP. The iron signal remains virtually unaffected in the presence of the second component of the sulfur $S(2p_{3/2})$ signal at 163.4 eV, and the shift of the $P(2p_{3/2})$ signal to lower binding energies (133.4 eV) (compared with the 133.8 eV found in the pure additive) may suggest that a change in the structure of the i-ZnDTP has occurred. The presence of the metallic signal in the iron spectra at 706.9 eV suggests that the oxy-hydroxide film thickness is less than the escape depth of the photoelectrons (i.e. ≈ 5 nm). Thus it can be concluded that in the non-contact area, an atmospherically formed oxy-hydroxide film is present, together with a very thin physically adsorbed organic layer.

Contact area: Chemical state of the elements and composition. The peak at 530.2 ± 0.1 eV in the O(1s) spectra taken in the contact area of the tribostressed sample arises from iron oxide and is present in all spectra (both on the steel prior to tribotesting and within the contact area for all loads). This shows that there is always some oxide present in the tribofilm. A second oxygen peak is found at 531.6 ± 0.1 eV. The peak position is close, within experimental error, to the position of the iron hydroxides (531.8 ± 0.1 eV) [43] and to the oxygen of iron phosphate (531.7 eV table 1), also in agreement with the literature [44]. The intensity of this second peak increases with applied load, as does the amount of phosphorus found in the contact area. The ratio between the oxygen in this chemical state and the total phosphorus signal, corrected with the appropriate sensitivity factors, was found to be close to 4 : 1 for high loads and increases at low loads (figure 6). The oxygen-to-phosphorus ratio of 4 : 1 would agree with a phosphate structure, while at lower loads some hydroxides might also be present, in addition to the phosphate. The presence of a phosphate might also be supported by the position of the $P(2p_{3/2})$ signal found at 133.6 ± 0.1 eV—close to the measured value of the reference iron phosphate at 133.7 eV and 0.2 eV lower than the value found for the free i-ZnDTP (table 1). Reference measurements on surfaces tribostressed in the presence of decane only (no additive) only showed small changes in the relative intensities of the different O1s signal contributions [31]. In this case the ratio of the hydroxide peak to the oxide peak was smaller in the tribostressed area than in the non-contact area, which substantiates our assertion that the O1s peak at 531.6 eV in the contact areas is due to oxygen bound to phosphorus.

As discussed earlier, the binding energies of bridging oxygen (BO) and non-bridging oxygen (NBO) in the O(1s) spectra of a polyphosphate are separated by approximately 1.5 eV [19–22]. The O(1s) peak positions range between 531.6 and 532.6 eV for NBO and 533.1 and 534.1 eV for BO. The differences found in the literature are probably due to different sample preparation methods. The peak positions of the second O(1s) peak found in the spectra of the contact area are close to the lower values reported in the literature for NBO oxygen. In addition, a small peak can be detected at 532.9 eV, a value which would be rather low for BO. Also, the binding energy of the $P(2p_{3/2})$ signal at 133.6 eV is in agreement with literature values for phosphate (133.4–133.8 eV [22,44]) but does not correspond to the literature values for the polyphosphates (134.0–135.0 eV [19–22]). The oxygen-to-phosphorus ratio would also be too high for a polyphosphate. This leads to the conclusion that only phosphates are formed under these experimental conditions at the surface and no polyphosphates are detected. This finding is in agreement with the literature, where it is reported that polyphosphates are formed thermally on the sample and modified with the applied tribological stress [5,12]. The tribological experiments in this work were performed at temperatures where no polyphosphates are thermally formed, leading to a different film composition. It is important to mention that a thin protective film is also formed under these conditions.

The peak position of the component of the S(2p) spectra at 161.9 eV is typical for sulfides. Possible species formed could be zinc sulfide or iron sulfide, but the peak positions of the Fe(2p) and the Zn(2p) signal do not provide any further evidence in this regard. The Zn(LMM) signal (not shown) indicates that the presence of zinc sulfide can be ruled out, and thus zinc might be available to participate in the formation of tribofilms in the phosphate layer. The use of the two-dimensional chemical-state plot, using the Zn(2p) and the Zn(LMM) signal, seems to be a powerful tool to obtain more information on the chemical bonds of zinc [26]. The second component of the sulfur peak at 163.2 eV can be assigned to organosulfur species (e.g. thiols). It can be assumed that after the reaction of the phosphorus present in the ZnDTP molecule to phosphate, part of the sulfur reacts with a metal to form sulfides. The remaining sulfur may react with alkyl chains to form organosulfur species. These species are soluble in the lubricant, and only a small amount would stay adsorbed at the surface or become incorporated in the surface film. This leads to the depletion of S in the contact area in comparison to P.

The XPS iron signals detected in the contact regions are different from those collected on the steel prior to tribotesting: only the component at 706.8 eV is always present. This signal is assigned to the metallic iron of the steel substrate. The differences in binding energies between the Fe(II) and Fe(III) components of the tribofilms and of the natural oxide film can be explained taking into account that in the case of the naturally formed film the aging of the film resulted in a magnetite film, whereas the peak position after tribotest suggests the formation of FeO and maghemite, as has already been found in model alloy studies [45]. Furthermore, the presence of the peak at 712.4 eV suggests at least a partial removal or reaction of the hydroxides, resulting in the formation of iron phosphate: this is also supported by the oxygen-to-phosphorus ratio, which is almost stoichiometric at higher loads (i.e. 4 : 1). Apparently all traces of hydroxides have been removed from the surface, since no hydroxide peak can be detected at 711.9 eV [46].

The approximate film thickness (including a carbon overlayer) can be calculated from the attenuation by the overlying film of the electrons emitted from the metallic iron present in the steel beneath the surface film. It was found to be 5.0 ± 0.2 nm after the 5 N test. This film thickness is significantly smaller than that reported in most other studies. This may be due to the fact that most of these studies have been carried out at temperatures above 80 °C. At these temperatures, but not below, a thick thermal film is formed on the surface. Under the conditions applied in this study, the contact temperature and the tribochemical reaction form a thin phosphate film, but the thermal activation is insufficient to form a thick polyphosphate film.

In order to ascertain the degree of homogeneity of the contact area a Scanning Auger Microscopy O(KLL) map of a 1 N contact area (figure 7) has also been collected on the tribologically stressed samples. It can be seen that the surface composition is not homogenous. The wear tracks are visible, with a regular spacing of 25 μm, arising probably from the 25 μm steps made during the tribotest. In Auger microscopy, it is difficult to distinguish between different chemical states, even though recently the O(KLL) signal has been applied to the chemical-state identification of the species present in a tribofilm [34], while in XPS we have to be aware that we are averaging over the analyzed area. It can be assumed that changes in the distribution of these areas, as reported in the present work, reflect the load dependency of the tribofilm.

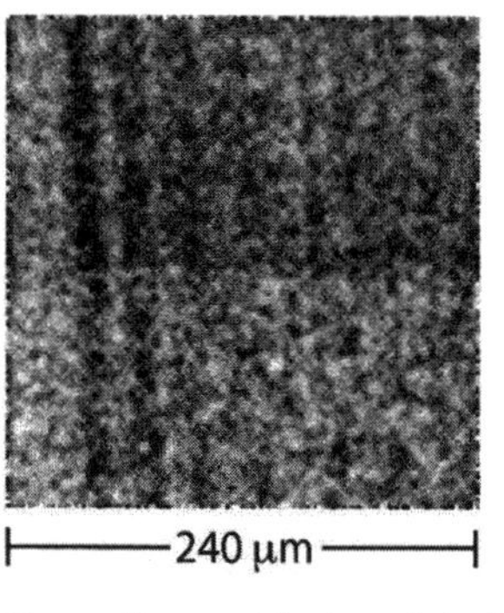

Figure 7. O (KLL) map of an area tribostressed at a load of 1 N. The intensity of the signal shows a periodicity of 25 μm, corresponding to the step size of the tribotest sequence.

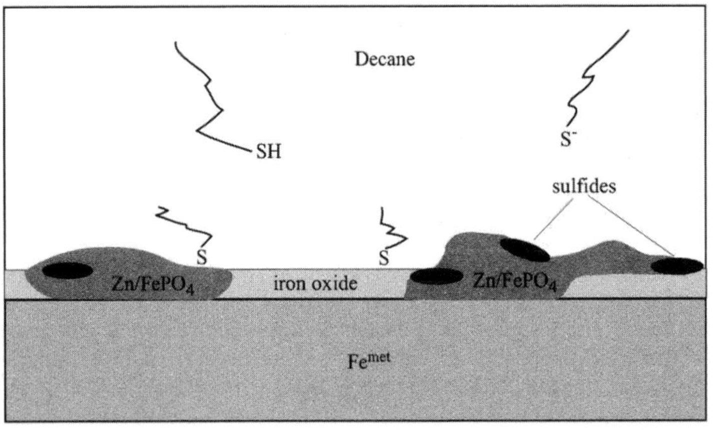

Figure 8. Model of the tribofilm: a mixed iron/zinc phosphate film is formed on the surface. Due to the absence of thermal activation, the protective film is only formed in areas with high tribological stress, occurring predominantly at the asperities. Some iron oxide remains in the less-stressed areas (valleys). Some sulfide is also present. Organosulfur species are formed and remain either dissolved in the oil or adsorbed on the surface. The total film thickness is approximately 5 nm.

4.3. Mechanism of film formation

The proposed mechanism of reaction under these experimental conditions is the following: in the non-contact area a thin, organic layer (including ZnDTP or LI-ZnDTP) is adsorbed. Due to the mechanical and thermal stress during the tribological contact, the adsorbed species and additive molecules from the solution react with the iron oxide to form a protective film. This takes place in the region of highest mechanical and thermal stress. The phosphorus in the ZnDTP reacts with oxygen from the oxide and dissolved oxygen in the oil to form an iron phosphate-containing film, while organosulfur species are released into the solution. Some of these species may react to form sulfides that are incorporated in the film or adsorbed on the surface. The phosphate film is thin (<5 nm) and inhomogeneous. The composition and amount of phosphate film formed on the surface depend on the applied load. The higher the load, the more phosphate is found on the surface. This may be explained by the fact that the higher the applied load, the higher the real contact area. The highest tribological and thermal stresses are found in the real contact area and therefore the tribofilm is formed preferentially in these areas. The amount and the composition of the tribofilm are therefore dependent on the size and the tribological stress in the real contact area. The proposed film structure is shown in figure 8.

5. Conclusion

The results presented here suggest that in the non-contact area, only physical adsorption of ZnDTP takes place on the existing oxide layer. In the contact region, a reaction occurs due to the tribostress and a thin tribofilm is formed. The tribofilm contains iron phosphate and other reaction products, such as sulfides and organosulfur species. The amount and composition of the tribofilm is dependent on load. The phosphate shows a greater increase with increasing applied load than the other species. No polyphosphate could be detected, due to the lack of thermal activation.

The combinatorial approach, especially when combined with imaging surface spectroscopies, has proven itself to be useful in characterizing the lubricant–additive interaction. It allows the tribochemical reaction occurring under a variety of tribological conditions to be characterized, and this approach shows promise as a tool for the screening of new antiwear additives.

Acknowledgments

Financial support of the ETH Zurich and Italian MURST (ex 60% grant to A.R.) is gratefully acknowledged. Prof. J.M. Martin (Ecole Centrale de Lyon, France) is thanked for supplying the i-ZnDTP. Prof. B. Elsener (University of Cagliari, Italy) is thanked for his help with the AES measurements.

References

[1] M. Eglin, A. Rossi and N.D. Spencer, Tribol. Lett. (this issue).
[2] C.H. Bovington, *Chemistry and Technology of Lubricants*, eds. R.M. Mortier and S.T. Orszulik (Blackie and Son, Glasgow and London, 1997).
[3] A.J. Gellman and N.D. Spencer, J. Eng. Tribol. 216 (2002) 443.
[4] D. Klamann, *Lubricants and Related Products* (Verlag Chemie, Weinheim, 1984).
[5] J.M. Martin, Tribol. Lett. 6 (1999) 1.
[6] Z. Yin, M. Kasrai, M. Fuller, G.M. Bancroft, K. Fyfe and K.H. Tan, Wear 202 (1997) 172.
[7] M.L.S. Fuller, M. Kasrai, G.M. Bancroft, K. Fyfe and K.H. Tan, Tribol. Int. 31 (1998) 627.
[8] C.H. Bovington and B. Dacre, ASLE Trans. 27 (1984) 252.
[9] R.B. Jones and R.C. Coy, ASLE Trans. 24 (1981) 91.

[10] S.H. Choa, K.C. Ludema, G.E. Potter, B.M. Dekoven, T.A. Morgan and K.K. Kar, Wear 177 (1994) 33.

[11] F.M. Piras, *In situ* Attenuated Total Reflection Tribometry. PhD Thesis, presented at the Department of Materials, ETH Zürich, 2002.

[12] F.M. Piras, A. Rossi and N.D. Spencer, in: *Proceedings of the 28th Leeds–Lyon Symposium on Tribology*, eds. D. Dowson *et al.*, Vienna, 2002, p. 199.

[13] F.M. Piras, A. Rossi and N.D. Spencer, Langmuir 18 (2002) 6606.

[14] J.S. Sheasby, T.A. Caughlin, A.G. Blahey and K.F. Laycock, Tribol. Int. 23 (1990) 301.

[15] R.J. Bird and G.D. Galvin, Wear 37 (1976) 143.

[16] T.P. Debies and W.G. Johnston, ASLE Trans. 23 (1980) 289.

[17] S. Jahanmir, J. Tribol. Trans. ASME 109 (1987) 577.

[18] Z.F. Yin, M. Kasrai, G.M. Bancroft, K.F. Laycock and K.H. Tan, Tribol. Int. 26 (1993) 383.

[19] E.C. Onyiriuka, J. Non-Crystal. Solids 163 (1993) 268.

[20] R.K. Brow, J. Non-Crystal. Solids 194 (1996) 267.

[21] M. Kasrai, M.L.S. Fuller, M. Scaini, Z. Yin, R.W. Brunner, G.M. Bancroft, M.E. Fleet, K. Fyfe and K.H. Tan, in: *Proceedings of the 21st Leeds–Lyon Symposium on Tribology*, eds. D. Dowson *et al.* Leeds, 1994, p. 659.

[22] F.M. Piras, A. Rossi and N.D. Spencer, Tribol. Lett. (in press) (2003).

[23] R. Gresch, W. Müller-Warmuth and H. Dutz, J. Non-Crystal. Solids 34 (1979) 127.

[24] S.W. Gaarenstroom and N. Winograd, J. Chem. Phys. 67 (1977) 3500.

[25] C.D. Wagner, L.H. Gale and R.H. Raymond, Anal. Chem. 51 (1979) 466.

[26] A. Rossi, K. Matsumoto, M. Eglin, F.M. Piras and N.D. Spencer, in preparation.

[27] J.S. Sheasby and Z. Nisenholz, Tribol. Trans. 36 (1993) 741.

[28] J.F. Graham, C. McCague and P.R. Norton, Tribol. Lett. 6 (1999) 149.

[29] N.D. Spencer, in: *Proceedings of the JAST Tribology Conference, Tokyo*, 1998, p. 396.

[30] F.M. Piras, A. Rossi and N.D. Spencer, submitted to Surf. Interface Anal. (2003).

[31] M. Eglin, A. Rossi and N.D. Spencer, in: *Proceedings of the 28th Leeds–Lyon Symposium* on Tribology, eds. D. Dowson *et al.* Vienna, 2002, p. 49.

[32] G.W. Canning, M.L.S. Fuller, G.M. Bancroft, M. Kasrai, J.N. Cutler, G. De Stasio and B. Gilbert, Tribol. Lett. 6 (1999) 159.

[33] T. Le Mogne, J.M. Martin and C. Grossiord, Tribol. Series 36 (1999) 413.

[34] J.M. Martin, C. Grossiord, T. Le Mogne, S. Bec and A. Tonck, Tribol. Int. 34 (2001) 523.

[35] W.A. Glaeser, D. Baer and M. Engelhardt, Wear 162–164 (1993) 132.

[36] M.P. Seah, Surf. Interface Anal. 31 (2001) 721.

[37] J.H. Scofield, J. Elect. Spect. Rel. Phen. (1976) 129.

[38] M.P. Seah and W.A. Dench, Surf. Interface Anal. 1 (1979) 2.

[39] K. Berresheim, M. Matternklosson and M. Wilmers, Fresenius J. Anal. Chem. 341 (1991) 121.

[40] R.P. Gupta and S.K. Sen, Phys. Rev. B 12 (1975) 15.

[41] M. Eglin, A. Rossi, F.M. Piras and N.D. Spencer, Surface Sci. Spectra 8 (2001) 97.

[42] G.W. Stachowiak and A.W. Batchelor, *Engineering Tribology* (Butterworth–Heinemann, Boston, 2001).

[43] A. Rossi and B. Elsener, Mat. Sci. Forum 185–188 (1995) 337.

[44] D. Schuetzle, R.O. Carter, J. Shyu, R.A. Dickie, J. Holubka and N.S. McIntyre, Appl. Spect. 40 (1986) 641.

[45] A. Rossi, B. Elsener and G. Puddu, *Materials and Corrosion* (in press) (2002).

[46] N.S. McIntyre and D.G. Zetaruk, Anal. Chem. 49 (1977) 1521.

Tribology Letters, Vol. 15, No. 3, October 2003 (© 2003) 181

Combined *in situ* (ATR FT-IR) and *ex situ* (XPS) study of the ZnDTP-iron surface interaction

Federica M. Piras[a], Antonella Rossi[a,b] and Nicholas D. Spencer[a]

[a]*Laboratory for Surface Science and Technology, Department of Materials, Swiss Federal Institute of Technology, ETH Zürich, CH-8092 Zürich, Switzerland*

[b]*Department of Inorganic and Analytical Chemistry, University of Cagliari, INSTM, Cagliari, Italy*

Received 15 September 2002; accepted 23 February 2003

Attenuated total reflection infrared (ATR FT-IR) and X-ray photoelectron spectroscopy (XPS) have been used for the *in situ* and *ex situ* characterization of thermal and tribological films formed on iron from a commercial zinc dialkyldithiophosphate (ZnDTP). From *in situ* ATR FT-IR analysis, information on the chemical changes occurring at the iron/lubricant additive interface was obtained during heating and sliding at high temperatures. Different mechanisms and chemical compositions have been found for the thermal and tribochemical reactions between the ZnDTP and the iron surface under the experimental conditions used in this work. Both the ATR FT-IR and the XPS results show the decomposition of ZnDTP with the formation of polyphosphates following thermal testing at 150 °C. However, after tribological testing at the same temperature an inorganic phosphate film has been detected on the iron surface instead.

KEY WORDS: tribochemistry, boundary lubrication, ZnDTP, attenuated total reflection, FT-IR, small-area X-ray photoelectron spectroscopy

1. Introduction

Zinc dialkyldithiophosphates (ZnDTPs) have been the most widely used lubricant additives in engine and industrial oil formulations since the 1940s, due to their multifunctional performance [1]. The ZnDTP family comprises primary and secondary aliphatic dithiophosphates, with chain lengths from C3 to C12, and alkylated phenoldithiophosphates. ZnDTPs were added first as antioxidants, but it was soon recognized that they can reduce or even prevent both mild and severe wear, acting as *antiwear* and *extreme-pressure* additives, respectively [2,3]. It is generally accepted that the antiwear and extreme-pressure performance of ZnDTP results from the formation of tribofilms, which show a shear strength that is sufficiently low to ensure that the shear plane is located within the protective tribofilm, while being sufficiently high to maintain its integrity [4,5]. In addition, the rate of formation of the ZnDTP tribofilms is low enough to avoid the corrosive wear of the substrate and sufficiently high to avoid its complete removal during sliding. ZnDTP has been described as a "smart" material [4] due to its ability to act, under different tribological conditions, as both antiwear and extreme-pressure additive, and to form films with different mechanical properties under different conditions.

The structure and the chemical composition of ZnDTP antiwear tribofilms have been mainly studied *ex situ*, with surface analytical techniques such as X-ray photoelectron spectroscopy (XPS) [6,7], Auger electron spectroscopy (AES) [8,9], and X-ray absorption near-edge spectroscopy (XANES) [10,11]. It has been found that the film composition depends on the temperature and tribological test conditions. Some authors have suggested that the films consist of inorganic amorphous phosphates, mainly orthophosphate (PO_4^{3-}) and pyrophosphate ($P_2O_7^{4-}$) associated with zinc and other metals, such as Ca^{2+}, derived from overbased detergent additives [12,13]. Other authors have proposed the formation of a mixture of short- and long-chain polyphosphates on the basis of XANES results obtained comparing spectra collected on tribostressed samples with those from sodium phosphate compounds of known chain length [14]. In addition, the presence of sulfides and oxides in the inner part of the tribofilm has been indicated [11] by nanoindentation measurements [15] and depth-profiling results obtained by SIMS [16].

Martin *et al.* recently proposed a two-layer structure of the ZnDTP tribofilms, where a thin, (~ 10 nm) long-chain zinc poly(thio)phosphate film is superimposed on a thicker (~ 100 nm) short-chain mixed Fe/Zn polyphosphate film, containing some embedded nanocrystallites of ZnO and ZnS [17]. A model for elucidating the mechanism of the tribochemical reaction of zinc dithiophosphate with a steel substrate has been based on the hard and soft acids and bases (HSAB) theory, developed by Pearson [18,19] and applied to the ZnDTP by Jones and Coy [20] and by Martin [21].

Despite the large number of investigations conducted on the formation, structure, and chemical composition of the ZnDTP tribofilms, further studies need to be

1023-8883/03/1000–0181/0 © 2003 Plenum Publishing Corporation carried out in order to truly understand the mechanism of the ZnDTP's antiwear action. This will be necessary in order to be able to replace it with new more environmentally friendly alternatives.

The use of modern analytical techniques has yielded important information concerning the ZnDTP antiwear action, but no single technique provides the whole picture of the ZnDTP tribofilm formation mechanism and structure. The combination of *in situ* methods with *ex situ* analyses may be the most powerful approach of all, since tribofilms surface-chemically analyzed during a friction test (i.e. *in situ*) are representative of the films in their active state, and monitoring changes with time, temperature, or other variables is possible. *Ex situ* analysis, on the other hand, generally includes air exposure during the transfer from the tribometer to the analytical instrument, as well as the solvent washing of the tribological surfaces before the analysis, which may render the analyzed tribofilms unrepresentative of the original surface material. The great advantage of *ex situ* analysis is the possibility of combining various complementary modern surface analytical techniques.

In this work, a combination of *in situ* attenuated total reflection infrared (ATR FT-IR) tribometry together with *ex situ* small-area X-ray photoelectron spectroscopy (SAXPS) has been used to obtain information on the chemical changes occurring during heating and sliding and to determine the spatial distribution of the elements and of their respective chemical states. The results obtained using a commercial ZnDTP, both pure and dissolved in poly-α-olefin (PAO), are presented and discussed. Three solutions of ZnDTP in PAO were tested in the ATR tribometer under tribological conditions at 150 °C, at concentration of 5, 10, and 20 wt%. Due to the low intensity of the peak assigned to the P–O–C stretching vibration in the ATR spectra of the 5 and 10 wt% solutions, spectral changes during the experiments were not clearly detectable. Significant and clear changes were observed only in the case of the 20 wt% solution.

2. Experimental

2.1. Materials

The ZnDTP additive investigated is a commercial secondary ZnDTP (C3 + C6, Hitec 7169, Ethyl Petroleum Additives International), purified by liquid chromatography [22]. A commercial poly-α-olefin (PAO, Durasyn 166, Tunap Industrie GmbH. & Co.) was used as a base oil. The viscosity of the PAO used in this work was 0.0256 Pa·s at 40 °C, and 0.00487 Pa·s at 100 °C. Solutions ranging from 5 to 20% of ZnDTP in PAO were investigated.

Before XPS analysis, the tribo-stressed iron-coated germanium ATR crystals were washed with cyclohexane (p.a. $\geq$ 99.5%, Fluka), in order to remove the residual additive and the base oil. The samples that had been subjected to thermal treatment only were apparently free of volatile materials and were analyzed without solvent washing to avoid removal of reaction products.

2.2. Methods

2.2.1. In situ *attenuated total reflection infrared (ATR FT-IR) tribometry*

ATR FT-IR spectra were obtained with a Nicolet Magna-IR System 550 Fourier Transform Spectrometer equipped with a Greasby-Specac advanced overhead (specaflow) ATR System. The spectra were measured using the experimental conditions listed in table 1. Trapezoidal ATR elements of monocrystalline germanium with an angle of incidence of 45°, dimensions 72 × 10 × 6 mm and 7 reflections have been used in this work. The iron coating of the crystals was performed by magnetron sputtering at the Paul Scherrer Institut (PSI, Villigen, Switzerland). Before each experiment, the thickness of the iron coating was checked by ellipsometry and found to be 12.0 ± 0.3 nm. The XPS analysis of the iron surface indicated the presence of a thin iron oxide film.

As a background spectrum, the single-beam spectrum of the iron-coated germanium ATR crystal was acquired before each experiment. The ATR FT-IR spectra presented in this work are reported after subtraction of the background spectrum, without any other correction.

The ATR tribological tests were performed under pure sliding conditions with an ATR tribometer, as previously described in detail [23] and shown schematically in figure 1. The tribological experimental conditions used in this work (table 1) were chosen so as to assure boundary lubrication conditions.

The ATR FT-IR spectra have been measured after cooling the ATR tribometer down to room temperature, due to the increasing absorbance of the germanium with temperature [23].

2.2.2. *X-ray photoelectron spectroscopy (XPS)*

XPS analyses were performed using a PHI 5700 spectrophotometer equipped with a concentric hemispherical analyzer in the standard configuration (Physical Electronics, Eden Prairie, MN, USA). The vacuum system consists of a turbo-molecular pump, ion pump, and a titanium sublimation pump. The base pressure before the analysis was better than 10^{-7} Pa. The X-ray source was AlKα (1486.6 eV), run at 300 watts. The incident angle was 54.7° and the emission angle was 45° with respect to the sample surface normal. All the spectra were obtained in digital mode. A constant energy of 23.50 eV was set across the hemispheres of the electron analyzer operated in the Fixed Analyzer Transmission (FAT) mode for the detailed spectra; the survey spectra have been acquired with 187.85 eV pass

Table 1
ATR FT-IR and tribological experimental conditions.

ATR FT-IR experimental conditions		Tribological experimental conditions	
Detector	MCT/A	Apparent contact area	0.1 mm^2
Spectral range	4000–650 cm^{-1}	Normal load	7 N
Number of scans	1024	Average contact pressure	34 MPa
Resolution	2 cm^{-1}	Average sliding velocity	24 mm/min (pseudo-sinusoidal)
Acquisition time	15 min	Additive	secondary ZnDTP
		Lubricant	poly-α-olefin
		Temperature	150 °C

energy. The instrument operated in minimum area mode and the aperture was 0.4 mm diameter. The instrument was calibrated using the inert-gas-ion-sputter-cleaned reference materials SCAA90 of Cu, Ag and Au [24]. The accuracy of the binding energy values was found to be ± 0.05 eV.

The binding-energy values reported in this work are the mean values over at least three independent measurements. The standard deviation is also reported.

To compensate for sample charging during the analysis, all the binding energies were referred to the aliphatic carbon, C1s, signal taken at 285.0 eV, according to [25]. The spectra were resolved into their Gaussian–Lorentzian components after background subtraction, according to [26].

The atomic concentration of the element j was calculated as: $X_j = (I_{ij}/S_{ij})/\sum_j (I_{ij}/S_{ij})$, where I_{ij} is the area of the peak i of the element j and S_{ij} is the sensitivity factor. The sensitivity factors were calculated from the Scofield photoionization cross sections [27], the attenuation length corrected for the emission angle, and the transmission function of the analyzer, as described in [28], assuming the sample to be homogeneous. The attenuation length corrected for the emission angle (ϑ) was calculated as: $\Lambda_{ij} = A/(KE) + B \times \sqrt{KE} \times \cos\vartheta$, where the values of A and B are 31 and 0.087, respectively, and Λ_{ij} is in nm. These values are valid for organic compounds, according to [29], and they have been used to calculate the sensitivity factors for ZnDTP frozen on gold (table 2).

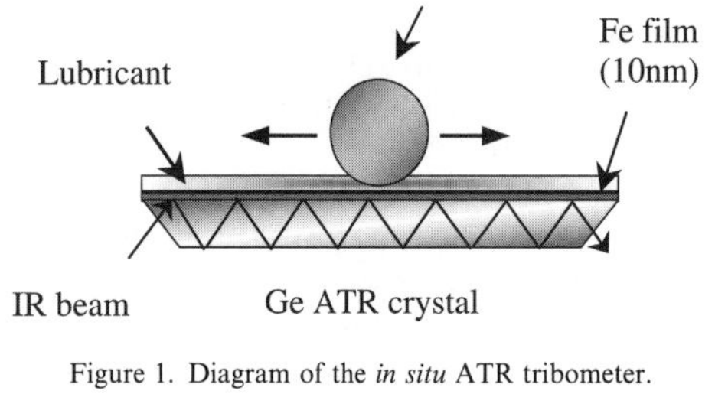

Figure 1. Diagram of the *in situ* ATR tribometer.

3. Results

In situ ATR thermal and tribological tests have been performed in the presence of both pure ZnDTP and a ZnDTP solution in poly-α-olefin (PAO). The XPS spectra were collected *ex situ* on thermal and tribological films formed from pure ZnDTP and from ZnDTP dissolved in PAO. The XPS results of pure ZnDTP and of the iron-coated germanium ATR crystal are shown for comparison. In the following, only the O1s and the P2p XPS spectra are presented, because these two signals allow the identification of non-bridging oxygen ($P\text{–}O^-$ and P=O) and bridging oxygen (P–O–P) [7,30,31]. A more detailed analysis of the XPS spectra of all the elements detected on the thermal and tribological films is reported elsewhere [28].

The ATR FT-IR spectra were measured *in situ* during both thermal and tribological tests.

3.1. Reference material: ZnDTP frozen on gold

One drop of the commercially pure ZnDTP was placed on a sputtered gold substrate and cooled down to liquid nitrogen temperature in the spectrometer introduction system and subsequently transferred to a liquid nitrogen cooling stage in the main chamber. The survey and the detailed spectra of C1s, O1s, P2p, S2p, and $Zn2p_{3/2}$ obtained on ZnDTP frozen on sputtered gold are shown in figure 2. The binding energy values are summarized in table 2. The C1s signal is asymmetric, containing a contribution at 285.0 eV and one at 286.7 ± 0.2 eV. The first component is assigned to the carbon of the aliphatic chains [32] and the second to the carbon covalently bonded to oxygen of the thiophosphate group **C**–O–P(S_2) [33,34]. The O1s signal is fitted with only one Gaussian/Lorentzian curve having a maximum at 533.0 ± 0.1 eV; this value is typical for a bridging oxygen [33,35]. The P2p signal is a doublet with $2p_{1/2}$ and $2p_{3/2}$ components: their energy separation (0.95 eV) and area ratio of 0.5 was fixed for the curve-fitting analysis of all P2p spectra. The $P2p_{3/2}$ peak maximum is at 133.6 ± 0.1 eV. The S2p is also asym-

Table 2
XPS binding energies (B.E.), full width at half-maximum height (FWHM) and experimental results of quantitative analysis of bulk ZnDTP (frozen on gold). Attenuation length [Λ] and sensitivity factors [S] calculated for the commercial ZnDTP frozen on gold.

	C1s (1)		C1s (2)	O1s	P2p	S2p	$Zn2p_{3/2}$
Assignment	**CH_2, CH_3**		**C–O–P**	**P–O–C**	**$(CO)_2$-P$(S)_2$-**	**P–S**	**Zn(II)**
B.E. (eV)	285.0		286.7 ± 0.2	533.0 ± 0.1	133.6 ± 0.1	162.6 ± 0.2	1022.5 ± 0.1
FWHM	1.7		1.7	2.0	1.8	2.0	2.1
Area ratio		3:1		–	–	–	–
Attenuation length Λ (nm)		2.1		1.9	2.3	2.3	1.4
Sensitivity factor, *S*		1.68		4.62	2.16	3.05	24.10
Conc. % calc.		62%		14%	7%	14%	3%
Conc. % exper.		60%		13%	8%	15%	4%

metric because of spin-orbit splitting, and thus it was fitted keeping the separation of the two components and their branching ratio always equal to 1.2 and 0.5 eV, respectively. The $S2p_{3/2}$ peak maximum is found at 162.6 ± 0.2 eV. A single peak at 1022.5 ± 0.1 eV characterizes the $Zn2p_{3/2}$ signal.

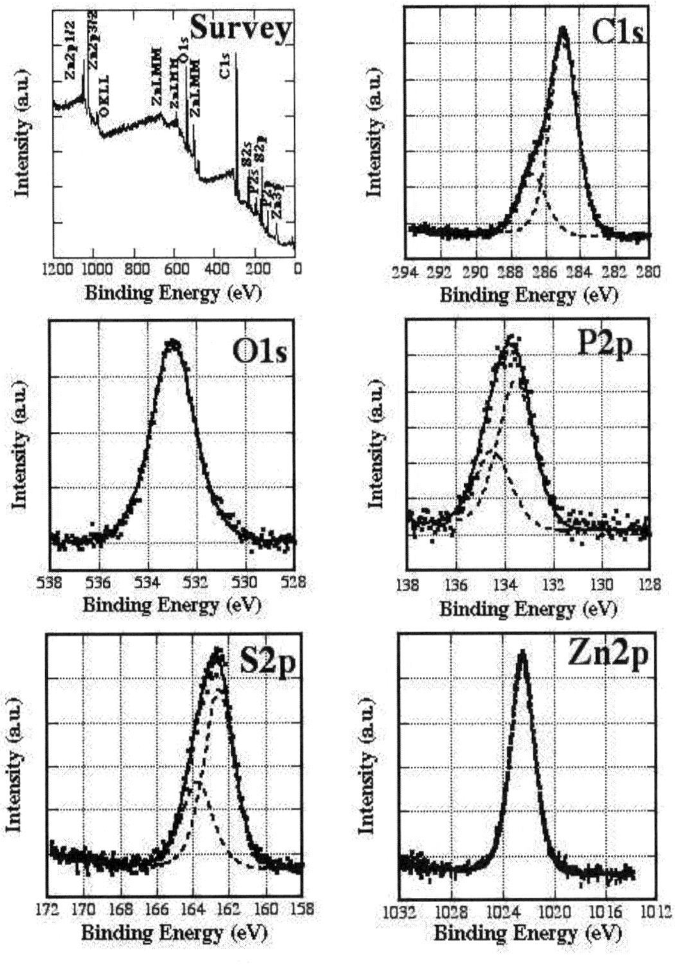

Figure 2. Survey and detailed XP spectra of C1s, O1s, P2p, S2p, and Zn2p of a commercial purified ZnDTP, frozen on sputtered gold and used as reference compound. In each detailed spectrum, the points are the original data; the line between the points is the envelope of the model Gaussian–Lorentzian product functions (dotted lines) used in the curve-fitting routine. The P2p and S2p signals are fitted with two components to account for the spin–orbit splitting.

The results of qualitative and quantitative analysis are summarized in table 2. The experimental atomic concentration percentages (see "Conc. % exper." in table 2) are found as expected from the result of the elemental analysis [22] and the stoichiometry (see "Conc. % calc." in table 2).

3.2. *Substrate: Iron film*

Germanium ATR crystals were coated with an iron film, 10 nm in nominal thickness, by magnetron sputtering [23] and analyzed "as received": only iron, oxygen, and carbon signals were detected. The carbon is due to a thin contamination layer present on the iron film.

The high-resolution XPS spectrum of $Fe2p_{3/2}$ after satellite [26] and background subtractions is shown in figure 3. The curve-fitting of the $Fe2p_{3/2}$ signal was resolved into four contributions: at 707.0 ± 0.1 eV, 708.8 ± 0.1 eV, 710.3 ± 0.1 eV and 711.8 ± 0.1 eV. The signal at 707.0 eV is assigned to iron in the metallic state. The signals at higher binding energies, 708.8 eV and 710.3 eV, are assigned to iron (II) and iron (III), respectively. The signal at the highest binding energy, 711.8 eV, is assigned to iron hydroxide (FeO**O**H), according to [36,37].

The high-resolution XPS spectrum of O1s, which is asymmetric and contains contributions at 530.2 ± 0.1 eV, 531.8 ± 0.1 eV and 533.1 ± 0.1 eV, is also shown in figure 3. The signals at low binding energies, 530.2 eV and 531.8 eV, are assigned to oxygen bonded to iron in the oxide (Fe–**O**) and the hydroxy-group (Fe–**O**–**H**), respectively. The third component, 533.1 eV, is assigned to adsorbed water [36,37].

Table 3 contains the results of qualitative analysis and the calculated thickness of the iron oxy-hydroxide layer detected on the iron film.

3.3. *Thermal tests*

3.3.1. In situ *ATR tribometry*

The typical ATR FT-IR spectrum acquired after thermal testing of the commercially pure ZnDTP at 150 °C on an iron-coated germanium ATR crystal is shown in figures 4(a) and (b) (see Thermal test). The spectrum collected at the beginning of the experiment (0 h) shows the characteristic IR peaks of the ZnDTP molecule [23]. The spectrum acquired after 38 h of heating at 150 °C (see figures 4(a) and (b), Thermal test) shows a broad band in the region around 1100 cm^{-1}, due to the overlap of the peaks assigned to the stretching vibration of the P=O and P–O groups, which are found in the region 1320–1140 cm^{-1} and 950–1060 cm^{-1}, respectively. The very intense peak at 916 cm^{-1} is assigned to the stretching vibration of the P–O–P group [23].

A typical ATR FT-IR spectrum collected after a 126-hour thermal test performed at 150 °C in the presence of the 20 wt% solution of ZnDTP in PAO on an iron-coated (10 nm) germanium ATR crystal is reported in figures 4(c) and (d). At the beginning of the experiment, the ATR spectrum (0 h) shows the characteristic peaks of the ZnDTP and PAO molecules on a germanium crystal coated with an iron film (10 nm) [22]. The spectrum recorded after 126 h of heating at 150 °C shows a broad band between 1300 and 900 cm^{-1}, assigned to the stretching vibrations of P–O–P, P–O, and P=O groups, thus indicating thermal decomposition of the ZnDTP molecule. According to the results obtained after the thermal test at 150 °C in the presence of pure ZnDTP (see figures 4(a) and (b)), the peak at 914 cm^{-1} is assigned to the asymmetric stretching vibration of the P–O–P bond (νP–O–P) and the band around 1100 cm^{-1} to an overlap of the peaks assigned to the stretching vibrations of the P=O and P–O group (νP=O, νP–O).

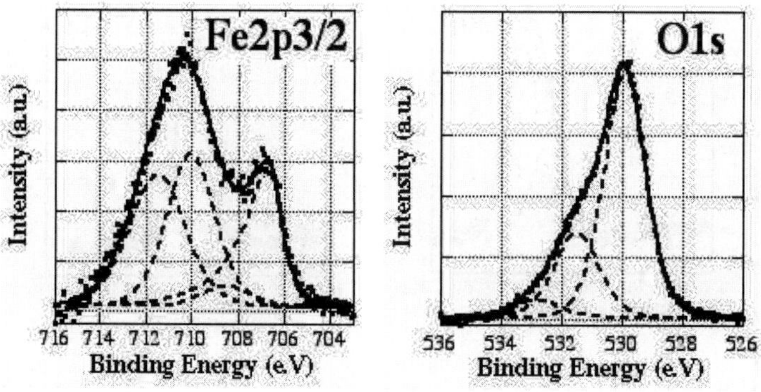

Figure 3. Detailed XP spectra of Fe2p3/2 and O1s regions of the iron film deposited on a germanium crystal. In each detailed spectrum, the points are the original data; the line between the points is the envelope of the model Gaussian–Lorentzian product functions (dotted lines) used in the curve-fitting routine.

Table 3
XPS binding energies (B.E. ±0.1 eV), full width at half-maximum height (FWHM) and calculated thickness of the iron oxy-hydroxide layer detected on the iron substrate.

	O1s (1)	O1s (2)	O1s (3)	$Fe2p_{3/2}$ (1)	$Fe2p_{3/2}$ (2)	$Fe2p_{3/2}$ (3)	$Fe2p_{3/2}$ (4)
Assignment	Fe–**O**	FeOOH	H_2**O**	**Fe**^{0}	**Fe(II)**	**Fe(III)**	**Fe**OOH
B.E. (eV)	530.2	531.8	533.1	707.0	708.8	710.3	711.8
FWHM	1.7	1.8	1.8	1.3	2.6	2.6	3.0

Oxy-hydroxide film thickness 2.6 ± 0.3 nm.

3.3.2. Ex situ *XPS analysis*

Commercially pure ZnDTP or the 20 wt% solution in PAO have been deposited onto iron-coated germanium ATR crystals, in the ATR tribometer. Subsequently, the temperature of the tribometer was raised to 150 °C. ATR FT-IR spectra were then collected periodically *in situ*, during the experiments, cooling the system down to room temperature during the short periods required to acquire the spectra. XPS analysis was subsequently performed *ex situ*, after no further changes were detected in the ATR FT-IR spectra, i.e. after 38 h of heating at 150 °C in the presence of the commercially pure ZnDTP and 126 h of heating at 150 °C in the 20 wt% solution of it in PAO.

In figure 5(a), the O1s and P2p XPS spectra measured after a thermal test in the presence of pure ZnDTP are reported. The O1s signal consists of four contributions. The first component, 530.1 ± 0.3 eV, is very small and it might be assigned to the formation of ZnO as side product. All three other signals are assigned to oxygen

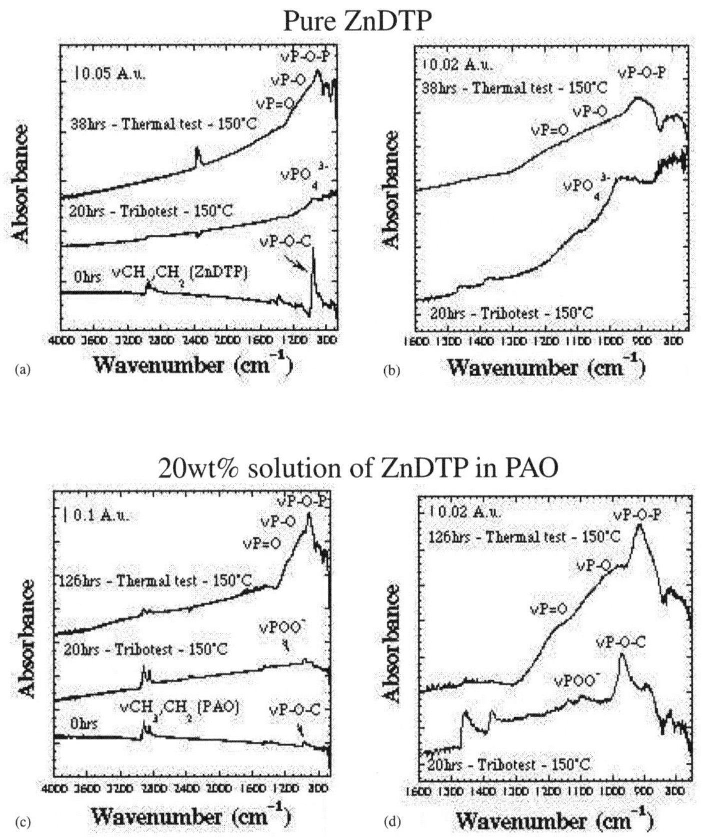

Figure 4. ATR FT-IR spectra of pure ZnDTP (a) and (b) and 20 wt% solution of ZnDTP in PAO (c) and (d) acquired before (0 h) and after thermal- and tribo-testing at 150 °C, using germanium ATR crystals coated with iron (10 nm). The absorbance of the two spectra named "20 h – Tribotest – 150 °C" reported in the enlarged regions (b) and (d) has been multiplied by 4.

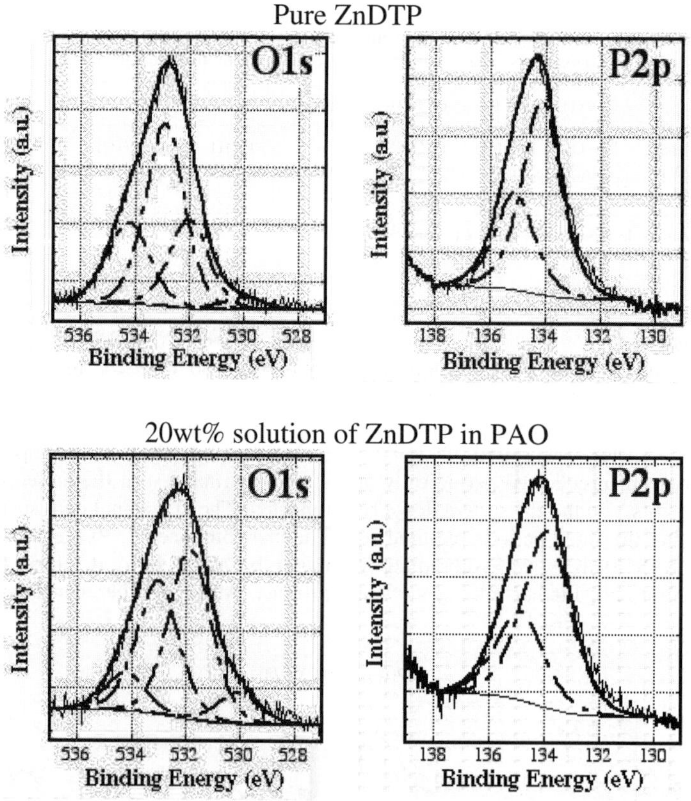

Figure 5. High-resolution XPS spectra of O1s and P2p of the iron-coated germanium ATR crystals after 38 h (a) and after 126 h (b) thermal-testing at 150 °C, in the presence of commercially pure ZnDTP (a) and 20 wt% solution of ZnDTP in PAO (b). In each detailed spectrum, the scattered lines are the original data; the line between the points is the envelope of the model Gaussian–Lorentzian product functions (dotted lines) used in the curve-fitting routine.

bonded to phosphorus: the first (532.2 ± 0.2 eV) to a non-bridging oxygen, involved in a P=**O** or P−**O**$^-$ bond [30], the second (533.1 ± 0.3 eV) to a P−**O**−C (see table 2) and the third (534.4 ± 0.3 eV) to a P−**O**−P group [7,30,31]. The P2p signal is shifted to higher B.E. after heating at 150 °C: from 133.6 (see table 2) to 134.1 ± 0.2 eV, which is closer to the value reported for polyphosphates [7,30,31]. Also the S2p signal (not shown) is shifted to higher B.E.: from 162.6 eV (see table 2) to 163.1 ± 0.2 eV, which may be assigned to a thiolate group [38]. The Zn2p$_{3/2}$ signal (not shown) was found at 1022.6 ± 0.1 eV. No Fe2p signal was detected.

The high-resolution XPS spectra of O1s and P2p, acquired after a thermal test in the presence of the 20 wt% solution of ZnDTP in PAO, are presented in figure 5(b). The O1s signal consists of four peaks at the same binding energies detected for the pure ZnDTP, i.e. 530.0 ± 0.3 eV, 531.9 ± 0.3 eV, 533.0 ± 0.3 eV, and 534.4± 0.2 eV. The P2p$_{3/2}$ was found at 133.8 ± 0.2 eV. The S2p$_{3/2}$ and Zn2p$_{3/2}$ signals (not shown) were detected at 162.9 ± 0.1 eV and 1022.5 ± 0.1 eV, respectively. The Fe2p photoelectron peaks were not detected.

3.4. *Tribological tests*

Tribological tests were performed with the ATR tribometer after depositing commercially pure ZnDTP or a solution of it in PAO on an iron-coated germanium ATR crystal. ATR FT-IR spectra were collected *in situ* for 20 h, and XPS spectra were acquired *ex situ*, after 20 h tribo-testing at 150 °C.

3.4.1. In situ *ATR tribometry*

A typical ZnDTP ATR FT-IR spectrum acquired after a 20 h tribological test at 150 °C, on an iron-coated germanium ATR crystal, is reported in figures 4(a) and (b) (see Tribotest). The spectrum shows two bands with maximum intensities at 1102 and 972 cm^{-1}, assigned to the stretching vibrations of the PO_4^{3-} group [23].

The ATR FT-IR spectrum typically measured after sliding at 150 °C in the presence of the 20 wt% solution of ZnDTP in PAO on a germanium ATR crystal coated with an iron film (10 nm) is shown in figures 4(c) and (d) (see Tribotest). The spectrum shows two peaks at 1138 and 1099 cm^{-1}, assigned to the asymmetric and symmetric stretching vibrations of $(POO)^-$ group, respectively [22].

3.4.2. Ex situ *XPS analysis*

Visual inspection of the iron surface, after tribotesting at 150 °C, revealed areas with different morphologies. A wear scar showing several distinct wear tracks was detected across the contact area. In figure 6, an XPS image of $Fe2p_{3/2}$ and an example of the distribution of the selected points for small-area XPS analysis across the contact area is shown.

A good agreement has been observed between the XPS data obtained from the high-resolution XPS spectra acquired at several points distributed throughout the contact area.

The high-resolution XPS spectra of O1s and P2p acquired in the contact area after tribo-testing in the presence of pure ZnDTP are presented in figure 7(a). The O1s signal consists of three contributions: the first at 530.4 ± 0.1 eV, assigned to a Fe–**O** bonding (see table 3), the second at 532.0 ± 0.1 eV to a non-bridging oxygen bonded to phosphorus [30], and the third at 533.2 ± 0.1 eV, to a C–**O**–P and adsorbed water (see tables 2 and 3). The binding energy of the $P2p_{3/2}$ component is at 133.4 ± 0.1 eV and is assigned to phosphorus in an orthophosphate group ($\mathbf{P}O_4^{3-}$) [34]. The $S2p_{3/2}$ (not shown) is found at 162.4 ± 0.1 eV. The $Zn2p_{3/2}$ signal has a single contribution at 1022.5 ± 0.1 eV. The $Fe2p_{3/2}$ signal shows three contributions: at 709.1 ± 0.1 eV, 710.5 ± 0.1 eV and 712.4 ± 0.1 eV. The signals at lower binding energies, 709.1 eV and 710.5 eV, are assigned to iron (II) and iron (III), respectively (see table 3), while the signal at the highest binding energy, 712.4 eV, is attributable to iron phosphate (**Fe**PO_4) [34]. The $Fe2p_{3/2}$ signal assigned to iron in the metallic state was not detected. Germanium signals were neither detected in the contact area nor inside the wear scar.

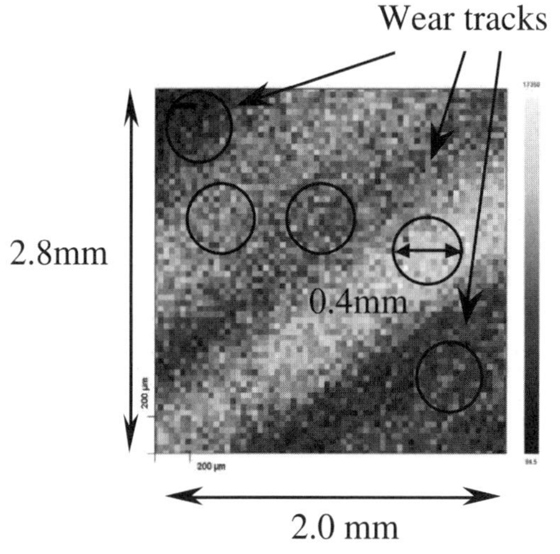

Figure 6. XPS image of $Fe2p_{3/2}$ of an Fe-coated Ge ATR crystal after tribotesting at 150 °C in the presence of the 20 wt% ZnDTP solution in PAO. The position of the wear tracks in the acquired XPS image and the distribution of the selected points for subsequent small-area XPS analysis are shown. The dimensions 2 × 2.8 mm indicate the real size of the analyzed area on the surface sample taking into account the emission angle of 45° and the consequent distortion of the image.

In figure 7(b), the high-resolution XPS spectra of O1s and P2p measured on a 20 wt% solution of ZnDTP in PAO after tribo-testing are shown. The O1s signal consists of only two components: the first at 532.0 ± 0.1 eV (non-bridging oxygen) assigned to a phosphate group, and the second at 533.4 ± 0.1 eV, assigned to a C–**O**–P group or adsorbed water. The peaks at 530.0 and 534.4 eV were not detected. The binding energy of the $P2p_{3/2}$ component is found at 133.9 ± 0.1 eV, and that of the $S2p_{3/2}$ at 162.4 ± 0.1 eV. The $Zn2p_{3/2}$ signal (not shown) has a single contribution at 1022.7 ± 0.1 eV. The $Fe2p_{3/2}$ signal showed a very low intensity and signal-to-noise ratio; thus it was difficult to analyze its components by curve-fitting. The maximum of the peak was found at 710.7 eV, indicating the presence of Fe(III).

4. Discussion

Despite the large number of studies on the tribochemistry of ZnDTP reported in the literature, several points regarding the antiwear action of ZnDTP are still unclear and under discussion. Two open questions concern the kinetics of the tribofilm formation and the relative roles of tribochemical and thermal processes. Film formation in the presence of ZnDTP also occurs at lower apparent temperatures due to flash temperatures induced by the metal–metal contact itself [39,41]. The dependence of the ZnDTP decomposition rate on the temperature has been also described in the literature and considered one of the most important features in the anti-wear film formation mechanism [10].

4.1. Chemical composition of commercially pure ZnDTP

The comparison of the atomic percentages obtained from quantitative XPS analysis with the atomic percentages calculated from the molecular formula $C_{18}H_{40}O_4P_2S_4Zn$ for the mixture of di-isopropyl ZnDTP and hexyl ZnDTP (C3 + C6 secondary ZnDTP) indicates that the ZnDTP investigated is pure and neutral, in agreement with the results obtained from elemental analysis [22]. After purification by liquid chromatography, the commercially pure ZnDTP contains neither diluting oil nor zinc oxide.

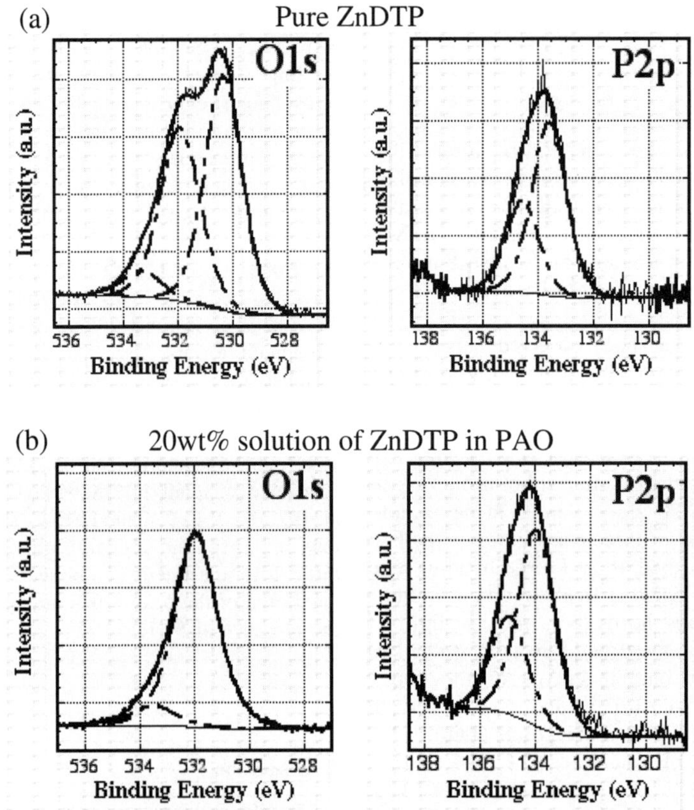

Figure 7. High-resolution XPS spectra of O1s and P2p of pure ZnDTP (a) and 20 wt% ZnDTP solution in PAO (b) on Fe-coated Ge ATR crystals, after tribotesting at 150 °C. In each detailed spectrum, the scattered lines are the original data; the line between the points is the envelope of the model Gaussian–Lorentzian product functions (dotted lines) used in the curve-fitting routine.

4.2. Chemical composition of thermal reaction products

Both the XPS and ATR results indicate that after heating at 150 °C, ZnDTP thermally decomposes to form polyphosphates from both pure ZnDTP and 20 wt% solution of ZnDTP in PAO. The P2p XPS signal (figure 5) shows a component at high binding energy of 134.4 ± 0.1 eV — characteristic of polyphosphates, according to previous studies [7,30,31]. The O1s peak at 534.4 ± 0.3 eV is assigned to the corresponding P−O−P group. In addition, O1s peaks assigned to non-bridging oxygens (P=O and P−O species), at 532.2 ± 0.2 eV, have been detected, in agreement with the ATR spectra (figure 4).

A comparison of the O1s and P2p XPS spectra acquired for the pure ZnDTP (figure 5(a)) with those measured for the 20 wt% ZnDTP solution in PAO (figure 5(b)) indicates that the only significant difference is the relative intensity of the O1s components. In the diluted ZnDTP, the non-bridging oxygen signal is more intense. This may be due to the involvement of solvent molecules in the thermally induced rearrangement mechanism. This difference in the mechanism seems to be supported by previous *in situ* ATR studies [22,23], which have shown that the rate of thermal decomposition of ZnDTP in the 20 wt% solution in PAO is lower than that of the pure additive. An induction period of about 26 h was observed for the pure ZnDTP [22] before any precipitate was visible and modifications of the ATR spectrum were detected; the thermal reaction was complete after only 38 h of heating at 150 °C (see figures 4(a) and (b), Thermal test). Only after 63 h of heating at 150 °C [23] did the ATR spectrum of the solution show any variations in the fingerprint region, and it was not until 126 h that the thermal reaction was complete (see figures 4(c) and (d), Thermal test).

The Fe2p XPS signal has not been detected in the thermal films formed from either pure ZnDTP and its solution in PAO on iron-coated germanium ATR crystals. This indicates there is no formation of polyphosphates associated with iron by thermochemical reaction of ZnDTP on iron surfaces, under the experimental conditions used in this work.

Both the XPS and ATR results presented in this work for the thermal reaction product formed from ZnDTP on iron surfaces are in good agreement with the mechanism of decomposition proposed in the literature [20,40].

4.3. Chemical composition of tribochemical films

Both the XPS and the ATR results reported in this work indicate the formation of inorganic phosphates (orthophosphates) following 20 h tribo-testing at 150 °C in the presence of the commercially pure ZnDTP. The binding energies of oxygen, phosphorus, and iron XPS signals measured after tribo-testing at 150 °C indicate the formation of iron orthophosphate in the contact area (figure 7(a)). After the tribological test at 150 °C, bands assigned to the stretching vibration of the PO_4^{3-} group were detected in the pure ZnDTP ATR spectrum (figures 4(a) and (b), Tribotest).

The XPS data obtained after the tribological test in the presence of the 20 wt% ZnDTP in PAO also show the presence of iron phosphate in the contact area, in agreement with the results obtained on the pure ZnDTP. Differences were detected in the oxygen signal, which does not show the component at low binding energy (530.4 eV, see figures 7(a) and (b)) and assigned to an Fe–O bond. This may suggest the formation of a thicker organic film on the sample surface in the presence of PAO, which may be adsorbed in the external part of the film (the C1s is a single signal at 285.0 eV).

The ATR spectrum acquired on the same sample (figures 4(c) and (d), Tribotest) shows two bands assigned to POO^- species, which may be present in the solution above an uneven film of phosphate, since this species is not detectable (after washing) by XPS. Since the ATR spectrum is obtained by an averaged spectrum obtained along the length of the germanium ATR crystal, the intensity of the νPO_4 peaks might be so weak compared to that of the νPOO^- peaks that it is not detected.

Furthermore, the S2p binding energy (162.4 ± 0.1 eV) indicates the formation of sulfides, in agreement with the model proposed by Martin *et al.* [17].

Germanium XPS signals were neither detected in the contact area nor inside the wear scar. The absence of germanium XPS signals and the disappearance of the metallic iron signal inside the wear scar is an indication of the formation of thick tribofilms. As reported in the literature, thicknesses of up to 120 nm have been measured for ZnDTP tribofilms formed on iron/steel surfaces [10,41].

As mentioned above, it is generally accepted that antiwear tribofilms of zinc/iron polyphosphates and poly(thio)phosphates are formed on the tribo-stressed surfaces from ZnDTP solution. Models proposed for ZnDTP tribofilms consist of a multilayered structure, where a polyphosphate film is responsible for the wear reduction [4,16,17].

Both the XPS and the ATR results reported in this work indicate the formation of iron/zinc orthophosphates following 20 h tribo-testing at 150 °C in the presence of the commercially pure ZnDTP. The only evidence for the P–O–P species, characteristic of polyphosphates, was the presence of the P2p XPS signal at 134.4 eV in the more severely worn area.

It has been previously pointed out that the tribological conditions used in this work are milder than those typically reported in the literature [23]. Under very severe conditions the formation of a bi-layered tribofilm, consisting of polyphosphates overlying mixed short chain phosphates, has been predicted, according to a model of the mechanism of formation of ZnDTP tribofilms on steel proposed by Martin and based on the Pearson's theory of hard and soft acids and bases (HSAB) [21]. The same model predicts the formation of short-chain iron/zinc phosphates under less severe conditions, such as those used in this work.

5. Conclusions

The utility and the power of combining *ex situ* XPS and *in situ* ATR studies for the characterization of thermal and tribological films formed on iron from ZnDTP have been demonstrated. By means of *ex situ* small-area XPS, combined with state-of-the-art data analysis methods, the spatial distribution of the elements and of their respective chemical states was investigated. *In situ* ATR analysis provided information on the chemical changes occurring at the iron/ZnDTP interface, and thus the mechanism of the action of the additive under these conditions was elucidated during heating and sliding at high temperatures.

The XPS and ATR results reported in this work show a good correlation. The chemical analysis of the ZnDTP (pure and dissolved in PAO) thermally decomposed on iron surfaces indicates the formation of zinc polyphosphates. The inorganic phosphates were detected by ATR analysis on the tribological films formed from ZnDTP on iron. XPS spectra acquired on the same samples suggest the formation of iron/zinc orthophosphate and show the presence of metal sulphides in the contact area and inside the wear scar. XPS also provided some evidence that polyphosphates were being formed in the most severely worn regions.

It can be concluded that, under the conditions used in this study and at 150 °C, the thermal and tribochemical reactions of ZnDTP with iron surfaces display quite different mechanisms.

Acknowledgments

The authors would like to thank Dr. H. Camenzind (Ciba Specialty Chemicals, Switzerland) for supplying the purified commercial ZnDTP and Mr. M. Horisberger (PSI, Villigen, Switzerland) for preparing the iron coatings by magnetron sputtering. Financial support from the ETH and Italian MURST (60% grant to A. Rossi) is gratefully acknowledged.

References

[1] A.J. Gellman and N.D. Spencer, J. Eng. Trib., Proc. Instn. Mech. Engrs. Part J, 216 (2002) 443 and A.M. Barnes, K. Bartle and V.R.A. Thibon, Tribol. Int. 34 (2001) 389.

[2] M. Rasberg, in: *Chemistry and Technology of Lubricants*, eds. R.M. Mortier and S.T. Orszulik (VCH Publishers, 1992) Ch. 4.

[3] A.R. Lansdown, in: *Chemistry and Technology of Lubricants*, eds. R.M. Mortier and S.T. Orszulik (VCH Publishers, 1992) Ch. 12.

[4] S. Bec, A. Tonck, J.M. Georges, R.C. Coy, J.C. Bell and G.W. Roper, *Proc. R. Soc. Lond. A* 455 (1999) 4181.

[5] A. Tonck, S. Bec, J.M. Georges, R.C. Coy, J.C. Bell and G.W. Roper, in: *Proc. 25th Leeds–Lyon Symp.*, eds. D. Dowson *et al.* (Elsevier Science, 1999) p. 39.

[6] R.J. Bird and G.D. Galvin, Wear 37 (1976) 143.

[7] M. Kasrai, M. Fuller, M. Scaini, Z. Yin, R.W. Brunner, G.M. Bancroft, M.E. Fleet, K. Fyfe and K.H. Tan, in: *Proc. 21st Leeds–Lyon Symp.*, eds. D. Dowson *et al.* (Elsevier Science, 1995) p. 659.

[8] T.P. Debies and W.G. Johnston, ASLE Trans. 23 (1980) 289.

[9] S. Jahanmir, J. Tribol. 109 (1987) 577.

[10] Z. Yin, M. Kasrai, M. Fuller, G.M. Bancroft, K. Fyfe and K.H. Tan, Wear 202 (1997) 172.

[11] M. Fuller, Z. Yin, M. Kasrai, G.M. Bancroft, E.S. Yamaguchi, P.R. Ryason, P.A. Willermet and K.H. Tan, Tribol. Int. 30 (1997) 305.

[12] G.M. Bancroft, M. Kasrai, M. Fuller, Z. Yin, K. Fyfe and K.H. Tan, Tribol. Lett. 3 (1997) 47.

[13] P.A. Willermet, R.O. Carter III and E.N. Boulos, Tribol. Int. 25 (1992) 371.

[14] P.A. Willermet, D.P. Dailey, R.O. Carter III, P.J. Schmitz, W. Zhu, J.C. Bell and D. Park, Tribol. Int. 28 (1995) 163.

[15] R.C. Watkins, Tribol. Int. 15 (1982) 523.

[16] J.C. Bell, K.M. Delargy and A.M. Seeney, in: *Proc. 18th Leeds–Lyon Symp.*, eds. D. Dowson *et al.* (Elsevier Science, 1992) p. 387.

[17] J.M. Martin, C. Grossiord, Th. Le Mogne, S. Bec and A. Tonck, Tribol. Int. 34 (2001) 523.

[18] R.G. Pearson, *Chemical Hardness* (Wiley-VCH, 1997).

[19] T.-L. Ho, Chem. Rev. 75 (1975) 1.

[20] R.C. Coy and R.B. Jones, ASLE Trans. 24 (1980) 77; and R.B. Jones and R.C. Coy, ASLE Trans. 24 (1981) 91.

[21] J.M. Martin, Tribol. Lett. 6 (1999) 1.

[22] F.M. Piras, A. Rossi and N.D. Spencer, in: *Proc. 28th Leeds–Lyon Symp.*, eds. D. Dowson *et al.* (Elsevier Science, 2002) p. 199.

[23] F.M. Piras, A. Rossi and N.D. Spencer, Langmuir 18 (2002) 17, 6606.

[24] M.P. Seah, Surf. Inter. Anal. 14 (1989) 488.

[25] M.P. Seah, in: *Practical Surface Analysis*, eds. D. Briggs and M. P. Seah (Wiley, Chichester, 1990) Appendix 2.

[26] P.M.A. Sherwood, in: *Practical Surface Analysis*, eds. D. Briggs and M. P. Seah (Wiley, Chichester, 1983) Appendix 3.

[27] J.H. Scofield, J. Elec. Spect. Rel. Phen. 8 (1976) 129.

[28] F.M. Piras, A. Rossi and N.D. Spencer, *Manuscript in preparation.*

[29] M.P. Seah and W.A. Dench, Surf. Inter. Anal. 1 (1979) 2.

[30] E.C. Onyiriuka, J. Non-Cryst. Solids 163 (1993) 268.

[31] R.K. Brow, J. Non-Cryst. Solids 194 (1996) 267.

[32] J.F. Moulder, W. F. Stickle, P.E. Sobol, K.D. Bomben, *Handbook of X-ray Photoelectron Spectroscopy*, ed. J. Chastain (Perkin Elmer Corporation, Eden Prairie, 1992).

[33] M. Eglin, A. Rossi, F.M. Piras and N.D. Spencer, Surf. Sci. Spec., 8 (2001) 2, 97.

[34] D. Schuetzle, R.O. Carter III, J. Shyu, R.A. Dickie, J. Holobka, and N.S. McIntyre, Appl. Spec. 40 (1986) 5.

[35] M. Textor, L. Ruiz, R. Hofer, A. Rossi, K. Feldman, G. Hähner and N.D. Spencer, Langmuir 16 (2000) 3257.

[36] N.S. McIntyre and D.G. Zetaruk, Anal. Chem. 49 (1977) 1521.

[37] A. Rossi, B. Elsener, G. Hähner, M. Textor and N.D. Spencer, Surf. Interface Anal. 29 (2000) 460.

[38] F. Bensebaa, Y. Zhou, Y. Deslandes, E. Kruus and T.H. Ellis, Surf. Sc. 405 (1998) L472.

[39] M. Eglin, A. Rossi and N.D. Spencer, in: *Proc. 28th Leeds–Lyon Symp.*, eds. D. Dowson *et al.* (Elsevier Science, 2002).

[40] J.J. Dickert and C.N. Rowe, J. Org. Chem. 32 (1967) 647; and C.N. Rowe and J.J. Dickert, ASLE Trans. 10 (1967) 85.

[41] L. Taylor, A. Dratva and H.A. Spikes, Tribol. Trans. 43 (2000) 469.

Available online at www.sciencedirect.com
SCIENCE DIRECT
Wear 256 (2004) 578–584
WEAR
www.elsevier.com/locate/wear

Surface analytical studies of surface-additive interactions, by means of in situ and combinatorial approaches

A. Rossi [a,b], M. Eglin [a], F.M. Piras [a,1], K. Matsumoto [a,2], N.D. Spencer [a,*]

[a] Laboratory for Surface Science and Technology, Department of Materials, Swiss Federal Institute of Technology, ETH-Zürich, Switzerland
[b] Department of Inorganic and Analytical Chemistry, University of Cagliari, INSTM, Cagliari, Italy

Abstract

The influence of tribological conditions on the surface reactions occurring between zinc dialkyldithiophosphate (ZnDTP) and steel surfaces has been studied by means of a combination of X-ray photoelectron spectroscopy (XPS), time-of-flight secondary ion mass spectroscopy (ToF-SIMS), in situ attenuated total reflection (ATR) infrared spectroscopy, and high-throughput combinatorial approaches. Purely thermal treatment at 150 °C appears to lead to the formation of zinc polyphosphates. However, in the presence of tribological stress, simple phosphates appear to dominate, with some indication that higher load conditions lead to an increase in the surface concentration of both phosphate and, at higher temperatures, polyphosphate.

Keywords: Surface analysis; X-ray photoelectron spectroscopy; Time-of-flight secondary ion mass spectroscopy; Attenuated total reflection infrared spectroscopy; Zinc dialkyldithiophosphate; Combinatorial methods; In situ methods; Tribochemistry

1. Introduction

Originally added to lubricating oils as an antioxidant [1], zinc dialkyldithiophosphate (ZnDTP) has become one of the most widespread and heavily investigated antiwear and extreme-pressure additives. Crucial to the functioning of ZnDTP is the formation of a protective film on rubbing surfaces in situations where elastohydrodynamic lubrication has broken down [2]. Despite innumerable studies that have sought to characterize such films, the literature has not, to date, reached a consensus on the chemical pathways involved in this process. One probable reason for this situation, recently expounded by Bec et al. [3], is that ZnDTP functions as a "smart" material, producing films with different chemical and mechanical properties, depending on the prevailing tribological conditions. In other words, since most laboratories have employed different sets of conditions for their investigations, the films that they have been scrutinizing have displayed correspondingly different properties.

In this study, we have employed a combination of imaging surface-analytical approaches: X-ray photoelectron spectroscopy (XPS) and time-of-flight secondary ion mass spectroscopy (ToF-SIMS), a recently developed in situ tribometer [4] and a combinatorial, high-throughput screening method [5], in order to determine the effects of temperature and load on the reaction pathways that occur during tribochemical activation of ZnDTP.

2. Experimental

2.1. Materials

A purified commercial secondary ZnDTP (C3 + C6, Hitec 7169, Ethyl Petroleum Additives International Ltd., UK. 30, Shell International Trading Company) and a commercial poly-α-olefin (PAO, Durasyn 166, Tunap Industrie GmbH & Co. Germany) were used as lubricant additive and base oil, respectively, unless stated otherwise. The purification of the commercial ZnDTP was carried out by liquid chromatography. 100 g of the additive was eluted from a column of 200 g of column-chromatography-grade silica gel (short column, Silica gel 60, 0.063-0.22, Merck, Switzerland) with hexane and ethyl acetate.

In the ball-on-disk experiments, a 1 wt.% solution of di-isopropyl zinc dithiophosphate (*i*-ZnDTP) in decane was used as a lubricant. To dissolve the additive in the lubricant, the solution was stirred at 60 °C for 30 min.

* Corresponding author. Tel.: +41-1-632-5850; fax: +41-1-633-1027.
E-mail address: nicholas.spencer@mat.ethz.ch (N.D. Spencer).
[1] Present address: Department of Chemical and Biosystem Sciences and Technologies, University of Siena, Siena, Italy.
[2] Present address: Sumitomo Metal Industries Ltd., Pipe & Tube R&D Department, Amagasaki, Hyogo, Japan.

0043-1648/$ – see front matter © 2003 Elsevier B.V. All rights reserved.
doi:10.1016/j.wear.2003.10.001

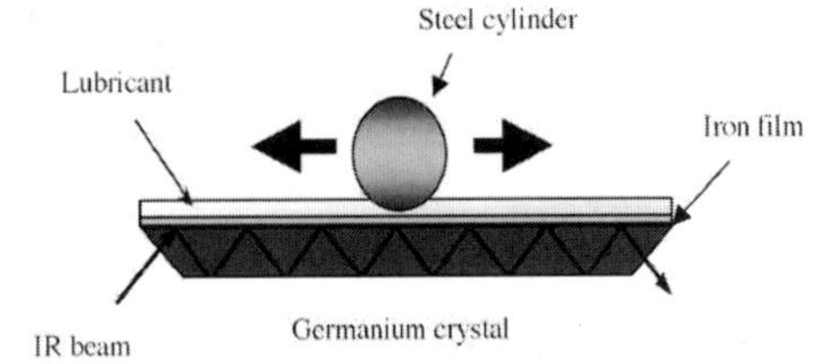

Fig. 1. Schematic diagram of the ATR FTIR tribometer.

2.2. Methods

2.2.1. Attenuated total reflection (ATR) infrared tribometry

ATR spectra were obtained in situ with a Nicolet Magna-IR System 550 Fourier Transform Spectrometer, equipped with a Greasby-Specac advanced overhead (specaflow) 1401 Series ATR System. The ATR spectra were collected in the spectral range 4000–650 cm^{-1}, with a resolution of 2 cm^{-1}. The ATR tribometer is equipped with a crystal holder that is heatable to 200 °C, so that both purely thermal reactions and tribochemical reactions can be studied at elevated temperatures (Fig. 1). During sliding, a normal load of 7 N (apparent contact pressure 34 MPa) was applied, the average sliding velocity (sinusoidal) being 24 mm/min. Further details may be found in [4].

2.2.2. X-ray photoelectron spectroscopy

The XPS analyses were performed on a PHI 5700 system with an Omni Focus IV lens system (Physical Electronics, Eden Prairie, MN, USA). The residual pressure in the spectrometer during the data acquisition was always below 5×10^{-7} Pa. The X-ray source was Al Kα (1486.6 eV), run at 350 W. The diameter of the analyzed area was either 400 or 120 μm in the case of the combinatorial approach experiments. The spectrometer was operated in the fixed analyzer transmission (FAT) mode [10,12]. The instrument was calibrated using the spectral lines of $Au(4f_{7/2})$ and $Cu(2p_{3/2})$ at 83.98 and 932.67 eV, respectively [316]. $Ag(3d_{5/2})$, Cu(3p), Cu(LMM) and Ag(MNN) peak energies at 368.26, 75.14, 567.96, 1128.78 eV, respectively, were used to verify the linearity of the binding-energy scale. The accuracy was ±0.05 eV.

Spectroscopic maps were acquired using the imaging capabilities of the Omni Focus IV lens system. The analyzed spot is electrostatically rastered over the sample (typically 64 × 64 pixels, 2 mm × 2 mm). For each pixel a full spectrum of the selected energy region is acquired. The acquired images were processed with the PHI Multipak (V6.0) software. Spectra can be extracted from the map by selecting a region of interest and can be used to reconstruct "Chemical State Maps" with a linear-least-squares (LLS) routine. The correlation between the extracted spectra and the spectra at each pixel generates a new map, which shows regions having similar chemical states of the given element.

From these maps, areas of interest were determined and analyzed by small-area XPS (SAXPS). For curve fitting, CASA XPS (V2.0) software (CASAXPS Software Ltd., UK) was used. The spectra were resolved into their components after an integrated Shirley background subtraction. Curve-fitting parameters (FWHM and Gaussian/Lorenzian ratio) were determined from reference spectra. The spectra were fitted with Gaussian/Lorenzian peaks, keeping the FWHM and the Gaussian/Lorenzian ratio constant and fitting the energy of the peaks and their heights using a LLS algorithm. Quantitative analysis was carried out correcting the intensities for the inelastic mean free path according to Seah and Dench [6], the photoionization cross-section according to Scofield [8] and the analyzer transmission function [9].

2.2.3. Time-of-flight secondary ion mass spectrometry

The spectra were acquired using a Physical Electronics PHI 7200 ToF-SIMS instrument (Physical Electronics, Eden Prairie, MN, USA), allowing parallel mass registration with high sensitivity and high mass resolution. A gallium liquid-metal ion (LMI) gun at 25 keV beam energy was used for spatially resolved ToF-SIMS analysis. The beam diameter was 0.25 μm, the pulse width 100 ns and the TDC bin size 10 ns. A mass resolution $(m/\Delta m)$ of 4500 at $m/z = 29$ was obtained.

Images of the same region of the sample were collected in positive and negative SIMS polarity over an area of typically 100 μm × 100 μm using the LMI gun at 25 keV beam energy. The total primary ion dose for each image was about 10^{13} ions/cm^2. A Sun workstation was used for spectral acquisition and processing using ToF-Pak software.

2.2.4. Tribometer experiments

Flat-on-disk tribological measurements were performed with a CSEM tribometer (CSEM, Neuchâtel, Switzerland) using steel/steel tribo-pairs (AISI 52100). The load was kept constant at 10 N (nominal pressure: 857 Pa) while the velocity was varied stepwise from 1 to 10,000 mm/min. During the tribological experiment the pin was maintained at a constant distance from the center of the disc. The steel was mechanically polished and the final roughness (R_a) was 0.01 μm for the flat and less than 0.01 μm for the disk. The steel pair was immersed in the lubricant during the tribological testing, which was carried out in air at 25 ± 1 °C, at a RH of 50%.

Ball-on-disk measurements were carried out with a CETR UMT-2 (CETR, Campbell, CA, USA) tribometer, which allows the programming of load, rotational velocity and duration of the tribotest. The tribometer is equipped with a load cell with a maximum capacity of 5 N and a resolution of ±5 mN in two axes (normal load and friction force). Normal load is applied via a spring, and is constantly monitored and adjusted via a feedback loop. The spring constant was determined to be 2.7 N/mm in the z-direction and 5.7 N/mm in the y-direction (friction force). Normal load, friction force,

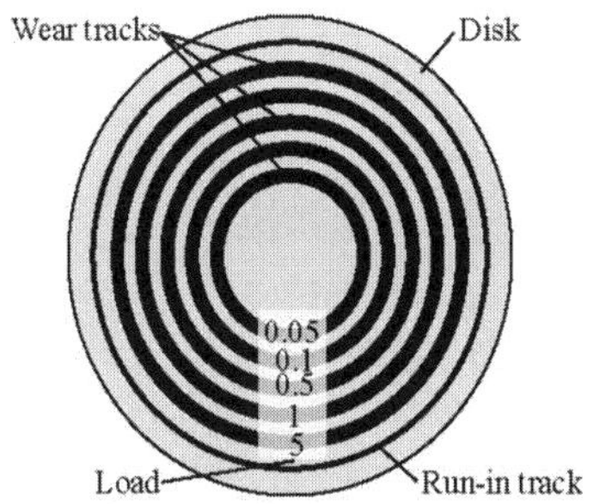

Fig. 2. Concentric 250 μm wide test regions with different loads in the ball-on-disk test.

rotational position of the disc and the *y*-position of the head versus time can all be digitally recorded.

The tests were carried at room temperature (24 ± 0.5 °C). The relative humidity of the air was recorded during each test and was between 22 and 38%. Prior to the actual tests, a running-in of the ball was performed at 5 N load with a speed of 31.4 mm/min for 2 h, outside the region that was later to be analyzed in order to create a flat area on the bottom of the ball.

Each ball-on-disk test consisted of five individual concentric test regions, each of which corresponded to a different load (Fig. 2). The relative velocity between the ball and the disc surface was kept constant at 31.4 mm/min for all the tribostressed regions, while the load was changed from 5 to 0.05 N. The number of turns of the ball on the disc was kept constant for each load, in order to reveal comparable tribostress. In order to create tribostressed areas that were sufficiently wide to be analyzed by XPS, the ball was moved after every fifth turn by 25 μm. Thus for each load, 11 overlapping wear tracks form a tribostressed annulus with a width of >250 μm. This allows a XPS analysis to be performed completely within the tribostressed area. After the tribotest, the samples were cleaned in cyclohexane in an ultrasonic bath for 30 s, dried under an argon stream and subsequently introduced into the surface analysis chamber.

3. Results

3.1. ATR infrared tribometry

The ATR FTIR spectra acquired after heating undiluted ZnDTP deposited on an iron-coated germanium crystal at 150 °C for 38 h and after sliding for 90 and 20 h at 25 and 150 °C, respectively, are shown in Fig. 3. The spectrum collected at the beginning of the experiment (0 h) at 25 °C is also given for comparison. This spectrum shows the characteristic IR peaks of the ZnDTP molecule [4]. The spectrum acquired after 38 h of heating at 150 °C shows a broad band in the region around 1100 cm^{-1}, due to the overlap of the peaks assigned to the stretching vibration of the P=O and P–O groups, which are found in the region 1320–1140 and

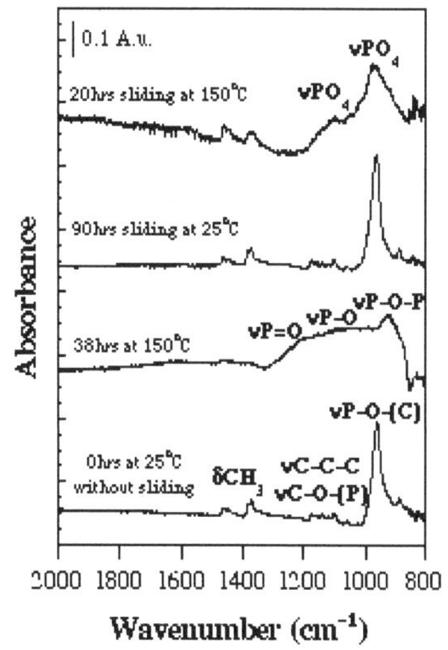

Fig. 3. In situ ATR FTIR spectra of thermal films formed at room temperature and at 150 °C and of tribofilms formed during tribotests at room temperature and at 150 °C. Iron-coated (10 nm) germanium ATR crystal. Baseline correction was applied. The spectrum collected on undiluted ZnDTP deposited on the iron-coated Ge crystal without sliding or heating is shown for comparison. The normal load was 7 N, the Hertzian contact area 0.1 mm^2, average apparent contact pressure 34 MPa, mean sliding velocity 24 mm/min.

1060–950 cm^{-1}. The more pronounced signal, at 916 cm^{-1}, is assigned to the stretching vibration of the P–O–P group [4]. The spectrum collected after prolonged sliding (90 h) at room temperature shows no changes when compared to the spectrum collected at the beginning of the experiment. Only those sliding experiments carried out at elevated temperatures show a detectable reaction product: the spectrum shows two bands with maximum intensities at 1102 and 972 cm^{-1}, assigned to the stretching vibrations of the PO_4^{3-} group in agreement with the spectra acquired on phosphates using the same analysis conditions [15].

3.2. Imaging X-ray photoelectron spectroscopy (i-XPS) and SAXPS

The XPS images shown in Fig. 4 were obtained from tribostressed samples (single track) subjected to running-in for 3 h and tribological testing over a wide range of velocities using a ball-on-disk configuration, with a constant load of 10 N, at room temperature in the presence of 2% ZnDTP in poly-alpha-olefin (PAO) (viscosity, 31 cS) as a base oil. The elemental distribution of carbon, oxygen, phosphorus, sulfur, zinc and iron around the contact region is shown. In the maps, yellow represents the highest signal intensity, black the lowest. It can be observed that the intensities of phosphorus, sulfur and zinc are higher in the contact region, whereas the iron intensity is lower in the wear scar with respect to the non-contact area. The spectra associated with the pixels in a defined area can be summed to create a

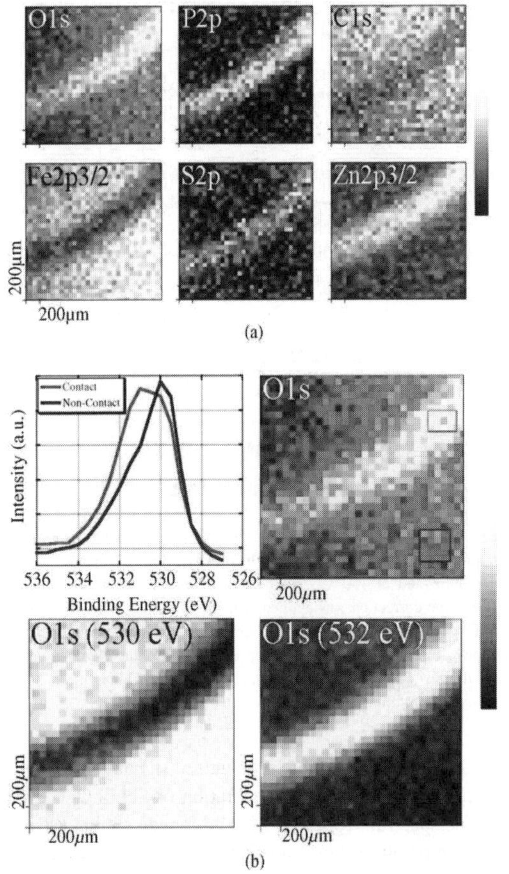

Fig. 4. (a) XPS elemental maps obtained from a tribostressed sample (ball-on-disk, 25 °C, load 10 N): worn and non-contact areas are clearly distinguished. (b) O1s spectra extracted from the contact area (red) and non-contact area (blue) of the O1s map. The O1s chemical state reconstructed maps are shown.

basis spectrum. A new map can subsequently be generated by fitting all spectra stored in the original map to the basis spectra by means of a LLS procedure. This second, derived map will then be the chemical state map for the given element. As an example in Fig. 4b, two spectra extracted from two areas of the oxygen map are shown together with the reconstructed chemical state maps.

The same i-XPS and SAXPS analyses were applied to the characterization of the thermal and tribological films formed after completing the in situ ATR tests [10]. After tribotesting, different areas were distinguishable, both in terms of morphology and composition (imaging XPS results, Fig. 5). After having distinguished the different regions of interest by imaging, SAXPS analysis was also performed. The detailed analysis of the high-resolution spectra allowed the identification of the different components of the elements. O1s could be fitted with signals assigned on the basis of reference compounds to metal oxide at 530.0 ± 0.2 eV and to phosphate group non-bridging oxygen (P–O–, POH, P=O)

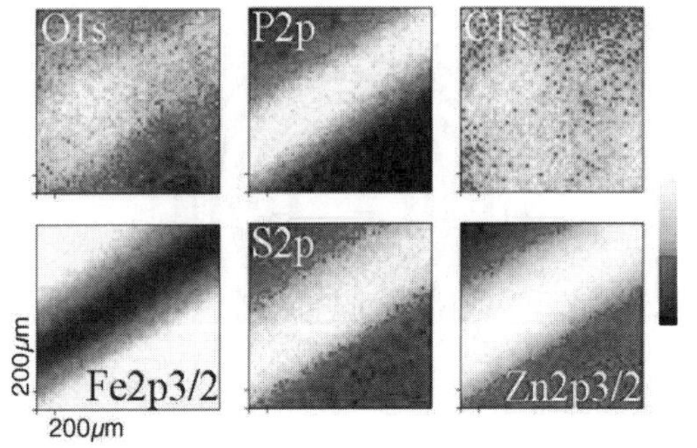

Fig. 5. XPS maps of a tribostressed iron-coated germanium ATR crystal after sliding with a steel cylinder at 150 °C for 20 h [10].

at 532.0 ± 0.2 eV. A third component assigned to the bridging oxygen at 534.4 ± 0.2 eV appeared only after the thermal test. After tribological testing at the same temperature the O1s was found at 533.2 eV [10]. The phosphorus P2p signal has been used to distinguish between phosphates and polyphosphates and the sulfur signal to differentiate thiolates, sulfides and sulfates (Fig. 4a and b).

SAXPS results confirmed the indication of the ATR FTIR spectra (Fig. 3): after purely thermal testing (150 °C) the product is a zinc polyphosphate, whereas sliding at the same temperature resulted in a phosphate film except in the center of the worn area, where polyphosphates were detected. These results show that there is a strong dependence of the reaction product on the temperature and on the tribological test conditions.

A combinatorial approach has been recently introduced to allow the characterization of the tribological films produced under a variety of conditions, spatially distributed on one sample [5,11,12]. An example of this new approach is shown in Fig. 6, where testing of *i*-ZnDTP in decane has been carried out under a series of different load conditions in concentric tracks on a single disk.

3.3. Time-of-flight secondary ion mass spectroscopy

The total positive ion map from a 100Cr6 sample after tribotesting in a ZnDTP-containing PAO solution at room

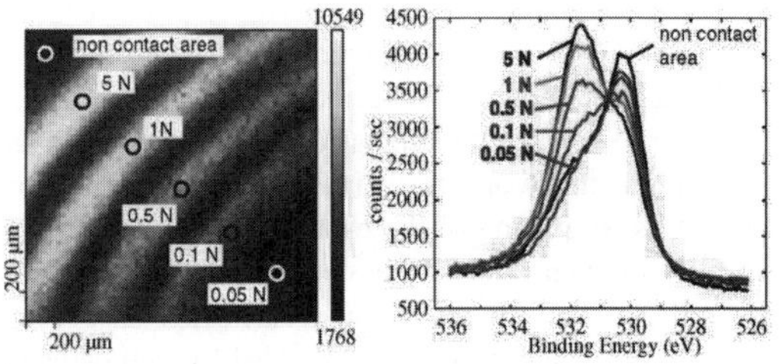

Fig. 6. O1s LLS map (left side) of a spectrum extracted from the tribostressed area (O1s = 532 eV); on the right the spectra extracted from the tested regions subjected to different loads [11,12].

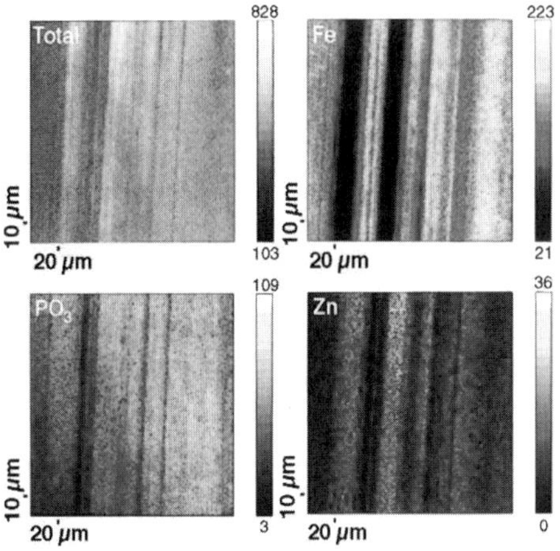

Fig. 7. ToF-SIMS images of a 100Cr6 steel disk after running-in and tribological tests over a wide velocity range, 25 °C, 50% RH. Contact area: fragment distribution in and outside scars.

temperature with a flat-on-disk configuration was collected, along with maps of Zn^+ and Fe^+ ions (Fig. 7).

In the same experiment, the maps of the negative fragments were also recorded, and the negative ion map of the PO_3^- fragment is also shown in Fig. 7. A higher intensity of the zinc signal is revealed where the iron intensity is lower, and the distribution of the PO_3^- fragments appears to be quite homogeneous within the contact area, even over wear-scar regions.

4. Discussion

Innumerable tribological research papers have demonstrated the effectiveness of the derivatives of thiophosphoric acids, such as the class of zinc dithiophosphates, as lubricant additives for ferrous materials. In the last 20 years, great progress has been made in determining the surface composition of the films formed—even if the mechanism of film formation from mixed-element additives is poorly understood and sometimes contradictory. This might be due to differences in temperature, test duration, applied load, and, last but not least, different in situ and ex situ analytical techniques and their individual effects on tribofilm composition. The aim of the present study is to assess the respective roles of physical and instrumental parameters.

4.1. Effect of temperature

One of the prerequisites for an antiwear additive is to form a surface film in the region where mechanical stress is applied. Some authors have reported that the thermal decomposition of ZnDTP is the key step in the mechanism of tribofilm formation [13], whereas other researchers have suggested that the tribofilm can also be formed at room temperature [14], implying a different mechanism of film formation at low and at high temperatures.

At room temperature in the absence of sliding, the ATR FTIR spectrum reported in Fig. 3 shows no changes in the peak position or in the peak shape, in comparison to the transmission spectrum [4]. Adsorption tests performed for up to 96 h confirmed these findings [15]. The spectrum recorded after 38 h at 150 °C indicates a rearrangement of the molecule with the formation of P–O–P bonds and the elimination of alkenes, as suggested by the decrease in the intensity of the alkyl peaks. The XPS analyses confirm the formation of polyphosphates. These results are in agreement with the literature [16–18].

Even after sliding tests at 25 °C for up to 90 h, no evidence of film formation is detectable in the ATR spectrum (Fig. 3). In contrast, the XPS results obtained after tribological testing at 25 °C (Fig. 4a and b) indicate the formation of a tribofilm in the mechanical stressed area. The tribofilm is constituted mainly of zinc/iron phosphates in the worn area and its elemental and chemical composition differs from the film present in the non-contact region, where the ZnDTP is probably only weakly adsorbed. Tribotests performed with constant velocity at a 10 N load in the same ball-on-disk configuration showed high intensity of the O1s component assigned to the phosphate non-bridging group [5]. From the presence of the iron signal at 706.9 ± 0.1 eV, assigned to metallic iron, either a film thickness lower than the XPS sampling depth (8 nm [7]) or a laterally highly inhomogeneous film has to be assumed.

After sliding tests carried out at 150 °C for 20 h, both in situ ATR FTIR spectra (Fig. 3) and ex situ XPS results (Fig. 5) support the formation of a tribofilm that is mainly constituted of zinc/iron phosphates [10]. The tribofilm is now thicker than that formed at room temperature, since no signal from the metallic substrate is detectable. The ATR spectra, which are averaged over the entire contact region, as described above, do not show P–O–P signals, in agreement with the XPS spectra collected in those regions of the contact area where the wear was not severe. Only in the center of a worn area does the photoelectron P2p signal found at 134 eV suggest the formation of P–O–P bonds. The formation of polyphosphate has been also reported by other authors [3,14,19,20] and seems thus to occur only at high temperatures and under severe wear conditions.

4.2. Effect of load

The chemical composition of the tribofilm can be expected to vary with the applied load [14]. Systematic investigations on the influence of the applied load are rare in the literature. The applied load probably affects the film-formation rate [21] by increasing the temperature in the contact area and may also change the length of the polyphosphate chains formed in the films [22]. In the present work, the effect of the applied load on the chemical composition of the surface layer in the contact region has been studied using a combinatorial approach [11]. A parameter library has been constructed, applying various tribological conditions at different lateral positions on a single disk, which is subsequently subjected to imaging surface analysis [12]. The results clearly show that the intensity of the O1s component assigned to the phosphate–oxygen of the non-bridging group is more pronounced at higher loads (Fig. 6). The O1s chemical state map (Fig. 6) was reconstructed using a LLS algorithm, as described in Section 2. This representation allows the contact area to be distinguished from the non-contact area, even when the wear is so mild that optical microscopic observations do not reveal morphological differences. The results of the SAXPS indicate that no changes in the binding energy of the elements (within the experimental error $\pm$ 0.2 eV) were detected as the applied load was systematically varied from 0.05 to 5 N with an apparent contact pressure ranging from 2.8 to 280 MPa (ball-on-disk) [5,11,12]. Similar binding energy values were observed applying an apparent contact pressure of 34 MPa (cylinder-on-flat, ATR crystal analyzed by XPS) [10].

Although it is clear that the tribofilm is inhomogeneous on a small scale (2–20 μm) and that the spot size of the XPS analysis is relatively large (120 μm, diameter), it is possible to estimate the average composition as a function of the applied load [12]. In particular, the O/P ratio was found to approach 4:1 at the highest load, thus supporting the hypothesis of the formation of a short-chain phosphate film at room temperature. Furthermore, the estimation of the average layer thickness by means of a model that also takes into account the attenuation of the emitted photoelectrons due to the organic reaction layer [23], indicated that the film thickness increases from 4.2 $\pm$ 0.2 nm at 0.05 N to 5.0 $\pm$ 0.2 nm at 10 N.

4.3. *Lateral resolution of the surface analytical techniques*

The highest lateral resolution achieved in this work was 1 μm, as displayed in the ToF-SIMS maps in Fig. 7. The results clearly show that the contact area after a chip-on-disk test is not homogeneous, but is composed of individual scars. In the region where the wear is more severe, there is evidence of zinc phosphate, whereas the less-worn areas of the contact region are covered by an iron phosphate layer. Similar information has also been obtained by means of photoelectron emission microscopy (micro-XAS) [24]: this technique allows spatially resolved elemental analysis to be performed, although the data are difficult to quantify.

The XPS maps (Figs. 4 and 5 with a lateral resolution of 400 μm, and Fig. 6 with a lateral resolution down to 120 μm) represent an average over both severely and mildly worn areas of the contact region. The observed dependence of the film composition and thickness on the applied load could be interpreted as an increase in the proportion of severely worn areas within the contact region.

4.4. *In situ versus ex situ techniques*

Most analytical studies currently performed to study the tribochemistry of additive-derived films formed in tribostressed systems consist of ex situ, post-mortem analyses, i.e. they are carried out outside the tribometer and following the tribological test. The advantage of ex situ analysis is the possibility of using the panoply of complementary modern surface analytical techniques, such as XPS, AES, ToF-SIMS and XAS. Almost all of these combine spectroscopic information with microscopy at a micron or sub-micron spatial resolution. The XPS maps and spectra reported here (Figs. 4–6) provide experimental evidence that not only differences in the elemental distribution are present between the contact and non-contact regions, as has already been demonstrated by AES mapping [25–27], but also chemical state variations in the elements that are participating in the tribofilm formation. It is apparent that in the contact area, the intensity of the iron signal is lower, whereas zinc, sulfur and phosphorus intensities are higher. Furthermore, it is shown (Figs. 4b and 5) that changes in the chemical state of the different elements and their relative concentrations vary on the micrometer scale, as a function of the mechanical conditions applied during the tribological tests.

In the non-contact area, a thin film of adsorbed organic thiophosphate and iron oxy-hydroxides is clearly identified. The iron oxide-related oxygen clearly dominates outside the contact area, whereas the contact region consists mostly of phosphate-type oxygen. These results illustrate the complexity of the surface chemical analysis of a tribostressed sample.

The major disadvantages of the ex situ techniques are that it is not possible to follow the kinetics of film formation and that changes in the surface composition may occur during the transfer to the spectrometer. In contrast, ATR spectroscopy readily yields information, in situ, on the kinetics of adsorption and reaction at liquid/solid interfaces, and can readily be adapted to the analysis of thin metal coatings. The results obtained under the conditions used in this work and at 150 °C, show that the thermal and the tribochemical reactions display different kinetics, the tribochemical reaction having a higher rate than the purely thermal one. However, the information from ATR FTIR experiments is averaged over the total area of the iron-coated germanium crystal. Changes in the film composition will be detected only when a sufficiently large area, including wear scars, is contributing to the signal.

5. Conclusions

The chemical reaction of zinc dithiophosphate has been studied using in situ ATR FTIR spectroscopy as well as with imaging XPS and ToF-SIMS. In situ ATR FTIR confirms the existence of a decomposition pathway that proceeds through the elimination of alkanes under pure thermal conditions. The reaction product is zinc polyphosphate. Under conditions of mild mechanical stress at high temperatures, as well as under more severe mechanical conditions at 25 °C, the data presented here suggest the formation of thin phosphate films. Sulfur is mainly present as sulfide and is depleted in comparison to the adsorbed ZnDTP film present in the non-contact region. The study of the effect of load by means of a combinatorial experiment has demonstrated that this is a powerful tool for the systematic study of the influence of physical parameters on the composition of tribochemical films.

The combination of modern surface analysis techniques with state-of-the art data interpretation provides detailed insight into the reaction mechanisms and thus a better understanding of the action of additives:

- i-XPS combined with SAXPS provide spatial distribution not only of elements but of their respective chemical states.
- ToF-SIMS combines monolayer surface sensitivity with excellent lateral resolution and allows reaction products on the surface to be identified.
- In situ test methods, such as ATR FTIR, provide further information on the chemical changes occurring during sliding.

Acknowledgements

This study was financially supported by the Council of the Swiss Federal Institute of Technology (PPM-Program) and by the Italian MURST (60% grant to A.R.). Prof. J.M. Martin (Ecole Centrale de Lyon, France) and Dr. H. Camenzind (Ciba Specialty Chemicals, Switzerland) are thanked for supplying the pure additives. The authors are also grateful to Mrs. Irene Klingenfuss for technical help in the ToF-SIMS measurements and to Mr. Horisberger (PSI, Villigen, Switzerland) for the coating of the ATR crystals.

References

[1] J.J. Habeeb, W.H. Stover, The role of hydroperoxides in engine oil and the effect of zinc dialkyldithiophosphates, ASLE Trans. 30 (1987) 419–426.

[2] G.W. Roper, J.C. Bell, Review and evaluation of lubricated wear in simulated valve train contact conditions (SAE 952473), in: Recent Snapshots and Insights into Lubricant Tribology SP-1116, Society of Automotive Engineers, Warrendale, PA, 1995, pp. 67–83.

[3] S. Bec, A. Tonck, J.M. Georges, R.C. Coy, J.C. Bell, G.W. Roper, Relationship between mechanical properties and structures of zinc dithiophosphate anti-wear films, Proc. Roy. Soc. London A 455 (1999) 4181–4203.

[4] F.M. Piras, A. Rossi, N.D. Spencer, Growth of tribological films: in situ characterization based on attenuated total reflection infrared spectroscopy, Langmuir 18 (2002) 6606–6613.

[5] M. Eglin, A. Rossi, N.D. Spencer, Additive-surface interaction in boundary lubrication: a combinatorial approach, in: D. Dowson, M. Priest, G. Dalmaz, A.A. Lubrecht (Eds.), Boundary and Mixed Lubrication, Proceedings of the 28th Leeds–Lyon Symposium on Tribology, Elsevier, Amsterdam, 2002, pp. 49–57.

[6] M.P. Seah, Post-1989 calibration energy for X-ray photoelectron spectrometers and the 1990 Josephson constant, Surf. Interf. Anal. 14 (1989) 488.

[7] M.P. Seah, W.A. Dench, Quantitative electron spectroscopy of surfaces: a standard data base for electron inelastic mean free pathsin solids, Surf. Interf. Anal. 1 (1979) 2–11.

[8] J.H. Scofield, Hartree-Slater subshell photo-ionisation cross-sections at 1254 and 1487 eV, J. Electr. Spectr. Relat. Phenom. 8 (1976) 129–137.

[9] K. Berresheim, M. Mattern-Klosson, M. Wilmers, A standard form of spectra for quantitative ESCA-analysis, Fresenius J. Anal. Chem. 341 (1991) 121–124.

[10] F.M. Piras, A. Rossi, N.D. Spencer, Combined in situ (ATR FTIR) and ex situ (XPS) study of the ZnDTP–iron surface interaction, Tribology Letters 15 (2003) 181–191.

[11] M. Eglin, A. Rossi, N.D. Spencer, A combinatorial approach to elucidating tribochemical mechanisms, Tribology Letters 15 (2003) 193–198.

[12] M. Eglin, A. Rossi, N.D. Spencer, X-ray photo-electron spectroscopy analysis of tribostressed samples in presence of *i*-ZnDTP using a combinatorial approach, Tribology Letters 15 (2003) 199–209.

[13] G.M. Bancroft, M. Kasrai, M. Fuller, Z. Yin, K. Fyfe, K.H. Tan, Mechanism of tribochemical film formation: stability of tribo- and thermally generated ZDDP films, Tribol. Lett. 3 (1997) 47–51.

[14] J.M. Martin, Antiwear mechanisms of zinc dithio-phosphate: a chemical hardness approach, Tribol. Lett. 6 (1999) 1–8.

[15] F.M. Piras, In situ total attenuated total reflection tribometry. Thesis No. 14638. Presented at the Department of Materials, ETH-Zurich, 2002.

[16] J.J. Dickert, C.N. Rowe, The thermal decomposition of metal *O,O*,-dialkylphosphorodithiolates, J. Org. Chem. 32 (1967) 647–653.

[17] R.C. Coy, R.B. Jones, The thermal degradation and EP performance of zinc dialkyldithiophosphate additives in white oil, ASLE Trans. 24 (1980) 77–90.

[18] R.B. Jones, R.C. Coy, The chemistry of the thermal degradation of zinc dialkyl-dithiophosphate additives, ASLE Trans. 24 (1981) 91–97.

[19] J.C. Bell, K.M. Delargy, A.M. Seeney, The removal of substrate material through zinc dithiophosphate anti-wear films, in: D. Dowson, et al. (Eds.), Proceedings of the 18th Leeds/Lyon Symposium on Wear Particles, Elsevier, Amsterdam, 1992, pp. 387–396.

[20] M.L. Suominen Fuller, M. Kasrai, G.M. Bancroft, K. Fyfe, K.H. Tan, Solution decomposition of zinc dialkyl-dithiophosphate and its effect on antiwear and thermal film formation studied by X-ray absorption spectroscopy, Tribol. Int. 31 (1998) 627–644.

[21] S.H. Choa, K.C. Ludema, G.E. Potter, B.M. Dekoven, T.A. Morgan, K.K. Kar, A model of the dynamics of boundary film formation, Wear 177 (1994) 33–45.

[22] Z. Yin, M. Kasrai, M. Fuller, G.M. Bancroft, K. Fyfe, K.H. Tan, Application of soft X-ray absorption spectroscopy in chemical characterization of antiwear films generated by ZDDP. Part I. The effects of physical parameters, Wear 202 (1997) 172–191.

[23] A. Rossi, B. Elsener, XPS analysis of passive film on the amorphous alloy Fe70Cr10P13C7: the effect of the applied potential, Surf. Interf. Anal. 18 (1992) 499–504.

[24] G.W. Canning, M.L. Suominen Fuller, G.M. Bancroft, M. Kasrai, J.N. Cutler, G. De Stasio, B. Gilbert, Spectromicroscopy of tribological films from engine oil additives. Part I. Films from ZDDP's, Tribol. Lett. 6 (1999) 159–169.

[25] T.P. Debies, W.G. Johnston, Surface chemistry of some antiwear additives as determined by electron spectroscopy, ASLE Trans. 23 (1980) 289–297.

[26] W.A. Glaeser, D. Baer, M. Engelhardt, In situ wear experiments in the scanning Auger spectrometer, Wear 162–164 (1993) 132–138.

[27] J.M. Martin, C. Grossiord, T. Le Mogne, S. Bec, A. Tonck, The two-layer structure of Zndtp tribofilms. Part I. AES, XPS and XANES analyses, Tribol. Int. 34 (2001) 523–530.

Tribol Lett (2007) 28:209–222
DOI 10.1007/s11249-007-9267-0

ORIGINAL PAPER

Pressure Dependence of ZnDTP Tribochemical Film Formation: A Combinatorial Approach

Roman Heuberger · Antonella Rossi · Nicholas D. Spencer

Received: 1 July 2007 / Accepted: 24 August 2007 / Published online: 15 September 2007

Abstract Combinatorial testing has been performed on zinc dialkyldithiophosphate (ZnDTP)-containing lubricants, to investigate the effects of contact pressure on the formation of tribochemical films. Contact pressures ranging from 25 to 500 MPa were applied in ball-on-disc tribotests with oscillating load. Both the ball and the disc were investigated by means of small-area and imaging X-ray photoelectron spectroscopy (XPS). The thickness and the composition of the reaction layer were estimated from the XPS data. The thickness of the reaction layer in the tribologically stressed areas of the ball and of the disc increased with both temperature and contact pressure. The reaction layer mainly consisted of short-chain poly(thio)phosphates, shorter chains being observed at higher contact pressures. At high pressures, the presence of a thick, high-toughness short-chain poly(thio)phosphate layer can explain the lower friction and dimensional wear coefficients observed. On the ball, similar anti-wear film formation mechanisms were observed as on the disc, zinc sulphide being deposited in the post-contact region.

Keywords X-ray photoelectron spectroscopy · XPS · ZnDTP · Boundary lubrication · Tribochemistry · Combinatorial testing · Load · Temperature · Friction · Wear

R. Heuberger · A. Rossi · N. D. Spencer (✉)
Laboratory for Surface Science and Technology, Department of Materials, ETH Zurich, Wolfgang-Pauli-Strasse 10, 8093 Zurich, Switzerland
e-mail: spencer@mat.ethz.ch

A. Rossi
Dipartimento di Chimica Inorganica ed Analitica, Università degli Studi di Cagliari, Cittadella Universitaria di Monserrato, 09100 Cagliari, Italy

1 Introduction

Combinatorial testing is frequently used in the pharmaceutical industry for rapid parallel screening. In tribology, friction and wear strongly depend on various experimental parameters such as the material of the tribopairs, their surface roughness and geometry, the lubricant composition, and the test conditions, such as contact pressure, sliding speed and temperature [1–3]. The investigation of the effect of all these parameters is both time consuming and laborious: the design of an experiment where a set of spatially separated areas are tested under different conditions, i.e. the development of a combinatorial approach to tribometry, allows high throughput in lubricant-additive testing. An example of a combinatorial test in tribology is the scratch test, where a diamond tip is slid over the surface with increasing load [4]. Hogmark et al. developed a crossed-sample configuration with load varying along the sample [4, 5]. Both tests have been used to screen load effects on hard coatings [4]. Eglin et al. developed two combinatorial approaches to ball-on-disc testing using a tribometer that allows the free programming of load, radius and rotational speed [6–9]. In the first approach, the effect of different loads were studied at different radii on a single disc [6–8]. The contact region was artificially expanded with closely spaced multiple radii to enable subsequent surface-analytical investigations of the tribotrack [7, 8]. In the second approach, the applied load was ramped up and down as a function of the geometrical position of the disc while different sliding times were applied at different radii; this approach focuses on the tribological results (friction, wear)

[9]. The application of different conditions at different radii seems to be a very effective and useful approach at low temperatures, while in some high-temperature experiments, the ongoing thermal reactions may influence the chemical composition of the tribofilm produced on another track created at an earlier stage of the experiment [10].

One of the aims of the present study was to set up a combinatorial test that could be effectively applied at both low and high temperatures without artificially expanding the contact region with closely spaced multiple radii. Earlier publications [7, 8, 11] relied on this approach in order to overcome the resolution limits of the XPS instruments employed. The newer generation of XPS spectrometers, however, is able to focus a monochromatic X-ray beam down to 5–10 μm, and this allows imaging and small-area XPS measurements to be performed on single wear tracks produced in a ball-on-disc experiment.

Zinc dialkyldithiophosphate (ZnDTP, formula is in Fig. 1)—a family of organometallic compounds—is still widely used as an anti-wear additive and exhibits excellent performance under boundary-lubrication conditions [12, 13]. The disadvantage of ZnDTPs is that they contain large amounts of phosphorus, sulphur and zinc, which impair the environment, both directly and indirectly, by poisoning exhaust catalysts. For this reason environmental legislation will be limiting the concentration of phosphorus and sulphur in engine oils: Replacements for ZnDTP, as well as an understanding of the working mechanisms of both ZnDTP and its possible alternatives, are badly needed.

Much effort has been expended in understanding the anti-wear mechanism of ZnDTP, especially its temperature dependence. Tribochemical films produced from ZnDTP have been investigated with numerous tools, such as atomic force microscopy (AFM) [14–17], nano-indentation [17–20], XPS [7, 11, 20–22], X-ray absorption spectroscopy (XANES) [22–24] and in situ attenuated total reflection infrared spectroscopy [25, 26]. The current understanding is that thermal degradation of ZnDTP begins with the migration of the carbon chain from the oxygen atom to the sulphur atom at temperatures as low as 60 °C [11, 27–29]. Some of the carbon chains and some thiols are released into the lubricant solution [11, 30]. Starting below 100 °C, phosphoryl groups react with each other to form a pyrophosphate $P_2O_7^{4-}$, also known as "short-chain polyphosphates". At higher temperatures (above ~150 °C), higher concentrations of fragments are available in the lubricant and the

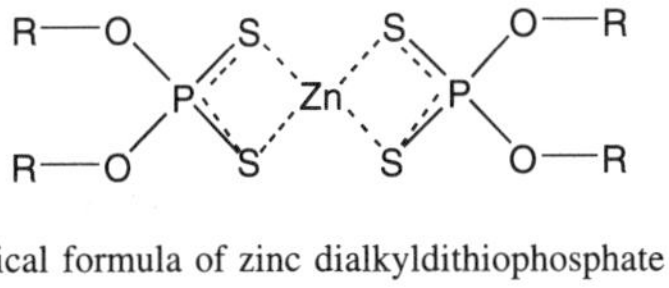

Fig. 1 Chemical formula of zinc dialkyldithiophosphate (ZnDTP)

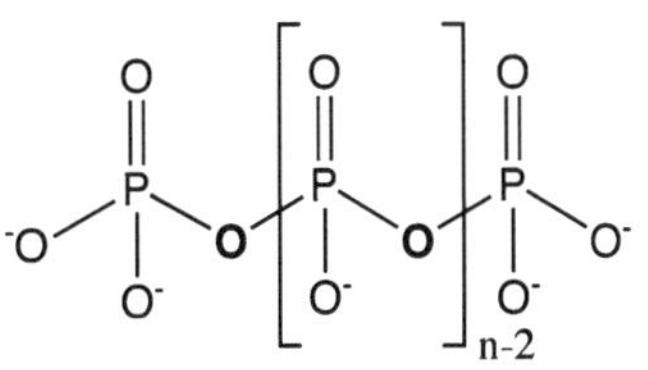

Fig. 2 Scheme of a polyphosphate with the chain length *n*. The oxygen atoms linking two phosphorus atoms together (*bold*) are denoted "bridging oxygen" (BO), while the terminal oxygen atoms in the phosphate groups are labelled "non-bridging oxygen" (NBO) [31]

formation of longer chains, known as polyphosphates (see Fig. 2), can occur [11–13, 24]. It has been suggested that during a tribological contact similar reactions occur on the surfaces of both counterparts, starting at room temperature [11, 32]. But while it has been proven that the tribofilm changes on the disc as a function of test temperature from a zinc phosphate at low temperature to a zinc polyphosphate at higher temperature [11, 20, 22, 25, 26] only very little analysis of the ball has been carried out [33, 34].

The influence of load on the tribofilm composition has been examined by only a few research groups to date. While Bird and Galvin found only small effects within the experimental uncertainties [21], Eglin et al. working at ambient temperature, observed the formation of orthophosphates at a load of 5 N [7, 8] and Yin et al. found higher formation rates and longer polyphosphate chains with higher loads at 80 °C [35]. Although much effort has been put into understanding the reaction mechanism, there remain many open questions concerning the reacting species, the film formation and its kinetics [12, 13].

In this study, a modified combinatorial approach has been applied for the first time to investigate the pressure dependence of the formation mechanism of tribofilms. Pressure was varied from 26 ± 5 to 515 ± 100 MPa on a single tribotrack. Temperatures of 30 °C, 80 °C and 150 °C were investigated, as they all typically occur under different conditions in an automobile engine.

2 Experimental

2.1 Tribological Testing

A CETR UMT-2 tribometer with temperature control (Center for Tribology, Campbell, CA, USA) was employed for the ball-on-disc test. The load cell could be moved up and down and the normal load was generated by a compressed spring. Both ball and disc (both hardened bearing steel 100Cr6, AISI 52100) were immersed in the lubricant oil. A 1 wt.% solution of a commercial secondary zinc dialkyldithiophosphate ZnDTP (C_3H_7 + C_6H_{13}, see Fig. 1, HiTEC®7169, Afton Chemical Corporation, Richmond,

VA, USA, purified by liquid chromatography) in poly-α-olefin (Durasyn 166, Tunap Industries GmbH. & Co., Mississauga, Canada) was used as a lubricant. Tests were performed at 30 °C, 80 °C and 150 °C (all ±3 °C) and each test was repeated at least three times. The relative humidity was between 20% and 60%. Prior to the actual oscillating-load test, a running-in of the ball applying a load of 10 N was performed for 1000 turns with a sliding speed of 0.005 m/s at a radius of 6.0 ± 0.2 mm in ZnDTP solution. The running-in of the ball yielded a measurable flat contact area that was used to determine the contact pressure by dividing the applied load by the measured apparent contact area.

During the oscillating-load test, the load was changed from 0.5 to 10 N over two cycles in each rotation (see scheme in Fig. 5). The sliding speed was 0.005 m/s and 1000 turns at a radius of 5.0 ± 0.2 mm were performed. Before further investigation, the samples were rinsed and ultrasonically cleaned in ethanol (p.a.) for 3 min.

The data were processed with Matlab 7.1 software (The MathWorks, Inc., Natick, MA, USA) to visualize the friction coefficient in colour plots and to determine the friction coefficients in dependence on the applied load (average of 2000 values/load).

2.2 Laser Profilometry

The topography of the discs was measured by means of laser profilometry (UBM, type UBC 14, UBM Messtechnik GmbH, Ettlingen, Germany), in order to determine the amount of wear. A Kr^{++}-laser beam (beam size 1 μm, wave length 407 and 413 nm) was dynamically focussed onto the sample and the z-value measured (accuracy ±10 nm) [36]. The sample table can be laterally moved and an area of 13×13 mm^2 was measured with a resolution of 120 points/mm. The data were stored in a z-value matrix.

The wear rate was calculated with Matlab Software (V6.5). Radial profiles were read out of the z-value matrix and averaged over sectors of 5° (50 lines) to improve the signal-to-noise ratio. Applying a linear background, the area below the background was summed and converted into wear rates.

2.3 X-Ray Photoelectron Spectroscopy

The surface chemistry of the tribostressed samples was investigated with X-ray photoelectron spectroscopy (XPS), analysing selected areas, both on the tribotrack and on the non-contact areas.

The X-ray source of the PHI Quantera SXM (ULVAC-PHI, Chanhassen, MN, USA) is a focussed and scanned monochromatic AlKα beam with a diameter that can be chosen between 5 and 200 μm. The emitted electrons are collected and retarded with an Omega lens system at an emission angle of 45°. After passing a spherical capacitor energy analyzer, the electrons are detected by a 32-channel detector. The system is equipped with a high-performance floating-column ion gun and an electron neutralizer for charge compensation. The residual pressure was always below 5×10^{-7} Pa. The system was calibrated according to ISO 15472:2001 and the accuracy was better than ±0.05 eV.

The unambiguous identification of the tribologically stressed areas was performed by means of the sample-positioning station (SPS), which allows photographs of the sample surface to be acquired outside the instrument. The SPS, in combination with a scanning X-ray image (SXI), facilitates rapid location of the desired analysis region and correlation with the coordinates calculated using the angular position from the tribological test. Two areas of the disc with each of the load values (10 N, 5 N and 0.5 N) were measured, as well as two non-contact areas. The arc length of the region tribostressed at any given load ±0.1 N was 165 μm and the width of the wear scar was ≤150 μm. On the ball, the centre of the tribostressed area, the non-contact areas 500 μm in front and behind the centre and, if present, the deposited material (approximately 250 μm behind the centre) were analysed. The positions of the points were double-checked with imaging XPS and XPS line scans.

Small-area XPS spectra were collected with a beam diameter of 20 μm and a power of 4.25 W in the constant-analyzer-energy (CAE) mode, using a pass energy of 69 eV and a step size of 0.125 eV. Under these conditions, the full width at half maximum height (fwhm) for $Ag3d_{5/2}$ is 1.3 eV. Survey spectra were acquired with 280 eV pass energy and a step size of 1 eV. The whole set of spectra (detailed and survey spectra) was acquired within 40–60 min/area. Imaging XPS and line scans were performed with a beam size of 10 μm, a power of 2.3 W and a pass energy of 140 eV in the snapshot mode, allowing the simultaneous acquisition of spectra with a binding-energy range of 15.5 eV. For imaging XPS, the spectra of all elements of interest were measured on an area of typically 600×600 μm^2 (lateral resolution normally 10 μm) within 6–8 h of acquisition.

2.3.1 Data Processing

Detailed spectra were processed with CasaXPS software (V2.3.12, Casa Software Ltd., UK). An iterated Shirley-Sherwood background subtraction was applied before peak fitting using a linear-least-squares algorithm. Minor charging was observed and corrected by referencing the aliphatic carbon to 285.0 eV. Details on the curve-fitting parameters, which have been measured on reference compounds, have been published elsewhere [10, 11].

2.3.2 Quantitative Analysis

Quantitative analysis was performed on the basis of a first-principles model [37, 38], assuming that the electrons from sublayers of a multilayer structure are attenuated by the overlayers according to the Beer-Lambert law. Further assumptions are that each layer is homogeneous in thickness and composition and that there are no gradients within the layers. The equations that correlate the area of the photoelectron signal with the concentration are written for each species of the substrate and for the multilayer structure according to [37, 38] and then solved numerically by iteration. Three-layer models based on these assumptions have been developed and applied by different groups to single- and multi-component substrates and overlayers [39–42] using a system of parametric equations.

In this work, since the substrate consists of only one component (i.e. iron), the system of equations can be simplified considerably, since the electrons from this single component are attenuated by the overlayer(s). Thus, in some cases, a film with four layers was proposed and the corresponding system of non-linear equations could still be solved using the Newton method. More details are provided in [10]. The presence of a four-layer structure—steel/oxide layer/reaction layer/organic layer—was proposed on the basis of angle-resolved profiles (not shown here). The peak areas were corrected with the sensitivity factor $S_i = \sigma_i \times L(\gamma) \times G(E_i) \times \rho/A_i \times \lambda(E_i) \times \cos(\theta)$, which includes the photoionization cross-section σ_i according to Scofield [43], the angular asymmetry function $L(\gamma)$ [44, 45], the étendue $G(E_i)$ as defined for Auger electron spectroscopy ($G(E_i)$ = *analysed area* × *transmission function* × *detector efficiency*) [37], the density of the material ρ divided by the atomic weight of the element A_i, the inelastic-mean-free path (IMFP) $\lambda(E_i)$ according to Tanuma et al. [46, 47] and $\cos(\theta)$, where θ is the emission angle. As model compounds for the densities of the layers and for calculating the IMFP, carbon, metallic iron, iron oxide Fe_3O_4, hydrated zinc orthophosphate for the reaction layer in the non-contact areas at 30 °C and 80 °C and zinc pyrophosphate for the tribofilms and the non-contact areas at 150 °C were chosen [11]. Details on the equations and factors mentioned above are given in [10].

3 Results

3.1 Tribological Tests

A colour plot of the friction coefficient μ of a typical tribological test performed at 80 °C is shown in Fig. 3. The load was cycled in such a way that the same angular position always experienced the same load. The friction coefficient changed only slightly during the first 500 turns and remained constant during the last 500 turns. Regions with lighter colour and therefore higher friction coefficient are to be found in the track at 0° (and 360°) and 180°, which correspond to the areas where the lowest load of 0.5 ± 0.1 N was applied. Slightly higher friction coefficients at lower loads were observed at all temperatures, as shown in Fig. 4. The friction coefficient increased with increasing temperature; for 10 N load it rose from 0.13 ± 0.01 at 30 °C to 0.15 ± 0.01 at 80 °C up to 0.20 ± 0.01 at 150 °C.

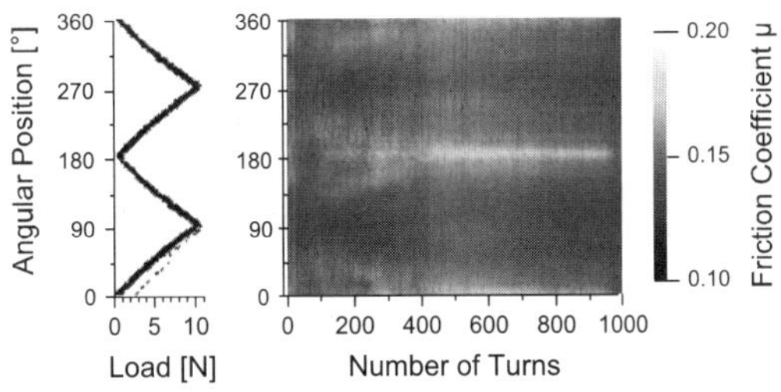

Fig. 3 Colour plot of the friction coefficient μ of a sample tribostressed at 80 °C in ZnDTP solution. Left: a plot of the applied load vs. the angular position of the rotating disc

3.2 Wear

The wear rate increased in a quasi-linear fashion with applied load (Fig. 5). High scatter of the wear rate due to the increased surface roughness caused by the wear was observed at high loads. The dimensional wear coefficients k (wear rates divided by the applied load, Fig. 6) showed lower average wear coefficients at higher loads. With increasing temperature, higher wear coefficients were obtained; for 10 N load, the wear coefficient increased from 1.1 ± 0.6 × 10^{-6} mm^3/Nm at 30 °C to 1.6 ± 0.3 × 10^{-6} mm^3/Nm at 80 °C to 1.8 ± 0.7 × 10^{-6} mm^3/Nm at 150 °C.

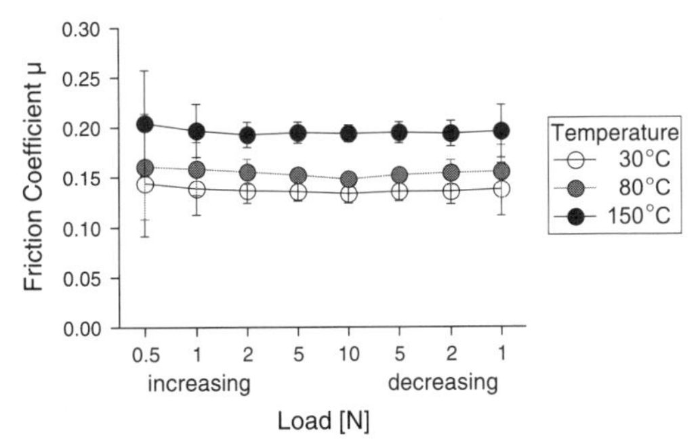

Fig. 4 Friction coefficient μ vs. load for experiments performed at different temperatures in ZnDTP solution

Tribol Lett (2007) 28:209–222

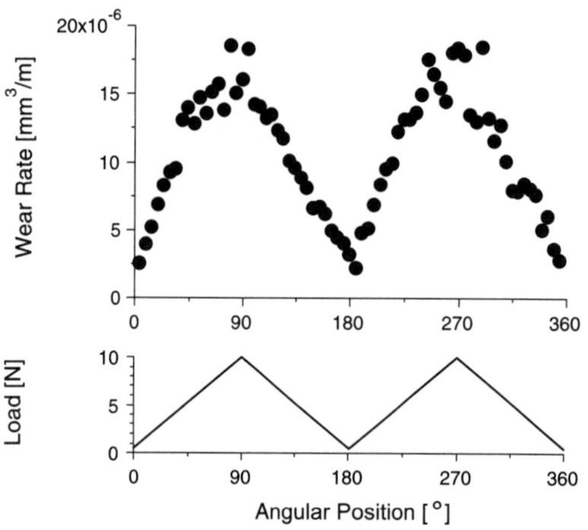

Fig. 5 Wear rate of a disc tribostressed at 80 °C in an oscillating-load test. The load dependence is shown as a function of the angular position of the disc in the scheme below

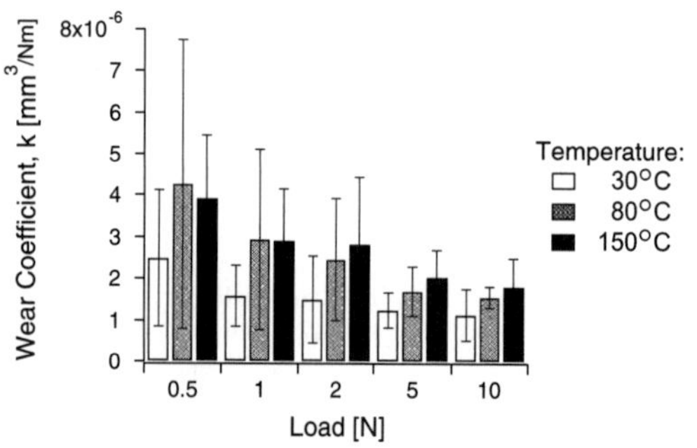

Fig. 6 Dimensional wear coefficient k on the disc as a function of the applied load for different temperatures of the ZnDTP solution

3.3 XPS Results

Survey and detailed high-resolution spectra were measured on both the disc and on the ball. Survey spectra were used for peak identification and to check the presence of contaminants. Detailed spectra of phosphorus 2p, together with zinc 3s, sulphur 2p, carbon 1s, oxygen 1s, iron 2p, zinc $2p_{3/2}$ and zinc LMM were recorded to identify the different chemical states of the species and to perform the quantitative analysis.

Spectra of a spot tribostressed with 10 N load at 80 °C are shown in Fig. 7. The *phosphorus 2p* signal was fitted with two peaks, $2p_{3/2}$ and $2p_{1/2}$, due to the spin-orbit splitting with an energy difference of 0.85 and an area ratio of 2:1. The phosphorus $2p_{3/2}$ peak was at 133.7 ± 0.1 eV. In the same spectrum the *zinc* 3s peak was at 140.5 ± 0.1 eV. This peak was used for the quantitative analysis instead of the more intense zinc $2p_{3/2}$ peak because the kinetic energy of the Zn3s electrons is closer to that of P2p and S2p and the IMFPs and therefore the sampling depths are similar.

The *sulphur $2p_{3/2}$* peak was found at 162.2 ± 0.1 eV, and can be assigned to a sulphur having a formal oxidation state of –2, as found in sulphides [48, 49], thiolates [50, 51] or when sulphur is substituting oxygen atoms in poly(thio)phosphates.

In the *carbon 1s* spectrum, the most intense peak was at 285.0 eV due to aliphatic carbon [52, 53]. Minor contributions were found at 286.8 eV due to carbon bound to oxygen and sulphur [52, 54, 55] and at 289.0 ± 0.2 eV indicating carbonate [49, 56] and/or carboxylic groups [52, 55].

The *oxygen 1s* signal was composed of three peaks: the first at 530.3 ± 0.1 eV being assigned to oxygen in iron and zinc oxides [7, 57, 58], the main peak at 531.7 ± 0.1 eV to non-bridging oxygen (NBO) in polyphosphates and to oxygen bound to carbon [7, 59, 60] and the high-binding-energy peak at 533.4 eV to bridging oxygen (BO) in polyphosphates [60, 61].

The *iron* signal (Fig. 7) presented two main peaks at 711 and 724 eV due to the spin-orbit splitting, assigned to $Fe2p_{3/2}$ and $Fe2p_{1/2}$ [53]; only the more prominent iron $2p_{3/2}$ signal was used for quantification. On this sample, very low-intensity iron signals were detected. Peaks of two oxidation states Fe(II) and Fe(III) in iron oxides at 709.6 ± 0.1 and 711.0 ± 0.1 eV [49, 57, 58] were detected, while no signal attributable to metallic iron (707 eV) was revealed. Both oxides exhibit high-binding-energy satellites but only the satellite of iron (II) at 715.1 ± 0.1 eV was within the fitted region: it was fixed at an area of 7% of the main peak [57, 58]. A small, high-binding-energy peak at 713.5 eV assigned to iron phosphate [62] was detected.

3.3.1 *Imaging XPS*

Imaging XPS pictures and line scans were taken to determine the position of the analysed area and to check for lateral inhomogeneities of the tribotrack and of the non-contact area. The total peak area is colour coded and plotted as a function of the lateral position. In the case of oxygen, the spectra collected for the maps were additionally fitted with three peaks at 530.3, 531.9 and 533.4 eV and the maps reconstructed with a linear-least-square fit routine (see bottom row of Fig. 8).

Disc An example of the images collected on a tribotrack produced with 10 N at 80 °C and on the ball is shown in Fig. 8 to compare the films formed on both sliding partners. The tribotrack on the disc was clearly visible in the phosphorus 2p image, where the tribotrack showed higher

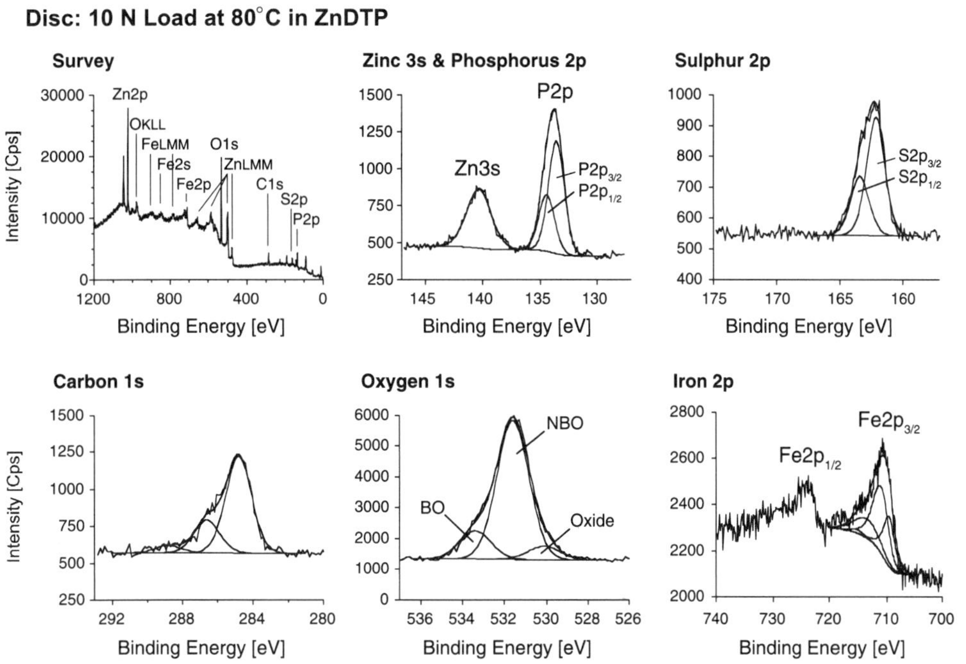

Fig. 7 XPS survey and detailed spectra of Zn3s and P2p, S2p, C1s, O1s and Fe2p measured on the tribotrack on the disc following an experiment under 10 N load at 80 °C. In the O1s spectrum the component at 531.8 eV is labelled as NBO, but this signal also contains contributions from sulphates and from oxygen bound to carbon

intensity than the non-contact area. The same distribution was found with the sulphur (II) peak at 162 eV binding energy. There was less contrast in the case of the peak at 168 eV, where slightly higher intensity was found in the non-contact area (not shown). To create the colour plots, the more intense zinc $2p_{3/2}$ peak was used instead of the Zn3s peak because of the better signal-to-noise ratio. Again the tribotrack shows a higher signal, and there was a region along the tribotrack with even higher intensity. The carbon image of the disc shows that there was slightly less carbon in the tribotrack than in the surrounding area, whereas the total oxygen map showed higher levels in the tribostressed region. The three different oxygen peaks showed that the oxide, which is more intense in the non-contact area, was covered with a film containing both non-bridging (NBO) and bridging oxygen (BO). The iron signal in the contact area appears lower than in the non-contact area due to the formation of a thicker reaction layer.

Ball In the tribostressed region on the ball (round area), the maps show higher intensities of the phosphorus, sulphur, zinc and NBO and BO signals; in this region, lower intensities of carbon, iron and O(oxide) signals were detected. In the region behind the contact area during the tribotest (left side of the contact area in the map), there was a tail with an enhanced intensity of both sulphur (II) and zinc.

3.3.2 Spectra at Different Loads

Disc Detailed spectra of P2p, Zn3s, S2p and O1s taken from small areas tribostressed with different loads, as well as from the non-contact area and of the centre of the ball are presented in Fig. 9. The spectra of areas tribostressed with 5 N were almost identical to those stressed with 10 N. A further decrease of the load to 0.5 N did not change the binding energy values of the photoelectron signals (uncertainty ±0.1 eV). The intensity of phosphorus, zinc and sulphur signals decreased with decreasing load, while the peak in the oxygen O1s spectrum assigned to the oxide became more pronounced.

In the *non-contact regions*, only low-intensity signals of phosphorus and zinc were detected at slightly lower binding energies of 133.4 ± 0.2 and 140.3 ± 0.1 eV. The sulphur peak at 162.1 ± 0.1 eV was much smaller compared to that measured in the 10 N area, but there was an additional noisy peak at 168.4 ± 0.3 eV that can be assigned to oxidized sulphur species [48, 49]. The non-contact region also revealed a more pronounced oxide peak at 530.0 ± 0.1 eV in the oxygen spectrum, while in the carbon spectrum, the high-binding energy peak (carbonates and/or carboxylic groups) at 288.8 ± 0.1 eV was more intense. The contributions to the O1s signals, labelled as

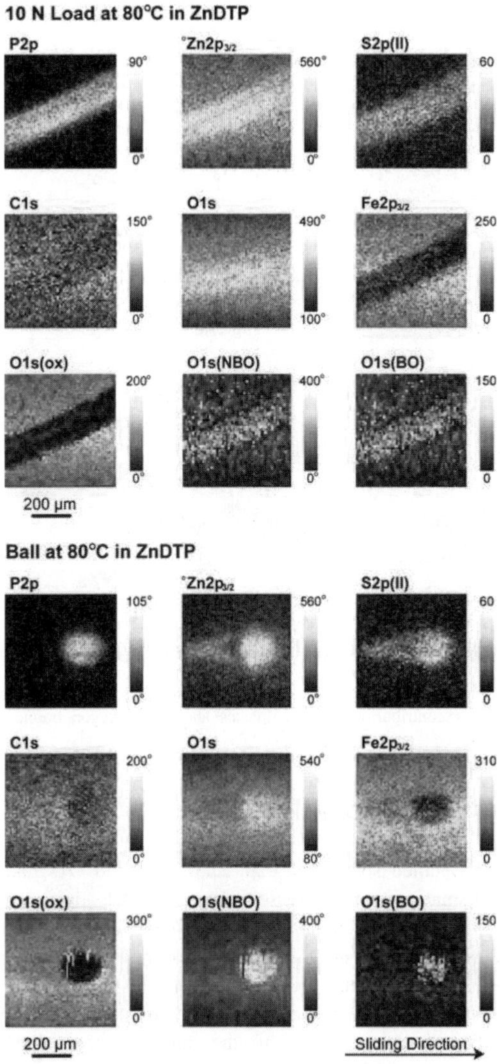

Fig. 8 Imaging-XPS of the tribotrack on the disc (tribostressed under 10 N load at 80 °C) and of the tribostressed region on the ball from the same experiment. Complete spectra were acquired at each point of the map and the area under the peak taken for the colour plot

NBO, also included those due to oxygen present in iron hydroxides and sulphate groups. In addition, adsorbed water has to be considered together with the BO-peak [58]. The iron peak in the non-contact region was very pronounced and metallic iron was detected at 706.7 eV. On some samples, small peaks at 711.8 eV assigned to iron hydroxide were found [7, 58].

Ball The peaks measured in the centre of the tribostressed region on the ball were similar to those collected on the disc after tribotesting with 5 and 10 N load, but a slightly higher intensity of the phosphorus and sulphur (II) peaks was observed.

3.3.3 Spectra at Different Temperatures

Disc Spectra of the 10 N area produced at 30 °C, 80 °C and 150 °C are presented in Fig. 10. At 30 °C, the binding energy of phosphorus $2p_{3/2}$ (133.6 ± 0.2 eV) was slightly lower than at 80 °C. A small sulphate peak was detected on the tribotrack and in the oxygen spectrum the peak of BO was less intense than that detected at 80 °C. The spectra from samples tribostressed at 150 °C were almost identical to those seen in the 10 N area produced at 80 °C; the binding energies of P2p, Zn3s and the low-binding energy peaks of S2p were each shifted by 0.2 eV, while the intensities of the peaks were similar. At this high temperature, minor contributions from sulphates were found in the tribotracks at 169.2 ± 0.1 eV. Almost no oxide peak in the oxygen signal and only weak iron signals were revealed. The spectra of the areas tribostressed with 0.5 N load showed intermediate characteristics between those of the high loads and the non-contact area.

The spectra of the non-contact areas at 30 °C were almost identical to those of the non-contact region of the 80 °C samples, while those recorded at 150 °C were similar to the tribostressed spectra, with slightly less intense phosphorus peaks but a higher intensity in the sulphate peaks. Only traces of iron were detected.

Ball On the tribostressed regions on the balls at all temperatures, similar spectra were observed as on the tribotracks on the discs stressed with high loads.

3.3.4 Thickness and Composition

Results of angle-resolved XPS profiles (not shown in this work) suggest that the surface was not uniform within the XPS sampling depth [10]. On the steel substrate an iron oxy-hydroxide layer was formed, covered by a reaction layer containing phosphorus, sulphur, zinc, iron phosphate and the two oxygen species NBO and BO. The outermost part of the surface presented an organic layer containing the contribution to the C1s signal due to aliphatic and high-binding energy carbons and some oxygen bound to carbon (contributing to the NBO peak). The same multilayer structure was found on similar samples in earlier work [9, 11]. The four-layer model was applied if the intensity of metallic iron was at least 5% of the overlayer intensity [63] and a three-layer model—assuming a semi-infinite oxide layer as a substrate—was used if this criterion was not met. Applying the multi-layer models, the thicknesses and compositions of the individual layers could be calculated simultaneously.

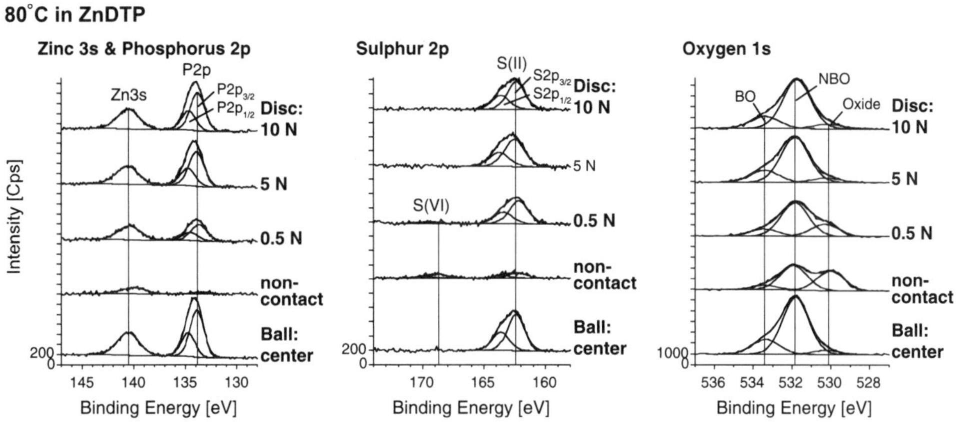

Fig. 9 XPS spectra of P2p with Zn3s, S2p and O1s measured in the tribotrack stressed under varying load, on the non-contact area on the disc and in the centre of the tribostressed region of the ball. The experiment was performed at 80 °C

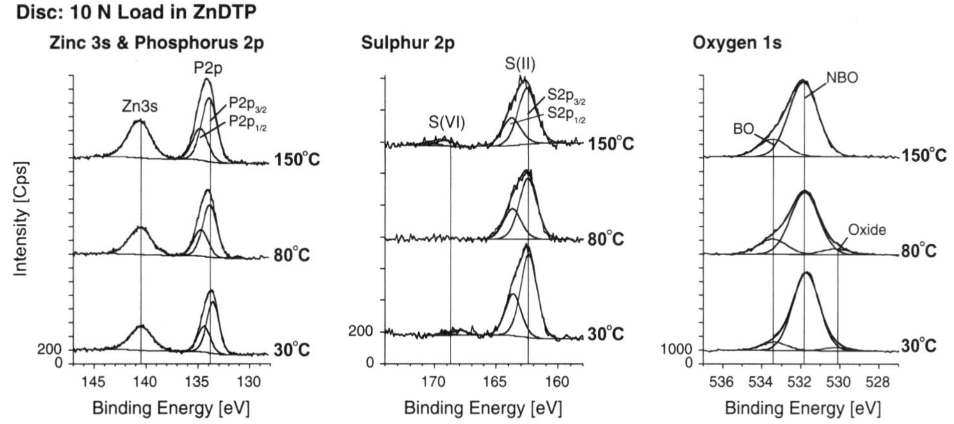

Fig. 10 XPS spectra of Zn3s and P2p, S2p and O1s measured in the tribotrack on the disc stressed with 10 N load at 30 °C, 80 °C and 150 °C

Thickness The thicknesses of the reaction layers are shown in Fig. 11 for different loads and temperatures. At both 30 °C and 80 °C, films of about 4 nm were formed at 5 N and 10 N loads, while in the non-contact areas of both the disc and the ball only very thin films of less than 1 nm were found. In the tribostressed region in the centre of the ball, similar thicknesses were measured as on the disc stressed at high loads. At 80 °C, the deposited material behind the contact region of the ball (compare with Fig. 8) led to an increased film thickness of 3.7 ± 0.4 nm compared to the non-contact areas of the same samples (1.4 ± 0.7 nm).

At 150 °C, tribofilms of 6 nm or thicker were formed on the disc. When the layers had a thickness higher than 6 nm no signal from the iron oxide layer was detected and the reaction layer has to be assumed to be semi-infinite and the thickness of the film is reported as being higher than 6 nm. Thinner films of 4 ± 1 nm were observed with 0.5 N load. In the non-contact areas, thick thermal films of more than 6 nm were produced. On the ball, thick reaction layers were found all over the surface.

Composition The composition of the reaction layer changed with the applied load and temperature (see Table 1). While on the tribotracks and the thermal films at 150 °C the amount of oxygen was about three times that of phosphorus, far more oxygen was present in the non-contact areas at 30 and 80 °C. Compared to the ZnDTP molecule, where a S:P ratio of 2:1 is present, a depletion of sulphur was found in the tribotracks and non-contact areas, while on the ball the deposited material behind the tribological contact at 80 °C was enriched in sulphur. Small amounts of oxidized sulphur species were found in the non-

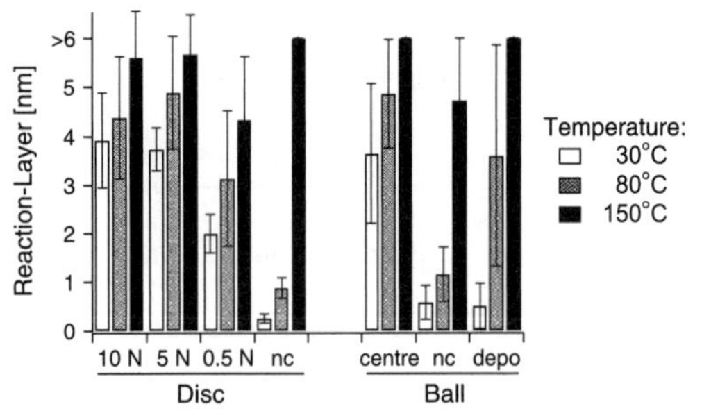

Fig. 11 Thickness of the reaction layer on the discs and balls as a function of the applied load and temperature. Please note that "nc" stands for the non-contact area, "centre" means the centre of the tribostressed region of the ball and "depo" is the deposited material on the ball. Reaction layers with thicknesses higher than 6 nm could not be calculated, therefore they are labelled as ">6"

contact areas only and in trace quantities in the tribotracks produced at 150 °C. In comparison to ZnDTP (Zn:P = 0.5:1), the reaction layers were enriched in zinc. Usually only small amounts of iron in the form of iron phosphate were to be found.

For the tribofilms at 30 °C, the oxygen-to-phosphorus ratio of 2.9 ± 0.1:1 is lower than expected for a short-chain polyphosphate. This might suggest that some sulphur (II) has most likely substituted oxygen in the phosphate chain. The tribofilm thus mainly consisted of a short-chain zinc poly(thio)phosphate, together with traces of zinc sulphide. At 30 °C, almost no changes in the film composition were observed with decreasing load. In the non-contact areas, there were small amounts of zinc oxide with traces of phosphate compounds and sulphides.

The 10 N tribofilm produced at 80 °C mainly contained short-chain zinc poly(thio)phosphate, with some sulphides. Applying a 5 N load led to the same film composition, and with 0.5 N there was slightly more sulphur present. The non-contact film was similar to that at 30 °C, but with higher amounts of zinc sulphide and zinc sulphate.

At 150 °C, the 10 N tribofilm was made of short-chain zinc poly(thio)phosphate, together with small amounts of zinc sulphate and sulphides. Again the same film composition was found at 5 N load. At the lowest load of 0.5 N, the main compound was still zinc poly(thio)phosphate, but the film contained more zinc sulphide and sulphate. Some iron substituting for zinc was detected as well. In the non-contact area there was a long-chain zinc poly(thio)phosphate with higher amounts of zinc sulphate and zinc sulphide.

The O/Fe ratio of the layer beneath the surface film was found to be equal to 1.3, in excellent agreement with the expected value, assuming the presence of an iron oxide (Fe_3O_4) layer on the steel surface [10]. This finding validates the model proposed here for the calculation of the

Table 1 Elemental ratio of the reaction layer normalised to phosphorus

			O	:	P	:	S(II)	:	S(VI)	:	Zn	:	Fe
30 °C	Disc	10 N	2.9 ± 0.1	:	1.0	:	0.5 ± 0.1	:	0.0	:	0.7 ± 0.1	:	0.1 ± 0.1
		5 N	2.8 ± 0.3	:	1.0	:	0.5 ± 0.1	:	0.0	:	0.7 ± 0.1	:	0.1 ± 0.0
		0.5 N	2.9 ± 0.5	:	1.0	:	0.5 ± 0.1	:	0.0	:	0.8 ± 0.1	:	0.0
		nc	8 ± 3	:	1.0	:	0.7 ± 0.7	:	0.4 ± 0.8	:	4.4 ± 1.8	:	0.0
	Ball	centre	2.8 ± 0.2	:	1.0	:	0.5 ± 0.1	:	0.0	:	0.6 ± 0.1	:	0.1 ± 0.0
		nc	6 ± 2	:	1.0	:	0.3 ± 0.1	:	1.0 ± 0.7	:	1.2 ± 0.5	:	0.0
80 °C	Disc	10 N	2.5 ± 0.3	:	1.0	:	0.5 ± 0.1	:	0.0	:	0.9 ± 0.1	:	0.0
		5 N	2.6 ± 0.2	:	1.0	:	0.5 ± 0.1	:	0.0	:	0.9 ± 0.1	:	0.0
		0.5 N	3.0 ± 0.4	:	1.0	:	0.7 ± 0.1	:	0.0	:	1.1 ± 0.1	:	0.0
		nc	10 ± 3	:	1.0	:	1.9 ± 0.5	:	1.0 ± 0.6	:	7 ± 2	:	0.0
	Ball	centre	2.8 ± 0.2	:	1.0	:	0.4 ± 0.0	:	0.0	:	0.7 ± 0.1	:	0.0
		nc	6 ± 3	:	1.0	:	1.4 ± 1.0	:	0.7 ± 0.8	:	2 ± 2	:	0.1 ± 0.1
		depo	4 ± 2	:	1.0	:	6 ± 1	:	0.6 ± 0.6	:	5 ± 2	:	0.0
150 °C	Disc	10 N	3.0 ± 0.2	:	1.0	:	0.3 ± 0.0	:	0.1 ± 0.1	:	0.9 ± 0.1	:	0.0
		5 N	2.9 ± 0.2	:	1.0	:	0.3 ± 0.1	:	0.0	:	0.9 ± 0.1	:	0.0
		0.5 N	3.3 ± 0.5	:	1.0	:	0.4 ± 0.1	:	0.1 ± 0.1	:	0.9 ± 0.1	:	0.1 ± 0.1
		nc	3.8 ± 0.2	:	1.0	:	0.9 ± 0.3	:	0.3 ± 0.1	:	1.5 ± 0.1	:	0.0
	Ball	centre	2.9 ± 0.0	:	1.0	:	0.3 ± 0.0	:	0.0	:	0.8 ± 0.0	:	0.0
		nc	4.3 ± 0.0	:	1.0	:	1.3 ± 0.1	:	0.4 ± 0.0	:	1.8 ± 0.1	:	0.0
		depo	3.4 ± 0.0	:	1.0	:	1.2 ± 0.0	:	0.3 ± 0.0	:	1.6 ± 0.0	:	0.0

In the case of oxygen, only the NBO and BO peaks were taken into account. For iron, only the peak assigned to iron phosphate was attributed to the reaction layer composition and the thickness of the surface layers and of the substrate.

The composition of the tribofilm on the *ball* was very similar to that of tribofilms produced on the disc at high loads; no significant differences were detected. The non-contact areas on the ball half a millimetre in front of and behind the contact were equivalent, within the usual uncertainty, and similar to the non-contact areas on the disc. The material deposited at 80 °C about 250 μm behind the sliding contact mainly consisted of zinc sulphide. At 150 °C, the composition of the deposited material was similar to that of the non-contact areas.

3.3.5 Chain Length

The binding energy of phosphorus $2p_{3/2}$ can be used to identify the kind of phosphate formed on the surface: i.e. between orthophosphates (single phosphate groups), or short- or long-chain polyphosphates [7, 59, 60, 62, 64]. The chain length of these polyphosphates can be calculated using the BO-to-NBO ratio (see Fig. 12). The ratios were corrected by subtracting the contributions to the NBO signal of oxygen in sulphate groups and oxygen bound to carbon in the organic layer.

At both 30 °C and 80 °C, shorter chain lengths were observed corresponding to higher contact pressures. In the non-contact areas of 30 °C and 80 °C, the binding energy of phosphorus $P2p_{3/2}$ of 133.3 ± 0.2 eV indicates that orthophosphates were present [59]. The BO-to-NBO ratio of the non-contact areas is not plotted for these temperatures, because adsorbed water was contributing to the BO peak of these samples.

At 150 °C, chains of ~4 phosphate groups were found at high contact pressures while shorter chains of three units were found at low pressures. This was confirmed by the binding energy of phosphorus being at 134.0 ± 0.2 eV at

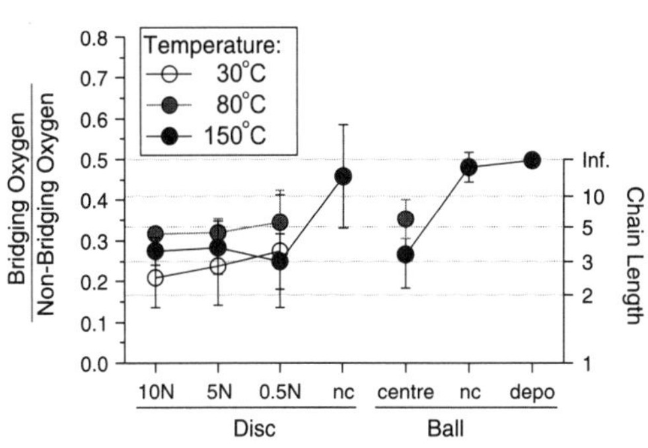

Fig. 12 Bridging to non-bridging oxygen ratio vs. the applied load and temperature. The ratio is corrected for the contributions to NBO originating from sulphate groups and carbon bound to oxygen

high pressures and 0.2 eV lower at low pressure. In the non-contact areas, the thermal film consisted of a long-chain polyphosphate, as indicated by the BO-to-NBO ratio of almost 0.5.

The chain length of the tribostressed region on the *ball* remained unchanged with the applied temperature and ranged between 3 and 6 phosphate units. In the non-contact areas and in the near contact area (deposited material) at 150 °C, long chains were calculated.

4 Discussion

4.1 Comparison of the Constant-Load and Oscillating-Load Approaches

In this paper, the modified combinatorial approach with oscillating load in combination with surface analysis using XPS is presented for the first time. The friction coefficients measured with the oscillating-load test (Fig. 4) are in excellent agreement with earlier work performed on the same system but with constant loads: applying the constant-load test, the friction coefficient increased from 0.14 ± 0.02 at room temperature to 0.20 ± 0.2 above 130 °C [11]. Only few data are to be found in the literature comparing the friction coefficient in a ZnDTP solution at different loads. For experiments with constant loads at 150 °C in ZnDTP solution, Eglin et al. [9] reported friction coefficients at 1 N and 5 N load of 0.19 ± 0.1, in agreement with this work. At low loads, 0.1 N, a higher value of 0.27 ± 0.3 was found.

The dimensional wear coefficients *k* obtained with the hardened steel samples stressed with the oscillating-load test were between 1 and 4×10^{-6} mm^3/Nm (see Fig. 6), which correspond to severe wear. This is unsurprising in the boundary-lubrication regime and the wear coefficients *k* would normally be expected to lie between 10^{-8} and 10^{-6} mm^3/Nm [1]. There are two reasons for the high wear rates observed in these experiments: First, the sliding speeds are very low and second, the contact areas are small, resulting in a high average contact pressure up to 500 MPa, even at relatively low applied loads.

The tribological films formed in this study were thicker than those formed in the constant-load test, while the thermal films had similar thicknesses [11]. This is likely to be due to the longer time of tribostress; only five turns per annulus were applied in the constant-load test, while 1000 turns were carried out with the oscillating-load test applied in this work. Short-chain poly(thio)phosphates with 2–6 phosphate units were found at all temperatures in the tribotracks, which corresponds to results obtained with constant loads [11, 65] and is in agreement with proposals made in the literature [22, 35]. At 150 °C, the thermal film was composed of long-chain poly(thio)phosphates, in agreement with previous work [11, 29].

A potential drawback of the oscillating-load test approach is that part of the film formed at high contact pressure might be transferred to areas at low contact pressures [33, 34]. However, no evidence for a possible transfer was found and the results are in agreement with results obtained with the constant-load test [11].

4.2 Influence of Contact Pressure on Thickness and Composition of the Tribological Films

It is difficult to compare the applied loads reported in different studies, since often only the load is given but neither the contact pressure nor the contact area are reported. With applied loads ranging from 0.5 to 10 N in this work, average contact pressures in the range of 25–500 MPa were probed.

The tribological stress leads to the formation of poly(thio)phosphate tribofilms at all temperatures investigated, while in the non-contact areas only very thin thermal films were formed at 30 °C and 80°C. The contact pressure presumably leads to local frictional heating, which accelerates a chemical reaction of the adsorbed additives (see Fig. 13). Also, a purely tribochemical effect is likely: at 30 °C, iron phosphate was found in the reaction layer, which is an indication that nascent iron, released because of wear, reacts with the phosphoryl groups and forms a component of the tribofilm.

At 150 °C, thick thermal layers (>6 nm) composed of long-chain poly(thio)phosphates were formed. Under mechanical stress, there are further potential pathways: Wear might have removed the thermal layer, formed prior to contact, and upon tribostress a similar film-formation mechanism as at lower temperatures takes place [66]. Alternatively iron oxide reacts with the poly(thio)phosphates causing a shortening of the chains [67]. Iron is known to function as a depolymerising element for phosphate glasses [62]. Yet a further possibility is that the contact pressure leads to a cleavage of the long chains [11]. Any combination of these mechanisms could explain the presence of short-chain poly(thio)phosphates in the tribotracks.

At higher contact pressures, shorter poly(thio)phosphate chains were formed with correspondingly less sulphur. Only at 0.5 N load (150 °C) were shorter chains found than at higher contact pressures. An explanation might be that on the thick tribofilms formed under high contact pressure, only the outer part, which generally contains longer chains [22, 35, 65, 68], was probed, while at 0.5 N load the entire film, including the shorter chains at the bottom, were analysed. Less sulphur at higher contact pressures was also found by both Yin et al. and Eglin et al. [7, 8, 35], while Yin found longer chain lengths at higher contact pressures in an experiment at much higher sliding speed (0.35 m/s) at 100 °C [35], under which conditions a different lubrication regime may have been operative. According to nanoindentation measurements, the hardness of the tribofilm correlates with the applied contact pressure [17, 20]. Therefore a softer viscous film of longer chains might be present at lower contact pressures, which would explain the incrementally higher friction coefficient observed under these conditions. High contact pressures cause more frictional heating, therefore thicker films were formed in these areas. Together with the increased hardness at higher contact pressures, thicker and tougher films [69] of shorter polyphosphates were formed, which exhibited higher shear resistance, resulting in lower dimensional wear coefficients (see Fig. 6).

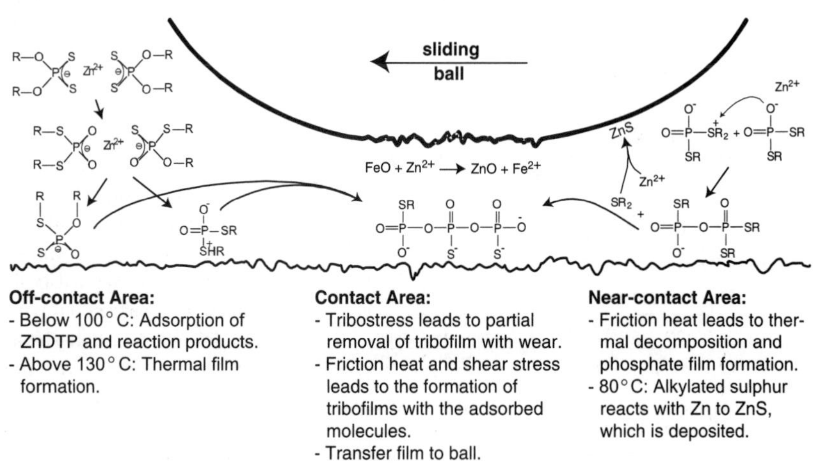

Fig. 13 Schematic of reactions occurring in the contact, near-contact and off-contact areas on samples tribostressed under ZnDTP solution

4.3 Influence of Temperature on Thickness and Composition of Thermal and Tribological Films

In the non-contact areas at 30 °C and 80 °C, only very thin reaction layers of zinc sulphide and zinc phosphate were formed, whereas the films at 80 °C were thicker. This is in agreement with earlier temperature-dependent studies [11] and according to Bovington and Dacre, the thermal decomposition starts at 60 °C in the presence of iron powder as a catalyst [67]. At 150 °C, thick thermal films made of long-chain poly(thio)phosphate were formed, due to thermo-oxidative processes [3, 11, 29, 70]. The thickness and the P/S ratio measured on the thermal film formed in this investigation, after approximately 4 h of exposure to the ZnDTP-containing solution, were found to be 6 nm and 0.8 respectively: these values correlate well with those estimated on the basis of PIXE measurements (11 nm and with a P/S ratio of 1.5) after 8 h of exposure assuming the film to be composed of $Zn_2P_2O_7$ [24]. XPS thus allows the thickness and the composition of each layer, and simultaneously the composition of the substrate, to be calculated with very high flexibility and accuracy, since it is possible to use the proper density value for each layer.

In the tribostressed areas, the average thickness of the reaction layer increased with higher temperatures because of the external temperature, which together with the frictional heat favours the polyphosphate film formation. On the tribofilms formed at 150 °C, a reduced sulphide content was found while species containing oxidized sulphur were present. This is likely due to a temperature-dependent oxidation of the tribofilm [3].

4.4 ZnDTP Reactivity: Ball vs. Disc Surface Composition and Thickness

During a ball-on-disc experiment, the ball is under permanent tribostress while a single spot in the centre of the tribotrack is stressed for about 40 ms every 6.3 s. Despite the different rubbing time at varying loads experienced by the ball and the disc, both the thickness and the compositions of the reaction layers were the same for the tribostressed region on the ball and for the tribotrack on the disc stressed with high contact pressures (see Figs. 9 and 11 and Table 1): a short-chain poly(thio)phosphate of 2–6 units was always found with the same amount of sulphur (II) and zinc.

The non-contact areas on the ball 0.5 mm in front and behind the tribostressed region were essentially similar in thickness and composition as the non-contact areas on the disc, thus indicating that the thermal film was not affected because of the tribological contact half a millimetre away.

On the ball tribostressed at 80 °C, there was a tail behind the tribostressed region composed mainly of zinc sulphides (see Fig. 8 and Table 1). Upon frictional heating, reactive sulphur compounds are probably produced, which then react with zinc to form zinc sulphide and deposit behind the tribological contact. These products are readily detectable on the ball because they are always collected on the same area during the entire experiment, while on the other hand they are thinly distributed over the entire disc. At 30 °C no deposited material was found on the ball and at 150 °C, a thicker deposit behind the tribological contact than in the non-contact area was formed, with the same composition as that of the non-contact film. Thus at 150 °C, more reactive species were present in the vicinity of the tribological contact, leading to the thicker film.

5 Conclusion

Real applications, as opposed to most tribological tests, are characterised by long lifetimes with many stress cycles, often at different loads. This modified combinatorial approach now allows the screening of the effect of different loads along a single track, while the permanent change of the load does not appear to affect the frictional behaviour, as discussed above. This approach is far faster than changing the radius for each load and provides the possibility of performing long-term combinatorial experiments, which allow for a better comparison with real systems.

X-ray photoelectron spectroscopic characterization permitted not only the assessment of contact-pressure effects on the chemical composition, but also the calculation of thicknesses and compositions of the films formed both on the ball and on the disc. These findings help to understand the anti-wear mechanism of ZnDTP and allow the proposal of a mechanism that takes into account the changes occurring on the two counterparts. The same composition of the tribofilm was found on the ball and on the disc whenever high loads were applied. This suggests that the same mechanism took place on the ball and on the disc tribostressed with high contact pressures. While on the disc the pressure dependence was investigated, the reactions behind the tribological contact could be followed on the ball.

A higher contact pressure produced more frictional heat, which causes the formation of thicker tribofilms of poly(thio)phosphates. The higher pressure has the net effect of reducing the chain length, resulting in poly(thio)phosphates of 3–5 phosphate units. These films are presumed to be harder and tougher, which results in lower wear.

References

1. Hutchings, I.M.: Tribology – Friction and Wear of Engineering Materials. Butterworth-Heinemann Ltd, London (1992)
2. Hähner, G., Spencer, N.D.: Rubbing and scrubbing. Phys. Today **51**, 22–27 (1998)
3. Gellman, A.J., Spencer, N.D.: Surface chemistry in tribology. Proc. Inst. Mech. Eng. J. **216**, 443–461 (2002)
4. Hogmark, S., Jacobson, S., Larsson, M.: Design and evaluation of tribological coatings. Wear **246**, 20–33 (2000)
5. Hogmark, S., Jacobson, S., Waenstrand, O.: Universal test for friction, wear and lubrication, used in tribological studies. Swedish Patent, SE9802017-5 (1998)
6. Eglin, M., Rossi, A., Spencer, N.D.: Additive-surface interaction in boundary lubrication: a combinatorial approach. In: Proc. 28th Leeds-Lyon Symp., Vienna, pp. 49–57. Elsevier Science B.V. (2001)
7. Eglin, M., Rossi, A., Spencer, N.D.: X-ray photoelectron spectroscopy analysis of tribostressed samples in the presence of ZnDTP: a combinatorial approach. Tribol. Lett. **15**, 199–209 (2003)
8. Eglin, M., Rossi, A., Spencer, N.D.: A combinatorial approach to elucidating tribochemical mechanisms. Tribol. Lett. **15**, 193–198 (2003)
9. Eglin, M.: Development of a combinatorial approach to lubricant additive characterization. Ph.D. Thesis, ETH Zurich, Zurich, Switzerland (2003)
10. Heuberger, R.: Combinatorial study of the tribochemistry of anti-wear lubricant additives. Ph.D. Thesis No. 17207, ETH Zurich (2007)
11. Heuberger, R., Rossi, A., Spencer, N.D.: XPS study of the influence of temperature on ZnDTP tribofilm composition. Tribol. Lett. **25**, 185–196 (2007)
12. Spikes, H.: The history and mechanisms of ZDDP. Tribol. Lett. **17**, 469–489 (2004)
13. Nicholls, M.A., Do, T., Norton, P.R., Kasrai, M., Bancroft, G.M.: Review of the lubrication of metallic surfaces by zinc dialkyl-dithiophosphates. Tribol. Int. **38**, 15–39 (2005)
14. Pidduck, A.J., Smith, G.C.: Scanning probe microscopy of automotive anti-wear films. Wear **212**, 254–264 (1997)
15. Aktary, M., McDermott, M.T., Torkelson, J.: Morphological evolution of films formed from thermooxidative decomposition of ZDDP. Wear **247**, 172–179 (2001)
16. Morina, A., Green, J.H., Neville, A., Priest, M.: Surface and tribological characteristics of tribofilms formed in the boundary lubrication regime with application to internal combustion engines. Tribol. Lett. **15**, 443–452 (2003)
17. Graham, J.F., McCague, C., Norton, P.R.: Topography and nano-mechanical properties of tribochemical films derived from zinc dialkyl and diaryl dithiophosphates. Tribol. Lett. **6**, 149–157 (1999)
18. Bec, S., Tonck, A., Georges, J.M., Georges, E., Loubet, J.L.: Improvements in the indentation method with a surface force apparatus. Philos. Mag. A **74**, 1061–1072 (1996)
19. Warren, O.L., Graham, J.F., Norton, P.R., Houston, J.E., Michalske, T.A.: Nanomechanical properties of films derived from zinc dialkyldithiophosphate. Tribol. Lett. **4**, 189–198 (1998)
20. Bec, S., Tonck, A., Georges, J.M., Coy, R.C., Bell, J.C., Roper, G.W.: Relationship between mechanical properties and structures of zinc dithiophosphate anti-wear films. Proc. R. Soc. London A **455**, 4181–4203 (1999)
21. Bird, R.J., Galvin, G.D.: Application of photoelectron-spectroscopy to the study of e.p. films on lubricated surfaces. Wear **37**, 143–167 (1976)
22. Martin, J.M., Grossiord, C., Le Mogne, T., Bec, S., Tonck, A.: The two-layer structure of zndtp tribofilms Part 1: AES, XPS and XANES analyses. Tribol. Int. **34**, 523–530 (2001)
23. Yin, Z.F., Kasrai, M., Bancroft, G.M., Laycock, K.F., Tan, K.H.: Chemical characterization of antiwear films generated on steel by zinc dialkyl dithiophosphate using X-ray-absorption spectroscopy. Tribol. Int. **26**, 383–388 (1993)
24. Fuller, M.L.S., Fernandez, L.R., Massoumi, G.R., Lennard, W.N., Kasrai, M., Bancroft, G.M.: The use of X-ray absorption spectroscopy for monitoring the thickness of antiwear films from ZDDP. Tribol. Lett. **8**, 187–192 (2000)
25. Piras, F.M., Rossi, A., Spencer, N.D.: In situ Attenuated Total Reflection (ATR) spectroscopic analysis of tribological phenomena. In: Proc. 28th Leeds-Lyon Symp., Vienna, pp. 199–207. Elsevier Science B.V. (2001)
26. Piras, F.M., Rossi, A., Spencer, N.D.: Growth of tribological films: In situ characterization based on attenuated total reflection infrared spectroscopy. Langmuir **18**, 6606–6613 (2002)
27. Coy, R.C., Jones, R.B.: The thermal-degradation and ep performance of zinc dialkyldithiophosphate additives in white oil. ASLE Trans. **24**, 77–90 (1981)
28. Jones, R.B., Coy, R.C.: The chemistry of the thermal-degradation of zinc dialkyldithiophosphate additives. ASLE Trans. **24**, 91–97 (1981)
29. Fuller, M.L.S., Kasrai, M., Bancroft, G.M., Fyfe, K., Tan, K.H.: Solution decomposition of zinc dialkyl dithiophosphate and its effect on antiwear and thermal film formation studied by X-ray absorption spectroscopy. Tribol. Int. **31**, 627–644 (1998)
30. Luther, H., Sinha, S.K.: Zur Reaktionskinetik der thermischen Zersetzung von Zinkdialkyldithiophosphaten. Erdoel Kohle, Erdgas Petrochem. **17**, 91–97 (1964)
31. Brow, R.K.: Review: the structure of simple phosphate glasses. J. Non-Cryst. Solids **263**, 1–28 (2000)
32. Fujita, H., Spikes, H.A.: The formation of zinc dithiophosphate antiwear films. Proc. Inst. Mech. Eng. J. **218**, 265–277 (2004)
33. Minfray, C., Le Mogne, T., Lubrecht, A.A., Martin, J.M.: Experimental simulation of chemical reactions between ZDDP tribofilms and steel surfaces during friction processes. Tribol. Lett. **21**, 67–78 (2006)
34. Martin, J.M., Grossiord, C., Le Mogne, T., Igarashi, J.: Transfer films and friction under boundary lubrication. Wear **245**, 107–115 (2000)
35. Yin, Z.F., Kasrai, M., Fuller, M., Bancroft, G.M., Fyfe, K., Tan, K.H.: Application of soft X-ray absorption spectroscopy in chemical characterization of antiwear films generated by ZDDP.1. The effects of physical parameters. Wear **202**, 172–191 (1997)
36. Windecker, R.: Optical autofocussing profilometer. Techn. Messen. **60**, 267–270 (1993)
37. Seah, M.P.: Quantification of AES and XPS. In: Briggs, D., Seah, M.P. (eds.) Practical Surface Analysis, pp. 201–255. John Wiley & Sons Ltd, Chichester, UK (1990)
38. Fadley, C.S.: Solid state – and surface – analysis by means of angular-dependent x-ray photoelectron spectroscopy. Prog. Solid State Chem. **11**, 265–343 (1976)
39. Asami, K., Hashimoto, K.: X-Ray photoelectron-spectra of several oxides of iron and chromium. Corr. Sci. **17**, 559–570 (1977)
40. Rossi, A., Elsener, B.: XPS analysis of passive films on the amorphous alloy Fe70Cr10P13C7 – effect of the applied potential. Surf. Interface Anal. **18**, 499–504 (1992)
41. Elsener, B., DeFilippo, D., Rossi, A.: Composition of passive films on stainless steels formed in neutral sulphate solutions. In: Marcus, P., Baroux, B., Keddam, M (eds.) Modification of Passive Films, pp. 6–11. The Institute of Materials, London (1994)
42. Matsumoto, K.: Surface chemical and tribological investigations of phosphorus-containing lubricant additives. Ph.D. Thesis, ETH Zurich, Zurich, Switzerland (2003)
43. Scofield, J.H.: Hartree-Slater subshell photoionization cross-sections at 1254 and 1487 eV. J. Electron. Spectrosc. **8**, 129–137 (1976)

44. Reilman, R.F., Msezane, A., Manson, S.T.: Relative intensities in photoelectron-spectroscopy of atoms and molecules. J. Electron. Spectrosc. **8**, 389–394 (1976)
45. Seah, M.P.: Quantification of AES and XPS. In: Briggs, D., Grant, J.T. (eds.) Surface Analysis by Auger and X-Ray Photoelectron Spectroscopy, pp. 345–375. IM Publications and Surface Spectra Limited, Chichester and Manchester, UK (2003)
46. Tanuma, S., Powell, C.J., Penn, D.R.: Calculations of electron inelastic mean free paths. 5. Data for 14 organic-compounds over the 50-2000 eV range. Surf. Interface Anal. **21**, 165–176 (1993)
47. Tanuma, S., Powell, C.J., Penn, D.R.: Calculation of electron inelastic mean free paths (IMFPs) VII. Reliability of the TPP-2M IMFP predictive equation. Surf. Interface Anal. **35**, 268–275 (2003)
48. de Donato, P., Mustin, C., Benoit, R., Erre, R.: Spatial distribution of iron and sulphur species on the surface of pyrite. Appl. Surf. Sci. **68**, 81–93 (1993)
49. Descostes, M., Mercier, F., Thromat, N., Beaucaire, C., Gautier-Soyer, M.: Use of XPS in the determination of chemical environment and oxidation state of iron and sulfur samples: constitution of a data basis in binding energies for Fe and S reference compounds and applications to the evidence of surface species of an oxidized pyrite in a carbonate medium. Appl. Surf. Sci. **165**, 288–302 (2000)
50. Zerulla, D., Chasse, T.: X-ray induced damage of self-assembled alkanethiols on gold and indium phosphide. Langmuir **15**, 5285–5294 (1999)
51. Heister, K., Zharnikov, M., Grunze, M., Johansson, L.S.O., Ulman, A.: Characterization of X-ray induced damage in alkanethiolate monolayers by high-resolution photoelectron spectroscopy. Langmuir **17**, 8–11 (2001)
52. Beamson, G., Briggs, D.: High Resolution XPS of Organic Polymers. John Wiley & Sons Ltd, Chichester, UK (1992)
53. Moulder, J.F., Stickle, W.F., Sobol, P.E., Bomben, K.D.: Handbook of X-Ray Photoelectron Spectroscopy. Physical Electronics, Inc., Eden Prairie, MN, USA (1995)
54. McCafferty, E., Wightman, J.P.: Determination of the concentration of surface hydroxyl groups on metal oxide films by a quantitative XPS method. Surf. Interface Anal. **26**, 549–564 (1998)
55. Huang, N.P., Michel, R., Voros, J., Textor, M., Hofer, R., Rossi, A., Elbert, D.L., Hubbell, J.A., Spencer, N.D.: Poly(L-lysine)-g-poly(ethylene glycol) layers on metal oxide surfaces: surface-analytical characterization and resistance to serum and fibrinogen adsorption. Langmuir **17**, 489–498 (2001)
56. Heuer, J.K., Stubbins, J.F.: An XPS characterization of FeCO3 films from CO2 corrosion. Corr. Sci. **41**, 1231–1243 (1999)
57. Brundle, C.R., Chuang, T.J., Wandelt, K.: Core and valence level photoemission studies of iron-oxide surfaces and oxidation of iron. Surf. Sci. **68**, 459–468 (1977)
58. Olla, M., Navarra, G., Elsener, B., Rossi, A.: Nondestructive in-depth composition profile of oxy-hydroxide nanolayers on iron surfaces from ARXPS measurement. Surf. Interface Anal. **38**, 964–974 (2006)
59. Pinna, R.: Synthesis and characterization of Zn and Fe polyphosphates glasses. Diploma Thesis, University of Cagliari, Cagliari, Italy (2001/2002)
60. Onyiriuka, E.C.: Zinc phosphate-glass surfaces studied by XPS. J. Non-Cryst. Solids **163**, 268–273 (1993)
61. Brückner, R., Chun, H.-U., Goretzki, H., Sammet, M.: XPS measurements and structural aspects of silicate and phosphate glasses. J. Non-Cryst. Solids **42**, 49–60 (1980)
62. Brow, R.K., Arens, C.M., Yu, X., Day, E.: An XPS study of iron phosphate-glasses. Phys. Chem. Glass. **35**, 132–136 (1994)
63. Cumpson, P.J.: Angle-resolved X-ray photoelectron spectroscopy. In: Briggs, D., Grant, J.T. (eds.) Surface Analysis by Auger and X-Ray Photoelectron Spectroscopy, pp. 651–675. IM Publications and Surface Spectra Limited, Chichester and Manchester, UK (2003)
64. Piras, F.M., Rossi, A., Spencer, N.D.: Combined in situ (ATR FT-IR) and ex situ (XPS) study of the ZnDTP-iron surface interaction. Tribol. Lett. **15**, 181–191 (2003)
65. Minfray, C., Martin, J.M., Esnouf, C., Le Mogne, T., Kersting, R., Hagenhoff, B.: Antiwear mechanisms of zinc dithiophosphate: a chemical hardness approach. Thin Solid Films **447**, 272–277 (2004)
66. Stachowiak, G.W., Batchelor, A.W.: Engineering Tribology. Elsevier Inc., Oxford, UK (2005)
67. Bovington, C.H., Dacre, B.: The adsorption and reaction of decomposition products of zinc di-isopropyldiophosphate on steel. ASLE Trans. **27**, 252–258 (1984)
68. De Barros, M.I., Bouchet, J., Raoult, I., Le Mogne, T., Martin, J.M., Kasrai, M., Yamada, Y.: Friction reduction by metal sulfides in boundary lubrication studied by XPS and XANES analyses. Wear **254**, 863–870 (2003)
69. Pawlak, Z., Yarlagadda, P.K.D.V., Frost, R., Hargreaves, D.: The mechanical strength of phosphates under friction-induced cross-linking. J. Achieve. Mater. Manuf. Eng. **17**, 201–204 (2006)
70. Martin, J.M.: Antiwear mechanisms of zinc dithiophosphate: a chemical hardness approach. Tribol. Lett. **6**, 1–8 (1999)

Tribol Lett (2009) 35:31–43
DOI 10.1007/s11249-009-9429-3

ORIGINAL PAPER

Reactivity of Triphenyl Phosphorothionate in Lubricant Oil Solution

Filippo Mangolini · Antonella Rossi · Nicholas D. Spencer

Received: 9 February 2009 / Accepted: 4 March 2009 / Published online: 27 March 2009

Abstract Investigating the thermo-oxidative reactivity of anti-wear additives in lubricant oil solution at high temperature can significantly contribute to an understanding of the mechanism of thermal film and tribofilm formation on metal surfaces. In this study, the reactivity of triphenyl phosphorothionate (TPPT) in lubricant oil solution at high temperature (423 and 473 K) has been studied by Fourier transform infrared spectroscopy (FT-IR) and nuclear magnetic resonance (NMR) spectroscopy. The results show that the TPPT molecule was highly thermally stable and did not completely decompose in oil solution even upon heating at 423 K for 168 h and at 473 K for 72 h. The degradation of the TPPT molecule, which turned out to be a first-order reaction, started taking place after 6 h at both temperatures, leading to the breakage of the P=S bond with the formation of triphenyl phosphate. During these heating experiments, no oil-insoluble compounds were detected. The oxidation of the base oil as a result of the prolonged heating demonstrated that the TPPT molecule did not effectively act as oxidation inhibitor.

Keywords Thermo-oxidative degradation · Triphenyl phosphorothionate · Ashless anti-wear additive · Low-SAPS additive · FT-IR · NMR

F. Mangolini · A. Rossi · N. D. Spencer (✉)
Laboratory for Surface Science and Technology, Department of Materials, ETH Zurich, HCI H 523, Wolfgang-Pauli-Strasse 10, 8093 Zurich, Switzerland
e-mail: spencer@mat.ethz.ch

A. Rossi
Dipartimento di Chimica Inorganica ed Analitica, Università degli Studi di Cagliari, INSTM Unit—Cittadella Universitaria di Monserrato, 09100 Cagliari, Italy

1 Introduction

Zinc dialkyldithiophosphates (ZnDTPs) have been used as antioxidants, corrosion inhibitors and anti-wear agents in engine oil formulations for decades [1–3]. This class of additives contains a large amount of phosphorus, sulfur and zinc, which can lead to a loss of emission-control catalyst effectiveness and block filters within the exhaust gas after-treatment system [4].

Engine lubricant specifications (e.g. European Automobile Manufacturers' Association ACEA, American Petroleum Institute API, International Lubricant Standardization and Approval Committee ILSAC and Japanese Automotive Standards Organization JASO) are progressively limiting the level of sulfated ash, phosphorus, and sulfur (SAPS) in oil formulations. Several low- or zero-SAPS additives, such as thiophosphates, amine phosphates, metal dithiocarbamates, organometallics and organoboron compounds, have recently been proposed to partially or fully replace ZnDTPs [5].

In contrast to the situation with ZnDTPs, relatively little fundamental research has been carried out with new ashless (i.e. metal-free) additives, such as organosulfur compounds [6], phosphates [6–16], amine phosphates [12], phosphites [11, 17–19], phosphorothionates [12–14, 20–26], and dithiophosphates [12, 21, 22, 24].

Most of the published work deals with films produced by these alternative additives to ZnDTPs under tribological and/or pure thermal conditions. Several analytical techniques have been employed to characterize the reaction layers, such as X-ray photoelectron spectroscopy (XPS) [7, 12, 14, 25, 26], X-ray absorption spectroscopy (XAS) [13, 21–24], X-ray photoelectron emission microscopy (X-PEEM) [24], scanning electron microscopy (SEM) [20, 21], Fourier transform infrared spectroscopy (FT-IR)

[7, 14, 16], Auger electron spectroscopy (AES) [8–10, 14, 17, 18], and atomic force microscopy (AFM) [22].

Najman et al. investigated the thermal films and tribofilms formed by phenyl phosphates on steel [13]. No thermal film formation was detected up to 423 K for triphenyl phosphate, while diphenyl phosphate reacted more easily with steel substrates: in this case the formation of iron(II) polyphosphate was suggested. In the tribofilms formed at 373 K, the data interpretation led the authors to propose the formation of short-chain iron(II) polyphosphate for both additives.

The same authors studied the thermal films produced by triphenyl phosphorothionate (TPPT) at 423 K. Also, in this case, the formation of short-chain iron(II) polyphosphate with a low amount of iron sulfate was suggested [22]. In the tribotracks, short-chain iron(II) polyphosphate, iron sulfides, and sulfates were detected. Increasing sliding time induced an increase in the amount of iron sulfate. After 6 h, comparable amounts of iron sulfate and phosphates were observed. These findings were explained by the authors with a decomposition temperature for triphenyl phoshorothionate of around 423 K. Following the adsorption, the thermal oxidative process took place: after the breakage of the P=S bond, the formation of iron phosphate and sulfate was proposed to occur.

Recently, the reactivity of alkylated TPPTs with air-oxidized steel was investigated in our group [25, 26]. In the absence of mechanical stress, the phosphorothionate molecules adsorbed on the substrate at low temperature (303–353 K), as described by Koyama et al. [27]. The activation temperature for the thermal decomposition of TPPT molecules was found to be around 423 K. The thermal films produced at this temperature consisted of short-chain polyphosphates and oxidized sulfur species, as indicated by the XPS results. The proposed reaction mechanism started with the P=S bond scission, followed by the cleavage of the C–O or P–O bond. The released sulfur was then oxidized to form sulfates.

The thermal films and tribofilms produced on air-oxidized iron surfaces at 373 K in the presence of tributyl thiophosphate were studied by in situ attenuated total reflection (ATR/FT-IR) tribometry, XPS, and temperature programmed reaction spectrometry (TPRS) [14]. Iron polyphosphate and sulfate were detected on the iron surface for both thermal films and tribofilms. After tribological tests, however, the amounts of sulfate and polyphosphate in the films were higher. The presence of sulfur in tributyl thiophosphate was found to favor the formation of long-chain iron polyphosphate and to lower the temperature of chemical and tribochemical reactions by around 50 K compared to the surface reaction of tributyl phosphate [7, 14]. The reaction between TBT and iron was proposed to occur via an initial P=S bond scission to give tributyl phosphite. In accordance with [3, 8, 11, 17–19], this compound successively produced butoxy groups through P–O bond cleavage.

In contrast with these findings, Sung and Gellman found that the decomposition of tributyl phosphate on iron foil produced butyl intermediates through C–O bond scission [9]. On the other hand, the initial step of the decomposition of tricresyl phosphate, a model compound for aryl phosphates, was the cleavage of the P–O bond with the formation of aryloxy intermediates, independent of the position of the methyl group [10].

While the reactivity of new ashless additives with metal surfaces has been investigated quite extensively, less is known about the thermal degradation of these compounds in oil solution. As was the case for ZnDTPs [1, 3], investigating the reactions that take place in oil solution at high temperature could significantly contribute to elucidating the mechanism of thermal film and tribofilm formation on metal surfaces.

Hilgetag and Teichman have accurately described the chemical behavior of organophosphates and organothiophosphates, pointing out that the key chemical feature of these compounds is their strong alkylating power [28, 29]. All the chemical reactions could be explained on the basis of Pearson's acid-basis concept [30–32], according to which oxygen is a "hard base" and reacts preferentially with "hard acids," such as protons, "A" metals (alkali metal cations, alkaline earth metal cations, and light transition metal cations in high oxidation states), carbonyl carbon and phosphoryl phosphorus. Sulfur, on the contrary, is a "soft base" and reacts mainly with "soft acids," such as "B" metals (heavy transition metal cations in low oxidation states), tetrahedral carbon, halogens.

This basic knowledge has been fundamental for understanding the reaction of ZnDTPs in mineral oil at temperatures above 423 K, i.e. the formation of polyphosphates and organic sulfides [1].

In the present work, the thermo-oxidative reaction of TPPT in poly-α-olefin (PAO) has been investigated at 423 and 473 K, which are temperatures typically experienced by automobile engines [33]. A combined investigation using Fourier transform infrared spectroscopy (FT-IR) and nuclear magnetic resonance (NMR) spectroscopy has been carried out for the first time to analyze the oil solutions after the heating experiments.

2 Experimental

2.1 Materials

Purified TPPT (Irgalube® TPPT, Ciba® Speciality Chemicals, Basel, Switzerland) (Fig. 1) and commercial poly-α-olefin (PAO, Durasyn® 166, Tunap Industrie GmbH & Co.,

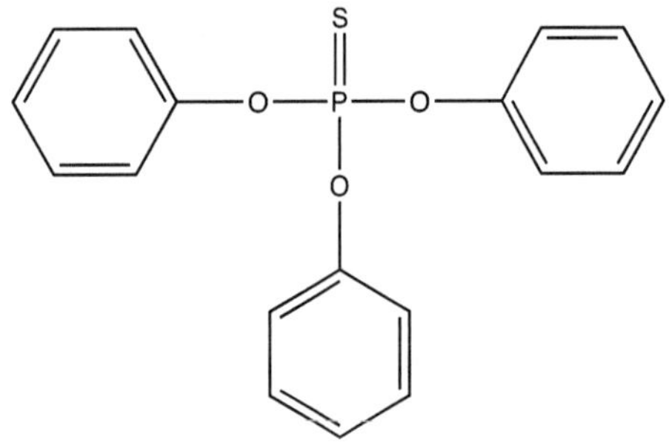

Fig. 1 Chemical formula of triphenyl phosphorothionate

Mississauga, Canada) were used as lubricant additive and base oil, respectively. The purification of TPPT was carried out by liquid chromatography.

For the heating experiments, a 0.044 mol dm^{-3} (1.82 wt%) solution of TPPT in PAO was prepared. For dissolving the crystalline TPPT in the base oil, an ultrasonic bath was employed six times for 10 min. The temperature of the lubricant solution was always maintained below 318 K during the ultrasonic bath treatment.

The vacuum filtration of the solutions at the end of the experiments was carried out using cellulose filter papers with a 2.5-μm particle retention in liquid (Grade 42, Whatman®, Maidstone, England).

2.2 Methods

2.2.1 Heating Experiments

The heating experiments were performed immersing a glass flask (40 cm^3, 95 × 27.5 mm^2, VWR International, Dietikon, Switzerland) containing 15 cm^3 of the lubricant solution in a silicone oil bath heated at 423 or 473 K (±1 K). During the experiment the solution, open to air, was neither stirred nor purged with any gas. The duration of the heating varied between 3 and 168 h at 423 K and between 3 and 72 h at 473 K. The relative humidity was between 22% and 30%.

After the test, the glass flask was removed from the silicone oil bath and left 45 min in air to cool down without any cover. The solution was then vacuum filtered in order to remove all the precipitates.

2.2.2 Microelemental Analysis

The microelemental analysis was carried out using a CHN-900 (Leco Corporation, St. Joseph, MI, USA) for carbon and hydrogen, a CHNS-932 (Leco Corporation, St. Joseph, MI, USA) for sulfur and a RO-478 (Leco Corporation, St. Joseph, MI, USA) for oxygen. The phosphorus content was determined by the molybdovanadate method after microwave digestion and leaching (2 × 50 min) in concentrated sulfuric acid and perchloric acid at 483–508 K.

Table 1 Transmission FT-IR experimental conditions

Detector	DGTS/KBr
Resolution	2 cm^{-1}
Number of scans	64
Scan velocity	0.6329 cm/s
Acquisition time	136 s
Gain control	1

2.2.3 Fourier Transform Infrared Spectroscopy

Transmission FT-IR spectra were acquired with a Nicolet™ 5700 Fourier Transform Infrared spectrometer (Thermo Electron Corporation, Madison, WI, USA). The experimental conditions are listed in Table 1. Sampling was performed by placing one drop of solution onto a KBr pellet. The crystalline, purified TPPT was ground with dried KBr and pressed into a pellet.

The spectra were processed with OMNIC™ software (V7.2, Thermo Electron Corporation, Madison, WI, USA). The single-beam spectrum of the KBr pellet was acquired before each measurement as a background spectrum. Normalization with respect to the methyl asymmetric deformation band, overlapped by the methylene scissor vibration band, of PAO at 1466 cm^{-1} [34] and to the C–O–(P) stretching vibration band at 1183 cm^{-1} [34] was performed for the lubricant solutions and for the purified TPPT, respectively.

2.2.4 Nuclear Magnetic Resonance Spectroscopy

The NMR spectra were recorded in $CDCl_3$ (99.8 at.% D, Armar Chemicals, Döttingen, Switzerland) at 300 K using a Bruker Avance 500 NMR spectrometer operating at 500.1 (1-H), 125.8 (13-C), and 202.5 (31-P) MHz. The chemical shifts, given as dimensionless δ values, were referred to TMS (1-H and 13-C) and H_3PO_4 (85%) following the IUPAC recommendation [35].

3 Results

3.1 Characterization of Purified TPPT

3.1.1 Microelemental Analysis

The microelemental analysis of TPPT after purification by liquid chromatography is reported in Table 2 together with

Table 2 Elemental analysis of purified triphenyl phosphorothionate

Element	Concentration (wt%)	
	Measured	Expected
C	63.16 ± 0.07	63.15
H	4.55 ± 0.02	4.42
O	14.1 ± 0.1	14.02
P	9.04 ± 0.05	9.05
S	9.47 ± 0.08	9.37

the expected weight percentages. The composition was found to correspond to the molecular formula $C_{18}H_{15}O_3PS$.

3.1.2 Fourier Transform Infrared Spectroscopy

The transmission FT-IR of TPPT after purification by liquid chromatography (Fig. 2) clearly revealed the characteristic peaks of the TPPT molecule. The IR peaks and the assigned functional groups are listed in Table 3. Three spectral regions could be distinguished:

(i) the region at high wavenumbers (3150–3000 cm^{-1})—aromatic CH stretching vibrations;
(ii) the region between 2100 and 1660 cm^{-1}—overtone and combination bands due to the CH out-of-plane deformation vibrations; and
(iii) the fingerprint region (1600–400 cm^{-1}).

All the reported vibration frequencies are in agreement with the literature [34, 36–39].

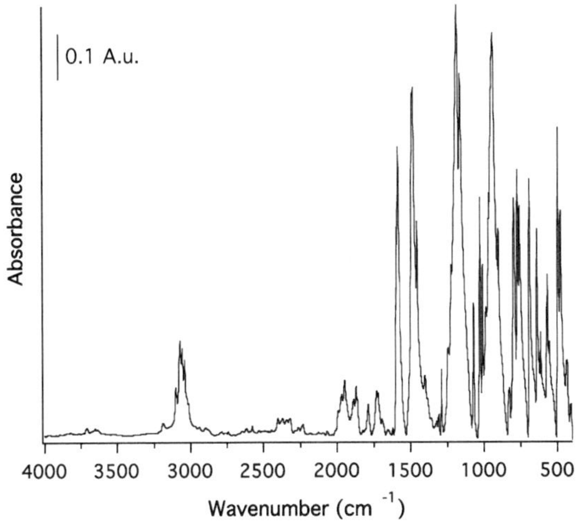

Fig. 2 Transmission FT-IR of purified TPPT. Three spectral regions can be distinguished: (i) 3150–3000 cm^{-1}: aromatic CH stretching vibrations, (ii) 2100–1660 cm^{-1}: overtone and combination bands due to the CH out-of-plane deformation vibrations, and (iii) 1600–400 cm^{-1}: fingerprint region

Table 3 IR frequencies (cm^{-1}) and functional groups for the transmission FT-IR spectrum of purified TPPT

Frequency (cm^{-1})	Functional group
3102 w, 3095, 3072, 3057, 3040, 3025 s, 3013 w	νCH
2067, 1992, 1967,1947, 1884, 1869, 1790, 1734, 1723, 1696	Overtone and combination bands due to γCH
1598, 1587 s, 1490, 1483 s	νPh
1455	ωP–O–Ar
1289, 1245, 1221	δCH
1183 vs, 1161	νC–O–(P)
1074, 1068, 1024	δCH
1007	νPh
984 w, 963 w	γCH
940 vs	νP–O–(C)
900, 830, 825	γCH
796 m	P=S (I)
771 m, 758	ν_sP–O–(C)
690	γCH
639 m	P=S (II)
614	δCH
573, 568, 555 w	δP–O–Ar
497, 479	γCH

ν: stretching; γ: out-of-plane deformation vibration; δ: in-plane deformation vibration; ω: wagging

3.1.3 Nuclear Magnetic Resonance Spectroscopy

The NMR spectra (1-H, 13-C, and 31-P) of TPPT after purification by liquid chromatography are shown in Fig. 3.

Both the 1-H and the 13-C spectra showed the characteristic signals of monosubstituted benzene [40, 41]: in the 1-H spectrum two multiplets are found at 7.40 and 7.28 ppm, whereas in the 13-C spectrum four signals are found in the 120–155 ppm region (150.8, 129.7, 125.7, and 121.3 ppm).

The 31-P spectrum showed a singlet at 54.3 ppm, in agreement with the literature [42–44].

3.2 Heating Experiments

During the heating experiments at 423 K, the solutions slowly changed their color, turning yellow-brown after 6–13 h and dark brown after 72–168 h. At 473 K, the color change was faster: the solutions appeared yellow even after 3 h and turned dark brown after 24 h.

After the experiments, all the solutions were vacuum filtered in order to remove all the precipitates. No insoluble compounds were separated out on the filter at either 423 or 473 K.

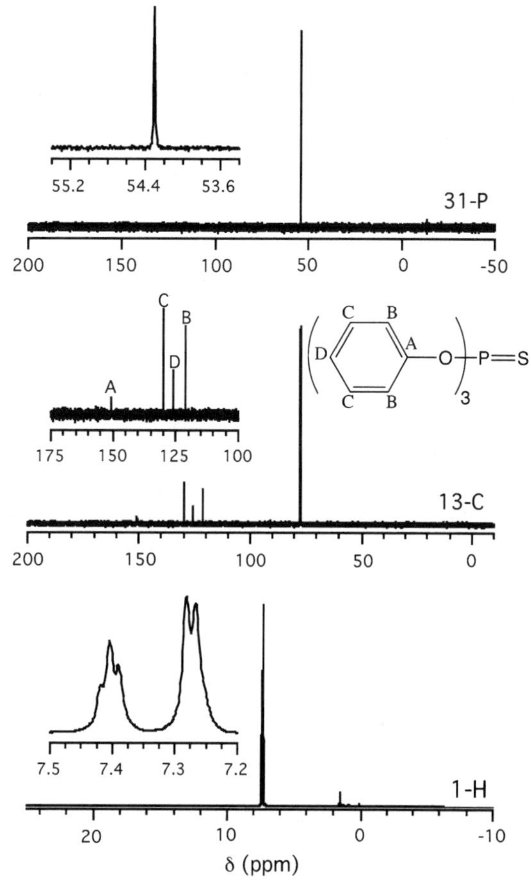

Fig. 3 NMR spectra (1-H, 13-C, and 31-P) of purified TPPT

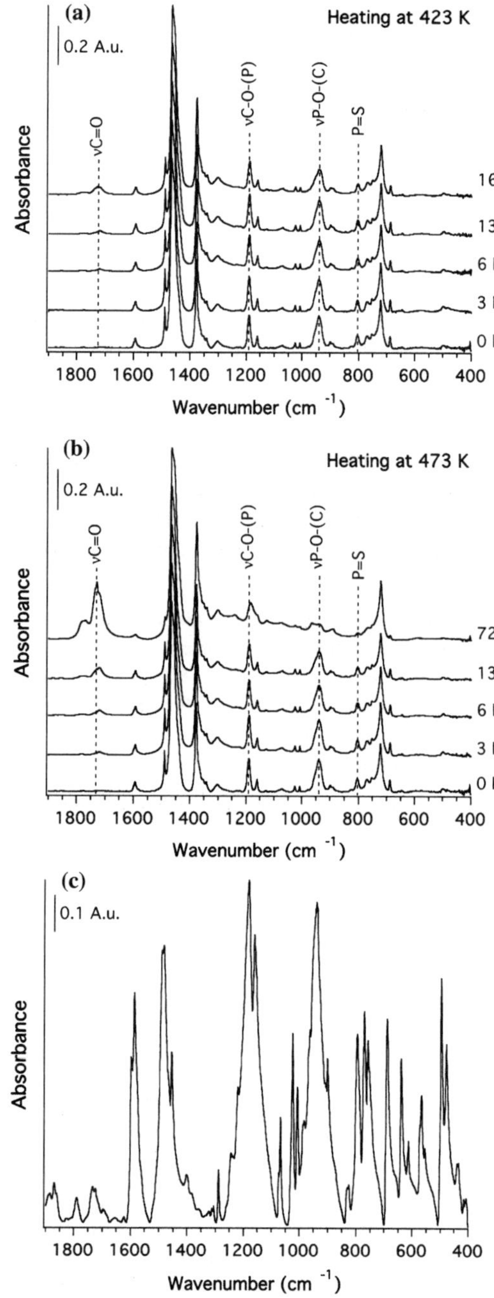

Fig. 4 Transmission FT-IR spectra of a 0.044 mol dm^{-3} solution of TPPT in PAO heated at 423 K (**a**) and 473 K (**b**) for different times and of purified TPPT (**c**)

3.2.1 Fourier Transform Infrared Spectroscopy

3.2.1.1 Heating at 423 K The transmission spectra of a 0.044 mol dm^{-3} solution of TPPT in PAO heated at 423 K for 3–168 h are reported in Figs. 4a and 5.

The spectrum of the unheated solution (0 h) showed the characteristic IR peaks of the TPPT (Table 3) and PAO (1466 cm^{-1} $\delta_{as}CH_3$ and δCH_2, 1378 cm^{-1} $\delta_s CH_3$, 890 cm^{-1} ρCH_3, 722 cm^{-1} ρCH_2) [34, 36, 45] molecules. The C–O–(P) stretching (1190 cm^{-1}) and the P=S(I) (803 cm^{-1}) vibration bands of TPPT in PAO shifted toward higher wavenumbers with respect to the transmission spectrum of purified TPPT (Table 3).

Upon heating at 423 K, no changes in peak position were detected up to 168 h. A slight decrease in peak intensity was observed after 168 h for the νPh (1593 and 1490 cm^{-1}), νC–O–(P) (1190 cm^{-1}), νP–O–(C) (940 cm^{-1}), and P=S (803 cm^{-1}) vibrations. The band at 940 cm^{-1} also exhibited a shoulder at higher wavenumbers after 13–168 h (arrow in Fig. 5c).

In the 1550–1850 cm^{-1} region, where the C=O stretching vibration is found [34, 45], a new weak and broad peak having a maximum at 1719 cm^{-1} was detected

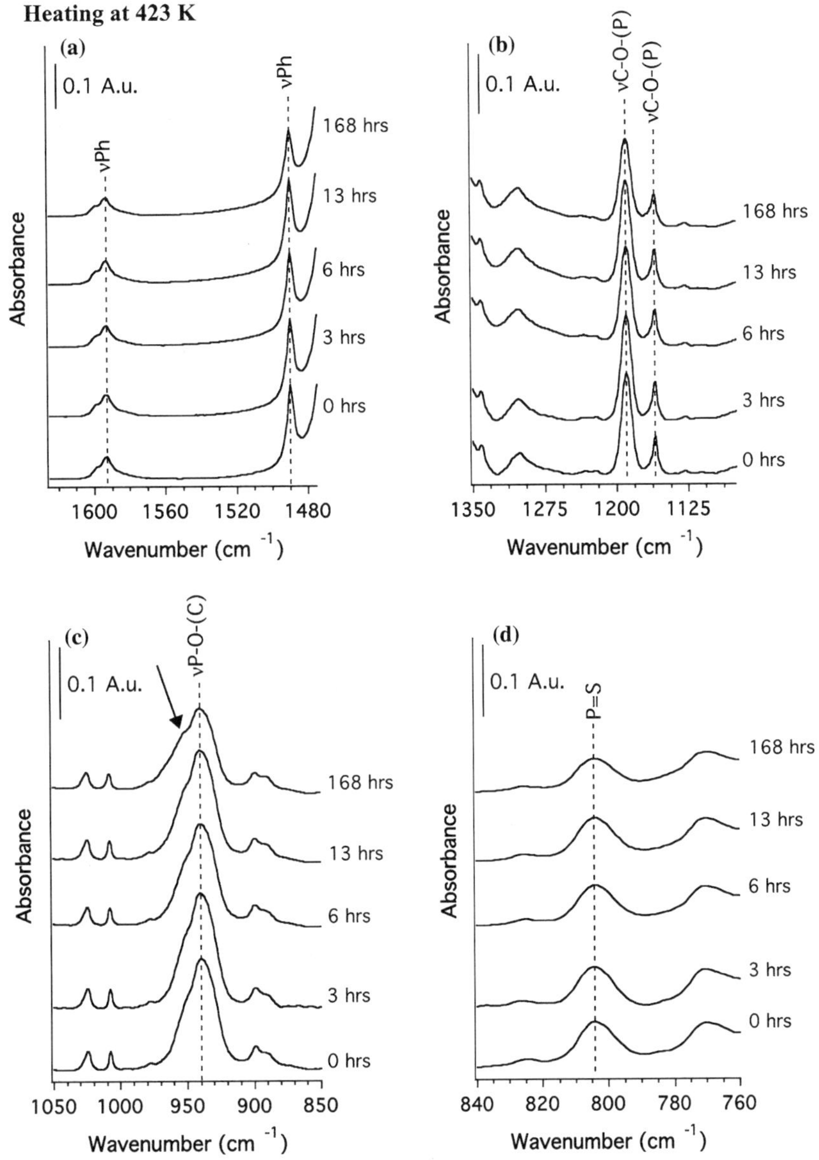

Fig. 5 Transmission FT-IR spectra of a 0.044 mol dm^{-3} solution of TPPT in PAO heated at 423 K for different times

after 13 h (Fig. 4a). After 168 h, this broad absorption band increased in intensity and showed two maxima at 1726 and 1774 cm^{-1}. Correspondingly, a slight increase of the baseline in the 850–1300 cm^{-1} region, where the C–O stretching vibration is found [34, 45], was also observed.

3.2.1.2 Heating at 473 K The absorbance of all the characteristic peaks of TPPT (*ν*Ph, *ν*C–O–(P), *ν*P–O–(C), and P=S(I)) progressively decreased upon heating at 473 K.

A shoulder at higher wavenumbers characterized the *ν*P–O–(C) absorption band after 13 h and strongly increased after 72 h, when a peak could be clearly observed at 965 cm^{-1} (arrow in Fig. 6c). A new weak peak at 1243 cm^{-1} was also detected after 72 h (arrow in Fig. 6b).

The transmission FT-IR spectrum of the solution heated at 473 K for 72 h exhibited a change in shape and a shift to lower wavenumbers (1186 cm^{-1}) of the band assigned to the C–O–(P) stretching vibration, when compared to the spectrum of the unheated solution (1190 cm^{-1}) (Figs. 4b

Fig. 6 Transmission FT-IR spectra of a 0.044 mol dm^{-3} solution of TPPT in PAO heated at 473 K for different times

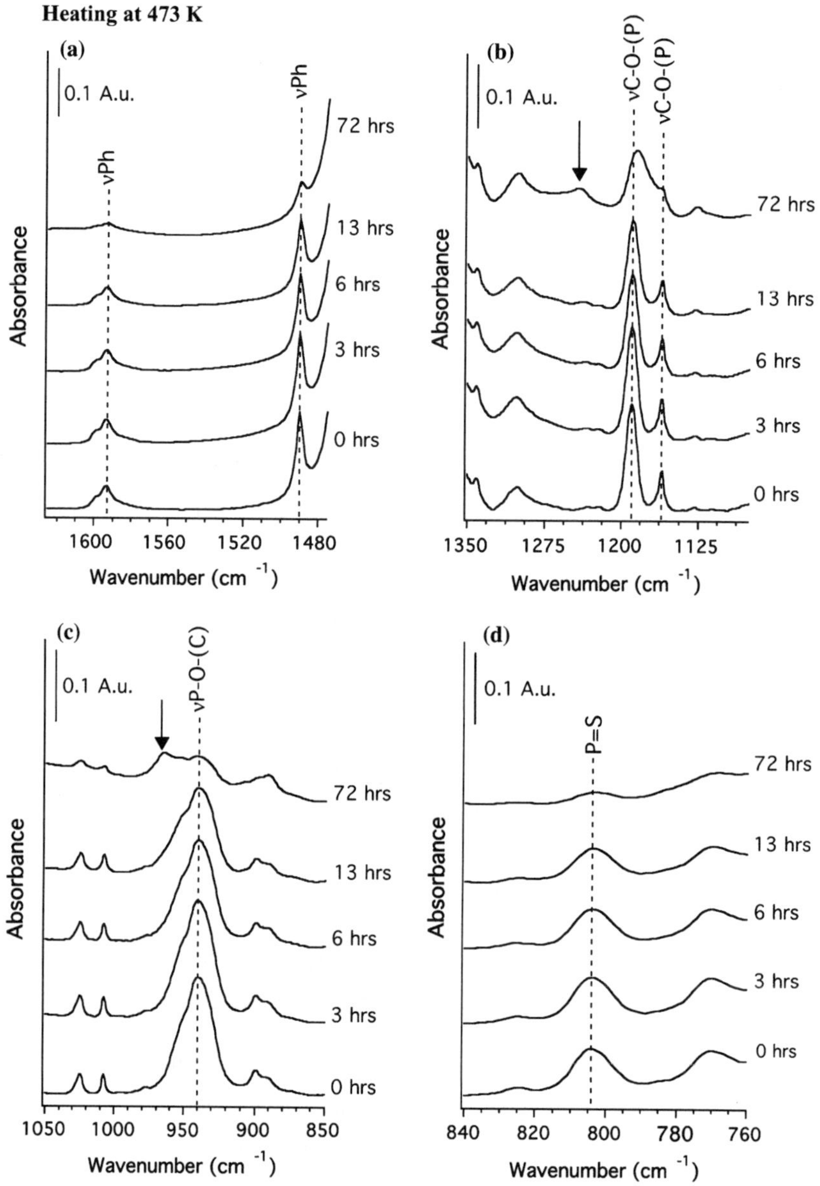

and 6b). This might be due to the superposition of the C–O stretching vibration band at ca. 1175 cm^{-1} as reported in [46].

In the 1550–1850 cm^{-1} region, a broad band was already observed after 3 h (Fig. 4b). Besides increasing in intensity, this band had a new small peak at higher wavenumbers (around 1775 cm^{-1}) after 13 h. After 72 h, a very intense absorption band with two maxima at 1731 and 1777 cm^{-1} could be seen. Correspondingly, an increase of the baseline in the 850–1300 cm^{-1} region, where the C–O stretching vibration is found [34, 45], was also observed.

3.2.1.3 Degradation Index In order to investigate the degradation kinetics, the Degradation Index (*DI*) has been defined as:

$$DI = \frac{A_{t,i}}{A_{0,i}} \times 100 \quad (1)$$

where $A_{t,i}$ is the absorbance of the *i*-th vibration at *t* heating hours, while $A_{0,i}$ is the absorbance of the *i*-th vibration for the unheated solution.

Figure 7 reports the *DI* of three characteristic vibrations of TPPT ($\nu Ph_{1593\ cm^{-1}}$, νP–O–(C), P=S(I)) for a

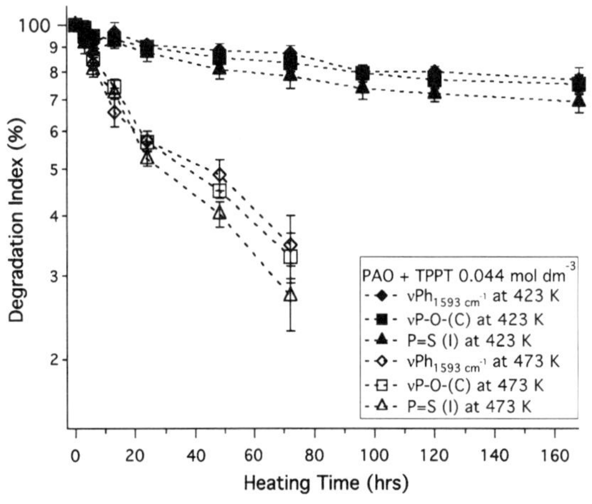

Fig. 7 Degradation Index versus heating time at 423 and 473 K for three characteristic vibrations of TPPT in PAO solution

0.044 mol dm^{-3} solution of TPPT in PAO as a function of the heating time at 423 and 473 K.

All three absorption bands showed a monotonic decrease of the *DI* values (logarithmic scale on the *DI* axis) as the solutions were heated at both temperatures. Comparing the experiments performed at 423 and at 473 K, an increase in the degradation rate was observed passing from 423 to 473 K. At both temperatures, the *DI* values of the $\nu Ph_{1593\ cm^{-1}}$ and νP–O–(C) bands were always comparable, whereas those of the P=S(I) vibration were slightly lower. The ratio of the *DI* values for the νPh, νP–O–(C) and P=S(I) absorption bands was 77 ± 5 : 75 ± 4 : 69 ± 4 at 423 K and 35 ± 5 : 33 ± 4 : 27 ± 4 at 473 K at the end of the experiments, i.e. after, respectively, 168 and 72 h.

As for the characteristic νC–O–(P) vibration of the TPPT molecule, its *DI* (not shown in Fig. 7) turned out to be always higher than that of the absorption bands considered above ($\nu Ph_{1593\ cm^{-1}}$, νP–O–(C), P=S(I)) as a result of the superposition of the C–O stretching vibration. However, as for the other vibrations of the TPPT molecule ($\nu Ph_{1593\ cm^{-1}}$, νP–O–(C), P=S(I)), a decrease of the *DI* values as the solutions were heated at high temperature and an increase in the degradation rate as the temperature passed from 423 to 473 K was observed for the νC–O–(P) absorption band.

3.2.2 Nuclear Magnetic Resonance Spectroscopy

The NMR spectra (1-H, 13-C, and 31-P) of a 0.044 mol dm^{-3} solution of TPPT in PAO heated at 423 and 473 K for, respectively, 168 and 72 h are shown in Fig. 8 together with the spectrum of the unheated solution.

In the case of the unheated solution (0 h), the characteristic peaks of the TPPT molecule were found (see Sect. 3.1.3). The intense peaks between 2 and 0 ppm in the 1-H spectrum and between 60 and 0 ppm in the 13-C spectrum correspond to the aliphatic functional groups of the base oil (PAO) [40, 47].

Upon heating at 423 K for 168 h and at 473 K for 72 h, the characteristic peaks of the TPPT molecule in the 1-H and 13-C spectra progressively decreased in intensity. Moreover, in the former new but weak peaks were found in the 6–2 ppm region, which can be assigned to the hydrogen of a methyl or methylene group bound to a carbonyl or ester group [48]. In the 13-C spectrum no peaks, which can be assigned to these species, can be found because of the small magnetic moment and low natural abundance of carbon-13 [41].

In the 31-P NMR spectrum of the unheated solution (0 h), a narrow peak at 54.4 ppm was found. The dissolution of TPPT in PAO produced a 0.1-ppm downfield shift with respect to the 31-P NMR spectrum of purified TPPT (Fig. 3).

Upon heating at 423 K for 168 h, a new small signal was detected at −16.3 ppm. The singlet at 54.4 ppm was still narrow and intense. Comparing the 31-P NMR spectrum of the solution heated at 423 K for 168 h with that of the solution heated at 473 K for 72 h, an upfield shift of 0.1 ppm was observed for both signals. Moreover, the peak at −16.4 ppm increased in intensity. The integrated intensity ratio of the peaks at 54.4 and −16.3 ppm was 100:18 for the solution heated at 423 K for 168 h and 100:37 for the solution heated at 473 K for 72 h.

4 Discussion

4.1 Oxidation of the Base Oil

The thermal degradation of lubricant base oils is known to occur at high temperature and to result in the formation of several carbonyl and hydroxy species (lactones, peroxy esters, peroxy acids, esters, ketones, etc.), which are formed by a free radical-initiated chain reaction [33].

FT-IR and NMR spectroscopy have already been used to identify the oxidation products and to investigate the oxidation kinetics [46, 49–54]. The appearance of two broad absorption bands in the 1850–1685 cm^{-1} and in the 1300–850 cm^{-1} regions of the FT-IR spectrum, corresponding, respectively, to the stretching vibration of the C=O and C–O bonds, has been used to follow the oil-aging process [46, 50, 53]. However, the complex carbonyl envelope made it difficult to give a definitive statement on the chemical nature and distribution of all the species involved [46, 50].

In the present work, the appearance of a broad band in the 1850–1550 cm^{-1} region and the increase of the baseline in the 1300–850 cm^{-1} region observed in the FT-IR

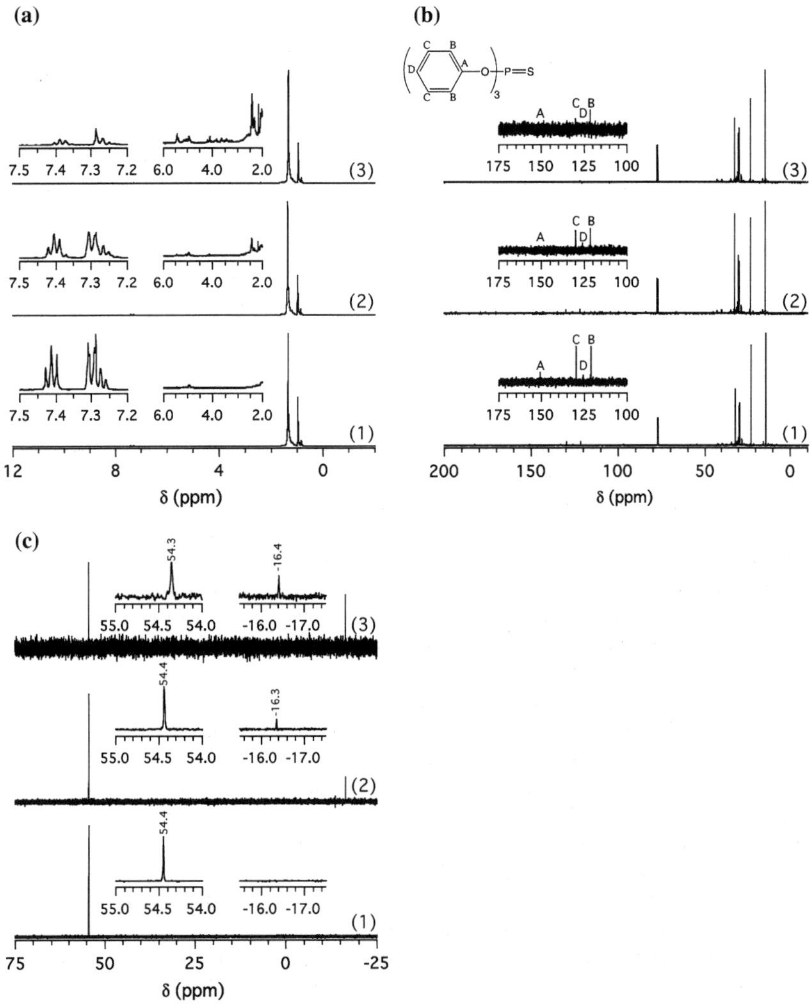

Fig. 8 NMR spectra (1-H (**a**); 13-C (**b**); 31-P (**c**)) of 0.044 mol dm^{-3} solutions of TPPT in PAO before (1) and after heating at 423 K for 168 h (2) and at 473 K for 72 h (3)

spectra, as well as the presence of new peaks between 6 and 2 ppm in the 1-H spectrum of the TPPT solutions heated at 423 and 473 K, clearly indicate that the base oil (PAO) was getting oxidized during the heating process.

Moreover, the shift of the peak maxima in the 1850–1550 cm^{-1} region as the heating time increased, i.e. as the base oil was getting oxidized, could be due to the different distributions of the oxidation products. In the final stages, i.e. after heating at 423 K for 168 h and at 473 K for 72 h, the bands at, respectively, 1774 and 1777 cm^{-1} are expected to be mainly from five-membered ring lactones and possibly from small amounts of peroxy esters, whereas the peaks at, respectively, 1726 and 1731 cm^{-1} might correspond to ester compounds [46, 50].

Since the solution was not completely depleted of the additive even after heating at 423 K for 168 h and at 473 K for 72 h (presence of the characteristic TPPT absorption bands in the FT-IR spectra), the TPPT molecule turned out not to be an effective oxidation inhibitor, as opposite to ZnDTPs, which acts as both primary (radical trapping) and secondary (peroxide-decomposing) antioxidants [1, 55].

Control heating experiments on pure PAO (not shown) have also been performed at the same temperatures used in the present work (i.e. 423 and 473 K). The comparison of the FT-IR spectra of pure PAO and solutions of TPPT in PAO heated at the same temperature for the same time showed that the TPPT molecule decreased the oxidation of the base oil (PAO) at 423, but not at 473 K. Moreover, the presence of TPPT in PAO was found to affect the distribution of the species formed during the oil-aging process (shift of the peak maximum in the 1850–1550 cm^{-1} region of the FT-IR spectra).

4.2 Temperature and Time Effect on TPPT Reactivity in Oil Solution

The transmission FT-IR and NMR spectra presented in this work showed the presence of the characteristic peaks of the TPPT molecule even after heating at 423 K for 168 h and at 473 K for 72 h. During the heating experiments, no oil-insoluble deposit was formed. These findings suggest that the TPPT molecule has a higher thermal stability than ZnDTPs, which are known to completely react in a few hours at high temperature (403–503 K) by a thermo-oxidative degradation route with the formation of a precipitate, whose composition, corresponding to a zinc poly(thio)phosphate compound, resembles that of the surface films obtained in tribological experiments with ZnDTP additives [1, 56–59].

Although the solutions started changing color after 6 h at 423 K and after 3 h at 473 K, the transmission FT-IR spectra exhibited a shoulder at higher wavenumbers (at ca. 965 cm^{-1}) in the νP–O–(C) peak only upon heating at 423 and 473 K for 13 h, indicating the activation of the molecule at these temperatures. Moreover, after heating at 473 K for 72 h a new but weak absorption band at 1243 cm^{-1}, which can be assigned to the P=O stretching vibration [34], appeared in the FT-IR spectrum. The presence of this band indicates the exchange of the sulfur in the phosphorothionate compound with a phosphoryl oxygen, as reported by Teichman and Hilgetag [29]. The formation of triphenyl phosphate also explains the appearance of a shoulder at higher wavenumbers (at ca. 965 cm^{-1}) in the νP–O–(C) peak [38, 60]. The absence of the P=O stretching vibration band in the FT-IR spectrum of the solution heated at 423 K for 168 h can be due to the increase of the baseline in the 1300–850 cm^{-1} region, where the C–O stretching vibration is found [34, 45], induced by the oxidation of the base oil.

These band assignments could be confirmed by the NMR spectra. The 31-P NMR spectra of the solutions heated at 423 K for 168 h and at 473 K for 72 h, in particular, showed the presence of a new peak at, respectively, −16.3 and −16.4 ppm, which corresponds to triphenyl phosphate [42, 44].

The scission of the P=S bond upon heating at 423 and 473 K is also suggested by the *DI* values of the corresponding absorption band in the FT-IR spectra (Fig. 7), which are always slightly lower than the values of the Ph and P–O–(C) stretching vibration peaks. The monotonic decrease of all the *DI* values in Fig. 7, where a logarithmic scale for the vertical axis was used, also suggests that the degradation of the TPPT molecule is a first-order reaction.

A definitive statement about the reaction step following the P=S bond scission, i.e. the cleavage of the P–O bond to form aryloxy groups or the scission of the C–O bond to give aryl group, cannot be done on the basis of the FT-IR spectra since the superposition of the C–O stretching vibration band of the ester compounds (at ca. 1175 cm^{-1}) produced by the oxidation of the base oil with the νC–O–(P) band results in *DI* values which are not representative of the decrease in absorbance of the latter.

4.3 Proposed Reaction Mechanism

Based on the findings mentioned above, the reaction mechanism depicted in Scheme 1a is suggested.

Upon heating at 423 and 473 K for more than 13 h, the TPPT molecule undergoes a nucleophilic attack at the phosphorus atom to cause P=S bond scission to give triphenyl phosphite. Volatile compounds might be also formed during this reaction step. A thermo-oxidative reaction then occurs, which leads to the oxidation of triphenyl phosphite to triphenyl phosphate. No statement can

Scheme 1 Reaction mechanism in lubricant oil solution of TPPT at 423 and 473 K (**a**) and of zinc dialkyldithiophosphates at 403–503 K [1, 56–58] (**b**)

(a)

$$(PhO)_3PS \xrightarrow{T} (PhO)_3P + S \xrightarrow{O_2} (PhO)_3PO$$

(b)

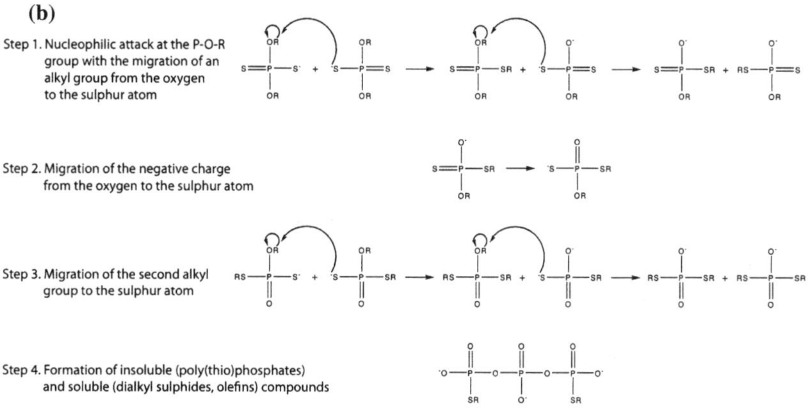

be made about the following step, i.e. the cleavage of the P–O or C–O bond.

The reaction rate is strongly accelerated as the temperature is increased from 423 to 473 K. However, the TPPT molecule does not completely react in oil solution even upon heating at 423 K for 168 h or at 473 K for 72 h.

It has to be emphasized that the chemical reaction of TPPT might not be the only reason for the decrease in intensity of its characteristic signals in the FT-IR and NMR spectra. The compound volatilization has been addressed as the main reason for the weight loss of triaryl monothiophosphates during thermogravimetric analyses (TGA) [61]. Although different experimental conditions were employed in this work and this makes it difficult to compare the present results with those reported in [61], some evaporation of the molecule cannot be ruled out as a contributing factor in the decrease in absorbance of the characteristic bands of the TPPT molecule observed in the present study.

The reaction mechanism described above significantly differs from the thermal and thermo-oxidative degradation route of ZnDTPs (Scheme 1b) [1, 56–58]. At temperatures ranging from 403 to 503 K depending on the alkyl groups present, ZnDTPs, being very strong alkylating agents [28–32], undergo a nucleophilic attack at the P–O–R group, which is often followed by an autocatalytic realkylation (Step 1 in Scheme 1b). After that, the negative charge migrates from the oxygen to the sulfur atom (Step 2 in Scheme 1b). The second stage of the isomerization reaction can then start resulting in the migration of the alkyl groups from the oxygen to the sulfur atom (Step 3 in Scheme 1b). Subsequent reactions lead to the formation of insoluble (poly(thio)phosphates) and soluble (diaklyl sulfides, and olefins) compounds (Step 4 in Scheme 1b).

Aromatic esters, having no tetrahedral carbon bound to oxygen, cannot undergo dearylation by the same S_N2-type process (i.e. reaction initiated by a nucleophilic attack on the α-carbon of the alkyl group) [58] and this results in a higher thermal stability than alkyl thiophosphate esters [33].

5 Conclusions

The following conclusions can be drawn from the results presented in this work:

(i) Upon heating solutions of TPPT in PAO at 423 and 473 K, the base oil (PAO) gets oxidized, indicating that the TPPT molecule is not an effective oxidation inhibitor;

(ii) The TPPT molecule, being not completely decomposed upon heating at 423 K for 168 h and at 473 K for 72 h, has a high thermal stability in lubricant oil solution;

(iii) The thermo-oxidative reaction of TPPT in lubricant oil solution starts with the scission of the P=S bond, leading to the formation of triphenyl phosphate;

(iv) The reaction rate is strongly accelerated as the temperature is increased from 423 to 473 K;

(v) No oil-insoluble compounds are formed during the heating experiments.

Acknowledgments The authors wish to express their gratitude to the ETH Research Committee for its support of this work. Dr. H. Camenzind (Ciba® Speciality Chemicals, Basel, Switzerland) is thanked for supplying the pure additive. Mrs. D. Sutter and Mr. M. Schneider kindly performed the NMR and the elemental analysis, respectively.

References

1. Spikes, H.: The history and mechanisms of ZDDP. Tribol. Lett. **17**, 469–489 (2004). doi:10.1023/B:TRIL.0000044495.26882.b5
2. Nicholls, M.A., Do, T., Norton, P.R., Kasrai, M., Bancroft, G.M.: Review of the lubrication of metallic surfaces by zinc dialkyl-dithiophosphates. Tribol. Int. **38**, 15–39 (2005). doi:10.1016/j.triboint.2004.05.009
3. Gellman, A.J., Spencer, N.D.: Surface chemistry in tribology. J. Eng. Tribol. **216**, 443–461 (2002)
4. Kubsh, J.: Three-way catalyst deactivation associated with oil-derived poisons. In: Bode, H. (ed.) Materials aspects in automotive catalytic converters, pp. 215–222. Wiley-VCH Verlag GmbH & Co, Weinheim (2003)
5. Spikes, H.: Low- and zero-sulphated ash, phosphorus and sulphur anti-wear additives for engine oils. Lubricat. Sci. **20**, 103–136 (2008). doi:10.1002/ls.57
6. Bovington, C.H.: Friction, wear and the role of additives in their control. In: Mortier, R.M., Orszulik, S.T. (eds.) Chemistry and Technology of Lubricants, pp. 320–348. Blackie Academic & Professional, London (1997)
7. Piras, F.M.: In situ attenuated total reflection tribometry. PhD thesis no. 14638, ETH Zurich, Zurich, Switzerland (2002)
8. Gao, F., Kotvis, P.V., Stacchiola, D., Tysoe, W.T.: Reaction of tributyl phosphate with oxidized iron: surface chemistry and tribological significance. Tribol. Lett. **18**, 377–384 (2005). doi:10.1007/s11249-004-2768-1
9. Sung, D., Gellman, A.J.: The surface chemistry of alkyl and arylphosphate vapor phase lubricants on Fe foil. Tribol. Int. **35**, 579–590 (2002). doi:10.1016/S0301-679X(02)00045-2
10. Sung, D., Gellman, A.J.: Thermal decomposition of tricresylphosphate isomers on Fe. Tribol. Lett. **13**, 9–14 (2002). doi:10.1023/A:1016599502098
11. Ren, D., Gellman, A.J.: Reaction mechanisms in organophosphate vapor phase lubrication of metal surfaces. Tribol. Int. **34**, 353–365 (2001). doi:10.1016/S0301-679X(01)00025-1
12. Matsumoto, K.: Surface chemical and tribological investigations of phosphorus-containing lubricant additives. PhD thesis no. 15150, ETH Zurich, Zurich, Switzerland (2003)
13. Najman, M.N., Kasrai, M., Bancroft, G.M., Miller, A.: Study of the chemistry of films generated from phosphate ester additives on 52100 steel using X-ray absorption spectroscopy. Tribol. Lett. **13**, 209–218 (2002). doi:10.1023/A:1020164127000
14. Rossi, A., Piras, F.M., Kim, D., Gellman, A.J., Spencer, N.D.: Surface reactivity of tributyl thiophosphate: effects of temperature and mechanical stress. Tribol. Lett. **23**, 197–208 (2006). doi:10.1007/s11249-006-9051-6

15. Saba, C.S., Forster, N.H.: Reactions of aromatic phosphate esters with metals and their oxides. Tribol. Lett. **12**, 135–146 (2002). doi:10.1023/A:1014081523491
16. Perez, J.M., Ku, C.S., Pei, P., Hegemann, B.E., Hsu, S.M.: Characterization of tricresylphosphate lubricating films by micro-Fourier transform infrared spectroscopy. Tribol. Trans. **33**, 131–139 (1990). doi:10.1080/10402009008981939
17. Gao, F., Furlong, O., Kotvis, P.V., Tysoe, W.T.: Reaction of tributyl phosphite with oxidized iron: surface and tribological chemistry. Langmuir **20**, 7557–7568 (2004). doi:10.1021/la049438t
18. Ren, D., Gellman, A.: Initial steps in the surface chemistry of vapor phase lubrication by organophosphorus compounds. Tribol. Lett. **6**, 191–194 (1999). doi:10.1023/A:1019184312290
19. Holbert, A.W., Batteas, J.D., Wong-Foy, A., Rufael, T.S., Friend, C.M.: Passivation of Fe(110) via phosphorus deposition: the reactions of trimethylphosphite. Surf. Sci. **401**, L437–L443 (1998). doi:10.1016/S0039-6028(98)00076-4
20. Gong, Q., He, W., Liu, W.: The tribological behavior of thiophosphates as additives in rapeseed oil. Tribol. Int. **36**, 733–738 (2003). doi:10.1016/S0301-679X(03)00053-7
21. Najman, M., Kasrai, M., Bancroft, G.M., Davidson, R.: Combination of ashless antiwear additives with metallic detergents: interactions with neutral and overbased calcium sulfonates. Tribol. Int. **39**, 342–355 (2006). doi:10.1016/j.triboint.2005.02.014
22. Najman, M.N., Kasrai, M., Bancroft, G.M.: Chemistry of antiwear films from ashless thiophosphate oil additives. Tribol. Lett. **17**, 217–229 (2004). doi:10.1023/B:TRIL.0000032448.77085.f4
23. Najman, M.N., Kasrai, M., Bancroft, G.M.: Investigating binary oil additive systems containing P and S using X-ray absorption near-edge structure spectroscopy. Wear **257**, 32–40 (2004). doi:10.1016/S0043-1648(03)00537-4
24. Najman, M.N., Kasrai, M., Bancroft, G.M., Frazer, B.H., De Stasio, G.: The correlation of microchemical properties to antiwear (AW) performance in ashless thiophosphate oil additives. Tribol. Lett. **17**, 811–822 (2004). doi:10.1007/s11249-004-8089-6
25. Heuberger, R.: Combinatorial study of the tribochemistry of antiwear lubricant additives. PhD thesis no. 17207, ETH Zurich, Zurich, Switzerland (2007)
26. Heuberger, R., Rossi, A., Spencer, N.D.: Reactivity of alkylated phosphorothionates with steel: a tribological and surface-analytical study. Lubricat. Sci. **20**, 79–102 (2008). doi:10.1002/ls.56
27. Koyama, M., Hayakawa, J., Onodera, T., Ito, K., Tsuboi, H., Endou, A., Kubo, M., Del Carpio, C.A., Miyamoto, A.: Tribochemical reaction dynamics of phosphoric ester lubricant additive by using a hybrid tight-binding quantum chemical molecular dynamics method. J. Phys. Chem. B **110**, 17507–17511 (2006). doi:10.1021/jp061210m
28. Hilgetag, G., Teichmann, H.: The alkylating properties of alkyl thiophosphates. Angew. Chem. Int. Ed. **4**, 914–922 (1965). doi:10.1002/anie.196509141
29. Teichmann, H., Hilgetag, G.: Nucleophilic reactivity of the thiophosphoryl group. Angew. Chem. Int. Ed. **6**, 1013–1023 (1967). doi:10.1002/anie.196710131
30. Pearson, R.G.: Hard and soft acids and bases. J. Am. Chem. Soc. **85**, 3533–3539 (1963). doi:10.1021/ja00905a001
31. Pearson, R.G., Songstad, J.: Application of the principle of hard and soft acids and bases to organic chemistry. J. Am. Chem. Soc. **89**, 1827–1836 (1967). doi:10.1021/ja00984a014
32. Pearson, R.G.: Chemical Hardness. Wiley, New York (1997)
33. Rasberger, M.: Oxidative degradation and stabilization of mineral oil based lubricants. In: Mortier, R.M., Orszulik, S.T. (eds.) Chemistry and Technology of Lubricants, pp. 98–143. Blackie Academic & Professional, London (1997)
34. Socrates, G.: Infrared and Raman Characteristic Group Frequencies. Wiley, Chichester (2001)
35. Harris, R.K., Becker, E.D., Cabral De Menezes, S.M., Goodfellow, R., Granger, P.: NMR nomenclature. nuclear spin properties and conventions for chemical shifts. Pure Appl. Chem. **73**, 1795–1818 (2001). doi:10.1351/pac200173111795
36. Roeges, N.P.G.: A Guide to the Complete Interpretation of Infrared Spectra of Organic Structures. Wiley, Chichester (1994)
37. Chittenden, R.A., Thomas, L.C.: Characteristic infra-red absorption frequencies of organophosphorus compounds—III. Phosphorus–sulphur and phosphorus–selenium bonds. Spectrochim. Acta **20**, 1679–1696 (1964). doi:10.1016/0371-1951(64)80173-9
38. Thomas, L.C., Chittenden, R.A.: Characteristic infrared absorption frequencies of organophosphorus compounds—II. P–O–(X) bonds. Spectrochim. Acta **20**, 489–502 (1964). doi:10.1016/0371-1951(64)80044-8
39. Lin-Vien, D., Colthup, N.B., Fateley, W.G., Grasselli, J.G.: The Handbook of Infrared and Raman Characteristic Frequencies of Organic Molecules. Academic Press, San Diego (1991)
40. Friebolin, H.: Basic One- and Two-Dimensional NMR Spectroscopy. Wiley, New York (2005)
41. Günther, H.: NMR Spectroscopy—Basic Principles, Concepts, and Applications in Chemistry. Wiley, New York (1995)
42. Berger, S., Braun, S., Kalinowski, H.O.: NMR Spectroscopy of the non-metallic elements. Wiley, New York (1997)
43. Quin, L.D.: A Guide to Organophosphorus Chemistry. Wiley, New York (2000)
44. Crutchfield, M.M., Dungan, C.H., Letcher, J.H., Mark, V., Van Wazer, J.R.: P31 nuclear magnetic resonance. In: Grayson, M., Griffith, E.J. (eds.) Topics in Phophorus Chemistry. Wiley, New York (1967)
45. Colthup, N.B., Daly, L.H., Wiberley, S.E.: Introduction to Infrared and Raman Spectroscopy. Academic Press, London (1990)
46. Adhvaryu, A., Perez, J.M., Singh, I.D., Tyagi, O.S.: Spectroscopic studies of oxidative degradation of base oils. Energy Fuels **12**, 1369–1374 (1998). doi:10.1021/ef980134m
47. Silverstein, R.M., Webster, F.X., Kiemle, D.J.: Spectroscopic Identification of Organic Compounds. Wiley, New York (2005)
48. Hesse, M., Meier, H., Zeeh, B.: Spektroskopische Methoden in der Organischen Chemie, 3. überarbeitete Auflage. Georg Thieme Verlag, Stuttgart (1987)
49. Barman, B.N.: Behavioral differences between group I and group II base oils during thermo-oxidative degradation. Tribol. Int. **35**, 15–26 (2002). doi:10.1016/S0301-679X(01)00073-1
50. Coates, J.P., Setti, L.C.: Infrared spectroscopic methods for the study of lubricant oxidation products. ASLE Trans. **29**, 394–401 (1986)
51. George, G.A., Celina, M., Vassallo, A.M., Cole-Clarke, P.A.: Real-time analysis of the thermal oxidation of polyolefins by FT-IR emission. Polym. Degrad. Stabil. **48**, 199–210 (1995). doi:10.1016/0141-3910(95)00035-K
52. Owrang, F., Mattsson, H., Olsson, J., Pedersen, J.: Investigation of oxidation of a mineral and a synthetic engine oil. Thermochim. Acta **413**, 241–248 (2004). doi:10.1016/j.tca.2003.09.016
53. Priéri, F., Gresser, E., Le Dréau, Y., Obiols, J., Kister, J.: New method of simulation to evaluate the sensitivity to oxidation of lubricating oils: an aging cell coupled with Fourier transform infrared spectroscopy. Appl. Spectrosc. **62**, 810–816 (2008). doi:10.1366/000370208784909571
54. Santos, J.C.O., Santos, I.M.G., Souza, A.G., Sobrinho, E.V., Fernandes, J.V.J., Silva, A.J.N.: Thermoanalytical and rheological characterization of automotive mineral lubricants after thermal degradation. Fuel **83**, 2393–2399 (2004). doi:10.1016/j.fuel.2004.06.016
55. Willermet, P.A., Carter, R.O., Schmitz, P.J., Everson, M., Scholl, D.J., Weber, W.H.: Formation, structure, and properties of

lubricant-derived antiwear films. Lubricat. Sci. **9**, 325–348 (1997). doi:10.1002/ls.3010090402

56. Coy, R.C., Jones, R.B.: The thermal degradation and EP performance of zinc dialkyldithiophosphate additives in white oil. ASLE Trans. **24**, 77–90 (1981)
57. Dickert, J.J.J., Rowe, C.N.: The thermal decomposition of metal *O*,*O*-dialkylphosphorodithioates. J. Org. Chem. **32**, 647–653 (1967). doi:10.1021/jo01278a031
58. Jones, R.B., Coy, R.C.: The chemistry and thermal degradation of zinc dialkyldithiophosphate additives. ASLE Trans. **24**, 91–97 (1981)
59. Spedding, H., Watkins, R.C.: The antiwear mechanism of zddp's. Part I. Tribol. Int. **15**, 9–12 (1982). doi:10.1016/0301-679X(82)90101-3
60. Mortimer, F.S.: Vibrational assignment and rotational isomerism in some simple organic phosphates. Spectrochim. Acta **9**, 270–281 (1957). doi:10.1016/0371-1951(57)80142-8
61. Ribeaud, M.: Volatility of phosphorus-containing anti-wear agents for motor oils. Lubricat. Sci. **18**, 231–241 (2006). doi:10.1002/ls.20

2e. Surface modification for lubrication of implants

Commentary

While synovial fluid works well as a lubricant for cartilage, there is no particular reason to suppose that it is an effective lubricant for man-made materials used in implants, such as polyethylene, CoCrMo, or alumina. This is an issue that we have explored using a number of different techniques. Our first effort in this area (2.40) involved focusing on albumin, which is the protein with the highest concentration in synovial fluid. We found that this protein is very susceptible to elevated temperatures, and that, even at 60°C, substantial degradation can take place quite rapidly. While such temperatures are unlikely to be encountered *in vivo*, this has consequences for implant testing, where the lubricants frequently become warm due to frictional heating. We were able to follow the level of denaturation by means of circular dichroism spectroscopy, and could correlate this with changes in friction measured in a pin-on-disk tester.

Our next foray into this field, in collaboration with Rowena Crockett at the Empa (Swiss Federal Laboratories for Materials Science and Technology), took a broader look at the proteins in synovial fluid. We came to the conclusion that albumin actually increases friction in comparison to saline, while glycoproteins, in general, lubricated implant-material sliding surfaces effectively. Despite the massive excess of albumin in synovial fluid, we could show, by fluorescence spectroscopy, that the glycoproteins made it to the surface and lubricated the materials (2.41). Fluorescence microscopy also proved useful for monitoring the transfer of polyethylene to inorganic countersurfaces, and it was found, contrary to conventional wisdom, that this transfer also occurs in the presence of proteins or even synovial fluid (2.42).

Biomaterials often require short-term lubrication for particular procedures, and this is the case for catheter insertion, for example. Silicone oil is frequently used for this purpose, but the use of brush-like systems could improve lubrication and, therefore, render the experience less painful for the patient. We started from the observation in Phil Messersmith's group (Northwestern University) that dopa- and lysine-containing molecules in the mussel appear to polymerize upon oxidation, leading to an adhesive bond to virtually any material. We then used PEG-dopa-lysine to coat poly(dimethylsiloxane), and tribologically tested the resulting coating (2.43). Friction coefficients of around 0.03 at 1 N load could be maintained for relatively long sliding distances (approx. 8m), without the need for the presence of polymer in the lubricant. The PEG brushes formed in this way also have the advantage of being resistant to protein adsorption.

Available online at www.sciencedirect.com
SCIENCE @ DIRECT•
Biomaterials 26 (2005) 1165–1173
Biomaterials
www.elsevier.com/locate/biomaterials

Protein-mediated boundary lubrication in arthroplasty

M.P. Heuberger[a,*], M.R. Widmer[a], E. Zobeley[b], R. Glockshuber[b], N.D. Spencer[a]

[a] *Laboratory for Surface Science and Technology, Department of Materials, ETH Zürich, CH-8092 Zürich, Switzerland*
[b] *Molecular Biology and Biophysics, Department of Biology, ETH Zürich, Switzerland*

Received 15 December 2003; accepted 26 May 2004

Available online 2 July 2004

Abstract

Wear of articulated surfaces can be a major lifetime-limiting factor in arthroplasty. In the natural joint, lubrication is effected by the body's natural synovial fluid. Following arthroplasty, and the subsequent reformation of the synovial membrane, a fluid of similar composition surrounds the artificial joint. Synovial fluid contains, among many other constituents, a substantial concentration of the readily adsorbing protein albumin. The ability of human serum albumin to act as a boundary lubricant in joint prostheses has been investigated using a pin-on-disc tribometer. Circular dichroism spectroscopy was employed to follow the temperature- and time-dependent conformational changes of human serum albumin in the model lubricant solution. Effects of protein conformation and polymer surface hydrophilicity on protein adsorption and the resulting friction in the boundary lubrication regime have been investigated. Unfolded proteins preferentially adsorb onto hydrophobic polymer surfaces, where they form a compact, passivating layer and increase sliding friction—an effect that can be largely suppressed by rendering the substrate more hydrophilic. A molecular model for protein-mediated boundary friction is proposed to consolidate the observations. The relevance of the results for in vivo performance and ex vivo hip-joint testing are discussed.

Keywords: Albumin; Protein adsorption; Protein conformation; Joint replacement; Arthroplasty; Friction; Interface; Polyethylene; Hydrophilicity

1. Introduction

By virtue of our modern lifestyle, more than one-third of the readers of this article will likely experience the failure of their native hip joint(s)—sooner or later [1]. Fortunately, hip-joint arthroplasty is on hand to deal with this condition. A major unsolved problem, however, is the mechanical wear of the artificial joint, which limits the lifetime to about 10–15 years [2]. While most lubrication of the healthy natural joint relies on a film of synovial fluid (SF), the artificial joint consists of synthetic materials and is mainly lubricated in the mixed and boundary regimes [3]. In a widely used arthroplastic design [4], the acetabular cup consists of a linear of ultra-high-molecular-weight poly(ethylene) (UHMWPE), and the femoral head is a polished metal or ceramic ball. The wear [2] of the UHMWPE lining is generally regarded as the key lifetime-limiting factor. Up to 100,000 μm-sized wear particles are released per footstep. Such wear particles can activate the immune system [5] as well as be deposited into the surrounding tissue [6]. These factors can ultimately lead to osteolysis and mechanical loosening of the implant [7]. In this study, we investigated boundary lubrication mediated by naturally occurring amphiphilic molecules—an aspect that has received recent attention [8–11].

From an engineering standpoint, SF can be regarded as an aqueous electrolyte solution containing proteins, lipids and hyaluronic acid [12]. SF is also important for implants since a few weeks into the healing process, it surrounds the artificial joint. The predominant lubrication mode is in the mixed and boundary regimes [13]. In this study, we focus on the role of the most abundant protein in SF—human serum albumin (HSA). A screening study involving proteins and lipids was used

*Corresponding author. Tel.: +41-1-632-6452; fax: +41-1-633-1027.
E-mail address: manfred.heuberger@mat.ethz.ch (M.P. Heuberger).

0142-9612/$ - see front matter © 2004 Elsevier Ltd. All rights reserved.
doi:10.1016/j.biomaterials.2004.05.020

to identify the molecules relevant for boundary lubrication. In agreement with previous studies [14], we identified HSA as a constituent with a significant effect on sliding friction. We wanted to focus on the boundary lubrication properties of the HSA component adsorbed onto poly(ethylene). Poly(ethylene) is considerably more hydrophobic than natural cartilage, and therefore proteins are expected to adsorb in very different ways onto the two surfaces. In a previous pin-on-disc study [15], we indeed found that the friction of UHMWPE versus ceramic was reduced in a HSA solution after the polymer was transiently rendered more hydrophilic. An oxygen RF-plasma treatment of the polymer surface was used to achieve the desired hydrophilicity. The use of more permanently hydrophilic surfaces by changing the bulk polymer chemistry was shown to be impractical, using available materials, due to their inferior mechanical properties [16]. Here, we extend our study to include the temperature-dependent protein conformation and clarify its behavior during adsorption and boundary lubrication in the UHMWPE/ceramic tribopair.

HSA consists of 585 amino acids and is comparatively rich in cysteine, which allow HSA to form a total of 17 internal disulfide bridges [17]. HSA has a rather high α-helical content [18], which can readily be detected using the circular dichroism (CD) technique. HSA consists of three domains with similar primary structure and is glycosylated at residues Asn-342 and Asp-518. The unfolding of HSA is known to proceed in several incremental steps [19].

In solution, proteins can be unfolded (i.e. denatured) by a change in chemical environment [20], pressure [21] or by changes in temperature [22].

Recently, it has been found that the temperature in articulated joints may rise up to 45°C below the cartilage surface in vivo [23,24] and up to 90°C in ex vivo hip-joint simulator tests [25,26]. Irreversible unfolding of proteins occurring under such conditions can significantly affect their boundary lubrication properties.

2. Experimental

2.1. Tribometer

A tribometer in pin-on-disc configuration was used to characterize the tribological properties of the test surfaces. Although this simple setup does not reproduce all clinical parameters relevant for the accurate simulation of wear (e.g. contact geometry, loading cycle, fluid entrainment or multi-directional sliding), it can give valuable fundamental information about the effects of interfacial protein adsorption on friction. The tribometer (CSEM, Switzerland) had a nominal loading range of 0.1–10 N (static weights) and a nominal velocity range of 10^{-4}–10^{-1} m/s. The sliding radius was set to either $r_1 = 3$ mm or $r_2 = 9$ mm, to extend the nominal velocity range. For comparison, the typical in vivo sliding speeds of the hip joint are known to be in the range of 0–10^{-1} m/s [27]. Friction was measured both in start–stop and steady-state sliding conditions. The friction data were recorded digitally at a sampling rate of 45 Hz. The nominal sensitivity of the friction–force measurement was 10 mN. A Plexiglas hood covered the tribometer, allowing physiological temperatures (37 ± 2°C) of pin, disc and lubricant to be maintained during all friction experiments. An IR heater with adjustable power and a DC fan inside the hood were used for this purpose. For some experiments, the model lubricant solution was independently heated to achieve higher temperatures: A peristaltic pump was used to continuously circulate the lubricant through a tubing system that passed through a heated water bath outside the tribometer. The temperature of the water bath could be adjusted in a range of 25°C to 90°C. Under typical circulation conditions, an individual protein molecule was thus repeatedly exposed to elevated temperatures by passage through the water bath. Typically, this amounted to $\approx 8\%$ of the total circulation time. Comparison between cyclic and steady tempering indeed revealed that the relevant parameter for irreversible protein unfolding was the integral exposure time to elevated temperature [28]. Multiplication of the time axis by a factor 0.08 was thus used to normalize the exposure times shown in Fig. 4 below.

2.2. Polymer and ceramic materials

We used flat-ended pins of ultra-high-molecular-weight poly(ethylene), (UHMWPE—medical-grade, DSM Stamylan, UH210). The dimensions of the square pins were 1.5 mm × 3.5 mm × 3.5 mm, as cut out of a larger block. The surface roughness of the pins was constant, $R_a = 1.5 \pm 0.5$ µm, after several hours of run-in, and a fixed pair of pin and disc was used for all measurements with each kind of lubricant composition. Rough pins were chosen because a statistical analysis of 20 independent friction measurements at 10 N load revealed superior reproducibility of the rough pins ($R_a = 1.5 \pm 0.5$ µm, $\sigma_{rough} = 50$ mN) as opposed to smooth pins ($R_a < 0.1$ µm, $\sigma_{smooth} = 180$ mN). Pins and disc were reused and thoroughly cleaned before each experiment (see below).

Medical-quality UHMWPE differs from the industrial polymer mainly in the smaller proportion of additives and trace elements. This is achieved by additional purification procedures and omission of antioxidants or stabilizers. This polymer is therefore susceptible to aging [29]. The ceramic discs used for tribotesting were made of surface-polished (roughness: $R_A = 5$–$8\,nm$) Al_2O_3 or ZrO_2 [30] discs (Metoxit AG, Switzerland). The outer diameter of the discs was 30 mm. Prior to insertion into the tribometer, all pins were sonicated for 10 min in an aqueous detergent solution, then thoroughly rinsed and sonicated again in acetone for an additional 10 min. The ceramic discs were sonicated in acetone for 15 min. During the experiment, the disc and part of the pin were submerged into the model lubricant inside an aluminum cup of volume $\approx 25\,ml$.

2.3. *The hydrophilic–hydrophobic balance*

The polymer pins were optionally treated in a 100 W radio frequency (RF) oxygen plasma cleaner (Harrick Scientific Corporation, NY (PDC-32G)) to render the surface temporarily more hydrophilic. Contact angle (CA) measurements were used to establish the dependence of the surface hydrophilicity with plasma treatment time for flat samples of the same material. The contact angle was found to decrease nearly exponentially with treatment time from $87 \pm 2°$ for untreated UHMWPE to $< 10 \pm 5°$ for plasma treatment times exceeding 100 s [15,16]. Unless otherwise stated, a treatment time of 10 s was chosen for all plasma treatments used in this study, resulting in a moderately hydrophilic contact angle of $41 \pm 5°$. Using profilometry, we could detect a reduction of $\approx 1\,nm$ polymer film thickness per second of plasma treatment. The surface roughness was, however, found not to change significantly during plasma treatment. Using a number of extended-period friction measurements, it was also confirmed that the lifetime ($> 30\,min$) of the hydrophilic surface modification was long enough to reliably perform the friction measurements on hydrophilic surfaces in solution.

2.4. *Chemical surface analysis*

X-ray photoelectron spectroscopy (XPS) (PHI 5700, Physical Electronics, USA) was used to characterize the chemical composition of the UHMWPE surface prior to and following RF-plasma treatment. We used Al-Kα radiation at 300 W and a takeoff angle of 90° (i.e. normal to surface). The base pressure was $< 5 \times 10^{-9}$ Torr. Low-resolution survey spectra were recorded in the binding-energy range 0–1200 eV at 0.4 eV intervals (pass energy: 187.85 eV). Detailed spectra were recorded around the peaks of C_{1s}, O_{1s}, N_{1s}, and Al_{2p} at 0.125 eV intervals (pass energy: 58.7 eV). The integration time was 150 ms for all measurements. Sample exposure to X-rays was kept at a minimum to reduce any degradation effects. Spectral evaluation was carried out by means of PHI MultiPack software.

2.5. *Aqueous solutions*

As a model lubricant, we used an aqueous solution of human serum albumin (HSA). The base solution was isotonic Ringer's solution (Merck, isotonic standard solution), with ultra-pure water (Easypure) as a basis. The HSA was obtained from Sigma (fraction V powder, No. A1653) in powder form and stirred into the solution prior to the experiment. Unless otherwise noted, the concentration of HSA was chosen to be 8 mg/ml, to mimic the natural albumin concentration in human SF [31].

2.6. *Protein conformation*

The conformation of albumin in solution was monitored using circular dichroism (CD) (spectropolarimeter, Jasco J-715, Omnilab). In an asymmetric chiral medium, circularly polarized light becomes elliptically polarized due to asymmetric absorption. CD is based on the measurement of the difference in absorbance between right- and left-circularly polarized light [32–34]. CD is particularly well suited to determine the α-helix content of proteins in solution [34]. As the protein unfolds (i.e. denatures), its α-helical domains unfold and the ellipticity (in degrees) of the transmitted light decreases towards zero (i.e. random-coil conformation). The UV wavelengths at $\lambda_1 = 208\,nm$ and $\lambda_2 = 222\,nm$ are particularly sensitive indicators of this phenomenon, since they represent the strongest signal in the CD spectrum, characteristic for α-helix. Since Beer's law applies to the CD technique, the signal intensities depend on spectral absorbance, light path length and absorber solution concentration. To obtain optimal signal conditions, the concentration of HSA in solution had to be reduced to 1 mg/ml, as opposed to 8 mg/ml (i.e. synovia) used in the friction measurements. The light path length was 1 mm inside a quartz cuvette. The sample solution (400–500 μl) was filled into the cuvette, which was then sealed with a plastic cap to prevent evaporation. The temperature was controlled using a Peltier element in a closed temperature feedback loop. Non-equilibrium (i.e. ramp) measurements were conducted at a rate of 1 °C/min. For quasi-equilibrium measurements at selected temperatures, an equilibration time of 10–200 min was allowed, as required.

2.7. *Protein adsorption*

The kinetics and quantity of HSA adsorption on poly(ethylene) were measured using optical waveguide lightmode spectroscopy (OWLS) [35–38]. In this technique, a laser beam (He–Ne, $\lambda = 632.8\,nm$) is coupled into an SiO_2–TiO_2 waveguide via an optical grating structure. The in-coupling angle is a sensitive function of the effective refractive index of the medium adjacent to the waveguide–liquid interface above the coupling grating. The total of the in-coupled light intensity was measured at the end of the waveguide (ASI 2400 μV, Microvacuum Ltd., Hungary) using a photosensitive element. Both TE and TM mode intensity peaks were recorded. The evanescent field above the grating detects small changes in adsorbed (protein) mass, at a distance up to about $\lambda/2$ from the interface. For the OWLS adsorption studies, 25 ± 15 nm thick films of low-density poly-(ethylene) (LDPE) (DSM, Stamylan, 2101) were spin-coated (Fairchild Convac Spincoater 1001) onto the TiO_2 optical surface of the waveguides from a 0.7% (w/w) solution of LDPE in *p*-xylene at 120°C. LDPE had to be used here as a substitute for UHMWPE because the latter is insoluble. The spin-coating protocol was set up in two steps. Namely, spinning at 5000 rpm for 120 s, followed by spinning at 5900 rpm for another 30 s. The spin-coated LDPE films were characterized using profilometry and atomic force microscopy (AFM) (roughness: $R_A < 1$ nm).

3. Results

As mentioned above, friction experiments were carried out while varying two main parameters: (i) the surface hydrophilicity and (ii) the conformation of HSA in the model lubricant solution. The results presented in this section are therefore organized as follows: first, we document the physical effects of the RF-plasma treatment with respect to the surface properties of UHMWPE; second, we report on temperature-dependent protein folding in solution before we describe the results of the friction measurements.

3.1. Polymer surface modification

A typical XPS spectrum of oxygen RF-plasma-treated UHMWPE is depicted in Fig. 1. The plasma treatment time was 10 s. The survey spectrum reveals incorporation of oxygen species into the surface that were not present before the treatment (not shown). The inset shows the detail spectrum of the carbon C_{1s} energy region. A deconvolution of the carbon peak, considering carbonyl (C=O) and hydroxyl (C–OH) species, yields the following atomic concentrations: of all carbon atoms, 19% are covalently bound to oxygen, most of which (85%) are hydroxyl-related and a minor part (15%) is of the carbonyl type. The incorporation of oxygen species is correlated with a significant decrease of the water contact angle as described in the experimental section above. After a plasma-treatment time of 10 s, the contact angle was found to be $41 \pm 5°$. These observations are in good agreement with previous studies of oxygen-plasma-treated poly(ethylenes) [39, 40].

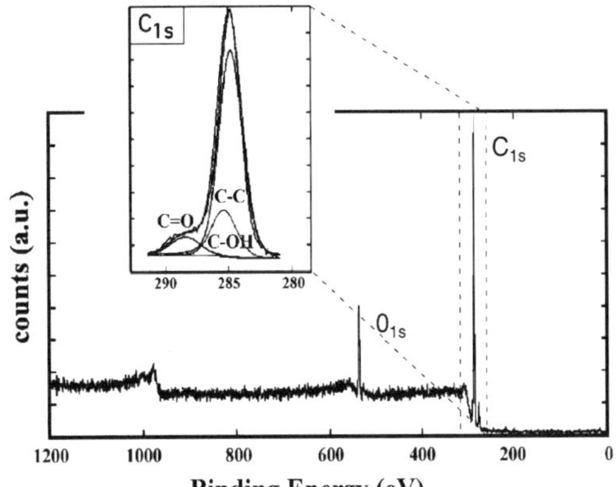

Fig. 1. XPS survey from 0–1200 eV binding energy after oxygen RF-plasma treatment; the inset shows a detail spectrum of the C_{1s} line with a three-component de-convolution fit.

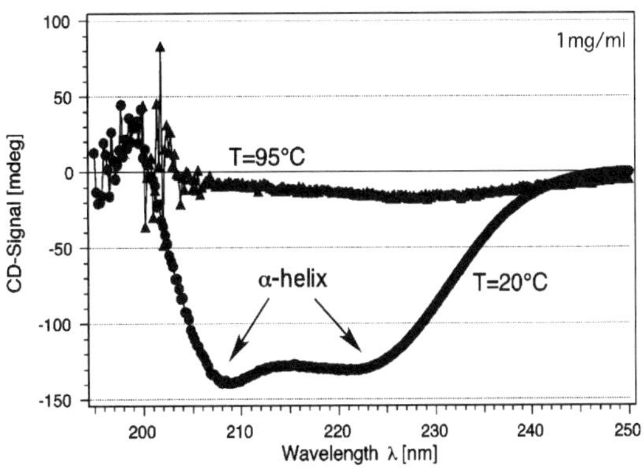

Fig. 2. CD spectrum measured in a 1 mg/ml solution of HSA. The curve taken at 20°C is typical for α-helix domains in (N)-HSA. At elevated temperatures of 95°C, the proteins denature and the CD spectrum suggests an unfolded, random coil structure.

3.2. Thermal protein unfolding

The far-UV CD spectrum of a 1 mg/ml HSA solution is illustrated in Fig. 2. The spectra measured at two different (quasi) equilibrium temperatures are shown. A protein concentration of 1 mg/ml was chosen for optimal signal intensity levels. The curve taken at a temperature of $T = 20°C$ shows two minima, namely around $\lambda_1 = 208$ nm and $\lambda_2 = 222$ nm, which are characteristic for the presence of the α-helix secondary structure in HSA. These typical features have completely vanished at a temperature of $T = 95°C$, indicating complete loss of secondary structure. HSA has no β-sheet domains. Monitoring the CD signal at one of the above wavelengths during a temperature ramp thus allows the kinetics of protein unfolding to be followed quantitatively and in situ. The outcome of a temperature sweep is shown in Fig. 3. We observe a continuous

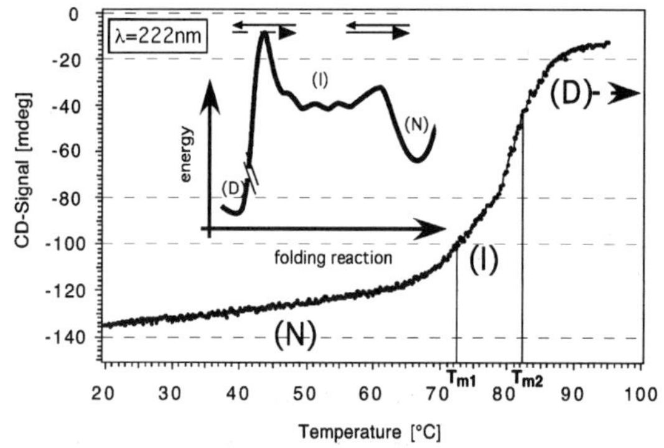

Fig. 3. CD measured at a fixed wavelength, $\lambda = 222$ nm, which is a measure for the α-helix content in the HSA solution. A temperature ramp was induced at a rate of 1°C/min. Three regions can be distinguished: (N), where HSA exists mainly in the native form, (I), region of multiple transitions involving reversibly unfolded intermediate states and (D), where the irreversible unfolding (i.e. denaturation) of HSA becomes important.

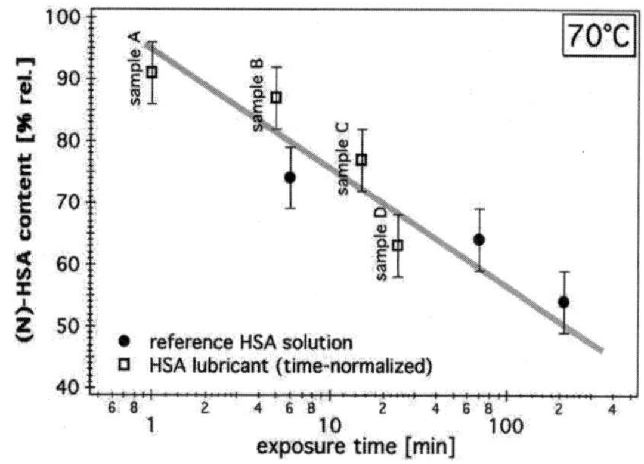

Fig. 4. Percentage of native (N)-HSA in solution after exposure to 70°C as calculated from CD measurements; the CD signal (all measured at 25°C) is plotted logarithmically against the effective exposure time. Open squares (labeled) represent measurements that were taken periodically from the lubricant solution during the course of the continuous pin-on-disc friction experiment also shown in Fig. 5. The lubricant samples were diluted at a ratio 1:8 to adjust CD signal levels. The time-normalization procedure is explained in the text. Filled circles stand for CD reference measurements obtained in situ (i.e. measured after exposing to 70°C for the indicated amount of time and subsequent cooling to 25°C). The gray line in the background is to guide the eye.

decrease of α-helix content between 20°C and 50°C, followed by a clear increase of unfolding, in several steps over a relatively wide temperature range, until the CD signal vanishes above $\approx$90°C. Comparison of repeated heating–cooling cycles (below $\approx$70°C) revealed that initially native (N)-HSA can reversibly unfold into intermediate (I) states and thus refold upon cooling. At temperatures above $\approx$70°C, however, we observed a noticeable amount of irreversibly denatured (D)-HSA, which aggregates. As shown in the sketched inset of Fig. 3, the unfolding process includes at least two steps: first, a reversible, partial unfolding, followed by an irreversible unfolding into a random coil configuration. A similar behavior has been previously reported [41,42]. A sharp division of the temperature scale into (N), (I) and (D) regimes in Fig. 3 is somewhat arbitrary, but simplifies the following discussion. The occurrence of irreversible unfolding can cause history effects, mainly in the (I)-regime (Fig. 3). In particular, the relative amount of unfolded HSA in the solution was found to depend exponentially on the exposure time. Fig. 4 (filled circles) depicts the corresponding decrease of native (N)-HSA in a reference solution as a function of the exposure time at 70°C. For the second data set (open squares), the lubricant was continuously circulated through a heated water bath at 70°C. While the steady-state temperature in the lubricant reservoir remained 37°C, the solution temperature rose transiently to 70°C in the tubing, causing a cumulative degree of protein denaturation. We note that both sets of data reveal a nearly exponential decrease of (N)-HSA with a characteristic time constant of 20 min. This is the characteristic time constant for irreversible unfolding at 70°C. Based on the good agreement between both experiments, we can also conclude that the tribo-mechanical action of the pin-on-disc experiment did not significantly affect the unfolding kinetics in the lubricant. In the light of boundary lubrication, it is important to note that completely unfolded (D)-HSA is effectively more hydrophobic and thus expected to adsorb differently, depending on the hydrophilic/hydrophobic balance at the UHMWPE surfaces.

3.3. Protein adsorption and boundary lubrication

We have also investigated the kinetics of protein adsorption from differently pretreated HSA solutions onto LDPE of different degrees of hydrophilicity using the OWLS technique. A 5 s oxygen RF-plasma treatment was used to render part of the investigated polymer surfaces more hydrophilic. It is known that proteins can denature upon adsorption, especially on hydrophobic surfaces. Here, we focus on the effect of the folding state in solution on the amount of adsorption, which is an important factor, as shown in Table 1. To obtain partially denatured HSA, the protein solution was heated to 70°C for a duration of 30 min, which resulted in a relative amount of 35% (D)-HSA in the solution. The solution was cooled to 25°C prior to injection into the OWLS flow cell. The smallest amount of adsorbed mass (80 ± 20 ng/cm^2) was detected on the hydrophobic LDPE surface when the solution was partially denatured. Comparing this amount with that from the native (N)-HSA solution (110 ± 20 ng/cm^2) adsorbing on a similar surface, we can conclude that the

Table 1
Typical friction coefficient, μ and mass of proteins on differently prepared polymer surfaces adsorbed from solutions with different content of native (N) and irreversibly denatured (D) albumin

polymer substrate \ HSA solution	100% (N)	65% (N) + 35% (D)
Hydrophilic	μ=0.09, m_{ad}=190 ng/cm^2	μ=0.09, m_{ad}=190 ng/cm^2
Hydrophobic	μ=0.10, m_{ad}=110 ng/cm^2	μ=0.13, m_{ad}=80 ng/cm^2

The adsorbed mass was determined at 25°C solution temperature at typical accuracy of ±20 ng/cm^2 using optical waveguide lightmode spectroscopy (OWLS). A short oxygen RF-plasma treatment was used to render the polymer surface temporarily more hydrophilic. The partially denatured HSA solution was prepared by heating it to 70°C for 30 min. A schematic illustration of the molecular model proposed on the basis of the presented results is also shown. All four investigated combinations are shown to clarify the role of protein conformation in solution and the morphology of the adsorbed boundary layer as a function of surface hydrophilicity. The interface morphology is nearly identical in the two cases with hydrophobic (p$^-$) and hydrophilic (p$^+$) polymer surfaces. The ceramic (c) surface is unchanged in all four cases.

presence of (D)-HSA in solution leads to a considerable reduction in the total adsorbed mass. The expected mechanism is that (D)-HSA adsorbs more readily onto the hydrophobic surface than (N)-HSA and thus occupies a larger surface area per molecule than is the case with the purely native protein solution. The adsorbed layer of denatured (D)-HSA effectively passivates the surface and prevents adsorption of further proteins from solution. OWLS reveals a significantly higher adsorbed mass on the *hydrophilized* polymer surface. Independent of the (D)-HSA content in the solution, we find an adsorbed mass of 190±20 ng/cm^2. This insensitivity of hydrophilic surfaces towards the folding state of HSA in solution suggests that it is (N)-HSA that is predominantly adsorbed from both types of solutions. In this case, we may also conclude that (N)-HSA occupies less surface area per molecule and forms a thicker adsorbate film. Fig. 5 shows the friction measured in a pin-on-disc setup using a lubricant containing thermally denatured (D)-HSA

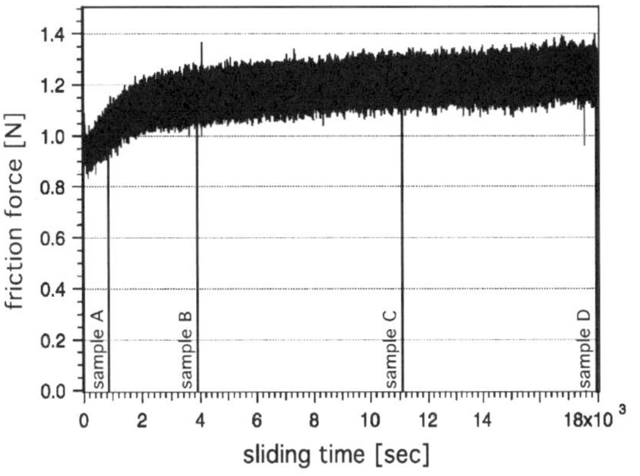

Fig. 5. Typical friction force recording of a pin-on-disc experiment as a function of sliding time at 40°C, the lubricant was constantly pumped through a pumping system that passed through a water bath at 70°C. The sliding velocity was v=2.07 mm/s and the normal load was L=10 N. Small samples of the lubricant solution were taken at four times, A, B, C, D. Vertical lines are added to indicate the respective times of sample extraction.

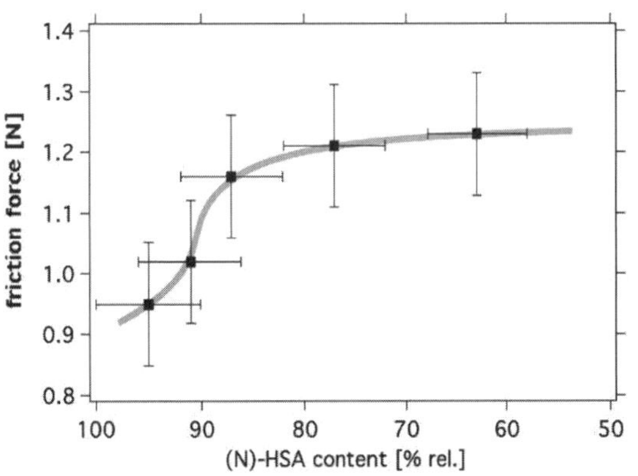

Fig. 6. Correlation between friction and conformation of HSA; a clear increase of friction is observed as the relative content of native protein declines below some 90%. The gray line in the background is to guide the eye.

(i.e. intermittent heating). The HSA lubricant solution was intermittently heated to 70°C in the external water bath as described in the experimental section above. Four liquid samples (A–D) were taken from the lubricant at different times to determine the relative (N)-HSA content in the CD spectrometer. For convenience, the same CD measurements are also included in Fig. 4 above. On hydrophobic (i.e. untreated) UHMWPE, we observed a 30% increase of the friction force with time (i.e. amount of denaturation). We note that the increase of friction correlates with the relative decrease of (N)-HSA (i.e. increased denaturation) in solution. This correlation is illustrated in Fig. 6. The existence of such a clear correlation suggests again that denatured proteins adsorb readily onto the hydrophobic surface and even replace the previously adsorbed native forms. The enrichment of the boundary layer with (D)-HSA changes the boundary lubrication conditions towards higher friction forces. Extended-time experiments with pure (N)-HSA solution (not shown) revealed no such increase of friction with time, which excludes the possibility of run-in or wear effects.

4. Discussion

Although the role of proteins [9] and lipids [8] has attracted some attention in joint tribology, the role of protein folding on boundary lubrication has not been systematically studied to date. Based on the CD measurements presented above, we can outline a picture of HSA conformation in solution that is in good agreement with previous studies of HSA [19,22]. While the current study deals with varying polymer hydrophilicity, the role of protein folding on adsorption onto the opposing ceramic surface was not included as a variable here. We can still propose a molecular model of protein-mediated boundary lubrication if we suppose that proteins adsorb onto the (hydrophilic) ceramic surface independently of both the polymer hydrophilicity and the protein conformation in solution (c.f. Table 1). Table 1 depicts schematic illustrations of the proposed molecular model. When the polymer surface is rendered hydrophilic, we observe a selective adsorption of (N)-HSA. A dense and consequently thick boundary layer is formed. The selectivity is clear since the adsorbed mass does not change if we add 35% (D)-HSA to the solution. Furthermore, in the two cases that included hydrophilic polymer surfaces (Table 1), we find identical friction coefficients. The ceramic counter-surface is also hydrophilic, and thus we expect a similar adsorption behavior, i.e. selective adsorption of (N)-HSA. This symmetry of the tribo-system is broken when the polymer surface is hydrophobic. Hydrophobic surfaces selectively adsorb hydrophobic (D)-HSA from the solution via strong hydrophobic interactions. For this reason, the molecular morphology of the adsorbed films is sensitive to even a small concentration of (D)-HSA in the lubricant solution. If the solution contains only native protein, the hydrophobic surface adsorbs only a relatively low total mass of (N)-HSA. It is thus reasonable to assume that a loosely adsorbed film of native proteins is formed [43–46]. As the concentration of (D)-HSA is increased, a monolayer of strongly denatured proteins is adsorbed, eventually replacing the native proteins. A large surface area per adsorbed molecule is present after this substitution, which also inhibits further adsorption from solution. We observe a higher friction coefficient on these latter films, presumably because adsorbed denatured proteins are less efficiently hydrated and thus function less well as low-shear-modulus boundary lubrication additives.

The relationship between friction force and wear is non-trivial [47]. It has been shown, for example, that a higher friction can lead to less wear in some cases [9,48]. More experiments are thus needed to clarify the effect of protein-mediated boundary lubrication for the known problem of wear. Another limitation of the present study is that the friction test was unidirectional. Multi-directional sliding is known to affect the wear rates [49]. The combination of experimental techniques presented here is thus a first step towards a more complete investigation of the molecular aspects in joint lubrication. In particular, it focuses on the relationship between protein folding, protein adsorption and boundary lubrication, as well as the significance of these effects in bio-tribological studies.

It is common practice to assess the performance of implant joints via repetitive articulation in more sophisticated mechanical simulators [50] over extended periods of time, in order to predict their in vivo reliability. A temperature rise in the lubricant of 60°C to 90°C has been observed [9,48] in such tests. The inevitable consequence of excessive temperatures is a progressive, irreversible unfolding of proteins. Unfolded proteins will preferentially adsorb onto hydrophobic surfaces or other unfolded (hydrophobic) proteins—the latter phenomenon leading to protein precipitation. In this case, both friction and wear results may be significantly affected by the adsorption of these denatured proteins, especially for hydrophobic surfaces. An appropriate temperature-control system for the lubricant solution is thus absolutely necessary for such joint-simulator tests. The situation may be less severe in the case of the artificial joint in vivo, since the temperature is usually between 43°C and 46°C [24]. But since these temperatures were measured below the cartilage surface, the lubricant could also reach higher temperatures at the interface, inducing the formation of a significant amount of (D)-HSA. More accurate in vivo temperature measurements are needed to fully resolve this issue. In addition, the SF in vivo is constantly being renewed. According to our results, we would thus not expect a progressively increasing amount of (D)-HSA in vivo, and thus little modification of the boundary lubrication on the hydrophobic polymer surface of the endoprosthesis is to be expected for relative (D)-HSA concentrations below the critical value of 10% (c.f. Fig. 6).

The adsorption of proteins from solution onto polymeric surfaces depends on protein–water, protein–surface and surface–water interactions [51,52]. Our optical adsorption study indicates that hydrophobic surfaces preferentially adsorb denatured proteins from a mixed solution. This finding underlines the amphiphilic nature of proteins and the important role of hydrophobic attraction for protein–surface interactions [45]. In contrast, hydrophilic surfaces selectively adsorb native proteins from a mixed solution, indicating that the driving force here is not the hydrophobic interaction but the ubiquitous Van der Waals force. When adsorbed, native proteins form a thicker film and remain essentially natively folded. Upon adsorption, native proteins maintain the more hydrophilic moieties at their surface, which means that they are more hydrated. Our results thus also suggest that the observed differences in boundary lubrication are linked to the different capacities of the adsorbed film to bind water. The use of hydration water bound to adsorbed hydrophilic molecules may be a promising line of attack for bio-lubricated systems beyond arthroplasty [53].

5. Conclusions

We have found that the conformational state of human serum albumin (HSA) in the lubricating solution can modify the efficacy of HSA-mediated boundary lubrication on a hydrophobic polymer surface. Furthermore, we have shown the importance of surface hydrophilicity on the lubricating ability of irreversibly denatured HSA. The thermal denaturation of HSA was followed by far-UV circular dichroism spectroscopy and at least two unfolding steps (one reversible, one irreversible) were discernable. On hydrophobic surfaces, a positive correlation exists between friction force and the concentration of denatured proteins in solution. A resistance towards the adsorption of denatured HSA can be achieved by rendering the polymer sliding partner more hydrophilic—more hydrophilic surfaces preferentially adsorb proteins of native conformation, which form thicker, denser films that have the potential to reduce boundary-lubricated friction. We conclude that the unfolding of proteins in the lubricant is of relevance for a number of practical situations. In particular, our results point out the need for rigorous control of lubricant temperature during joint-simulator tests, if they are to provide information that is useful for the design of implants.

Acknowledgements

This work has profited from scientific discussions with J. Vörös, C. Rieker (Centerpulse) and W. Rieger (Metoxit AG). We acknowledge a tribological feasibility study carried out by G. Franzolin, and T. Künzler for his help with the surface roughness measurements. Furthermore, we would like to thank M. Elsener and J. Vanicek for their technical support. Financial support was provided by the Commission for Technology and Innovation, and the Research Committee of the ETH Zurich.

References

[1] Hench LL. Medical materials for the next millenium. MRS Bull 1999;13–19.
[2] Lewis G. Polyethylene wear in total hip and knee arthroplasties. J Biomed Mater Res 1997;38(1):55–75.
[3] Unsworth A. Tribology of human and artificial joints. J Eng Med 1991;205:163–72.
[4] Charnley J. Low friction arthroplasty of the hip. Berlin: Springer; 1979.
[5] McGee MA, Howie DW, Neale SD, Haynes DR, Pearcy MJ. The role of polyethylene wear in joint replacement failure. Proc Inst Mech Eng Part H J Eng Med 1997;211(1):65–72.
[6] Mathiesen EB, Lindgren JU, Reinholt FP, Sudmann E. Tissue reactions to wear products from polyacetal (Delrin®) and UHMWPE in total hip replacement. J Biomed Mater Res 1987; 21:459–66.
[7] McGee MA, Howie DW, Costi K, Haynes DR, Wildenauer CI, Pearcy MJ, et al. Implant retrieval studies of the wear and loosening of prosthetic joints: a review. Wear 2000;241:158–65.
[8] Williams PF, Iwasaki Y, Ishihara K, Powell GL, Gilbert JA, Nakabayashi N, et al. Evaluation of the frictional properties of an elastomer with enhanced lipid-adsorbing ability. Proc Inst Mech Eng Part H J Eng Med 1997;211(5):359–68.
[9] Liao Y-S, Benya PD, McKellop HA. Effect of protein lubrication on the wear properties of materials for prosthetic joints. J Biomed Mater Res 1999;48:465–73.
[10] Sawae Y, Murakami T, Chen J. Effect of synovia constituents on friction and wear of ultra high molecular weight polyethylene (UHMWPE) sliding against prosthetic joint materials. Wear 1998; 216:213–9.
[11] Bos MA, van Vliet T. Interfacial rheological properties of adsorbed protein layers and surfactants: a review. Adv Colloid Interface Sci 2001;91:437–71.
[12] Kirk P. Physical properties of synovial fluid: composition, rheology and lubrication properties. Report No. M5874. Winterthur: Sulzer Orthopedics; 1997.
[13] Jalali-Vahid D, Jagatia M, Jin ZM, Dowson D. Prediction of lubricating film thickness in UHMWPE hip joint replacements. J Biomech 2001;34(2):261–6.
[14] Saikko VO, Ahlroos T. Wear simulation of UHMWPE for total hip replacement with a multidirectional pin-on-disk device: effects of counterface material, contact area and lubricant. J Biomed Mater Res 2000;49:147–54.
[15] Widmer MR, Heuberger M, Voros J, Spencer ND. Influence of polymer surface chemistry on frictional properties under protein-lubrication conditions: implications for hip-implant design. Tribol Lett 2001;10(1–2):111–6.
[16] Widmer MR, Heuberger M, Spencer ND. Boundary lubrication and friction of polyethylene and polyamides under protein-containing solutions. In: 28th Leeds-Lyon Symposium on Tribology 2002. Vienna: Elsevier Sci; 2002. p. 361–366.
[17] Sugio S, Kashima A, Mochizuki S, Noda M, Kobayashi K. Crystal structure of human serum albumin at 2.5 Å resolution. Protein Eng 1999;12(6):439–46.
[18] Carter DC, He XM, Munson SH, Twigg PD, Gernert KM, Broom MB, et al. Three-dimensional structure of human serum albumin. Science 1989;244:1195–8.
[19] Pico GA. Thermodynamic features of the thermal unfolding of human serum albumin. Int J Biol Macromol 1997;20(1):63–73.
[20] Muralidhara BK, Prakash V. Molten globule state of human serum albumin in urea. Curr Sci 1997;72(11):831–4.

[21] Tanaka N, Nishizawa H, Kunugi S. Structure of pressure-induced denatured state of human serum albumin: a comparison with the intermediate in urea-induced denaturation. Biochim Biophys Acta 1997;1338(1):13–20.
[22] Farruggia B, Pico GA. Thermodynamic features of the chemical and thermal denaturations of human serum albumin. Int J Biol Macromol 1999;26(5):317–23.
[23] Kloss H, Willmann G, Woydt M. Temperaturberechnung an den Artikulationsflächen beim künstlichen Hüftgelenk. Mat-wiss u Werkstofftechnik 2001;32:200–10.
[24] Bergmann G, Graichen F, Rohlmann A, Verdonschot N, van Lenthe GH. Frictional heating of total hip implants. Part 1: measurements in patients. J Biomech 2001;34(4):421–8.
[25] Lu Z, McKellop H. Frictional heating of bearing materials tested in a hip joint wear simulator. Proc Inst Mech Engrs 1997;211(Part H):101–8.
[26] Sawyer G. Temperature modeling in a total knee joint replacement using patient specific kinematics. Tribol Lett 2003;16(4): 343–52.
[27] Dawihl W, Mittelmeier H, Dörre E, Altmeyer G, Hanser U. Zur Tribologie von Hüftgelenk-Endoprothesen aus Aluminiumoxidkeramik. Medizinisch-Orthopädische Technik 1979;99:114–8.
[28] Widmer M. Modified molecular friction in artificial hip joints; PhD Thesis; Zürich: Swiss Federal Institute of Technology (ETH); 2002.
[29] Streicher RM. Ultra-high-molecular polyethylene as material for hip-joint cups. Stuttgart: Georg Thieme Verlag; 1986.
[30] Piconi C, Maccauro G. Zirconia as a ceramic biomaterial. Biomaterials 1999;20:1–25.
[31] Ciba-Geigy. Synovialflüssigkeit: Ciba-Geigy AG, Basel; 1977.
[32] Johnson Jr WC. Secondary structure of proteins through circular dichroism spectroscopy. Annual Rev Biophy Biophys Chem 1988;17:145–66.
[33] Johnson Jr WC. Protein secondary structure and circular dichroism: a practical guide. Proteins 1990;7(3):205–14.
[34] Johnson Jr WC. Analysis of circular dichroism spectra. Methods Enzymol 1992;210:426–47.
[35] Ramsden JJ. Review of new experimental techniques for investigating random sequential adsorption. J Stat Phys 1993; 73(5/6):853–77.
[36] Csucs G, Ramsden JJ. Interaction of phospholipid vesicles with smooth metal-oxide surfaces. Biochim Biophys Acta 1998; 1369:61–70.
[37] Ramsden JJ. OWLS: a versatile technique for sensing with bioarrays. Chimia 1999;53(3):67–71.
[38] Vörös J, Ramsden JJ, Csucs G, Szendro I, De Paul SM, Textor M. Optical grating coupler biosensors. Biomaterials 2002;23: 3699–710.
[39] Wheale SH, Barker CP, Badyal JPS. Chemical reaction pathways at the plasma-polymer interface. Langmuir 1998;14:6699–704.
[40] Lee S-G, Kang T-J, Yoon T-H. Enhanced interfacial adhesion of UHMWPE fibres by oxygen plasma treatment. J Adhes Sci Technol 1998;12(7):731–48.
[41] Krishnakumar SS, Panda D. Spatial relationship between the Prodan site, Trp-214, and Cys-34 residues in human serum albumin and loss of structure through incremental unfolding. Biochemistry 2002;41(23):7443–52.
[42] Muzammil S, Kumar Y, Tayyab S. Anion-induced stabilization of human serum albumin prevents the formation of intermediate during urea denaturation. PROTEINS: Struct Funct Genet 2000;40:29–38.
[43] Absolom DR, Zingg W, Neumann AW. Protein adsorption to polymer particles: role of surface properties. J Biomed Mater Res 1987;21:161–71.
[44] Young BR, Pitt WG, Cooper SL. Protein adsorption on polymeric biomaterials II. Adsorption kinetics. J Colloid Interface Sci 1988;125:246–60.
[45] Young BR, Pitt WG, Cooper SL. Protein adsorption on polymeric biomaterials I. Adsorption isotherms. J Colloid Interface Sci 1988;124:28–43.
[46] Underwood PA, Steel JG. Practical limitations of estimation of protein adsorption to polymer surfaces. J Immunol Methods 1991;142:83–94.
[47] Streicher RM, Schön R. Tribological behaviour of various materials and surfaces against polyethylene. In: 17th Annual Meeting of the Society for biomaterials, 1–5 May 1991. p. 289.
[48] Lu Z, McKellop H, Liao P, Benya P. Potential thermal artifacts in hip joint wear simulators. J Biomed Mater Res (Appl Biomater) 1999;48:458–64.
[49] Bragdon CR, O'Connor DO, Lowenstein Jr JD, Syniuta WD. The importance of multidirectional motion for the wear of polyethylene in the hip. In: World Tribology Congress, Proceedings of the Institution of Mechanical Engineers; London: 1997. p. 735.
[50] Saikko VO. Tribology of total replacement hip joints studied with new hip joint simulators and a materials-screening apparatus. Acta Polytech Scand 1993;110:44.
[51] van Straaten J, Peppas NA. Modelling of protein adsorption on polymeric surfaces. J Biomater Sci Polymer Edn 1991;2:91–111.
[52] Malmsten M. Formation of adsorbed protein layers. J Colloid Interf Sci 1998;207:186–99.
[53] Lee S, Müller M, Ratoi-Salagean M, Voros J, Pasche S, De Paul SM, et al. Boundary lubrication of oxide surfaces by poly(L-lysine)-g-poly(ethylene glycol) (PLL-g-PEG) in aqueous media. Tribol Lett 2003;15(3):231–9.

Biomaterials 30 (2009) 2072–2078

Contents lists available at ScienceDirect

Biomaterials

journal homepage: www.elsevier.com/locate/biomaterials

The adsorption and lubrication behavior of synovial fluid proteins and glycoproteins on the bearing-surface materials of hip replacements

Marcella Roba [a,b], Marco Naka [a], Emanuel Gautier [c], Nicholas D. Spencer [a,*], Rowena Crockett [b]

[a] *Laboratory for Surface Science and Technology, Department of Materials, ETH Zurich, Wolfgang-Pauli-Strasse 10, CH-8093 Zürich, Switzerland*
[b] *Empa, Swiss Federal Institute for Material Science and Technology, Ueberlandstrasse 129, CH-8600 Dübendorf, Switzerland*
[c] *Orthopaedic Clinic, Cantonal Hospital, CH-1708 Fribourg, Switzerland*

ARTICLE INFO

Article history:
Received 24 October 2008
Accepted 25 December 2008
Available online 19 January 2009

Keywords:
Synovial fluid
Hip implants
Lubrication
Glycoproteins
Adsorption
Alumina

ABSTRACT

The selectivity of synovial fluid protein adsorption onto ultra-high molecular weight polyethylene (UHMWPE) and alumina (Al_2O_3), and in particular the ability of glycoproteins to adsorb in the presence of all the other synovial fluid proteins, was investigated by means of fluorescence microscopy and gel electrophoresis (SDS-PAGE). The non-specific nature of protein adsorption from synovial fluid indicated that the lubrication of artificial hip-joint materials may not be attributable to a single protein as has been frequently suggested. The friction behavior of polyethylene (PE) sliding against Al_2O_3 in solutions of bovine serum albumin (BSA), alpha-1-acid glycoprotein (AGP) and alpha-1-antitrypsin (A1AT) was investigated by means of colloidal probe atomic force microscopy. BSA was shown to be a poorer boundary lubricant than the phosphate buffered saline used as a control. This was attributed to denaturation of the BSA upon adsorption, which provided a high-shear-strength layer at the interface, impairing the lubrication. Interestingly, both the glycoproteins AGP and A1AT, despite their low concentrations, improved lubrication. The lubricating properties of AGP and A1AT were attributed to adsorption via the hydrophobic backbone, allowing the hydrophilic carbohydrate moieties to be exposed to the aqueous solution, thus providing a low-shear-strength fluid film that lubricated the system. The amount of glycoprotein adsorbed on hydrophobic surfaces was determined by means of optical waveguide lightmode spectroscopy (OWLS), allowing conclusions to be drawn about the conformation of the glycan residues following adsorption.

1. Introduction

Under physiological conditions, synovial fluid, an aqueous electrolyte solution rich in proteins, lipids and hyaluronan, in combination with articular cartilage, is responsible for the low friction coefficients of hip-joints [1]. Following artificial hip-implant surgery (arthroplasty), the synovial membrane is reformed and the artificial materials are lubricated by pseudo-synovial fluid [2], which is believed to be similar in composition to the synovial fluid present before the surgery. The tribological performance of the artificial joint is significantly poorer than that of the natural joint [3]. For ultra-high molecular weight polyethylene (UHMWPE)–metal and UHMWPE–ceramic pairings, for example, the production of UHMWPE wear debris, with subsequent failure of the prosthesis due to osteolysis, provides an impetus for research into understanding and improving the materials used in artificial joints [4,5].

Investigating the adsorption of synovial fluid components onto artificial surfaces is of great importance in understanding the lubrication mechanisms occurring in artificial hip implants [6–8]. However, the extreme complexity of synovial fluid, characterized by large concentrations of hyaluronan and the presence of many different proteins, most of which are derived from plasma, is often not accounted for in such adsorption studies [9].

Protein adsorption studies on artificial hip-joint materials have focused mainly on albumin, as this is the most abundant protein in synovial fluid. Studies to evaluate albumin adsorption on ceramics and metals have been carried out using XPS and radiolabeling techniques [10,11] as well as sodium dodecyl sulfate polyacrylamide gel electrophoresis (SDS-PAGE) and chromatography [12,13]. Fluorescence microscopy has been applied to evaluate albumin adsorption on UHMWPE [6,14].

Both synovial fluid and solutions of albumin have been used to study the friction behavior of the UHMWPE–CoCrMo pairing, with

* Corresponding author. Tel.: +41 44 632 58 50.
E-mail address: spencer@mat.ethz.ch (N.D. Spencer).

0142-9612/$ – see front matter © 2008 Elsevier Ltd. All rights reserved.
doi:10.1016/j.biomaterials.2008.12.062

the conclusion that albumin is not acting as the boundary lubricant in this system despite being the most abundant protein [7]. Other studies have focused attention on lubrication of UHMWPE–alumina (Al_2O_3) pairings by salt solutions containing albumin and hyaluronan [15].

To date, no attempt has been made to understand the relationship between the adsorption of biomolecules from the complex synovial fluid and the friction between the surfaces of artificial joint materials on a molecular level. There have, however, been a number of studies on the mechanisms of aqueous lubrication in the presence of synthetic macromolecules. A particularly efficient mechanism of lubrication involves the formation of hydrophilic polymer brushes at the surface, which maintain a fluid film between the opposing surfaces [16,17]. The low friction between the surfaces has been attributed to the fluidity of the hydration layers around the polymers and resistance to interpenetration between the polymer brushes [18].

In this study, we have investigated to what extent specific, less abundant biomolecules in synovial fluid can influence the lubrication of artificial joint materials. The adsorption and friction behavior of two glycoproteins, alpha-1-acid glycoprotein (AGP) and alpha-1-antitrypsin (A1AT) were compared to those of albumin in order to determine the influence of glycosylation on friction. These glycoproteins were chosen as AGP has a very high degree of glycosylation at approximately 45–50%, and A1AT, with a lower amount of carbohydrate, is one of the most abundant glycoproteins in synovial fluid. SDS-PAGE was used to investigate the adsorption of proteins from bovine synovial fluid onto UHMWPE. This technique proved to be useful in evaluating the selectivity of protein adsorption. However, due to the small total mass of proteins adsorbing, detection of adsorbed proteins that were present in low concentrations in synovial fluid was extremely difficult. Therefore, a fluorescence-labeling technique was applied to confirm the adsorption of glycoproteins onto the materials analyzed in this study. Optical waveguide lightmode spectroscopy (OWLS) was used to quantify glycoprotein adsorption onto hydrophobic surfaces. Friction between a polyethylene (PE) and Al_2O_3 tribopair in the presence of albumin and glycoproteins was determined by atomic force microscopy (AFM).

2. Materials and methods

2.1. Materials

UHMWPE and Al_2O_3 samples were provided by Mathys Ltd. Bettlach (Switzerland) and were the same materials as those used in the production of artificial hip-joints. The UHMWPE samples had been γ-irradiated with a dose of 25–30 kGy in a nitrogen atmosphere. Medical quality Al_2O_3 specimens were cleaned in an ultrasonic bath with, consecutively, isopropanol, ethanol and pure water for 10 min each followed by drying with a N_2 gas jet. The roughness (Ra) of UHMWPE samples was 0.1–0.2 μm and that of Al_2O_3 was 4–8 nm.

2.2. Labeling of proteins

Fluorescent labeling of albumin, alpha-1-acid glycoprotein and alpha-1-antitrypsin was carried out with ATTO 488 NHS ester (Sigma–Aldrich, Switzerland). The structure of ATTO 488 is the undisclosed property of Atto-Tec GmbH, Siegen, Germany (www.atto-tec.com).

Fatty-acid-free bovine serum albumin (BSA, Sigma–Aldrich, Switzerland) was dissolved in a sodium bicarbonate buffer (0.1 M, pH 8.3, Merck, Switzerland) to a concentration of 5 mg/ml. ATTO 488 NHS ester (2 mg/ml) was dissolved in amine-free, anhydrous dimethylsulfoxide (DMSO, Sigma–Aldrich, Switzerland) immediately before conjugation. The conjugation reaction was carried out by adding a two-fold molar excess of dye to the protein solution, followed by incubation for 60 min at 24 °C with stirring. Labeled proteins were separated from unreacted dye by dialysis overnight against phosphate buffered saline (PBS, Sigma–Aldrich, Switzerland) using 10 kDa dialysis membranes (Slide-A-Lyzer, Perbio Science N.V., Belgium). Aliquots of labeled albumin were stored at −20 °C. The same protocol was used to label alpha-1-acid glycoprotein from bovine serum (AGP, Sigma–Aldrich, Switzerland) and alpha-1-antitrypsin from human plasma (A1AT, Sigma–Aldrich, Switzerland).

2.3. Adsorption studies via fluorescence microscopy

Solutions for incubation tests were obtained by mixing bovine synovial fluid (BSF) with labeled BSA, labeled AGP and labeled A1AT to concentrations of 12.5 mg/ml for BSA and of 0.3 mg/ml for the glycoproteins. BSF was removed from the knee joint of cows not older than 18 months at the Hinwil slaughterhouse in Switzerland. All samples were clear, yellow, particle-free liquids and were stabilized with protease inhibitor (Sigma–Aldrich, Switzerland) and 0.1% sodium azide.

UHMWPE and Al_2O_3 samples were incubated, using a drop of each solution, for 60 min in a dark environment, to avoid photobleaching of the labeled proteins. After rinsing with PBS and pure water, fluorescence images at the border of the droplets were obtained with an AX10 Imager M1m (Zeiss, Germany).

2.4. Nanoscale friction measurements

Friction was measured between a colloidal probe of PE and a disc of Al_2O_3. The measurements were carried out using an AFM apparatus (Dimension 3000, Digital Instruments, Santa Barbara, CA) at room temperature (24 °C). The AFM colloidal probes, with a particle of PE (5 μm of diameter) on the tip were provided by Novascan Technologies (Ames, IA, USA). The stiffness of the Si_3N_4 cantilevers used was 0.58 N/m. The scanning velocity was 10 μm/s. Friction forces were measured in arbitrary units (a.u.) and an average of 10 TMRs (trace minus retrace) was calculated. Each trace and retrace was measured on a different line over a distance of 40 μm. Varying loads were applied for each specimen in the presence of individual and combined proteins. The solutions used were PBS (control), and BSA, AGP and A1AT were added gradually to the PBS solution. The concentrations were 0.1, 0.5 and 1 mg/ml for BSA and AGP, and 0.1, 0.2 and 0.5 mg/ml for A1AT. In addition, BSA solutions containing AGP or A1AT were measured. Each set of tests included a measurement of PBS to allow a comparison between the different solutions and involved a single, new colloidal probe. Results presented in each figure correspond to a single set.

2.5. Evaluation of protein adsorption using SDS-PAGE

UHMWPE samples were incubated in a 20% BSF solution in PBS at 37 °C for 1 h and 24 h. After the incubation time, samples were rinsed with PBS and pure water and incubated at 24 °C for 1 h in 2 ml of 80% acetonitrile solution, to allow protein desorption. Blank samples, not incubated in BSF, were used as controls. Subsequently the acetonitrile solutions were concentrated under nitrogen flow, and proteins were dissolved in sample buffer containing Tris–HCl, SDS, glycerol, mercaptoethanol and bromophenol blue. Samples were then heated at 95 °C for 5 min before being loaded in the gel.

The gel used for SDS-PAGE was a 13% acrylamide, self-cast gel. It had a total of ten lanes, two of which were loaded with desorption solutions from blank samples. Four lanes were used for the desorbed protein solutions and one lane was loaded with protein standards.

2.6. Quantitative evaluation of protein adsorption using OWLS

AGP and A1AT adsorption was evaluated on fluorosilane-functionalized SiO_2/TiO_2 waveguides ($Si_{0.75}Ti_{0.25}O_2$ on glass, 1.2 × 0.8 cm^2, Microvacuum, Budapest, Hungary). Waveguides were rinsed with 0.1% HCl and isopropanol, and subsequently cleaned with O_2 plasma (Plasma Cleaner/Sterilizer, PDC-32G instrument, Harrick, Ossining, NY). It was determined that a hydrophobic surface could be fabricated on the OWLS waveguide with high reproducibility by means of perfluorooctyltrichlorosilane. Functionalization with 1H,1H,2H,2H-perfluorooctyltrichlorosilanes (ABCR GmbH & Co. KG, Germany) was carried out from the gas phase in a desiccator. The static water contact angle of the functionalized waveguides was determined to be 100°.

Adsorption in the OWLS instrument (OWLS 110, Microvacuum, Hungary) was carried out at 37 °C, using PBS as buffer. After having recorded the baseline, either 0.5 mg/ml AGP solution or 0.5 mg/ml A1AT solution were injected into the flow chamber. Protein adsorption was allowed to proceed until a plateau in the adsorption curve was observed. At this stage, non- or weakly adsorbed proteins were washed away by rinsing the flow chamber with PBS.

2.7. Evaluation of synovial fluid protein composition by means of 2D gel electrophoresis

The 2D gel shown in Fig. 1 was measured as part of a DIGE (differential in-gel electrophoresis) experiment (to be published). Human synovial fluid samples (obtained with ethics approval of the University Hospital Fribourg) were dialyzed against PBS overnight to eliminate protease inhibitor and sodium azide, which had been added to stabilize the fluids. Subsequently, the samples were incubated in hyaluronidase solution (Hyaluronidase VI-S, Sigma–Aldrich, Switzerland) at 37 °C for 48 h to digest hyaluranon. After sample clean up and quantification with a 2-D

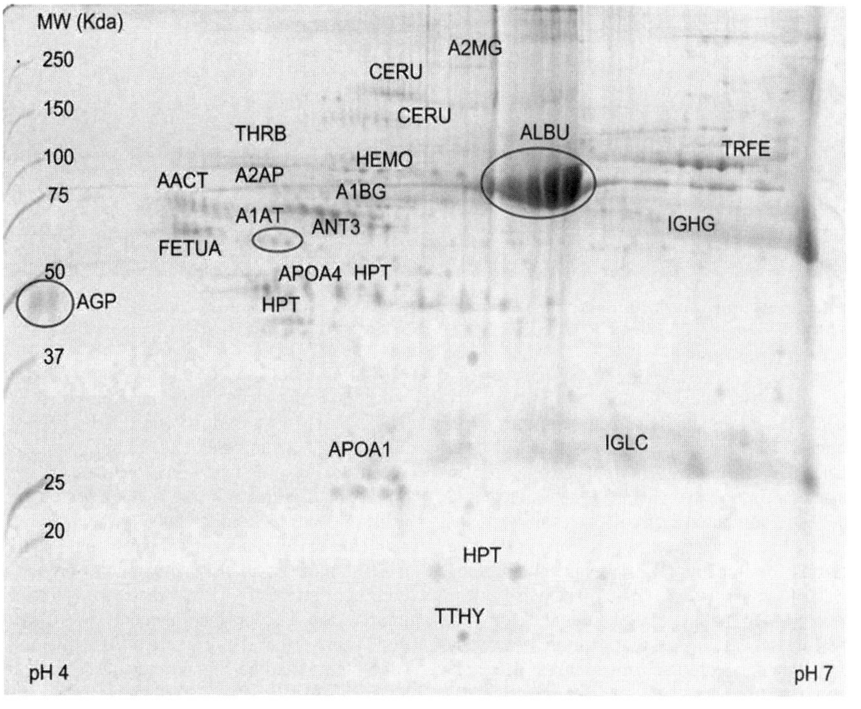

Fig. 1. 2D gel electrophoresis of human synovial fluid. The proteins circled in the picture are the proteins used in this work: albumin (ALBU), AGP and A1AT. Other proteins present in synovial fluid are also indicated. Alpha-1-antichymotrypsin (AACT), Alpha-1B-glycoprotein (A1BG), Alpha-2-antiplasmin (A2AP), Alpha-2-HS-glycoprotein (FETUA), Alpha-2-macroglobulin (A2MG), Antithrombin-III (ANT3), Apolipoprotein A-I (APOA1), Apolipoprotein A-IV (APOA4), Ceruloplasmin (CERU), Haptoglobin (HPT), Hemopexin (Beta-1B-glycoprotein) (HEMO), Immunoglobulin heavy chain gamma (IGHG), Immunoglobulin light chain (IGLC), Prothrombin (THRB), Serotransferrin (Transferrin) (TRFE), Transthyretin (Prealbumin) (TTHY).

Quant Kit (Ettan™ Sample Preparation Kit and Reagents, Amersham Biosciences, UK), labeling was carried out with CyDye DIGE fluor (Amersham Biosciences, UK) fluorescent dyes. For the isoelectric focusing, a 24 cm ReadyStrip™ IPG strip, with p*I* range 4–7 (BioRad, Hercules, USA), was used. The protein samples were loaded onto the strip and isoelectric focusing was carried out overnight by means of an IGPphor™ Isoelectric Focusing System (Ettan™, Amersham Biosciences, UK). After equilibration, the strip was applied onto one end of a self-cast, 11% acrylamide gel. Separation of the proteins according to molecular weight was then carried out by means of Ettan™ DALTtwelve system (Amersham Biosciences, UK). The gel was fixed using a 10% methanol, 7% acetic acid solution and stained using SYPRO® Ruby protein gel stain (Molecular Probes, Holland). Scanning of the gel was carried out using a Typhoon 9400 scanner (Molecular Dynamics, Amersham Biosciences, UK).

3. Results

3.1. Synovial fluid composition analysis

The 2D gel electrophoresis of human synovial fluid closely corresponded to the typical protein distribution found in plasma (Fig. 1) [19]. Albumin and immunoglobulin (heavy and light chains) dominated the gel in terms of concentration. A1AT could be observed in the low isoelectric point region, at around p*I* 5, together with other plasma glycoproteins.

3.2. Protein adsorption analysis with SDS-PAGE and OWLS

UHMWPE samples were incubated in BSF, and investigated for adsorbed proteins in the range of plasma proteins by means of protein desorption and gel electrophoresis. Bands of adsorbed proteins corresponding to albumin, immunoglobulin heavy chain gamma (IGHG) and immunoglobulin light chain (IGLC) were observed at all incubation times (Fig. 2). Fainter bands in the range of 30–60 KDa could also be detected on the gel. At molecular weights lower than 31 KDa, prealbumin was observed.

Quantitative glycoprotein adsorption measurements carried out with OWLS showed that the amount of AGP adsorbed onto a hydrophobic surface in the absence of other proteins was 110 ± 14 ng/cm^2, and the corresponding amount of A1AT was 152 ± 13 ng/cm^2.

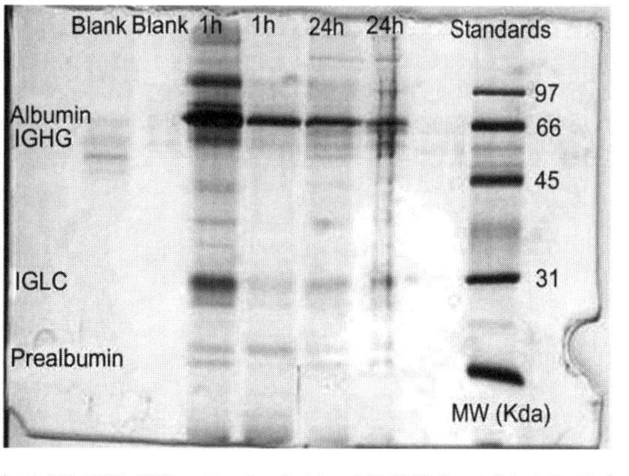

Fig. 2. SDS-PAGE of BSF proteins desorbed from UHMWPE. From left to right: blank sample lanes, 1 h BSF incubation lanes, 24 h BSF incubation lanes, protein standard lane.

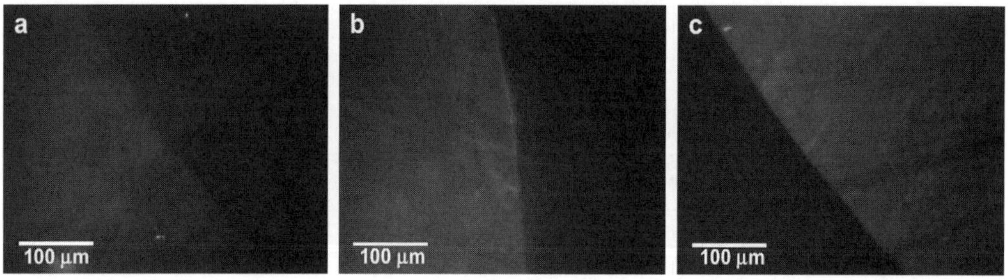

Fig. 3. Fluorescence images of UHMWPE samples after incubation in synovial fluid containing (a) labeled BSA, (b) labeled AGP, (c) labeled A1AT.

3.3. Fluorescence imaging

UHMWPE and Al_2O_3 samples for fluorescence imaging were incubated with droplets of bovine synovial fluids that had been "spiked" with labeled BSA, AGP and A1AT. Droplets were used, in order to highlight any contrast between material that was exposed to protein and the clean, unexposed surface. Fluorescence microscope images of UHMWPE samples, taken at the border of the droplets, show that all the three proteins investigated adsorbed onto the surface from synovial fluid (Fig. 3). There is a clear difference in the fluorescence intensity at the border of the droplet, indicating the difference between the blank and the area wetted by the droplet, where protein adsorption had occurred. All three proteins also adsorbed from synovial fluid onto the Al_2O_3 substrates (Fig. 4).

3.4. Nanofriction measurements

Friction measurements were carried out with the same colloidal probe for each set of experiments and different colloidal probes between sets. PBS was measured in all sets as a reference to allow a comparison between the different groups of experiments.

Fig. 5a shows friction curves in relation to normal load, measured with AFM, for one PE colloidal probe sliding against Al_2O_3 substrates in a solution of BSA in PBS. When using a concentration of 0.1 mg/ml BSA, a significant increase in friction force, for loads higher than 65 nN, could be observed in comparison to PBS alone. Increasing the concentration of BSA up to 1.0 mg/ml led to a further increase in the friction force.

Friction curves for solutions of AGP showed a different behavior (Fig. 5b) in comparison to BSA. The friction force curve for a concentration of 0.1 mg/ml was very similar to that of PBS. Increasing the concentration of the AGP lubricant up to 0.5 mg/ml and 1.0 mg/ml resulted in a considerable reduction in friction. A1AT in PBS behaved in a similar way to AGP (Fig. 5c). At 0.1 mg/ml, the lubricating properties of A1AT solution were similar to those of PBS. At concentrations of 0.2 mg/ml and 0.5 mg/ml, a significant reduction in the friction force was observed.

Fig. 5d shows a comparison of friction measurements carried out using 1.0 mg/ml BSA solution and mixtures of BSA and AGP. When adding AGP to the BSA solution, at a concentration of 0.1 mg/ml, a decrease in the friction was observed at the lower loads. At loads higher than 200 nN, the friction force values were similar to those obtained for the BSA solution. Increasing the AGP concentration up to 0.5 mg/ml in the presence of BSA led to a further decrease of friction that was maintained above 200 nN normal load. Friction measurements with mixtures of BSA and A1AT showed that the presence of A1AT in a BSA solution at low concentration did not significantly affect the lubrication (Fig. 5e). When increasing the A1AT concentration up to 0.5 mg/ml, a decrease in the friction could be observed at lower loads, whereas above 200 nN, the lubricating behavior was similar to that of BSA.

4. Discussion

From the large number of proteins present in synovial fluid (Fig. 1), BSA and the glycoproteins AGP and A1AT were chosen for the investigation of their role in the lubrication of PE vs. Al_2O_3, for reasons of both their abundance and their chemical structure. BSA, a globular protein characterized by a hydrophobic core and hydrophilic surface, is the most abundant protein in synovial fluid [20]. The chemical composition of glycoproteins is very different from that of BSA, as it is characterized by carbohydrate moieties covalently linked to a polypeptide backbone [21]. Apart from albumin, all other plasma proteins are potentially glycoproteins. AGP is negatively charged and has a high carbohydrate content, comprising 45–50% of the total weight [22], while the peptide backbone itself is mainly characterized by hydrophobic domains. AGP was an interesting target for investigation due to its high carbohydrate concentration, despite its low concentration. A1AT was chosen since it is one of the most abundant glycoprotein in

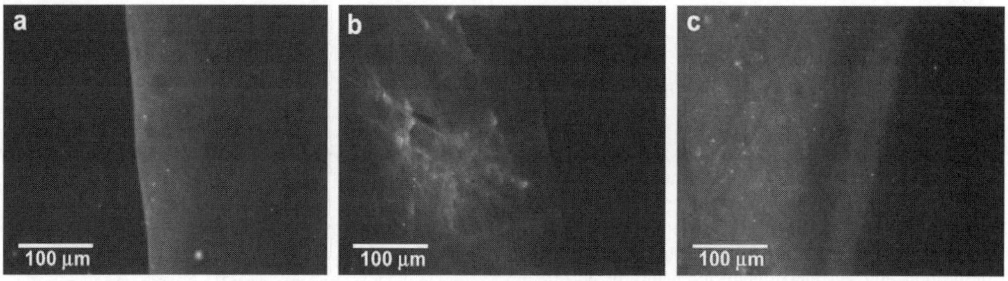

Fig. 4. Fluorescent images of Al_2O_3 samples after incubation in synovial fluid containing (a) labeled BSA, (b) labeled AGP, (c) labeled A1AT.

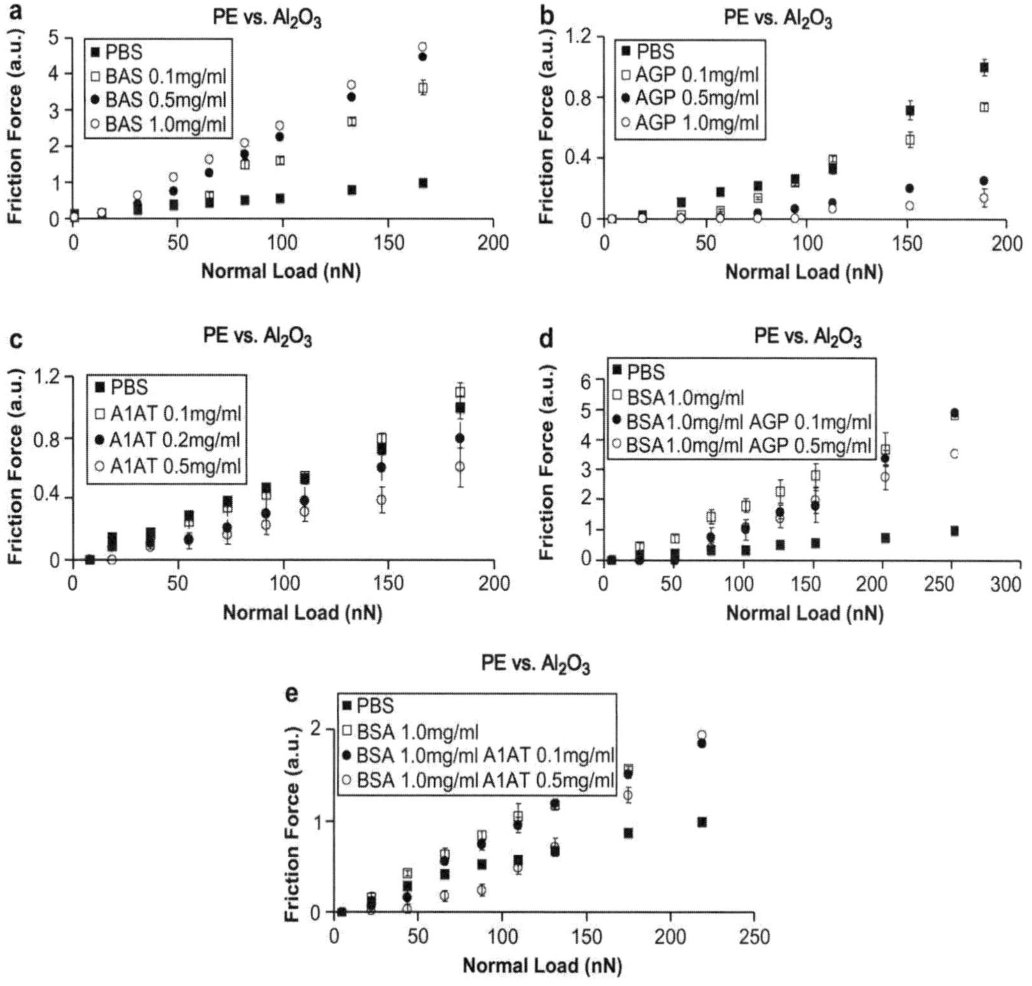

Fig. 5. Friction force versus the normal load for PE sliding against Al_2O_3 in (a) BSA at increasing concentrations, (b) AGP at increasing concentrations, (c) A1AT at increasing concentrations, (d) BSA and AGP mixtures, (e) BSA and A1AT mixtures. PBS was always used as control in each set of tests. Error bars are not always visible due to a small standard deviation. Each experiment was repeated three times.

synovial fluid, with a carbohydrate content of approximately 12% and a predominantly hydrophilic protein backbone [23].

BSA was shown to be a less effective boundary lubricant than PBS, AGP and A1AT. Upon adsorption on hydrophobic surfaces, albumin unfolds, exhibiting the hydrophobic core and adsorbing via hydrophobic interactions [24,25]. This behavior leads to the presence of denatured adsorbed BSA on the PE surface, resulting in a high-shear-strength layer with strong hydrophobic interactions between the albumin and the surface and between adjacent albumin molecules [26]. AGP and A1AT behaved in a very different way, enhancing the lubrication of PE vs. Al_2O_3 in comparison to PBS, as the concentration was increased. The adsorption of AGP and A1AT on PE occurred via hydrophobic interactions with the protein moiety and also onto Al_2O_3 via electrostatic interactions. Although the peptidic component in A1AT is hydrophilic overall, it does contain hydrophobic amino acid residues [27].

The natural lubricant present after arthroplasty is a complex mixture containing proteins, lipids and hyaluronan. However, the literature on natural lubrication has focused on single-component studies rather than on synovial fluid, mostly for reasons of availability and convenience. As albumin is by far the most abundant protein and readily adsorbs onto artificial materials, it has widely been considered to be the most important in lubrication studies. However, SDS-PAGE of the proteins desorbed from UHMWPE showed that the whole range of plasma proteins can adsorb onto polyethylene (Fig. 2). Over 1 and 24 h periods of incubation, no evidence of adsorption selectivity could be detected. Additionally, fluorescence studies on the adsorption of BSA, AGP and A1AT showed that all three adsorbed onto the UHMWPE (Fig. 3) and alumina (Fig. 4), even in the presence of all other biomolecules in synovial fluid. These results imply that, given the absence of selective adsorption, many of synovial fluid proteins or glycoproteins could be implicated in the boundary lubrication of the UHMWPE/Al_2O_3 hip-implant pairing.

AFM friction measurements showed that AGP, both at lower and higher concentrations, maintained its lubricating properties in the presence of BSA (Fig. 5d). However, at low AGP concentration and high normal loads the friction force was similar to BSA alone. The structure of BSA is characterized by the presence of more adsorption sites in comparison to AGP, making it capable of binding more

M. Roba et al. / Biomaterials 30 (2009) 2072–2078 2077

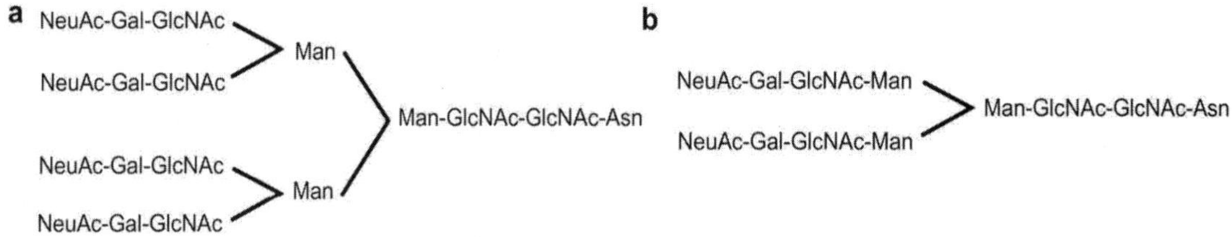

Fig. 6. Examples of the carbohydrates found on AGP and A1AT, (a) tetraantennary and (b) biantennary glycans. NeuAc: *N*-acetyl neuramic acid (sialic acid), Gal: galactose, GlcNAc: *N*-acetylglucosamine, Man: mannose, Asn: asparagine.

strongly to the surfaces [28,29]. AGP, at a low concentration, may be squeezed out of the interface at high loads, and the effect of BSA on lubrication would prevail resulting in a similar friction behavior to that of BSA alone (Fig. 5d). However, it is also possible that the mechanism by which the adsorbed AGP lowers friction cannot withstand the higher loads. A1AT behaved in a similar way to AGP in the presence of BSA (Fig. 5e). The temperature during the AFM measurements was 24 °C and the relative adsorption of the albumin and glycoproteins may be influenced by this. However, as the adsorption studies have shown, there was no indication of specific adsorption from synovial fluid at 24 °C or 37 °C.

The low friction observed for AGP and A1AT can be explained by the structure and hydrophilic nature of the glycans. AGP contains 5 glycan residues that are mainly tetraantennary and triantennary, whereas A1AT contains 3, mainly biantennary, glycan residues (Fig. 6) [23,30]. The structures of the glycans on AGP in solution have been determined [30–32]. These residues do not extend out into the aqueous solution, but instead are forced into a conformation parallel to the protein by electrostatic interactions between the negatively charged sialic acid groups on the glycan and the positively charged lysine groups on the peptide backbone [31]. A model constructed with the software Hyperchem (Hypercube Inc., USA) showed that such conformations could also account for the amount of AGP and A1AT adsorbed onto hydrophobic surfaces. Structures extended parallel to the surface for the tetraantennary and biantennary glycans in AGP and A1AT, based on those in [31], are shown in Fig. 7a and b, respectively. If it is assumed that these can rotate freely then they will occupy an area given by a circle with a radius of 2.05 nm for AGP and of 2.40 nm for A1AT (Fig. 8). The mass of AGP and A1AT on the surface, as measured with OWLS, can be converted into the number of glycan residues per nanometer using molecular weights of 44 kDa and 54 kDa, respectively. Assuming that the glycans occupy close-packed circles and taking into account the spaces between the circles, the radius of the glycans in AGP calculated from the OWLS measurements corresponded to 1.98 ± 0.14 nm and the radius of the glycans in A1AT was 2.39 ± 0.10 nm. It should be noted that these calculations only give the average size occupied by the glycans. The two closest glycan residues on AGP are separated by only 10 amino acid residues [29], and, therefore, the maximum separation between the asparagine residues that these glycans are attached to is 3.7 nm. Two tetraantennary glycans with the conformation shown in Fig. 7a require a separation between asparagines of 4.10 nm. All other glycans on AGP and A1AT are sufficiently far apart on the peptide backbone to accommodate the structures shown in Fig. 7a and b respectively.

These calculations lead to the conclusion that the coverage of pure AGP or A1AT on a hydrophobic surface is determined by the size and packing of the glycan residues. The packed glycoproteins form a highly hydrophilic layer capable of trapping water at the surface to provide a lubricious, low-shear-strength, fluid film. Adsorption of AGP leads to a lower friction than A1AT, presumably due to the larger amount of carbohydrate per unit area (Fig. 8). In the presence of BSA, the glycoproteins inevitably adsorb further apart on the surface, reducing their ability to influence friction but not eliminating it. It should be emphasized here that AGP and A1AT were investigated to determine the influence of glycosylation on friction behavior. Bovine AGP and human A1AT were used, as these have been well characterized. The structure of the glycans and the degree of glycosylation varies from species to species and as AGP is an acute-phase protein, the structure and concentration will also depend on the health of the patient. As no specific adsorption behavior of the synovial fluid proteins could be detected, it is possible that other glycoproteins play a role in determining the friction behavior. However, these experiments show that the glycoproteins behave very differently from albumin and that they should be considered when studying the lubrication of artificial hip-joint materials.

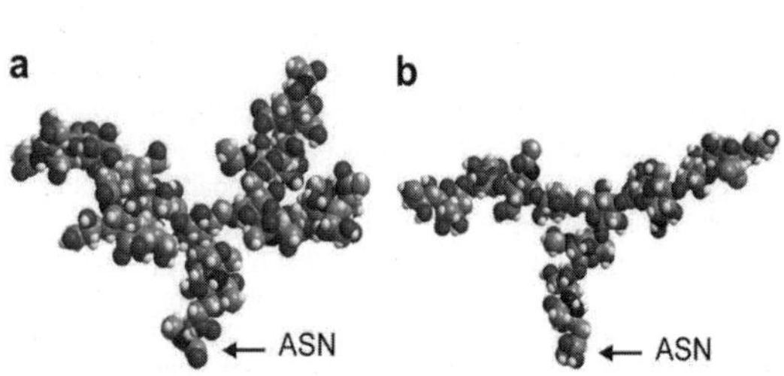

Fig. 7. Conformations of (a) tetraantennary and (b) biantennary glycans. The amino acid residue asparagine (ASN) is at the bottom in both cases.

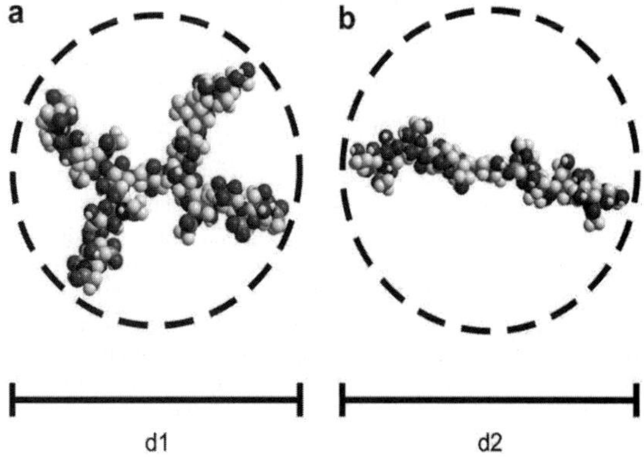

Fig. 8. The structures shown from above are: (a) tetraantennary glycan, the diameter d1 is 4.10 nm when calculated from the structure in (a) and 3.96 nm when calculated from the OWLS measurements; (b) biantennary glycans, d2 is 4.80 nm from the structure in (b) and 4.78 nm from the OWLS measurements.

5. Conclusions

SDS-PAGE and fluorescence microscopy have shown that protein adsorption from synovial fluid onto UHMWPE and Al_2O_3 is non-specific. Not only were abundant proteins such as albumin and gamma globulins found to adsorb, but also the less abundant ones such as AGP and A1AT, indicating that any protein present in synovial fluid can, in principle, play a role, positive or negative, in boundary lubrication. The glycoproteins AGP and A1AT were shown to enhance the lubrication of PE vs. Al_2O_3 on the molecular level, both on their own and in the presence of BSA in solution. The lubricating properties of AGP and A1AT were attributed to the presence of hydrophilic carbohydrate chains, which possess the ability to form a hydrophilic, low-shear-strength fluid layer upon adsorption. The possibility of reducing friction in artificial joints by tailoring the surface chemistry to selectively adsorb certain glycoproteins may potentially have an impact on the development of new prosthetic materials.

Acknowledgments

This work was financially supported by the Swiss Confederation's innovation promotion agency (CTI), Mathys AG Bettlach, and the Empa research commission. We also gratefully acknowledge the assistance and advice of Dr. Beat Gasser (Robert Mathys Foundation) and Dr. Daniel Delfosse (Mathys AG Bettlach). We would also like to thank Dr. Heinz Troxler and Peter Kleinert (University of Zurich Children's Hospital) and Dr. Peter Gehrig and Dr. Mike Scott (Functional Genomics Center Zurich) for their indispensable assistance in carrying out the protein analyses.

Appendix

Figures with essential colour discrimination. Certain figures in this article, in particular Figures 3, 4, 7 and 8 may be difficult to interpret in black and white. The full colour image can be found in the on-line version, at doi:10.1016/j.biomaterials.2008.12.062.

References

[1] Prete PE, Gurakar-Osborne A, Kashyap ML. Synovial fluid lipids and apolipoproteins: a contemporary perspective. Biorheology 1995;32(1):1–16.
[2] Hallab NJ, Messina C, Skipor A, Jacobs JJ. Differences in the fretting corrosion of metal–metal and ceramic–metal modular junctions of total hip replacements. Journal of Orthopaedic Research 2004;22(2):250–9.
[3] Jin ZM, Stone M, Ingham E, Fisher J. Biotribology. Current Orthopaedics 2006;20(1):32–40.
[4] Ingham E, Fisher J. The role of macrophages in osteolysis of total joint replacement. Biomaterials 2005;26(11):1271–86.
[5] Lewis G. Polyethylene wear in total hip and knee arthroplasties. Journal of Biomedical Materials Research 1997;38:55–75.
[6] Karuppiah K, Sundararajan S, Xu Z-H, Li X. The effect of protein adsorption on the friction behavior of ultra-high molecular weight polyethylene. Tribology Letters 2006;22(2):181–8.
[7] Mazzucco D, Spector M. The John Charnley Award Paper: the role of joint fluid in the tribology of total joint arthroplasty. Clinical Orthopaedics and Related Research 2004;429:17–32.
[8] Scholes SC, Unsworth A. The effects of proteins on the friction and lubrication of artificial joints. Proceedings of the Institution of Mechanical Engineers, Part H 2006;220(6):687–93.
[9] Smith MA, Bains SK, Betts JC, Choy EHS, Zanders ED. Use of two-dimensional gel electrophoresis to measure changes in synovial fluid proteins from patients with rheumatoid arthritis treated with antibody to CD4. Clinical and Diagnostic Laboratory Immunology January 1, 2001;8(1):105–11.
[10] Serro AP, Gispert MP, Martins MCL, Brogueira P, Colaco R, Saramago B. Adsorption of albumin on prosthetic materials: implication for tribological behavior. Journal of Biomedical Materials Research Part A 2006;78(3):581–9.
[11] Williams RL, Williams DF. Albumin adsorption on metal surfaces. Biomaterials 1988;9(3):206–12.
[12] Rosengren A, Pavlovic E, Oscarsson S, Krajewski A, Ravaglioli A, Piancastelli A. Plasma protein adsorption pattern on characterized ceramic biomaterials. Biomaterials 2002;23(4):1237–47.
[13] Takami Y, Yamane S, Makinouchi K, Otsuka G, Glueck J, Benkowski R, et al. Protein adsorption onto ceramic surfaces. Journal of Biomedical Materials Research 1998;40(1):24–30.
[14] Crockett R, Roba M, Naka M, Gasser B, Delfosse D, Frauchiger V, et al. Friction, lubrication, and polymer transfer between UHMWPE and CoCrMo hip-implant materials: a fluorescence microscopy study. Journal of Biomedical Materials Research 2008. doi:10.1002/jbm.a.32036.
[15] Gispert MP, Serro AP, Colaco R, Saramago B. Friction and wear mechanisms in hip prosthesis: comparison of joint materials behaviour in several lubricants. Wear 2006;260(1–2):149–58.
[16] Lee S, Muller M, Ratoi-Salagean M, Voros J, Pasche S, De Paul SM, et al. Boundary lubrication of oxide surfaces by poly(L-lysine)-g-poly(ethylene glycol) (PLL-g-PEG) in aqueous media. Tribology Letters 2003;15(3):231–9.
[17] Muller M, Lee S, Spikes HA, Spencer ND. The influence of molecular architecture on the macroscopic lubrication properties of the brush-like co-polyelectrolyte poly(L-lysine)-g-poly(ethylene glycol) (PLL-g-PEG) adsorbed on oxide surfaces. Tribology Letters 2003;15(4):395–405.
[18] Raviv U, Giasson S, Kampf N, Gohy J-F, Jerome R, Klein J. Lubrication by charged polymers. Nature 2003;425(6954):163–5.
[19] Anderson NL, Anderson NG. The human plasma proteome: history, character, and diagnostic prospects. Molecular & Cellular Proteomics 2002;1(11):845–67.
[20] He XM, Carter DC. Atomic-structure and chemistry of human serum-albumin. Nature 1992;358(6383):209–15.
[21] Kornfeld R, Kornfeld S. Comparative aspects of glycoprotein structure. Annual Review of Biochemistry 1976;45(1):217–38.
[22] Kleinert P, Kuster T, Arnold D, Jaeken J, Heizmann CW, Troxler H. Effect of glycosylation on the protein pattern in 2-D-gel electrophoresis. Proteomics 2007;7(1):15–22.
[23] Kolarich D, Weber A, Turecek PL, Schwarz HP, Altmann F. Comprehensive glyco-proteomic analysis of human alpha(1)-antitrypsin and its charge isoforms. Proteomics 2006;6(11):3369–80.
[24] Kleijn M, Norde W. The adsorption of proteins from aqueous solution on solid surfaces. Heterogeneous Chemistry Reviews 1995;2(3):157–72.
[25] Nakanishi K, Sakiyama T, Imamura K. On the adsorption of proteins on solid surfaces, a common but very complicated phenomenon. Journal of Bioscience and Bioengineering 2001;91(3):233–44.
[26] Heuberger MP, Widmer MR, Zobeley E, Glockshuber R, Spencer ND. Protein-mediated boundary lubrication in arthroplasty. Biomaterials 2005;26(10): 1165–73.
[27] Carrell RW, Jeppsson JO, Laurell CB, Brennan SO, Owen MC, Vaughan L, et al. Structure and variation of human alpha-1-antitrypsin. Nature 1982; 298(5872):329–34.
[28] Meloun B, Moravek L, Kostka V. Complete amino-acid sequence of human-serum albumin. FEBS Letters 1975;58(1):134–7.
[29] Schmid K, Kaufmann H, Isemura S, Bauer F, Emura J, Motoyama T, et al. Structure of alpha1-acid glycoprotein: complete amino-acid sequence, multiple amino-acid substitutions, and homology with immunoglobulins. Biochemistry 1973;12(14):2711–24.
[30] Montreuil J. Spatial conformation of glycans and glycoproteins. Biology of the Cell 1984;51(2):115–31.
[31] Li ZQ, Perkins SJ, Loucheuxlefebvre MH. Alpha-1 acid glycoprotein: a small-angle neutron-scattering study of a human-plasma glycoprotein. European Journal of Biochemistry 1983;130(2):275–9.
[32] Montreuil J. Structure and conformation of glycans and glycoproteins (a review): their relationship with normal and pathological metabolism and function. Biology of the Cell 1982;45:303.

Friction, lubrication, and polymer transfer between UHMWPE and CoCrMo hip-implant materials: A fluorescence microscopy study

Rowena Crockett,[1] Marcella Roba,[1,2] Marco Naka,[2] Beat Gasser,[3] Daniel Delfosse,[4] Vinzenz Frauchiger,[3] Nicholas D. Spencer[2]
[1]*Empa, Swiss Federal Institute for Material Science and Technology, Ueberlandstrasse 129, 8600 Duebendorf, Switzerland*
[2]*Laboratory for Surface Science and Technology, Department of Materials, ETH Zurich, Wolfgang-Pauli-Strasse 10, 8093 Zürich, Switzerland*
[3]*Dr. H. C. Robert Mathys Foundation, Bischmattstrasse 12, 2544 Bettlach, Switzerland*
[4]*Mathys AG Bettlach, Gueterstrasse 5, 2544 Bettlach, Switzerland*

Received 8 July 2007; revised 25 January 2008; accepted 1 February 2008
Published online 13 May 2008 in Wiley InterScience (www.interscience.wiley.com). DOI: 10.1002/jbm.a.32036

Abstract: The friction coefficients of CoCrMo sliding against UHMWPE and CoCrMo were measured in solutions of albumin and synovial fluid containing fluorescently labeled albumin. No fluorescence could be observed on the CoCrMo disc following incubation in labeled albumin or after sliding against CoCrMo. This was due to quenching of the fluorophore by the metal and indicated that a protein film thicker than 10 nm was not formed on the surface. A more complicated behavior was observed for UHMWPE sliding against CoCrMo. For each lubricating solution and at each load, a bimodal distribution of steady-state friction values was observed, the friction coefficient either remaining constant or decreasing during the early stages of the measurement. As no quenching of the fluorophores occurred on the UHMWPE surface, the fluorescence labeling method could be used to reveal polyethylene (PE) transfer and to show that it correlates with the friction coefficient: Low friction coefficients corresponded to a low density of PE spots on the CoCrMo surface. In addition, it was found that the friction coefficients for UHMWPE sliding against CoCrMo in synovial fluid were not significantly different from those in phosphate-buffered saline (PBS), but that the addition of albumin to PBS did cause a significant increase in the friction coefficient. © 2008 Wiley Periodicals, Inc. J Biomed Mater Res 89A: 1011–1018, 2009

Key words: artificial hip joints; albumin; friction; UHMWPE; CoCrMo; fluorescence

INTRODUCTION

The need to develop new materials for artificial hip joints is driven, in part, by the local and systemic biological consequences of wear debris arising from the currently used materials.[1] As a result, most studies of artificial joint materials, such as alumina, cobalt-chromium-molybdenum alloys (CoCrMo), and ultrahigh molecular weight polyethylene (UHMWPE), concentrate on wear analyses, most reliably carried out with a hip-joint simulator and involving wear-rate measurements and wear-particle characterization.[2] Although these studies are essential before a new material can be implanted, bench tests have been useful in investigations on the mechanisms of wear, friction, and lubrication between implant materials. Such studies are important in the development of new materials.[3]

Although there is no simple relationship between wear and friction, a low friction can generally be expected to enhance the performance of an artificial hip.[4,5] Mazzuco and Spector investigated the effect of fluid composition on the static and kinetic friction coefficients between polyethylene (PE) and CoCrMo or oxidized zirconium alloys.[6] By comparing synthetic preparations with serum and synovial fluid, they concluded that hyaluronan, phospholipids, albumin, and γ-globulin do not act as boundary lubricants in implants. Interestingly, they found large variations in the lubricating ability of human syno-

Correspondence to: R. Crockett; e-mail: rowena.crockett@empa.ch

Contract grant sponsor: Swiss Commission for Technology and Innovation
Contract grant sponsor: Mathys AG Bettlach
Contract grant sponsor: Empa Research Commission

vial fluid samples from different subjects, suggesting that this is an important factor in the failure of artificial joints.[6]

A difference in the behavior of bovine and human synovial fluids has been observed in the average friction coefficients of a variety of polymers sliding against CoCrMo.[7] In the same study, there appeared to be no relationship between protein concentration and average friction coefficient. However, the variation in the protein concentrations for bovine and human synovial fluid was not discussed. The unusually low protein concentration in the bovine synovial fluid (BSF) suggests that these samples were not stabilized and that the high concentration in the human samples may be due to contamination with blood. Both of these factors can be expected to cause large variations in protein concentration. Gispert et al. also observed that the concentration of albumin had no effect on the coefficient of friction.[8]

Friction studies were carried out with a hip-joint simulator by Scholes et al. They compared the friction behavior between a number of implant materials using both carboxy methyl cellulose (CMC) solutions and biological fluids as lubricants.[9–11] They found that friction coefficient for alumina/alumina was higher for biological fluids than for CMC, and attributed this to the inhibition of fluid-film formation by a protein film on the surfaces. For metal/metal joints, the friction coefficients were lower in biological fluids, leading to the conclusion that the formation of a protein film assisted boundary lubrication.

The adsorption of albumin, the most abundant protein in synovial fluid, on implant materials was studied in detail by means of XPS and I^{125}-labelled albumin.[12] These techniques showed the formation of a monolayer of albumin on alumina but multilayered islands of protein on the CoCrMo surface.[12] No PE transfer was observed in AFM images of the surface of CoCrMo after tribological tests when an albumin solution was used as the lubricant. However, contrary to the studies described earlier, the addition of albumin to a solution of Hank's balanced salt solution caused a dramatic drop in the friction coefficient for UHMWPE sliding against CoCrMo.[8]

It has been accepted by many that no PE transfer takes place in the presence of proteins since McKellop et al. studied wear debris in 1978.[13–15] The transfer film from an UHMWPE pin onto a stainless steel plate was observed visibly and with optical microscopy when the lubricant was pure water. When serum was used as the lubricant, a transfer film was not observed.

In this study, fluorescent labeling of albumin was used to investigate the adsorption of this protein onto the surfaces of artificial implant materials before and during tribological measurements. The labeled protein was either used together with unlabeled albumin or used to "spike" a synovial fluid sample. The fluorescence from a protein layer of less than 10 nm cannot be detected on CoCrMo as fluorescence quenching occurs.[16] However, there is no quenching of the fluorophore by UHMWPE or a UHMWPE transfer layer on CoCrMo. Therefore, this method has proved useful in relating the friction coefficient of UHMWPE sliding against CoCrMo to the behavior of the protein and the transfer of polymer to the metal surface.

MATERIALS AND METHODS

Materials

Pins and discs were provided by Mathys Bettlach (Switzerland), and the materials used were CoCrMo and UHMWPE. CoCrMo was forged according to ISO 5832-12 with a high carbon content ($0.20 < C < 0.25$ wt %). UHMWPE was grade 1020 according to ISO 5834-2 and supplied by Quadrant (Switzerland). The pins were cylindrical, with one spherically shaped end of radius 35 mm for UHMWPE and 1200 mm for CoCrMo. This approach yielded contact pressures in the range of 8–20 MPa for loads of 0.5–5 N. The maximum contact pressure in CoCrMo *versus* CoCrMo hip joints is ~70 MPa, however, pressure is not distributed evenly over the contact area and this high pressure is only experienced at a small part.[17,18] In addition, an unrealistically large amount of adhesive wear occurred when the contact pressure was increased above 20 MPa in the pin-on-disc measurements.

The specimens were cleaned in an ultrasonic bath with, consecutively, hexane, isopropanol, and pure water for 15 min each followed by drying with an N_2 gas jet. This procedure was carried out twice. The roughness (Ra) of the CoCrMo discs was 5 ± 2 nm and that of the UHMWPE discs was 0.1–0.2 μm. The UHMWPE pins and discs were γ-irradiated with a dose of 25–30 kGy in a nitrogen atmosphere prior to use.

Labeling of proteins

Albumin labeling was carried out using ATTO 488 NHS ester and ATTO 565 NHS ester (Sigma-Aldrich, Switzerland) as fluorescent dyes. The structure of ATTO 488 is the undisclosed property of Atto-Tec GmbH, Siegen, Germany (www.atto-tec.com). ATTO 565 is a rhodamine-based dye.

Fatty acid-free bovine serum albumin (BSA, Sigma-Aldrich) was dissolved in a sodium bicarbonate buffer (0.1*M*, pH 8.3, Merck, Switzerland) to a concentration of 5 mg/mL. ATTO NHS ester (2 mg/mL) was dissolved in amine-free, anhydrous dimethylsulfoxide (DMSO, Sigma-Aldrich) immediately before conjugation. The conjugation reaction was carried out by adding a twofold molar excess of dye to the protein solution followed by incubation for 60 min at 24°C with stirring. Labeled proteins were separated from unreacted dye by dialysis overnight against phosphate-buffered saline (PBS, Sigma-Aldrich) using 10-kDa dialysis membranes (Slide-A-Lyzer, Perbio Science N. V., Belgium). Aliquots of labeled albumin were stored at −20°C.

Lubricating solutions

BSF was removed from the knee joints of cows no older than 18 months at the Hinwil slaughterhouse in Switzerland. All samples were clear, yellow, particle-free liquids, and were stabilized with protease inhibitor (Sigma-Aldrich) and 0.1% sodium azide. BSF was mixed with labeled BSA solution (5 mg/mL) to give a final total protein concentration of 12.5 mg/mL. The BSA solutions used in the tribological experiments were produced by mixing BSA in PBS (20 mg/mL) with the labeled BSA solution (5 mg/mL) to give a final total protein concentration of 12.5 mg/mL.

Pre- and post-incubation of UHMWPE discs were carried out with labeled BSA solutions (0.1 mg/mL), preincubation of CoCrMo discs was carried out with unlabeled BSA solutions (0.1 mg/mL).

Tribological measurements

Coefficients of friction were measured with a pin-on-disc apparatus (CSEM S.A., Switzerland). The CoCrMo pins were run-in against CoCrMo in PBS at a load of 2 N and a sliding speed of 5 mm/s for 60 laps. The UHMWPE pins were run-in against CoCrMo in PBS at a load of 2 N and a sliding speed of 5 mm/s for 40 laps.

For measurements of friction coefficient, the normal load was varied in the range of 0.5–5 N at a sliding velocity of 10 mm/s. The steady-state friction was taken as the average of the friction coefficient in the 20th lap. All values are the average of at least four measurements. After the friction tests, the samples were washed with PBS followed by deionized water and dried with nitrogen.

To avoid the photobleaching of labeled BSA, the specimens and lubricants were kept in a dark environment during all steps of the experiments. Fluorescence images were obtained with an AX10 Imager M1m (Zeiss, Germany).

The discs were examined by means of AFM (Dimension 3000, Santa Barbara, USA) with Si_3N_4 probes (Veeco Instruments, USA) with a nominal spring constant of 0.12 N/m. Phase imaging was carried out with AFM (Multimode IIIA, Santa Barbara, USA) and silicon AFM probes (BS Multi75Al, NanoAndMore, Germany) with a nominal spring constant of 3 N/m and resonance frequency of 75 kHz.

Force–distance curves were obtained with a silicon probe, (BS Multi75Al, NanoAndMore GmbH, Germany) with a spring constant of 4.8 N/m, calibrated with a probe of known spring constant. The constant compliance was calculated from the force–distance curves of the tip against a clean CoCrMo surface. The force–distance curves of the silicon tip against spots on the surface were fitted with the Hertz equation for an infinitely stiff, spherical tip against a soft material.[19]

Environmental scanning electron microscopy (ESEM, Philips ESEM-FEG XL30, FEI, Netherlands) was used to image the AFM tip used for force–distance measurements and the surface of the CoCrMo discs.

RESULTS

Friction measurements

Friction coefficients were measured for CoCrMo sliding against CoCrMo in three solutions: PBS, diluted BSA, and BSF. At a load of 0.5 N, the friction coefficient was lower than at higher loads. For all the three lubricants, there was no significant change in the friction coefficient from 1 to 5 N, consistent with boundary lubrication, and, therefore, the average μ over these loads was calculated. The highest friction coefficient of 0.71 ± 0.12 was measured for PBS. Addition of BSA to the solution caused a >50% drop in the friction coefficient to 0.33 ± 0.06 and the lowest value was found for BSF at 0.25 ± 0.03.

When UHMWPE was slid against CoCrMo, a far more complicated behavior of the friction coefficient was observed (Fig. 1). Although there was no significant difference in the average friction coefficient upon increasing the load from 1 to 5 N, there was a decrease in the range of friction coefficients observed for BSF and BSA solutions. There was no significant difference between the mean friction coefficients for UHMWPE sliding against CoCrMo in PBS and BSF. However, higher mean friction coefficients were observed when a solution of BSA was used as the lubricant (Fig. 1).

A closer examination of the friction-coefficient measurements revealed that, for each load, two well-distinguished classes of behavior could be identified. In some cases, the friction coefficient remained con-

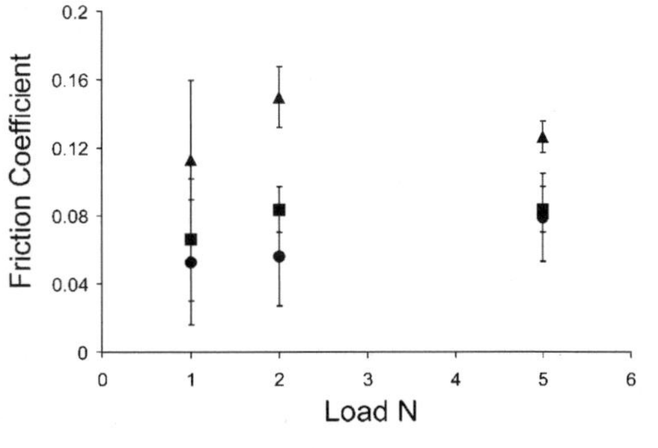

Figure 1. Friction coefficients for UHMWPE sliding against CoCrMo at 1, 2, and 5 N in PBS (●), BSF (■), and BSA (▲).

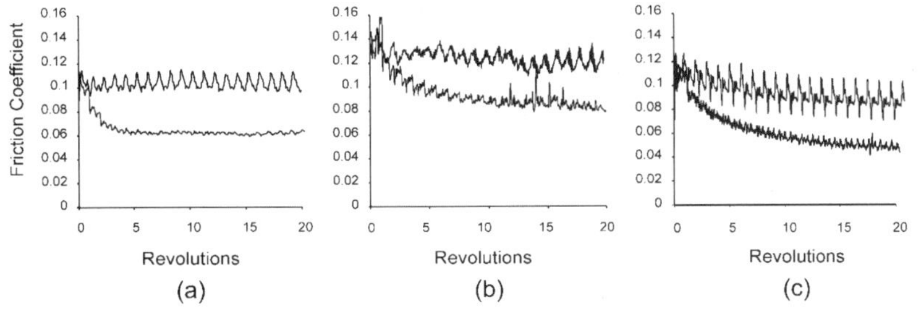

Figure 2. Change in friction coefficient with the number of laps at 10 mm/s and 1 N for the lubricating solutions: (a) BSF, (b) BSA, and (c) PBS.

stant throughout the test, whereas in others there was an initial decrease followed by a constant friction coefficient [Fig. 2(a) for BSF]. The initial friction coefficient in both cases was the same. Sufficient measurements were carried out using PBS to distinguish two significantly different friction coefficients for loads of 1 N (0.021 ± 0.002 and 0.085 ± 0.008), 2 N (0.03 ± 002 and 0.081 ± 0.007), and 5 N (0.039 ± 0.004 and 0.092 ± 0.009). When BSA was added to PBS [Fig. 2(b)], the ranges of friction coefficients observed were 0.07–0.21 (1 N), 0.13–0.17 (2 N), and 0.11–0.14 (5 N). When BSF was the lubricant [Fig. 2(a)], the ranges of friction coefficient were 0.03–0.1 (1 N), 0.06–0.11 (2 N), and 0.07–0.10 (5 N).

Fluorescence imaging

No fluorescence was observed inside or outside the wear track on the CoCrMo surface after sliding against CoCrMo in either BSA or BSF solutions containing albumin labeled with atto-488, a green fluorescent dye (not shown). The fluorescent images on the CoCrMo disc, obtained after sliding against an UHMWPE pin in BSF and BSA, are shown in Figure 3(a,b), respectively. Images with a similar morphology were visible on the disc after sliding in BSA and BSF at loads of 1, 2, and 5 N. In all cases, small, bright spots were observed in the sliding track and no fluorescence was observed outside the track.

To determine whether these structures were also formed during sliding in the absence of protein, the CoCrMo disc was incubated in a labeled albumin solution after sliding against UHMWPE in PBS [Fig. 3(c)]. These images were similar to those obtained after sliding in the labeled BSF and BSA solutions.

Variations in the density of the fluorescent spots on the CoCrMo disc were observed for measurements carried out under identical conditions for both BSF and BSA. These differences were most apparent at loads of 1 and 2 N for both solutions. A lower density of spots was observed on tracks where the friction coefficient fell during the measurement, whereas tracks on which the friction coefficient remained at a higher value had a correspondingly higher density of fluorescent spots (Fig. 4).

Since PE does not quench the fluorescence signal, discs of UHMWPE were used to follow the exchange between albumin in solution and albumin adsorbed on the surface. The discs were incubated in a solution of albumin that had been labeled with atto-565, a red fluorescent dye. Typical fluorescent images following sliding of a CoCrMo pin against a UHMWPE disc under a load of 2 N in BSA and BSF containing albumin labeled with atto-488 are shown in Figure 5(a,b), respectively. Exchange of the red-labeled albumin with green-labeled albumin was observed

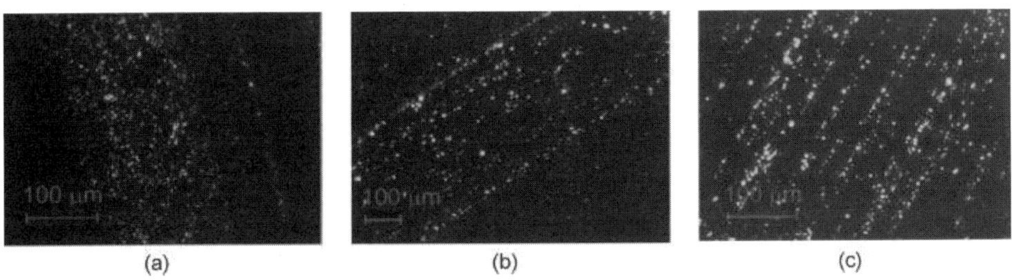

Figure 3. Fluorescent images on CoCrMo after sliding against a UHMWPE pin at a load of 2 N in (a) BSF, (b) BSA, and (c) PBS.

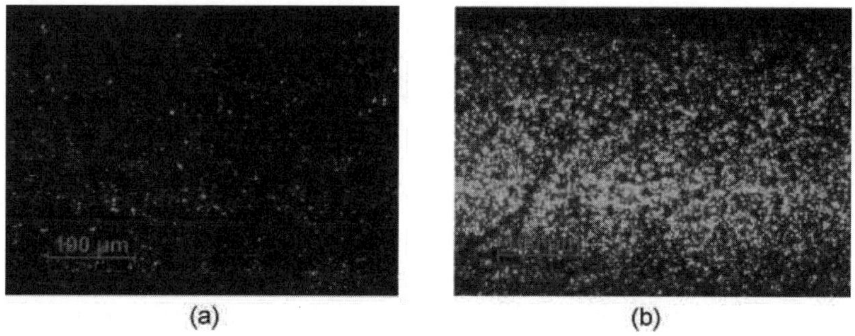

Figure 4. Fluorescent images on CoCrMo disc after sliding against UHMWPE at 10 mm/s and 1 N in BSF: (a) friction coefficient 0.06 and (b) friction coefficient 0.11.

inside the sliding track, but no exchange or adsorption of the green-labeled albumin was observed outside the track.

Surface morphology

Figure 6(a) shows a typical AFM image of the spots observed with fluorescence imaging obtained after sliding UHMWPE against CoCrMo in BSF at 2 N. The features observed in the AFM image were of the same size, shape, and distribution as those observed with fluorescence imaging. The height of the spots was determined by measuring profiles on the images [Fig. 6(b)]. The average height was 150 ± 70 nm. AFM imaging also showed the presence of protrusions from the surface of the CoCrMo that were identified by back-scattered electron microscopy as carbide grains. The average height of these carbides was only 8 ± 4 nm.

Phase imaging was also carried out with AFM [Fig. 7(a)] and showed a contrast between the spots and the substrate. Phase imaging detects changes in adhesion and in the mechanical properties of the surface.[20] Therefore, force–distance measurements were carried out on the spots and the CoCrMo substrate to determine the contribution of these two properties to the phase image [Fig. 7(b)]. All measurements were carried out on spots with a height of more than 200 nm. The force–distance curves were fitted with the Hertz equation for a sphere, and a total of 35 measurements on seven spots gave an average effective modulus of 1 ± 0.2 GPa. SEM images of the AFM probe showed that the tip was conical in shape, although the very small indentation depth is likely to have resulted in behavior similar to that of a sphere.

DISCUSSION

The friction coefficients observed for CoCrMo sliding against CoCrMo in the three solutions were consistent with those reported in the literature.[3] That no fluorescence image was detected on the CoCrMo surfaces either after incubation in labeled protein or after the CoCrMo-CoCrMo measurements indicated that the fluorescence of the adsorbed albumin was

Figure 5. Fluorescent images on UHMWPE disc after sliding against CoCrMo at a load of 2 N in (a) BSA and (b) BSF. The discs were incubated in red-labeled albumin and the green-labeled albumin was in the lubricating solutions.

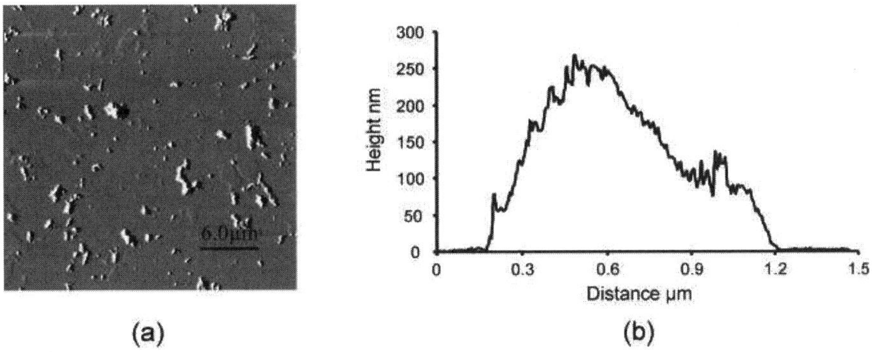

Figure 6. (a) AFM deflection image of CoCrMo surface after sliding against UHMWPE pin in BSF at a load of 2 N. (b) Profile of one of the larger spots taken from the AFM height image.

quenched by the metal and a protein film thicker than 10 nm was not formed.

After sliding UHMWPE against CoCrMo in all solutions, fluorescence was detected on the surface of the metal. The fluorescent spots showed similar sizes and shapes when they were generated in PBS, and visualized by postincubation with labeled albumin, as when they were generated in BSA and BSF containing labeled albumin. It could, therefore, be concluded that these spots were due to transfer of UHMWPE to the CoCrMo surface. This postulation was supported by AFM phase imaging, which showed a contrast between the CoCrMo surface, and the measurement of force–distance curves that were consistent with the effective modulus of UHMWPE. The measured effective modulus of 1 ± 0.2 GPa lay within the range of bulk UHMWPE (0.85) and the surface of γ-irradiated UHMWPE (up to 2.6 GPa).[21]

The transfer of UHMWPE to CoCrMo has been previously observed when the tribopairs slide against each other in pure water or salt solution. The transfer of UHMWPE in the presence of protein has not been reported previously. Clinical studies have not shown the presence of a transfer film on explanted artificial joints, however, this may simply be because no technique that could reasonably be expected to detect UHMWPE has been used to measure the surfaces. This discrepancy is also encountered in the literature for UHMWPE sliding against alumina in protein-containing solutions. Some studies have shown that there is no transfer of polymer to alumina in the presence of albumin, whereas other groups have observed such transfer.[8,22–24]

Such differences in studies on tribological behavior would generally be attributed to the use of different tribological parameters, such as load, contact pressure, or speed. In this study, tribological measurements were carried out in the boundary regime, whereas in the studies of Gispert et al, measurements were carried out in the mixed regime.[8] This difference may be an explanation for the observation here of increasing friction coefficient upon adding albumin to PBS and the transfer of polymer to the metal surface.

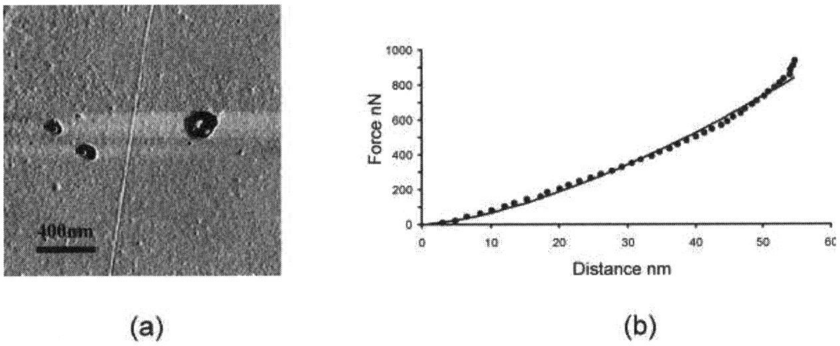

Figure 7. (a) AFM phase image at the surface of a CoCrMo disc after sliding against UHMWPE. (b) Force–distance curve measured on one spot shown in Figure 6(b) (●), the line was calculated from the Hertz equation for a sphere with an effective modulus (E') of 0.9 GPa.

However, the results presented here demonstrate that very different behaviors can also be observed under identical tribological conditions, with and without protein in the lubricating solution. The average surface roughness does not give an indication of the number of contacting asperities at the beginning of each measurement, and this may vary between samples sufficiently to cause variations in the tribological behavior. However, the observation that the friction coefficient was, within experimental error, the same at the beginning of each measurement suggests that differences in the topography of the pin surfaces were small.

The mechanical properties at the surface of the UHMWPE will also play an important role in determining the tribological behavior.[21,25–27] The UHMWPE pins were produced and machined in the same way and γ-irradiated with the same dose before use to ensure the same surface properties as UHMWPE used in artificial joints. It can, therefore, be assumed that there is little variation in the mechanical properties. However, small variations in the shear strength at the surface may be sufficient to determine whether lumpy transfer of polymer occurs or a smooth transfer film is formed. If the transfer film was thinner than 10 nm it would not be detected in the fluorescence imaging but may have been formed in those cases where the friction coefficient dropped to a lower value.

UHMWPE tends to form transfer layers on metal counterfaces in water through interlamellar shear of the polymer in the same way as other low-friction polymers, such as polytetrafluorethylene (PTFE).[25,28] The low friction coefficients of these materials are attributed to the ease with which the polymer molecules shear against each other. Most other polymers show poor friction properties due to lumpy transfer of material to the metal surface. However, lumpy transfer, in which debris adheres to the metal surface, can also occur for PTFE or UHMWPE under certain conditions. For example, lumpy transfer of PTFE occurs at low sliding speeds and was shown to give a friction coefficient that was approximately twice that of the thin transfer film.[29]

It has been postulated that the formation of thick, multilayer protein films plays an important role in determining the friction coefficient of artificial hip implants.[8,12,23,30–33] Albumin is sensitive to many factors including heat, protein concentration, salt concentration, and pH, and aggregates through hydrophobic interactions in solution over time to form solid, insoluble particles.[34,35] These particles are then deposited on the surface. Under the conditions used in this study, no deposition of protein to form a thick, aggregated film could be observed, either with AFM or with fluorescence imaging. It has been shown elsewhere that protein films with an average thickness of 43 nm can be deposited onto CoCrMo from BSA solutions.[12] A film of this thickness would have been visible in the fluorescence image in our studies. It may be the case, therefore, that while polymer transfer is inhibited by the deposition of a thick, aggregated protein film, it is not inhibited by an adsorbed protein monolayer. Therefore, it appears that the determining factors for the tribological behavior of an artificial hip joint *ex vivo* are the stability of the protein solution and the length of time over which the measurements were carried out.

Fluorescence imaging of UHMWPE discs following incubation in solutions containing red-labeled albumin showed that albumin adsorbed onto the polymer as a homogeneous monolayer. Following sliding against a CoCrMo pin, some of the red-labeled albumin was replaced by the green-labeled albumin from the solution, but the remaining red fluorescence outside of the contact area indicated that these areas were not coated with albumin from the solution. Therefore, as with the CoCrMo, the UHMWPE was apparently coated with a protein monolayer. No protein aggregates were deposited from the solution onto the UHMWPE.

CONCLUSION

The use of fluorescently labeled albumin in BSF and BSA solutions allowed the transfer of PE from the UHMWPE pin to the CoCrMo disc to be observed under the conditions used here. The transfer of UHMWPE was also shown to correlate with the friction coefficient. In those measurements where the friction coefficient dropped during sliding, a lower density of PE was observed on the metal surface. A higher density of spots was observed in those cases where the friction coefficient remained constant at a high value.

In this study, neither on CoCrMo nor on UHMWPE could the deposition of a thick protein film be observed. It is postulated that the transfer of PE is not inhibited by monolayers of adsorbed protein, as expected in the *in vivo* situation, but could be inhibited by the adhesion of aggregated proteins, brought about by experimental artifacts.

The authors are indebted to the Hinwil slaughterhouse, Switzerland, for collecting and providing bovine synovial fluid.

References

1. American Academy of Orthopaedic Surgeons. Implant Wear in Total Joint Replacement: Clinical and Biologic Issues, Material and Design Considerations. American Academy of Orthopaedic Surgeons: Rosemont, IL; 2001.

2. Essner A, Schmidig G, Wang A. The clinical relevance of hip joint simulator testing: In vitro and in vivo comparisons. Wear 2005;259:882–886.
3. Jin ZM, Stone M, Ingham E, Fisher J. Biotribology Curr Orthop 2006;20:32–40.
4. Saikko V, Calonius O. Simulation of wear rates and mechanisms in total knee prostheses by ball-on-flat contact in a five-station, three-axis test rig. Wear 2002;253:424–429.
5. Wang A, Essner A, Schmidig G The effects of lubricant composition on in vitro wear testing of polymeric acetabular components. J Biomed Mater Res Part B: Appl Biomater 2004; 68:45–52.
6. Mazzucco D, Spector M. The John Charnley Award Paper—The role of joint fluid in the tribology of total joint arthroplasty. Clin Orthop Relat Res 2004;429:17–32.
7. Yao JQ, Laurent MP, Johnson TS, Blanchard CR, Crowninshield RD. The influences of lubricant and material on polymer/CoCr sliding friction. Wear 2003;255:780–784.
8. Gispert MP, Serro AP, Colaco R, Saramago B. Friction and wear mechanisms in hip prosthesis: Comparison of joint materials behaviour in several lubricants. Wear 2006;260:149–158.
9. Scholes SC, Unsworth A. Comparison of friction and lubrication of different hip prostheses. Proc Inst Mech Eng Part H: J Eng Med 2000;214:49–57.
10. Scholes SC, Unsworth A, Goldsmith AAJ. A frictional study of total hip joint replacements. Phys Med Biol 2000;45:3721–3735.
11. Scholes SC, Unsworth A, Hall RM, Scott R. The effects of material combination and lubricant on the friction of total hip prostheses. Wear 2000;241:209–213.
12. Serro AP, Gispert MP, Martins MCL, Brogueira P, Colaco R, Saramago B. Adsorption of albumin on prosthetic materials: Implication for tribological behavior. J Biomed Mater Res Part A 2006;78:581–589.
13. McKellop H, Clarke IC, Markolf KL, Amstutz HC. Wear characteristics of UHMW polyethylene—Method for accurately measuring extremely low wear rates. J Biomed Mater Res 1978;12:895–927.
14. Saikko V. Effect of lubricant protein concentration on the wear of ultra-high molecular weight polyethylene sliding against a CoCr counterface. J Tribol-Trans ASME 2003;125: 638–642.
15. Saikko V. Effect of contact pressure on wear and friction of ultra-high molecular weight polyethylene in multidirectional sliding. Proc Inst Mech Eng Part H: J Eng Med 2006;220:723–731.
16. Perez-Luna VH, Yang SP, Rabinovich EM, Buranda T, Sklar LA, Hampton PD, Lopez GP. Fluorescence biosensing strategy based on energy transfer between fluorescently labeled receptors and a metallic surface. Biosens Bioelectron 2002; 17:71–78.
17. Fialho JC, Fernandes PR, Eca L, Folgado J. Computational hip joint simulator for wear and heat generation. J Biomech 2007;40:2358–2366.
18. Muller O, Parak WJ, Wiedemann MG, Martini F. Three-dimensional measurements of the pressure distribution in artificial joints with a capacitive sensor array. J Biomech 2004;37:1623–1625.
19. Dedkov GV. Experimental and theoretical aspects of the modern nanotribology. Physica Status Solidi a-Appl Res 2000;179:3–75.
20. Lemieux M, Usov D, Minko S, Stamm M, Shulha H, Tsukruk VV. Reorganization of binary polymer brushes: Reversible switching of surface microstructures and nanomechanical properties. Macromolecules 2003;36:7244–7255.
21. Benson RS. Use of radiation in biomaterials science. Nucl Instrum Methods Phys Res Sect B-Beam Interact Mater Atoms 2002;191:752–757.
22. Kumar P, Oka M, Ikeuchi K, Shimizu K, Yamamuro T, Okumura H, Kotoura Y. Low wear rate of uhmwpe against zirconia ceramic (Y-Psz) in comparison to alumina ceramic and Sus 316l alloy. J Biomed Mater Res 1991;25:813–828.
23. Sawae Y, Murakami T, Chen J. Effect of synovia constituents on friction and wear of ultra-high molecular weight polyethylene sliding against prosthetic joint materials. Wear 1998; 216:213–219.
24. Widmer MR, Heuberger M, Voros J, Spencer ND. Influence of polymer surface chemistry on frictional properties under protein-lubrication conditions: Implications for hip-implant design. Tribol Lett 2001;10:111–116.
25. Barrett TS, Stachowiak GW, Batchelor AW. Effect of roughness and sliding speed on the wear and friction of ultra-high-molecular-weight polyethylene. Wear 1992;153:331–350.
26. Gibbs C, Bender JW. A study of the nanotribological fatigue of ultra-high molecular weight polyethylene. Tribol Lett 2006;22:85–93.
27. Maszybrocka J, Cybo J, Frackowiak JE, Krzemien K. Change of micromechanical properties of polyethylene induced by a tribological process in polymer/metal system. Adv Mater Tech 2006;513:75–84.
28. Biswas SK, Vijayan K. Friction and wear of Ptfe—A review. Wear 1992;158:193–211.
29. Makinson KR, Tabor D. Friction + transfer of polytetrafluoroethylene. Proc R Soc Lond A Math Phys Sci 1964;281:49–61.
30. Chandrasekaran M, Loh NL. Effect of counterface on the tribology of UHMWPE in the presence of proteins. Wear 2001;250:237–241.
31. Karuppiah KSK, Sundararajan S, Xu ZH, Li XD. The effect of protein adsorption on the friction behavior of ultra-high molecular weight polyethylene. Tribol Lett 2006;22:181–188.
32. Kitano T, Ateshian GA, Mow VC, Kadoya Y, Yamano Y. Constituents and pH changes in protein rich hyaluronan solution affect the biotribological properties of artificial articular joints. J Biomech 2001;34:1031–1037.
33. Liao YS, Benya PD, McKellop HA. Effect of protein lubrication on the wear properties of materials for prosthetic joints. J Biomed Mater Res 1999;48:465–473.
34. Bier M, Nord FF. Aggregation phenomena in egg albumin solutions as determined by light scattering measurements. Proc Natl Acad Sci USA 1949;35:17–23.
35. Heuberger MP, Widmer MR, Zobeley E, Glockshuber R, Spencer ND. Protein-mediated boundary lubrication in arthroplasty. Biomaterials 2005;26:1165–1173.

A novel low-friction surface for biomedical applications: Modification of poly(dimethylsiloxane) (PDMS) with polyethylene glycol(PEG)-DOPA-lysine

Kanika Chawla,[1] Seunghwan Lee,[2] Bruce P. Lee,[1] Jeffrey L. Dalsin,[1] Phillip B. Messersmith,[1] Nicholas D. Spencer[2]

[1]*Department of Biomedical Engineering, Northwestern University, Evanston, Illinois*

[2]*Department of Materials, Laboratory for Surface Science and Technology, ETH Zurich, Wolfgang-Pauli-Strasse 10, CH-8093, Zurich, Switzerland*

Received 24 January 2008; revised 9 April 2008; accepted 28 April 2008
Published online 20 June 2008 in Wiley InterScience (www.interscience.wiley.com). DOI: 10.1002/jbm.a.32141

Abstract: Aqueous biocompatible tribosystems are desirable for a variety of tissue-contacting medical devices. L-3,4-dihydroxyphenylalanine (DOPA) and lysine (K) peptide mimics of mussel adhesive proteins strongly interact with surfaces and may be useful for surface attachment of lubricating polymers in tribosystems. Here, we describe a significant improvement in lubrication properties of poly (dimethylsiloxane) (PDMS) surfaces when modified with PEG-DOPA-K. Surfaces were characterized by optical and atomic force microscopy, contact angle, PM-IRRAS, and X-ray photoelectron spectroscopy. Such surfaces, tested over the course of 200 rotations (~8 m in length), maintained an extremely low friction coefficient (μ) (0.03 ± 0.00) compared to bare PDMS (0.98 ± 0.02). These results indicate the potential applications of PEG-DOPA-K for the modification of device surfaces. Extremely low μ values were maintained over relatively long length scales and a range of sliding speeds without the need for substrate pre-activation and in the absence of excess polymer in aqueous solution. These results were only obtained when DOPA was bound to lysine (modification with PEG-DOPA did not have an effect on μ) suggesting the critical role of lysine in obtaining a lowered friction coefficient. © 2008 Wiley Periodicals, Inc. J Biomed Mater Res 90A: 742–749, 2009

Key words: aqueous lubrication; surface modification; DOPA; lysine; tribology

INTRODUCTION

Surface-grafted poly(ethylene glycol) (PEG)-based polymers have demonstrated several useful biointerfacial properties including protein and cell resistance,[1,2] and suppression of immunogenic and antigenic activity.[3,4] Such properties are essential in surface-modifying polymers for biomedical applications.

**Present address*: Jeffrey L. Dalsin and Bruce P. Lee: Nerites Corporation, Madison, WI, USA.

Correspondence to: N. D. Spencer; e-mail: spencer@mat.ethz.ch or P. B. Messersmith, Biomaterials Group, Department of Biomedical Engineering, Northwestern University, 2145 Sheridan Rd., Evanston, IL 60208, USA; e-mail: philm@northwestern.edu

Contract grant sponsor: NIH and the International Institute for Nanotechnology

Lubricity is also a desirable property for biomedical applications involving moving parts, such as the artificial joint, as well as tubular devices, for example catheters and endoscopes.[5] End-grafting of hydrophilic polymer chains through surface-initiated polymerization of monomers, known as the "grafting from" approach[6] has been extensively investigated as an effective means to impart surface hydrophilicity and lubricity to various polymeric materials that are used for tissue-contacting devices.[5,7–9] Although remarkable lubricating properties have been achieved, these methods typically require surface activation by means of chemical or physical (high energy) methods, such as UV irradiation, plasma, or corona discharge, because of the initially nonreactive, that is, hydrophobic and/or nonpolar, surface properties of polymeric materials.[7,10–12]

PEG is very attractive for surface modification since it can simultaneously impart biocompatibility and lubricity to materials. The grafting of PEG chains onto polymeric surfaces can be achieved through noncovalent interaction between hydrophobic anchoring groups of PEG-based copolymers and the surfaces in aqueous solvents,[9,13,14] and the application of the aforementioned "grafting-from" approach has also been reported.[15–17] Previous tribological studies involving grafting of PEG chains, including poly(L-lysine)-*graft*-poly(ethylene glycol) (PLL-*g*-PEG)[9] or Pluronic[TM,13,14] onto polymeric surfaces indicated a dramatic reduction of friction forces under aqueous conditions.[9,13,14,18] Such lubricating effects are mediated by the presence of excess polymer in the bulk, whereby the region encountering tribostress and wear of the polymer layer can be rapidly replenished by adsorption of polymer from solution, and thus a "self-healing" mechanism can be activated.[13] Although this characteristic could be advantageous for conditions where continuous and cyclic tribological contacts are expected, such as bearing systems,[19] it makes them unsuitable for tissue-contacting devices, since the presence of "excess polymer" *in vivo* is not a viable option. Nevertheless, tribological contacts for tubular medical devices are expected to be very mild, and do not necessitate either cyclic or long-term service. Thus, a PEG layer, stably attached onto the surface, may be sufficient for the efficient lubrication of such devices.

Synthetic polymers containing DOPA and lysine (K) have a strong affinity for many surfaces[20–23] and may be useful in biomaterial tribosystems. The presence of L-3,4-dihydroxyphenylalanine (DOPA) in mussel adhesive proteins is believed to be critical for their impressive interfacial properties.[24] Thus, the objective of this study was to determine if application of PEG-DOPA-K (Fig. 1) to PDMS surfaces results in an improvement in the grafting of PEG chains, and hence aqueous lubrication properties with biomedical relevance. We have chosen PDMS as the tribopair for three reasons. Firstly, PDMS can represent silicone-based polymeric materials that have already found a broad range of biomedical applications,[25] and can further represent the broad class of hydrophobic polymeric materials used for tissue-contacting devices. Secondly, tribological interactions involving elastomers, either on one or both sides of the contact, typically yield mild contact pressures, analogously to the application of tissue-contacting devices. Thirdly, many previous PEG-based copolymers have been tested with tribopairs involving PDMS,[13,14] and thus these can be directly compared with PEG-DOPA-K for their lubricating efficacy. To this end, we have selected two other PEG-based copolymers for comparison, PEG-DOPA and Pluronic[TM], both of which are known to adsorb readily onto hydrophobic surfaces. The comparison with PEG-DOPA, which is identical to PEG-DOPA-K except for the absence of lysine is, in particular, expected to reveal the role of lysine in the lubricating behavior.

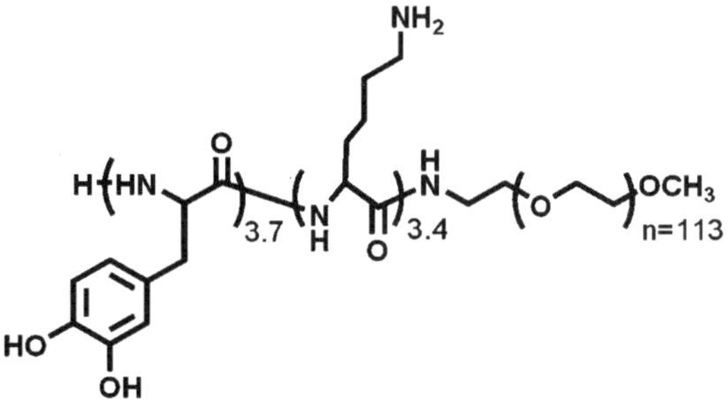

Figure 1. Chemical structure of PEG-DOPA-K. The average number of DOPA units and lysine units is 3.7 and 3.4, respectively. It should be noted that DOPA and K residues are statistically distributed.

MATERIALS AND METHODS

Tribopairs

Poly(dimethylsiloxane) (PDMS) elastomer was used for both pin and disk, as previously described.[14] Briefly, the base and curing agent were thoroughly mixed in a 10:1 ratio (w/w) and ensuing bubbles generated from mixing were removed by vacuum. Tribopairs consisting of hemispherical pins (6 mm diameter) and flat disks (30 mm diameter, 5 mm thickness) were fabricated in polystyrene 96-well cell-culture plates and a custom-machined aluminum mold, respectively.[14]

Synthesis of PEG-DOPA-K and surface modification

N-carboxyanhydrides (NCAs) of DOPA (diacetyl-DOPA-NCA) and lysine (Fmoc-K-NCA) were prepared.[26] Briefly, methoxy-PEG-NH_2 (PEG-NH_2, MW 5000 Da) was dried by azeotropic evaporation with benzene and further dried in a desiccator for 3 h. Ring-opening polymerization of NCA was performed by dissolving PEG-NH_2 in anhydrous THF at 100 mg/mL, purging with argon, and adding a 6*M* excess of undiluted diacetyl-DOPA-NCA and Fmoc-K-NCA. The reaction mixture was stirred at room temperature for five days under exclusion of water vapor. The peptide-modified block copolymers were purified in succession with diethyl ether, cold methanol, and again diethyl ether. Peptide-coupled PEG was dissolved in anhydrous DMF at a concentration of 50 mg/mL and sparged with argon for 10 min. Pyridine was added to make a 5% solution and stirred for 15 min with argon bubbling. The mixture was rotary-evaporated to remove excess pyridine and precipitated in diethyl ether. The crude polymer was further purified by dialysis (MWCO > 3400 Da) for 4 h and lyophilized to yield PEG-DOPA-K (Fig. 1).

Lysine content was verified by ^1H NMR. The DOPA content of the block copolymers was determined using UV absorbance of polymer solutions in 12.1 m*M* HCl at the maximum absorbance wavelength of the catechol (λ_{max} = 280 nm).[27] Solutions containing known concentrations of free DOPA amino acid were used to construct the calibration curve.

Tribopairs were incubated in 1 mg/mL PEG-DOPA-K in 0.6*M* K_2SO_4, 0.1*M* N-morpholinopropanesulfonic acid (MOPS), pH 9.0 for 18 h at 50°C. Following modification, tribopairs were rinsed with ultrapure water and blown dry with nitrogen.

For some comparison tests, samples were incubated overnight in other PEG-containing copolymers, 1 mg/mL PEG-DOPA(MW of PEG is 5000 Da)[20] or 1 mg/mL Pluronic™ P105 (molecular formula EO_{37}-PO_{56}-EO_{37}, that is, MW of PEG chains are 3250 Da, according to the manufacturer)[13] under identical conditions to those described previously.

Contact angle

Static water contact angle (Ramé-Hart, Netcong, NJ) was measured before and after modification with PEG-DOPA-K.

Atomic force microscopy

Topographic images of PDMS surfaces were obtained in air by tapping-mode atomic force microscopy (AFM) (Asylum MFP-3D, Santa Barbara, CA) with silicon cantilevers (f = 280 kHz). The film thickness of PEG-DOPA-K on PDMS was measured from the height difference between modified and unmodified regions on the same sample.

Gel Permeation Chromatography

Multimer formation in solution was detected by overnight incubation of PEG-DOPA-K and PEG-DOPA in 0.6M K_2SO_4, 0.1M MOPS, pH 9.0 at 50°C. Gel permeation chromatography (GPC) analysis was performed on this solution using multi-angle laser light scattering (Wyatt Technology, Santa Barbara, CA) in a mobile phase consisting of 0.1*M* NaCl, 50 m*M* PO_4^-, and 0.05 % NaN_3.

Optical microscopy

Modified and unmodified PDMS surfaces were imaged *en face* by optical microscopy using a Zeiss Axiovert 135 microscope equipped with a CCD camera (ORCA-ER, Hamamatsu, Japan).

X-ray photoelectron spectroscopy

Survey and high-resolution X-ray photoelectron spectroscopy (XPS) spectra were collected on an Omicron ESCALAB (Omicron, Taunusstein, Germany) configured with a monochromated Al Kα (1486.8 eV) 300W X-ray source, 1.5 mm circular spot size, a flood gun to counter charging effects, and operating under ultrahigh vacuum ($<10^{-8}$ Torr). The takeoff angle, defined as the angle between the substrate normal and the detector, was fixed at 45°. Substrates were mounted on sample studs by means of double-sided adhesive tape. All binding energies were calibrated using the C(1s) carbon peak (284.6 eV). Analysis consisted of a broad survey scan (50.0 eV pass energy) and a 10-min high-resolution scan (22.0 eV pass energy) at 90–110 eV for Si(2p), 275–295 eV for C(1s), 390–410 eV for N(1s), and 525–545 eV O(1s). High-resolution spectra were acquired and used to calculate atomic composition.

Polarization-modulation infrared reflection-absorption spectroscopy

A thin film of PDMS (ca. 30 nm thickness) was spin coated onto gold substrates.[28] The thin film was then modified with 1 mg/ml PEG-DOPA-K in buffer. After 1h or 18h of modification at 50°C, samples were rinsed with ultrapure water and dried with nitrogen. High-resolution polarization-modulation infrared reflection-absorption spectroscopy (PM-IRRAS)on a Bruker IFS 66v IR spectrometer, equipped with a PMA37 polarization-modulation accessory (Bruker Optics, Germany), was employed to determine the presence of PEG-DOPA-K films on the PDMS surface. The interferogram from the spectrometer's external beam port was passed through a KRS-5 wire-grid polarizer and a ZnSe photoelastic modulator before reflecting off the sample surface at an angle of 80° and being detected with a liquid-nitrogen-cooled MCT detector. Typically, 1024 scans of multiplexed interferograms were collected with 8 cm^{-1} resolution and processed with OPUS software (Bruker Optics, Germany).

Pin-on-disk tribometry

Lubricating properties of PEG-DOPA-K on PDMS were characterized by testing apposed pairs in a pin-on-disk geometry (CSM, Neuchâtel, Switzerland) in HEPES buffer.[13,14] Briefly, the load was controlled by dead weight and the sliding speed by a motor underneath the disk. Frictional forces generated during sliding contact were monitored by a strain gauge and measured as a function of speed (0.00025–0.1 m/s) at fixed load (1 N, unless otherwise mentioned) or rotations (up to 200) at fixed speed (0.005 m/s) and load (1 N, mean Hertzian contact pressure = 0.36 MPa).[14] For these measurements, the average friction over a defined number of rotations (20) was obtained at each speed. Generally, the friction forces in the initial few rotations showed a characteristic change ("running in" behavior) yet exhibited a steady kinetic friction force, F_k, after no more than five rotations. For μ-versus-speed plots, the latter half of the total number of rotations (11th–20th) was averaged in order to eliminate the "running-in" effect. Long-term friction measurements for 200 rotations or more were also conducted at fixed speed (0.005 m/s) and load (1 N).

RESULTS

Film thickness, wettability, and gross morphology

Surface modification of PDMS with PEG-DOPA-K was confirmed on both macro and micro scales.

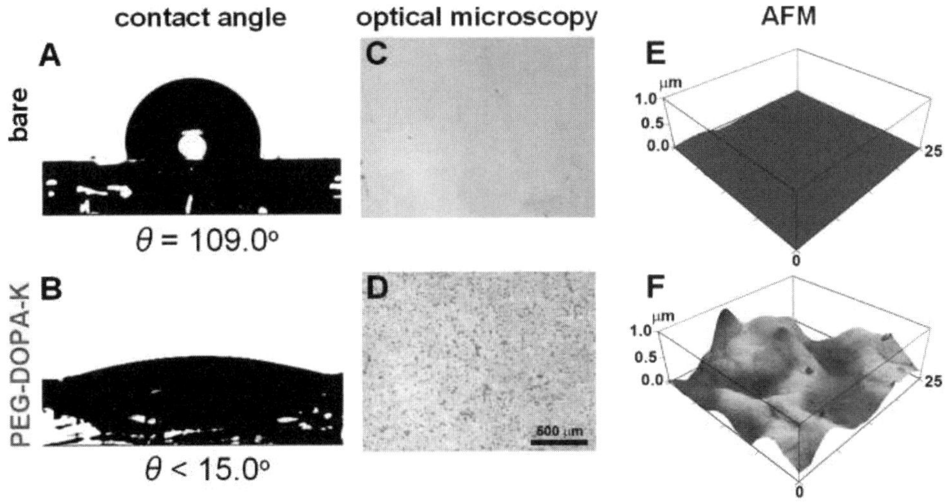

Figure 2. Wettability and gross morphology of (A, C, E) bare and (B, D, F) PEG-DOPA-K modified PDMS surfaces by (A),(B) contact angle, (C),(D) optical microscopy, and (E),(F) atomic force microscopy (AFM). [Color figure can be viewed in the online issue, which is available at www.interscience.wiley.com.]

Wettability (water contact angle) measurements showed increased hydrophilicity of PDMS after PEG-DOPA-K modification [<15°, Fig. 2(B)] compared to before [109°, Fig. 2(A)]. Optical micrographs [Fig. 2(c,d)] indicated a fairly uniform layer of PEG-DOPA-K [Fig. 2(f)] whereas AFM micrographs showed some heterogeneity, with a layer thickness on the order of 0.5–1 μm compared to unmodified samples [Fig. 2(e)].

XPS

Differences in chemical composition between bare and PEG-DOPA-K modified PDMS surfaces were evident by XPS analysis. XPS spectra indicated the presence of increased N(1s) (+2.1%), increased C(1s) (+4.8%), and decreased Si(2p) (−4.1%) after adsorption of PEG-DOPA-K onto the PDMS surface (Fig. 3 and Table I). Spectra of modified samples showed the presence of a N(1s) peak at 399.7 eV, which was not observed in the spectra of unmodified (control) PDMS surfaces and was attributed to the peptide in adsorbed PEG-DOPA-K. Further, a diminished Si(2p) signal, representative of the silicon present in PDMS, was also noted after modification.

PM-IRRAS

Reflective PM-IRRAS was applied to qualitatively determine the presence and chemical composition of the PEG-DOPA-K layer on thin-layer PDMS. PEG-DOPA-K modification on thin-film PDMS (on a Au substrate) was initially performed for 18 h; however, PM-IRRAS spectra indicated saturated reflection [Fig. 4(A)], that is, the polymer layer was too thick to allow measurements by PM-IRRAS. Bands associated with Si—O (1100 cm^{-1}) and Si—CH_3 (1265 cm^{-1}) chemical species from PDMS were not observable when high amounts of PEG-DOPA-K were deposited [Fig. 4(A)]. When the modification time was reduced to 1 h, a thinner film of PEG-DOPA-K accumulated on the PDMS thin film surface. PM-IRRAS analysis revealed distinct amide bands at 1650 and

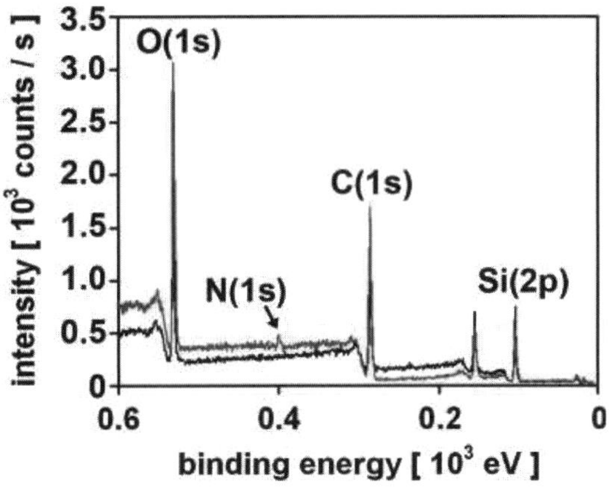

Figure 3. XPS chemical composition analysis of bare (black) and PEG-DOPA-K (red) modified PDMS samples. [Color figure can be viewed in the online issue, which is available at www.interscience.wiley.com.]

TABLE I
Quantitative XPS Analysis of Substrata

Substratum	Atomic Composition (%)			
	Si	C	N	O
Unmodifed PDMS	28.8	44.8	0.0	26.4
PEG-DOPA-K modified PDMS	24.7	49.6	2.1	23.6

1540 cm^{-1} in the polymer-associated spectrum. [Fig. 4(B)].

Pin-on disk tribometry

PEG-DOPA-K coating resulted in extremely low friction at self-mated sliding contacts between coated PDMS surfaces when tested in a pin-on-disk geometry. PEG-DOPA-K-modified PDMS tested over the course of 200 rotations (~8 m in total distance) [Fig. 5(A)], at fixed speed (0.005 m/s) and load (1 N), maintained an extremely low average friction coefficient (0.03 ± 0.00) compared to that of bare PDMS (0.98 ± 0.02) [Fig. 5(A)]. The lubricating effect by PEG-DOPA-K surface modification represents a 33-fold reduction in μ values compared to bare PDMS. While the remarkable lubricating effect of PEG-DOPA-K ($\mu \leq 0.03$) was reproducibly observed in sliding contacts up to 200 rotations, extended measurements up to 1000 rotations revealed a gradual increase in μ, commencing between the 300th and 500th rotation, with μ reaching 0.1 to 0.3 by the end of measurements (data not shown). During shorter experiments carried out as a function of speed (20 rotations at 0.00025–0.1 m/s, 1 N), PEG-DOPA-K was found to reduce μ by 42-fold compared to bare PDMS [Fig. 5(B)] with a slight increasing trend with increasing sliding speed. Similarly effective lubricating properties were observed from the measurements carried out under 2 N and 5 N (data not shown). Meanwhile, the other PEG-containing polymers, such as PEG-DOPA and Pluronic™ P105, did not reveal any noticeable lubricating effect, and resulted in higher friction forces compared to bare PDMS.

GPC

Formation of PEG-DOPA-K multimers in solution was monitored by performing GPC of PEG-DOPA-K solutions incubated under conditions identical to the surface modification reactions (0.6*M* K_2SO_4, 0.1*M* MOPS, pH 9.0 at 50°C). The presence of multimers at elution times about 60–62 min was clearly evident in the PEG-DOPA-K sample incubated at pH 9.0 (Fig. 6). Multimer formation was notably lower for PEG-DOPA and PEG-DOPA-K at pH 6.0, suggesting that both alkaline pH and the presence of lysine residues are important in forming multimers in solution under these conditions.

DISCUSSION

As was addressed in the Introduction, the aim of this work was to develop an approach to modify polymeric surfaces with PEG chains, with particular interest in improving the lubricating properties of tissue-contacting medical devices. Surface grafting of PEG chains for this particular purpose is not a trivial task since many other established methods require surface preactivation[6,29] or presence of excess

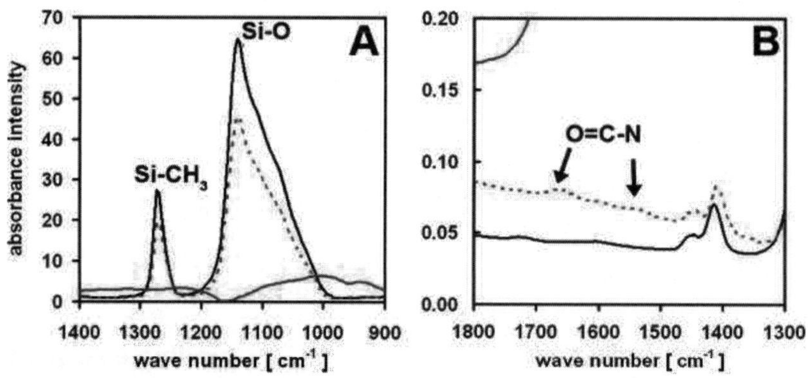

Figure 4. PM-IRRAS chemical composition analysis of lower (-) and higher (−) amounts of PEG-DOPA-K deposited on PDMS-Au substrate (black). Bands associated with (A) Si—O (1110 cm^{-1}) and Si—CH_3 (1265 cm^{-1}) chemical species from PDMS were not observable when (B) higher amounts of PEG-DOPA-K were deposited and peaks associated with amide bonds were displayed. [Color figure can be viewed in the online issue, which is available at www.interscience.wiley.com.]

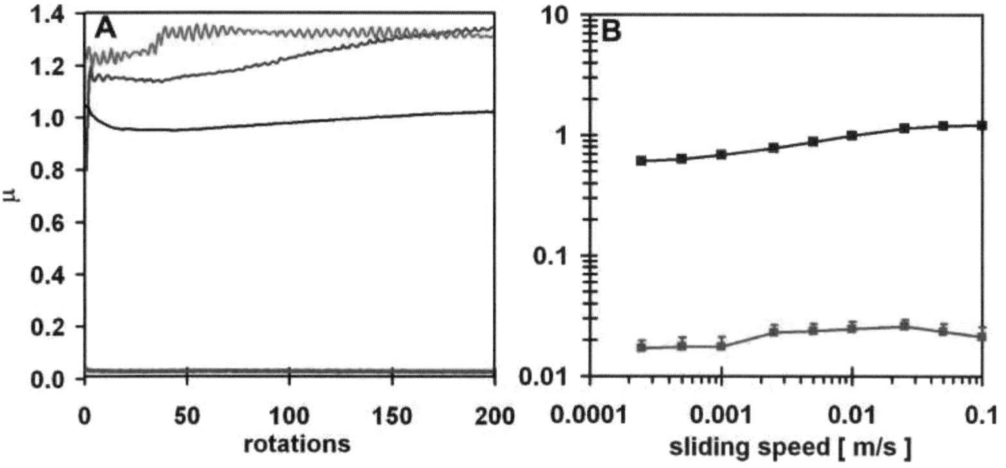

Figure 5. Effects of surface modification on friction coefficient (μ) as a function of (A) rotations and (B) sliding speed. For sliding speed measurements, the average friction over a defined number of rotations (20) was obtained at each speed. PDMS surface were left bare (black), or were modified with PEG-DOPA-K (red), PEG-DOPA (green), or Pluronic^{TM} P105 (blue). [Color figure can be viewed in the online issue, which is available at www.interscience.wiley.com.]

polymer,[19] which may not be possible for *in vivo* situations.

These results indicate substantial improvement in lubrication properties of PDMS after surface modification with PEG-DOPA-K. Modified surfaces were qualitatively characterized by contact angle, optical microscopy, and AFM [Fig. 2(A–E)]. Chemical composition of the modification layer was further verified by XPS (Fig. 3) and PM-IRRAS (Fig. 4) analyses. In comparison to other PEG-containing copolymers, such as PEG-DOPA and Pluronic^{TM} P105, PEG-DOPA-K modification resulted in a 42-fold reduction in the friction coefficient ($\mu = 0.03$) over long-term tests [Fig. 5(A)]. Additionally, an extremely low friction coefficient was maintained as a function of sliding speed, over the whole speed range employed in this work [Fig. 5(B)]. Taken together, these results demonstrate the possibility of biomedical applications of PEG-DOPA-K surface modification for medical devices, such as catheters. In contrast, the other PEG-containing copolymers, such as PEG-DOPA and Pluronic^{TM} P105, revealed no noticeable lubricating effect, and in contrast, slightly higher friction forces compared to bare PDMS surfaces. In addition, in preliminary studies, PEG-DOPA-K surfaces have indicated anti-fouling properties as well as decreased protein and cell attachment (unpublished data) similar to those previously reported for PEG-DOPA.[20,23]

Very high frictional forces between sliding contacts of two PDMS surfaces in an aqueous environment [Fig. 5(A)] are ascribed to the strong hydrophobic adhesive forces,[14] and surface-grafted PEG chains on PDMS surface are known to provide surface hydrophilicity necessary for aqueous lubrication.[13,14] In addition, the load-carrying capacity may be improved by the repulsion between two opposing surfaces bearing PEG chains in good solvents (water), arising from the osmotic pressure developed within the solvent-laden, brush-like polymer chains.[14,25] The Pluronic^{TM} block copolymer

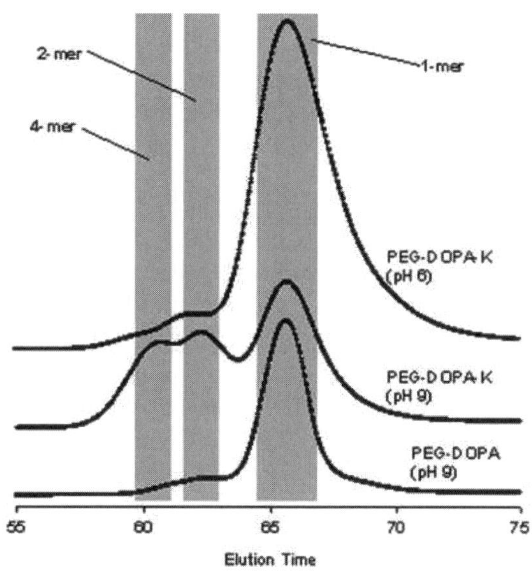

Figure 6. Gel permeation chromatogram of DOPA-functionalized PEGs after overnight incubation in 0.6*M* K_2SO_4, 0.1*M* MOPS, pH 9.0 at 50°C.

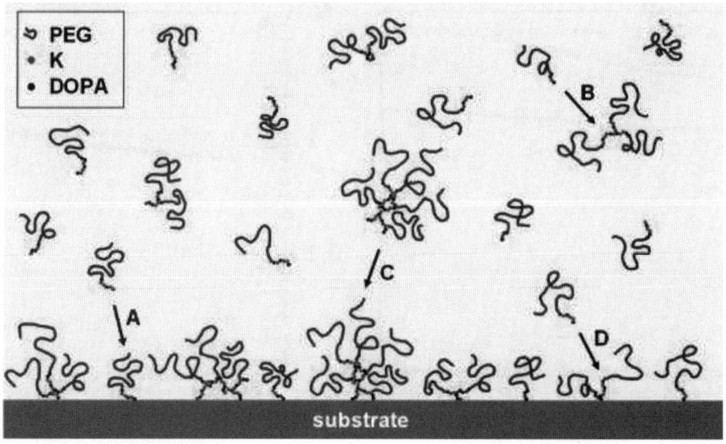

Figure 7. Potential mechanism for PEG-DOPA-K modification of PDMS (not drawn to scale). The formation of a coating by immersion of substrate in a solution of PEG-DOPA-K can occur through several possible pathways as illustrated by arrows in the figure. Individual PEG-DOPA-K molecules can directly adsorb (graft-to) onto the substrate surface (A) or polymerize first with other molecules in solution (B) followed by adsorption of polymer clusters onto the substrate (C). Alternatively, individual PEG-DOPA-K molecules may become immobilized through polymerization with surface bound molecules (D) in a process that resembles graft from approaches. [Color figure can be viewed in the online issue, which is available at www.interscience.wiley.com.]

employed in this work is expected to be immobilized through the hydrophobic interaction between the PPO block and PDMS surfaces. However, the exact nature of the interaction between PEG-DOPA, PEG-DOPA-K, and PDMS is unknown at this time. Since all tribological measurements in this work have been performed under aqueous buffer solution with no excess polymers, the lubricating performance directly indicates the stability of each PEG-based polymer layer.[19] Experimental results suggest that interactions between PEG-DOPA and Pluronic™ P105 and PDMS surface are too weak to withstand the tribological stress under the given conditions, while previous experiments revealed effective aqueous lubrication properties in the presence of excess polymers in the bulk solution.[13]

The molecular structure of the PEG-DOPA-K coating is likely to be comprised of both adsorbed single chains of PEG-DOPA-K and oligomers that have polymerized through their peptide endgroups (Fig. 7). Under the alkaline conditions employed during surface modification, catechol side chains of DOPA residues readily oxidize to yield quinones, which are capable of further reacting with other DOPA residues[30,31] and with primary amines[24,32] to form oligomers of PEG-DOPA-K, both in solution (Fig. 6) and on the surface (Fig. 7). Although further studies will be needed to fully understand the impact of PEG-DOPA-K polymerization on tribological properties, we can speculate that polymerization of the peptide anchors could enhance the stability of the polymer coating towards shear between the sliding surfaces. An interesting observation was that PEG-DOPA-K coatings performed much better than PEG-DOPA coatings (Fig. 5), suggesting a role for the lysine residues in lowering the friction coefficient. This effect could be manifested during formation of the coating, anchoring of the coating to PDMS, or possibly in altering tribology-relevant chemical characteristics (charge, hydrophilicity, etc.) of the coating.

Lubricious biocompatible aqueous tribosystems are desirable for tissue-contacting medical devices, such as catheters, endoscopes, and angioplasty balloons. The modification of PDMS surfaces described here is a simple, thermally activated dip-coating procedure, which results in highly effective lubrication properties. Although the lubricating effects of the coating were eventually reduced after many rotations (~1000, equivalent to 25 m in length), in its current form the coating may be suitable for single-use medical devices. In the future, it would be useful to further elucidate the lubricating properties of PEG-DOPA-K on other relevant biomaterial surfaces and to study the biological response of modified surfaces. Incorporation of other PEG-based copolymers and/or varying the DOPA:K ratio in future studies may provide further insight into the mechanistic basis for the reduction in friction.

The authors acknowledge Dr. Venkataraman Nagaiyanallur for assistance with PM-IRRAS and Dr. Haeshin Lee and Dr. Andrea Statz for their technical assistance.

References

1. Harris JM, editor. Poly(ethylene glycol) chemistry, biotechnical, and biomedical applications. New York: Plenum Press; 1992.
2. Alcantar N, Aydil ES, Israelachvili JN. Polyethylene glycol-coated biocompatible surfaces. J Biomed Mater Res 2000;51: 343–351.
3. Holmberg K, Bergstrom K, Stark MB. Immobilization of proteins via PEG chains. In: Harris JM, editor. Poly(ethylene glycol) Chemistry, Biotechnical, and Biomedical Applications. New York: Plenum Press; 1992. p 303–324.
4. Lee JH, Lee HB, Andrade JD. Blood compatibility of polyethylene oxide surfaces. Prog Polym Sci 1995;20:1043–1079.
5. Ikada Y. Lubricating Polymer Surfaces: Technomic Publishing; Lancaster, Basal, 1993.
6. Uyama Y, Kato K, Ikada Y. Surface modification of polymers by grafting. Graft Character Tech Kinet Model 1998;137:1–39.
7. Uyama Y, Tadokoro H, Ikada Y. Low-frictional catheter materials by photoinduced graft-polymerization. Biomaterials 1991; 12:71–75.
8. Ikeuchi K, Takii T, Norikane H, Tomita N, Ohsumi T, Uyama Y, Ikada Y. Water lubrication of polyurethane grafted with dimethylacrylamide for medical use. Wear 1993;161(1/2):179–185.
9. Lee S, Spencer ND. Poly(L-lysine)-*graft*-Poly(ethylene glycol) (PLL-*g*-PEG): A versatile aqueous lubricant additive for tribosystems involving thermoplastics. Lubricat Sci 2008;20:21–34.
10. Ikeuchi K, Kouchiyama M, Tomita N, Uyama Y, Ikada Y. Friction control with a graft layer of a thermo-sensing polymer. Wear 1996;199:197–201.
11. Gupta B, Anjum N. Plasma and radiation-induced graft modification of polymers for biomedical applications. Advances in Polymer Science. Berlin: Springer; 2003. pp 35–61.
12. Kato K, Uchida E, Kang E-T, Uyama Y, Ikada Y. Polymer surface with graft chains. Prog Polym Sci 2003;2003:209–259.
13. Lee S, Iten R, Muller M, Spencer ND. Influence of molecular architecture on the adsorption of poly(ethylene oxide)-poly (propylene oxide)-poly(ethylene oxide) (PEO-PPO-PEO) on PDMS surfaces and implications for aqueous lubrication. Macromolecules 2004;37:8349–8356.
14. Lee S, Spencer ND. Aqueous lubrication of polymers: Influence of surface modification. Tribol Int 2005;38:922–930.
15. Fan XW, Lin LJ, Messersmith PB. Cell fouling resistance of polymer brushes grafted from Ti substrates by surface-initiated polymerization: Effect of ethylene glycol side chain length. Biomacromolecules 2006;7:2443–2448.
16. Tugulu S, Arnold A, Sielaff I, Johnsson K, Klok HA. Protein-functionalized polymer brushes. Biomacromolecules 2005;6: 1602–1607.
17. Ma HW, Hyun JH, Stiller P, Chilkoti A. "Non-fouling" oligo (ethylene glycol)-functionalized polymer brushes synthesized by surface-initiated atom transfer radical polymerization. Adv Mater 2004;16:338–341.
18. Lee S, Spencer ND. Achieving ultralow friction by aqueous, brush-assisted lubrication. In: Erdemir A, Martin J-M, editors. Superlubricity. Amsterdam: Elsevier; 2007. p 365–396.
19. Lee S, Muller M, Heeb R, Zurcher S, Tosatti S, Heinrich M, Amstad F, Pechmann S, Spencer ND. Self-healing behavior of a polyelectrolyte-based lubricant additive for aqueous lubrication of oxide materials. Tribol Lett 2006;24:217–223.
20. Dalsin JL, Hu BH, Lee BP, Messersmith PB. Mussel adhesive protein mimetic polymers for the preparation of nonfouling surfaces. J Am Chem Soc 2003;125:4253–4258.
21. Statz AR, Barron AE, Messersmith PB. Protein, cell, and bacterial fouling resistance of polypeptoid-modified surfaces: Effect of side-chain chemistry. Soft Matter 2008;4:131–139.
22. Statz AR, Meagher RJ, Barron AE, Messersmith PB. New peptidomimetic polymers for antifouling surfaces. J Am Chem Soc 2005;127:7972–7973.
23. Dalsin JL, Lin L, Tosatti S, Voros J, Textor M, Messersmith PB. Protein resistance of titanium oxide surfaces modified by biologically inspired mPEG-DOPA. Langmuir 2005;21: 640–646.
24. Lee H, Dellatore SM, Miller WM, Messersmith PB. Mussel-inspired surface chemistry for multifunctional coatings. Science 2007;318:426–430.
25. Klein J. Shear, friction, and lubrication forces between polymer-bearing surfaces. Annu Rev Mater Sci 1996;26:581–612.
26. Fuller WD, Verlander MS, Goodman M. DOPA-containing polypeptides. I. Improved synthesis of high-molecular-weight poly(L-DOPA) and water-soluble copolypeptides. Biopolymers 1978;17: 2939–2943.
27. Lee BP, Chao C-Y, Nunalee FN, Motan E, Shull KR, Messersmith PB. Rapid gel formation and adhesion in photocurable and biodegradable block copolymers with high DOPA content. Macromolecules 2006;39:1740–1748.
28. Lee S, Vörös J. An aqueous-based surface modification of poly (dimethylsiloxane) (PDMS) to prevent biofouling. Langmuir 2005;21:11957–11962.
29. Uyama Y, Tadokoro H, Ikada Y. Surface lubrication of polymer-films by photoinduced graft-polymerization. J Appl Polym Sci 1990;39:489–498.
30. Burzio LA, Waite JH. Cross-linking in adhesive quinoproteins: Studies with model decapeptides. Biochemistry 2000;39: 11147–11153.
31. Lee BP, Dalsin JL, Messersmith PB. Synthesis and gelation of DOPA-modified poly(ethylene glycol) hydrogels. Biomacromolecules 2002;3:1038–1047.
32. Lee H, Scherer NF, Messersmith PB. Single-molecule mechanics of mussel adhesion. Proc Natl Acad Sci USA 2006;103: 12999–13003.

CHAPTER 3

Effects of Surface Morphology and Structure

3a. The influence of atomic-scale structure on catalytic activity

Commentary

It has been known for many decades that different crystal faces of the same element display subtly different chemical properties. This could be for a number of reasons, including differences in the electronic band structure at the surface, or due to the fact that different structures stabilize different intermediates or transition states to different extents for geometrical reasons. Clearly this is an important issue in catalysis, but maintaining the crystal faces in a clean, well-defined state before the reaction (which might be at high pressure) has been a challenge to many researchers in this field.

In this study of ammonia synthesis over iron single crystals — which I carried out in the Somorjai group in Berkeley in collaboration with Richard Schoonmaker, from Oberlin College — we were able to use the newly designed high-pressure-low-pressure apparatus to carry out catalytic reactions at high pressure (20 atm.), cleaning and analyzing the surface in UHV before and the reaction, and analyzing it afterwards without removing the sample into the air (3.1, 3.2). Moreover, we were able to measure the catalytic activity over single-crystal samples, noting the difference in activity between different crystal faces. The literature had already pointed to some particle-size dependence of ammonia synthesis, so we had a hint that we might find a crystal-face dependence, since certain surface orientations are favored on small particles, for energetic reasons. The structure sensitivity we found, however, was larger than any that had been observed before: the ammonia synthesis rate varied in the ratio 418 : 25 :1 for the faces Fe(111) : Fe(100) : Fe(110).

Nature Vol. 294 17 December 1981

Structure sensitivity in the iron single-crystal catalysed synthesis of ammonia

N. D. Spencer, R. C. Schoonmaker* & G. A. Somorjai

Materials and Molecular Research Division,
Lawrence Berkeley Laboratory, and Department of Chemistry,
University of California, Berkeley, California 94720, USA

There have been many investigations of the kinetics of the iron-catalysed synthesis of ammonia and the details of the mechanism of its reaction[1-3]. However, we report here the first direct and, therefore, unambiguous measurement of the surface structure sensitivity of the iron-catalysed ammonia synthesis reaction using the (111), (100) and (110) crystal faces of iron and operating at pressures approaching those used industrially (20 atmospheres). The Fe(111) face was found to be the most active, the activity ratio being 418:25:1 for the Fe(111), Fe(100) and Fe(110) planes. We have also observed the formation of a strongly bound nitrogen species after the reaction, in the presence of which ammonia continues to be produced.

The apparatus[4] consisted of a stainless steel ultrahigh vacuum chamber equipped with a four-grid retarding field analyser for Auger and low-energy electron diffraction measurements of the surface composition and structure, a residual gas analyser, an ion gun for specimen cleaning and a high-pressure cell connected to an external loop. Reactant and product gases at a total pressure of 20 atmospheres were continuously circulated over the heated iron crystal specimen by a positive displacement pump; samples, periodically withdrawn using a gas chromatograph type sampling valve, were passed by a nitrogen carrier through a photoionization detector to determine the amount of ammonia in the loop. Ammonia was the only species present which could be detected by the photoionization detector obviating the need for a separating column.

Circular iron specimens with crystal faces (111), (110) and (100) were spark-cut from a high-purity (4N) single-crystal rod and diamond polished, the final dimensions being ~6 mm diameter and 1 mm thickness. Sulphur and carbon were found to be the principal contaminants after this procedure, but could be removed by several days of heating at 800°C in 1 atmosphere of hydrogen, followed by many cycles of argon ion bombardment and annealing at 700 °C. To minimize the catalysis occurring on ill-defined iron surfaces, platinum foil was spot-welded to the rough edges of the specimens.

Results of measurement of structure sensitivity are shown in Table 1 from which it can be seen that the (111) surface of iron catalyses the synthesis of ammonia from hydrogen and nitrogen at least 418 times faster than the ordered (110) surface and 16 times faster than the ordered (100) surface. Our result is consistent with the low pressure chemisorption studies of Ertl and co-workers[2] where it was reported that the rate of dissociative chemisorption of nitrogen is greatest on the (111) plane of iron

* Permanent address: Department of Chemistry, Oberlin College, Oberlin, Ohio 44074, USA.

0028-0836/81/510643—02$01.00

644 *Nature Vol. 294 17 December 1981*

Table 1 Surface structure sensitivity of ammonia synthesis

Catalyst surface	Area (cm^2)	Rate* (mols NH_3 cm^{-2} s^{-1})
Fe(111)	0.63	4.6×10^{-8}
Fe(100), ordered	0.80	2.8×10^{-9}
Fe(110), ordered	0.57	0†
Fe(100), disordered‡	0.80	4.5×10^{-9}
Fe(110), disordered‡	0.57	9.7×10^{-10}

* Total pressure, 20 atm.; $P_{H_2}/P_{N_2}=3/1$; temperature = 798 K.
† Detection limit $\sim1.1\times10^{10}$ mol NH_3 cm^{-2} s^{-1}.
‡ Argon ion bombardment (1.5×10^{-5}) A cm^{-2} for 10 min at 500 V).

and least on the (110) plane, and also with previous suggestions[1-3,5] that this step is rate-determining in the ammonia synthesis reaction.

The (111) plane of iron contains both seven-coordinated (C_7) and four-coordinated (C_4) atoms, whereas the (110) and (100) planes contain only six-coordinated (C_6) and C_4 atoms, respectively. The large variation in rates of ammonia synthesis on the various planes (Table 1) and the anomalously high catalytic activity that we have also detected on the (110) and (100) planes when deliberately disordered by ion bombardment, support the proposals of Brill[6] and others[3,7] regarding the importance of clusters and C_7 iron atoms as components of the active site on the catalyst surface.

Although bulk nitrides of iron are thermodynamically unstable in the conditions of these experiments, we have observed the existence of a long-lived, strongly-bound nitrogen species after the synthesis reaction on both Fe(111) and Fe(100) crystal faces in the presence of which ammonia continued to be produced at an undiminished rate. The catalytic significance of this surprising observation has yet to be determined.

This work was supported by the Director, Office of Energy Research, Office of Basic Energy Sciences, Chemical Sciences Division of the US Department of Energy under contract W-7405-ENG-48. We also thank the Dow Chemical Co. for financial help.

Received 17 August; accepted 8 October 1981.

1. Emmett, P. H. in *The Physical Basis for Heterogeneous Catalysis* (eds Drauglis, E. & Jaffe, R.) 3 (Plenum, New York, 1975).
2. Ertl, G. *Catal. Rev.-Sci. Engng* **21**, 201 (1980).
3. Boudart, M. *Catal. Rev.-Sci. Engng* **23**, 1 (1981).
4. Blakely, D. W., Kozak, E., Sexton, B. A. & Somorjai, G. A. *J. Vac. Sci. Technol.* **13**, 1091 (1976).
5. Ozaki, A., Aika, K. & Morikawa, Y. in *Proc. 5th int. Congr. Catalysis* Vol. 2 (ed. Hightower, J.) 1251 (North Holland, Amsterdam, 1973).
6. Brill, R., Richter, E.-L. & Ruch, E. *Angew. Chem. (int. Edn)* **6**, 882 (1967).
7. Huang, K. H. in *Proc. 7th Int. Congr. on Catalysis*, Vol. A, 554 (Elsevier, New York, 1981).

0028-0836/81/510644—02$01.00

JOURNAL OF CATALYSIS 74, 129–135 (1982)

Iron Single Crystals as Ammonia Synthesis Catalysts: Effect of Surface Structure on Catalyst Activity

N. D. SPENCER, R. C. SCHOONMAKER,[1] AND G. A. SOMORJAI

Materials and Molecular Research Division, Lawrence Berkeley Laboratory, and Department of Chemistry, University of California, Berkeley, California 94720

Received September 23, 1981; revised November 25, 1981

The activity of three iron single-crystal faces with respect to their ability to catalyze ammonia synthesis was investigated. At 798 K and a total pressure of 20 atm of a stoichiometric mixture of hydrogen and nitrogen, the relative rates of ammonia formation were found to be 418 : 25 : 1 for Fe(111), Fe(100), and Fe(110) samples, respectively. The activation energy for ammonia synthesis on the Fe(111) surface was determined as 19.4 kcal mole^{-1}. Significant quantities of nitrogen were found to be present in the bulk and on the surface of the iron samples after catalytic runs. This is the first study in which ammonia synthesis activity has been directly correlated with surface crystallographic structure under well-defined, high-pressure conditions.

INTRODUCTION

Until recently, studies of the kinetics of ammonia synthesis have been carried out either under industrial conditions in reactors, operating at upwards of 1 atm (1 atm = 101.3 kN m^{-2}) total pressure (*1*, *2*), or indirectly in the clean environment of the ultrahigh-vacuum chamber (*3*) where pressures rarely exceed 10^{-4} Torr (1 Torr = 133.3 N m^{-2}). While many useful results have come from both these approaches, the former method suffers from a lack of direct information on the nature of the catalyst surface and the latter, low-pressure, technique does not allow the synthesis of ammonia to occur to any measurable extent, since the thermodynamic equilibrium is unfavorable. The present investigation, using an apparatus which allows both UHV and high-pressure conditions to be obtained within the same chamber, bridges the pressure gap and enables the rate of formation of ammonia to be directly related to the state of the catalyst surface.

Reactor studies (*1*) have, with few exceptions, shown that the rate-determining step in iron-catalyzed ammonia synthesis is the dissociative adsorption of nitrogen. Using field emission microscopy, Brill *et al.* (*4*) found that the (111) plane of iron preferentially adsorbed nitrogen and brief exposures to the gas resulted in the faceting of (110) and (100) planes to a (111) orientation. Furthermore, Brill and Kurzidim (*5*) made the suggestion, based on indirect evidence, that the Fe(111) plane of iron was the only one active in ammonia synthesis. Dumesic *et al.* (*6*), working with small iron particles in a reactor experiment using Mössbauer spectroscopy and CO chemisorption, came to the conclusion that seven-coordinated (C_7) iron atoms were active in ammonia synthesis. This is consistent with the work of Brill and Kurzidim (*5*), since the Fe(111) plane has the highest C_7 concentration. Ertl (*3*) and Bozso *et al.* (*7*, *8*), using UHV techniques, reported initial rates of nitrogen adsorption at 550 K to be in the ratio 60 : 3 : 1 for Fe(111), Fe(100), and Fe(110), respectively, but because of thermodynamic limitations they were unable to investigate actual ammonia synthesis under the low-pressure conditions ($P_{N_2} \leq 10^{-4}$ Torr) that were employed.

The present study is the first which directly and unambiguously probes the iron

[1] Permanent address: Department of Chemistry, Oberlin College, Oberlin, Ohio 44074.

catalyst surface structure sensitivity for the ammonia synthesis reaction under conditions of high temperature and pressure which approach those used in industry.

METHODS

The apparatus consisted of a bakeable stainless-steel ultrahigh-vacuum chamber of base pressure 10^{-9} Torr equipped with a four-grid retarding field analyzer for Auger electron spectroscopy, a quadrupole mass spectrometer for residual gas analysis, a sputter ion gun for specimen cleaning, and a hydraulically operated high-pressure cell (*9*) connected to an external stainless-steel loop. After closing the high-pressure cell, reactant and product gases at a total pressure of 20 atm were circulated over the heated iron crystal specimen by a positive displacement pump. Samples were periodically withdrawn using a gas chromatography-type sampling valve and swept by a nitrogen carrier gas into a photoionization detector (PID, lamp energy = 10.2 eV) in order to determine the amount of ammonia in the loop. With the photon energy employed, ammonia was the only species present which could be detected by the PID, rendering a separating column unnecessary. The detection system was calibrated by introducing known amounts of ammonia into the loop followed by pressurizing to 20 atm with a stoichiometric (1 : 3) mixture of nitrogen and hydrogen. The output signal of the PID was found to be a linear function of ammonia concentration in the regime encountered in our synthesis experiments: the partial pressure of ammonia produced in a synthesis run typically did not exceed 10 Torr, which at 723 K would be about 1.5% of the equilibrium value.

The high-pressure loop consisted primarily of $\frac{1}{4}$-in (6.4 mm) stainless-steel tubing and had a total volume of 170 cm^3. Initially, the loop was contaminated with oxygen and chlorine, which were removed by cooling an isolable condenser section with liquid nitrogen while circulating hydrogen gas at 6 atm pressure for several hours and by subsequently flushing the loop several times with 15 atm of hydrogen before each synthesis run. The usual flow rate of gas in the loop during a run was ~157 $cm^3\ sec^{-1}$ providing some laminar flow in the straight sections of loop (N_{Re} = 200), but turbulence and mixing at corners and constrictions. Reactants were mixed *in situ* by slowly bleeding nitrogen into circulating hydrogen, and product ammonia was allowed to build up in the high-pressure loop during the run. Pumping of the loop was achieved by a combination of mechanical and liquid nitrogen sorption pumps, and after most runs the crystal was maintained at reaction temperature until the loop was pumped out and the cell reopened to the UHV chamber.

Iron is a metal with a body-centered cubic lattice. The three low-index planes used in the present study are shown in Fig. 1. The Fe(110) plane is the closest-packed iron surface with only six-coordinated (C_6) atoms exposed. The Fe(100) plane is less closely packed and only has four-coordinated (C_4) atoms in its surface. The Fe(111) plane was the most open surface studied and has both C_4 and seven-coordinated (C_7) atoms exposed. Three approximately circular iron specimens were cut from a high-purity (4*N*) single-crystal rod and diamond polished, the final dimensions being approximately 6 mm diameter and 1 mm thickness.

The specimens were spot-welded to 0.25-mm platinum support wires which, in turn,

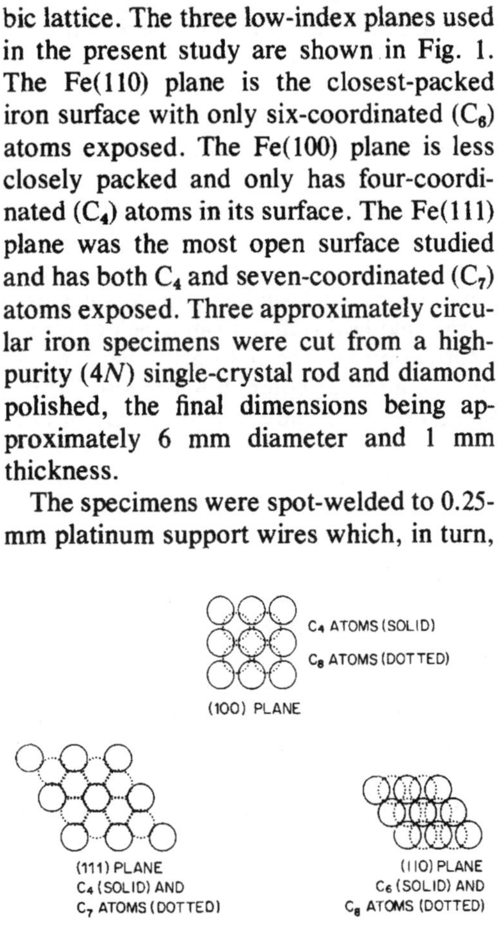

FIG. 1. Low-index planes of iron.

were spot-welded to 2-mm platinum rods on the manipulator. Platinum was chosen as a support material because of its inactivity as an ammonia synthesis catalyst. The crystal temperature was controlled to ±0.5 K and monitored by a Chromel–Alumel thermocouple, spot-welded to the edge of the specimen. Sulfur and carbon were the principal contaminants in the iron, but the surface concentrations of these could be substantially reduced by several days of heating at 1070 K in 1 atm of hydrogen, prior to insertion into the chamber, followed by many cycles of argon ion bombardment (on both sides of the specimens) and annealing at 973 K. In order to minimize catalysis occurring on ill-defined iron surfaces, platinum foil was spot-welded to the rough edges of the specimens.

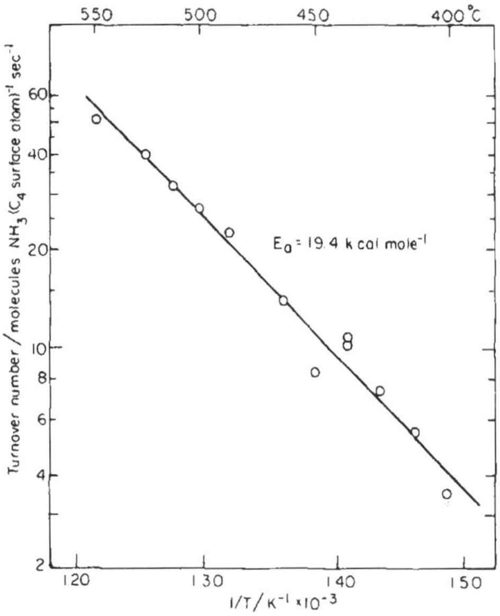

FIG. 3. Arrhenius plot for ammonia synthesis on Fe(111).

RESULTS

1. The (111) Crystal Face of Iron

The gradual buildup of ammonia in the high-pressure loop during a typical synthesis run at 736 K is shown in Fig. 2. The shape of the initial portion of the curve is the result of slow diffusion and adsorption of ammonia on the walls of the loop. Although the initial increase in ammonia concentration nearly corresponds to the total rate of synthesis, there follows a period during which ammonia starts to adsorb on the walls of the loop and the detected rate of formation is equal to the actual rate only after the walls have become saturated. In order to test this hypothesis, enough ammonia (~1 Torr) was added to the initial reactant mixture to saturate the walls, with the consequence that the detected ammonia increased from the outset at a constant rate. A turnover number for the catalyst can be derived from the part of the curve where ammonia concentration increases linearly. In the case of Fe(111) at 736 K (Fig. 2), the turnover number is 13.75 molecules NH_3 per C_4 surface atom per second. The turnover numbers (based on the number of C_4 atoms) for the Fe(111)-catalyzed reaction were determined at several temperatures within a 150 K range and plotted vs reciprocal temperature in Arrhenius form (Fig. 3). The activation energy derived from the straight line so obtained was 19.4 kcal mole^{-1}. The gross kinetic effects of varying

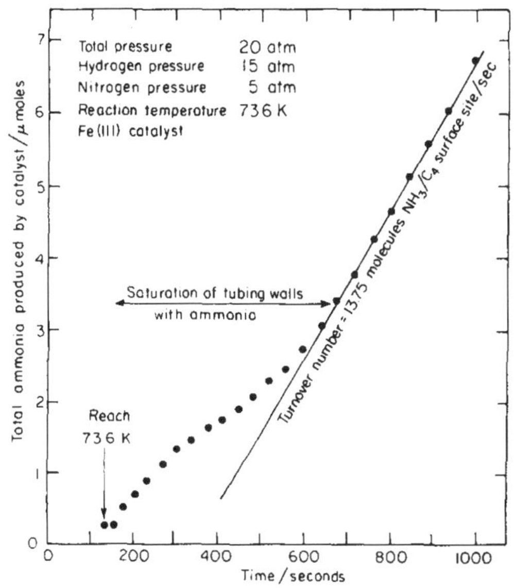

FIG. 2. Increase in ammonia concentration in loop during run.

TABLE 1

Effect of Changing Reactant Partial Pressures

Run temperature (K)	P_{N_2} (atm)	P_{H_2} (atm)	P_{Ar} (atm)	P_{total} (atm)	Turnover number (molecules NH_3 (C_4 atom)$^{-1}$ sec^{-1})
748	15	5	0	20	11.1 ± 2.0
748	5	15	0	20	12.7 ± 2.0
748	5	5	10	20	3.2 ± 1.0
748	5	5	0	10	5.0 ± 1.0

the partial pressure of the reactants at a crystal temperature of 748 K were also briefly investigated and the results are summarized in Table 1, where it can be seen that the turnover number is sensitive to both hydrogen and nitrogen partial pressures under these conditions, although extensive data which would be necessary to derive an exact kinetic expression were not obtained.

As is evident from the Auger spectra (Fig. 4) taken before and after the run shown in Fig. 2, the catalyst surface was clean prior to the synthesis reaction apart from a small amount of oxygen. This was removed during the run, presumably by the hydrogen as water. After the reaction, apart from a small amount of sulfur which had diffused out of the bulk iron (a process which appears to be enhanced by the presence of hydrogen) the only noniron surface species detected was nitrogen. This was always present after catalytic runs, even if they were terminated at a point where the first small ammonia production was observed. This nitrogen could not be removed by heating alone, which suggests that it is present not simply as an adsorbed species (7). Moreover, argon ion bombardment at room temperature, while cleaning the surface (as monitored by Auger spectroscopy), did not remove *all* nitrogen, which continued to be evolved at elevated temperatures. The nitrogen could, however, be removed by circulating 15 atm of hydrogen over the sample while it was being maintained at a temperature similar to that used in the synthesis run. This procedure led to the formation of some ammonia which, although too low in concentration to be detected by the PID, was found in the background gas mass spectrum at a low level after the loop was evacuated and the high-pressure cell reopened to UHV. When a synthesis reaction was carried out immediately on the N-contaminated surface from a previous run, ammonia was produced at a rate identical to that obtained on the clean surface. When an

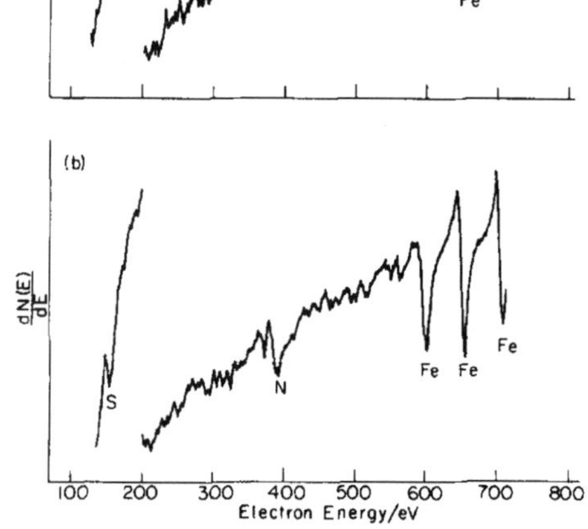

FIG. 4. Auger spectra: (a) before synthesis run; (b) after synthesis run.

initially clean crystal was heated at 673 K in 5×10^{-4} Torr of ammonia for 10 h to produce several layers of Fe_4N (*10*), the subsequent Auger spectrum of the sample strongly resembled Fig. 4b, which suggests that a similar amount of nitrogen was deposited during a normal catalytic run. A synthesis run carried out on the ammonia-treated sample showed catalytic activity which was comparable to that of the clean crystal.

If the sample was heated *in vacuo* for several hours at 973 K, large quantities of sulfur were found to segregate to the surface, presumably forming FeS (*11*). When treated in this way, the sample was totally inactive as an ammonia synthesis catalyst, as might be expected from industrial observations that sulfur is a serious poison of the real catalyst.

An attempt was made to study the catalytic activity of a heavily oxidized iron surface, but the oxide was found to be so short-lived under the reducing conditions of the reactant mixture that its activity could not be distinguished from that of clean iron. Addition of 5 Torr of water to the reaction mixture, however, completely suppressed catalytic activity, possibly suggesting that if it is continually replenished, the oxidized surface is inactive as a catalyst.

There was no effect on the turnover number (Table 2) when the Fe(111) sample was argon ion bombarded on both sides for varying lengths of time before some of the catalytic runs. In order to test for the possibility that the stainless-steel high-pressure cell was catalyzing the reaction, the Fe(111) crystal was replaced by a platinum sample of similar dimensions in order to reproduce, as closely as possible, the heat transfer between the crystal and the cell walls. When subjected to the usual synthesis reaction conditions at 773 K for many minutes, the platinum sample and cell walls produced no ammonia either as detected in the PID or as background gas in the mass spectrum after the cell was reopened to the UHV chamber.

2. *The (110) (Close-Packed) Crystal Face of Iron*

After thorough cleaning and annealing of the Fe(110) specimen, it did not produce enough ammonia to be detected by the PID under the catalytic conditions used in the studies of ammonia synthesis on the Fe(111) surface. It did catalyze the synthesis reaction to a small extent, however, since ammonia could be seen in the background mass spectrum after reopening the cell to the UHV chamber. If the detection limit of the PID is taken to be the upper limit on ammonia production for the Fe(110) sample, the catalytic activity of the (111) specimen is at least 418 times larger than that of the (110) crystal. Brief argon ion bombardment (1.5×10^{-5} A cm^{-2} for 10 min at 500 V) without subsequent annealing of the Fe(110) surface increased its catalytic activity by nearly an order of magnitude although it was still far below that of the Fe(111) surface. Exposure to catalytic conditions produced a significant nitrogen signal in the Auger spectrum taken after the cell was reopened to UHV.

3. *The (100) Crystal Face of Iron*

The rate of ammonia synthesis on this crystal face lies between those of the (111) and (110) faces. It was found to have an activity of approximately $\frac{1}{18}$th that of the Fe(111) face, and its activity increased by nearly a factor of 2 after deliberate disordering of the surface by argon ion bombardment (Table 2). Using the PID limit of detection for activity on the (110) face, we obtain activity ratios of ~418 : 25 : 1 for Fe(111) : Fe(100) : Fe(110). As with the other two surfaces, nitrogen was present in the Auger spectrum after all synthesis runs on this plane.

DISCUSSION

The results of the present study show clearly that Fe(111) is by far the most active and Fe(110) the least active plane in ammonia synthesis catalysis, Fe(100) having in-

TABLE 2

Ammonia Synthesis Surface Structure Sensitivity

Catalyst surface	Area (cm^2)	Rate[a] (moles NH_3 cm^{-2} sec^{-1})
Fe(111)	0.63	4.6×10^{-8}
Fe(100), ordered	0.80	2.8×10^{-9}
Fe(110), ordered	0.57	1.1×10^{-10} [b]
Fe(100), disordered[c]	0.80	4.5×10^{-9}
Fe(110), disordered[c]	0.57	9.7×10^{-10}

[a] Total pressure, 20 atm (2.03 MNm^{-2}), $P_{H_2}/P_{N_2} = 3/1$, temp. = 798 K.

[b] Detection limit of photoionization detector.

[c] Argon ion bombardment (1.5×10^{-5} A cm^{-2} for 600 sec at 500 V).

termediate activity and all three surfaces being partially covered with nitrogen after the reaction. The activation energy for ammonia synthesis has also been determined for the Fe(111) face as being 19.4 kcal $mole^{-1}$. The presence of surface nitrogen and oxygen did not appear to affect the performance of the single crystals as catalysts, the oxygen being rapidly removed as soon as the samples were heated in the reaction mixture. Both surface sulfur and small quantities of water in the reactant mixture were found to suppress all catalytic activity in the samples. Deliberate surface disordering by argon ion bombardment of the samples, while not noticeably affecting the catalytic activity of the Fe(111) surface, was found to increase those of the Fe(100) and Fe(110) specimens, the activity of the latter being enhanced by nearly an order of magnitude over the ordered surface value.

The actual activity ratio of Fe(111): Fe(100): Fe(110) at 798 K was found to be 418:25:1 for catalysis of ammonia synthesis, ammonia being formed at a rate of 4.6×10^{-8} moles cm^{-2} sec^{-1} on the Fe(111) face at this temperature. This is in broad agreement with the ratio obtained (*3*, *7*, *8*) for nitrogen dissociative chemisorption (60:3:1), the discrepancy probably being due to slight differences in crystal orientation (which would have a great effect in the case of the least active (110) crystal), temperature differences (the chemisorption results were obtained at 550 K, whereas the present work was carried out at 798 K), and differences in nitrogen coverage (the chemisorption results refer to rates at zero coverage). The activation energy for the synthesis reaction is consistent with that obtained by Scholten and Zwietering (*12*) for nitrogen chemisorption on the real catalyst (10–30 kcal $mole^{-1}$, depending on the coverage). This, together with the similarity of the relative nitrogen adsorption and ammonia synthesis activities for the different crystal planes adds further support to the now generally accepted conclusion that nitrogen dissociative chemisorption is the rate-determining step in iron-catalyzed ammonia synthesis.

The state of the surface during the reaction and the nature of the catalytically active site are still not certain. Although a considerable amount of surface nitrogen appeared in the Auger spectrum *after* all synthesis runs in the present study, we may not infer that nitrogen is necessarily present in significant quantities on the catalyst surface while ammonia is actually being synthesized. The fact that the nitrogen can be removed by hydrogen treatment implies that the surface nitrogen seen in the Auger spectrum is similar in nature to that which reacts to form ammonia during the synthesis run, and it should not be regarded as inactive. Nitrogen is known to dissolve in the bulk of iron (*7*), and during the reaction a considerable amount of atomic nitrogen presumably diffuses away from the catalyst surface into the bulk before it can react with surface hydrogen. It is possible that the postrun surface nitrogen (which may be a thin layer of Fe_4N as described by Ertl *et al.* (*10*) diffuses back from the bulk and segregates at the surface while the crystal is still at reaction temperature and the reactants are being pumped away. As has been suggested by Bozso *et al.* (*7*), the outward-diffusing nitrogen may cause a reconstruction at the surface to form a surface nitride. Such a surface may also be present during synthesis, but at present there is insufficient evidence to substantiate its existence.

It would appear (Table 2) that the close-packed Fe(110) plane contains few, if any, active sites, but that active sites are created by argon ion bombardment, where the term "active sites" is taken to mean sites upon which the catalytic reaction occurs *or* regions of the surface which may be converted to such sites (e.g., by the formation of a surface compound). Furthermore, it is clear that the total number of active sites on the Fe(111) plane is insensitive to argon ion bombardment. These observations are consistent with the hypothesis of Dumesic *et al.* (*6*) that C_7 iron atoms are components of the active site, since the inclusion of surface vacancies by argon ion bombardment can be envisioned as causing a substantial increase in the number of C_7 atoms on the close-packed Fe(110) surface (which initially has none) while not dramatically affecting the number on the open Fe(111) surface. However, owing to our inability to probe directly the nature of the catalyst surface *during* reaction, we cannot verify the details of the active site on the basis of the present data alone.

ACKNOWLEDGMENTS

We gratefully acknowledge Professors P. H. Emmett, G. Ertl, and M. Boudart, and Dr. M. Grunze for useful discussions during the preparation of this paper. This work was supported by the Director, Office of Energy Research, Office of Basic Energy Sciences, Chemical Sciences Division of the U.S. Department of Energy under Contract W-7405-ENG-48. We are also indebted to the Dow Chemical Company for financial support.

REFERENCES

1. Emmett, P. H., *in* "The Physical Basis for Heterogeneous Catalysis" (E. Drauglis and R. I. Jaffe, Eds.), p. 3. Plenum, New York, 1975.
2. Nielsen, A., *Catal. Rev. Sci. Eng.* **4,** 1 (1970).
3. Ertl, G., *Catal. Rev. Sci. Eng.* **21,** 201 (1980).
4. Brill, R., Richter, E.-L., and Ruch, R., *Angew. Chem. Int. Ed. Engl.* **6,** 882 (1967).
5. Brill, R., and Kurzidim, J., *in* "Colloque int. sur la structure et les propriétés des surfaces des solides," p. 99. Paris, 1969. CNRS, Paris, 1970.
6. Dumesic, J. A., Topsøe, H., and Boudart, M., *J. Catal.* **37,** 513 (1975).
7. Bozso, F., Ertl, G., Grunze, M., and Weiss, M., *J. Catal.* **49,** 18 (1977).
8. Bozso, F., Ertl, G., and Weiss, M., *J. Catal.* **50,** 519 (1977).
9. Blakely, D. W., Kozak, E., Sexton, B. A., and Somorjai, G. A., *J. Vac. Sci. Technol.* **13,** 1091 (1976).
10. Ertl, G., Huber, M., and Thiele, N., *Z. Naturforsch. A* **34,** 30 (1979).
11. Margot, E., Oudar, J., and Benard, J., *Compt. Rend. (Paris) C* **270,** 1261 (1970).
12. Scholten, J. J. F., and Zwietering, P., *Trans. Faraday Soc.* **53,** 1363 (1957).

3b. Surface structure and wetting

Commentary

Structure plays a major role in wetting, and many studies have been carried out on the so-called "Lotus Effect". We wanted to get beyond the existing studies by independently varying both structure and chemistry. This was achieved by means of the self-assembled monolayer chemistry that is so frequently applied in our group, and we were able to construct diagrams that relate wetting on rough surfaces to wetting on the chemically equivalent flat surface over a broad range of surface energies. The shape of the resulting curves can indicate the presence of superhydrophobicity, pinning effects, and the degree of conformity with existing wetting theories (3.3).

Combining our interests in wetting and gradients, we used a simple form of photolithography and elastomer replication to create a series of hydrophobic hole-to-pillar-density gradients. In this way we could vary the ratio of air to solid underneath a water droplet in a controlled way. The data obeyed the Cassie and Baxter equation well up to an air fraction of 70%, and we were able to observe major influences on contact and roll-off angles of the type of surface feature that was causing the roughness (3.4).

Langmuir **2008,** *24,* 5411–5417 5411

Beyond the Lotus Effect: Roughness Influences on Wetting over a Wide Surface-Energy Range

Doris M. Spori,† Tanja Drobek,† Stefan Zürcher,†,|| Mirjam Ochsner,† Christoph Sprecher,‡ Andreas Mühlebach,§ and Nicholas D. Spencer*,†

Laboratory for Surface Science and Technology, Department of Materials, ETH Zurich, Zürich, Wolfgang-Pauli-Strasse 10, 8093 Zürich, Switzerland, AO Research Institute, Clavadelerstrasse, 7270 Davos, Switzerland, Ciba Specialty Chemicals, Klybeckstrasse 141, 4002 Basel, Switzerland, SuSoS AG, Lagerstrasse 14, 8600 Dübendorf, Switzerland

Received January 22, 2008. Revised Manuscript Received February 24, 2008

To enhance our understanding of liquids in contact with rough surfaces, a systematic study has been carried out in which water contact angle measurements were performed on a wide variety of rough surfaces with precisely controlled surface chemistry. Surface morphologies consisted of sandblasted glass slides as well as replicas of acid-etched, sandblasted titanium, lotus leaves, and photolithographically manufactured golf-tee shaped micropillars (GTMs). The GTMs display an extraordinarily stable, Cassie-type hydrophobicity, even in the presence of hydrophilic surface chemistry. Due to pinning effects, contact angles on hydrophilic rough surfaces are shifted to more hydrophobic values, unless roughness or surface energy are such that capillary forces become significant, leading to complete wetting. The observed hydrophobicity is thus not consistent with the well-known Wenzel equation. We have shown that the pinning strength of a surface is independent of the surface chemistry, provided that neither capillary forces nor air enclosure are involved. In addition, pinning strength can be described by the axis intercept of the cosine–cosine plot of contact angles for rough versus flat surfaces with the same surface chemistries.

Introduction

For many practical applications, such as coating or fluid handling, the wettability of a surface plays a crucial role. The contact angle, θ^Y, that a liquid drop makes with an ideally flat surface corresponds to a minimum in the energy of the liquid–solid–ambient gas system[1] (see Figure 1). The prediction of contact angles for real surfaces presents a significant challenge, however, since it is well known that roughness exerts a significant influence over wetting phenomena.[2–5]

Only on ideally flat, uniform surfaces does θ^Y have a unique value. On real surfaces, depending on how the drop is deposited, the contact angle θ can vary between the so-called advancing and receding contact angles. This hysteresis can be ascribed to inhomogeneities in the distribution of adsorbates or the presence of contaminants, to surface roughness, or to time-dependent surface rearrangements.[2] On rough surfaces, the surface morphology strongly influences the value of θ.

On rough, hydrophobic surfaces, the liquid can either follow the surface topography and show strong pinning or can bridge from asperity to asperity while enclosing air beneath and showing almost no hysteresis in the contact angle. For the first case, Wenzel[6] introduced a roughness factor, r, to describe the roughness influence on θ (eq 1).

$$\cos\theta^W = r\cdot\cos\theta^Y \qquad (1)$$

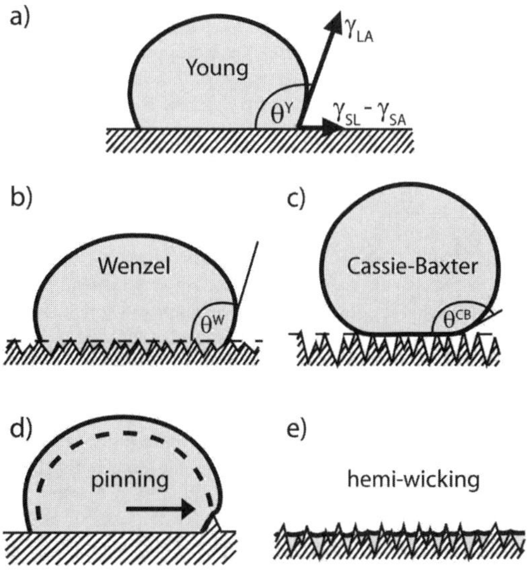

Figure 1. Scheme of wetting phenomena: (a) definition of contact angle on an ideally flat surface; (b) Wenzel-type wetting; (c) Cassie-type wetting; (d) pinning, a growing drop pinned by one obstacle; and (e) hemiwicking.

where r is calculated by dividing the actual, roughness-enhanced surface area by its projection. This behavior is often referred to as Wenzel-type wetting. If the cosines in eq 1 are plotted versus each other, the effect of roughness is evident in the deviation from a straight line with slope 1.

For the second case, Cassie and Baxter[7] modified Wenzel's equation by introducing the fractions f_1 and f_2, where f_1 corresponds to the area in contact with the liquid divided by the projected area, and f_2, to the area in contact with the air trapped

* To whom correspondence should be addressed. E-mail: spencer@mat.ethz.ch.
† ETH Zurich.
|| SuSoS AG.
‡ AO Research Institute.
§ Ciba Specialty Chemicals.

(1) Young, T. *Philos. Trans. R. Soc. London* **1805**, *95*, 65–87.
(2) Degennes, P. G. *Rev. Mod. Phys.* **1985**, *57*, 827–863.
(3) Dettre, R. H.; Johnson, R. E. *Adv. Chem.* **1964**, 43.
(4) Genzer, J.; Efimenko, K. *Biofouling* **2006**, *22*, 339–360.
(5) Quere, D. *Physica A* **2002**, *313*, 32–46.

Published on Web 04/29/2008

beneath the drop, also divided by the projected area (Supporting Information discussion 1):

$$\cos \theta^{CB} = f_1 \cdot \cos \theta^{Y} - f_2 \quad (2)$$

Since the introduction of these equations, wetting on rough surfaces has been the subject of intensive research,[2,3] which was significantly boosted by the discovery of the "self-cleaning" properties of the superhydrophobic lotus leaves ("lotus effect").[8] Although the many publications on this topic have increased our knowledge of superhydrophobic behavior,[4,9–13] showing, for example, that the most stable topographies to achieve superhydrophobicity are "undercut",[14–16] many aspects of this field still remain controversial.[17–20]

Structured surfaces that exhibit superhydrophobicity can also show an effect known as *hemiwicking*[5] or *superwetting* if they are surface-chemically functionalized to be hydrophilic. Hemiwicking is complete wetting due to the presence of capillary forces in two dimensions.[21,22]

We have chosen to examine wetting-property changes upon significant variation in the surface chemistry of samples with four different surface topographies. This is of both fundamental and practical relevance, since many surface coatings change their surface chemistry over time due to contamination or oxidation. Surfaces have been analyzed by scanning electron microscopy and roughness factors evaluated by means of white-light profilometry. Static (θs) and dynamic (θa, θr) contact angles have been measured. To vary surface chemistry over a wide range of surface energies, the surfaces were coated with gold and subsequently functionalized by means of mixed, self-assembled monolayers of methyl- and hydroxyl-terminated alkane thiols. In this way, contact angles between 20° and 105° can be obtained on a flat surface. Surfaces examined were sandblasted glass microscope slides (SBG), as well as replicas of sandblasted (large grit), acid-etched titanium (SLA); lotus leaves (LLR); and golf-tee-shaped micropillars of photoresist (GTM) on a silicon wafer. The GTM pillars show an extraordinarily stable, Cassie-type hydrophobicity. All four surface topographies are uniformly rough, such that it does not make a difference where a drop is placed. This precondition, as emphasized by McHale,[19] has to be met to be able to compare contact angle data with eqs 1 and 2. By applying the Wenzel equation over this wide range of surface chemistries, the roughness factor fails to predict the data. If the θ data are presented in a cosine−cosine plot of rough vs smooth θ for the same surface chemistries, the three major classes of behavior ("hemiwicking", "pinned", "Wenzel- or Cassie-type" wetting (see Figure 1)) can be readily distinguished. For the surface-energy range where pinning has the most profound influence, it is suggested that the axis intercept be surface-morphology-sensitive.

Methods

Substrates. Glass microscope slides were sandblasted with an air pressure of 8 bar. The sand jet was passed over the microscope slide twice for about 15 s in perpendicular directions to achieve a homogeneously roughened surface. Subsequently, these SBG substrates were ultrasonicated in ethanol for 10 min, rinsed with ethanol, and dried under a stream of nitrogen.

GTM masters were produced by standard photolithography on 4-in. silicon wafers upon which, after cleaning, SU-8 2025 negative photoresist (Microchem) was spin-coated for 60 s at 2000 rpm to a thickness of 30 μm and soft-baked on a hotplate for 2 min at 65 °C and 3 min at 95 °C. Subsequently, the wafer was exposed to UV light (MA6, Karl Süss) with constant intensity (total energy 180 mJ/cm^2) through a chromium mask having circular patterns of 20 μm diameter and 70 μm pitch distance (center to center). After postbaking for 2 min at 65 °C and 3 min at 95 °C, the wafers were developed for 5 min in SU-8 developer. Finally, the wafer was hard-baked at 190 °C for 10 min. The prepared wafer was used as a master to make replicas.

Replicas of acid-etched, large-grit-sandblasted (SLA) titanium (Straumann, Switzerland), the SU-8 pillar-structured silicon wafer, and lotus leaves (*Nelumbo nucifera*) obtained from the Botanical Garden in Zürich were prepared according to Wieland et al.[23] In short, a low-viscosity and fast-curing silicone (PROVILnovo, Light C.D. 2, fast set; Heraeus Kulzer GmbH, Germany) was cast in a mold affixed to the surface to be replicated. This silicone sample, exhibiting the original's negative structure was then used in turn as a mold to cast positive replicas with an epoxy blend (EPO-TEK 302-3; Epoxy Technology, Billerica, MA). The epoxy was then cured for 5 h at 60 °C, removed from the mold, and subsequently postcured for 1 h at 150 °C. The lotus leaf replicas (LLR) were sawn into 5 × 15 mm^2 pieces. The SLA replicas were cast in a circular mold of 12 mm diameter. All replicas were cleaned in a 2 vol % solution of Hellmanex (Hellma, Germany) in ultrapure water (resistance 18.2 MΩ, EASYpure by Barnstead, Dubuque, IA) and subsequently rinsed five times with ultrapure water.

Single-side-polished silicon wafers (Si-Mat Silicon Materials, Germany) were cut into 10 × 10 mm^2 pieces. To remove glue residues from the cutting step, they were sonicated for 10 min in toluene and 10 min in ethanol.

Gold Coating. The rough surfaces and single-side-polished silicon wafers were cleaned for 2 min in air plasma and then coated by resistance evaporation (MED 020 coating system, BALTEC, Liechtenstein) with 10−15 nm of Cr and 50 nm of Au (purity >99.99%, Unaxis, Liechtenstein). The rough surfaces were rotated during evaporation, and the pillars were also tilted by 25° to achieve a homogeneous coating. The gold-coated silicon wafers were used in every experiment as a flat reference.

Surface Modification. To achieve a wide surface-energy range, self-assembled monolayers (SAMs) of 11-mercaptoundecanol and dodecanethiol (Aldrich Chemicals, USA) were chemisorbed in both pure and mixed forms on freshly gold-coated samples. The samples were immersed in 0.1 mM ethanolic thiol solutions for 20 min. For the solution preparation, the total thiol concentration of 0.1 mM was held constant, and the composition of the two compounds was varied in terms of dodecylthiol molar ratio from 100, 70, 50, 45, 30, 15, 10 (GTM only) to 0%. After assembly, the samples were rinsed with ethanol (purity >99.8%, Merck, Germany) and dried under a stream of nitrogen.

Contact Angle Measurements. Static water contact angle (θs) measurements were performed on a Ramé-Hart contact-angle goniometer on freshly prepared surfaces. A drop of 6−8 μL was produced and then gently placed on the surface. For superhydrophobic

(6) Wenzel, R. N. *Ind. Eng. Chem.* **1936**, *28*, 988–994.
(7) Cassie, A. B. D.; Baxter, S. *Trans. Faraday Soc.* **1944**, *40*, 0546–0550.
(8) Barthlott, W.; Neinhuis, C. *Planta* **1997**, *202*, 1–8.
(9) Barbieri, L.; Wagner, E.; Hoffmann, P. *Langmuir* **2007**, *23*, 1723–1734.
(10) Bhushan, B.; Chae Jung, Y. *Ultramicroscopy* **2007**, *107*, 1033–1041.
(11) Blossey, R. *Nat. Mater.* **2003**, *2*, 301–306.
(12) Callies, M.; Quere, D. *Soft Matter* **2005**, *1*, 55–61.
(13) Patankar, N. A. *Langmuir* **2003**, *19*, 1249–1253.
(14) Ahuja, A.; Taylor, J. A.; Lifton, V.; Sidorenko, A. A.; Salamon, T. R.; Lobaton, E. J.; Kolodner, P.; Krupenkin, T. N. *Langmuir* **2008**, *24*, 9–14.
(15) Li, X. M.; Reinhoudt, D.; Crego-Calama, M. *Chem. Soc. Rev.* **2007**, *36*, 1350–1368.
(16) Tuteja, A.; Choi, W.; Ma, M. L.; Mabry, J. M.; Mazzella, S. A.; Rutledge, G. C.; McKinley, G. H.; Cohen, R. E. *Science* **2007**, *318*, 1618–1622.
(17) Gao, L. C.; McCarthy, T. J. *Langmuir* **2007**, *23*, 3762–3765.
(18) Gao, L. C.; McCarthy, T. J. *Langmuir* **2007**, *23*, 13243–13243.
(19) McHale, G. *Langmuir* **2007**, *23*, 8200–8205.
(20) Panchagnula, M. V.; Vedantam, S. *Langmuir* **2007**, *23*, 13242–13242.
(21) Extrand, C. W.; Moon, S. I.; Hall, P.; Schmidt, D. *Langmuir* **2007**, *23*, 8882–8890.
(22) Martines, E.; Seunarine, K.; Morgan, H.; Gadegaard, N.; Wilkinson, C. D. W.; Riehle, M. O. *Nano Lett.* **2005**, *5*, 2097–2103.

(23) Wieland, M.; Chehroudi, B.; Textor, M.; Brunette, D. M. *J. Biomed. Mater. Res.* **2002**, *60*, 434–444.

surfaces, the drop had to be enlarged up to 12 μL for the drop to be able to detach from the syringe. The contact angle is then defined as depicted in Figure 1.

Dynamic water contact angle (advancing (θa) and receding (θr)) measurements were performed on a Krüss contact-angle-measuring system (G2/G40 2.05-D, Krüss GmbH, Germany) with a speed of 15 μL/min. Two movies with 40 images were recorded for the advancing contact angle and only one for the receding. Analysis was carried out by means of the *tangent method 2* routine of the Krüss Drop-Shape Analysis program (DSA version 1.80.0.2 for Windows 9x/NT4/2000, 1997–2002 Kruess). The movies were evaluated as follows: If the drop exhibited a stick-jump behavior, but moved over the entire period of recording, all contact angles were evaluated. In this way, the stick-jump led to a high standard deviation. If one side of the drop did not move at all during the recording, only the other, moving side was evaluated. In the superhydrophobic case, since the drop is confined between the syringe and the surface while being very strongly repelled, it becomes squeezed out at one side. In this case, only the mobile, squeezed-out side was evaluated. The tangent method 2 routine, a fourth-order polynomial function, has difficulty in fitting drops in the Cassie-type wetting regime, leading to a systematic underestimation of a few degrees. For many experiments, the receding contact angle could not be defined, since the drop was so strongly pinned to the surface. A value of 0° was assumed for these cases. Several exemplary movies can be found in the Supporting Information.

Scanning Electron Microcopy. Replicas and SBGs were analyzed in a Zeiss Gemini 1530 FEG SEM at 3−5 kV, at room temperature, gold-coated, as described above.

White Light Profilometry. For the evaluation of the roughness factor, an optical profilometer was used (FRT MicroGlider, Fries Research & Technology GmbH, Germany). *X* and *Y* resolution were 1 μm, *Z* resolution was better than 10 nm. Areas of 2 × 2 mm^2 were measured with a resolution of 1000 × 1000 pixels. The data points were assembled into adjacent, nonoverlapping triangles, and their areas were summed up to achieve the actual surface area. The roughness factor was then obtained by dividing this actual surface area by the analyzed area (2 mm^2).

Results and Discussion

Surface-Morphology Characterization. SEM analysis of the investigated surfaces reveals considerable differences in surface morphology (Figure 2). The SBG (Figure 2a and b) shows a very profound microroughness with very rough patches, nanosized features, and very flat conchoidal fracture areas. The SLA replica surface (Figure 2c and d) also contains two major roughness scales. The microscale roughness originates from the sandblasting step, leading to troughs. The superimposed nanoscale roughness was created by the acid-etching process.[24–26] The LLR (Figure 2e and f) shows the wavy structure of the lotus leaf cells topped with papillae. The papillae exhibit a diameter of 9 ± 2 μm, a spacing of 21 ± 7 μm, and a height of around 20 μm. The nanoscale roughness is an approximate replica of the leaf's wax structure, which was compressed during the replication process (Supporting Information 1). The GTM pillars (Figure 2g and f) measure 24 μm in diameter on top and thin out to a diameter of 15 μm toward the bottom. Their height slightly exceeds 30 μm, and the pitch distance (center to center) is 70 μm. The pillar tops constitute 9% of the projected geometric surface area. The pillar surface is highly ordered, whereas the other three samples show nonperiodic structures.

(24) Kunzler, T. P.; Drobek, T.; Sprecher, C. M.; Schuler, M.; Spencer, N. D. *Appl. Surf. Sci.* **2006**, *253*, 2148–2153.
(25) Tosatti, S.; Michel, R.; Textor, M.; Spencer, N. D. *Langmuir* **2002**, *18*, 3537–3548.
(26) Wieland, M.; Textor, M.; Spencer, N. D.; Brunette, D. M. *Int. J. Oral Maxillofacial Implants* **2001**, *16*, 163–181.

Table 1. Relevant Factors for SBG (sand-blasted glass), SLA (replica of sand-blasted, acid-etched titanium), LLR (lotus leaf replica), GTM (replica of golf-tee shaped micropillars) Extracted From Data Analysis: Roughness Factors, *r*, From White-Light Profilometry (WLP) Data and SEM Image[a]

	SBG	SLA	LLR	GTM
r	2.3 (WLP)	1.5 (WLP)	1.7 (WLP)	1.3 (SEM)
k	1.18 ± 0.07	1.11 ± 0.08	1.09 ± 0.04	–
d_S	−0.03 ± 0.03	−0.86 ± 0.04	−0.31 ± 0.01	–

[a] k and d_S are the corresponding slopes and axis intercepts from the linear fits in Figure 4a.

Table 1 contains roughness factors for the studied surfaces, as well as corresponding k and d_S factors, as described below in Figure 4 and eq 3.

Water Contact Angle Data. Figure 3a) is an explanatory graph. It illustrates the effects of roughness on a surface chemistry defined by the contact angle found on the flat reference. The maximum θ that can be ordinarily achieved on a flat surface is 120°. Lower surface energies are thus excluded, and no data can be obtained beyond 120° on the flat surface (area A). With a hydrophobic coating (90−120° on the flat surface), the drop shows either Wenzel- or Cassie-type wetting (area B). Below 90° on a flat surface, even though the surface chemistry is intrinsically hydrophilic, the drop on rough surfaces shows a higher θ, and even hydrophobic values are possible (area C). In some cases, surface topography can be such that Cassie-type wetting persists in the hydrophilic regime[27] (area F). In the high surface energy range (low θ), hemiwicking can occur (area D). Surface topography determines the surface energy at which capillary forces come into play.

Presenting the data in a cosine−cosine plot facilitates comparison with the Wenzel predictions (area E), a linear function of surface energy (cosine of Young contact angle) with the roughness factor as slope and axis intercept of zero.

Water contact angles were measured on both rough and flat surfaces that had been exposed to the same thiol mixtures. In Figure 3, the θ data are plotted such that the cosines of the static and dynamic contact angles on the rough surfaces are plotted versus the cosine of the static contact angles measured on flat surfaces with the corresponding surface chemistry. Thus, each data point corresponds to one comparative experiment. With these diagrams, a direct comparison with the Wenzel theory[6] (eq 1) is possible.

Figure 3b−e shows the data obtained on SBG, SLA, LLR, and the GTM surfaces. The static and the advancing θ values for SBG (Figure 3b) display two linear trends, changing from slope 1 to slope 2.3 (corresponding to a Wenzel roughness factor, Table 1) at 90°. The surface cannot be made superhydrophobic, but becomes superwetting when functionalized with a pure OH-terminated SAM. As long as the surface chemistry remains hydrophilic, the receding contact angle is zero because the contact line is pinned to the surface.

The static and advancing θ values on the SLA replica (Figure 3c) show an extremely strong shift toward hydrophobicity compared to the θ predicted by the Wenzel equation. All data points in the lower right-hand quadrant correspond to hydrophilic surface chemistry, whereas the apparent contact angle on the SLA surface is hydrophobic. This phenomenon can be explained by pinning of the contact line. If the intrinsic surface energy is hydrophobic, the surface is capable of achieving contact angles near 150° but remains essentially in a Wenzel-type wetting regime.

(27) Liu, J. L.; Feng, X. Q.; Wang, G. F.; Yu, S. W. *J. Phys.: Condens. Matter* **2007**, *19*, 12.

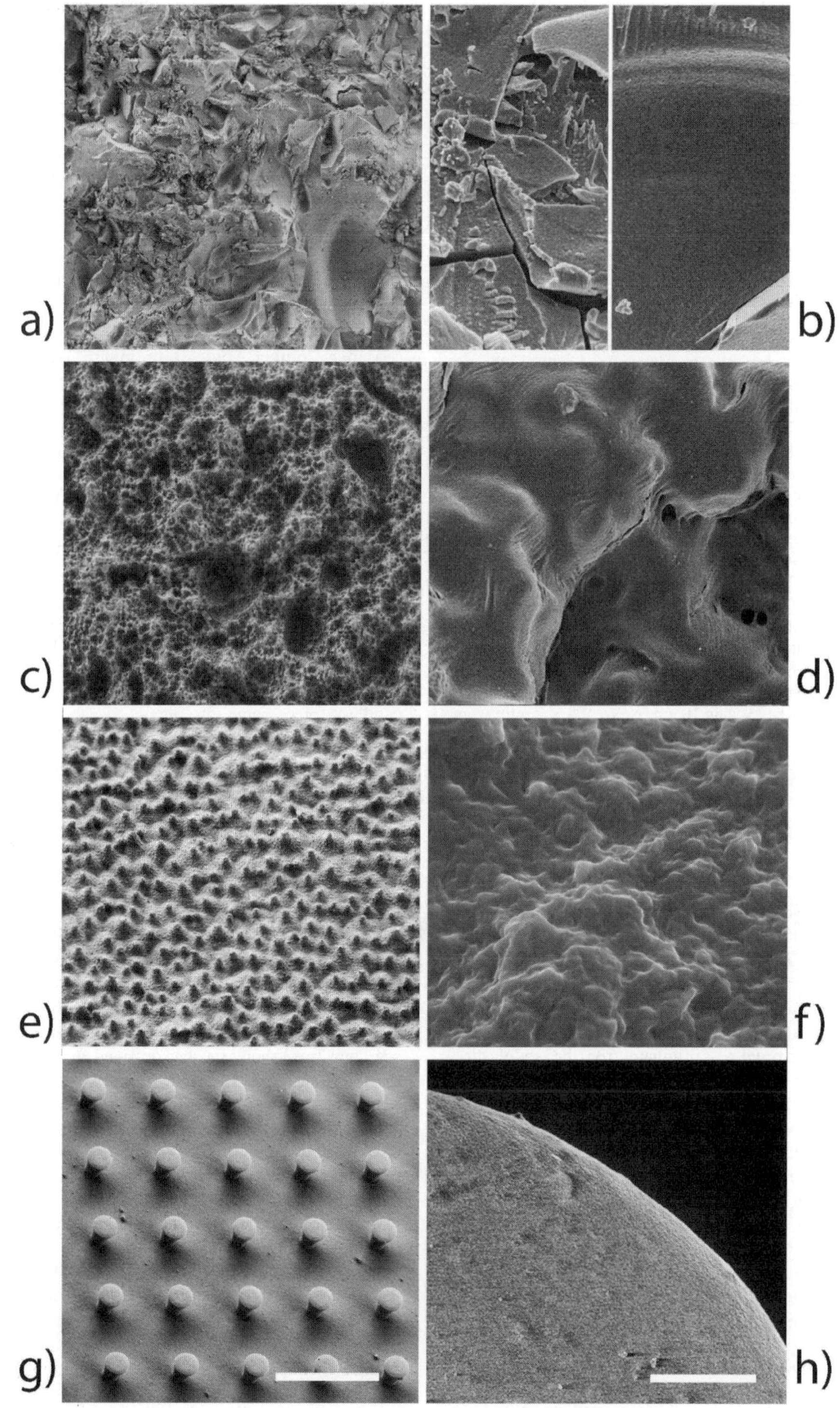

Figure 2. SEM images at two different magnifications of the examined surfaces. The scale bar on the left side corresponds to 100 μm, and the one on the right side, to 2 μm. (a, b) SBG (rough and flat regions enlarged), (c, d) SLA, (e, f) LLR, and (g, h) GTM.

The pinning strength is still sufficiently strong that the drop exhibits no roll-off, even though the receding θ is high. Similarly to SBG, SLA surfaces can also be rendered hemiwicking when coated with OH-terminated thiols.

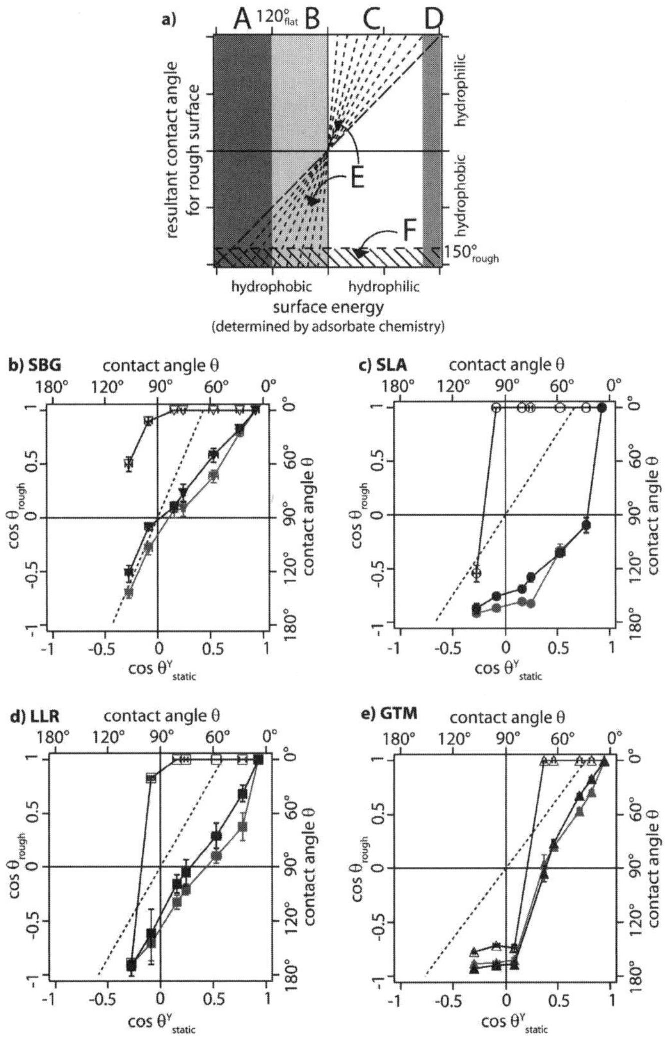

Figure 3. (a) Model graph clarifying the meaning of the axes in b−e and the effects that can be observed in the plot. (A) surface energies leading to contact angles above 120° on a flat surface are physically not possible, (B) hydrophobic region (Wenzel- or Cassie-type wetting occurs), (C) pinning effects cause the drop to adopt (considerably) higher contact angles as compared to the flat surface, (D) capillary forces occur, (E) series of Wenzel predictions, (F) contact angles above 150°, Cassie-type wetting. (b−e) Water contact angle data for four different surfaces; static (solid symbol), advancing (grey symbol), and receding (open symbol) θ. The x axis corresponds to the cosine of the static contact angles of the flat reference; the y values correspond to the cosine of the advancing, static, and receding contact angles on the corresponding rough surface. The dashed line corresponds to the prediction by Wenzel when calculated with the roughness factors derived from white light profilometry (displayed in Table 1): (b) SBG, (c) SLA, (d) LLR, (e) GTM.

The LLR (Figure 3d) shows a shift to hydrophobic contact angles due to pinning effects, which is greater than that of the SBG, but less than the case of the SLA. In contrast to SBG and SLA, the LLR functionalized with a pure CH_3-terminated SAM shows extremely high dynamic and static contact angles, an indication of Cassie-type wetting. The papillae are able to support the drop, even in the absence of the tubular wax of the original leaf. As soon as there is a hydrophilic contribution present in the SAM, Wenzel-type wetting occurs. Note the high standard deviations, which arise from the fact that in some measurements, the Cassie state still persisted, whereas in others, the drop penetrated into the structures. Both types of behavior were sometimes even observed on the same sample, which is an indication of the presence of metastable states. An explanation for this metastable Cassie regime might be found in the naturally grown leaf topography, which is inhomogeneous in pillar density, top perimeter, and height. On some areas of the leaf, the papillae can support the drop, whereas on others, they cannot. Again, the LLR shows superwetting when coated with a pure OH-terminated SAM.

The GTM surface (Figure 3e) behaves somewhat differently from the others. Thanks to the golf-tee, undercut shape, Cassie-type wetting is moderately stable, even at higher surface energies, as predicted by Liu et al.[27] Since the cross section of the pillar becomes smaller toward the bottom, the drop would need to form a larger liquid–air interface to follow the topography, which would be energetically unfavorable. Therefore, the surface energy of the solid must be quite high to overcome this energy barrier. Tilting the sample also helps to overcome this energy barrier.

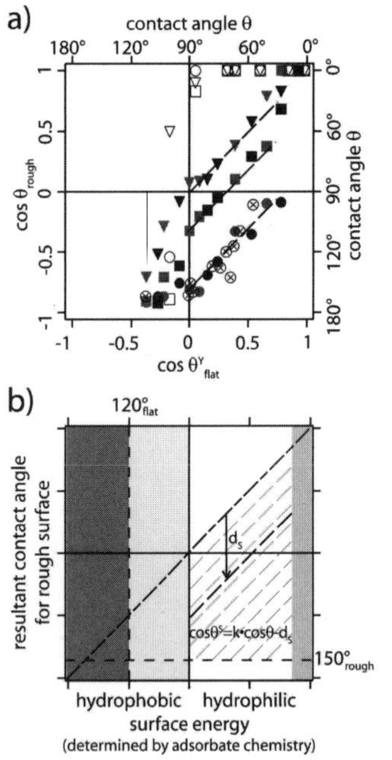

Figure 4. Water contact angle data and the derivation of an empirical description. (a) Water contact angle data of the three nonperiodical topographies. Here, the *x* axis consists of not only the static but also the dynamic values on the flat surface. Open symbols correspond to receding; solid, to static; and grey, to advancing contact angles. SBG, ▼; LLR, ■; SLA, ●. The circles with a cross are advancing contact angle data found by Tosatti et al.[25] by applying mixed SAMs of OH- and CH_3-terminated alkylphosphates on SLA titanium surfaces. The dashed line has slope 1. (b) Empirical description of the mechanisms found in the intermediate hydrophilic regime where pinning effects are predominant.

The hysteresis of dynamic measurements on the GTM is quite high, as compared to that of the LLR with the pure hydrophobic coating, for example. This is commonly observed for micrometer-sized pillars[28] in the absence of a second, smaller-scale structure. If the GTM surface is coated with a pure OH-terminated SAM, water undergoes spreading until it completely fills the volume between the pillars. At this point, it forms a drop coexisting with a liquid film within the structure. Although the water film fully spreads over the whole patterned area and is visible by eye, a contact angle of around 6° can still be measured (Supporting Information 2).

If static θ_{flat} vs static θ_{rough} and advancing θ_{flat} vs advancing θ_{rough} are presented in the same cosine−cosine plot, as shown in Figure 4a, differences between advancing and static measurements are canceled out. The difference between the advancing and the static contact angles on the rough surface is equal to that on the flat surface. This means that it does not matter whether θs or θa is measured, since the effects observed are governed by the environment at the contact line and describe the same tendency. In the intermediate surface-energy region, additional effects, such as air enclosure or capillary forces, can be excluded. Thus, the slope for all three nonperiodical surfaces tested (SBG, SLA, LLR) is close to unity (see *k*, Table 1) and no significant effect on the slope, originating from surface roughness, can be distinguished. Roughness mainly influences the axis intercept (see d_S, Table 1), which is an indication of the magnitude of the energy barrier pinning the contact line. We propose the following empirical description to characterize nonperiodical surface topographies in the intermediate hydrophilic regime (θ^Y, ranging approximately from 40° to 90°). θ^S is the contact angle measured on the rough surface, $k \approx 1$ is the slope, and d_S is the axis intercept found in the linear regression through the data.

$$\cos \theta^S = k \cdot \cos \theta^Y - d_S \quad (3)$$

The explanation of why SBG, the surface with the highest roughness factor, has the lowest hydrophobic shift, can be ascribed to the presence of conchoidal fracture areas on the surface, which present the energetically most favorable route for the contact line to move forward.[2] As soon as a step forward is made, the energy barrier is overcome, and the rest of the contact line will follow. This behavior resembles that found in dislocation movement in metals and in the special case of kink pairs (Seeger's dislocation mechanism).[29] Pinning of the contact line is therefore not very effective in this case. The LLR has no flat patches but a pillarlike structure, which presents fewer pinning sites than the dimpled SLA replica surface. It therefore seems that, for an understanding of roughness effects, the energy barrier is a more fruitful avenue to pursue than the effect of increased surface area. The receding contact angle remains close to zero, as long as the corresponding angle on the chemically equivalent flat surface remains below 90°. The detachment of the contact line is facilitated by a hydrophobic coating. The energy barrier governing this detachment is significantly lowered by the presence of air on Cassie-type wetted surfaces, leading to extremely low hysteresis. The advancing contact angle data of Tosatti et al.,[25] measured on SLA titanium with SAMs of mixed OH- and CH_3-terminated alkylphosphate monolayers, coincide extremely well with the mixed thiol data taken on the SLA replica in the present study. This demonstrates that the approach is valid, independent of the specific adsorbate–substrate system.

White-Light Profilometry. The white-light profilometry data was used to calculate the actual surface area, in order to be able to define the roughness factor, *r*. The roughness factor for the GTM surface was calculated by measuring the relevant lengths of a detached pillar in a SEM image (see Table 1) Additionally, "roughness factors" *k* from Figure 4a were extracted by applying a linear regression through the data found in the surface energy region corresponding to the contact angles from 40° to 90° on the flat surface.

Contrary to Wenzel's predictions, all tested surfaces show at least a small axis intercept. Additionally, slopes for SBG, SLA, and LLR in the region corresponding to 40−90° on the flat surface lie between 1 and 1.2; that is, far from the roughness values extracted from the white-light profilometry data in Table 1. Due to the pillar surface's strong and persistent Cassie-type wetting, the data show a transition from composite behavior directly to hemiwicking, leaving little room for roughness-dependent pinning.

Conclusions

Four different, heavily structured surfaces have been analyzed over a wide range of surface energies via water contact angle measurements. The data show three wetting regimes: If the surface energy is high, wettability is indeed enhanced by the surface roughness, causing hemiwicking in many cases due to capillary forces. At lower surface energies, pinning of the contact line results in a shift to more hydrophobic θ values. It was found that the surface topography defines the pinning strength and with it

(28) Bico, J.; Marzolin, C.; Quere, D. *Europhys. Lett.* **1999**, *47*, 220–226.

(29) Hull, D. *Introduction to Dislocations*, 2nd ed.; Pergamon Press: Oxford, 1975; Vol. 16.

the energy barrier counteracting the wetting behavior of the drop. An indication of this barrier is the axis intercept seen in the cosine−cosine plot (see Figure 4). This plot is surface-topography-sensitive, and the behavior of all tested surfaces can be readily distinguished. The topographical influence on θ cannot simply be predicted via a roughness factor. With the exception of the hydrophobic data in the case of the SBG surface, none of the measured contact angles could have been predicted by the Wenzel equation.

With regard to superhydrophobic surfaces, the golf-tee-shaped (GTM) pillars show stable superhydrophobicity over a wide range of surface energies. This topography seems to be a very effective design for microstructured, superhydrophobic surfaces.

Acknowledgment. The authors thank Ciba Speciality Chemicals for their generous financial assistance; the Botanical Garden of Zurich for providing the lotus leaves; and Electron Microscopy ETH Zurich, EMEZ, for their support.

Supporting Information Available: Supporting discussion 1 emphasizes that in the initial equation (eq 2) by Cassie and Baxter, $f_1 + f_2 \geq 1$, and only in special cases (e.g., flat-top pillars) does it equal unity. Discussion 2 shows that eq 2 can also be applied to calculate the contact angle in the very hydrophilic wetting regime. As soon as all cavities are filled with water, a finite contact angle can be observed, as predicted by eq 2. Additionally, a SEM analysis shows the different steps of the lotus leaf replica process, and five illustrative movies are shown for the evaluation of the dynamic contact angles: advancing and receding in the supherhydrophobic case, the stick-jump behavior, and a one-side pinned drop, as well as a movie of a receding drop showing no finite contact-angle value. This information is available free of charge via the Internet at http://pubs.acs.org.

LA800215R

Langmuir Article
pubs.acs.org/Langmuir

Cassie-State Wetting Investigated by Means of a Hole-to-Pillar Density Gradient

Doris M. Spori,[†] Tanja Drobek,[‡] Stefan Zürcher,[†,§] and Nicholas D. Spencer*,[†]

[†]*Laboratory for Surface Science and Technology, Department of Materials, ETH Zurich, Wolfgang-Pauli-Strasse 10, 8093 Zurich, Switzerland,* [‡]*Department Earth and Environmental Sciences, Section Crystallography, LMU Munich, Theresienstrasse 41/III, 80333 Munich, Germany, and* [§]*SuSoS AG, Lagerstrasse 14, 8600 Dübendorf, Switzerland*

Received December 15, 2009. Revised Manuscript Received March 7, 2010

The superhydrophobicity of rough surfaces owes its existence to heterogeneous wetting. To investigate this phenomenon, density gradients of randomly placed holes and pillars have been fabricated by means of photolithography. On such surfaces, drops can be observed in the Cassie state over the full range of f_1 (fraction of the drop's footprint area in contact with the solid). The gradient was produced with four different surface chemistries: native PDMS (polydimethylsiloxane), perfluorosilanized PDMS, epoxy, and CH_3-terminated thiols on gold. It was found that f_1 is the key parameter influencing the static water contact angle. Advancing and receding contact angles at any given position on the gradient are sensitive to the type of surface feature—hole or pillar—that is prevalent. In addition, roll-off angles have been measured and found to be influenced not only by the drop weight but also by suction events, edge pinning, and f_1.

Introduction

Since the superhydrophobic, self-cleaning properties of the lotus leaf[1] were attributed to its rough surface structure, many different approaches have been developed to produce similar superhydrophobic surfaces artificially (see e.g. the review by Roach et al.[2]). While both natural and industrially produced superhydrophobic surfaces generally have a stochastic distribution of surface features, most investigations that have sought to determine which parameters lead to stable self-cleaning properties have been performed on micrometer-sized, photolithographically fabricated, periodically arranged pillars.[3−6] This approach does not take into account the possibility that the inherently strong anisotropy of periodic structures, or even the periodicity itself, may have an effect on the wettability of a surface. It is therefore of interest to examine controlled, aperiodic surfaces, which are also better analogues for technically relevant systems. Gradients are very helpful tools[7] for the investigation of a system's sensitivity to specific parameters over a given range, since they constitute an inherently high-throughput approach. On surface gradients, one parameter is changed along the length of the sample, and therefore all other conditions, such as temperature, pressure, or surface energy, can be maintained constant while multiple measurements are made simultaneously over the parameter range selected. The parameter investigated in the present study was stochastically distributed pillar/hole density. This substrate allows models from the literature to be tested as to their validity on surfaces that resemble technically relevant systems.

*Corresponding author. E-mail: nspencer@ethz.ch.

(1) Barthlott, W.; Neinhuis, C. *Planta* **1997**, *202*, 1–8.
(2) Roach, P.; Shirtcliffe, N. J.; Newton, M. I. *Soft Matter* **2008**, *4*, 224–240.
(3) Barbieri, L.; Wagner, E.; Hoffmann, P. *Langmuir* **2007**, *23*, 1723–1734.
(4) Bico, J.; Marzolin, C.; Quéré, D. *Europhys. Lett.* **1999**, *47*, 220–226.
(5) Callies, M.; Chen, Y.; Marty, F.; Pepin, A.; Quéré, D. *Microelectron. Eng.* **2005**, *78–79*, 100–105.
(6) Quéré, D. *Annu. Rev. Mater. Res.* **2008**, *38*, 71–99.
(7) Morgenthaler, S.; Zink, C.; Spencer, N. D. *Soft Matter* **2008**, *4*, 419–434.

Generally, on a rough, somewhat hydrophobic substrate two different wetting states are possible: Either the drop completely conforms to the surface topography, or it bridges from asperity to asperity, spanning the air gaps beneath. The first case can be denoted as the Wenzel[8] state and the second case as the Cassie and Baxter (CB) state.[9] Both states can be associated with very high contact angles, but only in the CB state are very low roll-off angles in evidence, along with the concomitant self-cleaning effect. In this work, most drops described are in the CB state, but only few have low roll-off angles (≤10°).

On rough surfaces, as a consequence of contact-line pinning, there is not a single contact angle that describes wetting. In fact, the contact angle can adopt any value between a minimum (receding contact angle, θ_r) and a maximum (advancing contact angle, θ_a) value. The difference between θ_a and θ_r is known as the hysteresis.

Findings on Regular Pillar Surfaces. Among other parameters, Barbieri et al.[3] have investigated the influence of pitch distance (in hexagonal, square, and honeycomb periodic patterns) and pillar-top perimeter on the static contact angle of water drops atop perfluorinated silicon pillars. They generally found that the drop resides in the CB state much longer than would be predicted by calculating the thermodynamic transition to the Wenzel state. Pattern parameters, such as symmetry, pitch distance, and the pillar perimeter, only influenced the stability of the CB drop near the transition point, hexagonal and long perimeters favoring the CB state.

Öner[10] and Dorrer[11] have shown that the advancing contact angle of a drop suspended on pillars in the CB state for a given f_1 (fraction of the area under the drop in contact with the solid) is independent of the shape or spacing of the pillars. Dorrer and Rühe[11] proposed that the receding motion of the contact line is

(8) Wenzel, R. N. *Ind. Eng. Chem.* **1936**, *28*, 988–994.
(9) Cassie, A. B. D.; Baxter, S. *Trans. Faraday Soc.* **1944**, *40*, 0546–0550.
(10) Öner, D.; McCarthy, T. *Langmuir* **2000**, *16*, 7777–7782.
(11) Dorrer, C.; Rühe, J. *Langmuir* **2006**, *22*, 7652–7657.

governed by a jumping motion from one post to another, starting at the outermost post that inhibits the drop in assuming its preferred spherical shape. By holding f_1 constant, they showed that increasing the pillar-top dimensions reduces the receding contact angle (thus increasing hysteresis), whereas increasing the pitch distance leads to an increase in the receding contact angle and reduced hysteresis. Öner[10] suggests that the contact-angle hysteresis on a surface of random roughness should be smaller than that on a regular surface with the same f_1 due to the higher distortion of the contact line. Callies et al.[5] measured constant advancing contact angles on perfluorinated silicon pillars for f_1 values from 1% up to 25% and additionally showed how, for low pillar densities, it is possible to have drops in both the Wenzel and CB state, depending on the way in which the drop was put on the surface.

Several frequently used models for wetting have been tested with the acquired data. The models investigated were the Cassie–Baxter approximation,[9] the Furmidge equation for roll-off angles,[12] and a proposition from Patankar[13] for receding contact angles.

Models Tested on the Pillar-Density Gradients. On an ideal surface, surface tensions γ_{XY} (where the subscripts X and Y refer to the phases at the corresponding X–Y interface: solid (S), liquid (L), or air (A)) in balance lead to one specific contact angle, the Young's[14] contact angle θ^Y: $\cos \theta^Y = (|\vec{\gamma}_{SA}| - |\vec{\gamma}_{SL}|)/|\vec{\gamma}_{LA}|$. However, most surfaces are not ideal and have a certain degree of roughness. Roughness generally increases the contact angle,[15] unless surface energy is high enough to induce hemiwicking.[16] Wenzel[8] introduced an area roughness factor r to modify Young's equation, which takes into account the increased surface area and its influence on contact angle (θ^W): $\cos \theta^W = r \cos \theta^Y$. Concerns about this equation have been expressed, from both a theoretical[17] and an experimental[15,18] standpoint, but Wenzel's description of the conformal wetting state is a useful one. On certain surfaces the roughness is so profound that the surface tension of the liquid is sufficiently high to bridge from asperity to asperity and enclose air beneath the drop. Cassie and Baxter[9] modified Wenzel's equation to take the behavior of this composite surface into account:

$$\cos \theta^{CB} = f_1 \cos \theta - f_2 \qquad (1)$$

f_1 and f_2 describe the area fractions of the drop's footprint in contact with the solid and in contact with air, respectively. f_1 corresponds to the solid surface area wetted by the drop and f_2 to the area where the drop spans over air enclosures. Since both parameters are normalized by the analyzed projected area, $f_1 + f_2 \geq 1$. For small drops where gravity can be neglected and the Laplace pressure in the liquid is assumed to be constant,[6] sagging of the drop can be neglected and the area fraction constituting f_2 is assumed to be flat. When working with flat-top pillars, the area fraction of f_1 can also be assumed to be flat, and then $f_1 + f_2 = 1$. There remains a controversy in the field[18,19] that can be summarized by the phrase "area vs line". The models of Wenzel and Cassie–Baxter are based on area considerations. If they are tested by free-energy calculations and free-energy barriers, they correctly predict the equilibrium contact angles for the noncomposite (Wenzel) and the composite (Cassie–Baxter) state.[20] Nevertheless, the shape of the drop is also determined by the interaction at the three-phase-contact line; sharp edges influence the contact angle via pinning phenomena that are not considered in the thermodynamic approach.[18,21] Surprisingly, the CB approximation has not prompted as many criticisms regarding its validity as the Wenzel equation, even though they are based on similar principles.

In 1962, Furmidge[12] was looking into the conditions necessary for drop retention on inclined surfaces and, by equating all the forces involved, defined a correlation between the inclination α of the surface with the drop weight m, drop diameter perpendicular to the sliding direction w, advancing and receding contact angles (θ_a, θ_r), and the surface tension $\vec{\gamma}_{LA}$ of the liquid.

$$mg \sin \alpha = w|\vec{\gamma}_{LA}|(\cos \theta_r - \cos \theta_a) \qquad (2)$$

Furmidge assumed the footprint of the drop to be rectangular during sliding, which may be a source of error, but the equation is sufficiently accurate to be able to account for sliding drops on a flat surface.

The Cassie–Baxter model was designed to describe drops in thermodynamic equilibrium. Advancing and receding contact angles deviate from the equilibrium contact angles. Therefore, Patankar[13] deduced, via energy considerations, a condition that would describe the case of the receding drop if it would not leave a dry, but a wet, surface behind. In this case, the θ_r should obey the following rule:

$$\cos \theta_r^P = 2f_1 - 1 \qquad (3)$$

In this work, a novel, rapid method has been developed to explore the influence of quasi-randomly placed pillars and holes on the wetting of surfaces over a controlled range of f_1 values. In order to investigate a large parameter range, hole-to-pillar density gradients have been prepared that span the range from isolated holes to isolated pillars, both of 20–30 μm diameter. A systematic study of the wetting mechanisms and the effect of the real contact area on static and dynamic contact angles and roll-off angles has been carried out on hole-to-pillar density gradients prepared with four different surface chemistries. In contrast to earlier systematic studies,[3,5,22] this gradient approach covers the full range of f_1 (0–100%) and is, by the choice of the pillar distribution, a useful model for technically relevant surfaces.

Materials and Methods

A variety of morphological gradients with identical topography and different surface chemistry was prepared, following the procedure summarized in Figure 1.

Masks. The design of the mask is based on a 900 dpi bitmap (28.2 μm/pixel) with a random distribution of black and white pixels, which was printed on a foil mask. Photoshop CS (Version 8.0 for Macintosh) was used to create an 8-bit grayscale image. The image of the black-to-white gradient was modified with the linear gradient tool, which was set to span from black (gray value of 0) to white (gray value of 255) for the whole range of pillar density. Afterwards, the image was transformed to a binary image of black and white pixels by using the option "diffusion dither" in the image mode menu. The resulting image was used for the printing process. For technical reasons the whole range of pillar densities was split onto two samples, thus giving a higher accuracy

(12) Furmidge, C. G. *J. Colloid Sci.* **1962**, *17*, 309–324.
(13) Patankar, N. A. *Langmuir* **2003**, *19*, 1249–1253.
(14) Young, T. *Philos. Trans. R. Soc. London* **1805**, *95*, 65–87.
(15) Spori, D. M.; Drobek, T.; Zürcher, S.; Ochsner, M.; Sprecher, C.; Mühlebach, A.; Spencer, N. D. *Langmuir* **2008**, *24*, 5411–5417.
(16) Quéré, D. *Physica A (Amsterdam, Neth.)* **2002**, *313*, 32–46.
(17) Gray, V. R. *Chem. Ind.* **1965**, *23*, 969.
(18) Gao, L.; McCarthy, T. J. *Langmuir* **2007**, *23*, 3762–3765.
(19) Gao, L.; McCarthy, T. J. *Langmuir* **2009**, *25*, 7249–7255.
(20) Li, W.; Amirfazli, A. *Adv. Colloid Interface Sci.* **2007**, *132*, 51–68.
(21) Extrand, C. W. *Langmuir* **2002**, *18*, 7991–7999.
(22) Priest, C.; Albrecht, T. W. J.; Sedev, R.; Ralston, J. *Langmuir* **2009**, *25*, 5655–5660.

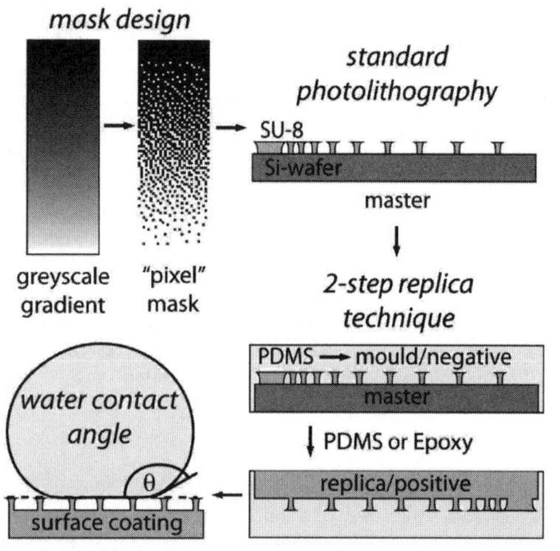

Figure 1. Scheme of experimental setup: A pixel image of a grayscale gradient going from black to white was prepared and then converted into a black-and-white bitmap. Printing this bitmap on a transparency foil leads to a "pixel" mask for standard photolithography. A topographical gradient master was prepared by using standard photolithography of SU-8 on silicon. In a two-step replication process, copies of the master were prepared from epoxy or PDMS. These substrates were functionalized with hydrophobic surface coatings. The wetting properties were analyzed by measuring dynamic and static water contact angles and roll-off angles.

for positioning the water drops. Each substrate was 42 mm long and 12 mm wide. The gradient covered 35 mm of the substrate, leaving 7 mm of microstructure-free area for handling of the substrate and measuring the reference contact angles (f_1 = 100%). The microstructure-free area is flat, only exhibiting the nanoscale roughness possibly created during the photolithographic step and thus having the same surface topography as the flat-top pillars.

Because of the presets of gamma correction in the color workspace of the computer used to generate the gradient mask, the distribution of black and white pixels along the gradient shows a distinct nonlinearity. A detailed analysis of the pixel distribution on the mask is shown in Supporting Information 1.

The mask manufacturing (Fotosatz Salinger AG, Zürich) was carried out with a drum exposure system (3810 dpi resolution) on a polyester foil of 0.1 mm thickness.

Masters. The photolithographic masks were used to prepare the masters for the replica technique. Standard photolithography was performed on 4 in. silicon wafers. After wafer cleaning, SU-8 2025 negative photoresist (MicroChem, USA) was spin-coated for 60 s at 2000 rpm to a thickness of 30 μm and soft-baked on a hot plate for 2 min at 65 °C and 3 min at 95 °C. Subsequently, the wafer was exposed to UV light (MA6, Süss MicroTec, Germany) with constant intensity (total energy 180 mJ/cm^2) through the photolithographic mask described above. After postbaking for 2 min at 65 °C and 3 min at 95 °C, the wafers were developed for 5 min in SU-8 developer (MicroChem, USA). Finally, the wafer was hard baked at 190 °C for 10 min.

Replicas. In order to facilitate the replica process, the masters were exposed to air plasma for 2 min and then functionalized with fluorosilanes (1*H*,1*H*,2*H*,2*H*-perfluorooctyltrichlorosilane, ABCR GmbH, Germany) by vapor phase deposition in a desiccator (rough vacuum) for 1 h. Then, a mixture of polydimethylsiloxane (PDMS, 1:10 curing agent to base, SYLGARD 184 silicone elastomer, base and curing agent, Dow Corning, USA) was cast over the master and allowed to cure overnight at 70 °C. The cured PDMS replica represented the negative of the master. After that, the PDMS negative replica was exposed to 2 min air plasma (RF level high, 0.1 Torr, PDC-32G, Harrick Plasma, USA) and functionalized with a layer of fluorosilanes. Again, a 10:1 mixture of PDMS was cast and cured overnight at 70 °C to yield a positive replica of the master. Contact angles were measured on the bare PDMS positive and on its surface following subsequent functionalization with fluorosilanes.

Epoxy positive replicas were produced as substrates for surface functionalization by self-assembled monolayers (SAMs) with thiol headgroups, similar to the procedure described in our previous publication.[15] The PDMS negatives were used as molds to cast an epoxy blend (EPO-TEK 302-3; Epoxy Technology, USA). The epoxy was mixed according to the protocol provided by the producer (4.5 g of Part A with 10 g of Part B), cast from the negative under vacuum, allowed to cure at 60 °C overnight, and postcured at 150 °C for 1 h. Later, these epoxy replicas were cleaned by ultrasonication for 10 min in a 2 vol % aqueous solution of Hellmanex (Hellma, Germany) and subsequently rinsed five times with ultrapure water (resistance 18.2 MΩ, TKA-GenPure, Huber & Co. AG, Switzerland). Contact angle measurements were performed after this step. By resistance evaporation (MED 020 coating system, BALTEC, Liechtenstein) the samples were coated with a layer of 10 nm Cr and 50 nm Au (purity > 99.99%, Umicore, Liechtenstein). During coating, the stage was rotated and tilted by 25°. Directly after coating, the samples were immersed in a 0.1 mM solution of dodecylthiol (Aldrich Chemicals, USA) in ethanol for 20 min, rinsed with ethanol, and blown dry under a stream of nitrogen. Contact angles were measured on the freshly prepared samples.

Contact Angle and Roll-Off Angle Measurements. Static and dynamic contact angle measurements were performed on a Krüss DSA 100 (Krüss, Germany). Static contact angles (θ_s) were usually measured with drop volumes of 6 or 9 μL on surfaces with contact angles above 140°. The drop was produced, still hanging on the syringe, and then the stage with the substrate was slowly lifted until the substrate touched the drop. Upon lowering the stage again, the drop detached, and after it came to rest, an image was taken. Thus, the history of the contact line was a purely advancing motion. The drop volume had to be increased to 9 μL for contact angles above 140°, since otherwise the drop would not detach from the syringe.

For dynamic contact angles (advancing (θ_a) and receding (θ_r)) measurements the drop volume was increased and decreased with a speed of 15 μL/min. This leads to low-rate contact-angle measurements with advancing contact-line speeds below 0.012 mm/s. Receding contact-line speeds are slightly higher on strongly pinning surfaces, at around 0.03 mm/s. For the advancing drop, two movies with 200 frames and, for the receding drop, one movie with 250 frames were recorded. In cases where the drop was pinned on one side, only the moving side of the drop was taken into account for the evaluation.[15]

All drops were fitted with the tangent method 2 routine, a fourth-order polynomial function, of the DSA3 software (Krüss, Germany). This routine has difficulty in fitting drops with large contact angles in the Cassie regime, leading to a systematic underestimation of a few degrees. Therefore, all drops with contact angles higher than 135° were fitted in ImageJ. Static contact angles were evaluated by means of the drop-analysis[23] plug-in and dynamic contact angles with the simple angle tool delivered with ImageJ. The reasons are more closely described in Supporting Information 2.

Roll-off angles were measured on a home-built device. It consists of a stage that is fixed on one side to a spindle and a goniometer that indicates the angle of the stage. The substrate is placed on the stage, and a drop of 6–9 μL is placed onto the sample surface. The stage is tilted until the drop starts to move, and the roll-off angle is recorded.

(23) Stalder, A. F.; Kulik, G.; Sage, D.; Barbieri, L.; Hoffmann, P. *Colloids Surf., A* **2006**, *286*, 92–103.

Scanning Electron Microscopy. The epoxy substrates were analyzed in a Gemini 1530 FEG SEM (Zeiss, Germany) at 3−5 kV, at room temperature, gold-coated as described above.

Extraction of f_1 and f_2. The grayscale SEM images were imported into the ImageJ program and then transformed into black-and-white bitmaps. These bitmaps can be analyzed with the "analyze particles" tool, yielding pillar-top area and perimeter for each pillar on the image. Summing this pillar-top area and then dividing the sum by the total analyzed (projected) area results in a value of f_1. For f_2 it is assumed that $f_1 + f_2 = 1$, since the pillar tops are flat, exhibiting no significant roughness.

The analyzed areas were 1.88×1.27 mm^2, except for the positions at 51, 56, 61, and 66 mm (corresponding to f_2 values of 80.2, 85.9, 88.2, and 94.4%), where the pillar density fell below 1000 pillars per analyzed area. Below this pillar density the contrast in gray values was too low for the program to distinguish between pillar and background. Therefore, the contrast had to be enhanced by hand by coloring the pillar tops white, and the analyzed area was reduced to 0.39×0.26 mm^2.

Results and Discussion

By the use of standard photolithography, the original bitmap from the Adobe Photoshop program was transferred into a morphological gradient by adding the third dimension (height) to the 2D pattern (see Figure 2). By using a negative photoresist, such as SU-8, all white pixels are cross-linked while the black pixels are etched away. This led to a morphology containing hole features on the "white" side of the gradient, gradually merging to larger holes, until islands were isolated, ending up as single pillars on the "black" side of the gradient. The same parameters for the photolithographic step were used as in our former publication,[15] leading to all features being flat on top and slightly undercut, similar to golf tees. The undercut shape helps to support the drop in the Cassie state[24] and ensures that the drop only wets the tops of the pillars.

By choosing different materials (PDMS and epoxy) and surface coatings (perfluorinated silanes and CH_3-terminated thiols on gold/epoxy), the surface chemistry of the substrates was varied.

In Figure 3a the distribution of the black pixels (black dots) along the gradient extracted from the bitmap is compared to the f_2 measured on SEM images on the positive (epoxy) replica (gray dots). The spots chosen for the SEM analysis were the same spots as used for the contact-angle measurements. The dotted line is a guide to the eye to show a gradient of black and white pixels having an entirely linear increase of black pixels along the distance. As already mentioned in the Materials and Methods section, the black pixels do not increase linearly along the gradient due to presets of the color workspace. The slight deviations between the bitmap and the topography occur during the printing and photolithography steps of the master fabrication. In the middle range (20−45 mm), where f_2 is higher than the bitmap data, color bleeding in the printing process led to larger black areas on the mask. At positions 0−20 mm, f_2 is actually lower than expected because of the cross-linking of the SU-8 epoxy. On an ideal substrate the corners of diagonally neighbored pixels would only touch, but on our master, the SU-8 actually forms narrow bridges across the corners. At the other end of the gradient (positions 45−70 mm) the larger printing of the black pixels compensates the bridges formed between the single, neighboring white pixels (f_1).

Figure 3b is a correlation graphic. It shows f_2 extracted from the SEM analysis plotted versus f_2 calculated from the contact

(24) Tuteja, A.; Choi, W.; Ma, M. L.; Mabry, J. M.; Mazzella, S. A.; Rutledge, G. C.; McKinley, G. H.; Cohen, R. E. *Science* **2007**, *318*, 1618–1622.

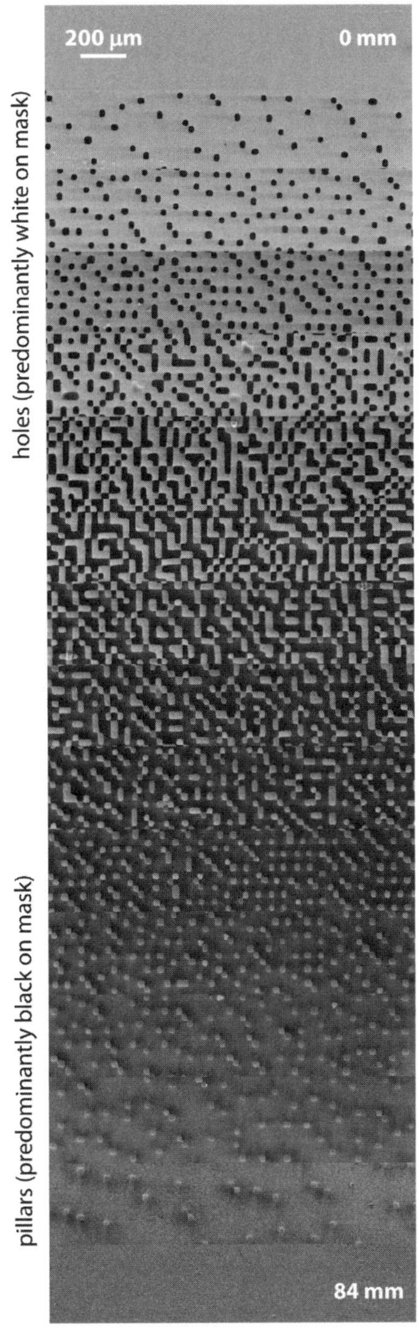

Figure 2. SEM-image composite of the gold-coated epoxy replica, containing 16 images that have been sampled every 5 mm along the gradient. The original white-to-black gradient on the photolithographic mask stretched from the top, white end, to the bottom, black end. The top end of the gradient contains isolated holes, whereas the bottom part is dominated by isolated pillars. In the central part of the gradient, the pixels grow together to yield more complex structures.

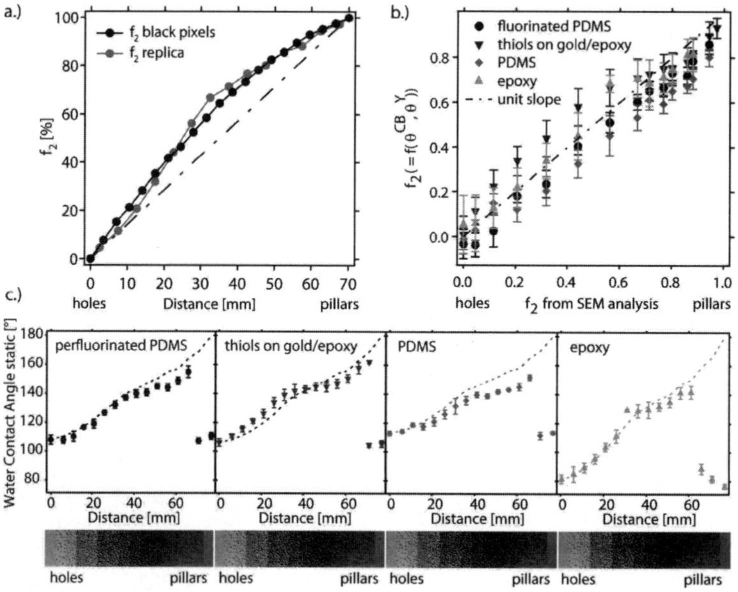

Figure 3. Statistical analysis and static contact angles. (a) f_2 distribution along the gradient. Black circle shows the data of a black pixel analysis on the initial gradient bitmap, and gray circle shows the extracted f_2 from the SEM analysis on the epoxy replica and the dotted line represents a linear increase of f_2 along the substrate. The error bars for the extraction of f_2 from the SEM images of the epoxy replica were smaller than the circles used for the data presentation. (b) Correlation between f_2 measured on the epoxy replica and f_2 calculated from the static contact angle data by solving the Cassie–Baxter equation. (c) The static water contact angle measurements on all four substrates and the corresponding Cassie–Baxter approximation (dashed lines): perfluorinated PDMS, gold- and dodecylthiol-coated epoxy, native PDMS, native epoxy. The contact angles at high f_2 values after the steep drop are in the Wenzel state. There, the pillars are set too far apart from each other to suspend the drop in the CB state.

angle data, solving the CB approximation (eq 1) for f_2:

$$f_2 = \frac{\cos\theta^{Y} - \cos\theta^{CB}}{\cos\theta^{Y} + 1} \tag{4}$$

This calculation was performed for all available surface chemistries and the obtained data entered into the graph. Since, in principle, the same parameter (f_2) is plotted on both axes, the data should exactly follow the bisector. The data show some scattering around the bisector but do not display a significant deviation that would imply another (or no) correlation. At f_2 values exceeding 70%, the scattering pattern changes slightly, which we suspect to be due to gravitational influences on the drop (see Supporting Information 3).

Influence of f_1/f_2 on Contact Angle. When plotting static contact angle data versus distance on the replicate substrates (Figure 3c), the data clearly show the same trend as depicted in the statistical analysis of the distribution of the initial black pixels (and f_2) versus distance (Figure 3a). The shoulder in the black and white distribution occurs at the same position in the contact angle data. This is an indication that on our quasi-random substrates f_1 and f_2 are indeed the significant parameters influencing the contact angle of drops in the CB state. Since the static measurement is closest to thermodynamic equilibrium, it is justifiable to compare the data with the predictions calculated according to the CB approximation (eq 1). In the first part of the gradient in Figure 3c, with f_2 going from 0 to 70% (equal to the distance of 0–41 mm) θ_s shows a good correlation to the Cassie–Baxter approximation. At the black end, where f_2 approaches 100%, θ_s drops down to the angle measured on a flat sample with the same surface chemistry. Here, the distance between the pillars gets so large that the drop cannot span from one pillar top to the next and wets in the Wenzel state. At the position of f_2 equal to 97.3% the drop on the sample coated with thiols on gold/epoxy is metastable. Drops were observed to be in both states: Wenzel and CB state. Just before this transition at f_2 values between 70 and ~90% (positions between 41 and ~61 mm), a deviation from the CB prediction occurs, which may be due to gravitational and discretization effects (see Supporting Information 3).

"Line vs Area". The question arises as to how the parameters f_1/f_2, which are area fractions, can be the main parameters influencing the contact angle in the CB state after it was shown[18,25] that the contact angle of a drop is only defined by what is in the vicinity of the contact line and not by what is to be found underneath the drop.

The explanation is quite simple: Since the topographical features are small compared to the base diameter of the drop (pillar side length of ~30 μm versus drop base diameter of > 1 mm), the value of f_1 determined as an area fraction is equivalent to the fraction of the contact line in contact with the solid. In this case, f_1 is actually a line parameter, not an area parameter. The question as to where the contact line truly contacts the pillars is irrelevant for this consideration, since the pillar tops are small compared to the footprint of the drop and deviations from the ideal circular footprint are negligible.

Dynamic Contact Angle Data. Many equations describing wetting phenomena on rough surfaces such as eqs 1 and 3 include a cosine function. Therefore, plotting the cosine rather than the pure contact angle data versus f_2 yields a straight line, if the data

(25) Extrand, C. W. *Langmuir* **2003**, *19*, 3793–3796.

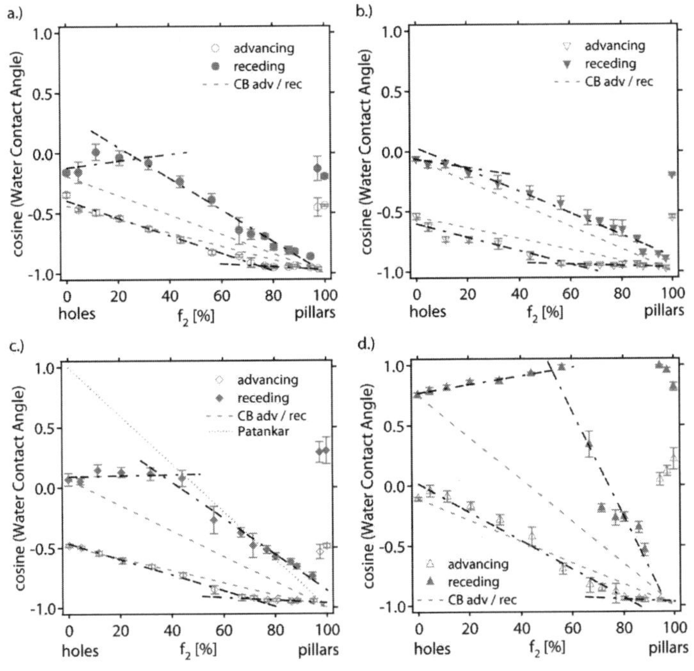

Figure 4. Comparison of dynamic-contact-angle data on different surface chemistries with the prediction by Cassie and Baxter (two lines: advancing slope always lower than receding slope) and Patankar: (a) perfluorinated PDMS surface, (b) gold and CH_3-terminated SAM on epoxy substrate, (c) native PDMS surface, and (d) native epoxy surface. Symbols: open symbols = advancing; full symbols = receding; dashed gray lines = corresponding Cassie predictions; dotted gray line = Patankar prediction (only for the graph of the native PDMS since its shape is the same for all other chemistries). The alternating dashed black lines are linear fits to the data.

follow the CB predictions. Thus, Figure 4 shows the cosine of the dynamic contact angle data measured on all four substrates versus f_2 (perfluorinated PDMS, PDMS; CH_3-terminated thiols and the pure epoxy surface). Additionally, the CB predictions for the dynamic case were calculated, using the f_2 extracted from the SEM images. The condition for receding contact angles as proposed by Patankar (eq 3) was also computed. Since eq 3 is only dependent on f_1 and not on surface chemistry, it looks the same for all data sets. In order to keep Figure 4 readable, it is only added to the graph of the native PDMS because there the strongest correlation between Patankar's prediction and the receding contact angle can be seen.

Supporting Information 4 shows the same figure with the raw measured contact-angle data rather than the cosines.

Similarly to the static contact angles, the advancing and receding contact angles show a distinct dependence on f_2. A detailed analysis shows that the gradients can be split into different regimes: For values of f_2 larger than 95%, the distance between neighboring pillars is too large to be spanned by the drop, and therefore the drop is in the Wenzel state, showing contact angles close to those found on a flat surface. For smaller f_2 values the drop only touches the pillar tops. In Figure 4, linear fits were added to selected data points (alternating dashed black lines). In the central region of the gradients, advancing as well as receding contact angles rise as air enclosure increases. For the perfluorinated PDMS surface, this region ranges from f_2 = 0% to 75% for advancing angles and f_2 = 28% to 95% for receding angles. Receding angles show a tendency to level off for low values of f_2, whereas for advancing angles, a plateau is visible for large f_2. Qualitatively, the behavior is the same for all the materials investigated, although the transitions from one regime to the other occur at different positions on the gradient.

Looking at the advancing contact angles, they rise with increasing hole density. The rise can be approximated by a linear trend line and in the case of the PDMS and the fluorinated PDMS even follows the CB prediction exactly. As the features increasingly resemble single pillars, the θ_a levels off at about 160° (150° for the more hydrophilic epoxy substrate), and no change is detectable with decreasing pillar density. The phenomenon of leveling off at high advancing contact angles (around 160°) has also been reported by others,[5,22] working with periodically distributed micrometer-sized pillars (4-fold symmetry). In this work, stagnation starts between 60% and 80% of air, and not at the same location for all substrates, even though they are replicates of the same master. Clearly, surface energy plays a role in the onset of stagnation. Additionally, since the sequence (CH_3-terminated thiols, perfluorinated PDMS, PDMS, epoxy) does not strictly follow surface energy, other properties of the material can apparently play a role. For example, PDMS, an elastomer, has a much lower elastic modulus than epoxy, which is a thermoset. The low modulus of PDMS is also the reason why the advancing contact angle on the flat part of PDMS is larger than that on the perfluorinated PDMS: the y-component of the liquid−air surface tension cannot be fully compensated and thus forms a rim around the base perimeter of the drop. This rim acts as surface roughness and slightly increases the contact angle.[26,27] During plasma

(26) Extrand, C. W.; Kumagai, Y. *J. Colloid Interface Sci.* **1996**, *184*, 191–200.
(27) Shanahan, M. E. R.; De Gennes, P. G. *Equilibrium of the Triple Line Solid/Liquid/Fluid of a Sessile Drop*; Elsevier Applied Science: New York, 1987.

treatment of the PDMS, which is necessary to functionalize the surface with perfluorinated silanes, a silica-like layer is formed.[28] This layer is stiffer than the bulk PDMS. Therefore, the perfluorinated PDMS has a surface coating that is both more hydrophobic and stiffer than the native PDMS; thus, no rim formation occurs on these samples (see Supporting Information 5).

The receding contact angles mirrored this behavior to some extent. On the holey side they first stagnate or even decrease, again until the holes become larger and the features resemble pillars, at which point the θ_r begins to increase. The rise starts at around 10% of air for the consistently hydrophobic substrates (CH_3-terminated thiols and perfluorinated PDMS) and at 50% of air for the more hydrophilic substrates (PDMS and epoxy). PDMS is known to change the orientation of its methyl groups upon contact with water, making it more hydrophilic in this situation.[29,30] This change of orientation and the aforementioned rim formation are the reason for the high contact angle hysteresis—even on flat PDMS; besides the roughness of the rim, the advancing contact line encounters a more hydrophobic surface than the receding contact line. The CB equation generally greatly overestimates θ_r, and the Patankar derivation strongly underestimates the measured θ_r of the hydrophobic substrates and clearly overestimates the θ_r of the epoxy surface. Only in the case of PDMS does the θ_r rise follow the Patankar prediction from 45% to 80% of air. This phenomenon is not yet a proof that a water film is actually left behind (the Patankar assumption), since other factors such as reorientation could be playing a role.

Priest et al.[22] showed in their investigations on substrates consisting only of holes or of pillars that the contact-angle hysteresis actually indicates on what type of feature the drop sits. Stagnation of the advancing and linear increasing of the receding contact angle with an increasing amount of air enclosure is a characteristic of the pillar surface. A linear increase of the advancing (a rather close agreement with the CB approximation) and stagnation or even decrease of the receding contact angle is found on holey surfaces. On our gradient surface where holes slowly merge together and pillars are formed, this behavior is also observed. However, although all four substrates exhibit the same topography, the transition point at which the behavior changes from hole contact to pillar contact cannot be found at the same f_2 value (see Figure 4, intercepts of the linear regressions (alternating dashed lines)). Most probably, between air percentages of 20–70% the effects of topography intermingle and surface energy starts to play a more important role.

In Figure 5, the cosine of the contact-angle hysteresis ($\Delta \cos \theta = \cos \theta_r - \cos \theta_a$) vs f_2 is shown. With increasing air content the hysteresis increases for all substrates up to about ~50% air. Beyond this point, the hysteresis decreases rapidly until it reaches its minimum, before the collapse into the Wenzel state (indicated by a sharp increase in hysteresis at large f_2 values). For the two consistently hydrophobic substrates (CH_3-terminated thiols and the perfluorinated PDMS) a plateau in hysteresis can be distinguished between 20% and 50% of air. Nevertheless, all curves show a minimum at around 45% of air. Dividing the data sets into two parts (0 to 45% and 45 to ~100% for all drops in CB state), the first part is predominantly governed by the mechanism originating from the hole structure and the second part is mostly governed by the presence of pillars. Each part can be fitted by a linear curve fit. According to Priest et al.,[22] the slope in this linear fit is a measure for the pinning energy E_{pin} and equal to E_{pin}/γ_{LA} (see eq 5):

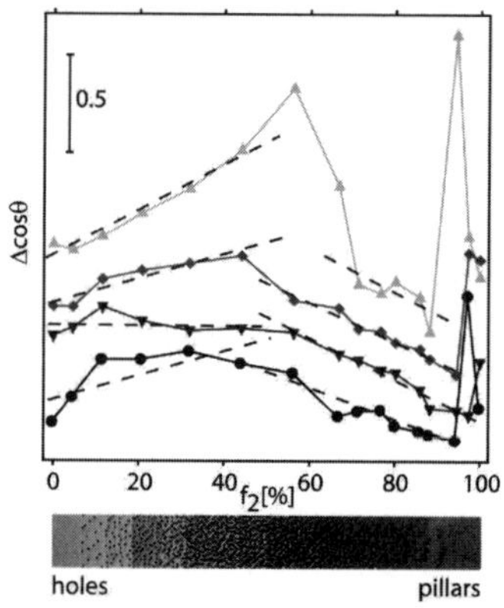

Figure 5. Contact-angle hysteresis ($\cos \theta_r - \cos \theta_a$) as a function of f_2 ($f_2 = 1 - f_1$). The curves have been separated for clarity. The slopes of the linear regressions are a measure for the pinning energy[22] (circles = perfluorinated PDMS; triangles = thiols on gold/epoxy; squares = native PDMS; pyramids = epoxy).

Table 1. Slopes of the Linear Fit in Figure 5 (Dashed Lines) Are Compared to the Values Found by Priest et al. on Homogeneous Substrates

E_{pin}/γ_{LA}	holes	pillars
Priest et al.[22]	0.4	1.2
perfluorinated PDMS	0.6	0.8
CH_3-terminated thiols	0	1.1
PDMS	0.6	1
epoxy	1.2	1.2

$$\Delta \cos \theta = \Delta \cos \theta_0 + f_D \frac{E_{pin}}{\gamma_{LA}} \quad (5)$$

f_D corresponds to the area fraction of the "defect". A surface consisting of holes can be considered as a matrix of substrate with very low-energy defect patches (air enclosures). Therefore, f_D corresponds to f_2 on the holey side of the substrate. Similarly, a pillar surface consists of a matrix of air with high-energy defects (pillars; even a hydrophobic surface has a higher interaction with the liquid than air). Thus, on the pillar side, f_D corresponds to f_1.

The slopes found on the substrates used in this work clearly show the same tendency: pinning energy of the pillars is significantly higher than the pinning energy of the holes (Table 1). On the holey surface the contact line is mostly in contact with the solid. In this way the energy barriers to adopting the contact angle corresponding to the lowest-free-energy state are lowered because of increased flexibility in the positioning of the contact line on the substrate. By being able to meander between the defects, the contact line does not have to follow the shape of the defects exactly, as it does on the pillar side, but can actually average over the defects on the surface, as was assumed in the CB approximation. This averaging might also be the reason why the CB approximation fits the data of the θ_a better on the holey side than on the pillar side. But, as mentioned above, the difference between the values of E_{pin}/γ_{LA} (Table 1, holes vs pillars) is not quite as profound as in the work of Priest et al., most probably due to the fact that both features are present in the middle of the gradient. The differences in E_{pin}/γ_{LA} between the present work (stochastic)

(28) Hillborg, H.; Sandelin, M.; Gedde, U. W. *Polymer* **2001**, *42*, 7349–7362.
(29) Chen, C.; Wang, J.; Chen, Z. *Langmuir* **2004**, *20*, 10186–10193.
(30) Morra, M.; Occhiello, E.; Marola, R.; Garbassi, F.; Humphrey, P.; Johnson, D. *J. Colloid Interface Sci.* **1990**, *137*, 11–24.

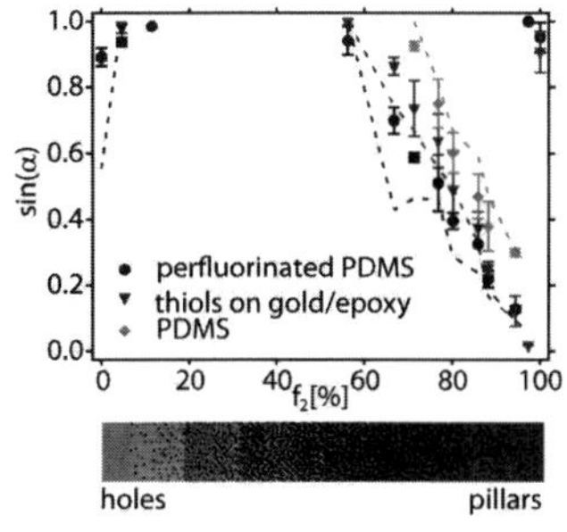

Figure 6. Sine of the roll-off angles of all analyzed surfaces. Only data points are shown where a roll-off angle could actually be measured. All positions where the drop did not move up until a tilt angle of 90° are not displayed. Drops on the pure epoxy surface did not roll off at any angle. The dotted line shows the prediction by the Furmidge equation for each surface.

and that of Priest et al. (periodic) for similar features and surface chemistries could also be an effect of the greater contact line distortion in the stochastic case, as was suggested by Öner.[10]

Despite the strong hysteresis, all drops evaluated were still in the CB state. In Supporting Information 6, examples of light microscopy images taken through a water drop are presented and clearly show the air enclosure over the full gradient. Despite its lower hydrophobicity, the epoxy surface is clearly in the CB state, since it shows qualitatively the same behavior as observed on the more hydrophobic substrates, which would not be the case if it were in the Wenzel state.[31]

Figure 6 shows the sine of the roll-off angles measured on the four different substrates together with the approximation calculated by the Furmidge equation (see eq 2, dashed lines). All parameters occurring in the Furmidge equation were measured independently from the roll-off angle (drop weight m, drop diameter w, advancing and receding contact angles θ_a, θ_r). Positions where no roll-off angles occurred are not displayed in that graph. For the pure epoxy surface no roll-off angle at any position could be observed for the given drop volumes. The native PDMS does not show any roll-off on the flat area due to the high hysteresis induced by the reorientation of the polymer side chains upon contact with water. For the two consistently hydrophobic substrates (perfluorinated PDMS and CH_3-terminated thiols on gold), on the holey side of the gradient the roll-off angle increases with increasing hole density, until no movement is observable even up to 90°. Movement is only possible again when the topography changes from holes to pillars. There, the sine of the roll-off decreases linearly with f_2 (R^2 of the linear regression > 97%).

As observed in Figure 5, the holey side shows a lower pinning energy than the pillar side. But in Figure 6 the drop rolls off far more easily on the pillar side. One reason for this is that the contact line has far more contact (low f_2) on the holey side than on the pillar side. Therefore, the inherent contact-angle hysteresis on the material plays a bigger role on the adhesion of the drop. At first this adhesion is increased by the introduction of pinning defects such as holes. Additionally, there is a range on the holey side of the gradient where the drop could not roll off at all. Two reasons may explain this phenomenon: (1) the receding contact line was strongly pinned, possibly aided by the finite curvature of the edges, at the front end of the holes;[22] (2) due to the suction induced by the sealed air cushions,[32,33] the drop was held on the surface. We therefore conclude that there are at least three different influences beyond f_2 that determine whether a drop can roll off a surface consisting of holes (sealed air cushions): one, the size and with it the weight of the drop defines the driving energy for downward motion (as seen by Reyssat et al.[34]); two, pinning at the front edge of the drop; and three, the magnitude of the suction effect.

On the pillar side the drop started moving at an air content of 56.2% (CH_3-terminated thiols and perfluorinated PDMS) and for the PDMS at 71.4%. The onset of drop movement for the consistently hydrophobic substrates coincides with the position chosen to divide the data for the linear regressions in Figure 5, probably indicating a change in the prevalent pinning mechanism. The Furmidge equation is able to predict the general behavior of the roll-off angles. It is an equation consisting only of measured values. Since no distinct deviation of the measured values can be observed, it can be concluded that all necessary parameters to describe the behavior were identified. The suction or pinning events mentioned above also influence the receding contact angle and are therefore taken into account in the equation.

Conclusions

A novel, rapid method for creating photolithographic masks has proven invaluable for the fabrication of gradients of 3D structures. Pseudo-random hole-to-pillar density gradients, covering the total f_1 range of 0−1, have been shown to be a useful tool for investigating the influence of structural effects on a variety of different wetting phenomena, such as static and dynamic contact angles and roll-off angle.

By the use of a gradient surface consisting of nearly randomly placed pillars slowly agglomerating into holes, different wetting mechanisms of drops in the CB state could be identified. Static contact angles increase more or less linearly with f_2, clearly indicating the importance of f_1 and f_2 for the wetting behavior of a drop in the CB state. Since the topographical structures here are small compared to the drop diameter, the area parameters f_1 and f_2 can also be considered as line parameters influencing the contact line.

When measuring advancing and receding contact angles, analyzing the hysteresis can distinguish whether holes or pillars are the predominant topographical feature. Dynamic measurements have shown that pinning energy is higher on pillared than on holey surfaces. Wetting behavior is consistent with that reported in previous (nongradient) studies,[22] which utilized periodic surface structures, and thus the degree of periodicity appears to be of secondary importance to that of the f_1 value.

Drops of 6 μL were found to be pinned so strongly on holey structures that they do not roll off at all. Two potential causes were identified: pinning at the front edge of the drop and suction events of sealed air cushions.

Acknowledgment. The authors thank Ciba Speciality Chemicals for their generous financial assistance and Electron Microscopy ETH Zurich, EMEZ, for their support. Additionally, we thank René Tölke for help with the photolithography, Cathrein Hückstädt for fruitful discussions, and Martin Elsener for building the roll-off angle goniometer. Financial support of TD by BMBF grant "Geotechnologien" 03G0709A is gratefully acknowledged.

(31) Fetzer, R.; Ralston, J. *J. Phys. Chem. C* **2009**, *113*, 8888–8894.
(32) Wang, S.; Jiang, L. *Adv. Mater.* **2007**, *19*, 3423–3424.
(33) Steinberger, A.; Cottin-Bizonne, C.; Kleimann, P.; Charlaix, E. *Nat. Mater.* **2007**, *6*, 665–668.
(34) Reyssat, M.; Quéré, D. *J. Phys. Chem. B* **2009**, *113*, 3906–3909.

Supporting Information Available: Supporting Information 1: more information on the mask characterization; Supporting Information 2: the difficulties in analyzing very high contact angles by testing different analysis tools; Supporting Information 3: discussion of the possibility that for drops in the CB state (showing very high contact angles) gravitational effects may occur at smaller volumes than is generally assumed by calculating the capillary length; Supporting Information 4: contact angle data acquired on the gradient vs f_2; Supporting Information 5: discussion of how the modulus of an elastomer can influence contact angle hysteresis; Supporting Information 6: light microscopy study showing that the drops were in a CB state over the full range of the topographical gradient. This material is available free of charge via the Internet at http://pubs.acs.org.

3c. Surface structural effects on cells

Commentary

It is well known that roughness affects the behavior of cells on surfaces, and this effect is put to use in implants, in order to influence whether or not bone is going to grow onto the surface. We wanted to study this in a controlled way, and to this end created roughness gradients that allowed us to follow the effect of roughness over a significant range. In the case of roughness on the micrometer scale (3.5), we found, as others have done, that osteoblasts proliferate preferentially on rougher surfaces, while fibroblasts proliferate preferentially on smooth surfaces. On the nanometer scale the situation is somewhat different, and we found that a high surface concentration of adherent nanoparticles leads to an almost total inhibition of cell attachment (3.6).

Available online at www.sciencedirect.com
ScienceDirect
ELSEVIER
Biomaterials 28 (2007) 2175–2182
Biomaterials
www.elsevier.com/locate/biomaterials

Systematic study of osteoblast and fibroblast response to roughness by means of surface-morphology gradients

Tobias P. Kunzler, Tanja Drobek, Martin Schuler, Nicholas D. Spencer*

Laboratory for Surface Science and Technology, Department of Materials, ETH Zurich, Wolfgang-Pauli-Strasse 10, CH-8093 Zurich, Switzerland

Received 15 October 2006; accepted 9 January 2007
Available online 14 January 2007

Abstract

The surface roughness of a medical implant is of great importance since the surface is in direct contact with the host tissue (e.g. bone, fibrous tissue). The response of cells to roughness is different depending on the cell type. However, the influence of roughness on cell behavior has only rarely been systematically studied. We have developed a surface-modification process to produce roughness gradients that cover a wide range of roughness values on one substratum. Such gradients allow for systematic investigations of roughness on cell behavior. Gradients were fabricated using a two-step roughening and smoothening process, involving sandblasting and a subsequent chemical polishing step. In order to produce a set of identical surfaces we applied a replica technique. Cell experiments were carried out with rat calvarial osteoblasts (RCO) and human gingival fibroblasts (HGF). RCOs showed a significantly increased proliferation rate with increasing surface roughness. The footprint of osteoblasts varied in size at different positions on the gradient, remaining small on the rough end of the gradient and increasing considerably as the roughness decreased. HGF showed the opposite proliferation behavior, proliferation decreasing with increasing roughness. The fibroblast morphology was found to be similar to that seen for osteoblasts.

Keywords: Osteoblast; Fibroblast; Cell proliferation; Cell morphology; Surface roughness; Surface modification

1. Introduction

The surface chemistry and topography of a biomaterial are both known to play a crucial role in success or failure upon placement in a biological environment [1]. In the case of dental implants, a primary requirement is the anchoring and attachment of bone with the implant, a process termed osseointegration [2]. A key issue for successful osseointegration is the topographical and chemical design of the implant surface [3]. It is complicated further by the fact that different tissue types than bone (epithelium and connective tissue) interface with the implant surface. Around the neck of a dental implant it is important that connective tissue rapidly forms a seal to prevent bacterial invasion [4] and the downgrowth of the overlaying epithelium [5]. For the bone-contacting region of the implant, rapid initial bone formation is required to permit early loading and conducive conditions for bone remodeling and long-term stability [6].

One approach currently employed with dental implants is to vary the surface topography [7], as certain cell types have been observed to show a preference for smooth or rough topography [8]. It is well known that osteoblastic cells prefer rougher surfaces, whereas fibroblasts, the most common cell type found in connective tissue, favor smooth surfaces [8,9]. Although the cell response to surface roughness has been extensively investigated, studies have often been limited to either using rough or smooth substrata, covering only a small range of roughness values. The influence of roughness has only rarely been *systematically* studied [10,11]. In addition, substrata are produced by different surface modification techniques, creating different types of surface features. Therefore, it is difficult to compare the cell behavior on these surfaces with each other.

Roughness gradients, which are fabricated with the same surface-modification technique, have similar surface features of varying roughness value on one surface. This can

*Corresponding author. Tel.: +41 44 632 5850; fax: +41 44 633 1027.
E-mail address: spencer@mat.ethz.ch (N.D. Spencer).

0142-9612/$ - see front matter © 2007 Elsevier Ltd. All rights reserved.
doi:10.1016/j.biomaterials.2007.01.019

greatly facilitate the systematic investigation of cell behavior, allowing studies to be performed with a continuously varying surface parameter within a single experiment. We have recently developed a method that allows the production of stochastic roughness gradients over a centimeter scale with topographical features in the micrometer and sub-micrometer range, which were created by a single surface-modification technique [12]. Subsequent application of a replica technique permits roughness gradients to be fabricated in large numbers, all with identical surface topography and well-defined surface chemistry [13,14].

There are only few reports in which roughness gradients have been applied in cell-culture studies [10,11]. Such gradients were produced by varying the crystallinity of a polymer film by means of a temperature-gradient stage. This method created roughness gradients with surface features in the low-nanometer range. It is well known that the biological response to surfaces roughness is also significantly affected by features in the micrometer range [15,16]. To our knowledge no systematic study has been performed on the influence of stochastic sub-micrometer to micrometer-scale roughness on cells. The present study provides systematic investigation of the morphological and proliferative behavior of osteoblasts and fibroblasts by means of roughness gradients.

2. Materials and methods

2.1. Gradient surfaces

2.1.1. Fabrication

Gradient surfaces were produced as described by Kunzler et al. [12]. In brief, a high-purity aluminum sheet was sandblasted and gradually immersed into a chemical polishing solution that preferentially removed features with a small radius of curvature as a function of time. In order to obtain gradients over a length of 10 mm, an immersion speed of 17.5 μm/s was used, resulting in an exposure time between 0 and 38 min.

Replicas of original aluminum roughness gradients were produced as described by Wieland et al. [13] and Schuler et al. [14]. An impression ("negative") of the aluminum gradient was produced using poly-vinyl-siloxane (PROVIL novo Light; Heraeus-Kulzer, Switzerland). Epoxy resin (EPO-TEK 302-3M; Epoxy Technology, Polyscience AG, Switzerland) was then cast onto these poly-vinyl-siloxane negatives and cured at 60 °C for 24 h. Cured, positive epoxy replicas were cleaned in a 2% solution of Helmanex (HELLMA GmbH & Co., Müllheim, Germany) and rinsed ten times with ultra pure water. After drying the replica substrata in air overnight, they were sputter-coated with a 60 nm layer of titanium using reactive magnetron sputtering (physical vapor deposition (PVD), Paul Scherrer Institute, Villigen, Switzerland).

2.1.2. Characterization

Gradients were characterized with a laser profilometer (UBM, Hilpert Electronics AG, Baden-Daettwil, Switzerland). Profiles were measured perpendicular to the long gradient axis with a resolution of 1000 points/mm. The data were evaluated with Matlab 7.0.4 R14. Profiles were fitted with a polynomial of degree 4 after removing outliers. The roughness parameters were calculated according to DIN EN ISO 4288-98. Qualitative characterization of the gradients was performed with a SEM (Hitachi S-4100, Hitachi, Japan) in the secondary-electron mode. Images were taken at a voltage of 5 kV and an emission current of 45 μA.

2.2. Cell culture

2.2.1. Osteoblasts (RCO)

Osteoblastic cells were obtained from new born rat calvariae, isolated and sub-cultured as described by Hasegawa et al. [17] and Chehroudi et al. [18]. Frontal, parietal and occipital bone were dissected, rinsed in phosphate-buffered saline (PBS, Fluka Chemicals, Buchs, Switzerland), placed in Dulbecco's modified Eagles medium (α-DMEM, Invitrogen, Basel, Switzerland), supplemented with 1% antibiotics (Penstrep, Invitrogen, Basel, Switzerland) and 10% foetal bovine serum (Invitrogen, Basel, Switzerland). The minced tissue was digested with a mixture of clostridial collagenase and trypsin (both from Sigma-Aldrich, Buchs, Switzerland) and then placed in tissue-culture flasks.

2.2.2. Fibroblasts (HGF)

Human gingival fibroblasts (HGF) were maintained according to the method of Elvin and Evans [19]. Stock cultures were recovered from liquid nitrogen and plated at 200,000 cells per 25 cm^2 tissue-culture flask in α-DMEM with 10% foetal bovine serum and 1% antibiotics.

2.3. Experimental design

2.3.1. Substrate preparation

Gradient epoxy replicas (five replica substrata per time period) were placed in 24-well plates together with epoxy replicas of SLA (sandblasted and acid etched) cpTi discs (Institut Straumann, Basel, Switzerland) [20] and titanium-coated Thermanox discs (Nunc, Wiesbaden, Germany) as reference surfaces. Substrata were sterilized and the surface activated by oxygen-plasma treatment (Harrick Plasma, Ithaca, USA) for 4 min. Immediately after the plasma treatment, the substrata were covered with media (α-DMEM with 10% foetal bovine serum and 1% antibiotics).

2.3.2. Cell incubation

After reaching confluency, cells were detached from the flask with trypsin-EDTA solution (Sigma-Aldrich, Buchs, Switzerland), centrifuged at 5000 rpm for 5 min, resuspended in media (α-DMEM with 10% foetal bovine serum and 1% antibiotics) and seeded at a concentration of 3500 cells/cm^2. The cells were incubated for 1, 2, 4 and 7 days at 37 °C in a humidified atmosphere of 93% air and 7% CO_2.

2.3.3. Cell staining

Cells were washed with prewarmed PBS (Fluka Chemicals, Buchs, Switzerland) and fixed in a 4% paraformaldehyde solution (freshly prepared from a 20% stock solution, paraformaldehyde powder from Fluka Chemicals, Buchs, Switzerland). After rinsing with PBS, the cells were permeabilized using 0.5% Triton X-100 (Fluka Chemicals, Buchs, Switzerland). The substrata were incubated with Alexa Fluor 488 phalloidin (1:100 dilution in PBS; Invitrogen AG, Basel Switzerland) for 25 min at room temperature and rinsed with PBS. The nucleus was stained with DAPI (1:1000 dilution in PBS; Invitrogen, Basel, Switzerland) for 15 min at room temperature. The substrata were rinsed again with PBS and kept in PBS in order not to dry off.

2.3.4. Evaluation

Substrata were observed with a fluorescent microscope (Axio IMAGER M1m, Zeiss, Oberkochen, Germany). In order to determine the cell number and morphology, three images at a time were taken at 11 positions (1 mm apart) along the gradient with a DAPI and FITC filter set (filter set no. 49 and 10, Zeiss, Oberkochen, Germany). Five substrata per time period were investigated. The number of cell nuclei was counted with ImageJ software (Version 1.36b for Mac OS X). Comparison between measurement series were made using non-parametric tests (SPSS 11.4 for Mac OS X) because variances were not homogeneous and observed values were not distributed normally even after appropriate transformations. All mean values are shown ± SE (standard error). We set our level of significance at 0.05.

T.P. Kunzler et al. / Biomaterials 28 (2007) 2175–2182 2177

3. Results

3.1. Roughness gradients

The replica technique enabled a large number of identical roughness gradient substrata to be produced from the original aluminum roughness gradient. By means of the titanium coating a homogeneous surface chemistry could be achieved. A photograph of a titanium-coated roughness gradient replica is shown in Fig. 1.

Characterization by laser profilometry showed that the roughness decreased monotonically along the gradient axis (Fig. 2). Maximum peak-to-valley values of about 45 μm were measured at the rough end. Using the chemical-polishing procedure it was possible to reduce these peak-to-valley values to minimum values of about 3 μm at the smooth end. Roughness values were calculated according to the DIN EN ISO 4288-98 standard. Table 1 shows an overview of different roughness values. The amplitude parameters Ra, Rq and Rz were found to decrease monotonically. The spacing parameter, Sm, increases with decreasing roughness. The skewness value Sk shows no trend. It oscillates around zero, indicating that the profile remains symmetric throughout the polishing process.

SEM investigations revealed the removal of small features even after a short polishing time. Fig. 3A shows SEM images of different positions on the gradient, corresponding to different polishing times. The first image in row A of Fig. 3 shows a section that was sandblasted only. Sharp edges, ridges and peaks created by the sandblasting procedure can be observed. These sharp edges, ridges and peaks were removed within minutes of the onset of chemical polishing (Fig. 3A, 1 mm). After about 30 min of polishing, features such as larger pits and protuberances were removed (Fig. 3A, 4 and 9 mm).

3.2. Cell response

3.2.1. Osteoblasts (RCO)

After 1 day of culture, no significant difference in the number of osteoblasts was found at different positions on the gradient (Kruskal–Wallis: $\chi^2_{10} = 13.46$, $p = 0.199$), indicating that the attachment was similar on the rough and smooth parts of the gradient (Fig. 4). After 2 days the number of osteoblasts was slightly higher towards the rougher end of the gradient (Fig. 4). Statistical analysis showed a significant difference (Kruskal–Wallis: $\chi^2_{10} = 67.96$, $p<0.001$). Further analysis using multiple comparisons showed that the cell number at the highest Ra value was significantly different compared to the cell number

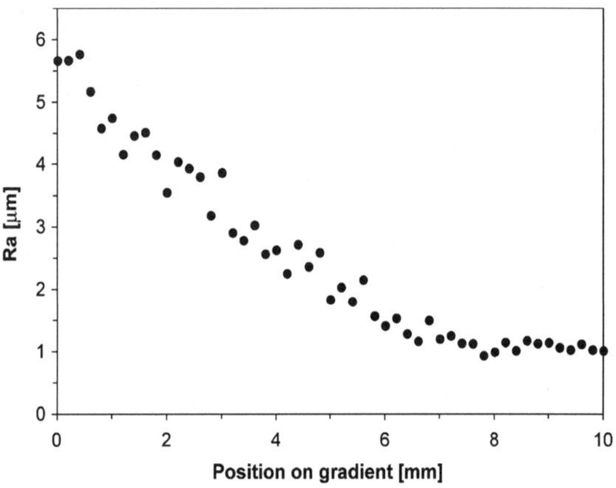

Fig. 2. Roughness value Ra along the gradient axis. Each point is calculated from a measured length perpendicular to the gradient axis. The Ra value was determined according to DIN EN ISO 4288-98. Measurements were carried out with a laser profilometer.

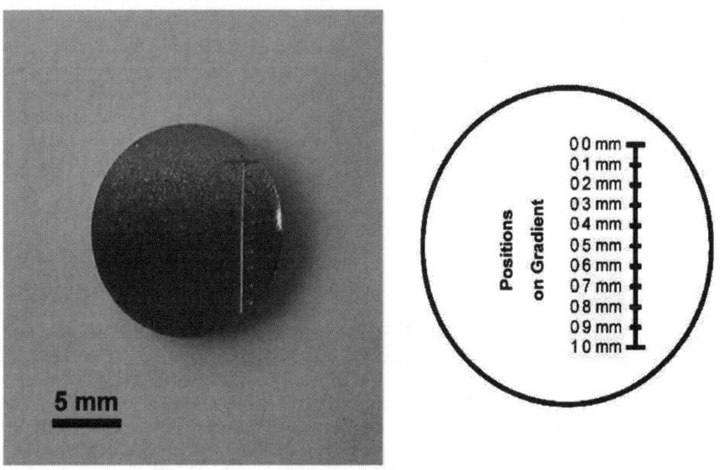

Fig. 1. Photograph of an epoxy replica roughness gradient of an original aluminum roughness gradient (left). Epoxy replica substratum is coated with 60 nm titanium. Schematic illustration of the epoxy replica gradient with the labeling of the different positions on the gradient (right).

Table 1
Standard roughness values of an SLA and a Thermanox surface and of different positions on the roughness gradient measured by laser profilometry

Roughness values (μm)	SLA	Thermanox	Position on sample			
			Sand blasted	1 mm	4 mm	9 mm
Ra	3.00	0.04	5.70	4.50	2.48	1.12
Rq	3.74	0.05	7.56	5.69	3.39	1.41
Rz	17.22	0.23	34.35	28.16	13.30	3.24
Sm	13.00	12.86	27.01	22.43	13.67	22.98
Sk	0.002	−0.035	−0.022	−0.011	0.061	0.016

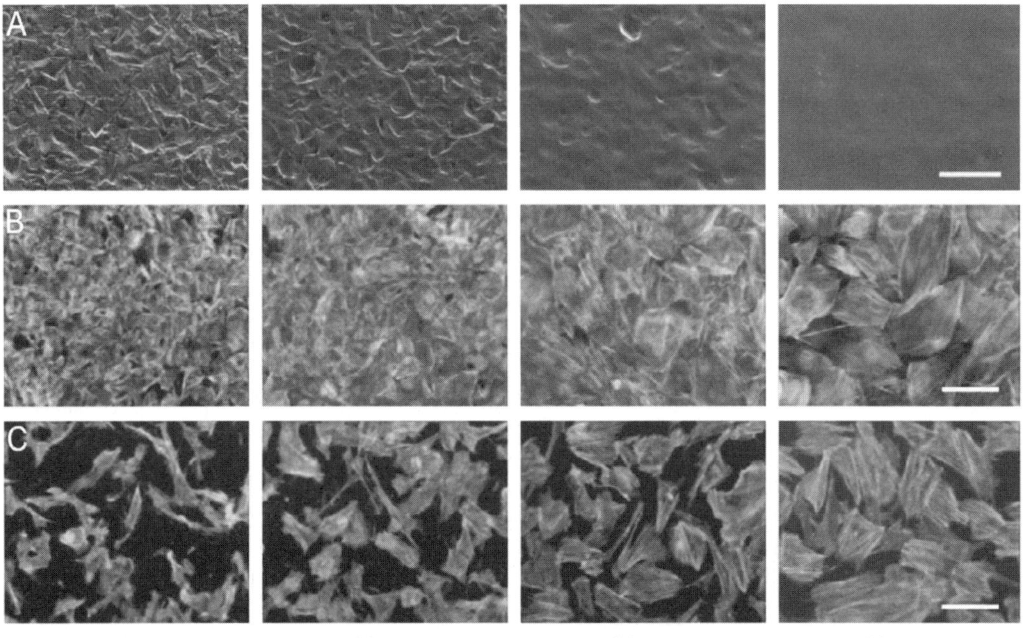

Fig. 3. (A) SEM images of different positions (sand blasted (sb), 1, 4 and 9 mm) and the according roughness value Ra on the roughness gradient. Morphology of rat calvarial osteoblasts (B) and human gingival fibroblasts (C) at different positions on the gradient. Cells were cultured for 7 days. The nucleus was stained with DAPI (blue) and the actin network with Alexa Fluor 488 phalloidin (green). Scale bar is 200 μm.

below Ra values of 4 μm (Tamhane test, $p<0.020$). A significant increase in the number of osteoblasts at the rough end could be observed after 4 days (Fig. 4) (Kruskal–Wallis: $\chi^2_{10} = 120.65$, $p<0.001$). At an Ra value of approximately 2 μm the number of osteoblasts continuously increased with increasing roughness to about two times the number found at Ra values below 2 μm. Multiple comparisons between different positions revealed that the cell number at the four positions with highest Ra values (Ra > 3.3 μm) was significantly different compared to positions with an Ra value below 2 μm (Tamhane test, $p<0.001$). After 7 days, osteoblasts showed similar behavior as already observed after 4 days (Fig. 4). The number of cells significantly increased with increasing roughness (Kruskal–Wallis: $\chi^2_{10} = 97.38$, $p<0.001$). The onset of the increase of number of cells is at an Ra value of about 2 μm. The number of osteoblasts was two times higher on the roughest part compared to the smooth part. The cell number at the three roughest positions (Ra > 3.9 μm) was statistically different from the cell number found at positions below an Ra value of 2 μm (Tamhane test, $p<0.018$). The control experiments on smooth Thermanox discs and on SLA surfaces showed a slightly increased initial attachment to SLA surfaces than to the Thermanox surface (Fig. 4). After 7 days the number of osteoblasts was 1.7 times higher on the Thermanox discs compared to the SLA surface.

The cell morphology was observed to change continuously following the gradient axis from rough to smooth. At the smoother end of the gradient, osteoblasts spread well and were roundish in shape with a distinctive actin network (Fig. 3B). Towards the rough end of the gradient, osteoblasts were less spread and had a much smaller diameter and a more cuboidal shape. On the Thermanox discs, osteoblasts showed similar behavior as seen for cells on the smooth part of the gradient. They were well spread with a distinctive, filamentous actin network (Fig. 5A).

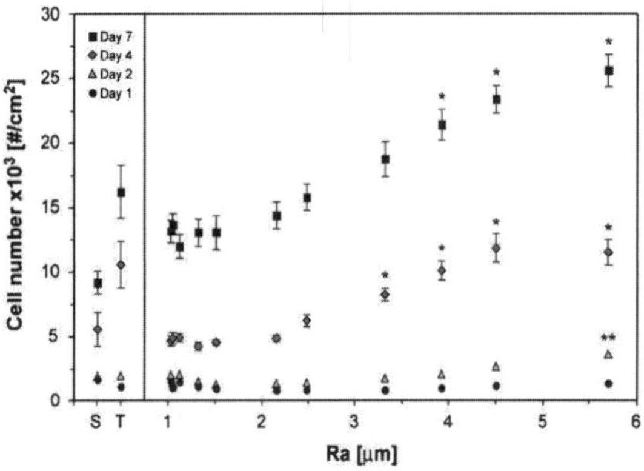

Fig. 4. Number of rat calvarial osteoblasts at different positions on the roughness gradient and on a SLA (S) and Thermanox (T) control surface. Cells were seeded at a density of 3500 cells/cm^2 and cultured for 1, 2, 4 and 7 days ($n = 5$). *Significant difference compared to cell number values below an Ra value of 2 µm (Tamhane test, $p<0.05$). **Significant difference compared to cell number values below an Ra value of 4 µm (Tamhane test, $p<0.05$).

3.2.2. Fibroblasts (HGF)

Fibroblasts showed a slightly higher initial attachment on the rough part of the gradient after 1 day of culture (Fig. 6). A statistical difference could be found (Kruskal–Wallis: $\chi^2_{10} = 19.14$, $p = 0.039$), although multiple comparison analysis revealed that the cell number only significantly differed at the roughest Ra value (5.9 µm) in comparison with that at an Ra value of 1.5 µm (Tamhane test, $p<0.022$). After 2 days the number of fibroblasts did not noticeably increase on any position of the gradient (Fig. 6), but statistically showed some difference (Kruskal–Wallis: $\chi^2_{10} = 26.74$, $p = 0.003$). A difference could only be found for the comparison between the cell number at an Ra value of 4.5 µm and the cell number at Ra values between 1.6 and 2.2 µm (Tamhane test, $p<0.018$). Four days post seeding the number of fibroblasts slightly increased for Ra values below 1.5 µm (Fig. 6), although statistically no difference could be found (Kruskal–Wallis: $\chi^2_{10} = 10.56$, $p = 0.393$). After 7 days the proliferation behavior of fibroblasts was contrary to that found for osteoblasts. The cell number for different roughness values was significantly different (Kruskal–Wallis: $\chi^2_{10} = 47.56$, $p<0.001$). Below an Ra value of 2 µm the number of cells gradually increased with decreasing roughness to about two times the number found on the roughest part of the gradient (Fig. 6). The two highest cell number values (at Ra values of 1.0 and 1.1 µm) were significantly different compared to cell number values at Ra values higher than 2 µm (Tamhane test, $p<0.037$). The control experiments on smooth Thermanox discs and

Fig. 5. Morphology of osteoblasts (A) and fibroblasts (B) on SLA and Thermanox surfaces with the according roughness value Ra. Cells were cultured for 7 days. The nucleus was stained with DAPI (blue) and the actin network with Alexa Fluor 488 phalloidin (green). Scale bar is 200 µm.

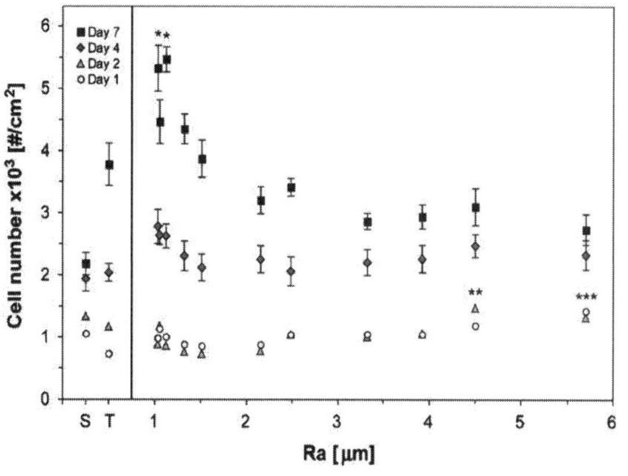

Fig. 6. Number of human gingival fibroblasts at different positions on the roughness gradient and on a SLA (S) and Thermanox (T) control surface. Cells were seeded at a density of 3500 cells/cm^2 and cultured for 1, 2, 4 and 7 days ($n = 5$). *Significant difference compared to cell number values above an Ra value of 2 µm (Tamhane test, $p<0.05$). **Two days post seeding, a significant difference was found only for the comparison between the cell number at an Ra value of 4.5 µm and the cell number values between Ra values of 1.6 and 2.2 µm (Tamhane test, $p<0.05$). ***One day post seeding, a significant difference could only be found comparing the cell number at the roughest Ra value of the gradient with the cell number at an Ra value of 1.5 µm (Tamhane test, $p<0.05$).

on SLA surfaces showed a slightly increased initial attachment to SLA surfaces than to the Thermanox surfaces (Fig. 6). After 7 days the number of fibroblasts was 1.7 times higher on the Thermanox discs compared to the SLA surface.

Fibroblasts exhibited a morphological behavior that was similar to that found for osteoblasts as a function of position on the gradient (Fig. 3C). Towards the smooth end of the gradient, fibroblasts were roundish and well spread with a diameter of up to 300 µm and a distinctive actin network. In contrast, at the rough end, fibroblasts spread less. They were elongated in shape with long filamentous extensions. Fibroblasts on SLA showed a thin, elongated shape with long filamentous extensions (Fig. 5B). On the Thermanox discs, fibroblasts showed similar behavior as seen for cells on the smooth part of the gradient. They were well spread with a distinctive, filamentous actin network (Fig. 5B).

4. Discussion

Typically, roughness values (Ra) of surfaces used in cell studies lie between 0.2 µm for polished, 3.4 µm for blasted surfaces and 4.9 µm for combined surface modifications (sandblasted and acid etched or plasma sprayed) [21–23]. Roughness gradients produced by the presented method cover most of the roughness values typically found in cell studies on a single surface. The advantage of a gradient surface is clearly that cell experiments can be performed over a wide range of roughness values under *identical* experimental conditions. In order to further decrease the variability in the cell experiments we developed a replica technique that allows for a production of identical substrata with well-defined surface chemistry. Previously, we have demonstrated the accuracy of this replication technique [14]. Surface features in the sub-micrometer range could exactly be reproduced over a large number of substrata and by means of the titanium coating, a controlled surface chemistry could be obtained, which at the same time shows excellent biocompatibility [24].

Osteoblastic cells are known to prefer rougher surfaces [5]. On our roughness gradients osteoblast proliferation increased in close correlation with increasing surface roughness (Fig. 4). The number of osteoblasts was almost two times higher on the rough end compared to the smooth end of the gradients. Increased proliferation of osteoblasts on rough substrata has been previously reported [22,25–30]. A direct comparison with these literature results is difficult since different methods of determining the roughness were used in the former studies, resulting in different roughness values. Often, the roughness was also created by different surface modification techniques. The roughness value of these substrata may be identical, but the surface characteristic is completely different. Therefore, we conclude that meaningful comparisons can only be drawn with studies that have used a similar surface-modification technique, i.e. in this case sandblasting. Mustafa et al. [27] used 300 µm TiO_2 particles to blast titanium discs and observed an increase of osteoblast proliferation on these discs of 55% compared to smooth ones. Degasne et al. [25] observed an increase in proliferation of 25% on discs blasted with SiO_2 particles between 225 and 500 µm in diameter and Rosa et al. [29] showed that the doubling rate of osteoblasts is 21% slower on smooth titanium surfaces compared to surfaces that were blasted with 250 µm Al_2O_3 particles. Although these studies were all carried out with different osteoblastic cells, they show a significant increase in proliferation for osteoblastic cells on blasted titanium surfaces. This increase in proliferation correlates well with the findings of this study. However, there are a number of reports of decreased proliferation on blasted titanium surfaces [21,31–34]. One possible explanation for this oppositional behavior is that osteoblasts in different maturation states were used. Lohman et al. showed that osteoblastic cells respond in a differential manner to changes in the surface roughness depending on their maturation state [23]. Some studies have also used transformed cells from osteosarcoma, and there is a concern that these cells may not respond to surface morphologies in a manner typical of normal osteoblasts [35]. Most of the studies where such cells were used showed decreased proliferation with increasing roughness, which is the opposite of our results. Another reason for these authors' results is the time period over which the cells are observed. Our results showed no significant difference in cell proliferation for osteoblasts during the first 2 days (Fig. 4). Only after 4 days were significantly more cells to be found on the rougher part of the gradient compared to the smooth part. Therefore, studies carried out for 2–3 days or less depend strongly on the initial attachment of the cells and may not be extrapolated to estimate the number of cells for longer periods. A further point to which attention has to be drawn is the level of confluency. It is known that cells behave differently once they reached confluency [36]. Since the osteoblasts on the smoother part of the gradients are more spread and have a large footprint they reach confluency faster than cells on the rougher part, which have a smaller footprint. It could be that the proliferation rate is slowed down on the smooth part and in that way contribute to the observed reduced cell number (Fig. 3B).

Osteoblasts on SLA showed a decreased proliferation compared to the smooth Thermanox disc, although the SLA surface is much rougher (Fig. 4). This observation has also been reported by others [37]. SLA surfaces consist of a coarse structure coming from the sandblasting with an overlaid fine structure formed by etching, which may cause a different cell response. Such a fine-etched structure in the nanometer range is not present on our roughness gradient surface, even though roughness with the same roughness value as SLA is covered by our gradient surface. The different behavior of osteoblasts on a blasted surface compared to an SLA surface is also evident in the morphology of the cells. Osteoblasts on the rough part of the gradients are small and compact (Fig. 3B, sb and 1 mm), whereas on the SLA surface they are larger with a lot of fine filamentous extensions that are probably attaching to the fine etched structure (Fig. 5A, SLA).

The results for the fibroblasts are in agreement with most other studies that have looked at HGFs on surfaces with different roughness (Fig. 6). Könönen et al. [38] showed that the number of fibroblasts is almost three times higher on polished surfaces compared to blasted titanium surfaces after 7 days. The investigations of Cochran et al. [39] revealed a decrease in the number of fibroblasts of a factor 1.4 after 7 days and a factor of two after 9 days for cells cultured on blasted and etched titanium discs. Other surface roughening treatments seem to have the same effect on fibroblasts, showing decreased proliferation on rougher substrata [40,41]. For fibroblasts on SLA and Thermanox surfaces it seems that the cells show a higher initial attachment on SLA, but after 4 days the proliferation slows down rapidly. The cell number on Thermanox exceeds that on SLA by almost a factor of two after 7 days (Fig. 6).

In terms of morphology, fibroblasts showed a flattened, well-spread appearance with an organized filamentous actin network on the smooth part of the gradients and were less spread with only occasional thin actin filaments on the rough part (Fig. 3C). A similar behavior for fibroblasts was also reported by Könönen et al. on smooth and rough substrata [38]. It appears to be a general tendency that cells spread less on rough and more on smooth substrata, since the same observation was made for the fibroblast as well as for the osteoblast morphology. One explanation could be found in the increased specific surface area of rough surfaces. Also, the mechanical stabilization produced by surface topography could play a role [42]. Cells supported by ridges and pitches do not need to express a distinctive actin network whereas cells on a smooth surface have to spread and develop a strong actin network in order to stabilize themselves on the surface.

5. Conclusions

We systematically studied the cell response of osteoblasts and fibroblasts to surface roughness by means of gradient substrata with a continuously varying roughness value and similar topographical features. Cell experiments performed on such roughness gradients showed a gradual change in the cell behavior along the gradient axis. Although cells were exposed to a variety of stimuli caused by changing surface roughness, cells still showed their type specific behavior, i.e. osteoblasts preferred the rougher part whereas fibroblasts favored the smoother part of the roughness gradient. In this sense roughness gradients serve as the ultimate control for cell response to surface roughness.

Acknowledgments

The authors would like to thank the AO Research Institute, Davos, Switzerland for access to the SEM and assistance with the SEM, and Dr. Gethin Owen, Dr. Douglas Hamilton and Prof. Don Brunette, University of British Columbia, for invaluable discussions. This work was financially supported by the Swiss National Science Foundation (SNF).

References

[1] Kasemo B. Biological surface science. Surf Sci 2002;500:656–77.
[2] Ellingsen JE, Lyngstadaas SP. In: Bio-implant interface: improving biomaterials and tissue reactions. Boca Raton, FL: CRC Press; 2003.
[3] Puleo DA, Nanci A. Understanding and controlling the bone–implant interface. Biomaterials 1999;20:2311–21.
[4] Quirynen M, De Soete M, van Steenberghe D. Infectious risks for oral implants: a review of the literature. Clin Oral Implant Res 2002;13:1–19.
[5] Chehroudi B, Gould TR, Brunette DM. Effects of a grooved epoxy substratum on epithelial cell behavior in vitro and in vivo. J Biomed Mater Res 1988;22:459–73.
[6] Cochran DL. A comparison of endosseous dental implant surfaces. J Periodontol 1999;70:1523–39.
[7] Scacchi M. The development of the ITI DENTAL IMPLANT SYSTEM. Clin Oral Implant Res 2000;11(Suppl.):8–11.
[8] Brunette DM. Principles of cell behavior on titanium surfaces and their application to implanted devices. In: Brunette DM, Tengvall P, Textor M, Thomsen P, editors. Titanium in medicine. Berlin and Heidelberg: Springer; 2001. p. 485–512.
[9] Rich A, Harris AK. Anomalous preferences of cultured macrophages for hydrophobic and roughened substrata. J Cell Sci 1981;50:1–7.
[10] Meredith JC, Sormana J-L, Keselowsky BG, Garcia AJ, Tona A, Karim A, et al. Combinatorial characterization of cell interactions with polymer surfaces. J Biomed Mater Res 2003;66:483–90.
[11] Washburn NR, Yamada KM, Simon CG, Kennedy SB, Amis E. High-throughput investigation of osteoblast response to polymer

cristallinity: influence of nanometer-scale roughness on proliferation. Biomaterials 2004;25:1215–24.

[12] Kunzler TP, Drobek T, Sprecher CM, Schuler M, Spencer ND. Fabrication of material-independent morphology gradients for high-throughput applications. Appl Surf Sci 2006;253:2148–53.

[13] Wieland M, Chehroudi B, Textor M, Brunette DM. Use of Ti-coated replicas to investigate the effects on fibroblast shape of surface with varying roughness and constant chemical composition. J Biomed Mater Res 2002;60:434–44.

[14] Schuler M, Kunzler TP, De Wild M, Sprecher CM, Textor M, Brunette DM, Tosatti SGP. Fabrication of TiO2-coated epoxy replicas with identical dual-type surface topographies for use in cell culture assays. J Biomed Mater Res 2006; submitted for publication.

[15] Abrams GA, Teixeira AI, Nealey PF, Murphy CJ. Effects of substratum topography on cell behvaior. In: Dillow AK, Lowman AM, editors. Biomimetic materials and design. New York and Basel: Marcel Dekker; 2002. p. 91–137.

[16] Flemming RG, Murphy CJ, Abrams GA, Goodman SL, Nealey PF. Effects of synthetic micro- and nano-structured surfaces on cell behavior. Biomaterials 1998;20:573–88.

[17] Hasegawa S, Sato S, Saito S, Suzuki Y, Brunette DM. Mechanical stretching increases the number of cultured bone-cells synthesizing DNA and alters their pattern of protein-synthesis. Calcified Tissue Int 1985;37:431–6.

[18] Chehroudi B, Ratkay J, Brunette DM. The role of implant surface geometry on mineralization in vivo and in vitro—a transmission and scanning electron-microscopy study. Cell Mater 1992;2:89–104.

[19] Elvin P, Evans CW. The adhesiveness of normal and Sv40-transformed Balb/C 3T3-cells—effects of culture density and shear rate. Eur J Cancer Clin Oncol 1982;18:669–75.

[20] Wieland M. Experimental determination and quantitative evaluation of the surface composition and topography of medical implant surfaces and their influence on osteoblastic cell–surface interactions. Ph.D. thesis. Zurich: ETH; 1999.

[21] Anselme K, Linez P, Bigerelle M, Le Maguer D, Hardouin P, Hildebrand HF, et al. The relative influence of the topography and chemistry of TiAl6V4 surfaces on osteoblastic cell behavior. Biomaterials 2000;21:1567–77.

[22] Hatano K, Inoue H, Kojo T, Matsunaga T, Tsujisawa T, Uchiyama C, et al. Effect of surface roughness on proliferation and alkaline phosphotase expression of rat calvarial cells cultured on polystyrene. Bone 1999;25:439–45.

[23] Lohmann CH, Bonewald LF, Sisk MA, Sylvia VL, Cochran DL, Dean DD, et al. Maturation state determine the response of osteogenic cells to surface roughness and 1,25-dihydroxyvitamin D3. J Bone Miner Res 2000;15:1169–80.

[24] Brunette DM, Tengvall P, Textor M, Thomsen P. In: Titanium in medicine. Berlin, Heidelberg: Springer; 2001.

[25] Degasne I, Baslé MF, Demais V, Lesourd M, Grolleau B, Mercier L, et al. Effects of roughness, fibronectin and vitronectin on attachment, spreading, and proliferation of human osteoblast-like (Saos-2) on titanium surfaces. Calcified Tissue Int 1999;64:499–507.

[26] Deligianni DD, Katsala N, Ladas S, Sotiropoulou D, Amedee J, Missirlis YF. Effect of surface roughness of titanium alloy Ti-6Al-4V on human bone marrow cell response and on protein adsorption. Biomaterials 2001;22:1241–51.

[27] Mustafa K, Wennerberg A, Wroblewski J, Hultenby K, Silva Lopez B, Arvidson K. Determining optimal surface roughness of TiO2 blasted titanium implant material for attachment, proliferation and differentiation of cells derived from human mandibular alveolar bone. Clin Oral Implant Res 2001;12:515–25.

[28] Nishimura N, Kawai T. Effect of microstructure of titanium surface on the behavior of osteogenic cell line MC3T3-E1. J Mater Sci-Mater Methods 1998;9:99–102.

[29] Rosa AL, Beloti MM. Effect of cpTi surface roughness on human bone marrow cell attachment, proliferation, and differentiation. Braz Dent J 2003;14:16–21.

[30] Zinger O, Anselme K, Denzer A, Habersetzer P, Wieland M, Jeanfils J, et al. Time-dependent morphology and adhesion of osteoblastic cells on titanium model surfaces featuring scale-resolved topography. Biomaterials 2004;25:2695–711.

[31] Anselme K, Bigerelle M, Noel B, Dufresne E, Judas D, Iost A, et al. Qualitative and quantitative study of human osteoblast adhesion on materials with various surface roughness. J Biomed Mater Res 2000; 49:155–66.

[32] Kieswetter K, Schwartz Z, Hummert TW, Cochran DL, Simpson J, Dean DD, et al. Surface roughness modulates the local production of growth factors and cytokines by osteoblast-like MG-63 cells. J Biomed Mater Res 1996;32:55–63.

[33] Kim HJ, Kim SH, Kim MS, Lee EJ, Oh HG, Oh WM, et al. Varying Ti-6Al-4V surface roughness induces different early morphologic and molecular responses in MG63 osteoblast-like cells. J Biomed Mater Res 2005;74A:366–73.

[34] Martin JY, Schwartz Z, Hummert TW, Schraub DM, Simpson J, Lankford J, et al. Effect of titanium surface roughness on proliferation, differentiation, and protein synthesis of human osteoblast-like cells (MG63). J Biomed Mater Res 1995;29: 389–401.

[35] Boyan BD, Lohmann CH, Dean DD, Sylvia VL, Cochran DL, Schwartz Z. Mechanism involved in osteoblast response to implant surface morphology. Annu Rev Mater Res 2001;31:357–71.

[36] Boyan BD, Schwartz Z, Bonewald LF, Swain LD. Localization of 1,25-(OH)2D3-responsive alkaline phosphatase in osteoblast-like cells (ROS 17/2.8, MG63, and MC 3T3) and growth cartilage cells in culture. J Biol Chem 1989;264:11879–966.

[37] Zinger O, Zhao G, Schwartz Z, Simpson J, Wieland M, Landolt D, et al. Differential regulation of osteoblasts by substrate microstructural features. Biomaterials 2005;26:1837–47.

[38] Könönen M, Hormia M, Kivilahti J, Hautaniemi J, Thesleff I. Effect of surface processing on attachment, orientation, and proliferation of human gingival fibroblasts on titanium. J Biomed Mater Res 1992; 26:1325–41.

[39] Cochran DL, Simpson J, Weber HP, Buser D. Attachment and growth of periodontal cells on smooth and rough titanium. Int J Oral Max Implant 1994;9:289–97.

[40] Ponsonnet L, Comte V, Othmane A, Lagneau C, Charbonnier M, Lissac M, et al. Effect of surface topography and chemistry on adhesion, orientation and growth of fibroblasts on nickel–titanium substrates. Mater Sci Eng C 2002;21:157–65.

[41] Wirth C, Comte V, Lagneau C, Exbrayat P, Lissac M, Jaffrezic-Renault N, et al. Nitinol surface roughness modulates in vitro cell response: a comparison between fibroblasts and osteoblasts. Mater Sci Eng C 2005;25:51–60.

[42] Brunette DM, Chehroudi B. The effects of the surface topography of micromachined titanium substrata on cell behavior in vitro and in vivo. J Biomech Eng 1999;121:49–57.

Available online at www.sciencedirect.com
ScienceDirect
Biomaterials 28 (2007) 5000–5006
Biomaterials
www.elsevier.com/locate/biomaterials

Systematic study of osteoblast response to nanotopography by means of nanoparticle-density gradients

Tobias P. Kunzler[a], Christoph Huwiler[b], Tanja Drobek[a], Janos Vörös[b], Nicholas D. Spencer[a,*]

[a]*Laboratory for Surface Science and Technology, Department of Materials, ETH Zurich, Wolfgang-Pauli-Strasse 10, 8093 Zurich, Switzerland*
[b]*Laboratory of Biosensors and Bioelectronics, Department of Information Technology and Electrical Engineering, Institute of Biomedical Engineering, ETH Zurich, Gloriastrasse 35, 8092 Zurich, Switzerland*

Received 16 May 2007; accepted 2 August 2007
Available online 27 August 2007

Abstract

Features over a wide range of length scales affect the biological response to a surface. While the influence of micro-features has been extensively studied, the effect of nano-features has only rarely been systematically investigated. We have developed a simple method to produce nano-featured gradients by kinetically controlled adsorption of negatively charged silica nanoparticles onto positively charged, poly(ethylene imine) (PEI)-coated silicon wafers. Subsequent sintering of the particles allowed a tuning of the particle morphology and resulted in a firm anchoring of the particles to the surface. Particle-density gradients were characterized by atomic force microscopy (AFM). Cell experiments with rat calvarial osteoblasts (RCO) on nano-featured gradients exhibited a significant decrease in proliferation at locations with higher particle coverage. Seven days post seeding, the number of osteoblasts was eight times higher at positions without particles compared to positions with maximum particle coverage. While cells spread well and developed a well-organized actin network in the absence of particles, spreading and formation of a strong actin network was considerably hindered at locations with maximum particle density.

Keywords: Cell proliferation; Cell morphology; Nanotopography; Nanoparticle; Osteoblast

1. Introduction

It is well known that micrometer-scale roughness has an influence on cell proliferation and morphology [1–3]. The roughness in this range provides the micro-environment for the cell, but at the interface between the cell and its extracellular matrix (ECM), interactions also take place on a smaller scale [4,5]. Looking at the adhesion sites of the cell (focal adhesions), which are in the range of 5–200 nm [6], it is clear that these very small components of a cell are influenced by nano-scale rather than micro-scale features. While the influence of micro-scale topography on cellular behavior has been recognized for decades [3,7], the nano-scale influence had remained virtually uninvestigated until about 15 years ago [8]. This was primarily due to the lack of techniques to produce controlled and accurate nano-topographies [1]. Recent progress in the production of nano-topographies, however, has permitted the fabrication of smaller and smaller features.

One of the first reports to describe the behavior of cells on substrata with nano-sized features was presented by Clark et al. [9]. They found that cells aligned to grooves (130 nm in width and separated by 130 nm) with a depth of 100 nm. Further work by others with decreasing groove dimensions [1,10] demonstrated that some cells even respond to grooves with a step size as low as 5 nm [11]. However, although grooves may represent the nanometer-scale in two dimensions (spacing/width and depth), they still extend over micrometers in one dimension (length), which may raise doubts as to whether the effects are caused by the nanometer topography alone or are also influenced by the larger microscopic extension of the features. Therefore, several studies have used substrata with features

*Corresponding author. Tel.: +41 44 632 5850; fax: +41 44 633 1027.
E-mail address: spencer@mat.ethz.ch (N.D. Spencer).

0142-9612/$ - see front matter © 2007 Elsevier Ltd. All rights reserved.
doi:10.1016/j.biomaterials.2007.08.009

such as pits [12–14] or pillars [13,15], which were nanometer-sized in all dimensions (width, length and depth). In contrast to grooved substrata, on which cells align to the ridges of the grooves, cells on nanopits or nanopillars show decreased adhesion, proliferation and a different morphological behavior [12,15]. The effect depends on the cell type, the shape of the nano-features, order and symmetry, and the total area over which the features are spread [5].

Compared to the number of publications investigating the cell response to micrometer topography, the work published on the influence of nano-sized features on cells is very limited and many questions remain unsolved. Systematic studies of the nano-topography would clearly contribute to resolving some of these questions. A nano-featured gradient surface with continuously changing surface-feature parameters is a powerful tool for such systematic studies. To our knowledge there has only been one report in the literature on the fabrication of nanometer-scale roughness gradients and their application in cell experiments [16]. This approach to gradient fabrication was based on changing the crystallinity of a polymer film by means of annealing on a temperature-gradient stage. The resulting roughness gradients showed continuously varying roughness parameters, but at the same time the shape of the nano-features also changed along the gradient axis. We recently developed a simple method to produce nano-featured gradients with continuously changing spacing between nanoparticles while the shape and size of the nano-features remains unchanged over the entire gradient [17]. In this work, proliferation and morphology of rat calvarial osteoblasts (RCO) have been systematically studied by means of these nanoparticle-density gradients.

2. Materials and methods

2.1. Nanoparticle-density gradients

Nanoparticle-density gradients were produced as described by Huwiler et al. [17]. In brief, silicon wafers (with a native oxide layer) were coated with poly(ethylene imine) (PEI) (Sigma-Aldrich, Germany) and lowered into an aqueous suspension of silica particles (average particle size: 73 nm in diameter; Klebosol, Clariant, France) by means of a linear-motion drive (Owis Staufen, Germany). The immersion profile was set to $x(t) = at^2$, where $x(t)$ is the position on the gradient at the time t, which was running from 0 to 1800 s and $a = -3.09 \times 10^{-6}\,\mathrm{m/s^2}$. This setup results in the creation of a gradient over 10 mm after a total immersion time of 30 min. For sintering of the particles to the surface, the substrates were heated at 10 °C/min to a final temperature of 1125 °C, held for 30 min and then cooled down at 1.5 °C/min to room temperature (RT). Particle density and spacing were calculated from scanning electron microscopy (SEM) images with ImageJ software (Version 1.37 for Windows). Atomic force microscopy (AFM) images were taken in Tapping Mode (DI Dimension 3000 AFM, Veeco Instruments GmbH, Mannheim, Germany) using a Pointprobe Plus beam-shaped silicon cantilever (nominal $T = 4\,\mu m$, $W = 28\,\mu m$, $L = 128\,\mu m$, $c = 38\,N/m$, $f_0 = 322\,kHz$). A scan rate of 1 Hz was used with an integral gain of 0.3 and proportional gain of 0. The images were evaluated with WSxM 4.0 software (Nanotec Electronica, Spain).

2.2. Cell culture

RCOs were obtained from newborn rat calvariae, isolated and subcultured as described by Hasegawa et al. [18] and Chehroudi et al. [19]. Frontal, parietal and occipital bone samples were dissected, rinsed in phosphate-buffered saline (PBS) (Fluka Chemicals, Buchs, Switzerland), placed in Dulbecco's modified Eagles medium (α-DMEM) (Invitrogen, Basel, Switzerland), supplemented with 1% antibiotics (Penstrep, Invitrogen, Basel, Switzerland) and 10% fetal bovine serum (FBS) (Invitrogen, Basel, Switzerland). The minced tissue was digested with a mixture of clostridial collagenase and trypsin (both from Sigma-Aldrich, Buchs, Switzerland) and then placed in tissue-culture flasks.

2.3. Experimental design

2.3.1. Substratum preparation

Sintered particle gradients were cleaned in a 2% solution of Hellmanex (HELLMA GmbH & Co., Müllheim, Germany) in an ultrasonic bath for 10 min and rinsed 10 times with ultra pure water. After oxygen-plasma treatment at RF level 'Hi' (PDC-002, Harrick Plasma, Ithaca, USA) for 4 min, substrata and titanium-coated Thermanox disks (Nunc, Wiesbaden, Germany), used as control surfaces, were immediately placed 30 min in a poly(L-lysine)-graft-poly(ethylene glycol) (PLL-*g*-PEG)-RGD (arginine–glycine–aspartic acid) solution (concentration: 0.5 mg/ml in 4-(2-hydroxyethyl)-1-piperazineethanesulfonic acid buffer with 150 mM added NaCl, pH 7.4 (HEPES 2), peptide density: 5.14 pmol/cm^2 (corresponding to a calculated ligand–ligand distance of 6 nm [20]); PLL-*g*-PEG-RGD was purchased from SuSoS SurfaceSolutionS GmbH, Zurich, Switzerland). The PLL-*g*-PEG-RGD solution had previously been filtered with a 0.22 μm filter in order to sterilize the solution. PLL-*g*-PEG-RGD-coated substrata were rinsed with HEPES 2 and covered with media (α-DMEM with 10% FBS and 1% antibiotics). The coating process was carried out in a sterile environment.

2.3.2. Cell harvesting and incubation

RCO between the 2nd and 3rd subculture were used for the experiments. After reaching confluency, cells were detached from the flask with trypsin-EDTA solution (Sigma-Aldrich, Buchs, Switzerland), centrifuged at 5000 rpm for 5 min (Rotofix 32 centrifuge, Hettich Laborapparate, Bäch, Switzerland), resuspended in media (α-DMEM with 10% FBS and 1% antibiotics) and seeded at a concentration of 3500 cells/cm^2. The cells were incubated for 2, 4 and 7 days at 37 °C in a humidified atmosphere of 93% air and 7% CO_2.

2.3.3. Cell staining

Cells were washed with PBS, which was pre-warmed to 37 °C (PBS, Fluka Chemicals, Buchs, Switzerland) and fixed in a 4% paraformaldehyde solution (freshly prepared from a 20% stock solution; paraformaldehyde powder was purchased from Fluka Chemicals, Buchs, Switzerland). After rinsing with PBS, the cells were permeabilized by exposure to a 0.5% Triton X-100 solution (Fluka Chemicals, Buchs, Switzerland) for 3 min at RT. In order to block non-specific binding sites, the substrata were immersed into a 3% BSA solution (BSA was dissolved in PBS; BSA was purchased from Sigma-Aldrich, Buchs, Switzerland) for 30 min at RT. Cells were incubated with the primary antibody (mouse monoclonal anti-human vinculin, 1:400 dilution in 0.1% BSA/PBS solution; Sigma-Aldrich, Oakville, Ontario, Canada) for 60 min at RT. After rinsing three times in 1% BSA/PBS solution, cells were incubated with a secondary antibody (Alexa 546 goat anti-mouse IgG, dilution 1:200 in 0.1% BSA/PBS solution; Invitrogen AG, Basel, Switzerland) for 90 min at RT. Substrata were rinsed again three times in 1% BSA/PBS solution. Afterwards, the substrata were incubated with Alexa Fluor 488 phalloidin (1:100 dilution in PBS; Invitrogen AG, Basel Switzerland) for 25 min at RT and rinsed three times with PBS. The nucleus was stained with DAPI (1:1000 dilution in PBS; Invitrogen, Basel, Switzerland) for 15 min at RT and rinsed three times with PBS. Subsequently, the substrata were immersed into a 4% paraformaldehyde solution for 15 min at RT. After

5002 T.P. Kunzler et al. / Biomaterials 28 (2007) 5000–5006

rinsing three times with PBS, substrata were kept in PBS, to prevent them from drying out.

2.3.4. Evaluation

Substrata were observed with a fluorescence microscope (Axio IMAGER M1m, Zeiss, Oberkochen, Germany). In order to determine the cell number and morphology, three images at a time were taken at 13 positions (1 mm apart) along the gradient with a DAPI, FITC and Rhodamine filter set (filter set no. 49 with excitation wavelength $\lambda = 365$ nm, no. 10 with excitation wavelength $\lambda = 488$ nm and no. 15 with excitation wavelength $\lambda = 546$ nm, Zeiss, Oberkochen, Germany). A total of four substrata per time period were investigated. The number of cell nuclei was counted with ImageJ software (Version 1.36b for Mac OS X).

Comparisons between measurement series were made using non-parametric tests (SPSS 11.4 for Mac OS X) because variances were not homogeneous and observed values were not distributed normally, even after appropriate transformations. All mean values are shown $\pm$SE (standard error). The level of significance p was set at 0.05.

3. Results

3.1. Nanoparticle-density gradients

AFM images at different positions along the gradient are shown in Fig. 1. The maximum particle coverage is 21% corresponding to a mean particle spacing of 190 nm ($\pm$20 nm) at position 0 mm on the gradient (position exposed to the particle solution for the longest time) (Fig. 1a). The particle coverage continuously decreases, and the particle spacing concomitantly increases, following the gradient axis towards shorter dipping times (Fig. 1b–f). At the end of the gradient that was exposed to the colloidal solution for a very short time only, almost no particles are present (Fig. 1g). There were a small number of clusters present at the highest coverages, but few of these consisted of >2 particles.

The heat-treatment temperature was set to 1125 °C, which proved to be an optimal temperature to establish a firm binding of the particles to the surface while at the same time still maintaining a nearly spherical particle shape. The contact angle between particle and surface was determined to be around 125°. AFM measurements showed that the average particle height above the substrate was 63.7($\pm$4.3) nm (Fig. 2).

3.2. Cell response

After 2 days of culture no significant difference in the number of RCO was found at different positions on the gradient (Kruskal–Wallis: $\chi^2_{11} = 18.32$, $p = 0.146$), indicating that the initial cell attachment was similar everywhere on the gradient (Fig. 3).

A significant difference in the cell number could be determined after 4 days of culture (Kruskal–Wallis: $\chi^2_{11} = 51.28$, $p<0.001$). At positions with the highest particle coverage (position 0 and 1 mm) the cell number had not increased significantly compared to that observed after culturing for 2 days (Fig. 3). Moving along the gradient to lower particle coverages, by position 2 mm on

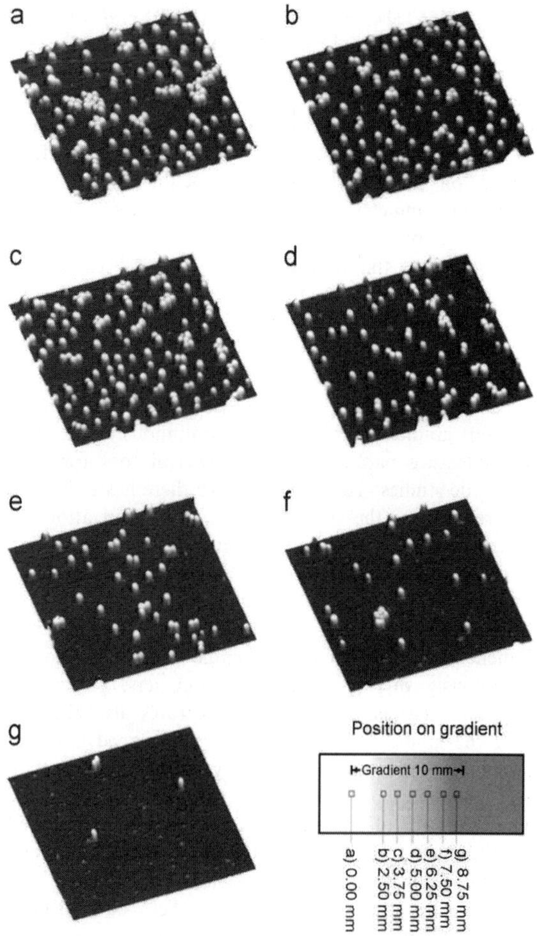

Fig. 1. AFM images of different positions on a particle-density gradient over 10 mm. AFM images were taken in Tapping Mode. The different positions on the gradient are indicated in the sketch in the lower-right corner of the figure. Gradients were produced by immersing a PEI-coated silicon wafer into a water-based colloidal solution with a concentration of 0.002 wt% 73 nm silica particles. The substratum was immersed into the colloid suspension with an immersion profile $x(t) = at^2$, with t running from 0 to 1800 s and $a = -3.09 \times 10^{-6}$. The side of an AFM image is 2 μm.

the gradient, the cell number had begun to continuously increase as a function of decreasing particle coverage. The cell number at the end of the gradient with no particles (position 10 mm) was found to be 11 times higher than that at the end with maximum particle coverage (position 0 mm). After position 10 mm (position 11 and 12 mm with no particles) the cell number remained constant.

After 7 days of culture, a statistically significant difference in the cell number at different positions on the gradient was observed (Kruskal–Wallis: $\chi^2_{11} = 99.27$, $p<0.001$). The cell number at positions between 7 and 12 mm was significantly higher compared to the number found at positions between 0 and 2 mm (Fig. 3). The number of osteoblasts monotonically increased as a function of decreasing particle coverage up to position 9 mm.

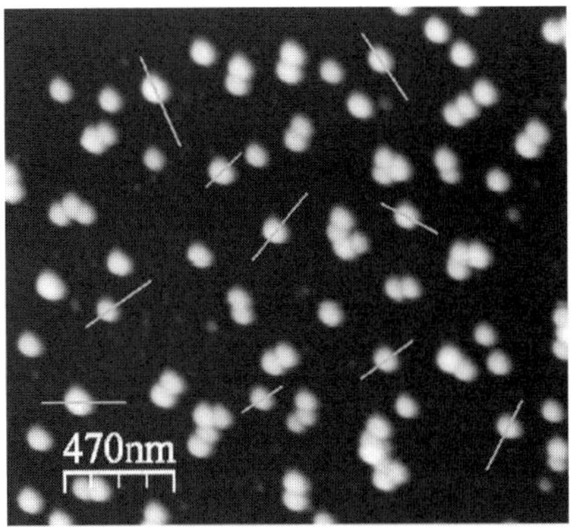

Fig. 2. Characterization of particle size in the nanomorphology gradient: AFM images of silica particles immobilized onto a silicon wafer substratum at three different positions along gradient (one example shown). Particles were sintered to the surface by a heat treatment at 1125 °C. Profiles were measured through the centers of 10 particles in each region. The mean height of the particles (originally 73 nm in diameter) above the substratum is 63.7(±4.3 nm). During the sintering process the particles are melted approximately 10 nm 'into' the substratum surface (thermal growth of the oxide layer on the substratum may also contribute to this).

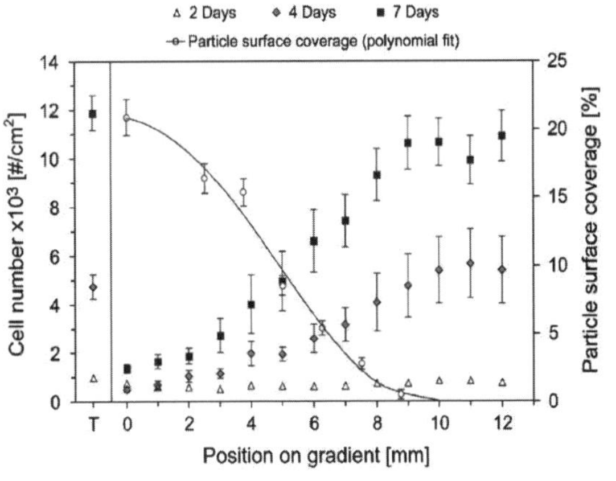

Fig. 3. Particle surface coverage and number of RCO at different positions on particle-density gradients. Cells were seeded at a density of 3500 cells/cm^2 and cultured for 2, 4 and 7 days ($n = 4$). The particle surface coverage was calculated from AFM images. T = Thermanox control surface.

At the end of the gradient with no particles (position 10 mm) the cell number was eight times higher than at the position with maximum particle coverage (position 0 mm).

The cell number on the Thermanox control surfaces was similar to that obtained at positions on the gradient without particles after culturing for 2 and 4 days. After 7 days of culture, the number of osteoblasts on Thermanox disks was slightly higher than on positions on the gradient with no particles (position 10–12 mm) (Fig. 3).

An overview of osteoblasts on a particle-density gradient is shown in Fig. 4. The cell number increases dramatically between position 7 and 8 mm on the gradient (Fig. 4a). The cell morphology also changes with decreasing particle coverage (Fig. 4b). At maximum nanoparticle coverage, osteoblasts were rather small in size with some long filamentous extensions (Fig. 5a). The cells showed no distinctive actin network, nor did they exhibit pronounced focal adhesions. With decreasing particle coverage the cells were more spread, some with filamentous extensions, some with a rounder footprint (Fig. 5b and c). The formation of an actin network with distinctive focal adhesions was only observed at positions with low particle coverage (Fig. 5d). At positions on the gradient with no particles, osteoblasts were well spread, reaching a diameter of up to 300 μm. The footprints of the osteoblasts were round in shape with a few small filamentous extensions at the edge of the cell membrane. The osteoblasts developed a distinctive actin network with many focal adhesions. On the Thermanox control surfaces, osteoblasts showed a similar morphological behavior as observed at positions on the gradient without particles (Fig. 5e).

4. Discussion

Experiments with RCO on nanoparticle-density gradients showed that the nano-topography has a distinctive influence on the cell behavior. The cell number after 4 and 7 days of culture was found to decrease significantly with increasing particle density (Fig. 3). Investigations of the cellular behavior on nano-featured gradients have also been carried out by Washburn et al. [16]. They found a decreasing number of MC3T3-E1 osteoblasts with increasing nanometer roughness, which correlates with the findings in this work. The roughness gradients used in their work were fabricated by different degrees of crystallization of a polymer film. As a result, size, shape and spacing of the features all change along the gradient axis. Since it is known that feature size, shape as well as spacing between the features influence the cell behavior [5], it is difficult to evaluate how much each parameter contributes to the overall effect of the observed cell response. The nano-featured gradients used in this paper have identically shaped and sized particles, with the spacing between particles as the only changing parameter along the gradient.

Previous studies with different types of fibroblasts on substrata with nano-features have shown a similar behavior to that observed for osteoblasts [12,14,21]. After several days of culture the number of fibroblasts was significantly lower than that found on flat control substrata. Drawing a conclusion from the observations made for osteoblasts and fibroblasts, it seems to be a general tendency that certain nano-features reduce cell attachment and proliferation, independently of the cell type. The influence of nano-features on the cell behavior can be explained by several different possible mechanisms.

Starting with the substratum surface in general, there are certain events to be considered that take place upon exposure of the surface to a biological milieu. The first

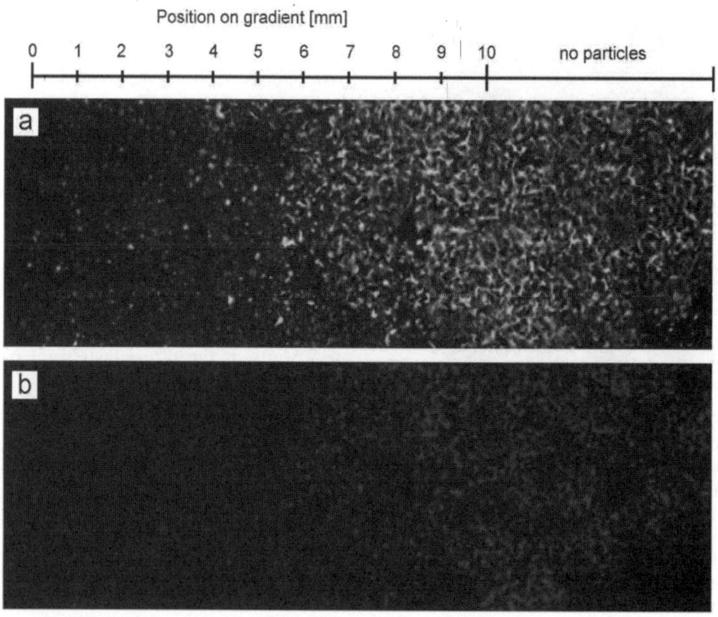

Fig. 4. Fluorescence images of RCO on a nanoparticle-density gradient. Images show an overview of a large part of the gradient. Cells were seeded at a density of 3500 cells/cm^2 and cultured for 7 days. After fixation, cells were stained for actin (green) and the nuclei (blue). Images are taken from the same sample: (a) once with a FITC filter and (b) with a DAPI filter.

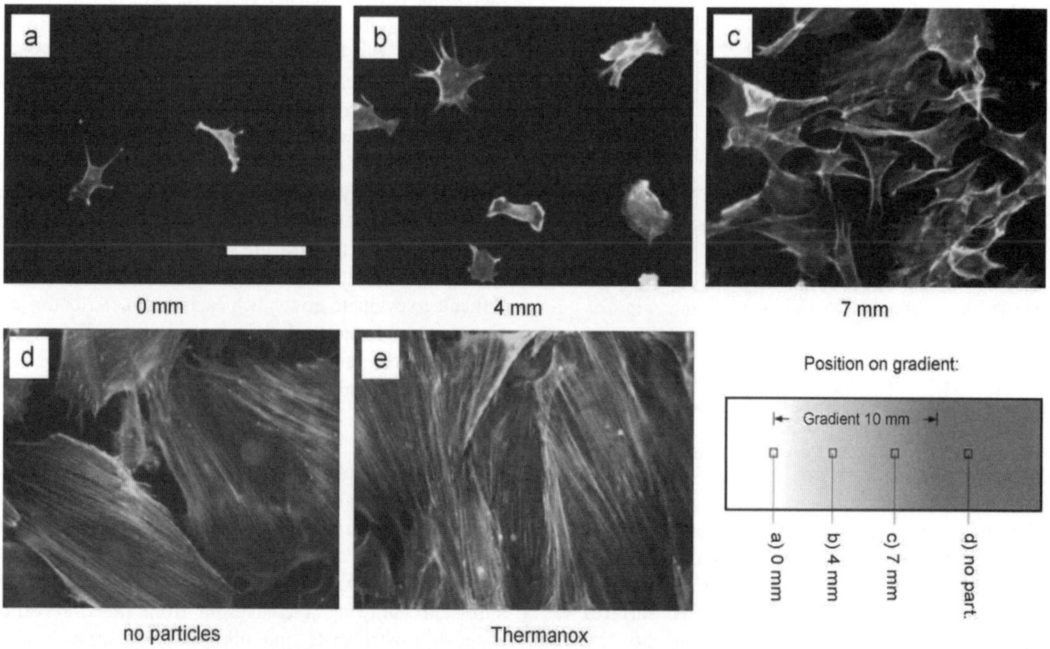

Fig. 5. Fluorescence images of cell morphology at different positions on a particle density gradient. RCOs were seeded at a density of 3500 cells/cm^2 and cultured for 7 days. After fixation cells were stained for vinculin (red), actin (green) and the nuclei (blue). With decreasing particle density, cells formed well-constituted focal adhesions (red) and a distinctive actin network (green). Image e is the Thermanox control surface. Scale bar is 100 μm.

molecules to reach a surface are water molecules, followed by the adsorption of proteins and finally the arrival of cells [22]. Hence, what the cells actually interact with is not the surface of a substratum itself, but a 'carpet' of proteins on top of it. The size of a protein is in the nanometer range, and it seems possible that protein adsorption or even conformation can be influenced by the nano-features on the surface. A change in protein adsorption or conformation will lead to a different cell behavior. That would mean that the cell response may not be caused by the nano-features directly, but by a change in the protein conformation. Whether nano-sized morphological features do really influence protein adsorption or conformation is controversial. There are reports demonstrating both that the nano-topography has an effect [23] and that nano-features have little or no effect on the protein adsorption [24,25].

In order to exclude the effect of possible changes in the protein adsorption, particle gradients used in this paper were coated with a monolayer of PLL-*g*-PEG-RGD. The PLL-*g*-PEG-RGD layer serves several purposes: PLL-*g*-PEG is known to be protein resistant, and therefore, proteins that could be modified by the nano-features are not present upon arrival of the cells on the surface. However, in order to provide binding sites for the cells, an RGD peptide sequence is linked to the PLL-*g*-PEG molecules. RGD is found within the ECM and has the ability to bind to cells via specific cell-surface receptors [26]. Furthermore, coating with PLL-*g*-PEG-RGD achieves a well-defined and homogeneous surface chemistry all over the gradient. This enables us to focus on geometric issues, as outlined in the following possible explanation for our results.

An important issue in cell adhesion is the formation of focal adhesions. Without adhesion, osteoblasts cannot spread, and spreading is necessary for cell division [4]. Focal adhesions are clusters of integrins, linking the cell to the ECM. Clustering of integrins has been shown to be essential in the formation of mature focal adhesions [27,28]. Spatz et al. [29] developed a method for the controlled deposition of gold nanodots to produce an array of specific adhesion sites for individual integrins with adjustable spacing. A separation of >73 nm between the integrins resulted in limited cell attachment and spreading [6].

On the particle gradients used in this work, the spacing between particles was determined to be 190 nm at a position with the highest particle density. With decreasing particle density the particle spacing is even higher. Assuming that a cell lies on top of the particles and integrins bind to the particles only, one would expect to find few cells all over the gradient, since the spacing of the binding sites is well above 73 nm, which was found to be a critical value for cell attachment [6].

Particles on gradients, which were used in this work, are 73 nm in diameter, and since integrins are about 8–12 nm in diameter [30,31], it is probable that more than one integrin binds to the top of a particle. However, the clustering of a few integrins on a particle is not enough to establish stable and mature focal adhesions. In the case of areas with higher particle density, no formation of distinctive focal adhesions is observed (Fig. 5a). Since the cytoplasmic domains of integrins act as sites of cytoskeletal nucleation [32], integrin clustering is also required in order to form a strong actin network. This is evident in Fig. 5a where the absence of mature focal adhesions inhibits the formation of a distinctive actin cytoskeleton.

On areas with lower particle coverage a higher number of cells with well-constituted focal adhesions and an organized actin network is found (Fig. 5c). As discussed above, a cell that settles down onto the particles encounters a limited number of integrin binding sites. In order to form stable focal adhesions, the cell attempts to expand to find more sites for integrins to bind to. Integrins are 30 nm in length (of which about 20 nm sticks out of the cell membrane) [4,33], and it is therefore impossible for them to reach the substrate between the particles. As a result, the cell has to distort its plasma membrane. The bending of the membrane requires energy, which increases with decreasing radius of curvature [34]. For particles that are close together the radius of curvature is smaller compared to larger spaced particles and hence, the cell needs to do more work to deform the membrane (Fig. 6).

At parts of the gradient with high particle density, a cell sees many particles, which are close together. In an effort to reach the bottom of the substratum between particles many convexities of the cell membrane, each with a small radius of curvature, are formed. This causes a stress situation in the cell. A possible reaction to reduce the stress is to minimize the contact area with the substratum. Such a behavior is indeed observed for areas with higher particle density (Fig. 5a). Another possibility for the cell is to move in order to find a location that provides more easily accessible attachment sites. Migrating cells are usually characterized by extended filopodia [6]: behavior that was also observed on areas with higher particle coverage (Fig. 5a and b). Either way, migration or reducing the contact area, the cells cannot spread, which is a requirement for the cell to proliferate. This could be a possible explanation for the low cell number found at positions with increasing particle density.

In contrast to areas with high particle coverage, cells are found to spread well on areas with few or no particles (Fig. 5c and d). At a certain particle distance the cell is no

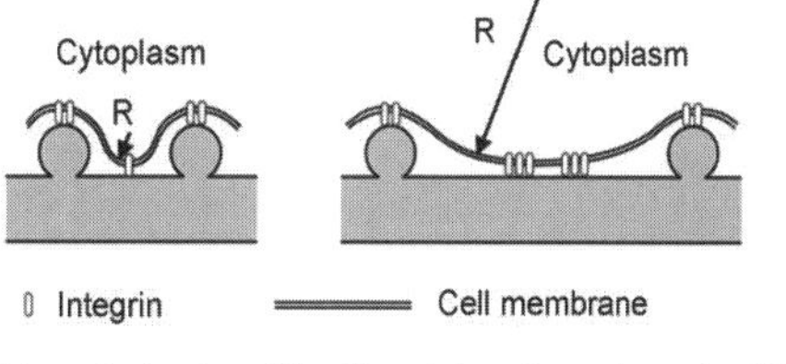

Fig. 6. Schematic drawing of bending of the cell membrane for different particle spacings. The space on top of the particles is limited for the attachment of integrins. In order to establish more integrin bindings the cell has to distort to reach the substratum between the particles. The energy needed to bend the membrane depends on the radius of curvature *R*. For a small radius of curvature (left) more work is needed to form such a protuberance than for a larger radius of curvature (right). In addition, there is more space to form integrin clusters between larger particles than between particles that are close together.

longer forced to bend its membrane into a small radius of curvature, which reduces the energy a cell has to generate and also the stress caused in the cell (Fig. 6). Furthermore, the cell finds enough space between the particles for integrin clustering and establishing well-constituted focal adhesions with a strong actin cytoskeleton.

5. Conclusion

Our results have shown that the nano-topography of particle gradients has a significant influence on cell proliferation and morphology. By keeping the size and shape of the nano-features constant (with the spacing as the only changing parameter along the gradient), we demonstrated that the distance between nano-features plays a crucial role in the response of cells to these nano-features. With the particle-density gradient we could show that the cell behavior can be continuously changed on one and the same surface. The observation of a dramatically decreased cell number on areas with high particle coverage is of particular interest in regard to many potential applications in cell engineering. It could be a simple approach for substantially reducing cell adhesion on a surface in a permanent way.

Acknowledgement

This work was financially supported by the Swiss National Science Foundation (SNF).

References

[1] Abrams GA, Teixeira AI, Nealey PF, Murphy CJ. Effects of substratum topography on cell behavior. In: Dillow AK, Lowman AM, editors. Biomimetic materials and design. New York, Basel: Marcel Dekker; 2002. p. 91–137.

[2] Brunette DM, Chehroudi B. The effects of the surface topography of micromachined titanium substrata on cell behavior in vitro and in vivo. J Biomech Eng 1999;121:49–57.

[3] Flemming RG, Murphy CJ, Abrams GA, Goodman SL, Nealey PF. Effects of synthetic micro- and nano-structured surfaces on cell behavior. Biomaterials 1998;20:573–88.

[4] Alberts A, Johnson A, Lewis J, Raff M, Roberts K, Walter P. In: Molecular biology of the cell. New York: Garland Science; 2002.

[5] Curtis AS. Tutorial on the biology of nanotopography. IEEE Trans Nanobiosci 2004;3:293–5.

[6] Arnold M, Calvacanti-Adam EA, Glass R, Blümmel J, Eck W, Kantlehner M, et al. Activation of integrin function by nanopatterned adhesive interface. Chem Phys Chem 2004;5:383–8.

[7] Harrison RG. On the stereotropism of embryotic cells. Science 1911;34:279–81.

[8] Curtis AS, Wilkinson C. Nanotechniques and approaches in biotechnology. Trends Biotechnol 2001;19:97–101.

[9] Clark P, Connolly P, Curtis AS, Dow JAT, Wilkinson CDW. Cell guidance by ultrafine topography in vitro. J Cell Sci 1991;99:73–7.

[10] Wojciak-Stothard B, Curtis AS, Monaghan W, Macdonald K, Wilkinson C. Guidance and activation of murine macrophages by nanometric scale topography. Exp Cell Res 1996;223:426–35.

[11] Rajnicek AM, McCaig CD. Contact guidance of CNS neurites on grooved quartz: influence of groove dimensions, neuronal age and cell type. J Cell Sci 1997;110:2915–24.

[12] Curtis AS, Gadegaard N, Dalby MJ, Riehle M, Wilkinson C, Aitchinson G. Cells react to nanoscale order and symmetry in their surroundings. IEEE Trans Nanobiosci 2004;3:61–5.

[13] Gadegaard N, Martines E, Riehle M, Seunarine K, Wilkinson C. Applications of nano-patterning to tissue engineering. Microelectron Eng 2006;83:1577–81.

[14] Gallagher JO, McGhee KF, Wilkinson C, Riehle M. Interactions of animal cells with ordered nanotopography. IEEE Trans Nanobiosci 2002;1:24–8.

[15] Wilkinson C, Curtis AS, Crossan J. Nanofabrication in cellular engineering. J Vac Sci Technol B 1998;16:3132–6.

[16] Washburn NR, Yamada KM, Simon CG, Kennedy SB, Amis E. High-throughput investigation of osteoblast response to polymer crystallinity: influence of nanometer-scale roughness on proliferation. Biomaterials 2004;25:1215–24.

[17] Huwiler C, Kunzler TP, Textor M, Vörös J, Spencer ND. Functionalizable nano-morphology gradients via colloidal self-assembly. Langmuir 2007;23:5929–35.

[18] Hasegawa S, Sato S, Saito S, Suzuki Y, Brunette DM. Mechanical stretching increases the number of cultured bone-cells synthesizing DNA and alters their pattern of protein-synthesis. Calcified Tissue Int 1985;37:431–6.

[19] Chehroudi B, Ratkay J, Brunette DM. The role of implant surface geometry on mineralization in vivo and in vitro—a transmission and scanning electron-microscopy study. Cell Mater 1992;2:89–104.

[20] Schuler M, Owen GR, Hamilton DW, De Wild M, Textor M, Brunette DM, et al. Biomimetic modification of titanium dental implant model surfaces using the RGDSP-peptide sequence: a cell morphology study. Biomaterials 2006;27:4003–15.

[21] Dalby MJ, Riehle M, Johnstone HJH, Affrossman S, Curtis AS. Polymer-demixing nanotopography: control of fibroblast spreading and proliferation. Tissue Eng 2002;8:1099–108.

[22] Kasemo B. Biological surface science. Surf Sci 2002;500:656–77.

[23] Webster TJ, Schadler LS, Siegel RW, Bizios R. Mechanisms of enhanced osteoblast adhesion on nanophase alumina involve vitronectin. Tissue Eng 2001;7:291–301.

[24] Cai K, Bossert J, Jandt KD. Does the nanometer scale topography of titanium influence protein adsorption and cell proliferation? Colloid Surface B 2006;49:136–44.

[25] Han M, Sethuraman A, Kane RS, Belfort G. Nanometer-scale roughness having little effect on the amount or structure of adsorbed protein. Langmuir 2003;19:9868–72.

[26] Pierschbacher MD, Ruoslahti E. Cell attachment activity of fibronectin can be duplicated by small synthetic fragments of the molecule. Nature 1984;309:30–3.

[27] Hersel U, Dahmen C, Kessler H. RGD modified polymers: biomaterials for stimulated cell adhesion and beyond. Biomaterials 2003;24:4385–415.

[28] Spatz PS. Cell-nanostructure interactions. Nanobiotechnology. Weinheim: Wiley-VCH; 2004. p. 53–65.

[29] Spatz PS, Mössmer S, Hartmann C, Möller M. Ordered deposition of inorganic clusters from micellar block copolymer films. Langmuir 2000;16:407–15.

[30] Erb E-M, Tangemann K, Bohrmann B, Müller B, Engel J. Integrin aIIbB3 reconstituted into lipid bilayers is nonclustered in its activated state but clusters after fibrinogen binding. Biochemistry 1997;36: 7395–402.

[31] Xiong JP, Stehle T, Diefenbach B, Zhang R, Dunker R, Scott DL, et al. Crystal structure of the extracellular segment of integrin aVb3. Science 2001;294:339–45.

[32] Juliano RL, Haskill S. Signal transduction from the extracellular matrix. J Cell Biol 1993;129:577–85.

[33] Xiong JP, Stehle T, Zhang R, Joachimiak A, Frech M, Goodman SL, et al. Crystal structure of the extracellular segment of integrin aVb3 in complex with an Arg-Gly-Asp ligand. Science 2002;296:151–5.

[34] Zimmerberg J, Kozlov MM. How proteins produce cellular membrane curvature. Nat Rev Mol Cell Biol 2006;7:9–19.

CHAPTER 4
Spatial Control of Surface Modification

4a. Surface gradients

Commentary

We were aware of surface-chemical gradients being used to carry out experiments in a high-throughput manner, but also saw several limitations in the ways this had been done in the literature (4.8). In particular, we wanted to develop a method that did not rely on diffusional processes, since these limit the shapes of gradients that can be fabricated. The result was the dipping method (4.1), in which a substrate is slowly lowered into a dilute solution of adsorbing molecules, such that the two ends of the substrate have substantially different surface concentrations, simply due to residence-time effects. A second component can then be added by backfilling the empty sites with a second adsorbate. The first system used for this process involved thiols on gold. We went on to examine the nanoscopic and molecular-level structure of these gradients and found that the process inherently led to the formation of islands of the first component and a "sea" of changing composition along the gradient (4.2). This can be avoided if a photocatalytic process is used to create the one-component gradient, for example (4.6). The degree of order in the two components seemed to be very dependent on the nature of the adsorbates (4.3). Thiol-based gradients were then extended to three components and two orthogonal directions (4.10), and also used in a study of the electronic effects of adsorbing highly electronegative species on a gold surface (4.9). This also turned out to be a way of tuning work function as a function of location on the surface.

We went on to apply polyelectrolyte-anchored PEG chains to an oxide surface in a similar dipping configuration, and used these gradients to follow the dependence of protein adsorption on PEG surface concentration. Also, by fabricating a gradient of biotin-terminated PEG within a PEG background, we created a platform that can be used (by means of the streptavidin-biotin interaction) to form a gradient of any biotinylated biomolecule on a surface (4.5). An alternative chemistry that appears suited to gradient fabrication on oxide surfaces is based on catechol chemistry, as described above (2.8).

Switching to morphology gradients, we developed two approaches to varying roughness along a surface. On the micrometer scale, we achieved this by sandblasting a surface initially, and then polishing it by means of a chemical polishing procedure, using the dipping/residence-time approach described above to create the gradient. The master produced in this way could then be replicated by means of an elastomeric intermediate, from which an epoxy positive could then be cast and coated with the material of interest by physical means, such as sputtering (4.4). On the nanometer scale, we adopted a different approach, using the time-dependent adsorption of silica nanoparticles onto a silicon wafer (coated with a positively charged polyelectrolyte) to create the gradient (4.7). Both microscale and nanoscale gradients were used in cell studies as described above (3.5, 3.6).

Langmuir
The ACS Journal of Surfaces and Colloids

DECEMBER 9, 2003
VOLUME 19, NUMBER 25

Letters

A Simple, Reproducible Approach to the Preparation of Surface-Chemical Gradients

Sara Morgenthaler, Seunghwan Lee, Stefan Zürcher, and Nicholas D. Spencer*

Laboratory for Surface Science and Technology, Department of Materials, Swiss Federal Institute of Technology (ETH), Zürich, Sonneggstrasse 5, CH-8092 Zürich, Switzerland

Received April 25, 2003. In Final Form: July 13, 2003

We demonstrate a very simple and reproducible preparative approach for the fabrication of surface-chemical gradients. A surface concentration gradient of adsorbed methyl- or hydroxyl-terminated thiolates was achieved upon gradually immersing a gold-coated substrate into a very dilute thiol solution (0.0033 mM) by means of a linear-motion drive. Subsequent immersion of the substrate into the complementary thiol solution provided a hydrophobicity gradient with a large range (50° of the water-contact angle) and over a significant distance (35 mm). The self-assembled monolayer gradient produced in this way also displayed a high packing density, as demonstrated by dynamic contact-angle and X-ray photoelectron spectroscopy measurements.

1. Introduction

The self-assembly of alkanethiols on gold is a well-known process that has been the subject of considerable research.[1-10] The mechanisms leading to the formation of single-component and mixed self-assembled monolayers (SAMs) have been studied extensively. The mixed systems investigated have often consisted of methyl- and hydroxyl-terminated thiols because the results of adsorption can be readily monitored by water contact-angle measurements. Such mixed monolayers were found to be stable and readily produced.

Chemical gradients are of great interest for numerous practical applications, such as investigating biomolecular interactions, cell-motility studies, diagnostics, nanotribology, or microfluidics, and naturally lend themselves to combinatorial studies because an entire spectrum of chemical properties can be covered in a single experiment. A number of gradient preparation techniques for various substrates have been described,[11-26] and such gradients have been used for further experiments and applications.[11,23-29] Several methods have been reported for the generation of thiol-based chemical gradients: the cross-diffusion of two thiol solutions through a polysac-

* To whom correspondence should be addressed. E-mail: nicholas.spencer@mat.ethz.ch. Fax: +41 1 633 10 27.

(1) Bain, C. D.; Troughton, E. B.; Tao, Y.-T.; Evall, J.; Whitesides, G. M.; Nuzzo, R. G. *J. Am. Chem. Soc.* **1989**, *111*, 321–335.
(2) Bain, C. D.; Whitesides, G. M. *J. Am. Chem. Soc.* **1988**, *110*, 6560–6561.
(3) Bain, C. D.; Evall, J.; Whitesides, G. M. *J. Am. Chem. Soc.* **1989**, *111*, 7155–7164.
(4) Bain, C. D.; Whitesides, G. M. *J. Am. Chem. Soc.* **1989**, *111*, 7164–7175.
(5) Ishida, T.; Nishida, N.; Tsuneda, S.; Hara, M.; Sasabe, H.; Knoll, W. *Jpn. J. Appl. Phys.* **1996**, *35*, L1710–L1713.
(6) Tamada, K.; Hara, M.; Sasabe, H.; Knoll, W. *Langmuir* **1997**, *13*, 1558–1566.
(7) Karpovich, D. S.; Blanchard, G. J. *Langmuir* **1994**, *10*, 3315–3322.
(8) Folkers, J. P.; Laibinis, P. E.; Whitesides, G. M. *Langmuir* **1992**, *8*, 1330–1341.
(9) Folkers, J. P.; Laibinis, P. E.; Whitesides, G. M. *J. Adhes. Sci. Technol.* **1992**, *6*, 1397–1410.
(10) Zerulla, D.; Uhlig, I.; Szargan, R.; Chassé, T. *Surf. Sci.* **1998**, *404*, 604–608.
(11) Ruardy, T. G.; Schakenraad, J. M.; van der Mei, H. C.; Busscher, H. J. *Surf. Sci. Rep.* **1997**, *29*, 1–30.
(12) Liedberg, B.; Tengvall, P. *Langmuir* **1995**, *11*, 3821–3827.
(13) Liedberg, B.; Wirde, M.; Tao, Y.-T.; Tengvall, P.; Gelius, U. *Langmuir* **1997**, *13*, 5329–5334.
(14) Lestelius, M.; Engquist, I.; Tengvall, P.; Chaudhury, M. K.; Liedberg, B. *Colloid Surf., B* **1999**, *15*, 57–70.
(15) Terrill, R. H.; Balss, K. M.; Zhang, Y.; Bohn, P. W. *J. Am. Chem. Soc.* **2000**, *122*, 988–989.
(16) Balss, K. M.; Fried, G. A.; Bohn, P. W. *J. Electrochem. Soc.* **2002**, *149*, C450–C455.

Published on Web 09/10/2003

charide matrix,[12-14] applying an electrochemical potential to a substrate during adsorption,[15,16] the use of microfluidic devices,[17,18] and scanning-tunneling-microscopy-based replacement lithography.[19] With the exception of the electrochemical potential approach, the gradients formed have been limited in physical size. The aim of our research was to develop a very simple, generally applicable method for fabricating chemical gradients of SAMs on gold on the millimeter–centimeter scale.

2. Materials and Methods

Materials. The two alkanethiols employed in this study were dodecanethiol [$CH_3(CH_2)_{11}SH$] and 11-mercapto-1-undecanol [$HO(CH_2)_{11}SH$], both purchased from Aldrich Chemicals (Milwaukee, WI). Ethanol (purity >99.8%, Merck, Darmstadt, Germany) was used as a solvent. The substrates for SAM films were prepared by evaporating gold (purity >99.99%, Unaxis, Balzers, Liechtenstein) onto silicon wafers (POWATEC, Cham, Switzerland), according to a standard method.[1] The silicon wafers were coated with a 6-nm-thick chromium adhesion layer, followed by an 80-nm gold film in an evaporation chamber (MED 020 coating system, BALTEC, Balzers, Liechtenstein) at a pressure of about 2×10^{-5} mbar. All glassware was cleaned with *piranha solution* (7:3 concentrated H_2SO_4/30% H_2O_2) for 20 min and rinsed copiously with deionized water and ethanol.

Preparation of Gradient SAM Films. The stock solutions were prepared by dissolving $CH_3(CH_2)_{11}SH$ or $HO(CH_2)_{11}SH$ in ethanol at a concentration of 1 mM. All other solutions were prepared by further dilution of the corresponding stock solution. The gradient SAM films were generated by varying the immersion time in alkanethiol-containing solutions along the longitudinal axis of the gold-coated silicon substrate (length 4 cm, width 1 cm). The immersion of the substrates was controlled by a computer-driven linear-motion drive (OWIS, Staufen, Germany). For comparison with static experiments, small pieces (1 × 1 cm) of the substrate were also immersed in the solutions for different lengths of time. All substrates were rinsed with ethanol, dried with nitrogen, and plasma-cleaned (30 s N_2, high power, Harrick Plasma Cleaner/Sterilizer PDC-32G instrument, Ossining, NY) before immersion. The speed of the linear-motion drive can be varied between 2.5 μm/s and 2.5 mm/s. Before characterization, the substrates were again rinsed with ethanol and dried with nitrogen.

Characterization of Gradient SAM Films. The hydrophobicity variation of the gradient SAM films was characterized by water contact-angle measurements as a function of position along the longitudinal axis of the sample. Both static and dynamic contact angles were measured employing a contact-angle goniometer (Ramé Hart model 100, Ramé Hart, Inc., Mountain Lakes, NJ, U.S.A., and G2/G40 2.05-D, Krüss GmbH, Hamburg, Germany, respectively). The results of dynamic contact-angle measurements were evaluated using digital image analysis. X-ray photoelectron spectroscopy (XPS) spectra were obtained using a PHI 5700 spectrometer with an Al Kα source (350 W, 15 kV) at a takeoff angle of 45°. We have chosen a pass energy of 46.95 and 0.1 eV per step to limit the exposure time and, therefore, keep X-ray damage to a minimum while having reasonable signal-to-noise ratios. The exposure time for each measurement of the four regions [C(1s), O(1s), S(2p), and Au(4f)] was 700 s.[30,31]

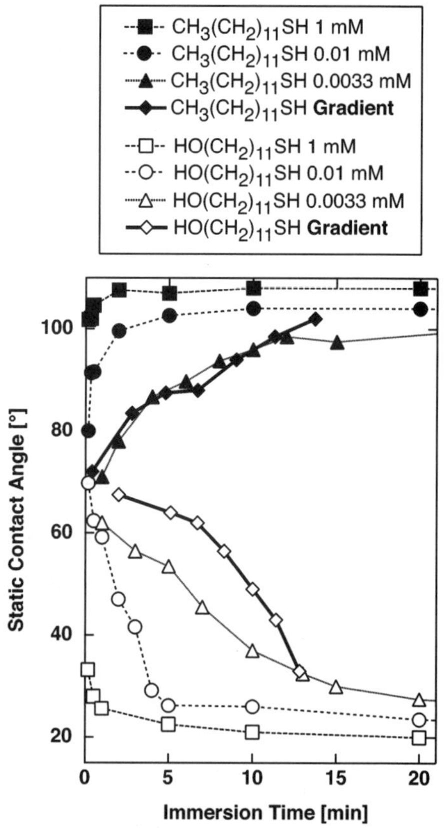

Figure 1. Gold substrates immersed in either $CH_3(CH_2)_{11}SH$- or $HO(CH_2)_{11}SH$-containing ethanol solutions. Static contact angles were measured after different immersion times. The lower the concentration of the solutions, the longer it takes to reach the saturation values. For comparison, two gradient samples are also shown: the gradient sample was prepared by immersing a gold-coated substrate (4 cm in length) in 0.003 mM $CH_3(CH_2)_{11}SH$ solution at a speed of 40 μm/s by means of a linear-motion drive.

3. Results and Discussion

In previous studies, the kinetics of alkanethiol adsorption on gold substrates has been investigated by varying the exposure time of a series of samples in solutions containing alkanethiol moieties.[1-7] In this work, this approach was reproduced for both $CH_3(CH_2)_{11}SH$ and $HO(CH_2)_{11}SH$ at different concentrations: 1, 0.01, and 0.0033 mM. In all cases, water contact angles (static) after 24 h of immersion reached the expected saturated values [110° for $CH_3(CH_2)_{11}SH$ and 26° for $HO(CH_2)_{11}SH$). As shown in Figure 1, however, the adsorption behavior in the initial stages (<30 min) showed a strong dependence on the solution concentration and the type of alkanethiol.

(17) Jeon, N. L.; Dertinger, S. K. W.; Chiu, D. T.; Choi, I. S.; Stroock, A. D.; Whitesides, G. M. *Langmuir* **2000**, *16*, 8311–8316.
(18) Dertinger, S. K. W.; Chiu, D. T.; Jeon, N. L.; Whitesides, G. M. *Anal. Chem.* **2001**, *73*, 1240–1246.
(19) Fuierer, R. R.; Carroll, R. L.; Feldheim, D. L.; Gorman, Ch. B. *Adv. Mater.* **2002**, *14*, 154–157.
(20) Jeong, B. J.; Lee, J. H.; Lee, H. B. *J. Colloid Interface Sci.* **1996**, *178*, 757–763.
(21) Efimenko, K.; Genzer, J. *Adv. Mater.* **2001**, *13*, 1560–1563.
(22) Wijesundara, M. B. J.; Fuoco, E.; Hanley, L. *Langmuir* **2001**, *17*, 5721–5726.
(23) Herbert, C. B.; McLernon, T. L.; Hypolite, C. L.; Adams, D. N.; Pikus, L.; Huang, C. C.; Fields, G. B.; Letourneau, P. C.; Distefano, M. D.; Hu, W. S. *Chem. Biol.* **1997**, *4*, 731–737.
(24) Spijker, H. T.; Bos, R.; van Oeveren, W.; de Vries, J.; Busscher, H. J. *Colloid Surf., B* **1999**, *15*, 89–97.
(25) Hypolite, C. L.; McLernon, T. L.; Adams, D. N.; Chapman, K. E.; Herbert, C. B.; Huang, C. C.; Distefano, M. D.; Hu, W.-S. *Bioconjugate Chem.* **1997**, *8*, 658–663.
(26) Wu, T.; Efimenko, K.; Vlcek, P.; Subr, V.; Genzer, J. *Macromolecules* **2003**, *36*, 2448–2453.
(27) Jayaraman, S.; Hillier, A. C. *Langmuir* **2001**, *17*, 7857–7864.
(28) Sehayek, T.; Vaskevich, A.; Rubinstein, I. *J. Am. Chem. Soc.* **2003**, *125*, 4718–4719.
(29) Houseman, B. T.; Mrksich, M. *Chem. Biol.* **2002**, *9*, 443–454.
(30) Zerulla, D.; Chassé, T. *Langmuir* **1999**, *15*, 5285–5294.
(31) Heister, K.; Zharnikov, M.; Grunze, M. *Langmuir* **2001**, *17*, 8–11.

For the highly concentrated solution (1 mM), both thiols [$CH_3(CH_2)_{11}SH$ and $HO(CH_2)_{11}SH$] reached water contact angles with less than 5% deviation from the saturated monolayer values immediately after immersion (<1 min), while systematically longer times were needed to reach these values for the dilute solutions (0.01 and 0.0033 mM). It was also noted that $HO(CH_2)_{11}SH$ displayed a slower adsorption behavior than $CH_3(CH_2)_{11}SH$, presumably as a result of its greater affinity for ethanol.

The different hydrophobicity values on separate samples shown in Figure 1 can be produced on a single substrate if the immersion time is spatially controlled, and, thus, a *gradient* can be generated. This was accomplished through a controlled immersion of a substrate, such that the position along the sample corresponds directly to a particular immersion time. Thus, the immersion speed has to be carefully selected according to the adsorption kinetics. In this study, a concentration of 0.0033 mM and a speed of 40 μm/s were chosen. The water contact angles (static) measured at several discrete points along the gradient films are also plotted in Figure 1, after the conversion of the spatial position to the corresponding immersion time. As expected, the contact-angle variation achieved along the gradient is in good agreement with that obtained from a series of homogeneously functionalized samples (70–100°) for a $CH_3(CH_2)_{11}SH$ gradient. For the $HO(CH_2)_{11}SH$ gradient, agreement was less good, presumably as a result of the higher intrinsic surface energy of the layer and, thus, a greater susceptibility to the adsorption of airborne contamination: A Au substrate that was rinsed with ethanol and briefly exposed to laboratory air showed a water contact angle of 65°.

A hydrophobicity gradient film composed of a single component, however, consists of coverage gradients of alkanethiols along the immersion axis of the substrate. Because partial monolayers are less ordered than full monolayers, this initial surface also displays a gradient in order. To remove this inhomogeneity in order and to promote the formation of a complete monolayer, while maintaining the hydrophobicity gradient, the sample was immersed in the complimentary thiol solution in a second step. This is also expected to provide an extended hydrophobicity gradient range.

Two approaches have been employed: (a) the sample was immersed in the same way as in the first step, allowing the end that was least exposed to the first component to be initially immersed in the complementary solution (*head-to-tail* method); (b) following the initial step, the sample was fully immersed in the complementary solution for a given time (*full-immersion* method). In addition, the issue of sequence, that is, whether a $CH_3(CH_2)_{11}SH$ or $HO(CH_2)_{11}SH$ gradient is generated first for each approach, provides four alternatives in total, denoted as *head-to-tail (CH_3-first)*, *head-to-tail (OH-first)*, *full-immersion (CH_3-first)*, and *full-immersion (OH-first)*. To facilitate the filling of vacant binding sites, a higher concentration (0.01 mM) was selected for the second solution. The samples were rinsed with ethanol and blown dry with a stream of nitrogen prior to their immersion into the second solution.

All four alternatives showed that the hydrophobicity gradient range is extended after immersion into the second solution: For both the *head-to-tail* and the *full-immersion* methods, after the second immersion step, *CH_3-first* showed a range of 40–100° and *OH-first* showed a range of 30–90° (static water contact angles). However, in terms of monolayer completion and reproducibility/stability, *full-immersion (CH_3-first)* provided the best results. The advancing and receding contact-angle measurements

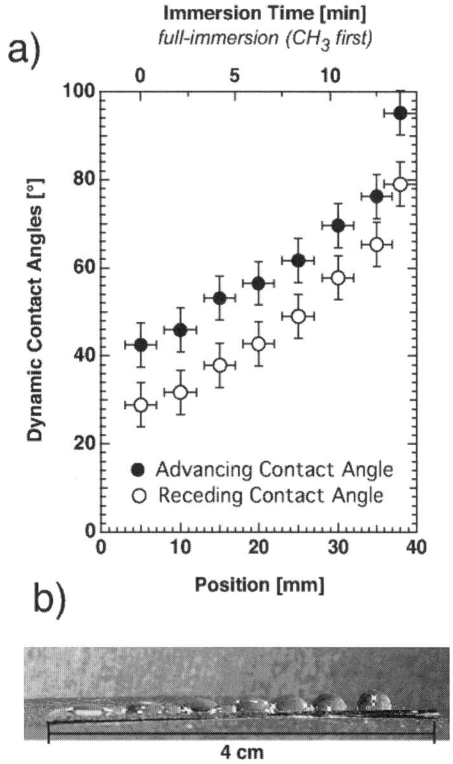

Figure 2. (a) Dynamic contact angles along a hydrophobicity gradient using *full-immersion (CH_3-first)* as the preparation method. The small hysteresis of less than 15° between the advancing and the receding contact angles is an indication for the formation of a full monolayer. (b) Water droplets along a hydrophobicity gradient using *full-immersion (CH_3-first)* as the preparation method.

obtained from *full-immersion (CH_3-first)* overnight are shown in Figure 2a. In this plot, the results obtained from five different gradient films are plotted to show their reproducibility (±5°). A fairly linear hydrophobicity gradient with an average water contact-angle slope of 1.5°/mm over 35 mm is obtained. The average hysteresis of 14° between advancing and receding contact angles indicates that the monolayer formation is nearly complete along the gradient.[1] The photograph in Figure 2b provides a more direct illustration of the hydrophobicity gradient generated by this method.

All the other approaches, including the *head-to-tail* methods and *full-immersion (OH-first)*, have the common drawback that the immersion time in the second solution is practically limited (<15 min under the experimental parameters selected in this work). For the *head-to-tail* method, this is because the immersion speed can only be varied between 2.5 μm/s to 2.5 mm/s. The *full-immersion (OH-first)* method has a limitation because an elongated immersion (>1 h) of the $HO(CH_2)_{11}SH$ precoated film in the second solution (0.01 mM $CH_3(CH_2)_{11}SH$) results in a loss of the hydrophobicity gradient as a result of the displacement of adsorbed $HO(CH_2)_{11}SH$ by $CH_3(CH_2)_{11}SH$. Thus, for the *full-immersion (OH-first)* method the immersion time in the second solution needs to be a compromise between the full saturation of sites and the maintenance of a hydrophobicity gradient (10 min was selected in this work). In contrast, *full-immersion (CH_3-first)* suffered virtually no corresponding displacement in

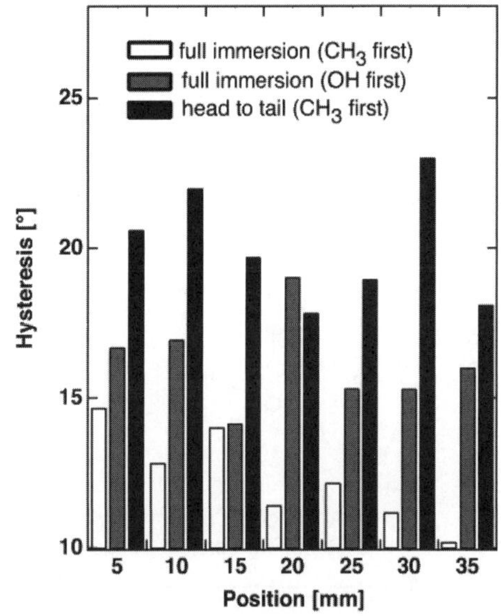

Figure 3. Comparison of the hysteresis of the advancing and the receding contact angles for three different preparation methods (see text). *Full-immersion (CH_3-first)* shows the smallest hysteresis values indicating the most complete monolayer formation.

the second step, even after 24 h of immersion, thus leading to an improved packing of the film. This behavior was confirmed by comparison with the behavior of single-component monolayers generated from either $HO(CH_2)_{11}SH$ or $CH_3(CH_2)_{11}SH$ moieties, immersed in the complementary solution. *Full-immersion (CH_3-first)* displayed the lowest contact-angle hysteresis of all the approaches tested, as shown in Figure 3.

The chemical composition of such a gradient [*full-immersion (CH_3-first)*] was also characterized by XPS immediately after preparation. For a full monolayer, a constant sulfur concentration (about 6 atom %) is expected across the whole gradient. At the same time, the normalized detected atomic concentration of oxygen is expected to increase from the hydrophobic to the hydrophilic side, while the amount of carbon should decrease somewhat because the terminal methyl groups are increasingly replaced by hydroxyl groups. An almost linear increase for the O(1s), with a concomitant decrease in the C(1s) signals, was found in the experiment, in agreement with the contact-angle results (see Figure 4). The comparison of the two extreme ends of the gradient with two control samples immersed for 24 h in either 0.003 mM $HO(CH_2)_{11}SH$ or 0.003 mM $CH_3(CH_2)_{11}SH$ demonstrates that the chemical composition is changing from an almost complete monolayer of $CH_3(CH_2)_{11}SH$ to an almost complete monolayer of $HO(CH_2)_{11}SH$ in a very smooth and almost linear way. The composition of the pure monolayers was compared with a theoretical model, where we corrected for instrumental functions and the attenuation effects of the monolayer at a takeoff angle of 45°.[32] In the case of the pure $CH_3(CH_2)_{11}SH$ sample, a perfect agreement was observed, whereas in the case of the pure $HO(CH_2)_{11}SH$ sample, an excess of carbon was found. This can be explained by a higher affinity to carbon contamination by the higher-surface-energy samples, compared

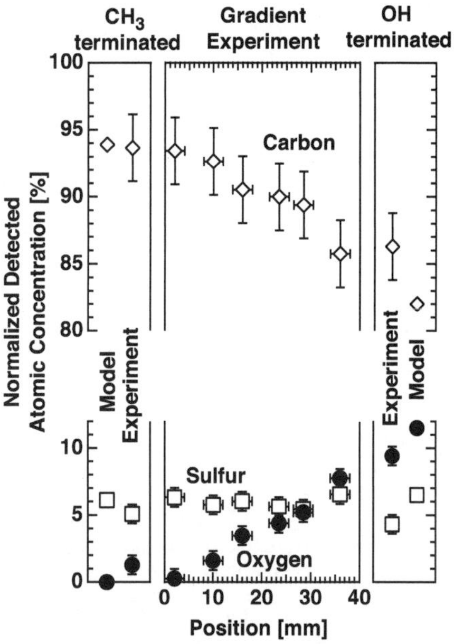

Figure 4. Normalized detected atomic concentrations along a hydrophobicity gradient [*full-immersion (CH_3-first)*] by XPS. Both ends are in good agreement with samples immersed in either $CH_3(CH_2)_{11}SH$ or $HO(CH_2)_{11}SH$. Theoretical values for a full CH_3-terminated or a full OH-terminated thiol film were calculated using a 15-Å-thick model [value from ellipsometry and modeling; electron takeoff angles of 45°: attenuation length of 0.085 × (kinetic energy)$^{0.5}$]. The discrepancy between the calculated and the experimental values in the case of the OH-terminated film can be explained by additional carbon contamination of the hydrophilic sample.

to the low-surface-energy, hydrophobic methyl-terminated surfaces. This explanation is in good agreement with results from ellipsometry, where films fabricated from $HO(CH_2)_{11}SH$ are always found to be a few angstroms thicker than $CH_3(CH_2)_{11}SH$ films.[12,13] If a monolayer of carbonaceous contamination is assumed to be present on the OH-terminated surface, the calculated normalized atomic concentrations match the experimental values within the error bars.

Gradients were stored in different media, such as air, water, ethanol, a vacuum, and nitrogen. The vacuum and nitrogen proved to be effective storage conditions for gradients for up to 5 days. In all other cases, considerable changes in the contact angle were detected after 5 days.

This very simple and reproducible method for the preparation of hydrophobicity gradients could be easily extended to the preparation of other gradients of various (bio)chemical functionalities. It can also be used to prepare gradients of tailored slopes and lengths, as determined by the concentrations of the adsorbing solutions and the velocity program of the linear-motion drive. While there is no general upper limit to the length of gradients that may be prepared in this way, a lower limit is presumably imposed by the finite size of the meniscus at the air–gold–thiol solution interface.

Acknowledgment. Our thanks go to Drs. Johan Ubbink and Prisca Schär-Zammaretti of the Nestlé Research Centre (Vers-chez-les-Blanc, Switzerland) for their support of this work.

LA034707L

(32) Laibinis, P. E.; Bain, C. D.; Whitesides, G. M. *J. Phys. Chem.* **1991**, *95*, 7017–7021.

2706 *Langmuir* **2006,** *22,* 2706–2711

Submicrometer Structure of Surface-Chemical Gradients Prepared by a Two-Step Immersion Method

Sara M. Morgenthaler, Seunghwan Lee, and Nicholas D. Spencer*

Laboratory for Surface Science and Technology, Department of Materials, Swiss Federal Institute of Technology, ETH Zurich, Wolfgang-Pauli-Strasse 10, CH-8093 Zürich, Switzerland

Received October 3, 2005. In Final Form: December 24, 2005

Lateral force microscopy and microdroplet density measurements have been used to examine the microstructure of surface-chemical gradients of thiols on gold, prepared by a two-step immersion method. A single-component coverage gradient, generated by gradual immersion of a gold surface into a solution of a single thiol, yielded islands of ~25 nm in diameter at the end that had only been briefly immersed, whereas an increasingly continuous film was formed along the gradient. After saturation with a second thiol with a different end group, the structure generated during the initial immersion step was found to persist.

1. Introduction

Self-assembled monolayers (SAMs) are often used as model systems to investigate surface interactions and are applied in such diverse fields as biological sensing, molecular-device fabrication, diagnostics, and nanotribology.[1−7] SAMs have attracted a lot of attention because of their spontaneous formation from solution, high degree of order due to van der Waals interchain interactions, stability, and well-defined surface properties. The most widely investigated SAMs have been prepared from alkanethiols or disulfides, adsorbed on gold surfaces.[2,8,9]

A full monolayer of chemisorbed, simple alkanethiol chains is known to display a $(\sqrt{3} \times \sqrt{3})$ R30° super-lattice structure on a Au(111) substrate.[10] The evolution of such phases during SAM growth has been studied in detail by STM. Differently packed, "lying-down" phases have been found in the initial growth stage before the final well-packed, "standing-up" phase is formed.[10−12] Submonolayer-coverage islands have also been employed as a model system to study the structure of alkanethiol SAMs under varying applied loads.[13,14] The islands were found to ripen over time with an increase in stability. Mixed monolayers are, however, more interesting for applications since their properties can be readily tailored by choosing appropriate mixing ratios of two different thiols. The structure of mixed SAMs prepared by coadsorption of two components depends on the mixing ratio, the solvent used for preparation, the nature of the molecular backbones and terminal groups, as well as the difference in length.[15] It is generally known, and has been shown by molecular dynamics simulations, that if alkanethiols with a chain-length difference of more than three CH_2 groups are coadsorbed, phase separation occurs.[16] This phase separation is due to increased van der Waals interaction between the longer chains and consequent preferential aggregation between them[16−21] and can be inhibited by performing the coadsorption at higher solution temperatures.[22] Differences in the chemical nature of backbone chains and end groups can also lead to phase separation,[23−25] usually due to increased hydrogen bonding between one of the two components. Most authors report that no phase separation has been observed for mixed SAMs composed of two components with different end groups, such as $-CH_3$, $-OH$, $-COOH$, or $-CN$,[9,26] as long as the backbone chain length remains the same, although Brewer et al. claimed phase separation in mixed OH- and CH_3-terminated SAMs of similar length by means of atomic force microscopy (AFM) pull-off force measurements.[27]

A mixed monolayer whose surface composition steadily changes along a sample constitutes a surface-chemical gradient. Several gradient preparation methods for alkanethiols have been

* To whom correspondence should be addressed. E-mail: nicholas.spencer@mat.ethz.ch. Fax: +41 44 633 10 27.

(1) Ulman, A. *Introduction to Ultrathin Organic Films: from Langmuir–Blodgett to Self-Assembly*; Academic Press: San Diego, 1991.
(2) Love, J. C.; Estroff, L. A.; Kriebel, J. K.; Nuzzo, R. G.; Whitesides, G. M. *Chem. Rev.* **2005**, *105*, 1103–1169.
(3) Mrksich, M. *Chem. Soc. Rev.* **2000**, *29*, 267–273.
(4) Flink, S.; van Veggel, F.; Reinhoudt, D. N. *Adv. Mater.* **2000**, *12*, 1315–1328.
(5) Overney, R. M.; Meyer, E.; Frommer, J.; Brodbeck, D.; Lüthi, R.; Howald, L.; Güntherodt, H. J.; Fujihira, M.; Takano, H.; Gotoh, Y. *Nature* **1992**, *359*, 133–135.
(6) Frisbie, C. D.; Rozsnyai, L. F.; Noy, A.; Wrighton, M. S.; Lieber, C. M. *Science* **1994**, *265*, 2071–2074.
(7) Noy, A.; Vezenov, D. V.; Lieber, C. M. *Annu. Rev. Mater. Sci.* **1997**, *27*, 381–421.
(8) Nuzzo, R. G.; Allara, D. L. *J. Am. Chem. Soc.* **1983**, *105*, 4481–4483.
(9) Bain, C. D.; Troughton, E. B.; Tao, Y. T.; Evall, J.; Whitesides, G. M.; Nuzzo, R. G. *J. Am. Chem. Soc.* **1989**, *111*, 321–335.
(10) Schreiber, F. *Prog. Surf. Sci.* **2000**, *65*, 151–256.
(11) Xu, S.; Cruchon-Dupeyrat, S. J. N.; Garno, J. C.; Liu, G. Y.; Jennings, G. K.; Yong, T. H.; Laibinis, P. E. *J. Chem. Phys.* **1998**, *108*, 5002–5012.
(12) Yamada, R.; Uosaki, K. *Langmuir* **1998**, *14*, 855–861.
(13) Barrena, E.; Ocal, C.; Salmeron, M. *J. Chem. Phys.* **1999**, *111*, 9797–9802.
(14) Barrena, E.; Palacios-Lidon, E.; Munuera, C.; Torrelles, X.; Ferrer, S.; Jonas, U.; Salmeron, M.; Ocal, C. *J. Am. Chem. Soc.* **2004**, *126*, 385–395.
(15) Folkers, J. P.; Laibinis, P. E.; Whitesides, G. M.; Deutch, J. *J. Phys. Chem.* **1994**, *98*, 563–571.
(16) Shevade, A. V.; Zhou, J.; Zin, M. T.; Jiang, S. Y. *Langmuir* **2001**, *17*, 7566–7572.
(17) Atre, S. V.; Liedberg, B.; Allara, D. L. *Langmuir* **1995**, *11*, 3882–3893.
(18) Folkers, J. P.; Laibinis, P. E.; Whitesides, G. M. *Langmuir* **1992**, *8*, 1330–1341.
(19) Hobara, D.; Ota, M.; Imabayashi, S.; Niki, K.; Kakiuchi, T. *J. Electroanal. Chem.* **1998**, *444*, 113–119.
(20) Shon, Y. S.; Lee, S.; Perry, S. S.; Lee, T. R. *J. Am. Chem. Soc.* **2000**, *122*, 1278–1281.
(21) Tamada, K.; Hara, M.; Sasabe, H.; Knoll, W. *Langmuir* **1997**, *13*, 1558–1566.
(22) Chen, S. F.; Li, L. Y.; Boozer, C. L.; Jiang, S. Y. *J. Phys. Chem. B* **2001**, *105*, 2975–2980.
(23) Nyquist, R. M.; Eberhardt, A. S.; Silks, L. A.; Li, Z.; Yang, X.; Swanson, B. I. *Langmuir* **2000**, *16*, 1793–1800.
(24) Smith, R. K.; Reed, S. M.; Lewis, P. A.; Monnell, J. D.; Clegg, R. S.; Kelly, K. F.; Bumm, L. A.; Hutchison, J. E.; Weiss, P. S. *J. Phys. Chem. B* **2001**, *105*, 1119–1122.
(25) Stranick, S. J.; Parikh, A. N.; Tao, Y. T.; Allara, D. L.; Weiss, P. S. *J. Phys. Chem.* **1994**, *98*, 7636–7646.
(26) Stranick, S. J.; Atre, S. V.; Parikh, A. N.; Wood, M. C.; Allara, D. L.; Winograd, N.; Weiss, P. S. *Nanotechnology* **1996**, *7*, 438–442.
(27) Brewer, N. J.; Leggett, G. J. *Langmuir* **2004**, *20*, 4109–4115.

Published on Web 02/07/2006

developed and gradients with compositional changes over a few micrometers to centimeters have been prepared.[28−33] The macroscopic properties of these gradients have been well characterized by X-ray photoelectron spectroscopy (XPS),[29,33] contact-angle measurements,[29,33] ellipsometry,[29] infrared spectroscopy,[29] scanning tunneling microscopy (STM),[32] and fluorescence microscopy.[31] To our knowledge, information on the microstructure of such gradients has not been reported. However, microstructure is of great importance when such gradients are applied in protein-adsorption studies or in biological sensing; Huang et al. have shown that the microstructure of mixed ω-functionalized monolayers strongly affects the adsorption of proteins.[34]

Atomic force microscopy, especially lateral force microscopy (LFM), has proven to be useful when mixed or patterned SAMs with different end functionalities need to be characterized.[35] Interactions between the AFM tip and the SAM strongly depend on the nature of the end group. Given that other structural parameters are identical, the more hydrophilic the end group of the SAM, the higher the friction force measured against a regular Si_3N_4 AFM tip under ambient conditions. This friction contrast may be enhanced by the use of a chemically modified tip,[6,7,36] although tip modifications may result in the broadening of the tip geometry, which itself leads to a degradation of resolution.

In this study, we present LFM measurements with a regular Si_3N_4 tip on surface-chemical gradients prepared from methyl- and hydroxyl- or carboxyl-terminated alkanethiols by a two-step immersion procedure.[33] During the first step, one type of alkanethiol is adsorbed with a gradient in surface concentration onto a gold surface by a gradual immersion process. This single-component gradient is saturated with a component of different end functionality in a second, rapid immersion process to create a densely packed, two-component monolayer. In the first section, we describe the island structure of single-component gradients and later show that this structure persists after saturation with the second component.

2. Experimental Section

Chemicals. 1-Dodecanethiol (98+%, Aldrich Chemicals, St. Louis, MO), 11-mercapto-1-undecanol (97%, Aldrich Chemicals, St. Louis, MO), 1-mercapto-undecanoic acid (95%, Aldrich Chemicals, St. Louis, MO), H_2SO_4 (95−97%, Merck, Germany), and H_2O_2 (30%, Merck, Germany) were used as received. Ethanol (analytical grade, Scharlau Chemicals SA, Spain) was used as a solvent for all experiments. Water was purified using a Milli-Q water system (Millipore, Billerica, MA).

Preparation of Gold Surfaces. Template-stripped gold surfaces were used for AFM measurements. They were prepared according to the procedure reported by Wagner et al.[37] A freshly cleaved mica sheet (50 mm × 80 mm) was mounted on a heating stage in a Bal-Tec BAE-370 vacuum coating system (Bal-Tec, Lichtenstein). The mica sheet was preheated for 12 h at 300 °C to outgas adsorbed water and other volatile contaminants. In a first step, a 5 nm layer of gold (99.99%, Umicore Materials AG, Lichtenstein) was deposited at a rate of 1 Å/s and a pressure of below 1×10^{-6} mbar. After the sample was annealed for 6 h at 300 °C, another 195 nm gold layer was deposited. The thin film was cooled at a rate of 10 °C/min and cut into 10 mm × 40 mm and 10 mm × 10 mm samples. Si wafers (POWATEC GmbH, Switzerland) were cut and precleaned twice by ultrasonication in toluene and ethanol and dried with pure nitrogen. The gold-coated mica sheets were then glued, gold face down, onto the Si slides with an epoxy resin (Epo-tek 377, Polyscience AG, Switzerland) and cured at 150 °C for more than 2 h. Immediately before use, the mica was detached mechanically from the gold by tweezers.

Microdroplet density measurements (μDD) were carried out on polycrystalline gold films. Si wafers were precleaned as described above. Additionally, they were cleaned for 10 min in piranha solution (7:3 concentrated H_2SO_4/30% H_2O_2), rinsed with plenty of water and dried with nitrogen. ***Attention:** Piranha solution reacts violently with all organics and should be handled with care!* The Si wafers were then coated with a 6 nm chromium adhesion layer followed by an 80 nm gold film in an evaporation chamber (MED 020 coating system, Bal-Tec, Liechtenstein) at a pressure of about 2×10^{-5} mbar. Before use, the polycrystalline substrates were cleaned by ultrasonication in ethanol followed by 30 s of air plasma treatment (PDC-001, Harrick Scientific Corporation, NY) and another 10 min immersion in ethanol to remove gold oxides that were produced during plasma treatment.[38]

Gradient Preparation and SAM Formation. Gradients were prepared on 10 mm × 40 mm samples according to the previously reported procedure.[33] The gold-coated substrates were mounted onto a computer-controlled, linear-motion drive (OWIS GmbH, Germany), which allowed them to be slowly immersed into a dilute ethanolic alkanethiol solution. The first gradual adsorption step was carried out in 0.005 mM dodecanethiol solutions. The substrate was immersed at a speed of 75 μm/s. Following total immersion, it was removed quickly from the solution, rinsed abundantly with ethanol, and dried with nitrogen. The substrate was then immersed overnight in a second 0.01 mM solution of either 11-mercapto-1-undecanol or 1-mercapto-undecanoic acid. Both single- (after the first step) and two-component (after the second step) gradients were then analyzed with AFM.

As a control, mixed monolayers were prepared in two different ways on 10 mm × 10 mm samples; either they were coadsorbed from a mixed 0.01 mM solution, or they were prepared in two subsequent immersion steps by mimicking the gradient preparation method described above. In the coadsorption case, the composition of the monolayers was controlled by varying the mixing ratio of the two components in solution. For the two-step case, the substrate was immersed completely into a 0.005 mM dodecanethiol solution for a given time, and then it was saturated overnight in a 0.01 mM solution of a different alkanethiol. As another control, a microcontact printed pattern composed of two alkanethiols was generated with a poly(dimethylsiloxane) (PDMS) stamp as described by Xia et al.[39]

Contact-Angle Measurements. Static contact angles were measured at room temperature and ambient humidity along the gradient samples. Single measurements of sessile drops were performed every 5 mm along the substrate with a contact-angle goniometer (Ramé Hart model 100, Ramé Hart, Inc., Mountain Lakes, NJ). One drop of Milli-Q water was 6 μL. All contact-angle measurements were averaged over several samples.

Lateral Force Microscopy. LFM measurements were performed with a commercial AFM (Dimension 3000, Veeco Instruments, Plainview, NY) equipped with a standard scanner. The images were obtained under ambient conditions using a regular V-shaped Si_3N_4 cantilever and tip (Veeco Instruments Inc., Plainview, NY) with a spring constant of 0.12 N/m. Images were recorded every 5 mm along the sample length, and at least three measurements were taken by changing the position across the sample width at a given length. Height and friction images were recorded simultaneously for each

(28) Ruardy, T. G.; Schakenraad, J. M.; van der Mei, H. C.; Busscher, H. J. *Surf. Sci. Rep.* **1997**, *29*, 3−30.
(29) Liedberg, B.; Tengvall, P. *Langmuir* **1995**, *11*, 3821−3827.
(30) Terrill, R. H.; Balss, K. M.; Zhang, Y. M.; Bohn, P. W. *J. Am. Chem. Soc.* **2000**, *122*, 988−989.
(31) Jeon, N. L.; Dertinger, S. K. W.; Chiu, D. T.; Choi, I. S.; Stroock, A. D.; Whitesides, G. M. *Langmuir* **2000**, *16*, 8311−8316.
(32) Fuierer, R. R.; Carroll, R. L.; Feldheim, D. L.; Gorman, C. B. *Adv. Mater.* **2002**, *14*, 154−157.
(33) Morgenthaler, S.; Lee, S.; Zürcher, S.; Spencer, N. D. *Langmuir* **2003**, *19*, 10459−10462.
(34) Huang, Y. W.; Gupta, V. K. *J. Chem. Phys.* **2004**, *121*, 2264−2271.
(35) Wilbur, J. L.; Biebuyck, H. A.; MacDonald, J. C.; Whitesides, G. M. *Langmuir* **1995**, *11*, 825−831.
(36) Headrick, J. E.; Berrie, C. L. *Langmuir* **2004**, *20*, 4124−4131.
(37) Wagner, P.; Hegner, M.; Güntherodt, H. J.; Semenza, G. *Langmuir* **1995**, *11*, 3867−3875.
(38) Ron, H.; Rubinstein, I. *Langmuir* **1994**, *10*, 4566−4573
(39) Xia, Y. N.; Whitesides, G. M. *Angew. Chem., Int. Ed.* **1998**, *37*, 551−575.

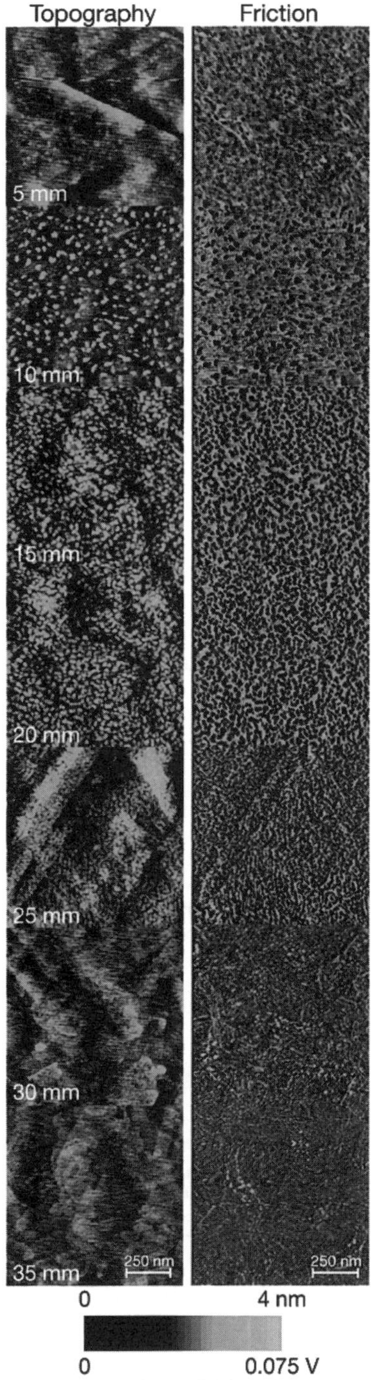

Figure 1. Topography (left) and friction (right) images (1 μm × 1 μm) from contact-mode AFM along a single-component (dodecanethiol) gradient from the briefly to the lengthily immersed end. Island structures are visible both in topography and friction images. The island coverage increases toward the end that was immersed for a longer time (bottom). Note that, while the same dynamic range is employed in each topographical or friction image, the offset has been automatically adjusted by the SPM software to preserve the same mean value. Thus, *absolute* colors cannot be compared from one end of the gradient to the other.

point. The images were flattened using installed software (Veeco Instruments ver. 5.12, Plainview, NY) before further analysis was performed.

Friction histograms were plotted for each image and were fitted with two peaks. Best fits were obtained with a 100% Lorentzian curve and a fixed full width at half-maximum (fwhm). Peak areas were used to estimate low- and high-friction areas.

Microdroplet Density. μDD measurements were performed as described by Hofer et al.[40] Briefly, the sample was placed onto a metal stage in a transparent humidity chamber. While the metal stage was cooled by iced water, the nucleation and growth of water-droplets was recorded using a CCD camera (model WV-BP 310/6, Matsushita Communication Deutschland GmbH, Germany) coupled to a microscope (Carl Zeiss (Schweiz) AG, Switzerland). Images at the point of droplet nucleation were saved and the number of droplets determined by means of digital image analysis.

3. Results and Discussion

Structure of Single-Component Gradients (after First Step). Single-component gradients were prepared as described in the Experimental Section. A single-component gradient consists of an incomplete methyl-terminated molecular film whose surface concentration changes along the sample. Dodecanethiol was selected as the first component to be adsorbed in order to minimize replacement occurring during the second step; dodecanethiol replaces mercapto-undecanol more effectively than the other way around. Figure 1 shows topographic and friction contrast images along a single-component gradient obtained by LFM. Islands can be distinguished from the underlying substrate on both the topography and the friction images. The triangular structures of the underlying Au(111) terraces appearing at 5 and 25 mm from the briefly immersed end can be detected through the organic layer. Thus, the friction images were employed for quantitative analysis. Individual islands with clear boundaries can be found at the low-concentration end, whereas these islands coalesce with increasing surface density of dodecanethiol. The low-friction (darker) areas are attributed to densely packed islands of alkanethiols, which display a much lower friction force than the underlying gold or a disordered film of similar thiols.[41] We assume that the regions around the islands are not bare gold but are covered with a layer of lying-down alkanethiol molecules, as has been reported by Barrena et al.[14] The islands showed an average height of around 10 Å above the surrounding region, which corresponds well with this picture.

The fraction of surface covered by islands was estimated from an analysis of friction-force histograms. The friction-force histograms were analyzed for their lower- and their higher-friction components, assuming that the underlying gold is completely covered with a layer of lying-down molecules. The peak areas for the lower friction contribution were normalized with the total peak intensity and are plotted in Figure 2. The inset in Figure 2 shows a fit of a histogram taken at 20 mm from the briefly immersed end. Two peaks can clearly be distinguished, best fits being obtained by keeping the peak shape (Lorentzian) and the fwhm constant. The amount of the lower-friction component increases toward the end that was immersed for longer, which agrees with an increased coverage at longer immersion times. A Langmuir-type adsorption isotherm (solid line) fits the curve well, and thus we can conclude that island growth follows Langmuir-type adsorption behavior. However, this growth behavior does not correspond directly to the overall adsorption kinetics of dodecanethiol from a 5 μM solution.

Structure of Two-Component Gradients (after Second Step). The single-component gradients are saturated with the

(40) Hofer, R.; Textor, M.; Spencer, N. D. *Langmuir* **2001**, *17*, 4123–4125.
(41) Carpick, R. W.; Salmeron, M. *Chem. Rev.* **1997**, *97*, 1163–1194.

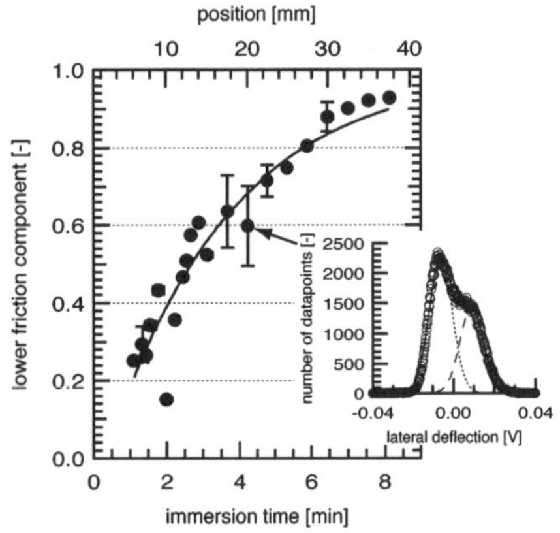

Figure 2. Fraction of lower friction component along single-component gradients determined by histogram analysis. The lower-friction contribution increases from the left to the right end showing an increasing coverage. The inset shows the peak fitting of the friction histogram at 20 mm from the end. A Langmuir-type isotherm fits the data well.

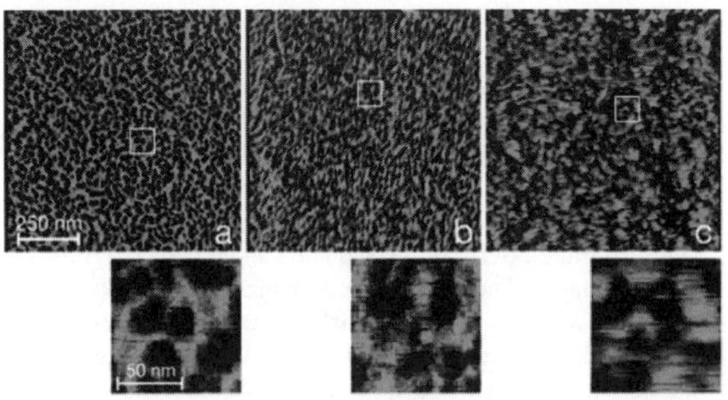

Figure 3. Images from lateral force microscopy displaying the structure of the gradients at 20 mm from the ends (a, single-component gradient; b, two-component gradient (OH-terminated); c, two-component gradient (COOH-terminated)). Overview images (top, 1 μm × 1 μm) and zoomed-in images (below, 0.1 μm × 0.1 μm) are presented. An island-like structure is found in all three images.

second component during a second adsorption step. Friction-force images from three gradients are compared in Figure 3. Images at 20 mm from the ends of a single-component gradient (a), a two-component gradient saturated with OH-terminated

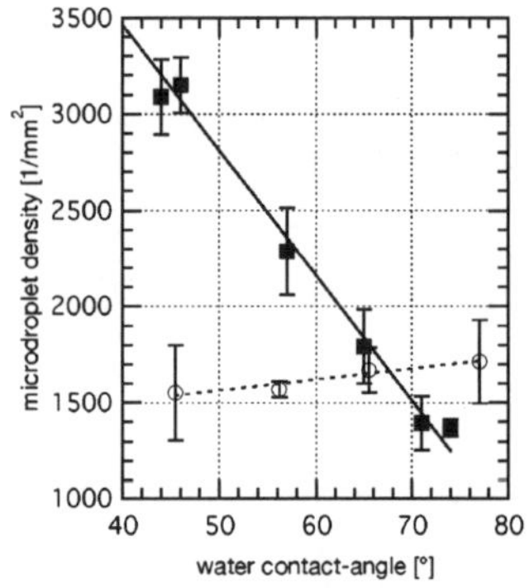

Figure 5. Microdroplet density measurements on gradients (squares) and mixed monolayers prepared by coadsorption (circles). The microdroplet density changes little with contact angle for homogeneously mixed monolayers; however, it changes significantly with contact angle for a gradient.

alkanethiols (b), and another two-component gradient saturated with COOH-terminated alkanethiols (c) are shown. A similarly inhomogeneous structure can be observed in all three images. Islands with clear boundaries are visible on a single-component gradient. These clear boundaries blur out when the gradient is saturated with a second component. However, the shape and size of the islands remain (Figure 3, zoomed-in images). Comparable behavior is observed in both hydroxyl- and carboxyl-terminated, two-component gradients.

In Figure 4a, a microcontact-printed pattern of dodecanethiol backfilled with mercapto-undecanol on polycrystalline gold is displayed to show the feasibility of using lateral-force images to distinguish between spatially segregated methyl- and hydroxyl-terminated regions. The hydroxyl-terminated region displays higher friction than the methyl-terminated region, as has been previously reported.[39] In Figure 4b, a two-component gradient is compared with mixed monolayers (Figure 4c−d). Mixed monolayers can be prepared either by two immersion steps (c), similar to the gradient preparation technique, or by coadsorption of the two components from a mixed solution (d). Contact angles were measured on all three samples (51° on average) and differ by only ±3°, which implies that the overall composition of all three monolayers is very similar. The inhomogeneity of a gradient sample (b) is comparable with the inhomogeneity of a mixed monolayer prepared by two steps (c). However, a mixed

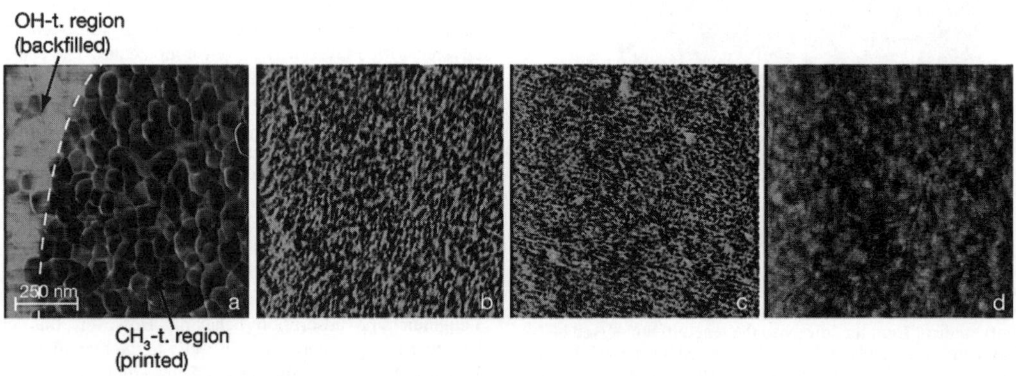

Figure 4. Comparison of friction force images of gradients and mixed monolayers prepared by different techniques. Image (a) shows a microcontact-printed pattern of dodecanethiol backfilled with mercapto-undecanol. Image (b) displays the friction force image for a two-component gradient (OH-terminated) at 20 mm from the end and images (c) and (d) show mixed monolayers (c, prepared by two successive adsorption steps; d, prepared by coadsorption of two thiols from a mixed solution). All images are 1 μm × 1 μm in size. Coadsorbed monolayers show more homogeneous friction images than monolayers prepared in two steps.

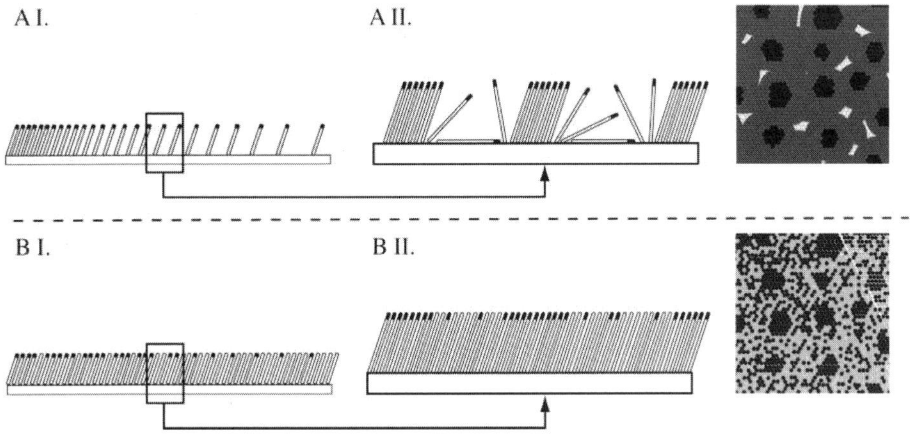

Figure 6. Sketch of the proposed structure for both a single-component (A) and a two-component gradient (B). A 4 cm-long, single-component gradient (a coverage gradient), transitions from a densely to a sparsely covered end (A, I). On a smaller scale, islands are observed (A, II). A layer of disordered alkanethiols surrounds the islands, sometimes leaving free gold crevices at grain boundaries (see clear gray spots in top view). When the incomplete gradient is saturated, a well-ordered SAM with a gradually changing end-functionality is generated (B, I). The alkanethiol islands remain mainly intact, while the flat-lying molecules arrange with the second component to form a well-ordered mixed monolayer (B, II).

monolayer prepared by coadsorption definitely shows a more homogeneous microstructure.

From Figure 3, we conclude that the island-type structure observed after the first immersion step persists during the saturation of the gradient, as the island size and shape remain unchanged. We have confirmed that a clear difference between coadsorbed methyl- and hydroxyl-terminated alkanethiols and mixed monolayers prepared by two steps or gradients exists (Figure 4b–d). A higher friction force contrast was observed on microcontact printed dodecanethiol monolayers, backfilled with mercapto-undecanol. We attribute the reduced friction force contrast in two-step or gradient monolayers to the fact that after the first adsorption step a phase of lying-down molecules is present between the islands. These molecules arrange themselves into a *mixed*, well-packed monolayer with the second component during the second adsorption step. Thus, a mixed monolayer of methyl- and hydroxyl-terminated alkanethiols forms the background to the methyl-terminated islands generated during the first immersion step.

Hofer et al. used microdroplet density measurements as a simple and powerful tool to determine the homogeneity of hydrocarbon-based SAMs.[40] They investigated mixed alkyl phosphate monolayers and found that the microdroplet density does not depend on the hydrophobicity of the substrate but on the microstructure. Incomplete monolayers showed a high microdroplet density, whereas for mixed monolayers prepared by coadsorption, no change in microdroplet density was found when varying the mixing ratios. We used microdroplet density to determine the microstructure of gradients and mixed monolayers with different contact angles prepared by coadsorption (Figure 5). Similarly to the studies on alkyl phosphate SAMs, mixed alkanethiol SAMs prepared from coadsorption revealed no significant change in the microdroplet density as a function of water contact angle. However, on a gradient, the microdroplet density increases toward the more hydrophilic end.

We have shown elsewhere by polarization-modulation infrared spectroscopy (PM-IRRAS) that gradient samples prepared by the two-step method exhibit a highly ordered structure along the gradient and do not show noticeable differences when compared with single or mixed, fully covered monolayers.[42] We thus conclude that the change in microdroplet density of the gradients must originate from inhomogeneities in the composition of topmost terminal groups. The microdroplet density at the more hydrophobic end is similar to that of a mixed monolayer of the same wettability, suggesting that the microstructure of a gradient at this position is similar to that of a mixed monolayer prepared by coadsorption. After the first immersion step into dodecanethiol solution, the gradient microstructure is almost homogeneous at the high concentration end (Figure 1, 35 mm). Thus, the second component only fills up defect sites and is distributed homogeneously over the substrate, the resulting surface resembling a homogeneously mixed monolayer of the same composition. However, islands of tens of nanometers in size are found toward the briefly immersed end (Figure 1, 10 mm), which remain when the sample is saturated with a second component. This structure clearly differs from a homogeneously mixed monolayer of the same composition since, for a coadsorbed monolayer, the lower friction component would be homogeneously distributed in the monolayer. The microdroplet density is thus much higher for the more inhomogeneous monolayer on the gradient sample. The difference between the structure of a coadsorbed monolayer and a gradient increases toward the more hydrophilic end, and thus the difference in microdroplet density also increases.

A sketch of the proposed structure for both the single-component (A) and the two-component gradient (B) is given in Figure 6. Side and top views are sketched on the molecular scale.

4. Conclusions

Lateral force microscopy and microdroplet density measurements have revealed that the structure of chemical gradients prepared by a two-step immersion procedure[33] is inhomogeneous on the <100 nm scale. Islands (a few tens of nanometers in diameter) surrounded by a layer of flat-lying molecules are found after the first adsorption step. These islands grow in a Langmuir-type fashion. After the second adsorption step, the islands remain intact, although their boundaries become less well defined due to a mixing of the lying-down phase with the second component.

(42) Venkataraman, N. V.; Zürcher, S.; Spencer, N. D. *Langmuir*, submitted for publication.

The $<$100 nm-scale phase separation on the gradient prepared in this approach is a unique feature that results from the two immersion steps, which involve a quenching of the adsorption of the first component followed by saturation with the second component. A mixed monolayer prepared in a similar way also revealed such phase separation, whereas a mixed monolayer prepared from coadsorption of two components from a mixed solution revealed homogeneously mixed monolayers on the same scale. These microstructural inhomogeneities will have an impact on the applicability of such gradients, especially when small particles, such as proteins, are involved. However, gradients prepared in this way may also find application as a means to produce functionalized nanoscale islands.

Acknowledgment. The authors acknowledge generous financial assistance for this project from the Swiss National Science Foundation.

LA0526840

4184 *Langmuir* **2006,** *22,* 4184–4189

Order and Composition of Methyl-Carboxyl and Methyl-Hydroxyl Surface-Chemical Gradients

Nagaiyanallur V. Venkataraman, Stefan Zürcher, and Nicholas D. Spencer*

Laboratory for Surface Science and Technology, Department of Materials, ETH Zurich, Wolfgang-Pauli-Strasse 10, CH-8093 Zurich, Switzerland

Received December 6, 2005. In Final Form: February 14, 2006

A detailed infrared and XPS characterization of surface-chemical gradients of dodecanethiol with 11-mercaptoundecanol or 11-mercaptoundecanoic acid self-assembled on gold, is reported. Gradients were prepared using a simple, two-step process previously reported from our laboratory, which involves a controlled immersion of a polycrystalline gold substrate in a dilute (5 μM) solution of one component and a subsequent back-filling with the other. FTIR measurements show that a single-component gradient of dodecanethiol is composed of disordered, liquidlike alkyl chain conformations. Such a gradient, when back-filled with a complementary thiol, having either a hydroxyl or carboxyl end-group, yields two-component gradients that show similar changes in wettability along their lengths. However, while gradients composed of methyl and hydroxyl end-groups show a well-ordered alkyl chain structure over their entire length, methyl-carboxyl gradients exhibit a greater conformational disordering toward the carboxyl-rich end.

Introduction

The self-assembly of long-chain amphiphilic molecules on solid surfaces has been extensively studied as a convenient route to surface modification;[1-5] of particular interest are self-assembled monolayers (SAMs) formed by long-chain alkanethiols on gold due to their high stability and ease of preparation.[6,7] Many applications of alkanethiol SAMs can be attributed to their utility in controlling macroscopic surface properties, such as wettability,[8,9] adhesion, or friction,[10,11] by a simple change of the chemical functional group at the alkyl chain termini. A systematic variation in surface properties can be achieved by coadsorption of two dissimilarly end-functionalized thiols from solution, and numerous investigations have been carried out on such two-component mixed SAMs of thiols on gold with a variety of combinations of end-groups including $-CH_3$, $-OH$, $-COOH$, $-NH_2$, $-CF_3$, $-CONH_2$, ferrocene, and oligo(ethylene glycol).[8,9,12-19] Surfaces with such well-defined surface-chemical compositions are useful for a systematic study of many interesting surface phenomena. Surfaces exhibiting changes in chemical composition within a single sample are of interest since they allow for an array of surface compositions to be explored within a single experiment. There have been several methods described previously for the preparation of such gradient surfaces with varying surface-chemical composition over a length of a few centimeters.[20] The earliest of such gradients reported include the wettability gradients based on silane adsorption on silica.[21-23] Surface-chemical gradients from SAMs of thiols on gold have also been prepared using several different methods such as cross-diffusion of two differently end-functionalized thiols in a polysaccharide matrix,[24,25] electrochemical desorption of thiol monolayers,[26] mass-transfer limited microcontact printing,[27] and UV/ozone treatment of SAMs.[28] The use of gradient surfaces to explore several surface compositions in a single measurement had been demonstrated in studies of surfactant and protein adsorption on a wettability gradient,[21,29] optimizing the immobilization of streptavidin and in relation to protein repellency of oligo(ethylene glycol)-terminated surfaces.[30] Surface-chemical gradients have also been used to prepare a gradient in surface

* To whom correspondence should be addressed. E-mail: nicholas.spencer@mat.ethz.ch. Fax: +41 44 633 10 27.

(1) Allara, D, L.; Nuzzo, R. G. *Langmuir* **1985**, *1*, 45.
(2) Allara, D. L.; Nuzzo, R. G. *Langmuir* **1985**, *1*, 52.
(3) Porter, M. D.; Bright, T. B.; Allara, D. L.; Chidsey, C. D. E. *J. Am. Chem. Soc.* **1987**, *109*, 3559.
(4) Bain, C. D.; Whitesides, G. M. *Angew. Chem., Int. Ed. Engl.* **1989**, *28*, 506.
(5) Bain, C. D.; Troughton, B. E.; Tao, Y.-T.; Evall, J.; Whitesides, G. M.; Nuzzo, R. G. *J. Am. Chem. Soc.* **1989**, *111*, 321.
(6) Ulman, A. *Chem. Rev.* **1996**, *96*, 1533.
(7) Love, J. C.; Estroff, L. A.; Kriebel, J. K.; Nuzzo, R. G.; Whitesides, G. M. *Chem. Rev.* **2005**, *105*, 1103.
(8) Nuzzo, R. G.; Dubois, L. H.; Allara, D. L. *J. Am. Chem. Soc.* **1990**, *112*, 558.
(9) Dubois, L. H.; Zergaski, B. R.; Nuzzo, R. G. *J. Am. Chem. Soc.* **1990**, *112*, 570.
(10) Kim, H. I.; Graupe, M.; Oloba, O.; Koini, T.; Imaduddin, S.; Lee, T. R.; Perry, S. S. *Langmuir* **1999**, *15*, 3179.
(11) Perry, S. S.; Lee, S.; Shon, Y.-S.; Colorado, R., Jr.; Lee, T. R. *Tribol. Lett.* **2001**, *10*, 81.
(12) Bain, C. D.; Whitesides, G. M. *J. Am. Chem. Soc.* **1988**, *110*, 6560.
(13) Bain, C. D.; Evall, J.; Whitesides, G. M. *J. Am. Chem. Soc.* **1989**, *111*, 7155.
(14) Ulman, A.; Evans, S. D.; Shnidman, Y.; Sharma, R.; Eilers, J. E. *Adv. Coll. Int. Sci.* **1992**, *39*, 175.
(15) Laibinis, P. E.; Whitesides, G. M. *J. Am. Chem. Soc.* **1992**, *114*, 1990.
(16) Balamurugan, S.; Ista, L. K.; Yan, J.; Lopez, G. P.; Fick, J.; Himmelhaus, M.; Grunze, M. *J. Am. Chem. Soc.* **2005**, *127*, 14548.
(17) Auletta, T.; van Veggel, F. C. J. M.; Reinhoudt, D. N. *Langmuir* **2002**, *18*, 1288.
(18) Li, S.; Cao, P.; Colorado, R. Jr.; Yan, X.; Wenzl, I.; Shmakova, O. E.; Graupe, M.; Lee, T. R.; Perry, S. S. *Langmuir* **2005**, *21*, 933.
(19) Tsao, W.-M.; Hoffmann, C. L.; Rabolt, J. F.; Johnson, H. E.; Castner, D. G.; Erdelen, C.; Ringsdorf, H. *Langmuir* **1997**, *13*, 4317.
(20) Busscher, H. J.; Elwing, H. *Colloids Surf., B* **1999**, *15*, 1 and references therein.
(21) Elwing, H.; Welin, S.; Askendal A.; Nilsson, U.; Lundstrom, I. *J. Colloid Interface Sci.* **1987**, *119*, 203.
(22) Welin-Klintstrom, S.; Askendal, A.; Elwing, H. *J. Colloid Interface Sci.* **1993**, *158*, 188.
(23) Golander, G. C.; Caldwell, K.; Lin, Y. S. *Colloids Surf.* **1989**, *42*, 165.
(24) Liedberg, B.; Tengvall, P. *Langmuir* **1995**, *11*, 3821.
(25) Lestelius, M.; Engquist, I.; Tengvall, P.; Chaudhury, M. K.; Liedberg, B. *Colloids Surf., B* **1999**, *15*, 57.
(26) Plummer, S. T.; Wang, Q.; Bohn, P. W.; Stockton, R.; Schwartz, M. A. *Langmuir* **2003**, *19*, 7528.
(27) Kraus, T.; Stutz, R.; Balmer, T, E.; Schmid, H.; Malaquin, L.; Spencer, N. D.; Wolf, H. *Langmuir* **2005**, *21*, 7796.
(28) Loos, K.; Kennedy, S. B.; Eidelman, N.; Tai, Y.; Zharnikov, M.; Amis, E. J.; Ulman, A.; Gross, R. A. *Langmuir* **2005**, *21*, 5237.
(29) Welin-Klintstrom, S.; Lestelius, M.; Liedberg, B.; Tengvall, P. *Colloids Surf., B* **1999**, *15*, 81.
(30) Riepl, M.; Ostblom, M.; Lundstrom, I.; Svensson, S. C. T.; Denier van der Gon, A. W.; Schaferling, M.; Liedberg, B. *Langmuir* **2005**, *21*, 1042.

Published on Web 03/23/2006

density of nanoparticles[31] and proteins[26] by utilizing the differences in chemical reactivity of the terminal functional group.

An alternative method of preparation of a two-component surface-chemical gradient of alkanethiols on gold was recently reported from our laboratory.[32] The method utilizes the kinetics of adsorption of dodecanethiol from very dilute solution (5 μM) by a controlled immersion step producing a gradient in concentration of dodecanethiol molecules on the surface, followed by back-filling with 11-mercaptoundecanol. The surface composition of hydroxyl- and methyl-terminated thiols has been shown to vary almost linearly along the length of a 4-cm-long polycrystalline gold surface. This method is simple and could, in principle, be extended to any combination of end-functional groups and sample size by appropriately controlling the thiol concentrations and immersion speeds. Here we demonstrate the ease of this method to produce chemical functionality gradients with different end-groups and report a detailed structural characterization of these gradients using contact angle measurements, X-ray photoelectron spectroscopy, and infrared spectroscopy. We report gradients composed of methyl-hydroxyl and methyl-carboxyl end-groups, which both show similar, linear changes in wettability with water. The deprotonation of the carboxyl groups allows one to produce surfaces with well-controlled gradients in surface charge, which could also be of interest in many surface-interaction studies, such as the field of protein adsorption.[30] Moreover, since carboxyl-terminated SAMs have been shown to be useful for surface immobilization of several interesting species ranging from simple bivalent metal ions[33] to magnetic nanoclusters,[34,35] polymers,[36] and proteins,[26] carboxyl-terminated gradients open the door to innumerable, more complex, and technologically relevant gradient systems.

Experimental Section

Dodecanethiol, 11-mercaptoundecanoic acid, and 11-mercaptoundecanol were obtained from Aldrich and used as received. Ethanol (Fluka) was used as solvent for all preparations. The substrates, 2 × 4 cm silicon wafers (POWATEC, Cham, Switzerland) were cleaned with oxidizing piranha solution, a 7:3 mixture of concentrated H_2SO_4 and 30% H_2O_2 (*Caution: piranha solution reacts violently when contacted with organic molecules and should be handled with extreme care*), thoroughly rinsed with copious amounts of MilliQ water (> 18 MΩ), subsequently cleaned with O_2 plasma (2 min O_2, high power, Harrick Plasma Cleaner/Sterilizer), and covered with a 6-nm adhesive layer of chromium before evaporation of 80-nm-thick layer of gold (MED020 coating system, BALTEC, Balzers, Lichtenstein). Prior to the immersion step, the substrates were cleaned for 30 s in N_2 plasma (30 s, high power) and left in pure ethanol for 10 min. Gradients were formed by controlled immersion of the substrate into a dilute (5 μM) ethanolic solution of dodecanethiol, the speed of immersion (75 μm/s) being accurately controlled by a linear-motion drive (OWIS, Staufen, Germany). The sample was quickly removed from the thiol solution and thoroughly rinsed with ethanol to remove physisorbed and loosely bound thiol molecules. The sample was then immersed in an ethanolic solution (0.01 mM) of 11-mercaptoundecanol or 11-mercaptoundecanoic acid overnight followed by thorough rinsing with ethanol and drying in a flow of high-purity N_2. Gradients made from 11-mercaptoundecanoic acid were, additionally, rinsed with acidified ethanol and dried in a flow of N_2.

Contact Angle Measurements. Dynamic contact angles were measured employing a contact angle goniometer (G2/G40 2.05-D, Kruss GmbH, Hamburg, Germany). Advancing and receding contact angles were determined using digital image analysis. Measurements were carried out every 5 mm along the substrate. All contact-angle measurements were averaged over several samples.

XPS Measurements. XPS analyses were performed using a VG Theta Probe spectrophotometer (Thermo Electron Corporation, West Sussex, UK) equipped with a concentric hemispherical analyzer and a two-dimensional channel plate detector with 112 energy and 96 angle channels. Spectra were acquired at a base pressure of 10^{-9} mbar or below using a monochromatic Al Kα source with a spot size of 300 μm. The instrument was run in the standard lens mode with electrons emitted at 53° to the surface normal and an acceptance angle of ±30°. The analyzer was used in the constant analyzer energy mode. Pass energies used for survey scans and detailed scans were 200 and 100 eV, respectively, for gold Au4f, carbon C1s, oxygen O1s, and sulfur S2p. Under these conditions, the energy resolution (full width at half-maximum height, fwhm) measured on gold Au4f$_{7/2}$ is 1.95 and 0.82 eV, respectively. Acquisition times were approximately 5 min for survey scans and 30 min (total) for high-energy-resolution elemental scans. These experimental conditions were chosen in order to obtain an adequate signal-to-noise ratio in a minimum time and to limit beam-induced damage. Under these conditions, sample damage was negligible, and reproducible analyzing conditions were obtained on all samples. All recorded spectra were referenced to the gold Au4f$_{7/2}$ signal at 83.96 eV. Data were analyzed using the program CasaXPS [Version 2.3.5, www.casaxps.com]. The signals were fitted using Gaussian–Lorentzian functions and Tauc asymmetry in the case of gold and least-squares-fit routines following Shirley iterative background subtraction. Sensitivity factors were calculated using published ionization cross sections[37] corrected for the angular asymmetry[38] and the attenuation length dependence with kinetic energy.

Infrared Spectroscopy. Polarization-modulation infrared reflection–absorption spectra (PM-IRRAS) were recorded on a Bruker IFS 66v IR spectrometer, equipped with a PMA37 polarization-modulation accessory (Bruker Optics, Germany). The interferogram from the spectrometer's external beam port was passed through a KRS-5 wire-grid polarizer and a ZnSe photoelastic modulator before reflecting off the sample surface at an angle of 80° and being detected with a liquid-nitrogen-cooled MCT detector. The frequency of polarization modulation was 50 kHz, with a maximum of polarization retardation set at 3000 cm^{-1}. The sample compartment was continuously purged with dry air. The sample holder was suitably modified to be able to record spectra of different regions of the gradient with a 2-mm aperture. Typically, 1024 scans of multiplexed interferograms were collected with 4-cm^{-1} resolution and processed with the OPUS software (Bruker Optics, Germany). The spectra were background-corrected with a polynomial.

Results and Discussion

The advancing and receding water-contact angles measured along a gradient of 1-dodecanethiol that had been back-filled with 11-mercaptoundecanoic acid (methyl-carboxyl gradient) are plotted in Figure 1. The adsorption parameters had been chosen, as described in our earlier publication,[32] such that the coverage varies almost linearly along the gradient axis, resulting in linear variation of the water-contact angles upon back-filling with the complementary thiol. The advancing contact angle at the methyl-rich end is about 80°, decreasing almost linearly along the length of the gradient, and reaches a value of about 40° at the carboxyl-rich end. The hysteresis in contact angle was between 10° and 15° along the length of the gradient, similar to those observed on a gradient prepared from dodecanethiol and 11-mercaptoundecanol[32] (methyl-hydroxyl gradient). For compari-

(31) Bhat, R. R.; Genzer, J.; Chaney, B. N.; Sugg, H. W.; Liebmann-Vinson, A. *Nanotechnology* **2003**, *14*, 1145.
(32) Morgenthaler, S.; Lee, S.; Zürcher, S.; Spencer, N. D. *Langmuir* **2003**, *19*, 10459.
(33) Yang, K.-L.; Cadwell, K.; Abbott, N. L. *Adv. Mater.* **2003**, *15*, 1819.
(34) Naitabdi, A.; Bucher, J.-P.; Gerbier, P.; Rabu, P.; Drillon, M. *Adv. Mater.* **2005**, *17*, 1612.
(35) Steckel, J. S.; Persky, N. S.; Martinez, C. R.; Barnes, C. L.; Fry, E. A.; Kulkarni, J.; Burgess, J. D.;. Pacheco, R. B.; Stoll, S. L. *Nano Lett.* **2004**, *4*, 399.
(36) Barreira, S. V. P.; Silva, F. *Langmuir* **2003**, *19*, 10324.
(37) Scofield, J. H. *J. Electron Spectrosc. Relat. Phenom.* **1976**, *8*, 129.
(38) Reilman, R. F.; Msezane, A.; Manson, S. T. *J. Electron Spectrosc. Relat. Phenom.* **1976**, *8*, 389.

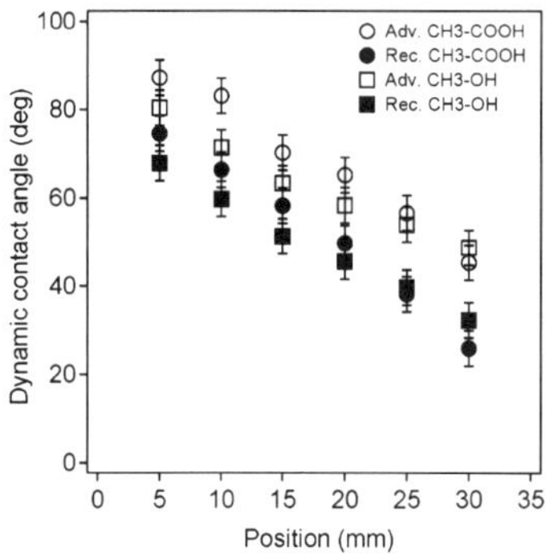

Figure 1. Advancing (open symbols) and receding (filled symbols) water contact angles on a methyl-carboxyl (circles) and a methyl-hydroxyl (squares) gradient measured along the length.

son, the contact angles measured on a methyl-hydroxyl gradient are also shown in Figure 1. It may be noted that the contact angles seen at the two extremities of the gradient are quite different from those observed on full monolayers of either of the two components (>110° for dodecanethiol and <10° for 11-mercaptoundecanoic acid), indicating that the gradients do not span the entire composition range. This is attributed to the fact that the single-component gradient formed after the first immersion step (as will be shown below from XPS measurements) never attains full coverage even at the highest-concentration end due to the low concentration of thiol used and the relatively short immersion times. Also, some replacement of the adsorbed dodecanethiol molecules by the 11-mercaptoundecanoic acid is expected during the second immersion step. However, substantial replacement of adsorbed thiol molecules by a second thiol is unlikely due to the low concentrations of the thiols employed during back-filling step.

XPS. X-ray photoelectron spectra in the C1s and O1s regions measured at 5 mm intervals on a methyl-carboxyl gradient are shown in Figure 2. The spectra in the C1s region show an intense peak at 285.0 eV (not shown in the spectra), corresponding to the C1s of the aliphatic chain with a less-intense component at 286.7 eV assigned to the C1s of the CH_2 group closest to either the S or COOH group. The peak at 289.7 eV is assigned to the C1s of the carboxyl group. The intensity of this peak increases along the gradient from the methyl-rich end to the carboxyl-rich end. The spectra shown here are for the samples prepared by rinsing with acidified ethanol after the second immersion step. This ensures the complete protonation of the carboxyl group, as shown by the O1s spectra in Figure 2b. Two oxygen species can be detected with an area ratio of nearly 1:1. These two peaks can be assigned to the C=**O** and the C−**O**H oxygens of the carboxyl group. When the gradients were rinsed with acid-free ethanol, an additional component for the carbon peak could be detected at 288.8 eV. This peak is assigned to a deprotonated carboxylate species, the intensity of which indicates that ~50% of all carboxyl groups had been deprotonated. Previous studies on full monolayers of carboxyl-terminated thiols on gold prepared from ethanol have shown that carboxyl groups are generally deprotonated and exist as carboxylates, often with the presence of trace amounts of metal ions, typically sodium, as counterion.[39,40] Often, in our measurements when no acid rinsing step was employed, contamination with low concentrations of bivalent metal ions, such as zinc, could be detected and a third oxygen species and a new carbon species that are attributed to carboxylate were observed.

In general, there is somewhat more oxygen and less sulfur detected than expected for the stoichiometric amount calculated for the adsorbed molecules. This is a consequence of the orientation of the carboxylic acids toward the vacuum−SAM interface, leading to the sulfur being buried underneath the SAM. Also, angle-resolved measurements show that this deviation is more pronounced for grazing-angle electron detection than for angles closer to the surface normal, as would be expected. The presence of other oxygen species, such as adsorbed water, H_3O^+, or oxidized gold cannot be completely excluded. The positions and widths of all the species are summarized in Table 1.

Quantification of the intensities of the various species was performed according to the procedure described above. These are shown in Figure 3a, for different positions along the gradient. The carbon intensity decreases, while that of oxygen increases linearly along the length of the gradient. A quantitative differentiation between the aliphatic thiol and the carboxy thiol is very difficult due to the weakness of the carboxylate signal in comparison to the aliphatic carbon and the uncertainty of the presence of other oxygen species on the samples. Also shown in Figure 3b are the intensity ratios of the overlayer over the substrate intensity calculated as $\sum(I_{S2p} + I_{C1s} + I_{O1s})/I_{Au4f}$. This ratio is directly related to the thickness of the overlayer[41] and therefore a measure for the coverage. As can be seen in Figure 3b, for a back-filled monolayer, this intensity ratio stays nearly constant over the entire length of the gradient and is comparable to the values obtained for single-component SAMs of either dodecanethiol or 11-mercaptoundecanoic acid (shown at the two extremes of the plot). Whereas in the case of a non-back-filled gradient, the ratio is decreasing, starting at ~80% going down to ~30% of the value of a full monolayer. Due to rapid contamination of the uncovered regions with adventitious carbon, the real thiol coverage on the low-coverage side could be even lower than the measured value. This could also be seen from the increase of the carbon-to-sulfur ratio with decreasing coverage (data not shown). Similar arguments of contamination with adventitious carbon could also explain the slightly larger values seen on the carboxyl side of the back-filled gradient (position 30 mm). Another feature of the plot shown in Figure 3b is the measure of compositional homogeneity of the gradients. For three of the positions along the single-component gradient (positions at 7.5, 19.5, and 32 mm), intensity ratios are also shown from measurements at five different locations along the direction perpendicular to the gradient direction. The variations in the intensity ratios are within the experimental uncertainties, indicating that the gradients are homogeneous along the perpendicular axis.

Infrared Spectroscopy. Infrared spectroscopy has been shown to be a useful method to study structure of self-assembled thiol monolayers, especially the orientation[42] and conformation[3] of the alkyl chains. PM-IRRA spectra of a single-component dodecanethiol gradient measured at every 5 mm along the length of the gradient are shown in Figure 4. The most prominent features in the spectra could be readily assigned to the C−H stretching modes arising from the $-CH_2-$ groups and the terminal CH_3 groups[2,3,5]. Peaks at ~ 2850 and ~ 2920 cm^{-1} are assigned to the CH_2 symmetric and antisymmetric stretching modes and the peaks at 2880 and 2960 cm^{-1} to the symmetric and asymmetric stretching modes of the CH_3 group, respectively. At the lowest-

(39) Arnold, R.; Azzam, W.; Terfort, A.; Woll, C. *Langmuir* **2002**, *18*, 3980.
(40) Willey, T. M.; Vance, A. L.; van Buuren, T.; Bostedt, C.; Nelson, A. J.; Terminello, L. J.; Fadley, C. S. *Langmuir* **2004**, *20*, 2746.
(41) Fadley, C. S. *Prog. Surf. Sci.* **1984**, *16*, 275.
(42) Parikh, A. N.; Allara, D. L. *J. Chem. Phys.* **1992**, *96*, 927.

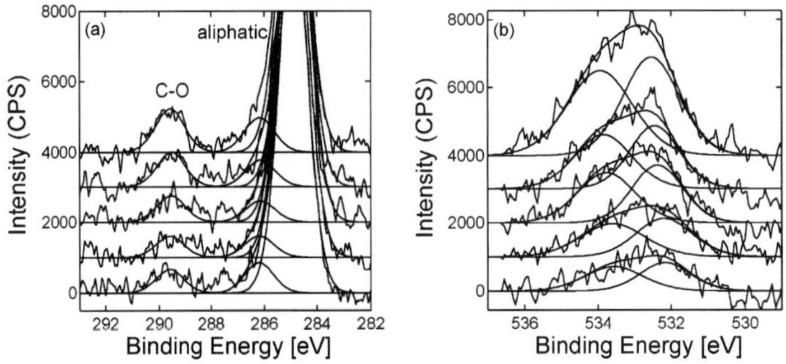

Figure 2. X-ray photoelectron spectra in the (a) C1s and (b) O1s regions measured on a methyl-carboxyl gradient. Spectra shown are measured from the methyl-rich end (bottom) to the carboxyl rich end (top) at 4, 11, 18, 25, and 34 mm from the methyl-rich end. The C1s region is expanded to highlight the change in the C−O intensity. The spectra are displaced along the *y* axis by 1000 CPS for clarity.

Table 1. Positions and Widths of Important XPS Peaks

	average binding energy (eV)	fwhm
$Au4f_{7/2}$	83.96	0.82
$Au4f_{5/2}$	87.69	0.91
C1s aliphatic	285.02	1.45
C1s CH_2−S/CH_2−COOH	286.72	1.45
C1s COOH	289.72	1.45
O1s C=O	532.41	1.61
O1s C−OH	533.81	1.86
$S2p_{3/2}$	162.12	1.00
$S2p_{1/2}$	163.30	1.00

concentration end (top trace) of the gradient, a shoulder around 2900 cm^{-1} could be seen, which could be due to a Fermi resonance involving the bending modes and the symmetric stretch or to a methylene antisymmetric stretch arising from the CH_2 groups nearest to the sulfur atoms.[43]

The symmetric and the antisymmetric stretching modes of the CH_2 groups appear at 2854 and 2924 cm^{-1} through the entire sample length. It is well known in the literature from a variety of studies on the structure of alkyl-chain assemblies, such as crystalline *n*-alkanes,[44,45] lipids, and SAMs,[46] that these vibrational modes are sensitive to conformational order of the alkyl chain. In a crystalline, all-trans conformation, the modes appear in the regions 2914−2918 and 2846−2848 cm^{-1} for the methylene antisymmetric and symmetric stretching modes,[45] respectively, shifting progressively to higher wavenumbers with increasing disorder and appearing in the region 2924−2926 cm^{-1} for the antisymmetric and 2854−2858 cm^{-1} for the symmetric stretching modes.[44] This indicates that, on the single-component dodecanethiol gradient, the alkyl chain possesses a disordered liquidlike conformation over the entire length of the gradient. This is expected, since at such low concentration (5 μM) and relatively short immersion times, of the order of a few minutes, the monolayer is expected to be composed of a mixture of lying-down and standing-up molecules with a considerable amount of disorder. Nevertheless, it should be noted that, even at the most concentrated end, the spectrum does not resemble that of a complete monolayer (shown as a dotted line in Figure 4) in terms of the intensity ratios of the methyl symmetric to methylene symmetric stretching modes, as well as in the position of the methylene stretching modes. This indicates that the structure of the monolayer, even at the highest-concentration end, is far from crystalline and ordered. The intensity of the CH_3 symmetric stretching mode at 2880 cm^{-1} further illustrates this fact. This vibrational mode has a transition moment along the C−C bond axis of the terminal $-CH_2-CH_3$ bond, and its intensity relative to the overall intensity of the C−H stretching modes increases rapidly toward the higher-concentration end of the gradient. Apart from changes in this intensity due to increasing concentration of thiols on the surface (as shown by the XPS intensity ratios in Figure 3b), this nonlinear increase toward the high-concentration end of the gradient could be due to a change in orientation of the adsorbed thiol molecules from a lying-down phase at the less-concentrated end to a standing-up phase. AFM measurements on such a gradient showed an islandlike structure with a typical height difference of 1 nm, consistent with the fact that the islands are surrounded by a disordered lying-down phase rather than bare gold.[47] This is not surprising considering the fact that the structure of thiol monolayers is known to evolve over long time scales and infrared spectral intensity changes in uniform SAMs have been reported over several hours/days.[48] The changes in intensity observed here are similar to ex situ infrared measurements on incomplete thiol monolayers prepared from dilute concentrations.[49] A detailed description of the structure of such a gradient at sub-micrometer length scales obtained with atomic force microscopy has been recently reported elsewhere.[47]

Infrared spectra of a two-component, methyl-hydroxyl gradient, prepared by back-filling a single-component dodecanethiol gradient with 11-mercaptoundecanol, are shown in Figure 5a. The most noticeable feature in the series of spectra, measured at different positions along the length of the gradient, is the change in the intensity of C−H stretching modes of the terminal CH_3 groups at 2960 cm^{-1}. The intensity of this band, as expected, decreases along the length of the gradient moving from the methyl-rich end to the hydroxyl-rich end. This correlates well with the changes observed in the contact angles and XPS measurements. The apparent increase in the intensity at 2880 cm^{-1} band toward the hydroxyl side of the gradient could be due to the presence of an additional band, often observed in full monolayers of 11-mercaptoundecanol, assigned to the stretching mode of $-CH_2$ group adjacent to −OH.[8,50] The most important difference in the series of spectra on a methyl-hydroxyl gradient shown above, in comparison to the single-component gradient, is the position

(43) Parikh, A. N.; Gilmor, S. D.; Beers, J. D.; Beardmore, K. M.; Cutts, R. W.; Swanson, B. I. *J. Phys. Chem. B* **1999**, *103*, 2850.
(44) Snyder, R. G.; Strauss, H. L.; Elliger, C. A. *J. Phys. Chem.* **1982**, *86*, 5145.
(45) MacPhail, R. A.; Strauss, H. L.; Snyder, R. G.; Elliger, C. A. *J. Phys. Chem.* **1984**, *88*, 334.
(46) Nuzzo, R. G.; Korenic, E. M.; Dubois, L. H. *J. Chem. Phys.* **1990**, *93*, 767.
(47) Morgenthaler, S. M.; Lee, S.; Spencer, N. D. *Langmuir* **2006**, *22*, 2706.
(48) Terrill, R. H.; Tanzer, T. A.; Bohn, P. W. *Langmuir* **1998**, *14*, 845.
(49) Bensebaa, F.; Voicu, R.; Huron, L.; Ellis, T. H.; Kruus, E. *Langmuir* **1997**, *13*, 5335.
(50) Atre, S. V.; Leidberg, B.; Allara, D. L. *Langmuir* **1995**, *11*, 3882.

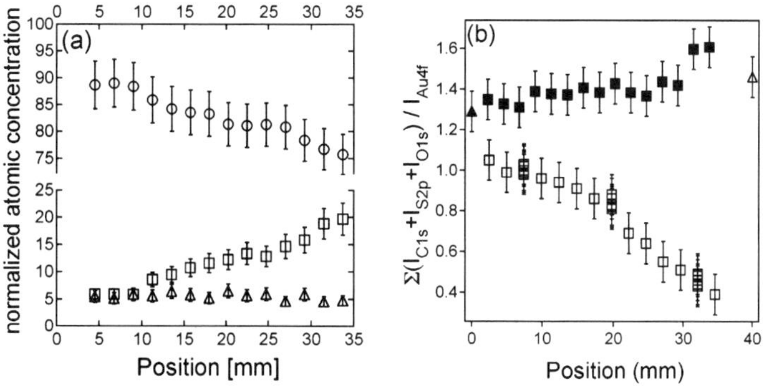

Figure 3. (a) Normalized atomic concentrations of the elements related to the SAM [carbon (circles), oxygen (squares) and sulfur (triangles)] along the length of a methyl-carboxyl gradient. The positions shown are the distances from the highest dodecanethiol concentration end. Error bars are estimated from the signal-to-noise ratios. (b) Film coverage determined by the ratio of the sum of S2p, C1s, and O1s intensities to substrate (Au4f) intensity along the length of a methyl-carboxyl (filled squares), single component (open squares) gradients compared to full monolayers of dodecanthiol (filled triangle) and 11-mercaptoundecanoic acid (open triangle). The intensity ratios at positions 7.5, 19.5, and 32 mm are also shown from measurements at five different locations along the perpendicular axis of the gradients to indicate the homogeneity of the gradients.

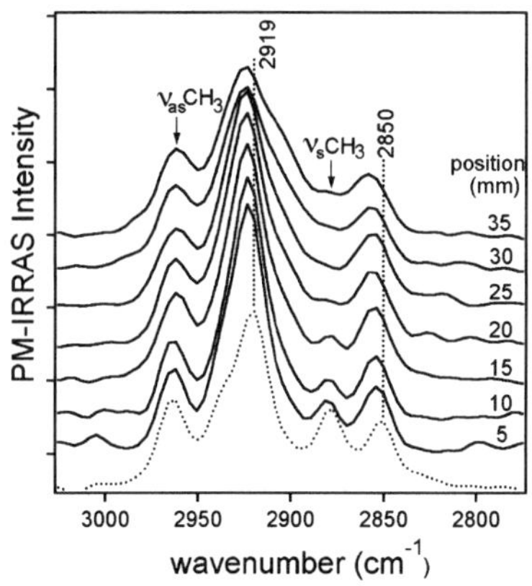

Figure 4. PM-IRRA spectra measured on a single-component 1-dodecanethiol gradient (solid lines) measured every 5 mm along the gradient from the high-concentration end (lower trace) to the low-concentration end (top). The spectra are displaced for clarity. The spectrum of a full monolayer of dodecanethiol (dotted line) is shown for comparison.

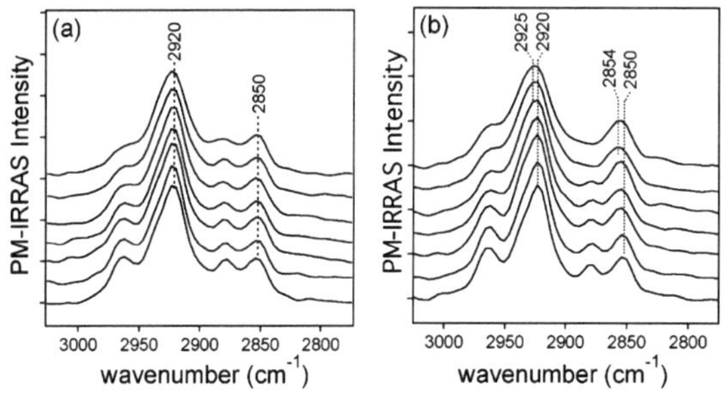

Figure 5. Infrared spectra of two-component gradients prepared from dodecanethiol back-filled with (a) 11-mercaptoundecanol and (b) 11-mercaptoundecanoic acid. The spectra shown are measured at 5 mm intervals along a 35-mm-long sample.

of the methylene stretching modes. Upon back-filling with 11-mercaptoundecanol, the symmetric and antisymmetric CH_2 stretching modes shift to 2850 and 2920 cm^{-1}, respectively. This indicates the establishment of a well-ordered monolayer throughout the entire length of the two-component gradient. Also, a slight decrease in the intensity of the methyl asymmetric stretching mode may be seen in the two-component gradients as compared to the single-component gradient. This is due to the replacement of dodecanethiol molecules by 11-mercaptoundecanol during the second immersion step. No quantitative analysis of the replacement kinetics from these intensities has been attempted here.

Infrared spectra in the C−H stretching region of a gradient back-filled with 11-mercaptoundecanoic acid are shown in Figure 5b. The intensities of the symmetric and asymmetric stretching modes of the methyl group decrease, as expected, from the CH_3-rich end to the COOH-rich end, in a similar way to that observed for a methyl-hydroxyl gradient. The positions of the methylene stretching modes also shift to lower wavenumbers compared to single-component gradients. This indicates a change in conformational ordering of the alkyl chains upon back-filling. However, in comparison to the methyl-hydroxyl gradient, the methylene stretching modes in the methyl-carboxyl gradient appear at higher wavenumbers at the carboxyl-rich end of the gradient. These modes appear at 2851 and 2921 cm^{-1} on the $-CH_3$ side of the sample, whereas they appear at 2854 and 2924 cm^{-1} at the −COOH end of the sample. The positions of the methylene stretching modes in all three gradients are plotted in Figure 6. This indicates that, while the methyl-hydroxyl gradient is conformationally well-ordered throughout its entire length, the methyl-carboxyl gradient shows an increasing conformational disorder toward the −COOH-rich end. The nature of the end functionality of thiol SAMs can significantly influence their packing, thereby modulating the conformational order of the alkyl chains.[51,52]

Carbonyl Stretching Region. Spectra in the 1900−1300 cm^{-1} region measured at different positions along a methyl-carboxyl gradient are shown in Figure 7. The spectra shown are those measured on an acid-rinsed sample. The spectra recorded on a sample prior to the acid-rinsing step are shown, for the same spectral region, as an inset. Two intense, broad bands centered around 1700 and 1450 cm^{-1} can be assigned to the carbonyl C═O stretching mode of a carboxylic acid and the symmetric stretching of carboxylate ion, respectively. The COO^- asymmetric

(51) Frey, S.; Shaporwnko, A.; Zharnikov, M.; Harder, P.; Allara, D. L. *J. Phys. Chem. B* **2003**, *107*, 7716.

(52) Dannenberger, O.; Weiss, K.; Himmel, H.-J.; Jager, B.; Buck, M.; Woll, C. *Thin Solid Films* **1997**, *307*, 183.

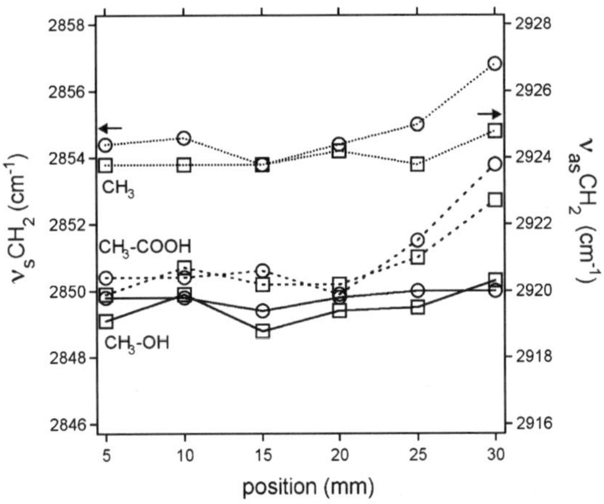

Figure 6. Positions of the methylene symmetric (circles) and antisymmetric (squares) stretching modes in single-component dodecanethiol gradients (dotted lines), and back-filled two-component gradients with 11-mercaptoundecanol (solid lines) and 11-mercapto-undecanoic acid (dashed line), along the length of the gradients.

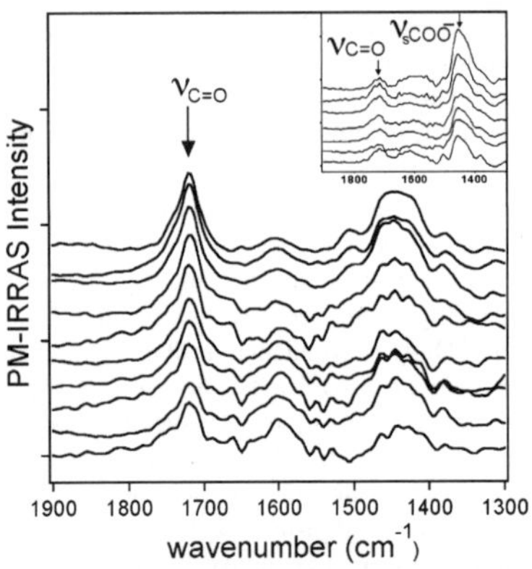

Figure 7. Infrared spectra of a methyl-carboxyl gradient in the carbonyl stretching region, showing the C=O stretching mode increasing in intensity along the length of the gradient moving to the carboxyl end. The inset shows spectra from the same region of a gradient prepared without the additional rinsing step with acidified ethanol, showing that the carboxyl groups are present predominantly as carboxylate species.

stretching band that appears in the region of 1570 cm^{-1} could not be clearly distinguished (inset spectra). From the inset spectra, it may thus be seen that the samples prepared from ethanol are almost exclusively composed of carboxylate species. This is in accordance with previous studies on −COOH-terminated monolayers and has been attributed to the presence of trace amounts of water and metal ions in the solvent.[39,40] A rinsing step with an ethanolic solution of HCl results in a predominantly carboxylic acid-terminated surface.[53] However, the intensity in the 1450−1600 cm^{-1} region does not go to zero upon rinsing with HCl.

(53) Methivier, C.; Beccard, B.; Pradier, C. M. *Langmuir* **2003**, *19*, 8807.

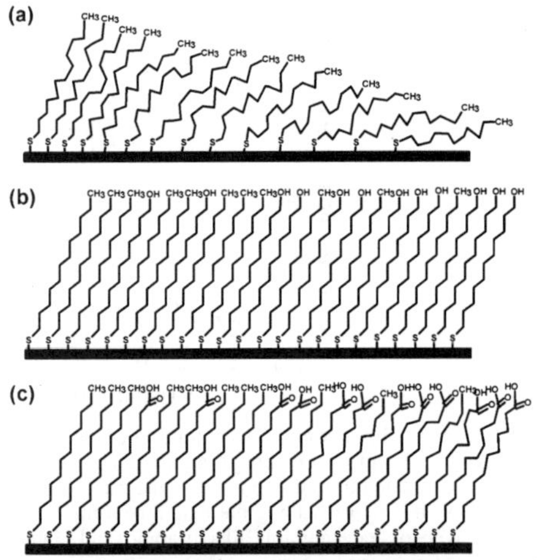

Figure 8. Schematic depiction of the structure of (a) a single-component dodecanethiol gradient and a two-component gradient with (b) 11-mercaptoundecanol or (c) 11-mercaptoundecanoic acid. The alkyl chains in the methyl-hydroxyl gradient (b) are well ordered, whereas in a methyl-carboxyl gradient (c) there is an increased disorder at the carboxyl end. The figure is meant to be illustrative and is not a quantitative description of conformational disorder.

This could be due to the contribution to this intensity from other species such as water molecules bound to the carboxyl groups. Also, the intensity around 1400 cm^{-1} is further complicated by the presence of several possible bands (e.g., deformation of COOH, scissoring modes of CH_2, stretching modes of COO^-). A quantitative differentiation of the ratio of carboxylic to carboxylate from these intensities was difficult, owing to the presence of multiple peaks in this region.

Conclusions

Surface-chemical gradients on centimeter length scales with different end-groups have been prepared by a simple two-step immersion method. The single-component gradient of dodecanethiol resulting after the first immersion step is composed of a disordered lying-down phase at the low-concentration end, gradually changing into a disordered standing-up phase at the higher-concentration end. Upon back-filling such a gradient with either 11-mercaptoundecanol or 11-mercaptoundecanoic acid, a complete monolayer, exhibiting almost linear variations in surface concentration of either hydroxyl or carboxyl functionality, is formed. The gradient composed of methyl- and hydroxyl-terminated thiols shows a well-ordered structure along the entire length of the sample, whereas the methyl-carboxyl gradient shows an increased conformational disorder at the carboxyl-rich end. The carboxyl groups in a methyl-carboxyl gradient prepared from ethanol are present as carboxylate ions, which could be converted to carboxylic acid by rinsing with acidified ethanol. The proposed structures of the single- and two-component gradients are shown as a schematic representation (Figure 8). Gradients of alkanethiols with other ionizable end-groups are currently being investigated.

LA053302T

Available online at www.sciencedirect.com

ScienceDirect

ELSEVIER

Applied Surface Science 253 (2006) 2148–2153

applied surface science

www.elsevier.com/locate/apsusc

Fabrication of material-independent morphology gradients for high-throughput applications

Tobias P. Kunzler [a], Tanja Drobek [a], Christoph M. Sprecher [b], Martin Schuler [a], Nicholas D. Spencer [a,*]

[a] *Laboratory for Surface Science and Technology, Department of Materials, ETH Zurich, Wolfgang-Pauli-Strasse 10, 8093 Zurich, Switzerland*

[b] *AO Research Institute, Clavadelerstrasse, 7270 Davos, Switzerland*

Received 14 March 2006; received in revised form 4 April 2006; accepted 5 April 2006
Available online 24 May 2006

Abstract

Gradient surfaces allow rapid, high-throughput investigations and systematic studies in many disparate fields, including biology, tribology and adhesion. We describe a novel method for the fabrication of material-independent morphology gradients, involving a two-step process of particle erosion followed by a chemical polishing procedure that preferentially removes features with a small radius of curvature as a function of time. Gradients are fabricated on aluminium surfaces, but they may be readily transferred to other materials via a replication technique, which allows for the production of identical roughness gradient samples with any chosen surface chemistry. The gradients have been characterized by means of scanning electron microscopy and optical profilometry. Standard roughness parameters (Ra, Rq, Rz, Sm and Sk) were calculated from optical profilometry data. The roughness has also been assessed over different wavelength windows by means of a fast Fourier transformation approach.

Keywords: Surface roughness; Gradient; FFT; Window roughness; High-throughput; Replicas

1. Introduction

Modern diagnostic and combinatorial studies in materials science, medicine and biotechnology require analysis techniques that are capable of performing as many tests as possible within minimum time. Gradient surfaces allow systematic studies to be performed with a continuously varying surface parameter within a single experiment. One such parameter can be the surface chemistry, and a number of methods for the fabrication of surface-chemical gradients have recently been published [1–4]. However, surface morphology is an equally significant parameter that influences both materials science and biological phenomena. It is well known that roughness over a wide range of length scales affects the biological response to surfaces, e.g. cell adhesion, proliferation and differentiation [5–7]. Further, in materials science, roughness on a micrometre and nanometre scale plays an important role in contact-related phenomena, such as tribology or adhesion [8–10].

Previous investigations of the effects of surface morphology have often been limited to either rough or smooth samples. The influence of roughness has only rarely been *systematically* studied [11–13]. Roughness gradients can greatly facilitate this kind of investigation. In the literature there are only very few reports of gradient fabrication by modification techniques that produce stochastic surface morphologies. These include porous silicon with a gradient in the pore-size distribution between 5 and 20 nm, made by using inhomogeneous electric fields in an etching process [14]. Another approach is to vary the crystallinity of a polymer film by means of a temperature-gradient stage [11,13]. However, the limited lateral dimensions and feature sizes of these gradients and the restriction to specific materials are disadvantageous in many applications.

We have developed a method that allows for the production of stochastic roughness gradients over a centimetre scale with topographical features in the nanometre and micrometre range. Subsequent application of a replica technique [15] permits

* Corresponding author. Tel.: +41 44 632 5850; fax: +41 44 633 1027.
E-mail address: nicholas.spencer@mat.ethz.ch (N.D. Spencer).

0169-4332/$ – see front matter © 2006 Published by Elsevier B.V.
doi:10.1016/j.apsusc.2006.04.014

morphological gradients to be transferred to different materials and surface chemistries.

2. Experimental

2.1. Gradient fabrication

Rolled aluminium sheets (purity 99.5%, dimensions: 15 mm × 70 mm × 1 mm (Alcan, Switzerland) were roughened in a conventional sand-blasting cabinet (Sablux, Switzerland) using fractured corundum particles (Sablux, Switzerland) with a diameter between 300 and 420 μm. The sand-blasting was performed at an air pressure of 3 bar under an impact angle of 70° and a working distance of 100 mm. Subsequently, the sheets were cleaned with ethanol for 10 min in an ultrasonic bath, followed by drying with a nitrogen jet. The subsequent chemical polishing step was carried out in a solution consisting of 77.5% (v/v) phosphoric acid (ρ = 1.74 g cm^{-3}), 16.5% (v/v) sulphuric acid (ρ = 1.84 g cm^{-3}) and 6% (v/v) nitric acid (ρ = 1.41 g cm^{-3})) at a temperature of 80 ± 1 °C. The polishing bath was kept slightly agitated during the procedure. For the polishing step, the sample was first dipped into the bath for 10 s in order to create a homogeneous aluminium oxide layer on the surface and then rinsed in hot distilled water. Afterwards, the sample was completely immersed into the bath again and then continuously withdrawn from the solution with a linear-motion drive (Owis Staufen, Germany) at a speed of 35 μm s^{-1}. Hence, a sample length of 20 mm corresponds to an exposure time to the polishing solution of between 0 and 38 min. In order to prevent gas bubbles from evolving from the aluminium sheet, a platinum electrode was immersed into the solution and a potential of 5 V was applied. During the polishing process this potential was gradually reduced to 2.5 V. After complete removal from the solution the sample was rinsed in hot distilled water.

Replicas of original aluminium roughness gradients were produced as described by Wieland et al. [15]. An impression ("negative") of the aluminium gradient was produced using polyvinylsiloxane (PROVIL novo Light; Heraeus-Kulzer, Switzerland). Epoxy-resin (EPO-TEK 302-3M; Epoxy Technology, Polyscience AG, Switzerland) was then cast onto these polyvinylsiloxane negatives and cured at 60 °C for 24 h. Cured, positive epoxy replicas were cleaned in a 2% solution of Helmanex (HELLMA GmbH & Co., Müllheim, Germany) and rinsed ten times with ultra pure water. After drying the replica samples in air overnight, they were sputter-coated with a 60 nm thick layer of gold or titanium.

2.2. Characterization

Qualitative characterization was carried out with an optical profilometer (FRT MicroGlider, Fries Research & Technology GmbH, Bergisch Gladbach, Germany). Samples were measured at a sampling rate of 30–1000 Hz and a resolution of 1000 points mm^{-1}. The data were evaluated with Matlab 7.0.4 R14. For analysis of sections of the gradient, outliers were removed and a plane fit was applied. Profiles were fitted with a polynomial of degree 5 after removing outliers. The wavelength-dependent evaluation was carried out using the fast Fourier and inverse fast Fourier transformation algorithm that is implemented in Matlab [16]. The roughness parameters were calculated according to DIN EN ISO 4288–98. For Ra values between 0.1 and 10 μm, DIN EN ISO 4288–98 requires the measurement of five consecutive lengths of 2.5 mm (total 12.5 mm). Each roughness value is the calculated average of these five measured values. Qualitative characterization of the gradients was performed with a SEM (Hitachi S-4100, Hitachi, Japan) in the secondary-electron mode. The original aluminium roughness gradient was coated with a 10 nm thin layer of carbon to minimize the negative charge accumulation on the surface. Replica samples were investigated without any further carbon coating. Images were taken at a voltage of 5 kV and an emission current of 45 μA.

3. Results and discussion

The method we developed to produce stochastic roughness gradients is a two-step process that involves homogeneous roughening of the surface by particle erosion followed by a kinetically controlled smoothing using a chemical polishing procedure. In the first step, an aluminium sheet is sand blasted to create a homogeneously rough surface. The sand-blasting step determines the maximum topographical variations of the sample. Several parameters of the blasting procedure, such as particle size and shape [17], angle of impact and air pressure [18,19], can be adjusted to alter the "high-end" roughness characteristics.

In a second step, a chemical polishing procedure is applied to reduce the roughness as a function of distance along the sample. Chemical polishing of aluminium is a well-known industrial technique [20,21] and the mechanisms of chemical polishing have been described extensively in the refs. [22–24]. In brief, the polishing solution contains a reagent that forms a passivation layer on the aluminium surface and another reagent that simultaneously dissolves this passivation layer. As a consequence of diffusion mechanisms, the reaction products are replaced more rapidly by fresh reagents at protuberances than in cavities, which leads to a smoothing out of the surface topography. The degree of smoothing depends on the reaction kinetics and therefore can be controlled by the temperature or by the residence time of a specific surface location in the polishing solution. The actual gradient is created by withdrawing the sand-blasted sample continuously from the solution using a linear-motion drive. In this way, the variation of the residence time leads to different degrees of polishing of the surface. Typical residence times are between seconds and 38 min.

A roughness gradient over a length of 20 mm is shown in Fig. 1. The gradient is presented together with a region that shows the initial surface prior to sand-blasting (Fig. 1a) and a region that is sand-blasted only (Fig. 1b). The roughness variation is visible due to the influence of the morphology on the light-scattering properties of the sample. At the rough end

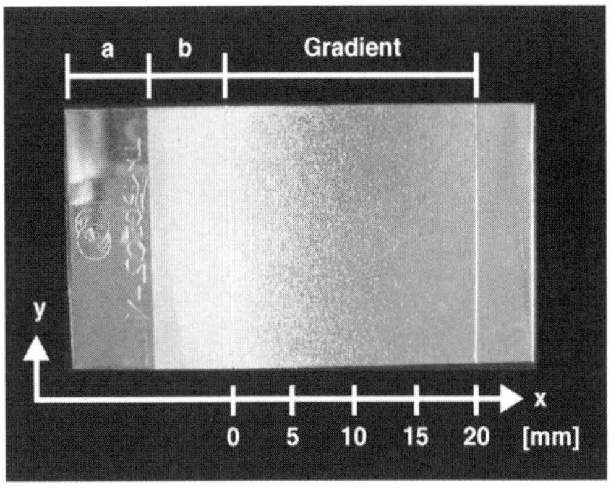

Fig. 1. Photograph of a roughness gradient on an aluminium surface over 20 mm. The section (a) shows the initial, non-blasted surface and (b) a section that was sand blasted only.

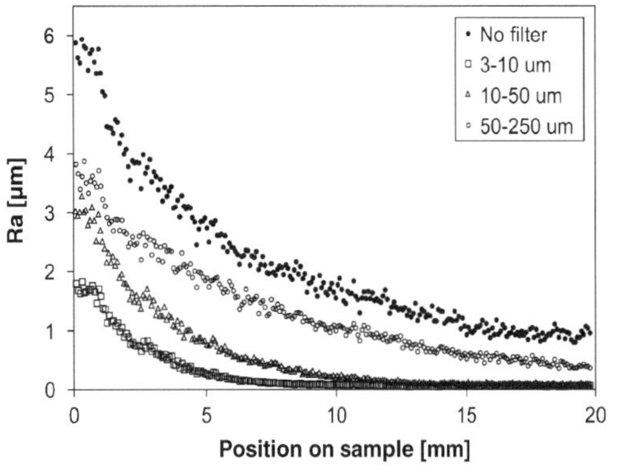

Fig. 2. Roughness value Ra calculated for different wavelength windows. Each point corresponds to a measured length in the y direction according to DIN EN ISO 4288–98. Data was acquired on an original aluminium roughness gradient by optical profilometry and evaluated applying fast Fourier transformation.

the light is scattered in all directions and thus leads to a dull appearance, whereas at the smooth end the light is reflected and the surface appears shiny.

While our method is, in principle, a simple and easy process for creating roughness gradients, some technical details need to be considered in order to produce well-defined morphology gradients. During the chemical reaction of the polishing process, gas bubbles (NO_x) are liberated, which can generate a texture when ascending along the aluminium sheet. Bubbles can also become pinned at peaks in the rough surface, leading to deep pit formation. These gas-induced defects have been observed and described by Clifford and Arrowsmith [25]. In addition to maintaining the homogeneity of the polishing solution, agitation can help to reduce such defects. However, agitation also leads to inhomogeneous metal removal due to turbulence induced at the edges of the sheet. By applying a small potential to the aluminium with respect to an additional electrode, however, the evolution of gas is transferred away from the sheet to the electrode. The potential has to be adjusted during the polishing procedure since the withdrawal of the sample results in a change of surface area and hence to a change of the potential. Further, the temperature of the polishing solution has to be carefully controlled because of its significant impact on reaction kinetics and thus on the overall shape of the gradient.

Gradients were characterized quantitatively by optical profilometry, measuring along and perpendicular to the gradient axis. Profiles along the gradient axis, in the x direction of the gradient, show that the roughness amplitude decreases monotonically. Maximum peak-to-valley values of about 50 µm were measured at the roughest end. With the chemical polishing procedure it was possible to reduce these peak-to-valley values to minimum values of about 3 µm.

A comparison of profiles at different y positions showed that the morphology is homogeneous across the surface, perpendicular to the gradient axis. This could be confirmed by profiles measured in the y direction of the gradient. From such profiles the roughness values were calculated according to the DIN EN ISO 4288–98 standard. Table 1 shows an overview of different roughness values. The amplitude parameters Ra, Rq and Rz were found to decrease monotonically. The spacing parameter, Sm, increases with decreasing roughness. The skewness value Sk shows no trend. It oscillates around zero, indicating that the profile remains symmetric throughout the polishing process.

Describing the topography of a surface in terms of standardized integral amplitude roughness values is often of limited value, since it does not adequately describe the surface characteristics that may be important in a particular application. On surfaces with large features and overlaid fine-roughness features, the fine features are often hidden by the coarser contributions to roughness. These fine-roughness features may be of particular importance for the performance of the surface in certain applications (e.g. cell–surface interactions [26,27] or superhydrophobicity [28]). Wieland et al. [29] have developed

Table 1
Standard roughness values

Roughness value (µm)	Position on sample					
	Sand blasted	3 mm	6 mm	9 mm	15 mm	20 mm
Ra	5.69	4.41	2.39	1.84	1.00	0.87
Rq	7.83	6.20	3.06	2.34	1.27	1.08
Rz	54.37	46.48	15.12	9.89	5.22	4.39
Sm	28.07	28.79	37.62	39.12	40.51	27.23
Sk	0.0071	−0.0028	−0.0044	−0.0150	0.0111	−0.0030

Roughness values were calculated from measurements with optical profilometry at different x positions on an original aluminium roughness gradient.

T.P. Kunzler et al. / Applied Surface Science 253 (2006) 2148–2153 2151

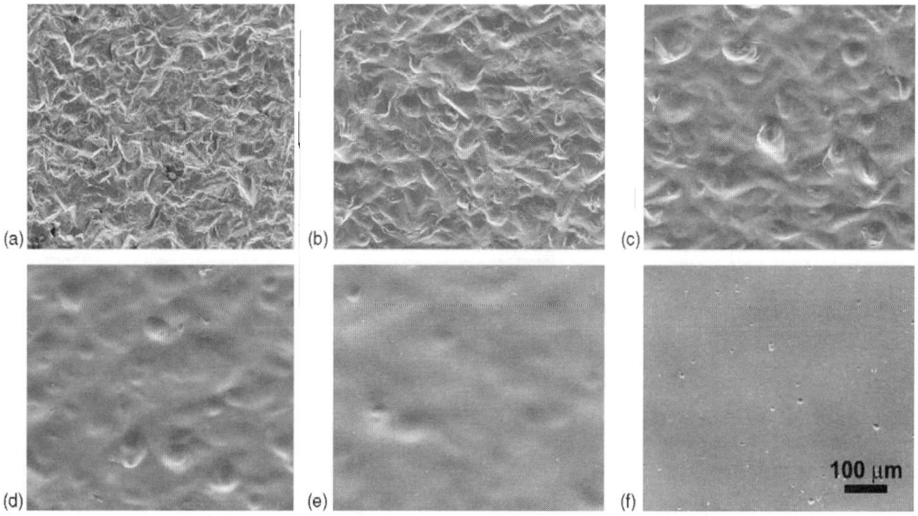

Fig. 3. SEM images of different x positions on a roughness gradient. (a) Sand-blasted only, (b) 3 mm, (c) 6 mm, (d) 9 mm, (e) 15 mm and (f) 20 mm. SEM images show the original aluminium surface of a roughness gradient. Topographical features were continuously removed from the surface as the polishing time elapsed.

a method for the wavelength-dependent evaluation of surface topography, which allows for a much more detailed description of the surface topographical properties. Using this approach, contributions in different wavelength windows can be individually evaluated. To determine wavelength-dependent roughness values of the gradient, the method was applied to the data obtained from laser profilometry. The roughness values were calculated for three different windows with wavelengths between 3–10 μm, 10–50 μm and 50–250 μm. Each wavelength window provides roughness information that was contained within the original profile. For every window, the roughness values can be calculated individually. The results for the arithmetic average Ra are shown in Fig. 2. A comparison of the roughness values in the different windows shows that features do not decrease at the same rate over different length scales. Features with short wavelengths are removed very fast at the beginning of the polishing procedure, whereas larger features disappear more gradually over the whole gradient. After about 8 mm from the rough end, the roughness in the wavelength window between 3 and 10 μm approaches a constant value of 0.06 μm. The behaviour of the small features could only be detected by means of a wavelength-dependent evaluation of the roughness data. Furthermore, the evaluation based on wavelength windows confirms the selectivity of the chemical polishing procedure, in which small features such as peaks and sharp ridges are initially attacked by the solution, large features being removed only after longer polishing times.

The removal of small features after a short polishing time can also be observed in SEM investigations. Fig. 3 shows SEM images of different positions of the gradient, corresponding to

Fig. 4. Optical profilometry measurements at different x positions on the gradient. (a) Sand-blasted only, (b) 3 mm, (c) 6 mm, (d) 9 mm, (e) 15 mm and (f) 20 mm. Measurements were carried out on an original aluminium roughness gradient. The side length of one square is 1 mm, the z scale is given in micrometres.

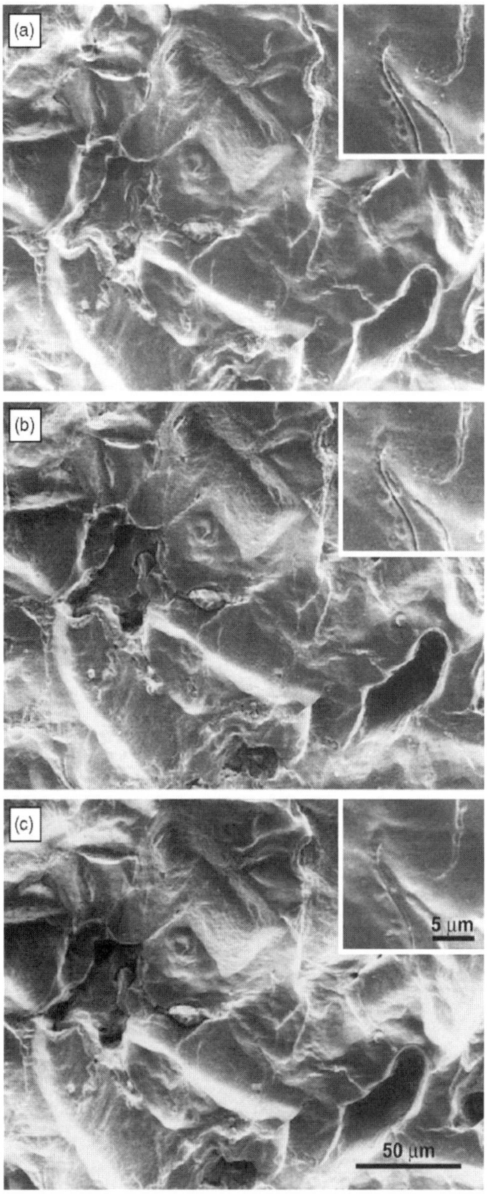

Fig. 5. SEM images of an original roughness gradient and its replicas. (a) Original aluminium surface of a roughness gradient, (b) replica (positive) of a roughness gradient, coated with 60 nm of titanium, (c) replica (positive) of a roughness gradient, coated with 60 nm of gold. All the three images show the same position on the gradient.

different polishing times. Fig. 3a shows an image of a section that was sand-blasted only. Sharp edges, ridges and peaks created by the sand-blasting procedure can be observed. These sharp edges, ridges and peaks were removed within minutes of the onset of chemical polishing (Fig. 3b). After about 30 min of polishing, features such as larger pits and protuberances were removed (Fig. 3c–f). The disadvantage of standard SEM is that quantitative information obtained from images is limited to two dimensions and that only relatively small areas can be investigated. In contrast, optical profilometry allows three-dimensional data to be collected over a larger area. Some regions of the gradient are shown in Fig. 4.

The initial gradient-fabrication method presented in this work has been limited to aluminium surfaces, because of its suitability for both the erosion and chemical polishing steps. In principle, the procedure could be extended to other metallic materials such as titanium or steel, if appropriate adjustments of the sand-blasting and polishing parameters and chemistry were made. Alternatively, high-resolution polymer replicas of gradients can be fabricated (via an elastomeric negative) from the aluminium gradients [15]. In order to create a well-defined surface chemistry, such replicas can be subsequently coated with thin films of a different material (e.g. by evaporation or sputtering of metals or oxides). Further manipulation of the replica surface can be obtained by surface functionalization with monolayers, e.g. by thiols on gold-coated surfaces [30] or by phosphates or phosphonates on titanium-coated surfaces [31]. The advantage of this replica technique is that a series of identical morphology gradients can be produced from only one master sample and that the surface chemistry can be defined in a subsequent functionalization step.

We produced a series of replicas of an aluminium gradient and coated these samples with gold and titanium. Fig. 5 shows SEM images of a section of the original aluminium gradient surface (Fig. 5a) and the same position on the replica, which was coated with titanium (Fig. 5b) and gold (Fig. 5c). A comparison between original and replica surfaces reveals no apparent differences in surface structure and confirms the fidelity of the replica process. Even very small and complex features in the sub-micrometer range are faithfully replicated.

We envision a variety of different applications for morphological gradients in the fields of biology, tribology and adhesion. For instance, surface structure plays an important role in cell growth, proliferation and attachment to the surface [5–7] and it is known that different kinds of cells favour surfaces with different roughnesses [32]. Gradient surfaces offer the possibility to systematically study the influence of morphology on cell–surface interactions of specific cell types by in vitro experiments that simultaneously probe a large morphological parameter space. In such tests, cell behaviour (e.g. cell adhesion, motility and differentiation) as well as the individual preferences of the different cell types can be addressed. Moreover, in contact-related phenomena, such as tribology or adhesion, the effects of roughness on a micrometre and nanometre scale are of great significance in mechanical as well as in biological systems [8–10]. By means of gradients, which cover a large parameter set on a single sample, it is possible to carry out combinatorial tribological experiments or systematically study the adhesion force between two surfaces in peel tests.

4. Conclusion

We have presented a simple method for the production of morphology gradients, which have the potential to facilitate high-throughput studies. A gradient was produced using the combination of particle erosion with a subsequent chemical polishing step. With this procedure it is possible to reproducibly fabricate morphology gradients over a length between 5 and 100 mm. We demonstrated that the gradient surface can be transferred to materials other than aluminium via replication and subsequent coating and/or surface-chemical functionalisation with monolayers. Such material-independent roughness gradients open up a very wide range of applications.

Acknowledgements

The authors would like to thank the AO Research Institute, Davos, Switzerland for access to the SEM and assistance with the SEM. This work was financially supported by the Swiss National Science Foundation (SNF).

References

[1] K.M. Balss, G.A. Fried, P.W. Bohn, J. Electrochem. Soc. 149 (2002) C450.
[2] B. Liedberg, P. Tengvall, Langmuir 11 (1995) 3821.
[3] S. Morgenthaler, S. Lee, S. Zürcher, N.D. Spencer, Langmuir 19 (2003) 10459.
[4] T.G. Ruardy, H.E. Moorlag, J.M. Schakenraad, H.C. Van der Mei, H.J. Busscher, J. Colloid Interface Sci. 188 (1997) 209.
[5] G.A. Abrams, A.I. Teixeira, P.F. Nealey, C.J. Murphy, Effects of Substratum Topography on Cell Behavior, Springer, 1999.
[6] D.M. Brunette, B. Chehroudi, J. Biomech. Eng. 121 (1999) 49.
[7] R.G. Flemming, C.J. Murphy, G.A. Abrams, S.L. Goodman, P.F. Nealey, Biomaterials 20 (1998) 573.
[8] D.A. Dillard, A.V. Pocius, M.K. Chaudhury, The Mechanics of Adhesion/Surfaces Chemistry Applications, Elsevier, Amsterdam, 2002.
[9] I.M. Hutchings, Tribology: Friction and Wear of Engineering Materials, Arnold, London, 1995.
[10] G.W. Stachowiak, A.W. Batchelor, Engineering Tribology, Elsevier, Amsterdam, 1993.
[11] J.C. Meredith, J.-L. Sormana, B.G. Keselowsky, A.J. Garcia, A. Tona, A. Karim, E. Amis, J. Biomed. Mater. Res. 66 (2003) 483.
[12] C.Y. Poon, R.S. Sayles, J. Phys. D Appl. Phys. 25 (1992) A249.
[13] N.R. Washburn, K.M. Yamada, C.G. Simon, S.B. Kennedy, E. Amis, Biomaterials 25 (2004) 1215.
[14] L.M. Karlsson, P. Tengvall, I. Lundström, H. Arwin, J. Electrochem. Soc. 149 (2002) C648.
[15] M. Wieland, B. Chehroudi, M. Textor, D.M. Brunette, J. Biomed. Mater. Res. 60 (2002) 434.
[16] P. Duhamel, M. Vetterli, Signal Process 19 (1990) 259.
[17] P.J. Slikkerveer, P.C.P. Bouten, F.H. int' Veld, H. Scholten, Wear 217 (1998) 237.
[18] S. Bouzid, N. Bouaouadja, J. Eur. Ceram. Soc. 20 (2000) 481.
[19] D.S. Park, M.-W. Cho, H. Lee, Int. J. Manuf. Tech. 23 (2004) 444.
[20] W.K. Bates, C.D. Coppard, Met. Finish. (1958) 5.
[21] R. Lattey, H. Neunzig, Metalloberfläche 9 (1955) 97.
[22] P. V. Shigolev, Electrolytic and Chemical Polishing of Metals, Freund, Tel-Aviv, 1974.
[23] H. Spindler, Metalloberfläche 10 (1956) 309.
[24] S. Wernick, R. Pinner, P.G. Sheasby, The Surface Treatment and Finishing of Aluminium and Its Alloys, ASM International, 1987.
[25] A.W. Clifford, D.J. Arrowsmith, Trans. Inst. Met. Finish. 56 (1978) 46.
[26] A.S. Curtis, P. Clark, Crit. Rev. Biocompat. 5 (1990) 343.
[27] P.S. Spatz, Cell-Nanostructure Interactions, Wiley-VCH, Weinheim, 2004
[28] W. Barthlott, C. Neinhuis, Planta 202 (1997) 1.
[29] M. Wieland, P. Haenggi, W. Hotz, M. Textor, B.A. Keller, N.D. Spencer, Wear 237 (1998) 231.
[30] C.D. Bain, E.B. Troughton, Y.-T. Tao, J. Evall, G.M. Whitesides, R.G. Nuzzo, J. Am. Chem. Soc. 111 (1989) 321.
[31] S. Tosatti, R. Michel, M. Textor, N.D. Spencer, Langmuir 18 (2002) 3537.
[32] D.M. Brunette, P. Tengvall, M. Textor, P. Thomsen, Titanium in Medicine, Springer, 2001.

Poly(L-lysine)-*grafted*-poly(ethylene glycol)-based surface-chemical gradients. Preparation, characterization, and first applications

Sara Morgenthaler and Christian Zink
Laboratory for Surface Science and Technology, Department of Materials, ETH Zurich, Wolfgang-Pauli-Strasse 10, CH-8093 Zurich, Switzerland

Brigitte Städler and Janos Vörös
Laboratory of Biosensors and Bioelectronics, Institute for Biomedical Engineering, Department of Information Technology and Electrical Engineering, ETH Zurich, Gloriastrasse 35, CH-8092 Zurich, Switzerland

Seunghwan Lee, Nicholas D. Spencer,[a] and Samuele G. P. Tosatti
Laboratory for Surface Science and Technology, Department of Materials, ETH Zurich, Wolfgang-Pauli-Strasse 10, CH-8093 Zurich, Switzerland

(Received 16 October 2006; accepted 11 December 2006; published 31 January 2007)

A simple dipping process has been used to prepare PEGylated surface gradients from the polycationic polymer poly(L-lysine), grafted with poly(ethylene glycol) (PLL-*g*-PEG), on metal oxide substrates, such as TiO_2 and Nb_2O_5. PLL-*g*-PEG coverage gradients were prepared during an initial, controlled immersion and characterized with variable angle spectroscopic ellipsometry and x-ray photoelectron spectroscopy. Gradients with a linear change in thickness and coverage were generated by the use of an immersion program based on an exponential function. These single-component gradients were used to study the adsorption of proteins of different sizes and shapes, namely, albumin, immunoglobulin G, and fibrinogen. The authors have shown that the density and size of defects in the PLL-*g*-PEG adlayer determine the amount of protein that is adsorbed at a certain adlayer thickness. In a second step, single-component gradients of functionalized PLL-*g*-PEG were backfilled with nonfunctionalized PLL-*g*-PEG to generate two-component gradients containing functional groups, such as biotin, in a protein-resistant background. Such gradients were combined with a patterning technique to generate individually addressable spots on a gradient surface. The surfaces generated in this way show promise as a useful and versatile biochemical screening tool and could readily be incorporated into a method for studying the behavior of cells on functionalized surfaces.
[DOI: 10.1116/1.2431704]

I. INTRODUCTION

Controlling the processes regulating the spontaneous adsorption of biomolecules onto artificial material surfaces is a critical consideration when designing and developing modern biomedical and bioanalytical devices.[1–4] Surfaces such as polymers and metal oxides, which are widely used in the biomaterials area, have indeed been shown to nonselectively adsorb large quantities of proteins in their native state. For applications in areas such as tissue engineering, implants, or biosensors, those surfaces need to be rendered "protein resistant," which means resistant towards nonspecific protein adsorption to minimize nonspecific biological response. Common features of nonfouling (i.e., protein-resistant) surfaces are their hydrophilicity, their charge neutrality, as well as the presence of hydrogen bond acceptors, but absence of hydrogen bond donor groups.[5]

Several ways to create protein-resistant surfaces have been proposed: poly(ethylene glycol) chemistry,[2,5,6] functionalized alkanethiols,[7–9] supported phospholipid bilayers,[10,11] polysaccharide chemistry,[12–15] and others. Among them, the most popular approach is based on the use of poly(ethylene glycol) (PEG). The protein resistance of PEG modified surfaces is attributed mainly to entropic repulsion and the high water content of the PEG chains. Among other methods, PEG chains may be immobilized on surfaces via covalent coupling, either by "grafting to"[16–19] or "grafting from,"[20] via the adsorption of PEG-containing block copolymers,[18,21–23] graft copolymers,[24,25] and interpenetrating polymer networks [26,27] and via functionalization with ethylene glycol–terminated alkanethiols[7,28–31] or silanes.[32,33] For the platforms dealing with PEG in a brushlike conformation, it was found that the most important parameter determining the protein resistance is the ethylene glycol monomer density on the surface n_{EG} expressed as EG units/surface unit.[34].

We have used the graft copolymer poly(L-lysine)-*graft*-poly(ethylene glycol) (PLL-*g*-PEG), because the latter system has several advantages over other PEG-based approaches [Fig. 1(a)].[24] The positively charged PLL backbone adsorbs electrostatically from an aqueous solution onto negatively charged surfaces, such as TiO_2 or Nb_2O_5, and the grafted PEG chains render these surfaces resistant towards nonspecific protein adsorption.[24,35] The architecture of the

[a]Author to whom correspondence should be addressed; electronic mail: spencer@mat.ethz.ch

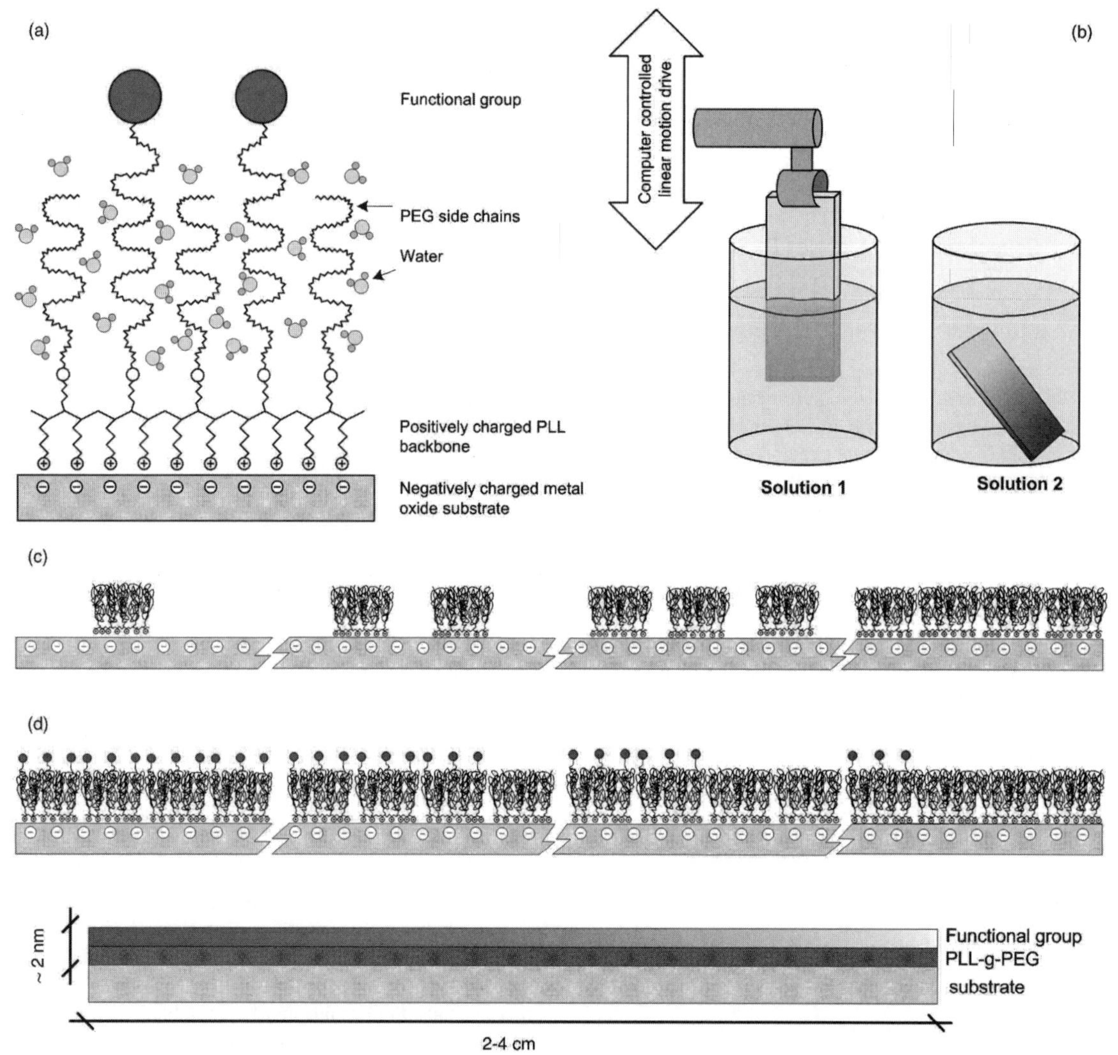

FIG. 1. (a) Schematic view of a poly(L-lysine)-*graft*-poly(ethylene glycol) (PLL-*g*-PEG/PEG-X) adlayer on a metal oxide surface. The positively charged PLL backbone attaches to the negatively charged metal oxide layer through electrostatic interactions. The grafted PEG chains are hydrated (represented by the H_2O molecules between the PEG chains) and extend into the aqueous environment (reproduced with permission from Tosatti *et al.*). (b) Schematic of the gradient preparation process. Gradients are prepared in two steps, a gradual immersion into one type of solution, followed by a full immersion into a second type of solution. Schematic of the two gradients prepared in this study: (c) PLL-*g*-PEG coverage gradient after a single immersion step (PLL-*g*-PEG on bare TiO_2/Nb_2O_5) and (d) functionalized PLL-*g*-PEG gradient after two immersion steps (PLL-*g*-PEG/PLL-*g*-PEG/PEG-biotin).

graft copolymer (molecular weight of the PEG chains and grafting ratio) and the adsorbed mass determine the ethylene glycol density on the surface.[34,36,37] Serum adsorption was found to decrease below the detection limit of optical waveguide lightmode spectroscopy (OWLS) measurements for ethylene glycol densities $\geqslant 20\ nm^{-2}$.[34] Specific (bio)functional groups can be attached at the ω position of the PEG side chains, e.g., biotin,[38] nitrilotriacetic acid,[39] or bioadhesive peptides such as Arg-Gly-Asp (RGD),[40,41] resulting in a surface that exposes a specific functionality in a protein-resistant background. In all those examples the degree of functionalization is of crucial importance when looking for optimal antibody immobilization[38] or cell response.[42]

Since a screening process is often expensive in terms of time and material, a surface gradient with a gradually changing antigen or receptor density can contribute to improving selection processes while screening a large range of properties on one single sample under the same experimental conditions. Additionally, since gradients are found to play a key role in understanding biological processes such as the growth of nerve cells, fabricated surface (bio)chemical gradients may be a powerful experimental tool to further investigate these mechanisms.[43] A variety of techniques to generate grafted polymer gradients is available, including the preparation of initiator gradients by corona discharge,[44,45] by vapor diffusion,[46,47] by controlling adsorption kinetics[48] or temperature,[49] the control of the polymerization conditions, e.g., time[50,51] or temperature,[52] and diffusion methods.[53,54] Some of these surfaces have been used to study cell growth. Other techniques have been used to gradually immobilize biomolecules, such as proteins, on surfaces, for example, microfluidics,[55–57] covalent coupling to an alkanethiol gradient,[58,59] covalent coupling by laser irradiation,[60] the use of stamping techniques and electrophoretics,[61] ink-jet printing,[62,63] drainage,[64] or by controlling the adsorption kinetics.[65]

In this study we present the generation and characterization of PLL-*g*-PEG gradients prepared by means of an immersion process originally developed for alkanethiols [Fig. 1(b)].[66] Two different types of PLL-*g*-PEG gradients with either nonfunctionalized [Fig. 1(c)] or biotinylated PEG chains [Fig. 1(d)] were prepared on titanium and niobium oxide surfaces, respectively, and characterized by means of variable angle spectroscopic ellipsometry (VASE) and x-ray photoelectron spectroscopy (XPS).

The gradients on TiO_2 were used to investigate the influence of PLL-*g*-PEG surface coverage, i.e., ethylene glycol density, on the adsorption behavior of different proteins that are relevant in terms of their occurrence and role in blood functions and wound healing processes, namely, human serum albumin, fibrinogen, and immunoglobulin G, and mixed protein solutions, such as blood serum and blood plasma. Since TiO_2 is a widely used implant material, a detailed knowledge of protein interaction with this surface is of great importance and gradient techniques should allow one to determine the minimum EG monomer surface density needed to prevent protein adsorption for a certain type of protein.[67] Additionally, the streptavidin/biotin interaction was monitored with biotinylated gradients by confocal laser scanning microscopy. These gradients were then combined with an in-house patterning technique termed molecular assembly patterning by lift-off (MAPL),[68] which creates micropatterned surfaces of functionalized spots in a protein-resistant background of PLL-*g*-PEG. A simple combination of gradients with a patterning technique offers the possibility for a quantitative comparison of different samples, as the protein-resistant background of the pattern enables us to calibrate the measurements. The gradual change in active-group concentration in successive patches makes these gradients interesting tools, likely to find manifold applications in the areas of biosensors or cell studies.

II. EXPERIMENTAL SECTION

A. Materials

All adsorption experiments were carried out from a "HEPES 2" buffer consisting of 10 mM 4-(2-hydroxyethyl)piperazine-1-ethane-sulfonic acid and 150 mM NaCl (both from MicroSelect, Fluka Chemie GmbH, Switzerland) in ultrapure water [MilliQ gradient A 10 system, resistance ≈ 18 MΩ/cm, total organic carbon <4 ppb (parts per 10^9), Millipore Corporation, USA]. The buffer was adjusted to *p*H 7.4 by the use of 6*M* NaOH and filtered through a 0.2 μm filter (Millex-GW, Millipore, Switzerland) prior to use.

PLL-*g*-PEG, a graft copolymer with a PLL backbone of 20 kDa, (including counterions, Br^-), PEG side chains of 2 kDa, and a grafting ratio of 3.5, was used for all experiments. The biotinylated polymer (PLL-*g*-PEG/PEG-biotin) had the same architecture, with 50% of its side chains biotinylated using PEG-biotin of 3.4 kDa. Both polymers were synthesized and characterized as previously described in detail.[34,38] Briefly, a 100 mM solution of PLL-HBr (Fluka, Switzerland) in 50 mM sodium tetraborate buffer (*p*H 8.5) was prepared and filter sterilized (0.22 μm pore size filter, Millex, Sigma-Aldrich, Switzerland). The grafting reaction was carried out by adding *N*-hydroxysuccinimidyl ester of methoxypoly(ethylene glycol) propionic acid (Nektar, USA) and allowing to react for 6 h at room temperature. Subsequently, the reaction mixture was dialyzed (Spectra-Por, molecular weight cutoff size of 6–8 kDa, Spectrum Laboratories Inc., USA) for 48 h against de-ionized water. The grafting ratio of the polymer was determined by ^{1}H NMR. The product was freeze dried and stored at −20 °C before use. Human serum albumin [molecular weight (MW) =66.4 kDa], rabbit Immunoglobulin G (IgG) (MW =150 kDa), fibrinogen (MW=340 kDa, all from Sigma Aldrich Chemie GmbH, Germany), blood serum (Precinorm U, Roche, Switzerland), and fresh frozen plasma were used for the protein-adsorption studies. Streptavidin alexa fluor488 (MW=52.8 kDa, Invitrogen, Switzerland) was used for confocal laser scanning microscopy.

B. Substrates

TiO_2 and Nb_2O_5 thin films (15 nm) were sputter coated onto silicon wafers (WaferNet GmbH, Germany) and Nb_2O_5 (6 nm) onto Pyrex wafers (SensorPrep Services, USA) using reactive magnetron sputtering (PSI Villigen, Switzerland). Prior to use, the oxide-coated substrates were cleaned by the following protocol: (i) 10 min sonication in 2-propanol and (ii) 2 min oxygen-plasma cleaning in a plasma cleaner/sterilizer PDC–32G instrument (Harrick, Ossining, NY, USA). The Nb_2O_5 coated Pyrex substrates were prepatterned with photoresist (S1818, Shipley, USA) according to the procedure described by Falconnet *et al.*[68] Finally, these samples were sonicated in water for 10 min and plasma cleaned for 5 s in an oxygen plasma prior to functionalization.

C. Surface modification

Gradients were prepared based on a procedure adapted from Morgenthaler *et al.*[66] The PLL-*g*-PEG was gradually adsorbed onto an oxide surface by an immersion process. A concentration of 0.02 mg/ml PLL-*g*-PEG in HEPES 2 and a total immersion time of 17 min were used. The substrate was dipped gradually with a linear motion drive (OWIS GmbH, Germany), according to a stepwise immersion program

(LABVIEW software V7.1, National Instruments). After this gradual coating step the substrates were rinsed immediately with HEPES 2 and ultrapure water and dried under a nitrogen stream. When biotinylated gradients were prepared, the gradual coating step was performed in a 0.02 mg/ml PLL-*g*-PEG/PEG-biotin solution with the same immersion program. After rinsing and drying the substrate was backfilled with nonfunctionalized PLL-*g*-PEG (0.1 mg/ml) for 40 min.

As references, homogeneously coated surfaces were prepared according to a previously published protocol[34] and bare, oxide-coated surfaces were immersed in HEPES 2 for 17 min.

D. Protein adsorption

All single proteins were adsorbed from a 0.1 mg/ml solution in HEPES 2. Gradient and reference samples were exposed to the protein solution for 15 min, then subjected to rinsing with HEPES 2 and ultrapure water, and finally dried under a stream of nitrogen. Serum and plasma were used as received without further dilution. Variable angle spectroscopic ellipsometry measurements were carried out in a dry state.

E. Variable angle spectroscopic ellipsometry (VASE)

The dry thicknesses of polymer and protein adlayers were determined by VASE (M-2000F, L.O.T. Oriel GmbH, Germany). Measurements were conducted under ambient conditions at three angles of incidence (65°, 70°, and 75°) in the spectral range of 370–1000 nm. Spectroscopic scans were taken after every step (after cleaning, after PLL-*g*-PEG adsorption, and after protein adsorption) every 3 or 5 mm along the sample. Three samples were analyzed for each type of protein. Measurements were fitted with the WVASE32 analysis software using a multilayer model for an oxide layer on silicon and an organic adlayer (polymer and protein). The n and k values for the oxide layers were fitted, and the adlayer thickness for both the PLL-*g*-PEG and the proteins was determined using a Cauchy model ($A=1.45$, $B=0.01$, and $C=0$).[23]

F. X-ray photoelectron spectroscopy (XPS)

XPS analysis was performed using a VG Theta Probe spectrophotometer (Thermo Electron Corporation, West Sussex, UK) equipped with a concentric hemispherical analyzer and a two-dimensional channel plate detector with 112 energy and 96 angle channels and a total aperture of 60°. Spectra at 10 or 20 different locations on the gradient sample (line scan with a point-to-point analysis spacing of 3.4 or 1.7 mm, respectively) were acquired at a base pressure of 10^{-9} mbar or below using a monochromatic Al $K\alpha$ source with a spot size of 300 μm. The measurements were repeated three times for each type of gradient. The instrument was run in the standard lens mode at 53° to the surface normal for survey spectra and in an eight-angle-channel mode for detailed spectra, each channel covering a sector of 7.5° and having the most grazing angle at 79.25° from the surface normal. The analyzer was used in the constant analyzer energy mode. Pass energies used for survey scans and detailed scans were 200 and 150 eV, respectively, for titanium Ti $2p$, carbon C $1s$, oxygen O $1s$, and nitrogen N $1s$. Acquisition times were approximately 30 min in total for high-energy-resolution elemental scans and 5 min for survey scans. These experimental conditions were chosen to obtain an adequate signal-to-noise ratio in a minimum time and to limit beam-induced damage. Under these conditions, sample damage was negligible, and reproducible analyzing conditions were obtained on all samples. All recorded spectra were referenced to the hydrocarbon C $1s$ signal at 285.0 eV. Data were analyzed using the program CASAXPS (Version 2.3.5, www.casaxps.com).

The signals were fitted using Gaussian-Lorentzian functions and least-squares-fit routines following a Shirley iterative background subtraction according to the protocol published by Huang *et al.*[35]

The different intensity ratios have been calculated by dividing the corresponding areas underneath the spectra. Since the main aim of the XPS measurements was to observe trends along the gradient, sensitivity factors were not employed.

G. Molecular assembly patterning by lift-off (MAPL)

The MAPL technique was applied to Nb_2O_5 coated Pyrex wafers (Sensor Prep Services, USA) as described by Falconnet *et al.*[68] Briefly, photolithography was used to create substrates with a patterned photoresist coating. After coating the prepatterned substrate with PLL-*g*-PEG/PEG-biotin in a concentration gradient, the photoresist was lifted off in 1-methyl-2-pyrrolidone [peptide-synthesis grade (⩾99.5%), Fluka Chemie GmbH, Switzerland]. Subsequently, the uncovered background was backfilled with nonfunctionalized PLL-*g*-PEG (0.1 mg/ml). The sample was placed in a polydimethylsiloxane-based flow cell (12 mm in length) and rinsed with buffer solution. The buffer solution was replaced with streptavidin alexa fluor488 (20 μg/ml) for 40 min. After rinsing with buffer, the sample was investigated using a confocal laser scanning microscope (Zeiss LSM 510, Germany) equipped with a 10× objective (0.3 numerical aperture Ph1 Plan-Neofluar, Zeiss, Germany), an argon laser, and the required filter sets. All images were taken with exactly the same instrument settings, allowing for a quantitative comparison.

H. Optical waveguide lightmode spectroscopy (OWLS)

OWLS measurements were carried out in OWLS 110 instruments (MicroVacuum, Hungary) using a laminar flow-through cell ($8\times2\times1$ mm^3). The formula of Defeijter *et al.* was applied to calculate the adsorbed mass[69] with a refractive index increment (dn/dc) of 0.139 cm^3/g for PLL-*g*-PEG and 0.182 cm^3/g for proteins.[38] The adsorbed mass was further converted into EG monomer surface density according to Pasche *et al.*[34] Waveguides were initially placed in the buffer immediately after the cleaning and allowed to soak

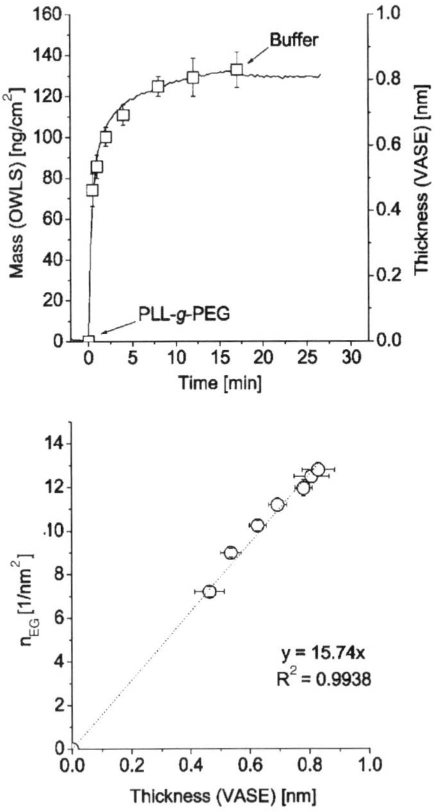

FIG. 2. Upper panel: adsorption kinetics for a 0.02 mg/ml PLL-*g*-PEG solution in HEPES 2 as measured by OWLS (adsorbed mass vs time, line) and VASE (layer thickness vs time, □). 17 min is needed to reach a plateau value in the adsorption curve. Lower panel: direct comparison of layer thickness from VASE and EG monomer surface density derived from OWLS measurements for different adsorption times. A good correlation was found between both techniques, allowing for a conversion of adsorbed thickness into EG monomer surface density n_{EG}.

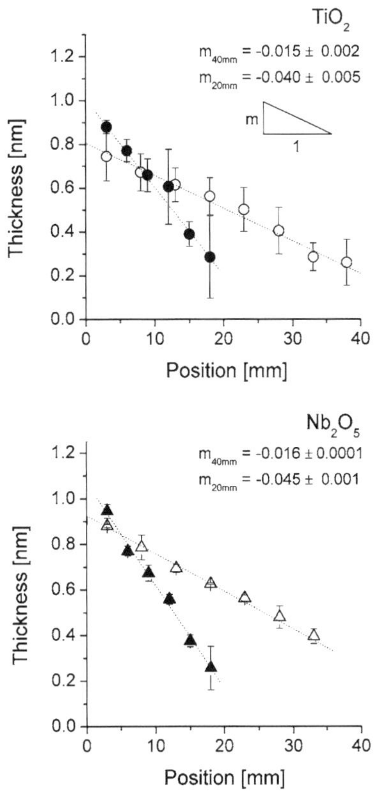

FIG. 3. Thickness of the adsorbed polymer adlayer measured by VASE as a function of position for different gradients. Upper panel: 20-mm-long, (●, R^2=−0.9973) and 40-mm-long (○, R^2=−0.9908) gradients on TiO_2 substrates; lower panel: 20-mm-long (▲, R^2=−0.9981) and 40-mm-long (△, R^2=−0.9966) gradients on Nb_2O_5 substrates. A linear increase in thickness was found for all types of gradients (see R^2 values). Similar slopes (similar *m* values) are obtained for the same length on both oxide substrates (both 40-mm- and 20-mm-long gradients were prepared by the same immersion program).

overnight. The samples were exposed *in situ* to the PLL-*g*-PEG solution at a concentration of 0.02 mg/ml. The adsorption was subsequently monitored for 17 min. Then, the polymer solution was replaced with buffer. Next the PLL-*g*-PEG modified samples were exposed to full human serum for 15 min before rinsing again with buffer solution.

III. RESULTS AND DISCUSSION

A. Gradient preparation

The gradient-preparation method used was based on controlling the adsorption kinetics. OWLS measurements were applied to determine the adsorption kinetics for different PLL-*g*-PEG concentrations (data not shown). For all concentrations a fast initial adsorption step was observed, followed by slow surface rearrangements that allow the adsorption of more polymer.[24] Huang *et al.* found that for a concentration of 1 mg/ml, 95% of the adlayer is formed in the first 5 min of the adsorption, while it takes 20 min to reach saturation.[35] The time needed to form a complete adlayer increases as the concentration of the solution decreases. We determined, by means of OWLS, that for a concentration of 0.02 mg/ml PLL-*g*-PEG, the saturation level is reached after an immersion time of 17 min (Fig. 2, upper panel). Serum adsorption was reduced by 99% compared to a bare TiO_2 coated substrate on such a coating (data not shown). This is in spite of the relatively low density of the PLL-*g*-PEG layer generated from the 0.02 mg/ml solution (7.8±0.4 Å). For comparison, a layer formed from 1 mg/ml solution for 30 min is 11.7±0.4 Å in thickness. The lower panel in Fig. 2 shows the correlation between EG monomer surface density, n_{EG}, derived from OWLS and adlayer thickness from VASE measurements for identical adsorption times. The correlation of these data allowed us to switch between the two methods using the conversion factor obtained by linear regression.

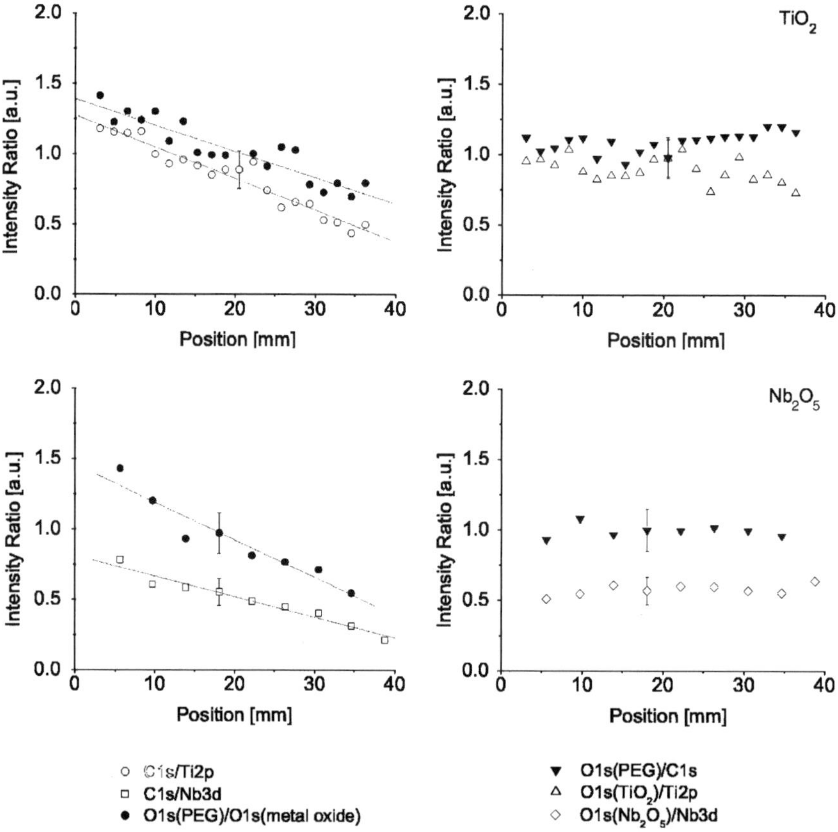

FIG. 4. Intensity ratios measured by XPS at 79.25° takeoff angle as a function of position for 40 mm gradients on TiO_2 (upper panel) and Nb_2O_5 (lower panel) substrates. Left column: both ratios that are sensitive to the surface coverage (C 1*s*/metal oxide) and O 1*s* (PEG)/O 1*s* (metal oxide) indicate the presence of a linear gradient composition [C 1*s*/Ti 2*p*: R^2=−0.97, O 1*s* (PEG)/O 1*s* (TiO_2): R^2=−0.92, C 1*s*/Nb 3*d*: R^2=−0.98, and O 1*s* (PEG)/O 1*s* (Nb_2O_5): R^2=−0.96], while the ratios relative to the composition of the substrate [O 1*s* (metal oxide)/metal oxide] or the adlayer [O 1*s* (PEG)/C 1*s*] remain constant (right column) as function of the position.

We observed that when a substrate is immersed at a constant speed for 17 min (for example, at 40 μm/s for a 4-cm-long sample), a nonlinear coverage gradient is formed, corresponding to the shape of the adsorption curve as measured by OWLS (Fig. 2, upper panel). The immersion program was therefore modified to generate a linear coverage gradient by gradually changing the immersion speed by means of a program based on an exponential function. This gradient preparation method was used to generate two types of gradients: PLL-*g*-PEG coverage gradients on bare oxide surfaces [Fig. 1(b)] and functionalized PLL-*g*-PEG/PEG-biotin gradients [Fig. 1(c)].

B. PLL-*g*-PEG versus oxide gradients

One-component PLL-*g*-PEG gradients were obtained by the immersion of either a TiO_2 or a Nb_2O_5 coated substrate according to a nonlinear speed program. Figure 3 presents the results obtained by means of variable angle spectroscopic ellipsometry for 20- and 40-mm-long gradients on TiO_2 (upper panel) and Nb_2O_5 (lower panel) substrates. The same immersion program was used for both oxide substrates, which leads to a very similar slope, indicating that the adsorption kinetics on both substrates are comparable. A linear increase in adlayer thickness can be found for all types of gradients (all R^2 values are higher than −0.99). Grazing-angle x-ray photoelectron spectroscopy measurements were also performed on such gradients (Fig. 4). We expect that the adlayer surface coverage decreases with decreasing immersion time (expressed as position on the gradient). One parameter that is highly sensitive to variations in the adlayer film thickness is the carbon/metal ratio, since these elements are found either only in the adlayer (carbon) or in the substrate (metal). The same information can be obtained when considering the ratio between the O 1*s* (PEG) and the O 1*s* (metal oxide) signals (left column). Finally, the fact that both the ratios O 1*s* (PEG) versus carbon and O 1*s* (metal oxide) versus metal oxide (Ti 2*p* or Nb 3*d*) remain constant along the gradient suggests that the adsorbed adlayer consists primarily of PLL-*g*-PEG.

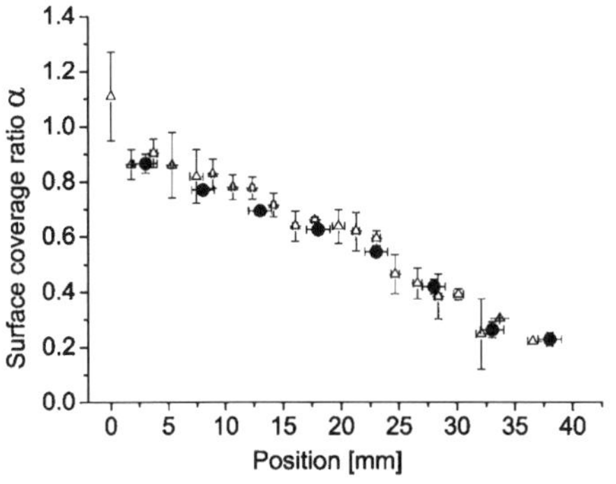

FIG. 5. Relative PLL-g-PEG surface coverage α for a single-component gradient on TiO_2 as determined by XPS (△) and VASE (●). A bare metal oxide surface exposed to HEPES 2 (α equal to zero) and a homogeneously coated sample (α equal to 1) were used as references. The relative surface coverage α decreases continuously towards the loosely covered end. VASE and XPS measurements correlate with each other.

To further compare VASE and XPS data a surface coverage parameter α was defined as the ratio between coated and uncoated surface areas at a point X on the gradient. A bare metal oxide surface exposed to HEPES 2 (α equal to zero) and a homogeneously PLL-g-PEG-coated sample (α equal to 1) were used as references. We chose to use the C/Ti ratio measured by XPS at an angle of 79.25° normal to the surface to calculate α, since at such a grazing angle, the signals are highly surface sensitive. α is then calculated from the adlayer thickness measured by VASE and the C/Ti ratio measured by XPS at individual positions on the gradient. The results show that both relative surface coverages are found to correspond well to each other and to decrease monotonically along the substrate length with decreasing immersion time (Fig. 5). The fact that zero coverage is not reached is mostly due to the adventitious contamination that adsorbs when exposing the gradient to ambient conditions during the deposition.

C. Protein adsorption on PLL-*g*-PEG coverage gradients

PLL-g-PEG coverage gradients on TiO_2 were exposed to a series of different protein solutions. The thickness of the PLL-g-PEG adlayer and the adsorbed protein layer was measured with VASE under dry conditions. Figure 6 represents the relative adsorbed amount of proteins that are deposited on a PLL-g-PEG coverage gradient compared to a bare TiO_2 substrate (value=1) and a homogeneously coated substrate as a function of the ethylene glycol monomer surface density n_{EG}. The measured dry thickness of the gradient polymer adlayer was thereby converted into EG monomer surface density, as proposed above. The amount of adsorbed protein was found to decrease towards higher EG monomer surface densities, which corresponds to the more densely covered end of the gradient. Albumin, IgG, and fibrinogen adsorption all decrease to below the detection limit of ellipsometry measurements (estimated to be 5–10 ng/cm^2)[23,41] at the PLL-g-PEG rich end as well as on the homogeneously coated substrates. However, the minimal PLL-g-PEG thickness, at which the protein adsorption falls below 5% (equals to at least a 95% reduction) in comparison with bare titanium, differs for each of the proteins: n_{EG} (IgG) $\geqslant 9.5$ nm^{-2}, n_{EG} (albumin) $\geqslant 9.9$ nm^{-2}, and n_{EG} (fibrinogen) $\geqslant 11.6$ nm^{-2}. This may be explained by the size, shape, and charge distribution of the proteins. While albumin is a triangular protein with a small molecular weight (MW of around 66 kDa), IgG (Y shaped, MW of around 150 kDa) and fibrinogen (elongated rod, MW of around 340 kDa) are substantially larger, however, at physiological pH they are all slightly negatively

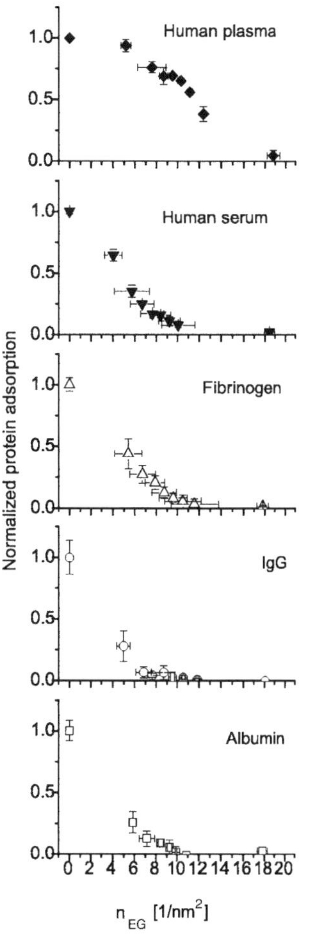

FIG. 6. Relative amount of protein adsorbed on PLL-g-PEG coverage gradients on TiO_2 plotted as a function of the EG monomer surface density, n_{EG}, calculated from OWLS/VASE adsorption data and NMR grafting ratios of the bulk polymers. The adsorbed amount of protein has been normalized by the values measured on the homogeneously coated and bare substrates. Protein adsorption is found to decrease towards the more densely covered end. The minimal PLL-g-PEG thickness at which protein adsorption is reduced to below 5% differs for each of the proteins (albumin, IgG, and fibrinogen). Mixed protein solutions, such as plasma and serum, require a higher EG monomer surface density to provide a protein-resistant coating.

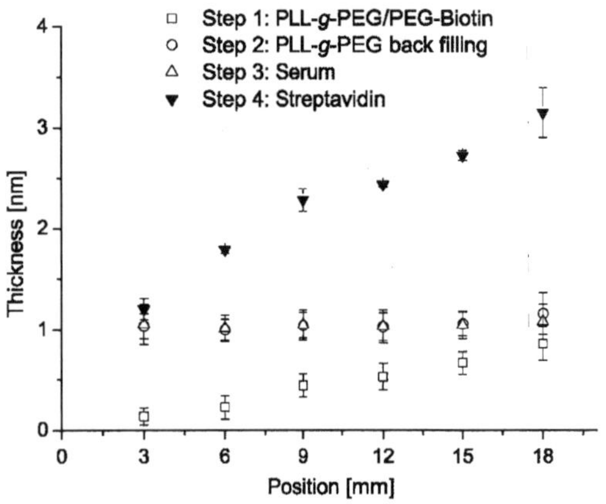

FIG. 7. Ellipsometric adlayer thickness of a biotinylated gradient as a function of the gradient position. Results are shown after four different protocol steps. A biotinylated coverage gradient was generated (step 1, linear increase in thickness), which was then backfilled with unmodified PLL-*g*-PEG (step 2, constant layer thickness). Negligible serum adsorption was found on such a functionalized gradient (step 3, no notable increase in layer thickness), while the amount of immobilized streptavidin gradually increased along the gradient (step 4, increasing thickness with increasing biotin density).

charged.[5] Fibrinogen adsorbs on denser PLL-*g*-PEG layers because it has the possibility to attach to the substrate via various binding sites due to its elongated shape and high molecular weight.[5] This suggests the existence of some very small uncovered patches (defects) present in the denser PLL-*g*-PEG that allow fibrinogen to reach the substrate-polymer interface and interact, whereas albumin and IgG need either larger or more closely spaced uncovered patches to interact with the surface, as suggested by the lower PLL-*g*-PEG coverage needed to reduce adsorption.

Adsorption from two mixed protein solutions, human serum and plasma, was not totally inhibited even on the high-EG-monomer-surface-density end of the gradient. Serum adsorption is reduced by 90%, whereas plasma adsorption is only reduced by 60%. There are two possible reasons, one being that both protein solutions were used at high concentrations, without further dilution. The chance for protein aggregation is higher at high concentrations. Such aggregates are less mobile than single proteins, which could possibly lead to an increased sedimentation that is difficult to remove with a short rinsing step. Another reason for the higher amount of protein adsorbed from plasma could be the presence of clotting factors, such as fibrinogen, that would lead to an increased aggregate formation.

D. PEG versus PEG-biotin gradients

Functionalized PLL-*g*-PEG gradients were prepared in two subsequent immersion steps, as described above. Figure 7 displays the adlayer thickness after the different steps of the protocol as a function of the position on the gradient. A one-component PLL-*g*-PEG/PEG-biotin gradient (Fig. 7, step 1) was backfilled with unmodified PLL-*g*-PEG, which gradually filled up the empty sites on the surface (Fig. 7, step 2). When exposing such a functionalized gradient to serum, a minimum adsorption (below the detection limit) was to be found all over the sample (Fig. 7, step 3), showing that the functionalized gradient coating is protein resistant. However, when the substrate was exposed to streptavidin, a gradient in adsorption along the substrate was found, being higher at the highly biotinylated end (Fig. 7, step 4). This demonstrates that the attached functional groups are active and available for specific immobilization.

E. Patterning of PLL-*g*-PEG/PEG-biotin gradients

Falconnet *et al.* have presented a simple method to pattern substrates by a combination of standard photolithography and molecular self-assembly termed MAPL.[68] The combination of the gradient approach with the MAPL technique allows for the generation of an array of discrete surface patches with variable bioligand concentrations along the direction of the gradient. This was demonstrated by the fabrication of a gradient of PLL-*g*-PEG and PLL-*g*-PEG/PEG-biotin on a Nb_2O_5 surface covered with prepatterned photoresist. After photoresist removal and backfilling with nonfunctionalized PLL-*g*-PEG, biotin-functionalized patches were created in a protein-resistant background. Such gradient patterns were visualized by adsorption of fluorescently labeled streptavidin and imaged by means of confocal laser scanning microscopy. Figure 8 depicts the fluorescence intensity of the surface-immobilized streptavidin, which decreases along the gradient when moving from the high towards the low biotin-density end of the sample. The size of the flow cell allowed us to image only 12 mm in the middle of the 20-mm-long gradient. This means that the highest measured fluorescent intensity (here set as 1) represents the most densely packed biotin spots within the flow cell but not on the full gradient, whereas the intensity at low biotin density, which corresponds to 15 mm along the gradient, is not equal to zero. Even if we lose a certain part of the information given by the gradient, it is important to use a flow cell to exchange the solutions, because drying effects, which might induce denaturation of certain proteins, would influence the outcome of our experiments. Two images taken at the extreme ends of the gradient in the flow cell are provided in addition to the fluorescence-intensity plot (at around 5 mm from the ends of the gradient). This array of discrete patches with variable biotin concentrations in a background that is resistant to streptavidin adsorption readily allows for the use of such gradients in the biosensor and molecular recognition area.

IV. CONCLUSIONS

We have presented a method to prepare chemical gradients from the polycationic graft copolymer PLL-*g*-PEG on oxide surfaces. First, PLL-*g*-PEG coverage gradients were prepared by controlling the adsorption time of the molecules during an immersion process. Single-component gradients from functionalized PLL-*g*-PEG may be backfilled with nonfunctionalized PLL-*g*-PEG to generate two-component gradients with functional groups such as biotin presented in a

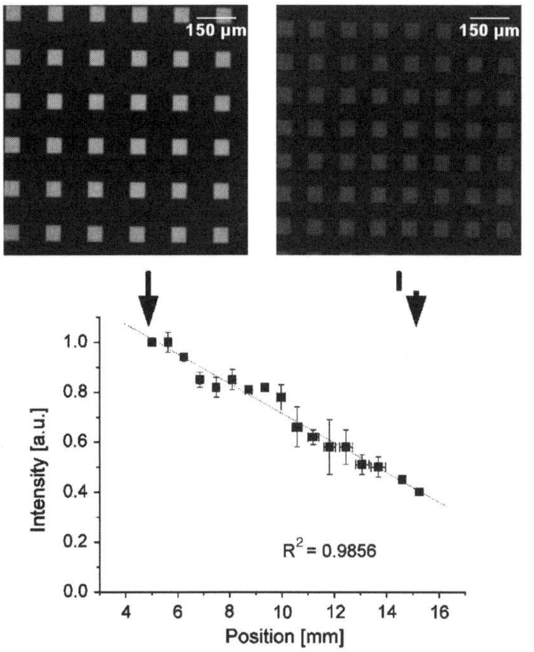

FIG. 8. Fluorescence intensity taken from confocal fluorescence microscopy images as a function of position on patterned biotinylated PLL-*g*-PEG gradients functionalized with fluorescently labeled streptavidin. The fluorescence signal decreases along the gradient, corresponding well to the trend found on unpatterned samples (Fig. 7). Images taken at a distance of 5 mm (left) and 15 mm (right) on the gradient, corresponding to both ends of the flow cell we used, are given as indicated by the arrows.

protein-resistant background. Proteins of different sizes and shapes were adsorbed onto single-component gradients. We have shown that the adsorption of smaller (albumin) and irregularly shaped (IgG) proteins can be inhibited by relatively less dense PLL-*g*-PEG layers than those required to prevent the adsorption of fibrinogen. Two-component gradients with biotinylated and nonfunctionalized PLL-*g*-PEG were also prepared and characterized. Such gradients, combined with patterning techniques, can be useful, high-throughput, and cost-effective tools in the study of biointerfaces, especially for probing the effect of bioligand concentration on specific interactions.

ACKNOWLEDGMENT

The authors acknowledge generous financial assistance for this project from the Swiss National Science Foundation.

[1]W. Senaratne, L. Andruzzi, and C. K. Ober, Biomacromolecules **6**, 2427 (2005).
[2]P. Kingshott and H. J. Griesser, Curr. Opin. Solid State Mater. Sci. **4**, 403 (1999).
[3]M. Tirrell, E. Kokkoli, and M. Biesalski, Surf. Sci. **500**, 61 (2002).
[4]B. Kasemo, Surf. Sci. **500**, 656 (2002).
[5]P. Vermette and L. Meagher, Colloids Surf., B **28**, 153 (2003).
[6]J. M. Harris, *Poly(ethylene glycol) Chemistry and Biological Applications* (American Chemical Society, Washington, DC, 1997), Vol. 680.
[7]C. Palegrosdemange, E. S. Simon, K. L. Prime, and G. M. Whitesides, J. Am. Chem. Soc. **113**, 12 (1991).
[8]Y. Y. Luk, M. Kato, and M. Mrksich, Langmuir **16**, 9604 (2000).
[9]S. F. Chen, L. Y. Liu, and S. Y. Jiang, Langmuir **22**, 2418 (2006).
[10]K. Glasmastar, C. Larsson, F. Hook, and B. Kasemo, J. Colloid Interface Sci. **246**, 40 (2002).
[11]M. Malmsten, J. Colloid Interface Sci. **168**, 247 (1994).
[12]E. Osterberg, K. Bergstrom, K. Holmberg, J. A. Riggs, J. M. Van Alstine, T. P. Schuman, N. L. Burns, and J. M. Harris, Colloids Surf., A **77**, 159 (1993).
[13]R. E. Marchant, S. Yuan, and G. Szakalasgratzl, J. Biomater. Sci., Polym. Ed. **6**, 549 (1994).
[14]S. L. McArthur, K. M. McLean, P. Kingshott, H. A. W. St. John, R. C. Chatelier, and H. J. Griesser, Colloids Surf., B **17**, 37 (2000).
[15]R. A. Frazier, G. Matthijs, M. C. Davies, C. J. Roberts, E. Schacht, and S. J. B. Tendler, Biomaterials **21**, 957 (2000).
[16]P. Kingshott, H. Thissen, and H. J. Griesser, Biomaterials **23**, 2043 (2002).
[17]J. L. Dalsin, L. J. Lin, S. Tosatti, J. Voros, M. Textor, and P. B. Messersmith, Langmuir **21**, 640 (2005).
[18]D. Lazos, S. Franzka, and M. Ulbricht, Langmuir **21**, 8774 (2005).
[19]O. H. Kwon, Y. C. Nho, K. D. Park, and Y. H. Kim, J. Appl. Polym. Sci. **71**, 631 (1999).
[20]H. W. Ma, J. H. Hyun, P. Stiller, and A. Chilkoti, Adv. Mater. (Weinheim, Ger.) **16**, 338 (2004).
[21]R. J. Green, M. C. Davies, C. J. Roberts, and S. J. B. Tendler, J. Biomed. Mater. Res. **42**, 165 (1998).
[22]M. Amiji and K. Park, Biomaterials **13**, 682 (1992).
[23]L. M. Feller, S. Cerritelli, M. Textor, J. A. Hubbell, and S. G. P. Tosatti, Macromolecules **38**, 10503 (2005).
[24]G. L. Kenausis *et al.*, J. Phys. Chem. B **104**, 3298 (2000).
[25]C. FreijLarsson, T. Nylander, P. Jannasch, and B. Wesslen, Biomaterials **17**, 2199 (1996).
[26]J. P. Bearinger, D. G. Castner, S. L. Golledge, A. Rezania, S. Hubchak, and K. E. Healy, Langmuir **13**, 5175 (1997).
[27]T. A. Barber, S. L. Golledge, D. G. Castner, and K. E. Healy, J. Biomed. Mater. Res. Part A **64**, 38 (2003).
[28]E. Ostuni, L. Yan, and G. M. Whitesides, Colloids Surf., B **15**, 3 (1999).
[29]K. L. Prime and G. M. Whitesides, J. Am. Chem. Soc. **115**, 10714 (1993).
[30]M. Mrksich, G. B. Sigal, and G. M. Whitesides, Langmuir **11**, 4383 (1995).
[31]K. Feldman, G. Hahner, N. D. Spencer, P. Harder, and M. Grunze, J. Am. Chem. Soc. **121**, 10134 (1999).
[32]S. W. Lee and P. E. Laibinis, Biomaterials **19**, 1669 (1998).
[33]N. Faucheux, R. Schweiss, K. Lutzow, C. Werner, and T. Groth, Biomaterials **25**, 2721 (2004).
[34]S. Pasche, S. M. De Paul, J. Voros, N. D. Spencer, and M. Textor, Langmuir **19**, 9216 (2003).
[35]N. P. Huang *et al.*, Langmuir **17**, 489 (2001).
[36]R. Michel, S. Pasche, M. Textor, and D. G. Castner, Langmuir **21**, 12327 (2005).
[37]M. S. Wagner, S. Pasche, D. G. Castner, and M. Textor, Anal. Chem. **76**, 1483 (2004).
[38]N. P. Huang, J. Voros, S. M. De Paul, M. Textor, and N. D. Spencer, Langmuir **18**, 220 (2002).
[39]G. L. Zhen, D. Falconnet, E. Kuennemann, J. Voros, N. D. Spencer, M. Textor, and S. Zurcher, Adv. Funct. Mater. **16**, 243 (2006).
[40]S. VandeVondele, J. Voros, and J. A. Hubbell, Biotechnol. Bioeng. **82**, 784 (2003).
[41]S. Tosatti, S. M. De Paul, A. Askendal, S. VandeVondele, J. A. Hubbell, P. Tengvall, and M. Textor, Biomaterials **24**, 4949 (2003).
[42]M. Schuler, G. R. Owen, D. W. Hamilton, M. De Wilde, M. Textor, D. M. Brunette, and S. G. P. Tosatti, Biomaterials **27**, 4003 (2006).
[43]J. Dodd and T. M. Jessell, Science **242**, 692 (1988).
[44]J. H. Lee, B. J. Jeong, and H. B. Lee, J. Biomed. Mater. Res. **34**, 105 (1997).
[45]Y. Iwasaki, S. Sawada, N. Nakabayashi, G. Khang, H. B. Lee, and K. Ishihara, Biomaterials **20**, 2185 (1999).
[46]T. Wu, K. Efimenko, and J. Genzer, J. Am. Chem. Soc. **124**, 9394 (2002).
[47]B. Zhao, Langmuir **20**, 11748 (2004).
[48]Y. Mei *et al.*, Langmuir **21**, 12309 (2005).
[49]Y. Liu, V. Klep, B. Zdyrko, and I. Luzinov, Langmuir **21**, 11806 (2005).

[50]M. R. Tomlinson and J. Genzer, Macromolecules **36**, 3449 (2003).
[51]C. Xu, T. Wu, C. M. Drain, J. D. Batteas, and K. L. Beers, Macromolecules **38**, 6 (2005).
[52]L. Ionov, B. Zdyrko, A. Sidorenko, S. Minko, V. Klep, I. Luzinov, and M. Stamm, Macromol. Rapid Commun. **25**, 360 (2004).
[53]K. Mougin, A. S. Ham, M. B. Lawrence, E. J. Fernandez, and A. C. Hillier, Langmuir **21**, 4809 (2005).
[54]Y.-S. Lin and V. Hlady, Colloids Surf., B **2**, 481 (1994).
[55]S. K. W. Dertinger, X. Y. Jiang, Z. Y. Li, V. N. Murthy, and G. M. Whitesides, Proc. Natl. Acad. Sci. U.S.A. **99**, 12542 (2002).
[56]I. Caelen, A. Bernard, D. Juncker, B. Michel, H. Heinzelmann, and E. Delamarche, Langmuir **16**, 9125 (2000).
[57]K. A. Fosser and R. G. Nuzzo, Anal. Chem. **75**, 5775 (2003).
[58]S. T. Plummer, Q. Wang, P. W. Bohn, R. Stockton, and M. A. Schwartz, Langmuir **19**, 7528 (2003).
[59]J. T. Smith, J. K. Tomfohr, M. C. Wells, T. P. Beebe, T. B. Kepler, and W. M. Reichert, Langmuir **20**, 8279 (2004).
[60]C. L. Hypolite, T. L. McLernon, D. N. Adams, K. E. Chapman, C. B. Herbert, C. C. Huang, M. D. Distefano, and W. S. Hu, Bioconjugate Chem. **8**, 658 (1997).
[61]R. A. Venkateswar, D. W. Branch, and B. C. Wheeler, Biomed. Microdevices **2**, 255 (2000).
[62]A. Y. Sankhe, B. D. Booth, N. J. Wiker, and S. M. Kilbey, Langmuir **21**, 5332 (2005).
[63]L. Pardo, W. C. Wilson, and T. J. Boland, Langmuir **19**, 1462 (2003).
[64]H. Baier and F. Bonhoeffer, Science **255**, 472 (1992).
[65]S. Kramer, H. Xie, J. Gaff, J. R. Williamson, A. G. Tkachenko, N. Nouri, D. A. Feldheim, and D. L. Feldheim, J. Am. Chem. Soc. **126**, 5388 (2004).
[66]S. Morgenthaler, S. Lee, S. Zürcher, and N. D. Spencer, Langmuir **19**, 10459 (2003).
[67]P. Tengvall, in *Titanium in Medicine: Material Science, Surface Science, Engineering, Biological Response and Medical Applications*, edited by D. M. Brunette, P. Tengvall, M. Textor, and P. Thomsen (Springer, Heidelberg, 2000).
[68]D. Falconnet, A. Koenig, T. Assi, and M. Textor, Adv. Funct. Mater. **14**, 749 (2004).
[69]J. A. Defeijter, J. Benjamins, and F. A. Veer, Biopolymers **17**, 1759 (1978).

Langmuir **2007**, *23*, 3489–3494 3489

Fabrication of Multiscale Surface-Chemical Gradients by Means of Photocatalytic Lithography

Nicolas Blondiaux,[†,‡] Stefan Zürcher,[‡] Martha Liley,[*,†] and Nicholas D. Spencer[‡]

Centre Suisse d'Electronique et de Microtechnique, SA, Jacquet Droz 1, CH-2000 Neuchâtel, Switzerland, and Laboratory for Surface Science and Technology, Department of Materials, ETH Zurich, Wolfgang-Pauli-Strasse 10, CH-8093 Zurich, Switzerland

Received November 1, 2006. In Final Form: January 19, 2007

We describe a new method for the fabrication of surface-chemical gradients. A film of titanium dioxide is brought into close proximity to a uniformly monolayer-covered surface and exposed to UV light to produce oxygen radicals. The use of a gradated grayscale mask between the UV source and the TiO_2 allows the production of surface-chemical gradients via oxidation of the monolayer. The technique is demonstrated on gold surfaces bearing alkanethiol SAMs. Oxidation and subsequent replacement of the oxidized thiols has been used to produce surface-chemical gradients with lengths on the submillimeter to centimeter scales. The oxidation, removal, and replacement of the thiols during the process have been demonstrated by means of XPS. This oxidative process may be applied to other surface chemistries. Similarly, other shapes and slopes of gradients may be produced, depending on the photomask employed.

Introduction

Surface-chemical gradient fabrication constitutes a very active field of research due to the utility of gradients in innumerable research applications. Surface gradients are useful tools for the combinatorial investigation of cell behavior or in biochemical assays because a vast number of surface-chemistry combinations can be screened for a single sample.[1,2] In a different field, wettability gradients can be used to control the motion and positioning of liquid drops on surfaces.[3–5] The design of gradients is, however, very dependent on the envisaged application. In high-throughput studies, the critical parameter is the number of conditions that can be screened on a single sample, whereas in the case of motion and positioning of liquids, emphasis is put on the slope of the gradients. In both situations, the length and slope of the gradient are important parameters.

Several approaches to creating chemical gradients have already been reported, such as the use of cross diffusion of thiol solutions through a polysaccharide matrix[6,7] or the gradual immersion of a gold-coated substrate in a dilute thiol solution.[8] Thiol composition gradients have also been generated by microcontact printing a gold surface with a PDMS stamp having a thickness gradient.[5] The length of the gradients produced with the techniques mentioned above is generally between a millimeter and a few centimeters. For the fabrication of submillimeter gradients, very different approaches have been reported. Choi et al. used microcontact printing to create gradients in silane surface concentration.[9] Gradients have also been created by selectively desorbing thiols using electrochemistry.[10] Fuierer et al. reported an STM-based technique permitting the controlled desorption of thiols by varying both the scan rate and the voltage applied.[11] The techniques based on microcontact printing have great flexibility with regard to the shape and length of the gradient.[9] Nevertheless, the use of PDMS as an elastomeric stamp may pose problems for some applications as a result of the possibility of surface contamination.[12] Approaches based on SPM techniques also offer good control over the properties of the gradients, but serial techniques such as these are inevitably limited in terms of throughput. Thus there is a need for new parallel techniques for the fabrication of short gradients.

We describe a technique that allows the generation of both long as well as short surface-chemical gradients. We have focused on the use of light to produce the gradients, using titanium dioxide (TiO_2) remote photocatalytic oxidation. In this technique, also called photocatalytic lithography, a titanium dioxide layer is held in close proximity to a monolayer-covered surface and irradiated with UV light. Radicals created in the TiO_2 layer diffuse across the air gap and react with organics adsorbed on the nearby surface. The degradation of silane and thiol self-assembled monolayers (SAMs) by means of photocatalytic lithography has already been demonstrated.[13,14] It has been used to pattern surfaces,[14,15] with a photomask being employed to illuminate the surface selectively and locally degrade the adsorbed species. In this study, we have combined photocatalytic lithography with grayscale lithography to degrade a thiol SAM gradually and produce a surface-chemical gradient.

The technique used is shown in Figure 1: a TiO_2-coated slide was placed above the substrate to be modified with a 60 μm intervening air gap. Upon UV illumination of the slide, oxidizing

* To whom correspondence should be addressed. E-mail: martha.liley@csem.ch. Tel: +41 32 720 51 84. Fax: +41 32 720 57 40.

† Centre Suisse d'Electronique et de Microtechnique.

‡ ETH Zurich.

(1) Meredith, J. C.; Karim, A.; Amis, E. J. *MRS Bull.* **2002**, *27*, 330–335.
(2) Herbert, C. B.; McLernon, T. L.; Hypolite, C. L.; Adams, D. N.; Pikus, L.; Huang, C. C.; Fields, G. B.; Letourneau, P. C.; Distefano, M. D.; Hu, W. S. *Chem. Biol.* **1997**, *4*, 731–737.
(3) Daniel, S.; Chaudhury, M. K.; Chen, J. C. *Science* **2001**, *291*, 633–636.
(4) Gallardo, B. S.; Gupta, V. K.; Eagerton, F. D.; Jong, L. I.; Craig, V. S.; Shah, R. R.; Abbott, N. L. *Science* **1999**, *283*, 57–60.
(5) Kraus, T.; Stutz, R.; Balmer, T. E.; Schmid, H.; Malaquin, L.; Spencer, N. D.; Wolf, H. *Langmuir* **2005**, *21*, 7796–7804.
(6) Liedberg, B.; Tengvall, P. *Langmuir* **1995**, *11*, 3821–3827.
(7) Liedberg, B.; Wirde, M.; Tao, Y. T.; Tengvall, P.; Gelius, U. *Langmuir* **1997**, *13*, 5329–5334.
(8) Morgenthaler, S.; Lee, S. W.; Zürcher, S.; Spencer, N. D. *Langmuir* **2003**, *19*, 10459–10462.
(9) Choi, S. H.; Newby, B. M. Z. *Langmuir* **2003**, *19*, 7427–7435.
(10) Bohn, P.; Plummer, S.; Wang, Q.; Coleman, B.; Swint, A.; Castle, P. *Abstr. Pap. Am. Chem. Soc.* **2003**, *225*, U687.
(11) Fuierer, R. R.; Carroll, R. L.; Feldheim, D. L.; Gorman, C. B. *Adv. Mater.* **2002**, *14*, 154–157.
(12) Csucs, G.; Kunzler, T.; Feldman, K.; Robin, F.; Spencer, N. D. *Langmuir* **2003**, *19*, 6104–6109.
(13) Tatsuma, T.; Tachibana, S.; Fujishima, A. *J. Phys. Chem. B* **2001**, *105*, 6987–6992.
(14) Notsu, H.; Kubo, W.; Shitanda, I.; Tatsuma, T. *J. Mater. Chem.* **2005**, *15*, 1523–1527.
(15) Kubo, W.; Tatsuma, T.; Fujishima, A.; Kobayashi, H. *J. Phys. Chem. B* **2004**, *108*, 3005–3009.

Published on Web 02/27/2007

3490 Langmuir, Vol. 23, No. 7, 2007 Letters

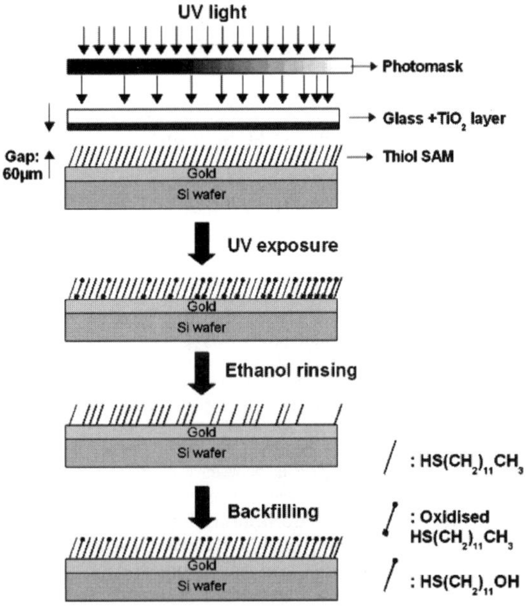

Figure 1. Remote photocatalytic oxidation of a thiol SAM under a gradient of UV illumination and subsequent backfilling with a second component.

radicals were created in the TiO_2 film as a result of its photocatalytic properties. The use of a photomask with a gray-tone gradient yielded a gradual variation of UV intensity along the TiO_2 film. Because the photocatalytic activity of TiO_2 depends on the intensity of the UV illumination,[16] the intensity variation was transposed into an oxygen radical concentration gradient, which led to a gradient in the oxidation of the substrate surface.

Materials and Methods

Formation of Thiol SAMs. The substrates used for the experiments were prepared by evaporating 30 nm of chromium followed by 120 nm of gold (99.99%, Umicore, Bösingen, Switzerland) onto silicon wafers (type P/Bor, ⟨100⟩, Siltronix, Archamps, France) using an e-beam evaporator (Leybold). The thiols used were $(CH_3(CH_2)_{11}$-SH (dodecanethiol 98+%) and $(HO(CH_2)_{11}SH$ (11-mercapto-1-undecanol 97+%), both purchased from Aldrich Chemicals (Milwaukee, WI). 1H,1H,2H,2H-Perfluorodecanethiol was purchased from Fluorous Technologies Inc. (Pittsburgh, PA). Ethanol (UVasol for spectroscopy) was obtained from Merck. The starting SAM was made by placing the gold-coated substrates overnight in a 1 mM dodecanethiol solution. The surface was then rinsed twice in ethanol and blown dry with nitrogen.

Photocatalytic Remote Oxidation. The photocatalytic layer was made by spin coating a TiO_2 nanoparticle suspension onto a clean glass slide. The TiO_2 nanoparticles were in the anatase form and were 7 nm in diameter. The TiO_2 suspension (STS-01) was kindly supplied by Ishihara Sangyo Gaisha, Ltd. (Yokkaichi, Japan). A small amount of poly(*N*-vinyl pyrrolidone) (K90 from Fluka (Buchs, Switzerland)) was added to the suspension to improve the quality of the film and avoid the formation of cracks during film formation. The final solution contained 10 wt % of TiO_2 and 0.2 mg/mL poly-(*N*-vinyl pyrrolidone). Glass slides were cleaned using piranha solution (4:1 v/v H_2SO_4/H_2O_2) for 10 min at 120 °C and rinsed in flowing water (MilliQ 185 plus, Millipore AG, Switzerland). ***Attention:*** *Piranha solution reacts violently with all organics and should be handled with care.* The solution was then spin coated at 6000 rpm onto the cleaned glass slide. To remove all traces of organic compounds present in the layer, the TiO_2-coated slides were then calcined at 400 °C for 1 h in an oven (Nabertherm, model L15/12/P320, Bremen, Germany). Because the anatase-rutile phase transition in TiO_2 occurs at temperatures above 600 °C,[17] we do not expect any phase transition due to this treatment. The cleaning of the TiO_2 layer proved to be an important issue. Before each experiment, the TiO_2 surfaces were cleaned in an oxygen plasma for 15 min using an rf power of 30 W and a pressure of 0.250 Torr (Plasmalab 80 plus, Oxford Instruments). The TiO_2-coated glass slides were then exposed to UV for 1 h. The irradiance of the lamp used (Oriel 300 W solar ultraviolet simulator, Lot-Oriel) was 17 mW/cm^2 at a wavelength of 365 nm. The photomask used for photocatalytic lithography was custom designed and fabricated by Photronics Ltd. (Manchester, U.K.). The lengths of the gray-tone gradients were 1.8 cm and 720 μm for the long and short gradients, respectively. The short gradients were repeated several times along the photomask. The grayscale gradients were not continuous but comprised 12 distinct gray tones.

After the remote photocatalytic oxidation, the samples were rinsed in ethanol. Backfilling was made by dipping the samples for 10 min in a 1 mM solution of 11-mercapto-1-undecanol or 1H,1H,2H,2H-perfluorodecanethiol dissolved in ethanol. The samples were then rinsed again in ethanol.

Contact-Angle Measurements. The wettability changes of the surfaces were characterized by measuring the contact angle of sessile water droplets as a function of their position along the gradient. Static water contact angles were determined using a DSA10 drop shape analysis system provided by Krüss (Hamburg, Germany).

Standard deviations were calculated using eight series of measurements (two series per sample on four distinct samples) and were used to calculate both the standard error of the mean and the confidence intervals. The confidence intervals were calculated using the Student's *t* distribution instead of the normal distribution because of the small number of samples considered. The error bars shown on the graph correspond to an 80% confidence interval.

XPS Measurements. XPS analysis was performed using a VG Theta Probe spectrophotometer (Thermo Electron Corporation, West Sussex, U.K.) equipped with a concentric hemispherical analyzer and a 2D channel plate detector with 112 energy channels and 96 angle channels. Spectra were acquired at a base pressure of 10^{-9} mbar or below using a monochromatic Al Kα source with a spot size of 300 μm. The instrument was run in the standard lens mode with electrons emitted at 53° with respect to the surface normal and an acceptance angle of ±30°. The analyzer was used in the constant-analyzer-energy mode. Pass energies used for survey scans and detailed scans were 200 and 100 eV, respectively for Au 4f, C 1s, O 1s, S 2p, and F 1s. Under these conditions, the energy resolutions (fwhm) measured on Au $4f_{7/2}$ are 1.95 and 0.82 eV, respectively. Acquisition times were approximately 5 min for survey scans and 30 min (total) for high-energy-resolution elemental scans. These experimental conditions were chosen to obtain an adequate signal-to-noise ratio in a minimum time and to limit beam-induced damage. Under these conditions, sample damage was negligible, and reproducible analysis conditions were obtained on all samples. All recorded spectra were referenced to the Au $4f_{7/2}$ signal at 83.96 eV. Data were analyzed using the program CasaXPS (version 2.3.5 www.casaxps.com). The signals were fitted using Gaussian–Lorentzian functions and Tauc asymmetry in the case of gold and least-squares-fit routines following Shirley iterative background subtraction. Sensitivity factors were calculated using published ionization cross sections[18] corrected for the angular asymmetry[19] and the attenuation-length dependence on kinetic energy.

(16) Tatsuma, T.; Kubo, W.; Fujishima, A. *Langmuir* **2002**, *18*, 9632–9634.

(17) Ranade, M. R.; Navrotsky, A.; Zhang, H. Z.; Banfield, J. F.; Elder, S. H.; Zaban, A.; Borse, P. H.; Kulkarni, S. K.; Doran, G. S.; Whitfield, H. J. *Proc. Natl. Acad. Sci. U.S.A.* **2002**, *99*, 6476–6481.

(18) Scofield, J. H. *J. Electron Spectrosc. Relat. Phenom.* **1976**, *8*, 129–137.

(19) Reilman, R. F.; Msezane, A.; Manson, S. T. *J. Electron Spectrosc. Relat. Phenom.* **1976**, *8*, 389–394.

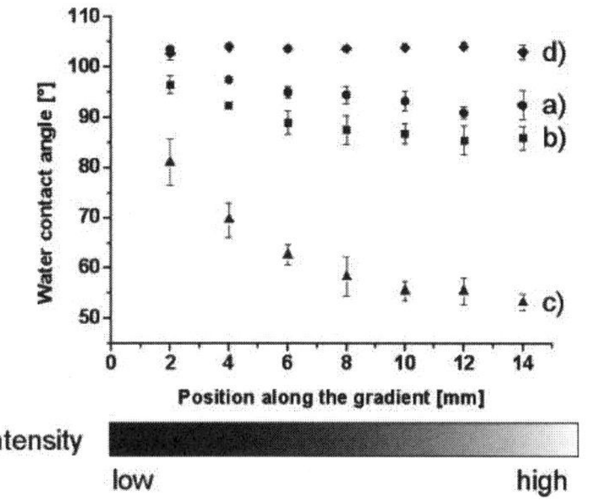

Figure 2. Water contact angles measured along the long gradient after different steps of the experiment. Results after (a) photocatalytic lithography, (b) photocatalytic lithography and rinsing with ethanol, (c) photocatalytic lithography, rinsing with ethanol, and backfilling with 11-mercapto-1-undecanol, and (d) rinsing a 1-dodecanethiol SAM with ethanol and backfilling with 11-mercapto-1-undecanol. The error bars correspond to an 80% confidence interval.

Results and Discussion

Gold-coated silicon wafers bearing a dodecanethiolate SAM were employed as substrates. A TiO_2-coated glass slide was placed on a 60 μm spacer on the gold surface, and the photomask was positioned above the assembly. Two kinds of gray-tone gradients were used: either a long 18 mm gray-tone gradient or a series of repeating short gradients, 720 μm long, as shown in Figure 4. The system was exposed to UV for 50 min, and the samples were then rinsed in ethanol. The SAM was backfilled by dipping the sample in a solution of 11-mercapto-1-undecanol. The long grayscale gradient was used initially to demonstrate the feasibility of the technique. The wettability gradients obtained were characterized by means of water contact angle measurements, performed after each step of the experiment. The results are presented in Figure 2.

Following the photocatalytic lithography step, the water contact angle was found to change only slightly along the gradient and was lower on the most exposed side of the sample. After the sample was rinsed with ethanol, a wettability gradient was observed, but the water-contact-angle changed only from 95 to 85°. The most hydrophilic part of the gradient corresponded to the most UV-exposed area. For comparison, a freshly prepared gold surface after rinsing in ethanol shows a water contact angle of 65°.[8] A much more pronounced gradient was obtained after backfilling the SAM with 11-mercapto-1-undecanol, when the water contact angle varied from 80 to 55°. The decrease in wettability was nonlinear, with the slope of the gradient decreasing toward the hydrophilic. A significant decrease was observed at distances between 2 and 8 mm from the low-UV-intensity end. At distances between 8 and 15 mm, the measurements were not significantly different, and a plateau was observed. To highlight the effect of photocatalysis, a control experiment was conducted where a homogeneous dodecanethiol SAM was not subjected to photocatalytic lithography but was rinsed with ethanol and backfilled with 11-mercapto-1-undecanol instead. As shown in Figure 2, no changes in contact angles were observed in this case.

XPS measurements were performed to gain a better understanding of the mechanism of photocatalytic degradation of the thiol SAM. Homogeneous samples were subjected to the process (without the gradient photomask) and XPS spectra measured at each step. The homogeneous sample thus corresponded to a position at the most hydrophilic end of the gradients. We mainly focused on the O 1s, C 1s, and S 2p regions of the spectrum. One critical step of the process was the backfilling of the oxidized SAM. The importance of this step was highlighted by backfilling the SAM with two different thiols: 11-mercapto-1-undecanol and 1H,1H,2H,2H-perfluorodecanethiol. When the SAM was backfilled with the perfluorinated thiol, the F 1s region of the spectrum was also measured.

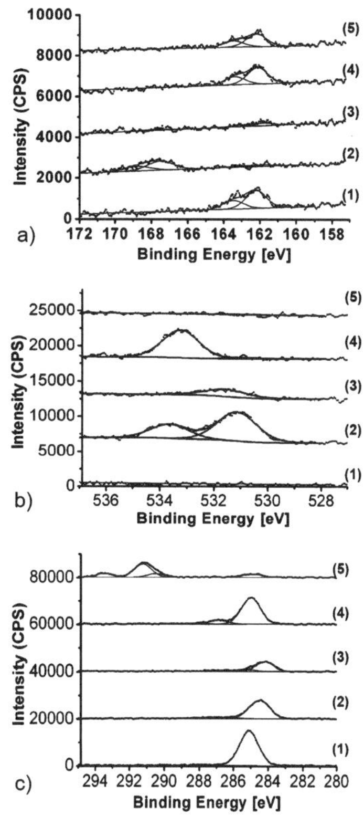

Figure 3. XPS spectra of (a) sulfur, (b) oxygen, and (c) carbon. The spectra were measured after each step of the process: (1) For the dodecanethiol SAM and following (2) photocatalytic lithography, (3) photocatalytic lithography and ethanol rinsing, (4) photocatalytic lithography, ethanol rinsing, and backfilling in 11-mercapto-1-undecanol, and (5) photocatalytic lithography, ethanol rinsing, and backfilling in 1H,H,2H,2H-perfluorodecanethiol. For clarity, the spectra are displaced along the *y* axis by 2000 CPS for S 2p, 6000 CPS for O 1s, and 20000 CPS for C 1s.

S2p spectra measured at each step of the process are shown in Figure 3a. For the dodecanethiol monolayer, one sulfur species was detected with a binding energy of the S $2p_{3/2}$ signal at 162.2 eV. After remote photocatalytic oxidation, a second sulfur peak appeared with a S $2p_{3/2}$ binding energy of 167.5 eV. Previous studies attributed this peak to the formation of sulfonate compounds.[20] After ethanol rinsing, the high-binding-energy sulfur peak disappeared as a result of the dissolution of the oxidized species in ethanol. Once the vacant sites were backfilled either with 11-mercapto-1-undecanol or 1H,1H,2H,2H-perfluorodecanethiol, the signal at 162.2 eV reappeared, confirming the adsorption of new thiols on the gold surface.

O 1s spectra are shown in Figure 3b for the same series of samples. As expected, no oxygen was detected for the dode-

(20) Brewer, N. J.; Janusz, S.; Critchley, K.; Evans, S. D.; Leggett, G. J. *J. Phys. Chem. B* **2005**, *109*, 11247−11256.

canethiol SAM. After remote photocatalytic oxidation, two peaks were observed: one at 531.1 eV and the other at 533.6 eV. Previous studies, focusing on the air oxidation of dodecanethiol, also reported the appearance of a peak at 531 eV.[21] The authors ascribed this peak to the formation of sulfonate species. As a comparison, the O 1s binding energy of 4-aminobenzenesulfonic acid was measured at 531.3 eV, and the S $2p_{3/2}$ binding energy was measured at 167.8 eV.[22] The second peak we observed may have arisen from a slight degree of oxidation of the alkyl chain on the thiol. Remote photocatalytic oxidation was already been reported to oxidize the alkyl chains of silane molecules efficiently.[23] Thus, there are two processes occurring in parallel during the experiment: the fast oxidation of the thiol headgroup to a sulfonate and the possibly slower oxidation of the alkyl chain. Slight oxidation of the gold surface also cannot be completely excluded. After ethanol rinsing, the amount of oxygen diminished significantly as a result of the dissolution of the sulfonate species. After backfilling with 11-mercapto-1-undecanol, a strong peak at 533.2 eV was observed again, confirming the adsorption of the hydroxyl-terminated thiol. When the SAM was backfilled with the perfluorinated thiol, no oxygen was detected.

C 1s spectra are shown in Figure 3c for the same series of samples. For the dodecanethiol SAM, a peak was observed at 285.1 eV, corresponding to the aliphatic carbon of dodecanethiol. After remote photocatalytic oxidation, a decrease in the peak intensity and a shift of 0.7 eV to lower binding energies were noticed. This effect was also reported by Willey et al. after air oxidation of dodecanethiol SAMs.[21] Ethanol rinsing led to a further decrease in the intensity of this peak. After backfilling with 11-mercapto-1-undecanol, the peak was observed again at 285 eV as a result of the adsorption of the new thiols. The shift in the C 1s peak is thought to be due to a slight shift in the surface potential of the gold as the electron-withdrawing S−Au bond is removed by oxidation and replaced by sulfonate species with little surface interaction (the molecules being subsequently held in place by interchain van der Waals interactions). The surface potential and C 1s peak return to the initial value following backfilling with the second thiol. Related effects have been observed in the case of fluorinated thiol gradients.[24]

Following backfilling with the perfluorinated thiol, a series of new peaks at 293.5 (CF_3), 291.2 (CF_2), 290.6 (CF_2 beside CH_2), and 284.9 eV (aliphatic carbon) were observed. This was confirmed by the presence of an intense peak at 688.5 eV in the F 1s spectrum of this sample (data not shown).

A summary of the normalized atomic elemental composition of each sample is shown in Table 1. The changes in the carbon signal after each step were analyzed in greater detail. In Figure 4, the ratio I_{C1s}/I_{Au4f} after each step are presented. As can be seen, this ratio strongly decreases after the UV illumination step. This may be linked to a minor degradation of the alkyl chains on the dodecanethiol during the process, which reduces the number of carbon atoms in the alkyl chains. A further decrease happens after rinsing with ethanol, meaning that some of the sulfonates species are removed, as described above in the discussion on the S 2p signals. After backfilling with 11-mercapto-1-undecanol or the perfluorinated thiol, the layer is restored and the ratio increases again. A smaller ratio is however observed when backfilling with the perfluorinated thiol. This lower carbon ratio can be partially explained by the smaller number of carbon atoms in the 1H,1H,2H,2H-perfluorodecanthiol than in 11-mercapto-1-undecanol.

Table 1. Elemental Composition Measured from the XPS Spectra of the SAM after Each Step of the Process[a]

	elemental composition (atom %)						
			O1s		S2p		
sample	Au 4f	C 1s	533.5 eV	531 eV	163 eV	168 eV	F 1s
1	44.2	52.7	0.0	0.0	3.1	0.0	0.0
2	52.4	35.2	3.2	6.8	0.3	2.0	0.0
3	76.9	20.1	0.1	2.1	0.6	0.2	0.0
4	45.4	45.9	5.7	0.2	2.8	0.0	0.0
5	29.7	24.0	0.0	0.0	1.5	0.0	44.8

[a] (1) Dodecanethiol SAM. (2) Following photocatalytic lithography. (3) Following photocatalytic lithography and ethanol rinsing. (4) Following photocatalytic lithography, ethanol rinsing, and backfilling in 11-mercapto-1-undecanol. (5) Following photocatalytic lithography, ethanol rinsing, and backfilling in 1H,H,2H,2H-perfluorodecanethiol.

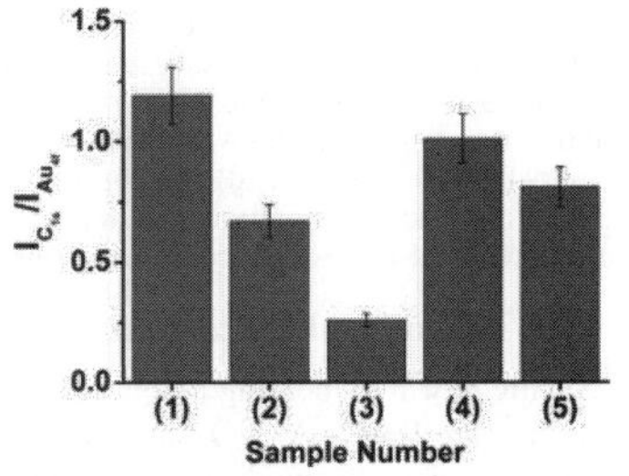

Figure 4. C 1s intensity divided by substrate intensity (Au 4f) measured after the five subsequent preparation steps. A clear reduction of the signal can be observed after (2) UV irradiation and (3) subsequent ethanol rinsing. The layers are restored by backfilling with (4) 11-mercaptoundecanol or (5) perfluorinated thiol, respectively. Error bars were estimated from the signal-to-noise ratios.

The results obtained from contact angle and XPS measurements suggest a potential degradation mechanism of the thiol monolayer. One clear aspect is the formation of sulfonate species due to the oxidation of thiols, which has been reported in many studies.[20,25,26] In parallel to thiol oxidation, some oxidation of the alkyl chains is happening. After the ethanol rinsing step, only the sulfonate species are removed from the SAM. This leads to an incomplete thiol SAM on the surface. During backfilling, the vacant sites exposed by the ethanol rinse are occupied by new thiol molecules, completing the SAM once more.

To fabricate a surface-chemical gradient, the TiO_2 layer was illuminated with a spatial gradient in UV intensity. Because the photocatalytic activity of TiO_2 depends on the intensity of UV light illumination,[16] the number of radicals created on the TiO_2 surface varied along the gradient. The SAM was therefore oxidized in a gradient. After the ethanol rinse step, the SAM was backfilled with the 11-mercapto-1-undecanol to obtain a more complete SAM exhibiting a composition gradient as illustrated in Figure 1. As shown in Figure 2, the gradients obtained were not linear, which may be undesirable for some practical applications. This issue is, however, readily addressed in our case by designing a nonlinear gray-tone gradient that would compensate for the nonlinearity of the process. The strength of this technique is

(21) Willey, T. M.; Vance, A. L.; van Buuren, T.; Bostedt, C.; Terminello, L. J.; Fadley, C. S. *Surf. Sci.* **2005**, *576*, 188−196.
(22) Harker, H.; Sherwood, P. M. *Philos. Mag.* **1973**, *27*, 1241−1244.
(23) Kubo, W.; Tatsuma, T. *J. Mater. Chem.* **2005**, *15*, 3104−3108.
(24) Zürcher, S.; et al. *J. Am. Chem. Soc.*, to be submitted for publication.
(25) Tarlov, M. J.; Burgess, D. R. F.; Gillen, G. *J. Am. Chem. Soc.* **1993**, *115*, 5305−5306.
(26) Dishner, M. H.; Feher, F. J.; Hemminger, J. C. *Chem. Commun.* **1996**, 1971−1972.

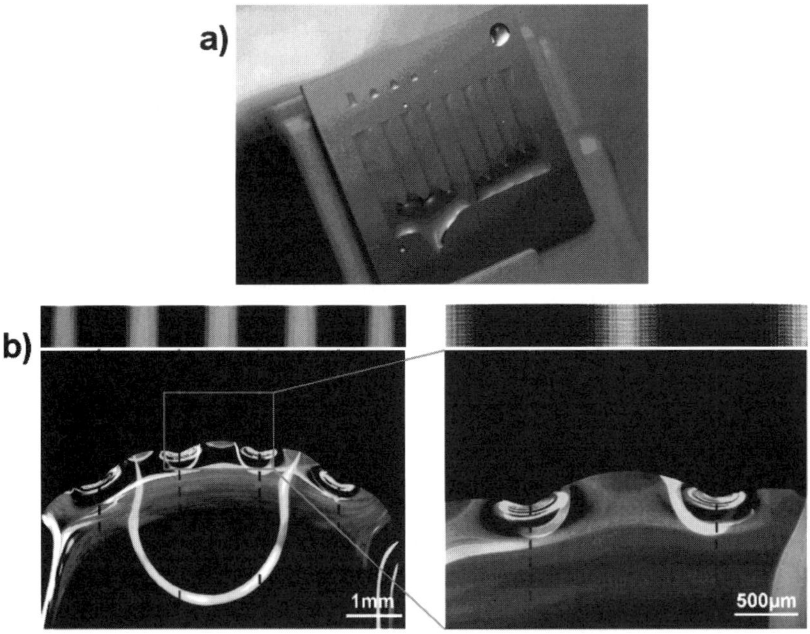

Figure 5. (a) Optical image of ethanol dewetting following photocatalytic lithography. (The sample was 1.5 cm wide.) (b) Dark-field images of a water droplet deposited on the surface bearing the repeating 720 μm gradients. The corresponding grayscale variations are shown above each image.

indeed its flexibility because the shapes, lengths, and slopes of the gradient may be adjusted at will by modifying the gray-tone gradients.

The final gradient of oxidized species will also depend on the photocatalysis parameters. Unfortunately, the underlying process of photocatalytic remote oxidation is not yet fully understood. The generation of radicals at the TiO_2 surface has been the subject of many studies.[27,28] It is well accepted that the main radicals created at the TiO_2 surface in the presence of oxygen are hydroxyl ($HO^•$) and superoxide radicals ($O_2^{•-}$ and $HO_2^•$). The generation of such radicals depends on the physical and chemical characteristics of the TiO_2 nanoparticles. As already reported in literature,[29] the size of the nanoparticles is important: smaller nanoparticles lead to a larger surface area, which increases the number of surface-active sites and thus the number of radicals generated. However, a decrease in particle size also means a higher rate of electron–hole recombination,[30] which is detrimental to the photocatalytic process. It has been found that there is actually an optimal nanoparticle size, corresponding to 10 nm, at which the highest photocatalytic efficiency is achieved.[29] Another critical aspect is the nanoparticle surface chemistry. Several studies have reported that the density of hydroxyl groups on the TiO_2 surface plays an important role in the photocatalytic process.[29] To have reproducible surface chemistry, the TiO_2 surfaces used in this study were subjected to oxygen plasma cleaning followed by UV exposure. One last important parameter is the crystallinity of the nanoparticles. It is well known that the brookite and rutile TiO_2 phases have lower photocatalytic efficiency than does the anatase form.[23] In our case, we used nanoparticles of anatase TiO_2 that were 7 nm in diameter.

The last part of this work concerns the fabrication of submillimeter wettability gradients. The method described above was repeated using the 720 μm gray-tone gradient. However, characterization could not be performed by means of standard contact-angle measurements because of the small gradient size. Instead, the wetting properties of ethanol and water on the gradients were used to obtain a qualitative view of the gradient (Figure 5). Upon dipping the samples in ethanol after photocatalytic remote oxidation, ethanol dewetted from the hydrophobic areas of the sample but wetted the hydrophilic, oxidized regions. Because the gradient was periodic, this led to stripes of ethanol. As mentioned above, no significant effect on the water contact angle was observed immediately following photocatalytic remote oxidation. However, after rinsing with ethanol and backfilling with 11-mercapto-1-undecanol, the shape of a sessile water drop revealed the wettability changes along the gradient (Figure 5c). On the hydrophobic regions, the drop dewetted more than on the hydrophilic areas. Because the short wettability gradients repeated along the sample, this resulted in the formation of a wavelike triple line. The shape of the sessile drop demonstrates that this technique allows the wettability of the substrate to be modified on the submillimeter scale.

The limits of lateral resolution in this technique will determine the minimal gradient length achievable. The gradient in chemistry is here obtained from a gradient in UV-light intensity. The length of the gray-tone gradient used for that was 720 μm. The first limitation is linked to the fabrication of a gray-tone gradient on a smaller scale. We used standard photolithography processes that may not be suitable for the fabrication of gradients smaller than 10 μm. To overcome this problem, the photomask could be fabricated, for instance, using an e-beam pattern generator and high-energy-beam-sensitive glass (HEBS).[31] The problems of resolution in photocatalytic lithography have already been

(27) Lee, N. C.; Choi, W. Y. *J. Phys. Chem. B* **2002**, *106*, 11818–11822.
(28) Ishibashi, K.; Nosaka, Y.; Hashimoto, K.; Fujishima, A. *J. Phys. Chem. B* **1998**, *102*, 2117–2120.
(29) Carp, O.; Huisman, C. L.; Reller, A. *Prog. Solid State Chem.* **2004**, *32*, 33–177.
(30) Zhang, Z.; Wang, C. C.; Zakaria, R.; Ying, J. Y. *J. Phys. Chem. B* **1998**, *102*, 10871–10878.

addressed by Kubo et al.[15] These authors used photocatalytic lithography to make surface patterning and reported resolutions of 10 μm. The fabrication of shorter gradients than described here can thus be envisaged. A final limitation arises from the lateral diffusion of the radicals through the 60 μm air gap. The resolution could readily be improved by decreasing the air gap between the TiO_2 layer and the substrate. Reducing the gap significantly below that used in the current work would, however, necessitate using clean-room conditions because of the possibility of dust lodging in the air gap and giving a nonreproducible and nonuniform gap.

Conclusions and Outlook

Both centimeter- and submillimeter-long surface-chemical gradients have been obtained using a simple technique that combines photocatalytic and grayscale lithography. The creation of surface-chemical gradients was demonstrated by gradually degrading alkanethiol SAMs and backfilling to produce wettability gradients. The technique may be applied to other combinations of thiol end-groups or, indeed, to other surface chemistry entirely. To date, the modification of surface wettability by photocatalytic lithography has also been shown for systems such as silane SAMs and some polymer surfaces,[23] correspondingly increasing the possible range of applications.

The shape and length of the gradients are controlled via the design of the grayscale gradient on the photomask. This approach may easily be extended to other gradient shapes (for instance, radial gradients) and other gradient slopes simply by using a different photomask. One interesting point is the size of the shortest gradient that can be obtained: previous studies using photocatalytic lithography to pattern surfaces reported resolutions of 10 μm[15], which suggests that gradients approaching such scales may be feasible.

The chemistry of the lithographic process has not yet been fully elucidated. However, several parameters are expected to influence the final gradients, such as the precise structure and chemistry of the TiO_2 layer, the humidity of the ambient air,[27,32] and the size of the air gap between the substrate and the photocatalytic layer.[33] Optimization of these parameters should allow better control of the final chemical gradient shape, slope, and resolution.

Acknowledgment. We thank Dr. Emmanuel Scolan and Dr. Rolf Steiger for useful scientific discussions. We also acknowledge Arrayon Biotechnology, Dr. Harry Heinzelmann, and Dr. Raphaël Pugin for their support of this work.

LA063186+

(31) Däschner, W.; Long, P.; Stein, R.; Wu, C.; Lee, S. H. *Appl. Opt. IP* **1997**, *36*, 4675–4680.

(32) Kozlov, D. V.; Panchenko, A. A.; Bavykin, D. V.; Savinov, E. N.; Smirniotis, P. G. *Russ. Chem. Bull.* **2003**, *52*, 1100–1105.

(33) Tatsuma, T.; Tachibana, S.; Miwa, T.; Tryk, D. A.; Fujishima, A. *J. Phys. Chem. B* **1999**, *103*, 8033–8035.

Langmuir **2007**, *23*, 5929–5935 5929

Functionalizable Nanomorphology Gradients via Colloidal Self-Assembly

Christoph Huwiler,[†] Tobias P. Kunzler,[‡] Marcus Textor,[‡] Janos Vörös,[†] and Nicholas D. Spencer*[,‡]

Laboratory of Biosensors and Bioelectronics, Institute of Biomedical Engineering, Department of Information Technology and Electrical Engineering, ETH Zürich, Gloriastrasse 35, CH-8092 Zurich, Switzerland, and Laboratory for Surface Science and Technology, Department of Materials and Materials Research Center, ETH Zürich, Wolfgang-Pauli-Strasse 10, CH-8093 Zürich, Switzerland

Received January 8, 2007. In Final Form: February 26, 2007

We present a novel approach for the fabrication of tailored nanomorphology gradients on metal oxide surfaces. We first show the direct formation of a nanocolloidal density gradient by a dip-coating process. The obtained silica nanoparticle gradients are then subjected to a heat treatment. Control of this sintering step allows the precise tailoring of the particle morphology on the surface. Both these processes together provide a new tool to form precise, tunable, and material-independent nanomorphology gradients.

Introduction

Surfaces with a continuously changing surface parameter (gradient surfaces) have distinct advantages for certain applications. Such surfaces, for example, allow rapid screening tests or combinatorial diagnostic studies to be performed on a single sample, which is of particular importance in biotechnology or medicine. The surface parameters of interest range from crystallinity[1] to porosity[2] and surface chemistry. Among surface-chemistry parameters that have been varied along gradients, wettability has been reported most frequently. In these gradients, the wettability changes gradually along the length of the sample because of a controlled change in surface chemistry. Various methods have been developed to produce such wettability gradients.[3−6]

Another parameter that plays a crucial role, especially in many bioscience applications, is surface morphology. It has been known for a long time that surface roughness affects the biological response to surfaces (e.g., cell adhesion, proliferation, and differentiation). Also, parameters related to surface-contact effects (such as tribological or adhesion phenomena) will be influenced by a varying surface topography. However, only little progress has been made in producing large-scale topographical gradients. Crystallinity gradients of a polymer film induced by a temperature-gradient stage have been described previously,[1] as have gradients of pore size in silicon, fabricated by etching in an anisotropic electric field.[2] Only very recently, an approach was presented by Kunzler et al. that allows the production of stochastic roughness gradients over centimeter length-scales with topographical features in the nanometer and micrometer range.[7]

A different approach to the production of surface morphology gradients is to capitalize on the self-assembly potential of colloidal particles. To date, the only particle-gradient structures produced have been colloidal crystals for use as photonic band gap materials. Most of these colloidal crystal gradients are fabricated by means of polymer-infiltration techniques, where a polymer with a gradually changing refractive index is forced into the interstices of a colloidal crystal.[8,9] It was not until recently that the first colloidal crystal gradients were produced.[10−12] An indirect way of producing a number-density gradient of colloidal particles was presented by Bhat et al. They decorated a molecular gradient (of amino groups) with colloidal gold particles to obtain nanocolloidal density gradients.[13] A similar approach was also presented by the same group to produce gold nanoparticle density gradients in two orthogonal directions.[14]

In this article, an alternative way of producing a particle-density gradient on a surface using nanosized colloidal particles is presented. In a simple dip-coating process, a gradient in nanoparticle density is achieved directly on the substrate. The number density can be varied from a high-coverage, random monolayer at one end of the gradient to only a few particles per square micrometer on the other. Not only is this method very universal as far as substrate and particle materials are concerned, but we also show how the shape of the colloidal particles on the substrate can be varied over a wide range using a heat-treatment step. This heat-treatment step simultaneously increases the adhesion of the particles to the substrate, providing the necessary robustness for such particle gradients to be used in replication techniques or in experiments that require a robust morphology gradient (e.g., in cell biology experiments).

* To whom correspondence should be addressed. E-mail: spencer@mat.ethz.ch.

[†] Laboratory of Biosensors and Bioelectronics, ETH Zürich.

[‡] Laboratory for Surface Science and Technology, ETH Zürich.

(1) Washburn, N. R.; Yamada, K. M.; Simon, C. G.; Kennedy, S. B.; Amis, E. *Biomaterials* **2004**, *25*, 1215.

(2) Karlsson, L. M.; Tengvall, P.; Lundström, I.; Arwin, H. *J. Electrochem. Soc.* **2002**, *149*, C648.

(3) Elwing, H.; Welin, S.; Askendal, A.; Nilsson, U.; Lundstroem, I. *J. Colloid Interface Sci.* **1987**, *119*, 203.

(4) Lee, J. H.; Kim, H. G.; Khang, G. S.; Lee, H. B.; Jhon, M. S. *J. Colloid Interface Sci.* **1992**, *151*, 563.

(5) Liedberg, B.; Tengvall, P. *Langmuir* **1995**, *11*, 3821.

(6) Morgenthaler, S.; Lee, S.; Zürcher, S.; Spencer, N. D. *Langmuir* **2003**, *19*, 10459.

(7) Kunzler, T. P.; Drobek, T.; Sprecher, C. M.; Schuler, M.; Spencer, N. D. *Appl. Surf. Sci.* **2006**, *253*, 2148.

(8) Park, J.-H.; Choi, W. S.; Koo, H. Y.; Kim, D.-Y. *Adv. Mater.* **2005**, *17*, 879.

(9) von Freymann, G.; John, S.; Kitaev, V.; Ozin, G. A. *Adv. Mater.* **2005**, *17*, 1273.

(10) Abkarian, M.; Nunes, J.; Stone, H. A. *Langmuir* **2004**, *126*, 5978.

(11) Li, J.; Han, Y. *Langmuir* **2006**, *22*, 1885.

(12) Zhang, J.; Xue, L.; Han, Y. *Langmuir* **2005**, *21*, 5667.

(13) Bhat, R. R.; Fischer, D. A.; Genzer, J. *Langmuir* **2002**, *18*, 5640.

(14) Bhat, R. R.; Tomlinson, M. R.; Genzer, J. *Macromol. Rapid Commun.* **2004**, *25*, 270.

Published on Web 04/21/2007

Experimental Section

Colloidal Gradient Preparation. Negatively charged silica particles will readily adsorb onto a positively charged surface by electrostatic attraction. To provide such a positively charged surface, a thin film of poly(ethylene imine) (PEI) (a branched, cationic polyelectrolyte polymer,[15] Sigma-Aldrich, Germany) was adsorbed onto a silicon wafer surface, previously cleaned by ultrasonication (10 min in isopropanol and then in Millipore water for another 10 min). The samples were then oxygen plasma treated at radio frequency (RF) level "Hi" (PDC-002, Harrick Plasma, Ithaca, United States) for 3 min. Cleaned samples were immersed in 1 mg/mL filtered PEI solution in ultrapure water for 30 min. After adsorption of the PEI electrolyte, the silicon wafer pieces were rinsed with ultrapure water, were blown dry in a nitrogen jet, and were stored in a closed container at room temperature. Measurements of PEI-coated particles exhibited a positive ζ-potential of around 30 mV at neutral pH, indicating that a PEI-coated surface is indeed positively charged at neutral pH values.

The nanoparticle gradients were produced using a suspension with silica particles with an average diameter of 73 nm (Klebosol, Clariant, France). The particle diameter was determined using scanning electron microscopy (SEM) and X-ray disc centrifuge methods. ζ-Potential experiments were conducted over a large pH range, and the surface charge (ζ-potential) was -60 mV at a pH of 7. The original, unfiltered 30 wt % stock suspension was homogenized for 5 min in an ultrasonic bath and was diluted to 1 wt % with ultrapure water, HEPES 1 (10 mM 4-(2-hydroxyethyl)piperazine-1-ethane-sulfonic acid, adjusted to pH 7.4 with 6 M NaOH) or HEPES 2 (HEPES 1 with 150 mM NaCl). This 1 wt % suspension was further diluted with ultrapure water, HEPES 1 or HEPES 2, to a final 0.002 wt % working suspension. Before use, the working suspension was homogenized again for 5 min in an ultrasonic bath. No ion exchange was performed prior to experiments.

The PEI-coated substrate was lowered into the colloidal suspension with a linear-motion drive (Owis Staufen, Germany). The immersion profile was set to $x(t) = a(t - u)^2$, where $x(t)$ is the position on the gradient at time t, which was running from 0 to 1800 s, $u = -100$ s (a parameter needed only for technical reasons), and $a = -3 \times 10^{-6}$ m/s^2. This setup results in the creation of a gradient over 10 mm after a total immersion time of 30 min. The suspension was kept slightly agitated during the procedure. Upon completion of the immersion process, the beaker containing the suspension was immediately flushed with about 1000 mL of distilled water to remove the particle suspension and to prevent further particles adsorbing onto the substrate. Afterward, the substrate was rinsed with ultrapure water and was blown dry with a nitrogen jet.

Sintering of Colloidal Gradients. Sintering of the gradient substrates was carried out in a high-temperature furnace under ambient atmosphere. Substrates were heated at 10 °C/min to final temperatures of either 1075, 1100, 1125, 1150, 1175, or 1200 °C, were held for 2 h, and then were cooled down at 1.5 °C/min to room temperature. To determine the "contact angles" shown in Figure 7, SEM images were analyzed using image analysis software (image J). The sintered gradient samples were tilted in the SEM to allow for a side view of the surface and to achieve a more reliable estimation of the contact angle.

Cell Culture. To achieve a more cell-compatible surface environment, the entire particle gradient was dipped into a PLL-*g*-PEG-RGD solution. The RGD-peptide sequence is end-grafted to the PLL-*g*-PEG copolymer and is known to promote cell attachment by interacting with cell integrins.[16,17] Rat calvarial osteoblasts (RCO) were cultivated for 7 days on a particle gradient with the following seeding conditions: 6000 cells/cm^2 in α-DMEM with 10% fetal bovine serum and 1% antibiotics medium, incubation at 37 °C, 7% CO_2, 100% humidity. After 7 days of culture, the cells were fixed, and the nucleus was stained with DAPI and the actin network with phalloidin as described by Kunzler et al.[18]

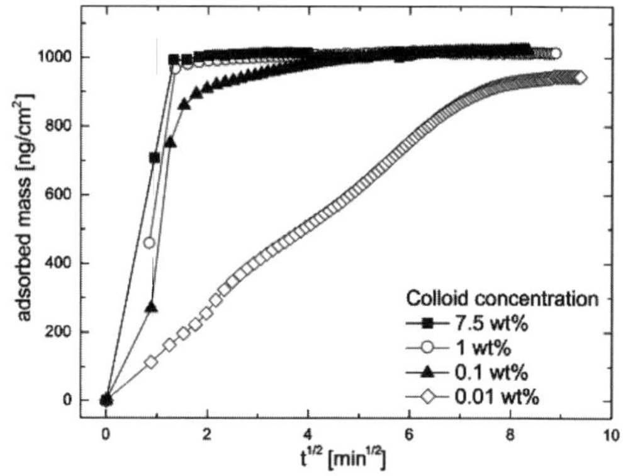

Figure 1. Influence of colloid concentration on the adsorption behavior of 73 nm silica colloids buffered in HEPES 1 (10 mmol salt concentration) at pH 7.4 on a poly(ethylene imine) coated silicon wafer observed by OWLS (optical waveguide light-mode spectroscopy). The negatively charged silica nano colloids adsorb electrostatically onto the positively charged PEI coated surface. OWLS is a surface-sensitive optical detection technique used to monitor adsorption processes in situ.

Results and Discussion

Silica particles, 73 nm in diameter, adsorb onto a PEI-coated silica surface in a controllable way, as shown in Figure 1. Depending on the concentration of the colloidal particles in the suspension, a monolayer of particles adsorbs onto the PEI-coated silica surface very rapidly, over a time span of several minutes. With a particle concentration of more than 1 wt %, adsorption of a monolayer takes place in less than a minute. For dilute suspensions, say 0.01 wt %, adsorption requires in excess of 1 h to produce a complete monolayer. The extent of this adsorption process is proportional to $\sqrt{t}$, as discussed later in this section.

The adsorption of colloidal particles by electrostatic interactions may be viewed as a one-step process, where a particle adsorbs electrostatically onto the substrate and may no longer be moved on the surface. Such processes are often referred to as irreversible "random sequential adsorption" (RSA),[19,20] where particles "stick" to a surface and cannot be moved after adsorption. As a consequence, a jamming limit exists for the particle coverage, which is found to be 0.547.[21] This means that the theoretical maximum surface coverage for the system used in this work may not exceed 54.7%. However, this value may, in reality, not be reached, since the particles are not only defined by their "physical" size, but also by the electric double layer around them. This double layer has a finite thickness, which must be added to the particle's physical diameter. Thus, the longer ranging the electric double layer, the lower the particle surface coverage at the jamming limit. The thickness and the numeric value of the electric double layer can readily be influenced by pH and ionic strength. In Figure 2, a schematic representation of the RSA model is given and corresponding SEM images are shown to illustrate the situation in our real system. At the jamming limit (Figure 2b), no more particles (or disks) may be deposited without moving particles on the surface (some of the forbidden particle positions

(15) Pericet-Camara, R.; Papastavrou, G.; Behrens, S. H.; Helm, C. A.; Bokovec, M. *J. Colliod Interface Sci.* **2006**, *296*, 496.
(16) Ruoslahti, E. *Annu. Rev. Cell Dev. Biol.* **1996**, *12*, 697.
(17) Schuler, M.; Owen, G. R.; Hamilton, D. W.; De Wild, M.; Textor, M.; Brunette, D. M.; Tosatti, S. *Biomaterials* **2006**, *27*, 4003.
(18) Kunzler, T. P.; Drobek, T.; Schuler, M.; Spencer, N. D. *Biomaterials* **2007**, *28*, 2175.
(19) Evans, J. W. *Rev. Mod. Phys.* **1993**, *65*, 1281.
(20) Hinrichsen, E. L.; Feder, J.; Jossang, T. *J. Stat. Phys.* **1986**, *44*, 793.
(21) Feder, J. *J. Theor. Biol.* **1980**, *87*, 237.

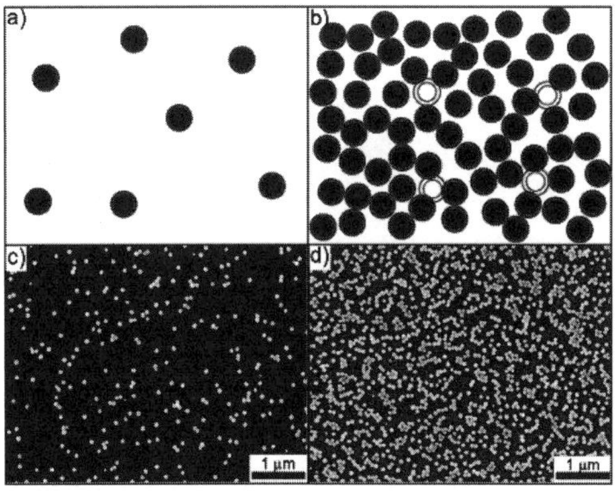

Figure 2. Random sequential adsorption model sketch and SEM images illustrating two stages of the process. (a) After short times (and depending on particle concentration), a few particles have irreversibly adsorbed to the surface. They cannot be moved on the surface, since they are strongly bound by electrostatic forces. (b) At or near the jamming limit of the RSA model: no further particles can be deposited, since they would overlap (some of the forbidden positions are indicated with red, hollow spheres). The effective "radius" of the particle consists not only of the particle itself but also of the ion cloud that it carries (symbolized by the gray ring around each black particle). This electric double layer (characterized by the Debye length) can be changed by varying pH or ionic strength of the solution. Increasing the Debye length therefore lowers the actual particle density on the surface. (c, d) SEM images of 73 nm silica particles adsorbing on a positively charged PEI surface. (c) After short times, a few particles have adsorbed on the surface. At this stage, some deviation of the RSA model can be observed, such as clustering of a few particles. (d) After longer adsorption times (30 min), a complete RSA "monolayer" forms as shown in b. Also in this image, deviations between the ideal situation in b and the system used in this work, d, can be seen. These deviations (clustering of particles and free space on the surface) are artifacts from the drying process, which modifies the pattern formed in suspension.

are indicated). In real systems, however, it is rarely the case that particles are immovable on the surface. This is the most common deviation from RSA models. For example, in the SEM images of Figure 2, some particles are clustered together, and as a consequence, some free space is produced on the surface. However, the particles can only move under the influence of capillary forces during drying. As long as the particles are in suspension, they will randomly adsorb, as predicted by the RSA theory, and will stick to the surface by electrostatic forces (however, many other factors might be responsible for deviations from a "pure" RSA model in this case, as mentioned later). During drying, capillary forces (stemming from water bridges that form between two particles) may be as high as the electrostatic forces holding the particles in place and therefore may move some particles together and form clusters of particles, as observed in Figure 2.

As mentioned above, two main parameters have long been identified to influence particle adsorption in cases where electrostatic interactions dominate the adsorption process: pH and ionic strength. These parameters were also tested in the silica nanoparticle−PEI system that was used in this work. It was found that particles near the isoelectric point (IEP) (at a pH of about 2.3), tend to agglomerate and form multilayers on the substrate, while with increasing pH, the particle concentration on the surface decreases because of the higher charge of the silica particles at higher pH values. Similar trends were found when lowering the ionic strength of the suspension from 160 mmol to a pure water suspension. Fewer and fewer particles adsorb to the positively charged surface because the electric double layer becomes decreasingly shielding as fewer ions are present in the suspension. This "shielding" effect of the double layer is often expressed by the Debye length, a parameter describing the "thickness" of the electric double layer. In our system, the Debye length is below 1 nm for a 160 mmol HEPES (pH 7.4 buffer) salt suspension (NaCl) and around 30 nm for a "pure water" suspension (we assume a minimum salt concentration of around 100 μmol for practical reasons (ions from stock solution, contamination from the air and handling, etc.))

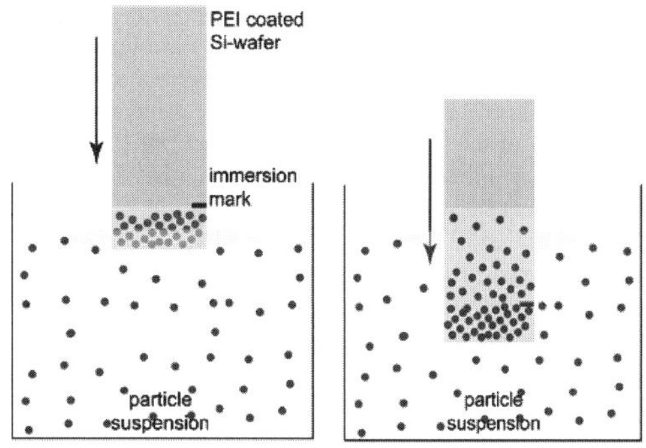

Figure 3. Schematic drawing of the situation at the commencement of the dip-coating process. The sample is immersed a few millimeters into the suspension to avoid unfavorable edge effects and to provide an area where a dense RSA monolayer can form. The immersion mark placed on the sample indicates the line from which on the actual dip-coating process starts.

Colloidal gradients were prepared by successively exposing a surface to a colloidal suspension with known properties for a specific time interval. If the surface and the particles interact electrostatically, particles will start to adsorb and form an incomplete particle layer, the number adsorbed depending on the time for which the particles were allowed to adsorb. This process is sketched in Figure 2, which (a) shows the situation after short adsorption times and (b) represents the jamming limit of the monolayer (which is reached after a longer time interval, depending on particle concentration).

The challenge to form a colloidal topographical gradient on a substrate can therefore be tackled by exposing different parts of the substrate for increasing times to a colloidal suspension. This is most conveniently done by simply dip-coating the substrate into the colloidal suspension at a distinct speed profile. By doing so, the end of the substrate that is immersed initially remains in the suspension much longer than the end of the substrate that is only immersed during the last part of the dipping process. Removing a sample from the suspension leads to the formation of a drying front, which both disturbs the colloidal structure itself and, more importantly, causes additional colloidal particles to be dragged to the three-phase contact line. In this case, particles are not only adsorbed by electrostatic interactions but also by capillary forces, an undesired and less controllable process. Therefore, the substrate sample has to be immersed into the nanocolloid suspension. In our experiments, the substrate was immersed about 5 mm into the suspension and was held there for 30 min. Only then was the sample immersed into the suspension at a given speed profile. Figure 3 shows a schematic of the situation prior to the start of the dip-coating process on the left and the situation at the end of the dip-coating process on the right. This procedure has two advantages: first, it minimizes the influence of edge effects that occur when a sample is immersed into the suspension. For example, a "jump" occurs when the sample first touches the suspension surface and contacts the substrate: the water film wets the sample very rapidly to form the water meniscus as depicted schematically in Figure 3. If the sample were not immersed a few millimeters into the suspension prior to the start of the experiment, such edge effects may affect the formation of the particle gradient. Second, the area where the sample was initially immersed provides a useful reference region to subsequently determine the jamming limit of the particle suspension, since this area is maintained for a significant length of time in the suspension, and thus a RSA monolayer of particles was allowed to form.

Another critical point is the end of the dip-coating process. To avoid drying effects as much as possible, the sample was not removed through the suspension−air interface. At the end of the dip-coating process, when the sample is immersed completely into the suspension (right-hand side of Figure 3), the beaker with the suspension is carefully flushed with extensive amounts of water. The beaker overflows and the suspension is rapidly diluted. After the suspension is diluted so much that ideally no more particles are present, the sample is removed from the suspension, is rinsed with more water, and is dried under a nitrogen flow. By means of this process, the influence of capillary forces during the drying process can be minimized and only a few particles are allowed to adsorb during the drying step.

Capillary forces are of great importance in such systems and are comparable in magnitude with the attractive electrostatic interactions between particle and surface. The drying step must therefore be controlled as described above. It is not an easy task to assign numbers to the capillary and electrostatic interaction energies because of assumptions that are made in the existing models, which are not always fulfilled in real systems. Nevertheless, the attractive capillary interaction energy between two particles was calculated to be about 4.4×10^6 kT for our system (on the basis of equations derived from refs 22−24). The electrostatic interaction energy between the surface (assuming a ζ-potential of 100 mV) and a particle (ζ-potential of −65 mV, as measured) is around 10^2 kT (upon contact of the two, rapidly decaying if the separation distance is increased).[25] This demonstrates that capillary forces are very strong compared to the electrostatic interaction energies. In fact, judging from the calculated values, they are several orders of magnitude higher than the electrostatic attraction of the particles to the surface. This is especially true if the particles are only separated by small distances. Capillary forces therefore gain in importance at high colloid densities on the surface, while they are relatively insignificant at low particle concentrations on the surface. These numbers also rationalize why very often in these experiments, doublets, triplets, or small clusters of particles are to be found on the surface instead of well-separated particles. As soon as two, three, or more particles are electrostatically adsorbed at relatively small separations, it is almost inevitable that these particles cluster together, since the capillary forces in these situations significantly exceed the electrostatic interactions.

Figure 4 shows an example of a colloidal gradient produced with 73 nm silica particles at a concentration of 0.002 wt % in pure water. An almost perfectly linear increase in particle concentration along the 1 cm gradient can be observed. Only a very few particles were adsorbed at one end, while almost a complete RSA-monolayer was formed on the other end. While there is virtually no visible influence of capillary forces on the low particle concentration images, clustering of particles increases

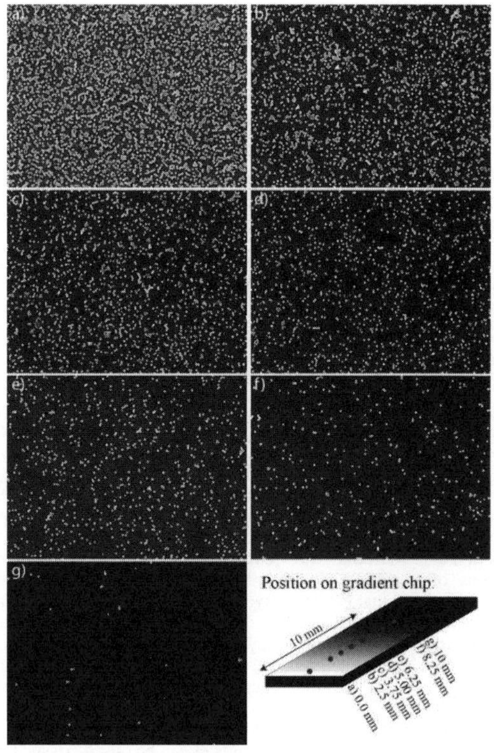

Figure 4. SEM images of a colloidal gradient and a sketch of the position of these images on the 1 cm gradient. Particle diameter was 73 nm and a particle concentration of 0.002 wt % was used in a pure water suspension. The suspension was slightly stirred during the dip-coating process, which lasted for 30 min with a designed immersion profile. The immersion profile was set to $x(t) = a(t - u)^2$, where $x(t)$ is the position on the gradient at time t, which was running from 0 to 1800 s, $u = -100$ s (a parameter needed only for technical reasons), and $a = -3 \times 10^{-6}$ m/s^2.

as the concentration of particles on the surface increases. This behavior is consistent with the force estimations made above, since as soon as the interparticle distance decreases, capillary forces gain in importance and tend to influence the resulting pattern more strongly. The immersion profile used (see figure caption in Figure 4) has a second-order polynomial shape to account for the $\sqrt{t}$ dependence of the diffusion-controlled particle adsorption kinetics observed in Figure 1. The combination of a second-order polynomial immersion profile and the $\sqrt{t}$ dependence of the particle adsorption kinetics would be expected to lead to a linear particle gradient. Stirring of the suspension during the process has a distinct influence on the particle gradient formation (causing far more particles to adsorb, in comparison to static conditions). Without stirring, particle coverage on the surface reached no more than 5%, while in the case of the gradient experiments, more then 30% coverage was obtained.

In the real system, this linearization of the particle gradient proved to work rather successfully, as observed in Figure 5. Here, SEM images were analyzed and statistical examination of these images yielded the particle concentration as a function of the position on the gradient (see Figure 5). It is observed that the particle gradients evolve nearly linearly over the entire 1 cm gradient. The gradient can be expanded to 2 cm by simply adapting the parameters such that instead of 10 mm, 20 mm will be dip-coated during the 1800 s. The value for the highest particle

(22) Aizenberg, J.; Braun, P. V.; Wiltzius, P. *Phys. Rev. Lett.* **2000**, *84*, 2997.
(23) Kralchevsky, P. A.; Nagayama, K. *Langmuir* **1994**, *10*, 23.
(24) Paunov, V. N.; Kralchevsky, P. A.; Denkov, N. D.; Nagayama, K. *J. Colloid Interface Sci.* **1993**, *157*, 100.
(25) Ninham, B. W. *Adv. Colloid Interface Sci.* **1999**, *83*, 1.

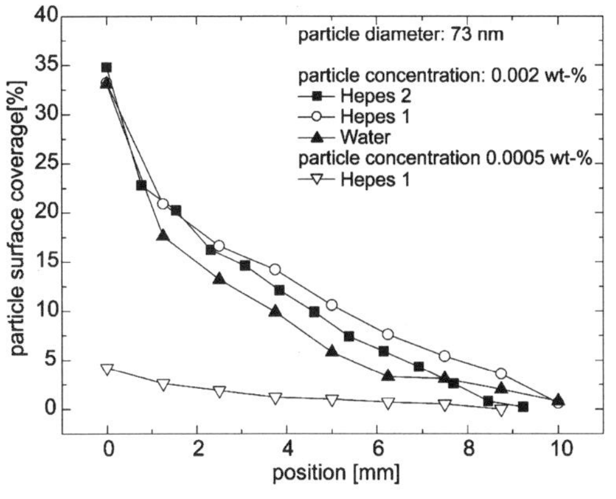

Figure 5. Colloid coverage of selected 1 cm gradients as a function of position on the gradient. Statistical evaluation of SEM images from different positions on the colloidal gradients yields the colloid coverage over the whole gradient. Particle suspensions with 0.002 wt % silica particles (73 nm in diamter) were used in different solvents. There is no significant difference between HEPES 1 and 2 buffered suspensions and ultrapure water suspensions (which were at pH 9.9). The end coverage reached values between 32 and 37%. The end coverage was measured in the area below the "immersion mark" indicated in Figure 3. This explains why this first point (at position 0 cm) is slightly higher than expected from the other values on the gradient, which lie on a rather straight line. If particle concentration was 4 times lower than for the first three experiments, the end coverage was significantly lower (around 5%), but the gradient still shows very linear behavior but in a narrower coverage range.

coverage (at position 0 cm) is taken within the area below the "immersion mark" (see Figure 3) and is therefore a little higher than one would expect from the extrapolation of the other values. This first value can be considered to be the particle coverage near the jamming limit and was found in this work to be in the range of 32−37%. This value is considerably lower than the theoretical value of 54.7% found in RSA-models. However, it must be remembered that in this theoretical model, no electric double layer is taken into account, which must be added to the particle diameter. It must also be stressed that the adsorption process of such small particles is influenced by many other factors, such as Brownian motion, the exact nature of the polymeric adlayer (bridging effects, mobility of particles on a polymeric adlayer), surface charge distribution, and others, and thus, it is not possible to attribute the adsorption process to a simple RSA-model. Also, capillary forces disturb the pattern formation to some extent near the jamming limit, such that statistical analysis underestimates the true particle coverage.

Another convenient way of changing the morphology of the produced particle gradients besides simply changing the particle size is to expose the colloidal gradient to a heat-treatment procedure. This will have two effects on the particle gradients. For one, all organic components used during the production of the gradient will be burned off, and second, the colloidal particles will start to sinter onto the substrate. The sintering process will thus increase mechanical stability of these gradients considerably, and by controlling the sintering conditions, the morphology of the particles may be changed in a predictable way. Figure 6 shows the influence of such a sintering treatment on the evolution of the particle morphology. During the heat-treatment step, the temperature was always maintained at the chosen sintering temperature for 2 h. Surface diffusion in the silica nanoparticles was activated at temperatures around 1100 °C. Below 1075 °C, there was no evidence of sintering and the particles remained in

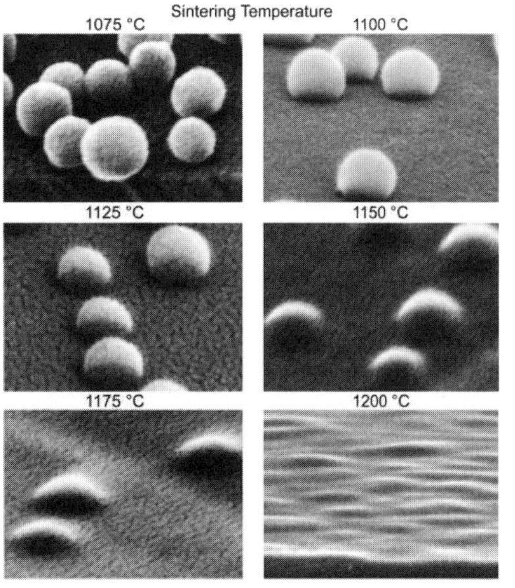

Figure 6. SEM images of heat-treated colloidal silica particles (73 nm in diameter) (tilted view). Heat treatment consisted of heating the samples to the desired temperature (with about 10 °C/min), holding at the end temperature for 2 h, and passive cooling (which took several hours). In the interesting temperature regime (1075−1200 °C), particle morphology changes continuously with increasing sintering temperature, as observed in the images. Below 1075 °C, the temperature is not high enough to induce surface diffusion on the silica particles and no change in particle morphology was seen. Temperatures above 1200 °C are sufficient to sinter the particles completely together with the oxidized silicon wafer surface, while at 1200 °C traces of the particles are still seen.

a spherical shape on the surface. With increasing temperature, the particles began to sinter to the surface. At 1100 and 1125 °C, the effects are not as dominant, but a slight "neck formation" was already evident. This neck formation is typical for the onset of the sintering process and is associated with the diffusion of surface atoms from the colloidal particle to the contact region of the particle with the substrate. The surface diffusion of atoms alone will not lead to shrinkage of the particles as observed for temperatures higher than 1150 °C. At these temperatures, other diffusion mechanisms are also active, such as volume diffusion (matter is transported from the bulk of the particle to the neck region) and diffusion along the particle−substrate interface. Both of these diffusion-transport mechanisms are responsible for the shrinkage of the particles at higher temperatures. The driving force for these sintering/diffusion processes is related to the minimization of the free surface energy that can be achieved as the spherical particles gradually minimize their surface area. The end point of this process is reached for temperatures above 1200 °C, where the particles lose their shape and completely diffuse into the surface. Figure 6 illustrates these steps and depending on the sintering conditions, the morphology of the particle gradient can be tailored. In Figure 7, the apparent contact angle of the particle with respect to the surface as determined from SEM image analysis is plotted versus the sintering temperature. A linear relation between the two is found, which allows the precise tuning of the morphology of our particle gradients. Corresponding to this change in contact angle, a change in the height of the colloidal particles is observed. The particle height changes in a similar way as the contact angle. At low temperatures, the particle height is equal to the particle diameter

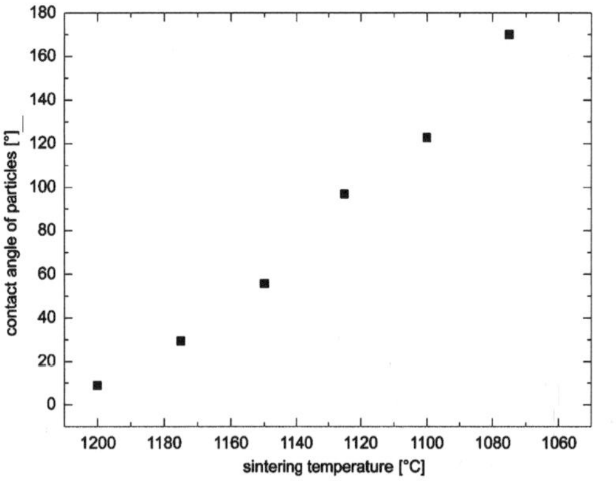

Figure 7. Graph of apparent contact angle of particle on a colloidal gradient versus sintering temperature. A linear relation between the contact angle (determined from SEM images as presented in Figure 6) and the sintering temperature is observed. Sintering at higher temperatures allows the spherical silica particles to diffuse much better into the surface and the spherical particles can decrease their large surface area more efficiently. This surface area reduction upon sintering is the driving force for the sintering process and as soon as surface diffusion processes are activated, the particle morphology starts to change. At lower sintering temperatures, fewer diffusion processes are activated and consequently only little influence on the particle morphology is observed.

and decreases with increasing sintering time. At high temperatures (above 1200 °C), the particle height decreases to 0 and the topography pattern vanishes. Thus, the heat treatment is an effective way of changing the particle topography on the gradient from large, spherical particles to small "hill-like" structures at higher temperatures.

Beside the adjustability of the morphology of the particle gradients that is offered by this sintering process, the mechanical stability is increased drastically: the particles are no longer simply adsorbed but are firmly sintered onto the surface. This fact makes these gradients also useful for several applications for which particle arrays were not previously suited, for example in adhesion and friction studies or as templates in replication techniques.[26] Choosing sintering conditions such that the contact angle of the particles is below 90° (particles appear as half-spheres on the sample) allows such samples to be used as templates since no "undercutting" during a replication process (casting in PDMS, for example) would be observed.

A first example of how such particle gradients can be useful is shown in Figure 8. Fluorescent labeling of the actin filaments and the nuclei of these cells is shown in the top two fluorescence microscopy images. It is observed that cell adhesion and spreading is largely reduced in areas where the density of particles on the surface is high (left side of fluorescence images). With decreasing particle density, the number of cells growing on the substrate increases significantly and reaches the highest number in the areas where no particles were on the surface (on the right-hand side of the fluorescence images). The graph in Figure 8 quantifies these results and indeed an inverse dependence of cell number with particle density is found. These are initial, promising results from cell biology experiments carried out with this kind of colloidal gradients and more detailed studies are currently underway.[27] This experiment is a good example of how such

(26) Wieland, M.; Chehroudi, B.; Textor, M.; Brunette, D. M. *J. Biomed. Mater. Res.* **2002**, *60*, 434.

(27) Kunzler, T. P. Diss. ETH No. 17049, ETH Zurich, Zurich, 2007.

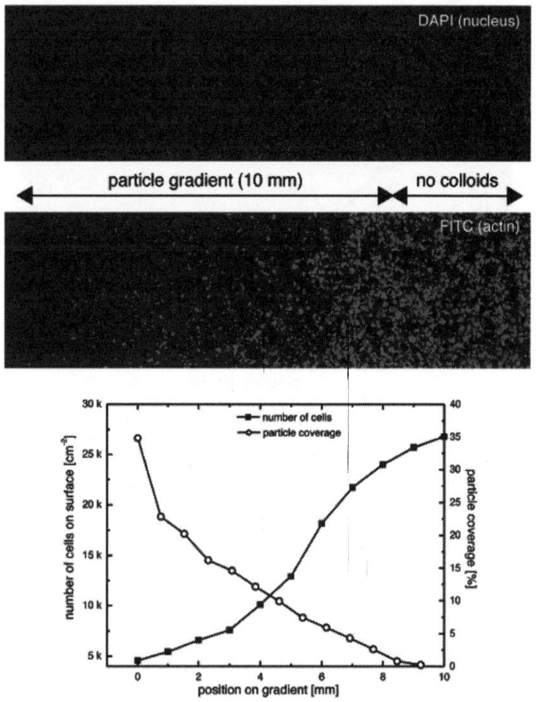

Figure 8. Fluorescence microscopy images of stained actin filaments (FITC) and cell nuclei (DAPI) on a particle gradient (0.002 wt % in H_2O). The particle gradient was uniformly coated with PLL-*g*-PEG-RGD, a polymer that introduces the RGD function, a peptide sequence known to induce cell growth, onto the particle surface. At high particle surface coverage (to the left of the images), only a few cells adhere and spread on the surface, while with decreasing particle density (moving to the right-hand side of the fluorescence microscopy images) the number of cells increases significantly. Quantitative analysis (graph below the images) shows how indeed a reciprocal relation exists between particle surface coverage and the number of rat calvarial osteoblast cells growing on the substrate.

nanomorphology gradient surfaces will help in rapid screening tests and how combinatorial and diagnostic studies can be performed on a single sample in the future. The techniques presented here may help in fabricating suitable particle gradients for these kind of applications in a straightforward and customizable way.

Conclusion

Combining the knowledge gained from particle-adsorption experiments as a function of particle concentration, pH, and ionic strength with a dip-coating process led to the development of a colloidal patterning method that allows the production of colloidal gradients with specifically adjustable parameters on a centimeter-length scale. Because of the nature of this process, almost any material combination could be used in a similar way to produce a similar kind of particle gradient provided that the adsorption kinetics are known and can be controlled as shown here. In this work, silica nanoparticles were assembled into a gradient on a silicon wafer, but no significant difference would be expected if the substrate material were changed to a different metal oxide, provided that appropriate modifications are made to account for differences in IEPs, adsorption kinetics, and sintering temperatures.

The subsequent sintering process proved to be an efficient way of altering the morphology of the colloidal gradients. The globular shape of the adsorbed particles can be continuously changed as a function of the exact sintering conditions. Furthermore, this heat treatment provides a route to stabilize the colloidal array on the surface. Thus, mechanically stable, topographical gradients were produced using a versatile technique that may find applications in various fields because of their tunable properties.

Acknowledgment. The authors thank Michael Horisberger for the metal oxide coatings (PSI, Switzerland) and Brandon Bürgler for SEM imaging (ETH Zürich). This work was financially supported by the ETH Council Nanotechnology funding program, Top Nano 21 (Project 5971.2), as well as by Nanocues (EU: FP6-NMP) and ETH Zurich.

LA0700422

Surface-chemical and -morphological gradients

Sara Morgenthaler, Christian Zink and Nicholas D. Spencer*

Received 9th October 2007, Accepted 17th December 2007
First published as an Advance Article on the web 30th January 2008
DOI: 10.1039/b715466f

Surface gradients of chemistry or morphology represent powerful tools for the high-throughput investigation of interfacial phenomena in the areas of physics, chemistry, materials science and biology. A wide variety of methods for the fabrication of such gradients has been developed in recent years, relying on principles ranging from diffusion to time-dependent irradiation in order to achieve a gradual change of a particular parameter across a surface. In this review we have endeavoured to cover the principal fabrication approaches for surface-chemical and surface-morphological gradients that have been described in the literature, and to provide examples of their applications in a variety of different fields.

1. Introduction

Surface gradients are surfaces with chemical or physical properties that gradually change over a given distance. A gradual change in a physical property, such as the wettability, can be induced by a change in surface chemistry, for example a gradually changing surface composition.

The motivation to prepare gradients is two-fold. Gradients are, on the one hand, ubiquitous in nature and thus biomimetic gradients allow us to gain a deeper insight into biological processes. Concentration gradients across membranes are central to energy generation in cells, and biomolecule gradients direct the haptotaxis of cells, for example, in axonal outgrowth. In order to understand these processes more clearly one needs to mimic the *in vivo* situation in the lab, which requires controlled chemical gradients. In addition, gradients can be a valuable materials research tool that allows high-throughput and cost-effective analysis of the influence of a wide range of parameters in the minimum amount of time.

The influence of surface properties, such as wettability, on the attachment and growth of cells and bacteria, the adhesion of proteins, as well as adhesion phenomena in general have traditionally been studied using series of individual samples. A large number of samples is typically prepared, each having a different surface property. This approach, however, has several disadvantages; different samples are prone to show a distribution of other properties, such as substrate roughness. It can therefore be difficult to attribute a particular effect to one single varying parameter, thus introducing a potential source of error, necessitating many repeat experiments. Furthermore, the handling of many samples means that experiments are time consuming. This is particularly problematic when working with biological specimens, since they may change their properties with time. Finally, multiple repeated studies extending over a long period of time may also encounter a range of different ambient conditions, which may influence the outcome of the experiments. By incorporating a range of surface properties into one single surface gradient, the need for lengthy repetitive procedures is avoided and many of these problems can be overcome.

In general, surface-chemical gradients can be created in two ways. Either the outermost surface layer of a substrate is gradually modified, for example by irradiation with an energetic beam,[1,2] or by chemical etching[3] or a surface coating, such as a self-assembled monolayer or a thin polymer film, is attached to the surface in a gradual manner. Whereas the former approach was more popular in early studies (see reviews by Elwing and Golander[4]

Laboratory for Surface Science and Technology, Department of Materials, ETH Zurich, Wolfgang-Pauli-Strasse 10, CH-8093 Zurich, Switzerland. E-mail: spencer@mat.ethz.ch; Tel: +41-44-632-5850

and Ruardy *et al.*[5]), the gradual application of surface coatings has become increasingly popular, since the bulk properties of the materials are less likely to be changed during this process.

Gradients have been applied in a large variety of studies, most often in those relevant to biomedical and biosensor applications, for example to investigate protein adsorption and cell adhesion[4,6–9] However, they have also been used in both fundamental and applied materials science studies, such as the growth of silane monolayers[10] or the generation of polymer libraries for sensor applications.[11]

We will first discuss the preparation methods that have been proposed in the literature and then cover the applications that have been addressed with surface gradients. The following section will be divided into the preparation of chemical gradients and the preparation of morphological gradients.

2. Preparation methods

2.1. Surface-chemical gradients

Surface-chemical gradients are mainly prepared *via* two systems: self-assembled monolayers (SAMs) or polymer coatings, especially brush-like polymer coatings—Table 1 provides an overview of the different preparation methods for the generation of surface-chemical gradients and the applications for which they have been used.

2.1.1. SAM-based techniques. SAM-based gradient preparation techniques can be divided into those developed for silanes on glass or silica surfaces and those applicable to alkanethiols on gold or silver.

Table 1 Surface-chemical gradient preparation techniques, systems and their applications

Technique		Adsorbate	Substrate	Application	References
Diffusion	Vapour phase	Silanes	Si, PDMS	Cell adhesion, protein adsorption, polymerisation template	10,13–21,55,56,62,63
	Solvent	Silanes	Si	Protein adsorption, polymerisation template	6,7,23–25,57,106, 107,109,110
	Through a matrix	Alkanethiols, polymer	Au	Protein adsorption, cell adhesion	7,34–38,67
Printing	Contact	Alkanethiols	Au		47,48
		Silanes	Si		26
		Biomolecules	Funct. Si		82
	Ink jet	Alkanethiols	Au	Polymerisation template	49,50
		Proteins	Si		49
Desorption	By potential	Alkanethiols, polymers	Au	Cell adhesion, protein adsorption	39,40–43,74,80
Advancing solution	Concentration gradient	Proteins	Glass, Si, Au	Cell adhesion, further functionalisation	8,86–88
		Monomers	Funct. Si		75
		Monomer solution	Funct. Si	Solvent effect in copolymer brushes	65,66,68
		Proteins, biomolecules	Funct. Si, filter	Cell growth	83–85
	Controlling reaction time	Silane	Si	Polymerisation template, protein adsorption, cell adhesion	33
		Alkanethiols	Au		51–53
		Copolymer	Oxide surfaces	Protein adsorption	70
		—	PVC	Cell adhesion	3
	Depletion	Proteins	PDMS, glass		89,90
Irradiation	Change intensity	—	PE	Cell adhesion and growth, protein adsorption, polymerisation template	1,9,54,58–61,64
		Polyatomic ions	PMMA		12
	Change exposure time	—	Polymer		2
		Monomer solution	Funct. Si		69,77
		Proteins	Funct. PS		78,79
		Silane	Si, nanoporous Si	Protein adsorption, cell adhesion	29–32
	Through mask	Silane	Glass	Enzyme adsorption	27,28
		Monomer	Funct. Si	Cell adhesion	77
Temperature		Polymer	Funct. Si	Solvent, pH effects in copolymer brushes, protein adsorption, microtubule motion	71–73
Irradiation and replacement	Through mask	Alkanethiols	Au		44,45
	Change exposure time	Alkanethiols	Au		46
Physically controlled polymerisation	Plasma polymerisation (mask)	Polymer	Any		76
	Electropolymerisation (pot. gradient)	Polymer	Au	Cell adhesion	80

The most commonly used technique for silanes was developed in 1992 by Chaudhury and Whitesides,[13] who dissolved decyltrichlorosilane in paraffin oil and let it evaporate next to a silicon surface. The silane diffuses along the surface, partially adsorbs and generates a gradual change in coverage. This method has been used extensively by the groups of Chaudhury and Genzer. Chaudhury's group and others have used this approach to study the parameters that govern droplet movement along a wettability gradient.[14–17] Genzer *et al.* have used the silane-diffusion method to generate a wide range of gradients in chemical functionality, including gradients of polymerisation initiators that were subsequently used to generate brush-copolymer gradients.[10,18] The details for these techniques are provided in the following subsection on polymer-based techniques. Genzer and coworkers also varied the substrate, using both PDMS[19,20] and porous silica.[21] Silane gradients have also been prepared by diffusion through solvents. Elwing *et al.* were among the first to develop a gradient-preparation technique that used silane diffusion in liquids.[22] A solution of a short silane ($Cl_2(CH_3)_2Si$) dissolved in tricholoroethylene is covered by xylene. The two solvents slowly diffuse into each other, allowing the creation of a wettability gradient on an immersed silicon or glass substrate. Wettability gradients created by this method were used to study protein and polymer adsorption.[6,7,23–25] Other methods to generate wettability gradients from silanes include the use of microcontact printing to generate a gradual silane coating[26] and the gradual oxidation of silanes by UV irradiation through a density filter[27,28] or by varying irradiation time.[29–31] Han *et al.* created a gradient from superhydrophobic to superhydrophilic wetting by combining the gradual oxidation of a silane SAM with a nanoporous substrate, prepared by layer-by-layer assembly of negatively charged silica nanoparticles and positively charged poly(allylamine hydrochloride).[32] Gradients of silane-bound initiators that were further used for polymerisation were also created by pumping silane solution slowly into a vessel. By controlling the adsorption time for the silanes, a coverage gradient is generated.[33]

The first technique for generating alkanethiol gradients was developed in 1995 by Liedberg and Tengvall,[34] who again used a diffusion method to generate a gradient from two differently functionalised alkanethiols. They covered a gold substrate with a polysaccharide matrix and added different alkanethiol solutions behind glass frits, fixed at the two ends of the substrate, as shown in Fig. 1a. The alkanethiols were then left to diffuse for several hours, allowing them to generate a densely packed monolayer with a gradually changing end-functionality. These gradients were thoroughly characterised by ellipsometry, IR spectroscopy and X-ray photoelectron spectroscopy,[34,35] Fig. 1b. Order–disorder gradients[36] using alkanethiols of different lengths, and wettability gradients have been used to study protein adsorption and cell adhesion.[7,37,38]

In 2000, Terrill *et al.* presented a method that relied on the electrochemical desorption of alkanethiols from a fully covered SAM by application of a potential.[39] The width of the potential window and the position of the electrodes thereby determine the width and slope of the gradients (see Fig. 2). Alkanethiol gradients prepared by this method were used in a variety of other experiments, for example to test new mass-spectroscopic techniques,[40] to study cell adhesion,[41] or to investigate nanoparticle attachment.[42] In addition to electrochemical desorption,

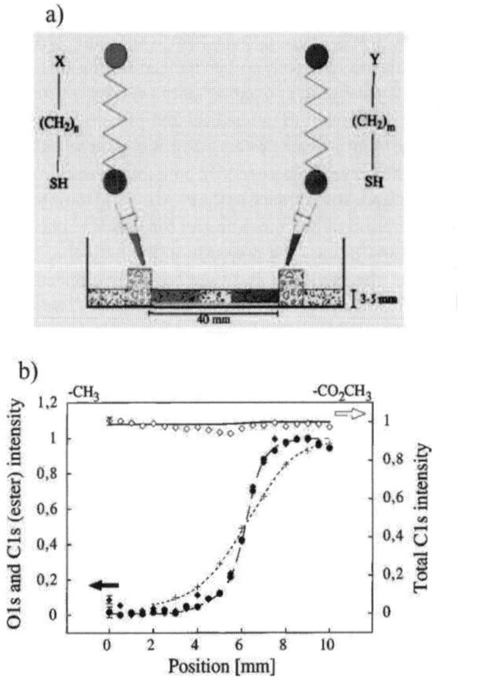

Fig. 1 (a) Schematic diagram of the cross-diffusion geometry used for the preparation of alkanethiol gradients on gold. Two alkanethiol solutions, placed behind glass filters, are left to cross-diffuse in a polysaccharide matrix.[34] (b) XPS intensity profiles obtained from a $HS(CH_2)_{15}CO_2CH_3$–$HS(CH_2)_{15}CH_3$ gradient prepared by this method. (●) O1s intensity, (◆) C1s (ester), obtained from the sum of the peak intensities of the four chemically shifted C1s peaks; (◇) total C1s intensity (282–292 eV). The profile clearly exhibits the characteristic shape of a diffusion process.[35]

alkanethiols have also been desorbed with a scanning tunneling microscope tip with a gradually increasing bias or scan speed.[43] One type of alkanethiol was desorbed from a full monolayer during the scanning process and replaced immediately by differently functionalised molecules from solution, which resulted in a two-component gradient in the submicrometre range.[43]

Photocatalytic oxidation, low-energy electrons or a focused X-ray beam have also been used to degrade alkanethiol SAMs in order to create surface-chemical gradients. Two-component gradients have been generated by Blondiaux *et al.*[44] Molecules from a full monolayer were gradually removed by oxygen radical oxidation and subsequently saturated with a second component. The oxygen radical concentration gradient was generated by UV irradiation of a thin TiO_2 film through a gray-tone photomask. In this way, repeating gradients in the micrometre range were created, which could be visualised *via* the contact line of a water droplet (see Fig. 3). A related technique presented by Ballav *et al.* used a variable dose of low-energy electrons to modify the SAM surface.[45] The exchange process during the subsequent immersion into a second component depended on the irradiation dose, thus leading to a higher exchange rate at the highly irradiated region. This allowed the creation of, for example, a wettability gradient. The desorption of alkanethiols by a focused X-ray

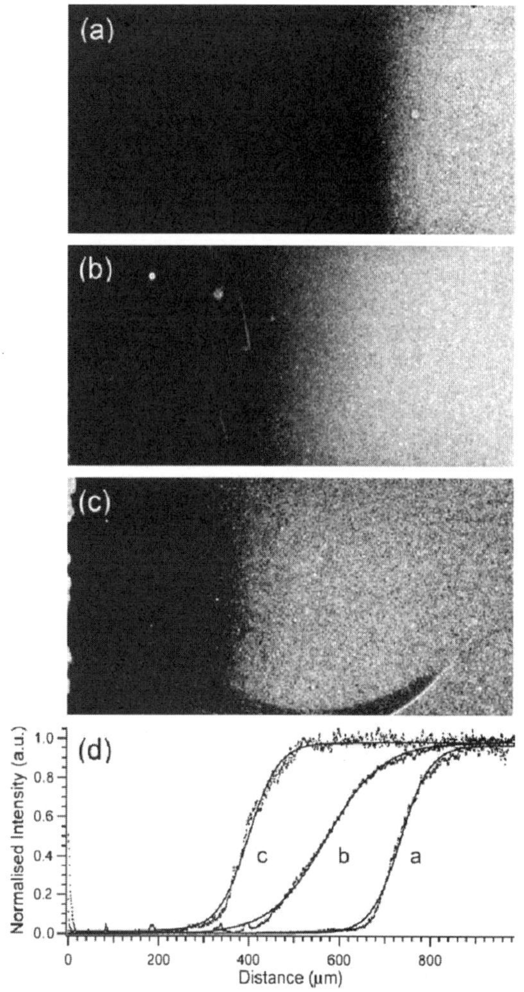

Fig. 2 Fluorescence images of nanosphere-tagged aminoethanethiol gradients prepared by electrochemical desorption with potential windows of (a) −1000 mV (left) ≤ $V(x)$ ≤ −200 mV (right), (b) −900 mV (left) ≤ $V(x)$ ≤ −200 mV (right), and (c) −800 mV (left) ≤ $V(x)$ ≤ −200 mV (right). (d) Normalised intensity as a function of position for each image, with dotted lines corresponding to data and solid lines to fits of the data.[41]

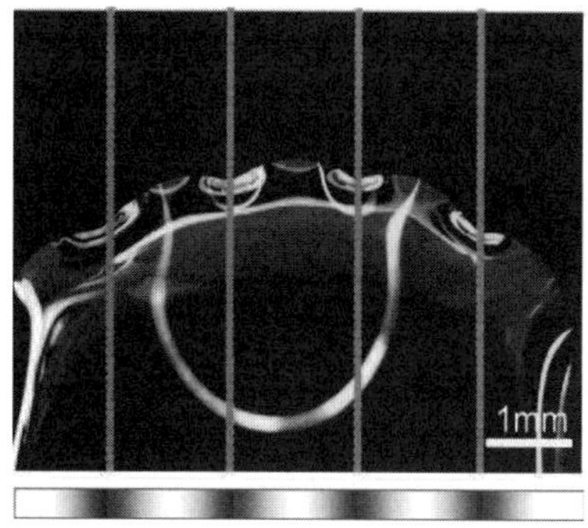

Fig. 3 Dark-field image of a single water droplet deposited on the surface bearing repeating 720 μm wettability (thiol–gold) gradients produced by photocatalytic lithography. The corresponding grayscale variation in the photomask is shown below the image.[44]

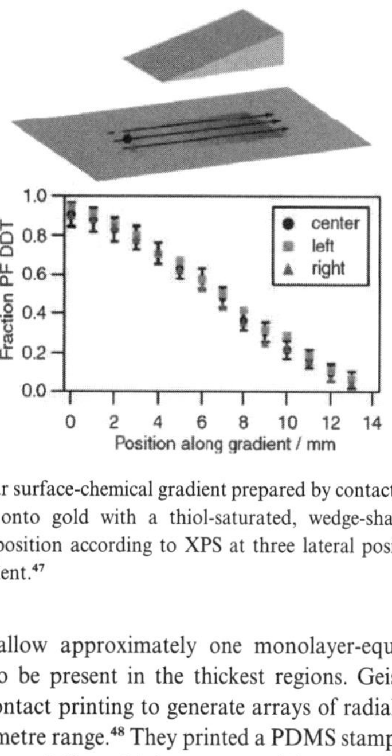

Fig. 4 Linear surface-chemical gradient prepared by contact-printing of alkanethiols onto gold with a thiol-saturated, wedge-shaped stamp. Surface composition according to XPS at three lateral positions along a single gradient.[47]

beam has been controlled by varying the exposure time.[46] However, only single-component gradients of a variety of alkanethiols have been created by this technique, *i.e.* without using a subsequent backfilling step.

Next to diffusion- and desorption-based techniques, another class of methods has been developed based on printing. Regular contact printing has been used to generate gradients with a wide variety of different shapes[47] by applying very thin, contoured, alkanethiol-saturated PDMS stamps to a substrate (see schematic in Fig. 4). More molecules are available in the thicker regions of the PDMS stamp than in the thinner areas, thus allowing the formation of a higher-surface-concentration layer in the thicker regions, whereas fewer alkanethiol molecules are adsorbed in the thinner areas. The thickness of the PDMS is chosen to allow approximately one monolayer-equivalent of molecules to be present in the thickest regions. Geissler *et al.* also used contact printing to generate arrays of radial gradients in the nanometre range.[48] They printed a PDMS stamp saturated with a mixture of alkanethiols onto a particle array, allowing the diffusion of alkanethiols along the particles onto the surface. A radial gradient forms because of the different diffusion constants of the two alkanethiols. Ink-jet printing was also used to generate alkanethiol gradients in the centimetre range by printing one component and backfilling with a second.[49,50]

Finally, surface-chemical gradients can also be prepared by a two-step immersion process.[51] During the first immersion step a gradual change in surface concentration of one type of molecule, for example a methyl-terminated alkanethiol, is achieved by slowly immersing the substrate into a dilute solution of the adsorbate. The conditions are chosen such that longer immersion times (*i.e.* at the leading edge of the substrate) result in a densely packed monolayer assembly, whereas only very few molecules adsorb during short immersion times (*e.g.* at the trailing edge). By choosing an appropriate combination of concentration

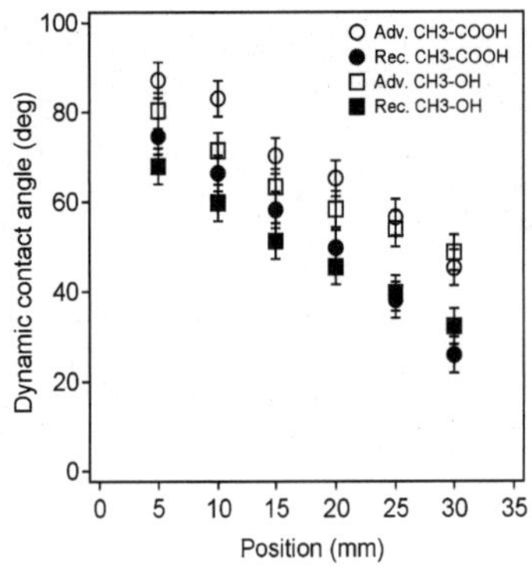

Fig. 5 Advancing (open symbols) and receding (filled symbols) water contact angles on a methyl-carboxyl (circles) and a methyl-hydroxyl (squares) gradient produced by the gradual immersion method (thiol–gold), measured along the major axis.[53]

and maximum immersion time, a coverage gradient of adsorbates can be created. This coverage gradient is mostly at submonolayer surface concentrations and can be saturated by subsequent total immersion of the substrate into a solution of a second type of adsorbate, for example a hydroxyl-terminated alkanethiol.[51] Such two-component gradients have an island-type submicrometre structure[52] but overall exhibit a high degree of molecular organisation.[53] Gradients with a linear change in wettability can readily be generated in this way, as shown in Fig. 5.

2.1.2. Polymer-based techniques. The polymer-based techniques can be again divided into two sections: those based on surface modification by, for example, irradiation and those based on the application of a gradually changing coating.

The review article by Ruardy *et al.* provides a good overview mainly of the creation of wettability gradients by surface modification.[5] Different types of surface irradiation, corona discharge or radio frequency plasma discharge, as well as etching solutions have been used to modify polymer surfaces. Gradients have been generated by either changing the time during which a surface is exposed to irradiation[2] or etching solution,[3] or by changing the irradiation power along the substrate.[1,54] During the exposure of the polymer substrate, a gradient in the concentration of activated oxygenated species is created on the surface, resulting in a gradient in wettability. However, all of these preparation methods have the drawback that the surface species generated are not well defined and that the substrates are also roughened during the process.

The number of techniques available to generate brush-like polymer gradients is large. The first part of this subsection will deal with the fabrication methods that employ silanes as grafting sites on the surface. The second part of the subsection covers all other techniques that allow brush-like polymer gradients to be created.

The silane-diffusion techniques developed by Chaudhury and Whitesides,[13] and Elwing *et al.*[22] have been used to generate gradients of grafting sites on silica surfaces. These grafting sites were used to either initiate the polymerisation of various monomers by atom transfer radical polymerisation (ATRP), a "grafting-from" technique,[55,56] or were used to attach PEG chains by the "grafting-to" technique.[57] Brush-like polymer gradients were also generated from gradients of alkanethiol initiator sites applied by ink-jet printing.[50] Functionality gradients prepared by corona treatment were also further used to graft polymer chains. The oxygenated species were used as grafting sites for various polymers, such as PEG[58] or others.[59–61] Brush-like polymer gradients prepared by these techniques have been used to immobilise nanoparticles[62,63] and to study protein adsorption and cell adhesion.[59,64]

Other techniques begin with a homogeneous coverage in initiator density and control the polymerisation conditions, for example the polymerisation time. Tomlinson and Genzer used gradual draining of a monomer solution from a vessel to generate brush-like polymer gradients with a gradually changing molecular weight,[65,66] Fig. 6a, step A. A second polymer can be grafted from this MW gradient for the generation of a block-copolymer gradient, either by a similar draining step, which yields a block-copolymer brush with a constant thickness and gradually varying block length (Fig. 6a, steps D,E), or by complete immersion, which yields a gradient with gradually changing thickness but similar block length of the second component (Fig. 6a, steps B,C). Such block-copolymer gradients can also be prepared in an orthogonal way, leading to complete block-copolymer composition libraries, as shown in Fig. 6b. These have been used to study the effect of solvent on chain conformation. Other groups have used diffusion to control the adsorption time of a polymer during grafting. Mougin *et al.* allowed an NHS-functionalised PEG chain to diffuse through an agarose matrix—a similar approach to that developed by Liedberg for alkanethiols[34,67] whereas Xu *et al.* pumped a monomer solution slowly through a microchannel, controlling the contact time to generate a gradient in molecular weight.[68] Finally, a moving shutter has also been used to control the irradiation time of a mixed PEG methacrylate solution, which resulted in the generation of a crosslinked PEG thickness gradient that could be used for biosensor applications.[69] Polymer brush coverage gradients can also be prepared by controlling the adsorption kinetics of a brush-forming graft copolymer, for example poly(L-lysine)-*g*-poly(ethylene glycol) by a simple dipping technique[70] originally developed for the generation of alkanethiol gradients. After the generation of a coverage gradient, the surface can be easily saturated with a differently functionalised copolymer, which allows the creation of gradients from different functional groups, such as peptides or biotin. The effect of temperature on the grafting rate was exploited by Ionov *et al.*[71–73] They first coated a silica substrate with an anchoring layer of either silanes or a polymer and then a second layer of polymer. The degree of anchoring of the top-layer polymer, which was to form a brush-like coating, to the layer below is controlled by applying a temperature gradient. More chains are grafted to the surface at a higher temperature, thus allowing for the fabrication of a gradient in grafting density. Such gradients were used, for example, to study the behaviour of polyelectrolyte brushes at different pH values.[72] Full alkanethiol-initiator SAMs were also used for the preparation of mixed polymer brush gradients.[74] A first monomer, *N*-isopropylacrylamide, was polymerised on the full SAM by ATRP, then the polymer chains were desorbed by application of an electrochemical potential, as developed by Terrill *et al.*[39] Subsequently the empty sites were re-saturated with the initiator and a second monomer, for example 2-hydroxyethylmethacrylate, was polymerised, which led to the formation of a mixed polymer brush gradient.

Not only have polymerisation conditions been gradually varied, but also the composition of the monomers used for polymerisation has been gradually adjusted. Xu *et al.* created a concentration gradient of (*n*-butylmethacrylate and 2-(*N*, *N*-dimethylamino)ethylmethacrylate in a microchannel by gradually changing the infusion rates of the two solutions.[75] Atom transfer radical polymerisation (ATRP) then led to the generation of a brush-copolymer gradient. Whittle *et al.* used plasma polymerisation to generate a gradient in chemical functionality by gradually changing the composition of a plasma and at the same time gradually shielding the substrate.[76]

Various groups have used photoimmobilisation or photopolymerisation to generate brush-like polymer gradients by changing the dose of light, for example, with a photomask,[77] or by changing the irradiation time.[78,79]

Also an electropolymerisation process in the presence of an in-plane electrochemical potential gradient was further used to generate poly(acrylic acid) and poly(acrylamide) thickness gradients.[80] Subsequent surface derivatisation of such thickness gradients could be used to generate peptide, fluorinated or nanoparticle gradients.

Finally, not only polymers but also proteins have been immobilised on surfaces as a gradient. The attachment and movement of cells depends strongly on the surface concentration and the confirmation of adsorbed proteins, and thus it is useful to study this behaviour in a high-throughput manner. Proteins have been covalently coupled to alkanethiol gradients by, for example, an amine group[37,81] or by laser irradiation[78] and they have been ink-jet printed[50] or stamped.[82] Other groups have controlled the adsorption kinetics by changing the contact time of a substrate with a protein solution.[83,84] Controlled adsorption kinetics were also used to immobilise single-stranded DNA onto an indium–tin oxide surface,[85] which allows the generation of a biosensor surface. Finally, microfluidics have also been used to generate concentration gradients of molecules in microchannels. Whitesides' group has created gradients in proteins and other biomolecules by controlling the laminar-flow conditions and have investigated cell attachment and growth[8,86–88] on these surfaces, while the groups of Caelen and Fosser have used the depletion of protein solutions along microchannels to generate coverage gradients.[89,90]

2.2. Morphological gradients

Within the last decade, various approaches have been described to fabricate different kinds of morphological gradients, varying in feature shape and size, length and in the fundamental principle that is used to create them. In order to compare these techniques, those fabricated by similar basic principles have been grouped together.

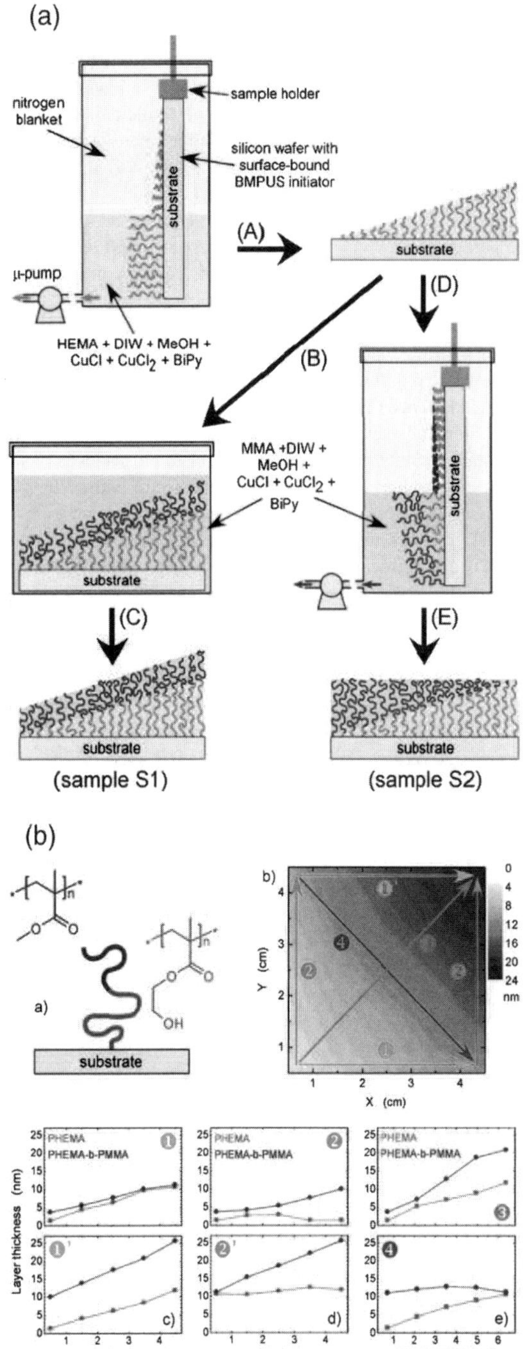

Fig. 6 (a) The preparation of surface-anchored PHEMA-*b*-PMMA gradients: a custom designed apparatus is used to decorate the sample surface with a grafted PHEMA having a gradient in molecular weight (arrow A). Surface-grafted PHEMA acts as a macroinitiator for the polymerisation of the PMMA block that has either a constant molecular weight (arrow B) or a variable molecular weight (arrow D). The overall process results in PHEMA-*b*-PMMA block copolymers with a constant PMMA length and a variable total length (arrow C) or a gradual PMMA length and a constant total length (arrow E).[65] (b) Dry thickness profile of PHEMA-*b*-PMMA MW1/MW2 orthogonal brush gradients as a function of the position on the substrate. (c–e) PHEMA (red squares) and total copolymer (blue circles) thicknesses along the directions depicted in the total thickness profile shown in part b.[66]

2.2.1. Particles. One widely used approach to create morphological gradients is to bind particles onto a smooth substrate. Genzer and his colleagues[62,66,91–93] have established a simple but highly efficient procedure to create nanofeature gradients by creating gradients of polymers and then adsorbing gold particles. They have employed two basic principles to create one-dimensional gradients or a combination of both to prepare two-dimensional gradients.

The first principle is the Chaudhury and Whitesides[13] procedure (see section 2.1.1), involving the evaporation of functionalised silanes next to a silicon substrate. If a silanised polymerisation initiator is used, this approach can produce a gradient in grafting density of polymer chains on the surface (see section 2.1.2). The second principle is simply withdrawing the polymerisation solution while the process is still running. The lower parts of the sample are thus grafted with longer chains, or in other words a gradient in molecular weight of the grafted polymer can be produced (see section 2.1.2). The particle gradient is produced by immersing a polymer gradient, such as poly(2-(dimethylamino) ethylmethacrylate) (PDMAEMA) into a gold nanoparticle suspension and letting the particles adsorb onto the polymer over several hours (see Fig. 7). Depending on the particle size, "quasi" 2D structures or 3D dispersions of nanoparticles can be achieved. Larger particles (diameter ≈ 16 nm) are not able to penentrate the polymer brush and they always remain in the top layer of the polymer. In order to form a 3D dispersion the particle size has to be reduced drastically (down to ≈ 3.5 nm) to allow the particles to penetrate throughout the thickness of the polymer brush.

Huwiler *et al.*[94] have developed a direct technique (*i.e.* not *via* a chemical gradient) for the fabrication of a nanoparticle density gradient by a simple dip-coating process. The principle was to adsorb negatively charged silica particles onto a positively charged surface (a poly(ethyleneimine) (PEI)-coated silicon wafer) by means of electrostatic interactions. Since the adsorption is a kinetically controlled process, a particle gradient can be created by exposing different parts of the substrate for increasing times to a colloidal suspension. This was achieved by slowly immersing the substrate into a colloidal suspension of silica nanoparticles, leading to a gradual increase in particle density along the substrate surface. In order to increase the mechanical stability of the particle array and to adjust their shape the gradients were partially sintered into the substrate at temperatures between 1075 and 1200 °C, at which temperature all organic compounds were burned out, leaving nothing behind but a bare SiO_2 nanomorphology gradient.

Another approach to the deposition of nanometre-sized particles on a surface has been described by Roth *et al.*[95] who thermally evaporated gold onto polystyrene in a vacuum chamber to form a thin film. Since the polymer–metal interaction is much weaker than the metal–metal interaction, the gold atoms form clusters in a self-assembly process. The gradient in cluster density was induced by using a blocking mask to shadow parts of the substrate.

2.2.2. Electrochemical etching (Si). If silicon is used as an anode in an electrochemical etching process, its surface is rendered porous. The mean pore size and the size distribution can be adjusted by varying the HF concentration in the electrolyte and the current density on the anode. A current density gradient along the substrate can be achieved by suitable design of the electrodes, leading to a porosity gradient in the silicon wafer. Different configurations of electrodes and compositions of the HF solution have been used to create such gradients on boron-doped silicon substrates (p+). Collins *et al.*[96] placed an off-centre Pt cathode 1 mm above the substrates (1.2 cm diameter discs). The pore diameters in the gradients ranged from 600 nm immediately below the counter electrode to 10 nm on the side furthest away from it.

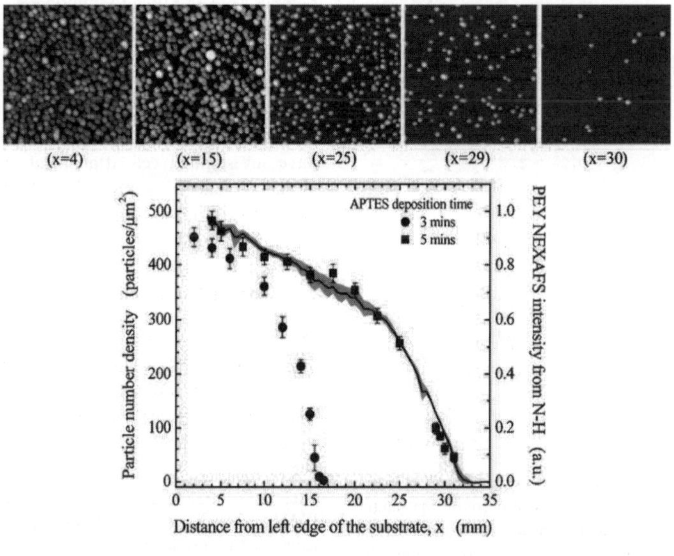

Fig. 7 (upper panel) AFM images of gold particles adsorbed along a substrate prepared by evaporating a mixture of (3-aminopropyl)triethoxysilane (APTES) and paraffin oil (PO) for 5 min followed by immersion in colloidal gold solution for 24 h (edge of each image = 1 μm). (lower panel) Particle number density profile (left) for two gradients prepared by evaporating APTES–PO mixtures for 3 and 5 min. The line represents the PEY NEXAFS profile (right) of N–H bonds from an ATEPS gradient prepared by evaporating the APTES–PO mixture for 5 min.[93]

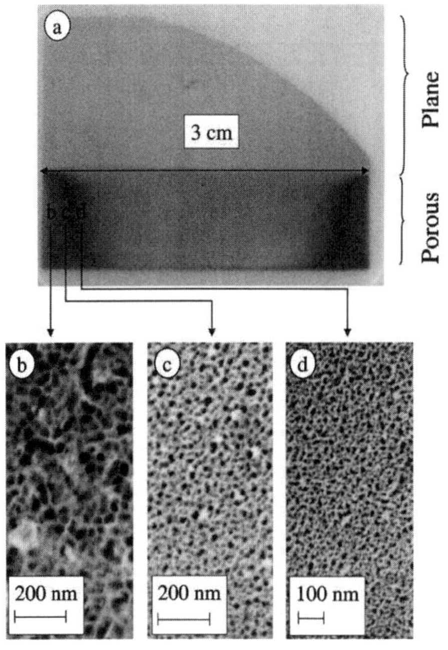

Fig. 8 (a) An etched porous silicon gradient. One gradient stretches from the left to the center and one from the right to the centre, as seen by the change in the gray scale. (b–d) SEM pictures taken of the wafer seen in (a) at different distances from the edge; (b) at the edge of the wafer (*i.e.*, 0 mm), (c) 2.5 mm from the edge of the wafer, and (d) 4 mm from the edge of the wafer.[97]

A different approach has been described by Karlsson *et al.*[97] They placed two similarly sized silicon wafers with their unpolished sides facing each other, separated by 2–15 cm. When applying a voltage between the wafers, the current density between them decreases from the edges towards the centre, yielding pore diameters in these gradients ranging from 20 nm to 3 nm from the edge to the centre (see Fig. 8).

2.2.3. Erosion/chemical polishing—replica methods. Another approach to the generation of morphological gradients is to gradually polish a rough surface. This was first described by Kunzler *et al.*,[98] who established a two-step process of particle erosion followed by chemical polishing. Pure aluminium sheets were sand-blasted with corundum particles and the substrate was subsequently fully immersed into the chemical polishing solution and then withdrawn in a controlled manner by means of a linear-motion drive, creating a roughness gradient on the centimetre scale (see Fig. 9a). A chemical-polishing solution was chosen that preferentially removed exposed features with a small radius of curvature. The composition of the polishing solution was 77.5% (v/v) phosphoric acid, 16.5% (v/v) sulfuric acid and 6% (v/v) nitric acid.

While aluminium is well suited to this approach, it may not be appropriate for a particular application. Also, it may be desirable to produce multiple identical samples for reproducibility reasons. These issues can be solved by replicating the gradients.[98] A negative impression can be made with a polyvinylsiloxane replica material, which is widely used in dentistry. Subsequently an epoxy positive can be cast from the negative, creating a replica of the roughness gradient. These replicas can then be coated with metals or oxides, and these coatings can, in turn, be functionalised, providing great flexibility in the final surface chemistry of the morphology gradient.

2.2.4. Polymers—temperature gradient. Polymers are very versatile, displaying a wide range of properties, and thus many polymer-based approaches have been developed. A method introduced by Meredith *et al.*[99] is based on the phase separation of a polymer blend upon heating. Since the authors intended to create a two-dimensional gradient it was necessary to create a composition gradient perpendicular to the temperature gradient. A continuously changing composition of two polymers in solution was obtained by pumping one polymer into a solution of the other. Simultaneously the mixture was withdrawn into a syringe, which thereby became filled with a composition gradient that could subsequently be used to extrude a stripe of changing composition onto a silicon substrate. A knife-edge coater was then used to spread the stripe in a thin film with a gradually changing composition perpendicular to a temperature gradient, which in turn led to a gradual phase separation of the polymer blend. The result is a two-dimensional gradient of feature size and distribution. The same group used a similar coater to create thickness gradients of thin polymer films.[100] A bead of polymer solution was placed between the blade and the substrate, which was firmly mounted onto the stage. Then, the blade was constantly accelerated with respect to the substrate causing the frictional drag to increase and thus increasing the amount of polymer deposited on the substrate. After the solvent was evaporated, a thickness gradient remained, whose steepness is defined by the acceleration of the blade (with thicknesses ranging from 50 to 250 nm).

Washburn *et al.*[101] used a temperature gradient to induce varying crystallinity and thus a roughness gradient on the nanometre length-scale. They created a thin film of poly(L-lactic acid) (PLLA) on a silanised silicon wafer, which was annealed on a temperature gradient ranging from 44 to 100 °C. The RMS values achieved with this method ranged from 0.54 to 13 nm.

Quite a different approach was used by Lu *et al.*[102] They produced a porosity gradient in a film of low-density polyethylene (LDPE) on a silicon wafer. The LDPE was dissolved in xylene at 90 °C and then recrystallised by slowly lowering the temperature while stirring. A clean silicon wafer was then dipped into this suspension to create a porous polymer film. By annealing the substrate on a temperature gradient reaching from 0 °C to well above 107 °C (Tm of the LDPE used) the porosity gradually changes along the gradient (see Fig. 10).

2.2.5. Polymers—solvents, spincoating. Recently a method was reported by Blondiaux *et al.*[103] that takes advantage of the phase separation of an immiscible polymer blend thin film in the presence of a surface-energy gradient. The gradually changing surface energy induces the phases of the polymer blend to separate into different morphologies along the gradient.

As a polymer blend a (50 : 50) w/w mixture of poly(methylmethacrylate) (PMMA) and poly(2-vinylpyridine) (P2VP) dissolved in MEK was used. It was then spin coated onto the substrates.

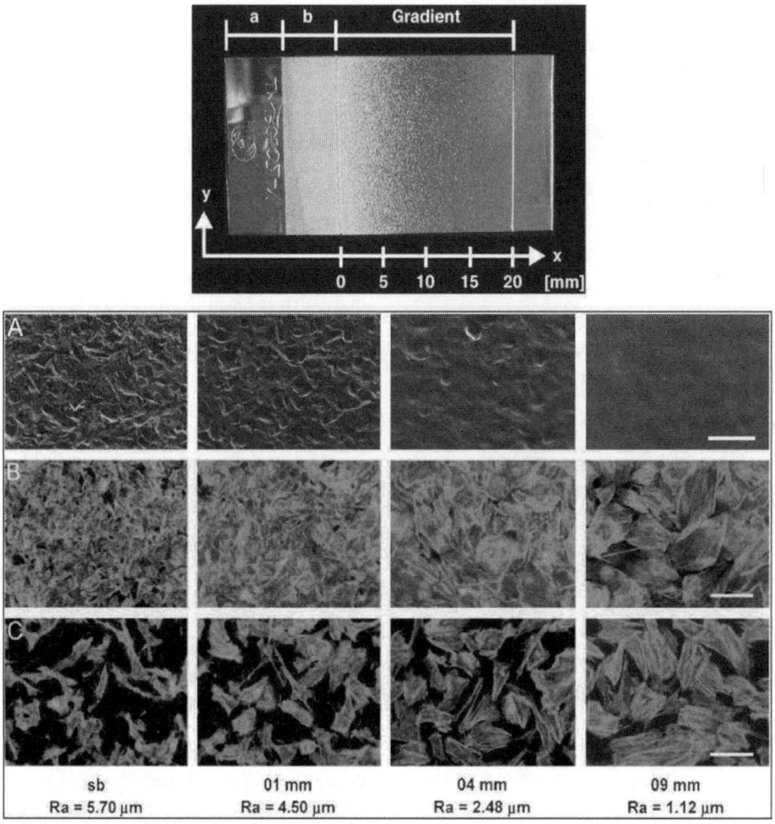

Fig. 9 (upper panel) Optical image of a roughness gradient on an aluminium surface over 20 mm. Section (a) shows the untreated aluminium surface and (b) a section that was sand blasted only.[98] (lower panel) (A) SEM images at different positions (sand blasted (sb), 1, 4 and 9 mm) and the corresponding roughness value Ra on a 10 mm roughness gradient of a similar type to that shown in Fig. 9a. Morphology of rat calvarial osteoblasts (B) and human gingival fibroblasts (C) at different positions on the gradient. Cells were cultured for 7 days. The nuclei were stained with DAPI (blue) and the actin network with Alexa Fluor 488 phalloidin (green). Scale bar is 200 μm.[117]

The morphology trend could be accentuated if one of the polymers was subsequently removed by means of a selective solvent.

2.2.6. Lithography. Photolithography can also be used to create morphological gradients on a surface. Rather than just creating a mask with gradually changing features, Cao *et al.*[104] used a regular photomask with uniform microfeatures. To create a gradient, an additional blocking mask was mounted just above the photoresist surface. While exposing the substrate, light diffraction occurs at the edge of the blocking mask, generating a gradient of light intensity on the low-contrast photoresist below. This leads to a gradient in dissolution by the developer that subsequently is transferred onto the Si substrate by reactive ion etching. The gradients created by these authors have been used to slowly narrow down the cross section in a fluidic channel, in order to allow long DNA molecules to be stretched so they can be inserted into nanochannels.

3. Applications

Chemical and morphological gradients have been used in a variety of applications, especially in the biomedical field and for studying wetting phenomena. These applications will be covered in more detail in the following section. In 1997 Ruardy *et al.* published a review[5] on the interaction of different types of proteins with wettability gradients, and a recent review by Genzer *et al.*[105] covers the use of polymer brush gradients for nanoparticle assembly and for studying protein adsorption.

3.1. Protein adsorption

Protein adsorption studies on gradient surfaces can be divided into roughly two categories. In the biomedical area one seeks to generate biocompatible implant surfaces to prevent foreign body response and facilitate the ingrowth of the implant. For this purpose it is important to understand the driving forces governing non-specific protein adsorption, for example on surfaces with different wettability, roughness or polymer-chain density. A different approach is needed for biosensors, where specific interactions, such as specific protein adsorption or antibody–antigen interactions are monitored.

Most of the early studies using surface gradients and proteins focused on their interaction with wettability gradients. The adsorption behaviour was found to vary for the same type of

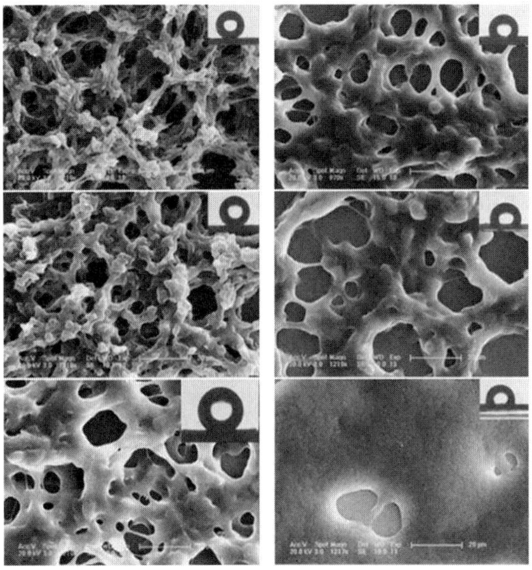

Fig. 10 FE-SEM micrographs of low-density polyethylene (LDPE) gradient surface along the major axis after processing in a temperature-gradient field. The micrographs correspond to the different location on the surface from low to high temperature. The insets are the corresponding profiles of a water drop on the surfaces.[102]

wettability gradient, depending on the adsorption conditions and the type of protein. The adsorption of fibrinogen and immunoglobulin G (IgG) was, for example, reduced when albumin was present in the solution due to the difference in adsorption kinetics between the proteins.[25] Surface wettability was also found to play a role in the Vroman effect.[106] While more human serum albumin (HSA), IgG, fibrinogen and lysozyme adsorbed on the hydrophobic end,[22,24,107,108] more human low-density lipoprotein (LDL) and high molecular weight kininogen was found to adsorb on the hydrophilic end.[22,109] This difference in adsorption behaviour can be attributed exclusively to the properties of the proteins. The following example, however, shows that this is not always the case. Spijker *et al.* reported more human serum albumin, fibrinogen and IgG adsorption on the hydrophilic end of a wettability gradient,[54] whereas Gölander *et al.* reported opposite behaviour.[25] The differences of the results in this case were attributed mainly to the different techniques used to prepare the gradients, and concomitantly to the different surface-chemical functionalities and the change in roughness. Similarly Loos *et al.* reported that the higher amount of enzyme (*Candida antarctica* lipase B) found on the hydrophilic end of a wettability gradient could not be attributed to the wettability alone, but was also influenced by surface roughness.[27] Not only the adsorption of proteins was studied, but also their desorption when exposed to nonionic detergents.[110] Welin-Klintström *et al.* compared wettability gradients prepared by two different methods and found that for one type of gradient not only the hydrophilicity but also the amount of negatively charged groups increased towards one end.[7] This additional effect was found to influence fibrinogen adsorption, which was higher on the charged surface. Finally, Riepl *et al.* have not only gradually varied surface wettability but also surface

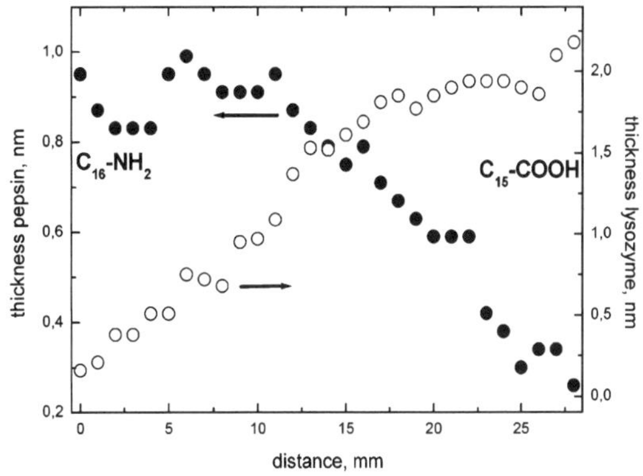

Fig. 11 Adsorption of pepsin and lysozyme onto a charge gradient prepared by alkanethiol cross-diffusion (C_{16}–NH_3^+–C_{15}–COO^-) studied by scanning ellipsometry. The adsorption time was 1.5 h, the concentration was 1 mg mL^{-1}, and the pH was 6.0.[38]

charge.[38] These charge gradients allowed the separation of lysozyme and pepsin protein mixtures, since the negatively charged pepsin adsorbs with higher probability onto the positively charged end, whereas the positively charged lysozyme behaves in the opposite way (see Fig. 11).

Riepl *et al.* have also used oligo(ethylene glycol) modified surfaces to study fibrinogen adsorption.[38] They found that the conformation of short oligo(ethylene glycol) chains controls protein adsorption. A helical chain conformation appeared to be resistant towards non-specific fibrinogen adsorption, whereas an all-*trans* orientation allowed protein adsorption. Brush-like poly(ethylene glycol) (PEG) coatings have been shown to render surfaces resistant towards non-specific protein adsorption.[111] PEG surface-density gradients have also been used to monitor protein adsorption.[57,64,70] Less protein (human serum albumin, IgG, fibrinogen, human serum and human plasma protein) was found to adsorb at the high-density end, as expected from studies with homogeneously covered substrates. Lin *et al.* reported contradictory results, where more fibrinogen was found to adsorb on the high PEG-chain density. However, this behaviour was exceptional and was attributed to interactions with additional, charged groups on the surface. Additionally, it has been shown that depending on the shape and size of the protein, different PEG densities are needed to reach a similar level of protein resistance.[70] Also, shorter PEG chains are less effective at reducing human plasma protein adsorption.[64] Other brush-like polymer coatings with gradually changing surface density, such as gradients from poly(2-hydroxyethylmethacrylate) (HEMA),[33,112] and poly(ω-methacryloyloxyalkylphosphorylcholine) (MAPC)[59] have also been used to gradually reduce fibronectin adsorption with increasing chain density.

The adsorption of proteins onto morphological gradients has not yet been studied extensively. Collins *et al.*.[96] and Karlsson *et al.*[97] both reported a higher albumin adsorption at the highly porous end of a morphology gradient, than at the gradient end with smaller pore sizes. However, the adsorption behaviour could not solely be attributed to the change in pore size, since the total thickness of the layer and the chemistry may also change with position.

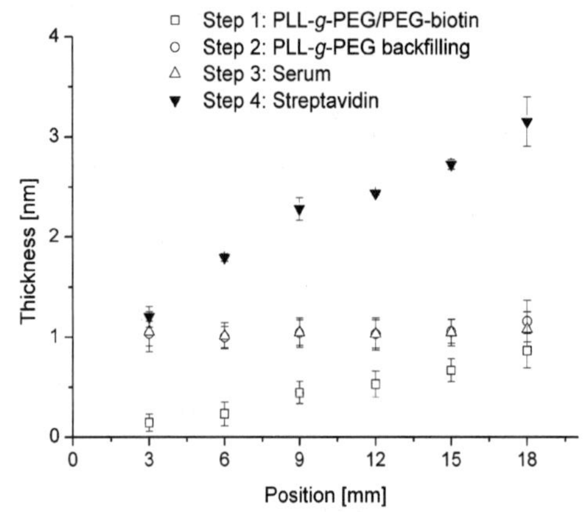

Fig. 12 Ellipsometric adlayer thickness of a biotinylated gradient as a function of the gradient position. A biotinylated coverage gradient was generated (step 1, linear increase in thickness), which was then backfilled with unmodified PLL-*g*-PEG (step 2, constant layer thickness). Negligible serum adsorption was found on such a functionalised gradient (step 3, no notable increase in layer thickness), while the amount of immobilised streptavidin gradually increased along the gradient (step 4, increasing thickness with increasing biotin density).[70]

Gradients have also been prepared directly from immunoglobulin G (IgG) by gradual immobilisation.[79,89] The efficiency of the antibody–antigen interaction on such IgG density gradients depends on the antigen surface density. An optimum surface density was found for both types of gradient. Steric hindrance and the confined space on the surface seem to prevent an efficient interaction at high antigen surface densities. Finally, biotin has also been gradually immobilised onto other gradients: streptavidin adsorption was found to reach saturation at a certain biotin density on a biotin gradient prepared by adsorption onto a wettability gradient.[60] Biotin gradients were also prepared by the saturation of a PEG-chain density gradient with a functionalised (biotinylated) brush-copolymer.[70] Fig. 12 shows that streptavidin binds gradually onto such substrates (step 4), whereas no unspecific serum adsorption occurs (step 3).

3.2. Cell adhesion

The adhesion of cells is important in many biomedical applications, for example for hip or dental root implants. A successful integration of these implants requires that bone cells adhere to the implant surface and proliferate. It is therefore important to understand the adhesion and proliferation of cells on surfaces with different properties, *e.g.* wettability or roughness. Different types of gradients, wettability, polymer-chain density, protein and morphology gradients, have therefore been used to study cell adhesion.

Depending on the type of cells and the presence of proteins in the solution, the adhesion of cells to wettability gradients has been observed to vary. Whereas endothelial cells adsorb, proliferate and grow more pronouncedly on the hydrophobic end of wettability gradients,[3] algal spores behave differently and adhere more strongly to the hydrophilic end.[113] On the other hand, Kennedy *et al.* found similar osteoblast adhesion along the entire wettability gradient, although more pronounced proliferation was observed at the hydrophobic end.[29] Finally, Lee *et al.* reported increased pheochromocytoma cell (PC-12) adhesion to an intermediate surface wettability, whereas neurite growth was enhanced at the hydrophilic end.[9] The results of all these studies, however, are influenced by the presence of proteins and other components in the cell media, which may pre-adsorb onto the wettability gradient and mediate cell adhesion, especially in those cases where enhanced cell adhesion corresponds to an enhanced protein adsorption.[9]

A more consistent picture is found for cells adhering to gradients in polymer-chain density. The protein-resistant coatings provided by high-density polyHEMA, polyMAPC, PEG coatings correspond to non-adhesive regions for cells. This means that fibronectin, as well as fibroblasts,[33,59] and osteoblasts[112] preferentially adhere to the low-density end of the gradients. Mougin *et al.* also found that the adhesion of endothelial cells was decelerated at the low-density end.[67] Not only the cells themselves, but also cell fragments, such as blood platelets, were found to adhere increasingly to the low PEG density end of a gradient.[64] Finally, a PEG-chain density gradient was backfilled with kinesin, a motor protein, which allows microtubules to be sorted according to their length[73] (see Fig. 13).

Since cell adhesion depends on the presence of certain proteins on a surface, protein density gradients have also been prepared and tested. Bovine serum albumin (BSA) is known to decrease cell adhesion, while other proteins, for example fibronectin, enhance the interaction. A fibronectin concentration gradient saturated with BSA leads to the selective adhesion of fibroblasts to the fibronectin-coated regions[81] (Fig. 14) and to endothelial

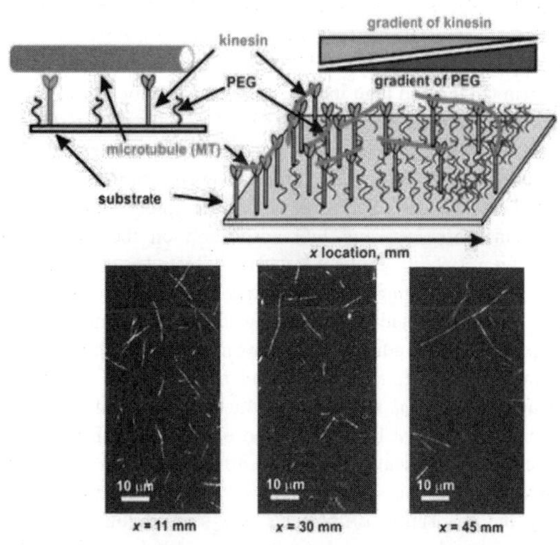

Fig. 13 Gliding motility of microtubules on a poly(ethylene glycol) (PEG)-gradient surface with immobilised kinesin. Upper part: schematic diagram of the motility system. Although the grafting density of PEG increases from left to right, the kinesin gradient is formed in the opposite direction. Lower part: fluorescence micrographs of gliding microtubules taken at three different locations along the gradient surface. At lower kinesin density, the number of microtubules per field of view decreases, whereas the average length of the microtubules increases.[73]

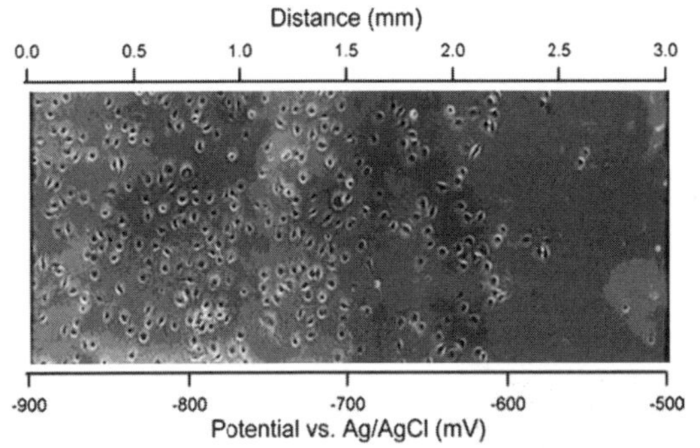

Fig. 14 Optical micrograph showing the adhesion of 3T3 fibroblast cells on a MUA–MUD-derived FN–BSA gradient prepared by electrochemical desorption. The top and bottom *x*-axes show the spatial/potential distribution of the cells.[81]

cell migration towards higher fibronectin densities.[37] The cell migration and directedness of endothelial cells was also found to increase with a higher laminin density.[8] Several protein gradients, *e.g.* laminin concentration gradients, were employed to direct the outgrowth of axons into the direction of a higher protein concentration, mimicking axon outgrowth in nature[83,114,115] (see Fig. 15).

Not only surface chemistry but also surface roughness is known to influence cell adhesion and proliferation.[116] Two different ranges of surface roughness gradients have been tested, on the one hand roughness gradients with microfeatures (see Fig. 9b), on the other hand gradients with a roughness change on the nanoscale (Fig. 16). Osteoblasts exhibited a different

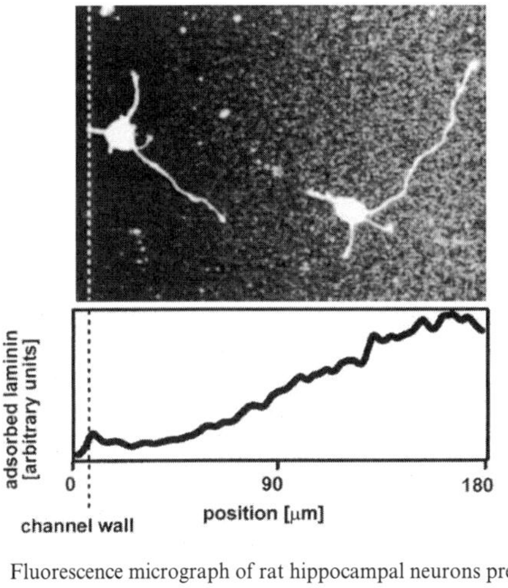

Fig. 15 Fluorescence micrograph of rat hippocampal neurons preferentially extending their presumptive axon (their longest process) in the direction of increasing surface density of laminin. Neurons were fixed after 24 h in culture and immunostained for laminin (to visualise the substrate-bound gradient in laminin) and tubulin (to picture the microtubules of the neurons). The shape of the immobilised gradient in laminin is shown in the graph below the micrograph. The dotted line indicates the left wall of the channel.[114]

behaviour on the two types of gradients. Whereas proliferation increased towards a higher surface roughness on the micrometre scale,[117] the opposite behaviour was found for nanofeature gradients.[101,118] In order to adhere to surfaces, cells need to form focal contacts. The spacing between particles is in the nanometre range, meaning much smaller than the size of the cells. If the spacing between particles is small, the cells need to deform their membrane substantially in order to attach. This unfavorable deformation leads to a decreased cell adhesion. Meredith *et al.* also found osteoblast adhesion and differentiation to differ on gradients with varying chemistry, microstructure and roughness.[99] Fibroblasts were not yet tested on both types of gradients, but contradictory behaviour was found on the micrometre scale. There, fibroblasts showed the opposite proliferation behaviour with the proliferation decreasing with increasing roughness.[117]

Finally, specific peptide sequences have also been gradually immobilised on surfaces, since certain peptide sequences are thought to trigger cell adhesion. Two gradient studies have confirmed this view, Herbert *et al.* having found an enhanced fibroblast adhesion on the high-peptide-density end,[119] and smooth muscle cells (SMC) also having displayed increased adhesion[30] (Fig. 17a–c).

3.3. Wettability effects

Wettability gradients are the ideal tool for studying various effects related to changing surface energy, such as superhydrophobicity or the movement of small water droplets.

In 1992 Chaudhury and Whitesides reported the movement of a water droplet along a 1 cm long wettability gradient, prepared by silane vapour diffusion[13] (Fig. 18). They demonstrated that a microlitre droplet will move along a wettability gradient, and even on an inclined surface, if the contact-angle hysteresis on the surface is small (<10°). The movement of droplets along wettability gradients was further studied in detail by Daniel *et al.*[13,15,120] They reported that the droplet movement is driven by the unbalanced contact angles at the droplet edges in the direction of the gradient. The contact-angle hysteresis could be overcome by vibration, thus causing droplet motion also on surfaces with higher contact-angle hysteresis.[15] They studied the influence of the amplitude and frequency of the vibration in detail and the influence of the viscosity of the liquid on the velocity of the drop. Several groups have also presented theoretical considerations concerning droplet movement and the prediction of droplet speed, and have compared the different models with experimental results.[17,121,122] Petrie *et al.* showed that the velocity of the droplet can not only be increased by reducing the surface hysteresis, but also by reducing friction through the combination of an etched porous silica surface and fluorinated silane SAMs.[21] Droplet movement was also studied in order to characterise the surface wettability by Choi and Newby, who combined polymer dewetting with droplet movement and extracted surface tension values.[26] Not only the movement of microlitre droplets, but also the condensation of very small droplets was observed on wettability gradients. Zhao and Beysens found no directed movement of such small droplets, but a different growth behaviour at the ends of the gradients.[16] With a growing droplet size, they observed that the centre of gravity of the droplet moved towards the more hydrophilic end of the gradient.

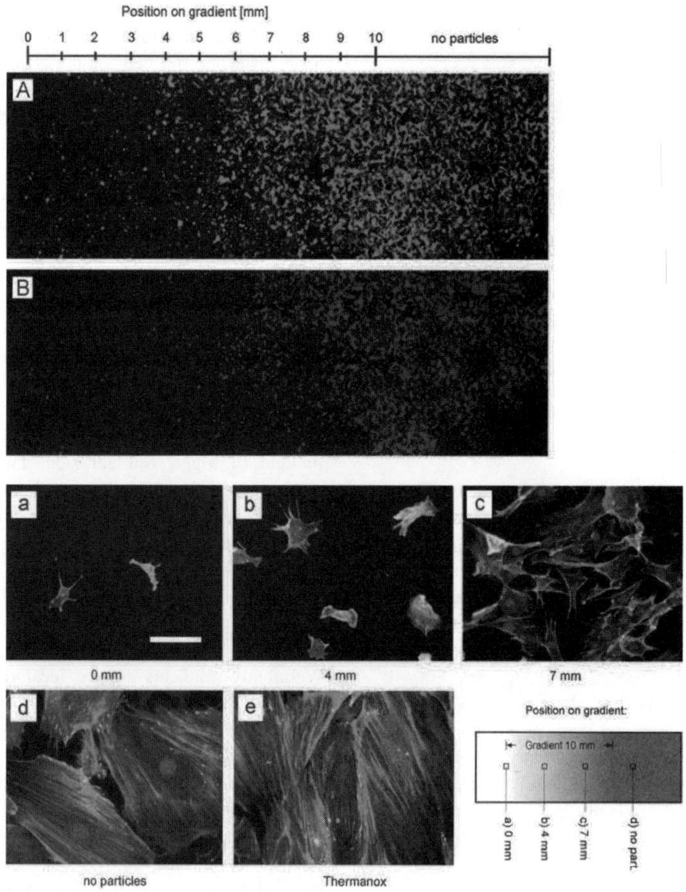

Fig. 16 (a) Fluorescence images of RCO on a nanoparticle-density gradient. Images show an overview of a large part of the gradient. Cells were seeded at a density of 3500 cells cm^{-2} and cultured for 7 days. After fixation, cells were stained with FITC for actin (green) and with DAPI for the nuclei (blue). (b) Fluorescence images of cell morphology at different positions on a nanoparticle-density gradient. RCOs were seeded at a density of 3500 cells cm^{-2} and cultured for 7 days. After fixation cells were stained for vinculin (red), actin (green) and the nuclei (blue). With decreasing particle density, cells formed well-constituted focal adhesions (red) and a distinctive actin network (green). Image e is the Thermanox control surface. Scale bar is 100 μm.[118]

Superhydrophobicity is an increasingly studied phenomenon since the easy generation of superhydrophobic surfaces would open up many new possibilities for industrial applications. Several groups have combined rough surfaces with a chemical gradient.[32,123] In this way they managed to create surfaces with a superhydrophilic to superhydrophobic transition with a high hysteresis. Lu *et al.* did not vary the chemistry, but only the surface topography in a gradual way and managed to generate a transition from superhydrophobic to hydrophobic (150° to 97°).[102]

Finally, composition gradients from two different polymer chains have also been used to study wettability effects. A polystyrene–poly(2-vinylpyridine) gradient allows a switchable wettability if submitted to a selective solvent treatment.[71] In the selective solvent, one of the two chains collapses and the properties of the surface are determined by mainly the other component. The same effect can be exploited when submitting a polyelectrolyte gradient (poly(*tert*-butylacrylate) *vs.* poly(2-vinylpyridine)) to a pH change,[72] Fig. 19.

3.4. Other applications

Surface chemical and morphological gradients have also been used for several other purposes, ranging from the study of fundamental phenomena to industrial applications.

Several chemical gradients have been used to carry out fundamental studies. Wu *et al.* have observed the mushroom-to-brush transition in polymer brush gradients,[55] while Genzer *et al.* studied the formation mechanism and structure of silane SAMs.[10] They found that the formation mechanism depends strongly on the type of head group and the humidity. Also two-component polymer brush gradients were submitted to a selective solvent treatment, which allowed the creation of a roughness gradient due to the collapse of one type of polymer chains.[56]

Gradients also simplify the development of new analytical techniques, as shown by Bhat *et al.*, since they allow the testing of very different surface compositions under the same environmental conditions, thus minimising the experimental error.[63]

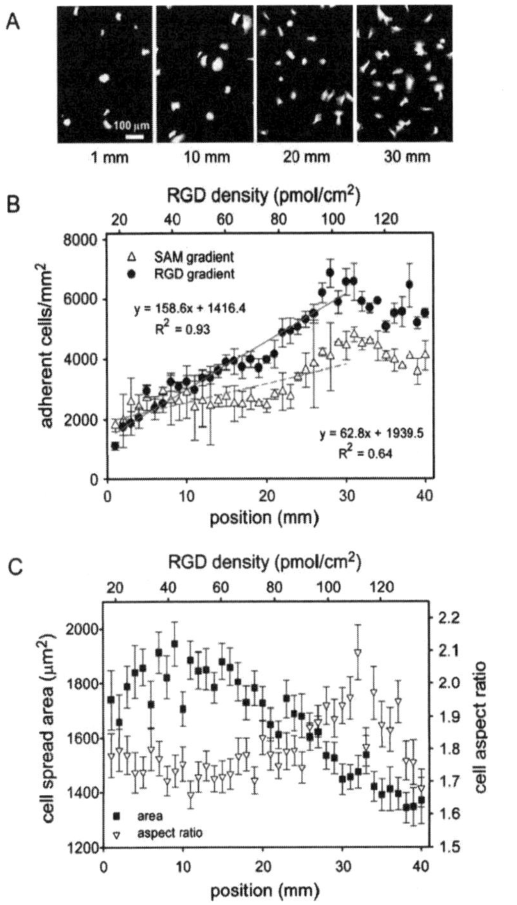

Fig. 17 Cell adhesion and morphology vary with surface-conjugated RGD peptide density. A) Cells were fluorescently labeled, and cell numbers, areas and aspect ratios quantified. B) The number of cells adhering to SAM (△) or RGD (●) conjugated gradients increases with position. A second axis (top) was added to indicate cell adhesion as a function of approximate RGD density. C) Cell areas (■) and aspect ratios (◆, mean ± S.E., $n > 45$) *versus* position and RGD concentration (top axis, derived from linear regression in Fig. 2) show different trends.[30]

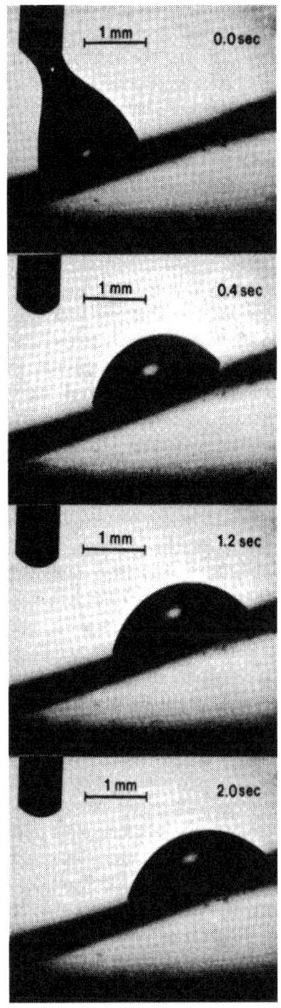

Fig. 18 Uphill motion of a drop of water on a gradient surface. The gradient surface was inclined by <15° from the horizontal plane. The volume of the drop was <1 µl. The moving drop was photographed with an automatic camera that exposed one frame every 0.4 s. The drop moved more rapidly on the initial part of the gradient than on the final part.[13]

A recent area of interest also lies in the field of liquid crystal displays (LCDs). Almost all liquid crystal devices rely on the control of liquid crystal orientation at the device interface. Surface-chemical gradients are a useful tool to study the alignment of LCs, since they facilitate the study of the systematic change in orientation from homeotropic to planar. Gradient studies have revealed that several factors seem to influence this change in orientation, for example the wettability of the surface, the thickness of the liquid crystal device[124] or the arrangement and tilt angle of the underlying anchoring layer.[125]

Surface-chemical as well as -morphological gradients have been used to study the adhesion of thin polymeric films. Chiang *et al.* carried out delamination tests on an orthogonal wettability *vs.* thickness gradient. They found that the delamination of the polymer film with increasing temperature depends on both

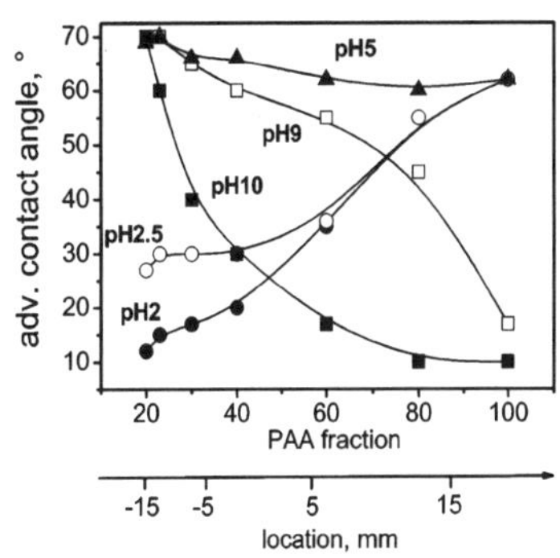

Fig. 19 Switching of the water contact angle of the gradient PAA-*mix*-PVP brush *vs.* composition (upper *X*-axis) and location of the probing drop on the sample (lower *X*-axis) upon exposure to water of different pHs ((●) pH = 2, (○) pH = 2.5, (▲) pH = 5, (□) pH = 9, (■) pH = 10).[72]

parameters.[126] The generation of such combined libraries allows the rapid screening of optimal surface parameters for different applications. Preliminary peel tests carried out in our laboratory have shown that not only the surface chemistry, but also the surface roughness can be explored in a high-throughput way on a gradient.[127] Tests with commercially available tape showed that the adhesion increases with surface roughness, although it also depends on the viscoelastic properties and thickness of the glue film. Morphology gradients have also been used to align DNA molecules.[104]

Finally, many other applications can also be envisaged, including optimising the surface chemistry for sensors, filters, or catalysts, as well as the use of gradients as templates for further experiments, such as the growth of nanostructures or crystals or for fundamental studies in tribology.

References

1 J. H. Lee, H. G. Kim, G. S. Khang, H. B. Lee and M. S. Jhon, *J. Colloid Interface Sci.*, 1992, **151**, 563–570.
2 W. G. Pitt, *J. Colloid Interface Sci.*, 1989, **133**, 223–227.
3 T. Uedayukoshi and T. Matsuda, *Langmuir*, 1995, **11**, 4135–4140.
4 H. Elwing and C. G. Golander, *Adv. Colloid Interface Sci.*, 1990, **32**, 317–339.
5 T. G. Ruardy, J. M. Schakenraad, H. C. vanderMei and H. J. Busscher, *Surf. Sci. Rep.*, 1997, **29**, 3–30.
6 V. Hlady and C. H. Ho, *Materialwiss. Werkstofftech.*, 2001, **32**, 185–192.
7 S. Welin-Klintstrom, M. Lestelius, B. Liedberg and P. Tengvall, *Colloids Surf., B*, 1999, **15**, 81–87.
8 R. C. Gunawan, J. Silvestre, H. R. Gaskins, P. J. A. Kenis and D. E. Leckband, *Langmuir*, 2006, **22**, 4250–4258.
9 S. J. Lee, G. Khang, Y. M. Lee and H. B. Lee, *J. Colloid Interface Sci.*, 2003, **259**, 228–235.
10 J. Genzer, K. Efimenko and D. A. Fischer, *Langmuir*, 2006, **22**, 8532–8541.
11 R. A. Potyrailo and L. Hassib, *Rev. Sci. Instrum.*, 2005, **76**, 062225.
12 M. B. J. Wijesundara, E. Fuoco and L. Hanley, *Langmuir*, 2001, **17**, 5721–5726.
13 M. K. Chaudhury and G. M. Whitesides, *Science*, 1992, **256**, 1539–1541.
14 S. Daniel, M. K. Chaudhury and J. C. Chen, *Science*, 2001, **291**, 633–636.
15 S. Daniel and M. K. Chaudhury, *Langmuir*, 2002, **18**, 3404–3407.
16 H. Zhao and D. Beysens, *Langmuir*, 1995, **11**, 627–634.
17 H. Suda and S. Yamada, *Langmuir*, 2003, **19**, 529–531.
18 J. Genzer, *J. Adhes.*, 2005, **81**, 417–435.
19 K. Efimenko and J. Genzer, *Adv. Mater.*, 2001, **13**, 1560.
20 J. Genzer, D. A. Fischer and K. Efimenko, *Adv. Mater.*, 2003, **15**, 1545.
21 R. J. Petrie, T. Bailey, C. B. Gorman and J. Genzer, *Langmuir*, 2004, **20**, 9893–9896.
22 H. Elwing, S. Welin, A. Askendal, U. Nilsson and I. Lundstrom, *J. Colloid Interface Sci.*, 1987, **119**, 203–210.
23 L. M. A. Vandesteeg and C. G. Golander, *Colloids Surf.*, 1991, **55**, 105–119.
24 V. Hlady, *Appl. Spectrosc.*, 1991, **45**, 246–252.
25 C. G. Golander, Y. S. Lin, V. Hlady and J. D. Andrade, *Colloids Surf.*, 1990, **49**, 289–302.
26 S. H. Choi and B. M. Z. Newby, *Langmuir*, 2003, **19**, 7427–7435.
27 K. Loos, S. B. Kennedy, N. Eidelman, Y. Tai, M. Zharnikov, E. J. Amis, A. Ulman and R. A. Gross, *Langmuir*, 2005, **21**, 5237–5241.
28 S. V. Roberson, A. J. Fahey, A. Sehgal and A. Karim, *Appl. Surf. Sci.*, 2002, **200**, 150–164.
29 S. B. Kennedy, N. R. Washburn, C. G. Simon and E. J. Amis, *Biomaterials*, 2006, **27**, 3817–3824.
30 N. D. Gallant, K. A. Lavery, E. J. Amis and M. L. Becker, *Adv. Mater.*, 2007, **19**, 965–969.
31 Y. Ito, M. Heydari, A. Hashimoto, T. Konno, A. Hirasawa, S. Hori, K. Kurita and A. Nakajima, *Langmuir*, 2007, **23**, 1845–1850.
32 J. T. Han, S. Kim and A. Karim, *Langmuir*, 2007, **23**, 2608–2614.
33 Y. Mei, T. Wu, C. Xu, K. J. Langenbach, J. T. Elliott, B. D. Vogt, K. L. Beers, E. J. Amis and N. R. Washburn, *Langmuir*, 2005, **21**, 12309–12314.
34 B. Liedberg and P. Tengvall, *Langmuir*, 1995, **11**, 3821–3827.
35 B. Liedberg, M. Wirde, Y. T. Tao, P. Tengvall and U. Gelius, *Langmuir*, 1997, **13**, 5329–5334.
36 M. Lestelius, I. Engquist, P. Tengvall, M. K. Chaudhury and B. Liedberg, *Colloids Surf., B*, 1999, **15**, 57–70.
37 J. T. Smith, J. K. Tomfohr, M. C. Wells, T. P. Beebe, T. B. Kepler and W. M. Reichert, *Langmuir*, 2004, **20**, 8279–8286.
38 M. Riepl, M. Ostblom, I. Lundstrom, S. C. T. Svensson, A. W. D. van der Gon, M. Schaferling and B. Liedberg, *Langmuir*, 2005, **21**, 1042–1050.
39 R. H. Terrill, K. M. Balss, Y. M. Zhang and P. W. Bohn, *J. Am. Chem. Soc.*, 2000, **122**, 988–989.
40 Q. Wang, J. A. Jakubowski, J. V. Sweedler and P. W. Bohn, *Anal. Chem.*, 2004, **76**, 1–8.
41 S. T. Plummer and P. W. Bohn, *Langmuir*, 2002, **18**, 4142–4149.
42 K. M. Balss, G. A. Fried and P. W. Bohn, *J. Electrochem. Soc.*, 2002, **149**, C450–C455.
43 R. R. Fuierer, R. L. Carroll, D. L. Feldheim and C. B. Gorman, *Adv. Mater.*, 2002, **14**, 154–157.
44 N. Blondiaux, S. Zurcher, M. Liley and N. D. Spencer, *Langmuir*, 2007, **23**, 3489–3494.
45 N. Ballav, A. Shaporenko, A. Terfort and M. Zharnikov, *Adv. Mater.*, 2007, **19**, 998–1000.
46 R. Klauser, C. H. Chen, M. L. Huang, S. C. Wang, T. J. Chuang and M. Zharnikov, *J. Electron Spectrosc. Relat. Phenom.*, 2005, **144**, 393–396.
47 T. Kraus, R. Stutz, T. E. Balmer, H. Schmid, L. Malaquin, N. D. Spencer and H. Wolf, *Langmuir*, 2005, **21**, 7796–7804.
48 M. Geissler, P. Chalsani, N. S. Cameron and T. Veres, *Small*, 2006, **2**, 760–765.
49 L. Pardo, W. C. Wilson and T. J. Boland, *Langmuir*, 2003, **19**, 1462–1466.
50 A. Y. Sankhe, B. D. Booth, N. J. Wiker and S. M. Kilbey, *Langmuir*, 2005, **21**, 5332–5336.
51 S. Morgenthaler, S. Lee, S. Zürcher and N. D. Spencer, *Langmuir*, 2003, **19**, 10459–10462.
52 S. Morgenthaler, S. Lee and N. D. Spencer, *Langmuir*, 2006, **22**, 2706–2711.
53 N. V. Venkataraman, S. Zürcher and N. D. Spencer, *Langmuir*, 2006, **22**, 4184–4189.
54 H. T. Spijker, R. Bos, W. van Oeveren, J. de Vries and H. Busscher, *Colloids Surf., B*, 1999, **15**, 89–97.
55 T. Wu, K. Efimenko and J. Genzer, *J. Am. Chem. Soc.*, 2002, **124**, 9394–9395.
56 B. Zhao, *Langmuir*, 2004, **20**, 11748–11755.
57 Y. S. Lin, V. Hlady and C. Golander, *Colloids Surf., B*, 1994, **3**, 49–62.
58 B. J. Jeong, J. H. Lee and H. B. Lee, *J. Colloid Interface Sci.*, 1996, **178**, 757–763.
59 Y. Iwasaki, S. Sawada, N. Nakabayashi, G. Khang, H. B. Lee and K. Ishihara, *Biomaterials*, 1999, **20**, 2185–2191.
60 M. S. Kim, K. S. Seo, G. Khang and H. B. Lee, *Langmuir*, 2005, **21**, 4066–4070.
61 J. H. Lee, H. W. Kim, P. K. Pak and H. B. Lee, *J. Polym. Sci., Part A: Polym. Chem.*, 1994, **32**, 1569–1579.
62 R. R. Bhat, J. Genzer, B. N. Chaney, H. W. Sugg and A. Liebmann-Vinson, *Nanotechnology*, 2003, **14**, 1145–1152.
63 R. R. Bhat, M. R. Tomlinson and J. Genzer, *J. Polym. Sci., Part B: Polym. Phys.*, 2005, **43**, 3384–3394.
64 J. H. Lee, B. J. Jeong and H. B. Lee, *J. Biomed. Mater. Res.*, 1997, **34**, 105–114.
65 M. R. Tomlinson and J. Genzer, *Macromolecules*, 2003, **36**, 3449–3451.
66 M. R. Tomlinson and J. Genzer, *Langmuir*, 2005, **21**, 11552–11555.
67 K. Mougin, A. S. Ham, M. B. Lawrence, E. J. Fernandez and A. C. Hillier, *Langmuir*, 2005, **21**, 4809–4812.
68 C. Xu, T. Wu, C. M. Drain, J. D. Batteas and K. L. Beers, *Macromolecules*, 2005, **38**, 6–8.
69 A. Larsson and B. Liedberg, *Langmuir*, 2007, **23**, 11319–11325.

70 S. Morgenthaler, C. Zink, B. Städler, J. Vörös, S. Lee, N. D. Spencer and S. G. P. Tosatti, *Biointerphases*, 2006, **1**, 156–165.
71 L. Ionov, A. Sidorenko, M. Stamm, S. Minko, B. Zdyrko, V. Klep and I. Luzinov, *Macromolecules*, 2004, **37**, 7421–7423.
72 L. Ionov, N. Houbenov, A. Sidorenko, M. Stamm, I. Luzinov and S. Minko, *Langmuir*, 2004, **20**, 9916–9919.
73 L. Ionov, M. Stamm and S. Diez, *Nano Lett.*, 2005, **5**, 1910–1914.
74 X. J. Wang and P. W. Bohn, *Adv. Mater.*, 2007, **19**, 515–520.
75 C. Xu, S. E. Barnes, T. Wu, D. A. Fischer, D. M. DeLongchamp, J. D. Batteas and K. L. Beers, *Adv. Mater.*, 2006, **18**, 1427–1430.
76 J. D. Whittle, D. Barton, M. R. Alexander and R. D. Short, *Chem. Commun.*, 2003, **14**, 1766–1767.
77 B. P. Harris and A. T. Metters, *Macromolecules*, 2006, **39**, 2764–2772.
78 C. L. Hypolite, T. L. McLernon, D. N. Adams, K. E. Chapman, C. B. Herbert, C. C. Huang, M. D. Distefano and W. S. Hu, *Bioconjugate Chem.*, 1997, **8**, 658–663.
79 I. Caelen, H. Gao and H. Sigrist, *Langmuir*, 2002, **18**, 2463–2467.
80 X. J. Wang and P. W. Bohn, *J. Am. Chem. Soc.*, 2004, **126**, 6825–6832.
81 S. T. Plummer, Q. Wang, P. W. Bohn, R. Stockton and M. A. Schwartz, *Langmuir*, 2003, **19**, 7528–7536.
82 R. A. Venkateswar, D. W. Branch and B. C. Wheeler, *Biomed. Microdevices*, 2000, **2**, 255–264.
83 H. Baier and F. Bonhoeffer, *Science*, 1992, **255**, 472–475.
84 S. Kramer, H. Xie, J. Gaff, J. R. Williamson, A. G. Tkachenko, N. Nouri, D. A. Feldheim and D. L. Feldheim, *J. Am. Chem. Soc.*, 2004, **126**, 5388–5395.
85 S. H. Park and U. Krull, *Anal. Chim. Acta*, 2006, **564**, 133–140.
86 N. L. Jeon, S. K. W. Dertinger, D. T. Chiu, I. S. Choi, A. D. Stroock and G. M. Whitesides, *Langmuir*, 2000, **16**, 8311–8316.
87 S. K. W. Dertinger, D. T. Chiu, N. L. Jeon and G. M. Whitesides, *Anal. Chem.*, 2001, **73**, 1240–1246.
88 X. Y. Jiang, Q. B. Xu, S. K. W. Dertinger, A. D. Stroock, T. M. Fu and G. M. Whitesides, *Anal. Chem.*, 2005, **77**, 2338–2347.
89 I. Caelen, A. Bernard, D. Juncker, B. Michel, H. Heinzelmann and E. Delamarche, *Langmuir*, 2000, **16**, 9125–9130.
90 K. A. Fosser and R. G. Nuzzo, *Anal. Chem.*, 2003, **75**, 5775–5782.
91 R. R. Bhat and J. Genzer, *Appl. Surf. Sci.*, 2006, **252**, 2549–2554.
92 R. R. Bhat and J. Genzer, *Surf. Sci.*, 2005, **596**, 187–196.
93 R. R. Bhat, D. A. Fischer and J. Genzer, *Langmuir*, 2002, **18**, 5640–5643.
94 C. Huwiler, T. P. Künzler, M. Textor, J. Vörös and N. D. Spencer, *Langmuir*, 2007, **23**, 5929–5935.
95 S. V. Roth, M. Burghammer, C. Rieckel, P. Müller-Bauschbaum, A. Diethert, P. Panagiotou and H. Walter, *Appl. Phys. Lett.*, 2003, **82**, 1935.
96 B. E. Collins, K. S. Dancil, G. Abbi and M. J. Sailor, *Adv. Funct. Mater.*, 2002, **12**, 187–191.
97 L. M. Karlsson, P. Tengval, I. Lundström and H. Arwin, *J. Electrochem. Soc.*, 2002, **149**, C648–C652.
98 T. P. Kunzler, T. Drobek, C. M. Sprecher, M. Schuler and N. D. Spencer, *Appl. Surf. Sci.*, 2006, **253**, 2148–2153.
99 J. C. Meredith, J. L. Sormana, B. G. Keselowsky, A. J. Garcia, A. Tona, A. Karim and E. J. Amis, *J. Biomed. Mater. Res.*, 2003, **66A**, 483.
100 C. M. Stafford, K. E. Roskov, T. H. Epps III and M. J. Fasolka, *Rev. Sci. Instrum.*, 2006, **77**, 23908.
101 N. R. Washburn, K. M. Yamada, C. G. Simon Jr, S. B. Kennedy and E. J. Amis, *Biomaterials*, 2004, **25**, 1215.
102 X. Lu, J. Zhang, C. Zhang and Y. Han, *Macromol. Rapid Commun.*, 2005, **26**, 637.
103 N. Blondiaux, PhD thesis, No. 16699, ETH Zurich, 2006.
104 H. Cao, J. O. Tegenfeldt, R. H. Austin and S. Y. Chou, *Appl. Phys. Lett.*, 2002, **81**, 3058–3060.
105 R. R. Bhat, M. R. Tomlinson, T. Wu and J. Genzer, *Adv. Polym. Sci.*, 2006, **198**, 51–124.
106 H. Elwing, A. Askendal and I. Lundstrom, *J. Biomed. Mater. Res.*, 1987, **21**, 1023–1028.
107 Y.-S. Lin and V. Hlady, *Colloids Surf., B*, 1994, **2**, 481–491.
108 H. Elwing, S. Welin, A. Askendahl and I. Lundstrom, *J. Colloid Interface Sci.*, 1988, **123**, 306–308.
109 P. Tengvall, A. Askendal, I. Lundstrom and H. Elwing, *Biomaterials*, 1992, **13**, 367–374.
110 H. Elwing, A. Askendal and I. Lundstrom, *J. Colloid Interface Sci.*, 1989, **128**, 296–300.
111 J. M. Harris, *Poly(ethylene glycol) Chemistry and Biological Applications*, American Chemical Society, Washington, DC, 1997, vol. 680.
112 R. R. Bhat, B. N. Chaney, J. Rowley, A. Liebmann-Vinson and J. Genzer, *Adv. Mater.*, 2005, **17**, 2802–2807.
113 M. K. Chaudhury, S. Daniel, M. E. Callow, J. A. Callow and J. A. Finlay, *Biointerphases*, 2006, **1**, 18–21.
114 S. K. W. Dertinger, X. Y. Jiang, Z. Y. Li, V. N. Murthy and G. M. Whitesides, *Proc. Natl. Acad. Sci. U. S. A.*, 2002, **99**, 12542–12547.
115 W. Halfter, *J. Neurosci.*, 1996, **16**, 4389–4401.
116 D. Brunette, Principles of Cell Behavior on Titanium Surfaces and Their Application to Implanted Devices, in *Titanium in Medicine*, ed. D. Brunette, P. Tengvall, M. Textor and P. Thomsen, Springer, Berlin, Heidelberg, 2001, pp. 485–512.
117 T. P. Kunzler, T. Drobek, M. Schuler and N. D. Spencer, *Biomaterials*, 2007, **28**, 2175–2182.
118 T. P. Kunzler, C. Huwiler, T. Drobek, J. Vörös and N. D. Spencer, *Biomaterials*, 2007, **28**, 5000–5006.
119 C. B. Herbert, T. L. McLernon, C. L. Hypolite, D. N. Adams, L. Pikus, C. C. Huang, G. B. Fields, P. C. Letourneau, M. D. Distefano and W. S. Hu, *Chem. Biol.*, 1997, **4**, 731–737.
120 S. Daniel, S. Sircar, J. Gliem and M. K. Chaudhury, *Langmuir*, 2004, **20**, 4085–4092.
121 R. S. Subramanian, N. Moumen and J. B. McLaughlin, *Langmuir*, 2005, **21**, 11844–11849.
122 N. Moumen, R. S. Subramanian and J. B. McLaughlin, *Langmuir*, 2006, **22**, 2682–2690.
123 X. Yu, Z. Q. Wang, Y. G. Jiang and X. Zhang, *Langmuir*, 2006, **22**, 4483–4486.
124 A. D. Price and D. K. Schwartz, *Langmuir*, 2006, **22**, 9753–9759.
125 B. H. Clare, K. Efimenko, D. A. Fischer, J. Genzer and N. L. Abbott, *Chem. Mater.*, 2006, **18**, 2357–2363.
126 M. Y. M. Chiang, R. Song, A. J. Crosby, A. Karim, C. K. Chiang and E. J. Amis, *Thin Solid Films*, 2005, **476**, 379–385.
127 T. Künzler, PhD Thesis Nr 17049, ETH Zurich, 2007.

5620 *J. Phys. Chem. C* **2009,** *113,* 5620–5628

Spatial Tuning of the Metal Work Function by Means of Alkanethiol and Fluorinated Alkanethiol Gradients

Nagaiyanallur V. Venkataraman,[†] Stefan Zürcher,[†] Antonella Rossi,[†,§] Seunghwan Lee,[†] Nicola Naujoks,[‡,||] and Nicholas D. Spencer*[,†]

Laboratory for Surface Science and Technology, Department of Materials, and Materials Research Center, ETH Zurich, Wolfgang-Pauli-Str. 10, CH-8093 Zurich, Switzerland, Nanotechnology Group, ETH Zurich, Tannenstrasse 3, CH-8092 Zurich, Switzerland, and Dipartimento di Chimica Inorganica ed Analitica, Università degli Studi di Cagliari, Cittadella Universitaria di Monserrato, I - 09100 Cagliari, Italy

Received: October 16, 2008; Revised Manuscript Received: January 30, 2009

Surface-chemical gradients composed of self-assembled monolayers (SAM) of decanethiol (DT) and a partially fluorinated decanethiol (PFDT) on gold, exhibiting gradual changes in surface concentration of one or both components, have been prepared by a simple, controlled-immersion process. Infrared spectroscopic studies on a single-component PFDT gradient indicate a change in average molecular orientation with increasing surface coverage, whereas on a two-component gradient, the orientation remains invariant over the entire length of the gradient. X-ray photoelectron spectroscopic measurements on a single-component PFDT gradient show a systematic decrease in the fluorine (F 1s) binding energy with increasing surface coverage, whereas a single-component DT gradient shows an increase in the carbon (C 1s) binding energy. In two-component (DT-PFDT) gradients, the molar ratios of the two components at any particular location on the sample surface determine the magnitude of the binding-energy shifts at that location. Such shifts, which are on the order of 1 eV, are shown to be a consequence of work-function changes in the underlying gold upon SAM formation. These results are discussed in light of the surface-potential measurements on a DT-PFDT gradient by Kelvin Probe Force Microscopy and XP spectra acquired on "floating" and grounded samples.

Introduction

Self-assembled monolayers (SAMs) of thiols on gold[1,2] have attracted great interest as a convenient route for surface modification in widely different fields, such as microelectronics, biosensing, single-molecule electronics, or photovoltaics.[3−6] Many of these applications can be attributed to the ability of SAMs to precisely modify interfacial properties by exposing a particular chemical functional group at the SAM/air or SAM/liquid interface. For example, a simple change of terminal group in an alkanethiol SAM from $-CH_3$ to $-CF_3$ is sufficient to bring about dramatic changes in the interfacial dipole moment, wetting, and friction of these films.[7−9] Being able to adjust the surface potential and thereby the work function of a metal is of great interest, since it allows tuning of charge- and hole-injection properties in semiconducting devices, such as organic light-emitting diodes and other molecular electronic systems that involve a metal-organic junction.[10] Although the presence of any adsorbed species alters the work function of a metallic surface, SAMs are particularly attractive since they can lead to a highly ordered array of molecular dipoles on the surface, resulting in a net nonzero dipole moment normal to the surface, whose magnitude and sign can be adjusted by changing the chemical nature of the assembling molecules.

* To whom correspondence should be addressed. E-mail: spencer@mat.ethz.ch. Fax: +41 44 633 10 27.

† Laboratory for Surface Science and Technology, Department of Materials, and Materials Research Center, ETH Zurich.

‡ Nanotechnology Group, ETH Zurich.

§ Dipartimento di Chimica Inorganica ed Analitica, Università degli Studi di Cagliari, Cittadella Universitaria di Monserrato.

|| Present Address: Applied Physics, Chalmers Institute of Technology, 412 96 Göteborg, Sweden.

Several recent experimental as well as theoretical studies have compared SAMs of alkanethiols and fluorinated alkanethiols on gold and silver with respect to their ability to modify the work function of the metal.[11−17] Campbell et al. have studied, by Kelvin-probe measurements, the change in the surface potential of a silver electrode upon SAM formation with different thiols and correlated them with the calculated molecular dipole moments.[17] Alloway et al. measured the work function of a series of partially fluorinated thiol SAMs on gold and compared them to their nonfluorinated analogs, assigning the differences to the changes in molecular dipoles.[16] While Ray et al. studied the organization-induced changes in the work function of a series of thiol monolayers with different end-groups,[18] Carbacos et al. have recently shown the effect of embedded dipoles at different locations along an alkyl thiol monolayer.[19] De Boer et al. have demonstrated that SAMs of fluoroalkyl or alkyl thiols dramatically alter the hole-injection barrier in a metal−polymer contact.[15] These studies have established a clear correlation between molecular properties of the adsorbed thiolate species with the electronic properties of the resulting interface.

Most of the above studies, however, have focused on single-component SAMs. The present study demonstrates that by mixing an alkanethiol with a highly fluorinated alkanethiol, it is possible to precisely modify the work function of the resulting interface. Mixed monolayers can be readily obtained by self-assembly from solutions containing the two different thiols. However, the resulting surface compositions are often different from the composition of the solution, and to achieve precise control, it is necessary to study multiple samples. An alternative, more convenient way to explore a range of surface compositions is to deliberately prepare a gradient in surface composition of the two components, since it circumvents the need to prepare

Published on Web 03/17/2009

multiple sets of samples and provides improved control and reliability in the measured experimental results, which are all obtained under identical conditions.[20−23] There is a variety of methods available for the preparation of such surface-chemical gradients of thiols on gold, exhibiting well-defined changes in surface composition along a sample surface.[24−33] Using a simple methodology developed in our laboratory, gradient surfaces of some tens of millimeters in length with virtually any two types of alkanethiol molecules can be easily fabricated.[34−36] Here, we report surface-chemical gradients that are composed of an alkanethiol and a highly fluorinated alkanethiol, of identical chain lengths. We describe a detailed structural characterization, using a variety of surface-analytical methods. In particular, we focus on the ability of the two different thiols to modulate the charge redistribution at the interface and thereby the work function of the resulting interface.

It is well known that the work function of a metallic surface can be shifted dramatically by gas-phase deposition of controlled amounts of an alkali metal.[37,38] Ahn et al. have recently shown that such a change in work function upon deposition of different amounts of potassium onto an alkanethiol-SAM covering a gold substrate results in systematic shifts in the binding energy of the C 1s peak.[39] Their results offer a detailed explanation of some of the previously observed shifts in binding energies of alkanethiol SAMs upon deposition of a metal.[40] We show that shifts in binding energies upon changing the metal work function are not necessarily restricted to such external perturbations, but are also inherent in SAMs, depending on the nature of the assembling molecules. We emphasize that a close examination of binding energies obtained by carefully calibrated XPS measurements on SAMs can reveal additional information concerning the electronic structure of the resulting interface.

Experimental Section

Decanethiol (DT) (Aldrich) and 1H,1H,2H,2H-perfluorodecanethiol (PFDT) (Fluorous Technologies Inc., Pittsburgh, USA) were used as received. Ethanol (Fluka, puriss >99%) was used as solvent for all preparations. Silicon (POWATEC, Cham, Switzerland) substrates of appropriate dimensions were cleaned with oxidizing *piranha solution,* a 7:3 mixture of concentrated H_2SO_4 and 30% H_2O_2 (*Caution: piranha solution reacts violently when contacted with organic material and should be handled with extreme care*), thoroughly rinsed with copious amounts of MilliQ water (>18 MΩ), subsequently cleaned with O_2 plasma (Harrick Plasma Cleaner/Sterilizer, PDC-32G instrument, Ossining, NY) and covered with a 6 nm adhesive layer of chromium before evaporation of a 100 nm layer of gold (MED020 coating system, BALTEC, Balzers, Lichtenstein). Prior to the immersion step, the substrates were cleaned for 30 s in air plasma (high power) and left in pure ethanol for 10 min. Gradients were formed by a controlled immersion of the substrate into a dilute (5 μM) ethanol solution of decanethiol, the speed of immersion (75 μm/sec) being accurately controlled by a linear-motion drive (OWIS, Staufen, Germany). The sample was subsequently immersed in ethanol solution (0.01 mM) of PFDT overnight, followed by thorough rinsing with ethanol and drying in a flow of high-purity N_2. Mixed monolayers were prepared using an identical substrate-cleaning protocol and leaving them in a mixed thiol solution containing different molar ratios of DT and PFDT in ethanol at a total concentration of 1 mM for 1 day, followed by thorough rinsing in pure ethanol.

Contact-Angle Measurements. Static contact angles were measured at room temperature and ambient humidity. On gradient samples, a single measurement of sessile drops was performed with one drop (4 μL) of hexadecane (Aldrich) at 5 mm intervals along the substrate with a contact-angle goniometer (Ramé Hart model 100, Ramé Hart Inc., USA). All contact-angle measurements reported are averages over at least 10 samples.

XPS Measurements. XPS analyses were performed using a VG Theta Probe spectrometer (Thermo Fisher Scientific, East Grinstead, UK) equipped with radian lens, a concentric hemispherical analyzer, and a two-dimensional channel-plate detector with 112 energy and 96 angle channels. Spectra were acquired at a base pressure of 10^{-7} Pa or below using a monochromatic Al-Kα source with a spot size of 300 μm. The instrument was run in the standard-lens mode with electrons emitted at 53° to the surface normal. The analyzer was used in the constant-analyzer-energy mode. Pass energies used for survey scans and detailed scans were 200 and 100 eV, respectively, for gold Au 4f, carbon C 1s, oxygen O 1s, and sulfur S 2p. Under these conditions, the energy resolution (full width at half-maximum height, fwhm) measured on sputtered clean gold Au $4f_{7/2}$ is 1.95 and 0.82 eV, respectively. Acquisition times were approximately 5 min for survey scans and 30 min (total) for high-energy-resolution elemental scans. These experimental conditions were chosen in order to obtain an adequate signal-to-noise ratio in a minimum time and to limit beam-induced damage. At the end of a series of experiments, the C 1s signal was acquired for a second time. Under these conditions, sample damage was negligible and reproducible analysis conditions were obtained on all samples. The composition of the SAMs was calculated following analysis and the stoichiometry was always found to be in agreement with the theoretical value, thereby ruling out the possibility of X-ray-induced damage. Unless specified, all reported binding energies were referenced to the conventional value of Au $4f_{7/2}$ at 83.96 eV. All of the binding energies are reported with an uncertainty of ± 0.1 eV. Data were analyzed using the CasaXPS program [Version 2.3.5 www.casaxps.com]. The signals were fitted using Gaussian−Lorentzian functions taking into account the asymmetry on the higher-binding-energy side due to multielectronic effects: in the case of gold GL(82) and T(5.5), T being a tail function.[41] Least-squares-fit routines following Shirley iterative background subtraction were applied. Quantitative analysis and calculation of the film thickness were carried out using the peak areas corrected with the published photoionization cross sections,[42] the angular asymmetry function[43] and the inelastic mean free path calculated according to ref 44. The spectrometer was calibrated according to international standard procedures.[45] Typical binding energy values found on sputter cleaned samples of pure gold, silver and copper are as follows: Au 4f7/2 at 83.8 ± 0.1 eV, Ag 3d5/2 at 368.2 ± 0.1 eV and Cu 2p3/2 at 932.5 ± 0.1 eV.

Infrared Spectroscopy. Polarization-modulation infrared reflection−absorption spectra (PM-IRRAS) were recorded on a Bruker IFS 66v FT-IR spectrometer, equipped with a PMA37 polarization-modulation accessory (Bruker Optics, Germany). The interferogram from the external beam port of the spectrometer was passed through a KRS-5 linear polarizer (Bruker Optics, Germany) and modulated using a ZnSe photoelastic modulator (PEM, Hinds Instruments, USA). The modulated beam was reflected off the sample at an incident angle of 80° and collected by a ZnSe lens onto a liquid-nitrogen-cooled MCT detector. The use of polarization modulation eliminates the need to measure a background spectrum.[46] Typically, 2048 multiplexed interferograms with 8 cm^{-1} resolution were coadded to improve the signal-to-noise ratio. The spectra were baseline-corrected with the spectrometer's OPUS software (Bruker

5622 *J. Phys. Chem. C, Vol. 113, No. 14, 2009* Venkataraman et al.

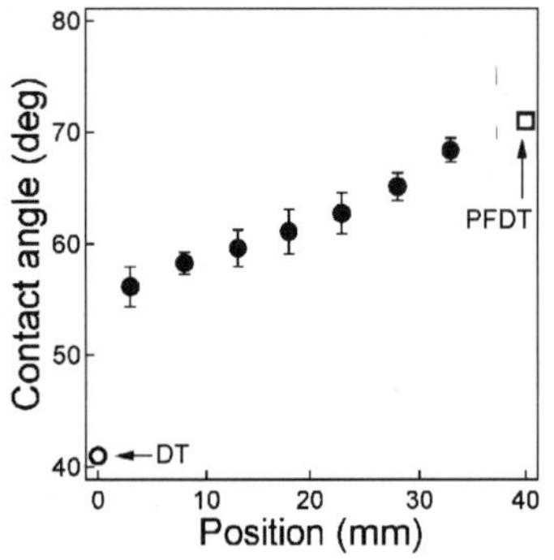

Figure 1. Static hexadecane contact angles measured at 5 mm intervals along a DT-PFDT gradient. Contact angles measured on pure DT and PFDT monolayers (open symbols) are displayed at the two extremes for comparison.

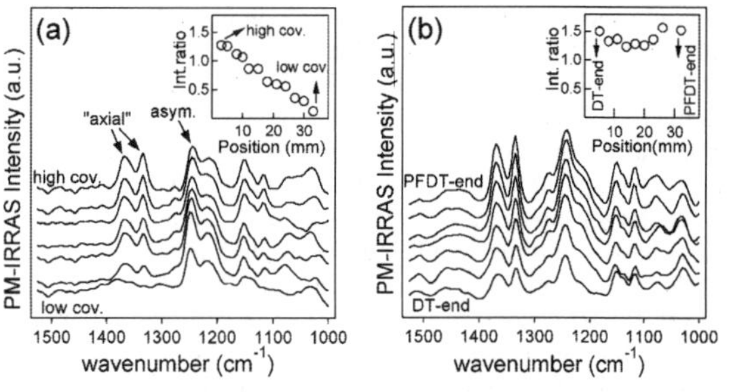

Figure 2. PM-IRRA Spectra in the C−F stretching region measured along the length (a) for a single-component PFDT gradient and (b) for a two-component DT-PFDT gradient. The spectra are measured at 5 mm intervals along the gradient sample arranged from bottom to top with increasing surface concentration of PFDT. Changes in the ratio of the "axial" stretching mode intensity to that of the CF_2 asymmetric stretch along length of the gradient are plotted as insets.

Optics, Germany) using a polynomial background that encompassed the featureless regions well outside the region of interest. The sample compartment was continuously purged with dry air during the measurements. Gradient samples of 4 × 2 cm dimensions were used for the measurements using a sample holder that was suitably modified to be able to record spectra of different regions of the gradient with a 2 mm slit.

Kelvin Probe Force Microscopy (KPFM). Kelvin probe force microscopy (KPFM) was employed to measure the surface potential along the surface of a gradient sample. The KPF Microscope used in this work is based on a modified commercial atomic force microscope (Nanoscope IIIa, MultiMode with Extender Electronics Module, Veeco Metrology Group, Santa Barbara, CA, USA) where topography and electric surface potential are measured sequentially using the lift-mode technique to minimize cross-talk.[47,48] To this end, the surface topography of a single line was first acquired in Tapping Mode and then the same line was immediately retraced at a set lift height (typically 15 nm) from the sample surface to measure the surface-potential distribution. Topographic and surface-potential images are thus obtained by repeating this procedure for each line along the slow-scan axis. Since the measured surface potential in this approach represents the potential difference (or offset) between tip and sample, all measurements were performed with the same tip/cantilever assembly on the same day. A commercial *n*-doped silicon tapping-mode tip (Nanosensors, Neuchatel, Switzerland, type PPP-NCHR) was employed.

Results and Discussion

Contact-Angle Measurements. Fluorinated thiol monolayers are both hydrophobic as well as oleophobic, leading to large contact angles with water as well as with liquid hydrocarbons. Initial characterization of the DT-PFDT gradients was achieved by measuring static hexadecane contact angles. The contact angles measured at 5 mm intervals along a 4 cm DT-PFDT gradient are shown in Figure 1. The data were averaged over at least 10 different samples. For measurements at different locations of the sample represented as position (in mm), the value 0 mm refers to the location of the sample with highest surface concentration of the thiol used in the first immersion step. A steady increase in the hexadecane contact angle is seen along the length of the gradients moving away from the DT-rich end, indicating an increase in surface concentration of the PFDT from one end of the sample to the other. Hexadecane contact angles measured on pure DT and PFDT SAMs are also shown in Figure 1 for comparison. Quantitative measurement of the surface concentrations of the two species was obtained from the XPS measurements described below. However, contact-angle measurements served as a quick check of the quality of duplicate gradient samples used in further spectroscopic characterization.

Infrared Spectroscopy. PM-IRRA Spectra in the region 1600−1000 cm^{-1} measured at different positions along a 4 cm long single-component PFDT gradient sample are shown in Figure 2a. Spectra from a two-component DT-PFDT gradient are displayed in Figure 2b. The spectra are shown displaced from bottom to top with increasing surface concentration of PFDT. This spectral region is characteristic of C−F stretching modes of a fluorinated chain molecule. The assignments of the bands given below are similar to those reported in the literature.[49−55] The strong band at around 1256 cm^{-1} is assigned to the asymmetric C−F stretching mode of the fluorinated part of the chain, while the weaker band centered at 1150 cm^{-1} is assigned to the symmetric C−F stretching mode. An important feature of the series of spectra shown in Figure 2 is the presence of intense bands at 1336 cm^{-1} and 1369 cm^{-1}. These vibrational modes, often referred to as "axial" stretching modes, are a characteristic feature of the infrared spectrum of long fluorinated chain assemblies.[50,54] These bands have their transition dipole moment along the fluorinated chain axis and therefore appear with much higher relative intensity in the reflection−absorption spectrum of ordered fluorinated monolayer assemblies measured on metallic substrates than in the corresponding bulk or transmission spectra.[54]

In the case of a single-component PFDT gradient, the relative intensity of the axial stretching modes compared to that of the CF_2 asymmetric stretch decreases with decreasing surface concentration of PFDT (top to bottom in Figure 2a). In the two-component DT-PFDT gradient (Figure 2b), however, both of these modes show an increase with increasing surface concentration of PFDT, leading to this intensity ratio remaining almost invariant over the entire length of the gradient. The intensity of the axial stretching mode at 1369 cm^{-1}, normalized to the CF_2 asymmetric stretch by the respective peak heights, are shown as insets in Figure 2, parts a and b. Reflection−absorption infrared spectral intensities measured on a metallic substrate are a consequence of the concentration of the surface species as well as their orientation on the surface, due to the surface selection rules.[56] As a first approximation, the contribution to the observed changes in intensities due to different surface concentrations of the PFDT may be taken to be equal in both single- and two-component gradients. The observed differences in the intensity ratios between the single component and two-component gradient can then be reconciled by considering changes in the average orientations of the PFDT chains in the two systems. The transition dipole moment for the CF_2 asymmetric stretching modes lies normal to the PFDT chain axis, whereas for the "axial" stretching modes, it lies parallel. In the case of the single-component gradient, the average molecular orientation changes continuously with increasing surface concentration of PFDT molecules along the length of the gradient. With increasing surface concentration of PFDT, there is an increasing tendency to adopt an upright orientation, leading to an increase in the intensity of the axial stretching modes at the expense of the CF_2 asymmetric stretch. On the other hand, in the two-component gradient composed of DT and PFDT molecules, although the surface concentration of PFDT molecules changes along the length of the gradient, the average molecular orientation remains almost invariant, leading to the intensity ratio of the two modes being constant.

It is also worth noting that the symmetric and antisymmetric C−H stretching modes of the DT molecules in a DT-PFDT gradient (in the C−H stretching region-spectra not shown) appeared at 2854 and 2924 cm^{-1}, respectively, along the entire length of the gradients, indicating that the alkyl chains of the decanethiol molecules adopt a disordered conformation throughout the entire length of the gradient. The overall intensity in the spectral region of 3000−2800 cm^{-1} decreases with increasing surface concentration of PFDT, since PFDT molecules do not have a measurable C−H stretching-mode intensity.

XPS. Photoelectron spectra of full monolayers of DT and PFDT were measured under conditions previously selected using a series of reference samples. Consistent with the previously reported values,[57] the measured binding energy of the sulfur 2p signals was found to be 162.1 ± 0.1 eV corresponding to a thiolate bound to gold. The binding energies of F 1s measured on the polymeric reference samples, poly(tetrafluoroethylene) (PTFE: $-(CF_2CF_2)_n-$), (689.7 ± 0.1 eV) and poly(vinylidene fluoride) (PVDF: $-(CH_2CF_2)_n-$) (688.2 ± 0.1 eV), are close to the published values.[58] Using Au $4f_{7/2}$ as a reference, the binding energy of F 1s in a full-coverage PFDT SAM is found at 688.2 ± 0.1 eV: this value of F 1s binding energy was rather surprising, since it is closer to that found on PVDF rather than the value for PTFE, which might be expected from the structure of the fluorinated PFDT molecule. This shift in the F 1s binding energy cannot be accounted for by a chemical-state effect, since no change in chemical state of the F atoms is expected upon SAM formation. The other possibility is that this shift is a consequence of referencing the binding energies to the underlying gold, disregarding the effect of SAM formation on the Au $4f_{7/2}$ photoemission, which leads to an apparent shift in the measured binding energies. It has been shown that the work function of a metal is drastically affected by SAM formation and depends on the nature of the assembling molecules.[17] A change in the metal work function in a SAM system has been shown to systematically change the measured binding energies of the monolayer atoms.[39]

To further test this hypothesis of the shift in binding energy of the F 1s signal being a consequence of the changes in gold work function, two single-component gradient samples with increasing surface coverage of either DT or PFDT were measured. If, indeed, the measured shifts in the binding energies are a consequence of changes in work function of gold upon SAM formation, then opposite trends in binding-energy shifts should be observed on the two gradients since the work function

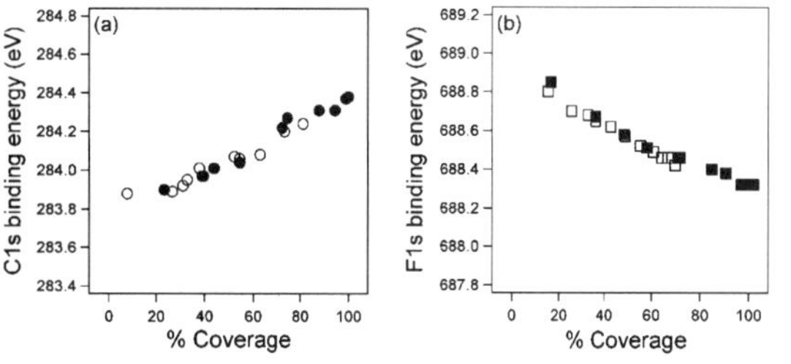

Figure 3. (a) Carbon (C 1s) and (b) Fluorine (F 1s) binding energies measured on a single-component DT gradient and single-component PFDT gradient, respectively. The open symbols represent gradient samples prepared by immersing at twice the speed as those represented by filled symbols, in order to expand the low coverage region and to ensure reliability of the measured binding energies in this region.

of gold shifts in an opposite direction for alkanethiol and fluorinated alkanethiol monolayers.[15] Therefore, in these measurements, it was chosen not to reference the measured binding energies to the Au $4f_{7/2}$ signal of the underlying gold substrate and rely on the instrumental calibration performed just prior to the sample measurements. The binding energies of aliphatic C 1s at different positions measured along a single-component DT-gradient and the F 1s binding energies on a PFDT-gradient are shown in Figure 3. The two gradients show systematic shifts in opposite directions with increasing surface coverage. With increasing DT coverage there is a systematic increase in the measured aliphatic C 1s binding energy (Figure 3a), whereas, with increasing coverage of PFDT there is a systematic decrease in F 1s (Figure 3b) as well as in the fluorinated carbon C 1s binding energies (not shown). The surface coverage was calculated from the film thickness obtained by means of a three-layer model adapted to this system.[59] In order to rule out the possibility of any instrumental artifacts being responsible for such shifts in the measured binding energies, a "blank gradient", prepared by slow immersion of a clean gold sample for the same duration of time into pure ethanol containing no thiol molecules, was measured under identical acquisition conditions. The C 1s binding energies measured at different locations on this sample showed no systematic trend, deviating only within a range of ± 0.1 eV from the mean value of 284.2 eV (see Supporting Information).

To enable a better correlation of the measured binding energy with the surface composition and to monitor the effect on the binding-energy shifts upon mixing of the two components, a series of mixed monolayers obtained by coadsorption of DT and PFDT from mixed solutions as well as two-component DT-PFDT gradients were thoroughly characterized by XPS. Two-component DT-PFDT gradients were obtained by back-filling a DT gradient with PFDT rather than the other way around, due to the greater tendency of replacement of PFDT by DT in ethanol during the second immersion step. The quantification of the surface species in both gradient and mixed-monolayer samples is presented in Figure 4. The values are plotted as a function of mole fraction of adsorbed PFDT to enable a direct comparison of the gradient and mixed monolayer samples. The mole-fraction of PFDT (n_{PFDT}) was calculated according to eq 1:

$$n_{PFDT} = \frac{A_{PFDT}}{A_{PFDT} + A_{DT}} \qquad (1)$$

where A_{PFDT} and A_{DT} are the areas of the PFDT and DT component in the C 1s spectra (see Figure 5c). The fluorinated

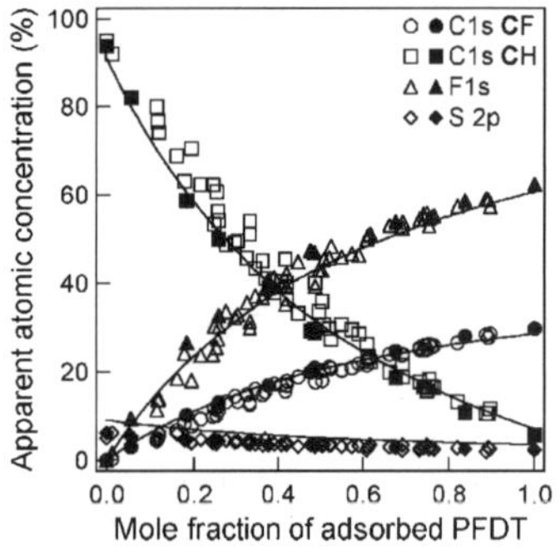

Figure 4. Apparent normalized atomic concentrations of the elements related to the SAM as a function of mole fraction of PFDT in two-component gradients (open symbols) and homogeneous mixed monolayers (closed symbols). The solid lines are the calculated values as a function of mole fraction of PFDT. The nonlinear behavior is due to the nondetectable hydrogen atoms, leading to a change in the total number of measured atoms from 11 when the mole fraction of PFDT is 0 to 28 when the mole fraction is 1.

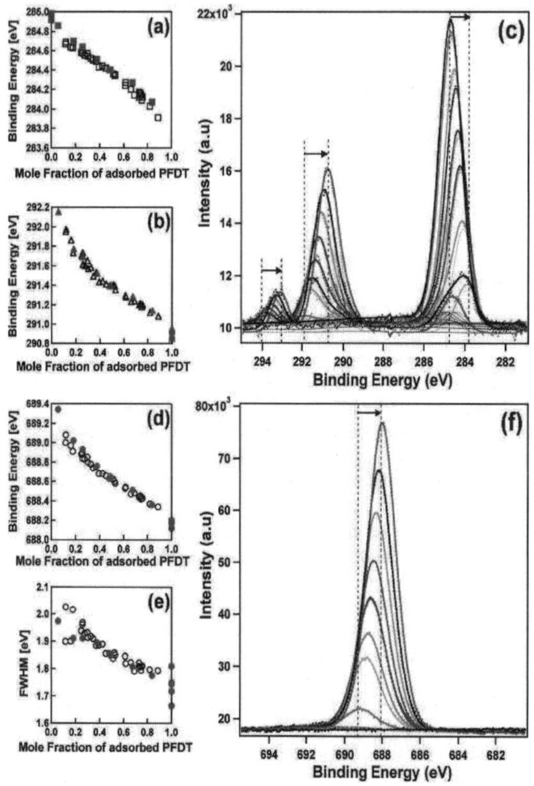

Figure 5. Binding energies of (a) aliphatic C 1s, (b) fluorinated C 1s, and (d) F 1s and (e) full-width at half-maximum (fwhm) of the F 1s signal are plotted as a function of the mole fraction of PFDT found in mixed monolayers (red) as well as two-component DT-PFDT gradient samples (open symbols). Representative XPS (c) C 1s and (f) F 1s signals measured on a series of mixed monolayers samples. The measured binding energies are indistinguishable for the gradient and mixed monolayer samples and depend solely on the surface molar ratio of the two thiols. The uncertainty in the measured binding energies is ± 0.1 eV.

and nonfluorinated C 1s peaks are completely separated on the binding-energy scale. The two aliphatic CH_2 groups belonging to the PFDT molecules have been constrained on the basis of the curve-fitting results obtained in the case of the pure full monolayer.

The amount of aliphatic carbon (C 1s **CH**) decreases continuously along the gradient, whereas the amounts of fluorinated carbon (C 1s **CF**) and fluorine (F 1s) increase with increasing adsorbed mole fraction of PFDT. The amount of sulfur (S 2p) remains nearly constant (Figure 4). The observed trends are consistent with the predicted values (plotted in the figure as solid lines). In the subsequent discussion, the mole fraction of adsorbed PFDT, rather than position, is used as a unique parameter to describe a specific location in a gradient sample (independent of gradient fabrication conditions) to afford direct comparison with mixed-monolayer samples.

Detailed XPS measurements on DT-PFDT gradients and mixed monolayers are presented in Figure 5. The data presented in Figure 5 are referenced to Au $4f_{7/2}$ by repositioning all of the binding energies to make the substrate gold signal appear at 83.96 eV. The typical value of the substrate gold signal was 83.8 ± 0.1 eV in all of the samples. The measured binding energies of aliphatic carbon (C 1s $\mathbf{CH_2}$), fluorinated carbon (C 1s $\mathbf{CF_2}$) and the fluorine (F 1s) signals are shown in panels (a), (c) and (d), respectively. Representative spectra showing the C 1s and F 1s signals measured on a series of DT-PFDT mixed monolayer sample are presented in panels (c) and (f), respectively. All measured binding energies are plotted as a function of mole fraction of adsorbed PFDT. The binding energies of all surface species show consistent shifts to lower binding energies, with the shifts scaling almost linearly with the mole fraction of adsorbed PFDT. Although the exact values of measured fractions of the two components differed slightly from sample to sample, the binding energies of all the signals plotted against the adsorbed mole fraction of PFDT follow the same trend, as evident from Figure 5.

The spread of the binding-energy shifts between the two extremes of coverage is about 1 eV for all components, such as the aliphatic carbon C 1s, fluorinated carbon C 1s or the fluorine F 1s signals. The shifts observable on the gradient samples span almost the entire range of the measured binding-energy shifts except in the two extremities of the concentration range. Also plotted in Figure 5(f) is the full-width at half-maximum (fwhm) of the F 1s signal as a function of mole fraction of adsorbed PFDT. The fwhm of the F 1s signal also shows a consistent shift to lower values with increasing PFDT concentration. A significant reduction in the fwhm of the F 1s signal of about 0.3 eV is observed between the two extremes of the gradient. The fwhm of the carbon 1s signals are not discernible due to their smaller natural line widths and different closely spaced and overlapping signals. To further substantiate the measured shifts in binding energies upon changing surface mole fraction of the two species, KPFM measurements were carried out on representative samples of DT-PFDT gradients.

Kelvin Probe Force Microscopy (KPFM). KPFM measures the contact potential difference of a sample surface with reference to the tip and affords a measure of surface potential with nanometer-scale resolution.[60] Figure 6 displays the relative surface potential of a DT-PFDT gradient, together with those of pure DT and PFDT SAMs measured by the KPFM approach. For this measurement, the gradient sample (1 cm (width) × 4 cm (length)) was cut into 4 pieces (1 cm (width) × 1 cm (length) each) due to geometric restrictions in the KPF-Microscope. KPFM measurements were performed at three different locations along the concentration gradient per sample piece. The surface potential values shown in Figure 6 were obtained from the

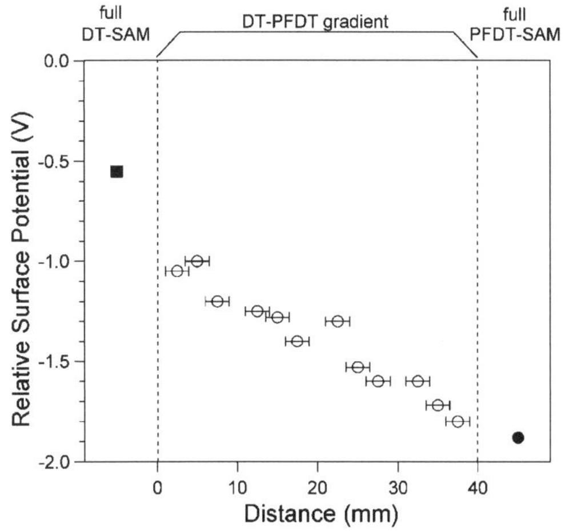

Figure 6. Surface potential of a DT-PFDT gradient and pure DT- and PFDT-SAM measured by means of Kelvin Probe Force Microscopy (KPFM) (The *x*-axis error bars represent the uncertainty of the position along the gradient sample).

section average across the middle line of each surface potential image (1 × 1 μm). However, values obtained at other locations on the same sample showed good correlation, with the trend shown in Figure 6; the maximum deviation was ± 0.01 V. The *x*-axis error bars represent the uncertainty of the position along the gradient sample.

Consistent with previous studies,[61−63] the full PFDT-SAM revealed a more negative surface potential (−1.88 V) than that of the full DT-SAM (−0.55 V). In addition, the surface potential along the gradient sample showed a gradual change in a virtually linear fashion toward more negative values from the DT-rich end to the PFDT-rich end. This observation supports the presence of a gradient in surface potential along the sample. Since KPFM measures the contact potential difference (CPD) between the tip and the sample, it sensitively depends on the tip conditions. Therefore, a careful calibration with a known standard is required to be able to extract absolute values of surface potential. However, the difference in surface potentials (ΔV) measured between two samples under identical experimental conditions with one and the same tip should provide a reasonable estimate of the relative shift in surface potential. In the present case, the difference in surface potential measured between full monolayers of DT and PFDT is 1.33 V. This value is in good agreement with the reported differences in work function measured on similar alkanethiol monolayer systems (1.4 eV).[15]

Origin of the Observed BE Shifts. Systematic shifts in carbon (C 1s) and fluorine (F 1s) binding energies were observed in XPS measurements at different positions along a gradient surface composed of decanethiol and partially fluorinated decanethiol, as well as in corresponding homogeneous mixed monolayer samples with different surface mole fractions of the two thiols. The binding energy of gold or the sulfur, however, did not show any significant shifts in measured binding energy with different surface mole fractions of the two components. The binding energy shifts of C 1s and F 1s are, at first, surprising considering the fact that no change in the chemical state of either the carbon or the fluorine atoms occurs upon monolayer formation. No change in chemical state of the two atoms is also likely to happen as a function of different surface molar ratios of the two thiols, thus ruling out *initial-state* effects as a possible cause of these shifts.

Binding energy in a typical XPS measurement is given by the following:

$$\mathrm{BE} = h\nu - \mathrm{KE} - \varphi_{\mathrm{sp}} \quad (2)$$

where $h\nu$ is the incoming photon energy, KE the measured kinetic energy, and ϕ_{sp} the spectrometer work function. The last term ensures that the binding energy of any particular chemical species reported from different measurements over several spectrometers is consistent. Binding energies are usually referenced to a known standard value. Clean metal surfaces with widely differing BEs (typically Cu, Ag, Au) are generally measured to ensure the linearity of the BE scale.[45] For a metallic substrate, the work function of the substrate does not influence the measured binding energy since it is "grounded" to the spectrometer, leading to the Fermi levels of the sample and the spectrometer being aligned and ensuring no spectrometer dependence of the measured BE. Therefore, for XPS measurements on metallic substrates, covered with thin organic overlayers of up to a few nanometers thickness, the core-level photoemission signal from the metal is generally used as a reference binding energy. For an insulating substrate, the photoemission process results in the build-up of positive charges on the surface, leading to an arbitrary shift in the measured BE and the need for a low-energy electron flux to neutralize the charge, thus allowing the measured BE to be used reliably. Nevertheless, as shown recently by Ahn et al.[39] changes in work function of a metallic substrate can also influence the measured BE, especially in the case of SAMs. With decreasing work function of the metal, in their experiment achieved by increasing amounts of a dosed potassium overlayer, there is an increase in the measured binding energy. A similar phenomenon could explain the observed shifts in binding energy seen on the DT-PFDT gradient samples. Before moving on to a discussion of the observed shifts, it is worth noting that either the referencing of all the observed signals to the substrate gold signal by appropriately repositioning them to make the Au $4f_{7/2}$ appear at 83.96 eV (as is practiced in literature) or carrying out measurements immediately following the calibration of the BE scale using sputter-cleaned metal samples without referencing to the substrate gold, only led to systematic shifts of the order of 0.2 eV. While the former approach provides precise values of BE to afford direct comparison with those values reported in literature, the latter is preferable when a more accurate value of the measured BE is desired. Nevertheless, the observed trends in BE shifts reported here were completely reproducible within the specified uncertainty of ± 0.1 eV as long as one or the other approach was adopted with consistency.

On a single-component DT-gradient, there is an increase in the measured C 1s BE with increasing surface coverage (Figure 3 a), consistent with the fact that the work function of gold decreases upon alkanethiol SAM formation. Similarly, the formation of a PFDT monolayer increases the work function of gold, leading to a decrease in the measured F 1s and C 1s binding energies with increasing surface coverage on a PFDT-gradient (Figure 3b). However, the exact magnitude of the shifts seen in single-component gradients does not entirely reflect the changes in the work function reported for these monolayers.[15] These differences could be rationalized by considering the fact that the change in work function ($\Delta\Phi$) upon SAM formation depends on the resulting net dipole moment change ($\Delta\mu\perp$) normal to the sample surface given by:[17]

5626 *J. Phys. Chem. C, Vol. 113, No. 14, 2009* Venkataraman et al.

$$\Delta\Phi = \frac{eN\Delta\mu_{\perp}}{\varepsilon} \quad (3)$$

where N is the surface density of the adsorbed species and ε the dielectric constant. In the case of single-component gradients, the net dipole moment change normal to the surface has two contributions, one due to the formation of an increasing number of Au–S bonds and the other from the net dipole moment normal to the surface arising from to the organic part of the monolayer. For simplicity, the former contribution may be taken to be approximately equal in magnitude for both DT and PFDT gradients disregarding the known differences in the surface densities of DT and PFDT SAMs arising from their different sizes.[52] The latter contribution, however, depends on the average orientation of the chains in the monolayer and the individual molecular dipole moments ($\mu_{\perp} = \mu_M \cos\theta$). Infrared spectroscopy indicates a change in average molecular orientation in single-component gradients of the PFDT (Figure 2) with increasing coverage, similar to that reported earlier for a single-component alkanethiol gradient.[34] However, it must be emphasized here that the infrared measurements report average properties of the films over a few square millimeters and do not reflect the details of the microstructure. In the single-component PFDT gradient, a change in overall tilt angle or an increase in the size of "islands" of ordered upright molecules surrounded by a "sea" of lying-down molecules would lead to similar changes in infrared spectral intensities. From previous atomic force microscopy (AFM) results on a single-component alkanethiol gradient,[35] the latter scenario is likely to be closer to the actual picture. Also, the presence of contaminants from the ambient in the nonbackfilled regions of the sample cannot be ruled out and could influence the observed shifts.

Two-component gradients provide a much clearer picture of the observed changes in binding energy. When single-component gradients are back-filled with the second component resulting in full monolayers, there is little or no change in average molecular orientation of the different components over the entire length of the gradient (Figure 2). Therefore, the net interfacial dipole moment at any particular location of the sample can almost entirely be ascribed to the molar ratio of the two thiols on the surface. The two-component gradient is also devoid of atmospheric contaminants, due to very low surface energies of the DT and PFDT monolayers, thus simplifying the interpretation of the observed shifts. The C 1s and F 1s XPS signals (Figure 5, parts a, b, d) show a steady decrease of about 1.2 eV between the two extremities of the measured concentration range. The DT-PFDT gradient shows a slightly smaller shift spanning 0.8 eV in the BEs between the two extremes of the sample. KPFM measurements on a DT-PFDT gradient indicates a change in work function (given by $\Delta\Phi = e\Delta V$) of about −1.0 eV on moving from the DT-rich end of the gradient to the PFDT-rich end of the sample and a difference $\Delta\Phi$ of −1.3 eV between the two full monolayers. The observed shifts in work function determined by KPFM and the shifts in BEs measured by XPS differ by 0.1−0.2 eV. It should be borne in mind that XPS measurements do not span the entire concentration range due to the low signal-to-noise ratio or the complete absence of a particular signal at the extremities of the concentration range, while no such restriction exists for KPFM measurements. A schematic depiction of the changes in work function induced by the opposite dipole moments of DT and PFDT in a DT-PFDT gradient sample is given in Figure 7. There is a continuous shift of the vacuum level (E_v)[64] due to the changing net dipole moment normal to the surface, determined by the surface molar ratio of the two thiols, resulting in a gradual change in the work function.

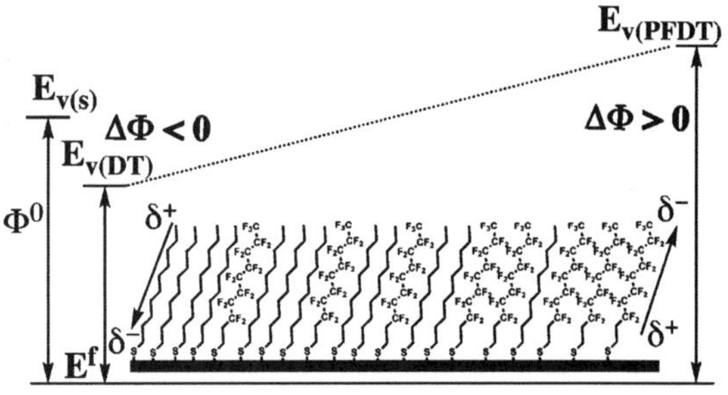

Figure 7. Schematic depiction of the changes in the work function along the length of a DT-PFDT gradient. The work function of SAM-covered gold, compared to the work function of pure gold (Φ_0), is lowered ($\Delta\Phi < 0$) upon DT adsorption, whereas it is raised ($\Delta\Phi > 0$) upon PFDT adsorption. The exact magnitude of the work function at any location along the gradient sample is, therefore, determined by the surface mole fraction of DT and PFDT.

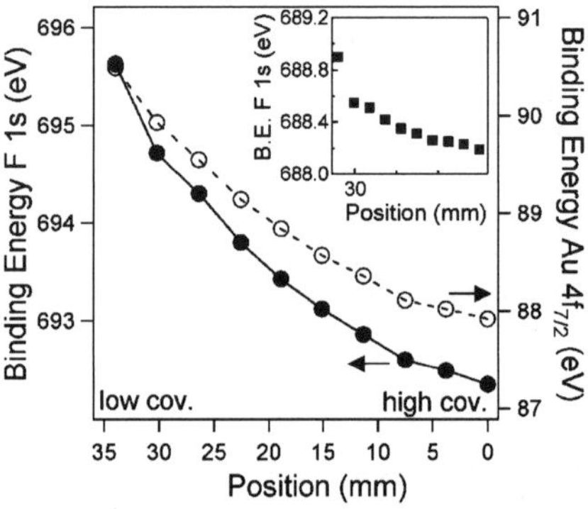

Figure 8. Binding energy of the Au $4f_{7/2}$ (open symbols) and F 1s (closed symbols) signals measured at different locations along a single-component PFDT gradient measured with the sample left "floating" with respect to the spectrometer (moving from position 35 to position 0 mm corresponds to increasing surface coverage of PFDT). Both Au $4f_{7/2}$ and F 1s signals show a shift toward lower values with increasing surface coverage. The inset shows BEs of the F 1s signals after referencing to the gold, by positioning the Au $4f_{7/2}$ signal at 83.96 eV. After referencing, the resulting shifts in the F 1s signal are similar to that seen in the measurements carried out with the sample grounded (Figure 3b).

To further substantiate this argument of the measured binding-energy shifts being a consequence of the shift in the work function of the underlying gold, XPS measurements were carried out on gradient samples with either the sample being grounded or left "floating" with respect to the spectrometer during the signal acquisition. The positions of the gold and fluorine signals measured at different locations along a single component PFDT gradient are shown in Figure 8. A single component PFDT-gradient rather than a DT-gradient was chosen due to the presence of an intense F 1s and distinctly different C 1s signals, well-separated from the adventitious carbon C 1s signal, making it relatively easier to measure the binding energies accurately. Unlike the measurement done on a single-component PFDT gradient with the sample grounded (Figure 3b), the samples measured "floating" showed shifts in the binding energies of all the acquired signals due to the sample charging, as shown for the gold and fluorine signals in Figure 8. These shifts varied from sample to sample and were of the order of 2.5 ± 0.2 eV.

$(\Delta(\delta BE) \approx \Delta\Phi_{Au})$.

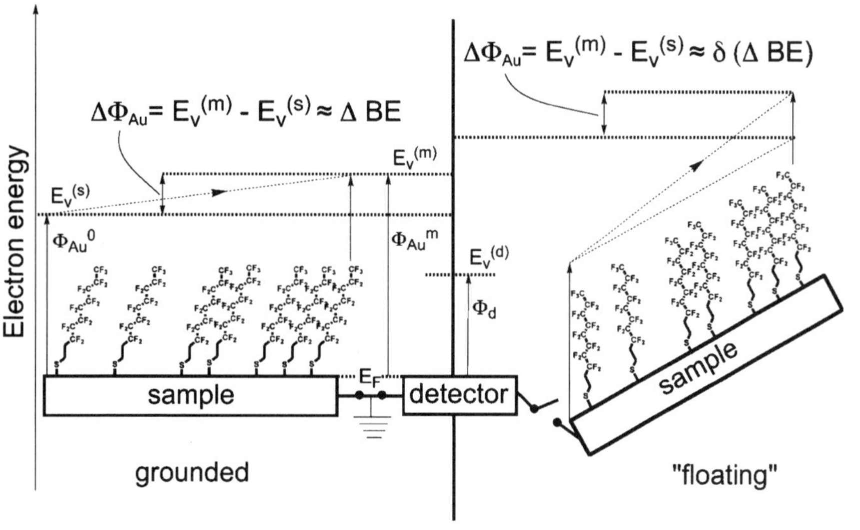

Figure 9. Schematic representation (adapted from ref 64) of the shifts in the gold work function along a single-component PFDT gradient and its effect on the measured binding energies. The left and right panels represent the situations when the sample is electrically grounded (left) or remains "floating" (right), with respect to the detector. In the left panel, since the Fermi level of the substrate is aligned to that of the detector, the relative positions of the vacuum levels of the pure and SAM-covered substrates (E_V^s and E_V^m) are well defined on the energy axis, allowing the observed BE shifts to be directly related to the shifts in the work function ($\Delta\Phi$) along the gradient sample. In the right panel, due to the lack of an energy-level alignment, the precise positions of these levels are arbitrary. The relative *differences* in the position of the levels (E_V^s and E_V^m) with respect to the vacuum level of the detector (E_v^d) are still directly related to the measured *shifts* in BE.

Even though the extent of this overall shift varied between different measurements, when the positions of all the signals were referenced to make the Au $4f_{7/2}$ appear at 83.96 eV (as in the case of grounded measurements), only the F 1s (Figure 8, inset) and C 1s (see Supporting Information) signals showed a shift to lower binding energy and the S 2p remained invariant (see Supporting Information). Furthermore, the difference between the change in binding energy of gold between the two extremes of the gradients and the shift in the F 1s signal between the high coverage and the low coverage regions of the gradient is 0.7 eV. These control measurements rule out the possibility of the measured shifts being a consequence of any instrumental artifacts and clearly establish that the nature of the shifts in binding energy is a consequence of the shifts in the work function of the underlying gold due to SAM formation.

A schematic representation of the two situations discussed above is shown in Figure 9. In the measurements carried out with the samples grounded, shifts in work function of the gold are reflected as shifts in binding energies of the C 1s and F 1s XPS signals with the Au 4f and S 2p remaining invariant ($\Delta BE \approx \Delta\Phi_{Au}$). In the "floating" case, however, the shifts are evident in all the measured binding energies due to the lifting of the degeneracy between the Fermi levels of the substrate and the detector, leading to shifts in the positions of all the measured signals. However, when the peak positions are referenced by appropriately shifting them to make the Au $4f_{7/2}$ signal appear at 83.96 eV, only the C 1s and F 1s signals show a shift in their binding energies, leaving the S 2p and the valence band of Au invariant (see Supporting Information), similar to that seen in the grounded case ($\Delta(\delta BE) \approx \Delta\Phi_{Au}$).

The above arguments clearly demonstrate the effect of changing metal work function on the measured binding energy. It is now clear why this difference in work function of gold does result in a similar shift in the measured sulfur 2p signal when measured "floating". The sulfur, being directly chemically bound to the gold atoms, therefore behaves similarly to the conducting substrate when measured grounded. It is also interesting to note that the fwhm of the fluorine 1s signal (Figure 5f) shows a consistent shift to lower values with increasing PFDT concentration. The reduction in fwhm of 0.3 eV is indeed non-negligible and could be due to a *final-state* effect. The fwhm (Γ) of a photoemission signal is inversely proportional to the lifetime (τ) of the created core hole ($\Gamma = h/\tau$). Preliminary data on mixed monolayer samples on silver exhibit a similar trend in the fwhm. This therefore implies that the lifetime of the core hole is significantly increased with increasing surface concentration of PFDT. Shifts in the work function of the metal might also contribute to the extent of stabilization of the created core hole.

Conclusions

A detailed XPS study of surface-chemical gradients of thiol SAMs on gold revealed structure-induced systematic shifts in binding energies of several constituent atoms of the monolayer when the BE scale is referenced to gold (Au $4f_{7/2}$), as is normally practiced in the literature. The shifts are shown to be due to interfacial charge redistribution that depends on the surface coverage, chemical nature, and molecular orientation of the organic constituents. Surface potential, measured by KPFM, indicates that this shift is a consequence of the shift in the work function of gold induced by SAM formation. These results are further substantiated by measurements carried out with the sample being left "floating" with respect to the spectrometer. The use of surface-chemical gradients simplifies the interpretation of these shifts by providing a self-consistent means to reliably explore several surface compositions on a single sample.

5628 *J. Phys. Chem. C, Vol. 113, No. 14, 2009* Venkataraman et al.

Although the measurements presented here are for the specific system of thiols adsorbing on gold, these effects can be generalized to any self-assembled-monolayer system.

Supporting Information Available: This material is available free of charge via the Internet at http://pubs.acs.org.

References and Notes

(1) Ulman, A. *Chem. Rev.* **1996**, *96*, 1533.

(2) Love, J. C.; Estroff, L. A.; Kriebel, J. K.; Nuzzo, R. G.; Whitesides, G. M. *Chem. Rev.* **2005**, *105*, 1103.

(3) Heimel, G.; Romaner, L.; Bredas, J.-L.; Zojer, E. *Surf. Sci.* **2006**, *600*, 4548.

(4) Yamada, H.; Imahori, H.; Nishimura, Y.; Yamazaki, I.; Ahn, T. K.; Kim, S. K.; Kim, D.; Fukuzumi, S. *J. Am. Chem. Soc.* **2003**, *125*, 9129.

(5) Chabinyc, M. L.; Chen, X.; Holmlin, R. E.; Jacobs, H.; Skulason, H.; Frisbie, C. D.; Mujica, V.; Ratner, M. A.; Rampi, M. A.; Whitesides, G. M. *J. Am. Chem. Soc.* **2002**, *124*, 11730.

(6) Maisch, S.; Buckel, F.; Effenberger, F. *J. Am. Chem. Soc.* **2005**, *127*, 17315.

(7) Barriet, D.; Lee, T. R. *Cur. Opin. Colloid Interface Sci.* **2003**, *8*, 236.

(8) Kim, H. I.; Koini, T.; Lee, T. R.; Perry, S. S. *Langmuir* **1997**, *13*, 7192.

(9) Graupe, M.; Takenaga, M.; Koini, T.; Colorado, R., Jr.; Lee, T. R. *J. Am. Chem. Soc.* **1999**, *121*, 3222.

(10) Ishii, H.; Sugiyama, K.; Ito, E.; Seki, K. *Adv. Mater.* **1999**, *11*, 605.

(11) Rusu, P. C.; Brocks, G. *Phys. Rev. B* **2006**, *74*, 073414.

(12) Rusu, P. C.; Brocks, G. *J. Phys. Chem. B* **2006**, *110*, 22628.

(13) Rousseau, R.; De Renzi, V.; Mazzarello, R.; Marchetto, D.; Biagi, R.; Scandolo, S.; del Pennino, U. *J. Phys. Chem. B* **2006**, *110*, 10862.

(14) Sun, Q.; Selloni, A. *J. Phys. Chem. A* **2006**, *110*, 11396.

(15) de Boer, B.; Hadipour, A.; Mandoc, M. M.; van Woudenbergh, T.; Blom, P. W. M. *Adv. Mater.* **2005**, *17*, 621.

(16) Alloway, D. M.; Hofmann, M.; Smith, D. L.; Gruhn, N. E.; Graham, A. L.; Colorado, Jr, R.; Wysocki, V. H.; Lee, T. R.; Lee, P. A.; Armstrong, N. R. *J. Phys. Chem. B* **2003**, *107*, 11690.

(17) Campbell, I. H.; Rubin, S.; Zawodzinski, T. A.; Kress, J. D.; Martin, R. L.; Smith, D. L.; Barashkov, N. N.; Ferraris, J. P. *Phys. Rev. B* **1996**, *54*, 14321.

(18) Ray, S. G.; Cohen, H.; Naaman, R.; Liu, H.; Waldeck, D. H. *J. Phys. Chem. B* **2005**, *109*, 14064.

(19) Cabarcos, O. M.; Shaporenko, A.; Weidner, T.; Uppili, S.; Dake, L. S.; Zharnikov, M.; Allara, D. L. *J. Phys. Chem. C.* **2008**, *112*, 10842.

(20) Elwing, H.; Welin, S.; Askendel, A.; Nilsson, U.; Lundstrom, I. *J. Colloid Interface Sci.* **1987**, *119*, 203.

(21) Golander, G. C.; Caldwell, K.; Lin, Y. S. *Colloids Surf.* **1989**, *42*, 165.

(22) Morgenthaler, S.; Zink, C.; Spencer, N. D. *Soft Mater.* **2008**, *4*, 419.

(23) Genzer, J.; Bhat, R. R. *Langmuir* **2008**, *24*, 2294.

(24) Tomlinson, M. R.; Genzer, J. *Macromolecules* **2003**, *36*, 3449.

(25) Genzer, J.; Efimenko, K.; Fischer, D. A. *Langmuir* **2006**, *22*, 8523.

(26) Ballav, N.; Shaporenko, A.; Terfort, A.; Zharnikov, M. *Adv. Mater.* **2007**, *19*, 998.

(27) Blondiaux, N.; Zurcher, S.; Liley, M.; Spencer, N. D. *Langmuir* **2007**, *23*, 3489.

(28) Kraus, T.; Stutz, R.; Balmer, T. E.; Schmid, H.; Malaquin, L.; Spencer, N. D.; Wolf, H. *Langmuir* **2005**, *21*, 7796.

(29) Plummer, S. T.; Wang, Q.; Bohn, P. W.; Stockton, R.; Schwartz, M. A. *Langmuir* **2003**, *19*, 7528.

(30) Weinstein, R. D.; Moriarty, J.; Cushnie, E.; Colorado, R., Jr.; Lee, T. R.; Patel, M.; Alesi, W. R.; Jennings, G. K. *J. Phys. Chem. B* **2003**, *107*, 11626.

(31) Lestelius, M.; Engquist, I.; Tengvall, P.; Chaudhury, M. K.; Liedberg, B. *Colloids Surf., B* **1999**, *15*, 57.

(32) Liedberg, B.; Tengvall, P. *Langmuir* **1995**, *11*, 3821.

(33) Chaudhury, M. K.; Whitesides, G. M. *Science* **1992**, *256*, 1539.

(34) Venkataraman, N. V.; Zurcher, S.; Spencer, N. D. *Langmuir* **2006**, *22*, 4184.

(35) Morgenthaler, S.; Lee, S.; Spencer, N. D. *Langmuir* **2006**, *22*, 2706.

(36) Morgenthaler, S.; Lee, S.; Zurcher, S.; Spencer, N. D. *Langmuir* **2003**, *19*, 10459.

(37) Spencer, N. D.; Lambert, R. M. *Surf. Sci.* **1981**, *104*, 63.

(38) Blaszczyszyn, R.; Blaszczyszyn, M.; Meclewski, R. *Surf. Sci.* **1975**, *51*, 396.

(39) Ahn, H.; Zharnikov, M.; Whitten, J. E. *Chem. Phys. Lett.* **2006**, *428*, 283.

(40) Tarlov, M. J. *Langmuir* **1992**, *8*, 80.

(41) Fairley, N.; Carrick, A. *The Casa Cookbook Part 1: Recipes for XPS Data Processing*; Acolyte Science: UK, 2005.

(42) Scofield, J. H. *J. Electron. Spectrosc. Relat. Phenom.* **1976**, *8*, 129.

(43) Reilman, R. F.; Msezane, A.; Manson, S. T. *J. Electron. Spectrosc. Relat. Phenom.* **1976**, *8*, 389.

(44) Tanuma, S.; Powell, C. J.; Penn, D. R. *Surf. Interface Anal.* **1993**, *165*, 21.

(45) ISO15472−2001. "Surface Chemical Analysis - X-ray Photoelectron Spectrometers - Calibration of energy scales" 2001.

(46) Buffetau, T.; Desbat, B.; Turlet, J. M. *Appl. Spectrosc.* **1991**, *45*, 380.

(47) Jacobs, H. O.; Stemmer, A. *Surf. Interface Anal.* **1999**, *27*, 361.

(48) Jacobs, H. O.; Knapp, H. F.; Stemmer, A. *Rev. Sci. Instrum.* **1999**, *70*, 1756.

(49) Pellerite, M. J.; Dunbar, T. D.; Boardman, L. D.; Wood, E. J. *J. Phys. Chem.* **2003**, *107*, 11726.

(50) Tsao, M.-W.; Rabolt, J. F.; Schonherr, H.; Castner, D. G. *Langmuir* **2000**, *16*, 1734.

(51) Tsao, M.-W.; Hoffmann, C. L.; Rabolt, J. F.; Johnson, H. E.; Castner, D. G.; Erdelen, C.; Ringsdorf, H. *Langmuir* **1997**, *13*, 4317.

(52) Alves, C. A.; Porter, M. D. *Langmuir* **1993**, *9*, 3507.

(53) Cho, H.-L.; Strauss, H. L.; Snyder, R. G. *J. Phys. Chem.* **1992**, *96*, 5290.

(54) Naselli, C.; Swalen, J. D.; Rabolt, J. F. *J. Chem. Phys.* **1989**, *90*, 3855.

(55) Rabolt, J. F.; Fanconi, B. *Macromolecules* **1978**, *11*, 740.

(56) Greenler, R. G. *J. Chem. Phys.* **1966**, *44*, 310.

(57) Hinds, K.; Grainger, D. W.; Castner, D. G. *Langmuir* **1996**, *12*, 5083.

(58) Beamson, G.; Briggs, D. *High Resolution XPS of Organic Polymers—The Scienta ESCA300 Database*; Wiley Interscience: New York, 1992.

(59) Rossi, A.; Elsener, B. *Surf. Interface Anal.* **1992**, *18*, 495.

(60) Palermo, V.; Palma, M.; Samorì, P. *Adv. Mater.* **2006**, *18*, 145.

(61) Hayashi, K.; Saito, N.; Sugimura, H.; Takai, O.; Nakagir, N. *Ultramicroscopy* **2002**, *91*, 151.

(62) Sugimura, H.; Hanji, T.; Hayashi, K.; Takai, O. *Adv. Mater.* **2002**, *14*, 524.

(63) Sugimura, H.; Hayashi, K.; Saito, N.; Takai, O.; Nakagiri, N. *Jpn. J. Appl. Phys.* **2001**, *40*, 4373.

(64) Cahen, D.; Kahn, A. *Adv. Mater.* **2003**, *15*, 271.

JP809156A

Langmuir Article
pubs.acs.org/Langmuir

Orthogonal, Three-Component, Alkanethiol-Based Surface-Chemical Gradients on Gold

Eva Beurer,[†] Nagaiyanallur V. Venkataraman,[†] Antonella Rossi,[†,‡] Florian Bachmann,[†] Roman Engeli,[†] and Nicholas D. Spencer*[,†]

[†]*Laboratory for Surface Science and Technology, Department of Materials, ETH Zurich, Wolfgang-Pauli-Strasse 10, CH-8093 Zurich, Switzerland, and* [‡]*Dipartimento di Chimica Inorganica ed Analitica, Università degli Studi di Cagliari, Cittadella Universitaria di Monserrato, I - 09100 Cagliari, Italy,*

Received December 22, 2009. Revised Manuscript Received February 3, 2010

An orthogonal surface-chemical gradient composed of self-assembled monolayers on gold has been prepared by successive, controlled immersions in orthogonal directions into dilute solutions of dodecanethiol and perfluorododecanethiol. The resulting two-component orthogonal gradient in surface coverage was backfilled with 11-mercaptoundecanol, leading to a two-directional, three-component surface-chemical gradient. Water and hexadecane show distinctly different wetting behaviors on the gradient surface because of the differences in the hydrophobic and oleophobic natures of the three different constituents. These results are correlated with the chemical composition maps of the surface obtained by X-ray photoelectron spectroscopy. The homogeneity and the ordering of the self-assembled monolayer were investigated by dynamic water contact angle measurements and polarization-modulation infrared reflection−absorption spectroscopy.

Introduction

The physicochemical characteristics of surfaces are central to a wide variety of applications ranging from tribology to cell biology. Self-assembled monolayers (SAMs) offer the possibility to tailor the surface chemistry precisely by exposing high concentrations of a certain chemical functional group (or groups) at the interface. In this regard, alkanethiol SAMs on gold have been heavily scrutinized as model systems because of their ease of preparation and high degree of monolayer ordering.[1] A host of thiols with different chemical or bioactive functional groups at their chain termini and their applications have been extensively studied and reviewed.[1,2] Mixed monolayers[3−7] comprising two or more components are as important as uniform SAMs because they allow the systematic tuning of surface characteristics such as the wettability, surface charge, work function, and exposed ligand density. Screening many individual mixed SAMs for optimal surface characteristics for a certain application is very time-consuming. However, surface gradients offer the possibility to optimize a desired property by testing a range of surface compositions on a single sample in a single high-throughput experiment.[8,9] The use of gradients also minimizes sample-to-sample variation and therefore enhances reliability. This is particularly important when two surface properties are to be simultaneously optimized, which can be achieved by combining two geometrically orthogonal gradients on the same sample. Such orthogonal gradients could be composed of either two different chemical functionalities or a combination of a morphological and a chemical gradient, for example. Clements et al.,[10] Zhang et al.,[11] and Yang et al.[12] have reported orthogonal gradients composed of surface roughness and surface chemistry, and orthogonal gradients combining two different grafted polymers were reported by Khire et al.[13] and Bhat et al.[14,15]

In this article, alkanethiol SAM-based orthogonal surface-chemical gradients are presented. The methodology used to generate such gradients is similar to that previously reported by our laboratory for the generation of a one-directional gradient.[16] The method is characterized by the gradual immersion of a gold-coated substrate into a dilute thiol solution, followed by backfilling with a complementary thiol to produce a continuous change in the surface composition of the two components. This simple methodology has since been utilized by our group and others[17] to generate a variety of thiol-based gradients on gold with different chemical functionalities. In this study, the methodology has been extended to create orthogonal surface-chemical gradients using a further immersion step in the orthogonal direction before backfilling with a third component. Experimental conditions have been optimized to minimize the replacement of

*To whom correspondence should be addressed. E-mail: nspencer@ethz.ch. Fax: +41 44 633 10 27.

(1) Love, J. C.; Estroff, L. A.; Kriebel, J. K.; Nuzzo, R. G.; Whitesides, G. M. *Chem. Rev.* **2005**, *105*, 1103–1169.
(2) Ulman, A. *Chem. Rev.* **1996**, *96*, 1533–1554.
(3) Bain, C. D.; Whitesides, G. M. *J. Am. Chem. Soc.* **1988**, *110*, 6560–6561.
(4) Laibinis, P. E.; Fox, M. A.; Folkers, J. P.; Whitesides, G. M. *Langmuir* **1991**, *7*, 3167–3173.
(5) Tamada, K.; Hara, M.; Sasabe, H.; Knoll, W. *Langmuir* **1997**, *13*, 1558–1566.
(6) Stranick, S. J.; Atre, S. V.; Parikh, A. N.; Wood, M. C.; Allara, D. L.; Winograd, N.; Weiss, P. S. *Nanotechnology* **1996**, *7*, 438–442.
(7) Lin, W. C.; Lee, S. H.; Karakachian, M.; Yu, B. Y.; Chen, Y. Y.; Lin, Y. C.; Kuo, C. H.; Shyue, J. J. *Phys. Chem. Chem. Phys.* **2009**, *11*, 6199–6204.
(8) Morgenthaler, S.; Zink, C.; Spencer, N. D. *Soft Matter* **2008**, *4*(3), 419–434.
(9) Genzer, J.; Bhat, R. R. *Langmuir* **2008**, *24*, 2294–2317.
(10) Clements, L. R.; Khung, Y. L.; Thissen, H.; Voelcker, N. H. *BioMEMS Nanotechnol.* **2008**, *6799*, U222–U230.
(11) Zhang, J. L.; Han, Y. C. *Langmuir* **2008**, *24*, 796–801.
(12) Yang, J.; Rose, F.; Gadegaard, N.; Alexander, M. R. *Adv. Mater.* **2009**, *21*, 300–304.
(13) Khire, V. S.; Benoit, D. S. W.; Anseth, K. S.; Bowman, C. N. *J. Polym. Sci., Part A: Polym. Chem.* **2006**, *44*, 7027–7039.
(14) Bhat, R. R.; Tomlinson, M. R.; Genzer, J. *J. Polym. Sci., Part B: Polym. Phys.* **2005**, *43*, 3384–3394.
(15) Bhat, R. R.; Tomlinson, M. R.; Wu, T.; Genzer, J. Surface-Grafted Polymer Gradients: Formation, Characterization, and Applications. In *Surface-Initiated Polymerization*; Springer-Verlag: Berlin, 2006; Vol. *198*, pp 51−124.
(16) Morgenthaler, S.; Lee, S. W.; Zürcher, S.; Spencer, N. D. *Langmuir* **2003**, *19*, 10459–10462.
(17) Burton, E. A.; Simon, K. A.; Hou, S.; Ren, D.; Luk, Y.-Y. *Langmuir* **2009**, *25*, 1547–1553.

preadsorbed thiols. The three different thiols have been chosen in order to result in very different wetting behavior by water and oil at different positions on the sample. In these orthogonal gradients, each location on the sample represents a unique combination of surface composition of three different components. These gradients are also relevant for fundamental studies on ternary SAMs because very few systematic studies have been carried out on such systems.[18] This is partially due to the fact that the number of samples required to study multicomponent SAMs increases exponentially with the number of components. Also, the surface composition of ternary SAMs is likely to be severely influenced by the different adsorption and replacement kinetics of each component, making it extremely difficult to control the exact surface composition precisely. Orthogonal gradients offer the possibility to study systematically the effects of varying the concentration of three different surface species independently.

Experimental Section

Chemicals. Dodecanethiol (98+%, Sigma-Aldrich), 11-mercaptoundecanol (97%, Sigma-Aldrich), and 1*H*,1*H*,2*H*,2*H*-perfluorododecanethiol (available from previous studies[19]) were used in this study. For all experiments, analytical-grade ethanol (Scharlau Chemie, Spain) was used as a solvent. Sulfuric acid (95–97%, Sigma-Aldrich), hydrogen peroxide (30%, p.a., Merck, Germany), chromium (99.6%, Balzers Materials, Liechtenstein), and gold (99.99%, Umicore Materials AG, Liechtenstein) were used as received. For contact angle measurements, ultrapure water (TKA GenPure, Germany) and hexadecane (≥98%, Fluka, Germany) were used.

Substrates. Glass slides (4 × 4 cm, Menzel GmbH, Germany) were cleaned for 10 min in hot piranha solution (7:3 sulfuric acid/hydrogen peroxide) and rinsed with copious amounts of water. ***Caution!*** *Piranha solution reacts violently when placed in contact with organic material and should be handled with care.* The glass slides were coated with a 10 nm chromium adhesion layer and 100 nm of gold by physical vapor deposition (MED 020 coating system, Bal-Tec, Liechtenstein). Prior to the immersion steps, the substrates were sonicated in ethanol for 10 min, plasma cleaned in air plasma (PDC-001, Harrick Scientific Corporation, NY) for 30 s, and subsequently immersed in ethanol for 10 min.

Gradient Preparation. The stock solutions were prepared by dissolving dodecanethiol, 11-mercaptoundecanol, or perfluorododecanethiol in ethanol to a concentration of 1 mM for dodecanethiol and mercaptoundecanol and 100 μM for perfluorododecanethiol. All other solutions were prepared by further dilution of the corresponding stock solution. In the first step, the sample was immersed into a 5 μM dodecanethiol solution by means of a linear-motion drive (OWIS GmbH, Germany) at a speed of 150 μm/s. It was removed immediately from the solution, rinsed with ethanol, and blown dry with nitrogen, followed by a second immersion in a 5 μM perfluorododecanethiol solution at the same immersion speed but with the immersion axis being perpendicular to the first. Subsequently, the sample was immersed for 30 min in 1 mM 11-mercaptoundecanethiol solution at 5 °C. Different immersion times and temperatures were tested, but the above conditions resulted in a reasonably well-ordered SAM without substantial replacement of the previously adsorbed thiol molecules, as judged from static contact angle measurements.

Contact Angle Measurements. Contact angle measurements were carried out with a contact angle goniometer (DSA 100, Krüss GmbH, Hamburg, Germany). Static contact angle measurements were performed with 6 μL droplets for water and 3 μL for hexadecane. Dynamic water contact angle measurements were carried out with drop volumes from 3 to 11 μL and at a dosing speed of 15 μL/min. Contact angles were measured every 5 mm along the orthogonal gradient, resulting in 49 measurements per sample. The presented contact angle values are an average over seven samples for water contact angles and over four samples for hexadecane contact angles. For the nonbackfilled samples, an average value over three samples is reported for both the water and the hexadecane contact angles.

Infrared Spectroscopy. Polarization-modulation infrared reflection–absorption spectra were recorded on a Bruker IFS 66v IR equipped with a PMA 37 polarization-modulation accessory. The incoming beam from the external beam port of the spectrometer, polarized by a KRS-5 wire-grid polarizer and modulated by a ZnSe photoelastic modulator, was reflected off the sample surface at an angle of 80° and detected with a liquid-nitrogen-cooled MCT detector. The maximum in polarization retardation was set at 3000 cm^{-1}, and the polarization was modulated with a frequency of 50 kHz. The sample compartment was continuously purged with dry air during the measurements. Spectra were acquired every 5 mm along the dodecanethiol gradient direction with a 2 mm aperture and a resolution of 4 cm^{-1} using 1024 scans of multiplexed interferograms. The data was processed with OPUS software (Bruker Optics, Germany) and background corrected with a polynomial.

XPS Measurements. X-ray photoelectron spectra were acquired with a VG Thetaprobe spectrophotometer (Thermo Electron Corporation, West Sussex, U.K.) equipped with a radian lens, a concentric hemispherical analyzer, and a 2D channel-plate detector with 112 energy and 96 angle channels. A monochromatic Al Kα source with a spot size of 400 μm was used. Electrons were emitted at 53° with respect to the surface normal, and the acceptance angle was ± 30°. The instrument was operated in standard lens mode, and the analyzer in constant-analyzer-energy mode. Pass energies used for survey scans and detailed scans were 200 and 100 eV, respectively, for Au 4f, C 1s, F 1s, O 1s, and S 2p. The energy resolution (fwhm) under these conditions measured on Au $4f_{7/2}$ is 1.55 and 0.95 eV, respectively. To obtain an adequate signal-to-noise ratio in a minimum time and to limit beam-induced damage, acquisition times of approximately 7 min for survey scans and 35 min (total) for high-energy-resolution elemental scans were chosen. These conditions provided reproducible XPS spectra, and sample damage was negligible. To analyze the data, CasaXPS (version 2.3.15, www.casaxps.com) was used. The signals were fitted using Gaussian–Lorentzian functions and least-squares-fit routines following Shirley iterative background subtraction. For the calculation of the sensitivity factors, published photoionization cross sections,[20] corrected for the attenuation-length dependence on kinetic energy and for the angular asymmetry,[21] were used. The total analysis time for one orthogonal gradient sample under these chosen conditions was about 5 days. (See below.)

Results and Discussion

Gradient Preparation. The experimental conditions, such as the solvent, concentration of the thiols, temperature, and substrate cleaning protocols, are similar to those used previously for one-directional gradients.[16] However, the preparation of orthogonal gradients requires some additional considerations because of the additional controlled-immersion step into a second thiol. First, the immersion speed during the first step was increased by a factor of 2 compared to the protocol for a linear, two-component gradient in order to result in only half the surface coverage. This was clearly necessary in order to leave a sufficient number of adsorption sites to allow the generation of an additional gradient

(18) Phong, P. H.; Ooi, Y.; Hobara, D.; Nishi, N.; Yamamoto, M.; Kakiuchi, T. *Langmuir* **2005**, *21*, 10581–10586.
(19) Kraus, T.; Stutz, R.; Balmer, T. E.; Schmid, H.; Malaquin, L.; Spencer, N. D.; Wolf, H. *Langmuir* **2005**, *21*, 7796–7804.
(20) Scofield, J. H. *J. Electron Spectrosc. Relat. Phenom.* **1976**, *8*, 129–137.
(21) Reilman, R. F.; Msezane, A.; Manson, S. T. *J. Electron Spectrosc. Relat. Phenom.* **1976**, *8*, 389–394.

along the orthogonal axis. The second consideration was the order in which the thiol immersions were carried out. From our earlier studies on alkyl-perfluoroalkyl gradients,[22] it is known that in ethanol, perfluoroalkanethiols are replaced more rapidly by alkyl thiols than the other way around; therefore, we chose the order of immersion to be alkyl followed by perfluoroalkyl. The third and most important set of parameters that needed to be reoptimized for the orthogonal gradients included the concentration, immersion time, and temperature during the backfilling with the third component. During this final step, the replacement of the previously adsorbed thiols needs to be minimized. However, such replacement-kinetics data is sparse in the literature for a single preadsorbed component[23-31] and nonexistent for two preadsorbed components. Therefore, several trials were carried out under different sets of conditions, and the static water contact angles were assessed every 5 mm along the 4 × 4 cm^2 surface (49 measurements per sample). In the first step, the backfilling process was performed at three different temperatures (5, 22, and 50 °C) because the replacement kinetics are significantly slowed by decreasing the temperature.[32] Because the highest difference in the contact angle was found on samples backfilled at 5 °C, the immersion time was varied for this backfilling temperature in a second step. After a backfilling time of 10 min, the difference in the contact angle between the two extremes started to drop. PMIRRAS measurements on samples backfilled for 10 min showed a high degree of disorder in the SAM. A backfilling time of 30 min was chosen as a trade off between the contact angle slope and degree of order in PMIRRAS. Thus, the conditions for the backfilling step have been chosen to be the following: 1 mM 11-mercaptoundecanol and an immersion time of 30 min at 5 °C. It is emphasized that this set of conditions is not necessarily unique and represents one of the several possible combinations of these parameters that lead to the desired wettability changes along the orthogonal axes.

Wetting Behavior. The hydrophobicity and oleophobicity of a surface can be determined by measuring the water and hexadecane contact angles, respectively. The water contact angle of a SAM-covered surface depends on the nature of the chemical termini and can be adjusted over a wide range of values by changing the ratio between different components. A full monolayer of hydrophilic mercaptoundecanol had a contact angle of 20°, and those of the two hydrophobic components, dodecanethiol and perfluorododecanethiol, had high water contact angles of 108 and 118°, respectively.

Figure 1 shows a 3D surface plot of the static water contact angles on the orthogonal gradient before (top panel) and after (middle panel) backfilling with mercaptoundecanol. The water contact angles on orthogonal gradients before backfilling show only a small increase along both immersion axes with increasing

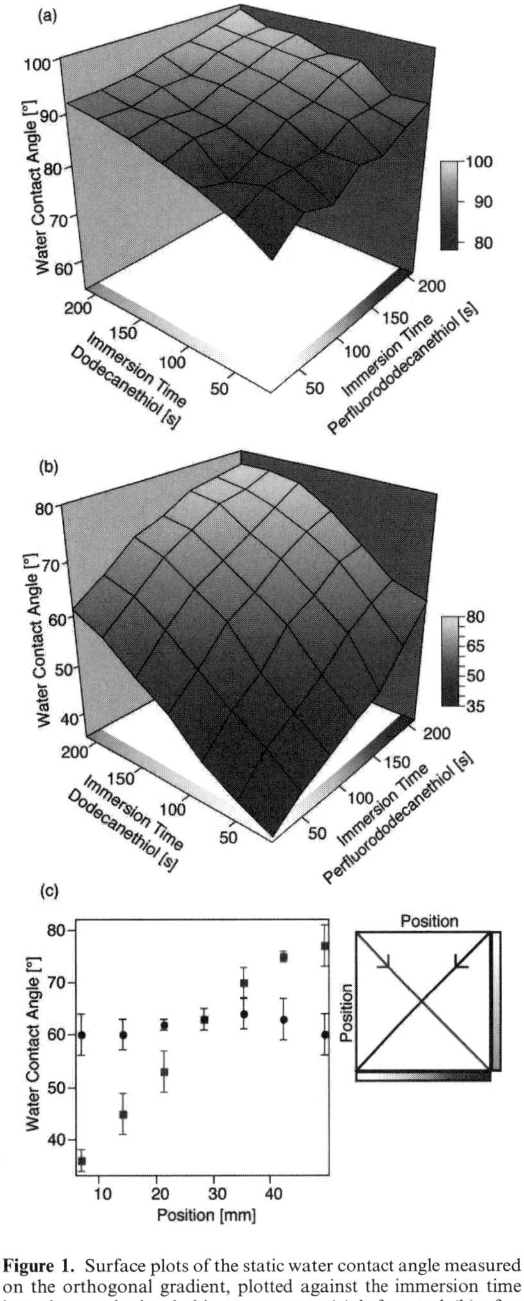

Figure 1. Surface plots of the static water contact angle measured on the orthogonal gradient, plotted against the immersion time into the two hydrophobic components (a) before and (b) after backfilling with mercaptoundecanol. Each intersection of the black gridlines marks a measured contact angle value. The more yellow the color, the more hydrophobic the location on the gradient sample. Color scales are displayed beside the plots. Additionally, for clarity, here and in subsequent Figures, the immersion times into dodecanethiol and perfluorododecanethiol solutions are indicated by green and a blue color intensity gradients, respectively. (c) Contact angles measured along the diagonals of a backfilled sample are displayed; the two diagonals correspond to the total immersion time in the hydrophobic solutions increasing (■) or remaining constant (●).

(22) Venkataraman, N. V.; Zürcher, S.; Rossi, A.; Lee, S.; Naujoks, N.; Spencer, N. D. *J. Phys. Chem. C* **2009**, *113*, 5620–5628.
(23) Patole, S. N.; Baddeley, C. J.; O'Hagan, D.; Richardson, N. V. *J. Phys. Chem. C* **2008**, *112*, 13997–14000.
(24) Cotton, C.; Glidle, A.; Beamson, G.; Cooper, J. M. *Langmuir* **1998**, *14*, 5139–5146.
(25) Imabayashi, S.; Gon, N.; Sasaki, T.; Hobara, D.; Kakiuchi, T. *Langmuir* **1998**, *14*, 2348–2351.
(26) Kakiuchi, T.; Sato, K.; Iida, M.; Hobara, D.; Imabayashi, S.; Niki, K. *Langmuir* **2000**, *16*, 7238–7244.
(27) Lin, P. H.; Guyot-Sionnest, P. *Langmuir* **1999**, *15*, 6825–6828.
(28) Chidsey, C. E. D.; Bertozzi, C. R.; Putvinski, T. M.; Mujsce, A. M. *J. Am. Chem. Soc.* **1990**, *112*, 4301–4306.
(29) Kajikawa, K.; Hara, M.; Sasabe, H.; Knoll, W. *Jpn J. Appl. Phys. Part 2* **1997**, *36*, L1116–L1119.
(30) Chung, C.; Lee, M. *J. Electroanal. Chem.* **1999**, *468*, 91–97.
(31) Kim, Y. K.; Koo, J. P.; Ha, J. S. *Appl. Surf. Sci.* **2005**, *249*, 7–11.
(32) Baralia, G. G.; Duwez, A. S.; Nysten, B.; Jonas, A. M. *Langmuir* **2005**, *21*, 6825–6829.

coverage of the hydrophobic components. This is expected because static water contact angle measurements do not sensitively distinguish between exposed methyl and methylene groups or between $-CF_2-$ and $-CF_3$ groups. Moreover, the presence of adventitiously deposited material on the nonbackfilled regions cannot be ruled out. However, the static water contact angles measured on the backfilled gradients (Figure 1b) show a clear increase along the two immersion axes. The presence of a gradient in water wettability is clearly visible. These values show the effect of backfilling with mercaptoundecanol along both alkyl and fluoroalkyl axes. The water contact angles measured on backfilled gradients reflect the extent of exposed $-OH$ and hydrophobic groups. A change in the contact angle to $>30°$ can be seen between the two extremes. Every position on the orthogonal surface-chemical gradient shows a different surface molar ratio of dodecanethiol, perfluorododecanethiol, and mercaptoundecanol. The ratio between hydrophilic and hydrophobic functional groups of the thiols is reflected in the static water contact angle measurements. This is more clearly shown in the lowest panel of Figure 1, wherein contact angles measured along the two diagonals of the sample are plotted. A wettability gradient with a linear increase of almost 40° can be observed along one of the sample diagonals (Figure 1c, ■). This line corresponds to the increasing total immersion time into the two hydrophobic components. Along the complementary diagonal (Figure 1c, ●), the contact angle does not show any significant variation. These data illustrate well the symmetric increase in the contact angle arising from dodecanethiol and perfluorododecanethiol. A small increase in the contact angle on moving from the dodecanethiol-rich region toward the perfluorododecanethiol-rich region, as might be expected from the pure-monolayer values, is not visible. Two effects could explain the absence of this increase. Either the adsorption of the perfluorododecanethiol component is slightly slower than the dodecanethiol adsorption or the replacement of perfluorododecanethiol with mercaptoundecanol is faster than the replacement of dodecanethiol. From our previous adsorption[22] and replacement kinetics measurements (data not shown), we conclude that explanation two is more likely. Moreover, it is to be noted that during the second immersion the number of available adsorption sites for perfluorothiol adsorption is reduced (almost linearly) along the dodecanethiol immersion axis.

Although the water contact angle data do not sensitively distinguish between the two hydrophobic components, the hexadecane contact angle is a measure of the oleophobicity of a surface.[33] Because the alkanethiols are oleophilic and fluoroalkanethiols oleophobic, this method is expected to reflect the compositional differences between the two hydrophobic components more sensitively. The single-component monolayers have very different hexadecane contact angles. The hexadecane contact angles for dodecanethiol SAM (45°) and mercaptoundecanol SAM (17°) are considerably lower than for the perfluorododecanethiol SAM (83°). Figure 2 shows the static hexadecane contact angle maps on nonbackfilled and backfilled orthogonal gradients. Unlike the water contact angles, the static hexadecane contact angles show an entirely different behavior on nonbackfilled gradients. The contact angle values increase rapidly along the perfluorothiol axis whereas along the dodecanethiol axis the contact angle is almost invariant. This is due to the different wetting behavior of hexadecane toward oleophilic or fluorinated surfaces. Along the perfluoro axis at low surface coverage a $-CF_2$-terminated surface is formed since the molecules are likely

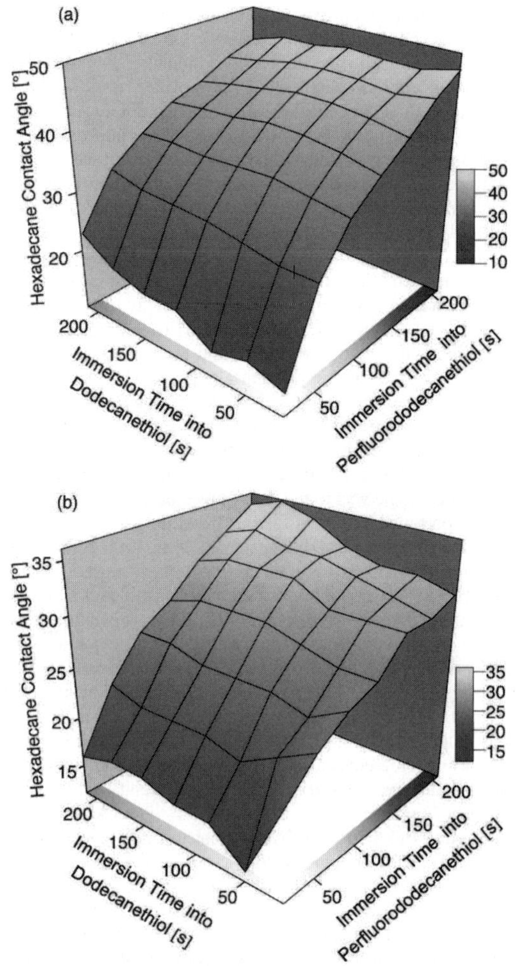

Figure 2. Surface plots of the static hexadecane contact angle measured on the orthogonal gradient plotted against immersion time into the two hydrophobic components (a) before and (b) after backfilling with mercaptoundecanol. Each intersection of the black gridlines marks a measured contact angle value. The more yellow the color on the surface plot, the higher the hexadecane contact angle on the surface of the orthogonal gradient. Color scales are displayed beside the plots.

to adopt a lying-down conformation. The wetting behavior of hexadecane is similar on surfaces exposing $-CF_2$ or $-CF_3$ groups. Therefore, the wetting behavior is mainly determined by the coverage and not by the ordering of the thiols. Along the dodecanethiol axis, at the low coverage of a nonbackfilled gradient, mainly the methylene groups of the alkyl chains are exposed, leading to much lower contact angles.[34] The surface coverage of both molecules is far below that of full SAMs, as evident from the measured values at the two extremes of the gradient. The hexadecane contact angles after backfilling (Figure 2b) show a gradient mainly along the perfluoro axis and only a small increase along the dodecanethiol axis. After the backfilling process, the surface density of thiols increases, thereby

(33) Graupe, M.; Koini, T.; Kim, H. I.; Garg, N.; Miura, Y. F.; Takenaga, M.; Perry, S. S.; Lee, T. R. *Mater. Res. Bull.* **1999**, *34*, 447–453.

(34) Bain, C. D.; Evall, J.; Whitesides, G. M. *J. Am. Chem. Soc.* **1989**, *111*, 7155–7164.

reducing the number of exposed methylene groups; however, the resulting surface exposes an increasing number of methyl groups together with the hydroxyl groups of mercaptoundecanol. This is one of the reasons that the static hexadecane contact angles, after backfilling, do not show an appreciable change in the dodecanethiol-rich regions of the sample compared to that for the perfluoroalkyl-rich regions. However, because of the short immersion time employed for the backfilling process the surface coverage of thiols is not maximal, leading to an increase in the hexadecane contact angle with the alkyl-rich end being smaller than expected. The slightly higher replacement of the perfluorothiol compared to alkylthiols by mercaptoundecanol is also reflected as a small increase toward the alkyl-rich end. The apparent lack of distinction between the mercaptoundecanol-rich and dodecanethiol-rich regions in the hexadecane contact angle may be understood by considering the disordered conformation of the alkyl chains leading to a considerable fraction of methylene groups still exposed at the interface. This is further evident from the dynamic contact angle and infrared spectral data discussed below.

Dynamic water contact angle measurements on backfilled gradients were performed to assess the homogeneity of the gradients. The difference between the advancing and receding contact angles (hysteresis) is a measure of the homogeneity of a surface. The larger the hysteresis, the less ordered and less homogeneous the surface. On the orthogonal gradient, the hysteresis of the dynamic water contact angle varies between 25 and 30° with no significant trend (Supporting Information). This indicates that the homogeneity of the SAM is comparable over the whole orthogonal gradient. However, this value is higher than that of the dodecanethiol-mercaptoundecanol two-component gradient system (around 14°) or those measured on full SAMs (10° for dodecanethiol SAMs and 15° for mercaptoundecanol SAMs). This could be due to the backfilling process not being complete, leading to less-well-organized SAMs. Increasing the backfilling time or the concentration of mercaptoundecanol did not lead to a significant improvement in the ordering without substantially increasing the fraction of replaced dodecanethiol and perfluorododecanethiol. The fact that the gradient components are not entirely homogeneously distributed but rather form nanoscopic islands rich in individual components during the two immersion steps could further contribute to this observed contact angle behavior.[35]

Chemical Composition. Although contact-angle measurements are only an indirect measure of the surface chemistry, X-ray photoelectron spectroscopy (XPS) allows the determination of the exact chemical composition of a surface. Unfortunately, this analysis is time-consuming for such orthogonal gradient samples. Whereas an orthogonal gradient is analyzed with static water contact angle measurements in about 20 min, the careful characterization of one orthogonal gradient demands an XPS acquisition time of more than 5 days. Therefore, only two sets of identically prepared samples were subjected to XPS analysis. Concerns about sample stability over such long acquisitions times and possible X-ray-induced sample degradation necessitated some precautions and led to severe restrictions on the number of repetitions carried out on such samples. To minimize the danger of sample degradation, a monochromatic X-ray source was used and the acquisition time for a single measuring spot was restricted to approximately 7 min for survey scans and 35 min (total) for high-energy-resolution elemental scans. After acquiring the spectra for all components (Au 4f, C 1s, F 1s, O 1s, S 2p, survey, and valence band) on one position, we re-recorded the C 1s peak to confirm that the sample was not degrading significantly during the measurements. For the quantification of the components' distribution, only the C 1s high-energy-resolution elemental scan has been analyzed. The high-energy-resolution elemental scans of F 1s and O 1s at selected positions close to the edges of the orthogonal gradient are shown in the Supporting Information. In the C 1s signal of the XPS spectra, different components of the carbon signal characteristic of the three components are observable because of their different chemical shifts. A typical C 1s XPS spectrum along with the fitted components is depicted in Figure 3.

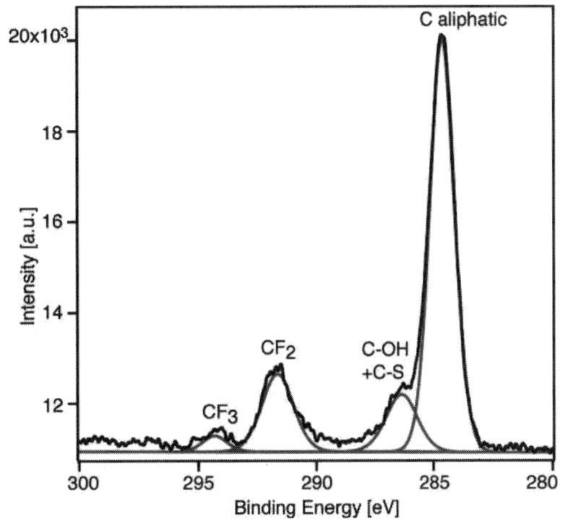

Figure 3. Typical C 1s XPS spectrum along with the fitted C aliphatic, C−OH, C−S, CF_2, and CF_3 components.

The aliphatic C 1s peak appears in the range between 284.5 and 284.7 eV binding energy, and this signal encompasses components arising from aliphatic dodecanethiol and the alkyl parts of mercaptoundecanol and the fluorinated thiol, together with any possible adventitiously deposited material. The C 1s (C−O) component arising from mercaptoundecanol gives rise to a carbon signal with peak maxima ranging from 286.1 to 286.6 eV. Characteristic of the fluorinated thiol are the C 1s, $C-F_2$, and the $C-F_3$ components, with peak maxima ranging from 291.5 to 291.9 and 294.0−294.5 eV, respectively. The signal of the C 1s (C−S) component is buried in the C 1s (C−O) component. In Figure 4, the contributions of the individual components to the total intensity are shown. To extract the true intensity of the C 1s (C−O) component, the contribution from the C 1s (C−S) signal has been subtracted in the following way: Assuming that no sulfur-containing contamination is present on the surface, the C 1s (C−S) contribution to this intensity can be calculated from the S 2p signal, correcting for the different sensitivity factors of carbon and sulfur signals. The contribution of the C 1s (C−S) signal is then simply subtracted from the total C 1s (C−O) signal.

For clarity, in Figure 4, in the top two representations the axes of the surface plots are turned by 180° with respect to the plots of the contact-angle measurements. In the first plot, the C 1s (C aliphatic) intensity is shown. It is decreasing mainly with increasing immersion time into the perfluorododecanethiol solution. Also, a minor drop is observable with decreasing immersion time into the dodecanethiol solution. The decrease in this intensity with increasing immersion time in perfluorododecanethiol arises because the perfluorododecanethiol component contains only 1 carbon atom bonded to hydrogens while dodecanethiol

(35) Morgenthaler, S. M.; Lee, S.; Spencer, N. D. *Langmuir* **2006**, *22*, 2706–2711.

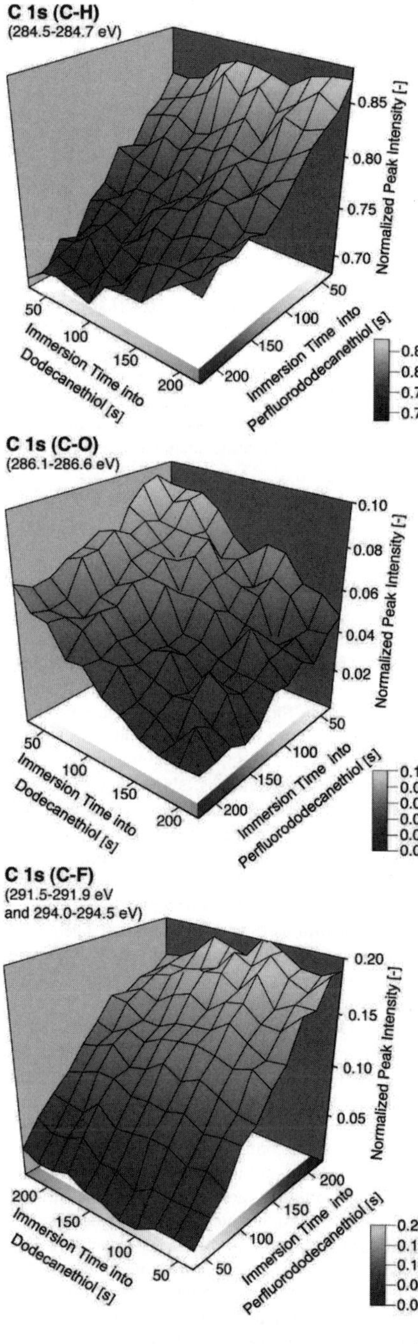

Figure 4. Components of the XPS C 1s signal, normalized to the total C 1s intensity. In the upper-most surface plot, the C 1s (C aliphatic) component is displayed, the middle panel shows the C 1s (C−O) component, and the bottom surface represents the C 1s (C−F) component. Each intersection of the black gridlines marks a measured value. Note that, for ease of display, the axes of the upper two surface plots are turned by 180° in comparison to the plot of the contact-angle measurements shown earlier. Color scales are displayed beside the plots.

and mercaptoundecanol have 11 and 10 such carbon atoms, respectively. The minor decrease in carbon intensity along the axis corresponding to decreasing immersion time into dodecanethiol arises from two sources: the reduction in chain length from 12 carbons in dodecanethiol to 11 carbons atoms in mercaptoundecanol, and the decrease in the perfluorododecanethiol concentration on the surface for longer immersion times into dodecanethiol solution due to saturation effects (i.e., nonlinear behavior in the adsorption kinetics at high coverage). In the middle panel of Figure 4, the contribution of C 1s (C−O) to the total C 1s peak is displayed. The maximum observed value of this signal normalized to the total C 1s signal is 0.103, which is slightly higher than the expected value of 0.091 for a full monolayer of mercaptoundecanol. The deviation from the expected value is observed even though the plotted C 1s (C−O) has been corrected for the C 1s (C−S) signal, as described above. This is due to the fact that these two signals (C−S and C−O) originate from different positions within the monolayer, leading to different amounts of attenuation: The C 1s (C−S) signal is attenuated by the entire monolayer above it, whereas the C 1s (C−O) signal is not attenuated at all because it arises from the outermost carbon, thus leading to an underestimation of the C 1s (C−S) signal and a consequent overestimation of the C 1s (C−O). Nevertheless, the C 1s (C−O) intensity, as expected, shows an increase with increasing concentration of the mercaptoundecanol. In the third part of Figure 4, the contribution of the C 1s (C−F) components to the total C 1s signal is shown. The C−F signals appear to be well separated from the other signals and therefore do not suffer from any of the above difficulties and clearly reflect the increasing concentration of perfluorododecanethiol with increasing immersion time. Almost no variation in this intensity is observed for any given immersion time along the dodecanethiol immersion axis, indicating the homogeneity of the gradients except for the longest immersion times into dodecanethiol, where the concentration of perfluorododecanethiol drops off slightly. This is entirely consistent with the contact-angle data presented earlier. This behavior can be explained by considering slightly modified Langmuir-type isotherm[36] absorption behavior and a progressively lower concentration of free binding sites for the perfluorothiols as a result of the previously adsorbed dodecanethiols. In the hexadecane contact angle measurement on backfilled gradients, this drop was not observable because the increase in the hexadecane contact angle due to increasing dodecanethiol concentration on the surface somewhat balanced out this effect. The mercaptoundecanol concentration, as evident from C 1s (C−O) on the surface, reflects well the water-wetting behavior shown in Figure 1. The higher its concentration, the lower the water contact angle. (Note that the axes of the surface plots are turned 180° with respect to each other.) This is more clearly seen in Figure 5, in which the intensity of the C−O signal is plotted as a function of the static water contact angle. A good correlation between the water contact angle and the C 1s (C−O) signal can be seen over the entire gradient.

In Figure 6, the correlation between the perfluorododecanethiol concentration on the surface and the hexadecane contact angle is shown. Here again, excellent correlation between the two quantities is evident. The increased scattering about the regression line for high hexadecane contact angles is due to the variation of the dodecanethiol concentration for constant perfluorododecanethiol concentration.

Organization. The crystallinity and organization of the alkyl chains of the SAM can be investigated by polarization

(36) Dannenberger, O.; Buck, M.; Grunze, M. *J. Phys. Chem. B* **1999**, *103*, 2202–2213.

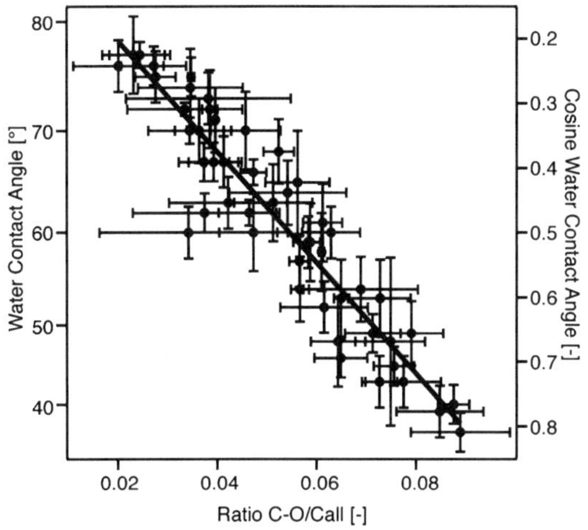

Figure 5. Static water contact angles are displayed against the normalized peak intensity of the XPS C 1s (C−O) signal.

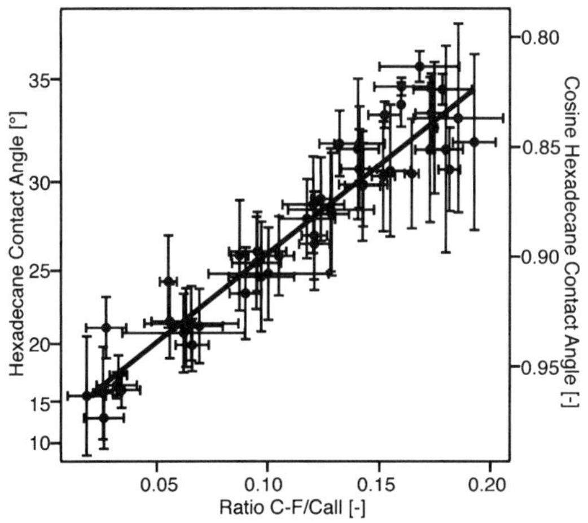

Figure 6. Correlation of the perfluorododecanethiol concentration on the surface (C 1s C−F) with the hexadecane contact angle.

modulation infrared reflection−absorption spectroscopy (PMIRRAS).[37]

Figure 7 shows the PMIRRA spectra of the orthogonal gradient every 5 mm along the dodecanethiol gradient axis. On the upper and lower ends of the gradient, the spectra of the uniform dodecanethiol and mercaptoundecanol SAM, respectively, are shown. The position of the peak maxima for the symmetric and antisymmetric CH_2 stretching mode with respect to the corresponding bands in the full SAMs shows that the alkyl chains are not well organized. On full monolayers, the symmetric and antisymmetric stretching modes are seen at 2850 and 2920 cm^{-1}, respectively. Although the positions of the symmetric and antisymmetric CH_2 stretching modes measured along the gradient decrease slightly with increasing immersion time into dodecanethiol, they do not reach the full-monolayer values. Nevertheless, the presence of a gradient in dodecanethiol concentration

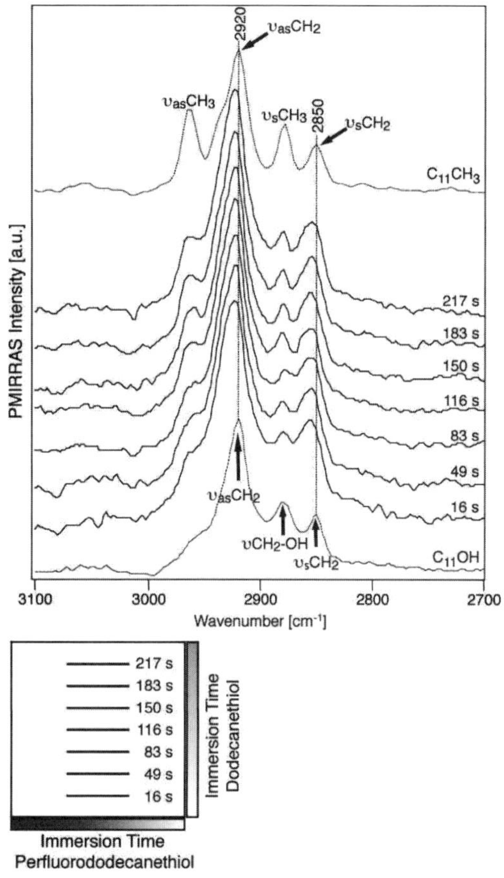

Figure 7. PMIRRA spectra of the orthogonal gradient measured every 5 mm along the dodecanethiol gradient axis. Each spectrum is labeled with the time the sample was immersed into dodecanethiol solution at this position. The spectra of pure dodecanethiol and the pure 11-mercaptoundecanol SAM are shown at the top and at the bottom of the graph, respectively. The lines indicate the approximate size and shape of the measured area.

can be seen from the increase in the asymmetric CH_3 stretching mode. Increasing the backfilling time or concentration of mercaptoundecanol did not lead to a significant improvement in the crystallinity and ordering of the SAM without substantially increasing the fraction of replaced dodecanethiol and perfluorododecanethiol. This is in contrast to the situation seen in the two-component gradients composed of dodecanethiol and mercaptoundecanol,[37] where the ordering and the crystallinity of the SAM after the backfilling step are not much different from those of the full monolayer, reflecting the complex nature of the three-component SAM systems. The spectra in the C−F stretching mode are not shown because the intensities of the peaks are quite low and the difference along the gradient was not observable. Unlike the two-component alkyl-perfluoroalkyl gradients, the typical helical modes of the CF_2 groups are not observable, even at the highest perfluorothiol coverage.[22] This is due to the lower concentration and poor organization of the perfluoro moieties on the orthogonal gradient rather than on the two-component gradients.

(37) Venkataraman, N. V.; Zürcher, S.; Spencer, N. D. *Langmuir* **2006**, *22*, 4184–4189.

Conclusions

A method to fabricate orthogonal surface-chemical gradients composed of three different thiols on gold is presented. Small deviations from orthogonal behavior occur in the most hydrophobic region because of saturation effects. The experimental conditions and the specific thiols have been chosen in such a way as to result in an orthogonal gradient that exhibits very different wetting behavior toward water and oil. The water contact angle shows a gradient along both immersion axes whereas the hexadecane contact angle varies only along one of the axes. This difference in wetting behavior can be switched by simply changing the order in which the immersions are carried out. For example, controlled immersions into dodecanethiol and mercaptoundecanol followed by backfilling with perfluorododecanethiol would lead to a complementary gradient in wetting behavior by the two liquids. The method can readily be extended to thiols with different functional groups (e.g., biologically active functional groups), providing a tool to investigate three-component mixed SAMs systematically, which can be utilized as a platform to optimize the effects of two surface properties simultaneously.

Acknowledgment. We thank the Swiss National Science Foundation for financial assistance.

Supporting Information Available: F 1s and O 1s XPS signals measured at four different locations close to the edges of the orthogonal gradient sample. Dynamic water contact angle hysteresis measured on the orthogonal gradient. Surface plots of the XPS F 1s and S 2p signal intensities measured on orthogonal gradients. This material is available free of charge via the Internet at http://pubs.acs.org.

4b. Surfaces patterns

Commentary

Surface-chemical patterns are often useful in sensor, diagnostic, or cell-biological applications, and there are many ways to create them. We found that one very flexible method was — by means of standard photolithography — to create a platform pattern of the right geometry based on two oxides with different adsorption chemistries (such as titania and silica), and then to use complementary adsorbates (such as PLL-g-PEG and alkylphosphates) to create the desired pattern in chemical functionality. In our paper (4.11), we describe the process, termed selective molecular assembly patterning or SMAP, and show how proteins can then be readily patterned on such a surface, and used, in turn as a template for the patterning of cells.

Microcontact printing (Figure 14) is a widely used approach to chemical patterning, but has been shown to have a number of drawbacks, in terms of both resolution and contamination issues, when used with commercially available elastomers, such as poly(dimethylsiloxane) (PDMS). We explored the use of an alternative elastomer — a commercially available copolymer of ethylene and octene — and were able to show improvements, both in cleanliness and in resolution when printing protein lines on a surface (4.12).

Microcontact printing has aroused much interest as a potential approach to the fabrication of devices on a sub-micron level, and in this context we have had the pleasure of some fruitful collaborations with Heiko Wolf's group at IBM Zurich. In an initial study, we measured the diffusion process of thiols in PDMS via microcontact printing experiments and were able to model the process with a Fickian approach. The knowledge gained is particularly significant when multiple printings with a single thiol inking are required.

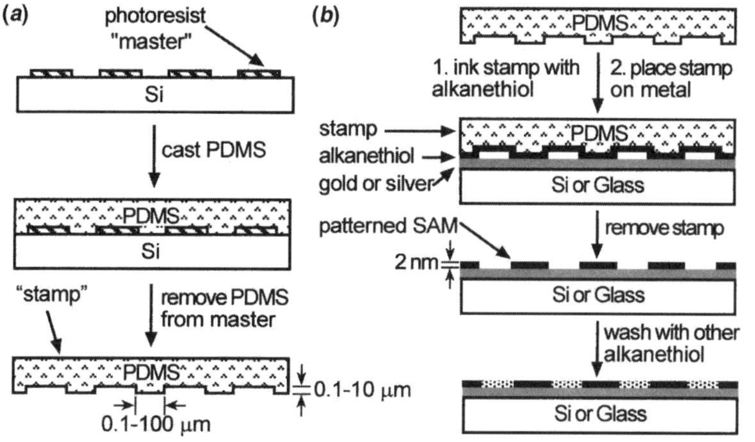

Figure 14: Principle of microcontact printing, in which a silicon or photoresist master is used to form a silicone stamp, which is in turn loaded with an "ink" such as an alkanethiol, which can then be used to transfer the original design as a chemical pattern onto a gold substrate, for example. From Reference 55 with kind permission.

Our further collaboration with IBM involved an extension of printing methods to small objects rather than molecules. This is of practical interest for device manufacture, since it could potentially allow the integration of high-resolution self-assembly procedures, physically separated from the main fabrication line, followed by a printing step to place the objects onto the device (an approach called Self-Assembly, Transfer, and Integration (SATI)) (4.14). This has been extended to the nanometer scale (4.15), and by means of specially structured surfaces, can also be used to place objects of different sizes in distinct patterns onto a surface (4.16).

Langmuir **2002,** *18,* 3281–3287 3281

Selective Molecular Assembly Patterning: A New Approach to Micro- and Nanochemical Patterning of Surfaces for Biological Applications

Roger Michel,[†] Jost W. Lussi,[‡] Gabor Csucs,[§] Ilya Reviakine,[†] Gaudenz Danuser,[§] Brigitte Ketterer,[‖] Jeffrey A. Hubbell,[‡] Marcus Textor,[*,†] and Nicholas D. Spencer[†]

Laboratory for Surface Science and Technology, Institute for Biomedical Engineering, and BioMicroMetricsGroup (BMMG), ETH Zurich, Wagistrasse 2, CH-8952 Schlieren, Switzerland, and Paul-Scherrer Institute, PSI Villigen, CH-5232 Villigen PSI, Switzerland

Received November 26, 2001. In Final Form: January 28, 2002

A novel patterning technique based on selective self-assembly of alkane phosphates on metal oxide surfaces is presented. Standard photolithography was used to create patterns of titanium dioxide within a matrix of silicon dioxide. Alkane phosphates were found to self-assemble on TiO_2, but not on SiO_2, surfaces. Subsequent adsorption of poly(L-lysine)-*g*-poly(ethylene glycol) (PLL-*g*-PEG) rendered the exposed SiO_2 surface resistant to protein adsorption. X-ray photoelectron spectroscopy and time-of-flight secondary ion mass spectrometry were employed to monitor the assembly processes. Protein-adsorption studies by means of fluorescence microscopy conclusively established that the resulting surfaces displayed protein-adhesive, alkyl phosphate modified TiO_2 features, arranged within a protein-resistant PLL-*g*-PEG-modified SiO_2 matrix. Human foreskin fibroblasts, incubated in a serum-containing medium, were found to selectively attach to the protein-adhesive areas, where they developed focal contacts. No interaction of cells with the PLL-*g*-PEG-coated SiO_2 areas was evident for at least 14 days. This patterning approach, termed selective molecular assembly patterning, is considered to be suitable for reproducible and cost-effective fabrication of biologically relevant chemical patterns over large areas.

1. Introduction

A variety of patterning techniques is currently available for studying the effects of chemical patterning and topographical microstructuring of surfaces on cell attachment, growth, differentiation, and death.[1–3] Interest in these effects is driven in part by the need to investigate, control, and improve implant–body interactions and implant integration. Moreover, the growing demand in biosensor technology for high-density, high-sensitivity, multianalyte chips can only be met with precise and reproducible patterning methodologies that allow a controlled juxtaposing of chemically distinct, active areas. The similarity of needs and constraints between the implant and the biosensor fields has led to the development of chemical,[2,4,5] topochemical,[6] and topographical[7,8] patterning methodologies that are applicable to both areas.

Surfaces can be chemically patterned using a number of techniques, such as microcontact printing,[2,9] microfluidic patterning,[10] photolithography,[11] and photodegradation[12,13] or photoactivation[14] of self-assembled monolayers (SAMs) that have been shown to present termini which resist protein adsorption[15] and cell attachment.[16] Although these patterning techniques have proven to be well suited for specific, mostly benchtop applications, they suffer from a number of limitations. While the nonphotolithographic techniques meet the criteria outlined above for spatial control, they are often incompatible with standard industrial processes, since the elastomeric stamps[2,10,17] are difficult to use reproducibly over large areas, transfer contaminants,[18] and degrade over time. Current photolithographic techniques that circumvent the use of polymeric stamps require complex chemistry,[11,14] while sol-

* Corresponding author: Dr. Marcus Textor, ETH Zürich, Laboratory for Surface Science and Technology, Wagistrasse 2, CH-8952 Schlieren, Switzerland. Phone: +41 (0)1 632 64 51. Fax: +41 (0)1 633 10 48. E-mail: textor@surface.mat.ethz.ch.

[†] Laboratory for Surface Science and Technology, ETH Zurich.

[‡] Institute for Biomedical Engineering, ETH Zurich.

[§] BioMicroMetricsGroup (BMMG), ETH Zurich.

[‖] Paul-Scherrer Institute, PSI Villigen.

(1) Blawas, A. S.; Reichert, W. M. *Biomaterials* **1998**, *19*, 595–609.

(2) Chen, C. S.; Mrksich, M.; Huang, S.; Whitesides, G. M.; Ingber, D. E. *Science* **1997**, *276*, 1425–1428.

(3) Ito, Y. *Biomaterials* **1999**, *20*, 2333–2342.

(4) Blawas, A. S.; Oliver, T. F.; Pirrung, M. C.; Reichert, W. M. *Langmuir* **1998**, *14*, 4243–4250.

(5) Whitesides, G. M.; Ostuni, E.; Takayama, S.; Jiang, X. Y.; Ingber, D. E. *Annu. Rev. Biom. Eng.* **2001**, *3*, 335–373.

(6) Shi, H. Q.; Tsai, W. B.; Garrison, M. D.; Ferrari, S.; Ratner, B. D. *Nature* **1999**, *398*, 593–597.

(7) Inoue, A.; Ishida, T.; Choi, N.; Mizutani, W.; Tokumoto, H. *Appl. Phys. Lett.* **1998**, *73*, 1976–1978.

(8) Brunette, D. M. *The effect of surface topography on cell migration and adhesion*; Ratner, B. D., Ed.; Elsevier Science Publishers B. V.: Amsterdam, 1988; pp 203–217.

(9) Bernard, A.; Delamarche, E.; Schmid, H.; Michel, B.; Bosshard, H. R.; Biebuyck, H. *Langmuir* **1998**, *14*, 2225–2229.

(10) Delamarche, E.; Bernard, A.; Schmid, H.; Michel, B.; Biebuyck, H. *Science* **1997**, *276*, 779–781.

(11) McFarland, C. D.; Thomas, C. H.; DeFilippis, C.; Steele, J. G.; Healy, K. E. *J. Biomed. Mater. Res.* **2000**, *49*, 200–210.

(12) Yang, X. M.; Peters, R. D.; Kim, T. K.; Nealey, P. F.; Brandow, S. L.; Chen, M. S.; Shirey, L. M.; Dressick, W. J. *Langmuir* **2001**, *17*, 228–233.

(13) Corey, J. M.; Wheeler, B. C.; Brewer, G. J. *IEEE Trans. Biomed. Eng.* **1996**, *43*, 944–955.

(14) Yang, Z. P.; Frey, W.; Oliver, T.; Chilkoti, A. *Langmuir* **2000**, *16*, 1751–1758.

(15) Ostuni, E.; Chapman, R. G.; Liang, M. N.; Meluleni, G.; Pier, G.; Ingber, D. E.; Whitesides, G. M. *Langmuir* **2001**, *17*, 6336–6343.

(16) Chapman, R. G.; Ostuni, E.; Liang, M. N.; Meluleni, G.; Kim, E.; Yan, L.; Pier, G.; Warren, H. S.; Whitesides, G. M. *Langmuir* **2001**, *17*, 1225–1233.

(17) Patel, N.; Bhandari, R.; Shakesheff, K. M.; Cannizzaro, S. M.; Davies, M. C.; Langer, R.; Roberts, C. J.; Tendler, S. J. B.; Williams, P. M. *J. Biomater. Sci., Polym. Ed.* **2000**, *11*, 319–331.

(18) Yang, Z. P.; Belu, A. M.; Liebmann-Vinson, A.; Sugg, H.; Chilkoti, A. *Langmuir* **2000**, *16*, 7482–7492.

Published on Web 03/09/2002

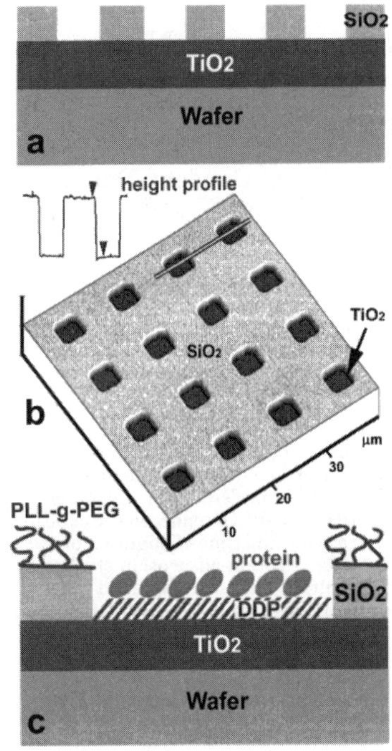

Figure 1. A schematic illustration of the SMAP methodology. (a) Sample exhibiting a material contrast, produced using common photolithographic techniques. TiO_2 squares within the SiO_2 matrix are shown. (b) An atomic force microscope image of the surface shown in (a). TiO_2 squares are located ~35 nm below the SiO_2 matrix (inset: a height profile of the surface). (c) Schematic view of the surface after the surface modification procedures: DDP on TiO_2, protein on DDP, and PLL-*g*-PEG on SiO_2, with the poly(lysine) backbone lying flat on the surface and PEG chains extending away from it.

vents used in conventional photolithography may denature or degrade deposited bio-organic layers.[1]

The preceding discussion indicates the demand for a technology that allows chemical patterning into areas that are distinctly protein and cell adhesive or nonadhesive. This technology should also be free of the limitations imposed by the use of polymeric stamps or photolithographic processes on organic overlayers. Selective molecular assembly patterning (SMAP) introduced in this work meets these criteria.

SMAP is based on a combination of lithographic structuring of metal oxide surfaces (Figure 1a,b) with the selective self-assembly of organic molecules, namely, alkyl phosphates[19,20] and poly(L-lysine)-*g*-poly(ethylene glycol) (PLL-*g*-PEG),[21] on distinct metal oxide surfaces. Alkyl phosphates are known to self-assemble on metal oxide substrates, such as TiO_2, from aqueous solutions,[22] rendering them hydrophobic. PLL-*g*-PEG, however, renders negatively charged metal oxide surfaces, such as SiO_2, resistant to protein adsorption and cell attachment.[21,23,24] Thus, by sequentially adsorbing alkyl phosphate and PLL-*g*-PEG onto a suitably structured surface consisting of a pattern of different adsorbing oxide regions (such as a TiO_2 pattern surrounded by a matrix of SiO_2), a surface exhibiting a pattern of protein-adhesive regions (alkyl phosphate) surrounded by a protein-resistant matrix (PLL-*g*-PEG) can be prepared (Figure 1c).

The spatial resolution of SMAP is limited by the lithographic approach used in the patterning of metal oxide substrates, which can easily reach nanometer length scales.[25] Recently introduced processes suitable for patterning metal oxide surfaces in the nanometer range, such as colloidal lithography,[26] hot embossing,[27] or template synthesis,[28] can also be applied in combination with SMAP, to generate large-scale, affordable, biologically relevant chemical patterns.

2. Materials and Methods

2.1. Substrate Preparation. Two kinds of substrates were employed for the investigation of the selective adsorption on SiO_2 and TiO_2 surfaces. Silicon wafer pieces (1 cm^2) or glass cover slips (Plano GmbH, Germany), coated half with SiO_2 and half with TiO_2, were used to quantitatively investigate the self-assembly processes. Additionally, whole 4 in. silicon ⟨110⟩ wafers were used to produce square patterns of 5×5 μm^2 and 60×60 μm^2, that were subsequently analyzed by imaging time-of-flight secondary ion mass spectrometry (ToF-SIMS) and used for protein-adhesion and cell-attachment experiments.

2.2. Substrate Coating. Four inch silicon ⟨110⟩ wafers (Wafernet GmbH, Germany) were sputter-coated with a 100 nm TiO_2 layer, followed by a 20 nm SiO_2 layer. The coated wafers were then used in subsequent photolithographic patterning steps (section 2.3). The 1 cm^2 silicon wafer pieces or glass coverslips were coated with a 12 nm SiO_2 layer, covered with aluminum foil to expose half of the slide, and coated with 12 nm of TiO_2, to be used in surface modification steps (section 2.4). All coating steps were carried out with a Leybold dc-magnetron Z600 sputtering plant. The deposition and characterization of these oxide coatings have been described previously.[29]

2.3. Patterning: Creating Material Contrast. A 30 nm aluminum hard mask was evaporated onto the TiO_2- and SiO_2-coated substrate surfaces with a Balzer BAK 600 coater.

A Shipley S1813 photoresist was spin-coated onto the aluminum layer with an STD5 Karl Süss spin coater at 4000 rpm for 25 s and baked at 90 °C for 60 s, resulting in a resist thickness of 1.3–1.5 μm. The resist was exposed to UV light through a suitable mask for 4–5 s using a MA6 Karl Süss mask aligner and developed with Shipley MF-84MX developer solution for 30 min.

After the photolithographic step, the resist pattern was transferred into the aluminum layer by wet etching[30] for 5 min. The resist was then stripped in a removal bath. After an additional cleaning step in acetone, the structured aluminum layer was used as a hard mask during the reactive ion etching (RIE) of silicon dioxide. The latter was carried out with an Oxford Plasma Lab 100, using a mixture of O_2 and CHF_3, in a ratio of 3:40 sccm, at 100 mTorr, 300 K, and 100 W, for 180 s. The temperature was controlled at 300 ± 5 K with a liquid-nitrogen-cooled cryotable. The depth of the etching was measured with a Tencor Alpha-step profilometer. After the RIE, the aluminum hard mask was

(19) Bram, C.; Jung, C.; Stratmann, M. *Fresenius' J. Anal. Chem.* **1997**, *358*, 108–111.
(20) Textor, M.; Ruiz, L.; Hofer, R.; Rossi, A.; Feldman, K.; Hahner, G.; Spencer, N. D. *Langmuir* **2000**, *16*, 3257–3271.
(21) Elbert, D. L.; Hubbell, J. A. *Chem. Biol.* **1998**, *5*, 177–183.
(22) Hofer, R.; Textor, M.; Spencer, N. D. *Langmuir* **2001**, *17*, 4014–4020.
(23) Kenausis, G. L.; Voros, J.; Elbert, D. L.; Huang, N. P.; Hofer, R.; Ruiz-Taylor, L.; Textor, M.; Hubbell, J. A.; Spencer, N. D. *J. Phys. Chem. B* **2000**, *104*, 3298–3309.
(24) Huang, N. P.; Michel, R.; Voros, J.; Textor, M.; Hofer, R.; Rossi, A.; Elbert, D. L.; Hubbell, J. A.; Spencer, N. D. *Langmuir* **2001**, *17*, 489–498.
(25) Xia, Y. N.; Rogers, J. A.; Paul, K. E.; Whitesides, G. M. *Chem. Rev.* **1999**, *99*, 1823–1848.
(26) Hanarp, P.; Sutherland, D.; Gold, J.; Kasemo, B. *Nanostruct. Mater.* **1999**, *12*, 429–432.
(27) Schift, H.; Heyderman, L. J.; Maur, M. A. D.; Gobrecht, J. *Nanotechnology* **2001**, *12*, 173–177.
(28) Schlottig, F.; Textor, M.; Spencer, N. D.; Sekinger, K.; Schnaut, U.; Paulet, J. F. *Fresenius' J. Anal. Chem.* **1998**, *361*, 684–686.
(29) Kurrat, R.; Textor, M.; Ramsden, J. J.; Boni, P.; Spencer, N. D. *Rev. Sci. Instrum.* **1997**, *68*, 2172–2176.
(30) Büttenbach, S. *Mikromechanik: Einführung in Technologie und Anwendung*; Teubner: Stuttgart, 1991.

removed by wet etching,[30] and wafers were cleaned and dried thoroughly, protected with 1.3 μm of Shipley S1813 photoresist, transferred to an adhesive backing, and sawn into 10 × 10 mm^2 pieces on a wafer-sawing machine (ESEC, Zug, Switzerland). Sawn samples were ultrasonically cleaned in hexane, ethanol, and water for 3 min each and stored until use. Immediately before the self-assembly processes, the samples were cleaned in an oxygen-plasma cleaner (Harrick Scientific Corp., Ossining, NY) for 2 min.

2.4. SMAP Patterning: Conversion of Material Contrast into Protein-Adhesion Contrast. It will be shown in the results section that the adsorption of the organic molecules described below is selective and creates a protein-adhesion contrast.

2.4.1. Selective Self-Assembly of Dodecyl Phosphate (DDP) on Titanium Oxide. Ammonium dodecyl phosphate, $DDPO_4(NH_4)_2$, was prepared as described previously.[22] It was dissolved at a concentration of 0.5 mM in high-purity water at 50 °C, cooled to room temperature, and stored (for a maximum of 14 days) until used. Samples (1 cm^2) exhibiting material contrast, prepared as described in section 2.1, were immersed in 1 mL of $DDPO_4(NH_4)_2$ solution and incubated for 24 h, after which they were removed, rinsed with high-purity water, and blown dry in a stream of nitrogen.

2.4.2. Assembly of PLL-*g*-PEG. The PLL(20 kD)-*g*[3.5]-PEG-(2 kD) graft copolymer (based on PLL of MW 20 000 and PEG of MW 2000, with a grafting ratio of PLL/PEG of 3.5) was synthesized and characterized as previously described.[23] Patterned, DDP-coated samples, prepared as described above, were immersed in 1 mg/mL copolymer solution in Hepes Z1 buffer (10 mM 4(2-hydroxyethyl)piperazine-1-ethanesulfonic acid, adjusted to pH 7.4 with 6 M NaOH, Fluka) for 15 min, washed with Hepes Z1 buffer and then with water, and dried with a stream of nitrogen. This step was performed in 24-well tissue-culture polystyrene plates for samples that were to be used in cell studies (section 2.5.2).

2.5. Protein Adsorption and Cell Attachment on Patterned Substrates. 2.5.1. Protein-Adsorption Experiments. The samples prepared as described in sections 2.4.1 and 2.4.2 were incubated in 40 μg/mL Oregon Green labeled streptavidin (Molecular Probes, Eugene, OR) for 1 h and rinsed with Hepes Z1 buffer, washed with water, and blown dry. For X-ray photoelectron spectroscopy (XPS) and ToF-SIMS experiments, SMAP samples were subjected to full serum (Human Control Serum N, Hoffmann-La Roche, Switzerland) for 40 min instead of the fluorescently labeled streptavidin.

2.5.2. Cell-Attachment Experiments. Cell attachment was measured inside 24-well tissue-culture test plates (TPP, Switzerland). Glass cover slips, coated half with TiO_2 and half with SiO_2, as well as Si wafers with 5 × 5 or 60 × 60 μm^2 protein-adhesive (TiO_2/DDP) squares in a protein-resistant (SiO_2/PLL-*g*-PEG) matrix (adhesion contrast, see sections 1 and 2) were used as substrates. Human foreskin fibroblasts (HFFs), kept under standard culture conditions, were plated onto substrates in Dulbecco's modified Eagle's medium (DMEM) containing Glutamax I (GIBCO Life Technologies) and sodium pyruvate, supplemented with 10% fetal bovine serum and a 1% antibiotic/antimyotic solution (GIBCO), at a seeding density of 5000 cells per cm^2. The medium was exchanged twice a week for the length of the experiment.

After 20 h of incubation, cells were fixed with 4% neutral-buffered formalin/0.01% glutaraldehyde solution in phosphate-buffered saline (PBS) for 2 h at 4 °C and permeabilized with 0.1% Triton X100 in PBS for 5 min. After washing and blocking (1 h in a mixture of 1.5% bovine serum albumin and 0.005% Tween20 (Aldrich) in PBS, referred to hereafter as the blocking buffer), cells were incubated with the primary monoclonal mouse anti-human vinculin clone hVIN-1 dissolved 1:400 in blocking buffer for 1 h, washed, and incubated with a solution containing a secondary FITC-labeled goat anti-mouse antibody, rhodamine-labeled phalloidin (which binds to f-actin), and Hoechst/DAPI nuclear stain.

Finally, the cells were washed with PBS and mounted on microscope slides with Vectashield (Vector Labs, Burlingame, CA) for observation in an optical microscope. All chemicals for immunostaining were obtained from Sigma (St. Louis, MO), unless specified otherwise.

2.6. Sample Characterization. 2.6.1. X-ray Photoelectron Spectroscopy. All XPS spectra were recorded on a SAGE 100 (SPECS, Berlin, Germany) using nonmonochromatic Mg Kα radiation with an energy of 240 W (12 kV, 20 mA), an electron takeoff angle of 90°, and electron-detector pass energies of 50 eV for survey and 14 eV for detailed spectra. For the high-resolution, detailed spectra, a reference Ag(3d$_{5/2}$) full width at half-maximum (fwhm) corresponds to 1.0 eV. During analysis, the base pressure remained below 5 × 10^{-8} mbar. All peaks were referenced to the C(1s) (hydrocarbon C−C, C−H) contribution at 285.0 eV.[31]

2.6.2. Time-of-Flight Secondary Ion Mass Spectrometry. Secondary ion mass spectra were recorded on a PHI 7200 time-of-flight secondary ion mass spectrometer in the mass range 0−200 *m/z*. Imaging ToF-SIMS was carried out with an indium liquid-metal primary ion gun at a current of ~2 mA. For imaging purposes, the gun was operated at 25 keV, at a pulse width of 10 ns. The mass resolution $\Delta M/M$ of the peak $C_2H_3^+$ at *m/z* 27 in positive ion mode remained around 500. All surfaces were scanned with a ~500 nm ion beam across 300 × 300 μm^2. No charge compensation was necessary for acquisition.

2.6.3. Fluorescence Microscopy. Fluorescence microscopy investigations of the modified surfaces and cell−surface interactions were carried out using a Zeiss LSM 510 confocal laser-scanning microscope. Three different laser lines were used in our experiments: Oregon Green and FITC probes were excited at 488 nm, and rhodamine phalloidin at 543 nm; the 633 nm line was used to visualize the surface contrast in reflection mode during the cell experiments. Either a 20× (0.4NA) LD Achroplan or a 40× (0.6NA) LD Achroplan objective was used for protein-adhesion experiments. Characterization of cell morphology was performed with a 63× (1.25NA) Plan-Neofluar oil-immersion objective.

3. Results

Samples, one half coated with SiO_2 and the other half with TiO_2, were used for quantitative surface analysis after each of the surface treatment steps (cleaning, self-assembly, and polymer and protein adsorption, section 2). These samples exhibit material contrast on a macroscopic scale and are discussed in section 3.1. Micropatterned surfaces were subjected to identical surface modification procedures and characterized qualitatively by imaging ToF-SIMS (section 3.2) and fluorescence microscopy (section 3.3) and were used in the cell experiments (section 3.4). In both types of samples, material contrast (on a macroscopic or microscopic scale, Figure 1a) is converted into contrast with respect to protein adhesion (Figure 1c) via a series of surface modification steps (self-assembly of DDP, adsorption of PLL-*g*-PEG; section 2).

3.1. Characterization of the Macroscopically Patterned Surfaces. 3.1.1. X-ray Photoelectron Spectroscopy. XPS is commonly used to analyze and quantify surface chemical composition.[32] Cleaned, oxygen-plasma-treated surfaces exhibited low C(1s) intensities (Figure 2a,b) on both SiO_2 and TiO_2, indicating only minor hydrocarbon contamination (<10 atomic %). After exposure of the surface to the DDP solution, the C(1s) intensity increased significantly on the TiO_2 surface to an amount typical of a DDP SAM.[22] No such increase was observed on the SiO_2 (Figure 2). Consistent with this observation, no phosphorus signal was detected on SiO_2 following exposure to the DDP solution, and the Si intensity remained unchanged (Figure 2d). In contrast, PLL-*g*-PEG was found to adsorb on both TiO_2/DDP and SiO_2 surfaces, as is indicated by the increase in both C(1s) and N(1s) intensities and the corresponding decrease in the substrate (Ti(2p) and Si(2p), respectively) intensities (Figure 2). On

(31) Wagner, C. D.; Riggs, W. M.; Davis, L. E.; Moulder, J. F.; Mulienberg, G. E. *Handbook of X-ray Photoelectron Spectroscopy*; Perkin-Elmer: Eden Prairie, MN, 1979.

(32) Vickerman, J. C. *Surface analysis − The principal techniques*; John Wiley: New York, 1997.

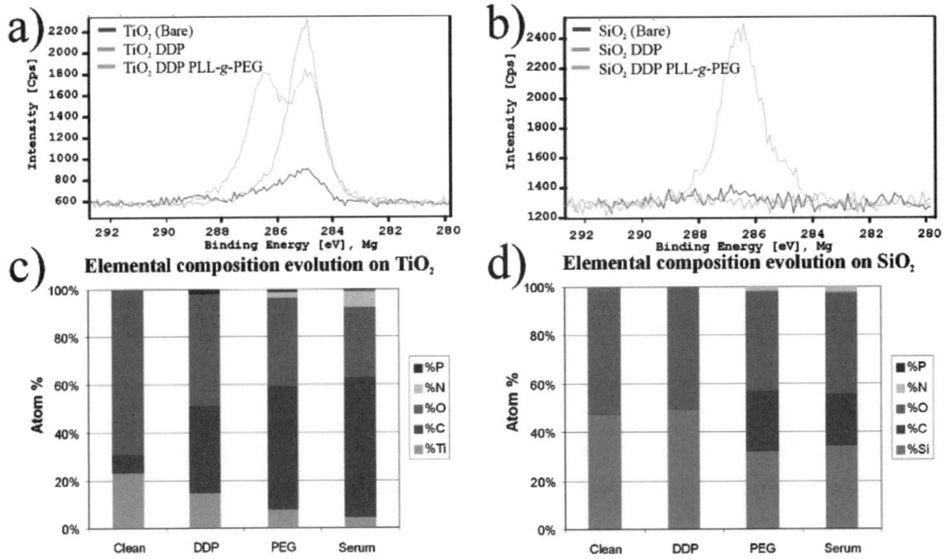

Figure 2. XPS spectra depicting the evolution of the C(1s) species on (a) TiO_2 and (b) SiO_2 areas of a sample. C(1s) is shown after cleaning and immersion in DDP and PLL-*g*-PEG. Elemental compositions of the surfaces are shown in (c) for TiO_2 and (d) for SiO_2, with serum immersion as an additional step.

SiO_2, deconvolution of the C(1s) peak revealed two components at 286.5 eV (C−O−C and C−O−H of PEG) and 285.0 eV (C−C and C−H of the lysine backbone), consistent with the chemical composition of the PLL-*g*-PEG molecule and an earlier detailed XPS study.[24] In the subsequent immersion step in full serum, protein replaced the PLL-*g*-PEG adsorbed onto the TiO_2/DDP (see below). The modified SiO_2/PLL-*g*-PEG surface remained unchanged (Figure 2d), while a decrease of the Ti(2p) as well as an increase in C(1s) and N(1s) (Figure 2c) signals indicated protein adsorption onto the TiO_2/DDP surface.

3.2. Characterization of the Patterned Surfaces. 3.2.1. Time-of-Flight Secondary Ion Mass Spectrometry. ToF-SIMS[32] is a valuable technique for qualitatively monitoring the adsorption of organic molecules onto surfaces, as it combines surface sensitivity with a high mass resolution. In addition, it can be used to investigate the localization of characteristic molecular ions with better than 1 μm lateral resolution, making it a valuable technique for monitoring surface modification steps on structured surfaces. When applied to the patterned substrates containing 60 × 60 μm^2 SiO_2 squares within a TiO_2 matrix, ToF-SIMS images (Figure 3) clearly indicated that Si^+ ions (*m/z* 28, Figure 3b) were found exclusively within the 60 μm squares, while the Ti^+ signal (*m/z* 47, Figure 3a) was only found in the surrounding matrix regions. The reverse was found on the samples with TiO_2 squares within a SiO_2 matrix (data not shown). To monitor the adsorption of DDP and PLL-*g*-PEG onto the substrate exhibiting material contrast, $PO_4H_4^+$ (*m/z* 99) and $C_3H_7O^+$ (*m/z* 59) ions were chosen. These ions are well separated from molecular ions originating from the metal oxide substrates and other organic fragments. A clear $PO_4H_4^+$ signal (Figure 3c), indicating phosphate adsorption onto the TiO_2 areas, was detected on the substrates incubated with DDP for 24 h. Hydrocarbon fragments occurring in the same mass range as the $PO_4H_4^+$ account for the slight increase in the intensity of the signal on the SiO_2 areas. On the other hand, PLL-*g*-PEG was found to adsorb onto both SiO_2 and TiO_2/DDP areas, as indicated by the presence of the $C_3H_7O^+$ ion in both areas (Figure 3d), consistent with the XPS findings (Figure 2a). Once exposed to serum, the amount of the $C_3H_7O^+$ ion on the TiO_2/DDP surface was found to decrease (Figure 3e). On the other hand, peaks due to amino acid ions with the generic structure $H_2N^+{=}CH{-}R$[33] appear on the TiO_2/DDP surface due to protein adsorption from the serum. These amino acid ions were not prominent on SiO_2. While the amount of the $C_3H_7O^+$ ion decreased by 1 order of magnitude after protein adsorption, the amount of Ti^+ ions detected remained constant. This suggests that the adsorbing proteins have replaced PLL-*g*-PEG on the DDP-functionalized regions.

3.3. Investigation of Protein Adsorption to Patterned Surfaces by Fluorescence Microscopy. Fluorescence microscopy was performed to obtain better insight into the adsorption behavior of proteins on the patterned TiO_2/SiO_2 oxide surfaces. Oregon Green labeled streptavidin was used as a model protein. Upon adsorption of the labeled streptavidin, the fluorescence signal (3.8 ± 0.09 × 10^4 au) was found to be localized on the hydrophobic, protein-adhesive DDP-coated TiO_2 5 × 5 μm^2 squares (Figure 4), while the fluorescence intensity on the PLL-*g*-PEG-coated SiO_2 matrix (6.4 ± 0.5 × 10^3 au) remained close to the background level (6.0 ± 0.5 × 10^3 au, determined by photobleaching of selected areas).

This is illustrated with a fluorescence intensity profile across several features (Figure 4, inset). It is known from measurements with optical waveguide lightmode spectroscopy (OWLS)[34] that below 1 ng/cm^2 of protein is adsorbed on PLL-*g*-PEG,[24] while ~50 ± 3 ng/cm^2 of streptavidin adsorbs on the DDP SAM (data not shown), resulting in a selectivity ratio (protein adsorbed on TiO_2/DDP over protein adsorbed on SiO_2/PLL-*g*-PEG) of at least ~50. The fluorescence intensity values quoted above yield a contrast ratio of ~100. The difference is likely to be caused by the lower sensitivity of OWLS.

Thus, the results of ToF-SIMS, XPS, and fluorescence measurements indicate that the material contrast present

(33) Mantus, D. S.; Ratner, B. D.; Carlson, B. A.; Moulder, J. F. *Anal. Chem.* **1993**, *65*, 1431−1438.
(34) Lukosz, W. *Sens. Actuators, B* **1995**, *29*, 37−50.

Figure 3. Positive ToF-SIMS images (300 × 300 μm^2) of 60 × 60 μm^2 SiO_2 areas, 60 μm apart, in a matrix of TiO_2. Positive ion maps are shown of (a) Ti^+ (*m/z* 47), (b) Si^+ (*m/z* 28), (c) $PO_4H_4^+$ (*m/z* 99), (d) $C_3H_7O^+$ (*m/z* 59) before serum adsorption, (e) $C_3H_7O^+$ (*m/z* 59) after serum adsorption, and (f) $C_3H_4NO^+$ (*m/z* 70), proline amino acid peak. The selectivity of the DDP onto the TiO_2 is shown as well as the PLL-*g*-PEG adsorption to the whole surface and its subsequent removal by protein on the TiO_2/DDP hydrophobic areas. A characteristic amino acid molecular ion, $C_3H_4NO^+$, was selected to show protein adsorption only to the TiO_2/DDP.

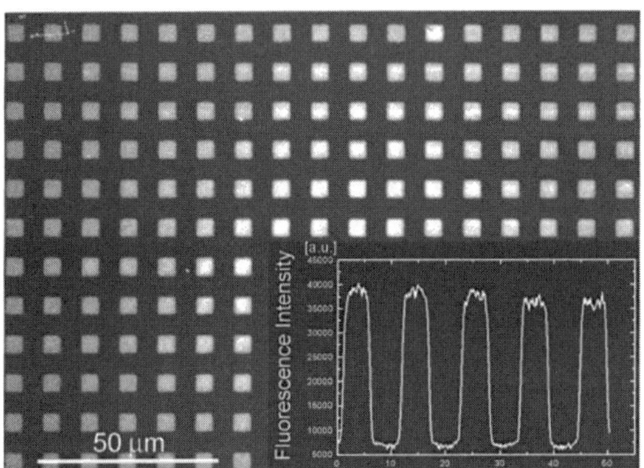

Figure 4. Fluorescence microscopy image on Oregon Green labeled streptavidin subjected to the SMAP-treated, 5 × 5 μm^2 TiO_2 in SiO_2 substrate. Streptavidin adsorption can only be observed on the TiO_2/DDP spots, while the SiO_2/PLL-*g*-PEG remains protein resistant. The inset shows the local distribution of fluorescence of the Oregon Green labeled streptavidin across the surface (in arbitrary units). A contrast of 100:1 was observed.

on the surface after the lithographic patterning steps (Figure 1a,b) was successfully converted into contrast with respect to protein adsorption (Figure 1c).

3.4. Cell Studies. In cell-culture assays, HFFs were incubated on three different adhesive/nonadhesive pattern geometries to test the feasibility of the SMAP technique for creating biologically relevant surface patterns: (a) half–half coated surfaces to judge attachment and motility of the cells on the adhesive (TiO_2/DDP/proteins) and the nonadhesive (SiO_2/PLL-*g*-PEG) areas; (b) adhesive patterns of 60 × 60 μm^2 size to test the feasibility of organizing cells on surfaces for applications in areas such as cell-based sensor chips; and (c) adhesive patterns of 5 × 5 μm^2 size to investigate the quality of subcellular-sized adhesive patches for localization of focal contacts and organization of cytoskeletal structures. In these cases, adhesion proteins mediating cell attachment spontaneously adsorbed to the DDP domains from the serum contained as a component in the culture medium.

(a) Surface-modified TiO_2 and SiO_2 half-coated samples were seeded with HFFs and incubated in a serum-containing medium. Cell attachment and spreading were monitored after various time intervals by means of phase-contrast microscopy. Cells attached and spread solely on the TiO_2/DDP surface. They were found to maintain a rounded shape above the SiO_2/PLL-*g*-PEG surface and were washed away upon medium removal. Furthermore, cells remained on the TiO_2/DDP side and were not observed to migrate onto the SiO_2/PLL-*g*-PEG-coated side within the experimental time scale (2 weeks). Tissue-culture polystyrene well plates, as well as untreated, oxide-coated glass cover slides, were used as controls. On both surfaces, cells were found to attach and spread.

(b) Transparent glass as well as Si wafer substrates with adhesive squares of 60 × 60 μm^2, separated by 60 μm, were used to test the suitability of the SMAP technique for immobilization of cells on regular patterns. Cells attached exclusively to the adhesive squares (of size in the order of a single cell) and remained geometrically confined and viable over the entire course of the experiments (2 weeks) (Figure 5,top).

(c) Si wafer substrates with a surface pattern of 5 × 5 μm^2 TiO_2 squares in a SiO_2 matrix, separated by 5 μm, were used in cell experiments to show the suitability of SMAP for creating adhesive patterns with subcellular feature sizes in a nonadhesive background. For visualization of the cell architecture on nontransparent Si wafer

3286 *Langmuir, Vol. 18, No. 8, 2002* *Michel et al.*

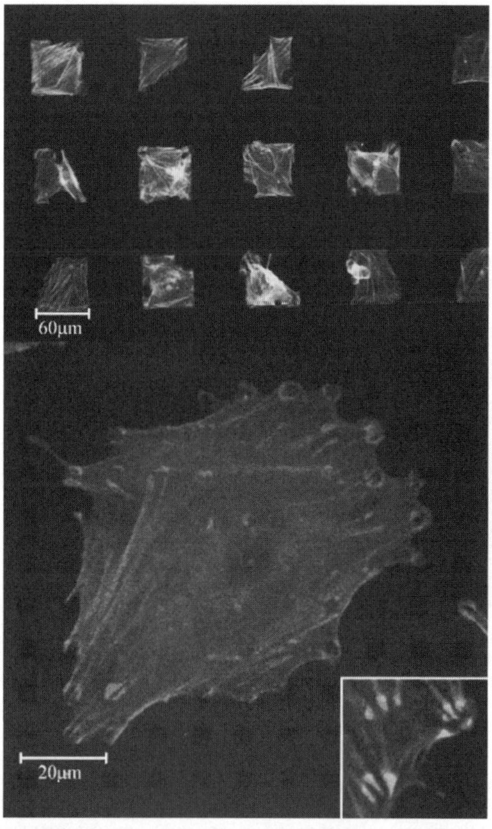

Figure 5. Top: 60 × 60 μm^2 of TiO_2 in SiO_2 substrate, SMAP treated. A representative array of fibroblast (HFF) cells attaching to the 60 × 60 μm^2 of TiO_2/DDP spots and spreading to the border of SiO_2/PLL-*g*-PEG is shown. Cells were visualized by immunostaining for f-actin. Bottom: HFFs spread on 5 × 5 μm^2 TiO_2/DDP cell-adhesive squares in a nonadhesive SiO_2/PLL-*g*-PEG background. Stress fibers were visualized by phalloidin staining for f-actin (red); vinculin was visualized by use of a monoclonal anti-vinculin primary antibody–fluorescently labeled secondary antibody combination (green). Substrate material contrast is visible in reflection mode (blue). The inset shows a magnification of focal contacts formed on the 5 × 5 μm^2 adhesive spots.

substrates, stress fibers were stained with rhodamine–phalloidin for f-actin. Vinculin, a protein present on the cytoplasmic side of focal adhesion complexes, was stained to visualize regions of focal contacts to the substrate. It has been previously shown that cells can spread on such surfaces, even when cell–surface contacts are established exclusively on adhesive islands, significantly smaller than that of the projected spread cell.[2]

HFFs were indeed able to attach and spread on these subcellular patterns but exhibited shapes different from those of cells incubated on nonpatterned substrates. Stress fibers were found to originate mainly above the 5 × 5 μm^2 adhesive patterns and often traversed several adhesive patches, while no interaction with the protein-resistant PLL-*g*-PEG background was evident. Immunostaining for vinculin showed that stress fibers were connected to the focal adhesion sites and that these focal adhesion sites were located exclusively on the 5 × 5 μm^2 adhesive features (Figure 5, bottom).

The results of these experiments clearly show that HFF cells consistently recognize the chemical contrast of the SMAP surfaces. We have strong evidence that this specificity is a direct consequence of the surface chemistry and is not influenced by the small surface topographical features. Due to the lithographic preparation, our substrates contained steps of 30–40 nm between the SiO_2 and TiO_2 areas. The effect of topography was tested in control experiments on SiO_2/TiO_2 patterned substrates, lacking the organic overlayers. Cells were found to spread freely on the different substrates with no sign of preferential attachment or orientation related to the nanosized step features (data not shown). This finding is consistent with those of earlier studies[35] where a topographic variation of 20 nm did not affect cell attachment.

4. Discussion and Conclusions

This report describes a new patterning technique for the preparation of biologically relevant chemical patterns with potential applications in biosensor and implant technologies, as well as in basic investigations of cell behavior. The method is based on selective self-assembly of DDP on TiO_2, but not SiO_2, from aqueous solution, in combination with the protein-resistant properties of PLL-*g*-PEG copolymer adsorbed onto negatively charged metal oxide surfaces. Due to the selective nature of the adsorption of the two adlayers, this technique is termed selective molecular assembly patterning. It relies on well-established sputter deposition, photolithography, and silicon etching techniques to create patterns composed of TiO_2 and SiO_2 areas with required dimensions, the attainable size of which is determined by the resolution of the specific patterning technique involved. The material contrast is then converted, in a series of simple dip-and-rinse processes involving aqueous solutions, into contrast with respect to protein and cell adhesion. This procedure was monitored with XPS and ToF-SIMS specifically used to analyze surface chemical composition, both of which showed unambiguously that DDP adsorbs selectively to the TiO_2 areas and PLL-*g*-PEG adsorbs to both the bare SiO_2 areas and the DDP-covered TiO_2 areas but proteins adsorb only to the DDP-covered TiO_2 areas, replacing the weakly bound PLL-*g*-PEG. The resulting contrast with respect to protein adsorption (defined in the results section) was estimated by fluorescence microscopy to be ~100-fold. Human foreskin fibroblasts, cultured in serum-containing media, were consistently found to adhere to the protein-adsorbing structures, where they developed focal contacts and survived for up to 14 days (the experimental time scale). No interaction between the cells and the protein-resistant (PLL-*g*-PEG) matrix was found within the time frame of the cell-culture experiments. The type of chemical contrast discussed in this report relies on the selective interaction of alkane phosphates with TiO_2 (or a number of other transition metal oxides) but not with SiO_2 surfaces. This is only one possible means of contrast formation within the frame of the SMAP technology. Strategies based on pH-dependent electrostatic contrast are currently being explored. Another area of interest is the extension of pattern size from the micrometer to the submicrometer or nanometer range, requiring substrates with finer oxide prepatterns that can be produced with the help of more sophisticated photolithographic techniques such as deep UV, X-ray, interference or colloidal particle lithography, electron beam lithography, or focused ion beam structuring.

SMAP presents a number of advantages, of particular relevance to an industrial environment, over the estab-

(35) Scotchford, C.; Winkelmann, M.; Gold, J.; Textor, M. *Biomaterials*, in preparation.

lished patterning techniques. First, the biologically relevant molecular adlayers form via a sequence of simple dip-and-rinse processes. No complex surface functionalization and immobilization schemes are required. Second, the production of required patterned oxide substrates is based on standard lithographic techniques. They are used to produce the required metal oxide patterns and do not cause any of the technological problems often encountered in alternative approaches based on photolithographic patterning of organic overlayers, where residual solvents can lead to inhomogeneities and degradation of the organic overlayers. These advantages of the SMAP technique, in turn, translate into cost-effectiveness, reproducibility, and compatibility with large-scale production of contamination-free chips bearing the desired biologically relevant patterns.

The SMAP technology has a number of potential applications, particularly in the area of microarray biosensor chips for DNA/RNA (genomics) or protein (proteomics) sensing, by providing patterns with defined physicochemical properties for improved spotting of recognition molecules and for cell-based sensor chips that require the provision of stable cell-adhesive areas on a highly protein- and cell-resistant background.

Acknowledgment. We thank Michael Horisberger for providing the sputter-coated oxide substrates and Paul Hug for help with the ToF-SIMS measurements and evaluation. Dr. Andreas Goessl, Dr. Janos Vörös, and Dr. Louis Tiefenauer are thanked for stimulating discussions.

LA011715Y

6104 *Langmuir* **2003,** *19,* 6104–6109

Microcontact Printing of Macromolecules with Submicrometer Resolution by Means of Polyolefin Stamps

Gabor Csucs,*[†,‡] Tobias Künzler,[†,§] Kirill Feldman,[||] Franck Robin,[⊥] and Nicholas D. Spencer[§]

BioMicroMetricsGroup, BMMG, ETH Zürich, Wagistrasse 4, CH-8952 Schlieren, Switzerland, Laboratory for Surface Science and Technology, Department of Materials, ETH Zürich, Sonneggstrasse 5, CH-8092 Zürich, Switzerland, Polymer Technology Group, Department of Materials, ETH Zürich, Universitätstrasse 41, CH-8092 Zürich, Switzerland, and Electromagnetic Fields and Microwave Electronics Laboratory, ETH Zürich, Gloriastrasse 35, CH-8092 Zürich, Switzerland

Received February 18, 2003. In Final Form: May 15, 2003

Microcontact printing (μCP) is a simple and cost-effective method to create micrometer-scale chemical patterns on surfaces. By careful modification of the conventionally used stamping material (poly(dimethylsiloxane) (PDMS)) and the stamping technique (e.g., "thin stamp μCP"), one can create surface chemical structures down to the submicrometer size range. In the present paper we report on the application of a new class of materials—polyolefin plastomers (POPs) for μCP applications. We show that the POP stamps are well suited to print proteins or block copolymers. Comparative studies on reproducibility, homogeneity, and quality of printing between POP and conventional PDMS stamps were also performed. The results show a superior performance of the POP stamps in the nanometer range and an identical performance in the micrometer range compared to PDMS. Further advantages of the POP-based μCP are faster stamp production, the lack of monomeric contamination (typical for PDMS stamps), and the possibility of recycling the POP stamps. We believe that POPs offer a useful alternative to PDMS for μCP and open new possibilities in submicrometer-range printing.

1. Introduction

During the past decade, microcontact printing (μCP) has become one of the most popular laboratory techniques for the fabrication of chemically microstructured surfaces. There are several reasons for this popularity: μCP is fast, is inexpensive, is simple, requires neither cleanroom instrumentation nor absolutely flat surfaces, and offers a way to create complex patterns, albeit with some geometrical constraints.[1] The achievable resolution is also remarkable—30 nm being the current limit (for thiol-based systems).[2] Although μCP was originally used to print self-assembled monolayers of alkanethiolates on gold surfaces,[1,3,4] it was soon extended to the stamping of proteins onto a variety of different surfaces.[5–7] The overwhelming majority of μCP studies have been carried out using poly(dimethylsiloxane) (PDMS) as a stamping material.[1] Although PDMS is well suited for many stamping applications, it has a number of serious drawbacks, which are partially connected to the softness (low mechanical stability) of the material. This softness sets serious geometrical constrains for the realizable structures and limits the achievable resolution of the standard PDMS-based technique.[1,8,9] To overcome these problems, two principal solutions have been introduced (and also combined with each other): (1) A supporting glass/plastic plate was used to increase the mechanical stability of the stamp.[6,10] (2) Special PDMS variants with better mechanical properties for high-resolution μCP were used.[11] Another (often neglected) drawback of PDMS-based μCP is the frequently observed low-molecular-weight ("monomer") PDMS contamination that is present on the stamped surface.[12,13] To solve these problems (mechanics/contamination), instead of creating new PDMS variants we have investigated the possibility of using a new class of materials—polyolefin plastomers (POPs) in μCP applications. In the present paper we describe the use of POPs for printing proteins (Alexa 488-fibrinogen) and block copolymers—fluorescein-poly-L-lysine-*g*-poly(ethylene glycol) (PLL-*g*-PEG-fl*).[13] We compare μCP with POPs to the conventional PDMS-based approach, in terms of both quality and reproducibility.

* To whom correspondence may be addressed. Fax: +41 1 633 1124. E-mail: csucs@biomech.mavt.ethz.ch.

† These authors contributed equally to this work.

‡ BioMicroMetricsGroup, BMMG, ETH Zürich.

§ Laboratory for Surface Science and Technology, Department of Materials, ETH Zürich.

|| Polymer Technology Group, Department of Materials, ETH Zürich.

⊥ Electromagnetic Fields and Microwave Electronics Laboratory, ETH Zürich.

(1) Xia, Y. N.; Whitesides, G. M. *Annu. Rev. Mater. Sci.* **1998**, *28*, 153–184.

(2) Biebuyck, H. A.; Larsen, N. B.; Delamarche, E.; Michel, B. *IBM J. Res. Dev.* **1997**, *41*, 159–170.

(3) Kumar, A.; Whitesides, G. M. *Appl. Phys. Lett.* **1993**, *63*, 2002–2004.

(4) Mrksich, M.; Chen, C. S.; Xia, Y. N.; Dike, L. E.; Ingber, D. E.; Whitesides, G. M. *Proc. Natl. Acad. Sci. U.S.A.* **1996**, *93*, 10775–10778.

(5) Bernard, A.; Delamarche, E.; Schmid, H.; Michel, B.; Bosshard, H. R.; Biebuyck, H. *Langmuir* **1998**, *14*, 2225–2229.

(6) James, C. D.; Davis, R. C.; Kam, L.; Craighead, H. G.; Isaacson, M.; Turner, J. N.; Shain, W. *Langmuir* **1998**, *14*, 741–744.

(7) Bernard, A.; Renault, J. P.; Michel, B.; Bosshard, H. R.; Delamarche, E. *Adv. Mater.* **2000**, *12*, 1067–1070.

(8) Delamarche, E.; Schmid, H.; Michel, B.; Biebuyck, H. *Adv. Mater.* **1997**, *9*, 741–746.

(9) Bietsch, A.; Michel, B. *J. Appl. Phys.* **2000**, *88*, 4310–4318.

(10) Rogers, J. A.; Paul, K. E.; Whitesides, G. M. *J. Vac. Sci. Technol., B* **1998**, *16*, 88–97.

(11) Schmid, H.; Michel, B. *Macromolecules* **2000**, *33*, 3042–3049.

(12) Yang, Z. P.; Belu, A. M.; Liebmann-Vinson, A.; Sugg, H.; Chilkoti, A. *Langmuir* **2000**, *16*, 7482–7492.

(13) Csucs, G.; Michel, R.; Lussi, J. W.; Textor, M.; Danuser, G. *Biomaterials* **2003**, *24*, 1713–1720.

Published on Web 06/27/2003

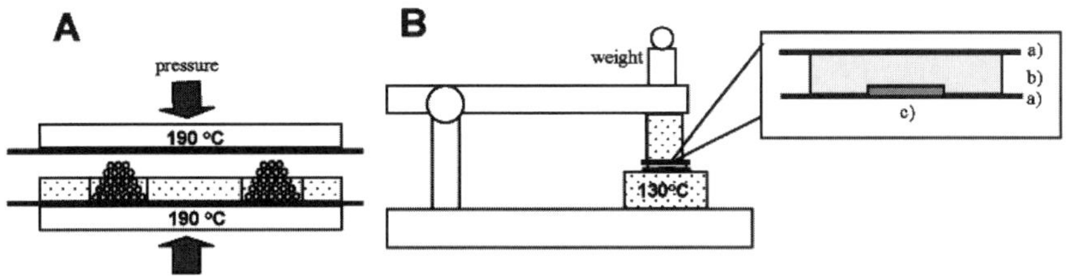

Figure 1. The two main steps of the POP stamp formation. (A) Melting the POP pellets into bars. The red line indicates the two polyimide foils. (B) Replicating the master in the POP bar: (a) silicon wafer (black line); (b) POP bar (yellow); (c) master (green). The sandwich was heated from both sides.

2. Experimental Section

2.1. Substrates. μCP was performed either on tissue-culture polystyrene (TCPS) well plates (Nunclon, NalgeNunc International, Rochester, NY) or on normal glass coverslips (Menzel Glass, Braunschweig, Germany). The precleaned coverslips were washed with ethanol and blown dry by N_2. TCPS plates were used without additional cleaning.

2.2. Stamp Masters. Masters for the micrometer-range structures were produced as previously described.[13]

The master contained lines of 1.5, 5, and 10 μm width and 1 × 1, 3 × 3, 5 × 5 μm squares. The separation distance between the structures was the same as their size.

The masters for the nanometer-range structures were fabricated using electron-beam lithography (EBL) by means of a Raith150 system (Raith GmbH, Dortmund, Germany): After being cleaned, the samples were coated with a 200 nm thick poly(methyl methacrylate) (PMMA) layer (MicroChem Corp., Newton, MA) and hard baked for 60 min at 180 °C. The resist was then exposed with an acceleration voltage of 25 kV and an electron dose of 330 μC/cm^2. The beam current was 13 pA. After exposure, the resist was developed in a methyl isobutyl ketone/isopropyl alcohol (MIBK/IPA) 1:3 solution for 60 s at 22 °C and rinsed with pure 2-propanol. After this lithographic step, the samples were transferred to a dry etching plant (Surface Technology Systems, STS T20, Ulm, Germany). The etching was performed with a mixture of 8 sccm CHF_3 and 20 sccm SF_6 at 41 mTorr, 300 K, and 55 W for 3 min.

2.3. Formation of the Stamps. PDMS (Sylgard 184 from Dow Corning, Midland, MI) stamps were formed as described previously.[13]

Affinity polyolefin plastomers were obtained as pellets from The Dow Chemical Company (Midland, MI). The plastomers used (Affinity VP8770, Affinity EG8200, Affinity EG8150) are a relatively new class of polymers that emerged from recent developments in metallocene polymerization catalysts. POPs are copolymers of ethylene and an α-olefin such as butene or octene. The metallocene catalyst selectively polymerizes the ethylene and comonomer sequences. Increasing the comonomer content will produce polymers with higher elasticity as the comonomer incorporation disrupts the polyethylene crystallinity. Stamps (from all three Affinity POP variants) were produced as follows (see also Figure 1): The POP pellets were first melted into blocks of 40 × 20 × 5 mm at 190 °C under a pressure of 4 bar using an appropriate metal template. To avoid the sticking of the polymers to the heated metal planes, thin polyimide foils were placed in between. After cooling to room temperature, the solid polymer bars were removed from the template, rinsed with ethanol, and dried under a stream of nitrogen. In the next step the bars were placed over the stamp masters (10 × 10 mm) and placed between two silicon wafers (50 × 30 mm). This "sandwich" was then placed on a heatable plate (heated to 130 °C from both sides). First, a weight of 200 g was put on the top of the sandwich for 5 min, which was afterward increased to 700 g for 4 min. After cooling, the master was easily peeled off from the POP bar, which was then cut to its proper size with a razor blade. Prior to usage, the POP stamps were cleaned with acetone for 5 min in an ultrasonic bath.

2.4. Printing of Proteins and Block Copolymers. Alexa-Green 488 fibrinogen (Molecular Probes, Eugene, OR) was dissolved in phosphate-buffered saline (PBS) (pH 7.4) at a concentration of 40 μg/mL. PLL-*g*-PEG-fluorescein was synthesized as described previously[13] and dissolved in 10 mM 4(2-hydroxyethyl)piperazine-1-ethanosulfonic acid (HEPES-Z1) (pH 7.4) at 1 mg/mL concentration. The clean stamps (both PDMS and POP) were incubated with the appropriate protein/block copolymer solution for 40 min in a laminar-flow hood, protected from light. After the incubation, the inking solution was removed and the stamps were blown dry with N_2. The stamps were then placed on the substrates for about 20 s. To ensure proper contact between the stamps and the surface, a small (stamp-surface-area-dependent) weight of 1 g/mm^2 was applied on top of the stamps. After the stamps were removed, the surfaces were washed with buffer. Optical microscopy was performed under buffer solution.

2.5. Fluorescence Microscopy. The fluorescence microscopy investigation of the printing quality was performed using a 20×/0.4 NA LD-Achroplan objective for the micrometer structures and a 100×/1.4 NA Plan-Apochromat for the nanometer structures (Carl Zeiss AG, Feldbach, Switzerland). Image acquisition was done by using a LSM510 laser-scanning confocal microscope (Carl Zeiss AG). The fluorophores were excited using the 488 nm line of the scanning unit.

2.6. X-ray Photoelectron Spectroscopy (XPS) Measurements. All XPS spectra were recorded on a SAGE 100 (SPECS, Berlin, Germany) using nonmonochromatic Al Kα radiation with an energy of 325 W (13 kV, 25 mA), electron takeoff angle of 90°, and electron-detector pass energies of 50 eV for survey and 14 eV for detailed spectra. During analysis, the base pressure remained below 5×10^{-8} mbar. All peaks are referenced to the C1s (hydrocarbon C–C, C–H) contribution set to 285.0 eV.

2.7. Contact Angle Measurements. Advancing contact angle measurements were performed on plane and PDMS/POP stamped surfaces by a Krüss G2 contact-angle-measurement system and the DSA1 drop-shape-analysis system (Krüss GmbH, Hamburg, Germany) using HPLC-quality water (Fluka, Buch, Switzerland). The experiments were performed at 22 °C and 44% relative humidity.

2.8. Scanning Electron Microscopy (SEM) Measurements. The SEM micrographs were acquired with the Raith150 EBL system with a LEO Gemini 1530 column (Raith). The acceleration voltage was 10 kV, and the sample current 100 pA.

2.9. Atomic Force Microscopy (AFM) Measurements. Imaging of the silicon master and the polymer replica stamps in PDMS and POP was performed with a NanoScope IIIa multimode scanning probe microscope (Veeco Instruments, Inc., Woodbury, NY). The silicon master was scanned in contact mode, while the PDMS and POP stamps were imaged in both contact mode and TappingMode. Oxide-sharpened Si_3N_4 probes (Veeco) with a nominal spring constant of 0.06 N/m were employed for scanning in contact mode at zero applied load (load being due to adhesion only) to ensure minimal damage to these low-modulus, highly compliant materials. TappingMode imaging was performed with TappingMode etched silicon probes (Veeco).

3. Results and Discussion

3.1. Analyzing the Stamping Quality. *Stamp Production.* Three different polyolefin elastomers, Affinity VP8770, Affinity EG8200, and Affinity EG8150, were tested for their suitability for μCP applications. The main difference between these elastomers is their crystallinity,

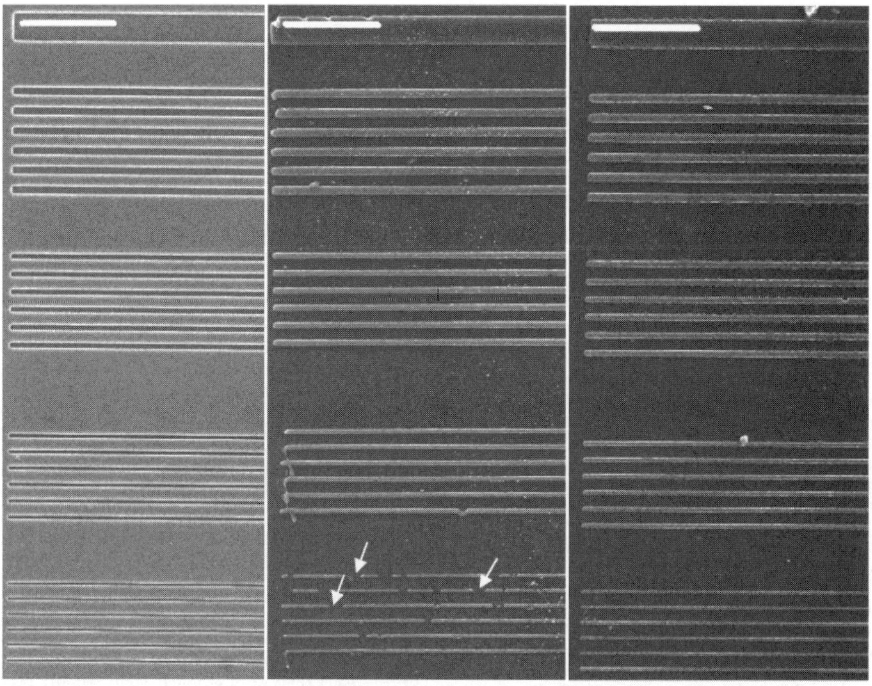

Figure 2. SEM images of the stamp master (left) a PDMS replica (middle), and a POP (VP8770, right) replica. The line thicknesses from the top to the bottom are 5000, 1000, 600, 300, and 100 nm. Scale bars in the corners are 20 μm.

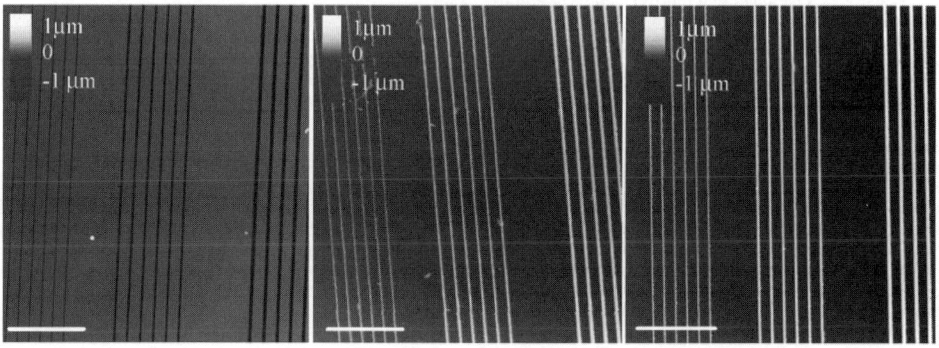

Figure 3. AFM images of the stamp master (left) a PDMS replica (middle), and a POP (VP8770) replica. The line thicknesses from the left to the right are 100, 300, and 600 nm, respectively. Scale bars in the corners are 20 μm.

VP 8770 being the most crystalline (with the highest melting point and mechanical stiffness) and EG 8150 being the most amorphous (with the lowest melting point and highest elasticity). The molding procedure for forming the stamps was the same for all materials. The production time for the stamps amounts to about 1 h—much faster than the approximately 24 h that is reported in the literature for PDMS stamp production.[11] The quality of the stamps was first checked by stamping fluorescently labeled proteins/polymers and using fluorescence microscopy to analyze the transferred structures. In further experiments, SEM and AFM were used to directly test the quality of the nanometer-scale stamps.

Scanning Electron Microscopy. Pattern replication from the master to the PDMS and POP (VP8770 variant) stamps was investigated (besides AFM) by SEM. The examinations concentrated on the nanometer range patterns, because this is the range where the occurrence of replication problems might be expected. Figure 2 shows a typical result of such an investigation showing 1000, 600, 300, and 100 nm thick lines (with 3 μm separations). The replication of the structures both by PDMS and POP seems to be excellent. However, some discontinuities of the 100 nm lines on the PDMS are also observable (see arrows on Figure 2B)—this may be due to wetting and viscoelastic properties of the PDMS, as well as its low bulk elastic modulus.[11]

Atomic Force Microscopy. In addition to SEM, AFM measurements were also performed, to further check the quality of the pattern replication by different stamp materials. Figure 3 illustrates such an experiment. The images nicely demonstrate that, qualitatively, the master patterns were indeed transferred to the stamps. In the case of the PDMS stamp, small defects are visible in the replication of the 100 nm lines. The existence of these discontinuities (similar those observed on the SEM images)

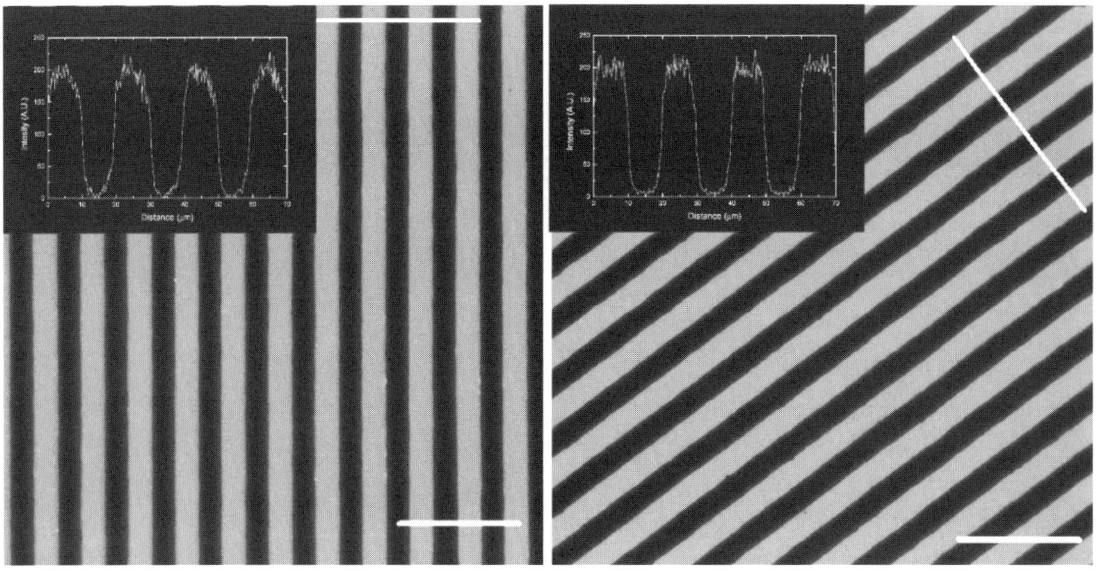

Figure 4. Typical printing assay, by means of fluorescence microscopy, of Alexa488 Fibrinogen using PDMS (left) and EG8150 (right) as stamp material and TCPS as substrate. The line thicknesses were 10 μm with 10 μm separations. The insets show cross sections of the fluorescence intensities measured perpendicular to the line directions. The lines across the stripes indicate the positions at which the cross sections were measured. Scale bars in the corners are 50 μm.

is probably again connected to both chemical and bulk mechanical properties of the PDMS.

Fluorescence Microscopy. The ultimate test for a new stamping material is whether it is possible to ink it and transfer molecules to another substrate, replicating the master structure. We have tested the performance of the POP stamps by printing proteins (Alexa488-fibrinogen) and block copolymers (PLL-*g*-PEG-fl*) onto both glass and tissue-culture polystyrene (TCPS) surfaces using two different masters containing micrometer or submicrometer-sized structures. In each case, the results were compared to those obtained with conventional PDMS stamps.

Figure 4 shows a typical example of a stamping experiment using the micrometer-range structures. There is no observable difference in the homogeneity and the quality of the printed structures between the POP (Affinity EG8150) and the PDMS printed surface. The protein-transfer ratio (from the stamp to the surface), as determined from the fluorescence intensities, was close to 100%—there was no measurable fluorescence left on the stamps as observed by microscopy. The insets on the images show that in both cases the edges are well-defined and there is no fluorescence present between the lines. The different POP variants all gave identical results (data not shown). After being cleaned with ethanol (10 min in ultrasonic bath), both the POP and the PDMS stamps were reusable. In our tests we have reused the POP stamps at least 10 times without observing any changes in the printing quality.

In a previous paper, we have shown that PDMS can be used to stamp the block copolymer PLL-*g*-PEG-fl* onto a variety of surfaces.[13] These observations could be replicated by using POP stamps. Again, there was no observable difference between PDMS- and the POP-stamped surfaces (data not shown). Both with POP and with PDMS stamps the printing quality seemed to be independent of the type of surface (glass or TCPS) or the type of micrometer-sized structure (10, 1.5 μm thick lines and 5 × 5, 3 × 3, 1 × 1 μm^2 squares with 10, 1.5, 5, 3, and 1 μm separations, respectively) (data not shown).

It is generally agreed in the literature that a significant challenge for μCP is to print submicrometer-sized structures with rather large (several micrometers) separations. One of the main problems when printing in this size range is the softness (small compression modulus) of the conventional PDMS stamps which results in the inability to replicate a thin line.[11] Different methods (special variants of PDMS and stabilizing solid substrates under the ultrathin PDMS layer) have been introduced to solve this problem. Since the POPs used in this work are much stiffer than PDMS, we hoped that their mechanical stability would be advantageous when printing in the submicrometer range. To test this hypothesis, we have produced a master containing different submicrometer-sized structures and used it to make POP and PDMS stamps. In these experiments we have used only the stiffest (VP7880) POP variant. Again, Alexa488-Fibrinogen was used for inking. Because the requirements of high-resolution optical microscopy, only glass coverslips were used as substrates. A typical result is shown in Figure 5. The stamped structures were 1000, 600, 300, and 100 nm thick lines (Figure 5A left to right), always with 3 μm separations. The separations between the groups of lines were even larger: 12, 17, and 12 μm, from the left to right. Analyzing the quality of the stamping, we can conclude that the thickness of the printed lines corresponds to that of the stamp master. Naturally, because of the limited resolution of optical microscopy, the 100 and 300 nm lines appear to be somewhat wider than their "real" size. It is remarkable that sagging of the stamps does not occur even at the large separations between the line groups. As with the micrometer structures, the stamps were again reusable and the reproducibility of the printing was good.

Figure 5B shows the best stamping result obtained by a "conventional" PDMS stamp using the same master. The difference is clear. Although the quality of the 1000, 600, and, to some extent, the 300 nm lines is still

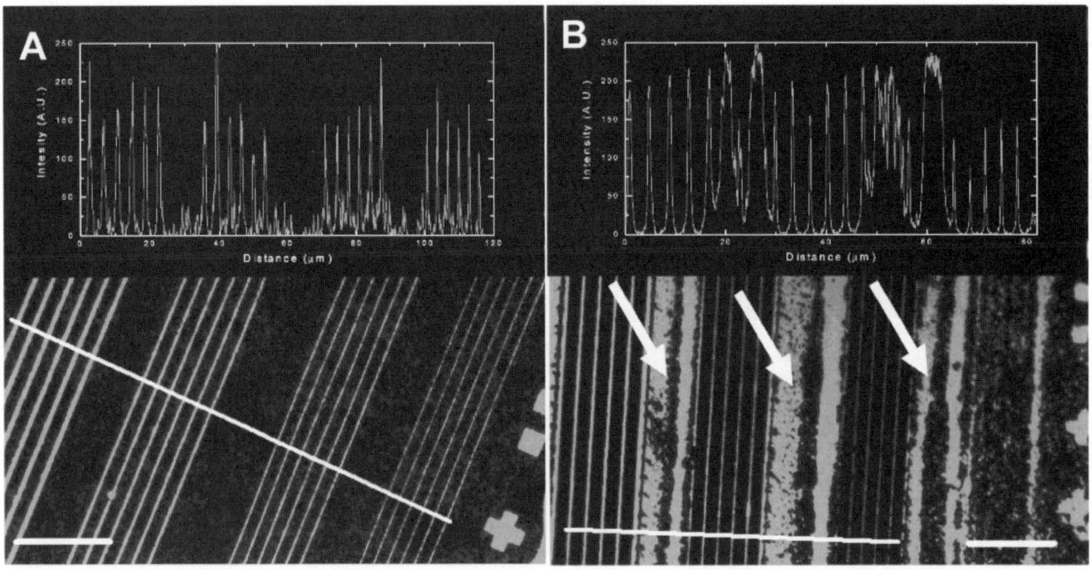

Figure 5. Nanometer-scale printing of Alexa488 Fibrinogen using VP8770 (A) and PDMS (B) as stamp materials, as determined by fluorescence microscopy. The line thicknesses are 1000, 600, 300, and 100 nm with 3 μm separation. The insets show cross sections of the fluorescence intensities measured perpendicular to the line directions. The lines across the stripes indicate the positions at which the cross sections were measured. Arrows on (B) indicate regions where sagging of the stamp occurred. Scale bars in the corners are 20 μm.

Table 1. Advancing Water Contact Angles Measured on the Surfaces of Stamp Materials and Glass Coverslips before and after Flat-Stamp Printing

	advancing water contact angle/deg		
substrate or stamp surface	surfaces before stamping	PDMS stamped	POE stamped
coverslip (plasma cleaned)	15 ± 5	87 ± 2	16 ± 5
coverslip (no plasma cleaning)	54 ± 2	76 ± 1	54 ± 2
PDMS	109 ± 2		
POP (Affinity EG8150)	105 ± 2		

acceptable, the 100 nm lines are totally distorted. Furthermore the thickness of the lines varies and does not always correspond to the master. Sagging occurred at all the larger separations (arrows).

3.2. Analyzing the Surface Contamination. *Contact-Angle Measurements.* To test whether stamping modifies the properties (hydrophilicity/hydrophobicity) of the stamped surface, we have measured the advancing contact angles on the stamps and on the stamped surfaces. Nonprinted substrates served as the reference. The stamps were in this case not inked, and nonstructured (flat) stamps were used, but otherwise the printing was performed as described for the proteins, on normal glass coverslips, cleaned either only with ethanol/distilled water washing or by (in some cases) a subsequent oxygen plasma treatment. The results are summarized in Table 1.

The values in the table show that both stamp materials are rather hydrophobic. Upon contacting the surface, however, POP (Affinity EG8150) has no measurable effect on the contact angle, as compared to the control values. PDMS, on the other hand, significantly modifies the properties of the surface, making it hydrophobic. This indicates a contamination of the substrate by the stamp. The effect is stronger on the highly reactive (clean) plasma-treated surface and less pronounced (but still present) on the slightly contaminated ethanol-cleaned surface. A possible explanation for this fact could be that the residual organic contamination present on the glass surface reduces the PDMS contamination.

Table 2. XPS Relative Detected Elemental Atomic Concentrations for Cleaned and POP and PDMS Printed TiO_2 Surfaces (n.d. = not detected)

	% Ti	% O	% C	% Si
TiO_2 clean	22	70	8	n.d.
TiO_2, POP printed	21	71	8	n.d.
TiO_2, PDMS printed	20	65	11	4

XPS Measurements. To obtain more information about possible surface contamination caused by the stamps, XPS measurements were also performed. The samples were prepared in a similar way to those used in contact angle measurements. Printing was again performed with flat/noninked POP (Affinity EG8150) and PDMS, but in this case sputter coated (20 nm) TiO_2 surfaces were employed as substrates. Prior to printing, all substrates were cleaned by oxygen plasma treatment. The cleaning time was 2 min in the case of TiO_2 surfaces and 10 s in the case of stamps. The results of the XPS measurements of stamped TiO_2 substrates were compared to those recorded on a nonstamped (clean) surface. Table 2 summarizes the results.

By analyzing the values, we can conclude the following: the POP printing has no detectable effect on the surface composition of the substrate. In the event of contamination by the POP one would expect an increased amount of carbon on the surface, but this was not observable. Si (as expected) is only detected in the case of PDMS stamping. This is a clear sign of monomeric PDMS contamination and has been previously reported.[12,13] The increased carbon content may be also explained by the monomeric contamination.

Conclusions

In the present paper we have demonstrated the feasibility of using polyolefin elastomers as stamp materials for microcontact printing of proteins and block copolymers. As shown by SEM and AFM investigations, the replication of the stamp master is effective with both POP (Affinity) and standard (Sylgard 184) PDMS materials, down to the nanometer range. Comparing the performance of POP stamps to those made from PDMS, we can conclude the following: (1) When printing micrometer-range structures, their performance is nearly identical. (2) In the submicrometer range (submicrometer structures with micrometer separations), much higher printing quality is achievable with the POP stamps. This fact is probably due to the higher bulk modulus of the POP stamps.

Comparing the possible surface contamination by the stamps using contact angle and XPS measurements, POP barely modifies the surface, while PDMS displays considerable contamination, presumably by low molecular weight material. The high processing speed and the recycling capability of the POP could be useful properties, when considering the technological applications of this method.

Acknowledgment. The authors wish to thank Dr. Thomas Allgeuer of The Dow Chemical Company, Horgen, Switzerland, for supplying the polyolefin plastomers.

LA0342823

622 *Langmuir* **2005,** *21,* 622−632

Diffusion of Alkanethiols in PDMS and Its Implications on Microcontact Printing (μCP)

Tobias E. Balmer,[†,‡] Heinz Schmid,[†] Richard Stutz,[†] Emmanuel Delamarche,[†] Bruno Michel,[†] Nicholas D. Spencer,[‡] and Heiko Wolf[*,†]

IBM Research GmbH, Zurich Research Laboratory, Säumerstrasse 4, CH-8803 Rüschlikon, Switzerland, and Laboratory for Surface Science & Technology, Department of Materials, Swiss Federal Institute of Technology, ETH Hoenggerberg, Wolfgang-Pauli-Strasse 10, CH-8093 Zürich, Switzerland

Received July 9, 2004. In Final Form: October 1, 2004

n-Alkanethiols $HS-(CH_2)_n-CH_3$ such as hexadecanethiol (HDT, $n = 15$), octadecanethiol (ODT, $n = 17$), and eicosanethiol (ECT, $n = 19$) have been shown to provide highly protective etch resists on microcontact-printed noble metals. As the quality of the printed pattern strongly depends on the mobility of the ink compound, we focused on understanding the diffusion behavior of HDT, ODT, and ECT in poly(dimethylsiloxane) (PDMS) stamps. We used a commercial PDMS material (Sylgard184), which is commonly used for microcontact printing (μCP), and a custom-synthesized one with a higher modulus. On the basis of linear-diffusion experiments, which maintained realistic printing conditions, we showed that the ink transport in the stamp follows Fick's law of diffusion. We then determined the diffusion coefficient by analytical and numerical modeling of the diffusion experiments. Numerical calculations were carried out with the finite-difference method applying more realistic boundary conditions (ink adsorption). Values for the diffusion coefficients of the three ink compounds in the two different PDMS materials all are on the order of $(4-7) \times 10^{-7}$ cm^2 s^{-1}. The scope and limits of the mathematical models are discussed. To demonstrate the potential of such models for microcontact printing, we simulate multiple printing cycles of an inked stamp and compare the results with experimental data.

1. Introduction

Microcontact printing (μCP) is a high-resolution technique used to transfer a pattern from an elastomeric stamp to a solid substrate by conformal contact. The method was introduced by Whitesides and co-workers,[1,2] who brought a polymer inked with alkanethiols into contact with a gold substrate to form a self-assembled monolayer (SAM) in the area of contact. This triggered a new field in scientific research that has been rapidly growing over the past 10 years. Today, μCP is known as one of several related techniques referred to as *soft lithography*. This term summarizes the common principle of these techniques of forming conformal contact between an elastomeric device and a substrate in order to obtain a chemical pattern.[3,4]

Microcontact printing of alkanethiols on gold using poly(dimethylsiloxane) (PDMS) stamps was the original demonstration of the technique, and it still is a widely used model system.[1,2] The alkanethiol SAM acts as a structured chemical resist in a subsequent etching process. The elastomeric stamp is obtained from liquid PDMS prepolymers, which are poured directly onto a master and cured. Specialty PDMS materials with high moduli have been developed for the printing of high-resolution patterns in the submicron range.[5−7] The hydrophobic properties of PDMS stamps are well suited for printing alkanethiols on gold and on other noble metals such as silver and copper, but other types of ink such as macromolecules, biological molecules, or catalysts may require a chemical treatment of the stamp's surface.[8−11] Examples of other self-assembling compounds or printable substances are alkoxysilanes or trichlorosilanes on silicon oxide (glass)[12−15] or alkyl phosphates on metal oxide surfaces,[16] reactants printed onto organic layers,[17] and proteins transferred to silicon or glass.[18]

† IBM Research GmbH, Zurich Research Laboratory.
‡ Swiss Federal Institute of Technology.

(1) Kumar, A.; Whitesides, G. M. *Appl. Phys. Lett.* **1993**, *63*, 2002−2004.
(2) Kumar, A.; Biebuyck, H. A.; Whitesides, G. M. *Langmuir* **1994**, *10*, 1498−1511.
(3) Xia, Y.; Whitesides, G. M. *Angew. Chem., Int. Ed.* **1998**, *37*, 550−575.
(4) Michel, B.; et al. *IBM J. Res. Dev.* **2001**, *45*, 697−719.
(5) Schmid, H.; Michel, B. *Macromolecules* **2000**, *33*, 3042−3049.
(6) Bietsch, A.; Michel, B. *J. Appl. Phys.* **2000**, *88*, 4310−4318.
(7) Odom, T. W.; Love, J. C.; Wolfe, D. B.; Paul, K. E.; Whitesides, G. M. *Langmuir* **2002**, *18*, 5314−5320.
(8) Ferguson, G. S.; Chaudhury, M. K.; Biebuyck, H. A.; Whitesides, G. M. *Macromolecules* **1993**, *26*, 5870−5875.
(9) Yan, L.; Huck, W. T. S.; Zhao, X. M.; Whitesides, G. M. *Langmuir* **1999**, *15*, 1208−1214.
(10) Donzel, C.; Geissler, M.; Bernard, A.; Wolf, H.; Michel, B.; Hilborn, J.; Delamarche, E. *Adv. Mater.* **2001**, *13*, 1164−1167.
(11) Delamarche, E.; Donzel, C.; Kamounah, F. S.; Wolf, H.; Geissler, M.; Stutz, R.; Schmidt-Winkel, P.; Michel, B.; Mathieu, H. J.; Schaumburg, K. *Langmuir* **2003**, *19*, 8749−8758.
(12) Sagiv, J. *J. Am. Chem. Soc.* **1980**, *102*, 92.
(13) St. John, P. M.; Craighead, H. G. *Appl. Phys. Lett.* **1996**, *68*, 1022−1024.
(14) Wang, D.; Thomas, S.; Wang, K.; Xia, Y.; Whitesides, G. *Appl. Phys. Lett.* **1997**, *70*, 1593−1595.
(15) Geissler, M.; Kind, H.; Schmidt-Winkel, P.; Michel, B.; Delamarche, E. *Langmuir* **2003**, *19*, 6283−6296.
(16) Hofer, R.; Textor, M.; Spencer, N. D. *Langmuir* **2001**, *17*, 4014−4020.
(17) Yan, L.; Zhao, X.-M.; Whitesides, G. M. *J. Am. Chem. Soc.* **1998**, *120*, 6179−6180.
(18) Bernard, A.; Delamarche, E.; Schmid, H.; Michel, B.; Bosshard, H. R.; Biebuyck, H. *Langmuir* **1998**, *14*, 2225−2229.

Published on Web 12/18/2004

In principle, it should be possible to use the stamps for multiple prints, but little is known about the ink distribution in a stamp, even after the first inking and printing cycle. The inking of the stamp with alkanethiols is a diffusion process in which the ink is transferred from a reservoir (ink source) into the stamp. Several methods for inking the stamp in microcontact printing have been introduced.[19,20] *Wet inking* is probably the most commonly used inking technique. In this case, the alkanethiols are dissolved in ethanol (EtOH) and the solution is applied to the entire surface of the stamp for a certain amount of time. A more controlled way to ink the stamp is to use a flat PDMS inker pad that has been impregnated or saturated in an ink solution (*contact inking*). A *permanently inked stamp* is obtained by placing an impregnated inker pad on the backside of the stamp and letting the ink diffuse through the stamp toward the patterned side. In an ideal case, the flux of ink through the bulk of the stamp reaches a steady state while the stamp is used for printing. This setup resembles a felt pen used for writing on paper with a reservoir of ink behind the tip.

Alkanethiol diffusion determines the quality of the microcontact-printed pattern.[19,21] *Bulk diffusion* through the PDMS is the limiting factor of the ink transport from stamp to substrate. It controls the completeness of the printed monolayer resist and thereby the selectivity in a subsequent etch. In addition, *surface diffusion* can play a role when ink molecules travel along the sidewalls of the stamp structure toward the substrate. This affects the quality of the pattern (blurring). Diffusion effects are mainly controlled by the concentration and mobility of the ink molecules. Low concentrations and a high molecular weight of the ink reduce the effects of surface diffusion.[21] This benefit is offset by longer printing times. In addition, compounds with high molecular weights tend to crystallize at the surface of the stamp, leading to contamination problems. To obtain the best possible quality of the pattern, a compromise between these parameters has to be found. So far, this has been achieved by using long-chain alkanethiols, such as hexadecanethiol (HDT), octadecanethiol (ODT), or eicosanethiol (ECT), for μCP on noble-metal surfaces.[3,4,21]

Efforts to evaluate the effects of ink diffusion when printing alkanethiols on metallic surfaces using PDMS stamps have been previously reported.[19,21] However, little is known about the magnitude of the diffusion coefficients that determine these effects. If μCP using alkanethiols is to become an industrial process, the quantitative characterization of ink diffusion in the stamp will be crucial in answering questions such as the following: What are optimal inking and printing times, or how much ink is taken up by the stamp and how many consecutive prints of desired quality are possible?

In this work, we demonstrate a method for characterizing the bulk diffusion of alkanethiol-based inks in PDMS stamps. The diffusion constants are obtained from mathematical models describing simple linear-diffusion experiments. The experiments are set up in such a way as to maintain realistic μCP conditions. We compare the diffusion of three different ink compounds (hexadecanethiol (HDT), octadecanethiol (OCT), and eicosanethiol (ECT)) in two different PDMS stamp materials (Sylgard184 and custom-synthesized *material C* with a higher cross-link density[22,23]). We then use the results to simulate the depletion of a stamp surface during multiple print cycles.

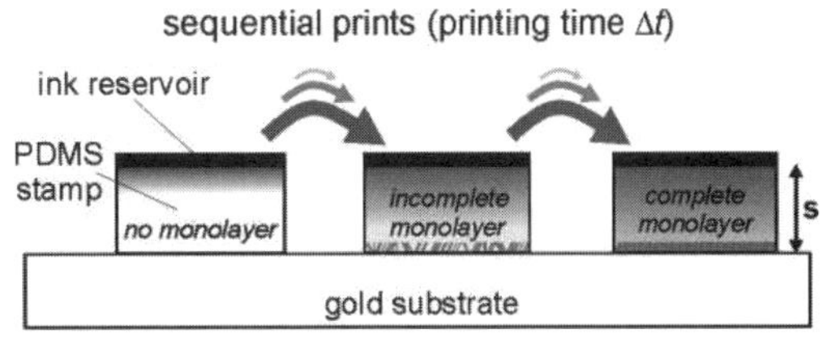

Figure 1. Principle of the diffusion experiment. A small cylindrical stamp is covered with an ink reservoir to provide linear diffusion. The stamp is then sequentially moved forward on a gold substrate with a printing time of Δt. With each print, more substance arrives at the bottom to be adsorbed onto the gold. After a certain amount of time (and number of prints), sufficient ink reaches the gold to form a complete monolayer within Δt.

2. Setup for Diffusion Experiments

Diffusion of a substance in a medium is quantitatively described by its diffusion coefficient. The common setup for determining this coefficient is to measure the flux of the substance through a plane of the medium (membrane), where the mass flux is induced by a concentration difference on either side of the plane. Our experiments differ from this setup because our aim is to characterize the ink diffusion under realistic printing conditions.

The principle of the linear-diffusion experiment used in this work is shown in Figure 1. A cylindrical, flat (i.e., nonpatterned) stamp is placed in contact with a gold substrate and inked from the upper face, which is not in contact with the gold. The alkanethiol has to diffuse through the entire stamp to reach the gold surface. As soon as the ink reservoir is added, timing is started and a sequence of prints carried out, with each print lasting for a defined period of time (Δt, printing time). After a certain number of prints, the ink starts to arrive on the printing side of the stamp, initially forming an incomplete monolayer on the gold. The flux of thiol increases with each print, leading to a more and more complete monolayer. Eventually, sufficient thiol will arrive on the bottom side of the stamp in the time interval (Δt) to form a complete monolayer on the gold substrate. The time at which this occurs is the parameter of interest in the diffusion experiment and is referred to as the *diffusion time* (t_d).

We expanded the ink source to cover the entire upper side of the cylindrical stamp to ensure linear-diffusion conditions. This was achieved in three different types of

(19) Libioulle, L.; Bietsch, A.; Schmid, H.; Michel, B.; Delamarche, E. *Langmuir* **1999**, *15*, 300–304.
(20) Geissler, M.; Bernard, A.; Bietsch, A.; Schmid, H.; Michel, B.; Delamarche, E. *J. Am. Chem. Soc.* **2000**, *122*, 6303–6304.
(21) Delamarche, E.; Schmid, H.; Bietsch, A.; Larsen, N. B.; Rothuizen, H.; Michel, B.; Biebuyck, H. *J. Phys. Chem. B* **1998**, *102*, 3324–3334.

(22) Geissler, M.; Wolf, H.; Stutz, R.; Delamarche, E.; Grummt, U. W.; Michel, B.; Bietsch, A. *Langmuir* **2003**, *19*, 6301–6311.
(23) Wolf, H.; et al. Manuscript in preparation.

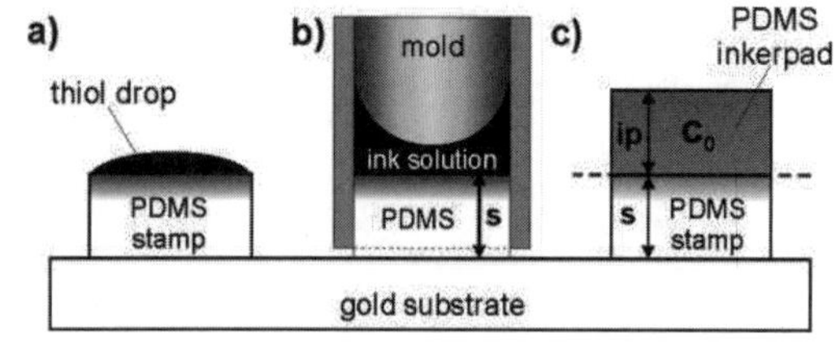

Figure 2. (a) Setup of the drop on top experiment. The ink reservoir is formed by a drop of liquid HDT to provide quasilinear diffusion. (b) Setup of a stamp for the mold experiment. The PDMS is cured inside a metal tube (mold), which provides an extended reservoir for an ink solution. (c) Principle of the inker pad experiment. The inker pad (ink source) of initial concentration C_0 provides true linear diffusion and is kept in conformal contact with the stamp during the entire experiment.

experiments shown in Figure 2. In part a, the surface of a small PDMS cylinder was covered with a drop of pure alkanethiol, leading to a constant concentration (saturation) in the upper boundary of the stamp ("drop on top"). In part b, a mold was designed in which the PDMS was cured and which provided a reservoir for pure thiol or a thiol solution ("mold"). In part c, a PDMS inker pad with known initial concentration was placed on the upper side of a stamp ("inker pad").

Knowing the (initial) concentration of the ink reservoir (C_0), the amount of thiol needed to form a complete monolayer on gold (c_{ml}), and the *diffusion time* (t_d) the diffusion coefficient (D) can be calculated by using an appropriate mathematical model (see the next section).

3. Mathematical Modeling of Diffusion Experiments

Fick's first law of diffusion describes the mass flux at a given point to be proportional to the magnitude of the concentration gradient at that point. The mass flux is defined as the amount of substance flowing through unit area per unit time. In one dimension, this relationship is

$$J_x = -D\frac{\partial C}{\partial x} \quad (1)$$

where J_x is the flux in the x direction, C the concentration of diffusing substance, and D the constant of proportionality called the *diffusivity* or *diffusion coefficient*. From eq 1 and the principle of mass conservation, *Fick's second law* can be derived.[24] Assuming that the diffusion coefficient (D) is constant, we obtain

$$\frac{\partial C}{\partial t} = D\frac{\partial^2 C}{\partial x^2} \qquad (D = \text{constant}) \quad (2)$$

This is also called *the linear-diffusion equation*, which in its most elementary form is a linear second-order partial differential equation (PDE). The assumption of a concentration-independent diffusion coefficient is generally true for diffusion in gases, liquids, and solutions. Polymers above the glass-transition temperature and, especially, rubbers such as PDMS can be expected to behave like liquids for small molecular diffusants.[25,26] Analytical solutions to the linear-diffusion equation, eq 2, for various problems and boundary conditions are to be found in the literature.[24,27] Here, we use the units g cm^{-3} for the concentration. The flux is given in mass per unit area per unit time (g cm^{-2} s^{-1}), and the diffusion coefficient is given in cm^2 s^{-1}. In the following, we derive the mathematical models to describe the linear-diffusion experiments introduced above.

Model 1 (Semi-Infinite Medium). For linear diffusion into a semi-infinite medium where the concentration at $x = 0$ is of the constant value C_0 for all times t, we find the solution for the concentration field[27] to be

$$C(x, t) = C_0 \operatorname{erfc}\left(\frac{x}{2\sqrt{Dt}}\right) \quad (3)$$

The error function (erf(z)) and its complement (erfc(z)), appearing in the above solution, are defined as

$$\operatorname{erfc}(z) = 1 - \operatorname{erf}(z) = 1 - \frac{2}{\sqrt{\pi}}\int_0^z e^{-\eta^2}\, d\eta \quad (4)$$

These functions appear frequently in the solutions of linear-diffusion problems and are usually found as built-in functions in commercially available mathematical software.

The boundary conditions of this solution basically correspond to the previously introduced drop on top and mold experiments (Figure 2a,b). For small diffusion times (little penetration of the concentration field), the stamp can be assumed to be of infinite extension. Also, as long the ratio between C_0 at $x = 0$ and the concentration at the stamp–gold interface ($C_{gold\text{-}if}$ at $x = s$) is sufficiently large, the model is considered to be accurate. This is expected in the case of pure alkanethiol compounds, where C_0 equals the saturation concentration of the ink in PDMS and is several orders of magnitude higher than the concentration needed to build up a monolayer within the printing interval.

Model 2-ip (Two Semi-Infinite Slabs in Contact). For the inker pad experiments, an analytical model can be derived analogously. Equation 3 only changes slightly to

$$C(x, t) = \frac{C_0}{2} \operatorname{erfc}\left(\frac{x}{2\sqrt{Dt}}\right) \quad (5)$$

This is the solution of two semi-infinite slabs being in contact at $x = 0$.[27] The concentration fields for $t = 0$, $t = T$, $t = 3T$, and $t = 9T$ are shown in Figure 3. Initially, the left slab (inker pad) has a homogeneous concentration of C_0 and the right slab (stamp) is empty. For $t > 0$, the ink starts to diffuse from the inker pad into the stamp, leveling off the initial concentration difference. These boundary conditions correspond to the inker pad experiment (Figure 2c). The same restrictions as those discussed above apply to ensure accuracy of this model (with assumption of infinite medium): small diffusion times and/or high concentration ratio between the interfaces $C_0/C_{gold\text{-}if}$ (as for $t < 3T$ in Figure 3).

Model 3-num (Numerical Approach). Finding analytical solutions for PDEs, such as eq 2, becomes difficult if complex geometries or special boundary conditions are involved. An alternative is to solve such equations numerically. We used the finite-difference method (FDM)

(24) Glicksman, M. E. *Diffusion in Solids: Field Theory, Solid-State Principles, and Applications*; John Wiley & Sons: New York, 2000.
(25) Crank, J. *Diffusion in Polymers*; Academic Press: London, 1968.
(26) Frisch, H. L.; Stern, S. A. *Crit. Rev. Solid State Mater. Sci.* **1983**, *11*, 124.
(27) Crank, J. *The Mathematics of Diffusion*; Oxford University Press: New York, 1975.

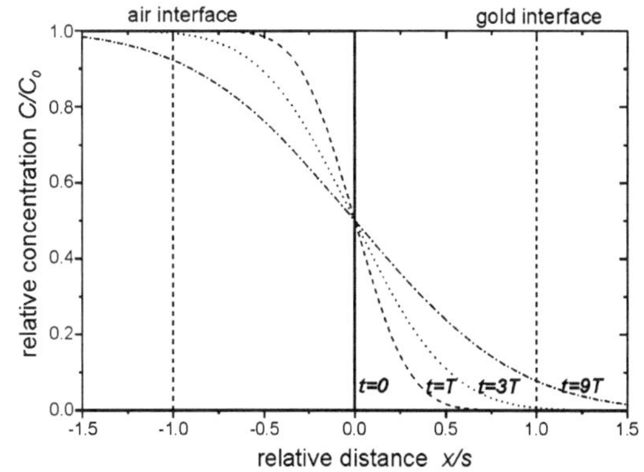

Figure 3. Calculated concentration field (model 2-ip) for two semi-infinite slabs at the initial concentration at $t = 0$ and three different times of $t > 0$. For small times ($t < 3T$), this model is able to describe the inker pad experiment with two finite slabs of thickness s in contact at $x = 0$.

to model the linear-diffusion experiments. In the FDM, eq 2 is solved by replacing the spatial and temporal derivatives with suitable difference equations (approximations). Instead of solving $C(x, t)$ with x and t continuously, $c_{i,j} = c(x_i, t_j)$ is solved by iteration with $x_i = i\,\mathrm{d}x$ and $t_j = j\,\mathrm{d}t$, where dx and dt are the corresponding spatial and temporal increments and i and j the indices in space and time.

An approximation for the time derivative is given by[27]

$$\left.\frac{\partial C}{\partial t}\right|_{x_i,t_j} \approx \frac{c_{i,j+1} - c_{i,j}}{\mathrm{d}t} \qquad (6)$$

and for the second spatial derivative at the position x_i by[27]

$$\left.\frac{\partial^2 C}{\partial x^2}\right|_{x_i} \approx \frac{c_{i+1,j} - 2c_{i,j} + c_{i-1,j}}{\mathrm{d}x^2} \qquad (7)$$

By substituting eqs 6 and 7 into eq 2 and rearranging, we obtain the explicit finite-difference formula

$$c_{i,j+1} = c_{i,j} + D\frac{\mathrm{d}t}{\mathrm{d}x^2}(c_{i+1,j} - 2c_{i,j} + c_{i-1,j}) \qquad (8)$$

The complete derivations for this equation and different methods of finite-difference equations are to be found in the literature (e.g., in refs 27 and 28).

Boundary Conditions. In the analytical solutions, no boundary effects at the air or gold interface have been considered because the slabs were approximated to be of infinite extension. With the FDM, it is now possible to consider these interfaces in order to use a more realistic model, referred to as model 3-num.

Alkanethiols such as HDT, ODT, and ECT have a very low vapor pressure under ambient conditions, and we can therefore neglect mass transport through the upper surface of the stamp. This represents a so-called no-flux boundary. Mathematically, this is enforced by setting the derivative of the concentration field at the interface to zero for all times:

$$\left.\frac{\partial c}{\partial x}\right|_{\text{air-if}} \overset{!}{=} 0 \qquad (\text{for all } t) \qquad (9)$$

On the bottom side of the stamp, chemisorption of the ink molecules occurs. The molecules close to the interface leave the diffusion system and bind irreversibly to the gold. This results in a sink boundary, which in this case is equal to a constant concentration set to zero at the interface.

$$c|_{\text{gold-if}} \overset{!}{=} 0 \qquad (\text{for all } t) \qquad (10)$$

There is no mass diffusing back in the direction of the stamp, and therefore, the net flow toward the gold side is increased compared to that when an infinite slab across the boundary (as in model 2-ip) is assumed. We used both the analytical and numerical models to determine the diffusion coefficient (D) and will compare these models in the results section.

Determination of the Diffusion Coefficient (D). To determine the diffusion coefficient (D) from linear-diffusion experiments, the following parameters must be known: (1) the initial concentration (C_0), (2) the surface concentration of a complete self-assembled thiol monolayer on a gold surface (c_{ml}), and (3) the diffusion time (t_d) to complete a monolayer within the printing time (Δt). With these three parameters, the mathematical model can be fitted to the experimental data using D as the fitting parameter. Other parameters such as the printing time (Δt) and the stamp thickness (s) are given by the experimental setup. t_d is determined by the linear-diffusion experiments, and C_0 is either given by the saturation concentration or set as the inker pad concentration (see the Experimental Section). Accordingly, only c_{ml} remains, which will be derived in the next paragraph.

The area per n-alkanethiol in a complete monolayer on a gold surface is 0.21 nm^2 per molecule.[29] This value does not depend on the chain length, at least not for the molecules HDT, ODT, and ECT used in this work. The packing density (ρ_{ml}) is obtained by inverting this value and equals 4.608×10^{14} molecules/cm^2. From atomic force microscopy (AFM) measurements on the sputtered gold of ~30 nm thickness used in this work, we assume a roughness factor of $R = 1.1$. The surface concentration of the complete monolayer in g cm^{-2} can now be calculated by

$$c_{ml} = \frac{R\rho_{ml}M_{th}}{N_A} \qquad (11)$$

where M_{th} is the molar mass of the thiol compound and $N_A = 6.022 \times 10^{23}$ Avogadro's number. For the compounds used, we obtain the following surface concentrations: HDT, 2.18×10^{-7} g cm^{-2}; ODT, 2.44×10^{-7} g cm^{-2}; and ECT, 2.65×10^{-7} g cm^{-2}.

The Flux Condition. During a linear-diffusion experiment, t_d is reached as soon as a complete monolayer is formed within the printing time (Δt). Knowing c_{ml}, the

(28) Epperson, J. F. *An Introduction to Numerical Methods and Analysis*; John Wiley & Sons: New York, 2002.

(29) Dubois, L. H.; Nuzzo, R. G. *Annu. Rev. Phys. Chem.* **1992**, *43*, 437.

average flux at the stamp–gold interface (during that specific print) is given by

$$\bar{J}(t_d)|_{x=s} = \frac{c_{ml}}{\Delta t} \quad (12)$$

With Fick's first law, eq 12 becomes

$$\bar{J}(t_d)|_{x=s} = \frac{c_{ml}}{\Delta t} = -D\frac{\partial C}{\partial x} \quad (13)$$

Forming the derivative at $x = s$ from eq 3 and substituting it into eq 13, we obtain

$$\left.\frac{\partial C(x, t_d)}{\partial x}\right|_{x=s} = \left.\frac{\partial}{\partial x}\left(C_0 \operatorname{erfc}\left(\frac{x}{2\sqrt{Dt}}\right)\right)\right|_{x=s} \stackrel{!}{=} \frac{-c_{ml}}{D\Delta t} \quad (14)$$

This is the *flux condition* for model 1 to be satisfied. All parameters except the diffusion coefficient are known, so eq 14 can be solved for D. We used mathematical software (MathCAD 8.0 Professional) to solve it by iteration in order to determine D. Only a slight change is necessary to obtain the flux condition for model 2-ip: eq 3 in eq 14 is substituted by eq 5:

$$\left.\frac{\partial C(x, t_d)}{\partial x}\right|_{x=s} = \left.\frac{\partial}{\partial x}\left(\frac{C_0}{2} \operatorname{erfc}\left(\frac{x}{2\sqrt{Dt_d}}\right)\right)\right|_{x=s} \stackrel{!}{=} \frac{-c_{ml}}{D\Delta t} \quad (15)$$

For the numerical solution, the flux at the stamp–gold interface can easily be approximated by

$$J(t)|_{x=s} \approx \text{flux}_j = \frac{-D(c_{n,j} - c_{n-1,j})}{dx} \quad (\text{for } t = j\ dt) \quad (16)$$

Then, the flux condition to be satisfied, using D as a fitting parameter, is

$$\frac{-D(c_{n,j} - c_{n-1,j})}{dx} \stackrel{!}{=} \frac{c_{ml}}{\Delta t} \quad (\text{for } j\ dt = t_d) \quad (17)$$

Note that, to solve the flux condition for the numerical model, the complete iteration using eq 8 must be performed until $j\ dt = t_d$, so that the flux condition at the end of the calculation is satisfied.

The Sum Condition. The *flux condition* is based on the approximation of an average flux value during the printing time (Δt). In reality, the flux at the stamp–gold interface increases continuously during the experiment,[30] but because Δt is chosen to be small compared with the diffusion time, it was appropriate to assume the flux to be nearly constant within the printing interval. However, a more precise way would be to calculate the amount of substance that passed the boundary at $x = s$ during the preceding interval (Δt) and compare it with the known amount needed to complete the monolayer (c_{ml}).

For model 3-num, the total sum of ink that has left the boundary at time t is the integration of the flux over the interval $[0, t]$:

$$\text{sum}|_t = \int_0^t C(x, \tau)\, d\tau \quad (18)$$

In the FDM, the integral is approximated by calculating

$$\text{sum}_j = \text{sum}_{j-1} + \text{flux}_j\, dt \quad \text{for} \quad j \in \left[1, \frac{t}{dt}\right] \quad (19)$$

The lower integration boundary only needs to be raised to $(t - \Delta t)$ to calculate the amount that passed through the stamp–gold interface during the preceding print. Then, the *sum condition* to be satisfied becomes

$$\text{sum}_{j-1} + \text{flux}_j\, dt \stackrel{!}{=} c_{ml} \quad \text{for} \quad j \in \left[\frac{t - \Delta t}{dt}, \frac{t}{dt}\right] \quad (20)$$

We used the *flux* as well as the *sum condition* to calculate D and found that, at least for the conditions used in this work, they yield the same values. In the following discussion, we therefore do not distinguish between the two methods.

4. Results and Discussion

(A) Determination of the Diffusion Time (t_d). We used two methods to determine the *diffusion time* (t_d) (Figure 4): selective etching and ellipsometry. In the first method, the consecutive prints of a linear-diffusion experiment are made visible by placing the gold substrate in an etching solution. Depending on the completeness of the thiol monolayer, the printed gold will be differently protected during etching. The first spot showing unetched gold (total monolayer coverage) determines t_d. The distinction can be done visually under an optical microscope. In general, the prints will lead to a broad gradient from no coverage to complete coverage, which makes it difficult to determine the exact spot on the etched sample visually (Figure 4b). Therefore, it is helpful to use a reference print from a saturated stamp that overlaps each spot.

A more quantitative way to determine t_d was achieved by using ellipsometry to measure the monolayer thickness (completeness) of the consecutive prints from a linear-diffusion experiment (Figure 4a). Ellipsometry allows the film thickness to be measured on a sub-monolayer level.[31] In this work, ellipsometry was used in combination with inker pad experiments using PDMS slabs of a size that provided a sufficient area for the ellipsometric measurement. An example of such a measurement from HDT prints is shown in Figure 4c. A linear fit is applied to the slope of the growing monolayer resulting from each consecutive print. The interception with the average value of the saturation level then corresponds to t_d. The linear estimation proved to be a good approximation of the slope,[32] and the two methods for determining t_d appeared to be equivalent.

(B) Diffusion Experiments. Diffusion from Pure Ink Compound. A series of drop on top experiments was performed using HDT and Sylgard184 stamps with thicknesses in the range of 0.7, 1.4, and 2.1 mm. If we plot the stamp thickness versus the square root of the determined diffusion time (t_d), a linear behavior with zero intercept is observed (Figure 5, open squares). This plot corresponds to a so-called *penetration-distance plot*, which can be derived from the mathematical solution for the semi-infinite slab (eq 3).[24] The penetration distance is defined as the distance at which a certain relative concentration (C/C_0) is reached as a function of the square root of time. The slope of the straight line represents the "penetration speed", the rate at which a certain concentration level (C/C_0) penetrates the bulk material. In our

(30) This is only true if ideal adsorption conditions are assumed, meaning that the possibility of adsorption is not reduced during the formation of the monolayer and that the concentration at the interface is zero for all times.

(31) Bain, C. D.; Troughton, E. B.; Tao, Y.; Evall, J.; Whitesides, G. M.; Nuzzo, R. G. *J. Am. Chem. Soc.* **1989**, *111*, 321–335.

(32) Note, that Figure 4c *does not* represent the monolayer growth kinetics at one single spot. Each value in the graph represents a thickness at the location of each of the consecutive prints.

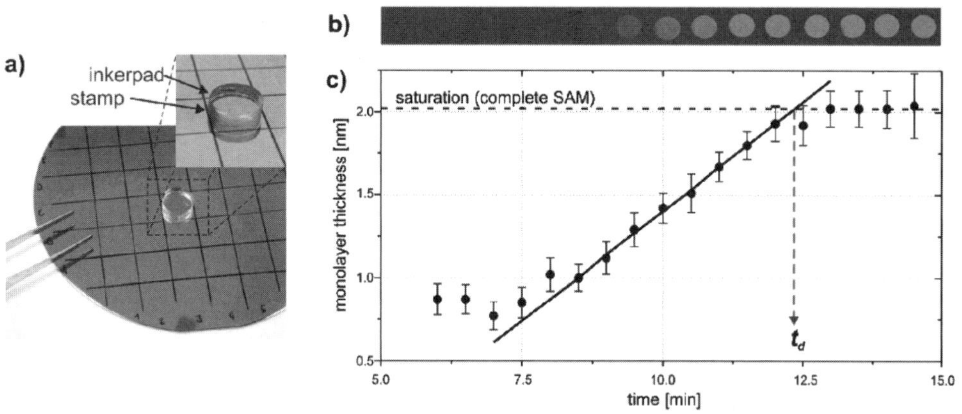

Figure 4. (a) Inker pad experiment on a 4 in. wafer coated with 30 nm Au and marked with a grid for consecutive printing. (b) Gradient of printed gold spots after selective etching. (c) Ellipsometry measurement to determine the diffusion time (t_d). The SAM thickness is measured for each printed spot, and a linear estimation is applied to the slope of the increasing monolayer to find the interception with the saturation level (complete SAM).

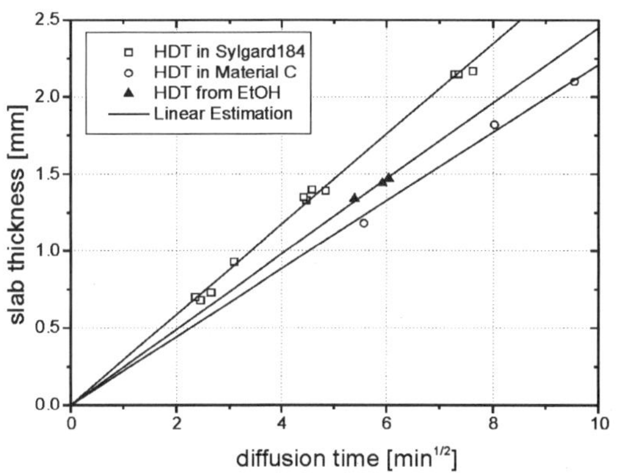

Figure 5. Results from the drop on top and mold experiments. All three data sets exhibit a linear trend with zero interception. This is expected for a diffusant obeying Fick's law in a system described by model 1. Thus, this result proves that the diffusion of the alkanethiols used in this work can be modeled with Fick's law and that model 1 is an appropriate model for the drop on top and mold experiments.

case, the constant concentration (C_0) at the top boundary corresponds to the saturation concentration of HDT in Sylgard184 and was determined as 12.7 wt %. We measured C_0 independently (see the Experimental Section), and the results are summarized in Table 1. The linear trend observed in the experimental data implies that (1) the diffusion of HDT in Sylgard184 is governed by Fick's laws of diffusion and (2) the drop on top experiment (and the mold experiment) can be described mathematically by model 1.

Analogous experiments were carried out with *material C*, and again, linearity is observed (Figure 5, open circles). The slope of the linear estimation of *material C* is significantly lower than that from Sylgard184. This indicates a reduced penetration speed of the ink in *material C*; however, it has to be taken into account that the C_0 value of the HDT is 50% lower than that of *material C*. Therefore, a qualitative difference of the diffusion coefficient cannot be determined solely from the different slopes in Figure 5.

Table 1. Overview of Saturation Concentration Determined and Swelling Due to HDT/ECT in PDMS

		saturation concentration			
ink	stamp material	(wt %)	(g cm^{-3})[a]	(mM)[b]	swelling (% lin.)
HDT	Sylgard184	12.7 ± 0.2[c]	0.126	486	6.1 ± 1.2
HDT	*material C*	6.4 ± 0.6[d]	0.064	246	3.5 ± 0.5
ECT	Sylgard184	≪0.25[e]	≪0.0025	≪8.2	

[a] PDMS density: $\rho = 1.03$ g cm^{-3} according to product data sheet (Sylgard184). [b] With respect to stamp volume. [c] Obtained after 5 h of saturation, representing a value of "raw" PDMS still containing 4–6 wt % extractables. For completely "extracted" Sylgard184, a saturation concentration of 11.5–12.0 wt % is assumed. [d] Obtained after 16 h of saturation. [e] Value at which no ECT surface crystallization was observed after 6 h of storage at 22 °C (however, it was observed after 75 h).

One might be tempted to derive the diffusion coefficient from the slope of the linear estimation in the penetration-distance plot if the relative concentration (C/C_0) is known. Unfortunately, this is not trivial because eq 7, containing the error function, cannot be solved for the distance x to obtain a linear equation, the slope of which can then be solved for D.

Diffusion from Ethanol Solution. In general, stamps and inker pads used for μCP are inked from diluted ethanol solutions. Even though the solvent is evaporated again prior to printing, it is of interest to evaluate the influence of the ethanol on the diffusion properties of the alkanethiol. Therefore, a 1.6 wt % HTD solution (~50 mM) was used in a series of mold experiments with Sylgard184 (filled triangles in Figure 5). The linear fit lies between the data of pure HDT in Sylgard184 and of *material C*. Again, this is not a qualitative representation of the diffusion coefficient because the different concentration level (C_0) on the top side of the stamp as compared with the other experiments has to be taken into account. It is assumed that the presence of the ethanol and the simultaneous swelling of the PDMS due to solvent penetration will increase the diffusion of HDT. However, we assume that the diffusion coefficient evaluated will not be of the same significance as the values obtained from pure HDT diffusion. The reason is the more complex experimental parameters: First, it has not been evaluated whether a concentration of 1.6 wt % HDT in EtOH truly corresponds to the same concentration in the top layer of the PDMS, assumed to be C_0. Second, evaporation of the solvent from

Table 2. Overview of Diffusion Coefficients Obtained from the Drop on Top, Mold, and Inker Pad Experiments along with the Calculated Concentrations at the Stamp–Gold Interface for $t = t_d$ and the Ratio $r = C_0/C_{gold\text{-}if}$

ink	stamp material	method	D (10^{-7} cm^2 s^{-1})	$C_{gold\text{-}if}$ (10^{-4} g cm^{-3})	r ($C_0/C_{gold\text{-}if}$)	model
HDT	Sylgard184	drop on top	5.9 ± 0.6	0.57 ± 0.17	2380	analytical
HDT in EtOH[a]	Sylgard184	mold	6.6 ± 0.1	0.93 ± 0.07	170	analytical
HDT	*material C*	drop on top	3.9 ± 0.3	0.71 ± 0.14	840	analytical
HDT	Sylgard184	inker pad	5.0 ± 0.2[b]	1.01 ± 0.22		numerical
ODT/ECT	Sylgard184	inker pad	~4[c]	~1.2/~1.7		numerical

[a] EtOH solution with 1.6 wt % HDT (50 mM). [b] Average value from measurements with C_0 between 0.4 and 6.3 wt % HDT. [c] Owing to experimental difficulties (precipitation), the diffusion coefficients of these two compounds could not be distinguished.

the reservoir cannot completely be prevented and results in a slight increase of C_0 during the experiment. Third, the model used to extract the diffusion coefficient does not imply the presence of a second diffusant that has an influence on the diffusion properties of the HDT.

(C) Diffusion Coefficients. Knowing the diffusion times, it is possible to calculate the diffusion coefficients, which is carried out for each set of data using the appropriate mathematical model. The values are then compared (in Table 2) and discussed below.

HDT in Sylgard184 from Drop on Top. Model 1 was used to calculate the diffusion coefficient for the drop on top experiments performed with HDT and Sylgard184 stamps. The concentration at $x = 0$ was set to $C_0 = 0.126$ g cm^{-3}, corresponding to the saturation concentration of HDT in Sylgard184 (see Table 1). The average flux is calculated from the printing time (Δt) and the surface concentration of the alkanethiol on the gold substrate, $c_{ml} = 2.18 \times 10^{-7}$ g cm^{-2}, according to eq 11. The average diffusion coefficient obtained from 11 measurements is $(5.9 \pm 0.6) \times 10^{-7}$ cm^2 s^{-1}.

A calculation of the thiol concentration in the stamp near the stamp–gold interface ($C_{gold\text{-}if}$) for t_d yields a value of $(0.5-1.0) \times 10^{-4}$ g cm^{-3}. This correlates with a molar concentration of 0.2–0.4 mM and indicates the concentration needed in the stamp to build up a monolayer of HDT on gold within 1 min.

HDT from EtOH Solution. An ethanol solution was used containing 1.6 wt % HDT (50 mM), and it was assumed that this is equivalent to C_0 at $x = 0$ in the Sylgard184 stamp. The diffusion coefficient determined using model 1 is $(6.6 \pm 0.1) \times 10^{-7}$ cm^2 s^{-1} (average value from four different measurements) and is slightly higher than that obtained for pure HDT diffusion from a concentrated source. The diffusion coefficient of pure ethanol in Sylgard184 was determined to be $(10.0 \pm 1.0) \times 10^{-7}$ cm^2 s^{-1} by Duineveld et al.,[33] and a list of other literature values is given, ranging from 10×10^{-7} to 60×10^{-7} cm^2 s^{-1}. Therefore, it can be expected that, owing to the swelling of the polymer network by the more rapidly diffusing ethanol, HDT transport is slightly faster than pure HDT diffusion. However, as mentioned above, there are other considerations to be made when dealing with a system containing two diffusants. It is assumed that under these conditions model 1 forfeits its accuracy, and therefore, the coefficient given is to be taken as specific to our experimental setup.

HDT in *Material C.* We determined the saturation concentration (C_0) of HDT in *material C* to 0.064 g cm^{-3} (see Table 1). The diffusion coefficient obtained with model 1 is $(3.9 \pm 0.2) \times 10^{-7}$ cm^2 s^{-1}, which is significantly smaller than that in the Sylgard material (Table 2). Because of the increased hardness (cross-link density) of *material C* as compared with Sylgard184, a smaller value is expected.

(33) Duineveld, P. C.; Lilja, M.; Johansson, T.; Inganäs, O. *Langmuir* **2002**, *18*, 9554–9559.

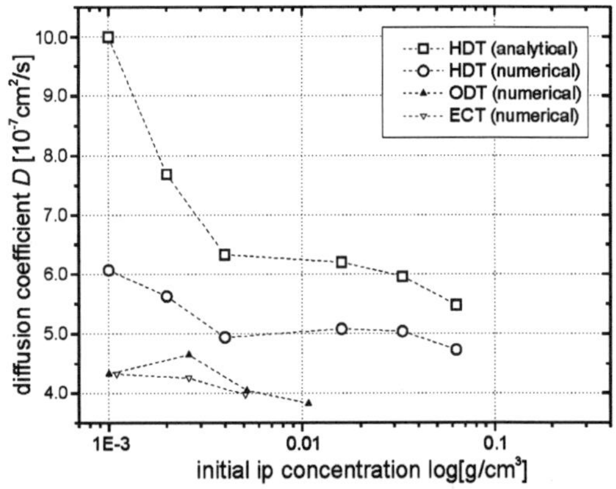

Figure 6. Diffusion coefficients for HDT, ODT, and ECT in Sylgard184 determined from inker pad experiments using reduced initial concentrations down to 0.001 g cm^{-3} (3–4 mM). The strong increase below 0.004 g cm^{-3} HDT of the coefficient calculated analytically is due to the failure of model 2-ip at low C_0. The shift in the value for HDT obtained from the numerical model is systematic and can be explained by the continuously increased flux at the stamp–gold interface due to the sink boundary. The coefficients for ODT and ECT are significantly lower, but the two values are hardly distinguishable.

Reduced Initial Concentration of HDT in Sylgard184. The saturation concentration involved in the experiments discussed above is on the order of some hundred millimolar and extremely high compared with the concentrations of 0.1–10 mM typically used in μCP.[19] Using homogenized inker pads with a defined initial concentration (C_0) as the ink source on top of Sylgard stamps, it was possible to monitor the course of the diffusion coefficient if the concentration was lowered toward "realistic" values, that is, to 0.001 g cm^{-3} (~4 mM). The analytical model 2-ip and the numerical model 3-num were used to calculate the diffusion coefficient from the corresponding experimental parameters. The results are shown in Figure 6. The diffusion coefficient for HDT in Sylgard184 determined with the analytical model slightly increases with decreasing C_0 between 0.063 and 0.004 g cm^{-3}. The values are on the order of $(6.0 \pm 0.5) \times 10^{-7}$ cm^2 s^{-1}, which is in good agreement with the value obtained from the drop on top experiments. Toward lower initial concentrations, however, the coefficient strongly increases up to 10×10^{-7} cm^2 s^{-1} for $C_0 = 0.001$ g cm^{-3} (Figure 6, open squares). Such an increase is not expected for a diffusant obeying Fick's law, the diffusion coefficient of which in a homogeneous medium is constant per definition.[27] We will show later that this strong increase is due to the "failure" of the analytical model 2-ip at low concentrations.

Looking at the diffusion coefficient of the same system (HDT in Sylgard from inker pad) but calculated numerically with model 3-num, it is seen that the values are shifted downward by roughly one digit to ~5×10^{-7} cm^2 s^{-1}.

The coefficient stays nearly constant down to 0.004 g cm^{-3} and then increases slightly to $\sim 6 \times 10^{-7}$ cm^2 s^{-1} at 0.001 g cm^{-3}; however, the increase is significantly smaller than that in the analytical model. The observed shift is systematic and can be explained by evaluating the flux at the stamp–gold interface: In the numerical model, the boundary condition at $x = s$ is set to $C(s, t) = 0$ (sink due to adsorption) for all values of t, whereas, in the analytical model, the slab is assumed to be of infinite extension and no actual boundary is considered. The sink boundary results in an increased "numerical flux" at the interface as compared with the "analytical flux" at any time during the calculation. Therefore, when using the flux condition, the diffusion coefficient in the analytical model 2-ip must be higher to reach the same flux level for $t = t_d$ at the end of the experiment.

In general, the numerical calculation with model 3-num is more realistic because its boundary conditions describe the system in a true physical way. Taking an average value over the complete concentration range, we obtain a diffusion coefficient of $(5.3 \pm 0.5) \times 10^{-7}$ cm^2 s^{-1} for HDT in Sylgard184. If applying the average to the plateau region between 0.004 and 0.063 g cm^{-3}, a coefficient of $(5.0 \pm 0.2) \times 10^{-7}$ cm^2 s^{-1} is found. So far it has not been possible to evaluate whether the slight increase of D from the numerical calculation toward concentrations below 0.004 g cm^{-3} is due to a simplification of the boundary conditions or whether it is a true physical phenomenon.

For an overall error estimation, we performed a calculation assuming that maximum errors of all input parameters[34] add up to result in the highest deviation of the diffusion coefficient. We find a maximum error of 15% in our values.

Comparison of the Analytical and Numerical Models. Even though the analytical model 2-ip lacks accuracy, its main advantage—the fast calculation time—should not be neglected. For example, it can be used to estimate the diffusion time to complete a monolayer for a given stamp thickness and initial concentration. As long as the scope of the model is taken into account, the result will be of useful accuracy. In the following, this scope shall be evaluated: We compared the concentration fields calculated with model 2-ip and model 3-num at $t = t_d$ for different initial concentrations (6.3, 0.4, and 0.1 wt % HDT; results not shown). For $C_0 = 6.3$ wt %, the two solutions are nearly identical. For $C_0 = 0.4$ wt %, quite a good match between the analytical and numerical models still exists; however, the concentration starts to differ slightly toward the boundaries. For $C_0 = 0.3$ wt %, finally, the difference in concentration fields becomes severe and the analytical model fails to predict the true course within the entire stamp. If evaluating the concentration ratio

$$r = \frac{C_0}{C_{\text{gold-if}}} \tag{21}$$

as a function of concentration, it drops from $r = 700$ (6.3 wt %) to $r = 8$ (0.1 wt %). For an initial concentration of 0.4 wt %, where the analytical solution starts to differ from the numerical one, a value of $r = 50$ is obtained. This point also correlates with the beginning of the strong increase of the diffusion coefficient determined using model 2-ip. Therefore, we state that, for $C_0/C_{\text{gold-if}} > 50$, model 2-ip can be used to perform calculations if a diffusion coefficient of $D = 6 \times 10^{-7}$ cm^2 s^{-1} (for HDT in Sylgard184) is used. For $C_0/C_{\text{gold-if}} < 50$, however, model 2-ip becomes inaccurate.

ODT/ECT in Sylgard184. Analogously to the experiments discussed above, the diffusion coefficients of ODT and ECT in Sylgard184 were determined. Given the strongly reduced solubility of these compounds in the Sylgard material at room temperature, the experimental range of possible initial inker pad concentrations was reduced. The diffusion coefficients obtained with model 3-num are shown in Figure 6. The values are on the order of 4×10^{-7} cm^2 s^{-1} and significantly lower than the values for HDT, as expected from the increased molecular weight. However, the two values are hardly distinguishable. There is only a slight trend for the ECT to have a lower coefficient than ODT. We assume that the lack of a clear distinction between the two diffusion coefficients is due to the more difficult experimental parameters such as the precipitation of ink substance from the inker pad. After the experiments performed with ECT, there were slight signs of precipitation on all inker pads used. On ODT inker pads, crystallization was also observed, but not until several hours after finishing the experiment. Because none of the models consider such precipitation, the diffusion coefficients obtained for ODT and ECT are probably of less precision than the value obtained for HDT.

5. Multiple Printing

It is desirable that a numerical model such as model 3-num can be applied to simulate realistic μCP processes, for example, multiple printing cycles. This would allow the printing parameters as well as stamp dimensions to be optimized on the basis of results from the calculation of the concentration field within the stamp during the printing process. To fulfill this task, we used model 3-num and extended it with variable boundary conditions so that printing and lift-off intervals can be simulated in the same calculation. As soon as contact with the gold substrate is made, a sink boundary is needed to simulate the adsorption of the ink. Once the monolayer is complete or the stamp lifted away from the surface, the interface changes to a no-flux boundary. It is relatively simple to implement such conditional statements in a numerical calculation. We simulated a multiprint experiment of a flat stamp containing a homogeneous initial concentration (C_0) and then compared the result with the corresponding experiment. The multiprint setup consists of several cycles with a printing interval and a subsequent lift-off interval of equal time. The numerical calculation was carried out with the MathCAD 8.0 Professional software.

Depletion of the Stamp's Surface. An experiment with 5 s printing and 5 s lift-off time (5/5 cycle) was calculated for 25 cycles. The following parameters were used: $s = 0.07$ cm (stamp thickness), $C_0 = 5.7 \times 10^{-3}$ g cm^{-3} ECT ($\sim$1.8 mM), $D = 4.0 \times 10^{-7}$ cm^2 s^{-1}, dt = 5 ms (time increment), and dx = 2.5×10^{-4} cm (spatial increment). Figure 7a shows the evolution of the simulated concentration field in steps of five cycles. The concentration at the surface of the stamp drops rapidly in the first 15 cycles, and the decrease slows down once the printing surface has been depleted. After 15 cycles, the extension of the depleted region within the PDMS bulk is only 100–200 μm. Therefore, the number of prints cannot be increased by increasing the thickness of the stamp when printing with such fast printing cycles. Note that at the end of a cycle the boundary is in its no-flux mode; therefore, the concentration at the stamp surface is partially recovered from the lift-off period and not at zero.

(34) We used the following errors of input parameters: initial concentration (C_0) $\pm 5 \times 10^{-5}$ g/cm^{-3}, stamp thickness (s) $\pm$0.02 mm, diffusion time (t_d) $\pm \Delta t$, printing time (Δt) $\pm$5 s, surface concentration of a complete monolayer (c_{ml}) $\pm$10%.

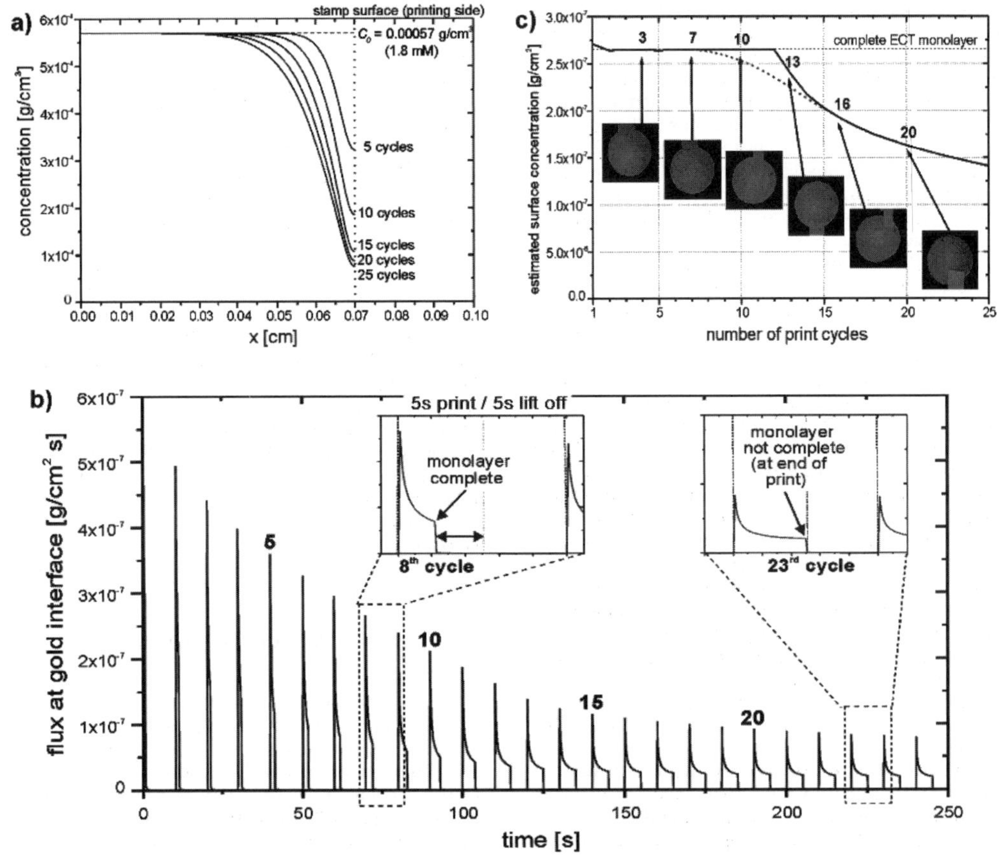

Figure 7. (a) Behavior of the concentration field during a multiprint experiment with 5/5 cycles (5 s print/5 s lift-off) and an initial concentration of 0.06 wt % (1.8 mM) in the stamp. After 15 cycles, the surface region (100−200 μm) of the stamp is depleted. (b) Calculated flux at the stamp surface during a multiprint experiment with 5/5 cycles. The insets show the effects of the variable boundary conditions, which are applied in the numerical calculation. (c) Calculated surface concentration during a multiprint experiment with 5/5 cycles and an initial concentration of 0.06 wt % ECT in the stamp (solid black line). The dashed red line shows the assumed course of the real surface concentration based on the observation made from the experiment.

Figure 7b shows the flux through the stamping surface, which is calculated in the numerical simulation according to eq 16. At the beginning of each print, there is a strong increase in the flux of ink molecules passing the interface and adsorbing onto the gold substrate. The flux then immediately decreases again because the ink transport to the surface is limited by the diffusion process. As soon as the monolayer is complete or when the stamp is lifted from the surface, the flux drops to zero again. This behavior shows the effect of the variable boundary condition applied in the numerical calculation. The criterion for a complete monolayer is fulfilled as soon as the amount of ink that left through the interface equals the surface concentration (c_{ml}) of an ECT monolayer. The difference between a print that results in a complete monolayer and one that results in a partial monolayer is shown in the two insets of Figure 7b. In the print interval of the 8th cycle, the monolayer is completed and the boundary condition changed from sink to no-flux mode. At the 22nd cycle, insufficient substance arrives to build up a complete monolayer during this print interval. In this case, the boundary condition is changed as soon as the stamp is lifted from the gold surface.

Comparison with Experiment. The amount of ink passing the stamp's surface can be calculated for each cycle using eq 20. This results in the estimation of the ECT surface concentration on the gold substrate as a function of cycles (prints). The result of the 5/5 multiprint simulation is shown in Figure 7c (solid black line). To evaluate the calculation, an experiment was carried out using a flat stamp of 4 mm in diameter and of corresponding initial ECT concentration. After multiple printings in 5/5 cycles, the sample was etched and the prints examined (see images of gold spots in Figure 7c). A reference print (rectangular feature) is used to characterize the degree of protection imparted by the ECT monolayer. The reference becomes clearly visible in the 13th print, after which the quality of the gold spot decreases with each further print. This indicates that the ECT coverage decreased as predicted by the simulation. However, under the optical microscope, a slight contrast to the reference is already visible starting with the 8th print. The dashed red line shows the assumed course of the real surface concentration based on the observation made from the experiment. The simulation fails to predict this early and smooth drop-off because it does not take into account the kinetics of monolayer formation in one printing cycle. However, overall, there is an excellent match between simulation and experiment. The agreement of simulation and experiment was reproduced for a 10 s/10 s multiprint (results not shown). In this case, the stamp surface was given more time to "recover" between prints, resulting in a higher number of complete prints. We calculated that 25 prints can be made before the stamp is depleted, at which point the quality of the prints rapidly decreases.

6. Conclusions and Outlook

The determination of the diffusion coefficient of alkanethiol ink in PDMS stamps is possible by means of simple linear-diffusion experiments, in which the basic parameters of μCP (ink concentration, printing time, and stamp geometry) are taken into account. Ink transport is monitored by direct adsorption on gold substrates from consecutive prints. We showed that the ink transport through the PDMS slab follows Fick's law of diffusion. A simplified analytical model was found to be accurate for experiments with high initial concentrations (saturation) but is likely to become inaccurate at low initial concentrations. Therefore, a more precise one-dimensional, numerical model based on the finite-difference method was developed, which also proved to be accurate at low concentrations.

The diffusion coefficients for all alkanethiols and PDMS materials investigated are on the order of 10^{-7} cm^2 s^{-1}. In addition, we simulated a multiprint process involving a flat stamp with a homogenized initial concentration and cycles of equal printing and lift-off times. The depletion of the stamp surface could be predicted and was in good agreement with the experiment.

The diffusion coefficients determined here are considered to build the basis of future simulations of arbitrary printing conditions. For this, the boundary conditions in the numerical model will have to be adjusted. For example, it must be taken into account that real stamps contain a patterned surface and therefore the amount of ink leaving the stamp with each print will be reduced. In addition, it is possible that the density of the pattern varies over the stamp surface, and therefore, the calculation needs to be expanded to two or three dimensions. Also, when printing high-resolution patterns, surface diffusion probably will start to play a more important role. Using experiments and calculations along the lines of this work, it will be possible to optimize the printing time, inking, and ink concentration as well as the stamp thickness for a given μCP application.

7. Experimental Section

PDMS Stamps and Inker Pads. PDMS stamps and inker pads were prepared as nonpatterned flat slabs by pouring the prepolymer mixture onto a planar polystyrene surface (Petri Dish, Falcon 1029, 1054, 1121, Becton Dickinson Labware, NJ). Two different types of PDMS material were used: Sylgard184 (Dow Corning, Midland, MI) and a 4−5 times harder custom-synthesized material described earlier[22] and referred to as *material C*.[23] The former was dispensed with an automatic mixing tool (DOPAG Micro-Mix E, Cham, Switzerland) and cured for at least 24 h at 60 °C; the latter was cured for at least 48 h at the same temperature. To prevent bubble formation, the mixture was degassed prior to curing. The actual thicknesses of the stamps and inker pads were measured with a micrometer (±0.01 mm), normally after the experiment to prevent unnecessary contamination of the slab surface. Slabs were cut in squares or punched out (using a very-thin-walled metallic tube) to obtain cylindrical stamps with a diameter between 4 and 10 mm. No slabs older than 3 weeks were used for experiments.

Thiols and Solutions. Hexadecanethiol (HDT, Fluka, M = 258.5 g/mol) was purified prior to its use by liquid column chromatography with heptane/ethyl acetate v/v = 20/1 (both obtained from Fluka) on silica gel (Kieselgel 60, 220−440 mesh, Fluka). Octadecanethiol (ODT, M = 286.6 g/mol) and eicosanethiol (ECT, M = 314.6 g/mol), both from Robinson Brothers Ltd., West Bromwich, U.K., were custom-synthesized and used as received without further purification. Ink solutions were prepared freshly with ethanol (Fluka, p.a.) prior to experiments.

Inker Pads with Defined C_0. Using the high precision of a microbalance (MT5 5.1 g/0.001 mg, Mettler Toledo), it was possible to expose PDMS slabs to a specific amount of ink to prepare inker pads with well-defined initial concentrations. To homogenize the concentrations, the inker pads were stored (under argon, in the dark) for at least 24 h prior to use. ECT/ODT inker pads were homogenized in the oven at 45 °C. Increasing the size of the inker pad leads to a higher precision of the final concentration. Therefore, mainly slabs of 8−10 mm in diameter and 0.7−1.4 mm in thickness were used. Using this method, inker pads with initial concentrations of as low as 0.05 wt % were prepared, which for ECT corresponds to a molar concentration of ~1.6 mM and lies in the range of concentration of EtOH solutions generally used for inking in μCP.

Substrates. Si/SiO_2 wafers (Siltronix, Geneva, Switzerland, 4 in. diameter) were primed with 2 nm of Ti (RF 100 W, 12 μbar, 20 sccm Ar, 30 s) and sputter-coated with ~30 nm of gold using Ar (20 sccm, DC 20 W) at a rate of 1 nm min^{-1} W^{-1} and a pressure of 12 μbar (LA 440 S, von Ardenne Anlagetechnik, Dresden, Germany). The gold surfaces were stored in Fluoroware wafer holders under an argon atmosphere and used as soon as possible. For ellipsometry measurements, the gold baseline was measured within 30 min after evaporation. As a guide for multiple printings, a 10 mm grid was marked on the entire wafer using a waterproof pen. For diffusion experiments with small stamps (4 mm in diameter), the wafer was cleaved into four even pieces and a fine grid scribed into the gold using a glass cutter.

Diffusion Experiments. *(i) Drop on Top.* Punched PDMS cylinders of 4 mm in diameter and various thicknesses (0.7−2.2 mm) were used. A drop of pure HDT (3−4 μL) was sufficient to cover the entire top surface of the stamp, providing linear diffusion through the stamp. The stamp was moved using tweezers, but special care was taken to prevent contact with the drop or the rim of the stamp, as otherwise the liquid would spill and ruin the experiment. In general, light pressure from the top was required after moving the stamp to induce conformal contact with the gold substrate. The set-off time between prints was kept as short as the handling of the stamp allowed (3−8 s), and printing times were chosen between 30 and 120 s, depending on the thickness of the slab, but kept constant during a single experiment. Drop on top experiments were performed with Sylgard184 and *material C* using pure HDT.

(ii) Mold. A special mold design allowed PDMS stamps to be produced with an extended ink reservoir. This ensured true linear-diffusion conditions and allowed the use of EtOH-based ink solutions. The PDMS was cured directly in a metal tube that was pretreated with primer (Dow Corning, 92-023) to provide good bonding between the metal surface and the polymer. To prevent evaporation of the ethanol, the reservoir was kept covered during the experiment. The mold was also used to form cylindrical stamps of *material C* for drop on top experiments. Because of its brittleness, *material C* cannot be cut or punched into shape.

(iii) Inker Pad. A homogenized inker pad (prepared as described above) acted as a reservoir from which the ink diffused into the stamp. Throughout the experiment, the inker pad was kept in conformal contact with the stamp. Using this method, experiments with reduced starting concentrations of as low as 0.01 wt % were carried out with HDT, ODT, and ECT and stamp thicknesses of ~0.7 mm. Linear-diffusion experiments using ODT and ECT were exclusively carried out as inker pad experiments in combination with ellipsometry to determine the diffusion time (t_d). Owing to possible precipitation of these compounds from the inker pad, concentrations below 1 wt % for ODT and 0.5 wt % for ECT were chosen.

All experiments were carried out at room temperature (22 ± 1 °C).

Etching. For the selective etching of printed gold substrates, a solution of 20 mM $Fe(NO_3)_3 \times 9H_2O$ (Fluka) and 30 mM thiourea (Fluka) in deionized water was used.[22] The etch bath reached its maximum activity 3−5 h after preparation (etch rate of ~4 min for 30 nm Au). The etching was performed at room temperature in a 500 mL container big enough to hold the wafer samples. The best results were obtained in a freshly made bath with slower but more homogeneous etching (~10 min for 30 nm Au). Light mechanical stirring (200 min^{-1}) and additional manual movement of the sample were necessary to achieve a homogeneous etch over the entire sample.

Ellipsometry. Measurements were carried out on a VASE instrument (variable angle spectroscopic ellipsometer, J. A. Woollam Co., Inc., Lincoln, NE) at an angle of 70° for a wavelength range of 400–800 nm. Stamps and inker pads of 10 mm in diameter were used for printing to provide sufficient area for the ellipsometric scan on the SAM. Printing was performed on a complete 4 in. wafer marked with a rectangular grid. For each wafer, a gold baseline was measured within 30 min after sputtering. A Cauchy dispersion relation for transparent films (extinction coefficient $k = 0$) was used for the parametrization of the refractive index of the thiol monolayer[35] using the Cauchy parameters $A = 1.48$, $B = 0.01$, and $C = 0$, which results in a physically reasonable index spectrum and $n = 1.50$ for $\lambda = 0.6328$ μm. In general, a value of 1.50 is suggested for organic monolayers, and a short discussion of this topic can be found in the work of Ulman.[36] From ellipsometry measurements using HDT, we determined a thickness of ~2 nm for a complete monolayer, which is 10% lower than the value found in the literature.[31] A plausible reason for this difference is that in our case the monolayer is not perfectly complete (ordered) as compared with one formed from solution over several hours. Within the relative short printing time and at low concentrations, it is not expected that the monolayer reaches its final density and order.[37] However, its density will be high enough to allow selective etching of the gold substrate.

(35) Tompkins, H. G.; McGahan, W. A. *Spectroscopic Ellipsometry and Reflectometry*; John Wiley & Sons: New York, 1999.

(36) Ulman, A. *An Introduction to Ultrathin Organic Films*; Academic Press, Inc.: San Diego, CA, 1991.

Saturation Concentration. The saturation concentration of the ink in the PDMS stamp was determined for HDT/ECT in Sylgard184 and *material C* (only HDT) by weighing "empty" and saturated stamps (MT5 5.1 g/0.001 mg, Mettler Toledo). The saturation concentration was calculated in weight % of ink contained in the PDMS, and the data were averaged from the values of at least five different samples. Saturation with ECT was carried out at 45 °C for 18 h. During cooling, the samples blurred immediately, owing to ECT crystallization in the bulk material. Within 2 h, the slabs cleared off again, starting from the edges, and strong precipitation on the surface could be observed. To determine a critical concentration value below which surface crystallization is suppressed, six slabs with a defined concentration lower than 1.1 wt % were prepared and homogenized at 45 °C for 68 h. Then, the slabs were checked for signs of ECT precipitation on their surface after 1, 6, and 75 h.

Acknowledgment. The authors would like to thank Matthias Geissler, Alexander Bietsch, Constance Rost, David Junker, and Urs Kloter for support and inspiring discussions. The authors thank P. F. Seidler for his continuous support. The partial support of the Swiss Federal Office for Education and Science (OFES) in the framework of the EC-funded project NaPa (Contract No. NMP4-CT-2003-500120) is gratefully acknowledged. The content of this work is the sole responsibility of the authors.

LA048273L

(37) Larsen, N. B.; Biebuyck, H.; Delamarche, E.; Michel, B. *J. Am. Chem. Soc.* **1997**, *119*, 3017–3026.

ADVANCED MATERIALS

Closing the Gap Between Self-Assembly and Microsystems Using Self-Assembly, Transfer, and Integration of Particles**

By *Tobias Kraus, Laurent Malaquin, Emmanuel Delamarche, Heinz Schmid, Nicholas D. Spencer,* and *Heiko Wolf**

Bulk processes allow efficient production of meso- to nanoscale particles with well-defined geometries and internal structures.[1] Various macroscopic applications rely on their advantageous optical,[2] catalytic,[3] mechanical,[4] and other properties, because random particle assemblies in suspensions, powders, or composites derive many useful properties from their constituent particles. Some properties, however, only become apparent at the individual particle level. Examples are size and geometry, which can be smaller and more regular than anything made by lithographic approaches,[5] and certain electronic or optical attributes.[6] Other properties, such as photonic bandgaps, only appear in specific particle arrangements.[7] Many of them are potentially useful in microsystems technology, but their exploitation requires integration of the particles at well-defined positions in a device. Microtechnology currently does not provide an established method for the integration of particles with dimensions below 100 μm. Pick-and-place robotic manipulation has been demonstrated for microparticles[8] and might be possible for nanoparticles, but is difficult and requires handling every particle individually. The large feature density in today's microsystems devices renders such serial fabrication economically unrealistic.

Self-assembly is a more efficient method of arranging large numbers of particles on a surface in parallel.[9] Order emerges either from particle–particle interactions, which limit the achievable particle arrangements to certain dense packings, or from particle–substrate interactions, which allow greater control of the final particle positions. Chemical or topographical patterning of the substrate is usually necessary to obtain the flexibility required in microfabrication. However, patterning requires an extra effort, is often incompatible with standard technology, or can even impede device functionality. Moreover, most self-assembly processes are based on suspensions, incurring all the contamination problems of wet processing. These shortcomings have so far prevented self-assembly from becoming a viable option in microtechnology. What is required is a process for handling particles that decouples the prerequisites of self-assembly from the actual integration on the device substrate, so that particle integration can be used as a modular step in microfabrication.

Here we show how controlled adhesion enables a microfabrication-compatible process to integrate particles on unstructured substrates through successive self-assembly, transfer, and integration (SATI). Self-assembly arranges the particles on a dedicated, reusable template without fixing them. An adhesive carrier transfers the entire particle arrangement to the target substrate. The final integration then creates a functional particle–substrate connection. SATI retains the advantages of self-assembly and is compatible with particles of different size ranges.

The separation of the different steps (Fig. 1) also alleviates some typical limitations of self-assembly. Conventional self-assembly processes provide strong particle adhesion during assembly to maintain particles in the assembled positions, which limits precision and requires a specific particle surface chemistry.[10] Processes in which long-range forces mediate the particle–surface interaction do not provide such strong adhesion, but allow greater alignment precision. This prompted some groups to use long-range forces for the assembly, and form the functional particle–substrate connection in later steps.[11–12] SATI spatially separates the assembly step from particle integration. This separation maintains the large throughput, high yield, and material versatility of self-assembly but allows greater flexibility. Moreover, a self-assembly method can be chosen that is particularly appropriate for the particles at hand (Fig. 2). For particles much larger than 1 μm, gravity is sufficiently strong to allow gravitational assembly with geometrical confinement of dry particles.[13] This resembles industrial techniques for ordering larger electronic components or tin beads. Smaller particles can be assembled from suspension through capillary assembly[14] or convective assembly[15] by forces in a moving meniscus that drive them into shallow recesses. Both processes are rapid and permit precise control of the particle positions, even for beads with small contact areas, yielding assemblies of dry particles on the template.

A parallel pick-and-place approach is used to transfer the ordered particles from the template to the target substrate. Adhesive forces hold particles firmly on a carrier, irrespective of their arrangement. In contrast to handling particles using aspirating nozzles, or electric or magnetic forces,[16] adhesion is a robust and technically simple approach that is compatible with very small particles. Well-defined particle surfaces allow the creation of an adhesion cascade, such that particles can be transferred

[*] Dr. H. Wolf, T. Kraus, Dr. L. Malaquin, Dr. E. Delamarche, H. Schmid
IBM Research GmbH, Zurich Research Laboratory
Säumerstrasse 4, CH-8803 Rüschlikon (Switzerland)
E-mail: hwo@zurich.ibm.com
T. Kraus, Prof. N. D. Spencer
Laboratory for Surface Science and Technology
Department of Materials, Swiss Federal Institute of Technology
ETH-Hönggerberg, Wolfgang-Pauli-Strasse 10
CH-8093 Zürich (Switzerland)

[**] A part of this project was funded by the Swiss Commission for Technology and Innovation. The partial support of the State Secretariat for Education and Research (SER) in the framework of the EC-funded project NaPa (contract no. NMP4-CT-2003-500120) is gratefully acknowledged. The content of this work is the sole responsibility of the authors. The authors thank Ute Drechsler and Richard Stutz for their assistance in microfabrication, and Walter Riess and Paul Seidler for their continuous support.

  DOI: 10.1002/adma.200501171

ADVANCED MATERIALS

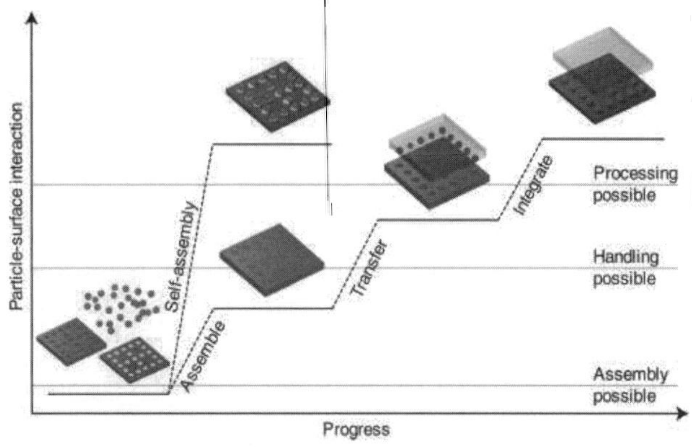

Figure 1. Single-step self-assembly versus the multistep SATI process. Conventional self-assembly processes order particles and establish a strong particle–substrate connection in a single step (assemble and integrate). In the SATI process, self-assembly is only used to assemble particles in a predefined order at a low level of adhesion. The particles are then transferred by a carrier at moderate adhesion levels that allow their handling without perturbing their order, and finally integrated on the target substrate at high adhesion levels so that further processing is possible.

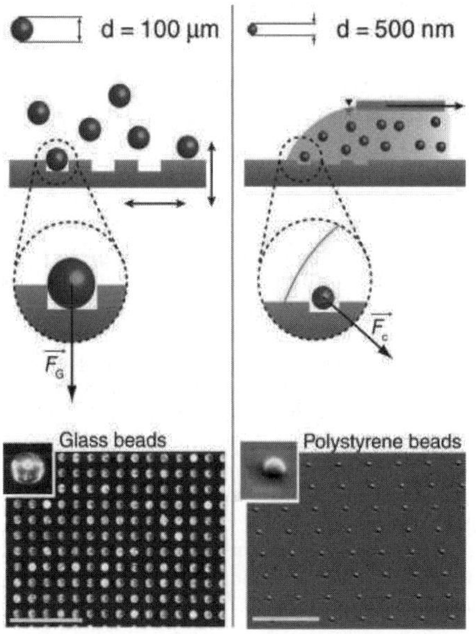

Figure 2. Self-assembly processes for different particle sizes. 100 μm diameter beads were assembled on a structured template (bottom left, scale bar: 1 mm) by gravitational assembly. Vibration of the template introduced sufficient mobility for the particles to move and be captured in holes in the template. Beads with 500 nm diameter were assembled from suspension in a meniscus that slowly moved over a structured carrier (bottom right, scale bar: 10 μm) during capillary assembly. The capillary forces dragged the particles into shallow recesses and maintained them in position during drying.

from surface to surface, exploiting the increasing adhesion. We minimized the initial particle adhesion to the gravitational assembly template by means of a silica surface covered by a low-energy self-assembled monolayer[17] (perfluorodecyltrichlorosilane with a surface energy of $\gamma \sim 6$ mJ m^{-2}). The carrier then provided adequate adhesion for picking the particles up from the template and holding them, but allowed their release in the integration step. Such a transfer has been demonstrated for thin gold layers;[18] for particles, however, both interfacial energies and contact areas influence adhesion. Standard Johnson–Kendall–Roberts theory[19] predicts a theoretical pull-off force F_A for a sphere with radius r in contact with a surface as

$$F_A = \frac{3}{2}\pi r W_A \qquad (1)$$

where W_A is the material-specific, areal work of adhesion. This adhesion force is usually very large. For a relatively low $W_A = 40$ mJ m^{-2} (far below typical bulk cohesion values), it already amounts to $F = -30$ μN, whereas gravity only exerts a force on the order of tens of nanonewtons. That would indicate that a carrier could readily support relatively large and heavy particles. Practical adhesion forces, however, are usually much weaker because of imperfections in the surfaces involved. A material combination with a larger W_A poses the risk of exceeding the cohesion on one side, leading to damage of the particles or the carrier[20] during separation. A compliant material that can deform upon contact with hard surfaces can compensate for particle imperfections[21] in conformal contact with an interfacial area close to the theoretical value. Here, the carriers were made from poly(dimethylsiloxane) (PDMS), a silicone rubber known for its mechanical compliance. Although its surface energy is relatively small[22] ($\gamma \sim 22$ mJ m^{-2}), the conformal contact creates a strong but reversible bond to particles. We found that a PDMS carrier can pick up comparatively heavy 100 μm diameter glass beads and hold them with sufficient strength to allow handling of the assembly without special precautions. PDMS also proved to be a suitable substrate for capillary assembly, allowing the direct ordering of small particles on the carrier.

For integration, the particles on the carrier were brought into contact with the target substrate. The flexible PDMS carrier allowed this even in the case of rough or wavy substrates. Because the particle–substrate interaction exceeded the particle–carrier interaction, all particles were printed onto the substrate. For certain materials, e.g., the polystyrene beads in Figure 3, an elevated temperature was sufficient to establish an irreversible connection. For other materials, or to establish electrical, optical, or thermal connectivity between particles and substrate, two-step processes were employed. Tin-coated particles were printed onto a thin polymer adhesion layer,

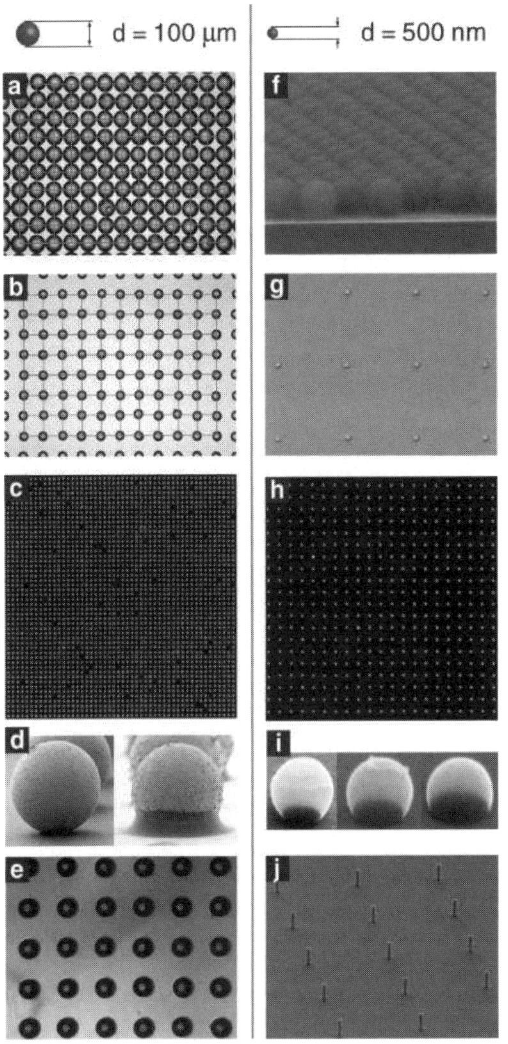

Figure 3. Particle monolayers and square grid arrays placed by SATI on unstructured substrates. Glass beads were printed onto glass substrates coated with a 400 nm thick spin-on glass adhesion layer without spacing (a) and with spacing (b) over an area of 1 cm × 1 cm (c). Tin-coated beads (d) that provided electrical contact after reflow were aligned with and integrated on gold contact pads (e). Both densely packed monolayers of polystyrene beads with hexagonal order (f) and arrays with 5 µm spacing (g) could be printed over large areas (h). The contact area, (i), was controlled via the process parameters. Arrays of this type were used in standard dry-etch processes to fabricate silicon rods (j).

which decomposed in the subsequent annealing step, establishing a conductive bridge. We printed glass particles onto a spin-on glass layer containing an organic component to provide conformal contact with the particles. After calcination, particles and substrates were connected by a glass bridge with good optical properties. Other possible modes of connection include the embedding of particles into a thick polymer layer[23] or the creation of covalent bonds between substrate and particle.[24]

We used SATI to integrate particles on glass and silicon substrates (Fig. 3). Dense hexagonal packings, as are used for photonic crystals, were assembled from polystyrene beads without a template and printed on silicon, whereas square packings were assembled from larger glass beads using a template. In applications where particles function individually, spaced particle arrays are required. Square spaced arrays of 100 µm diameter glass beads on glass substrates can act as optical elements, whereas 100 µm diameter tin-coated beads provide electrical connections to a silicon substrate with gold contacts similar to existing flip-chip assemblies.[25] Nanoscale particles can provide similar mechanical spacings or electrical connections. In addition, their small size and controllable contact area render them suitable as resists. Figures 3g–h show square arrays of 500 nm polystyrene beads printed in a 10 µm grid on silicon. The beads protected circular areas of the substrate in a subsequent etch step to pattern a metal layer or even etch posts from the underlying silicon (Fig. 3j). We were able to fabricate highly regular particle arrays of up to 1 cm × 1 cm with excellent overall regularity. Alignment and yield for the larger particles was limited (to about 98 %, see Fig. 3c) by deviations of sphericity and polydispersity. For the small particles, the limiting parameter was the template quality. The final integration step did not limit either yield or alignment precision.

Besides positioning individual particles, we assembled them into lines, planes, and pyramidal frusta (Fig. 4). Multiple 500 nm particles were assembled in a single step and transferred in parallel, forming (depending on temperature) continuous entities on the surface. The order in the three-dimensional (3D) structures assembled in a single step is limited to certain naturally occurring arrangements (i.e., dense hexagonal packing). However, because in the SATI process the assembly is separate from integration, fully controlled 3D structures were possible by stacking multiple planar assemblies. The adhesion concept allows particles to be stacked without intermediate planarization if sufficient adhesion is present. In a two-step experiment, a spin-on glass adhesion layer held 100 µm diameter glass beads aligned on top of an existing array in the standard printing process (Fig. 4a). It was also possible to shift the second array and perform an off-axis print so that the top particles sat on the side of the base particle (Fig. 4b). Similarly, 500 nm particles could be printed onto a particle monolayer (Fig. 4e).

All steps of the SATI process are compatible with very small particles. Self-assembly methods have been demonstrated for isotropic and anisotropic crystals of 50 nm at high resolution;[26] adhesion approaches can even transfer proteins from one surface to another,[27] and full monolayers of 5 nm gold particles have been transferred from a Langmuir–Schaefer trough via a stamp to a substrate.[28] The full control of

ADVANCED MATERIALS

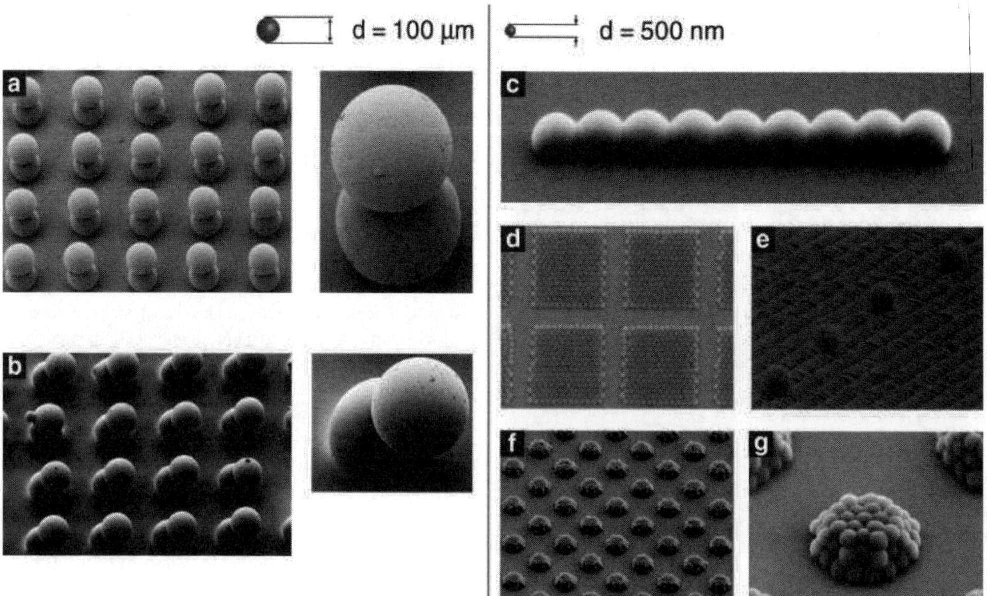

Figure 4. One-, two-, and three-dimensional particle assemblies fabricated using SATI. Arrays of glass microbeads were stacked on top of each other with full positional control both aligned (a) and off-axis (b) so that the beads were held in place only by the adhesion to a spin-on glass layer on the bottom array. Polystyrene particles were either assembled in a single step and then fused to lines (c), planes (d), or pyramidal frusta (f,g); or assembled in two different patterns and printed on top of each other (e).

placement achievable with SATI might therefore be used to integrate such promising components as nanoparticle transistors,[29] memory cells,[30] or sensors[31] into practical devices.

Experimental

Particles: Glass beads (98.9±5 μm diameter, 3.1 % variation coefficient, with a density of 2.45 $g\,cm^{-3}$) were obtained as a powder from Duke Scientific (Palo Alto, CA). Tin-coated polymer beads (100 μm) were from Sekisui Chemicals (Düsseldorf, Germany). Aqueous dispersions of polystyrene beads (499±5 nm, 1.3 % variation coefficient, 1 % solid content) were purchased from Duke Scientific and adjusted to 0.2–5 % solid content prior to use.

Self-Assembly: Gravitational assembly was performed on silicon templates with 60 μm deep holes defined through standard photolithography that were coated with a perfluorodecyltrichlorosilane monolayer. Particles were deposited on the coated template, which was then placed in a sealed cell. The entire setup was excited at 60–100 Hz so that the amplitude was just sufficient to fluidize the particles. When all holes in the template were filled with particles, the template was removed, and excess particles were picked up by a carrier that did not touch the assembled particles. Convective assembly [15] was used to order 500 nm polystyrene particles into 2D continuous layers. It was performed on flat PDMS templates previously oxidized by oxygen plasma. Capillary assembly [14] provided 500 nm polystyrene particle assemblies with full control of the positions on patterned PDMS templates without any surface modification. PDMS templates were fabricated by injection molding on flat or patterned silicon masters [32]. They consisted of a 200 μm thick layer of custom-synthesized PDMS [32] on a 175 μm thick glass backplane about 1 cm × 1 cm in size. Both convective and capillary assembly were performed using a dedicated setup. The PDMS template was fixed on the surface of a moving stage and placed below a fixed glass slide with a separation distance of about 500 μm. A defined volume (40 μL) of particle suspension was injected between the template and the slide. The liquid meniscus was moved over the template at a constant velocity of 1 $\mu m\,s^{-1}$ for convective assembly or between 1 and 10 $\mu m\,s^{-1}$ for capillary assembly. The entire setup was installed on the stage of an optical microscope for direct observation and control of the assembly process. A Peltier element allowed the template to be cooled to temperatures ranging from 10 to 18 °C to prevent evaporation of the suspension and allow control of the particle motion.

Transfer and Integration: Carriers were fabricated from 3 to 6 mm thick PDMS pieces that were bonded through plasma activation onto pieces of quartz photomask blanks. A vacuum chuck held the carriers in a setup that allowed alignment of the carrier and substrate parallel to each other, horizontal alignment of the particles with a pattern on the substrate, and well-controlled vertical movement for the integration. The substrate was held by a heatable vacuum chuck at the bottom of the setup. A stereomicroscope allowed optical alignment and control of the integration process. Adhesion layers were deposited onto silicon and glass substrates via spin-coating. For 100 μm glass beads, we used a 400 nm film of spin-on glass (type 400 F, Filmtronics, Butler, PA). The particles were integrated at a temperature of 80 °C so that the adhesion was sufficient for an immediate transfer and then calcined at 400 °C following the instructions of the manufacturer. For tin-coated 100 μm polymer beads, a polyisobutylene film (average molecular weight 500 000) with a thickness of about 5 μm provided strong adhesion throughout the tin reflow at 230 °C. Most of the polymer decomposed during the reflow; final traces were removed in toluene. Transfer of 500 nm polystyrene particles was performed by bring-

ADVANCED MATERIALS

ing the processed template into contact with a 2 cm × 2 cm silicon wafer piece (with native oxide) or a gold-coated silicon wafer piece, placed on a hot plate and heated to 120 °C in both cases. The PDMS template was pressed slightly onto the substrate until conformal contact was visibly established. Immediately after contact was achieved, the template and substrate were removed from the hot plate, allowed to cool down, and then separated.

Received: June 07, 2005
Final version: July 25, 2005
Published online: September 5, 2005

[1] F. Caruso, *Colloids and Colloidal Assemblies*, Wiley-VCH, Weinheim, Germany **2004**.
[2] W. Caseri, *Macromol. Rapid Commun.* **2000**, *21*, 705.
[3] A. Bell, *Science* **2003**, *299*, 1688.
[4] P. K. Mallick, *Fiber-Reinforced Composites: Materials, Manufacturing, and Design*, Dekker, New York **1993**.
[5] C. Murray, C. Kagan, M. Bawendi, *Annu. Rev. Mater. Sci.* **2000**, *30*, 545.
[6] T. Trindade, P. O'Brien, N. L. Pickett, *Chem. Mater.* **2001**, *13*, 3843.
[7] C. Lopez, *Adv. Mater.* **2003**, *15*, 1679.
[8] F. Garcia-Santamaria, H. T. Miyazaki, A. Urquila, M. Ibisate, M. Belmonte, N. Shinya, F. Meseguer, C. Lopez, *Adv. Mater.* **2002**, *14*, 1144.
[9] S. Maenosono, T. Okubo, Y. Yamaguchi, *J. Nanopart. Res.* **2003**, *5*, 5.
[10] G. Whitesides, M. Boncheva, *Proc. Natl. Acad. Sci.* **2002**, *99*, 4769.
[11] H. O. Jacobs, A. R. Tao, A. Schwartz, D. H. Gracias, G. M. Whitesides, *Science* **2002**, *296*, 323.
[12] X. R. Xiong, Y. Hanein, J. D. Fang, Y. B. Wang, W. H. Wang, D. T. Schwartz, K. F. Böhringer, *J. Microelectromech. Sys.* **2003**, *12*, 117.
[13] H.-J. J. Yeh, J. S. Smith, *IEEE Photonics Technol. Lett.* **1994**, *6*, 706.
[14] Y. Xia, Y. Yin, Y. Lu, J. McLellan, *Adv. Funct. Mater.* **2003**, *13*, 907.
[15] N. D. Denkov, O. D. Velev, P. A. Kralchevsky, O. B. Ivanov, H. Yoshimura, K. Nagayama, *Nature* **1993**, *361*, 26.
[16] J. Cecil, D. Vasquez, D. Powell, *Int. J. Prod. Res.* **2005**, *43*, 819.
[17] U. Srinivasan, M. Houston, R. Howe, R. Maboudian, *J. Microelectromech. Sys.* **1998**, *7*, 252.
[18] S. Hur, D. Khang, C. Kocabas, J. Rogers, *Appl. Phys. Lett.* **2004**, *85*, 5730.
[19] K. L. Johnson, K. Kendall, A. D. Roberts, *Proc. R. Soc. Lond. A* **1971**, *324*, 301.
[20] J. J. Bikerman, *The Science of Adhesive Joints*, Academic Press, New York **1968**.
[21] A. Bietsch, B. Michel, *J. Appl. Phys.* **2000**, *88*, 4310.
[22] C. Drummond, D. Chan, *Langmuir* **1997**, *13*, 3890.
[23] Y. Sun, D.-Y. Khang, F. Hua, K. Hurley, R. G. Nuzzo, J. A. Rogers, *Adv. Funct. Mater.* **2005**, *15*, 30.
[24] H. Schmid, H. Wolf, R. Allenspach, H. Riel, S. Karg, B. Michel, E. Delamarche, *Adv. Funct. Mater.* **2003**, *13*, 145.
[25] K. Gilleo, *Area Array Packaging Handbook: Manufacturing and Assembly*, McGraw-Hill Professional, New York **2001**.
[26] Y. Cui, M. T. Björk, J. A. Liddle, C. Sönnichsen, B. Boussert, A. P. Alivisatos, *Nano Lett.* **2004**, *4*, 1093.
[27] A. Bernard, D. Fitzli, P. Sonderegger, E. Delamarche, B. Michel, H. R. Bosshard, H. Biebuyck, *Nat. Biotechnol.* **2001**, *19*, 866.
[28] V. Santhanam, R. P. Andres, *Nano Lett.* **2004**, *4*, 41.
[29] D. L. Klein, R. Roth, A. K. L. Lim, A. P. Alivisatos, P. L. McEuen, *Nature* **1997**, *389*, 699.
[30] S. Kolliopoulou, P. Dimitrakis, P. Normand, H. L. Zhang, N. Cant, S. D. Evans, S. Paul, C. Pearson, A. Molloy, M. C. Petty, D. Tsoukalas, *J. Appl. Phys.* **2003**, *94*, 5234.
[31] A. B. Kharitonov, A. N. Shipway, I. Willner, *Anal. Chem.* **1999**, *71*, 5441.
[32] M. Geissler, H. Wolf, R. Stutz, E. Delamarche, U. W. Grummt, B. Michel, A. Bietsch, *Langmuir* **2003**, *19*, 6301.

  DOI: 10.1002/adma.200500170

ARTICLES

Nanoparticle printing with single-particle resolution

TOBIAS KRAUS[1,2], LAURENT MALAQUIN[1†], HEINZ SCHMID[1], WALTER RIESS[1], NICHOLAS D. SPENCER[2] AND HEIKO WOLF[1*]

[1]IBM Research GmbH, Zurich Research Laboratory, Säumerstrasse 4, 8803 Rüschlikon, Switzerland
[2]Laboratory for Surface Science and Technology, Department of Materials, ETH Zurich, Wolfgang-Pauli-Strasse 10, CH-8093 Zürich, Switzerland
†Present address: Laboratoire de Photonique et de Nanostructures, Route de Nozay, 91460 Marcoussis, France
*e-mail: hwo@zurich.ibm.com

Published online: 2 September 2007; doi:10.1038/nnano.2007.262

Bulk syntheses of colloids efficiently produce nanoparticles with unique and useful properties. Their integration onto surfaces is a prerequisite for exploiting these properties in practice. Ideally, the integration would be compatible with a variety of surfaces and particles, while also enabling the fabrication of large areas and arbitrarily high-accuracy patterns. Whereas printing routinely meets these demands at larger length scales, we have developed a novel printing process that enables positioning of sub-100-nm particles individually with high placement accuracy. A colloidal suspension is inked directly onto printing plates, whose wetting properties and geometry ensure that the nanoparticles only fill predefined topographical features. The dry particle assembly is subsequently printed from the plate onto plain substrates through tailored adhesion. We demonstrate that the process can create a variety of particle arrangements including lines, arrays and bitmaps, while preserving the catalytic and optical activity of the individual nanoparticles.

Today, printing is the most widespread technology to deposit small particles onto various surfaces. In book and fine-art printing, submicrometre pigment particles carry different colours, thus creating brilliant and bleach-resistant images. Small particles (often <100 nm in diameter) are increasingly used in other fields, too. In electronics, optics and biology, small particles are the target of intense research and are used in a number of commercial applications[1]. Such applications exploit the particle's confined electronic systems[2], their strong interaction with light, with their matrix and with each other[3], their well-defined surface properties[4], their high catalytic activity[5] and—in sufficiently small particles—their quantum confinement properties[6].

A prerequisite for future applications using particles as functional entities is often control of their arrangement on a surface, between electrodes or in a device. Doing so with standard microfabrication techniques is difficult, and it is often time-consuming and inefficient to create sparse patterns of small particles using subtractive top-down processing[7]. Printing techniques, in contrast, scale well to nanoscopic particle sizes. Figure 1 shows how gravure printing can be scaled to handle particles less than 100 nm in diameter. Using self-assembly processes to ink nanofabricated 'printing plates' and controlled adhesion for the transfer, we print nanocrystals at single-particle resolution. The process maintains the efficient parallel nature of self-assembly without requiring specially patterned substrates or elaborate surface treatments. Printing has been demonstrated on various targets, including soft and flexible materials that are challenging to handle by other microfabrication processes[8].

In conventional printing, the ink is treated as a continuous medium. Its wetting properties (as in offset printing) or the geometry of a template (as in gravure printing) define in which areas a printing plate is inked and from where the ink will later be transferred to the substrate. The arrangement of the final, embedded pigment particles in the print is stochastic. This principle holds even for molecular inks—many researchers have used micro-contact printing to transfer reactive molecules[9], proteins[10] and catalytic colloids[11] according to stamp geometries. The arrangement of the printed species itself is not imposed by a template, but is either irregular or imposed by the adsorption process[12].

In the process proposed here, discrete nanoparticles are arranged deterministically on structured surfaces using 'directed assembly'. In contrast to conventional inking, directed assembly does not merely fill predefined structures with randomly dispersed pigment particles, but arranges nanoparticles at positions that are defined by the geometry of a template. These silicone templates are moulded with total areas in the order of 1 cm^2 from hard masters with feature dimensions between 40 nm and tens of micrometres. The template geometries adopted (Fig. 2) include lines, producing close-packed nanoparticle wires as used in molecular electronics[13], spaced regular arrays for nanowire growth and arbitrary arrangements, such as the sun designed by the seventeenth-century alchemist Robert Fludd[14], which is composed of about 20,000 single particles.

HIGH-PRECISION INKING

For inking, the meniscus of a colloidal suspension is moved over the patterned polymer layer so that the dispersed particles arrange inside the features[15]. When the particles have assumed their desired positions, the liquid must be removed because brownian motion is sufficiently strong in nanoparticle suspensions

ARTICLES

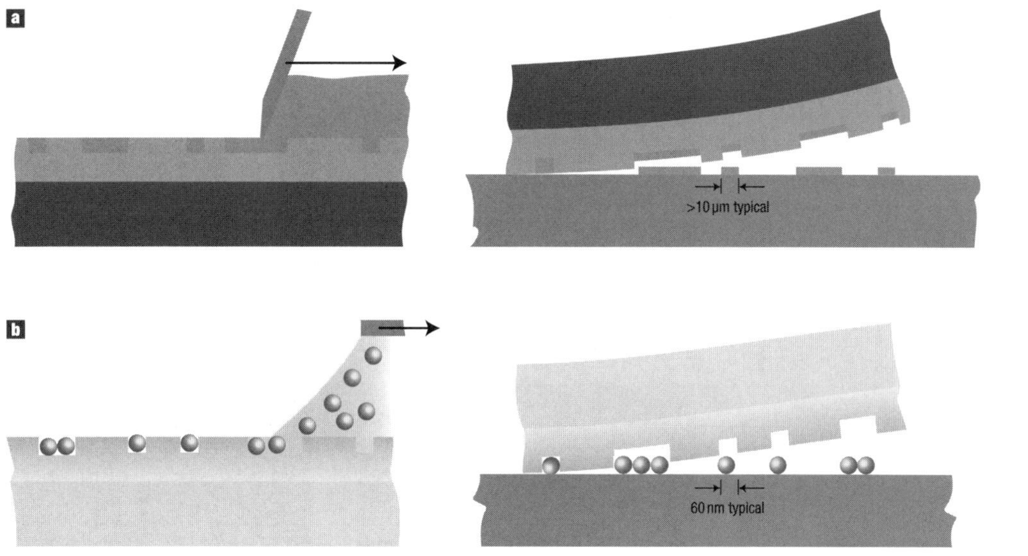

Figure 1 From traditional gravure printing to high-resolution particle printing. a, In gravure printing, a doctor blade fills the recessed features of a printing plate with ink. Pigment particles are randomly dispersed in the ink, which is transferred from the plate onto the substrate. **b**, In high-resolution particle printing, a self-assembly process controls the arrangement of nanoparticles on the printing plate. The entire assembly is printed onto the substrate, whereby the particle positions are precisely retained at a resolution that is three orders of magnitude higher than in conventional printing.

to destroy the order. The dry, filled printing plates (Fig. 2) are then stable and can be stored until the particles are transferred onto substrates.

In capillary assembly, as the colloid moves over the surface, particles are deposited in the desired regions. Excess particles are removed through the Stokes drag exerted by the liquid meniscus (Fig. 3a), which fulfils a similar function as the doctor blade in gravure printing (Fig. 1). High yield, precise arrangement and high contrast require control of particle transport, which is intimately connected with the wetting behaviour of the colloidal suspension.

Providing sufficient particles from the bulk colloid to the assembly region near the contact line is a prerequisite for particle deposition. This mass transport can be achieved by the convective flux of liquid towards the meniscus caused by water evaporating from the meniscus[16]. For 60-nm gold particles, the timescale of inertial response is very short—on the order of nanoseconds—and the brownian timescale is very long compared with convective timescales, so that in the bulk part of the fluid, particles can be regarded as solutes that travel along the streamlines (see Supplementary Information, page 2). The streamlines depend on experimental parameters, notably the temperature, the humidity and the substrate velocity, and can be simulated. Figure 3b shows computational fluid dynamics simulations, taking into account the substrate motion, which drags liquid towards the meniscus, and the evaporation from the meniscus, which removes liquid that has to be replenished from the bulk.

Two main features appear in the laminar flow of the colloid according to our simulations (see Supplementary Information, page 5): a flow component that is directed towards the meniscus and a recirculation flow in the upper part of the liquid. As the evaporation rate increases, the recirculation zone moves away from the meniscus. Higher colloid temperature and lower outside humidity increase particle flux to the assembly region both by increasing flow rates from the bulk and by decreasing the particle recirculation. For the liquid flow rate $\dot{V}_{\text{colloid}}$ that occurs at a colloid temperature of 27 °C and a bulk particle concentration of $c_{\text{p}} = 2 \times 10^{14}\ \text{l}^{-1}$, a flux of $\dot{N}_{\text{p}} = c_{\text{p}}\dot{V}_{\text{colloid}} \approx 1.5 \times 10^{8}\ \text{m}^{-1}\ \text{s}^{-1}$ is transported to the assembly region. This particle flux is larger than the flux required to form a full particle monolayer,

$$\dot{N}_{\text{ML}} = v_{\text{substr}} \frac{\pi/(3\sqrt{2})}{r_{\text{p}}^2 \pi} \approx 1.96 \times 10^7 \text{m}^{-1}\ \text{s}^{-1}$$

at a substrate velocity of $v_{\text{substr}} = 0.3\ \mu\text{m s}^{-1}$. Thus, sufficient particles to fill any feature on the printing plate are supplied. However, to ink with high yield and good contrast, additional prerequisites need to be satisfied.

In addition to the particle flux, the particle concentration in the assembly zone has to be sufficient. We can analyse the requirements for high-yield assembly using a microscopic model of particle transport. Whereas gaussian statistics describe the transport of particles to the accumulation zone, Poisson statistics apply when small numbers of particles need to move into specific positions. The immobilization of a single particle in a capture site of the template can be modelled by assuming that any particle from a certain volume V_{c} above this site will be captured (Fig. 3c). If no additional force exists, the average number $\langle n \rangle$ of particles in V_{c} depends solely on the particle concentration, $\langle n \rangle = c_{\text{p}} V_{\text{c}}$. Thus, the probability of capturing at least one particle becomes

$$P(n \geq 1) \approx \sum_{n=1}^{\infty} P(n) = 1 - e^{-\langle n \rangle}$$

ARTICLES

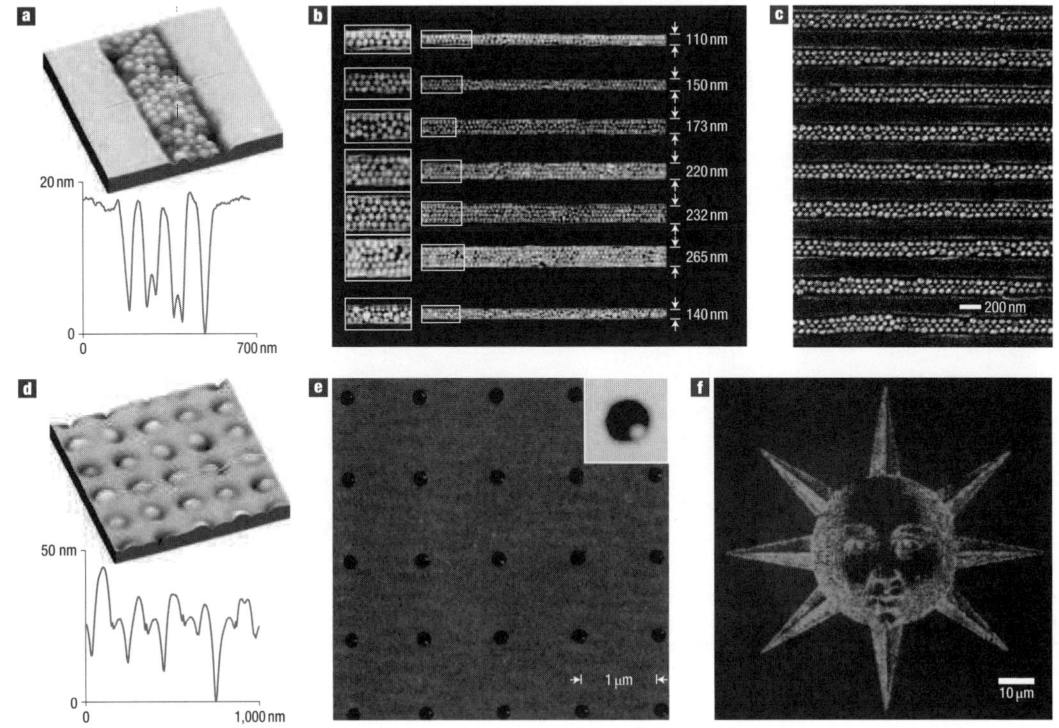

Figure 2 Particles arranged on printing plates after inking. Gold particles were assembled in poly(dimethylsiloxane) templates by hydrodynamic and surface forces. **a**, AFM topography and **b**, SEM micrographs of 60-nm Au monodispersed particles arranged in u-grooves of different widths, and one u-groove (bottom) filled with polydispersed particles. **c**, Larger areas of multiple, densely spaced lines have also been filled, for example 150-nm lines with 60-nm Au particles. **d**, AFM topography of 200-nm spaced array of 100-nm Au particles. **e**, 1-μm-spaced array of 60-nm Au particles. **f**, Bright-field optical micrograph of a sun composed of approximately 20,000 individual 60-nm Au particles.

so that the minimum concentration required to achieve a 90% yield across many capture sites is $c_p \geq 2.3/V_c$, which is large when compared with the original colloidal concentration. We therefore have to induce local particle accumulation without destabilizing the colloid to provide high assembly yield.

When contact angle and particle stabilization are adjusted properly, the excess particles that flow into the assembly region form a stable zone of high concentration close to the contact line (Fig. 3c). This 'accumulation zone' is self-limiting; it does not grow in size even if the meniscus is moved over an unstructured substrate, where no particles are deposited. A steady state between leaving and entering particles results when the accumulation zone has reached a certain size. The zone is reminiscent of the well-known concentration polarization in ultrafiltration[17], where the Stokes drag on the particles is balanced by interparticle forces (mainly hydrodynamic forces, electrostatic and entropic repulsion). This dynamic diffusion–convection equilibrium is tunable through temperature, which strongly affects the evaporation rate and mainly changes the Stokes drag on the particles, and through colloid composition, which changes the particle interaction forces.

More accurate descriptions of the assembly mechanism need to take into account the dewetting dynamics, particle–surface interaction and particle–particle forces. It was observed that the template geometry not only defines the particle arrangement, but also governs the mechanism leading to this arrangement. Larger features, for example the lines shown in Fig. 3d, create microscale flows. Particles can flow into the features and the actual arrangement often takes place at some distance from the main assembly front. In contrast, arrays that capture single particles pin the meniscus only very slightly, and the assembly takes place very close to the contact line. In all cases, the wetting behaviour is critical—the transition from the mobile to the assembled state takes place only upon dewetting.

Both the microscopic dewetting and the bulk liquid motion depend on the contact angle; inking at high yield and high precision requires its tuning. It is possible to modify the template surface to this end[15]. However, in a printing process, higher-energy surfaces are not an option because particles would strongly adhere to them, inhibiting the transfer step. Instead, we use surfactant systems to tune the wetting behaviour with more flexibility, as is frequently done in conventional printing, and to tune particle–surface interactions simultaneously. A surfactant system for particle inking has to change the contact angle of the colloidal suspension with the hydrophobic poly(dimethylsiloxane) (PDMS) to an angle between 50° and 60°, where we find the highest assembly yield. The surfactants should not, however,

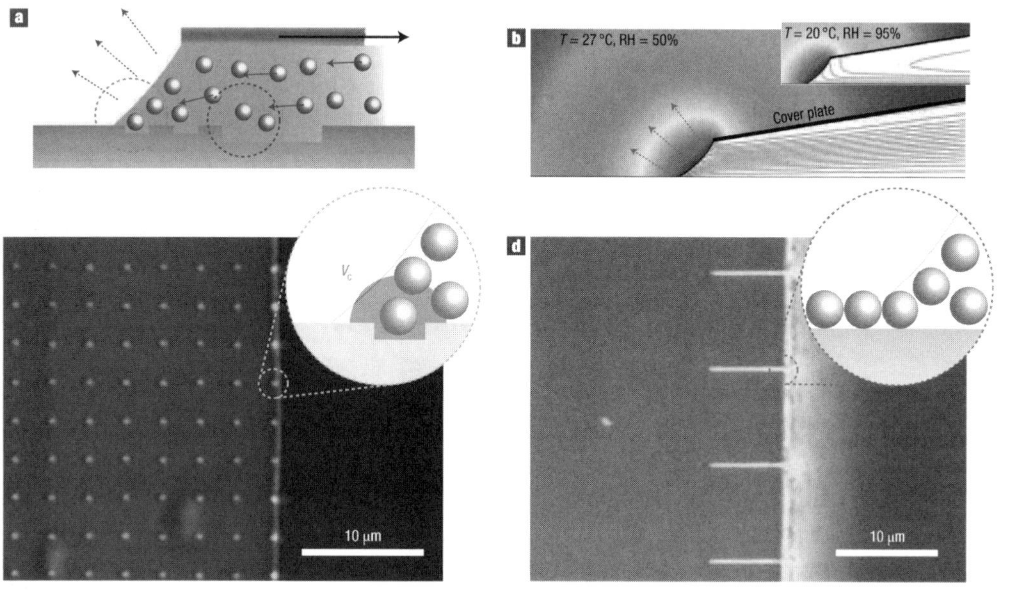

Figure 3 The inking process that arranges the particles. a, The meniscus of a colloidal suspension containing 60-nm Au particles is moving over the patterned PDMS surface of a printing plate. **b**, Simulation of the evaporation (colours indicate concentrations) and the streamlines (red lines) of the laminar flow that drags along particles towards the three-phase contact line, where they form an accumulation zone. The main panel is a simulation at a relative humidity (RH) of 50% and the inset for a RH of 95%. **c**,**d**, The accumulation zone, visible in the micrographs as a bright yellow line, moves over the printing plate. Depending on the geometry they encounter, particles at the contact line form sparse patterns, for example, arrays (**c**) or dense arrangements such as lines (**d**).

destabilize the colloidal particles, not even in the accumulation zone with its high particle concentrations, and it should only leave a very thin film on the inked plate. Less obvious are the effects of the surfactant on the particle–substrate interactions. We find that some surfactants hinder the assembly, possibly by creating a repulsive force between substrate and particles, whereas others (such as polyvinylpyrrolidone) form deposits that embed the particles on the substrate. A mixture of a non-ionic, rather hydrophobic surfactant—namely Triton X-45, an octylphenol ethoxylate with a short polyethyleneglycol chain—and the anionic dodecyl sulphonate appears to avoid all these issues, leaving the colloid stable even at high concentrations.

The surfactant system ensures a proper receding contact angle on an unstructured surface, but the local contact angle during assembly also depends on the pinning caused by the structures on the template's surface[16]. To achieve high yields without unspecific deposition, both an appropriate geometry and a sufficient depth of the capture sites have to be chosen. The results shown in Fig. 2 stem from arrays of single particles assembled in 40- to 45-nm deep holes and lines of particles assembled in 70- to 100-nm deep lines. The deeper the lines are, the more robust the assembly process, at the cost of a lower transfer yield. To produce sufficiently precise templates, masters patterned through electron-beam or optical lithography and etched into silicon or silicon oxide were used. They can be used to cast many templates in a high-elastic-modulus polydimethylsiloxane rubber[18,19]. First, a prepolymer mixture is poured onto the silicon master, then a flexible glass backplane is placed on top, and the polymer is cured. The backplane and polymer layer are peeled off, and the template is finally extracted with ethanol to remove residues before the inking of the printing plate (see Supplementary Information for details of printing plate fabrication).

PARTICLE TRANSFER

Inking is followed by transfer in the printing process. During transfer, the printing plate and substrate are brought into conformal contact, facilitated by the elastomer layer that adjusts to the topography of the substrate surface. On removal of the printing plate, adhesive forces hold the particles on the substrate, thereby creating the desired arrangement (Fig. 4). Traditional printing technologies use liquid inks, and ink transfer is based on wetting differences between the printing plate and the substrate (for example, paper). In this work, the ink solely consists of dry nanoparticles, which are far less mobile than liquids. Particles therefore can be positioned more precisely, but they are harder to print. Their transfer is based on the different levels of particle adhesion on the printing plate and the substrate. Such adhesion differences have already been used to transfer larger particles[20] and large molecules[10].

Wetting differences are caused by differences in interfacial energy. Likewise, the adhesion of small particles strongly depends on the interfacial energy of the particle–surface joint[21]. In contrast to solid–liquid interfaces, however, the particle–substrate interface is often not conformal, and its geometry can be complex[22]. When surfaces with different energies are used to transfer particles, it is also necessary to control the interfacial area, in particular when working with individual nanocrystals[20]. Such crystals often have irregular surfaces with sharp crystal edges. The adhesive force acting on one particle with a planar geometry can be much stronger than that acting on spherical ones. Moreover, solutes from the colloidal suspension frequently form adlayers on the metal surface while drying, thereby changing the chemical identity of the interface and increasing the particle–substrate distance. We

ARTICLES

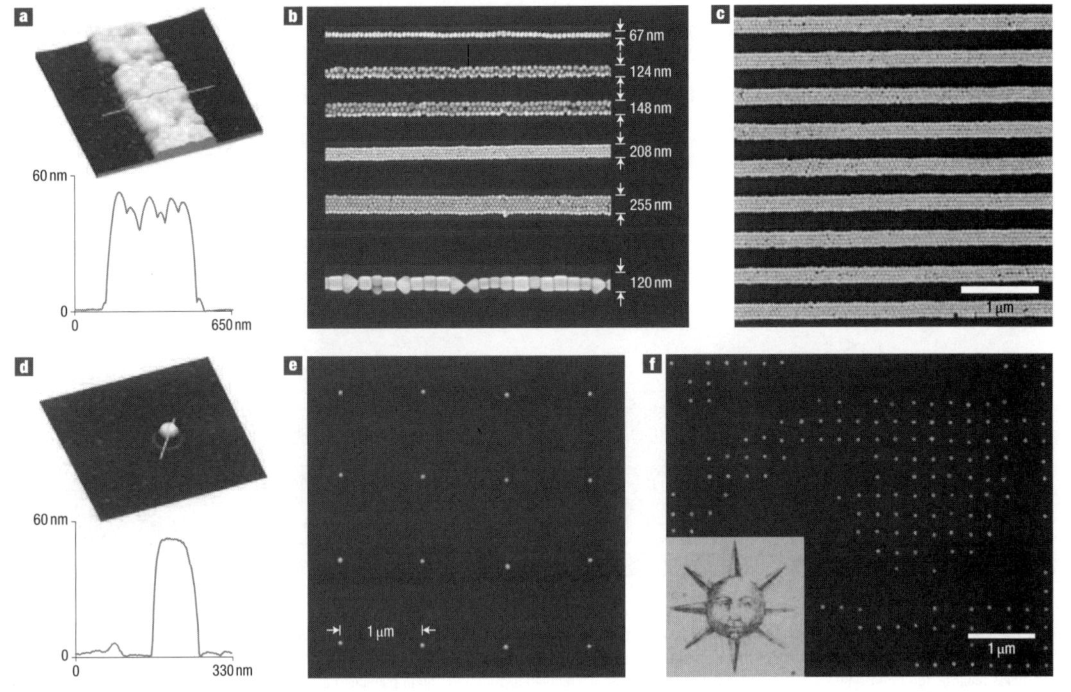

Figure 4 Particle structures printed on unpatterned Si substrates. Lines are directly printed onto the native oxide layer of Si wafers, and single-particle arrays are printed onto additional 30-nm PMMA adhesion layers. **a**, AFM and **b**, SEM images of lines from 60-nm Au particles. The bottom row in **b** is made from 100-nm Ag particles, which are mostly cubical in shape. **c**, Larger area of 200-nm-wide lines from 60-nm Au particles. **d**, AFM detail and **e**, SEM overview of 1-μm-spaced array of 60-nm Au particles. **f**, Detail (left eye) from a printed sun pattern composed of 60-nm-diameter Au particles with 280 nm pitch.

find that dry transfer of isolated spherical gold nanoparticles with 60 nm diameters leads to very low yields on silicon, quartz, and even fresh gold layers with strong van der Waals attraction to gold particles. The transfer of cubes and plates leads to much larger yields. Likewise, lines and layers of 60-nm gold spheres could be printed with good yield onto a hard silicon surface with a clean native oxide layer (Fig. 4). It appears that both a large crystal face in contact with the hard substrate and multiple smaller contact points increase adhesion and transfer yield, whereas organic adlayers on the particles decrease adhesion and yield. This is consistent with standard microscale models of adhesion, which stress the importance of the actual interfacial area and of contact splitting for adhesive strength[23]. Inside the lines, particles strongly adhere to each other, and each particle adheres to the substrate. Thus, to break the adhesive bond to the substrate, multiple particle–substrate contacts have to be broken, which requires a larger force than for each single adhesive contact[24].

A reliable transfer process thus requires both surface-energy differences and defined interfacial areas. The PDMS layer on the printing plate that we use provides a low-energy surface, well known for its capacity to release biomolecules onto more hydrophilic substrates[10]. The geometry of the PDMS layer was designed such that the particles come into contact with the substrate when the printing plate is placed on the target. This sometimes conflicts with the inking process, which for some particle arrangements (for example, for lines) requires that the template be deeper than the particle's diameter. In such cases, the gap was kept as small as possible. A height difference of 5–10 nm can be overcome when pressure is applied on the stamp during the transfer process. Figure 4a shows particle lines of different widths printed on oxidized silicon using this geometry.

In contrast to lines, arrays of individual spherical nanocrystals cannot be transferred onto hard substrates, even if they protrude significantly from the template (Fig. 2e,f). However, transfer with high yields is possible onto thin polymer layers, for example, spin-coated resists. A polymethylmethacrylate (PMMA) layer with a thickness of 10–30 nm increases the transfer yield to above 95%. Transfer takes place at a temperature of 110–120 °C, slightly above the glass transition temperature of PMMA. The polymer embeds the protruding segment of the crystal, thus providing a sufficiently large area of contact to create the adhesion necessary for transfer. Figure 4 shows arrays of individual nanocrystals printed on PMMA. If necessary, the PMMA can then be removed in hydrogen plasma, where it cleanly decomposes.

The position of the nanocrystals is preserved during the transfer with a local accuracy better than 100 nm (see also Fig. 5b). The long-range accuracy of the transfer is comparable to that of microcontact printing with thin-film stamps, for which relative average positioning errors below 20 p.p.m. have been demonstrated across 4-inch wafers[9]. Our printing plates feature the same combination of hard backplanes with soft elastomer layers to reduce the deformations of the stamp that cause long-range inaccuracies in soft lithography techniques[18].

ARTICLES

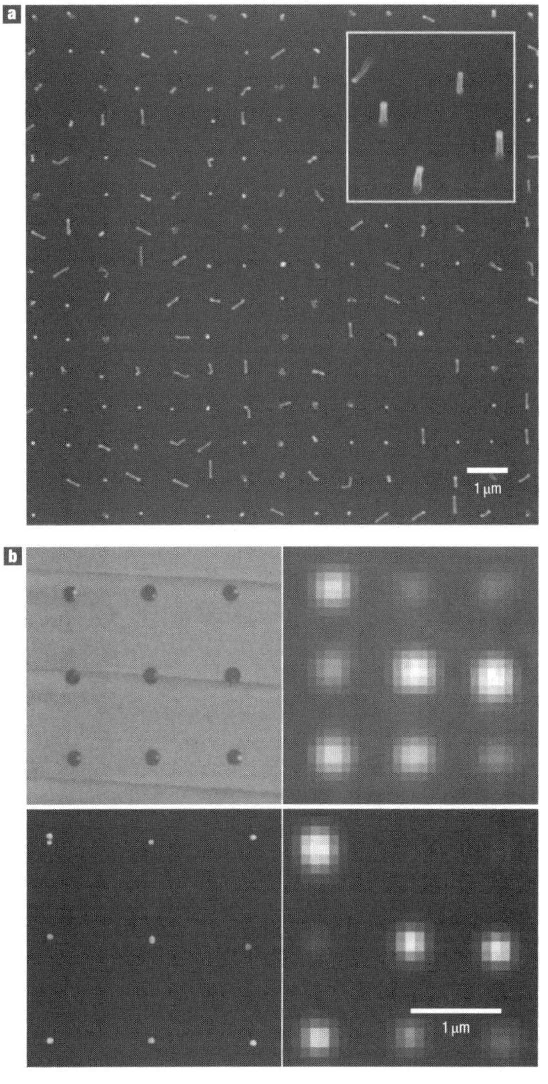

Figure 5 Demonstration of the activity of printed particles. In printed arrays, 60-nm Au particles retain both their catalytic and optical activities. **a**, Silicon nanowires grown through a vapour–liquid–solid process from a printed array of Au particles (inset is tilted). **b**, AFM topography and the corresponding dark-field image of particles on the printing plate (top row) and an SEM image and the corresponding dark-field micrograph on a silicon substrate (bottom row: the images were mirrored for convenience).

FUNCTIONAL ARRAYS

It is crucial that printed nanoparticles retain their useful properties during integration. Here, a surface-sensitive application was chosen to demonstrate that the individual gold nanocrystals are still catalytically active. Figure 5a shows silicon nanowires (SiNW) that were grown from an array of 60-nm particles using a vapour–liquid–solid (VLS) process[25]. The particles were printed onto a 30-nm-thick PMMA layer on top of a hydrogen-passivated silicon ⟨111⟩ substrate. The PMMA adhesion layer was removed by thermal decomposition at 300 °C in vacuum, assisted by hydrogen plasma. SiNW growth was then initiated at 460 °C in a reactor using silane as the precursor gas. Nanowires nucleated and grew from the gold particles, and their arrangement was preserved. The epitaxial relationship of the SiNW with the substrate is evident from the wires growing along both the vertical and the three inclined ⟨111⟩ growth directions. Such nanowires are regarded as novel building blocks for future transistors, memory cells and sensors, for example[26].

Metal nanocrystals also interact strongly with electromagnetic waves[27], a property that is exploited in applications such as glass staining, surface-enhanced Raman spectroscopy[28] (SERS) and surface plasmon resonance-based agglomeration assays[29]. It has been noticed that the characteristics of single particles can vary considerably[30], and the printing technique adopted here can create samples that allow investigating such single-particle effects in detail. When the particles are arranged in regular arrays, it is easy to find extraordinarily active particles and analyse them using a variety of methods. Figure 5b shows how some particles scatter visible light more strongly than others, both when immobilized in the PDMS template and when printed onto a silicon substrate. Comparison of the dark-field micrographs with the atomic force microscopy (AFM) and scanning electron micrograph (SEM) images reveals that this is not simply a function of size or overall shape (although doublets of particles often scatter the light strongly), and it will be interesting to apply single-particle SERS and high-resolution electron microscopy to identify the features that distinguish strongly scattering particles or SERS 'hot spots'.

CONCLUSIONS AND OUTLOOK

In conclusion, we have developed an efficient process to fabricate lines, arrays and complex arrangements of nanoparticles with high accuracy and single-particle resolution that retains individual particle functionality. Our printing approach is compatible with different particles, including not only metals, but also polymers, semiconductors and oxides[20]. It can handle bulk-synthesized particles directly in their original colloidal state even when they have been modified with functional molecules. Thanks to its high resolution, nanoparticle printing can efficiently define the critical dimensions for nanoscale devices. The long-range accuracy is similar to that of microcontact printing methods and reaches about 12 p.p.m. Reaching higher accuracies of 1 p.p.m., as required for large-scale integration in microelectronics, is a challenge that remains to be addressed.

Printing processes are adaptable to continuous processing and large areas[31]. Particle handling is based on adhesion, which scales favourably with particle size, so that the printing step can handle much smaller particles, and both self-assembly and directed assembly techniques have been reported to arrange particles down to 2 nm in diameter[15]. Many such techniques, including dry processes based on surface charging[32] and processes based on biological systems[33] could be combined with nanoparticle printing. This should enable not only the integration of much smaller objects but also the development of printing plates that can be programmed for any desired particle arrangement without requiring new templates.

METHODS

PARTICLE AND TEMPLATE PREPARATION

Metal nanocrystals were either purchased (gold colloid, 60 nm nominal diameter, British Biocell International) or synthesized through reduction from their respective salts. The polydispersed gold particles shown in Fig. 2 were obtained following a method from Brown *et al.*[34] in three steps, starting with small crystals produced according to Frens' method[35], which were used to seed the next stages. The silver nanocubes shown in Fig. 4 were obtained through polyol reduction of silver nitrate in a process adapted from Xia's group[36]. Experimental details can be found in the Supplementary Information. In brief, silver nitrate and polyvinylpyrrolidone were injected into ethylene glycol at 140 °C under acidic conditions, and 100-nm cubes were obtained after 17 h of growth. Those were then transferred into water and used for inking.

Templates for assembly were cast in PDMS using nanostructured silicon masters. See Supplementary Information, for details of the electron-beam writing and the processing of these masters. The masters were used to cast large numbers of silicone templates. The PDMS used is a variation of the commercially available product. It has a higher elastic modulus and can be patterned at higher fidelity. All templates were fabricated on backplanes cut from display glass (175 μm, Schott AG).

HIGH-PRECISION INKING

Of the many surfactants we tested, the best results were obtained with the following mixture, which was therefore adopted for all printing experiments. An aqueous 0.1 wt% emulsion of Triton X-45 (Fluka) was mixed with a 10 mM aqueous solution of sodiumdodecylsulphonate (Fluka), mixed and filtered through a PTFE 200-nm-pore syringe filter (Rotilabo). Then 50 μl of the mixture were added to 200 μl of the colloid, yielding a final particle concentration of $2 \times 10^{14}\ l^{-1}$.

The inking was performed in a homemade tool (the capillarity-assisted particle-assembly or 'CAPA' tool) described previously[16], which enables control of the temperature and substrate velocity while the template is being moved under a drop of the colloidal suspension. The substrate temperature was chosen such that a clearly visible accumulation zone formed, for example, 27 °C at the usual environmental humidity of 50%. A microscope camera recorded the contact angle of the meniscus, which was used to ensure good quality of the substrate and the colloid, indicated by a contact angle of 50–60°. During the inking, a microscope mounted on top of the setup was used to monitor and record the process.

PARTICLE TRANSFER

Particle transfer was carried out on both bare silicon surfaces (lightly doped, *p*-type, polished silicon wafers, Siltronic) and on PMMA-coated silicon surfaces (spin-coated with a 10–30 nm PMMA layer, having a molecular weight of 950 kDa, Microchem). In both cases, the printing plates with the particles were brought into contact with the substrates either manually or in a printing tool that allows microscopic inspection and alignment with existing structures as well as force and temperature control during the transfer. In the case of the uncoated substrates, the printing plate was pressed firmly onto the target and removed immediately after. In the case of the substrates with PMMA adhesion layers, the printing plate was placed on the target, the temperature was increased to 120 °C, pressure was applied on the backplane, the temperature was decreased, and the printing plate was removed as the temperature had dropped to 40 °C.

BULK COLLOID-FLOW SIMULATION

The laminar flows occurring in the inking process were simulated with a standard finite element modelling discretization of the Navier–Stokes equations using a commercial code (COMSOL, FEMLAB GmbH), in which the evaporation was calculated simultaneously to produce boundary conditions for the flow into the meniscus. (See Supplementary Information for details and results for a range of simulation parameters).

Received 24 May 2007; accepted 25 July 2007; published 2 September 2007.

References

1. Fendler, J. H. & Meldrum, F. C. The colloid chemical approach to nanostructured materials. *Adv. Mater.* **7,** 607–632 (1995).
2. Klein, D. L. *et al.* An approach to electrical studies of single nanocrystals. *Appl. Phys. Lett.* **68,** 2574–2576 (1996).
3. Maier, S. A. *et al.* Local detection of electromagnetic energy transport below the diffraction limit in metal nanoparticle plasmon waveguides. *Nature Mater.* **2,** 229–232 (2003).
4. Englebienne, P. Use of colloidal gold surface plasmon resonance peak shift to infer affinity constants from the interactions between protein antigens and antibodies specific for single or multiple epitopes. *Analyst* **123,** 1599–1603 (1998).
5. Bell, A. T. The impact of nanoscience on heterogeneous catalysis. *Science* **299,** 1688–1691 (2003).
6. Lee, T. H. & Dickson, R. M. Discrete two-terminal single nanocluster quantum optoelectronic logic operations at room temperature. *Proc. Natl Acad. Sci. USA* **100,** 3043–3046 (2003).
7. Parviz, B. A., Ryan, D. & Whitesides, G. M. Using self-assembly for the fabrication of nano-scale electronic and photonic devices. *IEEE Trans. Adv. Packag.* **26,** 233–241 (2003).
8. Ahn, J. H. *et al.* Heterogeneous three-dimensional electronics by use of printed semiconductor nanomaterials. *Science* **314,** 1754–1757 (2006).
9. Michel, B. *et al.* Printing meets lithography: Soft approaches to high-resolution printing. *IBM J. Res. Dev.* **45,** 697–719 (2001).
10. Bernard, A. *et al.* Microcontact printing of proteins. *Adv. Mater.* **12,** 1067–1070 (2000).
11. Hidber, P. C. *et al.* Microcontact printing of palladium colloids: Micron-scale patterning by electroless deposition of copper. *Langmuir* **12,** 1375–1380 (1996).
12. Larsen, N. B. *et al.* Order in microcontact printed self-assembled monolayers. *J. Am. Chem. Soc.* **119,** 3017–3026 (1997).
13. Liao, J. *et al.* Reversible formation of molecular junctions in 2D nanoparticle arrays. *Adv. Mater.* **18,** 2444–2447 (2006).
14. Fludd, R. & de Bry, J. T. Utriusque cosmi maioris scilicet et minoris metaphysica, physica atque technica historia, in duo volumina secundum cosmi differentiam diuisa. (Aere J. T. de Bry, typis H. Galleri, Oppenhemii, 1617).
15. Cui, Y. *et al.* Integration of colloidal nanocrystals into lithographically patterned devices. *Nano Lett.* **4,** 1093–1098 (2004).
16. Malaquin, L. *et al.* Controlled particle placement through convective and capillary assembly. *Langmuir* (in the press).
17. Peppin, S. S. L. & Elliott, J. A. W. Non-equilibrium thermodynamics of concentration polarization. *Adv. Colloid Interface Sci.* **92,** 1–72 (2001).
18. Schmid, H. & Michel, B. Siloxane polymers for high-resolution, high-accuracy soft lithography. *Macromolecules* **33,** 3042–3049 (2000).
19. Geissler, M. *et al.* Fabrication of metal nanowires using microcontact printing. *Langmuir* **19,** 6301–6311 (2003).
20. Kraus, T. *et al.* Closing the gap between self-assembly and microsystems using self-assembly, transfer, and integration of particles. *Adv. Mater.* **17,** 2438–2442 (2005).
21. Rimai, D. S., Ezenyilimba, M. C. & Quesnel, D. J. Effects of electrostatic and van der Waals interactions on the adhesion of spherical 7 μm particles. *J. Adhesion* **81,** 245–269 (2005).
22. Farshchi-Tabrizi, M. *et al.* On the adhesion between fine particles and nanocontacts: An atomic force microscope study. *Langmuir* **22,** 2171–2184 (2006).
23. Kendall, K. *Molecular Adhesion and its Applications: The Sticky Universe* vol. xix (Kluwer Academic/Plenum Publishers, New York, 2001).
24. Arzt, E., Gorb, S. & Spolenak, R. From micro to nano contacts in biological attachment devices. *Proc. Natl Acad. Soc. USA* **100,** 10603–10606 (2003).
25. Wagner, R. S. & Ellis, W. C. Vapor-liquid-solid mechanism of single crystal growth. *Appl. Phys. Lett.* **4,** 89–90 (1964).
26. Li, Y. *et al.* Nanowire electronic and optoelectronic devices. *Mater. Today* **9,** 18–27 (2006).
27. Hutter, E. & Fendler, J. H. Exploitation of localized surface plasmon resonance. *Adv. Mater.* **16,** 1685–1706 (2004).
28. Freeman, R. G. *et al.* Self-assembled metal colloid monolayers—an approach to sers substrates. *Science* **267,** 1629–1632 (1995).
29. Storhoff, J. J. *et al.* What controls the optical properties of DNA-linked gold nanoparticle assemblies? *J. Am. Chem. Soc.* **122,** 4640–4650 (2000).
30. Krug, J. T. *et al.* Efficient Raman enhancement and intermittent light emission observed in single gold nanocrystals. *J. Am. Chem. Soc.* **121,** 9208–9214 (1999).
31. Kipphan, H. *Handbook of Print Media* (Springer, Berlin, 2004).
32. Kim, H. *et al.* Parallel patterning of nanoparticles via electrodynamic focusing of charged aerosols. *Nature Nanotech.* **1,** 117–121 (2006).
33. Sleytr, U. B. *et al.* Crystalline bacterial cell surface layers (S layers): From supramolecular cell structure to biomimetics and nanotechnology. *Angew. Chem. Int. Edn* **38,** 1035–1054 (1999).
34. Brown, K. R., Walter, D. G. & Natan, M. J. Seeding of colloidal Au nanoparticle solutions. 2. Improved control of particle size and shape. *Chem. Mater.* **12,** 306–313 (2000).
35. Frens, G. Controlled nucleation for regulation of particle-size in monodisperse gold suspensions. *Nature Phys. Sci.* **241,** 20–22 (1973).
36. Im, S. H. *et al.* Large-scale synthesis of silver nanocubes: The role of HCl in promoting cube perfection and monodispersity. *Angew. Chem. Int. Edn* **44,** 2154–2157 (2005).

Acknowledgements

We thank U. Drechsler for her support in microfabrication as well as R. Stutz and M. Tschudy for their technical support. A part of this project was funded by the Swiss Commission for Technology and Innovation. The partial support of the State Secretariat for Education and Research (SER) in the framework of the EC-funded project NaPa (Contract No. NMP4-CT-2003-500120) is gratefully acknowledged. The content of this work is the sole responsibility of the authors. We thank P. Seidler for his continuous support.
Correspondence and requests for material should be addressed to H.W.
Supplementary information accompanies this paper on www.nature.com/naturenanotechnology.

Competing financial interests

The authors declare no competing financial interests.

ADVANCED MATERIALS
www.advmat.de

www.MaterialsViews.com

COMMUNICATION

Selective Assembly of Sub-Micrometer Polymer Particles

By *Cyrill Kuemin, K. Cathrein Huckstadt, Emanuel Lörtscher, Antje Rey, Andrea Decker, Nicholas D. Spencer,* and *Heiko Wolf**

Colloid syntheses provide us with a wealth of micro- and nanoscale particles.[1] Tight control over reaction conditions allows the engineering of the size, shape, morphology, composition, and surface chemistry of the colloidal particles synthesized, which, in turn, determine the mechanical,[2] electronic,[3] optical,[4] or biological properties[5] of the particles or their assemblies. The colloidal route thus offers broad synthetic flexibility, making such particles intriguing building blocks. Besides pushing the limits of colloid synthesis, current research efforts concomitantly investigate the efficient integration of particles from colloidal suspensions into materials and devices by self-assembly strategies.[6] Self-assembly requires finely balanced particle–particle (self-assembly per se) or particle–template interactions (directed self-assembly) for a desired particle architecture to arise from the disordered state.[7] Progress in particle self-assembly is clearly linked to the development of strategies that enable the organization of multiple different types of functional particles into arbitrarily complex, 1D, 2D, or even 3D architectures.

The present study focuses on the template-mediated, controlled placement of particles having different sizes onto surfaces by directed self-assembly. Hitherto, apart from a limited number of attempts using topographical templates,[8] mainly chemical templates have been utilized to self-assemble multiparticle patterns on surfaces. In these latter studies, specific interactions,[9] wettability,[10] or charge contrast[11] were harnessed for directing the assembly. Such assembly strategies are typically applicable to a limited subset of particles that match specific surface chemistry requirements. Here we present a more versatile approach, based on capillary assembly, to accurately place sub-micrometer-sized polymer particles of different diameters at size-selective assembly sites on dedicated topographical templates. In addition, we show that assembled particles can readily be transferred to a target substrate in a single printing step. By virtue of this approach, multiparticle patterns on surfaces can be fabricated with a degree of complexity that is difficult to achieve with other particle-assembly strategies.

In capillary assembly, a drop of colloid is moved mechanically over a topographically structured surface, and particles are assembled from the three-phase contact line of the receding meniscus (**Figure 1a**).[12] The horizontal component of the capillary forces, $F_{//}$, exerted on particles protruding through the liquid–air interface keeps particles dispersed and thus prevents non-specific deposition as the meniscus moves over flat regions. If the meniscus approaches a topographical feature, localized pinning and deformation of the meniscus redirect capillary forces so that the vertical component, $F_{\perp}$, becomes relevant. Upon capillary breakup, capillary and confinement forces consolidate and thus jointly assemble particles at appropriately designed topographical features of the template. As shown here, these forces have to be tuned by engineering the geometry of the topographical features (i.e., the assembly site) to selectively allow particles within a narrow size range to be assembled, whereas smaller and larger particles both pass by.

The feasibility of the approach described is demonstrated by the size-selective assembly of 200-nm, 350-nm, and 500-nm polystyrene particles on custom-designed elastomeric assembly templates. Figure 1b–g illustrates the three different types of size-selective assembly sites. Comprising arrangements of raised posts with appropriate spacing, they were conceived to selectively capture 200-nm, 350-nm, and 500-nm polystyrene particles. Raised posts rather than recessed features of different diameters (holes) were considered to facilitate subsequent particle transfer by printing. The assembly site for 200-nm particles (small trap, Figure 1b and 1c) was designed as an arrangement consisting of a front post and a trapping post placed aligned behind the former in the traveling direction of the meniscus. The front post will pin the approaching meniscus and serves as obstacle for larger particles, and the corner-shaped trapping post is the actual assembly site. The assembly sites for 350-nm particles (medium trap, Figure 1d and 1e) and 500-nm particles (large trap, Figure 1f and 1g) were designed as a pair of trapping posts behind a front post. To achieve selectivity for a specific particle size, the spacing between the posts of the different traps had to be adjusted, thereby tuning the forces at work in capillary assembly. For ease of template fabrication, all posts were of the same height (200 nm).

Figure 2 shows top-view scanning electron microscopy (SEM) images of real assembly templates comprising arrays of traps fabricated in high-modulus poly(dimethlysiloxane) PDMS and containing particles assembled in the traps. In capillary assembly experiments with monomodal colloids, the three trap types proved to capture the designated particles with high selectivity. Figure 2a–c shows SEM images of the templates after capillary assembly of 200-nm (P200), 350-nm (P350), and

[*] C. Kuemin, K. C. Huckstadt, Dr. E. Lörtscher, A. Rey, Dr. A. Decker, Dr. H. Wolf
IBM Research – Zurich
Säumerstrasse 4, 8803 Rüschlikon (Switzerland)
E-mail: hwo@zurich.ibm.com
C. Kuemin, K. C. Huckstadt, Prof. N. D. Spencer
Laboratory for Surface Science and Technology
Department of Materials
ETH Zurich, Wolfgang-Pauli-Strasse 10, 8093 Zürich (Switzerland)
A. Rey
Nanotechnology Group
Department of Mechanical and Process Engineering
ETH Zurich, Tannenstrasse 3, 8092 Zürich (Switzerland)

DOI: 10.1002/adma.201000396

ADVANCED MATERIALS
www.advmat.de
www.MaterialsViews.com

COMMUNICATION

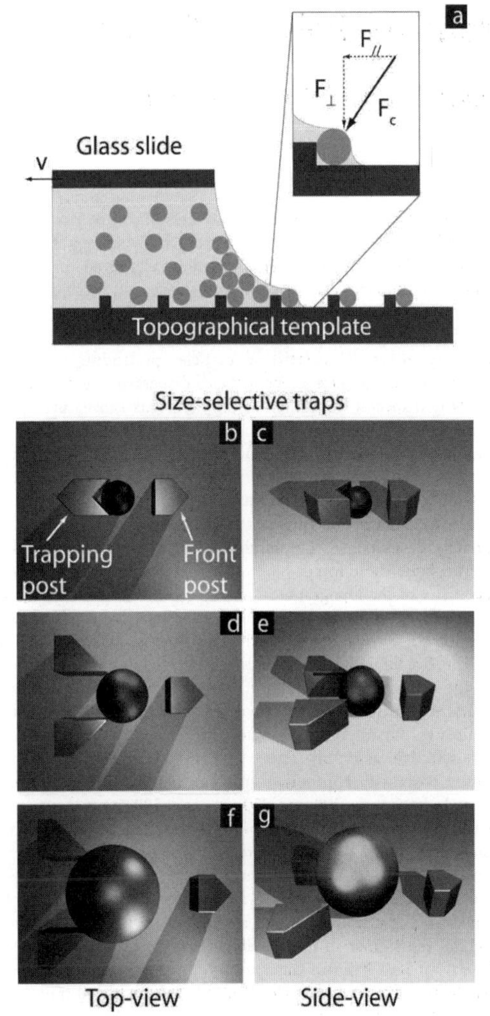

Figure 1. a) Schematic illustration of the capillary assembly process on a simplified template. A drop of colloid between a glass slide and the topographical template is moved at constant velocity ν. Capillary forces F_c ($F_{//}$ horizontal component; $F_{\perp}$: vertical component) act on a particle as the meniscus approaches a topographical feature. b–g) Size-selective assembly sites in top (left) and side view (right). Small traps (b,c) were conceived to capture 200-nm particles. Medium (d,e) and large traps (f,g) were designed for selective assembly of 350-nm and 500-nm particles, respectively. The assembly direction along which the meniscus and thus the particles move is from right to left.

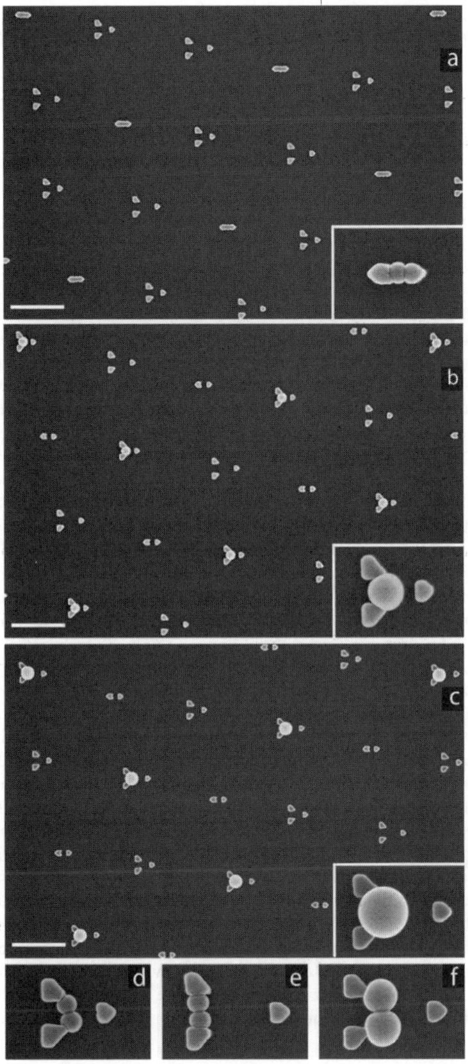

Figure 2. Particles selectively placed at designated traps on dedicated assembly templates by capillary assembly: a) 200-nm particles in small traps, b) 350-nm particles in medium traps, c) 500-nm in large traps. Traps other than the designated ones remained empty. The assembly of 200-nm particle dimers in d) medium and e) large traps, and f) 350-nm particle dimers in large traps constitute assembly errors. The assembly direction is from right to left. Samples were coated with 10 nm gold for imaging enhancement using SEM. The scale bars are 2 μm.

500-nm particles (P500). Complementary, quantitative data of these assembly experiments are displayed in the histograms in **Figure 3**, where the assembly yields $\gamma_{\text{Traptype/Particles}}$ (percentage of corresponding traps that captured a particle when colloids of the designated particles were moved over the assembly templates) and error rates $\varphi_{\text{Traptype/Particles}}$ (percentage of corresponding traps that captured a particle when colloids of non-designated particles were moved over the templates, differentiating between the two non-designated particle sizes) are plotted separately for small, medium, and large traps. The histograms

COMMUNICATION

ADVANCED MATERIALS
www.advmat.de

www.MaterialsViews.com

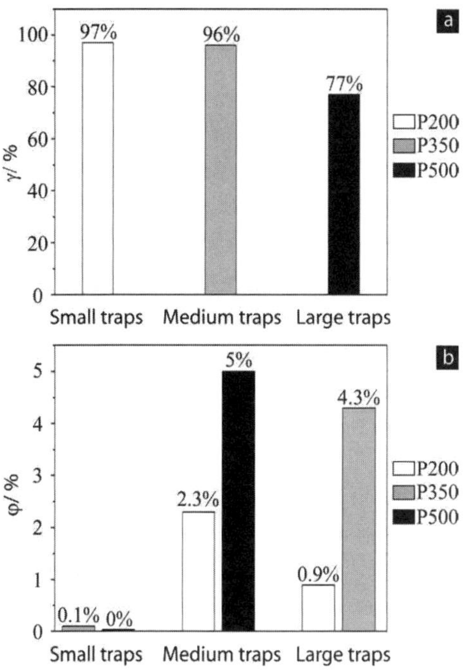

Figure 3. a) Yields γ and b) error rates φ of capillary assembly experiments with monomodal colloids plotted versus the trap type. Note the different scales of the y axes (a: 100%, b: 5%). 200-nm (P200), 350-nm (P350), and 500-nm particles (P500) were predominantly assembled in their corresponding trap types.

demonstrate that the three particle types were indeed assembled with good yields and low error rates, i.e., predominantly in their designated traps. As discussed below, small changes in the trap geometry, such as of the post shape, spacing between posts, or post height, were found to strongly affect the yield and error rates because of the geometry-dependent capillary and confinement forces that govern the assembly.

Owing to the low surface energy of PDMS (~22 mJ m^{-2}), the colloidal suspensions only partially wetted the template surfaces.[13] This resulted in receding wetting angles of 50°–55°, at which the lateral capillary forces (horizontal component) experienced by particles at the three-phase contact line were sufficiently strong to preclude non-specific deposition on the flat parts of the template surface. Thus, the meniscus acted as a doctor blade, dragging particles along. In this way, 200-nm particles could be assembled from the corresponding colloid with very high yields $\gamma_{SmallTraps/P200} = 97 \pm 1\%$, with the yield limited mainly by defects in the template. The highest selectivity for 200-nm particles was obtained with small traps in which the back of the front post and the foremost edge of the trapping post were separated by 205 ± 5 nm. The narrow gap guaranteed that exclusively 200-nm particles were assembled. Vertical capillary forces induced by pinning of the meniscus at the front post as well as confinement of the particles by the trapping post and by adjacent particles (particles near the three-phase contact line were densely packed) ensured assembly with both high yield and accurate placement in a self-limiting manner (Figure 2a). Tapering of the posts, thus avoiding post side walls that are parallel to the meniscus, eliminated deposition of 200-nm particles at the outer side of the traps. Moreover, 350-nm and 500-nm particles were dragged around small traps, resulting in very low error rates ($\varphi_{SmallTraps/P350} = 0.1\%$, $\varphi_{SmallTraps/P500} = 0\%$). As evidenced by SEM imaging, particles that were assembled in the small traps had diameters that were significantly smaller than the certified mean diameters (350 and 500 nm).

Medium traps with posts arranged as shown in Figure 2c and 2d (separation between trapping posts: 285 ± 5 nm; separation between trapping and front post: 325 ± 5 nm) captured 350-nm particles with the highest selectivity. No individual 200-nm particles were assembled in medium traps because they experienced confinement forces only transiently while bouncing against the posts: they were effectively being dragged along by lateral capillary forces, leaving the traps empty after capillary breakup. The error rate, $\varphi_{MediumTraps/P200} = 2.3\%$, resulted mainly from the assembly of particle dimers (Figure 2d). Those were formed directly in the traps from two particles that mutually hindered each other from passing through the gap. The high yields achieved, $\gamma_{MediumTraps/P350} = 96 \pm 1\%$, suggest that once 350-nm particles had been dragged into the medium traps by the meniscus, confinement forces exerted by the tips of the two trapping posts were sufficient to counter the lateral capillary forces. The vertical component of the capillary forces induced by pinning at the front post kept particles from rolling over the traps. Particles with an effective diameter of 500 nm were not assembled in medium traps. The error rate, $\varphi_{MediumTraps/P500} = 5.2\%$, results from the limited monodispersity of the colloid; captured particles were significantly smaller than the specified average diameter of 500 nm.

Large traps captured 500-nm particles with an assembly yield of $\gamma_{LargeTraps/P500} = 77 \pm 5\%$ (Figure 2c). Interestingly, large traps consisting solely of two trapping posts without a front post did not trap any 500-nm particles at all. This finding highlights the importance of a front post to induce pinning. Analogous to the case discussed for medium traps, 200-nm and 350-nm particles that were assembled as dimers (Figure 2e and 2f) accounted for the error rates $\varphi_{LargeTraps/P200} = 0.9\%$ and $\varphi_{LargeTraps/P350} = 4.3\%$. No individual 200- or 350-nm particles were captured because of the weak confinement they experienced as they were being dragged through the large traps. From the range of post spacings tested, we found that large traps were most selective for 500-nm particles when the two trapping posts were placed 360 ± 5 nm apart. Larger spacings reduced the assembly yield $\gamma_{LargeTraps/P500}$ and increased the error rate $\varphi_{LargeTraps/P350}$ (more dimers).

The assembly yield of $\gamma_{LargeTraps/P500} = 77 \pm 5\%$ for 500-nm particles is significantly lower than those for 200-nm and 350-nm particles. This result may be ascribed to two main effects: i) At the contact line, 500-nm particles may roll over traps because of the low posts height (200 nm), which in this case is much smaller than the particle diameter. This is especially true if defects are present at the traps. However, increasing the post height to counter this effect was not an option because even at a height of 230 nm, many 200-nm particles were found to assemble non-specifically at small and medium traps, leading to a loss of selectivity and placement accuracy. ii) Particles smaller than the specified average size (500 nm) may preferentially accumulate at the contact line because the meniscus constrains larger particles more strongly than smaller ones. Thus, smaller particles occupy positions that are favorable for assembly but without getting assembled, whereas particles with an effective diameter of 500 nm or larger remain at the back, where vertical capillary forces are insufficient to induce assembly. To test the latter hypothesis, colloid mixtures of 500/400-nm particles with a number ratio of 98:2 were prepared and moved over an array with large traps. Whereas 400-nm particles were too small to be assembled in large traps, the presence of as little as 2% of 400-nm particles in the mixture reduced the yield of 500-nm particles by as much as 17% (from 77% to 60%), which strongly supports hypothesis (ii).

The accumulation of smaller particles at the contact line probably also affected the error rates $\varphi_{SmallTraps/P350}$ and $\varphi_{MediumTraps/P500}$. Furthermore, this effect became manifest in the results of assembly experiments with bimodal colloid mixtures of 200/350-nm particles, simultaneously assembling two particle types at a time. For mixtures with a 1:1 number ratio, yields were significantly smaller ($\gamma_{SmallTraps/P200} < 90\%$, $\gamma_{MediumTraps/P350} < 50\%$) than in assembly experiments with monomodal colloids (cf. data in histograms), in particular for the larger particle type in the mixture. In this case, the favorable positions for assembly at the very front of the meniscus were occupied to a greater extent by 200-nm particles. The same trend was observed in assembly experiments with bimodal colloid mixtures of 350/500-nm. Generally, the presence of a second population of particles adversely affects the assembly of the targeted population.

Sequential assembly of particles from monomodal colloids on a single template enabled the fabrication of multiparticle patterns with assembly yields and error rates in the same range as presented in Figure 3 (showing one assembled particle type per template). This finding implies that previously assembled particles remained in place when a second colloid was moved over the same template. Furthermore, surface properties of the templates were not altered significantly in the course of previous assembly runs, for example, by surfactant deposition, which in turn could have led to enhanced wetting and thus convective assembly.[12c] **Figure 4a** shows an SEM image of sequentially assembled particles (a larger area view is available in the Supporting Information). Note that 200-nm particles were assembled first, followed by 350-nm and 500-nm particles, in succession. This order was imposed by the fact that assembled 350-nm particles in the medium traps as well as 500-nm particles in large traps would act as assembly sites for 200-nm particles, thus, rendering the traps non-selective.

Potential applications of a size-selective assembly process may require a multiparticle pattern to be integrated on substrates other than the PDMS template used here for the assembly. To accommodate that requirement, microcontact printing was used to transfer multiparticle arrays to a silicon model surface.[14] Figure 4b shows a detail of a multiparticle pattern printed on a plain silicon surface. Inspection of the assembly template after printing confirmed high transfer yield across the entire 250 μm × 250 μm large array area, leading to the conclusion that a majority of the assembled particles, irrespective of their size, was successfully transferred. Although 500-nm particles came into contact with the substrate first, the elastomeric nature of PDMS also enabled 350-nm and 200-nm particles to establish contact. In addition, partial deformation of the 500-nm and 350-nm particles due to the fact that printing was done at 90 °C substrate temperature certainly also facilitated the transfer. Note that the regularity of the array was not altered by the printing step. In traditional printing techniques, different components are typically printed sequentially. Whereas sequential printing faces issues related to the alignment of the individual components, in the present study the mutual alignment of the three individual particle arrays was defined by the trap pattern fabricated with high precision by lithography.

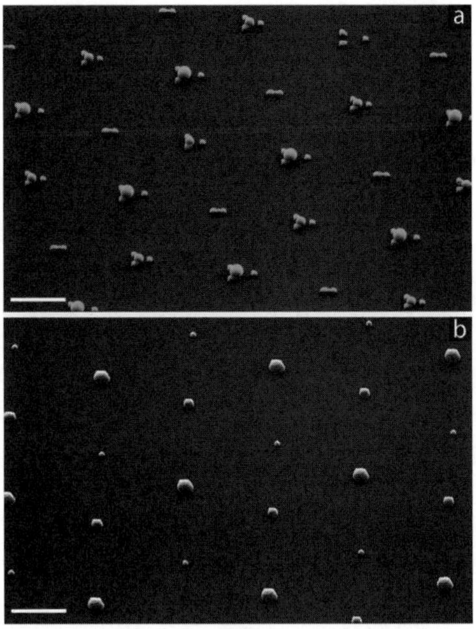

Figure 4. Sequentially assembled particles in their designated traps (a) and an array of the same particles printed on plain silicon (b). All particles, including the smallest, were transferred. The flat top of the particles is due to partial deformation during printing. Samples were coated with 10 nm gold for imaging enhancement using SEM. The scale bars are 2 μm.

This Communication presents a scalable approach to create multiparticle patterns in an efficient manner. By harnessing particle-size-dependent forces acting at the menisci of colloids, capillary assembly was proposed for the patterning of three differently sized polymer particles on one template, which could then be transferred by a single printing step. The feasibility of this approach was demonstrated with polystyrene particles of 200, 350, and 500 nm diameter. Because polymer particles are available in a large size range with different chemical compositions, colors, and also various surface chemical (or biological) functionalities, this particle-patterning process is a versatile approach for the fabrication of functional multiparticle patterns. Colloidal lithography, the fabrication of bead-based sensor platforms, security

ADVANCED MATERIALS
www.advmat.de
www.MaterialsViews.com

COMMUNICATION

features, and photonic devices are examples of technologies that can benefit from the particle-patterning method presented.

Experimental Section

Template Fabrication: Topographical templates were fabricated by molding liquid poly(dimethylsiloxane) (PDMS) against patterned silicon masters. Masters were fabricated by electron-beam lithography and deep reactive ion etching. First, the footprints of the small, medium, and large traps and their arrangement into arrays were defined in a design layout. The design comprised four separate arrays, one for each trap type and one with all three trap types placed adjacent to each other. Each array was 250 μm × 250 μm in size, comprising a total of 5400 traps. For patterning, a 220 nm thick polymethylmethacrylate (PMMA) film (950 kDa, 4% in anisol, microresist technology, Berlin, Germany) was spun onto a 4 in. *n*-doped silicon wafer and baked at 185 °C for 90 s. This film was exposed in an electron-beam lithography system (e line system from Raith GmbH, Dortmund, Germany) equipped with a Gemini electron column (gun voltage 20 kV, aperture 7.5 μm, nominal area dose 240 $\mu C\ cm^{-2}$). The area dose at which the traps were exposed was locally adjusted to correct for proximity effects. Development was done in a mixture of methylisobutylketone and isopropyl alcohol (1:3 by volume) for 55 s in an ultrasonic bath, followed by a 200 W oxygen plasma etch for 7 s (TePla plasma reactor, TePla AG, Asslar, Germany) to remove any residual PMMA at the bottom of the exposed regions. Pattern transfer into silicon was done by deep reactive ion etching with a plasma etcher (AMS200, Alcatel, Annecy, France). The PMMA mask was subsequently dissolved in acetone in an ultrasonic bath, followed by a rinsing step in isopropyl alcohol and exposure to oxygen plasma (200 W, 60 s, 1 mbar; 1 mbar = 100 Pa). Patterned silicon masters were rendered non-sticking by adsorption of 1*H*,1*H*,2*H*,2*H*-perfluorodecyltrichlorosilane (97 %, ABCR, Karlsruhe, Germany) from the gas phase in a desiccator (2 min, 25 mbar). They were then placed in an oven (80 °C, 50 mbar, 1 h) to ensure completion of the monolayer formation. Molding of PDMS was carried out as previously detailed.[15] Cured PDMS templates were immersed in ethanol for at least one day to extract low-molecular-weight silicone oils. Finally, templates were dried at 20 mbar for 1 h directly prior to use in assembly experiments.

Particle Assembly: Aqueous suspensions of like-charged polystyrene particles (Nanosphere size standards, 1% solids content, containing trace amounts of surfactants; certified mean diameters and coefficient of variations: 200 ± 4 nm/1.7%, 350 ± 6 nm/1.5%, 400 ± 5 nm/1.8%, and 498 ± 9 nm/1.6%), were purchased from Thermo Fisher Scientific Inc. (Waltham, MA, USA). Capillary assembly experiments were performed using a dedicated setup, as described previously.[14] For each assembly experiment, colloid (typically 50 μL) was injected between the assembly template and a stationary glass slide mounted approximately 500 μm above the assembly template. The colloid temperature was set by a Peltier element to approximately 15 °C above the dew point. At these temperatures, particles accumulated within a few minutes to form a high-concentration zone at the meniscus with an iridescent appearance, indicative of dense packing. The meniscus was moved over an assembly template with the help of a linear translation stage at a velocity of 0.5–1 $\mu m\ s^{-1}$. Image-analysis software (National Instruments Vision Assistant) with a custom-written script based on intensity thresholding of binarized dark-field optical micrograph images (recorded at the center of arrays with assembled particles, magnification: 100x) was used to determine the yield γ and error rate φ of the assemblies. Four assembly experiments per colloid were made to obtain the data shown in Figure 3. Yields γ and their standard deviations were rounded to whole percentage values. Error rates φ were rounded to one decimal point of a percent, thus avoiding that values between 0.05 and 0.5% were rounded to 0%.

Particle Printing: To transfer particles, a dedicated printing tool was employed.[14] Oxygen-plasma-cleaned plain silicon wafers were used as print substrates. The templates supporting the assembled multiparticle arrays were brought into conformal contact (pressure: ~1 bar) with the substrate (at 90 °C), followed by lifting off of the template.

Acknowledgements

Partial financial support from the Swiss National Science Foundation (National Research Program 47 "Supramolecular Functional Materials") is gratefully acknowledged. The authors thank Richard Stutz and Ute Drechsler for their assistance and Michel Despont and Walter Riess for their continuous support.

Supporting Information

Supporting Information is available online from Wiley InterScience or from the author.

Received: February 2, 2010
Revised: March 16, 2010
Published online: April 28, 2010

[1] a) S. E. Skrabalak, Y. Xia, *ACS Nano* **2009**, *3*, 10; b) D. Dendukuri, P. S. Doyle, *Adv. Mater.* **2009**, *21*, 1.
[2] L. J. Bonderer, A. S. Studart, L. J. Gauckler, *Science* **2008**, *319*, 1069.
[3] D. L. Klein, R. Roth, A. K. L. Lim, A. P. Alivisatos, P. L. McEuen, *Nature* **1998**, *389*, 699.
[4] X. Wang, X. Ren, K. Kahen, M. A. Hahn, M. Rajeswaran, S. Maccagnano-Zacher, J. Silcox, G. E. Cragg, A. L. Efros, T. D. Krauss, *Nature* **2009**, *459*, 686.
[5] K. L. Michael, L. C. Taylor, S. L. Schultz, D. R. Walt, *Anal. Chem.* **1998**, *70*, 1242.
[6] a) S. Mann, *Nat. Mater.* **2009**, *8*, 781; b) Y. Zhao, K. Thorkelsson, A. J. Mastroianni, T. Schilling, J. M. Luther, B. J. Rancatore, K. Matsunaga, H. Jinnai, Y. Wu, D. Poulsen, J. M. J. Fréchet, A. P. Alivisatos, T. Xu, *Nat. Mater.* **2009**, *8*, 979; c) A. N. Shipway, E. Katz, I. Willner, *Chem. Phys. Chem.* **2000**, *1*, 18; d) O. D. Velev, S. Gupta, *Adv. Mater.* **2009**, *21*, 1.
[7] a) Y. Min, M. Akbulut, K. Kristiansen, Y. Golan, J. Israelachvili, *Nat. Mater.* **2008**, *7*, 5277; b) B. A. Grzybowski, C. E. Wilmer, J. Kim, K. P. Browne, K. J. M. Bishop, *Soft Matter* **2009**, *5*, 1110.
[8] a) B. Varghese, F. C. Cheong, S. Sindhu, T. Yu, C. Lim, S. Valiyaveettil, C. Sow, *Langmuir* **2006**, *22*, 8248; b) Y. Yin, Y. Lu, B. Gates, Y. Xia, *J. Am. Chem. Soc.* **2001**, *123*, 8718.
[9] U. Plutowski, S. S. Jester, S. Lenhert, M. M. Kappes, C. Richert, *Adv. Mater.* **2007**, *19*, 1951.
[10] F. Fan, K. J. Stebe, *Langmuir* **2005**, *21*, 1149.
[11] a) I. Lee, H. Zheng, M. F. Rubner, P. T. Hammond, *Adv. Mater.* **2002**, *14*, 569; b) G. Gotesman, R. Naaman, *Langmuir* **2008**, *24*, 5981.
[12] a) Y. Yin, Y. Lu, B. Gates, Y. Xia, *J. Am. Chem. Soc.* **2001**, *123*, 8718; b) Y. Cui, M. T. Björk, J. Liddle, C. Sönnichsen, B. Boussert, A. P. Alivisatos, *Nano Lett.* **2004**, *4*, 1093; c) L. Malaquin, T. Kraus, H. Schmid, E. Delamarche, H. Wolf, *Langmuir* **2007**, *23*, 11513; d) M. Rycenga, P. H. C. Camargo, Y. Xia, *Soft Matter* **2009**, *5*, 1129.
[13] C. J. Drummond, D. Y. C. Chan, *Langmuir* **1997**, *13*, 3890.
[14] T. Krauss, L. Malaquin, H. Schmid, W. Riess, N. D. Spencer, H. Wolf, *Nat. Nanotech.* **2007**, *2*, 570.
[15] M. Geissler, H. Wolf, R. Stutz, E. Delamarche, U. Grummt, B. Michel, A. Bietsch, *Langmuir* **2003**, *19*, 6301.

CHAPTER 5
Methods for Characterizing Surface Modifications

5a. Roughness characterization

Commentary

Roughness is a concept that we think we all understand, but it is notoriously difficult to describe in a quantitative way. We have found that a very valuable approach is to take the Fourier transform of roughness profiles, to filter the transform between specific wavelength ranges and then to transform it back and calculate roughness values, such as R_a and R_q in the normal way. This wavelength-dependent roughness is then much more useful than the normally used integral parameters, since specific effects of roughness in fields such as biology or tribology are typically wavelength dependent. Moreover, the discrete wavelength-dependent roughness effects of individual steps (such as polishing or lacquering) on a total process can be usefully monitored in this way (5.1).

ELSEVIER

Wear 237 (2000) 231–252

WEAR

www.elsevier.com/locate/wear

Wavelength-dependent measurement and evaluation of surface topographies: application of a new concept of window roughness and surface transfer function

M. Wieland [a,b], P. Hänggi [c], W. Hotz [c], M. Textor [b], B.A. Keller [a], N.D. Spencer [b,*]

[a] *Swiss Federal Laboratories for Materials Testing and Research (EMPA), CH-8600 Dübendorf, Switzerland*

[b] *Department of Materials, Laboratory for Surface Science and Technology, ETH Zürich, CH-8092 Zürich, Switzerland*

[c] *Alusuisse Technology and Management AG, CH-8212 Neuhausen am Rheinfall, Switzerland*

Received 6 December 1998; received in revised form 23 July 1999; accepted 25 October 1999

Abstract

The technological performance of surfaces in fields such as tribology, biocompatibility and optics is often highly dependent on surface topography and roughness. The common practice of applying 'integral' roughness parameters, however, is often an incomplete and unsatisfactory way to describing surface topographies. Wavelength-dependent roughness evaluation is shown to be a successful method for the description of surface topographies in various characteristic roughness ranges, as well as being a useful indicator of the effect of surface treatment processes. This study examined the effects of common cut-off (COF) (high- and low-frequency) and average filtering (AFT) as well as fast Fourier transformation (FFT) techniques, as applied to synthetic and experimental profiles. Furthermore, the relationship between the roughness value R_q and the amplitudes C_n of the FFT power spectrum is demonstrated. To characterise a particular surface treatment process consisting of several consecutive processes, a surface treatment transfer function is defined using individual FFT coefficients C_n^x for each surface treatment step. To illustrate the application for industrial surfaces, two-dimensional (2-D) profiles on a micromachined steel surface (calibration sample) as well as on lacquered car body sheet and titanium implant surfaces were measured with a non-contact laser profilometer (LPM) and evaluated. FFT is shown to be a powerful method for the calculation of the wavelength-dependent roughness, as well as for the back-transformation of partial data sets into profiles in predefined wavelength ranges (''window roughness''). The more conventional filter methods give similar results, but some information are lost due to the fact that they do not correspond to a set of orthonormal functions. The concept of wavelength-dependent (''window'') roughness parameters is demonstrated to be a concept that allows for a much more detailed description of surface topographical properties with two main merits: (a) it allows macroscopic physico-technological properties to be correlated with roughness contributions in selected wavelength ranges; (b) the overall effect of consecutive surface treatment processes can be separated into wavelength-dependent contributions from each treatment step. The concept is demonstrated for consecutive surface treatment processes in the pretreatment and lacquering of aluminium car body sheet and for blasting and etching processes in the case of titanium medical implants.

Keywords: Surface engineering; Surface roughness; Surface topography; Fast Fourier transformation; Low- and high-frequency filter; Average filter; Surface transfer function; Surface wavelength-dependent roughness; Window roughness; Non-contact laser profilometer

1. Introduction

The properties of the surfaces of materials are generally described in terms of chemical composition, morphology and topography. Although all three surface aspects are important for the quality and functionality of materials and products, the influence of topographical properties are frequently underestimated. They are of particular relevance to applications such as formability, adhesion, tribology (wear and lubrication), optics and biocompatibility [1].

Two-dimensional (2-D) surface roughness parameters are often divided into three groups based on the characteristics of the surface that they quantify [1–3] (Table 1). (1) Amplitude parameters: these are solely height descriptive, for examples: R_a, R_q, R_{max} or R_{zDIN}. (2) Spacing parameters: these describe the spacing between the topographical

* Corresponding author. Tel.: +41-1-632-5850; telefax: +41-1-633-1027; e-mail: nspencer@surface.mat.ethz.ch

0043-1648/00/$ - see front matter
PII: S0043-1648(99)00347-6

232 M. Wieland et al. / Wear 237 (2000) 231–252

Table 1
Definition and description of the roughness parameters

Roughness parameter	Definition	Description
Amplitude parameters		
R_a	$R_a = (1/n)\Sigma_{i=1}^{n}\lvert y_i\rvert$	· the arithmetic average of the absolute values of all points of the profile; also called CLA (centre line average height)
R_q	$R_q = \sqrt{(1/n)\Sigma_{i=1}^{n} y_i^2}$	· the root mean square of the values of all points of the profile; also called RMS (root mean square)
R_{zDIN}	$R_{zDIN} = (1/5)\Sigma_{i=1}^{5} Y_i$	· the arithmetic average of the maximum peak to valley height of the roughness values Y_1 to Y_5 of five consecutive sampling sections over the filtered profile
R_{max}		· the maximum individual roughness depth encountered as determining R_{zDIN}
Spacing parameters		
S_m	$S_m = (1/n)\Sigma_{i=1}^{n} S_i$	· arithmetic average spacing between the falling flanks of peaks on the mean line
Hybrid parameters		
S_k	$S_k = (1/n)\Sigma_{i=1}^{n}(y_i^3/R_q^3)$	· amplitude distribution skew
K	$K = (1/n)\Sigma_{i=1}^{n}(y_i^4/R_q^4)$	· kurtosis is the comparison of the profile with a Gaussian amplitude and characterises the density as 'smooth' or 'peaky'
L_r	$L_r = L_0/L_m$	· the relation of the stretched length of the profile L_0 to the scanned length L_m

irregularities, for example: S_m. (3) Hybrid parameters: these include information about height as well as space, for example: S_k, K or L_r.

Roughness parameters are scale-dependent; the values will depend on the measurement scale and the sampling interval. Sayles and Thomas [4] demonstrated that the variance of the height distribution or R_q of a profile is linearly related to the measured distance along the surface. Furthermore, it is well-known that many types of surfaces used in engineering practice have a random structure. Such surface topographies can completely be defined by two characteristics: one related to the roughness height distribution or amplitude of the waveform and the other related to the spacing or wavelength [5,6]. Functions, which give information related to wavelength are the power spectra or the autocorrelation function of the surface whether periodic or random [3–7]; both are based on the Fourier transform. In recent years, a lot of effort has gone into investigating alternatives to the Fourier transform. The new transforms introduced fall into two main categories: (1) faster and simpler ways of producing comparable results; (2) transforms giving more information. In the first class there are the Walsh [8,9], Hartley [10,11], Hadamard [10–12], and Haar [13] transformation. They are all orthogonal transforms. In the second case there are the Wigner [14–16], ambiguity [14,16,17], wavelet [18] and Gabor [19,20] transforms, which are space-frequency transforms. Whitehouse [21] discusses the algorithm, advantages and disadvantages for all transformations. The discrete one-dimensional (1-D) Hartley transform offers advantages over the fast Fourier transformation (FFT) for numerical spectral analysis and therefore has great potential in communications, but there has been difficulty in carrying the advantage over to more than one dimension. Mulvaney [22] came to the general conclusion that after all the Fourier transform is best for surface characterisation. Although the orthogonal binary transforms such as Walsh, Haar, and so forth were faster they did not give spectral estimates which converged rapidly as a function of record length. The phase-invariant ones, such as Fourier, were slower to evaluate yet converged quickly. Also there is the point that the Fourier transform is very well-known and understood and is standard in many existing instruments. Displacing it would pose many educational and usage problems.

The wavelength-dependence can also be shown using cut-off filtering in the form of an infinitely sharp cut-off or one-stage or two-stage RC or Gauss filters [3,6], which are used in surface metrology for separating the waviness from the roughness of the surface. A complete overview on different kinds of filters has been given by Whitehouse [3].

Fractal analysis is another method used to characterise surface topographies. The essential difference to the approach discussed in this paper is the fact that while fractal analysis is a *scale-independent* evaluation technique, here, it is precisely the *scale-dependence* (the scale being a spatial frequency or wavelength dimension) that is of prime interest. For surfaces of fractal nature [23], a single *scale-independent* parameter — the fractal dimension — can be defined. Since our main interest is the extraction of wavelength-dependent topographical data from experimental profiles and the correlation of roughness data from different wavelength ranges with physico-chemical surface properties, the single *scale-independent* describing using the fractal analysis is not our preferred approach. Since the introduction of the concept of fractal dimensionality, by Mandelbrot [23,24], a large number of analytical strategies have been developed to allow the measurement of fractal dimensions [25–29].

There are three major difficulties related to the experimental measurement and evaluation of surface topographies.

(1) The experimental measurement of topographies across the whole scale-size of interest, often extending from the nanometer to the millimeter range, is generally impaired by distortion of the true surface profiles [30] (e.g., envelope system [31], limited lateral and/or vertical resolution and accessibility of surface features) and by artefacts (e.g., deformation of surface in contact-type measurements [32], optical artefacts in laser non-contact type measurements because of microgeometry, inclination and reflectivity of the surface [33,34]). The conclusion is that often a combination of different experimental techniques with different advantages/limitations is mandatory in order to get the best possible data [35].

(2) 2-D surface profiles $Z(x)$, although easier to determine, often do not adequately reflect the technological surface properties. Examples are found in the areas of formability and lubrication, where the presence of pit-like cavities favour the presence of lubricant-filled pockets that enhance lubrication [36]. Three-dimensional (3-D) surface data $Z(x, y)$ may therefore be a more realistic approach to describe topography-related surface functionality [2].

(3) The evaluation of topographical data in terms of standardised 'integral' amplitude roughness parameters such as R_a, R_q or R_{zDIN} is often of limited value for the description of real surfaces [37]. For example, two technical surfaces behaving very differently in a given situation may have the same R_a value [1]. In addition on random surfaces, fine-roughness features in the low micron- or submicron-range that may be important for performance in a given application, are often hidden by the coarser contributions to roughness. This is often a major limitation in conventional topography evaluation, and therefore a general interest exists in describing surface profiles with wavelength-dependent functions [3–7], rather than with integral topographical parameters.

The present paper focuses on this third aspect, i.e., a comparison of common COF (high- and low-frequency) and average filtering techniques (AFT) [38] as well as FFT techniques [39] to extract wavelength-dependent surface topography information from 2-D profiles $Z(x)$ in different wavelength ranges. The obtained roughness of these different ranges will be called "window roughness". For illustrative purposes, 'synthetic' surface profiles are comparatively evaluated by both mathematical procedures. Experimentally determined profiles are treated to illustrate applications on a micromachined calibration sample as well as for a lacquer system on car body sheets, and for titanium implant surfaces. Another application of the FFT is the surface treatment transfer function, which allows the effects of each consecutive surface treatment step to be separated. In subsequent papers, the extension of this formalism to 3-D surface data $Z(x, y)$ and a comparative study of different experimental methods for topography characterisation will be reported.

2. Material surfaces and experimental methods

2.1. Material surfaces

The surfaces listed in Table 2 served as calibration and reference samples to illustrate the application of the concepts described above to the areas of lacquered car body sheets and titanium implant surfaces.

The micromachined steel surface is a calibration sample (Rauhnormal RNDH3 Series No.: 0325) produced by Thyssen Hommelwerke GmbH (German calibration service), Germany. The characteristic roughness parameters are: $R_{max} = 9.99$ μm, $R_{zDIN} = 9.87$ μm and $R_a = 3.049$ μm, based on a cut-off of 0.8 mm.

The aluminium alloy Ac-120 samples of 50 × 200 mm gauge were produced with a EDT (electron discharge texturing) surface finish by Alusuisse Sierre AG, Sierre, Switzerland [40]. The surface topography was evaluated following each of four fabrication steps: (a) AR, (b) pretreated in a film-forming zinc phosphate process (PP), (c) after cataphoretic lacquer BC and (d) after application of the TC lacquer.

The cpTi surfaces were investigated in the form of discs, 15 mm in diameter and 1 mm in thickness. A stamping procedure was used to produce the Ti discs out of a grade 2 cpTi sheet (ASTM F67) in an annealed condition. Afterwards, one series was grit-blasted with alumina beads (B) under industrial particle-blasting conditions (average particle size: 250 μm), a second series was etched (E) in a solution of HCl/H_2SO_4 and a third

Table 2
List of the materials and surfaces investigated using a non-contact laser profilometer (LPM)

Application	Base material	Surface treatment
Calibration sample for reference measurements	steel	mechanically micromachined
Car body sheet surfaces after pretreatment and lacquering	aluminum alloy AlMgSi (Ac-120)	as-rolled (AR)
		after phosphate pretreatment (PP)
		with lacquer base coat (BC)
		with lacquer top coat (TC)
Titanium implant surfaces	commercially pure titanium (cpTi grade 2)	blasted (B)
		etched (E)
		blasted + etched (BE)

combination series (BE). The cpTi samples were surface-treated by Institut Straumann, Waldenburg, Switzerland.

2.2. Methods

2.2.1. Non-contact laser profilometry

Experimental 2-D roughness profiles were investigated with a non-contact LPM developed by UBM Messtechnik GmbH, Ettlingen, Germany, using a Microfocus sensor (vertical resolution 10 nm) [34]. The nominal lateral and vertical resolution of the system are 1 μm and 50 nm, respectively. Line profiles were obtained over a distance of 4.096 mm for all cpTi surfaces and the mechanically micromachined surface, with a lateral resolution of 1000 measurement points/mm. The cpTi surfaces were also measured according to DIN 4768 [41] (5.6 mm, 500 points/mm for surface E and 12.5 mm, 150 points/mm for surfaces B and BE). Line profiles of the aluminium car body sheet surfaces were measured after each coating step over a distance of 48.0 mm, with a lateral resolution of 250 points/mm.

2.2.2. Scanning electron microscopy (SEM)

The topographies of the micromachined calibration sample, the B, E and BE surfaces were investigated by scanning electron microscopy (Philips XL30).

2.3. Calculation procedures

COF, AFT and the calculation of standard roughness parameters were applied within the software provided with the LPM (UBM version 1.5). For standard roughness calculations a Gauss filter and an attenuation factor of 50% at the cut-off wavelength were chosen. The cut-off was set at 1/7 of the scanned profile (according to DIN 4768 and DIN 4777). The calculation of the roughness parameters after different cut-off settings, as well as average filtering steps were also carried out with a Gauss filter and an attenuation factor of 50%. The FFT and the subsequent ''window roughness'' calculations were performed using the software program 'Maple' (version Maple V Release 5). The profiles obtained by the LPM were exported as ASCII files and read in by 'Maple'. The profiles calculated using the ''window roughness'' concept were again imported in the UBM software and the roughness parameters calculated.

3. Theoretical background

The basic aim of this section is to outline and compare procedures for the evaluation of wavelength-dependent surface topography data. The introduction of an additional dimension (the scale, being a spatial frequency or wavelength dimension) requires a certain amount of computational effort, but it enables a particular surface topography to be described in more detail and wavelength-dependent surface changes taking place following a specific surface treatment such as etching, blasting, coating, etc., to be quantified.

Three different computational methods, the *FFT*, the *COF* and the *AFT*, are outlined and discussed in the following sections.

3.1. Fourier transformation procedure

Following a procedure discussed in detail by Beck and Gutsch [42], the profile $Z(x)$ can be written as a Fourier series expansion of periodic sine and cosine functions

$$F(x/L) = \frac{A_0}{2} + \sum_{n=1}^{N/2} \left[A_n \sin\left(f_n \frac{x}{L} \right) + B_n \cos\left(f_n \frac{x}{L} \right) \right] \tag{1}$$

where N is the number of points within the profile and L is the profile length normalised to unity and therefore dimensionless. Alternatively, by introducing phase-shifts φ_n for each wave considered:

$$F(x/L) = \frac{A_0}{2} + \sum_{n=1}^{N/2} \left[C_n \sin\left(f_n \frac{x}{L} + \varphi_n \right) \right] \tag{2}$$

where $C_n = \sqrt{A_n^2 + B_n^2}$ and $A_n/B_n = \tan(\varphi_n)$ in Eq. (2) are the amplitude- and phase-shift of each periodic sine function, respectively. The coefficients C_n correspond to the amplitude contribution of the n-th periodic wave to the surface profile and represent the normalised wavelength-dependent contributions to the profile. The spatial frequencies f_n are linked to the wavelengths λ_n through the Eq. (3):

$$\frac{1}{f_n} = \frac{\lambda_n}{2\pi} \quad \text{or dimensionless:} \quad \frac{L}{f_n} = \frac{\lambda_n^*}{2\pi} \tag{3}$$

with λ_n^* as the normalised wavelength. The transformations are mathematically strictly correct only for line profiles of infinite extension and of infinitesimal step size. In real situations (finite profile length and resolution), limits to the choice of λ_n, as well as edge boundary conditions, have to be observed.

1. According to the Nyquist theorem [43], the smallest λ_n (highest spatial frequency) that can be defined corresponds to twice the lateral distance between experimental points ('step size'). It is however safer to use four points for the smallest wavelength of interest.

Fig. 1. (a) First part of a synthetic profile based on four sinus waves with different wavelengths, phase-shifts and amplitudes. (b) The corresponding square of the FFT power spectrum $C_n^2 = A_n^2 + B_n^2$ vs. the logarithm of the profile wavelength. (c) R_q^2 as a function of the profile wavelength λ. According to Eq. (6), R_q^2 of the profile corresponds to the sum of the individual roughness contributions R_{qi}^2 from the different resolution steps.

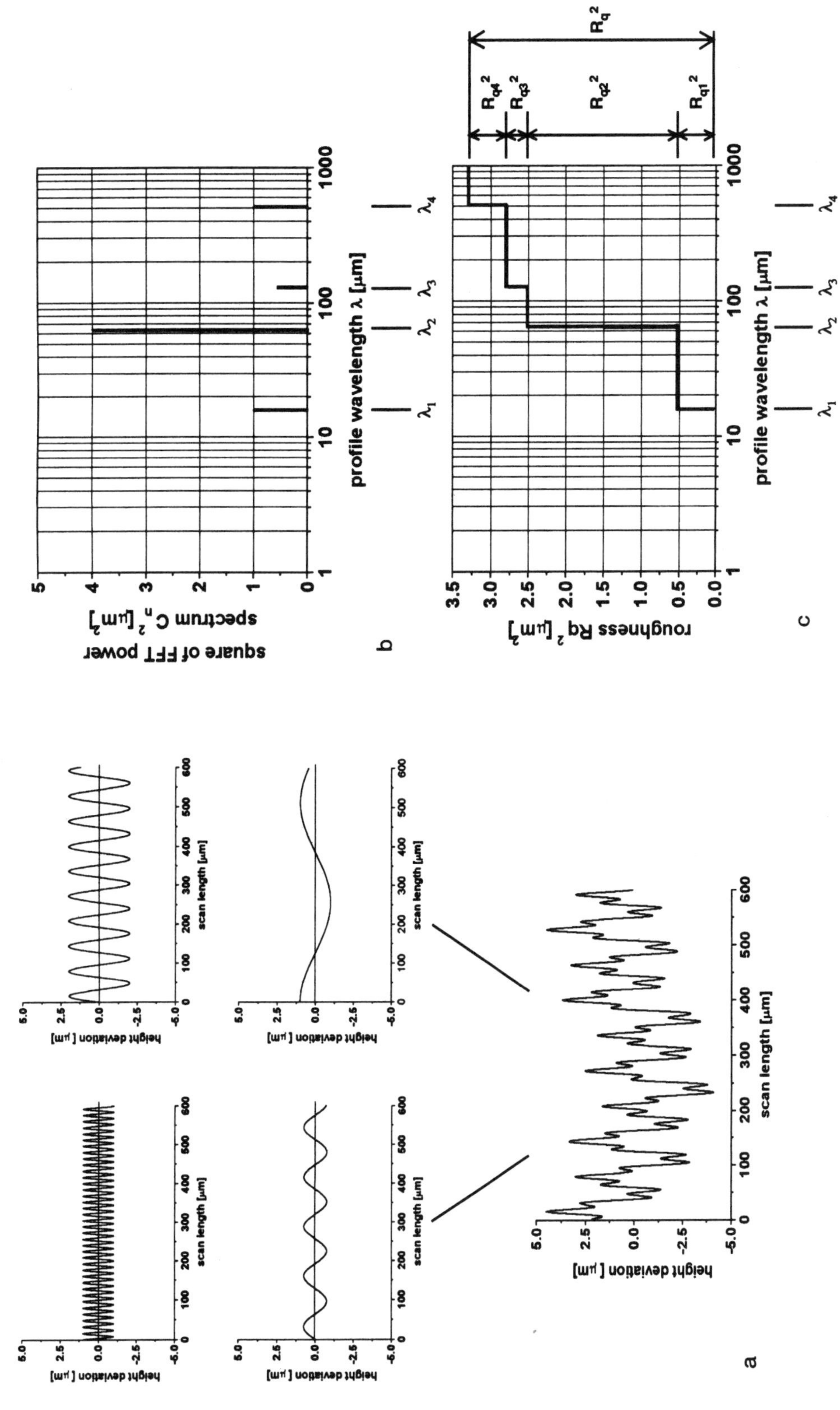
height deviation [μm]
scan length [μm]
a
square of FFT power spectrum C_n^2 [μm²]
profile wavelength λ [μm]
b
roughness Rq^2 [μm²]
profile wavelength λ [μm]
R_q^2
R_{q4}^2
R_{q3}^2
R_{q2}^2
R_{q1}^2
λ_1
λ_2
λ_3
λ_4
c

2. The largest meaningful wavelength λ_n (lowest spatial frequency) is approximately 0.2 times the scan length L of the profile. Despite the fact that larger λ_n can be theoretically be detected, their amplitudes are statistically very uncertain.

3.2. Relation between coefficients C_n and roughness parameter R_q

The root mean square roughness value, R_q, is defined for a 2-D surface profile $Z(x)$ according to Eq. (4), where L denotes the length of the profile in x direction. The waviness $\bar{Z}(x)$ has to be subtracted; alternatively $Z(x)$ in the Eqs. (4)–(6) has to be replaced by $Z(x) - \bar{Z}(x)$:

$$R_q^2 = L^{-1} \int_0^L Z(x)^2 \mathrm{d}x. \quad (4)$$

Expressing $Z(x)$ as a Fourier series expansion according to Eq. (2) results in:

$$R_q^2 = L^{-1} \int_0^L \left(\frac{A_0}{2} + \sum_{n=1}^{N/2} \left[C_n \sin\left(f_n \frac{x}{L} + \varphi_n \right) \right] \right)^2 \mathrm{d}x. \quad (5)$$

If L approaches infinity, Eqs. (4) and (5) can be converted to:

$$R_q^2 = \lim(L \to \infty) \left\{ L^{-1} \int_0^L Z(x)^2 \mathrm{d}x \right\}$$

$$= \frac{A_0^2}{2} + \frac{1}{2} \sum_{n=1}^{\infty} C_n^2 = \frac{A_0^2}{2} + \frac{1}{2} \sum_{n=1}^{\infty} (A_n^2 + B_n^2). \quad (6)$$

Eq. (6) proves that the mean square roughness parameter R_q^2 is proportional to the integral of the terms $C_n^2 = A_n^2 + B_n^2$. The series of $C_n = \sqrt{A_n^2 + B_n^2}$ terms is called the power spectrum. In other words, R_q^2, also called the variance of the height distribution, is proportional to the area under the power spectrum [4,44,45]. R_q^2 is a function of the wavelength and therefore a function of the resolution [4,46]. This relationship is true for both 2-D line profiles $Z(x)$ as well as 3-D surface topography images $Z(x,y)$ (see Section 3.4). Fig. 1 illustrates the relation.

In Fig. 1a, the first part of a synthesised profile based on four sine waves with different wavelengths, phase-shifts and amplitudes (Table 3) is shown. The whole profile length was 4.096 mm. In Fig. 1b, the square of the FFT power spectrum ($C_n^2 = A_n^2 + B_n^2$) of the profile is plotted vs. the logarithm of the profile wavelength, i.e., the inverse of the spatial frequency. Fig. 1c illustrates the roughness R_q^2 as a function of the profile wavelength, λ. The higher cut-off wavelength, λ_h, steadily decreases from 0.2 times the profile length to three times the step size of adjacent data points, while the lower resolution length, λ_l, is fixed at three times the resolution limit. R_q^2 was calculated using Eq. (6) for $n = N/2$ to $n = 1$. The result demonstrates the dependence of the roughness R_q^2 on the profile wavelength, λ, and the relationship to the square of the power spectrum C_n^2. The plot of R_q^2 vs. profile wavelengths falls in four steps, whose positions corresponding to the wavelengths of the four sine wave of the synthetic profile.

The coefficients C_n^2 are proportional to the weighting factors characterising the changes in roughness values R_q^2 as a function of the changes in resolution. The relationship between R_q^2 and the Fourier coefficients (Eq. (6)) allows the roughness of a profile to be understood as the sum of individual roughness contributions R_{qi}^2 from the different resolution steps (Fig. 1c).

The important advantage of the FFT formalism is the fact that individual contributions to R_q from any wavelength (or spatial frequency) range (λ_h to λ_l) of interest can be accurately and easily calculated.

3.2.1. The concept of the "window roughness"

It is possible to define a "differential roughness" as the limiting value of the change of roughness as the resolution step approaches zero, i.e., the first derivative of the roughness of the analysed profile vs. the resolution. In every practical case, however, one has as limited number of data points. In such a case, the roughness defined as R_q^2 corre-

Table 3
List of all synthetic profiles, based on four sinus waves, with their standardised roughness value R_q and their graphical results after calculation of R_q vs. λ using Eq. (7)

Figure	Synthetic profile	Amplitude (μm)				Wavelength λ (μm)				Phase-shift ϕ (°)				Standardised R_q value (μm)	FFT amplitude- and phase-plot	R_q vs. λ curve
Fig. 1a	I	1	0.75	2	1	16	64	128	512	90	0	0	90	1.811	–	–
Fig. 3a	I	1	1	1	1	8	16	32	64	90	90	90	90	1.414	I to IV are different	I to IV are different
	II	1	1	1	1	16	32	64	256	90	90	90	90	1.414		
	III	1	1	1	1	8	32	128	512	90	90	90	90	1.414		
	IV	1	1	1	1	16	64	128	512	90	90	90	90	1.414		
Profiles not shown	I	1	1	1	1	16	32	64	256	90	90	90	90	1.414	for I to IV only phase-plots are different	I to IV are equal
	II	1	1	1	1	16	32	64	256	90	0	0	90	1.414		
	III	1	1	1	1	16	32	64	256	0	0	90	0	1.414		
	IV	1	1	1	1	16	32	64	256	0	45	45	90	1.414		

sponds to the sum of the squared FFT coefficients in the corresponding range:

$$R_q^2(n,m) = \frac{A_0^2}{2} + \frac{1}{2}\sum_{i=n}^{i=m} C_i^2 \text{ or approximately}$$

$$R_q^2(n,m) = \frac{1}{2}\sum_{i=n}^{i=m} C_i^2 \quad (7)$$

since the term $A_0^2/2$ is generally negligible.

Eq. (7) will be defined as ''window roughness'', where n and m are the range of the Fourier coefficients, corresponding to the range of wavelengths and defining a certain ''window''. The lowest meaningful value n is the resolution length, the highest value is the cut-off length for the waviness.

Using the philosophy of a Fourier-transformed, resolution-dependent roughness curve as outlined above, the roughness is no longer a single value but a function of the resolution and of the waviness. In the ''window roughness'' concept, the roughness R_q^2 is proportional to the integral (Eq. (6)) with a lower limit λ_l (resolution) and an upper limit λ_h (waviness) of the Fourier coefficients. In the discrete notation (Eq. (7)) it is the sum of the squares of the FFT coefficients between a lower limit n and an upper limit m.

This concept allows roughness coefficients to be defined for different ''windows'' in the wavelength domain of interest and to separate contributions to the roughness parameters arising from different resolution ranges. Individual R_q values can be calculated for defined regions of interest by calculating the corresponding integrals or sums of the Fourier coefficients. This can also be done graphically (Fig. 2) in the plot of R_q vs. the logarithm of the profile wavelength, choosing the two limit wavelengths (λ_l and λ_h) and reading off the contribution of this region to R_q.

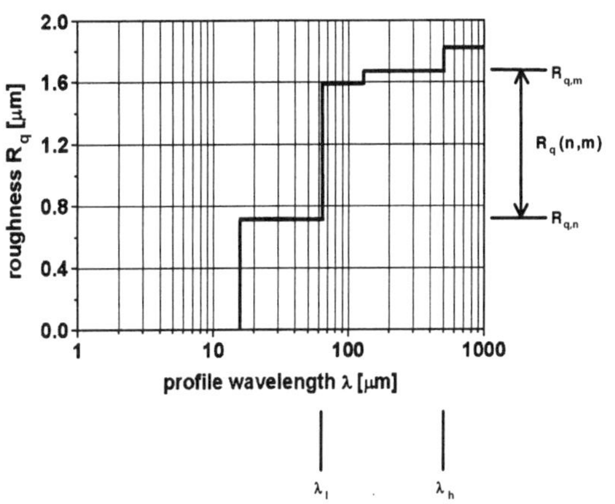

Fig. 2. ''Window roughness'' concept: $R_q(n,m)$ calculated between the profile wavelength limits λ_l and λ_h.

This formalism is closely related to the distinction of roughness, waviness and shape of surfaces. The high-pass (low spatial frequency) cut-off that is chosen sets the limit separating roughness and waviness. Using fixed cut-off values (such as defined in DIN4777 and DIN4768), a part of the information contained in experimental surface profiles, which may be relevant in practice, is completely lost (see Section 3.3).

3.2.2. FFT applied to synthetic roughness profiles: influence of wavelength and phase-shift

The first and second illustrative examples are based on profiles synthesised through the superposition of periodic sine wave functions with different wavelengths or phases but with equal amplitudes. The FFT of such profiles simply corresponds to the amplitudes for the given sine waves, with R_q being equal to the sum of the power spectrum. This means that with a constant number of amplitudes present in the spectrum, the wavelengths and phase-shifts of the different sine waves (correlated to the Fourier coefficients) do not influence the standardised R_q value.

To illustrate the influence of the wavelength, four profiles were synthesised, each of them using four sine waves with different wavelengths but with equal amplitude (1 μm) and phase-shift (90°) (Table 3). The profile length was 4.096 mm and the resolution 1 μm. The synthetic profiles have different aspects, but the standardised R_q values are the same for all of them ($R_q = 1.414$ μm, see Table 3). Each of the four profiles has been transformed by FFT technique and R_q calculated as a function of the wavelength. Fig. 3a shows the four profiles together with the corresponding FFT amplitude- and phase- plot, which are also different. Simulating a low-pass filter, the lower wavelength was steadily increased from three times the step size of adjacent data points (3 μm, $n = 3$) to 0.2 times the profile length (0.820 mm, $n = 820$), while the upper resolution length (waviness) was fixed (0.820 mm, $m = 820$). After each filtering step, R_q was calculated using Eq. (7). Fig. 3b shows the variation of R_q. The curves $R_q = f(\lambda)$ are clearly different reflecting the different wavelengths in the profiles. The plots start from the lowest wavelength at the standardised R_q value and fall in four steps at positions corresponding to the wavelengths of the sine wave periodicity of the synthetic profiles. In the same figure the dependence on R_q is indicated using the average and the low-pass filtering technique (discussed in Section 3.3).

In the second example (data not shown) four profiles were synthesised with a length of 4.096 mm and a resolution of 1 μm, each of them using four sine waves with constant wavelength of 16, 32, 64 and 256 μm, respectively, and equal amplitude of 1 μm, changing only the phases of the single waves (Table 3). The synthesised profiles are again different, but have the same standardised R_q values ($R_q = 1.414$ μm). Since R_q is only a function of the square of the amplitudes but does not depend on the

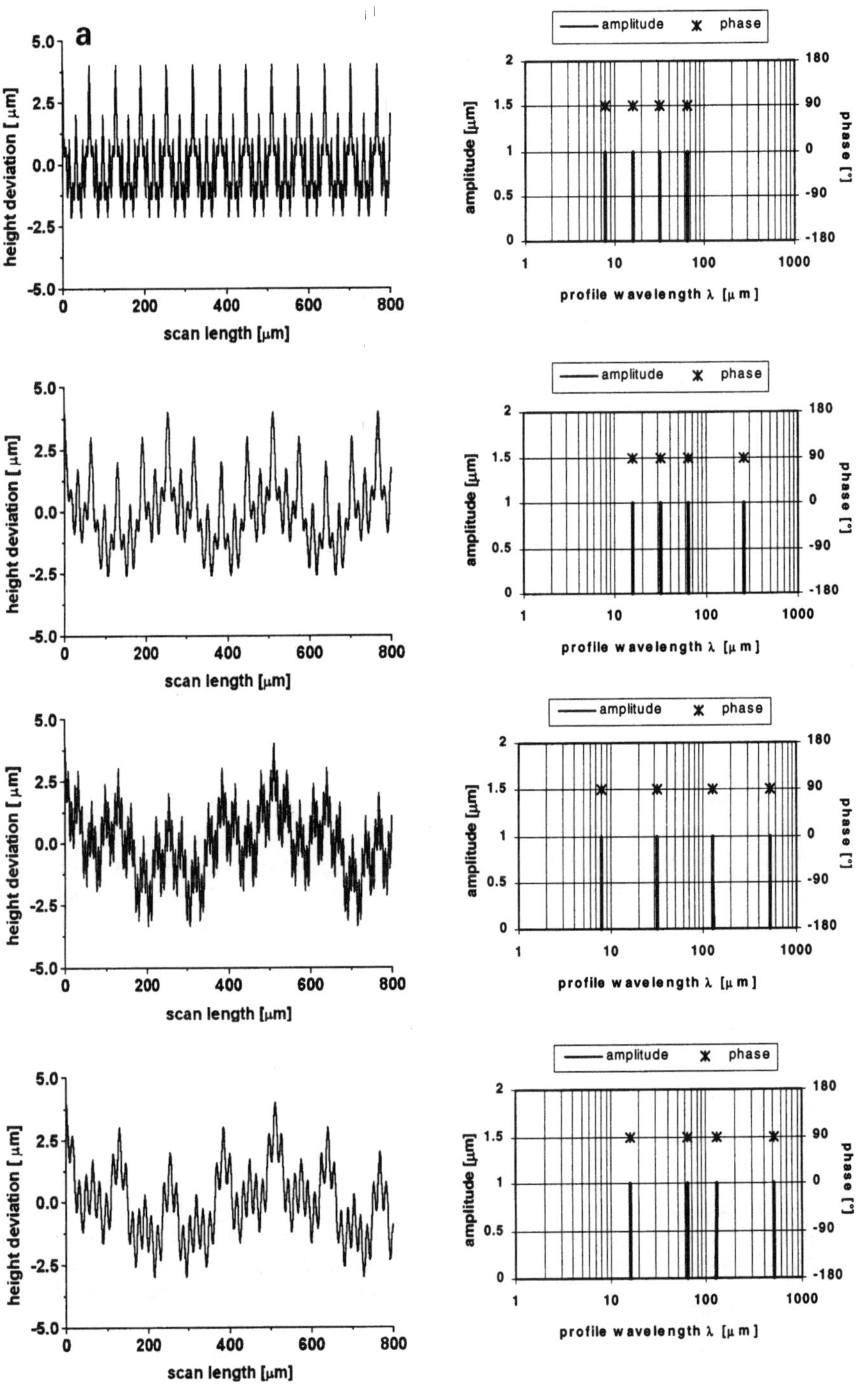

Fig. 3. (a) The profiles I to IV were each synthesised using four sinus waves with different wavelengths but equal amplitude- and phase-shifts. On the right hand side the corresponding FFT amplitude- and phase-shift plots. (b) The comparison of the dependence of the roughness R_q on the profile wavelength for the four profiles I to IV using the FFT, COF and AFT.

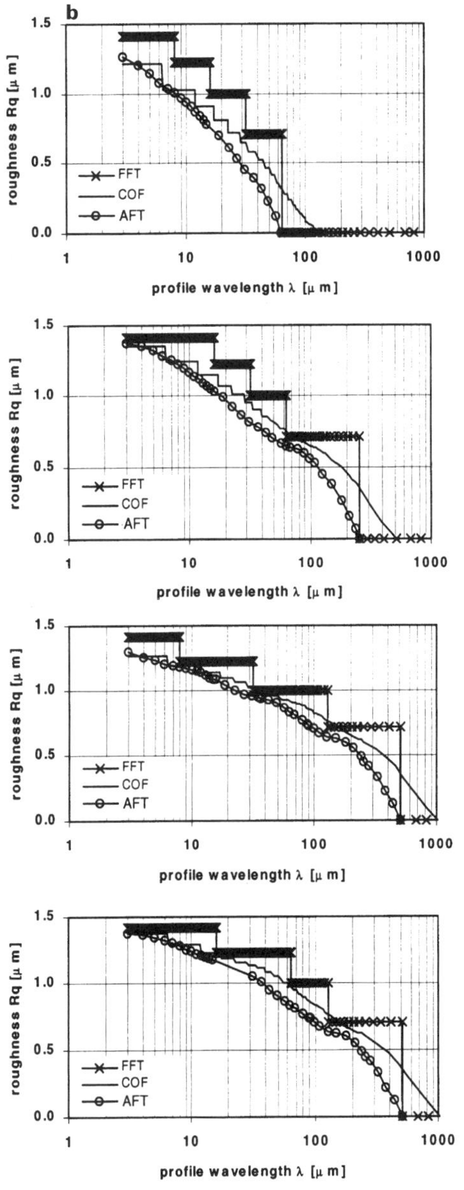

Fig. 3 (continued).

phases, all these profiles have the same dependence on R_q vs. wavelength. Such profiles can only be distinguished from each other by plotting their Fourier coefficients with the corresponding phases. In this case the plot of R_q vs. profile wavelengths is not sufficient to distinguish the four profiles from each other.

In Fig. 1a, not only wavelengths and phases are different, but also amplitudes. The standardised R_q value (R_q = 1.811 μm) is consequently different (Table 3).

In most cases, the plot of R_q vs. wavelength λ is well-suited to the adequate characterisation of a profile. However, such a plot may in certain practical cases be too complicated or time-consuming for a simple comparison of specific, topographical characteristics of a surface. In such cases, standardised roughness values may therefore be preferred.

Based on the Fourier transformation technique it is possible to demonstrate mathematically the correlation between the Fourier coefficients and the roughness R_q. Roughness values other than R_q can also be expressed as a function of wavelength. The mathematical procedure, however, is more complex and not further discussed here.

3.2.3. The inverse FFT, an application to the "window roughness"

An application of the "window roughness" approach is the inverse FFT. FFT-transformed profiles are inverse transformed in the wavelength domain of interest. The result is a low-, high- or band-pass filter applied to the original profile. Fig. 4 demonstrates the application of inverse transformation of the FFT to the synthetic profile in Fig. 1a cutting off the lower wavelength range ($\lambda_{cut\text{-}off}$ = 2, 16, 32, 64, 128, 256, 512 and 1024 μm). Table 4 shows the R_q values for the different inverse transformed profiles calculated using Eq. (7). Filtering within the ranges of 2–2048 and 16–2048, 32–2048 and 64–2048, 256–2048 and 512–2048 μm, respectively, showed no change in the inverse transformed profiles and roughness. This corresponds to the power spectrum in Fig. 1b and the dependence of the roughness R_q on the wavelength in Fig. 1c.

3.3. COF techniques

COF is a standardised procedure [37] applied before roughness parameters are calculated from experimental profiles. Low-frequency cut-off filters are used to cut off the range considered to be waviness, generally corresponding to 1/7 of the measured length of the profile. Valuable information may be lost by this standard technique, however.

An alternative to FFT, albeit a mathematically less accurate procedure, band-pass filtering may be applied to various spatial frequency/wavelength ranges of particular interest. To calculate wavelength-dependent roughness values similar to the FFT procedure discussed in Section 3.2, the effect of a varying low- or high-end cut-off is applied and the corresponding roughness values determined from the individual manipulated profiles.

Another more pragmatic way to handle the problem is to use filtering techniques such as those used for the COF (gliding Gauss or average filter), but with an intention other than just eliminating waviness. The procedure uses a continuously changing filter width starting from three times the data step size and ending at the cut-off for the waviness correction (1/7 of the profile length) or at any other predefined cut-off length shorter than half of the profile length. The result is a set of filtered profiles suitable for roughness calculations with cut-offs ranging from the pre-

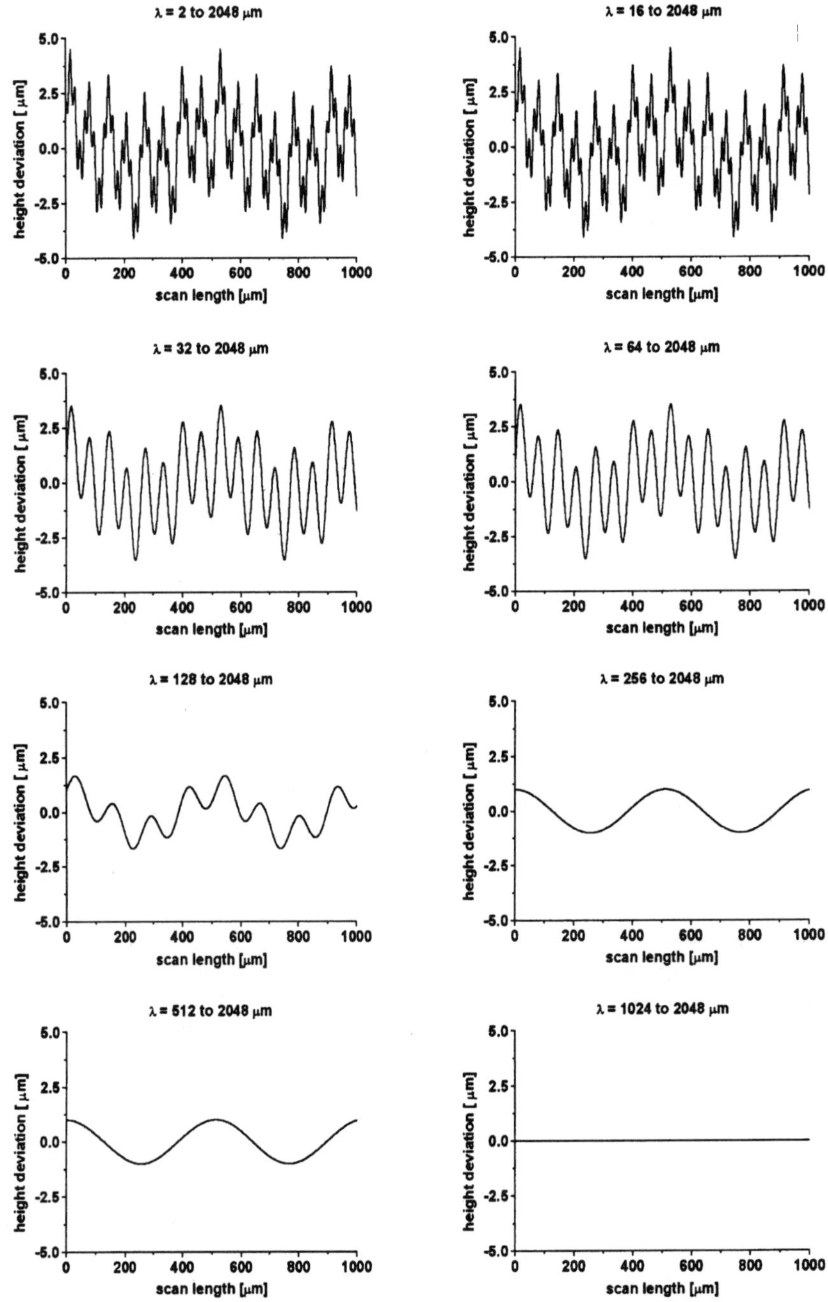

Fig. 4. The inverse FFT-transformed profiles in different ranges of ''window roughness'' from the profile I in Fig. 1 (see also Table 3).

defined cut-off length to three times the measuring step size, and a second set of filtered profiles with waviness information and cut-offs ranging from three times the measuring step size to the predefined cut-off length. At the higher limit the roughness profile is a straight (zero) line because the whole profile is interpreted as waviness. The lower the cut-off length, the larger the part of the profile that is considered to be roughness and the smaller the part considered to be waviness. From such profiles it is possible to calculate roughness parameters, not as a single

Table 4
List of R_q values calculated for the different inverse FFT-transformed profiles of the synthetic profile in Fig. 1a using Eq. (7)

Wavelength range (μm)	R_q value (μm)
2 to 2048	1.811
4 to 2048	1.811
8 to 2048	1.811
16 to 2048	1.811
32 to 2048	1.668
64 to 2048	1.668
128 to 2048	0.884
256 to 2048	0.707
512 to 2048	0.707
1024 to 2048	0

value, but as a function of the resolution or of the waviness.

In the case of the *roughness profile*, the curve $R(\lambda)$ starts at zero with a positive slope and a sigmoidal shape up to the R_a or R_q values, characteristic for the integral roughness values including the standard cut-off.

In the case of the *waviness profile*, the curve $W(\lambda)$ starts at R_a or R_q with a negative slope and a sigmoidal shape and ends at a value of zero for the standard waviness cut-off. $R(\lambda)$ and $W(\lambda)$ are simply related via the Eq. (8)

$$R(\lambda) + W(\lambda) = \text{const.} = R_a \text{ or } R_q. \quad (8)$$

Fig. 3b shows the application of the AFT to the four synthetic profiles in Fig. 3a. The filter width was continuously varied from three times the resolution limit (3 μm, i.e., the data step size between adjacent points) to the cut-off length (0.82 mm = 0.2 times the scan length). In case of the cut-off roughness, the resulting curves decreases with increasing average filter length. Compared with the curves obtained using the FFT procedure (Fig. 3b), the R_q values are, however, lower for the same wavelength and show no abrupt steps, since AFT technique does not correspond to a set of orthonormal functions like the sine and cosines curves of the Fourier transformations. R_q becomes zero at the highest profile wavelength. Although mathematically not strictly correct, the AFT technique is less cumbersome and time-consuming than the FFT technique. Additional examples with the low-pass COF technique are shown in Fig. 3b. The low-pass cut-off was raised from 3 μm to the upper limit of 0.82 mm. The result depends on the choice of the filter form (average, Gauss, rectangular, etc.). With COF techniques, it is also essential to cut-off at the lower and higher limit with a filtering dimension corresponding to half the filter size used; otherwise, boundary-related artefacts are produced. In this case, a Gauss filter and an attenuation factor of 50% were used. The resulting curves decrease in smaller steps than those obtained using the FFT technique, but no abrupt steps were observed at the highest wavelengths of the synthesised profiles. The R_q values become zero at twice the highest profile wavelength. A comparison of the R_q values at the same wavelength calculated by the three filtering techniques indicate the highest value for the FFT technique and the lowest for the AFT.

Furthermore, it is possible to apply a band-pass filter on the profile, having a lower limit corresponding to the resolution and an upper limit corresponding to the cut-off for the waviness. The limits can be chosen in such a way as to reflect ranges of interest for specific physical, mechanical, optical or other types of properties.

The philosophy discussed above for the Fourier transformation and the ''window roughness'' concept holds in the same way for the different filtering techniques. In the limit case (band-pass filter width → 0), both ''window values'' (FFT and COF) should be equal and define a correct ''window roughness'' curve vs. wavelength. The defined integral of this function between two chosen limits is equal to the roughness for the analysed profile determined between the boundaries of the lower limit (resolution) and the upper limit (waviness).

3.4. Evaluation of 3-D profiles

The formalism discussed above can be extended to the 3-D case. In the Fourier transform concept the resolution can be represented as concentric circles with the centre at $\lambda_x = \lambda_y = 0$, in the x/y wavelength or spatial frequency plane. The wavelength-dependent roughness is then proportional to the integral of the coefficients inside this circle. The weighting factor of the resolution changes in the ''window roughness'' concept corresponds to the integral over the annulus centred on zero in the x/y wavelength or spatial frequency plane with radius r to $r + \mathrm{d}r$. This corresponds to all vectors in the x/y wavelength or spatial frequency plane with length equal to the wavelength or spatial frequency range with the origin at zero.

Using the COF and AFT filtering techniques, a 2-D filter with a dimension that changes after each gliding filtering step is applied. After each filtering step, the surface roughness parameter has to be calculated. It is straightforward to use a square matrix filter, although a ''round'' kernel filter would be more appropriate. In the band-pass case, the surface has to be treated first with the lower and then with the upper surface filter and the obtained filtered topographies subtracted from each other.

It is also possible in the 3-D case to apply the filtering in the FFT space and to inverse transform the filtered FFT into real space. The result is a filtered topography that depends on the filter used to treat the FFT coefficients. From this filtered surface the roughness parameters for different ''windows'' can be calculated. The details of these formalisms and typical applications will be discussed in a second paper.

3.5. Surface treatment transfer functions

In the area of surface engineering, several consecutive surface treatment processes are often applied to achieve the required surface properties of the material or product. The surface topography may thereby be modified by each treatment, requiring careful control of the parameters and quality in each individual step. Examples are the following.

(1) Coating by electroplating, plasma spraying or lacquering onto surface-structured substrates, resulting in topographic changes related to both the substrate surface structure and the (macro and micro) throwing power of the applied surface coating technique.

(2) combinations of different surface roughening techniques, such as machining, shot blasting, etching, providing surface topographic effects in different dimensional ranges.

During such surface treatments the surface topography is often transformed in a ''resolution-dependent'' manner. It is proposed to use a new ''transfer function'' concept to follow quantitatively, and in a resolution-dependent way, the effect of each subsequent surface modification step. This may be helpful, both for the quality control of product surfaces and for the systematic development of specific surface structures based on combinations of existing surface treatment methods.

Referring to Eq. (5), we can define sets of FFT C_n^x coefficients for the surfaces after each surface modification step x.

C_n^{A} = set of FFT coefficients for surface produced in surface treatment step A.

C_n^{B} = set of FFT coefficients for surface produced in surface treatment steps A + B.

C_n^{C} = set of FFT coefficients for surface produced in surface treatment steps A + B + C etc.

''Multiplicative transfer functions'' $M^{y/x}$ are defined based on the following ratios of FFT coefficients:

$$M^{\mathrm{B/A}} = C_n^{\mathrm{B}}/C_n^{\mathrm{A}}$$

$$M^{\mathrm{C/B}} = C_n^{\mathrm{C}}/C_n^{\mathrm{B}}$$

$$M^{\mathrm{C/A}} = C_n^{\mathrm{C}}/C_n^{\mathrm{A}} = M^{\mathrm{C/B}}M^{\mathrm{B/A}} \text{ etc.} \tag{9}$$

Such transfer functions can be used to characterise the resolution-dependent ''topographical fingerprint'' of a particular surface treatment process, and, once known, can be used to back-calculate the effect of this surface treatment on starting surfaces with different topographies. For example, the effect of micro- and macro-throwing power of coating processes, such as electroplating or lacquering, in gradually levelling off substrate surface roughness can be followed in a quantitative way. A practical example will be shown and discussed in Section 4.2.

In general, however, consecutive surface treatments are unlikely to merely level off or amplify the surface topography of the base material, but will add their own specific surface roughness characteristics with different amplitude and spatial frequency/wavelength contributions. In such a

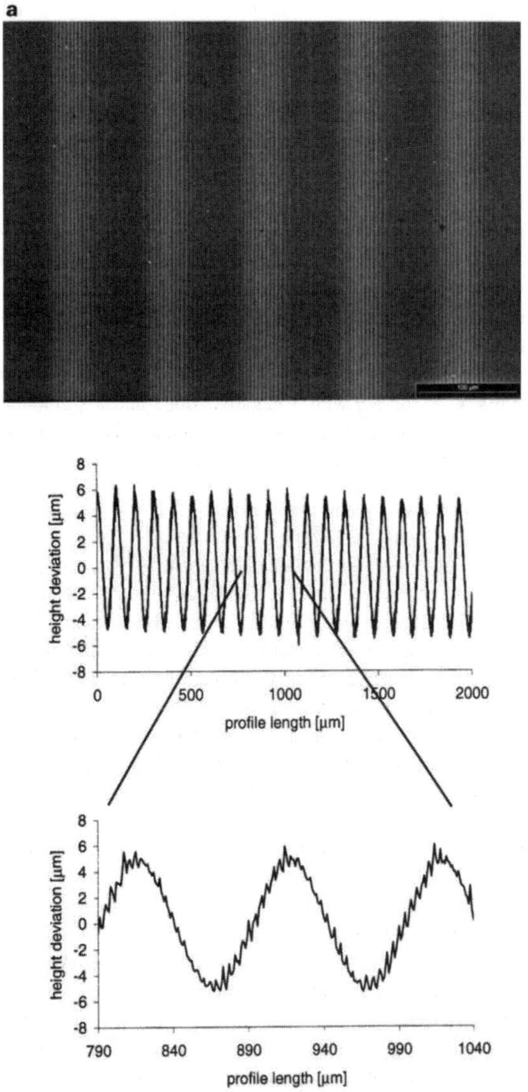

Fig. 5. (a) The SEM image of the micromachined steel surface and a 2-D LPM profile taken perpendicular to the machining direction. (b) The square of the FFT power spectrum vs. the logarithm of the profile wavelength of the 2-D LPM profile. (c) The comparison of the dependence of the roughness R_q on the profile wavelength for the micromachined steel surface using the FFT and COF technique. (d) The inverse FFT-transformed profiles in the wavelength ranges of A (85.3 to 113.8 µm) and B (3.81 to 4.31 µm), respectively. (e) The COF profiles in the same wavelength ranges A and B.

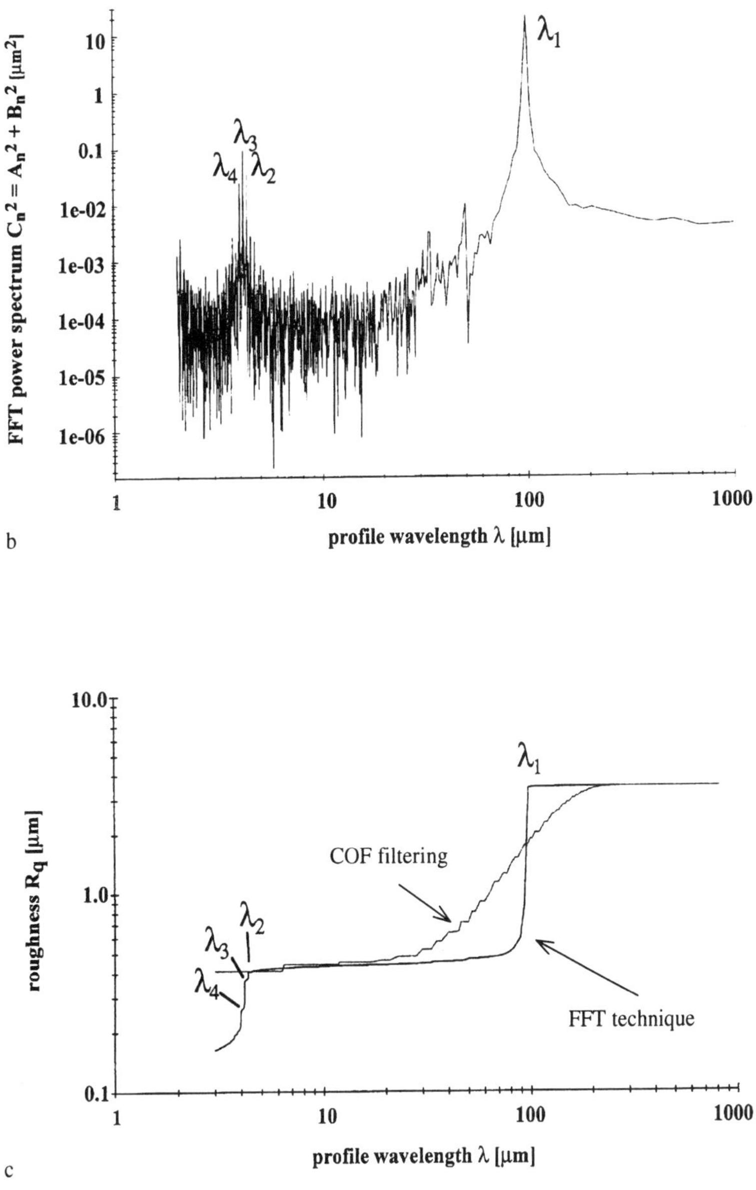

Fig. 5 (continued).

case — in addition to multiplicative factors — additive transfer terms (c_n^{B}, c_n^{C}, etc.) have to be considered as well.

$$M^{\mathrm{B/A}} = \left[C_n^{\mathrm{B}} / C_n^{\mathrm{A}} + c_n^{\mathrm{B}} \right]$$

$$M^{\mathrm{C/B}} = \left[C_n^{\mathrm{C}} / C_n^{\mathrm{B}} + c_n^{\mathrm{C}} \right]$$

$$M^{\mathrm{C/A}} = \left[C_n^{\mathrm{C}} / C_n^{\mathrm{B}} \left\{ C_n^{\mathrm{B}} / C_n^{\mathrm{A}} + c_n^{\mathrm{B}} \right\} + c_n^{\mathrm{C}} \right]$$
$$= C_n^{\mathrm{C}} / C_n^{\mathrm{B}} M^{\mathrm{B/A}} + c_n^{\mathrm{C}} \qquad (10)$$

To determine the additive transfer terms for a particular surface treatment process, starting substrates as flat as possible have to be used and the topography determined individually before and after each particular surface treatment step.

Two drawbacks of the transfer function concept should be mentioned.

(1) Since the transfer function involves ratios of coefficients, statistical scatter in the data are amplified. For statistically significant transfer function values, rather large

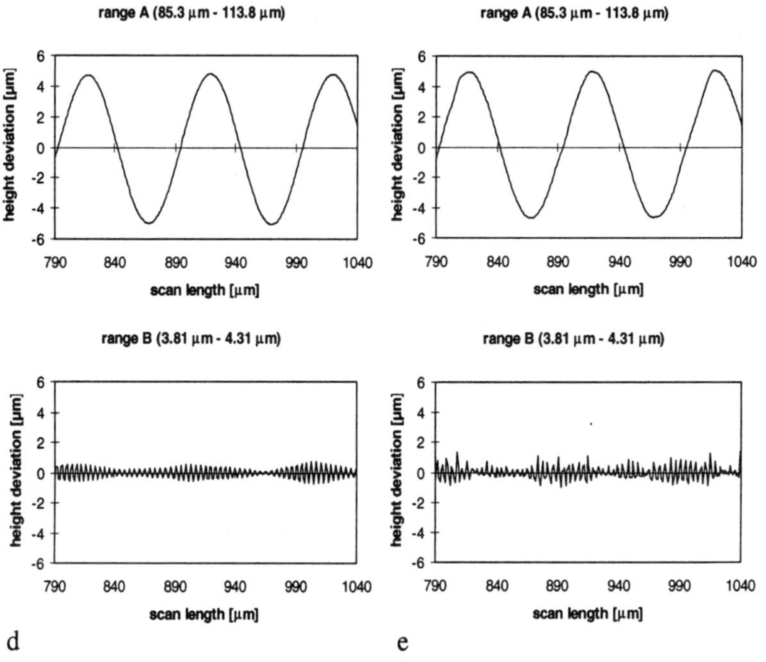

Fig. 5 (continued).

surface areas, characteristic of the chosen surface treatment process, have to be measured with adequate lateral and vertical resolution.

(2) Large data sets have to be handled with the consequence of increased computation time, particularly in the case of 3-D surface topographies.

The techniques described can be very useful, but have to be carefully evaluated in the context of the specific applications and questions raised.

4. Application to real surfaces

4.1. Micromachined reference steel surface

The micromachined steel surface (Fig. 5a) used as a reference is characterised by two superimposed surface structures with periodicities of the order of 100 and 4 μm,

Table 5

List of the roughness values R_a, R_q, R_{zDIN}, S_m, S_k, and L_r calculated in the wavelength ranges A (113.8–85.3 μm) and B (4.31–3.81 μm) with a Gauss filter and an attenuation factor of 50% at the cut-off of 0.82 mm as well as the parameters determined according to DIN4768

Wavelength range	Roughness parameters	Roughness values (FFT technique)	Roughness values (COF technique)	Wavelength range	Roughness parameters	Roughness values (standard norm DIN4768)
A (113.8–85.3 μm)	R_a (μm)	3.10	3.01	Whole (3–2048 μm)	R_a (μm)	3.10
	R_q (μm)	3.45	3.36		R_q (μm)	3.47
	R_{zDIN} (μm)	9.83	9.71		R_{zDIN} (μm)	11.49
	S_m (mm)	0.105	0.100		S_m (mm)	0.086
	S_k	−0.04	0.04		S_k	0.05
	L_r	1.02	1.02		L_r	1.18
B (4.31–3.81 μm)	R_a (μm)	0.265	0.34			
	R_q (μm)	0.323	0.44			
	R_{zDIN} (μm)	1.45	2.45			
	S_m (mm)	0.004	0.004			
	S_k	0.00	0.30			
	L_r	1.09	1.16			

respectively. This is clearly visible in the 2-D laser scan profile taken perpendicular to the machining direction over a length of 4.096 mm with a lateral resolution of 1 μm (Fig. 5a) as well as from the FFT power spectrum vs. the profile wavelength, which is dominated by the two basic wavelengths (Fig. 5b). Furthermore, the FFT power spectrum shows three distinct peaks around the wavelength of 4 μm. Fig. 5c shows the dependence of the R_q value on the profile wavelength λ using the FFT technique (λ_l is fixed at three times the resolution limit). The R_q value starts from the highest wavelength distribution at the standardised R_q value ($R_q = 3.47$ μm) and falls at the position corresponding to the first basic wavelength (97.5 μm) and again in three steps at the positions corresponding to the lower wavelength distribution.

If the high-pass filter is applied (Fig. 5c) the curve (COF) starts at the same R_q value ($R_q = 3.47$ μm) as obtained using the FFT technique and falls off continuously in small steps over a large range (approximately 250 to 20 μm). In the range between 20 and 3 μm, the curve falls over three steps and ends at the same R_q value ($R_q = 0.41$ μm) at which the FFT-based curve drops off. Using the COF technique, no abrupt steps are indicated around 100 and 4 μm. It follows that in the case of the high-pass filter, the most characteristic information of the profile is lost. Another way to characterise the profile is the "window roughness" concept. Fig. 5d and e illustrate the results of the inverse FFT and COF techniques when separately treating the two surface structures by dividing up the calculated wavelength range into A (113.8–85.3 μm) and B (4.31–3.81 μm). In the case of the COF technique a wavelength range A^* (250.0 to 12.0 μm) and B^* (9 to 0.5 μm) had to be chosen for the two ranges A and B described above in order to get correct results. Both methods allow the substructure-specific roughness parameters R_a, R_q, R_{zDIN}, S_m, S_k, and L_r to be determined with a Gauss filter, an attenuation factor of 50% and a cut-off of 0.82 mm (Table 5). In the range A (113.8–85.3 μm), the inverse FFT and the COF-filtered profile are very similar and the corresponding roughness values show only small differences between the two mathematical procedures for R_a, R_q, R_{zDIN}, S_m and L_r. However, the differences are larger in the range B (4.31–3.81 μm) for R_a, R_q, R_{zDIN}, S_k and L_r due to two reasons. (1) The band-pass filter has to be chosen at approximately 1/7 of the lower wavelength and twice the higher wavelength to filter in the predefined ranges A and B. (2) The band-pass filter does not correspond to a set of orthonormal functions like the sine and cosines curves of the Fourier transformation.

The differential roughness parameter sets are also compared to the conventional, standard 'integral' roughness values (Table 5). The standard amplitude parameters R_a, R_q, R_{zDIN} and the spacing parameter S_m are mostly determined by the roughness in the range A. The superimposed structure characterised by the range B shows an additive effect of the amplitude roughness values R_q^2, R_a^2 and

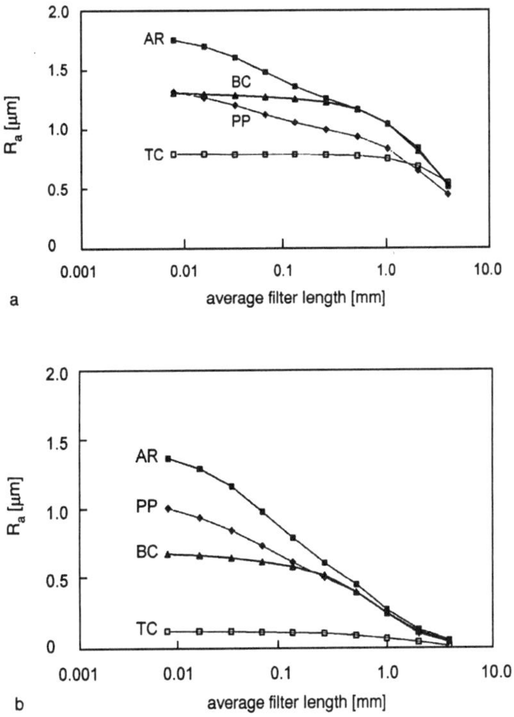

Fig. 6. (a) The resolution dependence of the roughness R_a on the average filtering length of the car body surface profiles after the different surface treatments: AR, after PP, after cataphoretic BC lacquer coating and after TC lacquering. The roughness R_a was calculated after each filtering step. (b) Before the AFT and the R_a calculation procedures were applied on the different surfaces AR, PP, BC and TC, the waviness of each profile was subtracted using a cut-off of 6.9 mm.

R_{zDIN}. The roughness value of L_r is dominated by the structures within range B.

4.2. Lacquered car body surfaces

The optical properties of lacquered car body sheet depend both on the properties of the lacquer and on the topography of the surface. The latter is a complex function of all surface-relevant fabrication steps, starting with the topography of the AR surface and ending with the final lacquering stage (TC). To judge the effect of pretreatment and lacquer systems on the topography of car body sheets, the AFT technique for the resolution-dependent roughness R_a and the FFT technique for the resolution-dependent surface treatment transfer functions were used. The line profiles were obtained with the non-contact LPM over a distance of 48.0 mm with a lateral resolution of 250 points/mm [47].

Fig. 6a shows the resolution-dependent roughness function of R_a for the four surfaces: AR, after PP, after cataphoretic lacquer BC and after TC lacquering. A sequential average filtering of the profiles starting from 4 μm to 4 mm was used. This treatment of the profiles corresponds to a simulated change in the measuring resolution from 4 μm to 4 mm. After each filtering step the R_a value was calculated using a Gauss filter and an attenuation factor of 50% (see Section 2.3). For surfaces AR and PP, the values of R_a steadily decrease with increasing filter length. In the case of surface BC there is a plateau at smaller resolution steps and the value of R_a does not decrease until higher resolution steps are reached. For surface TC this plateau is larger. Beside this tendency the curves cross each other and there seems to be no apparent correlation between the four surfaces.

However, a correlation between the four treatments is obvious when a band-pass filter is applied. First, the waviness of all profiles was subtracted using a cut-off length of 6.9 mm followed by applying the same average filtering and R_a calculation procedures as described above. Now a clear trend for all four resolution-dependent roughness curves is obvious in Fig. 6b. The COF has a normalising effect on the curves. The subtraction of the waviness from the measured profiles corresponds to an elimination of ''external'' factors, such as macroscopic bending, to the profile that have nothing to do with the roughness to be measured. The highest R_a value is obtained for surface AR, starting from 1.36 μm at a resolution of 4 μm. It steadily decreases to 0.07 μm at a resolution of 4 mm. Surface PP shows a similar curve, starting from an R_a value of 1.01 μm. For surface BC the calculated starting value of R_a is 0.68 μm. It is nearly constant up to a resolution of about 0.14 mm, followed by a steady decrease to 0.05 μm at a resolution of 4 mm. For surface TC the tendency is the same. The starting value of R_a is at 0.12 μm. The R_a value is nearly constant below a resolution of about 0.6 mm at which point it slowly decreases to 0.03 μm at a resolution of 4 mm.

The different surface treatment steps correspond to a kind of average filtering of the surface. Fig. 7 shows this effect. On the left hand side, the profiles of the surfaces AR, BC and TC are plotted using the same scale with a measuring length of 48.0 mm and a resolution of 250 points/mm. The roughness values are calculated as described in Section 2.3. The three profiles as well as their roughness values R_a and R_q are completely different (Fig. 7 and Table 6). On the right hand side, the same profiles are plotted to the same scale after average filtering over 2 mm. The three profiles and the roughness values R_a and R_q are roughly the same (Fig. 7 and Table 6). This demonstrates the ''filtering'' effect on surface roughness by lacquer treatment.

The FFT analysis allows a more detailed quantitative analysis of the effect of the lacquer system on the surface of the car body sheet (Fig. 8a). The wavelength-dependent contribution values are the highest for surface AR. They increase with increasing profile wavelength. The curve of surface PP is very similar. For surface BC, the values are typically one third as large below a wavelength of about 0.4 mm above which they match the values of curve AR. Surface TC contains additional lower amplitude contributions up to a wavelength of about 6 mm.

More detailed insight into the influence of each treatment step can be obtained in calculating the transfer functions between the different surfaces (Fig. 8b). There are three first-order transfer functions (PP/AR; BC/PP; TC/BC), two second-order functions (BC/AR; TC/PP)

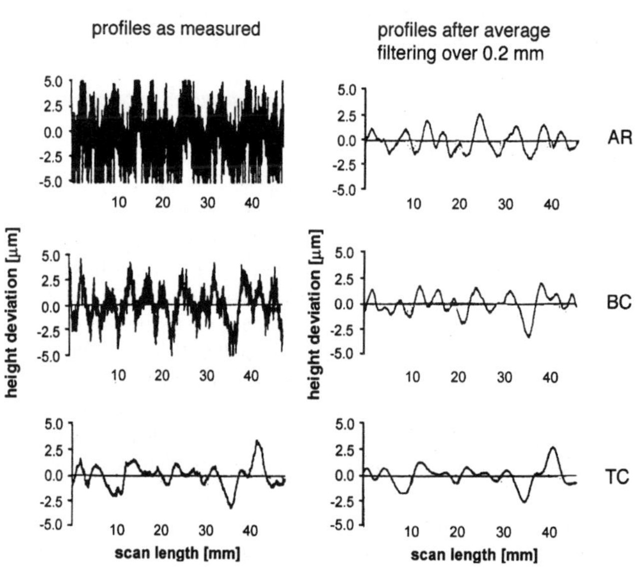

Fig. 7. The average filtering effect of the lacquer treatment process on a surface. On the left hand side the 2-D LPM profiles of the surfaces AR, BC and TC. On the right hand side the same profiles after the application of the average filter (filter length: 0.2 mm).

Table 6
List of the roughness values R_a and R_q from 2-D LPM profiles of the AR surface, after cataphoretic BC lacquering and after TC lacquering. Calculation according to DIN4768 with a cut-off of 6.9 mm

Type of car body sheet surface	Roughness parameters	Roughness values as measured	Roughness values after average filtering (0.2 mm)
AR	R_a (μm)	1.81	0.83
	R_q (μm)	2.38	1.00
After cataphoretic BC lacquer coating	R_a (μm)	1.39	0.84
	R_q (μm)	1.89	1.07
After TC lacquering	R_a (μm)	0.84	0.74
	R_q (μm)	1.12	0.99

and one third-order function (TC/AR). The first-order transfer functions characterise the effects of each single treatment step, the third order transfer function characterises the effect of the overall surface treatment process.

The PP process has only a minor effect, while the cataphoretic BC lacquer coating reduces the amplitude of the contributions to the roughness by a factor of three to four up to a wavelength of 300 μm, by a factor of two between 300 μm and 1 mm and no reduction above. The TC lacquering reduces the amplitude by a factor of five to six up to a wavelength of 1 mm and a factor of two between 1 and 6–10 mm.

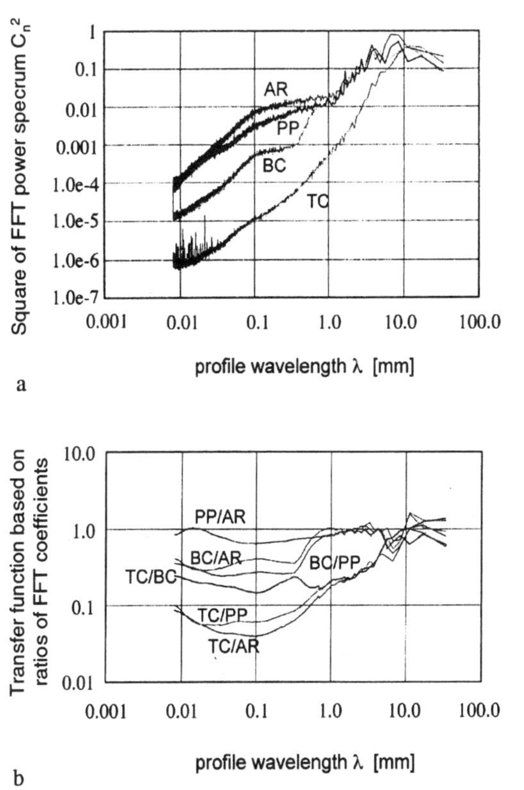

Fig. 8. (a) The square of the FFT power spectra of the car body sheet surfaces (AR, PP, BC and TC) after each treatment step vs. the logarithm of the profile wavelength. (b) The "multiplicative transfer functions" between the different surface treatment steps of the car body surface.

The total effect of the lacquer system is the multiplication of the individual first-order transfer functions. The resolution-dependent contributions to the roughness from surface AR are reduced through the used lacquer system as shown by the lowest curve drawn in Fig. 8b.

The calculated transfer function is characteristic for a particular surface treatment system such as the applied lacquer system. Within certain limits it is independent of the surface to which it is applied.

Once the transfer function for a system is defined, it can be applied to different starting surfaces to predict the final result. If, on the other hand, the final result is specified, one can define topography limits for the starting surface.

In the present lacquer system the concept of a simple "multiplicative transfer function" is applied, since, as a first approximation, only a proportional reduction of the roughness is expected. In the more general case for which roughness contributions are added in one or several steps, an "additive term" should be added to the transfer function. This term could be experimentally determined as discussed in Section 3.5.

4.3. Titanium implant surfaces

Surface topographies of bone-related prostheses and implants such as hip or knee joints, dental implants or bone-fracture plates are highly relevant for the behaviour of the biomaterial surface in contact with adjacent tissue [48]. Increased surface roughness in different wavelength ranges, achieved through processes such as particle blasting, plasma spraying or chemical/electrochemical etching can be used on titanium implants to promote bone integration and long-term stability of the implant in the patient [2,49].

A BE surface was investigated to illustrate the effect on cpTi surfaces of two consecutive surface-structuring processes resembling those used in dental implants, and which are known to lead to particularly effective bone integration and stability [50]. In this sequence, the titanium surface is first treated by particle-blasting with alumina beads, followed by a chemical etch process in a hot solution of HCl/H_2SO_4 that superimposes an approximately 10-times finer structure on top of the B surface. Following surface treatment, seven random 2-D profiles of each surface type were measured with the LPM according to DIN4768 (Section 2.2). Fig. 9 shows the SEM images of the B, E and BE surfaces and part of their 2-D LPM profiles. No major differences are immediately visible between the profiles B and BE. However, the standard roughness values R_a and R_q of the two surfaces are significantly different. Table 7 summarises the results of the standard roughness values R_a, R_q, R_{zDIN}, S_m, S_k, and L_r for the three surfaces B, E and BE. No significant differences could be observed for the standard roughness values R_{zDIN}, S_m, S_k, and L_r between the surfaces B and BE. The effect of the etching process is hidden by the coarser roughness contribution. Except for the roughness parameter S_k, all roughness values for surface E are significantly different to those of surfaces B and BE. An additive effect for the etching treatment of surface B could be observed for the amplitude parameters R_a and R_q, but not for R_{zDIN}, the spacing parameter S_m and the hybrid parameters S_k and L_r.

In order to get more specific information of each treatment step, the dependence of the roughness R_q on the profile wavelength was evaluated using Eq. (7) (FFT) and the high-pass filter (COF). Fig. 10 shows for surfaces B, E and BE both curves $R_q = f(\lambda)$. For surfaces B and E, the values of R_q (calculated by FFT) start at the standard roughness values and steadily decrease with decreasing higher wavelength limit. The two curves cross each other at a wavelength of 4 μm. In the case of surface BE, the

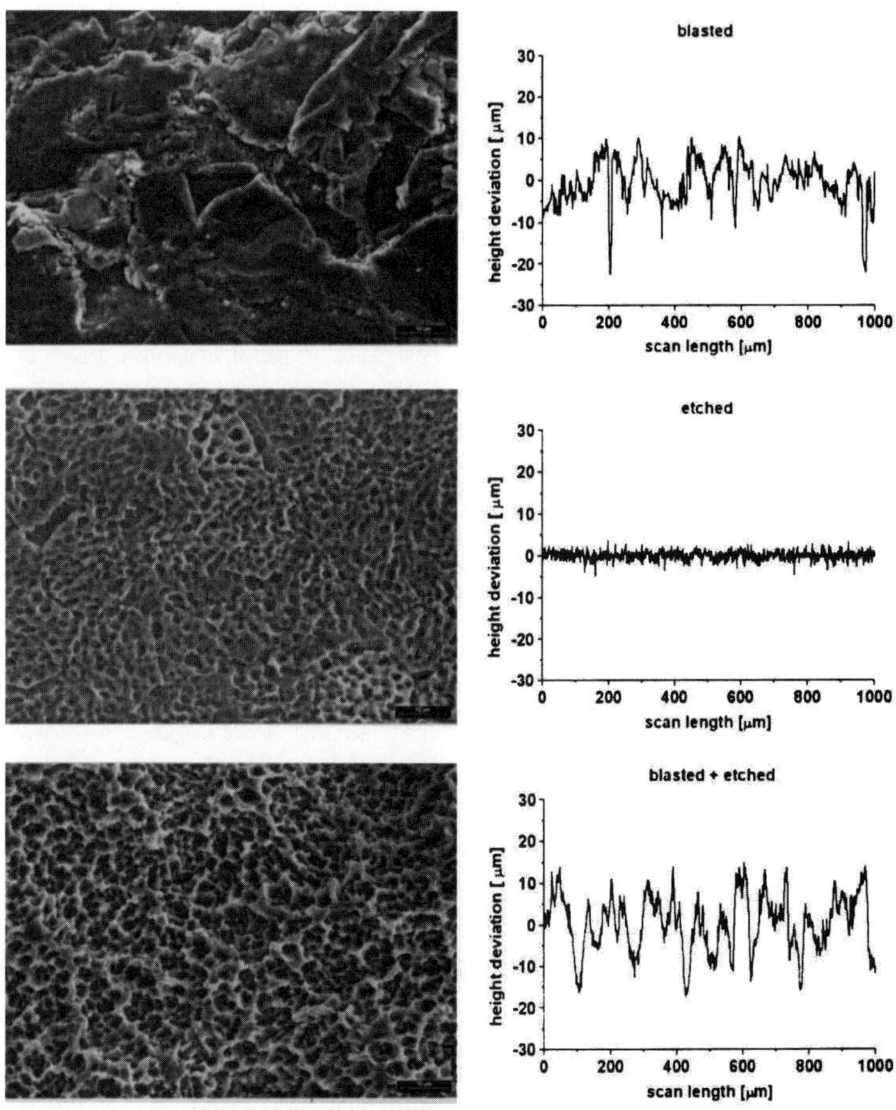

Fig. 9. SEM images of B, E and BE surfaces and their 2-D LPM profiles.

Table 7
List of the roughness values R_a, R_q, R_{zDIN}, S_m, S_k and L_r from 2-D laser scan profiles of the B, E and BE surfaces calculated according to DIN4768 ($n=7$ for each surface; mean values ± standard deviations)

Type of cpTi surfaces	Roughness parameters	Roughness value (standard norm DIN4768)
B surface	R_a (μm)	4.40 ± 0.18
	R_q (μm)	5.66 ± 0.23
	R_{zDIN} (μm)	33.97 ± 1.70
	S_m (mm)	0.070 ± 0.004
	S_k	−0.29 ± 0.20
	L_r	1.16 ± 0.01
E surface	R_a (μm)	0.96 ± 0.02
	R_q (μm)	1.22 ± 0.02
	R_{zDIN} (μm)	7.90 ± 0.29
	S_m (mm)	0.013 ± 0.000
	S_k	−0.16 ± 0.08
	L_r	1.26 ± 0.01
BE surface	R_a (μm)	5.50 ± 0.30
	R_q (μm)	6.91 ± 0.39
	R_{zDIN} (μm)	36.70 ± 1.63
	S_m (mm)	0.075 ± 0.005
	S_k	−0.14 ± 0.17
	L_r	1.17 ± 0.01

curve evaluated with the FFT technique also decreases, starting at the standard roughness value and crosses the

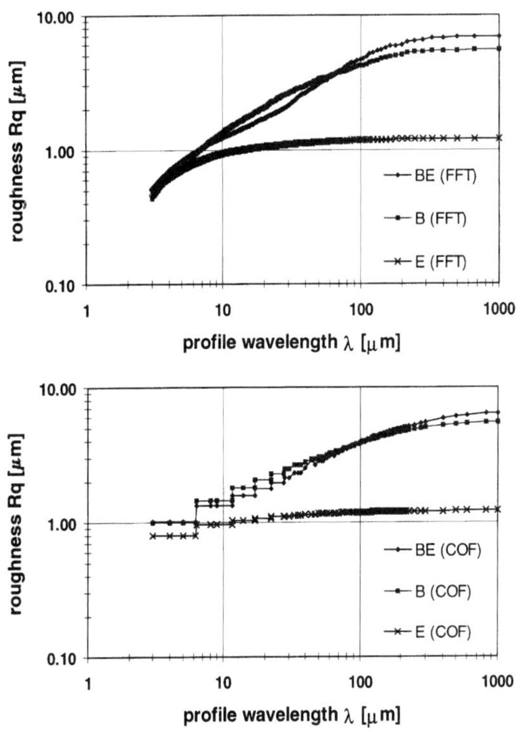

Fig. 10. The comparison of the dependence of the roughness R_q on the profile wavelength for B, E and BE cpTi surfaces calculated using the FFT and below COF technique.

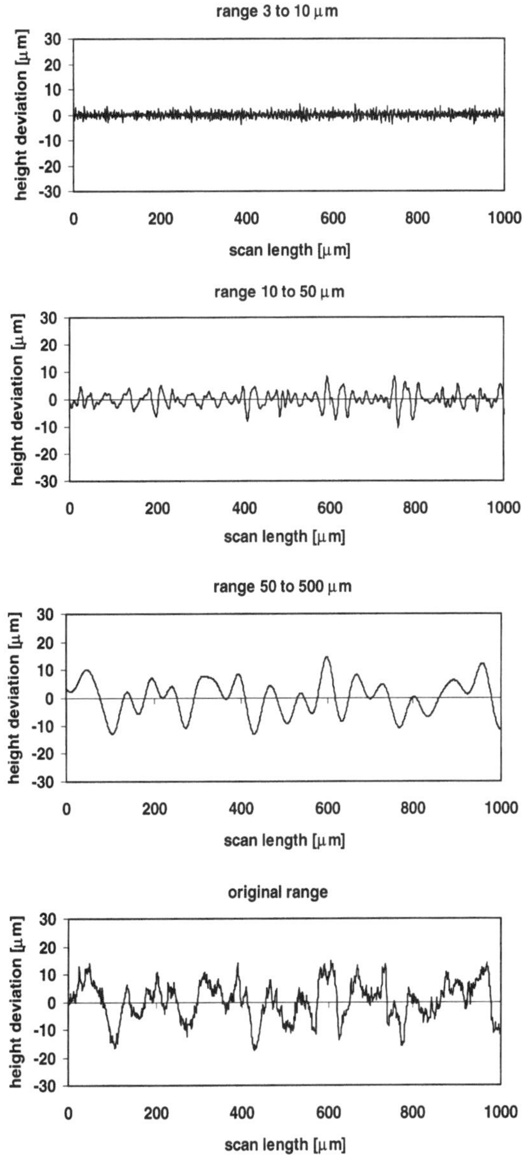

Fig. 11. The inverse FFT profiles of the BE surface in the three wavelength ranges of 3–10, 10–50 and 50–500 μm together with the original profile.

curve of surface B twice at wavelengths of 65 and 7.4 μm, respectively. The wavelength-dependent roughness within that window (65.0 and 7.4 μm) shows decreasing of R_q with decreasing λ for surface BE compared to surface B. This may be due to the fact that the etching process removes the alumina beads from the blasting process and smoothes sharp edges of the surface at the same time (see Figs. 9 and 10). Below 7.4 μm there is again an increase of R_q due to the superimposed structure produced by the etching process. For surface E the curves of the FFT and the COF technique are very similar above a wavelength of 30 μm, and below this value the COF curve falls stepwise. In the case of surface B, the two curves are equal above a wavelength of 700 μm, below the COF curve steadily decreases to a wavelength of approximately 90 μm and then falls off stepwise. For surface BE the curve calculated by the COF technique is always lower compared to the curve evaluated by the FFT technique and starts to decrease stepwise in the same range as the curve for surface B. The two curves cross each other at a wavelength of approximately 260 μm. Below a wavelength of 10 μm, the COF technique will be uncertain.

Our results illustrate that standard roughness parameter sets are of very limited value in describing the complex surface structures present on surface treated titanium implants, since the surfaces often show a variety of topographical features in different dimensional ranges that are all believed to be relevant to the interaction of the surface with biomolecules such as proteins, with cells and with tissue [51–53].

The λ range may be divided up into different ''windows'' of biological interest:

- 500–50 μm: roughness range important for the mechanical interlocking of tissue with the biomaterial surface,
- 50–10 μm: size range typical of the dimension of cells, important for the interaction of individual cells with surface cavities features,
- 10–3 μm: range of sub-cell features such as focal contact points; also relevant for the organisation of adsorbed biomolecule layers.

Fig. 11 shows the inverse FFT profiles of surface BE in the ranges of 3–10, 10–50 and 50–500 μm and the original 2-D LPM profile. The LPM profiles were randomly measured over a distance of 4.096 mm with a resolution of 1 μm. The profile in the range of 3–10 μm looks similar to the profile of surface E in Fig. 9. Table 8 lists the corresponding differential roughness parameters R_a, R_q, R_{zDIN}, S_m, S_k and L_r for the three ''windows'' defined above and the original profile. Seven profiles have been determined for each surface via the inverse FFT spectrum and back-calculation of differential $z(x)$ profiles using the UBM software. The cut-off was set at 0.82 mm using a Gauss filter and an attenuation factor of 50%. The amplitude parameters R_a and R_q, the spacing parameter S_m and the relative length L_r are significantly different in each range for each surface. The R_{zDIN} values in the three ranges of surfaces E and BE are also significantly different. The hybrid parameter S_k is not significantly different for the different ranges. A comparison of the roughness values R_q^2 and R_a^2 in the different ranges with those determined for the original profile indicate an additive effect for each range (see Section 3.2). In the case of surfaces B and BE, the spacing parameter S_m is dominated by the range of 10–50 μm, whereas for surface E the range of 3–10 μm is dominant. The highest relative length was observed in the range of 3–10 μm for all three surfaces. Although a comparison of the roughness values for the original etched profile with those for the lowest range of surface BE indicates no significant differences for the amplitude and hybrid parameters, the relative lengths L_r are clearly different. A comparison of the three surfaces within the ranges of 3–10 and 10–50 μm indicates no

Table 8
List of the roughness values R_a, R_q, R_{zDIN}, S_m, S_k, and L_r from 2-D laser scan profiles of the B, E and BE surfaces for the original profile and the three different size ranges of 500–50, 50–10 and 10–3 μm calculated with a Gauss filter and an attenuation factor of 50% at the cut-off of 0.82 mm ($n = 7$ for each surface; mean values ± standard deviations)

Type of cpTi surfaces	Roughness parameters	Roughness values			
		Range 3–10 μm	Range 10–50 μm	Range 50–500 μm	Original range
B surface	R_a (μm)	0.97 ± 0.04	2.08 ± 0.16	3.31 ± 0.31	4.19 ± 0.18
	R_q (μm)	1.30 ± 0.06	2.85 ± 0.24	4.27 ± 0.38	5.45 ± 0.19
	R_{zDIN} (μm)	10.48 ± 0.77	18.91 ± 2.21	18.33 ± 1.36	34.39 ± 3.27
	S_m (mm)	0.007 ± 0.000	0.027 ± 0.001	0.105 ± 0.006	0.029 ± 0.003
	S_k	0.05 ± 0.11	−0.24 ± 0.27	−0.09 ± 0.12	−0.38 ± 0.17
	L_r	1.62 ± 0.03	1.25 ± 0.02	1.04 ± 0.01	1.73 ± 0.03
E surface	R_a (μm)	0.73 ± 0.01	0.50 ± 0.02	0.265 ± 0.02	0.93 ± 0.04
	R_q (μm)	0.92 ± 0.02	0.64 ± 0.03	0.34 ± 0.02	1.21 ± 0.10
	R_{zDIN} (μm)	6.00 ± 0.35	3.84 ± 0.39	1.51 ± 0.14	8.80 ± 0.77
	S_m (mm)	0.006 ± 0.000	0.027 ± 0.001	0.40 ± 0.13	0.008 ± 0.001
	S_k	0.03 ± 0.07	−0.14 ± 0.10	−0.05 ± 0.34	0.11 ± 0.54
	L_r	1.53 ± 0.02	1.03 ± 0.01	1.00 ± 0.00	1.54 ± 0.01
BE surface	R_a (μm)	0.98 ± 0.02	2.01 ± 0.22	4.47 ± 0.39	5.34 ± 0.21
	R_q (μm)	1.25 ± 0.02	2.56 ± 0.28	5.66 ± 0.45	6.72 ± 0.31
	R_{zDIN} (μm)	8.06 ± 0.14	14.17 ± 2.27	23.25 ± 1.97	34.44 ± 2.11
	S_m (mm)	0.006 ± 0.000	0.028 ± 0.003	0.127 ± 0.018	0.033 ± 0.003
	S_k	0.07 ± 0.05	0.06 ± 0.14	−0.11 ± 0.35	−0.15 ± 0.30
	L_r	1.71 ± 0.01	1.18 ± 0.02	1.05 ± 0.01	1.82 ± 0.01

significant differences between the R_a, R_q, S_m and S_k values of surfaces B and BE and no differences between any of the all three surfaces for S_m. Within the range of 50–500 μm, no significant differences were observed between S_k and L_r of surfaces B and BE. Finally, significant differences were indicated between S_m and L_r of the original profiles of all surfaces measured and calculated according to DIN4768, measured over a length of 4.096 mm with a resolution of 1000 points/mm and calculated according to DIN4768, but with a cut-off of 0.82 mm. In general the calculation of the roughness parameters is a function of the filter type, attenuation factor, cut-off, scan length and resolution. Therefore, the roughness values should only be published together with all measurement and calculation parameters.

This set of results provides the opportunity to correlate in vitro and/or in vivo biological performance data with surface topographical data in the various size ranges of biological relevance. However, due to the importance of surface features in the low micron- (and submicron-) range, experimental techniques with increased lateral resolution (e.g., AFM, stereo-SEM) have to be used to complement the present study.

5. Conclusion and outlook

The concept of wavelength-dependent roughness evaluation is shown to be a useful approach in describing surface topographies, with the advantage that contributions in different wavelength ranges can be separately estimated. This possibility allows technological surface properties to be correlated with roughness contributions in different wavelength ranges.

2-D profiles do not, however, adequately describe 3-D properties, particularly in the case of anisotropic surfaces. The next step in the application of the wavelength-dependent roughness concept will therefore be an extension to 3-D evaluation, as discussed in Section 3.5. Another frequent limitation in applications is related to the limits of lateral and/or vertical resolution and to instrumental artefacts. In comparison to the laser profilometry technique, interference microscopy, AFM and stereo-SEM are able to resolve finer structures and surface features, although — in the case of AFM — problems in case of contacting envelope may be critical for strongly corrugated surfaces. A comparative study of different techniques will be published separately.

Acknowledgements

The authors would like to thank Dr. D. Snétivy of Straumann Institute AG, CH-4437 Waldenburg, Alusuisse Sierre AG, CH-3960 Sierre and Dr. D. Delfosse of Swiss Federal Laboratories for Materials Testing and Research, CH-3600 Thun for support and supply of the materials and V. Frauchiger, ETH Zürich, CH-8092 Zürich for LPM studies. This study was financially supported by the Swiss Priority Program on Materials Research (Program of the Board of the Swiss Federal Institutes of Technology).

References

[1] T.R. Thomas, Rough Surfaces, Longman, London, 1982.
[2] A. Wennerberg, On Surface Roughness and Implant Incorporation, Dissertation, Gothenburg, 1996.
[3] D.J. Whitehouse, Handbook of Surface Metrology, Institute of Physics Publishing, Bristol, 1994, pp. 10–20.
[4] R.S. Sayles, T.R. Thomas, Nature 271 (1978) 431–434.
[5] D.J. Whitehouse, Proc. R. Soc. London, Ser. A. 316 (1970) 97–121.
[6] R.C. Spragg, D.J. Whitehouse, Proc. Inst. Mech. Eng. 185 (1970) 697–707.
[7] T.R. Thomas, R.S. Sayles, Prog. Astronaut. Aeronaut. 39 (1975) 3–20.
[8] E.H. Smith, W.M. Walmsley, Wear 57 (1979) 157–166.
[9] M.I. Yolles, E.H. Smith, W.M. Walmsley, Wear 83 (1982) 151–164.
[10] W.K. Pratt, J. Kane, H.C. Andrews, Proc. IEEE 57 (1969) 58–69.
[11] J.A. Decker, M. Harwit, Appl. Opt. 8 (1969) 2552–2554.
[12] H.F. Harmuth, Transmission of Information by Orthogonal Functions, Springer, Vienna, 1972.
[13] R.S. Choras, Proc. Soc. Photo-Opt. Instrum. Eng. 359 (1982) 336–342.
[14] D.J. Whitehouse, K.G. Zheng, Proc. Inst. Mech. Congr. 206 (1991) 249–264.
[15] P.M. Woodward, Probability and Information Theory with Applications to Radar, Pergamon, London, 1953.
[16] D.J. Whitehouse, K.G. Zheng, Meas. Sci. Technol. 3 (9) (1992) 796–808.
[17] E. Wigner, Phys. Rev. 40 (1932) 749–759.
[18] I. Daubechies, Commun. Pure Appl. Math. 41 (1953) 909–996.
[19] S. Qian, D. Chen, Proc. 26th. Conf. on Inf. Sci. and Systems, Princeton University, March 1992, pp. 1–5.
[20] J. Wexler, S. Raz, Signal Process. 21 (1990) 207–221.
[21] D.J. Whitehouse, Handbook of Surface Metrology, Institute of Physics Publishing, Bristol, 1994, pp. 91–96 and 321–328.
[22] D.J. Mulvaney, Identification of surface texture signals by orthogonal transform analysis, PhD Thesis, Leeds University, 1983.
[23] B.B. Mandelbrot, The Fractal Geometry of Nature, Freeman, San Francisco, CA, 1982.
[24] B.B. Mandelbrot, Science 155 (1967) 636–638.
[25] P. Podsiadlo, G.W. Stachowiak, Wear 217 (1998) 24–34.
[26] C.A. Brown, Mater. Process. Technol. 44 (1994) 337–344.
[27] S. Chesters, H.Y. Wen, M. Lundin, G. Kasper, Appl. Surf. Sci. 40 (1989) 185–192.
[28] F.F. Ling, Wear 136 (1990) 141–156.
[29] M. Allen, G.J. Brown, N.J. Miles, Powder Technol. 84 (1995) 1–14.
[30] J.M. Bennett, L. Mattsson, Introduction to Surface Roughness and Scattering, Optical Society of America, Washington, DC, 1989.
[31] T.R. Thomas, Rough Surfaces, Longman, London, 1982, p. 21.
[32] C.A. Brown, G. Savary, Wear 141 (1991) 211–226.
[33] L. Mattsson, P. Wågberg, Precis. Eng. 15 (3) (1993) 141–149.
[34] R. Windecker, Tech. Mess. 60 (1993) 267–270.
[35] K.J. Stout, Three-Dimensional Surface Topography: Measurement, Interpretation and Applications, Penton Press, London, 1994, p. 59.
[36] M. Geiger, M. Pfestorf, U. Engel, Stahl Eisen 115 (7) (1995) 47–53.
[37] D.J. Whitehouse, Handbook of Surface Metrology, Institute of Physics Publishing, Bristol, 1994, p. 16.
[38] DIN 4777, Metrology of Surfaces; Profile Filters for Electrical Contact Stylus Instruments; Phase-Corrected Filters, Beuth Verlag, Berlin, 1990.

[39] E.O. Brigham, Schnelle Fourier-Transformation, München, Oldenburg, 1995.
[40] J. Timm, K. Armbruster, M. Osterhold, W. Hotz, Baender, Bleche, Rohre 35 (9) (1994) 110–114.
[41] DIN 4768, Determination of Surface Roughness Values of the Parameters R_a, R_z, R_{max} by Means of Electrical Contact (Stylus) Instruments; Terminology, Measuring Conditions, Beuth Verlag, Berlin, 1990.
[42] C. Beck, R. Gutsch, Feingeraetetechnik 24 (1977) 224–227.
[43] D.J. Whitehouse, Handbook of Surface Metrology, Institute of Physics Publishing, Bristol, 1994, p. 212.
[44] Y. Ju, T.N. Farris, ASME Tribol. 115 (1996) 320–328.
[45] I.A. Polonsky, T.P. Chsng, L.M. Keer, W.D. Sproul, Wear 208 (1994) 204–219.
[46] D.L. Jordan, R.C. Hollins, E. Jakeman, Wear 109 (1986) 127–134.
[47] M. Osterhold, W. Hotz, J. Timm, K. Armbruster, Baender, Bleche, Rohre 35 (10) (1994) 44–50.
[48] D.M. Brunette, Surface Characterisation of Biomaterials, Elsevier, Amsterdam, 1988, pp. 203–217.
[49] D. Buser, R.K. Schenk, S. Steinemann, J.P. Fiorellini, C.H. Fox, H. Stich, Biomed. Mater. Res. 25 (1991) 889–902.
[50] M. Wong, J. Eulenberger, R. Schenk, E. Hunziker, Biomed. Mater. Res. 29 (1995) 1567–1575.
[51] B. Kasemo, J. Lausmaa, CRC Crit. Rev. Biocompat. 2 (1986) 335–380.
[52] B. Chehroudi, T.R.L. Gould, D.M. Brunette, Biomed. Mater. Res. 23 (1989) 1067–1085.
[53] J.Y. Martin, Z. Schwartz, T.W. Hummert, D.M. Schraub, J. Simpson, J. Lankford, D.D. Dean, D.L. Cochran, B.D. Boyan, Biomed. Mater. Res. 29 (1995) 389–401.

5b. Chemical characterization by scanning-probe methods

Commentary

Atomic force microscopy (AFM) was a breakthrough technique in the late 1980s, since it brought nanoscale morphological imaging into the hands of many scientists who did not have the means to acquire vacuum equipment. Furthermore, it allowed imaging and force measurements in liquid media, which was of particular interest to biologists and colloid scientists. We felt that the force-measuring capabilities of AFM could readily be utilized as a chemically sensitive imaging method, by utilizing forces that were dependent on the chemical species present. Our first efforts in this direction relied upon the fact that different oxides under electrolytes have different isoelectric points. This means that their charges will flip at different pH values as the pH is varied and therefore they can be identified via the Coulombic interaction force with a charged AFM tip. These effects can be monitored by measuring both normal and lateral (friction) forces, the latter being particularly useful as a means to imaging the location of different oxides in a surface (5.2, 5.3).

When it comes to polymers, the situation is a little different, since the surfaces are rarely charged to the same extent as oxides. However, we were able to capitalize on the different van der Waals interactions between the AFM tip and different polymers (which have different refractive indices, leading to different Hamaker constant values, via the Lifshitz equation). In this way, the surface composition of immiscible polymer blends could be imaged in lateral-force mode (5.4). This work was a collaboration with our colleagues Paul Smith and Theo Tervoort at the ETH, who made the clever suggestion of using perfluorodecalin as an imaging medium. This liquid has a very low refractive index and therefore leads to a much greater Hamaker constant, which dramatically improves signal-to-noise ratio in the force measurements.

We used AFM force measurements in a collaboration with Michael Grunze's group at the University of Heidelberg, to examine the differences in protein-adhesion resistance of SAMs of oligoethylene glycols on silver and gold surfaces. We used alkanethiol-functionalized, gold-coated tips, to mimic the hydrophobic components of proteins interacting with the surfaces. We found that on the functionalized silver surfaces (which are not protein-resistant), an attractive interaction was observed with the hydrophobic tips, whereas on the functionalized gold surfaces (which are protein resistant), a repulsion was observed, which was ionic-strength dependent and therefore presumably electrostatic in origin (5.5).

4632 *Langmuir* **1995,** *11,* 4632–4635

Sensitivity of Frictional Forces to pH on a Nanometer Scale: A Lateral Force Microscopy Study

A. Marti, G. Hähner,* and N. D. Spencer*

Laboratory for Surface Science and Technology, Swiss Federal Institute of Technology, ETH-Zentrum, Sonneggstr. 5, CH-8092 Zürich, Switzerland

Received August 7, 1995. In Final Form: October 23, 1995®

An atomic force microscope (AFM) operating in the lateral (frictional) force mode (LFM) was used to investigate the pH dependence of kinetic friction between a silicon nitride tip and an oxidized silicon surface under aqueous electrolyte solutions. It was found that, besides the known pH dependence of the normal force curves, the frictional force is highly sensitive to the pH value of the solution. A linear relationship between frictional force and adhesion hysteresis (i.e., the dissipated energy), as recently shown to be valid for some systems by Israelachvili et al., is supported. The implication is that many chemical species on surfaces can be readily differentiated using AFM-LFM by varying the pH value of the surrounding medium and that a possible nanochemical imaging method could be developed, based upon this principle.

Introduction

Surface chemical imaging has great potential for the analysis of complex, multifunctional systems. However, despite the plethora of new scanning probe methods developed in the last decade, the ability to image surfaces chemically on a nanometer scale has not been significantly advanced until very recently. A promising approach seems to be the combination of the unique lateral resolution of the scanning probe methods with designed chemical interactions between tip and sample.[1–3]

Several different approaches to this problem have been investigated by others. These include functionalizing the scanning tip[1–4] or performing measurements in an electrochemical cell.[5] Most of these experiments have focused chiefly on identifying different functional groups within thin organic layers[1–3] although there also exists a significant technological interest in obtaining high-lateral-resolution chemical images of inorganic surfaces, such as the oxide systems used in semiconducting devices or catalysts.[6]

Oxide surfaces (and many adsorbed organic species) in an aqueous solution are generally electrically charged due to the interaction of surface $-OH$ groups with H^+ or OH^-. These surfaces display a *characteristic* isoelectric point (IEP), corresponding to the pH at which the net surface charge is zero (if there is no specific adsorption of ions other than H^+ or OH^-). A spatial mapping of the isoelectric point can therefore yield spatially resolved chemical information about the surface species present.

In this communication we describe a new approach to mapping the chemical nature of surface species via determination of local IEP, involving an atomic force microscope (AFM), fitted with a fluid cell, operating in both normal and lateral (frictional) force mode (LFM). The crucial point of these experiments is the variation in electrolyte pH between AFM-LFM measurements, while maintaining constant ionic strength. Although it is known that on the millimeter scale the pH value has an influence on the classical friction coefficient "μ"[7] and that the normal force between a silicon nitride tip and an oxide substrate sample depends strongly on the pH value,[8–10] to our knowledge this is the first published investigation dealing with the relation between the pH value and the frictional force on nanometer scale. Additionally, by comparing lateral force measurements with normal force curves (i.e. loading–unloading cycles that display adhesion hysteresis) taken under the same conditions, we have been able to support observations by Israelachvili et al.[11,12] that frictional force and adhesion hysteresis are linearly correlated.

Experimental Section

All samples were cut from a Si(100) wafer (Faselec AG). Prior to the measurements, the substrates were cleaned for 2 min by immersion in an acid mixture (H_2O_2 and H_2SO_4) in order to eliminate organic contaminants,[13] followed by an extensive rinse with ultrapure water (EASYpure RF, Barnstead, 18.2 MΩ cm). After the cleaning procedure, the samples were completely wettable by ultrapure water, demonstrating the hydrophilicity of the surfaces.

The measurements were performed with a commercially available AFM-LFM (Nanoscope III, Digital Instruments, Santa Barbara, CA), fitted with a liquid cell. With this instrument, both normal and lateral forces can be measured between a sample and a tip as a function of the displacement of the sample. The nominal spring constants of the commercial Si_3N_4 cantilevers used (Nanoprobes, Digital Instruments) were 0.12 and 0.58 N/m. The frictional force measurements were performed with both a 0.7 and 12 μm scanner and a scan size of 100 nm in single-scan-line mode. The scanning velocity was set to 0.8 μm/s. The lateral force was not observed to be dependent on the scan velocity, within the velocity range 0.2–8 μm/s, confirming that the nature

* To whom correspondence should be addressed.

® Abstract published in *Advance ACS Abstracts,* December 1, 1995.

(1) Frisbie, C. D.; Rozsnyai, L. F.; Noy, A.; Wrighton, M. S.; Lieber, C. M. *Science* **1994** *265,* 2071.
(2) Wilbur, J. L.; Biebuyck, H. A.; MacDonald, J. C.; Whitesides, G. M. *Langmuir* **1995**, *11,* 825.
(3) Akari, S.; Horn, D.; Keller, H.; Schrepp, W. *Adv. Mater.* **1995**, *7,* 549.
(4) Pierce, M.; Stuart, J.; Pungor, A.; Dryden, P.; Hlady, V. *Langmuir* **1994**, *10,* 3217.
(5) Binggeli, M.; Christoph, R.; Hintermann, H.-E. *Tribol. Lett.* **1995**, *1,* 13.
(6) Somorjai, G. A. *Introduction to Surface Chemistry and Catalysis*; John Wiley: New York, 1994.
(7) Zhu, Y. Y.; Kelsall, G. H.; Spikes, H. A. *Tribol. Trans.* **1994**, *37,* 811.
(8) Butt, H.-J. *Biophys. J.* **1991**, *60,* 1438.
(9) Lin, X.-Y.; Creuzet, F.; Arribart, H. *J. Phys. Chem.* **1993**, *97,* 7272.
(10) Senden, T. J.; Drummond, C. J.; Kekicheff, P. *Langmuir* **1994**, *10,* 358.
(11) Yoshizawa, H.; Chen, Y.-L.; Israelachvili, J. N. *J. Phys. Chem.* **1993**, *97,* 4128.
(12) Israelachvili, J. N.; Chem, Y.-L.; Yoshizawa, H. *J. Adhesion Sci. Technol.* **1994**, *8,* 1231.
(13) Ulman, A. *Ultrathin Organic Films*; Academic Press: London, 1991.

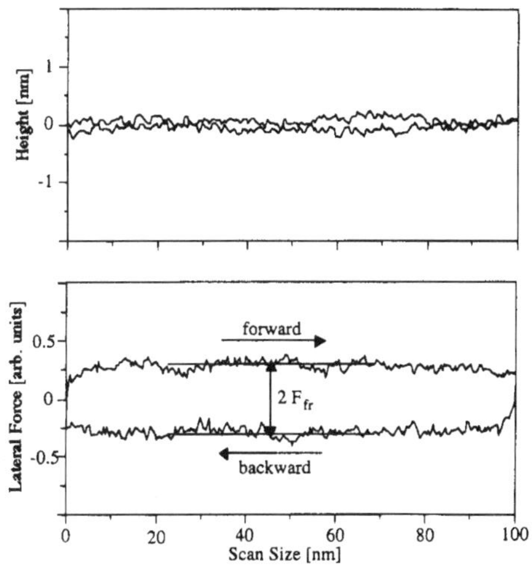

Figure 1. Topographical and lateral force line scans of 100 nm, recorded with an applied load of 9.5 nN and pH = 4. The mean roughness of the sample is below 0.1 nm. The difference between forward and backward scans in the lateral force scan corresponds to twice the frictional force.

of the observed lateral force is frictional rather than viscous. The applied load was varied between 2 and 40 nN. To check the constancy of the applied load during the frictional experiments, it was checked again immediately after the measurement. The observed deviation in all cases was below 1 nN.

The electrolyte used was a 1 mM aqueous NaCl solution (Fluka Chemie AG). The pH values were adjusted with HCl in the acid range and NaOH (Merck AG) in the alkaline range, choosing an appropriate acid or base concentration in order to maintain a nearly constant ionic strength. The H_3O^+ concentration of the aqueous electrolyte solution was monitored with a pH meter (Mettler, Delta 340). The pH values investigated in this work ranged from 3.5 to 9. It is known that at pH values greater than 9, SiO_2 starts dissolving appreciably.[14]

The force curves, i.e., loading–unloading cycles, were recorded with a velocity of 0.4 μm/s. Hysteresis in the force–distance curves is known to occur under certain conditions of scan velocity.[15] However, hysteresis was minimal for the velocity used in the present work. Following measurement, the tip displacement versus distance curves recorded with the AFM-LFM were transformed to force versus tip–sample separation curves, following the procedure described in refs 16 and 17.

Results and Discussion

The experimental results in Figure 1 display a single line scan over a range of 100 nm for an applied load of 9.5 nN and a pH of 4. The upper panel shows the topographical signal, while the lower panel shows the signal due to lateral forces. To determine the frictional force, we have calculated the difference between a forward and a backward scan. Since the samples are very smooth (roughness R_A below 0.1 nm over a scan range of 100 nm), a topographical contribution to the lateral force signal is expected to be small (see below). To improve the signal-to-noise ratio, up to four single scans were averaged to determine the kinetic frictional force.

(14) Iler, R. K. *The Chemistry of Silica*; Wiley-Interscience: New York, 1979.
(15) Hoh, J. H.; Engel, A. *Langmuir* **1993**, *9*, 3310.
(16) Ducker, W. A.; Senden, T. J.; Pashley, R. M. *Nature* **1991**, *353*, 239.
(17) Ducker, W. A.; Senden, T. J.; Pashley, R. M. *Langmuir* **1992**, *8*, 1831.

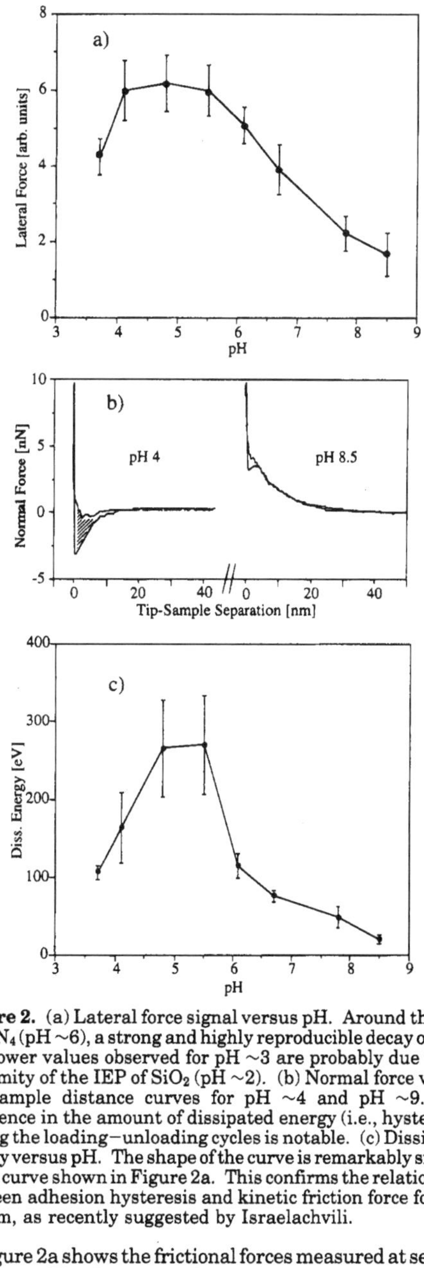

Figure 2. (a) Lateral force signal versus pH. Around the IEP of Si_3N_4 (pH ~6), a strong and highly reproducible decay occurs. The lower values observed for pH ~3 are probably due to the proximity of the IEP of SiO_2 (pH ~2). (b) Normal force versus tip–sample distance curves for pH ~4 and pH ~9. The difference in the amount of dissipated energy (i.e., hysteresis) during the loading–unloading cycles is notable. (c) Dissipated energy versus pH. The shape of the curve is remarkably similar to the curve shown in Figure 2a. This confirms the relationship between adhesion hysteresis and kinetic friction force for this system, as recently suggested by Israelachvili.

Figure 2a shows the frictional forces measured at several different pH values, together with their standard deviations. Below the isoelectric point of silicon nitride (pH ~ 6[9]), the tip and sample surfaces are oppositely charged (IEP of SiO_2 pH ~ 2[9,18]). In this region the surface charge (and hence the surface potential) shows a gradual change as the pH is moved toward the IEP of SiO_2. Around the IEP of Si_3N_4, a strong decay in frictional force is observed as the surface charge on the tip moves from positive to negative and the electrostatic interaction with the surface moves from attractive to repulsive.

Figure 2b shows the normal force (i.e., applied load) versus tip–sample distance curves for pH values of 4 and 8.5. The change in adhesion hysteresis (area enclosed by

(18) Parks, G. A. *Chem. Rev.* **1965**, *65*, 177.

the approaching and retracting curves), which corresponds to different amounts of dissipated energy, is very pronounced. We have calculated these energy values by integrating over loading–unloading cycles and have plotted these data versus the pH together with their standard deviations (Figure 2c). The amounts of dissipated energy are in the 100 eV range, which is most likely consumed as vibrational energy of the cantilever. It is unlikely that significant mechanical wear occurs during a loading–unloading cycle; i.e., the tip remains effectively unchanged during the experiments. This view is supported by the reproducibility of the measurements.

Both the kinetic frictional force and the adhesion hysteresis curves show remarkably similar behavior as the pH is varied. For pH values below the IEP of Si_3N_4, the lateral force and the amount of dissipated energy are both high. Above the IEP they both drop to low values, showing a characteristic and highly reproducible decrease around the IEP. In all cases a close relationship between the adhesion hysteresis and the kinetic frictional force was observed, with a marked sensitivity of both entities to pH at values close to the IEP.

The close relationship between the dissipated energy and the kinetic friction force can be understood in terms of a model due to Israelachvili et al. that was recently proposed and shown to be valid for several systems.[11,12] According to this model the kinetic friction force can be written as

$$F = C_1A + C_2L \quad (1)$$

where A is the adhesion hysteresis (energy dissipated) and L the applied load. The coefficient C_1 depends on the mutual adhesion of the two surfaces, while C_2 depends on the surface topography. While the second term corresponds to the "μL" term in Amontons' law, the first term varies in a more complex way with applied load.[12,19]

Since the silicon surface is very smooth, the coefficient C_2 is very small and hence the dominating contribution to the kinetic friction force comes from the adhesional component. Figure 3 displays the dependence of the lateral force on the applied load for two different pH values. Over a range of 40 nN, the slope of the $F-L$ curve (the "classical" friction coefficient) changes slightly with increasing L, the slope of both curves being somewhat steeper in the lower normal force range, changing to a gentler slope for higher normal forces. This observation suggests that the increase in the frictional force with load, L, comes principally from the first term in eq 1, which is nonlinear in L, and not from the second term. We have used the JKR model[20] to describe our data. The solid line in Figure 3 is the best fit of the JKR model to the measured data. Although the friction force is not calibrated in our experiment, the product RW of the radius of the tip and the adhesion energy can be determined. With an assumed radius of $R = 30$ nm,[21] the adhesion energy is on the order of 10 mJ m^{-2} for pH = 4 and 0.1 mJ m^{-2} for pH = 9. This is in excellent agreement with the values derived from the force curves, since RW can also be determined from the pull-off force,[19] yielding the same values. We interpret the rather small values of the adhesion energies as resulting from the creation of unit area in the *ionic* layer, i.e., energies connected with charge density fluctuations in the ionic layer surrounding tip and surface.

(19) Israelachvili, J. N. *Intermolecular and Surface Forces*, 2nd ed.; Academic Press: London, 1992.
(20) Johnson, K. L.; Kendall, K.; Roberts, A. D. *Proc. R. Soc. London* **1971**, *A324*, 301.
(21) Nanoscope III, Digital Instruments, Santa Barbara, CA.

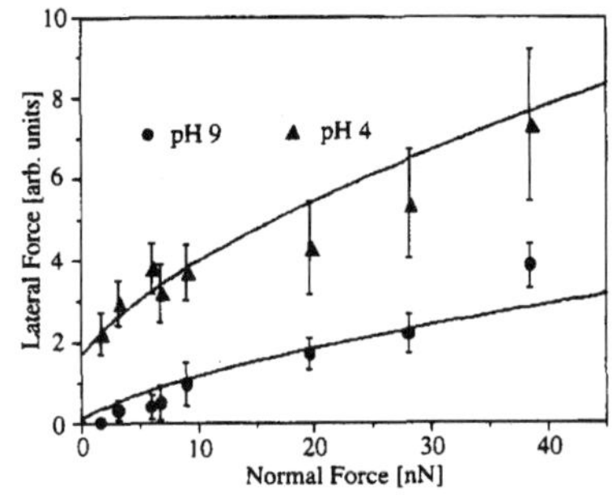

Figure 3. Lateral force versus applied load for pH ~4 and pH ~9. The solid curves are the best fit JKR model. Note that although the lateral force signal is not calibrated in our experiment, the adhesion energy can be determined from this fit, if the radius of the tip is known. The same entity can be obtained from the pull-off forces from loading–unloading cycles and the values are in exellent agreement.

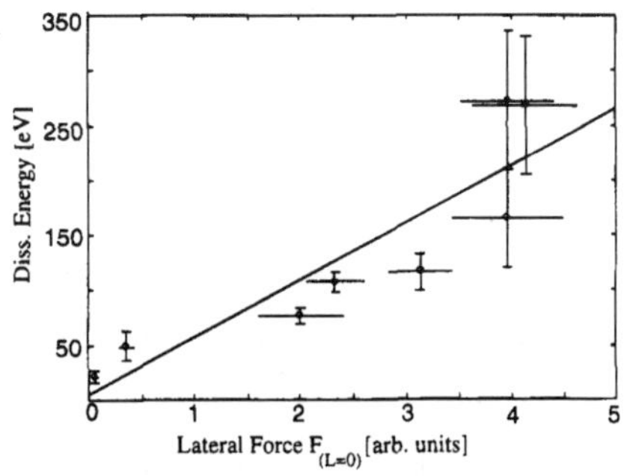

Figure 4. Dissipated energy versus lateral force that has been extrapolated to zero applied load. Also shown is the best fit from linear regression. The data support a linear relationship between both entities; however, the signal to noise ratio is insufficient to extract the exact relationship.

In order to examine the relationship between the dissipated energies and the lateral forces in more detail, we have extrapolated the force values measured with an applied load of 9.5 nN to zero load, assuming that the JKR model is applicable, based on Figure 3. Figure 4 shows the resulting data points together with the best fit straight line. Although the noise in the data is too great for the exact relationship to be extracted, the plot supports the proposed linear relationship between the two entities.

On the basis of the force curves shown in Figure 2b, it can be predicted that for different chemical species (with different IEPs) on a surface, a pH-dependent topographical contrast due to chemical differences should be detectable, provided that the applied load is in the appropriate range. In other words, the changes observed in a series of AFM images taken at different pH values can yield chemical information about the system. Similarly, chemical information may be derived from the pH dependence of LFM images due to the electrochemical component of the frictional force.

By variation of the pH, the surface charge distribution and therefore the electrostatic contributions to both normal and lateral forces are also changed. Our results suggest that for a given surface, the pH value likely to yield the highest chemical contrast in *lateral* force maps can be determined solely by determining the maximum adhesional hysteresis from *normal* force–distance curves, even if the exact state of the tip is not known. AFM and LFM measurements under programmed pH and constant ionic strength conditions can therefore help elucidate the surface chemistry, allowing surface species to be imaged according to their surface electrochemical properties. This method therefore shows potential for nanochemical imaging of both organic and inorganic surfaces.

A corollary of the behavior described above is that caution needs to be exercised when interpreting AFM scans carried out under electrolytes, since apparent topography may be due to electrochemically induced differences in the characteristic force curves for different surface species.

Acknowledgment. We thank J. Israelachvili for a fruitful discussion during the conference on Physics of Sliding Friction, held in Trieste, Italy, in June 1995.

LA950660V

Tribology Letters 3 (1997) 359–365

The influence of pH on friction between oxide surfaces in electrolytes, studied with lateral force microscopy: application as a nanochemical imaging technique

Georg Hähner, Andreas Marti and Nicholas D. Spencer

Laboratory for Surface Science and Technology, Department of Materials, ETH Zürich, ETH-Zentrum, Sonneggstrasse 5, CH-8092 Zurich, Switzerland

Received 10 February 1997; accepted 19 May 1997

We have investigated the lateral force (frictional) signal between the Si_3N_4 tip of an AFM and SiO_2 and AlO_x surfaces in a 1 mM NaCl solution, as a function of pH. It was found that the frictional force depends strongly on both the pH and the isoelectric points (IEP) of the materials under investigation. A simple linear model describing the dependence of the lateral force on the total normal force has been used to account qualitatively for the observed lateral force signal from the single-asperity contact between tip and surface on a nanometer scale. The observed pH-dependence of the lateral force signal on different oxide surfaces has been applied to reverse the chemical contrast on an appropriate sample composed of two different oxides, thus demonstrating the potential of this method for "chemical imaging".

Keywords: AFM/LFM, lateral force (friction) signal, oxide (amphoteric) surfaces, pH, nanochemical imaging

1. Introduction

The significance of friction has been realized since ancient times [1] since it plays a crucial role in many everyday processes. Most investigations in the area of tribology have been motivated by engineering problems, i.e., they were of practical interest and hence can be classified as "macrotribological". However, it has also been recognized for some time that a prerequisite for a thorough understanding of frictional phenomena on a macroscopic scale is an understanding of frictional processes on the microscopic level [2]. With the invention of surface analytical tools, such as the atomic force microscope (AFM), that allow single asperity contacts to be studied, experimental conditions are now available for tackling relevant problems at the nanometer scale [3]. Based on the potential of the AFM and other techniques, the area of "nanotribology" has grown significantly over the last few years [4–7]. Beside a fundamental interest in understanding the tribological properties of a single-asperity contact, significant effort has also been put in the direction of "chemical imaging", where attempts have been made to combine the frictional (chemical) sensitivity of the AFM with its unique lateral resolution [8–11]. Most of the work done in this area has been concerned with producing a contrast between chemically different organic species on surfaces, based on the different interactions of functional groups [8–11]. However, there are no standard recipes available to unambigously identify different chemical species on surfaces in general. In particular there is still a need for chemical imaging on inorganic surfaces, such as oxides, which play a vital role in many technical systems in such areas as catalysis and biocompatibility [12,13].

In the present paper we report on the sensitivity of the lateral force between an AFM tip and oxide surfaces to the pH of the surrounding electrolyte, and we demonstrate its potential for the chemical imaging of inorganic surfaces by capitalizing on the sensitivity of the lateral forces to both pH and the isoelectric points of the materials.

2. Experimental

The materials used in the pH-dependence experiments were pieces of oxidized silicon cut from a SiO_2-covered wafer (Faselec AG) and pieces of aluminum oxide from sapphire disks (Saphirwerke Brügg). For the imaging studies a regular pattern of surface-oxidized aluminum squares in the micrometer range was produced by means of electron-beam lithography on a Si wafer covered with a native oxide layer [14].

Prior to the friction measurements, the substrates were cleaned for 2 min by immersion in an acid mixture (H_2O_2 and H_2SO_4) in order to eliminate organic contaminants [15], followed by an extensive rinse with ultrapure water (EASYpure RF, Barnstead, 18.2 MΩ cm). After the cleaning procedure, the samples were completely wettable by ultrapure water, demonstrating the hydrophilicity of the surfaces.

The force measurements were performed by means of a commercially available AFM/LFM (Nanoscope III, Digital Instruments, Santa Barbara, CA), fitted with a liquid cell. With this instrument, both normal and lateral forces can be measured between a sample and a tip as a function of the displacement of the sample. The nominal spring constants of the commercial Si_3N_4 cantilevers used (Nanoprobes, Digital Instruments) were 0.12 N/m and 0.58 N/m. The frictional force measurements were performed both with a 0.7 μm and a 12 μm scanner and a scan size of 100 nm in single-scan-line mode. The frictional force was determined by calculating the difference of the lateral signal between forward and backward scanning [5]. The scanning velocity was set to 0.8 μm/s. The lateral force was not observed to be dependent on the scan velocity, within the velocity range of 0.2 to 8 μm/s, confirming that the nature of the observed lateral force is frictional rather than viscous. The applied load was typically below 10 nN. The load was checked immediately following the measurement in order to ensure its constancy during the frictional experiments. The observed deviation was below 1 nN in all cases.

For the imaging studies the samples were cleaned in an ultrasonic bath of a microdeposit remover [14] for 2 min followed by rinsing with the same liquid. Subsequently they were treated ultrasonically with ultrapure water, also followed by an extensive rinse. Finally they were blown dry in a nitrogen stream. Si_3N_4 tips (Park Scientific Instruments, Sunnyvale, CA) with a nominal cantilever force constant of 0.03 N/m were used. Prior to the measurement they were plasma-cleaned (Harrick PDC-32G) in an O_2 atmosphere for 5 min. The applied load during scanning was approximately 6 nN with a scanning velocity of 6 μm/s. The scanner allowed a maximal lateral range of 125 μm.

The electrolyte used in all experiments was a 1 mM aqueous NaCl solution (Fluka Chemie AG). The pH values were adjusted with HCl in the acid range and NaOH (Merck AG) in the alkaline range, choosing appropriate acid or base concentrations in order to maintain a nearly constant ionic strength. The H_3O^+ concentration of the aqueous electrolyte solution was monitored with a pH-meter (Mettler, Delta 340). The pH values investigated in this work ranged from 3.5 to 10.5.

3. Results and discussion

Figure 1 displays the experimentally observed lateral force (frictional) signal between the Si_3N_4 tip and a SiO_2 surface in a 1 mM NaCl solution for various pH values. The curve shows a maximum, with the strongest dependence on pH occurring between the values of 3.5 and 6.5.

Figure 2 shows the same measurement performed on an AlO_x surface. Similarly to the SiO_2 substrate, the frictional signal shows a maximum between two particular

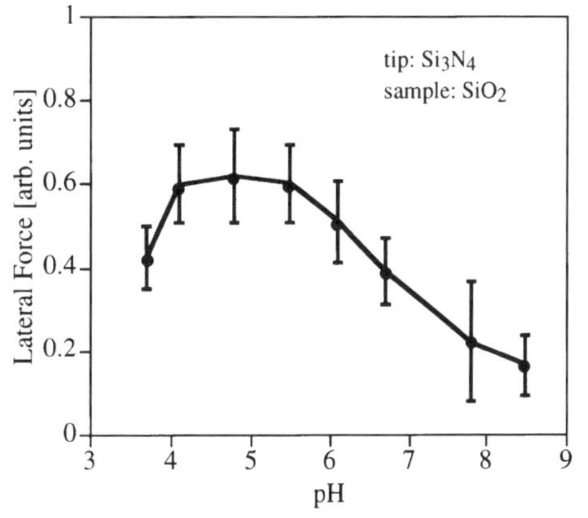

Figure 1. Lateral force signal between a Si_3N_4 tip and a SiO_2 surface in a 1 mM NaCl solution as a function of pH. The applied load during the measurement was on the order of 10 nN. The strongest dependence of the lateral signal is observed between pH values of 3.5 and 6.5, which are close to the isoelectric points of the materials in contact.

pH values, in this case 6 and 9, and lower, nearly constant values outside this range.

In general, our experimental observations on oxide surfaces indicate that the lateral signal, and hence frictional force, are strongly pH-dependent.

Oxide surfaces are typically hydroxyl terminated. In aqueous electrolytes these Me–OH groups can react with the OH^- and H_3O^+ ions from solution either by establishing Me–OH_2^+ (acidic conditions) or Me–O^- (basic conditions), hence forming a positively or negatively charged surface, depending on the pH. The isoelectric point (IEP) of these so-called amphoteric surfaces is the

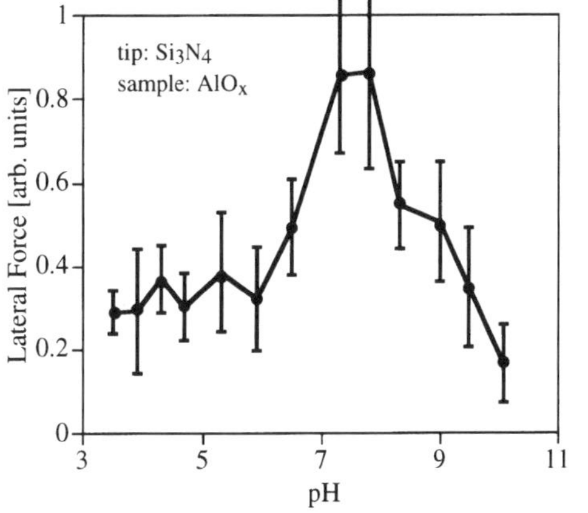

Figure 2. Same as figure 1 but for an AlO_x surface. The applied load during the measurement was on the order of 10 nN. The strongest dependence of the lateral signal in this case is observed between pH values of 6 and 9, the IEPs of tip and sample surface, while outside this interval a low and nearly constant signal is measured.

pH value at which the overall net surface charge is zero, provided that there is no specific adsorption of other ions. The IEP is characteristic for the material. However, although the *overall* surface charge is zero at the IEP, there are still charges present. The absolute number of charged sites at the IEP is also characteristic for the material [12,16].

The IEPs of the materials used are ~ 2 for SiO_2 [17] and typically between 5.5 and 7 for the Si_3N_4 tip [19], while the IEP of AlO_x is ~ 9 [17]. The limiting values of the pH range where the strongest change of the experimentally observed frictional signal occurs are thus close to the IEPs of the materials involved. Hence the observed lateral forces are sensitive to both pH and the IEPs of the materials under investigation.

3.1. Lateral force

Since we are concerned with a single-asperity contact in the submicrometer range and at low loads, Amontons' law is not applicable and adhesion plays a significant role [19]. This has previously been demonstrated for a Si_3N_4 tip and a SiO_2 surface in an electrolyte environment [20]. A simple linear equation describing lateral forces in situations where Amontons' law is not appropriate was suggested some 40 years ago [21]. Beside the externally applied load, an additional term due to "internal pressure" must be taken into consideration. In general, lateral forces, F_{lat}, in our experiment can be affected by the applied load, L, a possible (constant) adhesion force, K_0, and an internal pressure due to both Van der Waals and electrostatic interactions:

$$F_{lat} = K_0 + \mu_1 AP + \mu L , \quad (1)$$

where AP adds to the external load, L, and the adhesion constant, K_0. μ_1 and μ are constants, with μ being the classical coefficient of friction. In our case, the internal pressure P has two contributions: one stems from the attractive Van der Waals forces and the other from the electrostatic pressure due to the surface charges. A is the (purely geometrical) contact area, which stays constant, since L is constant. To first order, Van der Waals attraction is not affected by pH. Hence, the only term that depends on pH is the electrostatic contribution to P. This electrostatic pressure, P_{el}, is determined by the potentials of the two surfaces, Ψ, that in turn are related to the surface charges and the pH. The functional dependence of the lateral force F_{lat}, on the potentials Ψ, of the two surfaces, respectively on the pH of the surrounding liquid is, of course, not linear.

3.2. Oxide surfaces in electrolytes

The relation between the potential Ψ, and the pH can often be adequately described by the Nernst equation [12], resulting in a linear functional dependence. However, for oxide systems this equation is not always appropriate since the activity of the potential-determining ions depends on the surface potential [12].

Since the surface charge in oxide systems is caused by the reaction of amphoteric hydroxyl groups which, in turn, is determined by the pH, the surface potential, Ψ, is a function both of pH and the bulk electrolyte concentration which results in a more complex (non-linear) relation between pH and potential.

The final charge on the surface is balanced by counterions from the solution, establishing the so-called diffuse double layer. The solution of the non-linear Poisson–Boltzmann (PB) equation yields the density of counterions, the potential and the electric field at any point [19]. The characteristic length or "thickness" of the diffuse double layer is the so-called Debye length, which determines the exponential decay of the counterion density away from the surface. It can also be viewed as a screening length and depends solely on the properties of the liquid such as the concentration of the electrolyte, and not on any property of the surface. For a 1 mM solution of a 1 : 1 electrolyte it is approximately 10 nm [19].

The PB-equation, however, cannot be solved analytically for all geometries and all boundary conditions. While for a flat surface an analytical expression can be derived, for other geometries, such as a sphere, approximations have to be made, in order to deduce relations between potential and charge density [12].

In our experiment the two double layers surrounding tip and sample interact during scanning, while the two surfaces are "in contact".

3.3. Interaction between two oxide surfaces in an electrolyte

The interaction between two oxide surfaces in an electrolyte environment has been extensively studied both theoretically and experimentally [12]. The DLVO theory includes both Van der Waals attraction and electrostatic interactions [19]. Horn et al. showed that simple DLVO theory can be used to describe the interaction between sapphire surfaces in a 1 mM NaCl solution quite accurately up to distances of 8 nm, with the potentials being rather moderate (<40 mV), even for high pH values [22].

When two surfaces in an electrolyte environment approach one another, their double layers overlap and several situations may arise. In the case of oxide surfaces, the interaction may itself influence the degree of dissociation of surface groups, such that neither the surface potential nor the surface charge remains constant. A charge regulation model may then be more appropriate [23]. The constant charge and the constant potential model allow, however, both upper and lower limits for the strength of the interaction to be estimated. For most of the model calculations performed, it is assumed that only one of these entities changes as a function of the separation between the two surfaces.

We need to evaluate the expected pH dependence of the (normal) force at short distances (~1 nm) and moderate potentials for the particular geometry of tip and surface. While the tip has a pyramidal shape, the substrate can be well described by a flat. Since only the tip apex is important it may also be modeled by a ball. However, for none of these geometries are appropriate analytical expressions for a solution of the PB equation available. Recently the ball-on-plane geometry has been studied theoretically, but only computer simulations have been reported [24].

The main region of interest for our experiment is that between the IEPs of the materials, since it is here, where both surfaces are oppositely charged, that we observed the highest frictional forces and the strongest dependence on pH. Lower values were observed outside this range, where tip and sample surface are similarly charged. It has been shown by Gregory that in the case of dissimilar potentials, the constant-potential approximation for the linear PB equation works quite well [25]. The PB equation can be linearized if the so-called Debye–Hückel approximation is valid, i.e., if the electrical energy $ze\Psi$ is small compared to thermal energy kT (~ 25 meV). In this case a formula valid for constant potential during interaction can be derived for a flat–flat geometry.

Ohshima et al. elaborated an expression for higher dissimilar potentials and various geometries, based on a Taylor-series expansion of the PB equation and taking higher powers into account [26–28]. However, the deviation compared to the linear approximation is small and we will use the latter in the present paper.

Since our lateral force signal is not calibrated, the experimental data can only be modeled qualitatively. For low and constant potentials and a flat–flat geometry, the electrostatic component of the pressure between tip and surface is given by [12,25]

$$P_{\rm el} \sim (2\Psi_1\Psi_2\cosh(\kappa D) - \Psi_1^2 - \Psi_2^2)/\sinh(\kappa D)^2 \quad (2)$$

with Ψ_1 and Ψ_2 being the surface potentials, κ the inverse of the Debye length (~ 1/10 nm^{-1}) and D the distance between tip and substrate (~ 1 nm).

The same equation (2) was also used by Spikes et al. to model the classical friction coefficient in the presence of electrochemical effects between two metal oxide surfaces in aqueous solutions on a *macroscopic* scale [29,30] and by Lin et al. to model force interactions between tip and surface in an electrolyte for larger separations [18].

We tested both a linear relationship between potential and pH (Nernst behavior) and a non-linear relationship, with a maximum saturation value of the potential reached for approximately $\Delta\text{pH} \approx 4$, which is reasonable for the investigated substrates [17]. In the latter case we used a functional dependence $\Psi = B\tanh(\Delta\text{pH})$,

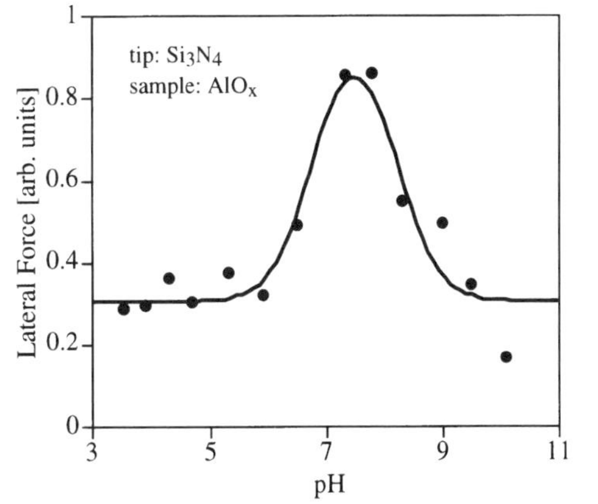

Figure 3. Experimental data from figure 2 together with a fit according to eq. (2). The simple linear model we have used qualitatively describes the observed dependence quite well, suggesting that the changes of the lateral force with pH are due to electrostatic interactions.

where B corresponds to the saturation value, i.e., maximum potential of the surface. We found empirically that the non-linear relationship led to better fit results. Figure 3 shows the experimental data observed for the AlO_x surface (from figure 2) together with a least-squares fit according to eq. (2), based on a non-linear dependence of the potential on pH.

The calculations we performed can only be reasonably interpreted on a qualitative basis, since we had to make several approximations. Moreover, the used models are continuum models and they break down at smaller distances, which are, however, important in our experiment. Ion-size effects and the discreteness of the charges have to be taken into consideration when discussing the interaction in more detail. It has been recog-

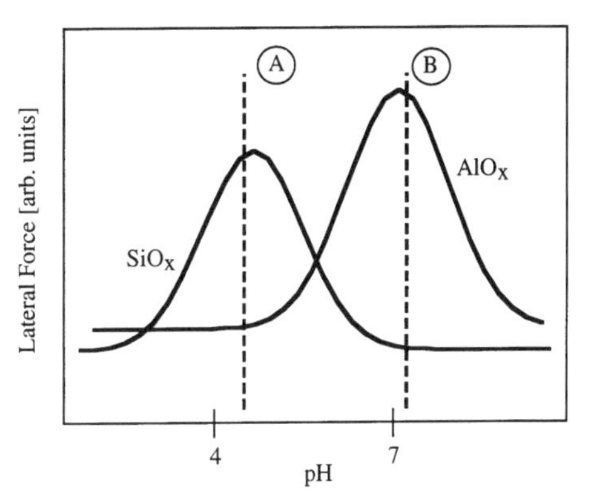

Figure 4. Schematic diagram of the friction force signal between an AFM tip and two different oxide surfaces. From the curves displayed, it becomes clear that the chemical contrast on oxide surfaces can be tailored by changing pH. On an appropriate sample even a reversal of the frictional contrast should be possible.

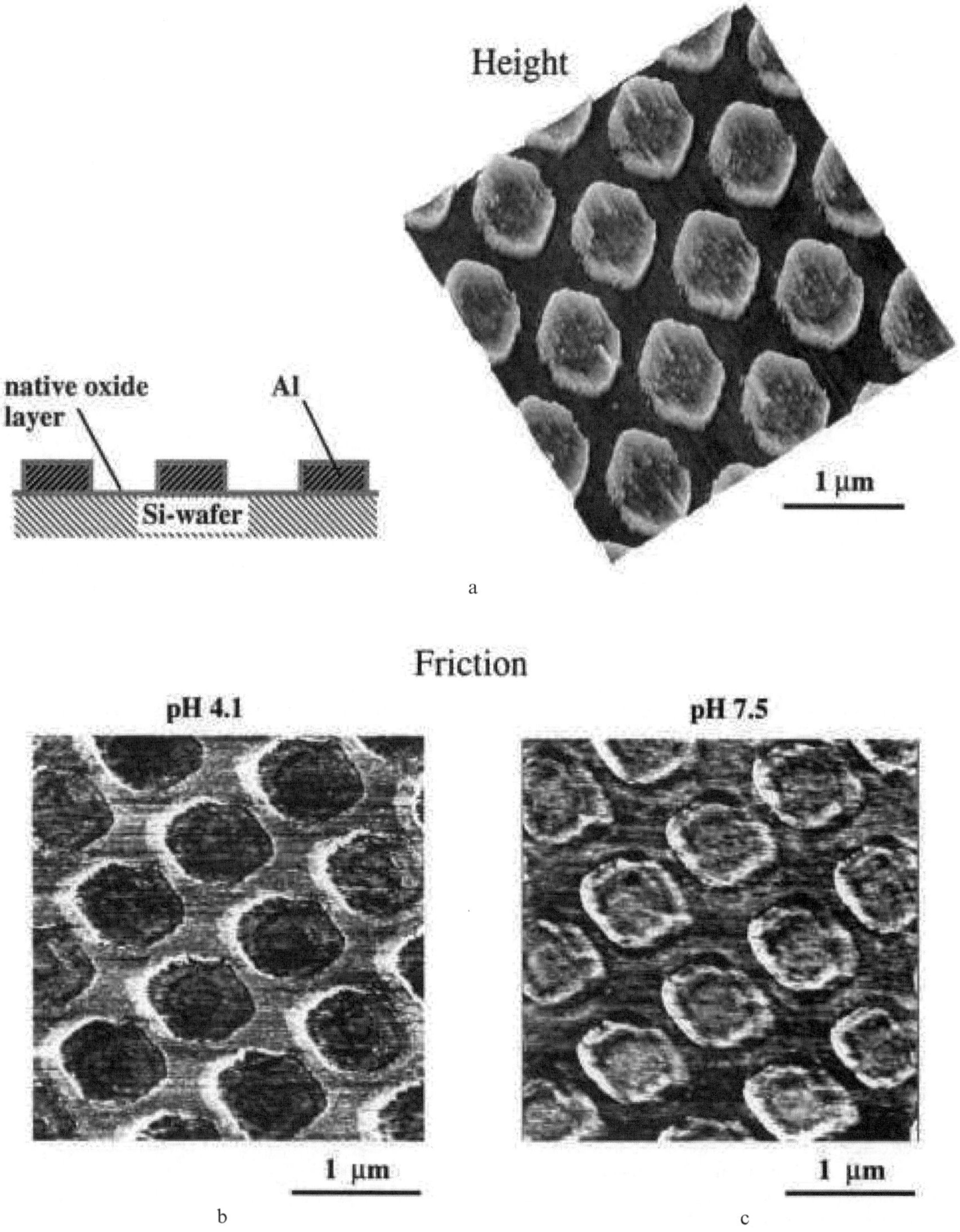

Figure 5. (a) Height image of a regular AlO_x pattern in the micrometer range on a SiO_2 substrate. The height of the AlO_x patches is approximately 5 nm. (b) Friction image of the sample measured at a pH of 4.1. The SiO_2 shows a higher lateral force signal. Note that the displayed image shows the difference between forward and backward scan and hence the frictional contrast, which is due to the chemical interaction between tip and sample. (c) Same as (b) but for a pH value of 7.5. Note that the friction contrast is reversed compared to (b), i.e., lateral force signal is higher on AlO_x in this case as explained in figure 4.

nized before, however, that DLVO theory works quite well, even for small separations [19]. In consequence, the dependence of the lateral force on pH is reproduced quite well by our model, i.e., the overall shape is likely to be mainly determined by the (pH-dependent) electrostatic interactions.

Although the empirical description of the lateral force works well, the dissipation mechanism is not clear. The ionic layers that are in direct contact with the surfaces are likely to be involved. Their destruction as well as an ordering of the ions in the gap between tip and surface may be responsible for the amount of energy dissipated during scanning. Further experiments are necessary to clarify this.

3.4. Nanochemical imaging

The experimentally determined frictional forces between a Si_3N_4 tip and SiO_2 and AlO_x surfaces (displayed in figures 1 and 2) show that frictional contrast between different chemical species is highly and characteristically pH-dependent. This, in turn, forms the basis of a novel, friction-based technique: From the observed pH-dependence of the friction force signal, it is evident that friction can be tailored to a certain extent by varying the pH of the surrounding electrolyte. Since the observed lateral forces are also sensitive to the IEPs and since the IEP is characteristic for a particular material, an analysis of different oxide surfaces and the identification of chemical species present becomes possible. On the basis of figures 1 and 2 it is clear that a reversal of the lateral force contrast of two different species on a surface is possible by changing pH. Figure 4 illustrates this concept: The relative positions of the characteristic pH-dependent frictional force curves measured for the SiO_2 and AlO_x surfaces are displayed schematically. To demonstrate this effect experimentally we used a sample consisting of a regular pattern of micrometer-range square, oxidized aluminum features on an oxidized silicon substrate. Figure 5a shows the height image of this sample, revealing the lateral dimension of the AlO_x islands to be approximately 0.7 μm, while their height is approximately 5 nm. Figures 5b and 5c show lateral force images obtained by subtracting the lateral signals from images recorded for forward and backward scanning [5] for pH values of 4.1 and 7.5, respectively. As shown schematically in figure 4, the lateral force contrast is indeed reversed upon pH-change and its pH-dependence is therefore chemically diagnostic. The results are promising for further studies of mixed oxide surfaces, which play an important role in biocompatibility, where the identification of different species with high lateral resolution as well as the determination of their isoelectric points is of fundamental importance.

4. Conclusions

We have demonstrated the sensitivity of the lateral force signal to both pH value and the IEPs of the materials involved for a nanometer-scale, single-asperity contact between two oxide surfaces in an electrolyte environment.

The resulting lateral force signal can be qualitatively described with a simple linear force model, where electrostatic interactions due to the influence of pH on surface hydroxyl groups are included in the total normal force.

For oxide systems, the frictional force can be tailored to a certain extent by pH variation. As a direct consequence, the chemical specificity of the IEP can be used to enhance chemical contrast and to identify different oxides. By means of changing pH, we were able to reverse the frictional contrast on an appropriate sample, and have thus demonstrated the potential of this approach as a chemical imaging technique.

Acknowledgement

We would like to thank J. Gold and B. Nilsson from Chalmers University, Göteborg, Sweden for the preparation of the samples we used for imaging and Professor J. Israelachvili for fruitful discussions concerning dissipation mechanisms.

References

[1] D. Dowson, *History of Tribology* (Longman, London, 1979).

[2] F.P. Bowden and D. Tabor, *The Friction and Lubrication of Solids* (Clarendon Press, Oxford, 1954).

[3] G. Binnig, C.F. Quate and C. Gerber, Phys. Rev. Lett. 56 (1986) 930.

[4] C.M. Mate, R. Erlandsson, G.M. McLelland and S. Chiang, Phys. Rev. Lett. 59 (1987) 1942.

[5] R. Overney and E. Meyer, MRS Bull. 18 (1993) 26.

[6] B. Bhushan, J.N. Israelachvili and U. Landman, Nature 374 (1995) 607.

[7] B. Bhushan, ed., *Micro/Nanotribology and Its Applications*, NATO ASI Series (Kluwer Academic, London, 1997).

[8] C.D. Frisbie, L.F. Rozsnyai, A. Noy, M.S. Wrighton and C.M Lieber, *Science* 265 (1994) 2071.

[9] A. Noy, C.D. Frisbie, L.F. Rozsnyai, M.S. Wrighton and C.M. Lieber, J. Am. Chem. Soc. 117 (1995) 7943.

[10] J.L. Wilbur, H.A. Biebuyck, J.C. MacDonald and G.M. Whitesides, Langmuir 11 (1995) 825.

[11] S. Akari, D. Horn, H. Keller and W. Schrepp, Adv. Mater. 7 (1995) 549.

[12] R.J. Hunter, *Foundations of Colloid Science*, Vol. 1 (Oxford University Press, London, 1989).

[13] G.A. Somorjai, *Introduction to Surface Chemistry and Catalysis* (Wiley, New York, 1994).

[14] J. Gold, PhD Thesis, Chalmers University Göteborg, Sweden (1996).

[15] A. Ulman, *Ultrathin Organic Films* (Academic Press, London, 1991).

[16] T.W. Healy and L.R. White, Adv. Coll. Interf. Sci. 9 (1978) 303.
[17] G.A. Parks, Chem. Rev. 65 (1965) 177.
[18] X.-Y. Lin, F. Creuzet and H. Arribart, J. Phys. Chem. 97 (1993) 7272.
[19] J.N. Israelachvili, *Intermolecular and Surface Forces*, 2nd Ed. (Academic Press, London, 1992).
[20] A. Marti, G. Hähner and N.D. Spencer, Langmuir 11 (1995) 4632.
[21] B.V. Deryaguin, V.V. Karassev, N.N. Zakhavaeva and V.P. Lazarev, Wear 1 (1957) 277.
[22] R.G. Horn, D.R. Clarke and M.T. Clarkson, J. Mater. Res. 3 (1988) 413.
[23] B.W. Ninham and V.A. Parsegian, J. Theor. Biol. 31 (1971) 405.
[24] J. Stankovich and S.L. Carnie, Langmuir 12 (1996) 1453.
[25] J. Gregory, J. Coll. Interf. Sci. 51 (1975) 44.
[26] H. Ohshima, T.W. Healy and L.R. White, J. Coll. Interf. Sci. 89 (1982) 484.
[27] H. Ohshima, D.Y.C. Chan, T.W. Healy and L.R. White, J. Coll. Interf. Sci. 92 (1983) 232.
[28] H. Ohshima, J. Coll. Interf. Sci. 162 (1994) 487.
[29] G.H. Kelsall, Y.Y. Zhu and H.A. Spikes, J. Chem. Soc. Faraday Trans. 89(2) (1993) 267.
[30] Y.Y. Zhu, G.H. Kelsall and H.A. Spikes, Tribol. Trans. 37 (1994) 811.

372 *Langmuir* **1998,** *14,* 372–378

Toward a Force Spectroscopy of Polymer Surfaces

Kirill Feldman, Theo Tervoort, Paul Smith, and Nicholas D. Spencer*

Department of Materials, Swiss Federal Institute of Technology, ETH-Zürich, CH-8092 Zürich, Switzerland

Received March 29, 1997. In Final Form: October 7, 1997

The adhesional forces between a series of polymer film surfaces and chemically well-defined atomic force microscopy tips have been measured and found to depend strongly on the chemical nature of both probe and sample surfaces. For a given series of polymers, the ranking in adhesion strength was markedly different for polar and nonpolar probes, irrespective of the precise chemical composition of those probes. In the case of nonpolar polymers, a correlation of adhesion force with calculations based on the Lifshitz theory of Van der Waals interactions was found. In the case of polar polymers, a reasonable correlation with water-contact angle was observed. The adhesional differences between different probe tips translate into reversals of chemical contrast in high-spatial-resolution lateral force images, when examining polymer blends using chemically different tips, demonstrating the potential of this approach for the nanometer-scale, friction-mediated surface-chemical imaging of polymers. Central to these experiments has been the use of perfluorodecalin as a medium for measuring interactions. Employment of this liquid greatly facilitates measurement of the forces between the probe tip and the polymer surface.

Introduction

Atomic and lateral force microscopy techniques (AFM and LFM) have, since their development in the 1980s,[1,2] shown considerable promise as methods for nanometer-scale, surface-chemical analysis, since they can provide quantitative, spatially resolved, chemically dependent information on interactions between the scanning probe and sample surfaces. This feature has been exploited by many researchers, using approaches such as chemical modification of probe tips for the recognition of specific surface groups[3–8] or monitoring the pH dependence of the tip–surface interaction.[9–11] The majority of such studies have involved self-assembled monolayers (SAMs) on flat gold surfaces,[12] which provide an idealized test surface, presenting a well-ordered, morphology-free, highly concentrated plane of functionality. The usefulness of SAMs as models for polymer surfaces is limited, however, since issues such as complex surface morphology, disorder, mechanical properties, and solvent interactions significantly complicate the issue with real polymers, making chemical imaging extremely challenging.[6,13,14]

Force–distance measurements with conventional (nonvacuum) scanning probe microscopes are often performed in a liquid environment in order to eliminate the contribution of capillary forces resulting from water adsorption from the air.[15] Moreover, the liquid environment can be used to tune the Van der Waals forces between the probe and the surface.[16a] This is a valuable approach that others have used for DNA imaging,[16b] for example. An important consideration here is the makeup of the van der Waals interaction, which can be calculated from the nonretarded Hamaker constant, A_{total}, and which, in turn, consists of the two terms $A_{\nu=0}$ and $A_{\nu>0}$, corresponding to the dipole–dipole and dipole–induced-dipole contributions and the dispersion (London) contributions, respectively, to the van der Waals interaction. According to Israelachvili's simplification[17] of the Lifshitz theory,[18] these contributions can be calculated for a system where two macroscopic phases interact across a third phase, from the respective static dielectric constants (ϵ_1, ϵ_2, ϵ_3) and optical refractive indexes (n_1, n_2, n_3) as follows:

$$A_{total} = A_{\nu=0} + A_{\nu>0} \approx \frac{3}{4}kT\left(\frac{\epsilon_1 - \epsilon_3}{\epsilon_1 + \epsilon_3}\right)\left(\frac{\epsilon_2 - \epsilon_3}{\epsilon_2 + \epsilon_3}\right) + \frac{3h\nu_e}{8\sqrt{2}} \times \frac{(n_1^2 - n_3^2)(n_2^2 - n_3^2)}{(n_1^2 + n_3^2)^{1/2}(n_2^2 - n_3^2)^{1/2}\{(n_1^2 + n_3^2)^{1/2} + (n_2^2 + n_3^2)^{1/2}\}} \quad (1)$$

where the electronic absorption frequency, ν_e, is assumed to be equal for all three components ($\nu_e = 3 \times 10^{15}$ Hz). The consequence of this relationship is that a close match between the dielectric constants of the tip, the sample, and the medium leads to a suppression of the first term, with the result that dispersion forces (determined by the optical refractive index) play the dominant role in determining the tip–sample adhesion. In fact, if the refractive index of the intervening medium is intermediate

(1) Binnig, G.; Quate, C. F.; Gerber, Ch. *Phys. Rev. Lett.* **1986**, *56*, 930.
(2) Mate, C. M.; Erlandsson, R.; McClelland, G. M.; Chiang, S. *Phys. Rev. Lett.* **1987**, *59*, 1942. Overney, R.; Meyer, E. *MRS Bull.* **1993**, *May*, 26.
(3) Frisbie, C. D.; Rozsnyai, L. F.; Noy, A.; Wrighton, M. S.; Lieber, C. M. *Science* **1994**, *265*, 2071.
(4) Lee, G. U.; Kidwell, D. A.; Colton, R. J. *Langmuir* **1994**, *10*, 354.
(5) Akari, S.; Horn, D.; Keller, H.; Schrepp, W. *Adv. Mater.* **1995**, *7*, 549.
(6) Sinniah, S. K.; Steel, A. B.; Miller, C. J.; Reutt-Robey, J. E. *J. Am. Chem. Soc.* **1996**, *118*, 8925.
(7) Green, J.-B. D.; McDermott, M. T.; Porter, M. D.; Siperko, L. M. *J. Phys. Chem.* **1995**, *99*, 10960.
(8) Han, T.; Williams, J. M.; Beebe, T. P. *Anal. Chim. Acta* **1995**, *307*, 365.
(9) Marti, A.; Hähner, G.; Spencer, N. D. *Langmuir* **1995**, *11*, 4632.
(10) Hähner, G.; Marti, A.; Spencer, N. D. *Trib. Lett.* **1997**, *3*, 359.
(11) Vezenov, D. V.; Noy, A.; Rozsnyai, L. F.; Lieber, C. M. *J. Am. Chem. Soc.* **1997**, *119*, 2006.
(12) Bain, C. D.; Evall, J.; Whitesides, G. M. *J. Am. Chem. Soc.* **1989**, *111*, 7155.
(13) Schönherr, H.; Vancso, G. J. *Polym. Prepr.* **1996**, *37* (2), 612.
(14) Aimé, J. P.; Elkaakour, Z.; Odin, C.; Bouhacina, T.; Michel, D.; Cuély, J.; Dautant, A. *J. Appl. Phys.* **1994**, *76*, 754.
(15) Grigg, D. A.; Russel, P. E.; Griffith, J. E. *J. Vac. Sci. Technol. A* **1992**, *10*, 680.
(16) (a) Hutter, J. L.; Bechhoefer, J. *J. Appl. Phys.* **1993**, *73*, 4123. (b) Hansma, H. G.; Sinsheimer, R. L.; Li, M.-Q.; Hansma, P. K. *Nucleic Acids Res.* **1992**, *20*, 3585.
(17) Israelachvili, J. *Intermolecular and Surface Forces*, 2nd ed.; Academic Press: London, 1992; Chapter 11.
(18) Lifshitz, E. M. *Sov. Phys.* **1956**, *JETP 2*, 73.

Published on Web 01/01/1998

between those of the other phases, a negative van der Waals interaction can result.[19]

The manipulation of surface forces by suitable choice of medium is central to the experiments in the present study. In the case of polymers, the choice of a suitable medium is greatly restricted by potential interactions, as a solvent, for example, with the polymer surface. In view of these constraints, we have chosen perfluorinated decalin ($C_{10}F_{18}$) as a measurement medium: Employment of this nonpolar liquid, which displays both a low dielectric constant and a low refractive index and is inert toward most polymers, greatly facilitates the measurement of the dispersion component of the Van der Waals forces between the probe tip and the polymer surface.

Below, we describe a systematic study of AFM-adhesion measurements on polymeric surfaces. We have endeavored to control as many of the complicating parameters on polymer surfaces as possible, such that our AFM adhesion measurements were primarily due to the dispersion component of the Van der Waals and H-bonding interactions between the tip and the polymer surface. Using an embryonic "force spectroscopy" approach, polymers could be distinguished by virtue of their differing hydrophobicities/hydrophilicities or, for purely hydrophobic systems, on the basis of their optical refractive index. By capitalizing on the intimate relationship between adhesion and friction,[20] these phenomena can be rendered visual by means of lateral force microscopy (LFM), enabling high-spatial-resolution chemical imaging of heterogeneous polymer systems—a challenging analytical task for conventional ultrahigh vacuum surface chemical imaging methods.

Experimental Section

Sample Preparation. A first series of thin (hydrophobic) polymer films consisted of polystyrene, PS (average MW = 250 000, Polysciences, Inc., Warrington, PA), isotactic polypropylene, i-PP (average MW = 250 000, Aldrich Chemical Co., Inc., Milwaukee, WI), poly(vinylidene fluoride), PVDF (average MW = 534 000, Aldrich Chemical Co., Inc., Milwaukee, WI), and poly(tetrafluoroethylene-*co*-hexafluoropropylene), FEP (Polymer Technology Group, ETH-Zürich).

A second series of thin films consisted of glassy polymers with different hydrophobicities/hydrophilicities: polystyrene, PS (average MW = 250 000, Polysciences, Inc.), polyacrylonitrile, PAN (average MW = 500 000, Polysciences, Inc.), poly(methyl methacrylate), PMMA (average MW = 350 000, Aldrich Chemical Co., Inc.), and poly(acrylic acid), PAA (average MW = 450 000, Aldrich Chemical Co., Inc.).

i-PP, PVDF, and FEP were first prepared from foils made by pressing powders between aluminum sheets above their corresponding melting temperatures. The films obtained were then pressed between plasma-cleaned silicon wafers (once again, above their melting temperatures) to achieve low surface roughness. It must be noted, however, that, due to the crystallization of isotactic polypropylene and PVDF upon cooling, the attainment of a comparable surface roughness to that of the silicon wafer was not anticipated; nevertheless, film roughnesses of ≈3 nm were attained. In addition, these polymers have glass transition temperatures of about −22 and −38 °C, respectively, meaning that additional chain rearrangements and an increase in crystallinity may take place during their subsequent storage at room temperature.

Thin films of the other polymers were prepared by spin casting 2 wt % solutions (PS and PMMA in toluene, PAN in *N,N*-dimethylformamide, and PAA in methanol) onto plasma-cleaned silicon wafers at 1000 rpm, followed by drying in a vacuum oven at 120 °C and 0.03 mbar for 24 h. All polymers of the second series are fully or highly amorphous and have glass transition temperatures of approximately 100 °C. Annealing above T_g ensured low surface roughness of the films and evaporation of the solvents. The thicknesses of the produced films were on the order of 50−100 μm, as measured by ellipsometry. The measured surface roughness of all films was close to that of the silicon substrates.

Probes. Two classes of AFM probe-tip surfaces were used in this work: polar and nonpolar. Polar tips were either oxygen-plasma-treated Si_3N_4 Microlevers (Park Scientific Instruments, Sunnyvale, CA) or Si_3N_4 cantilevers (Digital Instruments, Santa Barbara, CA) with attached COOH-functionalized glass spheres (Bioforce Laboratory, Ames, IA). The nonpolar tips were either gold-coated Si_3N_4 Microlevers or Si_3N_4 cantilevers with attached ≈7.5 μm diameter polystyrene beads (Bioforce Laboratory). Sharpened Microlevers were chosen because of their low spring constants (down to 0.007 N/m), providing high sensitivity to the measured pull-off forces and a small (10−20 nm, according to manufacturer's specifications) tip radius. Spring constants of the oxygen-plasma-treated and gold-coated Microlevers were calibrated by the method developed by Cleveland *et al.*[21] Oxygen-plasma treatment of the Microlevers was carried out in a radiofrequency plasma cleaner (Harrick Scientific Corp., Ossining, NY) operated at 40 W with an oxygen feed. The isoelectric point of the resulting tips was at approximately pH ≈ 3 (measured by the method of Marti *et al.*[9]), suggesting that the surface consisted chiefly of SiO_x. The gold-coated tips were prepared by thermal deposition of a 4 nm chromium adhesion layer, followed by 20 nm of gold in a Balzers (Liechtenstein) MED 010 coater operated at 2×10^{-5} mbar. A single tip of each type was used to measure all polymer surfaces within a given series, to ensure that the spring constant and tip radius were kept constant between samples.

Measurements. Force−distance measurements and lateral force imaging were performed with a scanning probe microscope (Nanoscope III Multimode, Digital Instruments) equipped with a liquid cell and enclosed in a thermally equilibrated environment. Up to 1280 force−distance curves at adjacent locations were collected for each sample. Prior to the measurements, the films were briefly placed under an α-radiation source, ^{210}Po (NRD, Inc., Grand Island, NY) to ensure that the static charge, which is likely to be present on the polymer surfaces, was removed and did not contribute to the overall forces measured. Following each set of experiments on a given series of polymers, the initial measurements were repeated, in order to check that the tip was unaltered and intact. Reproducibility was found to be within 15%.

In order to determine the influence of time dependent effects on the pull-off force for the polymer systems investigated, both load-dependent and frequency-dependent experiments were carried out (0.25−5.5 nN applied load and 0.1−5 Hz for loading−unloading cycles). No significant load dependence was detected over the investigated range, while the pull-off force was found to decrease by ≈25% over the measured frequency range, largely due to an increasing loading−unloading hysteresis. We decided to perform measurements at low frequency (0.5 Hz), below the onset of hysteresis.

Refractive index and film thicknesses were measured by ellipsometry (Type L-116 C, Gaertner Science Corp., Chicago, IL) using a 70° angle of incidence and a He−Ne laser. Static water-contact angle measurements were carried out using a contact angle goniometer (Ramé-Hart, Inc., Mountain Lakes, NJ).

We chose perfluorodecalin (PFD), $C_{10}F_{18}$, (Fluorochem, U.K.) as a medium for all AFM and LFM experiments, both for the reasons discussed in detail above and because PFD is not a solvent for any of the materials examined in this study. Additionally, PFD is convenient to use in AFM experiments because of its relatively low vapor pressure (P_v = 0.88 kPa), high boiling point (T_b = 141 °C), and nontoxicity.

(19) Burnham, N. A.; Colton, R. J. Force Spectroscopy. In *Scanning Tunneling Microscopy and Spectroscopy: Theory, Techniques, and Applications*; Bonnell, D. A., Ed.; VCH Publishers, Inc.: New York, 1993; Chapter 7.

(20) Israelachvili, J. N.; Chen, Y.-L.; Yoshizawa, H. *J. Adhesion Sci. Technol.* **1994**, *8*, 1234.

(21) Cleveland, J. P.; Manne, S.; Bocek, D.; Hansma, P. K. *Rev. Sci. Instrum.* **1993**, *64*, 403.

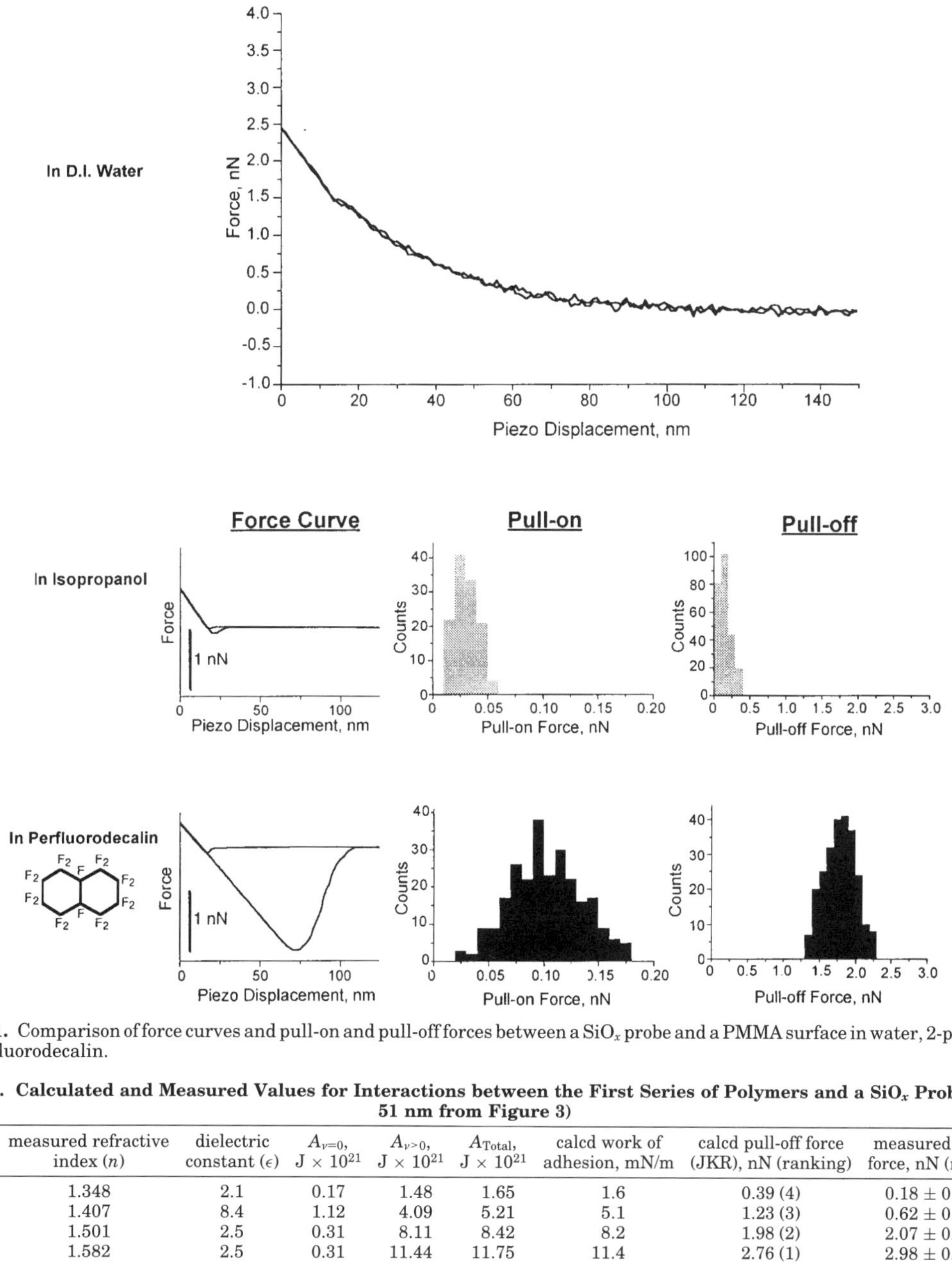

Figure 1. Comparison of force curves and pull-on and pull-off forces between a SiO_x probe and a PMMA surface in water, 2-propanol, and perfluorodecalin.

Table 1. Calculated and Measured Values for Interactions between the First Series of Polymers and a SiO_x Probe (R = 51 nm from Figure 3)

polymer	measured refractive index (n)	dielectric constant (ϵ)	$A_{\nu=0}$, J $\times$ 10^{21}	$A_{\nu>0}$, J $\times$ 10^{21}	A_{Total}, J $\times$ 10^{21}	calcd work of adhesion, mN/m	calcd pull-off force (JKR), nN (ranking)	measured pull-off force, nN (ranking)
FEP	1.348	2.1	0.17	1.48	1.65	1.6	0.39 (4)	0.18 $\pm$ 0.08 (4)
PVDF	1.407	8.4	1.12	4.09	5.21	5.1	1.23 (3)	0.62 $\pm$ 0.20 (3)
iPP	1.501	2.5	0.31	8.11	8.42	8.2	1.98 (2)	2.07 $\pm$ 0.15 (2)
PS	1.582	2.5	0.31	11.44	11.75	11.4	2.76 (1)	2.98 $\pm$ 0.16 (1)

Results and Discussion

An example of the effect of different media on force–distance measurements is shown in Figure 1 for the case of a PMMA surface scanned with a SiO_x probe in water, 2-propanol, and PFD. Since both the PMMA surface and the SiO_x probe are negatively charged at neutral pH, the force–distance measurements in water exhibit strong double-layer interaction forces: Most polymer surfaces acquire a charge in water,[22] and this is a common feature for force–distance measurements for polymers under water using a SiO_x probe. From Figure 1 it is also clear that PFD (n = 1.317) leads to a far greater tip–sample interaction (both pull-on and pull-off force) than 2-propanol (n = 1.378), when used as the intervening medium between a SiO_x (n = 1.480) tip and a PMMA (n = 1.482) surface. The higher pull-off forces and therefore greater signal-to-noise ratio obtained in force–distance measurements under PFD thus facilitate comparisons between different polymer samples. Calculations (Tables 1–3) show the interaction to be overwhelmingly dominated by the dispersion component for all polymers used in this study, due to the low refractive index, n, of PFD and the similarity

(22) Garbassi, F.; Morra, M.; Occhiello, E. *Polymer Surfaces: From Physics to Technology*; John Wiley & Sons Ltd: Chichester, 1994; Chapter 1.

Table 2. Calculated and Measured Values for Interactions between the Second Series of Polymers and a SiO_x Probe (*R* assumed to be 20 nm)

polymer	measured water contact angle, deg	measured refractive index (*n*)	dielectric constant[33] (ϵ)	$A_{\nu=0}$, J × 10^{21}	$A_{\nu>0}$, J × 10^{21}	A_{total}, J × 10^{21}	calcd work of adhesion, mN/m	calcd pull-off force (JKR), nN (ranking)	measured pull-off force, nN (ranking)
PAN	64 ± 2	1.356	6.5	0.99	1.84	2.83	2.8	0.26 (4)	1.32 ± 0.15 (3)
PMMA	68 ± 2	1.482	3.6	0.60	7.31	7.91	7.7	0.73 (3)	1.84 ± 0.16 (2)
PAA	H_2O soluble	1.506	5.0	0.83	8.32	9.15	8.9	0.84 (2)	2.13 ± 0.14 (1)
PS	91 ± 2	1.582	2.5	0.31	11.44	11.75	11.4	1.08 (1)	0.66 ± 0.10 (4)

Table 3. Calculated and Measured Values for Interactions between the Second Series of Polymers and a PS Particle (*R* ≈ 3.75 μm) Probe

polymer	measured refractive index (*n*)	dielectric constant (ϵ)	$A_{\nu=0}$, J × 10^{21}	$A_{\nu>0}$, J × 10^{21}	A_{total}, J × 10^{21}	calcd work of adhesion, mN/m	calcd pull-off force (JKR), nN (ranking)	measured pull-off force, nN (ranking)
PAN	1.356	6.5	0.33	2.91	3.24	3.2	55.7 (4)	54.4 ± 20.6 (4)
PMMA	1.482	3.6	0.20	11.57	11.77	11.5	202.6 (3)	112.8 ± 3.9 (2)
PAA	1.506	5.0	0.27	13.17	13.44	13.1	231.4 (2)	75.4 ± 4.5 (3)
PS	1.582	2.5	0.10	18.11	18.21	17.7	313.6 (1)	134.7 ± 7.5 (1)

of its dielectric constant, ϵ, to those of the surrounding solid phases.

To explore the use of PFD as a contrast-enhancing medium in force–distance measurements on polymer surfaces, two series of polymers were chosen for the experiments: a first series, consisting of apolar polymers with different refractive indexes, and a second series, in which polymers of different hydrophobicities/hydrophilicities were selected.

First Polymer Series. The first series of polymers was investigated, in order to limit the interaction between an AFM probe and polymer surfaces to only the *dispersion* (London) component of the Van der Waals force. Adhesional forces were measured between a SiO_x probe and a set of nonpolar polymers that provided a range of refractive indexes (as measured): polystyrene (1.582), isotactic polypropylene (1.501), poly(vinylidene fluoride) (1.407), and poly(tetrafluoroethylene-*co*-hexafluoropropylene) (1.348). The histograms of the pull-off forces, measured with a SiO_x probe, are shown in Figure 2 and tabulated with the calculated values for adhesion energy in Table 1.

The Hamaker constants for these systems were derived from eq 1, as described above. The work of adhesion, *W*, was calculated from the following approximation,[17] $W \approx A_{total}/12\pi D_0^2$, where $D_0 = 0.165$ nm is the commonly used value for the cutoff separation.[17] Plotting measured pull-off forces, *F*, vs the calculated work of adhesion values, *W*, should yield a slope of $1.5\pi R$, if the JKR[25] theory holds for our system (or $2\pi R$ if DMT[23] is a more appropriate model[24]). The results (Figure 3) indicate that, for this first polymer series, *F* scales linearly with *W*, with a corresponding tip radius (assuming JKR) of ≈50 nm. In other words, in the case of nonpolar polymers, AFM pull-off force results obtained under PFD scale quite well with adhesion energies predicted from Lifshitz theory.

Second Polymer Series. Our second series of experiments involved measurements of the pull-off forces in PFD between AFM probes and the surfaces of polymer films with varying degrees of hydrophobicity/hydrophilicity: PS, PAN, PMMA, and PAA. The histograms of the distributions of pull-off forces measured between AFM probes and polymer surfaces in PFD (Figures 4–6) clearly indicate two trends: one for the nonpolar tips (virtually identical results were obtained for both the gold and the polystyrene probes), where the adhesion is strongest for the polystyrene sample and weakest for poly(acrylonitrile), and the other for the polar probes (again, very similar results being obtained for the SiO_x- and COOH-coated tips), where the strongest adhesion is observed with the poly(acrylic acid) surface and the weakest with polystyrene.

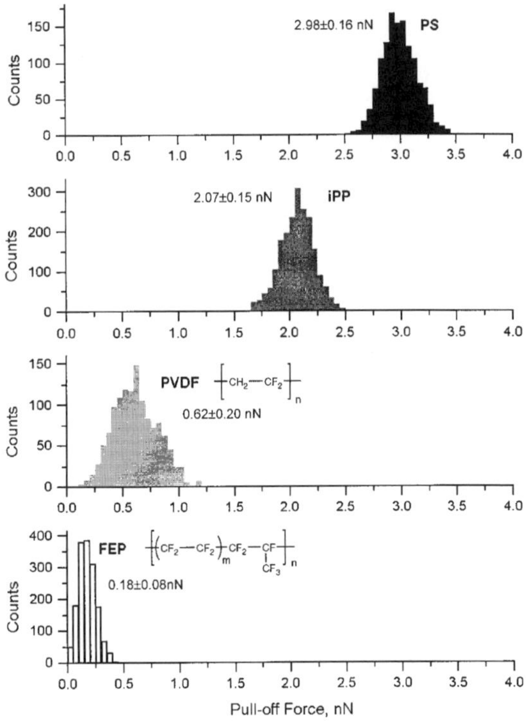

Figure 2. Histograms of pull-off forces measured between a SiO_x probe and PS, i-PP, PVDF, and FEP surfaces in perfluorodecalin.

The Hamaker constants and the work of adhesion for the second series were calculated as described above. Pull-off forces were calculated using the JKR theory,[25] $F = -1.5\pi WR$, where, for this second set of polymer films, the effective radius, *R*, of a sharpened tip was given a typical

(23) Derjaguin, B. V.; Muller, V. M.; Toporov, Yu. P. *J. Colloid Interface Sci.* **1975**, *53*, 314.

(24) Burnham, N. A.; Kulik, A. J.; Oulevey, F.; Mayencourt, C.; Gourdon, D.; Dupas, E.; Gremaud, G. In *Micro/Nanotribology and Its Applications*; Bhushan, B., Ed.; NATO ASI Series E: Applied Sciences; Kluwer Academic Publishers: Dordrecht, 1997; Vol. 330, p 421.

(25) Johnson, K. L.; Kendall, K.; Roberts, A. D. *Proc. R. Soc. London A* **1971**, *324*, 301.

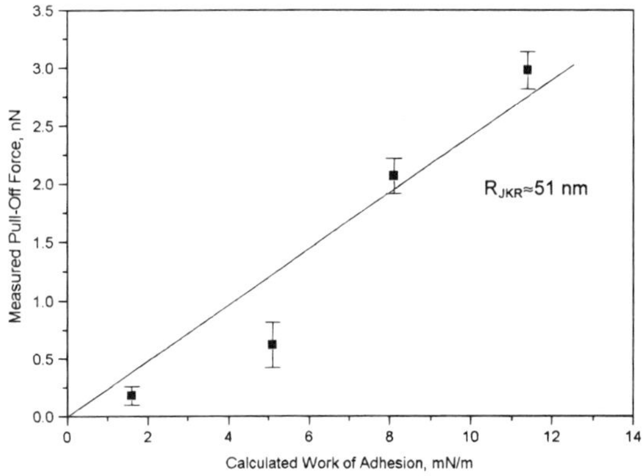

Figure 3. Comparison of measured pull-off forces and the calculated work of adhesion for interactions between a SiO_x probe and PS, i-PP, PVDF, and FEP surfaces in perfluorodecalin.

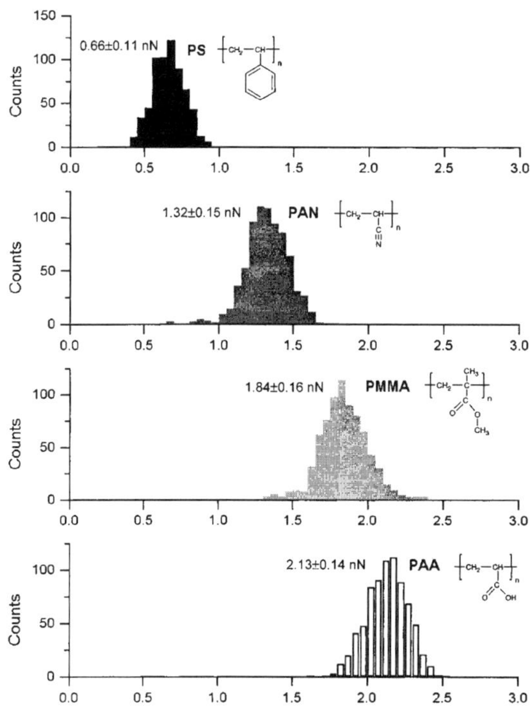

Figure 4. Histograms of pull-off forces measured between a SiO_x probe and PS, PAN, PMMA, and PAA surfaces in perfluorodecalin.

value of 20 nm.[26] These results highlight both the promise and the difficulties of this approach to polymer surface characterization. In the case of the nonpolar probes, the measured ranking in adhesion is roughly similar to that calculated for the PS-sphere probe (Table 3). Presumably the surface roughness of the PS-sphere is at least partially responsible for the significant disparities in absolute values. In the case of the polar probes, the measured order of adhesion is entirely different from the values calculated from the Lifshitz theory (Table 2). It must be

(26) Manufacturer's specifications: Park Scientific Instruments, Sunnyvale, CA.

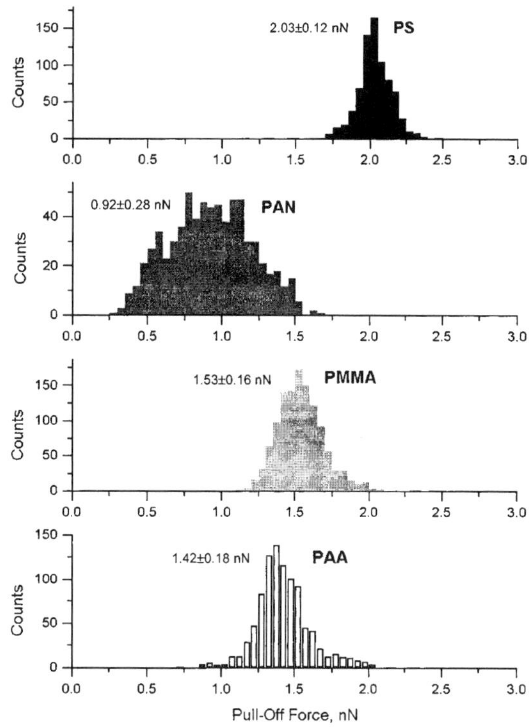

Figure 5. Histograms of pull-off forces measured between a gold probe and PS, PAN, PMMA, and PAA surfaces in perfluorodecalin.

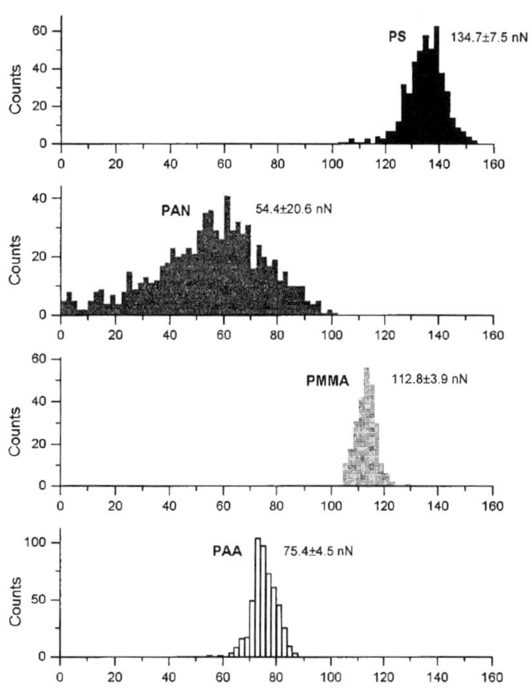

Figure 6. Histograms of pull-off forces measured between a polystyrene sphere (radius 3.75 μm) probe and PS, PAN, PMMA, and PAA surfaces in perfluorodecalin.

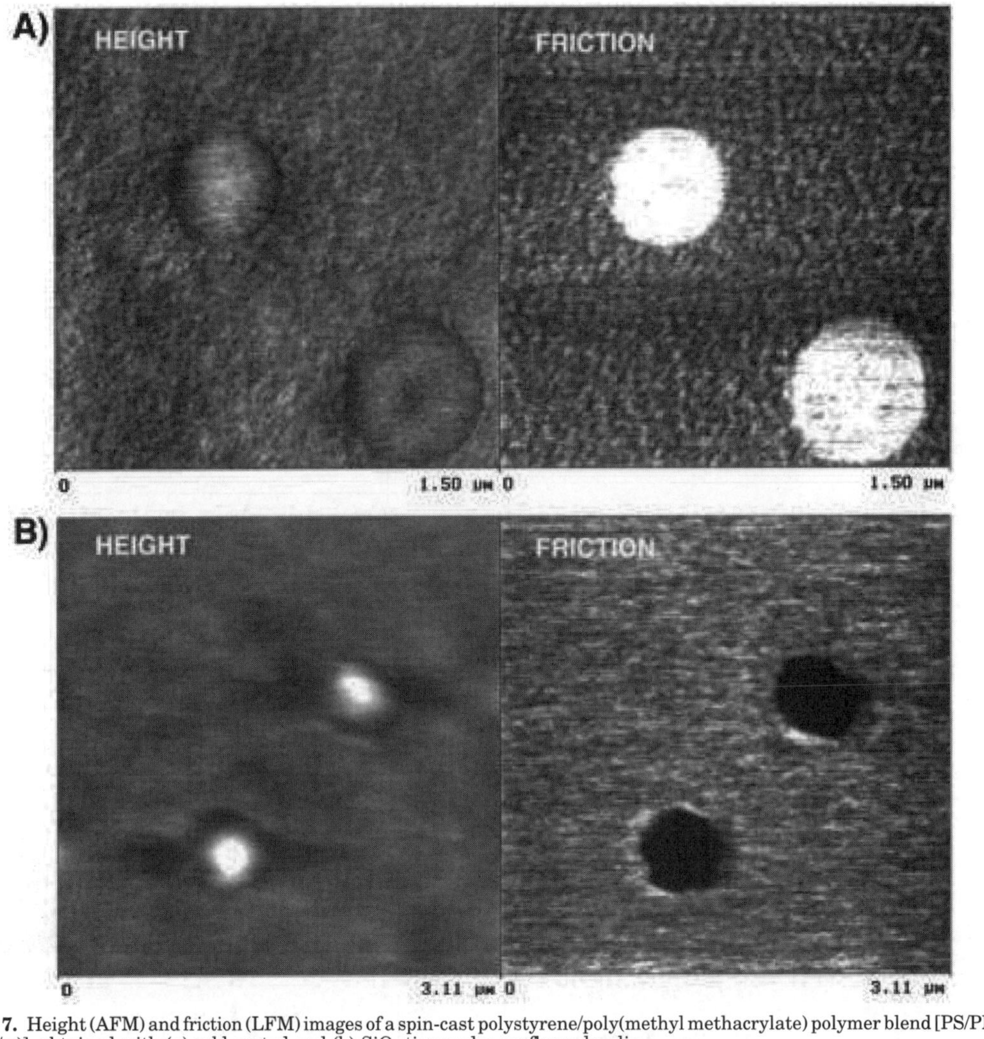

Figure 7. Height (AFM) and friction (LFM) images of a spin-cast polystyrene/poly(methyl methacrylate) polymer blend [PS/PMMA (1:10 w/w)], obtained with (a) gold-coated and (b) SiO_x tips under perfluorodecalin.

borne in mind, however, that many important surface properties of polymers are not taken into account by the calculation: PAA, with its free carboxyl groups, might reasonably be expected to form hydrogen bonds with hydroxyl species on the SiO_x tip. This effect would not, of course, be accounted for in the Lifshitz formalism. In the case of PMMA, which generally displayed strong interactions with both polar and nonpolar probes, surface rearrangement[27] could lead to a preferential orientation of either methyl or methacrylate groups toward the interface, depending on the nature of the approaching probe, thus increasing the strength of the interaction in both situations. Although our experiments were carried out at room temperature, some 80 °C below the bulk T_g, simulations by Mansfield and Theodorou[28] and experimental work by Kambour[29] (rapid craze healing of PMMA at room temperature) show that mobility at the surface is greatly enhanced over the corresponding bulk value, thus increasing the feasibility of rearrangement. Furthermore, the Van der Waals interaction of PFD with the polymer surfaces (all of which are wetted by PFD) presumably enhances surface mobility. The correspondence between the calculations and the measured interactions between PAN and the nonpolar probes is quite good, providing that the measured (ellipsometric, thin film), rather than the literature (bulk), values for the refractive index are used. The measured refractive index value is lower (1.356) than that in the literature (1.518). Frank *et al.*[30] have shown that the physical properties of spin-cast thin polymer films can differ substantially from those of corresponding bulk samples. In particular, these authors observed solvent incorporation to be higher in spin-cast films. This could account for the deviations of our refractive index measurements from literature values, in particular for PAN, where the presence of a small oxygen signal in XPS analysis of the film suggests that traces of solvent (*N,N*-dimethylformamide) were indeed present.

(27) Garbassi, F.; Morra, M.; Occhiello, E. *Polymer Surfaces: From Physics to Technology*; John Wiley & Sons Ltd: Chichester, 1994; Chapter 2.

(28) Mansfield, K. F.; Theodorou, D. N. *Macromolecules* **1991**, *24*, 6283.

(29) Kambour, R. P. *J. Polym. Sci.* **1964**, *2*, 4165.

(30) Frank, C. W.; Rao, V.; Despotopoulou, M. M.; Pease, R. F. W.; Hinsberg, W. D.; Miller, R. D.; Rabolt, J. F. *Science* **1996**, *273*, 912.

By reducing the density, solvent incorporation lowers the refractive index, which, in turn, affects the Van der Waals interaction, as shown in eq 1, and therefore leads to a decrease in the Hamaker constant.

The Lifshitz theory is insufficient to account for the behavior of the second polymer series, in particular when examined with the polar probes. However, since the SiO_x probe is known to contain hydroxyl groups, it was attempted to correlate the adhesion measurements with the behavior of water on the surface as measured by water-contact angle (see Table 2). The water-contact angle measurements were observed to correlate reasonably well with pull-off forces for the second polymer series and the polar probes, suggesting that the AFM–van der Waals approach could potentially be used to provide local hydrophobicity/hydrophilicity information and thus to differentiate between polymer surfaces, as an alternative to the method described by Sinniah *et al.*[6]

Differences in tip radius would account for the difference in the SiO_x–PS pull-off forces derived from the two series of measurements (Figures 2 and 4), given that the spring constants for the cantilevers appeared to be very similar. Imaging the tip used for the first polymer series (tip radius 50 nm) by field-emission scanning electron microscopy showed that it was, indeed, somewhat flattened. The tips for the second polymer series measurements had been taken from a different area of the wafer and were presumably closer to specifications (20 nm). This illustrates another difficulty with attempts to obtain quantitative analytical data with AFM and the need for independent measurements of tip properties.

Frictional Measurements. The histograms in Figures 4 and 5 demonstrate that the chemical nature of the probe tip determines, for example, whether it is PS or PMMA that exhibits the greater adhesional force between the tip and the polymer surface. Given that frictional forces are generally commensurate with adhesion hysteresis[20] (which usually varies monotonically with pull-off force), we would also expect to see a reversal of contrast between the two tip classes in LFM (frictional) images of PS–PMMA blends. It is worth noting that, while the bulk elastic moduli of the two polymers are very similar, 3200 and 3300 MPa for PS and PMMA, respectively, frictional contrast between these two polymers has been reported[31] for an applied load of 10 nN and found to be lower for PS, suggesting that surface nanomechanical properties may vary from those of the bulk. The same authors observed the reversal of contrast upon decreasing the applied load.[32]

We prepared a blend by spin-coating a 2 wt % solution of PS and PMMA (1:10 weight ratio) onto a Si wafer; the resulting film was annealed at 140 °C overnight to ensure substantial phase separation. The film was LFM-imaged in PFD with both SiO_x- and gold-coated tips at zero applied load (the load being due to the work of adhesion only). Both height and frictional (loop-subtracted) images are shown in Figure 7. These images clearly reveal the tip-dependent reversal of the frictional contrast for the two polymers: friction was lower for PS–SiO_x than for PMMA–SiO_x and higher for PS–Au than for PMMA–Au.

Conclusions

The surface nanochemical imaging of polymers with AFM is clearly an analytical challenge, in that the chemically induced properties are convoluted with many other factors, such as disorder, mechanical properties, surface dynamics, and morphology. Nevertheless, when these other factors are carefully controlled, it appears that AFM may be used to distinguish, in the case of nonpolar systems, between areas of differing optical refractive index, as manifested by the London component of the van der Waals force. In the case of polar systems, the approach can distinguish between regions of different hydrophilicity. By imaging in frictional mode (LFM), this same information could be used to provide a high-spatial-resolution chemical map of many heterogeneous polymer systems, especially in cases where the mechanical properties of the components are similar. It was found that, due to its low refractive index, the use of perfluorinated decalin as a medium in AFM experiments significantly enhanced the differences in pull-off forces measured on various polymer surfaces.

Acknowledgment. N.S. and K.F. are grateful for funding from the Council of the Swiss Federal Institute of Technology through their Priority Program on Materials and MINAST program. The authors also wish to acknowledge useful discussions with our colleague, Dr. Nancy Burnham, Ecole Polytechnique Fédérale de Lausanne, and some provocative observations by our anonymous reviewers.

LA9703353

(31) Krausch, G.; Hipp, M.; Böltau, M.; Marti, O.; Mlynek, J. *Macromolecules* **1995**, *28*, 260.

(32) The use of nanomechanical measurements is, of course, an alternative approach to that described in this paper for the chemical imaging of polymers. We believe, however, that it is likely to be more prone to ambiguity, due to the dependence of mechanical properties on processing conditions.

(33) *Polymer Handbook*, 3rd ed.; Brandrup, J., Immergut, E. H., Eds.; J. Wiley & Sons, Inc.: New York, 1989.

10134 *J. Am. Chem. Soc.* **1999,** *121,* 10134–10141

Probing Resistance to Protein Adsorption of Oligo(ethylene glycol)-Terminated Self-Assembled Monolayers by Scanning Force Microscopy

K. Feldman,[†] G. Hähner,*[,†] N. D. Spencer,[†] P. Harder,[‡] and M. Grunze*[,‡]

Contribution from the Laboratory for Surface Science and Technology, Department of Materials, ETH Zürich, Sonneggstrasse 5, CH-8092 Zürich, Switzerland, and Angewandte Physikalische Chemie, Universität Heidelberg, INF 253, 69120 Heidelberg, Germany

Received April 1, 1999

Abstract: Functionalized scanning force microscope (SFM) probes were used to investigate and to mimic the interaction between fibrinogen and self-assembled monolayers (SAMs) of methoxytri(ethylene glycol) undecanethiolates $-S(CH_2)_{11}(OCH_2CH_2)_3OCH_3$ (EG3-OMe) on gold and silver surfaces. The SAMs on gold are resistant to protein adsorption, whereas the films on silver adsorb variable amounts of fibrinogen. Experiments were performed with both charged and hydrophobic tips as models for local protein structures to determine the influence of these parameters on the interaction with the SAMs. A striking difference between the two monolayers was established when the forces were measured in an aqueous environment with hydrophobic probes. While a long-range attractive hydrophobic interaction was observed for the EG3-OMe on silver, a repulsive force was measured for EG3-OMe on gold. The strong dependence of the repulsive force for the EG3-OMe-gold system upon the solution ionic strength suggests that this interaction has a significant electrostatic contribution. The observed differences are attributed to the distinct molecular conformations of the oligo-(ethylene glycol) tails on the gold-supported (helical) and silver-supported ("all-trans") monolayers. A comparison of the force/distance curves for the EG3-OMe SAMs with those measured under identical conditions on end-grafted poly(ethylene glycol) (PEG 2000) on gold further emphasizes that the nature of the repulsive forces originating from the short-chain oligomers is unique and not related to a "steric repulsion" effect.

Introduction

Self-assembled monolayers (SAMs) of alkanethiolates on metal surfaces constitute a class of molecular assemblies formed by the spontaneous chemisorption of long-chain functionalized molecules on the surface of solid substrates.[1,2] Due to their ease of preparation, long-term stability, controllable surface chemical functionality, and high, crystal-like, two-dimensional order, SAMs represent suitable model surfaces to study molecular adsorption, adhesion, wetting, lubrication, and the interaction of proteins and cells with artificial organic surfaces. The latter phenomena are of crucial importance to the fields of biomaterials, biosensors, and medical devices.

The outstanding protein-resistant properties of poly(ethylene glycol) (PEG)-containing surfaces have been recognized for a long time and extensive experimental and theoretical work has been carried out to elucidate the physics underlying these properties.[3-5] When interpreted in terms of the "steric repulsion" theory,[3-5] the protein resistance of PEG is associated with a high conformational freedom and hence entropy of its solvated chains in the near-surface region. This theory is not appropriate, however, to explain the observation that protein resistance is also inherent in oligo(ethylene glycol)-terminated alkanethiol SAMs,[6] where the conformational freedom of the OEG tails is restricted by packing forces.

The protein adsorption characteristics of the methoxytri-(ethylene glycol) undecanethiolate (EG3-OMe) SAMs are strongly sensitive to the substrate used.[7] While the monolayers self-assembled on Au are protein resistant, those self-assembled on Ag are not. Fourier transform infrared reflection–absorption spectroscopic (FTIRAS) experiments show that the SAMs on Au and Ag differ in the conformation of the OEG tail.[7] On gold a conformation with spectral characteristics typical of helical and amorphous PEG is found, whereas the spectra observed on the Ag substrates resemble those for the planar all-trans conformation of stretched PEG samples (Figure 1). The different conformation and higher packing density in the Ag-supported monolayer is due to the formation of an incommensurate solid phase with nearly upright chains and a perceptibly smaller lattice spacing, as confirmed by calculation of the lowest-energy monolayer configurations on Au and Ag.[8] Good agreement is found between the experimental and calculated vibrational spectra of the lowest-energy SAM configurations on Au and Ag using our structural models.[9]

* Corresponding authors.
[†] ETH Zürich.
[‡] Universität Heidelberg.

(1) Ulman, A. *Chem. Rev.* **1996**, *96*, 1533–1554.
(2) Ulman, A. *An Introduction to Ultrathin Organic Films: From Langmuir–Blodgett to Self-Assembly*; Academic Press: Boston, 1991.
(3) Harris, J. M. *Poly(Ethylene Glycol) Chemistry*; Plenum: New York, 1992.
(4) Jeon, S. I.; Andrade, J. D. *J. Colloid Interface Sci.* **1991**, *142*, 159–166.
(5) Taunton, H. J.; Toprakcioglu, C.; Fetters, L. J.; Klein, J. *Nature* **1988**, *332*, 712–714.
(6) Prime, K. L.; Whitesides, G. M. *J. Am. Chem. Soc.* **1993**, *115*, 10714–21.
(7) Harder, P.; Grunze, M.; Dahint, R.; Whitesides, G. M.; Laibinis, P. E. *J. Phys. Chem. B* **1998**, *102*, 426–436.
(8) Pertsin, A. J.; Grunze, M.; Garbuzova, I. A. *J. Phys. Chem. B* **1998**, *102*, 4918–4926.

Published on Web 10/15/1999

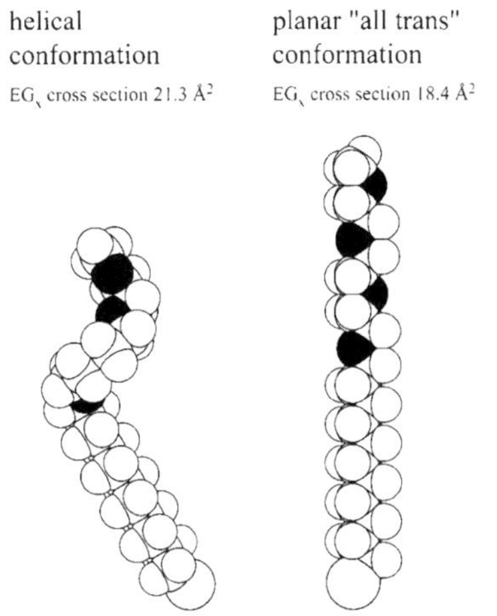

Figure 1. Molecular cross-sections for the helical EG3-OMe on gold with a ~30° tilt of the alkyl chain and a perpendicular orientation of "all-trans" EG3-OMe on silver. For details see ref 7.

To interpret the protein-resistance properties of the SAMs containing the helical conformers, the presence of a stable interphase water layer preventing the SAM from direct contact with protein molecules was postulated from *ab initio* 6-31G SCF calculations of the microscopic structure of the SAM/water interphase region.[10] The observed difference in protein adsorption between the helical and planar SAM phases is assumed to be caused by a difference in the structural organization of water near the SAM surface. Calculations show that the helical SAM phase easily accommodates water molecules, which act as a template for further adsorption of water via a hydrogen bridge-bonding network. The resulting interphase water layer is tightly bound to the helical or amorphous SAM surface, thereby preventing direct contact between the surface and the protein. In contrast, the denser SAM phase on silver does not allow water molecules to form strong hydrogen bonds with oxygen atoms in the EG3-OMe strands. As a result, no interphase water layer capable of hindering protein adsorption is formed. Similar explanations have been given in the literature[11] to correlate the inertness of some hydrophilic surfaces toward protein adsorption with the contact angles of water.

Force measurements have been performed with scanning probe techniques to study protein properties and protein–surface interactions. Apart from specific recognition and specific interactions,[12–18] single-molecule force spectroscopy,[19] adhesion forces between ligand–receptor pairs,[20,21] and protein adsorption onto polymer surfaces[22] have also been reported in the literature. However, the importance of the different chemical regions (neutral or charged) in a protein with respect to the net interaction of the macromolecule with a surface is still being debated.[23]

In the following, we first describe force–distance measurements between fibrinogen-modified probes and the surfaces of tri(ethylene glycol) (EG3-OMe)-terminated alkanethiolate monolayers on gold and silver, obtained with a commercial AFM to demonstrate the good correlation between these experiments and fibrinogen adsorption data obtained by FTIR. Subsequently, results obtained with charged (oxidized/hydroxylated) or hydrophobized (hexadecanethiol-derivatized) probes are reported. These experiments are conducted to elucidate qualitatively the different contributions and relevance of charged or hydrophobic patches in fibrinogen (the protein used in this study) to the resulting interactions with tri(ethylene glycol)-terminated SAMs on gold or silver.

Experimental Section

Materials. EG3-OMe (1-mercaptoundec-11-yl)tri(ethylene glycol) methyl ether was prepared according to a general procedure developed by Prime and Whitesides.[6] 11-Bromoundec-1-ene was added dropwise to a solution of tri(ethylene glycol) methyl ether in THF with 2 equiv of NaH and the solution was stirred overnight. Nonconsumed NaH was reacted with 2-propanol. The solvent was removed and the remaining product was purified by column chromatography. A solution of the olefin in methanol containing 4 equiv of thiolacetic acid and 10 mg of AIBN was irradiated for 6 h under an atmosphere of nitrogen with a 450 W, medium-pressure mercury lamp. Concentration of the reaction mixtures by rotary evaporation at reduced pressure followed by purification by chromatography on silica gel gave the thioacetate. The latter was refluxed overnight in 0.5 M HCl in methanol and the concentrated solution was purified by column chromatography. The purity of the thiol was checked with NMR and mass spectroscopy. The syntheses of EG6-OH and EG[3,1]-OMe have been described elsewhere.[24] (1-Mercaptoundec-11-yl)poly(ethylene glycol) methyl ether (MW 2000) was prepared analogous to a general procedure for mercaptoundecyl oligo(ethylenglycol) described by Prime and Whitesides.[24] The following changes were made: PEG 2000 monomethyl ether as starting material was dried over molecular sieve 48 h before use. For the coupling reaction to the alkyl chain, 2 equiv of 11-bromoundec-1-ene was added to a solution of PEG 2000 monomethyl ether in dry THF with 2 equiv of NaH and the mixture was stirred overnight. Purification of the different synthetic steps was carried out by column chromatography on silica gel with chloroform/methanol 1:3 as eluent. Purity and molecular weight distribution of the thiol were checked by NMR and MALDI mass spectrometry (M_n (number average) = 2199, M_w (weight average) = 2224; M_w/M_n = 1.01).

SAM Preparation. Polycrystalline gold (99.99%, Balzers Materials, Liechtenstein) and silver (99.99+%, Aldrich Chem. Co., Milwaukee, WI) substrates were prepared by thermal evaporation of these metals onto plasma-cleaned pieces of singly polished silicon (100) wafers (MEMC Electronic Materials, Inc., St. Peters, MO) in a BAL-TEC (Balzers, Liechtenstein) MED-020 coating system operated at (3–5) × 10^{-5} mbar. Evaporation of a 5-nm chromium adhesion layer was followed by deposition of a 100-nm layer of gold or silver at a rate of 0.5 nm/s. Coated substrates were immediately immersed into 2 mmol thiol solutions in ethanol. Upon removal from the thiol solution, the SAMs were rinsed with pure ethanol and dried with nitrogen.

(9) Pertsin, A. J.; Grunze, M. *Langmuir* **1994**, *10*, 3668–74.
(10) Wang, R. L. C.; Kreuzer, H. J.; Grunze, M. *J. Phys. Chem. B* **1997**, *101*, 9767–9773.
(11) Vogler, E. A. *Adv. Colloid Interface Sci.* **1998**, *74*, 69–117.
(12) Florin, E. L.; Moy, V. T.; Gaub, H. E. *Science* **1994**, *264*, 415.
(13) Ludwig, M.; Moy, V. T.; Rief, M.; Florin, E. L.; Gaub, H. E. *Microsc. Microanal. Microstrucrt.* **1994**, *5*, 321.
(14) Lee, G. U.; Kidwell, D. A.; Colton, R. J. *Langmuir* **1994**, *10*, 354–357.
(15) Florin, E. L.; Rief, M.; Lehmann, H.; Ludwig, M.; Dornmair, C.; Moy, V. T.; Gaub, H. E. *Biosens. Bioelectron.* **1995**, *10*, 895–901.
(16) Allen, S.; Chen, X. Y.; Davies, J.; Davies, M. C.; Dawkes, A. C.; Edwards, J. C.; Roberts, C. J.; Sefton, J.; Tendler, S. J. B.; Williams, P. M. *Biochemistry* **1997**, *36*, 7457–7463.
(17) Chowdhury, P. B.; Luckham, P. F. *Colloids Surf. A* **1998**, *143*, 53–57.
(18) Bowen, W. R.; Hilal, N.; Lovitt, R. W.; Wright, C. J. *J. Colloid Interface Sci.* **1998**, *197*, 348–352.
(19) Rief, M.; Oesterhelt, F.; Heymann, B.; Gaub, H. E. *Science* **1997**, *275*, 1295–1297.
(20) Lee, G. U.; Chrisey, L. A.; Colton, R. J. *Science* **1994**, *266*, 771.
(21) Moy, V. T.; Florin, E. L.; Gaub, H. E. *Colloids Surf. A* **1994**, *93*, 343.
(22) Chen, X.; Davies, M. C.; Roberts, C. J.; Tendler, S. J. B.; Williams, P. M.; Davies, J.; Dawkes, A. C.; Edwards, J. C. *Langmuir* **1997**, *13*, 4106–4111.
(23) Horbett, T. A., Brash, J. L.; Horbett, T. A.; Brash, J. L., Eds. American Chemical Society: Washington, DC, 1995; pp 11–14.
(24) Prime, K. L.; Whitesides, G. W. *J. Am. Chem. Soc.* **1991**, *113*, 12.

Characterization of SAMs involved static water-contact-angle measurements with a contact-angle goniometer (Ramé-Hart, Inc., Mountain Lakes, NJ) and film-thickness measurements by ellipsometry (Type L-116C, Gaertner Science Corp., Chicago, IL) with a 70° angle of incidence of a He−Ne laser. Results of water-contact-angle measurements were in agreement with those reported earlier:[7] 63° ± 2° for the EG3-OMe SAMs on *both* gold and silver. Representative films were checked by FTIR with respect to their overall quality and molecular conformation.

A sample of PEG 2000-thiolate was prepared by immersion of a gold substrate into an ethanolic 1 mmol thiol solution for 48 h; characterization of the sample revealed that the packing density of the molecules (distance between the binding sites) is slightly less than the unperturbed radius of gyration for the given molecular weight distribution ($R_g \sim 3.6$ nm).

Probes for SFM Measurements. It is known that formation of fibrinogen clusters on a hydrophobic surface is due to strong intermolecular interactions. These appear to be less significant in the presence of a mica surface.[25,26] Hence, to avoid clustering, we decided to mainly use oxygen-plasma-treated Si_3N_4 probes to adsorb fibrinogen. Similarly to mica, these probes acquire a negative charge at neutral pH. It should be noted, however, that similar results to those reported in the following were also obtained using hydrophobic (C16-functionalized) probes with adsorbed fibrinogen.

"Hydrophobic" probes (referred to below as C16-probes) were prepared by vapor-phase deposition of hexadecanethiol (Aldrich Chemical Co., Inc., Milwaukee, WI) onto gold-coated Si_3N_4 probes (Digital Instruments, Inc., Santa Barbara, CA) with a nominal radius of curvature of 30 nm. Both sides of the plasma-cleaned probes were coated with a 5-nm chromium layer followed by 50 nm of gold. After gold deposition, the probes were placed in a metal desiccator containing 200 mL of hexadecanethiol, the desiccator was then evacuated to ~0.1 mbar and kept under vacuum overnight. To ensure the quality of SAMs, a piece of a silicon wafer was gold coated and functionalized together with the probes and analyzed by ellipsometry, static contact angle, and XPS. The analyses did not reveal any notable differences between the SAMs of hexadecanethiol prepared via vapor-phase deposition and those prepared by immersion into a 5 mM solution in ethanol and therefore we assume a similar quality of the films on the Au-coated Si_3N_4 tips used in our force measurements.

Adsorption of fibrinogen onto SFM probes was conducted according to the following protocol: Si_3N_4 probes, previously cleaned with water-enriched oxygen plasma, or C16-probes (Digital Instruments, Inc., Santa Barbara, CA) were placed in phosphate-buffered saline solution (PBS) containing 0.01 M phosphate buffer, 0.0027 M KCl, and 0.137 M NaCl (solution obtained by dissolving PBS tablets, P-4417, Sigma Chem. Co., St. Louis, MO). Fraction I fibrinogen from human plasma (F-4883, Sigma Chem. Co., St. Louis, MO) was dissolved in PBS solution at a concentration of 2 mg/mL. This solution was then added to that containing SFM probes to achieve a final concentration of fibrinogen of approximately 1 mg/mL. After 1 h of adsorption, more PBS solution was added, followed by aspirating the liquid with a vacuum line to remove any fibrinogen film that may have formed at the liquid−air interface. This procedure was repeated several times, to ensure that upon removal of the probes from the solution no fibrinogen film was transferred from the air/water interface onto the probes.

Each set of samples was measured with only one probe to ensure that observed changes in the tip−surface interaction were not due to variability in cantilever stiffness or probe-tip radius, although the latter might be affected by the number of loading−unloading cycles. The force−distance measurements with the fibrinogen-preadsorbed probes were conducted in the following manner: at least 64 force−distance curves were collected in PBS solution, both at the same point and at adjacent locations.

To gain further insight into the nature of the distinctly different protein-resistance behavior of the two monolayers, we employed probes with better-defined surface compositions to "mimic" the nonspecific interaction of the fibrinogen macromolecule with monolayers of EG3-OMe. Since proteins contain both hydrophobic and charged domains,[27] we used hydrophobic (C16) and oxygen-plasma-treated Si_3N_4 probes, which acquire a net negative charge at biologically relevant pH values.

To obtain charged SFM probes we used "sharpened" Si_3N_4 Microlevers (Park Scientific Instruments, Sunnyvale, CA) with a nominal radius of curvature of 20 nm, and treated them for 45−60 s in a RF-plasma cleaner (Harrick Scientific Corp., Ossining, NY), operated at 40 W with a water-enriched oxygen feed to ensure both removal of contaminants and hydroxylation of the probe-tip surface. The plasma-treated tips were stored either in deionized water or in PBS buffer solution.

The largest dimension of a fibrinogen molecule in the native state is on the order of 470 Å.[25,28] The size of the charged αC-domains of this protein and of its hydrophobic D-domains is quite comparable to the contact area between the SFM probes and the surface.[28]

SFM Measurements. Force−distance measurements and imaging were performed with a Nanoscope IIIa scanning force microscope (Digital Instruments, Inc., Santa Barbara, CA) equipped with a liquid cell. We monitored the temperature inside the cell with a K-type thermocouple. The temperature was in the range of 27−30 °C during our measurements. Force−distance curves were collected with a cycle frequency of 0.3−0.5 Hz. All liquids introduced into the liquid cell were filtered with 0.22 μm "millex-GV", low-protein-binding filters (Millipore, Bedford, MA). Manufacturer-provided nominal values of the cantilevers' spring constants, k_0, and frequencies, ω_0, were used to calculate the actual spring constant, k, by measuring resonant frequencies, ω, of the probes according to the equation:

$$k = k_0*(\omega/\omega_0)^2$$

assuming that the effective mass of the cantilevers is constant. Prior to force-versus-distance measurements, 100- and 500-nm z-calibration gratings (TGZ-type, NT-MDT, Zelenograd, Russia) were scanned in contact mode to ensure proper calibration of the z-piezo. Piezo-displacement cantilever-deflection curves were converted into force−distance curves according to the procedure described in ref 29. Zero separation corresponds to a hard-wall potential, i.e., there is no absolute measure for the distance between tip and surface. Force-versus-distance measurements of each series of samples were performed with the same probe to minimize the error in distance due to variability in the spring constant value. It is important to note that only semiquantitative analysis of the force−distance data is possible due to the unknown precision of spring-constant and probe radii values. In addition, the probe radius may change due to blunting caused by many repetitions of the measurements.

Results

In establishing possible differences between the EG3-OMe SAMs on gold and silver, we first recorded 3 × 3 μm² SFM images of the two surfaces in TappingMode with a 5% reduction in the set-point amplitude. A surface-roughness analysis revealed no significant differences between the root-mean-square (RMS) roughness values of the two surfaces, which were both found to be around 1.6 nm. Hence, the differences in the force−distance curves on gold and silver described below cannot be related to differences in surface topography.

Figure 2 shows representative force−separation curves measured with a fibrinogen-modified probe in pure PBS solution on a series of surfaces including mica (Figure 2a), hexadecanethiol-covered gold (Figure 2b), a protein-resistant EG3-OMe monolayer on gold (Figure 2c), and a protein-adsorbing film on silver (Figure 2d). According to FTIR measurements, the latter adsorbed about 20% of a fibrinogen monolayer.

(25) Erlandsson, R.; Olsson, L.; Bongrand, P.; Claesson, P. M.; Curtis, A. S. G., Eds.; Springer: Berlin, 1994, Chapter 4.

(26) Ta, T. C.; Sykes, M. T.; McDermott, M. T. *Langmuir* **1998**, *14*, 2435−2443.

(27) Andrade, J. D.; Horbett, T. A.; Brash, J. L., Eds. ACS Symp. Series No. 343; American Chemical Society: Washington, DC, 1987; Chapter 1.

(28) Feng, L.; Andrade, J. D.; Horbett, T. A.; Brash, J. L., Eds. ACS Symp. Series No. 602; American Chemical Society: Washington, DC, 1995; Chapter 5.

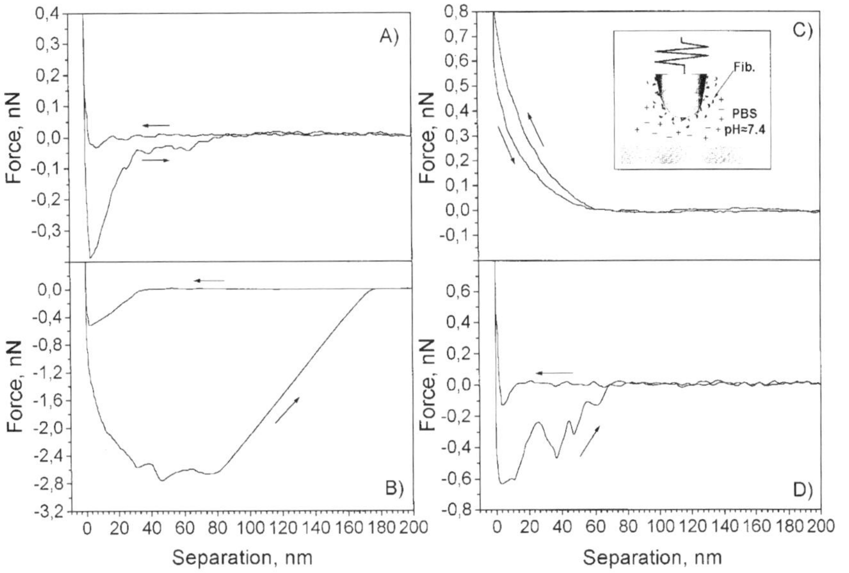

Figure 2. Representative force-versus-separation curves obtained with a fibrinogen-modified Si_3N_4 probe ($k = 0.03$ N/m) in PBS solution on (a) freshly cleaved mica surface, (b) hexadecanethiol SAM on gold, (c) EG3-OMe SAM on gold, and (d) EG3-OMe on silver showing attractive interaction and multiple pull-offs (representative of 20% of the force−distance data taken).

The hydrophilic (mica) and the hydrophobic (C16) surfaces are intended to serve as a reference for strong fibrinogen binding. Both surfaces show an attractive interaction with the fibrinogen-modified probe upon approach and strong adhesion upon retraction, indicating that the proteins establish contact to both probe and sample surfaces. Note the difference in the extent of the attractive interaction of fibrinogen between the two surfaces. Clearly, the attractive force upon approach, the pull-off forces, and the adhesion hysteresis (the area between the approaching and retracting curves) were all greatest for the fibrinogen-C16 system, which is in good agreement with results reported in the literature.[26]

As can be seen in Figure 2c, there are no attractive forces between a fibrinogen-modified probe and an EG3-OMe monolayer on gold. However, there is a reproducible loading−unloading hysteresis, which did not change when pure buffer was replaced by fibrinogen solution. Two hundred and fifty six force curves consecutively recorded in pure PBS and fibrinogen solution gave no evidence of attractive or adhesive forces. This was reproduced on several samples.

Similar observations have been reported by Sheth and Leckband by means of a surface forces apparatus.[30] These authors, however, measured an attractive interaction after *pressing* streptavidin onto PEG. The very small attractive forces found in their experiment might also be present here, but the effect is too small to be unambiguously detected.

For the EG3-OMe monolayer on silver, we observed multiple pull-off events indicating adhesion of the fibrinogen-coated probe, in agreement with the FTIR protein-adsorption measurements (Figure 2d).

Measurements with oxygen-plasma-treated probes were performed in air, deionized water, and the same PBS buffer as used in the protein-adsorption experiments. Force-versus-distance curves in air with an oxygen-plasma-treated probe ($k = 0.03$ N/m) revealed no significant differences between the EG3-OMe SAMs on gold and silver. A large adhesion hysteresis primarily due to capillary forces[31] was found for both surfaces. Pull-off forces for the SAM were determined to be 11.7 ± 1.4 nN on gold and 10.6 ± 0.3 nN on silver. The broader distribution of the pull-off forces from the EG3-OMe on gold might be due to the variations of the elastic modulus of the SAM on gold, which would, in turn, depend on the degree of crystalline order in the monolayer. The similarity of the pull-off forces on gold and silver in air measured with hydrophilic, largely hydroxylized probes is consistent with the similar water contact angles measured for the two monolayers.[7] Similar water-contact angles correspond to similar surface interfacial energies with water and, therefore, to similar values of the work of adhesion which, in turn, is reflected in the pull-off forces.

Force-versus-distance measurements were also performed with an oxygen-plasma-treated Si_3N_4 probe ($k = 0.01$ N/m) in deionized (DI) water (18.2 MΩ·cm) and PBS buffer (Figure 3). We previously found[32] that probes treated with water-rich oxygen plasma have an isoelectric point at pH ≈3, which implies that at higher pH values the probes acquire a negative surface charge due to deprotonation of hydroxyl groups. Our measurements of the EG3-OMe SAMs on gold and silver in DI water with a negatively charged probe showed a long-range repulsion followed by a short-range attraction, similar to a DLVO-type interaction.[33] The repulsive force was found to be stronger for the SAM on gold, while the pull-off forces were greater for that on silver. Measurements in PBS buffer did not reveal any significant differences between the two surfaces. As expected, the ion concentration had a significant effect on the observed interaction: for both surfaces attractive forces are observed in

(29) Ducker, W. A.; Senden, T. J.; Pashley, R. M. *Nature* **1991**, *353*, 239.

(30) Sheth, S. R.; Leckband, D. *Proc. Natl. Acad. Sci.* **1997**, *94*, 8399−8404.

(31) Grigg, D. A.; Russel, P. E.; Griffith, J. E. *J. Vac. Sci. Technol. A* **1992**, *10*, 680.

(32) Feldman, K.; Tervoort, T.; Smith, P.; Spencer, N. D. *Langmuir* **1998**, *14*, 372−378.

(33) Israelachvili, J. *Intermolecular and Surface Forces*, 2nd ed.; Academic Press: San Diego, CA, 1992.

10138 *J. Am. Chem. Soc., Vol. 121, No. 43, 1999* *Feldman et al.*

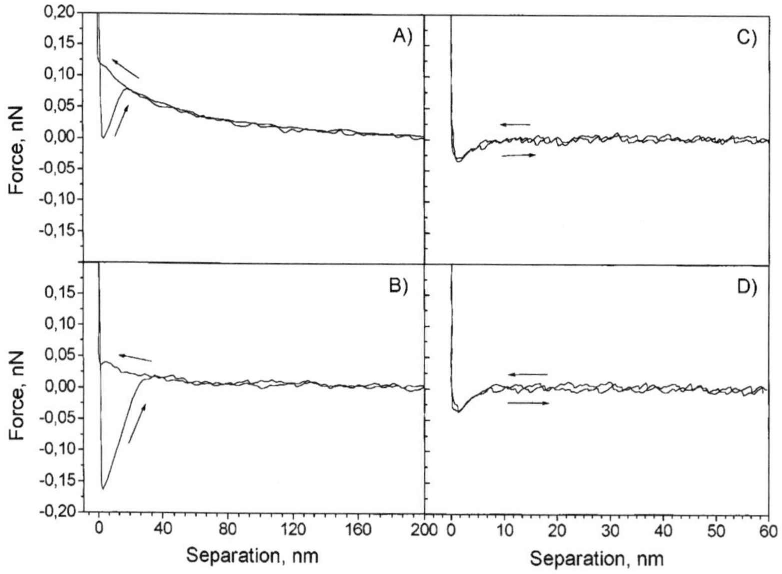

Figure 3. Representative force-versus-separation curves obtained with an oxygen plasma-treated Si_3N_4 probe (k = 0.02 N/m) measured on (a) EG3-OMe on gold in deionized water, (b) EG3-OMe on silver in deionized water, (c) EG3-OMe on gold in PBS solution, and (d) EG3-OMe on silver in PBS solution.

the PBS experiment, with a significantly smaller range of interaction compared to measurements in DI water.

The most interesting results were obtained with hydrophobic probes. Measurements in pure water with C16-probes revealed a long-range repulsion for the EG3-OMe monolayer on gold and a long-range attraction for that on silver (Figure 4). When DI water was replaced by PBS solution, the extent of the repulsive interaction diminished on gold, but did not change sign. For silver, however, only very small changes were found in PBS buffer as compared to pure water (Figure 4). The experiments were repeated many times with different samples of EG3-OMe and several batches of C16-probes to confirm their correlation with the FTIR protein-adsorption measurements. We also investigated two other 1-undecylthiols, either with a hydroxyl-terminated hexa(ethylene glycol) (EG6-OH) or a methoxy-terminated tri(ethylene glycol) (EG3-OMe) end group and a $-CH_2OCH_3$ side chain at the C-12 atom (EG[3,1]-OMe). Force−distance measurements on monolayers of either thiol on gold or silver established the same correlation for all investigated films. Whenever a repulsive force was observed upon approach of a *hydrophobic* probe in PBS solution, that surface resisted fibrinogen adsorption as determined by the FTIR measurements.

The repulsive curves measured upon approach to EG3-OMe on gold showed some variability with respect to the range of interaction. Some of them were purely repulsive, while occasionally a jump-to-contact at distances of ~5 nm before contact appeared. Whenever a jump-to-contact occurred, this was accompanied by a hysteresis in the loading−unloading cycle, while no hysteresis was found in the purely repulsive curves. This variability of the SFM data in terms of the *range* of the forces and the presence of an adhesion hysteresis might be attributed to the structural inhomogeneity of the monolayers, i.e., the presence of different phases with dissimilar mechanical properties in the EG3-OMe layer.

Force−distance measurements with a C16-probe and an EG3-OMe SAM on gold in DI water and PBS solution revealed a strong dependence of the range of the repulsive force upon the ionic strength of the solution; the same measurements for the

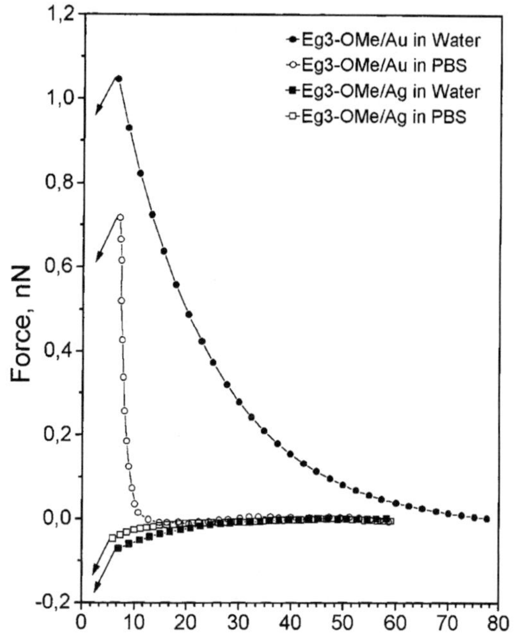

Figure 4. Advancing force-versus-separation curves measured with a hydrophobic C16-probe (k = 0.12N/m) in deionized water and PBS solution on the SAMs of EG3-OMe on gold and silver.

monolayer on silver showed a long-range attractive interaction with very little dependence on the ionic strength. These results suggest that, due to different conformations of the EG3-OMe molecules on gold and silver, the observed interactions may well be of a distinct physical nature, i.e., electrostatic for EG3-OMe on gold and hydrophobic for that on silver. To further elucidate this hypothesis, we performed force−distance mea-

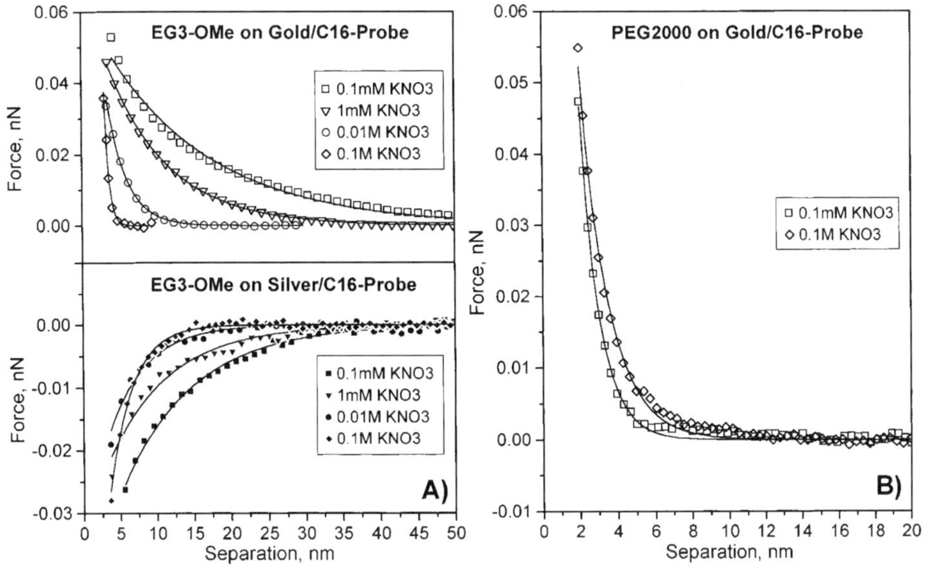

Figure 5. Advancing force-versus-separation curves measured with a hydrophobic C16-probe (k = 0.06 N/m) in aqueous solutions of KNO_3 on (a) the SAMs of EG3-OMe on gold and silver and (b) PEG2000-thiolate on gold. Solid lines represent fitted data.

surements with a C16-probe on the two monolayers in aqueous solutions of KNO_3 of various ionic strengths (Figure 5a), and, once again, detected repulsive forces for the EG3-OMe on gold that displayed a strong dependence upon the ionic strength of the solution and attractive long-range forces for that on silver that were less influenced by the presence and concentration of ions.

We also performed measurements on the end-grafted PEG 2000 brush on gold with a C16-probe (Figure 5b) to demonstrate that the long-range effects observed on packed, ordered monolayers of EG3-OMe are not detected in grafted polymer brushes. Also, in marked contrast to the behavior of the shorter molecules, end-grafted PEG scarcely revealed any ionic-strength dependence.

According to previous theoretical work on these systems,[10] strong interaction of water molecules with the helical conformer of the EG3-OMe SAM on gold is a necessary condition for protein resistance, because oligo(ethylene glycol) per se is not intrinsically protein resistant. Water cannot associate with the all-trans structure of the EG3-OMe SAM on silver, and hence this system exhibits protein-adsorption behavior that is characteristic for a slightly hydrophobic surface. To demonstrate unambiguously that the solvent plays a crucial role in explaining the difference in the adsorption characteristic of the helical and all-trans conformers, force–distance measurements with a hydrophobic probe were performed in perfluorodecalin $C_{10}F_{18}$ (Fluorochem, U.K.), which is a nonpolar, non-hydrogen-bonding liquid with a low dielectric constant ($\epsilon \sim 2.0$). In agreement with the theoretical arguments, we did not observe any long-range forces in perfluorodecalin on the two SAMs (Figure 6). Instead we found attractive interactions on both surfaces.

Discussion

The images recorded for EG3-OMe on gold and silver did not reveal any differences in roughness. Since the protein-adsorption experiments were conducted using fibrinogen, which is known to have an extremely high surface activity,[28] one would not expect roughness to be a factor influencing the surface affinity of fibrinogen. However, it is helpful to be able to exclude surface roughness as a possible variable in our experiments.

The presence of a hysteresis in the repulsive part of the force curve measured with a fibrinogen-coated C16-probe for EG3-OMe on gold (Figure 2c) indicates the presence of energy-dissipating processes, which can be attributed to viscoelastic conformational changes in the protein layer induced by the loading pressure. Although pressure-induced changes in protein conformation during loading and unloading might lead to aggregation of protein in the contact area, giving rise to long-range repulsive forces,[34] successive measurements at a single surface location showed no changes in the interaction. This also confirms that the protein layer recovers upon the retraction of the probe and that no plastic deformation is induced.

The interaction between charged bodies in electrolyte solutions can be described by the DLVO theory, which takes into account both electrostatic and van der Waals forces.[33] For low surface potentials (ψ < 25 mV) the electrostatic force per unit area between the two bodies bearing charge densities σ_1 and σ_2 can be described by the following equation:[35,36]

$$F_{\mathrm{el}} = \frac{2}{\epsilon\epsilon_0}[(\sigma_1^2 + \sigma_2^2)e^{-2\kappa D} + \sigma_1\sigma_2 e^{-\kappa D}] \qquad (1)$$

For larger separations, electrostatic force between a sphere of radius R and a flat surface becomes:[36,37]

$$F_{\mathrm{el}} = \frac{4\pi R\sigma_1\sigma_2}{\epsilon\epsilon_0\kappa}e^{-\kappa D} \qquad (2)$$

where $1/\kappa$ is the so-called Debye length and D is the separation between the two bodies. The Debye length, being a solution property, can be calculated from the following equation:[33]

(34) Claesson, P. M.; Blomberg, E.; Froberg, J. C.; Nylander, T.; Arnebrant, T. *Adv. Colloid Interface Sci.* **1995**, *57*, 161–227.
(35) Parsegian, V. A.; Gingell, D. *Biophys. J.* **1972**, *12*, 1192–1204.
(36) Butt, H.-J. *Biophys. J.* **1991**, *60*, 777–785.
(37) Butt, H.-J.; Jaschke, M.; Ducker, W. *Bioelectrochem. Bioenerg.* **1995**, *38*, 191.

$$\kappa = \left(\sum_i \frac{\rho_{\infty i} e^2 z_i^2}{\epsilon\epsilon_0 k_B T}\right)^{1/2} m^{-1} \qquad (3)$$

where $\rho_{\infty i}$ is the bulk ionic concentration, z_i is the ionic charge, k_B is Boltzmann's constant, e is the electronic charge, and T is the absolute temperature.

One important implication of eq 1 is that the overall electrostatic interaction is not eliminated when one of the surface charge densities is set to zero, e.g. $\sigma_1 = 0$ and $\sigma_2 \neq 0$. It is therefore not necessary for the both surfaces to carry charges.

The issue of surface charges on the EG3-OMe monolayer on gold is also important. The tails of EG3-OMe on gold can easily accommodate water molecules and seem to be present in both helical and amorphous states; for such "soft", permeable interfaces with polar tails it has been shown that the *effective* surface charge density, σ_{eff}, depends on the surface charge density, σ, and the dipole density, ν:[38,39,40]

$$\sigma_{eff} = [\sigma \cosh(\kappa l) + \nu\kappa \sinh(\kappa l)]e^{-\kappa l} \qquad (4)$$

where l is the thickness of the "soft" polar region. Therefore, if there are no surface charges (or the net surface charge is zero) but there is a dipole field, a surface would bear an effective non-zero charge, which would give rise to an electrostatic interaction.

Long-range repulsive forces followed by short-range attraction were measured in DI water for the EG3-OMe monolayers on gold and silver substrates with a negatively charged oxygen-plasma-treated Si_3N_4 probe (Figure 3). When comparing the force curves in DI water and PBS buffer it becomes clear that the ion concentration has a pronounced effect on the tip/substrate interaction. The two substrates differ only in the strength of the repulsive force. This difference may be explained if we now assume that, due to conformational differences, the EG3-OMe monolayer on gold carries an effective surface charge (eq 4), while that on silver does not. There will still be a repulsive electrostatic interaction on silver (see eq 1, with $\sigma_2 = 0$) but it will be weaker than that for gold where $\sigma_1 \neq 0$ and $\sigma_2 \neq 0$. Upon introduction of the PBS solution, the range of the electrostatic force is, as expected, dramatically reduced. The screening of the electrostatic repulsion is due to the high ionic strength of the PBS solution (we calculated the Debye length of the PBS solution to be $1/\kappa = 0.76$ nm at room temperature compared to $\sim$1 μm for DI water) and only the attractive van der Waals forces are clearly observed (Figure 3). In addition, the type of interaction changed from repulsive in pure water to attractive in PBS buffer, the range was found to be much shorter in the case of high ion concentrations, and almost no hysteresis was found for measurements performed in PBS buffer, although the hysteresis might also scale with the Debye length and hence be too small to be detectable here.

The EG3-OMe monolayer on silver consists of molecules that are tightly packed and expose their hydrophobic methyl groups at the film/water interphase. The observed long-range attractive forces between the hydrophobic C16-probe and the EG3-OMe SAM on silver in water and PBS solution (Figure 4) point to the presence of hydrophobic interaction.[41,42] Indeed, these force curves are well reproduced by exponential fits with decay lengths of 11 nm in pure water and 6 nm in PBS buffer (Figure 4). These are typical decay lengths for hydrophobic forces. Although the molecular origin of hydrophobic forces is not well understood[43] and the extent of the interaction is often questioned,[44] there is substantial experimental evidence of a long-range attractive interaction in an aqueous environment between hydrophobic surfaces with the attractive forces sometimes being sensed at separations of 50 nm and more.[41,45]

In contrast to the attractive hydrophobic interaction between EG3-OMe film on silver and the C16-probe, a strong dependence of the *repulsive* force on the ion concentration observed on the EG3-OMe monolayer on gold in DI water and PBS buffer suggests that this interaction involves a significant electrostatic contribution. Furthermore, since the helical or amorphous structure of the OEG tails of the SAM on gold allows a strong interaction of water molecules via hydrogen bonding with oxygen atoms of the OEG, there might be an additional repulsive contribution of hydration or steric-protrusion forces, which decay roughly exponentially with a distance of 0.2−0.4 nm.[42,45] This short-range, exponentially decaying force observed in PBS solution looks similar to data obtained on lipid bilayers in pure water with SFM[46] and the surface forces apparatus (SFA),[47] and is assigned to steric/hydration forces there.[33,46,47]

Study of the ionic concentration effect on the range of forces measured between the EG3-OMe SAMs and C16-probes (Figure 5a) provided further evidence that the overall interaction with the monolayer on gold had a significant electrostatic component. Data for EG3-OMe on gold shown in Figure 5a were fitted with an electrostatic force law for a case where the second surface (in our case, that of the C16-probe) does not carry a surface charge (see eq 1). Measurements of the symmetric system C16-probe/C16-surface did not show any repulsive interaction in water and salt solutions indicating that there were no charges present on these surfaces and the interaction was found to be strongly attractive and occurred at separations of about 20 nm. Note that traces of ions (in particular, HCO_3^- from dissolved CO_2) present in the water may reduce the Debye length of dilute solutions significantly.

Although for the case of charge on one side only, eq 1 reduces to $F_{el} \propto e^{-2\kappa D}$, we found that for most measurements and higher ionic strengths the experimental data were best fitted with $F_{el} \propto e^{-\kappa D}$. We performed measurements using a C16-probe and a bare SiO_x surface in aqueous solutions of KNO_3—also a "charge-on-one-side-only" system—and also observed in this case that the data were best fitted with the $e^{-\kappa D}$ coefficient. We believe that this is an image force effect, which comes into play for counterions in water that are surrounded by surfaces of higher dielectric constant.[33]

Results of the fits to the EG3-OMe data are shown in Table 1. The radius of the probe used was checked by scanning over the ridges of a $SrTiO_3$ single crystal[48] and was found to be $\sim$110 nm. Dipole moments per molecule were calculated from dipole density, ν, assuming $l = 1$ nm in eq 4 and packing density of 21.3 Å^2/molecule.[7] Data for EG3-OMe on silver were fitted with

(38) Bell, G. M.; Levine, P. L. J. *J. Colloid Interface Sci.* **1980**, *74*, 530−548.
(39) Belaya, M.; Levadny, V.; Pink, D. A. *Langmuir* **1994**, *10*, 2010−2014.
(40) Xu, W.; Blackford, B. L.; Cordes, J. G.; Jericho, M. H.; Pink, D. A.; Levadny, V. G.; Beveridge, T. *Biophys. J.* **1997**, *72*, 1404−1413.
(41) Yoon, R. H.; Flinn, D. H.; Rabinovich, Y. I. *J. Colloid Interface Sci.* **1997**, *185*, 363−370.
(42) Israelachvili, J.; Wennerström, H. *Nature* **1996**, *379*, 219−225.
(43) Tsao, Y.-H.; Evans, D. F.; Wennerström, H. *Science* **1993**, *262*, 547−550.
(44) Wood, J.; Sharma, R. *Langmuir* **1995**, *11*, 4797−4802.
(45) Claesson, P. M.; Bongrand, P.; Claesson, P. M.; Curtis, A. S. G., Eds. Springer: Berlin, 1994, Chapter 2.
(46) Dufrêne, Y. F.; Barger, W. R.; Green, J. B. D.; Lee, G. U. *Langmuir* **1997**, *13*, 4779−4784.
(47) Marra, J.; Israelachvili, J. *Biochemistry* **1985**, *24*, 4608.
(48) Sheiko, S. S.; Möller, M.; Reuvekamp, E. M. C. M.; Zandbergen, H. W. *Phys. Rev. B* **1993**, *48*, 5675.

Table 1. Fitting Results of the Data Shown in Figure 5a

	EG3-OMe on gold			
KNO_3 concn, M	$1/\kappa$ calcd, nm	$1/\kappa$ fitted, nm	dipole moments/ molecule, D	EG3-OMe on Ag $1/\lambda$, nm
10^{-4}	30.7	14.5	4.6	9.6
10^{-3}	9.7	9.3	2.5	8.3
10^{-2}	3.1	2.7	2.2	5.1
10^{-1}	1.0	0.7	1.4	3.1

empirical formula[41] of the form $F_h \propto \exp(-\lambda D)$ to demonstrate that, unlike the situation on gold, the constant $1/\lambda$ does not correlate with the calculated Debye length of the solution. Apparently, while the molecular conformers of OEG on the gold surface generate sufficiently strong dipolar fields to cause a screenable electrostatic interaction with the C16-probe, the interaction on silver is dominated by nonelectrostatic forces. A detailed discussion on the difference in dipolar moments for the helical, amorphous, and planar "all trans" conformers in contact with water will be presented elsewhere.[49]

Experiments with end-grafted PEG 2000 show a clear distinction between PEG brush behavior and that of EG3-OMe monolayers. The brushes exhibit *no long-range* interactions and little ionic strength dependence (Figure 5b). In fact, the force was found to become slightly more repulsive with increasing ionic strength, since the 0.1 M KNO_3 solution represents good solvent conditions.[33] The lack of long-range interactions for the polymer brush demonstrates that the repulsive interactions found for the short-chain oligomers is unique and not related to the steric repulsion effect responsible for the protein resistance for the end-grafted polymers.

Finally, the data recorded for EG3-OMe on Au and Ag under perfluorodecalin (Figure 6) illustrate the fact that the difference in the interfacial force does not only depend on the molecular conformation, but also on the solvent, and is significant only in an aqueous environment. Hence, to understand resistance to protein adsorption of biological or organic surfaces, it is important that the complete system, i.e., film and solvent, be considered.

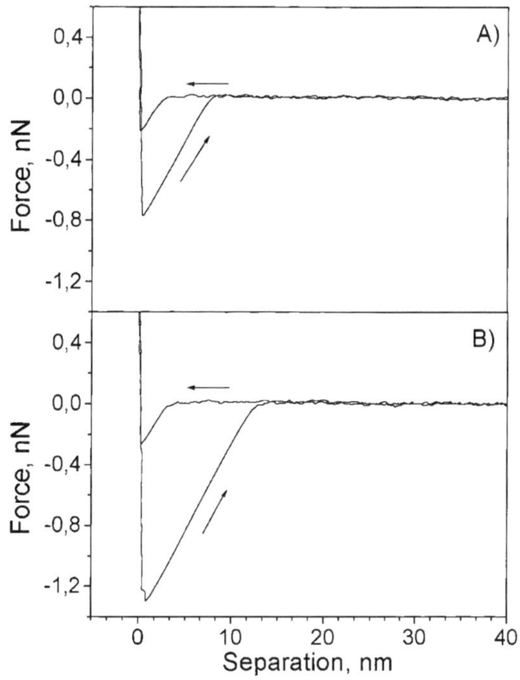

Figure 6. Representative force-versus-separation curves measured with the C16-probe (k = 0.12 N/m) in perfluorodecalin on (a) EG3-OMe on gold and (b) EG3-OMe on silver.

Conclusions

By means of a scanning force microscope with well-defined, functionalized tips, we were able to elucidate the contributions that are relevant to protein adsorption of fibrinogen on SAMs of EG3-OMe on gold and silver.

When using a hydrophobic tip to mimic hydrophobic regions of the protein, interactions of opposite sign were observed on the EG3-OMe-Au and EG3-OMe-Ag systems in aqueous media. While the attractive force observed on the silver system appeared to be hydrophobic in nature, the long-range repulsive force measured on the gold system displayed a strong ionic strength dependence, implying an important electrostatic component. The difference in sign of the forces can be correlated with the distinct molecular conformations of the OEG tails and the resulting difference in their dipolar fields. This effect dominates the interaction between the OEG-derivatized SAM and the SPM probe in the case of the helical and amorphous conformers on gold but is too small to contribute significantly to the total interaction in the case of the planar, "all trans" conformers on silver.

Acknowledgment. We thank the Deutsche Forschungsgemeinschaft, the Office of Naval Research, and the Council of the Swiss Federal Institutes of Technology (PPM and MINAST Program) for their financial support of this work. We thank W. Eck and S. Herrwerth for the EG3-OME and PEG 2000 materials.

JA991049B

(49) Zolk, M.; Buck, M.; Eisert, F.; Grunze, M.; Wang, R.; Kreuzer, H. J. In preparation.

5c. Following surface reactions in a UHV chamber

Commentary

Science relies on the development of new techniques in order to make new measurements and gain new insights. While the measurements have always been my priority, rather than the techniques, there was often the need to develop or improve certain methods along the way. The high-pressure-low-pressure apparatus, as described above (3.1, 3.2) enabled catalytic reactions to be carried out at pressures up to a hundred atmospheres, and to be followed by UHV surface characterization, by methods such as XPS, HREELS, LEED and mass spectroscopic measurements, without exposure to the air. This was a significant achievement within the Somorjai group, since it at least partially addressed the "pressure gap", between UHV surface science and experiments carried out in catalytic reactors (5.6).

Halogens are important as catalyst promoters, significant in a number of airborne pollution mechanisms, and generally of interest as reactive species. Carrying out halogen adsorption experiments in UHV chambers turned out to be challenging, however, since they were generally found to be too reactive to be simply introduced into the chamber as gas-phase elements. A number of authors resorted to the use of decomposition of organic halides, but this had drawbacks in terms of carbon-containing contamination. The solution was the development in the Lambert group of heated, highly collimated, solid-state electrochemical cells, which could produce halogens in small quantities in a local beam aimed directly at the sample, thus minimizing the adverse effects on the vacuum hardware (5.7). By modifying the control electronics to the electrochemical cell, an intensity waveform could be imparted to the halogen beam, which could then be used, in conjunction with mass spectrometric detection, for molecular-beam-reactive-scattering experiments, to determine, for example, the kinetics of reaction of the halogens with a surface (5.8).

Improved instrumentation to carry out surface analysis and to monitor chemical surface reactions *in situ* on small area catalysts over a wide pressure range (10^{-8}–10^{5} Torr)

A. L. Cabrera,[a] N. D. Spencer,[b] E. Kozak, P. W. Davies,[c] and G. A. Somorjai

Materials and Molecular Research Division, Lawrence Berkeley Laboratory and the Department of Chemistry, University of California, Berkeley, California 94720

(Received 11 May 1982; accepted for publication 2 August 1982)

Improved one-tier and two-tier designs for a high-pressure–low-pressure (HPLP) apparatus are described together with some of their recent applications. These instruments are used for surface analysis and *in situ* monitoring of catalytic reactions on small samples with a geometrical surface area of 1 cm^2. Catalytic reactions are studied over a wide range of pressures (10^{-8}–10^{5} Torr) and temperatures (273–1273 K). The apparatus incorporates surface analysis methods such as Electron Spectroscopy for Chemical Analysis (ESCA), High Resolution Electron Energy Loss Spectrometry (HREELS), and ^{14}C-radiotracer labeling in addition to Low Energy Electron Diffraction (LEED), Auger Electron Spectroscopy (AES), and mass spectrometry. Facilities are also available for argon sputter-ion cleaning, deposition of adsorbates by evaporation, and coverage calibration by means of a thickness monitor.

PACS numbers: 81.60. – j

INTRODUCTION

We have successfully studied a variety of reactions over single-crystal and polycrystalline surfaces of low surface area (1 cm^2)[1–8] using our previously reported high-pressure–low-pressure apparatus.[9] The basic design philosophy was to enable high-pressure reaction monitoring and catalyst surface characterization to be carried out in the same apparatus. Thus, since the sample is not removed from the UHV system, contamination of the catalyst by ambient gases can be kept to a minimum. Nevertheless, in a number of these studies, LEED and AES did not provide sufficient information to characterize the surface completely and the ability to combine many available techniques including ESCA, HREELS, ^{14}C-radiotracer labeling, among others, with high-pressure catalysis appears highly desirable. The combination of many surface analysis techniques provides a better understanding of a catalytic process. LEED is used for surface and adsorbate structure determinations and AES for surface composition measurements. ESCA gives information about the oxidation state and electronic structure of the surface and chemisorbed species, whereas HREELS allows the investigation of the bonds of chemisorbed molecules at the surface. Finally, radiotracer methods are among the most powerful techniques for measuring very low concentrations of materials in a transient environment.

The design of the improved HPLP apparatus allows for the easy exchange of surface analysis techniques and has a suitable geometry for rapid sample turnaround. Sensitivity to products has been maximized by making the high-pressure cell as small as possible. Two designs were needed (one tier and two tier) to satisfy the compatibility requirements of the electron spectrometers, evaporation stations, and other instruments.

I. ONE-TIER-HPLP SYSTEM

A photograph and a schematic diagram of this system are shown in Figs. 1 and 2, respectively. It consists of a conventional stainless-steel UHV chamber within which an isolation cell for high-pressure studies is situated. The chamber is cylindrical in shape and has a diameter of 12 in. Ultrahigh vacuum is obtained by a mechanically backed Varian 1 200 l/s diffusion pump with a liquid nitrogen trap, in combination with a water-cooled titanium sublimation pump. The background pressure of the system is lower than 1×10^{-9} Torr.

The sample is placed on the central axis of the chamber and is accessible to LEED, AES, ESCA, ion bombardment, mass spectrometry, deposition by evaporation, and beta-radiation detection facilities. Moreover, the sample can be rotated through 360°, thus allowing both front and back faces to be cleaned and inspected with the various surface science techniques.

Some care must be given to the choice of materials used in sample mounting. Materials which behave well under conditions of UHV will not necessarily stand up to the rigors of catalytic conditions, or may even catalyze unwanted reactions. Similarly, some substances which can be useful in constructing catalytic reactors may have a vapor pressure which is incompatible with UHV operation. In fact, a very limited selection of the elements finds application in a HPLP apparatus, and the final choice must be determined by the catalytic reactions under scrutiny. For the investigation of the properties of

FIG. 1. Photograph of a one-tier-HPLP system.

single crystals of platinum as hydrocarbon conversion catalysts, the crystals were spot welded to stout gold wires which, in turn, were held in copper rods. The copper was separated from the platinum in this way, so as to maintain it at a lower, less reactive temperature. The main source of crystal heating was resistive, within the crystal. For the investigation of the face specificity of iron-catalyzed ammonia synthesis, however, the iron crystals were spot welded to thin platinum wires, which, themselves, were spot welded to stout platinum rods. The main source of crystal heating in the latter case was conduction from the platinum wires. The cooling effect of the high hydrogen partial pressures used in this reaction necessitates a large heating current, which can lead to outgassing of the entire heating system. This conductive heating arrangement was chosen because it requires a lower current than crystal-resistive heating to achieve a given temperature. Platinum was chosen because of its inactivity as an ammonia synthesis catalyst.

The sample may be enclosed in a small high-pressure cell. This cell is displaced up and down by a hydraulic piston. To close the cell, a sealing pressure of 2 000 lb/in.^{2} is applied via the piston, the seal being made by an oxygen-free copper gasket. The gasket was annealed by heating to red heat, plunging into cold water, removing the oxide formed with concentrated HNO_3, and washing in water and acetone. It was found that plating the annealed gasket with 2/1000 in. gold improved both the efficiency and durability of the seal. Gaskets prepared in the above fashion can be used to seal the high-pressure cell effectively for at least 50 separate high-pressure runs. The cell is externally (to the UHV system) connected to a circulation pump, a pressure gauge, a gas manifold, and a gas sampling valve of a gas chromatograph. These units are connected in series forming a loop that starts at one end of the cell and is completed at the other end. The cell can be pressurized up to 100 atm with no significant increase in the background pressure of the vacuum chamber. Sulfur-containing lubricants must be vigorously excluded from all parts of the circulation loop since, especially when high-pressure hydrogen is being used, these substances can be readily vapor-transported onto the sample surface.

In the high-pressure mode, the isolation cell is closed and the reactants are admitted to the reaction loop. The reactants are continuously pumped by a mechanical circulation pump, and flow through the reaction loop and repeatedly over the sample. The gas mixture is sampled and analyzed by the gas chromatograph. The sample can be heated resistively by a high-current ac power supply and its temperature monitored by a chromel-alumel thermocouple.

One of several innovations in this apparatus consists of a LEED optics and a cylindrical mirror analyzer (CMA) mounted on hydraulically driven bellows, thus allowing these instruments to move back and forth relative to the sample. The optimum performance of the CMA is achieved when the end of the instrument is at a distance of 0.4 in. from the sample. Since the radius of the high-pressure cell is greater than this distance, the CMA would obstruct the cell while it was moving towards the closed position, unless the CMA were withdrawn. The LEED

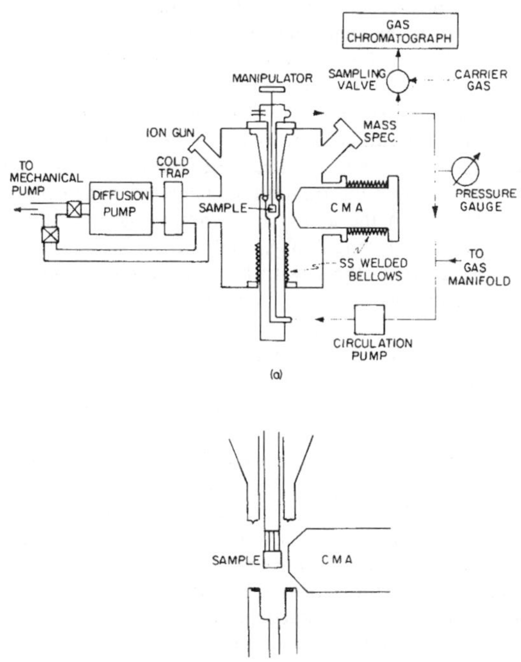

FIG. 2. (a) Schematic diagram of a one-tier-HPLP system with high-pressure cell closed. (b) Detail with high-pressure cell open.

FIG. 3. Photograph of two-tier-HPLP system.

optics are also mounted on bellows, since otherwise they would obstruct the CMA. The operation of ESCA/AES, LEED and the closing of the high-pressure cell are thus all mutually exclusive. Withdrawal of the LEED optics and the CMA is carried out by the same hydraulic system that controls the high-pressure cell.

II. TWO-TIER-HPLP SYSTEM

A photograph of this system is shown in Fig. 3. It consists of a stainless-steel ultrahigh vacuum (UHV) chamber built in two levels with an isolation cell for high-pressure studies (a schematic diagram of the system is displayed in Fig. 4). The upper portion of the UHV chamber contains the standard single-crystal surface analysis equipment (four grid LEED/Auger optics, glancing incidence electron gun, sputter ion gun for specimen cleaning and quadrupole mass spectrometer). The lower part of the UHV system contains the electron energy loss spectrometer (HREELS). The samples are lowered from the upper to the lower level by an extended travel precision manipulator. A similar system without the high-pressure isolation cell has been described in detail elsewhere.[10]

The high-pressure cell is identical to that in the one-tier–HPLP system but in this case both sides of the cell can be moved up and down. The top section of the cell has x and y displacement facilities, and, therefore, the sample can be precisely positioned in the x, y, and z directions and rotated through 360°. Nevertheless, when the cell is closed, this upper part of the cell must be locked in one position to withstand the sealing force. Due to the size of the top section, the LEED/Auger optics are mounted on bellows to facilitate easy withdrawal when the sample is lowered to the HREELS position. This system contains all the other features mentioned in the description of the one-tier–HPLP system.

III. SURFACE ANALYSIS

As we mentioned previously, the UHV chamber of the HPLP apparatus can incorporate combinations of

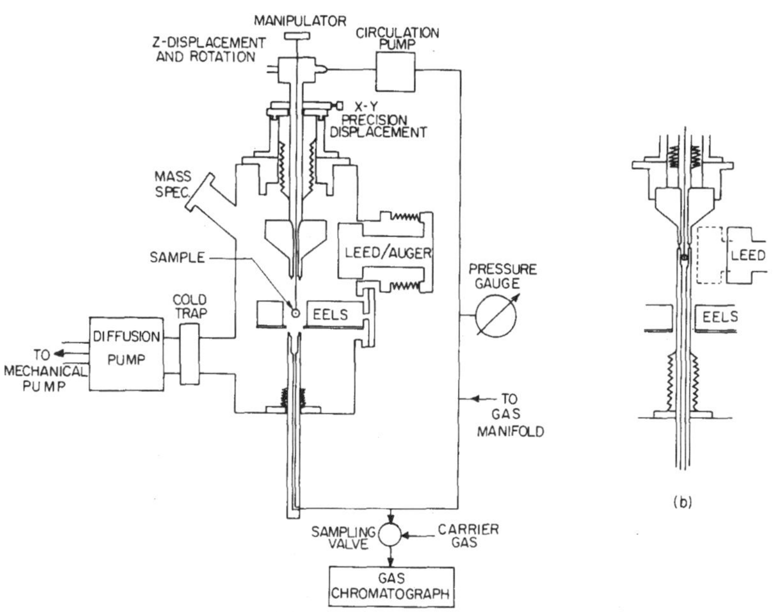

FIG. 4. (a) Schematic diagram of a two-tier-HPLP system with high-pressure cell open. (b) Detail with high-pressure cell closed.

many kinds of surface analysis techniques, thus providing very valuable information about surface composition and structure. As an example we list some of them which we have used and others that could be easily adapted: low-energy electron diffraction (LEED), Auger electron spectroscopy (AES), x-ray photoelectron spectroscopy (XPS), ultraviolet photoemission spectroscopy (UPS), high-resolution electron energy loss spectroscopy (HREELS), secondary ion mass spectroscopy (SIMS), and ^{14}C-radiotracer labeling. Moreover, instruments such as mass spectrometer, sputter-ion gun, evaporators, and thickness monitor can be easily fitted.

IV. USE OF A HPLP APPARATUS AS A CATALYTIC REACTOR

During high-pressure operation, the reaction cell can either be attached to a closed loop and run as a batch reactor, or used in a one-pass manner as a flow reactor. By removing the reaction products with a suitable condenser, the closed-loop system may also function as a flow reactor, and an example of this will be described below.

High-pressure photocatalytic reaction studies can be undertaken by incorporating a sapphire window into the reaction cell. This allows the sample to be illuminated by an external light source (UV or visible), when the sapphire window is aligned with another window in the UHV chamber. Product gases may be detected and analyzed by the use of a gas chromatograph attached to the sample loop. By the use of suitable chromatographic columns and detectors (such as flame ionization, thermal conductivity, or photoionization), most likely product combinations may be separated and analyzed. Moreover, a more direct analysis can be made if the product stream is throttled (by means of a leak valve) into a separately pumped mass spectrometer.

V. APPLICATION OF THE HIGH-PRESSURE–LOW-PRESSURE APPARATUS TO CATALYTIC SYSTEMS

The principal advantage of the type of apparatus discussed in the present paper over the more traditional catalytic reactor, is that the structure and composition of the catalyst surface can be readily ascertained both before and after the catalytic reaction has taken place. Thus, it is possible to investigate the catalytic properties of well-defined, ultraclean samples and observe the effects of the catalytic reaction on the sample surface.

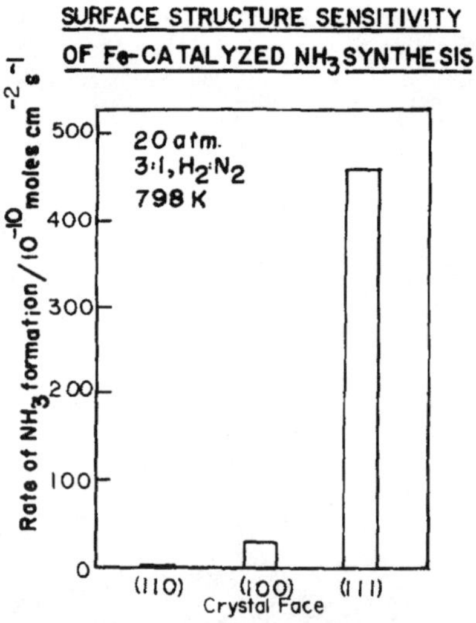

FIG. 6. The surface structure sensitivity of ammonia synthesis catalyzed by single-crystal iron under high-pressure conditions.

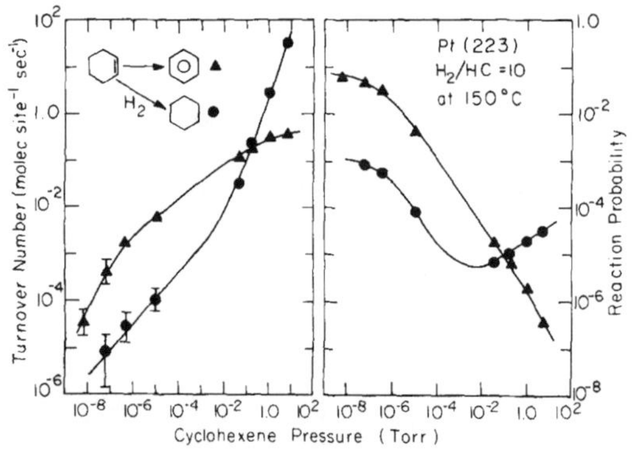

FIG. 5. Correlation of cyclohexene rates and reaction probabilities over a 10 orders of magnitude pressure range. The reactions were performed at 425 K over the stepped Pt(223) crystal surface with H_2/HC = 10.

The HPLP apparatus has been used to study the hydrogenation of carbon monoxide on iron and rhodium polycrystalline specimens[2,6] as well as on single crystals[4] and various rhodium compounds.[11] Postreaction surface analysis revealed the presence of a catalytically active carbonaceous layer on all the samples investigated. In addition, precise control of the rhodium surface oxidation state by oxygen pretreatment in the UHV chamber was found to have a marked effect on the product distribution of the rhodium catalyzed reaction.

An investigation of the catalytic effects of adsorbing gold on platinum surfaces has recently been undertaken.[12] A Knudsen oven and quartz crystal thickness monitor[13] have been attached to a HPLP system, enabling accurate, reproducible dosing of submonolayer quantities of gold under the ultraclean conditions of the UHV chamber. Results indicate that atoms of gold inhibit the catalytic activity of the platinum in heptane isomerization. However, annealing to form a 2-D alloy (and the consequent separation of surface platinum into small ensembles), actually led to an increase in isomerization activity over the clean platinum value.

The HPLP apparatus has enabled the study of catalytic reactions over a pressure range covering many orders of magnitude. The reaction of cyclohexene with hydrogen has been studied between 10^{-7} and 10^{2} Torr.[5] The changes in rate and reaction probability over this wide pressure range are displayed in Fig. 5. It was found that at low pressure the principal product was benzene, and at high pressure the reaction predominantly formed cyclohexane.

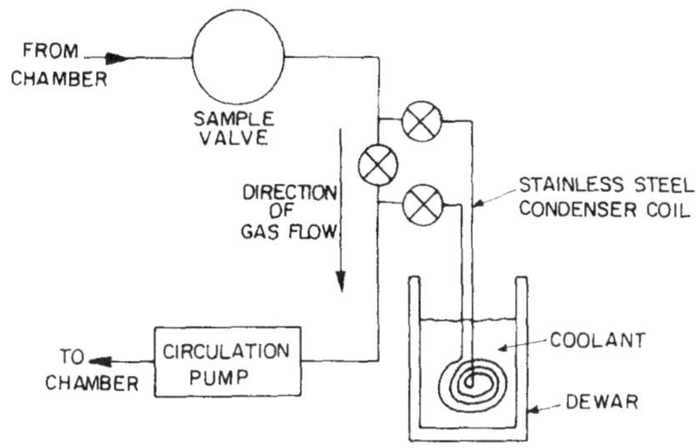

FIG. 7. Schematic diagram of a condenser used in studies of ammonia synthesis, that was attached to the external reaction loop.

It was also observed that while the low-pressure reaction took place on an essentially clean surface, the high-pressure reaction occurred in the presence of a near monolayer of carbonaceous species. In an effort to investigate this overlayer more fully, a ^{14}C-radioisotope technique was employed.[14] It was observed that when the platinum surface was dosed with ^{14}C-labeled hydrocarbons, the lability of the deposited radioactive carbonaceous layer varied with deposition temperature.[15]

At temperatures below 300 K, hydrocarbons were found to adsorb reversibly on platinum (i.e., with no carbon buildup). At temperatures greater than 750 K, multilayer carbon buildup was observed, with poisoning of the platinum's catalytic activity by graphitic carbon. However, at temperatures between these extremes, a mixture of hydrocarbon fragments appeared to exist on the platinum surface. This deposit was found to play an important role in hydrocarbon rearrangement reactions when they were carried out on the platinum surface under these conditions. The surface radioactivity was measured with a surface barrier detector mounted inside the UHV chamber on a rotatable feedthrough.

Ammonia synthesis from nitrogen and hydrogen on iron single crystal surfaces has been recently studied in a HPLP apparatus and yielded the interesting result that Fe(111) appears to be some 2½ orders of magnitude more active at catalyzing this reaction than the close-packed Fe(110) plane (at 798 K and 20 atm total pressure).[16,17] The experimental results are displayed in Fig. 6. This work has been the highest-pressure application of the HPLP apparatus to date, ammonia synthesis being thermodynamically favored at high pressures. During catalytic runs on the HPLP apparatus at 20–25 atm, the base pressure in the UHV chamber did not exceed 10^{-8} Torr, even after many closings of the cell on the same copper gasket. Indeed, it is clear that this seal would function satisfactorily at much higher pressures. Although the same can be said for most of the external circulation loop of the apparatus, some care must be exercised before subjecting electrical feedthroughs to high-pressure operation. Most of these feedthroughs are designed for use on UHV chambers, and are not capable of withstanding high pressures on what is normally their vacuum side. This situation is exacerbated by the large heating current required to maintain the catalyst temperature at high-gas pressures, since this leads to a warming and consequent weakening of the electrical feedthrough seals. This problem has been partially solved by directing a cooling fan onto the feedthroughs during a catalytic run. Feedthroughs capable of withstanding high internal pressure have been developed recently, however, and it is hoped that these will soon be readily available.

It was found during studies of ammonia synthesis on iron that the incorporation of a condenser downstream of the sample valve in the external circulation loop of the HPLP apparatus (Fig. 7), enabled the system to be run as a flow rather than a batch reactor. This is true for any reaction system where the reactants are more volatile than the products, since the condenser temperature can be adjusted to trap the products almost exclusively, allowing a nearly pure stream of reactants to impinge on the catalyst. In the case of ammonia synthesis, (where, next to the product, nitrogen at a partial pressure of 5 atm was the most condensable species) a slurry of isopentane (−159.9 °C) was found to be the ideal condenser medium. During a study of rhenium-catalyzed ammonia synthesis[18] the isopentane condenser was switched in periodically to reduce the ammonia partial pressure to below that at which it appeared to poison the catalyst. In this way, the rhenium was able to produce ammonia in excess of the amount usually leading to poisoning.

The ability to predose catalysts with controlled, submonolayer quantities of promoters is clearly an advantage of the HPLP apparatus over more conventional reactors. The effects of potassium on iron-catalyzed ammonia synthesis and CO hydrogenation are currently being investigated and there is evidence to suggest that the alkali metal plays a role in maintaining a clean catalyst surface. Potassium and other alkali metals are readily dispensed as ions from heated aluminosilicates.[19] The use of submonolayer quantities of potassium as a catalyst itself has also been recently studied by Cabrera *et al.*,[8] who produced methane by reacting water (20 Torr) with graphite in the presence of potassium at 200 °C.

Sulfur and halogens are important in catalysis, being capable of behaving as poisons and promoters in different catalytic systems. A study of the effect of submonolayer quantities of these elements on the performance of several catalysts is currently in progress. Both sulfur[20] and halogens[21] may be conveniently and reproducibly deposited on the catalyst surface by the *in situ* solid-state electrolysis of the respective silver compound.

[a] Present address: Corporate Science Center, Air Products and Chemicals, Inc., P.O. Box 538, Allentown, Pennsylvania 18105.
[b] Present address: W. R. Grace & Co., WRC, 7379 Route 32, Columbia, Maryland 21044.
[c] Present address: Bendix Corporation, ATC, 9140 Old Annapolis Road, Route 108, Columbia, Maryland 21045.

[1] B. A. Sexton and G. A. Somorjai, J. Catal. **46**, 167 (1977).
[2] D. J. Dwyer and G. A. Somorjai, J. Catal. **52**, 291 (1978).
[3] D. J. Dwyer, K. Yoshida, and G. A. Somorjai, *Proceedings of the*

Petrochemical Division of the ACS, 1978, Energy Research Abstracts **3**, No. 30498; *Adv. Chem. Ser.* **178**, edited by Kugler and Steffgen (ACS, Washington, 1979) Chap. 7.
[4] D. Dwyer and G. A. Somorjai, J. Catal. **56**, 249 (1979).
[5] S. M. Davis and G. A. Somorjai, J. Catal. **65**, 78 (1980).
[6] D. G. Castner, R. L. Blackadar, and G. A. Somorjai, J. Catal. **66**, 257 (1980).
[7] W. D. Gillespie, R. K. Herz, E. E. Petersen, and G. A. Somorjai, J. Catal. **70**, 147 (1981).
[8] A. L. Cabrera, H. Heinemann, and G. A. Somorjai, Chem. Phys. Lett. **81**, 402 (1981).
[9] D. W. Blakely, E. Kozak, B. A. Sexton, and G. A. Somorjai, J. Vac. Sci. Technol. **13**, 1091 (1976).
[10] L. H. Dubois, P. K. Hansma, and G. A. Somorjai, Appl. Surf. Sci. **6**, 173 (1980).
[11] P. R. Watson and G. A. Somorjai, J. Catal. **72**, 347 (1981).
[12] J. W. A. Sachtler and G. A. Somorjai (submitted to J. Catal.).
[13] J. W. A. Sachtler, J. P. Biberian, and G. A. Somorjai, Surf. Sci. **110**, 43 (1981).
[14] S. M. Davis, B. E. Gordon, M. Press, and G. A. Somorjai, J. Vac. Sci. Technol. **19**, 231 (1981).
[15] S. M. Davis, F. Zaera, and G. A. Somorjai (submitted to J. Catal.).
[16] N. D. Spencer, R. C. Schoonmaker, and G. A. Somorjai, Nature **294**, 643 (1981).
[17] N. D. Spencer, R. C. Schoonmaker, and G. A. Somorjai, J. Catal. **74**, 129 (1982).
[18] N. D. Spencer and G. A. Somorjai, J. Phys. Chem. **86**, 3493 (1982).
[19] R. K. Feeney, W. E. Sayle, and J. W. Hooper, Rev. Sci. Instrum. **47**, 964 (1976).
[20] W. Heegemann, K. H. Meister, E. Bechtold, and K. Hayek, Surf. Sci. **49**, 161 (1975).
[21] R. A. Marbrow and R. M. Lambert, Surf. Sci. **61**, 329 (1976).

A simple, controllable source for dosing molecular halogens in UHV

N. D. Spencer,[a] P. J. Goddard,[b] P. W. Davies,[c] M. Kitson,[d] and R. M. Lambert

Department of Physical Chemistry, University of Cambridge, Lensfield Road, Cambridge, CB2 1EP, England

(Received 4 October 1982; accepted 2 February 1983)

An electrochemical source for the reproducible generation and dosing of halogens in UHV systems is described. This approach minimizes corrosion damage to vacuum hardware and lends itself to the study of both submonolayer and multilayer halogen adsorption. The corrosive action of halogens on ultrahigh vacuum components has necessitated the development of molecular halogen beam sources, which allow *in situ* generation and dosing of fluorine, chlorine, bromine, and iodine, with a negligible increase in vacuum chamber background pressure. This communication describes the chlorine and bromine sources developed and routinely used in our laboratory. Iodine dosing may be accomplished using a source of similar design. Fluorine, however, because of its highly reactive nature, requires a somewhat different design of source, such as that recently described in detail by Bechtold.[1]

PACS numbers: 81.60. – j, 07.30.Kf, 68.45.Da

The basis of our halogen source is the solid-state electrochemical cell $Ag/AgX/Pt, X_2$, in which silver halide is electrolyzed to silver and halogen. This involves the electric field induced motions of Ag^+ ions through defects in the AgX lattice. A similar cell, using Ag_2S as an electrolyte[2] has been used for the UHV deposition of molecular sulphur (S_2).[3,4] In order to increase the mobility of silver ions in the AgX (and thereby lower the cell resistance), it was found necessary to heat the silver halide to 550 K. However, increasing the defect concentration by the addition of 4–8 wt.% of the corresponding cadmium halide,[5] permitted lower cell operating temperatures (420 K) and also facilitated steady electrolysis over extended periods of time.

Since the source is electrolytic, its halogen output may be easily switched on and off and regulated by controlling the current passing through the cell. In many applications of these sources, it has been necessary to collimate the emerging halogen molecules. It has been possible to dose samples reproducibly with precisely controlled numbers of halogen molecules.[9–12] This was done by using Faraday's law[6] (to obtain the number of generated molecules) and calculating the flow characteristics (given the cell-collimator-sample geometry and assuming Clausing flow.)[7,8] The cell electrodes consisted of a platinum anode and a silver wire cathode. The latter was bent into a flat spiral and a Pt/Pt10%Rh thermocouple was tied onto it. The CdX_2-doped AgX was melted in a crucible and poured into a mould (made of 10 mm o.d. Pyrex tubing), at the base of which the cathode had already been positioned. The platinum gauze anode was spot-welded to a platinum wire; it was then attached, while at red heat, to the other end of the solidified AgX pellet. The cell heating system consisted of a tin oxide-coated Pyrex tube (o.d. 10 mm, Fig. 1), which also served as mechanical support for the pellet and as beam collimator. The tin oxide coating was applied to the glass tube by passing a mixture of oxygen, tin (IV) chloride, and phenylhydrazine over the outside surface while it was being heated to 600 K. The resulting coating had a resistance of approximately 10 Ω/cm of tubing. The tube was clamped at the top and bottom by aluminum alloy supports, which were in turn attached to carry the feedthroughs of a 70 mm UHV flange. The supports were used to carry the heating current and to provide the source with mechanical stability (Fig. 2). At operating temperatures, the cell typically had a resistance of 1 kΩ.

The electrolyzing current for the cell was supplied by a constant current source, which conveniently provided a cell

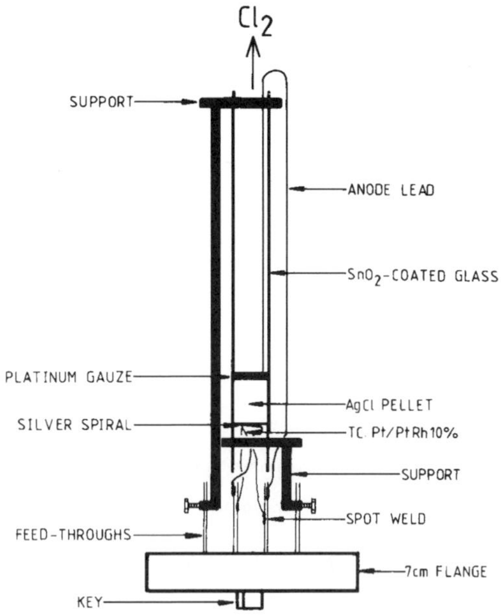

FIG. 1. Schematic of halogen source.

current proportional to a supplied programming voltage. Cell currents used were typically between 10 and 200 μA which, in the geometries of our experiments, corresponded to monolayer dosing times of between 1–100 s. For some applications a computer was used to program the constant current source, and it was found that there was relatively little harmonic distortion in the beam flux up to nearly 50 Hz. Such a computer controlled halogen source has recently been used to study the interaction of bromine with palladium.[13]

Some problems were initially encountered due to Ag*X* evaporation from the cell pellet onto the sample being dosed. However, at 420 K and dosing periods shorter than 10 min, this has not proved a problem. When collimation was used, it was found necessary to turn on the cell for a few seconds before starting a set of experiments, in order to ensure reproducibility. The collimating tube presumably becomes saturated with halogen during the start-up period.

After many hours of halogen generation, a dead short was sometimes observed through the cell, even at room temperature. This was an indication that a silver dendrite had grown within the Ag*X* pellet. The brief passage of a very high current through the heated cell was occasionally found to break the dendrite. However, it appears that it is, in fact, dendrite growth that limits the lifetime of these halogen sources.

Acknowledgments: We gratefully acknowledge Dr. J. N. Agar and Dr. K. Schwaha for useful discussions concerning the construction of these sources.

[a] W. R. Grace and Co., Washington Research Center, 7379 Route 32, Columbia, Maryland, 21044.
[b] ICI Agricultural Division, Billingham, Cleveland, TS23 1LD, U.K.
[c] Bendix Corporation, Advanced Technology Center, 9140 Old Annapolis Road, Route 108, Columbia, Maryland, 21045.
[d] BP Research Centre, Chertsey Road, Sunbury, TW16 7LN U.K.

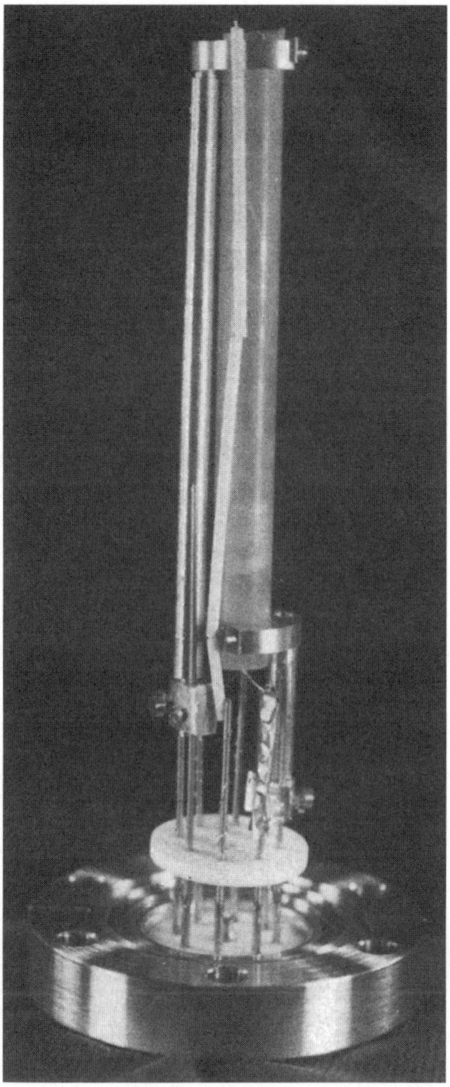

FIG. 2. Photograph of chlorine source.

[1] E. Bechtold, Appl. Surf. Sci. **7**, 231 (1981).
[2] C. Wagner, J. Chem. Phys. **21**, 1819 (1953).
[3] W. Heegeman, K. H. Meister, E. Bechtold, and K. Hayek, Surf. Sci. **49**, 161 (1975).
[4] K. Schwaha, N. D. Spencer, and R. M. Lambert, Surf. Sci. **81**, 273 (1974).
[5] J. Teltow, Ann. Phys. **6**, 68 (1949).
[6] M. Faraday, Philos. Trans. **77** (1834).
[7] B. A. Gottwald, Vak. Tech. **22**, 106 (1973).
[8] B. B. Dayton, Vak. Tech. **7**, 7 (1958).
[9] P. J. Goddard, K. Schwaha, and R. M. Lambert, Surf. Sci. **71**, 351 (1978).
[10] P. W. Davies and R. M. Lambert, Surf. Sci. **95**, 571 (1980).
[11] M. Kitson and R. M. Lambert, Surf. Sci. **100**, 368 (1980).
[12] N. D. Spencer and R. M. Lambert, Surf. Sci. **107**, 237 (1981).
[13] W. T. Tysoe, N. D. Spencer, and R. M. Lambert, Surf. Sci. **120**, 413 (1982).

Surface Science 120 (1982) 413–426 413
North-Holland Publishing Company

MOLECULAR BEAM REACTIVE SCATTERING OF Br_2 FROM Pd(111) USING AN ELECTROCHEMICAL EFFUSIVE SOURCE

W.T. TYSOE, N.D. SPENCER * and R.M. LAMBERT

Department of Physical Chemistry, University of Cambridge, Lensfield Road, Cambridge CB2 1EP, UK

Received 28 January 1982; accepted for publication 28 May 1982

A computer-controlled modulated molecular beam source is used to investigate the kinetics of the surface reactions which occur when bromine is reactively scattered by Pd(111). The reaction products are atomic bromine and molecular bromine: the latter species arises from an adatom recombination process and gives rise to a product vector modulated at twice the frequency of the incident beam (2ω.) By making suitable measurements of the temperature dependence of the product vector phase shifts at ω and 2ω, the four kinetic parameters which characterise the first-order and second-order rate processes are obtained. These are: $A_1 = 2.5 \times 10^9$ s^{-1}, $E_1 = 177$ kJ mol^{-1}, $A_2 = 3.6 \times 10^{-10}$ m^2 s^{-1}, $E_2 = 131$ kJ mol^{-1}. The significance of these values is discussed in terms of the properties of the transition state to desorption.

1. Introduction

It has recently been inferred that bromine overlayers on Pd(111) can desorb by either one of two pathways: a kinetically first-order process yields Br atoms while a second-order process yields molecular Br_2 [1]. The only other metal–halogen system which is found to desorb molecular halogen is Au(111)–Cl_2 [2]. In this case Cl_2 evolution was attributed to decomposition of $AuCl_3$, a view which was consistent with the observed Cl:Cl_2 desorption yield and which is supported by thermochemical and other evidence. Although $PdBr_2$ *is* formed on Pd(111) after very large Br_2 doses [1], the decomposition of a surface compound *cannot* be invoked to explain Br/Br_2 desorption in the low to intermediate exposure regime because Br desorption commences only after a critical coverage has been exceeded. This led us to propose that halogen evaporation was dominated by the processes

$$Br(a) \xrightarrow{E_1} Br(g), \tag{1}$$

* Present address: Department of Chemistry, University of California, Berkeley, California 94720, USA.

$$2\ Br(a) \xrightarrow{E_2} Br_2(g), \tag{2}$$

which requires the thermochemical balance

$$E_2 = 2E_1 - D(Br_2), \tag{3}$$

where $D(Br_2) = 194$ kJ is the dissociation energy of gaseous Br_2 [3]. While the LEED, thermal desorption and other evidence presented in ref. [1] is consistent with reactions (1) and (2), that evidence does not firmly establish the kinetic pathways for overlayer evaporation. Thus the Br_2 desorption spectra do not unambiguously identify the corresponding surface process as second-order atom recombination. Furthermore, the Br desorption spectra, when analysed with a "normal" first-order value of $10^{13}\ s^{-1}$ for the pre-exponential term, yield a value for E_1 of ~ 250 kJ mol^{-1}. This means that (3) fails to balance by about 170 kJ. The discrepancy can be resolved if one assumes that the pre-factor is several orders of magnitude lower than $10^{13}\ s^{-1}$ [4].

The present paper seeks to clarify these aspects of the Pd–Br_2 system by a modulated molecular beam reactive scattering investigation of the surface reaction kinetics. The object is to identify the kinetic order of all overlayer evaporation pathways and to obtain values for all rate parameters which characterise them.

The basic experimental technique has been described in detail by Madix (ref. [5] and references therein). As normally applied it requires the production of a well-collimated, high intensity amplitude modulated molecular beam. This is a difficult procedure [6] made more difficult when the beam gas is corrosive although McKinley has successfully employed DC halogen beams for surface studies [7].

Solid state electrochemical halogen sources [8] provide a convenient way of obtaining accurate doses of very pure halogens; they have also been used in modulated beam experiments [9]. The flux output of these sources is limited necessitating inconveniently long detection times in any AC experiment using conventional analogue signal processing. This difficulty can be overcome by using a microprocessor both to drive the beam source at very low frequency and to collect, store and Fourier analyse the scattered signal after averaging over many cycles [10]. This procedure yields reliable and accurate measurements of the phase and amplitude of the product vectors. In the following section we describe this improved experimental technique; data are presented which show the source's performance over a wide range of conditions in order to establish that the chemical results are free from artefacts.

2. Experimental

The UHV chamber, method of specimen preparation and the construction of the bromine source have been described in earlier publications [8,11]. The forward peaked Br_2 flux was collimated by a tube with $1/r = 20$ [12] which terminated 20 mm away from the Pd(111) crystal, and the system base pressure rose to $\sim 6 \times 10^{-10}$ Torr during operation of the halogen source. The angle of incidence was 12° and the scattered flux was detected by a quadrupole mass spectrometer located in the plane of incidence and at the specular angle. The specimen manipulator was fitted with an auxiliary target of borosilicate glass, which could be moved into the scattering position in substitution for the Pd crystal. This provided one of the two means of establishing the zero residence time phase shift due to finite molecular transit times and all other instrumental parameters.

The experiment was controlled by a North Star Z-80 microprocessor which was used to output a pre-programmed waveform via a D/A converter and a voltage programmable current source to the electrochemical Br_2 source. This arrangement permits the Br_2 flux to be modulated according to any desired complex waveform (e.g. a sum of odd harmonics). The product signal from the

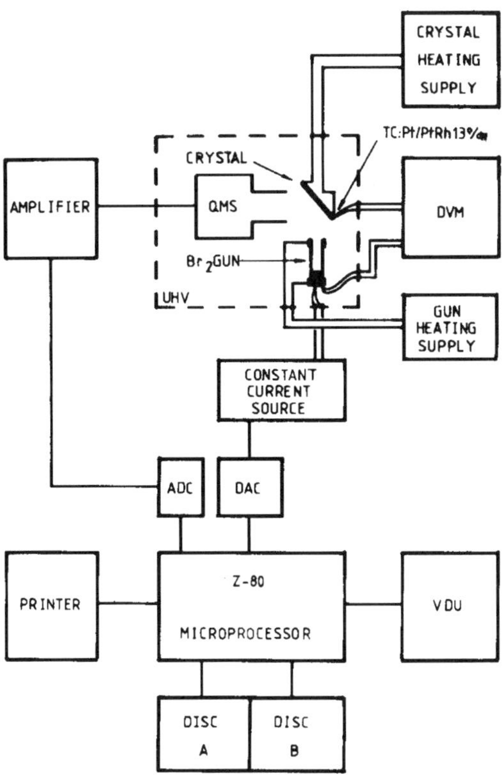

Fig. 1. Block schematic of the computer controlled molecular beam apparatus.

mass spectrometer was amplified, digitised and stored in memory; this waveform could be accumulated over many cycles (typically several hundred) in order to provide a very good signal-to-noise ratio before further processing. This signal processing was conveniently carried out by the microprocessor and consisted of scaling, removal of DC components, 5-point smoothing and Fourier transformation. A block schematic of the experimental arrangement is shown in fig. 1.

3. Results

3.1. Beam source characteristics

Modulation frequencies were in the range 0.5 to 22.5 Hz; stable, reproducible low-frequency modulation is readily obtained with such a computer driven beam source, the low-frequency operation itself being selected for two reasons: (i) The electrolyte response is likely to manifest itself as a significant complex impedance at high frequencies. (ii) Transit time phase shift and demodulation effects (arising from Maxwellian velocity distribution) become more important at higher frequencies. The relatively long specimen-to-detector distance of 120 mm would tend to emphasise such effects. Within the chosen frequency range molecular transit time effects in our apparatus produce a phase shift of $< 1°$ and a demodulation of $< 1\%$ over the temperature range of interest [13]. The source driving waveform is synthesised in 256 steps (8-bit system). In order to ensure that this did not introduce significant errors, the driving waveform was

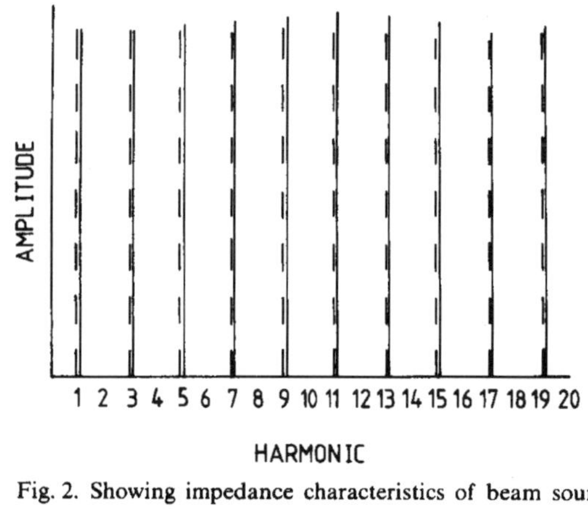

Fig. 2. Showing impedance characteristics of beam source. Fourier transform of source current driven by a voltage waveform containing 10 equal amplitude odd harmonics of 2.5 Hz. Solid line: crystal temperature 300 K; dashed line: crystal temperature 1250 K.

itself Fourier analysed: first harmonic amplitude was always $< 0.04\%$ that of the fundamental. Two further questions remain. Is the source performance affected by accidental radiative heating from the specimen? [14]. Does the halogen beam flux incident on the crystal actually follow the driving waveform? The first question is answered by the results illustrated in fig. 2 which

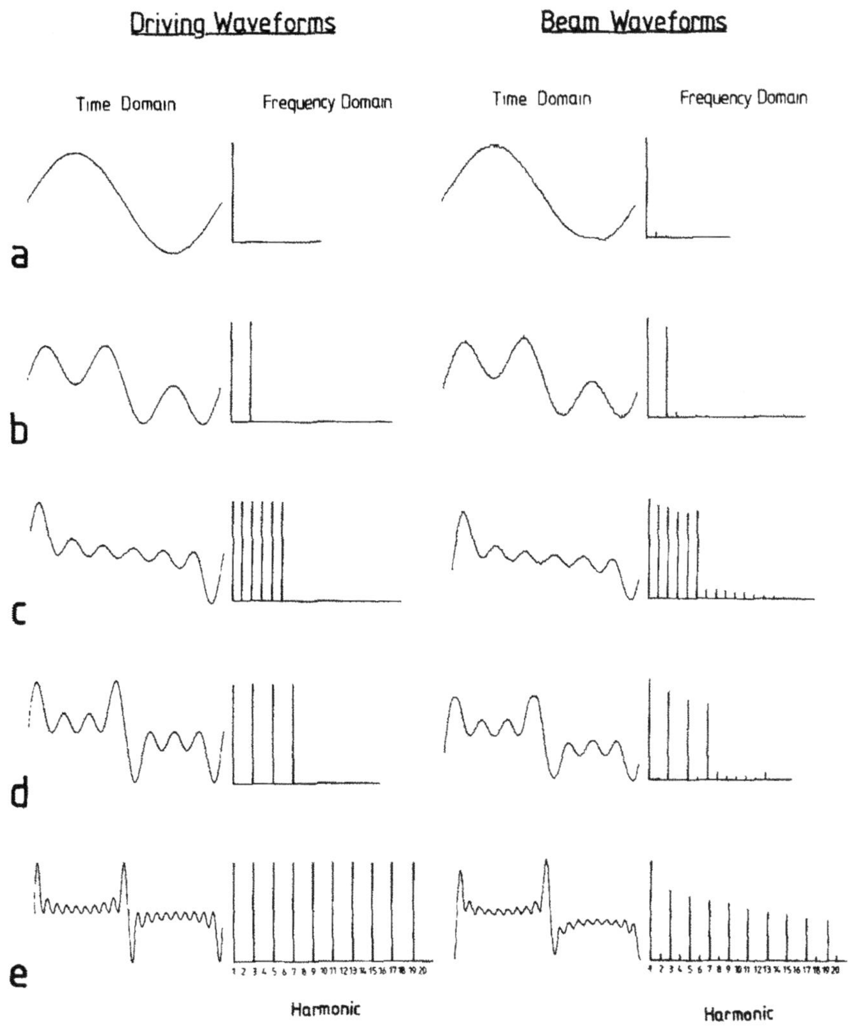

Fig. 3. Showing correspondence between driving waveform and *incident* molecular beam flux for a variety of synthesised waveforms. (The small 2ω component visible in beam waveform (a) is an experimental artefact – see text.)

shows that the source current accurately follows the driving waveform at a crystal temperature of 300 K. More importantly, its performance remains unchanged at a crystal temperature of 1250 K; all phase angles are within 0.5° of the driving frequency corresponding to a linear response between 2.5 and 47.5 Hz. The second question is answered by removing the specimen and placing the mass spectrometer opposite the halogen source. The fundamental frequency in these experiments was 2.5 Hz; it can be seen that relatively little demodulation occurs. (The small apparent 2ω component in the beam waveform which is visible in fig. 3a is an experimental artefact which arises from slightly incorrect adjustment of the D/A converter in this particular experiment. It can be seen that the negative excursion of the signal has been slightly "cut off".)

3.2. Reactive scattering of bromine from Pd(111)

Our earlier work [1] demonstrated that atomic Br and molecular Br_2 are the relevant product species. In order to discover the appropriate temperature regime for their investigation in beam experiments the following "DC" beam measurements were made. The mass spectrometer signal at 79 amu (Br^+) was monitored as a function of crystal temperature while switching the halogen source on and off at a frequency of *0.5 Hz*. The amplitude of the resulting signal is proportional to the scattered flux from the crystal at any given temperature. This procedure was adopted in order to circumvent the problem caused by a slowly rising background bromine signal which would have interfered with a true DC measurement. The results of this experiment are

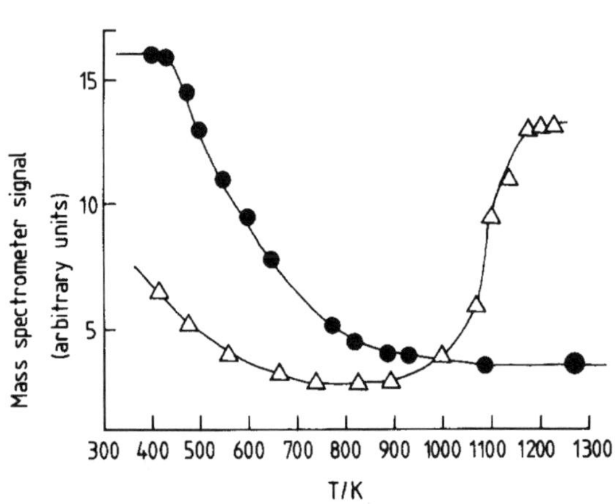

Fig. 4. Reflected intensity as a function of crystal temperature during a "DC" beam experiment (see text): (●) Br_1; (△) Br.

illustrated in fig. 4 from which it is evident that Br_2 desorption is most usefully studied between ~600 K and ~750 K while the optimum regime for Br is between ~1050 K and ~1300 K. Fig. 4 will be discussed in more detail in section 3.6.

3.3. Atomic bromine desorption

Modulated beam experiments were carried out using a single sinusoidal frequency of 2.5 Hz, and the scattered Br product was detected for surface temperatures in the range 1050–1330 K. The temperature dependence of the phase shift of the Br product vector is shown in fig. 5; these phase shifts have not been corrected for the constant phase shift due to all instrumental factors, and this latter quantity was determined in the following manner:
(1) The glass "dummy" target was placed in the scattering position in place of the Pd crystal, and the phase shift of the scattered bromine signal was measured. Bromine is only very weakly adsorbed on glass which should result in a very short residence time for the physically adsorbed molecules. The phase shift due to molecular flight times is negligible (see above) so that the observed shift in this experiment provides the zero residence time reference phase shift with respect to which all reactively scattered product phase shifts can be measured. The result so obtained is $33 \pm 1°$.
(2) This reference instrumental phase shift can also be measured by a completely different technique using the Br product data in fig. 5. At very high temperatures, the surface residence time of Br will be very small and tending to zero with increasing temperature. Consequently, the high temperature limiting phase shift will be due to instrumental factors alone. Fig. 5 yields a value of ~32° for this quantity – in good agreement with the result of method 1.

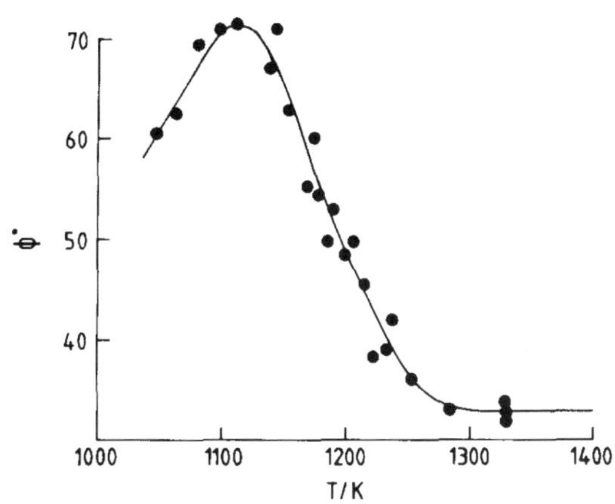

Fig. 5. Phase variation of the Br(ω) signal as a function of crystal temperature.

3.4. Molecular bromine desorption

Our earlier work [1] suggested that Br_2 desorption resulted from second-order Br(a) + Br(a) recombination reaction: this conclusion was based on an inspection of the Br_2 thermal desorption spectra. The present approach permits one to characterise the kinetic order of the surface reaction and allows a determination of the kinetic parameters. When products are formed by a non-linear surface process such as

$$Br(a) + Br(a) \rightarrow Br_2(g),$$

the scattered Br_2 signal should then contain a component modulated at 2ω, i.e. at twice the modulation frequency of the incident beam.

Fig. 6 shows the results of an experiment in which $\omega = 0.5$ Hz. Large signals at 2ω *are* indeed observed at 150 amu (Br_2^+) and may be taken as indicative of a non-linear surface process such as $Br(a) + Br(a) \rightarrow Br_2(g)$. The amplitude of these $Br_2(2\omega)$ signals was about 5% of the $Br_2(\omega)$ signal: this corresponds to a very large 2ω content because it should be kept in mind that a large proportion of the detected Br_2 flux represents *unreactively* scattered bromine which will necessarily turn up at ω. By contrast the first-order process $Br(a) \rightarrow Br(g)$ shows a $Br(2\omega)$ amplitude which is only 0.2% that of $Br(\omega)$ and in this case *all* the detected Br flux is due to reactive scattering. Clearly, Br_2 and Br desorption are dominated by second-order and first-order processes respectively. The analysis of the $Br_2(2\omega)$ data which follows is possibly *only* because the incident beam flux is accurately sinusoidally modulated: it depends on the absence of 2ω components in the non-reactively scattered Br_2 flux which would otherwise

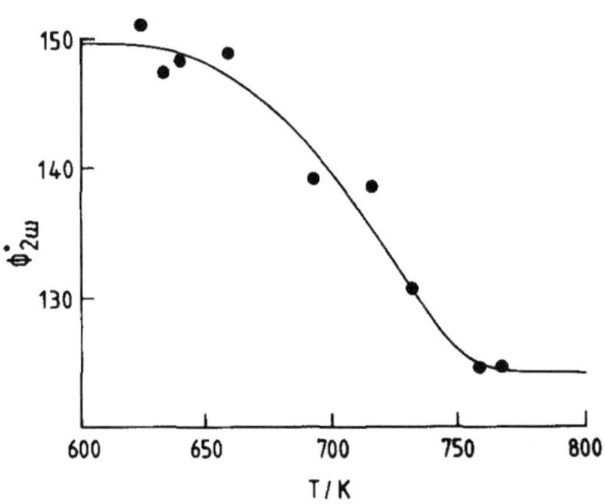

Fig. 6. Phase variation of the 2ω component for the scattering of Br_2 as a function of crystal temperature.

swamp the desired signal. Furthermore, it is because Br_2 and Br desorption occur in markedly different temperature regimes that we can achieve the simple separation in the data which is employed in the analysis. That is, the 79 amu signal *is* simply a measure of Br flux at higher temperatures because Br_2 desorption is now negligible: therefore the Br(ω) signal does not contain a significant component from Br_2 fragmentation in the mass spectrometer. A quantitative analysis of the 2ω data requires a knowledge of the absolute flux incident on the crystal. This was obtained with reference to LEED calibrated surface coverages [1] which showed that the incident flux was 2.12×10^{17} m^{-2} s^{-1}.

3.5. Analysis of results

The surface processes which occur during the reactive scattering of bromine may be summarised as follows:

$$Br_2(g) \underset{Pd(111)}{\overset{\lambda}{\rightarrow}} Br(a) \overset{k_2}{\rightarrow} \tfrac{1}{2}Br_2(g)$$
$$Br(a) \overset{k_1}{\rightarrow} Br(g)$$

where λ is the sticking probability. If $n(t)$ is the instantaneous concentration of Br(a) and $I(t)$ the incident flux, then

$$dn/dt = 2\lambda I(t) - k_1 n - 2k_2 n^2, \tag{4}$$

where k_1 and k_2 are first- and second-order rate constants. In the present case $I(t)$ is a pure sine wave, so

$$I(t) = \tfrac{1}{2}(1 + e^{i\omega t})I_0, \tag{5}$$

following the formalism of Olander and Ullman [15], we may calculate the amplitude ε and the phase lag ϕ of the Br atom signal at ω

$$\varepsilon = k_1 \lambda I_0 / \left[(k_1 + 4k_2 n_0)^2 + \omega^2\right]^{1/2}, \tag{6a}$$

$$\tan\phi = \omega/(k_1 + 4k_2 n_0). \tag{6b}$$

The expressions for the amplitude and phase of the $Br_2(2\omega)$ signal can be similarly derived, and they are fairly complex. They can be simplified by remembering that the Br atom desorption rate is negligible compared to the second-order Br_2 rate [1] in the temperature regime for which the Br_2 data are taken: i.e. $k_1 n_0 \ll k_2 n_0^2$. We then obtain for the phase of the Br_2 vector

$$\tan(\phi + \pi/2) = \left(32k_2^2 n_0^2 \omega - \omega^3\right) / \left(32k_2^3 n_0^3 - 10k_2 n_0 \omega^2\right), \tag{7}$$

where $n_0 = (\lambda I_0/2k_2)^{1/2}$. The limits of this equation are $\tan(\phi + \pi/2) \rightarrow \omega/10k_2 n_0$ for $\omega > k_2 n_0$ and $\tan(\phi + \pi/2) \rightarrow \omega/k_2 n_0$ for $\omega < k_2 n_0$. Although the modulation frequency is low, it is appropriate to choose the first limit since the incident beam flux I_0 is small enough for the former condition to apply. Using this approximation, eq. (7) simplifies to yield

$$\tan(\phi + \pi/2) = \omega/10 k_2 n_0, \tag{8}$$

which with the definition of n_0 yields

$$\tan(\phi + \pi/2) = \frac{\omega}{10}\left(\frac{2}{k_2 \lambda I_0}\right)^{1/2}. \tag{9}$$

Writing the rate constant k_2 as $k_2 = A_2 \exp(-E_2/RT)$, where A_2 and E_2, are the pre-exponential factor and activation energy for second-order Br_2 desorption, one obtains:

$$\ln \tan(\phi + \pi/2) = \ln\left[\left(\frac{2}{A_2 \lambda I_0}\right)^{1/2} \frac{\omega}{10}\right] + \frac{E_2}{2RT}.$$

3.6. *Intensity data*

Fig. 4 shows that the reflected Br_2 flux decreases monotonically with increasing temperature: the surface Br concentration varies with temperature such that the surface is saturated at $\lesssim 400$ K whereas at ~ 1250 K the coverage is vanishingly small. Thus at low temperature the large reflected Br_2 signal is the result of essentially specular reflection of Br_2 from a saturated overlayer (the detector is at the specular angle). As the temperature is raised Br(a) + Br(a) recombination begins to contribute to the scattered Br_2 flux, and these species leave the surface with a more isotropic angular distribution. Accordingly, the Br_2 flux reaching the detector begins to fall.

It can be seen that the detected Br_2 signal does *not* fall to zero even at the highest temperatures: this residual signal is not the result of reflection of the incident beam by the chamber walls because removal of the Pd crystal from the beam *does* cause it to vanish. The most likely explanation seems to be that it is due to the specularly reflected non-reactively scattered Br_2 molecules. A knowledge of the experimental geometry enables us to calculate that in the case of specular reflection $\sim 0.4\%$ of the incident flux arrives at the detector (i.e. at low temperatures). In the case of complete isotropic reactive scattering (high temperatures) this figure would be reduced by a factor of ~ 40. Inspection of fig. 4 shows that the actual ratio between the limiting low-temperature and high-temperature Br_2 signal amplitudes is about a factor of 10 greater than this. This implies that at high temperatures and low coverages $\sim 20\%$ of the incident molecules are not reactively scattered which corresponds approximately to a trapping probability of ~ 0.8 [16] in accordance with our earlier results [1].

No $PdBr_2$ *evaporation* was detected in the present work; this is unsurprising given the long interaction times which are necessary for the production of detectable amounts of dihalide [1].

3.7. Phase shift data

It should be noted that the Br(ω) data are taken over a temperature regime (fig. 5) in which the contribution of Br_2 to the Br^+ signal is relatively small (fig. 4). The relevant phase change was shown to be

$$\tan\phi = \omega/(k_1 + 4k_2 n_0). \tag{6b}$$

This equation can be simplified by noting that the "DC beam" results show that for temperatures $\gtrsim 1100$ K Br *atom* desorption is the dominant rate process. Using this to simplify (6b) by writing $k_1 \gg k_2 n_0$ and expressing k_1 in the form $k_1 = A_1 \exp(-E_1/RT)$ reduces (6b) to

$$\ln(\omega/\tan\phi(\omega)) = \ln A_1 - E_1/RT. \tag{10}$$

Using the measured zero residence time reference phase shift of $33 \pm 1°$, the data in fig. 5 may then be plotted in accordance with eq. (10). The result is displayed in fig. 7 from which the slope and intercept yield values of

$$E_1 = 177 \text{ kJ mol}^{-1}, \quad A_1 = 2.5 \times 10^9 \text{ s}^{-1},$$

for the desorption parameters of atomic Br.

The A_1 value is *several orders of magnitude lower* than the values which are often assumed in the analysis of first-order thermal desorption spectra (viz. 10^{13}–10^{15} s^{-1}). It is however in much closer agreement with the corresponding

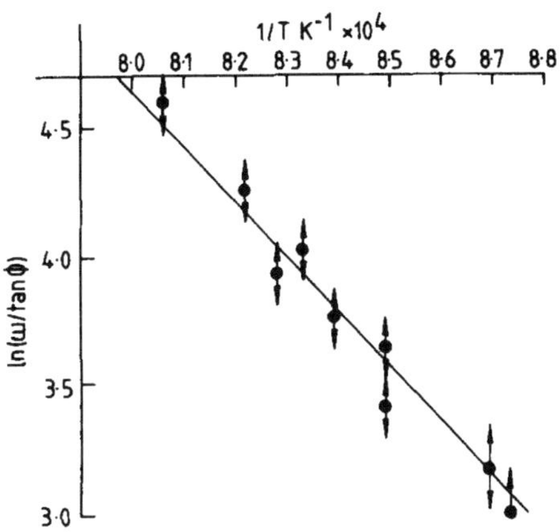

Fig. 7. Plot of $\ln(\omega/\tan\phi)$ versus $1/T$ for Br signal. The error bars show the extent of the contribution from the fragmentation of Br_2 in the mass spectrometer.

and even lower figure of $\sim 2 \times 10^7$ s^{-1} which was obtained by a detailed analysis of atomic Cl desorption spectra from Rh(111) [4]. (Such an analysis is not possible for the Pd(111)–Br spectra [1] due to their greater complexity.) These results suggest that atomic halogen desorption from transition metal surfaces takes place from a layer of relatively mobile atoms which desorb via a transition state which is localised at a step or some other surface defect. The hypothesis does not seem too unreasonable given that crossing a step leads to loss of surface atom coordination thus providing a low energy evaporation path. If the above value for A_1 is used in conjunction with the low-coverage Br desorption spectra of ref. [1] one obtains a desorption energy of 179 kJ mol^{-1} in good agreement with the above E_1 value.

A simple determination of a zero residence time reference phase cannot be easily carried out in the case of 2ω bromine molecule scattering since it contains contributions from both the electrochemical source driven at ω and from the detection electronics at 2ω.

Consideration of eq. (8) shows that as the crystal temperature is lowered, the ratio $\omega/k_2 n_0$ becomes larger, so that as $T \to 0$, $\phi \to 180°$. It is thus assumed that the phase shift at 600 K (i.e. 149°) corresponds to this low-temperature limit. A further check on the validity of this assumption is that considerable deviations from linearity will be observed in the Arrhenius plot if it is incorrect.

Fig. 8 shows a plot of the $Br_2(2\omega)$ data according to eq. (9). The slope yields $E_2 = 131$ kJ $mole^{-1}$ for the activation energy to associative desorption of Br as Br_2. The intercept yields a value of 6.3×10^7 s^{-1} for the quantity $A_2 \lambda I_0$. Knowing I_0 and λ (see above) gives us $A_2 = 3.6 \times 10^{-10}$ m^2 s^{-1} for the pre-exponential factor to second-order desorption. Calculation of $k_2 n_0$ with these data shows that the condition $\omega \gg k_2 n_0$ does indeed apply.

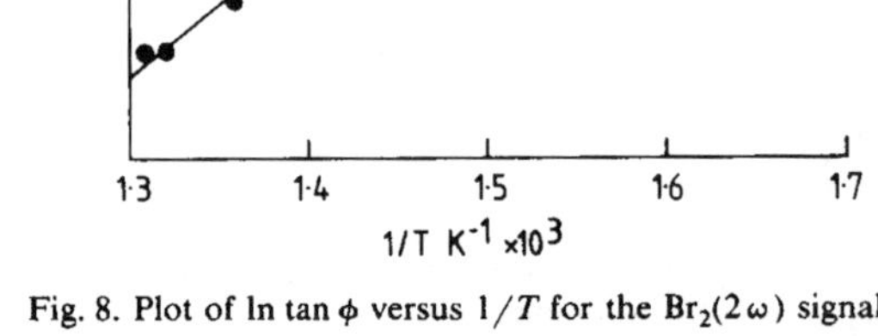

Fig. 8. Plot of ln tan ϕ versus $1/T$ for the $Br_2(2\omega)$ signal.

Once again these results can be interpreted in terms of transition state theory. For the limiting case of a rigid rotor confined to a plane with mobile adsorbate and transition state, the value of A_2 calculated on this basis is:

$$A_2(\text{calc}) = \eta \frac{kT}{h} 2\pi r_0^2,$$

where η is a transmission coefficient and r_0 the Br–Br internuclear separation in the transition state. Application of this formula to desorption of Br_2 yields a value of $A_2 = 4.8 \times 10^{-6}\ \eta\ \text{m}^2\ \text{s}^{-1}$. The discrepancy between the two values suggests that the rotation is hindered or $\eta \ll 1$. The assumption that the bromine molecular rotation in the transition state is completely hindered gives $A_2 = 1.6 \times 10^{-9}\ \text{m}^2\ \text{s}^{-1}$ which considerably improves the agreement.

A simple check on the consistency of E_1 and E_2 is provided by the thermochemical balance

$$E_2 = 2E_1 - D(Br_2),$$

where $D(Br_2)$ is the dissociation energy of Br_2 and is equal to 194 kJ mol^{-1}. Given E_1, this yields $E_2 = 144$ kJ in satisfactory agreement with the experimental value of 131 kJ.

Acknowledgements

W.T.T. thanks the SERC for financial support and N.D.S. acknowledges financial support by ICI Ltd. We are grateful to Johnson Matthey Ltd. for the loan of precious metals used in this work.

References

[1] W.T. Tysoe and R.M. Lambert, Surface Sci. 115 (1982) 37.
[2] N.D. Spencer and R.M. Lambert, Surface Sci. 107 (1981) 237.
[3] R.C. Weast, Ed., Handbook of Chemistry and Physics, 60th ed. (Chemical Rubber Co., Cleveland, OH, 1980).
[4] M.P. Cox and R.M. Lambert, Surface Sci. 107 (1981) 547.
[5] R.J. Madix, in: Physical Chemistry of Fast Reactions, Vol II. Ed. D.O. Hayward (Plenum, New York, 1976).
[6] N.F. Ramsey, Molecular Beams (Oxford University Press, Oxford, 1956).
[7] J.D. McKinley, J. Chem. Phys. 40 (1964) 120, 576; 45 (1966) 1690.
[8] P.J. Goddard and R.M. Lambert, Surface Sci. 67 (1977) 180.
[9] R.H. Prince and R.M. Lambert, Chem. Phys. Letters 67 (1979) 388.
[10] C.T. Foxon, M.R. Boudry and B.A. Joyce. Surface Sci. 44 (1974) 69.
[11] R.A. Marbrow and R.M. Lambert, Surface Sci. 61 (1976) 329.
[12] P. Clausing. Ann. Physik 12 (1932) 961.
[13] H. Halstead, D.G. Hummer and W.L. Fite, J. Chem. Phys. 41 (1964) 2567; J.A. Schwarz and R.J. Madix, J. Catalysis 12 (1968) 140.

[14] A.M. Kramer and E. Bauer. Surface Sci. 107 (1981) 1.
[15] D.R. Olander and A. Ullman, Intern. J. Chem. Kin. 8 (1976) 625.
[16] P. Kisliuk, J. Phys. Chem. Solids 3 (1957) 95; 5 (1958) 78;
D.A. King and M.G. Wells, Proc. Roy. Soc. (London) A339 (1974) 245.